IET CONTROL, ROBOTICS AND SENSORS SERIES 133

Control and Estimation of Dynamical Nonlinear and Partial Differential Equation Systems

Other volumes in this series:

Volume 68 **Flexible Robot Manipulators: Modelling, simulation and control** M.O. Tokhi and A.K.M. Azad (Editors)
Volume 69 **Advances in Unmanned Marine Vehicles** G. Roberts and R. Sutton (Editors)
Volume 70 **Intelligent Control Systems Using Computational Intelligence Techniques** A. Ruano (Editor)
Volume 71 **Advances in Cognitive Systems** S. Nefti and J. Gray (Editors)
Volume 72 **Control Theory: A guided tour, 3rd Edition** J. R. Leigh
Volume 73 **Adaptive Sampling with Mobile WSN** K. Sreenath, M.F. Mysorewala, D.O. Popa and F.L. Lewis
Volume 74 **Eigenstructure Control Algorithms: Applications to aircraft/rotorcraft handling qualities design** S. Srinathkumar
Volume 75 **Advanced Control for Constrained Processes and Systems** F. Garelli, R.J. Mantz and H. De Battista
Volume 76 **Developments in Control Theory towards Glocal Control** L. Qiu, J. Chen, T. Iwasaki and H. Fujioka (Editors)
Volume 77 **Further Advances in Unmanned Marine Vehicles** G.N. Roberts and R. Sutton (Editors)
Volume 78 **Frequency-Domain Control Design for High-Performance Systems** J. O'Brien
Volume 80 **Control-oriented Modelling and Identification: Theory and practice** M. Lovera (Editor)
Volume 81 **Optimal Adaptive Control and Differential Games by Reinforcement Learning Principles** D. Vrabie, K. Vamvoudakis and F. Lewis
Volume 83 **Robust and Adaptive Model Predictive Control of Nonlinear Systems** M. Guay, V. Adetola and D. DeHaan
Volume 84 **Nonlinear and Adaptive Control Systems** Z. Ding
Volume 86 **Modeling and Control of Flexible Robot Manipulators, 2nd edition** M. O. Tokhi and A.K.M. Azad
Volume 88 **Distributed Control and Filtering for Industrial Systems** M. Mahmoud
Volume 89 **Control-based Operating System Design** A. Leva *et al.*
Volume 90 **Application of Dimensional Analysis in Systems Modelling and Control Design** P. Balaguer
Volume 91 **An Introduction to Fractional Control** D. Valério and J. Costa
Volume 92 **Handbook of Vehicle Suspension Control Systems** H. Liu, H. Gao and P. Li
Volume 93 **Design and Development of Multi-Lane Smart Electromechanical Actuators** F.Y. Annaz
Volume 94 **Analysis and Design of Reset Control Systems** Y.Guo, L. Xie and Y. Wang
Volume 95 **Modelling Control Systems Using IEC 61499, 2nd Edition** R. Lewis & A. Zoitl
Volume 96 **Cyber-Physical System Design with Sensor Networking Technologies** S. Zeadally and N. Jabeur (Editors)
Volume 99 **Practical Robotics and Mechatronics: Marine, Space and Medical Applications** I. Yamamoto
Volume 100 **Organic Sensors: Materials and Applications** E Garcia-Breijo and P Cosseddu (Editors)
Volume 102 **Recent Trends in Sliding Mode Control** L. Fridman JP. Barbot and F. Plestan (Editors)
Volume 104 **Control of Mechatronic Systems** L. Guvenc, B.A Guvenc, B. Demirel, M.T Emirler
Volume 105 **Mechatronic Hands: Prosthetic and Robotic Design** P.H. Chappell

Volume 107 **Solved Problems in Dynamical Systems and Control** D. Valério, J. T Machado, A. M. Lopes and A. M. Galhano

Volume 108 **Wearable Exoskeleton Systems: Design, Control and Applications** S. Bai, G.S. Virk and T.G.Sugar

Volume 111 **The Inverted Pendulum in Control Theory and Robotics: From Theory to New Innovations** O. Boubaker and R. Iriarte (Editors)

Volume 112 **RFID Protocol Design, Optimization, and Security for the Internet of Things** A. X. Liu, M. Shahzad, X Liu and K. Li

Volume 113 **Design of Embedded Robust Control Systems Using MATLAB®/Simulink®** P.H. Petkov, T.N. Slavov and J.K. Kralev

Volume 114 **Signal Processing and Machine Learning for Brain-Machine Interfaces** T. Tanaka and M. Arvaneh (Editor)

Volume 116 **Imaging Sensor Technologies and Applications** W Yang (Editor)

Volume 117 **Data Fusion in Wireless Sensor Networks** D. Ciuonzo and P.S. Rossi (Editors)

Volume 118 **Modeling, Simulation and Control of Electrical Drives** M. F. Rahman, S. K. Dwivedi (Editors)

Volume 119 **Swarm Intelligence Volumes** 1 – 3 Y. Tan (Editor)

Volume 120 **Imaging and Sensing for Unmanned Aircraft Systems Vol1 and 2** V. Estrela, J. Hemanth, O. Saotome, G. Nikolakopoulos and R. Sabatini (Editors)

Volume 121 **Integrated Fault Diagnosis and Control Design of Linear Complex Systems** M. Davoodi, N. Meskin and K.Khorasani

Volume 123 **Data-Driven Modeling, Filtering and Control: Methods and applications** C. Novara and S. Formentin (Editors)

Volume 125 **Short-Range Micro-Motion Sensing with Radar Technology** C. Gu and J.Lien (Editors)

Volume 126 **Fault Diagnosis and Fault-tolerant Control of Robotic and Autonomous Systems** Monteriù, Longhi and Freddi (Editors)

Volume 127 **Sensors, Actuators, and Their Interfaces: A multidisciplinary introduction 2nd Edition** N. Ida

Volume 128 **IoT Technologies in Smart Cities: From sensors to big data, security and trust** F. Al-Turjman and M.Imran (Editors)

Volume 129 **Signal Processing to Drive Human-Computer Interaction: EEG and eye-controlled interfaces** I. Kompatsiaris, C. Kumar and S. Nikolopoulos (Editors)

Volume 130 **Transparency for Robots and Autonomous Systems: Fundamentals, technologies and applications** R.Wortham

Volume 131 **Space Robotics and Autonomous Systems: Technologies, advances and applications** Yang Gao (Editor)

Control and Estimation of Dynamical Nonlinear and Partial Differential Equation Systems

Theory and applications

Gerasimos Rigatos, Masoud Abbaszadeh and Pierluigi Siano

The Institution of Engineering and Technology

Published by The Institution of Engineering and Technology, London, United Kingdom

The Institution of Engineering and Technology is registered as a Charity in England & Wales (no. 211014) and Scotland (no. SC038698).

First published 2022

The Institution of Engineering and Technology
Futures Place
Kings Way, Stevenage
Hertfordshire, SG1 2UA, United Kingdom

www.theiet.org

British Library Cataloguing in Publication Data
A catalogue record for this product is available from the British Library

ISBN 978-1-83953-426-3 (hardback)
ISBN 978-1-83953-427-0 (PDF)

Typeset in India by MPS Limited

In memory of my mother Diamantina Rigatou (1939–2018)
Gerasimos Rigatos

To my wife Elham, and my sons, Arman and Ario
Masoud Abbaszadeh

In memory of my father Giuseppe and dedicated to my mother Maddalena, my wife Anna Maria, my children Marta and Francesco
Pierluigi Siano

Contents

About the authors

Gerasimos Rigatos is a research director at the Industrial Systems Institute, Greece. He has led research cooperation projects in the areas of nonlinear control, nonlinear filtering and control of distributed parameter systems. His results have appeared in eight research monographs. His research, being highly cited, has been also published in more than 120 journal articles where he is the first author. Since 2007, he has been awarded visiting professor positions at several academic institutes (Université Paris XI France; Harper-Adams University College, UK; University of Northumbria, UK; University of Salerno, Italy, Ecole Centrale de Nantes, France). He is a senior member of the IEEE and a member and CEng of the IET. He holds an associate editor's position for the *Journal of Advanced Robotic Systems*. He received both his diploma in Electrical and Computer Engineering and his PhD from the Department of Electrical and Computer Engineering of the National Technical University of Athens (NTUA), Greece.

Masoud Abbaszadeh is a principal research engineer at GE Research Center, Niskayuna, NY, USA. He received BSc, MSc and PhD degrees from Amirkabir University of Technology, Tehran, Iran, Sharif University of Technology, Tehran, Iran, and University of Alberta, Edmonton, Alberta, Canada, in 2000, 2002, and 2008, respectively, all in Electrical and Computer Engineering. From 2001 to 2003, he was working as a control and automation engineer in automotive manufacturing. From 2008 to 2011, he was with Maplesoft, Waterloo, Ontario, Canada, as a research engineer. He was the principal developer of the MapleSim Control Design Toolbox and was a member of a research team working on the Maplesoft-Toyota joint projects. From 2011 to 2013, he was a senior research engineer at United Technologies Research Center, East Hartford, CT, USA, working on advanced control systems, and complex systems modeling and simulation. He has been with GE Research Center, Niskayuna, NY, USA, since 2013, where he is currently a principal research engineer. He has also held an adjunct professor position at the ECSE department of the Rensselaer Polytechnic Institute, Troy, NY, USA. His research interests include estimation and detection theory, robust and nonlinear control, and machine learning with applications in diagnostics, cyber-physical resilience and autonomous systems. He has over 110 peer-reviewed papers, 4 book chapters, and holds 25 issued US patents, with 30+ more patents pending. He is an associate editor of *IEEE Transactions on Control Systems Technology*, and a member of IEEE CSS Conference editorial board.

Pierluigi Siano is a professor and scientific director of the Smart Grids and Smart Cities Laboratory with the Department of Management & Innovation Systems, University of Salerno, Italy. He received his MSc degree in electronic engineering and his

Ph.D. degree in information and electrical engineering from the University of Salerno, Salerno, Italy, in 2001 and 2006, respectively. Since 2021, he has been Distinguished Visiting Professor in the Department of Electrical & Electronic Engineering Science, University of Johannesburg.

His research activities are centered on demand response, energy management, the integration of distributed energy resources in smart grids, electricity markets, and on planning and management of power systems. In these research fields, he has co-authored more than 650 articles including more than 370 international journal papers that received in Scopus more than 12,700 citations with an H-index equal to 55. In 2019, 2020, and 2021, he has been awarded as Highly Cited Researcher in Engineering by the Web of Science Group. He has been a member of the IEEE since 2009 and a senior member since 2014. He has been the Chair of the IES TC on Smart Grids. He is the Editor for the Power & Energy Society Section of *IEEE Access, IEEE Transactions on Power Systems, IEEE Transactions on Industrial Informatics, IEEE Transactions on Industrial Electronics*, and *IEEE Systems*.

Preface

The monograph presents advances in applied control of nonlinear and PDE dynamical systems, and provides both theoretical analysis of the proposed control methods and case studies about their use in robotics, mechatronics, electric power generation, power electronics, micro-electronics, industrial production processes and cyberphysical systems comprising communication and computer networks. The monograph covers thoroughly the area of automatic control for complex nonlinear dynamical systems, including also applications to distributed parameter systems which are described by partial differential equations. The monograph has a meaningful contribution in the areas of automatic control and systems science. Its results can be classified in the following main approaches for the control of complex nonlinear dynamical systems: (i) control with methods of approximate (local) linearization being associated with the solution of the nonlinear optimal control problem, (ii) control with methods of exact (global) linearization comprising also adaptive control methods, (iii) control of distributed parameter systems (systems which are described by partial differential equations) and stochastic estimation methods.

With reference to approach (i) which is control methods based on approximate linearization, one can distinguish results towards extending H-infinity control to nonlinear dynamical systems and towards solving the associated nonlinear optimal control problem. The methods which are developed for nonlinear control problems rely on linearization of the systems' dynamics around local operating points while the designed feedback controllers make use of the approximately linearized state-space models. Such controllers exhibit also robustness to external perturbations, as well as to modelling errors, and achieve asymptotically (as time advances) the compensation of the nonlinear dynamics of the controlled systems. In this area, one can note an important research result which is a new method of H-infinity control. This approach makes use of an approximately linearized model of the system that is obtained through the computation of Jacobian matrices. In contrast to problems of linear control and the method of the linear quadratic regulator, it is far more difficult to achieve a solution of the optimal control problem in the case of nonlinear dynamical systems under model uncertainties and external disturbances. The nonlinear optimal control problem is usually treated with iterative computational methods that are not always of assured convergence to the optimum. Actually, one comes against a differential game where the control signal tries to minimize the system's cost function so as to achieve the convergence of the state vector to the designated reference values, whereas the disturbance inputs try to maximize this cost function. For such problems, the monograph comes to propose a novel H-infinity (optimal) control method. At each time-step of the optimization algorithm, approximate linearization takes place around local operating points, with the use of Taylor series expansion and through the computation of Jacobian matrices. The linearization error is considered to be an additional perturbation affecting the system. Next, for the linearized equivalent model of the system, an optimal H-infinity controller can be applied, while to compute this controller's gains an algebraic Riccati equation has to be iteratively solved at each time-step of the control algorithm. This control scheme is also shown to be sufficiently robust, thus assuring the compensation of modelling errors, parametric uncertainty or external disturbances that affect the control loop. With the use of Lyapunov analysis, the global stability properties of the control loop are proven and the convergence of the system's state vector to the designated reference setpoints is demonstrated. The new solutions for the nonlinear optimal (H-infinity) control problem are computationally efficient since they require the solution of only one Riccati equation.

With reference to approach (ii) that is control based on global linearization approaches, the monograph elaborates on transformations of the initial nonlinear dynamics of the controlled systems into equivalent linear state-space descriptions where finally the design of feedback controllers is performed and the solution of the related stochastic estimation (filtering) problems is also accomplished. In this approach belong the monograph's results on differential flatness theory-based control, which rely on the transformation of the state-space description of the system into the canonical (Brunovsky) form. Through the global linearization-based control approach one can avoid the modelling errors that follow approximate linearization methods and consequently control of high precision and robustness can be achieved. Flatness-based control relies on differential flatness theory and consists of state-variables transformations that finally bring the system's state-space model into an equivalent linear form where the application of the standard linear control and stabilization methods is enabled. The flatness-based method comprises also inverse transformations that allow to compute estimates of the state variables of the initial nonlinear system. In this area, one can also classify the monograph's results on a new nonlinear estimation method which is known as derivative-free nonlinear Kalman Filter and which contributes towards solving the filtering problem for nonlinear dynamical systems in an optimal manner.

In the above-noted area (ii) of global linearization-based control the monograph analyzes also flatness-based adaptive fuzzy control for a wide class of nonlinear dynamical systems. In the flatness-based adaptive fuzzy control approach, one performs first an initial transformation (diffeomorphism) of the system's state-space model into an equivalent linear form. In this new description, the transformed control inputs contain unknown nonlinear functions which can be identified with the use of nonlinear regressors (e.g. neurofuzzy networks, wavelet networks or other networks that comprise nonlinear kernel functions). Learning in such networks takes place with the use of gradient algorithms where the learning rate is regulated through conditions for the minimization of the system's Lyapunov function and for assuring that this cumulative energy function of the system will have always a negative first-order derivative and will be persistently decreasing. At each time-step of the adaptive control algorithm, the estimated values for the nonlinear functions that constitute the system's dynamics are used to compute the feedback control inputs. It is proven that this approach achieves the minimization of the system's Lyapunov function and that the control loop becomes globally asymptotically stable.

With reference to approach (iii), that is control for dynamical systems which are described by nonlinear partial differential equations (PDEs) the monograph advances towards boundary control of the dynamics of the partial differential equations. Actually, in the proposed PDE control methods the control inputs are related only to the boundary conditions of the PDE, thus one arrives at a PDE boundary control problem. The methods are based on semi-discretization of the PDE (only about its spatial dimension) and this allows to substitute the PDE with an equivalent set of ordinary differential equations (ODEs). The solution of the stabilization problem for the PDE with the use of control inputs which are applied through the boundary conditions can be achieved by exploiting differential flatness theory. The design of a stabilizing controller for the PDE is based on the proof that (i) the state-space model of the PDE is a differentially flat system, (ii) each row of the state-space model is also a differentially flat subsystem. Next, for each subsystem (row) being associated with an ordinary differential equation, one can compute a virtual control input that can stabilize the subsystem's dynamics and which can also eliminate the tracking error of the subsystem's output. The virtual control input for the i-th subsystem becomes a reference setpoint for the (i+1)-th subsystem. From the last row of the state-space description one can compute the control input (boundary condition) which should be finally applied to the partial differential equation. This control input comprises in a recursive manner all virtual control inputs which were computed for the rows (subsystems) that constitute the PDE's state-space description. Thus, by tracing backwards the rows of the state-space model of the PDE (from the last to the first row) one can finally compute the control input that should be applied in the form of a boundary condition so as all individual state variables to converge to their designated setpoints. The global stability of this PDE control method is proven through Lyapunov analysis. Finally, in the above-noted research area (iii), that is dynamical systems described by PDEs, it is also of worth to include several results on estimation for fault diagnosis. About this topic, one can note results on fault detection and isolation methods which rely on Kalman Filtering and on the statistical properties of the χ^2 distribution that is followed by the state vector's estimation error (residuals).

Regarding the application part of the control and estimation methods for nonlinear and PDE dynamical systems, the monograph has examined the following:

(a) Control and estimation based on approximate linearization for robotic systems: (i) Nonlinear control of the cart and double-pendulum overhead crane, (ii) Nonlinear control of the underactuated offshore crane, (iii) Nonlinear control of the inertia wheel and pendulum system, (iv) Nonlinear control of the torsional oscillator with rotational actuator, (v) Nonlinear control of robotic exoskeletons, (vi) Nonlinear control of brachiation robots, (vii) Nonlinear control of power line inspection robots, (viii) Nonlinear control of robots with electrohydraulic actuators, (ix) Nonlinear control of robots with electropneumatic actuators, (x) Nonlinear control of flexible joint robots, (x) Nonlinear control of redundant robotic manipulators and (xi) Nonlinear control of parallel closed-chain robotic manipulators.

(b) Control and estimation based on approximate linearization for autonomous vehicles: (i) Nonlinear control of tracked autonomous vehicles, (ii) Nonlinear control of the autonomous fire-truck, (iii) Nonlinear control of the truck and N-trailer system, (iv) Nonlinear control of the ball-bot autonomous robot, (v) Nonlinear control of the ball-and-plate dynamical system, (vi) Nonlinear control of 3-DOF unmanned surface vessels, (vii) Nonlinear control of the 3-DOF autonomous underwater vessel, (viii) Nonlinear control of the Vertical Take-off and Landing Aircraft, (ix) Nonlinear control of aerial manipulators, (x) Nonlinear control of the 6-DOF autonomous octocopter and (xi) Nonlinear control of hypersonic aerial vehicles.

(c) Control and estimation based on approximate linearization for energy conversion systems: (i) Nonlinear control of the VSI-fed three-phase PMSM, (ii) Nonlinear control of the VSI-fed six-phase PMSM, (iii) Nonlinear control of the DC electric microgrids, (iv) Nonlinear control of distributed marine-turbine power generation units, (v) Nonlinear control of PMLSGs (permanent magnet linear synchronous generators) in wave energy conversion systems, (vi) Nonlinear control of Permanent Magnet Brushless DC motors, (vii) Nonlinear optimal control of Hybrid Electric Vehicles powertrains, (viii) Nonlinear control of shipboard AC/DC microgrids and (ix) Nonlinear control of power generation in hybrid AC/DC microgrids.

(d) Control and estimation based on approximate linearization for mechatronic systems: (i) Nonlinear control of electrohydraulic actuators, (ii) Nonlinear control of electropneumatic actuators, (iii) Nonlinear control of hot-steel rolling mills, (iv) Nonlinear control of paper mills, (v) Nonlinear control of the injection moulding machine, (v) Nonlinear control of the slosh-container system dynamics, (vi) Nonlinear control of micro-satellites' attitude dynamics and (vi) Nonlinear control of the industrial crystallization process.

(e) Control and estimation based on global linearization for industrial and PDE systems: (i) Control of a robotic exoskeleton subject to time-delays, (ii) Adaptive control of synchronous reluctance machines (iii) Control of a mobile robotic manipulator, (iv) SoC (state-of-charge) estimation in Electric Vehicles with a Kalman Filter-based disturbance observer, (iv) Control of partial differential equations, and in particular, (v) Control of nonlinear wave PDE dynamics, (vi) Control of a data flow PDE for bandwidth allocation in internet routes, (vii) Control of a diffusion PDE describing data flow in communication networks, (viii) Control of the diffusion PDE in Li-ion batteries, (ix) Control of a diffusion PDE in industrial assets' management, (x) Estimation of PDE dynamics of the highway traffic and (xi) Estimation of the PDE dynamics of a cable-suspended bridge and use of the obtained estimates for fault diagnosis.

Through the above-noted developments and the methods proposed for control and estimation of nonlinear and PDE dynamical systems this monograph has a useful contribution in the area of nonlinear dynamical systems and control theory. Although being computationally and algorithmically simple, the presented control schemes assure precise tracking of setpoints and stabilization for complicated nonlinear and PDE dynamical systems. Besides, in several cases they ensure optimal performance of the control loop, as for instance with the solution of the nonlinear optimal control or with the solution of the nonlinear optimal state estimation problem. The application field is wide and comprises primarily what is considered to be industrial systems technology, that is robotic systems (robotic manipulators and autonomous robotic vehicles), several mechatronic systems, electric power generation and power electronics, industrial production processes and cyber-physical systems that include the complex dynamics of communication and computer networks. These results are undoubtedly of interest for the engineering and academic community and can be used for teaching-related courses at the late undergraduate or post-graduate level. Certainly, research in the field of control and estimation for nonlinear and PDE dynamical systems has the potential for arriving at many more significant and exploitable findings in the years to come.

Gerasimos Rigatos	Masoud Abbaszadeh	Pierluigi Siano
Athens, Greece	Clifton Park, NY, USA	Fisciano, Salerno, Italy
May 2022	May 2022	May 2022

Acknowledgement

The authors of the monograph appreciate the review remarks of colleagues working in the research area of control of nonlinear and PDE dynamical systems. These comments have enabled us to arrive at a concise, meaningful and clear presentation of the analyzed control and estimation methods for nonlinear and PDE dynamical systems and have contributed to developing a research document that will be useful for the engineering and academic community.

Chapter 1

Principles of non-linear control

Abstract

The chapter analyzes established concepts in the control of non-linear and PDE dynamical systems. The following main approaches for the control of complex non-linear dynamical systems are considered (i) control with methods of approximate (local) linearization being associated with the solution of the non-linear optimal control problem, (ii) control with methods of exact (global) linearization comprising also adaptive control methods and (iii) control of distributed parameter systems (systems which are described by partial differential equations) and stochastic estimation methods. In (i) one can distinguish a novel non-linear optimal (H-infinity) control method. In (ii) one can distinguish flatness-based control for non-linear dynamical systems with known model and adaptive flatness-based control in the case of systems with the unknown model. In (iii) one can distinguish a flatness-based boundary control approach for PDEs which is implemented in successive loops.

1.1 Control based on approximate linearization

1.1.1 Overview of the optimal control concept

1.1.1.1 Optimal control for continuous-time systems with Bellman's principle

The monograph presents a novel solution to the non-linear optimal control problem for dynamical systems. The method is of proven stability and remains computationally tractable. It retains the advantages of linear optimal control, that is fast and accurate tracking of reference setpoints under moderate variations of the control inputs. One can also point out the advantages of this non-linear optimal control method against Non-linear Model Predictive Control (NMPC) (in NMPC the stability properties of the control scheme remain unproven and the convergence of the iterative search for an optimum often depends on initialization and parameter values' selection). It is also noteworthy that the new non-linear optimal control method is applicable to a wider class of dynamical systems than approaches based on the solution of State Dependent Riccati Equations (SDRE) (the SDRE approaches can be applied only to dynamical systems which can be transformed to the Linear Parameter Varying – LPV – form). Besides, the non-linear optimal control method performs better than non-linear optimal control schemes which use an approximation of the solution of the Hamilton–Jacobi–Bellman equation by Galerkin series expansions (actually, the stability properties of the Galerkin series expansion-based optimal control approaches are still unproven).

Optimal control and its application exhibit a rapid development in the last years (Wei, Song *et al.*, 2016; Liu, Wei *et al.*, 2017; Löber, 2017; Borrelli *et al.*, 2017). However, non-linear optimal control problems remain difficult to handle and a conclusive solution about them has not been achieved so far (Grüne and Pannek, 2017; Aschepkov *et al.*, 2017; Berkovitz and Medhin, 2012; Sanchez and Ornelas-Tellez, 2017). In the area of optimal control, Bellman's optimality principle stands for one of the most efficient approaches (Crassidis and Junkins, 2012). The following continuous-time dynamical system is considered:

$$\dot{x} = f(x, u, t), \quad x(t_0) = x_0 \tag{1.1}$$

The objective is to select the control input $u(t),\ t \in [t_0, t_f]$ so as to achieve minimization of the following cost function

$$J(u) = \int_{t_0}^{t_f} L(x, u, t)dt \tag{1.2}$$

To this end, one defines

$$J^0(x, t) = \min_{u \in [t, t_f]} \int_t^{t_f} L(x, u, t)dt \tag{1.3}$$

where $J^0(x, t_0) = J^0(x_0)$. According to the principle of optimality, optimization can take place in two stages: (i) from t to $t + \Delta t$, (ii) from $t + \Delta t$ to t_f. Consequently, the previous cost function is decomposed as follows:

$$\begin{aligned} J^0(x, t) &= \min_{u \in [t, t_f]} \left\{ \int_t^{t+\Delta t} L(x, u, t)dt + \int_{t+\Delta t}^{t_f} L(x, u, t)dt \right\} \\ &= \min_{u \in [t, t_f]} \left\{ \int_t^{t+\Delta t} L(x, u, t)dt + J^0(x + \Delta x, t + \Delta t) \right\} \end{aligned} \tag{1.4}$$

After taking the Taylor series expansion of $J^O(x + \Delta x, t + \Delta t)$ one obtains

$$J^0(x + \Delta x, t + \Delta t) = J^0(x, t) + \frac{\partial J^0}{\partial x}\Delta x + \frac{\partial J^0}{\partial t}\Delta t + O(\Delta) \tag{1.5}$$

where $O(\Delta)$ stands for higher-order terms to be omitted from the Taylor series expansion. By substituting (1.5) in (1.4), one obtains

$$0 = \min_{u \in [t, t_f]} \left\{ L\Delta t + \frac{\partial J^0}{\partial x}\Delta x + \frac{\partial J^0}{\partial t}\Delta t + O(\Delta) \right\} \tag{1.6}$$

and after dividing by Δt and taking the limit $\Delta t \to 0$ one arrives at the relation

$$-\frac{\partial J^0}{\partial t} = \min_{u \in [t, t_f]} \mathrm{H} \tag{1.7}$$

$$H = L(x,u,t) + \frac{\partial J^0}{\partial x}\frac{dx}{dt} = L(x,u,t) + \frac{\partial J^0}{\partial x}f(x,u,t) \tag{1.8}$$

In the above relations, (1.7) is known as the Hamilton–Jacobi–Bellman equation, whereas (1.8) is the Hamiltonian function of the optimal control problem. By solving the Hamilton–Jacobi–Bellman equation one obtains both the control input $u(t)$ $t\in [t, t_f]$ that minimizes the objective function $J^0(x,t)$, as well as the minimized value of $J^0(x,t)$.

The solution of the Hamilton–Jacobi–Bellman equation is a non-trivial problem and in the case of non-linear dynamical systems, it may become a complicated or even non-tractable problem. If the system is a linear one, and the cost function comprises quadratic terms, then the solution of the Hamilton–Jacobi–Bellman equation is substituted by the solution of a Riccati differential equation.

1.1.1.2 Solution of the optimal control problem in closed form

The solution of the optimal control problem in closed form can be achieved in the case of linear continuous-time systems with a quadratic cost function. For a linear system of the form:

$$\dot{x}(t) = A(t)x(t) + B(t)u(t) \tag{1.9}$$

and for a cost function that comprises quadratic terms in the form (Crassidis and Junkins, 2012):

$$J = \int_{t_0}^{t_f} Ldt \tag{1.10}$$

with

$$L = \frac{1}{2}x^T Q(t)x + \frac{1}{2}u^T Ru \tag{1.11}$$

then the optimal value of the cost function is

$$J^0(x,t) = \frac{1}{2}x^T(t)P(t)x(t) \tag{1.12}$$

with $P(t)$ being a symmetric positive definite matrix. For such a system the Hamiltonian function is (Crassidis and Junkins, 2012)

$$H = \frac{1}{2}x^T Qx + \frac{1}{2}u^T Ru + \frac{1}{2}x^T P(Ax + Bu) + \frac{1}{2}(Ax + Bu)^T Px \tag{1.13}$$

By deriving the Hamiltonian function with respect to the control input u and by taking its extremum condition one gets

$$\frac{\partial H}{\partial u} = Ru + B^T Px = 0 \tag{1.14}$$

By solving the above equation with respect to $u(t)$ one finds the optimal control input that minimizes the quadratic cost function of the system. This is given by

$$u^0(t) = -R^1(t)B^T(t)P(t)x(t) \tag{1.15}$$

By substituting the optimal control input to the Hamiltonian of (1.13) one obtains

$$\begin{aligned} H^0 = & \min_{u\in[t,t_f]} H = \frac{1}{2}x^T Qx + \frac{1}{2}x^T PBR^{-1}B^T Px + \frac{1}{2}x^T A^T Px \\ & + \frac{1}{2}X^T PAx - x^T PBR^{-1}B^T Px \\ & = \frac{1}{2}x^T (Q + A^T P + PA - PBR^{-1}B^T P)x \end{aligned} \tag{1.16}$$

Next, by differentiating the cost function of (1.12) one obtains

$$\frac{\partial J^0(x,t)}{\partial t} = \frac{1}{2}x^T(t)\frac{dP(t)}{dt}x(t) \tag{1.17}$$

Next, by equating (1.16) to (1.17) one has the following formulation of the Hamilton–Jacobi–Bellman equation

$$\begin{aligned} -\frac{1}{2}x^T(t)\frac{dP(t)}{dt}x(t) = & \min_{u\in[t,t_f]} H = \frac{1}{2}x^T Qx + \frac{1}{2}x^T PBR^{-1}B^T Px + \frac{1}{2}x^T A^T Px \\ & + \frac{1}{2}X^T PAx - x^T PBR^{-1}B^T Px \\ & = \frac{1}{2}x^T (Q + A^T P + PA - PBR^{-1}B^T P)x \end{aligned} \tag{1.18}$$

Through intermediate computations, one arrives at the following differential Riccati equation (Crassidis and Junkins, 2012)

$$-\frac{dP(t)}{dt} = A^T P + PA + Q - PBR^{-1}B^T P \tag{1.19}$$

The boundary condition for the solution of this differential equation comes from the relation

$$J^0(x,t_f) = \frac{1}{2}x(t_f)P(t_f)x(t_f) = 0 \tag{1.20}$$

having also $P(t_f) = 0$. By integrating (1.19) in inverse time from t_f to t_0 one can find matrix $P(t) : t\in[t_0, t_f]$. Besides, using (1.15) one can find the optimal control input to be applied to the system. Moreover, using (1.12) one can find the minimum value of the system's cost function $J^0(x,t)$.

The above procedure for finding a solution to the optimal control problem of a linear dynamical system arrives at the design of a Linear Quadratic Regulator. If matrices A and B which appear in the state-space model of the system of (1.9) are time invariant and the weight matrices Q and R which appear in the quadratic cost function of (1.10) and (1.11) are time-invariant too, and the finite-time $t_f = \infty$ (infinite time horizon) then the solution of the differential Riccati equation $P(t)$ arrives at a steady-state value, that is $\frac{dP(t)}{dt} = 0$ for $t\to\infty$. In such a case, the differential Riccati equation of (1.19) becomes an algebraic Riccati equation of the form:

$$A^T P + PA + Q - PBR^{-1}B^T P = 0 \tag{1.21}$$

In such a case the optimal feedback control becomes

$$u = R^{-1}B^T P \tag{1.22}$$

where P now is the solution of the algebraic Riccati equation of (1.21). Thus, the optimized cost function becomes $J^0 = \frac{1}{2}x^T(0)Px(0)$.

1.1.1.3 Stability conditions for optimal control

The stability of a linear optimal control loop is related to the stabilizability properties of such a system. The linear system $\dot{x}(t) = Ax(t) + Bu(t)\ \ x\in R^n\ \ u\in R^m$ is stabilizable if the control input $u(t)$ can be chosen such that all closed-loop poles are stable. The stabilizability of a system is a more generic property than controllability. It actually means that some of the poles of the system can be placed at the left complex semi-plane through state feedback, while the rest of the system poles which cannot be relocated with feedback are inherently found in the left complex semi-plane. The conditions that assure global stability for the closed-loop of an optimal linear quadratic control scheme are given in the following theorem:

Theorem: *Assume C to be a $n\times n$ square root of matrix Q where Q is the weight matrix appearing in the quadratic cost function of* (1.10) *and* (1.11). *This signifies that $C^TC = Q$. Assume next, that the matrices pair (A,B) is stabilizable, and the matrices pair (A,C) is observable. Then the following hold*:

(i) there exists a unique symmetric positive-definite steady-state solution P of the differential Riccati equation given in (1.19), *which is not dependent on the final condition $P(t_f)$, (ii) the aforementioned matrix P is also the unique solution of the algebraic Riccati equation given in* (1.21) *and (iii) the closed-loop system is globally asymptotically stable.*

According to this theorem, the stability properties of an optimal linear quadratic control loop depend on the structure of the open-loop system, as defined by matrices A and B, and on the structure of the weight matrix Q which appears as design parameters in the quadratic cost function. The stabilizability of the pair (A,B) assures that through state-feedback all poles of the closed-loop system can be finally found in the left complex semi-plane. The observability of the pair $(A,\sqrt{Q})$ signifies that all state variables of the system appear finally in the cost function J which is subject to minimization. The poles of the closed-loop system, and consequently the transient characteristics of the optimal control loop are determined by the selection of matrices $Q\geq 0$ and $R > 0$.

1.1.2 Design of an H-infinity non-linear optimal controller

1.1.2.1 Equivalent linearized dynamics of the system

The proposed non-linear optimal control method is addressed to non-linear dynamical systems in the form:

$$\dot{x} = f(x) + g(x)u \tag{1.23}$$

where $x\in R^n$, $f(x)\in R^n$, $g(x)\in R^n$ and $u\in R^m$. It can be also applied to non-linear systems having an even more generic state-space description in the form $\dot{x} = \tilde{f}(x,u)$. After linearization of (1.23) around its current operating point, the system's dynamic model is also written as

$$\dot{x} = Ax + Bu + d_1 \tag{1.24}$$

Parameter d_1 stands for the linearization error in the system's dynamic model appearing in (1.24).

The linearization approach which has been followed for implementing the nonlinear optimal control scheme is based on Taylor series expansion and on the computation of Jacobian matrices for the state-space model of the system. The linearization takes place at each sampling instance around the time-varying operating point (x^*, u^*) where x^* is the present value of the system's state vector and u^* is the last sampled value of the control inputs vector. Consider for instance the following affine-in-the-input state-space model

$$\begin{aligned} \dot{x} &= f(x) + g(x)u \Rightarrow \\ \dot{x} &= [f(x^*) + \nabla_x f(x) \mid_{x^*} (x - x^*)] + [g(x^*) + \nabla_x g(x) \mid_{x^*} (x - x^*)]u^* \\ &\quad + g(x^*)u^* + g(x^*)(u - u^*) + \tilde{d}_1 \Rightarrow \\ \dot{x} &= [\nabla_x f(x) \mid_{x^*} + \nabla_x g(x) \mid_{x^*} u^*]x + g(x^*)u \\ &\quad - [\nabla_x f(x) \mid_{x^*} + \nabla_x g(x) \mid_{x^*} u^*]x^* + f(x^*) + g(x^*)u^* + \tilde{d}_1 \end{aligned} \tag{1.25}$$

where $\tilde{d}_1$ is the modelling error due to truncation of higher-order terms in the Taylor series expansion of $f(x)$ and $g(x)$. Next, by defining $A = [\nabla_x f(x) \mid_{x^*} + \nabla_x g(x) \mid_{x^*} u^*]$, $B = g(x^*)$ one obtains

$$\dot{x} = Ax + Bu - Ax^* + f(x^*) + g(x^*)u^* + \tilde{d}_1 \tag{1.26}$$

Moreover by denoting $d_1 = -Ax^* + f(x^*) + g(x^*)u^* + \tilde{d}_1$ about the cumulative modelling error term in the Taylor series expansion procedure one has

$$\dot{x} = Ax + Bu + d_1 \tag{1.27}$$

which is the approximately linearized model of the dynamics of the system of (1.24). The term $f(x^*) + g(x^*)u^*$ is the derivative of the state vector at (x^*, u^*) which is almost annihilated by $-Ax^*$. The Jacobian matrices A and B are indeed updated at each sampling instance by substituting the state vector elements by their values at the linearization point. As proven by Lyapunov stability analysis, the proposed optimal (H-infinity) controller achieves the compensation of the above-noted modelling error and the precise tracking of reference setpoints by the system's state variables.

The reference setpoints for the system's state vector are denoted by $\mathbf{x_d} = [x_1^d, \dots, x_n^d]$. Tracking of this trajectory is achieved after applying the control input u^*. At every time instant the control input u_d is assumed to differ from the control input u appearing in (1.24) by an amount equal to Δu, that is $u_d = u + \Delta u$

$$\dot{x}_d = Ax_d + Bu^* + d_2 \tag{1.28}$$

The dynamics of the controlled system described in (1.24) can be also written as

$$\dot{x} = Ax + Bu + Bu_d - Bu_d + d_1 \tag{1.29}$$

and by denoting $d_3 = -Bu_d + d_1$ as an aggregate disturbance term one obtains

$$\dot{x} = Ax + Bu + Bu_d + d_3 \tag{1.30}$$

By subtracting (1.28) from (1.30) one has

$$\dot{x} - \dot{x}_d = A(x - x_d) + Bu + d_3 - d_2 \tag{1.31}$$

By denoting the tracking error as $e = x - x_d$ and the aggregate disturbance term as $\tilde{d} = d_3 - d_2$, the tracking error dynamics becomes

$$\dot{e} = Ae + Bu + \tilde{d} \tag{1.32}$$

The above-linearized form of the system's model can be efficiently controlled after applying an H-infinity feedback control scheme.

1.1.2.2 The non-linear H-infinity control

Consider the case in which the non-linear model of the system is written in the form

$$\dot{x} = \tilde{f}(x, u) \;\; x \in R^n, \;\; u \in R^m \tag{1.33}$$

Linearization of the system is performed at each iteration of the control algorithm around its present operating point $(x^*, u^*) = (x(t), u(t - T_s))$, where T_s is the sampling period. The linearized equivalent model of the system is described by

$$\dot{x} = Ax + Bu + L\tilde{d} \;\; x \in R^n, \;\; u \in R^m, \;\; \tilde{d} \in R^q \tag{1.34}$$

where matrices A and B are obtained from the computation of the Jacobians

$$A = \begin{pmatrix} \frac{\partial \tilde{f}_1}{\partial x_1} & \frac{\partial \tilde{f}_1}{\partial x_2} & \cdots & \frac{\partial \tilde{f}_1}{\partial x_n} \\ \frac{\partial \tilde{f}_2}{\partial x_1} & \frac{\partial \tilde{f}_2}{\partial x_2} & \cdots & \frac{\partial \tilde{f}_2}{\partial x_n} \\ \cdots & \cdots & & \cdots \\ \frac{\partial \tilde{f}_n}{\partial x_1} & \frac{\partial \tilde{f}_n}{\partial x_2} & \cdots & \frac{\partial \tilde{f}_n}{\partial x_n} \end{pmatrix} |_{(x^*, u^*)} \quad B = \begin{pmatrix} \frac{\partial \tilde{f}_1}{\partial u_1} & \frac{\partial \tilde{f}_1}{\partial u_2} & \cdots & \frac{\partial \tilde{f}_1}{\partial u_m} \\ \frac{\partial \tilde{f}_2}{\partial u_1} & \frac{\partial \tilde{f}_2}{\partial u_2} & \cdots & \frac{\partial \tilde{f}_2}{\partial u_m} \\ \cdots & \cdots & & \cdots \\ \frac{\partial \tilde{f}_n}{\partial u_1} & \frac{\partial \tilde{f}_n}{\partial u_2} & \cdots & \frac{\partial \tilde{f}_n}{\partial u_m} \end{pmatrix} |_{(x^*, u^*)} \tag{1.35}$$

and vector $\tilde{d}$ denotes disturbance terms due to linearization errors. The problem of disturbance rejection for the linearized model that is described by

$$\begin{aligned} \dot{x} &= Ax + Bu + L\tilde{d} \\ y &= Cx \end{aligned} \tag{1.36}$$

where $x \in R^n$, $u \in R^m$, $\tilde{d} \in R^q$ and $y \in R^p$ cannot be treated efficiently if the classical LQR control scheme is applied. This is because of the existence of the perturbation term $\tilde{d}$. The disturbance term $\tilde{d}$ apart from modeling (parametric) uncertainty and external perturbation terms, can also represent noise terms of any distribution.

In the H_∞ control approach, a feedback control scheme is designed for trajectory tracking by the system's state vector and simultaneous disturbance rejection, considering that the disturbance affects the system in the worst possible manner. The disturbances' effects are incorporated in the following quadratic cost function:

$$J(t) = \frac{1}{2}\int_0^T [y^T(t)y(t) + ru^T(t)u(t) - \rho^2\tilde{d}^T(t)\tilde{d}(t)]dt, \quad r, \rho > 0 \tag{1.37}$$

The significance of the negative sign in the cost function's term that is associated with the perturbation variable $\tilde{d}(t)$ is that the disturbance tries to maximize the cost function $J(t)$ while the control signal $u(t)$ tries to minimize it. The physical meaning of the relation given above is that the control signal and the disturbances compete to each other within a min–max differential game. This problem of mini–max optimization can be written as

$$\min_u \max_{\tilde{d}} J(u, \tilde{d}) \tag{1.38}$$

The objective of the optimization procedure is to compute a control signal $u(t)$ which can compensate for the worst possible disturbance, that is externally imposed to the system. However, the solution to the min–max optimization problem is directly related to the value of parameter ρ. This means that there is an upper bound in the disturbances magnitude that can be annihilated by the control signal.

1.1.2.3 Computation of the feedback control gains

For the linearized system given by (1.36) the cost function of (1.37) is defined, where coefficient r determines the penalization of the control input and weight coefficient ρ determines the reward of the disturbances' effects.

It is assumed that (i) the energy that is transferred from the disturbances signal $\tilde{d}(t)$ is bounded, that is $\int_0^\infty \tilde{d}^T(t)\tilde{d}(t)dt < \infty$, (ii) matrices $[A, B]$ and $[A, L]$ are stabilizable, (iii) matrix $[A, C]$ is detectable. Then, the optimal feedback control law is given by

$$u(t) = -Kx(t) \tag{1.39}$$

with

$$K = \frac{1}{r}B^T P \tag{1.40}$$

where P is a positive definite symmetric matrix that is obtained from the solution of the Riccati equation

$$A^T P + PA + Q - P\left(\frac{2}{r}BB^T - \frac{1}{\rho^2}LL^T\right)P = 0 \tag{1.41}$$

where Q is also a positive semi-definite symmetric matrix. The worst-case disturbance is given by

$$\tilde{d}(t) = \frac{1}{\rho^2}L^T Px(t) \tag{1.42}$$

The computation of the worst-case disturbance that this controller can sustain comes from the superposition of Bellman's optimality principle when considering that the system is affected by two separate inputs (i) the control input u (ii) the cumulative disturbance input $\tilde{d}(t)$. Solving the optimal control problem for u that is for the minimum variation (optimal) control input that achieves the elimination of the state vector's tracking error gives $u = -\frac{1}{r}B^T Pe$. Equivalently, solving the optimal control problem for $\tilde{d}$, that is for the worst-case disturbance that the control loop can sustain gives $\tilde{d} = \frac{1}{\rho^2}L^T Pe$.

1.1.2.4 Riccati equation coefficients in H-infinity control robustness

Parameter ρ in (1.37) is an indication of the closed-loop system robustness. If the values of $\rho > 0$ are excessively decreased with respect to r, then the solution of the Riccati equation is no longer a positive definite matrix. Consequently, there is a lower bound $\rho_{\min}$ of ρ for which the H_∞ control problem has a solution. The acceptable values of ρ lie in the interval $[\rho_{\min}, \infty)$. If $\rho_{\min}$ is found and used in the design of the H_∞ controller, then the closed-loop system will have elevated robustness. Unlike this, if a value $\rho > \rho_{\min}$ is used, then an admissible stabilizing H_∞ controller will be derived but it will be a suboptimal one. The Hamiltonian matrix

$$H = \begin{pmatrix} A & -(\frac{1}{r}BB^T - \frac{1}{\rho^2}LL^T) \\ -Q & -A^T \end{pmatrix} \tag{1.43}$$

provides a criterion for the existence of a solution of the Riccati equation (1.41). A necessary condition for the solution of the algebraic Riccati equation to be a positive definite symmetric matrix is that H has no imaginary eigenvalues (Rigatos, 2011; Lublin and Athans, 1995).

1.1.2.5 Lyapunov stability analysis.

Through Lyapunov stability analysis, it will be shown that the proposed non-linear control scheme assures H_∞ tracking performance for a nonlinear system, and that in the case of bounded disturbance terms asymptotic convergence to the reference setpoints is achieved.

The tracking error dynamics for the non-linear system is written in the form

$$\dot{e} = Ae + Bu + L\tilde{d} \tag{1.44}$$

where in the case of a non-linear system with a state vector of dimension equal to n, one can define $L \in R^n$ to be the identity matrix. Variable $\tilde{d}$ denotes model uncertainties and external disturbances of the system's model. The following Lyapunov equation is considered

$$V = \frac{1}{2}e^T Pe \tag{1.45}$$

where $e = x - x_d$ is the tracking error. By differentiating with respect to time one obtains

$$\begin{aligned} \dot{V} &= \frac{1}{2}\dot{e}^T Pe + \frac{1}{2}e^T P\dot{e} \Rightarrow \\ \dot{V} &= \frac{1}{2}[Ae + Bu + L\tilde{d}]^T Pe + \frac{1}{2}e^T P[Ae + Bu + L\tilde{d}] \Rightarrow \end{aligned} \tag{1.46}$$

$$\dot{V} = \frac{1}{2}[e^T A^T + u^T B^T + \tilde{d}^T L^T]Pe \\ + \frac{1}{2}e^T P[Ae + Bu + L\tilde{d}] \Rightarrow \tag{1.47}$$

$$\dot{V} = \frac{1}{2}e^T A^T Pe + \frac{1}{2}u^T B^T Pe + \frac{1}{2}\tilde{d}^T L^T Pe \\ \frac{1}{2}e^T PAe + \frac{1}{2}e^T PBu + \frac{1}{2}e^T PL\tilde{d} \tag{1.48}$$

The previous equation is rewritten as

$$\dot{V} = \frac{1}{2}e^T(A^T P + PA)e + \left(\frac{1}{2}u^T B^T Pe + \frac{1}{2}e^T PBu\right) \\ + \left(\frac{1}{2}\tilde{d}^T L^T Pe + \frac{1}{2}e^T PL\tilde{d}\right) \tag{1.49}$$

Assumption: For given positive definite matrix Q and coefficients r and ρ there exists a positive definite matrix P, which is the solution of the following matrix equation

$$A^T P + PA = -Q + P\left(\frac{2}{r}BB^T - \frac{1}{\rho^2}LL^T\right)P \tag{1.50}$$

Moreover, the following feedback control law is applied to the system

$$u = -\frac{1}{r}B^T Pe \tag{1.51}$$

By substituting (1.50) and (1.51) one obtains

$$\dot{V} = \frac{1}{2}e^T\left[-Q + P\left(\frac{2}{r}BB^T - \frac{1}{\rho^2}LL^T\right)P\right]e \\ + e^T PB\left(-\frac{1}{r}B^T Pe\right) + e^T PL\tilde{d} \Rightarrow \tag{1.52}$$

$$\dot{V} = -\frac{1}{2}e^T Qe + \frac{1}{r}e^T PBB^T Pe - \frac{1}{2\rho^2}e^T PLL^T Pe \\ - \frac{1}{r}e^T PBB^T Pe + e^T PL\tilde{d} \tag{1.53}$$

which after intermediate operations gives

$$\dot{V} = -\frac{1}{2}e^T Qe - \frac{1}{2\rho^2}e^T PLL^T Pe + e^T PL\tilde{d} \tag{1.54}$$

or, equivalently

$$\dot{V} = -\frac{1}{2}e^T Qe - \frac{1}{2\rho^2}e^T PLL^T Pe \\ + \frac{1}{2}e^T PL\tilde{d} + \frac{1}{2}\tilde{d}^T L^T Pe \tag{1.55}$$

Lemma: *The following inequality holds*

$$\frac{1}{2}e^T PL\tilde{d} + \frac{1}{2}\tilde{d}L^T Pe - \frac{1}{2\rho^2}e^T PLL^T Pe \leq \frac{1}{2}\rho^2\tilde{d}^T\tilde{d} \tag{1.56}$$

Proof: The binomial $(\rho\alpha - \frac{1}{\rho}b)^2$ is considered. Expanding the left part of the above inequality one gets

$$\begin{aligned} &\rho^2 a^2 + \frac{1}{\rho^2}b^2 - 2ab \geq 0 \Rightarrow \frac{1}{2}\rho^2 a^2 + \frac{1}{2\rho^2}b^2 - ab \geq 0 \Rightarrow \\ &ab - \frac{1}{2\rho^2}b^2 \leq \frac{1}{2}\rho^2 a^2 \Rightarrow \frac{1}{2}ab + \frac{1}{2}ab - \frac{1}{2\rho^2}b^2 \leq \frac{1}{2}\rho^2 a^2 \end{aligned} \tag{1.57}$$

The following substitutions are carried out: $a = \tilde{d}$ and $b = e^T PL$ and the previous relation becomes

$$\frac{1}{2}\tilde{d}^T L^T Pe + \frac{1}{2}e^T PL\tilde{d} - \frac{1}{2\rho^2}e^T PLL^T Pe \leq \frac{1}{2}\rho^2 \tilde{d}^T \tilde{d} \tag{1.58}$$

Equations (1.58) is substituted in (1.55) and the inequality is enforced, thus giving

$$\dot{V} \leq -\frac{1}{2}e^T Qe + \frac{1}{2}\rho^2 \tilde{d}^T \tilde{d} \tag{1.59}$$

Equation (1.59) shows that the H_∞ tracking performance criterion is satisfied. The integration of $\dot{V}$ from 0 to T gives

$$\begin{aligned} &\int_0^T \dot{V}(t)dt \leq -\frac{1}{2}\int_0^T ||e||_Q^2 dt + \frac{1}{2}\rho^2 \int_0^T ||\tilde{d}||^2 dt \Rightarrow \\ &2V(T) + \int_0^T ||e||_Q^2 dt \leq 2V(0) + \rho^2 \int_0^T ||\tilde{d}||^2 dt \end{aligned} \tag{1.60}$$

Moreover, if there exists a positive constant $M_d > 0$ such that

$$\int_0^\infty ||\tilde{d}||^2 dt \leq M_d \tag{1.61}$$

then one gets

$$\int_0^\infty ||e||_Q^2 dt \leq 2V(0) + \rho^2 M_d \tag{1.62}$$

Thus, the integral $\int_0^\infty ||e||_Q^2 dt$ is bounded. Moreover, $V(T)$ is bounded and from the definition of the Lyapunov function V in (1.45) it becomes clear that $e(t)$ will be also bounded since $e(t) \in \Omega_e = \{e|e^T Pe \leq 2V(0) + \rho^2 M_d\}$. According to the above and with the use of Barbalat's Lemma one obtains $lim_{t\to\infty} e(t) = 0$.

Elaborating on the above, it can be noted that the proof of global asymptotic stability for the control loop of the non-linear system is based on (1.59) and on the application of Barbalat's Lemma. It uses the condition of (1.61) about the boundedness of the square of the aggregate disturbance and modelling error term $\tilde{d}$ that affects the model. However, the proof of global asymptotic stability is not restricted by this condition. By selecting the attenuation coefficient ρ to be sufficiently small and in particular to satisfy $\rho^2 < ||e||_Q^2/||\tilde{d}||^2$ one has that the first derivative of the Lyapunov function is upper bounded by 0. Therefore, for the ith time interval, it is proven that the Lyapunov function defined in Eq (1.45) is a decreasing one. This also assures that the Lyapunov function of the system defined in (1.45) will always have a negative first-order derivative.

1.1.2.6 Robust state estimation with the use of the H_∞ Kalman Filter

The control loop has to be implemented with the use of information provided by a small number of sensors and by processing only a small number of state variables. To reconstruct the missing information about the state vector of the non-linear system, it is proposed to use a filtering scheme and based on it to apply state estimation-based control (Rigatos, 2015; Gibbs, 2012; Simon, 2006). By denoting as $A(k), B(k), C(k)$ the discrete-time equivalents of matrices A, B, C which appear in Eq. (1.36), the recursion of the H_∞ Kalman Filter, for the model of the non-linear system, can be formulated in terms of a *Measurement update* and a *Time update* part

Measurement update:

$$\begin{aligned} D(k) &= [I - \theta W(k)P^-(k) + C^T(k)R(k)^{-1}C(k)P^-(k)]^{-1} \\ K(k) &= P^-(k)D(k)C^T(k)R(k)^{-1} \\ \hat{x}(k) &= \hat{x}^-(k) + K(k)[y(k) - C\hat{x}^-(k)] \end{aligned} \tag{1.63}$$

Time update:

$$\begin{aligned} \hat{x}^-(k+1) &= A(k)x(k) + B(k)u(k) \\ P^-(k+1) &= A(k)P^-(k)D(k)A^T(k) + Q(k) \end{aligned} \tag{1.64}$$

where it is assumed that parameter θ is sufficiently small to assure that the covariance matrix $P^-(k)^{-1} - \theta W(k) + C^T(k)R(k)^{-1}C(k)$ will be positive definite. When $\theta = 0$ the H_∞ Kalman Filter becomes equivalent to the standard Kalman Filter. One can measure only a part of the state vector of the non-linear system and can estimate through filtering the rest of the state vector elements.

1.1.3 Optimal state estimation with the H-infinity Kalman Filter

1.1.3.1 Outline of the filter

Throughout this monograph, the H-infinity Kalman Filter is often used as a robust optimal state estimator (Gibbs, 2012; Simon, 2006). This is a variant of the typical Kalman Filter. This filter is the result of an optimization procedure that maximizes the robustness of estimation against noise and against errors about the initial value of the system's state vector. The functioning and the performance of the filter is assessed as follows:

(i) Due to the non-linearity of the systems under control, state estimation should be normally performed with a non-linear filter or state-observer. This problem cannot be treated with the use of the linear Kalman Filter or with the use of linear state observers. However, the monograph's approach allows the use of the H-infinity Kalman Filter, although this is a linear filter too. This is achieved by applying approximate linearization around a time-varying operating point (equilibrium). The linearization relies on Taylor series expansion and on the computation of the associated Jacobian matrices.

(ii) Usually the problem of state estimation for non-linear systems is treated with non-linear filtering techniques. However, the related Extended Kalman Filtering methods are not particularly robust against linearization modelling errors and against measurement noise. Furthermore, Sigma Point Kalman Filtering methods are not of proven convergence and stability. Additionally, Particle Filtering techniques demand high computation power and their slow convergence shows that they are not the recommended approach for real-time state estimation purposes (Rigatos, 2016b; Rigatos, 2015). Compared to the aforementioned non-linear filtering schemes, the H-infinity Kalman Filtering approach exhibits elevated robustness against noise and is of proven convergence and stability.

(iii) To elaborate on the matrices which appear in the *Measurement update* part and in the *Time update part* of the H-infinity Kalman Filter, the following can be noted: Matrix $R(k)$ is the measurement noise covariance matrix, that is the covariance matrix of the measurement error vector of the system. Matrix $P^{-}(k)$ is the a priori state vector estimation error covariance matrix of the system, that is the covariance matrix of the state vector estimation error prior to receiving the updated measurement of the system's outputs. Matrix $W(k)$ is a weight matrix which defines the significance to be attributed by the H-infinity Kalman Filter to minimizing the state vector's estimation error, relatively to the effects that the noise affecting the system may have. Finally, matrix $D(k)$ stands for a modified a posteriori state vector estimation error covariance matrix, that is the covariance matrix of the state vector estimation error after receiving the updated measurement of the system's outputs. Conclusively, the H-infinity Kalman Filter retains the structure of the typical Kalman Filter, that is a recursion in discrete-time comprising a Time update part (computation of variables prior to receiving measurements) and a Measurement update part (computation of variables after measurements have been received). There is a modified a posteriori state vector estimation error covariance matrix, which in turn takes into account a weight matrix that defines the accuracy of the state estimation under the effects of elevated noise.

(iv) The stability of the H-infinity Kalman Filter and the convergence of the state estimation procedure that is performed by it relies on the detectability/observability properties of the non-linear systems treated in the monograph and not on initialization of the matrices that appear in the filter's recursion. The monograph's concept on performing non-linear state estimation with the use of the H-infinity Kalman filter is widely applicable (Rigatos, 2016b; Rigatos, 2015). The non-linear state-space model of the monitored system undergoes approximate linearization around a temporary operating point, which is recomputed at each time-step of the estimation algorithm. For the approximately linearized model, state estimation is performed with the use of the H-infinity Kalman Filter. The effects of measurement noise, as well as the modelling error which is due to truncation of higher-order terms in the Taylor series expansion is considered to be a perturbation that is asymptotically compensated by the robustness of the H-infinity Kalman Filter.

1.1.3.2 Implementation stages of the filter

The H-infinity Kalman Filter is often used in this monograph for the state estimation-based (sensorless) implementation of the proposed optimal control scheme. Filters designed to minimize a weighted norm of state errors are called H_∞ or minimax filters (Gibbs, 2012; Simon, 2006). The discrete-time H_∞ filter uses the same state-space model as the Kalman Filter, which has the form

$$\begin{aligned} x(k+1) &= A(k)x(k) + B(k)u(k) + w(k) \\ z(k) &= C(k)x(k) + v(k) \end{aligned} \tag{1.65}$$

$E[w(k)] = 0,\ E[w(k)w(k)^T] = Q(k)\delta ij,\ E[v(k)] = 0,\ E[v(k)v(k)^T] = R(k)\delta_{ij}$ and $E(w(k)v(k)^T) = 0$. The update of the state estimate is again given by

$$\hat{x}(k) = \hat{x}^-(k) + K(k)(z(k) - C(k)\hat{x}^-(k)) \tag{1.66}$$

and minimizes the trace of the covariance matrix of the state vector estimation error

$$J = \frac{1}{2}E\{\tilde{x}(k)^T \cdot \tilde{x}(k)\} = \frac{1}{2}tr(P^-(k)) \tag{1.67}$$

where $\tilde{x}^-(k) = x(k) - \hat{x}^-(k)$ and $P^-(k) = E[\tilde{x}^-(k)^T \cdot \tilde{x}^-(k)]$. The H_∞ filtering approach defines first a transformation $d(k) = L(k)x(k)$, where $L(k) \in R^{n\times n}$ is a full rank matrix. The use of the aforementioned transformation allows certain combinations of states to be given more weight than others. Next, defining the estimation error variable $\tilde{d}_1(i) = d(i) - \hat{d}(i)$, the cost function of the H_∞ filter is initially formulated as (Gibbs, 2012; Simon, 2006)

$$\begin{aligned} J(k) &= \sum_{i=0}^{k-1} \tilde{d}(i+1)^T S(i)\tilde{d}(i+1)/b \\ b &= \tilde{x}^-(0)^T P^-(0)^{-1}\tilde{x}^-(0) + \sum_{i=0}^{k-1} w^T(i+1)Q(i+1)^{-1}w(i+1) \\ &\quad + \sum_{i=0}^{k-1} v^T(i)R(i)^{-1}v(i) \end{aligned} \tag{1.68}$$

where S_i is a positive-definite symmetric weighting matrix. It can be observed that both matrices $S(k)$ and $L(k)$ appear in the cost function and thus affect the solution $\hat{x}^-(k+1)$ of the optimization problem. The objective is to find state vector estimates $\hat{x}^-(k)$ and $\hat{x}(k)$ that keep the cost function below a given value $1/\theta$ for worst-case conditions, i.e.

$$J(k) < \frac{1}{\theta} \tag{1.69}$$

By rewriting (1.68) and substituting (1.65) a modified cost function is obtained

$$\begin{aligned} J_a(k) &= -\frac{1}{\theta}\tilde{x}^-(0)^T P^-(0)\tilde{x}^-(0) + \sum_{i=0}^{k-1} \Gamma(i) \\ \Gamma(i) &= (x(i+1) - \hat{x}^-(i+1))^T W_i(x(i+1) - \hat{x}^-(i+1)) \\ &\quad - \frac{1}{\theta}(w^T(i+1)Q(i+1)^{-1}w(i+1) \\ &\quad + (y(i) - C(i)x^-(i))^T R(i)^{-1}(y(i) - C(i)x^-(i))) \end{aligned} \tag{1.70}$$

and

$$W(i) = L(i)^T S(i) L(i) \tag{1.71}$$

This cost function does not include the dynamic model of the system given in (1.65) and this cost is added by using a vector of Lagrange multipliers $\lambda(i+1)$. This gives

$$\begin{aligned} J(k) = & -\frac{1}{\theta}\tilde{x}^-(0)^T P^-(0)\tilde{x}^-(0) \\ & + \sum_{i=0}^{k-1}(\Gamma_i + 2\frac{\lambda(i+1)^T}{\theta})(A(i)\hat{x}(i) + B(i)u(i) \\ & + w(i) - x(i+1)) + \frac{2\lambda(0)^T}{\theta}x(0) - \frac{2\lambda(0)^T}{\theta}x(0) \end{aligned} \tag{1.72}$$

The cost function of the filter given in (1.72) can be used as the basis for the solution. It is aimed to find equations defining $\hat{x}^-(k+1)$, or equivalently a measurement weighting matrix (similar to the Kalman gain matrix), that minimize the cost for worst case assumptions about $x(0)$, $w(i)$ and $y(i)$. Thus, the optimization objective is formulated as

$$J^*(k) = \begin{matrix} \min \\ x_i \end{matrix} \quad \begin{matrix} \max \\ x(0), w(i), y(i) \end{matrix} J(k) \tag{1.73}$$

It is noted that the estimation algorithm uses knowledge of the output measurement $y(i)$ but receives no information about the initial conditions of the system $x(0)$ and the process noise $w(i)$. Under this assumption, the estimation should be able to compensate for worst-case values for the unknown parameters. This is a game-theoretic problem, that is solved in two steps (Gibbs, 2012; Simon, 2006).

In the first step of optimization, partial derivatives of $J(k)$ with respect to $x(0)$, $w(i)$ and $\lambda(i)$ are set to zero so as to maximize the cost function of (1.72), now being dependent only on the terms $\hat{x}^-(k+1)$ and $y(k)$ which are included in Γ_i. In the second step of optimization, the partial derivatives of $J(k)$ with respect to $\hat{x}^-(k+1)$ and $y(k)$ are set to zero, to obtain a condition for the filter's gain matrix that minimizes this cost function. From the optimization conditions $\partial J(k)/\partial x_0 = 0^T$, $\partial J(k)/\partial w(i) = 0^T$, $\partial J(k)/\partial \lambda(i) = 0^T$ ones obtains an expression of $J(k)$ as function of $\hat{x}^-(k+1)$ and $y(k)$. Next, from the optimization conditions $\partial J(k)/\partial \hat{x}^-(i+1) = 0^T$, and $\partial J(k)/\partial y(i) = 0^T$ one obtains the filter's equations.

The recursion of the H-infinity Kalman Filter can be formulated again in terms of a *Measurement update* and a *Time update* part (Gibbs, 2012; Simon, 2006)

Measurement update:

$$\begin{aligned} D(k) &= [I - \theta W(k)P^-(k) + C^T(k)R(k)^{-1}C(k)P^-(k)]^{-1} \\ K(k) &= P^-(k)D(k)C^T(k)R(k)^{-1} \\ \hat{x}(k) &= \hat{x}^-(k) + K(k)[y(k) - C\hat{x}^-(k)] \end{aligned} \tag{1.74}$$

Time update:

$$\begin{aligned} \hat{x}^-(k+1) &= A(k)x(k) + B(k)u(k) \\ P^-(k+1) &= A(k)P^-(k)D(k)A^T(k) + Q(k) \end{aligned} \tag{1.75}$$

where it is assumed that parameter θ is sufficiently small to maintain

$$P^{-}(k)^{-1} - \theta W(k) + C^{T}(k)R(k)^{-1}C(k) \tag{1.76}$$

positive definite. When $\theta = 0$ the H_∞ Kalman Filter becomes equivalent to the standard Kalman Filter. It is noted that apart from the process noise covariance matrix $Q(k)$ and the measurement noise covariance matrix $R(k)$ the H_∞ Kalman filter requires tuning of the weight matrices L and S, as well as of parameter θ.

1.2 Global linearization-based control concepts

1.2.1 Foundations of global linearization-based control

1.2.1.1 The differential geometric approach

In the area of global linearization-based control, one distinguishes primarily differential geometric approaches (Lie algebra-based control) and differential flatness theory approaches. Before proceeding to the analysis of the differential geometric approach for the control of non-linear systems, the following definitions are given (Isidori, 1995; Khalil, 1996):

Vector Field: A mapping $f : D \rightarrow R^n$ where $D \subset R^n$ is a domain, is said to be a vector field on D. A vector field is a n-dimensional column.

Covector Field: The transpose of a vector field is said to be a co-vector field. A covector field is a n-dimensional row.

Inner Product: If f and w are respectively a vector field and a covector field on D then, the inner product $< w,f >$ is defined as

$$< w,f >= w(x)f(x) = \sum_{i=1}^{n} w_i(x)f_i(x) \tag{1.77}$$

Differential: Let $h : D \rightarrow R$. The differential of h is a covector field defined as

$$dh = \frac{\partial h}{\partial x} = \left[\frac{\partial h}{\partial x_1}, \frac{\partial h}{\partial x_1}, \cdots, \frac{\partial h}{\partial x_n}\right] \tag{1.78}$$

Lie Derivatives: Let $h : D \rightarrow R$ and $f : D \rightarrow R^n$. The Lie derivative of h with respect to f or along f is denoted as $L_f h$ and is defined as

$$L_f h = \frac{\partial h}{\partial x} f(x) \tag{1.79}$$

This is the meaning of computing the derivatives of h along the trajectories of the system $\dot{x} = f(x)$. The previous notation can be generalized toward the computation of

Lie derivatives of higher order with respect to the same vector field, or with respect to a new vector field. Thus one has

$$L_g L_f h(x) = \frac{\partial L_f h}{\partial x} g(x) \tag{1.80}$$

while it also holds

$$\begin{aligned} &L_f^0 h(x) = h(x) \\ &L_f h(x) = \frac{\partial h_1}{\partial x_1} f_1 + \frac{\partial h_2}{\partial x_2} f_2 + \cdots + \frac{\partial h_n}{\partial x_n} f_n \\ &\cdots \\ &L_f^k h(x) = L_f L_f^{k-1} = \frac{\partial L_f^{k-1}}{\partial x} f(x) \end{aligned} \tag{1.81}$$

Using the Lie derivative notation one can arrive at the definition of the relative degree for non-linear dynamical systems. Consider for example the following single-input single-output non-linear system

$$\begin{aligned} &\dot{x} = f(x) + g(x)u \\ &y = h(x) \end{aligned} \tag{1.82}$$

has relative degree r in a domain D_0 if

$$L_g L_f^i h(x) = 0 \text{ for } 0 \leq i \leq r-2 \text{ and } L_g L_f^{r-1} h(x) \neq 0 \; \forall \, x \in D_0 \tag{1.83}$$

Lie Bracket: Let f and g be two vector fields on $D \subset R^n$. The Lie Bracket of f and g is written as $[f, g]$ and is a third vector field which is defined as

$$[f, g](x) = \frac{\partial g}{\partial x} f - \frac{\partial f}{\partial x} g \tag{1.84}$$

where $\frac{\partial g}{\partial x}$ and $\frac{\partial f}{\partial x}$ are Jacobian matrices. The definition of the Lie Brackets can be repetitively used as follows

$$\begin{aligned} &ad_f^0 g(x) = g(x) \\ &ad_f g(x) = [f, g](x) \\ &\cdots \\ &ad_f^k g(x) = [f, ad_f^{k-1} g](x) \end{aligned} \tag{1.85}$$

For Lie Brackets the following properties hold

Bilinear: Let f_1, f_2 and g_1, g_2 be vector fields and r_1, r_2 be real numbers. Then,

$$\begin{aligned} &[r_1 f_1 + r_2 f_2, g_1] = r_1 [f_1, g_1] + r_2 [f_2, g_2] \\ &[f_1, r_1 g_1 + r_2 g_2] = r_1 [f_1, g_1] + r_2 [f_1, g_2] \end{aligned} \tag{1.86}$$

Skew commutative: $[f, g] = -[g, f]$

Jacobi identity: If f and g are vector fields and h is a real-valued function, then

$$L_{[f,g]} h(x) = L_f L_g h(x) = L_g L_f h(x) \tag{1.87}$$

Diffeomorphism: A mapping $T : D \rightarrow R^n$ is a diffeomorphism on D if (i) it is invertible on D, that is there exists a function $T^{-1}(x)$ such that $T^{-1}(T(x)) = x$ for all $x \in D$ and (ii) both $T(x)$ and $T^{-1}(x)$ are continuously differentiable. If the Jacobian matrix $[\frac{\partial T}{\partial x}]$ is non-singular at a point $x_0 \in D$, then from the inverse function theorem it is concluded that there is a neighborhood N of x_0, such that T restricted to N is a diffeomorphism on N, T is said to be a global diffeomorphism if it is a diffeomorphism on R^n and $T(R^n) = R^n$. T is a global diffeomorphism if and only if

(i) $[\partial T/\partial x]$ is non-singular for all $x \in R^n$
(ii) T is proper, that is $lim_{||x|| \rightarrow \infty} ||T(x)|| = \infty$

Distribution: Let $f_1, f_2, \ldots, f_k$ be vector fields on $D \subset R^n$. At any fixed point $x \in D$, $f_1(x), f_2(x), \ldots, f_k(x)$ are vectors in R^n and

$$\Delta(x) = span\{f_1(x), f_2(x), \ldots, f_n(x)\} \tag{1.88}$$

is a subspace on R^n. To each point $x \in R^n$ a subspace $\Delta(x)$ is assigned. One can refer to this assignment using

$$\Delta = span(f_1, f_2, \ldots, f_n) \tag{1.89}$$

and this is called a distribution.This implies that, Δ is the collection of all vector spaces $\Delta(x)$ for $xinD$. The dimension of $\Delta(x)$ which is denoted as

$$dim(\Delta(x)) = rank[f_1(x), f_2(x), \ldots, f_k(x)] \tag{1.90}$$

may vary with x. However, when $\Delta = span\{f_1, f_2, \ldots, f_r\}$, where $\{f_1(x), \ldots, f_r(x)\}$ are linearly independent for all $x \in D$, then $dim(\Delta(x)) = r$ for all $x \in D$. In that case Δ is said to be a non-singular distribution on D, generated by $f_1, f_2, \ldots, f_r$. It follows from the smoothness of the vector fields that every $y \in \Delta$ can be expressed as

$$g(x) = \sum_{i=1}^{r} c_i(x) f_i(x) \tag{1.91}$$

where $c_i(x)$ are smooth functions defined on D.

Involutive distribution: A distribution Δ is involutive if

$$g_1 \in \Delta \text{ and } g_2 \in \Delta \Rightarrow [g_1, g_2] \in \Delta \tag{1.92}$$

If Δ is a non-singular distribution on Δ, generated by $f_1, f_2, \ldots, f_r$ then it can be verified that Δ is an involutive distribution if and only if

$$[f_i, f_j] \in \Delta \;\forall\; 1 \leq i, j \leq r \tag{1.93}$$

or equivalently

$$[f_i, f_j](x) = \sum_{k=1}^{r} \alpha_{ijk} f_k(x) \;\; \forall\; i, j \tag{1.94}$$

Equivalently, involutivity means that if one forms the Lie bracket of any pairs of vector fields from the set $[f_1, f_2, \ldots, f_m]$, the produced vector field can be expressed as a linear combination of the original set of vector fields. It is noted that checking if a set of vector fields $[f_1, \ldots, f_m]$ is involutive and is equivalent to checking if

$$rank(f_1(x), \ldots, f_m(x)) = rank(f_1(x), \ldots, f_m(x),\ [f_i, f_j](x)) \tag{1.95}$$

for all x and all i, j.

Codistribution: Codistributions are dual objects to distributions. They are defined using covector fields and possess properties in analogy to those of distributions. A particular codistribution is the one that annihilates a distribution Δ, which is denoted as $\Delta^{\perp}$, and which is defined as

$$\Delta^{\perp}(x) = \{w \in (R^n)^* | < w, v >= 0, \forall\ v \in \Delta(x)\} \tag{1.96}$$

where $(R^n)^*$ is the n-dimensional space of new vectors.
Complete integrability: Let Δ be a non-singular distribution on D, generated by $f_1, \ldots, f_n$. Then Δ is completely integrable if for each $x_0 \in D$, there exists a neighborhood N of x_0 and $n - r$ real-valued smooth functions $h_1(x), \ldots, h_{n-r}(x)$, such that $h_1(x), \ldots, h_{n-r}(x)$ satisfy the partial differential equations

$$\frac{\partial h_j}{\partial x} f_i(x) = 0,\ \forall\ 1 \leq i \leq r \text{ and } 1 \leq j \leq n - r \tag{1.97}$$

and the covector fields $dh_j(x)$ are linearly independent for all $x \in D$. Equivalently, one can write

$$\Delta^{\perp} = span\{dh_1, \ldots, dh_{n-r}\} \tag{1.98}$$

Finally a key result in differential geometry is *Frobenius' theorem* which states that: A non-singular distribution is completely integrable if it is involutive.

1.2.1.2 Elaboration on the Frobenius theorem

The Frobenius theorem is important for the formal treatment of feedback linearization of n-order non-linear systems. It provides a necessary and sufficient condition for the solvability of a special class of partial differential equations. Consider for example the set of first-order partial differential equations

$$\begin{aligned} &\frac{\partial h}{\partial x_1} f_1 + \frac{\partial h}{\partial x_2} f_2 + \frac{\partial h}{\partial x_3} f_3 = 0 \\ &\frac{\partial h}{\partial g_1} g_1 + \frac{\partial h}{\partial x_2} g_2 + \frac{\partial h}{\partial x_3} g_3 = 0 \end{aligned} \tag{1.99}$$

where $f_i(x_1, x_2, x_3)$ and $g_i(x_1, x_2, x_3)$, $i = 1, 2, 3$ are scalar functions of x_1, x_2, x_3 and $h(x_1, x_2, x_3)$ is an unknown function. If a solution $h(x_1, x_2, x_3)$ exists for the above set of partial differential equations, then it is said that the set of vector fields $[f, g]$ where $f = (f_1, f_2, f_3)^T$ and $g = (g_1, g_2, g_3)^T$ is completely integrable .

It has to be determined when this set of partial differential equations is solvable. The Frobenius theorem provides a relatively simple condition about this.

Equation (1.99) has a solution $h(x_1, x_2, x_3)$ if and only if there exist scalar functions $\alpha_1(x_1, x_2, x_3)$ and $\alpha_2(x_1, x_2, x_3)$, such that

$$[f, g] = \alpha_1 f + \alpha_2 g \tag{1.100}$$

which means that the Lie bracket of f and g can be expressed as a linear combination of f and g. This is the involutivity condition on the vector fields $[f, g]$. Thus, the Frobenius theorem states that the set of vector fields $[f, g]$ is completely integrable, if and only if, it is involutive.

Next, the Frobenius theorem is re-stated, according to the previous definitions of *complete integrability* and *involutivity*.

Frobenius theorem: Let $f_1, f_2, \ldots, f_m$ be a set of linearly independent vector fields. The set is completely integrable, if and only if it is involutive.

1.2.1.3 Input–output linearization

Input–output linearization is analyzed for non-linear dynamical systems of the form

$$\begin{aligned} \dot{x} &= f(x) + g(x)u \\ y &= h(x) \end{aligned} \tag{1.101}$$

where $f : D \to R^n$, $g : D \to R^n$ and $h : D \to R$ are smooth, $D \subset R^n$ is a domain. Moreover, y is the system's output. By input–output linearization, one means the generation of the linear differential relation between the output y and a new equivalent input v (Slotine, 1991). The basic approach of input–output linearization is to differentiate the output function y repeatedly until the input u appears.

The single-input single-output system of (1.101) is considered (Khalil, 1996)

$$\begin{aligned} \dot{x} &= f(x) + g(x)u \\ y &= h(x) \end{aligned} \tag{1.102}$$

The system has relative degree r on D, which means

$$L_g L_f^i h(x) = 0 \text{ for } 0 \leq i \leq r - 2, \text{ and } L_g L_f^{r-1} h(x) \neq 0, \ \forall x \in D \tag{1.103}$$

The following two lemmas are applied to the system (Khalil, 1996):

Lemma 1. *For all $x \in D$ and all integers k and j such that $k \geq 0$ and $0 \leq r - 1$ one has*

$$L_{ad_f^j g} L_f^k h(k) = \begin{cases} 0 \ \text{ if } 0 \leq j + k < r - 1 \\ (-1)^j L_g L_f^{r-1} h(x) \neq 0 \ \text{ if } j + k = r - 1 \end{cases} \tag{1.104}$$

Proof: The proof is based on induction. For $j = 0$ (1.104) holds because it becomes equivalent to the definition of the relative degree of the system of (1.102). Next, it is assumed that (1.104) holds for some j and it is proven for $j + 1$. Using the Jacobi identity one has

$$L_{[f, \beta]} \lambda(x) = L_f L_\beta \lambda(x) - L_\beta L_f \lambda(x) \tag{1.105}$$

for any real-valued function λ and any vector fields f and β. Taking $\lambda = L_f^k$, $\beta = ad_f^j g$ one obtains

$$L_{ad_f^{j+1}g}L_f^k h(k) = L_{[f,ad_f^i g]}L_f^k h(x) = L_f L_{ad_f^i g}L_f^k h(x) - L_{ad_j^i g}L_f^{k+1}h(x) \tag{1.106}$$

In the previous equation, the first term of the right hand side is equal to 0, that is $L_f L_{ad_f^i g}L_f^k h(x) = 0$ because $j+k+1 \leq r-1$ or $j+k < r-1$.

Moreover, using that (1.104) holds for j one has also

$$L_{ad_f^i g}L_f^{k+1}h(k) = \begin{cases} 0 \ \text{if}\ 0 \leq j+k+1 < r-1 \\ (-1)^j L_g L_f^{r-1}h(x) \neq 0 \ \ \text{if}\ j+k+1 = r-1 \end{cases} \tag{1.107}$$

Consequently, one obtains

$$L_{ad_f^{i+1}g}L_f^k h(k) = \begin{cases} 0 \ \text{if}\ 0 \leq j+k+1 < r-1 \\ (-1)^{j+1} L_g L_f^{r-1}h(x) \neq 0 \ \ \text{if}\ j+k+1 = r-1 \end{cases} \tag{1.108}$$

which completes the proof of the Lemma.

Lemma 2. *For all $x \in D$*

- *the row vectors $dh(x, d_{L_f})h(x), \ldots, dL_f^{r-1}(x)$ are linearly independent*
- *the column vectors $g(x), ad_f g(x), \ldots, ad_f^{r-1}g(x)$ are linearly independent*

Proof: One has

$$\begin{pmatrix} dh(x) \\ \cdots \\ dL_f^{r-1}h(x) \end{pmatrix} \left(g(x) \ \cdots \ ad_f^{r-1}g(x)\right)$$

$$= \begin{pmatrix} L_g h(x) & L_{ad_f g}h(x) & \cdots & \cdots & L_{ad_f^{r-1}g}h(x) \\ L_g L_f h(x) & \cdots & \cdots & L_{ad_f^{r-2}g}h(x) & * \\ \cdots & \cdots & \cdots & \cdots & \cdots \\ L_g L_f^{r-1}h(x) & * & \cdots & \cdots & * \end{pmatrix} \tag{1.109}$$

From the previous Lemma, the right-hand side matrix is written as

$$\begin{pmatrix} 0 & \cdots\cdots & 0 & \diamond \\ 0 & \cdots\cdots & \diamond & * \\ \cdots & \cdots & \cdots & \cdots \\ 0 & \diamond & \cdots\cdots & * \\ \diamond & * & \cdots\cdots & * \end{pmatrix} \tag{1.110}$$

where $\diamond$ denotes a non-zero elements. Thus, the matrix is non-singular. This provides a proof for the Lemma, because if any of the two matrices appearing in the left side of the equation had rank less than n, their product should be a singular matrix.

The previous lemma proves that $r \leq n$. Next, the main result of input–output linearization will be formulated: A single-input single-output system of relative degree

$r < n$ can be locally transformed into the canonical (normal) form. This is given in the following theorem:

Theorem: The system of (1.102) is considered and it is assumed that it has relative degree $r < n$ in D. Then, for every $x_0 \in D$, there exist a neighborhood N of x_0 and smooth functions $\phi_1(x), \ldots, \phi_{n-r}(x)$, such that

$$L_g\phi_i(x) = 0, \ \forall\, 1 \leq i \leq n-r, \ \forall\, x \in N \tag{1.111}$$

and the mapping

$$T(x) = \begin{pmatrix} \phi_1(x) \\ \cdots \\ \phi_{n-r}(x) \\ h(x) \\ \cdots \\ L_f^{r-1}h(x) \end{pmatrix} \tag{1.112}$$

restricted to N, is a diffeomorphism on N.

Proof: The distribution $\Delta = span\{g\}$ is non-singular, involutive and has dimension 1. By Frobenius' theorem Δ is completely integrable. Consequently, for every $x_0 \in D$ there exist a neighborhood N_1 of x_0 and $n-1$ smooth functions $\phi_1(x), \ldots, \phi_{n-1}(x)$ with linearly independent differentials such that

$$L_g\phi_i(x) = 0 \text{ for } 1 \leq i \leq n-1 \ \forall\, x \in N_1 \tag{1.113}$$

Equivalently

$$\Delta^{\perp} = span\{d\phi_1, \ldots, d\phi_{n-1}\} \tag{1.114}$$

on N_1. Since

$$L_gL_f^ih(x) = 0, \ \text{ for } 0 \leq i \leq r-2 \tag{1.115}$$

and $dh_1, \ldots, dL_f^{r-2}h(x)$ are linearly independent, one can use $h, \ldots, L_f^{r-2}h$ as part of these $n-1$ functions. In particular they can be denoted as $\phi_{n-r+1}, \ldots, \phi_{n-1}$. Since $L_gL_f^{r-1}h(x) \neq 0$, one has

$$\begin{aligned} &dL_f^{r-1}h(x_0) \notin \Delta^{\perp}(x_0) \Rightarrow rank\left[\frac{\partial T}{\partial x}(x_0)\right] = n \Rightarrow \\ &\frac{\partial T}{\partial x}(x_0) \text{ is non-singular} \end{aligned} \tag{1.116}$$

This shows that there is a neighborhood N_2 of x_0 such that $T(x)$, restricted to N_2, is a diffeomorphism on N_2. Taking $N = N_1 \bigcap N_2$ completes the theorem's proof.

1.2.2 Elaborating on input–output linearization

1.2.2.1 The case of a well-defined relative degree

First a region Ω_x of the state-space is considered. Using the notations of differential geometry, the process of repeated differentiation gives

$$\dot{y} = \nabla h(f + gu) = L_f h(x) + L_g h(x) u = v \tag{1.117}$$

If $L_g h(x) \neq 0$ for some $x = x_0$ in Ω_x, then by continuity this relation is verified in a finite neighborhood Ω of x_0. In Ω the control input that is finally exerted on the system is given by

$$u = \frac{1}{L_g h}(-L_f h + v) \tag{1.118}$$

results in a linear relation between u and v, which is $\dot{y} = v$.
If however, $L_g h(x) = 0$ for all $x \in \Omega_x$ one can attempt to differentiate $\dot{y}$ once more with respect to time, and this gives

$$\ddot{y} = L_f^2 h(x) + L_g L_f h(x) u \tag{1.119}$$

If $L_g L_f h(x)$ is again zero for all $x \in \Omega_x$, additional differentiations with respect to time can be attempted

$$y^{(i)} = L_f^i h(x) + L_g L_f^{i-1} h(x) u \tag{1.120}$$

until for some integer r one arrives at

$$L_g L_f^{i-1} h(x) \neq 0 \tag{1.121}$$

for some $x = x_0$ in Ω_x. If the above relation indeed holds then one can compute the stabilizing control input that is finally applied to the system

$$u = \frac{1}{L_g L_f^{i-1} h(x)}(-L_f^r h + v) \tag{1.122}$$

By applying the control input of (1.122) to the equivalent description of the system's dynamics one gets

$$y^{(r)} = L_f^r h(x) + L_g L_f^{r-1} h(x) u \tag{1.123}$$

one arrives at the input–output linearized description of the system

$$y^{(r)} = v \tag{1.124}$$

The number r of the differentiations required for the input u to appear is the relative-degree of the system. One has $r \leq n$, where n is the system's order. If $r = n$ then input–output linearization becomes equivalent to the input-state linearization of the system.

1.2.2.2 Input–output linearization and normal forms

When the relative degree r is defined and $r < n$, the non-linear system of (1.101) can be transformed to the normal form using the new state vector $[y, \dot{y}, \ldots, y^{(r-1)}]$. Let

$$\mu = [\mu_1, \mu_2, \ldots, \mu_r]^T = [y, \dot{y}, \ldots, y^{(r-1)}]^T \tag{1.125}$$

In a neighborhood Ω of a point x_0, the normal form of the system can be written as

$$\begin{pmatrix} \dot{\mu}_1 \\ \dot{\mu}_2 \\ \dot{\mu}_3 \\ \cdots \\ \cdots \\ \dot{\mu}_{r-1} \\ \dot{\mu}_r \end{pmatrix} = \begin{pmatrix} 0 & 1 & 0 & \cdots & \cdots & 0 & 0 \\ 0 & 0 & 1 & \cdots & \cdots & 0 & 0 \\ 0 & 0 & 0 & \cdots & \cdots & 0 & 0 \\ \cdots & \cdots & \cdots & \cdots & \cdots & \cdots & \cdots \\ \cdots & \cdots & \cdots & \cdots & \cdots & \cdots & \cdots \\ 0 & 0 & 0 & \cdots & \cdots & 0 & 1 \\ 0 & 0 & 0 & \cdots & \cdots & 0 & 0 \end{pmatrix} \begin{pmatrix} \mu_1 \\ \mu_2 \\ \mu_3 \\ \cdots \\ \cdots \\ \mu_{r-1} \\ \mu_r \end{pmatrix} + \begin{pmatrix} 0 \\ 0 \\ 0 \\ \cdots \\ \cdots \\ 0 \\ 1 \end{pmatrix} v \tag{1.126}$$

where $v = (\alpha(\mu, \psi) + \underline{(}\mu, \psi)u)$, with internal dynamics defined by

$$\dot{\psi} = w(\mu, \psi) \tag{1.127}$$

while the output function is defined by

$$y = \mu_1 \tag{1.128}$$

Variables μ_i and ψ_j are referred to as *normal coordinates* or *normal states* in Ω (or at x_0). It is noted that the internal dynamics of the system do not depend on the control input u.

To demonstrate that the non-linear system of (1.101) can indeed be transformed into the normal form of (1.126), it has to be shown not only that a suitable transformation exists. Equivalently, a diffeomorphism has to be defined

$$\phi(x) = [\mu_1, \ldots, \mu_r, \psi_1, \ldots, \psi_{n-r}]^T \tag{1.129}$$

such that (1.126) and (1.127) are verified. To show that a transformation ϕ is a diffeomorphism it suffices to show that its Jacobian is invertible, i.e. that the gradients $\nabla\mu_i$ and $\nabla\psi_j$ are linearly independent.

First it will be shown that the gradients $\nabla\mu_i$ are linearly independent which means that the components of μ can serve as state variables. This result states that the system's output and its first $r-1$ derivatives can serve as new state variables. Next, it will be shown that other $n-r$ vector fields ψ_j can be found to complete the new state vector, which means that the gradients $\nabla\mu_i$ and $\nabla\psi_j$ are also linearly independent.

The following Lemma holds about the linear independence of the gradients $\nabla\mu_i$ (Slotine, 1991):

Lemma: *If the relative degree of the system of Eq.* (1.101) *is r in the region* Ω, *then the gradients* $\nabla\mu_1, \nabla\mu_2, \cdots, \nabla\mu_r$ *are also linearly independent in* Ω.

Proof: It is first noted, that in terms of μ (1.121) can be written as

$$\begin{aligned} &\nabla\mu_i g = 0 \quad 1 \le i < r \\ &\nabla\mu_r g \neq 0 \end{aligned} \tag{1.130}$$

Next, the proof is provided for the case that $r = 3$, however it can be generalized to higher values of r. It is assumed that there exist smooth functions $\alpha_i(x)$, such that everywhere in Ω

$$a_1 \nabla \mu_1 + a_2 \nabla \mu_2 + a_3 \nabla \mu_3 = 0 \tag{1.131}$$

Multiplying (1.131) by g yields

$$[a_1 \nabla \mu_1 + a_2 \nabla \mu_2 + a_3 \nabla \mu_3] g = 0 \tag{1.132}$$

which from (1.131) implies that $\alpha_3 = 0$ (everywhere in Ω). Replacing $a_3 = 0$ in (1.132) gives

$$[a_1 \nabla \mu_1 + a_2 \nabla \mu_2] = 0 \tag{1.133}$$

Multiplying (1.133) by the Lie bracket $ad_f g$, and using Jacobi's identity, as well as the definition of the relative degree one has

$$\begin{aligned} 0 &= a_1 L_{ad_f g} \mu_1 + a_2 L_{ad_f g} \mu_2 \\ &= \alpha_1 [L_f L_g h - L_g L_f h] + \alpha_2 [L_f L_g - L_g L_f] L_f h \\ &\quad - a_2 L_g L_f^2 h \end{aligned} \tag{1.134}$$

which from the definition of the relative degree implies that $a_2 = 0$, everywhere in Ω. Replacing $a_2 = 0$ in (1.133) implies that

$$a_1 \nabla \mu_1 = 0 \tag{1.135}$$

Multiplying (1.135) by $ad_f^2 g$ and using the definition of the relative degree, as well as the following result from Jacobi's identity

$$\begin{aligned} L_{ad_f^2 g} h &= L_{ad_f ad_f g} h = L_f L_{ad_f g} h - L_{ad_f g} L_f h \\ &= L_f [L_f L_g h - L_g L_f h] - [L_f L_g - L_g L_f] L_f h \\ &= L_f^2 L_g h - 2 L_f L_g L_f h + L_g L_f^2 h \end{aligned} \tag{1.136}$$

one gets

$$0 = a_1 L_{ad_f^2 g} \mu_1 = a_1 [L_f^2 L_g h - 2 L_f L_g L_f h + L_g L_f^2 h] = a_1 L_g L_f^2 h \tag{1.137}$$

which again from the definition of the relative degree implies that $a_1 = 0$, everywhere in Ω. This finally confirms that the gradients $\nabla \mu_i$ are linearly independent. This completes the Lemma's proof $\diamond$.

Next, it is shown that there exist $n - r$ more functions ψ_j to complete the coordinates transformation. The proof about this is as follows:

It is needed to show that the gradient vectors $\nabla \psi_j$ can be found such that the $\nabla \mu_i$ and $\nabla \psi_j$ are all linearly independent. At x_0 (1.130) indicates that the first $r - 1$ vectors $\nabla \mu_i$, $(i = 1, 2, \ldots, r - 1)$ which have been shown to be linearly independent, are all within a hyperplane orthogonal to g. Since the dimension of such hyperplane is $n - 1$, one can find

$$n - r = (n - 1) - (r - 1) \tag{1.138}$$

vectors in that hyperplane that are linearly independent of the $\nabla\mu_i$ and linearly independent of each other. These vectors can be called $\nabla\psi_j$ $(j = 1, 2, \ldots, n - r)$ and by definition they verify

$$\nabla\psi_j g = 0 \quad 1 \leq j \leq n - r \tag{1.139}$$

Through Frobenius theorem, it can be also assured that there exist $n - 1$ linearly independent gradient functions ∇h_k $k = 1, \ldots, n - 1$ which satisfy $\nabla h_k g = 0$. Furthermore from (1.130) it is assured that μ_r is not in the hyperplane orthogonal to g. Thus the gradients $\nabla\mu_i$ $(i = 1, \ldots, r)$ and $\nabla\psi_j$ $(j = 1, \ldots, n - r)$ are all linearly independent. Thus the Jacobian matrix of the transformation given in (1.129) is invertible.

By continuity, the Jacobian remains invertible in a neighborhood Ω_1 of x_0. Redefining Ω as the intersection of Ω_1 and Ω, the transformation ϕ defines a diffeomorphism in Ω. Hence, in Ω this transformation is a state transformation that brings the non-linear system in the form of (1.126), with

$$\begin{aligned} \alpha(\mu, \psi) &= L_f^r h(x) = L_f^r h[\psi^{-1}(\mu, \psi)] \\ \beta(\mu, \psi) &= L_g L_f^{r-1} h(x) = L_g L_f^{r-1} h[\psi^{-1}(\mu, \psi)] \end{aligned} \tag{1.140}$$

while in (1.127) the input u does not appear, since the ψ_j verify

$$L_g \psi_j(x) = 0 \; \forall x \in \Omega \tag{1.141}$$

From a practical point of view, to find the vector field ψ that completes the transformation into a normal form requires the non-trivial step of solving the set of partial differential equations are given in (1.139).

1.2.3 Input-state linearization

A different approach for obtaining a linearized equivalent description of a non-linear dynamical system is input-state linearization.

Definition: A non-linear system

$$\dot{x} = f(x) + g(x)u \tag{1.142}$$

where $f : D_r \rightarrow R^n$ and $G : D_r \rightarrow R^{n \times p}$ are sufficiently smooth on a domain $D_x \subset R^n$, is said to be input-state linearizable if there exists a diffeomorphism $T : D_x \Rightarrow R^n$ such that $D_x = T(D_x)$ contains the origin, and the change of variables $z = T(x)$ transforms the system of (1.142) into the form

$$\dot{z} = Az + B\beta^{-1}(x)[u - \alpha(x)] \tag{1.143}$$

The following, single-input system is considered

$$\dot{x} = f(x) + g(x)u \tag{1.144}$$

where $f : D \rightarrow R^n$ and $g : D \rightarrow R^n$ are smooth functions on a domain $D \subset R^n$. The following Theorem gives necessary and sufficient conditions for the system of (1.144) to be input-to-state linearizable near a point $x_0 \in D$ (Khalil, 1996).

Theorem: The system of (1.144) is input-state linearizable if and only if there exists a domain $D_x \subset D$ such that

1. the matrix $G(x) = [g(x), ad_f g(x), \ldots, ad_f^{n-1} g(x)]$ has rank n for all $x \in D_x$
2. the distribution $D = span\{g, ad_f g, \ldots, ad_f^{n-2} g\}$ is involutive in D_x

Proof: The system of (1.144) is input-to-state linearizable if and only if there is a real-valued function $h(x)$ such that the system

$$\begin{aligned} \dot{x} &= f(x) + g(x)u \\ y &= h(x) \end{aligned} \tag{1.145}$$

has relative degree n in D_x. That is $h(x)$ satisfies

$$L_g L_f^i h(x) = 0 \text{ for } 0 \le i \le n-2 \text{ and } L_g L_f^{n-1} h(x) \neq 0 \; \forall x \in D_x \tag{1.146}$$

The "only if" part follows from the definition of input-to-state linearization while the "if part" follows from applying the change of variables which is given by $z = T(x) = [h(x), L_f h(x), \ldots, L_f^{n-1} h(x)]^T$. The previously given Lemma 2 confirms that the Jacobian matrix $[\frac{\partial T}{\partial x}](x)$ is non-singular for all $x \in D_x$. Hence, for each $x_0 \in D_x$, there is a neighborhood N of x_0 such that $T(x)$ is restricted to N, is a diffeomorphism. For the theorem's proof it is also necessary to show that the existence of $h(x)$ satisfying (1.146) is equivalent to the previous stated conditions 1 and 2.

Necessity: Suppose there is $h(x)$ satisfying (1.146) shows that $rank(G) = n$. Then D is non-singular and has dimension $n-1$. From (1.104), with $k = 0$ and $r = n$ one has

$$L_g h(x) = L_{ad_f g} h(x) = \cdots = L_{ad_f^{(n-2)} g} h(x) = 0 \tag{1.147}$$

which can be also written as

$$dh(x)[g(x), ad_f g(x), \ldots, ad_f^{n-2} g(x)] = 0 \tag{1.148}$$

The above equation implies that $D^{\perp} = span(dh)$. Hence, D is completely integrable and it follows from Frobenius theorem that D is involutive.

Sufficiency: It is assumed that conditions 1 and 2 hold. Then D is non-singular and has dimension $n-1$. By Frobenius' theorem, there exists $h(x)$ satisfying

$$L_g h(x) = L_{ad_f g} h(x) = \cdots = L_{ad_f^{n-2} g} h(x) = 0 \tag{1.149}$$

Using the Jacobi identity, it can be verified that

$$L_g h(x) = L_g L_f h(x) = \cdots = L_g L_f^{n-2} h(x) = 0 \tag{1.150}$$

Furthermore,

$$\begin{aligned} dh(x)G(x) &= dh(x)[g(x), ad_f g(x), \ldots, ad_f^{n-1} g(x)] \\ &= [0, \ldots, 0, L_{ad_f^{n-1} g} h(x)] \end{aligned} \tag{1.151}$$

Since $rank(G) = n$ and $dh(x) \neq 0$, it must be true that $L_{ad_f^{n-1}g}h(x) \neq 0$. Using the Jacobi identity, it can be verified that $L_g L_f^{n-1} h(x) \neq 0$, which completes the proof of the theorem.

1.2.3.1 Relation between input-state and input–output linearization

The formal definition of input-to-state-linearization is as follows:

Using the transformed state-space description of the system of (1.143) and by defining the new control input $v = \beta^{-1}(x)[u - \alpha(x)]$, one obtains

$$\dot{z} = Az + Bv \tag{1.152}$$

where

$$A = \begin{pmatrix} 0 & 1 & 0 & \cdots & 0 \\ 0 & 0 & 1 & \cdots & 0 \\ \cdots & \cdots & \cdots & \cdots & \cdots \\ \cdots & \cdots & \cdots & \cdots & \cdots \\ 0 & 0 & 0 & \cdots & 1 \\ 0 & 0 & 0 & \cdots & 0 \end{pmatrix} \quad B = \begin{pmatrix} 0 \\ 0 \\ \cdots \\ \cdots \\ 0 \\ 1 \end{pmatrix} \tag{1.153}$$

The new state vector z is called linearizing state and the control input of (1.152) is the linearizing control law. The transformed system dynamics given in (1.152) and (1.153) is in the so-called linear canonical form, which is both controllable and observable.

The canonical form of (1.152) stands for a special case of the input–output linearization, where the output function leads to a relative degree n. This means that if a system is input–output linearizable with relative degree n, it is also input-state linearizable. On the other hand, if a system is input-state linearizable with a first new state z_1 representing the output, the system is also input–output linearizable with relative degree n.

Consequently, the relation between input-state linearization and input–output linearization can be summarized as follows:

Lemma: An nth order non-linear system is input-state linearizable, if and only if, there exists a scalar function $z_1(x)$ such that the system's input–output linearization with $z_1(x)$ as output has relative degree n.

However, the above lemma provides no guidance about how to find the desirable output function $z_1(x)$.

1.2.3.2 Conditions for input-state linearization

It should be specified which are the non-linear systems of the form of (1.142) or (1.273) can be subjected to input-state linearization. The above question stands for one of the most important issues in feedback linearization theory. The necessary and sufficient conditions for input–output linearization of a dynamical system have been stated in the theorem given in Section 1.2.3. Next, the proof of this theorem is revisited (Slotine, 1991).

Theorem: The non-linear system of (1.273) with $f(x)$ and $g(x)$ being smooth vector fields, is input-state linearizable, if and only if, there exists a region Ω such that the following conditions hold:

- the vector fields $[g, ad_f g, \ldots, ad_f^{n-1} g]$ are linearly independent in Ω
- the set $[g, ad_f g, \ldots, ad_f^{n-2} g]$ is involutive in Ω

Prior to proving the theorem some remarks about the aforementioned conditions are given:

The first condition can be interpreted as a controllability condition for the non-linear system of (1.145). For linear systems, the vector fields $[g, ad_f g, \ldots, ad_f^{n-1} g]$ become $[b, Ab, \ldots, A^{n-1} b]$ and independence implies that the associated matrix is a full-rank one which is also invertible. It can be also shown that if a system's linear approximations in a closed connected region $\Omega \in R$ are all controllable, then under some mild smoothness assumptions, the system can be driven from any point in Ω to any point in Ω (the inverse does not hold because a non-linear system can be controllable whereas its linear approximation may not be).

The involutivity condition is less intuitive. It is trivially satisfied for linear systems (which have constant vector fields) but does not hold in general in the non-linear case.

Next, to proceed to the Theorem's proof, the following Lemma is analyzed:

Lemma: *Let $z(x)$ be a smooth function in a region Ω. Then, in Ω, the set of equations*

$$L_g z = L_g L_f z = \cdots = L_g L_f^k z = 0 \tag{1.154}$$

is equivalent to

$$L_g z = L_g L_{ad_f} g z = \cdots = L_g L_{ad_f^k} g z = 0 \tag{1.155}$$

for any positive integer k.

Proof: The Lemma's proof is based on induction. When $k = 0$ the result is obvious. When $k = 1$, then from Jacobi's identity one obtains

$$L_{ad_f g} z = L_f L_g z - L_g L_f z = 0 - 0 = 0 \tag{1.156}$$

When $k = 2$, the using again Jacobi's identity then one obtains

$$L_{ad_f^2 g} z = L_f^2 L_g z - 2 L_f L_g L_f z + L_g L_f^2 z = 0 - 0 + 0 = 0 \tag{1.157}$$

Repeating the procedure, it can be shown by induction that (1.156) implies (1.157) for any k. Finally, using Jacobi's identity in an inverse manner it can be shown that from (1.157) one can arrive at (1.156).

Proof of Theorem: First the necessity condition is proven. It is assumed that there exists a state transformation $z = z(x)$ and an input transformation $u = \alpha(x) + \beta(x)v$, such that z and v satisfy (1.152). The first row of (1.152) can be rewritten as

$$\dot{z}_1 = \frac{\partial z_1}{\partial x}(f + gu) = z_2 \tag{1.158}$$

Continuing, in a similar manner with the rest of the elements of the state vector one has the following differential equations

$$\frac{\partial z_1}{\partial x}f + \frac{\partial z_1}{\partial x}gu = z_2 \tag{1.159}$$

$$\frac{\partial z_2}{\partial x}f + \frac{\partial z_2}{\partial x}gu = z_3 \tag{1.160}$$

$$\cdots\;\cdots\;\cdots$$

$$\frac{\partial z_n}{\partial x}f + \frac{\partial z_n}{\partial x}gu = v \tag{1.161}$$

Since $z_1, \ldots, z_n$ are independent of u, while v is not, from the above equations it can be concluded that

$$L_g z_1 = L_g z_2 = \cdots = L_g z_{n-1} = 0 \text{ and } L_g z_n \neq 0 \tag{1.162}$$

$$L_f z_i = z_{i+1} \quad i = 1, 2, \cdots, n-1 \tag{1.163}$$

The above equations on z_i can be compressed into one single equation concerning only z_1. Using the Lemma proven above and (1.162) one gets

$$\nabla z_1 ad_f^k g = 0 \quad k = 0, 1, 2, \ldots, n-2 \tag{1.164}$$

Moreover, by continuing as in the proof of the previous Lemma it can be shown that

$$\nabla z_1 ad_f^{n-1} g = (-1)^{n-1} L_g z_n \tag{1.165}$$

which implies that

$$\nabla z_1 ad_f^{n-1} g \neq 0 \tag{1.166}$$

From (1.164), (1.165) and (1.166) one can conclude that the state vector fields $[g, ad_f g, \ldots, ad_f^{n-1} g]$ are linearly independent. Indeed, if for some number $i (i \leq n-1)$ there existed scalar functions $\alpha_1(x), \ldots, \alpha_{i-1}(x)$, such that

$$ad_f^i g = \sum_{k=0}^{i-1} \alpha_k ad_f^k g \tag{1.167}$$

one would also have

$$ad_f^{n-1} g = \sum_{k=0}^{n-2} \alpha_k ad_f^k g \tag{1.168}$$

Equation (1.168) together with (1.165) would imply that

$$\nabla z_1 ad_f^{n-1} g = \sum_{k=0}^{n-2} \alpha_k ad_f^k g = 0 \tag{1.169}$$

which is in contradiction to (1.166).

From (1.164), (1.165) and (1.166) it is also possible to conclude that the vector fields are involutive. This follows from the existence of a scalar function z_1 satisfying the $n-1$ partial differential equations of (1.164) and from the necessity part of the Frobenius theorem. This completes the necessity part of the present Theorem proof.

Next, it will be proven that the two conditions in the present Theorem are also sufficient for the state linearizability of the system of (1.145), which means that it is possible to find a state transformation and an input transformation such that (1.152) is satisfied. The stages of the proof are as follows: Since the involutivity condition is satisfied, from the Frobenius theorem there exists a non-zero scalar function $z_1(x)$ satisfying

$$L_g z_1 = L_{ad_f g} z_1 = \cdots = L_{ad_f^{n-2} g} z_1 = 0 \tag{1.170}$$

According to the previous Lemma, the above equation can be written as

$$L_g z_1 = L_g L_f z_1 = \cdots = L_g L_f^{n-2} z_1 = 0 \tag{1.171}$$

Consequently, if one uses as new state variables the elements of the vector $z = [z_1, L_f z_1, \cdots, L_f^{n-1} z_1]^T$, the first $n-1$ state equations give

$$\dot{z}_k = z_{k+1} \quad k = 1, \ldots, n-1 \tag{1.172}$$

while the last equation becomes

$$\dot{z}_n = L_f^n z_1 + L_g L_f^{n-1} z_1 u \tag{1.173}$$

The next question that arises is if the gain function $L_g L_f^{n-1} z_1$ can be equal to zero. Since the vector fields $[g, ad_f g, \ldots, ad_f^{n-1} g]$ are linearly independent in Ω, and using (1.171), one obtains in accordance to the proof of the previous Lemma that

$$L_g L_f^{n-1} z_1 = (-1)^{n-1} L_{ad_f^{n-1} g} z_1 \tag{1.174}$$

and consequently, it should hold

$$L_{ad_f^{n-1} g} z_1(x) \neq 0 \quad \forall x \in \Omega \tag{1.175}$$

Otherwise, the non-zero vector ∇z_1 should satisfy

$$\nabla z_1 [g, ad_f g, \ldots, ad_f^{n-1} g] = 0 \tag{1.176}$$

which implies that ∇z_1 would be orthogonal to n linearly independent vectors, which is a contradiction. For the system's dynamics described in (1.173), and by denoting as new control input $v = L_f^n z_1 + L_g L_f^{n-1} z_1 u$, the stabilizing feedback control law that is finally applied to the system is

$$u = (L_g L_f^{n-1} z_1)^{-1} (-L_f^n z_1 + v) \tag{1.177}$$

1.2.4 Stages in the implementation of input-state linearization

The stages for implementing input-state linearization of dynamical systems of the form of (1.145) and for designing the associated feedback control law, are as follows:

(i) the vector fields $[g, ad_f g, \ldots, ad_f^{n-1} g]$ for the given system is constructed.
(ii) it is checked if the controllability and involutivity conditions are satisfied.

If both conditions are satisfied, then one should find the first state z_1 (which is the output function that leads to input–output linearization of relative degree n) from (1.170) and (1.171), that is

$$\begin{aligned} \nabla z_1 ad_f^i &= 0 \quad i = 0, 1, \ldots, n-2 \\ \nabla z_1 ad_f^{n-1} &= 0 \end{aligned} \tag{1.178}$$

Next, the state transformation $z(x) = [z_1, L_f z_1, \ldots, L_f^{n-1} z_1]^T$ and the input transformation $u = \alpha(x) + \beta(x)v$ are computed where

$$\begin{aligned} \alpha(x) &= -\frac{L_f^n z_1}{L_g L_f^{n-1} z_1} \\ \beta(x) &= \frac{1}{L_g L_f^{n-1} z_1} \end{aligned} \tag{1.179}$$

The transformed control input that makes the linearized system $z_1^{(n)} = v$ be a Hurwitz stable one, i.e. to have all its poles in the left complex semi-plane is given by

$$\begin{aligned} v = z_{1,d}^{(n)} &- K_1(z_1^{(n-1)} - z_{1,d}^{(n-1)}) - K_2(z_1^{(n-2)} - z_{1,d}^{(n-2)}) - \cdots \\ &- K_{n-1}(\dot{z}_1 - \dot{z}_{1,d}) - K_n(z_1 - z_{1,d}) \end{aligned} \tag{1.180}$$

1.2.5 Input–output and input-state linearization for MIMO systems

The previously analyzed approaches to input–output and input-state linearization which were formulated for SISO systems can be also extended to MIMO systems. Input-output linearization of MIMO systems is obtained similarly to the SISO case, by differentiating the selected outputs y_i until the inputs re-appear. It is assumed that r_i stands for the smallest order of output derivations, such that at least one of the inputs appear in $y_i^{(ri)}$, that is

$$y_i^{(ri)} = L_f^{r_i} h_i + \sum_{j=1}^{m} L_{g_j} L_f^{r_i - 1} h_i u_j \tag{1.181}$$

with $L_{g_j} L_f^{r_i - 1} h_i(x) \neq 0$, for at least one j, in a neighborhood Ω of point x_0. Following this procedure of successive derivations for all chosen outputs of the system, one obtains the input–output linearized description of the system's dynamics in matrix form

$$\begin{pmatrix} y_1^{r_i} \\ \cdots \\ \cdots \\ y_m^{r_m} \end{pmatrix} = \begin{pmatrix} L_f^{r_1} h_1(x) \\ \cdots \\ \cdots \\ L_f^{r_m} h_m(x) \end{pmatrix} + \begin{pmatrix} L_{g_1^1} L_f^{r_1-1} h_1(x) & \cdots & L_{g_1^m} L_f^{r_m-1} h_m(x) \\ \cdots & \cdots & \cdots \\ \cdots & \cdots & \cdots \\ L_{g_1^m} L_f^{r_1-1} h_1(x) & \cdots & L_{g_m^m} L_f^{r_m-1} h_m(x) \end{pmatrix} \begin{pmatrix} u_1 \\ \cdots \\ \cdots \\ u_m \end{pmatrix} \tag{1.182}$$

In the above description, the linearized dynamical system is a square one, that is the number of its outputs is equal to the number of control inputs. Therefore, through the followed procedure one succeeds at the same time linearization and decoupling of the system's dynamics. Thus, finally, the linearized system is written in the form

$$\begin{aligned}\tilde{Y} &= \tilde{F} + \tilde{G}u \Rightarrow \\ \tilde{Y} &= v\end{aligned} \tag{1.183}$$

and the feedback control law that makes the elements of the output vector $\tilde{Y}$ converge to the desirable setpoints is $u = \tilde{G}^{-1}(v - \tilde{F})$. The above system is said to have relative degree $(r_1, \ldots, r_m)$ and the scalar $r = r_1 + \cdots + r_m$ is the total relative degree.

1.2.6 Dynamic extension

In certain cases, and despite successive derivations of the system's outputs y_i, $i = 1, 2, \ldots, m$ the control inputs may not re-appear and thus matrix $\tilde{G}$ given in (1.183) may be singular. To handle this case, the so-called dynamic extension or dynamic feedback linearization is performed. This means that the system's state vector is extended by considering some of the control inputs and their derivatives as additional state vector elements. For the extended state-space description it becomes possible to perform input–output linearization and to obtain an equivalent linearized and decoupled form of it. This control approach has as a result to generate a control input that contains integral terms of the initial state vector of the system.
Without loss of generality, one can consider dynamic extension in the case of a system with two inputs and two outputs and that matrix $\tilde{G}$ given in (1.183) is of rank 1. Assuming that one column of matrix $\tilde{G}$ becomes equal to zero, we get a description of the system (after successive differentiation) in the form

$$\begin{pmatrix} y_1^{(r_1)} \\ y_2^{(r_2)} \end{pmatrix} = \begin{pmatrix} L_f^{r_1} h_1 \\ L_f^{r_2} h_2 \end{pmatrix} + \tilde{g}_1 u_1 \tag{1.184}$$

By considering the control input as additional state variable, and by differentiating (1.184) once more with respect to time while also using the initial system's dynamics, one arrives at the description

$$\begin{pmatrix} y_1^{(r_1+1)} \\ y_2^{(r_2+1)} \end{pmatrix} = \begin{pmatrix} b_1(x, u_1) \\ b_2(x, u_1) \end{pmatrix} + \begin{pmatrix} \tilde{e}_{1,1}(x, u_1) & \tilde{e}_{1,2}(x, u_1) \\ \tilde{e}_{2,1}(x, u_1) & \tilde{e}_{2,2}(x, u_2) \end{pmatrix} \begin{pmatrix} \dot{u}_1 \\ u_2 \end{pmatrix} \tag{1.185}$$

which is also written as

$$\begin{pmatrix} y_1^{(r_1+1)} \\ y_2^{(r_2+1)} \end{pmatrix} = \tilde{B}(x, u_1) + \tilde{E}(x, u_1) \begin{pmatrix} \dot{u}_1 \\ u_2 \end{pmatrix} = \tilde{v} \tag{1.186}$$

The control input $\tilde{v}$ is chosen to place the poles of the input–output linearized system to the left complex semiplane. The control input $[\dot{u}_1, u_2]^T$ is given by

$$\begin{pmatrix} \dot{u}_1 \\ u_2 \end{pmatrix} = \tilde{E}^{-1}(\tilde{v} - \tilde{B}(x, u_1)) \tag{1.187}$$

After computing $\dot{u}_1$ from the above relation, the control input that is finally applied to the dynamical system is given by $[\int \dot{u}_1, u_2]^T$. The term $\int \dot{u}_1$ implies that the integral of the initial state vector of the system is used for implementing the feedback control.

1.3 Global linearization-based control using differential flatness theory

1.3.1 The background of differential flatness theory

Differential flatness theory and flatness-based control were introduced in the late 1980 by Michel Fliess and co-researchers and since then they keep on being developed and on providing efficient solutions to advanced control and state estimation problems (Fliess *et al.*, 1999).

The definition of a differentially flat system is as follows: A system $\dot{x} = f(x, u)$ with state vector $x \in R^n$, input vector $u \in R^m$ where f is a smooth vector field, is differentially flat if there exists a vector $y \in R^m$ in the form

$$y = h(x, u, \dot{u}, \ldots, u^{(r)}) \tag{1.188}$$

such that

$$\begin{aligned} x &= \phi(y, \dot{y}, \ldots, y^{(q)}) \\ u &= \alpha(y, \dot{y}, \ldots, y^{(q)}) \end{aligned} \tag{1.189}$$

where h, ϕ and α are smooth functions. This means that the new system's description is given by the m algebraic variables y_i, $i = 1, 2, \ldots, m$. The definition of the flat output given above was $y = h(x, u, \dot{u}, \ldots, u^{(r)})$. If the flat output is exclusively a function of the state vector x then the system is a 0-flat one. However, there may be need to express the flat output as a function of not only the state vector x but also as a function of the control u and of its derivatives. For instance, the latter holds in the case of dynamic feedback linearization and in the application of the so-called dynamic extension. This means that the state vector of the system is extended by considering as additional state variables the control inputs and its derivatives.

Equation (1.189) shows that the state vector of the differentially flat system and its control inputs can be expressed as a function of the flat output and of the flat output's derivatives. The basic question that arises in the study of differential flatness is whether given the differential equations that describe the non-linear system dynamics $\dot{x} = f(x, u)$, there exists a function $y = h()$ given by $y = h(x, u, \dot{u}, \ldots, u^{(r)}$, such that the state vector of the system x and the control input u can be expressed as functions of y and of its derivatives, as in (1.189).

This problem was initially set by D. Hilbert in 1912, for the second-order Monge's equation

$$\frac{d^2y}{dx^2} = F\left(x, y, z, \frac{dy}{dx}, \frac{dz}{dx}\right) \tag{1.190}$$

which is an under-determined differential-equation system (there is one differential equation with respect to x, connecting two functions of x, namely $y(x)$ and $z(x)$).

Hilbert speaks about a solution that is not based on the computation of integrals. He has shown that this problem is related to the classification of under-determined differential-equation systems through a group of transformations called "invertible without integral".

Soon afterward, Èlie Cartan, reworks on the question set by Hilbert and shows how the calculations on the Pfaff systems, permit to classify second-order Monge equations that admit a solution without integral. He deduced an explicit description for all curves in the Euclidean space R^3 for which the curvature ratio and the torsion are constant. The results of Èlie Cartan are complete for Pfaff systems of codimension 2, which are under-determined differential-equation systems of co-dimension 1, i.e. systems in which only one arbitrary function intervenes (otherwize stated, systems having only one control input). Èlie Cartan also suggested the notion of absolute equivalence, however he did not define it with precision. He also noted that for Pfaff systems of codimension equal or greater than 2 (which means systems having at least two control inputs) the solution of the above problem becomes extremely complicated.

Next, a formal definition will be given about the analogy in terms of the control theory, of what was described above as under-determined differential-equation systems which can be solved without integration.

1.3.2 Differential flatness for finite-dimensional systems

As noted in (1.188) and (1.189) differential flatness is a structural property of a class of non-linear dynamical systems, denoting that all system variables (such as state vector elements and control inputs) can be written in terms of a set of specific variables (the so-called flat outputs) and their derivatives. The following non-linear system is considered:

$$\dot{x}(t) = f(x(t), u(t)) \tag{1.191}$$

The time variable is $t \in R$, the state vector is $x(t) \in R^n$ with initial conditions $x(0) = x_0$, and the input variable is $u(t) \in R^m$. Next, the main principles of differentially flat systems are given (Rudolph, 2003; Villagra *et al.*, 2007):

The finite-dimensional system of (1.191) is in the general form of an ordinary differential equation (ODE), i.e. $S_i(w, \dot{w}, \ddot{w}, \ldots, w^{(i)})$, $i = 1, 2, \ldots, q$. The entity w is a generic notation for the system variables (these variables are the elements of the system's state vector $x(t)$ and the elements of the control input $u(t)$) while $w^{(i)}$, $i = 1, 2, \ldots, q$ are the associated derivatives. Such a system is said to be differentially flat if there is a collection of m functions $y = (y_1, \ldots, y_m)$ of the system variables and of their time-derivatives, i.e. $y_i = \phi(w, \dot{w}, \ddot{w}, \ldots, w^{(\alpha_i)})$, $i = 1, \ldots, m$ satisfying the following two conditions (Fliess and Mounier, 1999; Martin and Rouchon, 1999; Mounier and Rudolph, 2001; Mounier, 2005; Rigatos, 2010):

1. There does not exist any differential relation of the form $R(y, \dot{y}, \ldots, y^{(\beta)}) = 0$ which implies that the derivatives of the flat output are not coupled in the sense of an ODE, or equivalently it can be said that the flat output is differentially independent.

2. All system variables (i.e. the elements of the system's state vector w and the control input) can be expressed using only the flat output y and its time derivatives $w_i = \psi_i(y, \dot{y}, \ldots, y^{(\gamma_i)}),\ i = 1, \ldots, s$. An equivalent definition of differentially flat systems is as follows:

Definition: The system $\dot{x} = f(x, u)$, $x \in R^n$, $u \in R^m$ is differentially flat if there exist relations

$$\begin{aligned} h &: R^n \times (R^m)^{r+1} \rightarrow R^m, \\ \phi &: (R^m)^r \rightarrow R^n \text{ and} \\ \psi &: (R^m)^{r+1} \rightarrow R^m \end{aligned} \tag{1.192}$$

such that

$$\begin{aligned} y &= h(x, u, \dot{u}, \ldots, u^{(r)}), \\ x &= \phi(y, \dot{y}, \ldots, y^{(r-1)}), \text{ and} \\ u &= \psi(y, \dot{y}, \ldots, y^{(r-1)}, y^{(r)}). \end{aligned} \tag{1.193}$$

This means that all system dynamics can be expressed as a function of the flat output and its derivatives, therefore the state vector and the control input can be written as

$$\begin{aligned} x(t) &= \phi(y(t), \dot{y}(t), \ldots, y^{(r-1)}(t)), \text{and} \\ u(t) &= \psi(y(t), \dot{y}(t), \ldots, y^{(r)}(t)) \end{aligned} \tag{1.194}$$

1.3.3 Equivalence and differential flatness

1.3.3.1 Representation of system dynamics as vector fields of infinite dimension

A basic property of differentially flat systems is that through a change of variables (diffeomorphism) they can be transformed to an equivalent description, which is the linear canonical (Brunovsky) form. This is analyzed next and stands for the Lie-Backlünd isomorphism to equivalence and differential flatness (Fliess *et al.*, 1999).

First a dynamical system of the form

$$\dot{x} = f(x)\ x \in X \subset R^n \tag{1.195}$$

is considered. This is described by the couple (X, f), where X is defined in R^n and f is a vector field on X. A *trajectory* is considered to be the function $t \rightarrow x(t)$ such that $\dot{x}(t) = f(x(t))\ \forall t \geq 0$. One can consider also the output mapping $x \rightarrow h(x)$ for which holds

$$\frac{d}{dt}h(x(t)) = \frac{dh}{dt}(x(t)) \cdot \dot{x}(t) = \frac{dh}{dt}(x(t)) \cdot f(x(t))\ \forall\ t \geq 0 \tag{1.196}$$

Equation (1.196) gives the total derivative, i.e. $\frac{dh}{dt}(x(t)) \cdot f(x(t))$ is also called *time derivative* of the function h, and is noted by $\dot{h}$.

Next, the notions of the total derivative and of the time derivative of function h can be generalized to the case of a non-linear system with control input

$$\dot{x} = f(x, u) \tag{1.197}$$

where $X\times U\in R^n \times R^m$. Here, f is no-longer a vector field but is an infinite collection of vector fields parameterized by u. Actually, for every u the function $x\rightarrow f_u(x) = f(x,u)$ is a vector field on X. One can also consider the case of dynamic feedback in which the system's state vector is extended by defining as state vector element the control input (and its derivatives), while the equations describing the system's dynamics are extended as follows:

$$\begin{aligned}\dot{x} &= f(x,u)\\ \dot{u} &= v\end{aligned} \tag{1.198}$$

In the latter case, in place of the state-space X one has the state-space $X\times U$. Assume now that the solution of $\dot{x} = f(x,u)$ is the function $t\rightarrow(x(t,u(t)))$ which has values in $X\times U$, such that $\dot{x}(t) = f(x(t),u(t))\ \forall\ t\geq 0$. Moreover, one can consider the infinite function

$$t\rightarrow\xi(t) = (x(t),u(t),\dot{u}(t),\ldots) \tag{1.199}$$

which takes values in $X\times U\times R_m^{\infty}$, where $R_m^{\infty} = R^m\times R^m\times\ldots$ represents an infinite product formed by vectors defined in R^m. A point of R_m^{∞} is of the form $u^{(1)},u^{(2)},\ldots$ with $u^i\in R^m$, using that $\dot{x} = f(x,u)$. This function satisfies also the relation

$$\dot{\xi}(t) = (f(x(t),u(t)),\dot{u}(t),\ddot{u}(t),\ldots)\ \forall\ t\geq 0 \tag{1.200}$$

and this can be interpreted as a trajectory of an infinite vectors field

$$(x,u,u^{(1)},\ldots)\rightarrow F(x,u,u^{(1)},\ldots) = (f(x,u),u^{(1)},u^{(2)},\ldots,\ldots) \tag{1.201}$$

According to the above, the dynamical system $\dot{x} = f(x,u)$ is defined by the space $X\times U\times R_m^{\infty}$ and an infinite number of vector fields F on that space. As in the case of an autonomous dynamical system $\dot{x} = f(x)$, for the non-autonomous dynamical system $\dot{x} = f(x,u)$ one can also define the time derivative of the smooth output function $(x,u,u^{(1)},\ldots)\rightarrow h(x,u,u^{(1)},\ldots,u^{(k)})$ which is written as

$$\begin{aligned}\dot{h}(x,u,u^{(1)},\ldots,u^{(k+1)}) &:= Dh\cdot F\\ &= \frac{\partial h}{\partial x}f(x,u) + \frac{\partial h}{\partial u}u^{(1)} + \frac{\partial h}{\partial u^1}u^{(2)} + \cdots\end{aligned} \tag{1.202}$$

If $h()$ depends on a finite number of variables then the above sum becomes finite too. For the above type of functions, one can define the so-called Fréchet topology which implies computations on smooth functions which are defined in k copies of R^m, for k being sufficiently large.

After a change of variables, a differentially flat system is written in the canonical Brunovsky form. Thus, it becomes equivalent to the trivial system R_m^{∞}, F_m with coordinates $y,y^{(1)},y^{(2)},\ldots,$ and the vector fields

$$(y^{(1)},y^{(2)},y^{(3)},\ldots) = F_m(y,y^{(1)},y^{(2)},\ldots) \tag{1.203}$$

which describes systems composed by m chains of integrators (this is the canonical Brunovsky form for linear controllable systems).

1.3.3.2 Equivalence of systems

Two dynamical systems are considered to be equivalent if there exists an invertible relationship that exchanges their trajectories. The definition of equivalent systems is as follows:

Definition: Two systems (M, F) and (N, G) (notation referring to state vector and vector field respectively) are equivalent in $(p, q) \in (M, N)$, if and only if, there exists an endogenous transformation from a neighborhood of p to a neighborhood of q. The two systems (M, F) and (N, G) are equivalent if the equivalence holds for all pairs of points (p, q) of the space $M \times N$.

Using coordinates, the previous notions are expressed as follows: Considering the two systems $(X \times U \times R_m^\infty, F)$ and $(Y \times V \times R_s^\infty, G)$ describing the initial system dynamics

$$\dot{x} = f(x, u) \text{ with } (x, u) \in X \times U \subset R^n \times R^m \tag{1.204}$$

and the equivalent system dynamics

$$\dot{y} = g(y, v) \text{ with } (y, v) \in Y \times V \subset R^r \times R^s \tag{1.205}$$

The vector fields F, G are defined as

$$\begin{aligned} F(x, u, u^{(1)}, \ldots) &= (f(x, u), u^{(1)}, u^{(2)}, \ldots) \\ G(y, v, v^{(1)}, \ldots) &= (g(y, v), v^{(1)}, v^{(2)}, \ldots) \end{aligned} \tag{1.206}$$

If the two systems are equivalent the endogenous transformation takes the form

$$\Psi(x, u, u^{(1)}, \ldots) = (\psi(x, \bar{u}), \beta(x, \bar{u}), \dot{\beta}(x, \bar{u}), \ldots) \tag{1.207}$$

where $\bar{u}$ is the abbreviated notation $\bar{u} = (u, u^{(1)}, \ldots, u^{(k)})$. Similarly, one can define the inverse endogenous transformation

$$\Phi(y, v, v^{(1)}, \ldots) = (\phi(y, \bar{v}), \alpha(y, \bar{v}), \dot{\alpha}(y, \bar{v}), \ldots) \tag{1.208}$$

Since Φ and Ψ are inverse to each other, one has

$$\begin{aligned} \psi(\phi(y, \bar{v}), \bar{\alpha}(y, \bar{v})) = y \quad &\text{and} \quad \phi(\psi(x, \bar{u}), \bar{\beta}(x, \bar{u})) = x \\ \beta(\phi(y, \bar{v}), \bar{\alpha}(y, \bar{v})) = v \quad &\text{and} \quad \alpha(\psi(x, \bar{u}), \bar{\beta}(x, \bar{u})) = u \end{aligned} \tag{1.209}$$

About vector fields F and G which are related to the derivatives of x and y, respectively one has

$$f(\phi(y, \bar{v}), \alpha(y, \bar{v})) = D\phi(y, \bar{v}) \cdot \bar{g}(y, \bar{v}) \tag{1.210}$$

where $\bar{g}$ corresponds to $(g, v^{(1)}, \ldots, v^{(k)})$

$$g(\psi(x, \bar{u}), \beta(x, \bar{u})) = D\psi(x, \bar{u}) \cdot \bar{f}(x, \bar{u}) \tag{1.211}$$

where $\bar{f}$ corresponds to $(f, u^{(1)}, \ldots, u^{(k)})$. In additional, it can be stated that if the trajectory of the first system is denoted as

$$t \to (x(t), u(t)) \tag{1.212}$$

then the trajectory of the second system becomes

$$t \rightarrow (y(t), v(t)) = (\psi(x(t), \bar{u}(t)), \beta(x(t), \bar{u}(t))) \tag{1.213}$$

An important property of the endogenous transformations is that the number of control inputs in the initial and the final description of the system remain the same.

Theorem: *If the two systems* $(X \times U \times R_m^\infty, F)$ *and* $Y \times V \times R_s^\infty, G$ *are equivalent, then they will have the same number of control inputs, that is* $m = s$.

Proof: Consider the truncation Φ_m of Φ on $X \times U \times (R^m)^\mu$

$$\begin{aligned} &\Phi_\mu : X \times U \times (R^{m+k})^\mu \rightarrow Y \times V \times (R^s)^\mu \\ &(x, u, u^{(1)}, \ldots, u^{(k+\mu)}) \rightarrow (\phi, \alpha, \dot{\alpha}, \ldots, \alpha^{(\mu)}) \end{aligned} \tag{1.214}$$

which means the first $\mu + 2$ blocks of the components of Ψ, while μ is a sufficiently large entity. Since Ψ is invertible, Ψ_{mu} is a submersion for all μ. Thus the dimension of the dimension of the space from which Ψ has emanated is superior or equal to the dimension of the space of arrival. This is written as

$$n + m(k + \mu + 1) \geq s(\mu + 1) \ \forall \mu > 0 \tag{1.215}$$

This implies $m \geq s$. In a similar manner, it is shown that with the mapping Ψ it holds $s \geq m$. Consequently, one finally has $m = s$ which means that the number of control inputs in the two descriptions remains the same.

This definition of the equivalence is adapted to the equivalence between two so-called *diffieties*. The concept of a diffiety is as follows: Variable (X, τ_x) comprising the state vector of the non-linear system and its successive derivatives with respect to time, as well as the Lie derivatives of the mapping transformation along a vector field defined by the state vector elements and their derivatives is called a *manifold of jets* or a *diffiety*.

Given two diffieties (M, CTM) and (N, CTN) it is said that a function Ψ from M to N is Lie-Backlünd or a C-morphism, if the tangent (projection) function $T\Psi$ satisfies $T\Phi(CTM) \subset CTN$. Moreover, if Ψ has an inverse function Φ, such that $T\Psi(CTN) \subset CTM$, then Ψ is a Lie-Backlünd isomorphism or a C isomorphism. When such an isomorphism exists, the diffieties are called equivalent. Therefore, an endogenous transformation is a special type of a Lie-Bakclünd transformation which preserves the parametrization in time of the integral curves. It is also possible to introduce a more general concept which is the orbital equivalence, about the Lie-Bakclünd isomorphisms which preserve only the geometric locus of the integral curves.

1.3.3.3 Differentially flat systems and equivalence to the trivial system

Definition: The dynamical system (M, F) where M: is the state vector and F: is a vector field, is differentially flat in p (or respectively flat), if and only if it is equivalent in p (respectively equivalent) with the trivial system (therefore it can be described in the Brunovsky canonical form, as shown in the examples of Section 1.3.2).

The definition of equivalent systems can be used, and one can consider a differentially flat system $(X \times U \times R_m^\infty, F)$ associated with

$$\dot{x} = f(x,u) \text{ where } (x,u) \in X \times U \subset R^n \times R^m \quad (1.216)$$

By definition, the above system is equivalent to the trivial system R_n^∞, F_s, where the endogenous transformation Ψ takes the form

$$\Psi(x,u,u^{(1)},\dots) = (h(x,u), \dot{h}(x,u), \ddot{h}(x,u), \dots) \quad (1.217)$$

In other terms Ψ is the infinite extension of function $h()$. The inverse transformation of Ψ is denoted as Φ and holds

$$\Psi(\bar{y}) = (\psi(\bar{\psi}), \beta(\bar{\psi}), \dot{\beta}(\bar{\psi}). \cdots) \quad (1.218)$$

Since Φ and Ψ are inverse applications it holds that

$$\phi(\bar{h}(x,\bar{u})) = x \quad \text{and} \quad \alpha(\bar{h}(x,\bar{u})) = u \quad (1.219)$$

Moreover, $t \rightarrow y(t)$ is a trajectory of $y = v$ and

$$t \rightarrow (x(t), u(t)) = (\psi(\bar{\psi}(t)), \beta(\bar{\psi}(t))) \quad (1.220)$$

is a trajectory of $\dot{x} = f(x,u)$.

Definition: Assume that $\{M, F\}$ is a system and that Ψ is the transformation that brings it to the canonical form. The first component of Ψ is considered to be the flat output.

Another important property of differentially flat systems is that the dimension of the flat output is equal to the number of the control inputs (and thus linearization and decoupling is achieved). However, it is noted that the flat output is not unique. If y is a flat output, then $\gamma(y)$ is also a flat output, provided that γ is a smooth diffeomorphism. For single-input systems, the flat output is a scalar $s = m = 1$. It is easy to show that one can pass from one flat output to another by using only a static change of variables. The multi-variable case is clearly more complicated. For example for $s = m = 2$ it is easy to see that if y_1, y_2 is a flat output then $(y_1, y_2 + \gamma(y_1^{(r)}))$ is also a flat output, provided that $\gamma()$ is a smooth function and $r \geq 0$. Since the flat output of a system is not unique, it is preferable to select the flat output which leads to simple computation for the linearization of the system and the design of the feedback controller.

1.3.4 Differential flatness and trajectory planning

1.3.4.1 Applications of differential flatness in trajectory planning

Next, it will be shown how differential flatness can be used to solve the problem of trajectory planning (setpoints definition) for dynamical systems control. Assume the following non-linear system in which the control input u does not implement feedback control

$$\dot{x} = f(x,u), \quad x \in R^n \quad u \in R^m \quad (1.221)$$

with flat output $y = h(x, u, \dot{u}, \ldots, u^{(r)})$. In such a manner, the system's trajectories $(x(t), u(t))$ are written as a function of the flat output y and its derivatives $x = \phi(y, \dot{y}, \ldots, y^{(q)})$ and $u = \alpha(y, \dot{y}, \ldots, y^{(q)})$. The trajectory planning problem contains a start and an arrival state. It is assumed that the flat output is a vector with elements y_i which are written as

$$y_i(t) = \sum_j A_{ij}\lambda_j(t) \tag{1.222}$$

where $\lambda_j, j = 1, \ldots, N$ are basis functions. The problem of definition of the flat output components y_i becomes equivalent to finding its projections in the space spanned by the basis functions λ_i

Next, it is assumed that the initial state of the system is x_0 at t_0, while the final state of the system is x_f at t_f. The coefficients A_{ij} should satisfy the following

$$\begin{array}{ll} y_i(t_0) = \sum_j A_{ij}\lambda_j(t_0) & y_i(t_f) = \sum_j A_{ij}\lambda_j(t_f) \\ \cdots \quad\quad \cdots & \cdots \\ y_i^{(q)}(t_0) = \sum_j A_{ij}\lambda_j^{(q)}(t_0) & y_i^{(q)}(t_f) = \sum_j A_{ij}\lambda_j^{(q)}(t_f) \end{array} \tag{1.223}$$

Without loss of generality it assumed that the vector of the flat output is of dimension $i = 1$, that is $y = [y_1]$ (these results can be also generalized to the multi-dimensional case). The following vectors are defined:

$$\begin{aligned} \tilde{y}_0 &= (y_1(\tau_0), \ldots, y_1^{(q)}(\tau_0)) \\ \tilde{y}_f &= (y_1(\tau_f), \ldots, y_1^{(q)}(\tau_f)) \\ \tilde{y} &= (\tilde{y}_0, \tilde{y}_f) \end{aligned} \tag{1.224}$$

Using (1.224) and (1.223) one has that

$$\tilde{y} = \begin{pmatrix} \Lambda(\tau_0) \\ \Lambda(\tau_j) \end{pmatrix} A = \Lambda A \tag{1.225}$$

The elements of A should satisfy (1.225). The only condition for (1.225) to have a solution, is that matrix Λ is a full rank one. This means that the space of functions λ_j has to be sufficiently rich. It can be concluded that the trajectories planning problem for the case of differentially flat systems ends up with linear algebra theory.

1.3.4.2 Planning under constraints

In this case, the objective is to find a trajectory going from point a to point b, which will satisfy the constraints $K(x, u, \ldots, u^{(r)}) \leq 0$ at each time instant. In the flat coordinates, this consists in finding $T > 0$ and a function $[0, T] \ni t \rightarrow y(t)$ with $(y, \ldots, y^{(q)})$ to be given at $t = 0$ and T and to verify that $\forall t \in [0, T]$ the constraint $K(y, \ldots, y^{(v)})(t) \leq 0$ will hold, for a certain v and a specific function K, obtained from k, ϕ and α, where k is the constraint condition and ϕ, α are the previously defined mappings $\dot{x} = \phi(y, \bar{v})$ and $u = \alpha(y, \bar{v})$. The difficulty of this problem increases when $q = v = 0$.

Next, it is simply assumed that the initial state Y_0 and the final state Y_f are equilibria. It is also assumed that the motion from the initial to the final state satisfies the following constraint: there exists a path $[0, 1]\ni\sigma\rightarrow Y(\sigma)$ such that $Y(0)$ and $Y(1)$ correspond to start and arrival equilibria , for all $\sigma\in[0, 1]$, $K(Y(\sigma), 0, \ldots, 0) < 0$. Therefore, there exists $T > 0$ and $[0, T]\ni t\rightarrow y(t)$ which is a solution to the problem. It suffices to take $Y(\eta(t/T))$ where T is sufficiently large, and with η to be a smooth increasing function $[0, 1]\ni s\rightarrow\eta(s)\in[0, 1]$ with $\eta(0) = 0$, $\eta(1) = 1$ and $\frac{d^i\eta}{ds^i} = 0$ for $i = 1, \ldots, \max(q, v)$.

The problem becomes more complicated in the case of trajectory planning for systems subjected to non-holonomic constraints.

1.3.4.3 Planning of trajectories with singularities

With the previous analysis, it has been shown that the endogenous transformation

$$\Psi(x, u, u^{(1)}, \ldots) = (h(x, \bar{u}), \dot{h}(x, \bar{u}), \ddot{h}(x, \bar{u}), \ldots) \tag{1.226}$$

associated with the flat output $y = h(x, \bar{u})$ is defined everywhere, is smooth and invertible in a manner that always enables state variable x and control input u to be expressed as functions of the flat output and its derivatives

$$(y, \dot{y}, \ldots, y^{(q)})\rightarrow(x, u) = \phi(y, \dot{y}, \ldots, y^{(q)}) \tag{1.227}$$

However, a point of singularity may exist in the area of the state-space into which the control tries to bring the system's state vector (y is no longer invertible and (1.227) cannot be solved). As function ϕ cannot be defined at such a point, the previous computations become meaningless. A manner to circumvent this problem and to avoid the singularity is to choose a trajectory for the system

$$t\rightarrow\phi(y(t), \dot{y}(t), \ldots, y^{(q)}(t)) \tag{1.228}$$

such that the point of potential singularity is bypassed. This procedure is depicted in the following example:

Example: The following system with differentially flat dynamics is considered:

$$\dot{x}_1 = u_1 \quad \dot{x}_2 = u_2u_1 \quad \dot{x}_3 = x_2u_1 \tag{1.229}$$

where the flat output is defined as $y = (x_1, x_3)$. In case that $u_1 = 0$ it holds that $\dot{x}_1 = \dot{y}_1 = 0$ and a singularity arises because it holds

$$(y, \dot{y}, \ddot{y}) \overset{\phi}{\rightarrow} (x_1, x_2, x_3, u_1, u_2) = (y_1, \frac{\dot{y}_2}{\dot{y}_1}, y_2, \dot{y}_1, \frac{\ddot{y}_2\dot{y}_1 - \ddot{y}_1\dot{y}_2}{\dot{y}_1^3}) \tag{1.230}$$

and the inverse transformation ϕ is not defined at the points of singularity. However, if one considers trajectories of the following form $t\rightarrow y(t) := (\sigma(t), p(\sigma(t)))$, with σ and π being smooth functions he obtains that

$$\frac{y_2(t)}{y_1(t)} = \frac{p(\sigma(t))}{\sigma(t)} \tag{1.231}$$

$$\frac{\dot{y}_2(t)}{\dot{y}_1(t)} = \frac{\frac{dp}{d\sigma}(\sigma(t)\cdot)\dot{\sigma}(t)}{\dot{\sigma}(t)} \tag{1.232}$$

and

$$\frac{\ddot{y}_2\dot{y}_1 - \ddot{y}_1\dot{y}_2}{\dot{y}_1^3} = \frac{\frac{d^2p}{d\sigma^2}(\sigma(t))\cdot\dot{\sigma}^3(t)}{\dot{\sigma}^3(t)} \tag{1.233}$$

Therefore, one can extend $t \rightarrow \phi(y(t), \dot{y}(t), \ddot{y}(t))$ everywhere in time using

$$t \rightarrow (\sigma(t), \frac{dp}{d\sigma}(\sigma(t)), p(\sigma(t)), \dot{\sigma}(t), \frac{d^2p}{d\sigma^2}(\sigma(t))) \tag{1.234}$$

Trajectory planning is performed in a similar manner as before: actually, functions σ and p and their derivatives are constrained at the start and at the arrival point of the trajectory, while elsewhere they are free. In this example, the avoidance of the singularity is internally related with the symmetry of the system. Equations

$$\dot{x}_1 = u_1, \quad \dot{x}_2 = u_2u_1, \quad \dot{x}_3 = x_2u_1 \tag{1.235}$$

are linear in u_1. The transformation, $t \rightarrow \tilde{t} = \sigma(t)$ and $u_1 \rightarrow \tilde{u}_1 = \frac{u_1}{\dot{\sigma}(t)}$, where only the time t and the control input u change, leaves the equations invariants.

1.3.5 Differential flatness, feedback control and equivalence

1.3.5.1 Closed-loop systems under state feedback control and equivalence

The analysis of Section 1.3.4 referred to a system's dynamics $\dot{x} = f(x, u)$ in which u was not specifically implementing a feedback control action. Next, the notion of equivalence will be extended to the case of closed-loop control. To this end, the initial system dynamics $\dot{x} = f(x, u)$ and its equivalent description $\dot{y} = g(y, v)$ are considered

$$\begin{aligned} &\dot{x} = f(x, u), \ (x, u) \in X \times U \subset R^n \times R^m \\ &\dot{y} = g(y, v), \ (y, v) \in Y \times V \subset R^r \times R^p \end{aligned} \tag{1.236}$$

These correspond to the systems $(X \times U \times R_m^\infty, F)$ and $(Y \times V \times R_s^\infty, G)$, where F and G are defined by

$$\begin{aligned} &F(x, u, u^{(1)}, \ldots) := (f(x, u), u^{(1)}, u^{(2)}, \ldots) \\ &G(y, v, v^{(1)}, \ldots) := (g(y, v), v^{(1)}, v^{(2)}, \ldots) \end{aligned} \tag{1.237}$$

Considering that these systems are equivalent, they have the same trajectories. A question that arises is if a transition is possible from the system $\dot{x} = f(x, u)$ to the system $\dot{y} = g(y, v)$ by a feedback loop (primarily dynamic or alternatively static), or inversely. For the transformed state vectors z and control inputs v one has

$$\begin{aligned} &\dot{z} = \alpha(x, z, v) \ \ z \in Z \subset R^q \\ &u = \kappa(x, z, v) \end{aligned} \tag{1.238}$$

Next, the following theorem is introduced (Martin and Rouchon, 1999):

Theorem: *Assume that the systems $\dot{x} = f(x, u)$ and $\dot{y} = g(y, v)$ are equivalent. Thus $\dot{x} = f(x, u)$ can be transformed by a dynamic loop and change of coordinates (dynamic feedback linearization) into*

$$\dot{y} = g(y, v),\ \dot{v} = v^{(1)},\ \dot{v}^{(1)} = v^{(2)}, \ldots, \dot{v}^{(\mu)} = w \tag{1.239}$$

for an entity μ being sufficiently large. Inversely, $\dot{y} = g(y, v)$ can be transformed by a feedback loop (dynamic) and change of coordinates into

$$\dot{x} = g(x, u),\ \dot{u} = u^{(1)},\ \dot{u}^{1} = u^{(2)}, \ldots, \dot{u}^{(\ni)} = w \tag{1.240}$$

for an entity v being sufficiently large.

Proof: The notation F and G is used for the infinite dimensional vector fields which represent the two systems. The equivalence shows that there exists an invertible function

$$\Phi(y, \bar{v}) = (\phi(y, \bar{v}), \alpha(y, \bar{v}), \dot{\alpha}(y, \bar{v}), \cdots) \tag{1.241}$$

such that

$$F(\Phi(y, \bar{v})) = D\Phi(y, \bar{v}) \cdot G(y, \bar{v}) \tag{1.242}$$

Next, one sets $\bar{y} = (y, v, v^{(1)}, \ldots, v^{(\mu)})$ and $w = v^{(\mu+1)}$. For μ sufficiently large, it holds that ϕ (respectively α) depends exclusively on $\bar{y}$ (respectively on $\bar{y}$ and w). With these notations Φ is written as

$$\phi(\tilde{y}, w) = (\phi(\tilde{y}), \alpha(\tilde{y}, \bar{w}), \dot{\alpha}(\tilde{y}, \bar{w}), \ldots) \tag{1.243}$$

and (1.242) implies in particular that

$$f(\phi(\tilde{y}, \alpha(\tilde{y}, w))) = D\phi(\tilde{y})\bar{g}(\tilde{y}, w) \tag{1.244}$$

where $\bar{g} := (g, v^{(1)}, \ldots, v^{(k)})$. Since Φ is invertible, and ϕ is of full rank it can thus be completed by a function π to obtain a change of coordinates

$$\tilde{y} \rightarrow \phi(\tilde{\psi}) = (\phi(\tilde{y}), \pi(\tilde{\psi})) \tag{1.245}$$

Considering now the dynamic feedback, one has

$$\begin{aligned} u &= \alpha(\phi^{-1}(x, z), w) \\ \dot{z} &= D\pi(\phi^{-1}(x, z)) \cdot \tilde{g}(\phi^{-1}(x, z), w) \end{aligned} \tag{1.246}$$

which transforms $\dot{x} = f(x, u)$ into

$$\begin{pmatrix} \dot{x} \\ z \end{pmatrix} = \tilde{f}(x, z, w) := \begin{pmatrix} f(x, \alpha(\phi^{-1}(x, z), w)) \\ D\pi(\phi^{-1}(x, z)) \cdot \tilde{g}(\phi^{-1}(x, z), w) \end{pmatrix} \tag{1.247}$$

Using the previous example one has

$$\tilde{f}(\phi(\tilde{y}, w)) = \begin{pmatrix} f(\phi(\tilde{y}, \alpha(\tilde{y}, w))) \\ D\pi(\tilde{y})\cdot\tilde{g}(\tilde{y}, w) \end{pmatrix} = \begin{pmatrix} D\phi(\tilde{y} \\ D\pi(\tilde{y})) \end{pmatrix} \tilde{g}(\tilde{y}, w) = D\phi(\tilde{y})\cdot\tilde{g}(\tilde{y}, w) \tag{1.248}$$

Thus, $\tilde{f}$ and $\tilde{g}$ are ϕ-conjugates that completes the proof.

A generic remark is that a differentially flat system is equivalent to a system in the linear canonical (Brunovsky) form, and thus one arrives immediately at the following conclusion: A flat dynamics is linearizable by dynamic feedback and change of coordinates.

1.3.5.2 Closed-loop systems and equivalence under endogenous feedback

The dynamical system $\dot{x} = f(x, u)$ is considered. The feedback loop

$$\begin{aligned} u &= \kappa(x, z, w) \\ \dot{z} &= \alpha(x, z, w) \end{aligned} \tag{1.249}$$

is called endogenous since the system's dynamics in open loop $\dot{x} = f(x, u)$ is equivalent to the dynamics of the system in closed loop, the latter given by

$$\begin{aligned} \dot{x} &= f(x, \kappa(x, z, w)) \\ \dot{z} &= \alpha(x, z, w) \end{aligned} \tag{1.250}$$

The term endogenous has the meaning that variables z and w are not included in the state vector of the system (case also known as "zero dynamics"). Using the above one arrives at a more refined definition of equivalence between systems and differential flatness.

Theorem: *Two dynamics $\dot{x} = f(x, u)$ and $\dot{y} = g(y, v)$ are equivalent, if and only if, $\dot{x} = f(x, u)$ can be transformed by endogenous feedback and change of coordinates.*

Corollary 1: A dynamics is flat, if and only if it is linearizable by dynamic endogenous feedback and change of coordinates.

Corollary 2: Consider the system description

$$\begin{aligned} \dot{x} &= f(x, \kappa(x, z, w)) \\ \dot{z} &= \alpha(x, z, w) \end{aligned} \tag{1.251}$$

where the constituents of the dynamic endogenous feedback are

$$\begin{aligned} \dot{u} &= \kappa(x, z, w) \\ \dot{z} &= \alpha(x, z, w) \end{aligned} \tag{1.252}$$

This system can be transformed by endogenous feedback into

$$\dot{x} = f(x, u), \text{ where } \dot{u} = u^{(1)}, \ldots, \dot{u}^{(\mu)} = w \tag{1.253}$$

for μ sufficiently large (thus finally the system's control input contains multiple integral terms). This shows clearly the properties which are preserved by equivalence.

The property which is maintained by introducing integral terms and change of coordinates is controllability. An endogenous feedback is actually reversible to the effect of multiple integrators.

1.3.5.3 Trajectory tracking in feedback control of differentially flat systems

The problem of trajectory tracking for the system $\dot{x} = f(x, u)$ consists in finding a controller u which will permit the system state variables to track the reference setpoint $t \rightarrow (x_r(t), u_r(t))$. It is convenient to add to the open-loop control a correction term Δu, which is a function of the state vector error $\Delta x = x - x_r$. For differentially flat systems there is a systematic method to calculate Δu from the tracking error Δx.

If the dynamics admit as flat output $y = h(x, \bar{u})$, one can use the corollary that a differentially flat system is linearizable by (dynamic) feedback and change of coordinates, into the linear canonical form $y^{(\mu+1)} = w$. For the linearized equivalent model the control input is defined as $y_r^{(\mu+1)} - K\Delta\tilde{y}$, where K is a gains vector and Δy is a vector having as elements the flat output tracking error and its derivatives up to order μ. By applying such a feedback control input, the tracking error dynamics becomes

$$\Delta y^{(\mu+1)} = -K\Delta\tilde{y} \tag{1.254}$$

where $y_r(t) := (x_r(t), \bar{u}_r(t))$ and $\tilde{y} := (y, \dot{y}, \ldots, y^{\mu})$, and Δy represents the vector of the flat output tracking error. Such a feedback control law assures asymptotic tracking. There exists an invertible transformation

$$\Phi(\bar{y}) = (\phi(\bar{y}), \alpha(\bar{y}), \dot{\alpha}(\bar{y}), \ldots) \tag{1.255}$$

which connects the infinite vector fields $F(x, \bar{u}) := (f(x, u), u, u^{(1)}, \ldots)$ and $G(\bar{y}) := (y, y^{(1)}, \ldots)$. From the proof of the theorem on the equivalence of two systems (theorem given in Section 1.3.5.1) the above means that for the state vector one has

$$\begin{aligned} x &= \phi(\tilde{y}_r(t) + \Delta\tilde{y}) \\ &= \phi(\tilde{y}_r(t)) + R_\phi(y_r(t), \Delta\tilde{y})\Delta\tilde{y} \\ &= x_r(t) + R_\phi(y_r(t), \Delta\tilde{y})\Delta\tilde{y} \end{aligned} \tag{1.256}$$

and for the control input one has

$$\begin{aligned} u &= \alpha(\tilde{y}_r(t) + \Delta\tilde{y}, -K\Delta\tilde{y}) \\ &= \alpha(\tilde{y}_r(t)) + R_\alpha(y_r^{(\mu+1)}(t), \Delta\tilde{y}) \begin{pmatrix} \Delta\tilde{y} \\ -K\Delta\tilde{y} \end{pmatrix} \\ &= u_r(t) + R_\alpha(\tilde{y}_r(t), y_r^{(\mu+1)}(t), \Delta\tilde{y}, \Delta w) \begin{pmatrix} \Delta\tilde{y} \\ -K\Delta\tilde{y} \end{pmatrix} \end{aligned} \tag{1.257}$$

where the following classical result of factorization has been used

$$\begin{aligned} R_\phi(Y, \Delta Y) &:= \int_0^1 D\phi(Y + t\Delta Y)dt \\ R_\alpha(Y, w, \Delta Y, \Delta w) &:= \int_0^1 D\alpha(Y + t\Delta Y, w + t\Delta w)dt \end{aligned} \tag{1.258}$$

Since $\Delta y \rightarrow 0$ for $t \rightarrow \infty$, this means that $x \rightarrow x_r(t)$ and $u \rightarrow u_r(t)$.

1.3.6 Flatness-based control and state feedback under the model uncertainties

Up to now, for the systems used for analyzing differential flatness properties and flatness-based control it has been assumed that an exact dynamic model was available. Flatness-based control can be also applied to systems characterized by model uncertainties and exogenous disturbances. Denoting the coefficients vector of the system as $\theta \in R^p$, the generalized system dynamics of (1.259) is written as

$$\dot{x}(t) = f(\theta, x(t), u(t)) \tag{1.259}$$

In the generic case, the system's coefficients θ are subject to uncertainty, i.e.

$$\theta = \theta_0 + \tilde{\theta},\ \tilde{\theta}_i \in [\underline{\theta}_i, \bar{\theta}_i],\ i = 1, \ldots, p \tag{1.260}$$

where θ_0 is the nominal value of the parameter vector. In (1.259) the vector field $f : R^p \times R^n \times R^m \rightarrow R^n$ is smooth. Denoting the flat output $y(t) \in R$ and parameters vector θ, the definition of differential flatness given in Section 1.3.2 is rewritten as (Hegenmeyer and Delaleau, 2003; Hegenmeyer and Delaleau, 2010)

$$\begin{aligned} y &= h(\theta, x, u, \dot{u}, \ldots, u^{(r)}) \\ x &= \phi(\theta, y, \dot{y}, \ldots, y^{(r-1)}) \\ u &= \psi(\theta, y, \dot{y}, \ldots, y^{(r)}) \end{aligned} \tag{1.261}$$

where h, ϕ and ψ are smooth functions defining mappings $h : R^p \times R^n \times (R^m)^{r+1} \rightarrow R^m$, $\phi : R^p \times (R^m)^r \rightarrow R^n$ and $\psi : R^p \times (R^m)^{r+1} \rightarrow R^m$, respectively. It is also assumed that the flat output $y(t)$ is independent of the coefficients (parameters) vector θ. Equation (1.261) shows that for every given trajectory of the flat output $t \rightarrow y(t)$ the evolution of all other variables of the system $t \rightarrow x(t)$ and $t \rightarrow u(t)$ are given without integration of any differential equation. Moreover, for a sufficiently smooth desired trajectory of the flat output $t \rightarrow y^*(t)$ (1.261) can be used to design the corresponding feed-forward control $u^*(t)$ directly for the nominal system parameters θ_0. The trajectory y^* is called the nominal trajectory, while the trajectory u^* is called the nominal control. Thus, knowing the desirable system's output, one can also find the associated flat output y^* and subsequently the control input that makes the system track the desirable trajectory is given by

$$u^*(t) = \psi(\theta_0, y^*(t), \dot{y}^*(t), \ldots, y^{*(r)}(t)) \tag{1.262}$$

which means that for each admissible nominal trajectory $y^*(t)$ there is a nominal control input u^*. The initial condition of the desired trajectory of the flat output $t \rightarrow y^*(t)$ is defined as the vector $y_0^* = [y^*(0), \dot{y}^*(0), \ldots, y^{*(r-1)}(0)]^T$. The system is *consistent* with respect to the initial conditions x_0 if $x_0 = \phi(\theta, y_0^*)$. The following theorem has been proven (Hegenmeyer and Delaleau, 2003):

Theorem: *If the desired trajectory of the flat output is consistent with the initial condition* x_0 *and* $\theta = \theta_0$, *then when applying the nominal control input of Eq.* (1.262) *the system becomes equivalent, by a change of coordinates, to a linear system in Brunovksy (canonical) form.*

Without loss of generality, a single-input single-output flat system is considered. It is easy to show that such a flat system can be represented as follows: Setting $y = [y, \dot{y}, \ldots, y^{(n-1)}]^T = [y_1, y_2, \ldots, y_n]^T$ the system can be transformed via the diffeomorphism

$$y = h(\theta, x) \tag{1.263}$$

where $h = \phi^{-1}$ with ϕ defined in (1.261), into the control normal (canonical) form

$$\begin{aligned} \dot{y}_i(t) &= y_{i+1}(t)\ i \in \{1, \ldots, n-1\} \\ \dot{y}_n(t) &= \alpha(\theta, y(t), u(t)) \end{aligned} \tag{1.264}$$

where $y^{(i)}$ is the $i-1$-th order derivative of the flat output. From the second row of (1.264) one can see that the n-th order derivative of the flat output is finally written as a non-linear function of the parameters vector θ, of the flat output's vector $y(t)$ and of the control input $u(t)$. The system in the canonical form described by (1.264) and the system of (1.259), (1.261) and (1.262) have the same solution $t \to y(t)$.
Defining the new control input $v(t) = \alpha(\theta, y(t), u(t))$ the Brunovsky form of the transformed initial system is written as

$$\begin{aligned} \dot{y}_i(t) &= y_{i+1}(t)\ i \in \{1, \ldots, n-1\} \\ \dot{y}_n(t) &= v(t) \end{aligned} \tag{1.265}$$

Once the system has been written in the Brunovksy (canonical) form of (1.265) a control input that can assure tracking of any desirable trajectory $\dot{y}_n^*$ is given by

$$v = \dot{y}_n^* + K^T \bar{e} \tag{1.266}$$

where the term $K^T \bar{e}$ corresponds to an error feedback controller with an integral term and thus processes the augmented error vector $\bar{e} = [\int_0^t e_1(\tau) d\tau, e_1, e_2, \ldots, e_n]$. The feedback control term can be also of the form of a proportional-derivative controller and in the latter case processes the error vector $e = [e_1, e_2, \ldots, e_n]$, thus taking the form

$$K^T e = -\sum_{i=1}^{n} k_{n+1-i} e_i(t),\ k_{n+1-i} > 0 \tag{1.267}$$

Thus, in case that the exact values of the model's parameters θ are known and these are equal to the nominal values θ_0, and no integration of the flat output error is used in the feedback loop, i.e. the feedback control signal is of the form $K^T e = -\sum_{i=1}^{n} k_{n+1-i} e_i(t)$, the closed-loop system error vector dynamics become

$$\begin{aligned} &\dot{y}_n = v \Rightarrow \dot{y}_n = \dot{y}_n^* + K^T e \Rightarrow \\ &\dot{y}_n = \dot{y}_n^* - k_n e_1 - k_{n-1} e_2 + \cdots - k_1 e_n \Rightarrow \\ &\dot{y}_n - \dot{y}_n^* = -k_n e_1 - k_{n-1} e_2 + \cdots - k_1 e_n \Rightarrow \\ &e^{(n)} + k_1 e^{(n-1)} + k_2 e^{(n-2)} + \cdots + k_{n-1} \dot{e} + k_n e = 0 \end{aligned} \tag{1.268}$$

where $e_i = y_i - y_i^* = y^{(i-1)} - y^{(i-1)*}$. The dynamics of the flat output tracking error can be also written in a matrix form

$$\begin{pmatrix} \dot{e}_1 \\ \dot{e}_2 \\ \cdots \\ \cdots \\ \dot{e}_n \end{pmatrix} = \begin{pmatrix} 0 & 1 & 0 & \cdots & 0 \\ 0 & 0 & 1 & \cdots & 0 \\ \cdots & \cdots & \cdots & \cdots & \cdots \\ 0 & 0 & 0 & \cdots & 1 \\ -k_n & -k_{n-1} & -k_{n-2} & \cdots & -k_1 \end{pmatrix} \begin{pmatrix} e_1 \\ e_2 \\ \cdots \\ \cdots \\ e_n \end{pmatrix} \tag{1.269}$$

In the case of the error dynamics of (1.269) to succeed elimination of the tracking error it suffices to select the controller gains $k_1, k_2, \ldots, k_n$ so as the polynomial $e^{(n)} + k_1 e^{(n-1)} + k_2 e^{(n-2)} + \ldots + k_{n-1}\dot{e} + k_n e = 0$ to be a Hurwitz one.

When that the model's parameters θ are not precisely known or have deviated from their nominal values θ_0, one can apply again a control input of the form described in (1.266) and (1.267), and for the specific magnitude of the exogenous disturbances or parametric uncertainties can expect again satisfactory tracking of the reference trajectory. In such a case, the robustness features of the closed loop can be analyzed with the use of results from interval polynomial stability theory (e.g. Kharitonov's theory) or other results on interval polynomial analysis (e.g. see Rigatos and Siano, 2011). The problem of robustness of a flatness-based controller in case of parametric uncertainties or exogenous perturbations has been studied in Hegenmeyer and Delaleau, 2003, Hegenmeyer and Delaleau, 2010 and also in Rigatos, 2012a, Rigatos, 2012b.

1.3.7 Classification of types of differentially flat systems

1.3.7.1 Criteria about the differential flatness of a system

Definition of 0-flat systems

The existence of a computable criterion so as to decide if the dynamical system $\dot{x} = f(x, u)$, $x \in R^n$, $u \in R^m$ is differentially flat remains open. This means that there is no systematic method to construct a flat output. This situation is somehow equivalent to the definition of Lyapunov functions for non-linear dynamical systems. For even though Lyapunov functions are both theoretically and practically important for the study of the stability features of non-linear dynamical systems there is no systematic manner to define them.

The main difficulty in the computation of the flat output $y = h(x, u, \ldots, y^{(r)})$ remains in its dependence to the derivatives of the control input u, up to order r which can be arbitrarily large. To know if the order of r admits an upper bound (which can be dependent on n and m) remains a question. It is not known if for a certain dimension of the state vector and for a certain dimension of the control inputs vector such a finite boundary exists. In the following, it will be noted that a dynamical system is r-flat if it admits a flat output depending up to the rth order derivative of the control input. Equivalently, in a 0-flat system its flat output depends exclusively on the elements of the system's state vector.

The next example shows that such an upper bound should depend, in the best case, linearly to the dimension of the state vector. Consider the dynamical system

$$x_1^{(a_1)} = u_1 \quad x_1^{(a_2)} = u_2 \quad \dot{x}_3 = u_1 u_2 \tag{1.270}$$

The flat output is taken to be

$$y_1 = x_3 + \sum_{i=1}^{\alpha_1} (-1)^i x_1^{(\alpha_1 - i)} u_2^{(i-1)}, \qquad y_2 = x_2 \tag{1.271}$$

This system is r-flat, with $r = min(\alpha_1, \alpha_2) - 1$. There is no flat output depending on the derivatives of u which has an order equal or less than $r - 1$.

Systems that can be linearized by static or dynamic state feedback

It has been shown that dynamical systems which can be linearized with static state feedback can be also transformed to the linear canonical (Brunovsky) form. Thus, such systems are differentially flat, with the flat outputs to be exclusively functions of the state vector x (0-flat system). It is noted that a differentially flat system is not always a system which can be linearized with static state feedback. There exist systems that can be linearized through dynamic feedback linearization, that is their state vector is extended by including as additional state variables the control inputs and their derivatives (see examples on the linearization of the VTOL and the hovercraft model (Rigatos, 2015)). Such systems can be also written in the linear canonical form, and thus are also differentially flat.

Systems with only one control input

Single input dynamical systems, that is

$$\dot{x} = f(x, u) \; x \in R^n, \; u \in R \tag{1.272}$$

can be linearized with static state feedback linearization and can be finally written in the linear canonical form. Such systems are differentially flat.

Systems affine-in-the-input of co-dimension 1

A system of the form

$$\dot{x} = f_0(x) + \sum_{j=1}^{n-1} g_j(x) u_j \; x \in R^n \tag{1.273}$$

that is systems in which the number of control inputs is by one less than the dimension of the system's state vector, are 0-flat, as long as they are controllable (strongly accessible in x). The problem becomes more complicated when the system is not an affine in the input one. However, the latter systems can be finally transformed into affine-in-the-input ones too.

Systems without drift term

These are systems of the form $\dot{x} = \sum_{i=1}^{m} f_i(x)u_i$. Within this class of systems, one can distinguish several cases. The following theorem holds (Martin and Rouchon, 1999).

Theorem 1. *Systems without drift and with two control inputs*

$$\dot{x} = f_1(x)u_1 + f_2(x)u_2 \tag{1.274}$$

are flat if and only if, the generic rank of E_k *is equal to* $k+2$, *for* $k = 0, \ldots, n-2$, *where* $E_0 = \{f_1, f_2\}$ *and* $E_{k+1} = \{E_k, [E_k, E_k]\}, \quad k > 0$. *A dynamical system without drift and with two control inputs is always 0-flat. Such a system can be written in the chained form (which is a precursor of the linear canonical form), through static state feedback and a change of variables. For example, for a system of the form*

$$\dot{x}_1 = u_1 \quad \dot{x}_2 = u_2 \quad \dot{x}_3 = x_2 u_1 \quad \cdots \quad \dot{x}_n = x_{n-1}u_1 \tag{1.275}$$

is a differentially flat one and the flat output is $y = (x_1, x_n)$.

Systems in triangular form

The state-space description of systems in triangular form is as follows:

$$\begin{aligned}
\dot{x}_1 &= f_1(x_1) + g_1(x_1)x_2 \\
\dot{x}_2 &= f_2(x_1, x_2) + g_2(x_1, x_2)x_3 \\
\dot{x}_3 &= f_3(x_1, x_2, x_3) + g_3(x_1, x_2, x_3)x_4 \\
&\ldots \\
\dot{x}_i &= f_i(x_1, x_2, \cdots, x_i) + g_i(x_1, x_2, \cdots, x_i)x_{i+1} \\
&\ldots \\
\dot{x}_{n-1} &= f_{n-1}(x_1, x_2, \cdots, x_{n-1}) + g_{n-1}(x_1, x_2, \cdots, x_{n-1})x_n \\
\dot{x}_n &= f_n(x_1, x_2, \cdots, x_n) + g_n(x_1, x_2, \cdots, x_n)u
\end{aligned} \tag{1.276}$$

The flat output for such systems is $y = x_1$.

Theorem 2. *Driftless systems with a state vector of dimension n and* $n-2$ *control inputs. The system of the form*

$$\dot{x} = \sum_{i=1}^{n-2} f_i(x)u_i \tag{1.277}$$

is a differentially flat one if it is controllable (strongly accessible for almost all x). More precisely, such a system is 0-flat if n is an odd number and is a 1-flat one if n is an even number.

1.3.7.2 A sufficient condition for showing that a system is not differentially flat

The following theorem provides a sufficient condition for showing that dynamical systems are not differentially flat (Martin and Rouchon, 1999):

Theorem: *The criterion of the regulated variety. Assume that the dynamical system1* $\dot{x} = f(x,u)$ *is a differentially flat one. Then, the projection to the p-dimensional space of its state-space equation* $p = f(x,u)$ *is a subvariety regulated for all* x.

This criterion signifies that the elimination of the control input u from the n equations $\dot{x} = f(x,u)$ leads to $n-m$ equations of the form $F(x,\dot{x}) = 0$ with the following property: For all (x,p) such that $F(x,p) = 0$ there exists $\alpha \in R^n$, $\alpha \neq 0$ such that

$$\forall \lambda \in R \;\; F(x, p + \lambda a) = 0 \tag{1.278}$$

Proof: Consider $(\bar{x}, \bar{u}, \bar{p})$ such that $\bar{p} = f(\bar{x}, \bar{u})$. Generally, f is of the order $m = dim(u)$ with respect to u. If x is removed from the state space description then one arrives at $n-m$ equations which are based exclusively on x and p, that is $F(x,p) = 0$. Thus one has the constraint about the trajectory around $(\bar{x}, \bar{u})$: $\;F(x,\dot{x}) = 0$. If the system is differentially flat it holds that

$$x = \phi(y, \dot{y}, \ldots, y^{(q)}) \tag{1.279}$$

where variable y is the flat output. Thus, there exists $(\bar{y}, \dot{\bar{y}}, \ldots, \bar{y}^{(q)})$ such that $\bar{x} = \phi(\bar{y}, \dot{\bar{y}}, \ldots, \bar{y}^{(q)})$. Then, the following identity holds

$$F(\phi(\bar{y}, \dot{\bar{y}}, \ldots, \bar{y}^{(q)}), \phi_y \dot{y} + \phi_{\dot{y}} \ddot{y} + \cdots + \phi_{y^{(q)}} y^{(q+1)}) = 0 \tag{1.280}$$

Taking now that the derivatives of function y at $t = 0$ up to order q are given by $y^{(r)}(0) = \bar{y}^{(r)}$ $\; r = 0, \ldots, q$ and $y^{\alpha}(0) = \bar{y}^{q} + \xi$ where ξ is an arbitrary vector in R^n, one has the following identity for all $\xi \in R^n$

$$F(\bar{x}, \bar{x} + \phi_{y^{\alpha}} \xi) = 0 \tag{1.281}$$

with $\phi_y^{(q)}$ to be evaluated at $\bar{y}, \dot{\bar{y}}, \ldots, \bar{y}^{(q)}$. From the point $(\bar{x}, \bar{p})$ of $F(x,p) = 0$ passes the affine space which is parallel to the image of $\phi_{y^{(q)}}$, that is a space of dimension at most equal to m and at least equal to 1.

Therefore by showing that there does not exist any flat output with derivatives which are not coupled in the sense of an ODE (or equivalently it can be said that there does not exist a flat output being differentially independent) one can prove that a dynamical system is not differentially flat. This result is particularly simple. It stands however for an efficient method to show that certain multi-input systems are not differentially flat.

Example 1: The system

$$\dot{x}_1 = u_1, \quad \dot{x}_2 = u_2, \quad \dot{x}_3 = (u_1)^2 + (u_2)^3 \tag{1.282}$$

is not a differentially flat one. This is because the sub-variety $p_3 = p_1^2 + p_2^3$ is not a regulated one. There does not exist an $\alpha \in R^3$, $\alpha \neq 0$ such that

$$\forall \lambda \in R \;\; p_3 + \lambda\alpha_3 = (p_1 + \lambda\alpha_1)^2 + (p_2 + \lambda\alpha_2)^3 \tag{1.283}$$

Actually, the cubic term implies that $\alpha_2 = 0$ and the quadratic term implies that $\alpha_1 = 0$ and thus one has $\alpha_3 = 0$.

1.3.7.3 Liouvillian and non-differentially flat systems

There are some non-flat systems that do not satisfy differential flatness properties. Such systems cannot be linearized by state feedback (neither by static feedback linearization nor by dynamic feedback linearization). For such cases the notion of Liouvillian systems has been introduced, which can be seen as an extension of differentially flat systems.

Non-flat systems are algebraically characterized by an integer called the *defect*. This integer is well defined by introducing the notion of maximal linearizing output. Using an informal definition, a set of variables $y = (y_1, \ldots, y_m)$ is a maximal linearizing output, if the number of system variables which can be expressed as a differential function of y is maximal. Therefore, the maximal linearizing output characterizes the largest subsystem which satisfies the differential flatness property. This largest subsystem is called flat subsystem.

The defect represents the number of the system's variables which do not belong to the flat subsystem. A non-linear system is said to be *Liouvillian* or integrable by quadrature's if the elements which do not belong to the flat subsystem can be obtained by elementary integrations of the elements of the flat subsystem. In the following examples, the notion of integrability by quadrature's and that of Liouvillian systems is explained Martin and Rouchon, 1999.

Example 1:

$$\dot{x}_1 = x_2 + x_3^2, \quad \dot{x}_2 = x_3, \quad \dot{x}_3 = u \tag{1.284}$$

This system is not linearizable by static or dynamic state feedback. The system is of defect equal to 1 and is found to be Liouvillian (there is a state variable that does not belong to the flat subsystem). A maximum linearizing output is given by $y = x_2$ and x_1 is the only variable that does not belong to the flat subsystem. However, x_1 can be obtained from the integration $x_1 = \int y + \dot{y}^2$, that is x_1 is given by a pure integral of a differential function of y.

Example 2: The following dynamic model is considered

$$\dot{x}_1 = u_1, \quad \dot{x}_2 = u_2, \quad \dot{x}_3 = x_1u_2 - x_2u_1, \quad \dot{x}_4 = x_3u_1 \quad \dot{x}_5 = x_3u_2 \tag{1.285}$$

This system is nilpotent and completely controllable since it is driftless and satisfies the Lie algebra rank condition. However, the system is not flat and cannot be transformed into the canonical form. Nevertheless, there is a subsystem within it defined by the first four equations, which is differentially flat and consequently the system's defect is equal to 1. The maximal linearizing output of the system is given by $y=(y_1,y_2)$, where $y_1=x_1$ and $y_2=(zx_1-x_4)/3$, with z to be defined by $z=(x_1x_2-x_3)/2$. Indeed $z=\dot{y}_2/\dot{y}_1$, thus $x_4=y_1z-3y_2=\alpha(y_1,\dot{y}_1,y_2,\dot{y}_2)$, $x_3=\dot{x}_4/\dot{y}_1=b(y_1,\dot{y}_1,\ddot{y}_1,y_2,\dot{y}_2,\ddot{y}_2)$, $x_2=(2z+x_3)/y_1=c(y_1,\dot{y}_1,\ddot{y}_1,y_2,\dot{y}_2,\ddot{y}_2)$, $u_2=\dot{x}_2=q(y_1,\dots,y_1^{(3)},y_2,\dots,y_2^{(3)})$. Since $x_5=\int x_3u_2=\int b(y_1,\dots,\ddot{y}_2)q(y_1,\dots,y_2^{(3)})$ one gets that the above system of (1.285) is a Liouvillian one.

1.4 Control of PDE dynamical systems

1.4.1 Distributed parameter systems and transformation in the canonical form

Up to now, the analysis on differential flatness properties was focused on lumped parameter systems, that is systems described by ordinary differential equations. By applying differential flatness theory it is also possible to transform into the canonical form distributed parameter systems which are described by non-linear partial differential equations of the parabolic, elliptic and hyperbolic types. This is particularly significant for developing efficient non-linear control and filtering methods for such a type of system. To demonstrate this, the model of non-linear heat diffusion PDE will be used as an example.

The solution of the stabilization problem for PDE systems with the use of control inputs that are applied through the boundary conditions can be achieved by exploiting differential flatness theory. The design of a stabilizing controller for the PDE is based on the proof that (i) the state-space model of the PDE is a differentially flat system, (ii) each row of the state-space model is also a differentially flat subsystem. Next, for each subsystem (row) being associated with an ordinary differential equation, one can compute a virtual control input that can stabilize the subsystem's dynamics and which can also eliminate the tracking error of the subsystem's output. The virtual control input for the ith subsystem becomes a reference setpoint for the $(i+1)$th subsystem. From the last row of the state-space description, one can compute the control input (boundary condition) which should be finally applied to the partial differential equation. This control input comprises in a recursive manner all virtual control inputs which were computed for the rows (subsystems) that constitute the PDE's state-space description. Thus, by tracing backwards the rows of the state-space model of the PDE (from the last to the first row) one can finally compute the control input that should be applied in the form of a boundary condition so that all individual state variables to converge to their designated setpoints. The global stability of this PDE control method is proven through Lyapunov analysis.

1.4.2 State-space description of a heat diffusion dynamics

The following non-linear heat diffusion equation is considered, describing the spatiotemporal variations of temperature in a surface

$$\frac{\partial\phi}{\partial t}=K\frac{\partial^2\phi}{\partial x^2}+f(\phi)+u(x,t) \tag{1.286}$$

Using the approximation for the partial derivative in the partial differential equation given in (1.286) one has

$$\frac{\partial^2\phi}{\partial x^2} \simeq \frac{\phi_{i+1} - 2\phi_i + \phi_{i-1}}{\Delta x^2} \tag{1.287}$$

and considering spatial measurements of variable ϕ along axis x at points $x_0 + i\Delta x$, $i = 1, 2, \ldots, N$ one has

$$\frac{\partial \phi_i}{\partial t} = \frac{K}{\Delta x^2}\phi_{i+1} - \frac{2K}{\Delta x^2}\phi_i + \frac{K}{\Delta x^2}\phi_{i-1} + f(\phi_i) + u(x_i, t) \tag{1.288}$$

By considering the associated samples of ϕ given by $\phi_0, \phi_1, \ldots, \phi_N, \phi_{N+1}$ one has

$$\begin{aligned}
\frac{\partial \phi_1}{\partial t} &= \frac{K}{\Delta x^2}\phi_2 - \frac{2K}{\Delta x^2}\phi_1 + \frac{K}{\Delta x^2}\phi_0 + f(\phi_1) + u(x_1, t) \\
\frac{\partial \phi_2}{\partial t} &= \frac{K}{\Delta x^2}\phi_3 - \frac{2K}{\Delta x^2}\phi_2 + \frac{K}{\Delta x^2}\phi_1 + f(\phi_2) + u(x_2, t) \\
\frac{\partial \phi_3}{\partial t} &= \frac{K}{\Delta x^2}\phi_4 - \frac{2K}{\Delta x^2}\phi_3 + \frac{K}{\Delta x^2}\phi_2 + f(\phi_3) + u(x_3, t) \\
&\cdots \\
\frac{\partial \phi_{N-1}}{\partial t} &= \frac{K}{\Delta x^2}\phi_N - \frac{2K}{\Delta x^2}\phi_{N-1} + \frac{K}{\Delta x^2}\phi_{N-2} + f(\phi_{N-1}) + u(x_{N-1}, t) \\
\frac{\partial \phi_N}{\partial t} &= \frac{K}{\Delta x^2}\phi_{N+1} - \frac{2K}{\Delta x^2}\phi_N + \frac{K}{\Delta x^2}\phi_{N-1} + f(\phi_N) + u(x_N, t)
\end{aligned} \tag{1.289}$$

By defining the following state vector

$$x^T = (\phi_1, \phi_2, \ldots, \phi_N) \tag{1.290}$$

one obtains the state-space description

$$\begin{aligned}
\dot{x}_1 &= \frac{K}{\Delta x^2}x_2 - \frac{2K}{\Delta x^2}x_1 + \frac{K}{\Delta x^2}\phi_0 + f(x_1) + u(x_1) \\
\dot{x}_2 &= \frac{K}{\Delta x^2}x_3 - \frac{2K}{\Delta x^2}x_2 + \frac{K}{\Delta x^2}x_1 + f(x_2) + u(x_2) \\
\dot{x}_3 &= \frac{K}{\Delta x^2}x_4 - \frac{2K}{\Delta x^2}x_3 + \frac{K}{\Delta x^2}x_2 + f(x_3) + u(x_3) \\
&\cdots \\
\dot{x}_{N-1} &= \frac{K}{\Delta x^2}x_N - \frac{2K}{\Delta x^2}x_{N-1} + \frac{K}{\Delta x^2}x_{N-2} + f(x_{N-1}) + u(x_{N-1}) \\
\dot{x}_N &= \frac{K}{\Delta x^2}\phi_{N+1} - \frac{2K}{\Delta x^2}x_N + \frac{K}{\Delta x^2}x_{N-1} + f(x_N) + u(x_N)
\end{aligned} \tag{1.291}$$

Next, the following state variables are defined

$$\begin{aligned}
y_{1,i} &= x_i \\
y_{2,i} &= \dot{x}_i
\end{aligned} \tag{1.292}$$

and the state-space description of the system becomes as follows

$$\begin{aligned}
\dot{y}_{1,1} &= \tfrac{K}{\Delta x^2}y_{1,2} - \tfrac{2K}{\Delta x^2}y_{1,1} + \tfrac{K}{\Delta x^2}\phi_0 + f(y_{1,1}) + u(y_{1,1}) \\
\dot{y}_{1,2} &= \tfrac{K}{\Delta x^2}y_{1,3} - \tfrac{2K}{\Delta x^2}y_{1,2} + \tfrac{K}{\Delta x^2}y_{1,1} + f(y_{1,2}) + u(y_{1,2}) \\
\dot{y}_{1,3} &= \tfrac{K}{\Delta x^2}y_{1,4} - \tfrac{2K}{\Delta x^2}y_{1,3} + \tfrac{K}{\Delta x^2}y_{1,2} + f(y_{1,3}) + u(y_{1,3}) \\
&\cdots \\
&\cdots \\
\dot{y}_{1,N-1} &= \tfrac{K}{\Delta x^2}y_{1,N} - \tfrac{2K}{\Delta x^2}y_{1,N-1} + \tfrac{K}{\Delta x^2}y_{1,N-2} + f(y_{1,N-1}) + u(y_{1,N-1}) \\
\dot{y}_{1,N} &= \tfrac{K}{\Delta x^2}\phi_{N+1} - \tfrac{2K}{\Delta x^2}y_{1,N} + \tfrac{K}{\Delta x}y_{1,N-1} + f(y_{1,N}) + u(y_{1,N})
\end{aligned} \tag{1.293}$$

The dynamical system described in (1.293) is a differentially flat one with flat output defined as the vector $\tilde{y} = [y_{1,1}, y_{1,2}, \ldots, y_{1,N}]$. Indeed all state variables can be written as functions of the flat output and its derivatives. Moreover, by defining the new control inputs

$$\begin{aligned}
v_1 &= \frac{K}{\Delta x^2}\phi_0 + f(y_{1,1}) + u(y_{1,1}) \\
v_2 &= f(y_{1,2}) + u(y_{1,2}) \\
v_3 &= f(y_{1,3}) + u(y_{1,3}) \\
&\cdots \\
v_{N-1} &= f(y_{1,N-1}) + u(y_{1,N-1}) \\
v_N &= \frac{K}{\Delta x^2}\phi_{N+1} + f(y_{1,N}) + u(y_{1,N})
\end{aligned} \tag{1.294}$$

the following state-space description is obtained

$$\begin{pmatrix} \dot{y}_{1,1} \\ \dot{y}_{1,2} \\ \cdots \\ \dot{y}_{1,N-1} \\ \dot{y}_{1,N} \end{pmatrix} = \begin{pmatrix} -\frac{2K}{\Delta x^2} & \frac{K}{\Delta x^2} & 0 & 0 & \cdots & 0 & 0 & 0 & 0 \\ \frac{K}{\Delta x^2} & -\frac{2K}{\Delta x^2} & \frac{K}{\Delta x^2} & 0 & \cdots & 0 & 0 & 0 & 0 \\ 0 & \frac{K}{\Delta x^2} & -\frac{2K}{\Delta x^2} & \frac{K}{\Delta x^2} & \cdots & 0 & 0 & 0 & 0 \\ \cdots & \cdots & \cdots & \cdots & \cdots & \cdots & \cdots & \cdots & \cdots \\ 0 & 0 & 0 & 0 & \cdots & 0 & \frac{K}{\Delta x^2} & -\frac{2K}{\Delta x^2} & \frac{K}{\Delta x^2} \\ 0 & 0 & 0 & 0 & \cdots & 0 & 0 & \frac{K}{\Delta x^2} & -\frac{2K}{\Delta x^2} \end{pmatrix} \begin{pmatrix} y_{1,1} \\ y_{1,2} \\ \cdots \\ y_{1,N-1} \\ y_{1,N} \end{pmatrix} + \begin{pmatrix} 1 & 0 & 0 & \cdots & 0 & 0 \\ 0 & 1 & 0 & \cdots & 0 & 0 \\ 0 & 0 & 1 & \cdots & 0 & 0 \\ \cdots & \cdots & \cdots & \cdots & & \\ 0 & 0 & 0 & \cdots & 1 & 0 \\ 0 & 0 & 0 & \cdots & 0 & 1 \end{pmatrix} \begin{pmatrix} v_1 \\ v_2 \\ v_3 \\ \cdots \\ v_{N-1} \\ v_N \end{pmatrix} \tag{1.295}$$

Assuming that all measurements from the set of points $x_j\, j\in[1,2,\ldots,m]$ are available, the associated observation (measurement) equation becomes

$$\begin{pmatrix} z_1 \\ z_2 \\ \cdots \\ z_m \end{pmatrix} = \begin{pmatrix} 1 & 0 & 0 & \cdots & 0 & 0 \\ 0 & 1 & 0 & \cdots & 0 & 0 \\ \cdots & \cdots & \cdots & \cdots & & \\ 0 & 0 & 0 & \cdots & 1 & 0 \\ 0 & 0 & 0 & \cdots & 0 & 1 \end{pmatrix} \begin{pmatrix} y_{1,1} \\ y_{1,2} \\ y_{1,3} \\ \cdots \\ y_{1,N-1} \\ y_{1,N} \end{pmatrix} \tag{1.296}$$

Thus, in matrix form one has the following state-space description of the system

$$\begin{aligned} \dot{\tilde{y}} &= A\tilde{y} + Bv \\ \tilde{z} &= C\tilde{y} \end{aligned} \tag{1.297}$$

Moreover, denoting $a = \frac{K}{Dx^2}$ and $b = -\frac{2K}{Dx^2}$, the initial description of the system given in (1.295) is rewritten as follows

$$\begin{pmatrix} \dot{y}_{1,1} \\ \dot{y}_{1,2} \\ \cdots \\ \dot{y}_{1,N-1} \\ \dot{y}_{1,N} \end{pmatrix} = \begin{pmatrix} b\,a\,0\,0\,0 & \cdots & 0\,0\,0\,0\,0\,0 \\ a\,b\,a\,0\,0 & \cdots & 0\,0\,0\,0\,0\,0 \\ 0\,0\,a\,b\,a & \cdots & 0\,0\,0\,0\,0\,0 \\ \cdots & \cdots & \cdots \\ 0\,0\,0\,0\,0\,0\,0 & \cdots & a\,b\,a\,0 \\ 0\,0\,0\,0\,0\,0\,0 & \cdots & 0\,0\,a\,b \end{pmatrix} \begin{pmatrix} y_{1,1} \\ y_{1,2} \\ \cdots \\ y_{1,N-1} \\ y_{1,N} \end{pmatrix} + \begin{pmatrix} 1 & 0 & 0 & \cdots & 0 & 0 \\ 0 & 1 & 0 & \cdots & 0 & 0 \\ 0 & 0 & 1 & \cdots & 0 & 0 \\ \cdots & \cdots & \cdots & \cdots & & \\ 0 & 0 & 0 & \cdots & 1 & 0 \\ 0 & 0 & 0 & \cdots & 0 & 1 \end{pmatrix} \begin{pmatrix} v_1 \\ v_2 \\ v_3 \\ \cdots \\ v_{N-1} \\ v_N \end{pmatrix} \tag{1.298}$$

1.4.3 Differential flatness of the non-linear heat-diffusion PDE

The previously defined state vector of the PDE model is considered without the effect of the external control input $u(x,t)$.

$$\frac{\partial\phi}{\partial t} = K\frac{\partial^2\phi}{\partial x^2} + f(\phi) \tag{1.299}$$

The state vector is again $\tilde{Y} = [y_{1,1}, y_{1,2}, \ldots, y_{1,i}, \ldots, y_{1,N-1}, y_{1,N}]$, where $y_{1,1} = V_1$, $y_{1,2} = V_2, \ldots, y_{1,i} = V_i, \ldots, y_{1,N-1} = V_{N-1}$ and $y_{1,N} = V_N$. It will be shown in a different manner that the state-space description of the non-linear PDE dynamics, which has as control input only the boundary condition ϕ_0 is a differentially flat one. One has

$$\dot{y}_{1,1} = -\frac{2K}{\Delta x^2}y_{1,1} + \frac{K}{\Delta x^2}\phi_0 + f(y_{1,1}) + \frac{K}{\Delta x^2}y_{1,2} \tag{1.300}$$

$$\dot{y}_{1,2} = -\frac{2K}{\Delta x^2}y_{1,2} + \frac{K}{\Delta x^2}y_{1,1} + f(y_{1,2}) + \frac{K}{\Delta x^2}y_{1,3} \tag{1.301}$$

$$\dot{y}_{1,3} = -\frac{2K}{\Delta x^2}y_{1,3} + \frac{K}{\Delta x^2}y_{1,2} + f(y_{1,3}) + \frac{K}{\Delta x^2}y_{1,4} \tag{1.302}$$

$\cdots\ \cdots$

$$\dot{y}_{1,i} = -\frac{2K}{\Delta x^2}y_{1,i} + \frac{K}{\Delta x^2}y_{1,i-1} + f(y_{1,i}) + \frac{K}{\Delta x^2}y_{1,i+1} \tag{1.303}$$

$\cdots\ \cdots$

$$\dot{y}_{1,N-1} = -\frac{2K}{\Delta x^2}y_{1,N-1} + \frac{K}{\Delta x^2}y_{1,N-2} + f(y_{1,N-1}) + \frac{K}{\Delta x^2}y_{1,N} \tag{1.304}$$

$$\dot{y}_{1,N} = -\frac{2K}{\Delta x^2}y_{1,N} + \frac{K}{\Delta x^2}y_{1,N-1} + f(y_{1,N}) + \frac{K}{\Delta x^2}\phi_{N+1} \tag{1.305}$$

The flat output is considered to be the state variable $y_{1,N}$, which is denoted as $y = y_{1,N}$. Next, it is shown that all state variables which stand also for virtual control inputs of the system $\alpha_i = y_{1,N-i}$, can be written as functions of the flat output $y = y_{1,N}$.

From (1.305) and considering that ϕ_{N+1} is constant one obtains

$$\begin{aligned} y_{1,N-1} = \alpha_1 &= \frac{1}{K/\Delta x^2}\left[\dot{y}_{1,N} + \frac{2K}{\Delta x^2}y_{1,N} - \frac{K}{2\Delta x^2}\phi_{N+1} - f(y_{1,N})\right] \\ &\Rightarrow y_{1,N-1} = h_1(y, \dot{y}, \ldots) \end{aligned} \tag{1.306}$$

and following a similar procedure, from (1.304) one gets

$$\begin{aligned} y_{1,N-2} = \alpha_2 &= \frac{1}{K/\Delta x^2}\left[\dot{y}_{1,N-1} + \frac{2K}{\Delta x^2}y_{1,N-1} - \frac{K}{2\Delta x^2}y_{1,N} - f(y_{1,N-1})\right] \\ &\Rightarrow y_{1,N-2} = h_2(y, \dot{y}, \cdots) \end{aligned} \tag{1.307}$$

Continuing in a similar manner, from (1.303) one obtains

$$\begin{aligned} y_{1,i-1} = \alpha_{N-i+1} &= \frac{1}{K/\Delta x^2}\left[\dot{y}_{1,i} + \frac{2K}{\Delta x^2}y_{1,i} - \frac{K}{2\Delta x^2}y_{1,i+1} - f(y_{1,i})\right] \\ &\Rightarrow y_{1,i-1} = h_{N-i+1}(y, \dot{y}, \ldots) \end{aligned} \tag{1.308}$$

From (1.302) one obtains

$$\begin{aligned} y_{1,2} = \alpha_{N-2} &= \frac{1}{K/\Delta x^2}\left[\dot{y}_{1,3} + \frac{2K}{\Delta x^2}y_{1,3} - \frac{K}{\Delta x^2}y_{1,4} - f(y_{1,3})\right] \\ &\Rightarrow y_{1,2} = h_{N-2}(y, \dot{y}, \ldots) \end{aligned} \tag{1.309}$$

From (1.301) one obtains

$$\begin{aligned} y_{1,1} = \alpha_{N-1} &= \frac{1}{K/\Delta x^2}\left[\dot{y}_{1,2} + \frac{2K}{\Delta x^2}y_{1,2} - \frac{K}{\Delta x^2}y_{1,3} - f(y_{1,2})\right] \\ &\Rightarrow y_{1,1} = h_{N-1}(y, \dot{y}, \ldots) \end{aligned} \tag{1.310}$$

Finally, from (1.300) one obtains

$$\begin{aligned} \phi_0 = \alpha_N &= \frac{1}{K/\Delta x^2}\left[\dot{y}_{1,1} + \frac{2K}{\Delta x^2}y_{1,1} - \frac{K}{\Delta x^2}y_2 - f(y_{1,1})\right] \\ &\Rightarrow \phi_0 = h_N(y, \dot{y}, \ldots) \end{aligned} \tag{1.311}$$

The above procedure confirms that all state variables of the model

$$\begin{matrix} y_{1,1} & y_{1,2} & y_{1,3} & \cdots \\ y_{1,i} & \cdots & y_{1,N-1} & y_{1,N} \end{matrix} \tag{1.312}$$

and the control input which is the boundary condition ϕ_0 can be written as functions of the flat output $y = y_{1,N}$ and of the flat output's derivatives. Consequently, differential flatness for the heat PDE is shown.

Additionally, one can consider decomposition of the PDE state-space equation into flat submodels, where at each submodel the virtual control input is $\alpha_i = y_{1,N-i}$

$$\dot{y}_{1,i} = -\frac{2K}{\Delta x^2} y_{1,i} + \frac{K}{\Delta x^2} y_{1,i-1} + f(y_{1,i}) + \frac{K}{\Delta x^2} y_{1,i+1} \qquad (1.313)$$
$$\ldots \ldots$$

1.4.4 Computation of a boundary conditions-based feedback control law

To implement boundary PDE feedback control one defines $K_1 = -2K$, $K_2 = K$, $K_3 = K$ and the non-linear diffusion-PDE model is rewritten as follows:

$$\dot{y}_{1,N} = \frac{K_1}{\Delta x^2} y_{1,N} + f(y_{1,N}) + \frac{K_2}{2\Delta x^2} \phi_{N+1} + \frac{K_1}{\Delta x^2} y_{1,N-1} \qquad (1.314)$$

$$\dot{y}_{1,N-1} = \frac{K_1}{\Delta x^2} y_{1,N-1} + f(y_{1,N-1}) + \frac{K_2}{2\Delta x^2} y_{1,N} + \frac{K_1}{\Delta x^2} y_{1,N-2} \qquad (1.315)$$

$$\dot{y}_{1,N-2} = \frac{K_1}{\Delta x^2} y_{1,N-2} + f(y_{1,N-2}) + \frac{K_2}{2\Delta x^2} y_{1,N-1} + \frac{K_1}{\Delta x^2} y_{1,N-1} \qquad (1.316)$$
$$\ldots \ldots$$

$$\dot{y}_{1,i} = \frac{K_1}{\Delta x^2} y_{1,i} + f(y_{1,i}) + \frac{K_2}{2\Delta x^2} y_{1,i+1} + \frac{K_1}{\Delta x^2} y_{1,i-1} \qquad (1.317)$$
$$\ldots \ldots$$

$$\dot{y}_{1,3} = \frac{K_1}{\Delta x^2} y_{1,3} + f(y_{1,3}) + \frac{K_2}{2\Delta x^2} y_{1,4} + \frac{K_1}{\Delta x^2} y_{1,2} \qquad (1.318)$$
$$\ldots \ldots$$

$$\dot{y}_{1,2} = \frac{K_1}{\Delta x^2} y_{1,2} + f(y_{1,2}) + \frac{K_2}{2\Delta x^2} y_{1,3} + \frac{K_1}{\Delta x^2} y_{1,1} \qquad (1.319)$$

$$\dot{y}_{1,1} = \frac{K_1}{\Delta x^2} y_{1,1} + f(y_{1,1}) + \frac{K_2}{2\Delta x^2} y_{1,2} + \frac{K_1}{\Delta x^2} \phi_0 \qquad (1.320)$$

The boundary condition ϕ_{N+1} is assumed to be known and remains steady. The boundary condition ϕ_0 stands for the control input (for example, it can be dependent on the interest rate). The feedback control law is designed as follows:

$$\begin{aligned} \alpha_1 = y^*_{1,N-1} = \frac{1}{(K_1/\Delta x^2)} \Big[& \dot{y}^d_{1,N} - k_{p,1}(y_{1,N} - y^d_{1,N}) \\ & - \frac{K_1}{\Delta x^2} y_{1,N} - f(y_{1,N}) - \frac{K_2}{2\Delta x^2} \phi_{N+1} \Big] \end{aligned} \qquad (1.321)$$

$$\begin{aligned}\alpha_2 = y_{1,N-2}^* &= \frac{1}{(K_1/\Delta x^2)}\left[\dot{y}_{1,N-1}^d - k_{p,2}(y_{1,N-1} - y_{1,N-1}^d)\right.\\ &\left.- \frac{K_1}{\Delta x^2}y_{1,N-1} - f(y_{1,N-1}) - \frac{K_2}{2\Delta x^2}y_N\right] \Rightarrow\\ \alpha_2 &= \frac{1}{(K_1/\Delta x^2)}\left[\dot{\alpha}_1 - k_{p,2}(y_{1,N-1} - \alpha_1)\right.\\ &\left.- \frac{K_1}{\Delta x^2}y_{1,N-1} - f(y_{1,N-1}) - \frac{K_2}{2\Delta x^2}y_N\right]\end{aligned} \tag{1.322}$$

$$\begin{aligned}\alpha_3 = y_{1,N-3}^* &= \frac{1}{(K_1/\Delta x^2)}\left[\dot{y}_{1,N-2}^d - k_{p,3}(y_{1,N-2} - y_{1,N-2}^d)\right.\\ &\left.- \frac{K_1}{\Delta x^2}y_{1,N-2} - f(y_{1,N-2}) - \frac{K_2}{2\Delta x^2}y_{N-1}\right] \Rightarrow\\ \alpha_3 &= \frac{1}{(K_1/\Delta X^2)}\left[\dot{\alpha}_2 - k_{p,3}(y_{1,N-2} - \alpha_2)\right.\\ &\left.- \frac{K_1}{\Delta x^2}y_{1,N-2} - f(y_{1,N-2}) - \frac{K_2}{2\Delta x^2}y_{N-1}\right]\end{aligned} \tag{1.323}$$

and continuing in a similar manner

$$\begin{aligned}\alpha_i = y_{1,N-i}^* &= \frac{1}{(K_1/\Delta x^2)}\left[\dot{y}_{1,N-i+1}^d - k_{p,i}(y_{1,N-i+1} - y_{1,N-i+1}^d)\right.\\ &\left.- \frac{K_1}{\Delta x^2}y_{1,N-i+1} - f(y_{1,N-i+1}) - \frac{K_2}{2\Delta x^2}y_{N-i+2}\right] \Rightarrow\\ \alpha_i &= \frac{1}{(K_1/\Delta x^2)}\left[\dot{\alpha}_{i-1} - k_{p,i}(y_{1,N-i+1} - \alpha_{i-1})\right.\\ &\left.- \frac{K_1}{\Delta x^2}y_{1,N-i+1} - f(y_{1,N-i+1}) - \frac{K_2}{2\Delta x^2}y_{N-i+2}\right]\end{aligned} \tag{1.324}$$

Following this procedure one arrives to compute the control inputs which are associated with the last two rows of the state-space model

$$\begin{aligned}\alpha_{N-2} = y_{1,2}^* &= \frac{1}{(K_1/\Delta x^2)}\left[\dot{y}_{1,3}^d - k_{p,N-1}(y_{1,3} - y_{1,3}^d)\right.\\ &\left.- \frac{K_1}{\Delta x^2}y_{1,3} - f(y_{1,3}) - \frac{K_2}{2\Delta x^2}y_{1,4}\right] \Rightarrow\\ \alpha_{N-2} &= \frac{1}{(K_1/\Delta x^2)}\left[\dot{\alpha}_{N-3} - k_{p,N-2}(y_{1,3} - \alpha_{N-3})\right.\\ &\left.- \frac{K_1}{\Delta x^2}y_{1,3} - f(y_{1,3}) - \frac{K_2}{2\Delta x^2}y_{1,4}\right]\end{aligned} \tag{1.325}$$

$$\begin{aligned}\alpha_{N-1} = y_{1,1}^* &= \frac{1}{(K_1/\Delta X^2)}\left[\dot{y}_{1,2}^d - k_{p,N-1}(y_{1,2} - y_{1,2}^d) - \frac{K_1}{\Delta x^2}y_{1,2} - f(y_{1,2}) - \frac{K_2}{2\Delta x^2}y_{1,3}\right] \Rightarrow\\ \alpha_{N-1} &= \frac{1}{(K_1/\Delta X^2)}\left[\dot{\alpha}_{N-2} - k_{p,N-1}(y_{1,2} - \alpha_{N-2}) - \frac{K_1}{\Delta x^2}y_{1,2} - f(y_{1,2}) - \frac{K_2}{2\Delta x^2}y_{1,3}\right]\end{aligned} \tag{1.326}$$

and finally

$$\begin{aligned}\alpha_N = \phi_0 &= \frac{1}{(K_1/\Delta x^2)}\left[\dot{y}_{1,1}^d - k_{p,N}(y_{1,1} - y_{1,1}^d) - \frac{K_1}{\Delta x^2}y_{1,1} - f(y_{1,1}) - \frac{K_2}{2\Delta x^2}y_{1,2}\right] \Rightarrow \\ \alpha_N &= \frac{1}{(K_1/\Delta x^2)}\left[\dot{\alpha}_{N-1} - k_{p,N}(y_{1,1} - \alpha_{N-1}) - \frac{K_1}{\Delta x^2}y_{1,1} - f(y_{1,1}) - \frac{K_2}{2\Delta x^2}y_{1,2}\right]\end{aligned} \tag{1.327}$$

Consequently, the computation of the aggregate control input $\alpha_N = \phi_0$ which is applied to the PDE model is performed backward, by substituting recursively into ϕ_0 the virtual control inputs a_{N-1}, a_{N-2},. . .,a_i,. . .,a_2,a_1.

1.4.5 Closed loop dynamics

By substituting (1.327) into (1.320) of the state-space model of the PDE dynamics, and using the definition $y_{1,1} - y_{1,1}^d = z_1$ one has

$$\begin{aligned}&\dot{y}_{1,1} = \dot{y}_{1,1}^d - k_{p,1}(y_{1,1} - y_{1,1}^d) \Rightarrow \\ &(\dot{y}_{1,1} - \dot{y}_{1,1}) + k_{p,1}(y_{1,1} - y_{1,1}^d) = 0 \Rightarrow \\ &\dot{z}_1 + k_{p,1}z_1 = 0\end{aligned} \tag{1.328}$$

Equivalently, by substituting (1.326) into (1.319), and using the definition $y_{1,2} - \alpha_1 = z_2$ one has

$$\begin{aligned}&\dot{y}_{1,2} = \dot{\alpha}_1 - k_{p,2}(y_{1,2} - \alpha_1) \Rightarrow \\ &(\dot{y}_{1,2} - \dot{\alpha}_1) + k_{p,2}(y_{1,2} - \alpha_1) = 0 \Rightarrow \\ &\dot{z}_2 + k_{d,2}\dot{z}_2 + k_{p,2}z_2 = 0\end{aligned} \tag{1.329}$$

Similarly, continuing with the rest of the equations of the state-space model and by substituting (1.325) into (1.318), while also using the definition $y_{1,3} - \alpha_2 = z_3$ one has

$$\begin{aligned}&\dot{y}_{1,3} = \dot{\alpha}_2 - k_{p,3}(y_{1,3} - \alpha_2) \Rightarrow \\ &(\dot{y}_{1,3} - \dot{\alpha}_2) + k_{p,3}(y_{1,3} - \alpha_2) = 0 \Rightarrow \\ &\ddot{z}_3 + k_{p,3}z_3 = 0\end{aligned} \tag{1.330}$$

Moving backwards, and by substituting (1.324) into (1.317) of the state-space model of the PDE dynamics, and using the definition $y_{1,i} - \alpha_{i-1} = z_i$ one has

$$\begin{aligned}&\dot{y}_{1,i} = \dot{\alpha}_{i-1} - k_{p,i}(y_{1,i} - \alpha_{i-1}) \Rightarrow \\ &(\dot{y}_{1,i} - \dot{\alpha}_{i-1}) + k_{p,i}(y_{1,i} - \alpha_{i-1}) = 0 \Rightarrow \\ &\dot{z}_i + k_{p,i}z_i = 0\end{aligned} \tag{1.331}$$

By substituting (1.322) into (1.315) of the state-space model of the PDE dynamics, and using the definition $y_{1,N-1} - \alpha_{N-2} = z_{N-1}$ one has

$$\begin{aligned}&\dot{y}_{1,N-1} = \dot{\alpha}_{N-2} - k_{p,N-1}(y_{1,N-1} - \alpha_{N-2}) \Rightarrow \\ &(\dot{y}_{1,N-1} - \dot{\alpha}_{N-2}) + k_{p,N-1}(y_{1,N-1} - \alpha_{N-2}) = 0 \Rightarrow \\ &\dot{z}_{N-1} + k_{p,N-1}z_{N-1} = 0\end{aligned} \tag{1.332}$$

Finally, by substituting (1.321) into (1.315), and using the definition $y_{1,N} - \alpha_{N-1} = z_N$ one obtains

$$\begin{aligned}
&\dot{y}_{1,N} = \dot{\alpha}_{N-1} - k_{p,N}(y_{1,N} - \alpha_{N-1}) \Rightarrow \\
&(\dot{y}_{1,N} - \dot{\alpha}_{N-1}) + k_{p,N}(y_{1,N} - \alpha_{N-1}) = 0 \Rightarrow \\
&\dot{z}_N + k_{p,N} z_N = 0
\end{aligned} \tag{1.333}$$

Thus, the dynamics of the closed-loop system becomes

$$\begin{aligned}
&\dot{z}_1 + k_{p,1} z_1 = 0 \\
&\dot{z}_2 + k_{p,2} z_2 = 0 \\
&\dot{z}_3 + k_{p,3} z_3 = 0 \\
&\cdots \cdots \\
&\dot{z}_i + k_{p,i} z_i = 0 \\
&\cdots \cdots \\
&\dot{z}_{N-1} + k_{p,N-1} z_{N-1} = 0 \\
&\dot{z}_N + k_{p,N} z_N = 0
\end{aligned} \tag{1.334}$$

The dynamics of the closed-loop system can be also written in matrix form

$$\dot{\tilde{Z}} + K_P \tilde{Z} = 0 \tag{1.335}$$

where

$$\begin{aligned}
&\tilde{Z} = [z_1, z_2, z_3, \ldots, z_i, \ldots, z_{N-1}, z_N] \\
&K_P = \mathrm{diag}[k_{p,1}, k_{p,2}, k_{p,3}, \ldots, k_{p,i}, \ldots, k_{p,N-1}, k_{p,N}].
\end{aligned} \tag{1.336}$$

After suitable selection of the coefficients $k_{P,i}$, $i = 1, 2, \ldots, N$ such that the monomials $p(s) = s + k_{p,i}$ to have negative roots, it can be assured that $lim_{t \to \infty} z_i(t) = 0$ and that the closed-loop system is asymptotically stable.

Moreover, to prove asymptotic stability for the closed-loop system, the following Lyapunov function can be used

$$V = \sum_{i=1}^{N} \frac{1}{2} z_i^2 \tag{1.337}$$

The derivative of this Lyapunov function with respect to time is

$$\begin{aligned}
&V = \sum_{i=1}^{N} \frac{1}{2} z_i^2 \Rightarrow \dot{V} = \sum_{i=1}^{N} z_i(-k_{p_i} z_i) \Rightarrow \\
&\dot{V} = -\sum_{i=1}^{N} k_{p,i} z_i^2 < 0
\end{aligned} \tag{1.338}$$

Thus, it is proven again that the closed-loop system is globally asymptotically stable.

Chapter 2

Control based on approximate linearization for robotic systems

Abstract

The chapter analyzes methods and presents results on control and estimation based on approximate linearization for robotic systems: (i) nonlinear control of the cart and double-pendulum overhead crane, (ii) nonlinear control of the underactuated offshore crane, (iii) nonlinear control of the inertia wheel and pendulum system, (iv) nonlinear control of the torsional oscillator with rotational actuator, (v) nonlinear control of robotic exoskeletons, (vi) nonlinear control of brachiation robots, (vii) Nonlinear control of power line inspection robots, (viii) nonlinear control of robots with electrohydraulic actuators, (ix) nonlinear control of robots with electropneumatic actuators, (x) nonlinear control of flexible joint robots, (x) nonlinear control of redundant robotic manipulators, and (xi) nonlinear control of parallel closed-chain robotic manipulators.

2.1 Nonlinear control of the cart and double-pendulum overhead crane

2.1.1 Outline

Overhead cranes are widely used in industry, in construction works as well as in loading and unloading of cargo ships and heavy-duty vehicles and in supply-chain, as in (Rigatos and Busawon, 2018; Sun, Wu, Liang *et al.*, 2019; and also in Sun, Wu, Fang *et al.*, 2018; Ouyang, Wang *et al.*, 2017; Giaccomelli *et al.*, 2018; and Qian, Tong *et al.*, 2016). In several cases, overhead cranes are modeled as a point mass suspended from the cart (moving trolley) through a hoisting cable, thus becoming equivalent to a single pendulum (Ouyang, Tian *et al.*, 2020a; Vaughan *et al.*, 2010; Zhang, Zhang *et al.*, 2019). However, in practice, the payload is connected first to a hook. Besides, the mass of the hook is often non-negligible and the payload is large enough such that its distance from the hook cannot be neglected either (Tian, Yu *et al.*, 2021; Jimenez-Cano et al., 2013; Ouyang, Tian *et al.*, 2020b). In the latter case, the dynamics of the crane system is better approximated by the one of the double pendulums (Ouyang, Tian *et al.*, 2020c; Ouyang, Tian *et al.*, 2020d; Wu, Wang *et al.*, 2021).

In such a case, one should take into account additional degrees of freedom, that is the hook swing with respect to the trolley and the mass swing with respect to the hook (Ouyang, Xu *et al.*, 2020b; Ouyang, Xu *et al.*, 2020a; Sun, Fu, Yang *et al.*, 2020). The degrees of freedom of the double-pendulum overhead crane are further raised by considering the displacement of the moving trolley along the horizontal axis (Chen, Cheng *et al.*, 2019; Chen and Sun, 2020; Lu, Fang *et al.*, 2019). An additional degree of freedom can be included when the rope (link) on which the hook is attached is of variable length (Lu, Fang *et al.*, 2019). Due to nonlinearities and underactuation, control of the double-pendulum overhead crane is a challenging task, as shown in Sun, Fang, Chen *et al.*, 2017, Liang, Zhang, Feng *et al.*, 2021, and Lu, Feng *et al.*, 2018.

In this section, a new nonlinear optimal control method is developed for a double-pendulum overhead crane, following the results of Rigatos and Busawon, 2018, Rigatos, Siano and Abbaszadeh, 2017, and Rigatos, Siano *et al.*, 2015. First the nonlinear dynamic model of the crane undergoes approximate linearization around a temporary operating point which is recomputed at each iteration of the control algorithm. The linearization point is re-defined at each sampling instant by the present value of the crane's state vector and by the last sampled value of the control inputs vector. The linearization process relies on first-order Taylor series expansion and on the computation of the associated Jacobian matrices (Rigatos and Tzafestas, 2007; Basseville and Nikiforov, 1993; Rigatos and Zhang, 2009). For the approximately linearized model of the crane's dynamics, a stabilizing optimal (H-infinity) feedback controller is defined. The modeling error which is due to the truncation of higher-order terms in the Taylor series expansion is deemed to be a perturbation which is asymptotically compensated by the robustness of the control algorithm.

The proposed H-infinity controller provides a solution to the optimal control problem of the double-pendulum overhead crane under model uncertainty and external perturbations (Rigatos and Busawon, 2018). Actually, the proposed controller implements a min-max differential game that takes place between (i) the control inputs of the crane which try to minimize a quadratic cost function of the state vector's tracking error and (ii) the model uncertainty and external perturbation terms which try to maximize this cost function. To compute the gains of the stabilizing H-infinity controller an algebraic Riccati equation is repetitively solved at each time-step of the control algorithm (Rigatos, 2011; Rigatos, 2015; Rigatos, 2016b). The global stability properties of the control scheme are proven through Lyapunov analysis. First, it is demonstrated that the control method satisfies the H-infinity tracking performance criterion (Rigatos and Busawon, 2018; Tousssaint, Basar and Bullo, 2000). This signifies robustness against model uncertainty and external perturbations. Besides, under moderate conditions, it is proven that the control loop of the double-pendulum overhead crane is globally asymptotically stable. The proposed control scheme retains the advantages of linear optimal control, that is fast and accurate tracking of reference setpoints under moderate variations of the control inputs (Rigatos and Busawon, 2018). Moreover, to implement state estimation-based control without the need to measure the entire state-vector of the double-pendulum overhead crane, the H-infinity Kalman Filter is used as a robust state estimator.

2.1.2 Dynamic model of the double-pendulum overhead crane

In the following, an overhead crane system is considered, comprising a cart and a double-pendulum suspended from it as shown in Figure 2.1. A first link connects a hook to the cart while a second link connects the payload to the hook.

Thus, one has the following degrees of freedom: (i) x: is the motion of the cart along the horizontal axis Ox, (ii) θ_1 is the angle between the first link of length l_1 and the vertical axis Oy, (iii) θ_2: is the angle between the second link of length l_2 which connects the payload's center of gravity to the hook and the vertical axis Oy. The state vector of the system is $q = [x, \dot{x}, \theta_1, \dot{\theta}_1, \theta_2, \dot{\theta}_2]^T$. The mass of the cart is denoted as m, the mass of the hook is denoted as m_1 while the mass of the payload is denoted as m_2. Using Euler–Lagrange analysis, the dynamic model of the double-pendulum overhead crane is obtained. The state-space model of this system is given in the form of a robotic model which comprises an inertia matrix $M(q)$, a Coriolis and centrifugal forces vector $C(q, \dot{q})\dot{q}$ and a gravitational forces vector $G(q)$. This gives (Rigatos and Busawon, 2018; Sun, Wu, Liang *et al.*, 2019; Sun, Wu, Fang *et al.*, 2018; Ouyang, Wang *et al.*, 2017; Giaccomelli *et al.*, 2018; Qian, Tong *et al.*, 2016)

$$M(q)\ddot{q} + C(q, \dot{q})\dot{q} + G(q) = u \tag{2.1}$$

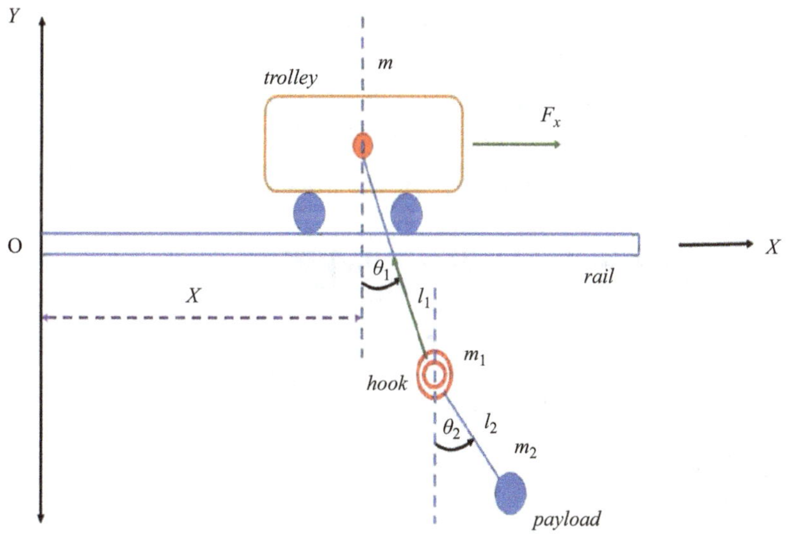

Figure 2.1 Diagram of the double-pendulum overhead crane with payload subject to hoisting and lowering

where the symmetric inertia matrix $M(q)$ is given by

$$M(q) = \begin{pmatrix} m_{11} & m_{12} & m_{13} \\ m_{21} & m_{22} & m_{23} \\ m_{31} & m_{32} & m_{33} \end{pmatrix} \tag{2.2}$$

the Coriolis and centrifugal forces matrix is given by

$$C(q,\dot{q}) = \begin{pmatrix} c_{11} & c_{12} & c_{13} \\ c_{21} & c_{22} & c_{23} \\ c_{31} & c_{32} & c_{33} \end{pmatrix} \tag{2.3}$$

while the gravitational forces vector and the control inputs vector are given by

$$G(q) = \begin{pmatrix} g_1 \\ g_2 \\ g_3 \end{pmatrix} \quad u = \begin{pmatrix} F_x \\ -d_1\dot{\theta}_1 \\ -d_2\dot{\theta}_2 \end{pmatrix} \tag{2.4}$$

where the elements of the above matrices are given in (Sun, Wu, Liang *et al.*, 2019; Qian, Tong *et al.*, 2016): $m_{11} = m + m_1 + m_2$, $m_{12} = (m_1 + m2)l_1\cos(\theta_1)$ $m_{13} = m_2l_2\cos(\theta_2)$, $m_{21} = (m_1 + m_2)l_1\cos(\theta_1)$, $m_{22} = (m_1 + m2)l_1$, $m_{23} = m_2l_1l_2\cos(\theta_1 - \theta_2)$, $m_{31} = m_2l_2\cos(\theta_2)$, $m_{32} = m_2l_1l_2\cos(\theta_1 - \theta_2)$, and $m_{33} = m_2l_2^2$.

While it also holds that (Sun, Wu, Liang *et al.*, 2019; Qian, Tong *et al.*, 2016): $c_{11} = 0, c_{12} = -(m_1 + m2)l_1\sin(\theta_1)\dot{x}, c_{13} = -m_2l_2\sin(\theta_2)\dot{\theta}_2, c_{21} = 0, c_{22} = 0, c_{23} = m_2l_1l_2\sin(\theta_1 - \theta_2)\dot{\theta}_2$, $c_{31} = 0$, $c_{32} = (m_1 + m_2)gl_1\sin(\theta_1)$, and $c_{33} = m_2gl_2\sin(\theta_2)$ and additionally (Sun, Wu, Liang *et al.*, 2019; Qian, Tong *et al.*, 2016): $g_1 = 0$, $g_2 = (m_1 + m_2)gl_1\sin(\theta_1)$ and $g_3 = m_2gl_2\sin(\theta)_2$. The terms $d_1\dot{\theta}_1$ and $d_2\dot{\theta}_2$ denote damping forces about the turn motion of the two links. It holds that

$$\begin{aligned} &M(q)\ddot{q} + C(q,\dot{q})\dot{q} + G(q) = u \Rightarrow \\ &\ddot{q} = -M^{-1}(q)C(q,\dot{q})\dot{q} - M^{-1}(q)G(q) + M^{-1}(q)u \end{aligned} \tag{2.5}$$

The inverse of the inertia matrix $M^{-1}(q)$ i computed as follows:

$$M^{-1}(q) = \frac{1}{detM}\begin{pmatrix} M_{11} & -M_{21} & M_{31} \\ -M_{12} & M_{22} & -M_{32} \\ M_{13} & -M_{23} & M_{33} \end{pmatrix} \tag{2.6}$$

where the sub-determinants M_{ij}, $i = 1,2,3$ and $j = 1,2,3$ are computed as follows:

$$M_{11} = m_{22}m_{33} - m_{32}m_{23}, M_{12} = m_{21}m_{33} - m_{31}m_{23}, M_{13} = m_{21}m_{32} - m_{31}m_{22}$$

$$M_{21} = m_{12}m_{33} - m_{32}m_{13}, M_{22} = m_{11}m_{33} - m_{31}m_{13}\ M_{23} = m_{11}m_{32} - m_{31}m_{12}$$

$$M_{31} = m_{12}m_{23} - m_{22}m_{13}, M_{32} = m_{11}m_{23} - m_{21}m_{13}, M_{33} = m_{11}m_{22} - m_{21}m_{12}$$

The determinant of the inertia matrix is

$$detM = m_{11}M_{11} - m_{12}M_{12} + m_{13}M_{13} \tag{2.7}$$

The Coriolis and centrifugal forces vector of the double-pendulum overhead crane is computed as follows:

$$\tilde{C}(q,\dot{q}) = C(q,\dot{q})\dot{q} \Rightarrow \tilde{C}(q,\dot{q}) = \begin{pmatrix} c_{11} & c_{12} & c_{13} \\ c_{21} & c_{22} & c_{23} \\ c_{31} & c_{32} & c_{33} \end{pmatrix} \begin{pmatrix} \dot{x} \\ \dot{\theta}_1 \\ \dot{\theta}_2 \end{pmatrix} \tag{2.8}$$

or equivalently

$$\tilde{C}(q,\dot{q}) = \begin{pmatrix} \tilde{c}_1 \\ \tilde{c}_2 \\ \tilde{c}_3 \end{pmatrix} = \begin{pmatrix} c_{11}\dot{x} + c_{12}\dot{\theta}_1 + c_{13}\dot{\theta}_2 \\ c_{21}\dot{x} + c_{22}\dot{\theta}_1 + c_{23}\dot{\theta}_2 \\ c_{31}\dot{x} + c_{32}\dot{\theta}_1 + c_{33}\dot{\theta}_2 \end{pmatrix} \tag{2.9}$$

while, the gravitational forces vector, after including in it the damping terms associated with the rate of change of angles θ_1 and θ_2, is given by

$$\tilde{G}(q) = \begin{pmatrix} \tilde{g}_1 \\ \tilde{g}_2 \\ \tilde{g}_3 \end{pmatrix} = \begin{pmatrix} 0 \\ (m_1 + m_2)gl_1 \sin(\theta)_1 + d_1\dot{\theta}_1 \\ m_2 gl_2 \sin(\theta_2) + d_2\dot{\theta}_2 \end{pmatrix} \tag{2.10}$$

Consequently, the dynamics of the double-pendulum overhead crane is written as

$$\begin{aligned} \begin{pmatrix} \ddot{x} \\ \ddot{\theta}_1 \\ \ddot{\theta}_2 \end{pmatrix} &= \frac{1}{detM} \begin{pmatrix} M_{11} & -M_{21} & M_{31} \\ -M_{12} & M_{22} & -M_{32} \\ M_{13} & -M_{23} & M_{33} \end{pmatrix} \begin{pmatrix} \tilde{c}_1 \\ \tilde{c}_2 \\ \tilde{c}_3 \end{pmatrix} \\ &- \frac{1}{detM} \begin{pmatrix} M_{11} & -M_{21} & M_{31} \\ -M_{12} & M_{22} & -M_{32} \\ M_{13} & -M_{23} & M_{33} \end{pmatrix} \begin{pmatrix} \tilde{g}_1 \\ \tilde{g}_2 \\ \tilde{g}_3 \end{pmatrix} \\ &+ \frac{1}{detM} \begin{pmatrix} M_{11} & -M_{21} & M_{31} \\ -M_{12} & M_{22} & -M_{32} \\ M_{13} & -M_{23} & M_{33} \end{pmatrix} \begin{pmatrix} u_1 \\ 0 \\ 0 \end{pmatrix} \end{aligned} \tag{2.11}$$

which can be also written in the form

$$\begin{pmatrix} \ddot{x} \\ \ddot{\theta}_1 \\ \ddot{\theta}_2 \end{pmatrix} = \begin{pmatrix} \dfrac{-M_{11}(\tilde{c}_1 + \tilde{g}_1) + M_{21}(\tilde{c}_2 + \tilde{g}_2) - M_{31}(\tilde{c}_3 + \tilde{g}_3)}{detM} \\ \dfrac{M_{13}(\tilde{c}_1 + \tilde{g}_1) - M_{22}(\tilde{c}_2 + \tilde{g}_2) + M_{32}(\tilde{c}_3 + \tilde{g}_3)}{detM} \\ \dfrac{-M_{13}(\tilde{c}_1 + \tilde{g}_1) + M_{23}(\tilde{c}_2 + \tilde{g}_2) - M_{33}(\tilde{c}_3 + \tilde{g}_3)}{detM} \end{pmatrix} + \begin{pmatrix} \dfrac{M_{11}}{detM} \\ \dfrac{-M_{12}}{detM} \\ \dfrac{M_{13}}{detM} \end{pmatrix} u \tag{2.12}$$

Next, by defining the state-vector of the crane as $x = [x_1, x_2, x_3, x_4, x_5, x_6]^T$ or $x = [x, \dot{x}, \theta_1, \dot{\theta}_1, \theta_2, \dot{\theta}_2]$, the state-space model of the double-pendulum overhead-crane system is written as:

$$\begin{pmatrix} \dot{x}_1 \\ \dot{x}_2 \\ \dot{x}_3 \\ \dot{x}_4 \\ \dot{x}_5 \\ \dot{x}_6 \end{pmatrix} = \begin{pmatrix} 0 \\ \dfrac{-M_{11}(\tilde{c}_1 + \tilde{g}_1) + M_{21}(\tilde{c}_2 + \tilde{g}_2) - M_{31}(\tilde{c}_3 + \tilde{g}_3)}{detM} \\ 0 \\ \dfrac{M_{13}(\tilde{c}_1 + \tilde{g}_1) - M_{22}(\tilde{c}_2 + \tilde{g}_2) + M_{32}(\tilde{c}_3 + \tilde{g}_3)}{detM} \\ 0 \\ \dfrac{-M_{13}(\tilde{c}_1 + \tilde{g}_1) + M_{23}(\tilde{c}_2 + \tilde{g}_2) - M_{33}(\tilde{c}_3 + \tilde{g}_3)}{detM} \end{pmatrix} + \begin{pmatrix} 0 \\ \dfrac{M_{11}}{detM} \\ 0 \\ \dfrac{-M_{12}}{detM} \\ 0 \\ \dfrac{M_{13}}{detM} \end{pmatrix} u \quad (2.13)$$

Equivalently, the state-space model of the double-pendulum overhead crane is written as

$$\begin{pmatrix} \dot{x}_1 \\ \dot{x}_2 \\ \dot{x}_3 \\ \dot{x}_4 \\ \dot{x}_5 \\ \dot{x}_6 \end{pmatrix} = \begin{pmatrix} f_1(x) \\ f_2(x) \\ f_3(x) \\ f_4(x) \\ f_5(x) \\ f_6(x) \end{pmatrix} + \begin{pmatrix} g_1(x) \\ g_2(x) \\ g_3(x) \\ g_4(x) \\ g_5(x) \\ g_6(x) \end{pmatrix} u \quad (2.14)$$

or in the concise nonlinear affine-in-the-input form

$$\dot{x} = f(x) + g(x)u \quad (2.15)$$

where $x \in R^{6\times1}$, $f(x) \in R^{6\times1}$, $g(x) \in R^{6\times1}$.

Using the previous notation of state variables, one has $m_{11} = m + m_1 + m_2$, $m_{12} = (m_1 + m_2)l_1 \cos(x_3)$, $m_{13} = m_2 l_2 \cos(x_5)$, $m_{21} = (m_1 + m_2)l_1 \cos(x_3)$, $m_{22} = (m_1 + m_2)l_1$, $m_{23} = m_2 l_1 l_2 \cos(x_3 - x_5)$, $m_{31} = m_2 l_2 \cos(x_5)$, $m_{32} = m_2 l_1 l_2 \cos(x_3 - x_5)$ and $m_{33} = m_2 l_2^2$ and also $c_{12} = 0$, $c_{12} = -(m_1 + m_2)l_1 \sin(x_3)x_2$, $c_{13} = -m_2 l_2 \sin(x_5)x_6$, $c_{21} = 0$, $c_{22} = 0$, $c_{23} = m_2 l_1 l_2 in(x_3 - x_5)x_6$, $c_{31} = 0$, $c_{32} = -m_2 l_1 l_2 \sin(x_3 - x_5)x_4$, and $c_{33} = 0$.

Additionally, one has $g_1 = 0$, $g_2 = (m_1 + m_2)g l_2 \sin(x_3)$ and $g_3 = m_2 g l_2 \sin(x_5)$.

2.1.3 Approximate linearization of the double-pendulum overhead crane

The nonlinear state-space model of the double-pendulum overhead crane being initially in the form

$$\dot{x} = f(x) + g(x)u \quad (2.16)$$

where $x \in R^{6\times1}$, $f(x) \in R^{6\times1}$, $g(x) \in R^{6\times1}$ and $u \in R$ undergo approximate linearization. The linearization process is based on the first-order Taylor series expansion and on the computation of the associated Jacobian matrices. The linearization is performed around the temporary operating point (x^*, u^*) which is recomputed at each sampling period, where x^* is the present value of the system's state vector and u^* is the last sampled value of the control inputs vector.

The equivalent linearized model of the double-pendulum overhead crane is given by

$$\dot{x} = Ax + Bu + \tilde{d} \tag{2.17}$$

where $\tilde{d}$ is the cumulative modeling error which is due to (i) truncation of higher-order terms in the Taylor series expansion, (ii) external perturbations, and (iii) sensors measurement noise of any distribution. Moreover, matrices A and B are the Jacobian matrices of the system which are computed as follows:

$$A = \nabla_x[f(x) + g(x)u \mid_{(x^*,u^*)}] \Rightarrow A = \nabla_x f(x) \mid_{x^*,u^*} + \nabla_x g(x)u \mid_{x^*,u^*} \tag{2.18}$$

$$B = \nabla_u[f(x) + g(x)u \mid_{(x^*,u^*)} \Rightarrow B = g(x) \mid_{(x^*,u^*)} \tag{2.19}$$

Computation of the elements of the Jacobian matrix $\nabla_x f(x) \mid_{x^*,u^*}$

First row of the Jacobian matrix $\nabla_x f(x) \mid_{x^*,u^*}$: $\frac{\partial f_1(x)}{\partial x_1} = 0$, $\frac{\partial f_1(x)}{\partial x_2} = 1$, $\frac{\partial f_1(x)}{\partial x_3} = 0$, $\frac{\partial f_1(x)}{\partial x_4} = 0$, $\frac{\partial f_1(x)}{\partial x_5} = 0$, $\frac{\partial f_1(x)}{\partial x_6} = 0$.

Second row of the Jacobian matrix $\nabla_x f(x) \mid_{x^*,u^*}$: For $i = 1, 2, \cdots, 8$, the numerator of function $f_2(x)$ is

$$f_{2n}(x) = -M_{11}(\tilde{c}_1 + \tilde{g}_1) + M_{21}(\tilde{c}_2 + \tilde{g}_2) - M_{31}(\tilde{c}_3 + \tilde{g}_3) \tag{2.20}$$

while the denominator of function $f_2(x)$ is

$$f_{2d}(x) = detM \tag{2.21}$$

For $i = 1, 2, \cdots, 6$, it holds that

$$\begin{aligned}\frac{\partial f_{2n}(x)}{\partial x_i} = & -\frac{\partial M_{11}(x)}{\partial x_i}(\tilde{c}_1(x) + \tilde{g}_1(x)) - M_{11}(x)\left(\frac{\partial \tilde{c}_1(x)}{\partial x_i} + \frac{\partial \tilde{g}_1(x)}{\partial x_i}\right) \\ & + \frac{\partial M_{21}(x)}{\partial x_i}(\tilde{c}_2(x) + \tilde{g}_2(x)) + M_{21}(x)\left(\frac{\partial \tilde{c}_2(x)}{\partial x_i} + \frac{\partial \tilde{g}_2(x)}{\partial x_i}\right) + \\ & - \frac{\partial M_{31}(x)}{\partial x_i}(\tilde{c}_3(x) + \tilde{g}_3(x)) - M_{31}(x)\left(\frac{\partial \tilde{c}_3(x)}{\partial x_i} + \frac{\partial \tilde{g}_3(x)}{\partial x_i}\right)\end{aligned} \tag{2.22}$$

$$\frac{\partial f_{2d}(x)}{\partial x_i} = \frac{\partial detM(x)}{\partial x_i} \tag{2.23}$$

while it also holds that for $i = 1, 2, \cdots, 6$

$$\frac{\partial f_2(x)}{\partial x_i} = \frac{\frac{\partial f_{2n}(x)}{\partial x_i} f_{2d}(x) - f_{2n}(x)\frac{\partial f_{2d}(x)}{\partial x_i}}{f_{2d}(x)^2} \tag{2.24}$$

Third row of the Jacobian matrix $\nabla_x f(x) \mid_{x^*,u^*}$: $\frac{\partial f_3(x)}{\partial x_1} = 0$, $\frac{\partial f_3(x)}{\partial x_2} = 0$, $\frac{\partial f_3(x)}{\partial x_3} = 0$, $\frac{\partial f_3(x)}{\partial x_4} = 1$, $\frac{\partial f_3(x)}{\partial x_5} = 0$, $\frac{\partial f_3(x)}{\partial x_6} = 0$.

Fourth row of the Jacobian matrix $\nabla_x f(x)\mid_{x^*,u^*}$: For $i = 1, 2, \cdots, 8$, the numerator of function $f_4(x)$ is

$$f_{4n}(x) = M_{12}(\tilde{c}_1 + \tilde{g}_1) - M_{22}(\tilde{c}_2 + \tilde{g}_2) + M_{32}(\tilde{c}_3 + \tilde{g}_3) \tag{2.25}$$

while the denominator of function $f_4(x)$ is

$$f_{4d}(x) = detM \tag{2.26}$$

For $i = 1, 2, \cdots, 6$, it holds that

$$\begin{aligned}\frac{\partial f_{4n}(x)}{\partial x_i} = {}& \frac{\partial M_{12}(x)}{\partial x_i}(\tilde{c}_1(x) + \tilde{g}_1(x)) + M_{12}(x)\left(\frac{\partial \tilde{c}_1(x)}{\partial x_i} + \frac{\partial \tilde{g}_1(x)}{\partial x_i}\right) \\ & - \frac{\partial M_{22}(x)}{\partial x_i}(\tilde{c}_2(x) + \tilde{g}_2(x)) - M_{22}(x)\left(\frac{\partial \tilde{c}_2(x)}{\partial x_i} + \frac{\partial \tilde{g}_2(x)}{\partial x_i}\right) \\ & + \frac{\partial M_{32}(x)}{\partial x_i}(\tilde{c}_3(x) + \tilde{g}_3(x)) + M_{32}(x)\left(\frac{\partial \tilde{c}_3(x)}{\partial x_i} + \frac{\partial \tilde{g}_3(x)}{\partial x_i}\right)\end{aligned} \tag{2.27}$$

$$\frac{\partial f_{4d}(x)}{\partial x_i} = \frac{\partial detM(x)}{\partial x_i} \tag{2.28}$$

while it also holds that for $i = 1, 2, \cdots, 6$

$$\frac{\partial f_4(x)}{\partial x_i} = \frac{\frac{\partial f_{4n}(x)}{\partial x_i} f_{4d}(x) - f_{4n}(x)\frac{\partial f_{4d}(x)}{\partial x_i}}{f_{4d}(x)^2} \tag{2.29}$$

Fifth row of the Jacobian matrix $\nabla_x f(x)\mid_{x^*,u^*}$: $\frac{\partial f_5(x)}{\partial x_1} = 0$, $\frac{\partial f_5(x)}{\partial x_2} = 0$, $\frac{\partial f_5(x)}{\partial x_3} = 0$, $\frac{\partial f_5(x)}{\partial x_4} = 0$, $\frac{\partial f_5(x)}{\partial x_5} = 0$, $\frac{\partial f_5(x)}{\partial x_6} = 1$.

Sixth row of the Jacobian matrix $\nabla_x f(x)\mid_{x^*,u^*}$: For $i = 1, 2, \cdots, 6$, the numerator of function $f_6(x)$ is

$$f_{6n}(x) = -M_{13}(\tilde{c}_1 + \tilde{g}_1) + M_{23}(\tilde{c}_2 + \tilde{g}_2) - M_{33}(\tilde{c}_3 + \tilde{g}_3) \tag{2.30}$$

while the denominator of function $f_6(x)$ is

$$f_{6d}(x) = detM \tag{2.31}$$

For $i = 1, 2, \cdots, 6$, it holds that

$$\begin{aligned}\frac{\partial f_{6n}(x)}{\partial x_i} = {}& -\frac{\partial M_{13}(x)}{\partial x_i}(\tilde{c}_1(x) + \tilde{g}_1(x)) - M_{13}(x)\left(\frac{\partial \tilde{c}_1(x)}{\partial x_i} + \frac{\partial \tilde{g}_1(x)}{\partial x_i}\right) \\ & + \frac{\partial M_{23}(x)}{\partial x_i}(\tilde{c}_2(x) + \tilde{g}_2(x)) + M_{23}(x)\left(\frac{\partial \tilde{c}_2(x)}{\partial x_i} + \frac{\partial \tilde{g}_2(x)}{\partial x_i}\right) \\ & - \frac{\partial M_{33}(x)}{\partial x_i}(\tilde{c}_3(x) + \tilde{g}_3(x)) - M_{33}(x)\left(\frac{\partial \tilde{c}_3(x)}{\partial x_i} + \frac{\partial \tilde{g}_3(x)}{\partial x_i}\right)\end{aligned} \tag{2.32}$$

$$\frac{\partial f_{6d}(x)}{\partial x_i} = \frac{\partial detM(x)}{\partial x_i} \tag{2.33}$$

while it also holds that for $i = 1, 2, \cdots, 6$

$$\frac{\partial f_6(x)}{\partial x_i} = \frac{\frac{\partial f_{6n}(x)}{\partial x_i} f_{6d}(x) - f_{6n}(x)\frac{\partial f_{6d}(x)}{\partial x_i}}{f_{6d}(x)^2} \tag{2.34}$$

Computation of the Jacobian matrix $\nabla_x g(x)\mid_{x^*,u^*}$: For $i = 1, 2, \cdots, 6$ it hold that

First row of the Jacobian matrix $\nabla_x g(x)\mid_{x^*,u^*}$: $\frac{\partial g_1(x)}{\partial x_i} = 0$ for $i = 1, 2, \cdots, 6$.

Second row of the Jacobian matrix $\nabla_x g(x)\mid_{x^*,u^*}$:

$$\frac{\partial g_2(x)}{\partial x_i} = \frac{\frac{\partial M_{11}(x)}{\partial x_i} detM(x) - M_{11}(x)\frac{\partial detM}{\partial x_i}}{detM(x)^2} \tag{2.35}$$

Third row of the Jacobian matrix $\nabla_x g(x)\mid_{x^*,u^*}$: $\frac{\partial g_3(x)}{\partial x_i} = 0$ for $i = 1, 2, \cdots, 6$.

Fourth row of the Jacobian matrix $\nabla_x g(x)\mid_{x^*,u^*}$:

$$\frac{\partial g_4(x)}{\partial x_i} = \frac{\frac{-\partial M_{12}(x)}{\partial x_i} detM(x) + M_{12}(x)\frac{\partial detM}{\partial x_i}}{detM(x)^2} \tag{2.36}$$

Fifth row of the Jacobian matrix $\nabla_x g(x)\mid_{x^*,u^*}$: $\frac{\partial g_5(x)}{\partial x_i} = 0$ for $i = 1, 2, \cdots, 6$.

Sixth row of the Jacobian matrix $\nabla_x g(x)\mid_{x^*,u^*}$:

$$\frac{\partial g_6(x)}{\partial x_i} = \frac{\frac{\partial M_{13}(x)}{\partial x_i} detM(x) - M_{13}(x)\frac{\partial detM}{\partial x_i}}{detM(x)^2} \tag{2.37}$$

Next, one computes the partial derivatives of the elements of the inertia matrix $M(x)$: for $i = 1, 2, \cdots, 6$, it holds that

$$\frac{\partial M_{11}}{\partial x_i} = \frac{dm_{22}}{dx_i}m_{33} + m_{22}\frac{dm_{33}}{dx_i} - \frac{dm_{32}}{dx_i}m_{23} - m_{32}\frac{dm_{23}}{dx_i} \tag{2.38}$$

$$\frac{\partial M_{12}}{\partial x_i} = \frac{dm_{21}}{dx_i}m_{33} + m_{21}\frac{dm_{33}}{dx_i} - \frac{dm_{31}}{dx_i}m_{23} - m_{31}\frac{dm_{23}}{dx_i} \tag{2.39}$$

$$\frac{\partial M_{13}}{\partial x_i} = \frac{dm_{21}}{dx_i}m_{32} + m_{21}\frac{dm_{32}}{dx_i} - \frac{dm_{31}}{dx_i}m_{22} - m_{31}\frac{dm_{22}}{dx_i} \tag{2.40}$$

$$\frac{\partial M_{21}}{\partial x_i} = \frac{dm_{12}}{dx_i}m_{23} + m_{12}\frac{dm_{23}}{dx_i} - \frac{dm_{32}}{dx_i}m_{13} - m_{32}\frac{dm_{13}}{dx_i} \tag{2.41}$$

$$\frac{\partial M_{22}}{\partial x_i} = \frac{dm_{11}}{dx_i}m_{33} + m_{11}\frac{dm_{33}}{dx_i} - \frac{dm_{31}}{dx_i}m_{13} - m_{31}\frac{dm_{13}}{dx_i} \tag{2.42}$$

$$\frac{\partial M_{23}}{\partial x_i} = \frac{dm_{11}}{dx_i}m_{22} + m_{11}\frac{dm_{22}}{dx_i} - \frac{dm_{31}}{dx_i}m_{12} - m_{31}\frac{dm_{12}}{dx_i} \tag{2.43}$$

$$\frac{\partial M_{31}}{\partial x_i} = \frac{dm_{12}}{dx_i}m_{23} + m_{12}\frac{dm_{23}}{dx_i} - \frac{dm_{22}}{dx_i}m_{13} - m_{22}\frac{dm_{13}}{dx_i} \tag{2.44}$$

$$\frac{\partial M_{32}}{\partial x_i} = \frac{dm_{11}}{dx_i}m_{23} + m_{11}\frac{dm_{23}}{dx_i} - \frac{dm_{21}}{dx_i}m_{13} - m_{21}\frac{dm_{13}}{dx_i} \tag{2.45}$$

$$\frac{\partial M_{33}}{\partial x_i} = \frac{dm_{11}}{dx_i}m_{22} + m_{11}\frac{dm_{22}}{dx_i} - \frac{dm_{21}}{dx_i}m_{12} - m_{21}\frac{dm_{12}}{dx_i} \tag{2.46}$$

About the partial derivative of $detM$ it holds that for $i = 1, 2, \cdots, 6$

$$\frac{\partial detM}{\partial x_i} = \frac{dm_{11}}{dx_i}M_{11} + m_{11}\frac{\partial M_{11}}{\partial x_i} - \frac{dm_{12}}{dx_i}M_{12} - m_{12}\frac{\partial M_{12}}{\partial x_i} + \frac{dm_{13}}{dx_i}M_{13} + m_{13}\frac{\partial M_{13}}{\partial x_i} \tag{2.47}$$

Next, one computes the partial derivatives of the elements of the Coriolis and centrifugal forces vector $\tilde{C}$

Partial derivatives of the first row of $\tilde{C}$: $\frac{\partial \tilde{c}_1}{\partial x_1} = \frac{dc_{11}}{dx_1}x_2 + \frac{dc_{12}}{dx_1}x_4 + \frac{dc_{13}}{dx_1}x_6$, $\frac{\partial \tilde{c}_1}{\partial x_2} = \frac{dc_{11}}{dx_2}x_2 + c_{11} + \frac{dc_{12}}{dx_2}x_4 + \frac{dc_{13}}{dx_2}x_6$, $\frac{\partial \tilde{c}_1}{\partial x_3} = \frac{dc_{11}}{dx_3}x_2 + \frac{dc_{12}}{dx_3}x_4 + \frac{dc_{13}}{dx_3}x_6$, $\frac{\partial \tilde{c}_1}{\partial x_4} = \frac{dc_{11}}{dx_4}x_2 + \frac{dc_{12}}{dx_4}x_4 + c_{12} + \frac{dc_{13}}{dx_4}x_6$, $\frac{\partial \tilde{c}_1}{\partial x_5} = \frac{dc_{11}}{dx_5}x_2 + \frac{dc_{12}}{dx_5}x_4 + \frac{dc_{13}}{dx_5}x_6$, $\frac{\partial \tilde{c}_1}{\partial x_6} = \frac{dc_{11}}{dx_6}x_2 + \frac{dc_{12}}{dx_6}x_4 + \frac{dc_{13}}{dx_6}x_6 + c_{13}$.

Partial derivatives of the second row of $\tilde{C}$: $\frac{\partial \tilde{c}_2}{\partial x_1} = \frac{dc_{21}}{dx_1}x_2 + \frac{dc_{22}}{dx_1}x_4 + \frac{dc_{23}}{dx_1}x_6$, $\frac{\partial \tilde{c}_2}{\partial x_2} = \frac{dc_{21}}{dx_2}x_2 + c_{21} + \frac{dc_{22}}{dx_2}x_4 + \frac{dc_{23}}{dx_2}x_6$, $\frac{\partial \tilde{c}_2}{\partial x_3} = \frac{dc_{21}}{dx_3}x_2 + \frac{dc_{22}}{dx_3}x_4 + \frac{dc_{23}}{dx_3}x_6$, $\frac{\partial \tilde{c}_2}{\partial x_4} = \frac{dc_{21}}{dx_4 1}x_2 + \frac{dc_{22}}{dx_4}x_4 + c_{22} + \frac{dc_{23}}{dx_4}x_6$, $\frac{\partial \tilde{c}_2}{\partial x_5} = \frac{dc_{21}}{dx_5}x_2 + \frac{dc_{22}}{dx_5}x_4 + \frac{dc_{23}}{dx_5}x_6$, $\frac{\partial \tilde{c}_2}{\partial x_6} = \frac{dc_{21}}{dx_6}x_2 + \frac{dc_{22}}{dx_6}x_4 + \frac{dc_{23}}{dx_6}x_6 + c_{23}$.

Partial derivatives of the third row of $\tilde{C}$: $\frac{\partial \tilde{c}_3}{\partial x_1} = \frac{dc_{31}}{dx_1}x_2 + \frac{dc_{32}}{dx_1}x_4 + \frac{dc_{33}}{dx_1}x_6$, $\frac{\partial \tilde{c}_3}{\partial x_2} = \frac{dc_{31}}{dx_2}x_2 + c_{31} + \frac{dc_{32}}{dx_2}x_4 + \frac{dc_{33}}{dx_2}x_6$, $\frac{\partial \tilde{c}_3}{\partial x_3} = \frac{dc_{31}}{dx_3}x_2 + \frac{dc_{32}}{dx_3}x_4 + \frac{dc_{33}}{dx_3}x_6$, $\frac{\partial \tilde{c}_3}{\partial x_4} = \frac{dc_{31}}{dx_4}x_2 + \frac{dc_{32}}{dx_4}x_4 + c_{32} + \frac{dc_{33}}{dx_4}x_6$, $\frac{\partial \tilde{c}_3}{\partial x_5} = \frac{dc_{31}}{dx_5}x_2 + \frac{dc_{32}}{dx_5}x_4 + \frac{dc_{33}}{dx_5}x_6$, $\frac{\partial \tilde{c}_3}{\partial x_6} = \frac{dc_{31}}{dx_6}x_2 + \frac{dc_{32}}{dx_6}x_4 + \frac{dc_{33}}{dx_6}x_6 + c_{33}$.

Next, one computes the partial derivatives of the gravitational forces vector $\tilde{G}$

Partial derivatives of the first row of $\tilde{G}$: $\frac{\partial \tilde{g}_1}{\partial x_1} = 0$, $\frac{\partial \tilde{g}_1}{\partial x_2} = 0$, $\frac{\partial \tilde{g}_1}{\partial x_3} = 0$, $\frac{\partial \tilde{g}_1}{\partial x_4} = 0$, $\frac{\partial \tilde{g}_1}{\partial x_5} = 0$, $\frac{\partial \tilde{g}_1}{\partial x_6} = 0$.

Partial derivatives of the second row of $\tilde{G}$: $\frac{\partial \tilde{g}_2}{\partial x_1} = 0$, $\frac{\partial \tilde{g}_2}{\partial x_2} = 0$, $\frac{\partial \tilde{g}_2}{\partial x_3} = (m_1 + m_2)gl_1\cos(x_3)$, $\frac{\partial \tilde{g}_2}{\partial x_4} = d_1$, $\frac{\partial \tilde{g}_2}{\partial x_5} = 0$, $\frac{\partial \tilde{g}_2}{\partial x_6} = 0$.

Partial derivatives of the third row of $\tilde{G}$: $\frac{\partial \tilde{g}_3}{\partial x_1} = 0$, $\frac{\partial \tilde{g}_3}{\partial x_2} = 0$, $\frac{\partial \tilde{g}_3}{\partial x_3} = 0$, $\frac{\partial \tilde{g}_3}{\partial x_4} = 0$, $\frac{\partial \tilde{g}_3}{\partial x_5} = m_2gl_2\cos(x_5)$, $\frac{\partial \tilde{g}_3}{\partial x_6} = d_2$.

Next, one computes the partial derivatives of the following parameters:

$\frac{\partial m_{11}}{\partial x_1} = 0$, $\frac{\partial m_{11}}{\partial x_2} = 0$, $\frac{\partial m_{11}}{\partial x_3} = 0$, $\frac{\partial m_{11}}{\partial x_4} = 0$, $\frac{\partial m_{11}}{\partial x_5} = 0$, $\frac{\partial m_{11}}{\partial x_6} = 0$.

$\frac{\partial m_{12}}{\partial x_1} = 0$, $\frac{\partial m_{12}}{\partial x_1} = 0$, $\frac{\partial m_{12}}{\partial x_3} = -(m_1 + m_2)l_2\sin(x_3)$, $\frac{\partial m_{12}}{\partial x_4} = 0$, $\frac{\partial m_{12}}{\partial x_5} = 0$, $\frac{\partial m_{12}}{\partial x_6} = 0$.

$\frac{\partial m_{13}}{\partial x_1}=0,\ \frac{\partial m_{13}}{\partial x_2}=0,\ \frac{\partial m_{13}}{\partial x_3}=-(m_1+m2)l_2\sin(x_3),\ \frac{\partial m_{13}}{\partial x_4}=0,\ \frac{\partial m_{13}}{\partial x_5}=0,\ \frac{\partial m_{13}}{\partial x_6}=0.$

$\frac{\partial m_{21}}{\partial x_1}=0,\ \frac{\partial m_{21}}{\partial x_2}=0,\ \frac{\partial m_{21}}{\partial x_3}=-(m_1+m_2)l_1\sin(x_3),\ \frac{\partial m_{21}}{\partial x_4}=0,\ \frac{\partial m_{21}}{\partial x_5}=0,\ \frac{\partial m_{21}}{\partial x_6}=0.$

$\frac{\partial m_{22}}{\partial x_1}=0,\ \frac{\partial m_{22}}{\partial x_2}=0,\ \frac{\partial m_{22}}{\partial x_3}=0,\ \frac{\partial m_{22}}{\partial x_4}=0,\ \frac{\partial m_{22}}{\partial x_5}=0,\ \frac{\partial m_{22}}{\partial x_6}=0.$

$\frac{\partial m_{23}}{\partial x_1}=0,\ \frac{\partial m_{23}}{\partial x_2}=0,\ \frac{\partial m_{23}}{\partial x_3}=-m_2l_1l_2\sin(x_3-x_5),\ \frac{\partial m_{23}}{\partial x_4}=0,\ \frac{\partial m_{23}}{\partial x_5}=m_2l_1l_2\sin(x_3-x_5),\ \frac{\partial m_{23}}{\partial x_6}=0.$

$\frac{\partial m_{31}}{\partial x_1}=0,\ \frac{\partial m_{31}}{\partial x_2}=0,\ \frac{\partial m_{31}}{\partial x_3}=0,\ \frac{\partial m_{31}}{\partial x_4}=0,\ \frac{\partial m_{31}}{\partial x_5}=-m_2l_2\sin(x_5),\ \frac{\partial m_{31}}{\partial x_6}=0.$

$\frac{\partial m_{32}}{\partial x_1}=0,\ \frac{\partial m_{32}}{\partial x_2}=0,\ \frac{\partial m_{32}}{\partial x_3}=-m_2l_1l_2\sin(x_3-x_5),\ \frac{\partial m_{32}}{\partial x_4}=0,\ \frac{\partial m_{32}}{\partial x_5}=m_2l_1l_2\sin(x_3-x_5),\ \frac{\partial m_{32}}{\partial x_6}=0.$

$\frac{\partial m_{33}}{\partial x_1}=0,\ \frac{\partial m_{33}}{\partial x_2}=0,\ \frac{\partial m_{33}}{\partial x_3}=0,\ \frac{\partial m_{33}}{\partial x_4}=0,\ \frac{\partial m_{33}}{\partial x_5}=0,\ \frac{\partial m_{33}}{\partial x_6}=0.$

In a similar manner one computes:

$\frac{\partial c_{11}}{\partial x_i}=0\ \frac{\partial c_{11}}{\partial x_1}=0,\ \frac{\partial c_{11}}{\partial x_2}=0,\ \frac{\partial c_{11}}{\partial x_3}=0,\ \frac{\partial c_{11}}{\partial x_4}=0,\ \frac{\partial c_{11}}{\partial x_5}=0,\ \frac{\partial c_{11}}{\partial x_6}=0.$

$\frac{\partial c_{12}}{\partial x_1}=0,\quad \frac{\partial c_{12}}{\partial x_2}=-(m_1+m_2)l_1\sin(x_3),\quad \frac{\partial c_{12}}{\partial x_3}=(m_1+m2)l_1\cos(x_3)x_2,\quad \frac{\partial c_{12}}{\partial x_4}=0,\ \frac{\partial c_{12}}{\partial x_5}=0,\ \frac{\partial c_{12}}{\partial x_6}=0.$

$\frac{\partial c_{13}}{\partial x_1}=0,\ \frac{\partial c_{13}}{\partial x_2}=0,\ \frac{\partial c_{13}}{\partial x_3}=0,\ \frac{\partial c_{13}}{\partial x_4}=0,\ \frac{\partial c_{13}}{\partial x_5}=-m_2l_2\cos(x_5)x_6,\ \frac{\partial c_{13}}{\partial x_6}=-m_2l_2\sin(x_5).$

$\frac{\partial c_{21}}{\partial x_1}=0,\ \frac{\partial c_{21}}{\partial x_2}=0,\ \frac{\partial c_{21}}{\partial x_3}=0,\ \frac{\partial c_{21}}{\partial x_4}=0,\ \frac{\partial c_{21}}{\partial x_5}=0,\ \frac{\partial c_{21}}{\partial x_6}=0.$

$\frac{\partial c_{22}}{\partial x_1}=0,\ \frac{\partial c_{22}}{\partial x_2}=0,\ \frac{\partial c_{22}}{\partial x_3}=0,\ \frac{\partial c_{22}}{\partial x_4}=0,\ \frac{\partial c_{22}}{\partial x_5}=0,\ \frac{\partial c_{22}}{\partial x_6}=0.$

$\frac{\partial c_{23}}{\partial x_1}=0,\ \frac{\partial c_{23}}{\partial x_2}=0,\ \frac{\partial c_{23}}{\partial x_3}=m_2l_1l_2\cos(x_3-x_5)x_6,\ \frac{\partial c_{23}}{\partial x_4}=0,\ \frac{\partial c_{23}}{\partial x_5}=-m_2l_1l_2\cos(x_3-x_5)x_6,\ \frac{\partial c_{23}}{\partial x_6}=0.$

$\frac{\partial c_{31}}{\partial x_1}=0,\ \frac{\partial c_{31}}{\partial x_2}=0,\ \frac{\partial c_{31}}{\partial x_3}=0,\ \frac{\partial c_{31}}{\partial x_4}=0,\ \frac{\partial c_{31}}{\partial x_5}=0,\ \frac{\partial c_{31}}{\partial x_6}=0.$

$\frac{\partial c_{32}}{\partial x_1}=0,\quad \frac{\partial c_{32}}{\partial x_2}=0,\quad \frac{\partial c_{32}}{\partial x_3}=-m_2l_1l_2\cos(x_1-x_5)x_4,\quad \frac{\partial c_{32}}{\partial x_4}=-m_2l_1l_2\sin(x_1-x_5),$ $\frac{\partial c_{32}}{\partial x_5}=m_2l_1l_2\cos(x_3-x_5)x_4$, and $\frac{\partial c_{32}}{\partial x_6}=0.$

$\frac{\partial c_{33}}{\partial x_1}=0,\ \frac{\partial c_{33}}{\partial x_2}=0,\ \frac{\partial c_{33}}{\partial x_3}=0,\ \frac{\partial c_{33}}{\partial x_4}=0,\ \frac{\partial c_{33}}{\partial x_5}=0,\ \frac{\partial c_{33}}{\partial x_6}=0.$

2.1.4 Computation of the feedback control gains

For the linearized system given by (1.36) the cost function of (1.37) is defined, where the coefficient r determines the penalization of the control input and the weight coefficient ρ determines the reward of the disturbances' effects. It is assumed that (i) the energy that is transferred from the disturbances signal $\tilde{d}(t)$ is bounded, that is $\int_0^\infty \tilde{d}^T(t)\tilde{d}(t)dt < \infty$, (ii) matrices $[A, B]$ and $[A, L]$ are stabilizable, and (iii) matrix $[A, C]$ is detectable. In the case of a tracking problem, the optimal feedback control law is given by

$$u(t) = -Ke(t) \tag{2.48}$$

with $e = x - x_d$ to be the tracking error, and $K = \frac{1}{r}B^T P$ where P is a positive definite symmetric matrix. Matrix P is obtained from the solution of the Riccati equation

$$A^T P + PA + Q - P\left(\frac{2}{r}BB^T - \frac{1}{\rho^2}LL^T\right)P = 0 \tag{2.49}$$

where Q is a positive semi-definite symmetric matrix. The worst case disturbance is given by

$$\tilde{d}(t) = \frac{1}{\rho^2}L^T Pe(t) \tag{2.50}$$

The diagram of the considered control loop for the double-pendulum overhead crane is depicted in Figure 2.2. The global stability properties of the control method can be proven following the procedure presented in Section 1.1.2.5.

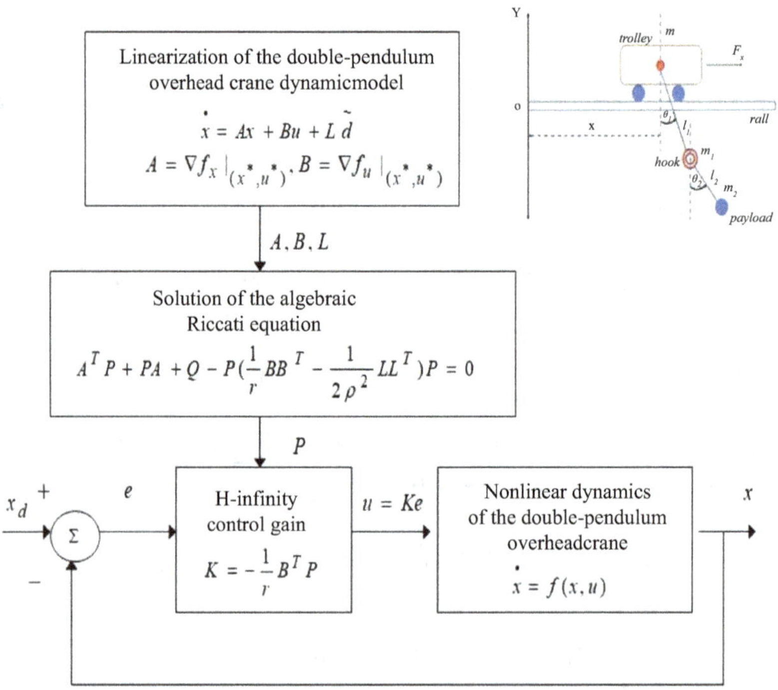

Figure 2.2 Diagram of the control scheme for the double-pendulum overhead crane

2.1.5 Simulation tests

The global stability properties of the control method and the elimination of the state vector's tracking error which were previously proven through Lyapunov analysis are further confirmed through simulation experiments. To compute the control inputs of the double-pendulum overhead crane, the algebraic Riccati equation of (2.49) had to

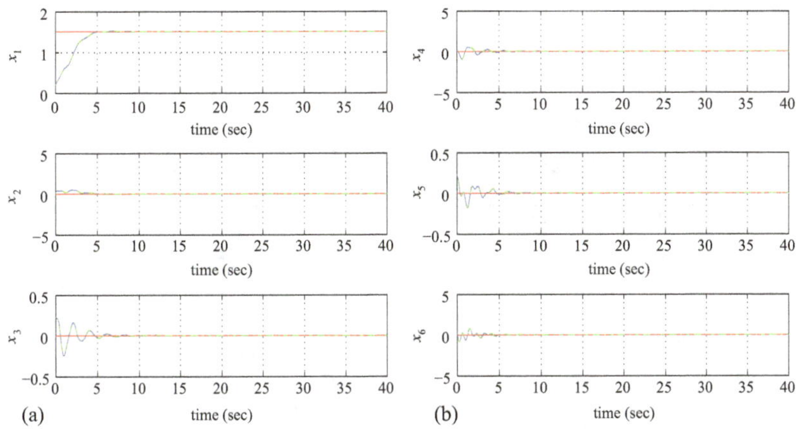

Figure 2.3 Tracking of setpoint 1 for the double-pendulum overhead crane: (a) convergence of state variables x_1 to x_3 to their reference setpoints (red line: setpoint, blue line: real value, green line: estimated value); (b) convergence of state variables x_4 to x_6 to their reference setpoints

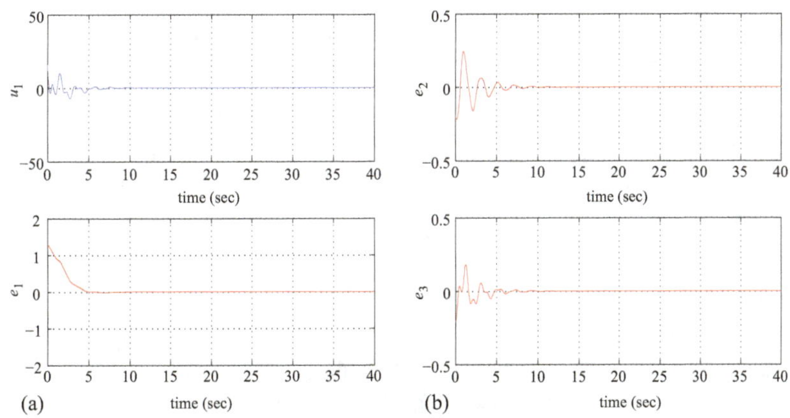

Figure 2.4 Tracking of setpoint 1 for the double-pendulum overhead crane: (a) control input u applied to crane and tracking error e_1 of the cart; (b) tracking error variables e_2 and e_3 of the double-pendulum

be solved at each time-step of the control algorithm. The tracking performance of the control scheme was tested using several reference setpoint that the state variables of the double-pendulum overhead crane had to follow. The obtained results are depicted in Figures 2.3 to 2.18. The real values of the state vector of the double-pendulum overhead crane are printed in blue, the estimated values which were provided by the

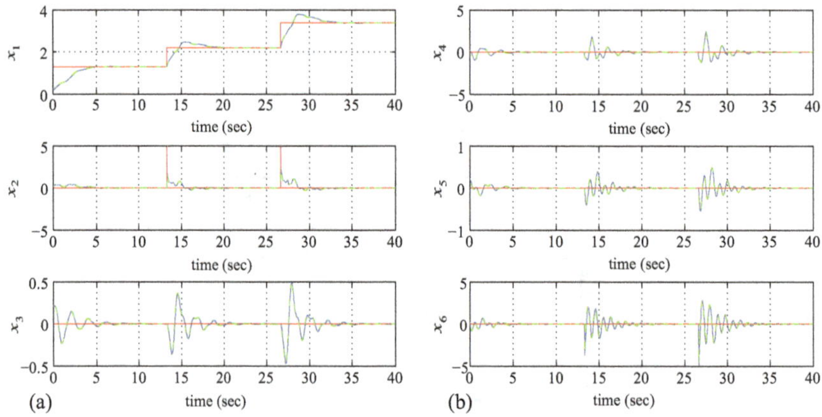

Figure 2.5 Tracking of setpoint 2 for the double-pendulum overhead crane: (a) convergence of state variables x_1 to x_3 to their reference setpoints (red line: setpoint, blue line: real value, green line: estimated value); (b) convergence of state variables x_4 to x_6 to their reference setpoints

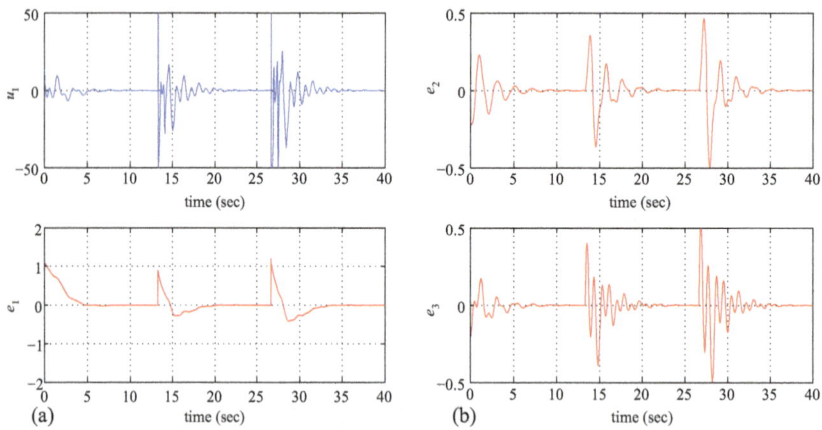

Figure 2.6 Tracking of setpoint 2 for the double-pendulum overhead crane: (a) control input u applied to crane and tracking error e_1 of the cart; (b) tracking error variables e_2 and e_3 of the double-pendulum

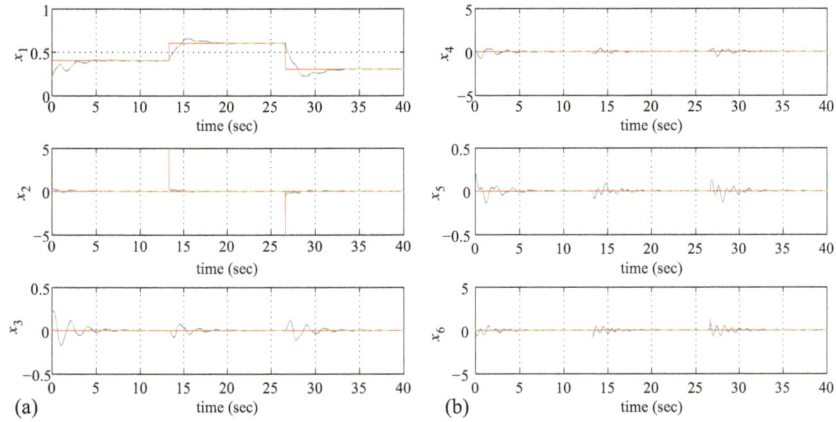

Figure 2.7 Tracking of setpoint 3 for the double-pendulum overhead crane: (a) convergence of state variables x_1 to x_3 to their reference setpoints (red line: setpoint, blue line: real value, green line: estimated value); (b) convergence of state variables x_4 to x_6 to their reference setpoints

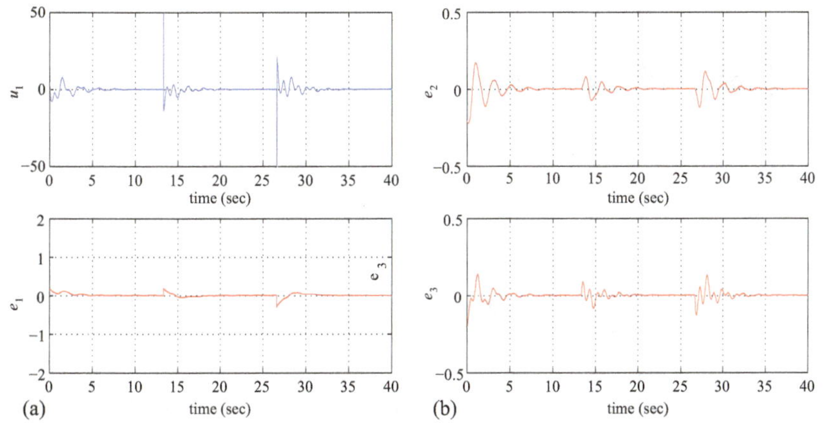

Figure 2.8 Tracking of setpoint 3 for the double-pendulum overhead crane: (a) control input u applied to crane and tracking error e_1 of the cart; (b) tracking error variables e_2 and e_3 of the double-pendulum

H-infinity Kalman Filter are plotted in green, while the related reference setpoints are shown in red. Distance is measured in meters (m), angles in rad, and time in sec. It can be noticed that in all cases, the proposed H-infinity controller achieved fast and accurate tracking of the reference paths under moderate variations of the control inputs.

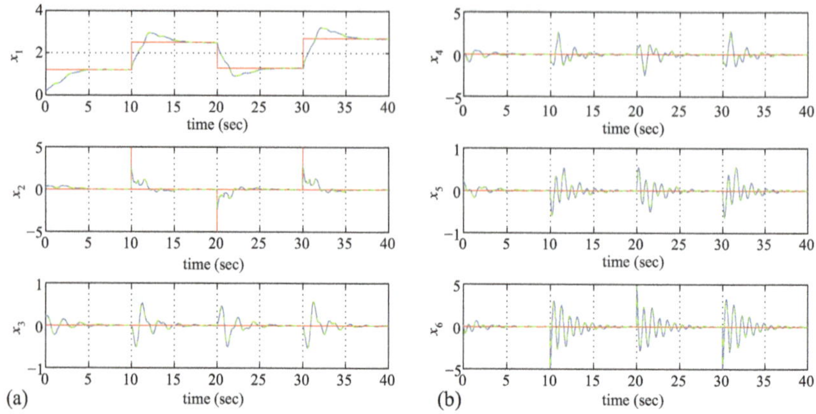

Figure 2.9 Tracking of setpoint 4 for the double-pendulum overhead crane: (a) convergence of state variables x_1 to x_3 to their reference setpoints (red line: setpoint, blue line: real value, green line: estimated value); (b) convergence of state variables x_4 to x_6 to their reference setpoints

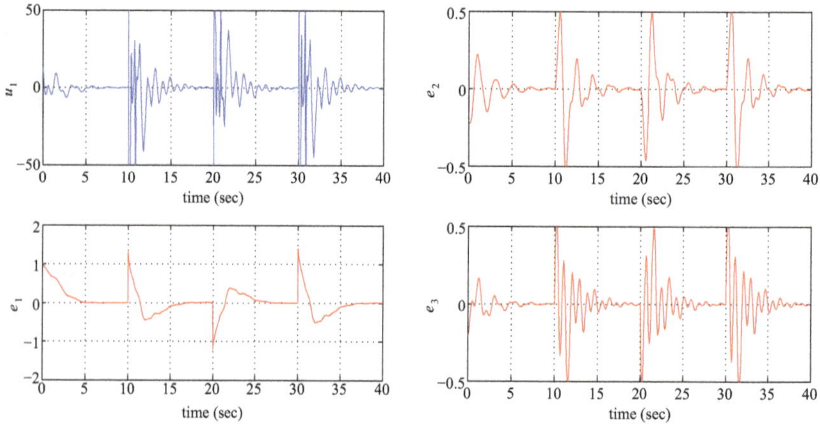

Figure 2.10 Tracking of setpoint 4 for the double-pendulum overhead crane: (a) control input u applied to crane and tracking error e_1 of the cart; (b) tracking error variables e_2 and e_3 of the double-pendulum

2.2 Nonlinear control of the underactuated offshore crane

2.2.1 Outline

The control problem of offshore boom cranes is important for minimizing the loading and unloading time of container ships and for executing such tasks with high accuracy

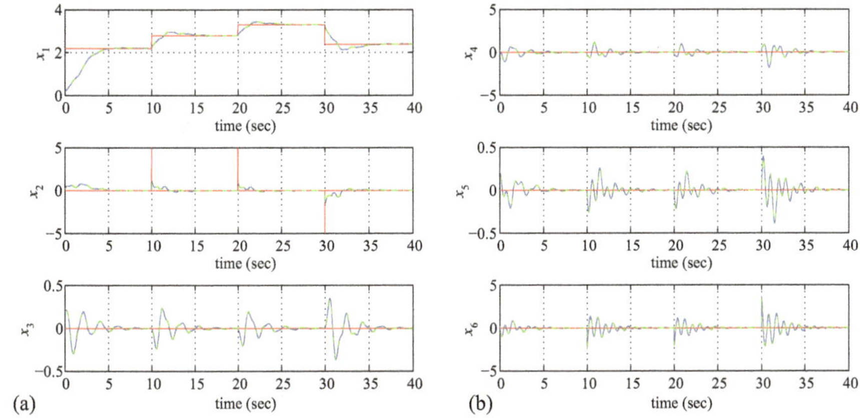

Figure 2.11 Tracking of setpoint 5 for the double-pendulum overhead crane: (a) convergence of state variables x_1 to x_3 to their reference setpoints (red line: setpoint, blue line: real value, green line: estimated value); (b) convergence of state variables x_4 to x_6 to their reference setpoints

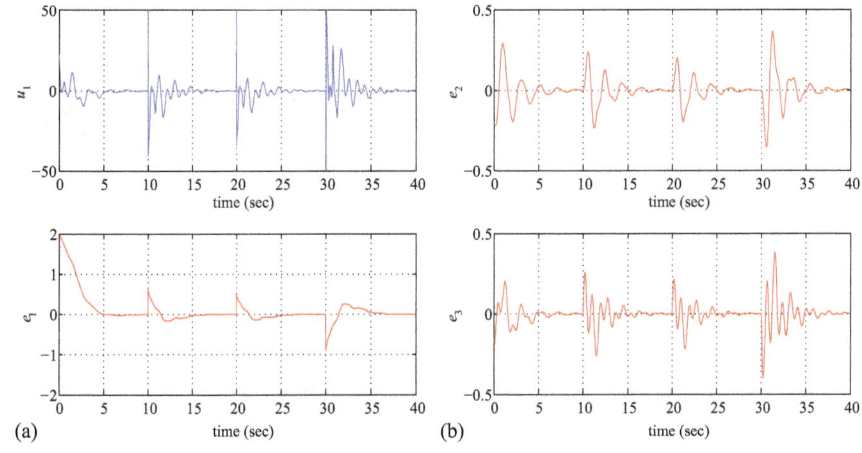

Figure 2.12 Tracking of setpoint 5 for the double-pendulum overhead crane: (a) control input u applied to crane and tracking error e_1 of the cart; (b) tracking error variables e_2 and e_3 of the double-pendulum

and safety, as noted in (Lu, Fang *et al.*, 2018; Fang, Wang *et al.*, 2014; Sun, Qiang, Xu *et al.*, 2017). This control problem is nontrivial due to the multivariable and nonlinear structure of the boom cranes' state-space model, as well as due to underactuation, as shown in (Neupert *et al.*, 2010; Kim and Park, 2015; Fang, Wang *et al.*, 2014; Yang, Sun *et al.*, 2019; Qian and Fang, 2018). Actually, a boom crane exhibits three degrees

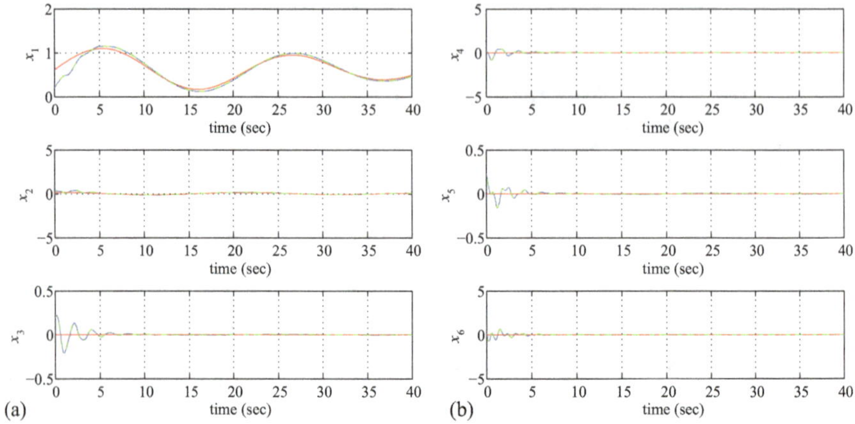

Figure 2.13 Tracking of setpoint 6 for the double-pendulum overhead crane: (a) convergence of state variables x_1 to x_3 to their reference setpoints (red line: setpoint, blue line: real value, green line: estimated value); (b) convergence of state variables x_4 to x_6 to their reference setpoints

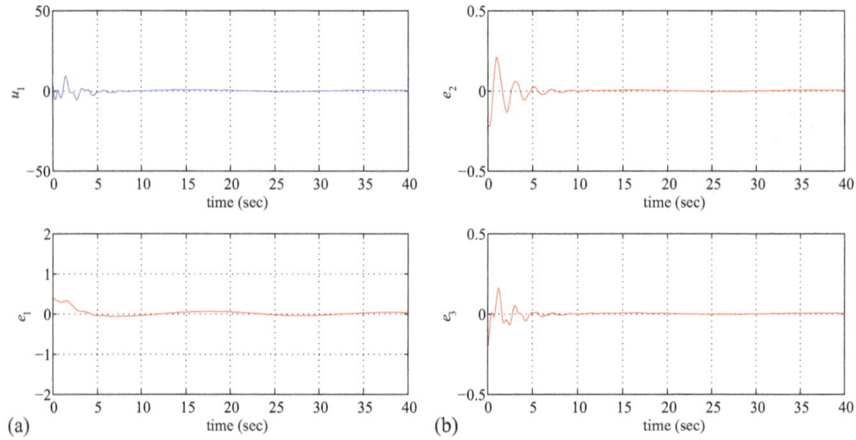

Figure 2.14 Tracking of setpoint 6 for the double-pendulum overhead crane: (a) control input u applied to crane and tracking error e_1 of the cart; (b) tracking error variables e_2 and e_3 of the double-pendulum

of freedom while receiving only two control inputs and this makes difficult the execution of precise manipulations, as well as trajectory tracking for the manipulated objects, while resulting also to unnecessary oscillations and high energy consumption by the boom crane's actuators, as shown in (Su, Fang *et al.*, 2016; Fang and Wang, 2012; Tuan, Cuang *et al.*, 2014; Ismail *et al.*, 2015; Leban *et al.*, 2015). So far,

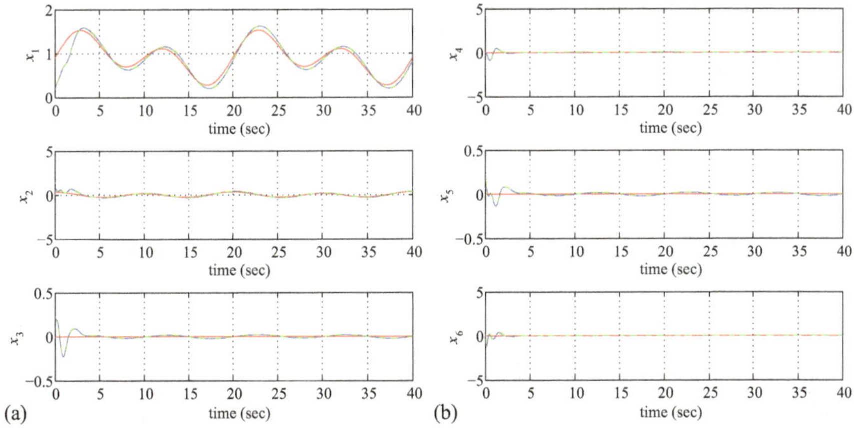

Figure 2.15 Tracking of setpoint 7 for the double-pendulum overhead crane: (a) convergence of state variables x_1 to x_3 to their reference setpoints (red line: setpoint, blue line: real value, green line: estimated value); (b) convergence of state variables x_4 to x_6 to their reference setpoints

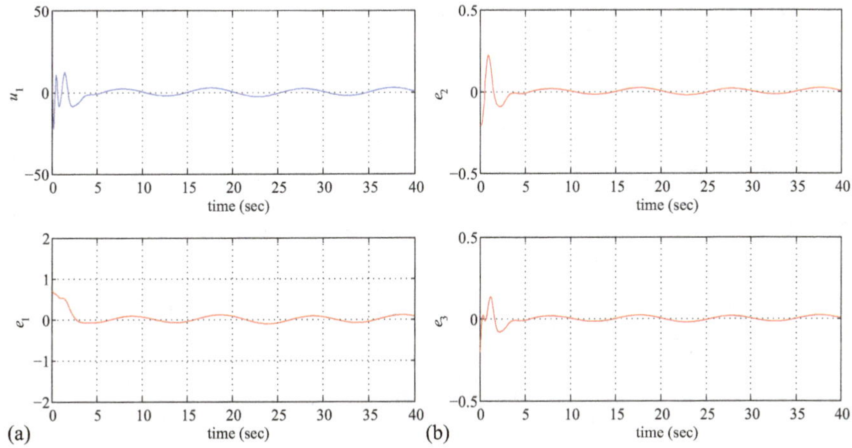

Figure 2.16 Tracking of setpoint 7 for the double-pendulum overhead crane: (a) control input u applied to crane and tracking error e_1 of the cart; (b) tracking error variables e_2 and e_3 of the double-pendulum

several approaches to the nonlinear control problem of offshore boom cranes have been attempted, comprising adaptive and robust control schemes, as well as various energy-based (Lyaponov theory-based) methods, as in (Tho and Terashima, 2019; Wu, Karkoub *et al.*, 2016; Raja Ismail and Ha, 2014; Fang, Wang *et al.*, 2014; Njo, Nguyen *et al.*, 2017). Besides, the application of popular industrial methods for optimal

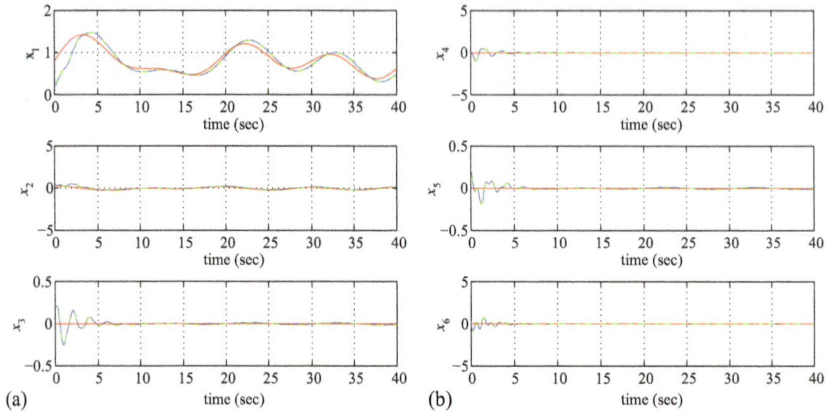

Figure 2.17 *Tracking of setpoint 8 for the double-pendulum overhead crane: (a) convergence of state variables x_1 to x_3 to their reference setpoints (red line: setpoint, blue line: real value, green line: estimated value); (b) convergence of state variables x_4 to x_6 to their reference setpoints*

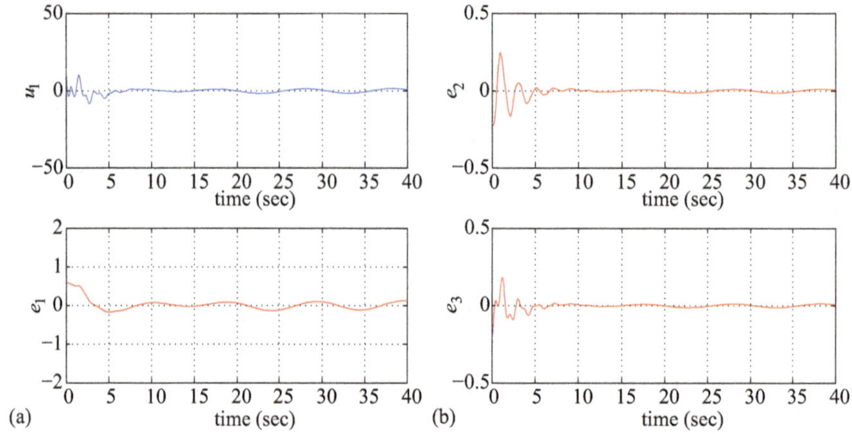

Figure 2.18 *Tracking of setpoint 8 for the double-pendulum overhead crane: (a) control input u applied to crane and tracking error e_1 of the cart; (b) tracking error variables e_2 and e_3 of the double-pendulum*

control, such as the MPC and the NMPC has been considered and efforts to overcome related limitations have been carried out, as in (Bock and Kugi, 2014; Chen, Fang *et al.*, 2016; Ho *et al.*, 2019; Devos and Levine, 2006). For instance, it is widely acknowledged that MPC is a linear control method and its use in the nonlinear dynamics of boom cranes will result into loss of the stability of the control loop. Additionally, in the case of NMPC, the convergence of the iterative search for an optimum depends on initialization and on parameters' values selection and cannot be always assured.

In the present section, a novel nonlinear optimal (H-infinity) control method is proposed for the dynamic model of offshore boom cranes (Rigatos and Busawon, 2018; Rigatos, Siano and Abbaszadeh, 2017; Rigatos, Siano *et al.*, 2015). This control scheme relies on approximate linearization of the boom crane's state-space model with the use of Taylor series expansion and the computation of the related Jacobian matrices (Rigatos and Tzafestas, 2007; Basseville and Nikiforov, 1993; Rigatos and Zhang, 2009). The linearization takes place around a time-varying operating point which is recomputed at each time-step of the control method. This point is defined by the present value of system's state vector and by the most recent value of the control inputs vector. The modeling error which is induced because of truncating higher-order terms in the Taylor series expansion is treated as a perturbation which is asymptotically eliminated by the robustness of the control algorithm. For the approximately linearized model of the boom crane a stabilizing H-infinity feedback controller is designed (Rigatos, 2011; Rigatos, 2015; Rigatos, 2016b).

The nonlinear optimal (H-infinity) controller stands for the solution of the optimal control problem for the boom crane's model under model uncertainty and external perturbations. Actually, it represents a min-max differential game in which the control inputs try to minimize a cost function that contains a quadratic term of the state vector's tracking error. On the other side, the disturbance and model uncertainty terms try to minimize this cost functional. For the computation of the controller's feedback gains an algebraic Riccati equation it repetitively solved at each time-step of the control algorithm. The stability and robustness properties of the control method are proven through Lyapunov analysis. First, it is proven that the H-infinity tracking performance criterion holds, and this signifies elevated robustness against model uncertainty and external perturbations (Tousssaint, Basar and Bullo, 2000; Lublin and Athans, 1995). Additionally, under moderate conditions, it is proven that the control scheme is globally asymptotically stable. Moreover, to implement state estimation-based control without the need to measure the entire state vector of the boom crane, the H-infinity Kalman Filter is proposed as a robust state estimator (Rigatos and Busawon, 2018). The proposed control scheme retains the advantages of linear optimal control, that is fast and accurate tracking of the reference setpoints under moderate variations of the control inputs.

2.2.2 Dynamic model of the boom crane

The diagram of the offshore boom crane is shown in Figure 2.19. There are two reference frames, namely the inertial frame OXY and the body-fixed frame $O_bY_bZ_b$. The body fixed frame is aligned with the inertial reference frame after rotation by an angle α. This angle appears because of the roll motion of the vessel which is induced by the waves' motion. The turn angle of the boom crane with respect to the horizontal axis of the body-fixed reference frame is denoted by ϕ. The moment of inertia of

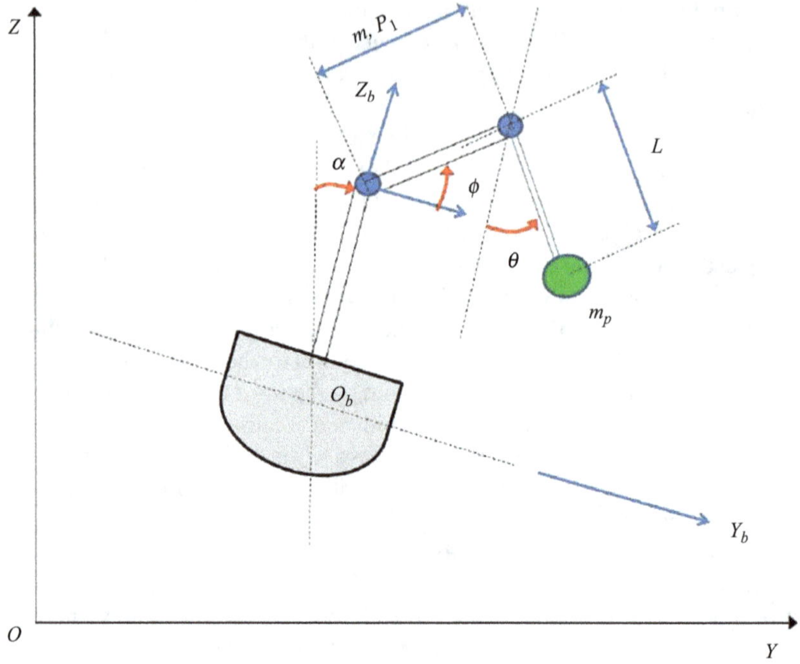

Figure 2.19 Body-fixed and inertial reference frame for the offshore boom crane

the boom is denoted as J. The mass of the boom crane is m and its length is P_l. The payload (e.g. lifted container) has mass m_p and is moved by a string of variable length L. The angle that is formed between the string and the vertical axis of the body-fixed reference frame of the vessel is denoted by θ. The state variables of this model are ϕ, L and θ. The state-space model of the boom crane is obtained through Euler–Lagrange analysis and after computing the potential and kinematic energy of the boom and of the load. This model is given by (Lu, Fang *et al.*, 2018)

$$M(q)\ddot{q} + V(q,\dot{q})\dot{q} = u + w \tag{2.51}$$

The inertial matrix $M(q)$ of the boom crane's model is given by (Lu, Fang *et al.*, 2018)

$$M(q) = \begin{pmatrix} m_{11} & m_{12} & m_{13} \\ m_{21} & m_{22} & m_{23} \\ m_{31} & m_{32} & m_{33}] \end{pmatrix} \tag{2.52}$$

where $m_{11} = J + m_p P_L^2, m_{12} = m_{21} = -m_P P_L \cos(\theta - \phi), m_{13} = m_{31} = m_P P_L \sin(\theta - \phi), m_{22} = m_p, m_{23} = m_{32} = 0$ and $m_{33} = m_p L^2$.

The Coriolis matrix $V(q,\dot{q})$ of the boom crane's model is given by (Lu, Fang *et al.*, 2018)

$$V(q,\dot{q}) = \begin{pmatrix} v_{11} & v_{12} & v_{13} \\ v_{21} & v_{22} & v_{23} \\ v_{31} & v_{32} & v_{33} \end{pmatrix} \tag{2.53}$$

where $v_{11} = 0$, $v_{12} = m_p P_L \sin(\theta - \phi)\dot{\phi}$, $v_{13} = m_p P_L \sin(\theta - \phi)\dot{L} + m_p P_L \cos(\theta - \phi)\dot{\theta}$, $v_{21} = -m_p P_L \sin(\theta - \phi)\dot{\phi}$, $v_{22} = 0$, $v_{23} = -m_p L\dot{\theta}$, $v_{31} = -m_p P_L L\cos(\theta - \phi)\dot{\phi}$, $v_{32} = m_p L\dot{\theta}$ and $v_{33} = m_p L\dot{L}$.

The state vector of the system is $= [\phi, L, \theta]^T$ and its first derivative is $\dot{q} = [\dot{\phi}, \dot{L}, \dot{\theta}]^T$. The control inputs vector $u \in R^{3\times1}$ is $u = [T_m, F_l, 0]^T$ where T_m is the torque applied to the basis joint of the boom, while F_l is the string force that lifts the payload. The disturbance inputs vector $w \in R^{3\times1}$ is obtained under the assumption that the roll angle of the vessel a varies slowly and consequently $\dot{\alpha} = 0$ and $\dot{\alpha} = 0$ (Lu, Fang *et al.*, 2018). The disturbances' vector is given by $w = [w_1, w_2, w_3]^T$, with $w_1 = -m_p P_l + m{\cdot}dg\cos(\phi - a)$, $w_2 = m_p g\cos(\theta - \alpha)$ and $w_3 = -m_p Lg\sin(\theta - \alpha)$. Variable d denotes the distance between the joint at the basis of the boom and its center of gravity.
The inverse of the inertia matrix $M(q)$ is

$$M(q) = \frac{1}{detM}\begin{pmatrix} M_{11} & -M_{21} & M_{31} \\ -M_{12} & M_{22} & -M_{32} \\ M_{13} & -M_{23} & M_{33} \end{pmatrix} \tag{2.54}$$

where the determinant $detM$ is given by

$$detM = Jm_p^2 L^2 \tag{2.55}$$

and the elements of the inverse inertia matrix are given by: $M_{11} = m_p^2 L^2$, $M_{12} = -m_p^2 L^2 P_L \cos(\theta - \phi)$, $M_{13} = -m_p^2 P_l L\sin(\theta - \phi)$, $M_{21} = -m_p^2 L^2 P_L \cos(\theta - \phi)$, $M_{22} = (J + m_p P_l^2)m_P L^2 - m_p^2 P_l^2 L^2 \sin^2(\theta - \phi)$, $M_{23} = m_p^2 P_l^2 L\sin(\theta - \phi)\cos(\theta - \phi)$, $M_{31} = -m_p^2 P_l L\sin(\theta - \phi)$, $M_{32} = m_p^2 P_l^2 L\sin(\theta - \phi)\cos(\theta - \phi)$ and $M_{33} = (J + m_p L^2)m_p - m_p^2 P_l^2 \cos^2(\theta - \phi)$.

By applying the inversion of the inertia matrix, the state-space model of the system becomes

$$\dot{q} = M^{-1}(q)\{-V(q,\dot{q})\dot{q} + w(q)\} + M^{-1}(q)u \tag{2.56}$$

Moreover, one can define the cumulative matrix $\tilde{C}(q,\dot{q}) = -V(q,\dot{q})\dot{q} + w(q)$ with $\tilde{C}(q,\dot{q}) = [\tilde{C}_1(q,\dot{q}), \tilde{C}_2(q,\dot{q}), \tilde{C}_3(q,\dot{q})]^T$, where it holds that

$$\begin{aligned} \tilde{C}_1(q,\dot{q}) = &-2m_p P_l \sin(\theta - \phi)\dot{\theta}\dot{L} - m_p P_l \cos(\theta - \phi)\dot{\theta}^2 \\ &+ (-m_p P_l + m{\cdot}d)g\cos(\phi - \alpha) \end{aligned} \tag{2.57}$$

$$\tilde{C}_2(q,\dot{q}) = m_p P_l \sin(\theta - \phi)\dot{\phi}^2 + m_p L\dot{\theta}^2 + m_p g\cos(\theta - \alpha) \tag{2.58}$$

$$\tilde{C}_3(q,\dot{q}) = m_p P_l L\cos(\theta - \phi)\dot{\phi}^2 - 2m_p L\dot{L}\dot{\theta} + m_p Lg\sin(\theta - \alpha) \tag{2.59}$$

Elaborating on (2.56), one arrives at the following state-space description:

$$\begin{pmatrix} \ddot{\phi} \\ \ddot{L} \\ \ddot{\theta} \end{pmatrix} = \frac{1}{detM} \begin{pmatrix} M_{11} & -M_{21} & M_{31} \\ -M_{12} & M_{22} & -M_{32} \\ M_{13} & -M_{23} & M_{33} \end{pmatrix} \begin{pmatrix} \tilde{C}_1(q,\dot{q}) \\ \tilde{C}_2(q,\dot{q}) \\ \tilde{C}_3(q,\dot{q}) \end{pmatrix} + \frac{1}{detM} \begin{pmatrix} M_{11} & -M_{21} & M_{31} \\ -M_{12} & M_{22} & -M_{32} \\ M_{13} & -M_{23} & M_{33} \end{pmatrix} \begin{pmatrix} T_m \\ F_l \\ 0 \end{pmatrix} \tag{2.60}$$

Continuing on the above, the state-space model of the offshore boom crane can be written as the following:

$$\begin{pmatrix} \ddot{\phi} \\ \ddot{L} \\ \ddot{\theta} \end{pmatrix} = \begin{pmatrix} \frac{M_{11}\tilde{C}_1 - M_{21}\tilde{C}_2 + M_{31}\tilde{C}_3}{detM} \\ \frac{-M_{12}\tilde{C}_1 + M_{22}\tilde{C}_2 - M_{32}\tilde{C}_3}{detM} \\ \frac{M_{13}1\tilde{C}_1 - M_{23}\tilde{C}_2 + M_{33}\tilde{C}_3}{detM} \end{pmatrix} + \begin{pmatrix} \frac{M_{11}}{detM} & \frac{-M_{21}}{detM} \\ \frac{-M_{12}}{detM} & \frac{M_{22}}{detM} \\ \frac{M_{13}}{detM} & \frac{-M_{22}}{detM} \end{pmatrix} \begin{pmatrix} T_m \\ F_l \end{pmatrix} \tag{2.61}$$

Next, the state vector $x = [x_1, x_2, x_3, x_4, x_5, x_6]^T = [\theta, \dot{\theta}, L, \dot{L}, \phi, \dot{\phi}]^T$ is defined, while the following functions are introduced:

$$\begin{aligned} & f_1(x) = x_2 && g_{11}(x) = 0 && g_{11}(x) = 0 \\ & f_2(x) = \frac{M_{11}\tilde{C}_1 - M_{21}\tilde{C}_2 + M_{31}\tilde{C}_3}{detM} && g_{21}(x) = \frac{M_{11}}{detM} && g_{22}(x) = \frac{-M_{21}}{detM} \\ & f_3(x) = x_4 && g_{31}(x) = 0 && g_{32}(x) = 0 \\ & f_4(x) = \frac{-M_{12}\tilde{C}_1 + M_{22}\tilde{C}_2 - M_{32}\tilde{C}_3}{detM} && g_{41}(x) = -\frac{M_{12}}{detM} && g_{42}(x) = \frac{M_{22}}{detM} \\ & f_5(x) = x_6 && g_{51}(x) = 0 && g_{52}(x) = 0 \\ & f_6(x) = \frac{M_{13}\tilde{C}_1 - M_{23}\tilde{C}_2 + M_{33}\tilde{C}_3}{detM} && g_{61}(x) = \frac{M_{13}}{detM} && g_{62}(x) = -\frac{M_{23}}{detM} \end{aligned} \tag{2.62}$$

Using (2.62) and by defining the control inputs vector $u = [u_1, u_2]^T = [T_m, F_L]^T$, one arrives at the following state-space description for the offshore boom crane's model

$$\begin{pmatrix} \dot{x}_1 \\ \dot{x}_2 \\ \dot{x}_3 \\ \dot{x}_4 \\ \dot{x}_5 \\ \dot{x}_6 \end{pmatrix} = \begin{pmatrix} f_1(x) \\ f_2(x) \\ f_3(x) \\ f_4(x) \\ f_5(x) \\ f_6(x) \end{pmatrix} + \begin{pmatrix} g_{11}(x) & g_{12}(x) \\ g_{21}(x) & g_{22}(x) \\ g_{31}(x) & g_{32}(x) \\ g_{41}(x) & g_{42}(x) \\ g_{51}(x) & g_{52}(x) \\ g_{61}(x) & g_{62}(x) \end{pmatrix} \begin{pmatrix} u_1 \\ u_2 \end{pmatrix} \tag{2.63}$$

The previous state-space model is in the affine-in-the-input form

$$\dot{x} = f(x) + g(x)u \tag{2.64}$$

where $x\in R^{6\times 1}$, $f(x)\in R^{6\times 1}$, $g(x)\in R^{6\times 2}$, and $u\in R^{2\times 1}$.

2.2.3 Approximate linearization of the dynamic model of boom cranes

The previously analyzed model of the offshore boom crane undergoes approximate linearization around the temporary operating point (x^*, u^*), where x^* is the present value of the system's state vector and u^* is the most recent value of the control inputs vectors. Starting from the nonlinear state-space model of (2.64), the linearization procedure gives

$$\dot{x} = Ax + Bu + \tilde{d} \tag{2.65}$$

where $\tilde{d}$ is the cumulative disturbances' vector and matrices A, B are given by the system's Jacobians

$$\begin{aligned} A &= \nabla_x[f(x) + g_1(x)u_1 + g_2(x)u_2] \mid_{(x^*,u^*)} \Rightarrow \\ A &= \nabla_x[f(x)] \mid_{(x^*,u^*)} + \nabla_x[g_1(x)]u \mid_{(x^*,u^*)} + \nabla_x[g_2(x)]u \mid_{(x^*,u^*)} \end{aligned} \tag{2.66}$$

$$\begin{aligned} B &= \nabla_u[f(x) + g_1(x)u_1 + g_2(x)u_2] \mid_{(x^*,u^*)} \Rightarrow \\ B &= [g_1(x), g_2(x)] \mid_{(x^*,u^*)} \Rightarrow B = g(x) \mid_{(x^*,u^*)} \end{aligned} \tag{2.67}$$

Computation of the Jacobian matrix $\nabla_x f(x)$:

$$\nabla_x f(x) = \begin{pmatrix} \frac{\partial f_1}{\partial x_1} & \frac{\partial f_1}{\partial x_2} & \frac{\partial f_1}{\partial x_3} & \frac{\partial f_1}{\partial x_4} & \frac{\partial f_1}{\partial x_5} & \frac{\partial f_1}{\partial x_6} \\ \frac{\partial f_2}{\partial x_1} & \frac{\partial f_2}{\partial x_2} & \frac{\partial f_2}{\partial x_3} & \frac{\partial f_2}{\partial x_4} & \frac{\partial f_2}{\partial x_5} & \frac{\partial f_2}{\partial x_6} \\ \frac{\partial f_3}{\partial x_1} & \frac{\partial f_3}{\partial x_2} & \frac{\partial f_3}{\partial x_3} & \frac{\partial f_3}{\partial x_4} & \frac{\partial f_3}{\partial x_5} & \frac{\partial f_3}{\partial x_6} \\ \frac{\partial f_4}{\partial x_1} & \frac{\partial f_4}{\partial x_2} & \frac{\partial f_4}{\partial x_3} & \frac{\partial f_4}{\partial x_4} & \frac{\partial f_4}{\partial x_5} & \frac{\partial f_4}{\partial x_6} \\ \frac{\partial f_5}{\partial x_1} & \frac{\partial f_5}{\partial x_2} & \frac{\partial f_5}{\partial x_3} & \frac{\partial f_5}{\partial x_4} & \frac{\partial f_5}{\partial x_5} & \frac{\partial f_5}{\partial x_6} \\ \frac{\partial f_6}{\partial x_1} & \frac{\partial f_6}{\partial x_2} & \frac{\partial f_6}{\partial x_3} & \frac{\partial f_6}{\partial x_4} & \frac{\partial f_6}{\partial x_5} & \frac{\partial f_6}{\partial x_6} \end{pmatrix} \tag{2.68}$$

About the first row of the Jacobian matrix $\nabla_x f(x)$, one has: $\frac{\partial f_1}{\partial x_1} = 0$, $\frac{\partial f_1}{\partial x_2} = 1$, $\frac{\partial f_1}{\partial x_3} = 0$, $\frac{\partial f_1}{\partial x_4} = 0$, $\frac{\partial f_1}{\partial x_5} = 0$, $\frac{\partial f_1}{\partial x_6} = 0$.

About the second row of the Jacobian matrix $\nabla_x f(x)$ and for $i = 1, 2, \cdots, 6$, one has:

$$\begin{aligned} \frac{\partial f_2}{\partial x_i} &= \frac{\left[\frac{\partial M_{11}}{\partial x_i}\tilde{C}_1 + M_{11}\frac{\partial \tilde{C}_1}{\partial x_i} - \frac{\partial M_{21}}{\partial x_i}\tilde{C}_2 - M_{21}\frac{\partial \tilde{C}_2}{\partial x_i} + \frac{\partial M_{31}}{\partial x_i}\tilde{C}_3 + M_{31}\frac{\partial \tilde{C}_1}{\partial x_i}\right]}{(detM)^2} \\ &\quad - \frac{[M_{11}\tilde{C}_1 - M_{21}\tilde{C}_2 + M_{31}\tilde{C}_3]\frac{\partial detM}{\partial x_i}}{(detM)^2} \end{aligned} \tag{2.69}$$

About the third row of the Jacobian matrix $\nabla_x f(x)$, one has: $\frac{\partial f_3}{\partial x_1} = 0$, $\frac{\partial f_3}{\partial x_2} = 0$, $\frac{\partial f_3}{\partial x_3} = 0$, $\frac{\partial f_3}{\partial x_4} = 1$, $\frac{\partial f_3}{\partial x_5} = 0$, $\frac{\partial f_3}{\partial x_6} = 0$.

About the fourth row of the Jacobian matrix $\nabla_x f(x)$ and for $i = 1, 2, \cdots, 6$, one has:

$$\frac{\partial f_4}{\partial x_i} = \frac{\left[\frac{-\partial M_{12}}{\partial x_i}\tilde{C}_1 - M_{12}\frac{\partial \tilde{C}_1}{\partial x_i} + \frac{\partial M_{22}}{\partial x_i}\tilde{C}_2 + M_{22}\frac{\partial \tilde{C}_2}{\partial x_i} - \frac{\partial M_{32}}{\partial x_i}\tilde{C}_3 - M_{32}\frac{\partial \tilde{C}_3}{\partial x_i}\right]}{(detM)^2} - \frac{[-M_{12}\tilde{C}_1 + M_{22}\tilde{C}_2 - M_{32}\tilde{C}_3]\frac{\partial detM}{\partial x_i}}{(detM)^2} \tag{2.70}$$

About the fifth row of the Jacobian matrix $\nabla_x f(x)$, one has: $\frac{\partial f_5}{\partial x_1} = 0$, $\frac{\partial f_5}{\partial x_2} = 0$, $\frac{\partial f_5}{\partial x_3} = 0$, $\frac{\partial f_5}{\partial x_4} = 0$, $\frac{\partial f_5}{\partial x_5} = 0$, $\frac{\partial f_5}{\partial x_6} = 1$.

About the sixth row of the Jacobian matrix $\nabla_x f(x)$ and for $i = 1, 2, \cdots, 6$, one has:

$$\frac{\partial f_6}{\partial x_i} = \frac{\left[\frac{\partial M_{13}}{\partial x_i}\tilde{C}_1 + M_{13}\frac{\partial \tilde{C}_1}{\partial x_i} - \frac{\partial M_{23}}{\partial x_i}\tilde{C}_2 - M_{23}\frac{\partial \tilde{C}_2}{\partial x_i} + \frac{\partial M_{33}}{\partial x_i}\tilde{C}_3 + M_{33}\frac{\partial \tilde{C}_1}{\partial x_i}\right]}{(detM)^2} - \frac{[M_{13}\tilde{C}_1 - M_{23}\tilde{C}_2 + M_{33}\tilde{C}_3]\frac{\partial detM}{\partial x_i}}{(detM)^2} \tag{2.71}$$

Computation of the Jacobian matrix $\nabla_x g_1(x)$:

$$\nabla_x g_1(x) = \begin{pmatrix} \frac{\partial g_{11}}{\partial x_1} & \frac{\partial g_{11}}{\partial x_2} & \frac{\partial g_{11}}{\partial x_3} & \frac{\partial g_{11}}{\partial x_4} & \frac{\partial g_{11}}{\partial x_5} & \frac{\partial g_{11}}{\partial x_6} \\ \frac{\partial g_{21}}{\partial x_1} & \frac{\partial g_{21}}{\partial x_2} & \frac{\partial g_{21}}{\partial x_3} & \frac{\partial g_{21}}{\partial x_4} & \frac{\partial g_{21}}{\partial x_5} & \frac{\partial g_{21}}{\partial x_6} \\ \frac{\partial g_{31}}{\partial x_1} & \frac{\partial g_{31}}{\partial x_2} & \frac{\partial g_{31}}{\partial x_3} & \frac{\partial g_{31}}{\partial x_4} & \frac{\partial g_{31}}{\partial x_5} & \frac{\partial g_{31}}{\partial x_6} \\ \frac{\partial g_{41}}{\partial x_1} & \frac{\partial g_{41}}{\partial x_2} & \frac{\partial g_{41}}{\partial x_3} & \frac{\partial g_{41}}{\partial x_4} & \frac{\partial g_{41}}{\partial x_5} & \frac{\partial g_{41}}{\partial x_6} \\ \frac{\partial g_{51}}{\partial x_1} & \frac{\partial g_{51}}{\partial x_2} & \frac{\partial g_{51}}{\partial x_3} & \frac{\partial g_{51}}{\partial x_4} & \frac{\partial g_{51}}{\partial x_5} & \frac{\partial g_{51}}{\partial x_6} \\ \frac{\partial g_{61}}{\partial x_1} & \frac{\partial g_{61}}{\partial x_2} & \frac{\partial g_{61}}{\partial x_3} & \frac{\partial g_{61}}{\partial x_4} & \frac{\partial g_{61}}{\partial x_5} & \frac{\partial g_{61}}{\partial x_6} \end{pmatrix} \tag{2.72}$$

About the first row of the Jacobian matrix $\nabla_x g_1(x)$, one has: $\frac{\partial g_{11}}{\partial x_1} = 0$, $\frac{\partial g_{11}}{\partial x_2} = 0$, $\frac{\partial g_{11}}{\partial x_3} = 0$, $\frac{\partial g_{11}}{\partial x_4} = 0$, $\frac{\partial g_{11}}{\partial x_5} = 0$, $\frac{\partial g_{11}}{\partial x_6} = 0$.

About the second row of the Jacobian matrix $\nabla_x g_1(x)$ and for $i = 1, 2, \cdots, 6$, one has:

$$\frac{\partial g_{21}}{\partial x_i} = \frac{\frac{\partial M_{11}}{\partial x_i}detM - M_{11}\frac{\partial detM}{\partial x_i}}{(detM)^2} \tag{2.73}$$

About the third row of the Jacobian matrix $\nabla_x g_1(x)$, one has: $\frac{\partial g_{21}}{\partial x_1}=0$, $\frac{\partial g_{21}}{\partial x_2}=0$, $\frac{\partial g_{21}}{\partial x_3}=0$, $\frac{\partial g_{21}}{\partial x_4}=0$, $\frac{\partial g_{21}}{\partial x_5}=0$, $\frac{\partial g_{21}}{\partial x_6}=0$.

About the fourth row of the Jacobian matrix $\nabla_x g_1(x)$ and for $i=1,2,\cdots,6$, one has:

$$\frac{\partial g_{41}}{\partial x_i}=-\frac{\frac{\partial M_{12}}{\partial x_i}detM-M_{12}\frac{\partial detM}{\partial x_i}}{(detM)^2} \tag{2.74}$$

About the fifth row of the Jacobian matrix $\nabla_x g_1(x)$, one has: $\frac{\partial g_{51}}{\partial x_1}=0$, $\frac{\partial g_{51}}{\partial x_2}=0$, $\frac{\partial g_{51}}{\partial x_3}=0$, $\frac{\partial g_{51}}{\partial x_4}=0$, $\frac{\partial g_{51}}{\partial x_5}=0$, $\frac{\partial g_{51}}{\partial x_6}=0$.

About the sixth row of the Jacobian matrix $\nabla_x g_1(x)$ and for $i=1,2,\cdots,6$, one has:

$$\frac{\partial g_{61}}{\partial x_i}=\frac{\frac{\partial M_{13}}{\partial x_i}detM-M_{13}\frac{\partial detM}{\partial x_i}}{(detM)^2} \tag{2.75}$$

Computation of the Jacobian matrix $\nabla_x g_2(x)$:

$$\nabla_x g_2(x)=\begin{pmatrix} \frac{\partial g_{12}}{\partial x_1} & \frac{\partial g_{12}}{\partial x_2} & \frac{\partial g_{12}}{\partial x_3} & \frac{\partial g_{12}}{\partial x_4} & \frac{\partial g_{12}}{\partial x_5} & \frac{\partial g_{12}}{\partial x_6} \\ \frac{\partial g_{22}}{\partial x_1} & \frac{\partial g_{22}}{\partial x_2} & \frac{\partial g_{22}}{\partial x_3} & \frac{\partial g_{22}}{\partial x_4} & \frac{\partial g_{22}}{\partial x_5} & \frac{\partial g_{22}}{\partial x_6} \\ \frac{\partial g_{32}}{\partial x_1} & \frac{\partial g_{32}}{\partial x_2} & \frac{\partial g_{32}}{\partial x_3} & \frac{\partial g_{32}}{\partial x_4} & \frac{\partial g_{32}}{\partial x_5} & \frac{\partial g_{32}}{\partial x_6} \\ \frac{\partial g_{42}}{\partial x_1} & \frac{\partial g_{42}}{\partial x_2} & \frac{\partial g_{42}}{\partial x_3} & \frac{\partial g_{42}}{\partial x_4} & \frac{\partial g_{42}}{\partial x_5} & \frac{\partial g_{42}}{\partial x_6} \\ \frac{\partial g_{52}}{\partial x_1} & \frac{\partial g_{52}}{\partial x_2} & \frac{\partial g_{52}}{\partial x_3} & \frac{\partial g_{52}}{\partial x_4} & \frac{\partial g_{52}}{\partial x_5} & \frac{\partial g_{52}}{\partial x_6} \\ \frac{\partial g_{62}}{\partial x_1} & \frac{\partial g_{62}}{\partial x_2} & \frac{\partial g_{62}}{\partial x_3} & \frac{\partial g_{62}}{\partial x_4} & \frac{\partial g_{62}}{\partial x_5} & \frac{\partial g_{62}}{\partial x_6} \end{pmatrix} \tag{2.76}$$

About the first row of the Jacobian matrix $\nabla_x g_2(x)$, one has: $\frac{\partial g_{12}}{\partial x_1}=0$, $\frac{\partial g_{12}}{\partial x_2}=0$, $\frac{\partial g_{12}}{\partial x_3}=0$, $\frac{\partial g_{12}}{\partial x_4}=0$, $\frac{\partial g_{12}}{\partial x_5}=0$, $\frac{\partial g_{12}}{\partial x_6}=0$.

About the second row of the Jacobian matrix $\nabla_x g_2(x)$ and for $i=1,2,\cdots,6$ one has:

$$\frac{\partial g_{22}}{\partial x_i}=-\frac{\frac{\partial M_{21}}{\partial x_i}detM-M_{21}\frac{\partial detM}{\partial x_i}}{(detM)^2} \tag{2.77}$$

About the third row of the Jacobian matrix $\nabla_x g_2(x)$, one has: $\frac{\partial g_{32}}{\partial x_1}=0$, $\frac{\partial g_{32}}{\partial x_2}=0$, $\frac{\partial g_{32}}{\partial x_3}=0$, $\frac{\partial g_{32}}{\partial x_4}=0$, $\frac{\partial g_{32}}{\partial x_5}=0$, $\frac{\partial g_{32}}{\partial x_6}=0$.

About the fourth row of the Jacobian matrix $\nabla_x g_2(x)$ and for $i=1,2,\cdots,6$ one has:

$$\frac{\partial g_{42}}{\partial x_i}=\frac{\frac{\partial M_{22}}{\partial x_i}detM-M_{22}\frac{\partial detM}{\partial x_i}}{(detM)^2} \tag{2.78}$$

About the fifth row of the Jacobian matrix $\nabla_x g_2(x)$, one has: $\frac{\partial g_{52}}{\partial x_1}=0$, $\frac{\partial g_{52}}{\partial x_2}=0$, $\frac{\partial g_{52}}{\partial x_3}=0$, $\frac{\partial g_{52}}{\partial x_4}=0$, $\frac{\partial g_{52}}{\partial x_5}=0$, $\frac{\partial g_{52}}{\partial x_6}=0$.

About the sixth row of the Jacobian matrix $\nabla_x g_2(x)$ and for $i=1,2,\cdots,6$ one has:

$$\frac{\partial g_{62}}{\partial x_i}=-\frac{\frac{\partial M_{23}}{\partial x_i}detM-M_{23}\frac{\partial detM}{\partial x_i}}{(detM)^2} \tag{2.79}$$

To compute the Jacobian matrices given above, one has to compute first the partial derivatives $\frac{\partial}{\partial x_i}$, $i=1,\ldots,6$ of the elements M_{ij} $i,j=1,\ldots,3$ of the inertial matrix, of the determinant $detM$ and finally of functions $\tilde{C}_i$, $i=1,2,3$.
(i) partial derivatives of the determinant $detM=Jm_p^2L^2$ or $detM=Jm_p^2x_3^2$.

It holds that $\frac{\partial detM}{\partial x_1}=0$, $\frac{\partial detM}{\partial x_2}=0$, $\frac{\partial detM}{\partial x_3}=2Jm_p^2x_3$, $\frac{\partial detM}{\partial x_4}=0$, $\frac{\partial detM}{\partial x_5}=0$ and $\frac{\partial detM}{\partial x_5}=0$,

(ii) partial derivatives of the element $M_{11}=m_p^2L^2$ or $M_{11}=m_p^2x_3^2$ of the inverse of the inertia matrix:

It holds that $\frac{\partial M_{11}}{\partial x_1}=0$, $\frac{\partial M_{11}}{\partial x_2}=0$, $\frac{\partial M_{11}}{\partial x_3}=2m_p^2x_3$, $\frac{\partial M_{11}}{\partial x_4}=0$, $\frac{\partial M_{11}}{\partial x_5}=0$ and $\frac{\partial M_{11}}{\partial x_6}=0$.

(iii) partial derivatives of the element $M_{12}=-m_p^2L^2P_l\cos(\theta-\phi)$ or $M_{12}=-m_p^2x_3^2P_l\cos(x(k,5)-x(k,1))$ of the inverse of the inertia matrix:

It holds that $\frac{\partial M_{12}}{\partial x_1}=-m_p^2x_3^2P_l\sin(x_5-x_1)$, $\frac{\partial M_{12}}{\partial x_2}=0$, $\frac{\partial M_{12}}{\partial x_3}=-2m_p^2x_3P_l\cos(x_5-x_1)$, $\frac{\partial M_{12}}{\partial x_4}=0$, $\frac{\partial M_{12}}{\partial x_5}=m_p^2x_3^2P_l\sin(x_5-x_1)$ and $\frac{\partial M_{12}}{\partial x_6}=0$.

(iv) partial derivatives of the element $M_{13}=-m_p^2P_L L\sin(\theta-\phi)$ of $M_{13}=-m_p^2P_Lx_3\sin(x_5-x_1)$ of the inverse of the inertia matrix:

It holds that $\frac{\partial M_{13}}{\partial x_1}=m_p^2P_lx_3\cos(x_5-x_1)$, $\frac{\partial M_{13}}{\partial x_2}=0$, $\frac{\partial M_{13}}{\partial x_3}=-m_p^2P_l\sin(x_5-x_1)$, $\frac{\partial M_{13}}{\partial x_4}=0$, $\frac{\partial M_{13}}{\partial x_5}=-m_p^2P_lx_3\cos(x_5-x_1)$, $\frac{\partial M_{13}}{\partial x_6}=0$

(v) partial derivatives of the element $M_{21}=-m_p^2L^2P_l\cos(\theta-\phi)$ or $M_{21}=-m_p^2x_3^2P_l\cos(x_5-x_1)$ of the inverse of the inertia matrix:

It holds that $\frac{\partial M_{21}}{\partial x_1}=-m_p^2x_3^2P_l\sin(x_5-x_1)$, $\frac{\partial M_{21}}{\partial x_2}=0$, $\frac{\partial M_{21}}{\partial x_3}=-m_p^22x_3P_l\cos(x_5-x_1)$, $\frac{\partial M_{21}}{\partial x_4}=0$, $\frac{\partial M_{21}}{\partial x_5}=m_p^2x_3^2P_l\sin(x_5-x_1)$ and $\frac{\partial M_{21}}{\partial x_6}=0$.

(vi) partial derivatives of the element $M_{22}=(J+m_pP_l^2)m_pL^2-m_p^2P_l^2L^2\sin^2(\theta-\phi)$ or $M_{22}=(J+m_pP_l^2)m_px_3^2-m_p^2P_l^2x_3^2\sin^2(x_5-x_1)$ of the inverse of the inertia matrix:

It holds that $\frac{\partial M_{22}}{\partial x_1} = m_p^2 P_l^2 x_3^2 2\sin(x_5 - x_1)\cos(x_5 - x_1)$, $\frac{\partial M_{22}}{\partial x_2} = 0$, $\frac{\partial M_{22}}{\partial x_3} = (J + m_p P_l^2)m_p 2x_3 - m_p^2 P_l^2 2x_3 \sin^2(x_5 - x_1)$, $\frac{\partial M_{22}}{\partial x_4} = 0$, $\frac{\partial M_{22}}{\partial x_5} = -m_p^2 P_l^2 x_3^2 2\sin(x_5 - x_1)$ $\cos(x_5 - x_1)$, and $\frac{\partial M_{22}}{\partial x_6} = 0$.

(vii) partial derivatives of the element $M_{23} = m_p^2 P_l^2 L\sin(\theta - \phi)\cos(\theta - \phi)$ or $M_{23} = m_p^2 P_l^2 x_3 \sin(x_5 - x_1)\cos(x_5 - x_1)$ of the inverse of the inertia matrix:

It holds that $\frac{\partial M_{23}}{\partial x_1} = -m_p^2 P_l^2 x_3$, $\frac{\partial M_{23}}{\partial x_2} = 0$, $\frac{\partial M_{23}}{\partial x_3} = -m_p^2 P_l^2 \sin(x_5 - x_1)\cos(x_5 - x_1)$, $\frac{\partial M_{23}}{\partial x_4} = 0$, $\frac{\partial M_{23}}{\partial x_5} = m_p^2 P_l^2 x_3$, $\frac{\partial M_{23}}{\partial x_6} = 0$

(viii) partial derivatives of the element $M_{31} = -m_p^2 P_l L\sin(\theta - \phi)$ or $M_{31} = -m_p^2 P_l x_3 \sin(x_5 - x_1)$ of the inverse of the inertia matrix:

It holds that $\frac{\partial M_{31}}{\partial x_1} = m_p^2 P_l x_3 \cos(x_5 - x_1)$, $\frac{\partial M_{31}}{\partial x_2} = 0$, $\frac{\partial M_{31}}{\partial x_3} = -m_p^2 P_l \sin(x_5 - x_1)$, $\frac{\partial M_{31}}{\partial x_4} = 0$, $\frac{\partial M_{31}}{\partial x_5} = -m_p^2 P_l^2 x_3 \cos(x_5 - x_1)$, and $\frac{\partial M_{31}}{\partial x_6} = 0$.

(ix) partial derivatives of the element $M_{32} = m_p^2 P_l^2 L\sin(\theta - \phi)\cos(\theta - \phi)$ or $M_{32} = m_p^2 P_l^2 x_3 \sin(x_5 - x_1)\cos(x_5 - x_1)$ of the inverse of the inertia matrix:

It holds that $\frac{\partial M_{32}}{\partial x_1} = -m_p^2 P_l^2 x_3 \cos^2(x_5 - x_1) + m_p^2 P_l^2 x_3 \sin^2(x_5 - x_1)$, $\frac{\partial M_{32}}{\partial x_2} = 0$, $\frac{\partial M_{32}}{\partial x_3} = m_p^2 P_l^2 \sin(x_5 - x_1)\cos(x_5 - x_1)$, $\frac{\partial M_{32}}{\partial x_4} = 0$, $\frac{\partial M_{32}}{\partial x_5} = m_p^2 P_l^2 x_3 \cos^2(x_5 - x_1) - m_p^2 P_l^2 x_3 \sin^2(x_5 - x_1)$, $\frac{\partial M_{32}}{\partial x_6} = 0$.

(x) partial derivatives of the element $M_{33} = (J + m_p P_l^2)m_p - m_p^2 P_l^2 \cos^2(\theta - \phi)$ or $M_{33} = (J + m_p P_l^2)m_p - m_p^2 P_l^2 \cos^2(x_5 - x_1)$ of the inverse of the inertia matrix:

It holds that $\frac{\partial M_{33}}{\partial x_1} = -m_p^2 P_l^2 2\cos(x_5 - x_1)\sin(x_5 - x_1)$, $\frac{\partial M_{33}}{\partial x_2} = 0$, $\frac{\partial M_{33}}{\partial x_3} = 0$, $\frac{\partial M_{33}}{\partial x_4} = 0$, $\frac{\partial M_{33}}{\partial x_5} = m_p^2 P_l^2 2\cos(x_5 - x_1)\sin(x_5 - x_1)$, and $\frac{\partial M_{33}}{\partial x_6} = 0$.

Moreover, about the elements of the extended Coriolis matrix $\tilde{C}_1(q, \dot{q}) = -2m_p P_l$ $\sin(\theta - \phi)\dot{\theta}\dot{L} - m_p P_l \cos(\theta - \phi)\dot{\theta}^2 + (-m_p P_l + m{\cdot}d)g\cos(\phi - \alpha)$ or $\tilde{C}_1(q, \dot{q}) = -2m_p P_l \sin(x_5 - x_1)x_6 x_4 - m_p P_l \cos(x_5 - x_1)x_6^2 + (-m_p P_l + m{\cdot}d)g\cos(x_5 - x_1)$.

About the partial derivatives of $\tilde{C}_1$ it holds that $\frac{\partial \tilde{C}_1}{\partial x_1} = 2m_p P_l \cos(x_5 - x_1)$ $x_6 x_4 - m_p P_l \sin(x_5 - x_1)x_6^2 - (m_p P_l + m{\cdot}d)g\sin(x_5 - \alpha)$, $\frac{\partial \tilde{C}_1}{\partial x_2} = 0$, $\frac{\partial \tilde{C}_1}{\partial x_3} = 0$, $\frac{\partial \tilde{C}_1}{\partial x_4} = -2m_p P_l \sin(x_5 - x_1)x_6$, $\frac{\partial \tilde{C}_1}{\partial x_5} = -2m_p P_l \sin(x_5 - x_1)x_6 - x_4 - m_p P_l \cos(x_5 - x_1)x_6^2$, and $\frac{\partial \tilde{C}_1}{\partial x_6} = 2m_p P_l \cos(x_5 - x_1)x_4 - 2m_p P_l \sin(x_5 - x_1)x_6$

Additionally, about the elements of the extended Coriolis matrix $\tilde{C}_2(q, \dot{q}) = -m_p P_l \sin(\theta - \phi)\dot{\phi}^2 - m_p L\dot{\theta}^2 + m_p g\cos(\theta - \alpha)$ or $\tilde{C}_2(q, \dot{q}) = -m_p P_l \sin(x_5 - x_1)$ $x_2^2 - m_p x_3 x_6^2 + m_p g\cos(x_5 - \alpha)$ one has

It holds that $\frac{\partial \tilde{C}_2}{\partial x_1} = -m_p P_l \cos(x_5 - x_1)x_2^2$, $\frac{\partial \tilde{C}_2}{\partial x_2} = m_p P_l \sin(x_5 - x_1)2x_2$, $\frac{\partial \tilde{C}_2}{\partial x_3} = m_p x_6^2$, $\frac{\partial \tilde{C}_2}{\partial x_4} = 0$, $\frac{\partial \tilde{C}_2}{\partial x_5} = m_p P_l \cos(x_5 - x_1)x_2^2 - m_p g \sin(x_5 - \alpha)$, $\frac{\partial \tilde{C}_2}{\partial x_6} = m_p x_3 2x_6$.

Furthermore, about the elements of the extended Coriolis matrix $\tilde{C}_3(q, \dot{q}) = m_p P_l L\cos(\theta - \phi)\dot{\phi}^2 - 2m_p x_3 x_4 x_6 + m_p Lg \sin(\theta - \alpha)$ or $\tilde{C}_3(q, \dot{q}) = m_p P_l L\cos(x_5 - x_1)x_2^2 - 2m_p x_3 x_4 x_6 + m_p x_3 g \sin(x_5 - \alpha)$ one has

It holds that $\frac{\partial \tilde{C}_3}{\partial x_1} = m_p P_l x_3 \sin(x_5 - x_1)x_2^2$, $\frac{\partial \tilde{C}_3}{\partial x_2} = 2m_p P_l x_3 \cos(x_5 - x_1)x_2$, $\frac{\partial \tilde{C}_3}{\partial x_3} = m_p P_l \cos(x_5 - x_1)x_2^2 - 2m_p x_4 x_6$, $\frac{\partial \tilde{C}_3}{\partial x_4} = -2m_p x_3 x_6$, $\frac{\partial \tilde{C}_3}{\partial x_5} = -m_p P_l x_3 \sin(x_5 - x_1)x_2^2 + m_p x_3 g \cos(x_5 - \alpha)$ and $\frac{\partial \tilde{C}_3}{\partial x_6} = -2m_p x_3 x_4$.

2.2.4 Computation of the feedback control gains

For the linearized system given by (1.36) the cost function of (1.37) is defined, where the coefficient r determines the penalization of the control input and the weight coefficient ρ determines the reward of the disturbances' effects. It is assumed that (i) the energy that is transferred from the disturbances signal $\tilde{d}(t)$ is bounded, that is $\int_0^\infty \tilde{d}^T(t)\tilde{d}(t)dt < \infty$, (ii) matrices $[A, B]$ and $[A, L]$ are stabilizable, (iii) matrix $[A, C]$ is detectable. Then, the optimal feedback control law is given by

$$u(t) = -Ke(t) \tag{2.80}$$

with $K = \frac{1}{r}B^T P$ where P is a positive definite symmetric matrix which is obtained from the solution of the Riccati equation

$$A^T P + PA + Q - P\left(\frac{2}{r}BB^T - \frac{1}{\rho^2}LL^T\right)P = 0 \tag{2.81}$$

and Q is a positive semi-definite symmetric matrix. The worst case disturbance is given by

$$\tilde{d}(t) = \frac{1}{\rho^2}L^T Pe(t) \tag{2.82}$$

The diagram of the considered control loop for the offshore boom crane is depicted in Figure 2.20. The global stability properties of the control method can be proven following the procedure presented in Section 1.1.2.5.

2.2.5 Simulation tests

The global stability properties of the control method and the elimination of the state vector's tracking error which were previously proven through Lyapunov analysis are further confirmed through simulation experiments. The feedback gains of the H-infinity controller are repetitively computed at each time-step of the control algorithm through the solution of the algebraic Riccati equation of (2.81). The obtained results are depicted in Figures 2.22 to 2.32. The real values of the state variables are depicted

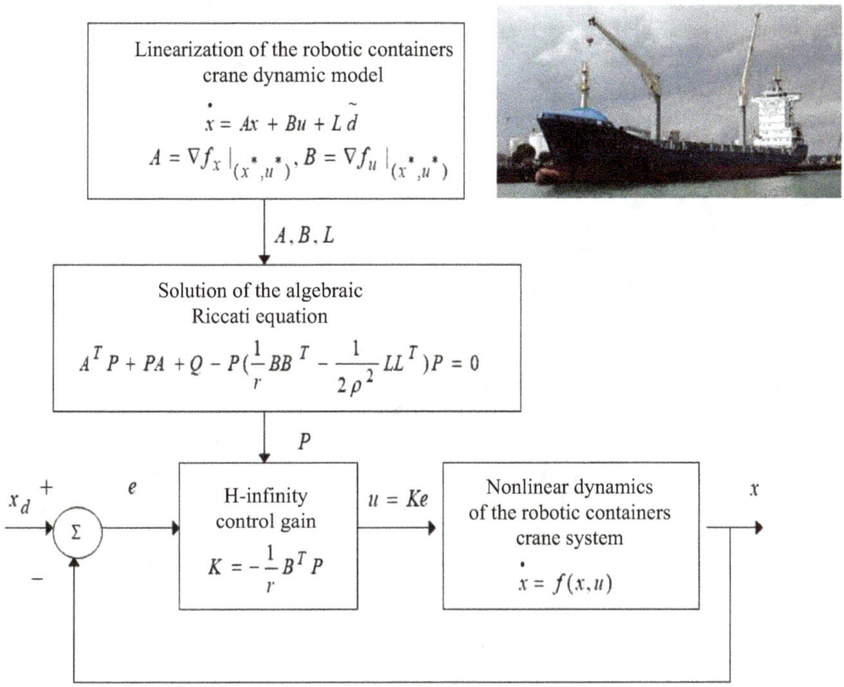

Figure 2.20 Diagram of the control scheme for the offshore boom crane

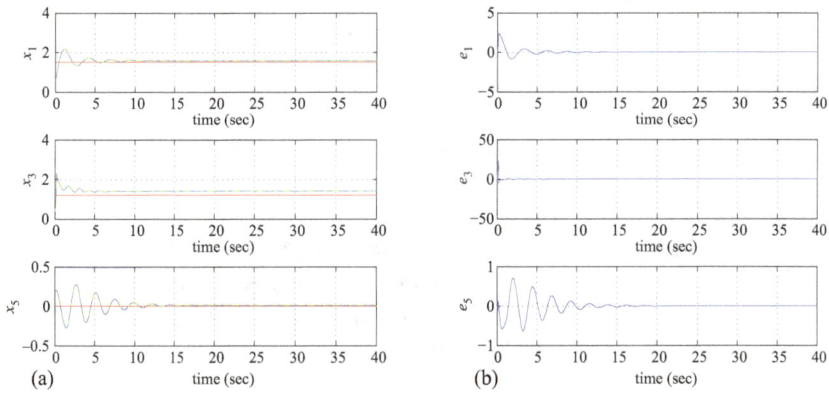

Figure 2.21 Tracking of setpoint 1 for the offshore boom crane: (a) convergence of state variables x_1, x_3, x_5 to their reference setpoints (red line: setpoint, blue line: real value, green line: estimated value); (b) tracking errors e_1, e_3, e_5 for state variables x_1, x_3, x_5

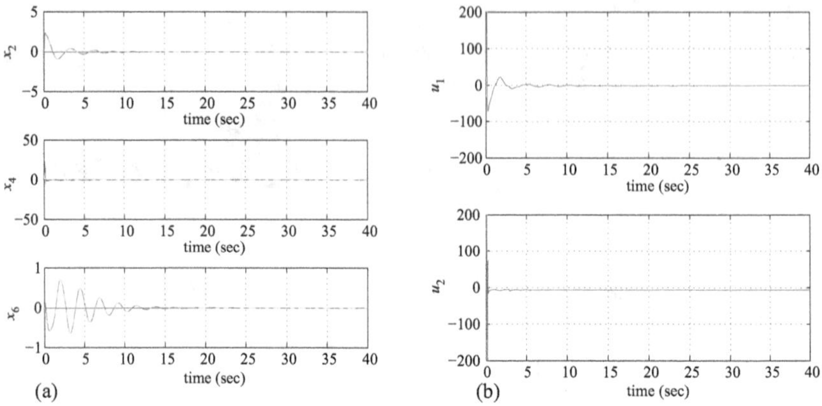

Figure 2.22 Tracking of setpoint 1 for the offshore boom crane: (a) convergence of state variables x_2, x_4, x_6 to their reference setpoints (red line: setpoint, blue line: real value, green line: estimated value); (b) control inputs u_1, u_2 of the underactuated crane

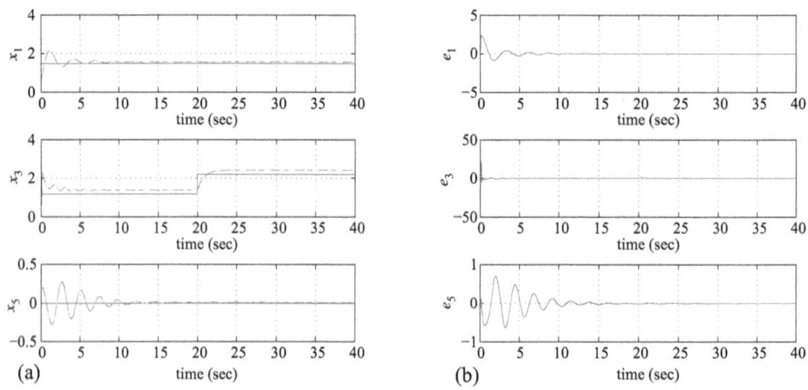

Figure 2.23 Tracking of setpoint 2 for the offshore boom crane: (a) convergence of state variables x_1, x_3, x_5 to their reference setpoints (red line: setpoint, blue line: real value, green line: estimated value); (b) tracking errors e_1, e_3, e_5 for state variables x_1, x_3, x_5

in blue, the estimated values are plotted in green, while the reference setpoints are printed in red. It can be confirmed that in all cases, the proposed control scheme achieved fast and accurate tracking of the reference setpoints under moderate variations of the control inputs.

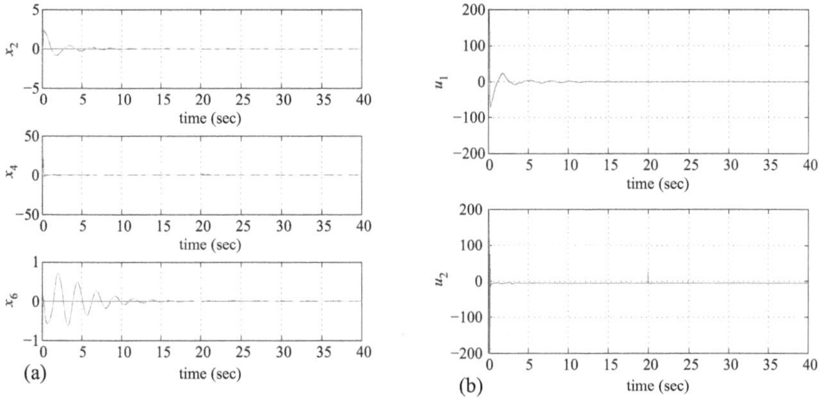

Figure 2.24 Tracking of setpoint 2 for the offshore boom crane: (a) convergence of state variables x_2, x_4, x_6 to their reference setpoints (red line: setpoint, blue line: real value, green line: estimated value); (b) control inputs u_1, u_2 of the underactuated crane

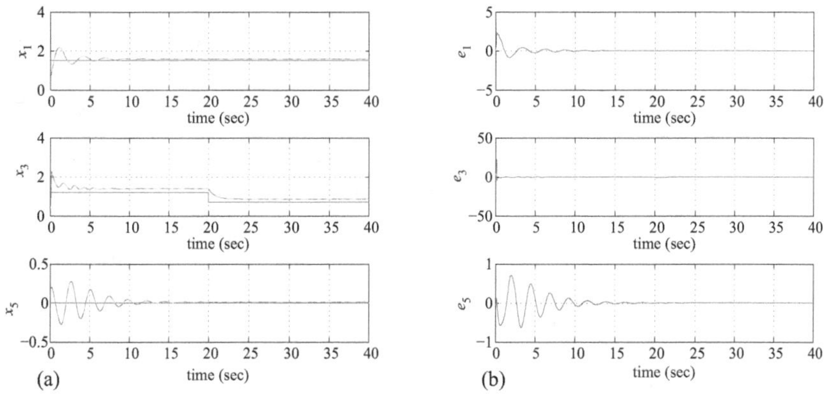

Figure 2.25 Tracking of setpoint 3 for the offshore boom crane: (a) convergence of state variables x_1, x_3, x_5 to their reference setpoints (red line: setpoint, blue line: real value, green line: estimated value); (b) tracking errors e_1, e_3, e_5 for state variables x_1, x_3, x_5

2.3 Nonlinear control of the inertia wheel and pendulum system

2.3.1 Outline

Control and stabilization of the dynamics of the inertia wheel inverted pendulum is a benchmark problem in the design of nonlinear control systems, as well as in

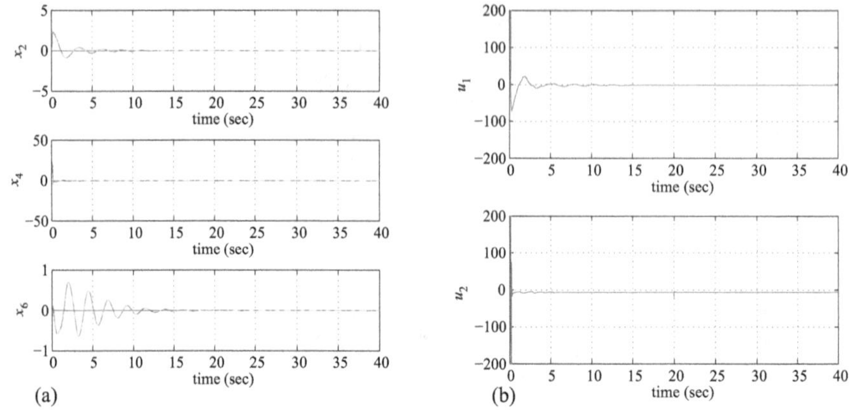

Figure 2.26 Tracking of setpoint 3 for the offshore boom crane: (a) convergence of state variables x_2, x_4, x_6 to their reference setpoints (red line: setpoint, blue line: real value, green line: estimated value); (b) control inputs u_1, u_2 of the underactuated crane

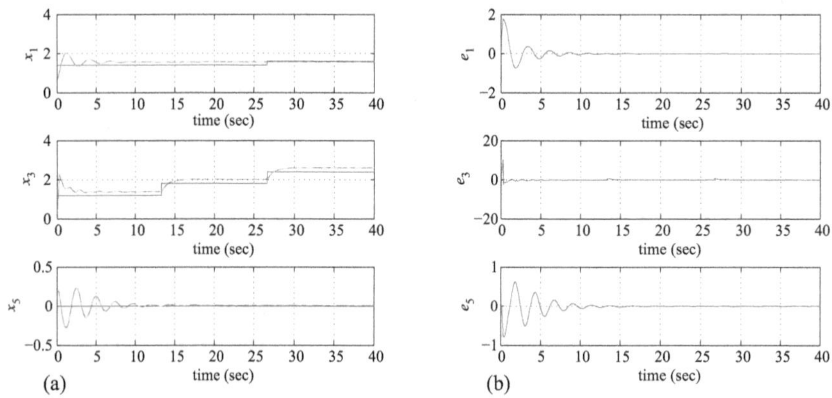

Figure 2.27 Tracking of setpoint 4 for the offshore boom crane: (a) convergence of state variables x_1, x_3, x_5 to their reference setpoints (red line: setpoint, blue line: real value, green line: estimated value); (b) tracking errors e_1, e_3, e_5 for state variables x_1, x_3, x_5

control systems education (Rigatos and Busawon, 2018; Kant and Mukherjee, 2018; Santibamez *et al.*, 2005; Ye, Wang *et al.*, 2007). Due to nonlinearities in the pendulum's dynamics and due to underactuation, this control problem is non-trivial, as shown in Haddad *et al.*, 2016, Audrey *et al.*, 2012, and Spong, Corke *et al.*, 2001. The inertia wheel inverted pendulum comprises a pendulum pole that rotates around a passive

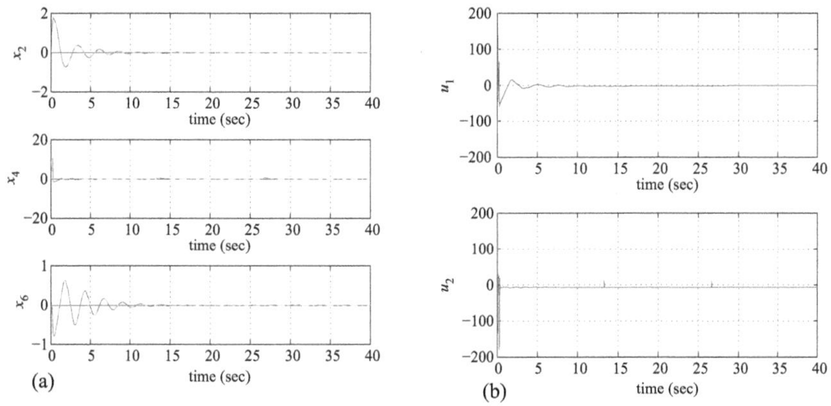

Figure 2.28 Tracking of setpoint 4 for the offshore boom crane: (a) convergence of state variables x_2, x_4, x_6 to their reference setpoints (red line: setpoint, blue line: real value, green line: estimated value); (b) control inputs u_1, u_2 of the underactuated crane

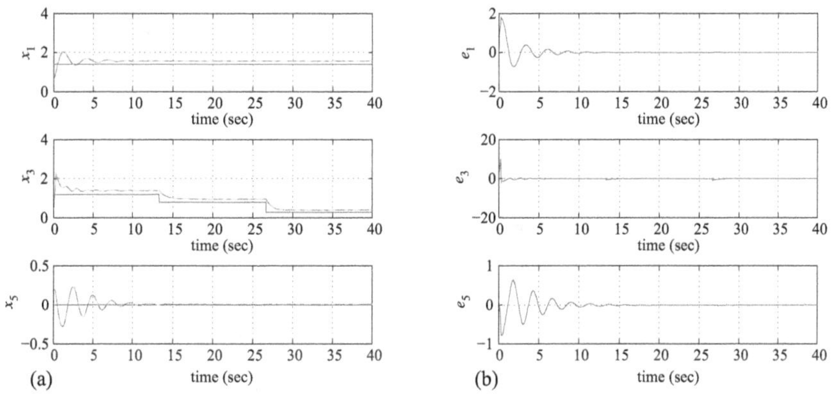

Figure 2.29 Tracking of setpoint 5 for the offshore boom crane: (a) convergence of state variables x_1, x_3, x_5 to their reference setpoints (red line: setpoint, blue line: real value, green line: estimated value); (b) tracking errors e_1, e_3, e_5 for state variables x_1, x_3, x_5

joint placed at its one end. At the other end of the pole, an active joint is placed and a wheel of specific mass and inertia performs rotation around it, as shown in (Aquilar *et al.*, 2012; Andary *et al.*, 2009; and also in Andary *et al.*, 2008; and Gritli and Belghith, 2018). There have been several attempts to solve the control problem for this two-mass pendulum system, as explained in (Mayr *et al.*, 2013; Ho and

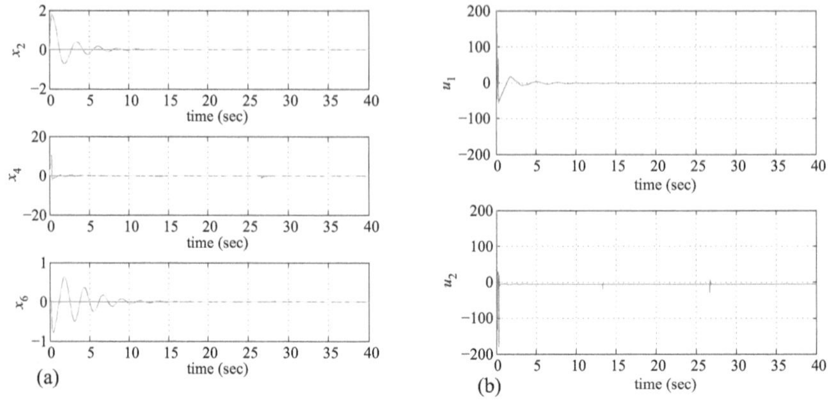

Figure 2.30 *Tracking of setpoint 5 for the offshore boom crane: (a) convergence of state variables* x_2, x_4, x_6 *to their reference setpoints (red line: setpoint, blue line: real value, green line: estimated value); (b) control inputs* u_1, u_2 *of the underactuated crane*

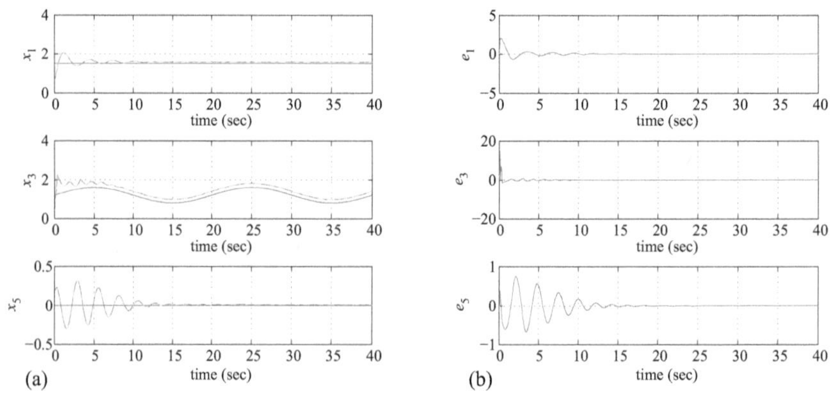

Figure 2.31 *Tracking of setpoint 1 for the offshore boom crane: (a) convergence of state variables* x_1, x_3, x_5 *to their reference setpoints (red line: setpoint, blue line: real value, green line: estimated value); (b) tracking errors* e_1, e_3, e_5 *for state variables* x_1, x_3, x_5

Pham, 2019; and Kant, Mukerjee *et al.*, 2017). Several results make use of Lyapunov methods for achieving stabilization (Montoya and Gil-Gonzalez, 2019; Alushev *et al.*, 2018; Franco, 2019). Moreover, one can distinguish results based on sliding-mode and model predictive control concepts as noted in (Iriarte *et al.*, 2013; Lu, Fang and Sun, 2016; Khalid and Memon, 2014; and Sowman *et al.*, 2015). There have been also attempts to define linearizing outputs for the pendulum's model and to apply

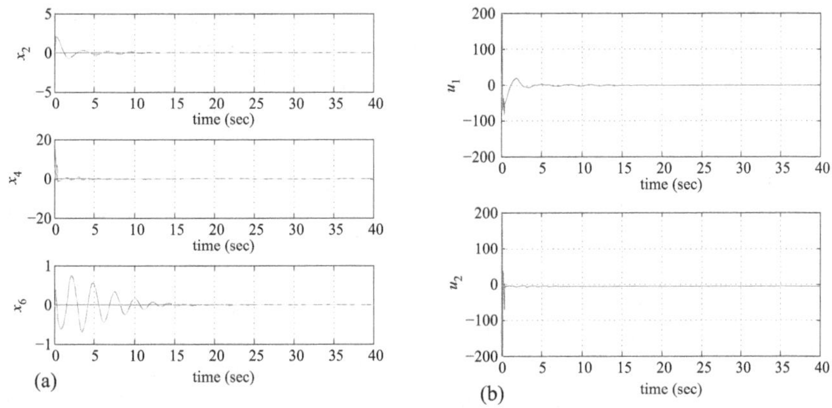

Figure 2.32 Tracking of setpoint 1 for the offshore boom crane: (a) convergence of state variables x_2, x_4, x_6 to their reference setpoints (red line: setpoint, blue line: real value, green line: estimated value); (b) control inputs u_1, u_2 of the underactuated crane

global linearization-based control (Enev, 2014). In many of the above-noted results, stability is claimed for the control loop (Ryalat and Laila, 2016; Zhang, Huang *et al.*, 2016; Scerri *et al.*, 2017). However, there exist hardly any results about the solution of the nonlinear optimal control problem for the system of the inertia wheel inverted pendulum.

In this section, a novel nonlinear optimal control method is developed for the inertia-wheel inverted pendulum while the global stability properties of the control scheme are also proven. First, the state-space model of the pendulum is subject to approximate linearization around a temporary operating point which is recomputed at each iteration of the control method (Rigatos and Siano, 2015; Rigatos, Siano *et al.*, 2015). This operating point is defined by the present value of the system's state vector and by the last value of the control inputs vector that was applied on it. The linearization procedure relies on first order Taylor series expansion and on the computation of the system's Jacobian matrices, as presented in (Rigatos and Tzafestas, 2007; Basseville and Nikiforov, 1993; and Rigatos and Zhang, 2009). The modeling error, which is due to the truncation of higher-order terms in the Taylor series expansion, is considered to be a perturbation which is asymptotically compensated by the robustness of the control loop. For the approximately linearized model of the pendulum, a stabilizing H-infinity feedback controller is designed.

Actually, the H-infinity feedback controller stands for the solution of the optimal control problem for the inertia wheel inverted pendulum under model uncertainty and external perturbations. The controller represents also the solution to a min–max differential game in which the control inputs try to minimize a cost function which comprises a quadratic term of the state vector's tracking error, whereas the model's uncertainty and other perturbation terms try to maximize this cost function. To compute the feedback gains of the H-infinity controller an algebraic Riccati equation is solved at iteration of the control algorithm (Rigatos, 2015;Rigatos, 2016b). The stability properties of the control scheme are proven through Lyapunov analysis (Rigatos and Busawon, 2018). At a first stage, it is proven that the control loop satisfies the H-infinity tracking performance criterion. This signifies elevated robustness against modeling errors and external disturbances. Moreover, it is proven that under moderate conditions the control loop is globally asymptotically stable . Furthermore, to implement state estimation-based control for the pendulum without the need to measure its entire state vector the H-infinity Kalman Filter is introduced as a robust state estimator (Rigatos, 2015; Rigatos, 2016b; Rigatos, 2015; Rigatos, 2016b). The proposed H-infinity control approach retains the advantages of the typical optimal control, that is fast convergence to the reference setpoints under moderate variations of the control inputs.

2.3.2 Dynamic model of the inertia wheel inverted pendulum

The inertia wheel inverted pendulum is an underactuated system, having two degrees of freedom and receiving one control input. Actually it comprises a pendulum point that rotates around a passive joint which is placed at its first end. At its second end, an active joint is mounted which generates torque and provides motion to a wheel. The diagram of this underactuated system is shown in Figure 2.33.

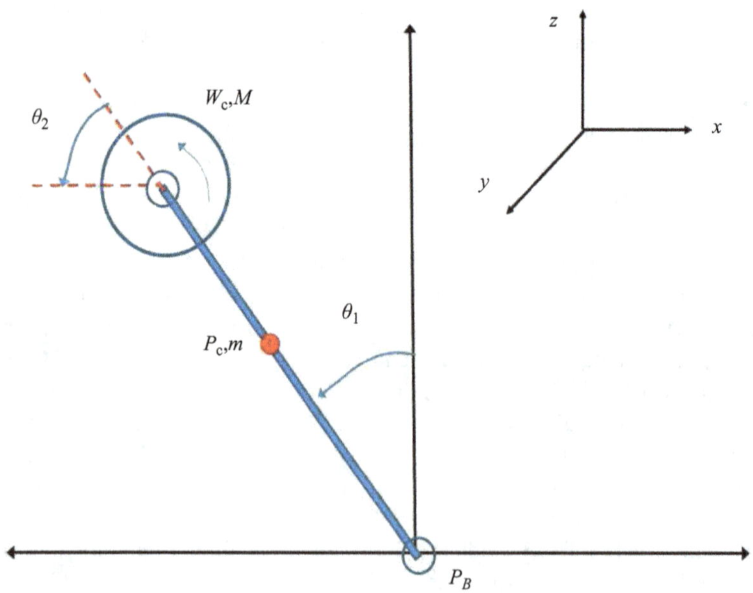

Figure 2.33 Diagram of the inertia wheel inverted pendulum system

The parameters of the model are defined as follows: θ_1 is the pendulum angle with respect to the vertical axis (rad), θ_2 is the wheel angle with respect to the pendulum axis (rad), m is the mass of the pendulum (kg), M is the mass of the wheel (kg), l is the length from the pendulum's base P_B to the pendulum's center of mass P_c (m), L is the length from the pendulum's base P_b to the wheel's center of mass W_c (m), I_{P_c} is the rotational inertia of the pendulum about its center of mass $kg{\cdot}m^2$, I_{P_B} is the rotational inertia of the pendulum about its base (kg·m^2), W_c is the rotational inertia of the wheel about its center of mass m^2, and g is gravity's acceleration constant m/sec^2. Moreover, the following parameters are defined $a = ML^2 + I_{P_B}$, $b = ml + ML$.

The system is underactuated and its state-space model is (Haddad *et al.*, 2016)

$$\begin{pmatrix} a + I_{W_C} & I_{W_C} \\ I_{W_C} & I_{W_C} \end{pmatrix} \begin{pmatrix} \ddot{\theta}_1 \\ \ddot{\theta}_2 \end{pmatrix} - \begin{pmatrix} b{\cdot}g\cdot\sin(\theta_1) \\ 0 \end{pmatrix} = \begin{pmatrix} 0 \\ u \end{pmatrix} \tag{2.83}$$

The following state variables are defined: $x_1 = \theta_1$, $x_2 = \dot{\theta}_1$, $x_3 = \theta_2$, $x_4 = \dot{\theta}_2$. The control input u is the torque at the wheel's active joint. The state-space description of the system is rewritten as

$$\begin{pmatrix} \ddot{\theta}_1 \\ \ddot{\theta}_2 \end{pmatrix} = \begin{pmatrix} a + I_{W_C} & I_{W_C} \\ I_{W_C} & I_{W_C} \end{pmatrix}^{-1} \cdot \begin{pmatrix} b{\cdot}g\cdot\sin(\theta_1) \\ u \end{pmatrix} \tag{2.84}$$

and by computing the inverse of the inertia matrix, one obtains

$$\begin{pmatrix} \ddot{\theta}_1 \\ \ddot{\theta}_2 \end{pmatrix} = \begin{pmatrix} \frac{1}{a} & -\frac{1}{a} \\ -\frac{1}{a} & \frac{a + I_{W_C}}{aI_{W_C}} \end{pmatrix}^{-1} \cdot \begin{pmatrix} b{\cdot}g\cdot\sin(\theta_1) \\ u \end{pmatrix} \tag{2.85}$$

Thus, one arrives at the following state-space description

$$\begin{aligned} \ddot{\theta}_1 &= \frac{b\cdot}{a} g\cdot\sin(\theta_1) - \frac{1}{a}u \\ \ddot{\theta}_2 &= -\frac{b\cdot}{a} g\cdot\sin(\theta_1) + \frac{a + I_{W_C}}{aI_{W_C}}u \end{aligned} \tag{2.86}$$

or using the previous state-variables notation, one obtains the following state-space description:

$$\begin{aligned} \dot{x}_1 &= x_2 \\ \dot{x}_2 &= \frac{b\cdot}{a} g\cdot\sin(x_1) - \frac{1}{a}u \\ \dot{x}_3 &= x_4 \\ \dot{x}_4 &= -\frac{b\cdot}{a} g\cdot\sin(x_1) + \frac{a + I_{W_C}}{aI_{W_C}}u \end{aligned} \tag{2.87}$$

In matrix form, the state-space description of the system becomes

$$\begin{pmatrix} \dot{x}_1 \\ \dot{x}_2 \\ \dot{x}_3 \\ \dot{x}_4 \end{pmatrix} = \begin{pmatrix} x_2 \\ \frac{b\cdot}{a} g \cdot \sin(x_1) \\ x_4 \\ -\frac{b\cdot}{a} g \cdot \sin(x_1) \end{pmatrix} + \begin{pmatrix} 0 \\ -\frac{1}{a} \\ 0 \\ \frac{a + I_{W_C}}{a I_{W_C}} \end{pmatrix} u \tag{2.88}$$

$$\dot{x} = f(x) + g(x)u \tag{2.89}$$

where $x \in R^{4\times 1}, f(x) \in R^{4\times 1}$, $g(x) \in R^{4\times 1}$ and $u \in R$.

2.3.3 Approximate linearization of the inertia wheel inverted pendulum

The state-space model undergoes approximate linearization around the time-varying operating point (x^*, u^*), where x^* is the present value of the system's state vector and u^* is the last value of the control inputs vector that was applied on it. The linearization procedure gives:

$$\dot{x} = Ax + Bu + \tilde{d} \tag{2.90}$$

where $\tilde{d}$ is a term that denotes the cumulative parametric uncertainty and the external perturbations of the model, while matrices A and B are computed from the system's Jacobians

$$\begin{aligned} A &= \nabla_x [f(x) + g(x)u] \mid_{(x^*,u^*)} \Rightarrow \\ A &= \nabla_x [f(x)] \mid_{(x^*,u^*)} \end{aligned} \tag{2.91}$$

$$\begin{aligned} B &= \nabla_u [f(x) + g(x)u] \mid_{(x^*,u^*)} \Rightarrow \\ B &= g(x) \mid_{(x^*,u^*)} \end{aligned} \tag{2.92}$$

The Jacobian matrix $\nabla_x [f(x)]$ is computed as follows:

$$\nabla_x [f(x)] \mid_{(x^*,u^*)} = \begin{pmatrix} 0 & 1 & 0 & 0 \\ \frac{b\cdot}{a} g \cdot \cos(x_1) & 0 & 0 & 0 \\ 0 & 0 & 0 & 1 \\ -\frac{b\cdot}{a} g \cdot \cos(x_1) & 0 & 0 & 0 \end{pmatrix} \mid_{(x^*,u^*)} \tag{2.93}$$

2.3.4 Computation of the feedback control gains

For the linearized system given by (1.36) the cost function of (1.37) is defined, where the coefficient r determines the penalization of the control input and the weight coefficient ρ determines the reward of the disturbances' effects.

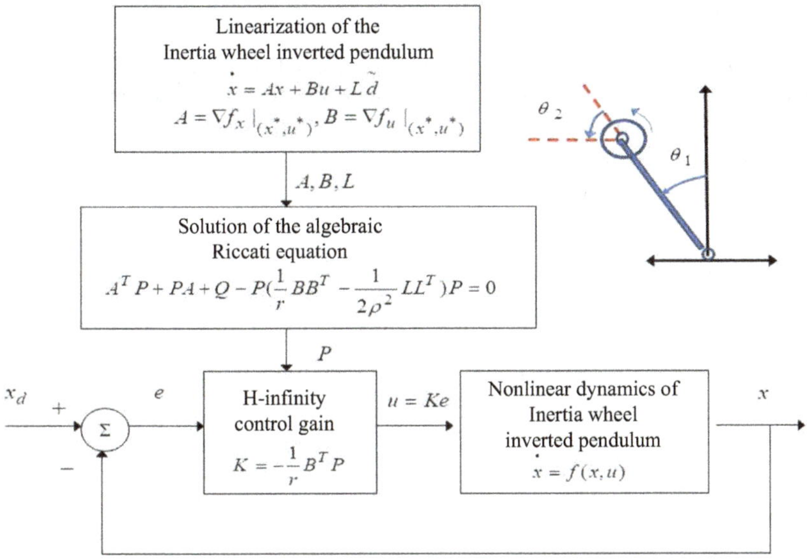

Figure 2.34 Nonlinear H-infinity control loop for the inertia wheel inverted pendulum: an H-infinity controller is applied to the approximately linearized model of the pendulum, which is obtained after Taylor series expansion of its state-space description

It is assumed that (i) the energy that is transferred from the disturbances signal $\tilde{d}(t)$ is bounded, that is $\int_0^\infty \tilde{d}^T(t)\tilde{d}(t)dt < \infty$, (ii) matrices $[A, B]$ and $[A, L]$ are stabilizable, and (iii) matrix $[A, C]$ is detectable. Then, the optimal feedback control law is given by

$$u(t) = -Ke(t) \tag{2.94}$$

with $K = \frac{1}{r}B^T P$ where P is a positive semi-definite symmetric matrix which is obtained from the solution of the Riccati equation

$$A^T P + PA + Q - P\left(\frac{1}{r}BB^T - \frac{1}{2\rho^2}LL^T\right)P = 0 \tag{2.95}$$

where Q is also a positive definite symmetric matrix. The worst case disturbance is given by $\tilde{d}(t) = \frac{1}{\rho^2}L^T Pe(t)$. This equation is obtained by solving the optimal control problem for the case that the system receives as input only the disturbance $\tilde{d}(t)$. The diagram of the considered control loop is depicted in Figure 2.34. The global stability properties of the control method can be proven following the procedure presented in Section 1.1.2.5.

2.3.5 Simulation tests

The performance of the considered nonlinear optimal control method for the inertia wheel inverted pendulum has been further confirmed through simulation experiments. The obtained results are depicted in Figures 2.35 to 2.42. The control scheme has been

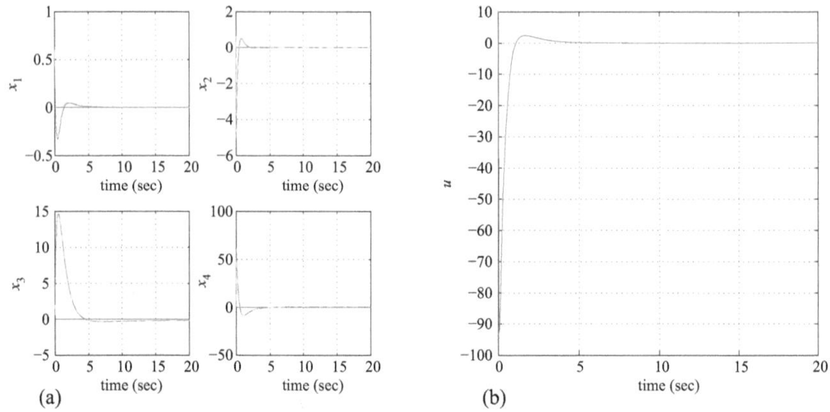

Figure 2.35 Nonlinear optimal control performance for the inertia wheel inverted pendulum in case of setpoint 1: (a) convergence of the state variables x_1 to x_4 to the reference setpoints and (b) control input u, applied to the inertia wheel inverted pendulum

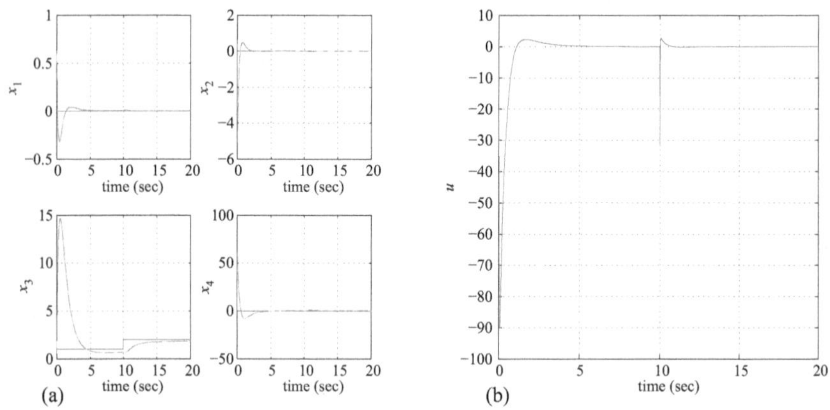

Figure 2.36 Nonlinear optimal control performance for the inertia wheel inverted pendulum in case of setpoint 2: (a) convergence of the state variables x_1 to x_4 to the reference setpoints and (b) control input u, applied to the inertia wheel inverted pendulum

implemented through partial measurement of the system's state vector, while the H-infinity Kalman Filter has been used to estimate the non-measurable state variables. In the provided diagrams, the real value of the state variables is printed in blue, the estimated value is plotted in green while the associated setpoints are printed in red. It can be observed that, in all cases, fast and accurate tracking of the reference setpoints

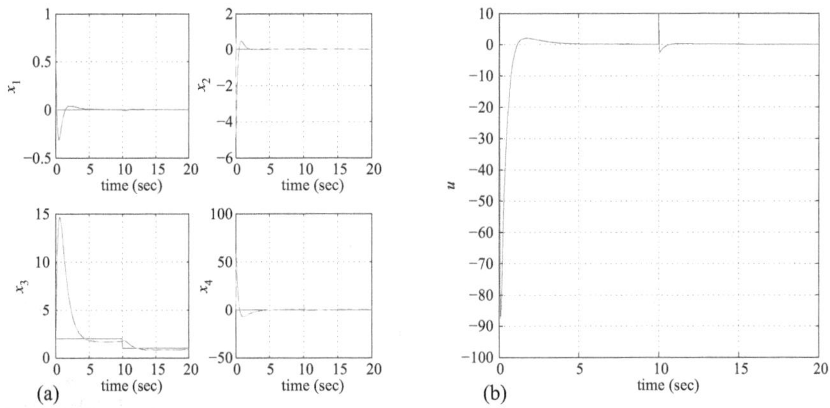

Figure 2.37 Nonlinear optimal control performance for the inertia wheel inverted pendulum in case of setpoint 3: (a) convergence of the state variables x_1 to x_4 to the reference setpoints; (b) control input u, applied to the inertia wheel inverted pendulum

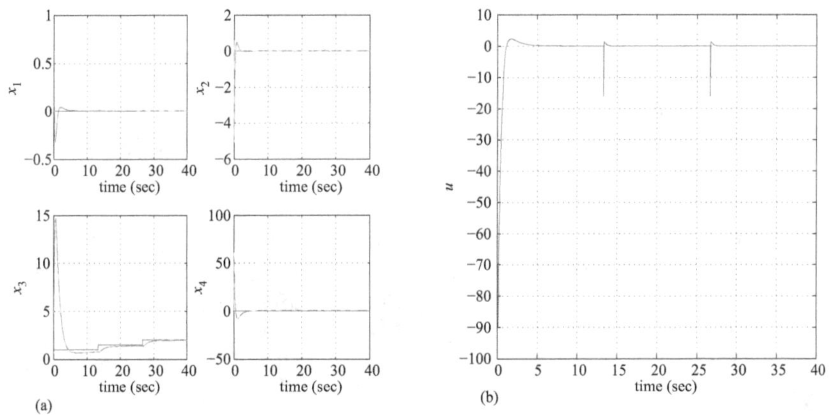

Figure 2.38 Nonlinear optimal control performance for the inertia wheel inverted pendulum in case of setpoint 4: (a) convergence of the state variables x_1 to x_4 to the reference setpoints; (b) control input u, applied to the inertia wheel inverted pendulum

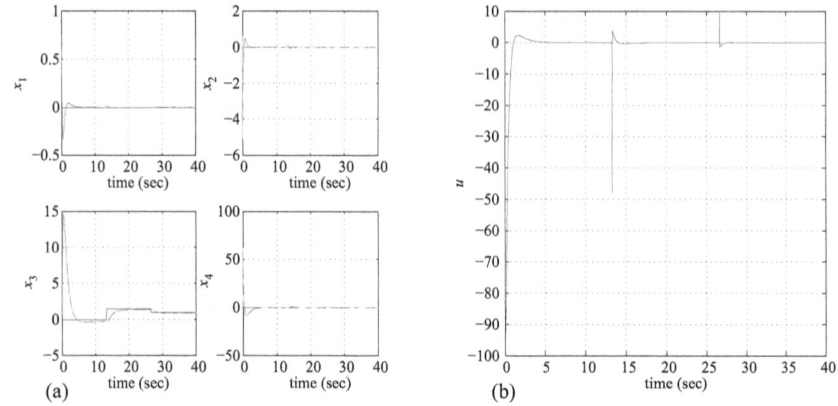

Figure 2.39 Nonlinear optimal control performance for the inertia wheel inverted pendulum in case of setpoint 5: (a) convergence of the state variables x_1 to x_4 to the reference setpoints; (b) control input u, applied to the inertia wheel inverted pendulum

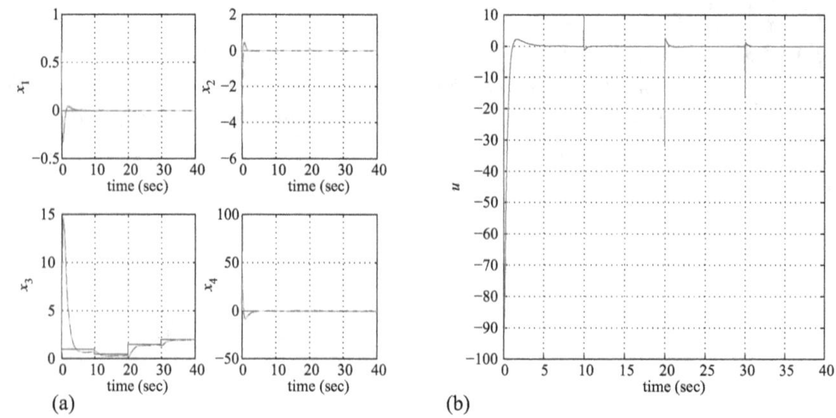

Figure 2.40 Nonlinear optimal control performance for the inertia wheel inverted pendulum in case of setpoint 6: (a) convergence of the state variables x_1 to x_4 to the reference setpoints; (b) control input u, applied to the inertia wheel inverted pendulum

has been achieved under moderate variations of the control inputs. The transient performance of the control method relies on the selection of the gains and matrices r, ρ, and Q that appear in the controller's algebraic Riccati equation. The robustness of the control method depends on the selection of ρ. Actually, the smallest value of this parameter that assures a solution for the algebraic Riccati equation of (2.95) is the one that provides maximum robustness to the control loop.

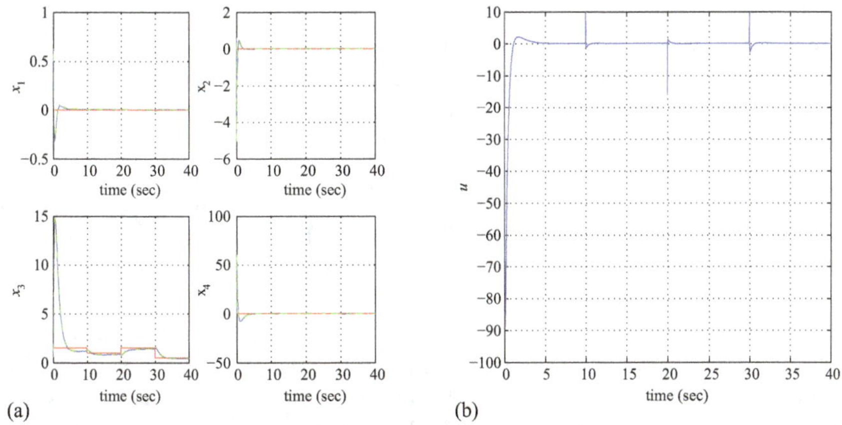

Figure 2.41 *Nonlinear optimal control performance for the inertia wheel inverted pendulum in case of setpoint 7: (a) convergence of the state variables x_1 to x_4 to the reference setpoints; (b) control input u, applied to the inertia wheel inverted pendulum*

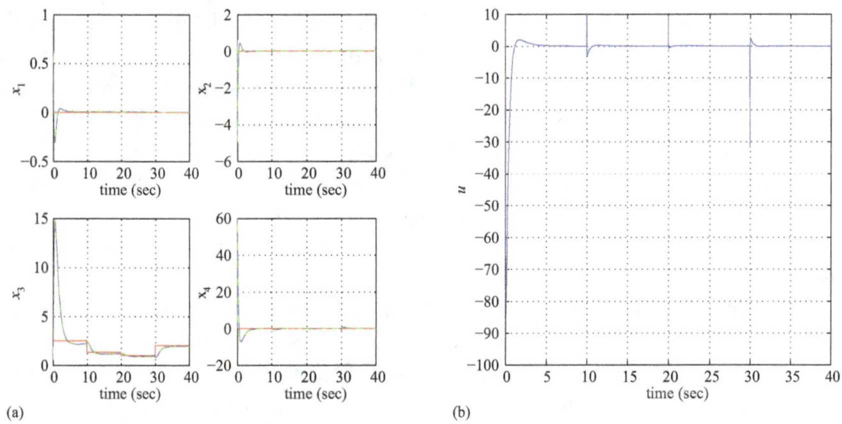

Figure 2.42 *Nonlinear optimal control performance for the inertia wheel inverted pendulum in case of setpoint 8: (a) convergence of the state variables x_1 to x_4 to the reference setpoints; (b) control input u, applied to the inertia wheel inverted pendulum*

2.4 Nonlinear control of the torsional oscillator with rotational actuator

2.4.1 Outline

The present section treats the nonlinear optimal control problem for the translational oscillator with rotating actuator (TORA system). This is an underactuated nonlinear mechatronic system exhibiting dynamics that is also met in specific satellite and aeronautics systems, as for instance the dual-spin spacecraft (Wu, Sun *et al.*, 2010; Guo and Liu, 2019; Huang, Khalil *et al.*, 2019; Sun, Wu, Fang *et al.*, 2017). This system is also known as Rotational-translational actuator (RTAC) (Pavlov *et al.*, 2007; Tadman, 2001; Avis *et al.*, 2010; Lan and Chen, 2006). It comprises a moving platform of mass M connected to a spring while in the interior of this platform there is a rod with a mass m at its one end being rotated by an electric motor which provides torque to its other end, as shown in Quan and Cai, 2013, Lan and Chen, 2006, Adlgostar *et al.*, 2006, and Rosales *et al.*, 2006. The control and stabilization of the considered system exhibits several difficulties, for instance due to its strongly nonlinear dynamic model as well as due to underactuation, as in (Gao and Zhao, 2014; Kawamoto *et al.*, 2001; and Kwon *et al.*, 2013). So far several nonlinear control approaches have been proposed for this system, including adaptive and robust control schemes as well as energy-based control (Wu, Gui *et al.*, 2018; Hung *et al.*, 2007; Lee and Chung, 2012; Avis *et al.*, 2008; Avis *et al.*, 2009). There have been also approaches towards state estimation-based control of the translational oscillator with rotating actuator (Celani, 2011; Burg and Dawson, 1997).

In this section, a nonlinear optimal (H-infinity) control method is proposed for the dynamic model of the translational oscillator with rotating actuator. The method relies on the approximate linearization of the state-space model of the system which takes place at each iteration of the control algorithm. The linearization process makes use of first-order Taylor series expansion and on the computation of the associated Jacobian matrices (Rigatos and Siano, 2015; Rigatos and Tzafestas, 2007; Basseville and Nikiforov, 1993; Rigatos and Zhang, 2009). The modeling error which is due to the truncation of higher-order terms in the Taylor series expansion is treated as a perturbation that is finally compensated by the robustness of the control algorithm. Thus the proposed method stands for the solution of the optimal control problem under model uncertainty and external perturbations.

Actually, the nonlinear optimal (H-infinity) control scheme represents the solution to a min-max differential game which takes place between the controller and the disturbances. The control inputs try to minimize a cost function that includes a quadratic term of the state vector's tracking error, whereas the model uncertainty and external perturbations try to maximize this cost functional. To compute the controller's feedback gains an algebraic Riccati equation is repetitively solved at each time-step of the control method (Rigatos, 2011; Rigatos, 2015; Rigatos and Busawon, 2018; Rigatos and Karapanou, 2020). The stability properties of the control scheme are proven through Lyapunov analysis. First, it is proven that the H-infinity tracking performance criterion holds and this signifies the method's robustness, as shown in Tousssaint, Basar and Bullo, 2000 and Lublin and Athans, 1995. Moreover, under moderate conditions it is proven that the control method is globally asymptotically stable . To implement state estimation-based control without taking velocity measurements out of this system, the H-infinity Kalman Filter is used as a robust state estimator (Rigatos and Busawon, 2018; Gibbs, 2012). Finally, through simulation experiments, it is confirmed that the new optimal control method achieves fast and accurate tracking of reference setpoints under moderate variations of the control inputs.

2.4.2 Dynamic model of the translational oscillator with rotating actuator

The system of the translational oscillator with rotating actuator comprises a platform of mass M and a rod with a mass m attached to its end, as shown in Figure 2.43. The rod is rotated by an actuator. The parameters of the dynamic model of the translational oscillator with rotating actuator are defined as follows: J is the moment of inertia of the rod, m is the mass which is attached at the end of the rod, r is the length of the rod, M is the mass of the platform, x is the horizontal displacement of the platform. An inertial reference frame is defined as shown in Figure 2.43 The position of mass m on the OX axis is $x + r\sin(\theta)$. The position of the mass m on the OY-axis is $r\sin(\theta)$. The potential energy of the system is given by:

$$P = \frac{1}{2}kx^2 + mgr(1 - \cos(\theta)) \tag{2.96}$$

The kinetic energy of the system is given by:

$$K = \frac{1}{2}M\dot{x}^2 + \frac{1}{2}m[(\dot{x} + r\cos(\theta)\dot{\theta})^2 + (-r\sin(\theta)\dot{\theta})^2] + \frac{1}{2}(mr^2 + J)\dot{\theta}^2 \tag{2.97}$$

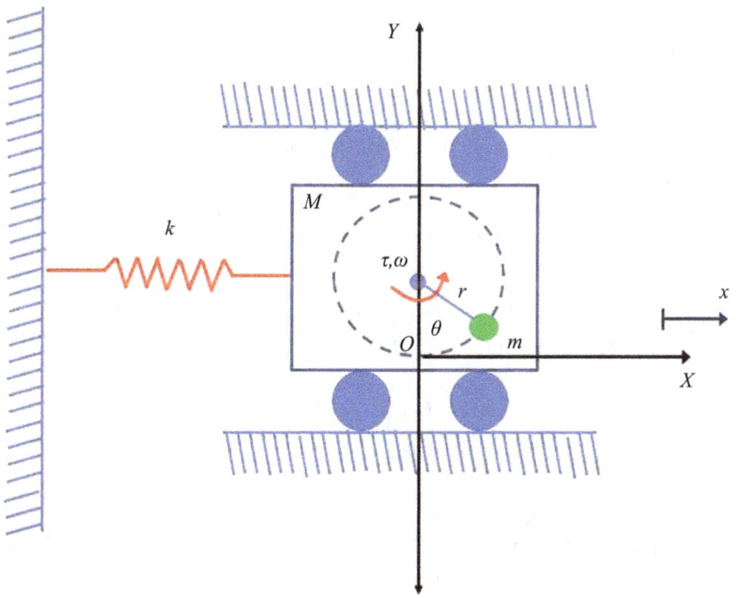

Figure 2.43 Diagram of the translational oscillator with rotating actuator (TORA system)

The Lagrangian of the system is $L = K - P$ which gives

$$\begin{aligned} L = &\frac{1}{2}M\dot{x}^2 + \frac{1}{2}m[(\dot{x} + r\cos(\theta)\dot{\theta})^2 + (-r\sin(\theta)\dot{\theta})^2] \\ &+ \frac{1}{2}(mr^2 + J)\dot{\theta}^2 - \frac{1}{2}kx^2 + mgr(1 - \cos(\theta)) \end{aligned} \tag{2.98}$$

The application of the Euler–Lagrange principle gives

$$\begin{aligned} &\frac{\partial}{\partial t}\frac{\partial}{\partial \dot{x}} - \frac{\partial L}{\partial x} = F_x \\ &\frac{\partial}{\partial t}\frac{\partial}{\partial \dot{\theta}} - \frac{\partial L}{\partial \theta} = \tau \end{aligned} \tag{2.99}$$

From the first row of the above Euler–Lagrange equations, and after intermediate operations, one obtains

$$(M + m)\ddot{x} + mr\cos(\theta)\ddot{\theta} - mr\sin(\theta)\dot{\theta}^2 + kx = 0 \tag{2.100}$$

From the second row of the above Euler–Lagrange equations, and after intermediate operations, one obtains

$$mr\cos(\theta)\ddot{x} + (2mr^2 + J)\ddot{\theta} + mgr\sin(\theta) = \tau \tag{2.101}$$

In matrix form, the dynamic model of the translational oscillator with rotating actuator becomes

$$\begin{pmatrix} (M+m) & mr\cos(\theta) \\ mr\cos(\theta) & 2mr^2 + J \end{pmatrix} \begin{pmatrix} \ddot{x} \\ \ddot{\theta} \end{pmatrix} = \begin{pmatrix} mr\sin(\theta)\dot{\theta}^2 - kx \\ -mgr\sin(\theta) \end{pmatrix} + \begin{pmatrix} 0 \\ 1 \end{pmatrix}\tau \tag{2.102}$$

Thus, the dynamics of the oscillator becomes

$$\begin{aligned} \begin{pmatrix} \ddot{x} \\ \ddot{\theta} \end{pmatrix} = &\begin{pmatrix} (M+m) & mr\cos(\theta) \\ mr\cos(\theta) & 2mr^2 + J \end{pmatrix}^{-1} \begin{pmatrix} mr\sin(\theta)\dot{\theta}^2 - kx \\ -mgr\sin(\theta) \end{pmatrix} \\ &+ \begin{pmatrix} (M+m) & mr\cos(\theta) \\ mr\cos(\theta) & 2mr^2 + J \end{pmatrix}^{-1} \begin{pmatrix} 0 \\ 1 \end{pmatrix}\tau \end{aligned} \tag{2.103}$$

The determinant of the inertia matrix is $detM = (M + m)(2mr^2 + J) - m^2r^2\cos^2(\theta)$. The inverse of the inertia matrix is

$$D^{-1} = \frac{1}{(M+m)(2mr^2 + J) - m^2r^2\cos^2(\theta)} \begin{pmatrix} 2mr^2 + J & -mr\cos(\theta) \\ -mr\cos(\theta) & (M+m) \end{pmatrix} \tag{2.104}$$

By performing additional operations, the state-space model of the system becomes

$$\begin{pmatrix}\ddot{x}\\ \ddot{\theta}\end{pmatrix}=\begin{pmatrix}\dfrac{(2mr^2+J)(mr\sin(\theta)\dot{\theta}^2-kx)+(mr\cos(\theta))(mgr\sin(\theta))}{(M+m)(2mr^2+J)-m^2r^2\cos^2(\theta)}\\ \dfrac{-(mr\cos(\theta))(mr\sin(\theta)\dot{\theta}^2-kx)-(M+m)(mgr\sin(\theta))}{(M+m)(2mr^2+J)-m^2r^2\cos^2(\theta)}\end{pmatrix}+\begin{pmatrix}\dfrac{-mr\cos(\theta)}{(M+m)(2mr^2+J)-m^2r^2\cos^2(\theta)}\\ \dfrac{((M+m)}{(M+m)(2mr^2+J)-m^2r^2\cos^2(\theta)}\end{pmatrix}\tau \tag{2.105}$$

By defining the state variables $x_1=x$, $x_2=\dot{x}$, $x_3=\theta$, and $x_4=\dot{\theta}$, the state-space description of the system becomes

$$\begin{pmatrix}\dot{x}_1\\ \dot{x}_2\\ \dot{x}_3\\ \dot{x}_4\end{pmatrix}=\begin{pmatrix}x_2\\ \dfrac{(2mr^2+J)(mr\sin(x_3)\dot{x}_3^2-kx_1)+(mr\cos(x_3))(mgr\sin(x_3))}{(M+m)(2mr^2+J)-m^2r^2\cos^2(x_3)}\\ x_4\\ \dfrac{-(mr\cos(x_3))(mr\sin(x_3)\dot{x}_3^2-kx)-(M+m)(mgr\sin(x_3))}{(M+m)(2mr^2+J)-m^2r^2\cos^2(x_3)}\end{pmatrix}+\begin{pmatrix}0\\ \dfrac{-mr\cos(x_3)}{(M+m)(2mr^2+J)-m^2r^2\cos^2(x_3)}\\ 0\\ \dfrac{((M+m)}{(M+m)(2mr^2+J)-m^2r^2\cos^2(x_3)}\end{pmatrix}\tau \tag{2.106}$$

or equivalently, the state-space model of the system can be written in the vector fields form

$$\dot{x}=f(x)+g(x)u \tag{2.107}$$

where $x\in R^{4\times 1}, f(x)\in R^{4\times 1}, g(x)\in R^{4\times 1}$ and $u\in R$, or

$$\begin{pmatrix}\dot{x}_1\\ \dot{x}_2\\ \dot{x}_3\\ \dot{x}_4\end{pmatrix}=\begin{pmatrix}f_1(x)\\ f_2(x)\\ f_3(x)\\ f_4(x)\end{pmatrix}+\begin{pmatrix}g_1(x)\\ g_2(x)\\ g_3(x)\\ g_4(x)\end{pmatrix}u \tag{2.108}$$

2.4.3 Approximate linearization of the state-space model

The dynamic model of the translational oscillator with rotating actuator undergoes linearization around the temporary operating point (x^*, u^*), where x^* is the present value of the system's state vector and u^* is the most recent control input that was applied to it. The linearization relies on Taylor series expansion and on the computation of the system's Jacobian matrices. The initial state-space model of the system is given in (2.107). By applying approximate linearization one arrives at the equivalent state-space description

$$\dot{x} = Ax + Bu + \tilde{d} \tag{2.109}$$

where $\tilde{d}$ is the cumulative disturbances vector while the Jacobian matrices A and B are given by

$$A = \nabla_x[f(x) + g(x)u]|_{(x^*,u^*)} \Rightarrow A = \nabla_x f(x)|_{(x^*,u^*)} + \nabla_x g(x)u|_{(x^*,u^*)} \tag{2.110}$$

$$B = \nabla_u[f(x) + g(x)u]|_{(x^*,u^*)} \Rightarrow B = g(x)|_{(x^*,u^*)} \tag{2.111}$$

The computation of the Jacobian matrix $\nabla_x f(x)|_{(x^*,u^*)}$ follows:

First row of the Jacobian matrix $\nabla_x f(x)$: $\frac{\partial f_1}{\partial x_1} = 0$, $\frac{\partial f_1}{\partial x_2} = 1$, $\frac{\partial f_1}{\partial x_3} = 0$, and $\frac{\partial f_1}{\partial x_4} = 0$,

Second row of the Jacobian matrix $\nabla_x f(x)$:

$$\frac{\partial f_2}{\partial x_i} = \frac{\frac{\partial f_{2n}}{\partial x_i} f_{2d} - f_{2n}\frac{\partial f_{2d}}{\partial x_i}}{f_{2d}^2} \tag{2.112}$$

for $i = 1, 2, 3, 4$ where $f_{2n} = (2mr^2 + J)(mr\sin(x_3))x_4^2 - kx_1 - (mr\cos(x_3))(mgr\sin(x_3))$. It holds that $\frac{\partial f_{2n}}{\partial x_1} = -(2mr^2 + J)$, $\frac{\partial f_{2n}}{\partial x_2} = 0$, $\frac{\partial f_{2n}}{\partial x_3} = (2mr^2 + J)(mr\cos(x_3)x_4^2) + (mr\sin(x_3))(mgr\sin(x_3)) - (mr\cos(x_3))(mgr\cos(x_3))$, $\frac{\partial f_{2n}}{\partial x_4} = (2mr^2 + J)(mr\sin(x_3)2x_4)$.

Besides $f_{2d} = (M + m)(2mr^2 + J) - m_2r^2\cos^2(x_3)$ while it also holds that $\frac{\partial f_{2d}}{\partial x_1} = 0$, $\frac{\partial f_{2d}}{\partial x_2} = 0$, $\frac{\partial f_{2d}}{\partial x_3} = 2m^2r^2\cos(x_3)\sin(x_3)$ and $\frac{\partial f_{2d}}{\partial x_4} = 0$.

Third row of the Jacobian matrix $\nabla_x f(x)$: $\frac{\partial f_3}{\partial x_1} = 0$, $\frac{\partial f_3}{\partial x_2} = 0$, $\frac{\partial f_3}{\partial x_3} = 0$, and $\frac{\partial f_3}{\partial x_4} = 1$.

Fourth row of the Jacobian matrix $\nabla_x f(x)$:

$$\frac{\partial f_4}{\partial x_i} = \frac{\frac{\partial f_{4n}}{\partial x_i} f_{4d} - f_{4n}\frac{\partial f_{4d}}{\partial x_i}}{f_{4d}^2} \tag{2.113}$$

for $i = 1, 2, 3, 4$, where $f_{4n} = -(mr\cos(x_3))(mr\sin(x_3)x_4^2 - kx_1) - (M + m)(mgr\sin(x_3))$ while it holds $\frac{\partial f_{4n}}{\partial x_1} = (mr\cos(x_3))k$, $\frac{\partial f_{4n}}{\partial x_2} = 0$, $\frac{\partial f_{4n}}{\partial x_3} = (mr\sin(x_3))(mr\sin(x_3)x_4^2 - kx_1) - (mr\cos(x_3))(mr\cos(x_3)) - (M + m)(mgr\cos(x_3))$ and also one has that $\frac{\partial f_{4n}}{\partial x_4} = -(mr\cos(x_3))(mr\sin(x_3)2x_4)$.

Besides $f_{4d} = (M + m)(2mr^2 + J) - m^2r^2\cos(x_3)$ while it also holds $\frac{\partial f_{4d}}{\partial x_1} = 0$, $\frac{\partial f_{4d}}{\partial x_2} = 0$, $\frac{\partial f_{4d}}{\partial x_3} = m^2r^2 2\cos(x_3)\sin(x_3)$ and $\frac{\partial f_{4d}}{\partial x_4} = 0$.

Next, the computation of the Jacobian matrix $\nabla_x g(x)|_{(x^*,u^*)}$ follows:

First row of the Jacobian matrix $\nabla_x g(x)$: $\frac{\partial g_1}{\partial x_1}=0$, $\frac{\partial g_1}{\partial x_2}=0$, $\frac{\partial g_1}{\partial x_3}=0$, and $\frac{\partial g_1}{\partial x_4}=0$,

Second row of the Jacobian matrix $\nabla_x g(x)$:

$$\frac{\partial g_2}{\partial x_i}=\frac{\frac{\partial g_{2n}}{\partial x_i}g_{2d}-g_{2n}\frac{\partial g_{2d}}{\partial x_i}}{g_{2d}^2} \tag{2.114}$$

for $i=1,2,3,4$ where $g_{2n}=-mr\cos(x_3)$. It holds that $\frac{\partial g_{2n}}{\partial x_1}=0$, $\frac{\partial g_{2n}}{\partial x_2}=0$, $\frac{\partial g_{2n}}{\partial x_3}=mr\sin(x_3)$, $\frac{\partial g_{2n}}{\partial x_4}=0$.

Besides $g_{2d}=(M+m)(2mr^2+J)-m^2r^2\cos^2(x_3)$ while it also holds that $\frac{\partial g_{2d}}{\partial x_1}=0$, $\frac{\partial g_{2d}}{\partial x_2}=0$, $\frac{\partial g_{2d}}{\partial x_3}=2m^2r^2\cos(x_3)\sin(x_3)$ and $\frac{\partial g_{2d}}{\partial x_4}=0$.

Third row of the Jacobian matrix $\nabla_x g(x)$: $\frac{\partial g_3}{\partial x_1}=0$, $\frac{\partial g_3}{\partial x_2}=0$, $\frac{\partial g_3}{\partial x_3}=0$, and $\frac{\partial g_3}{\partial x_4}=0$.

Fourth row of the Jacobian matrix $\nabla_x g(x)$:

$$\frac{\partial g_4}{\partial x_i}=\frac{\frac{\partial g_{4n}}{\partial x_i}g_{4d}-g_{4n}\frac{\partial g_{4d}}{\partial x_i}}{g_{4d}^2} \tag{2.115}$$

for $i=1,2,3,4$, where $g_{4n}=(M+m)$ while it holds $\frac{\partial g_{4n}}{\partial x_1}=0$, $\frac{\partial g_{4n}}{\partial x_2}=0$, $\frac{\partial g_{4n}}{\partial x_3}=0$ and $\frac{\partial g_{4n}}{\partial x_4}=0$.

Besides $g_{4d}=(M+m)(2mr^2+J)-m^2r^2\cos(x_3)$ while it also holds $\frac{\partial g_{4d}}{\partial x_1}=0$, $\frac{\partial g_{4d}}{\partial x_2}=0$, $\frac{\partial g_{4d}}{\partial x_3}=m^2r^2 2\cos(x_3)\sin(x_3)$ and $\frac{\partial g_{4d}}{\partial x_4}=0$.

$$\begin{aligned}\dot{x}&=Ax+Bu+L\tilde{d}\\ y&=Cx\end{aligned} \tag{2.116}$$

2.4.4 Computation of the feedback control gains

For the linearized system given by (1.36) the cost function of (1.37) is defined, where the coefficient r determines the penalization of the control input and the weight coefficient ρ determines the reward of the disturbances' effects. It is assumed that (i) The energy that is transferred from the disturbances signal $\tilde{d}(t)$ is bounded, that is $\int_0^\infty \tilde{d}^T(t)\tilde{d}(t)dt<\infty$, (ii) matrices $[A,B]$ and $[A,L]$ are stabilizable, and (iii) matrix $[A,C]$ is detectable. By denoting the state-vector's tracking error as $e=x-x_d$, the optimal feedback control law is given by

$$u(t)=-Ke(t) \tag{2.117}$$

with $K=\frac{1}{r}B^TP$ where P is a positive definite symmetric matrix which is obtained from the solution of the Riccati equation

$$A^TP+PA+Q-P\left(\frac{2}{r}BB^T-\frac{1}{\rho^2}LL^T\right)P=0 \tag{2.118}$$

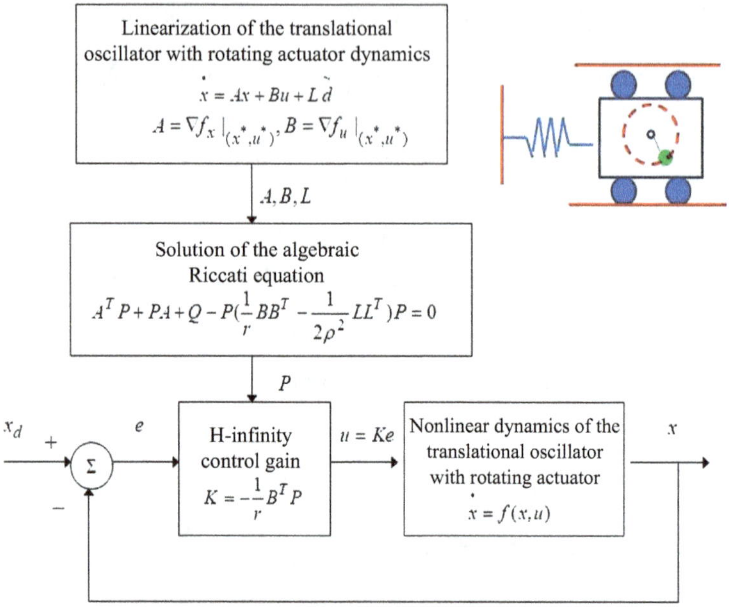

Figure 2.44 Diagram of the control scheme for the translational oscillator with rotating actuator

and Q is also a positive definite symmetric matrix. The worst case disturbance is given by $\tilde{d}(t) = \frac{1}{\rho^2} L^T Pe(t)$. The diagram of the considered control loop for the translational oscillator with rotating actuator is depicted in Figure 2.44. The global stability properties of the control method can be proven following the procedure presented in Section 1.1.2.5.

2.4.5 Simulation tests

The performance of the proposed nonlinear optimal (H-infinity) control scheme for the model of the translational oscillator with rotating actuator has been tested through simulation experiments. The obtained results are depicted in Figures 2.45 to 2.52. The real values of the state variables of the system are depicted in blue, the estimated values are printed in green, while the related setpoints are plotted in red. It can be considered that in all experiments fast and accurate tracking of the reference setpoints has been achieved. For the implementation of sensorless feedback control the H-infinity Kalman Filter has been used as a robust state estimator. By using the H-infinity Kalman Filter a state estimation-based implementation of the control method has been achieved. This allows the reliable functioning of the control loop after receiving measurements from a small number of sensors.

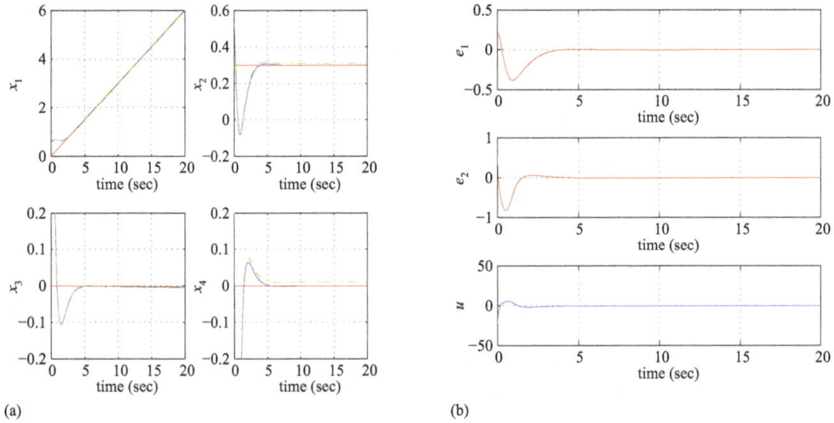

Figure 2.45 *Tracking of setpoint 1 for the translational oscillator with rotating actuator (TORA system): (a) convergence of state variables x_1 to x_4 to their reference setpoints (red line: setpoint, blue line: real value, green line: estimated value) and (b) tracking errors e_i $i = 1, 2$ for state variables x_2 and x_4 and control input u*

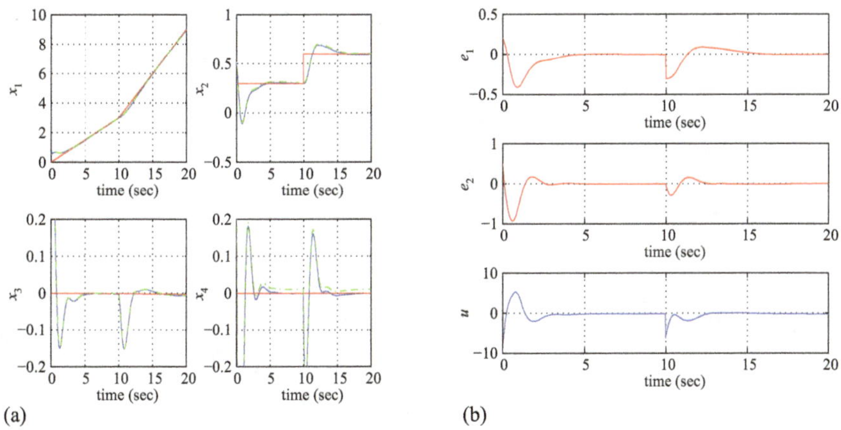

Figure 2.46 *Tracking of setpoint 2 for the translational oscillator with rotating actuator (TORA system): (a) convergence of state variables x_1 to x_4 to their reference setpoints (red line: setpoint, blue line: real value, green line: estimated value); (b) tracking errors e_i $i = 1, 2$ for state variables x_2 and x_4 and control input u*

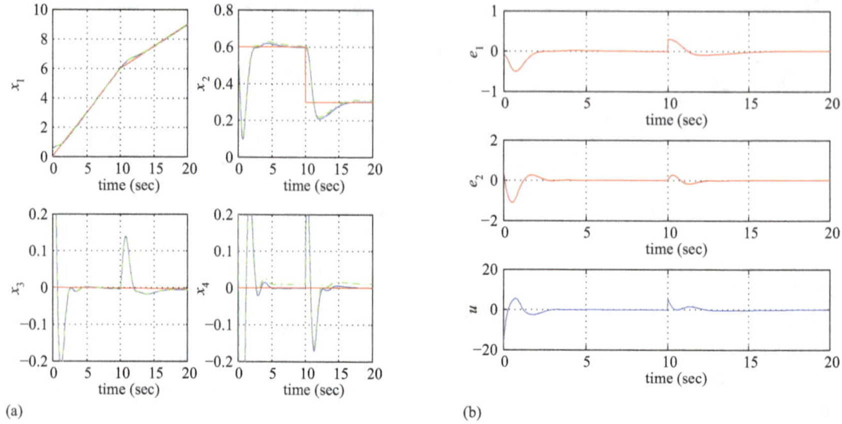

Figure 2.47 *Tracking of setpoint 3 for the translational oscillator with rotating actuator (TORA system): (a) convergence of state variables x_1 to x_4 to their reference setpoints (red line: setpoint, blue line: real value, green line: estimated value); (b) tracking errors e_i $i = 1, 2$ for state variables x_2 and x_4 and control input u*

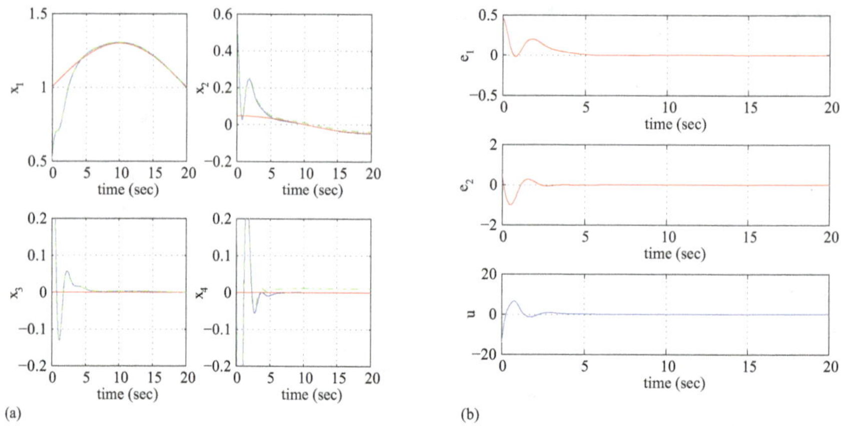

Figure 2.48 *Tracking of setpoint 4 for the translational oscillator with rotating actuator (TORA system): (a) convergence of state variables x_1 to x_4 to their reference setpoints (red line: setpoint, blue line: real value, green line: estimated value); (b) tracking errors e_i $i = 1, 2$ for state variables x_2 and x_4 and control input u*

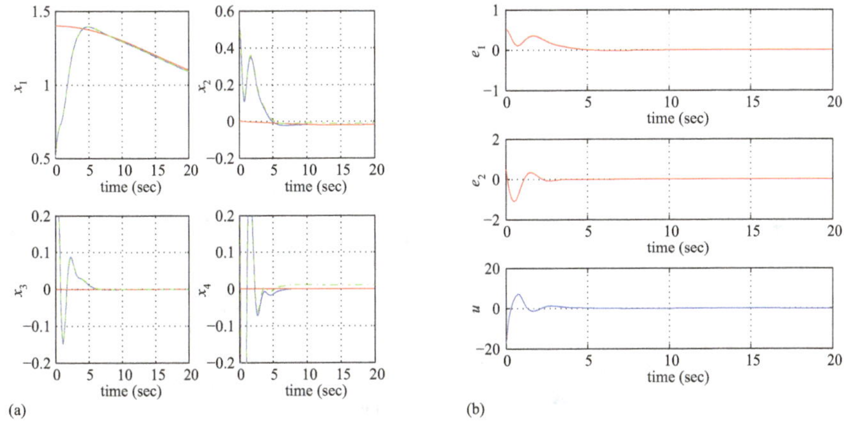

Figure 2.49 *Tracking of setpoint 5 for the translational oscillator with rotating actuator (TORA system): (a) convergence of state variables x_1 to x_4 to their reference setpoints (red line: setpoint, blue line: real value, green line: estimated value); (b) tracking errors e_i $i = 1, 2$ for state variables x_2 and x_4 and control input u*

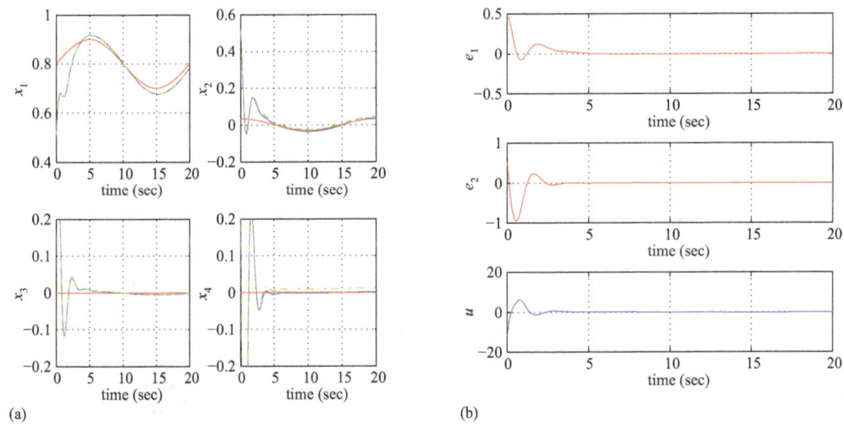

Figure 2.50 *Tracking of setpoint 6 for the translational oscillator with rotating actuator (TORA system): (a) convergence of state variables x_1 to x_4 to their reference setpoints (red line: setpoint, blue line: real value, green line: estimated value); (b) tracking errors e_i $i = 1, 2$ for state variables x_2 and x_4 and control input u*

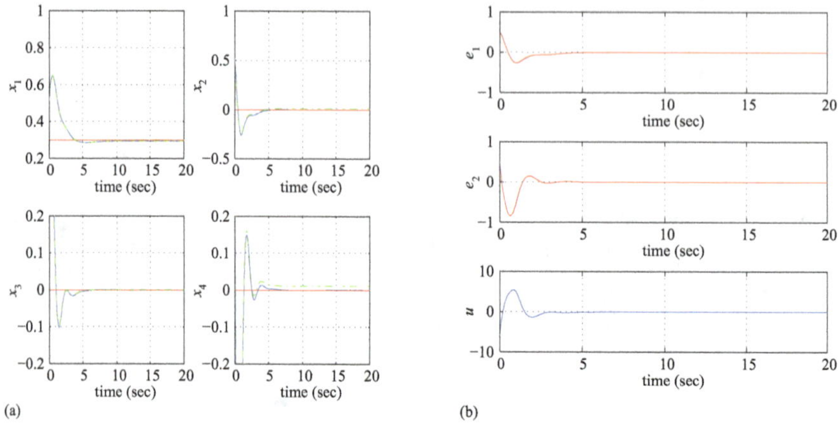

Figure 2.51 Tracking of setpoint 7 for the translational oscillator with rotating actuator (TORA system): (a) convergence of state variables x_1 to x_4 to their reference setpoints (red line: setpoint, blue line: real value, green line: estimated value); (b) tracking errors e_i $i = 1, 2$ for state variables x_2 and x_4 and control input u

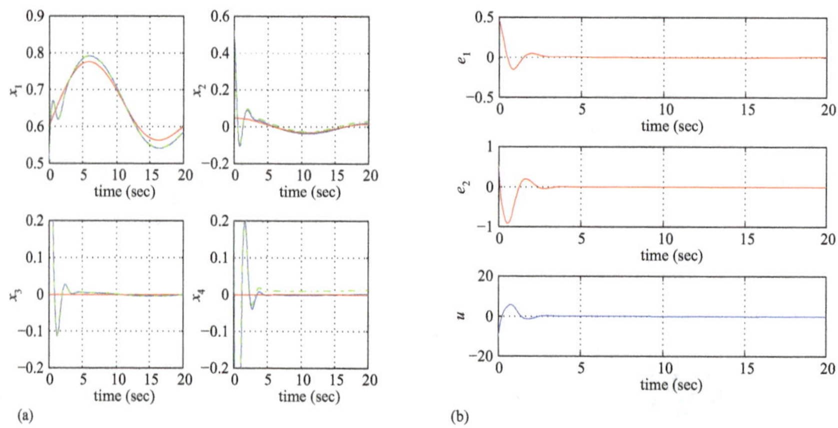

Figure 2.52 Tracking of setpoint 8 for the translational oscillator with rotating actuator (TORA system): (a) convergence of state variables x_1 to x_4 to their reference setpoints (red line: setpoint, blue line: real value, green line: estimated value); (b) tracking errors e_i $i = 1, 2$ for state variables x_2 and x_4 and control input u

2.5 Nonlinear control of robotic exoskeletons

2.5.1 Outline

Robotic exoskeletons find ample use in rehabilitation engineering for helping impaired people to regain their gait and mobility as well as their limbs' functionality after serious injuries, as in (Rigatos and Busawon, 2018; Pehlivan *et al.*, 2016; Mefoued, Mohammad *et al.*, 2013; and Yang, Ma *et al.*, 2019), and also analyzed in (Grun and Konogorski, 2012 and Yu, Lee *et al.*, 2016). Besides, they can be used in defense and security tasks by providing to humans in military or police corps the ability to lift and transfer heavy loads or to walk for longer distances without fatigue, as shown in (Huang, Li *et al.*, 2019; Chen, Chen *et al.*, 2018; and Lv, Zhou *et al.*, 2018). The control problem of limb exoskeletons is a nontrivial one due to the nonlinear and multivariable structure of the related state-space model (Chen, Li *et al.*, 2012; Zhang, Yang *et al.*, 2018; Thatte *et al.*, 2019). So far there have been attempts to solve this control problem with the use model-based and adaptive control techniques (Lu, Li *et al.*, 2014; Kang and Wa, 2013; Jabbari Asl *et al.*, 2018; Riani, Madani *et al.*, 2018). Moreover, one can note efforts to solve the control problem with the use of robust control methods, that may result in abrupt variations of the control inputs, as explained in (Cao *et al.*, 2018; Chen, Du *et al.*, 2019; Mohammad, Huo *et al.*, 2016; and Madani, Doachi *et al.*, 2016). Optimality concepts are usually not taken into account in the design of such controllers. On the one side, optimality signifies fast and accurate tracking of reference setpoints by the exoskeleton's state variables, thus improving the dexterity and precision of the performed motion. On the other side, optimality signifies minimal variations of the control inputs and minimal consumption of energy stored in batteries. Thus, extending the operational capacity and autonomy of exoskeletons without need to connect often to a power source (Ajjanaramvat and Parnichkan, 2018; Rodriguez *et al.*, 2017).

In this section, a novel solution to the nonlinear optimal control problem of robotic lower-limb exoskeletons is introduced. The method relies on approximate linearization of the exoskeleton's state-space model through first-order Taylor series expansion and through the computation of the related Jacobian matrices, as explained in (Rigatos and Tzafestas, 2007; Basseville and Nikiforov, 1993; and also presented in Rigatos and Zhang, 2009). The linearization takes place around a temporary operating point which is updated at each sampling period of the control algorithm and which is defined by the present value of the exoskeleton's state vector and by the most recent value of the control inputs vector. The modeling error, which is due to the truncation of higher-order terms in the Taylor series expansion, is considered to be a perturbation that is asymptotically compensated by the robustness of the control scheme. For the approximately linearized model of the exoskeleton an optimal (H-infinity) feedback controller is designed (Rigatos, 2016b; Rigatos, Siano and Cecati, 2015; Rigatos, Siano *et al.*, 2015).

Actually the proposed H-infinity controller stands for the solution of the optimal control problem for the robotic exoskeleton under model uncertainty and external perturbations. The H-infinity controller represents a min–max differential game in which the control inputs try to minimize a cost function which comprises quadratic terms of the state vector's tracking error. On the other side, the model uncertainty and disturbance terms try to maximize this cost function. To select the stabilizing gains of the H-infinity controller, an algebraic Riccati equation is solved at each sampling period of the control algorithm (Rigatos, 2011; Rigatos, 2015). The stability properties of the control method are proven through Lyapunov analysis. First, it is shown that the control loop satisfies the H-infinity tracking performance criterion, which signifies elevated robustness against model imprecision and exogenous disturbances, as analyzed in (Rigatos and Busawon, 2018; Tousssaint, Basar and Bullo, 2000; and Lublin and Athans, 1995). Moreover, under mild conditions it is proven that the control loop is globally asymptotically stable. Finally, to perform state estimation-based control without the need to measure the entire state vector of the exoskeleton, the H-infinity Kalman Filter is used as a robust state estimator, as previously shown in (Rigatos and Busawon, 2018 and Gibbs, 2012).

2.5.2 Dynamic model of the 2-DOF lower-limb exoskeleton

2.5.2.1 Dynamic model of the robotic exoskeleton

A 2-DOF lower-limb exoskeleton is considered as shown in Figure 2.53. Mass m_1 is the cumulative mass of the first link and of the leg's part from hip to knee. The related cumulative moment of inertia is I_1. Mass m_2 is the cumulative mass of the second link and of the leg's part from knee to ankle. The related cumulative moment of inertia is I_2. Mass m_1 is considered to be concentrated at a distance d_1 from the first joint. The length of the first link is l_1. Mass m_2 is considered to be concentrated at a distance d_2 from the second joint. The length of the second link is l_2.

The inertial reference frame shown in Figure 2.53 is considered. In this coordinates frame, the position of mass m_1 is

$$\begin{aligned} x_{m_1} &= d_1 \sin(\theta_1) \\ y_{m_1} &= l_1 + l_2 - d_1 \cos(\theta_1) \end{aligned} \tag{2.119}$$

The position of mass m_2 is given by:

$$\begin{aligned} x_{m_2} &= l_1 \sin(\theta_1) + d_2 \sin(\theta_1 - \theta_2) \\ y_{m_2} &= (l_1 + l_2) - l_1 \cos(\theta_1) - d_2 \cos(\theta_1 - \theta_2) \end{aligned} \tag{2.120}$$

The translational velocity of mass m_1 is

$$\begin{aligned} \dot{x}_{m_1} &= d_1 \cos(\theta_1)\dot{\theta}_1 \\ \dot{y}_{m_1} &= d_1 \sin(\theta_1)\dot{\theta}_1 \end{aligned} \tag{2.121}$$

The translational velocity of mass m_2 is

$$\begin{aligned} \dot{x}_{m_2} &= l_1 \cos(\theta_1)\dot{\theta}_1 + d_2 \cos(\theta_1 - \theta_2)(\dot{\theta}_1 - \dot{\theta}_2) \\ \dot{y}_{m_2} &= l_1 \sin(\theta_1)\dot{\theta}_1 + d_2 \sin(\theta_1 - \dot{\theta}_2)(\dot{\theta}_1 - \dot{\theta}_2) \end{aligned} \tag{2.122}$$

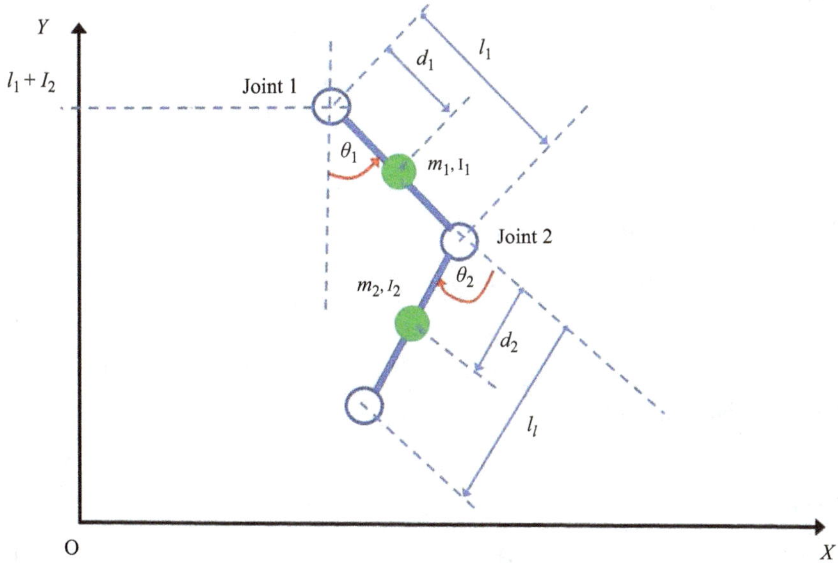

Figure 2.53 Diagram of a 2-DOF limb robotic exoskeleton performing the walking motion

The kinetic energy of link 1 is

$$K_1 = \frac{1}{2}m_1(\dot{x}_{m_1}^2 + \dot{y}_{m_1}^2) + \frac{1}{2}I_1\dot{\theta}_1^2 \Rightarrow$$
$$K_1 = \frac{1}{2}m_1[d_1^2\cos^2(\theta_1)\dot{\theta}_1^2 + d_1^2\sin^2(\theta_1)\dot{\theta}_1^2] + \frac{1}{2}I_1\dot{\theta}_1^2 \Rightarrow \quad (2.123)$$
$$K_1 = \frac{1}{2}m_1 d_1^2\dot{\theta}_1^2 + \frac{1}{2}I_1\dot{\theta}_1^2$$

The potential energy of link 1 is

$$P_1 = m_1 g[(l_1 + l_2) - d_1\cos(\theta_1)] \quad (2.124)$$

The kinetic energy of link 2 is

$$K_2 = \frac{1}{2}m_2(\dot{x}_{m_2}^2 + \dot{y}_{m_2}^2) + \frac{1}{2}I_2(\dot{\theta}_1 - \dot{\theta}_2)^2 \Rightarrow$$
$$K_2 = \frac{1}{2}m_2[l_1^2\cos^2(\theta_1)\dot{\theta}_1^2 + d_2^2\cos^2(\theta_1 - \theta_2)(\dot{\theta}_1 - \dot{\theta}_2)^2$$
$$+ 2l_1 d_2\cos(\theta_1)\cos(\theta_1 - \theta_2)\dot{\theta}_1(\dot{\theta}_1 - \dot{\theta}_2)] \quad (2.125)$$
$$+ \frac{1}{2}m_2[l_1^2\sin^2(\theta_1)\dot{\theta}_1^2 + d_2^2\sin^2(\theta_1 - \theta_2)(\dot{\theta}_1 - \dot{\theta}_2)^2$$
$$+ 2l_1 d_2\sin(\theta_1)\sin(\theta_1 - \theta_2)\dot{\theta}_1(\dot{\theta}_1 - \dot{\theta}_2)] + \frac{1}{2}I_2(\dot{\theta}_1 - \dot{\theta}_2)^2$$

which is also written as

$$\begin{aligned} K_2 = & \frac{1}{2} m_2 [l_1^2 \dot{\theta}_1^2 + d_2^2 (\dot{\theta}_1 - \dot{\theta}_2)^2 \\ & + 2 l_1 d_2 [\cos(\theta_2)\cos(\theta_1 - \theta_2) + \sin(\theta_2)\sin(\theta_1 - \theta_2)] \dot{\theta}_1 (\dot{\theta}_1 - \dot{\theta}_2)] \\ & + \frac{1}{2} I_2 (\dot{\theta}_1 - \dot{\theta}_2)^2 \end{aligned} \tag{2.126}$$

After intermediate operations one obtains

$$\begin{aligned} K_2 = & \frac{1}{2} m_2 [l_1^2 \dot{\theta}_1^2 + d_2^2 (\dot{\theta}_1 - \dot{\theta}_2^2) + 2 l_1 d_2 \cos(\theta_2) \dot{\theta}_1 (\dot{\theta}_1 - \dot{\theta}_2)] \\ & + \frac{1}{2} I_2 (\dot{\theta}_1 - \dot{\theta}_2)^2 \end{aligned} \tag{2.127}$$

The potential energy of link 2 is

$$P_2 = m_2 g [(l_1 + l_2) - l_1 \cos(\theta)_1 - d_2 \cos(\theta_1 - \theta_2)] \tag{2.128}$$

The Lagrangian of the robotic exoskeleton is given by

$$L = (K_1 + K_2) - (P_1 + P_2) \tag{2.129}$$

Thus, the Lagrangian of the exoskeleton is given by

$$\begin{aligned} L = & \left[\frac{1}{2} m_1 d_1^2 \dot{\theta}_1^2 + \frac{1}{2} I_1 \dot{\theta}_1^2 \right] + [\frac{1}{2} m_2 [l_1^2 \dot{\theta}_1^2 \\ & + d_2^2 (\dot{\theta}_1 - \dot{\theta}_2)^2 + 2 l_1 d_2 \cos(\theta_2) \dot{\theta}_1 (\dot{\theta}_1 - \dot{\theta}_2)] + \frac{1}{2} [I_2 (\dot{\theta}_1 - \dot{\theta}_2)^2] \\ & - [m_1 g [(l_1 + l_2) - d_1 \cos(\theta_1)] \\ & - [m_2 g [(l_1 + l_2) - l_1 \cos(\theta_1) - d_2 \cos(\theta_1 - \theta_2)] \end{aligned} \tag{2.130}$$

The equations of motion of the exoskeleton are obtained from the application of the Euler–Lagrange principle

$$\frac{\partial}{\partial t} \frac{\partial L}{\partial \dot{\theta}_1} - \frac{\partial L}{\partial \theta_1} = \tau_1 \tag{2.131}$$

$$\frac{\partial}{\partial t} \frac{\partial L}{\partial \dot{\theta}_2} - \frac{\partial L}{\partial \theta_2} = \tau_2 \tag{2.132}$$

where τ_1 and τ_2 are the torques which are generated by the motors that make the two joints of the exoskeleton rotate. About the first link it holds that

$$\begin{aligned} \frac{\partial L}{\partial \dot{\theta}_1} = & m_1 d_1^2 \dot{\theta}_1 + I_1 \dot{\theta}_1 \\ & + m_2 l_1^2 \dot{\theta}_1 + m_2 d_2^2 (\dot{\theta}_1 - \dot{\theta}_2) + m_2 l_1 d_2 \cos(\theta_2)(2\dot{\theta}_1 - \dot{\theta}_2) + I_2 (\dot{\theta}_1 - \dot{\theta}_2) \end{aligned} \tag{2.133}$$

and also

$$\begin{aligned} \frac{\partial}{\partial t} \frac{\partial L}{\partial \dot{\theta}_1} = & m_1 d_1^2 \ddot{\theta}_1 + I_1 \ddot{\theta}_1 \\ & + m_2 l_1^2 \ddot{\theta}_1 + m_2 d_2^2 (\ddot{\theta}_1 - \ddot{\theta}_2) - m_2 l_1 d_2 \sin(\theta_2) \dot{\theta}_2 (2\dot{\theta}_1 - \dot{\theta}_2) \\ & + m_2 l_1 d_2 \cos(\theta_2)(2\ddot{\theta}_1 - \ddot{\theta}_2) + I_2 (\ddot{\theta}_1 - \ddot{\theta}_2) \end{aligned} \tag{2.134}$$

while one has

$$\frac{\partial L}{\partial \theta_1} = -m_1 g d_1 \sin(\theta_1) - m_2 g l_1 \sin(\theta_1) - m_2 g d_2 \sin(\theta_1 - \theta_2) \tag{2.135}$$

Consequently, the equation of motion of the first link becomes

$$\begin{aligned}&[m_1 d_1^2 + I_1 + m_1 l_1^2 + m_1 d_1^2 + 2m_2 l_1 d_2 \cos(\theta_2) + I_2]\ddot{\theta}_1 \\ &\quad +[-m_2 d_2^2 - m_2 l_1 d_2 \cos(\theta_2) - I_2]\ddot{\theta}_2 - m_2 l_1 d_2 \sin(\theta_2)\dot{\theta}_2(2\dot{\theta}_1 - \dot{\theta}_2) \\ &\quad +(m_1 g d_1 + m_2 g l_1)\sin(\theta_1) + m_2 g d_2 \sin(\theta_1 - \theta_2) = \tau_1\end{aligned} \tag{2.136}$$

About the second link it holds that

$$\frac{\partial L}{\partial \dot{\theta}_2} = -m_2 d_2^2(\dot{\theta}_1 - \dot{\theta}_2) - m_2 l_1 d_2 \cos(\theta_2)\dot{\theta}_1 - I_2(\dot{\theta}_1 - \dot{\theta}_2) \tag{2.137}$$

and also

$$\begin{aligned}\frac{\partial}{\partial t}\frac{\partial L}{\partial \dot{\theta}_2} &= -m_2 d_2^2(\ddot{\theta}_1 - \ddot{\theta}_2) + m_2 l_1 d_2 \sin(\theta_2)\dot{\theta}_1\dot{\theta}_2 \\ &\quad -m_2 l_1 d_2 \cos(\theta_2)\ddot{\theta}_1 - I_1(\ddot{\theta}_1 - \ddot{\theta}_2)\end{aligned} \tag{2.138}$$

Moreover, one has

$$\frac{\partial L}{\partial \theta_2} = -m_2 l_1 d_2 \sin(\theta_2)\dot{\theta}_1(\dot{\theta}_1 - \dot{\theta}_2) - m_2 g d_2 \sin(\theta_1 - \theta_2) \tag{2.139}$$

Thus, the equation of motion of the second link becomes

$$\begin{aligned}&[-m_2 d_2^2 - m_2 l_1 d_2 \sin(\theta_2) - I_2]\ddot{\theta}_1 + [m_2 d_2^2 + I_2]\ddot{\theta}_2 \\ &\quad +m_2 l_1 d_2 \sin(\theta_2)\dot{\theta}_1\dot{\theta}_2 + 2m_2 l_1 d_2 \sin(\theta_2)\dot{\theta}_1(\dot{\theta}_! - \dot{\theta}_2) \\ &\quad +m_2 g d_2 \sin(\theta_1 - \theta_2) = \tau_2\end{aligned} \tag{2.140}$$

which is finally written as

$$\begin{aligned}&[-m_2 d_2^2 - m_2 l_1 d_2 \sin(\theta_2) - I_2]\ddot{\theta}_1 + [m_2 d_2^2 + I_2]\ddot{\theta}_2 \\ &\quad +m_2 l_1 d_2 \sin(\theta_2)\dot{\theta}_1^2 + m_2 g d_2 \sin(\theta_1 - \theta_2) = \tau_2\end{aligned} \tag{2.141}$$

In matrix form, the dynamic model of the exoskeleton is written as

$$\begin{pmatrix} m_1 d_1^3 + I_1 + m_2 l_1^2 + m_2 d_1^2 + 2m_2 l_1 d_2 \cos(\theta_2) + I_2 & -m_2 d_2^2 - m_2 l_1 d_2 \cos(\theta_2) - I_2 \\ -m_2 d_2^2 - m_2 l_1 d_2 \cos(\theta_2) - I_2 & m_2 d_2^2 + I_2 \end{pmatrix} \begin{pmatrix} \ddot{\theta}_1 \\ \ddot{\theta}_2 \end{pmatrix}$$

$$+ \begin{pmatrix} -m_2 l_1 d_2 \sin(\theta_2)\dot{\theta}_2(2\dot{\theta}_1 - \dot{\theta}_2) \\ m_2 l_1 d_2 \sin(\theta_2)\dot{\theta}_1^2 \end{pmatrix}$$

$$+ \begin{pmatrix} (m_1 g d_1 + m_2 g l_1)\sin(\theta_1) + m_2 g d_2 \sin(\theta_1 - \theta_2) \\ m_2 g d_2 \sin(\theta_1 - \theta_2) \end{pmatrix} = \begin{pmatrix} \tau_1 \\ \tau_2 \end{pmatrix}$$

2.5.2.2 State-space model of the robotic exoskeleton

The dynamic model of the robotic exoskeleton can be also written in the concise form

$$M(\theta)\ddot{\theta} + C(\theta, \dot{\theta}) + G(\theta) = \tau \tag{2.142}$$

where the state-vector is $\theta = [\theta_1, \theta_2]^T$, the control inputs vector is $\tau = [\tau_1, \tau_2]^T$, while the inertia matrix $M(\theta)$, the Coriolis matrix $C(\theta, \dot{\theta})$ and the gravitational forces matrix $G(\theta)$ are defined as follows:

$$M(\theta) = \begin{pmatrix} m_1 d_1^3 + I_1 + m_2 l_1^2 + m_2 d_1^2 + 2m_2 l_1 d_2 \cos(\theta_2) + I_2 & -m_2 d_2^2 - m_2 l_1 d_2 \cos(\theta_2) - I_2 \\ -m_2 d_2^2 - m_2 l_1 d_2 \cos(\theta_2) - I_2 & m_2 d_2^2 + I_2 \end{pmatrix}$$

$$C(\theta, \dot{\theta}) = \begin{pmatrix} -m_2 l_1 d_2 \sin(\theta_2)\dot{\theta}_2(2\dot{\theta}_1 - \dot{\theta}_2) \\ m_2 l_1 d_2 \sin(\theta_2)\dot{\theta}_1^2 \end{pmatrix}$$

$$G(\theta) = \begin{pmatrix} (m_1 g d_1 + m_2 g l_1)\sin(\theta_1) + m_2 g d_2 \sin(\theta_1 - \theta_2) \\ m_2 g d_2 \sin(\theta_1 - \theta_2) \end{pmatrix}$$

Using that he inertia matrix $M(\theta)$ is

$$M = \begin{pmatrix} M_{11} & M_{12} \\ M_{21} & M_{22} \end{pmatrix} \tag{2.143}$$

and that the determinant of $M(\theta)$ is $detM = M_{11}M_{22} - M_{12}^2$ the inverse of $M(\theta)$ is

$$M^{-1} = \begin{pmatrix} \frac{M_{22}}{M_{11}M_{22} - M_{12}^2} & \frac{-M_{12}}{M_{11}M_{22} - M_{12}^2} \\ \frac{-M_{21}}{M_{11}M_{22} - M_{12}^2} & \frac{M_{11}}{M_{11}M_{22} - M_{12}^2} \end{pmatrix} \tag{2.144}$$

Thus the dynamic model of the exoskeleton robot is written as

$$\ddot{\theta} = -M^{-1}(\theta)[C(\theta, \dot{\theta}) + G(\theta)] + M^{-1}(\theta)\tau \tag{2.145}$$

By performing operations between the individual matrices, one arrives at the state-space description

$$\begin{pmatrix} \ddot{\theta}_1 \\ \ddot{\theta}_2 \end{pmatrix} = \begin{pmatrix} \frac{-M_{22}(C_1 + G_1) + M_{12}(C_2 + G_2)}{M_{11}M_{22} - M_{12}^2} \\ \frac{M_{12}(C_1 + G_1) - M_{11}(C_2 + G_2)}{M_{11}M_{22} - M_{12}^2} \end{pmatrix} + \begin{pmatrix} \frac{M_{22}}{M_{11}M_{22} - M_{12}^2} & \frac{-M_{12}}{M_{11}M_{22} - M_{12}^2} \\ \frac{-M_{21}}{M_{11}M_{22} - M_{12}^2} & \frac{M_{11}}{M_{11}M_{22} - M_{12}^2} \end{pmatrix} \begin{pmatrix} \tau_1 \\ \tau_2 \end{pmatrix}$$

By defining the state-vector of the exoskeleton robot as $x_1 = \theta_1$, $x_2 = \dot{\theta}_1$, $x_3 = \theta_2$, and $x_4 = \dot{\theta}_2$, the state-space model of the exoskeleton becomes

$$\begin{pmatrix} \dot{x}_1 \\ \dot{x}_2 \\ \dot{x}_3 \\ \dot{x}_4 \end{pmatrix} = \begin{pmatrix} f_1(x) \\ f_2(x) \\ f_3(x) \\ f_4(x) \end{pmatrix} + \begin{pmatrix} 0 & 0 \\ g_{21}(x) & g_{22}(x) \\ 0 & 0 \\ g_{41}(x) & g_{42}(x) \end{pmatrix} \begin{pmatrix} u_1 \\ u_2 \end{pmatrix} \tag{2.146}$$

where $f_1(x) = 0$, $f_2(x) = \frac{-M_{22}(C_1+G_1)+M_{12}(C_2+G_2)}{M_{11}M_{22}-M_{12}^2}$, $f_3(x) = 0$ and also it holds that $f_4(x) = \frac{M_{12}(C_1+G_1)-M_{11}(C_2+G_2)}{M_{11}M_{22}-M_{12}^2}$

while $g_{11}(x)=0$, $g_{12}(x)=0$, $g_{21}(x)=\frac{M_{22}}{M_{11}M_{22}-M_{12}^2}$, $g_{22}(x)=\frac{-M_{12}}{M_{11}M_{22}-M_{12}^2}$, $g_{31}(x)=0$, $g_{31}(x)=0$, $g_{41}(x)=\frac{-M_{21}}{M_{11}M_{22}-M_{12}^2}$ and $g_{42}(x)=\frac{M_{11}}{M_{11}M_{22}-M_{12}^2}$

2.5.2.3 Differential flatness properties of the exoskeleton's model

It can be proven that the previously defined state-space model of the lower-limb robotic exoskeleton is differentially flat with flat outputs vector $Y=[x_1,x_3]^T$. This signifies that the flat outputs are the turn angles θ_1, θ_2 of the exoskeleton's joints. Actually, the differential flatness property means that (i) all state variables of the exoskeleton can be written as differential functions of the above noted flat outputs and (ii) the flat outputs are differentially independent, that is they are not connected through a relation in the form of a linear homogeneous differential equation.

From the first row of (2.146) one has $x_2=\dot{x}_1$, while from the third row of the same state-space model one gets $x_4=\dot{x}_3$. Consequently all state variables of the exoskeleton are expressed as differential functions of its flat outputs. From the second and the fourth row of the state-space description, one gets

$$\begin{pmatrix}u_1\\u_2\end{pmatrix}=\begin{pmatrix}g_{21}(x) & g_{22}(x)\\g_{41}(x) & g_{42}(x)\end{pmatrix}^{-1}\left[\begin{pmatrix}\ddot{x}_1\\\ddot{x}_3\end{pmatrix}-\begin{pmatrix}f_2(x)\\f_4(x)\end{pmatrix}\right] \tag{2.147}$$

Equation (2.147) confirms that the control inputs of the robotic exoskeleton are also differential functions of the flat outputs $Y=[x_1,x_3]^T$. This completes the proof about the differential flatness of the robotic model. The differential flatness property of the lower-limb exoskeleton can be used for defining setpoints for the related control loop.

2.5.3 Approximate linearization of the exoskeleton's dynamic model

2.5.3.1 Approximate linearization of the state-space model

The dynamic model of the robotic exoskeleton is initially written in the form

$$\dot{x}=f(x)+g(x)u \tag{2.148}$$

The model undergoes approximate linearization around the temporary operating point (x^*,u^*) where x^* is the present value of the exoskeleton's state vector and u^* is the last sampled value of the control inputs vector. The linearization is performed through first-order Taylor series expansion and through the computation of the associated Jacobian matrices.

The linearized state-space description of the system is written as

$$\dot{x}=Ax+Bu+\tilde{d} \tag{2.149}$$

where $\tilde{d}$ is the cumulative vector of disturbances and modeling errors while matrices A and B are defined by the system's Jacobians

$$A=\nabla_x[f(x)+g(x)u]\mid_{(x^*,u^*)}\Rightarrow A=\nabla_x f(x)\mid_{(x^*,u^*)}+\nabla_x g(x)u\mid_{(x^*,u^*)} \tag{2.150}$$

$$B=\nabla_u[f(x)+g(x)u]\mid_{(x^*,u^*)}\Rightarrow B=g(x)]\mid_{(x^*,u^*)} \tag{2.151}$$

The computation of the Jacobian matrix $\nabla_x f(x)\mid_{(x^*,u^*)}$ is as follows:

First row of the Jacobian matrix $\nabla_x f(x) \mid_{(x^*,u^*)}$: $\frac{\partial f_1}{\partial x_1} = 0$, $\frac{\partial f_1}{\partial x_2} = 1$, $\frac{\partial f_1}{\partial x_3} = 0$ and $\frac{\partial f_1}{\partial x_4} = 0$.

Second row of the Jacobian matrix $\nabla_x f(x) \mid_{(x^*,u^*)}$: for $i = 1, 2, 3, 4$

$$\frac{\partial f_2}{\partial x_i} = \frac{-\frac{\partial M_{22}}{\partial x_i}(C_1+G_1) - M_{22}(\frac{\partial C_1}{\partial x_i} + \frac{\partial G_1}{\partial x_i}) + \frac{\partial M_{12}}{\partial x_i}(C_2+G_2) + M_{12}(\frac{\partial C_2}{\partial x_i} + \frac{\partial G_2}{\partial x_i})}{[M_{11}M_{22} - M_{12}^2]} - \frac{[-M_{22}(C_1+G_1) + M_{12}(C_2+G_2)][\frac{\partial M_{11}}{\partial x_i}M_{22} + M_{11}\frac{\partial M_{22}}{\partial x_i} - 2M_{12}\frac{\partial M_{12}}{\partial x_i}]}{[M_{11}M_{22} - M_{12}^2]^2} \tag{2.152}$$

Third row of the Jacobian matrix $\nabla_x f(x) \mid_{(x^*,u^*)}$: $\frac{\partial f_3}{\partial x_1} = 0$, $\frac{\partial f_3}{\partial x_2} = 0$, $\frac{\partial f_3}{\partial x_3} = 0$ and $\frac{\partial f_3}{\partial x_4} = 1$.

Fourth row of the Jacobian matrix $\nabla_x f(x) \mid_{(x^*,u^*)}$: for $i = 1, 2, 3, 4$

$$\frac{\partial f_4}{\partial x_i} = \frac{\frac{\partial M_{12}}{\partial x_i}(C_1+G_1) + M_{12}(\frac{\partial C_1}{\partial x_i} + \frac{\partial G_1}{\partial x_i}) - \frac{\partial M_{11}}{\partial x_i}(C_2+G_2) - M_{11}(\frac{\partial C_2}{\partial x_i} + \frac{\partial G_2}{\partial x_i})}{[M_{11}M_{22} - M_{12}^2]} - \frac{[M_{12}(C_1+G_1) - M_{12}(C_2+G_2)][\frac{\partial M_{11}}{\partial x_i}M_{22} + M_{11}\frac{\partial M_{22}}{\partial x_i} - 2M_{12}\frac{\partial M_{12}}{\partial x_i}]}{[M_{11}M_{22} - M_{12}^2]^2} \tag{2.153}$$

Considering that the control inputs gain matrix consists of two column vectors, that is $g(x) = [g_1(x) \;\; g_2(x)]$, the computation of the Jacobian matrix $\nabla_x g_1(x) \mid_{(x^*,u^*)}$ is as follows:

First row of the Jacobian matrix $\nabla_x g_1(x) \mid_{(x^*,u^*)}$: $\frac{\partial g_{11}}{\partial x_1} = 0$, $\frac{\partial g_{11}}{\partial x_2} = 0$, $\frac{\partial g_{11}}{\partial x_3} = 0$ and $\frac{\partial g_{11}}{\partial x_4} = 1$.

Second row of the Jacobian matrix $\nabla_x g_1(x) \mid_{(x^*,u^*)}$: for $i = 1, 2, 3, 4$

$$\frac{\partial g_{21}}{\partial x_i} = \frac{\frac{\partial M_{22}}{\partial x_i}}{[M_{11}M_{22} - M_{12}^2]} - \frac{M_{22}[\frac{\partial M_{11}}{\partial x_i}M_{22} + M_{11}\frac{\partial M_{22}}{\partial x_i} - 2M_{12}\frac{\partial M_{12}}{\partial x_i}]}{[M_{11}M_{22} - M_{12}^2]^2} \tag{2.154}$$

Third row of the Jacobian matrix $\nabla_x g_1(x) \mid_{(x^*,u^*)}$: $\frac{\partial g_{31}}{\partial x_1} = 0$, $\frac{\partial g_{31}}{\partial x_2} = 0$, $\frac{\partial g_{31}}{\partial x_3} = 0$ and $\frac{\partial g_{31}}{\partial x_4} = 1$.

Fourth row of the Jacobian matrix $\nabla_x g_1(x) \mid_{(x^*,u^*)}$: for $i = 1, 2, 3, 4$

$$\frac{\partial g_{41}}{\partial x_i} = \frac{-\frac{\partial M_{21}}{\partial x_i}}{[M_{11}M_{22} - M_{12}^2]} - \frac{-M_{21}[\frac{\partial M_{11}}{\partial x_i}M_{22} + M_{11}\frac{\partial M_{22}}{\partial x_i} - 2M_{12}\frac{\partial M_{12}}{\partial x_i}]}{[M_{11}M_{22} - M_{12}^2]^2} \tag{2.155}$$

The computation of the Jacobian matrix $\nabla_x g_2(x) \mid_{(x^*,u^*)}$ is as follows:

First row of the Jacobian matrix $\nabla_x g_2(x) \mid_{(x^*,u^*)}$: $\frac{\partial g_{12}}{\partial x_1} = 0$, $\frac{\partial g_{12}}{\partial x_2} = 0$, $\frac{\partial g_{12}}{\partial x_3} = 0$ and $\frac{\partial g_{12}}{\partial x_4} = 1$.

Second row of the Jacobian matrix $\nabla_x g_2(x)\mid_{(x^*,u^*)}$: for $i = 1,2,3,4$

$$\frac{\partial g_{22}}{\partial x_i} = \frac{-\frac{\partial M_{12}}{\partial x_i}}{[M_{11}M_{22} - M_{12}^2]} - \frac{-M_{12}[\frac{\partial M_{11}}{\partial x_i}M_{22} + M_{11}\frac{\partial M_{22}}{\partial x_i} - 2M_{12}\frac{\partial M_{12}}{\partial x_i}]}{[M_{11}M_{22} - M_{12}^2]^2} \quad (2.156)$$

Third row of the Jacobian matrix $\nabla_x g_2(x)\mid_{(x^*,u^*)}$: $\frac{\partial g_{32}}{\partial x_1} = 0$, $\frac{\partial g_{32}}{\partial x_2} = 0$, $\frac{\partial g_{32}}{\partial x_3} = 0$ and $\frac{\partial g_{32}}{\partial x_4} = 1$.

Fourth row of the Jacobian matrix $\nabla_x g_2(x)\mid_{(x^*,u^*)}$: for $i = 1,2,3,4$

$$\frac{\partial g_{42}}{\partial x_i} = \frac{\frac{\partial M_{11}}{\partial x_i}}{[M_{11}M_{22} - M_{12}^2]} - \frac{M_{11}[\frac{\partial M_{11}}{\partial x_i}M_{22} + M_{11}\frac{\partial M_{22}}{\partial x_i} - 2M_{12}\frac{\partial M_{12}}{\partial x_i}]}{[M_{11}M_{22} - M_{12}^2]^2} \quad (2.157)$$

Next, the partial derivatives of the individual elements of the inertia, Coriolis and gravitational matrices are computed. It holds that:

Inertia matrix $M(\theta)$:

$\frac{\partial M_{11}}{\partial x_1} = 0$, $\frac{\partial M_{11}}{\partial x_2} = 0$, $\frac{\partial M_{11}}{\partial x_3} = -2m_2l_1d_2\sin(x_3)$, $\frac{\partial M_{11}}{\partial x_4} = 0$.

$\frac{\partial M_{12}}{\partial x_1} = 0$, $\frac{\partial M_{12}}{\partial x_2} = 0$, $\frac{\partial M_{12}}{\partial x_3} = m_2l_1d_2\sin(x_3)$, $\frac{\partial M_{12}}{\partial x_4} = 0$.

$\frac{\partial M_{21}}{\partial x_1} = 0$, $\frac{\partial M_{21}}{\partial x_2} = 0$, $\frac{\partial M_{21}}{\partial x_3} = m_2l_1d_2\sin(x_3)$, $\frac{\partial M_{21}}{\partial x_4} = 0$.

$\frac{\partial M_{22}}{\partial x_1} = 0$, $\frac{\partial M_{22}}{\partial x_2} = 0$, $\frac{\partial M_{22}}{\partial x_3} = 0$, $\frac{\partial M_{22}}{\partial x_4} = 0$.

Coriolis matrix $C(\theta,\dot{\theta})$:

$\frac{\partial C_1}{\partial x_1} = 0$, $\frac{\partial C_1}{\partial x_2} = -2m_2l_1d_2\sin(x_3)x_4$, $\frac{\partial C_1}{\partial x_3} = -m_2l_1d_2\cos(x_3)x_4(2x_2 - x_4)$, and also $\frac{\partial C_1}{\partial x_4} = -m_2l_1d_2\sin(x_3)(2x_2 - x_4) + m_2l_1d_2\sin(x_3)x_4$.

$\frac{\partial C_2}{\partial x_1} = 0$, $\frac{\partial C_2}{\partial x_2} = 2m_2l_1d_2\sin(x_3)x_2$, $\frac{\partial C_2}{\partial x_3} = 2m_2l_1d_2\cos(x_3)x_2^2$, $\frac{\partial C_2}{\partial x_4} = 0$

Gravitational matrix $G(\theta)$:

$\frac{\partial G_1}{\partial x_1} = (m_1gd_1 + m_2gl_1)\cos(x_1) + m_2gd_2\sin(x_1 - x_3)$, $\frac{\partial G_1}{\partial x_2} = 0$, $\frac{\partial G_1}{\partial x_3} = -m_2gd_2\cos(x_1 - x_3)$, $\frac{\partial G_1}{\partial x_4} = 0$.

$\frac{\partial G_2}{\partial x_1} = m_2gd_2\cos(x_1 - x_3)$, $\frac{\partial G_1}{\partial x_2} = 0$, $\frac{\partial G_1}{\partial x_3} = -m_2gd_2\cos(x_1 - x_3)$, $\frac{\partial G_1}{\partial x_4} = 0$.

2.5.3.2 Stabilizing feedback control

After linearization around its current operating point, the dynamic model of the lower-limb exoskeleton robot is written as

$$\dot{x} = Ax + Bu + d_1 \quad (2.158)$$

Parameter d_1 stands for the linearization error in the lower-limb exoskeleton robot's dynamic model appearing previously in (2.149). The reference setpoints for the lower-limb exoskeleton robot's state vector are denoted by $\mathbf{x_d} = [x_1^d, \cdots, x_4^d]$. Tracking of this trajectory is achieved after applying the control input u_d. At every time instant the control input u_d is assumed to differ from the control input u appearing in (2.158) by an amount equal to Δu, that is $u_d = u + \Delta u$

$$\dot{x}_d = Ax_d + Bu_d + d_2 \tag{2.159}$$

The dynamics of the controlled system described in (2.158) can be also written as

$$\dot{x} = Ax + Bu + Bu_d - Bu_d + d_1 \tag{2.160}$$

and by denoting $d_3 = -Bu_d + d_1$ as an aggregate disturbance term one obtains

$$\dot{x} = Ax + Bu + Bu_d + d_3 \tag{2.161}$$

By subtracting (2.159) from (2.161), one has

$$\dot{x} - \dot{x}_d = A(x - x_d) + Bu + d_3 - d_2 \tag{2.162}$$

By denoting the tracking error as $e = x - x_d$ and the aggregate disturbance term as $\tilde{d} = d_3 - d_2$, the tracking error dynamics becomes

$$\dot{e} = Ae + Bu + \tilde{d} \tag{2.163}$$

For the approximately linearized model of the system a stabilizing feedback controller is developed. The controller has the form

$$u(t) = -Ke(t) \tag{2.164}$$

with $K = \frac{1}{r}B^TP$ where P is a positive definite symmetric matrix which is obtained from the solution of the Riccati equation (Rigatos and Busawon, 2018)

$$A^TP + PA + Q - P\left(\frac{2}{r}BB^T - \frac{1}{\rho^2}LL^T\right)P = 0 \tag{2.165}$$

where Q is a positive semi-definite symmetric matrix. The diagram of the considered control loop is depicted in Figure 2.54. The global stability properties of the control method can be proven following the procedure presented in Section 1.1.2.5.

2.5.4 Simulation tests

The tracking performance of the proposed nonlinear optimal (H-infinity) control scheme for the robotic exoskeleton's model has been further confirmed through simulation experiments. The obtained results are depicted in Figures 2.55 to 2.60 It can be noticed that in all cases fast and accurate tracking of the reference setpoints was achieved under moderate variations of the control inputs. The real values of the state-variables of the exoskeleton are depicted in blue, their estimated values are printed in green, while the associated setpoints are plotted in red. The tracking accuracy of the control method as well as its transient performance depends on the selection of the gains r, ρ, and of matrix Q which appears in the algebraic Riccati equation of (2.165). The necessary and sufficient condition for the application of the proposed

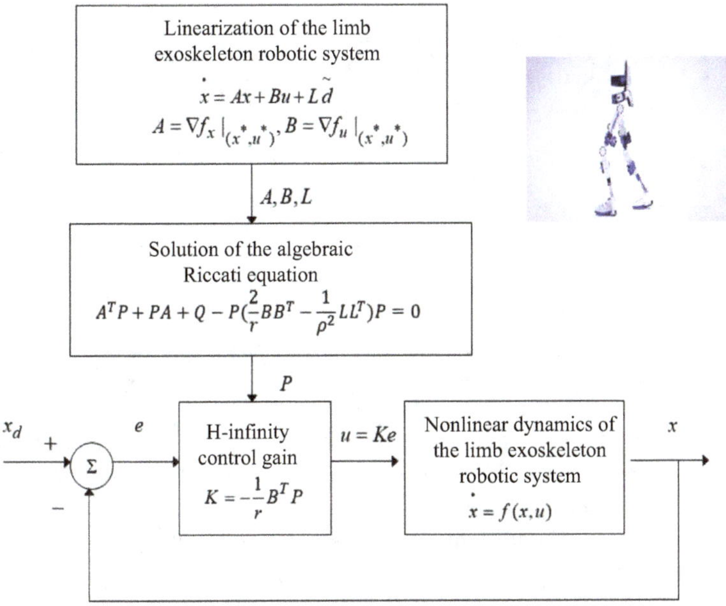

Figure 2.54 Diagram of the control scheme for the lower-limb robotic exoskeleton

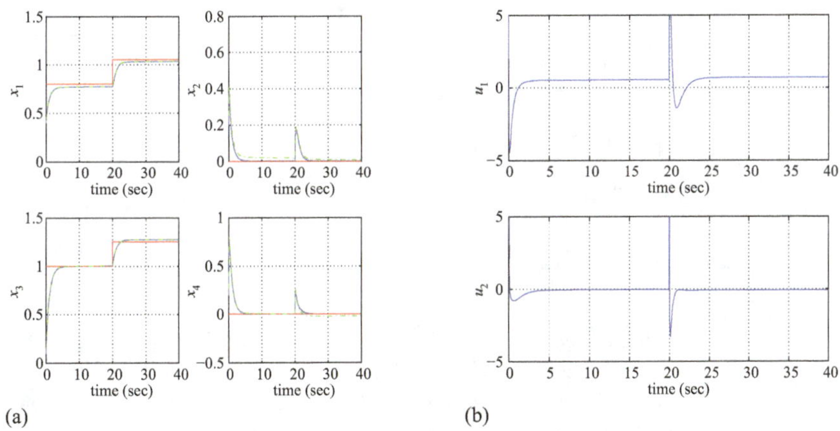

Figure 2.55 Tracking of setpoint 1 by the robotic exoskeleton: (a) convergence of the state variables of the exoskeleton x_i, $i = 1, \cdots, 4$ (blue line) to the associated reference values (red line) and estimated values (green line); (b) control inputs u_i, $i = 1, 2$ (motor torques) applied to the exoskeleton

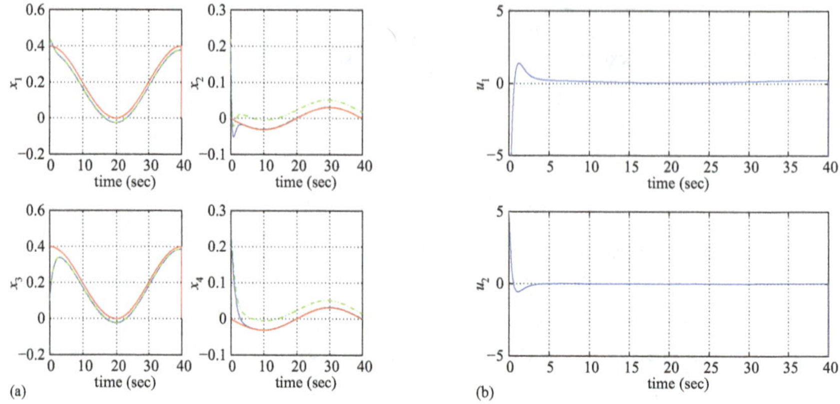

Figure 2.56 Tracking of setpoint 2 by the robotic exoskeleton: (a) convergence of the state variables of the exoskeleton x_i, $i = 1, \cdots, 4$ *(blue line) to the associated reference values (red line) and estimated values (green line); (b) control inputs* u_i, $i = 1, 2$ *(motor torques) applied to the exoskeleton*

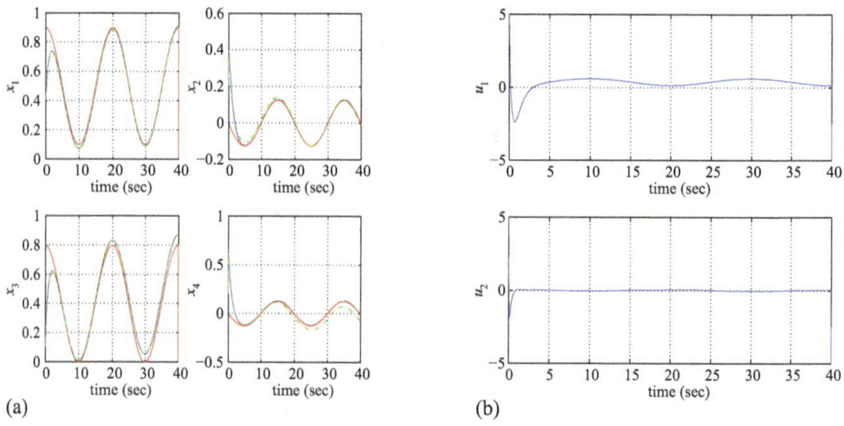

Figure 2.57 Tracking of setpoint 3 by the robotic exoskeleton: (a) convergence of the state variables of the exoskeleton x_i, $i = 1, \cdots, 4$ *(blue line) to the associated reference values (red line) and estimated values (green line); (b) control inputs* u_i, $i = 1, 2$ *(motor torques) applied to the exoskeleton*

nonlinear optimal control scheme is the existence of a solution of the aforementioned Riccati equation. The robustness of the control scheme depends on the selection of the attenuation coefficient ρ. Actually, the smallest value of ρ for which a solution of this Riccati equation can be obtained is the one that provides maximum robustness to the control loop.

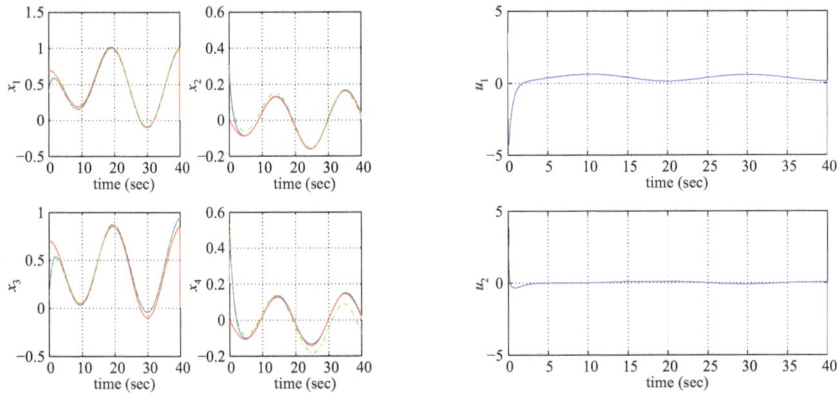

Figure 2.58 Tracking of setpoint 4 by the robotic exoskeleton: (a) convergence of the state variables of the exoskeleton x_i, $i = 1, \cdots, 4$ (blue line) to the associated reference values (red line) and estimated values (green line); (b) control inputs u_i, $i = 1, 2$ (motor torques) applied to the exoskeleton

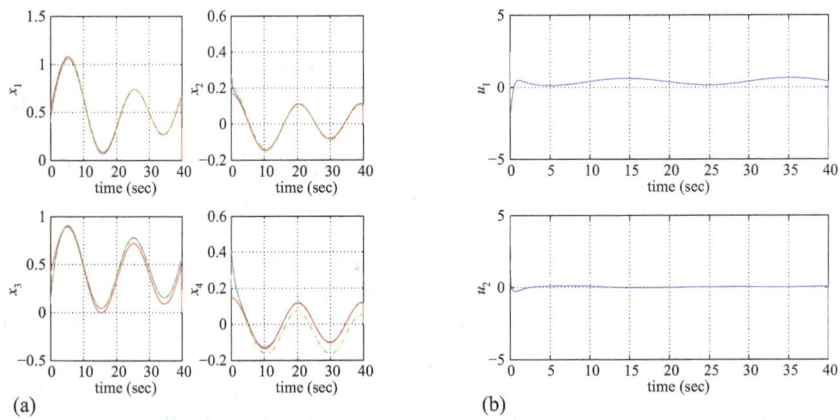

Figure 2.59 Tracking of setpoint 5 by the robotic exoskeleton: (a) convergence of the state variables of the exoskeleton x_i, $i = 1, \cdots, 4$ (blue line) to the associated reference values (red line) and estimated values (green line); (b) control inputs u_i, $i = 1, 2$ (motor torques) applied to the exoskeleton

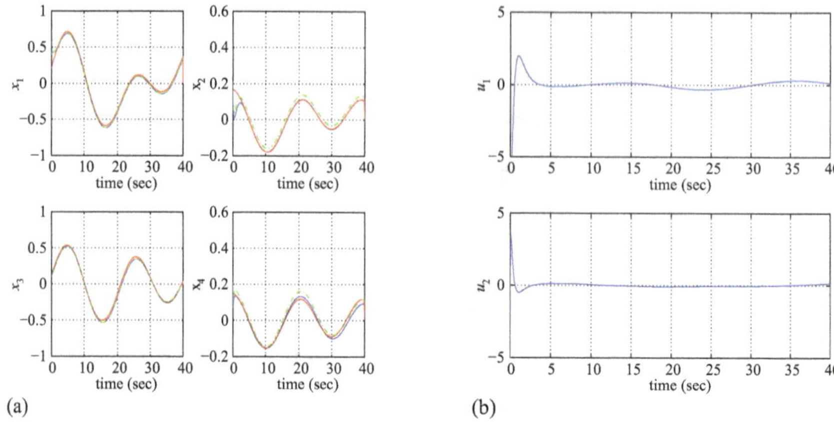

Figure 2.60 Tracking of setpoint 6 by the robotic exoskeleton: (a) convergence of the state variables of the exoskeleton x_i, $i = 1, \cdots, 4$ *(blue line) to the associated reference values (red line) and estimated values (green line); (b) control inputs* u_i, $i = 1, 2$ *(motor torques) applied to the exoskeleton*

2.6 Nonlinear control of brachiation robots

2.6.1 Outline

The control problem of brachiation robots is important because these robots can be used in inspection and maintenance tasks for critical infrastructure, such as bridges, dams, buildings or the electric power transmission and distribution system, as in (Wu, Huang *et al.*, 2015; Lu *et al.*, 2015; and Menendez, Auat Cheein *et al.*, 2017). This control problem exhibits significant difficulty because of the nonlinear and multivariable structure of the brachiation robots' state-space model, as presented in (Meghdari *et al.*, 2013; Menezes de Oliveira and Lages, 2006b; and Menezes de Oliveira and Lages, 2010). Control is more complicated when one considers that brachiation robots are often underactuated, as analyzed in (Meghdari *et al.*, 2011; Rigatos and Busawon, 2018; and Farzen *et al.*, 2019). On the one side, underactuation may be pursued in the design of these robotic systems in an aim to reduce their weight and energy consumption after removing specific actuators from them. On the other side, functioning of these robotic systems in an underactuation mode signifies that these robots are fault tolerant in the case of actuators' failures and that they continue to perform reliably their tasks even if specific actuators are subjected to fault (Nguyen and Liu, 2017; Gomes and Ruino, 2005; Ebrahimi *et al.*, 2016). So far several nonlinear control methods have been proposed for these robotic systems (Nakanishi *et al.*, 2000; Nguyen and Liu, 2019; Satel *et al.*, 2019). There have been several attempts to treat the related control problem with the use of global linearization-based techniques or with the use of robust control methods such as sliding-mode control. There have been also several results on the use of Model Predictive Control on these robotic systems (Menezes de Oliveira and Lages, 2007; Menezes de Oliveira and Lages, 2006b; Menezes de Oliveira and Lages, 2009). The confirmation of the global stability and of convergence to an optimum for such control methods remains a challenge (Yamukawa, 2018; Lin and Yang, 2020; Nakanishi and Vijayakumar, 2012). This research area remains open for further advancements and additional results on the control of brachiate and bi-brachiate robots can be found in (Liu and Yan, 2020; Yang, Fan *et al.*, 2020; Omoto, Dei *et al.*, 2019; Moro, 2016; Li, Guo *et al.*, 2018; Yang, Li *et al.*, 2010a; and Yang, Li *et al.*, 2010b).

In this section, a novel nonlinear optimal (H-infinity) control method is developed for multi-DOF underactuated brachiation robots. To this end, the dynamic model of the brachiation robots undergoes first approximate linearization around a temporary operating point which is updated at each time-step of the control method (Rigatos, Siano and Cecati, 2015; Rigatos, Siano, Wira *et al.*, 2019; Rigatos, Siano *et al.*, 2015; Rigatos, Busawon, Pomares *et al.*, 2019). This linearization point is defined by the present value of the state vector and by the last sampled value of the control inputs vector. The linearization process is based on first-order Taylor series expansion and on the computation of the related Jacobian matrices (Rigatos and Tzafestas, 2007; Basseville and Nikiforov, 1993; Rigatos and Zhang, 2009). The modeling error which is due to the truncation of higher-order terms from the Taylor series is considered to be a perturbation that is asymptotically compensated by the robustness of the control scheme. For the approximately linearized model of the brachiation robots, a stabilizing H-infinity feedback controller is designed. The proposed H-infinity control achieves the solution of the optimal control problem for the nonlinear dynamics of the brachiation robots, under model uncertainty and external perturbations.

Actually, this controller represents a min–max differential game in which the control inputs try to minimize a cost function that comprises a quadratic term of the state vector's tracking error, while the model uncertainty and exogenous disturbance terms try to maximize this cost function. To compute the feedback gains of the H-infinity controller an algebraic Riccati equation is repetitively solved at each iteration of the control algorithm (Rigatos, 2016b; Rigatos, 2015). The stability properties of the control method are proven through Lyapunov analysis. First, it is proven that the control system satisfies the H-infinity tracking performance criterion, which signifies elevated robustness against modeling errors and disturbances (Rigatos and Busawon, 2018; Tousssaint, Basar and Bullo, 2000). Moreover, under moderate conditions it is proven that the control system of the brachiation robots is globally asymptotically stable (Rigatos and Busawon, 2018). To implement state estimation-based control without the need to measure the entire state vector of the brachiation robot, the H-infinity Kalman Filter is used as a robust state estimator (Rigatos and Busawon, 2018).

The section's nonlinear optimal control method is a novel and significant result. Preceding approaches about the use of H-infinity control in nonlinear dynamical systems were limited to the case of affine-in-the-input systems with drift-only dynamics and considered that the control inputs gain matrix is not dependent of the values of the system's state vector. Moreover, in these approaches the linearization was performed around points of the desirable trajectory whereas in the present section's control method the linearization points are related with the value of the state vector at each sampling instant as well as with the last sampled value of the control inputs vector. The Riccati equation which has been proposed for computing the feedback gains of the controller is novel, so is the presented global stability proof through Lyapunov analysis.

2.6.2 Dynamic model of the multi-DOF brachiation robot

2.6.2.1 Overview of the dynamic model of the brachiation robot

The diagram of a brachiation robot which is used for inspection and maintenance of cables of the power grid is shown in Figure 2.61. The dynamic model of brachiation robot is

$$M(\theta)\ddot{\theta} + C(\theta,\dot{\theta}) + G(\theta) = \bar{B}\tau - F_v(\dot{\theta}) \tag{2.166}$$

where $\theta = [\theta_1, \theta_2, \theta_3]^T$ is the vector of the joints' angles, $M(\theta)$ is the inertia matrix, $C(\theta,\dot{\theta})$ is the Coriolis and Centrifugal terms vector, $G(\theta)$ is the gravitational terms vector, $\bar{B}\tau$ is the control input torques vector with matrix $\bar{B}$ to denote the specific type of underactuation, and $F_v(\dot{\theta})$ is a vector of viscous torques.

The lengths of the three links are denoted as l_i $i = 1, 2, 3$, while it is considered that l_1 and l_2 are greater than or equal to l_3. It is assumed that the masses of the three links are concentrated on the links' center of gravity. I_1, I_2, and I_3 are the moments of inertia of the three links of the considered brachiation robot, while b_1, b_2, and b_3 are friction coefficients defining the friction torques that appear in the joints of the above-noted links.

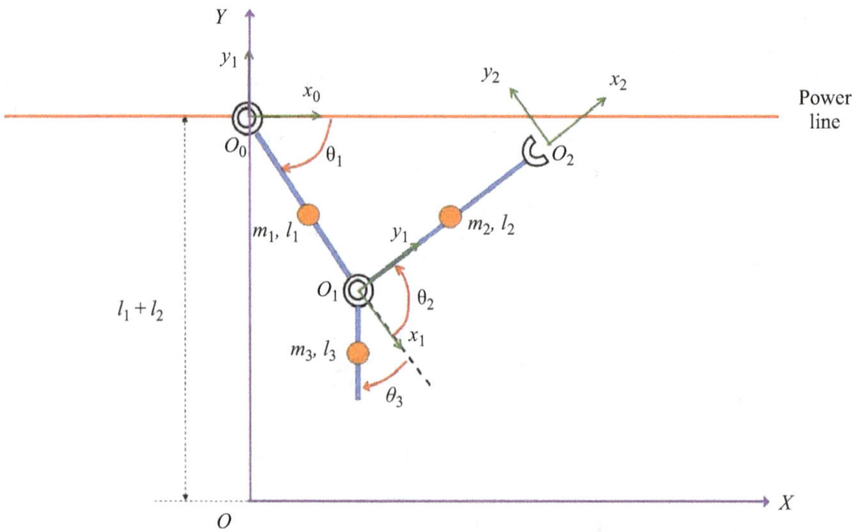

Figure 2.61 Diagram of the multi-DOF brachiation robot

Using Euler–Lagrange analysis one obtains the dynamic model of the brachiation robot which comprises the following three differential equations:

$$\begin{aligned}
&\left[m_1\frac{l_1^2}{2} + I_1 + m_2 l_1^2 \quad + m_2\frac{l_2^2}{4} + I_2 + m_3 l_1^2\right.\\
&\left.+m_3\frac{l_3^2}{4} + m_2 l_1 l_2\cos(\theta_2) + I_3 + m_3 l_1 l_3\cos(\theta_3)\right]\ddot{\theta}_1\\
&+\left[m_2\frac{l_2^2}{4} + I_2 + m_2 l_1 l_2\cos(\theta_2)\right]\ddot{\theta}_2 + \left[m_3\frac{l_3^2}{4} + I_3 + m_3 l_1 l_3\cos(\theta_3)\right]\ddot{\theta}_3\\
&+\left[-m_2 l_1 l_2\sin(\theta_2)\dot{\theta}_2\left(\dot{\theta}_1 + \frac{\dot{\theta}_2}{2}\right) - m_3 l_1 l_3\sin(\theta_3)\dot{\theta}_3\left(\dot{\theta}_1 + \frac{\dot{\theta}_3}{2}\right)\right]\\
&+\left[m_1 g\frac{l_2}{2}\cos(\theta_1) + m_2 g l_1\cos(\theta_2) + m_2 g\frac{l_2}{2}\cos(\theta_1+\theta_2)\right.\\
&\left.+m_3 g l_1\cos(\theta_1) + [m_3]g\frac{l_2}{2}\cos(\theta_1+\theta_3)\right] + b_1\dot{\theta}_1 = T_1
\end{aligned} \tag{2.167}$$

$$\begin{aligned}
&\left[m_2\frac{l_2^2}{4} + m_2 l_1 l_2\cos(\theta_2) + I_2\right]\ddot{\theta}_1 + \left[m_2\frac{l_2^2}{4} + I_2\right]\ddot{\theta}_2\\
&-m_2 l_1 l_2\sin(\theta_1)\dot{\theta}_1\dot{\theta}_2 - m_2 g\frac{l_2}{2}\cos(\theta_1+\theta_2) + b_2\dot{\theta}_2 = T_2
\end{aligned} \tag{2.168}$$

$$\begin{aligned}
&\left[m_3\frac{l_3^2}{4} + m_3 l_1 l_3\cos(\theta_3) + I_3\right]\ddot{\theta}_1 + \left[m_3\frac{l_3^2}{4} + I_3\right]\ddot{\theta}_3\\
&-m_3 l_1 l_3\sin(\theta_1)\dot{\theta}_1\dot{\theta}_3 - m_3 g\frac{l_3}{2}\cos(\theta_1+\theta_3) + b_3\dot{\theta}_3 = 0
\end{aligned} \tag{2.169}$$

Equivalently, the dynamic model of the brachiation robot can be written in the following matrix form:

$$\begin{pmatrix} m_{11} & m_{12} & m_{13}\\ m_{21} & m_{22} & m_{23}\\ m_{31} & m_{32} & m_{33}\end{pmatrix}\begin{pmatrix}\ddot{\theta}_1\\ \ddot{\theta}_2\\ \ddot{\theta}_3\end{pmatrix} + \begin{pmatrix} C_1(\theta,\dot{(\theta)})\\ C_2(\theta,\dot{(\theta)})\\ C_3(\theta,\dot{(\theta)})\end{pmatrix} + \begin{pmatrix} G_1(\theta)\\ G_2(\theta)\\ G_3(\theta)\end{pmatrix} + \begin{pmatrix} F_{v_1}\\ F_{v_2}\\ F_{v_3}\end{pmatrix} = \begin{pmatrix} T_1\\ T_2\\ 0\end{pmatrix} \tag{2.170}$$

where $m_{11} = [m_1\frac{l_1^2}{2} + I_1 + m_2 l_1^2 + m_2\frac{l_2^2}{4} + I_2 + m_3 l_1^2 + m_3\frac{l_3^2}{4} + m_2 l_1 l_2\cos(\theta_2) + I_3 + m_3 l_1 l_3\cos(\theta_3)]$, $m_{12} = [m_2\frac{l_2^2}{4} + I_2 + m_2 l_1 l_2\cos(\theta_2)]$, $m_{13} = [m_3\frac{l_3^2}{4} + I_3 + m_3 l_1 l_3\cos(\theta_3)]$, $m_{21} = [m_2\frac{l_2^2}{4} + I_2 + m_2 l_1 l_2\cos(\theta_2)]$, $m_{22} = [m_2\frac{l_2^2}{4} + I_2]$, $m_{23} = 0$, $m_{31} = [m_3\frac{l_3^2}{4} + m_3 l_1 l_3\cos(\theta_3) + I_3]$, $m_{32} = 0$, $m_{33} = [m_3\frac{l_3^2}{4} + I_3]$.

Besides, the elements of the Coriolis forces matrix $C(\theta,\dot{\theta})$ are given by $C_1(\theta,\dot{\theta}) = [-m_2 l_1 l_2\sin(\theta_2)\dot{\theta}_2(\dot{\theta}_1 + \frac{\dot{\theta}_2}{2}) - m_3 l_1 l_3\sin(\theta_3)\dot{\theta}_3(\dot{\theta}_1 + \frac{\dot{\theta}_3}{2})]$, $C_2(\theta,\dot{\theta}) = -m_2 l_1 l_2\sin(\theta_1)\dot{\theta}_1\dot{\theta}_2$, and $C_3(\theta,\dot{\theta}) = -m_3 l_1 l_3\sin(\theta_1)\dot{\theta}_1\dot{\theta}_3$.

About the elements of the gravitational forces vector one has $G_1(\theta) = [m_1 g\frac{l_2}{2}\cos(\theta_1) + m_2 g l_1\cos(\theta_2) + m_2 g\frac{l_2}{2}\cos(\theta_1+\theta_2) + m_3 g l_1\cos(\theta_1) + [m_3]g\frac{l_2}{2}\cos(\theta_1+\theta_3)]$, $G_2(\theta) = -m_2 g\frac{l_2}{2}\cos(\theta_1+\theta_2)$ and $G_3(\theta) = -m_3 g\frac{l_3}{2}\cos(\theta_1+\theta_3)$.

while the elements of the friction torques vector $F_v(\theta)$ are given by $F_{v_1}(\dot{\theta}) = b_1\dot{\theta}_1$, $F_{v_2}(\dot{\theta}) = b_2\dot{\theta}_2$ and $F_{v_3}(\dot{\theta}) = b_3\dot{\theta}_3$.

Finally, the control input torques vector τ is given by $\tau = [T_1, T_2, 0]^T$, which denotes that the brachiation robot functions in the underactuation mode. Next, the inverse of the inertia matrix M is computed as follows:

$$M^{-1} = \frac{1}{detM}\begin{pmatrix} M_{11} & -M_{21} & M_{31} \\ -M_{12} & M_{22} & -M_{32} \\ M_{13} & -M_{23} & M_{33} \end{pmatrix} \tag{2.171}$$

where using that $m_{23} = 0$ and $m_{32} = 0$, the determinant $detM$ is

$$\begin{aligned} detM &= m_{11}(m_{22}m_{33} - m_{32}m_{23}) - m_{12}(m_{21}m_{33} \\ &-m_{23}m_{31}) + m_{13}(m_{21}m_{32} - m_{31}m_{22}) \Rightarrow \\ detM &= m_{11}m_{22}m_{33} - m_{12}m_{21}m_{33} - m_{13}m_{31}m_{22} \end{aligned} \tag{2.172}$$

while it also holds that $M_{11} = m_{22}m_{33}$, $M_{12} = m_{21}m_{33}$, $M_{13} = -m_{31}m_{22}$, $M_{21} = m_{12}m_{33}$, $M_{22} = m_{11}m_{33} - m_{31}m_{13}$, $M_{23} = -m_{31}m_{12}$, $M_{31} = -m_{22}m_{13}$, $M_{32} = -m_{21}m_{13}$, $M_{33} = m_{11}m_{22} - m_{21}m_{12}$.

The state-space model of the robotic system is written as

$$\ddot{\theta} = -M^{-1}(\theta)C(\theta, \dot{\theta}) - M^{-1}(\theta)G(\theta) - M^{-1}(\theta)F_v(\dot{\theta}) + M^{-1}(\theta)\tau \tag{2.173}$$

After intermediate operations, the state-space model of the brachiation robot takes the form

$$\begin{pmatrix} \ddot{\theta}_1 \\ \ddot{\theta}_2 \\ \ddot{\theta}_3 \end{pmatrix} = \begin{pmatrix} \dfrac{\begin{matrix} -(M_{11}C_1 - M_{21}C_2 + M_{31}C_3) - (M_{11}G_1 - M_{21}G_2 + M_{31}G_3) \\ -(M_{11}F_{v1} - M_{21}F_{v2} + M_{31}F_{v3}) \end{matrix}}{detM} \\ \dfrac{\begin{matrix} -(-M_{12}C_1 + M_{22}C_2 - M_{32}C_3) - (-M_{12}G_1 + M_{22}G_2 - M_{32}G_3) \\ -(-M_{12}F_{v1} + M_{22}F_{v2} - M_{32}F_{v3}) \end{matrix}}{detM} \\ \dfrac{\begin{matrix} -(M_{13}C_1 - M_{23}C_2 + M_{33}C_3) - (M_{13}G_1 - M_{23}G_2 + M_{33}G_3) \\ -(M_{13}F_{v1} - M_{23}F_{v2} + M_{33}F_{v3}) \end{matrix}}{detM} \end{pmatrix} + \begin{pmatrix} \dfrac{M_{11}}{detM} \\ -\dfrac{M_{12}}{detM} \\ \dfrac{M_{13}}{detM} \end{pmatrix} T_1 + \begin{pmatrix} \dfrac{-M_{21}}{detM} \\ \dfrac{M_{22}}{detM} \\ \dfrac{-M_{23}}{detM} \end{pmatrix} T_2 \tag{2.174}$$

Next, the following state variables are defined: $x_1 = \theta_1$, $x_2 = \dot{\theta}_1$, $x_3 = \theta_2$, $x_4 = \dot{\theta}_2$, $x_5 = \theta_3, x_6 = \dot{\theta}_3$. Moreover, the following control input variables are defined: $u_1 = T_1$ and $u_2 = T_2$.

Thus, about the elements of the inertia matrix M it holds: where $m_{11}(x) = [m_1\frac{l_1^2}{2} + I_1 + m_2 l_1^2 + m_2\frac{l_2^2}{4} + I_2 + m_3 l_1^2 + m_3\frac{l_3^2}{4} + m_2 l_1 l_2\cos(x_3) + I_3 + m_3 l_1 l_3\cos(x_5)]$, $m_{12}(x) =$ / $[m_2\frac{l_2^2}{4} + I_2 + m_2 l_1 l_2\cos(x_3)]$, $m_{13}(x) = [m_3\frac{l_3^2}{4} + I_3 + m_3 l_1 l_3\cos(x_5)]$, $m_{21}(x) = [m_2\frac{l_2^2}{4} + I_2 + m_2 l_1 l_2\cos(x_3)]$, $m_{22}(x) = [m_2\frac{l_2^2}{4} + I_2]$, $m_{23}(x) = 0$, $m_{31}(x) = [m_3\frac{l_3^2}{4} + m_3 l_1 l_3\cos(x_5) + I_3]$, $m_{32}(x) = 0$, $m_{33}(x) = [m_3\frac{l_3^2}{4} + I_3]$.

where the determinant of M is

$$detM = m_{11}(x)m_{22}(x)m_{33}(x) - m_{12}(x)m_{21}(x)m_{33}(x) - m_{13}(x)m_{31}(x)m_{22}(x) \qquad (2.175)$$

In particular, it holds that $M_{11} = m_{22}(x)m_{33}(x)$, $M_{12} = m_{21}(x)m_{33}(x)$, $M_{13} = -m_{31}(x)m_{22}(x)$, $M_{21} = m_{12}(x)m_{33}(x)$, $M_{22} = m_{11}(x)m_{33}(x) - m_{31}(x)m_{13}(x)$, $M_{23} = -m_{31}(x)m_{12}(x)$, $M_{31} = -m_{22}(x)m_{13}(x)$, $M_{32} = -m_{21}(x)m_{13}(x)$, $M_{33} = m_{11}(x)m_{22}(x) - m_{21}(x)m_{12}(x)$.

Besides, the elements of the Coriolis forces matrix C are given by $C_1 = [-m_2 l_1 l_2\sin(x_3)x_4(x_2 + \frac{x_4}{2}) - m_3 l_1 l_3\sin(x_5)x_6(x_2 + \frac{x_6}{2})]$, $C_2 = -m_2 l_1 l_2\sin(x_3)x_2x_4$, and $C_3 = -m_3 l_1 l_3\sin(x_5)x_2x_6$.

About the elements of the gravitational forces vector, one has $G_1 = [m_1 g\frac{l_2}{2}\cos(x_1) + m_2 g l_1\cos(x_3) + m_2 g\frac{l_2}{2}\cos(x_1 + x_3) + m_3 g l_1\cos(x_1) + [m_3]g\frac{l_2}{2}\cos(x_1 + x_5)]$, $G_2 = -m_2 g\frac{l_2}{2}\cos(x_1 + x_3)$ and $G_3 = -m_3 g\frac{l_3}{2}\cos(x_1 + x_5)$.

While the elements of the friction torques vector $F_v(\theta)$ are given by $F_{v_1} = b_1x_2$, $F_{v_2} = b_2x_4$, and $F_{v_3} = b_3x_5$.

Finally, the state-space model of the brachiation robot is written in a form that comprises the following equations:

$$\dot{x}_1 = x_2 \qquad (2.176)$$

$$\dot{x}_2 = \frac{\begin{array}{c}-(M_{11}C_1 - M_{21}C_2 + M_{31}C_3) - (M_{11}G_1 - M_{21}G_2 + M_{31}G_3)\\ -(M_{11}F_{v1} - M_{21}F_{v2} + M_{31}F_{v3})\end{array}}{detM} + \frac{M_{11}}{detM}T_1 + \frac{-M_{21}}{detM}T_2 \qquad (2.177)$$

$$\dot{x}_3 = x_4 \qquad (2.178)$$

$$\dot{x}_4 = \frac{\begin{array}{c}-(-M_{12}C_1 + M_{22}C_2 - M_{32}C_3) - (-M_{12}G_1 + M_{22}G_2 - M_{32}G_3)\\ -(-M_{12}F_{v1} + M_{22}F_{v2} - M_{32}F_{v3})\end{array}}{detM} + \frac{-M_{12}}{detM}T_1 + \frac{M_{22}}{detM}T_2 \qquad (2.179)$$

$$\dot{x}_5 = x_6 \qquad (2.180)$$

$$\dot{x}_6 = \frac{\begin{array}{c}-(M_{13}C_1 - M_{23}C_2 + M_{33}C_3) - (M_{13}G_1 - M_{23}G_2 + M_{33}G_3) \\ -(M_{13}F_{v1} - M_{23}F_{v2} + M_{33}F_{v3})\end{array}}{detM} + \frac{M_{13}}{detM}T_1 + \frac{-M_{23}}{detM}T_2 \tag{2.181}$$

Thus, the state-space model of the brachiation robot is written in the following concise matrix form:

$$\dot{x} = f(x) + g_1(x)u_1 + g_2(x)u_2 \tag{2.182}$$

where $x \in R^{6\times 1}$, $f(x) \in R^{6\times 1}$, $g(x) \in R^{6\times 1}$, and $u \in R^{2\times 1}$. where vectors $f(x)$ and $g(x)$ are given by:

$$f(x) = \begin{pmatrix} x_2 \\ \dfrac{\begin{array}{c}-(M_{11}C_1 - M_{21}C_2 + M_{31}C_3) - (M_{11}G_1 - M_{21}G_2 + M_{31}G_3) \\ -(M_{11}F_{v1} - M_{21}F_{v2} + M_{31}F_{v3})\end{array}}{detM} \\ x_4 \\ \dfrac{\begin{array}{c}-(-M_{12}C_1 + M_{22}C_2 - M_{32}C_3) - (-M_{12}G_1 + M_{22}G_2 - M_{32}G_3) \\ -(-M_{12}F_{v1} + M_{22}F_{v2} - M_{32}F_{v3})\end{array}}{detM} \\ x_6 \\ \dfrac{\begin{array}{c}-(M_{13}C_1 - M_{23}C_2 + M_{33}C_3) - (M_{13}G_1 - M_{23}G_2 + M_{33}G_3) - (M_{13}F_{v1} \\ -M_{23}F_{v2} + M_{33}F_{v3})\end{array}}{detM} \end{pmatrix} \tag{2.183}$$

$$g_1(x) = \begin{pmatrix} 0 \\ \dfrac{M_{11}}{detM} \\ 0 \\ -\dfrac{M_{12}}{detM} \\ 0 \\ \dfrac{M_{13}}{detM} \end{pmatrix} \quad g_2(x) = \begin{pmatrix} 0 \\ \dfrac{-M_{21}}{detM} \\ 0 \\ \dfrac{M_{22}}{detM} \\ 0 \\ \dfrac{-M_{23}}{detM} \end{pmatrix} \tag{2.184}$$

2.6.2.2 Euler–Lagrange analysis for modeling of the brachiation robot's dynamics

In the inertial reference frame, the coordinates of the centers of gravity of the three links are:

$$x_{c_1} = \frac{l_1}{2}\cos(\theta_1), \qquad y_{c_1} = (l_1 + l_2) - \frac{l_1}{2}\sin(\theta_1)$$
$$x_{c_2} = \frac{l_1}{2}\cos(\theta_1) + \frac{l_2}{2}\cos(\theta_1 + \theta_2), \quad y_{c_2} = (l_1 + l_2) - \frac{l_1}{2}\sin(\theta_1) - \frac{l_2}{2}\sin(\theta_1 + \theta_2)$$
$$x_{c_3} = \frac{l_1}{2}\cos(\theta_1) + \frac{l_3}{2}\cos(\theta_1 + \theta_3), \quad y_{c_3} = (l_1 + l_2) - \frac{l_1}{2}\sin(\theta_1) - \frac{l_3}{2}\sin(\theta_1 + \theta_3)$$

The kinetic energy of the first link of the robot is:

$$\begin{aligned} K_1 &= \frac{1}{2}m_1(\dot{x}_{c_1}^2 + \dot{y}_{c_1}^2) + \frac{1}{2}I_1\dot{\theta}_1^2 \Rightarrow \\ K_1 &= \frac{1}{2}m_1\left[-\left(\frac{l_1}{2}\sin(\theta_1)\dot{\theta}_1\right)^2\right] + \left(\frac{l_1}{2}\cos(\theta_1)\dot{\theta}_1\right)^2 + \frac{1}{2}I_1\dot{\theta}_1^2 \Rightarrow \\ K_1 &= \frac{1}{2}m_1\frac{l_1^2}{4}\dot{\theta}_1^2 + \frac{1}{2}I_1\dot{\theta}_1^2 \end{aligned} \tag{2.185}$$

The potential energy of the first link of the robot is:

$$P_1 = m_1 g\left(l_1 + l_2 - \frac{l_1}{2}\sin(\theta_1)\right) \tag{2.186}$$

The kinetic energy of the second link of the robot is:

$$\begin{aligned} K_2 &= \frac{1}{2}m_2(\dot{x}_{c_2}^2 + \dot{y}_{c_2}^2) + \frac{1}{2}I_2(\dot{\theta}_1 + \dot{\theta}_2)^2 \Rightarrow \\ K_2 &= \frac{1}{2}m_2[(-l_1\sin(\theta_1)\dot{\theta}_1 - \frac{l_2}{2}\sin(\theta_1 + \theta_2)(\dot{\theta}_1 + \dot{\theta}_2))^2 \\ &\quad + (-l_1\cos(\theta_1)\dot{\theta}_1 - \frac{l_2}{2}\cos(\theta_1 + \theta_2)(\dot{\theta}_1 + \dot{\theta}_2))^2] + \frac{1}{2}I_2(\dot{\theta}_1 + \dot{\theta}_2) \Rightarrow \\ K_2 &= \frac{1}{2}m_2[l_1^2\dot{\theta}_1^2 + \frac{l_2^2}{4}(\dot{\theta}_1 + \dot{\theta}_2{}^2) + l_1 l_2\cos(\theta_2)\dot{\theta}_1(\dot{\theta}_1 + \dot{\theta}_2)] + \frac{1}{2}I_2(\dot{\theta}_1 + \dot{\theta}_2) \end{aligned} \tag{2.187}$$

The potential energy of the second link of the robot is:

$$P_2 = m_2 g\left(l_1 + l_2 - \frac{l_1}{2}\sin(\theta_1) - \frac{l_2}{2}\sin(\theta_1 + \theta_2)\right) \tag{2.188}$$

The kinetic energy of the third link of the robot is:

$$\begin{aligned} K_3 &= \frac{1}{2}m_3(\dot{x}_{c_3}^2 + \dot{y}_{c_3}^2) + \frac{1}{2}I_2(\dot{\theta}_1 + \dot{\theta}_3)^2 \Rightarrow \\ K_3 &= \frac{1}{2}m_3[(-l_1\sin(\theta_1)\dot{\theta}_1 - \frac{l_2}{2}\sin(\theta_1 + \theta_3)(\dot{\theta}_1 + \dot{\theta}_3))^2 \\ &\quad + (-l_1\cos(\theta_1)\dot{\theta}_1 - \frac{l_2}{2}\cos(\theta_1 + \theta_2)(\dot{\theta}_1 + \dot{\theta}_2))^2] + \frac{1}{2}I_3(\dot{\theta}_1 + \dot{\theta}_3) \Rightarrow \\ K_3 &= \frac{1}{2}m_3[l_1^2\dot{\theta}_1^2 + \frac{l_3^2}{4}(\dot{\theta}_1 + \dot{\theta}_3{}^2) + l_1 l_3\cos(\theta_3)\dot{\theta}_1(\dot{\theta}_1 + \dot{\theta}_3)] + \frac{1}{2}I_2(\dot{\theta}_1 + \dot{\theta}_2) \end{aligned} \tag{2.189}$$

The potential energy of the third link of the robot is:

$$P_3 = m_3 g\left(l_1 + l_2 - \frac{l_1}{2}\sin(\theta_1) - \frac{l_3}{2}\sin(\theta_1 + \theta_3)\right) \tag{2.190}$$

The Lagrangian of the brachiation robot is

$$\begin{aligned}
L &= K_1 + K_2 + K_3 - P_1 - P_2 - P_3 \Rightarrow \\
L &= \frac{1}{2}m_1\frac{l_1^2}{4}\dot{\theta}_1^2 + \frac{1}{2}I_1\dot{\theta}_1^2 \\
&\quad + \frac{1}{2}m_2\left[l_1^2\dot{\theta}_1^2 + \frac{l_2^2}{4}(\dot{\theta}_1 + \dot{\theta}_2{}^2) + l_1 l_2\cos(\theta_2)\dot{\theta}_1(\dot{\theta}_1 + \dot{\theta}_2)\right] + \frac{1}{2}I_2(\dot{\theta}_1 + \dot{\theta}_2) \\
&\quad + \frac{1}{2}m_3\left[l_1^2\dot{\theta}_1^2 + \frac{l_3^2}{4}(\dot{\theta}_1 + \dot{\theta}_3{}^2) + l_1 l_3\cos(\theta_3)\dot{\theta}_1(\dot{\theta}_1 + \dot{\theta}_3)\right] + \frac{1}{2}I_2(\dot{\theta}_1 + \dot{\theta}_2) \\
&\quad - m_1 g\left(l_1 + l_2 - \frac{l_1}{2}\sin(\theta_1)\right) - m_2 g\left(l_1 + l_2 - \frac{l_1}{2}\sin(\theta_1) - \frac{l_2}{2}\sin(\theta_1 + \theta_2)\right) \\
&\quad - m_3 g\left(l_1 + l_2 - \frac{l_1}{2}\sin(\theta_1) - \frac{l_3}{2}\sin(\theta_1 + \theta_3)\right)
\end{aligned} \tag{2.191}$$

It is considered that actuation is applied only to the first and second link while the third link of this robotic system is unactuated. Consequently, the brachiation robot of Figure 2.61 is underactuated. By applying the Euler–Lagrange method one obtains the equations of the dynamics of the brachiation robot (Rigatos and Busawon, 2018):

$$\frac{\partial}{\partial t}\frac{\partial L}{\partial \dot{\theta}_1} - \frac{\partial L}{\partial \theta_1} = T_1 - F_{v_1} \tag{2.192}$$

$$\frac{\partial}{\partial t}\frac{\partial L}{\partial \dot{\theta}_2} - \frac{\partial L}{\partial \theta_2} = T_2 - F_{v_2} \tag{2.193}$$

$$\frac{\partial}{\partial t}\frac{\partial L}{\partial \dot{\theta}_3} - \frac{\partial L}{\partial \theta_3} = -F_{v_3} \tag{2.194}$$

From (2.192), and by considering that $F_{v_1} = b_1\dot{\theta}_1$, one obtains

$$\begin{aligned}
&\left[m_1\frac{l_1^2}{2} + I_1 + m_2 l_1^2 + m_2\frac{l_2^2}{4} + I_2 + m_3 l_1^2 \right.\\
&\quad \left. + m_3\frac{l_3^2}{4} + m_2 l_1 l_2\cos(\theta_2) + I_3 + m_3 l_1 l_3\cos(\theta_3)\right]\ddot{\theta}_1\\
&\quad + \left[m_2\frac{l_2^2}{4} + I_2 + m_2 l_1 l_2\cos(\theta_2)\right]\ddot{\theta}_2 + \left[m_3\frac{l_3^2}{4} + I_3 + m_3 l_1 l_3\cos(\theta_3)\right]\ddot{\theta}_3\\
&\quad + \left[-m_2 l_1 l_2\sin(\theta_2)\dot{\theta}_2\left(\dot{\theta}_1 + \frac{\dot{\theta}_2}{2}\right) - m_3 l_1 l_3\sin(\theta_3)\dot{\theta}_3\left(\dot{\theta}_1 + \frac{\dot{\theta}_3}{2}\right)\right]\\
&\quad + \left[m_1 g\frac{l_2}{2}\cos(\theta_1) + m_2 g l_1\cos(\theta_2) + m_2 g\frac{l_2}{2}\cos(\theta_1 + \theta_2)\right.\\
&\quad \left. + m_3 g l_1\cos(\theta_1) + [m_3] g\frac{l_2}{2}\cos(\theta_1 + \theta_3)\right] + +b_1\dot{\theta}_1 = T_1
\end{aligned} \tag{2.195}$$

From (2.193), and by considering that $F_{v_2} = b_2\dot{\theta}_2$, one obtains

$$\begin{aligned}
&\left[m_2\frac{l_2^2}{4} + m_2 l_1 l_2\cos(\theta_2) + I_2\right]\ddot{\theta}_1 + \left[m_2\frac{l_2^2}{4} + I_2\right]\ddot{\theta}_2\\
&\quad - m_2 l_1 l_2\sin(\theta_1)\dot{\theta}_1\dot{\theta}_2 - m_2 g\frac{l_2}{2}\cos(\theta_1 + \theta_2) + b_2\dot{\theta}_2 = T_2
\end{aligned} \tag{2.196}$$

From (2.194), and by considering that $F_{v_3} = b_3\dot{\theta}_3$, one obtains

$$\begin{aligned}
&\left[m_3\frac{l_3^2}{4} + m_3 l_1 l_3\cos(\theta_3) + I_3\right]\ddot{\theta}_1 + \left[m_3\frac{l_3^2}{4} + I_3\right]\ddot{\theta}_3\\
&\quad - m_3 l_1 l_3\sin(\theta_1)\dot{\theta}_1\dot{\theta}_3 - m_3 g\frac{l_3}{2}\cos(\theta_1 + \theta_3) + b_3\dot{\theta}_3 = 0
\end{aligned} \tag{2.197}$$

The above three equations constitute the dynamic model of the 3-link brachiation robot, which can be written next in matrix form.

2.6.3 Approximate linearization of the dynamic model of the brachiation robot

2.6.3.1 Approximate linearization of the robot's model

The dynamic model of the brachiation robot undergoes approximate linearization around the time-varying operating point (x^*, u^*) which is updated at each iteration of the control algorithm, while x^* is the present value of the system's state vector and u^* is the last sampled value of the control inputs vector. The linearization process relies on first-order Taylor series expansion and on the computation of the related

Jacobian matrices. Thus, the state-space model of the robot being initially in the form $\dot{x} = f(x) + g(x)u$, where $g(x) = [g_1(x), g_2(x)]$ is written in the linearized form

$$\dot{x} = Ax + Bu + \tilde{d} \tag{2.198}$$

where $\tilde{d}$ is the cumulative vector of disturbances which may comprise (i) modeling error terms due to the truncation of higher-order terms in the Taylor series expansion, (ii) exogenous perturbations, and (iii) measurement noise of any distribution. Matrices A and B are obtained from the system's Jacobians

$$\begin{aligned} A &= \nabla_x[f(x) + g(x)u] \mid_{(x^*,u^*)} \Rightarrow \\ A &= \nabla_x[f(x)] \mid_{(x^*,u^*)} + \nabla_x[g_1(x)]u_1 \mid_{(x^*,u^*)} + \nabla_x[g_2(x)]u_2 \mid_{(x^*,u^*)} \end{aligned} \tag{2.199}$$

$$\begin{aligned} B &= \nabla_u[f(x) + g(x)u] \mid_{(x^*,u^*)} \Rightarrow \\ B &= g(x) \mid_{(x^*,u^*)} \end{aligned} \tag{2.200}$$

Next, the computation of the Jacobian matrix $A = \nabla_x f(x) \mid_{(x^*,u^*)}$ is performed

First row of the Jacobian matrix $\nabla_x f(x) \mid_{(x^*,u^*)}$: $\frac{\partial f_1}{\partial x_1} = 0$, $\frac{\partial f_1}{\partial x_2} = 1$, $\frac{\partial f_1}{\partial x_3} = 0$, $\frac{\partial f_1}{\partial x_4} = 0$, $\frac{\partial f_1}{\partial x_5} = 0$, $\frac{\partial f_1}{\partial x_6} = 0$, $\frac{\partial f_1}{\partial x_7} = 0$, and $\frac{\partial f_1}{\partial x_8} = 0$.

Second row of the Jacobian matrix $\nabla_x f(x) \mid_{(x^*,u^*)}$:

The numerator of $f_2(x)$ is

$$\begin{aligned} f_{2n} = &-(M_{11}C_1 - M_{21}C_2 + M_{31}C_3) - (M_{11}G_1 - M_{21}G_2 + M_{31}G_3) \\ &- (M_{11}F_{v1} - M_{21}F_{v2} + M_{31}F_{v3}) \end{aligned} \tag{2.201}$$

while the denominator of $f_2(x)$ is $f_{2d} = detM$. It holds that for $i = 1, 2, \cdots, 6$

$$\begin{aligned} \frac{\partial f_{2n}}{\partial x_i} = &-\left(\frac{\partial M_{11}}{\partial x_i}C_1 + M_{11}\frac{\partial C_1}{\partial x_i} - \frac{\partial M_{21}}{\partial x_i}C_2 - M_{21}\frac{\partial C_2}{\partial x_i} + \frac{\partial M_{31}}{\partial x_i}C_3 + M_{31}\frac{\partial C_3}{\partial x_i}\right) \\ &-\left(\frac{\partial M_{11}}{\partial x_i}G_1 + M_{11}\frac{\partial G_1}{\partial x_i} - \frac{\partial M_{21}}{\partial x_i}G_2 - M_{21}\frac{\partial G_2}{\partial x_i} + \frac{\partial M_{31}}{\partial x_i}G_3 + M_{31}\frac{\partial G_3}{\partial x_i}\right) \\ &-\left(\frac{\partial M_{11}}{\partial x_i}F_{v1} + M_{11}\frac{\partial F_{v1}}{\partial x_i} - \frac{\partial M_{21}}{\partial x_i}F_{v2} - M_{21}\frac{\partial F_{v2}}{\partial x_i} + \frac{\partial M_{31}}{\partial x_i}F_{v3} + M_{31}\frac{\partial F_{v3}}{\partial x_i}\right) \end{aligned} \tag{2.202}$$

Moreover, for $i = 1, 2, \cdots, 6$ one obtains that

$$\frac{\partial f_{2d}}{\partial x_i} = \frac{\partial detM}{\partial x_i} \tag{2.203}$$

Therefore, for the second row of the Jacobian matrix

$$\frac{\partial f_2}{\partial x_i} = \frac{\frac{\partial f_{2n}}{\partial x_i}f_{2d} - f_{2n}\frac{\partial f_{2d}}{\partial x_i}}{detM^2} \tag{2.204}$$

Third row of the Jacobian matrix $\nabla_x f(x) \mid_{(x^*,u^*)}$: $\frac{\partial f_3}{\partial x_1}=0$, $\frac{\partial f_3}{\partial x_2}=0$, $\frac{\partial f_3}{\partial x_3}=0$, $\frac{\partial f_3}{\partial x_4}=0$, $\frac{\partial f_3}{\partial x_5}=0$, $\frac{\partial f_3}{\partial x_6}=0$, $\frac{\partial f_3}{\partial x_7}=0$, and $\frac{\partial f_3}{\partial x_8}=0$.

Fourth row of the Jacobian matrix $\nabla_x f(x) \mid_{(x^*,u^*)}$:

The numerator of $f_4(x)$ is

$$\begin{aligned} f_{4n} = &-(-M_{12}C_1 + M_{22}C_2 - M_{32}C_3) - (-M_{12}G_1 + M_{22}G_2 - M_{32}G_3) \\ &- (-M_{12}F_{v1} + M_{22}F_{v2} - M_{32}F_{v3}) \end{aligned} \tag{2.205}$$

while the denominator of $f_4(x)$ is $f_{4d} = detM$. It holds that for $i = 1, 2, \cdots, 6$

$$\begin{aligned} \frac{\partial f_{4n}}{\partial x_i} = &-\left(-\frac{\partial M_{12}}{\partial x_i}C_1 - M_{12}\frac{\partial C_1}{\partial x_1} + \frac{\partial M_{22}}{\partial x_i}C_2 + M_{22}\frac{\partial C_2}{\partial x_i} - \frac{\partial M_{32}}{\partial x_i}C_3 - M_{32}\frac{\partial C_3}{\partial x_i}\right) \\ &-\left(-\frac{\partial M_{12}}{\partial x_i}G_1 - M_{12}\frac{\partial G_1}{\partial x_i} + \frac{\partial M_{22}}{\partial x_i}G_2 + M_{22}\frac{\partial G_2}{\partial x_i} - \frac{\partial M_{32}}{\partial x_i}G_3 - M_{32}\frac{\partial G_3}{\partial x_i}\right) \\ &-\left(\frac{\partial M_{12}}{\partial x_i}F_{v1} - M_{12}\frac{\partial F_{v1}}{\partial x_i} + \frac{\partial M_{22}}{\partial x_i}F_{v2} + M_{22}\frac{\partial F_{v2}}{\partial x_i} - \frac{\partial M_{32}}{\partial x_i}F_{v3} - M_{32}\frac{\partial F_{v3}}{\partial x_i}\right) \end{aligned} \tag{2.206}$$

Moreover, for $i = 1, 2, \cdots, 6$ one obtains that

$$\frac{\partial f_{4d}}{\partial x_i} = \frac{\partial detM}{\partial x_i} \tag{2.207}$$

Therefore, for the fourth row of the Jacobian matrix

$$\frac{\partial f_4}{\partial x_i} = \frac{\frac{\partial f_{4n}}{\partial x_i}f_{4d} - f_{4n}\frac{\partial f_{4d}}{\partial x_i}}{detM^2} \tag{2.208}$$

Fifth row of the Jacobian matrix $\nabla_x f(x) \mid_{(x^*,u^*)}$: $\frac{\partial f_5}{\partial x_1}=0$, $\frac{\partial f_5}{\partial x_2}=0$, $\frac{\partial f_5}{\partial x_3}=0$, $\frac{\partial f_5}{\partial x_4}=0$, $\frac{\partial f_5}{\partial x_5}=0$, $\frac{\partial f_5}{\partial x_6}=0$, $\frac{\partial f_5}{\partial x_7}=0$, and $\frac{\partial f_5}{\partial x_8}=0$.

Sixth row of the Jacobian matrix $\nabla_x f(x) \mid_{(x^*,u^*)}$:

The numerator of $f_6(x)$ is

$$\begin{aligned} f_{6n} = &-(M_{13}C_1 - M_{23}C_2 + M_{33}C_3) - (M_{13}G_1 - M_{23}G_2 + M_{33}G_3) \\ &- (M_{13}F_{v1} - M_{23}F_{v2} + M_{33}F_{v3}) \end{aligned} \tag{2.209}$$

while the denominator of $f_4(x)$ is $f_{4d} = detM$. It holds that for $i = 1, 2, \cdots, 6$

$$\frac{\partial f_{6n}}{\partial x_i} = -\left(-\frac{\partial M_{13}}{\partial x_i}C_1 - M_{13}\frac{\partial C_1}{\partial x_1} + \frac{\partial M_{23}}{\partial x_i}C_2 + M_{23}\frac{\partial C_2}{\partial x_i} - \frac{\partial M_{33}}{\partial x_i}C_3 - M_{33}\frac{\partial C_3}{\partial x_i}\right)$$
$$-\left(-\tfrac{\partial M_{13}}{\partial x_i}G_1 - M_{13}\tfrac{\partial G_1}{\partial x_i} + \tfrac{\partial M_{23}}{\partial x_i}G_2 + M_{23}\tfrac{\partial G_2}{\partial x_i} - \tfrac{\partial M_{33}}{\partial x_i}G_3 - M_{33}\tfrac{\partial G_3}{\partial x_i}\right)$$
$$-\left(\tfrac{\partial M_{13}}{\partial x_i}F_{v1} - M_{13}\tfrac{\partial F_{v1}}{\partial x_i} + \tfrac{\partial M_{23}}{\partial x_i}F_{v2} + M_{23}\tfrac{\partial F_{v2}}{\partial x_i} - \tfrac{\partial M_{33}}{\partial x_i}F_{v3} - M_{33}\tfrac{\partial F_{v3}}{\partial x_i}\right) \tag{2.210}$$

Moreover, for $i = 1, 2, \cdots, 6$ one obtains that

$$\frac{\partial f_{6d}}{\partial x_i} = \frac{\partial detM}{\partial x_i} \tag{2.211}$$

Therefore, for the sixth row of the Jacobian matrix

$$\frac{\partial f_6}{\partial x_i} = \frac{\frac{\partial f_{6n}}{\partial x_i}f_{6d} - f_{6n}\frac{\partial f_{6d}}{\partial x_i}}{detM^2} \tag{2.212}$$

Next, the computation of the Jacobian matrix $\nabla_x g_1(x) \mid_{(x^*,u^*)}$ is performed

For the first row of the Jacobian matrix and for $i = 1, 2, \cdots, 6$, it holds

$$\frac{\partial g_{11}}{\partial x_i} = 0 \tag{2.213}$$

For the second row of the Jacobian matrix and for $i = 1, 2, \cdots, 6$, it holds

$$\frac{\partial g_{21}}{\partial x_i} = \frac{\frac{\partial M_{11}}{\partial x_i}detM - M_{11}\frac{\partial detM}{\partial x_i}}{detM^2} \tag{2.214}$$

For the third row of the Jacobian matrix and for $i = 1, 2, \cdots, 6$, it holds

$$\frac{\partial g_{31}}{\partial x_i} = 0 \tag{2.215}$$

For the fourth row of the Jacobian matrix and for $i = 1, 2, \cdots, 6$, it holds

$$\frac{\partial g_{41}}{\partial x_i} = \frac{\frac{\partial M_{12}}{\partial x_i}detM - M_{12}\frac{\partial detM}{\partial x_i}}{detM^2} \tag{2.216}$$

For the fifth row of the Jacobian matrix and for $i = 1, 2, \cdots, 6$, it holds

$$\frac{\partial g_{51}}{\partial x_i} = 0 \tag{2.217}$$

For the sixth row of the Jacobian matrix and for $i = 1, 2, \cdots, 6$, it holds

$$\frac{\partial g_{61}}{\partial x_i} = \frac{\frac{\partial M_{13}}{\partial x_i}detM - M_{13}\frac{\partial detM}{\partial x_i}}{detM^2} \tag{2.218}$$

Additionally, the computation of the Jacobian matrix $\nabla_x g_2(x) \mid_{(x^*,u^*)}$ is performed

For the first row of the Jacobian matrix and for $i = 1, 2, \cdots, 6$ it holds

$$\frac{\partial g_{12}}{\partial x_i} = 0 \tag{2.219}$$

For the second row of the Jacobian matrix and for $i = 1, 2, \cdots, 6$, it holds

$$\frac{\partial g_{22}}{\partial x_i} = \frac{\frac{\partial M_{21}}{\partial x_i} detM - M_{21}\frac{\partial detM}{\partial x_i}}{detM^2} \tag{2.220}$$

For the third row of the Jacobian matrix and for $i = 1, 2, \cdots, 6$, it holds

$$\frac{\partial g_{33}}{\partial x_i} = 0 \tag{2.221}$$

For the fourth row of the Jacobian matrix and for $i = 1, 2, \cdots, 6$, it holds

$$\frac{\partial g_{42}}{\partial x_i} = \frac{\frac{\partial M_{22}}{\partial x_i} detM - M_{22}\frac{\partial detM}{\partial x_i}}{detM^2} \tag{2.222}$$

For the fifth row of the Jacobian matrix and for $i = 1, 2, \cdots, 6$, it holds

$$\frac{\partial g_{52}}{\partial x_i} = 0 \tag{2.223}$$

For the sixth row of the Jacobian matrix and for $i = 1, 2, \cdots, 6$, it holds

$$\frac{\partial g_{62}}{\partial x_i} = \frac{\frac{\partial M_{23}}{\partial x_i} detM - M_{33}\frac{\partial detM}{\partial x_i}}{detM^2} \tag{2.224}$$

Next, one computes the partial derivatives of the elements of the inertia matrix M, of the Coriolis and centrifugal terms matrix C, of the gravitational terms matrix G and of the viscous friction terms matrix F_v. It holds that

$\frac{\partial M_{11}}{\partial x_i} = 0$ for $i = 1, \cdots, 6$

$\frac{\partial M_{12}}{\partial x_3} = [-m_2 l_1 l_2 \sin(x_3)][m_3\frac{l_3^2}{4} + I_3]$ and $\frac{\partial M_{12}}{\partial x_i} = 0$ for $i = 1, 2, 4, 5, 6$

$\frac{\partial M_{13}}{\partial x_5} = [m_3 l_1 l_3 \sin(x_5)][m_2\frac{l_2^2}{4} + I_2]$ and $\frac{\partial M_{13}}{\partial x_i} = 0$ for $i = 1, 2, 3, 4, 6$

$\frac{\partial M_{21}}{\partial x_3} = [-m_2 l_1 l_2 \sin(x_3)][m_3\frac{l_3^2}{4} + I_3]$ and $\frac{\partial M_{21}}{\partial x_i} = 0$ for $i = 1, 2, 4, 5, 6$

$\frac{\partial M_{22}}{\partial x_i} = 0$ for $i = 1, 2, 4, 6$, $\frac{\partial M_{22}}{\partial x_3} = [-m_2 l_1 l_2 \sin(x_3)][m_3\frac{l_3^2}{4} + I_3]$,

$\frac{\partial M_{22}}{\partial x_5} = [-m_3 l_1 l_3 \sin(x_6)][m_3\frac{l_3^2}{4} + I_3] - 2[m_3\frac{l_3^2}{4} + I_3 + m_3 l_1 l_3 \cos(x_5)][-m_3 l_1 l_3 \sin(x_5)]$

$\frac{\partial M_{23}}{\partial x_3} = [m_3\frac{l_3^2}{4} + I_3 + m_3 l_1 l_3 \cos(x_5)][m_2 l_1 l_2 \sin(x_3)]$,

$\frac{\partial M_{23}}{\partial x_5} = [m_2\frac{l_2^2}{4} + I_2 + m_2 l_1 l_2 \cos(x_3)][m_3 l_1 l_3 \sin(x_5)]$, and $\frac{\partial M_{23}}{\partial x_i} = 0$ for $i = 1, 2, 4, 6$.

$\frac{\partial M_{31}}{\partial x_5} = [m_3 l_1 l_3 \sin(x_5)][m_2\frac{l_2^2}{4} + I_2]$ and $\frac{\partial M_{31}}{\partial x_i} = 0$ for $i = 1, 2, 3, 4, 6$

$\frac{\partial M_{32}}{\partial x_3} = [m_3\frac{l_3^2}{4} + I_3 + m_3l_1l_3\cos(x_5)][m_2l_1l_2\sin(x_3)]$,

$\frac{\partial M_{32}}{\partial x_5} = [m_2\frac{l_2^2}{4} + I_2 + m_2l_1l_2\cos(x_3)][m_3l_1l_3\sin(x_5)]$ and $\frac{\partial M_{32}}{\partial x_i} = 0$ for $i = 1, 2, 4, 6$.

$\frac{\partial M_{33}}{\partial x_i} = 0$ for $i = 1, 2, 4, 6$,

$\frac{\partial M_{33}}{\partial x_3} = [-m_2l_1l_2\sin(x_3)][m_2\frac{l_2^2}{4} + I_2] - 2[m_2\frac{l_2^2}{4} + I_2 + m_2l_1l_2\cos(x_3)][-m_2l_1l_2\sin(x_3)]$

$\frac{\partial M_{33}}{\partial x_5} = [-m_3l_1l_3\sin(x_5)][m_2\frac{l_2^2}{4} + I_2]$.

$$\frac{\partial detM}{\partial x_3} = [-m_2l_1l_2\sin(x_3)]\left[m_2\frac{l_2^2}{4} + I_2\right]\left[m_3\frac{l_3^2}{4} + I_3\right] + 2\left[m_2\frac{l_2^2}{4} + I_2 + m_2l_1l_2\cos(x_3)\right][m_2l_1l_2\sin(x_3)]\left[m_3\frac{l_3^2}{4} + I_3\right].$$

$$\frac{\partial detM}{\partial x_5} = [-m_3l_1l_3\sin(x_5)]\left[m_2\frac{l_2^2}{4} + +I_2\right]\left[m_2\frac{l_2^2}{4} + I_2\right] + 2\left[m_3\frac{l_3^2}{4} + I_3 + m_3l_1l_3\cos(x_5)\right][m_3l_1l_3\sin(x_5)]\left[m_2\frac{l_2^2}{4} + I_2\right].$$

$\frac{\partial detM}{\partial x_i} = 0$ for $i = 1.2.4.6$

Moreover, it holds that

$\frac{\partial C_1}{\partial x_1} = 0$, $\frac{\partial C_1}{\partial x_2} = -m_2l_1l_2\sin(x_3)x_4 - m_3l_1l_3\sin(x_5)x_6$, $\frac{\partial C_1}{\partial x_3} = -m_2l_1l_2\cos(x_3)x_4(x_2 + \frac{x_4}{2})$,
$\frac{\partial C_1}{\partial x_4} = -m_2l_1l_2\sin(x_3)(x_2 + \frac{x_4}{2}) - m_2l_1l_2\sin(x_3)\frac{x_4}{2}$, $\frac{\partial C_1}{\partial x_5} = -m_3l_1l_3\cos(x_5)x_6(x_2 + \frac{x_6}{2})$,
$\frac{\partial C_1}{\partial x_6} = -m_3l_1l_3\sin(x_5)(x_2 + \frac{x_6}{2}) - m_3l_1l_3\sin(x_5)\frac{x_6}{2}$

$\frac{\partial C_2}{\partial x_1} = -m_2l_1l_2\sin(x_3)x_4$, $\frac{\partial C_2}{\partial x_2} = 0$, $\frac{\partial C_2}{\partial x_3} = -m_2l_1l_2\cos(x_3)x_2x_4$, $\frac{\partial C_2}{\partial x_4} = -m_2l_1l_2\sin(x_3)x_2$,
$\frac{\partial C_1}{\partial x_5} = 0$, and $\frac{\partial C_2}{\partial x_6} = 0$.

$\frac{\partial C_3}{\partial x_1} = 0$, $\frac{\partial C_3}{\partial x_2} = -m_3l_1l_3\sin(x_5)x_6$, $\frac{\partial C_3}{\partial x_3} = 0$, $\frac{\partial C_3}{\partial x_4} = 0$, $\frac{\partial C_3}{\partial x_5} = -m_3l_1l_3\cos(x_5)x_2x_6$, and
$\frac{\partial C_3}{\partial x_6} = -m_3l_1l_3\sin(x_5)x_2$.

Furthermore, it holds that

$\frac{\partial G_1}{\partial x_1} = -m_1g\frac{l_2}{2}\sin(x_1) - m_3gl_1\sin(x_1) - m_2g\frac{l_2}{2}\sin(x_1 + x_3) - m_3gl_1\sin(x_1) - m_3g\frac{l_3}{2}$
$\sin(x_1 + x_3)$,

$\frac{\partial G_1}{\partial x_2} = 0$, $\frac{\partial G_1}{\partial x_3} = -m_2 g \frac{l_2}{2} \sin(x_1 + x_3)$, $\frac{\partial G_1}{\partial x_4} = 0$, $\frac{\partial G_1}{\partial x_5} = -m_3 g \frac{l_3}{2} \sin(x_1 + x_5)$, $\frac{\partial G_1}{\partial x_6} = 0$.

$\frac{\partial G_2}{\partial x_1} = m_2 g \frac{l_2}{2} \sin(x_1 + x_3)$, $\frac{\partial G_2}{\partial x_2} = 0$, $\frac{\partial G_2}{\partial x_3} = m_2 g \frac{l_2}{2} \sin(x_1 + x_3)$, $\frac{\partial G_2}{\partial x_4} = 0$, $\frac{\partial G_2}{\partial x_5} = 0$, and $\frac{\partial G_2}{\partial x_6} = 0$.

$\frac{\partial G_3}{\partial x_1} = m_3 g \frac{l3}{2} \sin(x_1 + x_5)$, $\frac{\partial G_3}{\partial x_3} = 0$, $\frac{\partial G_3}{\partial x_4} = 0$, $\frac{\partial G_3}{\partial x_5} = m_3 g \frac{l3}{2} \sin(x_1 + x_5)$, and $\frac{\partial G_3}{\partial x_6} = 0$.

Additionally, it holds that

$\frac{\partial F_{v1}}{\partial x_2} = b_1$ and $\frac{\partial F_{v1}}{\partial x_i} = 0$ for $i = 1, 3, 4, 5, 6$.

$\frac{\partial F_{v2}}{\partial x_4} = b_2$ and $\frac{\partial F_{v2}}{\partial x_i} = 0$ for $i = 1, 2, 3, 5, 6$.

$\frac{\partial F_{v3}}{\partial x_6} = b_3$ and $\frac{\partial F_{v2}}{\partial x_i} = 0$ for $i = 1, 2, 3, 4, 5$.

2.6.3.2 Stabilizing feedback control

After linearization around its current operating point, the dynamic model of the multi-DOF brachiation robot is written as (Rigatos and Busawon, 2018)

$$\dot{x} = Ax + Bu + d_1 \tag{2.225}$$

Parameter d_1 stands for the linearization error in the multi-DOF brachiation robot dynamic model appearing above in (2.225). The reference setpoints for the state vector of the multi-DOF brachiation robots are denoted by $\mathbf{x_d} = [x_1^d, \cdots, x_6^d]$. Tracking of this trajectory is achieved after applying the control input u^*. At every time instant the control input u^* is assumed to differ from the control input u appearing in (2.225) by an amount equal to Δu, that is $u^* = u + \Delta u$

$$\dot{x}_d = Ax_d + Bu^* + d_2 \tag{2.226}$$

The dynamics of the controlled system described in (2.225) can be also written as

$$\dot{x} = Ax + Bu + Bu^* - Bu^* + d_1 \tag{2.227}$$

and by denoting $d_3 = -Bu^* + d_1$ as an aggregate disturbance term one obtains

$$\dot{x} = Ax + Bu + Bu^* + d_3 \tag{2.228}$$

By subtracting (2.226) from (2.228) one has

$$\dot{x} - \dot{x}_d = A(x - x_d) + Bu + d_3 - d_2 \tag{2.229}$$

By denoting the tracking error as $e = x - x_d$ and the aggregate disturbance term as $\tilde{d} = d_3 - d_2$, the tracking error dynamics becomes

$$\dot{e} = Ae + Bu + \tilde{d} \tag{2.230}$$

For the approximately linearized model of the system, a stabilizing feedback controller is developed. The controller has the form

$$u(t) = -Ke(t) \tag{2.231}$$

with $K = \frac{1}{r}B^T P$ where P is a positive definite symmetric matrix which is obtained from the solution of the Riccati equation (Rigatos and Busawon, 2018)

$$A^T P + PA + Q - P\left(\frac{2}{r}BB^T - \frac{1}{\rho^2}LL^T\right)P = 0 \tag{2.232}$$

where Q is a positive semi-definite symmetric matrix. The diagram of the considered control loop is depicted in Figure 2.62. The global stability properties of the control method can be proven following the procedure presented in Section 1.1.2.5.

The solution of the H-infinity feedback control problem for the multi-DOF brachiation robot and the computation of the worst case disturbance that the related controller can sustain, comes from superposition of Bellman's optimality principle when considering that the robot is affected by two separate inputs (i) the control input u and (ii) the cumulative disturbance input $\tilde{d}(t)$. Solving the optimal control problem for u, that is for the minimum variation (optimal) control input that achieves elimination of the state vector's tracking error, gives $u = -\frac{1}{r}B^T Pe$. Equivalently, solving the optimal control problem for $\tilde{d}$, that is for the worst case disturbance that the control loop can sustain gives $\tilde{d} = \frac{1}{\rho^2}L^T Pe$.

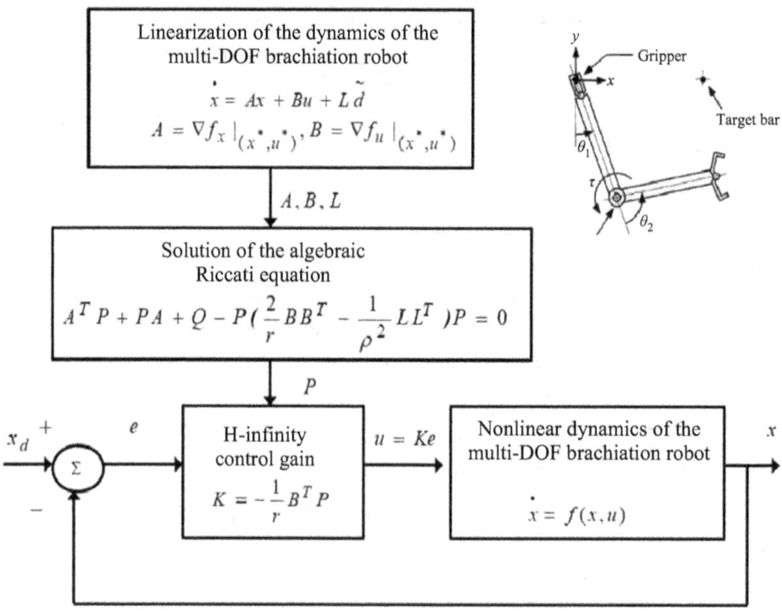

Figure 2.62 Diagram of the nonlinear optimal (H-infinity) control loop for the multi-DOF brachiation robot

2.6.4 Simulation tests

The performance of the proposed nonlinear optimal control scheme for the multi-DOF brachiation robot has been tested and confirmed through simulation experiments. The obtained results are depicted in Figures 2.63 to 2.78. Indicative values for the simulation experiments have been $m_1 = 1\,\text{kg}$, $l_1 = 1\,\text{m}$, $m_2 = 1\,\text{kg}$, $l_2 = 1\,\text{m}$ and

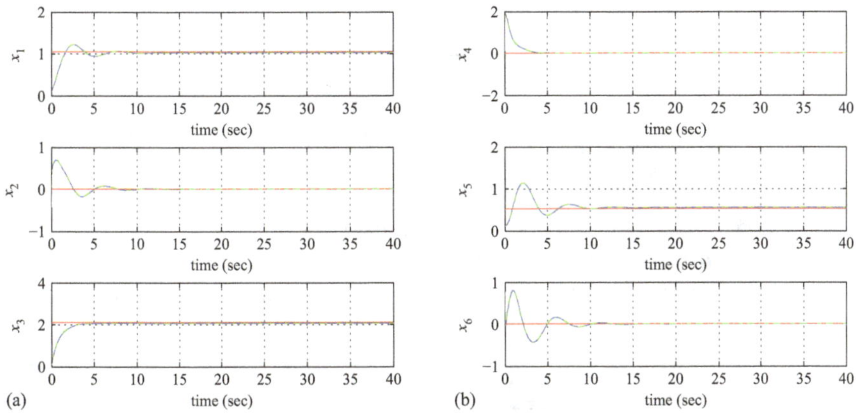

Figure 2.63 Tracking of setpoint 1 by the multi-DOF brachiation robot: (a) convergence of state variables x_1 to x_3, to the reference setpoints; (b) convergence of state variables x_4 to x_6, to the reference setpoints

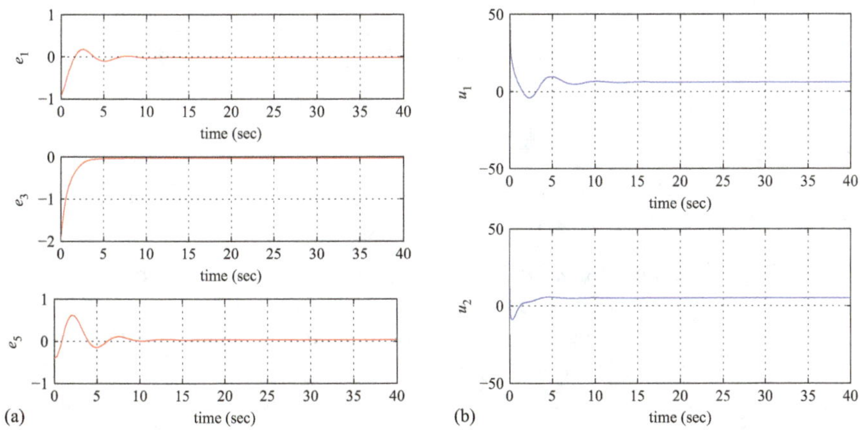

Figure 2.64 Tracking of setpoint 1 by the multi-DOF brachiation robot: (a) tracking error e_1, e_3 and e_5 of the joint angles of the brachiation robot; (b) control inputs u_1 and u_2 generated by the actuators of the brachiation robot

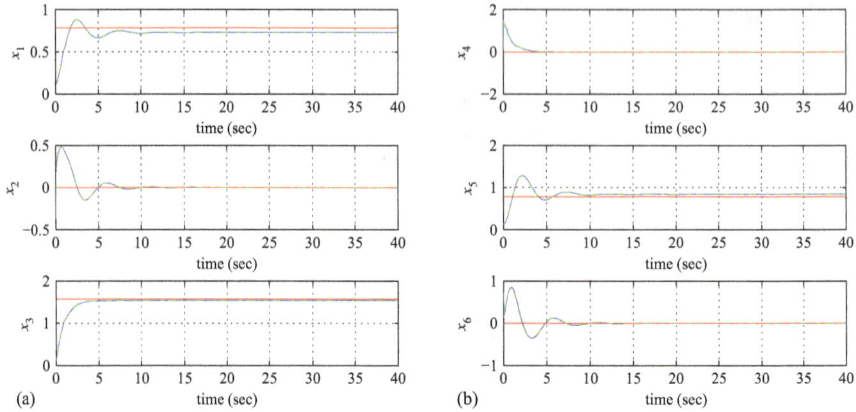

Figure 2.65 Tracking of setpoint 2 by the multi-DOF brachiation robot: (a) convergence of state variables x_1 to x_3, to the reference setpoints; (b) convergence of state variables x_4 to x_6, to the reference setpoints

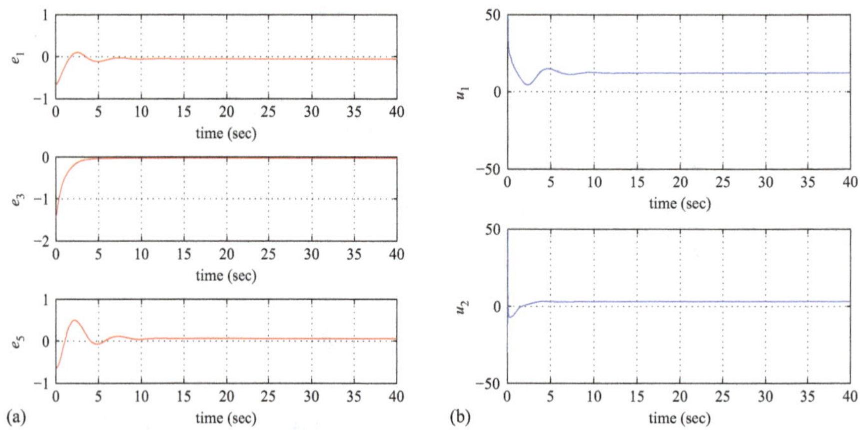

Figure 2.66 Tracking of setpoint 2 by the multi-DOF brachiation robot: (a) tracking error e_1, e_3 and e_5 of the joint angles of the brachiation robot; (b) control inputs u_1 and u_2 generated by the actuators of the brachiation robot

$m_3 = 0.5$ kg, $l_3 = 0.5$ m. It can be noticed that under the proposed nonlinear optimal control scheme, fast and accurate tracking of reference setpoints was achieved for all state variables of the underactuated brachiation robot. Besides it can be noticed that the variations of the control inputs remained moderate, which signifies that the

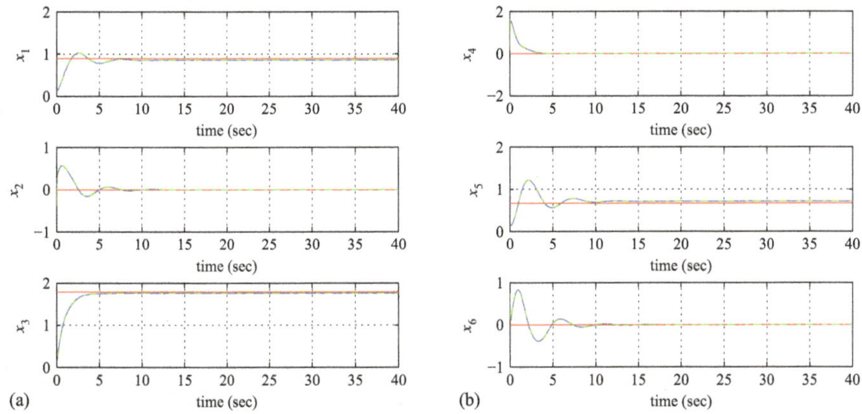

Figure 2.67 Tracking of setpoint 3 by the multi-DOF brachiation robot: (a) convergence of state variables x_1 to x_3, to the reference setpoints; (b) convergence of state variables x_4 to x_6, to the reference setpoints

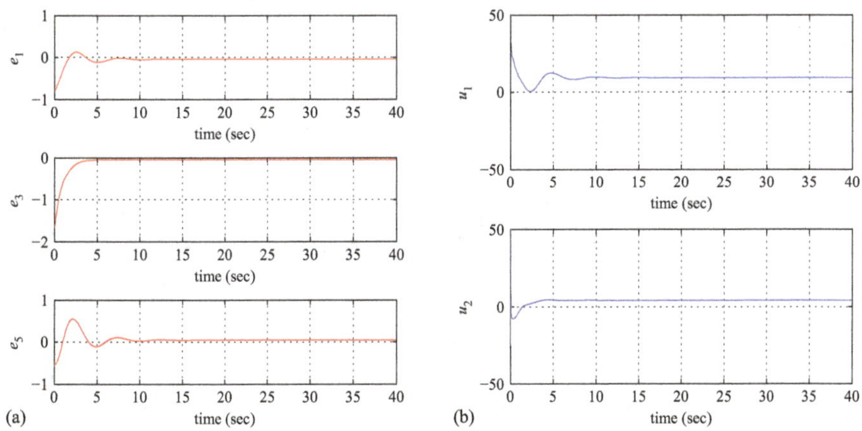

Figure 2.68 Tracking of setpoint 3 by the multi-DOF brachiation robot: (a) tracking error e_1, e_3 and e_5 of the joint angles of the brachiation robot; (b) control inputs u_1 and u_2 generated by the actuators of the brachiation robot

scheduled tasks can be accomplished by the brachiation robot under minimal energy consumption. This raises the autonomy and operational capacity of this robotic system. By limiting energy dispensing from the actuators of such robots one can ensure their uninterrupted functioning without the need for frequent re-charging of batteries.

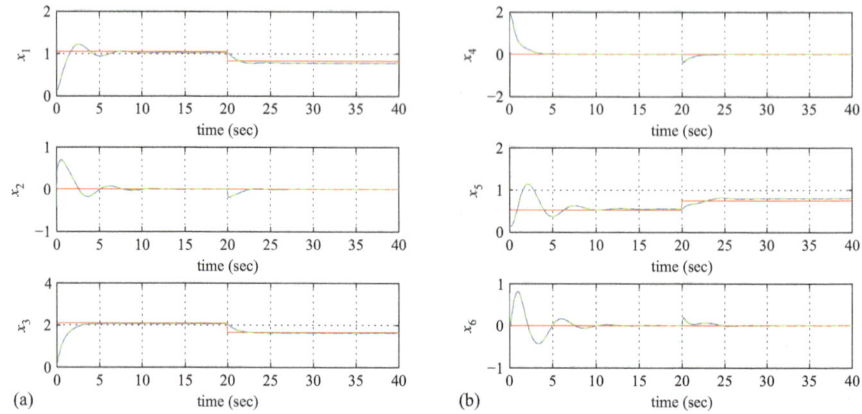

Figure 2.69 Tracking of setpoint 4 by the multi-DOF brachiation robot: (a) convergence of state variables x_1 to x_3, to the reference setpoints; (b) convergence of state variables x_4 to x_6, to the reference setpoints

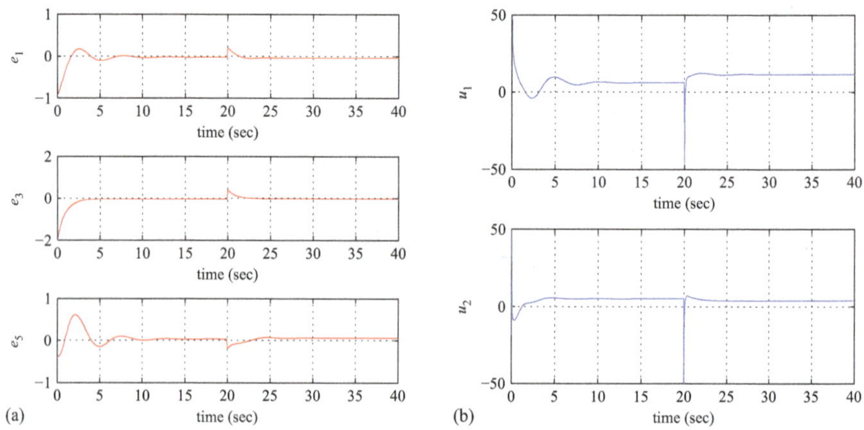

Figure 2.70 Tracking of setpoint 4 by the multi-DOF brachiation robot: (a) tracking error e_1, e_3 and e_5 of the joint angles of the brachiation robot; (b) control inputs u_1 and u_2 generated by the actuators of the brachiation robot

The aggregate simulation time was $T = 40$ sec. The adjustment time of each simulation experiment (time needed for the state variables of the robot to convergence to the targeted setpoints) did not exceed 10 sec. The variation ranges of the control inputs in the individual simulation experiments (maximum and minimum values of the torques generated by the actuators of the robot) were

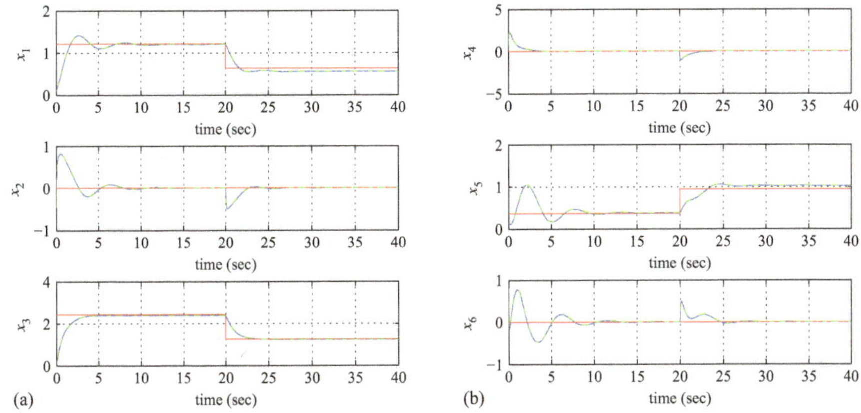

Figure 2.71 Tracking of setpoint 5 by the multi-DOF brachiation robot: (a) convergence of state variables x_1 to x_3, to the reference setpoints; (b) convergence of state variables x_4 to x_6, to the reference setpoints

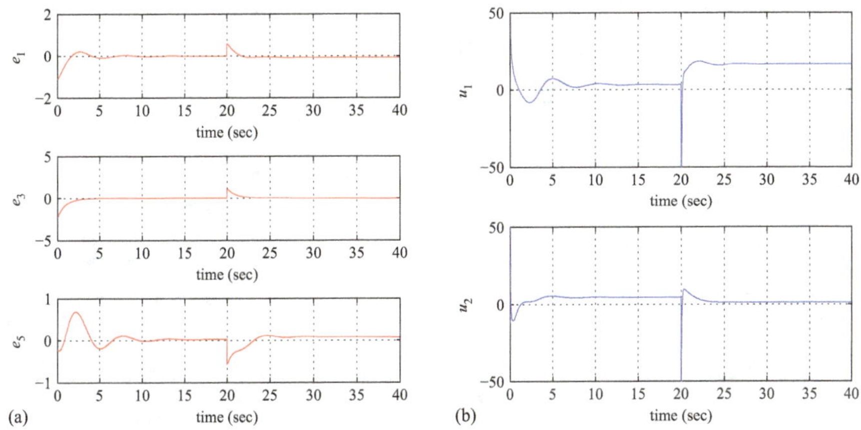

Figure 2.72 Tracking of setpoint 5 by the multi-DOF brachiation robot: (a) tracking error e_1, e_3 and e_5 of the joint angles of the brachiation robot; (b) control inputs u_1 and u_2 generated by the actuators of the brachiation robot

$[v_1^{min}, v_1^{max}] = [-200, 200]$, and $[v_2^{min}, v_2^{max}] = [-300, 300]$. Besides, the reference trajectories (setpoints) of the state variables of the brachiation robot were: (i) setpoint 1: $[\theta_1^d, \theta_2^d, \theta_3^d] = [\frac{\pi}{5}, \frac{2\pi}{5}, \frac{\pi}{2} - \frac{\pi}{5}]$, (ii) setpoint 2: $[\theta_1^d, \theta_2^d, \theta_3^d] = [\frac{\pi}{4}, \frac{2\pi}{4}, \frac{\pi}{2} - \frac{\pi}{4}]$, (iii) setpoint 3: $[\theta_1^d, \theta_2^d, \theta_3^d] = [\frac{\pi}{3.5}, \frac{2\pi}{3.5}, \frac{\pi}{2} - \frac{\pi}{3.5}]$, (iv) setpoint 4: $[\theta_1^d, \theta_2^d, \theta_3^d] = [\frac{\pi}{3}, \frac{2\pi}{3}, \frac{\pi}{2} - \frac{\pi}{3}]$ if $t \leq \frac{T}{2}$ and $[\theta_1^d, \theta_2^d, \theta_3^d] = [\frac{\pi}{3.8}, \frac{2\pi}{3.8}, \frac{\pi}{2} - \frac{\pi}{3.8}]$ if $t > \frac{T}{2}$, (v) setpoint 5:

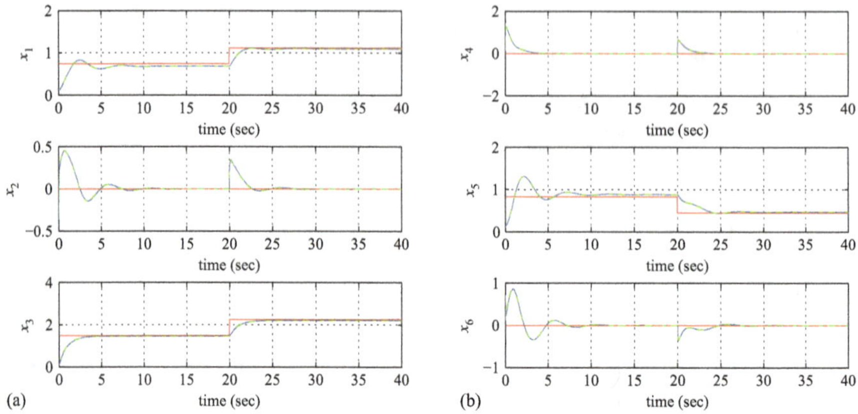

Figure 2.73 Tracking of setpoint 6 by the multi-DOF brachiation robot: (a) convergence of state variables x_1 to x_3, to the reference setpoints; (b) convergence of state variables x_4 to x_6, to the reference setpoints

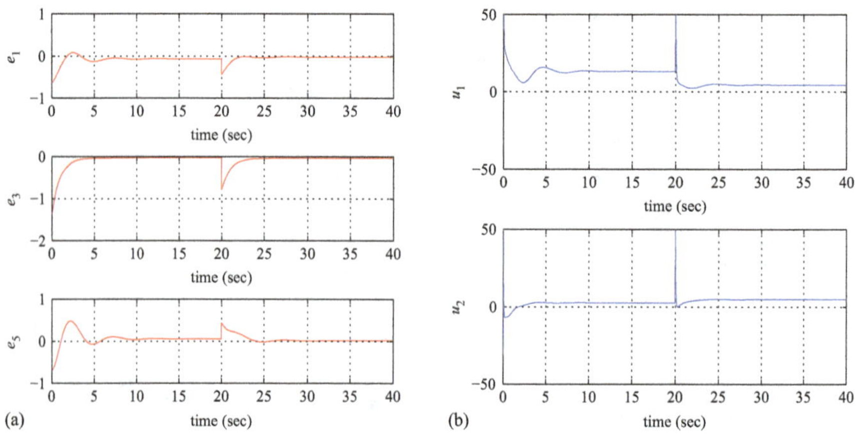

Figure 2.74 Tracking of setpoint 6 by the multi-DOF brachiation robot: (a) tracking error e_1, e_3 and e_5 of the joint angles of the brachiation robot; (b) control inputs u_1 and u_2 generated by the actuators of the brachiation robot

$[\theta_1^d,\theta_2^d,\theta_3^d]=[\frac{\pi}{2.6},\frac{2\pi}{2.6},\frac{\pi}{2}-\frac{\pi}{2.6}]$ if $t\leq\frac{T}{2}$ and $[\theta_1^d,\theta_2^d,\theta_3^d]=[\frac{\pi}{5},\frac{2\pi}{5},\frac{\pi}{2}-\frac{\pi}{5}]$ if $t>\frac{T}{2}$, (vi) setpoint 6: $[\theta_1^d,\theta_2^d,\theta_3^d]=[\frac{\pi}{4.2},\frac{2\pi}{4.2},\frac{\pi}{2}-\frac{\pi}{4.2}]$ if $t\leq\frac{T}{2}$ and $[\theta_1^d,\theta_2^d,\theta_3^d]=[\frac{\pi}{2.8},\frac{2\pi}{2.8},\frac{\pi}{2}-\frac{\pi}{2.8}]$ if $t>\frac{T}{2}$, (vii) setpoint 7: $[\theta_1^d,\theta_2^d,\theta_3^d]=[\frac{\pi}{4},\frac{2\pi}{4},\frac{\pi}{2}-\frac{\pi}{4}]$ if $t\leq\frac{T}{2}$ and $[\theta_1^d,\theta_2^d,\theta_3^d]=[\frac{\pi}{3},\frac{2\pi}{3},\frac{\pi}{2}-\frac{\pi}{3}]$ if $t>\frac{T}{2}$, and (viii) setpoint 8: $[\theta_1^d,\theta_2^d,\theta_3^d]=[\frac{\pi}{3},\frac{2\pi}{3},\frac{\pi}{2}-\frac{\pi}{3}]$ if $t\leq\frac{T}{2}$ and $[\theta_1^d,\theta_2^d,\theta_3^d]=[\frac{\pi}{5},\frac{2\pi}{5},\frac{\pi}{2}-\frac{\pi}{5}]$ if $t>\frac{T}{2}$.

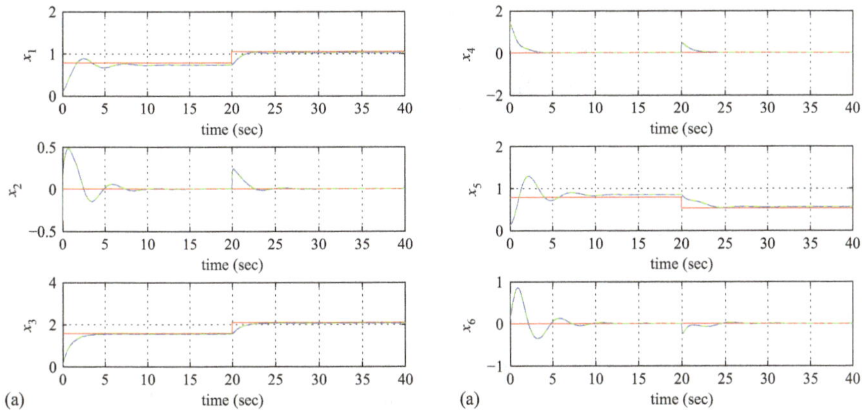

Figure 2.75 Tracking of setpoint 7 by the multi-DOF brachiation robot: (a) convergence of state variables x_1 to x_3 , to the reference setpoints; (b) convergence of state variables x_4 to x_6, to the reference setpoints

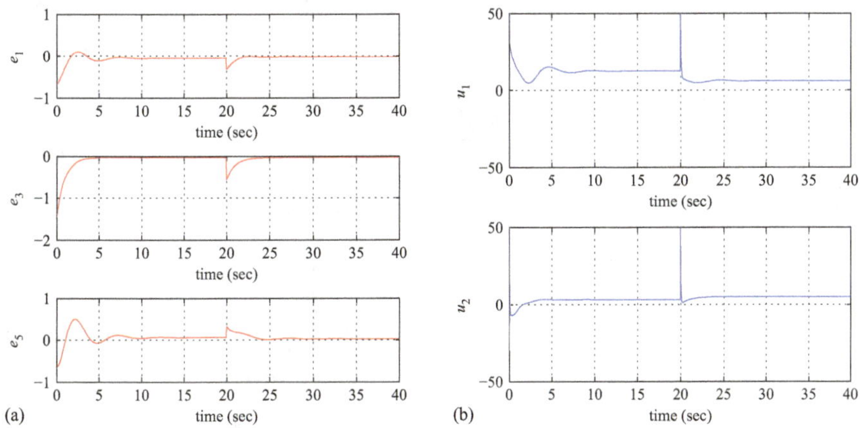

Figure 2.76 Tracking of setpoint 7 by the multi-DOF brachiation robot: (a) tracking error e_1, e_3 and e_5 of the joint angles of the brachiation robot; (b) control inputs u_1 and u_2 generated by the actuators of the brachiation robot

2.7 Nonlinear control of power line inspection robots

Inspection for faults of power grid components such as cables of the transmission and distribution system and insulators, as well as cleaning and de-icing of these components are often carried out by humans (Dian *et al.*, 2019; Li and Yuan, 2010;

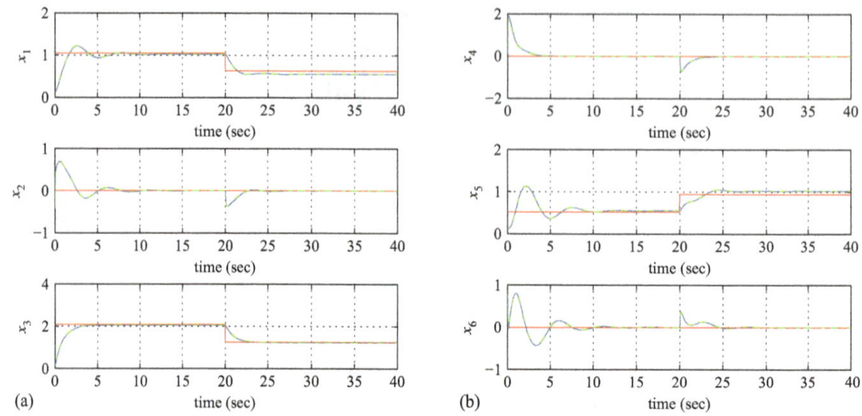

Figure 2.77 Tracking of setpoint 8 by the multi-DOF brachiation robot: (a) convergence of state variables x_1 to x_3, to the reference setpoints; (b) convergence of state variables x_4 to x_6, to the reference setpoints

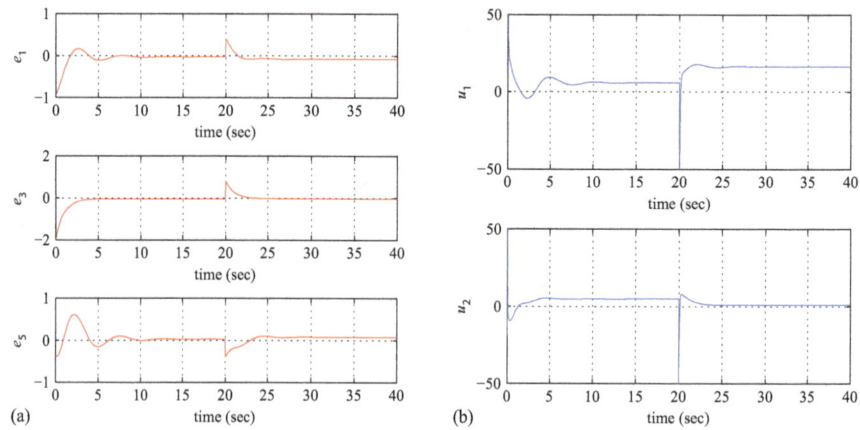

Figure 2.78 Tracking of setpoint 8 by the multi-DOF brachiation robot: (a) tracking error e_1, e_3 and e_5 of the joint angles of the brachiation robot; (b) control inputs u_1 and u_2 generated by the actuators of the brachiation robot

Chen, Liu *et al.*, 2016; He *et al.*, 2019; Wang and Wang, 2016). These are dangerous tasks, taking place in harsh conditions and exposing the life of the maintenance personnel to the risk of fatal accidents. To mitigate such risks and to perform faster and more efficient faults finding and repair in the power grid infrastructure, power line inspection robots can be used (Park, Lee *et al.*, 2012; Pouliot *et al.*, 2015;

Pouliot and Montamabault, 2009; Debenest *et al.*, 2008; Notomista *et al.*, 2019). The control problem of these robots remains a challenge because of the nonlinearities that characterize the related dynamic model and because of underactuation (Yang, Feng *et al.*, 2018; Seok and Kim, 2016; Chang *et al.*, 2017; Wang, Yan *et al.*, 2019). So far, several approaches have been proposed for solving the nonlinear control problem of power line inspection robots (Jiang, Yan *et al.*, 2019; Boje, 2016; Shen, Zhang *et al.*, 2018; Morozovsky and Bewley, 2019; Wang *et al.*, 2010). Stability proof has become a prerequisite in the development of controllers for such robotic mechanisms (Rigatos and Busawon, 2018; Yue *et al.*, 2017; Dian *et al.*, 2018; Guo, Mo *et al.*, 2015). Backstepping control and sliding mode control are among the nonlinear control methods one can consider for the power line inspection robots. However, such methods require the state-space model of the robots to be found in a specific form (Rigatos and Busawon, 2018). Furthermore, there exist few results on the nonlinear optimal control of power line inspection robots and on how energy dissipation by their actuators can be reduced (Menendez, Auat Cheein *et al.*, 2017; Zhang, Liu *et al.*, 2012).

In this section, a novel nonlinear optimal (H-infinity) control method is developed for the dynamic model of an underactuated power line inspection robot (Rigatos and Busawon, 2018). To this end, the robot's dynamic model undergoes first approximate linearization around a time-varying operating point which is updated at each time-step of the control method. Actually, at each sampling period this operating point is defined by the present value of the robot's state vector and by the last sampled value of the control inputs vector. The linearization relies on first-order Taylor series expansion and on the computation of the related Jacobian matrices (Rigatos and Tzafestas, 2007; Basseville and Nikiforov, 1993; Rigatos and Zhang, 2009). The truncation of higher-order terms from the Taylor series expansion is considered to be a perturbation that is asymptotically compensated by the robustness of the control method. For the approximately linearized model of the power line inspection robot an optimal (H-infinity) feedback controller is designed.

The H-infinity controller provides the solution to the optimal control problem for the power line inspection robots under model uncertainty and external disturbances (Rigatos, 2016b; Rigatos, Siano and Cecati, 2015; Rigatos, Siano *et al.*, 2015). It represents a min–max differential game taking place between (i) the control inputs which try to minimize a cost function that comprises a quadratic term of the state vector's tracking error and (ii) the model uncertainty and external perturbations which try to maximize this cost function. To compute the stabilizing feedback gains of the H-infinity controller an algebraic Riccati equation is also solved at each sampling period of the control method (Rigatos and Busawon, 2018). The stability properties of the control loop are proven through Lyapunov analysis. First, it is demonstrated that the control scheme satisfies the H-infinity tracking performance criterion (Rigatos, 2015; Tousssaint, Basar and Bullo, 2000). This signifies elevated robustness against model imprecision and exogenous perturbations. Moreover, under moderate conditions, it is proven that the control loop is globally asymptotically stable. Finally, to implement state estimation-based feedback control of the robot, without the need to measure its entire state vector, the H-infinity Kalman Filter is used as a robust state estimator.

2.7.1 Dynamic model of the power line inspection robot

2.7.1.1 Computation of the robotic system's Lagrangian

The specific type of the power line inspection robot to be analyzed next comes from (Dian *et al.*, 2019). The diagram of the underactuated power line inspection robot is depicted in Figure 2.79 and in Figure 2.80. Actually, in Figure 2.80 it is considered that the power line inspection robot has been rotated about its vertical position by an angle θ_1. Then, in the *OXYZ* inertial reference frame, the coordinates of mass m_1 are $x_{m_1} = 0$, $y_{m_1} = -h_1\cos(\theta_1)$ and $z_{m_1} = -h_1\sin(\theta_1)$. In the same reference frame, the coordinates of the counterweight mass m_2 are given by $x_{m_2} = l\cos(\theta_2)$, $y_{m_2} = (h_{20} - l\sin(\theta_2))\cos(\theta_1) + d_1\sin(\theta_1)$ and $z_{m_2} = (h_{20} - l\sin(\theta_2))\sin(\theta_1) - d_1\cos(\theta_1)$. The velocity of mass m_1 is given by:

$$\dot{x}_{m_1} = 0 \quad \dot{y}_{m_1} = h_1\sin(\theta_1))\dot{\theta}_1 \quad \dot{z}_{m_1} = -h_1\cos(\theta_1))\dot{\theta}_1 \tag{2.233}$$

The velocity of mass m_2 is given by:

$$\begin{aligned}
\dot{x}_{m_2} &= -l\sin(\theta_2)\dot{\theta}_2 \\
\dot{y}_{m_2} &= (-l\cos(\theta_2)\dot{\theta}_2)\cos(\theta_1) - (h_{20} - l\sin(\theta_2))\sin(\theta_1)\dot{\theta}_1 + d_1\cos(\theta_1)\dot{\theta}_1 \\
\dot{z}_{m_2} &= (-l\cos(\theta_2)\dot{\theta}_2)\sin(\theta_1) + (h_{20} - l\sin(\theta_2))\cos(\theta_1)\dot{\theta}_1 + d_1\sin(\theta_1)\dot{\theta}_1
\end{aligned} \tag{2.234}$$

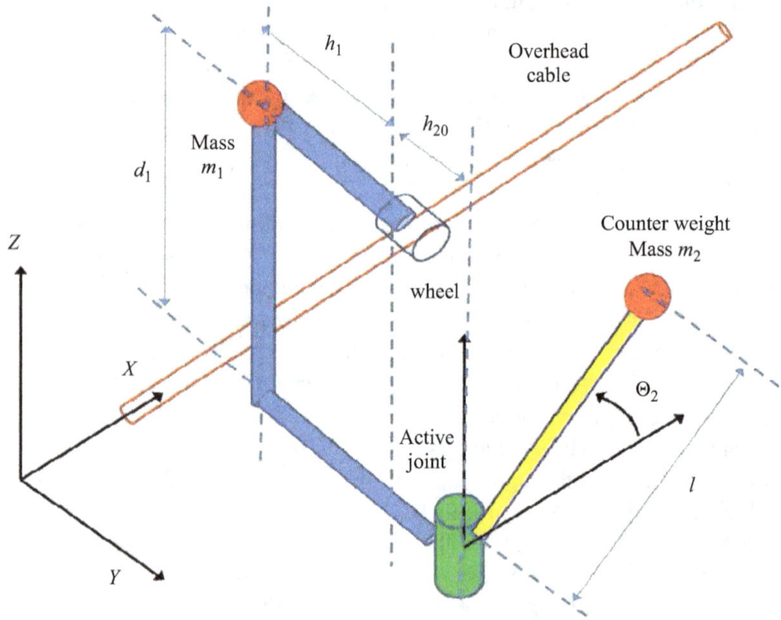

Figure 2.79 Diagram of the power line inspection robot

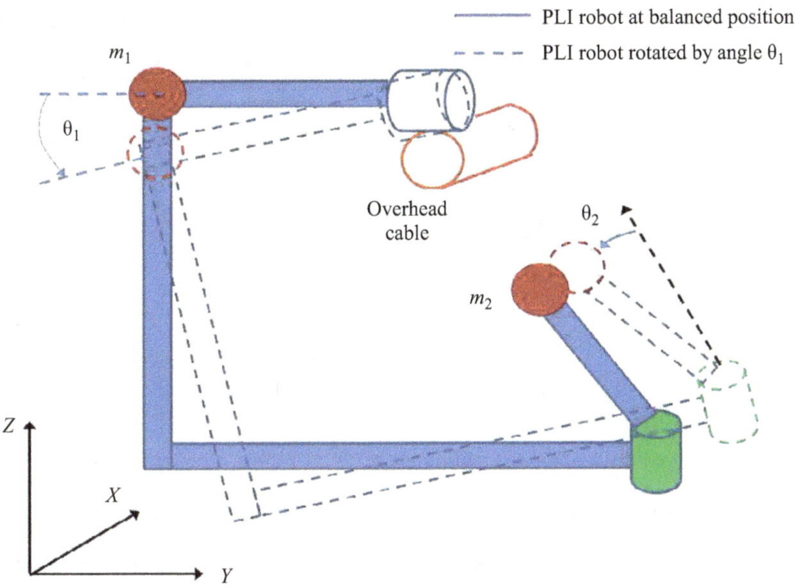

Figure 2.80 Rotation of the power line inspection robot by an angle θ_1 from its balanced position, and rotation of its counterweight mass m_2 by an angle θ_2

The kinetic energy due to translation of mass m_1 is

$$\begin{aligned} K_1 &= \frac{1}{2}m_1(\dot{x}_{m_1}^2 + \dot{y}_{m_1}^2) + \dot{z}_{m_1}^2 \Rightarrow \\ K_1 &= \frac{1}{2}m_1(h_1^2\sin^2(\theta_1)\dot{\theta}_1^2 + h_1^2\cos^2(\theta_1)\dot{\theta}_1^2) \Rightarrow \\ K_1 &= \frac{1}{2}m_1h_1^2\dot{\theta}_1^2 \end{aligned} \tag{2.235}$$

The kinetic energy due to translation of mass m_2 is

$$\begin{aligned} K_2 &= \frac{1}{2}m_2(\dot{x}_{m_2}^2 + \dot{y}_{m_2}^2) + \dot{z}_{m_2}^2 \Rightarrow \\ K_2 &= \frac{1}{2}m_2\{l^2\sin^2(\theta_2)\theta_2^2 + l^2\cos^2(\theta_1)\cos^2(\theta_2)\dot{\theta}_2^2\} \\ &\quad + [(h_{20} - l\sin(\theta_1)\sin(\theta_1)) - d_1\cos(\theta_1)]^2\dot{\theta}_1^2 \\ &\quad - 2[(-l\cos(\theta_2))\cos(\theta_1)][(h_{20} - l\sin(\theta_2))\sin(\theta_1) + d_1\cos(\theta_1)]\dot{\theta}_1\dot{\theta}_2 \\ &\quad + l^2\cos^2(\theta_2)\sin^2(\theta_1)\dot{\theta}_2^2 + [(h_{20} - l\sin(\theta_2))\cos(\theta_1) + d_1\sin(\theta)_1]^2\dot{\theta}_1^2 \\ &\quad + 2[(-l\cos(\theta_2))\sin(\theta_1)][(h_{20} - l\sin(\theta_2))\cos(\theta_1) + d_1\sin(\theta_1)]\dot{\theta}_1\dot{\theta}_2 \end{aligned} \tag{2.236}$$

After regrouping terms, one obtains

$$\begin{aligned} K_2 = \frac{1}{2}m_2\{&l^2\sin^2(\theta_2)\dot{\theta}_2^2 + l^2\cos^2(\theta_2)\dot{\theta}_2^2 + (h_{20} - l\sin(\theta_2))^2\dot{\theta}_1^2 + d_1^2\dot{\theta}_1^2 \\ &- 2(-l\cos(\theta_2))\cos(\theta_1)(h_{20} - l\sin(\theta_2))\sin(\theta_1)\dot{\theta}_1\dot{\theta}_2 \\ &- 2(l\cos(\theta_2))\cos(\theta_1)d_1\cos(\theta_1)\dot{\theta}_1\dot{\theta}_2 \\ &+ 2(l\cos(\theta_2))\sin(\theta_1)(h_{20} - l\sin(\theta_2))\cos(\theta_1)\dot{\theta}_1\dot{\theta}_2 \\ &+ 2(l\cos(\theta_2))\sin(\theta_1)d_1\cos(\theta_1)\dot{\theta}_1\dot{\theta}_2\} \end{aligned} \tag{2.237}$$

Through additional operations one obtains

$$K_2 = \frac{1}{2}m_2\{l^2\dot{\theta}_2^2 + (h_{20} - l\sin(\theta_1))^2\dot{\theta}_1^2 + d_1^2\dot{\theta}_1^2\} \tag{2.238}$$

Therefore, the total kinetic energy of the robotic mechanism is

$$\begin{aligned} &K = K_1 + K_2 \Rightarrow \\ &K = \frac{1}{2}m_1h_1^2\dot{\theta}_1^2 + \frac{1}{2}m_2l^2\dot{\theta}_2^2 + \frac{1}{2}m_2[(h_{20} - l\sin(\theta_1))^2 + d_1^2]\dot{\theta}_1^2 \end{aligned} \tag{2.239}$$

Next, the potential energy of the robotic mechanism is computed. The potential energy of mass m_1 is

$$\begin{aligned} &P_1 = m_1gz_1 \Rightarrow P_1 = m_1g(-h_1\sin(\theta_1)) \\ &\Rightarrow P_1 = -m_1gh_1\sin(\theta_1) \end{aligned} \tag{2.240}$$

The potential energy of mass m_2 is

$$\begin{aligned} &P_2 = m_2gz_2 \Rightarrow P_2 = m_2g[(h_{20} - l\sin(\theta_2))\sin(\theta_1) - d_1\cos(\theta_1)] \\ &\Rightarrow P_1 = -m_1gh_1\sin(\theta_1) \end{aligned} \tag{2.241}$$

The aggregate potential energy of the robotic system is given by:

$$\begin{aligned} &P = P_1 + P_2 \Rightarrow \\ &P = -m_1gh_1\sin(\theta_1) - m_2g[(h_{20} - l\sin(\theta_2))\sin(\theta_1) - d_1\cos(\theta_1)] \end{aligned} \tag{2.242}$$

The Lagrangian of the robotic system is (Dian *et al.*, 2019; Rigatos and Busawon, 2018)

$$\begin{aligned} &L = K - P \Rightarrow \\ &L = \frac{1}{2}m_1h_1^2\dot{\theta}_1^2 + \frac{1}{2}m_2l^2\dot{\theta}_2^2 + \frac{1}{2}m_2[(h_{20} - l\sin(\theta_1))^2 + d_1^2]\dot{\theta}_1^2 \\ &\quad + m_1gh_1\sin(\theta_1) + m_2g[(h_{20} - l\sin(\theta_2))\sin(\theta_1) - d_1\cos(\theta_1)] \end{aligned} \tag{2.243}$$

2.7.1.2 State-space model of the power line inspection robot

The equations that constitute the dynamic model of the power line inspection robot are obtained from the related Euler–Lagrange equations:

$$\begin{aligned} &\frac{\partial}{\partial t}\frac{\partial L}{\partial \dot{\theta}_1} - \frac{\partial L}{\partial \theta_1} = 0 \\ &\frac{\partial}{\partial t}\frac{\partial L}{\partial \dot{\theta}_2} - \frac{\partial L}{\partial \theta_2} = u \end{aligned} \tag{2.244}$$

where u is the torque applied to the actuator of the joint that rotates the counterweight of mass m_2. It holds that:

$$\frac{\partial L}{\partial \dot{\theta}_1} = m_1 h_1^2 \dot{\theta}_1 + m_2[(h_{20} - l\sin(\theta_2))^2 + d_1^2]\dot{\theta}_1 \tag{2.245}$$

Moreover, it holds that

$$\begin{aligned}\frac{d}{dt}\frac{\partial L}{\partial \dot{\theta}_1} &= m_1 h_1^2 \ddot{\theta}_1 + m_2[2(h_{20} - l\sin(\theta_2))(-l\cos(\theta_2))\dot{\theta}_2]\dot{\theta}_1 \\ &\quad + m_2[(h_{20} - l\sin(\theta_2))^2 + d_1^2]\ddot{\theta}_1 \Rightarrow \\ \frac{d}{dt}\frac{\partial L}{\partial \dot{\theta}_1} &= \{m_1 h_1^2 + m_2[(h_{20} - l\sin(\theta_2))^2 + d_1^2]\}\ddot{\theta}_1 \\ &\quad + m_2[2(h_{20} - l\sin(\theta_2))(-l\cos(\theta_2))]\dot{\theta}_1\dot{\theta}_2\end{aligned} \tag{2.246}$$

Additionally, one has that

$$\frac{\partial L}{\partial \theta_1} = m_1 g h_1 \cos(\theta_1) - m_2 g[(h_{20} - l\sin(\theta_2))\cos(\theta_1) + d_1 \sin(\theta_1)] \tag{2.247}$$

Consequently, the first equation of motion becomes:

$$\begin{aligned}&\frac{\partial}{\partial t}\frac{\partial L}{\partial \dot{\theta}_1} - \frac{\partial L}{\partial \theta_1} = 0 \Rightarrow \\ &\{m_1 h_1^2 + m_2[(h_{20} - l\sin(\theta_2))^2 + d_1^2]\}\ddot{\theta}_1 \\ &+ m_2[2(h_{20} - l\sin(\theta_2))(-l\cos(\theta_2))]\dot{\theta}_1\dot{\theta}_2 \\ &- m_1 g h_1 \cos(\theta_1) + m_2 g[(h_{20} - l\sin(\theta_2))\cos(\theta_1) + d_1 \sin(\theta_1)] = 0\end{aligned} \tag{2.248}$$

Furthermore, one has

$$\frac{\partial L}{\partial \dot{\theta}_2} = m_2 l^2 \dot{\theta}_2 \tag{2.249}$$

and also

$$\frac{d}{dt}\frac{\partial L}{\partial \dot{\theta}_2} = m_2 l^2 \ddot{\theta}_2 \tag{2.250}$$

Moreover, it holds that

$$\frac{\partial L}{\partial \theta_2} = m_2(h_{20} - l\sin(\theta_2))(-l\cos(\theta_2))\dot{\theta}_1^2 - m_2 g(-l\cos(\theta_2))\sin(\theta_1) \tag{2.251}$$

Thus, the second equation of motion of the robotic system becomes

$$\begin{aligned}&\frac{\partial}{\partial t}\frac{\partial L}{\partial \dot{\theta}_2} - \frac{\partial L}{\partial \theta_2} = u \Rightarrow \\ &m l^2 \ddot{\theta}_2 + m_2(h_{20} - l\sin(\theta_2))(l\cos(\theta_2))\dot{\theta}_1^2 - m_2 g(l\cos(\theta_2))\sin(\theta_1) = u\end{aligned} \tag{2.252}$$

The following state variables are defined: $x_1 = \theta_1$, $x_2 = \dot{\theta}_1$, $x_3 = \theta_2$ and $x_4 = \dot{\theta}_2$. Consequently, one can obtain the following state-space description:

$$\begin{aligned}&\{m_1 h_1^2 + m_2[(h_{20} - l\sin(x_3))^2 + d_1^2]\}\ddot{x}_1 \\ &+ m_2[2(h_{20} - l\sin(x_3))(-l\cos(x_3))]x_2 x_4 \\ &- m_1 g h_1 \cos(x_1) + m_2 g[(h_{20} - l\sin(x_3))\cos(x_1) + d_1 \sin(x_1)] = 0\end{aligned} \tag{2.253}$$

$$m_2 l^2 \ddot{x}_3 + m_2(h_{20} - l\sin(x_3)) l\cos(x_3) x_2^2 - m_1 g(l\cos(x_3))\sin(x_1) = u \tag{2.254}$$

The previous state-space model can be also written as:

$$\begin{aligned}\ddot{x}_1 = & \frac{m_2[2(h_{20} - l\sin(x_3))(l\cos(x_3))]x_2 x_4}{\{m_1 h_1^2 + m_2[(h_{20} - l\sin(x_3))^2 +]\}} \\ & + \frac{m_1 g h_1 \cos(x_1) + m_2 g[(h_{20} - l\sin(x_3))\cos(x_1) + d_1 \sin(x_1)]}{\{m_1 h_1^2 + m_2[(h_{20} - l\sin(x_3))^2]\}}\end{aligned} \tag{2.255}$$

$$\ddot{x}_3 = \frac{-m_2(h_{20} - l\sin(x_3)) l\cos(x_3) x_2^2 + m_1 g(l\cos(x_3))\sin(x_1)}{m_2 l^2} + \frac{u}{m_2 l^2} \tag{2.256}$$

Additionally, the state-space model of the system can be written as

$$\dot{x}_1 = x_2 \tag{2.257}$$

$$\begin{aligned}\dot{x}_2 = & \frac{m_2[2(h_{20} - l\sin(x_3))(l\cos(x_3))]x_2 x_4}{\{m_1 h_1^2 + m_2[(h_{20} - l\sin(x_3))^2 +]\}} \\ & + \frac{m_1 g h_1 \cos(x_1) + m_2 g[(h_{20} - l\sin(x_3))\cos(x_1) + d_1 \sin(x_1)]}{\{m_1 h_1^2 + m_2[(h_{20} - l\sin(x_3))^2]\}}\end{aligned} \tag{2.258}$$

$$\dot{x}_3 = x_4 \tag{2.259}$$

$$\dot{x}_4 = \frac{-m_2(h_{20} - l\sin(x_3)) l\cos(x_3) x_2^2 + m_1 g(l\cos(x_3))\sin(x_1)}{m_2 l^2} + \frac{u}{m_2 l^2} \tag{2.260}$$

Furthermore, one can consider the following concise description of the system in vector fields form

$$\dot{x} = f(x) + g(x)u \tag{2.261}$$

where $x \in R^{4\times1}, f(x) \in R^{4\times1}$ and $g(x) \in R^{4\times1}$ while $u \in R$, or equivalently

$$\begin{pmatrix}\dot{x}_1\\ \dot{x}_2\\ \dot{x}_3\\ \dot{x}_4\end{pmatrix} = \begin{pmatrix}f_1(x)\\ f_2(x)\\ f_3(x)\\ f_4(x)\end{pmatrix} + \begin{pmatrix}g_1(x)\\ g_2(x)\\ g_3(x)\\ g_4(x)\end{pmatrix} u \tag{2.262}$$

where $f_1(x) = x_2$,

$$f_2(x) = \frac{m_2[2(h_{20} - l\sin(x_3))(l\cos(x_3))]x_2 x_4}{\{m_1 h_1^2 + m_2[(h_{20} - l\sin(x_3))^2 +]\}}$$

$$+ \frac{m_1 g h_1 \cos(x_1) + m_2 g[(h_{20} - l\sin(x_3))\cos(x_1) + d_1 \sin(x_1)]}{\{m_1 h_1^2 + m_2[(h_{20} - l\sin(x_3))^2]\}}$$

$f_3(x) = x_4$,

$$f_4(x) = \frac{f_{4n}(x)}{f_{4d}(x)} = \frac{-m_2(h_{20} - l\sin(x_3)) l\cos(x_3) x_2^2 + m_1 g(l\cos(x_3))\sin(x_1)}{m_2 l^2}$$

and $g_1(x) = 0$, $g_2(x) = 0$, $g_3(x) = 0$ and $g_4(x) = \frac{1}{m_2 l^2}$.

2.7.2 Approximate linearization of the power line inspection robot

2.7.2.1 Approximately linearized model

The dynamic model of the power line inspection robot $\dot{x}=f(x)+g(x)u$ undergoes approximate linearization around the time-varying operating point (x^*,u^*), where x^* is the present value of the system's state vector and u^* is the most recent value of the control inputs vector that was applied to it. The linearization relies on first order Taylor series expansion and on the computation of the related Jacobian matrices.

This concept results into the approximately linearized model of the robotic system in the form:

$$\dot{x}=Ax+Bu+\tilde{d} \tag{2.263}$$

where $\tilde{d}$ is the cumulative perturbations' vector which comprises (i) the modeling error due to truncation of higher-order terms in the Taylor series expansion, (ii) external disturbances, and (iii) measurement noise of any distribution.

The Jacobian matrices of the robotic system are

$$\begin{aligned}&A=\nabla_x[f(x)+g(x)u]\mid_{(x^*,u^*)}\Rightarrow A=\nabla_x[f(x)]\mid_{(x^*,u^*)}+\nabla_x[g(x)]u\mid_{(x^*,u^*)}\Rightarrow\\&A=\nabla_x[f(x)]\mid_{(x^*,u^*)}\end{aligned} \tag{2.264}$$

$$B=\nabla_u[f(x)+g(x)u]\mid_{(x^*,u^*)}\Rightarrow B=g(x)\mid_{(x^*,u^*)} \tag{2.265}$$

Next, the elements of the Jacobian matrix $\nabla_x[f(x)]\mid_{(x^*,u^*)}$ are computed.

First row of the Jacobian matrix $\nabla_x[f(x)]\mid_{(x^*,u^*)}$: $\frac{\partial f_1}{\partial x_1}=0$, $\frac{\partial f_1}{\partial x_2}=1$, $\frac{\partial f_1}{\partial x_3}=0$ and $\frac{\partial f_1}{\partial x_4}=0$.

Second row of the Jacobian matrix $\nabla_x[f(x)]\mid_{(x^*,u^*)}$: for $i=1,\cdots,4$

$$\frac{\partial f_2}{\partial x_i}=\frac{\frac{\partial f_{2n}(x)}{\partial x_i}f_{2d}(x)-f_{2n}(x)\frac{\partial f_{2d}(x)}{\partial x_i}}{f_{2d}(x)^2} \tag{2.266}$$

where

$\frac{\partial f_{2n}}{\partial x_1}=0$, $\frac{\partial f_{2n}}{\partial x_2}=m_2[2(h_{20}-l\sin(x_3))(l\cos(x_3))]x_4$,

$$\begin{aligned}\frac{\partial f_{2n}}{\partial x_3}&=2m_2[(-l\cos(x_3))(l\cos(x_3))]x_2x_4\\&\quad-2m_2(h_{20}-l\sin(x_3))(l\sin(x_3))x_2x_4+m_2g[(l\cos(x_3))\cos(x_1)],\end{aligned}$$

$\frac{\partial f_{2n}}{\partial x_4}=m_2[2(h_{20}-l\sin(x_3))(l\cos(x_3))]x_2$.

and also $\frac{\partial f_{2d}}{\partial x_1}=0$, $\frac{\partial f_{2d}}{\partial x_12}=0$, $\frac{\partial f_{2d}}{\partial x_3}=m_22(h_{20}-l\sin(x_3))(-l\cos(x_3))$, $\frac{\partial f_{2d}}{\partial x_4}=0$.
Third row of the Jacobian matrix $\nabla_x[f(x)]\mid_{(x^*,u^*)}$: $\frac{\partial f_3}{\partial x_1}=0$, $\frac{\partial f_3}{\partial x_2}=0$, $\frac{\partial f_3}{\partial x_3}=0$ and $\frac{\partial f_3}{\partial x_4}=1$.

Fourth row of the Jacobian matrix $\nabla_x[f(x)]\mid_{(x^*,u^*)}$: $\frac{\partial f_4(x)}{\partial x_1}=\frac{1}{m_2 l^2}(m_2 g(l\cos(x_3))\cos(x_1))$, $\frac{\partial f_4(x)}{\partial x_2}=-\frac{2x_2}{m_2 l^2}$,

$$\begin{aligned}\frac{\partial f_4(x)}{\partial x_3}=&\frac{1}{m_2 l^2}[-m_2(-l\cos(x_3))(l\cos(x_3))x_2^2\\&-m_2(h_{20}-l\sin(x_3))(-l\sin(x_3))x_2^2-m_2 g(l\sin(x_3))\sin(x_1)],\end{aligned}$$

and $\frac{\partial f_4(x)}{\partial x_3}=0$

2.7.2.2 Stabilizing feedback control

After linearization around its current operating point, the dynamic model of the power line inspection robot is written as

$$\dot{x}=Ax+Bu+d_1 \tag{2.267}$$

Parameter d_1 stands for the linearization error in the power line inspection robot's dynamic model appearing previously in (2.263). The reference setpoints for the power line inspection robot's state vector are denoted by $\mathbf{x_d}=[x_1^d,\cdots,x_4^d]$. Tracking of this trajectory is achieved after applying the control input u^*. At every time instant the control input u_d is assumed to differ from the control input u appearing in (2.267) by an amount equal to Δu, that is $u_d=u+\Delta u$

$$\dot{x}_d=Ax_d+Bu_d+d_2 \tag{2.268}$$

The dynamics of the controlled system described in (2.267) can be also written as

$$\dot{x}=Ax+Bu+Bu_d-Bu_d+d_1 \tag{2.269}$$

and by denoting $d_3=-Bu_d+d_1$ as an aggregate disturbance term one obtains

$$\dot{x}=Ax+Bu+Bu_d+d_3 \tag{2.270}$$

By subtracting (2.268) from (2.270) one has

$$\dot{x}-\dot{x}_d=A(x-x_d)+Bu+d_3-d_2 \tag{2.271}$$

By denoting the tracking error as $e=x-x_d$ and the aggregate disturbance term as $\tilde{d}=d_3-d_2$, the tracking error dynamics becomes

$$\dot{e}=Ae+Bu+\tilde{d} \tag{2.272}$$

For the approximately linearized model of the system a stabilizing feedback controller is developed. The controller has the form

$$u(t)=-Ke(t) \tag{2.273}$$

with $K=\frac{1}{r}B^TP$ where P is a positive definite symmetric matrix which is obtained from the solution of the Riccati equation Rigatos and Busawon, 2018

$$A^TP+PA+Q-P\left(\frac{2}{r}BB^T-\frac{1}{\rho^2}LL^T\right)P=0 \tag{2.274}$$

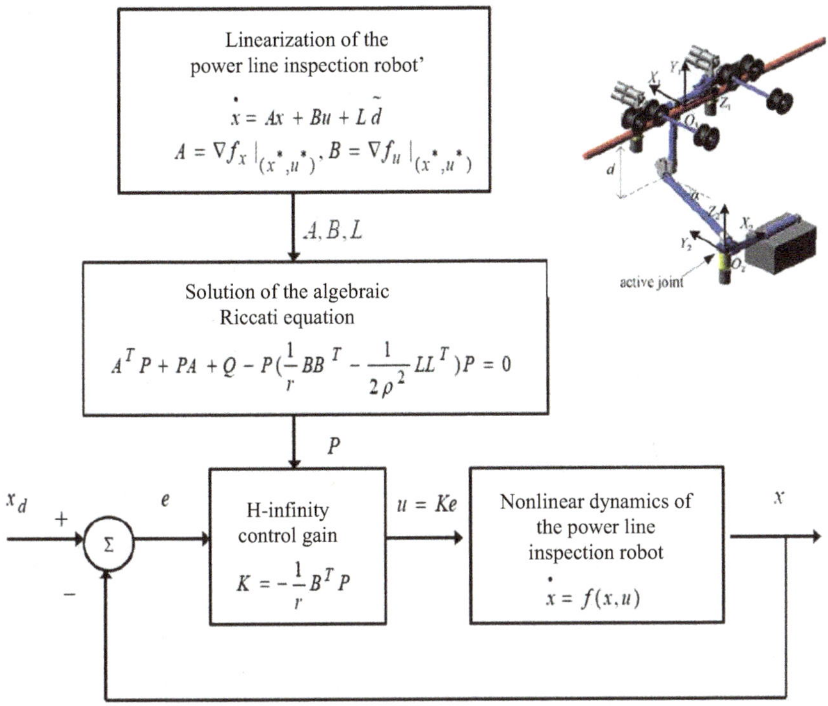

Figure 2.81 Diagram of the control scheme for the power line inspection robot (the figure of the PLI robot comes from Dian et al., 2019)

where Q is a positive semi-definite symmetric matrix. The diagram of the considered control loop is depicted in Figure 2.81. The global stability properties of the control method can be proven following the procedure presented in Section 1.1.2.5.

2.7.3 Simulation tests

The tracking performance of the proposed nonlinear optimal (H-infinity) control method for power line inspection robots was tested through simulation experiments. Indicative values for the parameters of the model of the robotic mechanism are: $l = 0.5$ m, $d_1 = 0.5$ m, $g = 10$ m/sec^2, $m_1 = 63$ kgr, $m_2 = 27$ kgr, $h_1 = 0.18$ m, and $h_{20} = 0.42$ m (Dian *et al.*, 2019). The obtained results are given in Figures 2.82 to 2.87. The state variables of the power line inspection robot are printed in blue color, the estimated state variables which were provided by the H-infinity Kalman Filter are printed in green, while the related setpoints plotted in red. The measurement units of angles θ_1 and θ_2 are in rad, while the measurement units of the angular velocities $\dot{\theta}_1$ and $\dot{\theta}_2$ are in rad/sec. The control input u is a torque measured in Ntm. It can be noticed that the tracking error for all state variables of the robotic mechanism has been eliminated, while the transient performance of the control loop was good. Actually, there were neither abrupt variations of the state variables nor excessive values of the

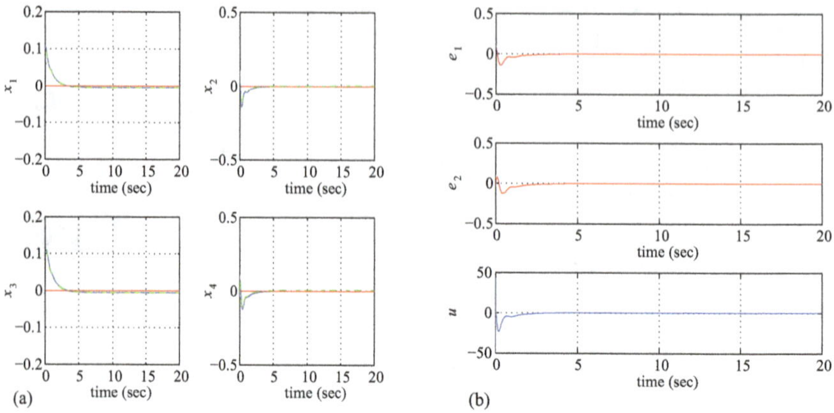

Figure 2.82 Tracking of setpoint 1 by the power line inspection robot: (a) state variables x_i $i = 1, \cdots, 4$ (blue lines), their estimated values (green lines) and the associated reference setpoints (red lines); (b) tracking errors e_1 (state variable x_1) and e_3 (state variable x_3) and control input u of the power line inspection robot

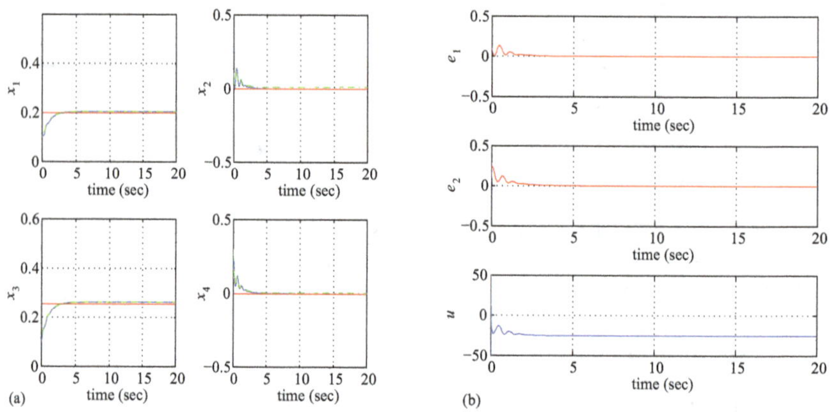

Figure 2.83 Tracking of setpoint 2 by the power line inspection robot: (a) state variables x_i $i = 1, \cdots, 4$ (blue lines), their estimated values (green lines) and the associated reference setpoints (red lines); (b) tracking errors e_1 (state variable x_1) and e_3 (state variable x_3) and control input u of the power line inspection robot

control inputs. The transient performance of the control loop depends on the selection of values for parameters r, ρ and Q which appear in the algebraic Riccati equation given in (2.274). Actually, the smallest value of ρ for which one can obtain a valid solution of the previously noted Riccati equation is the one that provides the control loop with maximum robustness.

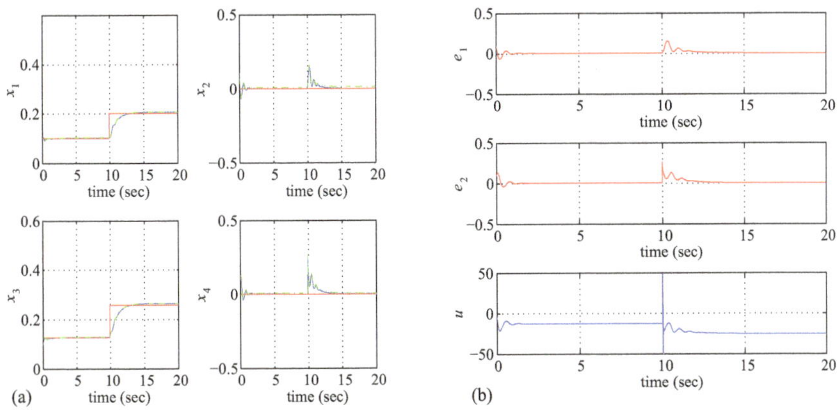

Figure 2.84 *Tracking of setpoint 3 by the power line inspection robot: (a) state variables x_i $i = 1, \cdots, 4$ (blue lines), their estimated values (green lines) and the associated reference setpoints (red lines); (b) tracking errors e_1 (state variable x_1) and e_3 (state variable x_3) and control input u of the power line inspection robot*

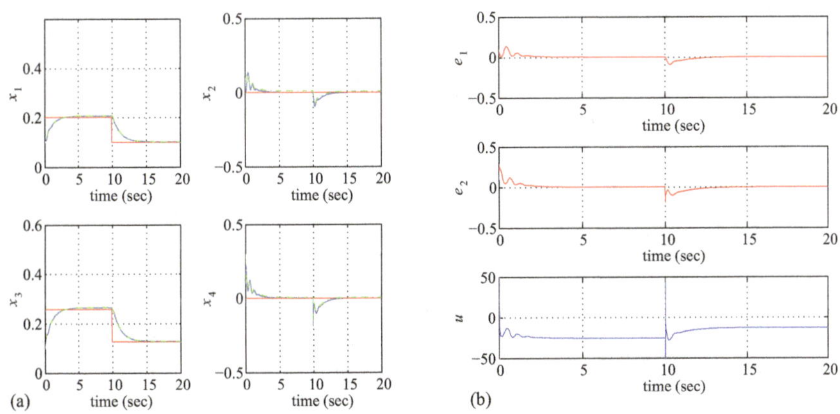

Figure 2.85 *Tracking of setpoint 4 by the power line inspection robot: (a) state variables x_i $i = 1, \cdots, 4$ (blue lines), their estimated values (green lines) and the associated reference setpoints (red lines); (b) tracking errors e_1 (state variable x_1) and e_3 (state variable x_3) and control input u of the power line inspection robot*

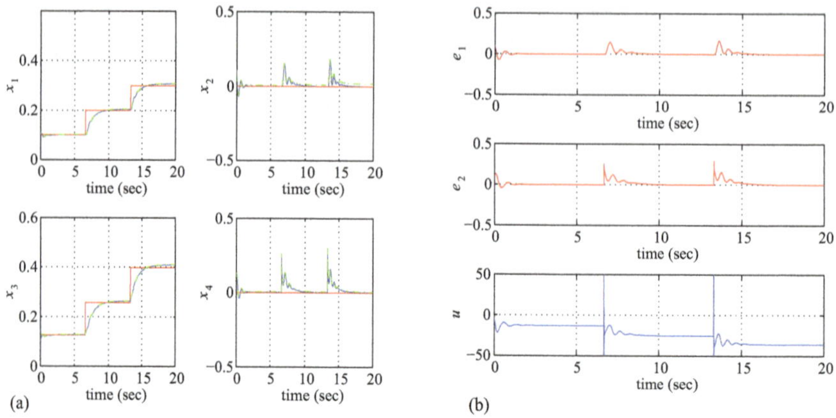

Figure 2.86 Tracking of setpoint 5 by the power line inspection robot: (a) state variables x_i $i = 1, \cdots, 4$ (blue lines), their estimated values (green lines) and the associated reference setpoints (red lines); (b) tracking errors e_1 (state variable x_1) and e_3 (state variable x_3) and control input u of the power line inspection robot

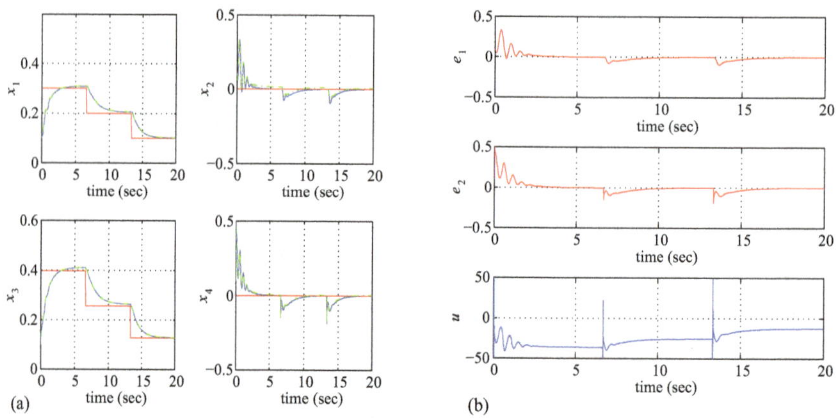

Figure 2.87 Tracking of setpoint 6 by the power line inspection robot: (a) state variables x_i $i = 1, \cdots, 4$ (blue lines), their estimated values (green lines) and the associated reference setpoints (red lines); (b) tracking errors e_1 (state variable x_1) and e_3 (state variable x_3) and control input u of the power line inspection robot

2.8 Nonlinear control of robots with electrohydraulic actuators

2.8.1 Outline

Electro-hydraulic actuation systems have been widely used in robotic manipulators, in construction machinery and in aircrafts (Rigatos and Busawon, 2018; Kim, Choi *et al.*, 2019; Nguyen, Tran *et al.*, 2018). Nowadays, manipulators and vehicle receiving hydraulic actuation constitute a huge global industrial sector (Mohanty and Yao, 2011b; Taylor and Robertson, 2013; Dinh *et al.*, 2018). Hydraulic manipulators are favored in industry, mining, agriculture and construction works where large actuation forces are required (Liang, Wan *et al.*, 2018; Guo, Sun *et al.*, 2016; Lu and Yao, 2014). For instance in 2016 the construction business alone sold 0.7 million units of construction machines, while in 2019 the sales of robotic manipulators, including also hydraulic manipulators, reached the level of 0.4 million units. Hydraulic actuation in robotic manipulators exhibits specific technical advantages, such as large force and torque, fast response, small size-to-power ratio, reliability and low manufacturing cost (Siroupour and Salcudean, 2001; Mohanty and Yao, 2011b; Tafazoli *et al.*, 2002). Hydraulic heavy-duty manipulators are irreplaceable for heavy workpiece handling. They improve the efficiency of fine manipulation of heavy loads, thus ensuring production, safety and reducing labor costs (Kovumaki and Mattila, 2015; Li, Xin *et al.*, 2019; Wang, Chi *et al.*, 2018). Comparing against electric actuators, hydraulic actuators have the advantage of producing high power while having a smaller size (Henicki *et al.*, 2016; Wind, Renner *et al.*, 2019). Minimization of energy consumption is an issue that has to be into taken account in the development of such robotic systems (Koivumaki, Zhu *et al.*, 2019; Maddahi *et al.*, 2016). Furthermore high precision motion is a vital functionality for electro-hydraulic robotic systems (Ranjan, Wrat *et al.*, 2020; Koivumaki and Mattila, 2015).

On the other side, the control of electro-hydraulic robotic manipulators remains a challenging problem that exhibits a higher degree of complexity than the control of electrically actuated manipulators. This is because electro-hydraulic manipulators have complicated nonlinear dynamics, and are subject to modeling uncertainties (such as friction, internal leakage, deadbands in valves operation and control saturation). Besides, the functioning of electro-hydraulic manipulators is subject to several parametric variations, such as uncertain or varying loads. Heavy loads lead to significant rigid–flexible coupling characteristics and mechanical deformation in multi-DOF hydraulic manipulators. For this reason, the design of reliable controllers for hydraulic manipulators is acknowledged to be a demanding task. Various control methods have been developed for electro-hydraulic robots. Most of the results on the control of these manipulators exclude actuator dynamics from the robots' dynamic behavior. However, actuator dynamics affects also the dynamics of the entire robotic system and the stability properties of the robots' control loop (La Hera *et al.*, 2009; Lee and Chung, 2019). The design of control and actuation schemes for hydraulic manipulators can be improved if rotary hydraulic actuators are used in place of linear hydraulic actuators Hedtmann and Brucker, 2005, Brucker and Heidtmann, 2004, Liang, Wang and Wang, 2019. Rotary vane actuators, functioning as rotational drives, provide rotational movements directly because they are constructed as joints and actuators in one. Nevertheless, these actuators may exhibit internal leakage and high friction. Therefore, the controller design for hydraulic manipulators should compensate for such perturbations too.

In this section, a novel nonlinear optimal control method is developed for multi-DOF electro-hydraulic robotic manipulators (Rigatos and Busawon, 2018). First, the dynamic model of the electro-hydraulic robotic manipulator undergoes approximate linearization around a temporary operating point which is updated at each time-step of the control method. This time-varying operating point is determined by the present value of the robot's state vector and by the last sampled value of the control inputs vector. The linearization process relies on first-order Taylor series expansion and on the computation of the related Jacobian matrices (Rigatos and Tzafestas, 2007; Basseville and Nikiforov, 1993; Rigatos and Zhang, 2009). The modeling error which is due to the truncation of higher-order terms in the Taylor series expansion is considered to be a perturbation which is asymptotically compensated by the robustness of the control method. For the approximately linearized model of the robot an H-infinity (optimal) feedback controller is designed.

There is ongoing research on the nonlinear control problem of hydraulic manipulators in several engineering tasks (Hyan *et al.*, 2020; Gao *et al.*, 2020; Li, Xie and Luo, 2019; Liang, Wan and Zhang, 2018; Kalmari *et al.*, 2017). The proposed H-infinity controller provides the solution of the optimal control problem for the dynamic model of the electro-hydraulic robotic manipulator under model uncertainties and external perturbations. Actually, it represents a min–max differential game taking place between (i) the controller which tries to minimize a quadratic cost function of the state vector's tracking error and (ii) the model uncertainty and external perturbation terms which try to maximize this cost function (Rigatos and Busawon, 2018; Rigatos, 2016b). To compute the feedback gains of the stabilizing feedback controller, an algebraic Riccati equation is being repetitively solved at each time-step of the control algorithm (Rigatos, 2011;Rigatos, 2015). The global stability properties of the control method are proven through Lyapunov analysis. First, it is demonstrated that the control scheme satisfies the H-infinity tracking performance criterion, which signifies elevated robustness against model uncertainty and external perturbations (Rigatos and Busawon, 2018; Tousssaint, Basar and Bullo, 2000). Moreover, under moderate conditions, it is proven that the control loop is globally asymptotically stable. Additionally, to implement state estimation-based control without the need to measure the entire state vector of the robotic system, the H-infinity Kalman Filter is used as a robust state estimator (Rigatos and Busawon, 2018; Rigatos, 2016b).

2.8.2 Dynamic model of the multi-DOF electrohydraulic manipulator

The model of a multi-DOF (2-DOF) rigid-link robot is depicted in Figure 2.88. Under the assumption that the masses of the links are concentrated at the links' end the dynamic model is given by (Rigatos and Busawon, 2018)

$$M(\theta)\ddot{\theta} + C(\theta,\dot{\theta}) + G(\theta) = \tau \tag{2.275}$$

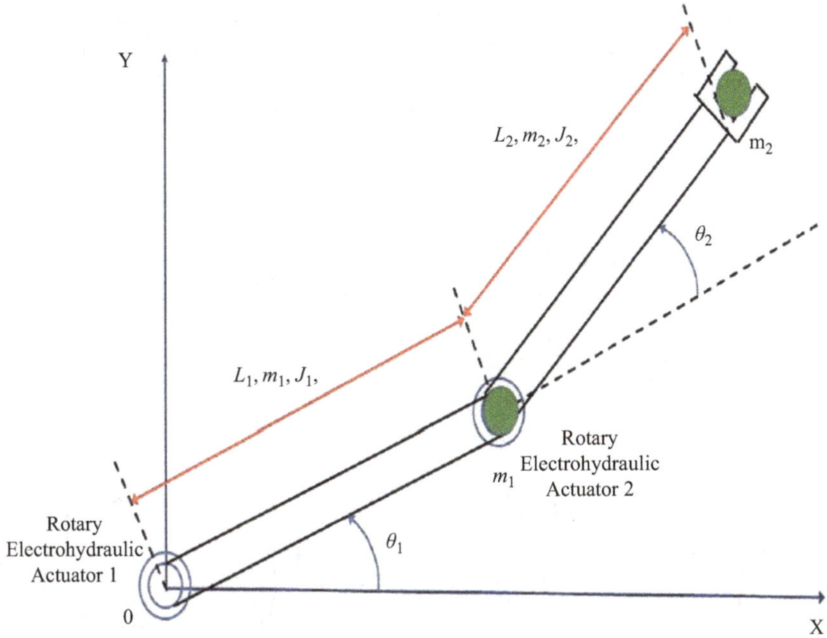

Figure 2.88 Diagram of the 2-DOF electrohydraulic manipulator

where $M(\theta)$ is the inertia matrix

$$M(\theta) = \begin{pmatrix} M_{11} & M_{12} \\ M_{21} & M_{22} \end{pmatrix} \tag{2.276}$$

where $M_{11} = (m_1 + m_2)l_1^2 + m_1 l_2^2 + 2m_2 l_1 l_2 \cos(\theta_2)$, $M_{12} = m_2 l_2^2 + m_2 l_1 l_2 \cos(\theta_2)$, $M_{21} = m_2 l_2^2 + m_2 l_1 l_2 \cos(\theta_2)$ and $M_{22} = m_2 l_2(l_1 + l_2)\cos(\theta_2)$. Moreover, $C(\theta, \dot{\theta})$ is the Coriolis and centrifugal forces vector

$$C(\theta, \dot{\theta}) = \begin{pmatrix} -m_2 l_1^2 \sin(\theta_2)\dot{\theta}_2^2 - 2m_2 l_1^2 \sin(\theta_2)\dot{\theta}_1\dot{\theta}_2 \\ m_2 l_1^2 \sin(\theta_2)\dot{\theta}_1^2 \end{pmatrix} \tag{2.277}$$

$G(\theta)$ is the gravitational forces vector

$$G(\theta) = \begin{pmatrix} (m_1 + m_2) g l_1 \cos(\theta_1) + m_2 g l_2 \cos(\theta_1 + \theta_2) \\ m_2 g l_2 \cos(\theta_1 + \theta_2) \end{pmatrix} \tag{2.278}$$

and $\tau(t)$ is the control inputs vector consisting of the torques that are generated by rotary electro-hydraulic actuators mounted on the robot's joints. The torques vector applied to the manipulator is $\tau = [\tau_1, \tau_2]^T$. Moreover, if one considers elasticity in the second joint, the torques vector becomes $\tau = [\tau_1, \tau_2 + k(\theta_1 - \theta_2)]^T$. Torques τ_1 and τ_2 are provided to the joints of the robot by rotary electro-hydraulic drives.

It holds that the determinant of the inertia matrix $M(\theta)$ is given by $detM = [M_{11}M_{22} - M_{12}^2]$. Moreover, after computing the inverse of the inertia matrix $M^{-1}(\theta)$ and the matrices products $M^{-1}(\theta)C(\theta,\dot{\theta})$, $M^{-1}(\theta)G(\theta)$ and $M^{-1}(\theta)\tau$, the following description of the dynamic model of the robot is obtained

$$\ddot{\theta} = -M^{-1}(\theta)C(\theta,\dot{\theta}) - M^{-1}(\theta)G(\theta) + M^{-1}(\theta)\tau \tag{2.279}$$

or analytically

$$\begin{pmatrix}\ddot{\theta}_1\\ \ddot{\theta}_2\end{pmatrix} = \frac{1}{detM}\left[\begin{pmatrix}M_{22}C_1 - M_{12}C_2\\ -M_{12}C_1 + M_{11}C_2\end{pmatrix} + \begin{pmatrix}M_{22}G_1 - M_{12}G_2\\ -M12G_1 + M_{11}G_2\end{pmatrix}\right.$$
$$\left. + \begin{pmatrix}M_{22} & -M_{12}\\ -M_{12} & M_{22}\end{pmatrix}\begin{pmatrix}\tau_1\\ \tau_2\end{pmatrix}\right] \tag{2.280}$$

or equivalently

$$\begin{pmatrix}\ddot{\theta}_1\\ \\ \ddot{\theta}_2\end{pmatrix} = \begin{pmatrix}\dfrac{M_{22}C_1 - M_{12}C_2 + M_{22}G_1 - M_{12}G_2}{detM}\\ \\ \dfrac{-M_{12}C_1 + M_{11}C_2 - M_{12}G_1 + M_{11}G_2}{detM}\end{pmatrix} + \begin{pmatrix}\dfrac{M_{22}}{detM} & -\dfrac{M_{12}}{detM}\\ \\ -\dfrac{M_{12}}{detM} & \dfrac{M_{11}}{detM}\end{pmatrix}\begin{pmatrix}\tau_1\\ \\ \tau_2\end{pmatrix} \tag{2.281}$$

Torques τ_1 and τ_2 are generated by rotary electro-hydraulic actuators and are given by

$$\tau_1 = K_{P_1}A_{P_1}P_{L_1} \quad \tau_2 = K_{P_2}A_{P_2}P_{L_2} \tag{2.282}$$

where K_{P_i}, $i = 1,2$ are force to torque conversion coefficients for the ith actuator, A_{P_i}, $i = 1,2$ is the effective area of the cylinder's chamber of the ith actuator and P_{L_i}, $i = 1,2$ is the differential load pressure of the cylinder of the ith actuator. The dynamics of the two hydraulic actuators are given by the following equations (Guo, Sun *et al.*, 2016; Lee and Chung, 2019; Hedtmann and Brucker, 2005):

$$\dot{P}_{L_1} = \frac{-4\beta_{e_1}A_{p1}}{V_{t_1}}r_1\omega_1 + \frac{4\beta_{e_1}C_{tl_1}}{V_{t_1}}P_{L_1} + \frac{4\beta_{e_1}C_{d_1}w_1}{V_{t_1}\sqrt{\rho_1}}\sqrt{P_{s_1} - \tanh(k_1x_{v_1})P_{L_1}}x_{v_1} \tag{2.283}$$

$$\dot{x}_{v_1} = -\frac{1}{T_{sv_1}}x_{v_1} + \frac{K_{sv_1}}{T_{sv_1}}u_1 \tag{2.284}$$

$$\dot{P}_{L_2} = \frac{-4\beta_{e_2}A_{p2}}{V_{t_2}}r_2\omega_2 + \frac{4\beta_{e_2}C_{tl_2}}{V_{t_2}}P_{L_2} + \frac{4\beta_{e_2}C_{d_2}w_2}{V_{t_2}\sqrt{\rho_2}}\sqrt{P_{s_2} - \tanh(k_2x_{v_2})P_{L_2}}x_{v_2} \tag{2.285}$$

$$\dot{x}_{v_2} = -\frac{1}{T_{sv_2}}x_{v_2} + \frac{K_{sv_2}}{T_{sv_2}}u_2 \tag{2.286}$$

where x_{v_i}, $i = 1,2$ is the position of the ith electrovalve.

Next, the following state-variables are defined: $x_1 = \theta_1$, $x_2 = \dot{\theta}_1$, $x_3 = \theta_2$, $x_4 = \dot{\theta}_2$, $x_5 = P_{L_1}$, $x_6 = x_{v_1}$, $x_7 = P_{L_2}$ and $x_8 = x_{v_2}$. With this notation of state-variables, the dynamic model of the system is written in the following form:

$$\dot{x}_1 = x_2 \tag{2.287}$$

$$\dot{x}_2 = \frac{M_{22}C_1 - M_{12}C_2 + M_{22}G_1 - M_{12}G_2}{detM} + \frac{M_{22}}{detM}K_{P_1}A_{P_1}x_5 - \frac{M_{12}}{detM}K_{P_2}A_{P_2}x_7 \tag{2.288}$$

$$\dot{x}_3 = x_4 \tag{2.289}$$

$$\dot{x}_4 = \frac{-M_{12}C_1 + M_{11}C_2 - M_{12}G_1 + M_{11}G_2}{detM} - \frac{M_{12}}{detM}K_{P_1}A_{P_1}x_5 + \frac{M_{11}}{detM}K_{P_2}A_{P_2}x_7 \tag{2.290}$$

$$\dot{x}_5 = \frac{-4\beta_{e_1}A_{p1}}{V_{t_1}}r_1x_2 + \frac{4\beta_{e_1}C_{tl_1}}{V_{t_1}}x_5 + \frac{4\beta_{e_1}C_{d_1}w_1}{V_{t1}\sqrt{\rho_1}}\sqrt{P_{s_1} - \tanh(k_1x_6)x_5}x_6 \tag{2.291}$$

$$\dot{x}_6 = -\frac{1}{T_{sv_1}}x_6 + \frac{K_{sv_1}}{T_{sv_1}}u_1 \tag{2.292}$$

$$\dot{x}_7 = \frac{-4\beta_{e_2}A_{p2}}{V_{t_2}}r_2x_4 + \frac{4\beta_{e_2}C_{tl_2}}{V_{t_2}}x_7 + \frac{4\beta_{e_2}C_{d_2}w_2}{V_{t_2}\sqrt{\rho_2}}\sqrt{P_{s_2} - \tanh(k_2x_8)x_7}x_8 \tag{2.293}$$

$$\dot{x}_8 = -\frac{1}{T_{sv_2}}x_{x_8} + \frac{K_{sv_2}}{T_{sv_2}}u_2 \tag{2.294}$$

Moreover, the state-space model of the electro-hydraulic manipulator can be written in the following matrix form:

$$\dot{x} = f(x) + g(x)u \tag{2.295}$$

with $x \in R^{8\times1}$, $f(x) \in R^{8\times1}$, $g(x) \in R^{8\times1}$, and $u \in R$, while vectors $f(x)$ and $g(x)$ are given by

$$f(x) = \qquad\qquad g(x=)$$

$$\begin{pmatrix} x_2 \\ \frac{M_{22}C_1 - M_{12}C_2 + M_{22}G_1 - M_{12}G_2}{detM} + \frac{M_{22}}{detM}K_{P_1}A_{P_1}x_5 - \frac{M_{12}}{detM}K_{P_2}A_{P_2}x_7 \\ x_4 \\ \frac{-M_{12}C_1 + M_{11}C_2 - M_{12}G_1 + M_{11}G_2}{detM} - \frac{M_{12}}{detM}K_{P_1}A_{P_1}x_5 + \frac{M_{11}}{detM}K_{P_2}A_{P_2}x_7 \\ \frac{-4\beta_{e_1}A_{p_1}}{V_{t_1}}r_1x_2 + \frac{4\beta_{e_1}C_{tl_1}}{V_{t_1}}x_5 + \frac{4\beta_{e_1}C_{d_1}w_1}{V_{t_1}\sqrt{\rho_1}}\sqrt{P_{s_1} - \tanh(k_1x_6)x_5}x_6 \\ -\frac{1}{T_{sv_1}}x_6 \\ \frac{-4\beta_{e_2}A_{p2}}{V_{t_2}}r_2x_4 + \frac{4\beta_{e_2}C_{tl_2}}{V_{t_2}}x_7 + \frac{4\beta_{e_2}C_{d_2}w_2}{V_{t_2}\sqrt{\rho_2}}\sqrt{P_{s_2} - \tanh(k_2x_8)x_7}x_8 \\ -\frac{1}{T_{sv_2}}x_{x_8} \end{pmatrix} \begin{pmatrix} 0 & 0 \\ 0 & 0 \\ 0 & 0 \\ 0 & 0 \\ 0 & 0 \\ \frac{K_{sv_1}}{T_{sv_1}} & 0 \\ 0 & 0 \\ 0 & \frac{K_{sv_2}}{T_{sv_2}} \end{pmatrix}$$

2.8.3 Differential flatness properties of the hydraulic robotic manipulator

It can be proven that the dynamic model of the electro-hydraulic robotic manipulator is differentially flat. The differential flatness property signifies that: (i) all state variables of the system and its control inputs can be written as differential functions of a subset of its state variables which are the system's flat outputs; (ii) the flat outputs are not connected through a relation in the form of an homogeneous differential equation, thus being differentially independent (Rigatos, 2015).

The differential flatness property allows to define setpoints for the control loop of the robotic manipulator. Actually, one can primarily define setpoints for the state variables which stand for the flat outputs of the robotic system, and through them can compute the setpoints' functions for the rest of the state variables of the robot.

The flat outputs of the electro-hydraulic robotic manipulator are taken to be $y_1 = x_1$ and $y_2 = x_3$, that is the two turn angles of the robot's joints. From the first row of the state-space model, one has

$$x_2 = \dot{x}_1 \Rightarrow x_2 = \dot{y}_1 \tag{2.296}$$

From the third row of the state-space model, one has

$$x_4 = \dot{x}_3 \Rightarrow x_4 = \dot{y}_2 \tag{2.297}$$

Thus, both state variables x_2 and x_4 are differential functions of the system's flat outputs. Next it is confirmed that the elements of the inertia matrix $M(\theta)$, of the Coriolis matrix $C(\theta,\dot{\theta})$ and of the gravitational matrix $G(\theta)$, as well as the determinant of the inertia matrix $detM$ are written as differential functions of the flat outputs of the system.

Additionally, from the second and the fourth row of the robot's state-space model, one has a system of equations with respect to state variables x_5 and x_7. Thus one obtains

$$\begin{pmatrix} \frac{M_{22}}{detM}K_{P_1}A_{P_1} & -\frac{M_{12}}{detM}K_{P_2}A_{P_2} \\ -\frac{M_{12}}{detM}K_{P_1}A_{P_1} & \frac{M_{11}}{detM}K_{P_2}A_{P_2} \end{pmatrix} \begin{pmatrix} x_5 \\ x_7 \end{pmatrix} = \begin{pmatrix} \dot{x}_2 - \frac{M_{22}C_1 - M_{12}C_2 + M_{22}G_1 - M_{12}G_2}{detM} \\ \dot{x}_4 - \frac{-M_{12}C_1 + M_{11}C_2 - M_{12}G_1 + M_{11}G_2}{detM} \end{pmatrix} \tag{2.298}$$

or equivalently

$$\begin{pmatrix} x_5 \\ x_7 \end{pmatrix} = \begin{pmatrix} \frac{M_{22}}{detM} K_{P_1} A_{P_1} & -\frac{M_{12}}{detM} K_{P_2} A_{P_2} \\ -\frac{M_{12}}{detM} K_{P_1} A_{P_1} & \frac{M_{11}}{detM} K_{P_2} A_{P_2} \end{pmatrix}^{-1} \cdot \begin{pmatrix} \dot{x}_2 - \frac{M_{22}C_1 - M_{12}C_2 + M_{22}G_1 - M_{12}G_2}{detM} \\ \dot{x}_4 - \frac{-M_{12}C_1 + M_{11}C_2 - M_{12}G_1 + M_{11}G_2}{detM} \end{pmatrix} \tag{2.299}$$

Consequently, it can be inferred that state variables x_5 and x_7 are also written as differential functions of the system's flat outputs. Moreover, from the fifth and the seventh row of the state-space model of the robotic manipulator one solves with respect to state variables x_6 and x_8. Actually, from the fifth row of the state-space model, one has

$$\sqrt{P_{s_1} - \tanh(k_1 x_6) x_5} x_6 = \frac{\dot{x}_5 + \frac{4\beta_{e_1} A_{p_1}}{V_{t_1}} r_1 x_2 - \frac{4\beta_{e_1} C_{tl_1}}{V_{t_1}} x_5}{\frac{4\beta_{e_1} C_{d_1} w_1}{V_{t_1} \sqrt{\rho_1}}} \tag{2.300}$$

Next, by defining as $h_1(x, \dot{x})$ the fraction which appears in the right-hand part of (2.300) and which does not include x_6 or x_8, one obtains

$$\sqrt{P_{s_1} - \tanh(k_1 x_6) x_5} x_6 = h_1(x, \dot{x}) \tag{2.301}$$

which signifies that state variable x_6 is a differential function of the system's flat output. Moreover, from the seventh row of the state-space model, one obtains

$$\sqrt{P_{s_2} - \tanh(k_2 x_8) x_7} x_8 = \frac{\dot{x}_7 + \frac{4\beta_{e_2} A_{p_2}}{V_{t_2}} r_2 x_4 - \frac{4\beta_{e_2} C_{tl_2}}{V_{t_2}} x_7}{\frac{4\beta_{e_2} C_{d_2} w_2}{V_{t_2} \sqrt{\rho_2}}} \tag{2.302}$$

Next, by defining as $h_2(x, \dot{x})$ the fraction which appears in the right-hand part of (2.300) and which does not include x_6 or x_8, one obtains

$$\sqrt{P_{s_2} - \tanh(k_2 x_8) x_7} x_8 = h_2(x, \dot{x}) \tag{2.303}$$

which signifies that state variable x_8 is a differential function of the system's flat output. Finally, from the sixth and the eighth row of the state-space model of the robotic manipulator one can solve with respect to the control inputs u_1 and u_2. This gives

$$u_1 = \frac{T_{sv_1} \dot{x}_6 + x_6}{K_{sv_1}} \tag{2.304}$$

$$u_2 = \frac{T_{sv_2} \dot{x}_8 + x_8}{K_{sv_2}} \tag{2.305}$$

These equations come to complete the proof of the differential flatness property of the state-space model of the multi-DOF (2-DOF) electro-hydraulic robotic manipulator.

By defining setpoints x_1^d and x_3^d for the state variables x_1 and x_3 which are also the flat outputs of the robotic model, one can compute next the functions of the setpoints of the rest of the state variables of the system, with the use of (2.296), (2.297), (2.299), (2.301), and (2.303).

2.8.4 *Approximate linearization of the electro-hydraulic manipulator*

2.8.4.1 Approximately linearized model

The dynamic model of the electro-hydraulic manipulator undergoes approximate linearization around the temporary operating point (x^*, u^*), where x^* is the present value of the state vector and u^* is the last sampled value of the control inputs vector. The linearization takes place at each time-step of the control algorithm and relies on first-order Taylor series expansion and on the computation of the related Jacobian matrices. The initial nonlinear model of the robotic manipulator is in the state-space form of (2.295), while after the linearization process, the equivalent state-space model becomes

$$\dot{x} = Ax + Bu + \tilde{d} \tag{2.306}$$

where $\tilde{d}$ is the cumulative disturbances vector which may comprise (i) modeling errors due to the truncation of higher-order terms from the Taylor series expansion, (ii) exogenous perturbations, and (iii) measurement noise of any distribution. Matrices A and B are computed from the system's Jacobians and are given by

$$A = \nabla_x[f(x) + g(x)u] \mid_{(x^*,u^*)} \Rightarrow A = \nabla_x[f(x)] \mid_{(x^*,u^*)} \tag{2.307}$$

$$B = \nabla_u[f(x) + g(x)u] \mid_{(x^*,u^*)} \Rightarrow B = g(x) \mid_{(x^*,u^*)} \tag{2.308}$$

Next, the elements of the Jacobian matrix $A = \nabla_x[f(x)] \mid_{(x^*,u^*)}$ are computed.

First row of the Jacobian matrix $A = \nabla_x[f(x)] \mid_{(x^*,u^*)}$: $\frac{\partial f_1}{\partial x_1} = 0$, $\frac{\partial f_1}{\partial x_2} = 1$, $\frac{\partial f_1}{\partial x_3} = 0$, $\frac{\partial f_1}{\partial x_4} = 0$, $\frac{\partial f_1}{\partial x_5} = 0$, $\frac{\partial f_1}{\partial x_6} = 0$, $\frac{\partial f_1}{\partial x_7} = 0$, $\frac{\partial f_1}{\partial x_8} = 0$.

Second row of the Jacobian matrix $A = \nabla_x[f(x)] \mid_{(x^*,u^*)}$: For $i = 1, 2, \cdots, 8$ and by defining $N_1 = K_{P_1}A_{P_1}x_5$, $N_2 = K_{P_2}A_{P_2}x_7$ and $f_{2n} = M_{22}C_1 - M_{12}C_2 + M_{22}G_1 - M_{12}G_2 + M_{22}N_1 - M_{12}N_2$ it holds that

$$\frac{\partial f_2}{\partial x_i} = \frac{\frac{\partial}{\partial x_i}[f_{2n}] - f_{2n}\frac{\partial detM}{\partial x_i}}{[detM]^2} \tag{2.309}$$

About $detM$ it holds that $detM = M_{11}M_{22} - M_{12}^2$, thus

$$\frac{\partial detM}{\partial x_i} = \frac{\partial M_{11}}{\partial x_i}M_{22} + M_{11}\frac{\partial M_{22}}{\partial x_i} - 2M_{12}\frac{\partial M_{12}}{\partial x_i} \tag{2.310}$$

with $M_{11} = (m_1 + m_2)l_1^2 + m_1 l_2^2 + 2m_2 l_1 l_2 \cos(x_3)$, thus $\frac{\partial M_{11}}{\partial x_1} = 0$, $\frac{\partial M_{11}}{\partial x_2} = 0$, $\frac{\partial M_{11}}{\partial x_3} = -2m_2 l_1 l_2 \sin(x_3)$, $\frac{\partial M_{11}}{\partial x_4} = 0$, $\frac{\partial M_{11}}{\partial x_5} = 0$, $\frac{\partial M_{11}}{\partial x_16} = 0$, $\frac{\partial M_{11}}{\partial x_7} = 0$ and $\frac{\partial M_{11}}{\partial x_8} = 0$.

and $M_{22}=m_2l_2(l_1+l_2)\cos(x_3)$, thus $\frac{\partial M_{22}}{\partial x_1}=0$, $\frac{\partial M_{22}}{\partial x_2}=0$, $\frac{\partial M_{22}}{\partial x_3}=-m_2l_2(l_1+l_2)\sin(x_3)$, $\frac{\partial M_{22}}{\partial x_4}=0$, $\frac{\partial M_{22}}{\partial x_5}=0$, $\frac{\partial M_{22}}{\partial x_16}=0$, $\frac{\partial M_{22}}{\partial x_7}=0$ and $\frac{\partial M_{22}}{\partial x_8}=0$.

and $M_{12}=m_2l_2^2+m_2l_1l_2\cos(x_3)$ thus $\frac{\partial M_{12}}{\partial x_1}=0$, $\frac{\partial M_{12}}{\partial x_2}=0$, $\frac{\partial M_{12}}{\partial x_3}=-m_2l_1l_2\sin(x_3)$, $\frac{\partial M_{12}}{\partial x_4}=0$, $\frac{\partial M_{12}}{\partial x_5}=0$, $\frac{\partial M_{12}}{\partial x_16}=0$, $\frac{\partial M_{12}}{\partial x_7}=0$ and $\frac{\partial M_{12}}{\partial x_8}=0$.

About the differentiation of the numerator's function $f_{2n}=[M_{22}C_1-M_{12}C_2+M_{22}G_1-M_{12}G_2+M_{22}N_1-M_{12}N_2]$ one has

$$\begin{aligned}
&\frac{\partial}{\partial x_i}[M_{22}C_1-M_{12}C_2+M_{22}G_1-M_{12}G_2]\\
&\quad=\frac{\partial M_{22}}{\partial x_i}C_1+M_{22}\frac{\partial C_1}{\partial x_i}-\frac{\partial M_{12}}{\partial x_i}C_2-M_{12}\frac{\partial C_2}{\partial x_i}\\
&\quad+\frac{\partial M_{22}}{\partial x_i}G_1+M_{22}\frac{\partial G_1}{\partial x_i}-\frac{\partial M_{12}}{\partial x_i}G_2-M_{12}\frac{\partial G_2}{\partial x_i}\\
&\quad+\frac{\partial M_{22}}{\partial x_i}N_1+M_{22}\frac{\partial N_1}{\partial x_i}-\frac{\partial M_{12}}{\partial x_i}N_2-M_{12}\frac{\partial N_2}{\partial x_i}
\end{aligned}\tag{2.311}$$

It holds that $C_1=-m_2l_1^2\sin(x_1)x_4^2-2m_2l_1^2\sin(x_3)x_2x_4$, thus $\frac{\partial C_1}{\partial x_1}=-m_2l_1^2os(x_2)x_4^2$, $\frac{\partial C_1}{\partial x_2}=-2m_2l_1^2\sin(x_3)x_4$, $\frac{\partial C_1}{\partial x_3}=-2m_2l_1^2\cos(x_3)x_2x_4$,
$\frac{\partial C_1}{\partial x_4}=-2m_2l_1^2\sin(x_1)x_4-2m_2l_1^2\sin(x_3)x_2$, $\frac{\partial C_1}{\partial x_5}=0$, $\frac{\partial C_1}{\partial x_6}=0$, $\frac{\partial C_1}{\partial x_7}=0$ and $\frac{\partial C_1}{\partial x_8}=0$.

It holds that $C_2=m_2l_1^2\sin(x_1)x_2^2$, thus $\frac{\partial C_2}{\partial x_1}=m_2l_1^2\cos(x_1)x_2^2$, $\frac{\partial C_2}{\partial x_2}=2m_2l_1^2\sin(x_1)x_2$, $\frac{\partial C_2}{\partial x_3}=0$, $\frac{\partial C_2}{\partial x_4}=0$, $\frac{\partial C_2}{\partial x_5}=0$, $\frac{\partial C_2}{\partial x_6}=0$, $\frac{\partial C_2}{\partial x_7}=0$, $\frac{\partial C_2}{\partial x_8}=0$ and $\frac{\partial C_2}{\partial x_9}=0$.

It holds that $G_1=(m_1+m_2)gl_1\cos(x_1)+m_2gl_2\cos(x_1+x_3)$ thus $\frac{\partial G_1}{\partial x_1}=-(m_1+m_2)gl_1\sin(x_1)-m_2gl_2\sin(x_1+x_2)$, $\frac{\partial G_1}{\partial x_2}=0$, $\frac{\partial G_1}{\partial x_3}=-m_2gl_2\sin(x_1+x_3)$, $\frac{\partial G_1}{\partial x_4}=0$, $\frac{\partial G_1}{\partial x_5}=0$, $\frac{\partial G_1}{\partial x_6}=0$, $\frac{\partial G_1}{\partial x_7}=0$ and $\frac{\partial G_1}{\partial x_8}=0$.

It holds that $G_2=m_2gl_2\cos(x_1+x_3)$ thus $\frac{\partial G_2}{\partial x_1}=-m_2gl_2\sin(x_1+x_3)$, $\frac{\partial G_2}{\partial x_2}=0$, $\frac{\partial G_2}{\partial x_3}=-m_2gl_2\sin(x_1+x_3)$, $\frac{\partial_2}{\partial x_4}=0$, $\frac{\partial G_2}{\partial x_5}=0$, $\frac{\partial G_2}{\partial x_6}=0$, $\frac{\partial G_2}{\partial x_7}=0$, $\frac{\partial G_2}{\partial x_8}=0$ and $\frac{\partial C_2}{\partial x_9}=0$.

It holds that $N_1=K_{P_1}A_{P_1}x_5$ thus $\frac{\partial N_1}{\partial x_1}=0$, $\frac{\partial N_1}{\partial x_2}=0$, $\frac{\partial N_1}{\partial x_3}=0$, $\frac{\partial N_1}{\partial x_4}=0$, $\frac{\partial N_1}{\partial x_5}=K_{P_1}A_{P_1}$, $\frac{\partial N_1}{\partial x_6}=0$, $\frac{\partial N_1}{\partial x_7}=0$ and $\frac{\partial N_1}{\partial x_8}=0$.

It holds that $N_2=K_{P_2}A_{P_2}x_7$ thus $\frac{\partial N_2}{\partial x_1}=0$, $\frac{\partial N_2}{\partial x_2}=0$, $\frac{\partial N_2}{\partial x_3}=0$, $\frac{\partial N_2}{\partial x_4}=0$, $\frac{\partial N_2}{\partial x_5}=0$, $\frac{\partial N_2}{\partial x_6}=0$, $\frac{\partial N_2}{\partial x_7}=K_{P_2}A_{P_2}$ and $\frac{\partial N_2}{\partial x_8}=0$.

Third row of the Jacobian matrix $A=\nabla_x[f(x)]\mid_{(x^*,u^*)}$: $\frac{\partial f_3}{\partial x_1}=0$, $\frac{\partial f_3}{\partial x_2}=0$, $\frac{\partial f_3}{\partial x_3}=0$, $\frac{\partial f_3}{\partial x_4}=1$, $\frac{\partial f_3}{\partial x_5}=0$, $\frac{\partial f_3}{\partial x_6}=0$, $\frac{\partial f_3}{\partial x_7}=0$, $\frac{\partial f_3}{\partial x_8}=0$.

Fourth row of the Jacobian matrix $A = \nabla_x[f(x)]\mid_{(x^*,u^*)}$: by defining again $N_1 = K_{P_1}A_{P_1}x_5$, $N_2 = K_{P_2}A_{P_2}x_7$ and the numerator function $f_{4n} = [-M_{12}C_1 + M_{11}C_2 - M_{12}G_1 + M_{11}G_2]$, then for $i = 1, 2, \cdots, 8$, one has that:

$$\frac{\partial f_4}{\partial x_i} = \frac{\partial f_{2n}}{\partial x_i}detM - f_{2n}\frac{\partial detM}{\partial x_i} \tag{2.312}$$

About the differentiation with respect to x_i of the numerator $f_{4n} = [-M_{12}C_1 + M_{11}C_2 - M_{12}G_1 + M_{11}G_2 - M_{12}N_1 + M_{11}N_2]]$, one has

$$\begin{aligned}&\frac{\partial}{\partial x_i}[-M_{12}C_1 - M_{11}C_2 + M_{22}G_1 + M_{11}G_2]\\ &\quad= -\frac{\partial M_{11}}{\partial x_i}C_1 - M_{22}\frac{\partial C_1}{\partial x_i} + \frac{\partial M_{11}}{\partial x_i}C_2 + M_{11}\frac{\partial C_2}{\partial x_i}\\ &\quad -\frac{\partial M_{12}}{\partial x_i}G_1 - M_{21}\frac{\partial G_1}{\partial x_i} + \frac{\partial M_{11}}{\partial x_i}G_2 = M_{11}\frac{\partial G_2}{\partial x_i}+\\ &\quad -\frac{\partial M_{12}}{\partial x_i}N_1 - M_{21}\frac{\partial N_1}{\partial x_i} + \frac{\partial M_{11}}{\partial x_i}N_2 = M_{11}\frac{\partial N_2}{\partial x_i}\end{aligned} \tag{2.313}$$

where the above noted partial derivatives were computed before.

Fifth row of the Jacobian matrix $A = \nabla_x[f(x)]\mid_{(x^*,u^*)}$: $\frac{\partial f_5}{\partial x_1} = 0$, $\frac{\partial f_5}{\partial x_2} = -\frac{4\beta_{e_1}A_{P_1}}{V_{t_1}}r_1$, $\frac{\partial f_5}{\partial x_3} = 0$, $\frac{\partial f_5}{\partial x_4} = 0$, $\frac{\partial f_5}{\partial x_5} = \frac{4\beta_{e_1}A_{P_1}}{V_{t_1}} + \frac{4\beta_{e_1}C_{d_1}w_1}{V_{t_1}\sqrt{\rho_1}}\frac{1}{2}\frac{1}{\sqrt{P_{s_1} - \tanh(k_1x_6)x_5}}(-\tanh(k_1x_6))x_6$, $\frac{\partial f_5}{\partial x_6} = \frac{4\beta_{e_1}C_{d_1}w_1}{V_{t_1}\sqrt{\rho_1}}\frac{1}{2}\cdot\frac{1}{\sqrt{P_{s_1} - \tanh(k_1x_6)x_5}}\cdot\{-k_1 + k_1[\tanh(k_1x_6)]^2\}x_5x_6 +$ $+\frac{4\beta_{e_1}C_{d_1}w_1}{V_{t_1}\sqrt{\rho_1}}\sqrt{P_{s_1} - \tanh(k_1x_6)x_5}$, $\frac{\partial f_5}{\partial x_7} = 0$, $\frac{\partial f_5}{\partial x_8} = 0$.

Sixth row of the Jacobian matrix $A = \nabla_x[f(x)]\mid_{(x^*,u^*)}$: $\frac{\partial f_6}{\partial x_1} = 0$, $\frac{\partial f_6}{\partial x_2} = 0$, $\frac{\partial f_6}{\partial x_3} = 0$, $\frac{\partial f_6}{\partial x_4} = 0$, $\frac{\partial f_6}{\partial x_5} = 0$, $\frac{\partial f_6}{\partial x_6} = -\frac{1}{T_{sv_1}}$, $\frac{\partial f_6}{\partial x_7} = 0$, $\frac{\partial f_6}{\partial x_8} = 0$.

Seventh row of the Jacobian $A = \nabla_x[f(x)]\mid_{(x^*,u^*)}$: $\frac{\partial f_7}{\partial x_1} = 0$, $\frac{\partial f_7}{\partial x_2} = 0$, $\frac{\partial f_7}{\partial x_3} = 0$, $\frac{\partial f_5}{\partial x_4} = -\frac{4\beta_{e_2}A_{P_2}}{V_{t_2}}r_2$, $\frac{\partial f_7}{\partial x_5} = 0$, $\frac{\partial f_7}{\partial x_6} = 0$, $\frac{\partial f_5}{\partial x_7} = \frac{4\beta_{e_2}A_{P_2}}{V_{t_2}} + \frac{4\beta_{e_2}C_{d_2}w_2}{V_{t_2}\sqrt{\rho_2}}\frac{1}{2}\cdot\frac{1}{\sqrt{P_{s_2} - \tanh(k_2x_8)x_7}}$ $(-\tanh(k_2x_8))x_8$ and $\frac{\partial f_5}{\partial x_8} = \frac{4\beta_{e_2}C_{d_2}w_2}{V_{t_2}\sqrt{\rho_2}}\frac{1}{2}\cdot\frac{1}{\sqrt{P_{s_2} - \tanh(k_2x_8)x_5}}\cdot\{-k_2 + k_2[\tanh(k_2x_8)]^2\}x_7x_8 +$ $+\frac{4\beta_{e_2}C_{d_2}w_2}{V_{t_2}\sqrt{\rho_2}}\sqrt{P_{s_2} - \tanh(k_2x_7)}$.

Eight row of the Jacobian matrix $A = \nabla_x[f(x)]\mid_{(x^*,u^*)}$: $\frac{\partial f_8}{\partial x_1} = 0$, $\frac{\partial f_8}{\partial x_2} = 0$, $\frac{\partial f_8}{\partial x_3} = 0$, $\frac{\partial f_8}{\partial x_4} = 0$, $\frac{\partial f_8}{\partial x_5} = 0$, $\frac{\partial f_8}{\partial x_6} = 0$, $\frac{\partial f_8}{\partial x_8} = -\frac{1}{T_{sv_2}}$.

2.8.4.2 Stabilizing feedback control

After linearization around its current operating point, the dynamic model of the electro-hydraulic robotic manipulator is written as (Rigatos and Busawon, 2018)

$$\dot{x} = Ax + Bu + d_1 \tag{2.314}$$

Parameter d_1 stands for the linearization error in the electro-hydraulic robotic manipulator's dynamic model appearing above in (2.314). The reference setpoints for the state vector of the electro-hydraulic robotic manipulator are denoted by $\mathbf{x_d} = [x_1^d, \cdots, x_8^d]$. Tracking of this trajectory is achieved after applying the control input u_d. At every time instant the control input u_d is assumed to differ from the control input u appearing in (2.314) by an amount equal to Δu, that is $u_d = u + \Delta u$

$$\dot{x}_d = Ax_d + Bu_d + d_2 \tag{2.315}$$

The dynamics of the controlled system described in (2.314) can be also written as

$$\dot{x} = Ax + Bu + Bu_d - Bu_d + d_1 \tag{2.316}$$

and by denoting $d_3 = -Bu_d + d_1$ as an aggregate disturbance term one obtains

$$\dot{x} = Ax + Bu + Bu_d + d_3 \tag{2.317}$$

By subtracting (2.315) from (2.317), one has

$$\dot{x} - \dot{x}_d = A(x - x_d) + Bu + d_3 - d_2 \tag{2.318}$$

By denoting the tracking error as $e = x - x_d$ and the aggregate disturbance term as $\tilde{d} = d_3 - d_2$, the tracking error dynamics becomes

$$\dot{e} = Ae + Bu + \tilde{d} \tag{2.319}$$

For the approximately linearized model of the system a stabilizing feedback controller is developed. The controller has the form

$$u(t) = -Ke(t) \tag{2.320}$$

with $K = \frac{1}{r}B^TP$ where P is a positive definite symmetric matrix which is obtained from the solution of the Riccati equation (Rigatos and Busawon, 2018)

$$A^TP + PA + Q - P\left(\frac{2}{r}BB^T - \frac{1}{\rho^2}LL^T\right)P = 0 \tag{2.321}$$

where Q is a positive semi-definite symmetric matrix. The diagram of the considered control loop is depicted in Figure 2.89. The global stability properties of the control method can be proven following the procedure presented in Section 1.1.2.5.

The solution of the H-infinity feedback control problem for the multi-DOF electro-hydraulic robotic manipulator and the computation of the worst case disturbance that this controller can sustain comes from superposition of Bellman's optimality principle when considering that the robot is affected by two separate inputs (i) the control input u and (ii) the cumulative disturbance input $\tilde{d}(t)$. Solving the optimal control problem for u that is for the minimum variation (optimal) control input that achieves elimination of the state vector's tracking error gives $u = -\frac{1}{r}B^TPe$. Equivalently, solving the optimal control problem for $\tilde{d}$, that is for the worst case disturbance that the control loop can sustain gives $\tilde{d} = \frac{1}{\rho^2}L^TPe$.

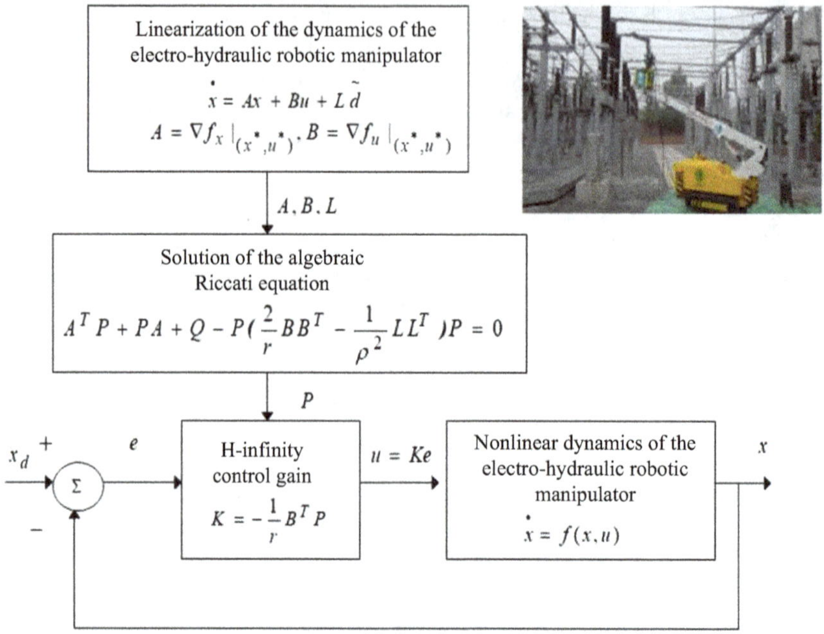

Figure 2.89 Diagram of the nonlinear optimal (H-infinity) control loop for the 2-DOF electro-hydraulic manipulator

2.8.5 Simulation tests

The performance of the proposed nonlinear optimal control scheme for the multi-DOF (2-DOF) electro-hydraulic manipulator has been tested and confirmed through simulation experiments. To select the feedback gains of the controller the algebraic Riccati equation of (2.321) had to be repetitively solve at each time-step of the control method. The obtained results are presented in Figures 2.90 to 2.105. The real values of the state variables of the robotic manipulator are plotted in blue, the estimated values which were provided by the H-infinity Kalman Filter are printed in green, while the associated reference setpoints are depicted in red. Through the provided results it can be noticed that fast and accurate tracking of the reference setpoints has been achieved for all state variables of the hydraulic manipulator, while the control inputs exhibited moderate variations.

2.9 Nonlinear control of robots with electropneumatic actuators

2.9.1 Outline

To provide high torques needed to move a robot's links, electric actuators are followed by a transmission system with a high transmission rate. For instance, gear ratios of

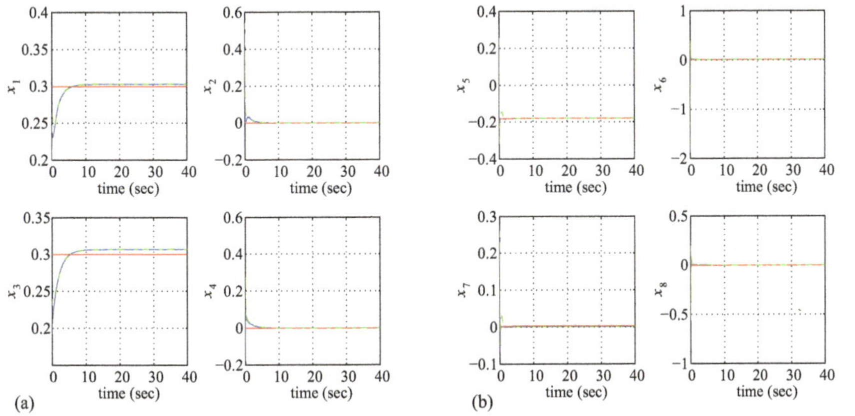

Figure 2.90 *Tracking of setpoint 1 by the electro-hydraulic robotic manipulator: (a) convergence of the state variables x_1 to x_4 (blue lines) to the reference setpoints (red lines) and KF-estimated state variables (green lines); (b) convergence of the state variables x_5 to x_8 to the reference setpoints*

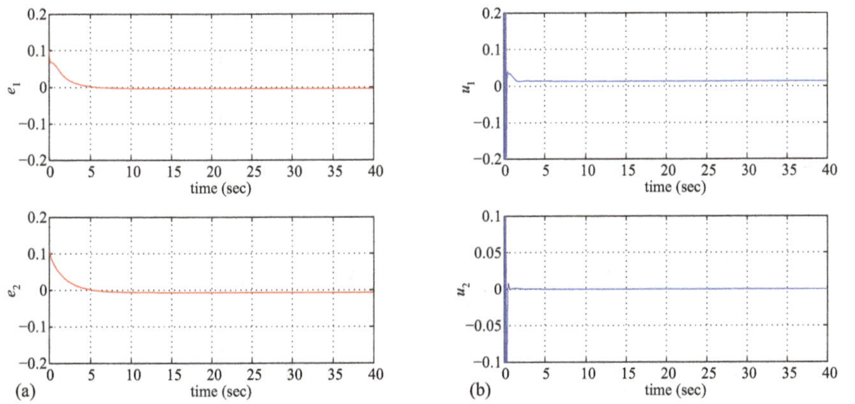

Figure 2.91 *Tracking of setpoint 1 by the electro-hydraulic robotic manipulator: (a) tracking error for state variables x_1 and x_3 (joint angles of the robot); (b) control inputs u_1 and u_2 generated by the electro-hydraulic actuators*

100:1 are often used in the joints of a robotic manipulator. This results into an actuator with large mechanical impedance (also known as non-back-drivable actuator). This in turn generates high contact forces when collision of the robotic mechanism occur and can cause humans' injury. Another disadvantage of electric actuators is that they can exhibit overheating when constant torques have to be provided. Comparing to electric

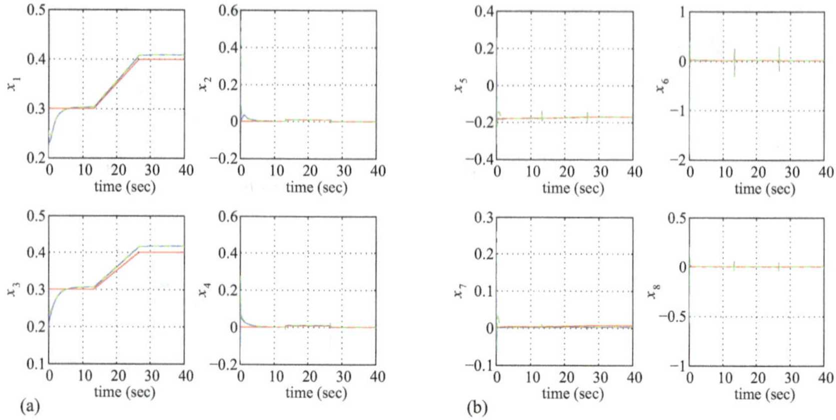

Figure 2.92 *Tracking of setpoint 2 by the electro-hydraulic robotic manipulator: (a) convergence of the state variables x_1 to x_4 (blue lines) to the reference setpoints (red lines) and KF-estimated state variables (green lines); (b) convergence of the state variables x_5 to x_8 to the reference setpoints*

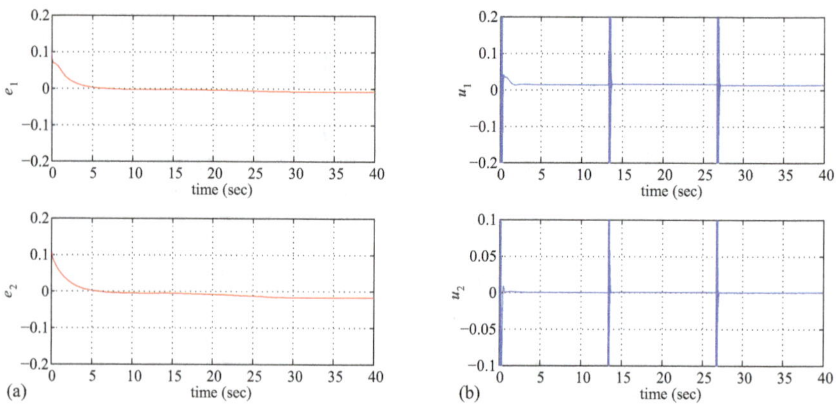

Figure 2.93 *Tracking of setpoint 2 by the electro-hydraulic robotic manipulator: (a) tracking error for state variables x_1 and x_3 (joint angles of the robot); (b) control inputs u_1 and u_2 generated by the electro-hydraulic actuators*

actuators, pneumatic actuators have promising properties for robotic applications, due to their low weight, simple mechanical design, low cost, and good power-to-weight ratio (Rigatos and Busawon, 2018; Chalaga and Plestan, 2021; Smaoui, Brun *et al.*, 2006a; Smaoui, Brun *et al.*, 2006b). Electropneumatically actuated robots usually have better friction properties (Le, Pham *et al.*, 2013; Behrad *et al.*, 2019;

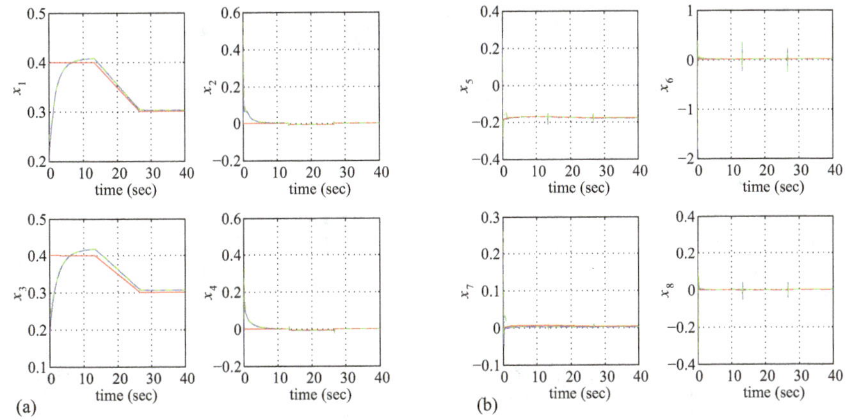

Figure 2.94 Tracking of setpoint 3 by the electro-hydraulic robotic manipulator: (a) convergence of the state variables x_1 to x_4 (blue lines) to the reference setpoints (red lines) and KF-estimated state variables (green lines); (b) convergence of the state variables x_5 to x_8 to the reference setpoints

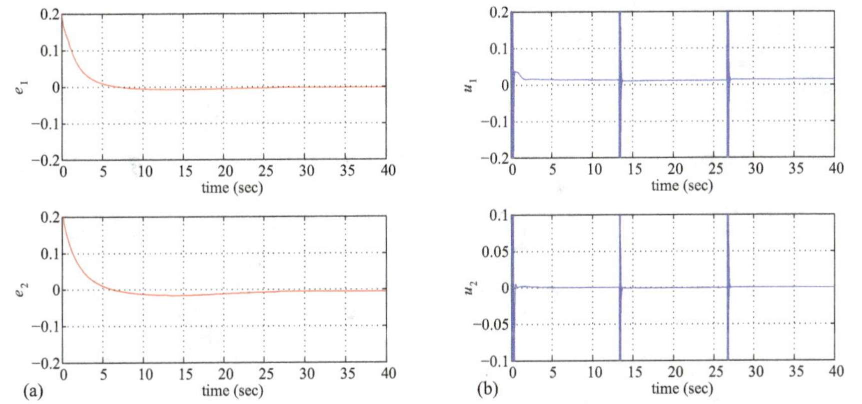

Figure 2.95 Tracking of setpoint 3 by the electro-hydraulic robotic manipulator: (a) tracking error for state variables x_1 and x_3 (joint angles of the robot); (b) control inputs u_1 and u_2 generated by the electro-hydraulic actuators

Rigatos, Zervos *et al.*, 2020; Van Damme *et al.*, 2009). Moreover, because of law mechanical impedance, pneumatic robots can provide moderate interaction forces which is important for robotic surgery and rehabilitation tasks (Toedtheide *et al.*, 2016; Kim, Chang *et al.*, 2008; Trumic *et al.*, 2020). Pneumatic actuators are also well suited for exoskeleton robots. Actuation in exoskeletons should have a fast and

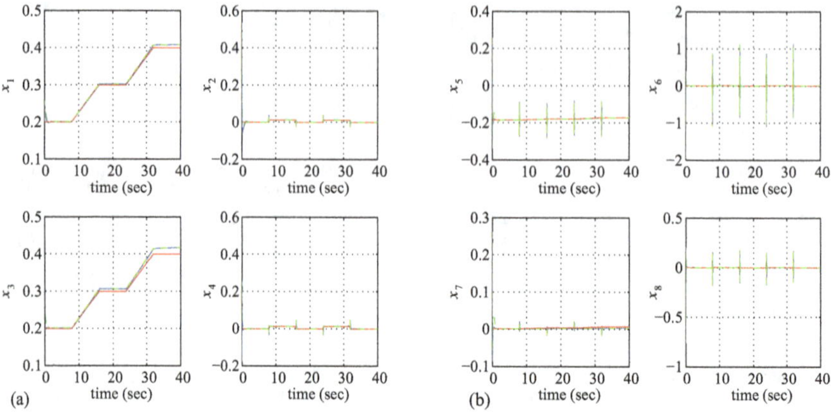

Figure 2.96 *Tracking of setpoint 4 by the electro-hydraulic robotic manipulator: (a) convergence of the state variables x_1 to x_4 (blue lines) to the reference setpoints (red lines) and KF-estimated state variables (green lines); (b) convergence of the state variables x_5 to x_8 to the reference setpoints*

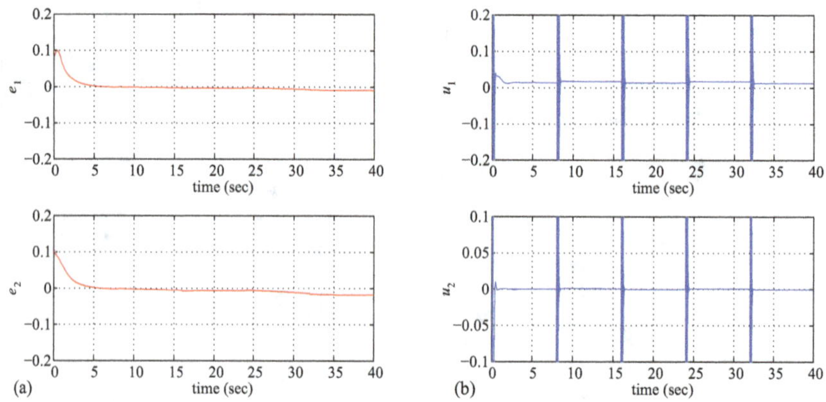

Figure 2.97 *Tracking of setpoint 4 by the electro-hydraulic robotic manipulator: (a) tracking error for state variables x_1 and x_3 (joint angles of the robot); (b) control inputs u_1 and u_2 generated by the electro-hydraulic actuators*

accurate response (Rigatos, Abbaszadeh *et al.*, 2020; Cao *et al.*, 2018; Varghese *et al.*, 2012; Cao and Huang, 2020). While electric motors come against high mechanical impedance and the risk of causing injuries, pneumatic actuators exhibit forces and torques which stay within moderate variation ranges (Hoang *et al.*, 2020; Hilerbrandth *et al.*, 2010; Liu, Guon *et al.*, 2018). Besides, unlike direct current electric motors,

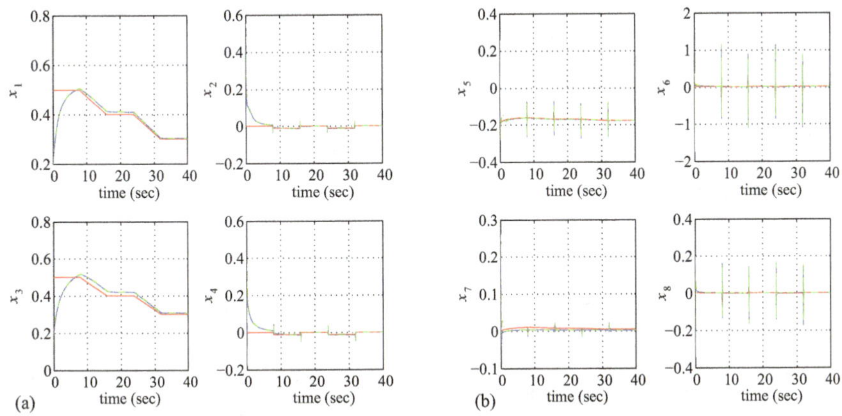

Figure 2.98 Tracking of setpoint 5 by the electro-hydraulic robotic manipulator: (a) convergence of the state variables x_1 to x_4 (blue lines) to the reference setpoints (red lines) and KF-estimated state variables (green lines); (b) convergence of the state variables x_5 to x_8 to the reference setpoints

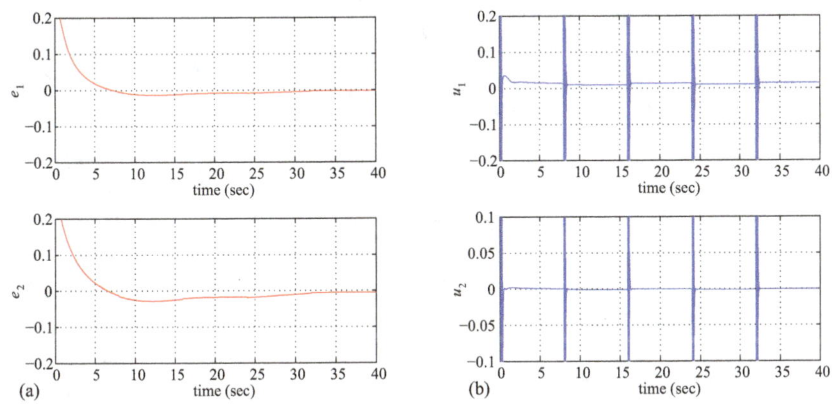

Figure 2.99 Tracking of setpoint 5 by the electro-hydraulic robotic manipulator: (a) tracking error for state variables x_1 and x_3 (joint angles of the robot); (b) control inputs u_1 and u_2 generated by the electro-hydraulic actuators

pneumatic actuators have an improved weight-to-power ratio and avoid overheating problems (Dong, Shi *et al.*, 2021; Sasaki and Kawasima, 2008; Zilic *et al.*, 2009).

However, pneumatic actuators are characterized by high nonlinearities which are due to air compressibility, thermodynamics and valves behavior and thus pneumatic robots require elaborated nonlinear control schemes to ensure their fast and precise

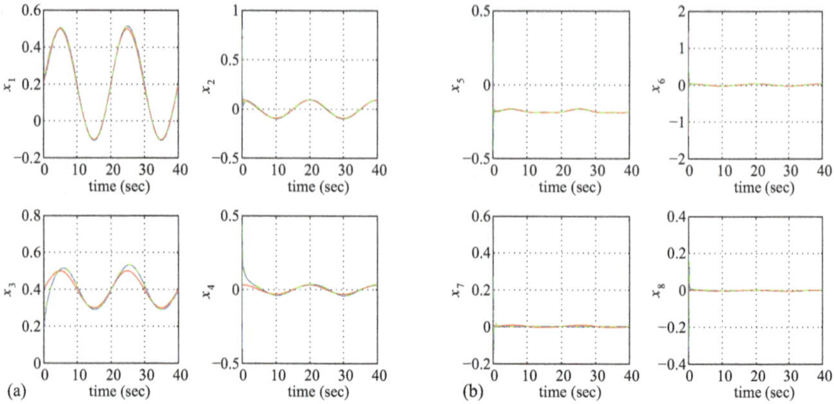

Figure 2.100 Tracking of setpoint 6 by the electro-hydraulic robotic manipulator: (a) convergence of the state variables x_1 to x_4 (blue lines) to the reference setpoints (red lines) and KF-estimated state variables (green lines); (b) convergence of the state variables x_5 to x_8 to the reference setpoints

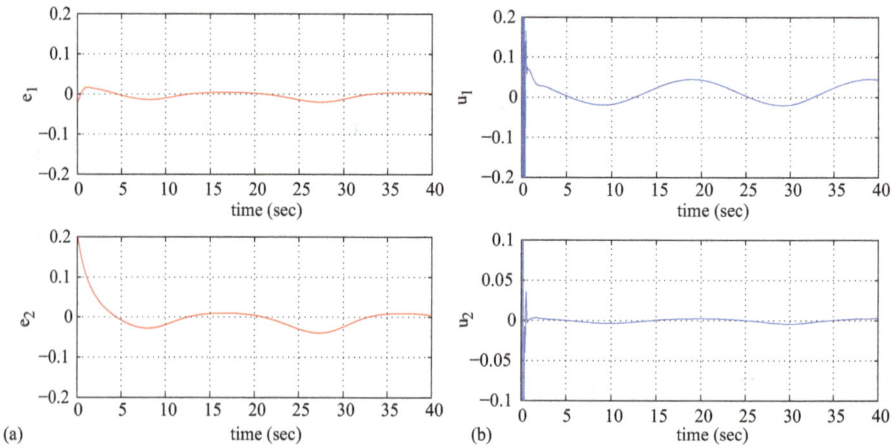

Figure 2.101 Tracking of setpoint 6 by the electro-hydraulic robotic manipulator: (a) tracking error for state variables x_1 and x_3 (joint angles of the robot); (b) control inputs u_1 and u_2 generated by the electro-hydraulic actuators

positioning (Schulter and Guelton, 2009; Bone *et al.*, 2015; Lofimejani *et al.*, 2016). Among the control methods which have been applied to pneumatic robots one can distinguish differential geometric approaches (Lie algebra-based control), differential flatness theory-based control, nonlinear model predictive control, sliding-mode control, backstepping control and multiple models-based fuzzy control (Tuvayonond

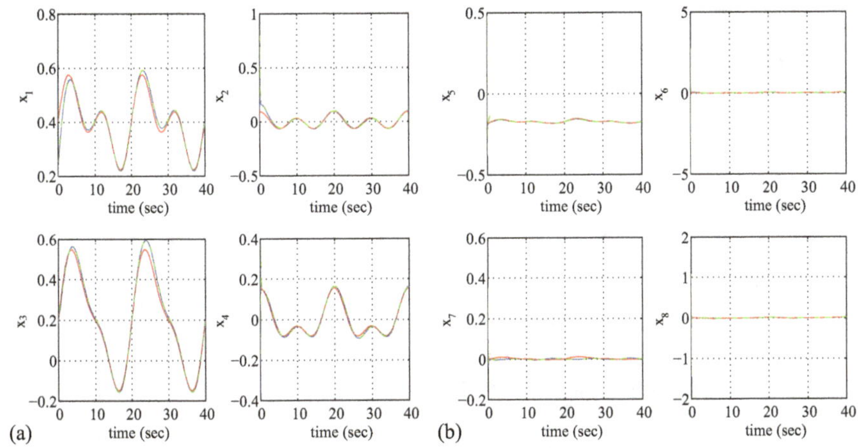

Figure 2.102 Tracking of setpoint 7 by the electro-hydraulic robotic manipulator: (a) convergence of the state variables x_1 to x_4 (blue lines) to the reference setpoints (red lines) and KF-estimated state variables (green lines); (b) convergence of the state variables x_5 to x_8 to the reference setpoints

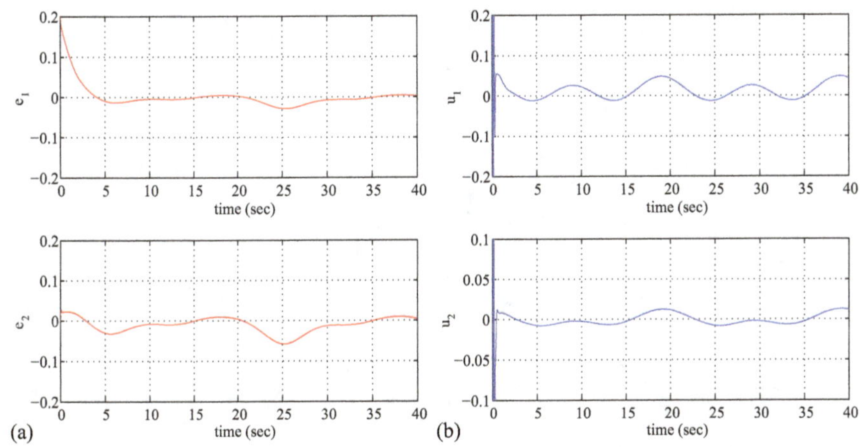

Figure 2.103 Tracking of setpoint 7 by the electro-hydraulic robotic manipulator: (a) tracking error for state variables x_1 and x_3 (joint angles of the robot); (b) control inputs u_1 and u_2 generated by the electro-hydraulic actuators

and Parnichkun, 2017; Pfeffer *et al.*, 2018; Best *et al.*, 2016). In this section, a novel solution has been proposed for the nonlinear optimal control problem of robotic exoskeletons with electropneumatic actuators. As a case study the dynamic model of a two-link lower-limb robotic exoskeleton with electropneumatic actuators has been considered. The dynamic model of this robotic system undergoes first approximate

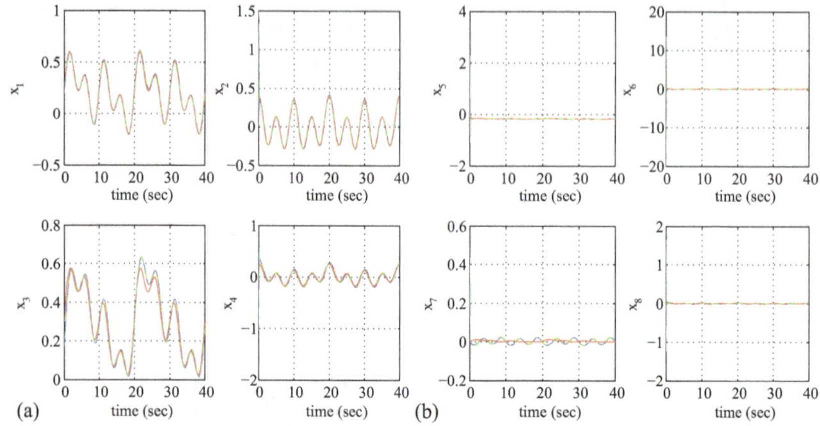

Figure 2.104 Tracking of setpoint 8 by the electro-hydraulic robotic manipulator: (a) convergence of the state variables x_1 to x_4 (blue lines) to the reference setpoints (red lines) and KF-estimated state variables (green lines); (b) convergence of the state variables x_5 to x_8 to the reference setpoints

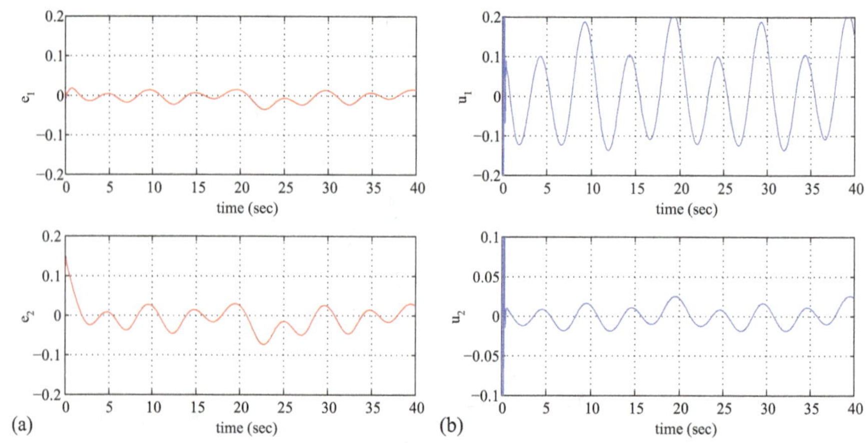

Figure 2.105 Tracking of setpoint 8 by the electro-hydraulic robotic manipulator: (a) tracking error for state variables x_1 and x_3 (joint angles of the robot); (b) control inputs u_1 and u_2 generated by the electro-hydraulic actuators

linearization at each iteration of the control algorithm around a temporary operating point. Within each sampling period this linearization point is defined by the present value of the robot's state vector and by the last sampled value of the control inputs vector. The linearization process relies on first-order Taylor series expansion and on the computation of the associated Jacobian matrices (Rigatos and Tzafestas, 2007; Basseville and Nikiforov, 1993; Rigatos and Zhang, 2009). The modeling error which is due to the truncation of higher-order terms from the Taylor series is considered to be a perturbation which is asymptotically compensated by the robustness of the control algorithm. To stabilize the dynamics of the electropneumatically actuated robot and to achieve precise tracking of reference setpoints an H-infinity (optimal) feedback controller is designed (Rigatos and Busawon, 2018).

Actually, the proposed H-infinity controller for the model of the 2-link electropneumatically actuated exoskeleton achieves the solution of the associated optimal control problem under model uncertainty and external disturbances. This controller implements a min–max differential game taking place between: (i) the control inputs which try to minimize a cost function which comprises a quadratic term of the state vector's tracking error; (ii) the model uncertainty and perturbation inputs which try to maximize this cost function Rigatos and Busawon, 2018. To select the stabilizing feedback gains of this H-infinity controller an algebraic Riccati equation is being repetitively solved at each time-step of the control method (Rigatos, 2016b, Rigatos, 2015). The global stability properties of the H-infinity control scheme are proven through Lyapunov analysis. First, it is proven that the robot's control loop achieves the H-infinity tracking performance criterion, which signifies elevated robustness against model imprecision and exogenous disturbances (Rigatos and Busawon, 2018; Tousssaint, Basar and Bullo, 2000). Moreover, under moderate conditions, the global asymptotic stability properties of the control scheme are proven (Rigatos and Busawon, 2018). The method achieves fast and accurate tracking of reference setpoints and despite strong nonlinearities in the electropneumatic actuators it keeps moderate the variations of the control inputs.

2.9.2 Dynamic model of a robotic manipulator with electropneumatic actuators

The dynamic model of a 2-link robotic exoskeleton with electropneumatic actuators is given in Figure 2.106. The dynamic model of the robotic exoskeleton can be also written in the concise form (Rigatos, Abbaszadeh *et al.*, 2020)

$$M(\theta)\ddot{\theta} + C(\theta,\dot{\theta}) + G(\theta) = \tau \tag{2.322}$$

where the state-vector is $\theta = [\theta_1, \theta_2]^T$, the joints' torques vector is $\tau = [\tau_1, \tau_2]^T$, while the inertia matrix $M(\theta)$, the Coriolis matrix $C(\theta,\dot{\theta})$ and the gravitational forces matrix $G(\theta)$ are defined as follows (Rigatos, Abbaszadeh *et al.*, 2020):

$$M(\theta) = \begin{pmatrix} m_1 d_1^2 + I_1 + m_2 l_1^2 + m_2 d_1^2 + 2m_2 l_1 d_2 \cos(\theta_2) + I_2 & -m_2 d_2^2 - m_2 l_1 d_2 \cos(\theta_2) - I_2 \\ -m_2 d_2^2 - m_2 l_1 d_2 \cos(\theta_2) - I_2 & m_2 d_2^2 + I_2 \end{pmatrix} \tag{2.323}$$

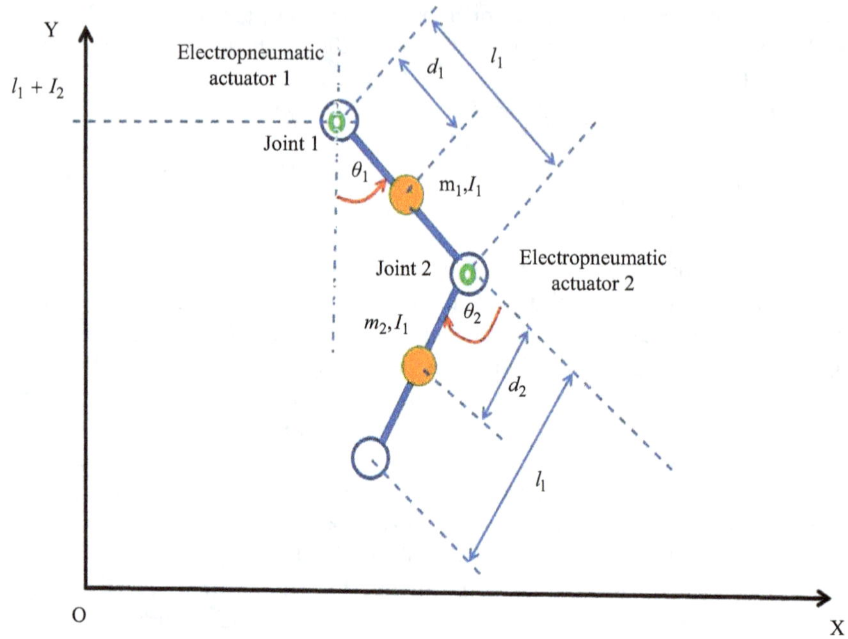

Figure 2.106 Diagram of a robotic exoskeleton with electropneumatic actuators

$$C(\theta,\dot{\theta}) = \begin{pmatrix} -m_2 l_1 d_2 \sin(\theta_2)\dot{\theta}_2(2\dot{\theta}_1 - \dot{\theta}_2) \\ m_2 l_1 d_2 \sin(\theta_2)\dot{\theta}_1^2 \end{pmatrix}$$
$$G(\theta) = \begin{pmatrix} (m_1 g d_1 + m_2 g l_1)\sin(\theta_1) + m_2 g d_2 \sin(\theta_1 - \theta_2) \\ m_2 g d_2 \sin(\theta_1 - \theta_2) \end{pmatrix} \tag{2.324}$$

Using that the symmetric inertia matrix $M(\theta)$ is

$$M = \begin{pmatrix} M_{11} & M_{12} \\ M_{21} & M_{22} \end{pmatrix} \tag{2.325}$$

and that the determinant of $M(\theta)$ is $detM = M_{11}M_{22} - M_{12}^2$ the inverse of $M(\theta)$ is

$$M^{-1} = \begin{pmatrix} \frac{M_{22}}{M_{11}M_{22} - M_{12}^2} & \frac{-M_{12}}{M_{11}M_{22} - M_{12}^2} \\ \frac{-M_{21}}{M_{11}M_{22} - M_{12}^2} & \frac{M_{11}}{M_{11}M_{22} - M_{12}^2} \end{pmatrix} \tag{2.326}$$

Thus the dynamic model of the exoskeleton robot is written as

$$\ddot{\theta} = -M^{-1}(\theta)[C(\theta,\dot{\theta}) + G(\theta)] + M^{-1}(\theta)\tau \tag{2.327}$$

By performing operations between the individual matrices, one arrives at the state-space description

$$\begin{pmatrix} \ddot{\theta}_1 \\ \ddot{\theta}_2 \end{pmatrix} = \begin{pmatrix} \frac{-M_{22}(C_1+G_1)+M_{12}(C_2+G_2)}{M_{11}M_{22}-M_{12}^2} \\ \frac{M_{12}(C_1+G_1)-M_{11}(C_2+G_2)}{M_{11}M_{22}-M_{12}^2} \end{pmatrix} + \begin{pmatrix} \frac{M_{22}}{M_{11}M_{22}-M_{12}^2} & \frac{-M_{12}}{M_{11}M_{22}-M_{12}^2} \\ \frac{-M_{21}}{M_{11}M_{22}-M_{12}^2} & \frac{M_{11}}{M_{11}M_{22}-M_{12}^2} \end{pmatrix} \begin{pmatrix} \tau_1 \\ \tau_2 \end{pmatrix} \tag{2.328}$$

Each electro-pneumatic actuator of the robotic exoskeleton uses two proportional servo-distributors (servo-valves or solenoid valves) which inject pressurized air in the compartments of a cylinder or release air out of these compartments. The main part of the actuator is this cylinder which comprises two compartments noted as P and N. These compartments have variable volume which is defined by the position of a sliding separator. The air pressure difference in the two chambers makes finally the free-end of each actuator move.

In such electropneumatic actuators, P_P is the pressure at compartment P of the cylinder, P_N is the pressure at compartment N of the cylinder (Chalaga and Plestan, 2021; Smaoui, Brun *et al.*, 2006b). Other parameters of the actuators' model are: S_P is the effective surface of compartment P of the cylinder, S_N is the effective surface of compartment N of the cylinder, b is the friction coefficient, K_r and r are coefficients in the ideal gas equation, T is the temperature variable in the ideal gas equation, V_P is the effective volume of the P chamber at the zero position, V_N is the effective volume of the N chamber at the zero position, V_{DP} and V_{DN} are the dead volumes which are present at each extremity of the cylinder, and l is the length of the cylinder.

Transmission of the linear motion x_i of the ith electropneumatic actuator's free-end to the rotary motion of the ith joint of the robotic exoskeleton is considered to be given by $\theta_i = k_i x_i$ where k_i is the associated transmission ratio. Equivalently, one has that $x_i = \frac{1}{k_i}\theta_i$ or $x_i = q_i\theta_i$ with $q_i = \frac{1}{k_i}$ for $i = 1, 2$. Using the above, the dynamics of the electropneumatic actuators of the 2-link exoskeleton is given in the following (Chalaga and Plestan, 2021; Smaoui, Brun *et al.*, 2006b):

First electropneumatic actuator:

$$\begin{aligned} m_1 q_1 \ddot{\theta}_1 &= [S_{P_1}P_{P_1} - S_{N_1}P_{N_1} - b_1 q_1 \dot{\theta}_1] \\ \dot{P}_{P_1} &= \frac{K_r T}{[V_{DP_1} + S_{P_1}\frac{l_1}{2} + S_{P_1}q_1\theta_1]}\left[\phi_{P_1}(q_1\theta_1) + \psi_{P_1}(q_1\theta_1)u_{11} - \frac{S_{P_1}}{rT}q_1\dot{\theta}_1 P_{P_1}\right] \\ \dot{P}_{N_1} &= \frac{K_r T}{[V_{DN_1} + S_{N_1}\frac{l_1}{2} + S_{N_1}q_1\theta_1]}\left[\phi_{N_1}(q_1\theta_1) + \psi_{N_1}(q_1\theta_1)u_{21} + \frac{S_{N_1}}{rT}q_1\dot{\theta}_1 P_{N_1}\right] \end{aligned} \tag{2.329}$$

Second electropneumatic actuator:

$$m_2q_2\ddot{\theta}_2 = [S_{P_2}P_{P_2} - S_{N_2}P_{N_2} - b_2q_2\dot{\theta}_2]$$

$$\dot{P}_{P_2} = \frac{K_rT}{[V_{DP_2} + S_{P_2}\frac{l_2}{2} + S_{P_2}q_2\theta_2]}\left[\phi_{P_2}(q_2\theta_2) + \psi_{P_2}(q_2\theta_2)u_{12} - \frac{S_{P_2}}{rT}q_2\dot{\theta}_2P_{P_2}\right]$$

$$\dot{P}_{N_2} = \frac{K_rT}{[V_{DN_2} + S_{N_2}\frac{l_2}{2} + S_{N_2}q_12\theta_2]}\left[\phi_{N_2}(q_2\theta_2) + \psi_{N_2}(q_2\theta_2)u_{22} + \frac{S_{N_2}}{rT}q_2\dot{\theta}_2P_{N_2}\right] \quad (2.330)$$

The torques which are applied from the electropneumatic actuators to the joints of the robotic exoskeleton are:

$$\begin{aligned}\tau_1 &= [S_{P_1}P_{P_1} - S_{N_1}P_{N_1} - b_1q_1\dot{\theta}_1]\\ \tau_2 &= [S_{P_2}P_{P_2} - S_{N_2}P_{N_2} - b_2q_2\dot{\theta}_2]\end{aligned} \quad (2.331)$$

Using the above, the aggregate dynamic model of the 2-link robotic exoskeleton with electropneumatic actuators is described by the following set of equations:

$$\begin{aligned}\ddot{\theta}_1 = &\frac{-M_{22}(C_1+G_1)+M_{12}(C_2+G_2)}{M_{11}M_{22}-M_{12}^2} + \frac{M_{22}[S_{P_1}P_{P_1} - S_{N_1}P_{N_1} - b_1q_1\dot{\theta}_1]}{M_{11}M_{22}-M_{12}^2}\\ &+\frac{-M_{12}[S_{P_2}P_{P_2} - S_{N_2}P_{N_2} - b_2q_2\dot{\theta}_2]}{M_{11}M_{22}-M_{12}^2}\end{aligned} \quad (2.332)$$

$$\begin{aligned}\ddot{\theta}_2 = &\frac{M_{12}(C_1+G_1)-M_{11}(C_2+G_2)}{M_{11}M_{22}-M_{12}^2} + \frac{-M_{21}[S_{P_1}P_{P_1} - S_{N_1}P_{N_1} - b_1q_1\dot{\theta}_1]}{M_{11}M_{22}-M_{12}^2}\\ &+\frac{M_{11}[S_{P_2}P_{P_2} - S_{N_2}P_{N_2} - b_2q_2\dot{\theta}_2]}{M_{11}M_{22}-M_{12}^2}\end{aligned} \quad (2.333)$$

$$\dot{P}_{P_1} = \frac{K_rT[\phi_{P_1}(q_1\theta_1) - \frac{S_{P_1}}{rT}q_1\dot{\theta}_1P_{P_1}]}{[V_{DP_1} + S_{P_1}\frac{l_1}{2} + S_{P_1}q_1\theta_1]} + \frac{K_rT\psi_{P_1}(q_1\theta_1)}{[V_{DP_1} + S_{N_1}\frac{l_1}{2} + S_{N_1}q_1\theta_1]}u_{11} \quad (2.334)$$

$$\dot{P}_{N_1} = \frac{K_rT[\phi_{N_1}(q_1\theta_1) - \frac{S_{N_1}}{rT}q_1\dot{\theta}_1P_{N_1}]}{[V_{DN_1} + S_{N_1}\frac{l_1}{2} + S_{N_1}q_1\theta_1]} + \frac{K_rT\psi_{N_1}(q_1\theta_1)}{[V_{DN_1} + S_{N_1}\frac{l_1}{2} + S_{N_1}q_1\theta_1]}u_{21} \quad (2.335)$$

$$\dot{P}_{P_2} = \frac{K_rT[\phi_{P_2}(q_2\theta_2) - \frac{S_{P_2}}{rT}q_2\dot{\theta}_2P_{P_2}]}{[V_{DP_2} + S_{P_2}\frac{l_2}{2} + S_{P_2}q_2\theta_2]} + \frac{K_rT\psi_{P_2}(q_2\theta_2)}{[V_{DP_2} + S_{N_2}\frac{l_2}{2} + S_{N_2}q_2\theta_2]}u_{12} \quad (2.336)$$

$$\dot{P}_{N_2} = \frac{K_rT[\phi_{N_2}(q_2\theta_2) - \frac{S_{N_2}}{rT}q_2\dot{\theta}_2P_{N_2}]}{[V_{DN_2} + S_{N_2}\frac{l_2}{2} + S_{N_2}q_2\theta_2]} + \frac{K_rT\psi_{N_2}(q_2\theta_2)}{[V_{DN_2} + S_{N_2}\frac{l_2}{2} + S_{N_2}q_2\theta_2]}u_{22} \quad (2.337)$$

The following state variables are defined for the electropneumatically actuated robotic system: $x_1=\theta_1$, $x_2=\dot{\theta}_1$, $x_3=\theta_2$, $x_4=\dot{\theta}_4$, $x_5=P_{P_1}$, $x_6=P_{N_1}$, $x_7=P_{P_2}$ and $x_8=P_{N_2}$. Using this notation, the dynamic model of the robot is written as follows:

$$\dot{x}_1 = x_2 \tag{2.338}$$

$$\begin{aligned}\dot{x}_2 = &\frac{-M_{22}(C_1+G_1)+M_{12}(C_2+G_2)}{M_{11}M_{22}-M_{12}^2}+\frac{M_{22}[S_{P_1}x_5-S_{N_1}x_6-b_1q_1x_2]}{M_{11}M_{22}-M_{12}^2}\\ &+\frac{-M_{12}[S_{P_2}x_7-S_{N_2}x_8-b_2q_2x_4]}{M_{11}M_{22}-M_{12}^2}\end{aligned} \tag{2.339}$$

$$\dot{x}_3 = x_4 \tag{2.340}$$

$$\begin{aligned}\dot{x}_4 = &\frac{M_{12}(C_1+G_1)-M_{11}(C_2+G_2)}{M_{11}M_{22}-M_{12}^2}+\frac{-M_{21}[S_{P_1}x_5-S_{N_1}x_6-b_1q_1x_2]}{M_{11}M_{22}-M_{12}^2}\\ &+\frac{M_{11}[S_{P_2}x_7-S_{N_2}x_8-b_2q_2x_4]}{M_{11}M_{22}-M_{12}^2}\end{aligned} \tag{2.341}$$

$$\dot{x}_5 = \frac{K_rT[\phi_{P_1}(q_1x_1)-\frac{S_{P_1}}{rT}q_1x_2x_5]}{[V_{DP_1}+S_{P_1}\frac{l_1}{2}+S_{P_1}q_1x_1]}+\frac{K_rT\psi_{P_1}(q_1x_1))}{[V_{DP_1}+S_{N_1}\frac{l_1}{2}+S_{N_1}q_1x_1]}u_{11} \tag{2.342}$$

$$\dot{x}_6 = \frac{K_rT[\phi_{N_1}(q_1x_1)-\frac{S_{N_1}}{rT}q_1x_2x_6]}{[V_{DN_1}+S_{N_1}\frac{l_1}{2}+S_{N_1}q_1x_1]}+\frac{K_rT\psi_{N_1}(q_1x_1)}{[V_{DN_1}+S_{N_1}\frac{l_1}{2}+S_{N_1}q_1x_1]}u_{21} \tag{2.343}$$

$$\dot{x}_7 = \frac{K_rT[\phi_{P_2}(q_2x_3)-\frac{S_{P_2}}{rT}q_2x_4x_7]}{[V_{DP_2}+S_{P_2}\frac{l_2}{2}+S_{P_2}q_2x_3]}+\frac{K_rT\psi_{P_2}(q_2x_3)}{[V_{DP_2}+S_{N_2}\frac{l_2}{2}+S_{N_2}q_2x_3]}u_{12} \tag{2.344}$$

$$\dot{x}_8 = \frac{K_rT[\phi_{N_2}(q_2x_3)-\frac{S_{N_2}}{rT}q_2x_4x_7]}{[V_{DN_2}+S_{N_2}\frac{l_2}{2}+S_{N_2}q_2x_3]}+\frac{K_rT\psi_{N_2}(q_2x_3)}{[V_{DN_2}+S_{N_2}\frac{l_2}{2}+S_{N_2}q_2x_3]}u_{22} \tag{2.345}$$

Thus, the state-space model of the electropneumatically actuated robotic exoskeleton is written as

$$\dot{x}_1 = f_1(x)+g_{11}u_{11}+g_{21}u_{21}+g_{31}u_{12}+g_{41}u_{22} \tag{2.346}$$

$$\dot{x}_2 = f_2(x)+g_{12}u_{11}+g_{22}u_{21}+g_{32}u_{12}+g_{42}u_{22} \tag{2.347}$$

$$\dot{x}_3 = f_3(x)+g_{13}u_{11}+g_{23}u_{21}+g_{33}u_{12}+g_{43}u_{22} \tag{2.348}$$

$$\dot{x}_4 = f_4(x)+g_{14}u_{11}+g_{24}u_{21}+g_{34}u_{12}+g_{44}u_{22} \tag{2.349}$$

$$\dot{x}_5 = f_5(x)+g_{15}u_{11}+g_{25}u_{21}+g_{35}u_{12}+g_{45}u_{22} \tag{2.350}$$

$$\dot{x}_6 = f_3(x)+g_{16}u_{11}+g_{26}u_{21}+g_{36}u_{12}+g_{46}u_{22} \tag{2.351}$$

$$\dot{x}_7 = f_7(x)+g_{17}u_{11}+g_{27}u_{21}+g_{37}u_{12}+g_{47}u_{22} \tag{2.352}$$

$$\dot{x}_8 = f_7(x)+g_{18}u_{11}+g_{28}u_{21}+g_{38}u_{12}+g_{48}u_{22} \tag{2.353}$$

or equivalently, the state-space model of the electropneumatic root takes the following matrix form:

$$\begin{pmatrix} \dot{x}_1 \\ \dot{x}_2 \\ \dot{x}_3 \\ \dot{x}_4 \\ \dot{x}_5 \\ \dot{x}_6 \\ \dot{x}_7 \\ \dot{x}_8 \end{pmatrix} = \begin{pmatrix} f_1(x) \\ f_2(x) \\ f_3(x) \\ f_4(x) \\ f_5(x) \\ f_6(x) \\ f_7(x) \\ f_8(x) \end{pmatrix} + \begin{pmatrix} 0 & 0 & 0 & 0 \\ 0 & 0 & 0 & 0 \\ 0 & 0 & 0 & 0 \\ 0 & 0 & 0 & 0 \\ g_{51} & 0 & 0 & 0 \\ 0 & g_{62} & 0 & 0 \\ 0 & 0 & g_{33} & 0 \\ 0 & 0 & 0 & g_{84} \end{pmatrix} \begin{pmatrix} u_{11} \\ u_{21} \\ u_{12} \\ u_{22} \end{pmatrix} \tag{2.354}$$

The model of the robotic exoskeleton with electropneumatic actuation can be also written in the concise form

$$\dot{x} = f(x) + g_1(x)u_{11} + g_2(x)u_{22} + g_3(x)u_{12} + g_4(x)u_{22} \tag{2.355}$$

where $x \in R^{8\times1}$, $f(x) \in R^{8\times1}$, $g_1(x) \in R^{8\times1}$, $g_2(x) \in R^{8\times1}$, $g_3(x) \in R^{8\times1}$ and $g_4(x) \in R^{8\times1}$, while the control inputs vector $u \in R^{4\times1}$ comprises the control variables $u = [u_{11}, u_{21}, u_{12}, u_{22}]^T$.

It holds that

$$\begin{aligned}
f_1(x) &= x_2 \\
f_2(x) &= \frac{-M_{22}(C_1+G_1)+M_{12}(C_2+G_2)}{M_{11}M_{22}-M_{12}^2} + \frac{M_{22}[S_{P_1}x_5 - S_{N_1}x_6 - b_1q_1x_2]}{M_{11}M_{22}-M_{12}^2} \\
&\quad + \frac{-M_{12}[S_{P_2}x_7 - S_{N_2}x_8 - b_2q_2x_4]}{M_{11}M_{22}-M_{12}^2} \\
f_3(x) &= x_4 \\
f_4(x) &= \frac{M_{12}(C_1+G_1)-M_{11}(C_2+G_2)}{M_{11}M_{22}-M_{12}^2} + \frac{-M_{21}[S_{P_1}x_5 - S_{N_1}x_6 - b_1q_1x_2]}{M_{11}M_{22}-M_{12}^2} \\
&\quad + \frac{M_{11}[S_{P_2}x_7 - S_{N_2}x_8 - b_2q_2x_4]}{M_{11}M_{22}-M_{12}^2} \\
f_5(x) &= \frac{K_rT[\phi_{P_1}(q_1x_1) - \frac{S_{P_1}}{rT}q_1x_2x_5]}{[V_{DP_1} + S_{P_1}\frac{l_1}{2} + S_{P_1}q_1x_1]} \quad f_6(x) = \frac{K_rT[\phi_{N_1}(q_1x_1) + \frac{S_{N_1}}{rT}q_1x_2x_6]}{[V_{DN_1} + S_{N_1}\frac{l_1}{2} + S_{N_1}q_1x_1]} \\
f_7(x) &= \frac{K_rT[\phi_{P_2}(q_2x_3) - \frac{S_{P_2}}{rT}q_2x_4x_7]}{[V_{DP_2} + S_{P_2}\frac{l_2}{2} + S_{P_2}q_2x_3]} \quad f_8(x) = \frac{K_rT[\phi_{N_2}(q_2x_3) + \frac{S_{N_2}}{rT}q_2x_4x_7]}{[V_{DN_2} + S_{N_2}\frac{l_2}{2} + S_{N_2}q_2x_3]}
\end{aligned} \tag{2.356}$$

It also holds that

$$\begin{aligned}
g_{51}(x) &= \frac{K_rT\psi_{P_1}(q_1x_1))}{[V_{DP_1} + S_{N_1}\frac{l_1}{2} + S_{N_1}q_1x_1]} \quad g_{62}(x) = \frac{K_rT\psi_{N_1}(q_1x_1)}{[V_{DN_1} + S_{N_1}\frac{l_1}{2} + S_{N_1}q_1x_1]} \\
g_{73}(x) &= \frac{K_rT\psi_{P_2}(q_2x_3)}{[V_{DP_2} + S_{N_2}\frac{l_2}{2} + S_{N_2}q_2x_3]} \quad g_{84}(x) = \frac{K_rT\psi_{N_2}(q_2x_3)}{[V_{DN_2} + S_{N_2}\frac{l_2}{2} + S_{N_2}q_2x_3]}
\end{aligned} \tag{2.357}$$

2.9.3 Approximate linearization of the robot with electropneumatic actuation

2.9.3.1 Approximate linearization of the robot

The dynamic model of the robotic system with pneumatic actuators undergoes approximate linearization around the temporary operating point (x^*, u^*). The linearization point is updated at each sampling instance, where x^* is the present value of the robot's state vector and u^* is the last sampled value of the control inputs vector. The linearization process is based on the first-order Taylor series expansion and on the computation of the associated Jacobian matrices (Rigatos and Busawon, 2018). The initial nonlinear model of the robot $\dot{x} = f(x, u)$ is transformed into the equivalent linear model

$$\dot{x} = Ax + Bu + \tilde{d} \tag{2.358}$$

where $\tilde{d}$ is the cumulative disturbances vector which represents: (i) modeling errors due to the truncation of higher-order terms in the Taylor series expansion, (ii) exogenous perturbations, and (iii) sensor measurement noise of any distribution. The linearization process relies on the Jacobian matrices

$$\begin{aligned} &A = \nabla_x[f(x) + g(x)] \mid_{(x^*,u^*)} \Rightarrow \\ &A = \nabla_x f(x) \mid_{(x^*,u^*)} + \nabla_x g_1(x) u_{11} \mid_{(x^*,u^*)} + \nabla_x g_2(x) u_{21} \mid_{(x^*,u^*)} \\ &\quad + \nabla_x g_3(x) u_{12} \mid_{(x^*,u^*)} + \nabla_x g_4(x) u_{22} \mid_{(x^*,u^*)} \end{aligned} \tag{2.359}$$

$$B = \nabla_u[f(x) + g(x)] \mid_{(x^*,u^*)} \Rightarrow B = g(x) \mid_{(x^*,u^*)} \tag{2.360}$$

Computation of the Jacobian matrix $\nabla_x f(x) \mid_{(x^*,u^*)}$.

First row of the Jacobian matrix $\nabla_x f(x) \mid_{(x^*,u^*)}$: $\frac{\partial f_1}{\partial x_1} = 0$, $\frac{\partial f_1}{\partial x_2} = 1$, $\frac{\partial f_1}{\partial x_3} = 0$, $\frac{\partial f_1}{\partial x_4} = 0$, $\frac{\partial f_1}{\partial x_5} = 0$, $\frac{\partial f_1}{\partial x_6} = 0$, $\frac{\partial f_1}{\partial x_7} = 0$ and $\frac{\partial f_1}{\partial x_8} = 0$.

Second row of the Jacobian matrix $\nabla_x f(x) \mid_{(x^*,u^*)}$: it holds that

$$f_2(x) = \frac{f_{2n}(x)}{f_{2d}(x)} \tag{2.361}$$

and thus one gets

$$\frac{\partial f_2}{\partial x_i} = \frac{\frac{\partial f_{2,num}}{\partial x_i} f[2, den] - f_{2,num} \frac{\partial f_{2,den}}{\partial x_i}}{f_{2,den}^2} \tag{2.362}$$

where

$$f_{2,num}(x) = -M_{21}(C_1 + G_1) + M_{12}(C_2 + G_2) + M_{22}\tau_1 - M_{22}\tau_2 \tag{2.363}$$

with the torques τ_1 and τ_2 to be given by $\tau_1 = S_{P_1}x_5 - S_{N_1}x_6 - b_1 q_1 x_2$, $\tau_2 = S_{P_2}x_7 - S_{N_2}x_8 - b_2 q_2 x_4$

$$f_{2,den}(x) = M_{11}M_{22} - M_{12}^2 \tag{2.364}$$

Thus, one gets for $i = 1, 2, \cdots, 8$

$$\begin{array}{l}\frac{\partial f_{2,num}(x)}{\partial x_i} = -\frac{\partial M_{22}}{\partial x_i}(C_1 + G_1) - M_{22}(\frac{\partial C_1}{\partial x_i} + \frac{\partial G_1}{\partial x_i})+ \\ \frac{\partial M_{12}}{\partial x_i}(C_2 + G_2) + M_{12}(\frac{\partial C_2}{\partial x_i} + \frac{\partial G_2}{\partial x_i})+ \\ \frac{\partial M_{22}}{\partial x_i}\tau_1 + M_{22}\frac{\partial \tau_1}{\partial x_i} - \frac{\partial M_{12}}{\partial x_i}\tau_2 - M_{12}\frac{\partial \tau_2}{\partial x_i}\end{array} \tag{2.365}$$

$$\frac{\partial f_{2,den}(x)}{\partial x_i} = \frac{\partial M_{11}}{\partial x_i}M_{22} + M_{11}\frac{\partial M_{22}}{\partial x_i} - 2M_{12}\frac{\partial M_{12}}{\partial x_i} \tag{2.366}$$

Third row of the Jacobian matrix $\nabla_x f(x)\mid_{(x^*,u^*)}$: $\frac{\partial f_3}{\partial x_1} = 0$, $\frac{\partial f_3}{\partial x_2} = 0$, $\frac{\partial f_3}{\partial x_3} = 0$, $\frac{\partial f_3}{\partial x_4} = 1$, $\frac{\partial f_3}{\partial x_5} = 0$, $\frac{\partial f_3}{\partial x_6} = 0$, $\frac{\partial f_3}{\partial x_7} = 0$ and $\frac{\partial f_3}{\partial x_8} = 0$.

Fourth row of the Jacobian matrix $\nabla_x f(x)\mid_{(x^*,u^*)}$: it holds that

$$f_4(x) = \frac{f_{4n}(x)}{f_{4d}(x)} \tag{2.367}$$

where

$$f_{4,num}(x) = M_{12}(C_1 + G_1) - M_{11}(C_2 + G_2) - M_{12}\tau_1 + M_{11}\tau_2 \tag{2.368}$$

with the torques τ_1 and τ_2 to be given by $\tau_1 = S_{P_1}x_5 - S_{N_1}x_6 - b_1q_1x_2$, $\tau_2 = S_{P_2}x_7 - S_{N_2}x_8 - b_2q_2x_4$

$$f_{4,den}(x) = M_{11}M_{22} - M_{12}^2 \tag{2.369}$$

and thus one gets for $i = 1, 2, \cdots, 8$

$$\frac{\partial f_4}{\partial x_i} = \frac{\frac{\partial f_{4,num}}{\partial x_i}f[4, den] - f_{4,num}\frac{\partial f_{4,den}}{\partial x_i}}{f_{4,den}^2} \tag{2.370}$$

Thus, one gets for $i = 1, 2, \cdots, 8$

$$\begin{aligned}\frac{\partial f_{4,num}(x)}{\partial x_i} &= \frac{\partial M_{12}}{\partial x_i}(C_1 + G_1) + M_{12}\left(\frac{\partial C_1}{\partial x_i} + \frac{\partial G_1}{\partial x_i}\right) + \\ &-\frac{\partial M_{11}}{\partial x_i}(C_2 + G_2) - M_{11}\left(\frac{\partial C_2}{\partial x_i} + \frac{\partial G_2}{\partial x_i}\right) + \\ &-\frac{\partial M_{12}}{\partial x_i}\tau_1 - M_{12}\frac{\partial \tau_1}{\partial x_i} + \frac{\partial M_{11}}{\partial x_i}\tau_2 + M_{11}\frac{\partial \tau_2}{\partial x_i}\end{aligned} \tag{2.371}$$

$$\frac{\partial f_{2,den}(x)}{\partial x_i} = \frac{\partial M_{11}}{\partial x_i}M_{22} + M_{11}\frac{\partial M_{22}}{\partial x_i} - 2M_{12}\frac{\partial M_{12}}{\partial x_i} \tag{2.372}$$

Fifth row of the Jacobian matrix $\nabla_x f(x)\mid_{(x^*,u^*)}$: it holds that

$$f_5(x) = \frac{f_{5,num}(x)}{f_{5,den}(x)} \tag{2.373}$$

where

$$f_{5,num}(x) = K_rT\left[\phi_{P_1}(q_1x_1) - \frac{S_{P_1}}{rT}q_1x_2x_5\right]$$
$$f_{5,den}(x) = \left[V_{DP_1} + S_{P_1}\frac{\tilde{l}_1}{2} + S_{P_1}q_1x_1\right] \tag{2.374}$$

and thus one gets for $i = 1, 2, \cdots, 8$

$$\frac{\partial f_5}{\partial x_i} = \frac{\frac{\partial f_{5,num}}{\partial x_i}f[5, den] - f_{5,num}\frac{\partial f_{5,den}}{\partial x_i}}{f_{5,den}^2} \tag{2.375}$$

Moreover, one has

$$\frac{\partial f_{5,num}}{\partial x_i} = K_rT\left[\frac{\partial \phi_{P_1}}{\partial x_i}(q_1x_1) - \frac{\partial}{\partial x_i}\frac{S_{P_1}}{rT}q_1x_2x_5\right] \tag{2.376}$$

$$\frac{\partial f_{5,den}}{\partial x_i} = \frac{\partial}{\partial x_i}\left[V_{DP_1} + S_{P_1}\frac{\tilde{l}_1}{2} + S_{P_1}q_1x_1\right] \tag{2.377}$$

Sixth row of the Jacobian matrix $\nabla_x f(x) \mid_{(x^*,u^*)}$: it holds that

$$f_6(x) = \frac{f_{6,num}(x)}{f_{6,den}(x)} \tag{2.378}$$

where

$$f_{6,num}(x) = K_rT\left[\phi_{N_1}(q_1x_1) + \frac{S_{N_1}}{rT}q_1x_2x_6\right]$$
$$f_{6,den}(x) = \left[V_{DN_1} + S_{N_1}\frac{\tilde{l}_1}{2} + S_{N_1}q_1x_1\right] \tag{2.379}$$

and thus one gets for $i = 1, 2, \cdots, 8$

$$\frac{\partial f_6}{\partial x_i} = \frac{\frac{\partial f_{6,num}}{\partial x_i}f_{6,den} - f_{5,num}\frac{\partial f_{6,den}}{\partial x_i}}{f_{6,den}^2} \tag{2.380}$$

Moreover, one has

$$\frac{\partial f_{6,num}}{\partial x_i} = K_rT\left[\frac{\partial \phi_{N_1}}{\partial x_i}(q_1x_1) + \frac{\partial}{\partial x_i}\frac{S_{N_1}}{rT}q_1x_2x_6\right] \tag{2.381}$$

$$\frac{\partial f_{6,den}}{\partial x_i} = \frac{\partial}{\partial x_i}\left[V_{DN_1} + S_{N_1}\frac{\tilde{l}_1}{2} + S_{N_1}q_1x_1\right] \tag{2.382}$$

Seventh row of the Jacobian matrix $\nabla_x f(x) \mid_{(x^*,u^*)}$: it holds that

$$f_7(x) = \frac{f_{7,num}(x)}{f_{7,den}(x)} \tag{2.383}$$

where

$$f_{7,num}(x) = K_r T\left[\phi_{P_2}(q_2x_3) - \frac{S_{P_2}}{rT}q_2x_4x_7\right]$$
$$f_{7,den}(x) = \left[V_{DP_2} + S_{P_2}\frac{\tilde{l}_2}{2} + S_{P_2}q_2x_3\right] \tag{2.384}$$

and thus one gets for $i = 1, 2, \cdots, 8$

$$\frac{\partial f_7}{\partial x_i} = \frac{\frac{\partial f_{7,num}}{\partial x_i} f[7, den] - f_{7,num}\frac{\partial f_{7,den}}{\partial x_i}}{f_{7,den}^2} \tag{2.385}$$

Moreover, one has

$$\frac{\partial f_{7,num}}{\partial x_i} = K_r T\left[\frac{\partial \phi_{P_2}}{\partial x_i}(q_2x_2) - \frac{\partial}{\partial x_i}\frac{S_{P_2}}{rT}q_2x4x_7\right] \tag{2.386}$$

$$\frac{\partial f_{7,den}}{\partial x_i} = \frac{\partial}{\partial x_i}\left[V_{DP_2} + S_{P_2}\frac{\tilde{l}_2}{2} + S_{P_2}q_2x_3\right] \tag{2.387}$$

Eighth row of the Jacobian matrix $\nabla_x f(x)\mid_{(x^*,u^*)}$: it holds that

$$f_8(x) = \frac{f_{8,num}(x)}{f_{8,den}(x)} \tag{2.388}$$

where

$$f_{8,num}(x) = K_r T\left[\phi_{N_2}(q_2x_3) + \frac{S_{N_2}}{rT}q_2x_4x_8\right]$$
$$f_{8,den}(x) = \left[V_{DN_2} + S_{N_2}\frac{\tilde{l}_2}{2} + S_{N_2}q_2x_3\right] \tag{2.389}$$

and thus one gets for $i = 1, 2, \cdots, 8$

$$\frac{\partial f_8}{\partial x_i} = \frac{\frac{\partial f_{8,num}}{\partial x_i} f_{8,den} - f_{8,num}\frac{\partial f_{8,den}}{\partial x_i}}{f_{8,den}^2} \tag{2.390}$$

Moreover, one has

$$\frac{\partial f_{8,num}}{\partial x_i} = K_r T\left[\frac{\partial \phi_{N_2}}{\partial x_i}(q_2x_3) + \frac{\partial}{\partial x_i}\frac{S_{N_2}}{rT}q_1x_4x_8\right] \tag{2.391}$$

$$\frac{\partial f_{8,den}}{\partial x_i} = \frac{\partial}{\partial x_i}\left[V_{DN_2} + S_{N_2}\frac{\tilde{l}_2}{2} + S_{N_2}q_2x_3\right] \tag{2.392}$$

Computation of the Jacobian matrix $\nabla_x g_1(x)\mid_{(x^*,u^*)}$: all rows of this Jacobian matrix have zero elements apart from the fifth row. It holds that:

$$g_{51}(x) = \frac{g_{51,num}(x)}{g_{51,den}(x)} \tag{2.393}$$

where

$$g_{51,num}(x) = K_r T \psi_{P_1}(q_1 x_1)$$
$$g_{51,den}(x) = \left[V_{DP_1} + S_{P_1}\frac{\tilde{l}_1}{2} + S_{P_1} q_1 x_1 \right] \tag{2.394}$$

and thus one gets for $i = 1, 2, \cdots, 8$

$$\frac{\partial g_{51}}{\partial x_i} = \frac{\frac{\partial g_{51,num}}{\partial x_i} g_{51,den} - g_{51,num}\frac{\partial g_{51,den}}{\partial x_i}}{g^2_{51,den}} \tag{2.395}$$

Moreover, one has

$$\frac{\partial g_{51,num}}{\partial x_i} = K_r T \frac{\partial \psi_{P_1}}{\partial x_i}(q_1 x_1) \tag{2.396}$$

$$\frac{\partial g_{51,den}}{\partial x_i} = \frac{\partial}{\partial x_i}\left[V_{DP_1} + S_{P_1}\frac{\tilde{l}_1}{2} + S_{P_1} q_1 x_1 \right] \tag{2.397}$$

Computation of the Jacobian matrix $\nabla_x g_2(x) \mid_{(x^*,u^*)}$: all rows of this Jacobian matrix have zero elements apart from the sixth row. It holds that:

$$g_{62}(x) = \frac{g_{62,num}(x)}{g_{62,den}(x)} \tag{2.398}$$

where

$$g_{62,num}(x) = K_r T \psi_{N_1}(q_1 x_1)$$
$$g_{62,den}(x) = \left[V_{DN_1} + S_{N_1}\frac{\tilde{l}_1}{2} + S_{N_1} q_1 x_1 \right] \tag{2.399}$$

and thus one gets for $i = 1, 2, \cdots, 8$

$$\frac{\partial g_{62}}{\partial x_i} = \frac{\frac{\partial g_{62,num}}{\partial x_i} g_{62,den} - g_{62,num}\frac{\partial g_{62,den}}{\partial x_i}}{g^2_{62,den}} \tag{2.400}$$

Moreover, one has

$$\frac{\partial g_{62,num}}{\partial x_i} = K_r T \frac{\partial \psi_{N_1}}{\partial x_i}(q_1 x_1) \tag{2.401}$$

$$\frac{\partial g_{62,den}}{\partial x_i} = \frac{\partial}{\partial x_i}\left[V_{DN_1} + S_{N_1}\frac{\tilde{l}_1}{2} + S_{N_1} q_1 x_1 \right] \tag{2.402}$$

Computation of the Jacobian matrix $\nabla_x g_3(x) \mid_{(x^*,u^*)}$: all rows of this Jacobian matrix have zero elements apart from the seventh row. It holds that:

$$g_{73}(x) = \frac{g_{71,num}(x)}{g_{71,den}(x)} \tag{2.403}$$

where

$$g_{73,num}(x) = K_r T \psi_{P_2}(q_2 x_3)$$
$$g_{73,den}(x) = \left[V_{DP_2} + S_{P_2}\frac{\tilde{l}_2}{2} + S_{P_2} q_2 x_3 \right] \tag{2.404}$$

and thus one gets for $i = 1, 2, \cdots, 8$

$$\frac{\partial g_{73}}{\partial x_i} = \frac{\frac{\partial g_{73,num}}{\partial x_i} g_{73,den} - g_{73,num}\frac{\partial g_{73,den}}{\partial x_i}}{g_{73,den}^2} \tag{2.405}$$

Moreover, one has

$$\frac{\partial g_{73,num}}{\partial x_i} = K_r T \frac{\partial \psi_{P_2}}{\partial x_i}(q_2 x_3) \tag{2.406}$$

$$\frac{\partial g_{73,den}}{\partial x_i} = \frac{\partial}{\partial x_i}\left[V_{DP_2} + S_{P_2}\frac{\tilde{l}_2}{2} + S_{P_2} q_2 x_3 \right] \tag{2.407}$$

Computation of the Jacobian matrix $\nabla_x g_4(x) \mid_{(x^*, u^*)}$: all rows of this Jacobian matrix have zero elements apart from the eighth row. It holds that:

$$g_{84}(x) = \frac{g_{84,num}(x)}{g_{84,den}(x)} \tag{2.408}$$

where

$$g_{84,num}(x) = K_r T \psi_{N_2}(q_2 x_3)$$
$$g_{84,den}(x) = \left[V_{DN_2} + S_{N_2}\frac{\tilde{l}_2}{2} + S_{N_2} q_2 x_3 \right] \tag{2.409}$$

and thus one gets for $i = 1, 2, \cdots, 8$

$$\frac{\partial g_{84}}{\partial x_i} = \frac{\frac{\partial g_{94,num}}{\partial x_i} g_{84,den} - g_{62,num}\frac{\partial g_{84,den}}{\partial x_i}}{g_{84,den}^2} \tag{2.410}$$

Moreover, one has

$$\frac{\partial g_{84,num}}{\partial x_i} = K_r T \frac{\partial \psi_{N_2}}{\partial x_i}(q_2 x_3) \tag{2.411}$$

$$\frac{\partial g_{84,den}}{\partial x_i} = \frac{\partial}{\partial x_i}[V_{DN_2} + S_{N_2}\frac{\tilde{l}_2}{2} + S_{N_2} q_2 x_3] \tag{2.412}$$

To complete the computation of the Jacobian matrices of the electropneumatically actuated robotic system, the following partial derivatives have to be computed:

(i) $M_{11} = m_1 d_1^2 + I_1 + m_2 l_1^2 + m_2 d_1^2 + 2m_2 l_1 \cos(x_3) + I_2$. It holds that $\frac{\partial M_{11}}{\partial x_i} = 0$ for $i = 1, 2, 4, 5, 6, 7, 8$ and $\frac{\partial M_{11}}{\partial x_3} = -2m_3 l_1 d_2 \sin(x_3)$.

$M_{12}=M_{21}=-m_2d_2^2-m_2l_1d_2\cos(x_3)+I_2$. It holds that $\frac{\partial M_{12}}{\partial x_i}=0$ for $i=1,2,4,5,6,7,8$ and $\frac{\partial M_{12}}{\partial x_3}=m_2l_1d_2\sin(x_3)$.

$M_{22}=m_2d_2^2+I_2$. It holds that $\frac{\partial M_{22}}{\partial x_i}=0$ for $i=1,2,3,4,5,6,7,8$.

(ii) $C_1(\theta,\dot{\theta})=-m_2l_1d_2\sin(x_3)x_4(2x_2-x_4)$. It holds that:
$\frac{\partial C_1}{\partial x_1}=0$, $\frac{\partial C_1}{\partial x_2}=-2m_1l_1d_2\sin(x_3)x_4$, $\frac{\partial C_1}{\partial x_3}=-m_2l_1d_2\cos(x_3)x_4(2x_2-x_4)$,
$\frac{\partial C_1}{\partial x_4}=-m_2l_1d_2\sin(x_3)x_4(2x_2-x_4)$, $\frac{\partial C_1}{\partial x_5}=0$, $\frac{\partial C_1}{\partial x_6}=0$, $\frac{\partial C_1}{\partial x_7}=0$, and $\frac{\partial C_1}{\partial x_8}=0$.

$C_2(\theta,\dot{\theta})=m_2l_1d_2\sin(x_3)x_2^2$.It holds that: $\frac{\partial C_2}{\partial x_1}=0$, $\frac{\partial C_2}{\partial x_2}=2m_2l_1d_2\sin(x_3)x_2$, $\frac{\partial C_2}{\partial x_3}=m_2l_1d_2\cos(x_3)x_2^2$, $\frac{\partial C_2}{\partial x_4}=0$, $\frac{\partial C_2}{\partial x_5}=0$, $\frac{\partial C_2}{\partial x_6}=0$, $\frac{\partial C_2}{\partial x_7}=0$, and $\frac{\partial C_2}{\partial x_8}=0$.

$G_1(\theta)=(m_1gd_1+m_2gl_1)\sin(x_1)+m_2gd_2\sin(x_1-x_3)$. It holds that $\frac{\partial G_1}{\partial x_1}=(m_1gd_1+m_2gl_1)\cos(x_1)+m_2gd_2\cos(x_1-x_3)$, $\frac{\partial G_1}{\partial x_2}=0$, $\frac{\partial G_1}{\partial x_3}=-m_2gd_2\cos(x_1-x_3)$, $\frac{\partial G_1}{\partial x_4}=0$, $\frac{\partial G_1}{\partial x_5}=0$, $\frac{\partial G_1}{\partial x_6}=0$, $\frac{\partial G_1}{\partial x_7}=0$, and $\frac{\partial G_1}{\partial x_8}=0$.
$G_2(\theta)=m_2gd_2\sin(x_1-x_3)$. It holds that $\frac{\partial G_2}{\partial x_1}=m_2gd_2\cos(x_1-x_3)$, $\frac{\partial G_2}{\partial x_2}=0$, $\frac{\partial G_2}{\partial x_3}=-m_2gd_2\cos(x_1-x_3)$, $\frac{\partial G_2}{\partial x_4}=0$, $\frac{\partial G_1}{\partial x_5}=0$, $\frac{\partial G_2}{\partial x_6}=0$, $\frac{\partial G_2}{\partial x_7}=0$, and $\frac{\partial G_2}{\partial x_8}=0$.

(iv) $\tau_1=S_{P_1}x_5-S_{N_1}x_6-b_1q_1x_2$. It holds that $\frac{\partial\tau_1}{\partial x_1}=0$, $\frac{\partial\tau_1}{\partial x_2}=-b_1q_1$, $\frac{\partial\tau_1}{\partial x_3}=0$, $\frac{\partial\tau_1}{\partial x_4}=0$, $\frac{\partial\tau_1}{\partial x_5}=S_{P_1}$, $\frac{\partial\tau_1}{\partial x_6}=-S_{N_1}$, $\frac{\partial\tau_1}{\partial x_7}=0$, and $\frac{\partial\tau_1}{\partial x_8}=0$.

$\tau_2=S_{P_2}x_7-S_{N_2}x_8-b_2q_2x_4$. It holds that $\frac{\partial\tau_2}{\partial x_1}=0$, $\frac{\partial\tau_2}{\partial x_2}=0$, $\frac{\partial\tau_2}{\partial x_3}=0$, $\frac{\partial\tau_2}{\partial x_4}=-b_2q_2$, $\frac{\partial\tau_2}{\partial x_5}=0$, $\frac{\partial\tau_2}{\partial x_6}=0$, $\frac{\partial\tau_2}{\partial x_7}=S_{P_2}$, and $\frac{\partial\tau_2}{\partial x_8}=-S_{N_2}$.

(v) $\phi_{P_1}(q_1x_1)=c_0^a+c_1^a(q_1x_1)+c_2^a(q_1x_1)^2+c_3^a(q_1x_1)^3+c_4^a(q_1x_1)^4+c_5^a(q_1x_1)^5$. It holds that: $\frac{\partial\phi(P_1)}{\partial x_1}=c_1^aq_1+c_2^a2(q_1x_1)+c_3^a3(q_1x_1)^2q_1+c_4^a4(q_1x_1)^3q_1+c_5^a5(q_1x_1)^4q_1$ and $\frac{\partial\phi(P_1)}{\partial x_i}=0$ for $i=2,3,4,5,6,7,8$.

$\phi_{P_2}(q_2x_3)=\bar{c}_0^a+\bar{c}_1^a(q_2x_3)+\bar{c}_2^a(q_2x_3)^2+\bar{c}_3^a(q_2x_3)^3+\bar{c}_4^a(q_2x_3)^4+\bar{c}_5^a(q_2x_3)^5$. It holds that: $\frac{\partial\phi(P_2)}{\partial x_3}=\bar{c}_1^aq_2+\bar{c}_2^a2(q_2x_3)+\bar{c}_3^a3(q_2x_3)^2q_2+\bar{c}_4^a4(q_2x_3)^3q_2+\bar{c}_5^a5(q_2x_3)^4q_2$ and $\frac{\partial\phi(P_2)}{\partial x_i}=0$ for $i=1,2,4,5,6,7,8$.

(vi) Computation of partial derivatives for $\psi_{P_1}(q_1x_1)$ and $\psi_{P_2}(q_2x_3)$

$$\psi_{P_1}(q_1x_1)=\begin{cases}c_0^b+c_1^b(q_1x_1)+c_2^b(q_1x_1)^2+c_3^b(q_1x_1)^3+c_4^b(q_1x_1)^4+c_5^b(q_1x_1)^5 & \text{if } u_{11}>0\\ c_0^c+c_1^c(q_1x_1)+c_2^c(q_1x_1)^2+c_3^c(q_1x_1)^3+c_4^c(q_1x_1)^4+c_5^c(q_1x_1)^5 & \text{if } u_{11}\leq 0\end{cases}$$

$$\frac{\partial\psi_{P_1}(q_1x_1)}{\partial x_1}=\begin{cases}c_1^bq_1+c_2^b2(q_1x_1)q_1+c_3^b3(q_1x_1)^2q_1+c_4^b4(q_1x_1)^3q_1+c_5^b5(q_1x_1)^4q_1 & \text{if } u_{11}>0\\ c_1^cq_1+c_2^c2(q_1x_1)q_1+c_3^c3(q_1x_1)^2q_1+c_4^c4(q_1x_1)^3q_1+c_5^c5(q_1x_1)^4q_1 & \text{if } u_{11}\leq 0\end{cases}$$

$\frac{\partial\psi_{P_1}(q_1x_1)}{\partial x_i}=0$ for $i=2,3,4,5,6,7,8$.

$$\psi_{P_2}(q_2x_3) = \begin{cases} \bar{c}_0^b + \bar{c}_1^b(q_2x_3) + \bar{c}_2^b(q_2x_3)^2 + \bar{c}_3^b(q_2x_3)^3 + \bar{c}_4^b(q_2x_3)^4 + \bar{c}_5^b(q_2x_3)^5 & \text{if } u_{12} > 0 \\ \bar{c}_0^c + \bar{c}_1^c(q_2x_3) + \bar{c}_2^c(q_2x_3)^2 + \bar{c}_3^c(q_2x_3)^3 + \bar{c}_4^c(q_2x_3)^4 + \bar{c}_5^c(q_2x_2)^5 & \text{if } u_{12} \le 0 \end{cases}$$

$$\frac{\partial \psi_{P_2}(q_2x_3)}{\partial x_3} = \begin{cases} \bar{c}_1^b q_2 + \bar{c}_2^b 2(q_2x_3)q_2 + \bar{c}_3^b 3(q_2x_3)^2 q_2 + \bar{c}_4^b 4(q_2x_3)^3 q_2 + \bar{c}_5^b 5(q_2x_3)^4 q_2 & \text{if } u_{12} > 0 \\ \bar{c}_1^c q_2 + \bar{c}_2^c 2(q_2x_3)q_2 + \bar{c}_3^c 3(q_2x_1)^2 q_2 + \bar{c}_4^c 4(q_2x_3)^3 q_2 + \bar{c}_5^c 5(q_2x_3)^4 q_2 & \text{if } u_{12} \le 0 \end{cases}$$

$\frac{\partial \psi_{P_2}(q_2x_3)}{\partial x_i} = 0$ for $i = 1, 2, 4, 5, 6, 7, 8$.

(vii) $\phi_{N_1}(q_1x_1) = d_0^a + d_1^a(q_1x_1) + d_2^a(q_1x_1)^2 + d_3^a(q_1x_1)^3 + d_4^a(q_1x_1)^4 + d_5^a(q_1x_1)^5$. It holds that: $\frac{\partial \phi(N_1)}{\partial x_1} = d_1^a q_1 + d_2^a 2(q_1x_1) + d_3^a 3(q_1x_1)^2 q_1 + d_4^a 4(q_1x_1)^3 q_1 + d_5^a 5(q_1x_1)^4 q_1$ and $\frac{\partial \phi(N_1)}{\partial x_i} = 0$ for $i = 2, 3, 4, 5, 6, 7, 8$.

$\phi_{N_2}(q_2x_3) = \bar{d}_0^a + \bar{d}_1^a(q_2x_3) + \bar{d}_2^a(q_2x_3)^2 + \bar{d}_3^a(q_2x_3)^3 + \bar{d}_4^a(q_2x_3)^4 + \bar{d}_5^a(q_2x_3)^5$. It holds that: $\frac{\partial \phi(N_2)}{\partial x_3} = \bar{d}_1^a q_2 + \bar{d}_2^a 2(q_2x_3) + \bar{d}_3^a 3(q_2x_3)^2 q_2 + \bar{d}_4^a 4(q_2x_3)^3 q_2 + \bar{d}_5^a 5(q_2x_3)^4 q_2$ and $\frac{\partial \phi(N_2)}{\partial x_i} = 0$ for $i = 1, 2, 4, 5, 6, 7, 8$.

(viii) Computation of partial derivatives for $\psi_{N_1}(q_1x_1)$ and $\psi_{N_2}(q_2x_3)$

$$\psi_{N_1}(q_1x_1) = \begin{cases} d_0^b + d_1^b(q_1x_1) + d_2^b(q_1x_1)^2 + d_3^b(q_1x_1)^3 + d_4^b(q_1x_1)^4 + d_5^b(q_1x_1)^5 & \text{if } u_{21} > 0 \\ d_0^c + d_1^c(q_1x_1) + d_2^c(q_1x_1)^2 + d_3^c(q_1x_1)^3 + d_4^c(q_1x_1)^4 + d_5^c(q_1x_1)^5 & \text{if } u_{21} \le 0 \end{cases}$$

$$\frac{\partial \psi_{N_1}(q_1x_1)}{\partial x_1} = \begin{cases} d_1^b q_1 + d_2^b 2(q_1x_1)q_1 + d_3^b 3(q_1x_1)^2 q_1 + d_4^b 4(q_1x_1)^3 q_1 + d_5^b 5(q_1x_1)^4 q_1 & \text{if } u_{21} > 0 \\ d_1^c q_1 + d_2^c 2(q_1x_1)q_1 + d_3^c 3(q_1x_1)^2 q_1 + d_4^c 4(q_1x_1)^3 q_1 + d_5^c 5(q_1x_1)^4 q_1 & \text{if } u_{21} \le 0 \end{cases}$$

$\frac{\partial \psi_{N_1}(q_1x_1)}{\partial x_i} = 0$ for $i = 2, 3, 4, 5, 6, 7, 8$.

$$\psi_{N_2}(q_2x_3) = \begin{cases} \bar{d}_0^b + \bar{d}_1^b(q_2x_3) + \bar{d}_2^b(q_2x_3)^2 + \bar{d}_3^b(q_2x_3)^3 + \bar{d}_4^b(q_2x_3)^4 + \bar{d}_5^b(q_2x_3)^5 & \text{if } u_{22} > 0 \\ \bar{d}_0^c + \bar{d}_1^c(q_2x_3) + \bar{d}_2^c(q_2x_3)^2 + \bar{d}_3^c(q_2x_3)^3 + \bar{d}_4^c(q_2x_3)^4 + \bar{d}_5^c(q_2x_2)^5 & \text{if } u_{22} \le 0 \end{cases}$$

$$\frac{\partial \psi_{N_2}(q_2x_3)}{\partial x_3} = \begin{cases} \bar{d}_1^b q_2 + \bar{d}_2^b 2(q_2x_3)q_2 + \bar{d}_3^b 3(q_2x_3)^2 q_2 + \bar{d}_4^b 4(q_2x_3)^3 q_2 + \bar{d}_5^b 5(q_2x_3)^4 q_2 & \text{if } u_{22} > 0 \\ \bar{d}_1^c q_2 + \bar{d}_2^c 2(q_2x_3)q_2 + \bar{d}_3^c 3(q_2x_1)^2 q_2 + \bar{d}_4^c 4(q_2x_3)^3 q_2 + \bar{d}_5^c 5(q_2x_3)^4 q_2 & \text{if } u_{22} \le 0 \end{cases}$$

$\frac{\partial \psi_{N_2}(q_2x_3)}{\partial x_i} = 0$ for $i = 1, 2, 4, 5, 6, 7, 8$.

Moreover, partial derivatives of the following functions are computed:

(ix) $\frac{\partial}{\partial x_2}[\frac{S_{P_1}}{rT} q_1x_2x_5] = \frac{S_{P_1}}{rT} q_1x_5$, $\frac{\partial}{\partial x_5}[\frac{S_{P_1}}{rT} q_1x_2x_5] = \frac{S_{P_1}}{rT} q_1x_2$ and $\frac{\partial}{\partial x_i}[\frac{S_{P_1}}{rT} q_1x_2x_5] = 0$ for $i = 1, 3, 4, 6, 7, 8$

$\frac{\partial}{\partial x_1}[V_{DP_1} + S_{P_1}\frac{\tilde{l}_1}{2} + S_{P_1}q_1x_1] = S_{P_1}q_1$

$\frac{\partial}{\partial x_i}[V_{DP_1} + S_{P_1}\frac{\tilde{l}_1}{2} + S_{P_1}q_1x_1] = 0$ for $i = 2,3,4,5,6,7,8$

(x) $\frac{\partial}{\partial x_2}[\frac{S_{N_1}}{rT}q_1x_2x_6] = \frac{S_{N_1}}{rT}q_1x_6$, $\frac{\partial}{\partial x_6}[\frac{S_{N_1}}{rT}q_1x_2x_6] == \frac{S_{N_1}}{rT}q_1x_2$ and $\frac{\partial}{\partial x_i}[\frac{S_{N_1}}{rT}q_1x_2x_6] = 0$ for $i = 1,3,4,5,7,8$

$\frac{\partial}{\partial x_1}[V_{DN_1} + S_{N_1}\frac{\tilde{l}_1}{2} + S_{N_1}q_1x_1] = S_{N_1}q_1$

$\frac{\partial}{\partial x_i}[V_{DN_1} + S_{N_1}\frac{\tilde{l}_1}{2} + S_{N_1}q_1x_1] = 0$ for $i = 2,3,4,5,6,7,8$.

(xi) $\frac{\partial}{\partial x_4}[\frac{S_{P_2}}{rT}q_2x_4x_7] = \frac{S_{P_2}}{rT}q_2x_7$, $\frac{\partial}{\partial x_7}[\frac{S_{P_2}}{rT}q_2x_4x_7] = \frac{S_{P_2}}{rT}q_2x_4$, and $\frac{\partial}{\partial x_i}[\frac{S_{P_2}}{rT}q_2x_4x_7] = 0$ for $i = 1,2,3,5,6,8$.

$\frac{\partial}{\partial x_3}[V_{DP_2+\frac{S_{P_2}\tilde{l}_2}{2}} + S_{P_2}q_2x_3] = S_{P_2}q_2 and$ $\frac{\partial}{\partial x_i}[V_{DP_2+\frac{S_{P_2}\tilde{l}_2}{2}} + S_{P_2}q_2x_3] = 0$ for $i = 1,2,4,$ $5,6,7,8$.

(xii) $\frac{\partial}{\partial x_4}[\frac{S_{N_2}}{rT}q_2x_4x_8] = \frac{S_{N_2}}{rT}q_2x_8$, $\frac{\partial}{\partial x_8}[\frac{S_{N_2}}{rT}q_2x_4x_8] = \frac{S_{N_2}}{rT}q_2x_8$, and $\frac{\partial}{\partial x_i}[\frac{S_{N_2}}{rT}q_2x_4x_8] = 0$ for $i = 1,2,3,5,6,7$.

$\frac{\partial}{\partial x_3}[V_{DN_2+\frac{S_{N_2}\tilde{l}_2}{2}} + S_{N_2}q_2x_3] = S_{N_2}q_2 and$ $\frac{\partial}{\partial x_i}[V_{DN_2+\frac{S_{N_2}\tilde{l}_2}{2}} + S_{N_2}q_2x_3] = 0$ for $i = 1,2,4,$ $5,6,7,8$.

2.9.3.2 Stabilizing feedback control

After linearization around its current operating point, the electropneumatically actuated robot is written in the form (Rigatos and Busawon, 2018; Rigatos, 2016b)

$$\dot{x} = Ax + Bu + d_1 \tag{2.413}$$

Parameter d_1 stands for the linearization error in the electropneumatically actuated robot appearing above in (2.413). The reference setpoints for the electropneumatically actuated robot's state vector are denoted by $\mathbf{x_d} = [x_1^d, \cdots, x_8^d]$. Tracking of this trajectory is achieved after applying the control input u_d. At every time instant, the control input u_d is assumed to differ from the control input u appearing in (2.413) by an amount equal to Δu, that is $u_d = u + \Delta u$

$$\dot{x}_d = Ax_d + Bu_d + d_2 \tag{2.414}$$

The dynamics of the controlled system described in (2.413) can be also written as

$$\dot{x} = Ax + Bu + Bu_d - Bu_d + d_1 \tag{2.415}$$

and by denoting $d_3 = -Bu_d + d_1$ as an aggregate disturbance term one obtains

$$\dot{x} = Ax + Bu + Bu_d + d_3 \tag{2.416}$$

By subtracting (2.414) from (2.416) one has

$$\dot{x} - \dot{x}_d = A(x - x_d) + Bu + d_3 - d_2 \tag{2.417}$$

By denoting the tracking error as $e = x - x_d$ and the aggregate disturbance term as $\tilde{d} = d_3 - d_2$, the tracking error dynamics becomes

$$\dot{e} = Ae + Bu + \tilde{d} \tag{2.418}$$

For the approximately linearized model of the system a stabilizing feedback controller is developed. The controller has the form

$$u(t) = -Ke(t) \tag{2.419}$$

with $K = \frac{1}{r}B^T P$ where P is a positive definite symmetric matrix which is obtained from the solution of the Riccati equation (Rigatos and Busawon, 2018; Rigatos, 2016b)

$$A^T P + PA + Q - P\left(\frac{2}{r}BB^T - \frac{1}{\rho^2}LL^T\right)P = 0 \tag{2.420}$$

where Q is a positive semi-definite symmetric matrix. The diagram of the considered control loop is depicted in Figure 2.107. The global stability properties of the control method can be proven following the procedure presented in Section 1.1.2.5.

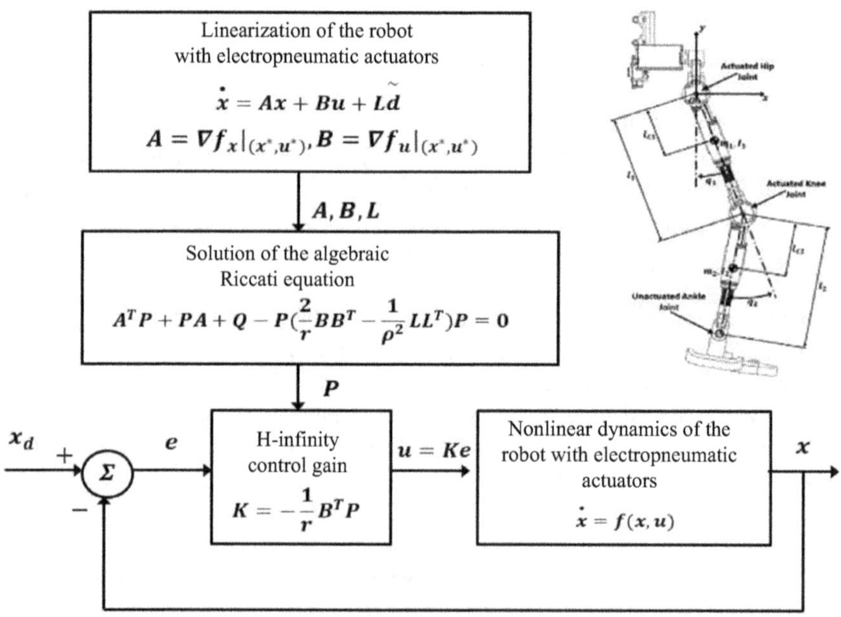

Figure 2.107 Diagram of the nonlinear optimal (H-infinity) control loop for the robotic exoskeleton with electropneumatic actuators

It is also noted that the solution of the H-infinity feedback control problem for the electropneumatically actuated robot and the computation of the worst case disturbance that this controller can sustain, comes from superposition of Bellman's optimality principle when considering that the actuator is affected by two separate inputs (i) the control input u and (ii) the cumulative disturbance input $\tilde{d}(t)$. Solving the optimal control problem for u that is for the minimum variation (optimal) control input that achieves elimination of the state vector's tracking error gives $u = -\frac{1}{r}B^T Pe$. Equivalently, solving the optimal control problem for $\tilde{d}$, that is for the worst case disturbance that the control loop can sustain gives $\tilde{d} = \frac{1}{\rho^2}L^T Pe$.

2.9.4 *Differential flatness of the robot with electropneumatic actuation*

It will be proven that the 2-link electropneumatically actuated robotic exoskeleton is a differentially flat system with flat outputs $y = [y_1, y_2, y_3, y_4]$ where

$$y_1 = x_1 = \theta_1 \quad y_2 = x_3 = \theta_2 \quad y_3 = x_5 = P_{P_1} \quad y_4 = x_7 = P_{P_2} \tag{2.421}$$

This signifies that all state variables and the control inputs of the robot can be written as differential functions of the robot's flat outputs. It also signifies that the flat outputs are not connected between them through a relation in the form of an homogeneous differential equation (Rigatos and Busawon, 2018; Rigatos, 2015). From the first row of the state-space model of the robot, one has

$$x_2 = \dot{x}_1 \Rightarrow x_2 = h_2(y, \dot{y}) \tag{2.422}$$

consequently, state variable x_2 is a differential function of the robot's flat outputs. Equivalently, from the third row of the state-space model, one has

$$x_4 = \dot{x}_3 \Rightarrow x_4 = h_4(y, \dot{y}) \tag{2.423}$$

consequently, state variable x_4 is a differential function of the robot's flat outputs. Next, from the second and fourth rows of the state-space model of the robot one has a system of equations with respect to state variables x_6 and x_8. Indeed:

$$\begin{aligned} &\dot{x}_2 - \frac{-M_{22}(C_1+G_1)+M_{12}(C_2+G2)}{M_{11}M_{22}-M_{12}^2} - \frac{M_{22}[S_{P_1}x_5 - b_1q_1x_2]}{M_{11}M_{22}-M_{12}^2} - \frac{M_2[S_{P_1}x_7 - b_2q_2x_4]}{M_{11}M_{22}-M_{12}^2} \\ &\quad = \frac{-M_{22}S_{N_1}}{M_{11}M_{22}-M_{12}^2}x_6 + \frac{M_{11}S_{N_2}}{M_{11}M_{22}-M_{12}^2}x_8 \end{aligned} \tag{2.424}$$

$$\begin{aligned} &\dot{x}_4 - \frac{M_{12}(C_1+G_1)-M_{11}(C_2+G2)}{M_{11}M_{22}-M_{12}^2} - \frac{-M_{21}[S_{P_1}x_5 - b_1q_1x_2]}{M_{11}M_{22}-M_{12}^2} - \frac{M_{11}[S_{P_1}x_7 - b_2q_2x_4]}{M_{11}M_{22}-M_{12}^2} \\ &\quad = \frac{M_{21}S_{N_1}}{M_{11}M_{22}-M_{12}^2}x_6 - \frac{M_{11}S_{N_2}}{M_{11}M_{22}-M_{12}^2}x_8 \end{aligned} \tag{2.425}$$

Equivalently, (2.424) and (2.425) can be written as

$$\begin{aligned} &\dot{x}_2[M_{11}M_{22} - M_{12}^2] - [-M_{22}(C_1+G_1) + M_{12}(C_2+G_2)] \\ &\quad -M_{22}[S_{P_1}x_5 - b_1q_1x_2] - M_{12}[S_{P_2}x_7 - b_2q_2x_4] = -M_{22}S_{N_1}x_6 + M_{12}S_{N_2}x_8 \end{aligned} \tag{2.426}$$

$$\begin{aligned} &\dot{x}_4[M_{11}M_{22} - M_{12}^2] - [M_{12}(C_1+G_1) - M_{11}(C_2+G_2)] \\ &\quad +M_{21}[S_{P_1}x_5 - b_1q_1x_2] - M_{11}[S_{P_2}x_7 - b_2q_2x_4] = M_{21}S_{N_1}x_6 - M_{11}S_{N_2}x_8 \end{aligned} \tag{2.427}$$

From (2.426) and (2.427), one can solve with respect to x_6 and x_8. This gives:

$$\begin{pmatrix} x_6 \\ x_8 \end{pmatrix} = \begin{pmatrix} -M_{22}S_{N_1} & M_{12}S_{N_2} \\ M_{21}S_{N_1} & -M_{11}S_{N_2} \end{pmatrix}^{-1} \cdot \begin{pmatrix} \dot{x}_2[M_{11}M_{22}-M_{12}^2]-[-M_{22}(C_1+G_1)+M_{12}(C_2+G_2)]- \\ M_{22}[S_{P_1}x_5-b_1q_1x_2]-M_{12}[S_{P_2}x_7-b_2q_2x_4] \\ \\ \dot{x}_4[M_{11}M_{22}-M_{12}^2]-[M_{12}(C_1+G_1)-M_{11}(C_2+G_2)]+ \\ M_{21}[S_{P_1}x_5-b_1q_1x_2]-M_{11}[S_{P_2}x_7-b_2q_2x_4] \end{pmatrix} \quad (2.428)$$

From (2.428) and taking into account that state variables x_1, x_3, x_5, and x_7 are also the flat outputs of the robot, while state variables x_2 and x_4 have already been expressed in the form of differential functions of the robot's flat outputs, one also obtains that state variables x_6 and x_8 are differential functions of the robotic system's flat outputs. Thus x_6 and x_8 can be rewritten as

$$x_6 = h_6(y,\dot{y},\ddot{y}) \quad x_8 = h_8(y,\dot{y},\ddot{y}) \quad (2.429)$$

Next, from the fifth row of the state-space model of the robot, one has

$$\begin{aligned} &\dot{x}_5[V_{DP_1}+S_{P_1}\frac{\tilde{l}_1}{2}+S_{P_1}q_1x_1] \\ &\quad = K_rT[\phi_{P_1}(q_1x_1)-\frac{S_{P_1}}{rT}q_1x_2x_5]+K_rT\psi_{P_1}(q_1x_1)u_{11}\Rightarrow \\ &u_{11} = \frac{\dot{x}_5[V_{DP_1}+S_{P_1}\frac{\tilde{l}_1}{2}+S_{P_1}q_1x_1]-K_rT[\phi_{P_1}(q_1x_1)-\frac{S_{P_1}}{rT}q_1x_2x_5]}{K_rT\psi_{P_1}(q_1x_1)} \end{aligned} \quad (2.430)$$

Equation (2.430) denotes that the control input u_{11} is a differential function of the flat outputs of the system, that is

$$u_{11} = h_{u_{11}}(y,\dot{y},\ddot{y}) \quad (2.431)$$

Besides, from the sixth row of the state-space model, one has that

$$\begin{aligned} &\dot{x}_6[V_{DN_1}+S_{N_1}\frac{\tilde{l}_1}{2}+S_{N_1}q_1x_1] = \\ &K_rT[\phi_{N_1}(q_1x_1)+\frac{S_{N_1}}{rT}q_1x_2x_6]+K_rT\psi_{N_1}(q_1x_1)u_{21}\Rightarrow \\ &u_{21} = \frac{\dot{x}_6[V_{DN_1}+S_{N_1}\frac{\tilde{l}_1}{2}+S_{N_1}q_1x_1]-K_rT[\phi_{N_1}(q_1x_1)+\frac{S_{N_1}}{rT}q_1x_2x_6]}{K_rT\psi_{N_1}(q_1x_1)} \end{aligned} \quad (2.432)$$

Equation (2.432) denotes that the control input u_{21} is a differential function of the flat outputs of the system, or

$$u_{21} = h_{u_{21}}(y,\dot{y},\ddot{y}) \quad (2.433)$$

Equivalently, from the seventh row of the state-space model of the robot, one has

$$\dot{x}_7[V_{DP_2}+S_{P_2}\frac{\tilde{l}_2}{2}+S_{P_2}q_2x_3]=$$
$$K_rT[\phi_{P_2}(q_2x_3)-\frac{S_{P_2}}{rT}q_2x_4x_7]+K_rT\psi_{P_2}(q_2x_3)u_{12}\Rightarrow \tag{2.434}$$
$$u_{12}=\frac{\dot{x}_7[V_{DP_2}+S_{P_2}\frac{\tilde{l}_2}{2}+S_{P_2}q_2x_3]-K_rT[\phi_{P_2}(q_2x_3)-\frac{S_{P_2}}{rT}q_2x_4x_7]}{K_rT\psi_{P_2}(q_2x_3)}$$

Equation (2.434) denotes that the control input u_{12} is a differential function of the flat outputs of the system, or

$$u_{12}=h_{u_{12}}(y,\dot{y},\ddot{y}) \tag{2.435}$$

Finally, from the eighth row of the state-space model, one has that

$$\dot{x}_8[V_{DN_2}+S_{N_2}\frac{\tilde{l}_2}{2}+S_{N_2}q_2x_3]=$$
$$K_rT[\phi_{N_2}(q_2x_3)+\frac{S_{N_2}}{rT}q_2x_4x_8]+K_rT\psi_{N_2}(q_2x_3)u_{22}\Rightarrow \tag{2.436}$$
$$u_{22}=\frac{\dot{x}_8[V_{DN_2}+S_{N_2}\frac{\tilde{l}_2}{2}+S_{N_2}q_2x_3]-K_rT[\phi_{N_2}(q_2x_3)+\frac{S_{N_2}}{rT}q_2x_4x_8]}{K_rT\psi_{N_2}(q_2x_3)}$$

Equation (2.436) denotes that the control input u_{22} is a differential function of the flat outputs of the system, or

$$u_{22}=h_{u_{22}}(y,\dot{y},\ddot{y}) \tag{2.437}$$

Thus, it has been proven that all state variables and the control inputs of the electropneumatically actuated robot are differential functions of the flat outputs of the system. Consequently, this is a differentially flat robotic system. On the one side the differential flatness properties allow the transformation of this robotic system into an input-output linearized form. On the other side, by using the relations that connect the state variable of the robot to the flat outputs one can solve the problem of setpoints selection. Actually, setpoints are defined first for the flat outputs of the robot, that is

$$y_1^d=x_1^d \quad y_2^d=x_2^d \quad y_3^d=x_3^d \quad y_4^d=x_4^d \tag{2.438}$$

and using the properties of differential flatness one defines setpoints also for the rest of the robot's state variables, that is x_2^d, x_4^d, x_6^d and x_8^d.

2.9.5 Simulation tests

The performance of the proposed nonlinear optimal (H-infinity) control method for electropneumatically actuated robots, and particularly for the model of the 2-link robotic exoskeleton, has been further confirmed through simulation experiments. To compute the feedback gains of the H-infinity controller the algebraic Riccati equation of (2.420) was being repetitively solved at each iteration of the control algorithm. The sampling period was chosen to be $T_s=0.01$ sec. The feasible setpoints for the state variables of the robot were computed by exploiting the differential flatness properties of its state-space model. The obtained results are depicted in Figures 2.108 to 2.123. It can be noticed that in all simulation experiments fast and accurate tracking of these setpoints by the state variables of the pneumatically actuated robots was achieved. It is noted that state estimation-based control has been implemented by using the H-infinity

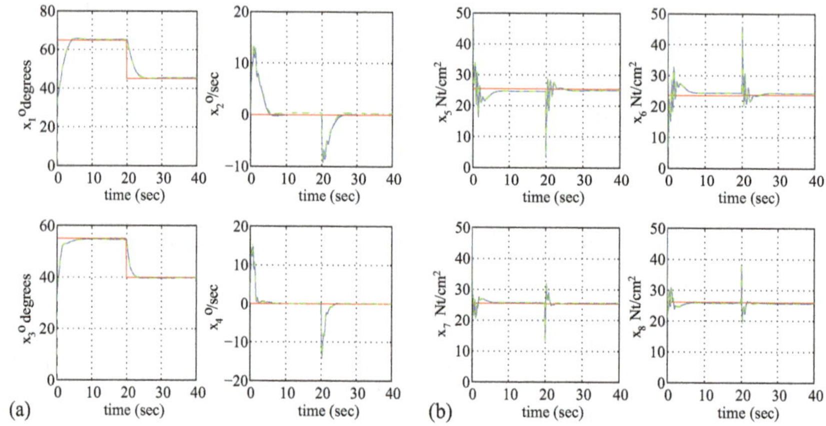

Figure 2.108 Tracking of setpoint 1 by the electropneumatically actuated robot: (a) convergence of state variables x_1 to x_4 (blue lines) to the reference setpoints (red lines) and their KF-based estimation (green line); (b) convergence of state variables x_5 to x_8 (blue lines) to the reference setpoints (red lines) and their KF-based estimation (green line)

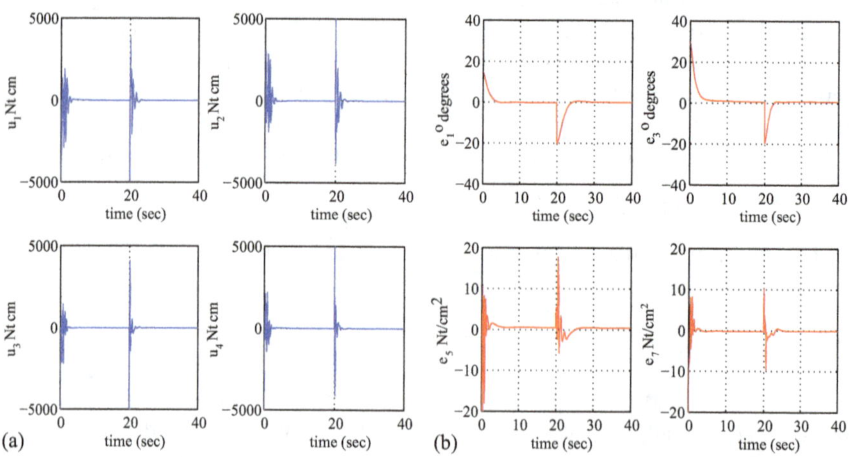

Figure 2.109 Tracking of setpoint 1 by the electropneumatically actuated robot: (a) variation of control inputs u_1 to u_4; (b) variation of tracking error variables e_1 to e_4 for the odd-numbered state variables of the robot

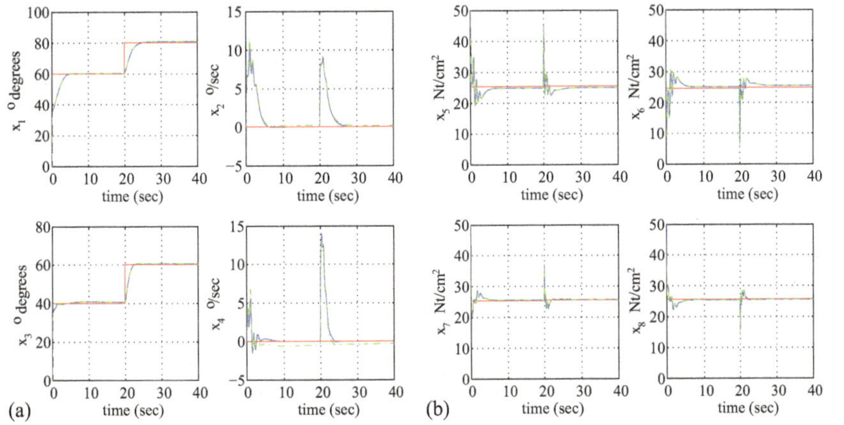

Figure 2.110 Tracking of setpoint 2 by the electropneumatically actuated robot: (a) convergence of state variables x_1 to x_4 (blue lines) to the reference setpoints (red lines) and their KF-based estimation (green line); (b) convergence of state variables x_5 to x_8 (blue lines) to the reference setpoints (red lines) and their KF-based estimation (green line)

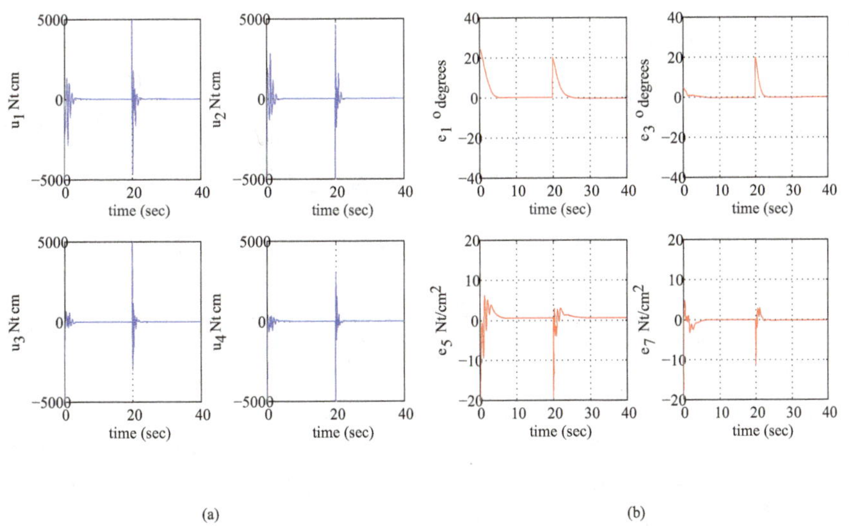

Figure 2.111 Tracking of setpoint 2 by the electropneumatically actuated robot: (a) variation of control inputs u_1 to u_4; (b) variation of tracking error variables e_1 to e_4 for the odd-numbered state variables of the robot

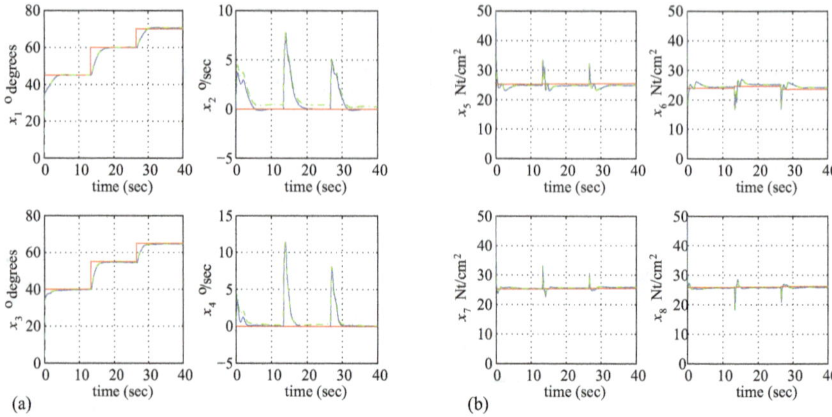

Figure 2.112 Tracking of setpoint 3 by the electropneumatically actuated robot: (a) convergence of state variables x_1 to x_4 (blue lines) to the reference setpoints (red lines) and their KF-based estimation (green line); (b) convergence of state variables x_5 to x_8 (blue lines) to the reference setpoints (red lines) and their KF-based estimation (green line)

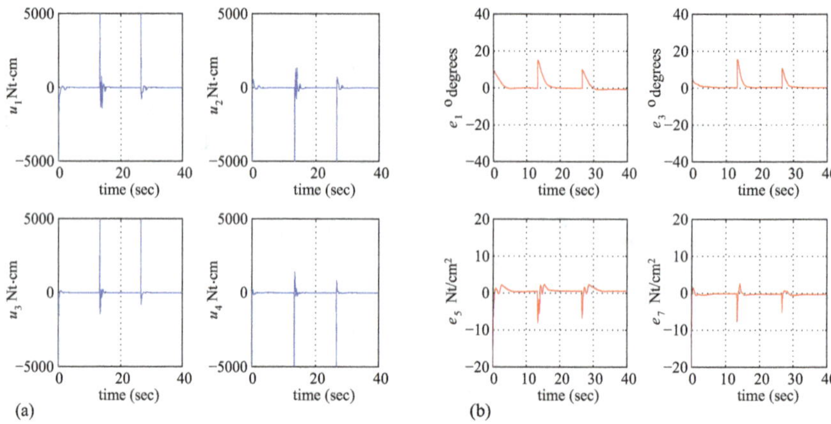

Figure 2.113 Tracking of setpoint 3 by the electropneumatically actuated robot: (a) variation of control inputs u_1 to u_4; (b) variation of tracking error variables e_1 to e_4 for the odd-numbered state variables of the robot

Kalman Filter as a robust state observer. Thus, in the performed experiments, the real values of the state variables of the robot have been printed in blue, the associated setpoints have been printed in red while the estimated values which were provided by the H-infinity Kalman Filter were plotted in green.

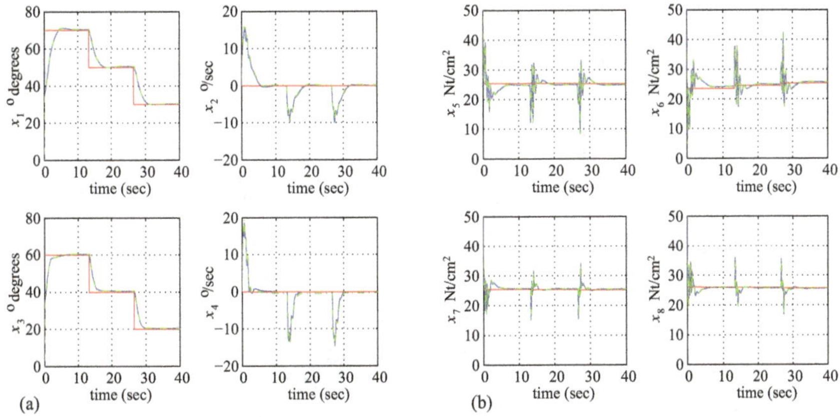

Figure 2.114 Tracking of setpoint 4 by the electropneumatically actuated robot: (a) convergence of state variables x_1 to x_4 (blue lines) to the reference setpoints (red lines) and their KF-based estimation (green line); (b) convergence of state variables x_5 to x_8 (blue lines) to the reference setpoints (red lines) and their KF-based estimation (green line)

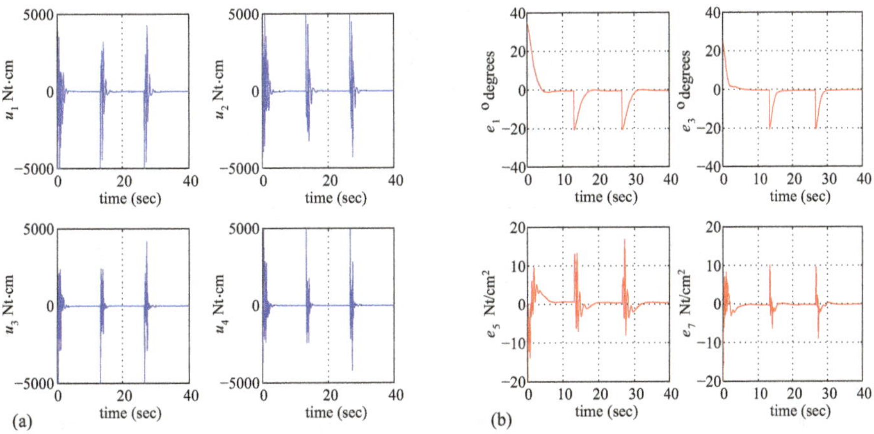

Figure 2.115 Tracking of setpoint 4 by the electropneumatically actuated robot: (a) variation of control inputs u_1 to u_4; (b) variation of tracking error variables e_1 to e_4 for the odd-numbered state variables of the robot

2.10 Nonlinear control of flexible joint robots

2.10.1 Outline

Synchronization between the turn angles of the robots' joints and the turn angles of the motors that provide actuation to these joints is not always direct to achieve

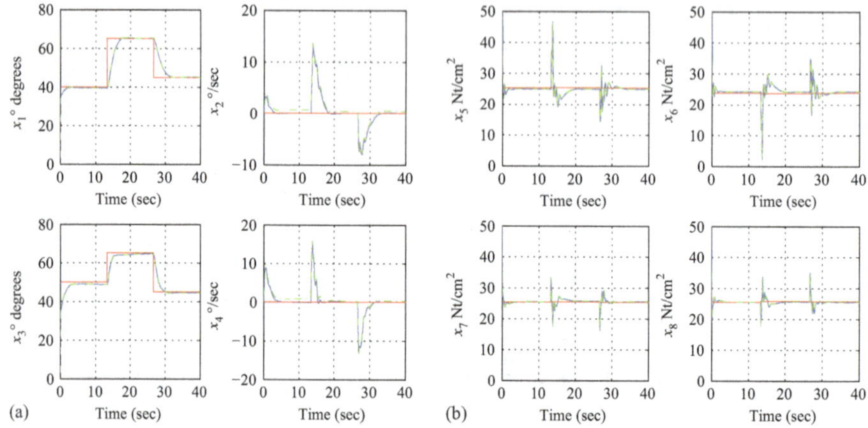

Figure 2.116 *Tracking of setpoint 5 by the electropneumatically actuated robot: (a) convergence of state variables* x_1 *to* x_4 *(blue lines) to the reference setpoints (red lines) and their KF-based estimation (green line); (b) convergence of state variables* x_5 *to* x_8 *(blue lines) to the reference setpoints (red lines) and their KF-based estimation (green line)*

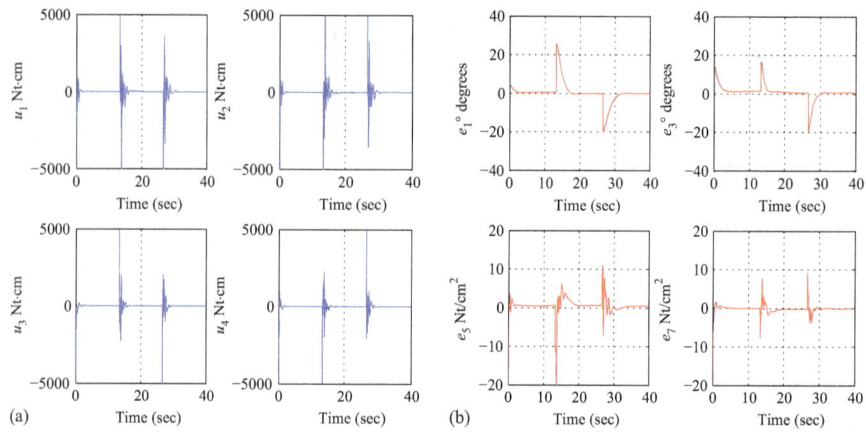

Figure 2.117 *Tracking of setpoint 5 by the electropneumatically actuated robot: (a) variation of control inputs* u_1 *to* u_4*; (b) variation of tracking error variables* e_1 *to* e_4 *for the odd-numbered state variables of the robot*

(Sun, Yu, Yang *et al.*, 2018; Kim, 2018; Lightcap and Banks, 2010; Ruderman, 2016). This happens for instance in the case of flexible-joint robots or in the case of robots with compliant actuators (Yu, Huang *et al.*, 2015; Li, Pan *et al.*, 2018). Flexibility in the joints of a robotic manipulator can affect negatively the precise

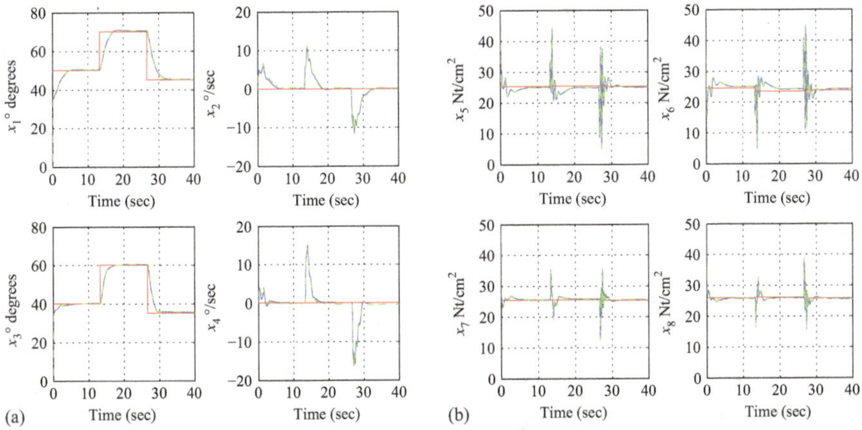

Figure 2.118 *Tracking of setpoint 6 by the electropneumatically actuated robot: (a) convergence of state variables x_1 to x_4 (blue lines) to the reference setpoints (red lines) and their KF-based estimation (green line); (b) convergence of state variables x_5 to x_8 (blue lines) to the reference setpoints (red lines) and their KF-based estimation (green line)*

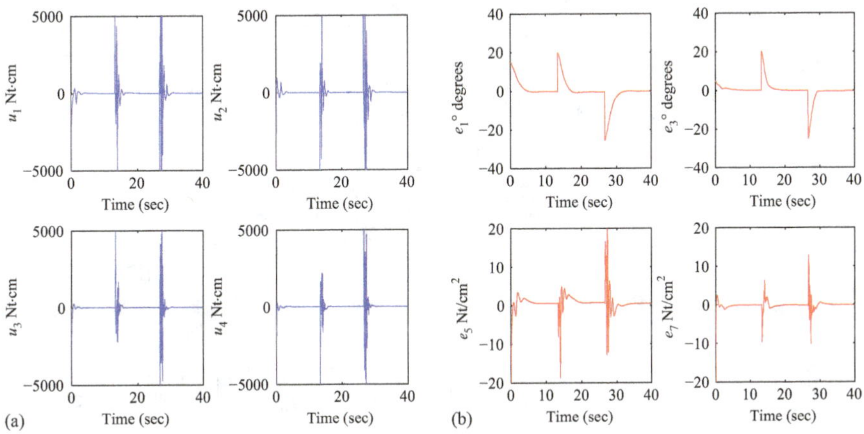

Figure 2.119 *Tracking of setpoint 6 by the electropneumatically actuated robot: (a) variation of control inputs u_1 to u_4; (b) variation of tracking error variables e_1 to e_4 for the odd-numbered state variables of the robot*

positioning of the robots' end-effector, and consequently can hinder the execution of several robotic tasks in which precise positioning is needed, as for instance robotic-guided radiotherapy, or robotic surgery (Niu, Shen *et al.*, 2019; Chen and Guo, 2019). On the other side, flexibility in the joints of a robotic manipulator or the use of compliant actuators can be useful, for instance when the appearance of abrupt

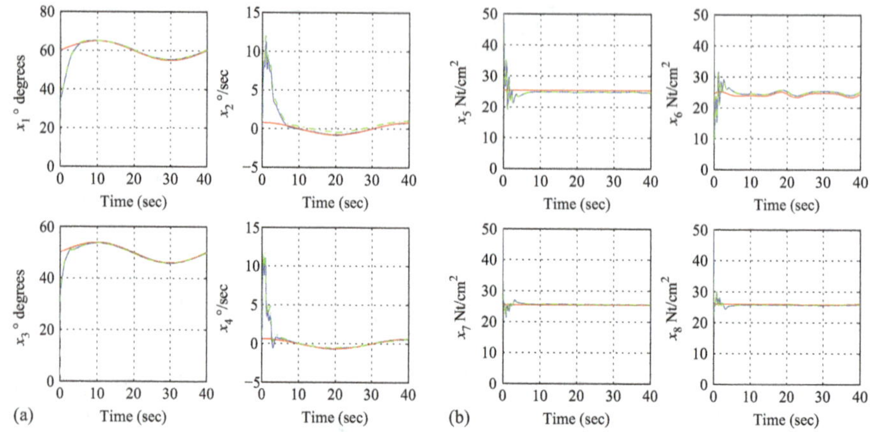

Figure 2.120 Tracking of setpoint 7 by the electropneumatically actuated robot: (a) convergence of state variables x_1 to x_4 (blue lines) to the reference setpoints (red lines) and their KF-based estimation (green line); (b) convergence of state variables x_5 to x_8 (blue lines) to the reference setpoints (red lines) and their KF-based estimation (green line)

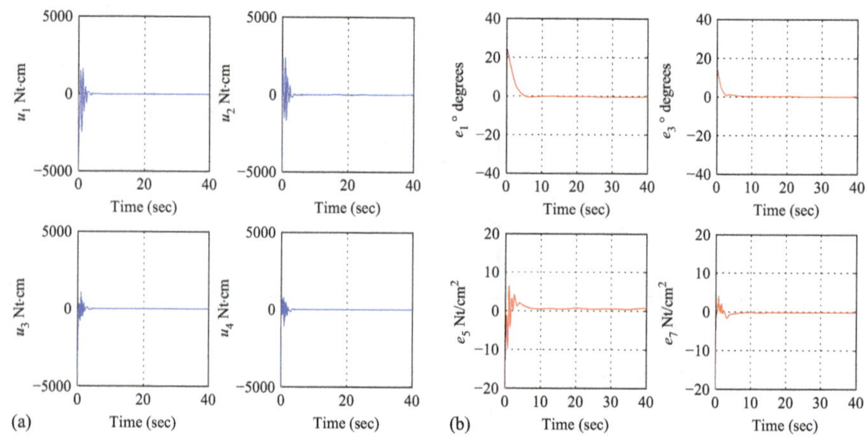

Figure 2.121 Tracking of setpoint 7 by the electropneumatically actuated robot: (a) variation of control inputs u_1 to u_4; (b) variation of tracking error variables e_1 to e_4 for the odd-numbered state variables of the robot

contact forces has to be avoided so as to assure the safe execution of several contact-based tasks (Kim and Song, 2012; Buandonno and de Luca, 2018). So far, several control approaches have been applied to robots with flexible joints and to robots with compliant actuators. One can note for instance sliding-mode control, sliding-mode control, backstepping control, or model predictive control (Yan, Lai *et al.*, 2020; Xiu,

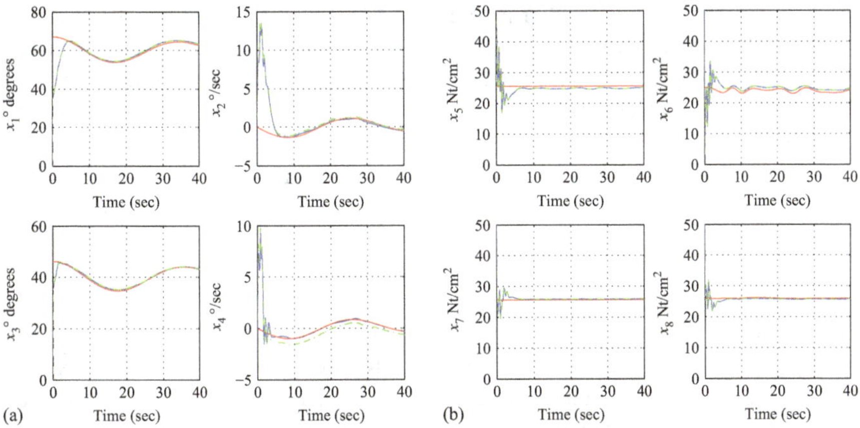

Figure 2.122 *Tracking of setpoint 8 by the electropneumatically actuated robot: (a) convergence of state variables x_1 to x_4 (blue lines) to the reference setpoints (red lines) and their KF-based estimation (green line); (b) convergence of state variables x_5 to x_8 (blue lines) to the reference setpoints (red lines) and their KF-based estimation (green line)*

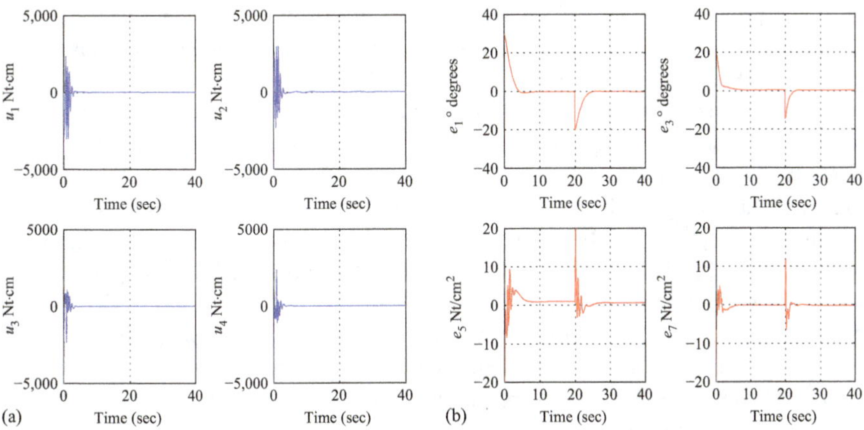

Figure 2.123 *Tracking of setpoint 8 by the electropneumatically actuated robot: (a) variation of control inputs u_1 to u_4; (b) variation of tracking error variables e_1 to e_4 for the odd-numbered state variables of the robot*

Xing *et al.*, 2019; Wang *et al.*, 2019; Yan, Zhang *et al.*, 2018; El Makrini *et al.*, 2017; Pan, Wang *et al.*, 2018). Often, nonlinear state-observers are used in such a type of robotic manipulators, so as to either estimate nonlinear state vector elements or to estimate disturbance forces and torques (Rsetam *et al.*, 2020; Talole, Kohle *et al.*, 2010; Farza *et al.*, 2013; Yun and Sun, 2014). On the other side, due to strong nonlinearities in the dynamic model of these robots and due to their multivariable state-space structure, little has been done towards the solution of the related nonlinear optimal control problem (Braun *et al.*, 2013; Order and Poznyak, 2012; Ordar and Poznyak, 2015; Sanchez, Cuvas *et al.*, 2020; Misgeld, Hewing *et al.*, 2019). Besides, the fact that flexible-joint robots are underactuated imposes additional difficulty in the implementation of optimal control schemes to them (Rigatos and Busawon, 2018; Coverly *et al.*, 2016).

In the present section, a novel nonlinear optimal control method is developed for robotic manipulators with flexible joints (Rigatos, 2016b; Rigatos, Siano and Cecati, 2015; Rigatos, Siano *et al.*, 2015). First, the state-space model of these robots undergoes approximate linearization around a time-varying operating point which is recomputed at each time-step of the control algorithm. This operating point is defined by the present value of the robot's state vector and by the last sampled value of its control inputs vector. The linearization process relies on the first-order Taylor series expansion and on the computation of the associated Jacobian matrices (Rigatos and Tzafestas, 2007; Basseville and Nikiforov, 1993; Rigatos and Zhang, 2009). The modeling inaccuracy which is due to truncation of higher-order terms in the Taylor series expansion is considered to be a perturbation which is asymptotically compensated by the robustness of the control method. For the approximately linearized model of the flexible-joint robots, a stabilizing (H-infinity) feedback controller is designed.

The proposed H-infinity controller achieves a solution to the optimal control problem for flexible-joint manipulators under model uncertainty and exogenous perturbations (Rigatos and Busawon, 2018). Actually, it represents a min–max differential game taking place between: (i) the control inputs, which try to minimize a cost function that comprises a quadratic term of the state vector's tracking error; (ii) the model uncertainty and disturbance inputs which try to maximize this cost function. To compute the stabilizing feedback gains of the H-infinity controller, as algebraic Riccati equation is repetitively solved at each iteration of the control algorithm (Rigatos and Busawon, 2018; Rigatos, 2011; Rigatos, 2015). The global stability properties of the control scheme are proven through Lyapunov analysis. First, it is shown that the control method achieves the H-infinity tracking performance criterion, which signifies elevated robustness against model uncertainty and external perturbations (Rigatos and Busawon, 2018; Tousssaint, Basar and Bullo, 2000). Moreover, under moderate conditions, it is proven that the control loop of the flexible-joint robotic manipulators is globally asymptotically stable (Rigatos and Busawon, 2018). Finally, to implement state estimation-based control without the need to measure the entire state vector of the flexible-joints robot, the H-infinity Kalman Filter is used as a robust state estimator (Rigatos and Busawon, 2018; Rigatos, 2015). It is remarkable that the proposed control method retains the advantages of linear optimal control, that is fast and accurate tracking of reference setpoints under moderate variations of the control inputs.

2.10.2 Dynamic model of a multi-DOF robotic manipulator with flexible joints

2.10.2.1 State-space model of the robotic system

The model of a multi-DOF flexible-joint robot is depicted in Figure 2.124. Under the assumption that the masses of the rigid links are concentrated at the links' end the dynamic model is given by (Sun, Yu, Yang *et al.*, 2018; Kim, 2018; Lightcap and Banks, 2010; Ruderman, 2016; Rigatos and Busawon, 2018)

$$\begin{aligned}&\tilde{M}(\theta)\ddot{\theta}+\tilde{C}(\theta,\dot{\theta})+\tilde{G}(\theta)=\tilde{K}(\theta-\theta_m)+\tilde{D}\dot{\theta}\\&\tilde{J}\ddot{\theta}_m+\tilde{B}\dot{\theta}_m+\tilde{K}(\theta-\theta_m)=\tau\end{aligned} \tag{2.439}$$

where $\theta\in R^{2\times 1}$ is the vector of the joints of the robotic manipulator, and $\theta_m\in R^{2\times 1}$ is the vector of the turn angles of the robot's actuators. Besides $\tilde{M}(\theta)$ is the inertia matrix

$$\tilde{M}(\theta)=\begin{pmatrix}M_{11} & M_{12}\\ M_{21} & M_{22}\end{pmatrix} \tag{2.440}$$

where $M_{11}=(m_1+m_2)l_1^2+m_1l_2^2+2m_2l_1l_2\cos(\theta_2)$, $M_{12}=m_2l_2^2+m_2l_1l_2\cos(\theta_2)$, $M_{21}=m_2l_2^2+m_2l_1l_2\cos(\theta_2)$ and $M_{22}=m_2l_2(l_1+l_2)\cos(\theta_2)$. Moreover, $C(\theta,\dot{\theta})$ is the Coriolis and centrifugal forces vector

$$\tilde{C}(\theta,\dot{\theta})=\begin{pmatrix}-m_2l_1^2\sin(\theta_2)\dot{\theta}_2^2-2m_2l_1^2\sin(\theta_2)\dot{\theta}_1\dot{\theta}_2\\ m_2l_1^2\sin(\theta_2)\dot{\theta}_1^2\end{pmatrix} \tag{2.441}$$

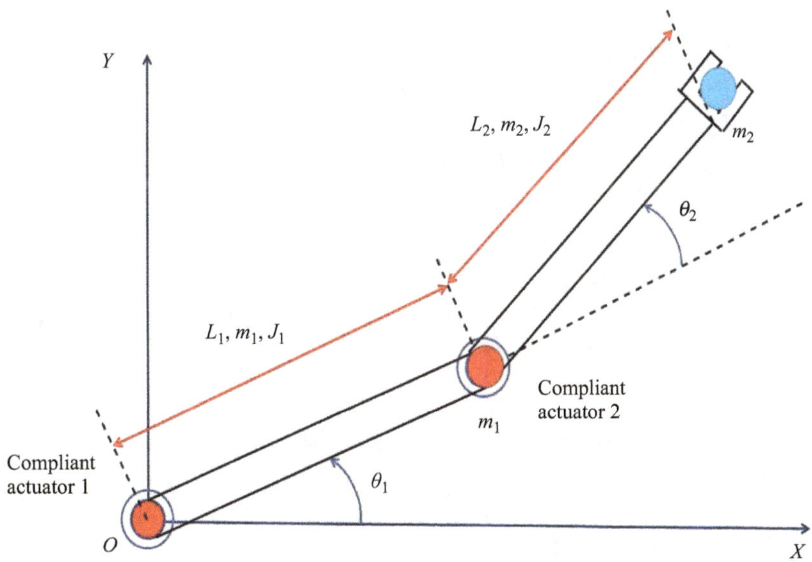

Figure 2.124 Diagram of the 2-link robotic manipulator with flexible joints

$\tilde{G}(\theta)$ is the gravitational forces vector

$$\tilde{G}(\theta) = \begin{pmatrix} (m_1 + m_2)gl_1 \cos(\theta_1) + m_2 gl_2 \cos(\theta_1 + \theta_2) \\ m_2 gl_2 \cos(\theta_1 + \theta_2) \end{pmatrix} \tag{2.442}$$

$\tilde{K}$ is an elasticity coefficients matrix defining the coupling between the joints and the actuators, $\tilde{D}$ is a damping coefficients matrix for the robots' joints, while $\tilde{J}$ is an inertia matrix of the robots' actuators, and B is a friction coefficients' matrix again for the robots' actuators, which are given by

$$\tilde{K} = \begin{pmatrix} k_1 & 0 \\ 0 & k_2 \end{pmatrix} \quad \tilde{D} = \begin{pmatrix} d_1 & 0 \\ 0 & d_2 \end{pmatrix} \quad \tilde{J} = \begin{pmatrix} J_1 & 0 \\ 0 & J_2 \end{pmatrix} \quad B = \begin{pmatrix} b_1 & 0 \\ 0 & b_2 \end{pmatrix} \tag{2.443}$$

The control input torques' vector is denoted as $\tau \in R^{2\times 1}$. Writing explicitly the differential equations of the two actuators, one obtains

$$\begin{aligned} J_1\ddot{\theta}_{m_1} + b_1\dot{\theta}_{m_1} + k_1(\theta_1 - \theta_{m_1}) = \tau_1 \\ J_2\ddot{\theta}_{m_2} + b_2\dot{\theta}_{m_2} + k_2(\theta_2 - \theta_{m_2}) = \tau_2 \end{aligned} \tag{2.444}$$

where τ_1 and τ_2 are the torques which drive the turn motion of the two actuators. The above two equations are also written in the following form:

$$\begin{aligned} \begin{pmatrix} \ddot{\theta}_1 \\ \ddot{\theta}_2 \end{pmatrix} &= -\begin{pmatrix} M_{11} & M_{12} \\ M_{21} & M_{22} \end{pmatrix}^{-1} \begin{pmatrix} C_1(\theta,\dot{\theta}) + G_1(\theta) \\ C_2(\theta,\dot{\theta}) + G_2(\theta) \end{pmatrix} \\ &\quad + \begin{pmatrix} M_{11} & M_{12} \\ M_{21} & M_{22} \end{pmatrix}^{-1} \begin{pmatrix} k_1(\theta_1 - \theta_{m_1}) + d_1\dot{\theta}_1 \\ k_2(\theta_2 - \theta_{m_2}) + d_2\dot{\theta}_2 \end{pmatrix} \end{aligned} \tag{2.445}$$

$$\begin{pmatrix} \ddot{\theta}_{m_1} \\ \ddot{\theta}_{m_2} \end{pmatrix} = \begin{pmatrix} -\frac{b_1}{J_1}\dot{\theta}_{m_1} - \frac{k_1}{J_1}(\theta_{m_1} - \theta_1) \\ -\frac{b_2}{J_2}\dot{\theta}_{m_2} - \frac{k_2}{J_2}(\theta_{m_2} - \theta_2) \end{pmatrix} + \begin{pmatrix} \frac{1}{J_1} & 0 \\ 0 & \frac{1}{J_2} \end{pmatrix} \begin{pmatrix} \tau_1 \\ \tau_2 \end{pmatrix} \tag{2.446}$$

About the inverse inertia matrix of the robot, one has

$$\begin{pmatrix} M_{11} & M_{12} \\ M_{21} & M_{22} \end{pmatrix}^{-1} = \frac{1}{detM} \begin{pmatrix} M_{22} & -M_{12} \\ -M_{21} & M_{11} \end{pmatrix} \tag{2.447}$$

where the determinant is $detM = M_{11}M_{22} - M_{12}M_{21}$. Consequently, the aggregate dynamic model of the 2-link robot with flexible links is written as:

$$\begin{pmatrix} \ddot{\theta}_1 \\ \ddot{\theta}_2 \end{pmatrix} = \begin{pmatrix} -\frac{M_{22}}{detM}(C_1(\theta,\dot{\theta}) + G_1(\theta)) + \\ \frac{M_{12}}{detM}(C_2(\theta,\dot{\theta}) + G_2(\theta)) + \\ \frac{M_{22}}{detM}(k_1(\theta_1 - \theta_{m_1}) + d_1\dot{\theta}_1) - \frac{M_{12}}{detM}(k_2(\theta_2 - \theta_{m_2}) + d_2\dot{\theta}_2) \\ \frac{M_{21}}{detM}(C_1(\theta,\dot{\theta}) + G_1(\theta)) - \\ \frac{M_{11}}{detM}(C_2(\theta,\dot{\theta}) + G_2(\theta)) - \\ \frac{M_{21}}{detM}(k_1(\theta_1 - \theta_{m_1}) + d_1\dot{\theta}_1) + \frac{M_{11}}{detM}(k_2(\theta_2 - \theta_{m_2}) + d_2\dot{\theta}_2) \end{pmatrix} \tag{2.448}$$

$$\begin{pmatrix} \ddot{\theta}_{m_1} \\ \ddot{\theta}_{m_2} \end{pmatrix} = \begin{pmatrix} -\frac{b_1}{J_1\dot{\theta}_{m_1}} - \frac{k_1}{J_1}(\theta_{m_1} - \theta_1) + \frac{1}{J_1}\tau_1 \\ -\frac{b_2}{J_2\dot{\theta}_{m_2}} - \frac{k_2}{J_2}(\theta_{m_2} - \theta_2) + \frac{1}{J_2}\tau_2 \end{pmatrix} \tag{2.449}$$

Next, the state variables of the system are defined as follows: $x_1 = \theta_1$, $x_2 = \dot{\theta}_1$, $x_3 = \theta_2$, $x_4 = \dot{\theta}_2$, $x_5 = \theta_{m_1}$, $x_6 = \dot{\theta}_{m_1}$, $x_7 = \theta_{m_2}$, and $x_8 = \dot{\theta}_{m_2}$. The state vector of the system is $x = [x_1, x_2, x_3, x_4, x_5, x_6, x_7, x_8]^T$ and the control inputs vector is denoted as $u = [u_1, u_2]^T$ or $u = [\tau_1, \tau_2]^T$. Thus, the state-space model of the system is written as:

$$\begin{aligned} \dot{x}_1 &= x_2 \\ \dot{x}_2 &= -\frac{M_{22}}{detM}(C_1(x_1, x_2, x_3, x_4) + G_1(x_1, x_3)) \\ &\quad + \frac{M_{12}}{detM}(C_2(x_1, x_2, x_3, x_4) + G_2(x_1, x_3)) \\ &\quad + \frac{M_{22}}{detM}(k_1(x_1 - x_5) + d_1x_2) - \frac{M_{12}}{detM}(k_2(x_3 - x_7) + d_2x_4) \\ \dot{x}_3 &= x_4 \\ \dot{x}_4 &= \frac{M_{21}}{detM}(C_1(x_1, x_2, x_3, x_4) + G_1(x_1, x_3)) \\ &\quad - \frac{M_{11}}{detM}(C_2(x_1, x_2, x_3, x_4) + G_2(x_1, x_3)) \\ &\quad - \frac{M_{21}}{detM}(k_1(x_1 - x_5) + d_1x_2) + \frac{M_{11}}{detM}(k_2(x_3 - x_7) + d_2x_4) \end{aligned}$$

$$\begin{aligned}
\dot{x}_5 &= x_6 \\
\dot{x}_6 &= -\frac{b_1}{J_1}x_6 - \frac{k_1}{J_1}(x_5 - x_1) + \frac{1}{J_1}u_1 \\
\dot{x}_7 &= x_8 \\
\dot{x}_8 &= -\frac{b_2}{J_2}x_8 - \frac{k_2}{J_2}(x_7 - x_2) + \frac{1}{J_2}u_2
\end{aligned} \tag{2.450}$$

Besides the following vector fields $f(x)\in R^{8\times 1}$ and $g(x)\in R^{8\times 2}$ are defined:

$$f(x) = \begin{pmatrix} x_2 \\ -\frac{M_{22}}{detM}(C_1(x_1,x_2,x_3,x_4) + G_1(x_1,x_3)) + \\ \frac{M_{12}}{detM}(C_2(x_1,x_2,x_3,x_4) + G_2(x_1,x_3)) + \\ \frac{M_{22}}{detM}(k_1(x_1 - x_5) + d_1x_2) - \frac{M_{12}}{detM}(k_2(x_3 - x_7) + d_2x_4) \\ x_4 \\ \frac{M_{21}}{detM}(C_1(x_1,x_2,x_3,x_4) + G_1(x_1,x_3)) - \\ \frac{M_{11}}{detM}(C_2(x_1,x_2,x_3,x_4) + G_2(x_1,x_3)) - \\ \frac{M_{21}}{detM}(k_1(x_1 - x_5) + d_1x_2) + \frac{M_{11}}{detM}(k_2(x_3 - x_7) + d_2x_4) \\ x_6 \\ -\frac{b_1}{J_1}x_6 - \frac{k_1}{J_1}(x_5 - x_1) \\ x_8 \\ -\frac{b_2}{J_2}x_8 - \frac{k_2}{J_2}(x_7 - x_2) \end{pmatrix} \tag{2.451}$$

$$g(x) = \begin{pmatrix} 0 & 0 & 0 & 0 & 0 & 0 & \frac{1}{J_1} & 0 \\ 0 & 0 & 0 & 0 & 0 & 0 & 0 & \frac{1}{J_2} \end{pmatrix}^T$$

and the robotic system with flexible joints is written in the following affine-in-the-input state-space form

$$\dot{x} = f(x) + g(x)u \tag{2.452}$$

where $x\in R^{8\times 1}, f(x)\in R^{8\times 1}, g(x)\in R^{8\times 2}$, and $u\in R^{2\times 1}$.

2.10.2.2 Differential flatness properties of the multi-DOF robot with flexible joints

It will be proven that the dynamic model of the multi-DOF robot with flexible joints is differentially flat. The flat outputs vector of the system is taken to be $y=[x_1,x_3]^T$. A dynamical system is considered to be differentially flat if the following two properties hold: (i) all state variables and the control inputs of the robotic system can be written as differential functions of a specific subset of its state variables, the latter standing for the flat outputs of the robotic system; (ii) the flat outputs of the robotic system are differentially independent, which means that they are not connected between them through a solution in the form of an homogeneous linear differential equation (Rigatos, 2015).

From the first row of the state-space model, one has

$$x_2=\dot{x}_1 \Rightarrow x_2=h_2(y,\dot{y}) \tag{2.453}$$

Besides, from the third row of the state-space model, one has

$$x_4=\dot{x}_3 \Rightarrow x_4=h_4(y,\dot{y}) \tag{2.454}$$

The above two equations signify that state variables x_2 and x_4 are differential functions of the system's flat outputs. Besides, it holds that the terms M_{11}, M_{12}, M_{21}, M_{22}, $detM$, C_1, C_2, G_1, and G_2 are differential functions of the flat outputs of the system $y_1=x_3$ and $y_2=x_3$.

Next, from the second and the fourth row of the state-space model of the system, one obtains a system of equations for state variables x_5 and x_7:

$$\begin{aligned}-\frac{M_{22}}{detM}k_1x_5+\frac{M_{12}}{detM}k_2x_7 &= \dot{x}_2+\frac{M_{22}}{detM}(C_1+G_1)-\frac{M_{12}}{detM}(C_2+G_2)\\ &\quad -\frac{M_{22}}{detM}(k_1x_1+d_1x_2)+\frac{M_{12}}{detM}(k_2x_3+d_2x_4)\end{aligned} \tag{2.455}$$

$$\begin{aligned}\frac{M_{21}}{detM}k_1x_5-\frac{M_{11}}{detM}k_2x_7 &= \dot{x}_4-\frac{M_{21}}{detM}(C_1+G_1)+\frac{M_{11}}{detM}(C_2+G_2)\\ &\quad +\frac{M_{21}}{detM}(k_1x_1+d_1x_2)-\frac{M_{11}}{detM}(k_2x_3+d_2x_4)\end{aligned} \tag{2.456}$$

The previous two equations are also written in matrix form:

$$\begin{aligned}&\begin{pmatrix}-\frac{M_{22}}{detM}k_1 & \frac{M_{12}}{detM}k_2\\ \frac{M_{21}}{detM}k_1 & -\frac{M_{11}}{detM}k_2\end{pmatrix}\begin{pmatrix}x_5\\ x_7\end{pmatrix}=\\ &\begin{pmatrix}\dot{x}_2+\frac{M_{22}}{detM}(C_1+G_1)-\frac{M_{12}}{detM}(C_2+G_2)-\frac{M_{22}}{detM}(k_1x_1+d_1x_2)+\frac{M_{12}}{detM}(k_2x_3+d_2x_4)\\ \dot{x}_4-\frac{M_{21}}{detM}(C_1+G_1)+\frac{M_{11}}{detM}(C_2+G_2)+\frac{M_{21}}{detM}(k_1x_1+d_1x_2)-\frac{M_{11}}{detM}(k_2x_3+d_2x_4)\end{pmatrix}\end{aligned} \tag{2.457}$$

or equivalently

$$\begin{pmatrix} x_5 \\ x_7 \end{pmatrix} = \begin{pmatrix} -\dfrac{M_{22}}{detM}k_1 & \frac{M_{12}}{detM}k_2 \\ \dfrac{M_{21}}{detM}k_1 & -\frac{M_{11}}{detM}k_2 \end{pmatrix}^{-1} \cdot$$

$$\begin{pmatrix} \dot{x}_2 + \dfrac{M_{22}}{detM}(C_1+G_1) - \dfrac{M_{12}}{detM}(C_2+G_2) - \dfrac{M_{22}}{detM}(k_1x_1+d_1x_2) + \dfrac{M_{12}}{detM}(k_2x_3+d_2x_4) \\ \dot{x}_4 - \dfrac{M_{21}}{detM}(C_1+G_1) + \dfrac{M_{11}}{detM}(C_2+G_2) + \dfrac{M_{21}}{detM}(k_1x_1+d_1x_2) - \dfrac{M_{11}}{detM}(k_2x_3+d_2x_4) \end{pmatrix} \tag{2.458}$$

The above equation demonstrates that state variables x_5 and x_7 can be written as differential functions of the flat outputs of the system. This is also written concisely as

$$x_5 = h_5(y,\dot{y}) \quad x_7 = h_7(y,\dot{y}) \tag{2.459}$$

Additionally, from the fifth row of the state-space model of the flexible joints robot, one has

$$x_6 = \dot{x}_5 \Rightarrow x_6 = h_6(y,\dot{y},\ddot{y}) \tag{2.460}$$

Moreover, from the seventh row of the state-space model of the flexible joints robot, one has

$$x_8 = \dot{x}_7 \Rightarrow x_8 = h_6(y,\dot{y},\ddot{y}) \tag{2.461}$$

The above two equations signify that state variables x_6 and x_8 are also differential functions of the flat outputs of the system. Furthermore, from the sixth row of the state-space model, one has

$$\begin{aligned} u_1 &= J_1[\dot{x}_6 + \frac{b_1}{J_1}x_6 + \frac{k_1}{J_1}(x_5 - x_1)] \Rightarrow \\ u_1 &= h_a(y,\dot{y},\ddot{y},y^{(3)}) \end{aligned} \tag{2.462}$$

Finally, from the eight row of the state-space model, one has

$$\begin{aligned} u_2 &= J_2[\dot{x}_8 + \frac{b_2}{J_2}x_8 + \frac{k_2}{J_2}(x_7 - x_3)] \Rightarrow \\ u_2 &= h_b(y,\dot{y},\ddot{y},y^{(3)}) \end{aligned} \tag{2.463}$$

Consequently, it has been shown that the control inputs of the system are also differential functions of its flat outputs. As a result of the above, it is confirmed that the dynamic model of the 2-link robotic manipulator with flexible joints is a differentially flat system.

By proving the differential flatness properties of the state-space model of the flexible-joints robot the selection of setpoints for its control loop can be performed in a systematic manner. First, one selects setpoints for the two flat outputs of the robot, that is the turn angles of its joints $x_1 = \theta_1$ and $x_3 = \theta_2$. Next, using the previously given relations about expressing the rest of the state variables of the robot as differential functions of its flat outputs one can compute setpoints' values for these state variables too.

2.10.3 Approximate linearization of the model of the flexible-joints robot

2.10.4 Jacobian matrices of the linearized model

In (2.451) giving vector field $f(x)$, the following functions are defined: $q_1 = k_1(x_1 - x_5) + d_1x_2$ and $q_2 = k_2(x_3 - x_7) + d_2x_4$. Thus $f(x)$ is rewritten as:

$$f(x) = \begin{pmatrix} x_2 \\ -\frac{M_{22}}{detM}(C_1+G_1) + \frac{M_{12}}{detM}(C_2+G_2) + \frac{M_{22}}{detM}q_1 - \frac{M_{12}}{detM}q_2 \\ x_4 \\ \frac{M_{21}}{detM}(C_1+G_1) - \frac{M_{11}}{detM}(C_2+G_2) - \frac{M_{21}}{detM}q_1 + \frac{M_{11}}{detM}q_2 \\ x_6 \\ -\frac{b_1}{J_1}x_6 - \frac{k_1}{J_1}(x_5 - x_1) \\ x_8 \\ -\frac{b_2}{J_2}x_8 - \frac{k_2}{J_2}(x_7 - x_2) \end{pmatrix} \tag{2.464}$$

The computation of the Jacobian matrix $\nabla_x f(x) \mid_{(x^*,u^*)}$ proceeds as follows:

First row of the Jacobian matrix $\nabla_x f(x) \mid_{(x^*,u^*)}$: $\frac{\partial f_1}{\partial x_1} = 0$, $\frac{\partial f_1}{\partial x_2} = 1$, $\frac{\partial f_1}{\partial x_3} = 0$, $\frac{\partial f_1}{\partial x_4} = 0$, $\frac{\partial f_1}{\partial x_5} = 0$, $\frac{\partial f_1}{\partial x_6} = 0$, $\frac{\partial f_1}{\partial x_7} = 0$, $\frac{\partial f_1}{\partial x_8} = 0$.

Second row of the Jacobian matrix $\nabla_x f(x) \mid_{(x^*,u^*)}$: for $i = 1, 2, \cdots, 8$, it holds

$$\begin{aligned} \frac{\partial f_2}{\partial x_1} = & -\frac{[\frac{\partial M_{22}}{\partial x_i}(C_1+G_1) + M_{22}(\frac{\partial C_1}{\partial x_i} + \frac{\partial G_1}{\partial x_i})]detM - [M_{22}(C_1+G_1)]\frac{\partial detM}{\partial x_i}}{detM^2} \\ & + \frac{[\frac{\partial M_{12}}{\partial x_i}(C_2+G_2) + M_{12}(\frac{\partial C_2}{\partial x_i} + \frac{\partial G_2}{\partial x_i})]detM - [M_{12}(C_2+G_2)]\frac{\partial detM}{\partial x_i}}{detM^2} \\ & + \frac{[\frac{\partial M_{22}}{\partial x_i}q_1 + M_{22}\frac{\partial q_1}{\partial x_i}]detM - M_{22}q_1\frac{\partial detM}{\partial x_i}}{detM^2} \\ & - \frac{[\frac{\partial M_{12}}{\partial x_i}q_2 + M_{12}\frac{\partial q_2}{\partial x_i}]detM - M_{12}q_2\frac{\partial detM}{\partial x_i}}{detM^2} \end{aligned} \tag{2.465}$$

Third row of the Jacobian matrix $\nabla_x f(x) \mid_{(x^*,u^*)}$: $\frac{\partial f_3}{\partial x_1} = 0$, $\frac{\partial f_3}{\partial x_2} = 0$, $\frac{\partial f_3}{\partial x_3} = 0$, $\frac{\partial f_3}{\partial x_4} = 1$, $\frac{\partial f_3}{\partial x_5} = 0$, $\frac{\partial f_3}{\partial x_6} = 0$, $\frac{\partial f_3}{\partial x_7} = 0$, $\frac{\partial f_3}{\partial x_8} = 0$.

Fourth row of the Jacobian matrix $\nabla_x f(x)\mid_{(x^*,u^*)}$: for $i = 1, 2, \cdots, 8$, it holds

$$\begin{aligned}\frac{\partial f_4}{\partial x_1} = & \frac{[\frac{\partial M_{21}}{\partial x_i}(C_1+G_1)+M_{21}(\frac{\partial C_1}{\partial x_i}+\frac{\partial G_1}{\partial x_i})]detM-[\frac{M_{21}}{detM}(C_1+G_1)]\frac{\partial detM}{\partial x_i}}{detM^2}\\ & -\frac{[\frac{\partial M_{11}}{\partial x_i}(C_2+G_2)+M_{11}(\frac{\partial C_2}{\partial x_i}+\frac{\partial G_2}{\partial x_i})]detM-[\frac{M_{11}}{detM}(C_2+G_2)]\frac{\partial detM}{\partial x_i}}{detM^2}\\ & -\frac{[\frac{\partial M_{21}}{\partial x_i}q_1+M_{21}\frac{\partial q_1}{\partial x_i}]detM-M_{21}q_1\frac{\partial detM}{\partial x_i}}{detM^2}\\ & +\frac{[\frac{\partial M_{11}}{\partial x_i}q_2+M_{11}\frac{\partial q_2}{\partial x_i}]detM-M_{11}q_2\frac{\partial detM}{\partial x_i}}{detM^2}\end{aligned} \tag{2.466}$$

Fifth row of the Jacobian matrix $\nabla_x f(x)\mid_{(x^*,u^*)}$: $\frac{\partial f_5}{\partial x_1} = 0$, $\frac{\partial f_5}{\partial x_2} = 0$, $\frac{\partial f_5}{\partial x_3} = 0$, $\frac{\partial f_5}{\partial x_4} = 0$, $\frac{\partial f_5}{\partial x_5} = 0$, $\frac{\partial f_5}{\partial x_6} = 1$, $\frac{\partial f_5}{\partial x_7} = 0$, $\frac{\partial f_5}{\partial x_8} = 0$.

Sixth row of the Jacobian matrix $\nabla_x f(x)\mid_{(x^*,u^*)}$: $\frac{\partial f_6}{\partial x_1} = \frac{k_1}{J_1}$, $\frac{\partial f_6}{\partial x_2} = 0$, $\frac{\partial f_6}{\partial x_3} = 0$, $\frac{\partial f_6}{\partial x_4} = 0$, $\frac{\partial f_6}{\partial x_5} = -\frac{k_1}{J_1}$, $\frac{\partial f_6}{\partial x_6} = -\frac{b_1}{J_1}$, $\frac{\partial f_6}{\partial x_7} = 0$, $\frac{\partial f_6}{\partial x_8} = 0$.

Seventh row of the Jacobian matrix $\nabla_x f(x)\mid_{(x^*,u^*)}$: $\frac{\partial f_7}{\partial x_1} = 0$, $\frac{\partial f_7}{\partial x_2} = 0$, $\frac{\partial f_7}{\partial x_3} = 0$, $\frac{\partial f_7}{\partial x_4} = 0$, $\frac{\partial f_7}{\partial x_5} = 0$, $\frac{\partial f_7}{\partial x_6} = 0$, $\frac{\partial f_7}{\partial x_7} = 0$, $\frac{\partial f_7}{\partial x_8} = 1$.

Eight row of the Jacobian matrix $\nabla_x f(x)\mid_{(x^*,u^*)}$: $\frac{\partial f_8}{\partial x_1} = 0$, $\frac{\partial f_8}{\partial x_2} = 0$, $\frac{\partial f_8}{\partial x_3} = \frac{k_2}{J_2}$, $\frac{\partial f_8}{\partial x_4} = 0$, $\frac{\partial f_8}{\partial x_5} = 0$, $\frac{\partial f_8}{\partial x_6} = -\frac{k_2}{J_2}$, $\frac{\partial f_8}{\partial x_7} = 0$, $\frac{\partial f_8}{\partial x_8} = -\frac{b_2}{J_2}$.

Next, one computes the partial derivatives of the individual components of $\tilde{M}$, $\tilde{C}$, $\tilde{G}$, q_1 and q_2.

One has that

$$\begin{aligned}&\tilde{M} = \begin{pmatrix} M_{11} & M_{12} \\ M_{21} & M_{22} \end{pmatrix} \Rightarrow \\ &M = \begin{pmatrix} (m_1+m_2)l_1^2 + m_1 l_2^2 + 2m_2 l_1 l_2 \cos(x_3) & m_2 l_2^2 + m_2 l_1 l_2 \cos(x_3) \\ m_2 l_2^2 + m_2 l_1 l_2 \cos(x_3) & m_2 l_2 (l_1 + l_2)\cos(x_3) \end{pmatrix}\end{aligned} \tag{2.467}$$

with $detM = M_{11}M_{22} - M12M_{22}$. The associated partial derivatives are:

$\frac{\partial M_{11}}{\partial x_i} = 0$, for all $i \neq 3$ and $\frac{\partial M_{11}}{\partial x_3} = -2m_2 l_1 l_2 \sin(x_3)$.

$\frac{\partial M_{12}}{\partial x_i} = 0$, for all $i \neq 3$ and $\frac{\partial M_{12}}{\partial x_3} = -2m_2 l_1 l_2 \sin(x_3)$.

$\frac{\partial M_{21}}{\partial x_i} = 0$, for all $i \neq 3$ and $\frac{\partial M_{21}}{\partial x_3} = -2m_2 l_1 l_2 \sin(x_3)$.

$\frac{\partial M_{22}}{\partial x_i} = 0$, for all $i \neq 3$ and $\frac{\partial M_{22}}{\partial x_3} = -m_2 l_1 l_2 \sin(x_3)$.

Additionally $\frac{\partial detM}{\partial x_i} = \frac{\partial M_{11}}{\partial x_i} M_{22} + M_{11} \frac{\partial M_{22}}{\partial x_i} - \frac{\partial M_{12}}{\partial x_i} M_{21} - \frac{\partial M_{21}}{\partial x_i} M_{12}$

Besides, it holds that

$$\tilde{C} = \begin{pmatrix} C_1 \\ C_2 \end{pmatrix} \Rightarrow \tilde{C} = \begin{pmatrix} -m_2 l_1^2 \sin(x_3) x_4^2 - 2m_2 l_1^2 \sin(x_3) x_2 x_4 \\ m_2 l_1^2 \sin(x_3) x_2^2 \end{pmatrix} \tag{2.468}$$

The associated partial derivatives are: $\frac{\partial C_1}{\partial x_1} = 0$, $\frac{\partial C_1}{\partial x_2} = -2m_2 l_1^2 \sin(x_3) x_4$, $\frac{\partial C_1}{\partial x_3} = -2m_2 l_1^2 \cos(x_3) x_4^2 - 2m_2 l_1^2 \cos(x_3) x_2 x_4$, $\frac{\partial C_1}{\partial x_4} = -m_2 l_1^2 \sin(x_3) 2x_4 - 2m_2 l_1^2 \sin(x_3) x_2$, $\frac{\partial C_1}{\partial x_5} = 0$, $\frac{\partial C_1}{\partial x_6} = 0$, $\frac{\partial C_1}{\partial x_7} = 0$, and $\frac{\partial C_1}{\partial x_8} = 0$.

and also: $\frac{\partial C_2}{\partial x_1} = 0$, $\frac{\partial C_2}{\partial x_2} = m_2 l_1^2 \sin(x_3) 2x_2$, $\frac{\partial C_2}{\partial x_3} = m_2 l_1^2 \cos(x_3) 2x_2$, $\frac{\partial C_2}{\partial x_4} = 0$, $\frac{\partial C_2}{\partial x_5} = 0$, $\frac{\partial C_2}{\partial x_6} = 0$, $\frac{\partial C_2}{\partial x_7} = 0$, and $\frac{\partial C_2}{\partial x_8} = 0$.

Moreover, one has that

$$\tilde{G} = \begin{pmatrix} G_1 \\ G_2 \end{pmatrix} \Rightarrow \tilde{G} = \begin{pmatrix} (m_1 + m_2) g l_1 \cos(x_1) + m_2 g l_2 \cos(x_1 + x_3) \\ m_2 g l_2 \cos(x_1 + x_3) \end{pmatrix} \tag{2.469}$$

The associated partial derivatives are: $\frac{\partial G_1}{\partial x_1} = -(m_1 + m_2) g l_1 \sin(x_1) - m_2 g l_2 \sin(x_1 + x_3)$, $\frac{\partial G_1}{\partial x_2} = 0$, $\frac{\partial G_1}{\partial x_3} = -m_2 g l_2 \sin(x_1 + x_3)$, $\frac{\partial G_1}{\partial x_4} = 0$, $\frac{\partial G_1}{\partial x_5} = 0$, $\frac{\partial G_1}{\partial x_6} = 0$, $\frac{\partial G_1}{\partial x_7} = 0$, and $\frac{\partial G_1}{\partial x_8} = 0$.

and also: $\frac{\partial G_2}{\partial x_1} = -m_2 g l_2 \sin(x_1 + x_3)$, $\frac{\partial G_2}{\partial x_2} = 0$, $\frac{\partial G_2}{\partial x_3} = -m_2 g l_2 \sin(x_1 + x_3)$, $\frac{\partial G_2}{\partial x_4} = 0$, $\frac{\partial G_2}{\partial x_5} = 0$, $\frac{\partial G_2}{\partial x_6} = 0$, $\frac{\partial G_2}{\partial x_7} = 0$, and $\frac{\partial G_2}{\partial x_8} = 0$.

Furthermore, it holds that $q_1 = k_1(x_1 - x_5) + d_1 x_2$.

The associated partial derivatives are: $\frac{\partial 1_1}{\partial x_1} = k_1$, $\frac{\partial q_1}{\partial x_2} = d_1$, $\frac{\partial q_1}{\partial x_3} = 0$, $\frac{\partial q_1}{\partial x_4} = 0$, $\frac{\partial q_1}{\partial x_5} = -k_1$, $\frac{\partial q_1}{\partial x_6} = 0$, $\frac{\partial q_1}{\partial x_7} = 0$, and $\frac{\partial q_1}{\partial x_8} = 0$.

and finally it holds that $q_2 = k_2(x_3 - x_7) + d_2 x_6$.

The associated partial derivatives are: $\frac{\partial q_2}{\partial x_1} = 0$, $\frac{\partial q_2}{\partial x_2} = 0$, $\frac{\partial q_2}{\partial x_3} = k_2$, $\frac{\partial q_2}{\partial x_4} = d_2$, $\frac{\partial q_2}{\partial x_5} = 0$, $\frac{\partial q_2}{\partial x_6} = 0$, $\frac{\partial q_2}{\partial x_7} = -k_2$, and $\frac{\partial q_2}{\partial x_8} = 0$.

2.10.4.1 Stabilizing feedback control

The nonlinear dynamic model of the multi-DOF robot with flexible joints, being initially in the state-space form $\dot{x} = f(x) + g(x)u$ undergoes approximate linearization around the temporary operating point (x^*, u^*), where x^* is the present value of the system's state vector and u^* is the last sampled value of the control inputs vector. This time-varying operating point is recomputed at each time-step of the control algorithm. The linearization process relies on first-order Taylor series expansion and on the computation of the related Jacobian matrices. The approximately linearized model of the robotic system is written as

$$\dot{x} = Ax + Bu + \tilde{d} \tag{2.470}$$

where $\tilde{d}$ is a cumulative disturbance term that may comprise (i) the modeling error due to the truncation of higher-order terms in the Taylor series expansion, (ii) the effects of exogenous perturbations, and (iii) measurement noise of any distribution. About the Jacobian matrices A and B it holds that

$$\begin{aligned} &A = \nabla_x[f(x) + g(x)u] \mid_{(x^*,u^*)} \\ &\Rightarrow A = \nabla_x f(x) \mid_{(x^*,u^*)} \end{aligned} \tag{2.471}$$

$$\begin{aligned} &B = \nabla_u[f(x) + g(x)u] \mid_{(x^*,u^*)} \\ &\Rightarrow B = g(x) \mid_{(x^*,u^*)} \end{aligned} \tag{2.472}$$

The Jacobians of the robotic system were computed explicitly above. After linearization around its current operating point (x^*, u^*), the dynamic model of the 2-link robotic manipulator with flexible joints is written as (Rigatos and Busawon, 2018)

$$\dot{x} = Ax + Bu + d_1 \tag{2.473}$$

Parameter d_1 stands for the linearization error in the robotic manipulator's model appearing previously in (2.473). The reference setpoints for the robotic manipulator's state vector are denoted by $\mathbf{x_d} = [x_1^d, \cdots, x_8^d]$. Tracking of this trajectory is achieved after applying the control input u_d. At every time instant, the control input u_d is assumed to differ from the control input u appearing in (2.473) by an amount equal to Δu, that is $u_d = u + \Delta u$

$$\dot{x}_d = Ax_d + Bu_d + d_2 \tag{2.474}$$

The dynamics of the controlled system described in (2.473) can be also written as

$$\dot{x} = Ax + Bu + Bu_d - Bu_d + d_1 \tag{2.475}$$

and by denoting $d_3 = -Bu_d + d_1$ as an aggregate disturbance term, one obtains

$$\dot{x} = Ax + Bu + Bu_d + d_3 \tag{2.476}$$

By subtracting (2.474) from (2.476), one has

$$\dot{x} - \dot{x}_d = A(x - x_d) + Bu + d_3 - d_2 \tag{2.477}$$

By denoting the tracking error as $e = x - x_d$ and the aggregate disturbance term as $\tilde{d} = d_3 - d_2$, the tracking error dynamics becomes

$$\dot{e} = Ae + Bu + \tilde{d} \tag{2.478}$$

To adhere to the description of the dynamics of the tracking error of the flexible-joints robot which has been given in (2.478), in which the control inputs vector u is applied through the control inputs gain matrix B, the cumulative disturbance term $\tilde{d}$ which appears in the right part of (2.478) can be also written as $L\tilde{d}$ where L is the disturbance inputs gain matrix. For the approximately linearized model of the

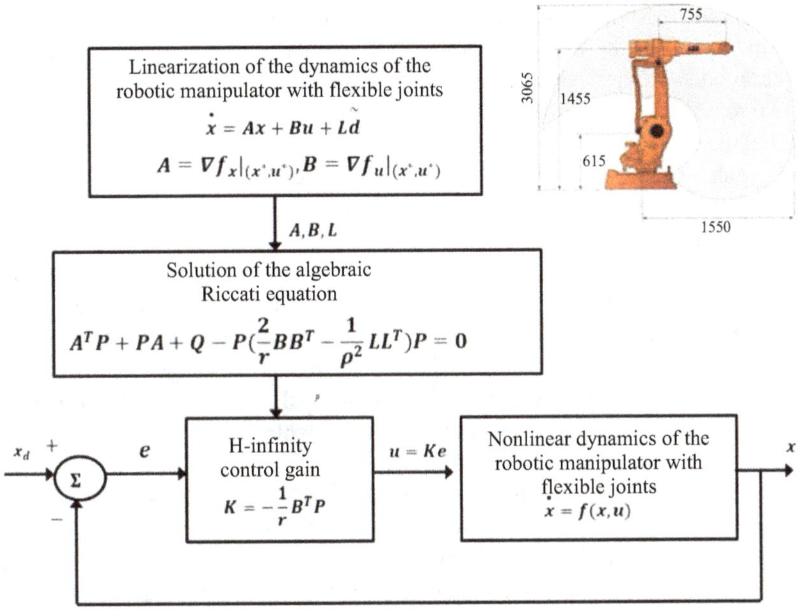

Figure 2.125 Diagram of the nonlinear optimal (H-infinity) control loop for the multi-DOF robotic manipulator with flexible joints

2-link robotic manipulator with flexible joints, a stabilizing feedback controller is developed. The controller has the form

$$u(t) = -Ke(t) \tag{2.479}$$

with $K = \frac{1}{r}B^T P$ where P is a positive definite symmetric matrix which is obtained from the solution of the Riccati equation (Rigatos and Busawon, 2018)

$$A^T P + PA + Q - P(\frac{2}{r}BB^T - \frac{1}{\rho^2}LL^T)P = 0 \tag{2.480}$$

where Q is a positive semi-definite symmetric matrix (Rigatos and Busawon, 2018). The previously analyzed concept about the nonlinear optimal control loop for the 2-link robotic manipulator with flexible joints is given in Figure 2.125. The global stability properties of the control method can be proven following the procedure presented in Section 1.1.2.5.

2.10.5 Simulation tests

The performance of the previously analyzed nonlinear optimal control method for the dynamic model of the 2-link robotic manipulator with flexible joints has been further confirmed through simulation experiments. To apply the optimal control method, the algebraic Riccati equation of Eq. (41) had to be solved at each time-step of the control algorithm, using Matlab's aresolv() function. State estimation-based control has been implemented with the use of the H-infinity Kalman Filter. The parameters of the robotic model which have been used in the simulation experiments were masses of the two robotic links $m_1 = 0.4$ kgr, and $m_2 = 0.2$ kgr, lengths of the two robotic links $l_1 = 0.3$ m and $l_2 = 0.4$ m gravity acceleration coefficient $g_c = 10$ m/sec^2, damping coefficients of the two links $d_1 = 10^{-3}$ Ntm/(rad/sec) and $d_2 = 10^{-3}$ Ntm/(rad/sec), elasticity coefficients of the two flexible joints $k_1 = 300$ Ntm/rad, $k_2 = 80$ Ntm/rad, damping coefficients of the two flexible joints $b_1 = 10^{-3}$ and $b_2 = 10^{-3}$. However, it is noted that (1) the implementation of the proposed control method does not rely on specific parameters values, since the method is robust and can tolerate parametric imprecision, (2) the coefficients that appear in the algebraic Riccati equation of the control system are $r = 0.001$, $\rho = 0.05$, $Q = 40I_{8\times8}$ and $L_m = 10^{-4}I_{8\times8}$, (3) the sampling period was $T_s = 0.01$s, and (4) initial values do not have to be defined because the proposed control method is globally stable and its performance does not depend on initial values.

The obtained results are depicted in Figures 2.126 to 2.141. The SI measurement system has been considered for the robot's state variables and control inputs. The turn angles of the robot's links and the turn angles of the motors of the flexible joints are measured in rads while the associated angular velocities are measured in rads/sec. The control inputs of the system, that is the torques of the motors are measured in Nt. The real values of the state vector elements are printed in blue, the estimated values are plotted in green while the associated reference setpoints are depicted in red. It can be noticed that under the proposed nonlinear optimal control scheme all

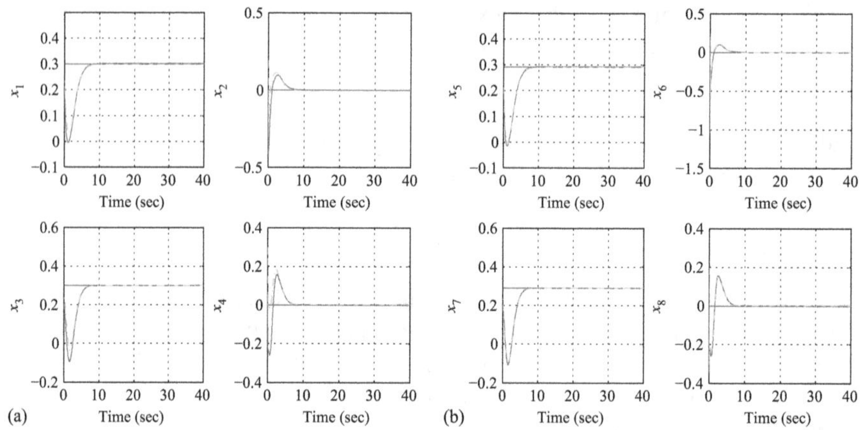

Figure 2.126 Tracking of setpoint 1 by the flexible-joint robot: (a) convergence of state variables x_1 to x_4 to their reference setpoints (red line: setpoint, blue line: real value, green line: estimated value); (b) convergence of state variables x_5 to x_8 to their reference setpoints (red line: setpoint, blue line: real value, green line: estimated value)

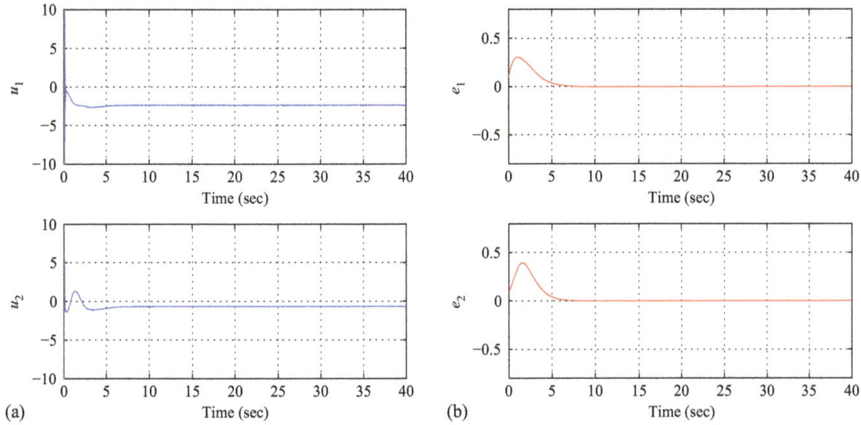

Figure 2.127 Tracking of setpoint 1 by the flexible-joint robot: (a) control inputs u_1 and u_2 applied to the actuators of the robot; (b) tracking error for the turn angles of the robot's links x_1, x_3

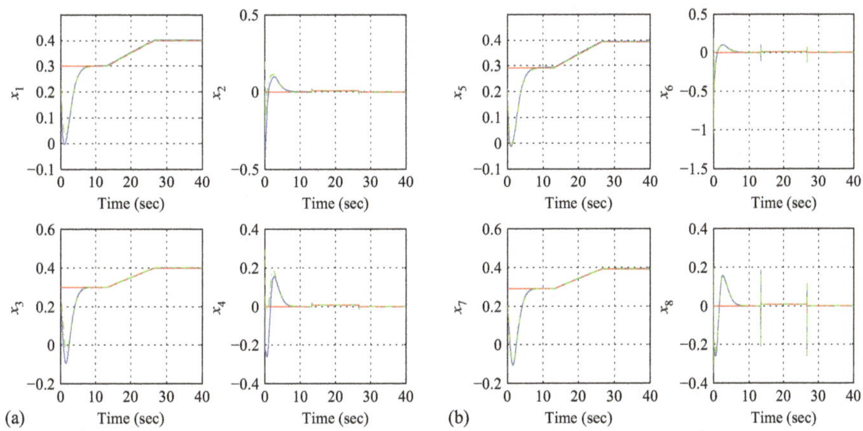

Figure 2.128 Tracking of setpoint 2 by the flexible-joint robot: (a) convergence of state variables x_1 to x_4 to their reference setpoints (red line: setpoint, blue line: real value, green line: estimated value); (b) convergence of state variables x_5 to x_8 to their reference setpoints (red line: setpoint, blue line: real value, green line: estimated value)

state variables converged rapidly to related setpoints and that the tracking error was minimized. Besides, it can be observed that the variations of the control inputs were moderate and consequently the energy consumption by the optimal control loop was minimal too.

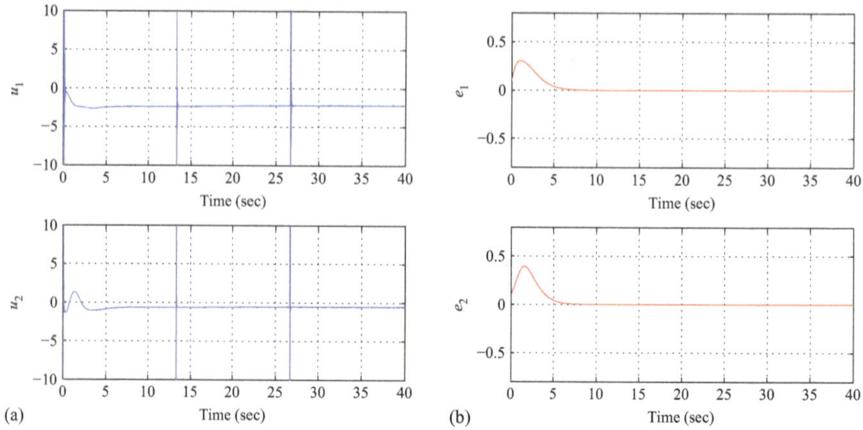

Figure 2.129 Tracking of setpoint 2 by the flexible-joint robot: (a) control inputs u_1 and u_2 applied to the actuators of the robot; (b) tracking error for the turn angles of the robot's links x_1, x_3

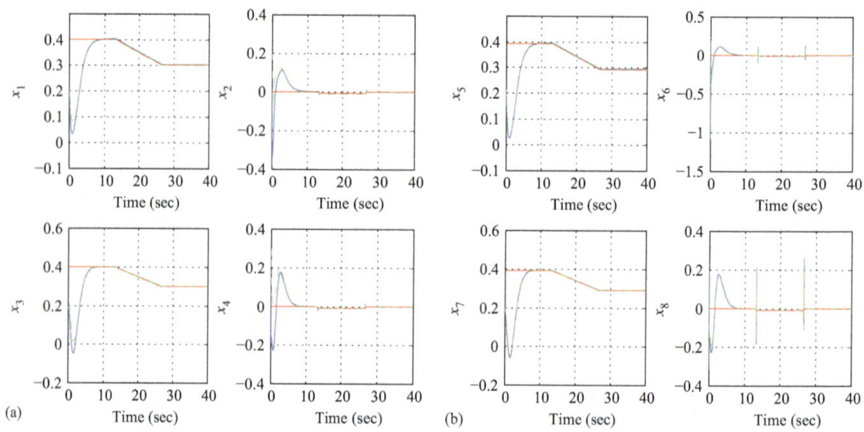

Figure 2.130 Tracking of setpoint 3 by the flexible-joint robot: (a) convergence of state variables x_1 to x_4 to their reference setpoints (red line: setpoint, blue line: real value, green line: estimated value); (b) convergence of state variables x_5 to x_8 to their reference setpoints (red line: setpoint, blue line: real value, green line: estimated value)

In the tracking of various setpoints by the state variables of the robot and for total simulation time equal to 40 sec the following can be noted: (i) in Figure 2.126, the convergence time is less than 8 sec, (ii) in Figure 2.128, the convergence time is about 5 sec (iii) in Figure 2.130 the convergence time is about 5 sec, (iv) in Figure 2.132 the convergence time is about 5 sec, (v) in Figure 2.134 the convergence time is about

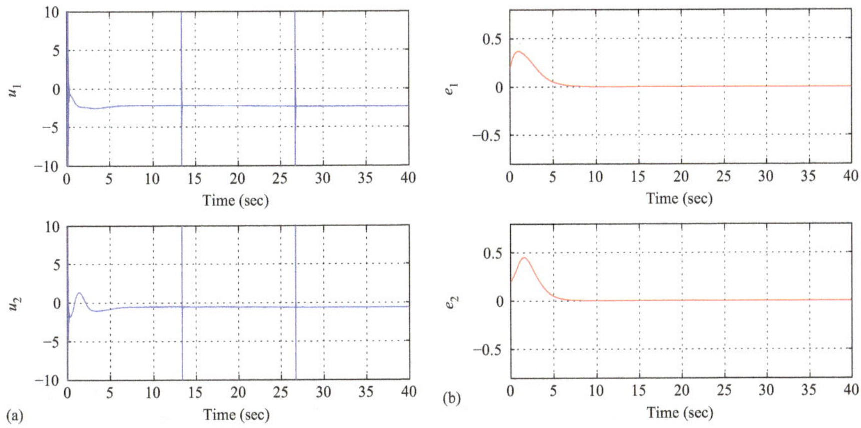

Figure 2.131 Tracking of setpoint 3 by the flexible-joint robot: (a) control inputs u_1 and u_2 applied to the actuators of the robot; (b) tracking error for the turn angles of the robot's links x_1, x_3

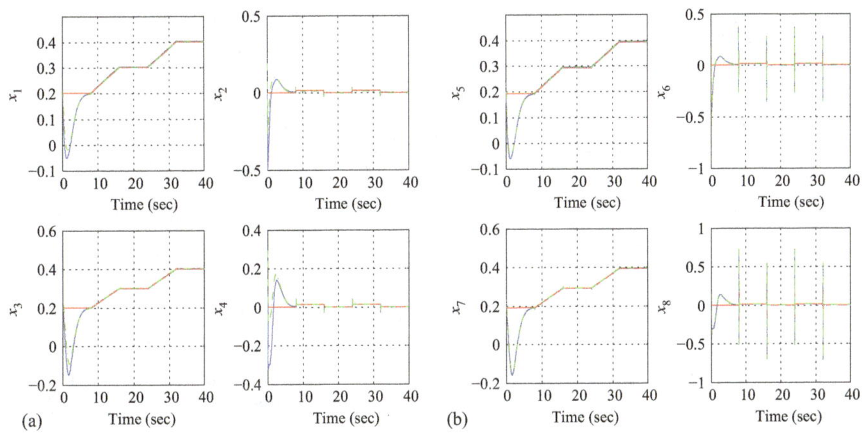

Figure 2.132 Tracking of setpoint 4 by the flexible-joint robot: (a) convergence of state variables x_1 to x_4 to their reference setpoints (red line: setpoint, blue line: real value, green line: estimated value); (b) convergence of state variables x_5 to x_8 to their reference setpoints (red line: setpoint, blue line: real value, green line: estimated value)

7 sec, (vi) in Figure 2.136 the convergence time is about 8 sec, (vii) in Figure 2.138 the convergence g time is about 8 sec, and (viii) in Figure 2.140 the convergence time is about 8 sec. In all cases, the overshoot was negligible. In Figures 2.127, 2.129, 2.131, 2.133, 2.135, 2.137, 2.139, and 2.141, the variations of the control inputs are moderate.

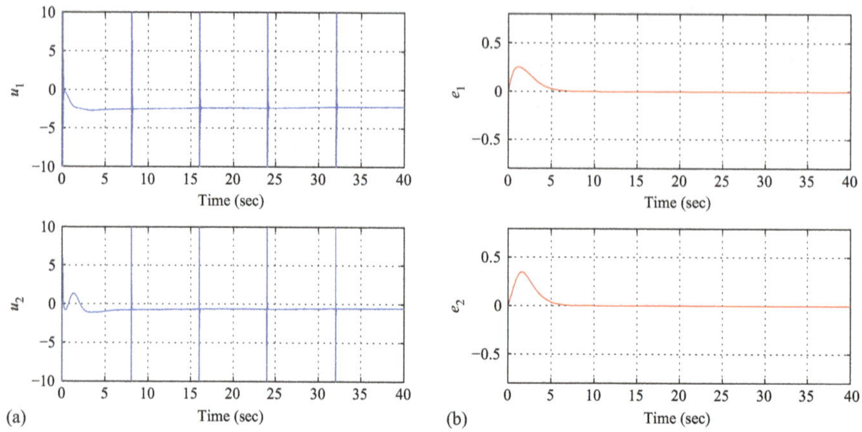

Figure 2.133 *Tracking of setpoint 4 by the flexible-joint robot: (a) control inputs u_1 and u_2 applied to the actuators of the robot; (b) tracking error for the turn angles of the robot's links x_1, x_3*

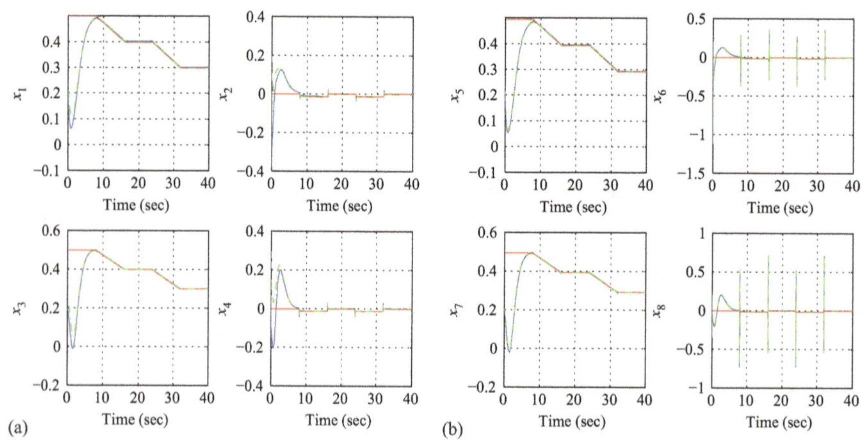

Figure 2.134 *Tracking of setpoint 5 by the flexible-joint robot: (a) convergence of state variables x_1 to x_4 to their reference setpoints (red line: setpoint, blue line: real value, green line: estimated value); (b) convergence of state variables x_5 to x_8 to their reference setpoints (red line: setpoint, blue line: real value, green line: estimated value)*

2.11 Nonlinear control of redundant robotic manipulators

2.11.1 Outline

Control of robotic manipulators with an end-effector that moves in a tasks-space with lower dimension than the robot's degrees of freedom comes against the problem of

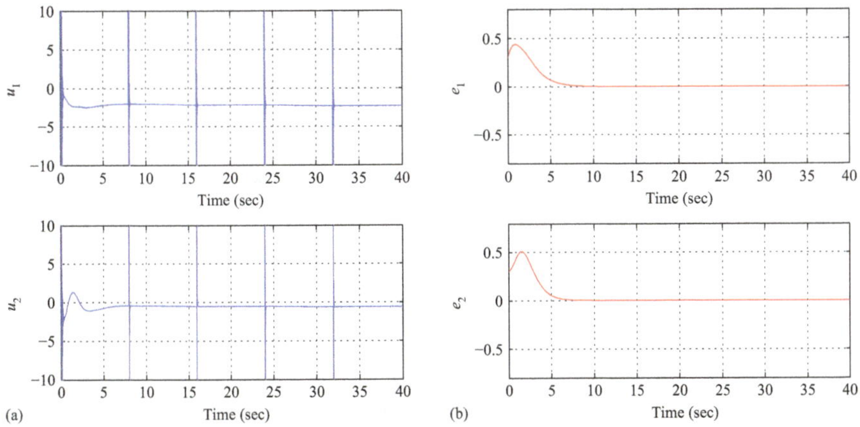

Figure 2.135 Tracking of setpoint 5 by the flexible-joint robot: (a) control inputs u_1 and u_2 applied to the actuators of the robot; (b) tracking error for the turn angles of the robot's links x_1, x_3

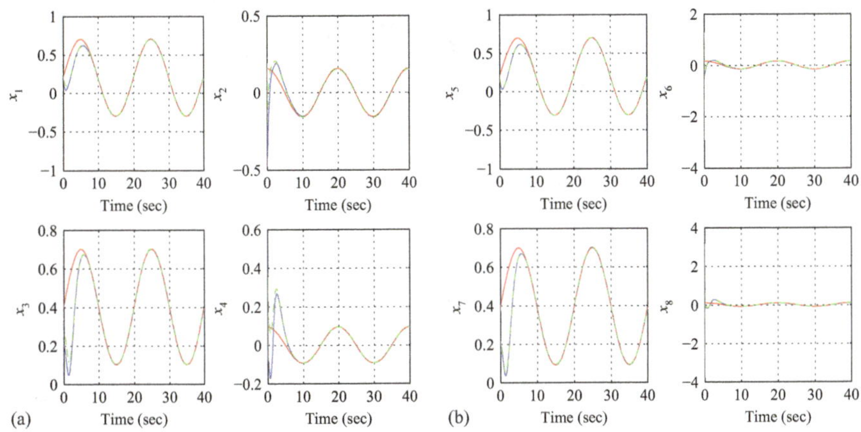

Figure 2.136 Tracking of setpoint 6 by the flexible-joint robot: (a) convergence of state variables x_1 to x_4 to their reference setpoints (red line: setpoint, blue line: real value, green line: estimated value); (b) convergence of state variables x_5 to x_8 to their reference setpoints (red line: setpoint, blue line: real value, green line: estimated value)

redundancy (Sadeghian *et al.*, 2014; Nicolis, Allewi *et al.*, 2020; Jin and Zhang, 2015). Kinematic redundancy occurs when a manipulator has more degrees of freedom than the minimum number required to execute a task. When a manipulator is redundant, it is anticipated that the inverse kinematic problem admits multiple solutions. This means that for a constant position of the end-effector, several configurations of the arm's joints can be obtained (Galicki, 2006; Galicki, 2014; Galicki, 2016). Redundant

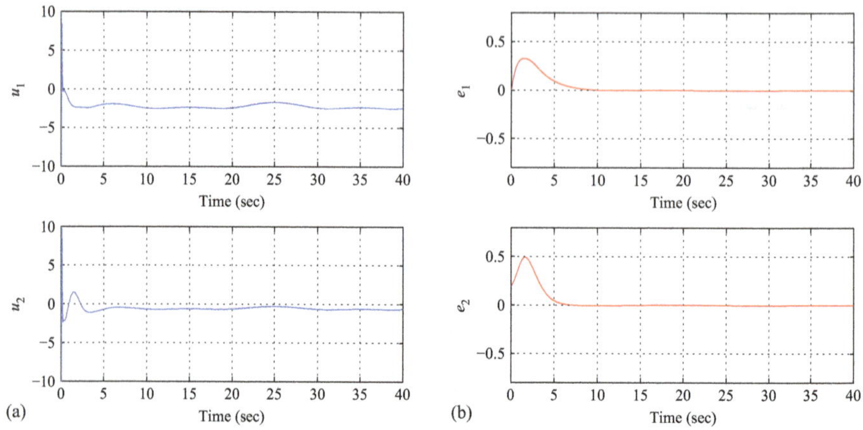

Figure 2.137 Tracking of setpoint 6 by the flexible-joint robot: (a) control inputs u_1 and u_2 applied to the actuators of the robot; (b) tracking error for the turn angles of the robot's links x_1, x_3

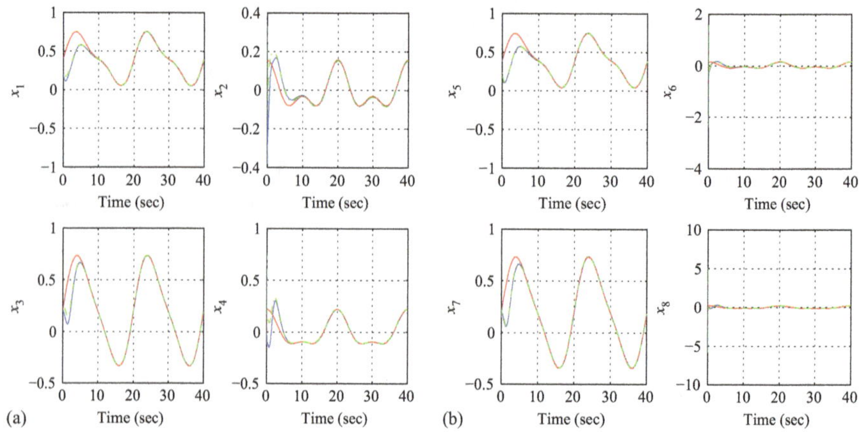

Figure 2.138 Tracking of setpoint 7 by the flexible-joint robot: (a) convergence of state variables x_1 to x_4 to their reference setpoints (red line: setpoint, blue line: real value, green line: estimated value); (b) convergence of state variables x_5 to x_8 to their reference setpoints (red line: setpoint, blue line: real value, green line: estimated value)

robots achieve a versatile motion in the joints space and dexterity in tasks execution (Galicki, 2008; Jiang, Huang *et al.*, 2019; Oriolo *et al.*, 2017). They can find use in robotic surgery as well as on several industrial tasks, because of offering multiple joints configurations that allow the end effector to reach easier the targeted position in the tasks' space (Mahindrakar *et al.*, 2005; Xiong, Lai *et al.*, 2016; Kayastha *et al.*, 2017).

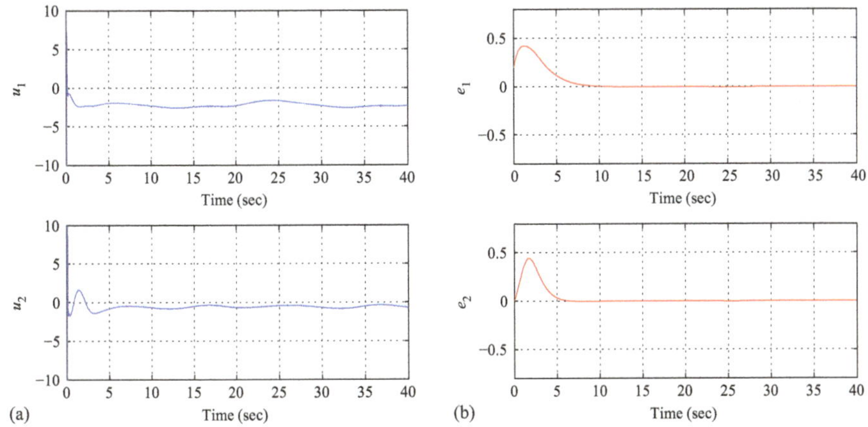

Figure 2.139 Tracking of setpoint 7 by the flexible-joint robot: (a) control inputs u_1 and u_2 applied to the actuators of the robot; (b) tracking error for the turn angles of the robot's links x_1, x_3

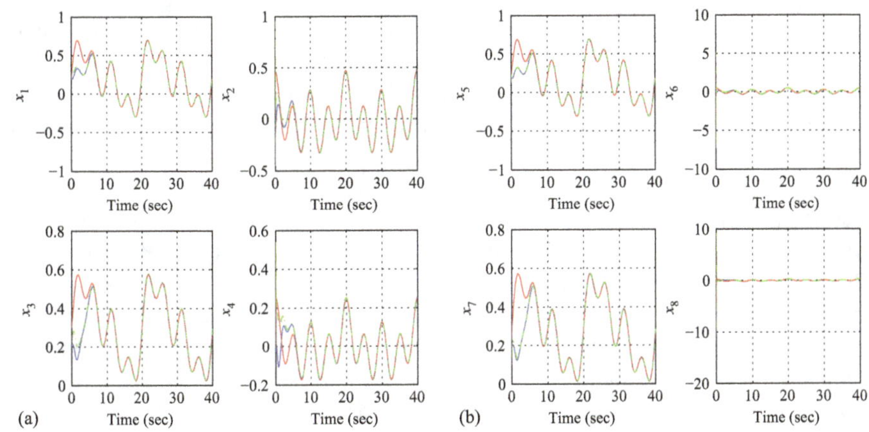

Figure 2.140 Tracking of setpoint 8 by the flexible-joint robot: (a) convergence of state variables x_1 to x_4 to their reference setpoints (red line: setpoint, blue line: real value, green line: estimated value); (b) convergence of state variables x_5 to x_8 to their reference setpoints (red line: setpoint, blue line: real value, green line: estimated value)

Generally, redundancy specifies a robot's internal movement without affecting the trajectory of the robot's end effector and permits the robot to react in a better way to the environment that surrounds. This in turn signifies that the robot can achieve obstacles' avoidance and can keep moderate the torques developed by its actuators (Kumar *et al.*, 2001; Sandeval *et al.*, 2018; Potkonjak, Popovic *et al.*, 1998). Control of redundant

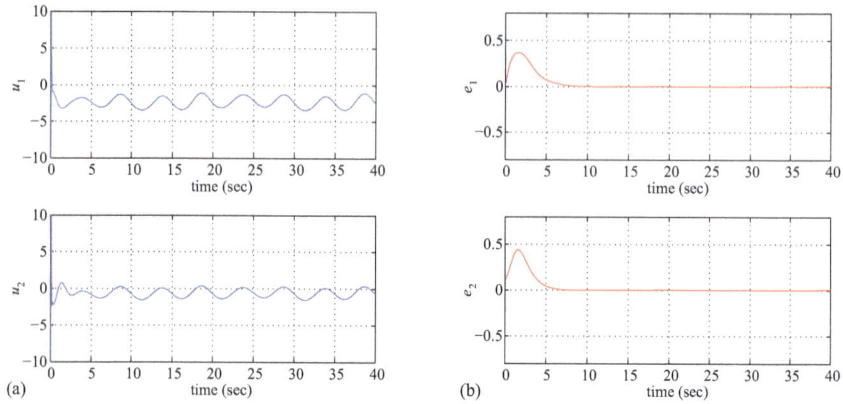

Figure 2.141 Tracking of setpoint 8 by the flexible-joint robot: (a) control inputs u_1 and u_2 applied to the actuators of the robot; (b) tracking error for the turn angles of the robot's links x_1, x_3

manipulators, which is also known as redundancy resolution, is to find a control action in the joints' space of a redundant robotic arm that leads to a desired motion of the end-effector in the tasks' space (Wang *et al.*, 2017; Jassour and Farrokhi, 2014; Xian, de Queiroz *et al.*, 2004). Several approaches have been proposed for the control of redundant robots (Nakanishi *et al.*, 2008; Petrovic and Lukic, 2017; Chiaverini *et al.*, 2016). The primary objective is to achieve stability properties for the redundant manipulator A related objective is the avoidance of singularities in the robot's inverse kinematic problem (Zhang, Li *et al.*, 2020; Karami *et al.*, 2019; Woolfrey *et al.*, 2019). So far the nonlinear optimal control problem for redundant manipulators has been little studied.

In the present section a new nonlinear optimal (H-infinity) method for control of redundant robotic manipulators is proposed (Rigatos and Busawon, 2018; Rigatos, 2016b; Rigatos, Zervos *et al.*, 2020). Without loss of generality a three-link planar redundant manipulator is considered (Mahindrakar *et al.*, 2005; Xiong, Lai *et al.*, 2016; Kayastha *et al.*, 2017; Rigatos and Busawon, 2018). First the joint kinematic and dynamic model of the robotic manipulator is defined . The first part of this model describes the system's kinematics, that the variation of the velocity of the end effector as a result of changes in the turn speed of the manipulator's joints. The second part of the model describes the dynamics of the manipulator that is the accelerations of its joints as a result of the torques that are developed by the joints' actuators. The joint kinematic and dynamic model of this manipulator, being strongly nonlinear, undergoes approximate linearization with the use of first-order Taylor series expansion and through the computation of the related Jacobian matrices (Rigatos and Tzafestas, 2007; Basseville and Nikiforov, 1993; Rigatos and Zhang, 2009). The linearization point is updated at each sampling instant and is defined by the present value of the system's state vector and by the last sampled value of the control inputs vector. The modeling error which is due to truncation of higher order terms in the Taylor series expansion is considered to be a perturbation which is asymptotically compensated by the robustness of the control method.

For the approximately linearized model of the redundant manipulator a stabilizing H-infinity feedback controller is designed. This controller solves the optimal control problem for the redundant manipulator under model uncertainty and external perturbations (Rigatos and Busawon, 2018; Rigatos, 2016b; Rigatos, Zervos *et al.*, 2020). Actually, this controller represents a min–max differential game, taking place between (i) the control inputs which try to minimize a quadratic function of the state vector's tracking error and (ii) the exogenous perturbations and model uncertainty terms which try to maximize this cost function. To select the feedback gains of the H-infinity controller an algebraic Riccati equation is repetitively solved at each time-step of the control method (Rigatos and Busawon, 2018; Rigatos, 2011; Rigatos, 2015). The global stability properties of the control scheme are proven through Lyapunov analysis. First, it is proven that the control loop satisfies the H-infinity tracking performance criterion, which signifies elevated robustness against model uncertainty and external disturbances (Rigatos and Busawon, 2018; Tousssaint, Basar and Bullo, 2000). Moreover, it is proven that under moderate conditions the control loop is globally asymptotically stable. The proposed nonlinear optimal control scheme for redundant manipulators achieves fast and accurate tracking of reference setpoints under moderate variations of the control inputs. Additionally, it is shown that it is possible to implement state estimation-based control for the redundant manipulator after using the H-infinity Kalman Filter as a robust state estimator (Rigatos and Busawon, 2018). The section's results can be also extended to the case of underactuated redundant robotic manipulators, where the robotic arm contains also passive joints.

2.11.2 Kinematic and dynamic model of the redundant manipulator

2.11.2.1 The kinematic and dynamic model for the 3-DOF redundant manipulator

Without loss of generality a 3-link redundant robotic manipulator is considered, as shown in Figure 2.142. The position of the end-effector of the robotic manipulator is given by (Mahindrakar *et al.*, 2005; Xiong, Lai *et al.*, 2016; Kayastha *et al.*, 2017; Rigatos and Busawon, 2018)

$$\begin{aligned} x &= l_1\cos(\theta_1) + l_2\cos(\theta_1+\theta_2) + l_3\cos(\theta_1+\theta_2+\theta_3) \\ y &= l_1\sin(\theta_1) + l_2\sin(\theta_1+\theta_2) + l_3\sin(\theta_1+\theta_2+\theta_3) \end{aligned} \tag{2.481}$$

The velocity of the end-effector of the robotic manipulator is given by

$$\begin{aligned} \dot{x} &= -l_1\sin(\theta_1)\dot{\theta}_1 - l_2\sin(\theta_1+\theta_2)(\dot{\theta}_1+\dot{\theta}_2) - l_3\sin(\theta_1+\theta_2+\theta_3)(\dot{\theta}_1+\dot{\theta}_2+\dot{\theta}_3) \\ \dot{y} &= l_1\cos(\theta_1)\dot{\theta}_1 + l_2\cos(\theta_1+\theta_2)(\dot{\theta}_1+\dot{\theta}_2) + l_3\mathrm{coss}(\theta_1+\theta_2+\theta_3)(\dot{\theta}_1+\dot{\theta}_2+\dot{\theta}_3) \end{aligned} \tag{2.482}$$

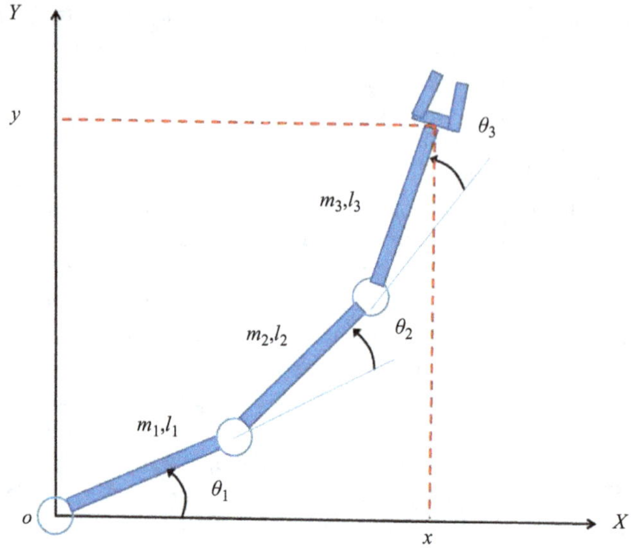

Figure 2.142 Diagram of the 3-DOF redundant robotic manipulator

Equivalently, the velocity of the end-effector of the robotic manipulator can be written as

$$\begin{aligned}\dot{x} &= [-l_1\sin(\theta_1) - l_2\sin(\theta_1+\theta_2) - l_3\sin(\theta_1+\theta_2+\theta_3)]\dot{\theta}_1\\&\quad + [-l_2\sin(\theta_1+\theta_2) - l_3\sin(\theta_1+\theta_2+\theta_3)]\dot{\theta}_2 + [-l_3\sin(\theta_1+\theta_2+\theta_3)]\dot{\theta}_3\\\dot{y} &= [l_1\cos(\theta_1) + l_2\cos(\theta_1+\theta_2) + l_3\cos(\theta_1+\theta_2+\theta_3)]\dot{\theta}_1\\&\quad + [l_2\cos(\theta_1+\theta_2) + l_3\cos(\theta_1+\theta_2+\theta_3)]\dot{\theta}_2 + [l_3\cos(\theta_1+\theta_2+\theta_3)]\dot{\theta}_3\end{aligned} \tag{2.483}$$

The following state variables are defined: $x_1 = x$, $x_2 = y$, $x_3 = \theta_1$, $x_4 = \dot{\theta}_1$, $x_5 = \theta_2$, $x_6 = \dot{\theta}_2$, $x_7 = \theta_3$, $x_8 = \dot{\theta}_3$. Thus, one obtains

$$\begin{aligned}\dot{x}_1 &= [-l_1\sin(x_3) - l_2\sin(x_3+x_5) - l_3\sin(x_3+x_5+x_7)]x_4\\&\quad + [-l_2\sin(x_3+x_5) - l_3\sin(x_3+x_5+x_7)]x_6 + [-l_3\sin(x_3+x_5+x_7)]x_8\\\dot{x}_2 &= [l_1\cos(x_3) + l_2\cos(x_3+x_5) + l_3\cos(x_3+x_5+x_7)]x_4\\&\quad + [l_2\cos(x_3+x_5) + l_3\cos(x_3+x_5+x_7)]x_6 + [l_3\cos(x_3+x_5+x_7)]x_8\end{aligned} \tag{2.484}$$

Besides, about motion in the joints space it holds that (Mahindrakar *et al.*, 2005; Kayastha *et al.*, 2017; Rigatos and Busawon, 2018)

$$M(\theta)\ddot{\theta} + C(\theta,\dot{\theta}) + G(\theta) = \tau \tag{2.485}$$

where $\theta \in R^{3\times 3}$ is the joints' vector, $M(\theta)\in R^{3\times 3}$ is the inertia matrix, $C(\theta,\dot{\theta})\in R^{3\times 1}$ is the Coriolis and centrifugal forces vector, $G(\theta)\in R^{3\times 1}$ is the gravitational forces vector and $\tau \in R^{3\times 1}$ is the control input torques vector. The previous relation can be also written as

$$\ddot{\theta} = -M(\theta)^{-1}C(\theta,\dot{\theta}) - M(\theta)^{-1}G(\theta) + M(\theta)^{-1}\tau \tag{2.486}$$

or in detailed form

$$\begin{pmatrix} \dot{\theta}_1 \\ \dot{\theta}_2 \\ \dot{\theta}_3 \end{pmatrix} = -\frac{1}{detM}\begin{pmatrix} M_{11} & -M_{21} & M_{31} \\ -M_{12} & M_{22} & -M_{32} \\ M_{13} & -M_{23} & M_{33} \end{pmatrix}\left\{\begin{pmatrix} C_1(\theta,\dot{\theta}) \\ C_2(\theta,\dot{\theta}) \\ C_3(\theta,\dot{\theta}) \end{pmatrix}\right.$$
$$\left. + \begin{pmatrix} G_1(\theta) \\ G_2(\theta) \\ G_3(\theta) \end{pmatrix}\right\} + \frac{1}{detM}\begin{pmatrix} M_{11} & -M_{21} & M_{31} \\ -M_{12} & M_{22} & -M_{32} \\ M_{13} & -M_{23} & M_{33} \end{pmatrix}\begin{pmatrix} u_1 \\ u_2 \\ u_3 \end{pmatrix} \tag{2.487}$$

Consequently, about the dynamics of the 3-link robotic manipulator one has that

$$\ddot{\theta}_1 = \frac{-(M_{11}C_1 - M_{21}C_2 + M_{31}C_3) - (M_{11}G_1 - M_{21}G_2 + M_{31}G_3)}{detM} + \frac{(M_{11}u_1 - M_{21}u_2 + M_{31}u_3)}{detM} \tag{2.488}$$

$$\ddot{\theta}_2 = \frac{-(-M_{12}C_1 + M_{22}C_2 - M_{32}C_3) - (-M_{12}G_1 + M_{22}G_2 - M_{32}G_3)}{detM} + \frac{(-M_{12}u_1 + M_{22}u_2 - M_{32}u_3)}{detM} \tag{2.489}$$

$$\ddot{\theta}_3 = \frac{-(M_{13}C_1 - M_{23}C_2 + M_{33}C_3) - (M_{13}G_1 - M_{23}G_2 + M_{33}G_3)}{detM} + \frac{(M_{13}u_1 - M_{23}u_2 + M_{33}u_3)}{detM} \tag{2.490}$$

Using the previous definition of state variables, the dynamics of motion in the tasks space becomes

$$
\begin{aligned}
\dot{x}_3 &= x_4\\
\dot{x}_4 &= \frac{-(M_{11}C_1 - M_{21}C_2 + M_{31}C_3) - (M_{11}G_1 - M_{21}G_2 + M_{31}G_3)}{detM}\\
&\quad + \frac{(M_{11}u_1 - M_{21}u_2 + M_{31}u_3)}{detM}\\
\dot{x}_5 &= x_6\\
\dot{x}_6 &= \frac{-(-M_{12}C_1 + M_{22}C_2 - M_{32}C_3) - (-M_{12}G_1 + M_{22}G_2 - M_{32}G_3)}{detM}\\
&\quad + \frac{(-M_{12}u_1 + M_{22}u_2 - M_{32}u_3)}{detM}\\
\dot{x}_7 &= x_8\\
\dot{x}_8 &= \frac{-(M_{13}C_1 - M_{23}C_2 + M_{33}C_3) - (M_{13}G_1 - M_{23}G_2 + M_{33}G_3)}{detM}\\
&\quad + \frac{(M_{13}u_1 - M_{23}u_2 + M_{33}u_3)}{detM}
\end{aligned}
\tag{2.491}
$$

Consequently, the state-space model of the redundant robotic manipulator can be written in the following state-space form:

$$
\begin{pmatrix}\dot{x}_1\\ \dot{x}_2\\ \dot{x}_3\\ \dot{x}_4\\ \dot{x}_5\\ \dot{x}_6\\ \dot{x}_7\\ \dot{x}_8\end{pmatrix} = \begin{pmatrix}f_1(x)\\ f_2(x)\\ f_3(x)\\ f_4(x)\\ f_5(x)\\ f_6(x)\\ f_7(x)\\ f_8(x)\end{pmatrix} + \begin{pmatrix}g_{11}(x) & g_{12}(x) & g_{13}(x)\\ g_{21}(x) & g_{22}(x) & g_{23}(x)\\ g_{31}(x) & g_{32}(x) & g_{33}(x)\\ g_{41}(x) & g_{42}(x) & g_{43}(x)\\ g_{51}(x) & g_{52}(x) & g_{53}(x)\\ g_{61}(x) & g_{62}(x) & g_{63}(x)\\ g_{71}(x) & g_{72}(x) & g_{73}(x)\\ g_{81}(x) & g_{82}(x) & g_{83}(x)\end{pmatrix} \begin{pmatrix}u_1\\ u_2\\ u_3\end{pmatrix}
\tag{2.492}
$$

or, in concise form the previous states-pace model becomes

$$
\dot{x} = f(x) + g(x)u \tag{2.493}
$$

with $x \in R^{8\times1}$, $u \in R^{3\times1}$, $f(x) \in R^{8\times1}$ and $g(x) \in R^{8\times3}$. About the elements of the above noted matrices it holds that

$$
\begin{aligned}
f_1(x) = {} & [-l_1 \sin(x_3) - l_2 \sin(x_3 + x_5) - l_3 \sin(x_3 + x_5 + x_7)]x_4\\
& +[-l_2 \sin(x_3 + x_5) - l_3 \sin(x_3 + x_5 + x_7)]x_6 + [-l_3 \sin(x_3 + x_5 + x_7)]x_8
\end{aligned}
\tag{2.494}
$$

$$
\begin{aligned}
f_2(x) = {} & [l_1 \cos(x_3) + l_2 \cos(x_3 + x_5) + l_3 \cos(x_3 + x_5 + x_7)]x_4\\
& +[l_2 \cos(x_3 + x_5) + l_3 \cos(x_3 + x_5 + x_7)]x_6 + [l_3 \cos(x_3 + x_5 + x_7)]x_8
\end{aligned}
\tag{2.495}
$$

$$f_3(x) = x_4 \tag{2.496}$$

$$f_4(x) = \frac{-(M_{11}C_1 - M_{21}C_2 + M_{31}C_3) - (M_{11}G_1 - M_{21}G_2 + M_{31}G_3)}{detM} \tag{2.497}$$

$$f_5(x) = x_6 \tag{2.498}$$

$$f_6(x) = \frac{-(-M_{12}C_1 + M_{22}C_2 - M_{32}C_3) - (-M_{12}G_1 + M_{22}G_2 - M_{32}G_3)}{detM} \tag{2.499}$$

$$f_7(x) = x_8 \tag{2.500}$$

$$f_8(x) = \frac{-(M_{13}C_1 - M_{23}C_2 + M_{33}C_3) - (M_{13}G_1 - M_{23}G_2 + M_{33}G_3)}{detM} \tag{2.501}$$

Moreover, about the input gains matrix $g(x)$, it holds that

$$g(x) = \begin{pmatrix} 0 & 0 & 0 \\ 0 & 0 & 0 \\ 0 & 0 & 0 \\ \frac{M_{11}}{detM} & -\frac{M_{21}}{detM} & \frac{M_{31}}{detM} \\ 0 & 0 & 0 \\ -\frac{M_{12}}{detM} & \frac{M_{22}}{detM} & -\frac{M_{32}}{detM} \\ 0 & 0 & 0 \\ \frac{M_{13}}{detM} & -\frac{M_{23}}{detM} & \frac{M_{33}}{detM} \end{pmatrix} \tag{2.502}$$

2.11.2.2 Dynamic model of the 3-DOF robotic manipulator in the joints' space

The dynamic model of the 3-DOF robotic manipulator, describing motion in the joints' space, is given by

$$M(\theta)\ddot{\theta} + C(\theta, \dot{\theta}) + G(\theta) = \tau(t) \tag{2.503}$$

where $\theta \in R^{3\times1}$ with $\theta = [\theta_1, \theta_2, \theta_3]^T$, $\dot{\theta} \in R^{3\times1}$ with $\dot{\theta} = [\dot{\theta}_1, \dot{\theta}_2, \dot{\theta}_3]^T$, $M(\theta) \in R^{3\times3}$ is the inertia matrix, $C(\theta, \dot{\theta}) \in R^{3\times1}$ is the Coriolis and centrifugal forces vector, $G(\theta) \in R^{3\times1}$ is the gravitational forces vector and $\tau(t) \in R^{3\times1}$ is the torques (control inputs) vector.

Using Euler-Lagrange analysis, one obtains the elements of matrices $M(\theta)$, $C(\theta, \dot{\theta})$ and $G(\theta)$ which are given as follows (Mahindrakar *et al.*, 2005; Kayastha *et al.*, 2017; Rigatos and Busawon, 2018)

$$M(\theta) = \begin{pmatrix} m_{11} & m_{12} & m_{13} \\ m_{21} & m_{22} & m_{23} \\ m_{31} & m_{32} & m_{33} \end{pmatrix} \tag{2.504}$$

where $m_{11} = a_1 + a_2 + a_4 + 2a_3\cos(\theta_2) + 2a_5\cos(\theta_2 + \theta_3) + 2a_6\cos(\theta_3)$, $m_{11} = m_{21} = a_2 + a_4 + a_3\cos(\theta_2) + a_5\cos(\theta_2 + \theta_3) + 2a_6\cos(\theta_3)$, $m_{13} = m_{31} = a_4 + a_5 \cos(\theta_2 + \theta_3) + a_6\cos(\theta_3)$, $m_{22} = a_2 + a_4 + a_6\cos(\theta_3)$, $m_{33} = a_4$

with $a_1 = m_1 l_1^2 + J_1 + (m_2 + m_3)L_1^2, a_2 = J_2 + m_2 l_2^2 + m_3 L_1^2, a_3 = (m_2 l_2 + m_3 L_2)L_1,$ $a_4 = J_3 + m_3 l_3^2, a_5 = m_3 l_3 L_1, a_6 = m_3 L_2 l_3$ (Xiong, Lai *et al.*, 2016).

$m_i,\ i = 1,2,3$ are the masses of the 3 links, $J_i\ i = 1,2,3$ are the moments of inertia of the three links for rotation performed around the joints found at their basis, $L_i\ i = 1,2,3$ are the lengths of the three links and $l_i\ i = 1,2,3$ are the distances between the basis of the links and their centers of gravity (Xiong, Lai *et al.*, 2016). With reference to matrix M, the following sub-matrices are defined

and the related determinant is $M_{11} = m_{22}m_{33} - m_{32}m_{23}$, $M_{12} = m_{21}m_{33} - m_{31}m_{23}$, $M_{13} = m_{12}m_{23} - m_{22}m_{13}, M_{21} = m_{12}m_{33} - m_{22}m_{13}, M_{22} = m_{11}m_{33} - m_{31}m_{13}, M_{23} =$ $m_{21}m_{32} - m_{31}m_{12}$, $M_{31} = m_{12}m_{23} - m_{22}m_{13}$, $M_{32} = m_{11}m_{23} - m_{21 m_{13}}$, $M_{33} =$ $m_{11}m_{22} - m_{21}m_{12}$.

while the determinant of the inertia matrix M is given by $detM = m_{11}(m_{22}m_{33} - m_{32}m_{23}) - m_{12}(m_2 m_{33} - m_{31}m_{23}) + m_{13}(m_{21}m_{32} - m_{31}m_{22})$.

About the Coriolis and centrifugal forces matrix, one has (Xiong, Lai *et al.*, 2016):

$$C(\theta,\dot{\theta}) = \begin{pmatrix} -a_3(2\dot{\theta}_1 + \dot{\theta}_2)\dot{\theta}_2 \sin(\theta_2) \\ -a_5(2\dot{\theta}_1 + \dot{\theta}_2 + \dot{\theta}_3)(\dot{\theta}_2 + \dot{\theta}_3)\sin(\theta_2 + \theta_3) \\ -a_6(2\dot{\theta}_1 + 2\dot{\theta}_2 + \dot{\theta}_3)\dot{\theta}_3 \sin(\theta_3) \\ \\ a_3\dot{\theta}_1^2 \sin(\theta_2) + a_5\dot{\theta}_1^2 \sin(\theta_2 + \theta_3) - \\ -a_6(2\dot{\theta}_1 + 2\dot{\theta}_2 + \dot{\theta}_3)\dot{\theta}_3 \sin(\theta_3) \\ \\ a_5\dot{\theta}_1^2 \sin(\theta_2 + \theta_3) + a_6(\dot{\theta}_1 + \dot{\theta}_2)^2 \sin(\theta_3) \end{pmatrix} \tag{2.505}$$

Since the manipulator is considered to be planar, the gravitational forces matrix is taken to be $0 \in R^{3\times 1}$, that is

$$G(\theta) = \begin{pmatrix} 0 \\ 0 \\ 0 \end{pmatrix} \tag{2.506}$$

The inverse of the inertia matrix $M(\theta)$ is computed as follows:

$$M^{-1}(\theta) = \frac{1}{detM}\begin{pmatrix} M_{11} & -M_{21} & M_{31} \\ -M_{12} & M_{22} & -M_{32} \\ M_{13} & -M_{23} & M_{33} \end{pmatrix} \tag{2.507}$$

Using the previous state variables notation $x_1 = x, x_2 = y, x_3 = \theta_1, x_4 = \dot{\theta}_1, x_5 = \theta_2,$ $x_6 = \dot{\theta}_2, x_7 = \theta_3, x_8 = \dot{\theta}_3$ one has that:

the elements of the inertia matrix are rewritten as $m_{11} = a_1 + a_2 + a_4 + 2a_3\cos(x_5) + 2a_5\cos(x_5 + x_7) + 2a_6\cos(x_7)$, $m_{12} = m_{21} = a_2 + a_4 + a_3\cos(x_5) + a_5\cos(x_5 + x_7) + 2a_6\cos(x_7)$, $m_{13} = m_{31} = a_4 + a_5\cos(x_5 + x_7) + a_6\cos(x_7)$,

$m_{22} = a_2 + a_4 + a_6\cos(x_7),\ m_{33} = a_4$

while the Coriolis matrix is rewritten as

$$C(\theta,\dot{\theta}) = \begin{pmatrix} -a_3(2x_4+x_6)x_6\sin(x_5) \\ -a_5(2x_4+x_6+x_8)(x_6+x_8)\sin(x_5+x_7) \\ -a_6(2x_4+2x_6+x_8)x_8\sin(x_7) \\ \\ a_3x_4^2\sin(x_5)+a_5x_4^2\sin(x_5+x_7)- \\ -a_6(2x_4+2x_6+x_8)x_8\sin(x_7) \\ \\ a_5x_4^2\sin(x_5+x_7)+a_6(x_4+x_6)^2\sin(x_7) \end{pmatrix} \tag{2.508}$$

2.11.3 Approximate linearization of the model of the redundant manipulator

2.11.3.1 Approximate linearization of the joint kinematic and dynamic model

The joint kinematic and dynamic model of the 3-DOF redundant robotic manipulator undergoes approximate linearization around the temporary operating point (x^*, u^*) where x^* is the present value of the robot's state vector and u^* is the last sampled value of the control inputs vector. The linearization point is updated at each iteration of the control algorithm. Thus, after applying Taylor series expansion, the initial model of the robot being in the form (Rigatos and Busawon, 2018)

$$\dot{x} = f(x) + g_1(x)u_1 + g_2(x)u_2 + +g_3(x)u_3 \tag{2.509}$$

is now written as

$$\dot{x} = Ax + Bu + \tilde{d} \tag{2.510}$$

where $\tilde{d}$ is the vector of cumulative disturbances which are due to (i) modeling errors because of truncating higher-order terms from the Taylor series expansion, (ii) exogenous perturbations, and (iii) measurement noise of any distribution. Matrices A and B are obtained from the robot's Jacobians:

$$\begin{aligned} &A = \nabla_x[f(x) + g_1(x)u_1 + g_2(x)u_2 + g_3(x)u_3]|_{(x^*,u^*)} \Rightarrow \\ &A = \nabla_x[f(x)]|_{(x^*,u^*)} + \nabla_x[g_1(x)u_1]|_{(x^*,u^*)} \\ &+\nabla_x[g_2(x)u_2]|_{(x^*,u^*)} + \nabla_x[g_3(x)u_3]|_{(x^*,u^*)} \end{aligned} \tag{2.511}$$

$$\begin{aligned} &B = \nabla_u[f(x) + g_1(x)u_1 + g_2(x)u_2 + g_3(x)u_3]|_{(x^*,u^*)} \Rightarrow \\ &B = [g_1(x)g_2(x)g_3(x)]|_{(x^*,u^*)} \Rightarrow B = g(x)|_{(x^*,u^*)} \end{aligned} \tag{2.512}$$

Computation of the Jacobian matrix $\nabla_x f(x)|_{x^*,u^*}$.

First row of the Jacobian matrix $\nabla_x f(x)|_{x^*,u^*}$: $\frac{\partial f_1}{\partial x_1} = 0$, $\frac{\partial f_1}{\partial x_2} = 0$, $\frac{\partial f_1}{\partial x_3} = [-l_1\cos(x_3) - l_2\cos(x_3+x_5) - l_3\cos(x_3+x_5+x_7)]x_4 + [-l_1\cos(x_3+x_5) - l_3]$, $\frac{\partial f_1}{\partial x_4} = [l_1\sin(x_3) - l_2\sin(x_3+x_5) - l_3\sin(x_3+x_5+x_7)]$. $\frac{\partial f_1}{\partial x_5} = [-l_2\cos(x_5+x_5) - l_3\cos(x_3+x_5+x_7)]x_4 + [-l_1\cos(x_5+x_5) - l_3\cos(x_3+x_5+x_7)]x_6 + [-l_3\cos(x_3+x_5+x_7)]x_8$, $\frac{\partial f_1}{\partial x_6} =$

$[-l_1\sin(x_3+x_5)-l_3\sin(x_3+x_5+x_7)]$, $\frac{\partial f_1}{\partial x_7}=[-l_3\cos(x_3+x_5+x_7)]x_4+[l_3\cos(x_3+x_5+x_7)]x_6+[-l_3\cos(x_3+x_5+x_7)]x_8$, $\frac{\partial f_1}{\partial x_8}=[-l_2\sin(x_3+x_5+x_7)]$.

Second row of the Jacobian matrix $\nabla_x f(x)|_{x^*,u^*}$: $\frac{\partial f_2}{\partial x_1}=0$, $\frac{\partial f_2}{\partial x_2}=0$, $\frac{\partial f_2}{\partial x_3}=[-l_1\sin(x_3)-l_2\sin(x_3+x_5)-l_3\sin(x_3+x_5+x_7)]x_4+[-l_2\sin(x_3+x_5)-l_3\sin(x_3+x_5+x_7)]x_6+[-l_3\sin(x_3+x_5+x_7)]x_8$, $\frac{\partial f_2}{\partial x_4}=l_1\cos(x_3)+l_2\cos(x_3+x_5)+l_3\cos(x_3+x_5+x_7)$, $\frac{\partial f_2}{\partial x_5}=[-l_2\sin(x_3+x_5)-l_3\sin(x_3+x_5+x_7)]x_4+[-l_2\sin(x_3+x_5)-l_3\sin(x_3+x_5+x_7)]x_6+[-l_3\sin(x_3+x_5+x_7)]x_8$, $\frac{\partial f_2}{\partial x_6}=[l_2\cos(x_3+x_5)+l_3\cos(x_3+x_5+x_7)]$, $\frac{\partial f_2}{\partial x_7}=[-l_3\sin(x_3+x_5+x_7)]x_4+[-l_3\sin(x_3+x_5+x_7)]x_6+[-l_3\sin(x_3+x_5+x_7)]x_8$, $\frac{\partial f_2}{\partial x_8}=[l_3\cos(x_3+x_5+x_7)]$.

Third row of the Jacobian matrix $\nabla_x f(x)|_{x^*,u^*}$: $\frac{\partial f_3}{\partial x_1}=0$, $\frac{\partial f_3}{\partial x_2}=0$, $\frac{\partial f_3}{\partial x_3}=0$, $\frac{\partial f_3}{\partial x_4}=1$, $\frac{\partial f_3}{\partial x_5}=0$, $\frac{\partial f_3}{\partial x_6}=0$, $\frac{\partial f_3}{\partial x_7}=0$, and $\frac{\partial f_3}{\partial x_8}=0$.

Fourth row of the Jacobian matrix $\nabla_x f(x)|_{x^*,u^*}$: it holds that $f_4(x)=\frac{f_{4n}(x)}{f_{4d}(x)}$, where $f_{4n}(x)=-(M_{11}C_1-M_{21}C_2+M_{31}C_3)-(M_{11}G_1-M_{21}G_2+M_{31}G_3)$ and $f_{4d}(x)=detM$. Thus, one has that for $i=1,2,\cdots,8$

$$\begin{aligned}\frac{\partial f_{4n}}{\partial x_i}=&-\left(\frac{\partial M_{11}}{\partial x_i}C_1+M_{11}\frac{\partial C_1}{\partial x_i}-\frac{\partial M_{21}}{\partial x_i}C_2-M_{21}\frac{\partial C_2}{\partial x_i}+\frac{\partial M_{31}}{\partial x_i}C_3+M_{31}\frac{\partial C_3}{\partial x_i}\right)\\&-\left(\frac{\partial M_{11}}{\partial x_i}G_1+M_{11}\frac{\partial G_1}{\partial x_i}-\frac{\partial M_{21}}{\partial x_i}G_2-M_{21}\frac{\partial G_2}{\partial x_i}+\frac{\partial M_{31}}{\partial x_i}G_3+M_{31}\frac{\partial G_3}{\partial x_i}\right)\end{aligned}\tag{2.513}$$

$$\frac{\partial f_{4d}}{\partial x_i}=\frac{\partial detM}{\partial x_i}\tag{2.514}$$

$$\frac{\partial f_4}{\partial x_i}=\frac{\frac{\partial f_{4n}}{\partial x_i}f_{4d}-f_{4n}\frac{\partial f_{4d}}{\partial x_i}}{f_{4d}^2}\tag{2.515}$$

Fifth row of the Jacobian matrix $\nabla_x f(x)|_{x^*,u^*}$: $\frac{\partial f_5}{\partial x_1}=0$, $\frac{\partial f_5}{\partial x_2}=0$, $\frac{\partial f_5}{\partial x_3}=0$, $\frac{\partial f_5}{\partial x_4}=0$, $\frac{\partial f_5}{\partial x_5}=0$, $\frac{\partial f_5}{\partial x_6}=1$, $\frac{\partial f_5}{\partial x_7}=0$, and $\frac{\partial f_5}{\partial x_8}=0$.

Sixth row of the Jacobian matrix $\nabla_x f(x)|_{x^*,u^*}$: it holds that $f_6(x)=\frac{f_{6n}(x)}{f_{6d}(x)}$, where $f_{6n}(x)=-(M_{12}C_1-M_{22}C_2+M_{32}C_3)-(M_{12}G_1-M_{22}G_2+M_{32}G_3)$ and $f_{6d}(x)=detM$. Thus, one has that for $i=1,2,\cdots,8$

$$\begin{aligned}\frac{\partial f_{6n}}{\partial x_i}=&-\left(\frac{\partial M_{12}}{\partial x_i}C_1+M_{12}\frac{\partial C_1}{\partial x_i}-\frac{\partial M_{22}}{\partial x_i}C_2-M_{22}\frac{\partial C_2}{\partial x_i}+\frac{\partial M_{32}}{\partial x_i}C_3+M_{32}\frac{\partial C_3}{\partial x_i}\right)\\&-\left(\frac{\partial M_{12}}{\partial x_i}G_1+M_{12}\frac{\partial G_1}{\partial x_i}-\frac{\partial M_{22}}{\partial x_i}G_2-M_{22}\frac{\partial G_2}{\partial x_i}+\frac{\partial M_{32}}{\partial x_i}G_3+M_{32}\frac{\partial G_3}{\partial x_i}\right)\end{aligned}\tag{2.516}$$

$$\frac{\partial f_{6d}}{\partial x_i}=\frac{\partial detM}{\partial x_i}\tag{2.517}$$

$$\frac{\partial f_6}{\partial x_i}=\frac{\frac{\partial f_{6n}}{\partial x_i}f_{6d}-f_{6n}\frac{\partial f_{6d}}{\partial x_i}}{f_{6d}^2}\tag{2.518}$$

Seventh row of the Jacobian matrix $\nabla_x f(x)|_{x^*,u^*}$: $\frac{\partial f_7}{\partial x_1}=0$, $\frac{\partial f_7}{\partial x_2}=0$, $\frac{\partial f_7}{\partial x_3}=0$, $\frac{\partial f_7}{\partial x_4}=0$, $\frac{\partial f_7}{\partial x_5}=0$, $\frac{\partial f_7}{\partial x_6}=0$, $\frac{\partial f_7}{\partial x_7}=0$, and $\frac{\partial f_7}{\partial x_8}=1$.

Eighth row of the Jacobian matrix $\nabla_x f(x)|_{x^*,u^*}$: it holds that $f_8(x)=\frac{f_{8n}(x)}{f_{8d}(x)}$, where $f_{8n}(x)=-(M_{13}C_1-M_{23}C_2+M_{33}C_3)-(M_{13}G_1-M_{23}G_2+M_{33}G_3)$ and $f_{8d}(x)=detM$. Thus, one has that for $i=1,2,\cdots,8$

$$\begin{aligned}\frac{\partial f_{8n}}{\partial x_i}=&-\left(\frac{\partial M_{13}}{\partial x_i}C_1+M_{13}\frac{\partial C_1}{\partial x_i}-\frac{\partial M_{23}}{\partial x_i}C_2-M_{23}\frac{\partial C_2}{\partial x_i}+\frac{\partial M_{33}}{\partial x_i}C_3+M_{33}\frac{\partial C_3}{\partial x_i}\right)\\&-\left(\frac{\partial M_{13}}{\partial x_i}G_1+M_{13}\frac{\partial G_1}{\partial x_i}-\frac{\partial M_{23}}{\partial x_i}G_2-M_{23}\frac{\partial G_2}{\partial x_i}+\frac{\partial M_{33}}{\partial x_i}G_3+M_{33}\frac{\partial G_3}{\partial x_i}\right)\end{aligned} \tag{2.519}$$

$$\frac{\partial f_{8d}}{\partial x_i}=\frac{\partial detM}{\partial x_i} \tag{2.520}$$

$$\frac{\partial f_8}{\partial x_i}=\frac{\frac{\partial f_{8n}}{\partial x_i}f_{8d}-f_{8n}\frac{\partial f_{8d}}{\partial x_i}}{f_{8d}^2} \tag{2.521}$$

Computation of the Jacobian matrix $\nabla_x g_1(x)|_{x^*,u^*}$.

First row of the Jacobian matrix $\nabla_x g_1(x)|_{x^*,u^*}$: $\frac{\partial g_{11}}{\partial x_1}=0$, $\frac{\partial g_{11}}{\partial x_2}=0$, $\frac{\partial g_{11}}{\partial x_3}=0$, $\frac{\partial g_{11}}{\partial x_4}=0$, $\frac{\partial g_{11}}{\partial x_5}=0$, $\frac{\partial g_{11}}{\partial x_6}=0$, $\frac{\partial g_{11}}{\partial x_7}=0$, and $\frac{\partial g_{11}}{\partial x_8}=0$.

Second row of the Jacobian matrix $\nabla_x g_1(x)|_{x^*,u^*}$: $\frac{\partial g_{21}}{\partial x_1}=0$, $\frac{\partial g_{21}}{\partial x_2}=0$, $\frac{\partial g_{21}}{\partial x_3}=0$, $\frac{\partial g_{21}}{\partial x_4}=0$, $\frac{\partial g_{21}}{\partial x_5}=0$, $\frac{\partial g_{21}}{\partial x_6}=0$, $\frac{\partial g_{21}}{\partial x_7}=0$, and $\frac{\partial g_{21}}{\partial x_8}=0$.

Third row of the Jacobian matrix $\nabla_x g_1(x)|_{x^*,u^*}$: $\frac{\partial g_{31}}{\partial x_1}=0$, $\frac{\partial g_{31}}{\partial x_2}=0$, $\frac{\partial g_{31}}{\partial x_3}=0$, $\frac{\partial g_{31}}{\partial x_4}=0$, $\frac{\partial g_{31}}{\partial x_5}=0$, $\frac{\partial g_{31}}{\partial x_6}=0$, $\frac{\partial g_{31}}{\partial x_7}=0$, and $\frac{\partial g_{31}}{\partial x_8}=0$.

Fourth row of the Jacobian matrix $\nabla_x g_1(x)|_{x^*,u^*}$, $i=1,2,\cdots,8$:

$$\frac{\partial g_{41}}{\partial x_i}=\frac{\frac{\partial M_{11}}{\partial x_i}detM-M_{11}\frac{\partial detM}{\partial x_i}}{detM^2} \tag{2.522}$$

Fifth row of the Jacobian matrix $\nabla_x g_1(x)|_{x^*,u^*}$: $\frac{\partial g_{51}}{\partial x_1}=0$, $\frac{\partial g_{51}}{\partial x_2}=0$, $\frac{\partial g_{51}}{\partial x_3}=0$, $\frac{\partial g_{51}}{\partial x_4}=0$, $\frac{\partial g_{51}}{\partial x_5}=0$, $\frac{\partial g_{51}}{\partial x_6}=0$, $\frac{\partial g_{51}}{\partial x_7}=0$, and $\frac{\partial g_{51}}{\partial x_8}=0$.

Sixth row of the Jacobian matrix $\nabla_x g_1(x)|_{x^*,u^*}$, $i=1,2,\cdots,8$:

$$\frac{\partial g_{61}}{\partial x_i}=-\frac{\frac{\partial M_{12}}{\partial x_i}detM-M_{12}\frac{\partial detM}{\partial x_i}}{detM^2} \tag{2.523}$$

Seventh row of the Jacobian matrix $\nabla_x g_1(x)|_{x^*,u^*}$: $\frac{\partial g_{71}}{\partial x_1}=0$, $\frac{\partial g_{71}}{\partial x_2}=0$, $\frac{\partial g_{71}}{\partial x_3}=0$, $\frac{\partial g_{71}}{\partial x_4}=0$, $\frac{\partial g_{71}}{\partial x_5}=0$, $\frac{\partial g_{71}}{\partial x_6}=0$, $\frac{\partial g_{71}}{\partial x_7}=0$, and $\frac{\partial g_{71}}{\partial x_8}=0$.

Eighth row of the Jacobian matrix $\nabla_x g_1(x)|_{x^*,u^*}$, $i = 1, 2, \cdots, 8$:

$$\frac{\partial g_{81}}{\partial x_i} = \frac{\frac{\partial M_{13}}{\partial x_i} detM - M_{13} \frac{\partial detM}{\partial x_i}}{detM^2} \tag{2.524}$$

Computation of the Jacobian matrix $\nabla_x g_2(x)|_{x^*,u^*}$.

First row of the Jacobian matrix $\nabla_x g_2(x)|_{x^*,u^*}$: $\frac{\partial g_{12}}{\partial x_1} = 0$, $\frac{\partial g_{12}}{\partial x_2} = 0$, $\frac{\partial g_{12}}{\partial x_3} = 0$, $\frac{\partial g_{12}}{\partial x_4} = 0$, $\frac{\partial g_{12}}{\partial x_5} = 0$, $\frac{\partial g_{12}}{\partial x_6} = 0$, $\frac{\partial g_{12}}{\partial x_7} = 0$, and $\frac{\partial g_{12}}{\partial x_8} = 0$.

Second row of the Jacobian matrix $\nabla_x g_2(x)|_{x^*,u^*}$: $\frac{\partial g_{22}}{\partial x_1} = 0$, $\frac{\partial g_{22}}{\partial x_2} = 0$, $\frac{\partial g_{22}}{\partial x_3} = 0$, $\frac{\partial g_{22}}{\partial x_4} = 0$, $\frac{\partial g_{22}}{\partial x_5} = 0$, $\frac{\partial g_{22}}{\partial x_6} = 0$, $\frac{\partial g_{22}}{\partial x_7} = 0$, and $\frac{\partial g_{22}}{\partial x_8} = 0$.

Third row of the Jacobian matrix $\nabla_x g_2(x)|_{x^*,u^*}$: $\frac{\partial g_{32}}{\partial x_1} = 0$, $\frac{\partial g_{32}}{\partial x_2} = 0$, $\frac{\partial g_{32}}{\partial x_3} = 0$, $\frac{\partial g_{32}}{\partial x_4} = 0$, $\frac{\partial g_{32}}{\partial x_5} = 0$, $\frac{\partial g_{32}}{\partial x_6} = 0$, $\frac{\partial g_{32}}{\partial x_7} = 0$, and $\frac{\partial g_{32}}{\partial x_8} = 0$.

Fourth row of the Jacobian matrix $\nabla_x g_2(x)|_{x^*,u^*}$, $i = 1, 2, \cdots, 8$:

$$\frac{\partial g_{42}}{\partial x_i} = \frac{\frac{\partial M_{21}}{\partial x_i} detM - M_{21} \frac{\partial detM}{\partial x_i}}{detM^2} \tag{2.525}$$

Fifth row of the Jacobian matrix $\nabla_x g_2(x)|_{x^*,u^*}$: $\frac{\partial g_{52}}{\partial x_1} = 0$, $\frac{\partial g_{52}}{\partial x_2} = 0$, $\frac{\partial g_{52}}{\partial x_3} = 0$, $\frac{\partial g_{52}}{\partial x_4} = 0$, $\frac{\partial g_{52}}{\partial x_5} = 0$, $\frac{\partial g_{52}}{\partial x_6} = 0$, $\frac{\partial g_{52}}{\partial x_7} = 0$, and $\frac{\partial g_{52}}{\partial x_8} = 0$.

Sixth row of the Jacobian matrix $\nabla_x g_2(x)|_{x^*,u^*}$, $i = 1, 2, \cdots, 8$:

$$\frac{\partial g_{62}}{\partial x_i} = -\frac{\frac{\partial M_{22}}{\partial x_i} detM - M_{22} \frac{\partial detM}{\partial x_i}}{detM^2} \tag{2.526}$$

Seventh row of the Jacobian matrix $\nabla_x g_2(x)|_{x^*,u^*}$: $\frac{\partial g_{72}}{\partial x_1} = 0$, $\frac{\partial g_{72}}{\partial x_2} = 0$, $\frac{\partial g_{72}}{\partial x_3} = 0$, $\frac{\partial g_{72}}{\partial x_4} = 0$, $\frac{\partial g_{72}}{\partial x_5} = 0$, $\frac{\partial g_{72}}{\partial x_6} = 0$, $\frac{\partial g_{72}}{\partial x_7} = 0$, and $\frac{\partial g_{72}}{\partial x_8} = 0$.

Eighth row of the Jacobian matrix $\nabla_x g_2(x)|_{x^*,u^*}$, $i = 1, 2, \cdots, 8$:

$$\frac{\partial g_{82}}{\partial x_i} = \frac{\frac{\partial M_{23}}{\partial x_i} detM - M_{23} \frac{\partial detM}{\partial x_i}}{detM^2} \tag{2.527}$$

Computation of the Jacobian matrix $\nabla_x g_3(x)|_{x^*,u^*}$.

First row of the Jacobian matrix $\nabla_x g_3(x)|_{x^*,u^*}$: $\frac{\partial g_{13}}{\partial x_1} = 0$, $\frac{\partial g_{13}}{\partial x_2} = 0$, $\frac{\partial g_{13}}{\partial x_3} = 0$, $\frac{\partial g_{13}}{\partial x_4} = 0$, $\frac{\partial g_{13}}{\partial x_5} = 0$, $\frac{\partial g_{13}}{\partial x_6} = 0$, $\frac{\partial g_{13}}{\partial x_7} = 0$, and $\frac{\partial g_{13}}{\partial x_8} = 0$.

Second row of the Jacobian matrix $\nabla_x g_3(x)|_{x^*,u^*}$: $\frac{\partial g_{23}}{\partial x_1} = 0$, $\frac{\partial g_{23}}{\partial x_2} = 0$, $\frac{\partial g_{23}}{\partial x_3} = 0$, $\frac{\partial g_{23}}{\partial x_4} = 0$, $\frac{\partial g_{23}}{\partial x_5} = 0$, $\frac{\partial g_{23}}{\partial x_6} = 0$, $\frac{\partial g_{23}}{\partial x_7} = 0$, and $\frac{\partial g_{23}}{\partial x_8} = 0$.

Third row of the Jacobian matrix $\nabla_x g_3(x)|_{x^*,u^*}$: $\frac{\partial g_{33}}{\partial x_1}=0$, $\frac{\partial g_{33}}{\partial x_2}=0$, $\frac{\partial g_{33}}{\partial x_3}=0$, $\frac{\partial g_{33}}{\partial x_4}=0$, $\frac{\partial g_{33}}{\partial x_5}=0$, $\frac{\partial g_{33}}{\partial x_6}=0$, $\frac{\partial g_{33}}{\partial x_7}=0$, and $\frac{\partial g_{33}}{\partial x_8}=0$.

Fourth row of the Jacobian matrix $\nabla_x g_3(x)|_{x^*,u^*}$, $i=1,2,\cdots,8$:

$$\frac{\partial g_{43}}{\partial x_i}=\frac{\frac{\partial M_{31}}{\partial x_i}detM-M_{31}\frac{\partial detM}{\partial x_i}}{detM^2} \tag{2.528}$$

Fifth row of the Jacobian matrix $\nabla_x g_3(x)|_{x^*,u^*}$: $\frac{\partial g_{53}}{\partial x_1}=0$, $\frac{\partial g_{53}}{\partial x_2}=0$, $\frac{\partial g_{53}}{\partial x_3}=0$, $\frac{\partial g_{53}}{\partial x_4}=0$, $\frac{\partial g_{53}}{\partial x_5}=0$, $\frac{\partial g_{53}}{\partial x_6}=0$, $\frac{\partial g_{53}}{\partial x_7}=0$, and $\frac{\partial g_{53}}{\partial x_8}=0$.

Sixth row of the Jacobian matrix $\nabla_x g_3(x)|_{x^*,u^*}$, $i=1,2,\cdots,8$:

$$\frac{\partial g_{63}}{\partial x_i}=-\frac{\frac{\partial M_{32}}{\partial x_i}detM-M_{32}\frac{\partial detM}{\partial x_i}}{detM^2} \tag{2.529}$$

Seventh row of the Jacobian matrix $\nabla_x g_3(x)|_{x^*,u^*}$: $\frac{\partial g_{73}}{\partial x_1}=0$, $\frac{\partial g_{73}}{\partial x_2}=0$, $\frac{\partial g_{73}}{\partial x_3}=0$, $\frac{\partial g_{73}}{\partial x_4}=0$, $\frac{\partial g_{73}}{\partial x_5}=0$, $\frac{\partial g_{73}}{\partial x_6}=0$, $\frac{\partial g_{73}}{\partial x_7}=0$, and $\frac{\partial g_{73}}{\partial x_8}=0$.

Eighth row of the Jacobian matrix $\nabla_x g_3(x)|_{x^*,u^*}$, $i=1,2,\cdots,8$:

$$\frac{\partial g_{83}}{\partial x_i}=\frac{\frac{\partial M_{33}}{\partial x_i}detM-M_{33}\frac{\partial detM}{\partial x_i}}{detM^2} \tag{2.530}$$

Next, one computes the partial derivatives of the following elements in the model of the redundant robotic manipulator: (i) elements of the inertial matrix M_{11}, M_{12}, M_{13}, M_{21}, M_{22}, M_{23}, and M_{31}, M_{32}, M_{33}, (ii) elements of the Coriolis and centrifugal forces vector C_1, C_2, C_3, and (iii) elements of the gravitational forces vector G_1, G_2, G_3.

About the partial derivatives of sub-matrices M_{ij}, $i=1,2,3$, $j=1,2,3$ of the inertia matrix M, it holds that

$$\frac{\partial M_{11}}{\partial x_i}=\frac{\partial m_{22}}{\partial x_i}m_{33}+m_{22}\frac{\partial m_{33}}{\partial x_i}-\frac{\partial m_{32}}{\partial x_i}m_{23}-m_{32}\frac{\partial m_{23}}{\partial x_i}\quad i=1,2,\cdots,8 \tag{2.531}$$

$$\frac{\partial M_{12}}{\partial x_i}=\frac{\partial m_{21}}{\partial x_i}m_{33}+m_{21}\frac{\partial m_{33}}{\partial x_i}-\frac{\partial m_{31}}{\partial x_i}m_{23}-m_{31}\frac{\partial m_{23}}{\partial x_i}\quad i=1,2,\cdots,8 \tag{2.532}$$

$$\frac{\partial M_{13}}{\partial x_i}=\frac{\partial m_{12}}{\partial x_i}m_{23}+m_{12}\frac{\partial m_{23}}{\partial x_i}-\frac{\partial m_{22}}{\partial x_i}m_{13}-m_{22}\frac{\partial m_{13}}{\partial x_i}\quad i=1,2,\cdots,8 \tag{2.533}$$

$$\frac{\partial M_{21}}{\partial x_i}=\frac{\partial m_{12}}{\partial x_i}m_{33}+m_{12}\frac{\partial m_{33}}{\partial x_i}-\frac{\partial m_{32}}{\partial x_i}m_{13}-m_{32}\frac{\partial m_{13}}{\partial x_i}\quad i=1,2,\cdots,8 \tag{2.534}$$

$$\frac{\partial M_{22}}{\partial x_i}=\frac{\partial m_{11}}{\partial x_i}m_{33}+m_{11}\frac{\partial m_{33}}{\partial x_i}-\frac{\partial m_{31}}{\partial x_i}m_{13}-m_{31}\frac{\partial m_{13}}{\partial x_i}\quad i=1,2,\cdots,8 \tag{2.535}$$

$$\frac{\partial M_{23}}{\partial x_i}=\frac{\partial m_{11}}{\partial x_i}m_{32}+m_{11}\frac{\partial m_{32}}{\partial x_i}-\frac{\partial m_{31}}{\partial x_i}m_{12}-m_{31}\frac{\partial m_{12}}{\partial x_i}\quad i=1,2,\cdots,8 \tag{2.536}$$

$$\frac{\partial M_{31}}{\partial x_i} = \frac{\partial m_{12}}{\partial x_i} m_{23} + m_{12}\frac{\partial m_{23}}{\partial x_i} - \frac{\partial m_{22}}{\partial x_i} m_{13} - m_{22}\frac{\partial m_{13}}{\partial x_i} \quad i = 1, 2, \cdots, 8 \quad (2.537)$$

$$\frac{\partial M_{32}}{\partial x_i} = \frac{\partial m_{11}}{\partial x_i} m_{23} + m_{11}\frac{\partial m_{23}}{\partial x_i} - \frac{\partial m_{21}}{\partial x_i} m_{13} - m_{21}\frac{\partial m_{13}}{\partial x_i} \quad i = 1, 2, \cdots, 8 \quad (2.538)$$

$$\frac{\partial M_{33}}{\partial x_i} = \frac{\partial m_{11}}{\partial x_i} m_{22} + m_{11}\frac{\partial m_{22}}{\partial x_i} - \frac{\partial m_{21}}{\partial x_i} m_{12} - m_{21}\frac{\partial m_{12}}{\partial x_i} \quad i = 1, 2, \cdots, 8 \quad (2.539)$$

About the determinant of the inertia matrix M the associated partial derivatives are (for $i = 1, 2, \cdots, 8$)

$$\begin{aligned} \frac{\partial detM}{\partial x_i} = & \frac{\partial m_{11}}{\partial x_i}(m_{22}m_{33} - m_{32}m_{23}) + m_{31}(\frac{\partial m_{22}}{\partial x_i} m_{33} + m_{22}\frac{\partial m_{33}}{\partial x_i} - \frac{\partial m_{32}}{\partial x_i} m_{23} - m_{32}\frac{\partial m_{23}}{\partial x_i}) \\ & - \frac{\partial m_{12}}{\partial x_i}(m_{21}m_{33} - m_{31}m_{23}) - m_{12}(\frac{\partial m_{21}}{\partial x_i} m_{33} + m_{21}\frac{\partial m_{33}}{\partial x_i} - \frac{\partial m_{31}}{\partial x_i} m_{23} - m_{31}\frac{\partial m_{23}}{\partial x_i}) \\ & + \frac{\partial m_{13}}{\partial x_i}(m_{21}m_{22} - m_{31}m_{22}) - m_{13}(\frac{\partial m_{21}}{\partial x_i} m_{22} + m_{24}\frac{\partial m_{23}}{\partial x_i} - \frac{\partial m_{31}}{\partial x_i} m_{22} - m_{31}\frac{\partial m_{22}}{\partial x_i}) + \end{aligned} \quad (2.540)$$

About the partial derivatives of the elements m_{ij}, $i = 1, 2, 3, j = 1, 2, 3$ of the inertia matrix, one has

For $m_{11} = a_1 + a_2 + a_4 + 2a_3\cos(x_5) + 2a_5\cos(x_3 + x_5) + 2a_6\cos(x_7)$: $\frac{\partial m_{11}}{\partial x_1} = 0$, $\frac{\partial m_{11}}{\partial x_2} = 0$, $\frac{\partial m_{11}}{\partial x_3} = 0$, $\frac{\partial m_{11}}{\partial x_4} = 0$, $\frac{\partial m_{11}}{\partial x_5} = -2a_3\sin x_5 - 2a_5\sin(x_5 + x_7)$, $\frac{\partial m_{11}}{\partial x_6} = 0$, $\frac{\partial m_{11}}{\partial x_7} = -2a_5\sin(x_5 + x_7) - 2a_6\sin(x_7)$, $\frac{\partial m_{11}}{\partial x_8} = 0$.

For $m_{12} = m_{21} = a_2 + a_4 + a_3\cos(x_5) + a_5\cos(x_5) + a_5\cos(x_5 + x_7) + 2a_6\cos(x_7)$: $\frac{\partial m_{12}}{\partial x_1} = 0$, $\frac{\partial m_{12}}{\partial x_2} = 0$, $\frac{\partial m_{12}}{\partial x_3} = 0$, $\frac{\partial m_{12}}{\partial x_4} = 0$, $\frac{\partial m_{12}}{\partial x_5} = -a_3\sin x_5 - a_5\sin(x_5 + x_7)$, $\frac{\partial m_{12}}{\partial x_6} = 0$, $\frac{\partial m_{12}}{\partial x_7} = -a_5\sin(x_5 + x_7) - 2a_6\sin(x_7)$, $\frac{\partial m_{12}}{\partial x_8} = 0$.

For $m_{13} = m_{31} = a_4 + a_5\cos(x_5 + x_7) + a_6\cos(x_7)$: $\frac{\partial m_{13}}{\partial x_1} = 0$, $\frac{\partial m_{13}}{\partial x_2} = 0$, $\frac{\partial m_{13}}{\partial x_3} = 0$, $\frac{\partial m_{13}}{\partial x_4} = 0$, $\frac{\partial m_{13}}{\partial x_5} = -a_5\sin(x_5 + x_7)$, $\frac{\partial m_{13}}{\partial x_6} = 0$, $\frac{\partial m_{13}}{\partial x_7} = -a_5\sin(x_5 + x_7)$, $\frac{\partial m_{13}}{\partial x_8} = -a_6\sin(x_7)$.

For $m_{22} = a_2 + a_4 + 2a_6\cos(x_7))$: $\frac{\partial m_{22}}{\partial x_1} = 0$, $\frac{\partial m_{22}}{\partial x_2} = 0$, $\frac{\partial m_{22}}{\partial x_3} = 0$, $\frac{\partial m_{22}}{\partial x_4} = 0$, $\frac{\partial m_{22}}{\partial x_5} = 0$, $\frac{\partial m_{22}}{\partial x_6} = 0$, $\frac{\partial m_{22}}{\partial x_7} = -2a_6\sin(x_7)$, $\frac{\partial m_{22}}{\partial x_8} = 0$.

For $m_{23} = m_{32} = a_4 + a_6\cos(x_7))$: $\frac{\partial m_{23}}{\partial x_1} = 0$, $\frac{\partial m_{23}}{\partial x_2} = 0$, $\frac{\partial m_{23}}{\partial x_3} = 0$, $\frac{\partial m_{23}}{\partial x_4} = 0$, $\frac{\partial m_{23}}{\partial x_5} = 0$, $\frac{\partial m_{23}}{\partial x_6} = 0$, $\frac{\partial m_{23}}{\partial x_7} = -a_6\cos(x_7)$, $\frac{\partial m_{23}}{\partial x_8} = 0$.

For $m_{33} = 0$: $\frac{\partial m_{33}}{\partial x_1} = 0$, $\frac{\partial m_{33}}{\partial x_2} = 0$, $\frac{\partial m_{33}}{\partial x_3} = 0$, $\frac{\partial m_{33}}{\partial x_4} = 0$, $\frac{\partial m_{33}}{\partial x_5} = 0$, $\frac{\partial m_{33}}{\partial x_6} = 0$, $\frac{\partial m_{33}}{\partial x_7} = 0$, $\frac{\partial m_{33}}{\partial x_8} = 0$.

Additionally, about the partial derivatives of the first element of the Coriolis and centrifugal forces vector C, one has $\frac{\partial C_1}{\partial x_1}=0$, $\frac{\partial C_1}{\partial x_2}=0$, $\frac{\partial C_1}{\partial x_3}=0$,

$$\frac{\partial C_1}{\partial x_4}=-a_1 2x_6\sin(x_5)-a_5 2(x_6+x_8)\sin(x_5+x_7)-a_6 2x_8\sin(x_7) \tag{2.541}$$

$$\begin{aligned}\frac{\partial C_1}{\partial x_5}&=-a_1(2x_4+x_6)x_6\cos(x_5)\\&\quad-a_5(2x_4+x_6+x_8)(x_6+x_8)\cos(x_5+x_7)\end{aligned} \tag{2.542}$$

$$\begin{aligned}\frac{\partial C_1}{\partial x_6}&=-a_1x_6\sin(x_5)-a_2(2x_4+x_6)\sin(x_5)-a_5(x_6+x_8)\sin(x_5+x_7)\\&\quad-a_5(2x_4+x_6+x_8)\sin(x_5+x_7)-a_6 2x_8\sin(x_7)\end{aligned} \tag{2.543}$$

$$\begin{aligned}\frac{\partial C_1}{\partial x_7}&=-a_5(2x_4+x_6+x_8)(x_6+x_8)\cos(x_5+x_7)\\&\quad-a_6(2x_4+2x_6+x_8)x_8\cos(x_7)\end{aligned} \tag{2.544}$$

$$\begin{aligned}\frac{\partial C_1}{\partial x_8}&=-a_5(x_6+x_8)\sin(x_5+x_7)-a_5(2x_4+x_6+x_8)\sin(x_5+x_7)\\&\quad-a_6x_8\sin(x_7)-a_6(2x_4+2x_6+x_8)\sin(x_7)\end{aligned} \tag{2.545}$$

Moreover, about the partial derivatives of the second element of the Coriolis and centrifugal forces vector C, one has $\frac{\partial C_2}{\partial x_1}=0$, $\frac{\partial C_2}{\partial x_2}=0$, $\frac{\partial C_2}{\partial x_3}=0$,

$$\frac{\partial C_2}{\partial x_4}=a_3 2x_4\sin(x_5)+a_5 2x_4\sin(x_5+x_7)-a_6 2x_8\sin(x_7) \tag{2.546}$$

$$\frac{\partial C_2}{\partial x_5}=a_3x_4^2\cos(x_5)+a_5x_4^2\cos(x_5+x_7) \tag{2.547}$$

$$\frac{\partial C_2}{\partial x_6}=-a_6 2x_8\sin(x_7) \tag{2.548}$$

$$\frac{\partial C_2}{\partial x_7}=a_5x_4^2\cos(x_5+x_7)-a_6(2x_4+2x_6+x_8)\cos(x_7) \tag{2.549}$$

$$\frac{\partial C_2}{\partial x_8}=-a_6x_8\sin(x_7)-a_6(2x_4+2x_6+x_8)\sin(x_7) \tag{2.550}$$

Furthermore, about the partial derivatives of the third element of the Coriolis and centrifugal forces vector C, one has $\frac{\partial C_3}{\partial x_1}=0$, $\frac{\partial C_3}{\partial x_2}=0$, $\frac{\partial C_3}{\partial x_3}=0$,

$$\frac{\partial C_3}{\partial x_4}=2a_1x_4\sin(x_5+x_7)+a_6 2(x_4+x_6)\sin(x_7) \tag{2.551}$$

$$\frac{\partial C_3}{\partial x_5}=a_1x_4^2\cos(x_5+x_7) \tag{2.552}$$

$$\frac{\partial C_3}{\partial x_6}=2a_6(x_4+x_6)\sin(x_7) \tag{2.553}$$

$$\frac{\partial C_3}{\partial x_7}=a_1x_4^2\cos(x_5+x_7)+a_6(x_4+x_6)^2\cos(x_7) \tag{2.554}$$

$$\frac{\partial C_3}{\partial x_8} = 0 \tag{2.555}$$

2.11.3.2 Stabilizing feedback control

After linearization around its current operating point, the dynamic model of the redundant three-link robotic manipulator is written as (Rigatos and Busawon, 2018)

$$\dot{x} = Ax + Bu + d_1 \tag{2.556}$$

Parameter d_1 stands for the linearization error in the redundant three-link robotic manipulator's dynamic model appearing above in (2.556). The reference setpoints for the state vector of the redundant three-link robotic manipulator are denoted by $\mathbf{x_d} = [x_1^d, \cdots, x_8^d]$. Tracking of this trajectory is achieved after applying the control input u_d. At every time instant the control input u_d is assumed to differ from the control input u appearing in (2.556) by an amount equal to Δu, that is $u_d = u + \Delta u$

$$\dot{x}_d = Ax_d + Bu_d + d_2 \tag{2.557}$$

The dynamics of the controlled system described in (2.556) can be also written as

$$\dot{x} = Ax + Bu + Bu_d - Bu_d + d_1 \tag{2.558}$$

and by denoting $d_3 = -Bu_d + d_1$ as an aggregate disturbance term, one obtains

$$\dot{x} = Ax + Bu + Bu_d + d_3 \tag{2.559}$$

By subtracting (2.557) from (2.559), one has

$$\dot{x} - \dot{x}_d = A(x - x_d) + Bu + d_3 - d_2 \tag{2.560}$$

By denoting the tracking error as $e = x - x_d$ and the aggregate disturbance term as $\tilde{d} = d_3 - d_2$, the tracking error dynamics becomes

$$\dot{e} = Ae + Bu + \tilde{d} \tag{2.561}$$

For the approximately linearized model of the system, a stabilizing feedback controller is developed. The controller has the form

$$u(t) = -Ke(t) \tag{2.562}$$

with $K = \frac{1}{r}B^T P$ where P is a positive definite symmetric matrix which is obtained from the solution of the Riccati equation (Rigatos and Busawon, 2018)

$$A^T P + PA + Q - P\left(\frac{2}{r}BB^T - \frac{1}{\rho^2}LL^T\right)P = 0 \tag{2.563}$$

where Q is a positive semi-definite symmetric matrix. The diagram of the considered control loop is depicted in Figure 2.143. The global stability properties of the control method can be proven following the procedure presented in Section 1.1.2.5.

2.11.4 Simulation tests

The performance of the proposed nonlinear optimal control scheme for redundant robotic manipulators has been further tested and confirmed through simulation experiments. To implement the control method the algebraic Riccati equation of (2.563)

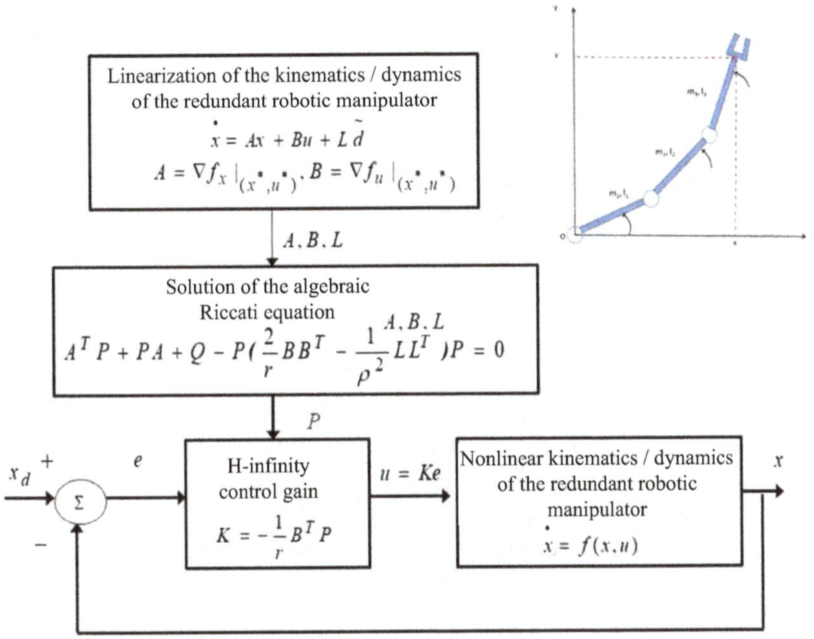

Figure 2.143 Diagram of the nonlinear optimal (H-infinity) control loop for the 3-link redundant robotic manipulator

had to be solved at each sampling period. The method is computationally efficient and time needed for solving the Riccati equation is significantly smaller than the sampling interval. The setpoints of the tests can be obtained in real-time from the solution of the manipulator's inverse kinematics problem. The obtained results are depicted in Figures 2.144 to 2.159. The real values of the state variables of the robot are depicted in blue, the estimated values (as provided by the H-infinity Kalman Filter) are printed in green while the associated setpoints are plotted in red. It can be noticed that fast and accurate tracking of all setpoints was achieved under moderate variations of the control inputs.

2.12 Nonlinear control of parallel closed-chain robotic manipulators

2.12.1 Outline

Parallel robots of closed kinematic chain are frequently used in industrial and humans assisting tasks (Rigatos and Busawon, 2018; Chang *et al.*, 2011; Ouyang, Li *et al.*, 2004). Such robotic mechanisms are met in applications where high stiffness, high

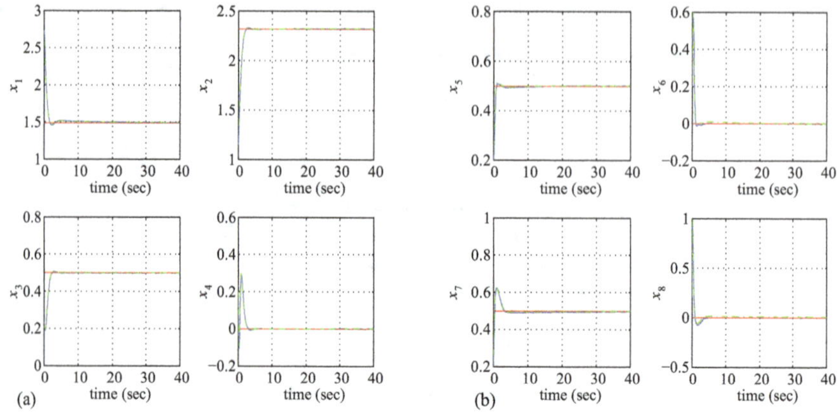

Figure 2.144 Tracking of setpoint 1 by the redundant 3-link planar robotic manipulator: (a) convergence of the state variables x_1 to x_4 (blue lines) to the reference setpoints (red lines) and KF-estimated state variables (green lines); (b) convergence of the state variables x_5 to x_8 to the reference setpoints

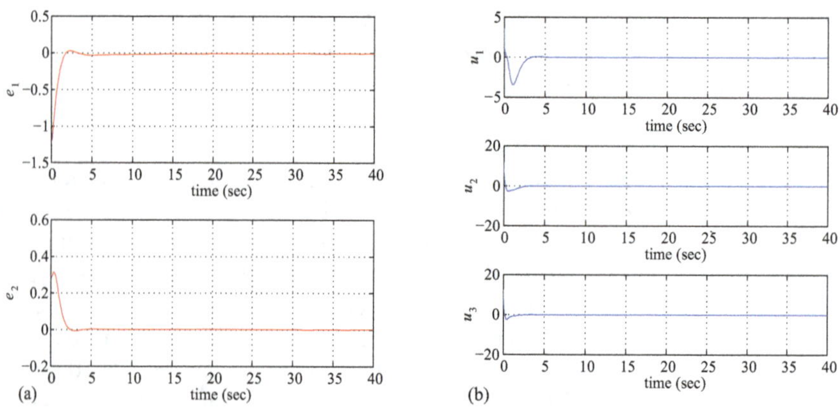

Figure 2.145 Tracking of setpoint 1 by the redundant 3-link planar robotic manipulator: (a) tracking error for state variables x_1 and x_3 (position error of the robot's end effector in the xy-plane); (b) control inputs u_1, u_2 and u_3 generated by the redundant manipulator

lift force and high velocity / acceleration of the joints and of the end-effector have to be developed (Ghorbel *et al.*, 2000; Muller and Hufnagel, 2012; Shang and Cong, 2020; Shang, Cong and Ge, 2011; Shang and Cong, 2014). Parallel manipulators achieve also precise positioning and for this reason they find use in manufacturing, in service robotics and in medical robotics (Wu, Zhang *et al.*, 2005; Chang *et al.*,

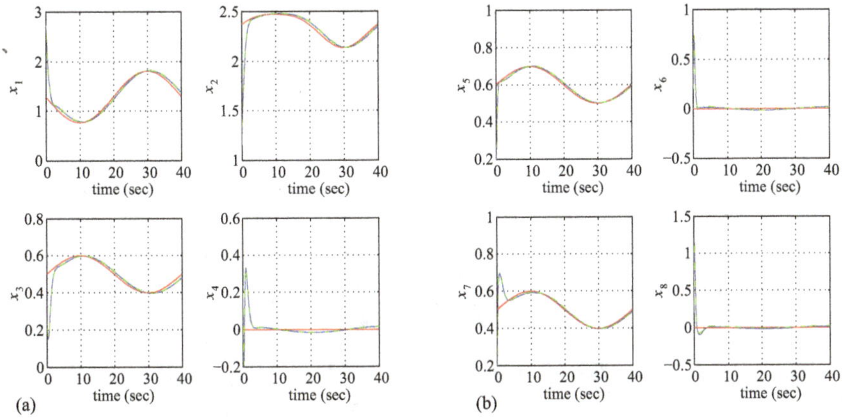

Figure 2.146 *Tracking of setpoint 2 by the redundant 3-link planar robotic manipulator: (a) convergence of the state variables x_1 to x_4 (blue lines) to the reference setpoints (red lines) and KF-estimated state variables (green lines); (b) convergence of the state variables x_5 to x_8 to the reference setpoints*

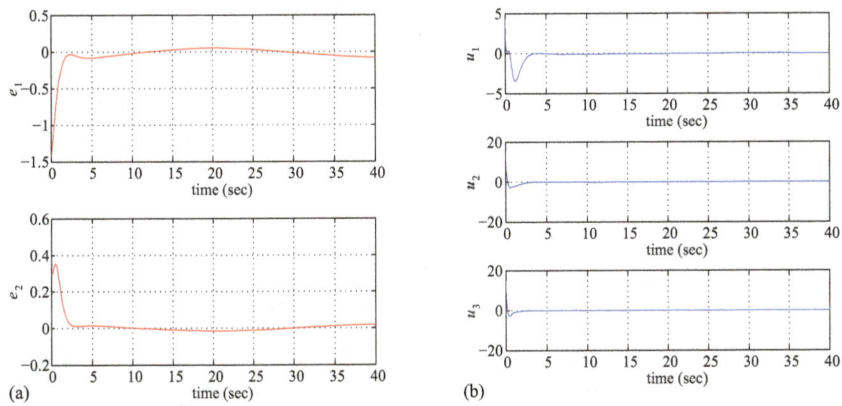

Figure 2.147 *Tracking of setpoint 2 by the redundant 3-link planar robotic manipulator: (a) tracking error for state variables x_1 and x_3 (position error of the robot's end effector in the xy-plane); (b) control inputs u_1, u_2 and u_3 generated by the redundant manipulator*

2012; Zi, Cao *et al.*, 2011). The present section examines the kinematic and dynamic model of a specific type of parallel robotic manipulators, which is the five-link closed kinematic chain robot (Chen, Wang *et al.*, 2011; Ouyang, Zhang *et al.*, 2006; Hou *et al.*, 2015). Five-bar mechanisms are typical parallel coupled mechanisms exhibiting relatively simple dynamics and large stiffness (Ouyang, Kong *et al.*, 2014; Ouyang,

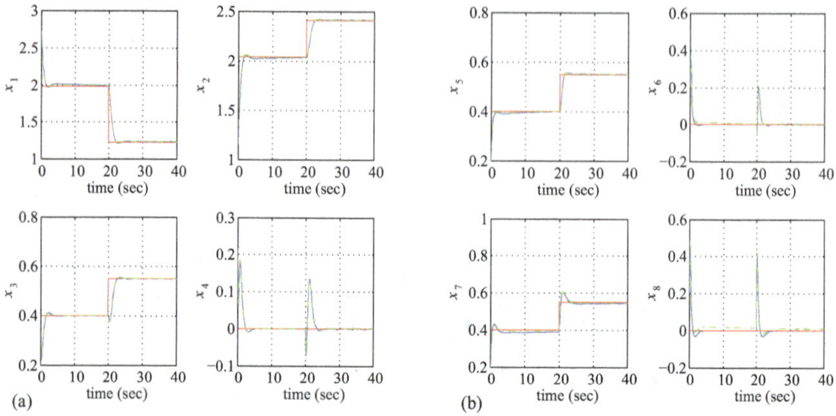

Figure 2.148 Tracking of setpoint 3 by the redundant 3-link planar robotic manipulator: (a) convergence of the state variables x_1 to x_4 (blue lines) to the reference setpoints (red lines) and KF-estimated state variables (green lines); (b) convergence of the state variables x_5 to x_8 to the reference setpoints

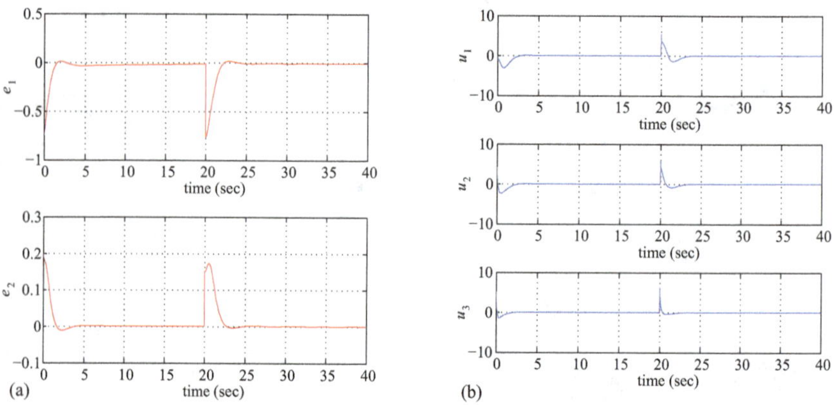

Figure 2.149 Tracking of setpoint 3 by the redundant 3-link planar robotic manipulator: (a) tracking error for state variables x_1 and x_3 (position error of the robot's end effector in the xy-plane); (b) control inputs u_1, u_2 and u_3 generated by the redundant manipulator

Kuang *et al.*, 2017; Le, Kung *et al.*, 2013). Such robots can be used in pick-and-place tasks and in performing tasks where their end-effector is either in contact with a surface or moves in contact-free mode (Cheng, Hu *et al.*, 2012; Cheung and Huang, 2005). In these robotic manipulators some joints may be passive thus resulting into an underactuated system (Yu, 2006; Bourbonnais *et al.*, 2015). The related control

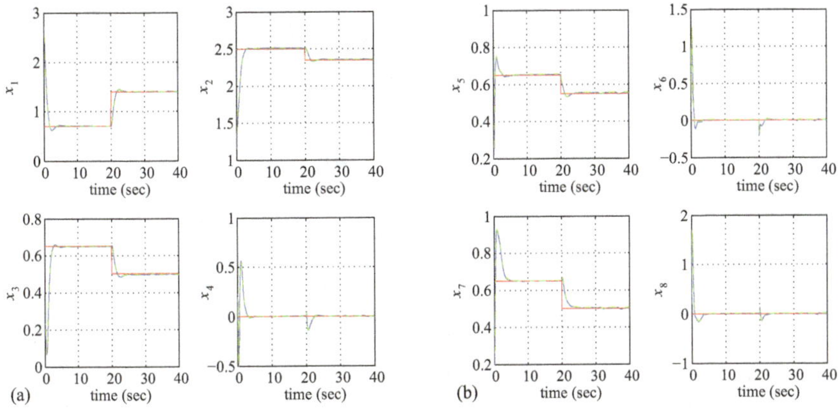

Figure 2.150 Tracking of setpoint 4 by the redundant 3-link planar robotic manipulator: (a) convergence of the state variables x_1 to x_4 (blue lines) to the reference setpoints (red lines) and KF-estimated state variables (green lines); (b) convergence of the state variables x_5 to x_8 to the reference setpoints

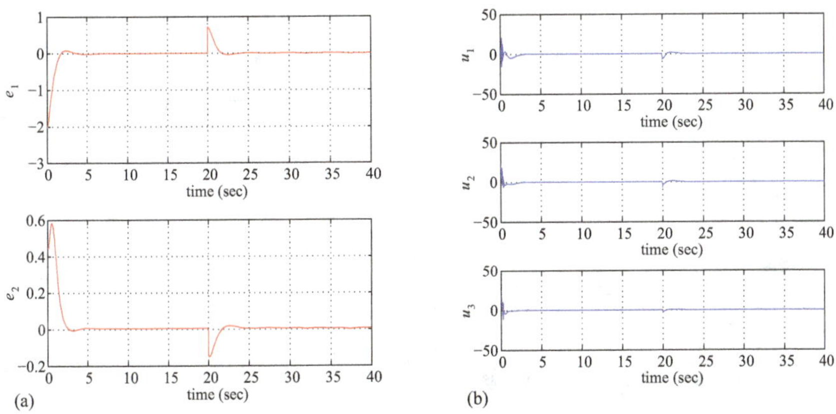

Figure 2.151 Tracking of setpoint 4 by the redundant 3-link planar robotic manipulator: (a) tracking error for state variables x_1 and x_3 (position error of the robot's end effector in the xy plane); (b) control inputs u_1, u_2 and u_3 generated by the redundant manipulator

problem is non-trivial, and several approaches require the transformation of the robot's dynamic model into a form where the computation of the stabilizing control inputs is finally enabled (Le, Kang and Soh, 2013; Lara-Molina *et al.*, 2016; Lara-Molina *et al.*, 2014; Le, Kung *et al.*, 2013). Indicative results on model-based and mode-free control of parallel robotic manipulators can be found in Harandi *et al.*, 2021, Xie

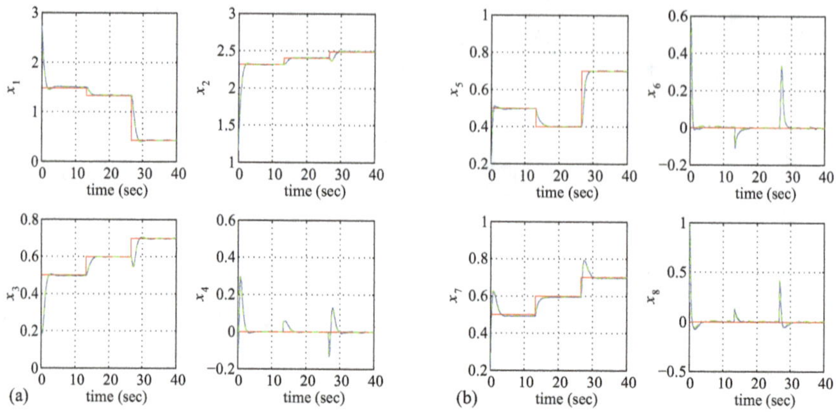

Figure 2.152 Tracking of setpoint 5 by the redundant 3-link planar robotic manipulator: (a) convergence of the state variables x_1 to x_4 (blue lines) to the reference setpoints (red lines) and KF-estimated state variables (green lines); (b) convergence of the state variables x_5 to x_8 to the reference setpoints

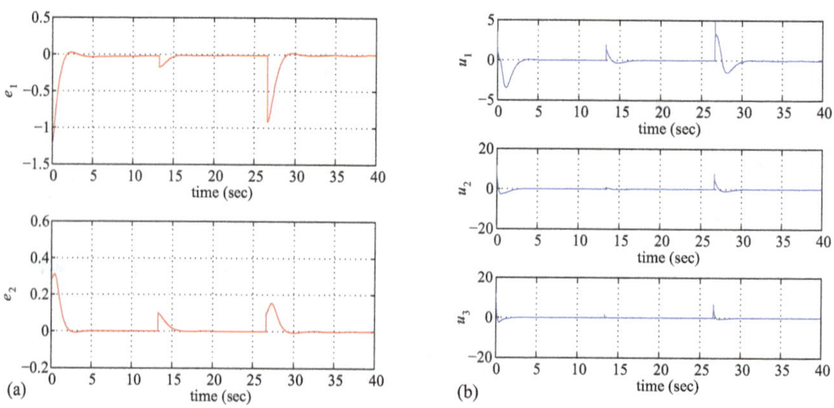

Figure 2.153 Tracking of setpoint 5 by the redundant 3-link planar robotic manipulator: (a) tracking error for state variables x_1 and x_3 (position error of the robot's end effector in the xy plane); (b) control inputs u_1, u_2 and u_3 generated by the redundant manipulator

et al., 2021, Morlock, Meyer *et al.*, 2021, Huang, Zhang *et al.*, 2020, and Jiang, Zhang *et al.*, 2021. However, little has been done so far for the solution of the optimal control problem for the dynamic model of the five-link closed-chain robotic mechanism.

In this section, a nonlinear optimal (H-infinity) controller is developed for the five-link closed-chain robotic manipulator, using the robot's dynamic model and

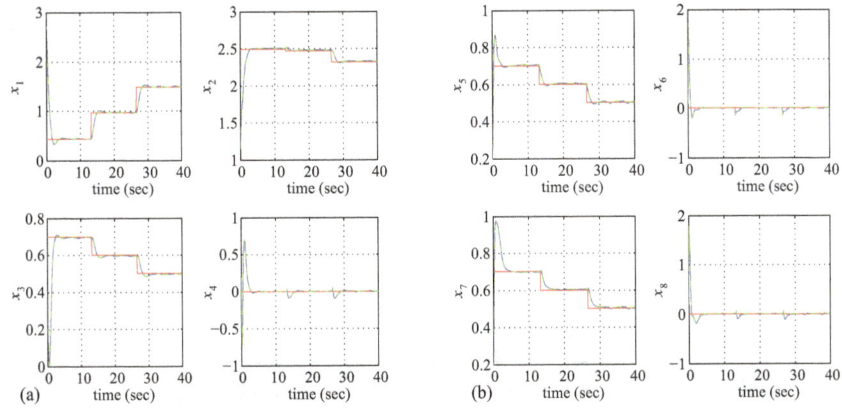

Figure 2.154 Tracking of setpoint 6 by the redundant 3-link planar robotic manipulator: (a) convergence of the state variables x_1 to x_4 (blue lines) to the reference setpoints (red lines) and KF-estimated state variables (green lines); (b) convergence of the state variables x_5 to x_8 to the reference setpoints

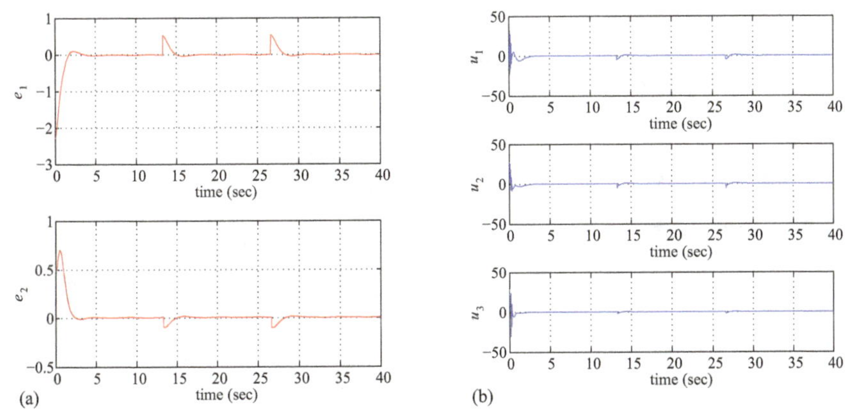

Figure 2.155 Tracking of setpoint 6 by the redundant 3-link planar robotic manipulator: (a) tracking error for state variables x_1 and x_3 (position error of the robot's end effector in the xy plane); (b) control inputs u_1, u_2 and u_3 generated by the redundant manipulator

taking into account the kinematic constraints in the motion of the links (Rigatos, 2016b; Rigatos, Zervos *et al.*, 2020). Unlike the popular computed torque method for robotic manipulators, the new control approach is characterized by optimality and is also applicable when the number of control inputs is not equal to the robot's number of DOFs. In the proposed method, the dynamic model of the robotic manipulator is

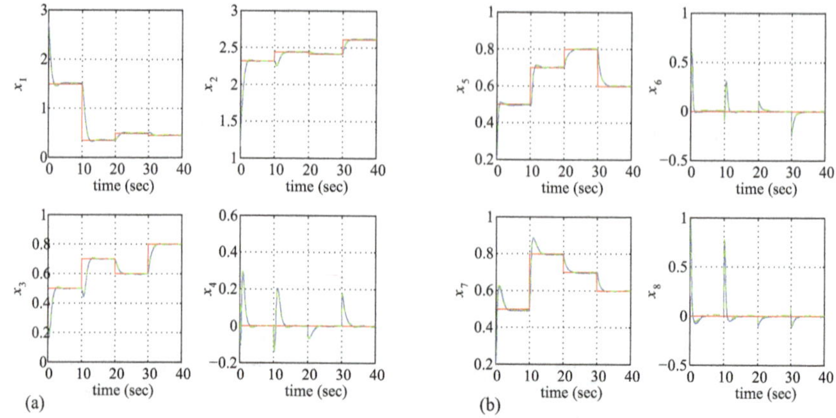

Figure 2.156 Tracking of setpoint 7 by the redundant 3-link planar robotic manipulator: (a) convergence of the state variables x_1 to x_4 (blue lines) to the reference setpoints (red lines) and KF-estimated state variables (green lines); (b) convergence of the state variables x_5 to x_8 to the reference setpoints

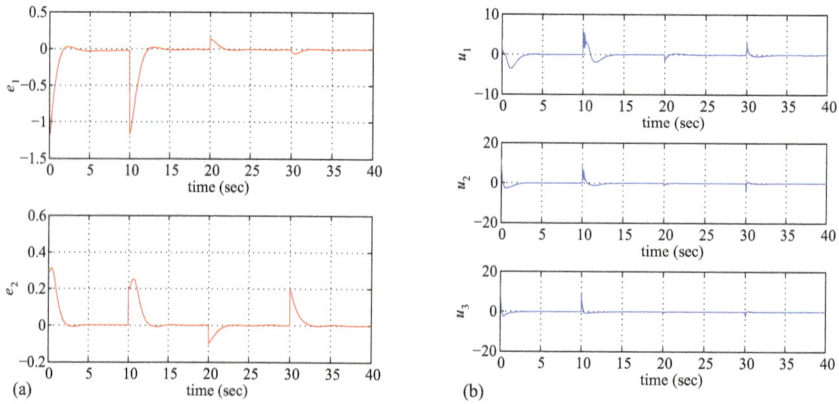

Figure 2.157 Tracking of setpoint 7 by the redundant 3-link planar robotic manipulator: (a) tracking error for state variables x_1 and x_3 (position error of the robot's end effector in the xy plane); (b) control inputs u_1, u_2 and u_3 generated by the redundant manipulator

first formulated in a state-space form, considering as state variables the turn angles of the joints and the related angular velocities. Next, the state-space model of the robot undergoes approximate linearization around a temporary operating point which is updated at each time-step of the control algorithm. The operating point is re-defined at each sampling instance by the present value of the system's state vector and by

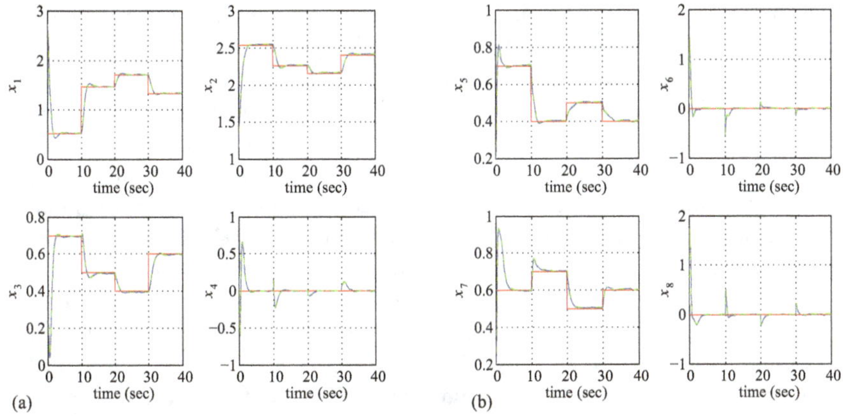

Figure 2.158 Tracking of setpoint 8 by the redundant 3-link planar robotic manipulator: (a) convergence of the state variables x_1 to x_4 (blue lines) to the reference setpoints (red lines) and KF-estimated state variables (green lines); (b) convergence of the state variables x_5 to x_8 to the reference setpoints

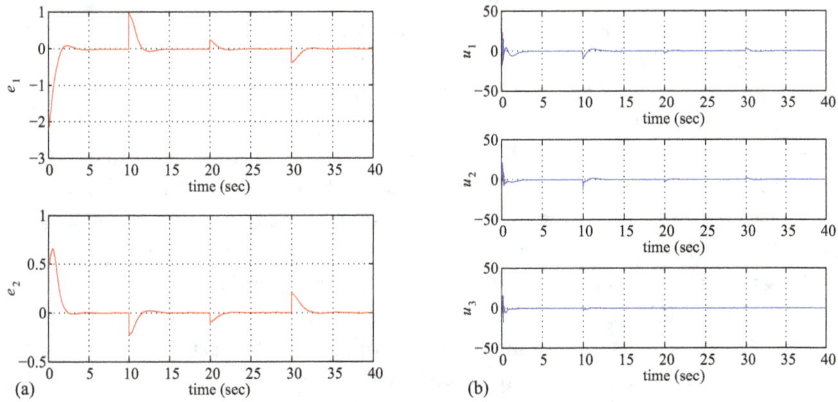

Figure 2.159 Tracking of setpoint 8 by the redundant 3-link planar robotic manipulator: (a) tracking error for state variables x_1 and x_3 (position error of the robot's end effector in the xy plane); (b) control inputs u_1, u_2 and u_3 generated by the redundant manipulator

the last sampled value of the control inputs vector. The linearization is based on the first-order Taylor series expansion and on the computation of the associated Jacobian matrices (Rigatos and Tzafestas, 2007; Basseville and Nikiforov, 1993; Rigatos and Zhang, 2009). The modeling error which is due to the truncation of higher-order terms from the Taylor series is considered to be a perturbation term that is asymptotically eliminated by the robustness of the control method.

For the approximately linearized model of the five-link closed kinematic-chain robot an H-infinity feedback controller is designed. This controller achieves the solution of the nonlinear optimal control problem for the robotic manipulator under model uncertainty and external perturbations. Actually, this controller represents a min–max differential game taking place between: (i) the control inputs which try to minimize a cost function that contains a quadratic term of the state vector's tracking error; (ii) the disturbance and model uncertainty inputs which try to maximize this cost function (Rigatos and Busawon, 2018). To select the feedback gains of the H-infinity controller an algebraic Riccati equation is repetitively solved at each time-step of the control algorithm (Rigatos, 2011; Rigatos, 2015). The global stability properties of the control method are proven through Lyapunov analysis. First it is shown that the control scheme satisfies the H-infinity tracking performance criterion which signifies elevated robustness against model uncertainty and external perturbations (Rigatos and Busawon, 2018; Tousssaint, Basar and Bullo, 2000). Next, under moderate conditions it is also shown that the control loop is globally asymptotically stable. The proposed control method retains the advantages of linear optimal control, that is fast and accurate tracking of reference setpoints under moderate variations of the control inputs. Finally, to perform state estimation-based control without the need to measure the entire state vector of the robotic system, the H-infinity Kalman Filter is used as a robust state estimator (Rigatos and Busawon, 2018).

Control and estimation for the nonlinear dynamics of parallel robotic manipulators remains a challenging research topic (Rigatos and Busawon, 2018). The present section provides the only existing solution to the nonlinear optimal control problem of the five-link parallel robotic system which is of proven global stability while also remaining computationally efficient. Preceding results on the use of H-infinity control in nonlinear dynamical systems were limited to the case of affine-in-the-input systems with drift-only dynamics and considered that the control inputs gain matrix is not dependent on the values of the system's state vector. Moreover, in these approaches, the linearization was performed around points of the desirable trajectory whereas in the present section's control method the linearization points are related with the value of the state vector at each sampling instant as well as with the last sampled value of the control inputs vector. The Riccati equation which has been proposed for computing the feedback gains of the controller is novel, so is the presented global stability proof through Lyapunov analysis.

The section's scientific contribution is outlined as follows: (i) the presented nonlinear optimal control method has improved performance when compared against other nonlinear control schemes that one can consider for the dynamic model of multi-DOF parallel robotic manipulators (such as Lie algebra-based control, differential flatness theory-based control, Model-based Predictive Control, Nonlinear Model-based Predictive Control, Sliding-mode control, Backstepping control), etc. (Rigatos and Busawon, 2018); (ii) it achieves fast and accurate tracking of all reference setpoints for the five-link parallel robot under moderate variations of the control inputs; (iii) it minimizes the consumption of energy by the actuators of the five-link parallel robots, thus improving the operational capacity and efficiency in tasks' execution for such robotic systems.

2.12.2 Dynamic model of the five-link parallel robot

2.12.2.1 Dynamics of the parallel robotic manipulator

Without loss of generality the proposed nonlinear optimal control method is applied to a closed-chain five-link parallel robot (Wu, Zhang *et al.*, 2005; Chang *et al.*, 2012; Chen, Wang *et al.*, 2011; Ouyang, Zhang *et al.*, 2006). The diagram of the parallel robot is given in Figure 2.160. The robotic mechanism comprises 5 links which are connected in closed chain. The lengths of the links are denoted as L_i $i = 1, \cdots, 5$. The fifth link is aligned with the Ox axis of the inertial reference frame The turn angles of the joints of the rest of the links are denoted as q_i $i = 1, \cdots 4$. Out of them joints q_1, q_2, q_3 and q_4 are active, whereas joint q_5 which is depicted at the top of the robot's diagram in Figure 2.160 is passive. The end-effector of the robot, which is located at the connection of the third and fourth link is denoted by Σ. The centers of gravity of the links are denoted as c_i $i = 1, \cdots, 5$. The distance between the basis joint of the i-th link and the link's center of gravity is denoted as r_i. The angle that is formed between the longitudinal axis of the ith link and the line that connects its basis joint to the link's center of gravity is denoted as δ_i $i = 1, \cdots, 5$. The end-effector Σ of the parallel robot has two degrees of freedom in the xy-plane. Such a type of robot can be used for performing dexterous operations in the 2D plane, as needed in medical

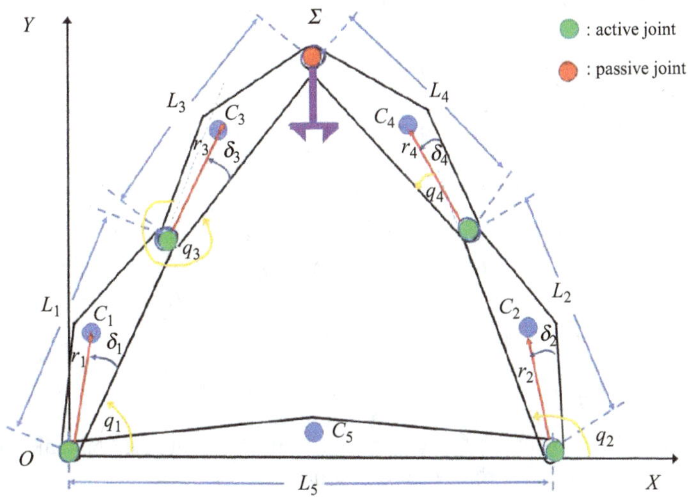

Figure 2.160 Diagram of the closed-chain five-link parallel robot with and 2-DOF end effector in the xy-plane

robotics, or in several pick-and-place tasks, and finally in industrial applications (laser cutting, welding, or CNC tasks).

The closed-chain five-link parallel robot is subject to the following kinematic constraints:

$$\begin{array}{l} L_1\cos(q_1)+L_3\cos(q_1+q_3)-L_5-L_2\cos(q_2)-L_4\cos(q_2+q_4)=0 \\ L_1\sin(q_1)+L_3\cos(q_1+q_3)-L_2\sin(q_2)-L_4\sin(q_2+q_4)=0 \end{array} \quad (2.564)$$

By differentiating the kinematic equations which are given in (2.564) one has

$$\begin{array}{l} -L_1\sin(q_1)\dot{q}_1-L_1\sin(q_1+q_3)\dot{q}_1-L_1\sin(q_1+q_3)\dot{q}_3+L_2\sin(q_2)\dot{q}_3 \\ +L_1\sin(q_2+q_4)\dot{q}_2+L_4\sin(q_2+q_4)\dot{q}_4=0 \end{array} \quad (2.565)$$

$$\begin{array}{l} L_1\cos(q_1)\dot{q}_1+L_3\cos(q_1+q_3)\dot{q}_1+L_3\cos(q_1+q_3)\dot{q}_3-L_2\cos(q_2)\dot{q}_2 \\ -L_4\cos(q_2+q_4)\dot{q}_2-L_4\cos(q_2+q_4)\dot{q}_4=0 \end{array} \quad (2.566)$$

Equivalently, by denoting the coordinates of the end effector Σ as (x,y), one has the following two conditions:

$$\begin{pmatrix} -L_1\sin(q_1)-L_3\sin(q_1+q_3) & -L_2\sin(q_1+q_3) \\ L_1\cos(q_1)+L_3\cos(q_1+q_3) & L_3\cos(q_1+q_3) \end{pmatrix}\begin{pmatrix} \dot{q}_1 \\ \dot{q}_3 \end{pmatrix}=\begin{pmatrix} \dot{x} \\ \dot{y} \end{pmatrix} \quad (2.567)$$

$$\begin{pmatrix} -L_2\sin(q_2)-L_4\sin(q_2+q_4) & -L_4\sin(q_2+q_4) \\ L_2\cos(q_2)+L_4\cos(q_2+q_4) & L_4\cos(q_2+q_4) \end{pmatrix}\begin{pmatrix} \dot{q}_2 \\ \dot{q}_4 \end{pmatrix}=\begin{pmatrix} \dot{x} \\ \dot{y} \end{pmatrix} \quad (2.568)$$

The dynamic model of the parallel robot is given by (Chang *et al.*, 2011; Ouyang, Li *et al.*, 2004; Cheng, Hu *et al.*, 2012; Cheung and Huang, 2005)

$$D(q)\ddot{q}+\tilde{C}(q,\dot{q})\dot{q}+G(q)=T \quad (2.569)$$

where the inertia matrix $D(q)$ is given by

$$D(q)=\begin{pmatrix} d_{11} & 0 & d_{13} & 0 \\ 0 & d_{22} & 0 & d_{24} \\ d_{31} & 0 & d_{33} & 0 \\ 0 & d_{42} & 0 & d_{44} \end{pmatrix} \quad (2.570)$$

and the elements of the inertia matrix are given by

$$\begin{array}{l} d_{11}=m_1r_1^2+m_3((L_1^2+r_3^2+2L_1r_3\cos(q_3+\delta_3))+J_1+J_3 \\ d_{13}=d_{31}=m_3(r_3^2+L_1r_3\cos(q_3+\delta_3))+J_3 \\ d_{22}=m_2r_2^2+m_4(L_2^2+r_4^2+2L_2r_4\cos(q_4+\delta_4))+J_2+J_4 \\ d_{24}=d_{42}=m_4(r_4^2+L_2r_4\cos(q_4+\delta_4))+J_4 \\ d_{33}=m_3r_3^2+J_3 \\ d_{44}=m_4r_4^2+J_4 \end{array} \quad (2.571)$$

where J_i, $i=1,2,3,4$ are the moments of inertia of the individual links. About the Coriolis and Centrifugal forces vector, one has

$$\tilde{C}(q,\dot{q})\dot{q}=\begin{pmatrix} h_1\dot{q}_3 & 0 & h_1(\dot{q}_1+\dot{q}_3) & 0 \\ 0 & h_2\dot{q}_4 & 0 & h_2(\dot{q}_2+\dot{q}_4) \\ -h_1\dot{q}_1 & 0 & 0 & 0 \\ 0 & -h_2\dot{q}_2 & 0 & 0 \end{pmatrix}\begin{pmatrix} \dot{q}_1 \\ \dot{q}_2 \\ \dot{q}_3 \\ \dot{q}_4 \end{pmatrix} \quad (2.572)$$

and by denoting $\tilde{C}(q,\dot{q})\dot{q}$ as $C(q,\dot{q})$ the previous relation can be written as

$$C(q,\dot{q}) = \begin{pmatrix} h_1\dot{q}_1\dot{q}_3 + h_1(\dot{q}_1+\dot{q}_3)\dot{q}_3 \\ h_2\dot{q}_2\dot{q}_2 + h_2(\dot{q}_2+\dot{q}_4)\dot{q}_4 \\ -h_1\dot{q}_1^2 \\ -h_2\dot{q}_2^2 \end{pmatrix} \tag{2.573}$$

where

$$\begin{aligned} h_1 &= -m_3L_1r_3\sin(q_3+\delta_3) \\ h_2 &= -m_4L_2r_4\sin(q_4+\delta_4) \end{aligned} \tag{2.574}$$

About the gravitational forces vector, one has

$$G(q) = g\begin{pmatrix} (m_1r_1+m_3L_1)\cos(q_1+\delta_1) + m_3r_3\cos(q_1+q_3+\delta_3) \\ (m_2r_2+m_4L_2)\cos(q_2+\delta_2) + m_4r_4\cos(q_2+q_4+\delta_4) \\ m_3r_3\cos(q_1+q_3+\delta_3) \\ m_4r_4\cos(q_2+q_4+\delta_4) \end{pmatrix} \tag{2.575}$$

where g is the acceleration of gravity. In concise form, the previously noted matrices and vectors are written as

$$D(q) = \begin{pmatrix} d_{11} & 0 & d_{13} & 0 \\ 0 & d_{22} & 0 & d_{24} \\ d_{31} & 0 & d_{33} & 0 \\ 0 & d_{42} & 0 & d_{44} \end{pmatrix} \quad C(q,\dot{q}) = \begin{pmatrix} c_1 \\ c_2 \\ c_3 \\ c_4 \end{pmatrix} \quad G(q) = \begin{pmatrix} g_1 \\ g_2 \\ g_3 \\ g_4 \end{pmatrix} \tag{2.576}$$

and the dynamic model of the robot that was previously given in (2.569) is now written as

$$\ddot{q} = -D^{-1}(q)C(q,\dot{q}) - D^{-1}(q)G(q) + D^{-1}(q)T \tag{2.577}$$

where the inverse of the inertia matrix $D(q)$ is given by

$$D^{-1}(q) = \frac{1}{detD}\begin{pmatrix} D_{11} & -D_{21} & D_{31} & -D_{41} \\ -D_{12} & D_{22} & -D_{32} & -D_{42} \\ D_{13} & -D_{23} & D_{33} & -D_{43} \\ -D_{14} & D_{24} & -D_{34} & D_{44} \end{pmatrix} \tag{2.578}$$

The determinant of the inertia matrix is given by

$$detD = d_{11}d_{22}d_{33}d_{44} - d_{11}d_{24}d_{42}d_{33} + d_{13}d_{24}d_{31}d_{42} \tag{2.579}$$

The sub-determinants of the inertia matrix D are defined as follows: $d_{11} = d_{22}d_{33}d_{44}$, $D_{12} = 0$, $D_{13} = -d_{31}(d_{22}d_{44} - d_{42}d_{44})$, $D_{14} = 0$, $D_{21} = 0$, $D_{22} = d_{44}(d_{11}d_{33} - d_{31}d_{13})$, $D_{23} = 0$, $D_{24} = -d_{42}(d_{11}d_{33} - d_{31}d_{13})$, $D_{31} = -d_{13}(d_{22}d_{44} - d_{42}d_{24})$, $D_{32} = 0$, $D_{33} = d_{11}(d_{22}d_{44} - d_{42}d_{24})$, $D_{34} = 0$, $D_{41} = 0$, $D_{42} = -d_{24}(d_{11}d_{33} - d_{31}d_{13})$, $D_{43} = 0$, $D_{44} = d_{22}(d_{11}d_{33} - d_{31}d_{13})$.

Thus, the inverse of the inertia matrix $D(q)$ becomes

$$D^{-1}(q) = \frac{1}{detD}\begin{pmatrix} D_{11} & 0 & D_{31} & 0 \\ 0 & D_{22} & 0 & D_{42} \\ D_{13} & 0 & D_{33} & 0 \\ 0 & D_{24} & 0 & D_{44} \end{pmatrix} \tag{2.580}$$

Using the above, the dynamic model of the robot given in (2.577) is written as

$$\ddot{q} = -D^{-1}(q)[C(q,\dot{q}) + G(q)] + D^{-1}(q)T \tag{2.581}$$

or equivalently

$$\begin{pmatrix} \ddot{q}_1 \\ \ddot{q}_2 \\ \ddot{q}_3 \\ \ddot{q}_4 \end{pmatrix} = -\frac{1}{detD}\begin{pmatrix} D_{11} & 0 & D_{31} & 0 \\ 0 & D_{22} & 0 & -D_{42} \\ D_{13} & 0 & D_{33} & 0 \\ 0 & D_{24} & 0 & D_{44} \end{pmatrix}\begin{pmatrix} c_1+g_1 \\ c_2+g_2 \\ c_3+g_3 \\ c_4+g_4 \end{pmatrix} + \frac{1}{detD}\begin{pmatrix} D_{11} & 0 & D_{31} & 0 \\ 0 & D_{22} & 0 & D_{42} \\ D_{13} & 0 & D_{33} & 0 \\ 0 & D_{24} & 0 & D_{44} \end{pmatrix}\begin{pmatrix} T_1 \\ T_2 \\ T_3 \\ T_4 \end{pmatrix} \tag{2.582}$$

After intermediate operations, one obtains

$$\begin{pmatrix} \ddot{q}_1 \\ \ddot{q}_2 \\ \ddot{q}_3 \\ \ddot{q}_4 \end{pmatrix} = -\frac{1}{detD}\begin{pmatrix} D_{11}(c_1+g_1)+D_{31}(c_3+g_3) \\ D_{22}(c_2+g_2)+D_{24}(c_4+g_4) \\ D_{13}(c_1+g_1)+D_{33}(c_3+g_3) \\ D_{24}(c_2+g_2)+D_{44}(c_4+g_4) \end{pmatrix} + \frac{1}{detD}\begin{pmatrix} D_{11}T_1+D_{31}T_3 \\ D_{22}T_2+D_{42}T_4 \\ D_{13}T_1+D_{33}T_3 \\ D_{24}T_2+D_{44}T_4 \end{pmatrix} \tag{2.583}$$

which finally gives

$$\begin{pmatrix} \ddot{q}_1 \\ \ddot{q}_2 \\ \ddot{q}_3 \\ \ddot{q}_4 \end{pmatrix} = \begin{pmatrix} \frac{-D_{11}(c_1+g_1)-D_{31}(c_3+g_3)}{detD} \\ \frac{D_{22}(c_2+g_2)+D_{24}(c_4+g_4)}{detD} \\ \frac{-D_{13}(c_1+g_1)-D_{33}(c_3+g_3)}{detD} \\ \frac{D_{24}(c_2+g_2)+D_{44}(c_4+g_4)}{detD} \end{pmatrix} + \begin{pmatrix} \frac{D_{11}}{detD} \\ 0 \\ \frac{D_{13}}{detD} \\ 0 \end{pmatrix}T_1 + \begin{pmatrix} 0 \\ \frac{D_{22}}{detD} \\ 0 \\ \frac{D_{24}}{detD} \end{pmatrix}T_2 + \begin{pmatrix} \frac{D_{31}}{detD} \\ 0 \\ \frac{D_{33}}{detD} \\ 0 \end{pmatrix}T_3 + \begin{pmatrix} 0 \\ \frac{D_{42}}{detD} \\ 0 \\ \frac{D_{44}}{detD} \end{pmatrix}T_4 \tag{2.584}$$

The following state variables are defined for the parallel robotic system:

$$x = [x_1,x_2,x_3,x_4,x_5,x_6,x_7,x_8]^T \Rightarrow x = [q_1,\dot{q}_1,q_2,\dot{q}_2,q_3,\dot{q}_3,q_4,\dot{q}_4]^T \tag{2.585}$$

The control inputs vector of the system is $u = [u_1, u_2, u_3, u_4]^T \Rightarrow u = [T_1, T_2, T_3, T_4]^T$. Consequently, the state-space model of the robot can be written as

$$\begin{pmatrix} \dot{x}_1 \\ \dot{x}_2 \\ \dot{x}_3 \\ \dot{x}_4 \\ \dot{x}_5 \\ \dot{x}_6 \\ \dot{x}_7 \\ \dot{x}_8 \end{pmatrix} = \begin{pmatrix} x_2 \\ \frac{-D_{11}(c_1+g_1) - D_{31}(c_3+g_3)}{detD} \\ x_4 \\ \frac{-D_{22}(c_2+g_2) - D_{24}(c_4+g_4)}{detD} \\ x_6 \\ \frac{-D_{13}(c_1+g_1) - D_{33}(c_3+g_3)}{detD} \\ x_8 \\ \frac{-D_{24}(c_2+g_2) - D_{44}(c_4+g_4)}{detD} \end{pmatrix}$$
$$+ \begin{pmatrix} 0 \\ \frac{D_{11}}{detD} \\ 0 \\ 0 \\ 0 \\ \frac{D_{13}}{detD} \\ 0 \\ 0 \end{pmatrix} u_1 + \begin{pmatrix} 0 \\ 0 \\ 0 \\ \frac{D_{22}}{detD} \\ 0 \\ 0 \\ 0 \\ -\frac{D_{24}}{detD} \end{pmatrix} u_2 + \begin{pmatrix} 0 \\ \frac{D_{31}}{detD} \\ 0 \\ 0 \\ 0 \\ \frac{D_{33}}{detD} \\ 0 \\ 0 \end{pmatrix} u_3 + \begin{pmatrix} 0 \\ 0 \\ 0 \\ \frac{D_{42}}{detD} \\ 0 \\ 0 \\ 0 \\ \frac{D_{44}}{detD} \end{pmatrix} u_4 \tag{2.586}$$

or in concise form, the state-space model of the robot is written as

$$\dot{x} = f(x) + G_1(x)u_1 + G_2(x)u_2 + G_3(x)u_3 + G_4(x)u_4 \tag{2.587}$$

where $x \in R^{8\times1}$, $f(x) \in R^{8\times1}$, $G_1(x) \in R^{8\times1}$, $u_1 \in R$, $G_2(x) \in R^{8\times1}$, $u_2 \in R$, $G_3(x) \in R^{8\times1}$, $u_3 \in R$, $G_4(x) \in R^{8\times1}$, $u_4 \in R$.

To complete the state-space model's definition, its constituent functions are expressed using the state-vector elements x_i, $i = 1, 2, \cdots, 8$. One has that

$$\begin{aligned} d_{11} &= m_1 r_1^2 + m_3((L_1^2 + r_3^2 + 2L_1 r_3 \cos(x_5 + \delta_3)) + J_1 + J_3 \\ d_{13} &= d_{31} = m_3(r_3^2 + L_1 r_3 \cos(x_5 + \delta_3)) + J_3 \\ d_{22} &= m_2 r_2^2 + m_4(L_2^2 + r_4^2 + 2L_2 r_4 \cos(x_7 + \delta_4)) + J_2 + J_4 \\ d_{24} &= d_{42} = m_4(r_4^2 + L_2 r_4 \cos(x_7 + \delta_4)) + J_4 \\ d_{33} &= m_3 r_3^2 + J_3 \\ d_{44} &= m_4 r_4^2 + J_4 \end{aligned} \tag{2.588}$$

$$\begin{aligned} h_1 &= -m_3 L_1 r_3 \sin(x_5 + \delta_3) \\ h_2 &= -m_4 L_2 r_4 \sin(x_7 + \delta_4) \end{aligned} \tag{2.589}$$

$$C(x, \dot{x}) = \begin{pmatrix} h_1 x_2 x_6 + h_1(x_2 + x_6)x_6 \\ h_2 x_4 x_8 + h_2(x_4 + x_8)x_8 \\ -h_1 x_2^2 \\ -h_2 x_4^2 \end{pmatrix} \tag{2.590}$$

$$G(x) = \begin{pmatrix} (m_1r_1 + m_3L_1)\cos(x_1 + \delta_1) + m_3r_3\cos(x_1 + x_5 + \delta_3) \\ (m_2r_2 + m_4L_2)\cos(x_3 + \delta_2) + m_4r_4\cos(x_3 + x_7 + \delta_4) \\ m_3r_3\cos(x_1 + x_5 + \delta_3) \\ m_4r_4\cos(x_3 + x_7 + \delta_4) \end{pmatrix} \tag{2.591}$$

Using the previous definitions, functions D_{ij} $i = 1, \cdots, 4, j = 1, \cdots, 4$ and the determinant $detD$ are also expressed using the state-vector elements x_i, $i = 1, \cdots, 8$. The state-space description is finally written in the following affine-in-the-input nonlinear state-space form

$$\dot{x} = f(x) + G_1(x)u_1 + G_2(x)u_2 + G_3(x)u_3 + G_4(x)u_4 \tag{2.592}$$

2.12.2.2 Differential flatness properties of the parallel robot

The considered 5-link parallel robotic manipulator is a differentially flat system, with flat outputs the turn angle variables of the robot's joints that is $x_1 = q_1, x_3 = q_2, x_5 = q_3$ and $x_7 = q_4$. The differential flatness properties of this robotic system signify that (i) all its state variables and its control inputs can be written as differential functions of a subset of its state vector elements that constitute the flat outputs vector, (ii) the flat outputs of the system are differentially independent which means that they are not connected between them through a relation in the form of an homogeneous linear differential equation (Rigatos and Busawon, 2018; Rigatos, 2015). The differential flatness properties of a system can be used to define setpoints about it.

From the first, third, fifth and seventh rows of the state-space model of the 5-link parallel robot given in (2.586), one has

$$x_2 = \dot{x}_1 \quad x_4 = \dot{x}_3 \quad x_6 = \dot{x}_5 \quad x_8 = \dot{x}_7 \tag{2.593}$$

which shows that state variables x_2, x_4, x_6 and x_8 are differential functions of the system's flat outputs. Besides, from the second, fourth, sixth, and eighth rows of the state-space model of (2.586), one arrives at the system of equations

$$\begin{pmatrix} \ddot{x}_1 \\ \ddot{x}_3 \\ \ddot{x}_5 \\ \ddot{x}_7 \end{pmatrix} = \begin{pmatrix} \frac{-D_{11}(c_1 + g_1) - D_{31}(c_3 + g_3)}{detD} \\ \frac{-D_{22}(c_2 + g_2) - D_{24}(c_4 + g_4)}{detD} \\ \frac{-D_{13}(c_1 + g_1) - D_{33}(c_3 + g_3)}{detD} \\ \frac{-D_{24}(c_2 + g_2) - D_{44}(c_4 + g_4)}{detD} \end{pmatrix} + \begin{pmatrix} \frac{D_{11}}{detD} & 0 & \frac{D_{13}}{detD} & 0 \\ 0 & \frac{D_{22}}{detD} & 0 & \frac{D_{42}}{detD} \\ \frac{D_{13}}{detD} & 0 & \frac{D_{33}}{detD} & 0 \\ 0 & \frac{D_{42}}{detD} & 0 & \frac{D_{44}}{detD} \end{pmatrix} \begin{pmatrix} T_1 \\ T_2 \\ T_3 \\ T_4 \end{pmatrix} \tag{2.594}$$

It holds that the terms D_{11}, D_{13}, D_{22}, D_{24}, D_{13}, D_{33}, D_{42}, D_{44} and $detD$ are functions of the flat outputs of the system x_1, x_3, x_5 and x_7. Thus, by solving the above relation with respect to the control inputs vector, one obtains

$$\begin{pmatrix} T_1 \\ T_2 \\ T_3 \\ T_4 \end{pmatrix} = \begin{pmatrix} \frac{D_{11}}{detD} & 0 & \frac{D_{13}}{detD} & 0 \\ 0 & \frac{D_{22}}{detD} & 0 & \frac{D_{42}}{detD} \\ \frac{D_{13}}{detD} & 0 & \frac{D_{33}}{detD} & 0 \\ 0 & \frac{D_{42}}{detD} & 0 & \frac{D_{44}}{detD} \end{pmatrix}^{-1} \left[\begin{pmatrix} \ddot{x}_1 \\ \ddot{x}_3 \\ \ddot{x}_5 \\ \ddot{x}_7 \end{pmatrix} - \begin{pmatrix} \frac{-D_{11}(c_1+g_1)-D_{31}(c_3+g_3)}{detD} \\ \frac{-D_{22}(c_2+g_2)-D_{24}(c_4+g_4)}{detD} \\ \frac{-D_{13}(c_1+g_1)-D_{33}(c_3+g_3)}{detD} \\ \frac{-D_{24}(c_2+g_2)-D_{44}(c_4+g_4)}{detD} \end{pmatrix} \right] \tag{2.595}$$

The above relation proves that the control inputs of the robotic system can be also written as differential functions of the system's flat outputs. Consequently, the five-link parallel robotic model is differentially flat.

2.12.3 Approximate linearization of the five-link parallel robot

2.12.3.1 Approximate linearization of the robot's dynamics

The previously described state-space model of the robot which has been given in (2.592) undergoes approximate linearization at each sampling instance around the temporary operating point (x^*, u^*), where x^* is the present value of the state vector of the robot and u^* is the last sampled value of the control inputs vector.

The linearization process relies on first-order Taylor series expansion and on the computation of the associated Jacobian matrices. This results into a linearized model of the form

$$\dot{x} = Ax + Bu + \tilde{d} \tag{2.596}$$

where $\tilde{d}$ is the cumulative disturbances vector which is due to (i) modeling errors coming from the truncation of higher-order terms in the Taylor series, (ii) external perturbations, and(iii) sensors measurement noise of any distribution. The Jacobian matrices of the model are given by

$$\begin{aligned} A &= \nabla_x[f(x) + G_1(x)u_1 + G_2(x)u_2 + G_3(x)u_3 + G_4(x)u_4] \mid_{(x^*,u^*)} \Rightarrow \\ A &= \nabla_x f(x) \mid_{(x^*,u^*)} + \nabla_x G_1(x)u_1 \mid_{(x^*,u^*)} + \nabla_x G_2(x)u_2 \mid_{(x^*,u^*)} \\ &\quad + \nabla_x G_3(x)u_3 \mid_{(x^*,u^*)} + \nabla_x G_4(x)u_4 \mid_{(x^*,u^*)} \end{aligned} \tag{2.597}$$

$$\begin{aligned} B &= \nabla_u[f(x) + G_1(x)u_1 + G_2(x)u_2 + G_3(x)u_3 + G_4(x)u_4] \mid_{(x^*,u^*)} \Rightarrow \\ B &= [G_1(x) \ \ G_2(x) \ \ G_3(x) \ \ G_4(x)] \mid_{(x^*,u^*)} \Rightarrow B = G(x) \mid_{(x^*,u^*)} \end{aligned} \tag{2.598}$$

The detailed computation of the elements of the Jacobian matrices A and B is given in the following subsection.

2.12.3.2 Computation of the Jacobian matrices of the five-link parallel robot

Computation of the Jacobian matrix $\nabla_x f(x) \mid_{(x^*,u^*)}$.

First row of the Jacobian matrix $\nabla_x f(x)\mid_{(x^*,u^*)}$: $\frac{\partial f_1}{\partial x_1}=0$, $\frac{\partial f_1}{\partial x_2}=1$, $\frac{\partial f_1}{\partial x_3}=0$, $\frac{\partial f_1}{\partial x_4}=0$, $\frac{\partial f_1}{\partial x_5}=0$, $\frac{\partial f_1}{\partial x_6}=0$, $\frac{\partial f_1}{\partial x_7}=0$, $\frac{\partial f_1}{\partial x_8}=0$.

Second row of the Jacobian matrix $\nabla_x f(x)\mid_{(x^*,u^*)}$: for $i=1,\cdots,8$

$$\frac{\partial f_2}{\partial x_i}=\frac{[-\frac{\partial D_{11}}{\partial x_i}(c_1+g_1)-D_{11}(\frac{\partial c_1}{\partial x_i}+\frac{\partial g_1}{\partial x_i})]detD}{[detD]^2}+\frac{[-\frac{\partial D_{31}}{\partial x_i}(c_3+g_3)-D_{31}(\frac{\partial c_3}{\partial x_i}+\frac{\partial g_3}{\partial x_i})]detD}{[detD]^2}-\frac{[-D_{11}(c_1+g_1)-D_{31}(c_3+g_3)]\frac{\partial detD}{\partial x_i}}{[detD]^2}$$

Third row of the Jacobian matrix $\nabla_x f(x)\mid_{(x^*,u^*)}$: $\frac{\partial f_3}{\partial x_1}=0$, $\frac{\partial f_3}{\partial x_2}=0$, $\frac{\partial f_3}{\partial x_3}=0$, $\frac{\partial f_3}{\partial x_4}=1$, $\frac{\partial f_3}{\partial x_5}=0$, $\frac{\partial f_3}{\partial x_6}=0$, $\frac{\partial f_3}{\partial x_7}=0$, $\frac{\partial f_3}{\partial x_8}=0$.

Fourth row of the Jacobian matrix $\nabla_x f(x)\mid_{(x^*,u^*)}$: for $i=1,\cdots,8$

$$\frac{\partial f_4}{\partial x_i}=\frac{[-\frac{\partial D_{22}}{\partial x_i}(c_2+g_2)+D_{22}(\frac{\partial c_2}{\partial x_i}-\frac{\partial g_2}{\partial x_i})]detD}{[detD]^2}+\frac{[\frac{\partial D_{34}}{\partial x_i}(c_4+g_4)+D_{34}(\frac{\partial c_4}{\partial x_i}-\frac{\partial g_4}{\partial x_i})]detD}{[detD]^2}-\frac{[D_{22}(c_2+g_2)+D_{34}(c_4+g_4)]\frac{\partial detD}{\partial x_i}}{[detD]^2}$$

Fifth row of the Jacobian matrix $\nabla_x f(x)\mid_{(x^*,u^*)}$: $\frac{\partial f_5}{\partial x_1}=0$, $\frac{\partial f_5}{\partial x_2}=0$, $\frac{\partial f_5}{\partial x_3}=0$, $\frac{\partial f_5}{\partial x_4}=1$, $\frac{\partial f_5}{\partial x_5}=0$, $\frac{\partial f_5}{\partial x_6}=1$, $\frac{\partial f_5}{\partial x_7}=0$, $\frac{\partial f_5}{\partial x_8}=0$.

Sixth row of the Jacobian matrix $\nabla_x f(x)\mid_{(x^*,u^*)}$: for $i=1,\cdots,8$

$$\frac{\partial f_6}{\partial x_i}=\frac{[-\frac{\partial D_{13}}{\partial x_i}(c_1+g_1)-D_{13}(\frac{\partial c_1}{\partial x_i}+\frac{\partial g_1}{\partial x_i})]detD}{[detD]^2}+\frac{[-\frac{\partial D_{33}}{\partial x_i}(c_3+g_3)-D_{33}(\frac{\partial c_3}{\partial x_i}+\frac{\partial g_3}{\partial x_i})]detD}{[detD]^2}-\frac{[-D_{13}(c_1+g_1)-D_{33}(c_3+g_3)]\frac{\partial detD}{\partial x_i}}{[detD]^2}$$

Seventh row of the Jacobian matrix $\nabla_x f(x)\mid_{(x^*,u^*)}$: $\frac{\partial f_7}{\partial x_1}=0$, $\frac{\partial f_7}{\partial x_2}=0$, $\frac{\partial f_7}{\partial x_3}=0$, $\frac{\partial f_7}{\partial x_4}=1$, $\frac{\partial f_7}{\partial x_5}=0$, $\frac{\partial f_7}{\partial x_6}=0$, $\frac{\partial f_7}{\partial x_7}=0$, $\frac{\partial f_7}{\partial x_8}=1$.

Eighth row of the Jacobian matrix $\nabla_x f(x)\mid_{(x^*,u^*)}$: for $i=1,\cdots,8$

$$\frac{\partial f_8}{\partial x_i}=\frac{[-\frac{\partial D_{24}}{\partial x_i}(c_2+g_2)+D_{24}(\frac{\partial c_2}{\partial x_i}-\frac{\partial g_2}{\partial x_i})]detD}{[detD]^2}+\frac{[\frac{\partial D_{44}}{\partial x_i}(c_4+g_4)+D_{44}(\frac{\partial c_4}{\partial x_i}-\frac{\partial g_4}{\partial x_i})]detD}{[detD]^2}-\frac{[D_{24}(c_2+g_2)+D_{44}(c_4+g_4)]\frac{\partial detD}{\partial x_i}}{[detD]^2}$$

Computation of the Jacobian matrix $\nabla_x G_1(x) \mid_{(x^*,u^*)}$.

First row of the Jacobian matrix $\nabla_x G_1(x) \mid_{(x^*,u^*)}$; $\frac{\partial G_{11}}{\partial x_i} = 0, \ i = 1, \cdots, 8$

Second row of the Jacobian matrix $\nabla_x G_1(x) \mid_{(x^*,u^*)}$: for $i = 1, \cdots, 8$: $\frac{\partial G_{21}}{\partial x_i} = \frac{\frac{\partial D_{11}}{\partial x_i} detD - D_{11} \frac{\partial D}{\partial x_i}}{[detD]^2}$

Third row of the Jacobian matrix $\nabla_x G_1(x) \mid_{(x^*,u^*)}$: $\frac{\partial G_{31}}{\partial x_i} = 0, \ i = 1, \cdots, 8$

Fourth row of the Jacobian matrix $\nabla_x G_1(x) \mid_{(x^*,u^*)}$: $\frac{\partial G_{41}}{\partial x_i} = 0, \ i = 1, \cdots, 8$

Fifth row of the Jacobian matrix $\nabla_x G_1(x) \mid_{(x^*,u^*)}$: $\frac{\partial G_{51}}{\partial x_i} = 0, \ i = 1, \cdots, 8$

Sixth row of the Jacobian matrix $\nabla_x G_1(x) \mid_{(x^*,u^*)}$: for $i = 1, \cdots, 8$: $\frac{\partial G_{61}}{\partial x_i} = \frac{\frac{\partial D_{13}}{\partial x_i} detD - D_{13} \frac{\partial D}{\partial x_i}}{[detD]^2}$

Seventh row of the Jacobian matrix $\nabla_x G_1(x) \mid_{(x^*,u^*)}$: $\frac{\partial G_{71}}{\partial x_i} = 0, \ i = 1, \cdots, 8$

Eighth row of the Jacobian matrix $\nabla_x G_1(x) \mid_{(x^*,u^*)}$: $\frac{\partial G_{81}}{\partial x_i} = 0, \ i = 1, \cdots, 8$

Computation of the Jacobian matrix $\nabla_x G_2(x) \mid_{(x^*,u^*)}$.

First row of the Jacobian matrix $\nabla_x G_2(x) \mid_{(x^*,u^*)}$: $\frac{\partial G_{12}}{\partial x_i} = 0, \ i = 1, \cdots, 8$

Second row of the Jacobian matrix $\nabla_x G_2(x) \mid_{(x^*,u^*)}$: $\frac{\partial G_{22}}{\partial x_i} = 0, \ i = 1, \cdots, 8$

Third row of the Jacobian matrix $\nabla_x G_2(x) \mid_{(x^*,u^*)}$: $\frac{\partial G_{32}}{\partial x_i} = 0, \ i = 1, \cdots, 8$

Fourth row of the Jacobian matrix $\nabla_x G_2(x) \mid_{(x^*,u^*)}$: for $i = 1, \cdots, 8$: $\frac{\partial G_{42}}{\partial x_i} = \frac{\frac{\partial D_{22}}{\partial x_i} detD - D_{22} \frac{\partial D}{\partial x_i}}{[detD]^2}$

Fifth row of the Jacobian matrix $\nabla_x G_2(x) \mid_{(x^*,u^*)}$: $\frac{\partial G_{52}}{\partial x_i} = 0, \ i = 1, \cdots, 8$

Sixth row of the Jacobian matrix $\nabla_x G_2(x) \mid_{(x^*,u^*)}$: $\frac{\partial G_{62}}{\partial x_i} = 0, \ i = 1, \cdots, 8$

Seventh row of the Jacobian matrix $\nabla_x G_2(x) \mid_{(x^*,u^*)}$: $\frac{\partial G_{72}}{\partial x_i} = 0, \ i = 1, \cdots, 8$

Eighth row of the Jacobian matrix $\nabla_x G_2(x) \mid_{(x^*,u^*)}$: for $i = 1, \cdots, 8$: $\frac{\partial G_{82}}{\partial x_i} = \frac{\frac{\partial D_{24}}{\partial x_i} detD - D_{24} \frac{\partial D}{\partial x_i}}{[detD]^2}$

Computation of the Jacobian matrix $\nabla_x G_3(x) \mid_{(x^*,u^*)}$.

First row of the Jacobian matrix $\nabla_x G_3(x)\mid_{(x^*,u^*)}$: $\frac{\partial G_{13}}{\partial x_i}=0,\ i=1,\cdots,8$

Second row of the Jacobian matrix $\nabla_x G_3(x)\mid_{(x^*,u^*)}$: for $i=1,\cdots,8$: $\frac{\partial G_{23}}{\partial x_i}=\frac{\frac{\partial D_{31}}{\partial x_i}detD-D_{31}\frac{\partial D}{\partial x_i}}{[detD]^2}$

Third row of the Jacobian matrix $\nabla_x G_3(x)\mid_{(x^*,u^*)}$: $\frac{\partial G_{33}}{\partial x_i}=0,\ i=1,\cdots,8$

Fourth row of the Jacobian matrix $\nabla_x G_3(x)\mid_{(x^*,u^*)}$: $\frac{\partial G_{43}}{\partial x_i}=0,\ i=1,\cdots,8$

Fifth row of the Jacobian matrix $\nabla_x G_3(x)\mid_{(x^*,u^*)}$: $\frac{\partial G_{53}}{\partial x_i}=0,\ i=1,\cdots,8$

Sixth row of the Jacobian matrix $\nabla_x G_3(x)\mid_{(x^*,u^*)}$: for $i=1,\cdots,8$: $\frac{\partial G_{63}}{\partial x_i}=\frac{\frac{\partial D_{33}}{\partial x_i}detD-D_{33}\frac{\partial D}{\partial x_i}}{[detD]^2}$

Seventh row of the Jacobian matrix $\nabla_x G_3(x)\mid_{(x^*,u^*)}$: $\frac{\partial G_{73}}{\partial x_i}=0,\ i=1,\cdots,8$

Eighth row of the Jacobian matrix $\nabla_x G_3(x)\mid_{(x^*,u^*)}$: $\frac{\partial G_{83}}{\partial x_i}=0,\ i=1,\cdots,8$

Computation of the Jacobian matrix $\nabla_x G_4(x)\mid_{(x^*,u^*)}$.

First row of the Jacobian matrix $\nabla_x G_4(x)\mid_{(x^*,u^*)}$: $\frac{\partial G_{14}}{\partial x_i}=0,\ i=1,\cdots,8$

Second row of the Jacobian matrix $\nabla_x G_4(x)\mid_{(x^*,u^*)}$: $\frac{\partial G_{24}}{\partial x_i}=0,\ i=1,\cdots,8$

Third row of the Jacobian matrix $\nabla_x G_4(x)\mid_{(x^*,u^*)}$: $\frac{\partial G_{34}}{\partial x_i}=0,\ i=1,\cdots,8$

Fourth row of the Jacobian matrix $\nabla_x G_4(x)\mid_{(x^*,u^*)}$: for $i=1,\cdots,8$: $\frac{\partial G_{44}}{\partial x_i}=\frac{\frac{\partial D_{42}}{\partial x_i}detD-D_{42}\frac{\partial D}{\partial x_i}}{[detD]^2}$

Fifth row of the Jacobian matrix $\nabla_x G_4(x)\mid_{(x^*,u^*)}$: $\frac{\partial G_{54}}{\partial x_i}=0,\ i=1,\cdots,8$

Sixth row of the Jacobian matrix $\nabla_x G_4(x)\mid_{(x^*,u^*)}$: $\frac{\partial G_{64}}{\partial x_i}=0,\ i=1,\cdots,8$

Seventh row of the Jacobian matrix $\nabla_x G_4(x)\mid_{(x^*,u^*)}$: $\frac{\partial G_{74}}{\partial x_i}=0,\ i=1,\cdots,8$

Eighth row of the Jacobian matrix $\nabla_x G_4(x)\mid_{(x^*,u^*)}$: for $i=1,\cdots,8$: $\frac{\partial G_{84}}{\partial x_i}=\frac{\frac{\partial D_{44}}{\partial x_i}detD-D_{44}\frac{\partial D}{\partial x_i}}{[detD]^2}$

Next, the partial derivatives of the elements of matrices $D(q)$, $D^{-1}(q)$, $C(q,\dot{q})$, and $G(q)$ are computed. It holds that

For $i = 1, \cdots, 8$

$$\begin{aligned}\frac{\partial D_{11}}{\partial x_i} &= \frac{\partial d_{22}}{\partial x_i} d_{33} d_{44} + d_{22} \frac{\partial d_{33}}{\partial x_i} d_{44} + d_{22} d_{33} \frac{\partial d_{44}}{\partial x_i} \\ &\quad - \frac{\partial d_{24}}{\partial x_i} d_{42} d_{33} - d_{24} \frac{\partial d_{42}}{\partial x_i} d_{23} - d_{24} d_{42} \frac{\partial d_{23}}{\partial x_i}\end{aligned} \tag{2.599}$$

For $i = 1, \cdots, 8$

$$\frac{\partial D_{12}}{\partial x_i} = 0 \tag{2.600}$$

For $i = 1, \cdots, 8$

$$\begin{aligned}\frac{\partial D_{11}}{\partial x_i} &= -\frac{\partial d_{31}}{\partial x_i} d_{22} d_{44} - d_{31} \frac{\partial d_{32}}{\partial x_i} d_{44} - d_{31} d_{33} \frac{\partial d_{44}}{\partial x_i} - \\ &\quad + \frac{\partial d_{31}}{\partial x_i} d_{42} d_{24} + d_{31} \frac{\partial d_{44}}{\partial x_i} d_{24} + d_{31} d_{42} \frac{\partial d_{24}}{\partial x_i}\end{aligned} \tag{2.601}$$

For $i = 1, \cdots, 8$

$$\frac{\partial D_{14}}{\partial x_i} = 0 \tag{2.602}$$

For $i = 1, \cdots, 8$

$$\frac{\partial D_{21}}{\partial x_i} = 0 \tag{2.603}$$

For $i = 1, \cdots, 8$

$$\begin{aligned}\frac{\partial D_{22}}{\partial x_i} &= \frac{\partial d_{44}}{\partial x_i} d_{11} d_{33} + d_{44} \frac{\partial d_{11}}{\partial x_i} d_{33} + d_{44} d_{11} \frac{\partial d_{33}}{\partial x_i} \\ &\quad - \frac{\partial d_{44}}{\partial x_i} d_{31} d_{13} - d_{44} \frac{\partial d_{31}}{\partial x_i} d_{13} - d_{44} d_{31} \frac{\partial d_{13}}{\partial x_i}\end{aligned} \tag{2.604}$$

For $i = 1, \cdots, 8$

$$\frac{\partial D_{23}}{\partial x_i} = 0 \tag{2.605}$$

For $i = 1, \cdots, 8$

$$\begin{aligned}\frac{\partial D_{24}}{\partial x_i} &= -\frac{\partial d_{42}}{\partial x_i} d_{11} d_{33} - d_{42} \frac{\partial d_{11}}{\partial x_i} d_{33} - d_{42} d_{11} \frac{\partial d_{33}}{\partial x_i} \\ &\quad + \frac{\partial d_{42}}{\partial x_i} d_{31} d_{13} + d_{42} \frac{\partial d_{31}}{\partial x_i} d_{13} + d_{42} d_{31} \frac{\partial d_{13}}{\partial x_i}\end{aligned} \tag{2.606}$$

For $i = 1, \cdots, 8$

$$\begin{aligned}\frac{\partial D_{31}}{\partial x_i} &= -\frac{\partial d_{13}}{\partial x_i} d_{22} d_{44} - d_{13} \frac{\partial d_{22}}{\partial x_i} d_{44} - d_{13} d_{22} \frac{\partial d_{44}}{\partial x_i} \\ &\quad + \frac{\partial d_{13}}{\partial x_i} d_{42} d_{24} + d_{13} \frac{\partial d_{42}}{\partial x_i} d_{24} + d_{13} d_{42} \frac{\partial d_{24}}{\partial x_i}\end{aligned} \tag{2.607}$$

For $i = 1, \cdots, 8$

$$\frac{\partial D_{32}}{\partial x_i} = 0 \tag{2.608}$$

For $i = 1, \cdots, 8$

$$\begin{aligned}\frac{\partial D_{33}}{\partial x_i} &= \frac{\partial d_{11}}{\partial x_i} d_{22} d_{44} + d_{11} \frac{\partial d_{22}}{\partial x_i} d_{44} + d_{11} d_{22} \frac{\partial d_{44}}{\partial x_i} \\ &\quad - \frac{\partial d_{11}}{\partial x_i} d_{42} d_{24} - d_{11} \frac{\partial d_{42}}{\partial x_i} d_{24} - d_{11} d_{42} \frac{\partial d_{24}}{\partial x_i}\end{aligned} \tag{2.609}$$

For $i = 1, \cdots, 8$

$$\frac{\partial D_{32}}{\partial x_i} = 0 \tag{2.610}$$

For $i = 1, \cdots, 8$

$$\begin{aligned}\frac{\partial D_{42}}{\partial x_i} &= -\frac{\partial d_{24}}{\partial x_i} d_{11} d_{33} - d_{24} \frac{\partial d_{11}}{\partial x_i} d_{33} - d_{24} d_{11} \frac{\partial d_{33}}{\partial x_i} \\ &\quad + \frac{\partial d_{24}}{\partial x_i} d_{31} d_{13} + d_{24} \frac{\partial d_{31}}{\partial x_i} d_{13} + d_{24} d_{31} \frac{\partial d_{13}}{\partial x_i}\end{aligned} \tag{2.611}$$

For $i = 1, \cdots, 8$

$$\frac{\partial D_{43}}{\partial x_i} = 0 \tag{2.612}$$

For $i = 1, \cdots, 8$

$$\begin{aligned}\frac{\partial D_{44}}{\partial x_i} &= \frac{\partial d_{22}}{\partial x_i} d_{11} d_{33} + d_{22} \frac{\partial d_{11}}{\partial x_i} d_{33} + d_{22} d_{11} \frac{\partial d_{33}}{\partial x_i} \\ &\quad - \frac{\partial d_{22}}{\partial x_i} d_{31} d_{13} - d_{22} \frac{\partial d_{31}}{\partial x_i} d_{13} - d_{22} d_{31} \frac{\partial d_{13}}{\partial x_i}\end{aligned} \tag{2.613}$$

For $i = 1, \cdots, 8$

$$\begin{aligned}\frac{\partial detD}{\partial x_i} = &\frac{\partial d_{11}}{\partial x_i} d_{22} d_{33} d_{44} + d_{11} \frac{\partial d_{22}}{\partial x_i} d_{33} d_{44} + d_{11} d_{22} \frac{\partial d_{33}}{\partial x_i} d_{44} + d_{11} d_{22} d_{33} \frac{\partial d_{44}}{\partial x_i} \\ &- \frac{\partial d_{11}}{\partial x_i} d_{34} d_{42} d_{33} - d_{11} \frac{\partial d_{34}}{\partial x_i} d_{42} d_{33} - d_{11} d_{34} \frac{\partial d_{42}}{\partial x_i} d_{33} - d_{11} d_{34} d_{42} \frac{\partial d_{33}}{\partial x_i} \\ &\frac{\partial d_{13}}{\partial x_i} d_{24} d_{31} d_{42} + d_{13} \frac{\partial d_{24}}{\partial x_i} d_{31} d_{42} + d_{13} d_{22} \frac{\partial d_{31}}{\partial x_i} d_{42} + d_{13} d_{22} d_{31} \frac{\partial d_{42}}{\partial x_i}\end{aligned} \tag{2.614}$$

Next, the following partial derivatives are computed

$$\frac{\partial d_{11}}{\partial x_i} = 0 \quad i = 1, 2, 3, 4, 6, 7, 8 \quad \frac{\partial d_{11}}{\partial x_5} = m_3(-2L_1 r_3 \sin(x_5 + \delta_3)) \tag{2.615}$$

$$\frac{\partial d_{13}}{\partial x_i} = 0 \quad i = 1, 2, 3, 4, 6, 7, 8 \quad \frac{\partial d_{13}}{\partial x_5} = m_3(-L_1 r_3 \sin(x_5 + \delta_3)) \tag{2.616}$$

$$\frac{\partial d_{22}}{\partial x_i}=0 \quad i=1,2,3,4,5,6,8 \quad \frac{\partial d_{22}}{\partial x_7}=m_4(-2L_2r_4\sin(x_7+\delta_4)) \tag{2.617}$$

$$\frac{\partial d_{24}}{\partial x_i}=0 \quad i=1,2,3,4,5,6,8 \quad \frac{\partial d_{24}}{\partial x_7}=m_4(L_2r_4\sin(x_7+\delta_4)) \tag{2.618}$$

$$\frac{\partial d_{33}}{\partial x_i}=0 \quad i=1,\cdots,8 \tag{2.619}$$

$$\frac{\partial d_{44}}{\partial x_i}=0 \quad i=1,\cdots,8 \tag{2.620}$$

$$\frac{\partial h_1}{\partial x_i}=0 \quad i=1,2,3,4,6,7,8 \quad \frac{\partial h_1}{\partial x_5}=-m_3(L_1r_3\cos(x_5+\delta_3)) \tag{2.621}$$

$$\frac{\partial h_2}{\partial x_i}=0 \quad i=1,2,3,4,5,6,8 \quad \frac{\partial h_2}{\partial x_7}=-m_4(L_2r_4\cos(x_7+\delta_4)) \tag{2.622}$$

Besides, one has that

$\frac{\partial c_1}{\partial x_1}=0$, $\frac{\partial c_1}{\partial x_2}=2h_1x_6$, $\frac{\partial c_1}{\partial x_3}=0$, $\frac{\partial c_1}{\partial x_4}=0$, $\frac{\partial c_1}{\partial x_5}=\frac{\partial h_1}{\partial x_5}x_2x_6+\frac{\partial h_1}{\partial x_5}(x_2+x_6)x_6$, $\frac{\partial c_1}{\partial x_6}=2h_1x_2+2h_1x_6$, $\frac{\partial c_1}{\partial x_7}=0$, and $\frac{\partial c_1}{\partial x_8}=0$.

$\frac{\partial c_2}{\partial x_1}=0$, $\frac{\partial c_2}{\partial x_2}=0$, $\frac{\partial c_2}{\partial x_3}=0$, $\frac{\partial c_2}{\partial x_4}=2h_2x_8$, $\frac{\partial c_2}{\partial x_5}=0$, $\frac{\partial c_2}{\partial x_6}=0$, $\frac{\partial c_2}{\partial x_7}=\frac{\partial h_2}{\partial x_7}x_4x_8+\frac{\partial h_2}{\partial x_7}(x_4+x_8)x_8$, $\frac{\partial c_2}{\partial x_8}=2h_2x_4+2h_2x_8$.

$\frac{\partial c_3}{\partial x_1}=0$, $\frac{\partial c_3}{\partial x_2}=-2h_1x_2$, $\frac{\partial c_3}{\partial x_3}=0$, $\frac{\partial c_3}{\partial x_4}=0$, $\frac{\partial c_3}{\partial x_5}=-\frac{\partial h_1}{\partial x_5}x_2^2$, $\frac{\partial c_3}{\partial x_6}=0$, $\frac{\partial c_3}{\partial x_7}=0$, and $\frac{\partial c_3}{\partial x_8}=0$.

$\frac{\partial c_4}{\partial x_1}=0$, $\frac{\partial c_4}{\partial x_2}=0$, $\frac{\partial c_4}{\partial x_3}=0$, $\frac{\partial c_4}{\partial x_4}=-2h_2x_4$, $\frac{\partial c_4}{\partial x_5}=0$, $\frac{\partial c_4}{\partial x_6}=0$, $\frac{\partial c_4}{\partial x_7}=-\frac{\partial h_3}{\partial x_7}x_4^2$, $\frac{\partial c_4}{\partial x_8}=0$.

Moreover, one has that

$\frac{\partial g_1}{\partial x_1}=g[-(m_1r_1+m_3L_1)\sin(x_1+\delta_1)]-m_2r_3\sin(x_1+x_5+\delta_3)$, $\frac{\partial g_1}{\partial x_2}=0$, $\frac{\partial g_1}{\partial x_3}=0$, $\frac{\partial g_1}{\partial x_4}=0$, $\frac{\partial g_1}{\partial x_5}=g[-m_2r_3\sin(x_1+x_5+\delta_3)]$, $\frac{\partial g_1}{\partial x_6}=0$, $\frac{\partial g_1}{\partial x_7}=0$ and $\frac{\partial g_1}{\partial x_8}=0$.

$\frac{\partial g_2}{\partial x_1}=0$, $\frac{\partial g_2}{\partial x_2}=0$, $\frac{\partial g_2}{\partial x_3}=g[-(m_2r_2+m_4L_2)\sin(x_3+\delta_2)-m_4r_4\sin(x_3+x_7+\delta_4)]$, $\frac{\partial g_2}{\partial x_4}=0$, $\frac{\partial g_2}{\partial x_5}=0$, $\frac{\partial g_2}{\partial x_6}=0$, $\frac{\partial g_2}{\partial x_7}=g[-m_4r_4\sin(x-3+x_7+\delta_4)]$, $\frac{\partial g_2}{\partial x_8}=0$.

$\frac{\partial g_3}{\partial x_1}=g[-m_3r_3\sin(x_1+x_5+\delta_3)]$, $\frac{\partial g_3}{\partial x_2}=0$, $\frac{\partial g_3}{\partial x_3}=0$, $\frac{\partial g_3}{\partial x_4}=0$, $\frac{\partial g_3}{\partial x_5}=g[-m_3r_3\sin(x_1+x_5+\delta_3)]$, $\frac{\partial g_3}{\partial x_6}=0$, $\frac{\partial g_3}{\partial x_7}=0$, $\frac{\partial g_3}{\partial x_8}=0$.

$\frac{\partial g_4}{\partial x_1}=0$, $\frac{\partial g_4}{\partial x_2}=0$, $\frac{\partial g_4}{\partial x_3}=g[-m_4r_3\sin(x_3+x_7+\delta_4)]$, $\frac{\partial g_4}{\partial x_4}=0$, $\frac{\partial g_4}{\partial x_5}=0$, $\frac{\partial g_4}{\partial x_6}=0$, $\frac{\partial g_4}{\partial x_7}=g[-m_4r_4\sin(x_3+x_7+\delta_4)]$, $\frac{\partial g_4}{\partial x_8}=0$.

2.12.3.3 Stabilizing feedback control

After linearization around its current operating point, the dynamic model of the five-link parallel robotic manipulator is written as (Rigatos and Busawon, 2018)

$$\dot{x} = Ax + Bu + d_1 \tag{2.623}$$

Parameter d_1 stands for the linearization error in the five-link parallel robotic manipulator's dynamic model appearing above in (2.623). The reference setpoints for the state vector of the five-link parallel robotic manipulator are denoted by $\mathbf{x_d} = [x_1^d, \cdots, x_8^d]$. Tracking of this trajectory is achieved after applying the control input u_d. At every time instant the control input u_d is assumed to differ from the control input u appearing in (2.623) by an amount equal to Δu, that is $u_d = u + \Delta u$

$$\dot{x}_d = Ax_d + Bu_d + d_2 \tag{2.624}$$

The dynamics of the controlled system described in (2.623) can be also written as

$$\dot{x} = Ax + Bu + Bu_d - Bu_d + d_1 \tag{2.625}$$

and by denoting $d_3 = -Bu_d + d_1$ as an aggregate disturbance term, one obtains

$$\dot{x} = Ax + Bu + Bu_d + d_3 \tag{2.626}$$

By subtracting (2.624) from (2.626), one has

$$\dot{x} - \dot{x}_d = A(x - x_d) + Bu + d_3 - d_2 \tag{2.627}$$

By denoting the tracking error as $e = x - x_d$ and the aggregate disturbance term as $\tilde{d} = d_3 - d_2$, the tracking error dynamics becomes

$$\dot{e} = Ae + Bu + \tilde{d} \tag{2.628}$$

In the H_∞ control approach, a feedback control scheme is designed for setpoints tracking by the system's state vector and simultaneous disturbance rejection, considering that the disturbance affects the system in the worst possible manner. Using that the system's output is $y = Cx$, the disturbances' effects are incorporated in the following quadratic cost function:

$$J(t) = \frac{1}{2}\int_0^T [y^T(t)y(t) + ru^T(t)u(t) - \rho^2\tilde{d}^T(t)\tilde{d}(t)]dt, \quad r, \rho > 0 \tag{2.629}$$

The significance of the negative sign in the cost function's term that is associated with the perturbation variable $\tilde{d}(t)$ is that the disturbance tries to maximize the cost function $J(t)$ while the control signal $u(t)$ tries to minimize it. The physical meaning of the relation given above is that the control signal and the disturbances compete to each other within a min–max differential game. This problem of min–max optimization can be written as $\min_u \max_{\tilde{d}} J(u, \tilde{d})$.

For the approximately linearized model of the system a stabilizing feedback controller is developed. The controller has the form

$$u(t) = -Ke(t) \tag{2.630}$$

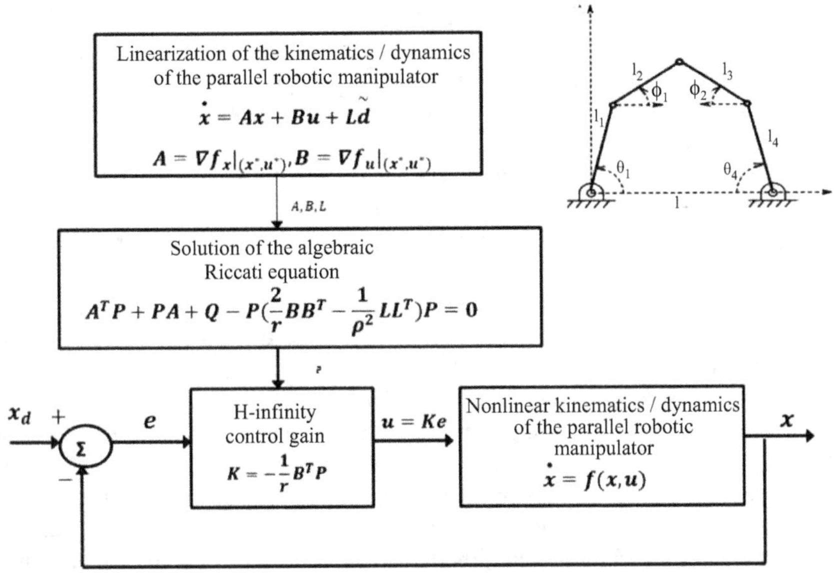

Figure 2.161 Diagram of the nonlinear optimal (H-infinity) control loop for the 5-link parallel robotic manipulator

with $K = \frac{1}{r}B^T P$ where P is a positive definite symmetric matrix which is obtained from the solution of the Riccati equation (Rigatos and Busawon, 2018)

$$A^T P + PA + Q - P\left(\frac{2}{r}BB^T - \frac{1}{\rho^2}LL^T\right)P = 0 \tag{2.631}$$

where Q is a positive semi-definite symmetric matrix. The diagram of the considered control loop is depicted in Figure 2.161. The global stability properties of the control method can be proven following the procedure presented in Section 1.1.2.5.

The solution of the H-infinity feedback control problem for the multi-DOF parallel robot and the computation of the worst case disturbance that the related controller can sustain, comes from superposition of Bellman's optimality principle when considering that the robot is affected by two separate inputs (i) the control input u and (ii) the cumulative disturbance input $\tilde{d}(t)$. Solving the optimal control problem for u, that is for the minimum variation (optimal) control input that achieves elimination of the state vector's tracking error, gives $u = -\frac{1}{r}B^T Pe$. Equivalently, solving the optimal control problem for $\tilde{d}$, that is for the worst case disturbance that the control loop can sustain gives $\tilde{d} = \frac{1}{\rho^2}L^T Pe$.

2.12.4 Simulation tests

The performance of the proposed nonlinear optimal control scheme for parallel robotic manipulators has been further tested and confirmed through simulation experiments.

To implement the control method the algebraic Riccati equation of (2.631) had to be solved at each sampling period. Indicative values for the parameters of the robotic model which were used in the simulation experiments were: $m_1 = 1.2\,\text{kgr}$, $m_2 = 1.2\,\text{kgr}$, $m_3 = 1.0\,\text{kgr}$, $m_4 = 1.0\,\text{kgr}$, $L_1 = 1.0\,\text{m}$, $L_2 = 1.0\,\text{m}$, $L_3 = 1.0\,\text{m}$, $L_4 = 1.0\,\text{m}$, $L_5 = 2.5\,\text{m}$, $r_1 = 0.5\,\text{m}$, $r_2 = 0.5\,\text{m}$, $r_3 = 0.5\,\text{m}$, $r_4 = 0.5\,\text{m}$, $\delta_1 = 0.3\,\text{rad}$, $\delta_2 = 0.3\,\text{rad}$, $\delta_3 = 0.3\,\text{rad}$, $\delta_4 = 0.3\,\text{rad}$, $J_1 = 1.2\,\text{kgr}\cdot \text{m}^2$, $J_2 = 1.2\,\text{kgr}\cdot \text{m}^2$, $J_3 = 1.0\,\text{kgr}\cdot \text{m}^2$, and $J_4 = 1.0\,\text{kgr}\cdot \text{m}^2$. Indicative values for the parameters of the H-infinity controller which were used in the simulation experiments were: $r = 0.001$, $\rho = 0.1$, $Q = diag[q_{ii}]\, i = 1, \cdots, 8$ with $q_{ii} = 20$ and $L = diag[l_{ii}]\, i = 1, \cdots, 8$ with $l_{ii} = 10^{-5}$.

The method is computationally efficient and the time needed for solving the Riccati equation is significantly smaller than the sampling interval. The setpoints of the tests have been obtained after taking into account the kinematic constraints of the parallel robot. The obtained results are depicted in Figures 2.162 to 2.177. The real values of the state variables of the robot are depicted in blue, the estimated values (as provided by the H-infinity Kalman Filter) are printed in green while the associated setpoints are plotted in red. It can be noticed that fast and accurate tracking of all setpoints was achieved under moderate variations of the control inputs. In the simulation experiments, the following convergence times for the state variables of the robot to their reference setpoints were observed: (i) Test 1 $\simeq$4 sec, (ii) Test 3 $\simeq$7 sec, (iii) Test 3 $\simeq$5 sec, (iv) Test 4 $\simeq$4 sec, (v) Test 5 $\simeq$5 sec, (vi) Test 6 $\simeq$5 sec, (vii) Test 7 $\simeq$3 sec, and (viii) Test 8 $\simeq$5 sec,

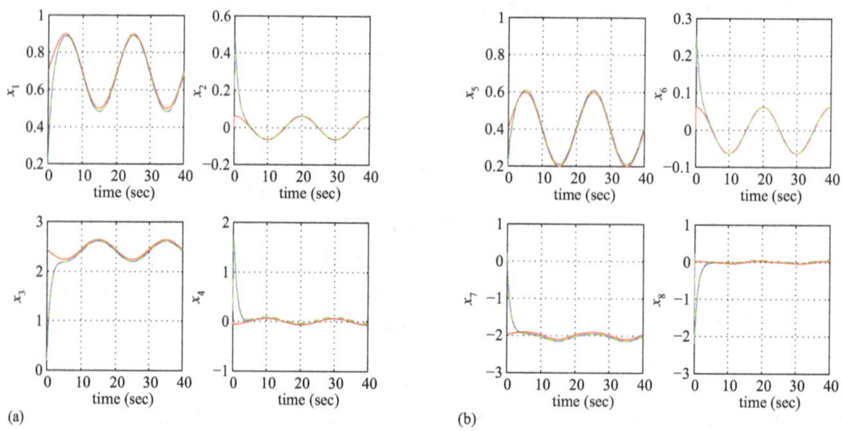

Figure 2.162 Tracking of setpoint 1 by the parallel 5-link robotic manipulator: (a) convergence of the state variables x_1 to x_4 (blue lines) to the reference setpoints (red lines) and KF-estimated state variables (green lines); (b) convergence of the state variables x_5 to x_8 to the reference setpoints

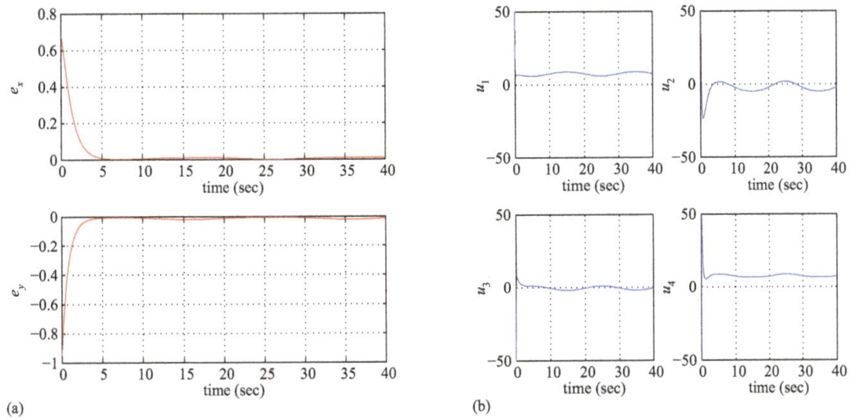

Figure 2.163 Tracking of setpoint 1 by the parallel 5-link planar robotic manipulator: (a) tracking error variables e_x and e_y (position error of the robot's end effector in the xy-plane); (b) control inputs u_1 to u_4 generated by the actuators of the parallel manipulator

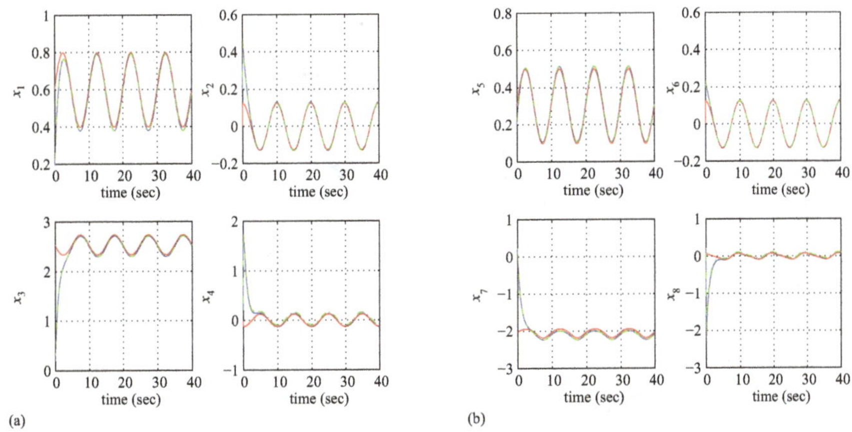

Figure 2.164 Tracking of setpoint 2 by the parallel 5-link robotic manipulator: (a) convergence of the state variables x_1 to x_4 (blue lines) to the reference setpoints (red lines) and KF-estimated state variables (green lines); (b) convergence of the state variables x_5 to x_8 to the reference setpoints

The control method has global (and not local) stability properties, as it can be confirmed by the ability to track changing setpoints. The transient performance of the control scheme depends on the parameters of the previously noted algebraic Riccati equation, such as gains r and ρ and matrix Q. Actually, for relatively small values

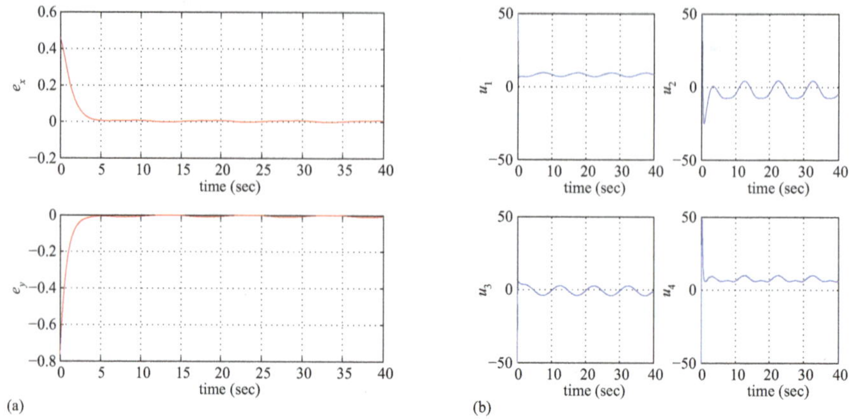

Figure 2.165 Tracking of setpoint 2 by the parallel 5-link planar robotic manipulator: (a) tracking error variables e_x and e_y (position error of the robot's end effector in the xy-plane); (b) control inputs u_1 to u_4 generated by the actuators of the parallel manipulator

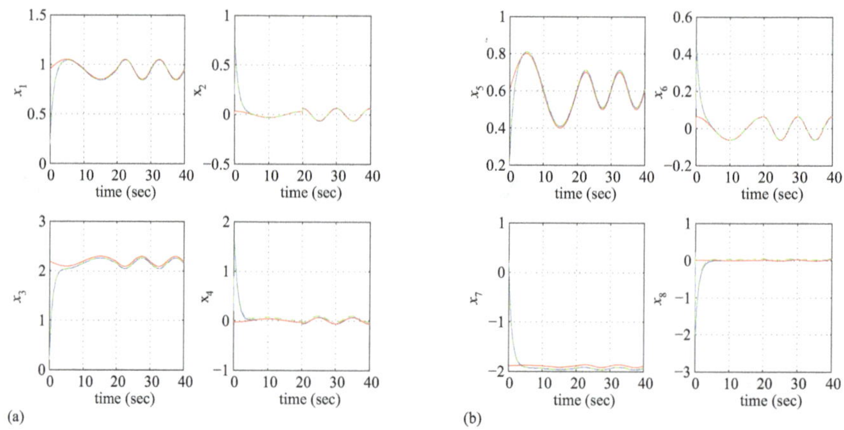

Figure 2.166 Tracking of setpoint 3 by the parallel 5-link robotic manipulator: (a) convergence of the state variables x_1 to x_4 (blue lines) to the reference setpoints (red lines) and KF-estimated state variables (green lines); (b) convergence of the state variables x_5 to x_8 to the reference setpoints

of r elimination of the state vector's tracking error is achieved, while for relatively high value of matrix Q, fast convergence of the state variables to setpoints is obtained. Besides, the smallest value of the attenuation coefficient ρ for which a positive definite and symmetric matrix P can be obtained from the solution of the aforementioned Riccati equation is the one that provides the control loop with maximum robustness.

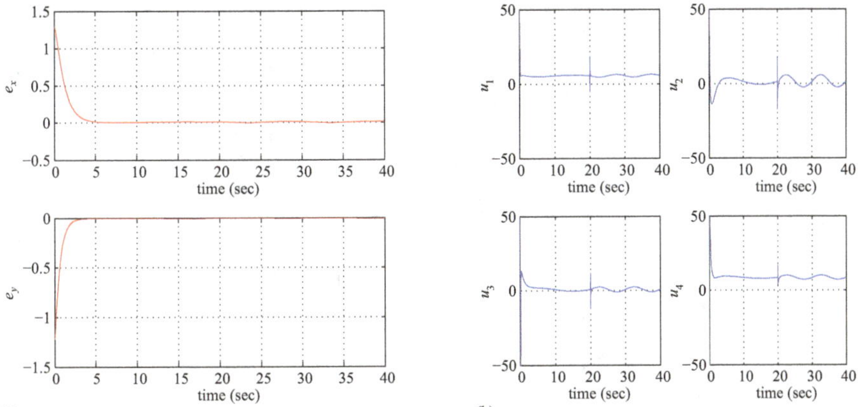

(a) (b)

Figure 2.167 Tracking of setpoint 3 by the parallel 5-link planar robotic manipulator: (a) tracking error variables e_x and e_y (position error of the robot's end effector in the xy-plane); (b) control inputs u_1 to u_4 generated by the actuators of the parallel manipulator

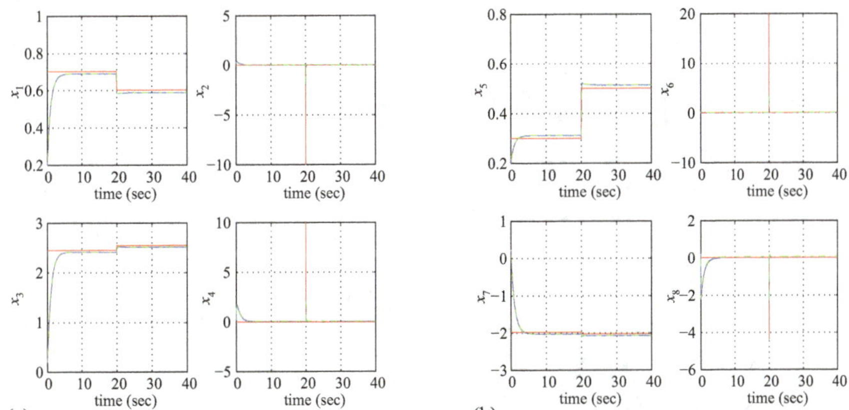

(a) (b)

Figure 2.168 Tracking of setpoint 4 by the parallel 5-link robotic manipulator: (a) convergence of the state variables x_1 to x_4 (blue lines) to the reference setpoints (red lines) and KF-estimated state variables (green lines); (b) convergence of the state variables x_5 to x_8 to the reference setpoints

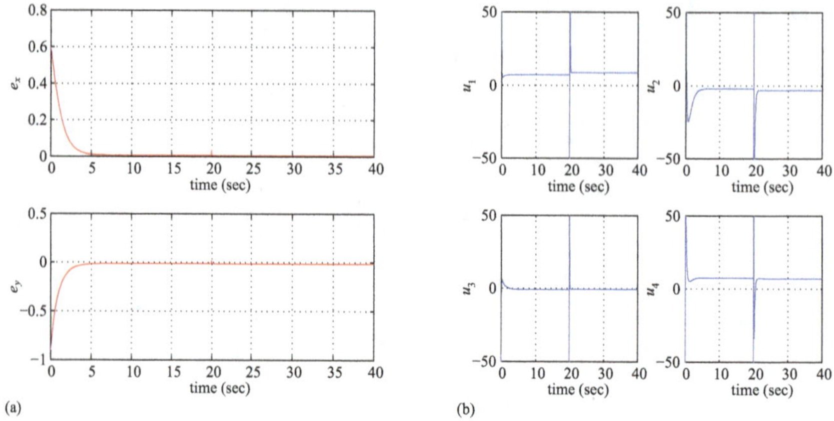

Figure 2.169 Tracking of setpoint 4 by the parallel 5-link planar robotic manipulator: (a) tracking error variables e_x and e_y (position error of the robot's end effector in the xy-plane); (b) control inputs u_1 to u_4 generated by the actuators of the parallel manipulator

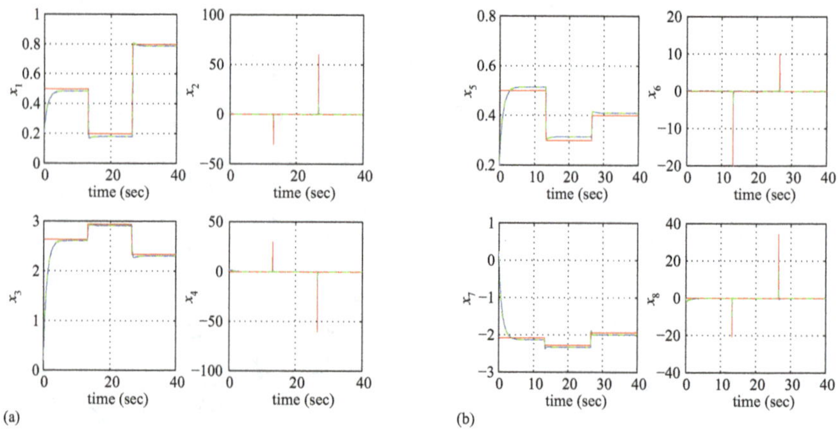

Figure 2.170 Tracking of setpoint 5 by the parallel 5-link robotic manipulator: (a) convergence of the state variables x_1 to x_4 (blue lines) to the reference setpoints (red lines) and KF-estimated state variables (green lines); (b) convergence of the state variables x_5 to x_8 to the reference setpoints

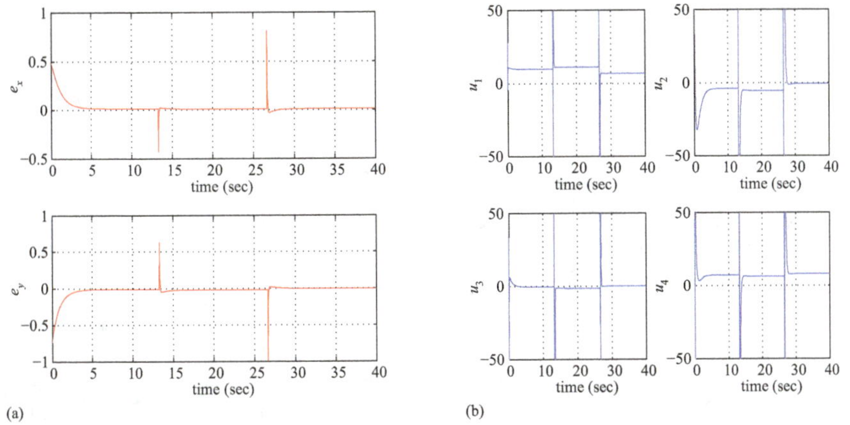

Figure 2.171 Tracking of setpoint 5 by the parallel 5-link planar robotic manipulator: (a) tracking error variables e_x and e_y (position error of the robot's end effector in the xy-plane); (b) control inputs u_1 to u_4 generated by the actuators of the parallel manipulator

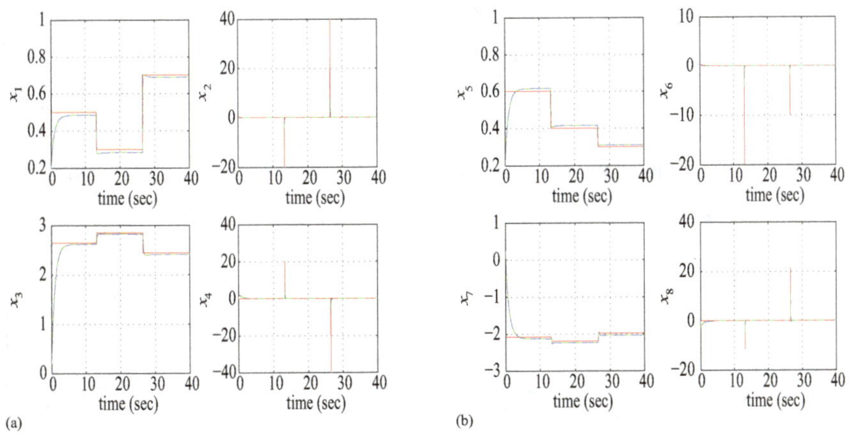

Figure 2.172 Tracking of setpoint 6 by the parallel 5-link robotic manipulator: (a) convergence of the state variables x_1 to x_4 (blue lines) to the reference setpoints (red lines) and KF-estimated state variables (green lines); (b) convergence of the state variables x_5 to x_8 to the reference setpoints

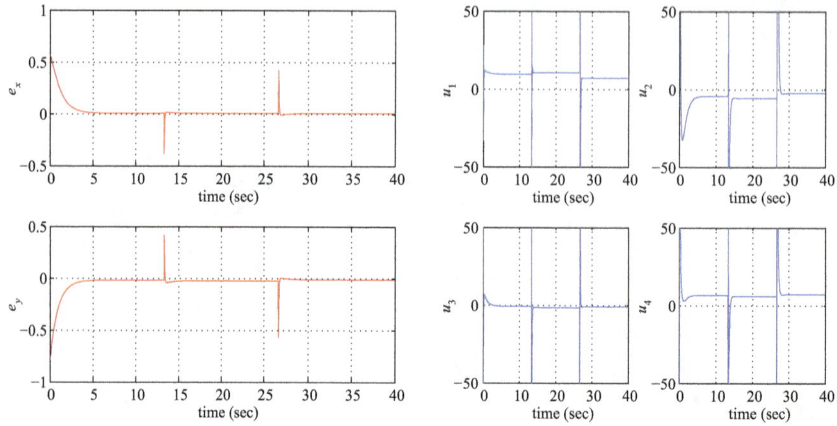

Figure 2.173 Tracking of setpoint 6 by the parallel 5-link planar robotic manipulator: (a) tracking error variables e_x and e_y (position error of the robot's end effector in the xy-plane); (b) control inputs u_1 to u_4 generated by the actuators of the parallel manipulator

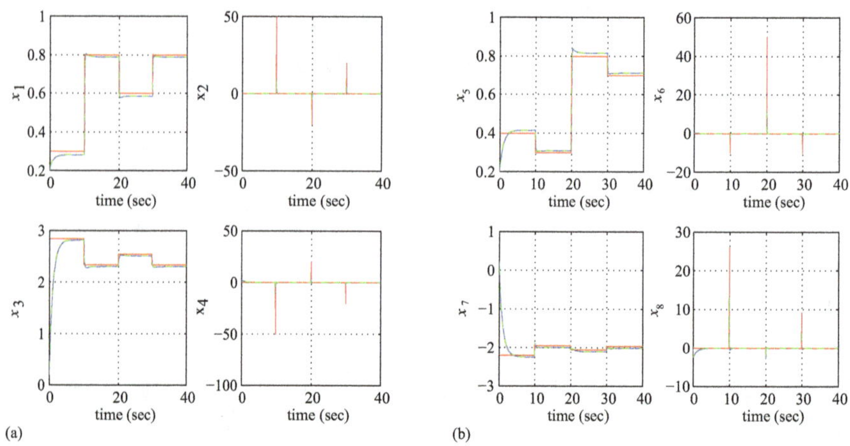

Figure 2.174 Tracking of setpoint 7 by the parallel 5-link robotic manipulator: (a) convergence of the state variables x_1 to x_4 (blue lines) to the reference setpoints (red lines) and KF-estimated state variables (green lines); (b) convergence of the state variables x_5 to x_8 to the reference setpoints

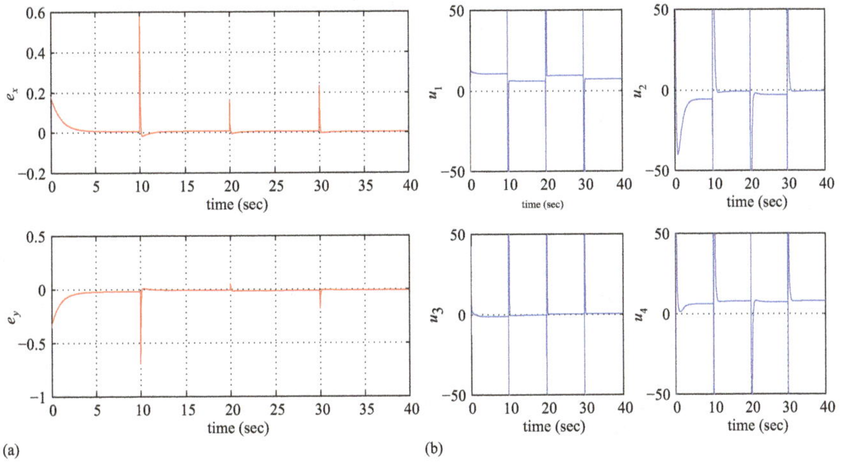

Figure 2.175 Tracking of setpoint 7 by the parallel 5-link planar robotic manipulator: (a) tracking error variables e_x and e_y (position error of the robot's end effector in the xy-plane); (b) control inputs u_1 to u_4 generated by the actuators of the parallel manipulator

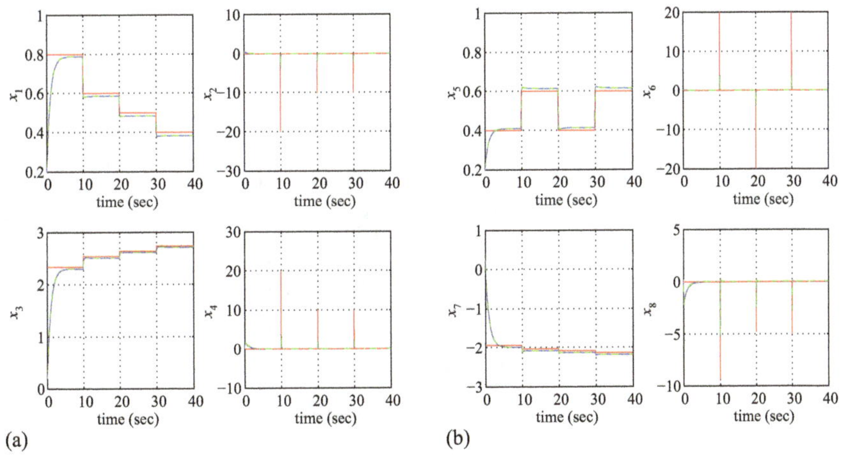

Figure 2.176 Tracking of setpoint 8 by the parallel 5-link robotic manipulator: (a) convergence of the state variables x_1 to x_4 (blue lines) to the reference setpoints (red lines) and KF-estimated state variables (green lines); (b) convergence of the state variables x_5 to x_8 to the reference setpoints

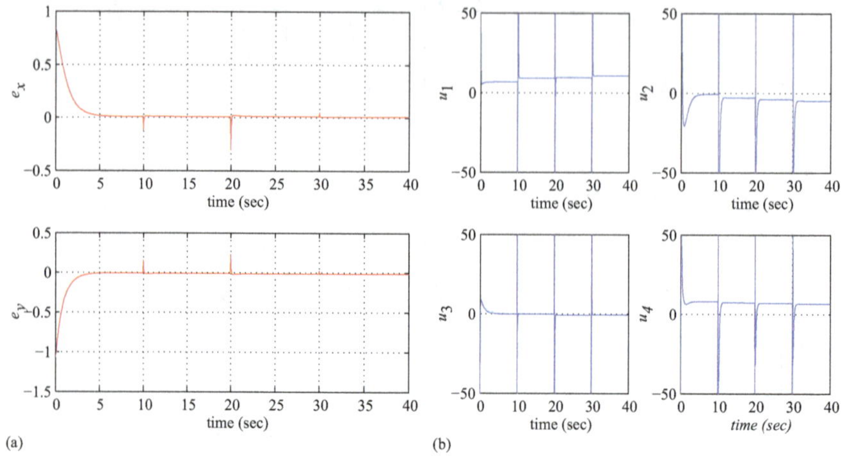

Figure 2.177 Tracking of setpoint 8 by the parallel 5-link planar robotic manipulator: (a) tracking error variables e_x and e_y (position error of the robot's end effector in the xy-plane); (b) control inputs u_1 to u_4 generated by the actuators of the parallel manipulator

Chapter 3
Control based on approximate linearization for autonomous vehicles

Abstract

The chapter analyzes methods and presents results on control and estimation based on approximate linearization for autonomous vehicles: (i) nonlinear control of tracked autonomous vehicles, (ii) nonlinear control of the autonomous fire-truck, (iii) nonlinear control of the truck and N-trailer system, (iv) nonlinear control of the ball-bot autonomous robot, (v) Nonlinear control of the ball-and-plate dynamical system, (vi) nonlinear control of 3-DOF unmanned surface vessels, (vii) nonlinear control of the 3-DOF autonomous underwater vessel, (viii) nonlinear control of the Vertical Take-off and Landing Aircraft, (ix) nonlinear control of aerial manipulators, (x) Nonlinear control of the 6-DOF autonomous octocopter, and (xi) nonlinear control of hypersonic aerial vehicles.

3.1 Nonlinear control of tracked autonomous vehicles

3.1.1 Outline

Tracked autonomous vehicles and tracked mobile robots can have numerous applications in construction works and mining (bulldozers and other construction machinery), in rescue and patrolling tasks where rough terrain has to be accessed, in defense (combat tanks and crawling vehicles used in defense operations) and certainly in agriculture where tracked tractors have found ample use (Rigatos and Busawon, 2018; Endo *et al.*, 2007; Guo, Yu *et al.*, 2019; Yamauchi *et al.*, 2017; Seron, Martinez *et al.*, 2014). Comparing to wheeled unmanned ground vehicles, tracked mobile robots and vehicles exhibit the capacity to function well in rough and difficultly accessible terrains and to accomplish off-road missions (Martinez, Mandow *et al.*, 2005; Yamauchi *et al.*, 2016; Cook *et al.*, 2016; Sebastian and Ben-Tzvi, 2019). Motion of such vehicles is provided by two tracks which are made to rotate by turning wheels being placed in the vehicle's sprockets (Tao, Ou and Feng, 2012; Zou, Angeles *et al.*, 2018; Liu and Liu, 2009a; Liu and Liu, 2009b). The kinematic model of such vehicles, in its essential form, can be considered as equivalent to the one of differential wheel-driven vehicles. However, due to the non-firm contact of the tracks with the ground, slippage effects are likely to appear (Saeed *et al.*, 2004; Fan, Singh *et al.*, 2010; Chen, Kamezaki *et al.*, 2018; Zweiri *et al.*, 2016). The motion of track-vehicles under slippage effects has been the subject of several research works during the last years (Mai and Wang, 2014; Sheng, Chen and Xi, 2008; Takai *et al.*, 2014; Peng, Yang *et al.*, 2019). Nevertheless, in such a case, the associated kinematic model becomes significantly more complicated and exhibits both multivariable structure and under-actuation, and consequently the solution of the related control problem has become a non-trivial task (Liu and Liu, 2010; Jiao *et al.*, 2014). Indicative results on the nonlinear control and autonomous navigation problem of tracked mobile robots can be found in (Zhou *et al.*, 2020; Wu, Wang *et al.*, 2019; Ziye, Haiou *et al.*, 2019; and Li, Jing *et al.*, 2009).

In this section, a novel solution is proposed for the nonlinear optimal control problem of tracked mobile robots and vehicles. First, the kinematic model of the tracked vehicles under slippage effects, undergoes approximate linearization around a temporary operating point (equilibrium) which is recomputed at each iteration of the control algorithm. This operating point is defined by the present value of the vehicle's state vector and by the last sampled value of the control inputs vector. The linearization process relies on the first-order Taylor series expansion and on the computation of the associated Jacobian matrices (Rigatos and Tzafestas, 2007; Basseville and Nikiforov, 1993; Rigatos and Zhang, 2009). The modeling error which is due to the truncation of higher-order terms in the Taylor series expansion is considered to be a perturbation that is asymptotically compensated by the robustness of the control method. Actually, for the linearized model of the tracked vehicle a stabilizing H-infinity feedback controller is designed (Rigatos and Siano, 2015; Rigatos, Siano *et al.*, 2015).

The H-infinity feedback controller is the solution of the optimal control problem for the autonomous tracked vehicle under model uncertainty and external perturbations. The H-infinity controller represents a min–max differential game in which the control inputs try to minimize a cost function which comprises a quadratic term of the vehicle's state vector error. On the other side, model uncertainty and external perturbation terms try to maximize this cost functional. To compute the controller's feedback gains an algebraic Riccati equation is repetitively solved at each time-step of the control algorithm (Rigatos, 2015; Rigatos, 2011; Rigatos, 2017). The stability properties of the control method are proven through Lyapunov analysis. At a first stage it is proven that the control loop satisfies the H-infinity tracking performance criterion. This signifies elevated robustness against model uncertainty and external perturbations (Rigatos and Busawon, 2018; Tousssaint, Basar and Bullo, 2000). At a second stage, and under moderate conditions, the global asymptotic stability of the control scheme is proven (Rigatos and Busawon, 2018). Finally, to implement state estimation-based control without measuring the entire state vector of the tracked vehicle, the H-infinity Kalman Filter is used as a robust state estimator (Rigatos and Busawon, 2018).

The robust stability properties of the proposed control method can be inferred by (a) the design concept of H-infinity control for the 3-DOF tracked mobile robot.

Actually, the computation of the worst-case disturbance that this controller can sustain, comes from superposition of Bellman's optimality principle when considering that the robot is affected by two separate inputs, that is the control input and the cumulative disturbance input; (b) the section's global stability proof and particularly the H-infinity tracking performance criterion which is shown to hold in (1.60). This criterion shows that by selecting the attenuation coefficient ρ to be sufficiently small then disturbance rejection can be achieved. It is also noted that apart from the design of a robust model-based control for the tracked mobile robot, it would be possible to pursue the development of completely model-free control for the kinematic or dynamic model of the autonomous tracked vehicle (Rigatos and Busawon, 2018; Yang, Sun *et al.*, 2019; Sun, Fu, Yang *et al.*, 2020).

Comparing to (Endo *et al.*, 2007 and Guo, Yu *et al.*, 2019), it can be noted that only the present section proposes a solution to the nonlinear optimal control problem of mobile robots and particularly of tracked robotic vehicles. The section's control approach differs from all control methods that one could have considered for the nonlinear kinematic model of tracked mobile robots. A large part of the competitive control methods is of known suboptimal performance (for instance global linearization-based control methods, sliding-mode control, back-stepping control, and PID control), while the rest of the control methods are not of proven global stability (MPC,NMPC, or PID control). Besides, multiple models-based optimal control approaches require the definition of multiple operating points, and the solution of multiple Riccati equations or LMIs and this incurs a significant computation effort comparing to the section's nonlinear optimal control approach. Finally, about LQR control it can be stated that this method is completely unsuitable for use in the case of nonlinear robotic systems because of not performing compensation of the modeling errors of the linearization procedure and consequently because of lacking a global stability proof. Comparing to (Rigatos and Siano, 2015), the latter section presents an application of the nonlinear optimal control method to the kinematic model of a wheeled unicycle-type robot. This is certainly a kinematic model of simpler nonlinearities than the model of the tracked autonomous ground vehicle. Finally, comparing to (Rigatos, Siano *et al.*, 2015), the latter section presents the application of the proposed nonlinear optimal control method to an induction motor which can be used for traction in electric vehicles. This is a radically different application area than the kinematic control of the tracked autonomous ground vehicles.

3.1.2 Kinematic model of the tracked mobile robot

3.1.2.1 The essential kinematic model of the tracked mobile robot

The coordinates of the center of gravity of the tracked vehicle are denoted as (x, y) while its heading angle is denoted by θ, as shown in Figure 3.1. Considering that the linear velocity of the left and the right track of the mobile robot is v_L and v_R

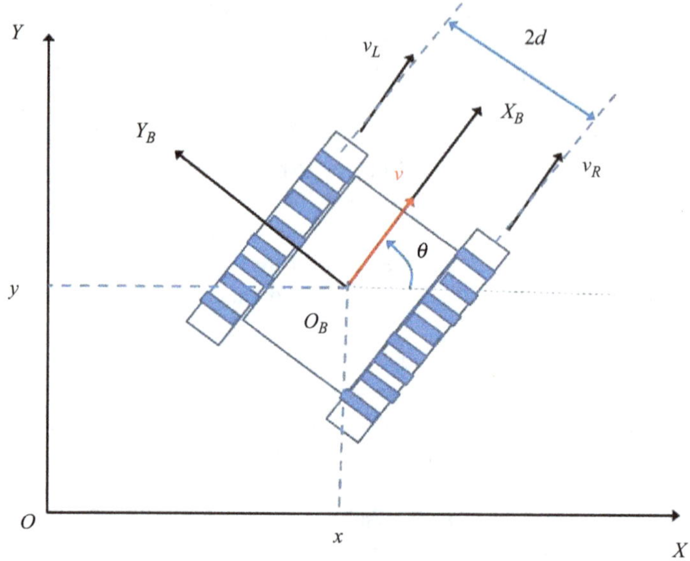

Figure 3.1 Reference axes and parameters of the essential kinematic model of the tracked mobile robot

respectively and that no slip effects appear on its tracks, the kinematic model of the vehicle is given by (Endo *et al.*, 2007)

$$\begin{aligned}\dot{x} &= \frac{v_R + v_L}{2}\cos(\theta)\\ \dot{y} &= \frac{v_R + v_L}{2}\sin(\theta)\\ \dot{\theta} &= \frac{v_R - v_L}{L}\end{aligned} \tag{3.1}$$

Consequently, one has that the state-space model of the tracked vehicle is written in the generic nonlinear state-space form $\dot{x} = f(x, u)$, where $x \in R^{3\times 1}$ and $f(x, u) \in R^{3\times 1}$.

3.1.2.2 The kinematic model of the tracked mobile robot under slip effects

Next, in the body-fixed reference frame of the tracked vehicle, the velocity of the tracked vehicle is given by the vector $V = [V_x, V_y]^T$. The existence of a non-zero velocity component along the y-axis is due to slippage effects. The slip angle is given by $\beta = \tan^{-1}(\frac{V_y}{V_x})$. To get the body-fixed frame aligned with the inertia reference frame its rotation by an angle θ is needed, as shown in Figure 3.2. As a result, the

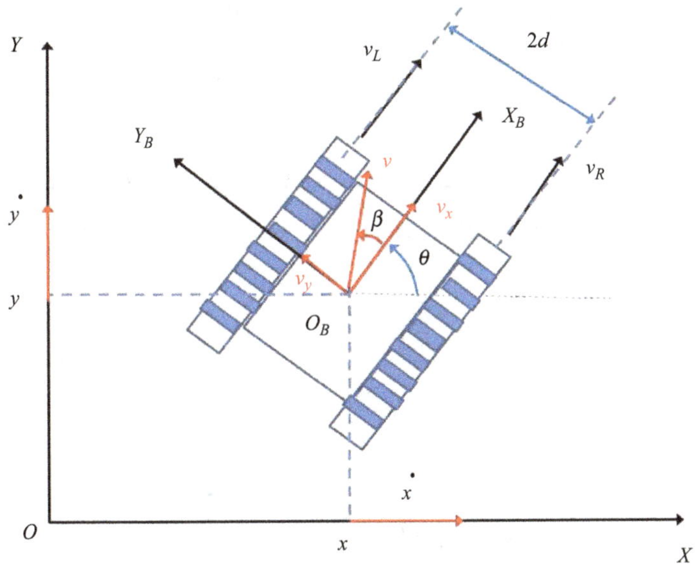

Figure 3.2 Reference axes and parameters of the slippage-affected kinematic model of the tracked mobile robot

kinematic model of the tracked vehicle, expressed in the inertial reference frame takes the following form (Yamauchi *et al.*, 2017):

$$\begin{aligned} \dot{x} &= V_x \cos(\theta) - V_y \sin(\theta) \\ \dot{y} &= V_x \sin(\theta) + V_y \cos(\theta) \\ \dot{\theta} &= \frac{v_R(1 - a_R) - v_L(1 - a_L)}{2d} \end{aligned} \tag{3.2}$$

where $2d$ is the distance between the tracks of the vehicle. It holds that (Yamauchi *et al.*, 2017)

$$\begin{aligned} V_x &= \frac{v_R(1 - a_R) + v_L(1 - a_L)}{2} \\ V_y &= V_x \tan(\beta) \end{aligned} \tag{3.3}$$

where a_R and a_L are the slip ratios of the right and the left track, respectively, while β is the slip angle. In particular about the slip ratios of the tracks, one has

$$a_R = 1 - \frac{v'_R}{v_R} \quad a_L = 1 - \frac{v'_L}{v_L} \tag{3.4}$$

Actually, v_R and v_L are the linear velocities of the tracks which are due to the rotational motion of the wheels in the vehicle's sprockets , while v'_R and v'_L are the velocities of the right and the left track at the contact point with the ground. When $v_R \neq v'_R$ and $v_L \neq v'_L$ slippage appears.

Using (3.2) and (3.3) the kinematic model of the tracked vehicle is written in the following form:

$$\begin{pmatrix} \dot{x} \\ \dot{y} \\ \dot{\theta} \end{pmatrix} = \begin{pmatrix} \frac{(1-a_r)\cos(\theta)-(1-a_r)\tan(\beta)\sin(\theta)}{2}v_R + \frac{(1-a_L)\cos(\theta)-(1-a_L)\tan(\beta)\sin(\theta)}{2}v_L \\ \frac{(1-a_r)\sin(\theta)+(1-a_r)\tan(\beta)\cos(\theta)}{2}v_R + \frac{(1-a_L)\sin(\theta)+(1-a_L)\tan(\beta)\cos(\theta)}{2}v_L \\ \frac{1-a_r}{2d}v_R - \frac{1-a_L}{2d}v_L \end{pmatrix} \quad (3.5)$$

Equivalently, the state-space model of the kinematics of the tracked mobile vehicle is written as

$$\begin{pmatrix} \dot{x} \\ \dot{y} \\ \dot{\theta} \end{pmatrix} = \begin{pmatrix} \frac{(1-a_r)\cos(\theta)-(1-a_r)\tan(\beta)\sin(\theta)}{2} & \frac{(1-a_L)\cos(\theta)-(1-a_L)\tan(\beta)\sin(\theta)}{2} \\ \frac{(1-a_r)\sin(\theta)+(1-a_r)\tan(\beta)\cos(\theta)}{2} & \frac{(1-a_L)\sin(\theta)+(1-a_L)\tan(\beta)\cos(\theta)}{2} \\ \frac{1-a_r}{2d} & -\frac{1-a_L}{2d} \end{pmatrix} \begin{pmatrix} v_R \\ v_L \end{pmatrix} \quad (3.6)$$

In the state-space representation of (3.5), the state vector is $x = [x_1.x_2, x_3]^T = [x, y, \theta]^T$, while the control inputs vector is $u = [u_1, u_2]^T = [u_R, u_L]^T$.

The dynamic model of the tracked robotic vehicle can be obtained by applying the Euler–Lagrange method. This requires the computation of the kinetic energy K and potential energy P of the vehicle and the related Lagrangian function $L = K - P$. The mass of the main body of the vehicle is M and the moment of inertia of the main body for rotation around the Oz axis is I_z. The masses of the left and the right track are M_L and M_R respectively. The kinetic energy of the vehicle is due to (i) its translational motion with velocity vector components $\dot{x}$ and $\dot{y}$, (ii) its rotational motion around the vertical axis with angular velocity $\dot{\theta}$, (iv) the translational motion of the left track with velocity v_L, and (v) the translational motion of the right track with velocity v_R. The previously noted velocity variables $\dot{x}$, $\dot{y}$ and $\dot{\theta}$ satisfy the equations of the kinematic model of the vehicle given in (5). The accelerations of the left and the right track are related with the traction forces F_L and F_R, through the relations $M_L\dot{v}_L = F_L$ and $M_R\dot{v}_R = F_R$. The state variables of the of the dynamic model are $[x, y, \theta, \dot{x}, \dot{y}, \dot{\theta}, v_L, v_R]$. The potential energy of the tracked vehicle is taken to be 0. The control inputs of the kinematic model are F_x which is the component of the vehicle's traction force along the Ox axis, F_y which is the component of the traction force of the vehicle along the Oy and T_z which is the torque due to uneven traction forces generated by the left and the right track of the vehicle. The proposed nonlinear optimal control method can be applied equally well to the kinematic model of the tracked vehicle which is of dimension 3, and to the vehicle's dynamic model which is of higher dimension.

3.1.3 Approximate linearization of the model of the tracked vehicle

3.1.3.1 Approximate linearization of the essential kinematic model of the tracked vehicle

The essential kinematic model of the tracked vehicle undergoes approximate linearization around its temporary operating point (x^*, u^*), where x^* is the present value of the system's state vector and u^* is the last value of the control inputs vector that was applied on it. The linearization relies on Taylor series expansion and on the computation of the associated Jacobian matrices. Using the state-space model of (3.1) in the form $\dot{x} = f(x, u)$, one has the following linearized state-space model:

$$\dot{x} = Ax + Bu + \tilde{d} \tag{3.7}$$

where $A = \nabla_x f(x, u)|_{(x^*,u^*)}$ and $B = \nabla_u f(x, u)|_{(x^*,u^*)}$. It holds that

$$A = \begin{pmatrix} 0 & 0 & -\frac{1}{2}\sin(\theta)v_R - \frac{1}{2}\sin(\theta)v_L \\ 0 & 0 & \frac{1}{2}\cos(\theta)v_R + \frac{1}{2}\cos(\theta)v_L \\ 0 & 0 & 0 \end{pmatrix} \tag{3.8}$$

$$B = \begin{pmatrix} \frac{1}{2}v_R\cos(\theta) & \frac{1}{2}v_R\cos(\theta) \\ \frac{1}{2}v_R\sin(\theta) & \frac{1}{2}v_R\sin(\theta) \\ \frac{1}{L} & -\frac{1}{L} \end{pmatrix} \tag{3.9}$$

3.1.3.2 Approximate linearization of the kinematic model of the tracked vehicle under slip effects

The kinematic model of the tracked vehicle under slip effects undergoes approximate linearization around its temporary operating point (x^*, u^*), where x^* is the present value of the system's state vector and u^* is the last value of the control inputs vector that was applied on it. The linearization relies on Taylor series expansion and on the computation of the associated Jacobian matrices. Using the state-space model of (3.5) in the form $\dot{x} = f(x, u)$, one has the following linearized state-space model:

$$\dot{x} = Ax + Bu + \tilde{d} \tag{3.10}$$

where $A = \nabla_x f(x, u)|_{(x^*,u^*)}$ and $B = \nabla_u f(x, u)|_{(x^*,u^*)}$. It holds that

$$A = \begin{pmatrix} 0 & 0 & \frac{-(1-a_r)\sin(\theta)-(1-a_r)\tan(\beta)\cos(\theta)}{2}v_R - \frac{(1-a_L)\sin(\theta)+(1-a_L)\tan(\beta)\cos(\theta)}{2}v_L \\ 0 & 0 & \frac{(1-a_r)\cos(\theta)-(1-a_r)\tan(\beta)\sin(\theta)}{2}v_R + \frac{(1-a_L)\cos(\theta)-(1-a_L)\tan(\beta)\sin(\theta)}{2}v_L \\ 0 & 0 & 0 \end{pmatrix} \tag{3.11}$$

$$B = \begin{pmatrix} \frac{(1-a_r)\cos(\theta)-(1-a_r)\tan(\beta)\sin(\theta)}{2} & \frac{(1-a_L)\cos(\theta)-(1-a_L)\tan(\beta)\sin(\theta)}{2} \\ \frac{(1-a_r)\sin(\theta)+(1-a_r)\tan(\beta)\cos(\theta)}{2} & \frac{(1-a_L)\sin(\theta)+(1-a_L)\tan(\beta)\cos(\theta)}{2} v_L \\ \frac{1-a_r}{2d} & -\frac{1-a_L}{2d} \end{pmatrix} \tag{3.12}$$

3.1.3.3 Stabilizing feedback control

After linearization around its current operating point, the dynamic model of the tracked vehicle is written as

$$\dot{x} = Ax + Bu + d_1 \tag{3.13}$$

Parameter d_1 stands for the linearization error in the tracked vehicle's dynamic model appearing previously in (3.10). The reference setpoints for the tracked vehicle's state vector are denoted by $\mathbf{x_d} = [x_1^d, x_2^d, x_3^d]$. Tracking of this trajectory is achieved after applying the control input u_d. At every time instant the control input u_d is assumed to differ from the control input u appearing in (3.13) by an amount equal to Δu, that is $u_d = u + \Delta u$

$$\dot{x}_d = Ax_d + Bu_d + d_2 \tag{3.14}$$

The dynamics of the controlled system described in (3.13) can be also written as

$$\dot{x} = Ax + Bu + Bu_d - Bu_d + d_1 \tag{3.15}$$

and by denoting $d_3 = -Bu_d + d_1$ as an aggregate disturbance term one obtains

$$\dot{x} = Ax + Bu + Bu_d + d_3 \tag{3.16}$$

By subtracting (3.14) from (3.16), one has

$$\dot{x} - \dot{x}_d = A(x - x_d) + Bu + d_3 - d_2 \tag{3.17}$$

By denoting the tracking error as $e = x - x_d$ and the aggregate disturbance term as $\tilde{d} = d_3 - d_2$, the tracking error dynamics becomes

$$\dot{e} = Ae + Bu + \tilde{d} \tag{3.18}$$

For the approximately linearized model of the system a stabilizing feedback controller is developed. The controller has the form

$$u(t) = -Ke(t) \tag{3.19}$$

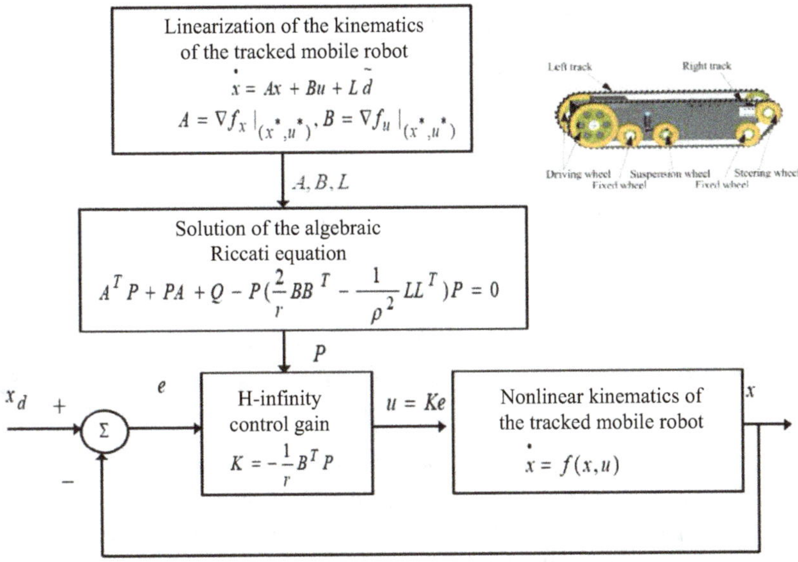

Figure 3.3 Diagram of the control scheme for the tracked mobile robot

with $K = \frac{1}{r}B^T P$ where P is a positive definite symmetric matrix which is obtained from the solution of the Riccati equation (Rigatos and Busawon, 2018)

$$A^T P + PA + Q - P\left(\frac{2}{r}BB^T - \frac{1}{\rho^2}LL^T\right)P = 0 \tag{3.20}$$

where Q is a positive semi-definite symmetric matrix. The diagram of the considered control loop is depicted in Figure 3.3. The global stability properties of the control method can be proven following the procedure presented in Section 1.1.2.5.

3.1.4 Simulation tests

The performance of the proposed nonlinear optimal (H-infinity) control method for the model of the tracked mobile robot under slippage effects has been tested through simulation tests. The simulation platform is Matlab and the algebraic Riccati equation which is needed for computing the stabilizing feedback control is solved with the use of function aresolv(). In particular the parameters of the robotic model which have been used in the simulation experiments were $a_R = 0.05$, $a_L = 0.05$, $\beta = 0.1$, $d = 1.2\,\text{m}$. It is noted that the implementation of the proposed control method does not rely on specific parameters values, since the method is robust and can tolerate parametric imprecision. The coefficients that appear in the algebraic Riccati equation of the control system are $r = 0.01$, $\rho = 0.02$, $Q = 2{\cdot}10^{-2}I_{3\times 3}$ and $L_m = 10^{-4}I_{3\times 3}$. The sampling period was $T_s = 0.01s$. Initial values do not have to be defined because

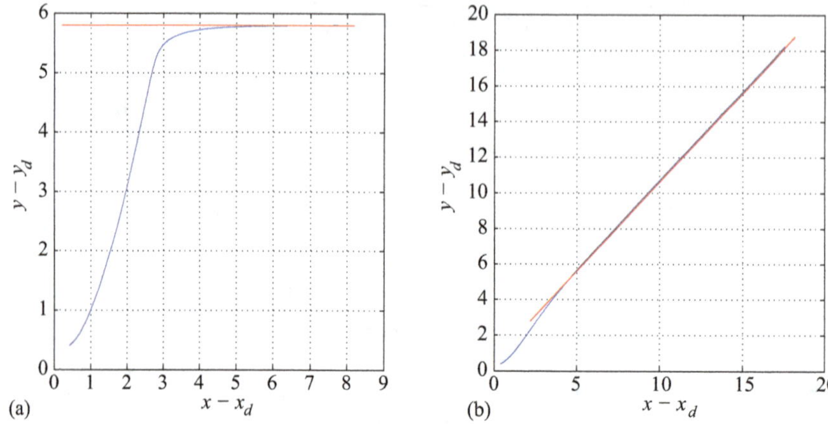

Figure 3.4 *(a) Tracking of the reference path (red line) in the xy-plane by the tracked mobile robot (blue line) in case of setpoint 1; (b) tracking of the reference path (red line) in the xy-plane by the tracked mobile robot in case of setpoint 2 (blue line)*

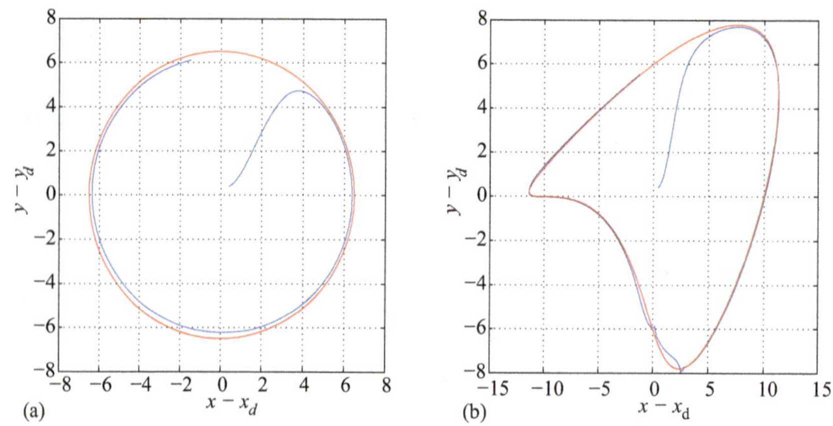

Figure 3.5 *(a) Tracking of the reference path (red line) in the xy-plane by the tracked mobile robot (blue line) in case of setpoint 3; (b) tracking of the reference path (red line) in the xy-plane by the tracked mobile robot in case of setpoint 4 (blue line)*

the proposed control method is globally stable and its performance does not depend on initial values. The Cartesian coordinates of the vehicle (x, y) are measured in meters (m) while its orientation angle is measured in radians (rad). The simulation tests come to confirm the satisfactory performance of the control method which exhibits fast and

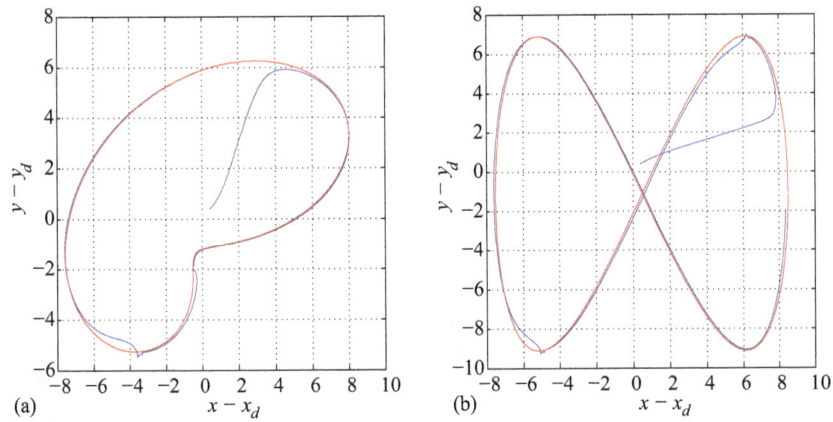

Figure 3.6 (a) Tracking of the reference path (red line) in the xy-plane by the tracked mobile robot (blue line) in case of setpoint 5; (b) tracking of the reference path (red line) in the xy-plane by the tracked mobile robot in case of setpoint 6 (blue line)

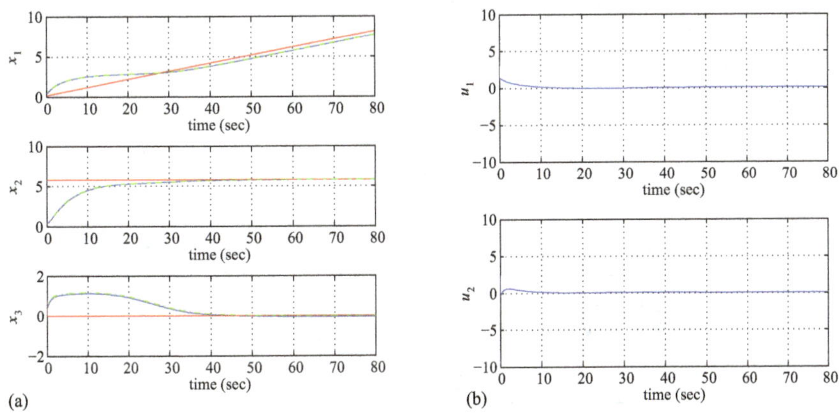

Figure 3.7 Tracking performance of the tracked mobile robot in case of setpoint 1: (a) convergence of state variables x_1 to x_3 to the reference setpoints; (b) control inputs u_1, u_2 applied to the tracked mobile robot

accurate tracking of the reference setpoints under moderate variations of the control inputs. The tracking of the reference paths in the *xy* plane that was achieved by the tracked mobile robots is shown in Figures 3.4 to 3.6. Besides, the variations of the state variables x_i, $i = 1, 2, 3$ of the tracked mobile robot and of the associated control inputs u_i, $i = 1, 2$ is shown in Figures 3.7 to 3.12.

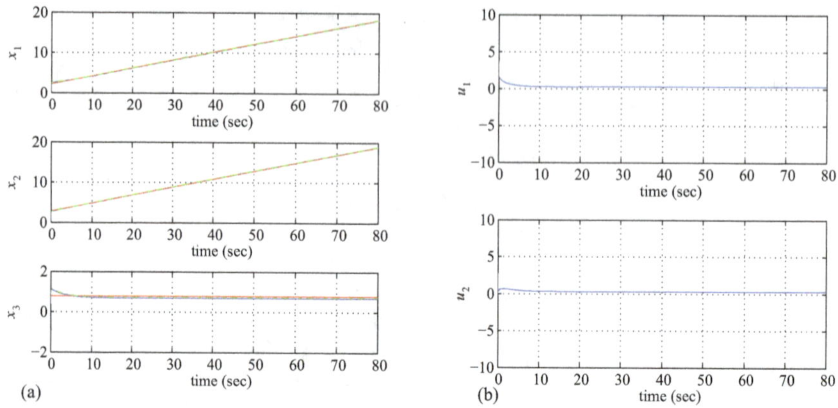

Figure 3.8 Tracking performance of the tracked mobile robot in case of setpoint 2: (a) convergence of state variables x_1 to x_3 to the reference setpoints; (b) control inputs u_1, u_2 applied to the tracked mobile robot

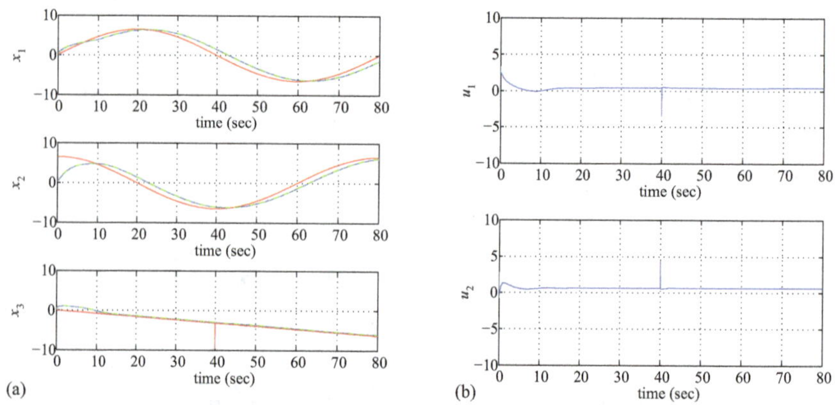

Figure 3.9 Tracking performance of the tracked mobile robot in case of setpoint 3: (a) convergence of state variables x_1 to x_3 to the reference setpoints; (b) control inputs u_1, u_2 applied to the tracked mobile robot

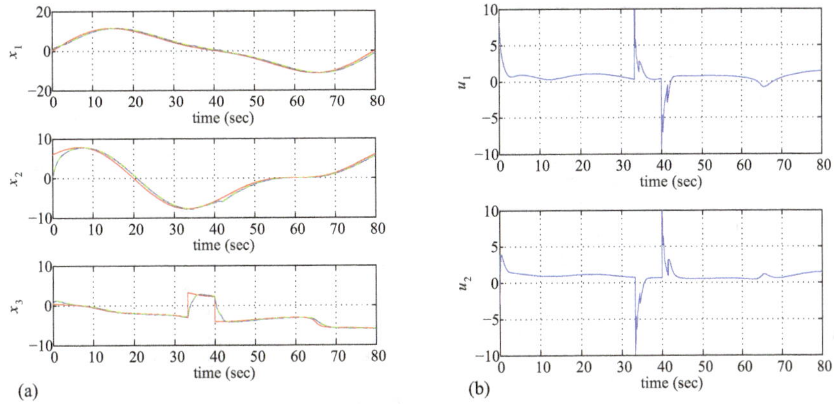

Figure 3.10 Tracking performance of the tracked mobile robot in case of setpoint 4: (a) convergence of state variables x_1 to x_3 to the reference setpoints; (b) control inputs u_1, u_2 applied to the tracked mobile robot

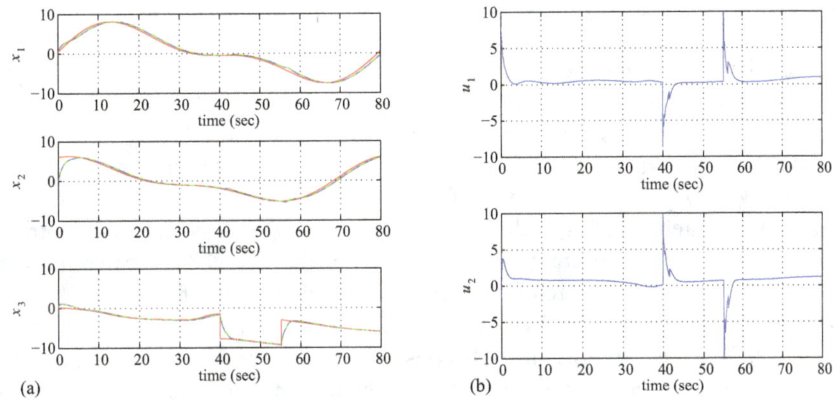

Figure 3.11 Tracking performance of the tracked mobile robot in case of setpoint 5: (a) convergence of state variables x_1 to x_3 to the reference setpoints; (b) control inputs u_1, u_2 applied to the tracked mobile robot

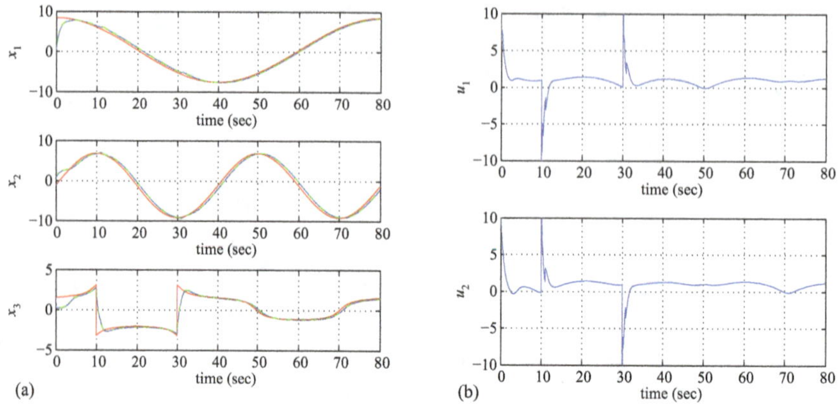

Figure 3.12 Tracking performance of the tracked mobile robot in case of setpoint 6: (a) convergence of state variables x_1 to x_3 to the reference setpoints; (b) control inputs u_1, u_2 applied to the tracked mobile robot

3.2 Nonlinear control of the autonomous articulated fire-truck

3.2.1 Outline

The problem of autonomous navigation of articulated fire-truck vehicles is treated with the use of the previously analyzed nonlinear optimal control method (Rigatos and Busawon, 2018). The problem of path following of articulated vehicles is a non-trivial one due to the nonlinearities and the multivariable structure of such systems (Rigatos and Busawon, 2018; Yuan, Sun *et al.*, 2015; Ritzen, Roebroek *et al.*, 2016; Kural *et al.*, 2017). Articulated vehicles of this type comprise two parts, the cab and the trailer. Usually the control inputs are applied exclusively on the cab and these can be the cab's forward velocity and the rate of change of the heading angle of its steering wheels (Watanabe, Ueda *et al.*, 2012; Jammazi, 2008; Bushnell *et al.*, 1995). To improve the vehicle's maneuverability it is also possible to apply an additional control input at the trailer, which is the rate of change of the heading angle of the wheels at the trailer's rear axle (Watanabe, Yamamoto *et al.*, 2010; Bushnell *et al.*, 1993; Tilbury *et al.*, 1994). Thus, in the fire-truck's case, one arrives at a truck and trailer vehicle with active trailer (Lamiraux *et al.*, 2005; Van de Wouw *et al.*, 2015; Jurnovich and Cebon, 2002). Consequently one can achieve more accurate path following by the trailer and the accomplishment of more precise maneuvers in the narrow streets of urban environments (Rangavajhula and Tsao, 2008; Abroshan *et al.*, 2014; Brun and van den Bosch, 1999).

There exist several attempts towards solving the nonlinear control problem for truck-and-trailer autonomous robotic vehicles when considering active control for the trailer (Wang *et al.*, 2018; Fletcher *et al.*, 2006; Ding, He *et al.*, 2012). These approaches either rely on simplified linear models of the kinematics of the articulated robotic vehicles, or make use of the vehicles' exact nonlinear model and subsequently apply global linearization transformations (Werner, Kormann *et al.*, 2013; Yamaguchi *et al.*, 2011; Lucet and Micaelli, 2019; Ni and He, 2018). In this section, a nonlinear optimal control method is developed for the kinematic model of the articulated autonomous fire-truck with active trailer. The section's approach is based on an approximately linearized model of the vehicle's kinematics (Rigatos, 2016b; Rigatos, Siano and Cecati, 2015; Rigatos, Siano *et al.*, 2015). Actually, at each time-instant, the state-space description of the vehicle's kinematics undergoes approximate linearization around a temporary operating. This linearization point is defined by the present value of the vehicle's state vector and by the last value of the vehicle's control inputs vector. This time-varying operating point is updated at each sampling period. The linearization procedure relies on first-order Taylor series expansion of the kinematic model of the vehicle and on the computation of the associated Jacobian matrices (Rigatos and Tzafestas, 2007; Basseville and Nikiforov, 1993; Rigatos and Zhang, 2009). The modeling error which is due to the truncation of higher order terms from the Taylor series is considered to be a perturbation which is asymptotically compensated by the robustness of the control method.

For the linearized model of the vehicle a stabilizing H-infinity feedback controller is designed. Actually, the H-infinity controller stands for the solution of the optimal control problem of the articulated fire-truck under model uncertainty and external perturbations. It represents a min–max differential game in which the control inputs try to minimize a quadratic cost function of the state vector's error whereas the model uncertainty and disturbance terms try to maximize this cost function. To compute the controller's feedback gains an algebraic Riccati equation is being repetitively solved at each time-step of the control algorithm (Rigatos, 2011; Rigatos, 2015). The stability properties of the control scheme are proven through Lyapunov analysis. At a first stage it is proven that the control loop satisfies the H-infinity tracking performance criterion (Rigatos and Busawon, 2018; Rigatos, 2016b; Tousssaint, Basar and Bullo, 2000). This signifies elevated robustness against model uncertainty and external perturbations. Moreover, under moderate conditions it is proven that the control loop is globally asymptotically stable. Finally to implement state estimation-based control without the need to measure the entire state vector of the robotic vehicle the H-infinity Kalman Filter is introduced as a robust state estimator (Rigatos and Busawon, 2018; Rigatos, 2016b; Gibbs, 2012).

3.2.2 Kinematic model of the autonomous fire-truck robot

3.2.2.1 State-space description of the kinematic model

The autonomous fire-truck comprises the cab and the trailer, as shown in Figure 3.13. The main parameters of the model of the autonomous fire-truck are as follows (Watanabe, Ueda *et al.*, 2012; Watanabe, Yamamoto *et al.*, 2010): (x_0, y_0): are the coordinates of the center of the front axle of the cab, (x_1, y_1) are the coordinates of the center of the rear axle of the cab, (ϕ_1) is the angle of the steering wheels of the cab with respect to the longitudinal axis of the cab, θ_1 is the angle between the longitudinal axis of the

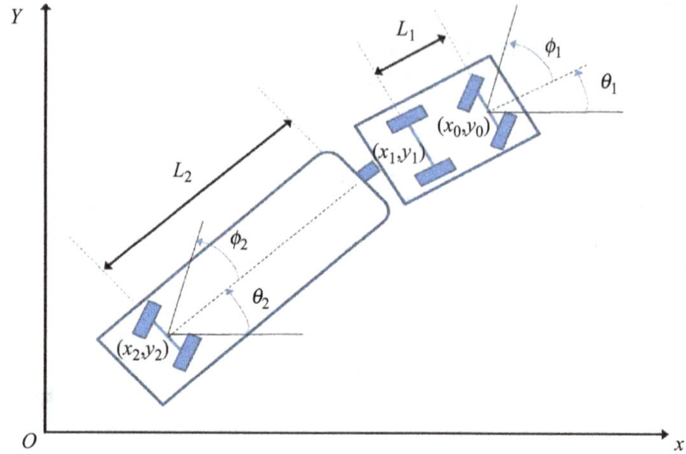

Figure 3.13 Diagram of the autonomous fire-truck robot which comprises the cab and the trailer

cab and the Ox axis of the inertial coordinates system, (x_2, y_2) are the coordinates of the center of the real axle of the trailer, ϕ_2 is the angle of the steering wheels of the trailer with respect to the longitudinal axis of the cab, θ_2 its the angle between the longitudinal axis of the trailer and the Ox axis of the inertial coordinates system. The length of the cab is denoted as L_0 and the length of the trailer is denoted as L_1.

The kinematic model of the autonomous fire-truck is given by (Watanabe, Ueda *et al.*, 2012; Watanabe, Yamamoto *et al.*, 2010):

$$\dot{x} = g_1(x)u_1 + g_2(x)u_2 + g_3(x)u_3 \tag{3.21}$$

where the control inputs of the vehicle are defined as u_1: is the forward driving velocity of the truck v_f, u_2 is the steering speed of the front wheels of the cab $\dot{\phi}_1$ and u_3 is the steering speed of the rear wheels of the trailer $\dot{\phi}_2$. Vector fields $g_1(x) \in R^{6\times 1}$, $g_2(x) \in R^{6\times 1}$ and $g_3(x) \in R^{6\times 1}$ are defined as:

$$g_1(x) = \begin{pmatrix} \cos(\theta_1) \\ \sin(\theta_1) \\ 0 \\ \dfrac{1}{L_o}\tan(\phi_2) \\ 0 \\ -\dfrac{1}{L_1}\dfrac{1}{\cos(\phi_2)}\sin(\phi_2 - \theta_1 + \theta_2) \end{pmatrix} \quad g_2(x) = \begin{pmatrix} 0 \\ 0 \\ 1 \\ 0 \\ 0 \\ 0 \end{pmatrix} \quad g_3(x) = \begin{pmatrix} 0 \\ 0 \\ 0 \\ 0 \\ 1 \\ 0 \end{pmatrix} \tag{3.22}$$

The state vector of the system is defined as $x = [x, y, \phi_1, \theta_1, \phi_2, \theta_2]^T$ or $x = [x_1, x_2, x_3, x_4, x_5, x_6]^T$. Using this state variables' notation and the model of (3.21) and (3.22) one obtains the following form of the state-space equations:

$$\dot{x}_1 = \cos(x_4)u_1 \tag{3.23}$$

$$\dot{x}_2 = \sin(x_4)u_1 \tag{3.24}$$

$$\dot{x}_3 = u_2 \tag{3.25}$$

$$\dot{x}_4 = \frac{1}{L_o}\tan(x_3) \tag{3.26}$$

$$\dot{x}_5 = u_3 \tag{3.27}$$

$$\dot{x}_6 = -\frac{1}{L_1}\frac{1}{\cos(x_5)}\sin(x_5 - x_4 + x_6) \tag{3.28}$$

3.2.2.2 Differential flatness properties of the kinematic model

The differential flatness properties of the model of the autonomous fire-truck are proven (Rigatos, 2015). The flat outputs vector of the system is taken to be $y = [y_1, y_2, y_3]^T = [x_1, x_2, x_6]^T$.

Then from (3.23) and (3.24), one has

$$\tan(x_4) = \frac{\dot{x}_2}{\dot{x}_1} \Rightarrow x_4 = \tan^{-1}\left(\frac{\dot{x}_2}{\dot{x}_1}\right) \tag{3.29}$$

Consequently x_4 is a differential function of the flat outputs of the system. From (3.26), one has

$$x_3 = \tan^{-1}(L_o\dot{x}_4) \tag{3.30}$$

Thus x_3 is a differential function of the flat outputs of the system. From (3.28), one has

$$\begin{aligned}
\dot{x}_6 &= -\frac{1}{L_1}\frac{1}{\cos(x_5)}[\sin(x_5)\cos(-x_4 + x_6) + \cos(x_5)\sin(-x_4 + x_6)] \Rightarrow \\
\dot{x}_6 &= -\frac{1}{L_1}\tan(x_5)\cos(-x_4 + x_6) - \frac{1}{L_1}\sin(-x_4 + x_6) \Rightarrow \\
&\dot{x}_6 + \frac{1}{L_1}\sin(-x_4 + x_6) = -\frac{1}{L_1}\tan(x_5)\cos(-x_4 + x_6) \Rightarrow \\
\tan(x_5) &= \frac{\dot{x}_6 + \frac{1}{L_1}\sin(-x_4 + x_6)}{-\frac{1}{L_1}\cos(-x_4 + x_6)} \Rightarrow \\
x_5 &= \tan^{-1}\left(\frac{\dot{x}_6 + \frac{1}{L_1}\sin(-x_4 + x_6)}{-\frac{1}{L_1}\cos(-x_4 + x_6)}\right)
\end{aligned} \tag{3.31}$$

As a result of the above, state variable x_5 is also a differential function of the flat outputs of the model of the autonomous fire-truck. Moreover, from (3.23) one has

$$u_1 = \frac{\dot{x}_1}{\cos(x_4)} \tag{3.32}$$

which signifies that for $x_4 \neq k\frac{\pi}{2}$ $k = 1, 2, 3, \cdots$, control input u_1 is a differential function of the flat outputs of the system. Besides, from (3.24), one has

$$u_2 = \dot{x}_3 \tag{3.33}$$

while from (3.27), one has

$$u_3 = \dot{x}_5 \tag{3.34}$$

Consequently, the control inputs u_2 and u_3 are also differential functions of the flat outputs of the system and the entire model of the autonomous fire-truck is a differentially flat system. By confirming the differential flatness properties of the articulated fire-truck one can define also setpoints for the vehicle's autonomous navigation problem (Rigatos, 2015).

3.2.3 Approximate linearization of the model of the autonomous fire-truck

By applying first-order Taylor series expansion in the kinematic model of the fire-truck

$$\dot{x} = g_1(x)u_1 + g_2(x)u_2 + g_3(x)u_3 \tag{3.35}$$

one obtains the following state-space description

$$\dot{x} = Ax + Bu + \tilde{d} \tag{3.36}$$

where matrices A and B rely on the computation of the system's Jacobians

$$A = \nabla_x g_1(x)u_1 \mid_{(x^*,u^*)} + \nabla_x g_2(x)u_2 \mid_{(x^*,u^*)} + \nabla_x g_3(x)u_3 \mid_{(x^*,u^*)} \tag{3.37}$$

$$B = [g_1(x)\; g_2(x)\; g_3(x)] \mid_{(x^*,u^*)} \tag{3.38}$$

First, the following Jacobian matrix is computed

$$\nabla_x g_1(x) = \nabla_x \begin{pmatrix} \cos(x_4) \\ \sin(x_4) \\ 0 \\ \frac{1}{L_0}\tan(x_3) \\ 0 \\ -\frac{1}{L_1}\frac{1}{\cos(x_5)}\sin(x_5 - x_4 + x_6) \end{pmatrix} \tag{3.39}$$

First row of the Jacobian matrix $\nabla_x g_1(x)$: $\frac{\partial g_{11}}{\partial x_1}=0$, $\frac{\partial g_{11}}{\partial x_2}=0$, $\frac{\partial g_{11}}{\partial x_3}=0$, $\frac{\partial g_{11}}{\partial x_4}=-\sin(x_4)$, $\frac{\partial g_{11}}{\partial x_5}=0$ and $\frac{\partial g_{11}}{\partial x_6}=0$.

Second row of the Jacobian matrix $\nabla_x g_1(x)$: $\frac{\partial g_{12}}{\partial x_1}=0$, $\frac{\partial g_{12}}{\partial x_2}=0$, $\frac{\partial g_{12}}{\partial x_3}=0$, $\frac{\partial g_{12}}{\partial x_4}=\cos(x_4)$, $\frac{\partial g_{12}}{\partial x_5}=0$ and $\frac{\partial g_{12}}{\partial x_6}=0$.

Third row of the Jacobian matrix $\nabla_x g_1(x)$: $\frac{\partial g_{13}}{\partial x_1}=0$, $\frac{\partial g_{13}}{\partial x_2}=0$, $\frac{\partial g_{13}}{\partial x_3}=0$, $\frac{\partial g_{13}}{\partial x_4}=0$, $\frac{\partial g_{13}}{\partial x_5}=0$ and $\frac{\partial g_{13}}{\partial x_6}=0$.

Fourth row of the Jacobian matrix $\nabla_x g_1(x)$: $\frac{\partial g_{14}}{\partial x_1}=0$, $\frac{\partial g_{14}}{\partial x_2}=0$, $\frac{\partial g_{14}}{\partial x_3}=\frac{1}{L_0}\frac{1}{cos^2(x_3)}$, $\frac{\partial g_{14}}{\partial x_4}=0$, $\frac{\partial g_{14}}{\partial x_5}=0$ and $\frac{\partial g_{14}}{\partial x_6}=0$.

Fifth row of the Jacobian matrix $\nabla_x g_1(x)$: $\frac{\partial g_{15}}{\partial x_1}=0$, $\frac{\partial g_{15}}{\partial x_2}=0$, $\frac{\partial g_{15}}{\partial x_3}=0$, $\frac{\partial g_{15}}{\partial x_4}=0$, $\frac{\partial g_{15}}{\partial x_5}=0$ and $\frac{\partial g_{15}}{\partial x_6}=0$.

Sixth row of the Jacobian matrix $\nabla_x g_1(x)$: $\frac{\partial g_{16}}{\partial x_1}=0$, $\frac{\partial g_{16}}{\partial x_2}=0$, $\frac{\partial g_{16}}{\partial x_3}=0$, $\frac{\partial g_{16}}{\partial x_4}=\frac{1}{L_1}\frac{1}{\cos(x_5)}\cos(x_5-x_4+x_6)$, $\frac{\partial g_{16}}{\partial x_5}=-\frac{1}{L_1}\frac{\sin(x_5)}{cos^2(x_5)}\sin(x_5-x_4+x_6)-\frac{1}{L_1}\frac{1}{\cos(x_5)}\cos(x_5-x_4+x_6)$ and $\frac{\partial g_{16}}{\partial x_6}=-\frac{1}{L_1}\frac{1}{\cos(x_5)}\cos(x_5-x_4+x_6)$.

Next, about the rest of the Jacobian matrices of the system, one has: $\nabla_x g_2(x)=0\in R^{6\times 6}$ and $\nabla_x g_3(x)=0\in R^{6\times 6}$.

3.2.4 The nonlinear H-infinity control

Next, a nonlinear optimal (H-infinity) controller will be developed for the autonomous fire-truck. The initial nonlinear model of the autonomous fire-truck is in the form

$$\dot{x}=f(x,u)\;\; x\in R^n,\;\; u\in R^m \tag{3.40}$$

Linearization is performed at each iteration of the control algorithm around its present operating point $(x^*,u^*)=(x(t),u(t-T_s))$. The linearized equivalent of the autonomous fire-truck is described by

$$\dot{x}=Ax+Bu+L\tilde{d}\;\; x\in R^n,\;\; u\in R^m,\;\; \tilde{d}\in R^q \tag{3.41}$$

Thus, after linearization around its current operating point, the model of the autonomous fire-truck is written as

$$\dot{x} = Ax + Bu + d_1 \tag{3.42}$$

Parameter d_1 stands for the linearization error in the model of the autonomous fire-truck appearing in (3.42). The reference setpoints for the system are denoted by $\mathbf{x_d} = [x_1^d, \cdots, x_6^d]$. Tracking of this trajectory is achieved after applying the control input u_d. At every time instant the control input u_d is assumed to differ from the control input u appearing in Eq. (3.42) by an amount equal to Δu, that is $u_d = u + \Delta u$. One can write

$$\dot{x}_d = Ax_d + Bu_d + d_2 \tag{3.43}$$

The dynamics of the controlled system described in (3.43) can be also written as

$$\dot{x} = Ax + Bu + Bu_d - Bu_d + d_1 \tag{3.44}$$

and by denoting $d_3 = -Bu_d + d_1$ as an aggregate disturbance term one obtains

$$\dot{x} = Ax + Bu + Bu_d + d_3 \tag{3.45}$$

By subtracting (3.43) from (3.45) one has

$$\dot{x} - \dot{x}_d = A(x - x_d) + Bu + d_3 - d_2 \tag{3.46}$$

By denoting the tracking error as $e = x - x_d$ and the aggregate disturbance term as $\tilde{d} = d_3 - d_2$, the tracking error dynamics becomes

$$\dot{e} = Ae + Bu + \tilde{d} \tag{3.47}$$

The above linearized form of the model of the autonomous fire-truck can be efficiently controlled after applying an H-infinity feedback control scheme. Then, the optimal feedback control law is given by

$$u(t) = -Ke(t) \tag{3.48}$$

with $K = \frac{1}{r}B^TP$, while P is a positive definite symmetric matrix which is obtained from the solution of the Riccati equation of the form

$$A^TP + PA + Q - P\left(\frac{2}{r}BB^T - \frac{1}{\rho^2}LL^T\right)P = 0 \tag{3.49}$$

The transients of the control algorithm are determined by matrix Q and also by gains r and ρ. The latter gain is the H-infinity attenuation coefficient and its minimum value that allows solution of (3.49) is the one that provides maximum robustness to the control algorithm for the autonomous fire-truck. The diagram of the control scheme

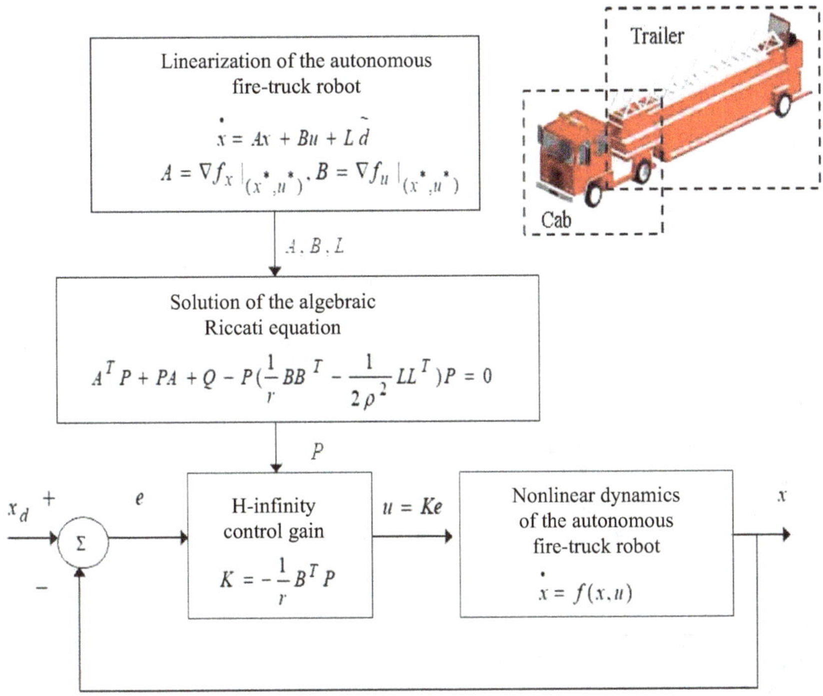

Figure 3.14 Diagram of the control scheme for the autonomous fire-truck vehicle

for the autonomous fire-truck vehicle is given in Figure 3.14. The global stability properties of the control method can be proven following the procedure presented in Section 1.1.2.5.

3.2.5 Simulation tests

The efficiency of the proposed nonlinear optimal control method for the model of the articulated autonomous fire-truck has been tested through simulation experiments. The obtained results are depicted in Figures 3.15 to 3.26.These demonstrate that fast and accurate tracking of the reference setpoints is achieved by all state vector elements of the robotic vehicle. The variations of the control inputs were moderate. For the computation of the control signal the algebraic Riccati equation of (3.49) had to be repetitively solved at each time-step of the control algorithm. The control inputs were applied on the initial nonlinear model of the vehicle and not on the equivalent linearized description of it that was obtained through the system's Jacobian matrices.

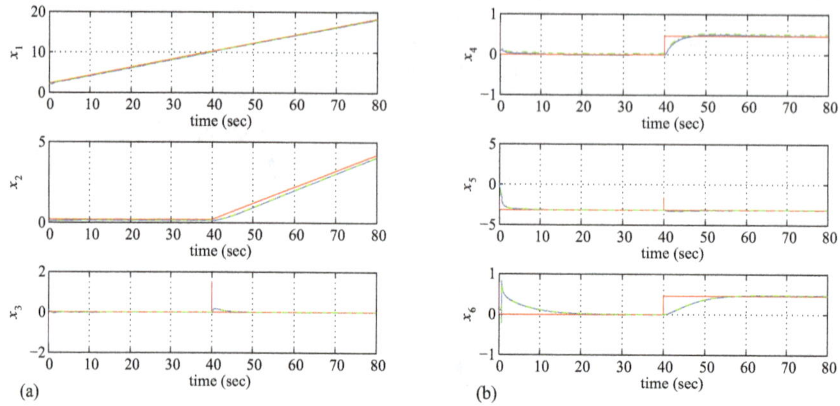

Figure 3.15 Tracking of setpoint 1 for the autonomous fire-truck robot: (a) convergence of state variables x_1 to x_3 to their reference setpoints; (b) convergence of state variables x_4 to x_6 to their reference setpoints (red line: setpoint, blue line: real value, green line: estimated value)

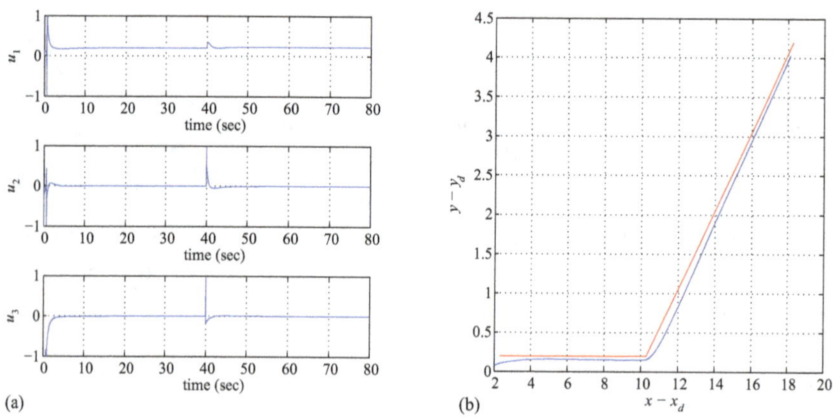

Figure 3.16 Tracking of setpoint 1 for the autonomous fire-truck robot: (a) variation of the control inputs u_1 to u_3; (b) path followed by the autonomous fire-truck robot on the xy-plane

3.3 Nonlinear control of the truck and N-trailer system

3.3.1 Outline

Autonomous navigation of multiple connected vehicles opens new directions for the development of intelligent transportation systems (Rigatos and Busawon, 2018; Felez

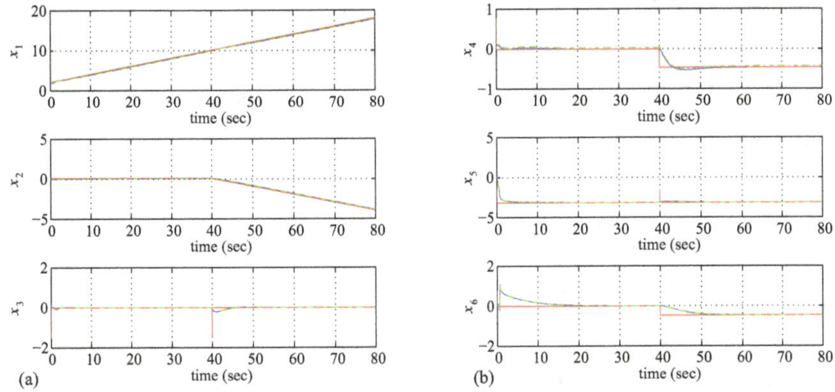

Figure 3.17 Tracking of setpoint 2 for the autonomous fire-truck robot: (a) convergence of state variables x_1 to x_3 to their reference setpoints; (b) convergence of state variables x_4 to x_6 to their reference setpoints (red line: setpoint, blue line: real value, green line: estimated value)

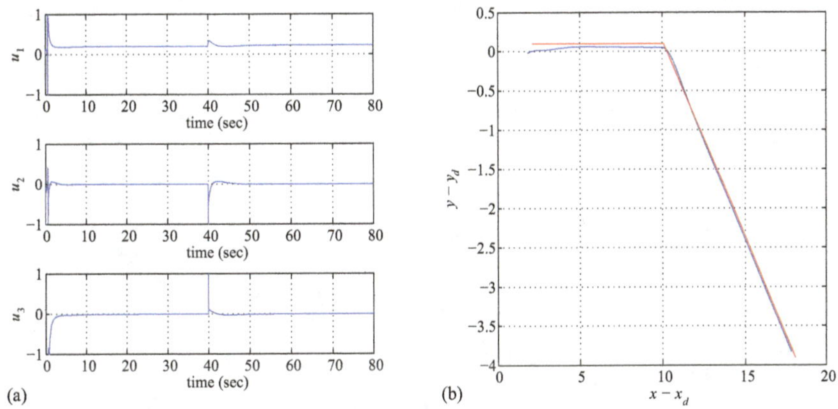

Figure 3.18 Tracking of setpoint 2 for the autonomous fire-truck robot: (a) variation of the control inputs u_1 to u_3; (b) path followed by the autonomous fire-truck robot on the xy-plane

et al., 2018; Miao and Cebon, 2012). As the volume of freight or agricultural products that has to be transported by heavy vehicles grows, and as the urban population that has to be commuted by public transportation means increases too, the use of multi-articulated buses and of vehicles of the truck and N-trailer type is expected to be deployed (Yuan, Yang *et al.*, 2018; Keymasi-Khalaji *et al.*, 2014; Chen, Peterson *et al.*, 2019). The control problem of the truck and multiple trailers robotic system

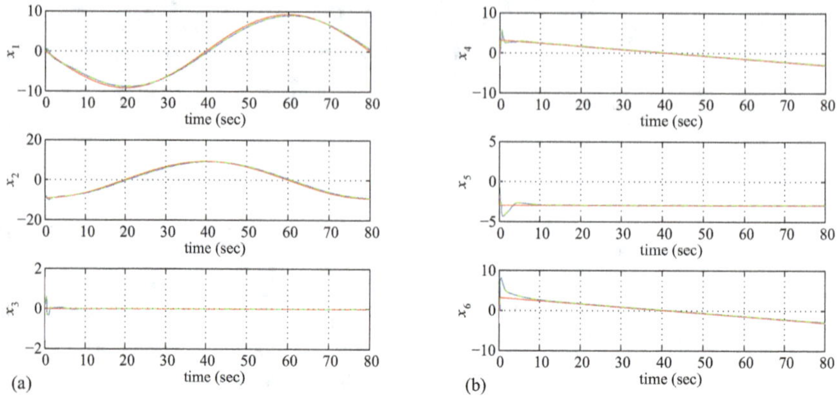

Figure 3.19 Tracking of setpoint 3 for the autonomous fire-truck robot: (a) convergence of state variables x_1 to x_3 to their reference setpoints; (b) convergence of state variables x_4 to x_6 to their reference setpoints (red line: setpoint, blue line: real value, green line: estimated value)

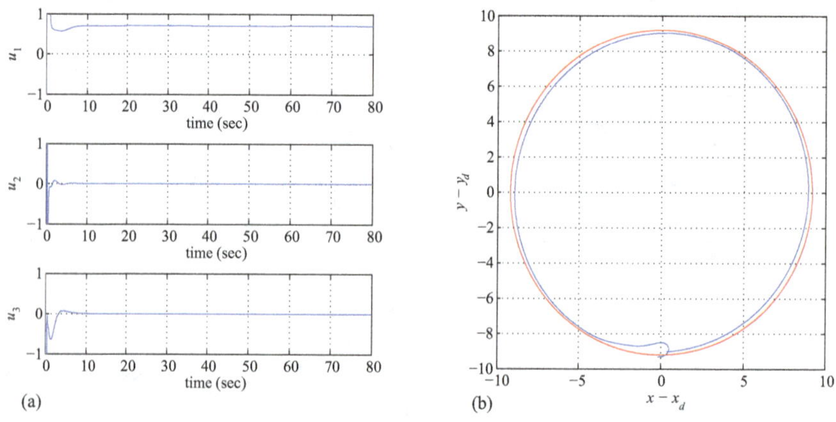

Figure 3.20 Tracking of setpoint 3 for the autonomous fire-truck robot: (a) variation of the control inputs u_1 to u_3; (b) path followed by the autonomous fire-truck robot on the xy-plane

is a nontrivial one, due to the underactuation and complex nonlinear structure of its kinematic model (Zhu and Xia, 2015; Michalek, 2019a; Michalek, 2019b). For a robotic system consisting of one truck and N trailers, there are $N+4$ degrees of freedom and only two control inputs. Actually, out of the $N+1$ connected vehicles one is considered to be a reference vehicle. Then, the degrees of freedom of the truck and N trailers robotic system are defined by the position and the orientation of the

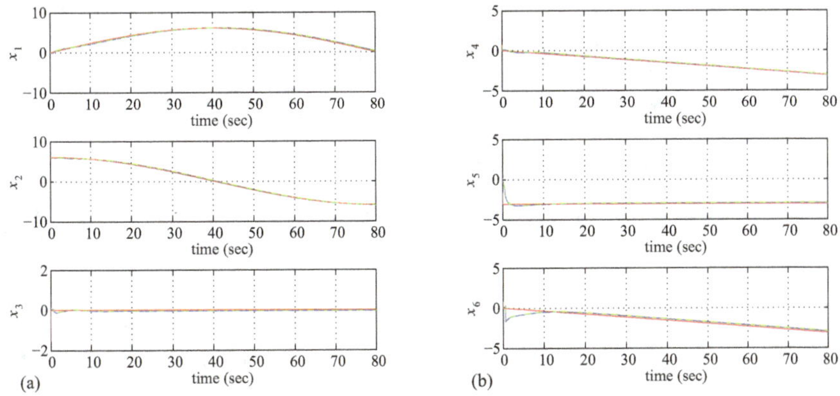

Figure 3.21 *Tracking of setpoint 4 for the autonomous fire-truck robot: (a) convergence of state variables x_1 to x_3 to their reference setpoints; (b) convergence of state variables x_4 to x_6 to their reference setpoints (red line: setpoint, blue line: real value, green line: estimated value)*

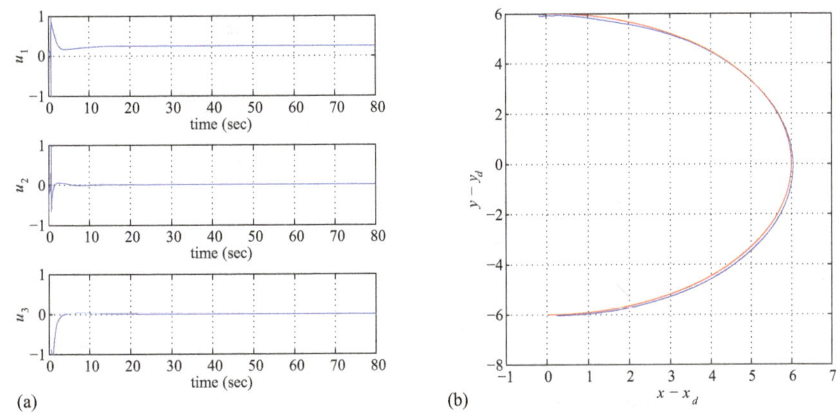

Figure 3.22 *Tracking of setpoint 4 for the autonomous fire-truck robot: (a) variation of the control inputs u_1 to u_3; (b) path followed by the autonomous fire-truck robot on the xy-plane*

reference vehicle on the *XY* plane, as well as by the steering angles of all $N + 1$ vehicles. At the same time there are only two control inputs, which are the velocity of the truck as well as the rate of change of the steering angle of the truck (Lizarraga *et al.*, 1998; Paszkowiack *et al.*, 2019; Bravo-Doddoli and Garcia-Naranjo, 2015). So far, there have been several attempts to solve the trajectory tracking problem for the truck and N-trailer robotic system (Morales, Martinez *et al.*, 2013; Guevara *et al.*, 2019; Yoo

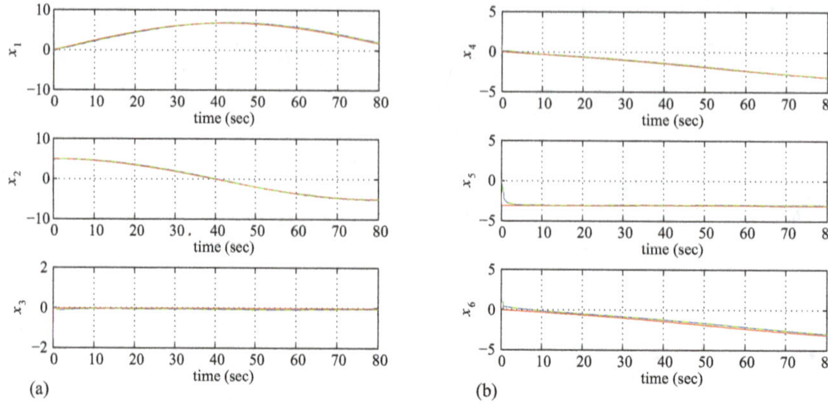

Figure 3.23 Tracking of setpoint 5 for the autonomous fire-truck robot: (a) convergence of state variables x_1 to x_3 to their reference setpoints; (b) convergence of state variables x_4 to x_6 to their reference setpoints (red line: setpoint, blue line: real value, green line: estimated value)

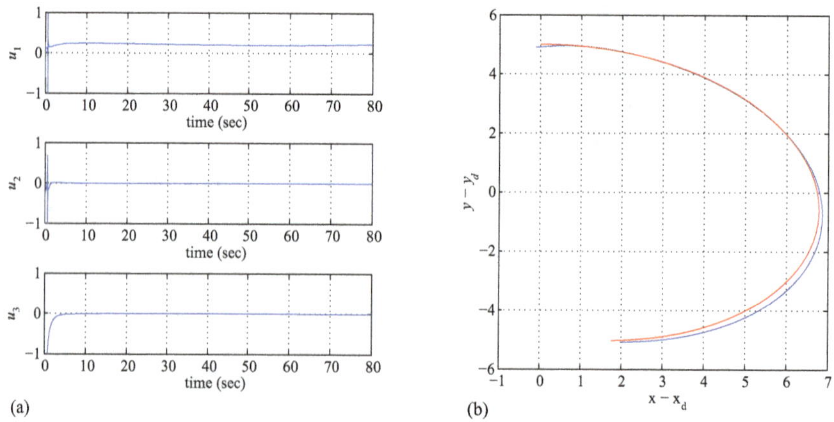

Figure 3.24 Tracking of setpoint 5 for the autonomous fire-truck robot: (a) variation of the control inputs u_1 to u_3; (b) path followed by the autonomous fire-truck robot on the xy-plane

and Chung, 2010; Pazderki *et al.*, 2012). For instance there exist several results on global linearization-based control methods (Lie algebra and differential flatness theory) which consider transformation of the state-space description of the robotic system through changes of state variables (diffeomorphisms) (Morin and Samson, 2008a; Morin and Samson, 2009; Morin and Samson, 2008b; Morin and Samson,

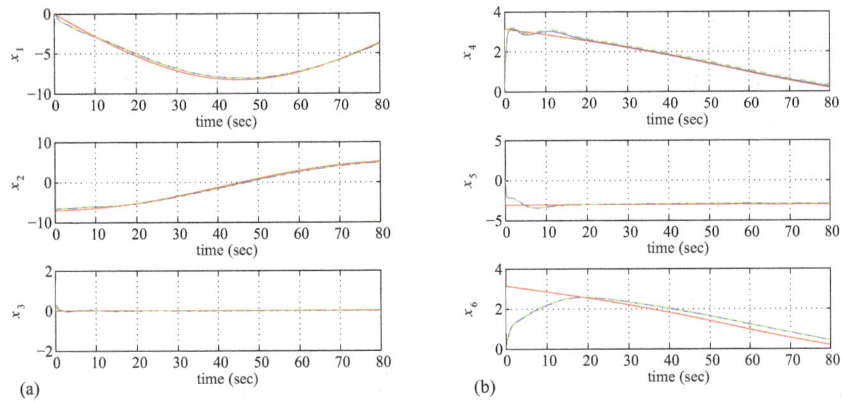

Figure 3.25 Tracking of setpoint 6 for the autonomous fire-truck robot: (a) convergence of state variables x_1 to x_3 to their reference setpoints; (b) convergence of state variables x_4 to x_6 to their reference setpoints (red line: setpoint, blue line: real value, green line: estimated value)

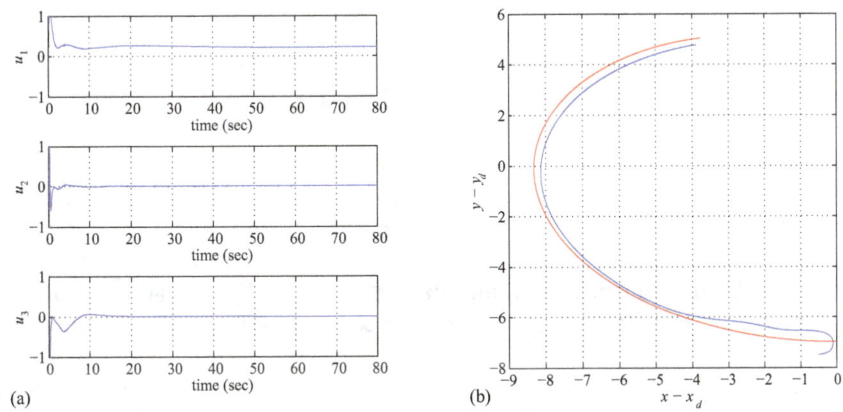

Figure 3.26 Tracking of setpoint 6 for the autonomous fire-truck robot: (a) variation of the control inputs u_1 to u_3; (b) path followed by the autonomous fire-truck robot on the xy-plane

2012). There exist also sporadic results on the use of backstepping control and sliding-mode control for this problem, which have to be further confirmed for truck and trailer formations of high dimensionality (Lashkari *et al.*, 2017; Lashkari *et al.*, 2018). Recent results analyzing challenges and potential applications for the truck and multi-trailer system can be found in Michalek, 2020, Michalek *et al.*, 2021, Ljungquist and Aschill, 2020, Guevara *et al.*, 2021, and Guevara *et al.*, 2020. It can be noted that there exist few developments on nonlinear optimal control approaches for trajectory tracking by this robotic system.

In this section, a novel solution to the nonlinear optimal control problem of the truck and N-trailer robotic system is developed (Rigatos and Busawon, 2018). To this end, the kinematic model of the robotic system undergoes approximate linearization around a temporary operating point which is recomputed at each iteration of the control algorithm. This operating point is defined at each sampling instance by the present value of the robotic system's state vector and by the last sampled value of the control inputs vector. The linearization process relies on the first-order Taylor series expansion and on the computation of the related Jacobian matrices (Rigatos and Tzafestas, 2007; Basseville and Nikiforov, 1993; Rigatos and Zhang, 2009). The truncation of higher-order terms from this Taylor series expansion is considered to be a perturbation that is asymptotically compensated by the robustness of the control algorithm. For the approximately linearized kinematic model of the truck and N-trailer system, an optimal (H-infinity) feedback controller is designed. The H-infinity controller stands for the solution of the nonlinear optimal control problem for the truck and N-trailer robotic system under model uncertainty and external perturbations (Rigatos, Siano and Cecati, 2015; Rigatos, Siano *et al.*, 2015; Rigatos, 2016b).

Actually, the H-infinity controller represents a min–max differential game taking place between (i) the control inputs which try to minimize a cost function which comprises a quadratic term of the state vector's tracking error and (ii) the model uncertainty terms and the exogenous perturbations which try to maximize this cost function (Rigatos, 2011; Rigatos, 2015). To define the feedback gains of the optimal (H-infinity) controller an algebraic Riccati equation is repetitively solved at each time-step of the control algorithm (Rigatos and Busawon, 2018). The stability properties of the control scheme are proven through Lyapunov analysis. It is shown that the control loop satisfies the H-infinity tracking performance criterion (Rigatos and Busawon, 2018; Tousssaint, Basar and Bullo, 2000; Lublin and Athans, 1995). This signifies elevated robustness against model uncertainty and external perturbations. Moreover, under moderate conditions, it is proven that the control loop is globally asymptotically stable. The proposed control method retains the advantages of typical (linear) optimal control, that is fast and accurate tracking of the reference setpoints under moderate variations of the control inputs. Finally, to implement state estimation-based feedback control through the measurement of a subset of the state variables of the truck and N-trailer robotic system, the H-infinity Kalman Filter is used as a robust observer (Rigatos, 2015; Gibbs, 2012).

3.3.2 Kinematic model of the truck and N trailer robotic system

The diagram of the truck and N-trailers robotic system in the OXY plane, as well as the related orientation angles, is given in Figure 3.27. The vehicles are numbered in a descending order, that is the trailer that is more distant from the truck gets No 1 while the trailer that is closest to the truck gets No. N. The tractor is numbered as vehicle $N+1$.

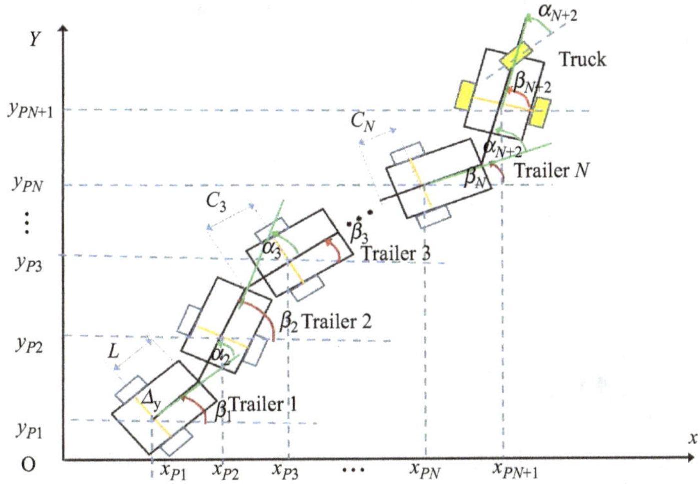

Figure 3.27 Diagram of the truck and N trailer robotic system in the XY plane and definition of the related orientation angles

The steering angle for the ith vehicle is denoted as a_{i+1} where $i = 1, 2, \cdots, N + 1$. Consequently, the steering angle of the truck (or equivalently the turn angle of the steering wheel) is denoted as a_{N+2}. The length of the ith vehicle along its longitudinal axis is L_i. The half of the length of the rear wheel axle of each vehicle is denoted as dy. The length of the drawbar of the ih vehicle is denoted as C_i (Lizarraga *et al.*, 1998; Morin and Samson, 2012).

A Frénet frame is considered next. This is defined by a curve C with curvature κ. One can use the tangential line at a point P_c of the curve to define a coordinates frame. The angle which is formed between the longitudinal axis of the i-th vehicle and the tangential line at point P_c is denoted as β_i. When this curve is a straight line, such as the OX axis of the inertial reference frame OXY, then the curvature is $\kappa = 0$. In the latter case, angles β_i are as shown in Figure 3.27 (Lizarraga *et al.*, 1998; Morin and Samson, 2012).

The motion of the i-th vehicle $i = 1, 2, \cdots, N, N + 1$ is defined by the state vector $x_1 = x_{P_i}$, $x_2 = y_{P_i}$, $x_3 = \beta_i$ and $x_4 = a_{i+1}$, where (x_{P_i}, y_{P_i}) are the Cartesian coordinates of the center of the rear axle of the last vehicle in the multi-vehicle chain (that is of trailer 1 which is the most distant trailer with respect to the truck and the front part of the vehicles' chain), β_i is the orientation of the last vehicle (trailer 1) in the inertial coordinates frame and a_{i+1} is its steering angle with reference to the preceding vehicle $i + 1$, including also the steering angle of the truck. Then, the kinematics of the truck and N-trailers system is given by the following driftless state-space equation (Lizarraga *et al.*, 1998; Morin and Samson, 2012)

$$\dot{x} = f(x)u_1 + g(x)u_2 \tag{3.50}$$

where $x \in R^{N+4}$, $f(x) \in R^{N+4}$, $g(x) \in R^{N+4}$ and $u \in R$. Vectors $f(x)$ and $g(x)$ are given by

$$f(x) = \begin{pmatrix} \frac{w(x,x_3)}{1-\kappa(x_1)x_2} \\ W(x, x_3 - \frac{\pi}{2}) \\ \frac{G_1(x)}{L_1} - \frac{\kappa(x_1)W(x,x_3)}{1-\kappa(x_1)x_2} \\ \frac{G_2(x)}{L_2} - \frac{G_1(x)}{L_1} \\ \frac{G_3(x)}{L_2} - \frac{G_2(x)}{L_2} \\ \vdots \\ \frac{G_{N+1}(x)}{L_{N+1}} - \frac{G_N(x)}{L_N} \\ 0 \end{pmatrix} \quad g(x) = \begin{pmatrix} 0 \\ 0 \\ 0 \\ 0 \\ 0 \\ \vdots \\ 1 \end{pmatrix} \tag{3.51}$$

In (3.50) and (3.51), the first three rows of the state-space model describe the motion of the trailer No. 1, that is the trailer which is more distant from the truck. The rest of the rows of the state-space model describe the change of the steering angles α_i of the individual vehicles, from trailer No. 1 up to the truck. The state vector x of the kinematic model is $x = [x_{P_1}, y_{P_1}, \beta_1, a_2, a_3, \cdots, a_{N+2}]^T$. Therefore, in case of a truck with N trailers, the state vector comprises $N+4$ state variables, that is $x = [x_1, x_2, x_3, x_4, x_5, \cdots, x_{N+4}]$.

In this kinematic model, the following functions and variables are defined (Lizarraga *et al.*, 1998; Morin and Samson, 2012):

$$W(x, x_3) = \cos(x_3)F_1(x) - [\sin(x_3)dx + \cos(x_3)dy]\frac{G_1(x)}{L_1} \tag{3.52}$$

Variables dx and dy describe the coordinates of the trailers' wheels in a Frénet reference frame placed at the center of rear axle of each vehicle. It holds that $dx = 0$. Besides functions $F_1(x)$ and $G_1(x)$ are defined in a recursive manner, using the relations (Lizarraga *et al.*, 1998; Morin and Samson, 2012)

$$\begin{aligned} F_1(x) &= \cos(x_4)F_2(x) + \frac{C_2}{L_2}\sin(x_4)G_2(x) \\ G_1(x) &= \sin(x_4)F_2(x) - \frac{C_2}{L_2}\cos(x_4)G_2(x) \end{aligned} \tag{3.53}$$

$$\begin{aligned} F_2(x) &= \cos(x_5)F_3(x) + \frac{C_3}{L_3}\sin(x_5)G_3(x) \\ G_2(x) &= \sin(x_5)F_3(x) - \frac{C_3}{L_3}\cos(x_5)G_3(x) \end{aligned} \tag{3.54}$$

$$\cdots, \cdots$$

$$\begin{aligned} F_N(x) &= \cos(x_{N+3})F_{N+1}(x) + \frac{C_{N+1}}{L_{N+1}}\sin(x_{N+3})G_{N+1}(x) \\ G_N(x) &= \sin(x_{N+3})F_{N+1}(x) - \frac{C_{N+1}}{L_{N+1}}\cos(x_{N+3})G_{N+1}(x) \end{aligned} \tag{3.55}$$

$$\begin{aligned} F_{N+1}(x) &= \cos(x_{N+4}) \\ G_{N+1}(x) &= \sin(x_{N+4}) \end{aligned} \tag{3.56}$$

Next, without loss of generality one considers the case of a mobile robotic system that comprises one truck and two trailers, that is $N = 2$. The state vector of the robotic system becomes $x = [x_1, x_2, x_3, x_4, x_5, x_6]^T$ or $x = [x_{P_1}, y_{P_1}, \beta_1, a_2, a_3, a_4]^T$ while it also holds:

For the truck

$$\begin{aligned} F_3(x) &= \cos(x_6) \\ G_3(x) &= \sin(x_6) \end{aligned} \tag{3.57}$$

For trailer No. 1

$$\begin{aligned} F_2(x) &= \cos(x_5)F_3(x) + \frac{C_3}{L_3}\sin(x_5)G_3(x) \\ G_2(x) &= \sin(x_5)F_3(x) - \frac{C_3}{L_3}\sin(x_5)G_3(x) \end{aligned} \tag{3.58}$$

or in detailed form

$$\begin{aligned} F_2(x) &= \cos(x_5)\cos(x_6) + \frac{C_3}{L_3}\sin(x_5)\sin(x_6) \\ G_2(x) &= \sin(x_5)\cos(x_6) - \frac{C_3}{L_3}\sin(x_5)\sin(x_6) \end{aligned} \tag{3.59}$$

For trailer No. 2

$$\begin{aligned} F_1(x) &= \cos(x_4)F_2(x) + \frac{C_2}{L_2}\sin(x_4)G_2(x) \\ G_1(x) &= \sin(x_4)F_2(x) - \frac{C_2}{L_2}\sin(x_4)G_2(x) \end{aligned} \tag{3.60}$$

or in detailed form

$$\begin{aligned} F_1(x) = {} & \cos(x_4)[\cos(x_5)\cos(x_6) + \frac{C_3}{L_3}\sin(x_5)\sin(x_6)] \\ & + \frac{C_2}{L_2}\sin(x_4)[\sin(x_5)\cos(x_6) - \frac{C_3}{L_3}\sin(x_5)\sin(x_6)] \\ G_1(x) = {} & \sin(x_4)[\cos(x_5)\cos(x_6) + \frac{C_3}{L_3}\sin(x_5)\sin(x_6)] \\ & - \frac{C_2}{L_2}\sin(x_4)[\sin(x_5)\cos(x_6) - \frac{C_3}{L_3}\sin(x_5)\sin(x_6)] \end{aligned} \tag{3.61}$$

Besides, one has that

$$\begin{aligned} W(x, x_3) &= \cos(x_3)F_1(x) - \cos(x_3)dy\frac{G_1(x)}{L_1} \\ W(x, x_3 - \frac{\pi}{2}) &= \sin(x_3)F_1(x) - \sin(x_3)dy\frac{G_1(x)}{L_1} \end{aligned} \tag{3.62}$$

Thus, the kinematic model of the truck and 2-trailer robotic system becomes

$$\begin{pmatrix} \dot{x}_1 \\ \dot{x}_2 \\ \dot{x}_3 \\ \dot{x}_4 \\ \dot{x}_5 \\ \dot{x}_6 \end{pmatrix} = \begin{pmatrix} \cos(x_3)F_1(x) - \cos(x_3)dy\dfrac{G_1(x)}{L_1} \\ \sin(x_3)F_1(x) - \sin(x_3)dy\dfrac{G_1(x)}{L_1} \\ \dfrac{G_1(x)}{L_1} \\ \dfrac{G_2(x)}{L_3} - \dfrac{G_1(x)}{L_2} \\ \dfrac{G_3(x)}{L_3} - \dfrac{G_2(x)}{L_2} \\ 0 \end{pmatrix} u_1 + \begin{pmatrix} 0 \\ 0 \\ 0 \\ 0 \\ 0 \\ 1 \end{pmatrix} u_2 \tag{3.63}$$

where u_1 is the velocity v of the truck and u_2 is the rate of change of the steering angle $\dot{a}_4$ of the truck.

3.3.3 Approximate linearization of the truck and N-trailer robotic system

3.3.3.1 Approximately linearized state-space model

The kinematic model of the truck and 2-trailer robotic system undergoes approximate linearization at each time-step of the control algorithm, around the temporary operating point (x^*, u^*), where x^* is the present value of the system's state vector and u^* is the last sampled value of the control inputs vector. The linearization relies on the first-order Taylor series expansion and on the computation of the associated Jacobian matrices. The linearized description is (Rigatos and Busawon, 2018)

$$\dot{x} = Ax + Bu + \tilde{d} \tag{3.64}$$

where $\tilde{d}$ is the cumulative disturbances' vector which comprises (i) modeling errors due to the truncation of higher-order terms from the Taylor-series expansion, (ii) exogenous perturbations, and (iii) measurement noises of any distribution. It holds that:

$$A = \nabla_x[f(x)u_1 + g(x)u_2] \mid_{(x^*,u^*)} \Rightarrow A = \nabla_x f(x)u_1 \mid_{(x^*,u^*)} \tag{3.65}$$

$$\begin{aligned} B_1 &= \nabla_{u_1}[f(x)u_1] \mid_{(x^*,u^*)} \Rightarrow B_1 = f(x) \mid_{(x^*,u^*)} \\ B_2 &= \nabla_{u_2}[g(x)u_2] \mid_{(x^*,u^*)} \Rightarrow B_2 = g(x) \mid_{(x^*,u^*)} \end{aligned} \tag{3.66}$$

Consequently, for the approximately linearized system of (3.64), the input gains matrix B is

$$B = [B_1 \; B_2] \Rightarrow B = [f(x) \; g(x)] \mid_{(x^*,u^*)} \tag{3.67}$$

The computation of the elements of the Jacobian matrix $\nabla_x f(x)u_1 \mid_{(x^*,u^*)}$ proceeds as follows:

First row of the Jacobian matrix $\nabla_x f(x)u_1 \mid_{(x^*,u^*)}$: $\frac{\partial f_1}{\partial x_i} = \cos(x_3)\frac{\partial F_1}{\partial x_i} - \cos(x_3)dy\frac{\partial G_1/\partial x_i}{L_1}$, for $i = 1, 2, 4, 5, 6$. Additionally, for $i = 3$ it holds that

$$\frac{\partial f_1}{\partial x_3} = -\sin(x_1)F_1(x) + \cos(x_3)\frac{\partial F_1}{\partial x_3} + \sin(x_3)dy\frac{G_1(x)}{L_1} - \cos(x_3)dy\frac{\partial G_1/\partial x_3}{L_1} \tag{3.68}$$

Second row of the Jacobian matrix $\nabla_x f(x)u_1 \mid_{(x^*,u^*)}$: $\frac{\partial f_2}{\partial x_i} = \sin(x_3)\frac{\partial F_1}{\partial x_i} - \sin(x_3)dy\frac{\partial G_1/\partial x_i}{L_1}$, for $i = 1, 2, 4, 5, 6$. Additionally, for $i = 3$ it holds that

$$\frac{\partial f_2}{\partial x_3} = \cos(x_1)F_1(x) + \sin(x_3)\frac{\partial F_1}{\partial x_3} - \cos(x_3)dy\frac{G_1(x)}{L_1} - \sin(x_3)dy\frac{\partial G_1/\partial x_3}{L_1} \tag{3.69}$$

Third row of the Jacobian matrix $\nabla_x f(x)u_1 \mid_{(x^*,u^*)}$: $\frac{\partial f_3}{\partial x_i} = \frac{\partial G_1/\partial x_i}{L_1}$ for $i = 1, 2, 3, 4, 5, 6$.

Fourth row of the Jacobian matrix $\nabla_x f(x)u_1 \mid_{(x^*,u^*)}$: $\frac{\partial f_4}{\partial x_i} = \frac{\partial G_2/\partial x_i}{L_2} - \frac{\partial G_1/\partial x_i}{L_1}$ for $i = 1, 2, 3, 4, 5, 6$.

Fifth row of the Jacobian matrix $\nabla_x f(x)u_1 \mid_{(x^*,u^*)}$: $\frac{\partial f_5}{\partial x_i} = \frac{\partial G_3/\partial x_i}{L_3} - \frac{\partial G_2/\partial x_i}{L_2}$ for $i = 1, 2, 3, 4, 5, 6$.

Sixth row of the Jacobian matrix $\nabla_x f(x)u_1 \mid_{(x^*,u^*)}$: $\frac{\partial f_6}{\partial x_i} = 0$ for $i = 1, 2, 3, 4, 5, 6$.

Next, the partial derivatives of functions $F_j(x), j = 1, 2, 3$ and $G_j(x), j = 1, 2, 3$ with respect to x_i, $i = 1, 2, 3, 4, 5, 6$ are computed. It holds that

$$\begin{aligned} &F_3(x) = \cos(x_6) && G_3(x) = \sin(x_6) \\ &\frac{\partial F_3}{\partial x_i} = 0,\ i = 1, 2, 3, 4, 5 && \frac{\partial F_3}{\partial x_6} = -\sin(x_6) \\ &\frac{\partial G_3}{\partial x_i} = 0,\ i = 1, 2, 3, 4, 5 && \frac{\partial F_3}{\partial x_6} = \cos(x_6) \end{aligned} \tag{3.70}$$

Equivalently, it holds that $F_2(x) = \cos(x_5)\cos(x_6) + \frac{C_3}{L_3}\sin(x_5)\sin(x_6)$, thus

$$\begin{aligned} &\frac{\partial F_2}{\partial x_i} = \cos(x_5)\frac{\partial F_3}{\partial x_i} + \frac{C_3}{L_3}\sin(x_5)\frac{\partial G_3}{\partial x_i} \\ &\Rightarrow \frac{\partial F_2}{\partial x_i} = 0,\ i = 1, 2, 3, 4 \end{aligned} \tag{3.71}$$

$$\begin{aligned} &\frac{\partial F_2}{\partial x_5} = -\sin(x_5)F_3 + \cos(x_5)\frac{\partial F_3}{\partial x_5} + \frac{C_3}{L_3}\cos(x_5)G_3 + \frac{C_3}{L_3}\sin(x_5)\frac{\partial G_3}{\partial x_5} \\ &\Rightarrow \frac{\partial F_2}{\partial x_5} = -\sin(x_5)F_3 + \frac{C_3}{L_3}\cos(x_5)G_3 \end{aligned} \tag{3.72}$$

$$\frac{\partial F_2}{\partial x_6} = \cos(x_5)\frac{\partial F_3}{\partial x_6} + \frac{C_3}{L_3}\sin(x_5)\frac{\partial G_3}{\partial x_6} \tag{3.73}$$

Additionally $G_2(x) = \sin(x_5)\cos(x_6) - \frac{C_3}{L_3}\sin(x_5)\sin(x_6)$, therefore

$$\begin{aligned}\frac{\partial G_2}{\partial x_i} &= \sin(x_5)\frac{\partial F_3}{\partial x_i} - \frac{C_3}{L_3}\cos(x_5)\frac{\partial G_3}{\partial x_i}\\ \Rightarrow \frac{\partial G_2}{\partial x_i} &= 0,\ i = 1,2,3,4\end{aligned} \tag{3.74}$$

$$\begin{aligned}\frac{\partial G_2}{\partial x_5} &= \cos(x_5)F_3 + \sin(x_5)\frac{\partial F_3}{\partial x_5} + \frac{C_3}{L_3}\sin(x_5)G_3 - \frac{C_3}{L_3}\cos(x_5)\frac{\partial G_3}{\partial x_5}\\ \Rightarrow \frac{\partial F_2}{\partial x_5} &= \cos(x_5)F_3 + \frac{C_3}{L_3}\sin(x_5)G_3\end{aligned} \tag{3.75}$$

$$\frac{\partial G_2}{\partial x_6} = \sin(x_5)\frac{\partial F_3}{\partial x_6} - \frac{C_3}{L_3}\cos(x_5)\frac{\partial G_3}{\partial x_6} \tag{3.76}$$

In a similar manner, one has $F_1(x) = \cos(x_4)F_2(x) + \frac{C_2}{L_2}\sin(x_4)G_2(x)$, therefore

$$\begin{aligned}\frac{\partial F_1}{\partial x_i} &= \cos(x_4)\frac{\partial F_2}{\partial x_i} + \frac{C_2}{L_2}\sin(x_4)\frac{\partial G_2}{\partial x_i}\\ \Rightarrow \frac{\partial F_1}{\partial x_i} &= 0,\ i = 1,2,3\end{aligned} \tag{3.77}$$

$$\begin{aligned}\frac{\partial F_1}{\partial x_4} &= -\sin(x_4)F_2 + \cos(x_4)\frac{\partial F_2}{\partial x_4} + \frac{C_2}{L_2}\cos(x_4)G_2 + \frac{C_2}{L_2}\sin(x_4)\frac{\partial G_2}{\partial x_4}\\ \Rightarrow \frac{\partial F_1}{\partial x_4} &= -\sin(x_4)F_2 + \frac{C_2}{L_2}\cos(x_4)G_2\end{aligned} \tag{3.78}$$

$$\frac{\partial F_1}{\partial x_5} = \cos(x_4)\frac{\partial F_2}{\partial x_5} + \frac{C_2}{L_2}\sin(x_4)\frac{\partial G_2}{\partial x_5} \tag{3.79}$$

$$\frac{\partial F_1}{\partial x_6} = \cos(x_4)\frac{\partial F_2}{\partial x_6} + \frac{C_2}{L_2}\sin(x_4)\frac{\partial G_2}{\partial x_6} \tag{3.80}$$

Moreover, one has $G_1(x) = \sin(x_4)F_2(x) - \frac{C_2}{L_2}\sin(x_4)G_2(x)$, therefore

$$\begin{aligned}\frac{\partial G_1}{\partial x_i} &= \sin(x_4)\frac{\partial F_2}{\partial x_i} - \frac{C_2}{L_2}\cos(x_4)\frac{\partial G_2}{\partial x_i}\\ \Rightarrow \frac{\partial G_1}{\partial x_i} &= 0,\ i = 1,2,3\end{aligned} \tag{3.81}$$

$$\begin{aligned}\frac{\partial G_1}{\partial x_4} &= \cos(x_4)F_2 + \sin(x_4)\frac{\partial F_2}{\partial x_4} + \frac{C_2}{L_2}\sin(x_4)G_2 - \frac{C_2}{L_2}\cos(x_4)\frac{\partial G_2}{\partial x_4}\\ \Rightarrow \frac{\partial G_1}{\partial x_4} &= \cos(x_4)F_2 + \frac{C_2}{L_2}\sin(x_4)G_2\end{aligned} \tag{3.82}$$

$$\frac{\partial G_1}{\partial x_5} = \sin(x_4)\frac{\partial F_2}{\partial x_5} - \frac{C_2}{L_2}\cos(x_4)\frac{\partial G_2}{\partial x_5} \tag{3.83}$$

$$\frac{\partial G_1}{\partial x_6} = \sin(x_4)\frac{\partial F_2}{\partial x_6} - \frac{C_2}{L_2}\cos(x_4)\frac{\partial G_2}{\partial x_6} \tag{3.84}$$

3.3.3.2 Stabilizing feedback control

After linearization around its current operating point, the kinematic model of the truck and N-trailer robotic system is written as

$$\dot{x} = Ax + Bu + d_1 \tag{3.85}$$

Parameter d_1 stands for the linearization error in the kinematic model of the truck and N-trailer robotic system appearing previously in (3.64). The reference setpoints for the state vector of the truck and N-trailer robotic system are denoted by $\mathbf{x_d} = [x_1^d, \cdots, x_6^d]$. Tracking of this trajectory is achieved after applying the control input u_d. At every time instant the control input u_d is assumed to differ from the control input u appearing in (3.85) by an amount equal to Δu, that is $u_d = u + \Delta u$ Rigatos and Busawon, 2018

$$\dot{x}_d = Ax_d + Bu_d + d_2 \tag{3.86}$$

The kinematics of the controlled system described in (3.85) can be also written as

$$\dot{x} = Ax + Bu + Bu_d - Bu_d + d_1 \tag{3.87}$$

and by denoting $d_3 = -Bu_d + d_1$ as an aggregate disturbance term one obtains

$$\dot{x} = Ax + Bu + Bu_d + d_3 \tag{3.88}$$

By subtracting (3.86) from (3.88), one has

$$\dot{x} - \dot{x}_d = A(x - x_d) + Bu + d_3 - d_2 \tag{3.89}$$

By denoting the tracking error as $e = x - x_d$ and the aggregate disturbance term as $\tilde{d} = d_3 - d_2$, the tracking error dynamics becomes

$$\dot{e} = Ae + Bu + \tilde{d} \tag{3.90}$$

For the approximately linearized model of the system a stabilizing feedback controller is developed. The controller, follows th optimal (H-infinity) control concept, and has the form

$$u(t) = -Ke(t) \tag{3.91}$$

with $K = \frac{1}{r}B^T P$ where P is a positive definite symmetric matrix which is obtained from the solution of the Riccati equation (Rigatos and Busawon, 2018)

$$A^T P + PA + Q - P\left(\frac{2}{r}BB^T - \frac{1}{\rho^2}LL^T\right)P = 0 \tag{3.92}$$

where Q is a positive semi-definite symmetric matrix. The diagram of the considered control loop is depicted in Figure 3.28. The global stability properties of the control method can be proven following the procedure presented in Section 1.1.2.5.

With reference to (3.64), the solution of the H-infinity feedback control problem for the truck and N-trailer system dynamics and the computation of the worst case disturbance that the related controller can sustain, comes from superposition of Bellman's optimality principle when considering that the truck and N-trailer system is affected by two separate inputs (i) the control input u; (ii) the cumulative disturbance input $\tilde{d}(t)$. Solving the optimal control problem for u, that is for the minimum variation

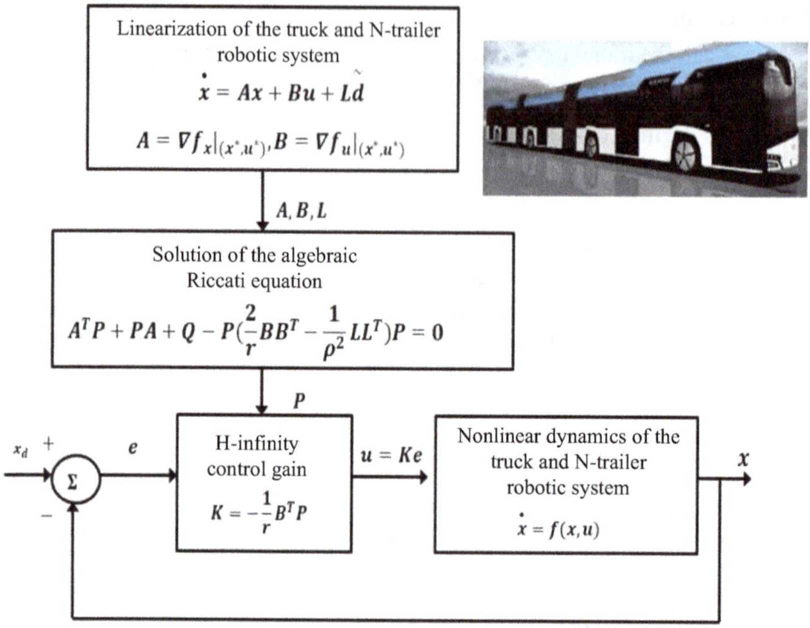

Figure 3.28 Diagram of the control scheme for the truck and N-trailer robotic system

(optimal) control input that achieves elimination of the state vector's tracking error, gives $u = -\frac{1}{r}B^TPe$. Equivalently, solving the optimal control problem for $\tilde{d}$, that is for the worst case disturbance that the control loop can sustain gives $\tilde{d} = \frac{1}{\rho^2}L^TPe$.

3.3.4 Simulation tests

The performance of the nonlinear optimal (H-infinity) control scheme for the model of the truck and N-trailer robotic system has been tested through simulation experiments. The sensors usually needed for the positioning and orientation estimation of the truck and N-trailer system were: (i) accelerometers or a GPS sensor that allow for finding at each sampling instance the position of the center of the rear axle of the truck and of the individual trailers; (ii) gyroscopic sensors that allow for finding at each sampling instance the steering angle of the individual vehicles that constitute this multi-vehicle system. It is noted though that by implementing a state estimation-based control scheme for the truck and N-trailer system it can be avoided to measure some of the aforementioned variables (the control method can use their estimated values instead).

A robotic system comprising a truck and two trailers has been considered. The following parameters of the kinematic model of the truck and N-trailer system have been used in the simulation experiments: (a) for trailer 1 (placed last in the multi-vehicle chain): the length of trailer 1 along its longitudinal axis was $L = 1.2$ m.

The half of the length of the rear wheel axle of trailer 1 was $dy = 0.5$ m. The length of the drawbar connecting trailer 1 with the center of the real axle of trailer 2 that was moving in front of it was $C = 0.3$ m. (b) For trailer 2 (placed second in the multi-vehicle chain): the length of trailer 2 along its longitudinal axis was $L = 1.2$ m. The half of the length of the rear wheel axle of trailer 2 was $dy = 0.5$ m. The length of the drawbar connecting trailer 2 with the center of the real axle of the truck that was moving in front of it was $C = 0.3$ m. (c) For the truck: the length along the truck's longitudinal axis was $L = 1.2$ m. The half of the length of the rear wheel axle of truck was $dy = 0.5$ m.

To select the controller's gains the algebraic Riccati equation of (3.92) was solved at each iteration of the control algorithm. The control scheme made use of the estimated state vector that was given by the H-infinity Kalman filter. The obtained results are depicted in Figures 3.29 to 3.38. Position variables are measured in meters (m) and angle variables are measured in radians (rad). The real values of the state variables are printed in blue, the estimated values are plotted in green while the related setpoints are depicted in red. The transient performance of the control scheme was determined by the selection of parameters r, ρ, and Q. Actually, the smallest value of ρ for which one can obtain a solution of the above noted Riccati equation is the one that provides the control scheme with maximum robustness. From the simulation experiments it can be noticed that the nonlinear optimal (H-infinity) control method achieved fast and accurate tracking of the reference setpoints under moderate variations of the control inputs.

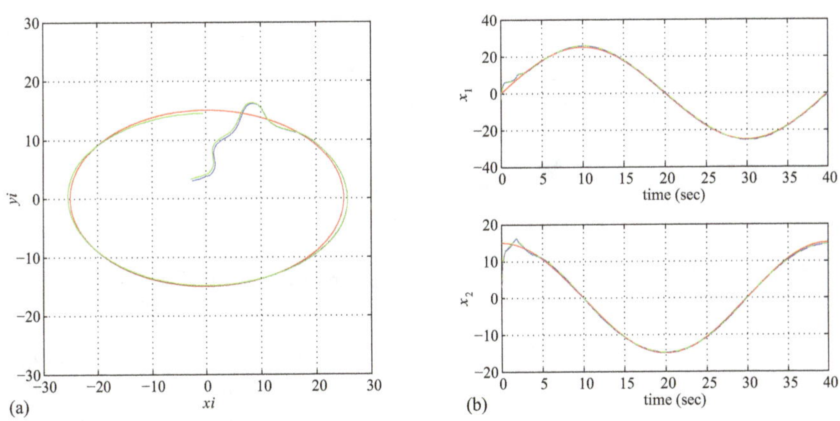

Figure 3.29 Tracking performance of the truck and N trailer robotic system in case of setpoint 1: (a) tracking of reference trajectory (red line) on the xy plane by the center of the rear axle of trailer No. 1 (blue line); (b) convergence of state variables x_1 and x_2 (blue lines) to reference setpoints (red lines) and estimated values provided by the KF (green lines)

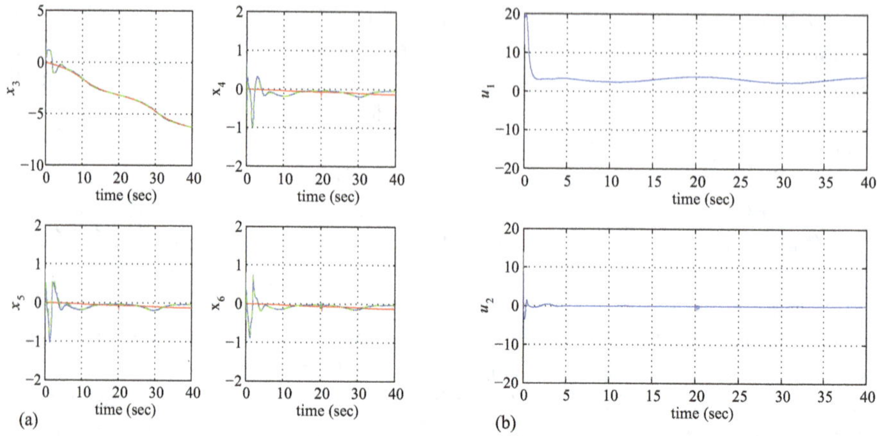

Figure 3.30 *Tracking performance of the truck and N trailer robotic system in case of setpoint 1: (a) convergence of state variables x_3 to x_6 (blue lines) to reference setpoints (red lines) and estimated values provided by the KF (green lines); (b) control inputs u_1 and u_2*

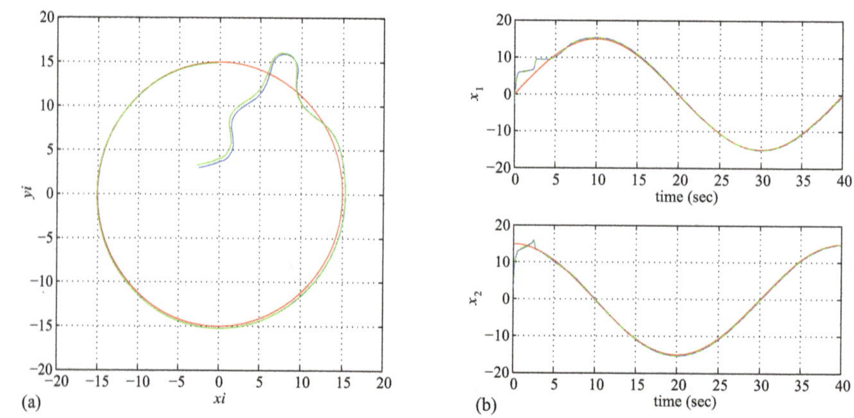

Figure 3.31 *Tracking performance of the truck and N trailer robotic system in case of setpoint 2: (a) tracking of reference trajectory (red line) on the xy plane by the center of the rear axle of trailer No. 1 (blue line); (b) convergence of state variables x_1 and x_2 (blue lines) to reference setpoints (red lines) and estimated values provided by the KF (green lines)*

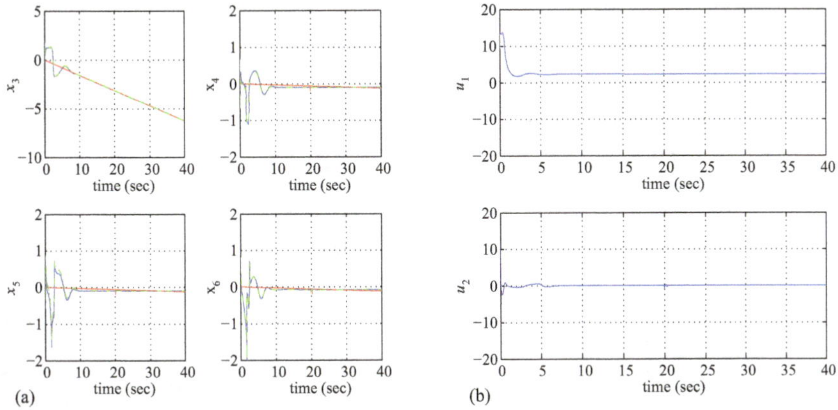

Figure 3.32 Tracking performance of the truck and N trailer robotic system in case of setpoint 2: (a) convergence of state variables x_3 to x_6 (blue lines) to reference setpoints (red lines) and estimated values provided by the KF (green lines); (b) control inputs u_1 and u_2

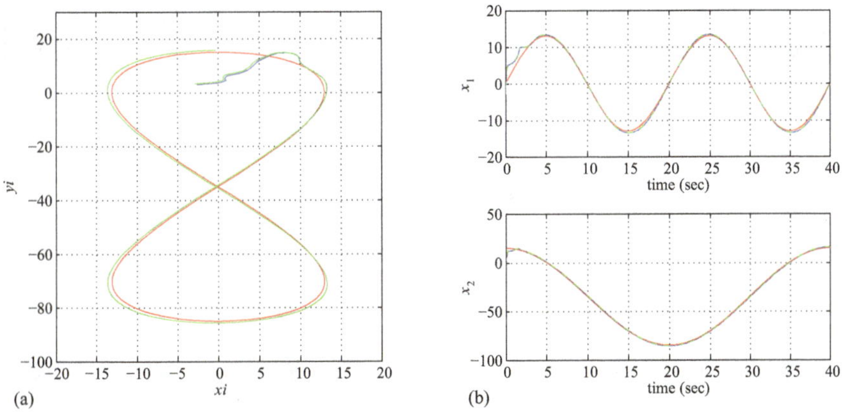

Figure 3.33 Tracking performance of the truck and N trailer robotic system in case of setpoint 3: (a) tracking of reference trajectory (red line) on the xy plane by the center of the rear axle of trailer No 1 (blue line); (b) convergence of state variables x_1 and x_2 (blue lines) to reference setpoints (red lines) and estimated values provided by the KF (green lines)

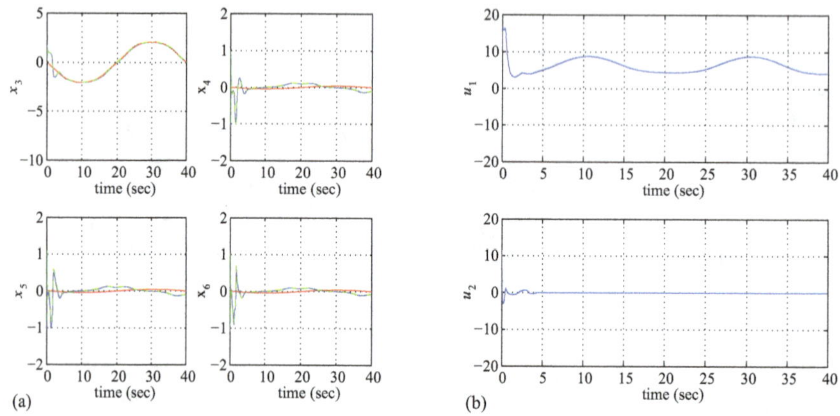

Figure 3.34 *Tracking performance of the truck and N trailer robotic system in case of setpoint 3: (a) convergence of state variables x_3 to x_6 (blue lines) to reference setpoints (red lines) and estimated values provided by the KF (green lines); (b) control inputs u_1 and u_2*

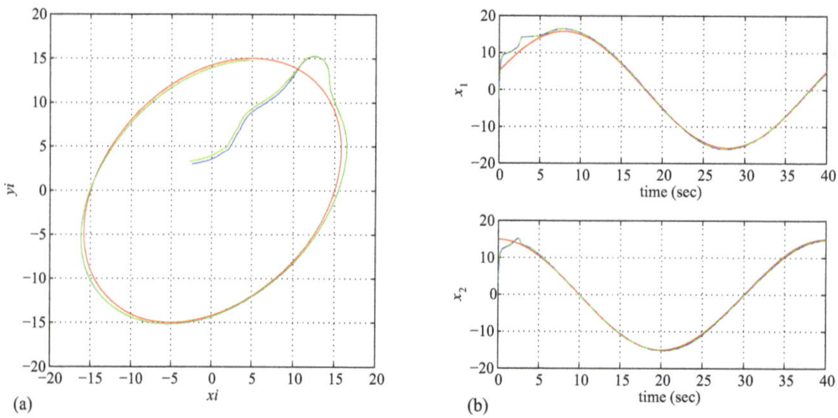

Figure 3.35 *Tracking performance of the truck and N trailer robotic system in case of setpoint 4: (a) tracking of reference trajectory (red line) on the xy plane by the center of the rear axle of trailer No 1 (blue line); (b) convergence of state variables x_1 and x_2 (blue lines) to reference setpoints (red lines) and estimated values provided by the KF (green lines)*

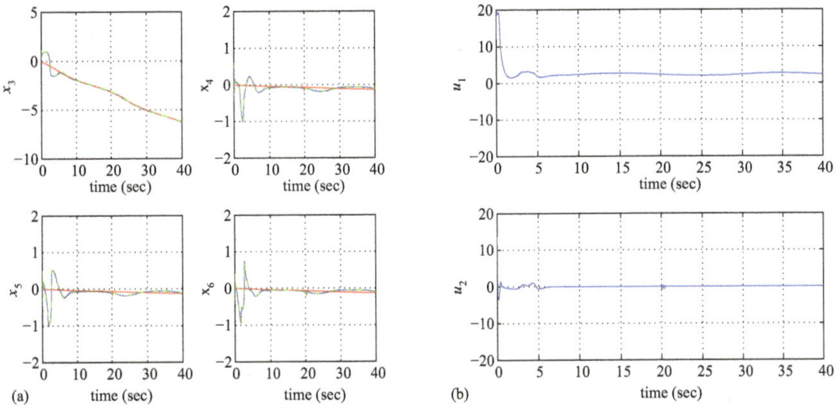

Figure 3.36 Tracking performance of the truck and N trailer robotic system in case of setpoint 4: (a) convergence of state variables x_3 to x_6 (blue lines) to reference setpoints (red lines) and estimated values provided by the KF (green lines); (b) control inputs u_1 and u_2

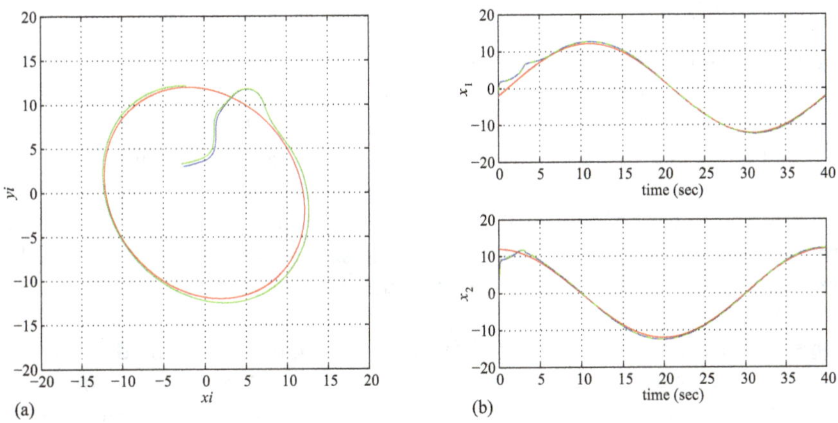

Figure 3.37 Tracking performance of the truck and N trailer robotic system in case of setpoint 5: (a) tracking of reference trajectory (red line) on the xy plane by the center of the rear axle of trailer No 1 (blue line); (b) convergence of state variables x_1 and x_2 (blue lines) to reference setpoints (red lines) and estimated values provided by the KF (green lines)

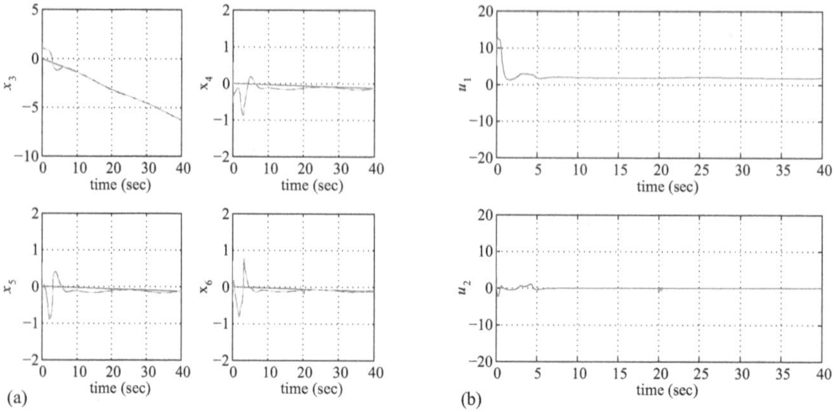

Figure 3.38 Tracking performance of the truck and N trailer robotic system in case of setpoint 5: (a) convergence of state variables x_3 to x_6 (blue lines) to reference setpoints (red lines) and estimated values provided by the KF (green lines); (b) control inputs u_1 and u_2

3.4 Nonlinear control of the ball-bot autonomous robot

3.4.1 Outline

Ballbots stand for a specific type of autonomous robotic vehicles comprising a rolling sphere and a rigid body that is mounted on it (Rigatos and Busawon, 2018; Chiu and Wang, 2019; Cai *et al.*, 2017; Thack and Lee, 2018; Tsai, Chiang *et al.*, 2015). Ballbots can be used for transportation of loads and humans in both civil and defense applications (Garcia-Garcia and Arias-Montiel, 2012; Pham, Kim *et al.*, 2019; Satici, Ruggiero *et al.*, 2016; Ba *et al.*, 2016; Thach, Ba *et al.*, 2017). The control problem of ballbots is a nontrivial one and so far several approaches have been proposed for it (Pham, Kim *et al.*, 2018; Satici, Donaire *et al.*, 2017; La Hera *et al.*, 2010; Voidya *et al.*, 2015; Su, Wang *et al.*, 2014). The solution of this control problem comes against several difficulties related with the nonlinear structure of the system's dynamics (Pham and Lee, 2019; Shomin and Hollis, 2013; Shamin, Fodicci *et al.*, 2015; Rawanasamy *et al.*, 2014). Additional complexity is imposed due to underactuation, since this robotic system receives one control input while having two degrees of freedom (Nagarajan *et al.*, 2013; Su and Gang, 2017; Herrera *et al.*, 2018; Sonleitner *et al.*, 2019; Nagarajan *et al.*, 2009). It is also remarkable about the difficulty of this control problem that the ballbot's dynamic model cannot be transformed directly into an input-output or input-state linearized form and consequently the application of global linearization-based control schemes is not straightforward. Besides, popular approaches for applying optimal control in industry such as MPC and NMPC have their own shortcomings related with global stability and convergence issues.

In this section, a novel nonlinear optimal (H-infinity) control method is developed for the dynamic model of the ballbot autonomous robotic system (Rigatos and

Busawon, 2018). To this end, the state-space description of the ballbot undergoes first approximate linearization around a time-varying operating point. The linearization relies on first-order Taylor series expansion and on the computation of the related Jacobian matrices (Rigatos and Tzafestas, 2007; Basseville and Nikiforov, 1993; Rigatos and Zhang, 2009). The time-varying operating point (temporary equilibrium) is updated at each iteration of the control method and is defined by the present value of the system's state vector and by the most recent value of the control inputs vector. The modeling error which is due to the truncation of higher-order terms in the Taylor series expansion is considered as a perturbation which is compensated by the robustness of the control algorithm. For the approximately linearized model of the ballbot, an H-infinity stabilizing feedback controller is designed.

The robot's H-infinity (optimal) controller achieves the solution of the related optimal control problem under model uncertainty and external perturbations (Rigatos, 2016b; Rigatos, Siano and Cecati, 2015; Rigatos, Siano *et al.*, 2015). Actually, it represents the solution to a min–max differential game in which the control inputs of the robotic vehicle try to minimize a cost function that comprises the state vector's tracking error whereas the model uncertainty terms and the exogenous disturbances try to maximize this cost function. To compute the controller's feedback gains, an algebraic Riccati equation is solved at each time-step of the control method. The stability properties of the control scheme are proven through Lyapunov analysis. First, it is proven that the control loop satisfies the H-infinity tracking performance criterion. This signifies elevated robustness against model uncertainties and external disturbances (Rigatos, 2011; Rigatos, 2015; Tousssaint, Basar and Bullo, 2000). Moreover, under moderate conditions it is proven that the control loop is globally asymptotically stable (Rigatos and Busawon, 2018). The control method retains the advantages of linear optimal control, that is fast and accurate tracking of reference setpoints under moderate variations of the control inputs. Finally, to achieve state estimation-based control without need to measure the entire state vector of the ballbot, the H-infinity Kalman Filter is used as a robust state estimator.

3.4.2 Dynamic model of the ballbot

The ballbot comprises a rolling sphere and a rigid body, in the form of an inverted pendulum, that is mounted on the top of the sphere. The system is underactuated and there is only a torque that makes the body rotate around its connection to the sphere. The main parameters of this robotic system are shown in Figure 3.39 and are defined as follows: ϕ is the angle between the vertical axis of the inertial reference frame OXY and the transversal axis of the body, θ is the angle between the vertical axis of the inertial reference frame OXY and the vertical axis of the ball-attached reference frame. M is the mass of the ball, m_n is the mass of the body, l is the distance between the center of gravity of the ball and the center of gravity of the frame, r is the radius of the ball, I_B is the moment of inertia of the ball, and I_b is the moment of inertia of the body (Chiu and Wang, 2019).

The dynamic model of the ballbot and the related equations of motion can be computed with the use of Euler–Lagrange analysis (Rigatos and Busawon, 2018).

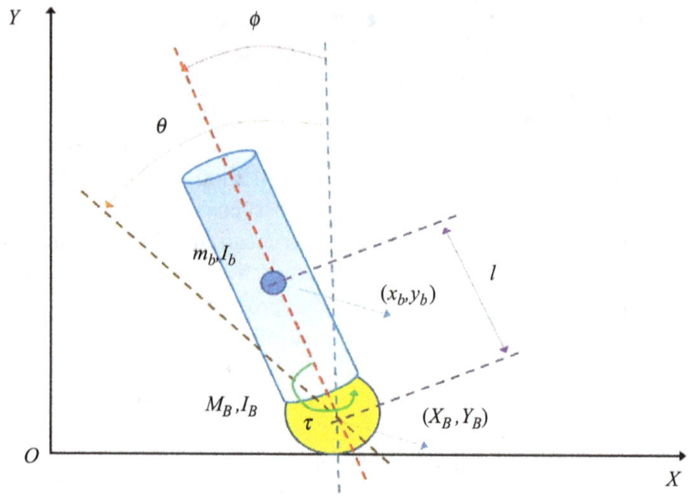

Figure 3.39 Diagram and reference frames for the ballbot robotic system

This requires to find first the kinetic and potential energy of the ball and of the body respectively, as well as the Lagrangian of the robotic system.

The kinetic energy of the ball is

$$K_B = \frac{1}{2}I_B\dot{\theta}^2 + \frac{1}{2}M(\dot{X}_B^2 + \dot{Y}_B^2) \tag{3.93}$$

where $\dot{X}_B = r\dot{\theta}$ and $\dot{Y}_B = 0$, thus

$$K_B = \frac{1}{2}I_B\dot{\theta}^2 + \frac{1}{2}Mr^2\dot{\theta}^2 \tag{3.94}$$

The potential energy of the ball is

$$P_B = Mgr \tag{3.95}$$

The kinetic energy of the body is

$$K_b = \frac{1}{2}I_b\dot{\phi}^2 + \frac{1}{2}m(\dot{x}_b^2 + \dot{y}_b^2) \tag{3.96}$$

where the coordinates of the center of mass of the body, expressed in the inertial reference frame are

$$\begin{aligned} x_m &= X_B - l\sin(\phi) \Rightarrow \dot{x}_m = \dot{X}_B - l\cos(\phi)\dot{\phi} \Rightarrow \dot{x}_m = r\dot{\theta} - l\cos(\phi)\dot{\phi} \\ y_m &= 2r + l\cos(\phi) \Rightarrow \dot{y}_m = -l\sin(\phi)\dot{\phi} \end{aligned} \tag{3.97}$$

Thus, the kinetic energy of the body is

$$K_b = \frac{1}{2}I_b\dot{\phi}^2 + \frac{1}{2}m(r^2\dot{\theta}^2 + l^2\cos^2(\phi)\dot{\phi}^2 - 2rl\cos(\phi)\dot{\theta}\dot{\phi} + l^2\sin^2(\phi)\dot{\phi}^2)\Rightarrow$$
$$K_b = \frac{1}{2}I_b\dot{\phi}^2 + \frac{1}{2}m(r^2\dot{\theta}^2 + l^2\dot{\phi}^2 - 2rl\dot{\theta}\dot{\phi}\cos(\phi)) \quad (3.98)$$

The potential energy of the body is

$$P_b = mg(2r + l\cos(\phi)) \quad (3.99)$$

Consequently, the Lagrangian of the ballbot is

$$L = (K_B + K_b) - (P_B + P_b)\Rightarrow$$
$$L = \frac{1}{2}I_b\dot{\phi}^2 + \frac{1}{2}m(\dot{x}_b^2 + \dot{y}_b^2) + \frac{1}{2}I_b\dot{\phi}^2 + \frac{1}{2}m(r^2\dot{\theta}^2 + l^2\dot{\phi}^2 - 2rl\dot{\theta}\dot{\phi}\cos(\phi))$$
$$- Mgr - \frac{1}{2}I_b\dot{\phi}^2 + \frac{1}{2}m(r^2\dot{\theta}^2 + l^2\dot{\phi}^2 - 2rl\dot{\theta}\dot{\phi}\cos(\phi)) \quad (3.100)$$

After intermediate operations, the Lagrangian of the robotic system is written in the following form

$$L = \frac{1}{2}(I_B + Mr^2 + mr^2)\dot{\theta}^2 + \frac{1}{2}(I_b + ml^2)\dot{\phi}^2$$
$$- mrl\dot{\theta}\dot{\phi}\cos(\phi) - (Mr + 2mr + ml\cos(\phi))g \quad (3.101)$$

By applying the Euler–Lagrange principle, one can find the equations of motion of the ballbot robotic system

$$\frac{\partial}{\partial t}\frac{\partial L}{\partial \dot{\theta}} - \frac{\partial L}{\partial \theta} = \tau \quad (3.102)$$

$$\frac{\partial}{\partial t}\frac{\partial L}{\partial \dot{\phi}} - \frac{\partial L}{\partial \phi} = -\tau \quad (3.103)$$

It holds that

$$\frac{\partial L}{\partial \dot{\theta}} = (I_B + M_b r^2 + mr^2)\dot{\theta} - mlr\cos(\phi)\dot{\phi} \quad (3.104)$$

and also

$$\frac{\partial}{\partial t}\frac{\partial L}{\partial \dot{\theta}} = (I_B + Mr^2 + mr^2)\ddot{\theta} + mlr\sin(\phi)\dot{\phi}^2 - mlr\cos(\phi)\ddot{\phi} \quad (3.105)$$

Moreover, one has

$$\frac{\partial L}{\partial \theta} = 0 \quad (3.106)$$

Additionally, it holds that

$$\frac{\partial L}{\partial \dot{\phi}} = (I_b + ml^2)\dot{\phi} - mlr\cos(\phi)\dot{\theta} \quad (3.107)$$

and also

$$\frac{\partial}{\partial t}\frac{\partial L}{\partial \dot{\phi}} = (I_b + ml^2)\ddot{\phi} + mlr\sin(\phi)\dot{\theta}\dot{\phi} - mrl\cos(\phi)\ddot{\theta} \quad (3.108)$$

Moreover, one has

$$\frac{\partial L}{\partial \phi} = mgl\sin(\phi) \tag{3.109}$$

Next, using (3.102) and (3.103) the equations of motion of the ballbot become

$$\begin{aligned} &(I_B + Mr^2 + mr^2)\ddot{\theta} + mlr\sin(\phi)\dot{\phi}^2 - mlr\cos(\phi)\ddot{\phi} = \tau \\ &-mrl\cos(\phi)\ddot{\theta} + (I_b + ml^2)\ddot{\phi} + mlr\sin(\phi)\dot{\theta}\dot{\phi} - mgl\sin(\phi) = -\tau \end{aligned} \tag{3.110}$$

Furthermore, in matrix form, the equations of motion of the ballbot are written as follows:

$$\begin{aligned} &\begin{pmatrix} (I_B + Mr^2 + mr^2) & -mlr\cos(\phi) \\ -mlr\cos(\phi) & (I_b + ml^2) \end{pmatrix} \begin{pmatrix} \ddot{\theta} \\ \ddot{\phi} \end{pmatrix} + \begin{pmatrix} 0 & mlr\dot{\phi}^2\sin(\phi) \\ 0 & mlr\dot{\theta}\sin(\phi) \end{pmatrix} \begin{pmatrix} \dot{\theta} \\ \dot{\phi} \end{pmatrix} \\ &+ \begin{pmatrix} 0 \\ -mgl\sin(\phi) \end{pmatrix} = \begin{pmatrix} \tau \\ -\tau \end{pmatrix} \end{aligned} \tag{3.111}$$

The dynamic model of the ballbot is also written in the concise matrix form:

$$M(q)\ddot{q} + C(q,\dot{q})\dot{q} + G(q) = U \tag{3.112}$$

where

$$q = \begin{pmatrix} \theta \\ \phi \end{pmatrix} \quad U = \begin{pmatrix} \tau \\ -\tau \end{pmatrix} \tag{3.113}$$

$$\begin{aligned} &M = \begin{pmatrix} I_B + mr^2 + Mr^2 & -mlr\cos(\phi) \\ -mlr\cos(\phi) & ml^2 + I_b \end{pmatrix} \\ &C = \begin{pmatrix} 0 & m\dot{\phi}lr\sin(\phi) \\ 0 & m\dot{\theta}lr\sin(\phi) \end{pmatrix} \qquad G = \begin{pmatrix} 0 \\ -mgl\sin(\phi) \end{pmatrix} \end{aligned} \tag{3.114}$$

Next, the dynamic model of the ballbot will be written into an affine-in-the-input state-space form. The determinant of the inertia matrix $M(q)$ is

$$detM = (I_B + (M+m)r^2)(ml^2 + I_b) - m^2l^2r^2\cos^2(\phi) \tag{3.115}$$

The inverse of the inertia matrix $M(q)$ is given by

$$M^{-1}(q) = \frac{1}{detM} \begin{pmatrix} ml^2 + I_b & ml\cos(\phi) \\ ml\cos(\phi) & I_B + (M+m)r^2 \end{pmatrix} \tag{3.116}$$

and the state-space description of the system is written as

$$\ddot{q} = -M^{-1}(q)[C(q,\dot{q})\dot{q} + G(q)] + M^{-1}(q)U \tag{3.117}$$

It holds that

$$C(q,\dot{q})\dot{q} + G(q) = \begin{pmatrix} m\dot{\phi}^2lr\sin(\phi) \\ m\dot{\phi}\dot{\theta}lr\sin(\phi) - mgl\sin(\phi) \end{pmatrix} \tag{3.118}$$

and also

$$-det^{-1}M(q)[C(q,\dot{q})\dot{q}+G(q)]$$

$$=-\begin{pmatrix} \frac{(ml^2+I_b)(m\dot{\phi}^2lr\sin(\phi))+(mlr\cos(\phi))(m\dot{\phi}\dot{\theta}lr\sin(\phi)-mgl\sin(\phi))}{detM} \\ \frac{(mlr\cos(\phi))(m\dot{\phi}^2lr\sin(\phi))+(I_B+(M+m)r^2)(m\dot{\phi}\dot{\theta}lr\sin(\phi)-mgl\sin(\phi))}{detM} \end{pmatrix} \tag{3.119}$$

Moreover, one has that

$$det^{-1}M(q)\cdot U=\begin{pmatrix} \frac{ml^2+I_b-mlr\cos(\phi)}{detM} \\ \frac{mlr\cos(\phi)-I_B-(M+m)r^2}{detM} \end{pmatrix}\tau \tag{3.120}$$

Consequently, the dynamic model of the ballbot is written as

$$\begin{pmatrix}\ddot{\theta}\\ \ddot{\phi}\end{pmatrix}=-\begin{pmatrix} \frac{(ml^2+I_b)(m\dot{\phi}^2lr\sin(\phi))+(mlr\cos(\phi))(m\dot{\phi}\dot{\theta}lr\sin(\phi)-mgl\sin(\phi))}{detM} \\ \frac{(mlr\cos(\phi))(m\dot{\phi}^2lr\sin(\phi))+(I_B+(M+m)r^2)(m\dot{\phi}\dot{\theta}lr\sin(\phi)-mgl\sin(\phi))}{detM} \end{pmatrix} + \begin{pmatrix} \frac{ml^2+I_b-ml\cos(\phi)}{detM} \\ \frac{mlr\cos(\phi)-I_B-(M+m)r^2}{detM} \end{pmatrix}\tau \tag{3.121}$$

Next, the following state variables are defined: $x_1=\theta$, $x_2=\dot{\theta}$, $x_3=\phi$ and $x_4=\dot{\phi}$. Besides, the control input is denoted as $u=\tau$. It holds that

$$\dot{x}_1=x_2 \tag{3.122}$$

$$\dot{x}_2=\frac{-(ml^2+I_b)(mx_4^2lr\sin(x_3))-(mlr\cos(x_3))(mx_4x_2lr\sin(x_3)-mgl\sin(x_3))}{detM} + \frac{ml^2+I_b-mlr\cos(x_3)}{detM}u \tag{3.123}$$

$$\dot{x}_3=x_4 \tag{3.124}$$

$$\dot{x}_4=\frac{-(mlr\cos(x_3))(mx_4^2lr\sin(x_3))-(I_B+(M+m)r^2)(mx_4x_2lr\sin(x_3)-mgl\sin(x_3))}{detM} + \frac{mlr\cos(x_3)-I_B-(M+m)r^2}{detM}u \tag{3.125}$$

Thus, one has an affine-in-the-input state-space description of the form

$$\dot{x} = f(x) + g(x)u \tag{3.126}$$

with $x \in R^{4\times 1}$, $u \in R$, $f(x) \in R^{4\times 1}$ and $g(x) \in R^{4\times 1}$. In particular, one can write

$$f(x) = \begin{pmatrix} x_2 \\ \dfrac{-(ml^2+I_b)(mx_4^2 lr\sin(x_3)) - (mlr\cos(x_3))(mx_4x_2 lr\sin(x_3) - mgl\sin(x_3))}{detM} \\ x_4 \\ \dfrac{-(mlr\cos(x_3))(mx_4^2 lr\sin(x_3)) - (I_B + (M+m)r^2)(mx_4x_2 lr\sin(x_3) - mgl\sin(x_3))}{detM} \end{pmatrix} \tag{3.127}$$

$$g(x) = \begin{pmatrix} 0 \\ \dfrac{ml^2 + I_b - mlr\cos(x_3)}{detM} \\ 0 \\ \dfrac{mlr\cos(x_3) - I_B - (M+m)r^2}{detM} \end{pmatrix} \tag{3.128}$$

The above noted vector fields $f(x)$ and $g(x)$ can be also written in the concise form

$$f(x) = \begin{pmatrix} x_2 \\ \dfrac{f_{2n}(x)}{detM} \\ x_4 \\ \dfrac{f_{4n}(x)}{detM} \end{pmatrix} \quad g(x) = \begin{pmatrix} 0 \\ \dfrac{g_{2n}(x)}{detM} \\ 0 \\ \dfrac{g_{4n}(x)}{detM} \end{pmatrix} \tag{3.129}$$

3.4.3 Approximate linearization of the ballbot's state-space model

The dynamic model of the ballbot undergoes approximate linearization around the temporary operating point (x^*, u^*), where x^* is the present value of the ballbot's state vector and u^* is the most recent value of the control inputs vector. The initial nonlinear state-space model of the robotic system was shown to be in the affine-in-the-input form $\dot{x} = f(x) + g(x)u$ that was given in (3.126), while after the proposed linearization procedure, it becomes

$$\dot{x} = Ax + bu + \tilde{d} \tag{3.130}$$

where $\tilde{d}$ is the cumulative perturbations term which comprises modeling errors due to approximate linearization, as well as external disturbances. Moreover, matrices A and B are related with the Jacobians of the robotic system, that is

$$A = \nabla_x[f(x) + g(x)u] \mid_{(x^*,u^*)} \Rightarrow A = \nabla_x[f(x)] \mid_{x^*,u^*} + \nabla_x[g(x)u] \mid_{x^*,u^*} \tag{3.131}$$

$$B = \nabla_u [f(x) + g(x)u] \mid_{(x^*,u^*)} \Rightarrow B = g(x) \mid_{(x^*),u^*} \tag{3.132}$$

The computation of the Jacobian matric $\nabla_x f(x) \mid_{(x^*,u^*)}$ of the robotic system proceeds as follows:

First row of the Jacobian matrix $\nabla_x f(x) \mid_{(x^*,u^*)}$: $\frac{\partial f_1(x)}{\partial x_1} = 0$, $\frac{\partial f_1(x)}{\partial x_2} = 1$, $\frac{\partial f_1(x)}{\partial x_3} = 0$ and $\frac{\partial f_1(x)}{\partial x_4} = 0$.

Second row of the Jacobian matrix $\nabla_x f(x) \mid_{(x^*,u^*)}$:

$$\frac{\partial f_2(x)}{\partial x_i} = \frac{\frac{\partial f_{2n}}{\partial x_i} detM - f_{2n} \frac{\partial detM}{\partial x_i}}{detM^2} \quad i = 1, \cdots, 4, \tag{3.133}$$

Third row of the Jacobian matrix $\nabla_x f(x) \mid_{(x^*,u^*)}$: $\frac{\partial f_3(x)}{\partial x_1} = 0$, $\frac{\partial f_3(x)}{\partial x_2} = 0$, $\frac{\partial f_3(x)}{\partial x_3} = 0$ and $\frac{\partial f_3(x)}{\partial x_4} = 1$.

Fourth row of the Jacobian matrix $\nabla_x f(x) \mid_{(x^*,u^*)}$:

$$\frac{\partial f_4(x)}{\partial x_i} = \frac{\frac{\partial f_{4n}}{\partial x_i} detM - f_{4n} \frac{\partial detM}{\partial x_i}}{detM^2} \quad i = 1, \cdots, 4, \tag{3.134}$$

The computation of the Jacobian matrix $\nabla_x g(x) \mid_{(x^*,u^*)}$ of the robotic system proceeds as follows:

First row of the Jacobian matrix $\nabla_x g(x) \mid_{(x^*,u^*)}$: $\frac{\partial g_1(x)}{\partial x_1} = 0$, $\frac{\partial g_1(x)}{\partial x_2} = 0$, $\frac{\partial g_1(x)}{\partial x_3} = 0$ and $\frac{\partial g_1(x)}{\partial x_4} = 0$.

Second row of the Jacobian matrix $\nabla_x g(x) \mid_{(x^*,u^*)}$:

$$\frac{\partial g_2(x)}{\partial x_i} = \frac{\frac{\partial g_{2n}}{\partial x_i} detM - g_{2n} \frac{\partial detM}{\partial x_i}}{detM^2} \quad i = 1, \cdots, 4, \tag{3.135}$$

Third row of the Jacobian matrix $\nabla_x g(x) \mid_{(x^*,u^*)}$: $\frac{\partial g_3(x)}{\partial x_1} = 0$, $\frac{\partial g_3(x)}{\partial x_2} = 0$, $\frac{\partial g_3(x)}{\partial x_3} = 0$ and $\frac{\partial g_3(x)}{\partial x_4} = 0$.

Fourth row of the Jacobian matrix $\nabla_x g(x) \mid_{(x^*,u^*)}$:

$$\frac{\partial g_4(x)}{\partial x_i} = \frac{\frac{\partial g_{4n}}{\partial x_i} detM - g_{4n} \frac{\partial detM}{\partial x_i}}{detM^2} \quad i = 1, \cdots, 4, \tag{3.136}$$

Next, one computes the partial derivatives of $f_{2n}(x)$ and $f_{4n}(x)$. It holds that:

$$\begin{aligned}
\frac{\partial f_{2n}}{\partial x_1} &= 0 \\
\frac{\partial f_{2n}}{\partial x_2} &= -mlr\cos(x_3)(mx_4 r\sin(x_3)) \\
\frac{\partial f_{2n}}{\partial x_3} &= -(ml^2 I_b)(mx_4^2)lr\cos(x_3) \\
&\quad -(-mlr\sin(x_3))(mx_2x_4 lr\sin(x_3) - mgl\sin(x_3)) \\
&\quad -(mlr\cos(x_3))(mx_2x_4 lr\cos(x_3) - mgl\cos(x_3)) \\
\frac{\partial f_{2n}}{\partial x_4} &= -(ml^2 + I_b)(2mx_4 lr\sin(x_3)) \\
&\quad -(mlrco(x_3))(mx_2 lr\sin(x_3))
\end{aligned} \tag{3.137}$$

Moreover, it holds that:

$$\begin{aligned}
\frac{\partial f_{4n}}{\partial x_1} &= 0 \\
\frac{\partial f_{4n}}{\partial x_2} &= -(I_B + (M+m)r^2)(mx_4 lr\sin(x_3)) \\
\frac{\partial f_{4n}}{\partial x_3} &= -(-mlr\sin(x_3))(mx_4^2 lr\sin(x_3)) - (mlr\cos(x_3))(mx_4^2 lr\cos(x_3)) \\
&\quad -(I_B + (M+m)r^2)(mx_2x_4 Lr\cos(x_3) - mglr\cos(x_3)) \\
\frac{\partial f_{4n}}{\partial x_4} &= -(mlr\cos(x_3))(m2x_4 lr\sin(x_3)) - (I_B + (M+m)r^2)(mx_2 L\sin(x_3))
\end{aligned} \tag{3.138}$$

Next, one computes the partial derivatives of g_{2n}, g_{4n}:

$$\begin{aligned}
&\frac{\partial g_{2n}}{\partial x_1} = 0 \qquad \frac{\partial g_{2n}}{\partial x_2} = 0 \\
&\frac{\partial g_{2n}}{\partial x_3} = ml\sin(x_3) \qquad \frac{\partial g_{2n}}{\partial x_4} = 0
\end{aligned} \tag{3.139}$$

$$\begin{aligned}
&\frac{\partial g_{4n}}{\partial x_1} = 0 \qquad \frac{\partial g_{4n}}{\partial x_2} = 0 \\
&\frac{\partial g_{4n}}{\partial x_3} = -ml\sin(x_3) \qquad \frac{\partial g_{4n}}{\partial x_4} = 0
\end{aligned} \tag{3.140}$$

Moreover, one computes the partial derivatives of *detM*

$$\begin{aligned}
&\frac{\partial det_M}{\partial x_1} = 0 \qquad \frac{\partial det_M}{\partial x_2} = 0 \\
&\frac{\partial det_M}{\partial x_3} = 2m^2l^2r^2\cos(x_3)\sin(x_3) \qquad \frac{\partial det_M}{\partial x_4} = 0
\end{aligned} \tag{3.141}$$

3.4.4 Computation of the feedback control gains

For the linearized system given by (1.36) the cost function of (1.37) is defined, where coefficient r determines the penalization of the control input and weight coefficient ρ determines the reward of the disturbances' effects. It is assumed that (i) the energy that is transferred from the disturbances signal $\tilde{d}(t)$ is bounded, that is $\int_0^{\infty}\tilde{d}^T(t)\tilde{d}(t)dt < \infty$, (ii) matrices $[A, B]$ and $[A, L]$ are stabilizable, and (iii) matrix $[A, C]$ is detectable. Then, by denoting the state vector tracking error as $e = x - x_d$, the optimal feedback control law is given by

$$u(t) = -Ke(t) \tag{3.142}$$

with $K = \frac{1}{r}B^TP$ where P is a positive definite symmetric matrix which is obtained from the solution of a Riccati equation of the form

$$A^TP + PA + Q - P\left(\frac{2}{r}BB^T - \frac{1}{1\rho^2}LL^T\right)P = 0 \tag{3.143}$$

where Q is also a positive semi-definite symmetric matrix. Moreover, by solving the dual optimal control problem after considering the disturbance input $d(t)$ in place of the control input $u(t)$, the worst case disturbance is found to be $\tilde{d}(t) = \frac{1}{\rho^2}L^TPe(t)$. The diagram of the considered control loop for the ballbot is depicted in Figure 3.40.

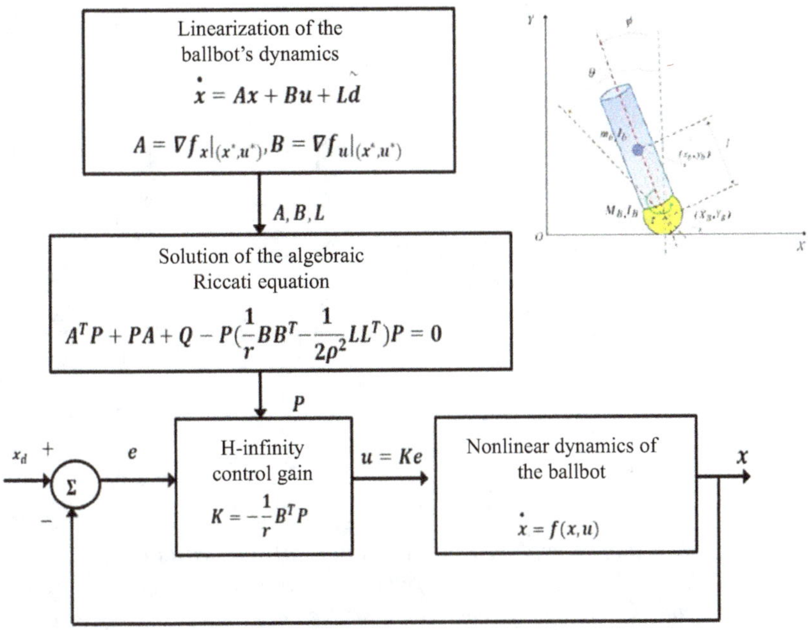

Figure 3.40 Diagram of the control scheme for the ballbot

The global stability properties of the control method can be proven following the procedure presented in Section 1.1.2.5.

3.4.5 Simulation tests

The tracking accuracy of the proposed nonlinear optimal (H-infinity) control method and its global stability properties were further confirmed through simulation experiments. Indicative parameter values for the ballbot robotic vehicle are $m = 4\,\text{kgr}$, $M = 4\,\text{kgr}$, $l = 1\,\text{m}$, $r = 0.2\,\text{m}$, $I_B = (2/3)\,\text{Mr}^2$, $I_b = (1/3)\,\text{mL}^2$ and $g = 10\,\text{m/sec}^2$. The obtained results are depicted in Figures 3.41 to 3.48. The real values of the ballbot's state variables are printed in blue, their estimated values (provided by the H-infinity Kalman Filter) are printed in green while the related reference setpoints are plotted in red. It can be noted that the proposed control method retains the advantages of linear optimal control, that is fast and accurate tracking of the reference setpoints under moderate variations of the control inputs. To select the controller's feedback gains the algebraic Riccati equation of (3.143) had to be solved at each time step of the control method. The computational effort to accomplish this was moderate and actually a solution of the Riccati equation could be obtained within a time interval that was significantly smaller than the sampling period ($f_s = 0.1\,\text{kHz}$). The transient performance of the control loop was determined by the selection of parameters and gains in the system's algebraic Riccati equation, such as r, ρ

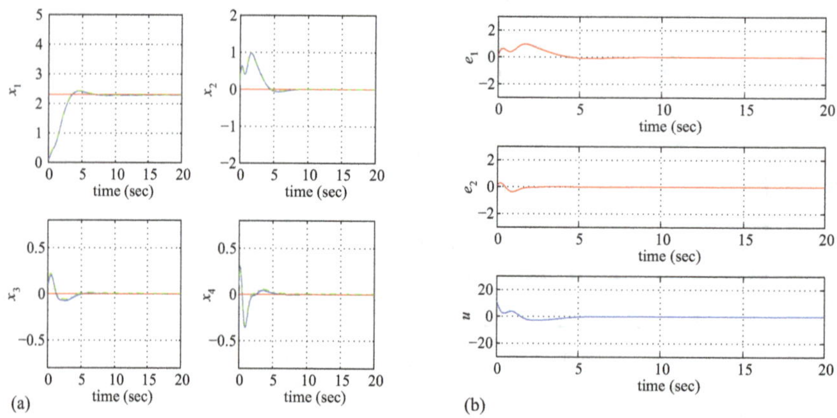

Figure 3.41 Tracking of setpoint 1 by the ballbot: (a) state variables x_i $i = 1, \cdots, 4$ (blue lines), their estimated values (green lines) and the associated reference setpoints (red lines); (b) tracking errors e_1 (state variable x_1) and e_3 (state variable x_3) and control input u of the ballbot

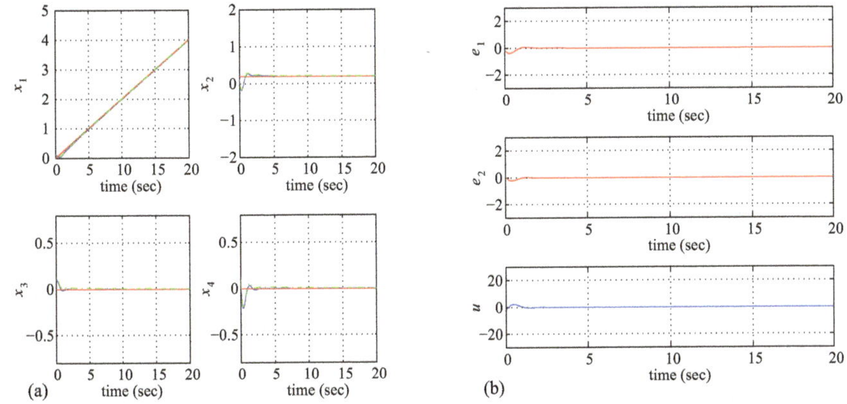

Figure 3.42 *Tracking of setpoint 2 by the ballbot: (a) state variables x_i $i = 1, \cdots, 4$ (blue lines), their estimated values (green lines) and the associated reference setpoints (red lines); (b) tracking errors e_1 (state variable x_1) and e_3 (state variable x_3) and control input u of the ballbot*

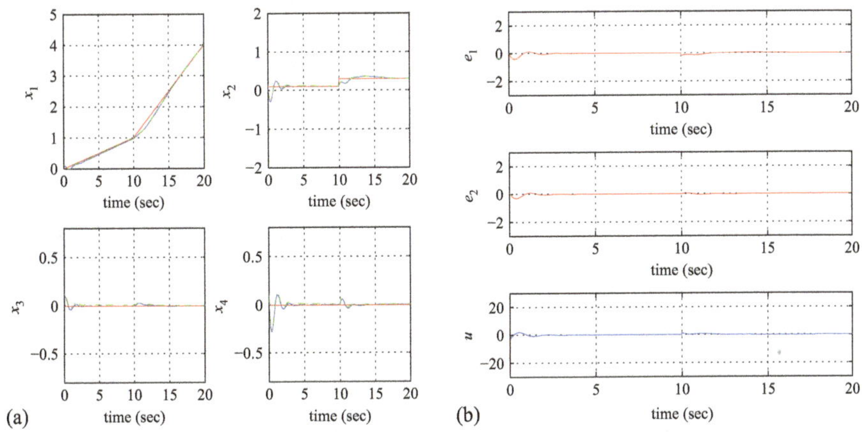

Figure 3.43 *Tracking of setpoint 3 by the ballbot: (a) state variables x_i $i = 1, \cdots, 4$ (blue lines), their estimated values (green lines) and the associated reference setpoints (red lines); (b) tracking errors e_1 (state variable x_1) and e_3 (state variable x_3) and control input u of the ballbot*

and Q. Actually, the smallest value of ρ for which one can obtain a valid solution of this Riccati equation is the one that provides to the control loop maximum robustness.

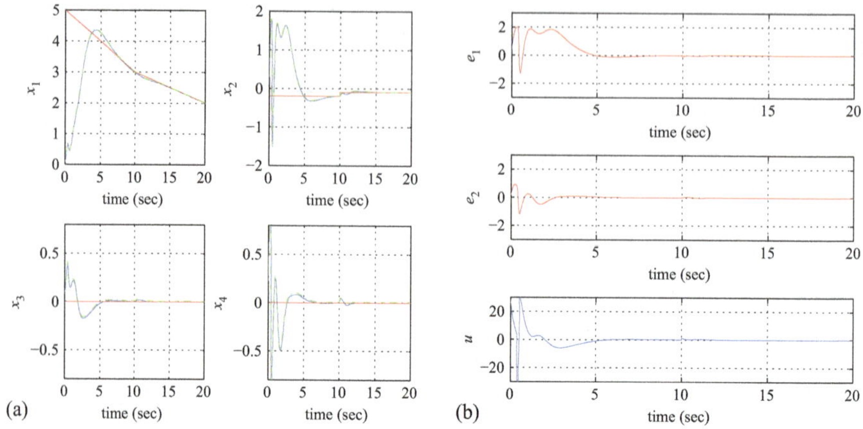

Figure 3.44 Tracking of setpoint 4 by the ballbot: (a) state variables x_i $i = 1, \cdots, 4$ (blue lines), their estimated values (green lines) and the associated reference setpoints (red lines); (b) tracking errors e_1 (state variable x_1) and e_3 (state variable x_3) and control input u of the ballbot

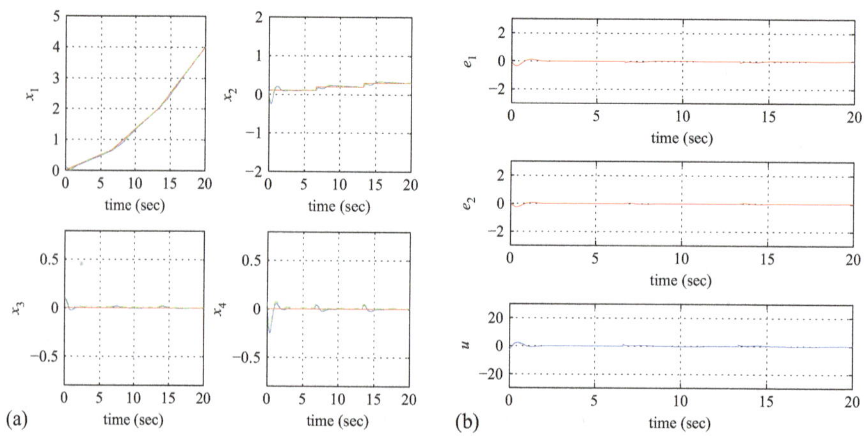

Figure 3.45 Tracking of setpoint 5 by the ballbot: (a) state variables x_i $i = 1, \cdots, 4$ (blue lines), their estimated values (green lines) and the associated reference setpoints (red lines); (b) tracking errors e_1 (state variable x_1) and e_3 (state variable x_3) and control input u of the ballbot

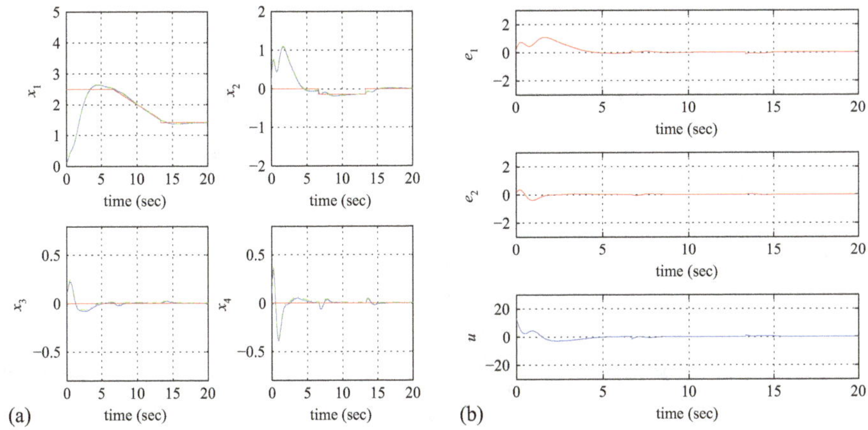

Figure 3.46 *Tracking of setpoint 6 by the ballbot: (a) state variables x_i $i = 1, \cdots, 4$ (blue lines), their estimated values (green lines) and the associated reference setpoints (red lines); (b) tracking errors e_1 (state variable x_1) and e_3 (state variable x_3) and control input u of the ballbot*

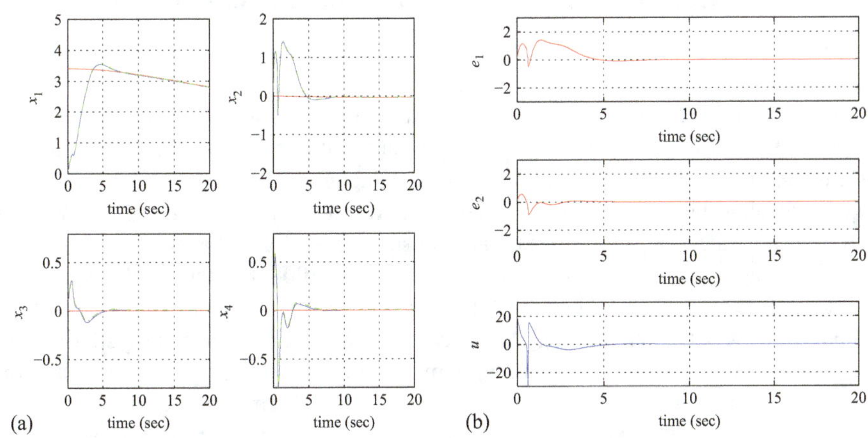

Figure 3.47 *Tracking of setpoint 7 by the ballbot: (a) state variables x_i $i = 1, \cdots, 4$ (blue lines), their estimated values (green lines) and the associated reference setpoints (red lines); (b) tracking errors e_1 (state variable x_1) and e_3 (state variable x_3) and control input u of the ballbot*

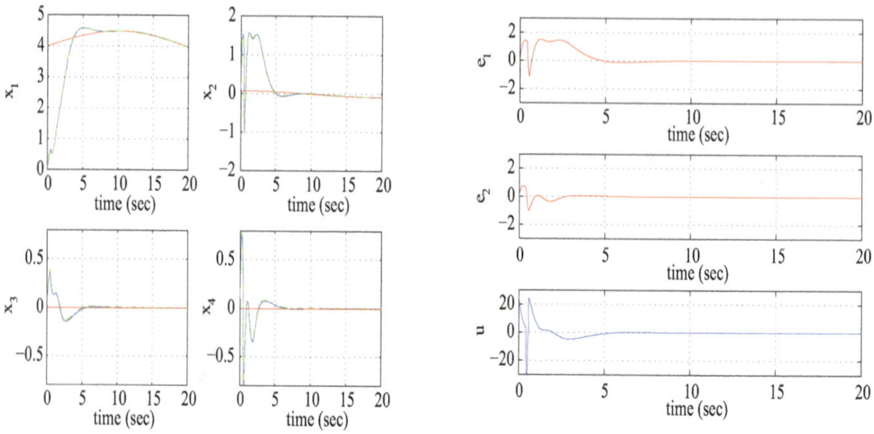

Figure 3.48 *Tracking of setpoint 8 by the ballbot: (a) state variables x_i $i = 1, \cdots, 4$ (blue lines), their estimated values (green lines) and the associated reference setpoints (red lines); (b) tracking errors e_1 (state variable x_1) and e_3 (state variable x_3) and control input u of the ballbot*

3.5 Nonlinear control of the ball-and-plate dynamical system

3.5.1 Outline

The ball and plate system consists of a plate that can rotate around its longitudinal and transversal axes of symmetry and of a ball that can roll on the plate's surface (Wang *et al.*, 2014; Fan, Zhang *et al.*, 2004; Wang *et al.*, 2008; Fu, Sui *et al.*, 2009). The control problem of the ball and plate system is non-trivial because this system in strongly nonlinear, and has a multivariable state-space form (Oriolo and Venditelli, 2005; Casagrande *et al.*, 2009; Morin and Samson, 2008c). Besides, this system is underactuated since it has four degrees of freedom while receiving only two control inputs (Huang, Zhao *et al.*, 2019; Mooref *et al.*, 2008). The ball and plate system is a 2D generalization of the ball and beam system and is a typical example of manipulation through rolling contacts (Zhenye and Zhang, 2015; Duan *et al.*, 2009). The main objective of the ball and plate system is to control the position of the ball by varying the inclination of the plate (Debono *et al.*, 2015; Liu, Tian *et al.*, 2009). Ball and plate systems are benchmark nonlinear systems on which the performance and efficiency of nonlinear control algorithms can be tested (Brezina *et al.*, 2019; Becker and Bretl, 2012; Fabrejas *et al.*, 2015). A large class of control schemes for the ball and plate system is model-based and includes methods such as sliding-mode control, back-stepping control, and Linear Quadratic Regulators (Rangel, Loukianov *et al.*, 2014; Hegenmeyer *et al.*, 2004; Burch *et al.*, 2018; Wang, Li *et al.*, 2014). Furthermore, Lyapunov theory-based control methods, Lie algebra-based control and differential flatness theory-based control have been applied to the ball and plate system (Kassem *et al.*, 2015; Silveira Castro *et al.*, 2014; Liao, Tu *et al.*, 2008). A second class of control approaches for the ball and plate system is to a great extent model-free and includes methods of estimation or neurofuzzy control (Moreno Armendaviz *et al.*, 2010; Hun and Liu, 2016).

In this section, the previously analyzed nonlinear optimal (H-infinity) control method is used for the control of the ball and plate system. First the dynamic model of the ball and plate system undergoes approximate linearization around a temporary operating point which is updated at each time-step of the control algorithm. The linearization point is defined at each sampling period by the present value of the system's state vector and by the last sampled value of the control inputs vector. The linearization is based on first-order Taylor series expansion and on the computation of the Jacobian matrices of the state-space model of the system (Rigatos and Tzafestas, 2007; Basseville and Nikiforov, 1993; Rigatos and Zhang, 2009). The modeling error which is induced by the truncation of higher-order terms in the Taylor series expansion is a perturbation that is asymptotically compensated by the robustness of the control algorithm. For the approximately linearized model of the ball and plate system a stabilizing H-infinity feedback controller is designed (Rigatos and Busawon, 2018; Rigatos, 2016b).

The proposed H-infinity controller provides a solution to the nonlinear optimal control problem for the ball and plate system. Actually, it represents a min–max differential game which takes place between (i) the control inputs which try to minimize a cost function that comprises a quadratic term of the state vector's tracking error, and (ii) the model uncertainty and exogenous perturbation inputs which try to maximize this cost function. To compute the feedback gains of the H-infinity controller an algebraic Riccati equation has to be repetitively solved at each iteration of the control algorithm (Rigatos, 2011; Rigatos, 2015). The global stability properties of the control scheme are proven through Lyapunov analysis. First, it is proven that the control loop of the ball and plate system satisfies the H-infinity tracking performance criterion, which signifies elevated robustness against model inaccuracies and external disturbances (Rigatos and Busawon, 2018; Tousssaint, Basar and Bullo, 2000). Moreover, under moderate conditions it is proven that the control scheme is globally asymptotically stable (Rigatos and Busawon, 2018; Rigatos, 2016b). Furthermore, to implement state estimation-based control the H-infinity Kalman Filter is used as a robust state estimator (Rigatos and Busawon, 2018; Rigatos, 2016b). The proposed control method retains the advantages of linear optimal control, that is fast and accurate tracking of reference setpoints by the state variables of the system and moderate variations for the system's control inputs.

The proposed H-infinity (optimal) control method for the ball and plate system exhibits several advantages when compared against other linear or nonlinear control schemes. For instance: (i) in contrast to global linearization-based control schemes (Lie algebra-based control and differential flatness theory-based control) it does not need complicated changes of state-variables (diffeomorphisms) and does not come against singularity problem in the computation of the control inputs; (ii) in contrast to popular approaches towards the implementation of optimal control in industrial systems, such as MPC or NMPC the performance of the new nonlinear optimal control method is assured. (iii) In contrast to sliding-mode control or to back-stepping control, the proposed nonlinear optimal control scheme does not require the state-space model of the system to be in a specific form (e.g. triangular, canonical, etc.). (iv) In contrast to PID control, the proposed nonlinear optimal control method is globally stable and functions well at changes of operating points. (v) In contrast to multi-models based control and linearization around multiple operating points, the nonlinear optimal control scheme requires linearization around one single operating point and thus it avoids the computational burden for solving multiple Riccati equations or LMIs.

3.5.2 Dynamic model of the ball and plate system

The ball and plate system comprises a plate that can rotate around its longitudinal and transversal axes of symmetry and of a ball that can roll on the plate's surface. The dynamic model of the ball and plate system is obtained from Euler–Lagrange analysis. The equations of the dynamic model of the ball and plate system are (Fan, Zhang *et al.*, 2004),

$$(J_p + J + mx^2)\ddot{\theta}_x + 2mx\dot{x}\dot{\theta}_x + mxy\ddot{\theta}_y + m\dot{x}y\dot{\theta}_y + mx\dot{y}\dot{\theta}_y + mgx\cos(\theta_x) = \tau_x \tag{3.144}$$

$$(J_p + J + my^2)\ddot{\theta}_y + 2my\dot{y}\dot{\theta}_y + mxy\ddot{\theta}_x + m\dot{x}y\dot{\theta}_x + mx\dot{y}\dot{\theta}_x + mgy\cos(\theta)_y = \tau_y \tag{3.145}$$

$$\left(m + \frac{J}{R^2}\ddot{x}\right) + mg\sin(\theta_x) - mx\dot{\theta}_x^2 + my\dot{\theta}_x\dot{\theta}_y = 0 \tag{3.146}$$

$$\left(m + \frac{J}{R^2}\ddot{y}\right) + mg\sin(\theta_y) - my\dot{\theta}_y^2 - mx\dot{\theta}_x\dot{\theta}_y = 0 \tag{3.147}$$

Equations (3.144) and (3.145) describe rotational motion of the plate around the x and y axes by angles θ_x and θ_y respectively. Equations (3.146) and (3.147) describe translational motion of the ball along the x and y axes respectively.

The main parameters of the model are defined as follows (Fan, Zhang *et al.*, 2004): m is the mass of the ball (0.11 kgr), R is the radius of the ball (0.02 m), S is the surface of the plate ($1.0 \times 1.0\,\text{m}^2$), x is the position of the ball on the x-axis (m), y is the position of the ball on the y-axis (m), $\dot{x}$ is the velocity of the ball along the x-axis, $\dot{y}$ is the velocity of the ball along the y-axis, ω is the rotational speed of the ball (rad/sec), $\dot{r}$ is the amplitude of the ball's velocity vector $r = \sqrt{\dot{x}^2 + \dot{y}^2}$ (m/sec), τ_x is the torque causing rotation of the plate around the x-axis (Nt·m), τ_y is the torque causing rotation of the plate around the y-axis (Nt·m), θ_x is the rotation angle of the plate around the x-axis, θ_y is the rotation angle of the plate around the y-axis, $\dot{\theta}_x$ is the rotation speed of the plate around the x-axis (rad/sec), $\dot{\theta}_y$ is the rotation speed of the plate around the y-axis (rad/sec), J_p is the moment of inertia of the plate ($0.5\,\text{kgr}\cdot\text{m}^2$), J is the moment of inertia of the ball ($\text{kgr}\cdot\text{m}^2$), and g is the acceleration of gravity ($9.8\,\text{m/sec}^2$).

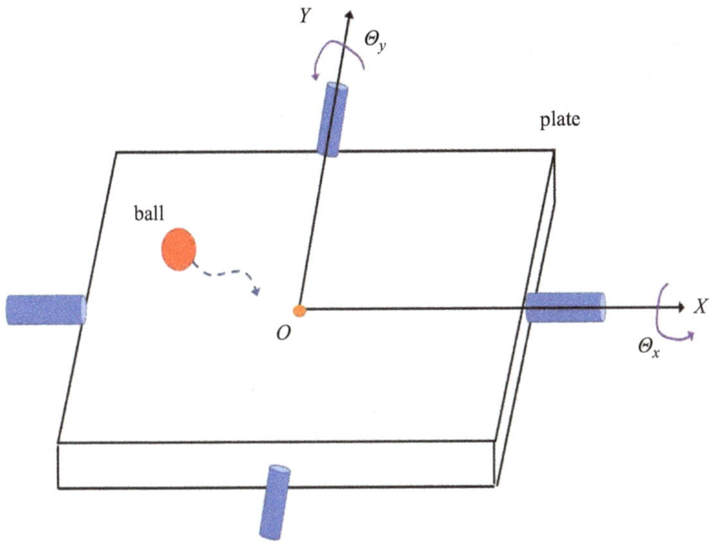

Figure 3.49 Diagram of the ball and plate dynamical system

Equations (3.144) and (3.145) can be written also in the following matrix form:

$$\begin{pmatrix} (J_P+J+mx^2) & mxy \\ mxy & (J_P+J+my^2) \end{pmatrix}\begin{pmatrix} \ddot{\theta}_x \\ \ddot{\theta}_y \end{pmatrix} = \begin{pmatrix} -2mx\dot{x}\dot{\theta}_x - m\dot{x}y\dot{\theta}_y - mx\dot{y}\dot{\theta}_y - mgx\cos(\theta)_x \\ -2my\dot{y}\dot{\theta}_y - m\dot{x}y\dot{\theta}_x - mx\dot{y}\dot{\theta}_x - mgy\cos(\theta)_y \end{pmatrix} + \begin{pmatrix} \tau_1 \\ \tau_2 \end{pmatrix} \tag{3.148}$$

Next, the following variables are defined $M_{11}=(J_P+J+mx^2)$, $M_{12}=mxy$, $M_{21}=mxy$, $M_{22}=(J_P+J+my^2)$, $detM=M_{11}M_{22}-M_{12}M_{21}$, $h_1=-2mx\dot{x}\dot{\theta}_x - m\dot{x}y\dot{\theta}_y - mx\dot{y}\dot{\theta}_y - mgx\cos(\theta)_x$ and $h_2=-2my\dot{y}\dot{\theta}_y - m\dot{x}y\dot{\theta}_x - mx\dot{y}\dot{\theta}_x - mgy\cos(\theta)_y$. Thus, after defining matrices

$$\theta=\begin{pmatrix} \theta_1 \\ \theta_2 \end{pmatrix} \quad M=\begin{pmatrix} M_{11} & M_{12} \\ M_{21} & M_{22} \end{pmatrix} \quad M^{-1}=\frac{1}{detM}\begin{pmatrix} M_{22} & -M_{12} \\ -M_{21} & M_{11} \end{pmatrix}$$
$$h=\begin{pmatrix} h_1 \\ h_2 \end{pmatrix} \qquad \tau=\begin{pmatrix} \tau_x \\ \tau_y \end{pmatrix} \tag{3.149}$$

the previous equation about the rotational motion of the plate is written in the form

$$\begin{aligned} &M\ddot{\theta}=h+\tau \Rightarrow \\ &\ddot{\theta}=M^{-1}h+M^{-1}\tau \end{aligned} \tag{3.150}$$

By performing multiplications between matrices in (3.150) one obtains the following description about the plate dynamics:

$$\begin{pmatrix} \ddot{\theta}_x \\ \\ \ddot{\theta}_y \end{pmatrix} = \begin{pmatrix} \dfrac{M_{22}h_1 - M_{12}h_2}{detM} \\ \\ \dfrac{-M_{12}h_1 + M_{11}h_2}{detM} \end{pmatrix} + \begin{pmatrix} \dfrac{M_{22}}{detM} & -\dfrac{M_{12}}{detM} \\ \\ -\dfrac{M_{21}}{detM} & \dfrac{M_{11}}{detM} \end{pmatrix} \begin{pmatrix} \tau_1 \\ \\ \tau_2 \end{pmatrix} \tag{3.151}$$

Besides, from (3.144) and (3.145) one has that

$$\ddot{x} = -\frac{m}{(m + \frac{J}{R^2})} g \sin(\theta_x) + \frac{m}{(m + \frac{J}{R^2})} x\dot{\theta}_x^2 + \frac{m}{(m + \frac{J}{R^2})} x\dot{\theta}_x\dot{\theta}_y \tag{3.152}$$

$$\ddot{y} = -\frac{m}{(m + \frac{J}{R^2})} g \sin(\theta_y) + \frac{m}{(m + \frac{J}{R^2})} y\dot{\theta}_y^2 + \frac{m}{(m + \frac{J}{R^2})} x\dot{\theta}_x\dot{\theta}_y \tag{3.153}$$

By defining coefficient $B = \frac{m}{(m + \frac{J}{R^2})}$, the above two equations become

$$\ddot{x} = -Bg \sin(\theta_x) + Bx\dot{\theta}_x^2 + Bx\dot{\theta}_x\dot{\theta}_y \tag{3.154}$$

$$\ddot{y} = -Bg \sin(\theta_y) + By\dot{\theta}_y^2 + Bx\dot{\theta}_x\dot{\theta}_y \tag{3.155}$$

Consequently, the dynamic model of the ball-plate system, is described by the following state equations:

$$\begin{aligned} \ddot{x} &= -Bg \cos(\theta_x) + Bx\dot{\theta}_x^2 + By\dot{\theta}_x\dot{\theta}_y \\ \ddot{y} &= -Bg \cos(\theta_y) + By\dot{\theta}_y^2 + Bx\dot{\theta}_x\dot{\theta}_y \\ \ddot{\theta}_x &= \frac{M_{22}h_1 - M_{12}h_2}{detM} + \frac{M_{22}}{detM}u_1 - \frac{M_{12}}{detM}u_2 \\ \ddot{\theta}_y &= \frac{-M_{12}h_1 + M_{11}h_2}{detM} - \frac{M_{21}}{detM}u_1 + \frac{M_{11}}{detM}u_2 \end{aligned} \tag{3.156}$$

Next, the following state variables are defined: $x_1 = x,\ x_2 = \dot{x},\ x_3 = y,\ x_4 = \dot{y}$, $x_5 = \theta_x,\ x_6 = \dot{\theta}_x,\ x_7 = \theta_y,\ x_8 = \dot{\theta}_y$. Following this definition it holds that $M_{11} = J_P + J + mx_1^2,\ \ M_{12} = mx_1x_3,\ \ M_{21} = mx_1x_3,\ \ M_{22} = J_P + J + mx_3^2,\ \ detM = (J_P + M)^2 + (J_P + J)(mx_1^2 + mx_3^2),\ \ h_1 = -2mx_1x_2x_6 - mx_2x_3x_8 - mx_1x_4x_8 - mgx_1 \cos(x_5)$,

$h_2 = -2mx_3x_4x_8 - mx_2x_3x_6 - mx_1x_4x_6 - mgx_3\cos(x_7)$. Consequently, the state-space model of the ball and plate system becomes

$$\begin{aligned}
\dot{x}_1 &= x_2\\
\dot{x}_2 &= -Bg\sin(x_5) + Bx_1x_6^2 + Bx_3x_6x_8\\
\dot{x}_3 &= x_4\\
\dot{x}_4 &= -Bg\sin(x_7) + Bx_3x_8^2 + Bx_1x_6x_8\\
\dot{x}_5 &= x_6\\
\dot{x}_6 &= \frac{M_{22}h_1 - M_{12}h_2}{detM} + \frac{M_{22}}{detM}u_1 - \frac{M_{12}}{detM}u_2\\
\dot{x}_7 &= x_8\\
\dot{x}_8 &= \frac{-M_{12}h_1 + M_{11}h_2}{detM} - \frac{M_{21}}{detM}u_1 + \frac{M_{11}}{detM}u_2
\end{aligned} \tag{3.157}$$

Equivalently, the state-space model of the ball and plate system is written in the following matrix form:

$$\begin{pmatrix}\dot{x}_1\\ \dot{x}_2\\ \dot{x}_3\\ \dot{x}_4\\ \dot{x}_5\\ \dot{x}_6\\ \dot{x}_7\\ \dot{x}_8\end{pmatrix} = \begin{pmatrix} x_2\\ -Bg\sin(x_5) + Bx_1x_6^2 + Bx_3x_6x_8\\ x_4\\ -Bg\sin(x_7) + Bx_3x_8^2 + Bx_1x_6x_8\\ x_6\\ \dfrac{M_{22}h_1 - M_{12}h_2}{detM}\\ x_8\\ \dfrac{-M_{12}h_1 + M_{11}h_2}{detM}\end{pmatrix} + \begin{pmatrix} 0 & 0\\ 0 & 0\\ 0 & 0\\ 0 & 0\\ 0 & 0\\ \dfrac{M_{22}}{detM} & -\frac{M_{12}}{detM}\\ 0 & 0\\ -\dfrac{M_{21}}{detM} & \frac{M_{11}}{detM}\end{pmatrix}\begin{pmatrix}u_1\\ u_2\end{pmatrix} \tag{3.158}$$

Moreover, the state-space model of the ball and plate system is written into the following concise form:

$$\dot{x} = f(x) + g(x)u \tag{3.159}$$

where $x \in R^{8\times 1}$, $f(x) \in R^{8\times 1}$, $g(x) \in R^{8\times 2}$ and $u \in R^{2\times 1}$.

3.5.3 *Approximate linearization of the model of the ball and plate system*

3.5.3.1 Approximately linearized model

The dynamic model of the ball and plate system has been written in the form of (3.159). The state-space model undergoes approximate linearization around the temporary operating point (x^*, u^*), where x^* is the present value of the system's state vector and u^* is the last sampled value of the control inputs vector. The linearization relies on the first-order Taylor series expansion and on the computation of the related Jacobian matrices. The linearization point is updated at each iteration of the control algorithm. The approximately linearized model is written as

$$\dot{x} = Ax + Bu + \tilde{d} \tag{3.160}$$

where $\tilde{d}$ is the cumulative disturbance vector which may comprise: (i) modeling errors due to the truncation of higher-order terms in the Taylor series expansion, (ii) exogenous perturbations, and (iii) noise terms of any distribution. Matrices A and B are the system's Jacobians and, and by considering that $g(x) = [g_1(x), g_2x()]$, they computed as follows:

$$\begin{aligned} A &= \nabla_x[f(x) + g_1(x)u_1 + g_2(x)u_2] \mid_{(x^*,u^*)} \Rightarrow \\ A &= \nabla_x[f(x)] \mid_{(x^*,u^*)} + \nabla_x[g_1(x)]u_1 \mid_{(x^*,u^*)} + + \nabla_x[g_2(x)]u_2 \mid_{(x^*,u^*)} \end{aligned} \tag{3.161}$$

$$\begin{aligned} B &= \nabla_u[f(x) + g_1(x)u_1 + g_2(x)u_2] \mid_{(x^*,u^*)} \Rightarrow \\ B &= \nabla_x[g_1(x) + g_2(x)] \mid_{(x^*,u^*)} \Rightarrow B = \nabla_x[g(x)] \mid_{(x^*,u^*)} \end{aligned} \tag{3.162}$$

Computation of the Jacobian matrix $\nabla_x[f(x)] \mid_{(x^*,u^*)}$.

First row of the Jacobian matrix $\nabla_x[f(x)] \mid_{(x^*,u^*)}$: $\frac{\partial f_1}{\partial x_1} = 0$, $\frac{\partial f_1}{\partial x_2} = 1$, $\frac{\partial f_1}{\partial x_3} = 0$, $\frac{\partial f_1}{\partial x_4} = 0$, $\frac{\partial f_1}{\partial x_5} = 0$, $\frac{\partial f_1}{\partial x_6} = 0$, $\frac{\partial f_1}{\partial x_7} = 0$, and $\frac{\partial f_1}{\partial x_8} = 0$.

Second row of the Jacobian matrix $\nabla_x[f(x)] \mid_{(x^*,u^*)}$: $\frac{\partial f_2}{\partial x_1} = Bx_6^2$, $\frac{\partial f_2}{\partial x_2} = 0$, $\frac{\partial f_2}{\partial x_3} = Bx_6x_8$, $\frac{\partial f_2}{\partial x_4} = 0$, $\frac{\partial f_2}{\partial x_5} = -Bg\cos(x_5)$, $\frac{\partial f_2}{\partial x_6} = 2Bx_1x_6 + Bx_3x_8$, $\frac{\partial f_2}{\partial x_7} = 0$, and $\frac{\partial f_2}{\partial x_8} = Bx_3x_6$,

Third row of the Jacobian matrix $\nabla_x[f(x)] \mid_{(x^*,u^*)}$: $\frac{\partial f_3}{\partial x_1} = 0$, $\frac{\partial f_3}{\partial x_2} = 0$, $\frac{\partial f_3}{\partial x_3} = 0$, $\frac{\partial f_3}{\partial x_4} = 1$, $\frac{\partial f_3}{\partial x_5} = 0$, $\frac{\partial f_3}{\partial x_6} = 0$, $\frac{\partial f_3}{\partial x_7} = 0$, and $\frac{\partial f_3}{\partial x_8} = 0$.

Fourth row of the Jacobian matrix $\nabla_x[f(x)] \mid_{(x^*,u^*)}$: $\frac{\partial f_4}{\partial x_1} = Bx_6x_8$, $\frac{\partial f_4}{\partial x_2} = 0$, $\frac{\partial f_4}{\partial x_3} = Bx_4^2$, $\frac{\partial f_4}{\partial x_4} = 0$, $\frac{\partial f_4}{\partial x_5} = 0$, $\frac{\partial f_4}{\partial x_6} = Bx_1x_8$, $\frac{\partial f_4}{\partial x_7} = -Bg\cos(x_7)$ and $\frac{\partial f_4}{\partial x_81} = 2Bx_3x_9 + Bx_1x_6$.

Fifth row of the Jacobian matrix $\nabla_x[f(x)] \mid_{(x^*,u^*)}$: $\frac{\partial f_5}{\partial x_1} = 0$, $\frac{\partial f_5}{\partial x_2} = 0$, $\frac{\partial f_5}{\partial x_3} = 0$, $\frac{\partial f_5}{\partial x_4} = 0$, $\frac{\partial f_5}{\partial x_5} = 0$, $\frac{\partial f_5}{\partial x_6} = 1$, $\frac{\partial f_5}{\partial x_7} = 0$, and $\frac{\partial f_5}{\partial x_8} = 0$.

Sixth row of the Jacobian matrix $\nabla_x[f(x)] \mid_{(x^*,u^*)}$: for $i = 1, 2, \cdots, 8$

$$\begin{aligned} \frac{\partial f_6}{\partial x_i} &= \frac{[\frac{\partial M_{22}}{\partial x_i}h_1 + M_{22}\frac{\partial h_1}{\partial x_i} - \frac{\partial M_{12}}{\partial x_i}h_2 - M_{12}\frac{\partial h_2}{\partial x_i}]detM}{[detM^2]} \\ &- \frac{[M_{22}h_1 - M_{12}h_2]\frac{\partial detM}{\partial x_i}}{[detM^2]} \end{aligned} \tag{3.163}$$

Seventh row of the Jacobian matrix $\nabla_x[f(x)]\mid_{(x^*,u^*)}$: $\frac{\partial f_7}{\partial x_1}=0$, $\frac{\partial f_7}{\partial x_2}=0$, $\frac{\partial f_7}{\partial x_3}=0$, $\frac{\partial f_7}{\partial x_4}=0$, $\frac{\partial f_7}{\partial x_5}=0$, $\frac{\partial f_7}{\partial x_6}=0$, $\frac{\partial f_7}{\partial x_7}=0$, and $\frac{\partial f_7}{\partial x_8}=1$.

Eighth row of the Jacobian matrix $\nabla_x[f(x)]\mid_{(x^*,u^*)}$: for $i=1,2,\cdots,8$

$$\frac{\partial f_8}{\partial x_i}=\frac{[\frac{-\partial M_{12}}{\partial x_i}h_1-M_{12}\frac{\partial h_1}{\partial x_i}+\frac{\partial M_{11}}{\partial x_i}h_2+M_{11}\frac{\partial h_2}{\partial x_i}]detM}{[detM^2]}-\\ -\frac{[-M_{12}h_1+M_{11}h_2]\frac{\partial detM}{\partial x_i}}{[detM^2]} \tag{3.164}$$

Computation of the Jacobian matrix $\nabla_x[g_1(x)]\mid_{(x^*,u^*)}$.

First, second, third, fourth, fifth and seventh rows of the Jacobian matrix $\nabla_x[g_1(x)]\mid_{(x^*,u^*)}$: for $i=1,2,3,4,5,7$ $\frac{\partial g_{i1}}{\partial x_1}=0$, $\frac{\partial g_{i1}}{\partial x_2}=0$, $\frac{\partial g_{i1}}{\partial x_3}=0$, $\frac{\partial g_{i1}}{\partial x_4}=0$, $\frac{\partial g_{i1}}{\partial x_5}=0$, $\frac{\partial g_{i1}}{\partial x_6}=0$, $\frac{\partial g_{i1}}{\partial x_7}=0$, and $\frac{\partial g_{i1}}{\partial x_8}=0$.

Sixth row of the Jacobian matrix $\nabla_x[g_1(x)]\mid_{(x^*,u^*)}$: for $i=1,2,\cdots,8$

$$\frac{\partial g_{61}}{\partial x_i}=\frac{\frac{\partial M_{22}}{\partial x_i}detM-M_{22}\frac{\partial detM}{\partial x_i}}{[detM^2]} \tag{3.165}$$

Eighth row of the Jacobian matrix $\nabla_x[g_1(x)]\mid_{(x^*,u^*)}$: for $i=1,2,\cdots,8$

$$\frac{\partial g_{81}}{\partial x_i}=\frac{\frac{-\partial M_{21}}{\partial x_i}detM+M_{22}\frac{\partial detM}{\partial x_i}}{[detM^2]} \tag{3.166}$$

Computation of the Jacobian matrix $\nabla_x[g_2(x)]\mid_{(x^*,u^*)}$.

First, second, third, fourth, fifth and seventh rows of the Jacobian matrix $\nabla_x[g_2(x)]\mid_{(x^*,u^*)}$: for $i=1,2,3,4,5,7$ $\frac{\partial g_{i2}}{\partial x_1}=0$, $\frac{\partial g_{i2}}{\partial x_2}=0$, $\frac{\partial g_{i2}}{\partial x_3}=0$, $\frac{\partial g_{i2}}{\partial x_4}=0$, $\frac{\partial g_{i2}}{\partial x_5}=0$, $\frac{\partial g_{i2}}{\partial x_6}=0$, $\frac{\partial g_{i2}}{\partial x_7}=0$, and $\frac{\partial g_{i2}}{\partial x_8}=0$.

Sixth row of the Jacobian matrix $\nabla_x[g_2(x)]\mid_{(x^*,u^*)}$: for $i=1,2,\cdots,8$

$$\frac{\partial g_{61}}{\partial x_i}=\frac{\frac{-\partial M_{12}}{\partial x_i}detM+M_{12}\frac{\partial detM}{\partial x_i}}{[detM^2]} \tag{3.167}$$

Eighth row of the Jacobian matrix $\nabla_x[g_2(x)]\mid_{(x^*,u^*)}$: for $i=1,2,\cdots,8$

$$\frac{\partial g_{81}}{\partial x_i}=\frac{\frac{0\partial M_{11}}{\partial x_i}detM-M_{11}\frac{\partial detM}{\partial x_i}}{[detM^2]} \tag{3.168}$$

Next, one computes the partial derivatives of terms M_{11}, M_{12}, M_{21}, M_{22}, $detM$, h_1, and h_2.

About M_{11}, it holds that: $\frac{\partial M_{11}}{\partial x_1} = 2mx_1$, $\frac{\partial M_{11}}{\partial x_i} = 0$ for $i = 2,3,4,5,6,7,8$.

About M_{12}, it holds that: $\frac{\partial M_{12}}{\partial x_1} = mx_3$, $\frac{\partial M_{12}}{\partial x_3} = mx_1$, $\frac{\partial M_{12}}{\partial x_i} = 0$ for $i = 2,4,5,6,7,8$

About M_{21}, it holds that: $\frac{\partial M_{21}}{\partial x_1} = mx_3$, $\frac{\partial M_{21}}{\partial x_3} = mx_1$, $\frac{\partial M_{21}}{\partial x_i} = 0$ for $i = 2,4,5,6,7,8$

About M_{22}, it holds that: $\frac{\partial M_{22}}{\partial x_3} = 2mx_3$, $\frac{\partial M_{22}}{\partial x_i} = 0$ for $i = 1,2,4,5,6,7,8$.

About $detM$, it holds that: $\frac{\partial detM}{\partial x_1} = (J_P + J)(2mx_1)$, $\frac{\partial detM}{\partial x_3} = (J_P + J)(2mx_3)$, $\frac{\partial detM}{\partial x_i} = 0$ for $i = 2,4,5,6,7,8$.

About h_1, it holds that: $\frac{\partial h_1}{\partial x_1} = -2mx_2x_6 - mx_4x_8 - mg\cos(x_5)$, $\frac{\partial h_1}{\partial x_2} = -2mx_1x_6 - mx_3x_8$, $\frac{\partial h_1}{\partial x_3} = -mx_2x_8$, $\frac{\partial h_1}{\partial x_4} = -mx_1x_8$, $\frac{\partial h_1}{\partial x_5} = mgx_1\sin(x_5)$, $\frac{\partial h_1}{\partial x_6} = -2mx_1x_2$, $\frac{\partial h_1}{\partial x_7} = 0$ and $\frac{\partial h_1}{\partial x_8} = -mx_2x_3 - mx_1x_4$.

About h_2, it holds that: $\frac{\partial h_2}{\partial x_1} = -mx_1x_6$, $\frac{\partial h_2}{\partial x_2} = -2mx_3x_6$, $\frac{\partial h_1}{\partial x_3} = -2mx_4x_8 - mx_2x_6 - mg\cos(x_7)$, $\frac{\partial h_2}{\partial x_4} = -2mx_3x_8 - mx_1x_6$, $\frac{\partial h_2}{\partial x_5} = 0$, $\frac{\partial h_2}{\partial x_6} = -mx_2x_3 - mx_1x_4$, $\frac{\partial h_2}{\partial x_7} = mgx_3\sin(x_7)$ and $\frac{\partial h_2}{\partial x_8} = -2mx_3x - 4$.

3.5.3.2 Stabilizing feedback control

After linearization around its current operating point (x^*, y^*), the dynamic model of the ball and plate system is written as (Rigatos and Busawon, 2018)

$$\dot{x} = Ax + Bu + d_1 \tag{3.169}$$

Parameter d_1 stands for the linearization error in the ball and plate system's dynamic model appearing above in (3.169). The reference setpoints for the state vector of the ball and plate system are denoted by $\mathbf{x_d} = [x_1^d, \cdots, x_8^d]$. Tracking of this trajectory is achieved after applying the control input u_d. At every time instant the control input u_d is assumed to differ from the control input u appearing in (3.169) by an amount equal to Δu, that is $u_d = u + \Delta u$

$$\dot{x}_d = Ax_d + Bu_d + d_2 \tag{3.170}$$

The dynamics of the controlled system described in (3.169) can be also written as

$$\dot{x} = Ax + Bu + Bu_d - Bu_d + d_1 \tag{3.171}$$

and by denoting $d_3 = -Bu_d + d_1$ as an aggregate disturbance term one obtains

$$\dot{x} = Ax + Bu + Bu_d + d_3 \tag{3.172}$$

By subtracting (3.170) from (3.172), one has

$$\dot{x} - \dot{x}_d = A(x - x_d) + Bu + d_3 - d_2 \tag{3.173}$$

By denoting the tracking error as $e = x - x_d$ and the aggregate disturbance term as $\tilde{d} = d_3 - d_2$, the tracking error dynamics becomes

$$\dot{e} = Ae + Bu + \tilde{d} \tag{3.174}$$

For the approximately linearized model of the system a stabilizing feedback controller is developed. The controller has the form

$$u(t) = -Ke(t) \tag{3.175}$$

with $K = \frac{1}{r}B^TP$ where P is a positive definite symmetric matrix which is obtained from the solution of the Riccati equation (Rigatos and Busawon, 2018)

$$A^TP + PA + Q - P\left(\frac{2}{r}BB^T - \frac{1}{\rho^2}LL^T\right)P = 0 \tag{3.176}$$

where Q is a positive semi-definite symmetric matrix. The diagram of the considered control loop is depicted in Figure 3.50. The global stability properties of the control method can be proven following the procedure presented in Section 1.1.2.5.

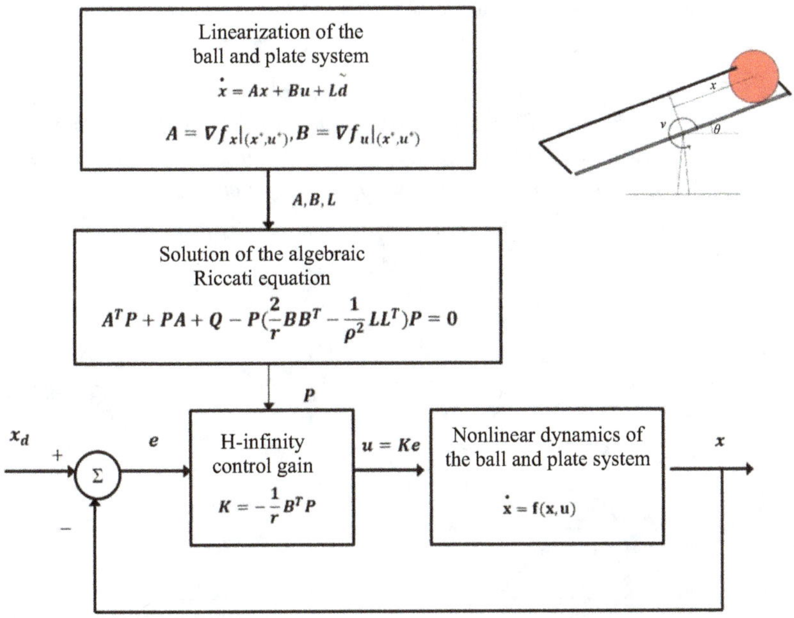

Figure 3.50 The nonlinear optimal (H-infinity) control loop of the ball and plate dynamical system

3.5.4 Simulation tests

The performance of the proposed nonlinear optimal control scheme for the ball and plate system has been tested and confirmed through simulation experiments. To implement this control method the algebraic Riccati equation of (3.176) had to be solved at each time-step of the algorithm. Besides, with the use of the H_∞ Kalman Filter state estimation-based control has been applied. State variables x_2 and x_4 were not measurable and had to be estimated with the use of the H-infinity Kalman Filter. The obtained results are depicted in Figures 3.51 to 3.66 The real values of the state

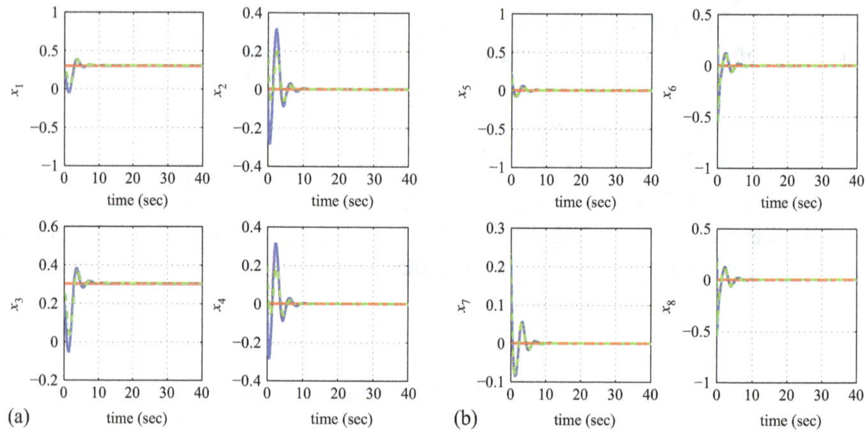

Figure 3.51 Tracking of setpoint 1 by the ball and plate system: (a) convergence of the state variables x_1 to x_4 (blue lines) to the reference setpoints (red lines) and KF-estimated state variables (green lines); (b) convergence of the state variables x_5 to x_8 to the reference setpoints

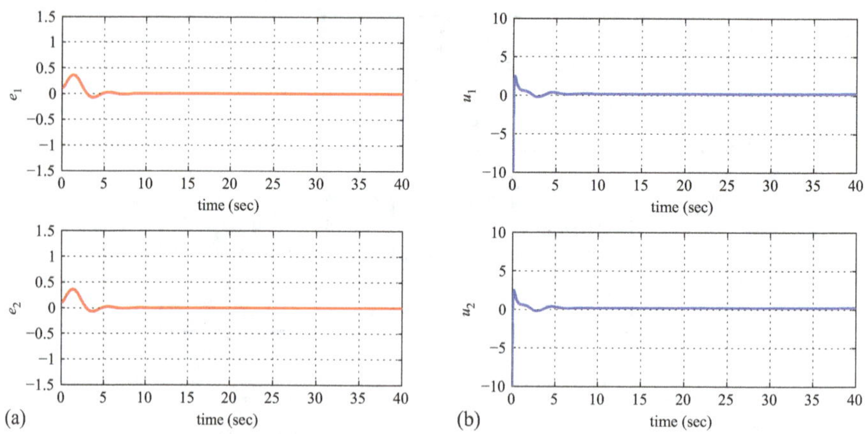

Figure 3.52 Tracking of setpoint 1 by the ball and plate system: (a) tracking error for state variables x_1 and x_3 (Cartesian coordinates of the ball); (b) control inputs u_1 and u_2 applied on the rotating plate

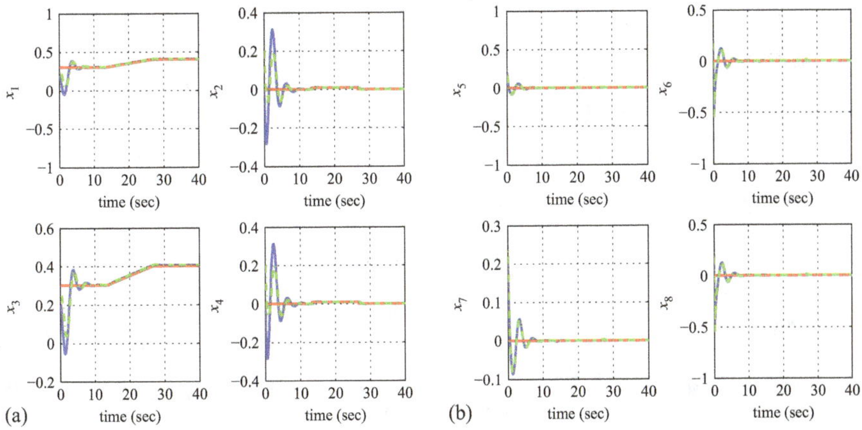

Figure 3.53 Tracking of setpoint 2 by the ball and plate system: (a) convergence of the state variables x_1 to x_4 (blue lines) to the reference setpoints (red lines) and KF-estimated state variables (green lines); (b) convergence of the state variables x_5 to x_8 to the reference setpoints

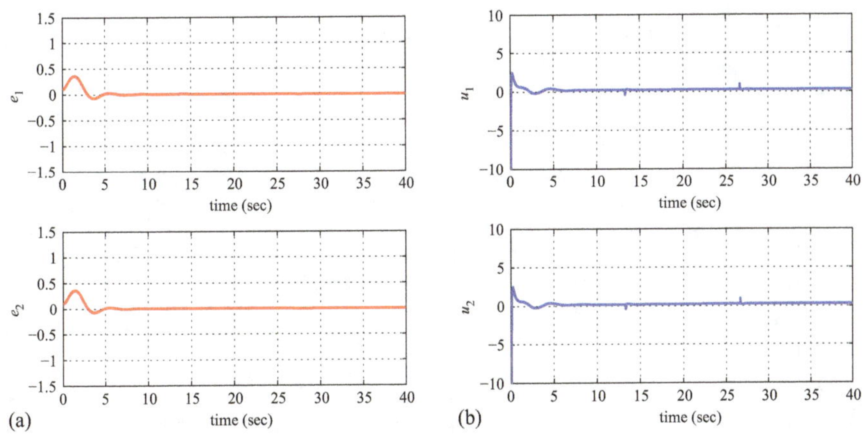

Figure 3.54 Tracking of setpoint 2 by the ball and plate system: (a) tracking error for state variables x_1 and x_3 (Cartesian coordinates of the ball); (b) control inputs u_1 and u_2 applied on the rotating plate

variables are depicted in blue, the estimated values of the state variables are printed in green while the related setpoints are plotted in red. It can be observed that in all test cases fast and accurate tracking of the reference setpoints was achieved under moderate variations of the control inputs.

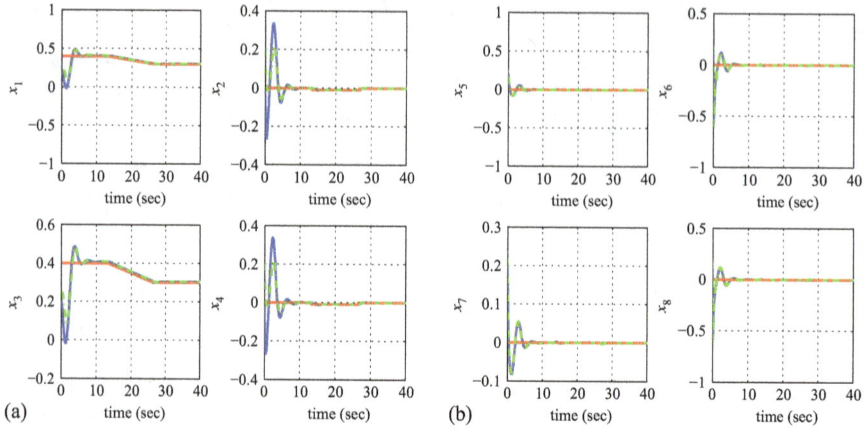

Figure 3.55 Tracking of setpoint 3 by the ball and plate system: (a) convergence of the state variables x_1 to x_4 (blue lines) to the reference setpoints (red lines) and KF-estimated state variables (green lines); (b) convergence of the state variables x_5 to x_8 to the reference setpoints

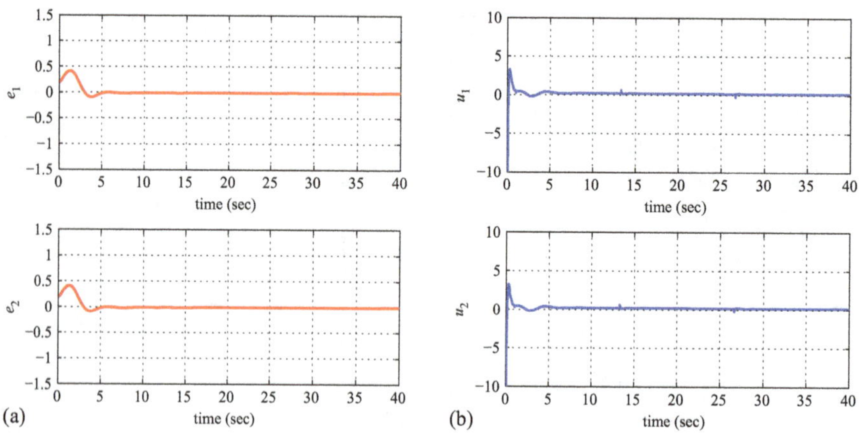

Figure 3.56 Tracking of setpoint 3 by the ball and plate system: (a) tracking error for state variables x_1 and x_3 (Cartesian coordinates of the ball); (b) control inputs u_1 and u_2 applied on the rotating plate

3.6 Nonlinear control of 3-DOF unmanned surface vessels

3.6.1 Outline

The solution of the nonlinear optimal control problem for 3-DOF unmanned surface vessels (USVs) remains always a technical challenge (Rigatos and Busawon,

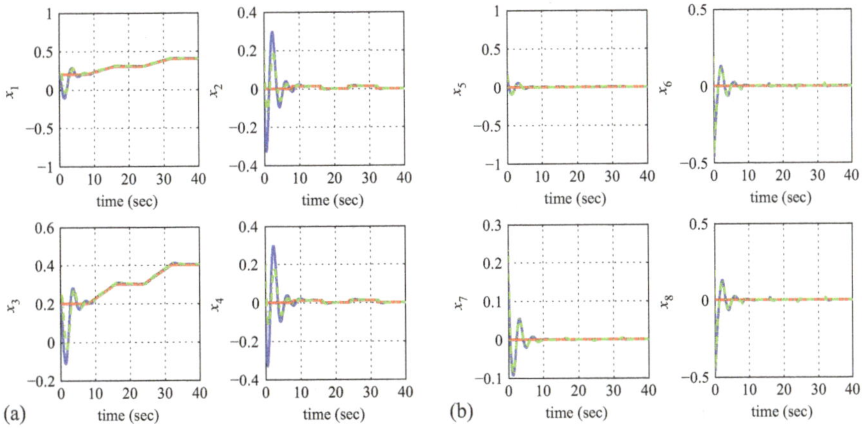

Figure 3.57 Tracking of setpoint 4 by the ball and plate system: (a) convergence of the state variables x_1 to x_4 (blue lines) to the reference setpoints (red lines) and KF-estimated state variables (green lines); (b) convergence of the state variables x_5 to x_8 to the reference setpoints

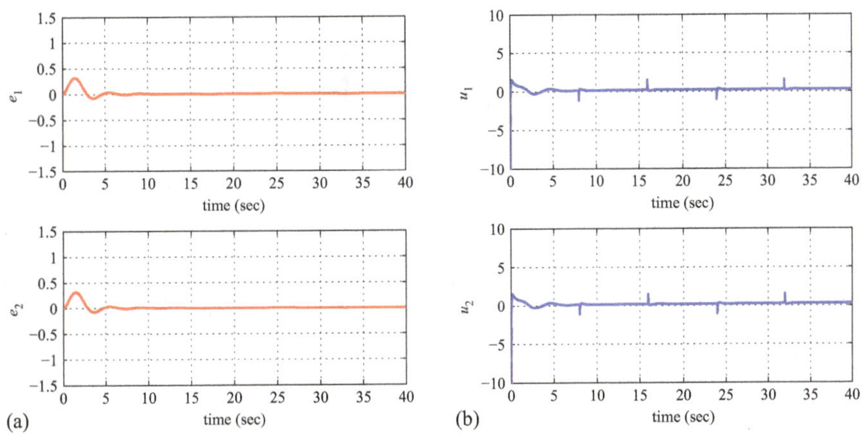

Figure 3.58 Tracking of setpoint 4 by the ball and plate system: (a) tracking error for state variables x_1 and x_3 (Cartesian coordinates of the ball); (b) control inputs u_1 and u_2 applied on the rotating plate

2018; Wang and Karimi, 2020; Fang, Zergeroglou *et al.*, 2004; Chen, Yang *et al.*, 2020). The dynamic model of 3-DOF USVs is a highly nonlinear and multivariable one (Huang, Gang *et al.*, 2019; Qin, Li *et al.*, 2020; Yang, Guo *et al.*, 2019). The vessel's model can be fully actuated (when receiving as control inputs the engine's forward propulsion, the rudder's torque and the thrusters' lateral propulsion. On the

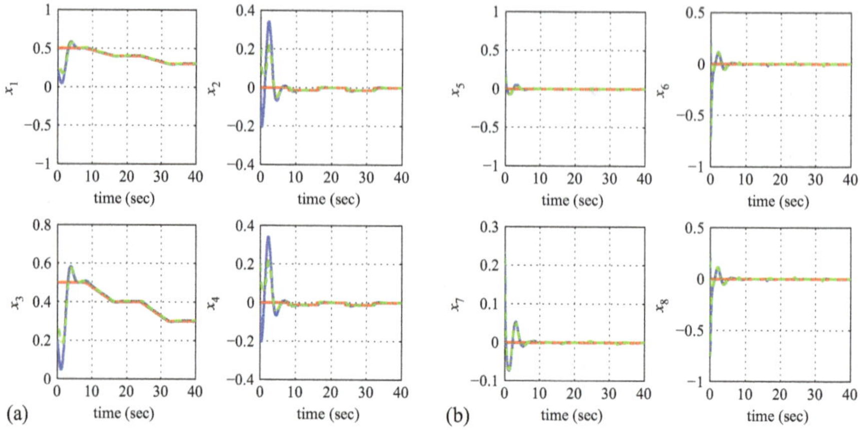

Figure 3.59 Tracking of setpoint 5 by the ball and plate system: (a) convergence of the state variables x_1 to x_4 (blue lines) to the reference setpoints (red lines) and KF-estimated state variables (green lines); (b) convergence of the state variables x_5 to x_8 to the reference setpoints

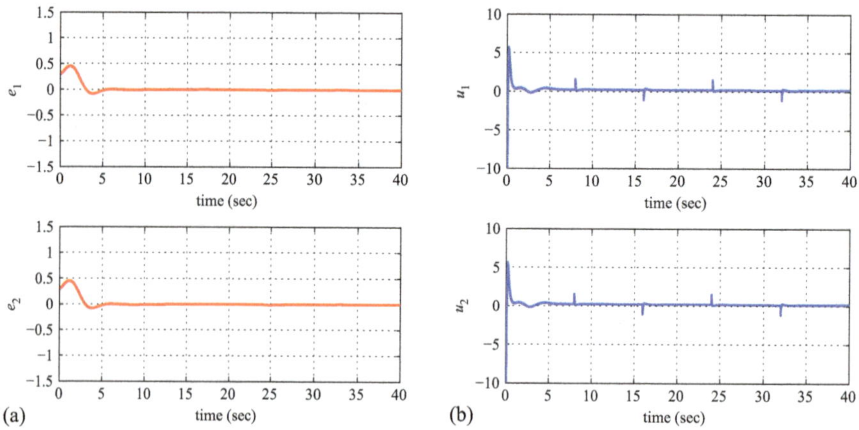

Figure 3.60 Tracking of setpoint 5 by the ball and plate system: (a) tracking error for state variables x_1 and x_3 (Cartesian coordinates of the ball); (b) control inputs u_1 and u_2 applied on the rotating plate

other side this dynamic model can be underactuated when thrusters' lateral propulsion is missing (Xiang, Yu *et al.*, 2020; Qiu, Wang *et al.*, 2020; Gao *et al.*, 2019). Solution of the control problem for USVs enables autonomous navigation of the individual vessels or coordination of multiple vessels' motion (Park and Yoo, 2019; Liang, Qu

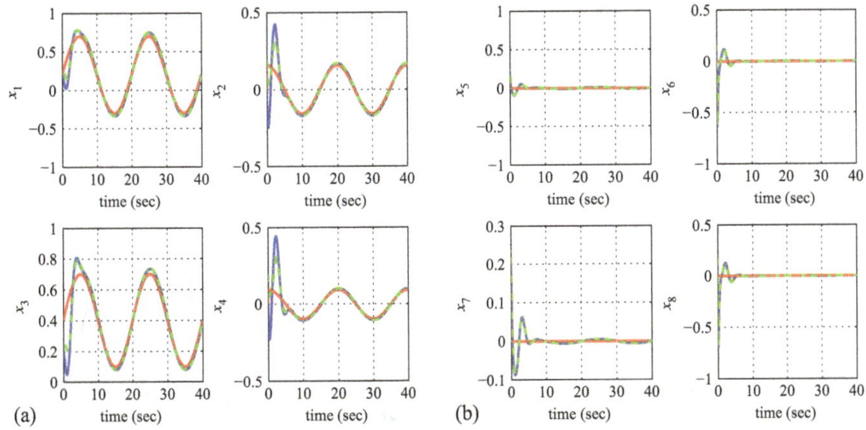

Figure 3.61 Tracking of setpoint 6 by the ball and plate system: (a) convergence of the state variables x_1 to x_4 (blue lines) to the reference setpoints (red lines) and KF-estimated state variables (green lines); (b) convergence of the state variables x_5 to x_8 to the reference setpoints

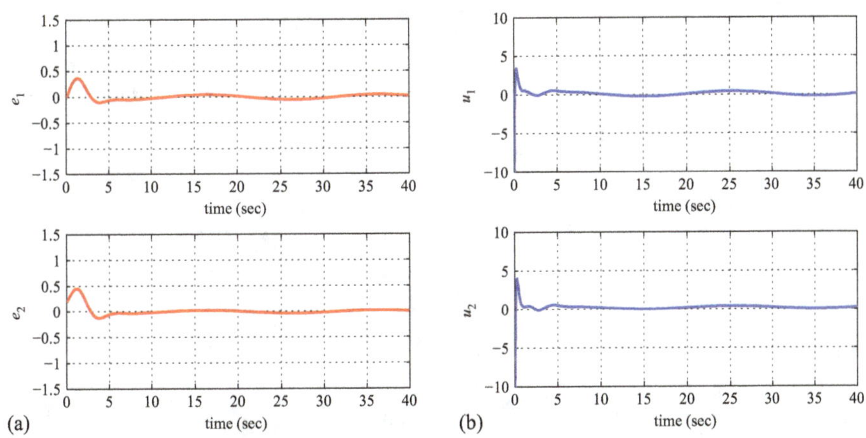

Figure 3.62 Tracking of setpoint 6 by the ball and plate system: (a) tracking error for state variables x_1 and x_3 (Cartesian coordinates of the ball); (b) control inputs u_1 and u_2 applied on the rotating plate

et al., 2020; Liang, Qu *et al.*, 2019). Several nonlinear control approaches, including robust and adaptive control schemes, have been developed. Precise tracking of reference setpoints and global stability properties of the USVs control loop are the design objectives in all related nonlinear control methods. Results on autonomous navigation of USVs with the use of adaptive control methods can be found in Chen,

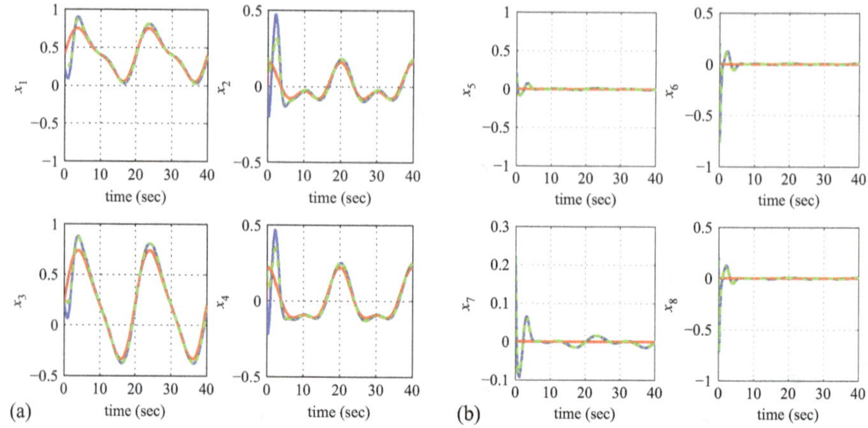

Figure 3.63 Tracking of setpoint 7 by the ball and plate system: (a) convergence of the state variables x_1 to x_4 (blue lines) to the reference setpoints (red lines) and KF-estimated state variables (green lines); (b) convergence of the state variables x_5 to x_8 to the reference setpoints

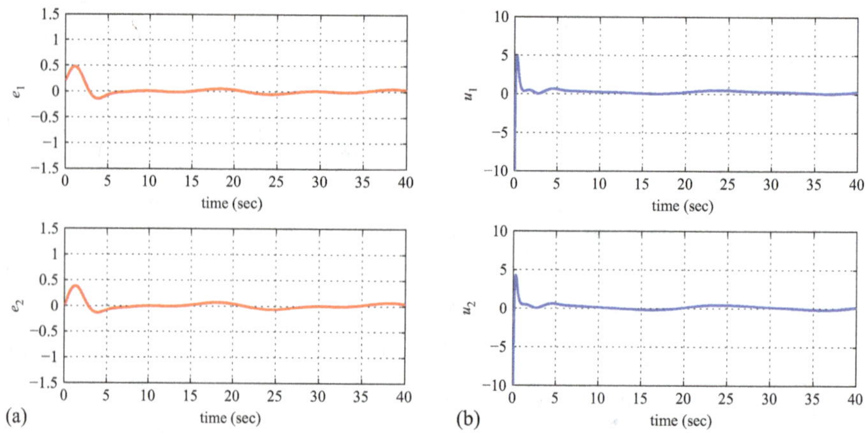

Figure 3.64 Tracking of setpoint 7 by the ball and plate system: (a) tracking error for state variables x_1 and x_3 (Cartesian coordinates of the ball); (b) control inputs u_1 and u_2 applied on the rotating plate

Zhang *et al.*, 2020, Ali-Hussain *et al.*, 2020, Chen, Liu *et al.*, 2018, and Rout, Cui *et al.*, 2020. Besides, results on the use of sliding-mode and robust control methods for solving the autonomous navigation problem of USVs are given in Liu, Hou *et al.*, 2020, Hu, Wang *et al.*, 2016, Zhou, Qi *et al.*, 2020, and Eshafani *et al.*, 2019. Additionally, the use of observers and disturbance estimators in control loops of

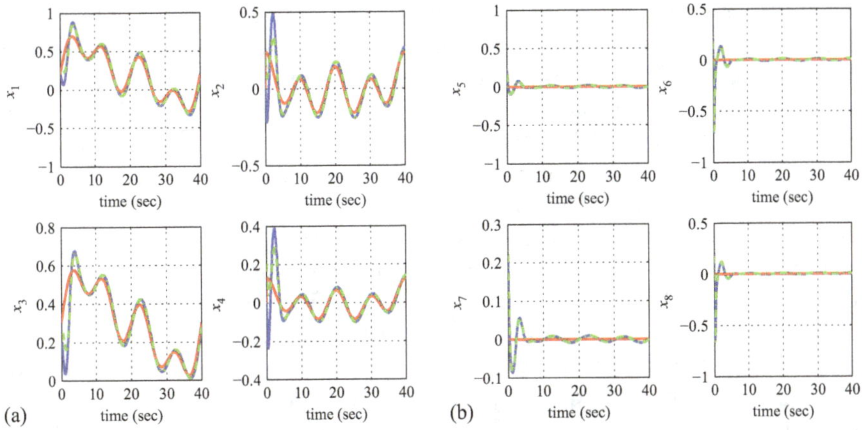

Figure 3.65 Tracking of setpoint 8 by the ball and plate system: (a) convergence of the state variables x_1 to x_4 (blue lines) to the reference setpoints (red lines) and KF-estimated state variables (green lines); (b) convergence of the state variables x_5 to x_8 to the reference setpoints

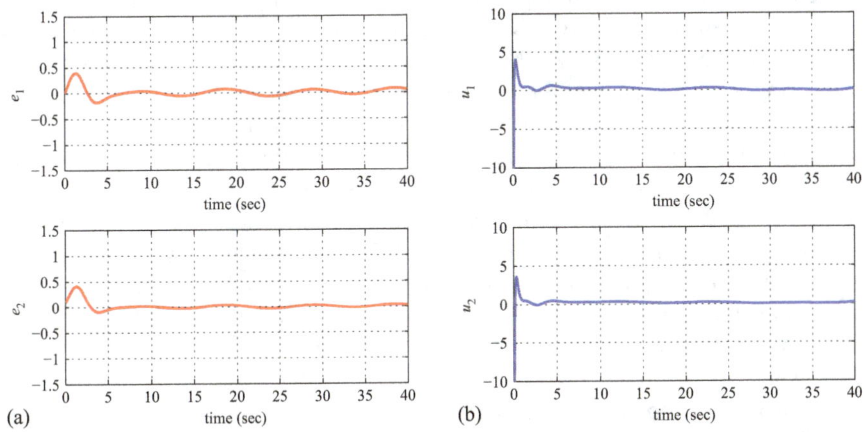

Figure 3.66 Tracking of setpoint 8 by the ball and plate system: (a) tracking error for state variables x_1 and x_3 (Cartesian coordinates of the ball); (b) control inputs u_1 and u_2 applied on the rotating plate

USVs can be found in Li, Guo *et al.*, 2019, Wang *et al.*, 2019, and Wang and Su, 2021.

In the present section, the previously analyzed nonlinear optimal control approach is developed for the dynamic model of 3-DOF unmanned surface vessels (USVs) (Rigatos and Busawon, 2018). This model is in the typical form of a robotic system, comprising an inertia matrix, a Coriolis and a centrifugal forces matrix, while receiving inputs from propulsion forces and rudder torques. It is shown that this model can be written in the affine-in-the-input nonlinear state-space form. Besides, it is proven that in the case of full actuation it satisfies differential flatness properties. Next, the dynamic model of the 3-DOF USV undergoes approximate linearization with the use of the first-order Taylor series expansion and through the computation of the associated Jacobian matrices (Rigatos, Siano and Cecati, 2015; Rigatos and Tzafestas, 2007; Basseville and Nikiforov, 1993; Rigatos and Zhang, 2009). Linearization takes place around a temporary operating point which is recomputed at each time-step of the control algorithm. This operating point is redefined at each sampling instance by the present value of the USV's state vector and by the last sampled value of the control inputs vector. The modeling error which is due to truncation of higher-order terms in the Taylor series expansion is viewed as a perturbation that is asymptotically compensated by the robustness of the control algorithm.

For the approximately linearized dynamic model of the 3-DOF USV, a stabilizing H-infinity feedback controller is designed (Rigatos and Busawon, 2018). This controller achieves the solution of the optimal control problem for the dynamics of the 3-DOF USV under model uncertainties and external disturbances. It can be stated that the H-infinity-controller represents a min–max differential game taking place between (i) the control inputs of the USV which try to minimize a quadratic cost function of the state vector's tracking error and (ii) the model uncertainty and external disturbance terms which try to maximize this cost function. To select the stabilizing feedback gains of the H-infinity controller, an algebraic Riccati equation is repetitively solved at each time-step of the control method (Rigatos, 2011; Rigatos, 2015; Rigatos and Karapanou, 2020). The global stability properties of the control scheme are proven through Lyapunov analysis. First, it is demonstrated that the control loop of the USV satisfies the H-infinity tracking performance criterion (Rigatos and Busawon, 2018; Tousssaint, Basar and Bullo, 2000). This signifies elevated robustness under model imprecision and external perturbations. Additionally, to implement state estimation-based control without the need to measure the entire state vector of the USV, the H-infinity Kalman Filter is used as a robust state estimator (Rigatos and Busawon, 2018; Gibbs, 2012).

The proposed nonlinear optimal control method retains the known advantages of linear optimal control, that is fast and accurate tracking of reference setpoints under moderate variations of the control inputs. Comparing to past attempts to solve the nonlinear optimal (H-infinity) control problem, the following advancements can be noted: (i) in past results the linearization of the system's dynamics is performed at points of the desirable trajectories (setpoints), whereas in the section's control method linearization is performed around a time-varying operating point which is redefined at each sampling instance by the present value of the system's state vector and by the last sampled value of the control inputs vector; (ii) in past results the system is considered to be exclusively in the affine-in-the-input state-space form with a control inputs gain matrix which is time-invariant. On the contrary, in the present section's approach there is no constraint to find the system in the affine-in-the-input state space form, while even in such a state-space description the control inputs gain matrix can be time-varying; (iii) in past approaches a different Lyapunov analysis procedure has been used to prove stability properties. In the present section, a novel Lyapunov function has been introduced to prove the global stability properties of the nonlinear optimal control scheme.

3.6.2 Dynamic model of the Unmanned Surface Vessel

3.6.2.1 Dynamics of the USV

The motion of the Unmanned Surface Vessel (USV) is described in both a body-fixed reference frame and in an inertial (earth-fixed) reference frame (Figure 3.67). In the body-fixed reference frame the joint kinematic and dynamic model of the USV comprises the following two-equations in matrix form (Rigatos and Busawon, 2018; Fang, Zergeroglou *et al.*, 2004; Rigatos, 2011):

$$\begin{aligned} &\dot{\eta} = R(\psi)v \Rightarrow \dot{v} = R^{-1}(\psi)\dot{\eta} \\ &M\dot{v} + Dv = \tau \end{aligned} \tag{3.177}$$

where $\eta(t) = [x(t)\, y(t)\, \psi(t)]^T \in R^{3\times 1}$ are the Cartesian coordinates and the orientation angle of the vessel in the inertial reference frame $OXYZ$. The associate velocities vector in the inertial reference frame is $\dot{\eta}(t) = [\dot{x}(t)\, \dot{y}(t)\, \dot{\psi}(t)]^T \in R^{3\times 1}$. The velocities vector in the body-fixed reference frame $\Sigma V_1 V_2 V_3$ is denoted as $v(t) = [v_1\ v_2\ v_3]^T$ where $v_!$ is the surge velocity, v_2 is the sway velocity and v_3 is the yaw turn velocity of the vessel.

In the above model, matrices M, D, and R are given by (Rigatos and Busawon, 2018; Fang, Zergeroglou *et al.*, 2004; Rigatos, 2011)

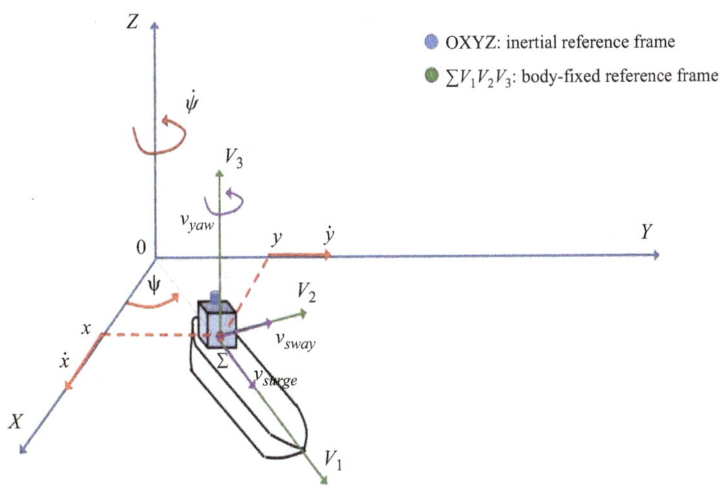

Figure 3.67 Monitoring of the motion of the USV in an inertial reference frame OXYZ as well as in a body-fixed reference frame $\Sigma V_1 V_2 V_3$

$$M = \begin{pmatrix} m_{11} & 0 & 0 \\ 0 & m_{22} & m_{33} \\ 0 & m_{23} & m_{33} \end{pmatrix}$$
$$D = \begin{pmatrix} d_{11} & 0 & 0 \\ 0 & d_{22} & d_{23} \\ 0 & d_{32} & d_{33} \end{pmatrix} \quad R = \begin{pmatrix} \cos(\psi) & -\sin(\psi) & 0 \\ \sin(\psi) & \cos(\psi) & 0 \\ 0 & 0 & 1 \end{pmatrix} \tag{3.178}$$

The dynamic model of the USV is expressed next exclusively in the inertial reference frame. Using the previous relations it holds that (Rigatos and Busawon, 2018; Fang, Zergeroglou *et al.*, 2004; Rigatos, 2011)

$$\tilde{J}\ddot{\eta} + \tilde{C}(\eta, \dot{\eta})\dot{\eta} + \tilde{F}\dot{\eta} = \tau^* \tag{3.179}$$

where $\tilde{J} = RMR^T \in R^{3\times3}$, $\tilde{C}(\eta, \dot{\eta}) = RM\dot{R}^T \in R^{3\times3}$, $\tilde{F} = RDR^T \in R^{3\times3}$ and $\tau^* = R\tau \in R^{3\times1}$. It holds that

$$\tilde{J}(\eta) = \begin{pmatrix} m_{11}\cos^2(\psi) + m_{22}\sin^2(\psi) & (m_{11} - m_{22})\sin(\psi)\cos(\psi) & -m_{23}\sin(\psi) \\ (m_{11} - m_{22})\sin(\psi)\cos(\psi) & m_{11}sin^2(\psi) + m_{22}\cos^2(\psi) & m_{23}\cos(\psi) \\ -m_{23}\sin(\psi) & m_{23}\cos(\psi) & m_{33} \end{pmatrix} \tag{3.180}$$

$$\tilde{C}(\eta, \dot{\eta}) = \begin{pmatrix} \dot{\psi}(m_{22} - m_{11})\sin(\psi)\cos(\psi) & \dot{\psi}(m_{11}\cos^2(\psi) + m_{22}\sin^2(\psi)) & 0 \\ -\dot{\psi}(m_{11}\sin^2(\psi) + m_{22}\cos^2(\psi)) & \dot{\psi}(m_{11} - m_{22})\sin(\psi)\cos(\psi) & 0 \\ -\dot{\psi}m_{23}\cos(\psi) & -\dot{\psi}m_{23}\sin(\psi) & 0 \end{pmatrix} \tag{3.181}$$

$$\tilde{F}(\eta) = \begin{pmatrix} d_{11}\cos^2(\psi) + d_{22}\sin^2(\psi) & (d_{11} - d_{22})\sin(\psi)\cos(\psi) & -d_{23}\sin(\psi) \\ (d_{11} - d_{22})\sin(\psi)\cos(\psi) & d_{11}\sin^2(\psi) + d_{22}\cos^2(\psi) & d_{23}\cos(\psi) \\ -d_{32}\sin(\psi) & d_{22}\cos(\psi) & d_{23} \end{pmatrix} \tag{3.182}$$

Using (3.179) the dynamic model of the USV can be written as (Rigatos and Busawon, 2018; Fang, Zergeroglou *et al.*, 2004; Rigatos, 2011)

$$\begin{aligned} &\tilde{J}(\eta)\ddot{\eta} + \tilde{C}(\eta, \dot{\eta})\dot{\eta} + \tilde{F}(\eta)\dot{\eta} = \tau^* \Rightarrow \\ &\ddot{\eta} = -\tilde{J}^{-1}(\eta)\tilde{C}(\eta, \dot{\eta})\dot{\eta} - \tilde{J}^{-1}(\eta)\tilde{F}(\eta)\dot{\eta} + \tilde{J}^{-1}(\eta)\tau^* \Rightarrow \\ &\ddot{\eta} = -\tilde{J}^{-1}(\eta)[\tilde{C}(\eta, \dot{\eta}) + \tilde{F}(\eta)]\dot{\eta} + \tilde{J}^{-1}\tau^* \end{aligned} \tag{3.183}$$

The inverse matrix $\tilde{J}^{-1}(\eta)$ is computed first. It holds that

$$\tilde{J}(\eta) = \begin{pmatrix} \tilde{J}_{11} & \tilde{J}_{12} & \tilde{J}_{13} \\ \tilde{J}_{21} & \tilde{J}_{22} & \tilde{J}_{23} \\ \tilde{J}_{31} & \tilde{J}_{32} & \tilde{J}_{33} \end{pmatrix} \tag{3.184}$$

with $\tilde{J}_{11} = m_{11}\cos^2(\psi) + m_{22}\sin^2(\psi)$, $\tilde{J}_{12} = (m_{11} - m_{22})\sin(\psi)\cos(\psi)$, $\tilde{J}_{13} = -m_{23}\sin(\psi)$, $\tilde{J}_{21} = (m_{11} - m_{22})\sin(\psi)\cos(\psi)$, $\tilde{J}_{22} = m_{11}sin^2(\psi) + m_{22}\cos^2(\psi)$, $\tilde{J}_{23} = m_{23}\cos(\psi)$, $\tilde{J}_{31} = -m_{23}\sin(\psi)$, $\tilde{J}_{32} = m_{23}\cos(\psi)$ and $J_{33} = m_{33}$.

The determinant of matrix $\tilde{J}(\eta,\dot{\eta})$ is

$$\begin{aligned} detJ &= \tilde{J}_{11}(\tilde{J}_{22}\tilde{J}_{33} - \tilde{J}_{32}\tilde{J}_{23}) - \tilde{J}_{12}(\tilde{J}_{21}\tilde{J}_{33} - \tilde{J}_{31}\tilde{J}_{23}) \\ &\quad + \tilde{J}_{13}(\tilde{J}_{21}\tilde{J}_{32} - \tilde{J}_{31}\tilde{J}_{22}) \end{aligned} \tag{3.185}$$

The sub-determinants of matrix $\tilde{J}(\eta,\dot{\eta})$ are defined as: $J_{11} = \tilde{J}_{22}\tilde{J}_{33} - \tilde{J}_{32}\tilde{J}_{23}$, $J_{12} = \tilde{J}_{21}\tilde{J}_{33} - \tilde{J}_{31}\tilde{J}_{23}$, $J_{13} = \tilde{J}_{21}\tilde{J}_{32} - \tilde{J}_{31}\tilde{J}_{22}$, $J_{21} = \tilde{J}_{12}\tilde{J}_{33} - \tilde{J}_{32}\tilde{J}_{13}$, $J_{22} = \tilde{J}_{11}\tilde{J}_{33} - \tilde{J}_{31}\tilde{J}_{33}$, $J_{23} = \tilde{J}_{11}\tilde{J}_{22} - \tilde{J}_{31}\tilde{J}_{12}$, $J_{31} = \tilde{J}_{12}\tilde{J}_{23} - \tilde{J}_{22}\tilde{J}_{13}$, $J_{32} = \tilde{J}_{11}\tilde{J}_{23} - \tilde{J}_{21}\tilde{J}_{13}$, $J_{33} = \tilde{J}_{11}\tilde{J}_{22} - \tilde{J}_{21}\tilde{J}_{12}$.

Thus, the inverse of matrix $\tilde{J}(\eta)$ is

$$\tilde{J}^{-1}(\eta) = \frac{1}{det\tilde{J}} \begin{pmatrix} J_{11} & -J_{21} & J_{31} \\ -J_{12} & J_{22} & -J_{32} \\ J_{13} & -J_{23} & J_{33} \end{pmatrix} \tag{3.186}$$

Moreover, the vector of the Coriolis and centrifugal forces $\tilde{C}(\eta,\dot{\eta})\dot{\eta}$ is written in the following form:

$$C(\eta,\dot{\eta}) = \begin{pmatrix} c_1 \\ c_2 \\ c_3 \end{pmatrix} = \begin{pmatrix} \tilde{C}_{11}\dot{\eta}_1 + \tilde{C}_{12}\dot{\eta}_2 + \tilde{C}_{13}\dot{\eta}_3 \\ \tilde{C}_{21}\dot{\eta}_1 + \tilde{C}_{22}\dot{\eta}_2 + \tilde{C}_{23}\dot{\eta}_3 \\ \tilde{C}_{31}\dot{\eta}_1 + \tilde{C}_{32}\dot{\eta}_2 + \tilde{C}_{33}\dot{\eta}_3 \end{pmatrix} \tag{3.187}$$

where $\tilde{C}_{11} = \dot{\psi}(m_{22} - m_{11})\sin(\psi)\cos(\psi)$, $\tilde{C}_{12} = \dot{\psi}(m_{11}\cos^2(\psi) + m_{22}\sin^2(\psi))$, $\tilde{C}_{13} = 0$, $\tilde{C}_{21} = -\dot{\psi}(m_{11}\sin^2(\psi) + m_{22}\cos^2(\psi))$, $\tilde{C}_{22} = \dot{\psi}(m_{11} - m_{22})\sin(\psi)\cos(\psi)$, $\tilde{C}_{23} = 0$, $\tilde{C}_{31} = -\dot{\psi}m_{23}\cos(\psi)$, $\tilde{C}_{32} = -\dot{\psi}m_{23}\sin(\psi)$, $\tilde{C}_{33} = 0$.

Additional, the damping forces vector $\tilde{F}(\eta)\dot{\eta}$ can be written in the form:

$$Q(\eta,\dot{\eta}) = \begin{pmatrix} q_1 \\ q_2 \\ q_3 \end{pmatrix} = \begin{pmatrix} \tilde{F}_{11}\dot{\eta}_1 + \tilde{F}_{12}\dot{\eta}_2 + \tilde{F}_{13}\dot{\eta}_3 \\ \tilde{F}_{21}\dot{\eta}_1 + \tilde{F}_{22}\dot{\eta}_2 + \tilde{F}_{23}\dot{\eta}_3 \\ \tilde{F}_{31}\dot{\eta}_1 + \tilde{F}_{32}\dot{\eta}_2 + \tilde{F}_{33}\dot{\eta}_3 \end{pmatrix} \tag{3.188}$$

where $\tilde{F}_{11} = d_{11}\cos^2(\psi) + d_{22}\sin^2(\psi)$, $\tilde{F}_{12} = (d_{11} - d_{22})\sin(\psi)\cos(\psi)$, $\tilde{F}_{13} = -d_{23}\sin(\psi)$, $\tilde{F}_{21} = (d_{11} - d_{22})\sin(\psi)\cos(\psi)$, $\tilde{F}_{22} = d_{11}sim^2(\psi) + d_{22}\cos^2(\psi)$, $\tilde{F}_{23} = d_{23}\cos(\psi)$, $\tilde{F}_{31} = -d_{32}\sin(\psi)$, $\tilde{F}_{32} = d_{22}\cos(\psi)$, $\tilde{F}_{33} = d_{23}$.

Consequently, the dynamic model of the autonomous surface vessel is written as

$$\ddot{\eta} = -\tilde{J}^{-1}(\eta)C(\eta,\dot{\eta}) - \tilde{J}^{-1}(\eta)Q(\eta,\dot{\eta}) + \tilde{J}^{-1}(\eta)\tau^* \quad (3.189)$$

or equivalently

$$\begin{pmatrix}\ddot{\eta}_1\\ \ddot{\eta}_2\\ \ddot{\eta}_3\end{pmatrix} = -\frac{1}{det\tilde{J}}\begin{pmatrix}J_{11} & -J_{21} & J_{31}\\ -J_{12} & J_{22} & -J_{32}\\ J_{13} & -J_{23} & J_{33}\end{pmatrix}\begin{pmatrix}c_1\\ c_2\\ c_3\end{pmatrix} - \frac{1}{det\tilde{J}}\begin{pmatrix}J_{11} & -J_{21} & J_{31}\\ -J_{12} & J_{22} & -J_{32}\\ J_{13} & -J_{23} & J_{33}\end{pmatrix}\begin{pmatrix}q_1\\ q_2\\ q_3\end{pmatrix}$$
$$+\frac{1}{det\tilde{J}}\begin{pmatrix}J_{11} & -J_{21} & J_{31}\\ -J_{12} & J_{22} & -J_{32}\\ J_{13} & -J_{23} & J_{33}\end{pmatrix}\begin{pmatrix}\tau_1\\ \tau_2\\ \tau_3\end{pmatrix} \quad (3.190)$$

or similarly

$$\begin{pmatrix}\ddot{\eta}_1\\ \ddot{\eta}_2\\ \ddot{\eta}_3\end{pmatrix} = -\frac{1}{det\tilde{J}}\begin{pmatrix}J_{11} & -J_{21} & J_{31}\\ -J_{12} & J_{22} & -J_{32}\\ J_{13} & -J_{23} & J_{33}\end{pmatrix}\begin{pmatrix}c_1+q_1\\ c_2+q_2\\ c_3+q_3\end{pmatrix}$$
$$+\frac{1}{det\tilde{J}}\begin{pmatrix}J_{11} & -J_{21} & J_{31}\\ -J_{12} & J_{22} & -J_{32}\\ J_{13} & -J_{23} & J_{33}\end{pmatrix}\begin{pmatrix}\tau_1\\ \tau_2\\ \tau_3\end{pmatrix} \quad (3.191)$$

Consequently, the dynamic model of the unmanned surface vessel can be written as

$$\begin{pmatrix}\ddot{\eta}_1\\ \\ \ddot{\eta}_2\\ \\ \ddot{\eta}_3\end{pmatrix} = \begin{pmatrix}\dfrac{J_{11}(c_1+q_1)+J_{21}(c_2+q_2)+J_{31}(c_3+q_3)}{detJ}\\ \\ \dfrac{J_{12}(c_1+q_1)+J_{22}(c_2+q_2)+J_{32}(c_3+q_3)}{detJ}\\ \\ \dfrac{J_{13}(c_1+q_1)+J_{23}(c_2+q_2)+J_{33}(c_3+q_3)}{detJ}\end{pmatrix}$$
$$+\begin{pmatrix}\dfrac{J_{11}}{detJ}\\ \\ -\dfrac{J_{12}}{detJ}\\ \\ \dfrac{J_{13}}{detJ}\end{pmatrix}\tau_1 + \begin{pmatrix}-\dfrac{J_{21}}{detJ}\\ \\ \dfrac{J_{22}}{detJ}\\ \\ -\dfrac{J_{23}}{detJ}\end{pmatrix}\tau_2 + \begin{pmatrix}\dfrac{J_{31}}{detJ}\\ \\ -\dfrac{J_{32}}{detJ}\\ \\ \dfrac{J_{33}}{detJ}\end{pmatrix}\tau_3 \quad (3.192)$$

Next, the following state vector is defined for the USV $x=[x_1,x_2,x_3,x_4,x_5,x_6]^T$, or $x=[\eta_1,\dot{\eta}_1,\eta_2,\dot{\eta}_2,\eta_3,\dot{\eta}_3]^T$, that is $x=[x,\dot{x},y,\dot{y},\psi,\dot{\psi}]^T$. Besides, the control inputs vector is defined as $u=[u_1,u_2,u_3]^T$ or $u=[\tau_1,\tau_2,\tau_3]^T$. Then, the dynamic model of the USV is written as

$$\begin{pmatrix}\dot{x}_1\\ \dot{x}_2\\ \dot{x}_3\\ \dot{x}_4\\ \dot{x}_5\\ \dot{x}_6\end{pmatrix}=\begin{pmatrix}x_2\\ \dfrac{J_{11}(c_1+q_1)+J_{21}(c_2+q_2)+J_{31}(c_3+q_3)}{detJ}\\ x_4\\ \dfrac{J_{12}(c_1+q_1)+J_{22}(c_2+q_2)+J_{32}(c_3+q_3)}{detJ}\\ x_6\\ \dfrac{J_{13}(c_1+q_1)+J_{23}(c_2+q_2)+J_{33}(c_3+q_3)}{detJ}\end{pmatrix}$$
$$+\begin{pmatrix}0\\ \dfrac{J_{11}}{detJ}\\ 0\\ -\dfrac{J_{12}}{detJ}\\ 0\\ \dfrac{J_{13}}{detJ}\end{pmatrix}u_1+\begin{pmatrix}0\\ -\dfrac{J_{21}}{detJ}\\ 0\\ \dfrac{J_{22}}{detJ}\\ 0\\ -\dfrac{J_{23}}{detJ}\end{pmatrix}u_2+\begin{pmatrix}0\\ \dfrac{J_{31}}{detJ}\\ 0\\ -\dfrac{J_{32}}{detJ}\\ 0\\ \dfrac{J_{33}}{detJ}\end{pmatrix}u_3 \tag{3.193}$$

Thus, the dynamic model of the USV is written in the following nonlinear affine-in-the-input state-space form:

$$\dot{x}=f(x)+g_1(x)u_1+g_2(x)u_2+g_3(x)u_3 \tag{3.194}$$

with $x\in R^{6\times 1}$, $f(x)\in R^{6\times 1}$, $g_1(x)\in R^{6\times 1}$, $g_2(x)\in R^{6\times 1}$, $g_3(x)\in R^{6\times 1}$ and $u\in R^{3\times 1}$. Next, the elements of the state-space description of the USV are written as functions of the state vector elements x_1 to x_6.

Thus $J_{11}=m_{11}\cos^2(x_5)+m_{22}\sin^2(x_5)$, $J_{12}=(m_{11}-m_{22})\sin(x_5)\cos(x_5)$, $J_{13}=-m_{23}\sin(x_5)$, $J_{21}=(m_{11}-m_{22})\sin(x_5)\cos(x_5)$, $J_{22}=m_{11}sin^2(x_5)+m_{22}\cos^2(x_5)$, $J_{23}=m_{23}\cos(x_5)$, $J_{31}=-m_{23}\sin(x_5)$, $J_{32}=m_{23}\cos(x_5)$ and $J_{33}=m_{33}$.

and $\tilde{C}_{11}=x_6(m_{22}-m_{11})\sin(x_5)\cos(x_5)$, $\tilde{C}_{12}=x_6(m_{11}\cos^2(x_5)+m_{22}\sin^2(x_5))$, $\tilde{C}_{13}=0$, $\tilde{C}_{21}=-x_6(m_{11}\sin^2(x_5)+m_{22}\cos^2(x_5))$, $\tilde{C}_{22}=x_6(m_{11}-m_{22})\sin(x_5)\cos(x_5)$, $\tilde{C}_{23}=0$, $\tilde{C}_{31}=-x_6m_{23}\cos(x_5)$, $\tilde{C}_{32}=-x_6m_{23}\sin(x_5)$, $\tilde{C}_{33}=0$.

and $\tilde{F}_{11}=d_{11}\cos^2(x_5)+d_{22}\sin^2(x_5)$, $\tilde{F}_{12}=(d_{11}-d_{22})\sin(x_5)\cos(x_5)$, $\tilde{F}_{13}=-d_{23}\sin(x_5)$, $\tilde{F}_{21}=(d_{11}-d_{22})\sin(x_5)\cos(x_5)$, $\tilde{F}_{22}=d_{11}sim^2(x_5)+d_{22}\cos^2(x_5)$, $\tilde{F}_{23}=d_{23}\cos(x_5)$, $\tilde{F}_{31}=-d_{32}\sin(x_5)$, $\tilde{F}_{32}=d_{22}\cos(x_5)$, $\tilde{F}_{33}=d_{23}$.

3.6.2.2 Differential flatness properties of the USV dynamic model

It will be proven that the dynamic model of the 3-DOF USV is differentially flat with flat outputs vector $y=[x,y,\psi]^T$ or $y=[x_1,x_3,x_5]^T$. From the first, third and fifth rows of the state-space model of the USV one obtains

$$x_2=\dot{x}_1 \quad x_4=\dot{x}_3 \quad x_6=\dot{x}_5 \tag{3.195}$$

This signifies that all state variables of the USV can be written as differential functions of the system's flat outputs. Besides, the state-space model of the USV can be written as

$$\begin{pmatrix}\ddot{x}_1\\ \ddot{x}_3\\ \ddot{x}_5\end{pmatrix}=\begin{pmatrix}\frac{J_{11}(c_1+q_1)+J_{21}(c_2+q_2)+J_{31}(c_3+q_3)}{detJ}\\ \frac{J_{12}(c_1+q_1)+J_{22}(c_2+q_2)+J_{32}(c_3+q_3)}{detJ}\\ \frac{J_{13}(c_1+q_1)+J_{23}(c_2+q_2)+J_{33}(c_3+q_3)}{detJ}\end{pmatrix}+\begin{pmatrix}\frac{J_{11}}{detJ} & -\frac{J_{21}}{detJ} & \frac{J_{31}}{detJ}\\ -\frac{J_{12}}{detJ} & \frac{J_{22}}{detJ} & -\frac{J_{32}}{detJ}\\ \frac{J_{13}}{detJ} & -\frac{J_{23}}{detJ} & \frac{J_{33}}{detJ}\end{pmatrix}\begin{pmatrix}u_1\\ u_2\\ u_3\end{pmatrix} \tag{3.196}$$

or equivalently

$$\begin{pmatrix}u_1\\ u_2\\ u_3\end{pmatrix}=\begin{pmatrix}\frac{J_{11}}{detJ} & -\frac{J_{21}}{detJ} & \frac{J_{31}}{detJ}\\ -\frac{J_{12}}{detJ} & \frac{J_{22}}{detJ} & -\frac{J_{32}}{detJ}\\ \frac{J_{13}}{detJ} & -\frac{J_{23}}{detJ} & \frac{J_{33}}{detJ}\end{pmatrix}^{-1} \cdot \left[\begin{pmatrix}\ddot{x}_1\\ \ddot{x}_3\\ \ddot{x}_5\end{pmatrix}-\begin{pmatrix}\frac{J_{11}(c_1+q_1)+J_{21}(c_2+q_2)+J_{31}(c_3+q_3)}{detJ}\\ \frac{J_{12}(c_1+q_1)+J_{22}(c_2+q_2)+J_{32}(c_3+q_3)}{detJ}\\ \frac{J_{13}(c_1+q_1)+J_{23}(c_2+q_2)+J_{33}(c_3+q_3)}{detJ}\end{pmatrix}\right] \tag{3.197}$$

Consequently, the control inputs of the USV are also differential functions of the flat output and the entire system is differentially flat. The differential flatness property of the USV can be used for defining setpoints for the system's state variables.

3.6.3 Approximate linearization of the USV state-space model

3.6.3.1 Approximately linearized dynamics of the USV

The dynamic model of the USV undergoes approximate linearization around the temporary operating point (x^*, u^*), where x^* is the present value of the USV's state vector and u^* is the last sampled value of the control inputs vector. The linearization relies on first-order Taylor series expansion and on the computation of the associated Jacobian matrices at each sampling instance. The initial nonlinear model of the USV

$$\begin{aligned}\dot{x} &= f(x,u) \;\; \text{or} \\ \dot{x} &= f(x) + g_1(x)u_1 + g_2(x)u_2 + g_3(x)u_3\end{aligned} \tag{3.198}$$

is substituted by its equivalent linearized description

$$\dot{x} = Ax + Bu + \tilde{d} \tag{3.199}$$

where $\tilde{d}$ is the cumulative disturbances vector. This may comprise (i) the modeling error due to truncation of higher-order terms in the Taylor series expansion, (ii) exogenous perturbations, and (iii) sensor measurement noise of any distribution.

Matrices A and B are Jacobians of the linearization process and are given by

$$\begin{aligned}A &= \nabla_x[f(x) + g_1(x)u_1 + g_2(x)u_2 + g_3(x)u_3] \mid_{(x^*,u^*)} \Rightarrow \\ A &= \nabla_x f(x) \mid_{(x^*,u^*)} + \nabla_x g_1(x)u_1 \mid_{(x^*,u^*)} + \\ &\quad + \nabla_x g_2(x)u_2 \mid_{(x^*,u^*)} + \nabla_x g_3(x)u_3 \mid_{(x^*,u^*)}\end{aligned} \tag{3.200}$$

$$\begin{aligned}B &= \nabla_u[f(x) + g_1(x)u_1 + g_2(x)u_2 + g_3(x)u_3] \mid_{(x^*,u^*)} \Rightarrow \\ B &= [g_1(x)\; g_2(x)\; g_3(x)] \mid_{(x^*,u^*)} \Rightarrow B = g(x) \mid_{(x^*,u^*)}\end{aligned} \tag{3.201}$$

Computation of the Jacobian matrix $\nabla_x f(x) \mid_{(x^*,u^*)}$.

First row of the Jacobian matrix $\nabla_x f(x) \mid_{(x^*,u^*)}$: $\frac{\partial f_1}{\partial x_1} = 0$, $\frac{\partial f_1}{\partial x_2} = 1$, $\frac{\partial f_1}{\partial x_3} = 0$, $\frac{\partial f_1}{\partial x_4} = 0$, $\frac{\partial f_1}{\partial x_5} = 0$, and $\frac{\partial f_1}{\partial x_6} = 0$.

Second row of the Jacobian matrix $\nabla_x f(x) \mid_{(x^*,u^*)}$: for $i = 1, 2, \cdots, 6$

$$\begin{aligned}\frac{\partial f_2}{\partial x_i} = &-\frac{[\frac{\partial J_{11}}{\partial x_i}(c_1+q_1) + J_{11}(\frac{\partial c_1}{\partial x_i} + \frac{\partial q_1}{\partial x_i})]detJ - [J_{11}(c_1+q_1)]\frac{\partial detJ}{\partial x_i}}{detJ^2} \\ &+ \frac{[\frac{\partial J_{21}}{\partial x_i}(c_2+q_2) + J_{21}(\frac{\partial c_2}{\partial x_i} + \frac{\partial q_2}{\partial x_i})]detJ - [J_{21}(c_2+q_2)]\frac{\partial detJ}{\partial x_i}}{detJ^2} \\ &- \frac{[\frac{\partial J_{31}}{\partial x_i}(c_3+q_3) + J_{31}(\frac{\partial c_3}{\partial x_i} + \frac{\partial q_3}{\partial x_i})]detJ - [J_{31}(c_3+q_3)]\frac{\partial detJ}{\partial x_i}}{detJ^2}\end{aligned} \tag{3.202}$$

Third row of the Jacobian matrix $\nabla_x f(x) \mid_{(x^*,u^*)}$: $\frac{\partial f_3}{\partial x_1} = 0$, $\frac{\partial f_3}{\partial x_2} = 0$, $\frac{\partial f_3}{\partial x_3} = 0$, $\frac{\partial f_3}{\partial x_4} = 1$, $\frac{\partial f_3}{\partial x_5} = 0$, and $\frac{\partial f_3}{\partial x_6} = 0$.

Fourth row of the Jacobian matrix $\nabla_x f(x)\mid_{(x^*,u^*)}$: for $i = 1, 2, \cdots, 6$

$$\begin{aligned}\frac{\partial f_4}{\partial x_i} &= \frac{[\frac{\partial J_{12}}{\partial x_i}(c_1+q_1)+J_{12}(\frac{\partial c_1}{\partial x_i}+\frac{\partial q_1}{\partial x_i})]detJ-[J_{12}(c_1+q_1)]\frac{\partial detJ}{\partial x_i}}{detJ^2}\\ &-\frac{[\frac{\partial J_{22}}{\partial x_i}(c_2+q_2)+J_{22}(\frac{\partial c_2}{\partial x_i}+\frac{\partial q_2}{\partial x_i})]detJ-[J_{22}(c_2+q_2)]\frac{\partial detJ}{\partial x_i}}{detJ^2}\\ &+\frac{[\frac{\partial J_{32}}{\partial x_i}(c_3+q_3)+J_{32}(\frac{\partial c_3}{\partial x_i}+\frac{\partial q_3}{\partial x_i})]detJ-[J_{32}(c_3+q_3)]\frac{\partial detJ}{\partial x_i}}{detJ^2}\end{aligned} \tag{3.203}$$

Fifth row of the Jacobian matrix $\nabla_x f(x)\mid_{(x^*,u^*)}$: $\frac{\partial f_5}{\partial x_1} = 0$, $\frac{\partial f_5}{\partial x_2} = 0$, $\frac{\partial f_5}{\partial x_3} = 0$, $\frac{\partial f_5}{\partial x_4} = 0$, $\frac{\partial f_5}{\partial x_5} = 0$, and $\frac{\partial f_5}{\partial x_6} = 1$.

Sixth row of the Jacobian matrix $\nabla_x f(x)\mid_{(x^*,u^*)}$: for $i = 1, 2, \cdots, 6$

$$\begin{aligned}\frac{\partial f_6}{\partial x_i} &= -\frac{[\frac{\partial J_{13}}{\partial x_i}(c_1+q_1)+J_{12}(\frac{\partial c_1}{\partial x_i}+\frac{\partial q_1}{\partial x_i})]detJ-[J_{13}(c_1+q_1)]\frac{\partial detJ}{\partial x_i}}{detJ^2}\\ &+\frac{[\frac{\partial J_{23}}{\partial x_i}(c_2+q_2)+J_{23}(\frac{\partial c_2}{\partial x_i}+\frac{\partial q_2}{\partial x_i})]detJ-[J_{23}(c_2+q_2)]\frac{\partial detJ}{\partial x_i}}{detJ^2}\\ &-\frac{[\frac{\partial J_{33}}{\partial x_i}(c_3+q_3)+J_{33}(\frac{\partial c_3}{\partial x_i}+\frac{\partial q_3}{\partial x_i})]detJ-[J_{33}(c_3+q_3)]\frac{\partial detJ}{\partial x_i}}{detJ^2}\end{aligned} \tag{3.204}$$

Computation of the Jacobian matrix $\nabla_x g_1(x)\mid_{(x^*,u^*)}$.

First row of the Jacobian matrix $\nabla_x g_1(x)\mid_{(x^*,u^*)}$: $\frac{\partial g_{11}}{\partial x_i} = 0$, for $i = 1, 2, \cdots, 6$.

Second row of the Jacobian matrix $\nabla_x g_1(x)\mid_{(x^*,u^*)}$ for $i = 1, 2, \cdots, 6$

$$\frac{\partial g_{21}}{\partial x_i} = \frac{\frac{\partial J_{11}}{\partial x_i}detJ - J_{11}\frac{\partial detJ}{\partial x_i}}{detJ^2} \tag{3.205}$$

Third row of the Jacobian matrix $\nabla_x g_1(x)\mid_{(x^*,u^*)}$: $\frac{\partial g_{31}}{\partial x_i} = 0$, for $i = 1, 2, \cdots, 6$.

Fourth row of the Jacobian matrix $\nabla_x g_1(x)\mid_{(x^*,u^*)}$ for $i = 1, 2, \cdots, 6$

$$\frac{\partial g_{41}}{\partial x_i} = \frac{\frac{\partial J_{12}}{\partial x_i}detJ - J_{12}\frac{\partial detJ}{\partial x_i}}{detJ^2} \tag{3.206}$$

Fifth row of the Jacobian matrix $\nabla_x g_1(x)\mid_{(x^*,u^*)}$: $\frac{\partial g_{51}}{\partial x_i} = 0$, for $i = 1, 2, \cdots, 6$.

Sixth row of the Jacobian matrix $\nabla_x g_1(x) \mid_{(x^*,u^*)}$ for $i = 1, 2, \cdots, 6$

$$\frac{\partial g_{61}}{\partial x_i} = \frac{\frac{\partial J_{13}}{\partial x_i} detJ - J_{13} \frac{\partial detJ}{\partial x_i}}{detJ^2} \tag{3.207}$$

Computation of the Jacobian matrix $\nabla_x g_2(x) \mid_{(x^*,u^*)}$.

First row of the Jacobian matrix $\nabla_x g_2(x) \mid_{(x^*,u^*)}$: $\frac{\partial g_{12}}{\partial x_i} = 0$, for $i = 1, 2, \cdots, 6$.

Second row of the Jacobian matrix $\nabla_x g_2(x) \mid_{(x^*,u^*)}$ for $i = 1, 2, \cdots, 6$

$$\frac{\partial g_{22}}{\partial x_i} = \frac{\frac{\partial J_{21}}{\partial x_i} detJ - J_{21} \frac{\partial detJ}{\partial x_i}}{detJ^2} \tag{3.208}$$

Third row of the Jacobian matrix $\nabla_x g_2(x) \mid_{(x^*,u^*)}$: $\frac{\partial g_{32}}{\partial x_i} = 0$, for $i = 1, 2, \cdots, 6$.

Fourth row of the Jacobian matrix $\nabla_x g_2(x) \mid_{(x^*,u^*)}$ for $i = 1, 2, \cdots, 6$

$$\frac{\partial g_{42}}{\partial x_i} = \frac{\frac{\partial J_{22}}{\partial x_i} detJ - J_{22} \frac{\partial detJ}{\partial x_i}}{detJ^2} \tag{3.209}$$

Fifth row of the Jacobian matrix $\nabla_x g_2(x) \mid_{(x^*,u^*)}$: $\frac{\partial g_{52}}{\partial x_i} = 0$, for $i = 1, 2, \cdots, 6$.

Sixth row of the Jacobian matrix $\nabla_x g_2(x) \mid_{(x^*,u^*)}$ for $i = 1, 2, \cdots, 6$

$$\frac{\partial g_{62}}{\partial x_i} = \frac{\frac{\partial J_{23}}{\partial x_i} detJ - J_{23} \frac{\partial detJ}{\partial x_i}}{detJ^2} \tag{3.210}$$

Computation of the Jacobian matrix $\nabla_x g_3(x) \mid_{(x^*,u^*)}$.

First row of the Jacobian matrix $\nabla_x g_3(x) \mid_{(x^*,u^*)}$: $\frac{\partial g_{13}}{\partial x_i} = 0$, for $i = 1, 2, \cdots, 6$.

Second row of the Jacobian matrix $\nabla_x g_3(x) \mid_{(x^*,u^*)}$ for $i = 1, 2, \cdots, 6$

$$\frac{\partial g_{23}}{\partial x_i} = \frac{\frac{\partial J_{31}}{\partial x_i} detJ - J_{31} \frac{\partial detJ}{\partial x_i}}{detJ^2} \tag{3.211}$$

Third row of the Jacobian matrix $\nabla_x g_3(x) \mid_{(x^*,u^*)}$: $\frac{\partial g_{33}}{\partial x_i} = 0$, for $i = 1, 2, \cdots, 6$.

Fourth row of the Jacobian matrix $\nabla_x g_3(x) \mid_{(x^*,u^*)}$ for $i = 1, 2, \cdots, 6$

$$\frac{\partial g_{43}}{\partial x_i} = \frac{\frac{\partial J_{32}}{\partial x_i} detJ - J_{32} \frac{\partial detJ}{\partial x_i}}{detJ^2} \tag{3.212}$$

Fifth row of the Jacobian matrix $\nabla_x g_3(x)\mid_{(x^*,u^*)}$: $\frac{\partial g_{53}}{\partial x_i}=0$, for $i=1,2,\cdots,6$.

Sixth row of the Jacobian matrix $\nabla_x g_3(x)\mid_{(x^*,u^*)}$ for $i=1,2,\cdots,6$

$$\frac{\partial g_{63}}{\partial x_i}=\frac{\frac{\partial J_{33}}{\partial x_i}detJ-J_{33}\frac{\partial detJ}{\partial x_i}}{detJ^2} \tag{3.213}$$

Next, the partial derivatives of the terms J_{ij} $i=1,2,3,\ \ j=1,2,3$ and of the determinant $detJ$ are computed. It holds that:

$$\frac{\partial J_{11}}{\partial x_i}=\frac{\partial\tilde{J}_{22}}{\partial x_i}\tilde{J}_{33}+\tilde{J}_{22}\frac{\partial\tilde{J}_{33}}{\partial x_i}-\frac{\partial\tilde{J}_{32}}{\partial x_i}\tilde{J}_{23}-\tilde{J}_{32}\frac{\partial\tilde{J}_{23}}{\partial x_i}\quad \text{for}\ \ i=1,2,\ldots,6 \tag{3.214}$$

$$\frac{\partial J_{12}}{\partial x_i}=\frac{\partial\tilde{J}_{21}}{\partial x_i}\tilde{J}_{33}+\tilde{J}_{21}\frac{\partial\tilde{J}_{33}}{\partial x_i}-\frac{\partial\tilde{J}_{31}}{\partial x_i}\tilde{J}_{23}-\tilde{J}_{31}\frac{\partial\tilde{J}_{23}}{\partial x_i}\quad \text{for}\ \ i=1,2,\ldots,6 \tag{3.215}$$

$$\frac{\partial J_{13}}{\partial x_i}=\frac{\partial\tilde{J}_{21}}{\partial x_i}\tilde{J}_{32}+\tilde{J}_{21}\frac{\partial\tilde{J}_{32}}{\partial x_i}-\frac{\partial\tilde{J}_{31}}{\partial x_i}\tilde{J}_{22}-\tilde{J}_{31}\frac{\partial\tilde{J}_{22}}{\partial x_i}\quad \text{for}\ \ i=1,2,\ldots,6 \tag{3.216}$$

$$\frac{\partial J_{21}}{\partial x_i}=\frac{\partial\tilde{J}_{12}}{\partial x_i}\tilde{J}_{33}+\tilde{J}_{12}\frac{\partial\tilde{J}_{33}}{\partial x_i}-\frac{\partial\tilde{J}_{32}}{\partial x_i}\tilde{J}_{13}-\tilde{J}_{32}\frac{\partial\tilde{J}_{13}}{\partial x_i}\quad \text{for}\ \ i=1,2,\ldots,6 \tag{3.217}$$

$$\frac{\partial J_{22}}{\partial x_i}=\frac{\partial\tilde{J}_{11}}{\partial x_i}\tilde{J}_{33}+\tilde{J}_{11}\frac{\partial\tilde{J}_{33}}{\partial x_i}-\frac{\partial\tilde{J}_{31}}{\partial x_i}\tilde{J}_{13}-\tilde{J}_{31}\frac{\partial\tilde{J}_{13}}{\partial x_i}\quad \text{for}\ \ i=1,2,\ldots,6 \tag{3.218}$$

$$\frac{\partial J_{23}}{\partial x_i}=\frac{\partial\tilde{J}_{11}}{\partial x_i}\tilde{J}_{32}+\tilde{J}_{11}\frac{\partial\tilde{J}_{32}}{\partial x_i}-\frac{\partial\tilde{J}_{31}}{\partial x_i}\tilde{J}_{12}-\tilde{J}_{31}\frac{\partial\tilde{J}_{12}}{\partial x_i}\quad \text{for}\ \ i=1,2,\ldots,6 \tag{3.219}$$

$$\frac{\partial J_{31}}{\partial x_i}=\frac{\partial\tilde{J}_{12}}{\partial x_i}\tilde{J}_{23}+\tilde{J}_{12}\frac{\partial\tilde{J}_{23}}{\partial x_i}-\frac{\partial\tilde{J}_{22}}{\partial x_i}\tilde{J}_{13}-\tilde{J}_{22}\frac{\partial\tilde{J}_{13}}{\partial x_i}\quad \text{for}\ \ i=1,2,\ldots,6 \tag{3.220}$$

$$\frac{\partial J_{32}}{\partial x_i}=\frac{\partial\tilde{J}_{11}}{\partial x_i}\tilde{J}_{23}+\tilde{J}_{11}\frac{\partial\tilde{J}_{23}}{\partial x_i}-\frac{\partial\tilde{J}_{21}}{\partial x_i}\tilde{J}_{13}-\tilde{J}_{21}\frac{\partial\tilde{J}_{13}}{\partial x_i}\quad \text{for}\ \ i=1,2,\ldots,6 \tag{3.221}$$

$$\frac{\partial J_{33}}{\partial x_i}=\frac{\partial\tilde{J}_{11}}{\partial x_i}\tilde{J}_{22}+\tilde{J}_{11}\frac{\partial\tilde{J}_{22}}{\partial x_i}-\frac{\partial\tilde{J}_{21}}{\partial x_i}\tilde{J}_{12}-\tilde{J}_{21}\frac{\partial\tilde{J}_{12}}{\partial x_i}\quad \text{for}\ \ i=1,2,\ldots,6 \tag{3.222}$$

Additionally, it holds that for $i=1,2,\ldots,6$

$$\begin{aligned}\frac{\partial detJ}{\partial x_i}=&\frac{\partial\tilde{J}_{11}}{\partial x_i}(\tilde{J}_{22}\tilde{J}_{33}-\tilde{J}_{32}\tilde{J}_{23})+\tilde{J}_{11}\left(\frac{\partial\tilde{J}_{22}}{\partial x_i}\tilde{J}_{33}+\tilde{J}_{22}\frac{\partial\tilde{J}_{33}}{\partial x_i}-\frac{\partial\tilde{J}_{32}}{\partial x_i}\tilde{J}_{23}-\tilde{J}_{32}\frac{\partial\tilde{J}_{23}}{\partial x_i}\right)\\&+\frac{\partial\tilde{J}_{12}}{\partial x_i}(\tilde{J}_{21}\tilde{J}_{33}-\tilde{J}_{31}\tilde{J}_{23})+\tilde{J}_{12}\left(\frac{\partial\tilde{J}_{21}}{\partial x_i}\tilde{J}_{33}+\tilde{J}_{21}\frac{\partial\tilde{J}_{33}}{\partial x_i}-\frac{\partial\tilde{J}_{31}}{\partial x_i}\tilde{J}_{23}-\tilde{J}_{31}\frac{\partial\tilde{J}_{23}}{\partial x_i}\right)\\&+\frac{\partial\tilde{J}_{13}}{\partial x_i}(\tilde{J}_{21}\tilde{J}_{32}-\tilde{J}_{31}\tilde{J}_{22})+\tilde{J}_{13}\left(\frac{\partial\tilde{J}_{21}}{\partial x_i}\tilde{J}_{32}+\tilde{J}_{21}\frac{\partial\tilde{J}_{32}}{\partial x_i}-\frac{\partial\tilde{J}_{31}}{\partial x_i}\tilde{J}_{22}-\tilde{J}_{31}\frac{\partial\tilde{J}_{22}}{\partial x_i}\right)\end{aligned} \tag{3.223}$$

Equivalently, one has for the partial derivatives of c_i, $i = 1,2,3$, for $i = 1,2\ldots,6$

$$\frac{\partial c_1}{\partial x_i} = \frac{\partial \tilde{C}_{11}}{\partial x_i}\dot{\eta}_1 + \tilde{C}_{11}\frac{\partial \dot{\eta}_1}{\partial x_i} + \frac{\partial \tilde{C}_{12}}{\partial x_i}\dot{\eta}_2 + \tilde{C}_{12}\frac{\partial \dot{\eta}_2}{\partial x_i} + \frac{\partial \tilde{C}_{13}}{\partial x_i}\dot{\eta}_3 + \tilde{C}_{13}\frac{\partial \dot{\eta}_3}{\partial x_i} \quad (3.224)$$

$$\frac{\partial c_2}{\partial x_i} = \frac{\partial \tilde{C}_{21}}{\partial x_i}\dot{\eta}_1 + \tilde{C}_{21}\frac{\partial \dot{\eta}_1}{\partial x_i} + \frac{\partial \tilde{C}_{22}}{\partial x_i}\dot{\eta}_2 + \tilde{C}_{22}\frac{\partial \dot{\eta}_2}{\partial x_i} + \frac{\partial \tilde{C}_{23}}{\partial x_i}\dot{\eta}_3 + \tilde{C}_{23}\frac{\partial \dot{\eta}_3}{\partial x_i} \quad (3.225)$$

$$\frac{\partial c_3}{\partial x_i} = \frac{\partial \tilde{C}_{31}}{\partial x_i}\dot{\eta}_1 + \tilde{C}_{31}\frac{\partial \dot{\eta}_1}{\partial x_i} + \frac{\partial \tilde{C}_{32}}{\partial x_i}\dot{\eta}_2 + \tilde{C}_{32}\frac{\partial \dot{\eta}_2}{\partial x_i} + \frac{\partial \tilde{C}_{33}}{\partial x_i}\dot{\eta}_3 + \tilde{C}_{33}\frac{\partial \dot{\eta}_3}{\partial x_i} \quad (3.226)$$

Additionally, one has for the partial derivatives of c_i, $i = 1,2,3$, for $i = 1,2\ldots,6$

$$\frac{\partial q_1}{\partial x_i} = \frac{\partial \tilde{F}_{11}}{\partial x_i}\dot{\eta}_1 + \tilde{F}_{11}\frac{\partial \dot{\eta}_1}{\partial x_i} + \frac{\partial \tilde{F}_{12}}{\partial x_i}\dot{\eta}_2 + \tilde{F}_{12}\frac{\partial \dot{\eta}_2}{\partial x_i} + \frac{\partial \tilde{F}_{13}}{\partial x_i}\dot{\eta}_3 + \tilde{F}_{13}\frac{\partial \dot{\eta}_3}{\partial x_i} \quad (3.227)$$

$$\frac{\partial q_2}{\partial x_i} = \frac{\partial \tilde{F}_{21}}{\partial x_i}\dot{\eta}_1 + \tilde{F}_{21}\frac{\partial \dot{\eta}_1}{\partial x_i} + \frac{\partial \tilde{F}_{22}}{\partial x_i}\dot{\eta}_2 + \tilde{F}_{22}\frac{\partial \dot{\eta}_2}{\partial x_i} + \frac{\partial \tilde{F}_{23}}{\partial x_i}\dot{\eta}_3 + \tilde{F}_{23}\frac{\partial \dot{\eta}_3}{\partial x_i} \quad (3.228)$$

$$\frac{\partial q_3}{\partial x_i} = \frac{\partial \tilde{F}_{31}}{\partial x_i}\dot{\eta}_1 + \tilde{F}_{31}\frac{\partial \dot{\eta}_1}{\partial x_i} + \frac{\partial \tilde{F}_{32}}{\partial x_i}\dot{\eta}_2 + \tilde{F}_{32}\frac{\partial \dot{\eta}_2}{\partial x_i} + \frac{\partial \tilde{F}_{33}}{\partial x_i}\dot{\eta}_3 + \tilde{F}_{33}\frac{\partial \dot{\eta}_3}{\partial x_i} \quad (3.229)$$

Moreover, about the partial derivatives of variables η_i, $i = 1,2,3$ one has that

$$\frac{\partial \dot{\eta}_1}{\partial x_2} = 1 \quad \frac{\partial \dot{\eta}_1}{\partial x_i} = 0 \text{ for } i = 1,3,4,5,6 \quad (3.230)$$

$$\frac{\partial \dot{\eta}_2}{\partial x_4} = 1 \quad \frac{\partial \dot{\eta}_2}{\partial x_i} = 0 \text{ for } i = 1,2,3,5,6 \quad (3.231)$$

$$\frac{\partial \dot{\eta}_3}{\partial x_6} = 1 \quad \frac{\partial \dot{\eta}_3}{\partial x_i} = 0 \text{ for } i = 1,2,3,4,5 \quad (3.232)$$

Next, one proceeds to the computation of the partial derivatives $\frac{\partial \tilde{J}_{ij}}{\partial x_i}$ $i = 1,2,3$, $j = 1,2,3$, $\frac{\partial \tilde{C}_{ij}}{\partial x_i}$ $i = 1,2,3, j = 1,2,3$ and $\frac{\partial \tilde{F}_{ij}}{\partial x_i}$ $i = 1,2,3, j = 1,2,3$. It holds that:

$$\frac{\partial \tilde{J}_{11}}{\partial x_i} = 0 \text{ for } i = 1,2,3,4,6 \text{ and } \frac{\partial \tilde{J}_{11}}{\partial x_5} = -2m_{11}\cos(x_5)\sin(x_5) + 2m_{22}\sin(x_5)\cos(x_5).$$

$$\frac{\partial \tilde{J}_{12}}{\partial x_i} = 0 \text{ for } i = 1,2,3,4,6 \text{ and } \frac{\partial \tilde{J}_{12}}{\partial x_5} = (m_{11} - m_{22})[\cos^2(x_5) - sin^2(x_5)].$$

$$\frac{\partial \tilde{J}_{13}}{\partial x_i} = 0 \text{ for } i = 1,2,3,4,6 \text{ and } \frac{\partial \tilde{J}_{13}}{\partial x_5} = -m_{23}\cos(x_5).$$

$$\frac{\partial \tilde{J}_{21}}{\partial x_i} = 0 \text{ for } i = 1,2,3,4,6 \text{ and } \frac{\partial \tilde{J}_{21}}{\partial x_5} = -(m_{11} - m_{22})[\cos^2(x_5) - sin^2(x_5)].$$

$$\frac{\partial \tilde{J}_{22}}{\partial x_i} = 0 \text{ for } i = 1,2,3,4,6 \text{ and } \frac{\partial \tilde{J}_{22}}{\partial x_5} = 2m_{11}\sin(x_5)\cos(x_5) - 2m_{22}\sin(x_5)\cos(x_5).$$

$$\frac{\partial \tilde{J}_{23}}{\partial x_i} = 0 \text{ for } i = 1,2,3,4,6 \text{ and } \frac{\partial \tilde{J}_{23}}{\partial x_5} = -m_{23}\sin(x_5).$$

$$\frac{\partial \tilde{J}_{31}}{\partial x_i} = 0 \text{ for } i = 1,2,3,4,6 \text{ and } \frac{\partial \tilde{J}_{31}}{\partial x_5} = -m_{23}\cos(x_5).$$

$$\frac{\partial \tilde{J}_{32}}{\partial x_i} = 0 \text{ for } i = 1,2,3,4,6 \text{ and } \frac{\partial \tilde{J}_{32}}{\partial x_5} = -m_{23}\cos(x_5).$$

$$\frac{\partial \tilde{J}_{33}}{\partial x_i} = 0 \text{ for } i = 1,2,3,4,4,6$$

Furthermore, it holds that

$$\frac{\partial \tilde{F}_{11}}{\partial x_i} = 0 \text{ for } i = 1,2,3,4,6 \text{ and } \frac{\partial \tilde{F}_{11}}{\partial x_5} = -2d_{11}\cos(x_5)\sin(x_5) + 2d_{22}\sin(x_5)\cos(x_5).$$

$$\frac{\partial \tilde{F}_{12}}{\partial x_i} = 0 \text{ for } i = 1,2,3,4,6 \text{ and } \frac{\partial \tilde{F}_{12}}{\partial x_5} = (d_{11} - d_{22})[\cos^2(x_5) - sin^2(x_5)].$$

$$\frac{\partial \tilde{F}_{13}}{\partial x_i} = 0 \text{ for } i = 1,2,3,4,6 \text{ and } \frac{\partial \tilde{F}_{13}}{\partial x_5} = -d_{23}\cos(x_5).$$

$$\frac{\partial \tilde{F}_{21}}{\partial x_i} = 0 \text{ for } i = 1,2,3,4,6 \text{ and } \frac{\partial \tilde{F}_{21}}{\partial x_5} = (d_{11} - d_{22})[\cos^2(x_5) - sin^2(x_5)].$$

$$\frac{\partial \tilde{F}_{22}}{\partial x_i} = 0 \text{ for } i = 1,2,3,4,6 \text{ and } \frac{\partial \tilde{F}_{22}}{\partial x_5} = 2d_{11}\sin(x_5)\cos(x_5) - 2d_{22}\sin(x_5)\cos(x_5).$$

$$\frac{\partial \tilde{F}_{23}}{\partial x_i} = 0 \text{ for } i = 1,2,3,4,6 \text{ and } \frac{\partial \tilde{F}_{23}}{\partial x_5} = -d_{22}\sin(x_5).$$

$$\frac{\partial \tilde{F}_{31}}{\partial x_i} = 0 \text{ for } i = 1,2,3,4,6 \text{ and } \frac{\partial \tilde{F}_{31}}{\partial x_5} = -d_{32}\cos(x_5).$$

$$\frac{\partial \tilde{F}_{32}}{\partial x_i} = 0 \text{ for } i = 1,2,3,4,6 \text{ and } \frac{\partial \tilde{F}_{32}}{\partial x_5} = -d_{32}\sin(x_5).$$

$$\frac{\partial \tilde{F}_{33}}{\partial x_i} = 0 \text{ for } i = 1,2,3,4,5,6.$$

3.6.3.2 Stabilizing feedback control

After linearization around its current operating point (x^*, u^*), the dynamic model of the Unmanned Surface Vessel (USV) is written as (Rigatos and Busawon, 2018)

$$\dot{x} = Ax + Bu + d_1 \tag{3.233}$$

Parameter d_1 stands for the linearization error in the USV's model appearing previously in (3.233). The reference setpoints for the USV's state vector are denoted by $\mathbf{x_d} = [x_1^d, \ldots, x_6^d]$. Tracking of this trajectory is achieved after applying the control input u_d. At every time instant the control input u_d is assumed to differ from the control input u appearing in (3.233) by an amount equal to Δu, that is $u_d = u + \Delta u$

$$\dot{x}_d = Ax_d + Bu_d + d_2 \tag{3.234}$$

The dynamics of the controlled system described in (3.233) can be also written as

$$\dot{x} = Ax + Bu + Bu_d - Bu_d + d_1 \tag{3.235}$$

and by denoting $d_3 = -Bu_d + d_1$ as an aggregate disturbance term one obtains

$$\dot{x} = Ax + Bu + Bu_d + d_3 \tag{3.236}$$

By subtracting (3.234) from (3.236) one has

$$\dot{x} - \dot{x}_d = A(x - x_d) + Bu + d_3 - d_2 \tag{3.237}$$

By denoting the tracking error as $e = x - x_d$ and the aggregate disturbance term as $\tilde{d} = d_3 - d_2$, the tracking error dynamics becomes

$$\dot{e} = Ae + Bu + \tilde{d} \tag{3.238}$$

For the approximately linearized model of the system a stabilizing feedback controller is developed. The controller has the form

$$u(t) = -Ke(t) \tag{3.239}$$

with $K = \frac{1}{r}B^T P$ where P is a positive definite symmetric matrix which is obtained from the solution of the Riccati equation (Rigatos and Busawon, 2018)

$$A^T P + PA + Q - P(\frac{2}{r}BB^T - \frac{1}{\rho^2}LL^T)P = 0 \tag{3.240}$$

where Q is a positive semi-definite symmetric matrix. The diagram of the considered control loop for the AUV is depicted in Figure 3.68. The global stability properties of the control method can be proven following the procedure presented in Section 1.1.2.5.

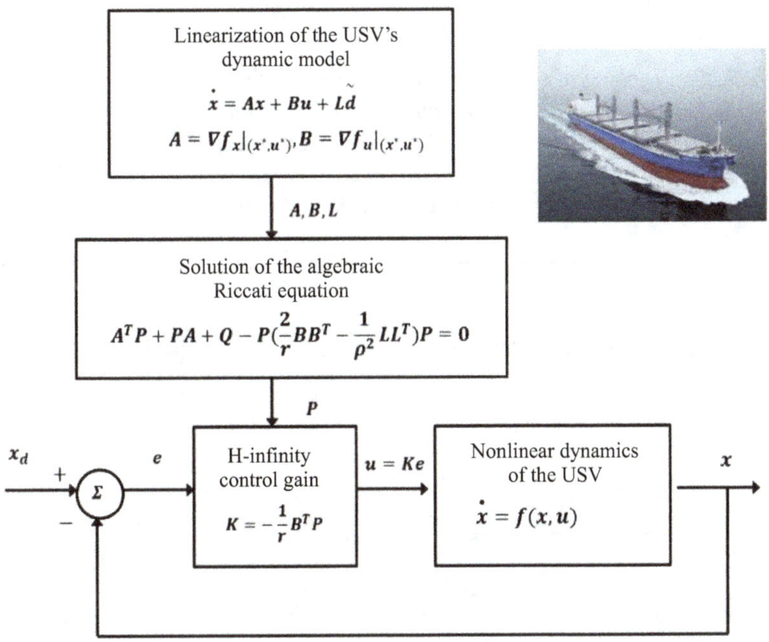

Figure 3.68 Diagram of the control scheme for the Unmanned Surface Vessel

3.6.4 Simulation tests

The performance of the proposed nonlinear optimal control scheme for the USV is further confirmed through simulation experiments. The model of the unmanned surface vessel that has been considered in these tests is the one which has been used in (Fang, Zergeroglou *et al.*, 2004). Certainly, the results can be generalized for underactuated USVs. To compute the control inputs of the USV, the algebraic Riccati equation of (3.240) had to be solved at each time-step of the control algorithm. The tracking performance of the control scheme was tested using several reference paths that the UAV had to follow. The obtained results are depicted in Figures 3.69 to 3.84.

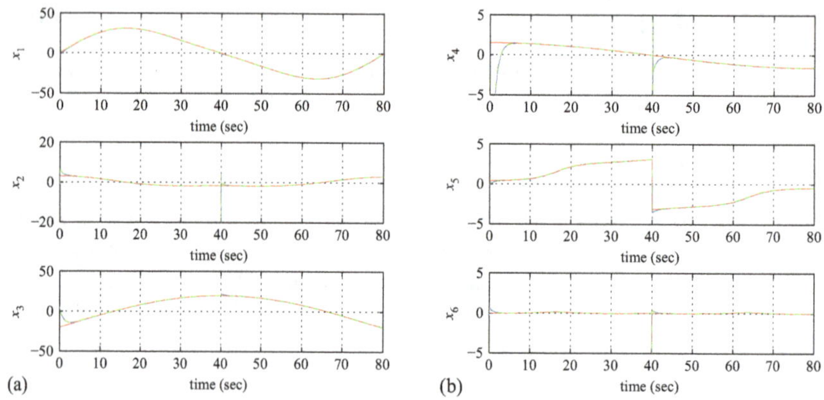

Figure 3.69 Tracking of setpoint 1 for the 3-DOF USV: (a) convergence of state variables x_1 to x_3 to their reference setpoints (red line: setpoint, blue line: real value, green line: estimated value); (b) convergence of state variables x_4 to x_6 to their reference setpoints (red line: setpoint, blue line: real value, green line: estimated value)

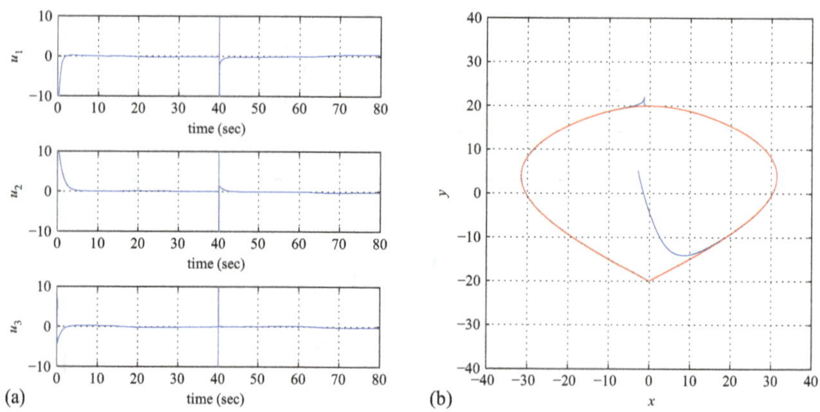

Figure 3.70 Tracking of setpoint 1 for the 3-DOF USV: (a) control inputs u_1 to u_3 applied to the propulsion and steering system of the USV; (b) tracking of the reference path (red line) by the USV (blue line) in the xy plane

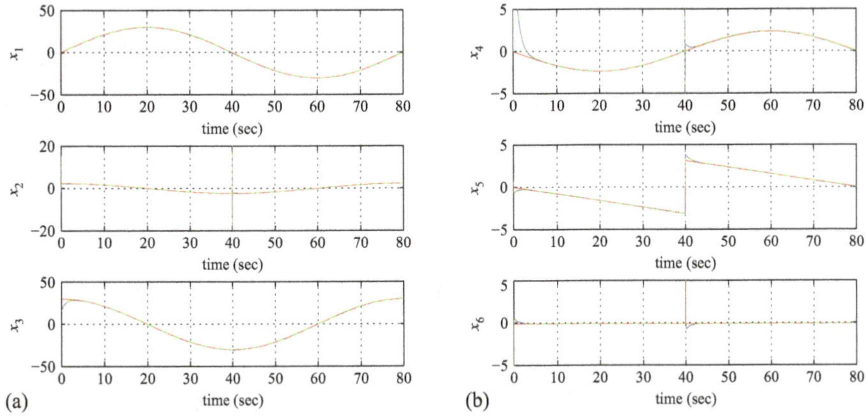

Figure 3.71 Tracking of setpoint 2 for the 3-DOF USV: (a) convergence of state variables x_1 to x_3 to their reference setpoints (red line: setpoint, blue line: real value, green line: estimated value); (b) convergence of state variables x_4 to x_6 to their reference setpoints (red line: setpoint, blue line: real value, green line: estimated value)

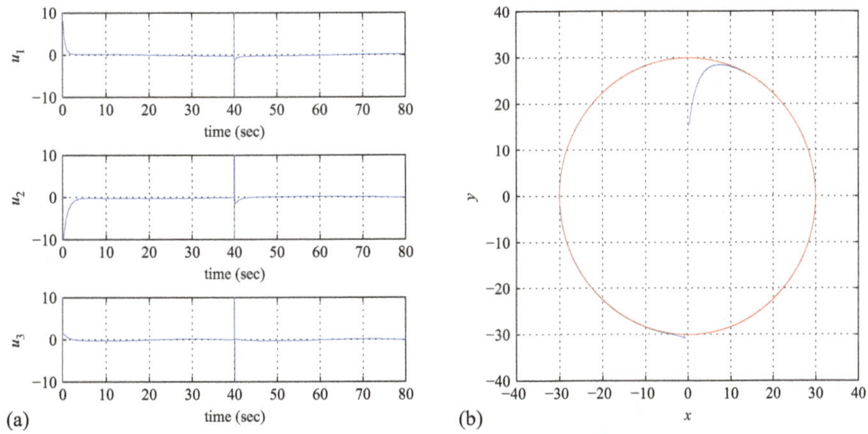

Figure 3.72 Tracking of setpoint 2 for the 3-DOF USV: (a) control inputs u_1 to u_3 applied to the propulsion and steering system of the USV; (b) tracking of the reference path (red line) by the USV (blue line) in the xy plane

The real values of the USV's state vector are printed in blue, the estimated values which were provided by the H-infinity Kalman Filter are plotted in green, while the related reference setpoints are shown in red. Distance is measured in meters (m), angles in rad and time in sec. It can be noticed that in all cases, the proposed H-infinity controller achieved fast and accurate tracking of the reference paths under moderate variations of the control inputs.

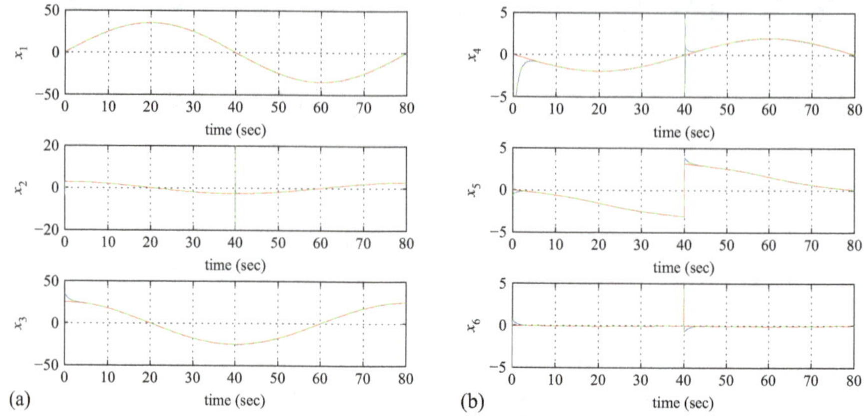

Figure 3.73 Tracking of setpoint 3 for the 3-DOF USV: (a) convergence of state variables x_1 to x_3 to their reference setpoints (red line: setpoint, blue line: real value, green line: estimated value); (b) convergence of state variables x_4 to x_6 to their reference setpoints (red line: setpoint, blue line: real value, green line: estimated value)

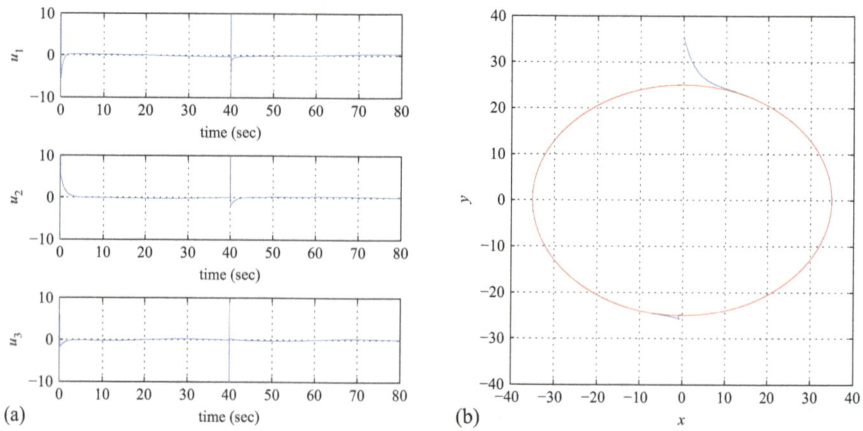

Figure 3.74 Tracking of setpoint 3 for the 3-DOF USV: (a) control inputs u_1 to u_3 applied to the propulsion and steering system of the USV; (b) tracking of the reference path (red line) by the USV (blue line) in the xy plane

3.7 Nonlinear control of the 3-DOF autonomous underwater vessel

3.7.1 Outline

The use of Autonomous Underwater Vessels (AUVs) is widely met in defense and security tasks, in missions for seabed exploration and exploitation, in the construction

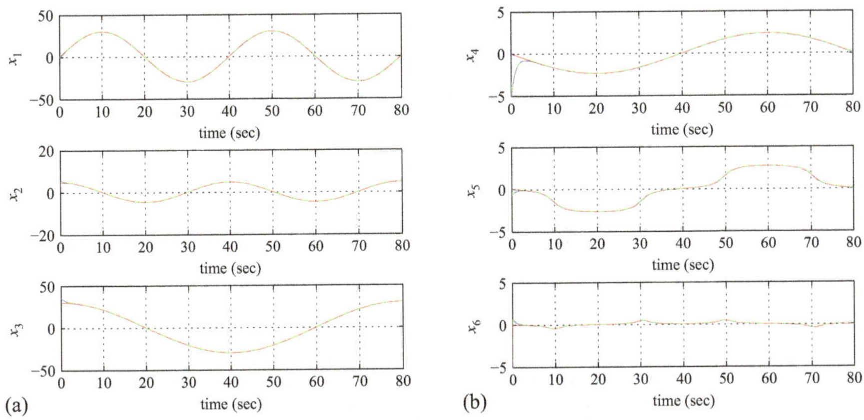

Figure 3.75 Tracking of setpoint 4 for the 3-DOF USV: (a) convergence of state variables x_1 to x_3 to their reference setpoints (red line: setpoint, blue line: real value, green line: estimated value); (b) convergence of state variables x_4 to x_6 to their reference setpoints (red line: setpoint, blue line: real value, green line: estimated value)

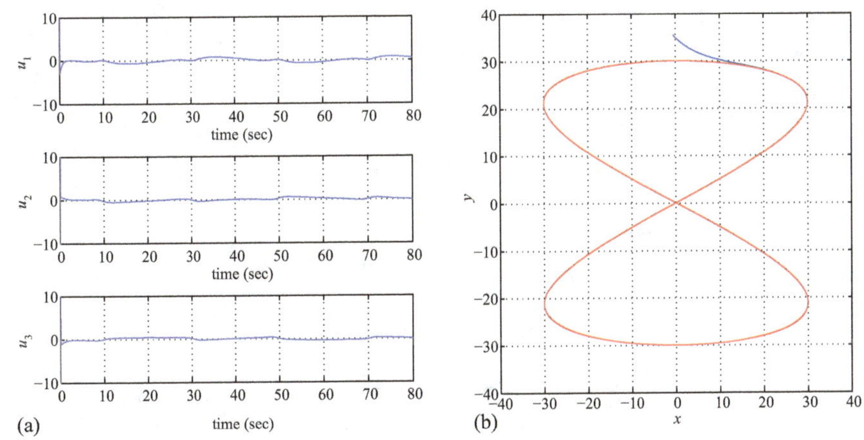

Figure 3.76 Tracking of setpoint 4 for the 3-DOF USV: (a) control inputs u_1 to u_3 applied to the propulsion and steering system of the USV; (b) tracking of the reference path (red line) by the USV (blue line) in the xy plane

of underwater pipelines and in the birthing of underwater cables, as well in scientific exploration missions (Rigatos and Busawon, 2018). The kinematic and dynamic model of these robotic systems is a strongly nonlinear one and receives multiple control inputs in the form of thrust forces from the AUVs' propulsion system (Shen,

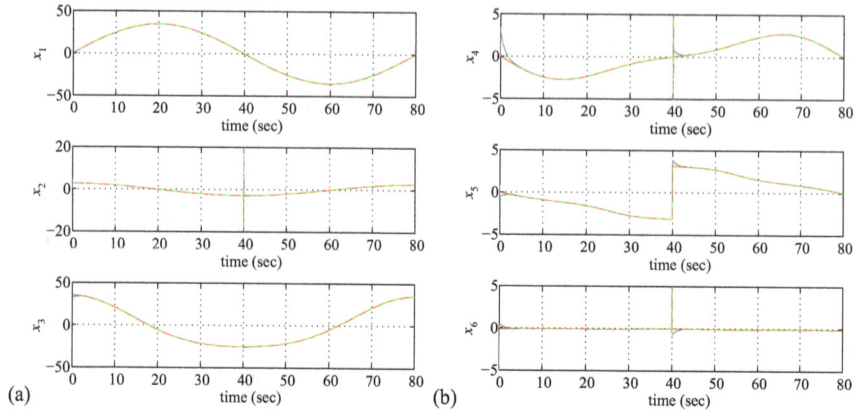

Figure 3.77 Tracking of setpoint 5 for the 3-DOF USV: (a) convergence of state variables x_1 to x_3 to their reference setpoints (red line: setpoint, blue line: real value, green line: estimated value); (b) convergence of state variables x_4 to x_6 to their reference setpoints (red line: setpoint, blue line: real value, green line: estimated value)

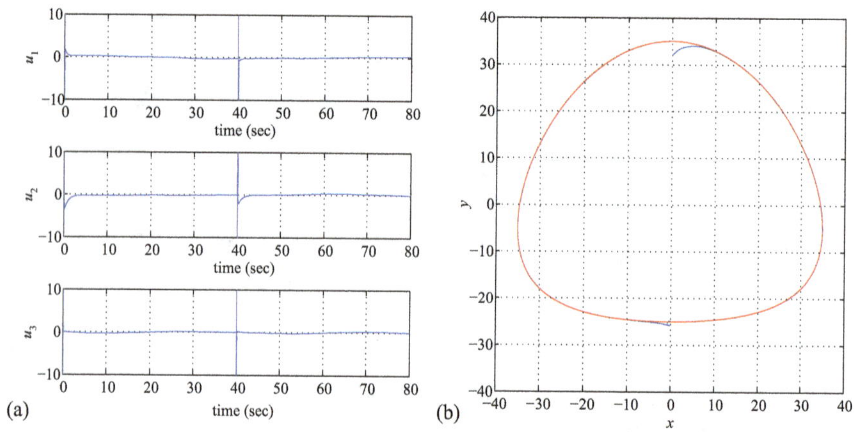

Figure 3.78 Tracking of setpoint 5 for the 3-DOF USV: (a) control inputs u_1 to u_3 applied to the propulsion and steering system of the USV; (b) tracking of the reference path (red line) by the USV (blue line) in the xy plane

Shi *et al.*, 2019; Shen, Shi *et al.*, 2018; Shen, Buckham *et al.*, 2017; Choukri and Qidan, 2014; Guerrero *et al.*, 2019). Depending on whether the number of degrees of freedom of such AUVs exceeds the number of their control inputs or not, AUVs can be classified into underactuated and fully actuated ones (Zacarrias Ferreira *et al.*,

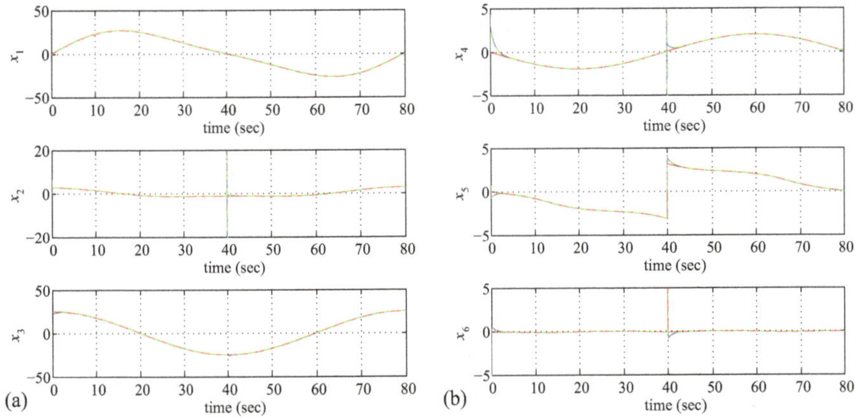

Figure 3.79 Tracking of setpoint 6 for the 3-DOF USV: (a) convergence of state variables x_1 to x_3 to their reference setpoints (red line: setpoint, blue line: real value, green line: estimated value); (b) convergence of state variables x_4 to x_6 to their reference setpoints (red line: setpoint, blue line: real value, green line: estimated value)

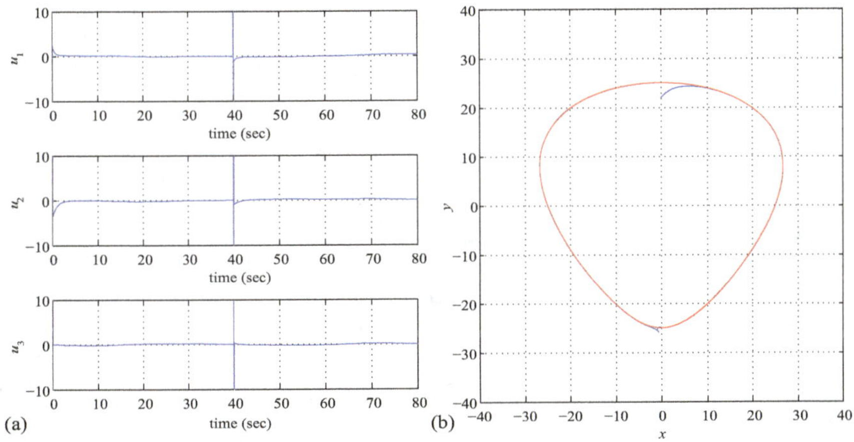

Figure 3.80 Tracking of setpoint 6 for the 3-DOF USV: (a) control inputs u_1 to u_3 applied to the propulsion and steering system of the USV; (b) tracking of the reference path (red line) by the USV (blue line) in the xy plane

2018; Makavita *et al.*, 2019; Ying, Yan *et al.*, 2020; Yan, Yang *et al.*, 2019). The control problem of AUVs is a non-trivial one and several nonlinear control methods have been proposed for it so far (Allibert *et al.*, 2019; Suarez Fernandez *et al.*, 2019; Xiang, Lapierre *et al.*, 2015; Fischer *et al.*, 2014; Xiang, Yu and Zhang, 2017). Precise path

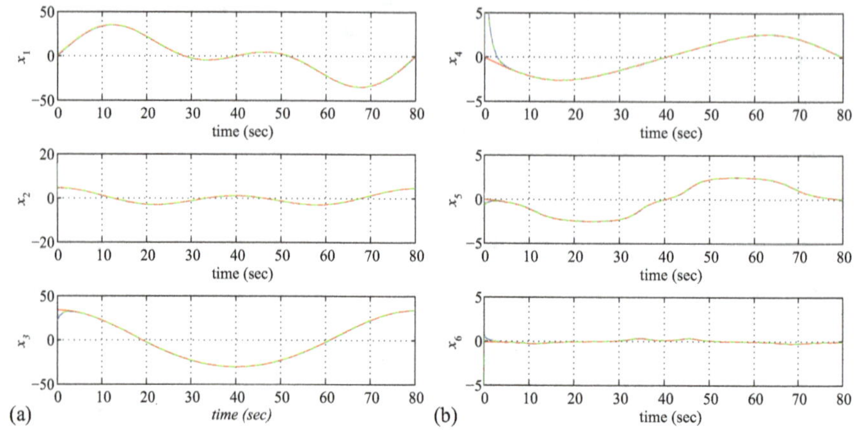

Figure 3.81 *Tracking of setpoint 7 for the 3-DOF USV: (a) convergence of state variables x_1 to x_3 to their reference setpoints (red line: setpoint, blue line: real value, green line: estimated value); (b) convergence of state variables x_4 to x_6 to their reference setpoints (red line: setpoint, blue line: real value, green line: estimated value)*

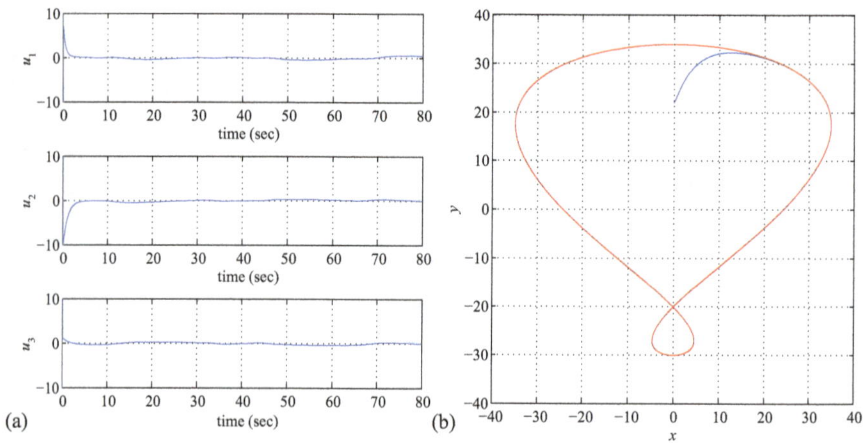

Figure 3.82 *Tracking of setpoint 7 for the 3-DOF USV: (a) control inputs u_1 to u_3 applied to the propulsion and steering system of the USV; (b) tracking of the reference path (red line) by the USV (blue line) in the xy plane*

following under model uncertainty and external perturbations is the main objectives of these control schemes (Qiao and Zhang, 2017; Ma, Hu *et al.*, 2019; Peng, Wang *et al.*, 2019; Yu, Xiang *et al.*, 2020; Xu, Tang *et al.*, 2019). Besides, minimization in energy consumption by the electric actuators of AUVs is another significant research

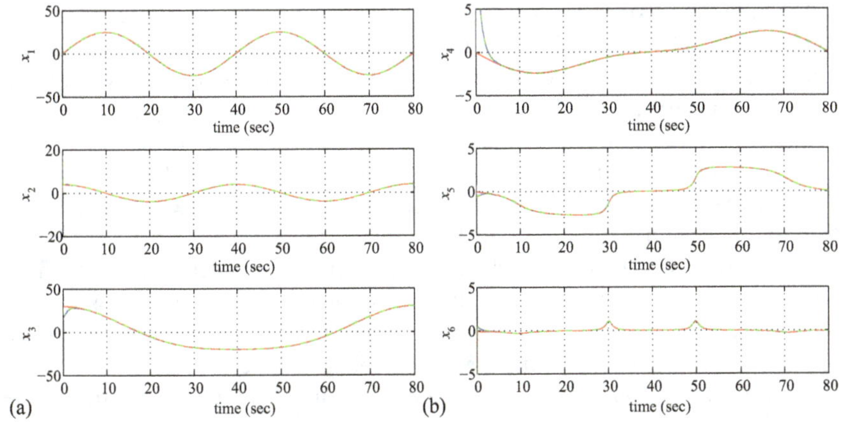

Figure 3.83 Tracking of setpoint 8 for the 3-DOF USV: (a) convergence of state variables x_1 to x_3 to their reference setpoints (red line: setpoint, blue line: real value, green line: estimated value); (b) convergence of state variables x_4 to x_6 to their reference setpoints (red line: setpoint, blue line: real value, green line: estimated value)

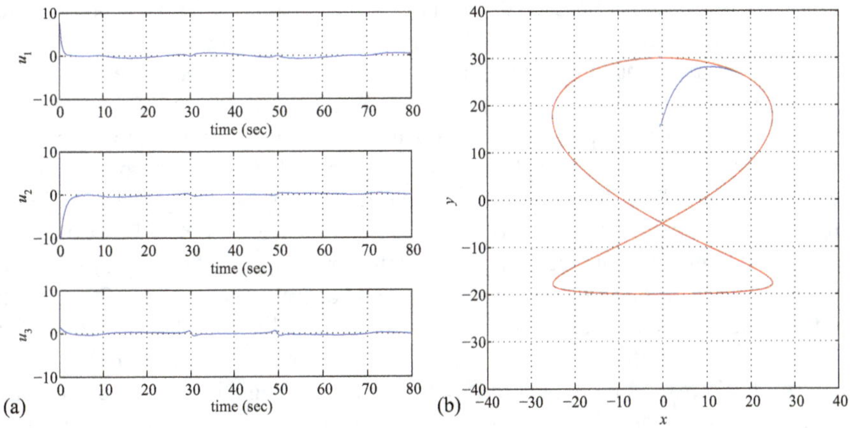

Figure 3.84 Tracking of setpoint 8 for the 3-DOF USV: (a) control inputs u_1 to u_3 applied to the propulsion and steering system of the USV; (b) tracking of the reference path (red line) by the USV (blue line) in the xy plane

target. To this end, MPC and NMPC control methods can be considered (Zhang, Liu *et al.*, 2019; Narayan Rath and Subudhi, 2019; Wei, She and Shi, 2020; Guo, Zhang *et al.*, 2019).

In this section, the previously analyzed nonlinear optimal (H-infinity) control method is proposed for fully actuated AUVs under model uncertainty and external perturbations (Rigatos and Busawon, 2018). At first, the joint kinematic and dynamic model of the fully actuated AUVs undergoes approximate linearization around a temporary operating point which is recomputed at each iteration of the control algorithm. This linearization point is defined by the present value of the system's state vector and by the last sampled value of the AUVs control inputs vector. The linearization process relies on first-order Taylor series expansion and on the computation of the related Jacobian matrices (Rigatos, Siano and Cecati, 2015; Rigatos and Tzafestas, 2007; Basseville and Nikiforov, 1993; Rigatos and Zhang, 2009). The induced modeling error, which is due to the truncation of higher-order terms in the Taylor series is considered to be a perturbation that is finally asymptotically compensated by the robustness of the control algorithm. For the approximately linearized model of the AUV an H-infinity stabilizing feedback controller is designed.

The H-infinity controller gives the solution of the AUVs optimal control problem under model imprecision and exogenous disturbances. Actually, it represents a min–max differential game taking place between (i) the control inputs which try to minimize a cost function that comprises a quadratic term of the AUVs state vector error; (ii) the model uncertainty and external perturbation terms which try to maximize this cost function. For the selection of the H-infinity controller's feedback gains an algebraic Riccati equation is being repetitively solved at each time-step of the control algorithm (Rigatos, 2011; Rigatos, 2015; Rigatos and Karapanou, 2020). The stability properties of the control method are proven through Lyapunov analysis. First, it is demonstrated that the control scheme satisfies the H-infinity tracking performance criterion, which signifies elevated robustness under model uncertainty and external perturbations (Rigatos and Busawon, 2018; Tousssaint, Basar and Bullo, 2000). Moreover, under moderate conditions it is proven that the control loop is globally asymptotically stable (Rigatos and Busawon, 2018). Finally, to implement feedback control without the need to measure the entire state vector of the AUVs the H-infinity Kalman Filter is proposed as a robust state estimator (Rigatos and Busawon, 2018; Gibbs, 2012).

The section's control method minimizes the consumption of energy by the actuators of the AUV thus improving the autonomy and operational capacity of this robotic system. The control of the AUVs' model is of elevated difficulty because the system is nonlinear and multivariable. The application of global linearization-based control methods to the AUVs model exhibits several shortcomings. For instance it requires complicated state variables transformations. Besides, to compute the control inputs that will be finally applied to the AUVs model, inverse transformations are needed. Furthermore, one cannot apply MPC because the AUVs system is a strongly nonlinear one and the use of MPC in its case would make the control loop unstable. Additionally, the convergence of NMPC schemes to an optimum, cannot be assured either since this method relies on the ad-hoc selection of parameters' values and initialization. It is also remarkable in the application of sliding-mode control, that to arrive at a well determined procedure for defining the sliding surface, transformation of the AUVs state-space model into an input–output linearized form is first needed. It is also noteworthy, that one cannot apply directly backstepping control to the AUVs model because this system is not found in the triangular (backstepping integral) state-space form. Finally, it would not be recommended to apply multiple local models-based control, through the linearization of the AUVs model around multiple operating points because this would require the solution of multiple Riccati equations or LMIs and would be a computationally intensive procedure.

3.7.2 Kinematic and dynamic model of the AUV

3.7.2.1 State-space description of the AUV

The joint kinematic and dynamic model of the Autonomous Underwater Vessel (AUV) is given by (Shen, Shi *et al.*, 2019; Shen, Buckham *et al.*, 2017):

$$\dot{x} = u\cos(\psi) - v\sin(\psi) \tag{3.241}$$

$$\dot{y} = u\sin(\psi) + v\cos(\psi) \tag{3.242}$$

$$\dot{\psi} = r \tag{3.243}$$

$$\dot{u} = \frac{M_{\dot{v}}}{M_{\dot{u}}}vr - \frac{X_u}{M_{\dot{u}}}u - \frac{D_u}{M_{\dot{u}}}u|u| + \frac{F_u}{M_{\dot{u}}} \tag{3.244}$$

$$\dot{v} = -\frac{M_{\dot{u}}}{M_{\dot{v}}}ur - \frac{Y_v}{M_{\dot{v}}}v - \frac{D_v}{M_{\dot{v}}}v|v| + \frac{F_v}{M_{\dot{v}}} \tag{3.245}$$

$$\dot{r} = \frac{M_{\dot{u}} - M_{\dot{v}}}{M_{\dot{r}}}uv - \frac{N_r}{M_{\dot{r}}}r - \frac{D_r}{M_{\dot{r}}}r|r| + \frac{F_r}{M_{\dot{r}}} \tag{3.246}$$

where (x,y) are the Cartesian coordinates of the center of gravity of the AUV, and ψ is its heading angle, expressed in an inertial coordinates system. Besides u and v are the surge and sway velocities of the AUV which are defined in a body-fixed reference frame, while r is the rate of change of its roll angle. The depth in which the AUV is found is considered to remain unchanged and thus its motion takes place in a xy horizontal plane. Propulsion to the AUV is provided by the vector of thrust forces $\tau = [F_u, F_v, F_r]$. In the previously given model, $M_{\dot{u}}$ $M_{\dot{v}}$ and $M_{\dot{r}}$ are inertial coefficients, X_u, Y_v, N_r are hydrodynamic coefficients for the linear terms of the model, and D_u, D_v, D_r are hydrodynamic coefficients for the quadratic terms of the model.

The following state variables are defined: $x = [x_1, x_2, x_3, x_4, x_5, x_6]^T$ or $x = [x, y, \psi, u, v, r]^T$ and the following control inputs are defined $u = [u_1, u_2, u_3]^T = [F_u, F_v, F_r]^T$. Thus, the joint kinematic and dynamic model of the AUV is written in the following state-space form

$$\dot{x}_1 = x_4\cos(x_3) - x_5\sin(x_3) \tag{3.247}$$

$$\dot{x}_2 = x_4\sin(x_3) + x_5\cos(x_3) \tag{3.248}$$

$$\dot{x}_3 = x_6 \tag{3.249}$$

$$\dot{x}_4 = \frac{M_{\dot{v}}}{M_{\dot{u}}}x_5x_6 - \frac{X_u}{M_{\dot{u}}}x_4 - \frac{D_u}{M_{\dot{u}}}x_4|x_4| + \frac{u_1}{M_{\dot{u}}} \tag{3.250}$$

$$\dot{x}_5 = -\frac{M_{\dot{u}}}{M_{\dot{v}}}x_4x_6 - \frac{Y_v}{M_{\dot{v}}}x_5 - \frac{D_v}{M_{\dot{v}}}x_5|x_5| + \frac{u_2}{M_{\dot{v}}} \tag{3.251}$$

$$\dot{x}_3 = \frac{M_{\dot{u}} - M_{\dot{v}}}{M_{\dot{r}}}x_4x_5 - \frac{N_r}{M_{\dot{r}}}x_6 - \frac{D_r}{M_{\dot{r}}}x_6|x_6| + \frac{u_3}{M_{\dot{r}}} \tag{3.252}$$

The state-space model of the AUV can be also written in the following affine-in-the-input matrix form:

$$\begin{pmatrix} \dot{x}_1 \\ \dot{x}_2 \\ \dot{x}_3 \\ \dot{x}_4 \\ \dot{x}_5 \\ \dot{x}_6 \end{pmatrix} = \begin{pmatrix} x_4\cos(x_3) - x_5\sin(x_3) \\ x_4\sin(x_3) + x_5\cos(x_3) \\ x_6 \\ \frac{M_{\dot{v}}}{M_{\dot{u}}}x_5x_6 - \frac{X_u}{M_{\dot{u}}}x_4 - \frac{D_u}{M_{\dot{u}}}x_4|x_4| \\ -\frac{M_{\dot{u}}}{M_{\dot{v}}}x_4x_6 - \frac{Y_v}{M_{\dot{v}}}x_5 - \frac{D_v}{M_{\dot{v}}}x_5|x_5| \\ \frac{M_{\dot{u}} - M_{\dot{v}}}{M_{\dot{r}}}x_4x_5 - \frac{N_r}{M_{\dot{r}}}x_6 - \frac{D_r}{M_{\dot{r}}}x_6|x_6| \end{pmatrix} + \begin{pmatrix} 0 & 0 & 0 \\ 0 & 0 & 0 \\ 0 & 0 & 0 \\ \frac{1}{M_{\dot{u}}} & 0 & 0 \\ 0 & \frac{1}{M_{\dot{v}}} & 0 \\ 0 & 0 & \frac{1}{M_{\dot{r}}} \end{pmatrix} \begin{pmatrix} u_1 \\ u_2 \\ u_3 \end{pmatrix} \tag{3.253}$$

Moreover, the AUV's model can be written in the concise form:

$$\dot{x} = f(x) + g(x)u \tag{3.254}$$

$x \in R^{6\times1}, f(x) \in R^{6\times1}$, $g(x) \in R^{6\times3}$, and $u \in R^{3\times1}$.

3.7.3 Differential flatness properties of the AUV's model

The AUV's joint kinematic and dynamic model is differentially flat, with flat outputs $y = [y_1, y_2, y_3]^T = [x, y, \psi]^T$ (Rigatos, 2015). The differential flatness property is concluded if (i) all state variables and the control inputs of a system can be expressed as differential functions of a subset of its state variables, where the latter stand for the system's flat outputs; (ii) the flat outputs of the system are differentially independent, which means that they are not connected through a relation in the form of a homogeneous differential equation.

From (3.247) and (3.248), one solves with respect to x_4 and x_5

$$\begin{aligned} &\begin{pmatrix} \dot{x}_1 \\ \dot{x}_2 \end{pmatrix} = \begin{pmatrix} \cos(x_3) & -\sin(x_3) \\ \sin(x_3) & \cos(x_3) \end{pmatrix} \begin{pmatrix} x_4 \\ x_5 \end{pmatrix} \Rightarrow \\ &\begin{pmatrix} x_4 \\ x_5 \end{pmatrix} = \begin{pmatrix} \cos(x_3) & -\sin(x_3) \\ \sin(x_3) & \cos(x_3) \end{pmatrix}^{-1} \begin{pmatrix} h_4(y, \dot{y}) \\ h_5(y, \dot{y}) \end{pmatrix} \end{aligned} \tag{3.255}$$

which signifies that x_4 and x_5 are differential functions of the system's flat outputs. Besides, from (3.249), one has

$$x_6 = \dot{x}_3 \Rightarrow x_6 = h_6(y, \dot{y}) \tag{3.256}$$

Additionally, from (3.250), (3.251), and (3.252), one can solve with respect to the control inputs F_u, F_v and F_r respectively, thus obtaining

$$F_u = M_{\dot{u}} \left\{ \dot{x}_4 - \frac{M_{\dot{v}}}{M_{\dot{u}}}x_5x_6 + \frac{X_u}{M_{\dot{u}}}x_4 + \frac{D_u}{M_{\dot{u}}}x_4|x_4| \right\} \tag{3.257}$$

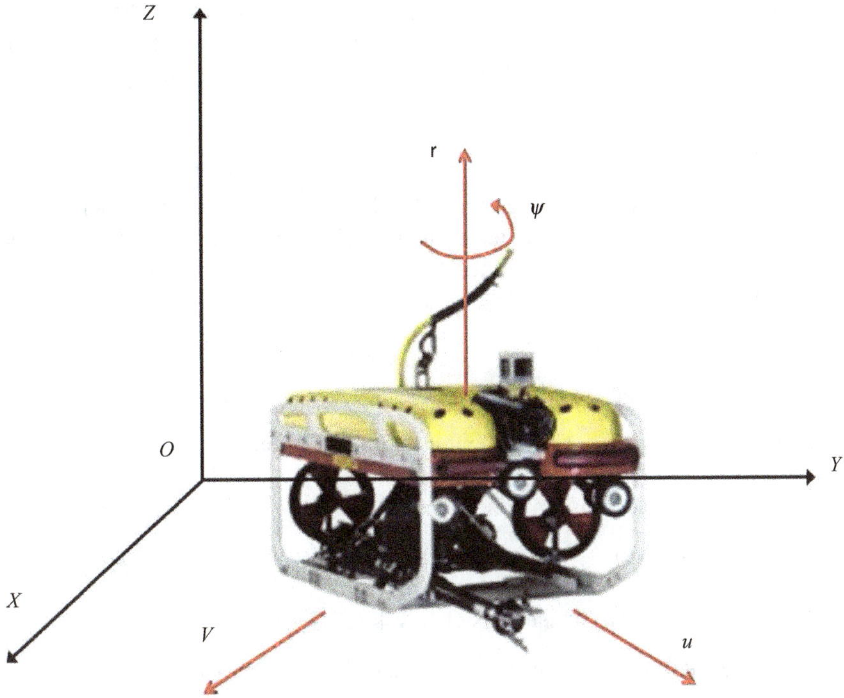

Figure 3.85 Diagram of the Autonomous Underwater Vessel and the related coordinates frames

$$F_v = M_{\dot{v}} \left\{ \dot{x}_5 + \frac{M_{\dot{u}}}{M_{\dot{v}}} x_4 x_6 + \frac{Y_v}{M_{\dot{v}}} x_5 + \frac{D_v}{M_{\dot{v}}} x_5 |x_5| \right\} \tag{3.258}$$

$$F_r = M_{\dot{r}} \left\{ \frac{\dot{x}_6 - M_{\dot{u}} - M_{\dot{v}}}{M_{\dot{r}} x_4 x_5 + \frac{N_r}{M_{\dot{r}}} x_6 + \frac{D_r}{M_{\dot{r}}} x_6 |x_6|} \right\} \tag{3.259}$$

The above equations signify that the AUV's model is differentially flat, because all its state variables and its control inputs can be written as differential functions of its flat outputs. Thus, by defining setpoints for $x_1 = x$, $x_2 = y$ and $x_3 = r$ one can compute also values for the setpoints of state variables x_4, x_5 and x_6 (surge, sway and roll velocities of the AUV).

3.7.4 Approximate linearization of the state-space model of the AUV

3.7.4.1 The approximately linearized model of the AUV

The nonlinear state-space model of the AUV being in the form $\dot{x} = f(x) + g(x)u$ undergoes approximate linearization around the temporary operating point (x^*, u^*), where

x^* is the present value of the state vector of the system and u^* is the most recent sampled value of the control inputs vector. The linearization process relies on first order Taylor series expansion and on the computation of the Jacobian matrices of the system. The linearization procedure gives

$$\dot{x} = Ax + Bu + \tilde{d} \tag{3.260}$$

where $\tilde{d}$ is the approximately linearized model which comprises: (i) modeling errors due to truncation of higher-order terms in the Taylor series expansion, (ii) the effects of external perturbations, and (iii) measurement noise of any distribution.

Matrices A and B are obtained from the computation of the system's Jacobians

$$A = \nabla_x[f(x) + g(x)u] \mid_{(x^*,u^*)} \Rightarrow A = \nabla_x[f(x)] \mid_{(x^*,u^*)} \tag{3.261}$$

$$B = \nabla_u[f(x) + g(x)u] \mid_{(x^*,u^*)} \Rightarrow B = g(x) \mid_{(x^*,u^*)} \tag{3.262}$$

Next, the Jacobian matrix $A = \nabla_x[f(x)] \mid_{(x^*,u^*)}$ is computed.

First row of the Jacobian matrix $A = \nabla_x[f(x)] \mid_{(x^*,u^*)}$: $\frac{\partial f_1}{\partial x_1} = 0$, $\frac{\partial f_1}{\partial x_2} = 0$, $\frac{\partial f_1}{\partial x_3} = -x_4\sin(x_3) - x_5\cos(x_3)$, $\frac{\partial f_1}{\partial x_4} = \cos(x_3)$, $\frac{\partial f_1}{\partial x_5} = -\sin(x_3)$ and $\frac{\partial f_1}{\partial x_6} = \cos(x_3)$.

Second row of the Jacobian matrix $A = \nabla_x[f(x)] \mid_{(x^*,u^*)}$: $\frac{\partial f_2}{\partial x_1} = 0$, $\frac{\partial f_2}{\partial x_2} = 0$, $\frac{\partial f_2}{\partial x_3} = x_4\cos(x_3) - x_5\sin(x_3)$, $\frac{\partial f_2}{\partial x_4} = \sin(x_3)$, $\frac{\partial f_2}{\partial x_5} = \cos(x_3)$, $\frac{\partial f_2}{\partial x_6} = 0$.

Third row of the Jacobian matrix $A = \nabla_x[f(x)] \mid_{(x^*,u^*)}$: $\frac{\partial f_3}{\partial x_1} = 0$, $\frac{\partial f_3}{\partial x_2} = 0$, $\frac{\partial f_3}{\partial x_3} = 0$, $\frac{\partial f_3}{\partial x_4} = 0$, $\frac{\partial f_3}{\partial x_5} = 0$, and $\frac{\partial f_3}{\partial x_6} = 0$.

Fourth row of the Jacobian matrix $A = \nabla_x[f(x)] \mid_{(x^*,u^*)}$: $\frac{\partial f_4}{\partial x_1} = 0$, $\frac{\partial f_4}{\partial x_2} = 0$, $\frac{\partial f_4}{\partial x_3} = 0$, $\frac{\partial f_4}{\partial x_4} = -\frac{X_u}{M_{\dot{u}}} - \frac{D_u}{M_{\dot{u}}}|x_4| - \frac{D_u}{M_{\dot{u}}}x_4 sgn(x_4)$, $\frac{\partial f_4}{\partial x_5} = \frac{M_{\dot{v}}}{M_{\dot{u}}}x_6$, and $\frac{\partial f_4}{\partial x_6} = \frac{M_{\dot{v}}}{M_{\dot{u}}}x_5$.

Fifth row of the Jacobian matrix $A = \nabla_x[f(x)] \mid_{(x^*,u^*)}$: $\frac{\partial f_5}{\partial x_1} = 0$, $\frac{\partial f_5}{\partial x_2} = 0$, $\frac{\partial f_5}{\partial x_3} = 0$, $\frac{\partial f_5}{\partial x_4} = -\frac{M_{\dot{u}}}{M_{\dot{v}}}x_6$, $\frac{\partial f_5}{\partial x_5} = -\frac{Y_v}{M_{\dot{v}}} - \frac{D_v}{M_{\dot{v}}}|x_5| - \frac{D_v}{M_{\dot{v}}}x_5 sgn(x_5)$, $\frac{\partial f_5}{\partial x_6} = - - \frac{M_{\dot{u}}}{M_{\dot{v}}}x_4$.

Sixth row of the Jacobian matrix $A = \nabla_x[f(x)] \mid_{(x^*,u^*)}$: $\frac{\partial f_6}{\partial x_1} = 0$, $\frac{\partial f_6}{\partial x_2} = 0$, $\frac{\partial f_6}{\partial x_3} = 0$, $\frac{\partial f_6}{\partial x_4} = \frac{M_{\dot{u}} - M_{\dot{v}}}{M_{\dot{r}}}x_5$, $\frac{\partial f_6}{\partial x_5} = \frac{M_{\dot{u}} - M_{\dot{v}}}{M_{\dot{r}}}x_4$, and $\frac{\partial f_6}{\partial x_6} = \frac{N_r}{M_{\dot{r}}} - \frac{D_r}{M_{\dot{r}}}|x_6| - \frac{D_r}{M_{\dot{r}}}x_6 sgn(x_6)$.

3.7.4.2 Stabilizing feedback control

After linearization around its current operating point (x^*, u^*), the dynamic model of the fully actuated Autonomous Underwater Vessel (AUV) is written as (Rigatos and Busawon, 2018)

$$\dot{x} = Ax + Bu + d_1 \tag{3.263}$$

Parameter d_1 stands for the linearization error in the AUV's model appearing previously in (3.263). The reference setpoints for the AUV's state vector are denoted by

$\mathbf{x_d} = [x_1^d, \ldots, x_6^d]$. Tracking of this trajectory is achieved after applying the control input u_d. At every time instant the control input u_d is assumed to differ from the control input u appearing in (3.263) by an amount equal to Δu, that is $u_d = u + \Delta u$

$$\dot{x}_d = Ax_d + Bu_d + d_2 \tag{3.264}$$

The dynamics of the controlled system described in (3.263) can be also written as

$$\dot{x} = Ax + Bu + Bu_d - Bu_d + d_1 \tag{3.265}$$

and by denoting $d_3 = -Bu_d + d_1$ as an aggregate disturbance term, one obtains

$$\dot{x} = Ax + Bu + Bu_d + d_3 \tag{3.266}$$

By subtracting (3.264) from (3.266), one has

$$\dot{x} - \dot{x}_d = A(x - x_d) + Bu + d_3 - d_2 \tag{3.267}$$

By denoting the tracking error as $e = x - x_d$ and the aggregate disturbance term as $\tilde{d} = d_3 - d_2$, the tracking error dynamics becomes

$$\dot{e} = Ae + Bu + \tilde{d} \tag{3.268}$$

For the approximately linearized model of the system a stabilizing feedback controller is developed. The controller has the form

$$u(t) = -Ke(t) \tag{3.269}$$

with $K = \frac{1}{r}B^T P$ where P is a positive definite symmetric matrix which is obtained from the solution of the Riccati equation (Rigatos and Busawon, 2018)

$$A^T P + PA + Q - P(\frac{2}{r}BB^T - \frac{1}{\rho^2}LL^T)P = 0 \tag{3.270}$$

where Q is a positive semi-definite symmetric matrix. The diagram of the considered control loop for the AUV is depicted in Figure 3.86. The global stability properties of the control method can be proven following the procedure presented in Section 1.1.2.5.

3.7.5 Simulation tests

The performance of the proposed nonlinear optimal control scheme for the fully actuated AUV is further confirmed through simulation experiments. The model of the fully actuated AUV that has been considered in these tests is the one which has been used in (Shen, Shi *et al.*, 2019 and Shen, Buckham *et al.*, 2017) and actually represents the joint kinematics and dynamics of SAAB's Falcon AUV. Certainly, the results can be generalized for any type of fully actuated AUV. To compute the control inputs of the AUV, the algebraic Riccati equation of (3.270) had to be solved at each time-step of the control algorithm. The tracking performance of the cntrol scheme was tested using several reference paths that the AUV had to follow. The obtained results are depicted in Figures 3.87 to 3.102. The real values of the AUV's state vector are printed in blue, the estimated values which were provided by the H-infinity Kalman

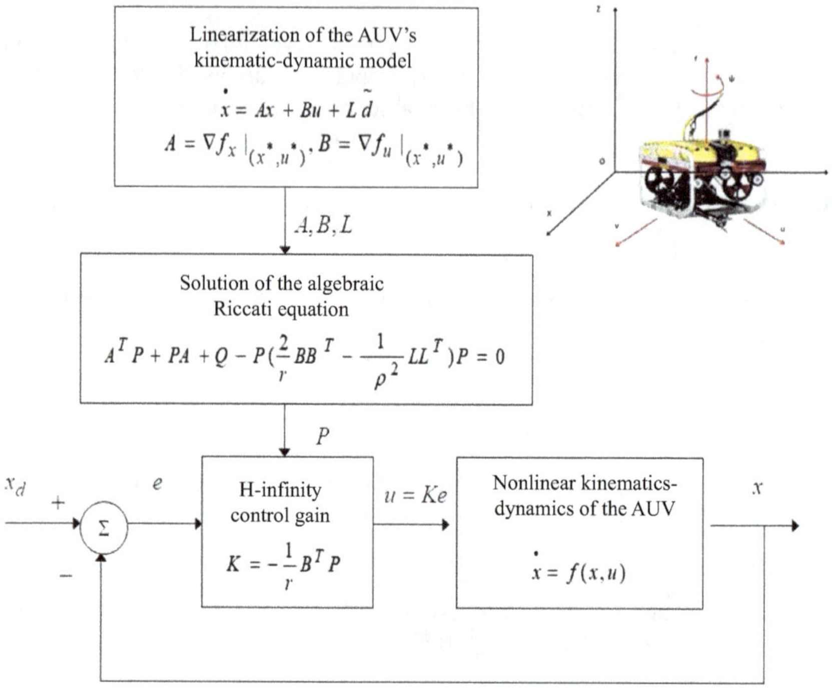

Figure 3.86 Diagram of the control scheme for the AUV that receives ful actuation

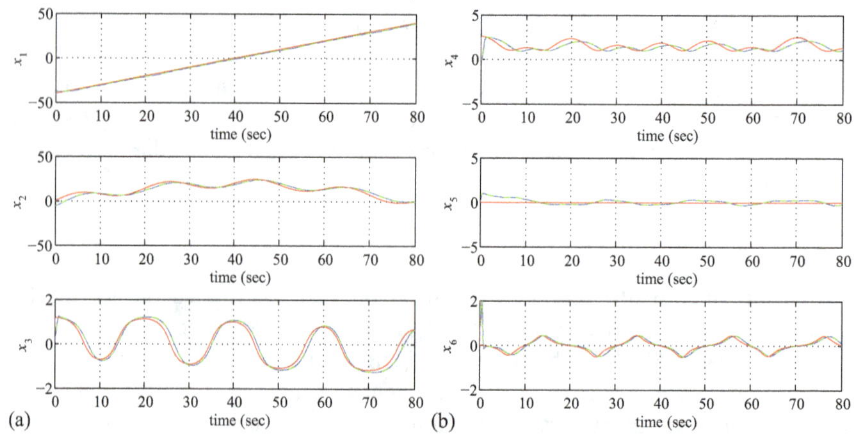

Figure 3.87 Tracking of setpoint 1 for the fully actuated AUV): (a) convergence of state variables x_1 to x_3 to their reference setpoints (red line: setpoint, blue line: real value, green line: estimated value); (b) convergence of state variables x_4 to x_6 to their reference setpoints (red line: setpoint, blue line: real value, green line: estimated value)

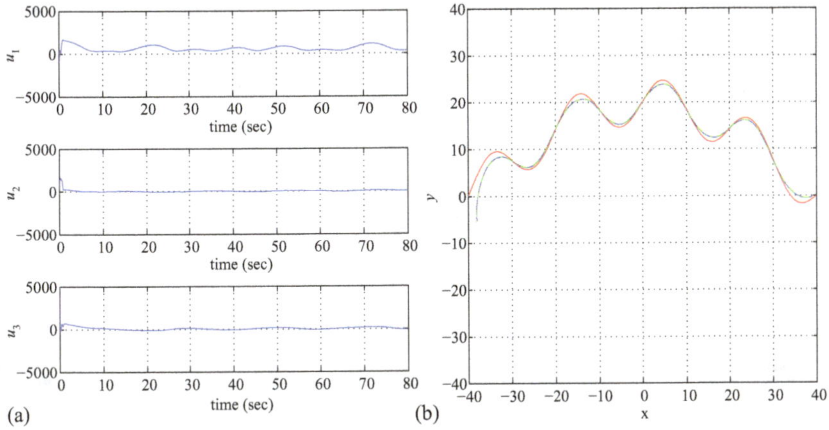

Figure 3.88 *Tracking of setpoint 1 for the fully actuated AUV: (a) control inputs u_1 to u_3 applied to the actuators of the AUV; (b) tracking of the reference path (red line) by the AUV (blue line) in the xy plane*

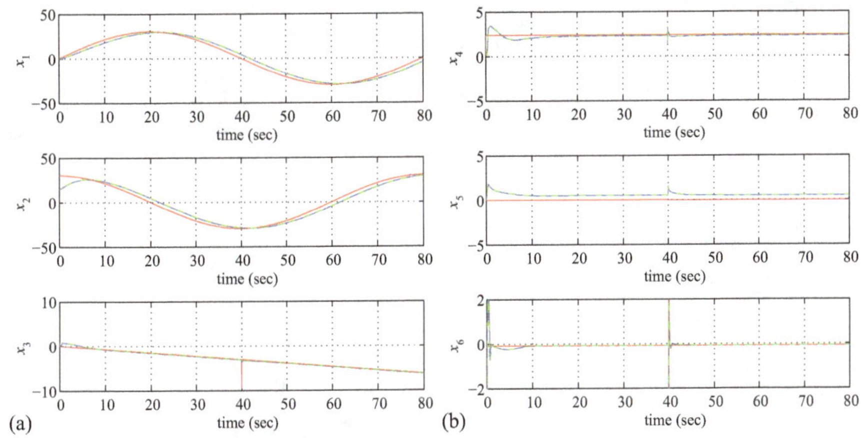

Figure 3.89 *Tracking of setpoint 2 for the fully actuated AUV: (a) convergence of state variables x_1 to x_3 to their reference setpoints (red line: setpoint, blue line: real value, green line: estimated value); (b) convergence of state variables x_4 to x_6 to their reference setpoints (red line: setpoint, blue line: real value, green line: estimated value)*

Filter are plotted in green, while the related reference setpoints are shown in red. Distance is measured in meters (m), angles in rad and time in sec. It can be noticed that in all cases, the proposed H-infinity controller achieved fast and accurate tracking of the reference paths under moderate variations of the control inputs.

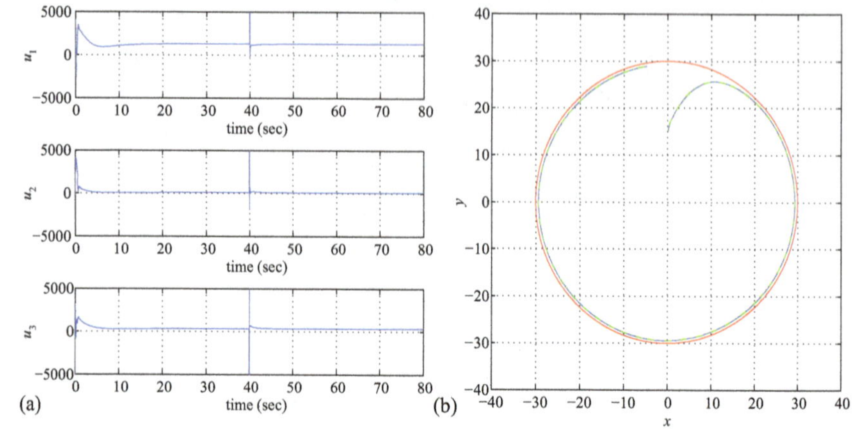

Figure 3.90 Tracking of setpoint 2 for the fully actuated AUV: (a) control inputs u_1 to u_3 applied to the actuators of the AUV; (b) tracking of the reference path (red line) by the AUV (blue line) in the xy plane

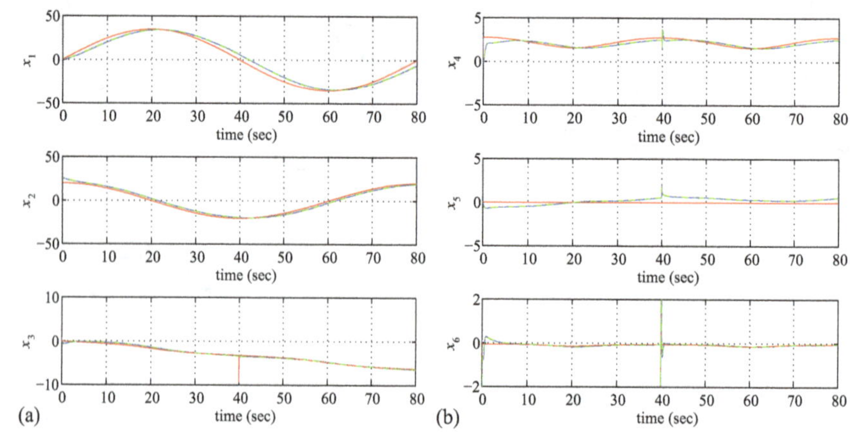

Figure 3.91 Tracking of setpoint 3 for the fully actuated AUV: (a) convergence of state variables x_1 to x_3 to their reference setpoints (red line: setpoint, blue line: real value, green line: estimated value); (b) convergence of state variables x_4 to x_6 to their reference setpoints (red line: setpoint, blue line: real value, green line: estimated value)

3.8 Nonlinear control of the vertical take-off and landing aircraft

3.8.1 Outline

The use of vertical take-off and landing aircrafts (VTOL) is deploying fast in several safety, surveillance and defense tasks (Rigatos and Busawon, 2018; Hua, Humel

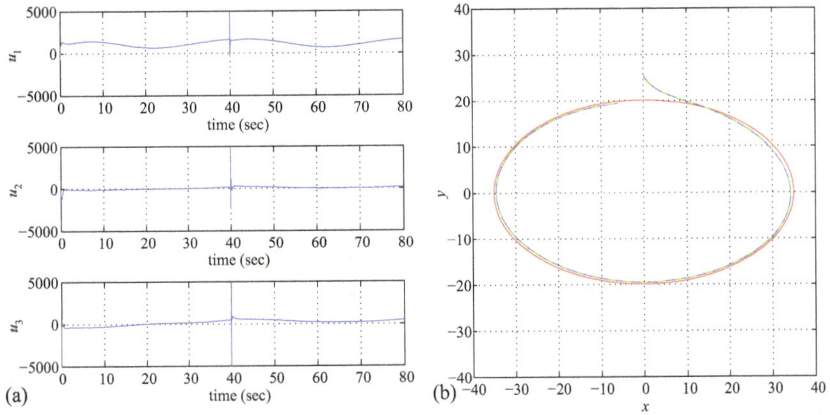

Figure 3.92 Tracking of setpoint 3 for the fully actuated AUV: (a) control inputs u_1 to u_3 applied to the actuators of the AUV; (b) tracking of the reference path (red line) by the AUV (blue line) in the xy plane

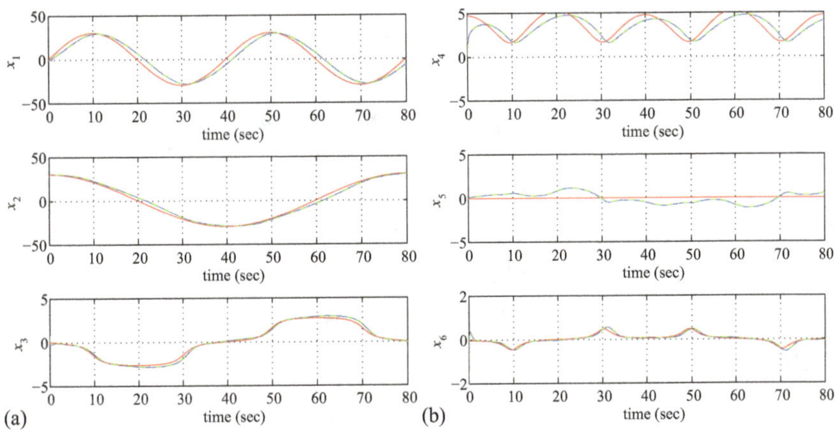

Figure 3.93 Tracking of setpoint 4 for the fully actuated AUV: (a) convergence of state variables x_1 to x_3 to their reference setpoints (red line: setpoint, blue line: real value, green line: estimated value); (b) convergence of state variables x_4 to x_6 to their reference setpoints (red line: setpoint, blue line: real value, green line: estimated value)

et al., 2013; Do *et al.*, 2003). The problem of nonlinear control and stabilization of autonomous vertical take-off and landing aircrafts is a non-trivial one, due to the nonlinear dynamics of such systems as well as due to their underactuation (Lyu, Zhou *et al.*, 2018; Peng, Guo *et al.*, 2019; Hua, Humel *et al.*, 2009). During take-off or landing such aircrafts exhibit three degrees of freedom while receiving only two

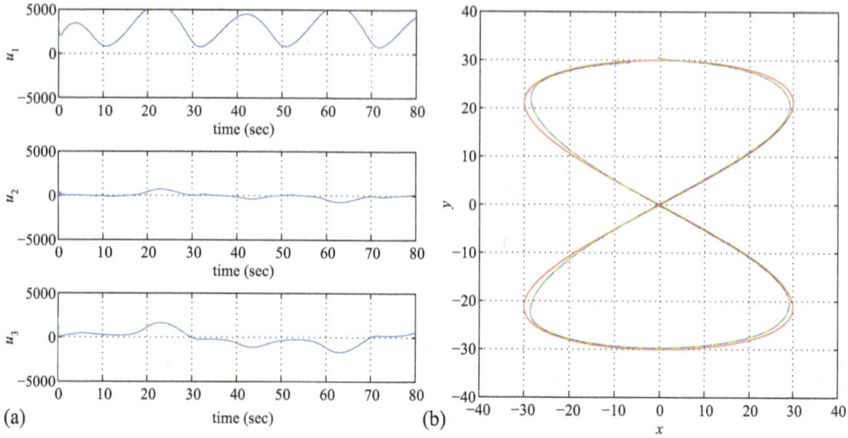

Figure 3.94 *Tracking of setpoint 4 for the fully actuated AUV: (a) control inputs u_1 to u_3 applied to the actuators of the AUV; (b) tracking of the reference path (red line) by the AUV (blue line) in the xy plane*

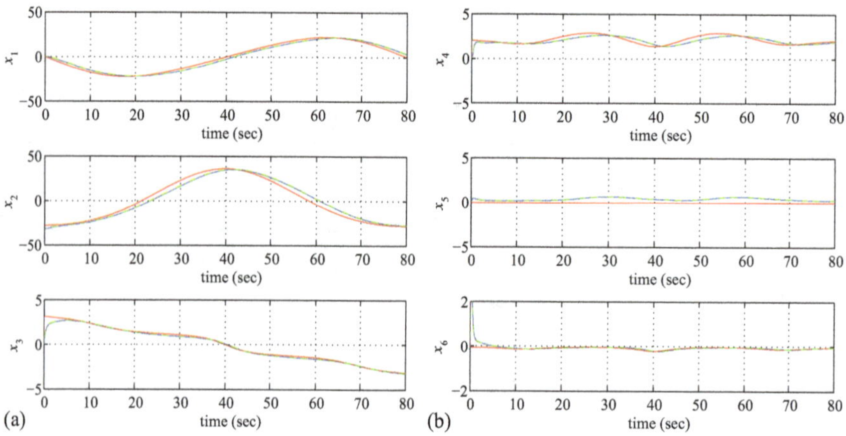

Figure 3.95 *Tracking of setpoint 5 for the fully actuated AUV: (a) convergence of state variables x_1 to x_3 to their reference setpoints (red line: setpoint, blue line: real value, green line: estimated value); (b) convergence of state variables x_4 to x_6 to their reference setpoints (red line: setpoint, blue line: real value, green line: estimated value)*

control inputs (Chwa, 2015; Su and Lin, 2011; Ailon *et al.*, 2010). Actually, their center of gravity can move along the two dimensions of the vertical plane, while they can also perform a roll motion (Pozo and Maggiore, 2014; Naidi *et al.*, 2017; Hua, Humel *et al.*, 2015). The control inputs they receive are a thrust force that can lift up the aircraft and forces applied on the wings tips that can change the position of

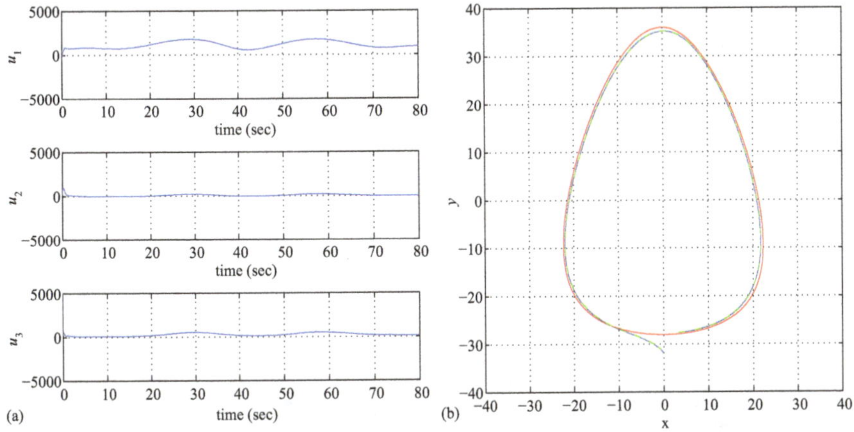

Figure 3.96 Tracking of setpoint 5 for the fully actuated AUV: (a) control inputs u_1 to u_3 applied to the actuators of the AUV; (b) tracking of the reference path (red line) by the AUV (blue line) in the xy plane

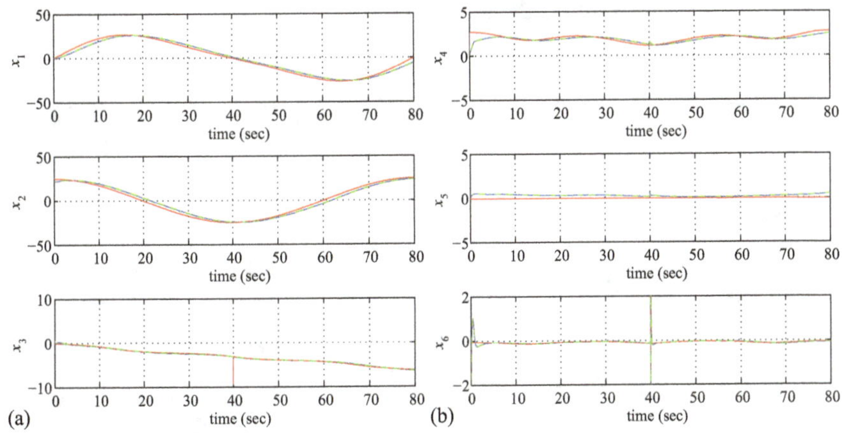

Figure 3.97 Tracking of setpoint 6 for the fully actuated AUV: (a) convergence of state variables x_1 to x_3 to their reference setpoints (red line: setpoint, blue line: real value, green line: estimated value); (b) convergence of state variables x_4 to x_6 to their reference setpoints (red line: setpoint, blue line: real value, green line: estimated value)

the aircraft's center of gravity or can change its roll angle during the take-off and landing stage (Olfati-Saber, 2002; Wang, 2013; Tayebi and McGilvray, 2006). The forces exerted on the wings tips are not perpendicular to the body axis of the aircraft but are tilted by a small angle (Chadli *et al.*, 2013; Hérisé *et al.*, 2012). There exist several attempts towards the solution of the nonlinear multivariable control problem

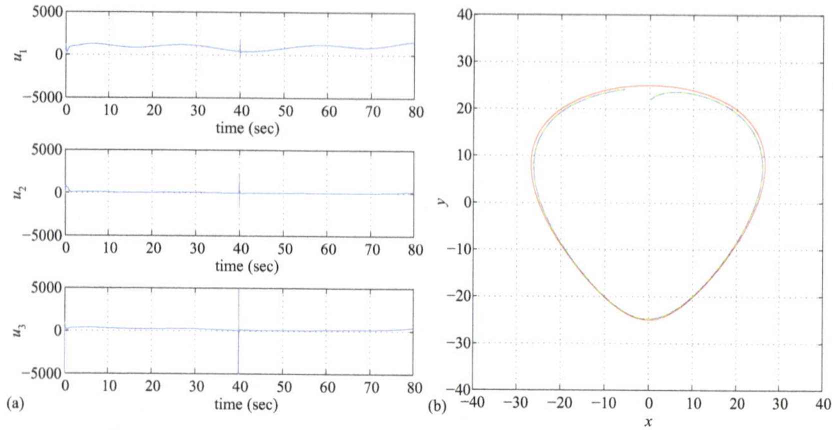

Figure 3.98 Tracking of setpoint 6 for the fully actuated AUV: (a) control inputs u_1 to u_3 applied to the actuators of the AUV; (b) tracking of the reference path (red line) by the AUV (blue line) in the xy plane

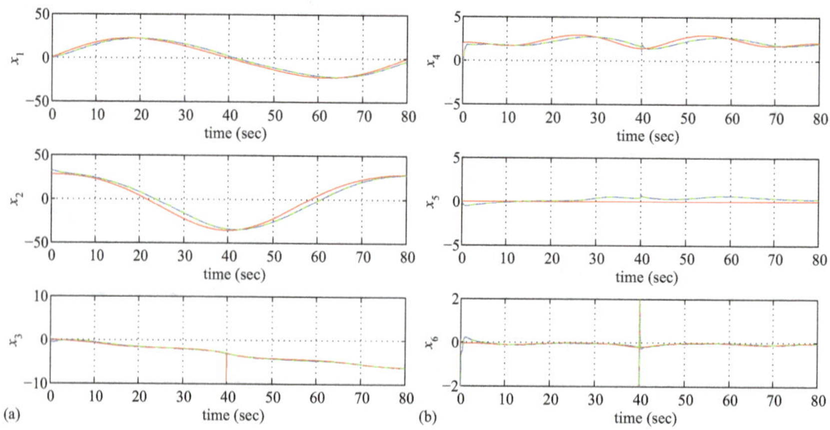

Figure 3.99 Tracking of setpoint 7 for the fully actuated AUV: (a) convergence of state variables x_1 to x_3 to their reference setpoints (red line: setpoint, blue line: real value, green line: estimated value); (b) convergence of state variables x_4 to x_6 to their reference setpoints (red line: setpoint, blue line: real value, green line: estimated value)

for the VTOL aircraft among which global linearization-based control schemes have a distinguished position (Hauser *et al.*, 1992; Fliess *et al.*, 1999; Martin, Devasia *et al.*, 1996). Global stability and robustness are the objectives that controllers' design for VTOL drones aims at achieving (Li, Zhang *et al.*, 2018; Abdessameud and Tayebi, 2009; Wang, Liu *et al.*, 2009).

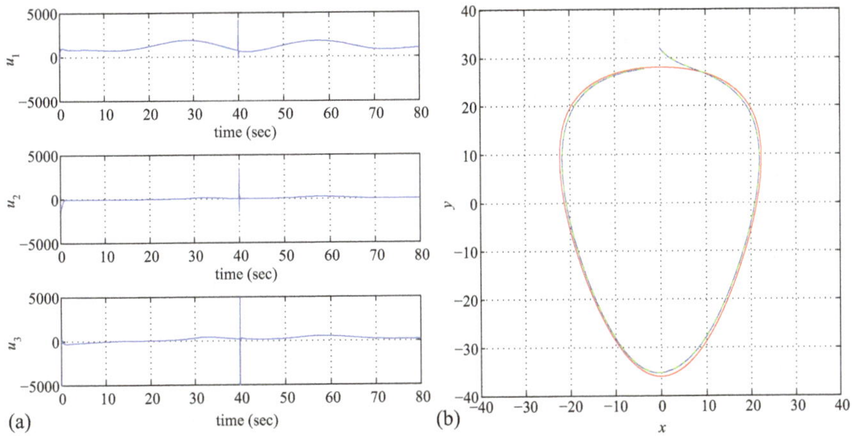

Figure 3.100 Tracking of setpoint 7 for the fully actuated AUV: (a) control inputs u_1 to u_3 applied to the actuators of the AUV; (b) tracking of the reference path by the AUV in the xy plane

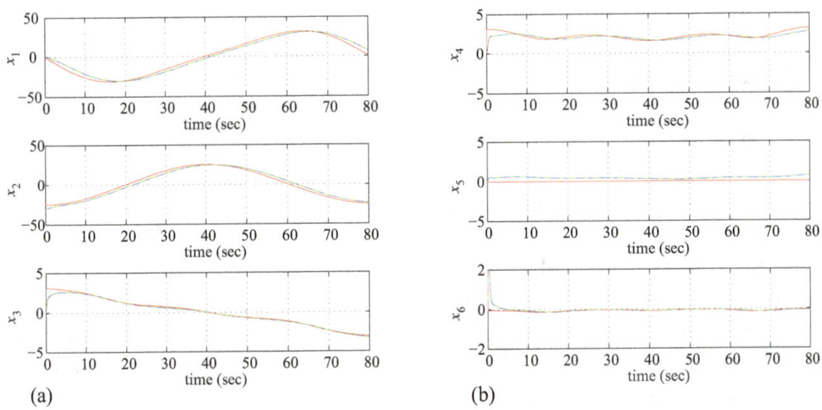

Figure 3.101 Tracking of setpoint 8 for the fully actuated AUV: (a) convergence of state variables x_1 to x_3 to their reference setpoints (red line: setpoint, blue line: real value, green line: estimated value); (b) convergence of state variables x_4 to x_6 to their reference setpoints (red line: setpoint, blue line: real value, green line: estimated value)

In this section, a nonlinear optimal control method is developed for the dynamical model of the VTOL aircraft (Rigatos, 2016b; Rigatos, Siano and Cecati, 2015; Rigatos, Siano *et al.*, 2015). To this end, the state-space model of the VTOL aircraft undergoes first approximate linearization around a temporary operating point which is updated at each time-step of the control algorithm (Rigatos and Tzafestas,

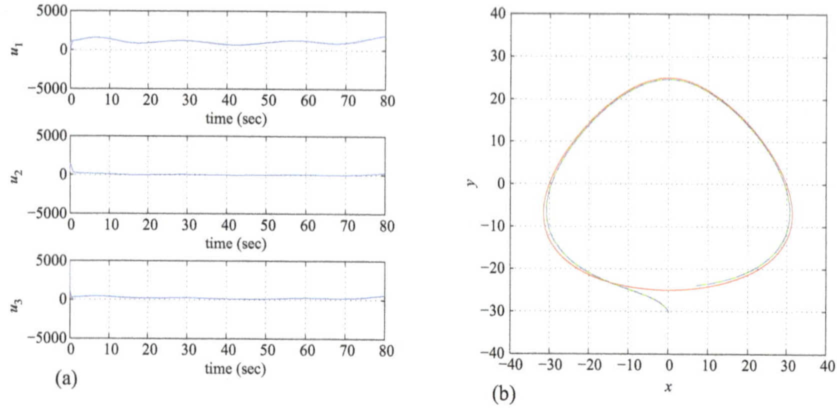

Figure 3.102 Tracking of setpoint 8 for the fully actuated AUV: (a) control inputs u_1 to u_3 applied to the actuators of the AUV; (b) tracking of the reference path by the AUV in the xy plane

2007; Basseville and Nikiforov, 1993; Rigatos and Zhang, 2009). The linearization point is defined by the present value of the VTOL aircraft's state vector and by the last value of the control inputs vector that was applied on it. The linearization relies on Taylor series expansion and on the computation of the system's Jacobian matrices. The modeling error which is due to the truncation of higher-order terms in the Taylor series expansion is considered to be a perturbation which is asymptotically compensated by the robustness of the control algorithm. For the approximately linearized model of the system an H-infinity feedback controller is designed.

Actually, the H-infinity controller stands for the solution of the optimal control problem of the VTOL aircraft under model uncertainties and external perturbations. It represents a min–max differential game taking place between (i) the control inputs that try to minimize a cost function which comprises a quadratic term of the state vector's tracking error and (ii) the modeling error and perturbation inputs which try to maximize this cost function. For the computation of the H-infinity controller's feedback gains an algebraic Riccati equation is repetitively being solved at each iteration of the control scheme (Rigatos, 2011; Rigatos, 2015). The robustness properties of the control method are proven through Lyapunov analysis. First, it is proven that the control scheme exhibits the H-infinity tracking performance (Rigatos and Busawon, 2018; Tousssaint, Basar and Bullo, 2000). This signifies elevated robustness against model uncertainties and external perturbations. Moreover, under moderate conditions, it is proven that the control loop is globally asymptotically stable. Finally, to implement state estimation-based control for the dynamic model of the VTOL aircraft without the need to measure its entire state-vector, the H-infinity Kalman Filter is used as a robust state estimator (Gibbs, 2012).

3.8.2 Dynamic model of the vertical take-off and landing aircraft

3.8.2.1 State-space form of the VTOL aircraft

Vertical take-off and landing aircrafts (VTOLS) are widely used for surveillance, patrolling and defense tasks. They receive as control input a directed thrust, as well as forces acting on their wing tips. The diagram of the VTOL is shown in Figure 3.103

The parameters of the model are defined as follows: m is the mass of the aircraft, J is the moment of inertia of the aircraft when performing roll motion, L is the wing's span, θ is the angle between the body-axis of the aircraft and the OX axis of the inertial coordinates frame. Forces exerted at the end of the wings are not perpendicular to the body axis of the aircraft but are tilted by a small angle a. Next, by defining $u_1 = \frac{T}{m}$, $u_2 = \frac{2F}{m}\cos(a)$, $\epsilon = \tan(a)$ and $\lambda = \frac{mL}{J}$ the dynamic model of the VTOL is given by Wang, 2013, Martin, Devasia *et al.*, 1996

$$\begin{aligned} \ddot{x} &= -u_1 \sin(\theta) + \epsilon u_2 \cos(\theta) \\ \ddot{z} &= u_1 \cos(\theta) + \epsilon u_2 \sin(\theta) - g \\ \ddot{\theta} &= \lambda u_2 \end{aligned} \tag{3.271}$$

where g is the acceleration of gravity. Next, the following control inputs variables are defined: $v_1 = u_1 \sin(\theta)$, $v_2 = u_1 \cos(\theta) - g$ and $v_3 = u_2$. After this control input variables transformation, the state-space model of the VTOL aircraft is written as

$$\begin{aligned} \ddot{x} &= -v_1 + \epsilon v_3 \cos(\theta) \\ \ddot{z} &= v_2 + \epsilon v_3 \sin(\theta) \\ \ddot{\theta} &= \lambda v_3 \end{aligned} \tag{3.272}$$

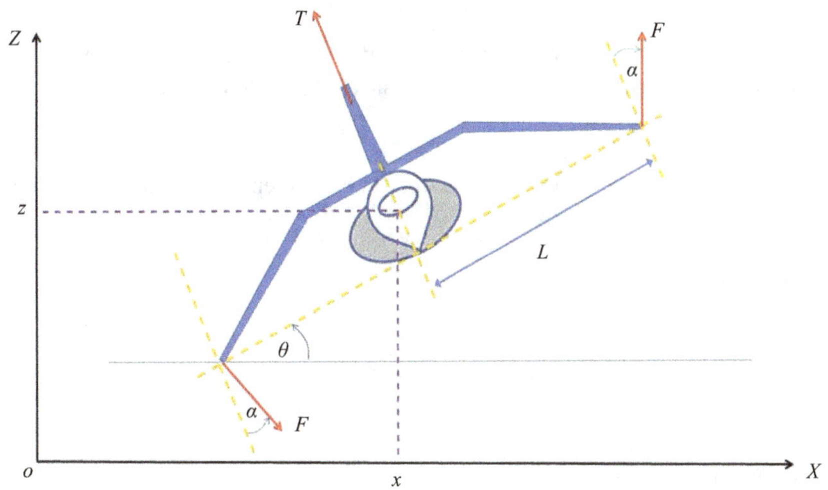

Figure 3.103 Diagram of the vertical take-off and landing aircraft

Considering that v_1, v_2 and v_3 are computed through the solution of the nonlinear control problem for the VTOL aircraft one has that $u_1 = \sqrt{v_1^2+(v_2+g)^2}$, while $u_2 = v_3$. Next, by defining the state-variables $x_1 = x$, $x_2 = \dot{x}$, $x_3 = z$, $x_4 = \dot{z}$, $x_5 = \theta$, and $x_6 = \dot{\theta}$ one obtains the following state-space description in matrix form

$$\begin{pmatrix} \dot{x}_1 \\ \dot{x}_2 \\ \dot{x}_3 \\ \dot{x}_4 \\ \dot{x}_5 \\ \dot{x}_6 \end{pmatrix} = \begin{pmatrix} x_2 \\ 0 \\ x_4 \\ 0 \\ x_6 \\ 0 \end{pmatrix} + \begin{pmatrix} 0 & 0 & 0 \\ -1 & 0 & \varepsilon\cos(x_5) \\ 0 & 0 & 0 \\ 0 & 1 & \varepsilon\sin(x_5) \\ 0 & 0 & 0 \\ 0 & 0 & \lambda \end{pmatrix} \begin{pmatrix} v_1 \\ v_2 \\ v_3 \end{pmatrix} \tag{3.273}$$

Using a vector fields notation, the above state-space model can be written as

$$\dot{x} = f(x) + g_1(x)v_1 + g_2(x)v_2 + g_3(x)v_3 \tag{3.274}$$

$f(x)\in R^{6\times1}$, $g_1(x)\in R^{6\times1}$, $g_2(x)\in R^{6\times1}$ and $g_3(x)\in R^{6\times1}$ are given by

$$f(x) = \begin{pmatrix} x_2 \\ 0 \\ x_4 \\ 0 \\ x_6 \\ 0 \end{pmatrix} \quad g_1(x) = \begin{pmatrix} 0 \\ -1 \\ 0 \\ 0 \\ 0 \\ 0 \end{pmatrix} \quad g_2(x) = \begin{pmatrix} 0 \\ 0 \\ 0 \\ 1 \\ 0 \\ 0 \end{pmatrix} \quad g_3(x) = \begin{pmatrix} 0 \\ \varepsilon\cos(x_5) \\ 0 \\ \varepsilon\sin(x_5) \\ 0 \\ 0 \end{pmatrix} \tag{3.275}$$

3.8.3 *Differential flatness properties of the VTOL aircraft*

The dynamic model of the VTOL aircraft is differentially flat. The flat outputs of the model are (Fliess *et al.*, 1999; Martin, Devasia *et al.*, 1996; Rigatos, 2015)

$$y_1 = x - \varepsilon\sin(\theta) \quad y_2 = y + \varepsilon\cos(\theta) \tag{3.276}$$

It holds that all state variables of the model can be written as differential functions of its flat outputs. Thus one has:

$$x = y_1 - \frac{\varepsilon\ddot{y}_1}{\sqrt{\ddot{y}_1^2+(\ddot{y}_2+g)^2}} \tag{3.277}$$

$$z = y_2 - \frac{\varepsilon\ddot{y}_2+g}{\sqrt{\ddot{y}_1^2+(\ddot{y}_2+g)^2}} \tag{3.278}$$

$$\theta = \tan^{-1}\left(\frac{\ddot{y}_1}{\ddot{y}_2+g}\right) \tag{3.279}$$

Moreover, using the above relations of (3.277), (3.278) and (3.279), as well as the model of (3.271), it can be shown that the control inputs u_1 and u_2 can be also written as differential functions of the flat outputs. Consequently, the model of the VTOL is differentially flat. This property can be used for defining reference setpoints for the associated control loop.

3.8.4 Approximate linearization of the VTOL aircraft dynamic model

The dynamic model of the VTOL aircraft undergoes approximate linearization around the temporary operating point (x^*, u^*). The linearization procedure is repeated at each time-step of the control algorithm and x^* is the present value of the system's state vector, while u^* is the last sampled value of the control inputs vector. The linearization is performed based on the first-order Taylor series expansion and on the computation of the Jacobian matrices of the state-space model. This results into the linearized description

$$\dot{x} = Ax + Bv + \tilde{d} \tag{3.280}$$

where matrices A and B are given by

$$\begin{aligned} A = \nabla_x f(x) \mid_{(x^*,u^*)} + \nabla_x g_1(x) v_1 \mid_{(x^*,u^*)} \\ + \nabla_x g_2(x) v_2 \mid_{(x^*,u^*)} + \nabla_x g_3(x) v_3 \mid_{(x^*,u^*)} \end{aligned} \tag{3.281}$$

$$B = \nabla_v [f(x) + g_1(x)v_1 + g_2(x)v_2 + g_3(x)v_3] \mid_{(x^*,u^*)} \Rightarrow B = g \mid_{(x^*,u^*)} \tag{3.282}$$

Next, the following Jacobian matrices are computed:

$$\nabla_x f(x) \mid_{(x^*,u^*)} = \begin{pmatrix} 0 & 1 & 0 & 0 & 0 & 0 \\ 0 & 0 & 0 & 0 & 0 & 0 \\ 0 & 0 & 0 & 1 & 0 & 0 \\ 0 & 0 & 0 & 0 & 0 & 0 \\ 0 & 0 & 0 & 0 & 0 & 1 \\ 0 & 0 & 0 & 0 & 0 & 0 \end{pmatrix} \tag{3.283}$$

and also $\nabla g_1(x) \mid_{(x^*,u^*)} = 0 \in R^{6\times 6}$, $\nabla g_2(x) \mid_{(x^*,u^*)} = 0 \in R^{6\times 6}$ and

$$\nabla_x g_3(x) \mid_{(x^*,u^*)} = \begin{pmatrix} 0 & 1 & 0 & 0 & 0 & 0 \\ 0 & 0 & 0 & 0 & -\varepsilon \sin(x_5) & 0 \\ 0 & 0 & 0 & 0 & 0 & 0 \\ 0 & 0 & 0 & 0 & \varepsilon \cos(x_5) & 0 \\ 0 & 0 & 0 & 0 & 0 & 0 \\ 0 & 0 & 0 & 0 & 0 & 0 \end{pmatrix} \tag{3.284}$$

3.8.5 H-infinity feedback control

For the linearized system given by (1.36) the cost function of (1.37) is defined, where coefficient r determines the penalization of the control input and weight coefficient ρ determines the reward of the disturbances' effects.

It is assumed that (i) the energy that is transferred from the disturbances signal $\tilde{d}(t)$ is bounded, that is $\int_0^\infty \tilde{d}^T(t)\tilde{d}(t)dt < \infty$, (ii) matrices $[A, B]$ and $[A, L]$ are stabilizable, and (iii) matrix $[A, C]$ is detectable. Then, by defining the state-vector's tracking error as $e = x - x_d$, the optimal feedback control law is given by

$$u(t) = -Ke(t) \tag{3.285}$$

with

$$K = \frac{1}{r} B^T P \tag{3.286}$$

where P is a positive definite symmetric matrix which is obtained from the solution of the Riccati equation

$$A^T P + PA + Q - P\left(\frac{2}{r} BB^T - \frac{1}{\rho^2} LL^T\right) P = 0 \tag{3.287}$$

where Q is also a positive semi-definite symmetric matrix. The worst case disturbance is given by

$$\tilde{d}(t) = \frac{1}{\rho^2} L^T Pe(t) \tag{3.288}$$

The above term has been computed after solving the dual H-infinity control problem for the linearized system in which the control input u is substituted by the disturbance input $\tilde{d}$. The previously analyzed concept about the nonlinear optimal control loop for the VTOL aircraft is given in Figure 3.104. The global stability

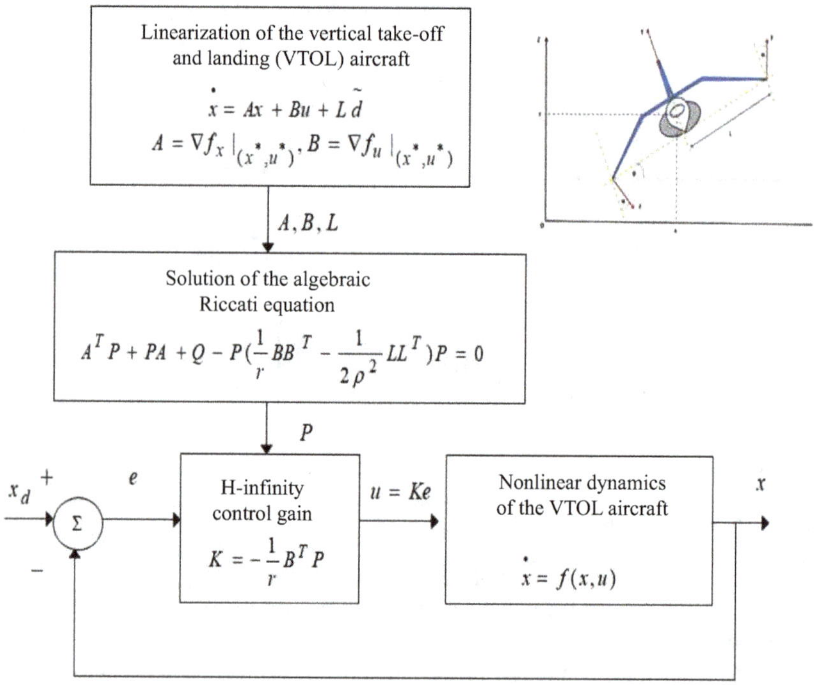

Figure 3.104 Diagram of the control scheme for the vertical take-off and landing aircraft

properties of the control method can be proven following the procedure presented in Section 1.1.2.5.

3.8.6 Simulation tests

The performance of the previously analyzed nonlinear optimal control method for the dynamic model of the VTOL aircraft has been further confirmed through simulation experiments. The presented results are depicted in Figures 3.105 to 3.116. For the computation of the H-infinity controller's feedback gains the algebraic Riccati equation appearing in (3.287) had to be solved at each time-step of the control method. The control inputs were applied directly on the initial nonlinear model of the VTOL aircraft. Through the obtained results, it can be noticed that fast and accurate tracking of the reference setpoints was achieved under moderate variations of the control inputs.

The simulation results show that the tracking error for all state variables of the VTOL aircraft has been eliminated. The transient performance of the control scheme relied on the control loop gains r and ρ and well as on the value of the diagonal elements of matrix Q. As explained above, the smallest value of the attenuation coefficient ρ for which the algebraic Riccati equation of (3.287) admits a solution, is the one that provides maximum robustness to the control system. It is also noted that by using the H-infinity Kalman Filter a state estimation-based implementation of the control method has been achieved. This allows the reliable functioning of the control loop after receiving measurements from a small number of sensors.

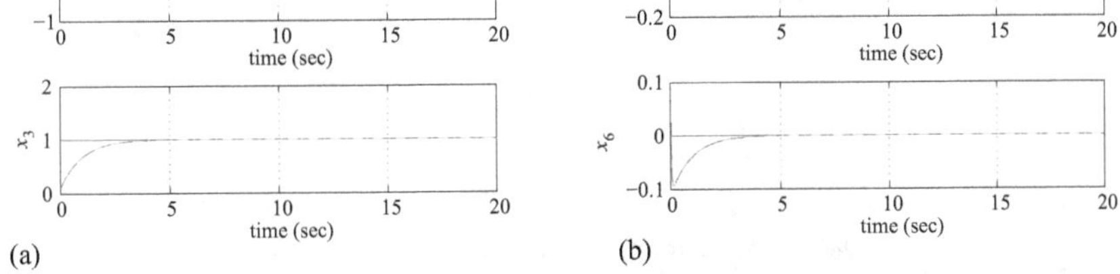

Figure 3.105 Tracking of setpoint 1 for the VTOL aircraft: (a) convergence of state variables x_1, x_2 and x_3 to their reference setpoints (red line: setpoint, blue line: real value, green line: estimated value); (b) convergence of state variables x_4, x_5 and x_6 to their reference setpoints (red line: setpoint, blue line: real value, green line: estimated value)

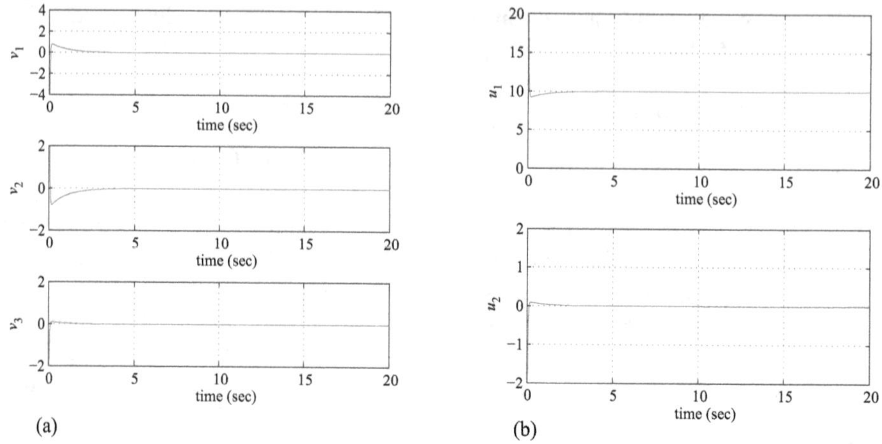

Figure 3.106 Tracking of setpoint 1 for the VTOL aircraft: (a) transformed control inputs u_1, u_2 and u_3; (b) real control inputs u_1 and u_2

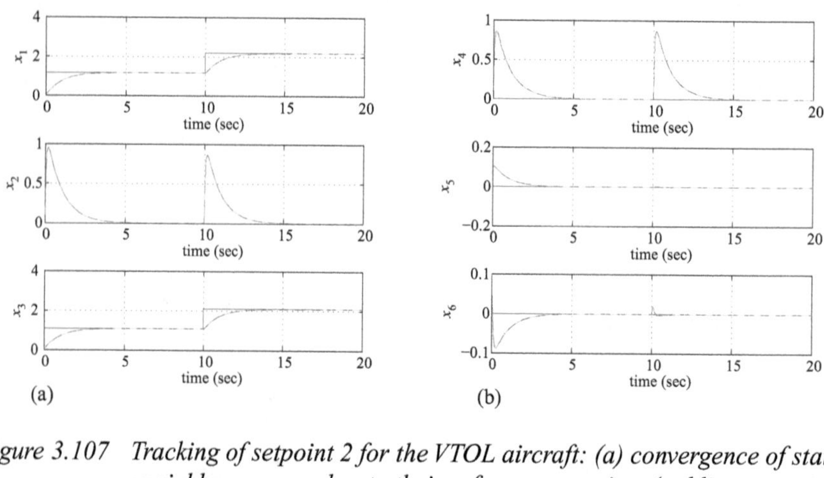

Figure 3.107 Tracking of setpoint 2 for the VTOL aircraft: (a) convergence of state variables x_1, x_2 and x_3 to their reference setpoints (red line: setpoint, blue line: real value, green line: estimated value); (b) convergence of state variables x_4, x_5 and x_6 to their reference setpoints (red line: setpoint, blue line: real value, green line: estimated value)

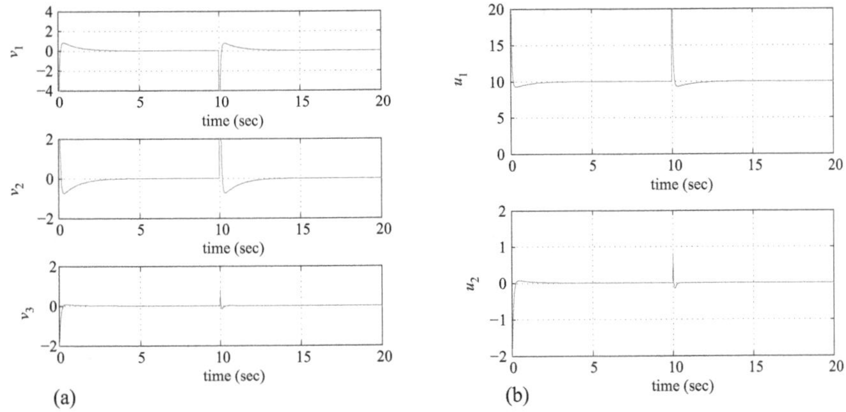

Figure 3.108 Tracking of setpoint 2 for the VTOL aircraft: (a) transformed control inputs u_1, u_2 and u_3; (b) real control inputs u_1 and u_2

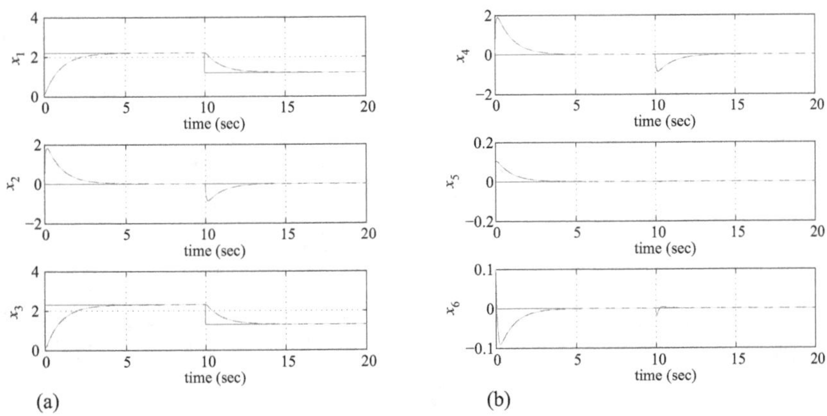

Figure 3.109 Tracking of setpoint 3 for the VTOL aircraft: (a) convergence of state variables x_1, x_2 and x_3 to their reference setpoints (red line: setpoint, blue line: real value, green line: estimated value); (b) convergence of state variables x_4, x_5 and x_6 to their reference setpoints (red line: setpoint, blue line: real value, green line: estimated value)

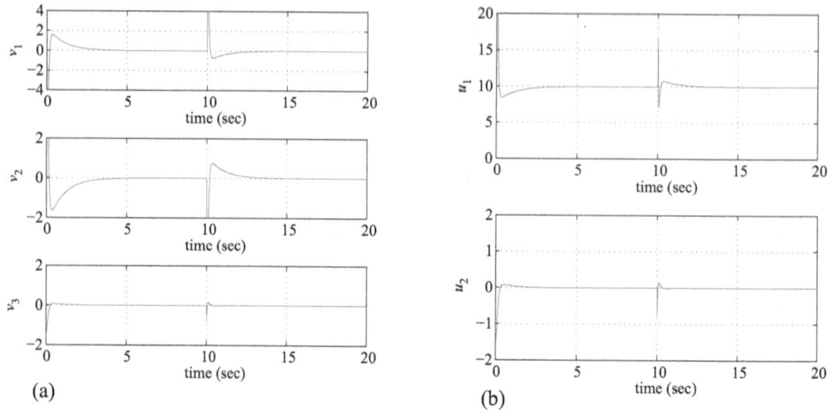

Figure 3.110 Tracking of setpoint 3 for the VTOL aircraft: (a) transformed control inputs u_1, u_2 and u_3; (b) real control inputs u_1 and u_2

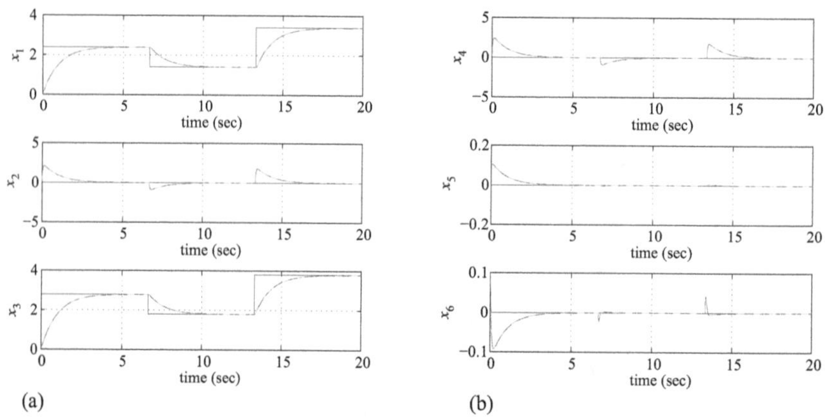

Figure 3.111 Tracking of setpoint 4 for the VTOL aircraft: (a) convergence of state variables x_1, x_2 and x_3 to their reference setpoints (red line: setpoint, blue line: real value, green line: estimated value); (b) convergence of state variables x_4, x_5 and x_6 to their reference setpoints (red line: setpoint, blue line: real value, green line: estimated value)

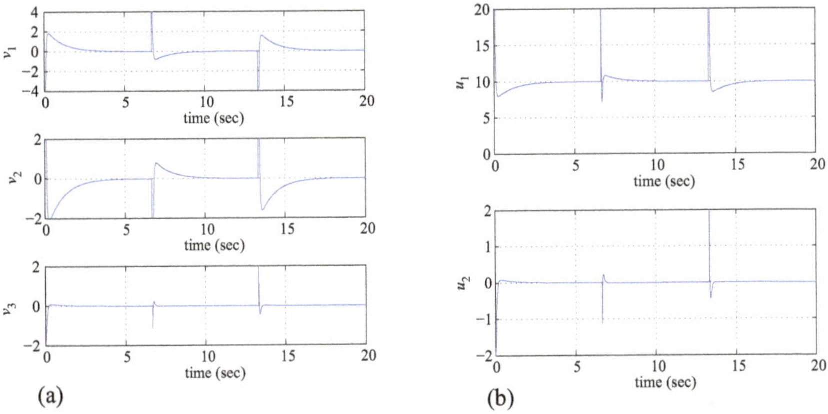

Figure 3.112 Tracking of setpoint 4 for the VTOL aircraft: (a) transformed control inputs u_1, u_2 and u_3; (b) real control inputs u_1 and u_2

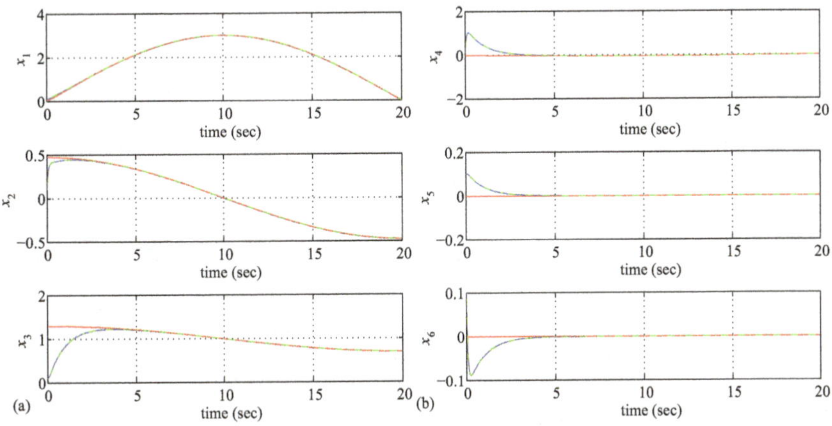

Figure 3.113 Tracking of setpoint 5 for the VTOL aircraft: (a) convergence of state variables x_1, x_2 and x_3 to their reference setpoints (red line: setpoint, blue line: real value, green line: estimated value); (b) convergence of state variables x_4, x_5 and x_6 to their reference setpoints (red line: setpoint, blue line: real value, green line: estimated value)

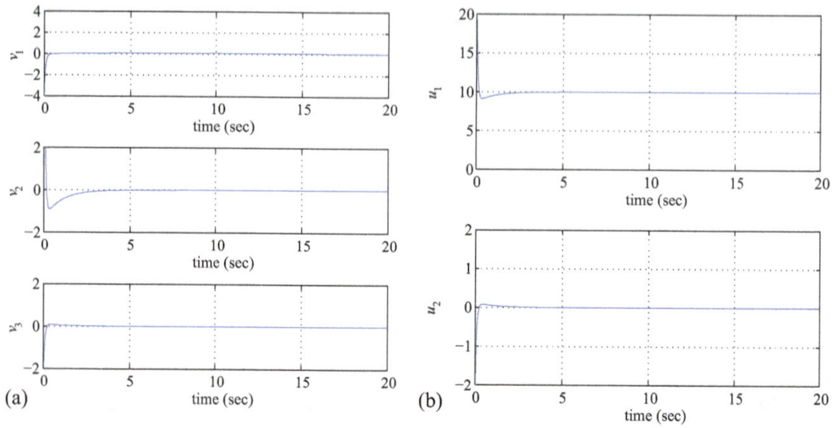

Figure 3.114 Tracking of setpoint 5 for the VTOL aircraft: (a) transformed control inputs u_1, u_2 and u_3; (b) real control inputs u_1 and u_2

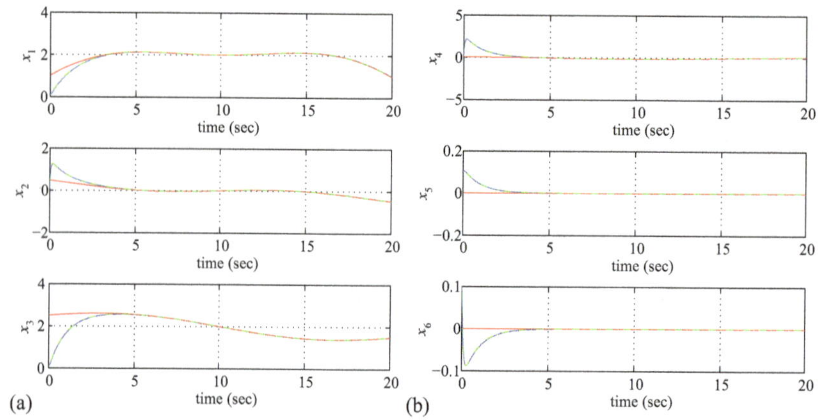

Figure 3.115 Tracking of setpoint 6 for the VTOL aircraft: (a) convergence of state variables x_1, x_2 and x_3 to their reference setpoints (red line: setpoint, blue line: real value, green line: estimated value); (b) convergence of state variables x_4, x_5 and x_6 to their reference setpoints (red line: setpoint, blue line: real value, green line: estimated value)

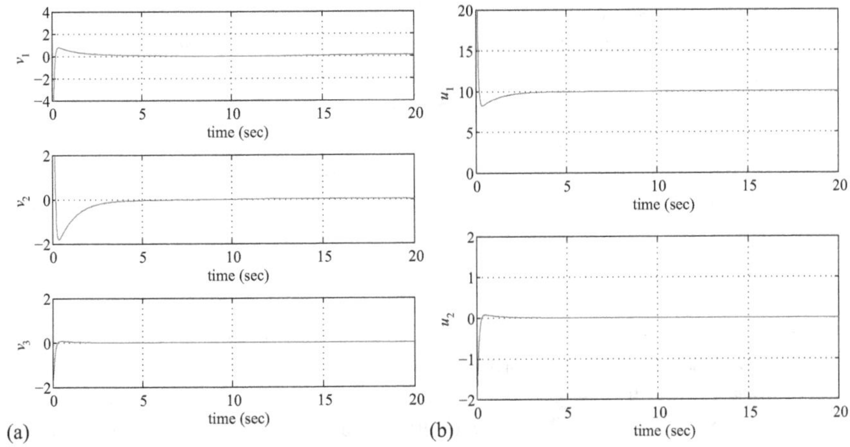

Figure 3.116 Tracking of setpoint 6 for the VTOL aircraft: (a) transformed control inputs u_1, u_2 and u_3; (b) real control inputs u_1 and u_2

3.9 Nonlinear control of aerial manipulators

3.9.1 Outline

The use of aerial robotic manipulators, also known as rotor flying manipulators, is rapidly deploying in rescue, defense and ammunition or transportation tasks (Rigatos and Busawon, 2018; Acosta *et al.*, 2016; de Cos *et al.*, 2017; Lumni *et al.*, 2017). Aerial robotic manipulators comprise a multi-rotor UAV and a robotic manipulator mounted on it (Beikzadeh and Liu, 2018; Jimenez-Cano et al., 2013; Forte *et al.*, 2012; Lippiello and Ruggiero, 2012a). The dynamic and kinematic model of such aerial robots accumulates the complexity of the models of its constituents, that is of the UAV and of the robotic manipulator. Actually, the dynamic model of such aerial robots is a multivariable and strongly nonlinear system and the solution of the related control problem is a non-trivial task (Suarez *et al.*, 2018; Acosta *et al.*, 2014; Rugierro *et al.*, 2015; Meng, He *et al.*, 2016). Besides, this system is underactuated, which means that it has more degrees of freedom than its control inputs and consequently the solution of the related control problem becomes far more complicated (Orsag *et al.*, 2017; Yang, He *et al.*, 2014; Caccavale *et al.*, 2014; Escareno *et al.*, 2017). The control forces and torques which are exerted on the rotor flying manipulator come from the thrust of the rotors and from the motors which rotate the links of the robotic arm (Suarez, Heredia *et al.*, 2018; Khalifa *et al.*, 2016; Garimella and Kobilarov, 2018; Lippiello and Ruggiero, 2012b). Usually, the manipulators in such robotic systems make use of rigid revolute joints. By substituting such joints with flexible ones the aerial manipulators can perform grasping and compliance tasks with more safety and the related damage risk can be reduced. On the other side the use of flexible joints introduces additional degrees of freedom in the dynamic model of the aerial manipulators and raises further the difficulty for solving the associated control problem (Yuksel, Mahboubi *et al.*, 2015; Yuksel, Buondonno *et al.*, 2016; Yuksel, Staub *et al.*, 2016).

In this section, the previously analyzed nonlinear optimal (H-infinity) control approach is proposed for the problem of control of aerial robotic manipulators with flexible joints (Rigatos, 2016b; Rigatos, Siano and Cecati, 2015; Rigatos, Siano *et al.*, 2015). To this end, the section develops first the dynamic model of the robotic system that comprises the rotorcraft UAV and the manipulator with flexible joints. This is done by computing the Lagrangian of the robotic system and by applying the Euler-Lagrange method. Thus, a nonlinear multi-DOF and underactuated state-space model of the aerial robot is obtained which is finally written in an affine-in-the-input form. Next, the state-space description of the aerial robot undergoes approximate linearization around a time-varying operating point, which is updated at each time-step of the control method. The linearization point is defined by the present value of the state vector of the robotic system and by the last value of the control inputs vector that was applied on it. The linearization relies on Taylor series expansion and on the computation of the associated Jacobian matrices (Rigatos and Tzafestas, 2007; Basseville and Nikiforov, 1993; Rigatos and Zhang, 2009). For the approximately linearized model of the aerial robot an optimal (H-infinity) feedback controller is designed. The modeling error which was due to truncation of higher-order terms in the Taylor series expansion is considered to be a perturbation that is finally compensated by the robustness of the control algorithm.

The proposed H-infinity control method stands for the solution of the optimal control problem for the nonlinear dynamics of the aerial robotic manipulator, under model uncertainties and external perturbations. Actually, it represents the solution to a min–max differential game, in which the controller tries to minimize a quadratic cost function of the system's state-vector error, whereas the model uncertainty and external disturbances terms try to maximize this cost functional. To compute the stabilizing feedback gains of the H-infinity controller, an algebraic Riccati equation is repetitively solved at each iteration of the control algorithm (Rigatos, 2011; Rigatos, 2015). The stability properties of the control scheme are proven through Lyapunov analysis. It is demonstrated that the control loop of the aerial robotic manipulator satisfies the H-infinity tracking performance criterion (Rigatos and Busawon, 2018; Tousssaint, Basar and Bullo, 2000). This signifies elevated robustness against model uncertainties and external perturbations. Besides, under moderate conditions it is proven that the control scheme is globally asymptotically stable. Finally, to implement feedback control without the need to measure the entire state vector of the aerial robotic system, the H-infinity Kalman Filter is used as a robust state estimator (Rigatos and Busawon, 2018; Gibbs, 2012).

3.9.2 Dynamic model of the aerial robotic manipulator

The aerial robotic manipulator comprises a quadrotor UAV and a robotic arm with flexible joints, as shown in Figure 3.117. The dynamic model of the aerial robotic system will be obtained with the use of the Euler–Lagrange method.

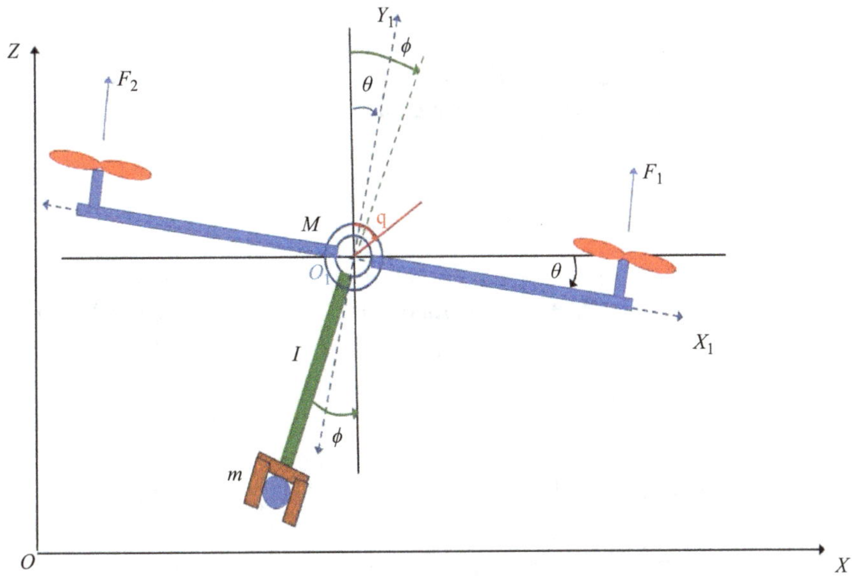

Figure 3.117 Reference frames for the aerial robotic manipulator with elastic joints

The aerial robot is considered to move on the *xz*-plane as shown in Figure 3.117. The inertial coordinates system is denoted as XOZ. The body-fixed coordinates system is denoted as $X_1O_1Z_1$. The rotation (roll) angle of the UAV is denoted as θ. The rotation angle of the link of the robotic manipulator with respect to the vertical axis of the inertial system is denoted as ϕ. The joint at the basis of the link exhibits flexibility. As a result, the turn angle of the motor that provides rotation to the joint it not equal to ϕ and is denoted as q.

The forces and torques which are exerted on the aerial manipulator system are as follows: (i) there are lift forces generated by the rotors of the UAV, (ii) when the lift forces are uneven, a torque T_θ is generated which makes the UAV turn by the roll angle θ, and (iii) the motor which is mounted on the basis of the robotic manipulator generates a torque T_m which makes the motor's rotor turn by an angle q. The transmission of this turn motion is performed through an elastic joint and the robotic link finally rotates by the aforementioned angle ϕ.

By denoting the mass of the UAV as M, its moment of inertia as I_U and the coordinates of its center of gravity in the inertial reference frame as (x, z), the kinetic energy of the UAV given by

$$K_u = \frac{1}{2}I_u\dot{\theta}^2 + \frac{1}{2}M\dot{x}^2 + \frac{1}{2}M\dot{z}^2 \tag{3.289}$$

It is considered that the moment of inertia of the robotic link is I_m and that its aggregate mass m is located at its end-effector, while the coordinates of the end-effector in the inertial frame are denoted as (x_m, y_m). Moreover, it is considered that the moment of inertia of the motor that rotates the flexible joint is J. Then, the kinetic energy of the robotic manipulator is given by

$$K_m = \frac{1}{2}I_m\dot{\theta}^2 + \frac{1}{2}m\dot{x}_m^2 + \frac{1}{2}m\dot{z}_m^2 + \frac{1}{2}J\dot{q}^2 \tag{3.290}$$

Considering that the length of the robotic link is l, the coordinates of the position and the velocity of the end-effector of the robotic manipulator in the inertial reference frame are given by

$$\begin{aligned} x_m &= x - l\sin(\phi) \Rightarrow \dot{x}_m = \dot{x} - l\cos(\phi)\dot{\phi} \\ z_m &= z - l\cos(\phi) \Rightarrow \dot{z}_m = \dot{z} + l\sin(\phi)\dot{\phi} \end{aligned} \tag{3.291}$$

Consequently, the aggregate kinetic energy of the aerial manipulator is:

$$\begin{aligned} K = &\frac{1}{2}I_u\dot{\theta}^2 + \frac{1}{2}m\dot{x}^2 + \frac{1}{2}m\dot{z}^2 + \\ &+ \frac{1}{2}I_m\dot{\phi}^2 + \frac{1}{2}m(\dot{x} - l\cos(\phi)\dot{\phi})^2 + \frac{1}{2}m(\dot{z} + l\sin(\phi)\dot{\phi})^2 + \frac{1}{2}J\dot{q}^2 \end{aligned} \tag{3.292}$$

The potential energy of UAV is

$$P_u = Mgz \tag{3.293}$$

Considering that all mass of the manipulator is concentrated at its end-effector, its potential energy is

$$P_m = Mgz + mgz_m + \frac{1}{2}k(\phi - q)^2 \Rightarrow P_m = mg(z - l\cos(\phi)) + \frac{1}{2}k(\phi - q)^2 \tag{3.294}$$

The aggregate potential energy of the aerial robotic manipulator is

$$P = Mgz + mg(z - l\cos(\phi)) + \frac{1}{2}k(\phi - q)^2 \tag{3.295}$$

As a result of the above, the Lagrangian of the aerial robotic manipulator is

$$\begin{aligned} L &= K - P \Rightarrow \\ L &= \left[\frac{1}{2}I_u\dot{\theta}^2 + \frac{1}{2}m\dot{x}^2 + \frac{1}{2}m\dot{z}^2 \right. \\ &\quad \left. + \frac{1}{2}I_m\dot{\phi}^2 + \frac{1}{2}m(\dot{x} - l\cos(\phi)\dot{\phi})^2 + \frac{1}{2}m(\dot{z} + l\sin(\phi)\dot{\phi})^2 + \frac{1}{2}J\dot{q}^2\right] \\ &\quad - \left[Mgz + mg(z - l\cos(\phi)) + \frac{1}{2}k(\phi - q)^2\right] \end{aligned} \tag{3.296}$$

The state variables of the system are $[x, z, \theta, \phi, q]$ while their time derivatives are $[\dot{x}, \dot{z}, \dot{\theta}, \dot{\phi}, \dot{q}]$. By applying the Euler–Lagrange method, one has

$$\frac{\partial}{\partial t}\frac{\partial L}{\partial \dot{x}} - \frac{\partial L}{\partial x} = F_x \quad \frac{\partial}{\partial t}\frac{\partial L}{\partial \dot{z}} - \frac{\partial L}{\partial z} = F_z$$

$$\frac{\partial}{\partial t}\frac{\partial L}{\partial \dot{\theta}} - \frac{\partial L}{\partial \theta} = T_\theta \quad \frac{\partial}{\partial t}\frac{\partial L}{\partial \dot{\phi}} - \frac{\partial L}{\partial \phi} = 0 \tag{3.297}$$

$$\frac{\partial}{\partial t}\frac{\partial L}{\partial \dot{q}} - \frac{\partial L}{\partial q} = T_m$$

where $F_x = (F_1 + F_2)\cos(\theta)$, $F_z = (F_1 + F_2)\sin(\theta)$, $T_\theta = (F_1 - F_2)d$, with d standing for the distance of the rotor from the UAV's center of gravity.

From the Euler–Lagrange equation with respect to x, and after intermediate operations one obtains

$$\begin{aligned}&\frac{\partial}{\partial t}\frac{\partial L}{\partial \dot{x}} - \frac{\partial L}{\partial x} = F_x \Rightarrow \\ &(M+m)\ddot{x} + ml\sin(\phi)\dot{\phi}^2 - ml\cos(\phi)\ddot{\phi} = F_x\end{aligned} \tag{3.298}$$

From the Euler–Lagrange equation with respect to z, and after intermediate operations, one obtains

$$\begin{aligned}&\frac{\partial}{\partial t}\frac{\partial L}{\partial \dot{z}} - \frac{\partial L}{\partial z} = F_z \Rightarrow \\ &(M+m)\ddot{z} + ml\cos(\phi)\dot{\phi}^2 + ml\sin(\phi)\ddot{\phi} - Mg - mg = F_z\end{aligned} \tag{3.299}$$

From the Euler–Lagrange equation with respect to θ, and after intermediate operations, one obtains

$$\begin{aligned}&\frac{\partial}{\partial t}\frac{\partial L}{\partial \dot{\theta}} - \frac{\partial L}{\partial \theta} = T_\theta \Rightarrow \\ &I_u\ddot{\theta} = T_\theta\end{aligned} \tag{3.300}$$

From the Euler–Lagrange equation with respect to ϕ, and after intermediate operations, one obtains

$$\begin{aligned}&\frac{\partial}{\partial t}\frac{\partial L}{\partial \dot{\phi}} - \frac{\partial L}{\partial \phi} = 0 \Rightarrow \\ &(I_m + ml^2)\ddot{\phi} - m\ddot{x}l\cos(\phi) + m\ddot{z}l\sin(\phi) - mgl\sin(\phi) - k(\phi - q) = 0\end{aligned} \tag{3.301}$$

From the Euler–Lagrange equation with respect to q, and after intermediate operations, one obtains

$$\begin{aligned}&\frac{\partial}{\partial t}\frac{\partial L}{\partial \dot{q}} - \frac{\partial L}{\partial q} = 0 \Rightarrow \\ &J\ddot{q} + k(\phi - q) = T_m\end{aligned} \tag{3.302}$$

Equation (3.300) is independent from the rest of the state-space equations of the aerial robotic manipulator with elastic joints. This signifies that the roll angle θ of the UAV can be directly controlled by a control input of the form $T_\theta = \frac{1}{I_u}[\ddot{\theta}_d - k_d(\dot{\theta} - \dot{\theta}_d) - k_p(\theta - \theta_d)]$, where the control gains $k_d > 0$, $k_p > 0$ are chosen so as the characteristic polynomial $p(s) = s^2 + k_d s + k_p$ to be a Hurwitz one.

Next, the state-space model of the aerial robotic system becomes

$$\begin{aligned} &(M+m)\ddot{x} + ml\sin(\phi)\dot{\phi}^2 - ml\cos(\phi)\ddot{\phi} = F_x \\ &(M+m)\ddot{z} + ml\cos(\phi)\dot{\phi}^2 + ml\sin(\phi)\ddot{\phi} - Mg - mg = F_z \\ &(I_m + ml^2)\ddot{\phi} - m\ddot{x}l\cos(\phi) + m\ddot{z}l\sin(\phi) - mgl\sin(\phi) - k(\phi - q) = 0 \\ &J\ddot{q} + k(\phi - q) = T_m \end{aligned} \tag{3.303}$$

By denoting the control inputs $u_1 = F_x \Rightarrow u_1 = (F_1 + F_2)\cos(\phi)$, $u_2 = F_Z + (M + m)g \Rightarrow u_2 = (F_1 + F_2)\sin(\phi) + (M + m)g$ and $u_3 = T_m$ it can be seen that by solving the control problem for the aerial robotic manipulator one arrives also at the following relation for the lift forces F_1, F_2:

$$(F_1 + F_2) = \{u_1^2 + [u_2 - (M+m)g]^2\}^{\frac{1}{2}} \tag{3.304}$$

Besides, using that the distance between the center of gravity of the UAV and the position of its rotors is equal to d, (3.300) gives

$$(F_1 - F_2) = \frac{I_u}{d}\ddot{\theta} \tag{3.305}$$

From (3.304) and (3.305), one can find the values of the lift forces of the rotors. Moreover, from Eq. $u_3 = T_m$ one can find the value of the torque provided by the motor of the robotic link.

Next, using (3.303) the state-space model of the aerial robotic manipulator is formulated in matrix form. Actually, one has:

$$\begin{aligned} &\begin{pmatrix} (M+m) & 0 & -ml\cos(\phi) & 0 \\ 0 & (M+m) & ml\sin(\phi) & 0 \\ -ml\cos(\phi) & ml\sin(\phi) & (I_m + ml^2) & 0 \\ 0 & 0 & 0 & J \end{pmatrix} \begin{pmatrix} \ddot{x} \\ \ddot{z} \\ \ddot{\phi} \\ \ddot{q} \end{pmatrix} + \\ &+ \begin{pmatrix} ml\sin(\phi)\dot{\phi}^2 \\ -ml\cos(\phi)\dot{\phi}^2 \\ -mgl\sin(\phi) - k(\phi - q) \\ k(\phi - q) \end{pmatrix} = \begin{pmatrix} 1 & 0 & 0 \\ 0 & 1 & 0 \\ 0 & 0 & 0 \\ 0 & 0 & 1 \end{pmatrix} \begin{pmatrix} u_1 \\ u_2 \\ u_3 \end{pmatrix} \end{aligned} \tag{3.306}$$

By denoting the system's state vector as $\tilde{q} = [x, z, \theta, \phi, q]^T$, the state-space description of the aerial robotic manipulator is written in the concise form

$$M(\tilde{q})\ddot{\tilde{q}} + C(\tilde{q}, \dot{\tilde{q}}) = G_m u \tag{3.307}$$

where $M(\tilde{q}) \in R^{4\times 4}$ is the inertia matrix, $C(\tilde{q},\dot{\tilde{q}}) \in R^{4\times 1}$ is the Coriolis and centrifugal forces matrix, and $G_m \in R^{4\times 3}$ is the control inputs gains matrix. Next the inverse of the inertia matrix $M(\tilde{q})^{-1}$ is computed as follows:

$$M(\tilde{q})^{-1} = \frac{1}{detM}\begin{pmatrix} M_{11} & -M_{21} & M_{31} & -M_{41} \\ -M_{12} & M_{22} & -M_{32} & M_{42} \\ M_{13} & -M23 & M_{33} & -M_{43} \\ -M_{14} & M_{24} & -M_{34} & M_{44} \end{pmatrix} \tag{3.308}$$

where $detM = J[(M+m)^2(I_m+ml^2)-(M+m)m^2l^2]$ and the sub-determinants of this matrix are defined as:

$M_{11} = J[(M+m)(I_m)+ml^2]-m^2l^2\sin^2(\phi)$, $M_{12} = J[m^2l^2\sin(\phi)\cos(\phi)]$, $M_{13} = J[(M+m)ml\cos(\phi)]$, $M_{14}=0$, $M_{21} = J[m^2l^2\sin(\phi)\cos(\phi)]$, $M_{22} = J[(M+m)(I_m+ml^2)-m^2l^2\cos^2(\phi)]$, $M_{23}=J[(M+m)ml\sin(\phi)]$ $M_{24}=0$, $M_{31}=J[(M+m)ml\cos(\phi)]$ $M_{32}=J[(M+m)ml\sin(\phi)]$, $M_{33}=J(M+m)^2$, $M_{34}=0$, $M_{41}=0$, $M_{42}=0$, $M_{43}=0$ and $M_{44}=(M+m)[(M+m)(I_m+ml^2)-m^2l^2]$.

Next, the state-space model of the aerial robotic manipulator is written as:

$$\begin{aligned} &M(\tilde{q})\ddot{\tilde{q}} + C(\tilde{q},\dot{\tilde{q}}) = G_m u \Rightarrow \\ &\ddot{\tilde{q}} = -M(\tilde{q})^{-1}C(\tilde{q},\dot{\tilde{q}}) + M(\tilde{q})^{-1}G_m u \end{aligned} \tag{3.309}$$

Using that $C(\tilde{q},\dot{\tilde{q}}) = [c_{11},c_{21},c_{31},c_{41}]^T$, the product term $-M(\tilde{q})^{-1}C(\tilde{q},\dot{\tilde{q}})$ becomes:

$$-M(\tilde{q})^{-1}C(\tilde{q},\dot{\tilde{q}}) = -\frac{1}{detM}\begin{pmatrix} M_{11}c_{11} - M_{21}c_{21} + M_{31}c_{31} - M_{41}c_{41} \\ -M_{12}c_{11} + M_{22}c_{21} - M_{32}c_{31} + M_{42}c_{41} \\ M_{13}c_{11} - M_{23}c_{21} + M_{33}c_{31} - M_{43}c_{41} \\ -M_{14}c_{11} + M_{24}c_{21} - M_{34}c_{31} + M_{44}c_{41} \end{pmatrix} \tag{3.310}$$

Moreover, about the product term $M(\tilde{q})^{-1}G_m$ one has

$$M(\tilde{q})^{-1}G_m = \frac{1}{detM}\begin{pmatrix} M_{11} & -M_{21} & -M_{41} \\ -M_{12} & M_{22} & M_{42} \\ M_{13} & -M_{23} & -M_{13} \\ -M_{14} & M_{24} & M_{44} \end{pmatrix} \tag{3.311}$$

Next, by defining the state variables of the aerial robotic system as $x_1 = x$, $x_2 = \dot{x}$, $x_3 = z, x_4 = \dot{z}, x_5 = \phi, x_6 = \dot{\phi}, x_7 = q$ and $x_8 = \dot{q}$ one obtains the following state-space description:

$$\dot{x}_1 = x_2 \tag{3.312}$$

$$\begin{aligned} \dot{x}_2 = &-\frac{1}{detM}(M_{11}c_{11} - M_{21}c_{21} + M_{31}c_{31} - M_{41}c_{41}) \\ &+ \frac{M_{11}}{detM}u_1 - \frac{M_{21}}{detM}u_2 - \frac{M_{41}}{detM}u_3 \end{aligned} \tag{3.313}$$

$$\dot{x}_3 = x_4 \tag{3.314}$$

$$\begin{aligned}\dot{x}_4 = & -\frac{1}{detM}(-M_{12}c_{11} + M_{22}c_{21} - M_{32}c_{31} + M_{42}c_{41}) \\ & \frac{-M_{12}}{detM}u_1 + \frac{M_{22}}{detM}u_2 + \frac{M_{42}}{detM}u_3\end{aligned} \tag{3.315}$$

$$\dot{x}_5 = x_6 \tag{3.316}$$

$$\begin{aligned}\dot{x}_6 = & -\frac{1}{detM}(M_{13}c_{11} - M_{23}c_{21} + M_{33}c_{31} - M_{43}c_{41}) \\ & + \frac{M_{13}}{detM}u_1 - \frac{M_{23}}{detM}u_2 + \frac{M_{43}}{detM}u_3\end{aligned} \tag{3.317}$$

$$\dot{x}_7 = x_8 \tag{3.318}$$

$$\begin{aligned}\dot{x}_8 = & -\frac{1}{detM}(-M_{14}c_{11} + M_{24}c_{21} - M_{34}c_{31} + M_{44}c_{41}) \\ & - \frac{M_{14}}{detM}u_1 + \frac{M_{24}}{detM}u_2 + \frac{M_{44}}{detM}u_3\end{aligned} \tag{3.319}$$

Equivalently, the state-space model of the aerial robotic manipulator can be written in the following affine-in-the-input form:

$$\begin{pmatrix}\dot{x}_1\\ \dot{x}_2\\ \dot{x}_3\\ \dot{x}_4\\ \dot{x}_5\\ \dot{x}_6\\ \dot{x}_7\\ \dot{x}_8\end{pmatrix} = \begin{pmatrix}f_1\\ f_2\\ f_3\\ f_4\\ f_5\\ f_6\\ f_7\\ f_8\end{pmatrix} + \begin{pmatrix}g_{11}\\ g_{21}\\ g_{31}\\ g_{41}\\ g_{51}\\ g_{61}\\ g_{71}\\ g_{81}\end{pmatrix}u_1 + \begin{pmatrix}g_{12}\\ g_{22}\\ g_{32}\\ g_{42}\\ g_{52}\\ g_{62}\\ g_{72}\\ g_{82}\end{pmatrix}u_2 + \begin{pmatrix}g_{13}\\ g_{23}\\ g_{33}\\ g_{43}\\ g_{53}\\ g_{63}\\ g_{73}\\ g_{83}\end{pmatrix}u_3 \tag{3.320}$$

which is also written concisely as

$$\dot{x} = f(x) + g_1(x)u_1 + g_2(x)u_2 + g_3(x)u_3 \tag{3.321}$$

with $x \in R^{8\times1}$, $u \in R^{3\times1}$, $f(x) \in R^{8\times1}$, $g_1(x) \in R^{8\times1}$, $g_2(x) \in R^{8\times1}$, and $g_3(x) \in R^{8\times1}$, while the elements of the aforementioned vector fields are defined as

$f_1(x) = x_2,\ g_{11}(x) = 0,\ g_{12}(x) = 0,\ g_{13}(x) = 0$

$f_2(x) = -\frac{1}{detM}(M_{11}c_{11} - M_{21}c_{21} + M_{31}c_{31} - M_{41}c_{41}),\ g_{21}(x) = \frac{M_{11}}{detM},\ g_{22}(x) = -\frac{M_{21}}{detM},$ $g_{23}(x) = -\frac{M_{41}}{detM}$

$f_3(x) = x_4,\ g_{31}(x) = 0,\ g_{32}(x) = 0,\ g_{33}(x) = 0$

$f_4(x) = -\frac{1}{detM}(-M_{12}c_{11} + M_{22}c_{21} - M_{32}c_{31} + M_{42}c_{41}),\quad g_{41}(x) = -\frac{M_{12}}{detM},\quad g_{42}(x) = \frac{M_{22}}{detM},\ g_{43}(x) = \frac{M_{42}}{detM}$

$f_5(x) = x_6, g_{51}(x) = 0, g_{52}(x) = 0, g_{53}(x) = 0$

$f_6(x) = -\frac{1}{detM}(M_{13}c_{11} - M_{23}c_{21} + M_{33}c_{31} - M_{43}c_{41}), g_{61}(x) = \frac{M_{13}}{detM}, g_{62}(x) = \frac{M_{23}}{detM}, g_{63}(x) = \frac{M_{43}}{detM}$

$f_7(x) = x_8, g_{71}(x) = 0, g_{72}(x) = 0, g_{73}(x) = 0$

$f_8(x) = -\frac{1}{detM}(-M_{14}c_{11} + M_{24}c_{21} - M_{34}c_{31} + M_{44}c_{41}), \quad g_{81}(x) = -\frac{M_{14}}{detM}, \quad g_{82}(x) = \frac{M_{24}}{detM}, g_{83}(x) = \frac{M_{44}}{detM}$

3.9.3 Approximate linearization of the model of the aerial robotic manipulator

The dynamic model of the aerial robotic manipulator, previously written in the state-space form

$$\dot{x} = f(x) + g_1(x)u_1 + g_2(x)u_2 + g_3(x)u_3 \tag{3.322}$$

undergoes approximate linearization around a time-varying operating point (x^*, u^*), where x^* is the present value of the system's state vector and u^* is the last sampled value of the control inputs vector. The linearization relies on Taylor series expansion and on the computation of the associated Jacobian matrices. The linearization procedure gives the model:

$$\dot{x} = Ax + Bu + \tilde{d} \tag{3.323}$$

where $\tilde{d}$ is the cumulative disturbance term which is due to the modeling error. This is induced by the truncation of higher-order terms in the Taylor series expansion. It may also comprise the effects of the external perturbations. About matrices A and B one has

$$\begin{aligned} A = [\nabla_x f(x)] \mid_{(x^*,u^*)} &+ [g_1(x)]u_1 \mid_{(x^*,u^*)} \\ &+ [g_2(x)]u_2 \mid_{(x^*,u^*)} + [g_3(x)]u_3 \mid_{(x^*,u^*)} \end{aligned} \tag{3.324}$$

$$B = [g_1(x)\; g_2(x)\; g_3(x)] \mid_{(x^*,u^*)} \tag{3.325}$$

To compute the aforementioned Jacobian matrices, the following partial derivatives are given first:

$\frac{\partial M_{11}}{\partial x_i} = 0$ for $i = 1,2,3,4,6,7,8$ and $\frac{\partial M_{11}}{\partial x_5} = -2x_i^2 l^2 \sin(\phi)\cos(\phi)$, $\frac{\partial M_{12}}{\partial x_i} = 0$ for $i = 1,2,3,4,6,7,8$ and $\frac{\partial M_{12}}{\partial x_5} = Jm^2l^2[\cos^2(\phi) - sin^(\phi)]$, $\frac{\partial M_{13}}{\partial x_i} = 0$ for $i = 1,2,3,4,6,7,8$ and $\frac{\partial M_{13}}{\partial x_5} = J(M+m)ml(-\sin(\phi))$, $\frac{\partial M_{14}}{\partial x_i} = 0$ for $i = 1,2,3,4,5,6,7,8$

$\frac{\partial M_{21}}{\partial x_i} = 0$ for $i = 1,2,3,4,6,7,8$ and $\frac{\partial M_{21}}{\partial x_5} = Jm^2l^2[\cos(\phi) - sin^2(\phi)]$, $\frac{\partial M_{22}}{\partial x_i} = 0$ for $i = 1,2,3,4,6,7,8$ and $\frac{\partial M_{22}}{\partial x_5} = Jm^2l^2 2\cos(\phi)\sin(\phi)$, $\frac{\partial M_{23}}{\partial x_i} = 0$ for $i = 1,2,3,4,6,7,8$ and $\frac{\partial M_{23}}{\partial x_5} = J(M+m)ml\cos(\phi)$, $\frac{\partial M_{24}}{\partial x_i} = 0$ for $i = 1,2,3,4,5,6,7,8$

$\frac{\partial M_{31}}{\partial x_i}=0$ for $i=1,2,3,4,6,7,8$ and $\frac{\partial M_{31}}{\partial x_5}=J(M+m)(ml)(-\sin(\phi))$, $\frac{\partial M_{32}}{\partial x_i}=0$ for $i=1,2,3,4,6,7,8$ and $\frac{\partial M_{32}}{\partial x_5}=J(M+m)(ml)\cos(\phi)$, $\frac{\partial M_{33}}{\partial x_i}=0$ for $i=1,2,3,4,5,6,7,8$, $\frac{\partial M_{34}}{\partial x_i}=0$ for $i=1,2,3,4,5,6,7,8$

$\frac{\partial M_{41}}{\partial x_i}=0$ for $i=1,2,3,4,5,6,7,8$, $\frac{\partial M_{42}}{\partial x_i}=0$ for $i=1,2,3,4,5,6,7,8$, $\frac{\partial M_{43}}{\partial x_i}=0$ for $i=1,2,3,4,5,6,7,8$, and $\frac{\partial M_{44}}{\partial x_i}=0$ for $i=1,2,3,4,5,6,7,8$

In a similar manner, one computes:

$\frac{\partial c_{11}}{\partial x_i}=0$ for $i=1,2,3,4,7,8$, $\dfrac{\partial c_{11}}{\partial x_5}=ml\cos(\phi)\dot{\phi}^2$ and $\dfrac{\partial c_{11}}{\partial x_6}=2ml\sin(\phi)\dot{\phi}$.

$\frac{\partial c_{21}}{\partial x_i}=0$ for $i=1,2,3,4,7,8$, $\dfrac{\partial c_{21}}{\partial x_5}=ml\sin(\phi)\dot{\phi}^2$ and $\dfrac{\partial c_{21}}{\partial x_6}=-2ml\cos(\phi)\dot{\phi}$.

$\frac{\partial c_{31}}{\partial x_i}=0$ for $i=1,2,3,4,6,8$, $\dfrac{\partial c_{31}}{\partial x_5}=-mgl\cos(\phi)-k$ and $\dfrac{\partial c_{31}}{\partial x_7}=k$.

$\frac{\partial c_{41}}{\partial x_i}=0$ for $i=1,2,3,4,6,8$, $\dfrac{\partial c_{41}}{\partial x_5}=k$ and $\dfrac{\partial c_{41}}{\partial x_7}=-k$.

The computation of the Jacobian matrices $\nabla_x f(x)\mid_{(x^*,u^*)}$, $\nabla_x g_1(x)\mid_{(x^*,u^*)}$, $\nabla_x g_2(x)\mid_{(x^*,u^*)}$, and $\nabla_x g_3(x)\mid_{(x^*,u^*)}$ is as follows:

First row of the Jacobian matrix $\nabla_x\mid_{(x^*,u^*)}$: $\frac{\partial f_1}{\partial x_1}=0$, $\dfrac{\partial f_1}{\partial x_2}=1$, $\frac{\partial f_1}{\partial x_3}=0$, $\dfrac{\partial f_1}{\partial x_4}=0$, $\frac{\partial f_1}{\partial x_5}=0$, $\dfrac{\partial f_1}{\partial x_6}=0$, $\frac{\partial f_1}{\partial x_7}=0$, $\dfrac{\partial f_1}{\partial x_8}=0$.

Second row of the Jacobian matrix $\nabla_x\mid_{(x^*,u^*)}$: for $i=1,2,\ldots,8$

$$\begin{aligned}\frac{\partial f_2}{\partial x_i}=-\frac{1}{detM}\cdot\Bigg[&\frac{\partial M_{11}}{\partial x_i}c_{11}+M_{11}\frac{\partial c_{11}}{\partial x_i}-\frac{\partial M_{21}}{\partial x_i}c_{21}-M_{21}\frac{\partial c_{21}}{\partial x_i}\\&+\frac{\partial M_{31}}{\partial x_i}c_{31}+M_{31}\frac{\partial c_{31}}{\partial x_i}-\frac{\partial M_{41}}{\partial x_i}c_{41}-M_{41}\frac{\partial c_{41}}{\partial x_i}\end{aligned}\tag{3.326}$$

Third row of the Jacobian matrix $\nabla_x\mid_{(x^*,u^*)}$: $\frac{\partial f_3}{\partial x_1}=0$, $\dfrac{\partial f_3}{\partial x_2}=0$, $\frac{\partial f_3}{\partial x_3}=0$, $\dfrac{\partial f_3}{\partial x_4}=1$, $\frac{\partial f_3}{\partial x_5}=0$, $\dfrac{\partial f_3}{\partial x_6}=0$, $\frac{\partial f_3}{\partial x_7}=0$, $\dfrac{\partial f_3}{\partial x_8}=0$.

Fourth row of the Jacobian matrix $\nabla_x\mid_{(x^*,u^*)}$: for $i=1,2,\ldots,8$

$$\begin{aligned}\frac{\partial f_4}{\partial x_i}=-\frac{1}{detM}\cdot\Bigg[&-\frac{\partial M_{12}}{\partial x_i}c_{11}-M_{12}\frac{\partial c_{11}}{\partial x_i}+\frac{\partial M_{22}}{\partial x_i}c_{21}+M_{22}\frac{\partial c_{21}}{\partial x_i}\\&-\frac{\partial M_{32}}{\partial x_i}c_{31}-M_{32}\frac{\partial c_{31}}{\partial x_i}+\frac{\partial M_{42}}{\partial x_i}c_{41}+M_{42}\frac{\partial c_{41}}{\partial x_i}\end{aligned}\tag{3.327}$$

Fifth row of the Jacobian matrix $\nabla_x \mid_{(x^*,u^*)}$: $\frac{\partial f_5}{\partial x_1}=0$, $\frac{\partial f_5}{\partial x_2}=0$, $\frac{\partial f_5}{\partial x_3}=0$, $\frac{\partial f_5}{\partial x_4}=0$, $\frac{\partial f_5}{\partial x_5}=0$, $\frac{\partial f_5}{\partial x_6}=1$, $\frac{\partial f_5}{\partial x_7}=0$, $\frac{\partial f_5}{\partial x_8}=0$.

Sixth row of the Jacobian matrix $\nabla_x \mid_{(x^*,u^*)}$: for $i=1,2,\ldots,8$

$$\begin{aligned}\frac{\partial f_6}{\partial x_i} &= -\frac{1}{detM}\cdot\left[\frac{\partial M_{13}}{\partial x_i}c_{11}+M_{13}\frac{\partial c_{11}}{\partial x_i}-\frac{\partial M_{23}}{\partial x_i}c_{21}-M_{23}\frac{\partial c_{21}}{\partial x_i}\right.\\ &= \left.+\frac{\partial M_{33}}{\partial x_i}c_{31}+M_{33}\frac{\partial c_{31}}{\partial x_i}-\frac{\partial M_{44}}{\partial x_i}c_{41}+M_{43}\frac{\partial c_{41}}{\partial x_i}\right.\end{aligned} \tag{3.328}$$

Seventh row of the Jacobian matrix $\nabla_x \mid_{(x^*,u^*)}$: $\frac{\partial f_7}{\partial x_1}=0$, $\frac{\partial f_7}{\partial x_2}=0$, $\frac{\partial f_7}{\partial x_3}=0$, $\frac{\partial f_7}{\partial x_4}=0$, $\frac{\partial f_7}{\partial x_5}=0$, $\frac{\partial f_7}{\partial x_6}=0$, $\frac{\partial f_7}{\partial x_7}=0$, $\frac{\partial f_7}{\partial x_8}=1$.

Eighth row of the Jacobian matrix $\nabla_x \mid_{(x^*,u^*)}$: for $i=1,2,\ldots,8$

$$\begin{aligned}\frac{\partial f_8}{\partial x_i} &= -\frac{1}{detM}\cdot\left[-\frac{\partial M_{14}}{\partial x_i}c_{11}-M_{14}\frac{\partial c_{11}}{\partial x_i}+\frac{\partial M_{24}}{\partial x_i}c_{21}+M_{24}\frac{\partial c_{21}}{\partial x_i}\right.\\ &\quad \left.-\frac{\partial M_{34}}{\partial x_i}c_{31}-M_{34}\frac{\partial c_{31}}{\partial x_i}+\frac{\partial M_{44}}{\partial x_i}c_{41}+M_{44}\frac{\partial c_{41}}{\partial x_i}\right.\end{aligned} \tag{3.329}$$

In a concise form, the Jacobian matrix $\nabla_x f(x) \mid_{(x^*,u^*)}$

$$\nabla_x f(x) = \begin{pmatrix} \frac{\partial f_1}{\partial x_1} & \frac{\partial f_1}{\partial x_2} & \cdots & \frac{\partial f_1}{\partial x_8} \\ \frac{\partial f_2}{\partial x_1} & \frac{\partial f_2}{\partial x_2} & \cdots & \frac{\partial f_2}{\partial x_8} \\ \cdots & \cdots & \cdots & \cdots \\ \frac{\partial f_8}{\partial x_1} & \frac{\partial f_8}{\partial x_2} & \cdots & \frac{\partial f_8}{\partial x_8} \end{pmatrix} \tag{3.330}$$

Computation of the Jacobian matrix $\nabla_x g_1(x) \mid_{(x^*,u^*)}$:

First row of the Jacobian matrix $\nabla_x g_1(x) \mid_{(x^*,u^*)}$: $\frac{\partial g_{11}(x)}{\partial x_i}=0$ for $i=1,2,\ldots,8$.

Second row of the Jacobian matrix $\nabla_x g_1(x) \mid_{(x^*,u^*)}$: $\frac{\partial g_{21}(x)}{\partial x_i}=\frac{\partial M_{11}}{\partial x_i}$ for $i=1,2,\ldots,8$

Third row of the Jacobian matrix $\nabla_x g_1(x) \mid_{(x^*,u^*)}$: $\frac{\partial g_{31}(x)}{\partial x_i}=0$ for $i=1,2,\ldots,8$.

Fourth row of the Jacobian matrix $\nabla_x g_1(x) \mid_{(x^*,u^*)}$: $\frac{\partial g_{41}(x)}{\partial x_i}=-\frac{\partial M_{12}}{\partial x_i}$ for $i=1,2,\ldots,8$

Fifth row of the Jacobian matrix $\nabla_x g_1(x) \mid_{(x^*,u^*)}$: $\frac{\partial g_{51}(x)}{\partial x_i}=0$ for $i=1,2,\ldots,8$.

Sixth row of the Jacobian matrix $\nabla_x g_1(x) \mid_{(x^*,u^*)}$: $\frac{\partial g_{61}(x)}{\partial x_i}=\frac{\partial M_{13}}{\partial x_i}$ for $i=1,2,\ldots,8$

Seventh row of the Jacobian matrix $\nabla_x g_1(x)\mid_{(x^*,u^*)}$: $\frac{\partial g_{71}(x)}{\partial x_i} = 0$ for $i = 1,2,\ldots,8$.

Eighth row of the Jacobian matrix $\nabla_x g_1(x)\mid_{(x^*,u^*)}$: $\frac{\partial g_{81}(x)}{\partial x_i} = -\frac{\partial M_{14}}{\partial x_i}$ for $i = 1,2,\ldots,8$

In a concise form, the Jacobian matrix $\nabla_x g_1(x)\mid_{(x^*,u^*)}$

$$\nabla_x g_1(x) = \begin{pmatrix} \frac{\partial g_{11}}{\partial x_1} & \frac{\partial g_{11}}{\partial x_2} & \cdots & \frac{\partial g_{11}}{\partial x_8} \\ \frac{\partial g_{21}}{\partial x_1} & \frac{\partial g_{21}}{\partial x_2} & \cdots & \frac{\partial g_{21}}{\partial x_8} \\ \cdots & \cdots & \cdots & \cdots \\ \frac{\partial g_{81}}{\partial x_1} & \frac{\partial g_{81}}{\partial x_2} & \cdots & \frac{\partial g_{81}}{\partial x_8} \end{pmatrix} \tag{3.331}$$

Computation of the Jacobian matrix $\nabla_x g_2(x)\mid_{(x^*,u^*)}$:

First row of the Jacobian matrix $\nabla_x g_2(x)\mid_{(x^*,u^*)}$: $\frac{\partial g_{12}(x)}{\partial x_i} = 0$ for $i = 1,2,\ldots,8$.

Second row of the Jacobian matrix $\nabla_x g_2(x)\mid_{(x^*,u^*)}$: $\frac{\partial g_{22}(x)}{\partial x_i} = -\frac{\partial M_{21}}{\partial x_i}$ for $i = 1,2,\ldots,8$

Third row of the Jacobian matrix $\nabla_x g_2(x)\mid_{(x^*,u^*)}$: $\frac{\partial g_{32}(x)}{\partial x_i} = 0$ for $i = 1,2,\ldots,8$.

Fourth row of the Jacobian matrix $\nabla_x g_2(x)\mid_{(x^*,u^*)}$: $\frac{\partial g_{42}(x)}{\partial x_i} = \frac{\partial M_{22}}{\partial x_i}$ for $i = 1,2,\ldots,8$

Fifth row of the Jacobian matrix $\nabla_x g_2(x)\mid_{(x^*,u^*)}$: $\frac{\partial g_{52}(x)}{\partial x_i} = 0$ for $i = 1,2,\ldots,8$.

Sixth row of the Jacobian matrix $\nabla_x g_2(x)\mid_{(x^*,u^*)}$: $\frac{\partial g_{62}(x)}{\partial x_i} = -\frac{\partial M_{23}}{\partial x_i}$ for $i = 1,2,\ldots,8$

Seventh row of the Jacobian matrix $\nabla_x g_2(x)\mid_{(x^*,u^*)}$: $\frac{\partial g_{72}(x)}{\partial x_i} = 0$ for $i = 1,2,\ldots,8$.

Eighth row of the Jacobian matrix $\nabla_x g_2(x)\mid_{(x^*,u^*)}$: $\frac{\partial g_{82}(x)}{\partial x_i} = \frac{\partial M_{24}}{\partial x_i}$ for $i = 1,2,\ldots,8$

In a concise form, the Jacobian matrix $\nabla_x g_2(x)\mid_{(x^*,u^*)}$

$$\nabla_x g_2(x) = \begin{pmatrix} \frac{\partial g_{12}}{\partial x_1} & \frac{\partial g_{12}}{\partial x_2} & \cdots & \frac{\partial g_{12}}{\partial x_8} \\ \frac{\partial g_{22}}{\partial x_1} & \frac{\partial g_{22}}{\partial x_2} & \cdots & \frac{\partial g_{22}}{\partial x_8} \\ \cdots & \cdots & \cdots & \cdots \\ \frac{\partial g_{82}}{\partial x_1} & \frac{\partial g_{82}}{\partial x_2} & \cdots & \frac{\partial g_{82}}{\partial x_8} \end{pmatrix} \tag{3.332}$$

Computation of the Jacobian matrix $\nabla_x g_3(x)\mid_{(x^*,u^*)}$:

First row of the Jacobian matrix $\nabla_x g_3(x)\mid_{(x^*,u^*)}$: $\frac{\partial g_{13}(x)}{\partial x_i} = 0$ for $i = 1,2,\ldots,8$.

Second row of the Jacobian matrix $\nabla_x g_3(x)\mid_{(x^*,u^*)}$: $\frac{\partial g_{23}(x)}{\partial x_i} = -\frac{\partial M_{41}}{\partial x_i}$ for $i = 1,2,\ldots,8$

Third row of the Jacobian matrix $\nabla_x g_3(x)\mid_{(x^*,u^*)}$: $\frac{\partial g_{33}(x)}{\partial x_i} = 0$ for $i = 1,2,\ldots,8$.

Fourth row of the Jacobian matrix $\nabla_x g_3(x)\mid_{(x^*,u^*)}$: $\frac{\partial g_{43}(x)}{\partial x_i} = \frac{\partial M_{42}}{\partial x_i}$ for $i = 1,2,\ldots,8$

Fifth row of the Jacobian matrix $\nabla_x g_3(x)\mid_{(x^*,u^*)}$: $\frac{\partial g_{53}(x)}{\partial x_i} = 0$ for $i = 1,2,\ldots,8$.

Sixth row of the Jacobian matrix $\nabla_x g_3(x)\mid_{(x^*,u^*)}$: $\frac{\partial g_{63}(x)}{\partial x_i} = -\frac{\partial M_{43}}{\partial x_i}$ for $i = 1,2,\ldots,8$

Seventh row of the Jacobian matrix $\nabla_x g_3(x)\mid_{(x^*,u^*)}$: $\frac{\partial g_{73}(x)}{\partial x_i} = 0$ for $i = 1,2,\ldots,8$.

Eighth row of the Jacobian matrix $\nabla_x g_3(x)\mid_{(x^*,u^*)}$: $\frac{\partial g_{83}(x)}{\partial x_i} = \frac{\partial M_{44}}{\partial x_i}$ for $i = 1,2,\ldots,8$

In a concise form, the Jacobian matrix $\nabla_x g_3(x)\mid_{(x^*,u^*)}$

$$\nabla_x g_3(x) = \begin{pmatrix} \frac{\partial g_{13}}{\partial x_1} & \frac{\partial g_{13}}{\partial x_2} & \cdots & \frac{\partial g_{13}}{\partial x_8} \\ \frac{\partial g_{23}}{\partial x_1} & \frac{\partial g_{23}}{\partial x_2} & \cdots & \frac{\partial g_{23}}{\partial x_8} \\ \cdots & \cdots & \cdots & \cdots \\ \frac{\partial g_{83}}{\partial x_1} & \frac{\partial g_{83}}{\partial x_2} & \cdots & \frac{\partial g_{83}}{\partial x_8} \end{pmatrix} \tag{3.333}$$

It should be noted that without loss of generality, in the preceding dynamic model of the aerial robotic system, the robotic manipulator consisted of 1 link and 1 flexible joint. The entire robotic system exhibits already 5 degrees of freedom, denoted by its state variables, that is $[x, y, \theta, \phi, q]$. It is straightforward to extend the modeling approach and the related design of the nonlinear optimal controller to the case of a robotic manipulator with more links and flexible joints.

3.9.4 *Differential flatness properties of the aerial robotic manipulator*

The flat outputs vector of the aerial robotic manipulator is taken to be $y = [y_1, y_2, y_3, y_4]^T$ or $y = [x.y.\phi, q]^T$. The control inputs vector, initially consisting of u_i, $i = 1,2,3$ is now extended with the inclusion of the joint elasticity parameter k as an additional control input. Thus, the extended control inputs vector is u_1, u_2, u_3, k. This allows to satisfy the condition that in a multivariable differentially flat system the number of flat outputs should be equal to the number of the control inputs (Rigatos, 2015).

From (3.312), one has

$$x_2 = \dot{x}_1 \Rightarrow x_2 = \dot{y}_1 \tag{3.334}$$

Consequently, state variable x_2 is a differential function of the system's flat outputs. From (3.314), one has

$$x_4 = \dot{x}_3 \Rightarrow x_2 = \dot{y}_2 \tag{3.335}$$

Consequently, state variable x_4 is a differential function of the system's flat outputs. From (3.316), one has

$$x_6 = \dot{x}_5 \Rightarrow x_6 = \dot{y}_3 \tag{3.336}$$

Consequently, state variable x_6 is a differential function of the system's flat outputs. From (3.318), one has

$$x_8 = \dot{x}_7 \Rightarrow x_8 = \dot{y}_4 \tag{3.337}$$

Consequently, state variable x_8 is a differential function of the system's flat outputs. Moreover, from (3.313), (3.315), (3.317), and (3.319), one has a system of four equations with unknowns the elements of the control inputs vector. As a result, the control inputs of the aerial robotic manipulator can be written also as differential functions of the system's flat outputs.

$$u_i = f_u(y, \dot{y}) \tag{3.338}$$

This signifies that the aerial robotic manipulator with flexible joints is a differentially flat system.

3.9.5 Computation of the feedback control gains

For the linearized system given by (1.36) the cost function of (1.37) is defined, where coefficient r determines the penalization of the control input and weight coefficient ρ determines the reward of the disturbances' effects. It is assumed that (i) the energy that is transferred from the disturbances signal $\tilde{d}(t)$ is bounded, that is $\int_0^\infty \tilde{d}^T(t)\tilde{d}(t)dt < \infty$, (ii) matrices $[A, B]$ and $[A, L]$ are stabilizable, and (iii) matrix $[A, C]$ is detectable. In the case of a tracking problem, the optimal feedback control law is given by

$$u(t) = -Ke(t) \tag{3.339}$$

with $e = x - x_d$ to be the tracking error, and $K = \frac{1}{r}B^T P$ where P is a positive definite symmetric matrix. Matrix P is obtained from the solution of the Riccati equation

$$A^T P + PA + Q - P\left(\frac{2}{r}BB^T - \frac{1}{\rho^2}LL^T\right)P = 0 \tag{3.340}$$

where Q is a positive semi-definite symmetric matrix. The worst case disturbance is given by

$$\tilde{d}(t) = \frac{1}{\rho^2}L^T Pe(t) \tag{3.341}$$

The solution of the H-infinity feedback control problem for the double-pendulum overhead crane and the computation of the worst case disturbance that the related

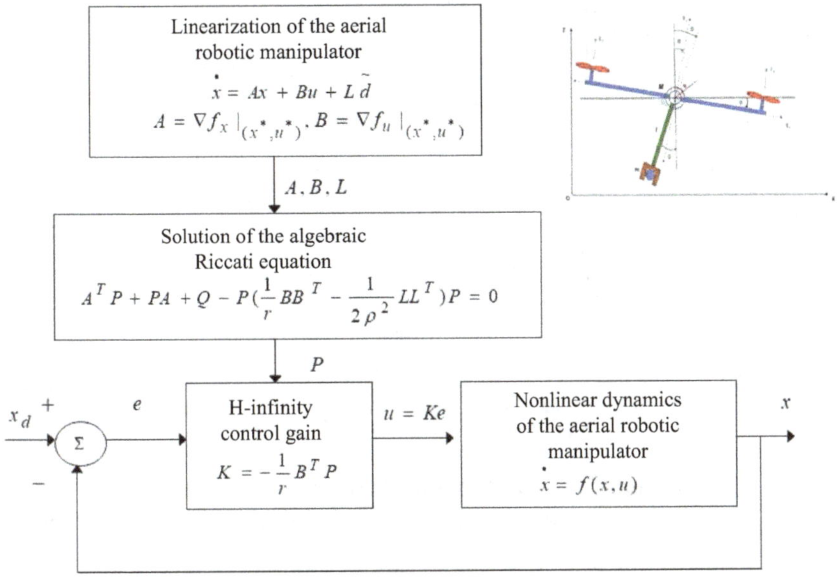

Figure 3.118 Diagram of the control scheme for the aerial robotic manipulator with elastic joints

controller can sustain, comes from superposition of Bellman's optimality principle when considering that the robot is affected by two separate inputs (i) the control input u and (ii) the cumulative disturbance input $\tilde{d}(t)$. Solving the optimal control problem for u, that is for the minimum variation (optimal) control input that achieves elimination of the state vector's tracking error, gives $u = -\frac{1}{r}B^T Pe$. Equivalently, solving the optimal control problem for $\tilde{d}$, that is for the worst case disturbance that the control loop can sustain gives $\tilde{d} = \frac{1}{\rho^2}L^T Pe$. The diagram of the control loop is depicted in Figure 3.118. The global stability properties of the control method can be proven following the procedure presented in Section 1.1.2.5.

3.9.6 Simulation tests

The performance of the proposed nonlinear optimal (H-infinity) control method for the model of the 5-DOF aerial robotic manipulator was tested through simulation experiments. For the computation of the controller's feedback gains the algebraic Riccati equation given in (3.340) had to be solved at each time-step of the control algorithm. The obtained results are depicted in Figures 3.119 to 3.130 and confirm fast and accurate tracking of the reference setpoints while also assuring that the control

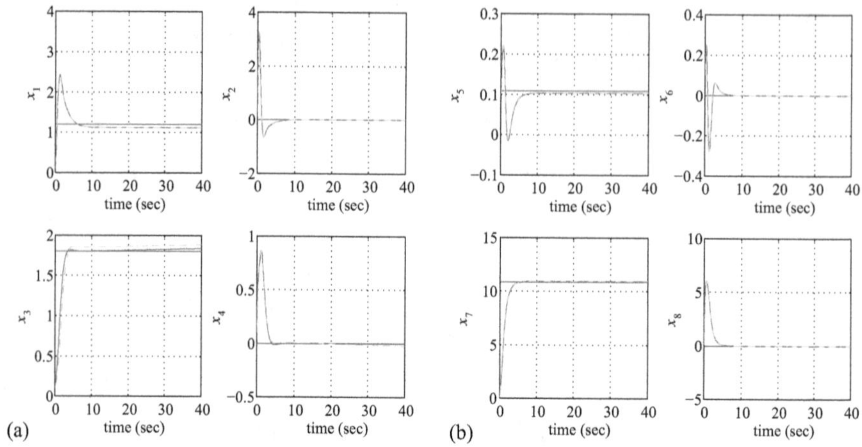

Figure 3.119 Tracking of setpoint 1 for the aerial robotic manipulator: (a) convergence of state variables x_1 , x_2, x_3 and x_4 to their reference setpoints; (b) convergence of state variables x_5, x_6, x_7 and x_8 to their reference setpoints

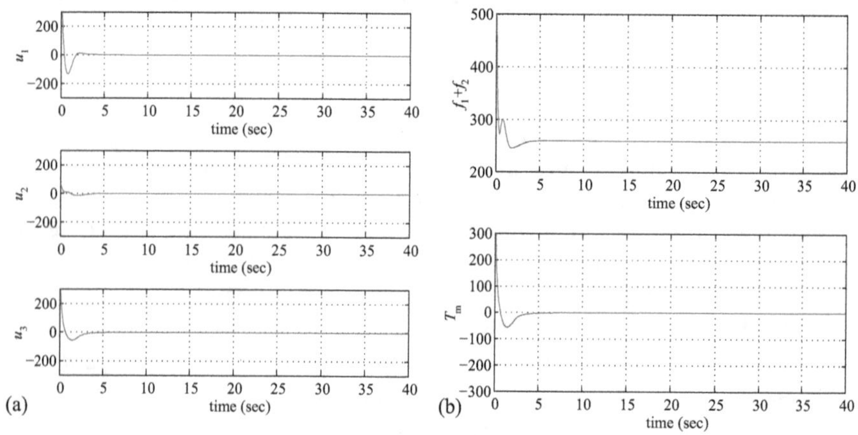

Figure 3.120 Tracking of setpoint 1 for the aerial robotic manipulator: (a) control inputs u_i $i = 1, 2, 3$ computed through the solution of the nonlinear optimal control problem; (b) control inputs $f_1 + f_2$ (aggregate lift force of the UAV) and T_m (torque of the motor mounted at the basis of the manipulator)

inputs will remain within the permitted ranges and that actuators saturation is avoided. In the obtained results, the real value of the system's state vector variables is depicted in blue, the estimated value provided by the H-infinity Kalman Filter is depicted in green, while the associated reference setpoints are plotted in red.

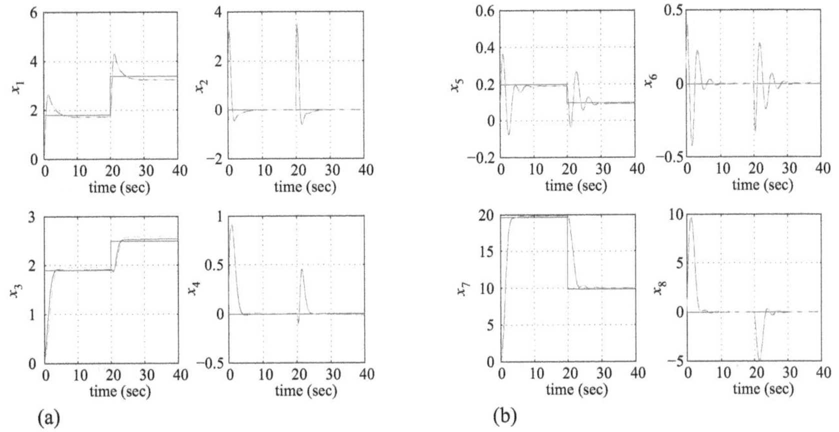

Figure 3.121 Tracking of setpoint 2 for the aerial robotic manipulator: (a) convergence of state variables x_1 , x_2, x_3 and x_4 to their reference setpoints; (b) convergence of state variables x_5, x_6, x_7 and x_8 to their reference setpoints

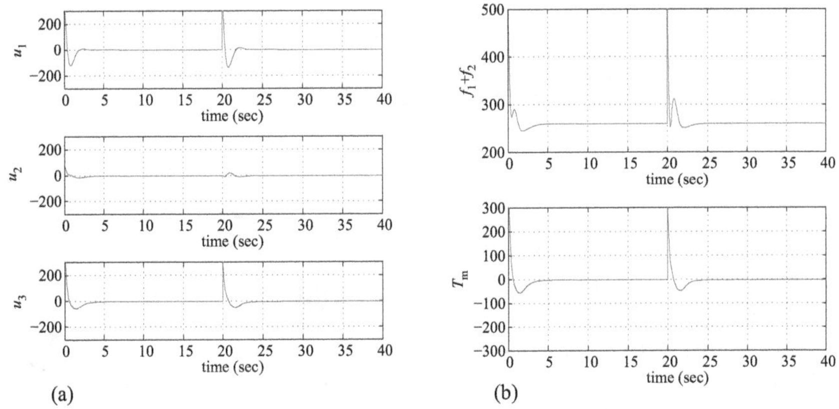

Figure 3.122 Tracking of setpoint 2 for the aerial robotic manipulator: (a) control inputs u_i $i = 1, 2, 3$ computed through the solution of the nonlinear optimal control problem; (b) control inputs $f_1 + f_2$ (aggregate lift force of the UAV's motors) and T_m (torque of the motor mounted at the basis of the manipulator)

3.10 Nonlinear control of the 6-DOF autonomous octocopter

3.10.1 Outline

Nonlinear control and autonomous navigation of UAVs is a nontrivial problem (Rigatos and Busawon, 2018). Octorotors are UAVs equipped with 8 rotors, which

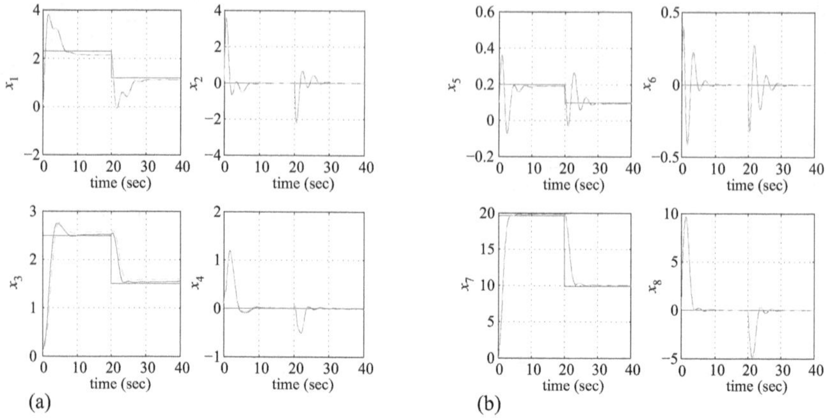

Figure 3.123 Tracking of setpoint 3 for the aerial robotic manipulator: (a) convergence of state variables x_1 , x_2, x_3 and x_4 to their reference setpoints; (b) convergence of state variables x_5, x_6, x_7 and x_8 to their reference setpoints

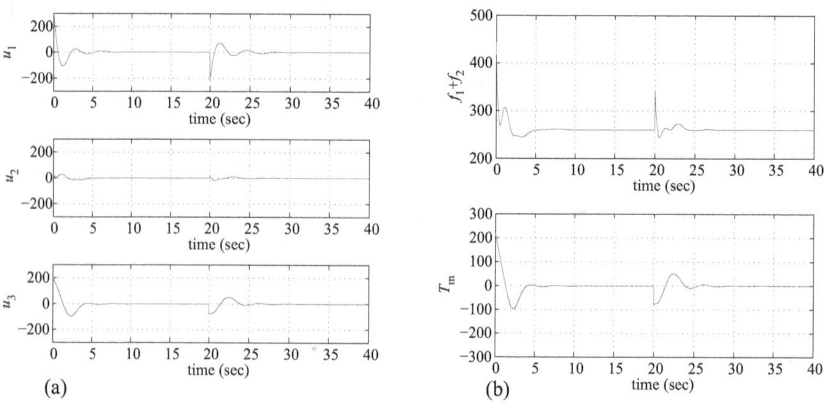

Figure 3.124 Tracking of setpoint 3 for the aerial robotic manipulator: (a) control inputs u_i $i = 1, 2, 3$ computed through the solution of the nonlinear optimal control problem; (b) control inputs $f_1 + f_2$ (aggregate lift force of the UAV's motors) and T_m (torque of the motor mounted at the basis of the manipulator)

are usually placed at equidistant positions on the vertices of an octagon (star-shaped octorotors). Alternatively, there can be octocopters where the rotors are connected on a basis that comprises two orthogonal rods (Alwi and Edwards, 2015; Bauer *et al.*, 2016; Zeglache *et al.*, 2019). Octorotors can be used for surveillance, patrolling and defense tasks. They can function reliably under rough conditions and in environments

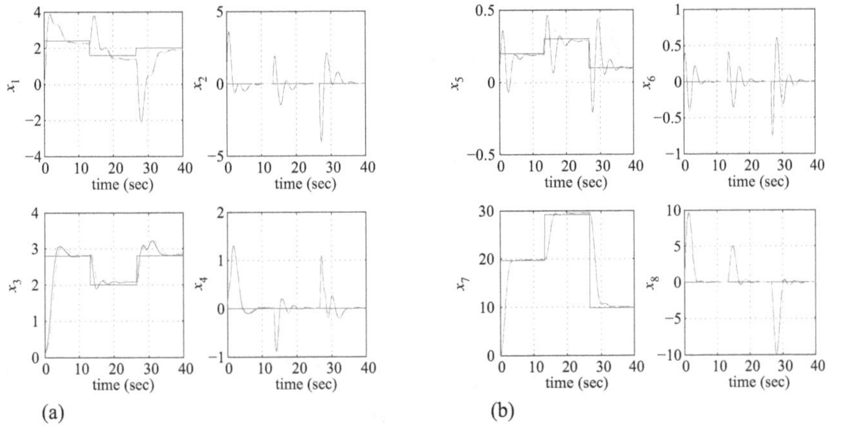

Figure 3.125 Tracking of setpoint 4 for the aerial robotic manipulator: (a) convergence of state variables x_1 , x_2, x_3 and x_4 to their reference setpoints; (b) convergence of state variables x_5, x_6, x_7 and x_8 to their reference setpoints

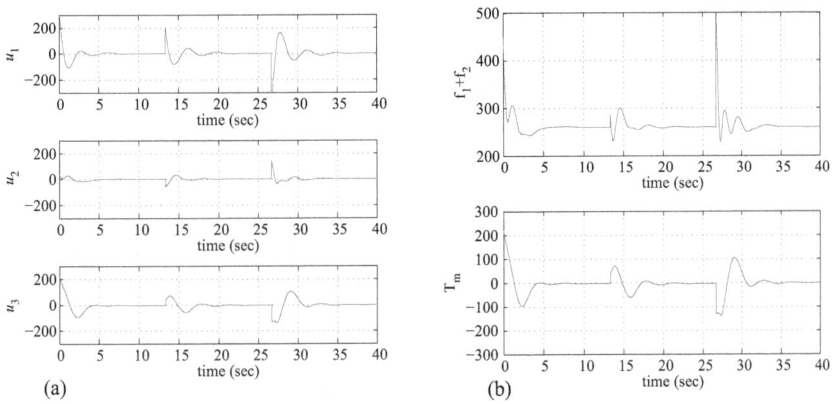

Figure 3.126 Tracking of setpoint 4 for the aerial robotic manipulator: (a) control inputs u_i $i = 1, 2, 3$ computed through the solution of the nonlinear optimal control problem; (b) control inputs $f_1 + f_2$ (aggregate lift force of the UAV's motors) and T_m (torque of the motor mounted at the basis of the manipulator)

which are difficult for humans to access (Bouzid *et al.*, 2016a; Marks, Whidborne *et al.*, 2012; Merheb, Noura *et al.*, 2014). Because of having a larger number of motors than tri-rotors and quadrotors they can transport heavier loads while also being more resilient against motors' failures (Alwi and Edwards, 2013; Alwi *et al.*, 2014; Saied,

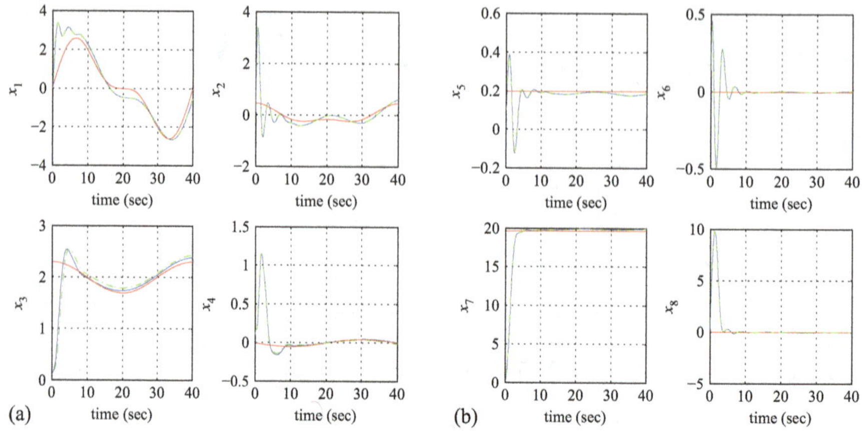

Figure 3.127 Tracking of setpoint 5 for the aerial robotic manipulator: (a) convergence of state variables x_1, x_2, x_3 and x_4 to their reference setpoints; (b) convergence of state variables x_5, x_6, x_7 and x_8 to their reference setpoints

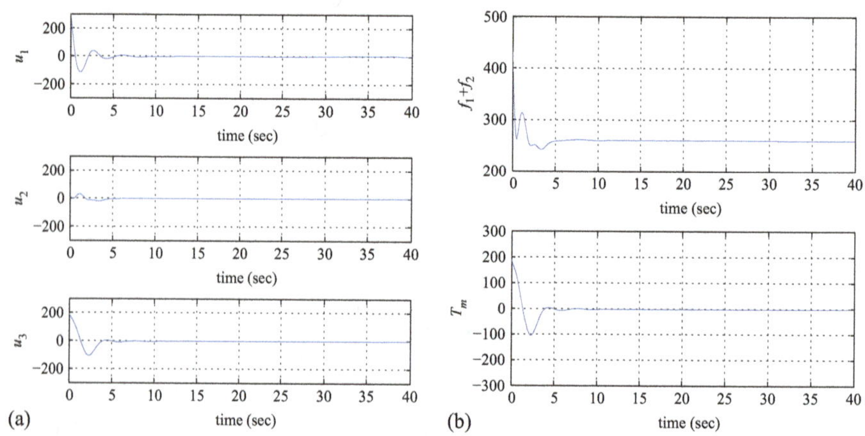

Figure 3.128 Tracking of setpoint 5 for the aerial robotic manipulator: (a) control inputs u_i $i = 1, 2, 3$ computed through the solution of the nonlinear optimal control problem; (b) control inputs $f_1 + f_2$ (aggregate lift force of the UAV's motors) and T_m (torque of the motor mounted at the basis of the manipulator)

Lussier *et al.*, 2016). The octorotor has six degrees of freedom, namely it can perform translational motion in the Cartesian space, while it can also perform rotations by a yaw, pitch and roll angle around the axes of an inertial coordinates reference frame (Chevet *et al.*, 2017; Salazar, Sanjuan *et al.*, 2017; Wang, Zhang *et al.*, 2017). An

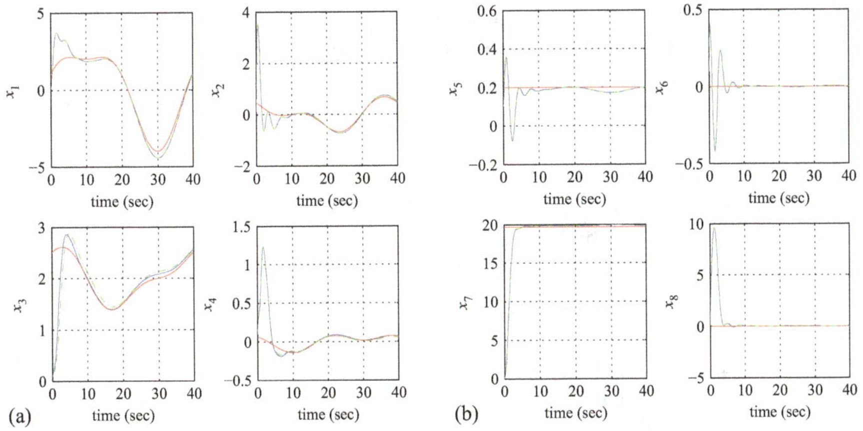

Figure 3.129 Tracking of setpoint 6 for the aerial robotic manipulator: (a) convergence of state variables x_1, x_2, x_3 and x_4 to their reference setpoints; (b) convergence of state variables x_5, x_6, x_7 and x_8 to their reference setpoints

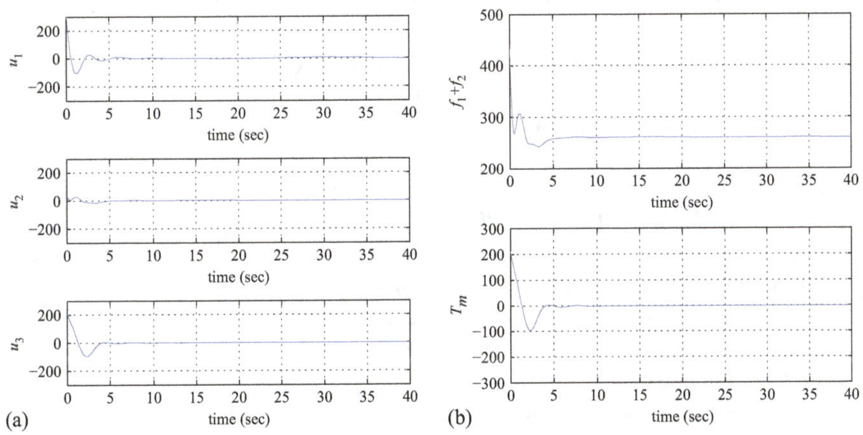

Figure 3.130 Tracking of setpoint 6 for the aerial robotic manipulator: (a) control inputs u_i $i = 1, 2, 3$ computed through the solution of the nonlinear optimal control problem; (b) control inputs $f_1 + f_2$ (aggregate lift force of the UAV's motors) and T_m (torque of the motor mounted at the basis of the manipulator)

octorotor receives four control inputs. These are the thrust provided by the rotors and three torques which result in the rotational motion of the UAV around the axes of the inertial reference system. The latter are the yaw, pitch and roll motion controls (Bouzid *et al.*, 2016b; Ali *et al.*, 2019; Li, Shin *et al.*, 2018). In turn these control inputs are determined by adjusting the power that is provided by the individual motors of the UAV (Niermeyer *et al.*, 2016; Yoon, Johnson *et al.*, 2018; Haus *et al.*, 2017). There exist several attempts towards the solution of the nonlinear multivariable control problem for octorotors. Global stability and robustness are the objectives that controllers' design for octorotors aims at achieving (Trachte *et al.*, 2014; Balsini *et al.*, 2018; Konrad and Abei, 2018; Yao *et al.*, 2018).

In this section, a nonlinear optimal control method is developed for a star-shaped octorotor (Rigatos, 2016b; Rigatos, Siano and Cecati, 2015; Rigatos, Siano *et al.*, 2015). To this end, the dynamic model of the octorotor undergoes first approximate linearization around a temporary operating point which is re-computed at each time-step of the control algorithm. This operating point is defined by the present value of the UAV's state vector and by the last sampled value of the control inputs vector. The linearization relies on first-order Taylor series expansion and on the computation of the Jacobian matrices of the state-space description of the UAV (Rigatos and Tzafestas, 2007; Basseville and Nikiforov, 1993; Rigatos and Zhang, 2009). The modeling inaccuracies which are due to the truncation of higher-order terms in the Taylor series expansion are considered to be disturbances which are asymptotically compensated by the robustness of the control method. For the linearized state-space description an H-infinity feedback controller is designed. Actually, the H-infinity controller stands for the solution of the octorotor's optimal control problem under model uncertainties and external perturbations.

The controller represents a min–max differential game taking place between (i) the control inputs which try to minimize a cost function that comprises a quadratic term of the state vector's tracking error and (ii) the model uncertainty and the external perturbation terms which try to maximize this cost function. To select the controller's feedback gains an algebraic Riccati equation is repetitively solved at each iteration of the control algorithm (Rigatos, 2011; Rigatos, 2015). The stability properties of the control method are proven with the use of Lyapunov analysis. First, it is proven that the control scheme satisfies the H-infinity tracking performance criterion. This signifies elevated robustness against model uncertainty and external perturbations (Rigatos and Busawon, 2018; Tousssaint, Basar and Bullo, 2000). Next, it is proven that under moderate conditions, the control loop is also globally asymptotically stable. Finally, to implement state estimation-based control without the need to measure the entire state vector of the octorotor, the H-infinity Kalman filter is used as a robust state estimator (Rigatos and Busawon, 2018; Gibbs, 2012).

The control problem of multicopters and in particular of octocopters remains always technically challenging (Zhang, Gu *et al.*, 2019; Chen, Zhang *et al.*, 2019; Ejaz *et al.*, 2019; Chebhi *et al.*, 2020; Jafarinasab *et al.*, 2019). The proposed nonlinear optimal control method for octocopters is a genuine and novel result that improves the functioning of such UAVs. Comparing to other nonlinear control approaches that could have been considered for this problem, the method exhibits several advantages. The section's control method minimizes the consumption of energy by the actuators of the octocopter thus improving the autonomy and operational capacity of this robotic system. The control of the octocopter's model is a demanding task because the system is nonlinear and multivariable. The application of global linearization-based control methods to the octocopter's model exhibits several shortcomings. For instance it requires complicated state variables transformations. Besides, to compute the control inputs that will be finally applied to this UAV's model, inverse transformations are needed. Furthermore, one cannot apply MPC because the octocopter's system is a strongly nonlinear one and the use of MPC in its case would make the control loop unstable. Additionally, the convergence of NMPC schemes to an optimum, cannot be assured either since this method relies on the ad-hoc selection of parameters' values and on initialization. It is also remarkable in the application of sliding-mode control, to arrive at a well determined procedure for defining the sliding surface, transformation of the octocopter's state-space model into an input-output linearized form is first needed. It is also noteworthy, that one cannot apply directly backstepping control to the octocopter's model because this system is not found in the triangular (backstepping integral) state-space form. Finally, it would not be recommended to apply multiple local models-based control, through the linearization of the octocopter's model around multiple operating points because this would require the solution of multiple Riccati equations or LMIs and would be a computationally intensive procedure.

3.10.2 Dynamic model of the octorotor

3.10.2.1 State-space description of the octorotor's model

The octorotor is a UAV receiving actuation from eight motors placed on the vertices of an octagon and is depicted in Figure 3.131. The Cartesian coordinates of the octorotor are denoted by the vector $[x, y, z]^T$, while the rotation (Euler) angles of the UAV with respect to the axes of the inertial coordinates system are given by the vector $[\phi, \theta, \psi]^T$.

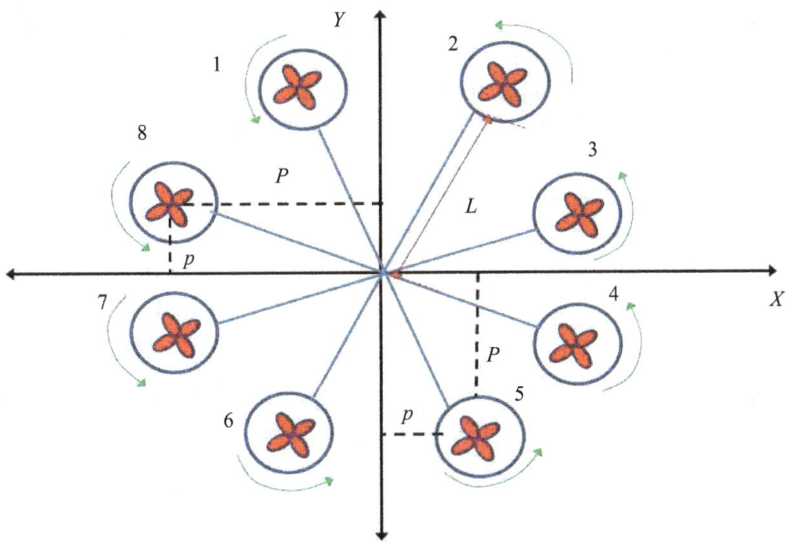

Figure 3.131 Diagram of the autonomous octorotor

The dynamic model of the octorotor, expressed in an inertial reference frame is given by the following set of equations (Merheb, Noura *et al.*, 2014)

$$\begin{aligned}
\ddot{x} &= \tfrac{U_1}{m}(\sin(\psi)\sin(\phi)+\cos(\psi)\sin(\theta)\cos(\phi)) - \tfrac{K_{ft,x}}{m}\dot{x} \\
\ddot{y} &= \tfrac{U_1}{m}(-\cos(\psi)\sin(\phi)+\sin(\psi)\sin(\theta)\cos(\phi)) - \tfrac{K_{ft,y}}{m}\dot{y} \\
\ddot{z} &= \tfrac{U_1}{m}(\cos(\theta)\cos(\psi)) - \tfrac{K_{ft,z}}{m}\dot{z} - g \\
\ddot{\phi} &= \tfrac{I_y-I_z}{I_x}\dot{\theta}\dot{\psi} - \tfrac{I_{rotor}}{I_x}\dot{\theta}\gamma - \tfrac{K_{fa,x}}{I_x}\dot{\phi}^2 + \tfrac{LU_2}{I_x} \\
\ddot{\theta} &= \tfrac{I_z-I_x}{I_y}\dot{\phi}\dot{\psi} + \tfrac{I_{rotor}}{I_y}\dot{\phi}\gamma - \tfrac{K_{fa,y}}{I_y}\dot{\theta}^2 + \tfrac{LU_3}{I_y} \\
\ddot{\psi} &= \tfrac{I_x-I_y}{I_z}\dot{\phi}\dot{\theta} - \tfrac{K_{fa,z}}{I_z}\dot{\psi}^2 + \tfrac{U_4}{I_y}
\end{aligned} \tag{3.342}$$

The parameters of the model are defined as follows: (i) the Cartesian coordinates of the octorotor in the inertial reference frame are denoted as x, y, and z respectively, (ii) the rotation angles of the octorotor (yaw, pitch and roll angles) around the axes of the inertial reference frame are denoted as ϕ, θ and ψ respectively, (iii) the moment of inertia of the octorotor for rotation around the axes of the inertial reference frame are denoted as I_x, I_y and I_z respectively, (iv) the mass of the octorotor is denoted as m, (v) the acceleration of gravity is g, (vi) the distance between the center of gravity of the octorotor and each motor is denoted by L, (vii) b is a coefficient measured in $Nt{\cdot}sec^2$ and d is a coefficient measured in $m{\cdot}sec^2$, (viii) $p = L\sin(a)$ and $P = L\cos(a)$ are distance coefficients, where a is the half of the angle formed by the line segments that connect the center of gravity of the UAV with two rotors placed on successive vertices of the octagon ($a = 22.5^o$). Moreover, aerodynamic forces and translation drag forces are taken into account, through the aerodynamic friction coefficients $K_{fa,x}$, $K_{fa,y}$ and $K_{fa,z}$ and through the coefficients of the translation drag forces $K_{ft,x}$, $K_{ft,y}$, and $K_{ft,z}$. The control inputs to the octorotor are denoted by U_1 which is the thrust provided by the rotor and U_2, U_3, U_4 which are torques resulting in the rotational motion of the UAV around the axes of the inertial reference system. The latter are the yaw, pitch and roll motion controls.

The control inputs U_1, U_2, U_3, and U_4 are related with the rotation speeds of the eight motors ω_1 to ω_8

$$\begin{pmatrix} U_1 \\ U_2 \\ U_3 \\ U_4 \end{pmatrix} = \begin{pmatrix} b & b & b & b & b & b & b & b \\ b\cdot p & -b\cdot p & -b\cdot P & -b\cdot P & -b\cdot p & b\cdot p & b\cdot P & b\cdot P \\ -b\cdot P & -b\cdot P & -b\cdot p & b\cdot p & b\cdot P & b\cdot P & b\cdot P & -b\cdot P \\ -d & -d & d & d & -d & -d & d & d \end{pmatrix} \begin{pmatrix} \omega_1^2 \\ \omega_2^2 \\ \omega_3^2 \\ \omega_4^2 \\ \omega_5^2 \\ \omega_6^2 \\ \omega_7^2 \\ \omega_8^2 \end{pmatrix} \tag{3.343}$$

About the definition of coefficient γ one has

$$\gamma = \omega_3^2 + \omega_4^2 + \omega_7^2 + \omega_8^2 - \omega_1^2 - \omega_2^2 - \omega_5^2 - \omega_6^2 \tag{3.344}$$

and stands for a measure of disequilibrium of the total thrust that is provided by the actuators to the octocopter. Next, the following state variables notation is used

$$\begin{aligned} x &= [x_1, x_2, x_3, x_4, x_5, x_6, x_7, x_8, x_9, x_{10}, x_{11}, x_{12}]^T \text{ or} \\ x &= [x, \dot{x}, y, \dot{y}, z, \dot{z}, \phi, \dot{\phi}, \theta, \dot{\theta}, \psi, \dot{\psi}]^T \end{aligned} \tag{3.345}$$

Using the previous state variables notation, the state-space model of the octorotor is written in the following form:

$$\begin{aligned}
\dot{x}_1 &= x_2 \\
\dot{x}_2 &= \frac{U_1}{m}(\sin(x_{11})\sin(x_7) + \cos(x_{11})\sin(x_9)\cos(x_7)) - \frac{K_{ft,x}}{m}x_2 \\
\dot{x}_3 &= x_4 \\
\dot{x}_4 &= \frac{U_1}{m}(-\cos(x_{11})\sin(x_7) + \sin(x_{11})\sin(x_9)\cos(x_7)) - \frac{K_{ft,y}}{m}x_4 \\
\dot{x}_5 &= x_6 \\
\dot{x}_6 &= \frac{U_1}{m}(\cos(x_9)\cos(x_{11})) - \frac{K_{ft,z}}{m}x_6 - g \\
\dot{x}_7 &= x_8 \\
\dot{x}_8 &= \frac{I_y - I_z}{I_x}x_{10}x_{12} - \frac{I_{rotor}}{I_x}x_{10}\gamma - \frac{K_{fa,x}}{I_x}x_8^2 + \frac{LU_2}{I_x} \\
\dot{x}_9 &= x_{10} \\
\dot{x}_{10} &= \frac{I_z - I_x}{I_y}x_8x_{12} + \frac{I_{rotor}}{I_y}x_8\gamma - \frac{K_{fa,y}}{I_y}x_{10}^2 + \frac{LU_3}{I_y} \\
\dot{x}_{11} &= x_{12} \\
\dot{x}_{12} &= \frac{I_x - I_y}{I_z}x_8x_{10} - \frac{K_{fa,z}}{I_z}x_{12}^2 + \frac{U_4}{I_y}
\end{aligned} \tag{3.346}$$

Next, the following control variables are defined

$$\begin{aligned}
v_1 &= \frac{U_1}{m}(\sin(x_{11})\sin(x_7) + \cos(x_{11})\sin(x_9)\cos(x_7)) \\
v_2 &= \frac{U_1}{m}(-\cos(x_{11})\sin(x_7) + \sin(x_{11})\sin(x_9)\cos(x_7)) \\
v_3 &= \frac{U_1}{m}(\cos(x_9)\cos(x_{11})) - g \qquad & v_4 = U_2 \\
v_5 &= U_3 \qquad & v_6 = U_4
\end{aligned} \tag{3.347}$$

Using (3.347) and after intermediate operations it can be proven that it holds

$$U_1 = \sqrt{v_1^2 + v_2^2 + (v_3 + g)^2} \tag{3.348}$$

Moreover, using (3.347), the state-space model of the octorotor is written as follows

$$\begin{aligned}
\dot{x}_1 &= x_2 \\
\dot{x}_2 &= -\frac{K_{ft,x}}{m}x_2 + v_1 \\
\dot{x}_3 &= x_4 \\
\dot{x}_4 &= -\frac{K_{ft,y}}{m}x_4 + v_2 \\
\dot{x}_5 &= x_6 \\
\dot{x}_6 &= -\frac{K_{ft,z}}{m}x_6 + v_3 \\
\dot{x}_7 &= x_8 \\
\dot{x}_8 &= \frac{I_y - I_z}{I_x}x_{10}x_{12} - \frac{I_{rotor}}{I_x}x_{10}\gamma - \frac{K_{fa,x}}{I_x}x_8^2 + \frac{Lv_4}{I_x} \\
\dot{x}_9 &= x_{10} \\
\dot{x}_{10} &= \frac{I_z - I_x}{I_y}x_8x_{12} + \frac{I_{rotor}}{I_y}x_8\gamma - \frac{K_{fa,y}}{I_y}x_{10}^2 + \frac{Lv_5}{I_y} \\
\dot{x}_{11} &= x_{12} \\
\dot{x}_{12} &= \frac{I_x - I_y}{I_z}x_8x_{10} - \frac{K_{fa,z}}{I_z}x_{12}^2 + \frac{v_6}{I_z}
\end{aligned} \tag{3.349}$$

Next, one obtains the following state-space description of the system in matrix form:

$$\begin{pmatrix} \dot{x}_1 \\ \dot{x}_2 \\ \dot{x}_3 \\ \dot{x}_4 \\ \dot{x}_5 \\ \dot{x}_6 \\ \dot{x}_7 \\ \dot{x}_8 \\ \dot{x}_9 \\ \dot{x}_{10} \\ \dot{x}_{11} \\ \dot{x}_{12} \end{pmatrix} = \begin{pmatrix} x_2 \\ -\frac{K_{ft,x}}{m}x_2 \\ x_4 \\ -\frac{K_{ft,y}}{m}x_4 \\ x_6 \\ -\frac{K_{ft,z}}{m}x_6 \\ x_8 \\ \frac{I_y - I_z}{I_x}x_{10}x_{12} - \frac{I_{rotor}}{I_x}x_{10}\gamma - \frac{K_{fa,x}}{I_x}x_8^2 \\ x_{10} \\ \frac{I_z - I_x}{I_y}x_8x_{12} + \frac{I_{rotor}}{I_y}x_8\gamma - \frac{K_{fa,y}}{I_y}x_{10}^2 \\ x_{12} \\ \frac{I_x - I_y}{I_z}x_8x_{10} - \frac{K_{fa,z}}{I_z}x_{12}^2 \end{pmatrix} + \begin{pmatrix} 0 & 0 & 0 & 0 & 0 & 0 \\ 1 & 0 & 0 & 0 & 0 & 0 \\ 0 & 0 & 0 & 0 & 0 & 0 \\ 0 & 1 & 0 & 0 & 0 & 0 \\ 0 & 0 & 0 & 0 & 0 & 0 \\ 0 & 0 & 1 & 0 & 0 & 0 \\ 0 & 0 & 0 & 0 & 0 & 0 \\ 0 & 0 & 0 & \frac{L}{I_x} & 0 & 0 \\ 0 & 0 & 0 & 0 & 0 & 0 \\ 0 & 0 & 0 & 0 & \frac{L}{I_y} & 0 \\ 0 & 0 & 0 & 0 & 0 & 0 \\ 0 & 0 & 0 & 0 & 0 & \frac{1}{I_z} \end{pmatrix} \begin{pmatrix} v_1 \\ v_2 \\ v_3 \\ v_4 \\ v_5 \\ v_6 \end{pmatrix} \tag{3.350}$$

Finally, the dynamic model of the octorotor is written in the following concise form

$$\dot{x} = f(x) + g(x)v \tag{3.351}$$

where $x \in R^{12\times 1}$, $f(x) \in R^{12\times 1}$, $g(x) \in R^{12\times 6}$ and $v \in R^{6\times 1}$.

3.10.2.2 Differential flatness properties of the octorotor's model

It will be shown that the previous state-space description of the octorotor's model given in (3.350) and (3.351) is a differentially flat system. Differential flatness for a dynamical system is concluded by the following two properties: (i) all state variables and the control inputs of the system can be expressed as differential functions of a subset of its state variables, which are known to stand for the system's flat outputs; (ii) the flat outputs of the system are differentially independent, which means that they are not connected between them through a relation of the form of an homogeneous ordinary differential equation. The flat outputs vector of the system is denoted as $Y=[x_1,x_3,x_5,x_7,x_9,x_{11}]^T=[x,y,z,\phi,\theta,\psi]^T$.

From the odd-numbered rows of the previous state-space model one obtains

$$\begin{array}{lll} x_2=\dot{x}_1 & x_4=\dot{x}_3 & x_6=\dot{x}_5 \\ x_8=\dot{x}_7 & x_{10}=\dot{x}_9 & x_{12}=\dot{x}_{11} \end{array} \tag{3.352}$$

From the even-numbered rows of the previous state-space model, one obtains

$$\begin{aligned} v_1&=\dot{x}_2+\frac{K_{ft,x}}{m}x_2\\ v_2&=\dot{x}_4+\frac{K_{ft,y}}{m}x_4\\ v_3&=\dot{x}_6+\frac{K_{ft,z}}{m}x_6\\ v_4&=\frac{I_x}{L}\left[\dot{x}_8-\frac{I_y-I_z}{I_x}x_{10}x_{12}+\frac{I_{rotor}}{I_x}x_{10}\gamma+\frac{K_{fa,x}}{I_x}x_8^2\right]\\ v_5&=\frac{I_y}{L}\left[\dot{x}_{10}-\frac{I_z-I_x}{I_y}x_8x_{12}-\frac{I_{rotor}}{I_y}x_8\gamma+\frac{K_{fa,y}}{I_y}x_{10}^2\right]\\ v_6&=\frac{I_z}{L}\left[\dot{x}_{12}-\frac{I_x-I_y}{I_z}x_8x_{10}+\frac{K_{fa,z}}{I_z}x_{12}^2\right] \end{aligned} \tag{3.353}$$

According to (3.352) and (3.353), all state variables and control inputs of the octorotor can be written as differential functions of the system's flat outputs. Besides, the flat outputs are differentially independent which means that they are not connected between them through a relation in the form of a linear homogeneous differential equation. Consequently, the octorotor is a differentially flat system. By proving that the model of the octorotor is differentially flat one can also define setpoints for all its state variables.

3.10.3 Approximate linearization of the octorotor's model

3.10.3.1 The approximately linearized model

The previously defined nonlinear state-space model of the octorotor, given in (3.351) undergoes approximate linearization around the time-varying operating point (x^*,v^*), where x^* is the present value of the system's state vector and v^* is the last value of the control input vectors that was applied on it. The linearization is performed based on first-order Taylor series expansion and on the computation of the Jacobian matrices of the state-space model. This results into the linearized description

$$\dot{x} = Ax + Bv + \tilde{d} \tag{3.354}$$

where $\tilde{d}$ is the modeling error due to truncation of higher order terms in the Taylor series expansion and due to external perturbations. Matrices A and B are given by

$$\begin{aligned} A &= \nabla_x[f(x) + g(x)v] \mid_{(x^*,u^*)} \Rightarrow A = \nabla_x[f(x)] \mid_{(x^*,u^*)} \\ B &= \nabla_v[f(x) + g(x)v] \mid_{(x^*,u^*)} \Rightarrow B = g \mid_{(x^*,u^*)} \end{aligned} \tag{3.355}$$

About the Jacobian matrix $A = \nabla_x[f(x)] \mid_{(x^*,u^*)}$ one has:
First row of the Jacobian matrix $\nabla_x[f(x)] \mid_{(x^*,u^*)}$: $\frac{\partial f_1}{\partial x_i} = 0 \; \forall \; i \neq 2$ and $\frac{\partial f_1}{\partial x_2} = 1$

Second row of the Jacobian matrix $\nabla_x[f(x)] \mid_{(x^*,u^*)}$: $\frac{\partial f_2}{\partial x_1} = 0$, $\frac{\partial f_2}{\partial x_2} = -\frac{K_{ft,x}}{m}$, $\frac{\partial f_2}{\partial x_3} = 0$, $\frac{\partial f_2}{\partial x_4} = 0$, $\frac{\partial f_2}{\partial x_5} = 0$, $\frac{\partial f_2}{\partial x_6} = 0$, $\frac{\partial f_2}{\partial x_7} = 0$, $\frac{\partial f_2}{\partial x_8} = 0$, $\frac{\partial f_2}{\partial x_9} = 0$, $\frac{\partial f_2}{\partial x_{10}} = 0$, $\frac{\partial f_2}{\partial x_{11}} = 0$, $\frac{\partial f_2}{\partial x_{12}} = 0$.

Third row of the Jacobian matrix $\nabla_x[f(x)] \mid_{(x^*,u^*)}$: $\frac{\partial f_3}{\partial x_i} = 0 \; \forall \; i \neq 4$ and $\frac{\partial f_3}{\partial x_4} = 1$

Fourth row of the Jacobian matrix $\nabla_x[f(x)] \mid_{(x^*,u^*)}$: $\frac{\partial f_4}{\partial x_1} = 0$, $\frac{\partial f_4}{\partial x_2} = 0$, $\frac{\partial f_4}{\partial x_3} = 0$, $\frac{\partial f_4}{\partial x_4} = -\frac{K_{ft,y}}{m}$, $\frac{\partial f_4}{\partial x_5} = 0$, $\frac{\partial f_4}{\partial x_6} = 0$, $\frac{\partial f_4}{\partial x_7} = 0$, $\frac{\partial f_4}{\partial x_8} = 0$, $\frac{\partial f_4}{\partial x_9} = 0$, $\frac{\partial f_4}{\partial x_{10}} = 0$, $\frac{\partial f_4}{\partial x_{11}} = 0$, $\frac{\partial f_4}{\partial x_{12}} = 0$.

Fifth row of the Jacobian matrix $\nabla_x[f(x)] \mid_{(x^*,u^*)}$: $\frac{\partial f_5}{\partial x_i} = 0 \; \forall \; i \neq 6$ and $\frac{\partial f_5}{\partial x_6} = 1$

Sixth row of the Jacobian matrix $\nabla_x[f(x)] \mid_{(x^*,u^*)}$: $\frac{\partial f_6}{\partial x_1} = 0$, $\frac{\partial f_6}{\partial x_2} = 0$, $\frac{\partial f_6}{\partial x_3} = 0$, $\frac{\partial f_6}{\partial x_4} = 0$, $\frac{\partial f_6}{\partial x_5} = 0$, $\frac{\partial f_6}{\partial x_6} = -\frac{K_{ft,z}}{m}$, $\frac{\partial f_6}{\partial x_7} = 0$, $\frac{\partial f_6}{\partial x_8} = 0$, $\frac{\partial f_6}{\partial x_9} = 0$, $\frac{\partial f_6}{\partial x_{10}} = 0$, $\frac{\partial f_6}{\partial x_{11}} = 0$, $\frac{\partial f_6}{\partial x_{12}} = 0$.

Seventh row of the Jacobian matrix $\nabla_x[f(x)] \mid_{(x^*,u^*)}$: $\frac{\partial f_7}{\partial x_i} = 0 \; \forall \; i \neq 8$ and $\frac{\partial f_7}{\partial x_8} = 1$

Eighth row of the Jacobian matrix $\nabla_x[f(x)] \mid_{(x^*,u^*)}$: $\frac{\partial f_8}{\partial x_1} = 0$, $\frac{\partial f_8}{\partial x_2} = 0$, $\frac{\partial f_8}{\partial x_3} = 0$, $\frac{\partial f_8}{\partial x_4} = 0$, $\frac{\partial f_8}{\partial x_5} = 0$, $\frac{\partial f_8}{\partial x_6} = 0$, $\frac{\partial f_8}{\partial x_7} = 0$, $\frac{\partial f_8}{\partial x_8} = -2\frac{K_{ft,x}}{I_x}x_8$, $\frac{\partial f_8}{\partial x_9} = 0$, $\frac{\partial f_8}{\partial x_{10}} = \frac{I_y - I_z}{I_x}x_{12} - \frac{I_{rotor}}{I_x}\gamma$, $\frac{\partial f_8}{\partial x_{11}} = 0$, $\frac{\partial f_8}{\partial x_{12}} = \frac{I_y - I_z}{I_x}x_{12}$.

Ninth row of the Jacobian matrix $\nabla_x[f(x)] \mid_{(x^*,u^*)}$: $\frac{\partial f_9}{\partial x_i} = 0 \; \forall \; i \neq 10$ and $\frac{\partial f_9}{\partial x_{10}} = 1$

Tenth row of the Jacobian matrix $\nabla_x[f(x)] \mid_{(x^*,u^*)}$: $\frac{\partial f_{10}}{\partial x_1} = 0$, $\frac{\partial f_{10}}{\partial x_2} = 0$, $\frac{\partial f_{10}}{\partial x_3} = 0$, $\frac{\partial f_{10}}{\partial x_4} = 0$, $\frac{\partial f_{10}}{\partial x_5} = 0$, $\frac{\partial f_{10}}{\partial x_6} = 0$, $\frac{\partial f_{10}}{\partial x_7} = 0$, $\frac{\partial f_{10}}{\partial x_{10}} = \frac{I_z - I_x}{I_y}x_{12} + \frac{I_{rotor}}{I_y}\gamma$, $\frac{\partial f_{10}}{\partial x_9} = 0$, $\frac{\partial f_{10}}{\partial x_{10}} = -2\frac{K_{ft,y}}{I_y}x_{10}$, $\frac{\partial f_{10}}{\partial x_{11}} = 0$, $\frac{\partial f_{10}}{\partial x_{12}} = \frac{I_z - I_x}{I_y}x_8$.

Eleventh row of the Jacobian matrix $\nabla_x[f(x)]\,|_{(x^*,u^*)}$: $\frac{\partial f_{11}}{\partial x_i}=0\ \forall\ i\neq 12$ and $\frac{\partial f_{11}}{\partial x_{12}}=1$

Twelfth row of the Jacobian matrix $\nabla_x[f(x)]\,|_{(x^*,u^*)}$: $\frac{\partial f_{12}}{\partial x_1}=0$, $\frac{\partial f_{12}}{\partial x_2}=0$, $\frac{\partial f_{12}}{\partial x_3}=0$, $\frac{\partial f_{12}}{\partial x_4}=0$, $\frac{\partial f_{12}}{\partial x_5}=0$, $\frac{\partial f_{12}}{\partial x_6}=0$, $\frac{\partial f_{12}}{\partial x_7}=0$, $\frac{\partial f_{12}}{\partial x_{12}}=\frac{I_x-I_y}{I_z}x_{10}$, $\frac{\partial f_{12}}{\partial x_9}=0$, $\frac{\partial f_{12}}{\partial x_{10}}=\frac{I_x-I_y}{I_z}x_8$, $\frac{\partial f_{12}}{\partial x_{11}}=0$, $\frac{\partial f_{12}}{\partial x_{12}}=-2\frac{K_{ft,z}}{I_z}x_{12}$.

3.10.3.2 Stabilizing feedback control

After linearization around its current operating point (x^*,u^*), the dynamic model of the octocopter is written as (Rigatos and Busawon, 2018)

$$\dot{x}=Ax+Bu+d_1 \tag{3.356}$$

Parameter d_1 stands for the linearization error in the octocopter's model appearing previously in (3.356). The reference setpoints for the octocopter's state vector are denoted by $\mathbf{x_d}=[x_1^d,\dots,x_{12}^d]$. Tracking of this trajectory is achieved after applying the control input u_d. At every time instant the control input u_d is assumed to differ from the control input u appearing in (3.356) by an amount equal to Δu, that is $u_d=u+\Delta u$

$$\dot{x}_d=Ax_d+Bu_d+d_2 \tag{3.357}$$

The dynamics of the controlled system described in (3.356) can be also written as

$$\dot{x}=Ax+Bu+Bu_d-Bu_d+d_1 \tag{3.358}$$

and by denoting $d_3=-Bu_d+d_1$ as an aggregate disturbance term one obtains

$$\dot{x}=Ax+Bu+Bu_d+d_3 \tag{3.359}$$

By subtracting (3.357) from (3.359), one has

$$\dot{x}-\dot{x}_d=A(x-x_d)+Bu+d_3-d_2 \tag{3.360}$$

By denoting the tracking error as $e=x-x_d$ and the aggregate disturbance term as $\tilde{d}=d_3-d_2$, the tracking error dynamics becomes

$$\dot{e}=Ae+Bu+\tilde{d} \tag{3.361}$$

For the approximately linearized model of the system, a stabilizing feedback controller is developed. The controller has the form

$$u(t)=-Ke(t) \tag{3.362}$$

with $K=\frac{1}{r}B^TP$ where P is a positive definite symmetric matrix which is obtained from the solution of the Riccati equation (Rigatos and Busawon, 2018)

$$A^TP+PA+Q-P\left(\frac{2}{r}BB^T-\frac{1}{\rho^2}LL^T\right)P=0 \tag{3.363}$$

where Q is a positive semi-definite symmetric matrix. The previously analyzed concept about the nonlinear optimal control loop for the octorotor's system is given in Figure 3.132. The global stability properties of the control method can be proven following the procedure presented in Section 1.1.2.5.

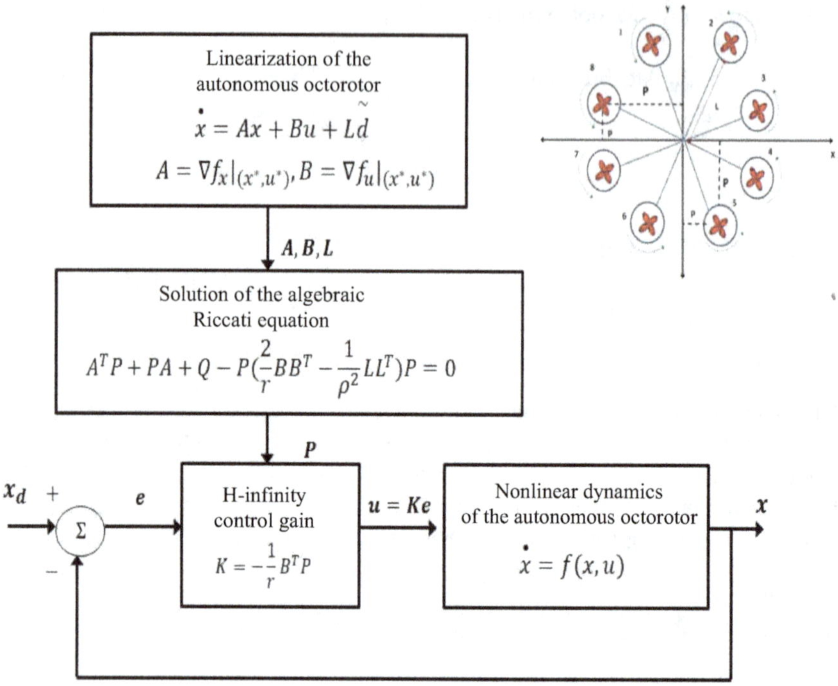

Figure 3.132 Diagram of the control scheme for the autonomous octorotor

3.10.4 Simulation tests

The performance of the previously analyzed nonlinear optimal control method for the dynamic model of the autonomous octorotor has been further confirmed through simulation tests. The presented results are depicted in Figures 3.133 to 3.144. The state variables of the octorotor were measured in SI units. For the computation of the H-infinity controller's feedback gains the algebraic Riccati equation appearing in (3.363) had to be solved at each time-step of the control method. The control inputs were applied directly on the initial nonlinear model of the autonomous octorotor. Through the obtained results it can be noticed that fast and accurate tracking of the reference setpoints was achieved under moderate variations of the control inputs.

The time-step in the algorithmic implementation of the control method is the sampling period, which in the octocopter's control loop has been chosen to be $T_s = 0.01$ s. This time interval is sufficient for solving the algebraic Riccati equation that finally provides matrix P which is used in the computation of the stabilizing feedback control gains. The implementation of the control method can be achieved either on-board, under an embedded control scheme or can be transmitted through a wireless communication channel to the UAV. In the latter case one has to consider also the

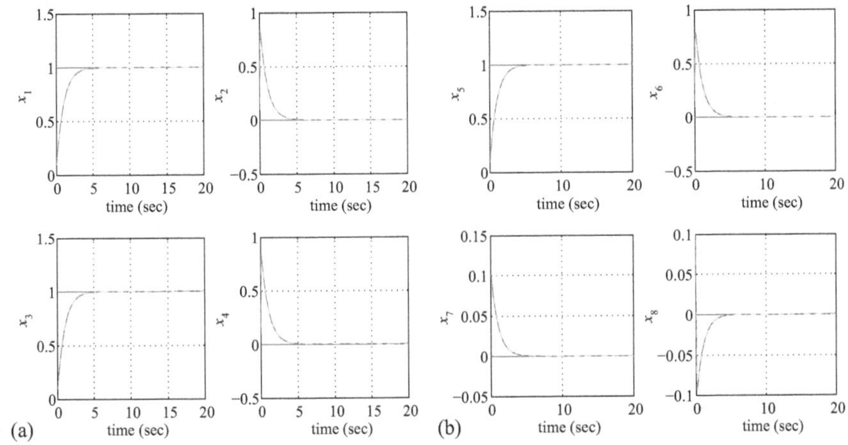

Figure 3.133 Tracking of setpoint 1 for the octorotor UAV, that is motion from initial position (x_A, y_A, z_A) to final position (x_B, y_B, z_B) and alignment of its rotation angles: (a) convergence of state variables x_1 to x_4 to their reference setpoints; (b) convergence of state variables x_5 to x_8 to their reference setpoints

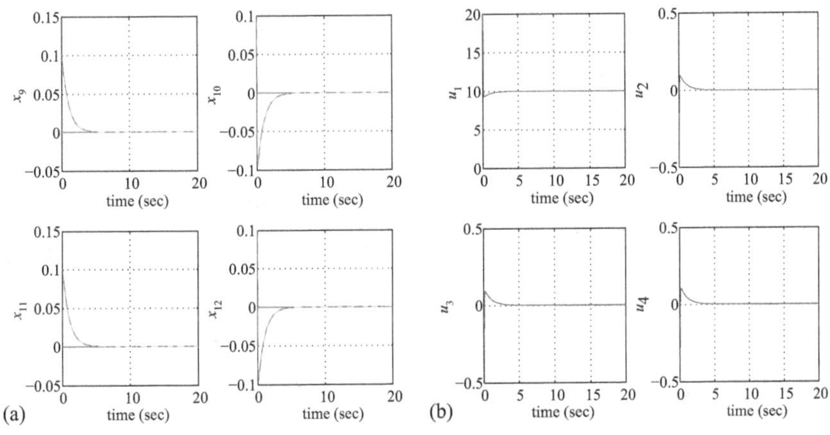

Figure 3.134 Tracking of setpoint 1 for the octorotor that is motion from initial position (x_A, y_A, z_A) to final position (x_B, y_B, z_B) and alignment of its rotation angles: (a) convergence of state variables x_9 to x_{12} to their reference setpoints; (b) real control inputs U_1 to U_4

effects of input-output time-delays in the dynamics of the octocopter and has to update the UAV's state-space model accordingly. In this manner the time-delays effects will appear in linearization process and will be taken into account in the design of the optimal stabilizing feedback controller.

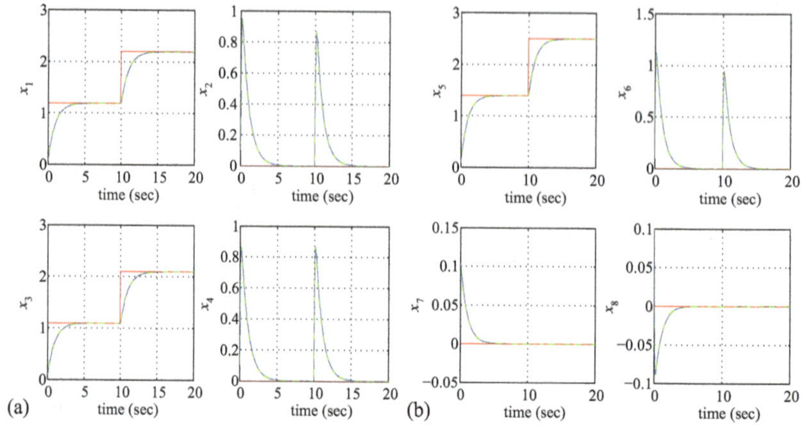

Figure 3.135 Tracking of setpoint 2 for the octorotor UAV, that is motion from initial position (x_A, y_A, z_A) to intermediate position (x_B, y_B, z_B) and to final position (x_C, y_C, z_C) and alignment of its rotation angles: (a) convergence of state variables x_1 to x_4 to their reference setpoints; (b) convergence of state variables x_5 to x_8 to their reference setpoints

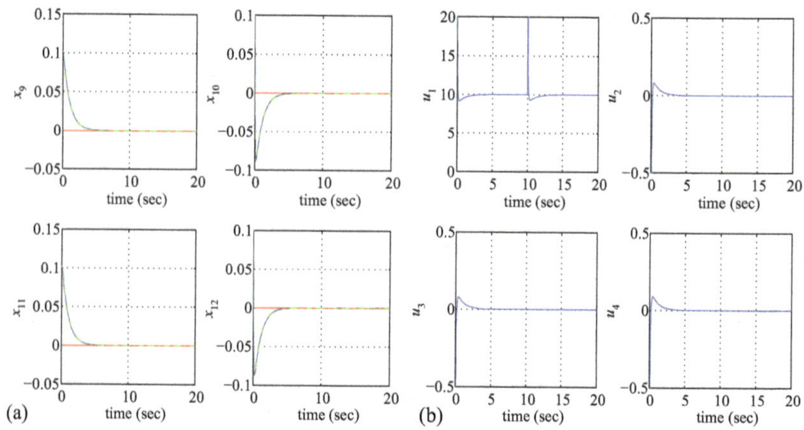

Figure 3.136 Tracking of setpoint 2 for the octorotor UAV, that is motion from initial position (x_A, y_A, z_A) to intermediate position (x_B, y_B, z_B) and to final position (x_C, y_C, z_C) and alignment of its rotation angles: (a) convergence of state variables x_9 to x_{12} to their reference setpoints; (b) real control inputs U_1 to U_4

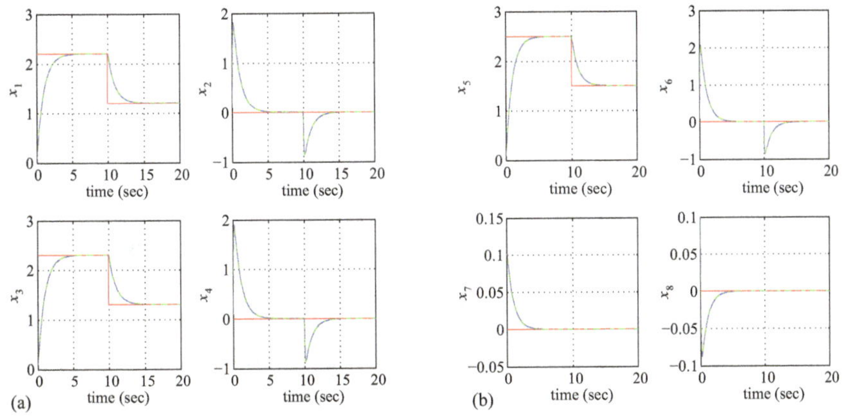

Figure 3.137 Tracking of setpoint 3 for the octorotor UAV, that is motion from initial position (x_A, y_A, z_A) to intermediate position (x_B, y_B, z_B) and to final position (x_C, y_C, z_C) and alignment of its rotation angles: (a) convergence of state variables x_1 to x_4 to their reference setpoints; (b) convergence of state variables x_5 to x_8 to their reference setpoints

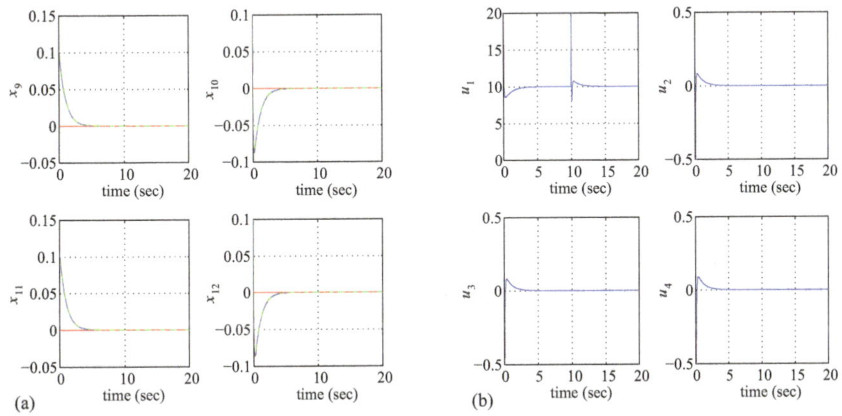

Figure 3.138 Tracking of setpoint 3 for the octorotor UAV, that is motion from initial position (x_A, y_A, z_A) to intermediate position (x_B, y_B, z_B) and to final position (x_C, y_C, z_C) and alignment of its rotation angles: (a) convergence of state variables x_9 to x_{12} to their reference setpoints; (b) real control inputs U_1 to U_4

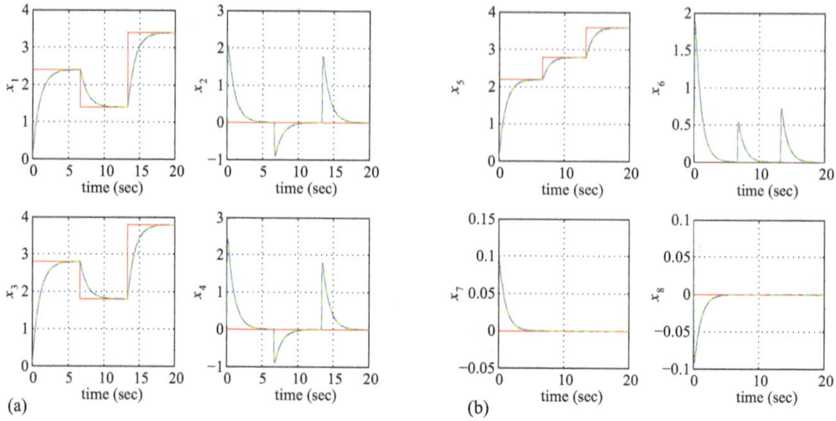

Figure 3.139 Tracking of setpoint 4 for the octorotor UAV, that is motion from initial position (x_A, y_A, z_A) to intermediate position (x_B, y_B, z_B), to subsequent position (x_C, y_C, z_C) and to final position (x_D, y_D, z_D) and alignment of its rotation angles: (a) convergence of state variables x_1 to x_4 to their reference setpoints; (b) convergence of state variables x_5 to x_8 to their reference setpoints

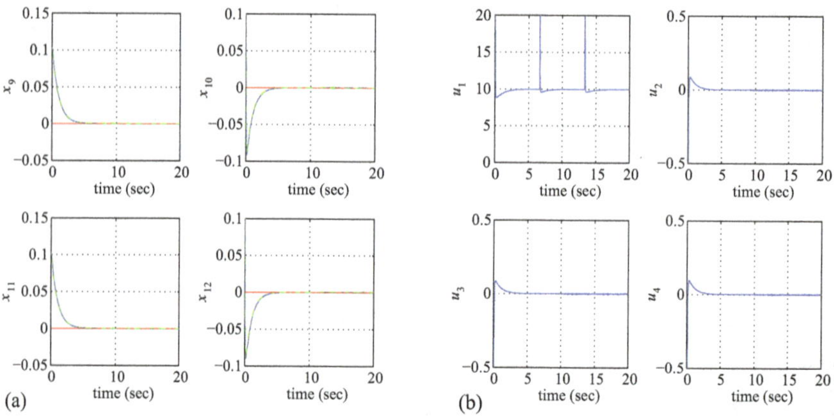

Figure 3.140 Tracking of setpoint 4 for the octorotor UAV, that is motion from initial position (x_A, y_A, z_A) to intermediate position (x_B, y_B, z_B), to subsequent position (x_C, y_C, z_C) and to final position (x_D, y_D, z_D) and alignment of its rotation angles: (a) convergence of state variables x_9 to x_{12} to their reference setpoints; (b) real control inputs U_1 to U_4

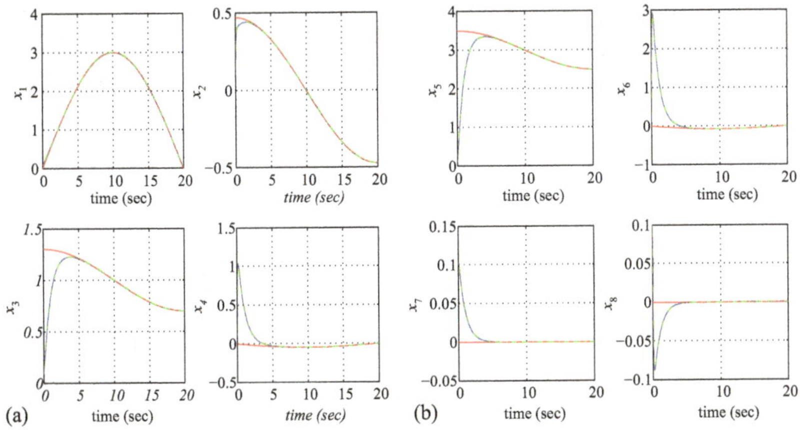

Figure 3.141 Tracking of setpoint 5 for the octorotor UAV, that is motion along a reference path in the [x, y, z] space and alignment of its rotation angles: (a) convergence of state variables x_1 to x_4 to their reference setpoints; (b) convergence of state variables x_5 to x_8 to their reference setpoints

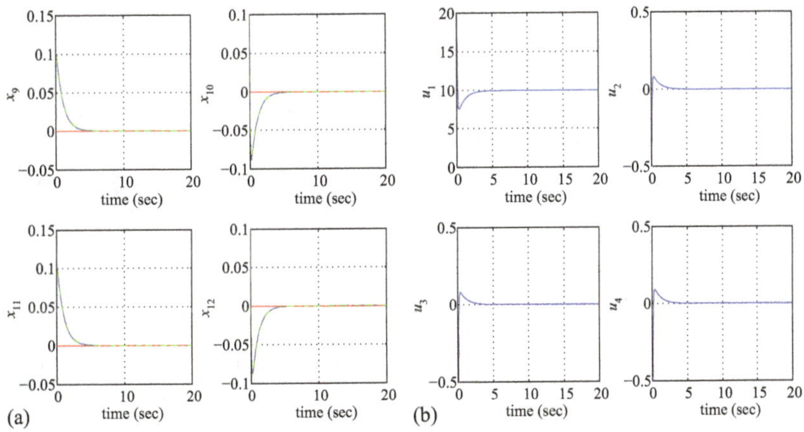

Figure 3.142 Tracking of setpoint 5 for the octorotor UAV, that is motion along a reference path in the [x, y, z] space and alignment of its rotation angles: (a) convergence of state variables x_9 to x_{12} to their reference setpoints; (b) real control inputs U_1 to U_4

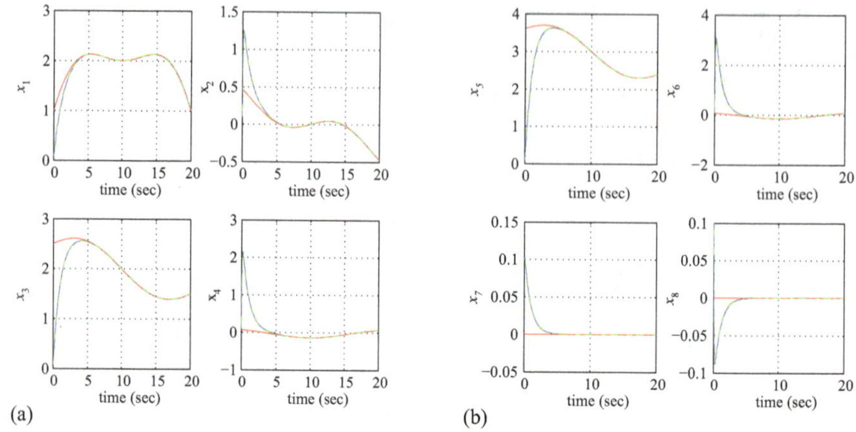

Figure 3.143 Tracking of setpoint 6 for the octorotor UAV, that is motion along a reference path in the $[x, y, z]$ space and alignment of its rotation angles: (a) convergence of state variables x_1 to x_4 to their reference setpoints; (b) convergence of state variables x_5 to x_8 to their reference setpoints

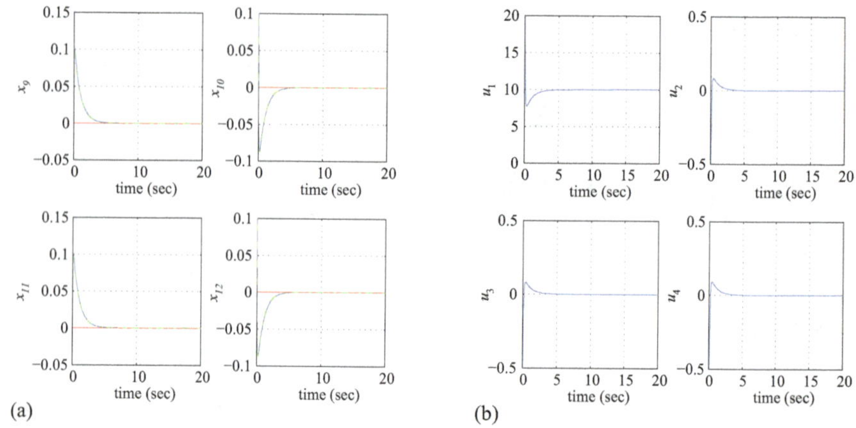

Figure 3.144 Tracking of setpoint 6 for the octorotor UAV, that is motion along a reference path in the $[x, y, z]$ space and alignment of its rotation angles: (a) convergence of state variables x_9 to x_{12} to their reference setpoints; (b) real control inputs U_1 to U_4

3.11 Nonlinear control of hypersonic aerial vehicles

3.11.1 Outline

Control of hypersonic aerial vehicles (HSVs) has emerged as a significant research topic during the last years with possible use for both transportation and defense purposes (Wu, Feng *et al.*, 2017; Wang, Bao *et al.*, 2017; Wamg, Chao *et al.*, 2014; Dou, Su *et al.*, 2018; An *et al.*, 2018). Hypersonic aerial vehicles can develop speeds of about 5–6 Mach, which allows for shortening the duration of long-distance flights (Wu, Zhuo *et al.*, 2017; Hu, Xu *et al.*, 2018; Chen, Niu *et al.*, 2018). Consequently HSVs can boost the operational capacity of airliners. On the other side hypersonic aerial vehicles are more difficult to be withheld by aerial defense and protection systems thus giving also a significant advantage in airforce operations (Xu, Wang *et al.*, 2017; Xu, Yang *et al.*, 2018; Yuan, Wang *et al.*, 2018; Ye, Zong *et al.*, 2018; Wang, Hou *et al.*, 2016). The control of hypersonic aerial vehicles is a non-trivial problem due to the multi-variable and strongly nonlinear model of such aircrafts (Wum, Shi *et al.*, 2018; Dou, Guo *et al.*, 2018; Gao *et al.*, 2018; Gruenwald *et al.*, 2018; Sun, Pu and Yi, 2019). Besides, such systems are underactuated because of having more degrees of freedom than control inputs, and this imposes additional difficulty in achieving flight stabilization and control (Zhang, Wang et. al., 2018; Zerar *et al.*, 2009; Xu and Pan, 2015; Moria *et al.*, 2009). Proof of stability is a main objective in the design of control methods for hypersonic vehicles during the last years. For instance one can note model-based Lyapunov approaches for hypersonic vehicles control (Xu, Shi *et al.*, 2019; Gao, Guo *et al.*, 2019; Tao, Yi *et al.*, 2019b; Zhang, Wang *et al.*, 2019; Dong, Liu *et al.*, 2019; Xia, Wu and Shao, 2019). Additionally one can note model-free Lyapunov approaches for hypersonic vehicles control (Wang, Yang *et al.*, 2019; Gao, Lin *et al.*, 2019; Basin *et al.*, 2018; Li and Li, 2019; Tao, Yi *et al.*, 2019b; Sun, Yi, Pu *et al.*, 2019).

In the present section, a nonlinear optimal (H-infinity) control method is developed for the dynamic model of hypersonic aerial vehicles (Rigatos and Busawon, 2018; Rigatos, 2016b; Rigatos, Siano and Cecati, 2015; Rigatos, Siano *et al.*, 2015). First, the dynamic model of the aerial vehicles undergoes approximate linearization around a temporary operating point which is recomputed at each time-step of the control method. This operating point consists of the present value of the system's state vector and the last value of the control inputs vector that was applied on the HSV. The linearization relies on Taylor series expansion and on the computation of the Jacobian matrices of the associated state-space model (Rigatos and Tzafestas, 2007; Basseville and Nikiforov, 1993; Rigatos and Zhang, 2009). For the approximately linearized model of the system, an H-infinity stabilizing feedback controller is designed. The modeling error which is due to the truncation of higher-order terms in the Taylor series expansion is treated as a perturbation that is compensated by the robustness of the controller.

The proposed H-infinity controller achieves the solution of the nonlinear optimal control problem for the model of the hypersonic vehicle under model uncertainties and external perturbations. It actually represents the solution of a min–max differential game in which the controller tries to minimize a quadratic cost functional of the state vector's error whereas the model uncertainty and perturbation terms try to maximize this cost functional. To compute the feedback gains of the control an algebraic Riccati equation has to be solved at each time-step of the control method (Rigatos, 2011, Rigatos, 2015). The stability properties of the control scheme are proven through Lyapunov analysis. First, it is demonstrated that the H-infinity tracking performance criterion is satisfied, which signifies elevated robustness against model uncertainties and external perturbations (Rigatos and Busawon, 2018; Tousssaint, Basar and Bullo, 2000). Besides, under moderate conditions, the global asymptotic stability of the control loop has been demonstrated. Additionally, to implement state estimation-based control without measuring the entire state vector of the system, the H-infinity Kalman Filter has been used as a robust state estimator (Rigatos and Busawon, 2018; Gibbs, 2012).

3.11.2 Dynamic model of the autonomous hypersonic aerial vehicle

3.11.2.1 Dynamics of the longitudinal motion of the hypersonic vehicle

The dynamic model of hypersonic aerial vehicles is a strongly nonlinear and multi-variable one. The model of the longitudinal dynamics of the autonomous hypersonic vehicle, shown in Figure 3.145 is given by (Wu, Feng *et al.*, 2017; Wang, Bao *et al.*, 2017; Wamg, Chao *et al.*, 2014; Dou, Su *et al.*, 2018; An *et al.*, 2018):

$$\dot{V} = \frac{T\cos(a) - D}{m} - \frac{\mu \sin(\gamma)}{r^2} \tag{3.364}$$

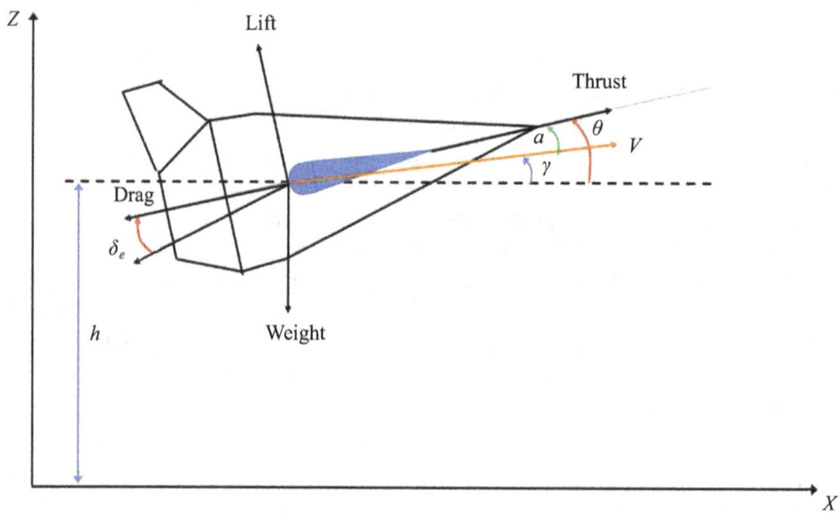

Figure 3.145 Diagram of the autonomous hypersonic vehicle

$$\dot{\gamma} = \frac{L + T\sin(a)}{mV} - \frac{(\mu - V^2 r)\cos(\gamma)}{Vr^2} \tag{3.365}$$

$$\dot{h} = V\sin(\gamma) \tag{3.366}$$

$$\dot{a} = q\dot{\gamma} \tag{3.367}$$

$$\dot{q} = \frac{M_{yy}}{I_{yy}} \tag{3.368}$$

where

$$\begin{array}{ll} L = \frac{1}{2}\rho V^2 S C_L(a) & D = \frac{1}{2}\rho V^2 S C_L(a) \\ T = \frac{1}{2}\rho V^2 S C_T(\beta) & M_{yy} = \frac{1}{2}\rho V^2 S\bar{c}[C_M(a) + C_M'(q) + C_M'(\delta_e, a)] \\ r = h + R_E & \end{array} \tag{3.369}$$

The state variables of the model are $x = [V, \gamma, h, a, q]^T$ where V is the velocity of the hypersonic aerial vehicle, γ is the flight path angle, h is the altitude, a is the angle of attack and q is the pitch rate. The control inputs of the model are δ_e which is the elevator's deflection and β which is the throttle's angle.

Besides in the state-space model of (3.364)– (3.368) L is a function describing lift effects, D is a function denoting drag effects, T is a function representing thrust effects, M_{yy} is a function providing the pitching moment and r is the function of the distance from earth's center.

Other functionals that appear in the previous state-space model are: $C_L(a)$ providing the lift coefficient, $C_D(a)$ providing the drag coefficient, $C_T(\beta)$ standing for the thrust coefficient, $C_M(a)$ which is the moment coefficient due to the angle of attack, $C_M(a)$ which is the moment coefficient due to the pitch rate and $C_M(\delta_e, a)$ which is the moment's coefficient due to elevator's deflection.

Indicative values for $C_L(a)$, $C_D(a)$, $C_T(a)$, $C_M(a)$, $C_M'(q)$, and $C_M''(\delta_e, a)$ are given in the following:

$$C_L(a) = 0.6203a \tag{3.370}$$

$$C_D(a) = 0.6450a^2 + 0.0043379a + 0.00372 \tag{3.371}$$

$$\begin{array}{l} C_T(\beta) = 0.02574\beta \ \text{ if } \ \beta < 1 \\ C_T(\beta) = 0.0224 + 0.00336\beta \ \text{ if } \ \beta \geq 1 \end{array} \tag{3.372}$$

$$C_M(a) = 0.035a^2 + 0.036617a + 5.3261 \cdot 10^{-6} \tag{3.373}$$

$$C_M'(V, a, q) = \frac{\ddot{c}}{2V} q(-6.796a^2 + 0.3015a - 0.2289) \tag{3.374}$$

$$C_M''(\delta_e, a) = C_M''(\delta_e) + C_M''(a) = 0.0292\delta_e - 0.0292a \tag{3.375}$$

Regarding the parameters of the model these are defined as follows: μ is the gravitational constant, ρ is the density of air, S is the reference area, and $\bar{c}$ is the mean aerodynamics cord and R_E is the radius of earth.

By using the previous state-space notation $x=[x_1,x_2,x_3,x_4,x_5]^T=[V,\gamma,h,a,q]^T$ and by defining the modified control inputs $u_1 = C_T(\beta)$ and $u_2 = C''_M(\delta_e)$ one obtains the following state-space description about the dynamics of the hypersonic vehicle:

$$\dot{x}_1 = \frac{1}{m}\frac{1}{2}\rho x_1^2 S u_1 \cos(x_4) - \frac{1}{m}\frac{1}{2}\rho x_1^2 S C_D(x_4) - \frac{\mu \sin(x_2)}{(x_3+R_E)^2} \tag{3.376}$$

$$\dot{x}_2 = \frac{1}{mx_1}\frac{1}{2}\rho x_1^2 S C_L(x_4) + \frac{1}{mx_1}\frac{1}{2}\rho x_1^2 S u_1 \sin(x_4) - \frac{\mu \sin(x_2)}{(x_3+R_E)^2} \tag{3.377}$$

$$\dot{x}_3 = x_1 \cos(x_2) \tag{3.378}$$

$$\dot{x}_4 = x_5 - \frac{1}{mx_1}\frac{1}{2}\rho x_1^2 S C_L(x_4) - \frac{1}{mx_1}\frac{1}{2}\rho x_1^2 S u_1 \sin(x_4) + \frac{\mu - x_1 r \sin(x_2)}{x_1(x_3+R_E)^2} \tag{3.379}$$

$$\dot{x}_5 = \frac{1}{I_{yy}}\frac{1}{2}\rho x_1^2 S\bar{c}[C_M(x_4) + u_2 - 0.0292x_4 + C'_M(x_1,x_4,x_5)] \tag{3.380}$$

Next, the following functions are defined:

$$\begin{aligned} f_1 &= -\frac{1}{m}\frac{1}{2}\rho x_1^2 S C_D(x_4) - \frac{\mu \sin(x_2)}{(x_3+R_E)^2} \\ g_{11} &= \frac{1}{m}\frac{1}{2}\rho x_1^2 S \cos(x_4) \qquad g_{12} = 0 \end{aligned} \tag{3.381}$$

$$\begin{aligned} f_2 &= \frac{1}{mx_1}\frac{1}{2}\rho x_1^2 S C_L(x_4) - \frac{(\mu - x_1^2(x_3+R_E))\cos(x_2)}{x_1(x_3+R_E)^2} \\ g_{21} &= \frac{1}{mx_1}\frac{1}{2}\rho x_1^2 S \sin(x_4) \qquad g_{22} = 0 \end{aligned} \tag{3.382}$$

$$\begin{aligned} f_3 &= x_1 \sin(x_2) \\ g_{31} &= 0 \qquad g_{32} = 0 \end{aligned} \tag{3.383}$$

$$\begin{aligned} f_4 &= x_5 - \frac{1}{mx_1}\frac{1}{2}\rho x_1^2 S C_L(x_4) + \frac{(\mu - x_1^2(x_3+R_E))\cos(x_2)}{x_1(x_3+R_E)^2} \\ g_{41} &= -\frac{1}{mx_1}\frac{1}{2}\rho x_1^2 S \sin(x_4) \qquad g_{42} = 0 \end{aligned} \tag{3.384}$$

$$\begin{aligned} f_5 &= \frac{1}{I_{yy}}\frac{1}{2}\rho x_1^2 S\bar{c}[C_M(x_4) - 0.0292x_4 + C'_M(x_1,x_4,x_5)] \\ g_{51} &= 0 \qquad g_{52} = \frac{1}{I_{yy}}\frac{1}{2}\rho x_1^2 S\bar{c} \end{aligned} \tag{3.385}$$

Consequently, the state-space description of the hypersonic vehicle is written in the affine-in-the-input form

$$\begin{pmatrix} \dot{x}_1 \\ \dot{x}_2 \\ \dot{x}_3 \\ \dot{x}_4 \\ \dot{x}_5 \end{pmatrix} = \begin{pmatrix} f_1(x) \\ f_2(x) \\ f_3(x) \\ f_4(x) \\ f_5(x) \end{pmatrix} + \begin{pmatrix} g_{11}(x) & g_{12}(x) \\ g_{21}(x) & g_{22}(x) \\ g_{31}(x) & g_{32}(x) \\ g_{41}(x) & g_{42}(x) \\ g_{151}(x) & g_{52}(x) \end{pmatrix} \begin{pmatrix} u_1 \\ u_2 \end{pmatrix} \tag{3.386}$$

By denoting $x \in R^{5\times 1}$, $u \in R^{5\times 2}$, $f(x) \in R^{5\times 1}$, and $G(x) \in R^{5\times 2}$, the previous state-space model can be also written in the concise form

$$\dot{x} = f(x) + G(x)u \tag{3.387}$$

3.11.2.2 Dynamics of the HSV under a small angle of attack

Under a small angle of attack a one has that $\cos(a) \simeq 1$, and $\sin(a) \simeq 0$. Then the first equation of the previous state-space model of the HSV becomes

$$\dot{x}_1 = -\frac{1}{m}\frac{1}{2}\rho x_1^2 S C_D(x_4) + \frac{1}{m}\frac{1}{2}\rho x_1^2 S u_1 \tag{3.388}$$

About the second state-space equation of the HSV, one has

$$\dot{x}_2 = \frac{1}{m x_1}\frac{1}{2}\rho x_1^2 S C_L(x_4) - \frac{(\mu - x_1^2(x_3 + R_E))\cos(x_2)}{x_1(x_3 + R_E)^2} \tag{3.389}$$

About the third state-space equation of the HSV, one has

$$\dot{x}_3 = x_1 \sin(x_2) \tag{3.390}$$

About the fourth state-space equation of the HSV, one has

$$\dot{x}_4 = x_5 - \frac{1}{m x_1}\frac{1}{2}\rho x_1^2 S C_L(x_4) + \frac{(\mu - x_1^2(x_3 + R_E))\cos(x_2)}{x_1(x_3 + R_E)^2} \tag{3.391}$$

Moreover, about the fifth state-space equation of the HSV, one obtains

$$\dot{x}_5 = \frac{1}{I_{yy}}\frac{1}{2}\rho x_1^2 S\bar{c}[C_M(x_4) - 0.0292 x_4 + C_M'(x_1, x_4, x_5)] + \frac{1}{I_{yy}}\frac{1}{2}\rho x_1^2 S\bar{c} u_2 \tag{3.392}$$

3.11.3 *Differential flatness properties of the hypersonic vehicle*

It is proven that the dynamic model of the hypersonic aerial vehicle is a differentially flat one (Rigatos and Busawon, 2018; Rigatos, 2015). The flat outputs of the system are defined as:

$$[y_1, y_2]^T = [V, h]^T \tag{3.393}$$

Under the assumption of a small angle of attack it holds: $\cos(a) \simeq 1$ and $\sin(a) \simeq 0$. Consequently, from the previous state-space model of (3.381)–(3.385)

$$\begin{aligned} f_1 &= -\frac{1}{m}\frac{1}{2}\rho x_1^2 S C_D(x_4) \\ g_{11} &= \frac{1}{m}\frac{1}{2}\rho x_1^2 S \qquad g_{12} = 0 \end{aligned} \tag{3.394}$$

$$f_2 = \frac{1}{mx_1}\frac{1}{2}\rho x_1^2 SC_L(x_4) - \frac{(\mu - x_1^2(x_3+R_E))\cos(x_2)}{x_1(x_3+R_E)^2}$$

$$g_{21} = 0 \qquad g_{22} = 0 \tag{3.395}$$

$$f_3 = x_1 \sin(x_2)$$

$$g_{31} = 0 \qquad g_{32} = 0 \tag{3.396}$$

$$f_4 = x_5 - \frac{1}{mx_1}\frac{1}{2}\rho x_1^2 SC_L(x_4) + \frac{(\mu - x_1^2(x_3+R_E))\cos(x_2)}{x_1(x_3+R_E)^2}$$

$$g_{41} = 0 \qquad g_{42} = 0 \tag{3.397}$$

$$f_5 = \frac{1}{I_{yy}}\frac{1}{2}\rho x_1^2 S\bar{c}[C_M(x_4) - 0.0292x_4 + C_M'(x_1, x_4, x_5)]$$

$$g_{51} = 0 \qquad g_{52} = \tfrac{1}{I_{yy}}\tfrac{1}{2}\rho x_1^2 S\bar{c} \tag{3.398}$$

Consequently, the state-space model of the hypersonic vehicle, under a small angle of attack, becomes:

$$\dot{x}_1 = f_1(x) + g_{11}(x)u_1 \tag{3.399}$$

$$\dot{x}_2 = f_2(x) \tag{3.400}$$

$$\dot{x}_3 = f_3(x) \tag{3.401}$$

$$\dot{x}_4 = f_4(x) \tag{3.402}$$

$$\dot{x}_5 = f_5(x) + g_{52}(x)u_2 \tag{3.403}$$

Using the previously defined flat outputs of the system y_1 and y_2 it holds that

$$x_1 = y_1 \quad x_2 = y_2 \tag{3.404}$$

From (3.401) one has:

$$\dot{x}_3 = x_1 \sin(x_2) \Rightarrow x_2 = sin^{-1}(\frac{\dot{x}_3}{x_1})$$
$$\Rightarrow x_2 = sin^{-1}(\frac{\dot{y}_2}{y_1}) \Rightarrow x_2 = h_2(y, \dot{y}) \tag{3.405}$$

which signifies that state variable x_2 is a differential function of the system's flat outputs. From (3.400) one has:

$$\dot{x}_2 = \frac{1}{mx_1}\frac{1}{2}\rho x_1^2 SC_L(x_4) - \frac{(\mu - x_1^2(x_3+R_E))}{x_1(x_3+R_E)^2} \Rightarrow$$
$$C_L(x_4) = \frac{\dot{x}_2 + \frac{(\mu - x_1^2(x_3+R_E))}{x_1(x_3+R_E)^2}}{\frac{1}{mx_1}\frac{1}{2}\rho x_1^2 S} \tag{3.406}$$

and using that $C_L(x_4) = 0.6203x_4$, one obtains

$$x_4 = \frac{1}{0.6203} \frac{\dot{x}_2 + \frac{(\mu - x_1^2(x_3+R_E))}{x_1(x_3+R_E)^2}}{\frac{1}{mx_1}\frac{1}{2}\rho x_1^2 S} \tag{3.407}$$

$$\Rightarrow x_4 = h_4(y, \dot{y})$$

which signifies that state variable x_4 is a differential function of the system's flat outputs. From (3.402), one has:

$$\dot{x}_4 = x_5 - \frac{1}{mx_1}\frac{1}{2}\rho x_1^2 SC_L(x_4) + \frac{(\mu - x_1^2(x_3 + R_E))}{x_1(x_3 + R_E)^2} \Rightarrow$$

$$x_5 = \dot{x}_4 + \frac{1}{mx_1}\frac{1}{2}\rho x_1^2 SC_L(x_4) - \frac{(\mu - x_1^2(x_3 + R_E))}{x_1(x_3 + R_E)^2} \Rightarrow \tag{3.408}$$

$$x_5 = h_5(y, \dot{y})$$

which signifies that state variable x_5 is a differential function of the system's flat outputs. From (3.399), one has:

$$u_1 = \frac{\dot{x}_1 - f_1(x)}{g_{11}(x)} \tag{3.409}$$

Since it has been proven that all state variables of the system are differential functions of its flat outputs, the previous relation signifies that control input u_1 is also a differential function of the flat outputs of the hypersonic vehicle's model. Moreover, from (3.403) one has:

$$u_2 = \frac{\dot{x}_5 - f_5(x)}{g_{52}(x)} \tag{3.410}$$

Since it has been proven that all state variables of the system are differential functions of its flat outputs, the previous relation signifies that control input u_2 is also a differential function of the flat outputs of the hypersonic vehicle's model.

As a result of the above, all state variables and the control inputs of the hypersonic aerial vehicle can be written as differential functions of its flat outputs, and consequently it is proven that this system is a differentially flat one. The differential flatness property of the system allows for solving the problem of reference trajectories (setpoints) selection for the HSV's state variables. Actually, by choosing reference values for the flat outputs of the system $y_1 = V$ (velocity of the HSV) and $y_2 = h$ (height of the USV) and by using the previous relations one can also compute the reference trajectories for the rest of the system's state variables that is $x_2 = \gamma$, $x_4 = a$ and $x_5 = q$.

3.11.4 Approximate linearization for the dynamic model of the hypersonic vehicle

The previously analyzed dynamic model of the hypersonic vehicle undergoes approximate linearization around the temporary operating point (x^*, u^*), where x^* is the present value of the system's state vector and u^* is the last sampled value of the control inputs vector. The concise model of the system was:

$$\dot{x} = f(x) + G(x)u \tag{3.411}$$

The linearization relies on Taylor-series expansion and on the computation of the associate Jacobian matrices. The equivalent linearized model that is obtained through this procedure

$$\dot{x} = Ax + Bu + \tilde{d} \tag{3.412}$$

where $\tilde{d}$ is the disturbance vector which is due to approximate linearization and the truncation of higher-order terms in the Taylor series expansion. This term can also comprise external perturbations. Besides, matrices A and B are given by:

$$\begin{gathered} A = \nabla_x[f(x) + G(x)u] \mid_{(x^*,u^*)} \Rightarrow \\ A = [\nabla_x[f(x)] \mid_{(x^*,u^*)} + \nabla_x[g_1(x)]u_1 \mid_{(x^*,u^*)} + \nabla_x[g_2(x)]u_2 \mid_{(x^*,u^*)} \end{gathered} \tag{3.413}$$

$$B = \nabla_u[f(x) + G(x)u] \mid_{(x^*,u^*)} \Rightarrow B = G(x) \mid_{(x^*,u^*)} \tag{3.414}$$

where $g_i(x)$ $i = 1,2$ denotes the i-th column of the control inputs gain matrix $G(x)$. About the computation of the Jacobian matrix $\nabla_x[f(x)] \mid_{(x^*,u^*)}$ one has:

First row of the Jacobian matrix $\nabla_x[f(x)] \mid_{(x^*,u^*)}$: $\frac{\partial f_1}{\partial x_1} = -\frac{1}{m}\rho x_1 S C_B(x_4)$, $\frac{\partial f_1}{\partial x_2} = -\frac{\mu\cos(x_2)}{(x_3+R_E)^2}$, $\frac{\partial f_1}{\partial x_3} = \frac{2\mu\sin(x_2)}{(x_3+R_E)^3}$, $\frac{\partial f_1}{\partial x_4} = -\frac{1}{m}\frac{1}{2}\rho x_1^2 S\frac{\partial C_D(x_4)}{\partial x_4}$, $\frac{\partial f_1}{\partial x_5} = 0$.

where $\frac{\partial C_D(x_4)}{\partial x_4} = 0.6450 \cdot 2 \cdot x_4 + 0.0043378$

Second row of the Jacobian matrix $\nabla_x[f(x)] \mid_{(x^*,u^*)}$:

$$\frac{\partial f_2}{\partial x_1} = -\frac{1}{m}\frac{1}{2}SC_L(x_4) - \frac{[(-2x_1x_3+R_E)(x_1x_3+R_E{}^2)\cos(x_2) - (\mu - x_1^2(x_3+R_E))(x_3+R_E)^2]}{(x_1(x_3+R_E)^2)^2},$$

$$\frac{\partial f_2}{\partial x_2} = \frac{(\mu - x_1^2(x_3+R_E))\sin(x_2)}{x_1(x_3+R_E)^2},$$

$$\frac{\partial f_2}{\partial x_3} = \frac{[x_1^2\cos(x_2)(x_1(x_3+R_E)^2) + (\mu - x_1^2(x_3+R_E))\cos(x_2)x_1 2(x_3+R_E)]}{(x_1(x_3+R_E)^2)^2}$$

$$\frac{\partial f_2}{\partial x_4} = \frac{1}{mx_1}\frac{1}{2}\rho x_1^2 S\frac{\partial C_L(x_4)}{\partial x_4}$$

$$\frac{\partial f_2}{\partial x_5} = 0$$

where $\frac{\partial C_L(x_4)}{\partial x_4} = 0.6203$.

Third row of the Jacobian matrix $\nabla_x[f(x)] \mid_{(x^*,u^*)}$: $\frac{\partial f_3}{\partial x_1} = \sin(x_2)$, $\frac{\partial f_3}{\partial x_2} = x_1 \cos(x_2)$, $\frac{\partial f_3}{\partial x_3} = 0$, $\frac{\partial f_2}{\partial x_4} = 0$ and $\frac{\partial f_3}{\partial x_5} = 0$.

Fourth row of the Jacobian matrix $\nabla_x[f(x)] \mid_{(x^*,u^*)}$

$$\frac{\partial f_4}{\partial x_1} = -\frac{1}{m}\frac{1}{2}SC_L(x_4) + \frac{[(-2x_1x_3+R_E)(x_1(x_3+R_E)^2)\cos(x_2) - (\mu - x_1^2(x_3+R_E))(x_3+R_E)^2]}{(x_1((x_3+R_E))^2)^2},$$

$$\frac{\partial f_4}{\partial x_2} = -\frac{(\mu - x_1^2(x_3+R_E))\sin(x_2)}{x_1(x_3+R_E)^2},$$

$$\frac{\partial f_4}{\partial x_3} = -\frac{[x_1^2\cos(x_2)(x_1(x_3+R_E)^2) + (\mu - x_1^2(x_3+R_E))\cos(x_2)x_1 2(x_3+R_E)]}{(x_1(x_3+R_E)^2)^2}$$

$$\frac{\partial f_4}{\partial x_4} = -\frac{1}{mx_2}\frac{1}{2}\rho x_1^2 S\frac{\partial C_L(x_4)}{\partial x_4}$$

$$\frac{\partial f_4}{\partial x_5} = 0$$

Fifth row of the Jacobian matrix $\nabla_x[f(x)] \mid_{(x^*,u^*)}$

$$\frac{\partial f_5}{\partial x_1} = \frac{1}{I_{yy}}\rho x_1 S\bar{C}[C_M(x_4) - 0.0292x_4 + C'_M(x_1,x_4,x_5)] + \frac{1}{I_{yy}}\rho x_1 S\bar{C}\frac{\partial C'_M(x_1,x_4,x_5)}{\partial x_1}$$

where $\frac{\partial C'_M(x_1,x_4,x_5)}{\partial x_1} = -\frac{\ddot{c}}{2x_1^2}x_5(-6.796x_4^2 + 0.3015x_4 - 0.2284)$.

$\frac{\partial f_5}{\partial x_2} = 0$ and $\frac{\partial f_5}{\partial x_3} = 0$.

$$\frac{\partial f_5}{\partial x_4} = \frac{1}{I_{yy}}\frac{1}{2}\rho x_1^2 S\bar{C}\left[\frac{\partial C_M(x_4)}{\partial x_4} - 0.0292 + \frac{\partial C'_M(x_1,x_4,x_5)}{\partial x_4}\right]$$

where $\frac{\partial C_M(x_4)}{\partial x_4} = -2\cdot 0.035x_4 + 0.036617$

and $\frac{\partial C'_M(x_1,x_4,x_5)}{\partial x_4} = \frac{\ddot{c}}{2x_1}x_5(-2\cdot 6.796x_4 + 0.3015)$

$$\frac{\partial f_5}{\partial x_5} = \frac{1}{I_{yy}}\rho x_1 S\bar{C}\frac{\partial C'_M(x_1,x_4,x_5)}{\partial x_5}$$

where $\frac{\partial C'_M(x_1,x_4,x_5)}{\partial x_5} = \frac{\ddot{c}}{2x_1}(-6.796x_4^2 + 0.3015x_4 - 0.2289)$

About the computation of the Jacobian matrix $\nabla_x[g_1(x)] \mid_{(x^*,u^*)}$, one has:

First row of the Jacobian matrix $\nabla_x g_1(x)$: $\frac{\partial g_{11}}{\partial x_1} = \frac{1}{m}\rho x_1 S\cos(x_4)$, $\frac{\partial g_{11}}{\partial x_2} = 0$, $\frac{\partial g_{11}}{\partial x_3} = 0$, $\frac{\partial g_{11}}{\partial x_4} = 0$, and $\frac{\partial g_{11}}{\partial x_5} = 0$.

Second row of the Jacobian matrix $\nabla_x g_1(x)$: $\frac{\partial g_{21}}{\partial x_1} = -\frac{1}{mx_1^2}\frac{1}{2}\rho S\cos(x_4)$, $\frac{\partial g_{21}}{\partial x_2} = 0$, $\frac{\partial g_{21}}{\partial x_3} = 0$, $\frac{\partial g_{21}}{\partial x_4} = \frac{1}{m}\frac{1}{2}\rho x_1 S\cos(x_4)$, and $\frac{\partial g_{11}}{\partial x_5} = 0$.

Third row of the Jacobian matrix $\nabla_x g_1(x)$: $\frac{\partial g_{31}}{\partial x_1} = 0$, $\frac{\partial g_{32}}{\partial x_2} = 0$, $\frac{\partial g_{33}}{\partial x_3} = 0$, $\frac{\partial g_{34}}{\partial x_4} = 0$, and $\frac{\partial g_{35}}{\partial x_5} = 0$.

Fourth row of the Jacobian matrix $\nabla_x g_1(x)$: $\frac{\partial g_{41}}{\partial x_1} = -\frac{1}{m}\frac{1}{2}\rho S\sin(x_4)$, $\frac{\partial g_{42}}{\partial x_2} = 0$, $\frac{\partial g_{43}}{\partial x_3} = 0$, $\frac{\partial g_{44}}{\partial x_4} = -\frac{1}{mx_1}\frac{1}{2}\rho x_1^2 S\cos(x_4)$, and $\frac{\partial g_{45}}{\partial x_5} = 0$.

Fifth row of the Jacobian matrix $\nabla_x g_1(x)$: $\frac{\partial g_{51}}{\partial x_1} = 0$, $\frac{\partial g_{52}}{\partial x_2} = 0$, $\frac{\partial g_{53}}{\partial x_3} = 0$, $\frac{\partial g_{54}}{\partial x_4} = 0$, and $\frac{\partial g_{55}}{\partial x_5} = 0$.

About the computation of the Jacobian matrix $\nabla_x[g_2(x)]\,|_{(x^*,u^*)}$ one has:

First row of the Jacobian matrix $\nabla_x g_2(x)$: $\frac{\partial g_{12}}{\partial x_1} = 0$, $\frac{\partial g_{12}}{\partial x_2} = 0$, $\frac{\partial g_{12}}{\partial x_3} = 0$, $\frac{\partial g_{12}}{\partial x_4} = 0$, and $\frac{\partial g_{12}}{\partial x_5} = 0$.

Second row of the Jacobian matrix $\nabla_x g_2(x)$: $\frac{\partial g_{22}}{\partial x_1} = 0$, $\frac{\partial g_{22}}{\partial x_2} = 0$, $\frac{\partial g_{22}}{\partial x_3} = 0$, $\frac{\partial g_{22}}{\partial x_4} = 0$, and $\frac{\partial g_{22}}{\partial x_5} = 0$.

Third row of the Jacobian matrix $\nabla_x g_2(x)$: $\frac{\partial g_{32}}{\partial x_1} = 0$, $\frac{\partial g_{32}}{\partial x_2} = 0$, $\frac{\partial g_{32}}{\partial x_3} = 0$, $\frac{\partial g_{32}}{\partial x_4} = 0$, and $\frac{\partial g_{32}}{\partial x_5} = 0$.

Fourth row of the Jacobian matrix $\nabla_x g_2(x)$: $\frac{\partial g_{42}}{\partial x_1} = 0$, $\frac{\partial g_{42}}{\partial x_2} = 0$, $\frac{\partial g_{42}}{\partial x_3} = 0$, $\frac{\partial g_{42}}{\partial x_4} = 0$, and $\frac{\partial g_{42}}{\partial x_5} = 0$.

Fifth row of the Jacobian matrix $\nabla_x g_2(x)$: $\frac{\partial g_{52}}{\partial x_1} = \frac{1}{I_{yy}}\rho x_1 S_{\bar{C}}$, $\frac{\partial g_{52}}{\partial x_2} = 0$, $\frac{\partial g_{52}}{\partial x_3} = 0$, $\frac{\partial g_{52}}{\partial x_4} = 0$, and $\frac{\partial g_{52}}{\partial x_5} = 0$.

3.11.5 Computation of the feedback control gains

For the linearized system given by (1.36) the cost function of (1.37) is defined, where the coefficient r determines the penalization of the control input and the weight coefficient ρ determines the reward of the disturbances' effects. It is assumed that (i) the energy that is transferred from the disturbances signal $\tilde{d}(t)$ is bounded, that is $\int_0^\infty \tilde{d}^T(t)\tilde{d}(t)dt < \infty$; (ii) matrices $[A, B]$ and $[A, L]$ are stabilizable; (iii) matrix $[A, C]$

is detectable. In the case of a tracking problem, the optimal feedback control law is given by

$$u(t) = -Ke(t) \tag{3.415}$$

with $e = x - x_d$ to be the tracking error, and $K = \frac{1}{r}B^T P$ where P is a positive definite symmetric matrix. Matrix P is obtained from the solution of the Riccati equation

$$A^T P + PA + Q - P(\frac{2}{r}BB^T - \frac{1}{\rho^2}LL^T)P = 0 \tag{3.416}$$

where Q is a positive semi-definite symmetric matrix. The worst case disturbance is given by

$$\tilde{d}(t) = \frac{1}{\rho^2}L^T Pe(t) \tag{3.417}$$

The diagram of the control loop for the hypersonic vehicle is depicted if Figure 3.146. The global stability properties of the control method can be proven following the procedure presented in Section 1.1.2.5.

3.11.6 Simulation tests

The performance of the proposed nonlinear optimal (H-infinity) control for the hypersonic aerial vehicle has been tested through simulation experiments, which were obtained with the use of the small-angle-of-attack HSV state-space model. To

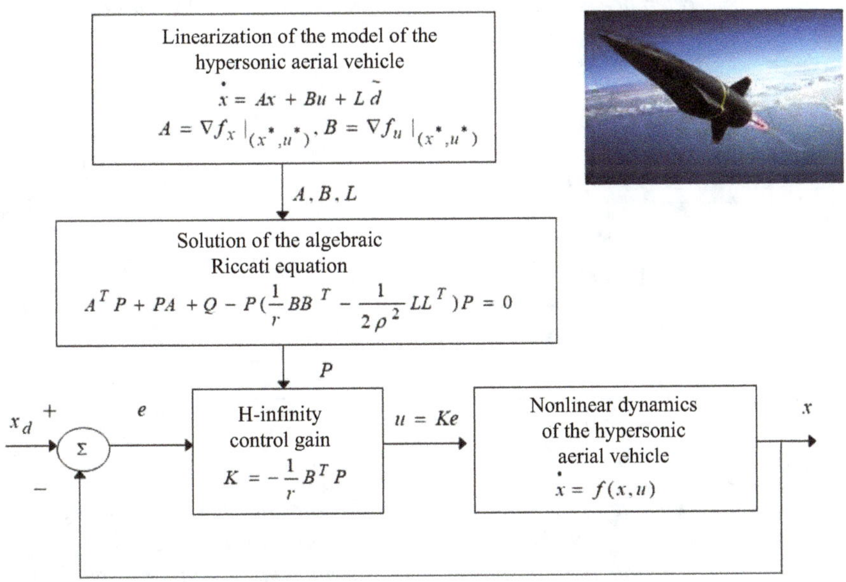

Figure 3.146 Diagram of the control scheme for the hypersonic aerial vehicle

implement the control method, the dynamic model of the hypersonic aerial vehicle was subject to iterative linearizations through Taylor series expansion that where taking place at each iteration of the control algorithm. Besides. to compute the feedback control gains, the algebraic Riccati equation of (3.416) had to be repetitively

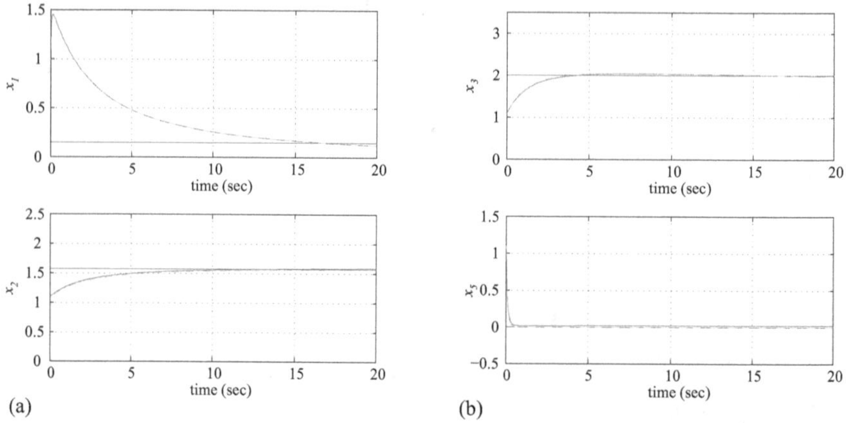

Figure 3.147 Tracking of setpoint 1 for the HSV state variables: (a) convergence of state variables $x_1 = V$ and $x_2 = \gamma$ (blue lines) to the reference setpoints (red lines) and their estimated values (green lines); (b) convergence of state variables $x_3 = h$ and $x_5 = q$ (blue lines) to the reference setpoints (red lines) and their estimated values (green lines)

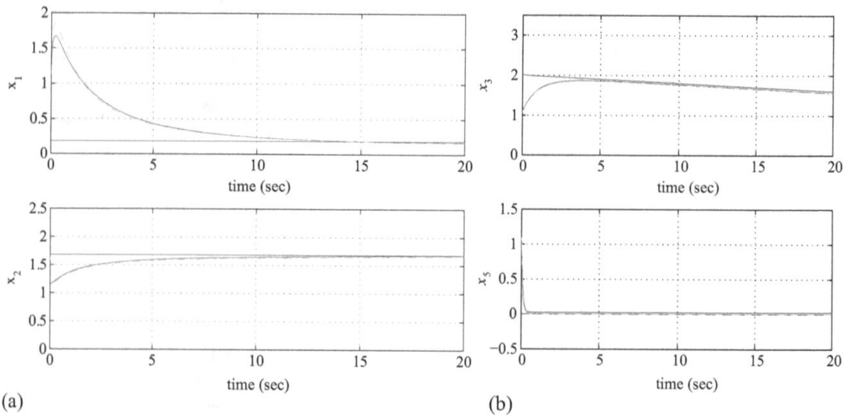

Figure 3.148 Tracking of setpoint 2 for the HSV state variables: (a) convergence of state variables $x_1 = V$ and $x_2 = \gamma$ (blue lines) to the reference setpoints (red lines) and their estimated values (green lines); (b) convergence of state variables $x_3 = h$ and $x_5 = q$ (blue lines) to the reference setpoints (red lines) and their estimated values (green lines)

solved at each time-step of the control scheme. As shown in Figure 3.147 to Figure 3.155, the proposed control approach achieved minimization of the tracking error for all its state variables while the control inputs varied within the allowed ranges.

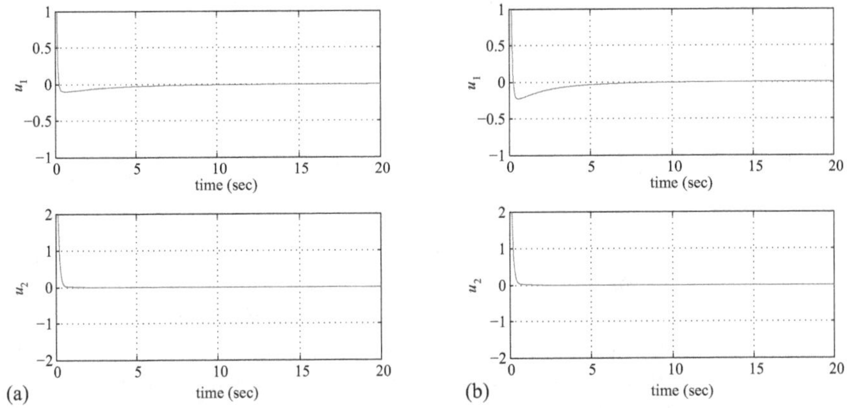

Figure 3.149 Variation of the control inputs $u_1 = \delta_e$ and $u_2 = \beta$ of the HSV: (a) when tracking of setpoint 1; (b) when tracking setpoint 2

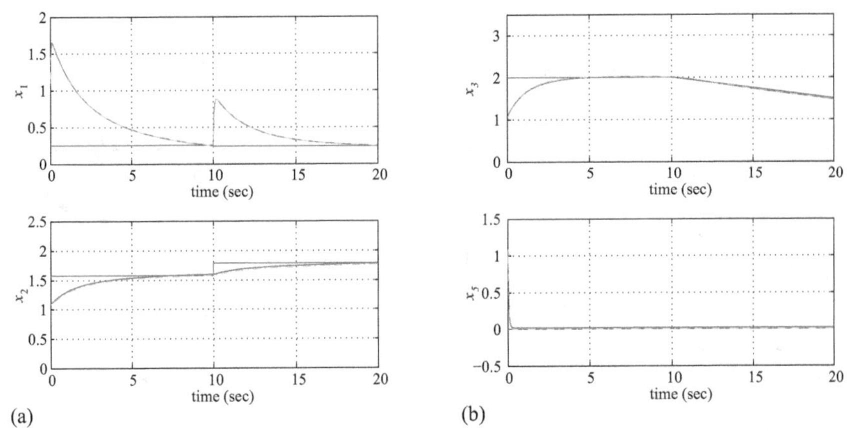

Figure 3.150 Tracking of setpoint 3 for the HSV state variables: (a) convergence of state variables $x_1 = V$ and $x_2 = \gamma$ (blue lines) to the reference setpoints (red lines) and their estimated values (green lines); (b) convergence of state variables $x_3 = h$ and $x_5 = q$ (blue lines) to the reference setpoints (red lines) and their estimated values (green lines)

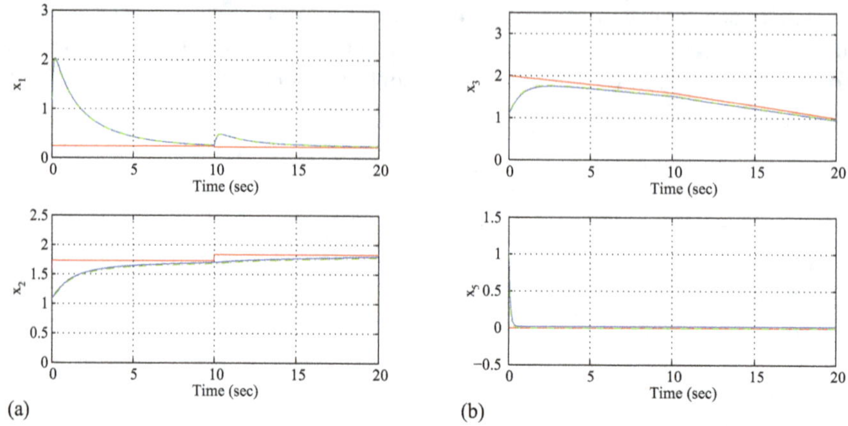

Figure 3.151 *Tracking of setpoint 4 for the HSV state variables: (a) convergence of state variables $x_1 = V$ and $x_2 = \gamma$ (blue lines) to the reference setpoints (red lines) and their estimated values (green lines); (b) convergence of state variables $x_3 = h$ and $x_5 = q$ (blue lines) to the reference setpoints (red lines) and their estimated values (green lines)*

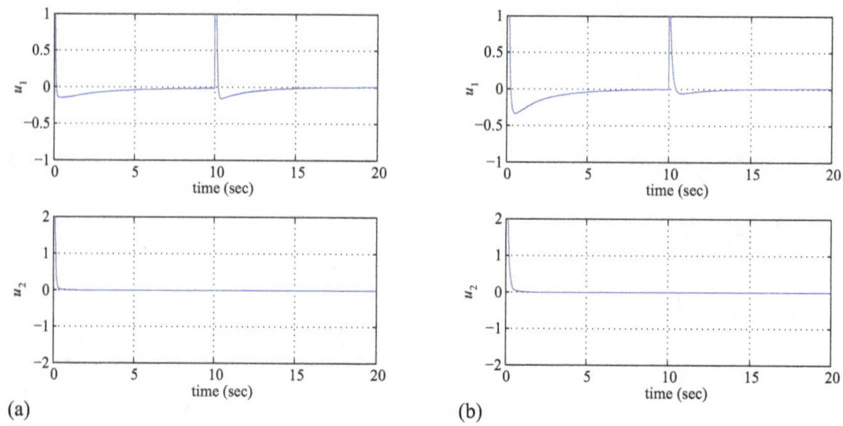

Figure 3.152 *Variation of the control inputs $u_1 = \delta_e$ and $u_2 = \beta$ of the HSV: (a) when tracking of setpoint 3; (b) when tracking setpoint 4*

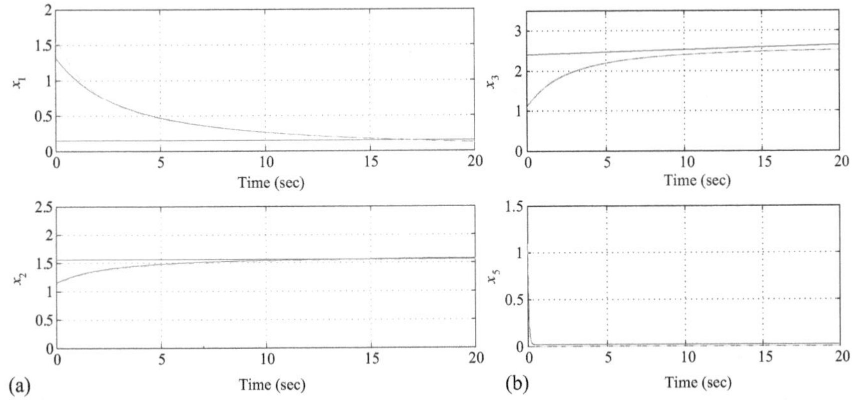

Figure 3.153 Tracking of setpoint 5 for the HSV state variables: (a) convergence of state variables $x_1 = V$ and $x_2 = \gamma$ (blue lines) to the reference setpoints (red lines) and their estimated values (green lines); (b) convergence of state variables $x_3 = h$ and $x_5 = q$ (blue lines) to the reference setpoints (red lines) and their estimated values (green lines)

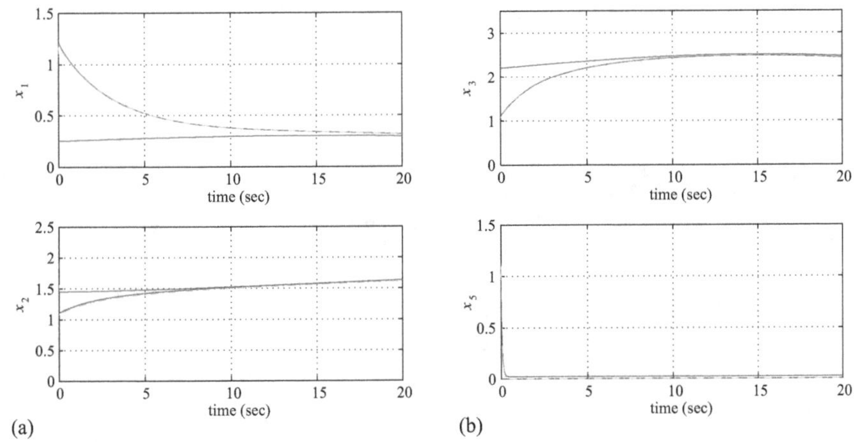

Figure 3.154 Tracking of setpoint 6 for the HSV state variables: (a) convergence of state variables $x_1 = V$ and $x_2 = \gamma$ (blue lines) to the reference setpoints (red lines) and their estimated values (green lines); (b) convergence of state variables $x_3 = h$ and $x_5 = q$ (blue lines) to the reference setpoints (red lines) and their estimated values (green lines)

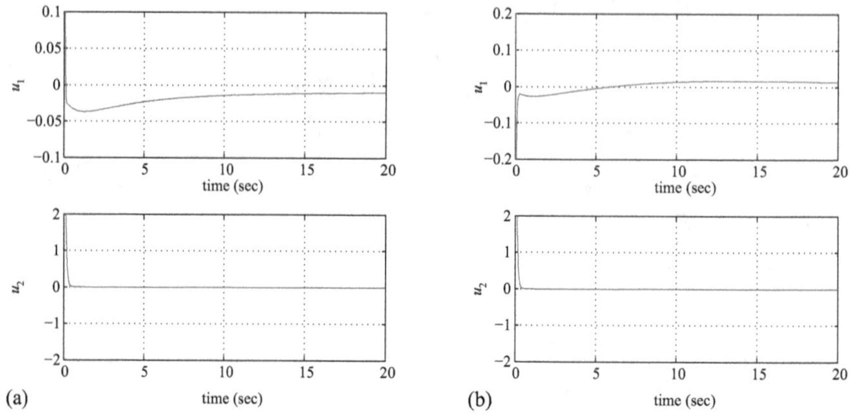

Figure 3.155 Variation of the control inputs $u_1 = \delta_e$ and $u_2 = \beta$ of the HSV: (a) when tracking of setpoint 5; (b) when tracking setpoint 6

Chapter 4
Control based on approximate linearization in energy conversion

Abstract

The chapter analyzes methods and presents results on control and estimation based on approximate linearization for energy conversion systems: (i) Nonlinear control of the VSI-fed three-phase PMSM, (ii) Nonlinear control of the VSI-fed six-phase PMSMs, (iii) Nonlinear control of DC electric microgrids, (iv) Nonlinear control of distributed marine power generation units, (v) Nonlinear control of PMLSGs in wave energy conversion systems, (vi) Nonlinear control of Permanent Magnet Brushless DC motors, (vii) Nonlinear optimal control of Hybrid Electric Vehicles powertrains, (viii) Nonlinear control of shipboard AC/DC microgrids and (ix) Nonlinear control of power generation in hybrid AC/DC microgrids.

4.1 Nonlinear control of the VSI-fed three-phase PMSM

4.1.1 Outline

Voltage Source Inverter-fed Permanent Magnet Synchronous Machines (VSI-PMSMs) are widely used in industrial actuation and mechatronic systems, in water pumping stations, as well as in the traction of transportation systems such as electric vehicles and electric trains (Rigatos, 2016b; Rigatos, 2015; Thounthong *et al.*, 2018; Kim, 2017; Kim and Lee, 2010; Wang, Chen *et al.*, 2016). The dynamic model of VSI-PMSMs is multivariable and exhibits complicated nonlinear dynamics (Morawiec, 2013; Xie, Lu *et al.*, 2016; Kim, Youn *et al.*, 2006; Chen, Yao *et al.*, 2019). The inverters' currents which are generated through a PWM process are used to control the stator currents of the PMSM which in turn control the rotational speed of this electric machine (Ge *et al.*, 2017; Zhou and Chen, 2018; Magri *et al.*, 2010; Madanzadeh *et al.*, 2020). So far, several nonlinear control schemes for VSI-PMSMs have been developed, having as primary objectives the precise tracking of setpoints by the system's state variables and the robustness to parametric changes or external perturbations (Cheema *et al.*, 2019; Li, Xia *et al.*, 2012; Wang, Chen *et al.*, 2015; Cai *et al.*, 2018; Linares-Flores *et al.*, 2015). To reduce dependence on sensor measurements for the VSI-PMSM control loops, sensorless control schemes have been also developed

(Wang, Zhang *et al.*, 2017; Raghavendra Gopinant and Shyama, 2019; Wang, Xu and Zou, 2019; Nahid-Mobarakeh *et al.*, 2007; Ramirez-Villalobos *et al.*, 2017). Optimal control schemes for VSI-PMSMs such as MPC and NMPC aim at minimizing also the variations of the control inputs, thus also reducing the consumption of energy by the control loop (Englert and Graicher, 2018; Wang, Fan *et al.*, 2018; Zhang, Xu *et al.*, 2018; Zhang, Zhang *et al.*, 2019; Nguyen and Jung, 2018). However, such control methods may come against limitations related with the nonlinear dynamics of the VSI-PMSM system and its functioning under variable operating conditions.

In the present section, the previously analyzed nonlinear optimal control method is proposed for Voltage Source Inverter-fed Permanent Magnet Synchronous Machines (Rigatos, Siano, Jovanovic *et al.*, 2017; Rigatos, Siano, Marignetti *et al.*, 2018; Rigatos, Siano and Cecati, 2015; Rigatos, Siano *et al.*, 2015). First, the dynamic model of the VSI-PMSM system undergoes approximate linearization around a temporary operating point which is recomputed at each time-step of the control method. This linearization point is defined by the present value of the system's state vector and the last sampled value of its control inputs vector. The linearization process is based on first-order Taylor series expansion and on the computation of the Jacobian matrices of the state-space model of the VSI-PMSM system (Rigatos and Tzafestas, 2007; Basseville and Nikiforov, 1993; Rigatos and Zhang, 2009). The resulting modelling error which is due to the truncation of higher-order terms in the Taylor series expansion is considered to be a perturbation which is asymptotically compensated by the robustness of the control method. For the approximately linearized model of the VSI-PMSM system, a stabilizing H-infinity feedback controller is designed.

The proposed H-infinity controller provides the solution to the optimal control problem for the VSI-PMSM system under model uncertainty and external perturbations (Rigatos, 2016b). Actually, this controller represents a min-max differential game taking place between (i) the control inputs which try to minimize a cost function that contains a quadratic term of the state vector's tracking error, (ii) the model uncertainty and exogenous disturbance terms which try to maximize this cost function. To select the feedback gains of the stabilizing feedback controller, an algebraic Riccati equation is repetitively solved at each time-step of the control algorithm (Rigatos, 2011; Rigatos and Busawon, 2018). To analyze the stability properties of the control scheme the Lyapunov method is used. It is proven that the VSI-PMSM loop has the H-infinity tracking performance property which signifies robustness against model uncertainty and disturbances (Rigatos, 2016b; Tousssaint, Basar and Bullo, 2000). Moreover, under moderate conditions the global asymptotic stability properties of this control scheme are proven (Rigatos, 2016b). The proposed control method achieves fast-tracking of reference setpoints by the VSI-PMSM state variables, while keeping also moderate the variations of the control inputs. The latter property indicates that energy consumption by the VSI-PMSM control loop can be minimized (Rigatos, 2016b; Gibbs, 2012).

The proposed nonlinear optimal control method for the VSI-PMSM system exhibits several advantages (Rigatos, 2016b; Rigatos and Busawon, 2018): Comparing to global linearization-based control methods, such as Lie algebra-based control or differential flatness theory-based control the nonlinear optimal control scheme avoids complicated state variables transformations (diffeomorphisms). Besides, its control inputs are applied directly to the initial nonlinear model of the VSI-PMSM system and thus inverse transformations and the related singularity problems are also avoided. Comparing to popular approaches for solving the optimal control problem, which are often met in the industry, such as Model Predictive Control (MPC) and Nonlinear Model Predictive Control (NMPC), the nonlinear optimal control method is of proven global stability and its convergence to an optimum is assured. On the other side, MPC is a linear control method which if applied to the nonlinear model of the VSI-PMSM system the stability of the control loop will be lost. Besides, the performance of NMPC's iterative search for an optimum is dependent on parameters values selection and initialization, and the method's convergence cannot be assured. Comparing to backstepping control, the nonlinear optimal control scheme does not require the state-space description of the controlled system to be found in the triangular (backstepping integral) form. Comparing to sliding-mode control there is no need to define in an often intuitive manner the sliding surfaces of the controlled system. It is noted that unless the controlled system is found in an input-output linearized form, there is no systematic procedure for defining its sliding surfaces. Finally, comparing to local models-based control the section's nonlinear optimal control method avoids linearization around multiple operating points and does not need the solution of multiple Riccati equations or LMIs. As a result of this, the nonlinear optimal control method is of less computational effort.

4.1.2 Dynamic model of the VSI-PMSM system

4.1.2.1 State-space model of the VSI-PMSM system

The diagram of the voltage Source Inverter-fed Permanent Magnet Synchronous Machine is shown in Figure 4.1. The dynamic model of the Permanent Magnet Synchronous Machine (PMSM) consists of the following equations (Rigatos, 2016b; Thounthong *et al.*, 2018)

$$\begin{aligned}
\dot{\theta} &= \omega_m \\
\dot{\omega}_m &= \frac{1}{J}(T_e - B\omega_m - T_m) \\
\frac{di_d}{dt} &= \frac{1}{L_d}(V_{L_d} - Ri_d + \omega_e L_q i_q) \\
\frac{di_q}{dt} &= \frac{1}{L_q}(v_q - Ri_q - \omega_e L_d i_d - \omega_m \Psi_m)
\end{aligned} \tag{4.1}$$

Variable θ is the turn angle of the machine's rotor and ω_m is the rotor's angular velocity. The electromagnetic torque which is generated by the machine is $T_e = pi_q(\Psi_m - (L_q - L_d)i_d)$ with p to denote the number of poles. The term $B\omega_m$ denotes friction effects, while the term T_m denotes torque due to an external mechanical load. Moreover, it holds that $\omega_e = p\omega_m$. The inductance coefficients in the dq reference frame are denoted as L_d and L_q, while the stator's resistance is R. The stator's currents expressed in the dq reference frame are i_d and i_q respectively. The term Ψ_m

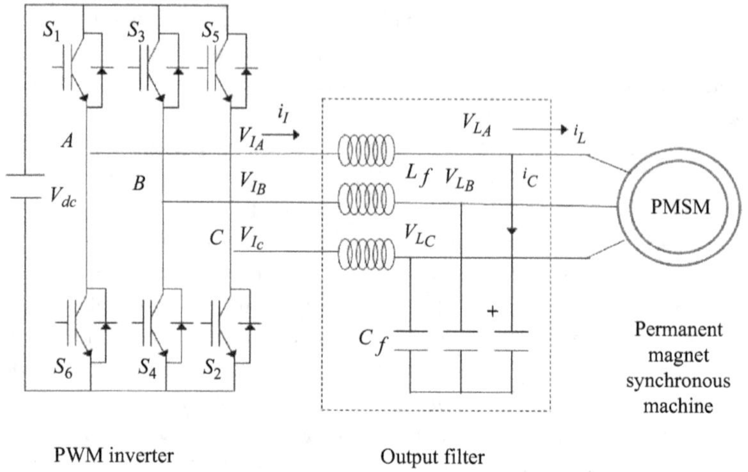

Figure 4.1 Diagram of the control scheme for the Voltage Source Inverter-fed Permanent Magnet Synchronous Machine (VSI-PMSM)

denotes magnetic flux due to the machine's permanent magnets. The external control voltages of the machine are v_d and v_q. The dynamics of the inverter are given by (Rigatos, 2016b; Kim and Lee, 2010)

$$\begin{aligned}
\frac{dV_{L,d}}{dt} &= \omega_0 V_{L,q} + \frac{1}{C_f} I_{I,d} - \frac{1}{C_f}\frac{p_f V_{L,d} + q_f V_{L,q}}{V_{L,d}^2 + V_{L,q}^2} + \omega_0 C_f V_{L,q} - \omega_0 L_f V_{L,q}\frac{(I_{I,d}^2 + I_{I,q}^2)}{V_{L,d}^2 + V_{L,q}^2} \\
\frac{dV_{L,q}}{dt} &= -\omega_0 V_{L,d} + \frac{1}{C_f} I_{I,q} - \frac{1}{C_f}\frac{p_f V_{L,q} - q_f V_{L,d}}{V_{L,d}^2 + V_{L,q}^2} - \omega_0 C_f V_{L,d} + \omega_0 L_f V_{L,d}\frac{(I_{I,d}^2 + I_{I,q}^2)}{V_{L,d}^2 + V_{L,q}^2} \\
\frac{dI_{L,d}}{dt} &= \omega_0 I_{L,q} - \frac{1}{L_f} V_{L,d} + \frac{1}{L_f} V_{I,d} \\
\frac{dI_{L,q}}{dt} &= \omega_0 I_{L,d} - \frac{1}{L_f} V_{L,q} + \frac{1}{L_f} V_{I,q}
\end{aligned} \tag{4.2}$$

Variables $V_{L,d}$, $V_{L,q}$ are voltages at the LC filter that is connected to the inverter and $I_{L,d}$, $I_{L,q}$ are the related currents, all expressed in the dq reference frame. The circuit elements of the filter are the inductance L_f and the capacitance C_f. The voltages and currents at the output of the inverter are denoted as $V_{I,d}$, $V_{I,q}$ and $I_{I,d}$, $I_{I,q}$, respectively. Coefficient ω_0 is related to the AC frequency. The active and reactive power of the inverter are given by (Rigatos, 2016b; Kim and Lee, 2010)

$$p_f = V_{L,d} I_{L,d} + V_{L,q} I_{L,q} \tag{4.3}$$

$$q_f = V_{L,q} I_{L,d} - V_{L,d} I_{L,q} - \omega_0 C_f (V_{L,d}^2 + V_{L,q}^2) + \omega_0 L_f (I_{I,d}^2 + I_{I,q}^2) \tag{4.4}$$

The state vector of the VSI-PMSM system is defined as: $x = [x_1, x_2, x_3, x_4, x_5, x_6, x_7, x_8]^T$ or $x = [\theta, \omega_m, i_d, i_q, V_{L,d}, V_{L,q}, I_{I,d}, I_{I,q}]^T$. The control input vector of the system is $u = [u_1, u_2]^T$ or $u = [V_{I,d}, V_{I,q}]^T$.

Considering that the torque of the external load is $T_L = mgl\sin(\theta)$ (that is the PMSM is used to lift a rod of mass m and length l) and using the above state variables notation, the state-space description of the system is:

$$
\begin{aligned}
\dot{x}_1 &= x_2 \\
\dot{x}_2 &= \frac{1}{J}[px_4(\Psi_m - (L_d - L_q)x_3)] - Bx_2 - mgl\sin(x_1) \\
\dot{x}_3 &= \frac{1}{L_d}(x_5 - Rx_3 + px_2L_qx_4) \\
\dot{x}_4 &= \frac{1}{L_q}(x_6 - Rx_4 - px_2L_dx_3 - x_2\Psi_m) \\
\dot{x}_5 &= \omega_0x_6 + \frac{1}{C_f}x_7 - \frac{1}{C_f}\frac{p_fx_5 + q_fx_6}{x_5^2 + x_6^2} + \omega_0C_fx_6 - \omega_0L_fx_6\frac{(x_7^2 + x_8^2)}{x_5^2 + x_6^2} \\
\dot{x}_6 &= -\omega_0x_5 + \frac{1}{C_f}x_8 - \frac{1}{C_f}\frac{p_fx_6 - q_fx_5}{x_5^2 + x_6^2} - \omega_0C_fx_5 + \omega_0L_fx_5\frac{(x_7^2 + x_8^2)}{x_5^2 + x_6^2} \\
\dot{x}_7 &= \omega_0x_8 - \frac{1}{L_f}x_5 + \frac{1}{L_f}u_1 \\
\dot{x}_8 &= \omega_0x_7 - \frac{1}{L_f}x_6 + \frac{1}{L_f}u_2
\end{aligned}
\tag{4.5}
$$

The state-space model of the VSI-PMSM system is also written in the following matrix form:

$$
\begin{pmatrix} \dot{x}_1 \\ \dot{x}_2 \\ \dot{x}_3 \\ \dot{x}_4 \\ \dot{x}_5 \\ \dot{x}_6 \\ \dot{x}_7 \\ \dot{x}_8 \end{pmatrix} =
\begin{pmatrix}
x_2 \\
\frac{1}{J}[px_4(\Psi_m - (L_d - L_q)x_3)] - Bx_2 - mgl\sin(x_1) \\
\frac{1}{L_d}(x_5 - Rx_3 + px_2L_qx_4) \\
\frac{1}{L_q}(x_6 - Rx_4 - px_2L_dx_3 - x_2\Psi_m) \\
\omega_0x_6 + \frac{1}{C_f}x_7 - \frac{1}{C_f}\frac{p_fx_5 + q_fx_6}{x_5^2 + x_6^2} + \omega_0C_fx_6 - \omega_0L_fx_6\frac{(x_7^2 + x_8^2)}{x_5^2 + x_6^2} \\
-\omega_0x_5 + \frac{1}{C_f}x_8 - \frac{1}{C_f}\frac{p_fx_6 - q_fx_5}{x_5^2 + x_6^2} - \omega_0C_fx_5 + \omega_0L_fx_5\frac{(x_7^2 + x_8^2)}{x_5^2 + x_6^2} \\
\omega_0x_8 - \frac{1}{L_f}x_5 \\
\omega_0x_7 - \frac{1}{L_f}x_6
\end{pmatrix}
+
\begin{pmatrix}
0 & 0 \\ 0 & 0 \\ 0 & 0 \\ 0 & 0 \\ 0 & 0 \\ 0 & 0 \\ \frac{1}{L_f} & 0 \\ 0 & \frac{1}{L_f}
\end{pmatrix}
\begin{pmatrix} u_1 \\ u_2 \end{pmatrix}
$$

where the active and reactive power coefficients of the system are given as

$$p_f = x_5I_{L,d} + x_6I_{L,q} \tag{4.6}$$

$$q_f = x_6I_{L,d} - x_5I_{L,q} - \omega_0C_f(x_5^2 + x_6^2) + \omega_0L_f(x_7^2 + x_8^2) \tag{4.7}$$

The state-space model of the inverter-PMSM system is thus written in the affine-in-the-input form

$$\dot{x} = f(x) + g(x)u \tag{4.8}$$

where $x \in R^{8\times 1}$, $f(x) \in R^{8\times 1}$, $g(x) \in R^{8\times 2}$ and $u \in R^{2\times 1}$.

4.1.2.2 Differential flatness properties of the VSI-PMSM model

It will be proven that the VSI-PMSM model is differentially flat, with flat outputs $y = [y_1, y_2]^T = [\theta, i_d]^T$. The differential flatness property for a dynamical system is proven by showing that (i) all state variables and their control inputs can be written as differential functions of a specific subset of its state variables, which stand for the system's flat outputs, (ii) the flat outputs of the system are differentially independent, which means that they are not connected between them through a relation in the form of a homogeneous linear differential equation (Rigatos, 2015).

From the 1st row of the state-space model, one has

$$x_2 = \dot{x}_1 \Rightarrow x_2 = h_2(y, \dot{y}) \tag{4.9}$$

which signifies that state variable x_2 is a differential function of the system's flat outputs. From the 2nd row of the state-space model, one has

$$\begin{aligned} \dot{x}_2 &= \frac{1}{J}[px_4(\Psi_m - (L_d - L_q)x_3)] - Bx_2 - mgl\sin(x_1) \Rightarrow \\ x_4 &= \frac{\dot{x}_2 + \frac{B}{J}x_2 + \frac{mgl}{J}\sin(x_1)}{\frac{p}{J}[\Psi_m - (L_d - L_q)x_3]} \Rightarrow \\ x_4 &= h_4(y, \dot{y}, \ddot{y}) \end{aligned} \tag{4.10}$$

because x_1, x_3 are elements of the flat outputs vector, and x_2 was proven to be a differential function of the system's flat outputs. The above relation signifies that state variable x_4 is a differential function of the system's flat outputs. From the third row of the state-space model one has:

$$\begin{aligned} \dot{x}_3 &= \frac{1}{L_d}(x_5 - Rx_3 + px_2L_qx_4) \Rightarrow \\ x_5 &= L_d\dot{x}_3 + Rx_3 - \mu x_2L_qx_4 \Rightarrow \\ x_5 &= h_5(y, \dot{y}, \ddot{y}) \end{aligned} \tag{4.11}$$

This signifies that state variable x_5 is also a differential function of the flat outputs of the system. From the fourth row of the state-space model one has:

$$\begin{aligned}
\dot{x}_4 &= \frac{1}{L_q}(x_6 - Rx_4 - px_2L_dx_3 - x_2\Psi_m) \Rightarrow \\
x_6 &= L_q\dot{x}_4 + Rx_4 + px_2L_dx_3 + x_2\Psi_m \Rightarrow \\
x_6 &= h_6(y, \dot{y}, \ddot{y}, y^{(3)})
\end{aligned} \tag{4.12}$$

This signifies also that state variable x_6 is a differential function of the system's flat outputs.

From the fifth row of the state-space model of the system one has:

$$\begin{aligned}
\dot{x}_5 &= \omega_0 x_6 + \frac{1}{C_f}x_7 - \frac{1}{C_f}\frac{p_f x_5 + q_f x_6}{x_5^2 + x_6^2} + \omega_0 C_f x_6 - \omega_0 L_f x_6 \frac{(x_7^2 + x_8^2)}{x_5^2 + x_6^2} \Rightarrow \\
x_5\dot{x}_5 &= \omega_0 x_5 x_6 + \frac{1}{C_f}x_5 x_7 - \frac{1}{C_f}x_5\frac{p_f x_5 + q_f x_6}{x_5^2 + x_6^2} + \omega_0 C_f x_5 x_6 - \omega_0 L_f x_5 x_6 \frac{(x_7^2 + x_8^2)}{x_5^2 + x_6^2}
\end{aligned} \tag{4.13}$$

From the sixth row of the state-space model of the system one has:

$$\begin{aligned}
\dot{x}_6 &= -\omega_0 x_5 + \frac{1}{C_f}x_8 - \frac{1}{C_f}\frac{p_f x_6 - q_f x_5}{x_5^2 + x_6^2} - \omega_0 C_f x_5 + \omega_0 L_f x_5 \frac{(x_7^2 + x_8^2)}{x_5^2 + x_6^2} \Rightarrow \\
x_6\dot{x}_6 &= -\omega_0 x_5 x_6 + \frac{1}{C_f}x_6 x_8 - \frac{1}{C_f}x_6\frac{p_f x_6 - q_f x_5}{x_5^2 + x_6^2} - \omega_0 C_f x_5 x_6 + \omega_0 L_f x_5 x_6 \frac{(x_7^2 + x_8^2)}{x_5^2 + x_6^2} \Rightarrow
\end{aligned} \tag{4.14}$$

By adding (4.13) and (4.14) one obtains

$$\begin{aligned}
x_5\dot{x}_5 + x_6\dot{x}_6 &= \frac{1}{C_f}x_5 x_7 + \frac{1}{C_f}x_6 x_8 - \frac{1}{C_f}x_5\frac{p_f x_5 + q_f x_6}{x_5^2 + x_6^2} - \frac{1}{C_f}x_6\frac{p_f x_6 - q_f x_5}{x_5^2 + x_6^2} \Rightarrow \\
x_5\dot{x}_5 + x_6\dot{x}_6 &= \frac{1}{C_f}x_5 x_7 + \frac{1}{C_f}x_6 x_8 - \frac{p_f}{C_f}
\end{aligned} \tag{4.15}$$

Thus one obtains the following two relations about x_7 and x_8

$$x_7 = \frac{x_5\dot{x}_5 + x_6\dot{x}_6 - \frac{1}{C_f}x_6 x_8 - \frac{p_f}{C_f}}{\frac{x_6}{C_f}} \tag{4.16}$$

$$x_8 = \frac{x_5\dot{x}_5 + x_6\dot{x}_6 - \frac{1}{C_f}x_5 x_7 + \frac{p_f}{C_f}}{\frac{x_5}{C_f}} \tag{4.17}$$

From (4.13) and (4.16) one has a binomial with respect to x_7. This proves that x_7 is a differential function of the flat output of the system y or

$$x_7 = h_7(y, \dot{y}, \ddot{y}, y^{(3)}, y^{(4)}) \tag{4.18}$$

From (4.14) and (4.17) one has a binomial with respect to x_8. This proves that x_8 is a differential function of the flat output of the system y or

$$x_8 = h_8(y, \dot{y}, \ddot{y}, y^{(3)}, y^{(4)}) \tag{4.19}$$

Besides, from the 7th row of the state-space model of the system one has that

$$\begin{aligned} \dot{x}_7 &= \omega_0 x_8 - \frac{1}{L_f}x_5 + \frac{1}{L_f}u_1 \Rightarrow \\ u_1 &= L_f[\dot{x}_7 - \omega_0 x_8 + \frac{1}{L_f}x_5] \end{aligned} \tag{4.20}$$

Moreover, from the 8th row of the state-space model of the system one has that

$$\begin{aligned} \dot{x}_8 &= -\omega_0 x_7 - \frac{1}{L_f}x_6 + \frac{1}{L_f}u_2 \Rightarrow \\ u_2 &= L_f[\dot{x}_8 + \omega_0 x_7 + \frac{1}{L_f}x_6] \end{aligned} \tag{4.21}$$

The above two relations signify that control inputs u_1 and u_2 are also differential functions of the flat outputs of the system. This completes the proof of the differential flatness properties of the VSI-PMSM model. The differential flatness property allows to define setpoints for all state variables, after selecting setpoints for the flat outputs x_1^d and x_3^d.

4.1.3 Approximate linearization of the inverter-PMSM dynamics

4.1.3.1 The approximately linearized model

The dynamic model of the inverter-PMSM system, being in the nonlinear form of (4.8) undergoes approximate linearization around the temporary operating point (x^*, u^*), where x^* is the present value of the system's state vector and u^* is the last sampled value of the control inputs vector. The linearization point is updated at each time-step of the control algorithm. This results in the linearized description

$$\dot{x} = Ax + Bu + \tilde{d} \tag{4.22}$$

where $\tilde{d}$ is the cumulative disturbances vector which comprises: (i) the modelling error due to the truncation of higher-order terms in the Taylor series expansion, (ii) the exogenous perturbations, and (iii) measurement noises of any distribution. Matrices A and B are computed from the system's Jacobians

$$\begin{aligned} A &= \nabla_x[f(x) + g(x)u] \mid_{(x^*,u^*)} \Rightarrow A = \nabla_x f(x) \mid_{(x^*,u^*)} \\ B &= \nabla_u[f(x) + g(x)u] \mid_{(x^*,u^*)} \Rightarrow B = g(x) \mid_{(x^*,u^*)} \end{aligned} \tag{4.23}$$

Computation of the elements of the Jacobian matrix $A = \nabla_x f(x) \mid_{(x^*,u^*)}$:

First row of the Jacobian matrix $A = \nabla_x f(x) \mid_{(x^*,u^*)}$: $\frac{\partial f_1}{\partial x_1} = 0$, $\frac{\partial f_1}{\partial x_2} = 0$, $\frac{\partial f_1}{\partial x_3} = 0$, $\frac{\partial f_1}{\partial x_4} = 0$, $\frac{\partial f_1}{\partial x_5} = 0$, $\frac{\partial f_1}{\partial x_6} = 0$, $\frac{\partial f_1}{\partial x_7} = 0$, and $\frac{\partial f_1}{\partial x_8} = 0$.

Second row of the Jacobian matrix $A = \nabla_x f(x) \mid_{(x^*,u^*)}$: $\frac{\partial f_2}{\partial x_1} = -\frac{1}{J}mgl\cos(x_1)$, $\frac{\partial f_2}{\partial x_2} = -\frac{B}{J}$, $\frac{\partial f_2}{\partial x_3} = -\frac{1}{J}px_4(L_d - L_q)$, $\frac{\partial f_2}{\partial x_4} = \frac{1}{J}p(\Psi_m - (L_d - L_q)x_3)$, $\frac{\partial f_2}{\partial x_5} = 0$, $\frac{\partial f_2}{\partial x_6} = 0$, $\frac{\partial f_2}{\partial x_7} = 0$, and $\frac{\partial f_2}{\partial x_8} = 0$.

Third row of the Jacobian matrix $A = \nabla_x f(x) \mid_{(x^*,u^*)}$: $\frac{\partial f_3}{\partial x_1} = 0$, $\frac{\partial f_3}{\partial x_2} = \frac{1}{L_d} p L_q x_4$, $\frac{\partial f_3}{\partial x_3} = -\frac{R}{L_d}$, $\frac{\partial f_3}{\partial x_4} = \frac{1}{L_d} p x_2 L_q$, $\frac{\partial f_3}{\partial x_5} = \frac{1}{L_d}$, $\frac{\partial f_3}{\partial x_6} = 0$, $\frac{\partial f_3}{\partial x_7} = 0$ and $\frac{\partial f_3}{\partial x_8} = 0$.

Fourth row of the Jacobian matrix $A = \nabla_x f(x) \mid_{(x^*,u^*)}$: $\frac{\partial f_4}{\partial x_1} = 0$, $\frac{\partial f_4}{\partial x_2} = \frac{-(pL_d x_3 + \Psi_m)}{L_q}$, $\frac{\partial f_4}{\partial x_3} = -\frac{p x_2 L_d}{L_q}$, $\frac{\partial f_4}{\partial x_4} = -\frac{R}{L_q}$, $\frac{\partial f_4}{\partial x_5} = 0$, $\frac{\partial f_4}{\partial x_6} = \frac{1}{L_q}$, $\frac{\partial f_4}{\partial x_7} = 0$, and $\frac{\partial f_4}{\partial x_8} = 0$.

Fifth row of the Jacobian matrix $A = \nabla_x f(x) \mid_{(x^*,u^*)}$: $\frac{\partial f_5}{\partial x_1} = 0$, $\frac{\partial f_5}{\partial x_2} = 0$, $\frac{\partial f_5}{\partial x_3} = 0$, $\frac{\partial f_5}{\partial x_4} = 0$,

$$\begin{aligned}
\frac{\partial f_5}{\partial x_5} &= -\frac{1}{C_f}\frac{p_f(x_5^2+x_6^2)-(p_f x_5+q_f x_6)2x_5}{(x_5^2+x_6^2)^2} + \frac{\omega_0 L_f x_6(x_7^2+x_8^2)2x_5}{x_5^2+x_6^{2^2}}, \\
\frac{\partial f_5}{\partial x_6} &= \omega_0 - \frac{1}{C_f}\frac{q_f(x_5^2+x_6^2)-(p_f x_5+q_f x_6)2x_6}{(x_5^2+x_6^2)^2} - \omega_0 C_f \\
&\quad + \omega_0 L_f \frac{[(x_7^2+x_8^2)(x_5^2+x_6^2) - x_6(x_7^2+x_8^2)2x_6]}{(x_5^2+x_6^2)^2} \\
\frac{\partial f_5}{\partial x_7} &= \frac{1}{C_f} - \frac{\omega_0 L_f x_6(2x_7)}{(x_5^2+x_6^2)} \text{ and } \frac{\partial f_5}{\partial x_8} = -\frac{\omega_0 L_f x_6(2x_5)}{(x_5^2+x_6^2)}.
\end{aligned}$$

Sixth row of the Jacobian matrix $A = \nabla_x f(x) \mid_{(x^*,u^*)}$: $\frac{\partial f_6}{\partial x_1} = 0$, $\frac{\partial f_6}{\partial x_2} = 0$, $\frac{\partial f_6}{\partial x_3} = 0$, $\frac{\partial f_6}{\partial x_4} = 0$,

$$\begin{aligned}
\frac{\partial f_6}{\partial x_5} &= -\omega_0 - \frac{1}{C_f}\frac{[(-q_f)(x_5^2+x_6^2)-(p_f x_6 - q_f x_5)(2x_5)]}{(x_5^2+x_6^2)^2} - \omega_0 C_f \\
&\quad + \frac{\omega_0 L_f[(x_7^2+x_8^2)(x_5^2+x_6^2) - x_5(x_7^2+x_8^2)(2x_5)]}{[(x_5^2+x_6^2)^2]}, \\
\frac{\partial f_6}{\partial x_6} &= -\frac{1}{C_f}\frac{[p_f(x_5^2+x_6^2)-(q_f x_6 - p_f x_5)(2x_6)]}{(x_5^2+x_6^2)^2} - \frac{\omega_0 L_f x_5(x_7^2+x_8^2)(2x_6)}{(x_5^2+x_6^2)^2}, \\
\frac{\partial f_6}{\partial x_7} &= \frac{\omega_0 L_f x_5(2x_7)}{(x_5^2+x_6^2)}, \text{ and } \frac{\partial f_6}{\partial x_8} = -\frac{1}{C_f} + \frac{\omega_0 L_f x_5(2x_5)}{(x_5^2+x_6^2)}
\end{aligned}$$

Seventh row of the Jacobian matrix $A = \nabla_x f(x) \mid_{(x^*,u^*)}$: $\frac{\partial f_7}{\partial x_1} = 0$, $\frac{\partial f_7}{\partial x_2} = 0$, $\frac{\partial f_7}{\partial x_3} = 0$, $\frac{\partial f_7}{\partial x_4} = 0$, $\frac{\partial f_7}{\partial x_5} = -\frac{1}{L_f}$, $\frac{\partial f_7}{\partial x_6} = 0$, $\frac{\partial f_7}{\partial x_7} = 0$, $\frac{\partial f_7}{\partial x_8} = \omega_0$.

Eighth row of the Jacobian matrix $A = \nabla_x f(x) \mid_{(x^*,u^*)}$: $\frac{\partial f_8}{\partial x_1} = 0$, $\frac{\partial f_8}{\partial x_2} = 0$, $\frac{\partial f_8}{\partial x_3} = 0$, $\frac{\partial f_8}{\partial x_4} = 0$, $\frac{\partial f_8}{\partial x_5} = 0$, $\frac{\partial f_8}{\partial x_6} = -\frac{1}{L_f}$, $\frac{\partial f_8}{\partial x_7} = \omega_0$, $\frac{\partial f_8}{\partial x_8} = \omega_0$.

4.1.3.2 Stabilizing feedback control

After linearization around its current operating point (x^*, u^*), the dynamic model of the VSI-PMSM system is written as (Rigatos, 2016b)

$$\dot{x} = Ax + Bu + d_1 \tag{4.24}$$

Parameter d_1 stands for the linearization error in the VSI-PMSM's model appearing previously in (4.24). The reference setpoints for the VSI-PMSM's state vector are denoted by $\mathbf{x_d} = [x_1^d, \cdots, x_8^d]$. Tracking of this trajectory is achieved after applying the control input u_d. At every time instant, the control input u_d is assumed to differ from the control input u appearing in (4.24) by an amount equal to Δu, that is $u_d = u + \Delta u$

$$\dot{x}_d = Ax_d + Bu_d + d_2 \tag{4.25}$$

The dynamics of the controlled system described in (4.24) can be also written as

$$\dot{x} = Ax + Bu + Bu_d - Bu_d + d_1 \tag{4.26}$$

and by denoting $d_3 = -Bu_d + d_1$ as an aggregate disturbance term one obtains

$$\dot{x} = Ax + Bu + Bu_d + d_3 \tag{4.27}$$

By subtracting (4.25) from (4.27) one has

$$\dot{x} - \dot{x}_d = A(x - x_d) + Bu + d_3 - d_2 \tag{4.28}$$

By denoting the tracking error as $e = x - x_d$ and the aggregate disturbance term as $\tilde{d} = d_3 - d_2$, the tracking error dynamics becomes

$$\dot{e} = Ae + Bu + \tilde{d} \tag{4.29}$$

For the approximately linearized model of the system, a stabilizing feedback controller is developed. The controller has the form

$$u(t) = -Ke(t) \tag{4.30}$$

with $K = \frac{1}{r}B^T P$ where P is a positive definite symmetric matrix which is obtained from the solution of the Riccati equation (Rigatos, 2016b)

$$A^T P + PA + Q - P\left(\frac{2}{r}BB^T - \frac{1}{\rho^2}LL^T\right)P = 0 \tag{4.31}$$

where Q is a positive semi-definite symmetric matrix. The previously analyzed concept about the nonlinear optimal control loop for the VSI-PMSM system is given in Figure 4.2. The global stability properties of the control method can be proven following the procedure presented in Section 1.10.2.5.

Whereas the Linear Quadratic Regulator (LQR) is the solution of the quadratic optimal control problem using Bellman's optimality principle, H-infinity control is the solution of the optimal control problem under model uncertainty and external perturbations. The cost function that is subject to minimization in the case of LQR

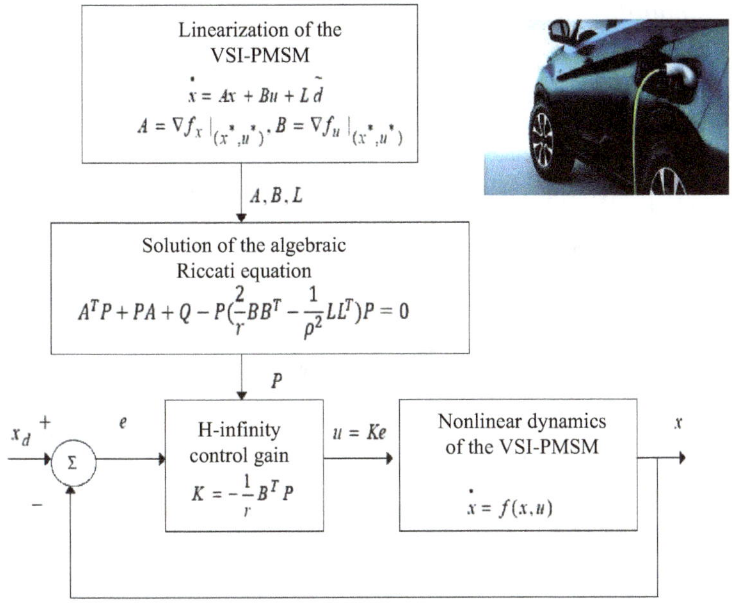

Figure 4.2 Diagram of the control scheme for the Voltage Source Inverter-fed PMSM (with potential application to an electric vehicle)

comprises a quadratic term of the state vector's tracking error, as well as a quadratic term of the variations of the control inputs. In the case of H-infinity control the cost function is extended with the inclusion of a quadratic term of the cumulative disturbance and model uncertainty inputs that affect the model of the controlled system. In the case of the VSI-PMSM, the model is nonlinear and is also affected by uncertainty and external perturbations. By applying approximate linearization to the VSI-PMSM's dynamics one obtains a linear state-space description which is subject to modelling imprecision and exogenous disturbances. Therefore, one can arrive at a solution of the related optimal control problem only by applying the H-infinity control approach.

It is also noted that the solution of the H-infinity feedback control problem for the VSI-PMSM and the computation of the worst-case disturbance that this controller can sustain, comes from superposition of Bellman's optimality principle when considering that the VSI-PMSM system is affected by two separate inputs (i) the control input u, (ii) the cumulative disturbance input $\tilde{d}(t)$. Solving the optimal control problem for u that is for the minimum variation (optimal) control input that achieves elimination of the state vector's tracking error gives $u = -\frac{1}{r}B^T Pe$. Equivalently, solving the optimal control problem for $\tilde{d}$, that is for the worst case disturbance that the control loop can sustain gives $\tilde{d} = \frac{1}{\rho^2}L^T Pe$.

4.1.4 Simulation tests

The performance of the proposed nonlinear optimal (H-infinity) control scheme for VSI-fed PMSMs has been further confirmed through simulation experiments. To compute the stabilizing H-infinity feedback control, the algebraic Riccati equation of (4.31) had to be iteratively solved at each time-step of the control algorithm. The obtained results are depicted in Figures 4.3 to 4.14 and verify that fast and accurate

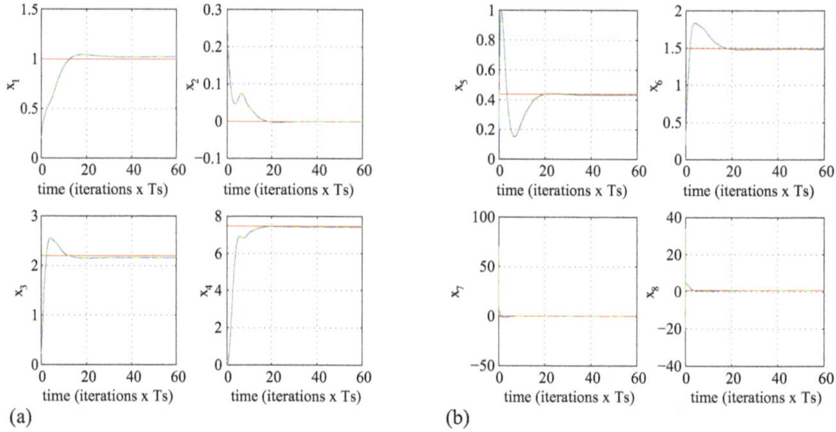

Figure 4.3 Tracking of setpoint 1 for the VSI-PMSM system (a) convergence of state variables x_1 to x_4 to their reference setpoints (red line: setpoint, blue line: real value, green line: estimated value), (b) convergence of state variables x_5 to x_8 to their reference setpoints (red line: setpoint, blue line: real value, green line: estimated value)

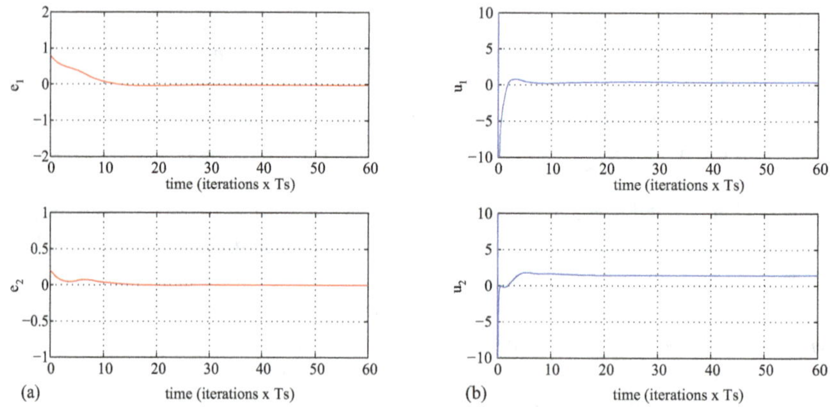

Figure 4.4 Tracking of setpoint 1 for the VSI-PMSM system (a) tracking error for state variables x_1 (turn angle of the PMSM) and x_2 (angular velocity of the PMSM), (b) control inputs u_1 (inverter's voltage $V_{I,d}$) and u_2 (inverter's voltage $V_{I,q}$)

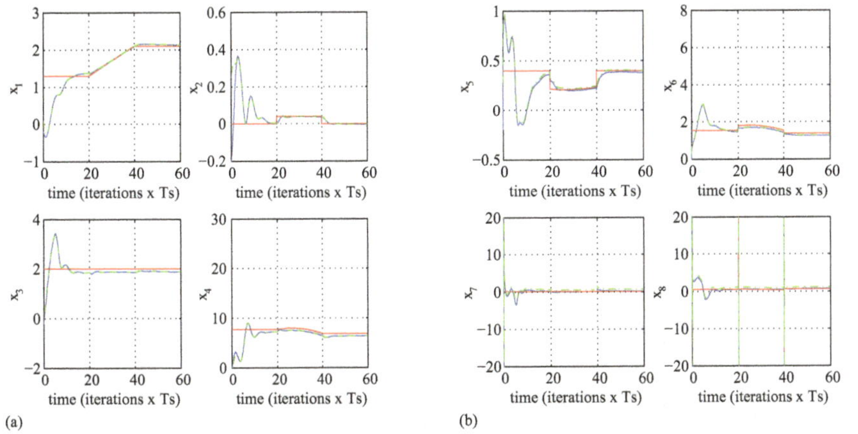

Figure 4.5 Tracking of setpoint 2 for the VSI-PMSM system (a) convergence of state variables x_1 to x_4 to their reference setpoints (red line: setpoint, blue line: real value, green line: estimated value), (b) convergence of state variables x_5 to x_8 to their reference setpoints (red line: setpoint, blue line: real value, green line: estimated value)

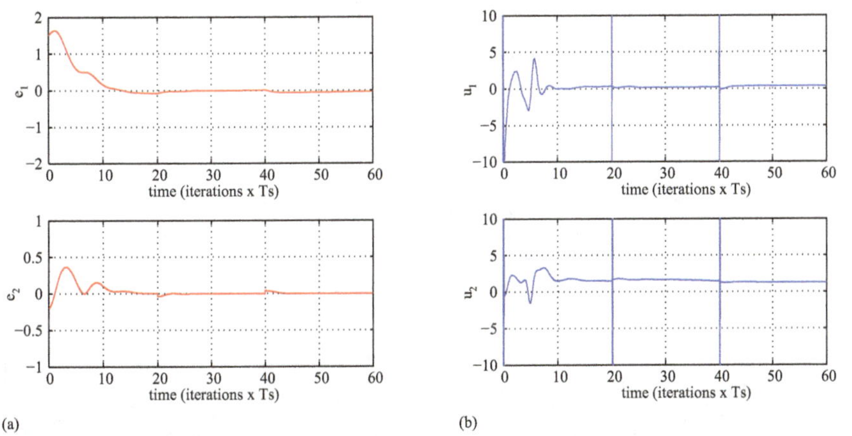

Figure 4.6 Tracking of setpoint 2 for the VSI-PMSM system (a) tracking error for state variables x_1 (turn angle of the PMSM) and x_2 (angular velocity of the PMSM), (b) control inputs u_1 (inverter's voltage $V_{I,d}$) and u_2 (inverter's voltage $V_{I,q}$))

tracking is achieved for all state variables of the VSI-PMSM system under moderate variations of the control inputs. The real values of the state variables are depicted in blue, the estimated values provided by the H-infinity Kalman Filter are printed in green, while the related setpoints are shown in red.

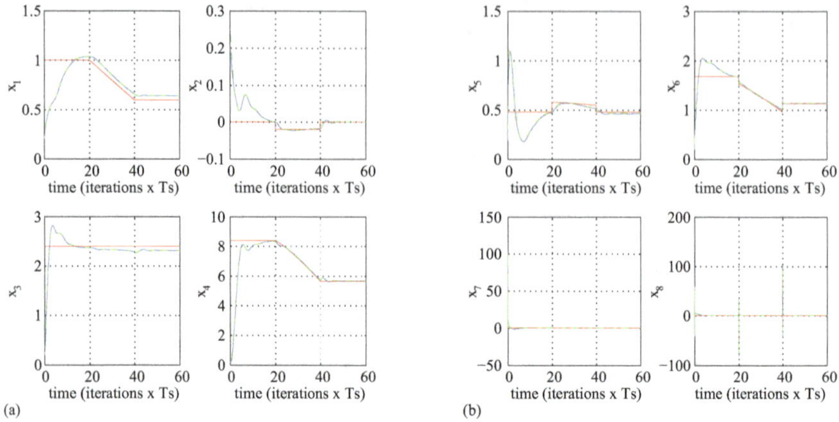

Figure 4.7 Tracking of setpoint 3 for the VSI-PMSM system (a) convergence of state variables x_1 to x_4 to their reference setpoints (red line: setpoint, blue line: real value, green line: estimated value), (b) convergence of state variables x_5 to x_8 to their reference setpoints (red line: setpoint, blue line: real value, green line: estimated value)

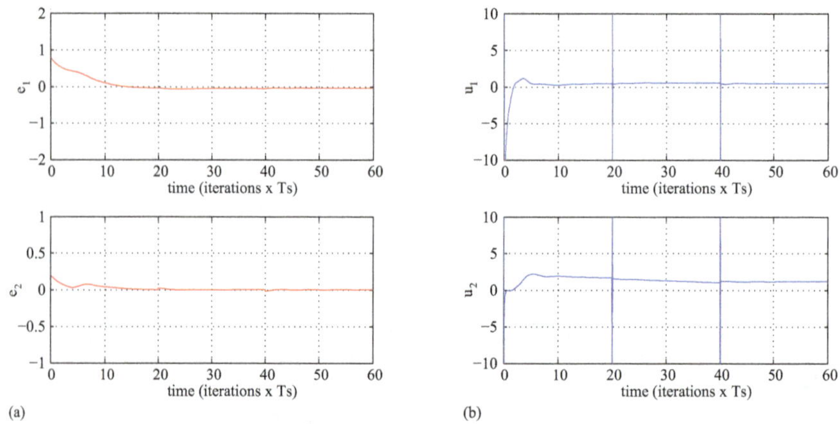

Figure 4.8 Tracking of setpoint 3 for the VSI-PMSM system (a) tracking error for state variables x_1 (turn angle of the PMSM) and x_2 (angular velocity of the PMSM), (b) control inputs u_1 (inverter's voltage $V_{I,d}$) and u_2 (inverter's voltage $V_{I,q}$))

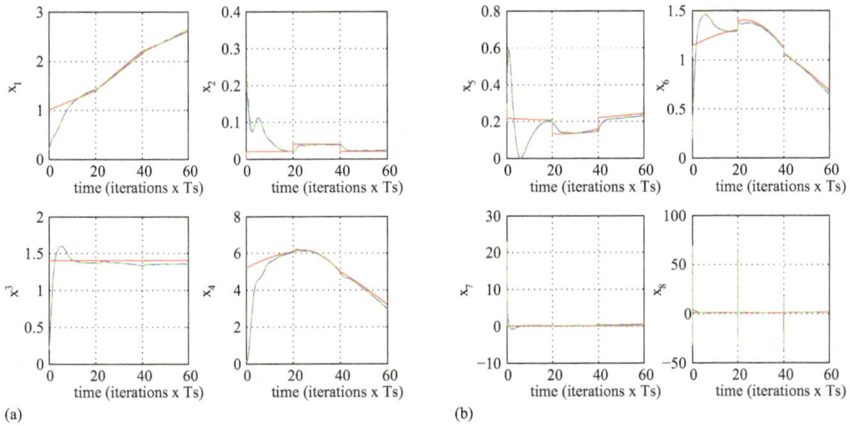

Figure 4.9 *Tracking of setpoint 4 for the VSI-PMSM system (a) convergence of state variables x_1 to x_4 to their reference setpoints (red line: setpoint, blue line: real value, green line: estimated value), (b) convergence of state variables x_5 to x_8 to their reference setpoints (red line: setpoint, blue line: real value, green line: estimated value)*

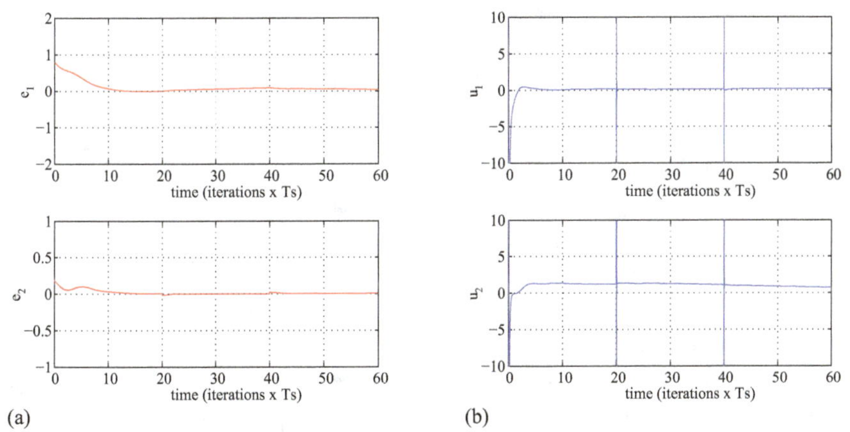

Figure 4.10 *Tracking of setpoint 4 for the VSI-PMSM system (a) tracking error for state variables x_1 (turn angle of the PMSM) and x_2 (angular velocity of the PMSM), (b) control inputs u_1 (inverter's voltage $V_{I,d}$) and u_2 (inverter's voltage $V_{I,q}$))*

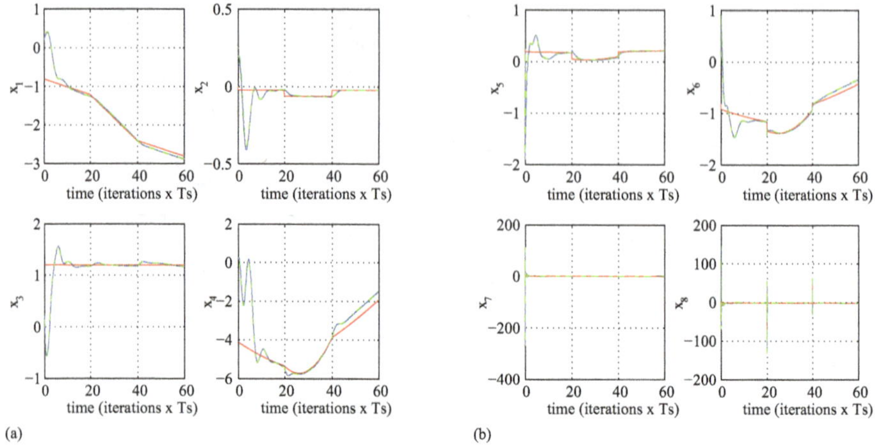

Figure 4.11 Tracking of setpoint 5 for the VSI-PMSM system (a) convergence of state variables x_1 to x_4 to their reference setpoints (red line: setpoint, blue line: real value, green line: estimated value), (b) convergence of state variables x_5 to x_8 to their reference setpoints (red line: setpoint, blue line: real value, green line: estimated value)

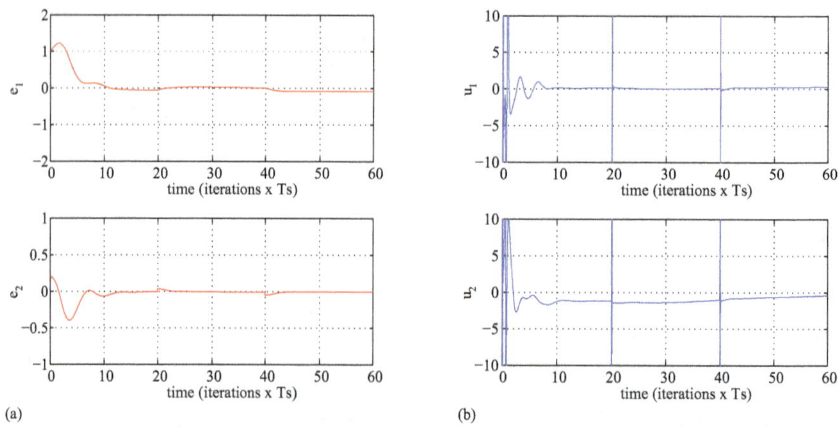

Figure 4.12 Tracking of setpoint 5 for the VSI-PMSM system (a) tracking error for state variables x_1 (turn angle of the PMSM) and x_2 (angular velocity of the PMSM), (b) control inputs u_1 (inverter's voltage $V_{I,d}$) and u_2 (inverter's voltage $V_{I,q}$)

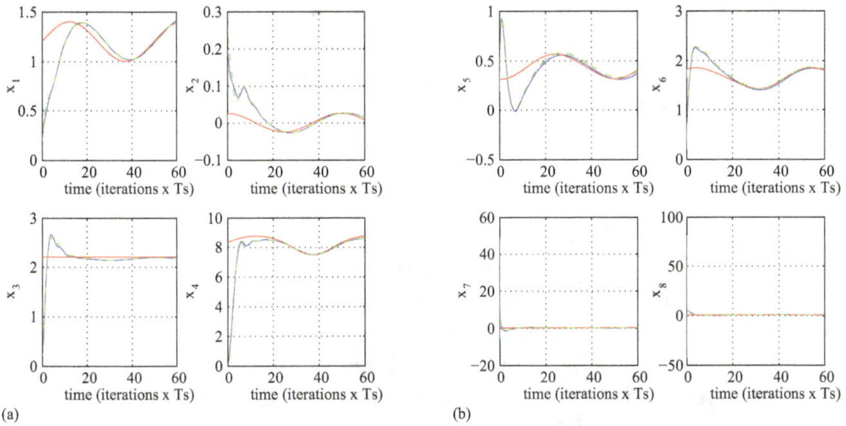

Figure 4.13 Tracking of setpoint 6 for the VSI-PMSM system (a) convergence of state variables x_1 to x_4 to their reference setpoints (red line: setpoint, blue line: real value, green line: estimated value), (b) convergence of state variables x_5 to x_8 to their reference setpoints (red line: setpoint, blue line: real value, green line: estimated value)

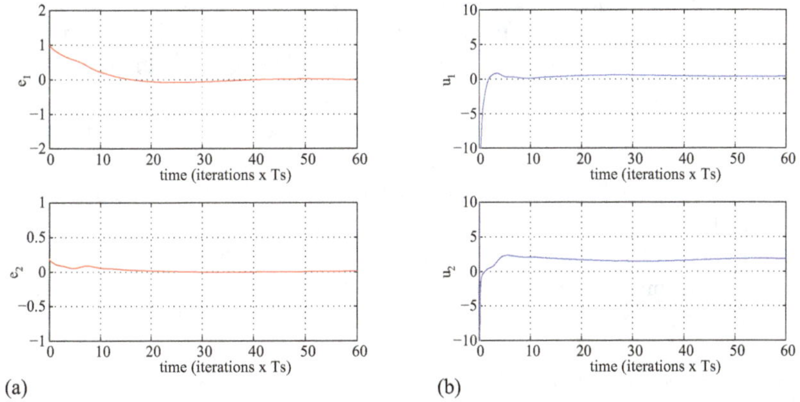

Figure 4.14 Tracking of setpoint 6 for the VSI-PMSM system (a) tracking error for state variables x_1 (turn angle of the PMSM) and x_2 (angular velocity of the PMSM), (b) control inputs u_1 (inverter's voltage $V_{I,d}$) and u_2 (inverter's voltage $V_{I,q}$)

4.2 Nonlinear control of VSI fed six-phase PMSMs

4.2.1 Outline

Multi-phase AC machines have specific advantages comparing to three-phase machines (Rigatos, 2016b). For instance, they have increased efficiency because of producing more power, they exhibit a balanced power distribution per phase, and have smooth electromagnetic torque pulsation (Liu, Guo *et al.*, 2017; Lin, Hang *et al.*, 2013; Zhou and Chen, 2018). Moreover, multi-phase motors have reduced losses and reduced magnetic flux harmonics. They exhibit also improved reliability and fault tolerance because they can remain functional even when faults affect a couple of their phases (Karttunen *et al.*, 2016; Yuan, Chen *et al.*, 2016; Lup and Liu, 2020). Efficient control of multi-phase machines through inverters can further improve their performance and this enables to avoid undesirable effects such as large harmonics of stator currents (Rigatos, 2015; Kim and Lee, 2010; Luo and Liu, 2019)

Multi-phase machines have been applied in many fields such as trains and electric vehicles traction, ships propulsion and actuation in aircrafts. They can also find use in high-speed elevators and in numerically controlled machining, while multi-phase generators find use in several renewable power installations (Bermudez *et al.*, 2021; Gocamves *et al.*, 2021; Duran *et al.*, 2016). In the last years, growing attention is being paid to six-phase Permanent Magnet Synchronous Machines (six-phase PMSM) under parallel control by two inverters (Geng, Lai *et al.*, 2019; Xian, Zhang *et al.*, 2020; Hu, Xiong *et al.*, 2020). Six-phase PMSMs are characterized by two windings in one stator, while the currents of each winding are controlled by separate inverters (Suzuki *et al.*, 2019; Xu, Zhang, Fang *et al.*, 2020). Actually, a frequent configuration is that of the asymmetrical six-phase windings which consists of two three-phase windings at the stator with a 30° shift between them (He *et al.*, 2020; Lin, Huang *et al.*, 2013). Recently, there has been an upsurge of the interest in six-phase PMSMs because of their potential use in Electric Vehicles (EVs) and Hybrid Electric Vehicles (HEVs) traction systems (Yung, Schofied *et al.*, 2019; Lu, Li *et al.*, 2019).

In this section, a nonlinear optimal (H-infinity) controller is developed for the dynamic model of the VSI-fed six-phase PMSM (Rigatos, 2016b; Rigatos, Siano, Jovanovic *et al.*, 2017; Rigatos, Siano, Marignetti *et al.*, 2018; Rigatos, Siano and Cecati, 2015; Rigatos, Siano *et al.*, 2015). At the first stage, the state-space model of the VSI-fed six-phase PMSM undergoes approximate linearization around a temporary operating point which is recomputed at each sampling instance. This linearization point is re-defined at each iteration of the control algorithm by the present value of the system's state vector and by the last sampled value of the control inputs vector. The linearization process is based on first-order Taylor series expansion and on the computation of the associated Jacobian matrices (Rigatos and Tzafestas, 2007; Basseville and Nikiforov, 1993; Rigatos and Zhang, 2009). The modelling error which is due to the truncation of higher-order terms in the Taylor series expansion is considered to be a perturbation that is asymptotically compensated by the robustness of the control method. For the approximately linearized model of the VSI-fed six-phase PMSM a stabilizing H-infinity (optimal) feedback controller is designed.

Actually, this H-infinity controller represents a min-max differential game taking place between (i) the control inputs which try to minimize a quadratic cost function containing a quadratic term of the state-vector's tracking error, (ii) the model uncertainty and exogenous perturbation terms which try to maximize this cost function. To compute the stabilizing feedback gains of this controller, an algebraic Riccati equation is repetitively solved at each sampling period of the control method (Rigatos, 2011; Rigatos and Busawon, 2018). The global stability properties of the control scheme are proven through Lyapunov analysis (Rigatos, 2016b; Rigatos, Siano, Jovanovic *et al.*, 2017; Rigatos, Siano, Marignetti *et al.*, 2018; Rigatos, Siano and Cecati, 2015; Rigatos, Siano *et al.*, 2015). First, it is demonstrated that the control loop satisfies the H-infinity tracking performance criterion. This signifies the control method's robustness against modelling imprecision and exogenous perturbations (Rigatos, 2016b; Tousssaint, Basar and Bullo, 2000). Additionally, under moderate conditions the global asymptotic stability of the control loop was proven. Moreover, by confirming that the VSI-fed six-phase PMSM system is differentially flat, a solution has been also reached for the setpoints definition problem for this system (Rigatos, 2016b; Rigatos, 2015). Furthermore, by using the H-infinity Kalman Filter as a robust state estimator it has been enabled to implement state estimation-based control for the VSI-fed six-phase PMSM without needing to measure the entire state vector of this system. As an outline of the performance of the control method, it can be stated that the nonlinear optimal controller achieves fast and accurate tracking of reference setpoints while it also avoids excessive variation of the control inputs.

Preceding results on the use of H-infinity control to nonlinear dynamical systems were limited to the case of affine-in-the-input systems with drift-only dynamics and considered that the control inputs gain matrix is not dependent on the values of the system's state vector. Moreover, in these approaches the linearization was performed around points of the desirable trajectory whereas in the present section's control method the linearization points are related with the value of the state vector at each sampling instance as well as with the last sampled value of the control inputs vector. The Riccati equation which has been proposed for computing the feedback gains of the controller is novel, so is the presented global stability proof through Lyapunov analysis. The section's approach enables a solution of the nonlinear optimal control problem without the need to solve State Dependent Riccati Equations (SDREs). It is noteworthy that the solution of SDREs is ensured only in the case of Linear Parameter Varying (LPV) systems, while the section's approach is not limited to a specific class of nonlinear dynamical systems.

The advantages from the application of the proposed control method are outlined as follows: (i) it relies on approximate linearization of the state-space model of the VSI-fed six-phase PMSM which makes use of the Taylor series expansion concept. Thus, complicated state variable transformations (diffeomorphisms) which appear in global linearization-based control approaches are avoided to a great extent, (ii) unlike popular methods for implementing optimal control in industry, such as MPC and NMPC, the nonlinear optimal control method is of proven global stability and of assured convergence to an optimum. By minimizing the variation of the control inputs, this control method achieves also minimization of energy consumption by the control loop of the VSI-fed six-phase PMSM system, (iii) unlike multiple-models-based control with a fuzzy decomposition of a system's nonlinear dynamics into multiple local linear models the section's approach does need to solve multiple Riccati equations or LMIs and remains computationally simple.

4.2.2 Dynamic model of the VSI-fed six-phase PMSM

4.2.2.1 State-space model of the six-phase PMSM

The diagram of the VSI-fed six-phase Permanent Magnet Synchronous Motor is shown in Figure 4.15. There are two three-phase systems, the first one is denoted as *abc* and the second one is denoted as *xyz* and results from the first system by a 30° rotation, as shown in Figure 4.16. To describe the dynamics of the 6-phase PMSM, the synchronously rotating *dq* reference frame is used (Rigatos, 2016b). The dynamic model of the six-phase PMSM (in a dual Y and shift by 30° windings configuration) consists of the stator voltage equation, the stator linkage equation, the electromagnetic torque equation and the mechanical rotation equation (Liu, Guo *et al.*, 2017).

To transform the dynamic model of the multi-phase PMSM from the *abc* and *xyz* reference frames into the $dq1$ and $dq2$ reference frames the following Park transformations are applied:

$$T_{dq1} = \frac{2}{3}\begin{pmatrix} \cos(\theta_e) & \cos(\theta_e - 120^\circ) & \cos(\theta_e + 120^\circ) \\ \sin(\theta_e) & \sin(\theta_e - 120^\circ) & \sin(\theta_e + 120^\circ) \\ \frac{1}{2} & \frac{1}{2} & \frac{1}{2} \end{pmatrix} \tag{4.32}$$

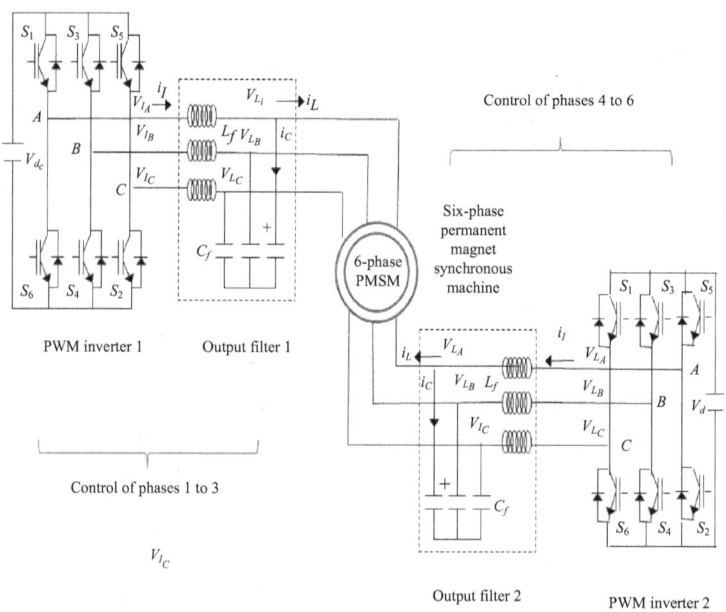

Figure 4.15 Diagram of the VSI-fed six-phase Permanent Magnet Synchronous Motor

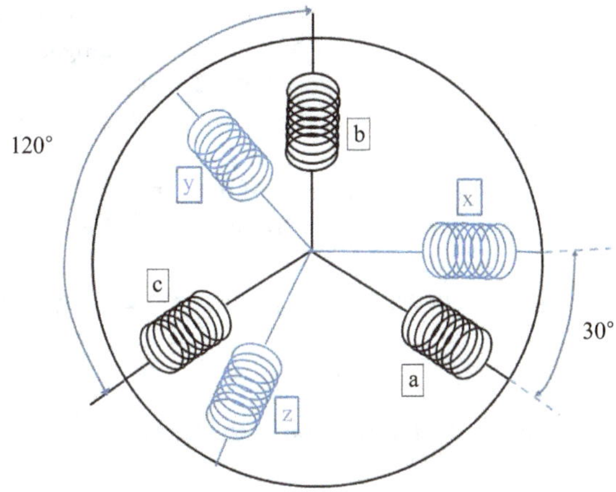

Figure 4.16 Three-phase frames in the stator of the six-phase PMSM

$$T_{dq2} = \frac{2}{3}\begin{pmatrix} \cos(\theta_e - 30^\circ) & \cos(\theta_e - 150^\circ) & \cos(\theta_e + 90^\circ) \\ \sin(\theta_e - 30^\circ) & \sin(\theta_e - 150^\circ) & \sin(\theta_e + 90^\circ) \\ \frac{1}{2} & \frac{1}{2} & \frac{1}{2} \end{pmatrix} \tag{4.33}$$

where T_{dq1} is the transformation leading from the *abc* reference frame to the $dq1$ reference frame, T_{dq2} is the transformation leading from the *xyz* reference frame to the $dq2$ reference frame, and θ_e is the rotor flux angle.

The stator voltage equations are (Liu, Guo *et al.*, 2017)

$$u_{d_1} = R_s i_{d_1} + \frac{d\psi_{d_1}}{dt} - n_p\omega\psi_{q_1} \tag{4.34}$$

$$u_{q_1} = R_s i_{q_1} + \frac{d\psi_{q_1}}{dt} + n_p\omega\psi_{d_1} \tag{4.35}$$

$$u_{d_2} = R_s i_{d_2} + \frac{d\psi_{d_2}}{dt} - n_p\omega\psi_{q_2} \tag{4.36}$$

$$u_{q_2} = R_s i_{q_2} + \frac{d\psi_{q_2}}{dt} + n_p\omega\psi_{d_2} \tag{4.37}$$

The stator linkage equation is (Liu, Guo *et al.*, 2017)

$$\psi_{d_1} = L_d i_{d_1} + L_{m_d} i_{d_2} + \psi_f \tag{4.38}$$

$$\psi_{q_1} = L_q i_{q_1} + L_{m_q} i_{q_2} \tag{4.39}$$

$$\psi_{d_2} = L_{m_d} i_{d_1} + L_d i_{q_2} + \psi_f \tag{4.40}$$

$$\psi_{q_2} = L_{m_q} i_{q_1} + L_q i_{q_2} \tag{4.41}$$

where u_{d_1}, u_{q_1} are the dq axis voltages of the first set of the stator windings, u_{d_2}, u_{q2} are the dq axis voltages of the second set of the stator windings, ψ_{d_1}, ψ_{q_1} are the dq axis magnetic flux components of the first set of the stator windings, ψ_{d_2}, ψ_{q_2} are the dq axis magnetic flux components of the second set of the stator windings, L_d, L_q are the dq axis components of the inductance of the windings circuits, L_{m_d}, L_{m_q} are the dq axis components of the mutual inductance of the windings circuits, R_s is the stator resistance, n_p is the number of pole pairs, ψ_f is the flux linkage of the permanent magnet.

The electromagnetic torque equation is given by (Liu, Guo *et al.*, 2017):

$$T_e = \frac{3n_p}{2}(\psi_{d_1} i_{q_1} + \psi_{d_2} i_{q_2} - \psi_{q_1} i_{d_1} - \psi_{q_2} i_{d_2}) \tag{4.42}$$

The mechanical equation of the turn motion of the motor is given by

$$J\frac{d\omega}{dt} = T_e - T_L - B_w\omega \tag{4.43}$$

where T_e is the electromagnetic torque, T_L is the load torque and $B_w\omega$ is a friction term. J is the rotor's moment of inertia. By substituting (4.38) to (4.41) into (4.34) to (4.37) and by defining the coefficients

$$\gamma_1 = \frac{1}{L^2 - L_m^2} \quad \gamma_2 = \frac{L_m}{L^2 - L_m^2} \quad \gamma_3 = \frac{1}{L + L_m} \tag{4.44}$$

the dynamic model of the 6-phase PMSM is written in the following form (Liu, Guo *et al.*, 2017):

$$\frac{d\theta}{dt} = \omega \tag{4.45}$$

$$\frac{d\omega}{dt} = \frac{3n_p}{2J}\psi_f(i_{q_1} + i_{q_2}) - \frac{T_L}{J} - \frac{B_m}{J}\omega \tag{4.46}$$

$$\frac{di_{d_1}}{dt} = \gamma_1 u_{d_1} - \gamma_2 u_{d_2} - \gamma_1 R_s i_{d_1} + \gamma_2 R_s i_{d_2} + n_p\omega i_{q_1} \tag{4.47}$$

$$\frac{di_{q_1}}{dt} = \gamma_2 u_{q_1} - \gamma_2 u_{q_2} - \gamma_1 R_s i_{q_2} - \gamma_2 R_s i_{q_2} - n_p\omega i_{d_1} - \gamma_2 n_p\omega\psi_f \tag{4.48}$$

$$\frac{di_{d_2}}{dt} = \gamma_1 u_{d_2} - \gamma_2 u_{d_1} - \gamma_1 R_s i_{d_2} + \gamma_2 R_s i_{d_1} + n_p\omega i_{q_2} \tag{4.49}$$

$$\frac{di_{q_2}}{dt} = \gamma_1 u_{q_2} - \gamma_2 u_{q_1} - \gamma_1 R_s i_{q_2} + \gamma_2 R_s i_{q_1} - n_p\omega i_{d_2} - \gamma_2 n_p\omega\psi_f \tag{4.50}$$

Next, using the following definitions for state and control input variables

$$x = [x_1, x_2, x_3, x_4 x_5, x_6]^T = [\theta, \omega, i_{d_1}, i_{q_1}, i_{d_2}, i_{q_2}]^T \tag{4.51}$$

$$u = [u_1, u_2, u_3, u_4]^T = [u_{d_1}, u_{q_1}, u_{d_2}, u_{q_2}]^T \tag{4.52}$$

the equations of the state-space model of the 6-phase PMSM become

$$\dot{x}_1 = x_2 \tag{4.53}$$

$$\dot{x}_2 = \frac{3n_p}{2J}\psi_f(x_4 + x_6) - \frac{T_L}{J} - \frac{B_m}{J}x_2 \tag{4.54}$$

$$\dot{x}_3 = \gamma_1 u_1 - \gamma_2 u_3 - \gamma_1 R_s x_3 + \gamma_2 R_s x_5 + n_p x_2 x_4 \tag{4.55}$$

$$\dot{x}_4 = \gamma_2 u_2 - \gamma_2 u_4 - \gamma_1 R_s x_4 - \gamma_2 R_s x_6 - n_p x_2 x_3 - \gamma_2 n_p x_2 \psi_f \tag{4.56}$$

$$\dot{x}_5 = \gamma_1 u_3 - \gamma_2 u_1 - \gamma_1 R_s x_5 + \gamma_2 R_s x_3 + n_p x_2 x_6 \tag{4.57}$$

$$\dot{x}_6 = \gamma_1 u_4 - \gamma_2 u_2 - \gamma_1 R_s x_6 + \gamma_2 R_s x_4 - n_p x_2 x_5 - \gamma_2 n_p \omega \psi_f \tag{4.58}$$

Consequently, the dynamic model of the six-phase PMSM can be written as $\dot{x} = f(x) + g(x)u$ $x \in R^{6\times 1}$, $f(x) \in R^{6\times 1}$, $g(x) \in R^{6\times 4}$ and $u \in R^{4\times 1}$. Analytically the state-space model of the six-phase PMSM is written as

$$\begin{pmatrix}\dot{x}_1\\ \dot{x}_2\\ \dot{x}_3\\ \dot{x}_4\\ \dot{x}_5\\ \dot{x}_6\end{pmatrix} = \begin{pmatrix} x_2 \\ \frac{3n_p}{2J}\psi_f(x_4 + x_6) - \frac{T_L}{J} - \frac{B_m}{J}x_2 \\ -\gamma_1 R_s x_3 + \gamma_2 R_s x_5 + n_p x_2 x_4 \\ -\gamma_1 R_s x_4 - \gamma_2 R_s x_6 - n_p x_2 x_3 - \gamma_2 n_p x_2 \psi_f \\ -\gamma_1 R_s x_5 + \gamma_2 R_s x_3 + n_p x_2 x_6 \\ -\gamma_1 R_s x_6 + \gamma_2 R_s x_4 - n_p x_2 x_5 - \gamma_2 n_p \omega \psi_f \end{pmatrix} + \begin{pmatrix} 0 & 0 & 0 & 0 \\ 0 & 0 & 0 & 0 \\ \gamma_1 & 0 & -\gamma_2 & 0 \\ 0 & \gamma_1 & 0 & -\gamma_2 \\ -\gamma_2 & 0 & \gamma_1 & 0 \\ 0 & -\gamma_2 & 0 & \gamma_1 \end{pmatrix} \begin{pmatrix} u_1 \\ u_2 \\ u_3 \\ u_4 \end{pmatrix} \tag{4.59}$$

4.2.2.2 State-space model of the inverters

The six-phase PMSM is controlled by two inverters, where the first inverter denoted as (a) controls phases 1 to 3 while the second inverter denoted as (b) controls phases 4 to 6.

The dynamics of the first inverter is given by (Rigatos, 2016b; Kim and Lee, 2010)

$$\begin{aligned}
\frac{dV_{L,d}}{dt} &= \omega_0 V_{L,q} + \frac{1}{C_f}I_{I,d} - \frac{1}{C_f}\frac{p_f V_{L,d} + q_f V_{L,q}}{V_{L,d}^2 + V_{L,q}^2} + \omega_0 C_f V_{L,q} - \omega_0 L_f V_{L,q}\frac{(I_{I,d}^2 + I_{I,q}^2)}{V_{L,d}^2 + V_{L,q}^2} \\
\frac{dV_{L,q}}{dt} &= -\omega_0 V_{L,d} + \frac{1}{C_f}I_{I,q} - \frac{1}{C_f}\frac{p_f V_{L,q} - q_f V_{L,d}}{V_{L,d}^2 + V_{L,q}^2} - \omega_0 C_f V_{L,d} + \omega_0 L_f V_{L,d}\frac{(I_{I,d}^2 + I_{I,q}^2)}{V_{L,d}^2 + V_{L,q}^2} \\
\frac{dI_{I,d}}{dt} &= \omega_0 I_{I,q} - \frac{1}{L_f}V_{L,d} + \frac{1}{L_f}V_{I,d} \\
\frac{dI_{I,q}}{dt} &= \omega_0 I_{I,d} - \frac{1}{L_f}V_{L,q} + \frac{1}{L_f}V_{I,q}
\end{aligned} \tag{4.60}$$

Variables $V_{L,d}$, $V_{L,q}$ are voltages at the LC filter that is connected to the first inverter and $I_{L,d}$, $I_{L,q}$ are the related currents, all expressed in the dq reference frame. The circuit elements of the filter are the inductance L_f and the capacitance C_f. The voltages and currents at the output of the inverter are denoted as $V_{I,d}$, $V_{I,q}$ and $I_{I,d}$, $I_{I,q}$, respectively. Coefficient ω_0 is related to the AC frequency. The active and reactive power of the inverter are given by (Rigatos, 2016b)

$$p_f = V_{L,d}I_{L,d} + V_{L,q}I_{L,q} \tag{4.61}$$

$$q_f = V_{L,q}I_{L,d} - V_{L,d}I_{L,q} - \omega_0 C_f(V_{L,d}^2 + V_{L,q}^2) + \omega_0 L_f(I_{I,d}^2 + I_{I,q}^2) \tag{4.62}$$

The dynamics of the second inverter is given by (Rigatos, 2016b; Kim and Lee, 2010)

$$\begin{aligned}
\frac{d\bar{V}_{L,d}}{dt} &= \omega_0\bar{V}_{L,q} + \frac{1}{\bar{C}_f}\bar{I}_{I,d} - \frac{1}{\bar{C}_f}\frac{\bar{p}_f\bar{V}_{L,d} + \bar{q}_f\bar{V}_{L,q}}{\bar{V}_{L,d}^2 + \bar{V}_{L,q}^2} + \omega_0\bar{C}_f\bar{V}_{L,q} - \omega_0\bar{L}_f\bar{V}_{L,q}\frac{(\bar{I}_{I,d}^2 + \bar{I}_{I,q}^2)}{\bar{V}_{L,d}^2 + \bar{V}_{L,q}^2}\\
\frac{d\bar{V}_{L,q}}{dt} &= -\omega_0\bar{V}_{L,d} + \frac{1}{\bar{C}_f}\bar{I}_{I,q} - \frac{1}{\bar{C}_f}\frac{\bar{p}_f\bar{V}_{L,q} - \bar{q}_f\bar{V}_{L,d}}{\bar{V}_{L,d}^2 + \bar{V}_{L,q}^2} - \omega_0\bar{C}_f\bar{V}_{L,d} + \omega_0\bar{L}_f\bar{V}_{L,d}\frac{(\bar{I}_{I,d}^2 + \bar{I}_{I,q}^2)}{\bar{V}_{L,d}^2 + \bar{V}_{L,q}^2}\\
\frac{d\bar{I}_{I,d}}{dt} &= \omega_0\bar{I}_{I,q} - \frac{1}{\bar{L}_f}\bar{V}_{L,d} + \frac{1}{\bar{L}_f}\bar{V}_{I,d}\\
\frac{d\bar{I}_{I,q}}{dt} &= \omega_0\bar{I}_{I,d} - \frac{1}{\bar{L}_f}\bar{V}_{L,q} + \frac{1}{\bar{L}_f}\bar{V}_{I,q}
\end{aligned} \tag{4.63}$$

Variables $\bar{V}_{L,d}$, $\bar{V}_{L,q}$ are voltages at the LC filter that is connected to the second inverter and $\bar{I}_{L,d}$, $\bar{I}_{L,q}$ are the related currents, all expressed in the dq reference frame. The circuit elements of the filter are the inductance $\bar{L}_f$ and the capacitance $\bar{C}_f$. The voltages and currents at the output of the inverter are denoted as $\bar{V}_{I,d}$, $\bar{V}_{I,q}$ and $\bar{I}_{I,d}$, $\bar{I}_{I,q}$, respectively. Coefficient ω_0 is related to the AC frequency. The active and reactive power of the inverter are given by (Rigatos, 2016b)

$$\bar{p}_f = \bar{V}_{L,d}\bar{I}_{L,d} + \bar{V}_{L,q}\bar{I}_{L,q} \tag{4.64}$$

$$\bar{q}_f = \bar{V}_{L,q}\bar{I}_{L,d} - \bar{V}_{L,d}\bar{I}_{L,q} - \omega_0\bar{C}_f(\bar{V}_{L,d}^2 + \bar{V}_{L,q}^2) + \omega_0 L_f(I_{I,d}^2 + I_{I,q}^2) \tag{4.65}$$

The following state variables and control inputs are defined for the first inverter: $x = [x_7, x_8, x_9, x_{10}]^T = [V_{L_d}, V_{L_q}, I_{I,d}, I_{I,q}]^T$ and $u = [u_1, u_2]^T = [V_{I,d}, V_{I,q}]^T$. Thus, the dynamic model of the first inverter becomes

$$\dot{x}_7 = \omega_0 x_8 + \frac{1}{C_f}x_9 - \frac{1}{C_f}\frac{p_f x_7 + q_f x_8}{x_7^2 + x_8^2} + \omega_0 C_f x_8 - \omega_0 L_f x_8\frac{(x_9^2 + x_{10}^2)}{x_7^2 + x_8^2} \tag{4.66}$$

$$\dot{x}_8 = -\omega_0 x_7 + \frac{1}{C_f}x_{10} - \frac{1}{C_f}\frac{p_f x_8 - q_f x_7}{x_7^2 + x_8^2} - \omega_0 C_f x_7 + \omega_0 L_f x_7 \frac{(x_9^2 + x_{10}^2)}{x_7^2 + x_8^2} \tag{4.67}$$

$$\dot{x}_9 = \omega_0 x_{10} - \frac{1}{L_f}x_7 + \frac{1}{L_f}u_1 \tag{4.68}$$

$$\dot{x}_{10} = \omega_0 x_9 - \frac{1}{L_f}x_8 + \frac{1}{L_f}u_2 \tag{4.69}$$

In matrix form, the state-space model of the first inverter is written as

$$\begin{pmatrix} \dot{x}_7 \\ \dot{x}_8 \\ \dot{x}_9 \\ \dot{x}_{10} \end{pmatrix} = \begin{pmatrix} \omega_0 x_8 + \frac{1}{C_f}x_9 - \frac{1}{C_f}\frac{p_f x_7 + q_f x_8}{x_7^2 + x_8^2} + \omega_0 C_f x_8 - \omega_0 L_f x_8 \frac{(x_9^2 + x_{10}^2)}{x_7^2 + x_8^2} \\ -\omega_0 x_7 + \frac{1}{C_f}x_{10} - \frac{1}{C_f}\frac{p_f x_8 - q_f x_7}{x_7^2 + x_8^2} - \omega_0 C_f x_7 + \omega_0 L_f x_7 \frac{(x_9^2 + x_{10}^2)}{x_7^2 + x_8^2} \\ \omega_0 x_{10} - \frac{1}{L_f}x_7 \\ \omega_0 x_9 - \frac{1}{L_f}x_8 \end{pmatrix} + \begin{pmatrix} 0 & 0 \\ 0 & 0 \\ \frac{1}{L_f} & 0 \\ 0 & \frac{1}{L_f} \end{pmatrix} \begin{pmatrix} u_1 \\ u_2 \end{pmatrix} \tag{4.70}$$

The following state variables and control inputs are defined for the second inverter: $x = [x_{11}, x_{12}, x_{13}, x_{14}]^T = [\bar{V}_{L_d}, \bar{V}_{L_q}, \bar{I}_{I,d}, \bar{I}_{I,q}]^T$ and $u = [u_1, u_2]^T = [V_{I,d}, V_{I,q}]^T$. Thus, the dynamic model of the second inverter becomes

$$\dot{x}_{11} = \omega_0 x_{12} + \frac{1}{\bar{C}_f}x_{13} - \frac{1}{\bar{C}_f}\frac{\bar{p}_f x_{11} + \bar{q}_f x_{12}}{x_{11}^2 + x_{12}^2} + \omega_0 \bar{C}_f x_{12} - \omega_0 \bar{L}_f x_{12} \frac{(x_{13}^2 + x_{14}^2)}{x_{11}^2 + x_{12}^2} \tag{4.71}$$

$$\dot{x}_{12} = -\omega_0 x_{11} + \frac{1}{\bar{C}_f}x_{14} - \frac{1}{\bar{C}_f}\frac{\bar{p}_f x_{12} - \bar{q}_f x_{11}}{x_{11}^2 + x_{12}^2} - \omega_0 \bar{C}_f x_{11} + \omega_0 \bar{L}_f x_{11} \frac{(x_{13}^2 + x_{14}^2)}{x_{11}^2 + x_{12}^2} \tag{4.72}$$

$$\dot{x}_{13} = \omega_0 x_{14} - \frac{1}{\bar{L}_f}x_{11} + \frac{1}{\bar{L}_f}u_3 \tag{4.73}$$

$$\dot{x}_{14} = \omega_0 x_{13} - \frac{1}{\bar{L}_f}x_{12} + \frac{1}{\bar{L}_f}u_4 \tag{4.74}$$

In matrix form, the state-space model of the second inverter is written as

$$\begin{pmatrix} \dot{x}_{11} \\ \dot{x}_{12} \\ \dot{x}_{13} \\ \dot{x}_{14} \end{pmatrix} = \begin{pmatrix} \omega_0 x_{12} + \frac{1}{\bar{C}_f} x_{13} - \frac{1}{\bar{C}_f} \frac{\bar{p}_f x_{11} + \bar{q}_f x_{12}}{x_{11}^2 + x_{12}^2} + \omega_0 \bar{C}_f x_{12} - \omega_0 \bar{L}_f x_{12} \frac{(x_{13}^2 + x_{14}^2)}{x_{11}^2 + x_{12}^2} \\ -\omega_0 x_{11} + \frac{1}{\bar{C}_f} x_{14} - \frac{1}{\bar{C}_f} \frac{\bar{p}_f x_{12} - \bar{q}_f x_{11}}{x_{11}^2 + x_{12}^2} - \omega_0 \bar{C}_f x_{11} + \omega_0 \bar{L}_f x_{11} \frac{(x_{13}^2 + x_{14}^2)}{x_{11}^2 + x_{12}^2} \\ \omega_0 x_{14} - \frac{1}{\bar{L}_f} x_{11} \\ \omega_0 x_{13} - \frac{1}{\bar{L}_f} x_{12} \end{pmatrix} + \begin{pmatrix} 0 & 0 \\ 0 & 0 \\ \frac{1}{\bar{L}_f} & 0 \\ 0 & \frac{1}{\bar{L}_f} \end{pmatrix} \begin{pmatrix} u_3 \\ u_4 \end{pmatrix} \tag{4.75}$$

4.2.2.3 A concise model of the inverters' dynamics

In this subsection the notation $z = [z_1, z_2, z_3, z_4]^T = [V_{L,q}, V_{L,q}, I_{I,d}, I_{I,q}]^T$ is used to describe the dynamics of the voltage source inverter, while the control inputs vector of the inverter is denoted as $v = [v_1, v_2]^T = [V_{I,d}, V_{I,q}]^T$. By considering p_f, q_f as piecewise constant and by deriving the first row of the states-space model of the individual inverter with respect to time, one has

$$\begin{aligned} \ddot{z}_1 &= \omega \dot{z}_2 + \frac{1}{C_f} \dot{z}_3 - \frac{1}{C_f} \frac{\{(p_f \dot{z}_1 + q_f \dot{z}_2)(z_1^2 + z_2^2) - (p_f z_1 + q_f z_2)(2z_1 \dot{z}_1 + 2z_2 \dot{z}_2)}{(z_1^2 + z_2^2)^2\}} \\ &- \omega \dot{z}_2 + \frac{1}{C_f} \left\{ \frac{\omega L_f \dot{z}_2 (z_3^2 + z_4^2)(z_1^2 + z_2^2) + \omega L_f z_2 (2z_3 \dot{z}_3 + 2z_4 \dot{z}_4)(z_1^2 + z_2^2)}{(z_1^2 + z_2^2)^2} \right\} \\ &- \frac{\omega L_f z_2 (z_3^2 + z_4^2)(2z_1 \dot{z}_1 + 2z_2 \dot{z}_2)}{(z_1^2 + z_2^2)^2} \end{aligned} \tag{4.76}$$

The time derivatives are substituted from the associated rows of the state-space equations. Next, substitutions of the terms $\dot{x}_3$ and $\dot{x}_4$ are performed. It holds that

$$\begin{aligned} \ddot{z}_1 &= \omega \dot{z}_2 + \frac{1}{C_f} (\omega z_4 - \frac{1}{L_f} z_1 + \frac{1}{L_f} u_1) \\ &- \frac{1}{C_f} \left\{ \frac{(p_f \dot{z}_1 + q_f \dot{z}_2)(z_1^2 + z_2^2) - (p_f z_1 + q_f z_2)(2z_1 \dot{z}_1 + 2z_2 \dot{z}_2)}{(x_1^2 + x_2^2)^2} \right\} \\ &- \omega \dot{z}_2 + \frac{1}{C_f} \left\{ \frac{\omega L_f \dot{z}_2 (z_3^2 + z_4^2)(z_1^2 + z_2^2) + \omega L_f z_2 (z_3^2 + z_4^2)(2z_1 \dot{z}_1 + 2z_2 \dot{z}_2)}{(z_1^2 + z_2^2)^2} \right\} \\ &+ \frac{\omega L_f z_2}{C_f} \frac{2z_3 (z_1^2 + z_2^2)}{(z_1^2 + z_2^2)^2} (\omega x z_4 - \frac{1}{L_f} z_1 + \frac{1}{L_f} v_1) + \frac{\omega L_f z_2}{C_f} \frac{2z_4 (z_1^2 + z_2^2)}{(z_1^2 + z_2^2)^2} (-\omega z_3 - \frac{1}{L_f} z_2 + \frac{1}{L_f} v_2) \end{aligned} \tag{4.77}$$

The previous relation is rewritten by regrouping terms that multiply the control inputs u_1 and u_2. Thus one obtains

$$\begin{aligned}\ddot{z}_1 = {} & \omega\dot{z}_2 + \frac{1}{C_f}(\omega z_4 - \frac{1}{L_f}z_1) - \frac{1}{C_f}\left\{\frac{(p_f\dot{z}_1 + q_f\dot{z}_2)(z_1^2+z_2^2) - (p_f z_1 + q_f z_2)(2z_1\dot{z}_1 + 2z_2\dot{z}_2)}{(z_1^2+z_2^2)^2}\right\} \\ & - \omega\dot{z}_2 + \frac{1}{C_f}\left\{\frac{\omega L_f\dot{z}_2(z_3^2+z_4^2)(z_1^2+z_2^2) - \omega L_f z_2(z_3^2+z_4^2)(2z_1\dot{z}_1 + 2z_2\dot{z}_2)}{(z_1^2+z_2^2)^2}\right\} \\ & + \frac{\omega L_f z_2}{C_f}\cdot\frac{2z_3(\omega z_4 - \frac{1}{L_f}z_1)}{(z_1^2+z_2^2)} + \frac{\omega L_f z_2}{C_f}\cdot\frac{2z_4(-\omega z_3 - \frac{1}{L_f}z_2)}{(z_1^2+z_2^2)} \\ & + \frac{1}{C_f}\{\frac{\omega L_f z_2(2z_3)}{(z_1^2+z_2^2)L_f} + \frac{1}{L_f}\}v_1 + \frac{\omega L_f z_2}{C_f}\frac{2z_4}{(z_1^2+z_2^2)}\frac{1}{L_f}v_2\end{aligned} \tag{4.78}$$

The previous relation can be also written in the form

$$\ddot{z}_1 = f_a(z) + g_{a_1}(z)v_1 + g_{a_2}(z)v_2 \tag{4.79}$$

In a similar manner, by differentiating the second row of the state-space equations of the individual inverters with respect to time one has

$$\begin{aligned}\ddot{z}_2 = {} & -\omega\dot{z}_1 + \frac{1}{C_f}\dot{z}_4 - \frac{1}{C_f}\frac{(p_f\dot{z}_2 - g_f\dot{z}_1)(z_1^2+z_2^2) - (p_f z_2 - g_f z_1(2z_1\dot{z}_1 + 2z_2\dot{z}_2))}{(z_1^2+z_2^2)^2} \\ & + \omega\dot{z}_1 - \frac{1}{C_f}\left\{\frac{\omega L_f\dot{z}_1(z_3^2+z_4^2)(z_1^2+z_2^2) + \omega L_f z_1(2z_3\dot{z}_3 + 2z_4\dot{z}_4)(z_1^2+z_2^2)}{(z_1^2+z_2^2)^2}\right. \\ & \left. - \frac{-\omega L_f z_1(z_3^2+z_4^2)(2z_1\dot{z}_1 + 2z_2\dot{z}_2)}{(z_1^2+z_2^2)^2}\right\}\end{aligned} \tag{4.80}$$

By substituting the derivative terms for z_3 and z_4, according to the third and fourth rows of the state-space equations one has

$$\begin{aligned}\ddot{z}_2 = {} & -\omega\dot{z}_1 + \frac{1}{C_f}(-\omega z_3 - \frac{1}{L_f}z_2 + \frac{1}{L_f}u_2) - \frac{1}{C_f} \\ & \cdot\left\{\frac{(p_f\dot{z}_2 - g_f\dot{z}_1)(z_1^2+z_2^2) - (p_f z_2 - g_f z_1)(2z_1\dot{z}_1 + 2x_2\dot{z}_2)}{(z_1^2+z_2^2)^2}\right\} \\ & +\omega\dot{z}_1 - \frac{1}{C_f}\left\{\frac{\omega L_f\dot{z}_1(z_3^2+z_4^2)(z_1^2+z_2^2) - \omega L_f z_1(z_3^2+z_4^2)(2z_1\dot{z}_1 + 2z_2\dot{z}_2)}{(z_1^2+z_2^2)^2}\right\} \\ & -\frac{1}{C_f}\frac{\omega L_f z_1 2z_3(z_1^2+z_2^2)}{(z_1^2+z_2^2)^2}(\omega z_4 - \frac{1}{L_f}x_1 + \frac{1}{L_f}v_1) \\ & -\frac{1}{C_f}\frac{\omega L_f z_1 2z_4(z_1^2+z_2^2)}{(z_1^2+z_2^2)^2}(-\omega z_3 - \frac{1}{L_f}z_2 + \frac{1}{L_f}v_2)\end{aligned} \tag{4.81}$$

By grouping the terms that multiply the control inputs u_1 and u_2 one obtains

$$\begin{aligned}
\ddot{z}_2 &= -\omega\dot{z}_1 + \frac{1}{C_f}(-\omega z_3) - \frac{1}{C_f}\left\{\frac{(p_f\dot{z}_2 - g_f\dot{}_1)(z_1^2+z_2^2) - (p_f z_2 - g_f z_1)(2z_1\dot{z}_1 + 2z_2\dot{z}_2)}{(z_1^2+z_2^2)^2}\right\} \\
&+ \omega\dot{z}_1 - \frac{1}{C_f}\left\{\frac{\omega L_f\dot{z}_1(z_3^2+z_4^2)(z_1^2+z_2^2) - \omega L_f z_1(z_3^2+z_4^2)(2z_1\dot{z}_1+2z_2\dot{z}_2)}{(z_1^2+z_2^2)^2}\right\} \\
&- \frac{1}{C_f}\frac{\omega L_f z_1 2z_3(\omega z_4 - \frac{1}{L_f}z_1)}{(z_1^2+z_2^2)} - \frac{1}{C_f}\frac{\omega L_f z_1 2z_4(-\omega z_3 - \frac{1}{L_f}z_2)}{(z_1^2+z_2^2)} \\
&- \frac{1}{C_f}\frac{\omega L_f z_1 2z_3}{(z_1^2+z_2^2)}\frac{1}{L_f}v_1 - \frac{1}{C_f}\{\frac{\omega L_f z_1 2z_4}{(z_1^2+z_2^2)}\frac{1}{L_f} + \frac{1}{L_f}\}v_2
\end{aligned} \tag{4.82}$$

The previous equation can be also written in the form

$$\ddot{z}_2 = f_b(z) + g_{b_1}(z)v_1 + g_{b_1}(z)v_2 \tag{4.83}$$

Thus, one obtains an input–output linearized description of the inverter in the form

$$\begin{aligned}
\ddot{z}_1 &= f_a(z) + g_{a_1}(z)v_1 + g_{a_2}(z)v_2 \\
\ddot{z}_2 &= f_b(z) + g_{b_1}(z)v_1 + g_{b_1}(z)v_2
\end{aligned} \tag{4.84}$$

The following control inputs are defined

$$\begin{aligned}
u_1 &= f_a(z) + g_{a_1}(z)v_1 + g_{a_2}(z)v_2 \\
u_2 &= f_b(z) + g_{b_1}(z)v_1 + g_{b_1}(z)v_2
\end{aligned} \tag{4.85}$$

Finally, the dynamics of each inverter is written in the form

$$\begin{aligned}
\ddot{z}_1 &= u_1 \\
\ddot{z}_2 &= u_2
\end{aligned} \tag{4.86}$$

4.2.2.4 State-space description of the VSI-fed six-phase PMSM

Next, in the state-space model of the VSI-fed PMSM the following substitutions are performed: $u_1 = V_{L,d} = x_7$, $u_2 = V_{L,q} = x_8$, $u_3 = \bar{V}_{L,d} = x_{11}$ and $u_4 = \bar{V}_{L,q} = x_{12}$. By substituting the control inputs of the six-phase PMSM with the dq output voltages of the two inverters, the state-space model of the motor is written as

$$\begin{pmatrix}\dot{x}_1\\ \dot{x}_2\\ \dot{x}_3\\ \dot{x}_4\\ \dot{x}_5\\ \dot{x}_6\end{pmatrix} = \begin{pmatrix} x_2 \\ \frac{3n_p}{2J}\psi_f(x_4+x_6) - \frac{T_L}{J} - \frac{B_m}{J}x_2 \\ -\gamma_1 R_s x_3 + \gamma_2 R_s x_5 + n_p x_2 x_4 + \gamma_1 x_7 - \gamma_2 x_{11} \\ -\gamma_1 R_s x_4 - \gamma_2 R_s x_6 - n_p x_2 x_3 - \gamma_2 n_p x_2 \psi_f + \gamma_1 x_8 - \gamma_2 x_{12} \\ -\gamma_1 R_s x_5 + \gamma_2 R_s x_3 + n_p x_2 x_6 - \gamma_2 x_7 + \gamma_1 x_{11} \\ -\gamma_1 R_s x_6 + \gamma_2 R_s x_4 - n_p x_2 x_5 - \gamma_2 n_p \omega \psi_f - \gamma_2 x_8 + \gamma_1 x_{12} \end{pmatrix} \tag{4.87}$$

By aggregating the state-space model of the 6-phase PMSM and of the two inverters one obtains the following state-space description

$$
\begin{pmatrix} \dot{x}_1 \\ \dot{x}_2 \\ \dot{x}_3 \\ \dot{x}_4 \\ \dot{x}_5 \\ \dot{x}_6 \\ \dot{x}_7 \\ \dot{x}_8 \\ \dot{x}_9 \\ \dot{x}_{10} \\ \dot{x}_{11} \\ \dot{x}_{12} \\ \dot{x}_{13} \\ \dot{x}_{14} \end{pmatrix} = \begin{pmatrix} x_2 \\ \frac{3n_p}{2J}\psi_f(x_4+x_6) - \frac{T_L}{J} - \frac{B_m}{J}x_2 \\ -\gamma_1 R_s x_3 + \gamma_2 R_s x_5 + n_p x_2 x_4 + \gamma_1 x_7 - \gamma_2 x_{11} \\ -\gamma_1 R_s x_4 - \gamma_2 R_s x_6 - n_p x_2 x_3 - \gamma_2 n_p x_2 \psi_f + \gamma_1 x_8 - \gamma_2 x_{12} \\ -\gamma_1 R_s x_5 + \gamma_2 R_s x_3 + n_p x_2 x_6 - \gamma_2 x_7 + \gamma_1 x_{11} \\ -\gamma_1 R_s x_6 + \gamma_2 R_s x_4 - n_p x_2 x_5 - \gamma_2 n_p \omega \psi_f - \gamma_2 x_8 + \gamma_1 x_{12} \\ \omega_0 x_8 + \frac{1}{C_f}x_9 - \frac{1}{C_f}\frac{p_f x_7 + q_f x_8}{x_7^2+x_8^2} + \omega_0 C_f x_8 - \omega_0 L_f x_8 \frac{(x_9^2+x_{10}^2)}{x_7^2+x_8^2} \\ -\omega_0 x_7 + \frac{1}{C_f}x_{10} - \frac{1}{C_f}\frac{p_f x_8 - q_f x_7}{x_7^2+x_8^2} - \omega_0 C_f x_7 + \omega_0 L_f x_7 \frac{(x_9^2+x_{10}^2)}{x_7^2+x_8^2} \\ \omega_0 x_{10} - \frac{1}{L_f}x_7 \\ \omega_0 x_9 - \frac{1}{L_f}x_8 \\ \omega_0 x_{12} + \frac{1}{\bar{C}_f}x_{13} - \frac{1}{\bar{C}_f}\frac{\bar{p}_f x_{11} + \bar{q}_f x_{12}}{x_{11}^2+x_{12}^2} + \omega_0 \bar{C}_f x_{12} - \omega_0 \bar{L}_f x_{12} \frac{(x_{13}^2+x_{14}^2)}{x_{11}^2+x_{12}^2} \\ -\omega_0 x_{11} + \frac{1}{\bar{C}_f}x_{14} - \frac{1}{\bar{C}_f}\frac{\bar{p}_f x_{12} - \bar{q}_f x_{11}}{x_{11}^2+x_{12}^2} - \omega_0 \bar{C}_f x_{11} + \omega_0 \bar{L}_f x_{11} \frac{(x_{13}^2+x_{14}^2)}{x_{11}^2+x_{12}^2} \\ \omega_0 x_{14} - \frac{1}{\bar{L}_f}x_{11} \\ \omega_0 x_{13} - \frac{1}{\bar{L}_f}x_{12} \end{pmatrix}
$$

$$
+ \begin{pmatrix} 0 & 0 & 0 & 0 \\ 0 & 0 & 0 & 0 \\ 0 & 0 & 0 & 0 \\ 0 & 0 & 0 & 0 \\ 0 & 0 & 0 & 0 \\ 0 & 0 & 0 & 0 \\ 0 & 0 & 0 & 0 \\ 0 & 0 & 0 & 0 \\ \frac{1}{L_f} & 0 & 0 & 0 \\ 0 & \frac{1}{L_f} & 0 & 0 \\ 0 & 0 & 0 & 0 \\ 0 & 0 & 0 & 0 \\ 0 & 0 & \frac{1}{\bar{L}_f} & 0 \\ 0 & 0 & 0 & \frac{1}{\bar{L}_f} \end{pmatrix} \begin{pmatrix} u_1 \\ u_2 \\ u_3 \\ u_4 \end{pmatrix} \quad (4.88)
$$

The state-space model of the VSI-fed six-phase PMSM can be also written in the following form:

$$\dot{x} = f(x) + g(x)u \tag{4.89}$$

where $x \in R^{14\times1}, f(x) \in R^{14\times1}$, $g(x) \in R^{14\times4}$, and $u \in R^{4\times1}$.

$$f(x) = \begin{pmatrix} x_2 \\ \frac{3n_p}{2J}\psi_f(x_4+x_6) - \frac{T_L}{J} - \frac{B_m}{J}x_2 \\ -\gamma_1 R_s x_3 + \gamma_2 R_s x_5 + n_p x_2 x_4 + \gamma_1 x_7 - \gamma_2 x_{11} \\ -\gamma_1 R_s x_4 - \gamma_2 R_s x_6 - n_p x_2 x_3 - \gamma_3 n_p x_2 \psi_f + \gamma_1 x_8 - \gamma_2 x_{12} \\ -\gamma_1 R_s x_5 + \gamma_2 R_s x_3 + n_p x_2 x_6 - \gamma_2 x_7 + \gamma_1 x_{11} \\ -\gamma_1 R_s x_6 + \gamma_2 R_s x_4 - n_p x_2 x_5 - \gamma_3 n_p \omega \psi_f - \gamma_2 x_8 + \gamma_1 x_{12} \\ \omega_0 x_8 + \frac{1}{C_f}x_9 - \frac{1}{C_f}\frac{p_f x_7 + q_f x_8}{x_7^2 + x_8^2} + \omega_0 C_f x_8 - \omega_0 L_f x_8 \frac{(x_9^2 + x_{10}^2)}{x_7^2 + x_8^2} \\ -\omega_0 x_7 + \frac{1}{C_f}x_{10} - \frac{1}{C_f}\frac{p_f x_8 - q_f x_7}{x_7^2 + x_8^2} - \omega_0 C_f x_7 + \omega_0 L_f x_7 \frac{(x_9^2 + x_{10}^2)}{x_7^2 + x_8^2} \\ \omega_0 x_{10} - \frac{1}{L_f}x_7 \\ \omega_0 x_9 - \frac{1}{L_f}x_8 \\ \omega_0 x_{12} + \frac{1}{\bar{C}_f}x_{13} - \frac{1}{\bar{C}_f}\frac{\bar{p}_f x_{11} + \bar{q}_f x_{12}}{x_{11}^2 + x_{12}^2} + \omega_0 \bar{C}_f x_{12} - \omega_0 \bar{L}_f x_{12} \frac{(x_{13}^2 + x_{14}^2)}{x_{11}^2 + x_{12}^2} \\ -\omega_0 x_{11} + \frac{1}{\bar{C}_f}x_{14} - \frac{1}{\bar{C}_f}\frac{\bar{p}_f x_{12} - \bar{q}_f x_{11}}{x_{11}^2 + x_{12}^2} - \omega_0 \bar{C}_f x_{11} + \omega_0 \bar{L}_f x_{11} \frac{(x_{13}^2 + x_{14}^2)}{x_{11}^2 + x_{12}^2} \\ \omega_0 x_{14} - \frac{1}{\bar{L}_f}x_{11} \\ \omega_0 x_{13} - \frac{1}{\bar{L}_f}x_{12} \end{pmatrix} \tag{4.90}$$

$$g(x) = \begin{pmatrix} 0 & 0 & 0 & 0 \\ 0 & 0 & 0 & 0 \\ 0 & 0 & 0 & 0 \\ 0 & 0 & 0 & 0 \\ 0 & 0 & 0 & 0 \\ 0 & 0 & 0 & 0 \\ 0 & 0 & 0 & 0 \\ 0 & 0 & 0 & 0 \\ \frac{1}{L_f} & 0 & 0 & 0 \\ 0 & \frac{1}{L_f} & 0 & 0 \\ 0 & 0 & 0 & 0 \\ 0 & 0 & 0 & 0 \\ 0 & 0 & \frac{1}{\bar{L}_f} & 0 \\ 0 & 0 & 0 & \frac{1}{\bar{L}_f} \end{pmatrix}$$

4.2.3 Differential flatness properties of the VSI-fed six-phase PMSM

It will be proven that the dynamic model of the VSI-fed six-phase PMSM is differentially flat with flat outputs $y_1 = x_1 = \theta$, $y_2 = x_3 = i_{d_1}$, $y_3 = x_5 = i_{d_2}$ and $y_4 = x_6 = i_{q_2}$. The state-space equations of the VSI-fed six-phase PMSM are:

$$\dot{x}_1 = x_2 \tag{4.91}$$

$$\dot{x}_2 = \frac{3n_p}{2J}\psi_f(x_4 + x_6) - \frac{T_L}{J} - \frac{B_m}{J}x_2 \tag{4.92}$$

$$\dot{x}_3 = -\gamma_1 R_s x_3 + \gamma_2 R_s x_5 + n_p x_2 x_4 + \gamma_1 x_7 - \gamma_2 x_{11} \tag{4.93}$$

$$\dot{x}_4 = -\gamma_1 R_s x_4 - \gamma_2 R_s x_6 - n_p x_2 x_3 - \gamma_2 n_p x_2 \psi_f + \gamma_1 x_8 - \gamma_2 x_{12} \tag{4.94}$$

$$\dot{x}_5 = -\gamma_1 R_s x_5 + \gamma_2 R_s x_3 + n_p x_2 x_6 - \gamma_2 x_7 + \gamma_1 x_{11} \tag{4.95}$$

$$\dot{x}_6 = -\gamma_1 R_s x_6 + \gamma_2 R_s x_4 - n_p x_2 x_5 - \gamma_3 n_p \omega \psi_f - \gamma_2 x_8 + \gamma_1 x_{12} \tag{4.96}$$

$$\dot{x}_7 = \omega_0 x_8 + \frac{1}{C_f}x_9 - \frac{1}{C_f}\frac{p_f x_7 + q_f x_8}{x_7^2 + x_8^2} + \omega_0 C_f x_8 - \omega_0 L_f x_8 \frac{(x_9^2 + x_{10}^2)}{x_7^2 + x_8^2} + \frac{1}{L_f}u_1 \tag{4.97}$$

$$\dot{x}_8 = -\omega_0 x_7 + \frac{1}{C_f}x_{10} - \frac{1}{C_f}\frac{p_f x_8 - q_f x_7}{x_7^2 + x_8^2} - \omega_0 C_f x_7 + \omega_0 L_f x_7 \frac{(x_9^2 + x_{10}^2)}{x_7^2 + x_8^2} + \frac{1}{L_f}u_2 \tag{4.98}$$

$$\dot{x}_9 = \omega_0 x_{10} - \frac{1}{L_f}x_7 \tag{4.99}$$

$$\dot{x}_{10} = \omega_0 x_9 - \frac{1}{L_f}x_8 \tag{4.100}$$

$$\dot{x}_{11} = \omega_0 x_{12} + \frac{1}{\bar{C}_f}x_{13} - \frac{1}{\bar{C}_f}\frac{\bar{p}_f x_{11} + \bar{q}_f x_{12}}{x_{11}^2 + x_{12}^2} + \omega_0 \bar{C}_f x_{12} - \omega_0 \bar{L}_f x_{12}\frac{(x_{13}^2 + x_{14}^2)}{x_{11}^2 + x_{12}^2} + \frac{1}{\bar{L}_f}u_3 \tag{4.101}$$

$$\dot{x}_{12} = -\omega_0 x_{11} + \frac{1}{\bar{C}_f}x_{14} - \frac{1}{\bar{C}_f}\frac{\bar{p}_f x_{12} - \bar{q}_f x_{11}}{x_{11}^2 + x_{12}^2} - \omega_0 \bar{C}_f x_{11} + \omega_0 \bar{L}_f x_{11}\frac{(x_{13}^2 + x_{14}^2)}{x_{11}^2 + x_{12}^2} + \frac{1}{\bar{L}_f}u_4 \tag{4.102}$$

$$\dot{x}_{13} = \omega_0 x_{14} - \frac{1}{\bar{L}_f}x_{11} \tag{4.103}$$

$$\dot{x}_{14} = \omega_0 x_{13} - \frac{1}{\bar{L}_f}x_{12} \tag{4.104}$$

From (4.91) one has

$$x_2 = \dot{x}_1 \Rightarrow x_2 = \dot{y}_1 \Rightarrow x_2 = h_2(y, \dot{y}) \tag{4.105}$$

which signifies that state variable x_2 is a differential function of the flat outputs of the system. From, (4.92) one solve for x_4. Thus gives

$$x_4 = \frac{\dot{x}_2 - \frac{3n_p\psi_f}{2J}x_6 + \frac{T_L}{J} + \frac{B_w}{J}x_2}{\frac{3n_p\psi_f}{2J}} \qquad (4.106)$$
$$\Rightarrow x_4 = h_4(y, \dot{y})$$

which signifies that state variable x_4 is a differential function of the system's flat outputs. Consequently, it has been proven that all state variables of the six-phase PMSM x_1 to x_6 are differential functions of the flat outputs of the integrated VSI six-phase PMSM system.

Besides, from (4.93) and (4.95) one has a system of equations with respect to x_7 and x_{11}. Thus

$$\begin{aligned} \gamma_1 x_7 - \gamma_2 x_{11} &= -\dot{x}_3 - \gamma_1 R_s x_3 + \gamma_2 R_s x_5 + n_p x_2 x_4 \\ -\gamma_2 x_7 + \gamma_1 x_{11} &= -\dot{x}_5 - \gamma_1 R_s x_5 + \gamma_3 R_s x_3 + n_p x_2 x_6 \end{aligned} \qquad (4.107)$$

By solving with respect to x_7 and x_{11} one obtains

$$\begin{pmatrix} x_7 \\ x_{11} \end{pmatrix} = \begin{pmatrix} \gamma_1 & -\gamma_2 \\ -\gamma_2 & \gamma_1 \end{pmatrix}^{-1} \begin{pmatrix} -\dot{x}_3 - \gamma_1 R_s x_3 + \gamma_2 R_s x_5 + n_p x_2 x_4 \\ -\dot{x}_5 - \gamma_1 R_s x_5 + \gamma_3 R_s x_3 + n_p x_2 x_6 \end{pmatrix} \qquad (4.108)$$

or equivalently

$$\begin{pmatrix} x_7 \\ x_{11} \end{pmatrix} = \begin{pmatrix} h_7(y, \dot{y}) \\ h_{11}(y, \dot{y}) \end{pmatrix} \qquad (4.109)$$

which signifies that state variables x_7 and x_{11} are differential functions of the flat outputs of the system. Moreever, from (4.94) and (4.96) one has a system of equations with respect to state variables x_8 and x_{12}, in the form

$$\begin{aligned} -\gamma_1 x_8 + \gamma_2 x_{12} &= -\dot{x}_4 - \gamma_1 R_s x_4 + \gamma_2 R_s x_6 - n_p x_2 x_3 - \gamma_3 n_p \psi_f x_2 \\ \gamma_2 x_8 - \gamma_1 x_{12} &= -\dot{x}_6 - \gamma_1 R_s x_6 + \gamma_2 R_s x_4 - n_p x_2 x_5 - \gamma_3 n_p \psi_f x_2 \end{aligned} \qquad (4.110)$$

By solving with respect to x_7 and x_{11} one obtains

$$\begin{pmatrix} x_8 \\ x_{12} \end{pmatrix} = \begin{pmatrix} \gamma_1 & -\gamma_2 \\ -\gamma_2 & \gamma_1 \end{pmatrix}^{-1} \begin{pmatrix} -\dot{x}_4 - \gamma_1 R_s x_4 + \gamma_2 R_s x_6 - n_p x_2 x_3 - \gamma_3 n_p \psi_f x_2 \\ -\dot{x}_6 - \gamma_1 R_s x_6 + \gamma_2 R_s x_4 - n_p x_2 x_5 - \gamma_3 n_p \psi_f x_2 \end{pmatrix} \qquad (4.111)$$

or equivalently

$$\begin{pmatrix} x_8 \\ x_{12} \end{pmatrix} = \begin{pmatrix} h_8(y, \dot{y}) \\ h_{12}(y, \dot{y}) \end{pmatrix} \qquad (4.112)$$

which signifies that state variables x_8 and x_{12} are differential functions of the flat outputs of the system. Next, by multiplying (4.97) with x_7 one obtains

$$x_7\dot{x}_7 = \omega_0 x_7 x_8 + \frac{1}{C_f} x_7 x_9 - \frac{1}{C_f} x_7 \frac{p_f x_7 + q_f x_8}{(x_7^2 + x_8^2)} + \omega_0 x_7 x_8 - \omega_0 L_f x_7 x_8 \frac{(x_9^2 + x_{10}^2)}{(x_7^2 + x_8^2)} \qquad (4.113)$$

Moreover, by multiplying (4.98) with x_8 one obtains

$$x_8\dot{x}_8 = -\omega_0 x_7 x_8 + \frac{1}{C_f} x_8 x_{10} - \frac{1}{C_f} x_8 \frac{p_f x_8 - q_f x_7}{(x_7^2 + x_8^2)} - \omega_0 x_7 x_8 + \omega_0 L_f x_7 x_8 \frac{(x_9^2 + x_{10}^2)}{(x_7^2 + x_8^2)} \qquad (4.114)$$

By adding (4.113) with (4.114) and after intermediate operations one obtains

$$\begin{aligned} x_7\dot{x}_7+x_8\dot{x}_8 &= \frac{1}{C_f}x_7x_9-\frac{1}{C_f}\frac{p_fx_7^2}{(x_7^2+x_8^2)}+\frac{1}{C_f}x_8x_{10}-\frac{1}{C_f}\frac{p_fx_8^2}{(x_7^2+x_8^2)}\Rightarrow \\ x_7\dot{x}_7+x_8\dot{x}_8 &= \frac{1}{C_f}x_7x_9+\frac{1}{C_f}x_8x_{10}-\frac{p_f}{C_f} \end{aligned} \tag{4.115}$$

Equation (4.115) can be solved with respect to x_{10} thus giving

$$x_{10}=\frac{C_fx_7\dot{x}_7+C_fx_8\dot{x}_9+p_f}{x_8}-\frac{x_7}{x_8}x_9 \tag{4.116}$$

Equation (4.116) can be also written in the more concise form

$$x_{10}=w_{10}(x_7,\dot{x}_7,x_8,\dot{x}_8)-\frac{x_7}{x_8}x_9 \tag{4.117}$$

By substituting (4.117) into (4.97) one obtains

$$\begin{aligned} x_7 &= \omega_0x_8-\frac{1}{C_f}x_9-\frac{1}{C_f}\frac{p_fx_7+q_fx_8}{x_7^2+x_8^2}+\omega_0C_fx_8 \\ &-\omega_0L_fx_8\frac{x_9^2}{(x_7^2+x_8^2)}-\omega_0L_fx_8\frac{w_{10}^2+(\frac{x_7}{x_8})^2x_9^2-2w_{10}(\frac{x_7}{x_8})x_9}{(x_7^2+x_8^2)} \end{aligned} \tag{4.118}$$

Equation (4.118) is a binomial with respect to x_9 and with coefficients depending on $x_7,\dot{x}_7,x_8,\dot{x}_8$. Solving this binomial with respect to x_9 one obtains that

$$x_9=h_9(x_7,\dot{x}_7,x_8,\dot{x}_8)\Rightarrow x_9=h_9(y,\dot{y}) \tag{4.119}$$

which means that x_9 is a differential function of the flat outputs of the system. Moreover, by substituting (4.119) into (4.117) one has

$$x_{10}=w_{10}(x_7,\dot{x}_7,x_8,\dot{x}_8)-\frac{x_7}{x_8}x_9\Rightarrow x_9=h_{10}(y,\dot{y}) \tag{4.120}$$

which signifies that state variable x_{10} is a differential function of the flat outputs of the system. Next, by multiplying (4.101) with x_{11} one obtains

$$\begin{aligned} x_{11}\dot{x}_{11} &= \omega_0x_{11}x_{12}+\frac{1}{\bar{C}_f}x_{11}x_{13}-\frac{1}{\bar{C}_f}x_{11}\frac{\bar{p}_fx_{11}+\bar{q}_fx_{12}}{(x_{11}^2+x_{12}^2)}+\omega_0x_{11}x_{12} \\ &-\omega_0\bar{L}_fx_{11}x_{12}\frac{(x_{13}^2+x_{14}^2)}{(x_{11}^2+x_{12}^2)} \end{aligned} \tag{4.121}$$

Moreover, by multiplying (4.102) with x_8 one obtains

$$x_{12}\dot{x}_{12} = -\omega_0 x_{11}x_{12} + \frac{1}{\bar{C}_f}x_{12}x_{14} - \frac{1}{\bar{C}_f}x_{12}\frac{\bar{p}_f x_{12} - \bar{q}_f x_{11}}{(x_{11}^2 + x_{12}^2)} - \omega_0 x_{11}x_{12} + \omega_0\bar{L}_f x_{11}x_{12}\frac{(x_{13}^2 + x_{14}^2)}{(x_{11}^2 + x_{12}^2)} \tag{4.122}$$

By adding (4.121) with (4.122) and after intermediate operations one obtains

$$\begin{aligned} x_{11}\dot{x}_{11} + x_{12}\dot{x}_{12} &= \frac{1}{\bar{C}_f}x_{11}x_{13} - \frac{1}{\bar{C}_f}\frac{\bar{p}_f x_{11}^2}{(x_{11}^2 + x_{12}^2)} + \frac{1}{\bar{C}_f}x_{12}x_{14} - \frac{1}{\bar{C}_f}\frac{\bar{p}_f x_{12}^2}{(x_{11}^2 + x_{12}^2)} \Rightarrow \\ x_{11}\dot{x}_{11} + x_{12}\dot{x}_{12} &= \frac{1}{\bar{C}_f}x_{11}x_{13} + \frac{1}{\bar{C}_f}x_{12}x_{14} - \frac{\bar{p}_f}{\bar{C}_f} \end{aligned} \tag{4.123}$$

Equation (4.123) can be solved with respect to x_{14} thus giving

$$x_{14} = \frac{\bar{C}_f x_{11}\dot{x}_{11} + \bar{C}_f x_{12}\dot{x}_{13} + \bar{p}_f}{x_{12}} - \frac{x_{11}}{x_{12}}x_{13} \tag{4.124}$$

Equation (4.124) can be also written in the more concise form

$$x_{14} = w_{14}(x_{11}, \dot{x}_{11}, x_{12}, \dot{x}_{12}) - \frac{x_{11}}{x_{12}}x_{13} \tag{4.125}$$

By substituting (4.125) into (4.101) one obtains

$$\begin{aligned} \dot{x}_{11} = \omega_0 x_{12} - \frac{1}{\bar{C}_f}x_{13} - \frac{1}{\bar{C}_f}\frac{\bar{p}_f x_{11} + \bar{q}_f x_{12}}{x_{11}^2 + x_{12}^2} + \omega_0\bar{C}_f x_{12} \\ - \omega_0\bar{L}_f x_{12}\frac{x_{13}^2}{(x_{11}^2 + x_{12}^2)} - \omega_0\bar{L}_f x_{12}\frac{w_{14}^2 + (\frac{x_{11}}{x_{12}})^2 x_{13}^2 - 2w_{14}(\frac{x_{11}}{x_{12}})x_{13}}{(x_{11}^2 + x_{12}^2)} \end{aligned} \tag{4.126}$$

Equation (4.126) is a binomial with respect to x_{13} and with coefficients depending on $x_{11}, \dot{x}_{11}, x_{12}, \dot{x}_{12}$. Solving this binomial with respect to x_{13} one obtains that

$$x_{13} = h_{13}(x_{11}, \dot{x}_{11}, x_{12}, \dot{x}_{12}) \Rightarrow x_{13} = h_{13}(y, \dot{y}) \tag{4.127}$$

which means that x_{13} is a differential function of the flat outputs of the system. Moreover, by substituting (4.127) into (4.125) one has

$$x_{14} = w_{14}(x_{11}, \dot{x}_{11}, x_{12}, \dot{x}_{12}) - \frac{x_{11}}{x_{12}}x_{13} \Rightarrow x_{14} = h_{14}(y, \dot{y}) \tag{4.128}$$

Thus, it has been proven that all state variables x_1 to x_{14} of the VSI-fed six-phase PMSM are differential functions of the flat outputs of the system. Next, using (4.99) one solves for the control input u_1, that is

$$u_1 = L_f[\dot{x}_9 - \omega_0 x_{10} + \frac{1}{L_f}x_9] \Rightarrow u_1 = h_{u_1}(y, \dot{y}) \tag{4.129}$$

which signifies that control input u_1 is a differential function of the flat outputs of the system. Equivalently, from (4.100) one solves for the control input u_2

$$u_2 = L_f[\dot{x}_{10} - \omega_0 x_9 + \frac{1}{L_f} x_8] \Rightarrow u_2 = h_{u_2}(y, \dot{y}) \tag{4.130}$$

which signifies that control input u_2 is also a differential function of the system's flat outputs. Equivalently, from (4.103) one solves for control input u_3

$$u_3 = \bar{L}_f[\dot{x}_{13} - \omega_0 x_{12} + \frac{1}{L_f} x_{11}] \Rightarrow u_2 = h_{u_2}(y, \dot{y}) \tag{4.131}$$

which signifies that control input u_3 is also a differential function of the flat outputs of the system. Finally, from (4.104) one solves for control input u_4, thus obtaining

$$u_4 = \bar{L}_f[\dot{x}_{14} - \omega_0 x_{14} + \frac{1}{\bar{L}_f} x_{12}] \Rightarrow u_4 = h_{u_4}(y, \dot{y}) \tag{4.132}$$

which signifies that control input u_4 is also a differentia; function of the flat outputs of the system. Consequently, control inputs u_1 to u_4 are differential functions of the flat outputs of the VSI-fed six-phase PMSM. As a result of the above, the entire system is differentially flat.

By knowing that the VSI-fed six-phase PMSM is a differentially flat system, it is confirmed that it can be transformed into the input–output linearized form. Besides, the associated setpoints definition problem is solved. One defines first setpoints for the flat outputs of the system $y_1 = x_1, y_2 = x_3, y_3 = x_5$ and $y_4 = x_6$ and based on them finds the setpoints for the rest of the state vector elements.

4.2.4 Approximate linearization of the model of the VSI-fed six-phase PMSM

4.2.4.1 Aggregate model of the VSI-fed 6-phase PMSM with concise inverter dynamics

Using the results of Subsection 4.2.2.3 the dynamic model of the first inverter is written in the concise form

$$\ddot{x}_7 = u_1 \quad \ddot{x}_9 = u_2 \tag{4.133}$$

while the dynamic model of the second inverter is written in the concise form

$$\ddot{x}_{11} = u_3 \quad \ddot{x}_{13} = u_4 \tag{4.134}$$

Using the concise dynamics of the inverter, given in (4.133) and in (4.134) as well as the notation $f_1(x) = x_2$, $f_2(x) = \frac{3n_p}{2J}\psi_f(x_4 + x_6) - \frac{T_L}{J} - \frac{B_m}{J}x_2$, $f_3(x) = -\gamma_1 R_s x_3 + \gamma_2 R_s x_5 + n_p x_2 x_4 + \gamma_1 x_7 - \gamma_2 x_{11}$, $f_4(x) = -\gamma_1 R_s x_4 - \gamma_2 R_s x_6 - n_p x_2 x_3 - \gamma_2 n_p x_2 \psi_f + \gamma_1 x_9 - \gamma_2 x_{13}, f_5(x) = -\gamma_1 R_s x_5 + \gamma_2 R_s x_3 + n_p x_2 x_6 - \gamma_2 x_7 + \gamma_1 x_{11}, f_6(x) = -\gamma_1 R_s x_6 + \gamma_2 R_s x_4 - n_p x_2 x_5 - \gamma_2 n_p \omega \psi_f - \gamma_2 x_9 + \gamma_1 x_{13}$, $f_7(x) = x_8$, $f_8(x) = 0$, $f_9(x) = x_{10}$, $f_{10}(x) = 0, f_{11}(x) = x_{12}, f_{12}(x) = 0, f_{13}(x) = x_{14}$, and $f_{14}(x) = 0$.

the aggregate state-space model of the VSI-fed six-phase PMSM is written as

$$\begin{pmatrix} \dot{x}_1 \\ \dot{x}_2 \\ \dot{x}_3 \\ \dot{x}_4 \\ \dot{x}_5 \\ \dot{x}_6 \\ \dot{x}_7 \\ \dot{x}_8 \\ \dot{x}_9 \\ \dot{x}_{10} \\ \dot{x}_{11} \\ \dot{x}_{12} \\ \dot{x}_{13} \\ \dot{x}_{14} \end{pmatrix} = \begin{pmatrix} f_1(x) \\ f_2(x) \\ f_3(x) \\ f_4(x) \\ f_5(z) \\ f_6(x) \\ f_7(x) \\ f_8(x) \\ f_9(x) \\ f_{10}(x) \\ f_{11}(x) \\ f_{12}(x) \\ f_{13}(x) \\ f_{14}(x) \end{pmatrix} + \begin{pmatrix} 0 & 0 & 0 & 0 \\ 0 & 0 & 0 & 0 \\ 0 & 0 & 0 & 0 \\ 0 & 0 & 0 & 0 \\ 0 & 0 & 0 & 0 \\ 0 & 0 & 0 & 0 \\ 0 & 0 & 0 & 0 \\ 1 & 0 & 0 & 0 \\ 0 & 0 & 0 & 0 \\ 0 & 1 & 0 & 0 \\ 0 & 0 & 0 & 0 \\ 0 & 0 & 1 & 0 \\ 0 & 0 & 0 & 0 \\ 0 & 0 & 0 & 1 \end{pmatrix} \begin{pmatrix} u_1 \\ u_2 \\ u_3 \\ u_4 \end{pmatrix} \tag{4.135}$$

4.2.4.2 Linearized model of the VSI-fed six-phase PMSM

The dynamic model of the VSI-fed six-phase PMSM undergoes approximate linearization around the temporary operating point (x^*, u^*) at each sampling instance. The linearization point is defined by the present value x^* of the system's state vector and by the last sampled value u^* of the control inputs vector. The linearization process is based on first-order Taylor series expansion and on the computation of the associated Jacobian matrices. Using the initial nonlinear state-space model of the VSI-fed six-phase PMSM that was given in (4.89) one arrives, through the above-noted linearization process, at the equivalent linearized description

$$\dot{x} = Ax + Bu + \tilde{d} \tag{4.136}$$

where $\tilde{d}$ is the cumulative disturbances vector which may comprise (a) the modelling error due to the truncation of higher-order terms in the Taylor-series expansion, (b) exogenous perturbations, and (c) sensors measurement noise of any distribution. Matrices A and B are the Jacobians of the 6-phase PMSM which are given by

$$A = \nabla_x[f(x) + g(x)u] \mid_{(x^*,u^*)} \Rightarrow A = \nabla_x f(x) \mid_{(x^*,u^*)} \tag{4.137}$$

$$B = \nabla_u[f(x) + g(x)u] \mid_{(x^*,u^*)} \Rightarrow B = g(x) \mid_{(x^*,u^*)} \tag{4.138}$$

Considering a load (torque) of the form $T_L = mgl\sin(x_1)$, the computation of the Jacobian matrix $\nabla_x f(x) \mid_{(x^*,u^*)}$ proceeds as follows:

First row of the Jacobian matrix $\nabla_x f(x) \mid_{(x^*,u^*)}$: $\frac{\partial f_1}{\partial x_1} = 0$, $\frac{\partial f_1}{\partial x_2} = 1$, $\frac{\partial f_1}{\partial x_3} = 0$, $\frac{\partial f_1}{\partial x_4} = 0$, $\frac{\partial f_1}{\partial x_5} = 0$, $\frac{\partial f_1}{\partial x_6} = 0$, $\frac{\partial f_1}{\partial x_7} = 0$, $\frac{\partial f_1}{\partial x_8} = 0$, $\frac{\partial f_1}{\partial x_9} = 0$, $\frac{\partial f_1}{\partial x_{10}} = 0$, $\frac{\partial f_1}{\partial x_{11}} = 0$, $\frac{\partial f_1}{\partial x_{12}} = 0$, $\frac{\partial f_1}{\partial x_{13}} = 0$, $\frac{\partial f_1}{\partial x_{14}} = 0$.

Second row of the Jacobian matrix $\nabla_x f(x) \mid_{(x^*,u^*)}$: $\frac{\partial f_2}{\partial x_1} = mgl\cos(x_1)$, $\frac{\partial f_2}{\partial x_2} = -\frac{B_w}{J}$, $\frac{\partial f_2}{\partial x_3} = 0$, $\frac{\partial f_2}{\partial x_4} = \frac{3n_p\psi_f}{2J}$, $\frac{\partial f_2}{\partial x_5} = 0$, $\frac{\partial f_2}{\partial x_6} = \frac{3n_p\psi_f}{2J}$, $\frac{\partial f_2}{\partial x_7} = 0$, $\frac{\partial f_2}{\partial x_8} = 0$, $\frac{\partial f_2}{\partial x_9} = 0$, $\frac{\partial f_2}{\partial x_{10}} = 0$, $\frac{\partial f_2}{\partial x_{11}} = 0$, $\frac{\partial f_2}{\partial x_{12}} = 0$, $\frac{\partial f_2}{\partial x_{13}} = 0$, $\frac{\partial f_2}{\partial x_{14}} = 0$.

Third row of the Jacobian matrix $\nabla_x f(x) \mid_{(x^*,u^*)}$: $\frac{\partial f_3}{\partial x_1}=0$, $\frac{\partial f_3}{\partial x_2}=0$, $\frac{\partial f_3}{\partial x_3}=-\gamma_1 R_s$, $\frac{\partial f_3}{\partial x_4}=0$, $\frac{\partial f_3}{\partial x_5}=\gamma_2 R_s$, $\frac{\partial f_3}{\partial x_6}=0$, $\frac{\partial f_3}{\partial x_7}=\gamma_1$, $\frac{\partial f_3}{\partial x_8}=0$, $\frac{\partial f_3}{\partial x_9}=0$, $\frac{\partial f_3}{\partial x_{10}}=0$, $\frac{\partial f_3}{\partial x_{11}}=-\gamma_2$, $\frac{\partial f_3}{\partial x_{12}}=0$, $\frac{\partial f_3}{\partial x_{13}}=0$, $\frac{\partial f_3}{\partial x_{14}}=0$.

Fourth row of the Jacobian matrix $\nabla_x f(x) \mid_{(x^*,u^*)}$: $\frac{\partial f_4}{\partial x_1}=0$, $\frac{\partial f_4}{\partial x_2}=-n_p x_3-\gamma)3n_p\psi_f$, $\frac{\partial f_4}{\partial x_3}=-n_p x_2$, $\frac{\partial f_4}{\partial x_4}=-\gamma_1 R_s$, $\frac{\partial f_4}{\partial x_5}=0$, $\frac{\partial f_4}{\partial x_6}=\gamma_2 R_s$, $\frac{\partial f_4}{\partial x_7}=0$, $\frac{\partial f_4}{\partial x_8}=0$, $\frac{\partial f_4}{\partial x_9}=\gamma_1$, $\frac{\partial f_4}{\partial x_{10}}=0$, $\frac{\partial f_4}{\partial x_{11}}=0$, $\frac{\partial f_4}{\partial x_{12}}=0$, $\frac{\partial f_4}{\partial x_{13}}=-gamma_2$, $\frac{\partial f_4}{\partial x_{14}}=0$.

Fifth row of the Jacobian matrix $\nabla_x f(x) \mid_{(x^*,u^*)}$: $\frac{\partial f_5}{\partial x_1}=0$, $\frac{\partial f_5}{\partial x_2}=n_p x_6$, $\frac{\partial f_5}{\partial x_3}=\gamma_3 R_s$, $\frac{\partial f_5}{\partial x_4}=0$, $\frac{\partial f_5}{\partial x_5}=-\gamma_1 R_s$, $\frac{\partial f_5}{\partial x_6}=n_p x_2$, $\frac{\partial f_5}{\partial x_7}=-\gamma_2$, $\frac{\partial f_5}{\partial x_8}=0$, $\frac{\partial f_5}{\partial x_9}=0$, $\frac{\partial f_5}{\partial x_{10}}=0$, $\frac{\partial f_5}{\partial x_{11}}=\gamma_1$, $\frac{\partial f_5}{\partial x_{12}}=$, $\frac{\partial f_5}{\partial x_{13}}=0$, $\frac{\partial f_5}{\partial x_{14}}=0$.

Sixth row of the Jacobian matrix $\nabla_x f(x) \mid_{(x^*,u^*)}$: $\frac{\partial f_6}{\partial x_1}=0$, $\frac{\partial f_6}{\partial x_2}=-n_p x_5-\gamma_3 n_p\psi_f$, $\frac{\partial f_6}{\partial x_3}=0$, $\frac{\partial f_6}{\partial x_4}=\gamma_2 R_s$, $\frac{\partial f_6}{\partial x_5}=-n_p x_2$, $\frac{\partial f_6}{\partial x_6}=-\gamma_1 R_s$, $\frac{\partial f_6}{\partial x_7}=0$, $\frac{\partial f_6}{\partial x_8}=0$, $\frac{\partial f_6}{\partial x_9}=-\gamma_2$, $\frac{\partial f_6}{\partial x_{10}}=0$, $\frac{\partial f_6}{\partial x_{11}}=\gamma_1$, $\frac{\partial f_6}{\partial x_{12}}=0$, $\frac{\partial f_6}{\partial x_{13}}=\gamma_1$, $\frac{\partial f_6}{\partial x_{14}}=0$.

Seventh row of the Jacobian matrix $\nabla_x f(x) \mid_{(x^*,u^*)}$: $\frac{\partial f_7}{\partial x_1}=0$, $\frac{\partial f_7}{\partial x_2}=0$, $\frac{\partial f_7}{\partial x_3}=0$, $\frac{\partial f_7}{\partial x_4}=0$, $\frac{\partial f_7}{\partial x_5}=0$, $\frac{\partial f_7}{\partial x_6}=0$, $\frac{\partial f_7}{\partial x_7}=0$, $\frac{\partial f_7}{\partial x_8}=1$, $\frac{\partial f_7}{\partial x_9}=0$, $\frac{\partial f_7}{\partial x_{10}}=0$, $\frac{\partial f_7}{\partial x_{11}}=0$, $\frac{\partial f_7}{\partial x_{12}}=0$, $\frac{\partial f_7}{\partial x_{13}}=0$, $\frac{\partial f_7}{\partial x_{14}}=0$.

Eighth row of the Jacobian matrix $\nabla_x f(x) \mid_{(x^*,u^*)}$: $\frac{\partial f_8}{\partial x_1}=0$, $\frac{\partial f_8}{\partial x_2}=0$, $\frac{\partial f_8}{\partial x_3}=0$, $\frac{\partial f_8}{\partial x_4}=0$, $\frac{\partial f_8}{\partial x_5}=0$, $\frac{\partial f_8}{\partial x_6}=0$, $\frac{\partial f_8}{\partial x_7}=0$, $\frac{\partial f_8}{\partial x_8}=0$, $\frac{\partial f_8}{\partial x_9}=0$, $\frac{\partial f_8}{\partial x_{10}}=0$, $\frac{\partial f_8}{\partial x_{11}}=0$, $\frac{\partial f_8}{\partial x_{12}}=0$, $\frac{\partial f_8}{\partial x_{13}}=0$, $\frac{\partial f_8}{\partial x_{14}}=0$.

Ninth row of the Jacobian matrix $\nabla_x f(x) \mid_{(x^*,u^*)}$: $\frac{\partial f_9}{\partial x_1}=0$, $\frac{\partial f_9}{\partial x_2}=0$, $\frac{\partial f_9}{\partial x_3}=0$, $\frac{\partial f_9}{\partial x_4}=0$, $\frac{\partial f_9}{\partial x_5}=0$, $\frac{\partial f_9}{\partial x_6}=0$, $\frac{\partial f_9}{\partial x_7}=0$, $\frac{\partial f_9}{\partial x_8}=0$, $\frac{\partial f_9}{\partial x_9}=0$, $\frac{\partial f_9}{\partial x_{10}}=1$, $\frac{\partial f_9}{\partial x_{11}}=0$, $\frac{\partial f_9}{\partial x_{12}}=0$, $\frac{\partial f_9}{\partial x_{13}}=0$, $\frac{\partial f_9}{\partial x_{14}}=0$.

Tenth row of the Jacobian matrix $\nabla_x f(x) \mid_{(x^*,u^*)}$: $\frac{\partial f_{10}}{\partial x_1}=0$, $\frac{\partial f_{10}}{\partial x_2}=0$, $\frac{\partial f_{10}}{\partial x_3}=0$, $\frac{\partial f_{10}}{\partial x_4}=0$, $\frac{\partial f_{10}}{\partial x_5}=0$, $\frac{\partial f_{10}}{\partial x_6}=0$, $\frac{\partial f_{10}}{\partial x_7}=0$, $\frac{\partial f_{10}}{\partial x_8}=0$, $\frac{\partial f_{10}}{\partial x_9}=0$, $\frac{\partial f_{10}}{\partial x_{10}}=0$, $\frac{\partial f_{10}}{\partial x_{11}}=0$, $\frac{\partial f_{10}}{\partial x_{12}}=0$, $\frac{\partial f_{10}}{\partial x_{13}}=0$, $\frac{\partial f_{10}}{\partial x_{14}}=0$.

Eleventh row of the Jacobian matrix $\nabla_x f(x) \mid_{(x^*,u^*)}$: $\frac{\partial f_{11}}{\partial x_1}=0$, $\frac{\partial f_{11}}{\partial x_2}=0$, $\frac{\partial f_{11}}{\partial x_3}=0$, $\frac{\partial f_{11}}{\partial x_4}=0$, $\frac{\partial f_{11}}{\partial x_5}=0$, $\frac{\partial f_{11}}{\partial x_6}=0$, $\frac{\partial f_{11}}{\partial x_7}=0$, $\frac{\partial f_{11}}{\partial x_8}=0$, $\frac{\partial f_{11}}{\partial x_9}=0$, $\frac{\partial f_{11}}{\partial x_{10}}=0$, $\frac{\partial f_{11}}{\partial x_{11}}=0$, $\frac{\partial f_{11}}{\partial x_{12}}=1$, $\frac{\partial f_{11}}{\partial x_{13}}=0$, $\frac{\partial f_{11}}{\partial x_{14}}=0$.

Twelfth row of the Jacobian matrix $\nabla_x f(x) \mid_{(x^*,u^*)}$: $\frac{\partial f_{12}}{\partial x_1}=0$, $\frac{\partial f_{12}}{\partial x_2}=0$, $\frac{\partial f_{12}}{\partial x_3}=0$, $\frac{\partial f_{12}}{\partial x_4}=0$, $\frac{\partial f_{12}}{\partial x_5}=0$, $\frac{\partial f_{12}}{\partial x_6}=0$, $\frac{\partial f_{12}}{\partial x_7}=0$, $\frac{\partial f_{12}}{\partial x_8}=0$, $\frac{\partial f_{12}}{\partial x_9}=0$, $\frac{\partial f_{12}}{\partial x_{10}}=0$, $\frac{\partial f_{12}}{\partial x_{11}}=0$, $\frac{\partial f_{12}}{\partial x_{12}}=0$, $\frac{\partial f_{12}}{\partial x_{13}}=0$, $\frac{\partial f_{12}}{\partial x_{14}}=0$.

Thirteenth row of the Jacobian matrix $\nabla_x f(x)\mid_{(x^*,u^*)}$: $\frac{\partial f_{13}}{\partial x_1}=0$, $\frac{\partial f_{13}}{\partial x_2}=0$, $\frac{\partial f_{13}}{\partial x_3}=0$, $\frac{\partial f_{13}}{\partial x_4}=0$, $\frac{\partial f_{13}}{\partial x_5}=0$, $\frac{\partial f_{13}}{\partial x_6}=0$, $\frac{\partial f_{13}}{\partial x_7}=0$, $\frac{\partial f_{13}}{\partial x_8}=0$, $\frac{\partial f_{13}}{\partial x_9}=0$, $\frac{\partial f_{13}}{\partial x_{10}}=0$, $\frac{\partial f_{13}}{\partial x_{11}}=0$, $\frac{\partial f_{13}}{\partial x_{12}}=0$, $\frac{\partial f_{13}}{\partial x_{13}}=0$, $\frac{\partial f_{13}}{\partial x_{14}}=1$.

Fourteenth row of the Jacobian matrix $\nabla_x f(x)\mid_{(x^*,u^*)}$: $\frac{\partial f_{14}}{\partial x_1}=0$, $\frac{\partial f_{14}}{\partial x_2}=0$, $\frac{\partial f_{14}}{\partial x_3}=0$, $\frac{\partial f_{14}}{\partial x_4}=0$, $\frac{\partial f_{14}}{\partial x_5}=0$, $\frac{\partial f_{14}}{\partial x_6}=0$, $\frac{\partial f_{14}}{\partial x_7}=0$, $\frac{\partial f_{14}}{\partial x_8}=0$, $\frac{\partial f_{14}}{\partial x_9}=0$, $\frac{\partial f_{14}}{\partial x_{10}}=0$, $\frac{\partial f_{14}}{\partial x_{11}}=0$, $\frac{\partial f_{14}}{\partial x_{12}}=0$, $\frac{\partial f_{14}}{\partial x_{13}}=0$, $\frac{\partial f_{14}}{\partial x_{14}}=0$.

4.2.4.3 Stabilizing feedback control

After linearization around its current operating point (x^*, u^*), the dynamic model of the VSI-fed six-phase PMSM is written as (Rigatos, 2016b)

$$\dot{x} = Ax + Bu + d_1 \tag{4.139}$$

Parameter d_1 stands for the linearization error in the VSI-fed six-phase PMSM 's model appearing previously in (4.139). The reference setpoints for the VSI-fed six-phase PMSM 's state vector are denoted by $\mathbf{x_d} = [x_1^d, \cdots, x_{14}^d]$. Tracking of this trajectory is achieved after applying the control input u_d. At every time instant, the control input u_d is assumed to differ from the control input u appearing in (4.139) by an amount equal to Δu, that is $u_d = u + \Delta u$

$$\dot{x}_d = Ax_d + Bu_d + d_2 \tag{4.140}$$

The dynamics of the controlled system described in (4.139) can be also written as

$$\dot{x} = Ax + Bu + Bu_d - Bu_d + d_1 \tag{4.141}$$

and by denoting $d_3 = -Bu_d + d_1$ as an aggregate disturbance term one obtains

$$\dot{x} = Ax + Bu + Bu_d + d_3 \tag{4.142}$$

By subtracting (4.140) from (4.142) one has

$$\dot{x} - \dot{x}_d = A(x - x_d) + Bu + d_3 - d_2 \tag{4.143}$$

By denoting the tracking error as $e = x - x_d$ and the aggregate disturbance term as $\tilde{d} = d_3 - d_2$, the tracking error dynamics becomes

$$\dot{e} = Ae + Bu + \tilde{d} \tag{4.144}$$

For the approximately linearized model of the system a stabilizing feedback controller, of the optimal (H-infinity) control-type, is developed. The controller has the form

$$u(t) = -Ke(t) \tag{4.145}$$

with $K = \frac{1}{r}B^T P$ where P is a positive definite symmetric matrix which is obtained from the solution of the Riccati equation (Rigatos, 2016b)

$$A^T P + PA + Q - P\left(\frac{2}{r}BB^T - \frac{1}{\rho^2}LL^T\right)P = 0 \tag{4.146}$$

where Q is a positive semi-definite symmetric matrix. The previously analyzed concept about the nonlinear optimal control loop for the VSI-fed six-phase PMSM system

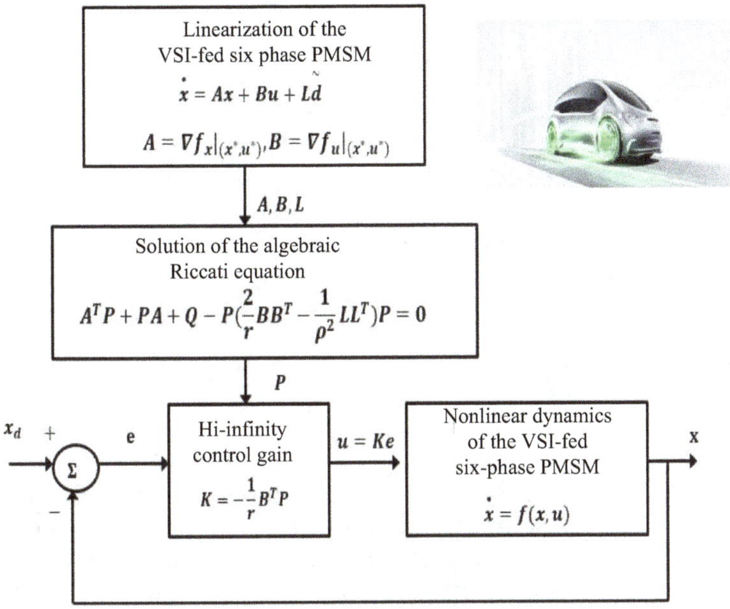

Figure 4.17 Diagram of the control scheme for the Voltage Source Inverter-fed six-phase PMSM (with potential application to an Electric Vehicle's traction system)

is given in Figure 4.17. The global stability properties of the control method can be proven following the procedure presented in Section 1.10.2.5.

It is noted that the solution of the H-infinity feedback control problem for the VSI-fed six-phase PMSM and the computation of the worst-case disturbance that this controller can sustain, comes from the superposition of Bellman's optimality principle when considering that the VSI-fed six-phase PMSM system is affected by two separate inputs (i) the control input u, (ii) the cumulative disturbance input $\tilde{d}(t)$. Solving the optimal control problem for u that is for the minimum variation (optimal) control input that achieves the elimination of the state vector's tracking error gives $u = -\frac{1}{r}B^T Pe$. Equivalently, solving the optimal control problem for $\tilde{d}$, that is for the worst case disturbance that the control loop can sustain gives $\tilde{d} = \frac{1}{\rho^2}L^T Pe$.

4.2.5 Simulation tests

The performance of the proposed nonlinear optimal (H-infinity) control scheme for VSI-fed six-phase PMSMs has been further confirmed through simulation experiments. To compute the stabilizing H-infinity feedback control, the algebraic Riccati equation of (4.146) had to be iteratively solved at each time-step of the control algorithm. The obtained results are depicted in Figures 4.18 to 4.33. The real values of the

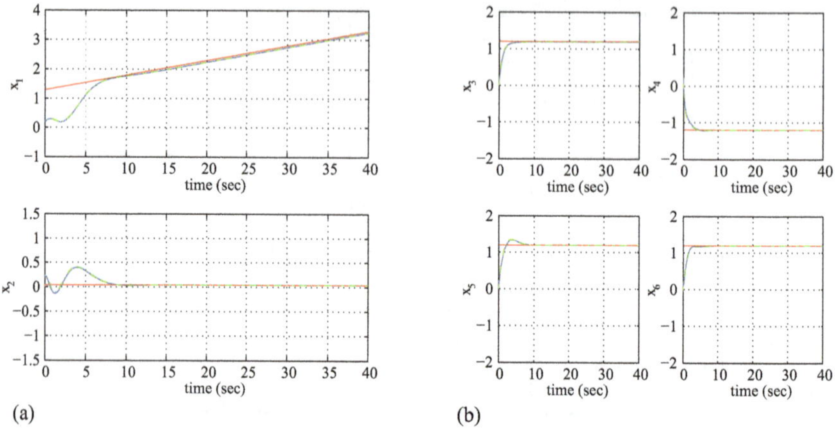

Figure 4.18 Tracking of setpoint 1 for the VSI-fed 6-phase PMSM system (a) convergence of state variables x_1 (turn angle of the motor θ) and x_2 (angular speed of the motor ω) to their reference setpoints (red line: setpoint, blue line: real value, green line: estimated value), (b) convergence of state variables x_3 to x_6 to their reference setpoints (red line: setpoint, blue line: real value, green line: estimated value)

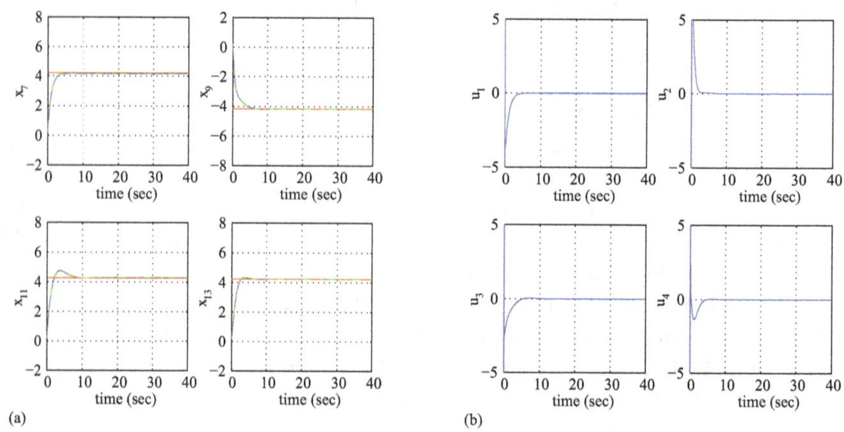

Figure 4.19 Tracking of setpoint 1 for the VSI-fed 6-phase PMSM system (a) tracking error for state variables x_7 to x_{14} (state variables of the Voltage Source Inverters), (b) control inputs u_1 to u_4 (used for calculating finally the inverters' control voltages $V_{I,d}$, $V_{I,q}$)

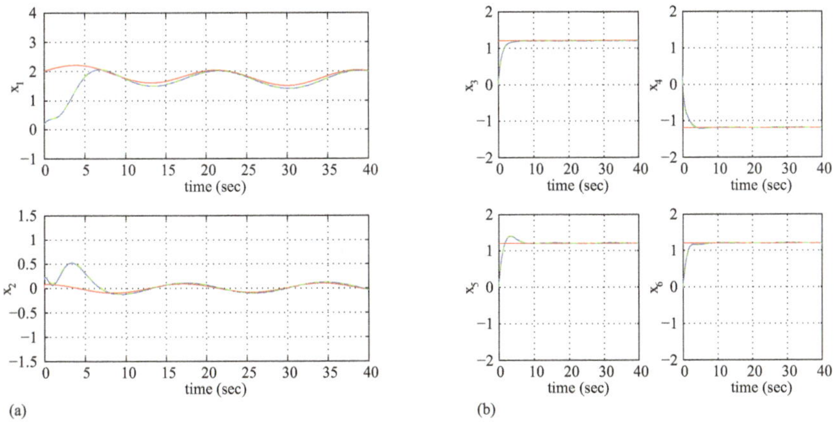

Figure 4.20 *Tracking of setpoint 2 for the VSI-fed 6-phase PMSM system (a) convergence of state variables x_1 (turn angle of the motor θ) and x_2 (angular speed of the motor ω) to their reference setpoints (red line: setpoint, blue line: real value, green line: estimated value), (b) convergence of state variables x_3 to x_6 to their reference setpoints (red line: setpoint, blue line: real value, green line: estimated value)*

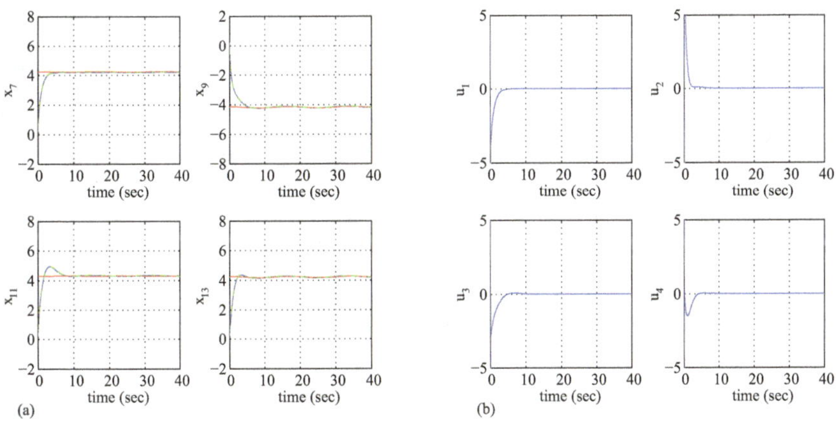

Figure 4.21 *Tracking of setpoint 2 for the VSI-fed 6-phase PMSM system (a) tracking error for state variables x_7 to x_{14} (state variables of the Voltage Source Inverters), (b) control inputs u_1 to u_4 (used for calculating finally the inverters' control voltages $V_{I,d}$, $V_{I,q}$)*

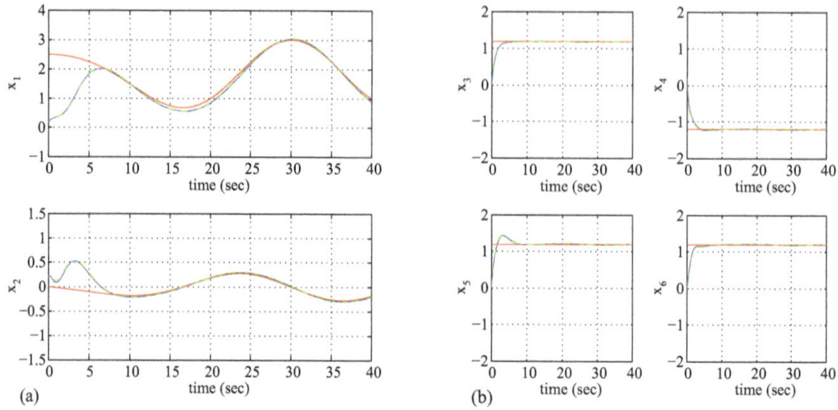

Figure 4.22 Tracking of setpoint 3 for the VSI-fed 6-phase PMSM system (a) convergence of state variables x_1 (turn angle of the motor θ) and x_2 (angular speed of the motor ω) to their reference setpoints (red line: setpoint, blue line: real value, green line: estimated value), (b) convergence of state variables x_3 to x_6 to their reference setpoints (red line: setpoint, blue line: real value, green line: estimated value)

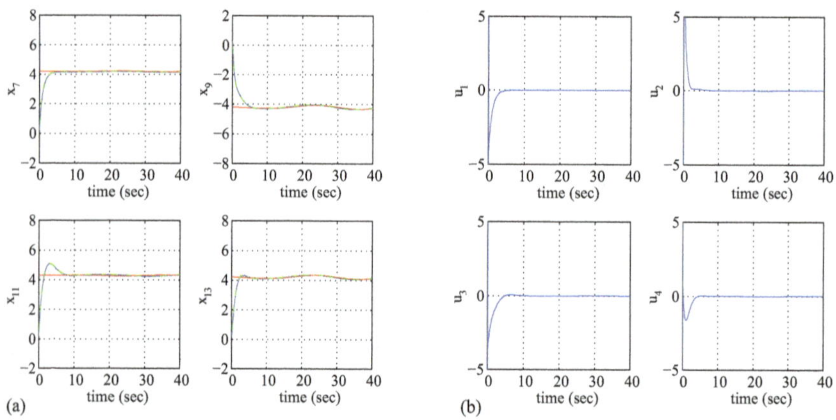

Figure 4.23 Tracking of setpoint 3 for the VSI-fed 6-phase PMSM system (a) tracking error for state variables x_7 to x_{14} (state variables of the Voltage Source Inverters), (b) control inputs u_1 to u_4 (used for calculating finally the inverters' control voltages $V_{1,d}$, $V_{1,q}$)

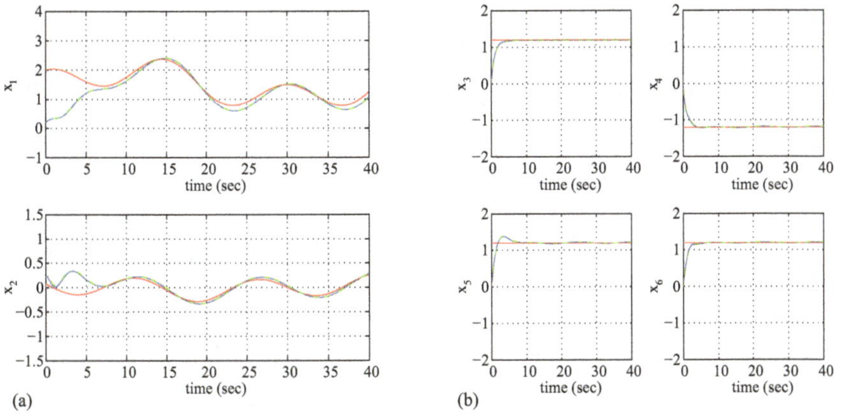

Figure 4.24 *Tracking of setpoint 4 for the VSI-fed 6-phase PMSM system (a) convergence of state variables x_1 (turn angle of the motor θ) and x_2 (angular speed of the motor ω) to their reference setpoints (red line: setpoint, blue line: real value, green line: estimated value), (b) convergence of state variables x_3 to x_6 to their reference setpoints (red line: setpoint, blue line: real value, green line: estimated value)*

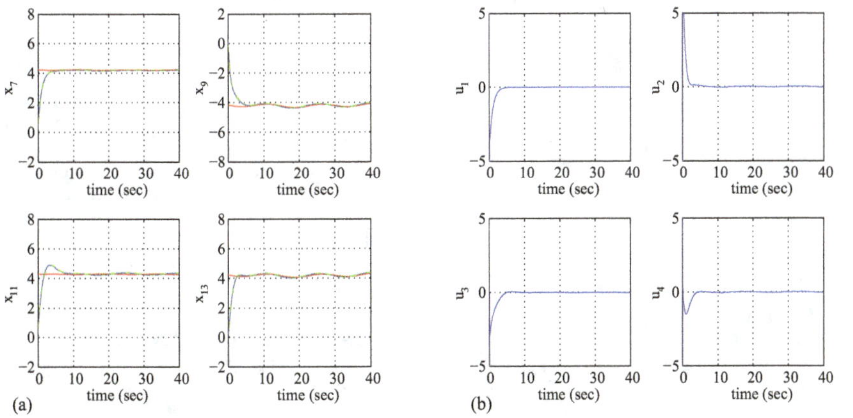

Figure 4.25 *Tracking of setpoint 4 for the VSI-fed 6-phase PMSM system (a) tracking error for state variables x_7 to x_{14} (state variables of the Voltage Source Inverters), (b) control inputs u_1 to u_4 (used for calculating finally the inverters' control voltages $V_{I,d}$, $V_{I,q}$)*

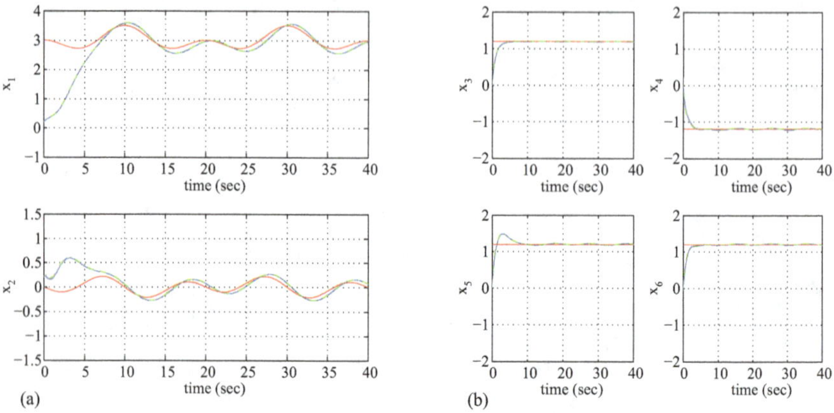

Figure 4.26 Tracking of setpoint 5 for the VSI-fed 6-phase PMSM system (a) convergence of state variables x_1 (turn angle of the motor θ) and x_2 (angular speed of the motor ω) to their reference setpoints (red line: setpoint, blue line: real value, green line: estimated value), (b) convergence of state variables x_3 to x_6 to their reference setpoints (red line: setpoint, blue line: real value, green line: estimated value)

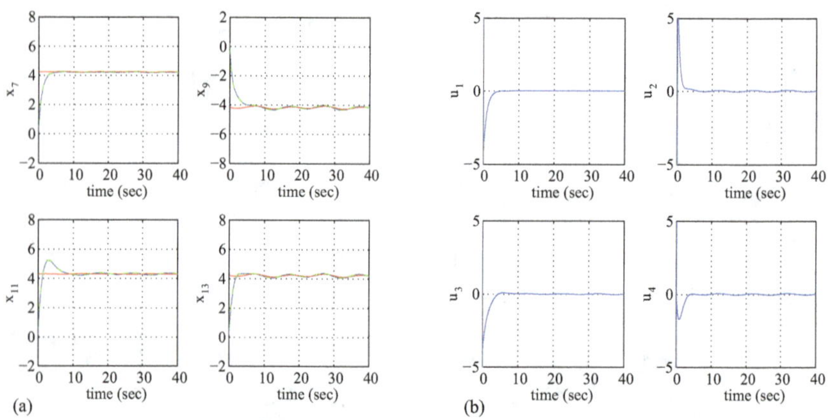

Figure 4.27 Tracking of setpoint 5 for the VSI-fed 6-phase PMSM system (a) tracking error for state variables x_7 to x_{14} (state variables of the Voltage Source Inverters), (b) control inputs u_1 to u_4 (used for calculating finally the inverters' control voltages $V_{I,d}$, $V_{I,q}$)

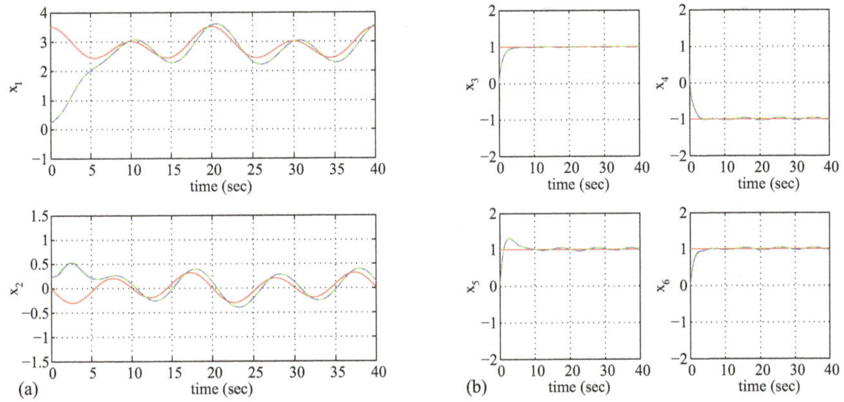

Figure 4.28 *Tracking of setpoint 6 for the VSI-fed 6-phase PMSM system (a) convergence of state variables x_1 (turn angle of the motor θ) and x_2 (angular speed of the motor ω) to their reference setpoints (red line: setpoint, blue line: real value, green line: estimated value), (b) convergence of state variables x_3 to x_6 to their reference setpoints (red line: setpoint, blue line: real value, green line: estimated value)*

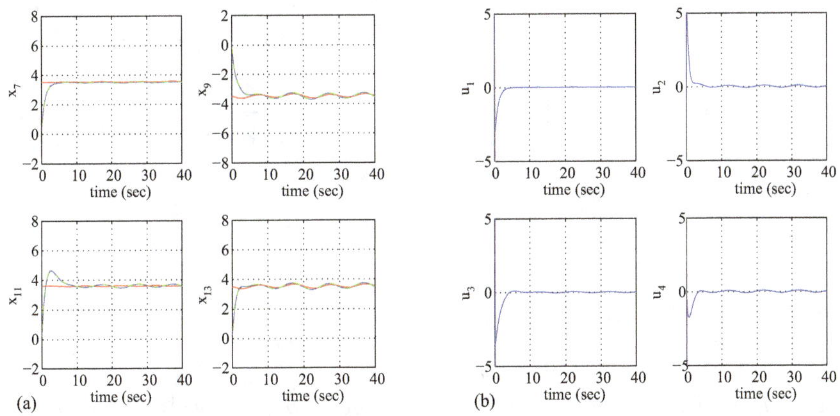

Figure 4.29 *Tracking of setpoint 6 for the VSI-fed 6-phase PMSM system (a) tracking error for state variables x_7 to x_{14} (state variables of the Voltage Source Inverters), (b) control inputs u_1 to u_4 (used for calculating finally the inverters' control voltages $V_{I,d}$, $V_{I,q}$)*

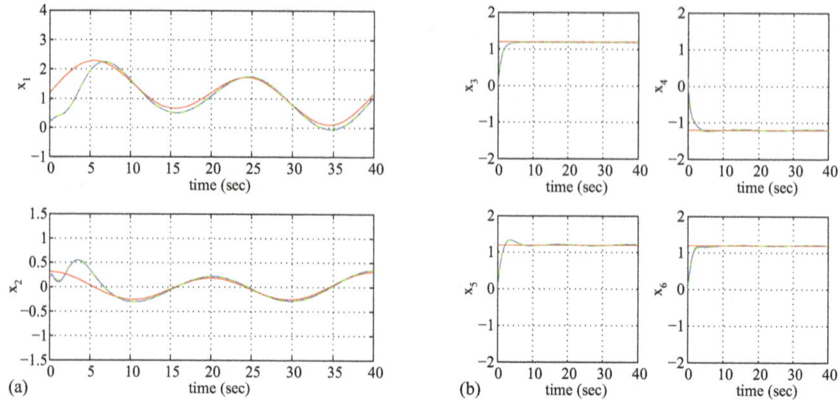

Figure 4.30 Tracking of setpoint 7 for the VSI-fed 6-phase PMSM system (a) convergence of state variables x_1 (turn angle of the motor θ) and x_2 (angular speed of the motor ω) to their reference setpoints (red line: setpoint, blue line: real value, green line: estimated value), (b) convergence of state variables x_3 to x_6 to their reference setpoints (red line: setpoint, blue line: real value, green line: estimated value)

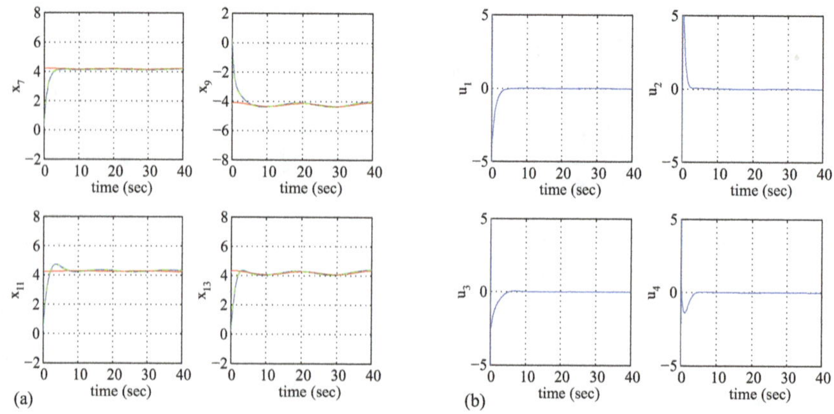

Figure 4.31 Tracking of setpoint 7 for the VSI-fed 6-phase PMSM system (a) tracking error for state variables x_7 to x_{14} (state variables of the Voltage Source Inverters), (b) control inputs u_1 to u_4 (used for calculating finally the inverters' control voltages $V_{I,d}$, $V_{I,q}$)

state variables are printed in blue, the estimated values are plotted in green while the related setpoints are depicted in red. The transient performance of the control scheme was determined by the selection of parameters r, ρ and Q. Actually, the smallest value of ρ for which one can obtain a solution of the above noted Riccati equation is

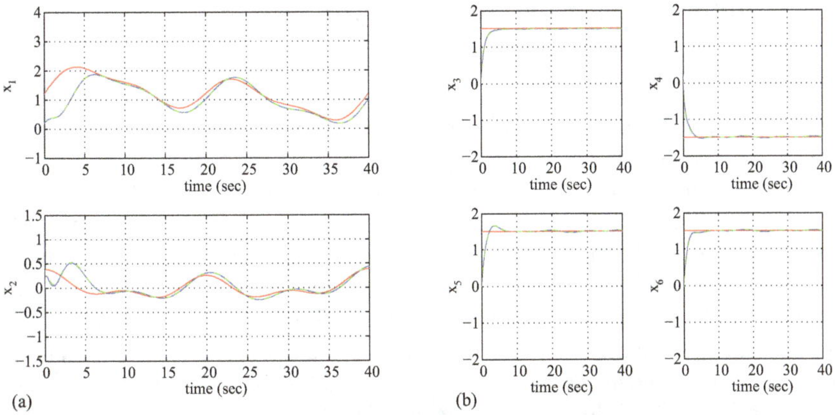

Figure 4.32 Tracking of setpoint 8 for the VSI-fed 6-phase PMSM system (a) convergence of state variables x_1 (turn angle of the motor θ) and x_2 (angular speed of the motor ω) to their reference setpoints (red line: setpoint, blue line: real value, green line: estimated value), (b) convergence of state variables x_3 to x_6 to their reference setpoints (red line: setpoint, blue line: real value, green line: estimated value)

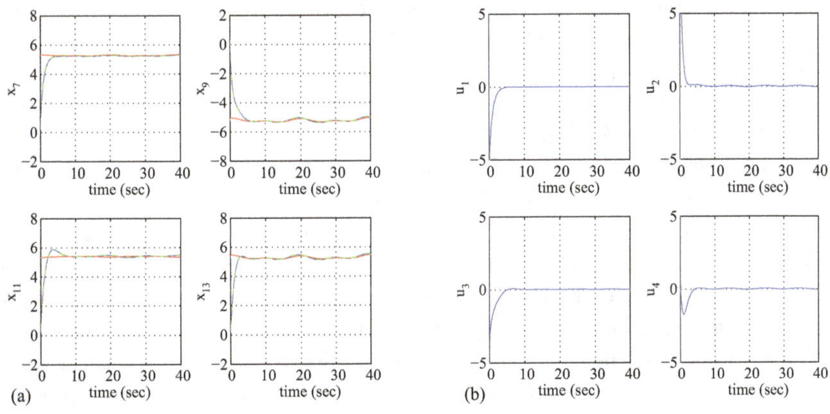

Figure 4.33 Tracking of setpoint 8 for the VSI-fed 6-phase PMSM system (a) tracking error for state variables x_7 to x_{14} (state variables of the Voltage Source Inverters), (b) control inputs u_1 to u_4 (used for calculating finally the inverters' control voltages $V_{I,d}$, $V_{I,q}$)

the one that provides the control scheme with maximum robustness. From the simulation experiments it can be noticed that the nonlinear optimal (H-infinity) control method achieved fast and accurate tracking of the reference setpoints under moderate variations of the control inputs.

4.3 Nonlinear control of DC electric microgrids

4.3.1 Outline

The use of DC industrial microgrids is rapidly deploying as a solution to the problem of uninterrupted power supply of industrial units and particularly of energy intensive ones (e.g. cement mills or steel rolling mills) (Rigatos, 2016b; Disqah *et al.*, 2015; Kardan *et al.*, 2019). Electric power accounts for a significant part of the functioning costs of energy intensive industries and the optimization of their power supply can contribute significantly towards raising the competitiveness and viability of such units (Orcajo *et al.*, 2016; Orcajo *et al.*, 2018; Chau *et al.*, 2018; Ghiasi *et al.*, 2014). Besides, it is important to keep the power supply uninterrupted and without perturbations so as to assure that the quality features of the scheduled production are reached (Iovine, Benamane *et al.*, 2015; Iovine, Rigaut *et al.*, 2019; Siad *et al.*, 2018; Iovine, Carrizosa *et al.*, 2019; Jamshidpour *et al.*, 2015). Additionally, by supplying electric power to industrial units through dedicated for this purpose microgrids, industrial production becomes less vulnerable to faults, perturbations and load variations in the main electricity grid (Alramlawi *et al.*, 2018; Xu, Liu *et al.*, 2019; Tucci *et al.*, 2018; Zhonglin *et al.*, 2016). To connect industrial units to DC power sources, such as photovoltaics, batteries and supercapacitors, DC to DC converters are used (Lenz *et al.*, 2014; Lemz and Pagano, 2013; Han *et al.*, 2017; Mane, Mejani *et al.*, 2017). To control power supply by DC microgrids, nonlinear control methods have been developed which in most cases do not follow optimal control concepts (Yang, Nahid-Mobarakeh *et al.*, 2016; Heymann *et al.*, 2018; Cavanagh *et al.*, 2018; Cucuzzela *et al.*, 2018). Furthermore by considering linear dynamics of the DC microgrid, model-predictive control has been also proposed (Lou, Gu *et al.*, 2018; Norrozi, Trip *et al.*, 2018; Gao, Li *et al.*, 2017; Mardani, Khooban *et al.*, 2019).

In the present section, a nonlinear optimal (H-infinity) control method is developed for an indicative model of a DC microgrid which comprises photovoltaic units, batteries and supercapacitors (Rigatos, 2016b; Rigatos and Busawon, 2018). The dynamic model of the DC microgrid undergoes first approximate linearization around a temporary operating point which is recomputed at each time-step of the control algorithm (Rigatos and Siano, 2015; Rigatos, Siano and Cecati, 2015). This time-varying operating point is defined by the present value of the DC microgrid's state vector and by the last sampled value of the control inputs vector. The linearization procedure relies on Taylor series expansion and on the computation of the associated Jacobian matrices (Rigatos and Tzafestas, 2007; Basseville and Nikiforov, 1993; Rigatos and Zhang, 2009). The modelling error which is due to truncation of higher-order terms in Taylor series expansion is considered to be a perturbation that is asymptotically compensated by the robustness of the control method. For the approximately linearized model of the system an H-infinity feedback controller is designed.

The proposed H-infinity controller stands for the solution of the optimal control problem for the nonlinear model of the DC microgrid under model uncertainty and external perturbations. Actually, it represents the solution to a min–max differential game in which the control inputs try to minimize a quadratic cost function of the state vector's tracking error, whereas the external perturbations and the model uncertainty terms try to maximize this cost functional. To select the controller's feedback gains an algebraic Riccati equation is repetitively solved at each iteration of the control algorithm (Rigatos, 2016b; Rigatos and Busawon, 2018; Rigatos, 2015). The stability properties of the control scheme are proven through Lyapunov analysis. First, it is demonstrated that the control loop of the DC microgrid satisfies the H-infinity tracking performance criterion. This signifies elevated robustness against model uncertainty and external disturbances (Rigatos, 2016b; Tousssaint, Basar and Bullo, 2000). Moreover, under moderate conditions it is also proven that this control loop exhibits global asymptotic stability. Finally, to implement state estimation-based control without the need to measure the entire state vector of the system, the H-infinity Kalman Filter is used as a robust state estimator (Rigatos, 2016b; Gibbs, 2012).

The present section offers a nonlinear optimal perspective to the nontrivial problem of control of DC microgrids (Lin *et al.*, 2019; Ornelas-Tellez *et al.*, 2018; Vafamand *et al.*, 2019). Popular approaches for industrial control such as MPC or NMPC may have questionable performance when applied to such control problems. Actually, MPC has been developed for linear dynamical systems and its use in the case of the nonlinear model of DC microgrids will risk the control loop's destabilization. Besides, the convergence of NMPC is not assured either. The convergence of the method's iterative search for an optimum depends on initialization and specific parameters' selection, therefore under NMPC one cannot always guarantee a solution for the nonlinear optimal control problem of DC microgrids. Finally, comparing to local models-based optimal control the section's approach exhibits specific advantages: (i) in the local-models based approach linearization is performed around multiple operating points (equilibria) which are selected off-line and which are not updated in time, whereas in the section's approach there is linearization only around one single operating point which is updated at each iteration of the control algorithm, (ii) in the local-models approach there is a need to find a solution of multiple Riccati equations associated with the individual models and this solution is performed offline. On the other side, in the section's approach there is a need to solve one single Riccati equation and this solution is repeated at each time-step of the control algorithm; (iii) in local models-based control there is a need to find a common solution for the individual Riccati equations, and one cannot assure that such a solution always exists. On the other side, in the section's approach there is need to obtain solution for one single Riccati equation and the existence of such a solution can be assured through suitable selection of gains and coefficients. In conclusion, comparing to local models-based control, the section's control method is computationally more efficient and is subject to less constraining assumptions.

4.3.2 Dynamic model of the DC microgrid

An indicative microgrid that can supply with power an industrial unit is shown in the following diagram (Figure 4.34) and is described by the following set of equations (Iovine, Benamane *et al.*, 2015):

$$\dot{x}_1 = -\frac{1}{R_1C_1}x_1 - \frac{1}{C_1}x_3 + \frac{1}{R_1C_1}V_{PV} \tag{4.147}$$

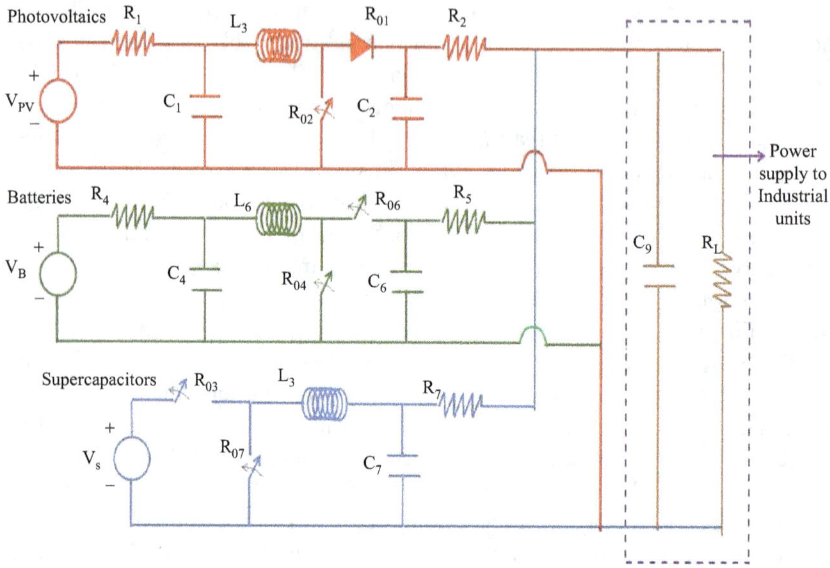

Figure 4.34 Use of a microgrid that comprises photovoltaics, batteries and supercapacitors to provide electric power to industrial units

$$\dot{x}_2 = -\frac{1}{R_2C_2}x_2 - \frac{1}{C_2}x_3 - \frac{1}{C_2}x_3u_1 + \frac{1}{R_2C_2}x_9 \tag{4.148}$$

$$\dot{x}_3 = \frac{1}{L_3}(x_1 - x_2 - R_{o_1}x_3) + \frac{1}{L_3}(x_2 + (R_{o_1} - R_{o_2})x_3)u_1 \tag{4.149}$$

$$\dot{x}_4 = -\frac{1}{R_4C_4}x_4 - \frac{1}{C_4}x_6 + \frac{1}{R_4C_4}V_B \tag{4.150}$$

$$\dot{x}_5 = -\frac{1}{R_5C_5}x_5 + \frac{1}{C_5}x_6 + \frac{1}{R_5C_5}x_9 - \frac{1}{C_5}x_6u_2 \tag{4.151}$$

$$\dot{x}_6 = \frac{1}{L_6}x_4 - \frac{1}{L_6}x_5 - \frac{R_{o4}}{L_6} + \frac{1}{L_6}x_5u_2 \tag{4.152}$$

$$\dot{x}_7 = -\frac{1}{R_7C_7}x_7 + \frac{1}{C_7}x_8 + \frac{1}{R_7C_7}x_9 \tag{4.153}$$

$$\dot{x}_8 = \frac{1}{L_8}V_su_3 - \frac{R_{08}}{L_8}x_8 - \frac{1}{L_8}x_7 \tag{4.154}$$

$$\dot{x}_9 = \frac{1}{C_9}\left(\frac{x_2 - x_9}{R_2} + \frac{x_5 - x_9}{R_5} + \frac{x_7 - x_9}{R_7} - x_9\frac{1}{R_L}\right) \tag{4.155}$$

The state variables of the dynamic model of the DC microgrid are defined as follows (Iovine, Benamane *et al.*, 2015): x_1 is voltage V_{c_1} at capacitor C_1, x_2 is voltage V_{c_2} at capacitor C_2, x_3 is current i_{L_3} at the inductor L_3, x_4 is voltage V_{C_4} at the capacitor C_4, x_5 is voltage V_{C_5} at the capacitor C_5, x_6 is current i_{L_6} at the inductor L_6, x_7 is voltage

V_{C_7} at the capacitor C_7, x_8 is current i_{L_8} at the inductor L_8 and x_9 is voltage V_{C_9} at the capacitor C_9.

Control is implemented with the use of the Pulse Width Modulation (PWM) approach. The control inputs of the dynamic model of the DC microgrid are: u_1 which stands for the duty cycle of the switch at the PV circuit, u_2 which is the duty cycle of the switch at the battery circuit and u_3 which is the duty cycle of the switch at the supercapacitor circuit.

Next, the state-space model of the DC microgrid is written in the following matrix form

$$\begin{pmatrix} \dot{x}_1 \\ \dot{x}_2 \\ \dot{x}_3 \\ \dot{x}_4 \\ \dot{x}_5 \\ \dot{x}_6 \\ \dot{x}_7 \\ \dot{x}_8 \\ \dot{x}_9 \end{pmatrix} = \begin{pmatrix} -\frac{1}{R_1C_1}x_1 - \frac{1}{C_1}x_3 + \frac{1}{R_1C_1}V_{PV} \\ -\frac{1}{R_2C_2}x_2 - \frac{1}{C_2}x_3 + \frac{1}{R_2C_2}x_9 \\ \frac{1}{L_3}(x_1 - x_2 - R_{o_1}x_3) \\ -\frac{1}{R_4C_4}x_4 - \frac{1}{C_4}x_6 + \frac{1}{R_4C_4}V_B \\ -\frac{1}{R_5C_5}x_5 + \frac{1}{C_5}x_6 + \frac{1}{R_5C_5}x_9 \\ \frac{1}{L_6}x_4 - \frac{1}{L_6}x_5 - \frac{R_{o4}}{L_6} \\ -\frac{1}{R_7C_7}x_7 + \frac{1}{C_7}x_8 + \frac{1}{R_7C_7}x_9 \\ -\frac{R_{08}}{L_8}x_8 - \frac{1}{L_8}x_7 \\ \frac{1}{C_9}\left(\frac{x_2 - x_9}{R_2} + \frac{x_5 - x_9}{R_5} + \frac{x_7 - x_9}{R_7} - x_9\frac{1}{R_L}\right) \end{pmatrix}$$
$$+ \begin{pmatrix} 0 & 0 & 0 \\ -\frac{1}{C_2}x_3 & 0 & 0 \\ \frac{1}{L_3}(x_2 + (R_{o_1} - R_{o_2})x_3) & 0 & 0 \\ 0 & 0 & 0 \\ 0 & -\frac{1}{C_5}x_6 & 0 \\ 0 & \frac{1}{L_6}x_5 & 0 \\ 0 & 0 & 0 \\ 0 & 0 & \frac{1}{L_8}V_s \\ 0 & 0 & 0 \end{pmatrix} \begin{pmatrix} u_1 \\ u_2 \\ u_3 \end{pmatrix} \qquad (4.156)$$

Consequently, the state-space model of DC microgrid is written in the following affine-in-the-input state-space form

$$\dot{x} = f(x) + g(x)u \tag{4.157}$$

where $x \in R^{9\times 1}$, $u \in R^{3\times 1}$, $f(x) \in R^{9\times 1}$ and $g(x) \in R^{9\times 3}$.

4.3.3 Approximate linearization of the state-space model of the DC microgrid

The dynamic model of the DC microgrid undergoes approximate linearization around the temporary operating point (x^*, u^*) which is recomputed at each time-step of the control method. This temporary operating point is defined by x^* which is the present value of the state vector of the DC microgrid and by u^* which is the last value of the control inputs vector that was applied to the system. The linearization procedure makes use of Taylor series expansion and requires the computation of the associated Jacobian matrices. The truncation of higher-order terms in the Taylor series results into modelling error which is finally compensated by the robustness of the control algorithm. The linearization procedure gives the new state-space model:

$$\dot{x} = Ax + Bu + \tilde{d} \tag{4.158}$$

where $\tilde{d}$ is the cumulative disturbance term while it holds that

$$A = \nabla_x[f(x) + g_1(x)u_1 + g_2(x)u_2 + g_3(x)u_3] \mid_{(x^*,u^*)} \Rightarrow$$

$$\begin{aligned} A = {} & \nabla_x[f(x)] \mid_{(x^*,u^*)} + \nabla_x[g_1(x)]u_1 \mid_{(x^*,u^*)} \\ & + \nabla_x[g_2(x)]u_2 \mid_{(x^*,u^*)} + \nabla_x[g_3(x)]u_3 \mid_{(x^*,u^*)} \end{aligned} \tag{4.159}$$

$$\begin{aligned} & B = \nabla_x[f(x) + g_1(x)u_1 + g_2(x)u_2 + g_3(x)u_3] \mid_{(x^*,u^*)} \\ & \Rightarrow B = g(x) \mid_{(x^*,u^*)} \end{aligned} \tag{4.160}$$

Computation of the Jacobian matrix $\nabla_x[f(x)]\,|_{(x^*,u^*)}$

$$\nabla_x[f(x)]\,|_{(x^*,u^*)} = \begin{pmatrix} -\frac{1}{R_1C_1} & 0 & -\frac{1}{C_3} & 0 & 0 & 0 & 0 & 0 & 0 \\ 0 & -\frac{1}{R_2C_2} & \frac{1}{C_3} & 0 & 0 & 0 & 0 & 0 & 0 \\ \frac{1}{L_3} & -\frac{1}{L_3} & -\frac{R_{o1}}{L_3} & 0 & 0 & 0 & 0 & 0 & 0 \\ 0 & 0 & 0 & -\frac{1}{R_4C_4} & 0 & -\frac{1}{C_4} & 0 & 0 & 0 \\ 0 & 0 & 0 & 0 & -\frac{1}{R_5C_5} & -\frac{1}{C_5} & 0 & 0 & \frac{1}{R_5c_5} \\ 0 & 0 & 0 & \frac{1}{L_6} & -\frac{1}{L_6} & -\frac{R_{o4}}{L_6} & 0 & 0 & 0 \\ 0 & 0 & 0 & 0 & 0 & 0 & -\frac{1}{R_7C_7} & \frac{1}{C_7} & \frac{1}{R_7C_7} \\ 0 & 0 & 0 & 0 & 0 & 0 & -\frac{1}{L_8} & -\frac{R_{o8}}{L_8} & 0 \\ 0 & \frac{1}{C_9R_2} & 0 & 0 & \frac{1}{C_9R_5} & 0 & \frac{1}{C_9R_7} & 0 & \frac{\partial f_9}{\partial x_9} \end{pmatrix} \tag{4.161}$$

where $\frac{\partial f_9}{\partial x_9} = -\frac{1}{C_9R_2} - \frac{1}{C_9R_5} - \frac{1}{C_9R_7} - \frac{1}{R_L}$

Computation of the Jacobian matrix $\nabla_x[g_1(x)]\,|_{(x^*,u^*)}$

$$\nabla_x g_1[x)]\,|_{(x^*,u^*)} = \begin{pmatrix} 0 & 0 & 0 & 0 & 0 & 0 & 0 & 0 & 0 \\ 0 & 0 & -\frac{1}{C_2} & 0 & 0 & 0 & 0 & 0 & 0 \\ 0 & \frac{1}{L_3} & \frac{R_{o1}-R_{o2}}{L_3} & 0 & 0 & 0 & 0 & 0 & 0 \\ 0 & 0 & 0 & 0 & 0 & 0 & 0 & 0 & 0 \\ 0 & 0 & 0 & 0 & 0 & 0 & 0 & 0 & 0 \\ 0 & 0 & 0 & 0 & 0 & 0 & 0 & 0 & 0 \\ 0 & 0 & 0 & 0 & 0 & 0 & 0 & 0 & 0 \\ 0 & 0 & 0 & 0 & 0 & 0 & 0 & 0 & 0 \\ 0 & 0 & 0 & 0 & 0 & 0 & 0 & 0 & 0 \end{pmatrix} \tag{4.162}$$

Computation of the Jacobian matrix $\nabla_x[g_2(x)]\mid_{(x^*,u^*)}$

$$\nabla_x g_2[x)]\mid_{(x^*,u^*)}=\begin{pmatrix} 0&0&0&0&0&0&0&0&0\\ 0&0&0&0&0&0&0&0&0\\ 0&0&0&0&0&0&0&0&0\\ 0&0&0&0&0&-\frac{1}{C_5}&0&0&0\\ 0&0&0&0&\frac{1}{C_6}&0&0&0&0\\ 0&0&0&0&0&0&0&0&0\\ 0&0&0&0&0&0&0&0&0\\ 0&0&0&0&0&0&0&0&0\\ 0&0&0&0&0&0&0&0&0 \end{pmatrix} \tag{4.163}$$

Computation of the Jacobian matrix $\nabla_x[g_3(x)]\mid_{(x^*,u^*)}$: $\nabla_x g_3[x)]\mid_{(x^*,u^*)}=0_{9\times 9}$.
where Q is also a positive definite symmetric matrix. The worst-case disturbance is given by $\tilde{d}(t)=\frac{1}{\rho^2}L^TPx(t)$. This equation is obtained by solving the optimal control problem for the case that the system receives as input only the disturbance $\tilde{d}(t)$.

4.3.4 Computation of the feedback control gains

For the linearized system given by (1.482) the cost function of (1.483) is defined, where coefficient r determines the penalization of the control input and weight coefficient ρ determines the reward of the disturbances' effects. It is assumed that (i) The energy that is transferred from the disturbances signal $\tilde{d}(t)$ is bounded, that is $\int_0^\infty \tilde{d}^T(t)\tilde{d}(t)dt<\infty$, (ii) matrices $[A,B]$ and $[A,L]$ are stabilizable, (iii) matrix $[A,C]$ is detectable. In the case of a tracking problem the optimal feedback control law is given by

$$u(t)=-Ke(t) \tag{4.164}$$

with $e=x-x_d$ to be the tracking error, and $K=\frac{1}{r}B^TP$ where P is a positive definite symmetric matrix. Matrix P is obtained from the solution of the Riccati equation

$$A^TP+PA+Q-P\left(\frac{2}{r}BB^T-\frac{1}{\rho^2}LL^T\right)P=0 \tag{4.165}$$

where Q is a positive semi-definite symmetric matrix. The worst-case disturbance is given by

$$\tilde{d}(t)=\frac{1}{\rho^2}L^TPe(t) \tag{4.166}$$

The previously analyzed concept about the nonlinear optimal control loop for the DC microgrid's system is given in Figure 4.35. The global stability properties of the control method can be proven following the procedure presented in Section 1.10.2.5.

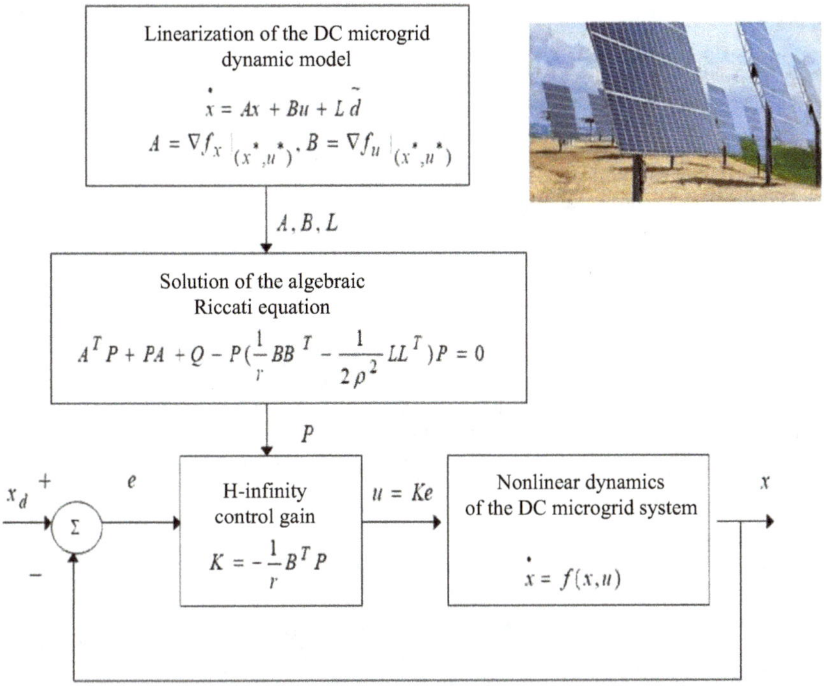

Figure 4.35 Diagram of the control scheme for the DC microgrid's system

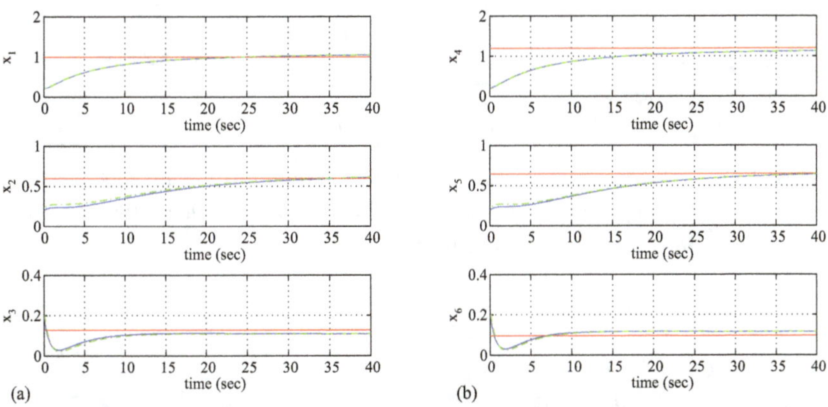

Figure 4.36 Test case 1: (a) tracking of reference setpoints by the state variables x_1 to x_3 of the DC microgrid; (b) tracking of reference setpoints by the state variables x_4 to x_6 of the DC microgrid

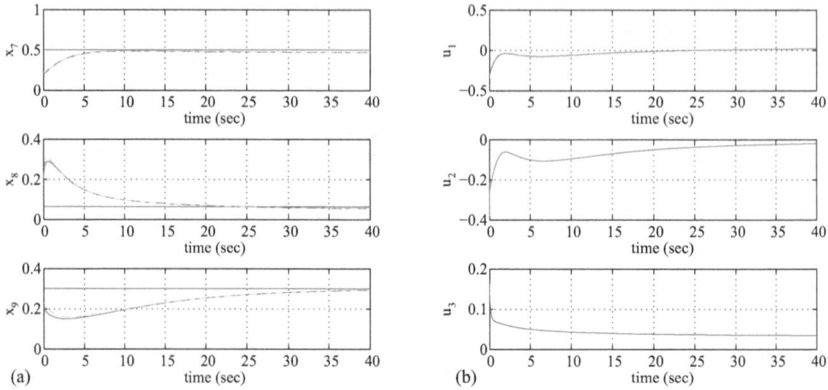

Figure 4.37 *Test case 1: (a) tracking of reference setpoints by the state variables x_7 to x_9 of the DC microgrid; (b) control inputs u_1 to u_3 of the DC microgrid*

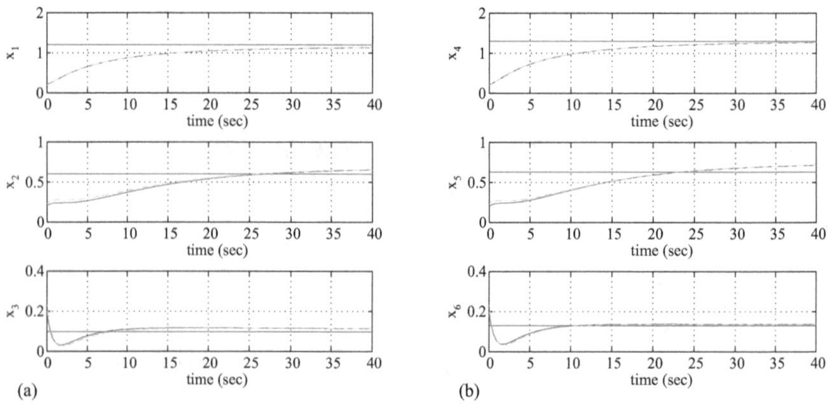

Figure 4.38 *Test case 2: (a) tracking of reference setpoints by the state variables x_1 to x_3 of the DC microgrid; (b) tracking of reference setpoints by the state variables x_4 to x_6 of the DC microgrid*

4.3.5 Simulation tests

The performance of the nonlinear optimal control for the dynamic model of the DC microgrid was tested through simulation experiments. To select the feedback gains of the H-infinity feedback controller the algebraic Riccati equation which appears in (1.496) had to be repetitively solved at each time-step of the control method. The obtained results are depicted in Figures 4.36 to 4.43. The real value of the state vector elements of the system is shown in blue, the estimated value as provided by the

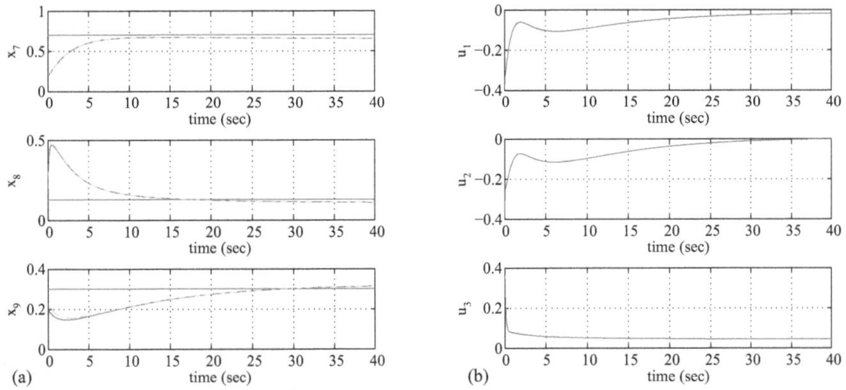

Figure 4.39 *Test case 2: (a) tracking of reference setpoints by the state variables x_7 to x_9 of the DC microgrid; (b) control inputs u_1 to u_3 of the DC microgrid*

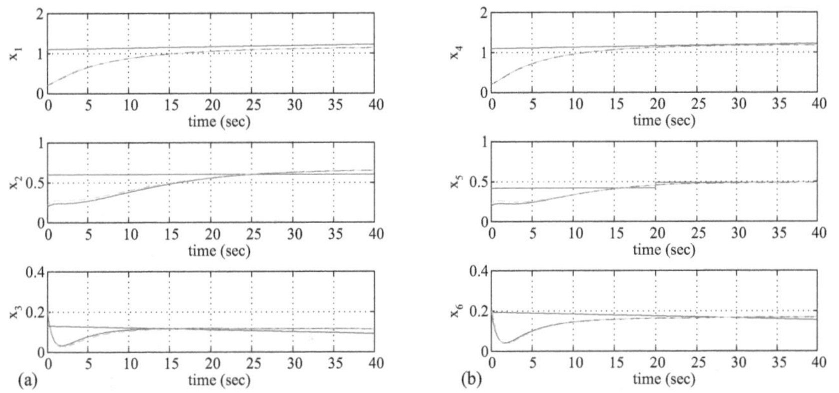

Figure 4.40 *Test case 3: (a) tracking of reference setpoints by the state variables x_1 to x_3 of the DC microgrid; (b) tracking of reference setpoints by the state variables x_4 to x_6 of the DC microgrid*

H-infinity Kalman Filter is shown in green, while the associated reference setpoints are printed in red. The simulation tests show that fast and accurate convergence of the state variables to their setpoints was achieved, under moderate variations of the control inputs.

The proposed control method achieves minimization of the tracking error for the state variables of the DC microgrid, and also minimizes the electric power consumed by it. The use of the H-infinity Kalman Filter has enabled to implement state estimation-based control without the need to measure the entire state vector of the DC

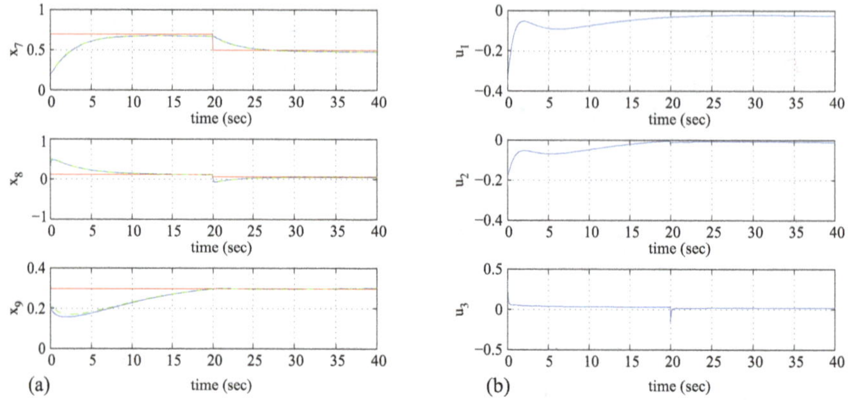

Figure 4.41 Test case 3: (a) tracking of reference setpoints by the state variables x_7 to x_9 of the DC microgrid; (b) control inputs u_1 to u_3 of the DC microgrid

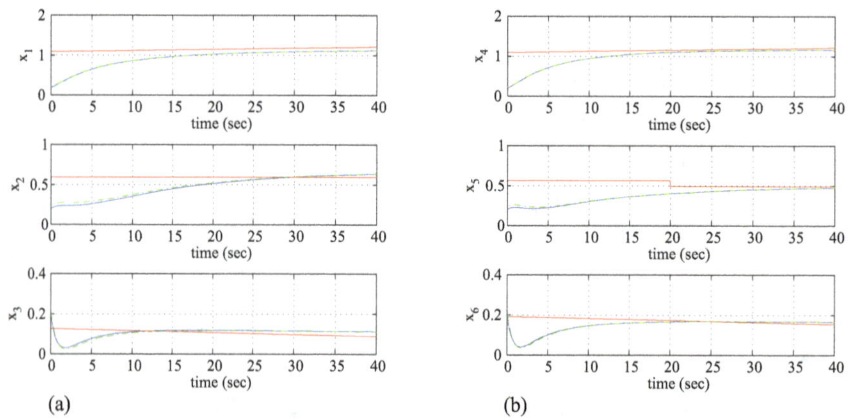

Figure 4.42 Test case 4: (a) tracking of reference setpoints by the state variables x_1 to x_3 of the DC microgrid; (b) tracking of reference setpoints by the state variables x_4 to x_6 of the DC microgrid

microgrid. Actually, stabilizing control could be applied to the system after measuring only state variables x_1, x_4 and x_7. Regarding the transient performance of the control method this is affected by the gains and parameters r, ρ and Q which appear in the Riccati equation of (4.165). Actually, the smallest value of the attenuation coefficient ρ for which a valid solution of the Riccati equation can be obtained, is the one that provides the control loop with maximum robustness.

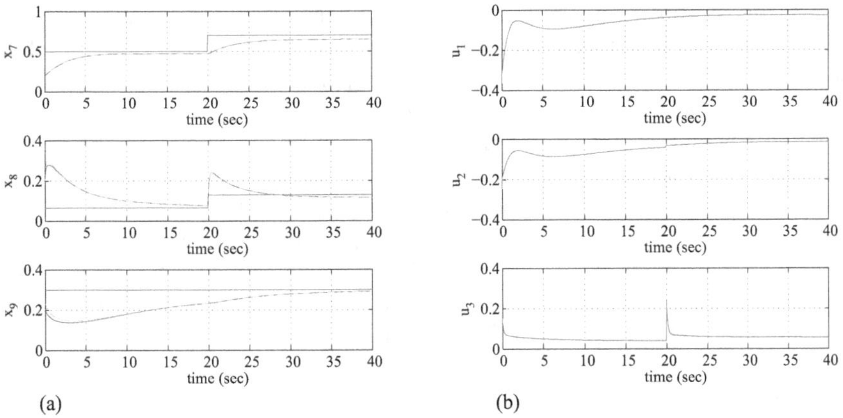

Figure 4.43 Test case 4: (a) tracking of reference setpoints by the state variables x_7 to x_9 of the DC microgrid; (b) control inputs u_1 to u_3 of the DC microgrid

4.4 Nonlinear control of distributed marine-turbine power generation units

4.4.1 Outline

Power generation from sea-currents and tidal streams offers a vast potential for raising the rates of renewable energy production and for covering a significant part of the energy consumption needs of the electricity grid (Rigatos, 2016b; Chabane *et al.*, 2018; Jahrami *et al.*, 2013; Chen, Tang *et al.*, 2018). To arrive at producing electric power for marine turbines in a reliable and uninterrupted manner, stabilization of the functioning of such power units and their synchronization to the grid's frequency is needed (Mahmud, Pota *et al.*, 2014; Galloway *et al.*, 2019; Song and Cameron, 2017; Jahrami *et al.*, 2013). To this end, nonlinear control schemes for marine-turbine power generation units have been developed (Colbio-Vega *et al.*, 2008; Dansoko *et al.*, 2018; Omkar *et al.*, 2019; Dansoko *et al.*, 2013). Control of distributed marine-turbine power systems is a non-trivial problem due to the nonlinearities and the multivariable structure of the associated state-space model (Ying, Lei *et al.*, 2018; Xu, Li *et al.*, 2016; Fall *et al.*, 2014; Mohanty *et al.*, 2016). Besides, such a control has to compensate for the interactions between the individual generators which appear when connecting the power generators to the grid's transmission lines (Dansoko *et al.*, 2014; Nachtane *et al.*, 2017; Aly, 2016). Optimality in the control of distributed marine-turbine power units has been little investigated so-far (Mekri *et al.*, 2017; Fan, Mu *et al.*, 2016; Benelgjali *et al.*, 2010). The solution of the related nonlinear optimal control problem is anticipated to assure for the individual power generators fast and accurate tracking of reference setpoints under minimal variations of the control inputs.

In this section, a solution is proposed for the nonlinear optimal control problem of distributed marine-turbine power units (Rigatos and Busawon, 2018; Rigatos, Siano and Cecati, 2015; Rigatos, Siano *et al.*, 2015). Each power unit is considered to comprise a marine-turbine that transmits, through a drivetrain, mechanical power and rotational motion to the rotor of a synchronous generator. The generators are connected next to a transmission line (infinite-bus) which provides power to electricity grid. First, the dynamic model of the distributed and interconnected power units undergoes approximate linearization around a temporary operating point which is recomputed at each time-step of the control algorithm. This operating point is defined at each time-instant by the present value of the power system's state vector and by the last sampled value of the control inputs vector. The linearization relies on first-order Taylor series expansion and on the computation of the associated Jacobian matrices (Rigatos and Tzafestas, 2007; Basseville and Nikiforov, 1993; Rigatos and Zhang, 2009). For the approximately linearized model of the distributed marine-turbines power system a stabilizing H-infinity (optimal) feedback controller is designed.

Actually, the H-infinity controller stands for the solution of the optimal control problem for the power system under model uncertainties and external perturbations. The controller represents a min-max differential game taking place between (i) the control inputs which try to minimize a quadratic cost function of the state vector error of the system, (ii) the model uncertainty and external perturbation terms which try to maximize this cost function. For the selection of the feedback gains of the H-infinity controller an algebraic Riccati equation is repetitively solved at each time-step of the control algorithm (Rigatos, 2011). The stability properties of the control method are proven through Lyapunov analysis. First, it is demonstrated that the H-infinity tracking performance criterion is satisfied. This signifies elevated robustness against model uncertainty and disturbances (Tousssaint, Basar and Bullo, 2000; Lublin and Athans, 1995). Moreover, under moderate conditions it is proven that the control scheme is globally asymptotically stable (Rigatos, 2016b). Finally, to implement state estimation-based control without the need to measure the entire state-vector of the distributed power system, the H-infinity Kalman Filter is used as a robust observer (Gibbs, 2012; Simon, 2006).

This section offers one of the very few existing approaches towards the solution of the nonlinear optimal control problem for distributed marine-turbine synchronous power generation, which is followed by a global stability proof. Preceding results on the use of H-infinity control to nonlinear dynamical systems were limited to the case of affine-in-the-input systems with drift-only dynamics and by considering that the control inputs gain matrix is not dependent on the values of the system's state vector. Moreover, in these approaches the linearization is performed around points of the desirable trajectory whereas in the present section's control method the linearization points are related to the value of the state vector at each sampling instant as well as with the last sampled value of the control inputs vector. The Riccati equation which has been proposed for computing the feedback gains of the controller is novel, so is the presented global stability proof through Lyapunov analysis.

Research on stabilization and synchronization of distributed power generation units is always a nontrivial and challenging topic (Sharma *et al.*, 2016; Nasiruddin *et al.*, 2014; Sharma, Nasiruddin *et al.*, 2015). Comparing to other nonlinear control approaches that could have been considered for the marine-turbine power generation system, the section's method exhibits several advantages. The section's control method minimizes the variations of the control inputs of the distributed power generation model, thus achieving the reduction of the functioning cost of this power system. The application of global linearization-based control methods to the marine-turbine power system dynamics has several shortcomings. For instance, it requires complicated state variables transformations. Besides, to compute the control inputs that will be finally applied to this distributed power generation system, inverse transformations are needed. Furthermore, one cannot apply MPC because the model of the marine turbines and synchronous generators power unit is a strongly nonlinear one and the use of MPC in its case would make the control loop unstable. Additionally, the convergence of NMPC schemes to an optimum, cannot be assured either since this method relies on the ad-hoc selection of parameters' values and on initialization. It is also remarkable in the application of sliding-mode control, to arrive at a well-determined procedure for defining the sliding surface, transformation of the distributed power system's state-space model into an input-output linearized form is first needed. It is also noteworthy, that one may not apply directly backstepping control to models of distributed power systems when these systems are not found in the triangular (backstepping integral) state-space form. Finally, it would not be recommended to apply multiple local models-based control, through the linearization of the model of the marine turbines and synchronous generators' power units around multiple operating points because this would require the solution of multiple Riccati equations or LMIs and would be a computationally intensive procedure.

The significance of the section's results and methods is summarized as follows: (1) the novelty/innovation of this research work is in presenting a novel and genuine nonlinear optimal control method for the distributed marine-turbine and synchronous generators electric power system, (2) the significance of the presented method is in achieving fast and accurate tracking of reference setpoints for the distributed power system under moderate variations of the control inputs. By minimizing energy consumption by the control system of the power generators the functioning cost of the considered distributed power system is also reduced, (3) the technical contribution of the section is in solving the nonlinear optimal control problem for distributed marine-turbine power generation units in an effective, but also computationally efficient manner.

4.4.2 Dynamic model of the distributed marine turbine power generation units

4.4.3 The dynamic model of the distributed power system

The diagram of the distributed marine turbine power generation units is given in Figure 4.44. Each power unit comprises a marine turbine, a drivetrain and a synchronous generator.

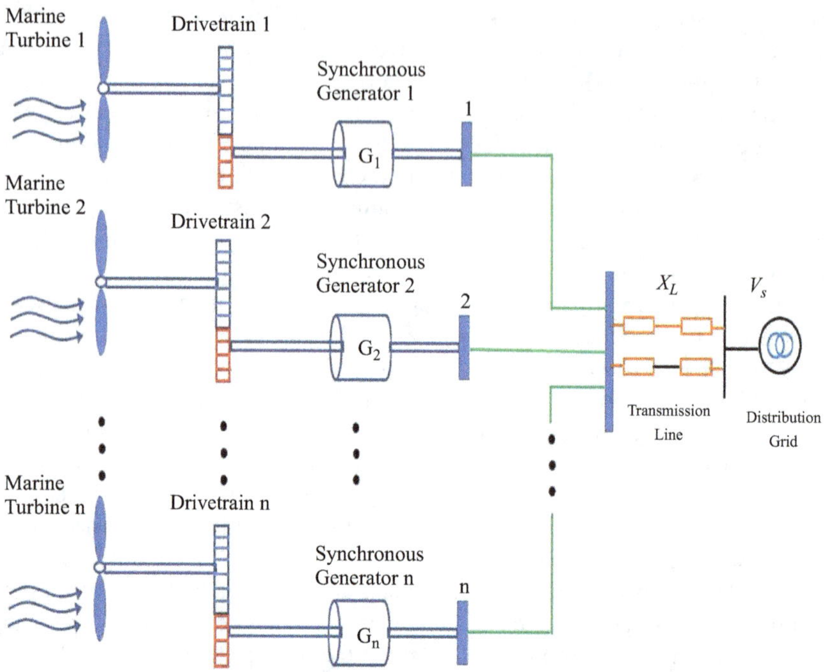

Figure 4.44 Diagram of the distributed marine turbine power generation units. Each unit comprises a marine turbine, a drivetrain and a synchronous generator

Considering that the drivetrain comprises: (i) a gear of n_t teeth at the side of the marine-turbine, (ii) a gear of n_g teeth at the side of the synchronous generator, one has that the dynamic model of the ith power generation unit is given by the following set of differential equations (Rigatos, 2016b):

$$\dot{\theta}_t^i = \omega_t^i \tag{4.167}$$

$$\dot{\omega}_t^i = \frac{1}{J_t^i}[c_{ba}^i T_m^i - K_1^i(\delta_t^i - \delta_g^i) - D_t^i \omega_t^i] \tag{4.168}$$

$$\dot{\theta}_g^i = \omega_g^i \tag{4.169}$$

$$\begin{aligned}\dot{\omega}_g^i = &\frac{1}{J_g^i}\{K_1^i(\delta_t^i - \delta_g^i)\frac{n_g^i}{n_t^i} - D_g^i(\omega_g^i - \omega_0) \\ &- \omega_0[G_{ii}E_{q_i}^{'2} + E_{q_i}^{'}\sum_{j=1, j\neq i}^{n} E_{q_j}^{'} G_{ij}\cos(\delta_g^i - \delta_g^j - a_{ij})]\}\end{aligned} \tag{4.170}$$

$$\dot{E}'_{q_i} = -\frac{1}{T'_{d_i}} E'_{q_i} + \frac{1}{T^i_{do}} \frac{x^i_d - x'^{\,i}_d}{x'^{\,i}_{d_\Sigma}} V_s \cos(\delta^i_g) + \frac{1}{T^i_{do}} E^i_f \tag{4.171}$$

The mechanical torque T^i_m which is provided to the ith marine turbine by sea-currents or tidal streams is given by

$$T^i_m = \frac{1}{2} \rho_i S_i V^3_{t_i} \tag{4.172}$$

where ρ_i is the density of the water, S_i is the cross-sectional area of the marine turbine and V_{t_i} is the velocity of the tidal stream at the ith turbine. Moreover, the following parameters of the model of the ith power generation unit are defined: K^i_1 is the elasticity of the drivetrain at the ith turbine resulting in torque (torsion) equal to $K^i_1(\theta^i_t - \theta^i_g)$, δ^i_t is the turn angle of the turbine, δ^i_g is the turn angle of the generator's rotor, D^i_t is the damping coefficient for the rotational motion of the turbine, D^i_g is the damping coefficient for the rotational motion of the generator, ω^i_t is the angular speed of the turbine, ω^i_g is the angular speed of the generator.

The electromagnetic torque that is applied at the ith generator, takes into account both the self-excitation at the generator and the excitation provided by the connection of other generators to the electricity grid. This is given by

$$T^i_e = \omega_0 [G_{ii} E'^{\,2}_{q_i} + E'_{q_i} \sum_{j=1, j \neq i}^{n} E'_{q_j} G_{ij} \cos(\delta^i_g - \delta^j_g - a_{ij})] \tag{4.173}$$

where E'_{q_i} is the quadrature axis transient voltage of the ith generator, E_{q_i} is the quadrature axis voltage, T^i_{do} is the direct axis open-circuit transient-time constant, V_s is the infinite bus (transmission line) voltage, $x'^{\,i}_{d_\Sigma} = x'^{\,i}_d + x^i_T + x^i_L$, $x'^{\,i}_d$ is the direct axis synchronous reactance, x^i_T is the reactance of the transformer, x^i_L is the reactance of the transmission line and E^i_f is the equivalent voltage at the excitation coil of the ith generator. The term $(\delta^i_g - \delta^j_g)$ stands for the difference in the turn angle of the ith and jth generator and a_{ij} denotes also a phase difference between these two generators. The terms G_{ij} denote cross-coupling between the ith and the jth generator, while the term G_{ii} denotes the strength attributed to the self-excitation of the ith generator.

Coefficient c^i_{ba} stands for a control input to the ith generator and is related primarily with the variable pitch angle of the generator's blades. It regulates the effect of the mechanical (tidal) torque that is exerted on the marine-turbine. Actually c_{ba} is a nonlinear function of the blades' pitch angle, of the rotor's speed and of the tidal stream's speed. This function may change by the water's speed. In practice the value of c_{ba} is part of the hydrodynamic maps of the turbine and is stored in look-up tables.

Next, without loss of generality, the case of three interconnected marine turbine power units is considered. The associated dynamic model is:

1st power unit

$$
\begin{aligned}
\dot{\theta}_t^1 &= \omega_t^1 \\
\dot{\omega}_t^1 &= \frac{1}{J_t^1}[c_{ba}^1 T_m^i - K_1^1(\delta_t^1 - \delta_g^1) - D_t^1 \omega_t^1] \\
\dot{\theta}_g^1 &= \omega_g^1 \\
\dot{\omega}_g^1 &= \frac{1}{J_g^1}\{K_1^1(\delta_t^1 - \delta_g^1)\}\frac{n_g^1}{n_t^1} - D_g^1(\omega_g^1 - \omega_0) \\
&\quad -\omega_0[G_{11}E_{q_1}^{'\,2} + E_{q_1}^{'}(E_q^{'\,2} G_{12}\cos(\delta_1 - \delta_2 - \alpha_{12}) + E_q^{'\,3} G_{13}\cos(\delta_1 - \delta_3 - \alpha_{13})] \\
\dot{E}_{q_1}^{'} &= -\frac{1}{T_{d_1}^{'}} E_{q_1}^{'} + \frac{1}{T_{do}^1}\frac{x_d^1 - x_d^{'\,1}}{x_{d_\Sigma}^{'\,1}} V_s\cos(\delta_g^1) + \frac{1}{T_{do}^1}E_f^1
\end{aligned}
\tag{4.174}
$$

2nd power unit

$$
\begin{aligned}
\dot{\theta}_t^2 &= \omega_t^2 \\
\dot{\omega}_t^2 &= \frac{1}{J_t^2}[c_{ba}^2 T_m^2 - K_1^2(\delta_t^2 - \delta_g^2) - D_t^2 \omega_t^2] \\
\dot{\theta}_g^2 &= \omega_g^2 \\
\dot{\omega}_g^2 &= \frac{1}{J_g^2}\{K_1^2(\delta_t^2 - \delta_g^2)\}\frac{n_g^2}{n_t^2} - D_g^2(\omega_g^2 - \omega_0) \\
&\quad -\omega_0[G_{22}E_{q_2}^{'\,2} + E_{q_2}^{'}(E_q^{'\,1} G_{21}\cos(\delta_2 - \delta_1 - \alpha_{21}) + E_q^{'\,3} G_{23}\cos(\delta_2 - \delta_3 - \alpha_{23})] \\
\dot{E}_{q_2}^{'} &= -\frac{1}{T_{d_2}^{'}} E_{q_2}^{'} + \frac{1}{T_{do}^2}\frac{x_d^2 - x_d^{'\,2}}{x_{d_\Sigma}^{'\,2}} V_s\cos(\delta_g^2) + \frac{1}{T_{do}^2}E_f^2
\end{aligned}
\tag{4.175}
$$

3rd power unit

$$
\begin{aligned}
\dot{\theta}_t^3 &= \omega_t^3 \\
\dot{\omega}_t^3 &= \frac{1}{J_t^3}[c_{ba}^3 T_m^3 - K_1^3(\delta_t^3 - \delta_g^3) - D_t^3 \omega_t^3] \\
\dot{\theta}_g^3 &= \omega_g^3 \\
\dot{\omega}_g^3 &= \frac{1}{J_g^3}\{K_1^3(\delta_t^3 - \delta_g^3)\}\frac{n_g^3}{n_t^3} - D_g^3(\omega_g^3 - \omega_0) \\
&\quad -\omega_0[G_{33}E_{q_3}^{'\,2} + E_{q_3}^{'}(E_q^{'\,1} G_{31}\cos(\delta_3 - \delta_1 - \alpha_{31}) + E_q^{'\,2} G_{32}\cos(\delta_3 - \delta_2 - \alpha_{32})] \\
\dot{E}_{q_3}^{'} &= -\frac{1}{T_{d_3}^{'}} E_{q_3}^{'} + \frac{1}{T_{do}^3}\frac{x_d^3 - x_d^{'\,3}}{x_{d_\Sigma}^{'\,3}} V_s\cos(\delta_g^3) + \frac{1}{T_{do}^3}E_f^3
\end{aligned}
\tag{4.176}
$$

Next, the following state variables and state vector are defined:

$$
\begin{aligned}
x &= [\Delta\delta_t^1, \Delta\omega_t^1, \Delta\delta_g^1, \Delta\omega_g^1, E_q^{'\,1}, \Delta\delta_t^2, \Delta\omega_t^2, \Delta\delta_g^2, \Delta\omega_g^2, E_q^{'\,2}, \\
&\quad \Delta\delta_t^3, \Delta\omega_t^3, \Delta\delta_g^3, \Delta\omega_g^3, E_q^{'\,3}]^T \text{ or} \\
x &= [x_1, x_2, x_3, x_4, x_5, x_6, x_7, x_8, x_9, x_{10}, x_{11}, x_{12}, x_{13}, x_{14}, x_{15}]^T
\end{aligned}
\tag{4.177}
$$

and the following control inputs vector is defined

$$u = [c_{ba}^1, E_f^1, c_{ba}^2, E_f^2, c_{ba}^3, E_f^3]^T \text{ or}$$
$$u = [u_1, u_2, u_3, u_4, u_5, u_6]^T \tag{4.178}$$

The distributed power generation system is in the nonlinear affine-in-the-input state space form

$$\dot{x} = f(x) + g(x)u$$

with state vector $x \in R^{15\times 1}$, drift vector $f(x) \in R^{15\times 1}$, control inputs gain matrix $g(x) \in R^{15\times 6}$ and control inputs vector $u \in R^{6\times 1}$.

The state-space description of the distributed marine power units takes the following form:

$$\begin{aligned}
\dot{x}_1 &= x_2\\
\dot{x}_2 &= \frac{1}{J_t^1}[u_1 T_m^1 - K_1^1(x_1 - x_3) - D_t^1 x_2]\\
\dot{x}_3 &= x_4\\
\dot{x}_4 &= \frac{1}{J_g^1}\{K_1^1(x_1 - x_3)\}\frac{n_g^1}{n_t^1} - D_g^1(x_4 - \omega_0)\\
&\quad -\omega_0[G_{11}{x_5}^2 + x_5(x_{10}G_{12}\cos(x_1 - x_3 - \alpha_{12}) + x_{15}G_{13}\cos(x_1 - x_3 - \alpha_{13})]\\
\dot{x}_5 &= -\frac{1}{T'_{d_1}}x_5 + \frac{1}{T_{do}^1}\frac{x_d^1 - {x'_d}^1}{{x'_{d_\Sigma}}^1}V_s\cos(x_1) + \frac{1}{T_{do}^1}u_2\\
\dot{x}_6 &= x_7\\
\dot{x}_7 &= \frac{1}{J_t^2}[u_3 T_m^2 - K_1^2(x_6 - x_8) - D_t^2 x_7]\\
\dot{x}_8 &= x_9\\
\dot{x}_9 &= \frac{1}{J_g^2}\{K_1^2(x_6 - x_8)\}\frac{n_g^2}{n_t^2} - D_g^2(x_9 - \omega_0)\\
&\quad -\omega_0[G_{22}x_10^2 + x_{10}(x_5 G_{21}\cos(x_6 - x_1 - \alpha_{21}) + x_{15}G_{23}\cos(x_8 - x_{11} - \alpha_{23})]\\
\dot{x}_{10} &= -\frac{1}{T'_{d_2}}x_{10} + \frac{1}{T_{do}^2}\frac{x_d^2 - {x'_d}^2}{{x'_{d_\Sigma}}^2}V_s\cos(x_6) + \frac{1}{T_{do}^2}u_4\\
\dot{x}_{11} &= x_{12}\\
\dot{x}_{12} &= \frac{1}{J_t^3}[u_5 T_m^3 - K_1^3(x_{11} - x_{13}) - D_t^3 x_{12}]\\
\dot{x}_{13} &= x_{14}\\
\dot{x}_{14} &= \frac{1}{J_g^3}\{K_1^3(x_{11} - x_{13})\}\frac{n_g^3}{n_t^3} - D_g^3(x_{14} - \omega_0)\\
&\quad -\omega_0[G_{33}x_{15}^2 + x_{15}(x_5 G_{31}\cos(x_{11} - x_1 - \alpha_{31}) + x_{10}G_{32}\cos(x_{11} - x_6 - \alpha_{32})]\\
\dot{x}_{15} &= -\frac{1}{T'_{d_3}}x_{15} + \frac{1}{T_{do}^3}\frac{x_d^3 - {x'_d}^3}{{x'_{d_\Sigma}}^3}V_s\cos(x_{11}) + \frac{1}{T_{do}^3}u_6
\end{aligned} \tag{4.179}$$

The state-space model of the system can be also written in matrix form:

$$\begin{pmatrix} \dot{x}_1 \\ \dot{x}_2 \\ \dot{x}_3 \\ \dot{x}_4 \\ \dot{x}_5 \\ \dot{x}_6 \\ \dot{x}_7 \\ \dot{x}_8 \\ \dot{x}_9 \\ \dot{x}_{10} \\ \dot{x}_{11} \\ \dot{x}_{12} \\ \dot{x}_{13} \\ \dot{x}_{14} \\ \dot{x}_{15} \end{pmatrix} = \begin{pmatrix} f_1(x) \\ f_2(x) \\ f_3(x) \\ f_4(x) \\ f_5(x) \\ f_6(x) \\ f_7(x) \\ f_8(x) \\ f_9(x) \\ f_{10}(x) \\ f_{11}(x) \\ f_{12}(x) \\ f_{13}(x) \\ f_{14}(x) \\ f_{15}(x) \end{pmatrix} + \begin{pmatrix} 0 & 0 & 0 & 0 & 0 & 0 \\ \frac{T_m^1}{J_t^1} & 0 & 0 & 0 & 0 & 0 \\ 0 & 0 & 0 & 0 & 0 & 0 \\ 0 & 0 & 0 & 0 & 0 & 0 \\ 0 & \frac{1}{T_{do}^1} & 0 & 0 & 0 & 0 \\ 0 & 0 & 0 & 0 & 0 & 0 \\ 0 & 0 & \frac{T_m^2}{J_t^2} & 0 & 0 & 0 \\ 0 & 0 & 0 & 0 & 0 & 0 \\ 0 & 0 & 0 & 0 & 0 & 0 \\ 0 & 0 & 0 & \frac{1}{T_{do}^2} & 0 & 0 \\ 0 & 0 & 0 & 0 & 0 & 0 \\ 0 & 0 & 0 & 0 & \frac{T_m^3}{J_t^3} & 0 \\ 0 & 0 & 0 & 0 & 0 & 0 \\ 0 & 0 & 0 & 0 & 0 & 0 \\ 0 & 0 & 0 & 0 & 0 & \frac{1}{T_{do}^3} \end{pmatrix} \begin{pmatrix} u_1 \\ u_2 \\ u_3 \\ u_4 \\ u_5 \\ u_6 \end{pmatrix} \tag{4.180}$$

where the elements of the drift matrix $f(x)$ are defined as:

$f_1(x) = x_2, \quad f_2(x) = \frac{1}{J_t^1}[-K_1^1(x_1 - x_3) - D_t^1 x_2], \quad f_3(x) = x_4, \quad f_4(x) = \frac{1}{J_g^1}\{K_1^1(x_1 - x_3)\}\frac{n_g^1}{n_t^1} - D_g^1(x_4 - \omega_0) - \omega_0[G_{11}x_5{}^2 + x_5(x_{10}G_{12}\cos(x_1 - x_3 - \alpha_{12}) + x_{15}G_{13}\cos(x_1 - x_3 - \alpha_{13})],\ f_5(x) = -\frac{1}{T'_{d_1}}x_5 + \frac{1}{T_{do}^1}\frac{x_d^1 - x_d^{\prime 1}}{x'_{d_\Sigma}{}^1} V_s\cos(x_1),\ f_6(x) = x_7,\ f_7(x) = \frac{1}{J_t^2}[-K_1^2(x_6 - x_8) - D_t^2 x_7], f_8(x) = x_9, f_9(x) == \frac{1}{J_g^2}\{K_1^2(x_6 - x_8)\}\frac{n_g^2}{n_t^2} - D_g^2(x_9 - \omega_0) -$
$-\omega_0[G_{22}x_10^2 + x_{10}(x_5G_{21}\cos(x_6 - x_1 - \alpha_{21}) + x_{15}G_{23}\cos(x_8 - x_{11} - \alpha_{23})], f_{10}(x) = -\frac{1}{T'_{d_2}}x_{10} + \frac{1}{T_{do}^2}\frac{x_d^2 - x_d^{\prime 2}}{x'_{d_\Sigma}{}^2} V_s\cos(x_6),\ \ f_{11}(x) = x_{12},\ \ f_{12}(x) = \frac{1}{J_t^3}[-K_1^3(x_{11} - x_{13}) - D_t^3 x_{12}],$
$f_{13}(x) = x_{14}, f_{14}(x) = \frac{1}{J_g^3}\{K_1^3(x_{11} - x_{13})\}\frac{n_g^3}{n_t^3} - D_g^3(x_{14} - \omega_0) -$
$-\omega_0[G_{33}x_{15}^2 + x_{15}(x_5G_{31}\cos(x_{11} - x_1 - \alpha_{31}) + x_{10}G_{32}\cos(x_{11} - x_6 - \alpha_{32})],\ f_{15}(x) = -\frac{1}{T'_{d_3}}x_{15} + \frac{1}{T_{do}^3}\frac{x_d^3 - x_d^{\prime 3}}{x'_{d_\Sigma}{}^3} V_s\cos(x_{11})$

4.4.4 Differential flatness of the distributed marine power generation units

It will be proven that the dynamic model of the distributed marine power generation system is differentially flat with flat outputs vector (Rigatos, 2016b)

$$Y = [x_1, x_3, x_6, x_8, x_{11}, x_{13}]^T \text{ or } \\ Y = [\Delta\delta_t^1, \Delta\delta_g^1, \Delta\delta_t^2, \Delta\delta_g^2, \Delta\delta_t^3, \Delta\delta_g^3] \tag{4.181}$$

From the first row of the state-space model of (4.179) one has

$$x_2 = \dot{x}_1 \Rightarrow x_2 = h_2(Y, \dot{Y}) \tag{4.182}$$

From the third row of the state-space model of (4.179) one has

$$x_4 = \dot{x}_3 \Rightarrow x_4 = h_4(Y, \dot{Y}) \tag{4.183}$$

From the sixth row of the state-space model of (4.179) one has

$$x_6 = \dot{x}_5 \Rightarrow x_6 = h_6(Y, \dot{Y}) \tag{4.184}$$

From the eighth row of the state-space model of (4.179) one has

$$x_8 = \dot{x}_7 \Rightarrow x_8 = h_8(Y, \dot{Y}) \tag{4.185}$$

From the eleventh row of the state-space model of (4.179) one has

$$x_{12} = \dot{x}_{11} \Rightarrow x_{12} = h_{12}(Y, \dot{Y}) \tag{4.186}$$

From the eighth row of the state-space model of (4.179) one has

$$x_{14} = \dot{x}_{13} \Rightarrow x_{14} = h_{14}(Y, \dot{Y}) \tag{4.187}$$

From the fourth row of the state-space model of (4.179) one has a first relation among x_5, x_{10} and x_{15}. From the ninth row of the state-space model of (4.179) one has a second relation among x_5, x_{10} and x_{15}. From the fourteenth row of the state-space model of (4.179) one has a third relation among x_5, x_{10} and x_{15}. Consequently, one arrives at a system of three equations with unknowns x_5, x_{10} and x_{15}. The rest of the parameters of this system of equations are differential functions of the flat outputs Y. Therefore, one can finally obtain

$$x_5 = h_5(Y, \dot{Y}) \quad x_{10} = h_{10}(Y, \dot{Y}) \quad x_{15} = h_{15}(Y, \dot{Y}) \tag{4.188}$$

Besides, from the second row of the state-space model of (4.179) one solves with respect to the control input u_1. This gives

$$u_1 = \frac{1}{T_m^1}\{J_t^1 \dot{x}_2 + K_1^1(x_1 - x_3) + D_t^1 x_2\} \Rightarrow u_1 = q_1(Y, \dot{Y}) \tag{4.189}$$

From the fifth row of the state-space model of (4.179) one solves with respect to the control input u_2. This gives

$$u_2 = T_{do}^1\{\dot{x}_5 + \frac{1}{T'_{d_1}} x_5 - \frac{1}{T_{do}^1} \frac{x_d^1 - x_d'^1}{x'_{d_{\Sigma}1}} V_s \cos(x_1)\} \Rightarrow u_2 = q_2(Y, \dot{Y}) \tag{4.190}$$

From the seventh row of the state-space model of (4.179) one solves with respect to the control input u_3. This gives

$$u_3 = \frac{1}{T_m^2}\{J_t^2\ddot{x}_7 + K_1^2(x_6 - x_8) + D_t^2 x_7\} \Rightarrow u_3 = q_3(Y, \dot{Y}) \tag{4.191}$$

From the tenth row of the state-space model of (4.179) one solves with respect to the control input u_4. This gives

$$u_4 = T_{do}^2\{\dot{x}_{10} + \frac{1}{T'_{d_2}}x_{10} - \frac{1}{T_{do}^2}\frac{x_d^2 - x_d'^2}{x'_{d_\Sigma}{}^2}V_s\cos(x_6)\}u_4 = q_4(Y, \dot{Y}) \tag{4.192}$$

From the twelfth row of the state-space model of (4.179) one solves with respect to the control input u_5. This gives

$$u_5 = \frac{1}{T_m^2}\{J_t^3\ddot{x}_{12} + K_1^3(x_{11} - x_{13}) + D_t^3 x_{10}\} \Rightarrow u_5 = q_5(Y, \dot{Y}) \tag{4.193}$$

From the fifteenth row of the state-space model of (4.179) one solves with respect to the control input u_6. This gives

$$u_6 = T_{do}^3\{\dot{x}_{15} + \frac{1}{T'_{d_3}}x_{15} - \frac{1}{T_{do}^3}\frac{x_d^3 - x_d'^3}{x'_{d_\Sigma}{}^3}V_s\cos(x_{11})\} \Rightarrow u_6 = q_6(Y, \dot{Y}) \tag{4.194}$$

Consequently, all state variables and the control inputs of the system can be written as differential functions of its flat outputs. This signifies that the system is differentially flat (Rigatos, 2016b).

4.4.5 Approximate linearization of the distributed marine power generators

The state-space model of the distributed marine power generation system has been written in the affine-in-the-input form of (4.180)

$$\dot{x} = f(x) + g(x)u \tag{4.195}$$

with $x \in R^{15\times 1}$, $u \in R^{6\times 1}$, $f(x) \in R^{15\times 1}$ and $g(x) \in R^{15\times 6}$. The state-space model undergoes approximate linearization around the temporary operating point (x^*, u^*), where x^* is the present value of the system's state vector and u^* is the last sampled value of the control inputs vector. The linearization relies on first-order Taylor series expansion and on the computation of the associated Jacobian matrices. The linearization procedure gives the state-space description

$$\dot{x} = Ax + Bu + \tilde{d} \tag{4.196}$$

where $\tilde{d}$ denotes the modelling error due to the truncation of higher-order terms in the Taylor series expansion and also the effects of external perturbations. Matrices A and B are obtained from the computation of the system's Jacobians

$$A = \nabla_x[f(x) + g(x)u] \mid_{(x^*, u^*)} \Rightarrow A = \nabla_x[f(x)] \mid_{(x^*, u^*)} \tag{4.197}$$

$$B = \nabla_u[f(x) + g(x)u] \mid_{(x^*, u^*)} \Rightarrow B = g(x) \mid_{(x^*, u^*)} \tag{4.198}$$

The elements of the Jacobian matrices of the nonlinear state-space model of the system are given in the Appendix.

4.4.5.1 Computation of the Jacobian matrices of the linearized model

The Jacobian matrix $A = \nabla_x[f(x)] \mid_{(x^*,u^*)}$ is computed.

For the first row of the Jacobian matrix $A = \nabla_x[f(x)] \mid_{(x^*,u^*)}$ one has: $\frac{\partial f_1}{\partial x_1} = 0$, $\frac{\partial f_1}{\partial x_2} = 1$, $\frac{\partial f_1}{\partial x_3} = 0$, $\frac{\partial f_1}{\partial x_4} = 0$, $\frac{\partial f_1}{\partial x_5} = 0$, $\frac{\partial f_1}{\partial x_6} = 0$, $\frac{\partial f_1}{\partial x_7} = 0$, $\frac{\partial f_1}{\partial x_8} = 0$, $\frac{\partial f_1}{\partial x_9} = 0$, $\frac{\partial f_1}{\partial x_{10}} = 0$, $\frac{\partial f_1}{\partial x_{11}} = 0$, $\frac{\partial f_1}{\partial x_{12}} = 0$, $\frac{\partial f_1}{\partial x_{13}} = 0$, $\frac{\partial f_1}{\partial x_{14}} = 0$, and $\frac{\partial f_1}{\partial x_{15}} = 0$.

For the second row of the Jacobian matrix $A = \nabla_x[f(x)] \mid_{(x^*,u^*)}$ one has: $\frac{\partial f_2}{\partial x_1} = \frac{1}{J_t^1}(-K_1^1)$, $\frac{\partial f_2}{\partial x_2} = \frac{1}{J_t^1}(-D_t^1)$, $\frac{\partial f_2}{\partial x_3} = \frac{1}{J_t^1}K_1^1$, $\frac{\partial f_2}{\partial x_4} = 0$, $\frac{\partial f_2}{\partial x_5} = 0$, $\frac{\partial f_2}{\partial x_6} = 0$, $\frac{\partial f_2}{\partial x_7} = 0$, $\frac{\partial f_2}{\partial x_8} = 0$, $\frac{\partial f_2}{\partial x_9} = 0$, $\frac{\partial f_2}{\partial x_{10}} = 0$, $\frac{\partial f_2}{\partial x_{11}} = 0$, $\frac{\partial f_2}{\partial x_{12}} = 0$, $\frac{\partial f_2}{\partial x_{13}} = 0$, $\frac{\partial f_2}{\partial x_{14}} = 0$, and $\frac{\partial f_2}{\partial x_{15}} = 0$.

For the third row of the Jacobian matrix $A = \nabla_x[f(x)] \mid_{(x^*,u^*)}$ one has: $\frac{\partial f_3}{\partial x_1} = 0$, $\frac{\partial f_3}{\partial x_2} = 0$, $\frac{\partial f_3}{\partial x_3} = 0$, $\frac{\partial f_3}{\partial x_4} = 1$, $\frac{\partial f_3}{\partial x_5} = 0$, $\frac{\partial f_3}{\partial x_6} = 0$, $\frac{\partial f_3}{\partial x_7} = 0$, $\frac{\partial f_3}{\partial x_8} = 0$, $\frac{\partial f_3}{\partial x_9} = 0$, $\frac{\partial f_3}{\partial x_{10}} = 0$, $\frac{\partial f_3}{\partial x_{11}} = 0$, $\frac{\partial f_3}{\partial x_{12}} = 0$, $\frac{\partial f_3}{\partial x_{13}} = 0$, $\frac{\partial f_3}{\partial x_{14}} = 0$, and $\frac{\partial f_3}{\partial x_{15}} = 0$.

For the fourth row of the Jacobian matrix $A = \nabla_x[f(x)] \mid_{(x^*,u^*)}$ one has: $\frac{\partial f_4}{\partial x_1} = -\frac{1}{J_g^1}\{K_1^1\frac{n_g^1}{n_t^1} + \omega_0 x_5 x_{10} G_{12}\sin(x_1 - x_6 - a_{12}) + \omega_0 x_5 x_{15} G_{13}\sin(x_1 - x_{11} - a_{13})\}$, $\frac{\partial f_4}{\partial x_2} = 0$, $\frac{\partial f_4}{\partial x_3} = \frac{1}{J_g^1}(-K_1^1\frac{n_g^1}{n_t^1})$, $\frac{\partial f_4}{\partial x_4} = \frac{1}{J_g^1}(= D_g^2)$, $\frac{\partial f_4}{\partial x_5} = \frac{1}{J_g^1}\{-\omega_0[2G_{11}x_5 - (x_{10}G_{12}\cos(x_1 - x_6 - a_{12}) + x_{15}G_{13}\cos(x_1 - x_{11} - a_{13}))]\}$, $\frac{\partial f_4}{\partial x_6} = \frac{1}{J_g^1}\{-\omega_0 x_5 x_{10} G_{12}\sin(x_1 - x_6 - a_{12})\}$, $\frac{\partial f_4}{\partial x_7} = 0$, $\frac{\partial f_4}{\partial x_8} = 0$, $\frac{\partial f_4}{\partial x_9} = 0$, $\frac{\partial f_4}{\partial x_{10}} = \frac{1}{J_g^1}\{-\omega_0 x_5 G_{12}\cos(x_1 - x_6 - a_{12})\}$, $\frac{\partial f_4}{\partial x_{11}} = \frac{1}{J_g^1}\{-\omega_0 x_5 x_{15} G_{13}\cos(x_1 - x_{11} - a_{13})\}$, $\frac{\partial f_4}{\partial x_{12}} = 0$, $\frac{\partial f_4}{\partial x_{13}} = 0$, $\frac{\partial f_4}{\partial x_{14}} = 0$, $\frac{\partial f_4}{\partial x_{15}} = \frac{1}{J_g^1}\{-\omega_0 x_5 G_{13}\cos(x_1 - x_{11} - a_{13})$.

For the fifth row of the Jacobian matrix $A = \nabla_x[f(x)] \mid_{(x^*,u^*)}$ one has:
$\frac{\partial f_5}{\partial x_1} = -\frac{1}{T_{do}^1}\frac{x_d^1 - x_d^{'1}}{x_{d_\Sigma}^{'1}}V_s\sin(x_1)$, $\frac{\partial f_5}{\partial x_2} = 0$, $\frac{\partial f_5}{\partial x_3} = 0$, $\frac{\partial f_5}{\partial x_4} = 0$, $\frac{\partial f_5}{\partial x_5} = -\frac{1}{T'_{d_1}}$, $\frac{\partial f_5}{\partial x_6} = 0$, $\frac{\partial f_5}{\partial x_7} = 0$, $\frac{\partial f_5}{\partial x_8} = 0$, $\frac{\partial f_5}{\partial x_9} = 0$, $\frac{\partial f_5}{\partial x_{10}} = 0$, $\frac{\partial f_5}{\partial x_{11}} = 0$, $\frac{\partial f_5}{\partial x_{12}} = 0$, $\frac{\partial f_5}{\partial x_{13}} = 0$, $\frac{\partial f_5}{\partial x_{14}} = 0$, and $\frac{\partial f_5}{\partial x_{15}} = 0$.

For the sixth row of the Jacobian matrix $A = \nabla_x[f(x)] \mid_{(x^*,u^*)}$ one has: $\frac{\partial f_6}{\partial x_1} = 0$, $\frac{\partial f_6}{\partial x_2} = 0$, $\frac{\partial f_6}{\partial x_3} = 0$, $\frac{\partial f_6}{\partial x_4} = 0$, $\frac{\partial f_6}{\partial x_5} = 0$, $\frac{\partial f_6}{\partial x_6} = 0$, $\frac{\partial f_6}{\partial x_7} = 1$, $\frac{\partial f_6}{\partial x_8} = 0$, $\frac{\partial f_6}{\partial x_9} = 0$, $\frac{\partial f_6}{\partial x_{10}} = 0$, $\frac{\partial f_6}{\partial x_{11}} = 0$, $\frac{\partial f_6}{\partial x_{12}} = 0$, $\frac{\partial f_6}{\partial x_{13}} = 0$, $\frac{\partial f_6}{\partial x_{14}} = 0$, and $\frac{\partial f_6}{\partial x_{15}} = 0$.

For the seventh row of the Jacobian matrix $A = \nabla_x[f(x)] \mid_{(x^*,u^*)}$ one has: $\frac{\partial f_7}{\partial x_1} = 0$, $\frac{\partial f_7}{\partial x_2} = 0$, $\frac{\partial f_7}{\partial x_3} = 0$, $\frac{\partial f_7}{\partial x_4} = 0$, $\frac{\partial f_7}{\partial x_5} = 0$, $\frac{\partial f_7}{\partial x_6} = \frac{1}{J_t^2}(-K_t^2)$, $\frac{\partial f_7}{\partial x_7} = \frac{1}{J_t^2}(-D_t^2)$, $\frac{\partial f_7}{\partial x_8} = \frac{1}{J_t^2}K_1^2$,

$\frac{\partial f_7}{\partial x_9} = 0$, $\frac{\partial f_7}{\partial x_{10}} = 0$, $\frac{\partial f_7}{\partial x_{11}} = 0$, $\frac{\partial f_7}{\partial x_{12}} = 0$, $\frac{\partial f_7}{\partial x_{13}} = 0$, $\frac{\partial f_7}{\partial x_{14}} = 0$, and $\frac{\partial f_7}{\partial x_{15}} = 0$.

For the eighth row of the Jacobian matrix $A = \nabla_x[f(x)] \mid_{(x^*,u^*)}$ one has: $\frac{\partial f_8}{\partial x_1} = 0$, $\frac{\partial f_8}{\partial x_2} = 0$, $\frac{\partial f_8}{\partial x_3} = 0$, $\frac{\partial f_8}{\partial x_4} = 0$, $\frac{\partial f_8}{\partial x_5} = 0$, $\frac{\partial f_8}{\partial x_6} = 0$, $\frac{\partial f_8}{\partial x_7} = 0$, $\frac{\partial f_8}{\partial x_8} = 0$, $\frac{\partial f_8}{\partial x_9} = 1$, $\frac{\partial f_8}{\partial x_{10}} = 0$, $\frac{\partial f_8}{\partial x_{11}} = 0$, $\frac{\partial f_8}{\partial x_{12}} = 0$, $\frac{\partial f_8}{\partial x_{13}} = 0$, $\frac{\partial f_8}{\partial x_{14}} = 0$, and $\frac{\partial f_8}{\partial x_{15}} = 0$.

For the ninth row of the Jacobian matrix $A = \nabla_x[f(x)] \mid_{(x^*,u^*)}$ one has: $\frac{\partial f_9}{\partial x_1} = \frac{1}{J_g^2}\{\omega_0 x_{10} x_5 G_{21} \sin(x_6 - x_1 - a_{21})\}$, $\frac{\partial f_9}{\partial x_2} = 0$, $\frac{\partial f_9}{\partial x_3} = 0$, $\frac{\partial f_9}{\partial x_4} = 0$, $\frac{\partial f_9}{\partial x_5} = \frac{1}{J_g^2}\{-\omega_0 x_{10} G_{21} \cos(x_6 - x_1 - a_{21})\}$, $\frac{\partial f_9}{\partial x_6} = \frac{1}{J_g^2}\{K_1^2 \frac{n_g^2}{n_t^2} + \omega_0 x_{10}[x_5 G_{21} \sin(x_6 - x_1 - a_{21}) + x_{15} G_{23} \sin(x_6 - x_{11} - a_{23})]\}$, $\frac{\partial f_9}{\partial x_7} = 0$, $\frac{\partial f_9}{\partial x_8} = \frac{1}{J_g^2}(-K_1^2)\frac{n_g^2}{n_t^2}$, $\frac{\partial f_9}{\partial x_9} = \frac{1}{J_g^2}(-D_t^2)$, $\frac{\partial f_9}{\partial x_{10}} = \frac{1}{J_g^2}\{-\omega_0[G_{22} 2x_{10} + (x_5 G_{21} \cos(x_6 - x_1 - a_{21}) + x_{15} G_{23} \cos(x_6 - x_{11} - a_{23}))]\}$, $\frac{\partial f_9}{\partial x_{11}} = \frac{1}{J_g^2}\{-\omega_0 x_{10} x_{15} G_{23} \sin(x_6 - x_{11} - a_{23})\}$, $\frac{\partial f_9}{\partial x_{12}} = 0$, $\frac{\partial f_9}{\partial x_{13}} = 0$, $\frac{\partial f_9}{\partial x_{14}} = 0$, and $\frac{\partial f_9}{\partial x_{15}} = 0$.

For the tenth row of the Jacobian matrix $A = \nabla_x[f(x)] \mid_{(x^*,u^*)}$ one has: $\frac{\partial f_{10}}{\partial x_1} = 0$, $\frac{\partial f_{10}}{\partial x_2} = 0$, $\frac{\partial f_{10}}{\partial x_3} = 0$, $\frac{\partial f_{10}}{\partial x_4} = 0$, $\frac{\partial f_{10}}{\partial x_5} = 0$, $\frac{\partial f_{10}}{\partial x_6} = -\frac{1}{T_{do}^2} \frac{x_d^2 - x_d'^2}{x'_{d_\Sigma}{}^2} V_s \sin(x_6)$, $\frac{\partial f_{10}}{\partial x_7} = 0$, $\frac{\partial f_{10}}{\partial x_8} = 0$, $\frac{\partial f_{10}}{\partial x_9} = 0$, $\frac{\partial f_{10}}{\partial x_{10}} = -\frac{1}{T'_{d_2}}$, $\frac{\partial f_{10}}{\partial x_{11}} = 0$, $\frac{\partial f_{10}}{\partial x_{12}} = 0$, $\frac{\partial f_{10}}{\partial x_{13}} = 0$, $\frac{\partial f_{10}}{\partial x_{14}} = 0$, and $\frac{\partial f_{10}}{\partial x_{15}} = 0$.

For the eleventh row of the Jacobian matrix $A = \nabla_x[f(x)] \mid_{(x^*,u^*)}$ one has: $\frac{\partial f_{11}}{\partial x_1} = 0$, $\frac{\partial f_{11}}{\partial x_2} = 0$, $\frac{\partial f_{11}}{\partial x_3} = 0$, $\frac{\partial f_{11}}{\partial x_4} = 0$, $\frac{\partial f_{11}}{\partial x_5} = 0$, $\frac{\partial f_{11}}{\partial x_6} = 0$, $\frac{\partial f_{11}}{\partial x_7} = 0$, $\frac{\partial f_{11}}{\partial x_8} = 0$, $\frac{\partial f_{11}}{\partial x_9} = 1$, $\frac{\partial f_{11}}{\partial x_{10}} = 0$, $\frac{\partial f_{11}}{\partial x_{11}} = 0$, $\frac{\partial f_{11}}{\partial x_{12}} = 1$, $\frac{\partial f_{11}}{\partial x_{13}} = 0$, $\frac{\partial f_{11}}{\partial x_{14}} = 0$, and $\frac{\partial f_{11}}{\partial x_{15}} = 0$.

For the twelfth row of the Jacobian matrix $A = \nabla_x[f(x)] \mid_{(x^*,u^*)}$ one has: $\frac{\partial f_{12}}{\partial x_1} = 0$, $\frac{\partial f_{12}}{\partial x_2} = 0$, $\frac{\partial f_{12}}{\partial x_3} = 0$, $\frac{\partial f_{12}}{\partial x_4} = 0$, $\frac{\partial f_{12}}{\partial x_5} = 0$, $\frac{\partial f_{12}}{\partial x_6} = 0$, $\frac{\partial f_{12}}{\partial x_7} = 0$, $\frac{\partial f_{12}}{\partial x_8} = 0$, $\frac{\partial f_{12}}{\partial x_9} = 1$, $\frac{\partial f_{12}}{\partial x_{10}} = 0$, $\frac{\partial f_{12}}{\partial x_{11}} = \frac{1}{J_t^3}(-K_1^3)$, $\frac{\partial f_{12}}{\partial x_{12}} = \frac{1}{J_t^3}(-D_t 1^3)$, $\frac{\partial f_{12}}{\partial x_{13}} = \frac{1}{J_t^3} K_1^3$, $\frac{\partial f_{12}}{\partial x_{14}} = 0$, and $\frac{\partial f_{12}}{\partial x_{15}} = 0$.

For the thirteenth row of the Jacobian matrix $A = \nabla_x[f(x)] \mid_{(x^*,u^*)}$ one has: $\frac{\partial f_{13}}{\partial x_1} = 0$, $\frac{\partial f_{13}}{\partial x_2} = 0$, $\frac{\partial f_{13}}{\partial x_3} = 0$, $\frac{\partial f_{13}}{\partial x_4} = 0$, $\frac{\partial f_{13}}{\partial x_5} = 0$, $\frac{\partial f_{13}}{\partial x_6} = 0$, $\frac{\partial f_{13}}{\partial x_7} = 0$, $\frac{\partial f_{13}}{\partial x_8} = 0$, $\frac{\partial f_{13}}{\partial x_9} = 1$, $\frac{\partial f_{13}}{\partial x_{10}} = 0$, $\frac{\partial f_{13}}{\partial x_{11}} = 0$, $\frac{\partial f_{13}}{\partial x_{12}} = 0$, $\frac{\partial f_{13}}{\partial x_{13}} = 0$, $\frac{\partial f_{13}}{\partial x_{14}} = 1$, and $\frac{\partial f_{13}}{\partial x_{15}} = 0$.

For the fourteenth row of the Jacobian matrix $A = \nabla_x[f(x)] \mid_{(x^*,u^*)}$ one has: $\frac{\partial f_{14}}{\partial x_1} = \frac{1}{J_g^3}\{-\omega_0 x_{15} x_5 G_{31} \sin(x_{11} - x_1 - a_{31})\}$, $\frac{\partial f_{14}}{\partial x_2} = 0$, $\frac{\partial f_{14}}{\partial x_3} = 0$, $\frac{\partial f_{14}}{\partial x_4} = 0$, $\frac{\partial f_{14}}{\partial x_5} = -\frac{1}{J_g^3}\{-\omega_0 x_{15} G_{31} \cos(x_{11} - x_1 - a_{31})\}$, $\frac{\partial f_{14}}{\partial x_6} = \frac{1}{J_g^3}\{-\omega_0 x_{15} x_{10} G_{32} \cdot \sin(x_{11} - x_6 - a_{32})\}$, $\frac{\partial f_{14}}{\partial x_7} = 0$, $\frac{\partial f_{14}}{\partial x_8} = 0$, $\frac{\partial f_{14}}{\partial x_9} = 0$, $\frac{\partial f_{14}}{\partial x_{10}} = \frac{1}{J_g^3}\{-\omega_0 x_{15} G_{32} \cos(x_{11} - x_6 - a_{32})\}$, $\frac{\partial f_{14}}{\partial x_{11}} = \frac{1}{J_g^3}\{K_1^3 \frac{n_g^3}{n_t^3} + \omega_0[x_{15}(x_5 G_{31} \sin(x_{11} - x_1 - a_{31}) + x_{10} G_{32} \sin(x_{11} - x_6 - a_{32}))]\}$,

$\frac{\partial f_{14}}{\partial x_{12}} = 1$, $\frac{\partial f_{14}}{\partial x_{13}} = \frac{1}{J_g^3}(-K_1^3)\frac{n_g^3}{n_t^3}$, $\frac{\partial f_{14}}{\partial x_{14}} = \frac{1}{J_g^3}(-D_g^3)$, and $\frac{\partial f_{14}}{\partial x_{15}} = \frac{1}{J_g^3}\{-\omega_0[G_{33}2x_{15} + (x_9 G_{31}\cos(x_{11} - x_1 - a_{31}) + x_{10}G_{32}\cos(x_{11} - x_6 - a_{32}))]\}$.

For the fifteenth row of the Jacobian matrix $A = \nabla_x[f(x)] \mid_{(x^*,u^*)}$ one has: $\frac{\partial f_{15}}{\partial x_1} = 0$, $\frac{\partial f_{15}}{\partial x_2} = 0$, $\frac{\partial f_{15}}{\partial x_3} = 0$, $\frac{\partial f_{15}}{\partial x_4} = 0$, $\frac{\partial f_{15}}{\partial x_5} = 0$, $\frac{\partial f_{15}}{\partial x_6} = 0$, $\frac{\partial f_{15}}{\partial x_7} = 0$, $\frac{\partial f_{15}}{\partial x_8} = 0$, $\frac{\partial f_{15}}{\partial x_9} = 0$, $\frac{\partial f_{15}}{\partial x_{10}} = 0$, $\frac{\partial f_{15}}{\partial x_{11}} = -\frac{1}{T_{do}^3}\frac{x_d^3 - x_d^{'3}}{x_{d_\Sigma}^{'3}}V_s\sin(x_{11})$, $\frac{\partial f_{15}}{\partial x_{12}} = 0$, $\frac{\partial f_{15}}{\partial x_{13}} = 0$, $\frac{\partial f_{15}}{\partial x_{14}} = 1$, and $\frac{\partial f_{15}}{\partial x_{15}} = -\frac{1}{T'_{d_3}}$.

4.4.6 Computation of the feedback control gains

For the linearized system given by (1.482) the cost function of (1.483) is defined, where coefficient r determines the penalization of the control input and with weight coefficient ρ determines the reward of the disturbances' effects. It is assumed that (i) The energy that is transferred from the disturbances signal $\tilde{d}(t)$ is bounded, that is $\int_0^\infty \tilde{d}^T(t)\tilde{d}(t)dt < \infty$, (ii) matrices $[A, B]$ and $[A, L]$ are stabilizable, (iii) matrix $[A, C]$ is detectable. In the case of a tracking problem the optimal feedback control law is given by

$$u(t) = -Ke(t) \tag{4.199}$$

with $e = x - x_d$ to be the tracking error, and $K = \frac{1}{r}B^T P$ where P is a positive definite symmetric matrix. Matrix P is obtained from the solution of the Riccati equation

$$A^T P + PA + Q - P(\frac{2}{r}BB^T - \frac{1}{\rho^2}LL^T)P = 0 \tag{4.200}$$

where Q is a positive semi-definite symmetric matrix. The worst-case disturbance is given by

$$\tilde{d}(t) = \frac{1}{\rho^2}L^T Pe(t) \tag{4.201}$$

It is noteworthy that the proposed control method provides a reliable and computationally efficient solution to the nonlinear optimal control problem for distributed marine-turbine power generators, which has not been solved in a conclusive manner up to now. The computation of the feedback control law that is applied to the power system is obtained by following Bellman's optimality principle. Other approaches to solve the nonlinear optimal control problem for the distributed power grid, such as MPC or NMPC have not been equally explored (Rigatos, 2016b). For instance, it is acknowledged that Model Predictive Control (MPC) has been primarily developed for linear dynamical systems and its application to the nonlinear distributed power generation units lacks a global asymptotic stability proof and risks a generators' tripping. On the other side, Nonlinear Model Predictive Control (NMPC) is based on an iterative search for the optimal solution that is not always of assured convergence while its performance is often dependent on initial parametrization. Consequently, the global stability properties of NMPC cannot be assured either. The diagram of the considered control loop is depicted in Figure 4.45. The global stability properties

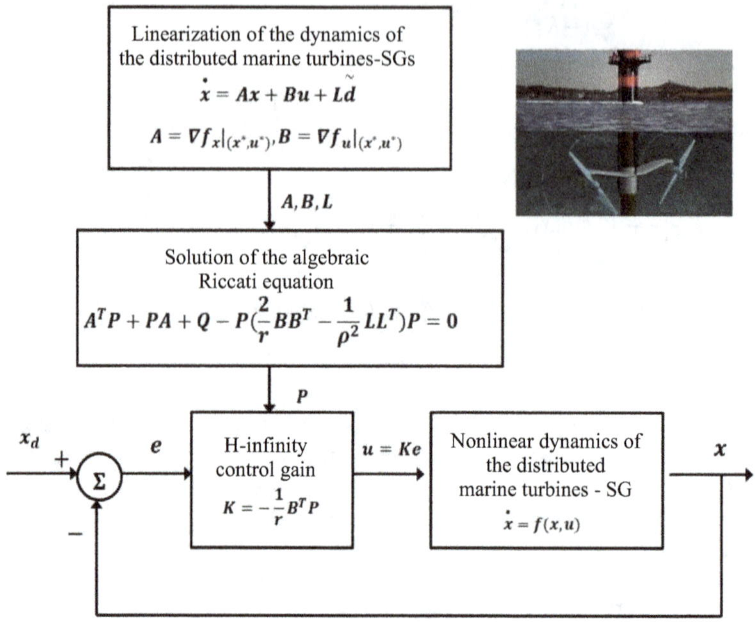

Figure 4.45 Diagram of the control scheme for the distributed marine-turbine power generators

of the control method can be proven following the procedure presented in Section 1.10.2.5.

4.4.7 Simulation tests

The performance of the control scheme is theoretically proven through Lyapunov analysis, and is practically confirmed through simulation experiments. Several reference setpoints have been considered for the state vector elements of the distributed marine turbines power generation system and the obtained results are depicted in Figures 4.46 to 4.60. To compute the controller's feedback gains, the algebraic Riccati equation that appears in (4.200) was repetitively solved at each time-step of the control method. The values of the parameters of the Riccati equation which have been used in the simulation experiments are $r = 0.001$, $\rho = 0.01$ while the diagonal elements of matrix Q have been given the value $q_{ii} = 10$, $i = 1, \cdots, 15$. To implement state estimation-based control without the need to measure the entire state vector of the system, the H-infinity Kalman Filter has been used as a robust state estimator. In the simulation diagrams, the real values of the state vector elements have been plotted in blue, the estimated values have been printed in green, and the setpoints have been marked in red. The state vector elements of the power system have been measured in the per-unit (p.u.) scale.

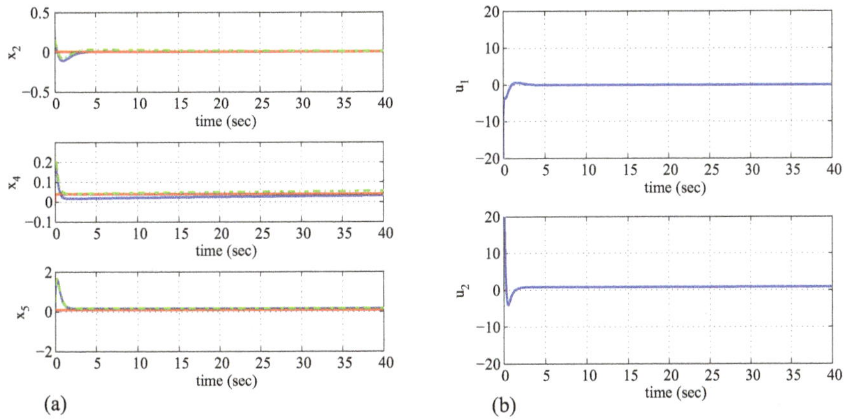

Figure 4.46 Tracking of setpoint 1 by marine-turbine power unit 1: (a) Convergence of the state variables of the marine-turbine generation unit x_2,x_4,x_5 (blue line) to the associated reference values (red line) and estimated values (green line); (b) Control inputs applied to the marine-turbine power unit 1

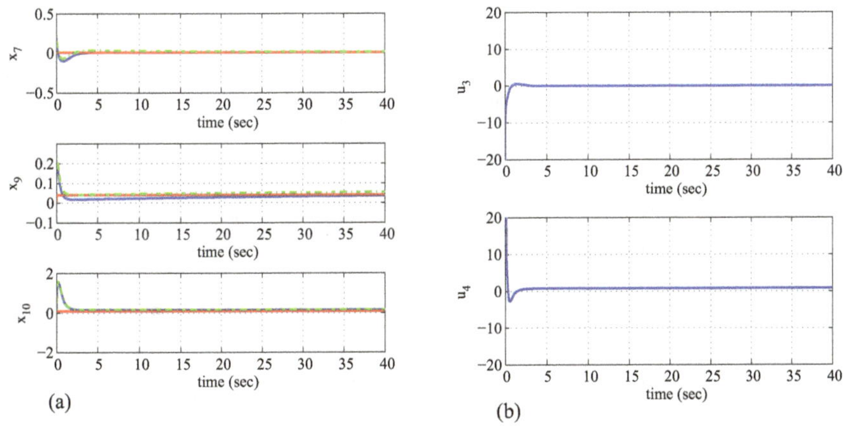

Figure 4.47 Tracking of setpoint 1 by marine-turbine power unit 2: (a) Convergence of the state variables of the marine-turbine generation unit x_7,x_9,x_{10} (blue line) to the associated reference values (red line) and estimated values (green line); (b) Control inputs applied to the marine-turbine power unit 2

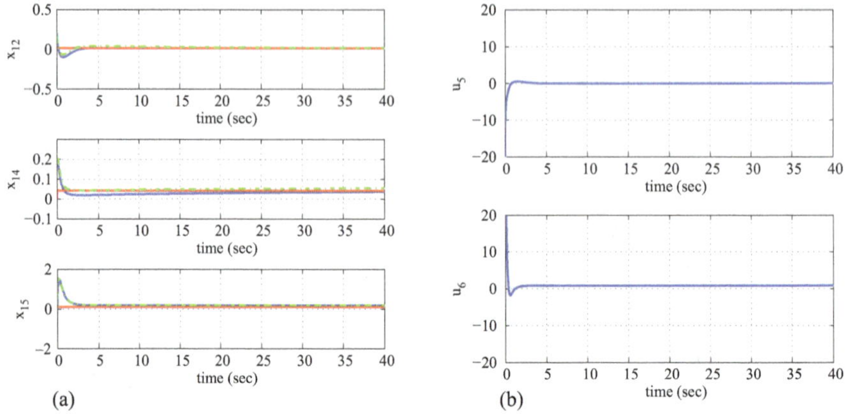

Figure 4.48 Tracking of setpoint 1 by marine-turbine power unit 3: (a) Convergence of the state variables of the marine-turbine generation unit x_{12},x_{14},x_{15} (blue line) to the associated reference values (red line) and estimated values (green line); (b) Control inputs applied to the marine-turbine power unit 3

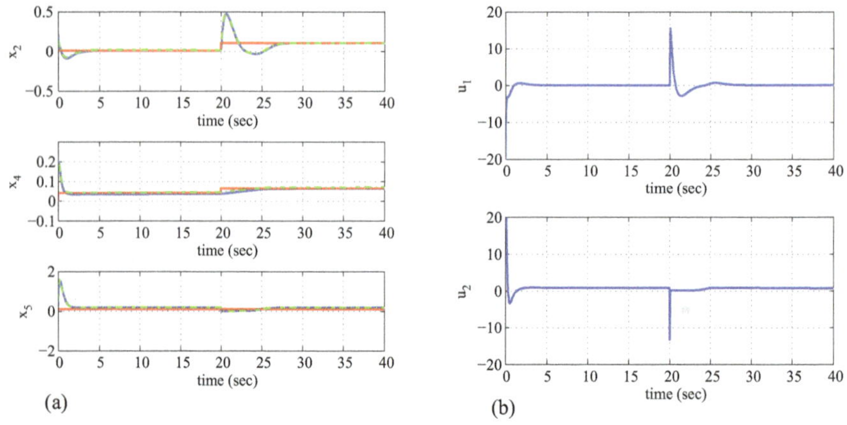

Figure 4.49 Tracking of setpoint 2 by marine-turbine power unit 1: (a) Convergence of the state variables of the marine-turbine generation unit x_2,x_4,x_5 (blue line) to the associated reference values (red line) and estimated values (green line); (b) Control inputs applied to the marine-turbine power unit 1

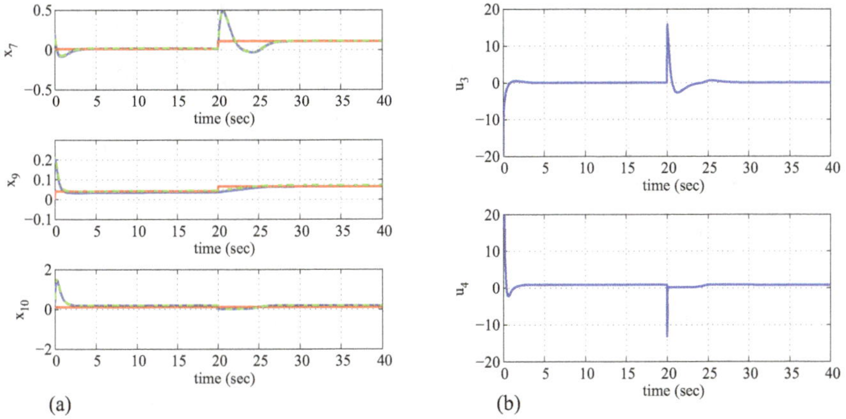

Figure 4.50 Tracking of setpoint 2 by marine-turbine power unit 2: (a) Convergence of the state variables of the marine-turbine generation unit x_7,x_9,x_{10} (blue line) to the associated reference values (red line) and estimated values (green line); (b) Control inputs applied to the marine-turbine power unit 2

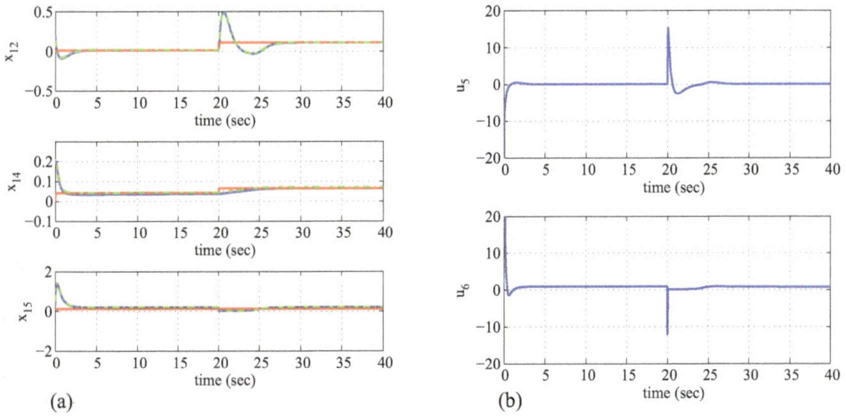

Figure 4.51 Tracking of setpoint 2 by marine-turbine power unit 3: (a) Convergence of the state variables of the marine-turbine generation unit x_{12},x_{14},x_{15} (blue line) to the associated reference values (red line) and estimated values (green line); (b) Control inputs applied to the marine-turbine power unit 3

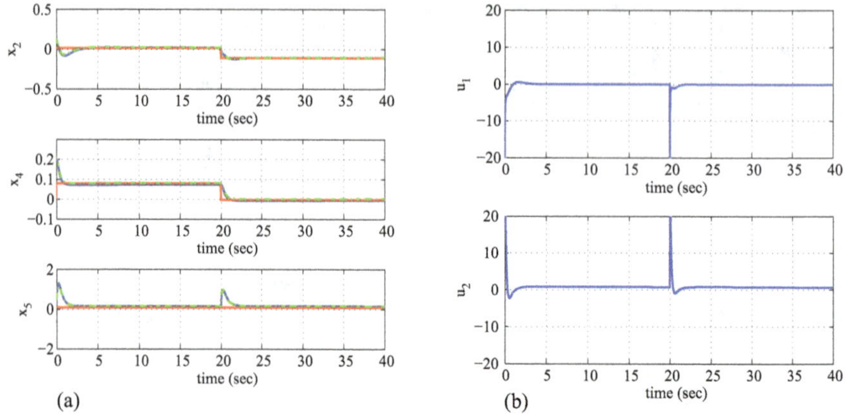

Figure 4.52 *Tracking of setpoint 3 by marine-turbine power unit 1: (a) Convergence of the state variables of the marine-turbine generation unit* x_2,x_4,x_5 *(blue line) to the associated reference values (red line) and estimated values (green line); (b) Control inputs applied to the marine-turbine power unit 1*

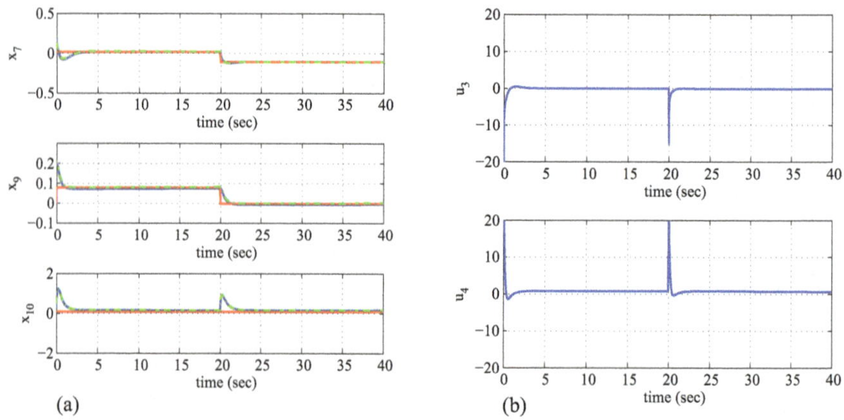

Figure 4.53 *Tracking of setpoint 3 by marine-turbine power unit 2: (a) Convergence of the state variables of the marine-turbine generation unit* x_7,x_9,x_{10} *(blue line) to the associated reference values (red line) and estimated values (green line); (b) Control inputs applied to the marine-turbine power unit 2*

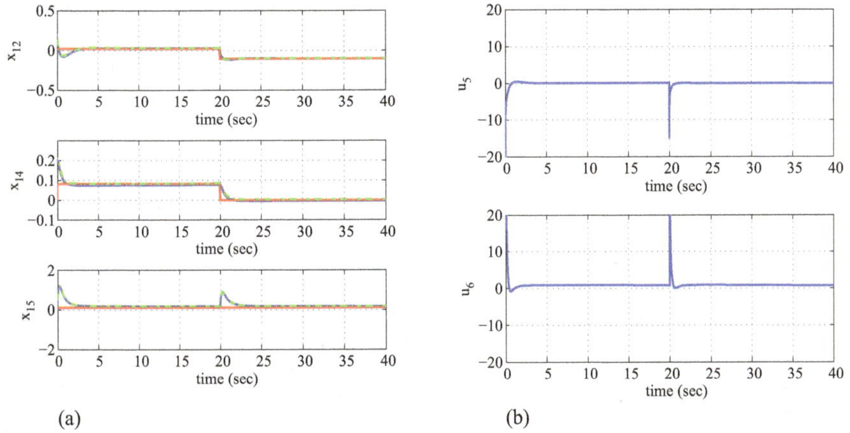

Figure 4.54 Tracking of setpoint 3 by marine-turbine power unit 3: (a) Convergence of the state variables of the marine-turbine generation unit x_{12},x_{14},x_{15} (blue line) to the associated reference values (red line) and estimated values (green line); (b) Control inputs applied to the marine-turbine power unit 3

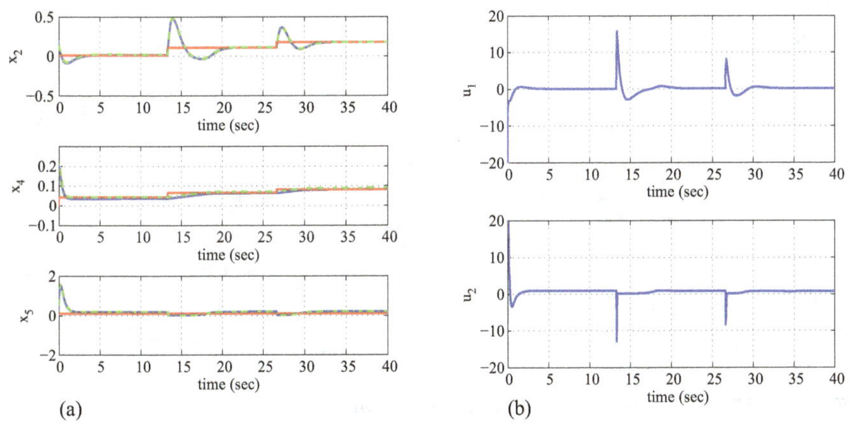

Figure 4.55 Tracking of setpoint 4 by marine-turbine power unit 1: (a) Convergence of the state variables of the marine-turbine generation unit x_2,x_4,x_5 (blue line) to the associated reference values (red line) and estimated values (green line); (b) Control inputs applied to the marine-turbine power unit 1

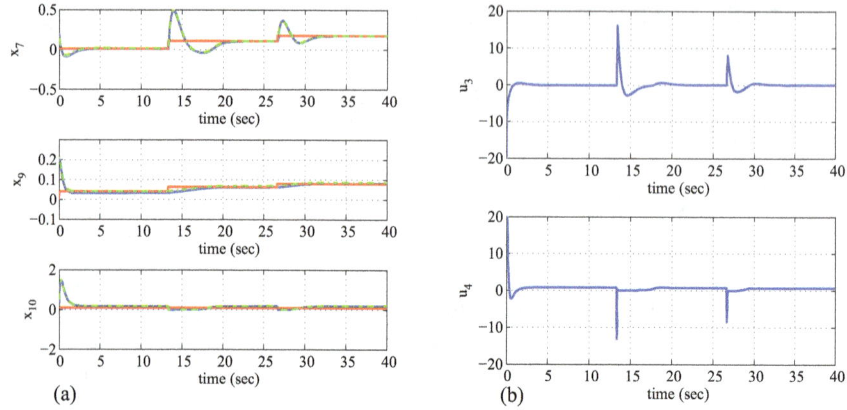

Figure 4.56 Tracking of setpoint 4 by marine-turbine power unit 2: (a) Convergence of the state variables of the marine-turbine generation unit x_7,x_9,x_{10} *(blue line) to the associated reference values (red line) and estimated values (green line); (b) Control inputs applied to the marine-turbine power unit 2*

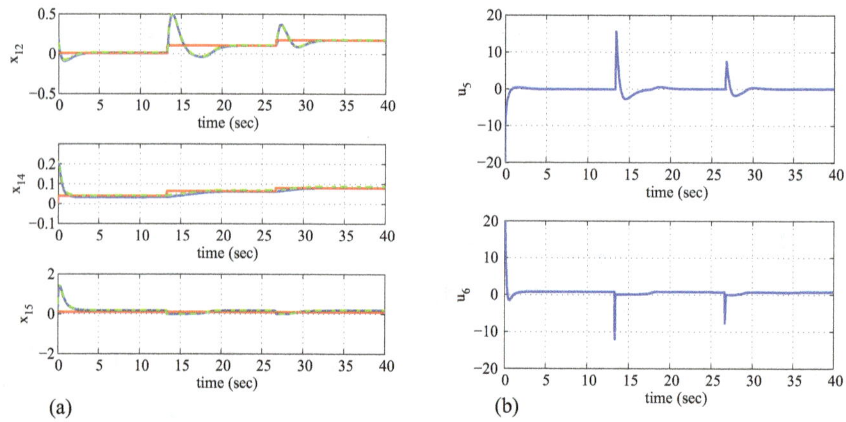

Figure 4.57 Tracking of setpoint 4 by marine-turbine power unit 3: (a) Convergence of the state variables of the marine-turbine generation unit x_{12},x_{14},x_{15} *(blue line) to the associated reference values (red line) and estimated values (green line); (b) Control inputs applied to the marine-turbine power unit 3*

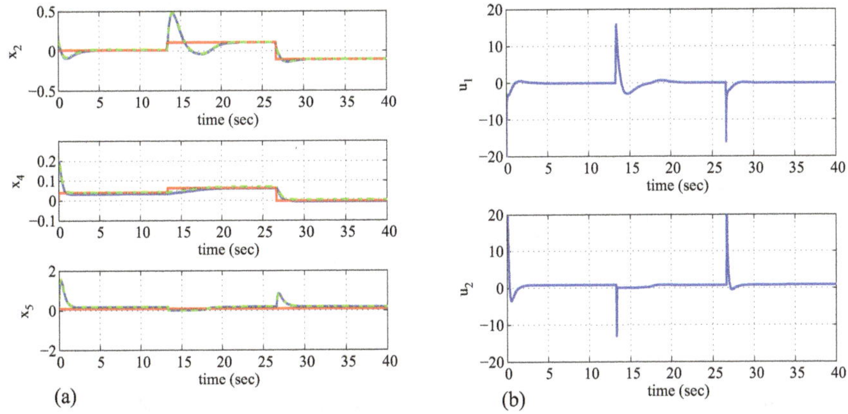

Figure 4.58 *Tracking of setpoint 5 by marine-turbine power unit 1: (a) Convergence of the state variables of the marine-turbine generation unit x_2,x_4,x_5 (blue line) to the associated reference values (red line) and estimated values (green line); (b) Control inputs applied to the marine-turbine power unit 1*

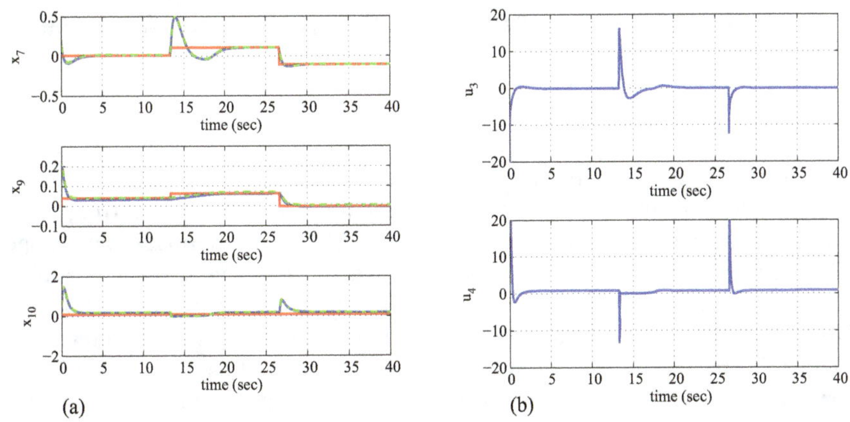

Figure 4.59 *Tracking of setpoint 5 by marine-turbine power unit 2: (a) Convergence of the state variables of the marine-turbine generation unit x_7,x_9,x_{10} (blue line) to the associated reference values (red line) and estimated values (green line); (b) Control inputs applied to the marine-turbine power unit 2*

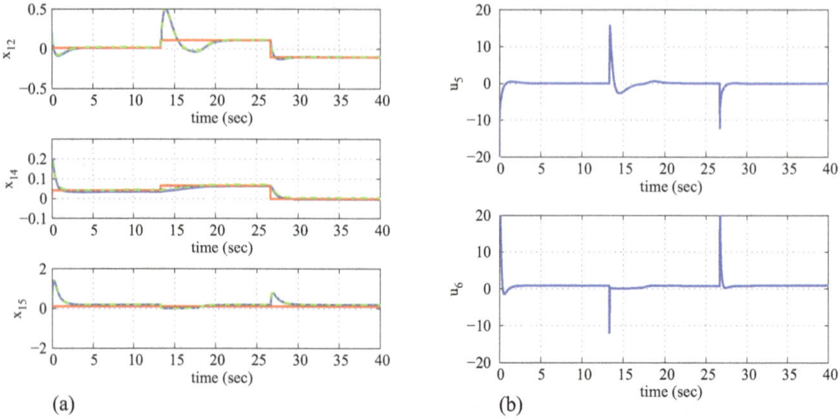

Figure 4.60 Tracking of setpoint 5 by marine-turbine power unit 3: (a) Convergence of the state variables of the marine-turbine generation unit x_{12},x_{14},x_{15} (blue line) to the associated reference values (red line) and estimated values (green line); (b) Control inputs applied to the marine-turbine power unit 3

4.5 Nonlinear control of PMLSGs in wave energy conversion systems

4.5.1 Outline

Conversion of sea-waves power into electricity can offer a vast renewable energy source capable of covering the growing demand for energy (Rigatos, 2016b; Prudell *et al.*, 2010; Dang *et al.*, 2015; Saeed, Wahydie *et al.*, 2017; Vermaak and Kampen, 2012; Rama Rao *et al.*, 2017). Therefore, the waves' power conversion control problem is a challenging task (Delli Colli *et al.*, 2006; Hodgins *et al.*, 2012; Jama and Wahudie, 2017; Wu, Zhang *et al.*, 2009). However, while attempting its solution one comes against several shortfalls and limitations. First, one cannot use conventional power generators in which the moving part performs a rotational motion (Ekstrom *et al.*, 2019; Temiz *et al.*, 2018; ang. Isberg *et al.*, 2018; Zou, Zhao *et al.*, 2012). On the other side, one has to use tubular linear power generators, with the tubular permanent magnet ones to stand for the most efficient type of electric machines for harvesting the waves' power (Oh, Park *et al.*, 2018; Polinder *et al.*, 2005; Lu and Ye, 2009). Second, one has to achieve a periodic motion of the moving part of the tubular permanent magnet linear synchronous generators so as to assure their synchronization with the electricity grid. To achieve this, control inputs which are generated by voltage source converters have to be applied to the generators (Maddugari *et al.*, 2019; Nie, Xiao *et al.*, 2013; Tom, 2014; Tom and Yeung, 2016). Actually, the control inputs should assure that the velocity of the moving part of the linear generators converges and stabilizes at desirable reference values, despite variations in the excitation force of the waves (Wu, Zhang *et al.*, 2008; Tedeschi and Molinas, 2012; Cappelli *et al.*, 2014). The solution of the related control problem is a nontrivial task due to the multivariable and nonlinear structure of the state-space model of the tubular permanent magnet linear synchronous generators.

In this section, a solution is developed for the nonlinear optimal (H-infinity) control problem of tubular Permanent Magnet Linear Synchronous power Generators (PMLSGs) (Rigatos and Siano, 2015; Rigatos, Siano *et al.*, 2015). To this end, the generators' state-space model undergoes first approximate linearization around a temporary operating point which is recomputed at each time-step of the control algorithm. This operating point is defined by the present value of the system's state vector and by the last value of the control inputs vector that was applied to it. The linearization procedure relies on first-order Taylor series expansion and on the computation of the system's Jacobian matrices (Rigatos and Tzafestas, 2007; Basseville and Nikiforov, 1993; Rigatos and Zhang, 2009). The modeling error which is due to the truncation of higher-order terms in the Taylor series expansion is considered to be a perturbation that is asymptotically compensated by the robustness of the control method. For the approximately linearized model of the tubular linear generator an optimal H-infinity stabilizing feedback controller is designed.

The H-infinity controller stands for the solution of the optimal control problem for the tubular linear generators under model uncertainty and external perturbations. It also represents the solution to a min–max differential game in which the control inputs of the generator try to minimize a cost function that comprises a quadratic term of the state vector error of the system. On the other side, the model uncertainty and external perturbation terms try to maximize this cost function. To select the feedback gains of the H-infinity controller an algebraic Riccati equation has to be repetitively solved at each iteration of the control algorithm (Rigatos, 2015; Rigatos, 2011; Rigatos and Busawon, 2018). The stability properties of the control loop are proven through Lyapunov analysis. First, it is proven that the control scheme satisfies the H-infinity tracking performance criterion (Rigatos, 2016b; Tousssaint, Basar and Bullo, 2000). This signifies elevated robustness against model uncertainties and external perturbations. Besides, under moderate conditions it is proven that the control loop is globally asymptotically stable. The proposed control method retains the advantages of linear optimal control in terms of accurate tracking of the reference setpoints by the state variables of linear synchronous generators, as well as in terms of minimization of the variations of the control inputs applied to the generator by voltage source converters.

4.5.2 Dynamics of the tubular permanent magnet linear synchronous generators

4.5.2.1 State-space description of the wave power conversion system

Wave energy conversion is performed with the use of tubular permanent magnet linear synchronous generators. The diagram of the wave power conversion system is shown in Figure 4.61. The generator comprises a stationary part consisting of permanent magnets and a moving part with windings. Due to the motion of the windings in the

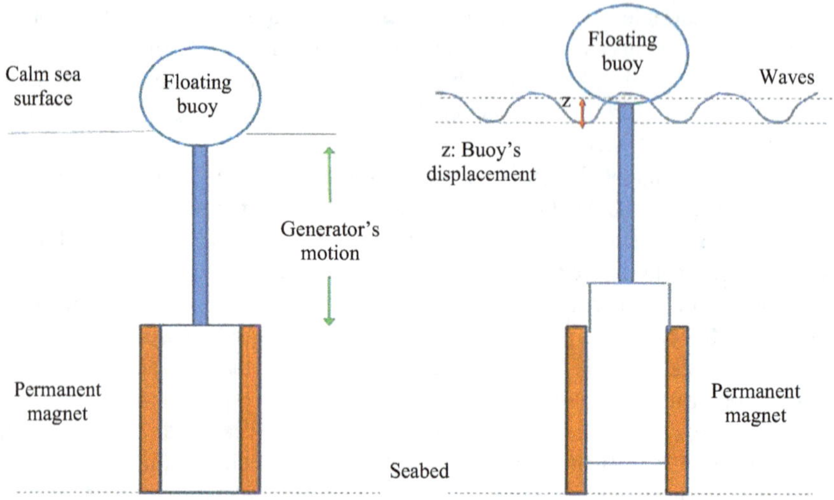

Figure 4.61 Diagram of the tubular permanent magnet linear synchronous generator used for wave energy conversion

magnetic field generated by the permanent magnets, induction currents and voltages appear. The dynamic model of these generators is analyzed next.

Dynamics of the mechanical part:

The simplified model of the motion of the mechanical part of the tubular linear generator is considered (Wu, Zhang *et al.*, 2009). By denoting as z the heave coordinate which represents the displacement of the center of gravity of the buoy from the calm free-water surface one has:

$$M\frac{d^2z}{dt^2} = -b_g\dot{z} - b_w\dot{z} - k_s{\cdot}z + F_e + F_{em} \tag{4.202}$$

where $-b_g\dot{z}$ is the friction force that resists to the motion of the moving part of the tubular generator, $-b_w\dot{z}$ is the hydrodynamic resistance force that resists to the motion of the buoy, $-k_s{\cdot}z$ is a spring-type force exerted on the buoy, F_e is the excitation force due to the incident wave on the buoy. One can enhance this model by including more forces in it, as for instance the Archimedes (lift) force F_b which is applied to the partially submerged buoy, or radiated forces F_r acting on the buoy because of the wave that is radiated as a result of the buoy's oscillations.

The electromagnetic force F_{em} which is applied to the moving part of the generator because of the generator's magnetic field is given by (Delli Colli *et al.*, 2006)

$$F_{em} = \frac{3}{2}\frac{\pi}{\tau_p}[i_d i_q(L_d - L_q) - \psi_{PM} i_q] \tag{4.203}$$

where τ_p is the poles pitch, i_d is the d-axis component at the moving part of the generator, i_q is the q-axis component at the moving part of the generator, L_d is the d-axis inductance at the primary (mover) part of the generator, L_q is the q-axis inductance at the moving part of the generator, ψ_{PM} is the magnetic flux due to the permanent magnets, and v_t is the linear speed of the generator.

Dynamics of the electrical part:

Considering decomposition of the electric circuit of the generator in the axes of the dq rotating frame, the dynamics of the electric part of the tubular permanent magnet linear synchronous generator is given by the following set of differential equations (Delli Colli *et al.*, 2006):

$$
\begin{aligned}
v_d &= -Ri_d - \frac{d\psi_d}{dt} + v_t\frac{\pi}{\tau_t}\psi_q \\
v_q &= -Ri_d - \frac{d\psi_q}{dt} - v_t\frac{\pi}{\tau_t}\psi_d
\end{aligned}
\tag{4.204}
$$

being complemented by the following algebraic equations:

$$
\begin{aligned}
\psi_d &= L_d i_d - \psi_{PM} \\
\psi_q &= L_q i_q
\end{aligned}
\tag{4.205}
$$

By considering term F_e as a disturbance that can be compensated by the robustness of the control algorithm one can omit F_d from the state-space description, thus arriving at the following state-space model

$$
\begin{aligned}
\ddot{z} &= -\frac{(b_g+b_w)}{M}\dot{z} - \frac{k_s}{M}z + \frac{3}{2M}\frac{\pi}{\tau_p}[i_d i_q(L_d-L_q) - \psi_{PM} i_q] \\
\frac{di_d}{dt} &= -\frac{R}{L_d}i_d + \frac{\dot{z}}{L_d}\frac{\pi}{\tau_t}(L_q i_q) - \frac{v_d}{L_d} \\
\frac{di_q}{dt} &= -\frac{R}{L_q}i_q - \frac{\dot{z}}{L_q}\frac{\pi}{\tau_t}(L_d i_d - \psi_{PM}) - \frac{v_q}{L_q}
\end{aligned}
\tag{4.206}
$$

Next, the following state variables are defined for the previous state-space description of the system $x_1 = z$, $x_2 = \dot{z}$, $x_3 = i_d$ and $x_4 = i_q$, while the control input variables are taken to be $u_1 = v_d$ and $v_2 = v_q$. Moreover, after setting $b = b_g + b_w$, this results into the following state-space description

$$
\begin{aligned}
\dot{x}_1 &= x_2 \\
\dot{x}_2 &= -\frac{k_s}{M}x_1 - \frac{b}{M}x_2 + \frac{3}{2M}\frac{\pi}{\tau_p}[x_3 x_4(L_d-L_q) - \psi_{PM}x_4] \\
\dot{x}_3 &= -\frac{R}{L_d}x_3 + \frac{x_2}{L_d}\frac{\pi}{\tau_t}(L_q x_4) - \frac{u_1}{L_d} \\
\dot{x}_4 &= -\frac{R}{L_q}x_4 - \frac{x_2}{L_q}\frac{\pi}{\tau_t}(L_d x_3) + \frac{x_2}{L_q}\frac{\pi}{\tau_t}\psi_{PM} - \frac{u_2}{L_q}
\end{aligned}
\tag{4.207}
$$

In matrix form, the state-space description of the wave power conversion system becomes

$$\begin{pmatrix} \dot{x}_1 \\ \dot{x}_2 \\ \dot{x}_3 \\ \dot{x}_4 \end{pmatrix} = \begin{pmatrix} x_2 \\ -\frac{k_s}{M}x_1 - \frac{b}{M}x_2 + \frac{3}{2M}\frac{\pi}{\tau_p}[x_3x_4(L_d - L_q) - \psi_{PM}x_4] \\ -\frac{R}{L_d}x_3 + \frac{x_2}{L_d}\frac{\pi}{\tau_t}(L_qx_4) \\ -\frac{R}{L_q}x_4 - \frac{x_2}{L_q}\frac{\pi}{\tau_t}(L_dx_3) + \frac{x_2}{L_q}\frac{\pi}{\tau_t}\psi_{PM} \end{pmatrix} + \begin{pmatrix} 0 & 0 \\ 0 & 0 \\ -\frac{1}{L_d} & 0 \\ 0 & -\frac{1}{L_q} \end{pmatrix} \begin{pmatrix} u_1 \\ u_2 \end{pmatrix} \tag{4.208}$$

4.5.2.2 Differential flatness properties of the wave-power conversion system

The wave-power conversion system which comprises the tubular permanent magnet linear synchronous generator is a differentially flat one, with flat outputs vector $Y = [x_1, x_3]^T = [z, i_d]^T$. It can be proven that the system satisfies differential flatness properties.

Indeed from the first row of its state-space model one has:

$$x_2 = \dot{x}_1 \Rightarrow x_2 = h_2(Y, \dot{Y}) \tag{4.209}$$

which signifies that state variable x_2 is a differential function of the flat outputs vector of the system. Besides the second row of the state-space model one solves with respect to x_4, thus obtaining

$$x_4 = \frac{\dot{x}_2 + \frac{k_s}{M}x_1 + \frac{b}{M}x_2}{\frac{3}{2M}\frac{\pi}{\tau_p}[x_3(L_d - L_q) - \psi_{PM}]} \Rightarrow x_4 = h_4(Y, \dot{Y}.\ddot{Y}) \tag{4.210}$$

which signifies that state variable x_4 is a differential function of the flat outputs vector of the system. Moreover, from the last two rows of the state-space model one solves with respect to the control inputs u_1 and u_2, thus obtaining

$$u_1 = -L_d[\dot{x_3} + \frac{R}{L_d}x_3 - \frac{x_2}{L_d}\frac{\pi}{\tau_t}(L_qx_4)] \Rightarrow u_1 = \tilde{h}_1(Y, \dot{Y}) \tag{4.211}$$

$$u_2 = -L_q[\dot{x_4} + \frac{R}{L_q}x_4 + \frac{x_2}{L_q}\frac{\pi}{\tau_t}(L_dx_3) - \frac{x_2}{L_q}\frac{\pi}{\tau_t}\psi_{PM}] \Rightarrow u_2 = \tilde{h}_2(Y, \dot{Y}, \ddot{Y}) \tag{4.212}$$

Consequently, the control input variables u_1 and u_2 are also differential functions of the flat outputs of the system.

Since (i) all state variables and the control inputs of the state-space model of the tubular permanent magnet linear synchronous generator are written as differential functions of the system's flat outputs, (ii) the flat outputs of the system are not connected to each other through a relation in the form of an homogenous linear differential equation, thus being differentially independent, it can be concluded that the system is differentially flat (Rigatos, 2015). The differential flatness properties of the system will be used for defining setpoints for the system's state variables.

4.5.3 Approximate linearization of the model of the tubular PMLSG

The state-space model of the wave power conversion system is the form

$$\dot{x} = f(x) + g(x)u \tag{4.213}$$

where $x \in R^{4\times1}$, $f(x) \in R^{4\times1}$, $g(x)x \in R^{4\times2}$ and $u \in R^{2\times1}$. The model undergoes approximate linearization around the temporary operating point (x^*, u^*), where x^* is the present value of the system's state vector and u^* is the last value of the control inputs vector that was applied to it.

The linearization procedure gives

$$\dot{x} = Ax + Bu + \tilde{d} \tag{4.214}$$

where $\tilde{d}$ is the cumulative disturbance term that may comprise the modeling errors due to the approximate linearization, as well as external perturbations. About matrices A and B one has

$$A = \nabla_x[f(x) + g(x)u] \mid_{(x^*,u^*)} \Rightarrow A = \nabla_x f(x) \mid_{(x^*,u^*)} \tag{4.215}$$

$$B = \nabla_u[f(x) + g(x)u] \mid_{(x^*,u^*)} \Rightarrow B = g(x) \mid_{(x^*,u^*)} \tag{4.216}$$

Next, the elements of the Jacobian matrix $A = \nabla_x f(x) \mid_{(x^*,u^*)}$ are computed:

First row of the Jacobian matrix $A = \nabla_x f(x) \mid_{(x^*,u^*)}$: $\frac{\partial f_1}{\partial x_1} = 0$, $\frac{\partial f_1}{\partial x_2} = 1$, $\frac{\partial f_1}{\partial x_3} = 0$ and $\frac{\partial f_1}{\partial x_4} = 0$.

Second row of the Jacobian matrix $A = \nabla_x f(x) \mid_{(x^*,u^*)}$: $\frac{\partial f_2}{\partial x_1} = -\frac{k_s}{M}$, $\frac{\partial f_2}{\partial x_2} = -\frac{b}{M}$, $\frac{\partial f_2}{\partial x_3} = \frac{3}{2M}\frac{\pi}{\tau_p}(L_d - L_q)x_4$ and $\frac{\partial f_2}{\partial x_4} = \frac{3}{2M}\frac{\pi}{\tau_p}(L_d - L_q)x_3 - \frac{3}{2M}\frac{\pi}{\tau_p}\psi_{PM}$.

Third row of the Jacobian matrix $A = \nabla_x f(x) \mid_{(x^*,u^*)}$: $\frac{\partial f_3}{\partial x_1} = 0$, $\frac{\partial f_3}{\partial x_2} = \frac{\pi}{L_d \tau_t} L_q x_4$, $\frac{\partial f_3}{\partial x_3} = -\frac{R}{L_d}$ and $\frac{\partial f_3}{\partial x_4} = \frac{x_2}{L_d}\frac{\pi}{\tau_t} L_q$.

Fourth row of the Jacobian matrix $A = \nabla_x f(x) \mid_{(x^*,u^*)}$: $\frac{\partial f_4}{\partial x_1} = 0$, $\frac{\partial f_4}{\partial x_2} = -\frac{\pi}{L_q \tau_t} L_d x_3 + \frac{\pi}{L_q \tau_t}\psi_{PM}$, $\frac{\partial f_4}{\partial x_3} = -\frac{x_2}{L_q}\frac{\pi}{\tau_t} L_d$ and $\frac{\partial f_4}{\partial x_4} = -\frac{R}{L_q}$.

For the approximately linearized model of the wave power conversion system an H-infinity stabilizing feedback controller can be designed.

4.5.4 Computation of the feedback control gains

For the linearized system given by (1.482) the cost function of (1.483) is defined, where the coefficient r determines the penalization of the control input and the weight coefficient ρ determines the reward of the disturbances' effects. It is assumed that (i) The energy that is transferred from the disturbances signal $\tilde{d}(t)$ is bounded, that is $\int_0^{\infty} \tilde{d}^T(t)\tilde{d}(t)dt < \infty$, (ii) matrices $[A, B]$ and $[A, L]$ are stabilizable, (iii) matrix $[A, C]$ is detectable. In the case of a tracking problem the optimal feedback control law is given by

$$u(t) = -Ke(t) \tag{4.217}$$

with $e = x - x_d$ to be the tracking error, and $K = \frac{1}{r}B^T P$ where P is a positive definite symmetric matrix. Matrix P is obtained from the solution of the Riccati equation

$$A^T P + PA + Q - P\left(\frac{2}{r}BB^T - \frac{1}{\rho^2}LL^T\right)P = 0 \tag{4.218}$$

where Q is a positive semi-definite symmetric matrix. The worst-case disturbance is given by

$$\tilde{d}(t) = \frac{1}{\rho^2}L^T Pe(t) \tag{4.219}$$

The diagram of the considered control loop is depicted in Figure 4.62. The global stability properties of the control method can be proven following the procedure presented in Section 1.10.2.5.

4.5.5 Simulation tests

The performance and tracking accuracy of the proposed nonlinear optimal control method for tubular permanent magnet linear synchronous generators used in wave power conversion has been demonstrated through simulation experiments. The excitation of the tubular linear generator by sea waves was considered to be a sinusoidal signal. To compute the controller's feedback gain the algebraic Riccati equation of (4.218) had to be repetitively solved at each iteration of the control algorithm. Besides to implement state estimation-based control the H-infinity Kalman Filter has been used for estimating the state variables of the tubular permanent magnet linear synchronous generator through the processing of measurements of the generator's currents $x_3 = i_d$ and $x_4 = i_q$. The obtained results are depicted in Figures 4.63 to 4.67. It can be noticed that under the proposed nonlinear optimal control scheme fast and accurate of the reference setpoints were achieved.

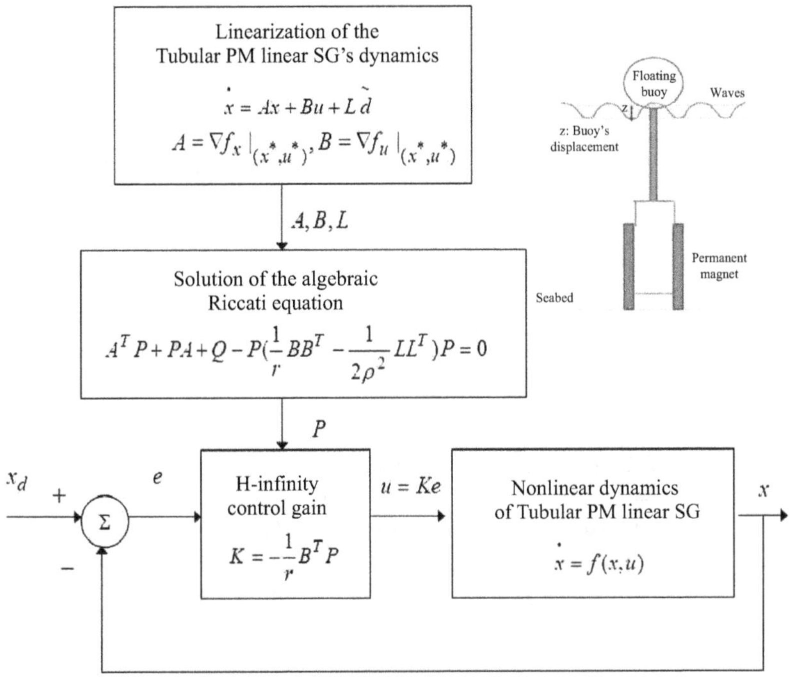

Figure 4.62 Diagram of the control scheme for the Tubular Permanent Magnet Linear Synchronous Generator

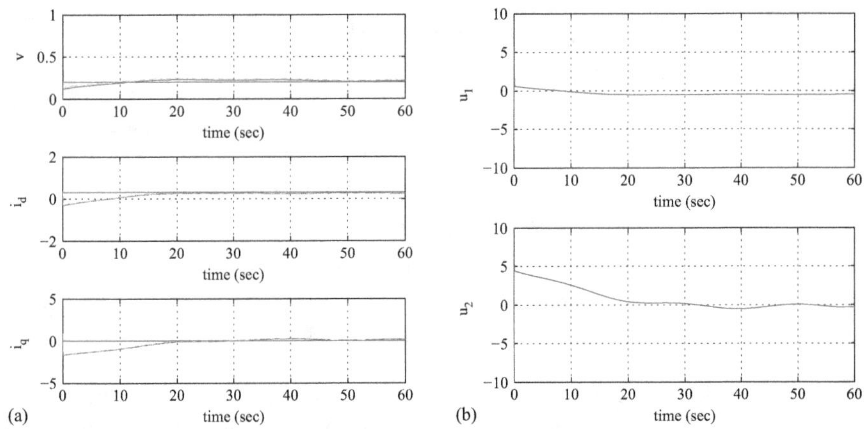

Figure 4.63 Tracking of reference setpoint 1: (a) convergence of state variables $x_2 = v$, $x_3 = i_d$ and $x_4 = i_q$ to their reference setpoints (red-lines) and estimated state variables provided by the Kalman Filter (green lines), (b) control inputs $u_1 = v_d$ and $u_2 = v_q$ applied to the tubular permanent magnet linear synchronous generator

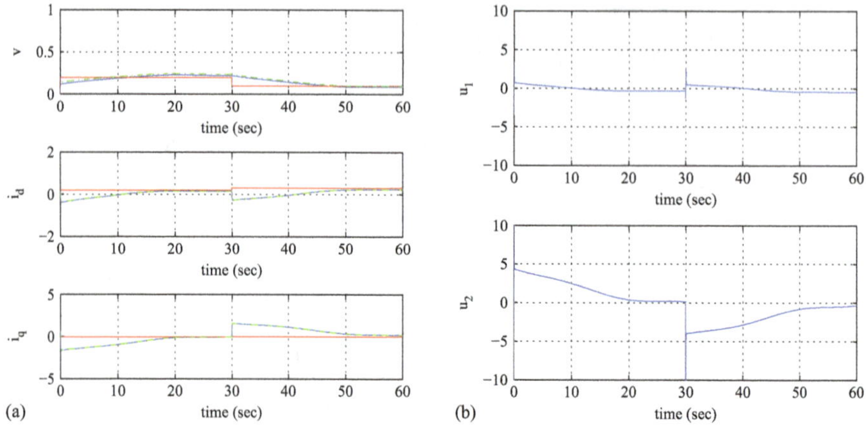

Figure 4.64 Tracking of reference setpoint 2: (a) convergence of state variables $x_2 = v$, $x_3 = i_d$ and $x_4 = i_q$ to their reference setpoints (red-lines) and estimated state variables provided by the Kalman Filter (green lines), (b) control inputs $u_1 = v_d$ and $u_2 = v_q$ applied to the tubular permanent magnet linear synchronous generator

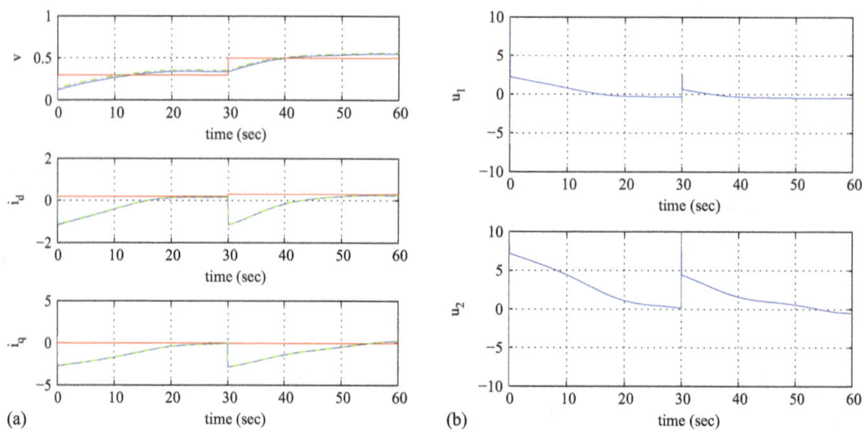

Figure 4.65 Tracking of reference setpoint 3: (a) convergence of state variables $x_2 = v$, $x_3 = i_d$ and $x_4 = i_q$ to their reference setpoints (red-lines) and estimated state variables provided by the Kalman Filter (green lines), (b) control inputs $u_1 = v_d$ and $u_2 = v_q$ applied to the tubular permanent magnet linear synchronous generator

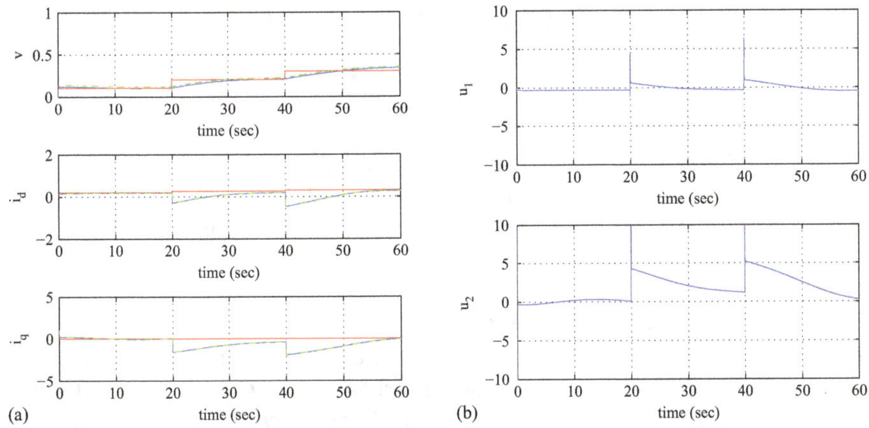

Figure 4.66 Tracking of reference setpoint 4: (a) convergence of state variables $x_2 = v$, $x_3 = i_d$ and $x_4 = i_q$ to their reference setpoints (red-lines) and estimated state variables provided by the Kalman Filter (green lines), (b) control inputs $u_1 = v_d$ and $u_2 = v_q$ applied to the tubular permanent magnet linear synchronous generator

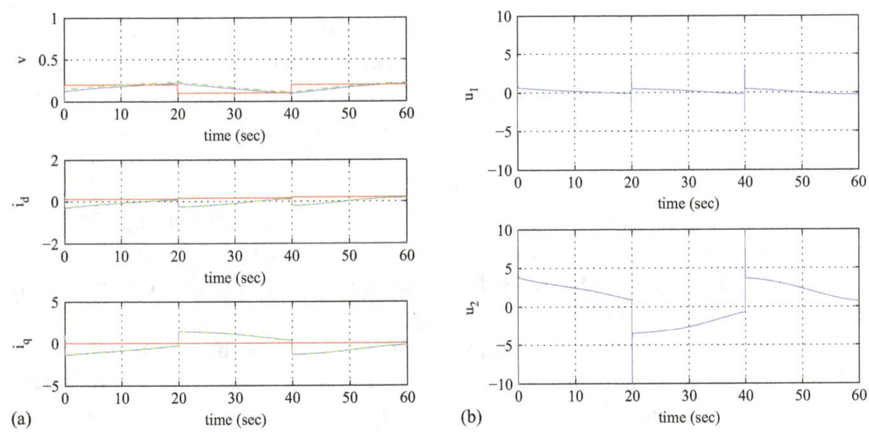

Figure 4.67 Tracking of reference setpoint 5: (a) convergence of state variables $x_2 = v$, $x_3 = i_d$ and $x_4 = i_q$ to their reference setpoints (red-lines) and estimated state variables provided by the Kalman Filter (green lines), (b) control inputs $u_1 = v_d$ and $u_2 = v_q$ applied to the tubular permanent magnet linear synchronous generator

4.6 Nonlinear control of Permanent Magnet Brushless DC motors

4.6.1 Outline

Brushless DC motors are synchronous three-phase motors with non-sinusoidal back electromotive force (back EMF) (Rigatos, 2016b; Bojec *et al.*, 2010; Kumar and Singh, 2016; Liu, Chen *et al.*, 2020). They are widely used in traction of electric vehicles as well as for providing actuation to several mechatronic systems (Joy and Ushakumari, 2021; Wang, Zhu *et al.*, 2020; Youssef, 2015; De *et al.*, 2019). The solution of the control problem for these motors in the abc reference frame is characterized by redundancy in the number of control inputs, which in turn hinders the definition of feasible setpoints for the stator currents and for the turn speed of the rotor (Aghili, 2011; Trivedi and Ku'Koshiri, 2020; Tian, An *et al.*, 2019; Chen and Liu, 2020). Whereas in the case of other synchronous motors (such as synchronous motors with windings at the rotor, permanent magnet synchronous motors and synchronous reluctance motors) a state-space model in the dq synchronously rotating frame is used for designing a feedback controller, such a model is not directly available in the case of brushless DC motors due to the non-sinusoidal form of the back EMF which makes inefficient the typical Park transformation (Kolano, 2019; Valle *et al.*, 2020; Lu, Wu *et al.*, 2020; Khan and Sueaz Miah, 2020). Actually, under the well established Park transformation, the model of the brushless DC motors in the dq reference still contains nonlinear non-constant terms which are related with back EMFs that impose difficulty to the design of a stabilizing feedback controller (Zhou, Chen *et al.*, 2018; Chen, Sun *et al.*, 2018; Vu, Nguyen *et al.*, 2014). To circumvent this shortcoming in the modelling and controller design for brushless DC motors elaborated Park transformations have been introduced (Munoz-Gomez *et al.*, 2018; Grenier and Louis, 1995; Ait-Gougen *et al.*, 2008; Kshirsagar and Krishnan, 2012). So far there have been attempts to solve the control problem for Permanent Magnet Brushless DC Motors using model-predictive controllers, sliding-mode controllers or global linearization-based methods (Lie algebra-based controllers) (Aeashloo *et al.*, 2018; Salehifar *et al.*, 2016; Salehifar *et al.*, 2018; Penz, Jiu *et al.*, 2018; Shiruai Bouroujani *et al.*, 2017) . On the other side, little has been done so far for the solution of the nonlinear optimal control problems for brushless DC motors.

In the present section, a solution is developed for the nonlinear optimal (H-infinity) problem of Permanent Magnet Brushless DC motors (Rigatos, 2016b; Rigatos and Busawon, 2018). First, after applying an elaborated Park transformation, the state-space description of the motor in the dq reference is obtained in which back EMF terms are removed from the rotor's dynamics. Next, this state-space model undergoes approximate linearization around a temporary operating point which is updated at each iteration of the control algorithm. This linearization point is defined by the present value of the motor's state vector and by the last sampled value of the control inputs vector. The linearization process relies on first-order Taylor series expansion and on computation of the related Jacobian matrices (Rigatos and Tzafestas, 2007; Basseville and Nikiforov, 1993; Rigatos and Zhang, 2009). The modelling error which is due to the truncation of higher-order terms in the related Taylor series expansion is viewed as a perturbation that is asymptotically compensated by the robustness of the control method. For the approximately linearized model of the Permanent Magnet Brushless DC motor, a stabilizing optimal (H-infinity) feedback controller is designed (Figure 4.68).

Actually, the proposed H-infinity control for Permanent Magnet Brushless DC motors achieves solution of the related nonlinear optimal control problem under model uncertainty and external perturbations (Rigatos, 2016b; Rigatos and Busawon, 2018; Tousssaint, Basar and Bullo, 2000). The proposed H-infinity controller represents a min–max differential game taking place between (i) the control inputs which try to minimize a quadratic function of the state vector's tracking error, and (ii) the model uncertainty and external perturbation terms which try to maximize this cost function. To select the stabilizing feedback gains of the H-infinity controller an algebraic Riccati equation has to be repetitively solved at each time-step of the control method (Rigatos, 2016b; Rigatos and Busawon, 2018). This Riccati equation uses the Jacobian matrices which are obtained from the linearization procedure for the motor's dynamics. The global stability properties of the control method are proven through Lyapunov analysis. First, it is demonstrated that the H-infinity tracking performance criterion is achieved which signifies elevated robustness against model uncertainty and external perturbations. Moreover, under moderate conditions, global asymptotic stability for the motor's control loop is proven (Rigatos, 2016b; Rigatos and Busawon, 2018; Tousssaint, Basar and Bullo, 2000). The proposed nonlinear optimal control method retains the advantages of typical linear optimal control, that is fast and accurate tracking of reference setpoints under moderate variations of the control inputs.

4.6.2 Dynamic model of the PMBLDC motor

4.6.2.1 Dynamic model of the PMBLDC motor in the *abc* reference frame

Permanent Magnet Brushless DC motors are a type of synchronous AC motors that exhibits non-sinusoidal back electro-motive force (back-EMF) (Rigatos, 2016b). By applying Kirchhoff's laws at the windings of the stator of this motor the following equation is obtained (Munoz-Gomez *et al.*, 2018)

$$\begin{pmatrix} v_{a_s} \\ v_{b_s} \\ v_{c_s} \end{pmatrix} = \begin{pmatrix} R_s & 0 & 0 \\ 0 & R_s & 0 \\ 0 & 0 & R_s \end{pmatrix} \begin{pmatrix} i_{a_s} \\ i_{b_s} \\ i_{c_s} \end{pmatrix} + \begin{pmatrix} 0 & L_s & 0 \\ 0 & 0 & l_s \\ 0 & 0 & L_s \end{pmatrix} \begin{pmatrix} \frac{d}{dt} i_{a_s} \\ \frac{d}{dt} i_{b_s} \\ \frac{d}{dt} i_{c_s} \end{pmatrix} + \begin{pmatrix} e_{a_s} \\ e_{b_s} \\ e_{c_s} \end{pmatrix} \tag{4.220}$$

where v_{a_s}, v_{b_s} and v_{c_s} are the stator phase voltages, i_{a_s}, i_{b_s} and i_{c_s} are the stator phase currents, R_s is the resistance at each phase of the stator, $L_s = L - M$ is the difference between the self-inductance L of each phase and the mutual inductance M between

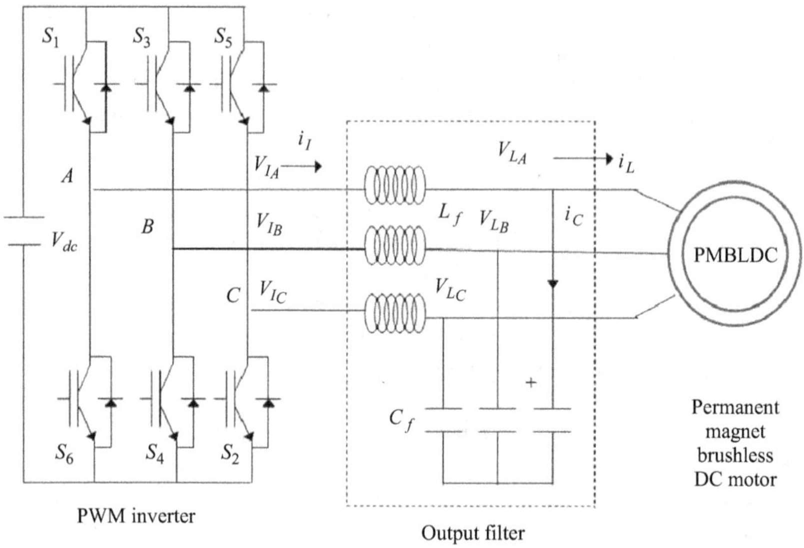

Figure 4.68 Equivalent circuits for the VSI-fed Permanent Magnet Brushless DC motor

phases, e_{a_s}, e_{b_s} and e_{c_s} are the stator back EMF (electro-motive force) about which it holds (Munoz-Gomez *et al.*, 2018)

$$\begin{pmatrix} e_{a_s} \\ e_{b_s} \\ e_{c_s} \end{pmatrix} = \omega_e \lambda_p \begin{pmatrix} f_{a_s}(\theta_e) \\ f_{b_s}(\theta_e) \\ f_{c_s}(\theta_e) \end{pmatrix} \tag{4.221}$$

where $\omega_e = \frac{d\theta_e}{dt}$ is the rotor's electrical velocity, being connected to the rotor's mechanical velocity with the relation $\theta_e = \frac{p}{2}\theta_m$ where p is the number of poles. Equivalently the relation between the electrical and the mechanical angular speed of the rotor is $\omega_e = \frac{p}{2}\omega_m$, and λ_p is the magnitude of the flux linkage generated by the robot's permanent magnets. Parameters $f_{a_s}(\theta_e)$, $f_{b_s}(\theta_e)$ and $f_{c_s}(\theta_e)$ are position-dependent trapezoidal functions with a magnitude equal to ± 11.

The electromagnetic torque of the 3-phase AC motor is (Munoz-Gomez *et al.*, 2018)

$$T_e = \frac{1}{\omega_e}[e_{a_s} i_{a_s} + e_{b_s} i_{b_s} + e_{c_s} i_{c_s}] \tag{4.222}$$

and after substituting the relations about the back-EMF one has

$$T_e = \frac{p\lambda_o}{2}[f_{a_s}(\theta_e) i_{a_s} + f_{b_s}(\theta_e) i_{b_s} + f_{c_s}(\theta_e) i_{c_s}] \tag{4.223}$$

About the mechanical torque of the rotor it holds that

$$J\dot{\omega}_m = T_e - T_L - B\omega_m \tag{4.224}$$

where T_L is the external load torque and $B\omega_m$ is the torque due to friction. Using the previous equations about the electrical the mechanical characteristics of the Permanent Magnet Brushless DC motor one has

$$\begin{aligned}
\dot{\omega}_m &= \frac{T_e}{J} - \frac{T_L}{J} - \frac{B\omega_m}{J} \\
\dot{i}_{a_s} &= -\frac{R_s}{L_s} i_{a_s} - \frac{p\omega_m\lambda_p}{2L_s} f_{a_s}(\theta_e) + \frac{1}{L_s} v_{a_s} \\
\dot{i}_{b_s} &= -\frac{R_s}{L_s} i_{b_s} - \frac{p\omega_m\lambda_p}{2L_s} f_{b_s}(\theta_e) + \frac{1}{L_s} v_{b_s} \\
\dot{i}_{c_s} &= -\frac{R_s}{L_s} i_{c_s} - \frac{p\omega_m\lambda_p}{2L_s} f_{c_s}(\theta_e) + \frac{1}{L_s} v_{c_s}
\end{aligned} \tag{4.225}$$

4.6.2.2 Dynamic model of the PMBLDC motor in the $\alpha\beta$ reference frame

First, Clarke's transform is applied to provide a description of the motor's dynamics in the $\alpha\beta$ frame. A balanced equation is assumed, that is $i_{a_s} + i_{b_s} + i_{c_s} = 0$. In the 2-phase description of the system one has (Munoz-Gomez *et al.*, 2018)

$$\begin{aligned}
v_\alpha &= R_s i_\alpha + L_s \frac{d}{dt} i_\alpha + \lambda_p \frac{p}{2} \omega_m f_\alpha(\theta_e) \\
v_\beta &= R_s i_\beta + L_s \frac{d}{dt} i_\beta + \lambda_p \frac{p}{2} \omega_m f_\beta(\theta_e)
\end{aligned} \tag{4.226}$$

The transformed electromagnetic torque is

$$T_{e(\alpha,\beta)} = \frac{3p\lambda_p}{4} [f_\alpha(\theta_e) i_\alpha + f_\beta(\theta_e) i_\beta] \tag{4.227}$$

The BLDC motor in the (α, β0 reference frame becomes

$$\begin{aligned}
\dot{\omega}_m &= \frac{3p\lambda_p}{4J} [f_\alpha(\theta_e) i_\alpha + f_\beta(\theta_e) i_\beta] - \frac{T_L}{J} - \frac{B\omega_m}{J} \\
\dot{i}_\alpha &= -\frac{R_s}{L_s} i_\alpha - \frac{p\omega_m\lambda_p}{2L_s} f_\alpha(\theta_e) + \frac{1}{L_s} v_\alpha \\
\dot{i}_\beta &= -\frac{R_s}{L_s} i_\beta - \frac{p\omega_m\lambda_p}{2L_s} f_\beta(\theta_e) + \frac{1}{L_s} v_\beta
\end{aligned} \tag{4.228}$$

4.6.2.3 Dynamics of the motor in the dq frame with standard Park's transformation

Next, transformation of the Permanent Magnet Brushless DC motor's model in the synchronously rotating dq reference frame is performed. The following rotation matrix is considered (Munoz-Gomez *et al.*, 2018)

$$T_{dq} = \begin{pmatrix} \cos(\theta_e) & \sin(\theta_e) \\ -\sin(\theta_e) & \cos\theta_e) \end{pmatrix} \tag{4.229}$$

The windings voltage equations expressed in dq reference frame are written as

$$\begin{aligned} v_d &= R_s i_d + L_s \frac{d}{dt} i_d - \omega_e L_s i_q + \lambda_p \frac{p}{2} \omega_m f_d(\theta_e) \\ v_q &= R_s i_q + L_s \frac{d}{dt} i_q - \omega_e L_s i_d + \lambda_p \frac{p}{2} \omega_m f_q(\theta_e) \end{aligned} \tag{4.230}$$

and the electromagnetic torque is given by

$$T_{e(dq)} = \frac{3p\kappa^2\lambda_p}{4}[f_d(\theta_e)i_d + f_q(\theta_e)i_q] \tag{4.231}$$

Thus, the dynamic model of the BLDC motor in dq reference frame is written as

$$\begin{aligned} \dot{\omega}_m &= \frac{3p\lambda_p}{4J}[f_d(\theta_e)i_d + f_q(\theta_e)i_q] - \frac{T_L}{J} - \frac{B\omega_m}{J} \\ \dot{i}_d &= -\frac{R_s}{L_s} i_d + \frac{p}{2}\omega_m i_q - \frac{p\omega_m\lambda_p}{2L_s} f_d(\theta_e) + \frac{1}{L_s} v_d \\ \dot{i}_q &= -\frac{R_s}{L_s} i_q + \frac{p}{2}\omega_m i_d - \frac{p\omega_m\lambda_p}{2L_s} f_q(\theta_e) + \frac{1}{L_s} v_q \end{aligned} \tag{4.232}$$

The particularity of the previous model of (4.232) is that because of the non-sinusoidal form of the back EMF terms of the motor, the model contains finally the non–DC (non-constant) variables $f_d(\theta_e)$ and $f_q(\theta_e)$ in the rotor's dynamics. Due to these terms the stabilizing control inputs for the motor which one may compute based on this model are expected to exhibit abrupt variations.

4.6.2.4 Dynamics of the motor in the *dq* reference with advanced Park's transformation

To remove non-sinusoidal (non-DC) terms from the dynamics of the rotor, a modification of Park's transformation is considered using a new set of variables κ and μ. These variables amend in this respect the conventional dq transformation and are defined as functions of the motor's electrical turn angle θ_e (Munoz-Gomez *et al.*, 2018)

$$\begin{aligned} \kappa &= \frac{\sqrt{f_\alpha(\theta_e)^2 + f_\beta(\theta_e)^2}}{\lambda_p} \\ \mu &= an^{-1}[-\frac{f_\alpha(\theta_e)}{f_\beta(\theta_e)}] - \theta_e \end{aligned} \tag{4.233}$$

The resulting transformation that will be applied to the (α, β) variables so as to provide an amended description of the motor's dynamics in the dq reference frame, without including non-constant terms $f_d(\theta_e)$ and $f_q(\theta_e)$ is given by

$$M_{dq} = \frac{1}{\kappa}\begin{pmatrix} \cos(\theta_e + \mu) & \sin(\theta_e + \mu) \\ -\sin(\theta_e + \mu) & \cos(\theta_e + \mu) \end{pmatrix} \tag{4.234}$$

By applying this amended Park's transformation to the circuits of the motor's windings one obtains:

$$\begin{aligned} v_{m_d} &= R_s i_{m_d} + L_s \frac{1}{\kappa}\frac{d\kappa}{dt} i_{md} + \frac{p}{2}\omega_m L_s i_{mq} + L_s \frac{d\mu}{dt} i_{mq} + L_s \frac{di_{md}}{dt} + \lambda_p \frac{p}{2}\omega_m f_{md}(\theta_e) \\ v_{m_q} &= R_s i_{m_q} + L_s \frac{1}{\kappa}\frac{d\kappa}{dt} i_{mq} - \frac{p}{2}\omega_m L_s i_{md} + -L_s \frac{d\mu}{dt} i_{md} - L_s \frac{di_{mq}}{dt} + \lambda_p \frac{p}{2}\omega_m f_{mq}(\theta_e) \end{aligned} \tag{4.235}$$

and the amended electromagnetic torque of the motor is given by

$$T_{e(md,mq)} = \frac{3\kappa^2 p \lambda_p}{4} = [f_{md}(\theta_e) i_{md} + f_{mq}(\theta_e) i_{mq}] \Rightarrow T_{e(md,mq)} = \frac{3p\lambda_p}{4} i_{mq} \tag{4.236}$$

where it has been used that after the amended transformation it holds that $f_{md}(\theta_e) = 0$ and $f_{mq}(\theta_e) = \frac{1}{\kappa^2}$. Finally, after using the amended Park's transformation the dynamics of the motor in the dq reference frame is written as

$$\begin{aligned} \dot{\omega}_m &= \frac{3p\lambda_p}{4J} i_{mq} - \frac{T_L}{J} - \frac{B\omega_m}{J} \\ \dot{i}_{md} &= -\frac{R_s}{L_s} i_{md} - \frac{1}{\kappa}\frac{d\kappa}{dt} i_{md} - \frac{p}{2}\omega_m i_{mq} - \frac{d\mu}{dt} i_{mq} + \frac{1}{L_s} v_{md} \\ \dot{i}_{mq} &= -\frac{R_s}{L_s} i_{mq} - \frac{1}{\kappa}\frac{d\kappa}{dt} i_{mq} + \frac{p}{2}\omega_m i_{md} + \frac{d\mu}{dt} i_{md} - \frac{p\omega_m \lambda_p}{2\kappa^2 L_s} + \frac{1}{L_s} v_{mq} \end{aligned} \tag{4.237}$$

4.6.2.5 State-space model of the motor under the amended *dq* transformation

Using the previous analysis and the dynamic model of the PMBLDC that has been given in (4.237) one can also write the state-space model of the motor in the affine-in-the-input nonlinear state-space form. The following state variables are defined $x_1 = \theta_m$, $x_2 = \omega_m$, $x_3 = i_{imd}$ and $x_4 = i_{mq}$. Thus, one obtains the state-space description (Munoz-Gomez *et al.*, 2018)

$$\dot{x} = f(x) + g(x)u \tag{4.238}$$

where $x \in R^{4\times1}$, $f(x) \in R^{4\times1}$, $g(x) \in R^{4\times2}$ and $u \in R^{2\times1}$. In matrix form the state-space model can be written as

$$\begin{pmatrix} \dot{x}_1 \\ \dot{x}_2 \\ \dot{x}_3 \\ \dot{x}_4 \end{pmatrix} = \begin{pmatrix} x_2 \\ \frac{3p\lambda_p}{4J} x_4 - \frac{T_L}{J} - \frac{Bx_2}{J} \\ -\frac{R_s}{L_s} x_3 - \frac{1}{\kappa}\frac{d\kappa}{dt} x_3 - \frac{p}{2}\omega_m x_4 - \frac{d\mu}{dt} x_4 \\ -\frac{R_s}{L_s} x_4 - \frac{1}{\kappa}\frac{d\kappa}{dt} x_4 + \frac{p}{2}\omega_m x_3 + \frac{d\mu}{dt} x_3 - \frac{p x_2 \lambda_p}{2\kappa^2 L_s} \end{pmatrix} + \begin{pmatrix} 0 & 0 \\ 0 & 0 \\ \frac{1}{L_s} & 0 \\ 0 & \frac{1}{L_s} \end{pmatrix} \begin{pmatrix} v_{md} \\ v_{mq} \end{pmatrix} \tag{4.239}$$

In the previous equation, the drift vector $f(x)$ and the control inputs gain matrix $g(x)$ are defined as follows:

$$f(x) = \begin{pmatrix} x_2 \\ \frac{3p\lambda_p}{4J}x_4 - \frac{T_L}{J} - \frac{Bx_2}{J} \\ -\frac{R_s}{L_s}x_3 - \frac{1}{\kappa}\frac{d\kappa}{dt}x_3 - \frac{p}{2}\omega_m x_4 - \frac{d\mu}{dt}x_4 \\ -\frac{R_s}{L_s}x_4 - \frac{1}{\kappa}\frac{d\kappa}{dt}x_4 + \frac{p}{2}\omega_m x_3 + \frac{d\mu}{dt}x_3 - \frac{px_2\lambda_p}{2\kappa^2 L_s} \end{pmatrix} \quad g(x) = \begin{pmatrix} 0 & 0 \\ 0 & 0 \\ \frac{1}{L_s} & 0 \\ 0 & \frac{1}{L_s} \end{pmatrix} \tag{4.240}$$

The torque due to external load is taken to be $T_L = mgl\sin(x_1)$, that is it is considered that the motor is used to lift (rotate) by an angle θ_m, a mass m which is attached to the end of a rod of length l. Considering that the measurable state variables of the motor are the rotor's turn angle θ_m and the stator currents i_{md}, i_{mq}, the output (measurement) equation for the model of the non-sinusoidal back EMF motor is given by

$$\begin{pmatrix} y_1 \\ y_2 \\ y_3 \end{pmatrix} = \begin{pmatrix} 1 & 0 & 0 & 0 \\ 0 & 0 & 1 & 0 \\ 0 & 0 & 0 & 1 \end{pmatrix} \begin{pmatrix} x_1 \\ x_2 \\ x_3 \\ x_4 \end{pmatrix} \tag{4.241}$$

4.6.2.6 Approximate linearization of the motor with non-sinusoidal back EMF

The dynamic model of the PMBLDC motor undergoes approximate linearization around the temporary operating point (x^*, u^*) which is updated at each sampling period of the control method. The linearization process relies on first-order Taylor series expansion and on the computation of the related Jacobian matrices. The linearization point is defined by the present value of the system's state vector x^* and by the last sampled value of the control inputs vector u^*. The initial nonlinear model of the motor, being in the form $\dot{x} = f(x) + g(x)u$, is turned into the equivalent linearized state-space description

$$\begin{aligned} \dot{x} &= Ax + Bu + \tilde{d} \\ y &= Cx \end{aligned} \tag{4.242}$$

where $\tilde{d}$ is the cumulative disturbances vector which comprises (i) the model uncertainty due to the truncation of higher-order terms in the Taylor series expansion, (ii) exogenous perturbations, and (iii) sensors measurement noise of any distribution. The detailed form of the Jacobian matrices A and B is given ny

$$A = \nabla_x[f(x) + g(x)u] \mid_{(x^*,u^*)} \Rightarrow A = \nabla_x[f(x)] \mid_{(x^*,u^*)} \tag{4.243}$$

$$B = \nabla_u[f(x) + g(x)u] \mid_{(x^*,u^*)} \Rightarrow B = g(x) \mid_{(x^*,u^*)} \tag{4.244}$$

Computation of the Jacobian matrix $\nabla_x[f(x)]\mid_{(x^*,u^*)}$.

First row of the Jacobian matrix $\nabla_x[f(x)]\mid_{(x^*,u^*)}$: $\frac{\partial f_1}{\partial x_1}=0$, $\frac{\partial f_1}{\partial x_2}=1$, $\frac{\partial f_1}{\partial x_3}=0$, and $\frac{\partial f_1}{\partial x_4}=0$,.

Second row of the Jacobian matrix $\nabla_x[f(x)]\mid_{(x^*,u^*)}$: $\frac{\partial f_2}{\partial x_1}=-\frac{mgl\cos(x_1)}{J}$, $\frac{\partial f_2}{\partial x_2}=-\frac{B}{J}$, $\frac{\partial f_2}{\partial x_3}=0$, and $\frac{\partial f_2}{\partial x_4}=\frac{3p\lambda_p}{4J}$

Third row of the Jacobian matrix $\nabla_x[f(x)]\mid_{(x^*,u^*)}$: $\frac{\partial f_3}{\partial x_1}=0$, $\frac{\partial f_3}{\partial x_2}=-\frac{p}{2}x_4$, $\frac{\partial f_3}{\partial x_3}=-\frac{R_s}{L_s}-\frac{1}{\kappa}\frac{d\kappa}{dt}$, and $\frac{\partial f_3}{\partial x_4}=-\frac{p}{2}x_3-\frac{d\mu}{dt}$

Fourth row of the Jacobian matrix $\nabla_x[f(x)]\mid_{(x^*,u^*)}$: $\frac{\partial f_4}{\partial x_1}=0$, $\frac{\partial f_4}{\partial x_2}=-\frac{p}{2}x_3-\frac{p\lambda_p}{2\kappa^2L_s}$, $\frac{\partial f_4}{\partial x_3}=\frac{p}{2}x_2+\frac{d\mu}{dt}$, $\frac{\partial f_4}{\partial x_4}=-\frac{R_s}{L_s}-\frac{1}{\kappa}\frac{d\kappa}{dt}$

4.6.3 *Differential flatness of the motor with non-sinusoidal back EMF*

The dynamic model of the PMBLDC motor is differentially flat with flat outputs $y_1=x_1=\theta_m$ and $y_2=x_3=i_{md}$ (Rigatos, 2016b; Rigatos and Busawon, 2018). From the first row of the state-space model of (4.240) one has

$$x_2=\dot{x}_1\Rightarrow x_2=h_a(y,\dot{y}) \tag{4.245}$$

Consequently, state variable x_2 is a differential function of the flat outputs vector of the system. From the second row of the state-space model of (4.240) and considering that the load torque is given by $T_L=mgl\cdot\sin(x_1)$, one has

$$\begin{aligned}\dot{x}_2&=-\frac{R_s}{L_s}x_3-\frac{1}{\kappa}\frac{d\kappa}{dt}x_3-\frac{p}{2}\omega_m x_4-\frac{d\mu}{dt}x_4\Rightarrow\\ x_4&=\frac{\dot{x}_2+\frac{ml\sin(x_1)}{J}+\frac{Bx_2}{J}}{\frac{3p\lambda_p}{2J}}\Rightarrow\\ x_4&=h_b(y,\dot{y},\ddot{y})\end{aligned} \tag{4.246}$$

Consequently, state variable x_4 is a differential function of the flat outputs vector of the system. Next, from the third row of the state-space model one solves with respect to the control input u_{md}

$$\begin{aligned}u_{md}&=l_s[\dot{x}_3+\frac{R_s}{L_s}x_3+\frac{1}{\kappa}\frac{d\kappa}{dt}x_3+\frac{p}{2}\omega_m x_4+\frac{d\mu}{dt}x_4]\Rightarrow\\ u_{md}&=h_c(y,\dot{y},\ddot{y},y^{(3)})\end{aligned} \tag{4.247}$$

Therefore, control input u_{md} is a differential function of the flat outputs vector of the system. Finally, from the fourth row of the state-space model one solves with respect to the control input u_{mq}

$$u_{mq} = L_s[\dot{x}_4 + \frac{R_s}{L_s}x_4 + \frac{1}{\kappa}\frac{d\kappa}{dt}x_4 - \frac{p}{2}\omega_m x_3 - \frac{d\mu}{dt}x_3 + \frac{px_2\lambda_p}{2\kappa^2 L_s}] \Rightarrow$$
$$u_{mq} = h_d(y, \dot{y}, \ddot{y}, y^{(3)}) \tag{4.248}$$

As a result of the above, all state variables and the control inputs of the Permanent Magnet Brushless DC motor are written as differential functions of the flat outputs vector of this system, and one concludes that the system is differentially flat. Actually one can select setpoints for state variables x_1 and x_3 and can use the previous relations about the differential flatness properties of the system to define setpoints about the rest of the state variables of the motor x_2 and x_4.

4.6.4 Computation of the feedback control gains

For the linearized system given by (4.242) the cost function of (1.483) is defined, where coefficient r determines the penalization of the control input and weight coefficient ρ determines the reward of the disturbances' effects. It is assumed that (i) The energy that is transferred from the disturbances signal $\tilde{d}(t)$ is bounded, that is $\int_0^\infty \tilde{d}^T(t)\tilde{d}(t)dt < \infty$, (ii) matrices $[A, B]$ and $[A, L]$ are stabilizable, (iii) matrix $[A, C]$ is detectable. In the case of a tracking problem the optimal feedback control law is given by

$$u(t) = -Ke(t) \tag{4.249}$$

with $e = x - x_d$ to be the tracking error, and $K = \frac{1}{r}B^T P$ where P is a positive definite symmetric matrix. Matrix P is obtained from the solution of the Riccati equation

$$A^T P + PA + Q - P\left(\frac{2}{r}BB^T - \frac{1}{\rho^2}LL^T\right)P = 0 \tag{4.250}$$

where Q is a positive semi-definite symmetric matrix. The worst-case disturbance is given by

$$\tilde{d}(t) = \frac{1}{\rho^2}L^T Pe(t) \tag{4.251}$$

The diagram of the considered control loop is depicted in Figure 4.69. The global stability properties of the control method can be proven following the procedure presented in Section 1.1.2.5.

4.6.5 Simulation tests

The previously proven global stability properties of the Permanent Magnet Brushless DC Motor are further confirmed through simulation experiments. The obtained results are depicted in Figures 4.70 to 4.75 In these diagrams the real values of the state variables of the Permanent Magnet Brushless DC Motor are printed in blue, the

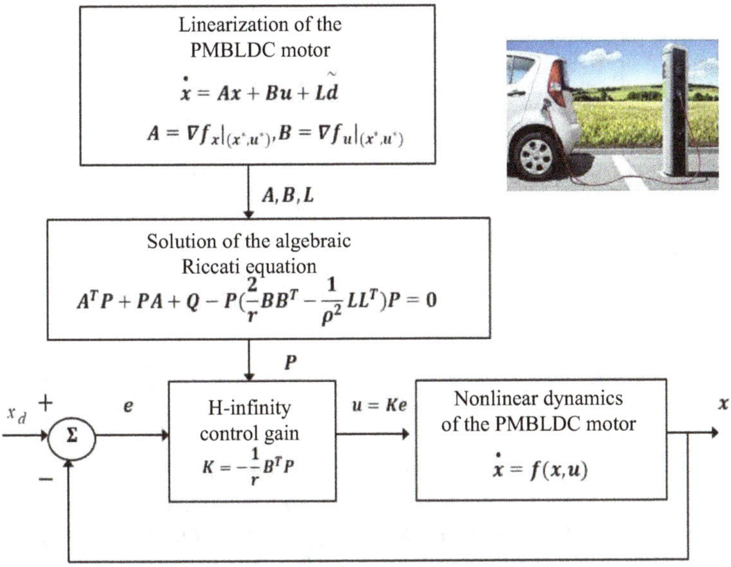

Figure 4.69 *Diagram of the Nonlinear Optimal Control loop for a Permanent Magnet Brushless DC motor*

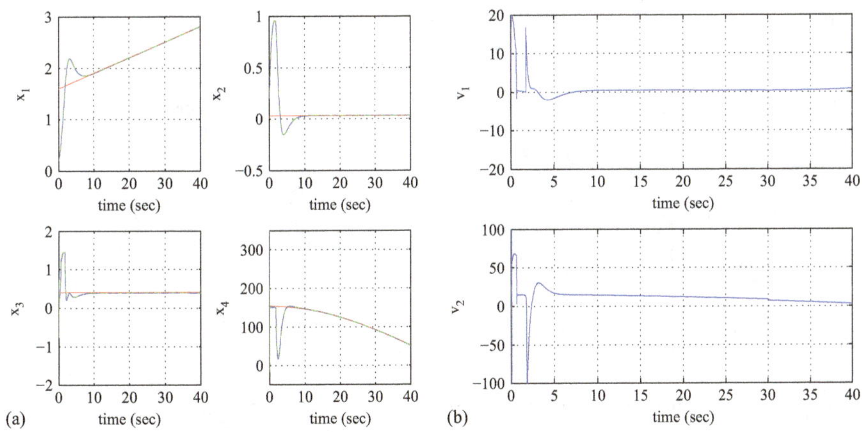

Figure 4.70 *Nonlinear optimal control of the Permanent Magnet Brushless DC Motor when tracking setpoint 1: (a) Convergence of state variables x_1 to x_4 to their reference setpoints, (a) Control inputs u_1 to u_2 (blue lines) which are applied to the PMBLDC Motor*

estimated values which are provided by the H-infinity Kalman Filter are printed in green while the associated setpoints are plotted in red. In all cases, it can be noticed that under the proposed nonlinear optimal control method fast and accurate tracking of the reference setpoints are achieved, under minimal variations of the control inputs.

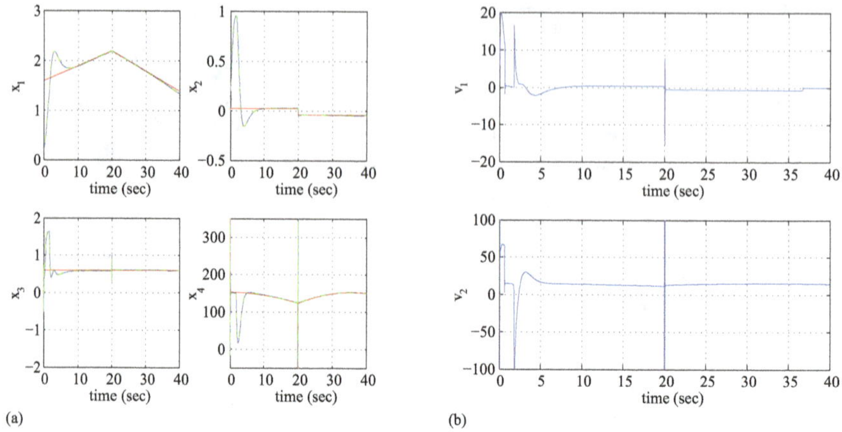

Figure 4.71 *Nonlinear optimal control of the Permanent Magnet Brushless DC Motor when tracking setpoint 2: (a) Convergence of state variables x_1 to x_4 to their reference setpoints, (a) Control inputs u_1 to u_2 (blue lines) which are applied to the PMBLDC Motor*

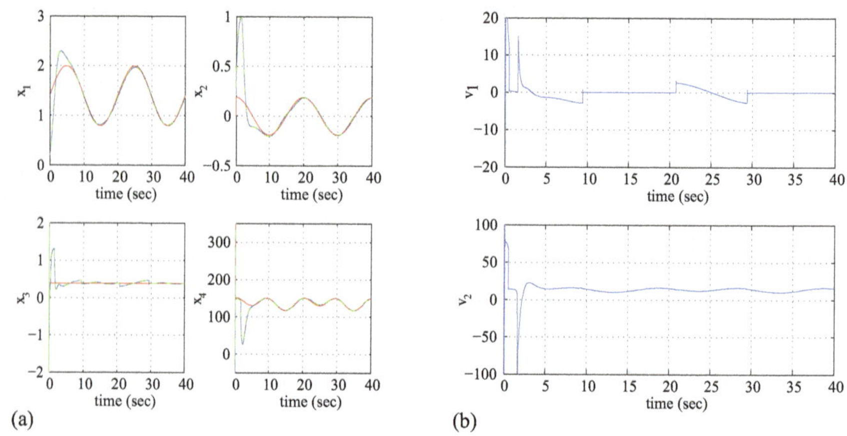

Figure 4.72 *Nonlinear optimal control of the Permanent Magnet Brushless DC Motor when tracking setpoint 3: (a) Convergence of state variables x_1 to x_4 to their reference setpoints, (a) Control inputs u_1 to u_2 (blue lines) which are applied to the PMBLDC Motor*

By keeping limited the control inputs of the motor it can be ensured that minimal energy is consumed for the implementation of the control method. Consequently, there is improvement in the autonomy of electric vehicles which receive traction by such a type of motors. Moreover, there is improvement in the operational capacity of mechatronic systems that receive actuation from such a type of Permanent Magnet Brushless DC motors.

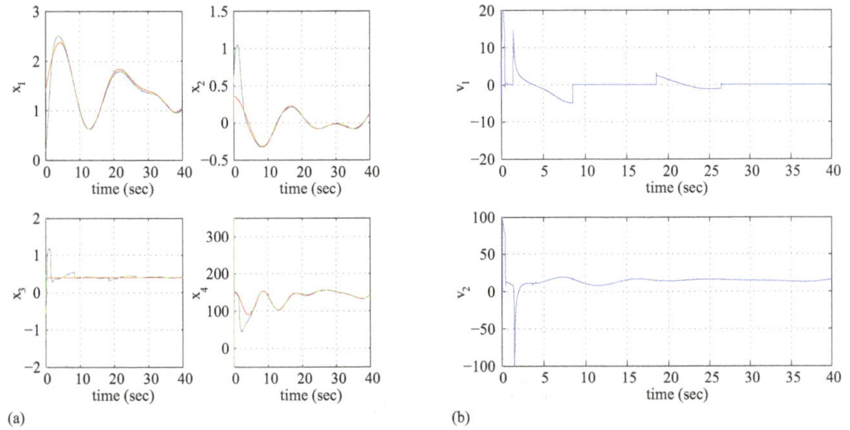

Figure 4.73 *Nonlinear optimal control of the Permanent Magnet Brushless DC Motor when tracking setpoint 4: (a) Convergence of state variables x_1 to x_4 to their reference setpoints, (a) Control inputs u_1 to u_2 (blue lines) which are applied to the PMBLDC Motor*

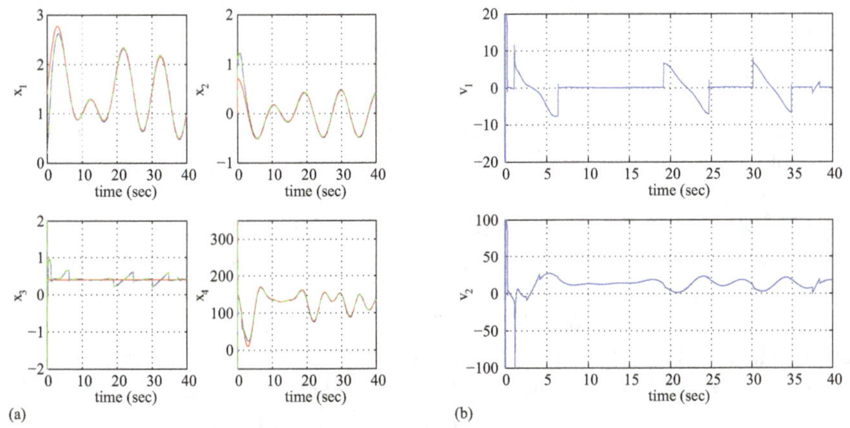

Figure 4.74 *Nonlinear optimal control of the Permanent Magnet Brushless DC Motor when tracking setpoint 5: (a) Convergence of state variables x_1 to x_4 to their reference setpoints, (a) Control inputs u_1 to u_2 (blue lines) which are applied to the PMBLDC Motor*

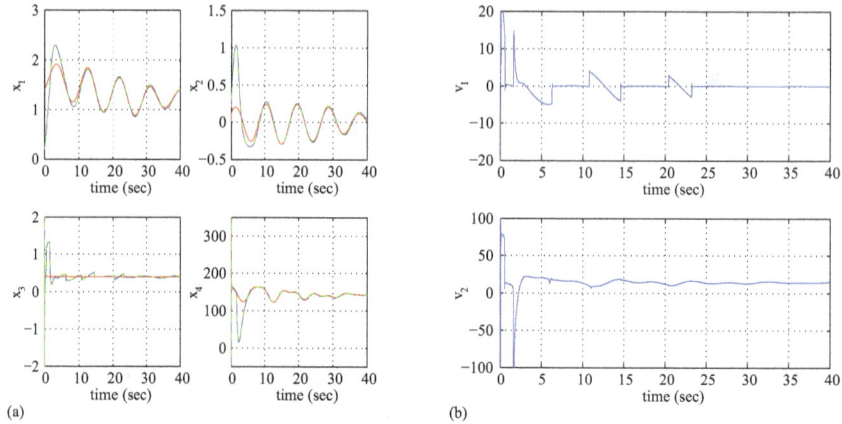

Figure 4.75 *Nonlinear optimal control of the Permanent Magnet Brushless DC Motor when tracking setpoint 6: (a) Convergence of state variables x_1 to x_4 to their reference setpoints, (a) Control inputs u_1 to u_2 (blue lines) which are applied to the PMBLDC Motor*

4.7 Nonlinear optimal control of Hybrid Electric Vehicles powertrains

4.7.1 Outline

Powertrains of hybrid electric vehicles usually comprise a combustion engine (diesel or gasoline engine) which provides torque for the turn motion of a synchronous or asynchronous electric machine that can be used as a generator or as a motor (Boldea *et al.*, 2014; Gokasan *et al.*, 2006; Zhuang *et al.*, 2020). In the generator functioning mode, the electric machine provides power to the vehicle's electric loads and to the batteries pack, while in the motor functioning mode it provides traction to the vehicle (Rigatos and Karapanou, 2020; Boldea, 2004; Chadir *et al.*, 2015; Tong, Cheng *et al.*, 2021). The alternating voltage of the generator is turned into DC voltage through an AC/DC converter thus allowing the charging of the vehicle's batteries (Lie, Chau *et al.*, 2010; Wang, Niu *et al.*, 2017; Zhuang, Zhang *et al.*, 2020; Hu, Song *et al.*, 2014). Control for the power supply and traction system of hybrid electric vehicles is a problem of elevated difficulty because of the nonlinearities and the multi-variable structure of this system's state-space model (Davis and Lorenz, 2003; Yong, Jiao *et al.*, 2014; Barba *et al.*, 2015; Zheng, Wu *et al.*, 2020). Actually, the integrated power-supply and traction system which comprises a diesel engine, a hybrid excited synchronous generator and an AC/DC converter addresses to a challenging control task that should be treated with methods of generic use (Emadi *et al.*, 2008; Ayaz and Mese, 2016; Zhou and Cai, 2019; Song, Wang *et al.*, 2021; Anton *et al.*, 2020). On the other side, it is necessary to solve the nonlinear control problem for HEVs' powertrains because this ensures improved performance and safety for such vehicles (Chen, Laio *et al.*, 2020; Zhang, Xu *et al.*, 2020; Yang, Yang *et al.*, 2020). Additionally, it is necessary to optimize energy consumption and to minimize losses in the HEVs' power-supply and traction system (Gao, Dai *et al.*, 2021; Broomhead *et al.*, 2017; Yang, Zha *et al.*, 2020).

In the present section, a nonlinear optimal (H-infinity) control method has been developed for the powertrain of hybrid electric vehicles which comprises a diesel engine, a hybrid excited synchronous machine (HESM) (having no windings at the rotor), and a three-phase AC/DC converter. The state-space model of the HEV's powertrain undergoes first approximate linearization through first-order Taylor series expansion and through the computation of the associated Jacobian matrices (Rigatos, Siano and Cecati, 2015; Rigatos and Tzafestas, 2007; Basseville and Nikiforov, 1993; Rigatos and Zhang, 2009). The linearization point is defined at each sampling instant by the present value of the system's state vector and by the last sampled value of the control inputs vectors. The truncation of higher-order terms in the Taylor series expansion introduces some model imprecision, however this is asymptotically compensated by the robustness of the control algorithm. For the approximately linearized model of the HEV's power-supply and traction system a stabilizing feedback H-infinity controller is developed.

The proposed H-infinity controller achieves a solution for the optimal control problem for the HEV's traction and propulsion system under model uncertainty and external perturbations. Actually, the H-infinity control approach implements a min-max differential game which takes place between (i) the control inputs which try to minimize a quadratic function of the state vector's tracking error, (ii) the model uncertainty and disturbance terms which try to maximize this cost function. To select the stabilizing feedback gains of the proposed H-infinity controller, an algebraic Riccati equation is repetitively solved at each time-step of the control algorithm (Rigatos, 2016b; Rigatos and Busawon, 2018). The global stability properties of the control scheme are proven through Lyapunov analysis. First, it is demonstrated that the control loop of the HEV's power-supply and traction system achieves the H-infinity tracking performance, which signifies robustness against model uncertainty and exogenous perturbations (Rigatos, 2016b; Tousssaint, Basar and Bullo, 2000). Moreover, under moderate conditions it is proven that the control method is globally asymptotically stable. Finally, to implement state estimation-based control without the need to measure the entire state vector of the HEV's powertrain, the H-infinity Kalman Filter is used as a robust state estimator. It is demonstrated that the H-infinity control scheme achieves fast and accurate tracking of reference setpoints under moderate variations of the control inputs. Besides, by minimizing the variations of the control inputs it reduces the dispersion of energy in the HEV's traction and propulsion system thus improving the vehicle's autonomy and operational capacity.

4.7.2 Dynamic model of the HEV power supply/traction system

The diagram of the HEV power train which comprises a Diesel engine, a hybrid excited synchronous machine and an AC/DC converter are depicted in Figure 4.76.

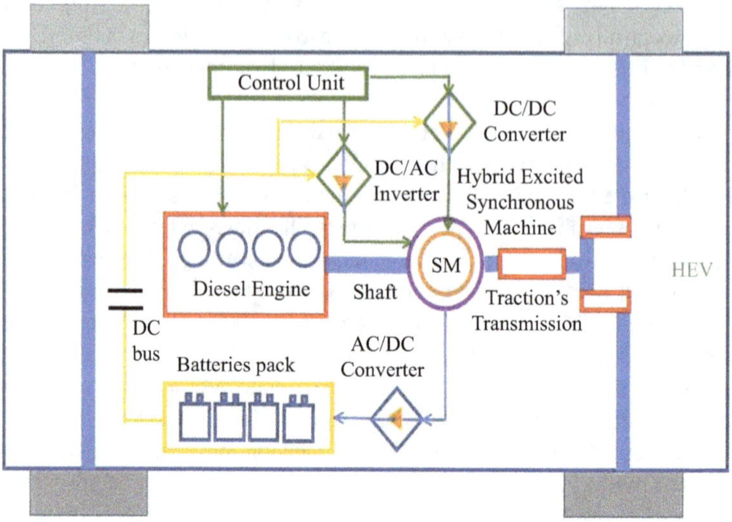

Figure 4.76 Diagram of the HEV's powertrain which comprises a Diesel engine, a hybrid excited synchronous machine and an AC/DC converter

The dynamic model of the AC/DC converter consists of the following equations (Rigatos and Karapanou, 2020; Rigatos, 2016b):

$$L\dot{i}_d = -Ri_d + L\omega_{dq}i_q + v_d - \frac{V_{dc}}{2}u_1 \tag{4.252}$$

$$L\dot{i}_q = -L\omega_{dq}i_d - Ri_q + v_q - \frac{V_{dc}}{2}u_2 \tag{4.253}$$

$$C_{dc}\dot{V}_{dc} = -\frac{1}{R_c}V_{dc} + \frac{3}{4}i_d u_1 + \frac{3}{4}i_q u_2 \tag{4.254}$$

where i_d, i_q are the line currents (i_a, i_b, i_c) after transformation in the dq reference frame, and equivalently v_d, v_q are the phase voltages v_a, v_b, v_c after transformation in the dq reference frame. Variable V_{dc} denotes the DC voltage output of the converter, $u_1 = \eta_d$ and $u_2 = \eta_q$ stand for control inputs. The line losses and any conduction losses are modelled by resistance R and the inverter switching losses are modeled by resistance R_c. Parameter L denotes inductance. Moreover, v_q is taken to be 0.

The dynamic model of the Hybrid Excited Synchronous Machine comprises the following equations (Rigatos and Karapanou, 2020)

$$\dot{\theta}_g = \omega_g \tag{4.255}$$

$$\dot{\omega}_g = -\frac{b}{J}\omega_g + \frac{P_n}{J}(L_d - L_q)i_{s_d}i_{s_q} + \frac{P_n\Psi_a}{J}i_{s_q} + \frac{P_nM_f}{J}i_{s_q}i_f - \frac{1}{J}T_m \tag{4.256}$$

$$\dot{i}_{s_d} = -L_fR_sKi_{s_d} + L_fL_qKP_n\omega_g i_{s_q} + M_fR_fKi_f + L_fKv_{s_d} - M_fKv_f \tag{4.257}$$

$$\dot{i}_{s_q} = -\frac{L_d}{L_q}P_n\omega_g i_{s_d} - \frac{R_s}{L_q}i_{s_q} - \frac{M_f}{L_q}P_n\omega_g i_f - \frac{\Psi_a}{L_q}P_n\omega_g + \frac{1}{L_q}v_{s_q} \tag{4.258}$$

$$\dot{i}_f = M_fR_sKi_{s_d} - M_fL_qKP_n\omega_g i_{s_q} - L_dR_fKi_f - M_fKv_{s_q} + L_dKv_f \tag{4.259}$$

The hybrid excited synchronous machine receives double excitation through (i) an AC/DC and DC/AC converter which is connected to the stator's windings, (ii) a DC/DC converter which is connected to an auxiliary winding of the stator. To describe the electric dynamics of the machine, the synchronously rotating dq reference frame is used. The parameters and state variables of the HESG model are defined as follows: v_{s_d} is the d-axis component of the stator voltage, v_{s_q} is the q-axis component of the stator voltage, v_f is the excitation (field) voltage at an auxiliary winding of the stator, i_{s_d} is the d-axis component of the stator current, i_{s_q} is the q-axis component of the stator current, i_f is the current at the auxiliary excitation circuit of the stator, R is the per-phase resistance of the stator, R_f is the resistance of the auxiliary excitation winding, L_d is the d-axis component of the stator inductance, L_q is the q-axis component of the stator inductance, and L_f is the inductance of the auxiliary excitation winding.

Besides, M_f is the mutual inductance between the auxiliary excitation winding and the d-axis winding of the stator, Ψ_a is the flux linkage of the rotor's permanent magnet, P_n is the number of poles of the machine, ω is the rotor's turn speed, J is the rotor's moment of inertia, b is the friction coefficient in the turn motion of the rotor and T_L is the mechanical torque exerted on the machine by the Diesel engine of the vehicle. Moreover, coefficient K is defined as $K = \frac{1}{L_dL_f - M_f^2}$.

The dynamic model of the Diesel Engine is given in a simplified form with respect to the model analyzed in (Rigatos and Karapanou, 2020):

$$\dot{T}_m = -\frac{T_m}{\bar{\tau}_s} + K_{en}f_{en} \tag{4.260}$$

The above model of the Diesel engine signifies that the produced torque is proportional to the opening of the fuel valve of the engine, subject to time-delay effects which are determined by the time constant $\bar{\tau}_s$.

For the aggregate dynamics of the HEV powertrain the following state variables are defined:

$$\begin{aligned} x &= [x_1, x_2, x_3, x_4, x_5, x_6, x_7, x_8, x_9]^T \Rightarrow \\ x &= [i_d, i_q, V_{dc}, \theta_g, \omega_g, i_{s_d}, i_{s_q}, i_f, T_m] \end{aligned} \tag{4.261}$$

The control inputs of the AC/DC converter are substituted by the stator currents of the HESM, that is $u_1 = K_1 i_{s_d}$ and $u_2 = K_2 i_{s_q}$, or equivalently $u_1 = K_1 x_6$ and $u_2 = K_2 x_7$.

The control inputs vector is defined as $u = [u_1, u_2, u_3, u_4]^T$ with $u_1 = v_{s_d}$, $u_2 = v_{s_q}$, $u_3 = v_f$ and $u_4 = f_{en}$.

Consequently one has the following state-space equations, which comprise the dynamics of the AC/DC converter, of the hybrid excited synchronous machine (HESM) and of the diesel engine:

(i) State-space dynamics of the converter

$$\dot{x}_1 = -\frac{R}{L}x_1 + \omega_{dq}x_2 + \frac{v_d}{L} - \frac{V_{dc}}{2L}K_1x_6 \tag{4.262}$$

$$\dot{x}_2 = -\omega_{dq}x_1 - \frac{R}{L}x_2 + \frac{v_q}{L} - \frac{V_{dc}}{2L}K_2x_7 \tag{4.263}$$

$$\dot{x}_3 = -\frac{1}{R_cC_{dc}}x_3 + \frac{3}{4C_{dc}}x_1K_1x_6 + \frac{3}{4C_{dc}}K_2x_7 \tag{4.264}$$

(ii) State-space dynamics of the hybrid excited synchronous machine

$$\dot{x}_4 = x_5 \tag{4.265}$$

$$\dot{x}_5 = -\frac{b}{J}x_5 + \frac{P_n}{J}(L_d - L_q)x_6x_7 + \frac{P_n\Psi_a}{J}x_7 + \frac{P_nM_f}{J}x_7x_8 - \frac{1}{J}x_9 \tag{4.266}$$

$$\dot{x}_6 = -L_fR_sKx_6 + L_fL_qKP_nx_5x_7 + M_fR_fKx_8 + L_fKu_1 - M_fKu_3 \tag{4.267}$$

$$\dot{x}_7 = -\frac{L_d}{L_q}P_nx_5x_6 - \frac{R_s}{L_q}x_7 - \frac{M_f}{L_q}P_nx_5x_8 - \frac{\Psi_a}{L_q}P_nx_5 + \frac{1}{L_q}u_2 \tag{4.268}$$

$$\dot{x}_8 = M_fR_sKx_6 - M_fL_qKP_nx_5x_7 - L_dR_fKx_8 - M_fKu_2 + L_dKu_3 \tag{4.269}$$

(iii) State-space dynamics of the Diesel engine

$$\dot{x}_9 = -\frac{x_9}{\bar{\tau}_s} + K_{en}u_4 \tag{4.270}$$

Consequently, one arrives at a nonlinear affine in-the-input state-space description of the form

$$\begin{pmatrix}\dot{x}_1\\ \dot{x}_2\\ \dot{x}_3\\ \dot{x}_4\\ \dot{x}_5\\ \dot{x}_6\\ \dot{x}_7\\ \dot{x}_8\\ \dot{x}_9\end{pmatrix}=\begin{pmatrix}-\frac{R}{L}x_1+\omega_{dq}x_2+\frac{v_d}{L}-\frac{V_{dc}}{2L}K_1x_6\\ -\omega_{dq}x_1-\frac{R}{L}x_2+\frac{v_q}{L}-\frac{V_{dc}}{2L}K_2x_7\\ -\frac{1}{R_cC_{dc}}x_3+\frac{3}{4C_{dc}}x_1K_1x_6+\frac{3}{4C_{dc}}K_2x_7\\ x_5\\ -\frac{b}{J}x_5+\frac{P_n}{J}(L_d-L_q)x_6x_7+\frac{P_n\Psi_a}{J}x_7+\frac{P_nM_f}{J}x_7x_8-\frac{1}{J}x_9\\ -L_fR_sKx_6+L_fL_qKP_nx_5x_7+M_fR_fKx_8\\ -\frac{L_d}{L_q}P_nx_5x_6-\frac{R_s}{L_q}x_7-\frac{M_f}{L_q}P_nx_5x_8-\frac{\Psi_a}{L_q}P_nx_5\\ M_fR_sKx_6-M_fL_qKP_nx_5x_7-L_dR_fKx_8\\ -\frac{x_9}{\bar{\tau}_s}\end{pmatrix}$$

$$+\begin{pmatrix}0&0&0&0\\0&0&0&0\\0&0&0&0\\0&0&0&0\\0&0&0&0\\L_fK&0&-M_fK&0\\0&\frac{1}{L_q}&0&0\\0&-M_fK&L_dK&0\\0&0&0&K_{en}\end{pmatrix}\begin{pmatrix}u_1\\u_2\\u_3\\u_4\end{pmatrix} \qquad (4.271)$$

or in concise form

$$\dot{x}=f(x)+g(x)u \qquad (4.272)$$

where $x\in R^{9\times 1}$, $f(x)\in R^{9\times 1}$, $g(x)\in R^{9\times 4}$, and $u\in R^{4\times 1}$.

4.7.3 Approximate linearization of the model of the HEV's powertrain

The state-space model of the HEV powertrain undergoes approximate linearization around the temporary operating point (x^*, u^*), where x^* is the present value of the state vector at each sampling instant and u^* is the last sampled value of the control inputs vector. The linearization relies on first-order Taylor series expansion of the nonlinear state-space model $\dot{x}=f(x)+g(x)u$ and on the computation of the associated Jacobian matrices. The modelling error which is due to the truncation of higher-order terms in the Taylor series expansion is deemed to be a perturbation which is asymptotically compensated by the robustness of the control algorithm. The linearized model is:

$$\dot{x} = Ax + Bu + \tilde{d} \tag{4.273}$$

where $\tilde{d}$ is a cumulative disturbance term that comprises: (i) the modelling error due to the truncation of higher-order terms in the Taylor series expansion, (ii) exogenous perturbations, (iii) sensor measurement noise of any distribution. Matrices A and B are the Jacobian matrices of the system which are given by

$$A = \nabla_x[f(x) + g(x)u] \mid_{(x^*,u^*)} \Rightarrow A = \nabla_x f(x) \mid_{(x^*,u^*)} \tag{4.274}$$

$$B = \nabla_u[f(x) + g(x)u] \mid_{(x^*,u^*)} \Rightarrow B = g(z) \mid_{(x^*,u^*)} \tag{4.275}$$

Computation of the Jacobian matrix $\nabla_x f(x) \mid_{(x^*,u^*)}$.

First row of the Jacobian matrix $\nabla_x f(x) \mid_{(x^*,u^*)}$: $\frac{\partial f_1}{\partial x_1} = -\frac{R}{L}$, $\frac{\partial f_1}{\partial x_2} = \omega_{dq}$, $\frac{\partial f_1}{\partial x_3} = -\frac{K_1 x_6}{2}$, $\frac{\partial f_1}{\partial x_4} = 0$, $\frac{\partial f_1}{\partial x_5} = 0$, $\frac{\partial f_1}{\partial x_6} = -\frac{K_1 x_3}{2}$, $\frac{\partial f_1}{\partial x_7} = 0$, $\frac{\partial f_1}{\partial x_8} = 0$, $\frac{\partial f_1}{\partial x_9} = 0$.

Second row of the Jacobian matrix $\nabla_x f(x) \mid_{(x^*,u^*)}$: $\frac{\partial f_2}{\partial x_1} = -\omega_{dq}$, $\frac{\partial f_2}{\partial x_2} = -\frac{R}{L}$, $\frac{\partial f_2}{\partial x_3} = -\frac{K_2 x_7}{2}$, $\frac{\partial f_2}{\partial x_4} = 0$, $\frac{\partial f_2}{\partial x_5} = 0$, $\frac{\partial f_2}{\partial x_6} = 0$, $\frac{\partial f_2}{\partial x_7} = -\frac{K_2 x_3}{2}$, $\frac{\partial f_2}{\partial x_8} = 0$, $\frac{\partial f_2}{\partial x_9} = 0$.

Third row of the Jacobian matrix $\nabla_x f(x) \mid_{(x^*,u^*)}$: $\frac{\partial f_3}{\partial x_1} = \frac{3K_1 x_6}{4C_{dc}}$, $\frac{\partial f_3}{\partial x_2} = \frac{3K_2 x_7}{4C_{dc}}$, $\frac{\partial f_3}{\partial x_3} = -\frac{1}{R_c C_{dc}}$, $\frac{\partial f_3}{\partial x_4} = 0$, $\frac{\partial f_3}{\partial x_5} = 0$, $\frac{\partial f_3}{\partial x_6} = \frac{3K_1 x_1}{4C_{dc}}$, $\frac{\partial f_3}{\partial x_7} = \frac{3K_2 x_2}{4C_{dc}}$, $\frac{\partial f_3}{\partial x_8} = 0$, $\frac{\partial f_3}{\partial x_9} = 0$.

Fourth row of the Jacobian matrix $\nabla_x f(x) \mid_{(x^*,u^*)}$: $\frac{\partial f_4}{\partial x_1} = 0$, $\frac{\partial f_4}{\partial x_2} = 0$, $\frac{\partial f_4}{\partial x_3} = 0$, $\frac{\partial f_4}{\partial x_4} = 0$, $\frac{\partial f_4}{\partial x_5} = 1$, $\frac{\partial f_4}{\partial x_6} = 0$, $\frac{\partial f_4}{\partial x_7} = 0$, $\frac{\partial f_4}{\partial x_8} = 0$, $\frac{\partial f_4}{\partial x_9} = 0$.

Fifth row of the Jacobian matrix $\nabla_x f(x) \mid_{(x^*,u^*)}$: $\frac{\partial f_5}{\partial x_1} = 0$, $\frac{\partial f_5}{\partial x_2} = 0$, $\frac{\partial f_5}{\partial x_3} = 0$, $\frac{\partial f_5}{\partial x_4} = 0$, $\frac{\partial f_5}{\partial x_5} = -\frac{b}{J}$, $\frac{\partial f_5}{\partial x_6} = \frac{P_n}{J}(L_d - L_q)x_7$, $\frac{\partial f_5}{\partial x_7} = \frac{P_n}{J}(L_d - L_q)x_6 + \frac{P_n \Psi_a}{J} + \frac{P_n M_f}{J} x_5$, $\frac{\partial f_5}{\partial x_8} = \frac{P_n M_L}{J} x_7$, $\frac{\partial f_5}{\partial x_9} = -\frac{1}{J}$.

Sixth row of the Jacobian matrix $\nabla_x f(x) \mid_{(x^*,u^*)}$: $\frac{\partial f_6}{\partial x_1} = 0$, $\frac{\partial f_6}{\partial x_2} = 0$, $\frac{\partial f_6}{\partial x_3} = 0$, $\frac{\partial f_6}{\partial x_4} = 0$, $\frac{\partial f_6}{\partial x_5} = L_f L_q K P_n x_2$, $\frac{\partial f_6}{\partial x_6} = -L_f R_s K$, $\frac{\partial f_6}{\partial x_7} = L_f L_q K P_n x_5$, $\frac{\partial f_6}{\partial x_8} = M_f R_f K$, $\frac{\partial f_6}{\partial x_9} = 0$.

Seventh row of the Jacobian matrix $\nabla_x f(x) \mid_{(x^*,u^*)}$: $\frac{\partial f_7}{\partial x_1} = 0$, $\frac{\partial f_7}{\partial x_2} = 0$, $\frac{\partial f_7}{\partial x_3} = 0$, $\frac{\partial f_7}{\partial x_4} = 0$, $\frac{\partial f_7}{\partial x_5} = \frac{L_d}{L_q} P_n x_6 - \frac{M_f}{L_q} P_n x_8 - \frac{\Psi_a}{L_q} P_n$, $\frac{\partial f_7}{\partial x_6} = -\frac{L_d}{L_q} P_n x_5 - \frac{R_s}{L_q}$, $\frac{\partial f_7}{\partial x_7} = 0$, $\frac{\partial f_7}{\partial x_8} = -\frac{M_f}{L_q} P_n x_5$, $\frac{\partial f_7}{\partial x_9} = 0$.

Eighth row of the Jacobian matrix $\nabla_x f(x) \mid_{(x^*,u^*)}$: $\frac{\partial f_8}{\partial x_1} = 0$, $\frac{\partial f_8}{\partial x_2} = 0$, $\frac{\partial f_8}{\partial x_3} = 0$, $\frac{\partial f_8}{\partial x_4} = 0$, $\frac{\partial f_8}{\partial x_5} = -M_f L_q K P_n x_7$, $\frac{\partial f_8}{\partial x_6} = M_f R_s K$, $\frac{\partial f_8}{\partial x_7} = -M_f L_q K P_n x_5$, $\frac{\partial f_8}{\partial x_8} = -L_d R_f K$, $\frac{\partial f_8}{\partial x_9} = 0$.

Ninth row of the Jacobian matrix $\nabla_x f(x) \mid_{(x^*,u^*)}$: $\frac{\partial f_9}{\partial x_1} = 0$, $\frac{\partial f_9}{\partial x_2} = 0$, $\frac{\partial f_9}{\partial x_3} = 0$, $\frac{\partial f_9}{\partial x_4} = 0$, $\frac{\partial f_9}{\partial x_5} = 0$, $\frac{\partial f_9}{\partial x_6} = 0$, $\frac{\partial f_9}{\partial x_7} = 0$, $\frac{\partial f_9}{\partial x_8} = 0$, $\frac{\partial f_9}{\partial x_9} = -\frac{1}{\tilde{\tau}_s}$.

4.7.4 Differential flatness properties of the HEV's powertrain

It will be proven that the dynamic model of the HEV's powertrain of (4.271), which comprises an AC/DC converter, a hybrid excited synchronous machine and a diesel engine, is differentially flat, with flat outputs: $y_1 = \frac{3L}{4}(x_1^2+x_2^2)+\frac{C_{dc}}{2}x_3^2$, $y_1 = x_2$, $y_3 = x_4$ and $y_4 = x_8$. This signifies that all state variables and the control inputs of the system can be written as differential functions of its flat outputs. It holds that:

$$\begin{aligned}\dot{y}_1 &= \frac{3L}{2}(x_1\dot{x}_1 + x_2\dot{x}_2) + C_{dc}x_3\dot{x}_3 \Rightarrow \\ \dot{y}_1 &= \frac{3L}{2}x_1\left[-\frac{R}{L}x_1 + \omega_{dq}x_2 + \frac{v_d}{L} - \frac{V_{dc}}{2L}K_1x_6\right] \\ &+ \frac{3L}{2}x_2\left[-\omega_{dq}x_1 - \frac{R}{L}x_2 + \frac{v_q}{L} - \frac{V_{dc}}{2L}K_2x_7\right] \\ &+ C_{dc}x_3\left[-\frac{1}{R_cC_{dc}}x_3 + \frac{3}{4C_{dc}}x_1K_1x_6 + \frac{3}{4C_{dc}}K_2x_7\right]\end{aligned} \tag{4.276}$$

which after intermediate operations and regrouping of terms give

$$\dot{y}_1 = -\frac{3R}{2}\left(x_1^2 + x_2^2 - \frac{1}{R_c}x_3^2 + \frac{3v_dx_1}{2} + \frac{3v_qx_2}{2}\right) \tag{4.277}$$

From the definition of the first flat output of the system one has

$$x_3^2 = \frac{y_1 - \frac{3L}{4}(x_1^2 + x_2^2)}{\frac{C_{dc}}{2}} \tag{4.278}$$

By substituting (4.278) into (4.277) one has

$$\dot{y}_1 = -\frac{3R}{2}x_1^2 - \frac{3R}{2}y_2^2 - \frac{1}{R_c}\frac{2}{C_{dc}}\left[y_1 - \frac{3L}{4}(x_1^2 + y_2^2) + \frac{3v_dx_1}{2} + \frac{3v_qy_2}{2}\right] \Rightarrow \tag{4.279}$$

$$\begin{aligned}\dot{y}_1 &= \left(-\frac{3R}{2} + \frac{3L}{2R_cC_{dc}}\right)x_1^2 + \left(-\frac{3R}{2} + \frac{3L}{2R_cC_{dc}}\right)y_2^2 - \frac{2}{R_cC_{dc}}y_1 \\ &+ \frac{3v_d}{2}x_1 + \frac{3v_d}{2}y_2 \Rightarrow\end{aligned} \tag{4.280}$$

and after regrouping of terms one obtains

$$\begin{aligned}&\left(-\frac{3R}{2} + \frac{3L}{2R_cC_{dc}}\right)x_1^2 + \frac{3v_d}{2}x_1 \\ &+ \left[-\dot{y}_1 + \left(-\frac{3R}{2} + \frac{3L}{2R_cC_{dc}}\right)y_2^2 - \frac{2}{R_cC_{dc}}y_1 + \frac{3v_q}{2}y_2\right] = 0\end{aligned} \tag{4.281}$$

The above relation is a binomial which is solved with respect to x_1. By retaining the maximum solution for x_1 one obtains

$$x_1 = h_1(y_1, \dot{y}_1, y_2, \dot{y}_2, y_3, \dot{y}_3, y_4, \dot{y}_4) \tag{4.282}$$

which signifies that state variable x_1 is a differential function of the system's flat outputs. By substituting (4.282) into (4.278) one has that

$$x_3^2 = \frac{y_1 - \frac{3L}{4}(h_1(y,\dot{y})^2 + y_2^2)}{\frac{C_{dc}}{2}} \tag{4.283}$$

which signifies that state variable x_3 is a differential function of the system's flat outputs, or

$$x_3 = h_3(y_1, \dot{y}_1, y_2, \dot{y}_2, y_3, \dot{y}_3, y_4, \dot{y}_4) \tag{4.284}$$

Next, from the first row of the state-space model one solves with respect to x_6. This gives

$$x_6 = \frac{\dot{x}_1 + \frac{R}{L}x_1 - \omega_{dq}x_2 - \frac{v_d}{L}}{-\frac{x_3}{2}K_1} \tag{4.285}$$

Since x_1, x_2, x_3 are differential functions of the system's flat outputs one can also infer that state variable x_6 is a differential function of the system's flat outputs, or

$$x_6 = h_6(y_1, \dot{y}_1, y_2, \dot{y}_2, y_2, \dot{y}_3, y_4, \dot{y}_4) \tag{4.286}$$

Additionally, from the second row of the state-space model one solves with respect to x_7. It holds that

$$x_7 = \frac{\dot{x}_2 + \omega_{dq}x_1 + \frac{R}{L}x_2 - \frac{v_q}{L}}{-\frac{x_3}{2}K_2} \tag{4.287}$$

Since x_1, x_2, x_3 are differential functions of the system's flat outputs one can also infer that state variable, x_7 is a differential function of the system's flat outputs, or

$$x_7 = h_7(y_1, \dot{y}_1, y_2, \dot{y}_2, y_2, \dot{y}_3, y_4, \dot{y}_4) \tag{4.288}$$

Besides, from the fourth row of the state-space model one has

$$x_5 = \dot{x}_4 \Rightarrow x_5 = \dot{y}_3 \tag{4.289}$$

Thus, x_5 is a differential function of the system's flat outputs. Additionally, from the fifth row of the state-space model one solves with respect to state variable x_9. This gives:

$$x_9 = \frac{\dot{x}_5 + \frac{b}{J}x_4 - \frac{P_n}{J}(L_d - L_q)x_6x_7 - \frac{P_n\Psi_a}{J}x_7 - \frac{P_nM_L}{J}x_7x_8}{-\frac{1}{J}} \tag{4.290}$$

Since state variables x_5, x_6, x_7 and $x_8 = y_4$ are differential functions of the system's flat outputs it also holds that x_9 is a differential function of the system's flat outputs, or

$$x_9 = h_9(y_1, \dot{y}_1, y_2, \dot{y}_2, y_3, \dot{y}_3, y_4, \dot{y}_4) \tag{4.291}$$

Thus, from the seventh row of the state-space model one solves with respect to the control input u_2. It holds that

$$u_2 = L_g\left[\dot{x}_7 - \frac{L_d}{L_q}P_nx_5x_6 + \frac{R_s}{L_q}x_6 + \frac{M_f}{L_q}P_nx_5x_8 - \frac{\Psi_a}{L_q}P_nx_5\right] \tag{4.292}$$

Since state variables x_5, x_6, x_7, x_8 are differential functions of the system's flat outputs it holds that control input u_2 is also a differential function of the system's flat outputs, or

$$u_2 = h_{u_2}(y_1, \dot{y}_1, y_2, \dot{y}_2, y_3, \dot{y}_3, y_4, \dot{y}_4) \tag{4.293}$$

Then, from the eighth row of the state-space model one solves with respect to the control input u_3. It holds that

$$u_3 = \frac{1}{L_d K}[\dot{x}_8 - M_f R_s K x_6 + M_f L_q K P_n x_5 x_7 + L_d R_f K x_8 + M_f K u_2] \tag{4.294}$$

Since state variables, x_5, x_6, x_7, x_8 and the control input u_2 are differential functions of the system's flat outputs, it holds that control input u_3 is also a differential function of the HEV's power supply and traction system flat outputs or

$$u_3 = h_{u_3}(y_1, \dot{y}_1, y_2, \dot{y}_2, y_3, \dot{y}_3, y_4, \dot{y}_4) \tag{4.295}$$

Equivalently, from the sixth row of the state-space model one solves with respect to the control input u_1. It holds that

$$u_1 = \frac{1}{L_f K}[\dot{x}_6 + L_f R_s K x_6 - L_f L_q K P_n x_5 x_6 - M_f R_f K x_8 + M K u_3] \tag{4.296}$$

Since state variables, x_5, x_6, x_8 and the control input u_3 are differential functions of the system's flat outputs, one has that control input u_1 can be also written as a differential function of the HEV's power and traction system flat outputs, or

$$u_1 = h_{u_1}(y_1, \dot{y}_1, y_2, \dot{y}_2, y_3, \dot{y}_3, y_4, \dot{y}_4) \tag{4.297}$$

Finally, from the ninth row of the state-space model one solves with respect to the control input u_4. It holds that

$$u_4 = \frac{1}{K_{en}}\left[\dot{x}_9 + \frac{1}{\bar{\tau}_s} x_9\right] \tag{4.298}$$

Since state variable, x_9 is a differential function of the system's flat outputs it holds that control input u_4 is also a differential function of the HEV's power and traction system flat outputs. Thus, one has

$$u_4 = h_{u_4}(y_1, \dot{y}_1, y_2, \dot{y}_2, y_3, \dot{y}_3, y_4, \dot{y}_4) \tag{4.299}$$

As a result of the above, all state variables and the control inputs of the HEV powertrain can be expressed as differential functions of the system's flat outputs. Consequently, the HEV powertrain is a differentially flat system.

The differential flatness property enables to define feasible setpoints for the control problem of the HEV powertrain. Actually, one can define first setpoints for the flat outputs y_1, y_2, y_3 and y_4 and next can use the previously given differential relations to obtain setpoints for state variables x_i, $i = 1, 2, \ldots, 9$.

4.7.5 Computation of the feedback control gains

For the linearized system given by (4.273) the cost function of (1.483) is defined, where the coefficient r determines the penalization of the control input and the weight coefficient ρ determines the reward of the disturbances' effects. It is assumed that (i) The energy that is transferred from the disturbances signal $\tilde{d}(t)$ is bounded, that is $\int_0^{\infty}\tilde{d}^T(t)\tilde{d}(t)dt < \infty$, (ii) matrices $[A, B]$ and $[A, L]$ are stabilizable, (iii) matrix $[A, C]$ is detectable. In the case of a tracking problem the optimal feedback control law is given by

$$u(t) = -Ke(t) \tag{4.300}$$

with $e = x - x_d$ to be the tracking error, and $K = \frac{1}{r}B^T P$ where P is a positive definite symmetric matrix. Matrix P is obtained from the solution of the Riccati equation

$$A^T P + PA + Q - P(\frac{2}{r}BB^T - \frac{1}{\rho^2}LL^T)P = 0 \tag{4.301}$$

where Q is a positive semi-definite symmetric matrix. The worst-case disturbance is given by

$$\tilde{d}(t) = \frac{1}{\rho^2}L^T Pe(t) \tag{4.302}$$

The diagram of the considered control loop for the HEV's powertrain is depicted in Figure 4.77. The global stability properties of the control method can be proven following the procedure presented in Section 1.1.2.5.

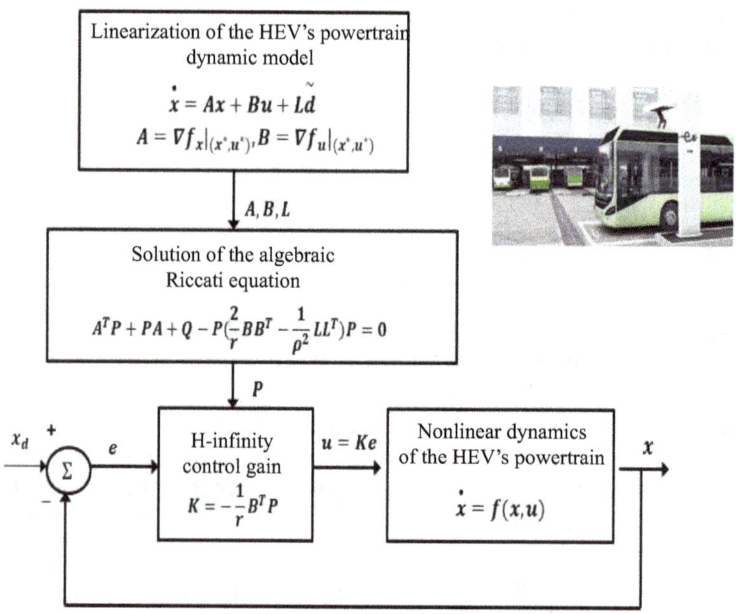

Figure 4.77 Diagram of the control scheme for the HEV's powertrain

4.7.6 Simulation tests

The global stability properties of the proposed nonlinear optimal control scheme for the HEV's powertrain, as well as the precise tracking of reference setpoints by the state variables of this powertrain has been further confirmed through simulation experiments. To implement the nonlinear optimal control scheme, the algebraic Riccati

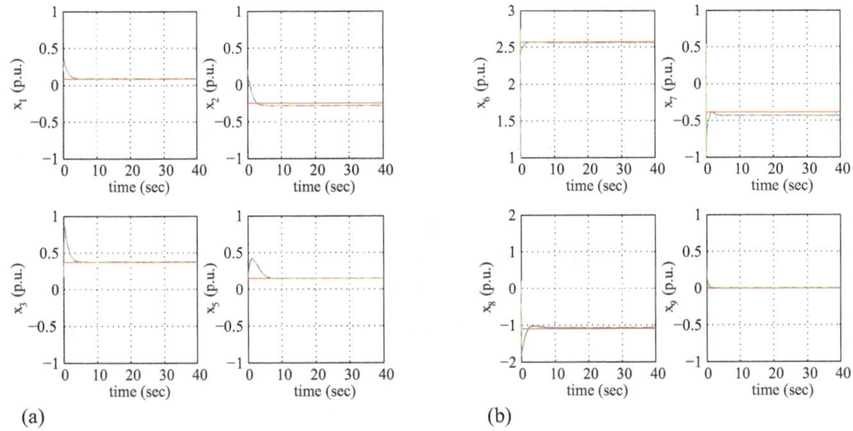

Figure 4.78 Tracking of setpoint 1 for the HEV powertrain (a) convergence of state variables x_1 to x_5 to their reference setpoints (red line: setpoint, blue line: real value, green line: estimated value), (b) convergence of state variables x_6 to x_9 to their reference setpoints

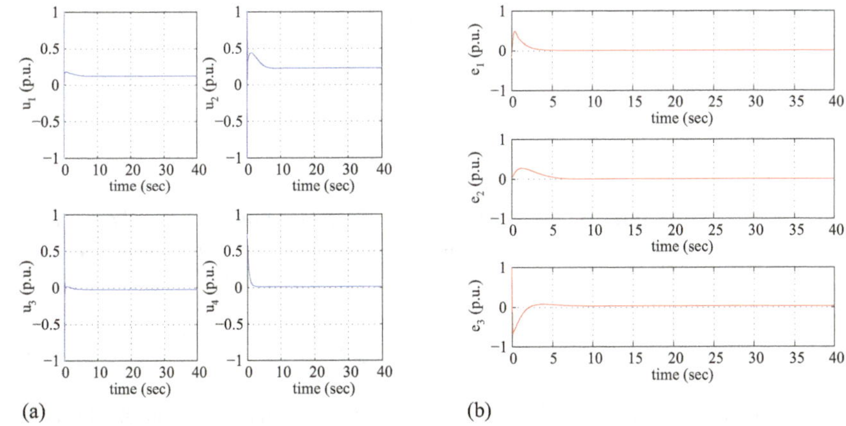

Figure 4.79 Tracking of setpoint 1 for the HEV powertrain (a) control inputs u_1 to u_4 applied to the HEV powertrain, (b) tracking error e_1, e_2, e_3, for state variables V_{dc}, ω_g, i_f of the HEV powertrain

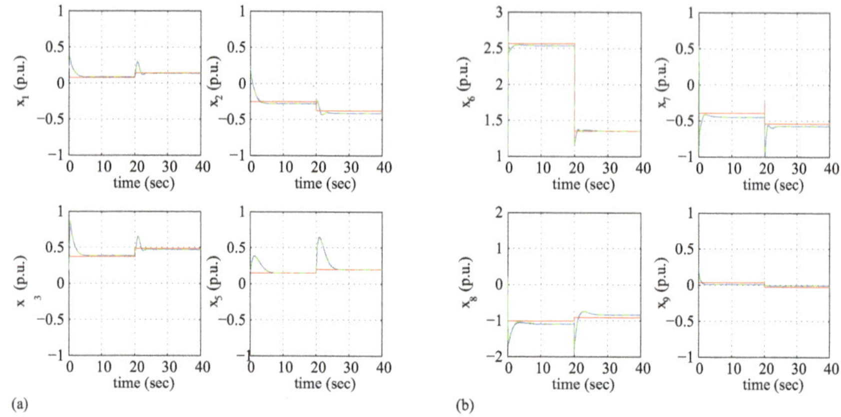

Figure 4.80 Tracking of setpoint 2 for the HEV power supply and traction system (a) convergence of state variables x_1 to x_5 to their reference setpoints (red line: setpoint, blue line: real value, green line: estimated value), (b) convergence of state variables x_6 to x_9 to their reference setpoints

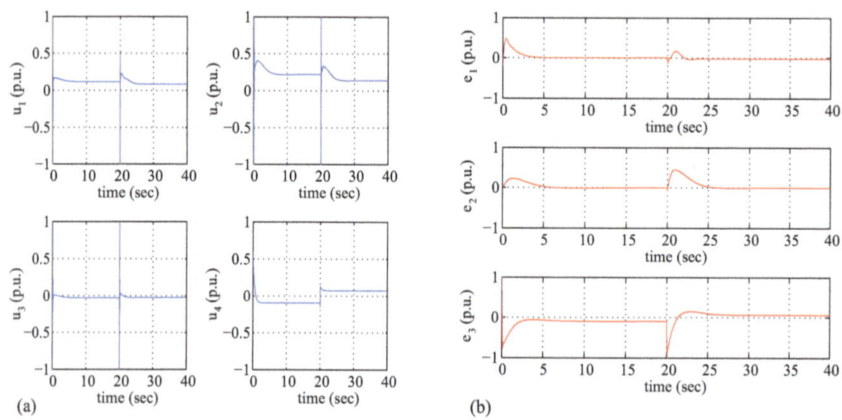

Figure 4.81 Tracking of setpoint 2 for the HEV powertrain (a) control inputs u_1 to u_4 applied to the HEV powertrain, (b) tracking error e_1, e_2, e_3, for state variables V_{dc}, ω_g, i_f of the HEV powertrain

equation which appears in (4.301) had to be solved at each time-step of the control algorithm. The obtained results are depicted in Figures 4.78 to 4.93. It can be noticed that in all cases fast and accurate tracking of reference setpoints was achieved under moderate variations of the control inputs. The real values of the state vector of the HEV's powertrain are printed in blue, the estimated state variables provided by the

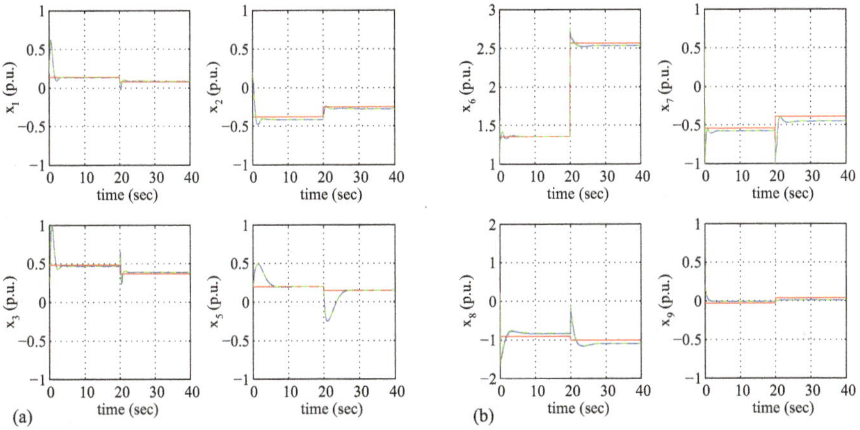

Figure 4.82 Tracking of setpoint 3 for the HEV powertrain (a) convergence of state variables x_1 to x_5 to their reference setpoints (red line: setpoint, blue line: real value, green line: estimated value), (b) convergence of state variables x_6 to x_9 to their reference setpoints

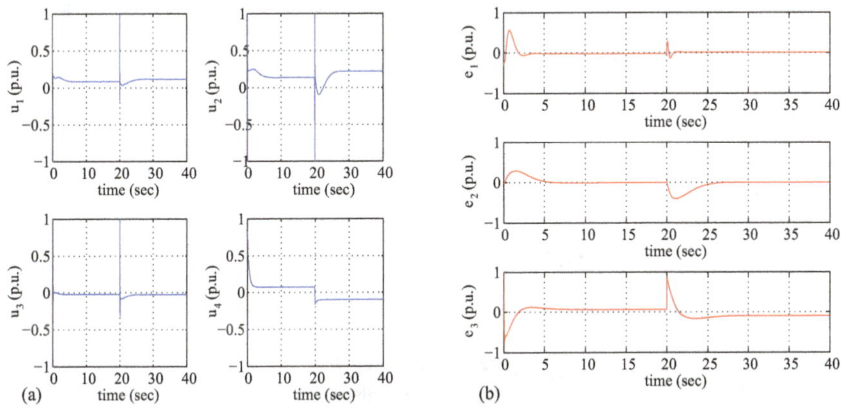

Figure 4.83 Tracking of setpoint 3 for the HEV powertrain (a) control inputs u_1 to u_4 applied to the HEV powertrain, (b) tracking error e_1, e_2, e_3, for state variables V_{dc}, ω_g, i_f of the HEV powertrain

H-infinity Kalman Filter are plotted in green, while the associated setpoints are shown in red colour.

The present section provides one of the few existing solutions to the nonlinear optimal control problem of HEV power supply and traction system which is of proven global stability while also remaining computationally efficient (Rigatos and

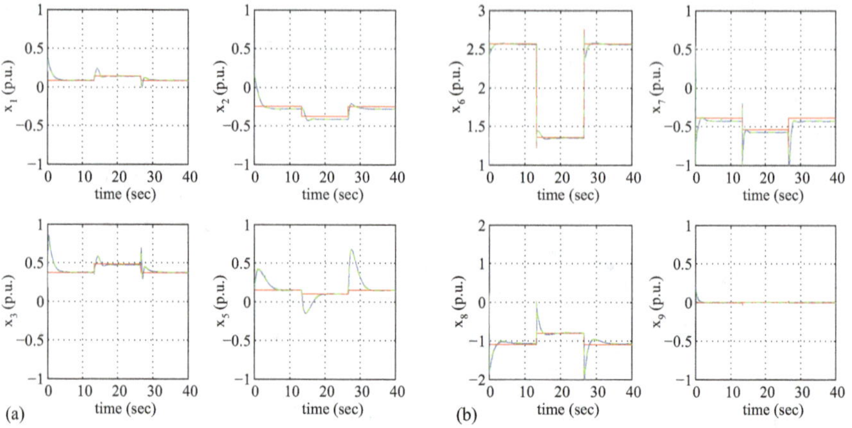

Figure 4.84 Tracking of setpoint 4 for the HEV powertrain (a) convergence of state variables x_1 to x_5 to their reference setpoints (red line: setpoint, blue line: real value, green line: estimated value), (b) convergence of state variables x_6 to x_9 to their reference setpoints

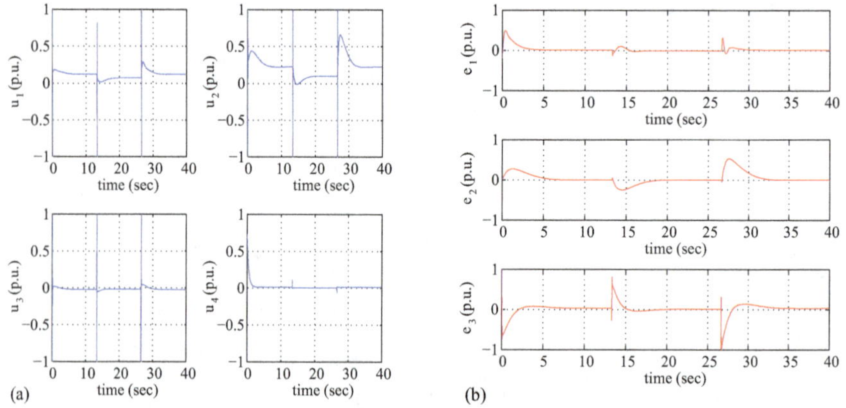

Figure 4.85 Tracking of setpoint 4 for the HEV powertrain (a) control inputs u_1 to u_4 applied to the HEV powertrain, (b) tracking error e_1, e_2, e_3, for state variables V_{dc}, ω_g, i_f of the HEV powertrain

Karapanou, 2020). Preceding results on the use of H-infinity control to nonlinear dynamical systems were limited to the case of affine-in-the-input systems with drift-only dynamics and considered that the control inputs gain matrix is not dependent on the values of the system's state vector. Moreover, in these approaches the linearization was performed around points of the desirable trajectory whereas in the present

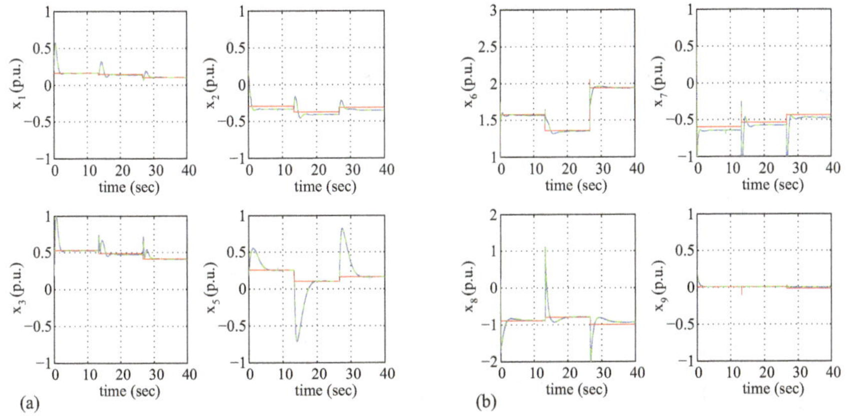

Figure 4.86 Tracking of setpoint 5 for the HEV powertrain (a) convergence of state variables x_1 to x_5 to their reference setpoints (red line: setpoint, blue line: real value, green line: estimated value), (b) convergence of state variables x_6 to x_9 to their reference setpoints

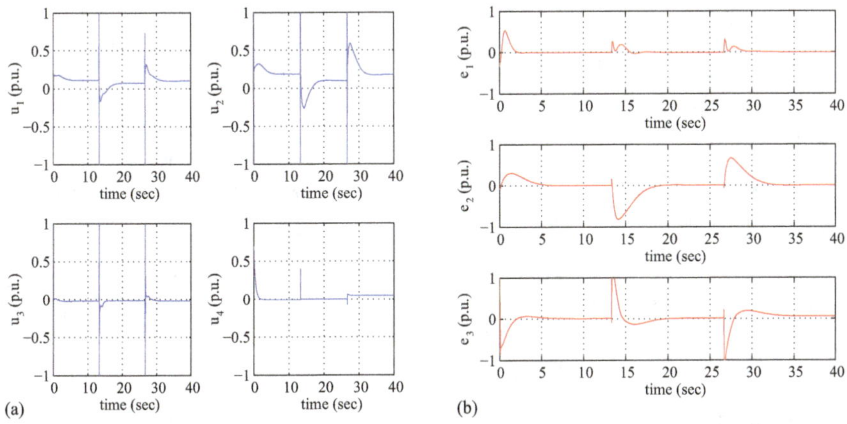

Figure 4.87 Tracking of setpoint 5 for the HEV powertrain (a) control inputs u_1 to u_4 applied to the HEV powertrain, (b) tracking error e_1, e_2, e_3, for state variables V_{dc}, ω_g, i_f of the HEV powertrain

section's control method the linearization points are related with the value of the state vector at each sampling instant as well as with the last sampled value of the control inputs vector. The Riccati equation which has been proposed for computing the feedback gains of the controller is novel, so is the presented global stability proof through Lyapunov analysis.

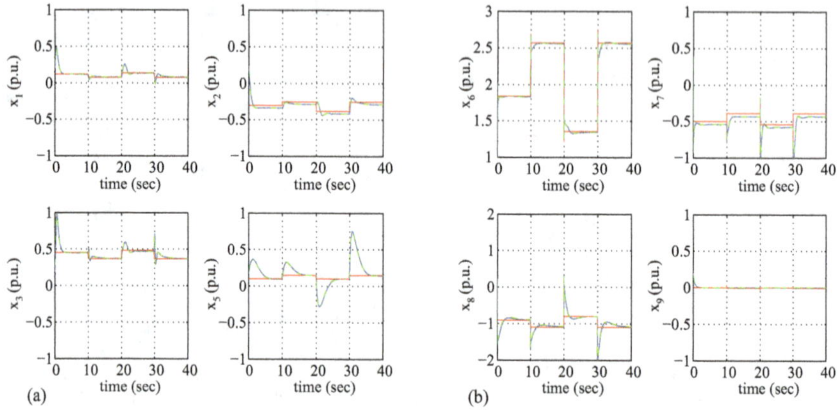

Figure 4.88 Tracking of setpoint 6 for the HEV powertrain (a) convergence of state variables x_1 to x_5 to their reference setpoints (red line: setpoint, blue line: real value, green line: estimated value), (b) convergence of state variables x_6 to x_9 to their reference setpoints

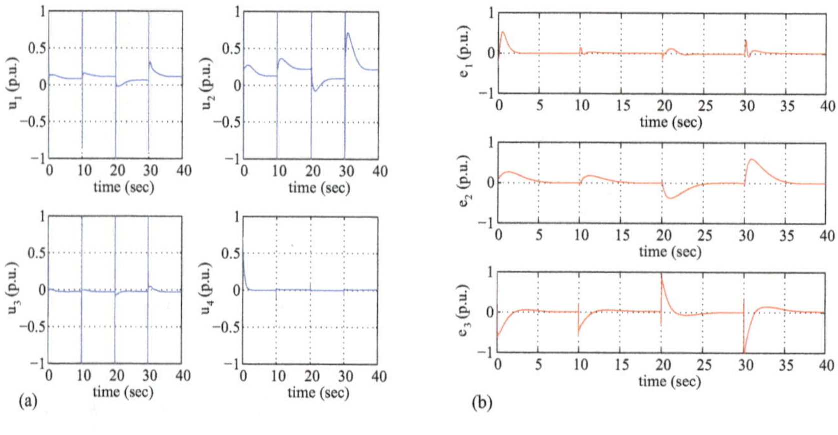

Figure 4.89 Tracking of setpoint 6 for the HEV powertrain (a) control inputs u_1 to u_4 applied to the HEV powertrain, (b) tracking error e_1, e_2, e_3, for state variables V_{dc}, ω_g, i_f of the HEV powertrain

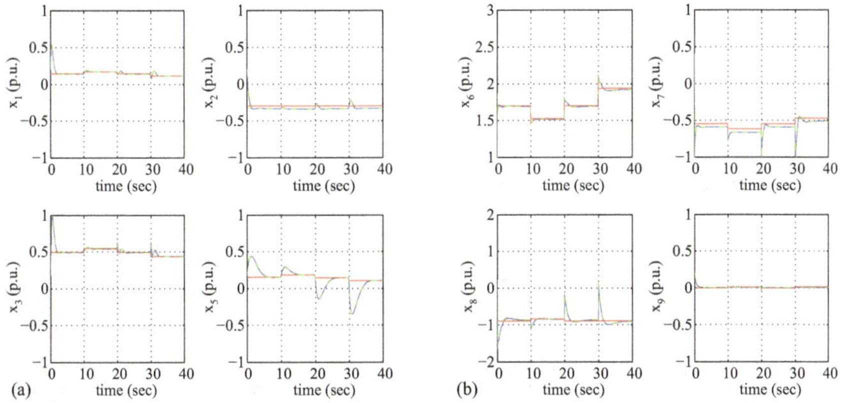

Figure 4.90 Tracking of setpoint 7 for the HEV powertrain (a) convergence of state variables x_1 to x_5 to their reference setpoints (red line: setpoint, blue line: real value, green line: estimated value), (b) convergence of state variables x_6 to x_9 to their reference setpoints

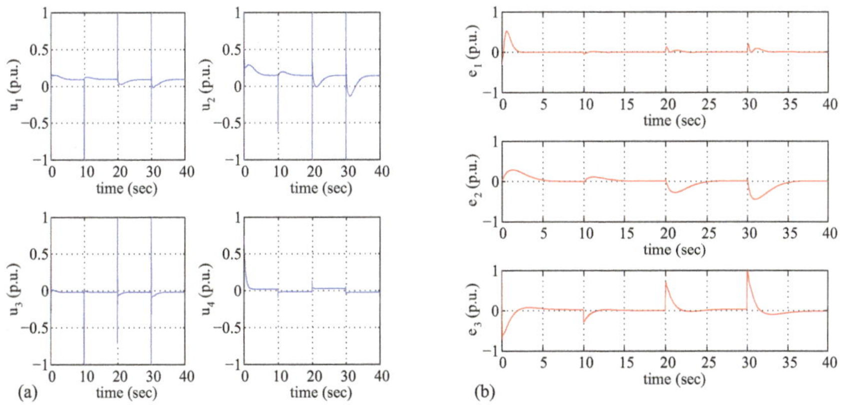

Figure 4.91 Tracking of setpoint 7 for the HEV powertrain (a) control inputs u_1 to u_4 applied to the HEV powertrain, (b) tracking error e_1, e_2, e_3, for state variables V_{dc}, ω_g, i_f of the HEV powertrain

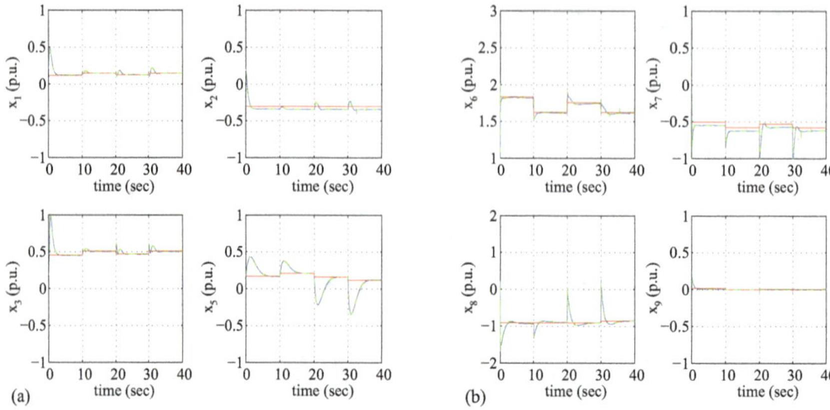

Figure 4.92 *Tracking of setpoint 8 for the HEV powertrain (a) convergence of state variables x_1 to x_5 to their reference setpoints (red line: setpoint, blue line: real value, green line: estimated value), (b) convergence of state variables x_6 to x_9 to their reference setpoints*

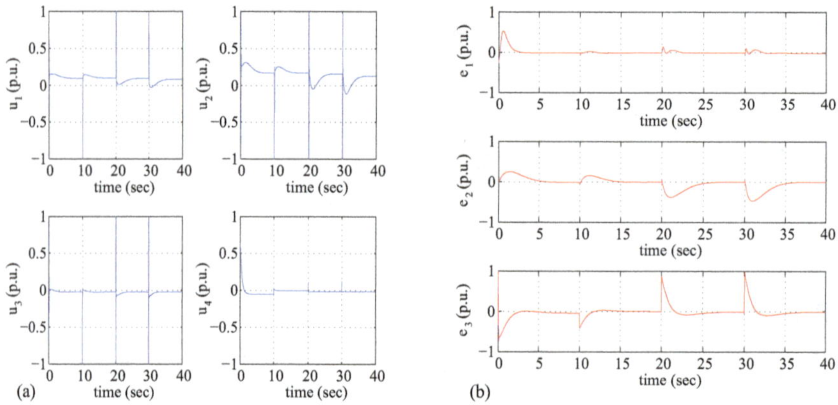

Figure 4.93 *Tracking of setpoint 8 for the HEV powertrain (a) control inputs u_1 to u_4 applied to the HEV powertrain, (b) tracking error e_1, e_2, e_3, for state variables V_{dc}, ω_g, i_f of the HEV powertrain*

4.8 Nonlinear control of shipboard AC/DC microgrids

4.8.1 Outline

During the last years, the concept of the all-electric-ship has gained much acceptance, because it exhibits several advantages (Rigatos, 2015; Mandani, Khooban *et al.*, 2017; Khooban *et al.*, 2020). Actually, it enables to develop highly reliable and powerful propulsion systems, that generate less polluting emissions while also keeping noise at low levels (Accetta and Pucci, 2019; Jin, Salligui *et al.*, 2016; Li, Xu *et al.*, 2020). The feasibility of the all-electric-ship approach depends on the implementation of automatic control methods that will ensure the accurate, synchronized and faultless functioning of power generators and power electronics (Xiao, Li *et al.*, 2020; Frances-Roger *et al.*, 2018; Fang, Gou *et al.*, 2019). Due to the complexity of the dynamics of prime-movers, power generators and power electronics that constitute AC/DC shipboard microgrids, the solution of the related nonlinear optimal control problem and of the associated power management problem is a non-trivial task (Jin, Meng *et al.*, 2018; Van Vu *et al.*, 2017; Hou *et al.*, 2019). Such a type of microgrids comprises both AC and DC elements (Rigatos, 2016b; Maqssod and Corzine, 2017; Satpathi *et al.*, 2019; Yousefizadeh *et al.*, 2021). For instance, diesel engines and gas turbines provide torque that causes the turn motion of the rotor of synchronous or asynchronous power generators. The generators provide voltage and power to the shipboard power grid. Often the AC voltage at the output of the generators is turned into DC voltage with the use of AC/DC converters, and next a DC voltage bus is used to distribute voltage to the ship's compartments, as well as to motors that can be used for the vessel's propulsion. Usually, there are three-phase motors of the synchronous or asynchronous type which have to be fed with alternating voltage. Consequently, they are connected to the previously noted DC bus through DC/AC inverters. So far there have been several results on linear and nonlinear control of shipboard AC/DC microgrids (Zhoxin, Tiamli *et al.*, 2019; Bu *et al.*, 2020). Under the linear or nonlinear dynamics assumption model-predictive control schemes of the MPC or NMPC type are respectively considered (Haseltalab and Negenborn, 2014; Haseltalab *et al.*, 2019; Haseltalab *et al.*, 2018). Moreover, elaborated methods aiming at achieving control and at optimizing energy management in shipboard AC/DC microgrids have been developed (Liu, Xu *et al.*, 2019; Li, Liu *et al.*, 2018; Hou *et al.*, 2021).

The present section proposes a nonlinear optimal control method for the dynamic model of a shipboard AC/DC microgrid that comprises Diesel engines, induction generators and AC/DC converters. This power system is used for supplying DC voltage to a DC bus that distributes power to the various compartments of the vessel. First, the dynamic model of the AC/DC microgrid undergoes approximate linearization around a temporary operating point which is recomputed at each time-step of the control algorithm, and which is defined by the present value of the system's state vector and by the last sampled value of the control inputs vector (Rigatos, 2016b). The linearization process relies on first-order Taylor-series expansion and on the computation of the associated Jacobian matrices (Rigatos, Siano and Cecati, 2015; Rigatos and Tzafestas, 2007; Basseville and Nikiforov, 1993; Rigatos and Zhang,

2009). The model inaccuracies which are due to the truncation of higher-order terms from the Taylor series are considered to be perturbations that are asymptotically compensated by the robustness of the control algorithm. For the linearized model of the shipboard AC/DC microgrid, a stabilizing feedback H-infinity controller is designed.

The proposed H-infinity controller implements a solution for the nonlinear optimal control problem of the AC/DC microgrid, under model uncertainty and external perturbations. Actually, it implements a min–max differential game taking place between the controller and the model uncertainty terms, in which (i) the controller tries to minimize a quadratic function of the state-vector's tracking error, (ii) the model uncertainty terms try to maximize this cost function. To compute the stabilizing feedback gains of the H-infinity controller, an algebraic Riccati equation is repetitively solved at each time-step of the control method (Rigatos, 2011; Rigatos and Busawon, 2018; Rigatos and Karapanou, 2020). The global stability properties of the control scheme are proven through Lyapunov analysis. First, it is demonstrated that the H-infinity tracking performance criterion holds. This signifies elevated robustness against model uncertainty and external perturbations (Rigatos, 2016b; Tousssaint, Basar and Bullo, 2000). Furthermore, under moderate conditions it is proven that the H-infinity control scheme is globally asymptotically stable. The proposed control approach retains the advantages of linear optimal control, that is fast and accurate tracking of reference setpoints under moderate variations of the control inputs (Rigatos, 2016b). To implement state estimation-based control, without the need to measure the entire state vector of the AC/DC microgrid, the H-infinity Kalman Filter is used as a robust state estimator (Rigatos, 2016b; Gibbs, 2012).

The present section provides one of the few existing solutions to the nonlinear optimal control problem of shipboard AC/DC microgrids which is of proven global stability while also remaining computationally efficient (Rigatos and Karapanou, 2020). Preceding results on the use of H-infinity control to nonlinear dynamical systems were limited to the case of affine-in-the-input systems with drift-only dynamics and considered that the control inputs gain matrix is not dependent on the values of the system's state vector. Moreover, in these approaches the linearization was performed around points of the desirable trajectory whereas in the present section's control method the linearization points are related with the value of the state vector at each sampling instant as well as with the last sampled value of the control inputs vector. The Riccati equation which has been proposed for computing the feedback gains of the controller is novel, so is the presented global stability proof through Lyapunov analysis.

A comparison of the proposed nonlinear optimal (H-infinity) control method against other linear and nonlinear control schemes for the shipboard AC/DC microgrid, shows the following: (1) Unlike global linearization-based control approaches, such as Lie algebra-based control and differential flatness theory-based control, the optimal control approach does not rely on complicated transformations (diffeomorphisms) of the system's state variables. Besides, the computed control inputs are applied directly on the initial nonlinear model of the shipboard AC/DC microgrid's dynamics and not on its linearized equivalent. The inverse transformations which are met in global linearization-based control are avoided and consequently one does not come against the related singularity problems. (2) Unlike Model Predictive Control (MPC) and Nonlinear Model Predictive control (NMPC), the proposed control method is of proven global stability. It is known that MPC is a linear control approach that if applied to the nonlinear dynamics of the shipboard AC/DC microgrid the stability of the control loop will be lost. Besides, in NMPC the convergence of its iterative search for an optimum depends on initialization and parameter values selection and consequently, the global stability of this control method cannot be always assured. (3) Unlike sliding-mode control and backstepping control the proposed optimal control method does not require the state-space description of the system to be found in a specific form. About sliding-mode control it is known that when the controlled system is not found in the input-output linearized form the definition of the sliding surface can be an intuitive procedure. About backstepping control it is known that it can not be directly applied to a dynamical system if the related state-space model is not found in the triangular (backstepping integral) form. (4) Unlike PID control, the proposed nonlinear optimal control method is of proven global stability, the selection of the controller's parameters does not rely on a heuristic tuning procedure, and the stability of the control loop is assured in the case of changes of operating points. (5) Unlike multiple local models-based control the nonlinear optimal control method uses only one linearization point and needs the solution of only one Riccati equation so as to compute the stabilizing feedback gains of the controller. Consequently, in terms of computation load the proposed control method for the shipboard AC/DC microgrid's dynamics is much more efficient.

4.8.2 Dynamic model of the shipboard AC/DC microgrid

4.8.2.1 Modelling of the ship's AC/DC microgrid

The ship's AC/DC microgrid comprises Diesel engines connected to induction generators while these asynchronous generators are connected in turn to an AC to DC converters. The converters provide DC voltage to a DC bus that distributes power to the vessel's compartments as well as to electric motor's which can be used for the ship's propulsion (Figure 4.94).

First, the equations of the AC–DC converter's dynamics are provided (Rigatos, 2016b):

$$L\dot{i}_d = -Ri_d + L\omega_{dq}i_q1 + v_d - \frac{V_{dc}}{2}\eta_1 \tag{4.303}$$

$$L\dot{i}_q = -L\omega_{dq}i_d - Ri_q + v_q - \frac{V_dC}{2}\eta_2 \tag{4.304}$$

$$C_{dc}\dot{V}_{dc} = -\frac{1}{R_c}V_{dc} + \frac{3}{4}i_d\eta_1 + \frac{3}{4}i_q\eta_2 \tag{4.305}$$

where i_d, i_q are the line currents (i_a, i_b, i_c) after transformation in the dq reference frame, and equivalently v_d, v_q are the phase voltages v_a, v_b, v_c after transformation in the dq reference frame. Variable V_{dc} denotes the DC voltage output of the converter, $u_1 = \eta_d$ and $u_2 = \eta_q$ stand for control inputs. The line losses and the transformer conduction losses are modelled by resistance R and the inverter switching losses are modeled by resistance R_c. Parameter L denotes inductance. Moreover, v_q is taken to be 0.

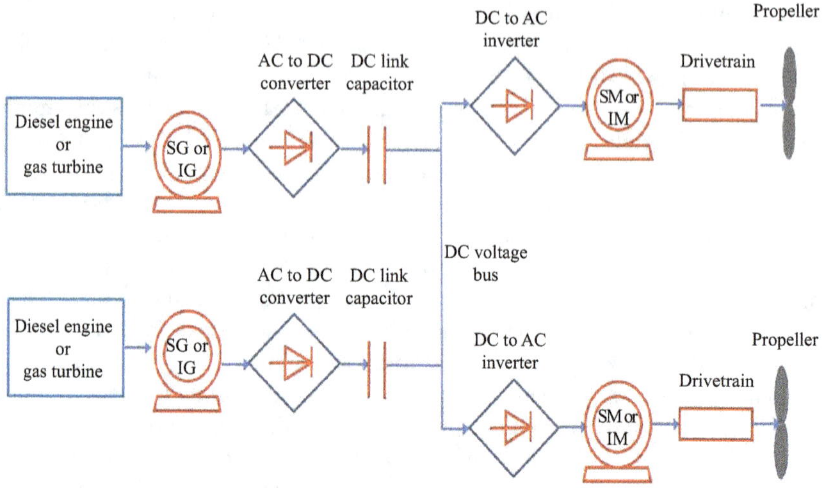

Figure 4.94 Diagram of the shipboard AC/DC microgrid

Next, the equations of the induction generator's dynamics are provided (Rigatos, 2016b)

$$\dot{\theta}_g = \omega_g \tag{4.306}$$

$$\dot{\omega}_g = \frac{1}{J}T_m - \frac{K_b}{J}\omega_g - \frac{n_p}{J}(i_{r_q}\psi_{s_d} - i_{r_d}\psi_{s_q}) \tag{4.307}$$

$$\dot{\psi}_{s_d} = \omega_{d_q}\psi_{r_q} - \frac{1}{\tau_s}\psi_{s_d} - \frac{M}{\tau_s}i_{r_d} + v_d \tag{4.308}$$

$$\dot{\psi}_{s_q} = -\frac{1}{\tau_s}\psi_{s_q} - \omega_{d_q}\psi_{s_d} + \frac{M}{\tau_s}i_{r_q} + v_q \tag{4.309}$$

$$\dot{i}_{r_d} = \beta\omega_g\psi_{s_q} + \frac{\beta}{\tau_s}\psi_{s_d} + (\omega_{d_q} - \omega_g)i_{r_q} - \gamma_2 i_{r_d} - \beta v_d + \frac{1}{\sigma L_r}v_{r_d} \tag{4.310}$$

$$\dot{i}_{r_q} = \frac{\beta}{\tau_s}\psi_{s_q} + \beta\omega_g\psi_{s_d} - \gamma_2 i_{r_q} - (\omega_{dq} - \omega_g)i_{r_d} - \beta v_q + \frac{1}{\sigma L_r}v_{r_q} \tag{4.311}$$

where ψ_{s_q}, ψ_{s_d}, i_{r_q}, i_{r_d} are the stator flux and the rotor currents variables, v_{s_q}, v_{s_d}, v_{r_q}, v_{r_d} are the stator and rotor voltages, L_s and L_r are the stator and rotor inductances, ω_r is the rotor's angular speed, M is the mutual inductance. Moreover, denoting as R_s and R_r the stator and rotor resistances the following parameters are defined

$$\begin{aligned} &\sigma = 1 - \frac{M^2}{L_r L_s} \quad \beta = \frac{1-\sigma}{M\sigma} \quad \tau_s = \frac{L_s}{R_s} \\ &\tau_r = \frac{L_r}{R_r} \quad \gamma_2 = \left(\frac{1-\sigma}{\sigma\tau_s} + \frac{1}{\sigma\tau_r}\right) \end{aligned} \tag{4.312}$$

The angle of the vectors that describe the magnetic flux $\psi_{s\alpha}$ and ψ_{sb} in the *ab* inertial reference frame is first defined for the stator, i.e. $\rho = tan^{-1}(\frac{\psi_{sb}}{\psi_{sa}})$. The angle between the inertial reference frame and the rotating reference frame is taken to be equal to ρ. Moreover, it holds that $\cos(\rho) = \frac{\psi_{sa}}{||\psi||}$, $\sin(\rho) = \frac{\psi_{sb}}{||\psi||}$, and $||\psi|| = \sqrt{\psi_{s\alpha}^2 + \psi_{s_b}^2}$. Therefore, in the rotating $d-q$ frame of the generator, and under the condition of field orientation, there will be only one non-zero component of the magnetic flux ψ_{s_d}, while the component of the flux along the q axis equals 0.

Further, the equations of the diesel engine's dynamics are given (Haseltalab and Negenborn, 2014; Haseltalab *et al.*, 2018; Rigatos and Karapanou, 2020)

$$\dot{T}_m = -\frac{T_m}{\bar{\tau}_s} + K_{en}f_{en} \tag{4.313}$$

This model of the Diesel engine signifies that the produced torque is proportional to the inflow of fuel to engine, subject to time-delay effects which are determined by the time constant $\bar{\tau}_s$.

4.8.2.2 State-space description of the shipboard AC/DC microgrid

The state-vector of the shipboard AC/DC microgrid is defined as

$$\begin{aligned} \dot{x} &= [x_1, x_2, x_3, x_4, x_5, x_6, x_7, x_8, x_9, x_{10}]^T \Rightarrow \\ \dot{x} &= [i_d, i_q, V_{dc}, \theta_g, \omega_g, \psi_{s_d}, \psi_{s_q}, i_{r_d}, i_{r_q}, T_m]^T \end{aligned} \tag{4.314}$$

Moreover, the control inputs vector is defined as $u_1 = v_{r_d}$, $u_2 = v_{r_q}$ and $u_3 = f_{en}$. Because of connecting the AC to DC converter to the rotor of the doubly-fed induction generator one has $\eta_1 = K_1 \cdot i_{r_d}$ and $\eta_2 = K_2 \cdot i_{r_q}$ where K_1 and K_2 are constants.

Using the previous state-space notation, the state-space equations of the AC-DC converter are re-written as:

$$\dot{x}_1 = -\frac{R}{L}x_1 + \omega_{dq}x_2 + \frac{v_d}{L} - \frac{x_3}{2L}K_1x_8 \tag{4.315}$$

$$\dot{x}_2 = -\omega_{dq}x_1 - \frac{R}{L}x_2 + \frac{v_q}{L} - \frac{x_3}{2L}K_2x_9 \tag{4.316}$$

$$\dot{x}_3 = -\frac{1}{R_cC_{dc}}x_3 + \frac{3K_1}{4C_{dc}}x_1x_8 + \frac{3K_2}{4C_{dc}}x_2x_9 \tag{4.317}$$

Next, the state-space equations of the induction generator are re-written as:

$$\dot{x}_4 = x_5 \tag{4.318}$$

$$\dot{x}_5 = \frac{1}{J}x_{10} - \frac{K_b}{J}x_5 - \frac{n_p}{J}(x_9x_6 - x_8x_7) \tag{4.319}$$

$$\dot{x}_6 = \omega_{d_q}x_7 - \frac{1}{\tau_s}x_6 - \frac{M}{\tau_s}x_8 + v_d \tag{4.320}$$

$$\dot{x}_7 = -\frac{1}{\tau_s}x_7 - \omega_{d_q}x_6 + \frac{M}{\tau_s}x_9 + v_q \tag{4.321}$$

$$\dot{x}_8 = \beta\omega_gx_7 + \frac{\beta}{\tau_s}x_6 + (\omega_{d_q} - \omega_g)x_9 - \gamma_2x_8 - \beta v_d + \frac{1}{\sigma L_r}u_1 \tag{4.322}$$

$$\dot{x}_9 = \frac{\beta}{\tau_s}x_7 + \beta\omega_g x_6 - \gamma_2 x_9 - (\omega_{dq} - \omega_g)x_8 - \beta v_q + \frac{1}{\sigma L_r}u_2 \tag{4.323}$$

Next, the state equations of the diesel engine are re-written as:

$$\dot{x}_{10} = -\frac{x_{10}}{\bar{\tau}_s} + K_{en}u_3 \tag{4.324}$$

Consequently, the model of the shipboard AC/DC microgrid is written as

$$\begin{pmatrix} \dot{x}_1 \\ \dot{x}_2 \\ \dot{x}_3 \\ \dot{x}_4 \\ \dot{x}_5 \\ \dot{x}_6 \\ \dot{x}_7 \\ \dot{x}_8 \\ \dot{x}_9 \\ \dot{x}_{10} \end{pmatrix} = \begin{pmatrix} -\frac{R}{L}x_1 + \omega_{dq}x_2 + \frac{v_d}{L} - \frac{x_3}{2L}K_1 x_8 \\ -\omega_{dq}x_1 - \frac{R}{L}x_2 + \frac{v_q}{L} - \frac{x_3}{2L}K_1 x_9 \\ -\frac{1}{R_c C_{dc}}x_3 + \frac{3K_1}{4C_{dc}}x_1 x_8 + \frac{3K_2}{4C_{dc}}x_2 x_9 \\ x_5 \\ \frac{1}{J}x_{10} - \frac{K_b}{J}x_5 - \frac{n_p}{J}(x_9 x_6 - x_8 x_7) \\ \omega_{d_q}x_7 - \frac{1}{\tau_s}x_6 - \frac{M}{\tau_s}x_8 + v_d \\ -\frac{1}{\tau_s}x_7 - \omega_{d_q}x_6 + \frac{M}{\tau_s}x_9 + v_q \\ \beta\omega_g x_7 + \frac{\beta}{\tau_s}x_6 + (\omega_{d_q} - \omega_g)x_9 - \gamma_2 x_8 - \beta v_d \\ \frac{\beta}{\tau_s}x_7 + \beta\omega_g x_6 - \gamma_2 x_9 - (\omega_{dq} - \omega_g)x_8 - \beta v_q \\ -\frac{x_{10}}{\bar{\tau}_s} \end{pmatrix} + \begin{pmatrix} 0 & 0 & 0 \\ 0 & 0 & 0 \\ 0 & 0 & 0 \\ 0 & 0 & 0 \\ 0 & 0 & 0 \\ 0 & 0 & 0 \\ 0 & 0 & 0 \\ \frac{1}{\sigma L_r} & 0 & 0 \\ 0 & \frac{1}{\sigma L_r} & 0 \\ 0 & 0 & K_{en} \end{pmatrix} \begin{pmatrix} u_1 \\ u_2 \\ u_3 \end{pmatrix} \tag{4.325}$$

Equivalently, the shipboard AC/DC microgrid is written in the concise state-space description

$$\dot{x} = f(x) + g(x)u \tag{4.326}$$

where $x \in R^{10\times 1}$, $f(x) \in R^{10\times 1}$, $g(x) \in R^{10\times 3}$ and $u \in R^{3\times 1}$.

4.8.2.3 Approximate linearization of the model of the shipboard AC/DC microgrid

The state-space model of the shipboard AC/DC microgrid undergoes approximate linearization around the temporary operating point (x^*, u^*), where x^* is the present value of the system's state vector and u^* is the last sampled value of the control inputs vector. The linearization process relies on first-order Taylor series expansion and on the computation of the associated Jacobian matrices (Rigatos, 2016b). The modelling error which is due to truncation of higher-order terms in the Taylor series is considered to be a perturbation which is asymptotically compensated by the robustness of the control method. The associated linearized model is:

$$\dot{x} = Ax + Bu + \tilde{d} \tag{4.327}$$

where matrices A and B are the system's Jacobian which are computed as follows:

$$A = \nabla_x[f(x) + g(x)u] \mid_{(x^*,u^*)} \Rightarrow A = \nabla_x f(x) \mid_{(x^*,u^*)} \tag{4.328}$$

$$B = \nabla_u[f(x) + g(x)u] \mid_{(x^*,u^*)} \Rightarrow B = g(x) \mid_{(x^*,u^*)} \tag{4.329}$$

The computation of the elements of the Jacobian matrix $\nabla_x f(x) \mid_{(x^*,u^*)}$ gives

First row of the Jacobian matrix $\nabla_x f(x) \mid_{(x^*,u^*)}$: $\frac{\partial f_1}{\partial x_1} = -\frac{R}{L}$, $\frac{\partial f_1}{\partial x_2} = \omega_{dq}$, $\frac{\partial f_1}{\partial x_3} = -\frac{K_1}{2L}x_8$, $\frac{\partial f_1}{\partial x_4} = 0$, $\frac{\partial f_1}{\partial x_5} = 0$, $\frac{\partial f_1}{\partial x_6} = 0$, $\frac{\partial f_1}{\partial x_7} = 0$, $\frac{\partial f_1}{\partial x_8} = -\frac{K_1}{2L}x_3$, $\frac{\partial f_1}{\partial x_9} = 0$, and $\frac{\partial f_1}{\partial x_{10}} = 0$.

Second row of the Jacobian matrix $\nabla_x f(x) \mid_{(x^*,u^*)}$: $\frac{\partial f_2}{\partial x_1} = -\omega_{dq}$, $\frac{\partial f_2}{\partial x_2} = -\frac{R}{L}$, $\frac{\partial f_2}{\partial x_3} = -\frac{K_1}{2L}x_9$, $\frac{\partial f_2}{\partial x_4} = 0$, $\frac{\partial f_2}{\partial x_5} = 0$, $\frac{\partial f_2}{\partial x_6} = 0$, $\frac{\partial f_2}{\partial x_7} = 0$, $\frac{\partial f_2}{\partial x_8} = 0$, $\frac{\partial f_2}{\partial x_9} = -\frac{K_2}{2L}x_3$, and $\frac{\partial f_2}{\partial x_{10}} = 0$.

Third row of the Jacobian matrix $\nabla_x f(x) \mid_{(x^*,u^*)}$: $\frac{\partial f_3}{\partial x_1} = \frac{3K_1}{4C_{dc}}x_8$, $\frac{\partial f_3}{\partial x_2} = \frac{3K_1}{4C_{dc}}x_9$, $\frac{\partial f_3}{\partial x_3} = -\frac{1}{C_{dc}R_c}$, $\frac{\partial f_3}{\partial x_4} = 0$, $\frac{\partial f_3}{\partial x_5} = 0$, $\frac{\partial f_3}{\partial x_6} = 0$, $\frac{\partial f_3}{\partial x_7} = 0$, $\frac{\partial f_3}{\partial x_8} = \frac{3K_1}{4C_{dc}}x_1$, $\frac{\partial f_3}{\partial x_9} = \frac{3K_1}{4C_{dc}}x_3$, and $\frac{\partial f_3}{\partial x_{10}} = 0$.

Fourth row of the Jacobian matrix $\nabla_x f(x) \mid_{(x^*,u^*)}$: $\frac{\partial f_4}{\partial x_1} = 0$, $\frac{\partial f_4}{\partial x_2} = 0$, $\frac{\partial f_4}{\partial x_3} = 0$, $\frac{\partial f_4}{\partial x_4} = 0$, $\frac{\partial f_4}{\partial x_5} = 1$, $\frac{\partial f_4}{\partial x_6} = 0$, $\frac{\partial f_4}{\partial x_7} = 0$, $\frac{\partial f_4}{\partial x_8} = 0$, $\frac{\partial f_4}{\partial x_9} = 0$, $\frac{\partial f_4}{\partial x_{10}} = 0$.

Fifth row of the Jacobian matrix $\nabla_x f(x) \mid_{(x^*,u^*)}$: $\frac{\partial f_5}{\partial x_1} = 0$, $\frac{\partial f_5}{\partial x_2} = 0$, $\frac{\partial f_5}{\partial x_3} = 0$, $\frac{\partial f_5}{\partial x_4} = -\frac{K_b}{J}$, $\frac{\partial f_5}{\partial x_5} = 0$, $\frac{\partial f_5}{\partial x_6} = -\frac{n_p}{J}x_9$, $\frac{\partial f_5}{\partial x_7} = \frac{n_p}{J}x_8$, $\frac{\partial f_5}{\partial x_8} = \frac{n_p}{J}x_7$, $\frac{\partial f_5}{\partial x_9} = -\frac{n_p}{J}x_6$, and $\frac{\partial f_5}{\partial x_{10}} = 0$.

Sixth row of the Jacobian matrix $\nabla_x f(x) \mid_{(x^*,u^*)}$: $\frac{\partial f_6}{\partial x_1} = 0$, $\frac{\partial f_6}{\partial x_2} = 0$, $\frac{\partial f_6}{\partial x_3} = 0$, $\frac{\partial f_6}{\partial x_4} = 0$, $\frac{\partial f_6}{\partial x_5} = 0$, $\frac{\partial f_6}{\partial x_6} = -\frac{1}{\tau_s}$, $\frac{\partial f_6}{\partial x_7} = \omega_{dq}$, $\frac{\partial f_6}{\partial x_8} = \frac{M}{\tau_s}$, $\frac{\partial f_6}{\partial x_9} = 0$, and $\frac{\partial f_6}{\partial x_{10}} = 0$.

Seventh row of the Jacobian matrix $\nabla_x f(x)\mid_{(x^*,u^*)}$: $\frac{\partial f_7}{\partial x_1}=0$, $\frac{\partial f_7}{\partial x_2}=0$, $\frac{\partial f_7}{\partial x_3}=0$, $\frac{\partial f_7}{\partial x_4}=0$, $\frac{\partial f_7}{\partial x_5}=0$, $\frac{\partial f_7}{\partial x_6}=-\omega_{dq}$, $\frac{\partial f_7}{\partial x_7}=-\frac{1}{\tau_s}$, $\frac{\partial f_7}{\partial x_8}=0$, $\frac{\partial f_7}{\partial x_9}=\frac{M}{\tau_s}$, and $\frac{\partial f_7}{\partial x_{10}}=0$.

Eighth row of the Jacobian matrix $\nabla_x f(x)\mid_{(x^*,u^*)}$: $\frac{\partial f_8}{\partial x_1}=0$, $\frac{\partial f_8}{\partial x_2}=0$, $\frac{\partial f_8}{\partial x_3}=0$, $\frac{\partial f_8}{\partial x_4}=0$, $\frac{\partial f_8}{\partial x_5}=-\beta x_7-x_9$, $\frac{\partial f_8}{\partial x_6}=\frac{\beta}{\tau_s}$, $\frac{\partial f_8}{\partial x_7}=-\beta x_5$, $\frac{\partial f_8}{\partial x_8}=-\gamma_2$, $\frac{\partial f_8}{\partial x_9}=\omega_{dq}-x_5$, and $\frac{\partial f_8}{\partial x_{10}}=0$.

Ninth row of the Jacobian matrix $\nabla_x f(x)\mid_{(x^*,u^*)}$: $\frac{\partial f_9}{\partial x_1}=0$, $\frac{\partial f_9}{\partial x_2}=0$, $\frac{\partial f_9}{\partial x_3}=0$, $\frac{\partial f_9}{\partial x_4}=0$, $\frac{\partial f_9}{\partial x_5}=\beta x_6+x_8$, $\frac{\partial f_9}{\partial x_6}=\beta x_5$, $\frac{\partial f_9}{\partial x_7}=\frac{\beta}{\tau_s}$, $\frac{\partial f_9}{\partial x_8}=-(\omega_{dq}-x_5)$, $\frac{\partial f_9}{\partial x_9}=-\gamma_2$, $\frac{\partial f_9}{\partial x_{10}}=0$,

Tenth row of the Jacobian matrix $\nabla_x f(x)\mid_{(x^*,u^*)}$: $\frac{\partial f_{10}}{\partial x_1}=0$, $\frac{\partial f_{10}}{\partial x_2}=0$, $\frac{\partial f_{10}}{\partial x_3}=0$, $\frac{\partial f_{10}}{\partial x_4}=0$, $\frac{\partial f_{10}}{\partial x_5}=0$, $\frac{\partial f_{10}}{\partial x_6}=0$, $\frac{\partial f_{10}}{\partial x_7}=0$, $\frac{\partial f_{10}}{\partial x_8}=0$, $\frac{\partial f_{10}}{\partial x_9}=0$, $\frac{\frac{\partial f_{10}}{\partial x_{10}}=1}{\tau_s}$.

4.8.3 Computation of the feedback control gains

For the linearized system given by (1.482) the cost function of (1.483) is defined, where coefficient r determines the penalization of the control input and weight coefficient ρ determines the reward of the disturbances' effects. It is assumed that (i) The energy that is transferred from the disturbances signal $\tilde{d}(t)$ is bounded, that is $\int_0^\infty \tilde{d}^T(t)\tilde{d}(t)dt<\infty$, (ii) matrices $[A,B]$ and $[A,L]$ are stabilizable, (iii) matrix $[A,C]$ is detectable. In the case of a tracking problem the optimal feedback control law is given by

$$u(t)=-Ke(t) \tag{4.330}$$

with $e=x-x_d$ to be the tracking error, and $K=\frac{1}{r}B^TP$ where P is a positive definite symmetric matrix. Matrix P is obtained from the solution of the Riccati equation

$$A^TP+PA+Q-P(\frac{2}{r}BB^T-\frac{1}{\rho^2}LL^T)P=0 \tag{4.331}$$

where Q is a positive semi-definite symmetric matrix. The worst-case disturbance is given by

$$\tilde{d}(t)=\frac{1}{\rho^2}L^TPe(t) \tag{4.332}$$

The diagram of the considered control loop for the shipboard AC/DC microgrid is depicted in Figure 4.95. The global stability properties of the control method can be proven following the procedure presented in Section 1.1.2.5.

4.8.3.1 Differential flatness properties of the shipboard AC/DC microgrid

It will be proven that the shipboard AC/DC microgrid model is differentially flat, with flat outputs $y_1=\frac{3L}{4}(x_1^2+x_2^2)+\frac{C_{dc}}{2}x_3^2$, $y_2=x_2$ and $y_3=x_4$. The differential flatness

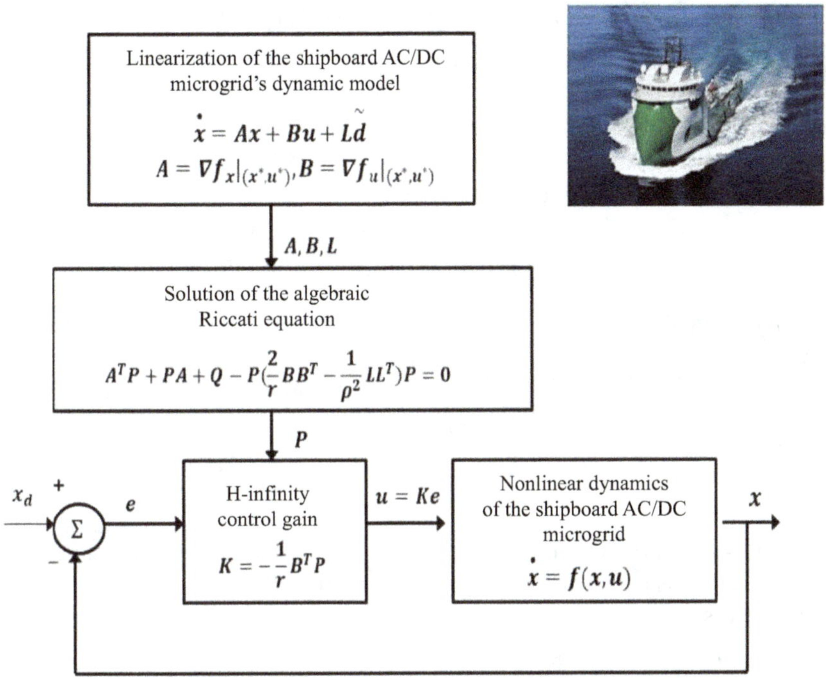

Figure 4.95 Diagram of the control scheme for the shipboard AC/DC microgrid

property signifies that all state variables and the control inputs of the system can be written as differential functions of the flat outputs. It holds that:

$$\begin{aligned}
\dot{y}_1 &= \frac{3L}{4}(x_1\dot{x}_1 + x_2\dot{x}_2) + \frac{C_{dc}}{2}x_3\dot{x}_3 \Rightarrow \\
\dot{y}_1 &= \frac{3L}{4}x_1\left(-\frac{R}{L}x_1 + \omega_{dq}x_2 + \frac{v_d}{L}\right) - \frac{K_1}{2L}x_3x_8 \\
&\quad + \frac{3L}{4}x_2\left(-\omega_{dq}x_1 - \frac{R}{L}x_3 + \frac{v_q}{L} - \frac{K_1}{2L}x_3x_4\right) \\
&\quad + \frac{C_{dc}}{2}x_3\left(-\frac{1}{C_{dc}R_c}x_3 + \frac{3K_1}{4C_{dc}}x_1x_8 + \frac{3K_2}{4C_{dc}}x_2x_9\right)
\end{aligned} \tag{4.333}$$

After intermediate operations and simplifications of terms, the previous relation gives

$$\dot{y}_1 = -\frac{3R}{4}(x_1^2 + x_2^2) - \frac{1}{2R_c}x_3^2 + \frac{3v_d}{4}x_1 + \frac{3v_d}{4}x_2 \tag{4.334}$$

Besides, from the definition of the flat output $y_1 = \frac{3L}{4}(x_1^2 + x_2^2) + \frac{C_{dc}}{2}x_3^2$ one can solve for x_3^2, thus obtaining

$$x_3^2 = \frac{y_1 - \frac{3L}{4}(x_1^2 + x_2^2)}{\frac{C_{dc}}{2}} \tag{4.335}$$

By substituting (4.335) into (4.334) one obtains

$$\dot{y}_1 = -\frac{3R}{4}(x_1^2 + x_2^2) - \frac{1}{2R_c}\frac{2}{C_{dc}}\left[y_1 - \frac{3L}{4}(x_1^2 + x_2^2)\right] + \frac{3v_d}{4}x_1 + \frac{3v_q}{4}x_2 \tag{4.336}$$

After intermediate operations and regrouping of terms, one arrives at the following binomial

$$\left(-\frac{3R}{4} + \frac{3L}{4R_cC_{dc}}\right)x_1^2 + \frac{3v_d}{4}x_1 + \left[\dot{y}_1 - \left(\frac{3R}{4} + \frac{3L}{4R_cC_{dc}}\right)y_1^2 + \frac{1}{R_cC_{dc}}y_1 - \frac{3v_q}{4}y_2\right] = 0 \tag{4.337}$$

Equation (4.337) is a binomial of x_1. The maximum solution for x_1 is retained. Thus, state variable x_1 is a differential function of the flat outputs of the system, which can be denoted as

$$x_1 = h_1(y_1, \dot{y}_1, y_2, \dot{y}_2, y_3, \dot{y}_3) \tag{4.338}$$

By substituting (4.338) into (4.335) one obtains

$$x_3^2 = \frac{y_1 - \frac{3L}{4}(h_1(y_1, \dot{y}_1, y_2.\dot{y}_2, y_3, \dot{y}_3)^2 + y_2^2)}{\frac{C_{dc}}{2}} \tag{4.339}$$

Therefore, state variable x_3 is also a differential function of the flat outputs of the system, or

$$x_3 = h_3(y_1, \dot{y}_1, y_2, \dot{y}_2, y_3, \dot{y}_3) \tag{4.340}$$

Next, from the first row of the state-space model of the shipboard AC/DC microgrid given in (4.315) one solves with respect to x_8, thus obtaining

$$x_9 = \frac{\dot{x}_1 + \frac{r}{L}x_1 - \omega_d q x_2 - \frac{v_d}{L}}{\frac{-K_1}{2L}x_3} \tag{4.341}$$

which signifies that state variable x_8 can be written as a differential function of the flat outputs of the system, or

$$x_8 = h_8(y_1, \dot{y}_1, y_2, \dot{y}_2, y_3, \dot{y}_3) \tag{4.342}$$

Besides, from the second row of the shipboard AC/DC microgrid given in (4.316) one solves with respect to x_9. This gives:

$$x_9 = \frac{\dot{x}_2 - \omega_d q x_1 + \frac{R}{L}x_2 - \frac{v_q}{L}}{\frac{-K_2}{2L}x_3} \tag{4.343}$$

which signifies that state variable x_8 can be written as a differential function of the flat outputs of the system, or

$$x_9 = h_8(y_1, \dot{y}_1, y_2, \dot{y}_2, y_3, \dot{y}_3) \tag{4.344}$$

Next, from the fourth row of the state-space model of the system one has

$$x_5 = \dot{x}_4 \Rightarrow x_5 = \dot{y}_2 \tag{4.345}$$

which signifies that state variable x_5 is also a differential function of the flat outputs of the system.

$$\dot{x}_5 = -\frac{1}{J}x_{10} - \frac{K_1}{J}x_2 - \frac{n_p}{J}(x_9x_6 - x_8x_7) \tag{4.346}$$

Besides, using the field orientation concept one obtains $x_7 = \psi_{s_q} = 0$, thus the fifth row of the state-space model of the system becomes

$$\dot{x}_5 = -\frac{1}{J}x_{10} - \frac{K_1}{J}x_2 - \frac{n_p}{J}x_9x_6 \tag{4.347}$$

Moreover, from the sixth row of the state-space model of the shipboard AC/DC microgrid, given in (4.320) one has

$$\dot{x}_6 = \omega_{dq}x_7 - \frac{1}{\tau_s}x_6 + \frac{M}{\tau_s}x_8 + v_d \tag{4.348}$$

while using again the field orientation concept $\psi_{s_q} = 0$ one obtains

$$\dot{x}_6 = -\frac{1}{\tau_s}x_6 + \frac{M}{\tau_s}x_8 + v_d \tag{4.349}$$

Additionally, from the seventh row of the state-space model given in (4.321) one has

$$\dot{x}_7 = -\frac{1}{\tau_s}x_7 + \omega_{dq}x_6 + \frac{M}{\tau_s}x_9 + v_q \tag{4.350}$$

Using again the field orientation concept $x_7 = \psi_{s_q} = 0$ one obtains

$$0 = \omega_{dq}x_6 + \frac{M}{\tau_s}x_9 + v_q \tag{4.351}$$

From (4.351) one can solve with respect to x_6 which gives

$$x_6 = \frac{-\frac{M}{\tau_s}x_8 - v_d}{\omega_{dq}} \tag{4.352}$$

The above relation signifies that state variable x_6 is a differential function of the system's flat outputs, or

$$x_6 = h_6(y_1, \dot{y}_1, y_2, \dot{y}_2, y_3, \dot{y}_3) \tag{4.353}$$

Returning to (4.347) and by solving with respect to x_{10} one obtains

$$x_{10} = \frac{\dot{x}_5 + \frac{K_b}{J}x_2 + \frac{n_p}{J}x_3x_6}{\frac{1}{J}} \tag{4.354}$$

which signifies that state variable x_{10} is a differential function of the system's flat outputs or

$$x_{10} = h_{10}(y_1, \dot{y}_1, y_2, \dot{y}_2, y_3, \dot{y}_3) \tag{4.355}$$

From the eighth row of the state-space model of the shipboard AC/DC microgrid, given in (4.322) one solves with respect to control input u_1. This gives

$$u_1 = \sigma L_r[\dot{x}_8 + \beta x_5 x_7 - \frac{\beta}{\tau_s} x_6 - (\omega_{dq} - x_5)x_9 + \gamma_2 x_8 + \beta v_d \tag{4.356}$$

which signifies that u_1 is a differential function of the system's flat outputs, or

$$u_1 = h_{u_1}(y_1, \dot{y}_1, y_2, \dot{y}_2, y_3, \dot{y}_3) \tag{4.357}$$

In a similar manner, from the ninth row of the state-space model of the shipboard AC/DC microgrid, given in (4.323) one solves with respect to the control input u_2. This gives

$$u_2 = \sigma L_r[\dot{x}_9 - \frac{\beta}{\tau_s} x_7 - \beta x_5 x_6 + \gamma_2 x_9 + (\omega_{dq} - x_5)x_8 + \beta v_q \tag{4.358}$$

which signifies that u_2 is a differential function of the system's flat outputs, or

$$u_2 = h_{u_2}(y_1, \dot{y}_1, y_2, \dot{y}_2, y_3, \dot{y}_3) \tag{4.359}$$

Finally, from the tenth row of the state-space model of the shipboard AC/DC microgrid, given in (4.324), one solves with respect to the control input u_3. Thus

$$u_3 = \frac{1}{K_{en}}\left[\dot{x}_{10} + \frac{x_{10}}{\bar{\tau}_s}\right] \tag{4.360}$$

which signifies that u_3 is a differential function of the system's flat outputs, or

$$u_3 = h_{u_3}(y_1, \dot{y}_1, y_2, \dot{y}_2, y_3, \dot{y}_3) \tag{4.361}$$

As a result of the previous analysis, all state variable and control inputs of the shipboard AC/DC microgrid can be written as differential functions of this system's flat outputs and the entire system is differentially flat. One can use the differential flatness property to define feasible setpoints for the state-variables of the shipboard AC/DC microgrid. Actually, one can define setpoints for the flat outputs of the system y_i $i = 1, 2, 3$ and using the relations that connect state variables x_i $i = 1, 2, \cdots, 10$ to the flat outputs and their derivatives can advance to the computation of setpoints x_i^d $i = 1, 2, \cdots, 10$.

4.8.4 Simulation tests

The global stability properties of the proposed nonlinear optimal control scheme for the shipboard AC/DC microgrid, as well as the precise tracking of reference setpoints by the state variables of this power system have been further confirmed through simulation experiments. To implement the nonlinear optimal control scheme, the algebraic Riccati equation which appears in (1.496) had to be solved at each time-step of the control algorithm. The obtained results are depicted in Figures 4.96 to 4.111. It can be noticed that in all cases fast and accurate tracking of reference setpoints was achieved under moderate variations of the control inputs. The real values of the state vector of the shipboard AC/DC microgrid are printed in blue, the estimated state variables provided by the H-infinity Kalman Filter are plotted in green, while the associated setpoints are shown in red colour.

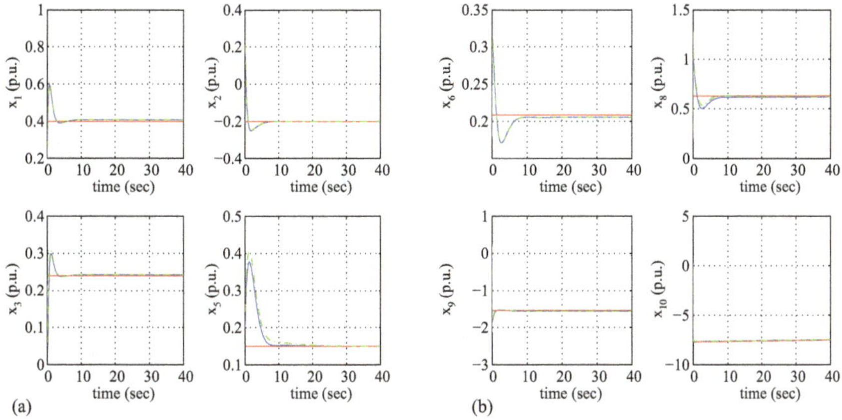

Figure 4.96 Tracking of setpoint 1 for the shipboard AC/DC microgrid (a) convergence of state variables x_1 to x_5 to their reference setpoints (red line: setpoint, blue line: real value, green line: estimated value), (b) convergence of state variables x_6 to x_{10} to their reference setpoints

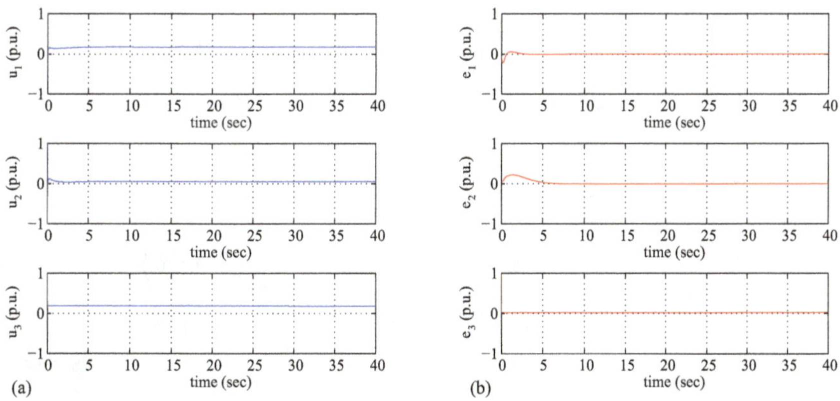

Figure 4.97 Tracking of setpoint 1 for the shipboard AC/DC microgrid (a) control inputs u_1 to u_3 applied to shipboard AC/DC microgrid, (b) tracking error e_1, e_2, e_3, for state variables V_{dc}, ω_g, T_m of the shipboard AC/DC microgrid

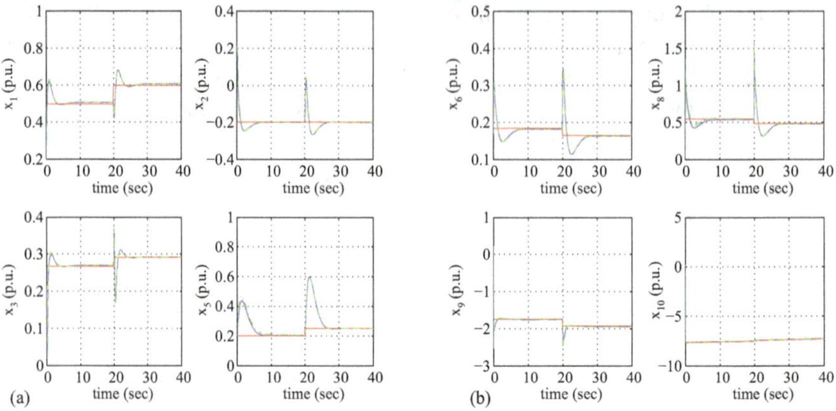

Figure 4.98 Tracking of setpoint 2 for the shipboard AC/DC microgrid (a) convergence of state variables x_1 to x_5 to their reference setpoints (red line: setpoint, blue line: real value, green line: estimated value), (b) convergence of state variables x_6 to x_{10} to their reference setpoints

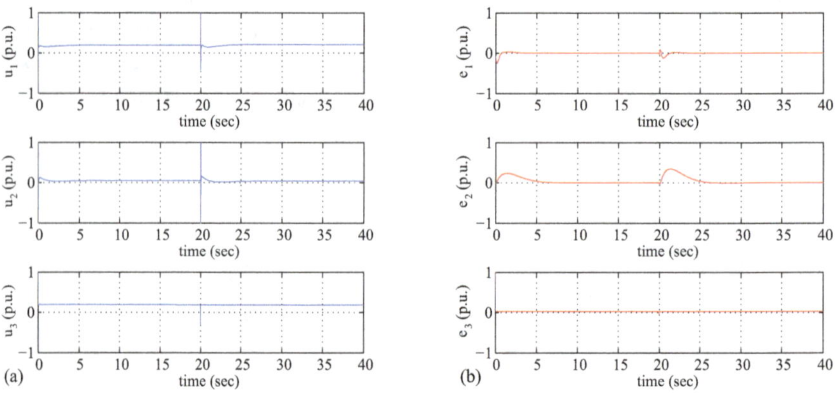

Figure 4.99 Tracking of setpoint 2 for the shipboard AC/DC microgrid (a) control inputs u_1 to u_3 applied to shipboard AC/DC microgrid, (b) tracking error e_1, e_2, e_3, for state variables V_{dc}, ω_g, T_m of the shipboard AC/DC microgrid

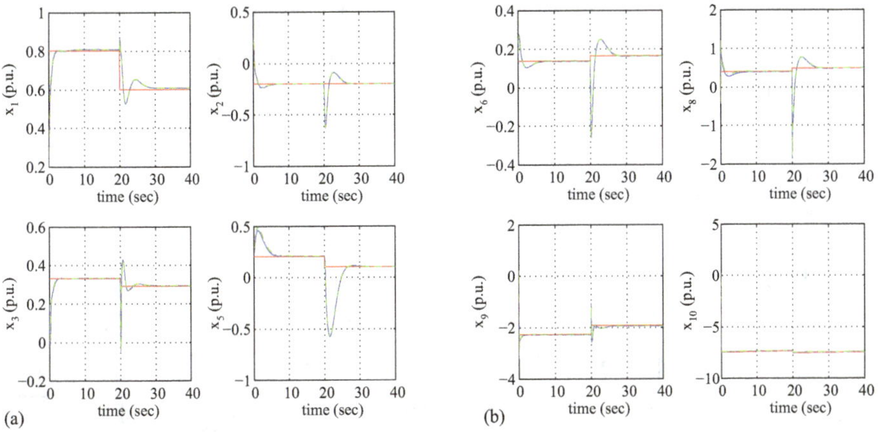

Figure 4.100 Tracking of setpoint 3 for the shipboard AC/DC microgrid (a) convergence of state variables x_1 to x_5 to their reference setpoints (red line: setpoint, blue line: real value, green line: estimated value), (b) convergence of state variables x_6 to x_{10} to their reference setpoints

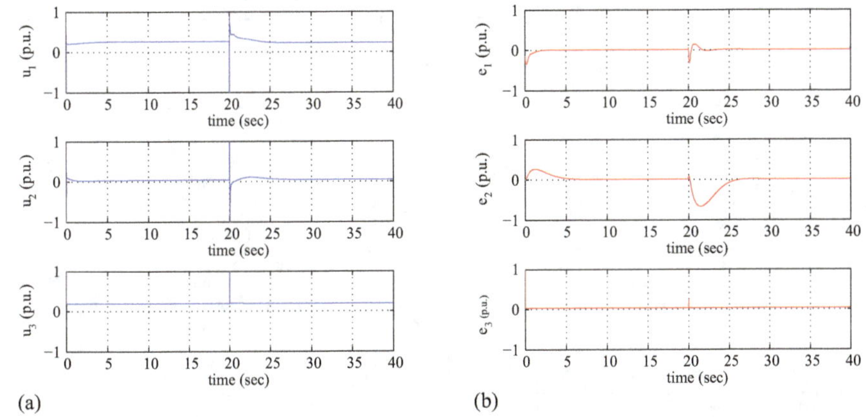

Figure 4.101 Tracking of setpoint 3 for the shipboard AC/DC microgrid (a) control inputs u_1 to u_3 applied to shipboard AC/DC microgrid, (b) tracking error e_1, e_2, e_3, for state variables V_{dc}, ω_g, T_m of the shipboard AC/DC microgrid

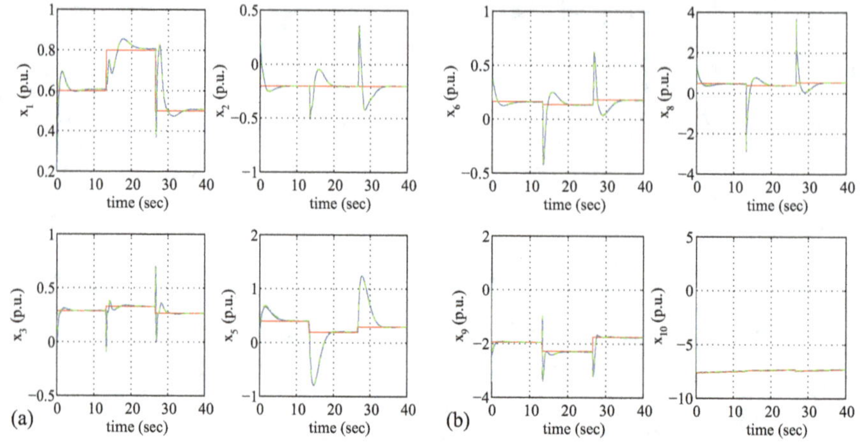

Figure 4.102 *Tracking of setpoint 4 for the shipboard AC/DC microgrid (a) convergence of state variables x_1 to x_5 to their reference setpoints (red line: setpoint, blue line: real value, green line: estimated value), (b) convergence of state variables x_6 to x_{10} to their reference setpoints*

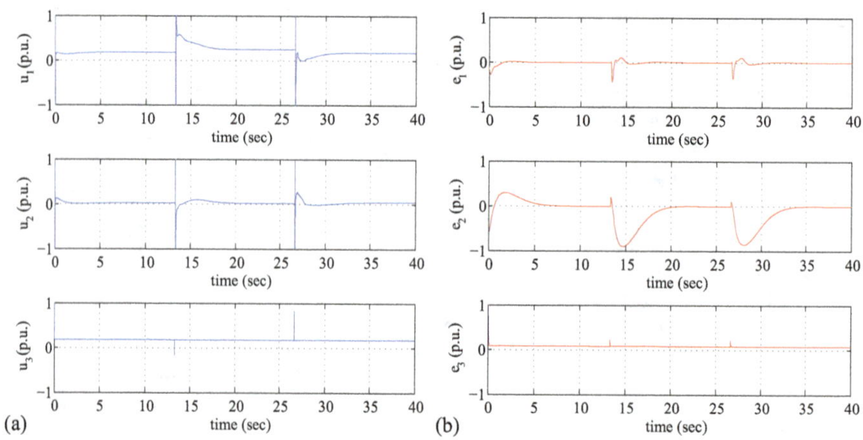

Figure 4.103 *Tracking of setpoint 4 for the shipboard AC/DC microgrid (a) control inputs u_1 to u_3 applied to shipboard AC/DC microgrid, (b) tracking error e_1, e_2, e_3, for state variables V_{dc}, ω_g, T_m of the shipboard AC/DC microgrid*

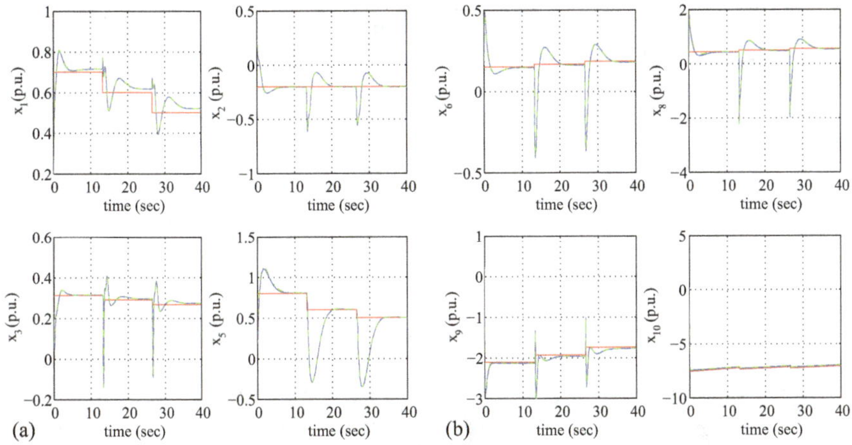

Figure 4.104 Tracking of setpoint 5 for the shipboard AC/DC microgrid (a) convergence of state variables x_1 to x_5 to their reference setpoints (red line: setpoint, blue line: real value, green line: estimated value), (b) convergence of state variables x_6 to x_{10} to their reference setpoints

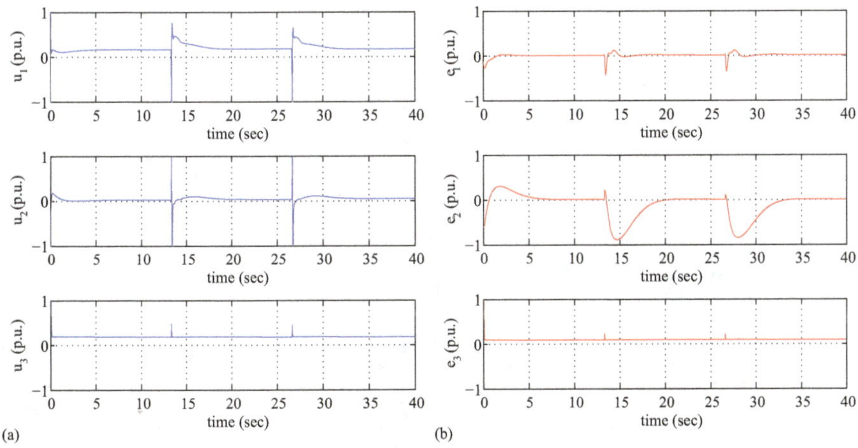

Figure 4.105 Tracking of setpoint 5 for the shipboard AC/DC microgrid (a) control inputs u_1 to u_3 applied to shipboard AC/DC microgrid, (b) tracking error e_1, e_2, e_3, for state variables V_{dc}, ω_g, T_m of the shipboard AC/DC microgrid

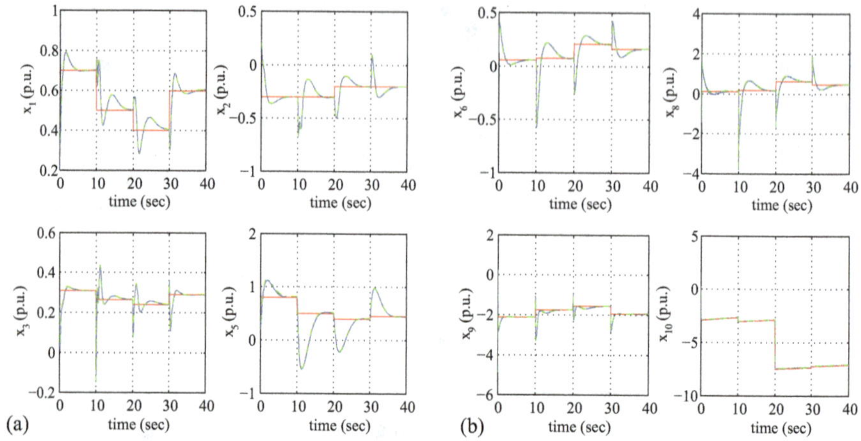

Figure 4.106 Tracking of setpoint 6 for the shipboard AC/DC microgrid (a) convergence of state variables x_1 to x_5 to their reference setpoints (red line: setpoint, blue line: real value, green line: estimated value), (b) convergence of state variables x_6 to x_{10} to their reference setpoints

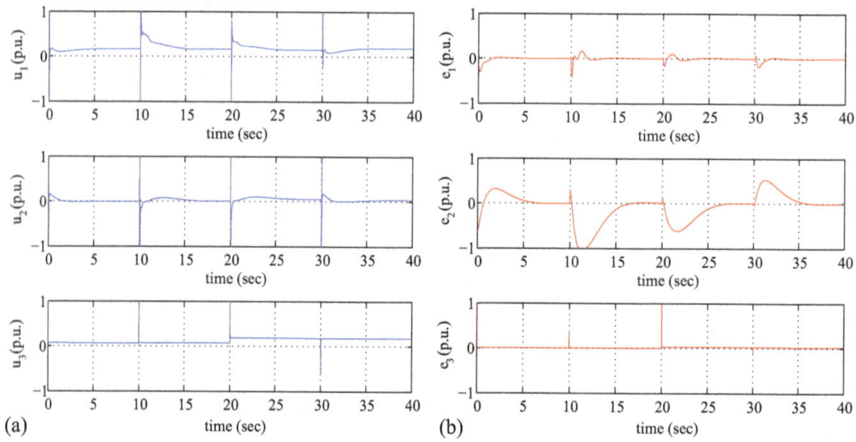

Figure 4.107 Tracking of setpoint 6 for the shipboard AC/DC microgrid (a) control inputs u_1 to u_3 applied to shipboard AC/DC microgrid, (b) tracking error e_1, e_2, e_3, for state variables V_{dc}, ω_g, T_m of the shipboard AC/DC microgrid

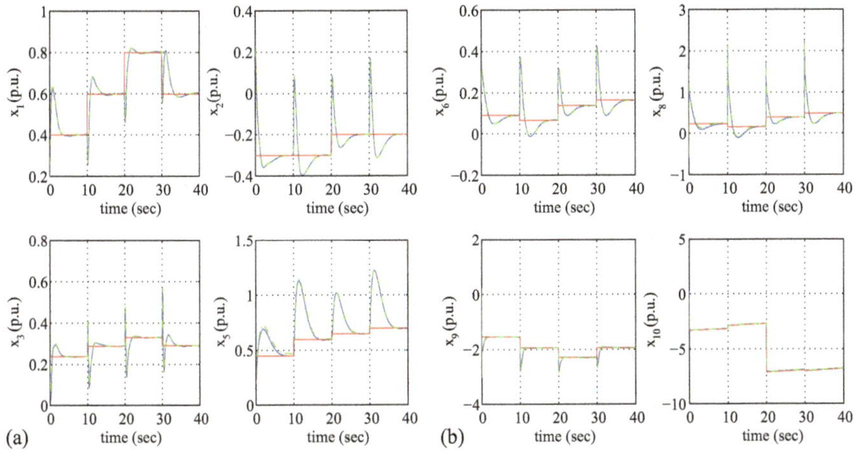

Figure 4.108 Tracking of setpoint 7 for the shipboard AC/DC microgrid (a) convergence of state variables x_1 to x_5 to their reference setpoints (red line: setpoint, blue line: real value, green line: estimated value), (b) convergence of state variables x_6 to x_{10} to their reference setpoints

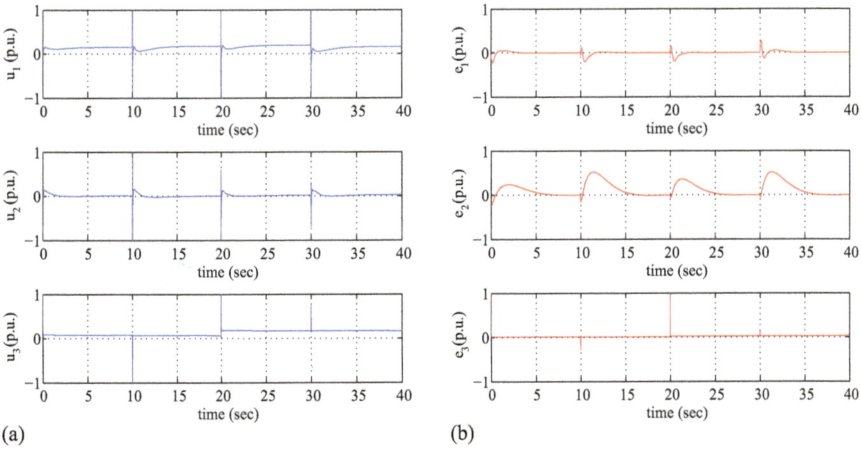

Figure 4.109 Tracking of setpoint 7 for the shipboard AC/DC microgrid (a) control inputs u_1 to u_3 applied to shipboard AC/DC microgrid, (b) tracking error e_1, e_2, e_3, for state variables V_{dc}, ω_g, T_m of the shipboard AC/DC microgrid

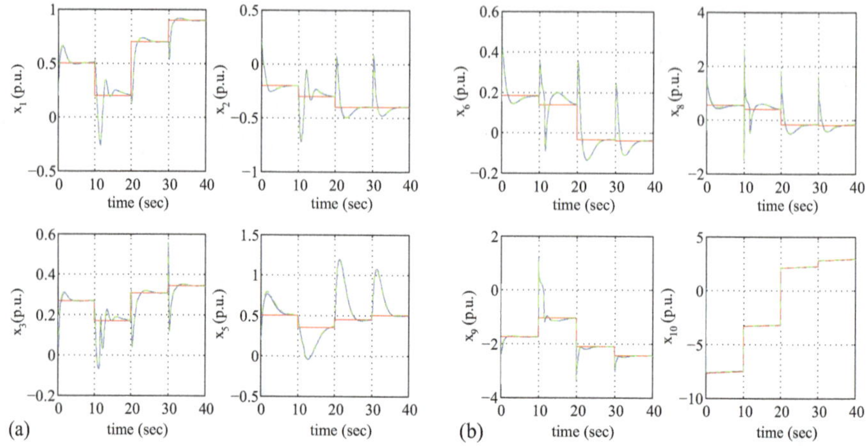

Figure 4.110 *Tracking of setpoint 8 for the shipboard AC/DC microgrid (a) convergence of state variables x_1 to x_5 to their reference setpoints (red line: setpoint, blue line: real value, green line: estimated value), (b) convergence of state variables x_6 to x_{10} to their reference setpoints*

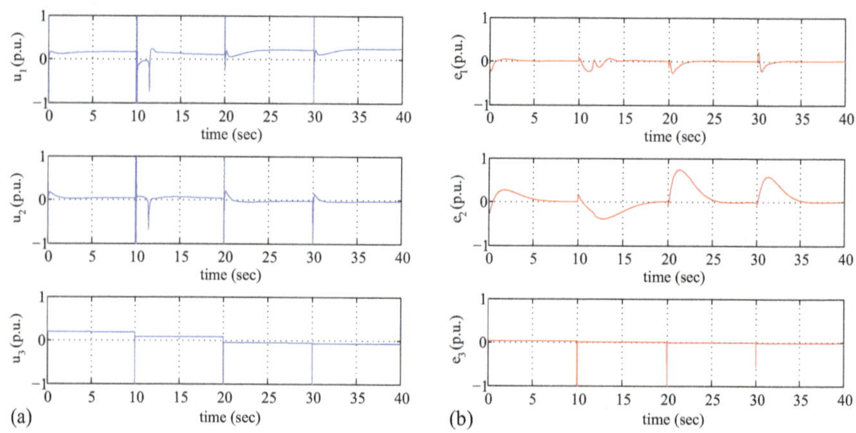

Figure 4.111 *Tracking of setpoint 8 for the shipboard AC/DC microgrid (a) control inputs u_1 to u_3 applied to shipboard AC/DC microgrid, (b) tracking error e_1, e_2, e_3, for state variables V_{dc}, ω_g, T_m of the shipboard AC/DC microgrid*

4.9 Nonlinear control of power generation in hybrid AC/DC microgrids

4.9.1 Outline

Microgrids incorporate AC or DC power generation units, energy storage elements and power electronics which enable to interconnect these power units, get them synchronized and finally, permit to aggregate and exploit the produced electric power (Espina *et al.*, 2020; Nehabbatkhah *et al.*, 2019; Ahmed *et al.*, 2020; Disqah *et al.*, 2015; Rigatos, 2016b). Thus, microgrids can be considered as controlled small-scale power systems which can be operated in an islanded or grid-connected mode (Sharma, Kewat *et al.*, 2019; Singh, Sarma *et al.*, 2021; Chishti *et al.*, 2020; Baghase *et al.*, 2021). They can provide supplementary power to the main electricity grid or can sustain the power needs of a specific area being exclusively fed by them (Garcia-Torres *et al.*, 2019; Liu, 2019; Ali *et al.*, 2020; Pourbehzadi *et al.*, 2019). Microgrids are becoming widely deployed and used because of their efficiency, reliability, flexibility and expansibility (Li, Guo, Zhou *et al.*, 2020, Li, Guo, Han *et al.*, 2021; Guan, Quintero *et al.*, 2015; Armghem *et al.*, 2020). According to the type of power generation units that microgrids incorporate they can be classified into (i) AC microgrids, that is microgrids that include AC power units, e.g. wind, hydro or marine power generation systems with various types of AC generators, as well as gas-turbine power generators and diesel-engine power generators (gensets), (ii) DC microgrids, that is microgrids that include DC power units such as photovoltaics and fuel cells or DC energy storage systems such as batteries and supercapacitors, (iii) hybrid AC/DC microgrids, that is microgrids which include both AC and DC power units of the aforementioned types (Wang, Fu *et al.*, 2021; Tian, Wen *et al.*, 2018; Wang, Li *et al.*, 2019; Zhang, Yue *et al.*, 2021). The main objectives in the functioning of microgrids are: (a) voltage and frequency control so as to assure stability and synchronization of the individual power units, (ii) improved energy harvesting and uninterrupted power production at desirable levels, (iii) robustness and reliable functioning under variable operating conditions and disturbances (Li, Guo, Li *et al.*, 2021; Yang, Xia *et al.*, 2018; ?; Li, Xu *et al.*, 2020).

In this section, a solution is proposed for the nonlinear optimal control problem of distributed hybrid microgrids that comprise both AC and DC power units. Actually, a microgrid is considered, consisting of (a) a wind power unit with a Doubly-fed induction generator (DFIG) being connected to a DC-bus through a 3-phase AC/DC converter, (b) a hydro-power unit with a synchronous reluctance generator (SRG) being connected to the previously-noted DC bus through a 3-phase AC/DC converter, (c) a photovoltaic power unit (PV) being connected to the above-noted DC bus through a DC/DC converter. To achieve control of the individual power units and to maintain the desirable levels of the aggregate power under variable operating conditions (variations in the wind, hydro or solar power) a novel nonlinear optimal control approach is proposed Rigatos, 2016b. First, the dynamic model of the hybrid microgrid undergoes approximate linearization around a time-varying operating point which is recomputed at each time-step of the control algorithm. This linearization point is defined at each sampling instance by the present value of the microgrid's state vector and by the last sampled value of the control inputs vector. The linearization process is based on first-order Taylor series expansion and on the computation of the associated Jacobian matrices (Rigatos, Siano and Cecati, 2015; Rigatos and Tzafestas, 2007; Basseville and Nikiforov, 1993; Rigatos and Zhang, 2009). The modelling error which is due to truncation of higher-order terms in the Taylor series expansion is deemed to be a perturbation that is asymptotically compensated by the robustness of the control method.

For the approximately linearized model of the hybrid microgrid a stabilizing H-infinity feedback controller is designed. The controller provides the solution to the optimal control problem for the hybrid microgrid, under model uncertainty and external perturbations. Actually, this controller represents a solution to a min-max differential game which takes place between (i) the control inputs which try to minimize a cost function that comprises a quadratic term of the state vector's tracking error, and (ii) the model uncertainty and exogenous disturbance terms which try to maximize this cost function. To compute the stabilizing feedback gains of the H-infinity controller an algebraic Riccati equation is repetitively solved at each iteration of the control algorithm (Rigatos, 2015; Rigatos and Busawon, 2018). The global stability properties of the control scheme are proven through Lyapunov analysis. First, it is demonstrated that the hybrid microgrid's control loop satisfies the H-infinity tracking performance criterion, which signifies elevated robustness against model imprecision and external perturbations (Tousssaint, Basar and Bullo, 2000; Rigatos and Karapanou, 2020). Next, under moderate conditions, the global asymptotic stability of the control loop is proven. Finally, to implement feedback control without measuring the entire state vector of the microgrid, the H-infinity Kalman Filter is used as a robust state estimator.

The present section provides one of the few existing solutions to the nonlinear optimal control problem of hybrid AC/DC microgrids which is of proven global stability while also remaining computationally efficient (Rigatos and Karapanou, 2020). Preceding results on the use of H-infinity control to nonlinear dynamical systems were limited to the case of affine-in-the-input systems with drift-only dynamics and considered that the control inputs gain matrix is not dependent on the values of the system's state vector. Moreover, in these approaches the linearization was performed around points of the desirable trajectory whereas in the present section's control method the linearization points are related to the value of the state vector at each sampling instant as well as with the last sampled value of the control inputs vector. The Riccati equation which has been proposed for computing the feedback gains of the controller is novel, so is the presented global stability proof through Lyapunov analysis. In brief, the control method which is proposed by the present section retains the advantages of linear optimal control, that is fast and accurate tracking of reference setpoints under moderate variations of the control inputs, while also remaining computationally simple and tractable.

4.9.2 Dynamic model of the hybrid distributed microgrid

The hybrid distributed microgrid of Figure 4.112 is considered. It comprises (i) a wind-power unit with Doubly-Fed induction Generator and its rotor being connected to the microgrid through an AC/DC converter, (ii) a hydro-power unit with synchronous reluctance generator and its stator being connected to the microgrid through an AC/DC converter and (iii) a photovoltaic power unit being connected to the microgrid through a DC/DC converter (Rigatos, 2016b)

The dynamic model of the Doubly-Fed Induction Generator (DFIG) which is the main component of the wind-power generation unit is given by (Rigatos, 2016b).

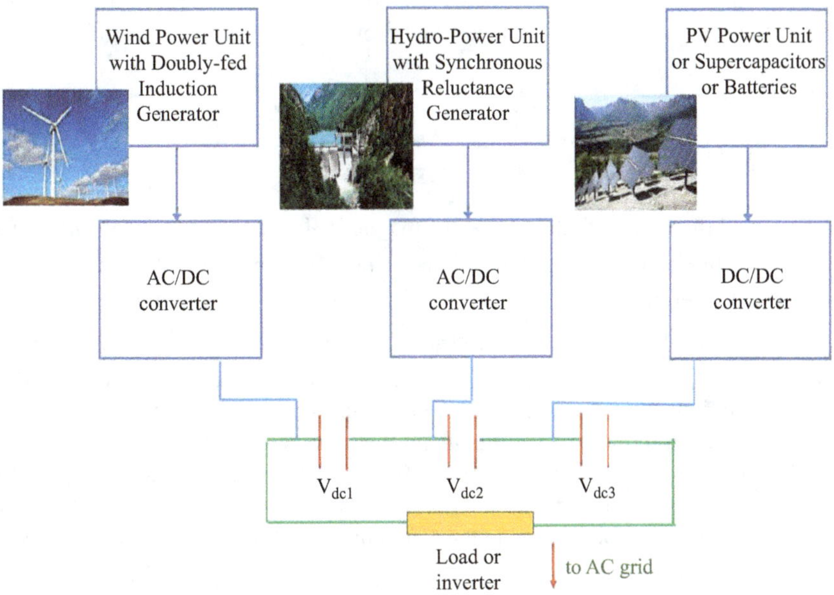

Figure 4.112 Hybrid distributed microgrid that comprises (i) a wind-power unit with Doubly-Fed induction Generator and its rotor being connected to the microgrid through an AC/DC converter, (ii) a hydro-power unit with synchronous reluctance generator and its stator being connected to the microgrid through an AC/DC converter, (iii) a photovoltaic power unit being connected to the microgrid through a DC/DC converter

A description of the DFIG's dynamics in the asynchronously rotating *dq* reference frame is used:

$$\begin{aligned}
\dot{\theta}_1 &= \omega_1 \\
\dot{\omega}_1 &= \frac{\beta}{J_1} - \frac{K_b}{J_1}\omega_1 - \frac{n_p}{J_1}(i_{r_d}\psi_{s_q} - i_{r_d}\psi_{s_q}) \\
\dot{\psi}_{s_d} &= \omega_{d_q}\psi_{s_q} - \frac{1}{\tau_s}\psi_{s_d} - \frac{M}{\tau_s}i_{r_d} + v_{s_d} \\
\dot{\psi}_{s_q} &= -\frac{1}{\tau_s}\psi_{s_q} - \omega_{d_q}\psi_{s_d} + \frac{M}{\tau_s}i_{r_q} + v_{s_q} \\
\dot{i}_{r_d} &= \beta\omega_1\psi_{s_q} + \frac{\beta}{\tau_s}\psi_{s_d} + (\omega_{dq} - \omega_1)i_{r_q} - \gamma_2 i_{r_d} - \beta v_d + \frac{1}{\sigma L_r}v_{r_d} \\
\dot{i}_{r_q} &= \frac{\beta}{\tau_s}\psi_{s_q} + \beta\omega_1\psi_{s_d} - \gamma_2 i_{r_q} - (\omega_{d_q} - \omega_1)i_{r_d} - \beta v_q + \frac{1}{\sigma L_r}v_{r_q}
\end{aligned} \tag{4.362}$$

where $\psi_{s_q}, \psi_{s_d}, i_{r_q}, i_{r_d}$ are the stator flux and the rotor currents variables, $v_{s_q}, v_{s_d}, v_{r_q}, v_{r_d}$ are the stator and rotor voltages, L_s and L_r are the stator and rotor inductances, ω_r is the rotor's angular speed, M is the mutual inductance. Moreover, denoting as R_s and R_r the stator and rotor resistances the following parameters are defined

$$\begin{aligned}
&\sigma = 1 - \frac{M^2}{L_r L_s} \quad \beta = \frac{1-\sigma}{M\sigma} \qquad \tau_s = \frac{L_s}{R_s} \\
&\tau_r = \frac{L_r}{R_r} \qquad \gamma_2 = \left(\frac{1-\sigma}{\sigma\tau_s} + \frac{1}{\sigma\tau_r}\right)
\end{aligned} \tag{4.363}$$

The angle of the vectors that describe the magnetic flux $\psi_{s\alpha}$ and ψ_{sb} in the *ab* inertial reference frame, is first defined for the stator, i.e. $\rho = tan^{-1}(\frac{\psi_{sb}}{\psi_{sa}})$. The angle between the inertial reference frame and the rotating reference frame is taken to be equal to ρ. Moreover, it holds that $\cos(\rho) = \frac{\psi_{sa}}{||\psi||}$, $\sin(\rho) = \frac{\psi_{sb}}{||\psi||}$, and $||\psi|| = \sqrt{\psi_{s\alpha}^2 + \psi_{s_b}^2}$. Therefore, in the rotating *dq* frame of the generator, and under the condition of the field orientation, there will be only one non-zero component of the magnetic flux ψ_{s_d}, while the component of the flux along the q axis equals 0.

The dynamic model of the AC/DC converter that connects the rotor of the DFIG to the microgrid is given by the following equations, after considering that the control inputs η_1 and η_2 are proportional to the rotor currents of the DFIG, that is $\eta_1 = k_{11}i_{r_d}$ and $\eta_2 = k_{21}i_{r_q}$ (Rigatos, 2016b):

$$\begin{aligned}
L_1\dot{i}_{d1} &= -R_1 i_d + L\omega_{dq}i_{q1} + v_d - \frac{V_{dc1}}{2}\eta_1 \\
L_1\dot{i}_{q1} &= -L_1\omega_{dq}i_d - Ri_{q1} + v_q - \frac{V_{dc1}}{2}\eta_2 \\
C_1\dot{V}_{dc1} &= -\frac{1}{Rc_1}V_{dc1} + \frac{3}{4}i_{d1}\eta_1 + \frac{3}{4}i_{q1}\eta_2
\end{aligned} \tag{4.364}$$

where i_d, i_q are the line currents (i_a, i_b, i_c) after transformation in the *dq* reference frame, and equivalently v_d, v_q are the phase voltages v_a, v_b, v_c after transformation in the dq reference frame. Variable V_{dc} denotes the DC voltage output of the converter, $u_1 = \eta_d$ and $u_2 = \eta_q$ stand for control inputs. The line losses and the transformer conduction losses are modelled by resistance R and the inverter switching losses are modeled by resistance R_c. Parameter L denotes inductance. Moreover, because of the field-orientation assumption, v_q is taken to be 0.

Next, by considering also a synchronously rotating dq reference frame, the dynamic model of the Synchronous Reluctance Generator (SRG), which is the main component of the hydro-power generation unit, jointly with the dynamics of the hydro-turbine, is given by (Rigatos, 2016b):

$$\begin{aligned}
\dot{\theta}_2 &= \omega_2 \\
\dot{\omega}_2 &= \frac{3P(L_d - L_q)}{4J_2} i_{sd} i_{s_q} - \frac{B_2}{J_2}\omega_2 - \frac{P}{2J_2}T_m \\
\dot{i}_{s_d} &= -\frac{R_s}{L_d} i_{s_d} + \frac{L_q}{L_d}\omega_2 i_{sq2} + \frac{1}{L_d} v_{s_d} \\
\dot{i}_{s_q} &= -\tfrac{R_s}{L_q} i_{s_q} - \tfrac{L_d}{L_q}\omega_2 i_{s_d} + \tfrac{1}{L_d} v_{s_q} \\
\dot{T}_m &= -\frac{1}{\tau_w} T_m + k_u u
\end{aligned} \tag{4.365}$$

where R_s is the stator's resistance, L_d is the direct axis inductance, L_q is the quadrature axis inductance, P is the number of poles of the machine, B_m is a friction coefficient and T_m is the mechanical torque. Moreover, $u = \alpha$ is the variable aperture that modifies the inflow of water towards the hydro-turbine. The electromagnetic torque which is developed by the synchronous reluctance machine is given by

$$\begin{aligned}
&T_e = K(\psi_d i_q - \psi_q i_d) \Rightarrow T_e = \frac{3}{2}\frac{P}{2}(i_d L_d i_q - i_q L_q i_d) \Rightarrow \\
&T_e = \frac{3P}{4}(L_d - L_q) i_d i_q
\end{aligned} \tag{4.366}$$

The dynamic model of the AC/DC converter that connects the stator of the Synchronous Reluctance Generator to the microgrid is given by the following equation, after considering that the control inputs η_1 and η_2 are proportional to the stator currents of the SRG this is $\eta_1 = k_{12} i_{s_d}$ and $\eta_2 = k_{22} i_{s_q}$ (Rigatos, 2016b):

$$\begin{aligned}
L\dot{i}_{d2} &= -R_2 i_{d2} + L_2 \omega_{dq} i_{q2} + v_d - \frac{V_{dc2}}{2}\eta_1 \\
L\dot{i}_{q2} &= -L_2 \omega_{dq} i_{d2} - R i_{q2} + v_q - \frac{V_{dc2}}{2}\eta_2 \\
C_2 \dot{V}_{dc2} &= -\frac{1}{Rc_2} V_{dc2} + \frac{3}{4} i_{d2}\eta_1 + \frac{3}{4} i_{q2}\eta_2
\end{aligned} \tag{4.367}$$

The dynamic model of the DC/DC converter that connects DC power sources (Photovoltaic power units, Supercapacitors or Batteries) to the microgrid is given by (Rigatos, 2016b):

$$\frac{dV_{dc3}}{dt} = \frac{1}{R_3C_3}V_{dc3} + \frac{1}{C_3}I_3$$
$$\frac{dI_3}{dt} = -\frac{V_{dc3}}{L_3} + \frac{1}{L_3}u \tag{4.368}$$

About the DC/DC converter one has that L_3 is the inductance at the circuit of the converter, C_3 is the capacitance at the circuit of the converter, R_3 is the resistance at the circuit of the converter, while $u = \delta$ is the PWM control input of the DC/DC converter. Moreover, by considering that the aggregate power which is provided to the microgrid by the wind-power unit, the hydro-power unit and the PV power unit is $P = (V_{dc1} + V_{dc2} + V_{dc3})^2/R_L$, then the dynamics of the power at the load which is connected to the DC microgrid is given by

$$\dot{P} = \frac{2}{R_L}(V_{dc1} + V_{dc2} + V_{dc3})(\dot{V}[dc1] + \dot{V}_{dc2} + \dot{V}_{dc3}) \Rightarrow$$
$$\dot{P} = \frac{2}{R_L}(V_{dc1} + V_{dc2} + V_{dc3}) \cdot \left\{\frac{1}{C_{dc1}}\left[-\frac{1}{RC}V_{dc_1} + \frac{3}{4}i_{d1}k_{11}i_{r_{d1}} + \frac{3}{4}i_{q1}k_{21}i_{r_{q1}}\right]\right. \tag{4.369}$$
$$\left. + \frac{1}{C_{dc2}}\left[-\frac{1}{RC}V_{dc_2} + \frac{3}{4}i_{d2}k_{12}i_{s_{d1}} + \frac{3}{4}i_{q2}k_{22}i_{s_{q2}}\right] + \left[\frac{1}{R_3C_3}V_{dc3} + \frac{1}{C_3}I_3\right]\right\}$$

The state vector of the microgrid is defined as

$$x = [x_1, x_2, x_3, x_4, x_5, x_6, x_7, x_8, x_9, x_{10}, x_{11}, x_{12}, x_{13}, x_{14}, x_{15}, x_{16}, x_{17}, x_{18}, x_{19}, x_{20}]^T \tag{4.370}$$

or equivalently

$$x = [\theta_1, \omega_1, \psi_{s_d}, \psi_{s_q}, i_{r_d}, i_{r_q}, i_{d1}, i_{q1}, V_{dc1}, \theta_2, \omega_2, i_{s_d}, i_{s_q}, T_m, i_{d2}, i_{q2}, V_{dc2}, I_3, V_{dc3}, P]^T \tag{4.371}$$

The control inputs vector of the microgrid is defined as

$$u = [u_1, u_2, u_3, u_4, u_5, u_6, u_7]^T \tag{4.372}$$

or equivalently

$$u = [\beta, v_{r_d}, v_{r_q}, v_{s_d}, v_{s_q}, \alpha, \delta]^T \tag{4.373}$$

Thus, the dynamics of the hybrid microgrid can be written in the following nonlinear affine-in-the-input state-space form

$$\dot{x} = f(x) + g(x)u \tag{4.374}$$

where $x \in R^{20\times 1}$, $f(x) \in R^{20\times 1}$, $g(x) \in R^{20\times 7}$, and $u \in R^{7\times 1}$. Moreover, using the previous state variables notation one arrives at the following description for the drift vector $f(x)$ and the control inputs matrix $g(x)$ of the state-space model of the microgrid

$$f(x) = \begin{pmatrix} x_2 \\ -\frac{K_b}{J_1}x_2 - \frac{n_p}{J_1}(x_5x_4 - x_6x_3 \\ \omega_{d_q}x_4 - \frac{1}{\tau_s}x_3 - \frac{M}{\tau_s}x_5 + v_{s_d} \\ -\frac{1}{\tau_s}x_4 - \omega_{d_q}x_3 + \frac{M}{\tau_s}x_6 + v_{s_q} \\ \beta x_2x_4 + \frac{\beta}{\tau_s}x_3 + (\omega_{dq} - x_2)x_6 - \gamma_2x_5 - \beta v_d \\ \frac{\beta}{\tau_s}x_4 + \beta x_2x_3 - \gamma_2x_6 - (\omega_{d_q} - x_2)x_5 - \beta v_q \\ -\frac{R_1}{L_1}x_7 + \omega_{dq}x_8 + \frac{v_d}{L_1} - \frac{x_9}{2L_1}k_{11}x_5 \\ -\omega_{dq}x_7 - \frac{R_1}{L_1}x_8 + \frac{v_q}{L_1} - \frac{x_9}{2L_1}k_{21}x_6 \\ -\frac{1}{Rc_1C_1}x_9 + \frac{3}{4C_1}x_7k_{11}x_5 + \frac{3}{4C_1}x_8k_{21}x_6 \\ x_{11} \\ \frac{3P(L_d - L_q)}{4J_2}x_{12}x_{13} - \frac{B_2}{J_2}x_{11} - \frac{P}{2J_2}x_{14} \\ -\frac{R_s}{L_d}x_{12} + \frac{L_q}{L_d}x_{11}x_{13} \\ -\frac{R_s}{L_q}x_{13} - \frac{L_d}{L_q}x_{11}x_{12} \\ -\frac{1}{\tau_w}x_{14} \\ -\frac{R_2}{L_2}x_{15} + \omega_{dq}x_{16} + \frac{v_d}{L_2} - \frac{x_{17}}{2L_2}k_{12}x_{12} \\ -\omega_{dq}x_{15} - \frac{R_2}{L_2}x_{16} + \frac{v_q}{L_2} - \frac{x_{17}}{2L_2}k_{22}x_{13} \\ -\frac{1}{Rc_2C_2}x_{17} + \frac{3}{4C_2}x_{15}k_{12}x_{12} + \frac{3}{4C_2}x_{16}k_{22}x_{13} \\ \frac{1}{R_3C_3}x_{18} + \frac{1}{C_3}x_{19} \\ -\frac{x_{18}}{L_3} \\ f_{20} \end{pmatrix} \tag{4.375}$$

where f_{20} is

$$f_{20} = \frac{2}{R_L}(x_9 + x_{17} + x_{18}) \cdot \left\{ \frac{1}{C_{dc1}} \left[-\frac{1}{RC}x_9 + \frac{3}{4}x_7 k_{11} x_5 + \frac{3}{4}x_8 k_{21} x_6 \right] + \frac{1}{C_{dc2}} \left[-\frac{1}{RC}x_{17} + \frac{3}{4}x_{15} k_{12} x_{12} + \frac{3}{4}x_{16} k_{22} x_{13} \right] + \left[\frac{1}{R_3 C_3}x_{18} + \frac{1}{C_3}x_{19} \right] \right\} \tag{4.376}$$

$$g(x) = \begin{pmatrix} 0 & 0 & 0 & 0 & 0 & 0 & 0 \\ \frac{T_{m1}}{J_1} & 0 & 0 & 0 & 0 & 0 & 0 \\ 0 & 0 & 0 & 0 & 0 & 0 & 0 \\ 0 & 0 & 0 & 0 & 0 & 0 & 0 \\ 0 & \frac{1}{\sigma L_r} & 0 & 0 & 0 & 0 & 0 \\ 0 & 0 & \frac{1}{\sigma L_r} & 0 & 0 & 0 & 0 \\ 0 & 0 & 0 & 0 & 0 & 0 & 0 \\ 0 & 0 & 0 & 0 & 0 & 0 & 0 \\ 0 & 0 & 0 & 0 & 0 & 0 & 0 \\ 0 & 0 & 0 & \frac{1}{L_d} & 0 & 0 & 0 \\ 0 & 0 & 0 & 0 & \frac{1}{L_q} & 0 & 0 \\ 0 & 0 & 0 & 0 & 0 & k_u & 0 \\ 0 & 0 & 0 & 0 & 0 & 0 & 0 \\ 0 & 0 & 0 & 0 & 0 & 0 & 0 \\ 0 & 0 & 0 & 0 & 0 & 0 & 0 \\ 0 & 0 & 0 & 0 & 0 & 0 & 0 \\ 0 & 0 & 0 & 0 & 0 & 0 & \frac{1}{L_3} \end{pmatrix} \tag{4.377}$$

4.9.3 Approximate linearization of the dynamic model of the hybrid microgrid

The state-space model of the hybrid microgrid which comprises wind/hydro turbines and PV-based power units undergoes approximate linearization around the temporary operating point (x^*, u^*), where x^* is the present value of the system's state vector and u^* is the last sampled value of the control inputs vector. The linearization process relies on first-order Taylor series expansion and on the computation of the associated Jacobian matrices (Rigatos, 2016b). The modelling error which is due to truncation of higher-order terms in the Taylor series is considered to be a perturbation that is asymptotically compensated by the robustness of the control method. The associated linearized model is:

$$\dot{x} = Ax + Bu + \tilde{d} \tag{4.378}$$

where matrices A and B are the system's Jacobian which are computed as follows:

$$A = \nabla_x[f(x) + g(x)u] \mid_{(x^*,u^*)} \Rightarrow A = \nabla_x f(x) \mid_{(x^*,u^*)} \tag{4.379}$$

$$B = \nabla_u[f(x) + g(x)u] \mid_{(x^*,u^*)} \Rightarrow B = g(x) \mid_{(x^*,u^*)} \tag{4.380}$$

The computation of the elements of the Jacobian matrix $\nabla_x f(x) \mid_{(x^*,u^*)}$ gives

First row of the Jacobian matrix $\nabla_x f(x) \mid_{(x^*,u^*)}$ gives: $\frac{\partial f_1}{\partial x_1} = 0$, $\frac{\partial f_1}{\partial x_2} = 1$, $\frac{\partial f_1}{\partial x_3} = 0$, $\frac{\partial f_1}{\partial x_4} = 0$, $\frac{\partial f_1}{\partial x_5} = 0$, $\frac{\partial f_1}{\partial x_6} = 0$, $\frac{\partial f_1}{\partial x_7} = 0$, $\frac{\partial f_1}{\partial x_8} = 0$, $\frac{\partial f_1}{\partial x_9} = 0$, $\frac{\partial f_1}{\partial x_{10}} = 0$, $\frac{\partial f_1}{\partial x_{11}} = 0$, $\frac{\partial f_1}{\partial x_{12}} = 0$, $\frac{\partial f_1}{\partial x_{13}} = 0$, $\frac{\partial f_1}{\partial x_{14}} = 0$, $\frac{\partial f_1}{\partial x_{15}} = 0$, $\frac{\partial f_1}{\partial x_{16}} = 0$, $\frac{\partial f_1}{\partial x_{17}} = 0$, $\frac{\partial f_1}{\partial x_{18}} = 0$, $\frac{\partial f_1}{\partial x_{19}} = 0$, and $\frac{\partial f_1}{\partial x_{20}} = 0$.

Second row of the Jacobian matrix $\nabla_x f(x) \mid_{(x^*,u^*)}$ gives: $\frac{\partial f_2}{\partial x_1} = 0$, $\frac{\partial f_2}{\partial x_2} = -\frac{K_b}{J_1}$, $\frac{\partial f_2}{\partial x_3} = \frac{n_p}{J_1}x_6$, $\frac{\partial f_2}{\partial x_4} = -\frac{n_p}{J_1}x_5$, $\frac{\partial f_2}{\partial x_5} = -\frac{n_p}{J_1}x_4$, $\frac{\partial f_2}{\partial x_6} = \frac{n_p}{J_1}x_3$, $\frac{\partial f_2}{\partial x_7} = 0$, $\frac{\partial f_2}{\partial x_8} = 0$, $\frac{\partial f_2}{\partial x_9} = 0$, $\frac{\partial f_2}{\partial x_{10}} = 0$, $\frac{\partial f_2}{\partial x_{11}} = 0$, $\frac{\partial f_2}{\partial x_{12}} = 0$, $\frac{\partial f_2}{\partial x_{13}} = 0$, $\frac{\partial f_2}{\partial x_{14}} = 0$, $\frac{\partial f_2}{\partial x_{15}} = 0$, $\frac{\partial f_2}{\partial x_{16}} = 0$, $\frac{\partial f_2}{\partial x_{17}} = 0$, $\frac{\partial f_2}{\partial x_{18}} = 0$, $\frac{\partial f_2}{\partial x_{19}} = 0$, and $\frac{\partial f_2}{\partial x_{20}} = 0$.

Third row of the Jacobian matrix $\nabla_x f(x) \mid_{(x^*,u^*)}$ gives: $\frac{\partial f_3}{\partial x_1} = 0$, $\frac{\partial f_3}{\partial x_2} = 0$, $\frac{\partial f_3}{\partial x_3} = -\frac{1}{\tau_s}$, $\frac{\partial f_3}{\partial x_4} = -\omega_{dq}$, $\frac{\partial f_3}{\partial x_5} = -\frac{M}{\tau_s}$, $\frac{\partial f_3}{\partial x_6} = 0$, $\frac{\partial f_3}{\partial x_7} = 0$, $\frac{\partial f_3}{\partial x_8} = 0$, $\frac{\partial f_3}{\partial x_9} = 0$, $\frac{\partial f_3}{\partial x_{10}} = 0$, $\frac{\partial f_3}{\partial x_{11}} = 0$, $\frac{\partial f_3}{\partial x_{12}} = 0$, $\frac{\partial f_3}{\partial x_{13}} = 0$, $\frac{\partial f_3}{\partial x_{14}} = 0$, $\frac{\partial f_3}{\partial x_{15}} = 0$, $\frac{\partial f_3}{\partial x_{16}} = 0$, $\frac{\partial f_3}{\partial x_{17}} = 0$, $\frac{\partial f_3}{\partial x_{18}} = 0$, $\frac{\partial f_3}{\partial x_{19}} = 0$, and $\frac{\partial f_3}{\partial x_{20}} = 0$.

Fourth row of the Jacobian matrix $\nabla_x f(x) \mid_{(x^*,u^*)}$ gives: $\frac{\partial f_4}{\partial x_1} = 0$, $\frac{\partial f_4}{\partial x_2} = 0$, $\frac{\partial f_4}{\partial x_3} = -\omega_{dq}$, $\frac{\partial f_4}{\partial x_4} = -\frac{1}{\tau_s}$, $\frac{\partial f_4}{\partial x_5} = 0$, $\frac{\partial f_4}{\partial x_6} = \frac{M}{\tau_s}$, $\frac{\partial f_4}{\partial x_7} = 0$, $\frac{\partial f_4}{\partial x_8} = 0$, $\frac{\partial f_4}{\partial x_9} = 0$, $\frac{\partial f_4}{\partial x_{10}} = 0$, $\frac{\partial f_4}{\partial x_{11}} = 0$, $\frac{\partial f_4}{\partial x_{12}} = 0$, $\frac{\partial f_4}{\partial x_{13}} = 0$, $\frac{\partial f_4}{\partial x_{14}} = 0$, $\frac{\partial f_4}{\partial x_{15}} = 0$, $\frac{\partial f_4}{\partial x_{16}} = 0$, $\frac{\partial f_4}{\partial x_{17}} = 0$, $\frac{\partial f_4}{\partial x_{18}} = 0$, $\frac{\partial f_4}{\partial x_{19}} = 0$, and $\frac{\partial f_4}{\partial x_{20}} = 0$.

Fifth row of the Jacobian matrix $\nabla_x f(x) \mid_{(x^*,u^*)}$ gives: $\frac{\partial f_5}{\partial x_1} = 0$, $\frac{\partial f_5}{\partial x_2} = \beta x_4 - x_6$, $\frac{\partial f_5}{\partial x_3} = \frac{\beta}{\tau_s}$, $\frac{\partial f_5}{\partial x_4} = \beta x_2$, $\frac{\partial f_5}{\partial x_5} = -\gamma_2$, $\frac{\partial f_5}{\partial x_6} = (\omega_{dq} - x_2)$, $\frac{\partial f_5}{\partial x_7} = 0$, $\frac{\partial f_5}{\partial x_8} = 0$, $\frac{\partial f_5}{\partial x_9} = 0$, $\frac{\partial f_5}{\partial x_{10}} = 0$, $\frac{\partial f_5}{\partial x_{11}} = 0$, $\frac{\partial f_5}{\partial x_{12}} = 0$, $\frac{\partial f_5}{\partial x_{13}} = 0$, $\frac{\partial f_5}{\partial x_{14}} = 0$, $\frac{\partial f_5}{\partial x_{15}} = 0$, $\frac{\partial f_5}{\partial x_{16}} = 0$, $\frac{\partial f_5}{\partial x_{17}} = 0$, $\frac{\partial f_5}{\partial x_{18}} = 0$, $\frac{\partial f_5}{\partial x_{19}} = 0$, and $\frac{\partial f_5}{\partial x_{20}} = 0$.

Sixth row of the Jacobian matrix $\nabla_x f(x) \mid_{(x^*,u^*)}$ gives: $\frac{\partial f_6}{\partial x_1} = 0$, $\frac{\partial f_6}{\partial x_2} = \beta x_3 + x_5$, $\frac{\partial f_6}{\partial x_3} = \frac{\beta}{\tau_s}$, $\frac{\partial f_6}{\partial x_4} = \beta x_2$, $\frac{\partial f_6}{\partial x_5} = -(\omega_{dq} - x_2)$, $\frac{\partial f_6}{\partial x_6} = -\gamma_2$, $\frac{\partial f_6}{\partial x_7} = 0$, $\frac{\partial f_6}{\partial x_8} = 0$, $\frac{\partial f_6}{\partial x_9} = 0$, $\frac{\partial f_6}{\partial x_{10}} = 0$, $\frac{\partial f_6}{\partial x_{11}} = 0$, $\frac{\partial f_6}{\partial x_{12}} = 0$, $\frac{\partial f_6}{\partial x_{13}} = 0$, $\frac{\partial f_6}{\partial x_{14}} = 0$, $\frac{\partial f_6}{\partial x_{15}} = 0$, $\frac{\partial f_6}{\partial x_{16}} = 0$, $\frac{\partial f_6}{\partial x_{17}} = 0$, $\frac{\partial f_6}{\partial x_{18}} = 0$, $\frac{\partial f_6}{\partial x_{19}} = 0$, and $\frac{\partial f_6}{\partial x_{20}} = 0$.

Seventh row of the Jacobian matrix $\nabla_x f(x) \mid_{(x^*,u^*)}$ gives: $\frac{\partial f_7}{\partial x_1} = 0$, $\frac{\partial f_7}{\partial x_2} = 0$, $\frac{\partial f_7}{\partial x_3} = 0$, $\frac{\partial f_7}{\partial x_4} = 0$, $\frac{\partial f_7}{\partial x_5} = -\frac{x_9}{2L_1}k_{11}$, $\frac{\partial f_7}{\partial x_6} = 0$, $\frac{\partial f_7}{\partial x_7} = -\frac{R_1}{L_1}$, $\frac{\partial f_7}{\partial x_8} = \omega_{dq}$, $\frac{\partial f_7}{\partial x_9} = -\frac{k_{11}}{P2L_1}x_5$, $\frac{\partial f_7}{\partial x_{10}} = 0$, $\frac{\partial f_7}{\partial x_{11}} = 0$, $\frac{\partial f_7}{\partial x_{12}} = 0$, $\frac{\partial f_7}{\partial x_{13}} = 0$, $\frac{\partial f_7}{\partial x_{14}} = 0$, $\frac{\partial f_7}{\partial x_{15}} = 0$, $\frac{\partial f_7}{\partial x_{16}} = 0$, $\frac{\partial f_7}{\partial x_{17}} = 0$, $\frac{\partial f_7}{\partial x_{18}} = 0$, $\frac{\partial f_7}{\partial x_{19}} = 0$, and $\frac{\partial f_7}{\partial x_{20}} = 0$.

Eighth row of the Jacobian matrix $\nabla_x f(x)\mid_{(x^*,u^*)}$ gives: $\frac{\partial f_8}{\partial x_1}=0$, $\frac{\partial f_8}{\partial x_2}=0$, $\frac{\partial f_8}{\partial x_3}=0$, $\frac{\partial f_8}{\partial x_4}=0$, $\frac{\partial f_8}{\partial x_5}=0$, $\frac{\partial f_8}{\partial x_6}=-\frac{x_9}{2L_1}k_{22}$, $\frac{\partial f_8}{\partial x_7}=-\omega_{dq}$, $\frac{\partial f_8}{\partial x_8}=-\frac{R_1}{L_1}$, $\frac{\partial f_8}{\partial x_9}=-\frac{k_{21}}{2L_1}x_6$, $\frac{\partial f_8}{\partial x_{10}}=0$, $\frac{\partial f_8}{\partial x_{11}}=0$, $\frac{\partial f_8}{\partial x_{12}}=0$, $\frac{\partial f_8}{\partial x_{13}}=0$, $\frac{\partial f_8}{\partial x_{14}}=0$, $\frac{\partial f_8}{\partial x_{15}}=0$, $\frac{\partial f_8}{\partial x_{16}}=0$, $\frac{\partial f_8}{\partial x_{17}}=0$, $\frac{\partial f_8}{\partial x_{18}}=0$, $\frac{\partial f_8}{\partial x_{19}}=0$, and $\frac{\partial f_8}{\partial x_{20}}=0$.

Ninth row of the Jacobian matrix $\nabla_x f(x)\mid_{(x^*,u^*)}$ gives: $\frac{\partial f_9}{\partial x_1}=0$, $\frac{\partial f_9}{\partial x_2}=0$, $\frac{\partial f_9}{\partial x_3}=0$, $\frac{\partial f_9}{\partial x_4}=0$, $\frac{\partial f_9}{\partial x_5}=\frac{3}{4C_1}x_7k_{11}$, $\frac{\partial f_9}{\partial x_6}=\frac{3}{4C_1}k_{21}x_8$, $\frac{\partial f_9}{\partial x_7}=\frac{3}{4C_1}k_{11}x_5$, $\frac{\partial f_9}{\partial x_8}=\frac{3}{4C_1}k_{21}x_6$, $\frac{\partial f_9}{\partial x_9}=-\frac{1}{Rc_1C_1}$, $\frac{\partial f_9}{\partial x_{10}}=0$, $\frac{\partial f_9}{\partial x_{11}}=0$, $\frac{\partial f_9}{\partial x_{12}}=0$, $\frac{\partial f_9}{\partial x_{13}}=0$, $\frac{\partial f_9}{\partial x_{14}}=0$, $\frac{\partial f_9}{\partial x_{15}}=0$, $\frac{\partial f_9}{\partial x_{16}}=0$, $\frac{\partial f_9}{\partial x_{17}}=0$, $\frac{\partial f_9}{\partial x_{18}}=0$, $\frac{\partial f_9}{\partial x_{19}}=0$, and $\frac{\partial f_9}{\partial x_{20}}=0$.

Tenth row of the Jacobian matrix $\nabla_x f(x)\mid_{(x^*,u^*)}$ gives: $\frac{\partial f_{10}}{\partial x_1}=0$, $\frac{\partial f_{10}}{\partial x_2}=0$, $\frac{\partial f_{10}}{\partial x_3}=0$, $\frac{\partial f_{10}}{\partial x_4}=0$, $\frac{\partial f_{10}}{\partial x_5}=0$, $\frac{\partial f_{10}}{\partial x_6}=0$, $\frac{\partial f_{10}}{\partial x_7}=0$, $\frac{\partial f_{10}}{\partial x_8}=0$, $\frac{\partial f_{10}}{\partial x_9}=0$, $\frac{\partial f_{10}}{\partial x_{10}}=0$, $\frac{\partial f_{10}}{\partial x_{11}}=1$, $\frac{\partial f_{10}}{\partial x_{12}}=0$, $\frac{\partial f_{10}}{\partial x_{13}}=0$, $\frac{\partial f_{10}}{\partial x_{14}}=0$, $\frac{\partial f_{10}}{\partial x_{15}}=0$, $\frac{\partial f_{10}}{\partial x_{16}}=0$, $\frac{\partial f_{10}}{\partial x_{17}}=0$, $\frac{\partial f_{10}}{\partial x_{18}}=0$, $\frac{\partial f_{10}}{\partial x_{19}}=0$, and $\frac{\partial f_{10}}{\partial x_{20}}=0$.

Eleventh row of the Jacobian matrix $\nabla_x f(x)\mid_{(x^*,u^*)}$ gives: $\frac{\partial f_{11}}{\partial x_1}=0$, $\frac{\partial f_{11}}{\partial x_2}=0$, $\frac{\partial f_{11}}{\partial x_3}=0$, $\frac{\partial f_{11}}{\partial x_4}=0$, $\frac{\partial f_{11}}{\partial x_5}=0$, $\frac{\partial f_{11}}{\partial x_6}=0$, $\frac{\partial f_{11}}{\partial x_7}=0$, $\frac{\partial f_{11}}{\partial x_8}=0$, $\frac{\partial f_{11}}{\partial x_9}=0$, $\frac{\partial f_{11}}{\partial x_{10}}=0$, $\frac{\partial f_{11}}{\partial x_{11}}=-\frac{B_2}{J_2}$, $\frac{\partial f_{11}}{\partial x_{12}}=\frac{3P(L_d-L_q)}{4J)2}x_{13}$, $\frac{\partial f_{11}}{\partial x_{13}}=\frac{3P(L_d-L_q)}{4J)2}x_{12}$, $\frac{\partial f_{11}}{\partial x_{14}}=-\frac{P}{2J_2}$, $\frac{\partial f_{11}}{\partial x_{15}}=0$, $\frac{\partial f_{11}}{\partial x_{16}}=0$, $\frac{\partial f_{11}}{\partial x_{17}}=0$, $\frac{\partial f_{11}}{\partial x_{18}}=0$, $\frac{\partial f_{11}}{\partial x_{19}}=0$, and $\frac{\partial f_{11}}{\partial x_{20}}=0$.

Twelfth row of the Jacobian matrix $\nabla_x f(x)\mid_{(x^*,u^*)}$ gives: $\frac{\partial f_{11}}{\partial x_1}=0$, $\frac{\partial f_{11}}{\partial x_2}=0$, $\frac{\partial f_{12}}{\partial x_3}=0$, $\frac{\partial f_{12}}{\partial x_4}=0$, $\frac{\partial f_{12}}{\partial x_5}=0$, $\frac{\partial f_{12}}{\partial x_6}=0$, $\frac{\partial f_{12}}{\partial x_7}=0$, $\frac{\partial f_{12}}{\partial x_8}=0$, $\frac{\partial f_{12}}{\partial x_9}=0$, $\frac{\partial f_{12}}{\partial x_{10}}=0$, $\frac{\partial f_{12}}{\partial x_{11}}=\frac{L_q}{L_d}x_{13}$, $\frac{\partial f_{12}}{\partial x_{12}}=-\frac{R_s}{L_d}$, $\frac{\partial f_{12}}{\partial x_{13}}=-\frac{L_q}{L_d}x_{11}$, $\frac{\partial f_{11}}{\partial x_{14}}=0$, $\frac{\partial f_{12}}{\partial x_{15}}=0$, $\frac{\partial f_{12}}{\partial x_{16}}=0$, $\frac{\partial f_{12}}{\partial x_{17}}=0$, $\frac{\partial f_{12}}{\partial x_{18}}=0$, $\frac{\partial f_{12}}{\partial x_{19}}=0$, and $\frac{\partial f_{12}}{\partial x_{20}}=0$.

Thirteenth row of the Jacobian matrix $\nabla_x f(x)\mid_{(x^*,u^*)}$ gives: $\frac{\partial f_{13}}{\partial x_1}=0$, $\frac{\partial f_{13}}{\partial x_2}=0$, $\frac{\partial f_{13}}{\partial x_3}=0$, $\frac{\partial f_{13}}{\partial x_4}=0$, $\frac{\partial f_{13}}{\partial x_5}=0$, $\frac{\partial f_{13}}{\partial x_6}=0$, $\frac{\partial f_{13}}{\partial x_7}=0$, $\frac{\partial f_{13}}{\partial x_8}=0$, $\frac{\partial f_{13}}{\partial x_9}=0$, $\frac{\partial f_{13}}{\partial x_{10}}=0$, $\frac{\partial f_{13}}{\partial x_{11}}=-\frac{L_d}{L_q}x_{12}$, $\frac{\partial f_{13}}{\partial x_{12}}=-\frac{L_d}{L_q}x_{11}$, $\frac{\partial f_{13}}{\partial x_{13}}=-\frac{R_s}{L_q}$, $\frac{\partial f_{13}}{\partial x_{14}}=0$, $\frac{\partial f_{13}}{\partial x_{15}}=0$, $\frac{\partial f_{13}}{\partial x_{16}}=0$, $\frac{\partial f_{13}}{\partial x_{17}}=0$, $\frac{\partial f_{13}}{\partial x_{18}}=0$, $\frac{\partial f_{13}}{\partial x_{19}}=0$, and $\frac{\partial f_{13}}{\partial x_{20}}=0$.

Fourteenth row of the Jacobian matrix $\nabla_x f(x)\mid_{(x^*,u^*)}$ gives: $\frac{\partial f_{14}}{\partial x_1}=0$, $\frac{\partial f_{14}}{\partial x_2}=0$, $\frac{\partial f_{14}}{\partial x_3}=0$, $\frac{\partial f_{14}}{\partial x_4}=0$, $\frac{\partial f_{14}}{\partial x_5}=0$, $\frac{\partial f_{14}}{\partial x_6}=0$, $\frac{\partial f_{14}}{\partial x_7}=0$, $\frac{\partial f_{14}}{\partial x_8}=0$, $\frac{\partial f_{14}}{\partial x_9}=0$, $\frac{\partial f_{14}}{\partial x_{10}}=0$, $\frac{\partial f_{14}}{\partial x_{11}}=0$, $\frac{\partial f_{14}}{\partial x_{12}}=0$, $\frac{\partial f_{14}}{\partial x_{13}}=0$, $\frac{\partial f_{14}}{\partial x_{14}}=-\frac{1}{\tau_w}$, $\frac{\partial f_{14}}{\partial x_{15}}=0$, $\frac{\partial f_{14}}{\partial x_{16}}=0$, $\frac{\partial f_{14}}{\partial x_{17}}=0$, $\frac{\partial f_{14}}{\partial x_{18}}=0$, $\frac{\partial f_{14}}{\partial x_{19}}=0$, and $\frac{\partial f_{14}}{\partial x_{20}}=0$.

Fifteenth row of the Jacobian matrix $\nabla_x f(x)\mid_{(x^*,u^*)}$ gives: $\frac{\partial f_{15}}{\partial x_1}=0$, $\frac{\partial f_{15}}{\partial x_2}=0$, $\frac{\partial f_{15}}{\partial x_3}=0$, $\frac{\partial f_{15}}{\partial x_4}=0$, $\frac{\partial f_{15}}{\partial x_5}=0$, $\frac{\partial f_{15}}{\partial x_6}=0$, $\frac{\partial f_{15}}{\partial x_7}=0$, $\frac{\partial f_{14}}{\partial x_8}=0$, $\frac{\partial f_{15}}{\partial x_9}=0$, $\frac{\partial f_{15}}{\partial x_{10}}=0$, $\frac{\partial f_{15}}{\partial x_{11}}=0$, $\frac{\partial f_{15}}{\partial x_{12}}=-\frac{x_{17}}{2L_2}k_{12}$, $\frac{\partial f_{15}}{\partial x_{13}}=0$, $\frac{\partial f_{15}}{\partial x_{14}}=0$, $\frac{\partial f_{15}}{\partial x_{15}}=-\frac{R_2}{L_2}$, $\frac{\partial f_{15}}{\partial x_{16}}=\omega_{dq}$, $\frac{\partial f_{15}}{\partial x_{17}}=-\frac{k_{12}}{2L_2}x_{12}$, $\frac{\partial f_{15}}{\partial x_{18}}=0$, $\frac{\partial f_{15}}{\partial x_{19}}=0$, and $\frac{\partial f_{15}}{\partial x_{20}}=0$.

Sixteenth row of the Jacobian matrix $\nabla_x f(x)\mid_{(x^*,u^*)}$ gives: $\frac{\partial f_{16}}{\partial x_1}=0$, $\frac{\partial f_{16}}{\partial x_2}=0$, $\frac{\partial f_{16}}{\partial x_3}=0$, $\frac{\partial f_{16}}{\partial x_4}=0$, $\frac{\partial f_{16}}{\partial x_5}=0$, $\frac{\partial f_{16}}{\partial x_6}=0$, $\frac{\partial f_{16}}{\partial x_7}=0$, $\frac{\partial f_{16}}{\partial x_8}=0$, $\frac{\partial f_{16}}{\partial x_9}=0$, $\frac{\partial f_{16}}{\partial x_{10}}=0$, $\frac{\partial f_{16}}{\partial x_{11}}=0$, $\frac{\partial f_{16}}{\partial x_{12}}=0$, $\frac{\partial f_{16}}{\partial x_{13}}=-\frac{x_{17}}{2L_2}k_{22}$, $\frac{\partial f_{16}}{\partial x_{14}}=0$, $\frac{\partial f_{16}}{\partial x_{15}}=-\omega_{dq}$, $\frac{\partial f_{16}}{\partial x_{16}}=-\frac{R_2}{L_2}$, $\frac{\partial f_{16}}{\partial x_{17}}=-\frac{k_{22}}{2L_2}x_{13}$, $\frac{\partial f_{16}}{\partial x_{18}}=0$, $\frac{\partial f_{16}}{\partial x_{19}}=0$, and $\frac{\partial f_{16}}{\partial x_{20}}=0$.

Seventeenth row of the Jacobian matrix $\nabla_x f(x)\mid_{(x^*,u^*)}$ gives: $\frac{\partial f_{17}}{\partial x_1}=0$, $\frac{\partial f_{17}}{\partial x_2}=0$, $\frac{\partial f_{17}}{\partial x_3}=0$, $\frac{\partial f_{17}}{\partial x_4}=0$, $\frac{\partial f_{17}}{\partial x_5}=0$, $\frac{\partial f_{17}}{\partial x_6}=0$, $\frac{\partial f_{17}}{\partial x_7}=0$, $\frac{\partial f_{17}}{\partial x_8}=0$, $\frac{\partial f_{17}}{\partial x_9}=0$, $\frac{\partial f_{17}}{\partial x_{10}}=0$, $\frac{\partial f_{17}}{\partial x_{11}}=0$, $\frac{\partial f_{17}}{\partial x_{12}}=\frac{3}{4C_2}k_{12}x_{15}$, $\frac{\partial f_{17}}{\partial x_{13}}=\frac{3}{4C_2}k_{22}x_{14}$, $\frac{\partial f_{17}}{\partial x_{14}}=\frac{3}{4C_2}k_{22}x_{13}$, $\frac{\partial f_{17}}{\partial x_{15}}=\frac{3}{4C_2}k_{12}x_{12}$, $\frac{\partial f_{17}}{\partial x_{16}}=0$, $\frac{\partial f_{17}}{\partial x_{17}}=-\frac{1}{Rc_2C_2}$, $\frac{\partial f_{17}}{\partial x_{18}}=0$, $\frac{\partial f_{17}}{\partial x_{19}}=0$, and $\frac{\partial f_{17}}{\partial x_{20}}=0$.

Eighteenth row of the Jacobian matrix $\nabla_x f(x)\mid_{(x^*,u^*)}$ gives: $\frac{\partial f_{18}}{\partial x_1}=0$, $\frac{\partial f_{18}}{\partial x_2}=0$, $\frac{\partial f_{18}}{\partial x_3}=0$, $\frac{\partial f_{18}}{\partial x_4}=0$, $\frac{\partial f_{18}}{\partial x_5}=0$, $\frac{\partial f_{18}}{\partial x_6}=0$, $\frac{\partial f_{18}}{\partial x_7}=0$, $\frac{\partial f_{18}}{\partial x_8}=0$, $\frac{\partial f_{18}}{\partial x_9}=0$, $\frac{\partial f_{18}}{\partial x_{10}}=0$, $\frac{\partial f_{18}}{\partial x_{11}}=0$, $\frac{\partial f_{18}}{\partial x_{12}}=0$, $\frac{\partial f_{18}}{\partial x_{13}}=0$, $\frac{\partial f_{18}}{\partial x_{14}}=0$, $\frac{\partial f_{18}}{\partial x_{15}}=0$, $\frac{\partial f_{18}}{\partial x_{16}}=0$, $\frac{\partial f_{18}}{\partial x_{17}}=0$, $\frac{\partial f_{18}}{\partial x_{18}}=\frac{1}{R_3C_3}$, $\frac{\partial f_{18}}{\partial x_{19}}=\frac{1}{C_3}$, and $\frac{\partial f_{18}}{\partial x_{20}}=0$.

Nineteenth row of the Jacobian matrix $\nabla_x f(x)\mid_{(x^*,u^*)}$ gives: $\frac{\partial f_{19}}{\partial x_1}=0$, $\frac{\partial f_{19}}{\partial x_2}=0$, $\frac{\partial f_{19}}{\partial x_3}=0$, $\frac{\partial f_{19}}{\partial x_4}=0$, $\frac{\partial f_{19}}{\partial x_5}=0$, $\frac{\partial f_{19}}{\partial x_6}=0$, $\frac{\partial f_{19}}{\partial x_7}=0$, $\frac{\partial f_{19}}{\partial x_8}=0$, $\frac{\partial f_{19}}{\partial x_9}=0$, $\frac{\partial f_{19}}{\partial x_{10}}=0$, $\frac{\partial f_{19}}{\partial x_{11}}=0$, $\frac{\partial f_{19}}{\partial x_{12}}=0$, $\frac{\partial f_{19}}{\partial x_{13}}=0$, $\frac{\partial f_{19}}{\partial x_{14}}=0$, $\frac{\partial f_{19}}{\partial x_{15}}=0$, $\frac{\partial f_{19}}{\partial x_{16}}=0$, $\frac{\partial f_{19}}{\partial x_{17}}=0$, $\frac{\partial f_{19}}{\partial x_{18}}=-\frac{1}{L_3}$, $\frac{\partial f_{19}}{\partial x_{19}}=0$, and $\frac{\partial f_{19}}{\partial x_{20}}=0$.

Twentieth row of the Jacobian matrix $\nabla_x f(x)\mid_{(x^*,u^*)}$ gives: $\frac{\partial f_{20}}{\partial x_i}=\frac{2}{R_L}(x_9+x_{17}+x_{18})(\frac{\partial f_9}{\partial x_i}+\frac{\partial f_{17}}{\partial x_i})+\frac{\partial f_{18}}{\partial x_i}$, if $i=1,2,3,4,5,6,7,8,10,11,12,13,14,15,16,19,20$ and $\frac{\partial f_{20}}{\partial x_9}=\frac{2}{R_L}(f_9+f_{17}+f_{18})+\frac{2}{R_L}(x_9+x_{17}+x_{18})(\frac{\partial f_9}{\partial x_9}+\frac{\partial f_{17}}{\partial x_9})+\frac{\partial f_{18}}{\partial x_9}$, $\frac{\partial f_{20}}{\partial x_{17}}=\frac{2}{R_L}(f_9+f_{17}+f_{18})+\frac{2}{R_L}(x_9+x_{17}+x_{18})(\frac{\partial f_9}{\partial x_{17}}+\frac{\partial f_{17}}{\partial x_{17}})+\frac{\partial f_{18}}{\partial x_{17}}$, $\frac{\partial f_{20}}{\partial x_{18}}=\frac{2}{R_L}(f_9+f_{17}+f_{18})+\frac{2}{R_L}(x_9+x_{17}+x_{18})(\frac{\partial f_9}{\partial x_{18}}+\frac{\partial f_{17}}{\partial x_{18}})+\frac{\partial f_{18}}{\partial x_{18}}$.

4.9.4 Computation of the feedback control gains

For the linearized system given by (1.482) the cost function of (1.483) is defined, where coefficient r determines the penalization of the control input and weight coefficient ρ determines the reward of the disturbances' effects. It is assumed that (i) The energy that is transferred from the disturbances signal $\tilde{d}(t)$ is bounded, that is $\int_0^{\infty}\tilde{d}^T(t)\tilde{d}(t)dt<\infty$, (ii) matrices $[A,B]$ and $[A,L]$ are stabilizable, (iii) matrix $[A,C]$ is detectable. In the case of a tracking problem the optimal feedback control law is given by

$$u(t)=-Ke(t) \tag{4.381}$$

with $e=x-x_d$ to be the tracking error, and $K=\frac{1}{r}B^TP$ where P is a positive definite symmetric matrix. Matrix P is obtained from the solution of the Riccati equation

$$A^TP+PA+Q-P(\frac{2}{r}BB^T-\frac{1}{\rho^2}LL^T)P=0 \tag{4.382}$$

where Q is a positive semi-definite symmetric matrix. The worst-case disturbance is given by

$$\tilde{d}(t) = \frac{1}{\rho^2} L^T P e(t) \tag{4.383}$$

The solution of the H-infinity feedback control problem for the hybrid microgrid and the computation of the worst-case disturbance that the related controller can sustain, comes from the superposition of Bellman's optimality principle when considering that the microgrid is affected by two separate inputs (i) the control input u, (ii) the cumulative disturbance input $\tilde{d}(t)$. Solving the optimal control problem for u, that is for the minimum variation (optimal) control input that achieves the elimination of the state vector's tracking error, gives $u = -\frac{1}{r}B^T P e$. Equivalently, solving the optimal control problem for $\tilde{d}$, that is for the worst-case disturbance that the control loop can sustain gives $\tilde{d} = \frac{1}{\rho^2}L^T P e$.

The diagram of the considered control loop for the hybrid microgrid is depicted in Figure 4.113. The global stability properties of the control method can be proven following the procedure presented in Section 1.10.2.5.

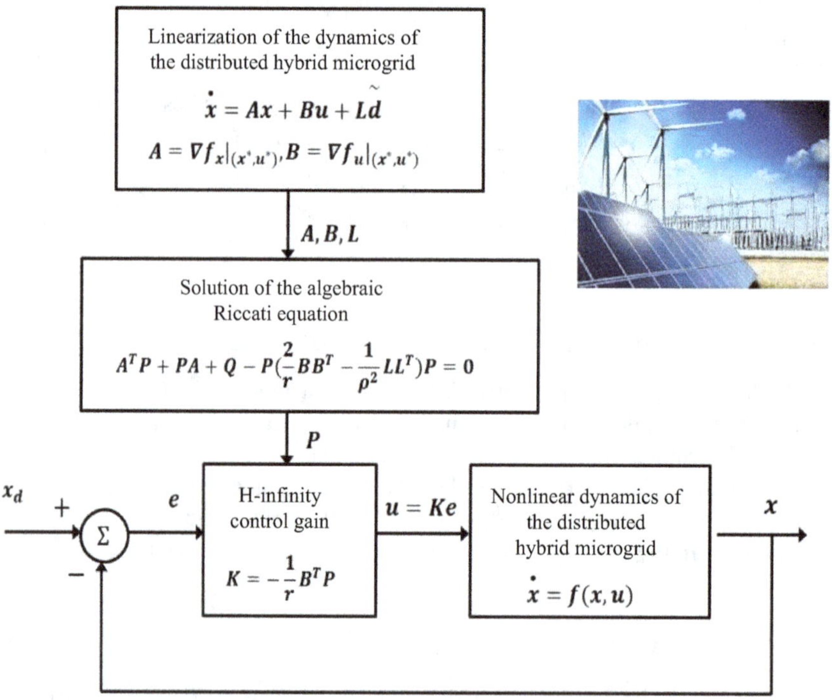

Figure 4.113 Diagram of the control scheme for the distributed hybrid microgrid, that comprises wind/hydro turbines and PV power units

4.9.5 *Differential flatness properties of the dynamic model of the microgrid*

Using (Rigatos, 2015), it will be proven that the dynamic model of the microgrid is a differentially flat system, with flat outputs: $y_1 = x_1$, $y_2 = \frac{3L_1}{4}(x_7^2 + x_8^2) + \frac{C_1}{2}x_3^2$, $y_3 = x_8, y_4 = x_{10}, y_5 = \frac{3L_2}{4}(x_{15}^2 + x_{16}^2) + \frac{C_2}{2}x_{17}^2, y_6 = x_{16}$ and $y_7 = x_{18}$. The state-space model of the microgrid comprises the following equations:

$$\dot{x}_1 = x_2 \tag{4.384}$$

$$\dot{x}_2 = -\frac{K_b}{J_1}x_2 - \frac{n_p}{J_1}(x_5x_4 - x_6x_3) + \frac{T_m}{J_1}u_1 \tag{4.385}$$

$$\dot{x}_2 = \omega_{dq}x_4 - \frac{1}{\tau_s}x_3 - \frac{M}{\tau_s}x_5 + v_{s_d} \tag{4.386}$$

$$\dot{x}_4 = -\frac{1}{\tau_s}x_4 - \omega_{dq}x_3 + \frac{M}{\tau_s}x_6 + v_{s_q} \tag{4.387}$$

$$\dot{x}_5 = \beta x_2x_4 + \frac{\beta}{\tau_s}x_3 + (\omega_{dq} - x_2)x_6 - \gamma_2x_5 - \beta v_d + \frac{1}{\sigma]L_r}u_2 \tag{4.388}$$

$$\dot{x}_6 = \frac{\beta}{\tau_s}x_7 + \beta x_2x_3 - \gamma_2x_6 - (\omega_{dq} - x_2)x_5 - \beta v_q + \frac{1}{\sigma]L_r}u_3 \tag{4.389}$$

$$\dot{x}_7 = -\frac{R_1}{L_1}x_7 + \omega_{dq}x_8 + \frac{v_d}{L_1} - \frac{x_9}{2L_1k_{11}}x_5 \tag{4.390}$$

$$\dot{x}_8 = -\omega_{dq}x_7 - \frac{R_1}{L_1}x_8 + \frac{v_q}{L_1} - \frac{x_9}{2L_1}k_{21}x_6 \tag{4.391}$$

$$\dot{x}_9 = -\frac{1}{Rc_1C_1}x_9 + \frac{3}{4C_1}x_7k_{11}x_5 + \frac{3}{4C_1}x_8k_{21}x_6 \tag{4.392}$$

$$\dot{x}_{10} = x_{11} \tag{4.393}$$

$$\dot{x}_{11} = \frac{3P(L_d - L_q)}{4J_2}x_{12}x_{13} - \frac{B_2}{J_2}x_{11} - \frac{P}{2J_2}x_{14} \tag{4.394}$$

$$\dot{x}_{12} = -\frac{R_s}{L_d}x_{12} + \frac{L_q}{L_d}x_{11}x_{13} + \frac{1}{L_d}u_4 \tag{4.395}$$

$$\dot{x}_{13} = -\frac{R_s}{L_q}x_{13} - \frac{L_d}{L_q}x_{11}x_{12} + \frac{1}{L_q}u_5 \tag{4.396}$$

$$\dot{x}_{14} = -\frac{1}{\tau_w}x_{14} + k_uu_6 \tag{4.397}$$

$$\dot{x}_{15} = -\frac{R_2}{L_2}x_{15} + \omega_{dq}x_{16} + \frac{v_d]}{L_2} - \frac{x_{17}}{3L_2}k_{12}x_{12} \tag{4.398}$$

$$\dot{x}_{16} = -\omega_{dq}x_{15} - \frac{R_2}{L_2}x_{16} + \frac{v_q}{L_2} - \frac{x_{17}}{3L_2}k_{22}x_{13} \tag{4.399}$$

$$\dot{x}_{17} = -\frac{1}{Rc_2C_2}x_{17} + \frac{3}{4C_2}x_{15}k_{12}x_{12} + \frac{3}{4C_2}x_{16}k_{22}x_{13} \tag{4.400}$$

$$\dot{x}_{18} = \frac{1}{R_3C_3}x_{18} + \frac{1}{C_3}x_{19} \tag{4.401}$$

$$\dot{x}_{19} - \frac{x_{18}}{L_3} + \frac{1}{L_3}u_7 \tag{4.402}$$

$$\dot{x}_{20} = \frac{2}{R_L}(x_9 + x_{17} + x_{18})(f_9 + f_{17} + f_{18}) \tag{4.403}$$

Equations (4.384) to (4.392) are examined first. The field orientation condition holds, that is $x_4 = \psi_{s_q} = 0$.

From the definition of the second flat output it holds that

$$\dot{y}_2 = \frac{3L_1}{4}(2x_7\dot{x}_7 + 2x_8\dot{x}_8) + \frac{C_1}{2}2x_9\dot{x}_9 \tag{4.404}$$

or equivalently

$$\begin{aligned}\dot{y}_2 = \frac{3L_1}{4}\left\{2x_7\left[-\frac{R_1}{L_1}x_7 + \omega_{dq}x_8 + \frac{v_d}{L_1} - \frac{x_9}{2L_1k_{11}}x_5\right]\right.\\ + 2x_8\left[-\omega_{dq}x_7 - \frac{R_1}{L_1}x_8 + \frac{v_q}{L_1} - \frac{x_9}{2L_1}k_{21}x_6\right]\\ \left. + C_1x_9\left[-\frac{1}{Rc_1C_1}x_9 + \frac{3}{4C_1}x_7k_{11}x_5 + \frac{3}{4C_1}x_8k_{21}x_6\right]\right\}\end{aligned} \tag{4.405}$$

which after intermediate operations gives

$$\dot{y}_2 = -\frac{3R_1}{2}(x_7^2 + x_8^2) - \frac{1}{Rc_1}x_9^2 + \frac{3v_d}{2}x_7 + \frac{3v_q}{2}x_8 \tag{4.406}$$

From the definition of the second flat output y_2 one solves with respect to x_9^2

$$x_9^2 = \frac{y_2 - \frac{3L_1}{4}(x_7^2 + x_8^2)}{\frac{C_1}{2}} \tag{4.407}$$

Equation (4.407) is substituted in (4.406) thus giving

$$\dot{y}_2 = -\frac{3R_1}{2}(x_7^2 + x_8^2) - \frac{1}{Rc_1}\frac{y_2 - \frac{3L_1}{4}(x_7^2 + x_8^2)}{\frac{C_1}{2}} + \frac{3v_d}{2}x_7 + \frac{3v_q}{2}x_8 \tag{4.408}$$

while after intermediate operations and by regrouping terms one obtains the following binomial

$$\left(\frac{3R_1}{2} - \frac{3L_1}{2Rc_1C_1}\right)x_7^2 - \frac{3v_d}{2}x_7 + \left[\dot{y}_2 + \frac{2}{Rc_1C_1}y_2 + \left(\frac{3R_1}{2} - \frac{3L_1}{2Rc_1C_1}\right)x_8^2 - \frac{3v_q}{2}x_8\right] = 0 \tag{4.409}$$

Moreover, using that the flat output y_3 coincides with state variable x_8, or $x_8 = y_3$, (4.409) becomes

$$\left(\frac{3R_1}{2} - \frac{3L_1}{2Rc_1C_1}\right)x_7^2 - \frac{3v_d}{2}x_7 + \left[\dot{y}_2 + \frac{2}{Rc_1C_1}y_2 + \left(\frac{3R_1}{2} - \frac{3L_1}{2Rc_1C_1}\right)y_3^2 - \frac{3v_q}{2}y_3\right] = 0 \tag{4.410}$$

The above given equation is a binomial with respect to x_7. By retaining the largest solution for x_7 one obtains

$$x_7 = h_7(y, \dot{y}) \tag{4.411}$$

where y is the flat outputs vector and $\dot{y}$ is its time-derivative. Consequently, x_7 is a differential function of the flat outputs of the system. Moreover, by substituting (4.411) into (4.407) one obtains

$$x_9^2 = \frac{y_2 - \frac{3L_1}{4}(h_7(y, \dot{y})^2 + y_3^2)}{\frac{C_1}{2}} \tag{4.412}$$

or equivalently,

$$x_9 = h_9(y, \dot{y}) \tag{4.413}$$

Consequently, x_9 is also a differential function of the flat outputs of the system. Next, from (4.390) one solves with respect to x_5 and this gives

$$x_5 = \frac{\dot{x}_7 + \frac{R_1}{L_1}x_7 - \omega_{dq}x_8 - \frac{v_d}{L_1}}{\frac{-x_9k_{11}}{2L_1}} \tag{4.414}$$

and using that x_7, x_8 and x_9 are differential functions of the flat outputs of the system it holds that x_5 is also a differential function of the system's flat outputs, or

$$x_5 = h_5(y, \dot{y}) \tag{4.415}$$

Moreover, from (4.391) one solves with respect to x_6. This gives

$$x_6 = \frac{\dot{x}_8 + \omega_{dq}x_7 + \frac{R_1}{L_1}x_8 - \frac{v_q}{L_1}}{\frac{-x_9k_{21}}{2L_1}} \tag{4.416}$$

and using that x_7, x_8 and x_9 are differential functions of the flat outputs of the system it holds that x_6 is also a differential function of the system's flat outputs, or

$$x_6 = h_6(y, \dot{y}) \tag{4.417}$$

From the field-orientation condition about the induction generator one has that $x_4 = \psi_{s_q} = 0$ and equivalently $x_4 = h_4(y, \dot{y})$. Using next (4.387), and solving with respect to x_3, one has that

$$x_3 = \frac{\dot{x}_4 + \frac{1}{\tau_s}x_4 - \frac{M}{\tau_s}x_6 - v_{s_q}}{-\omega_{dq}} \tag{4.418}$$

while using that x_6 is a differential function of the system's flat outputs, one gets that x_3 is also a differential function of the flat outputs of the microgrid, or

$$x_3 = h_3(y, \dot{y}) \tag{4.419}$$

Besides, from (4.384) one has that

$$x_2 = \dot{y}_1 \Rightarrow x_2 = h_2(y, \dot{y}) \tag{4.420}$$

which signifies that x_2 is a differential function of the flat outputs of the microgrid. After having proven that state variables x_1 to x_9 are functions of the flat outputs of the microgrid, then from (4.385) one can solve for control input u_1 which gives

$$\begin{aligned} u_1 &= \frac{J_1}{T_m}\left[\dot{x}_2 + \frac{K_1}{J_1}x_2 + \frac{K_1}{J_1} + \frac{n_p}{J_1}(x_5x_4 - x_6x_3)\right] \Rightarrow \\ u_1 &= h_{u_2}(y, \dot{y}) \end{aligned} \tag{4.421}$$

which signifies that control input u_1 is a differential function of the microgrid. Besides, from (4.388) one can solve for control input u_2 which gives

$$\begin{aligned} u_2 &= \sigma L_r\left[\dot{x}_5 - \beta x_2x_4 - \frac{\beta}{\tau_s}x_3 - (\omega_{dq} - x_2)x_6 + \gamma_2x_5 + \beta v_d\right] \\ &\Rightarrow u_2 = h_{u_2}(y, \dot{y}) \end{aligned} \tag{4.422}$$

which signifies that control input u_2 is a differential function of the microgrid. Besides, from (4.389) one can solve for control input u_3 which gives

$$\begin{aligned} u_3 &= \sigma L_r[\dot{x}_6 - \frac{\beta}{\tau_s}x_4 - \beta x_2x_3 + \gamma_2x_6 + (\omega_{dq} - x_2)x_5 + \beta v_d] \\ &\Rightarrow u_3 = h_{u_3}(y, \dot{y}) \end{aligned} \tag{4.423}$$

which signifies that control input u_2 is a differential function of the microgrid.

Next, the proof of differential flatness is concerned with state variables x_{10} to x_{17}. From the equation about flat output y_5 one has

$$\dot{y}_5 = \frac{3L_2}{4}(2x_{15}\dot{x}_{15} + 2x_{16}\dot{x}_{16}) + \frac{C_2}{2}2x_{17}\dot{x}_{17} \tag{4.424}$$

which is also written as

$$\begin{aligned} \dot{y}_5 = \frac{3L_2}{4}&(2x_{15}\left[-\frac{R_2}{L_2}x_{15} + \omega_{dq}x_{16} + \frac{v_d}{L_2} - \frac{x_{17}}{3L_2}k_{12}x_{12}\right] \\ &+ 2x_{16}\left[-\omega_{dq}x_{15} - \frac{R_2}{L_2}x_{16} + \frac{v_q}{L_2} - \frac{x_{17}}{3L_2}k_{22}x_{13}\right] \\ &+ C_2x_{17}\left[-\frac{1}{Rc_2C_2}x_{17} + \frac{3}{4C_2}x_{15}k_{12}x_{12} + \frac{3}{4C_2}x_{16}k_{22}x_{13}\right] \end{aligned} \tag{4.425}$$

By performing intermediate operations one has

$$\dot{y}_5 = -\frac{3R_2}{2}(x_{15}^2 + x_{16}^2) - \frac{1}{Rc_2}x_{17}^2 + \frac{3v_d}{2}x_{15} + 3v_q]over2x_{16} \tag{4.426}$$

From the relation about flat output y_5 one solves for x_{17}^2. Thus

$$x_{17}^2 = \frac{y_5 - \frac{3L_2}{4}(x_{15}^2 + x_{16}^2)}{\frac{C_2}{2}} \tag{4.427}$$

By substituting (4.427) into (4.426) one after performing intermediate operations and regrouping terms one obtains the binomial

$$\left(\frac{3R_2}{2} - \frac{3L_2}{2Rc_2C_2}\right)x_{15}^2 - \frac{3v_d}{2}x_{15} + \left[\dot{y}_5 + \frac{2}{Rc_2C_2}y_6 + \left(\frac{3R_2}{2} - \frac{3L_2}{2Rc_2C_2}\right)x_{16}^2 - \frac{3v_q}{2}x_{16}\right] = 0 \tag{4.428}$$

Moreover, using that the flat output y_6 coincides with state variable x_{16}, or $x_{16} = y_6$, (4.429) becomes

$$\left(\frac{3R_2}{2} - \frac{3L_2}{2Rc_2C_2}\right)x_{15}^2 - \frac{3v_d}{2}x_{15} + \left[\dot{y}_5 + \frac{2}{Rc_2C_2}y_6 + \left(\frac{3R_2}{2} - \frac{3L_2}{2Rc_2C_2}\right)y_6^2 - \frac{3v_q}{2}y_6\right] = 0 \tag{4.429}$$

By retaining the largest solution for x_{15} one has

$$x_{15} = h_{15}(y, \dot{y}) \tag{4.430}$$

which means that state variable x_{15} is a differential function of the system's flat output. Moreover, by substituting (4.430) into (4.427)

$$x_{17}^2 = \frac{y_5 - \frac{3L_2}{4}(h_{15}(y, \dot{y})^2 + y_6^2)}{\frac{C_2}{2}} \tag{4.431}$$

or equivalently

$$x_{17} = h_{15}(y, \dot{y}) \tag{4.432}$$

which signifies that state variable x_{17} is also a differential function of the system's flat outputs.

Next, by knowing that x_{15}, x_{16} and x_{17} are differential functions of the flat outputs of the microgrid, (4.398) can be used to solve with respect to x_{12}. Thus

$$x_{12} = \frac{\dot{x}_{15} + \frac{R_2}{L_2}x_{15} + \omega_{dq}x_{16} - \frac{v_d}{L_2}}{\frac{-x_{17}k_{12}}{2L_2}} \tag{4.433}$$

or equivalently

$$x_{12} = h_{12}(y, \dot{y}) \tag{4.434}$$

which means that state variable x_{12} is a differential function of the system's flat outputs. Next, knowing that x_{15} x_{16}, x_{17} are functions of the flat outputs of the microgrid and of their derivatives one can use (4.399) to solve with respect to x_{13}. Thus one obtains

$$x_{13} = \frac{\dot{x}_{16} + \omega_{dq}x_{15} + \frac{R_2}{L_2}x_{16} + \frac{v_q}{L_2}}{\frac{-x_{17}k_{22}}{2L_2}} \tag{4.435}$$

or equivalently

$$x_{13} = h_{13}(y, \dot{y}) \tag{4.436}$$

which means that state variable x_{13} is a differential function of the flat outputs of the microgrid. Moreover, from (4.393) one has

$$x_{11} = \dot{x}_{10} \Rightarrow x_{11} = h_{11}(y, \dot{y}) \tag{4.437}$$

which signifies that state-variable x_{11} is a differential function of the flat outputs of the microgrid. Moreover, from (4.394) it is allowed to solve for x_{14} thus giving

$$x_{14} = \frac{\dot{x}_{11} - \frac{3P(L_d-L_q)}{4J_2}x_{12}x_{13} + \frac{B_2}{J_2}x_{11}}{\frac{-P}{2J_2}} \tag{4.438}$$

or equivalently

$$x_{14} = h_{14}(y, \dot{y}) \tag{4.439}$$

which signifies that state variable x_{14} is a differential function of the flat outputs of the microgrid. Moreover, having proven that state variables x_{10} to x_{17} are either flat outputs or are differential functions of the flat outputs of the microgrid one can use (4.395) to solve for control input u_4. Thus, it holds

$$\begin{aligned} u_4 &= L_d\left[\dot{x}_{12} + \frac{R_s}{L_d}x_{12} - \frac{L_q}{L_d}x_{11}x_{13}\right] \Rightarrow \\ u_4 &= h_{u_4}(y, \dot{y}) \end{aligned} \tag{4.440}$$

Consequently, control input u_4 is a differential function of the flat outputs of the microgrid. Additionally, one can use (4.396) to solve for control input u_5. Thus, it holds

$$\begin{aligned} u_5 &= L_q\left[\dot{x}_{13} + \frac{R_s}{L_q}x_{13} - \frac{L_d}{L_1}x_{11}x_{12}\right] \Rightarrow \\ u_5 &= h_{u_5}(y, \dot{y}) \end{aligned} \tag{4.441}$$

Therefore, control input u_5 is a differential function of the flat outputs of the microgrid.

Next, differential flatness is proven for state-variables x_{18} and x_{19} of the DC/DC converter. From (4.401) one solves for state variable x_{19}, thus obtaining

$$x_{19} = C_3\left[\dot{x}_{18} - \frac{1}{R_3C_3}x_{18}\right] \tag{4.442}$$

and using that $x_{18} = y_7$, that is x_{18} is a flat output of the system, one obtains equivalently

$$x_{19} = h_{14}(y, \dot{y}) \tag{4.443}$$

which signifies that x_{19} is a differential function of the flat outputs of the system. Besides (4.402) one solves for the control input u_7. This gives

$$u_7 = L_3\dot{x}_{19} + x_{18} \tag{4.444}$$

which signifies that the control input u_7 is a differential function of the microgrid's flat outputs, or equivalently

$$u_7 = h_{u_7}(y, \dot{y}) \tag{4.445}$$

Finally, about state variable x_{20} it holds that

$$x_{20} = \frac{(x_9^2 + x_{17}^2 + x_{18}^2)}{R_L} \tag{4.446}$$

and having proven that state variables x_9, x_{17} and x_{18} are differential functions of the flat outputs of the microgrid it can be concluded that state variable x_{20} is a differential function of the flat outputs of the system. As a result of the previous analysis, it has been proven that state variables $x_1, x_2, \ldots, x_{20}$ and control inputs $u_1, u_2, \cdots, u_7$ can be expressed as functions of the flat outputs of the microgrid and of the associated time-derivatives. Consequently, the entire system is differentially flat.

By knowing that the microgrid is differentially flat it is confirmed that it is input–output linearizable (Rigatos, 2015). Besides, by using the differential flatness properties one can solve the setpoints' definition problem. Actually, one can select first setpoints for the state variables which are associated with the flat outputs of the system. Next, these reference values can be used to define setpoints for the rest of the state variables of the microgrid. In this regard, using (4.446) the setpoints for the power provided by the microgrid are given as functions of the setpoints x_9^d, x_{17}^d and x_{18}^d. Alternatively, one can use the amounts of power which are produced by the individual renewable power units as free variables that finally allow for injecting to the electricity grid the targeted power P_d. At the same time, the global stability of the microgrid is ensured.

4.9.6 Simulation tests

The global stability properties and the tracking performance of the nonlinear optimal control method for hybrid AC/DC microgrids have been tested and further confirmed through simulation experiments. To implement the nonlinear optimal control scheme, the algebraic Riccati equation which appears in (4.382) had to be solved at each time-step of the control algorithm. The obtained results are depicted in Figures 4.114 to 4.119. In these diagrams the following state variables are depicted: x_2: angular speed of the generator in the wind-power unit, x_9: DC output voltage of the AC/DC converter of the wind-power unit, x_{11}: angular speed of the generator of the hydro-power unit, x_{17}: DC output voltage of the AC/DC converter of the hydro-power unit, x_{18}: DC output voltage of the DC/DC converter of the PV-power unit, x_{20}: aggregate electric power produced by the hybrid microgrid. Besides the following control inputs are shown: u_1: blades' angle in the turbine of the DFIG generator, u_2, u_3: voltages applied to the rotor of the DFIG generator, u_4, u_5: voltages applied to the stator of the SRG power generator, u_7 PWM control input voltage of the DC/DC converter. It can be

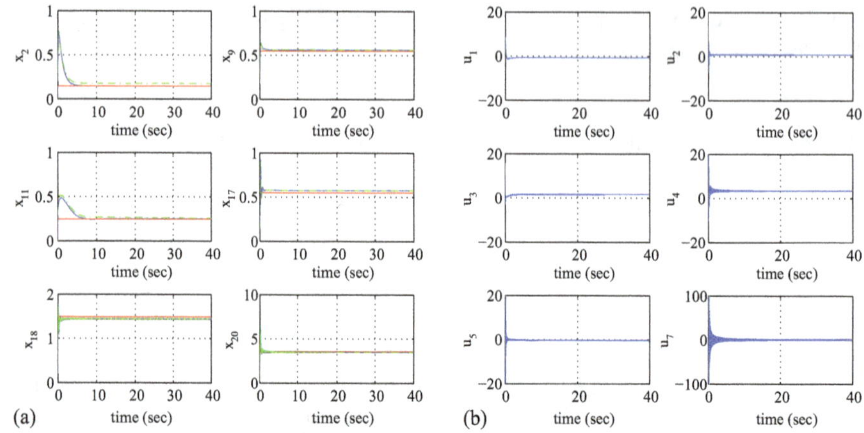

Figure 4.114 Tracking of setpoint 1 for the hybrid microgrid (a) convergence of state variables x_1 to x_{20} to their reference setpoints, (b) control inputs u_1 to u_7 of the hybrid microgrid

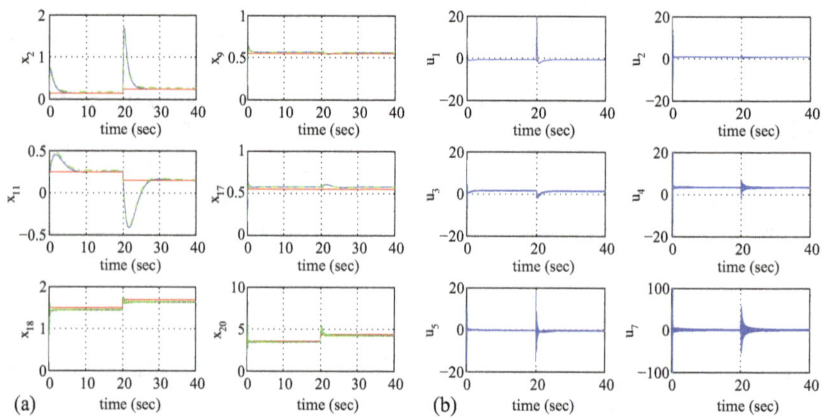

Figure 4.115 Tracking of setpoint 2 for the hybrid microgrid (a) convergence of state variables x_1 to x_{20} to their reference setpoints, (b) control inputs u_1 to u_7 of the hybrid microgrid

noticed that in all cases fast and accurate tracking of reference setpoints were achieved under moderate variations of the control inputs. The real values of the state vector elements of the hybrid microgrid are printed in blue, the estimated state variables provided by the H-infinity Kalman Filter are plotted in green, while the associated setpoints are shown in red colour.

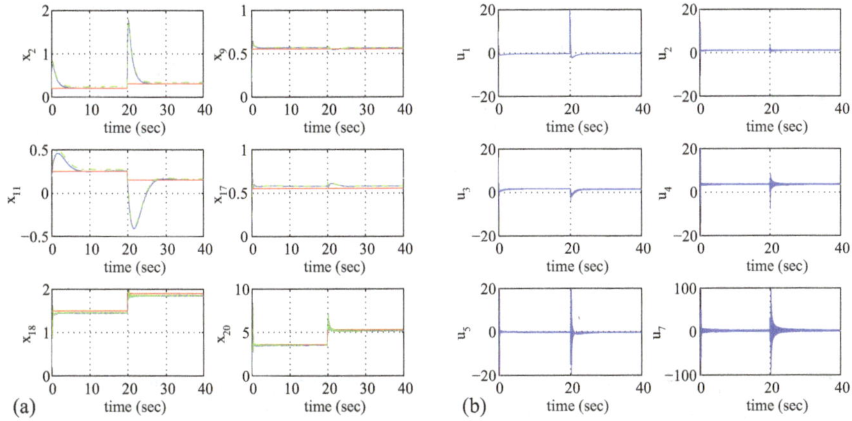

Figure 4.116 Tracking of setpoint 3 for the hybrid microgrid (a) convergence of state variables x_1 to x_{20} to their reference setpoints, (b) control inputs u_1 to u_7 of the hybrid microgrid

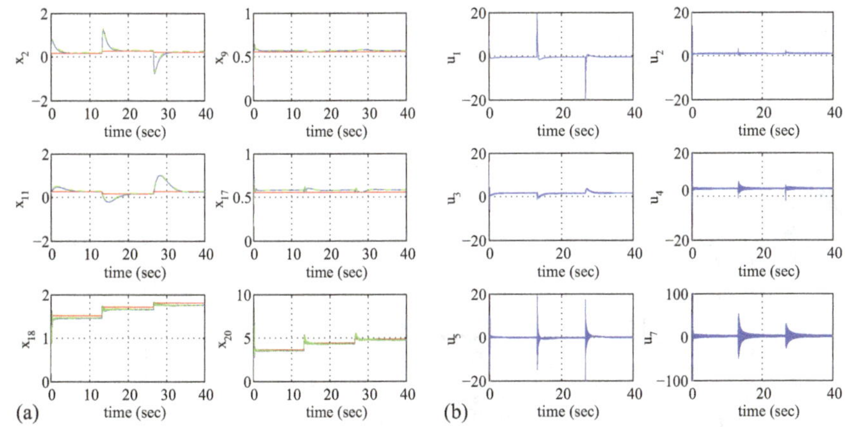

Figure 4.117 Tracking of setpoint 4 for the hybrid microgrid (a) convergence of state variables x_1 to x_{20} to their reference setpoints, (b) control inputs u_1 to u_7 of the hybrid microgrid

It can be noted that the transient performance of the control method depends on the selection of parameters r, ρ and Q which appear in the Riccati equation of (1.496). Actually, relatively small values of r result in the elimination of the tracking error while relatively large values of matrix Q result in fast convergence to the reference setpoints. Moreover, coefficient ρ affects the robustness of the control loop. The smallest value of ρ for which one can obtain a valid solution from the algebraic Riccati equation of

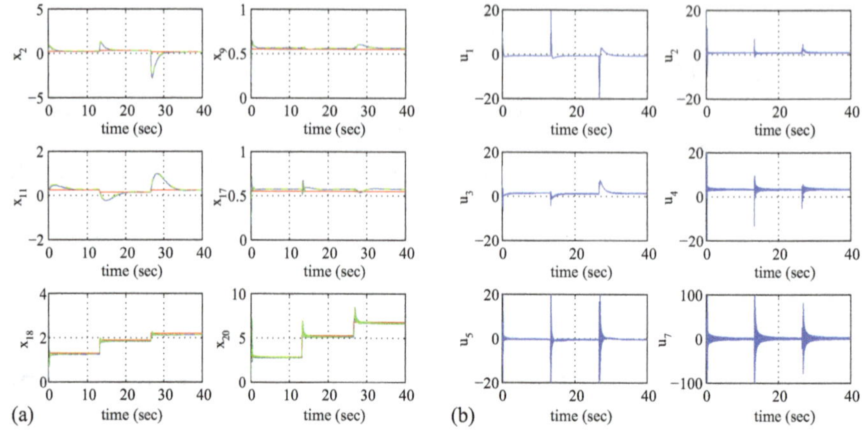

Figure 4.118 Tracking of setpoint 5 for the hybrid microgrid (a) convergence of state variables x_1 to x_{20} to their reference setpoints, (b) control inputs u_1 to u_7 of the hybrid microgrid

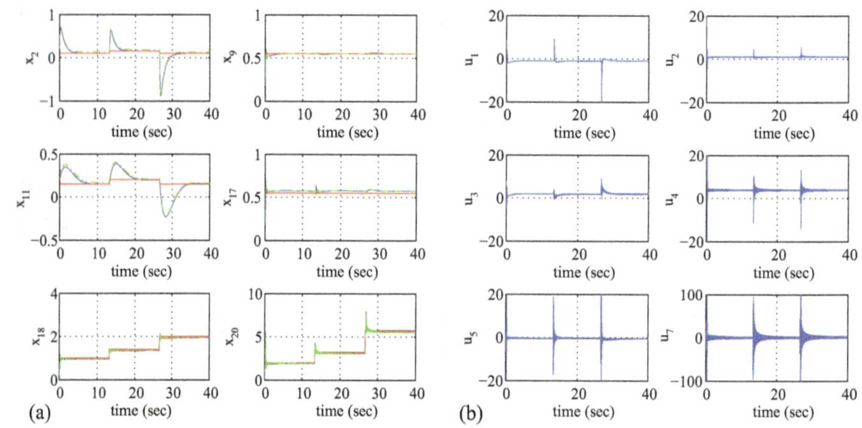

Figure 4.119 Tracking of setpoint 6 for the hybrid microgrid (a) convergence of state variables x_1 to x_{20} to their reference setpoints, (b) control inputs u_1 to u_7 of the hybrid microgrid

(1.496) is the one that provides the control loop with maximum robustness. It is also pointed out that the use of the H-infinity Kalman Filter as a robust state estimator has allowed to implement feedback control by using measurement of a subset of the state vector elements of the hybrid microgrid.

The section's scientific contribution is outlined as follows: (i) the presented nonlinear optimal control method has improved performance when compared against other nonlinear control schemes that one can consider for the dynamic model of hybrid microgrids (such as Lie algebra-based control, differential flatness theory-based control, Model-based Predictive Control, Nonlinear Model-based Predictive Control, Sliding-mode control, Backstepping control, etc.), (ii) it achieves fast and accurate tracking of all reference setpoints for the hybrid microgrid under moderate variations of the control inputs, (iii) it minimizes the dispersion of energy in the implementation of control and in the functioning of the hybrid microgrid.

Chapter 5

Control based on approximate linearization for mechatronic systems

Abstract

The chapter analyzes methods and presents results on control and estimation based on approximate linearization for mechatronic systems: (i) nonlinear control of electrohydraulic actuators, (ii) nonlinear control of electropneumatic actuators, (iii) nonlinear control of hot-steel rolling mills, (iv) nonlinear control of paper mills, (v) nonlinear control of the injection moulding machine, (v) nonlinear control of the slosh-container system dynamics, (vi) nonlinear control of micro-satellites' attitude dynamics and (vi) nonlinear control of the industrial crystallization process.

5.1 Nonlinear control of electrohydraulic actuators

5.1.1 Outline

Electrohydraulic actuators are extensively used in industry, in robotics and in avionics, because of their capability to produce high power and torque (Guo, Zhang *et al.*, 2018; Guo, Wang and Liu, 2018; Guo, Sun *et al.*, 2016, Lyn, Chen *et al.*, 2019). Electrohydraulic actuators have a small size-to-power ratio, large force output and high power-to-weight ratio. In particular, they are used in positioning, transfer of heavy loads and lifting tasks (Guan and Pan, 2008; Kim, Won *et al.*, 2012; Guom, Yin *et al.*, 2017). The dynamic model of electrohydraulic actuators comprises three parts, namely the electrical, the hydraulic and the mechanical part (Kaddissi *et al.*, 2011; Mintsa *et al.*, 2012; Ding, Deng *et al.*, 2017). Their control problem exhibits particular difficulties due to nonlinearities of the related state-space model, as well as due to underactuation (Yang, Yao *et al.*, 2016; Sofiane, 2015; Dinh *et al.*, 2018). So far several control methods have been proposed for this type of actuators, such as model predictive control, sliding-mode control and Lyapunov theory-based adaptive control (Yang and Yao, 2019; Guo, Wang and Li, 2013; Pisaturo and Senatore, 2015; Shi, Wang *et al.*, 2015). Another problem that arises is the difficulty in measuring the complete set of state variables of the actuator, as for instance pressures in its forward and return chamber (Ali *et al.*, 2016; Guo, Wei *et al.*, 2016; Kim, Shin *et al.*, 2013; Sakain and Tsuji, 2017). Precision in the positioning of the load as well as fast convergence to the targeted positions are the primary objectives in the design of the actuator's control loop (Ren, Deng *et al.*, 2019; Li, Bai *et al.*, 2018; Li and Wang, 2018; Karpenko and Sepehri, 2008).

In the present section, the previously analyzed nonlinear optimal (H-infinity) control scheme is proposed for electrohydraulic actuators. First, the dynamic model of the actuator undergoes approximate linearization around a time-varying operating point which is recomputed at each time-step of the control algorithm. The temporary operating point is defined by the present value of the actuator's state vector and by the last sampled value of the control inputs vector. The linearization process makes use of first-order Taylor series expansion and of the computation of the related Jacobian matrices (Rigatos and Tzafestas, 2007; Basseville and Nikiforov, 1993; Rigatos and Zhang, 2009). For the approximately linearized model of the system an optimal (H-infinity) feedback controller is designed. This achieves the solution of the optimal control problem for the nonlinear actuator, under model uncertainty and external perturbations.

The H-infinity (optimal) controller represents also a min–max differential game taking place between (i) the control inputs which try to minimize a cost function that contains a quadratic term of the state-vector's error and (ii) the model uncertainty and external perturbation terms which try to maximize this cost function (Rigatos and Busawon, 2018; Rigatos, 2016b). For the computation of the feedback gains of the controller an algebraic Riccati equation is repetitively solved at each iteration of the control method (Rigatos, 2011; Rigatos, 2015). The stability properties of the control scheme are proven through Lyapunov analysis. First, it is demonstrated that the control loop satisfies the H-infinity tracking performance criterion, and this signifies elevated robustness against model uncertainty and external disturbances (Rigatos and Busawon, 2018; Tousssaint, Basar and Bullo, 2000). Moreover, under moderate conditions it is proven that the control scheme is globally asymptotically stable (Rigatos and Busawon, 2018; Rigatos, 2015). Finally, to implement state estimation-based feedback control without the need to measure the entire state vector of the electrohydraulic actuator, the H-infinity Kalman Filter is introduced as a robust state estimator (Rigatos and Busawon, 2018; Rigatos, 2015).

Due to the nonlinearity of electrohydraulic actuators the solution of the related control and estimation problems and the proof of global stability and robustness for such methods remains always a challenging task (Yu, Song *et al.*, 2020; Heilan *et al.*, 2020; Ding and Yao, 2020; Huang, An, Yang *et al.*, 2020; Huang, An, Liang *et al.*, 2020). The section presents one of the few existing approaches to the nonlinear optimal control problem of electrohydraulic actuation systems. The concept of the method is novel and no equivalent approaches can be found in the entire bibliography. Comparing to previous attempts to solve the H-infinity (optimal) control problem for nonlinear dynamical systems the submitted section's approach exhibits the following points of novelty: Preceding results on the use of H-infinity control to nonlinear dynamical systems were limited to the case of affine-in-the-input systems with drift-only dynamics and considered that the control inputs gain matrix is not dependent of the values of the system's state vector. Moreover, in these approaches the linearization was performed around points of the desirable trajectory whereas in the present section's control method the linearization points are related with the value of the state vector at each sampling instant as well as with the last sampled value of the control inputs vector. The Riccati equation which has been proposed for computing the feedback gains of the controller is novel, so is the presented global stability proof through Lyapunov analysis (Rigatos and Busawon, 2018).

5.1.2 Dynamic model of the electrohydraulic actuator

5.1.2.1 State-space model of the actuator

The dynamic model of the electrohydraulic actuator is depicted in Figure 5.1. With the use of a pump, hydraulic fluid flows under pressure into the cylinder's chambers through cylinder ports which are controlled by servo-valves. The dynamic model of the electrohydraulic actuator comprises the electrical, the mechanical and the hydraulic part. The electrical part is the servo-valve which controls the fluid dynamics into the cylinder's chambers. In turn, the valve is driven by an electrical input current. The displacement of the valve, together with the load's pressure, control the fluid dynamics in the two chambers of the cylinder which stand for the hydraulic part of the system. The mechanical part of the actuator is the moving piston and the load attached to it (Guo, Sun *et al.*, 2016).

To obtain the dynamic model of the one-cylinder electrohydraulic actuator, the following procedure is followed: The load flow Q_L of a servo-valve is described by

$$Q_L = C_d w x_v \sqrt{\frac{1}{\rho}(P_s - \mathrm{sign}(x_v))P_L} \tag{5.1}$$

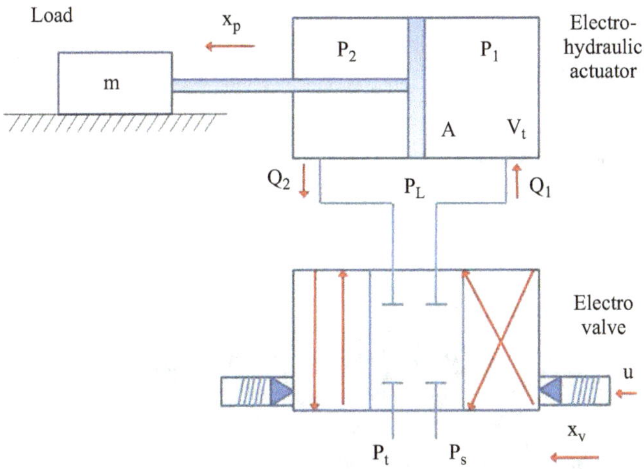

Figure 5.1 Diagram of the electrohydraulic actuator which is controlled by a servo-valve

where P_L is the load pressure of the cylinder, x_v is the position of the servo-valve, P_s is the supply pressure of the pump, C_d is the discharge coefficient, w is the area gradient of the valve spools and ρ is the viscosity of the flowing fluid.

The continuous-time flow-pressure equation of the hydraulic cylinder is

$$Q_L = A_P \frac{dx_P}{dt} + C_{tl} P_L + \frac{V_t}{4\beta_e} \frac{dP_L}{dt} \tag{5.2}$$

where x_P is the position of the cylinder's piston, C_{tl} is the coefficient of the total leakage of the cylinder, β_e is the effective bulk modulus, A_P is the effective area of the cylinder's chamber and V_t is the half volume of the cylinder.

By simplifying the main viscous function of the hydraulic oil and Coulomb friction, the mechanical dynamic equation driven by an hydraulic actuator can be written in the following form:

$$A_P P_L = m\ddot{x}_P + bx_P + kx_P + F_L \tag{5.3}$$

where m is the load mass, K is the load spring constant, b is the viscous function coefficient and F_L is the external load of the hydraulic actuator. Moreover, the relation between the valve's spool position and the input control voltage is described by

$$T_{sv}\dot{x}_v + x_v = k_{sv}u \tag{5.4}$$

k_{sv} is the gain of the control voltage u and T_{sv} is a response-time constant of the servo-valve. About the sign function which appears in (5.1) one can use the following approximation

$$\text{sign}(x_v) \simeq \tanh(kx_v) = \frac{e^{kx_v} - e^{-kx_v}}{e^{kx_v} + e^{-kx_v}} \tag{5.5}$$

By defining the state variables $\mathbf{x} = [x_1, x_2, x_3, x_4]^T = [x_P, \dot{x}_P, P_L, x_v]^T$ the state-space model of the electrohydraulic actuator becomes (Guo, Sun *et al.*, 2016)

$$\begin{aligned}
\dot{x}_1 &= x_2 \\
\dot{x}_2 &= \frac{1}{m}[-kx_1 - bx_2 + A_P x_3 - F_L] \\
\dot{x}_3 &= -\frac{4\beta_e A_P}{V_t} x_2 - \frac{4\beta_e C_{tl}}{V_t} x_3 + \frac{4\beta_e C_d w}{V_t\sqrt{\rho}} \sqrt{P_s - \tanh(kx_4)x_3 x_4} \\
\dot{x}_4 &= -\frac{1}{T_{sv}} x_4 + \frac{k_{sv}}{T_{sv}} u
\end{aligned} \tag{5.6}$$

where F_L is the force due to the load that is moved by the actuator. The previous state-space model can be also written in a vector-fields form

$$\begin{pmatrix}\dot{x}_1\\ \dot{x}_2\\ \dot{x}_3\\ \dot{x}_4\end{pmatrix}=\begin{pmatrix} x_2 \\ \frac{1}{m}[-kx_1-bx_2+A_Px_3-F_L] \\ \frac{4\beta_eA_P}{V_t}x_2-\frac{4\beta_eC_{tl}}{V_t}x_3+\frac{4\beta_eC_dw}{V_t\sqrt{\rho}}\sqrt{P_s-\tanh(kx_4)x_3x_4} \\ -\frac{1}{T_{sv}}x_4\end{pmatrix}+\begin{pmatrix}0\\0\\0\\ \frac{k_{sv}}{T_{sv}}\end{pmatrix}u \tag{5.7}$$

Equivalently, the model of the electrohydraulic actuator is written in the following concise form

$$\dot{x}=f(x)+g(x)u \tag{5.8}$$

where $x\in R^{4\times 1}, f(x)\in R^{4\times 1}, g(x)\in R^{4\times 1}$ and $u\in R$.

5.1.2.2 Differential flatness properties of the electrohydraulic actuator

The state-space model of the electrohydraulic actuator was defined in (5.6). It will be proven that this model is differentially flat, which signifies that (i) all its state variables and its control inputs can be written as differential functions of a subset of its state variables which constitute the flat outputs and (ii) the flat outputs of the system and their derivatives are differentially independent, which means that they are not connected through a relation in the form of a homogeneous differential equation (Rigatos, 2015).

In the previously defined model of the electrohydraulic actuator, the flat output is chosen to be the position of the piston, that is $y=x_P$. From the first row of the state-space model one has

$$x_2=\dot{x}_1\Rightarrow x_2=\dot{y} \tag{5.9}$$

which means that x_2 is a differential function of the flat output. Next from the second row of the state-space model one has

$$\begin{aligned} x_3&=\frac{1}{A_P}[m\dot{x}_2+kx_1+bx_2+F_L]\Rightarrow \\ x_3&=\frac{1}{A_P}[m\ddot{y}+ky+b\dot{y}+F_L]\end{aligned} \tag{5.10}$$

and considering that the load force F_L is constant one has that state variable x_3 is also a differential function of the flat output. From the third row of the state-space model one obtains that

$$\sqrt{P_s-\tanh(kx_4)x_3x_4}=\frac{1}{4\beta_eC_dw}[V_t\sqrt{\rho}\dot{x}_3+4\beta_eA_P\sqrt{\rho}x_2+4\beta_eC_{tl}\sqrt{\rho}x_3] \tag{5.11}$$

This is an equation that can be solved with respect to x_4 thus denoting that x_4 is finally a differential function of the system's flat output. Finally from the fourth row of the state-space model one has

$$u = \frac{T_{sv}\dot{x}_4 + x_4}{K_{sv}} \tag{5.12}$$

which signifies that the control input u is also a differential function of the system's flat output. This comes to complete the proof of the differential flatness properties of the electrohydraulic actuator. The differential flatness property allows to compute setpoints for the control loop of the actuator. Actually, one can define the setpoints for the flat output $y = x_1$ and next using (5.9)–(5.11) he can also define setpoints for the rest of the state variables of the system.

5.1.3 Approximate linearization of the electrohydraulic actuator's model

5.1.3.1 Approximate linearization of the state-space model

The dynamic model of the electrohydraulic actuator, being in the affine-in-the-input form $\dot{x} = f(x) + g(x)u$ undergoes approximate linearization after applying first-order Taylor series expansion. The linearization process requires the computation of the system's Jacobian matrices. It takes place around the temporary operating point (x^*, u^*), where x^* is the present value of the system's state vector and u^* is the last value of the control inputs vector that was applied on the actuator. The modelling error which is due to the omission of higher-order terms from the Taylor series is viewed as a perturbation that is asymptotically compensated by the robustness of the control algorithm. The linearization process results into a model of the form

$$\dot{x} = Ax + Bu + \tilde{d} \tag{5.13}$$

where $\tilde{d}$ is the cumulative vector of disturbances that may comprise: (i) the modelling error due to approximate linearization, (ii) exogenous perturbations and (iii) measurement noise of any distribution. Matrices A and B, stand for the Jacobian matrices of the Taylor series expansion, that is

$$\begin{aligned} A &= \nabla_x[f(x) + g(x)u] \mid_{(x^*,u^*)} \Rightarrow \\ A &= \nabla_x[f(x)] \mid_{(x^*,u^*)} + \nabla_x[g(x)]u \mid_{(x^*,u^*)} \Rightarrow \\ A &= \nabla_x[f(x)] \mid_{(x^*,u^*)} \end{aligned} \tag{5.14}$$

$$\begin{aligned} B &= \nabla_u[f(x) + g(x)u] \mid_{(x^*,u^*)} \Rightarrow \\ B &= g(x) \mid_{(x^*,u^*)} \end{aligned} \tag{5.15}$$

The computation of the Jacobian matrix $\nabla_x[f(x)] \mid_{(x^*,u^*)}$ proceeds as follows

First row of the Jacobian matrix $\nabla_x[f(x)] \mid_{(x^*,u^*)}$: $\frac{\partial f_1}{\partial x_1} = 0$, $\frac{\partial f_1}{\partial x_2} = 0$, $\frac{\partial f_1}{\partial x_3} = 0$ and $\frac{\partial f_1}{\partial x_4} = 0$.

Second row of the Jacobian matrix $\nabla_x[f(x)]\,|_{(x^*,u^*)}$: $\frac{\partial f_2}{\partial x_1}=-\frac{k}{m}$, $\frac{\partial f_2}{\partial x_2}=-\frac{b}{m}$, $\frac{\partial f_2}{\partial x_3}=\frac{A_P}{m}$, and $\frac{\partial f_2}{\partial x_4}=0$.

Third row of the Jacobian matrix $\nabla_x[f(x)]\,|_{(x^*,u^*)}$: $\frac{\partial f_3}{\partial x_1}=0$, $\frac{\partial f_3}{\partial x_2}=-\frac{4\beta_e A_P}{V_t}$, $\frac{\partial f_3}{\partial x_3}=-\frac{4\beta_e C_{tl}}{V_t}+\frac{4\beta_e C_d w}{V_t\sqrt{\rho}}(\frac{1}{2})\cdot\frac{1}{\sqrt{P_s-\tanh(kx_4)x_3}}x_4\cdot(-\tanh(kx_4))$ and $\frac{\partial f_3}{\partial x_4}=\frac{4\beta_e C_d w}{V_t\sqrt{\rho}}[(\frac{1}{2})\frac{1}{\sqrt{P_s-\tanh(kx_4)x_3}}x_4\cdot\frac{\partial\tanh(kx_4)}{\partial x_4}x_3+\sqrt{P_s-\tanh(kx_4)x_3}]$, where $\frac{\partial\tanh(kx_4)}{\partial x_4}=k(1-\tanh^2(kx_4))$.

Fourth row of the Jacobian matrix $\nabla_x[f(x)]\,|_{(x^*,u^*)}$: $\frac{\partial f_4}{\partial x_1}=0$, $\frac{\partial f_4}{\partial x_2}=0$, $\frac{\partial f_4}{\partial x_3}=0$ and $\frac{\partial f_4}{\partial x_4}=-\frac{1}{T_{sv}}$.

5.1.3.2 Stabilizing feedback control

After linearization around its current operating point, the electrohydraulic actuator's model is written in the form (Rigatos and Busawon, 2018)

$$\dot{x}=Ax+Bu+d_1 \tag{5.16}$$

Parameter d_1 stands for the linearization error in the electrohydraulic actuator's model appearing previously in (5.16). The reference setpoints for the electrohydraulic actuator's state vector are denoted by $\mathbf{x_d}=[x_1^d,\dots,x_4^d]$. Tracking of this trajectory is achieved after applying the control input u_d. At every time instant, the control input u_d is assumed to differ from the control input u appearing in (5.16) by an amount equal to Δu, that is $u_d=u+\Delta u$

$$\dot{x}_d=Ax_d+Bu_d+d_2 \tag{5.17}$$

The dynamics of the controlled system described in (5.16) can be also written as

$$\dot{x}=Ax+Bu+Bu_d-Bu_d+d_1 \tag{5.18}$$

and by denoting $d_3=-Bu_d+d_1$ as an aggregate disturbance term one obtains

$$\dot{x}=Ax+Bu+Bu_d+d_3 \tag{5.19}$$

By subtracting (5.17) from (5.19) one has

$$\dot{x}-\dot{x}_d=A(x-x_d)+Bu+d_3-d_2 \tag{5.20}$$

By denoting the tracking error as $e=x-x_d$ and the aggregate disturbance term as $\tilde{d}=d_3-d_2$, the tracking error dynamics becomes

$$\dot{e}=Ae+Bu+\tilde{d} \tag{5.21}$$

For the approximately linearized model of the system, a stabilizing feedback controller is developed. The controller has the form

$$u(t)=-Ke(t) \tag{5.22}$$

with $K = \frac{1}{r}B^T P$ where P is a positive definite symmetric matrix which is obtained from the solution of the Riccati equation (Rigatos and Busawon, 2018)

$$A^T P + PA + Q - P\left(\frac{2}{r}BB^T - \frac{1}{\rho^2}LL^T\right)P = 0 \tag{5.23}$$

where Q is a positive semi-definite symmetric matrix. The diagram of the considered control loop is depicted in Figure 5.2. The global stability properties of the control method can be proven following the procedure presented in Section 1.1.2.5.

It is also noted that the solution of the H-infinity feedback control problem for the electrohydraulic actuator and the computation of the worst-case disturbance that this controller can sustain, comes from superposition of Bellman's optimality principle when considering that the actuator is affected by two separate inputs (i) the control input u and (ii) the cumulative disturbance input $\tilde{d}(t)$. Solving the optimal control problem for u that is for the minimum variation (optimal) control input that achieves elimination of the state vector's tracking error gives $u = -\frac{1}{r}B^T Pe$. Equivalently, solving the optimal control problem for

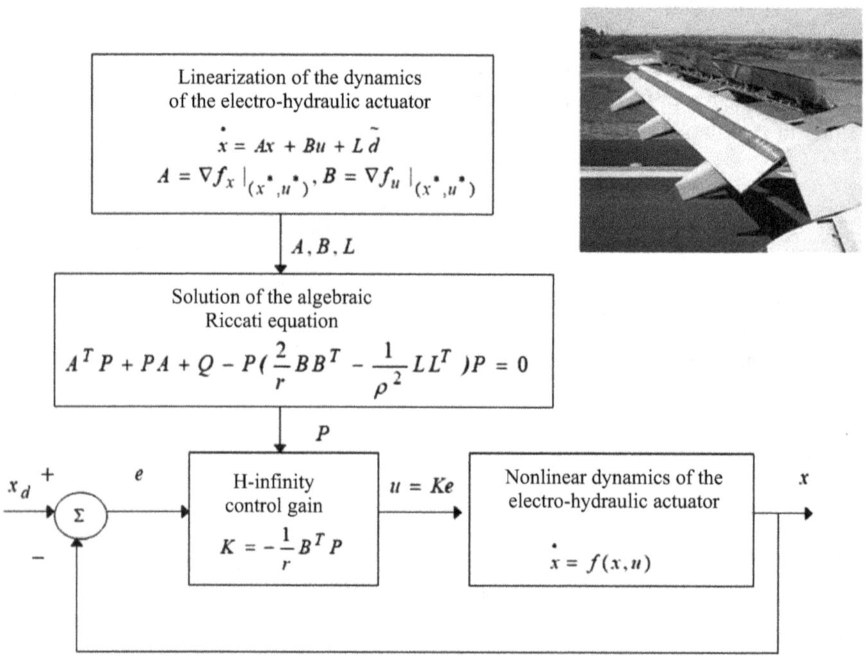

Figure 5.2 Diagram of the electrohydraulic actuator's nonlinear optimal (H-infinity) control loop

$\tilde{d}$, that is for the worst-case disturbance that the control loop can sustain gives $\tilde{d} = \frac{1}{\rho^2} L^T P e$.

5.1.4 Simulation tests

The performance of the proposed nonlinear optimal control method for the electro-hydraulic actuator has been tested through simulation experiments. To compute the H-infinity, controller's feedback gains the algebraic Riccati equation of (5.23) had to be solved at each iteration of the control method. The measurement units of the state variables of the model of the electrohydraulic actuator are x_1(m), x_2(m/s), x_3(bar) and x_4(mm). Indicative values for the parameters of this model are given in (Guo, Sun *et al.*, 2016). The ranges of variation of the state variables of the actuator are subject to scaling according to the values of the parameters of the system's model. The obtained results are depicted in Figures 5.3–5.8. It can be noticed that in all experiments fast convergence of the state variables to the related setpoints has been achieved, under moderate variations of the control inputs. Since only measurements of the actuator's position x_1 were considered to be available state estimation-based control has been applied. The H-infinity Kalman Filter has been used as a robust state estimator and in the simulation diagrams the real values of the state variables are printed in blue, the estimated values are plotted in green while the reference setpoints are depicted in red colour.

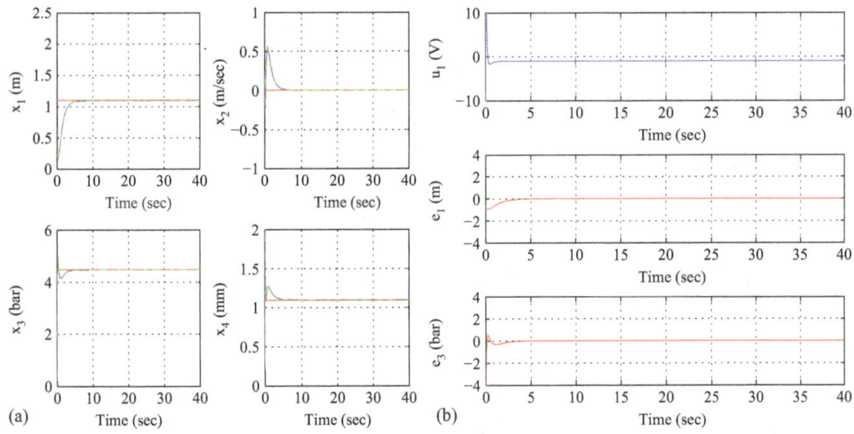

Figure 5.3 Tracking of setpoint 1 by the electrohydraulic actuator (a) state variables x_i i = 1,...,4 (blue lines), their estimated values (green lines) and the associated reference setpoints (red lines) and (b) control input u and tracking errors e_1 (state variable x_1) and e_3 (state variable x_3) for the electrohydraulic actuator

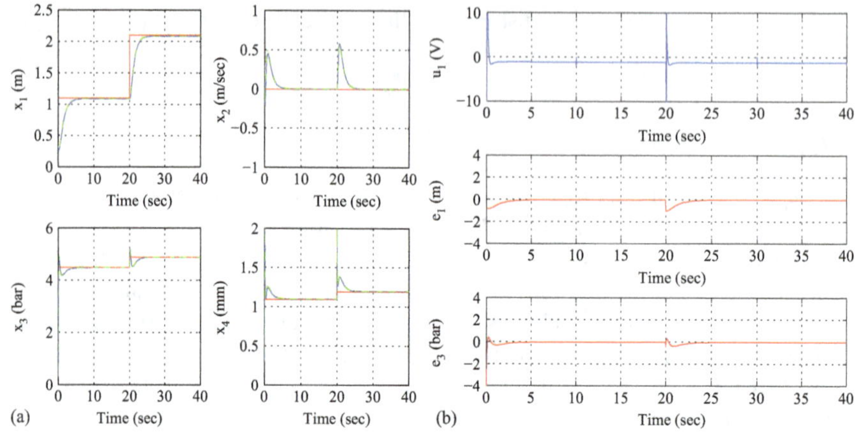

Figure 5.4 Tracking of setpoint 2 by the electrohydraulic actuator (a) state variables x_i $i = 1,\ldots,4$ (blue lines), their estimated values (green lines) and the associated reference setpoints (red lines) and (b) control input u and tracking errors e_1 (state variable x_1) and e_3 (state variable x_3) for the electrohydraulic actuator

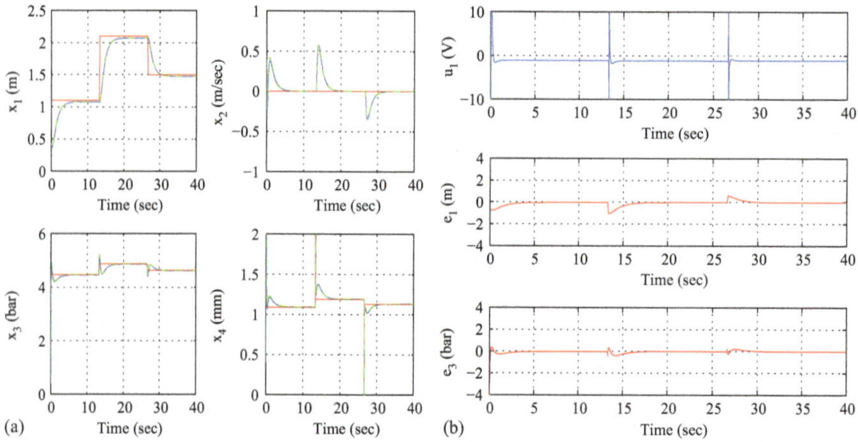

Figure 5.5 Tracking of setpoint 3 by the electrohydraulic actuator (a) state variables x_i $i = 1,\ldots,4$ (blue lines), their estimated values (green lines) and the associated reference setpoints (red lines) and (b) control input u and tracking errors e_1 (state variable x_1) and e_3 (state variable x_3) for the electrohydraulic actuator

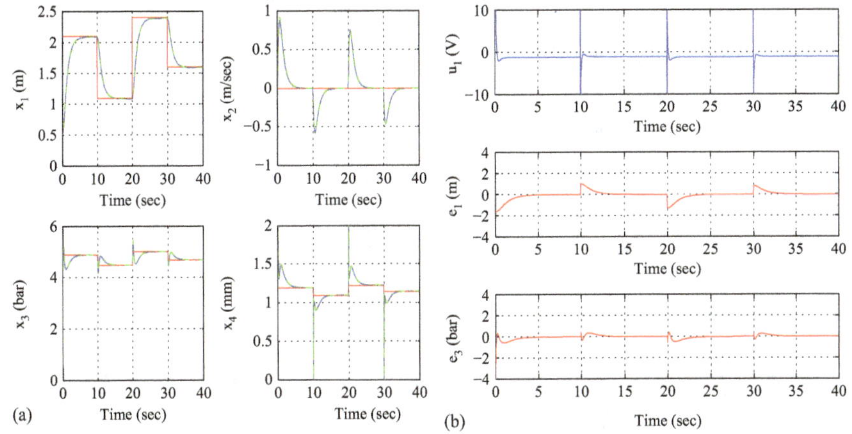

Figure 5.6 *Tracking of setpoint 4 by the electrohydraulic actuator (a) state variables x_i $i = 1,...,4$ (blue lines), their estimated values (green lines) and the associated reference setpoints (red lines) and (b) control input u and tracking errors e_1 (state variable x_1) and e_3 (state variable x_3) for the electrohydraulic actuator*

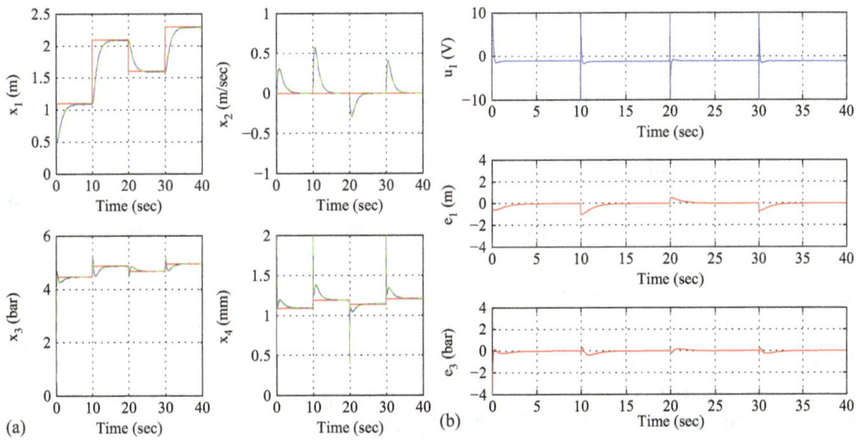

Figure 5.7 *Tracking of setpoint 5 by the electrohydraulic actuator (a) state variables x_i $i = 1,...,4$ (blue lines), their estimated values (green lines) and the associated reference setpoints (red lines) and (b) control input u and tracking errors e_1 (state variable x_1) and e_3 (state variable x_3) for the electrohydraulic actuator*

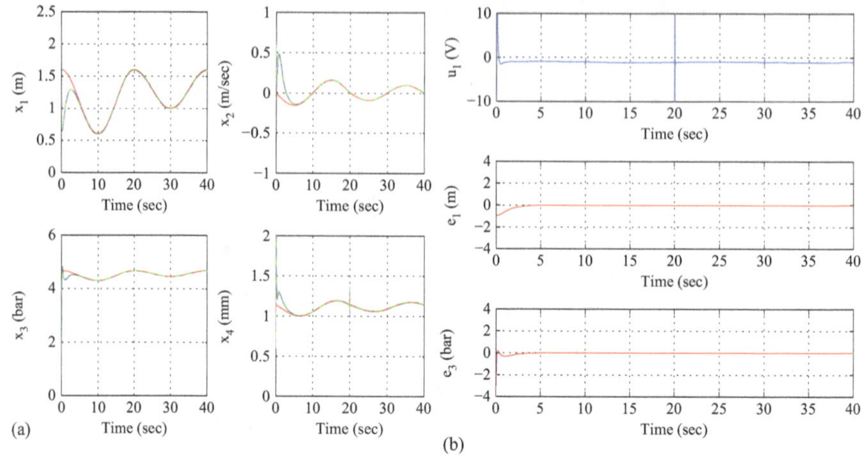

Figure 5.8 Tracking of setpoint 6 by the electrohydraulic actuator (a) state variables x_i $i = 1,\ldots,4$ (blue lines), their estimated values (green lines) and the associated reference setpoints (red lines) and (b) control input u and tracking errors e_1 (state variable x_1) and e_3 (state variable x_3) for the electrohydraulic actuator

5.2 Nonlinear control of electropneumatic actuators

5.2.1 Outline

Electro-pneumatic actuators are widely used in industrial actuation systems and in many mechatronics applications (Rigatos and Busawon, 2018; Rigatos, 2016b). They are frequently met in automotive systems as for instance in vehicles' clutches and brakes. They are also used in vehicles' combustion engines, for example in the control of turbochargers, exhaust gas recirculation and air intake manifolds (Ma, Zhu *et al.*, 2013; Schnidele *et al.*, 2012). Electro-pneumatic actuators exhibit several advantages when compared against electric or electro-hydraulic actuators (Yan, Estrada *et al.*, 2016; Plestan, Shtseel *et al.*, 2013). They are characterized by high output power to weight and size ratio (they use only air under pressure), easy maintenance, low cost and simple structure (Chalaga and Plestan, 2021; Smaoui, Brun *et al.*, 2006a; Smaoui, Brun *et al.*, 2006b). They are often used in rough and dusty environments, as well as in high force and pressure tasks (Shtessel *et al.*, 2012; Huber *et al.*, 2018; Zhou, Sun *et al.*, 2019). Their use can be also deployed in robotics and in precise positioning applications provided that their nonlinear dynamics and the associated nonlinear control problem are efficiently treated (Hodgsan *et al.*, 2012; Herzig *et al.*, 2018; Girin and Plestan, 2009). Linear control techniques such as MPC have been applied to electro-pneumatic actuators after considering locally linear dynamics and functioning around specific operating points (Tahoumi, Plestan *et al.*, 2021; Bone *et al.*, 2015; Chikh *et al.*, 2010). These techniques have been progressively overtaken by nonlinear control methods (Bouri and Thomasset, 2001; Ameur *et al.*, 2016; Michel, Selvarajan *et al.*, 2021). Thus one can distinguish the frequent use of feedback linearization (global linearization) methods, sliding mode control, backstepping control and nonlinear model predictive control techniques (Girin *et al.*, 2009; Abry *et al.*, 2016). State estimation-based control techniques have been also developed with the inclusion of state observers in the control loop of electropneumatic actuators (Lagrouche *et al.*, 2014; Estrada and Plestan, 2014). The implementation of nonlinear optimal control techniques for electro-pneumatic actuators is still an open and challenging problem (Chikh, Poignet *et al.*, 2010; Lagrouche *et al.*, 2006).

In this section, a nonlinear optimal control approach is proposed for electro-pneumatic actuators (Rigatos and Busawon, 2018; Rigatos, 2016b). The method relies on approximate linearization of the state-space model of these actuators around a temporary operating point which is updated at each sampling instance. The linearization is based on first-order Taylor series expansion and the computation of the Jacobian matrices of the state-space model of the system (Rigatos and Tzafestas, 2007; Basseville and Nikiforov, 1993; Rigatos and Zhang, 2009). The linearization point is defined at each time-step of the control algorithm by the present value of the actuator's state vector and by the last sampled value of the control inputs vector. The model imprecision which is induced by the truncation of higher-order terms in the Taylor series expansion is considered to be a perturbation that is asymptotically compensated by the robustness of the control method. For the approximately linearized model of the actuators a stabilizing H-infinity feedback controller is designed.

The proposed H-infinity controller provides a solution to the optimal control problem for electro-pneumatic actuators under model uncertainty and external perturbations (Rigatos and Busawon, 2018; Rigatos, 2016b). Actually, it represents a min–max differential game which takes place between (i) the control inputs that try to minimize a quadratic cost function of the state vector's tracking error and (ii) the model uncertainty and external perturbation terms which try to maximize this cost function. To compute the controller's feedback gains an algebraic Riccati equation is being repetitively solved at each iteration of the control algorithm (Rigatos, 2011; Rigatos, 2015). The global stability properties of the control scheme are proven through Lyapunov analysis. First, it is demonstrated that the control loop satisfies the H-infinity tracking performance criterion. This signifies the method's elevated robustness against model uncertainty and exogenous disturbances (Rigatos and Busawon, 2018; Rigatos, 2016b; Tousssaint, Basar and Bullo, 2000). Moreover, under moderate conditions the global asymptotic stability properties of the control scheme are proven. The proposed method retains the advantages of linear optimal control, that is fast and accurate tracking of reference setpoints under moderate variations of the control inputs. To implement state estimation-based control for electro-pneumatic actuators, without the need to measure the entire state vector of this system (for instance the velocity of moving load) the H-infinity Kalman Filter is used as a robust state estimator (Rigatos and Busawon, 2018; Rigatos, 2016b).

It is of worth to point out that preceding results on the use of H-infinity control to nonlinear dynamical systems were limited to the case of affine-in-the-input systems with drift-only dynamics and considered that the control inputs gain matrix is not dependent on the values of the system's state vector. Moreover, in these approaches the linearization was performed around points of the desirable trajectory whereas in the present section's control method the linearization points are related to the value of the state vector at each sampling instance as well as with the last sampled value of the control inputs vector. The Riccati equation which has been proposed for computing the feedback gains of the controller is novel, so is the presented global stability proof through Lyapunov analysis. The section's scientific contribution is outlined as follows: (i) the presented nonlinear optimal control method has improved or equally satisfactory performance when compared against other nonlinear control schemes that one can consider for the dynamic model of electro-hydraulic actuators (such as Lie algebra-based control, differential flatness theory-based control, Nonlinear Model-based Predictive Control, Sliding-mode control, Backstepping control), etc., (ii) it achieves fast and accurate tracking of all reference setpoints for the electro-pneumatic actuator under moderate variations of the control input and (iii) it minimizes the consumption of energy during the actuator's functioning.

5.2.2 Dynamic model of the electropneumatic actuator

An indicative diagram of an electropneumatic actuator is depicted in Figure 5.9. This electro-pneumatic actuator uses two proportional servo-distributors (servo-valves or solenoid valves) which inject pressurized air in the compartments of the cylinder or release air out of these compartments. The main part of the actuator is a cylinder that comprises two compartments (noted as P and N). These compartments have variable volume which is defined by the position of a sliding separator with a rod on it. The air pressure difference in the two chambers makes finally the rod move and in this manner actuation is given to the load which is attached to the free-end of the rod. The motion of the load's mass is defined by Newton's equations of motion subject to friction. The dynamic model of the electropneumatic actuator comprises the following equations (Chalaga and Plestan, 2021; Smaoui, Brun *et al.*, 2006a; Smaoui, Brun *et al.*, 2006b):

$$\begin{aligned}
\dot{y} &= v \\
\dot{v} &= \frac{1}{M}[S_p P_p - S_N P_N - bv - F_{ext}] \\
\dot{P}_p &= \frac{K_r T}{V_P(y)}\left[q_{mP}(u_P, P_p) - \frac{S_p}{rT} P_p v\right] \\
\dot{P}_N &= \frac{K_r T}{V_N(y)}\left[q_{mN}(u_N, P_N) + \frac{S_N}{rT} P_N v\right]
\end{aligned} \tag{5.24}$$

The state variables of the model are: y is the position of the piston, v is the velocity of the piston, P_P is the pressure at compartment P of the cylinder, P_N is the pressure at compartment N of the cylinder (Chalaga and Plestan, 2021; Smaoui, Brun *et al.*, 2006a; Smaoui, Brun *et al.*, 2006b).

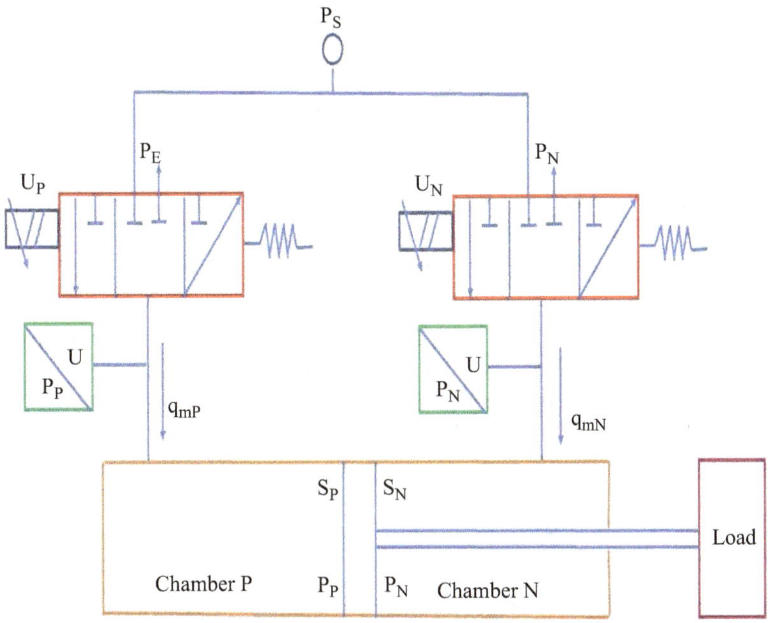

Figure 5.9 Diagram of the electropneumatic actuator

Other parameters of the model are: S_P is the effective surface of compartment P of the cylinder, S_N is the effective surface of compartment N of the cylinder, b is the friction coefficient, F_{ext} is the external force due to moving the load, K_p is a coefficient in the ideal gas equation, T is the temperature variable in the ideal gas equation, V_P is the effective volume of the P chamber at the zero position, V_N is te effective volume of the N chamber at the zero position, V_{DP} and V_{DN} are the dead volumes which are present at each extremity of the cylinder.

Defining the length of the cylinder as l, the following relations hold (Chalaga and Plestan, 2021; Smaoui, Brun *et al.*, 2006a; Smaoui, Brun *et al.*, 2006b):

$$\begin{aligned} V_P(y) &= V_P(0) + S_P y \ \text{ with } \ V_P(0) = V_{DP} + S_P \frac{l}{2} \\ V_N(y) &= V_N(0) + S_N y \ \text{ with } \ V_N(0) = V_{DN} + S_N \frac{l}{2} \end{aligned} \tag{5.25}$$

Besides $q_{mP}(u_P, P_P)$ is the mass flow rate in the P compartment of the cylinder and $q_{mN}(u_N, P_N)$ is the mass flow rate in the N compartment of the cylinder. Moreover, u_P is the input voltage of the electrovalve at the P compartment of the cylinder and u_N is the input voltage of the electrovalve at the N compartment of the cylinder. The analytical equations about the mass flow rates are (Chalaga and Plestan, 2021; Smaoui, Brun *et al.*, 2006b):

$$\begin{aligned} q_{mP}(u_P, P_P) &= \phi_P(P_P) + \psi_P(P_P, sgn(u_P))v \\ q_{mN}(u_N, P_N) &= \phi_N(P_N) + \psi_N(P_N, sgn(u_N))v \end{aligned} \tag{5.26}$$

$\phi_P(P_P)$ is a fifth order polynomial of pressure P_P, for instance $\phi_P(P_P) = c_0^a + c_1^a P_P + c_2^a P_P^2 + c_3^a P_P^3 + c_4^a P_P^4 + c_5^a P_P^5$. Besides $\psi_P(P_P, sgn(u_P))$ is a fifth-order polynomial of pressure P_P with coefficients which change with the sign of the control input u_P, for instance when $u_P > 0$, one has $\psi_P(P_P, sgn(u_P)) = c_0^b + c_1^b P_P + c_2^b P_P^2 + c_3^b P_P^3 + c_4^b P_P^4 + c_5^b P_P^5$ while when $u_P < 0$ one has $\psi_O(P_N, sgn(u_P)) = c_0^c + c_1^c P_P + c_2^c P_P^2 + c_3^c P_P^3 + c_4^c P_P^4 + c_5^c P_P^5$.

Equivalently $\phi_N(P_N)$ is a fifth order polynomial of pressure P_N, for instance $\phi_N(P_N) = d_0^a + d_1^a P_N + d_2^a P_N^2 + d_3^a P_N^3 + d_4^a P_N^4 + d_5^a P_N^5$. Besides $\psi_N(P_N, sgn(u_N))$ is a fifth-order polynomial of pressure P_N with coefficients which change with the sign of the control input u_N, for instance when $u_N > 0$, one has $\psi_N(P_N, sgn(u_N)) = d_0^b + d_1^b P_N + d_2^b P_N^2 + d_3^b P_N^3 + d_4^b P_N^4 + d_5^b P_N^5$ while when $u_N < 0$ one has $\psi_N(P_N, sgn(u_N)) = d_0^c + d_1^c P_N + d_2^c P_N^2 + d_3^c P_N^3 + d_4^c P_N^4 + d_5^c P_N^5$.

Using the state variables notation $x_1 = y, x_2 = v, x_3 = P_P$ and $x_4 = P_N$ the state-space model of the system becomes

$$\begin{aligned}
\dot{x}_1 &= x_2 \\
\dot{x}_2 &= \frac{1}{M}[S_p x_3 - S_N x_4 - b x_2 - F_{ext}] \\
\dot{x}_3 &= \frac{K_r T}{\left[V_{DP} + S_P \frac{l}{2} + S_P x_1\right]} \left[\phi_P(x_3) + \psi_P(x_3) u_1 - \frac{S_P}{rT} x_2 x_3\right] \\
\dot{x}_4 &= \frac{K_r T}{\left[V_{DN} + S_N \frac{l}{2} + S_N x_1\right]} \left[\phi_N(x_4) + \psi_N(x_4) u_1 + \frac{S_N}{rT} x_2 x_4\right]
\end{aligned} \tag{5.27}$$

or equivalently, in matrix form the state-space model is written as

$$\begin{pmatrix} \dot{x}_1 \\ \dot{x}_2 \\ \dot{x}_3 \\ \dot{x}_4 \end{pmatrix} = \begin{pmatrix} x_2 \\ \frac{1}{M}[S_P x_3 - S_N x_4 - b x_2 - F_{ext}] \\ \frac{K_r T[\phi_P(x_3) - \frac{S_P}{rT} x_2 x_3]}{[V_{DP} - S_P \frac{l}{2} + S_P x_1]} \\ \frac{K_r T[\phi_N(x_4) + \frac{S_N}{rT} x_2 x_4]}{[V_{DN} + S_N \frac{l}{2} + S_N x_1]} \end{pmatrix} + \begin{pmatrix} 0 & 0 \\ 0 & 0 \\ \frac{K_r T \psi_P(x_3, sgn(u_1))}{[V_{DP} + S_P \frac{l}{2} + S_P x_1]} & 0 \\ 0 & \frac{K_r T \psi_N(x_4, sgn(u_1))}{[V_{DN} + S_N \frac{l}{2} + S_N x_1]} \end{pmatrix} \begin{pmatrix} u_1 \\ u_2 \end{pmatrix} \tag{5.28}$$

Consequently, the model of the electropneumatic actuator is written in the affine-in-the-input state-space form

$$\dot{x} = f(x) + g(x)u \tag{5.29}$$

where $x \in R^{4\times1}$, $f(x) \in R^{4\times1}$, $g(x) \in R^{4\times2}$ and $u \in R^{2\times1}$. In the following, it will be considered that $F_{ext} = k_1 x_1$.

5.2.3 Approximate linearization of the model of the electropneumatic actuator

5.2.3.1 Approximately linearized model of the actuator

The dynamic model of the electropneumatic actuator undergoes approximate linearization around the temporary operating point (x^*, u^*) with the use of first-order Taylor series expansion and through the computation of the associated Jacobian matrices (Rigatos and Busawon, 2018; Rigatos, 2016b). The temporary operating point is re-defined at each sampling instance by the present value of the system's state vector and by the last sampled value of the control inputs vector. The linearization process gives

$$\dot{x} = Ax + Bu + \tilde{d} \tag{5.30}$$

where $\tilde{d}$ is the modelling error which can be due to: (i) truncation of higher-order terms in the Taylor series expansion, (ii) exogenous perturbations and (iii) sensor measurement noise of any distribution. Matrices A and B are the model's Jacobian matrices

$$\begin{aligned} A &= \nabla_x[f(x) + g(x)u] \mid_{(x^*,u^*)} \Rightarrow \\ A &= \nabla_x f(x) \mid_{(x^*,u^*)} + \nabla_x g_1(x) u_1 \mid_{(x^*,u^*)} + \nabla_x g_2(x) u_2 \mid_{(x^*,u^*)} \end{aligned} \tag{5.31}$$

$$B = \nabla_u[f(x) + g(x)u] \mid_{(x^*,u^*)} \Rightarrow B = g(x) \mid_{(x^*,u^*)} \tag{5.32}$$

Computation of the Jacobian matrix $\nabla_x f(x) \mid_{(x^*,u^*)}$:

First row of the Jacobian matrix $\nabla_x f(x) \mid_{(x^*,u^*)}$: $\frac{\partial f_1}{\partial x_1} = 0$, $\frac{\partial f_1}{\partial x_2} = 1$, $\frac{\partial f_2}{\partial x_3} = 0$, $\frac{\partial f_2}{\partial x_4} = 0$,.

Second row of the Jacobian matrix $\nabla_x f(x) \mid_{(x^*,u^*)}$: $\frac{\partial f_2}{\partial x_1} = -k_1$, $\frac{\partial f_2}{\partial x_2} = \frac{b}{M}$, $\frac{\partial f_2}{\partial x_3} = \frac{S_P}{M}$, $\frac{\partial f_2}{\partial x_4} = -\frac{S_N}{M}$.

Third row of the Jacobian matrix $\nabla_x f(x) \mid_{(x^*,u^*)}$:

The third row $f_3(x)$ is a fraction with numerator $f_{3,num} = K_r T[\phi_P(x_3) - \frac{S_P}{rT} x_2 x_3]$ and denominator $f_{3,den} = [V_{DP} - S_P \frac{l}{2} + S_P x_1]$. Thus:

$$\frac{\partial f_3}{\partial x_i} = \frac{\frac{\partial f_{3,num}}{\partial x_i} f_{3,den} - f_{3,num} \frac{\partial f_{3,den}}{\partial x_i}}{f_{3,den}^2} \quad i = 1, 2, 3, 4 \tag{5.33}$$

where it holds that

$$\frac{\partial f_{3,num}}{\partial x_1} = K_rT\left[\frac{\partial \phi_P(x_3)}{\partial x_1} - \frac{S_P}{rT}\frac{\partial}{\partial x_1}(x_2x_3)\right] = 0 \tag{5.34}$$

$$\frac{\partial f_{3,num}}{\partial x_2} = K_rT\left[\frac{\partial \phi_P(x_3)}{\partial x_2} - \frac{S_P}{rT}\frac{\partial}{\partial x_2}(x_2x_3)\right] = -K_rT\frac{S_P}{nT}x_3 \tag{5.35}$$

$$\begin{aligned}\frac{\partial f_{3,num}}{\partial x_3} &= K_rT\left[\frac{\partial \phi_P(x_3)}{\partial x_3} - \frac{S_P}{rT}\frac{\partial}{\partial x_3}(x_2x_3)\right]\\ &= K_rT\left[(c_1^a + 2c_2^a x_3 + 3c_3^a)x_3^2 + 4c_4^a x_3^3 + 5c_5^a x_5^4 - \frac{S_P}{rT}x_2\right]\end{aligned} \tag{5.36}$$

$$\frac{\partial f_{3,num}}{\partial x_4} = K_rT\left[\frac{\partial \phi_P(x_3)}{\partial x_4} - \frac{S_P}{rT}\frac{\partial}{\partial x_4}(x_2x_3)\right] = 0 \tag{5.37}$$

Additionally, it holds that $\frac{\partial f_{3,den}}{\partial x_1} = S_P$, $\frac{\partial f_{3,den}}{\partial x_2} = 0$, $\frac{\partial f_{3,den}}{\partial x_3} = 0$ and $\frac{\partial f_{3,den}}{\partial x_4} = 0$.

Fourth row of the Jacobian matrix $\nabla_x f(x) \mid_{(x^*,u^*)}$:

The fourth row $f_4(x)$ is a fraction with numerator $f_{4,num} = K_rT[\phi_N(x_4) + \frac{S_N}{rT}x_2x_4]$ and denominator $f_{3,den} = [V_{DN} + S_N\frac{l}{2} + S_Nx_1]$. Thus:

$$\frac{\partial f_4}{\partial x_i} = \frac{\frac{\partial f_{4,num}}{\partial x_i}f_{4,den} - f_{4,num}\frac{\partial f_{4,den}}{\partial x_i}}{f_{4,den}^2} \quad i = 1,2,3,4 \tag{5.38}$$

where it holds that

$$\frac{\partial f_{4,num}}{\partial x_1} = K_rT\left[\frac{\partial \phi_N(x_4)}{\partial x_1} - \frac{S_N}{rT}\frac{\partial}{\partial x_1}(x_2x_4)\right] = 0 \tag{5.39}$$

$$\frac{\partial f_{3,num}}{\partial x_2} = K_rT\left[\frac{\partial \phi_N(x_4)}{\partial x_2} - \frac{S_N}{rT}\frac{\partial}{\partial x_2}(x_2x_4)\right] = -K_rT\frac{S_N}{nT}x_4 \tag{5.40}$$

$$\frac{\partial f_{4,num}}{\partial x_3} = K_rT\left[\frac{\partial \phi_N(x_4)}{\partial x_3} - \frac{S_N}{rT}\frac{\partial}{\partial x_3}(x_2x_4)\right] = 0 \tag{5.41}$$

$$\begin{aligned}\frac{\partial f_{4,num}}{\partial x_4} &= K_rT\left[\frac{\partial \phi_N(x_4)}{\partial x_4} - \frac{S_N}{rT}\frac{\partial}{\partial x_4}(x_2x_4)\right]\\ &= K_rT\left[(d_1^a + 2d_2^a x_3 + 3d_3^a)x_3^2 + 4d_4^a x_3^3 + 5d_5^a x_5^4 + \frac{S_P}{rT}x_2\right]\end{aligned} \tag{5.42}$$

Additionally it holds that $\frac{\partial f_{4,den}}{\partial x_1} = S_N$, $\frac{\partial f_{4,den}}{\partial x_2} = 0$, $\frac{\partial f_{4,den}}{\partial x_3} = 0$ and $\frac{\partial f_{4,den}}{\partial x_4} = 0$.

Computation of the Jacobian matrix $\nabla_x g_1(x) \mid_{(x^*,u^*)}$:

First row of the Jacobian matrix $\nabla_x g_1(x)\mid_{(x^*,u^*)}$: $\frac{\partial g_{11}}{\partial x_1}=0$, $\frac{\partial g_{11}}{\partial x_2}=0$, $\frac{\partial g_{11}}{\partial x_3}=0$, and $\frac{\partial g_{11}}{\partial x_4}=0$.

Second row of the Jacobian matrix $\nabla_x g_1(x)\mid_{(x^*,u^*)}$: $\frac{\partial g_{21}}{\partial x_1}=0$, $\frac{\partial g_{21}}{\partial x_2}=0$, $\frac{\partial g_{21}}{\partial x_3}=0$, and $\frac{\partial g_{21}}{\partial x_4}=0$.

Third row of the Jacobian matrix $\nabla_x g_1(x)\mid_{(x^*,u^*)}$:

The third row $g_3(x)$ is a fraction with numerator $g_{3,num}=K_rT\psi_P(x_3,sgn(u_1))$ and denominator $g_{3,den}=[V_{DP}+S_P\frac{l}{2}+S_Px_1]$. Thus:

$$\frac{\partial g_{31}}{\partial x_i}=\frac{\frac{\partial g_{31,num}}{\partial x_i}g_{31,den}-g_{31,num}\frac{\partial g_{31,den}}{\partial x_i}}{g_{31,den}^2}\quad i=1,2,3,4 \tag{5.43}$$

It holds that

$$\frac{\partial g_{31,num}}{\partial x_1}=\frac{\partial}{\partial x_1}[K_rT\psi_P(x_3)]=0 \tag{5.44}$$

$$\frac{\partial g_{31,num}}{\partial x_2}=\frac{\partial}{\partial x_2}[K_rT\psi_P(x_3)]=0 \tag{5.45}$$

$$\begin{aligned}\frac{\partial g_{31,num}}{\partial x_3}&=\frac{\partial}{\partial x_3}[K_rT\psi_P(x_3)]\\&=K_rT[c_1^b+2c_2^bx_3+3d_3^bx-3^2+4c_4^bx_3^3+5c_5^bx_3^4]\ \text{if}\ u_P>=0\\&\quad K_rT[c_1^c+2c_2^cx_3+3d_3^cx-3^2+4c_4^cx_3^3+5c_5^cx_3^4]\ \text{if}\ u_P<0\end{aligned} \tag{5.46}$$

$$\frac{\partial g_{31,num}}{\partial x_4}=\frac{\partial}{\partial x_4}[K_rT\psi_P(x_3)]=0 \tag{5.47}$$

while it also holds $\frac{\partial g_{31,den}}{\partial x_1}=S_P$, $\frac{\partial g_{31,den}}{\partial x_2}=0$, $\frac{\partial g_{31,den}}{\partial x_3}=S_P$, and $\frac{\partial g_{31,den}}{\partial x_4}=0$.

Computation of the Jacobian matrix $\nabla_x g_2(x)\mid_{(x^*,u^*)}$:

First row of the Jacobian matrix $\nabla_x g_2(x)\mid_{(x^*,u^*)}$: $\frac{\partial g_{12}}{\partial x_1}=0$, $\frac{\partial g_{12}}{\partial x_2}=0$, $\frac{\partial g_{12}}{\partial x_3}=0$, and $\frac{\partial g_{12}}{\partial x_4}=0$.

Second row of the Jacobian matrix $\nabla_x g_2(x)\mid_{(x^*,u^*)}$: $\frac{\partial g_{22}}{\partial x_1}=0$, $\frac{\partial g_{22}}{\partial x_2}=0$, $\frac{\partial g_{22}}{\partial x_3}=0$, and $\frac{\partial g_{22}}{\partial x_4}=0$.

Third row of the Jacobian matrix $\nabla_x g_2(x)\mid_{(x^*,u^*)}$:

$$\frac{\partial g_{32}}{\partial x_i}=\frac{\frac{\partial g_{32,num}}{\partial x_i}g_{32,den}-g_{32,num}\frac{\partial g_{32,den}}{\partial x_i}}{g_{32,den}^2}\quad i=1,2,3,4 \tag{5.48}$$

Fourth row of the Jacobian matrix $\nabla_x g_2(x)\mid_{(x^*,u^*)}$:

The fourth row $g_4(x)$ is a fraction with numerator $g_{4,num} = K_r T\psi_N(x_4, sgn(u_1))$ and denominator $g_{4,den} = [V_{DN} + S_N\frac{l}{2} + S_N x_1]$. Thus

$$\frac{\partial g_{42}}{\partial x_i} = \frac{\frac{\partial g_{42,num}}{\partial x_i} g_{42,den} - g_{42,num}\frac{\partial g_{42,den}}{\partial x_i}}{g_{42,den}^2} \quad i = 1,2,3,4 \tag{5.49}$$

It holds that

$$\frac{\partial g_{42,num}}{\partial x_1} = \frac{\partial}{\partial x_1}[K_r T\psi_N(x_4)] = 0 \tag{5.50}$$

$$\frac{\partial g_{42,num}}{\partial x_2} = \frac{\partial}{\partial x_2}[K_r T\psi_N(x_4)] = 0 \tag{5.51}$$

$$\frac{\partial g_{42,num}}{\partial x_3} = \frac{\partial}{\partial x_3}[K_r T\psi_N(x_4)] = 0 \tag{5.52}$$

$$\begin{aligned}\frac{\partial g_{42,num}}{\partial x_4} &= \frac{\partial}{\partial x_4}[K_r T\psi_N(x_4)] \\ &= K_r T[d_1^b + 2d_2^b x_3 + 3d_3^b x_3^2 + 4d_4^b x_3^3 + 5d_5^b x_3^4] \ \text{if} \ u_N >= 0 \\ & \quad K_r T[c_1^c + 2c_2^c x_3 + 3d_3^c x_3^2 + 4d_4^c x_3^3 + 5d_5^c x_3^4] \ \text{if} \ u_N < 0\end{aligned} \tag{5.53}$$

while it also holds $\frac{\partial g_{42,den}}{\partial x_1} = S_N$, $\frac{\partial g_{42,den}}{\partial x_2} = 0$, $\frac{\partial g_{42,den}}{\partial x_3} = 0$, and $\frac{\partial g_{42,den}}{\partial x_4} = 0$.

5.2.3.2 Stabilizing feedback control

After linearization around its current operating point, the electropneumatic actuator's model is written in the form (Rigatos and Busawon, 2018; Rigatos, 2016b)

$$\dot{x} = Ax + Bu + d_1 \tag{5.54}$$

Parameter d_1 stands for the linearization error in the electropneumatic actuator's model appearing above in (5.54). The reference setpoints for the electropneumatic actuator's state vector are denoted by $\mathbf{x_d} = [x_1^d, \ldots, x_4^d]$. Tracking of this trajectory is achieved after applying the control input u_d. At every time instant, the control input u_d is assumed to differ from the control input u appearing in (5.54) by an amount equal to Δu, that is $u_d = u + \Delta u$

$$\dot{x}_d = Ax_d + Bu_d + d_2 \tag{5.55}$$

The dynamics of the controlled system described in (5.54) can be also written as

$$\dot{x} = Ax + Bu + Bu_d - Bu_d + d_1 \tag{5.56}$$

and by denoting $d_3 = -Bu_d + d_1$ as an aggregate disturbance term one obtains

$$\dot{x} = Ax + Bu + Bu_d + d_3 \tag{5.57}$$

By subtracting (5.55) from (5.57) one has

$$\dot{x} - \dot{x}_d = A(x - x_d) + Bu + d_3 - d_2 \tag{5.58}$$

By denoting the tracking error as $e = x - x_d$ and the aggregate disturbance term as $\tilde{d} = d_3 - d_2$, the tracking error dynamics becomes

$$\dot{e} = Ae + Bu + \tilde{d} \tag{5.59}$$

For the approximately linearized model of the system, a stabilizing feedback controller is developed. The controller has the form

$$u(t) = -Ke(t) \tag{5.60}$$

with $K = \frac{1}{r}B^TP$ where P is a positive definite symmetric matrix which is obtained from the solution of the Riccati equation (Rigatos and Busawon, 2018; Rigatos, 2016b)

$$A^TP + PA + Q - P(\tfrac{2}{r}BB^T - \tfrac{1}{\rho^2}LL^T)P = 0 \tag{5.61}$$

where Q is a positive semi-definite symmetric matrix. The diagram of the considered control loop is depicted in Figure 5.10. The global stability properties of the control method can be proven following the procedure presented in Section 1.1.2.5.

It is also noted that the solution of the H-infinity feedback control problem for the electropneumatic actuator and the computation of the worst-case disturbance that this controller can sustain, comes from superposition of Bellman's optimality principle when considering that the actuator is affected by two separate inputs (i) the control input u (ii) the cumulative disturbance input $\tilde{d}, (t)$. Solving the optimal control problem for u that is for the minimum variation (optimal) control input that achieves

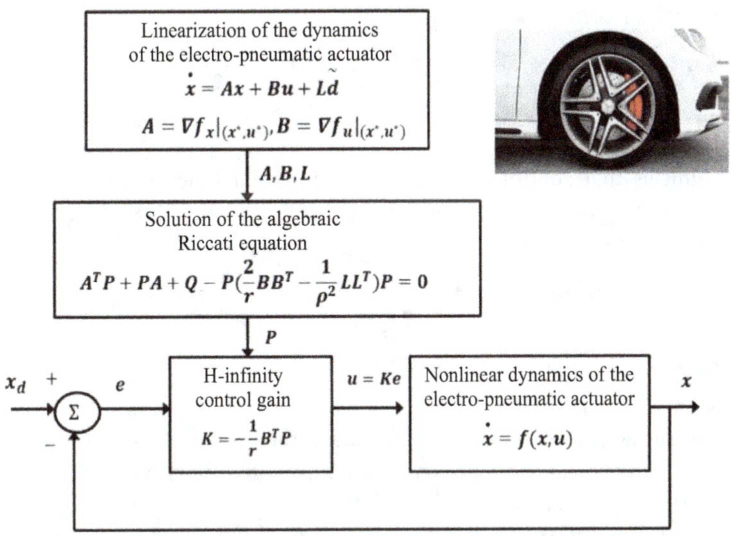

Figure 5.10 Diagram of the electropneumatic actuator's nonlinear optimal (H-infinity) control loop

elimination of the state vector's tracking error gives $u = -\frac{1}{r}B^T Pe$. Equivalently, solving the optimal control problem for $\tilde{d}$, that is for the worst-case disturbance that the control loop can sustain gives $\tilde{d} = \frac{1}{\rho^2}L^T Pe$.

5.2.4 Differential flatness properties of the electropneumatic actuator

5.2.4.1 Proof of differential flatness properties

It will be proven that the electropneumatic actuator is a differentially flat system with flat outputs $y_1 = x_1$ and $y_2 = x_3$ (Rigatos, 2015). From the first row of the state-space model, one has

$$\dot{x}_1 = x_2 \Rightarrow x_2 = \dot{y}_1 \Rightarrow x_2 = h_2(y, \dot{y}) \tag{5.62}$$

which demonstrates that x_2 is a differential function of the system's flat outputs. From the second row of the state-space model and considering that $F_{ext} = kx_1$ one has

$$\begin{aligned}
\dot{x}_2 &= \frac{1}{M}[S_P x_3 - S_N x_4 - bx_2 - kx_1] \Rightarrow \\
x_4 &= -\frac{1}{S_N}[M\dot{x}_2 - S_P x_3 - bx_2 + kx_1] \Rightarrow \\
x_4 &= -\frac{1}{S_N}[M\dot{h}_2(y, \dot{y}) - S_P y_2 - bh_2(y, \dot{y}) - ky_1] \Rightarrow \\
x_4 &= h_4(y, \dot{y}, \ddot{y})
\end{aligned} \tag{5.63}$$

Consequently, state variable x_4 is a differential function of the system's flat outputs. Next, from the third row of the state-space model of the system one solves with respect to the control input u_1. It holds that

$$\begin{aligned}
u_1 &= \frac{\frac{1}{K_r T}\dot{x}_3[V_{DP} + S_P\frac{l}{2} + S_P x_1] - \psi_P(x_3) + \frac{S_P}{rT}x_2 x_3}{\psi_P(x_3)} \Rightarrow \\
u_1 &= h(y, \dot{y})
\end{aligned} \tag{5.64}$$

which signifies that control input u_1 is a differential function of the system's flat outputs. From the fourth row of the state-space model one solves with respect to the flat output u_2. It holds that

$$\begin{aligned}
u_2 &= \frac{\frac{1}{K_r \cdot T}\dot{x}_4[V_{DP} + S_N\frac{l}{2} + S_M x_1] - \phi_N(x_4) - \frac{S_N}{rT}x_2 x_4}{\psi_N(x_4)} \Rightarrow \\
u_2 &= h_{u_2}(y, \dot{y}, \ddot{y}, y^{(3)})
\end{aligned} \tag{5.65}$$

Consequently, the control input u_2 is a differential function of the system's flat outputs. As a result of the previous analysis, all state variables and the control inputs of the electropneumatic actuator can be written as differential functions of the system's flat outputs and the system is differentially flat.

By using the differential flatness properties of the system one can define setpoints for the flat outputs x_1 and x_3 and next can use the relations $x_2 = h_2(y, \dot{y})$ and $x_4 = h_4(y, \dot{y}, \ddot{y})$ to compute setpoints for state variables x_2 and x_4 as well.

5.2.4.2 Flatness-based stabilization of the electropneumatic actuator

Besides, by knowing that the system is differentially flat, it is confirmed that it can be written in the input–output linearized form and consequently into the canonical Brunovsky form. By differentiating successively the flat output $y_1 = x_1$ one has

$$\begin{aligned}
&\dot{x}_1 = x_2 \Rightarrow \ddot{x}_1 = \dot{x}_2 \Rightarrow \\
&\ddot{x}_1 = \frac{1}{M}[S_P x_3 - S_N x_4 - b x_2 - K x_1] \Rightarrow \\
&x_1^{(3)} = \frac{1}{M}[S_P \dot{x}_3 - S_N \dot{x}_4 - b\dot{x}_2 - K\dot{x}_1]
\end{aligned} \tag{5.66}$$

thus one obtains

$$\begin{aligned}
x_1^{(3)} = &\frac{1}{M}\left\{\frac{S_P K_r T}{[V_{DP} + S_P \frac{l}{2} + S_P x_1]}\left[\phi_P(x_3) + \psi_P(x_3)u_1 - \frac{S_P}{nT}x_2 x_3\right]\right\} \\
&-\frac{1}{M}\left\{\frac{S_N K_r T}{[V_{DN} + S_N \frac{l}{2} + S_N x_1]}\left[\phi_N(x_4) + \psi_N(x_4)u_2 + \frac{S_N}{nT}x_2 x_4\right]\right\} \\
&-\frac{b}{M^2}\left[S_P x_3 - S_N x_4 - b x_2 - k x_1 - \frac{k}{M}x_2\right]
\end{aligned} \tag{5.67}$$

By regrouping terms, the previous relation is rewritten as

$$\begin{aligned}
x_1^{(3)} = &\frac{1}{M}\left\{\frac{S_P K_r T}{[V_{DP} + S_P \frac{l}{2} + S_P x_1]}\left[\phi_P(x_3) - \frac{S_P}{nT}x_2 x_3\right]\right\} \\
&-\frac{1}{M}\left\{\frac{S_N K_r T}{[V_{DN} + S_N \frac{l}{2} + S_N x_1]}\left[\phi_N(x_4) + \frac{S_N}{nT}x_2 x_4\right]\right\} \\
&-\frac{b}{M^2}\left[S_P x_3 - S_N x_4 - b x_2 - k x_1 - \frac{k}{M}x_2\right] \\
&+\frac{1}{M}\left\{\frac{S_P K_r T}{[V_{DP} + S_P \frac{l}{2} + S_P x_1]}[\psi_P(x_3)u_1]\right\} \\
&-\frac{1}{M}\left\{\frac{S_N K_r T}{[V_{DN} + S_N \frac{l}{2} + S_N x_1]}[\psi_N(x_4)u_2]\right\}
\end{aligned} \tag{5.68}$$

Consequently, it holds that

$$x_1^{(3)} = f_a(x) + g_{a_1}(x)u_1 + g_{a_2}(x)u_2 \tag{5.69}$$

Additionally, from the third row of the state-space model one gets

$$\dot{x}_3 = \frac{K_r T[\phi_P(x_3) - \frac{S_P}{rT} x_2 x_3]}{[V_{DP} + S_P \frac{l}{2} + S_P x_1]} + \frac{K_r T \psi_P(x_3, sgn(u_1))}{[V_{DP} + S_P \frac{l}{2} + S_P x_1]} u_1 \tag{5.70}$$

which is also written as

$$\dot{x}_3 = f_b(x) + g_{b_1}(x)u_1 + g_{b_2}(x)u_2 \tag{5.71}$$

Therefore, the state-space model of the electropneumatic actuator can be written in the input–output linearized form

$$\begin{aligned} x_1^{(3)} &= f_a(x) + g_{a_1}(x)u_1 + g_{a_2}(x)u_2 \\ \dot{x}_3 &= f_b(x) + g_{b_1}(x)u_1 + g_{b_2}(x)u_2 \end{aligned} \tag{5.72}$$

or in matrix form

$$\begin{pmatrix} x_1^{(3)} \\ \dot{x}_3 \end{pmatrix} = \begin{pmatrix} f_a(x) \\ f_b(x) \end{pmatrix} + \begin{pmatrix} g_{a_1}(x) & g_{a_2}(x) \\ g_{b_1}(x) & g_{b_2}(x) \end{pmatrix} \begin{pmatrix} u_1 \\ u_2 \end{pmatrix} \tag{5.73}$$

Next, by defining the virtual control inputs

$$\begin{aligned} v_1 &= f_1(x) + g_{a_1}(x)u_1 + g_{a_2}(x)u_2 \\ v_2 &= f_2(x) + g_{b_1}(x)u_1 + g_{b_2}(x)u_2 \end{aligned} \tag{5.74}$$

the input–output linearized form of the system is written as

$$x_1^{(3)} = v_1 \quad \dot{x}_3 = v_2 \tag{5.75}$$

The following stabilizing feedback control can be applied:

$$\begin{aligned} v_1 &= x_{1,d}^{(3)} - k_{11}(\ddot{x}_1 - \ddot{x}_{1,d}) - k_{21}(\dot{x}_1 - \dot{x}_{1,d}) - k_{31}(x_1 - x_{1,d}) \\ v_2 &= \ddot{x}_{3,d} - k_{12}(x_3 - x_{3,d}) \end{aligned} \tag{5.76}$$

Consequently, the dynamics of the closed-loop system becomes

$$\begin{aligned} &(x_1^{(3)} - x_{1,d}^{(3)}) + k_{11}(\ddot{x}_1 - \ddot{x}_{1,d}) + k_{21}(\dot{x}_1 - \dot{x}_{1,d}) + k_{31}(x_1 - x_{1,d}) = 0 \\ &(\dot{x}_3 - \dot{x}_{3,d}) + k_{12}(x_3 - x_{3,d}) = 0 \end{aligned} \tag{5.77}$$

Therefore, by defining the tracking error variables as $e_1 = x_1 - x_{1,d}$ and $e_3 = x_3 - x_{3,d}$ the tracking error dynamics becomes

$$\begin{aligned} &e_1^{(3)} + k_{11}\ddot{e}_1 + k_{21}\dot{e}_1 + k_{31}e_1 = 0 \\ &\dot{e}_3 + k_{12}e_3 = 0 \end{aligned} \tag{5.78}$$

Consequently, by selecting the above given feedback gains k_{11}, k_{21}, k_{31} and k_{12} so as the associated characteristic polynomials

$$\begin{aligned} p_1(s) &= \lambda_1^3 + k_{11}\lambda_1^2 + k_{21}\lambda + k_{31} \\ p_2(s) &= \lambda_2 + k_{12} \end{aligned} \tag{5.79}$$

to be Hurwitz stable, one has $\lim_{t\to\infty} e_1(t) = 0$ and $\lim_{t\to\infty} e_3(t) = 0$ thus, the closed-loop system is globally asymptotically stable. Using (5.74), the control inputs that should be applied to the initial nonlinear dynamics of the electropneumatic actuator, in the flatness-based control approach are

$$\begin{pmatrix} u_1 \\ u_2 \end{pmatrix} = \begin{pmatrix} g_{a_1}(x) & g_{a_2}(x) \\ g_{b_1}(x) & g_{b_2}(x) \end{pmatrix}^{-1} \cdot \left\{ \begin{pmatrix} v_1 \\ v_2 \end{pmatrix} - \begin{pmatrix} f_1(x) \\ f_2(x) \end{pmatrix} \right\} \tag{5.80}$$

5.2.5 Simulation tests

The efficiency of the proposed nonlinear optimal (H-infinity) control scheme for electropneumatic actuators has been tested through simulation experiments. The simulation tests have been carried out in Matlab and the method's algebraic Riccati equation has been solved using function aresolv(). The parameters of the model of the electropneumatic actuator have been chosen according to (Chalaga and Plestan, 2021; Smaoui, Brun *et al.*, 2006a; Smaoui, Brun *et al.*, 2006b). The sampling period was $T_s = 0.01$ s. The proposed control method is computationally efficient since the time interval which is needed for solving this Riccati equation and for computing the stabilizing feedback is much shorter than the sampling period. To implement this control method the algebraic Riccati equation of (5.61) had to be solved at each time-step of the algorithm. Besides, with the use of the H_∞ Kalman Filter, state estimation-based control has been applied. State variables x_2 was not directly measurable and had to be estimated with the use of this filter. The obtained results are depicted in Figure 5.11–5.16. The real values of the state variables are depicted in blue, the estimated values of the state variables are printed in green while the related setpoints are plotted in red. It can be observed that in all test cases fast and accurate tracking of the reference setpoints was achieved under moderate variations of the control inputs.

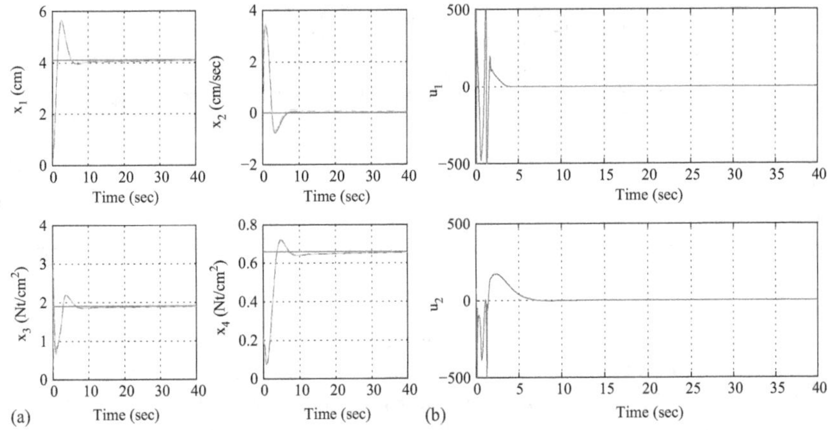

Figure 5.11 Tracking of setpoint 1 by the electropneumatic actuator (a) convergence of the state variables x_1 to x_4 (blue lines) to the reference setpoints (red lines) and their KF-based estimation (green line) and (b) control inputs u_1 and u_2 generated by the servo-distributors

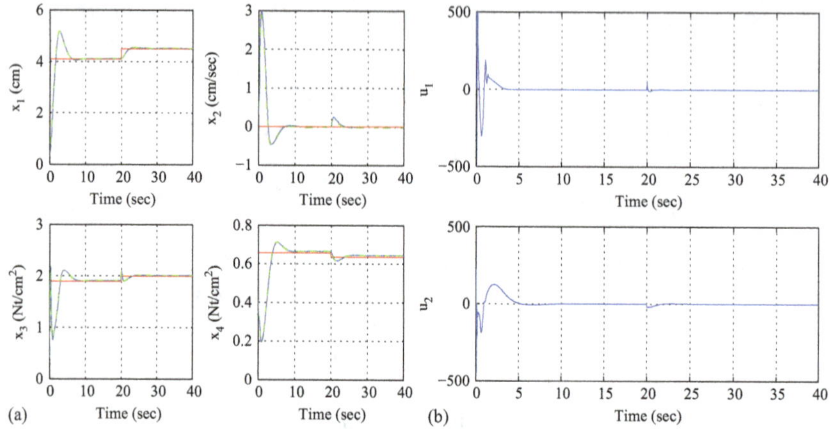

Figure 5.12 *Tracking of setpoint 2 by the electropneumatic actuator (a) convergence of the state variables* x_1 *to* x_4 *(blue lines) to the reference setpoints (red lines) and their KF-based estimation (green line) and (b) control inputs* u_1 *and* u_2 *generated by the servo-distributors*

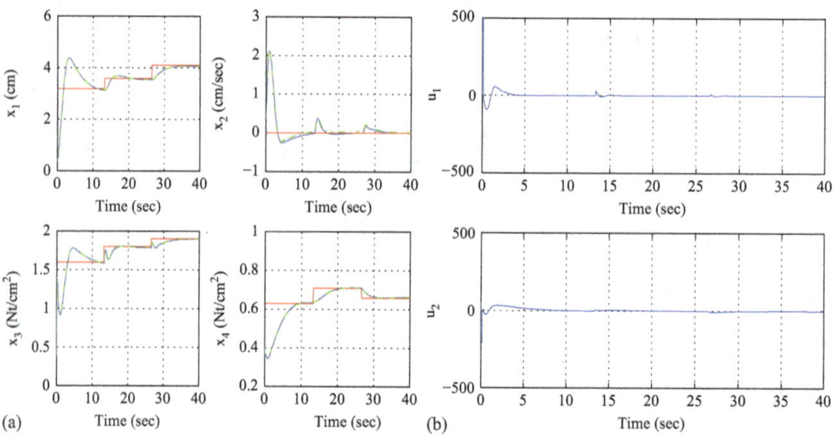

Figure 5.13 *Tracking of setpoint 3 by the electropneumatic actuator (a) convergence of the state variables* x_1 *to* x_4 *(blue lines) to the reference setpoints (red lines) and their KF-based estimation (green line) and (b) control inputs* u_1 *and* u_2 *generated by the servo-distributors*

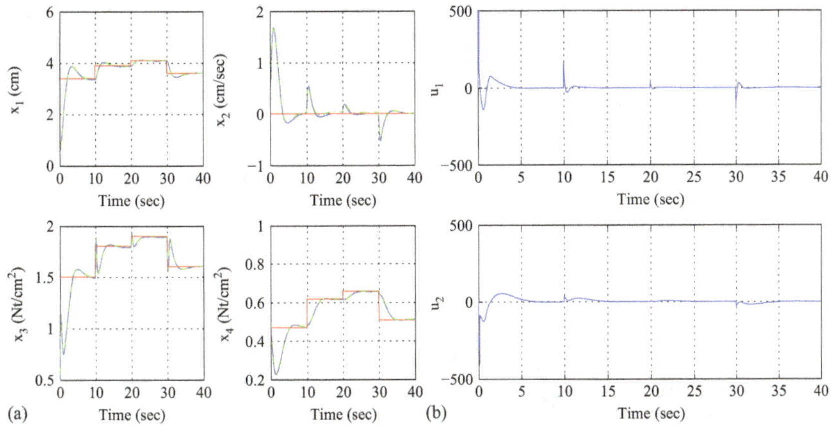

Figure 5.14 Tracking of setpoint 4 by the electropneumatic actuator (a) convergence of the state variables x_1 *to* x_4 *(blue lines) to the reference setpoints (red lines) and their KF-based estimation (green line) and (b) control inputs* u_1 *and* u_2 *generated by the servo-distributors*

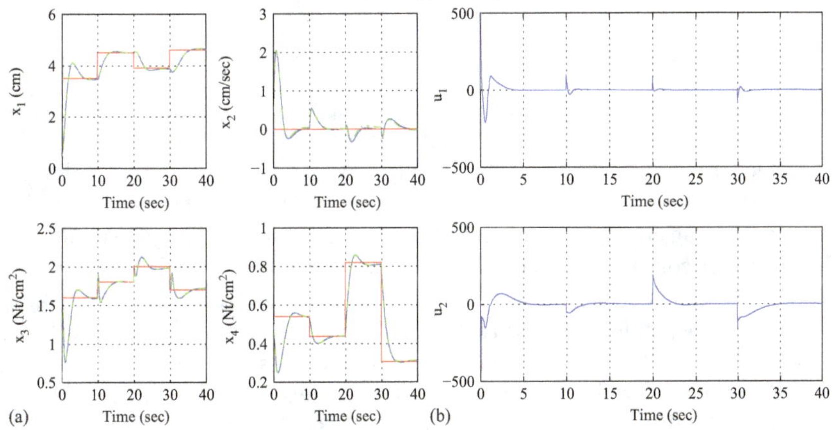

Figure 5.15 Tracking of setpoint 5 by the electropneumatic actuator (a) convergence of the state variables x_1 *to* x_4 *(blue lines) to the reference setpoints (red lines) and their KF-based estimation (green line) and (b) control inputs* u_1 *and* u_2 *generated by the servo-distributors*

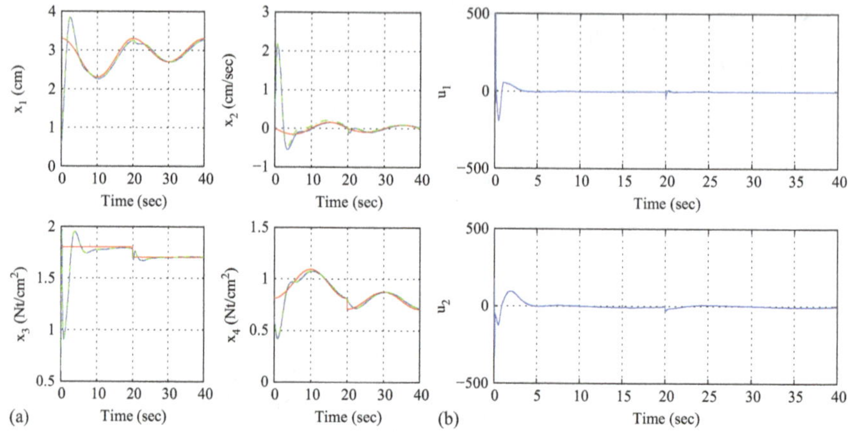

Figure 5.16 Tracking of setpoint 6 by the electropneumatic actuator (a) convergence of the state variables x_1 to x_4 (blue lines) to the reference setpoints (red lines) and their KF-based estimation (green line) and (b) control inputs u_1 and u_2 generated by the servo-distributors

5.3 Nonlinear control of hot-steel rolling mills

5.3.1 Outline

Control of hot steel rolling mills is of major importance for heavy industries (e.g. the automotive and ship-building industry) or for construction works since it allows to produce metal strips or plates of variable thickness and high strength (Zhong and Wang, 2011; Orcajo *et al.*, 2016; Orcajo *et al.*, 2018). Hot-steel rolling mills are classified among energy intensive industries and a major objective is to minimize the electric power consumed by them. The aim of the hot strip finishing mill process is to roll transfer bars, produced in a rough milling process (around 30 mm thick), into steel strips of variable thickness (between 0.8 mm and 20 mm) (Steinboeck *et al.*, 2014; Hou *et al.*, 2018). The reduction of the thickness is achieved by moving the metal strips through rolling strands where pressure is exerted (Ma, Li *et al.*, 2018; Yang and Zhang, 2009; Pedersen and Wittenmark, 1998). Besides, it is possible to apply interstrand pressure on the metal strips through their contact with a stiff pole which is called looper and which rotates around a pivot position (Hesketh *et al.*, 1998; Furlan *et al.*, 2008; Noh, Wan *et al.*, 2012). Actually, by varying the looper's angle with respect to the baseline the motion of the metal strip is uplifted and variable tension is applied on it (Hashimoto *et al.*, 2012; Geddes and Postlethwaite, 1998; Park, Noh and Wen, 2009). As a result, one can control more precisely the finishing of the processed strip. Control of hot steel rolling mills with a variable angle looper is a nontrivial nonlinear control problem (Yildiz, Forbes *et al.*, 2003; Fuchshumer *et al.*, 2005; Ozaki, Ohesuka *et al.*, 2009). The difficulties one has to overcome are related with the strongly nonlinear multivariable model of the looper's system and the inability to obtain directly measurements of the tension (Zhang, Zhou *et al.*, 2014; Grabmaix *et al.*, 2003; Schausberger *et al.*, 2018).

The outputs of the system are taken to be the turn angle of the looper and the tension that is exerted by the looper on the metal strips. The control inputs of the system are the speed (mass flow rate) of the metal strips in the interstand area and the torque that is developed by the actuator which makes the looper rotate. Model-based approaches to the related control problem make use of global or approximate linearization methods (Galicki, 2017; Dong, Liu *et al.*, 2010; Yildiz, Huang *et al.*, 2008). Moreover, in certain cases model-free control of the rolling mill system has been attempted (Xing and Chen, 2013; Liu, Chai *et al.*, 2011; Sbarbaro-Hofer *et al.*, 1993). In this section a nonlinear optimal control method is developed for the hot-steel roll milling system (Rigatos and Busawon, 2018; Rigatos, 2016b; Rigatos, 2015). The nonlinear dynamic model of the system undergoes first approximate linearization around a temporary operating point that is updated at every time-step for the control method (Rigatos and Siano, 2015; Rigatos, Siano *et al.*, 2015; Rigatos, Siano and Cecati, 2015). This linearization point consists of the present value of the state vector of the rolling mill and of the last sampled value of the control inputs vector. The linearization relies on Taylor series expansion and on the computation of the associated Jacobian matrices. For the linearized model of the rolling mill a stabilizing H-infinity feedback controller is designed (Rigatos and Tzafestas, 2007; Basseville and Nikiforov, 1993; Rigatos and Zhang, 2009).

The proposed H-infinity controller stands for the solution of the optimal control problem under model uncertainty and external perturbations. The controller represents the solution of a min–max differential game in which the control inputs try to minimize a cost function that comprises quadratic terms of the system's state vector error, whereas the model uncertainty and disturbance terms try to minimize this cost function. To compute the controller's feedback gains an algebraic Riccati equation has to be repetitively solved at each iteration of the control algorithm. The stability properties of the control scheme are proven through Lyapunov analysis. First, it is demonstrated that the control method satisfies the H-infinity tracking performance criterion, which signifies elevated robustness against model uncertainty and external perturbations (Tousssaint, Basar and Bullo, 2000; Lublin and Athans, 1995). Moreover, under moderate conditions is also proven that the control loop is globally asymptotically stable. This confirms that the tracking error for all state variables of the system is asymptotically eliminated. Finally, to implement state estimation-based control without the need to measure the entire state vector of the system, the H-infinity Kalman Filter is used as a robust state estimator (Gibbs, 2012; Simon, 2006).

The topic of control and optimization of the hot-steel rolling mill process is of interest for several industrial sectors and so far several efforts have been done towards its solution (Shi, Li *et al.*, 2012; Park, Jun and Won, 2010; Li and Janabi-Sharifi, 2009; Rossiter and Fleming, 2008; Choi *et al.*, 2007; Yin, Zhang *et al.*, 2017). The section offers one of the most effective solutions to the nonlinear optimal control problem of the hot-steel rolling mill system. Popular approaches for industrial control such as MPC or NMPC may have questionable performance when applied to this control problem. Actually, MPC has been developed for linear dynamical systems and its use in the case of the nonlinear model of the hot-steel rolling mill will risk the control loop's destabilization. Besides, the convergence of NMPC is not assured either. The convergence of the method's iterative search for an optimum depends on initialization and specific parameters' selection, therefore under NMPC one cannot always guarantee a solution for the nonlinear optimal control problem of the hot-steel rolling mill process. Finally, comparing to local models-based optimal control the section's approach exhibits specific advantages: (i) in the local-models based approach linearization is performed around multiple operating points (equilibria) which are selected off-line and which are not updated in time, whereas in the section's approach there is linearization only around one single operating which is updated at each iteration of the control algorithm, (ii) in the local-models approach there is a need to perform solution of multiple Riccati equations associated with the individual models and this solution is performed offline. On the other side, in the section's approach there is a need to solve one single Riccati equation and this solution is repeated at each time-step of the control algorithm and (iii) in local models-based control there is a need to find a common solution for the individual Riccati equations, and one cannot assure that such a solution always exists. On the other side, in the section's approach there is a need to obtain solution for one single Riccati equation and the existence of such a solution can be assured through suitable selection of gains and coefficients. In conclusion, comparing to local models-based control, the section's control method is computationally more efficient and is subject to less constraining assumptions.

5.3.2 Dynamic model of the hot-steel rolling mill

The diagram of the hot-steel rolling mill system is given in Figure 5.17. The following parameters can be defined (Zhong and Wang, 2011; Hesketh *et al.*, 1998): θ is the looper angle, ω is the looper angular speed, R_l is the looper arm length, h is the strip exit thickness, H is the strip entry thickness, w is the strip width, R_w^i is the work roll radius of the i-th stand, R_w^{i+1} is the work roll radius of the $(i+1)$th stand, M_L is the looper mass, R_G is the distance between the pivoting point of the looper and the center of gravity of the looper, R_r is the radius of the looper roll, L is the distance between two stands, L_1 is the distance between the i-th stand and the looper pivot, L_2 is the distance between the looper pivot and the $(i+1)$-th stand, ρ is the steel density, $l_1(\theta)$ is the distance between the i-th stand and the looper roll, $l_2(\theta)$ is the distance between the $i+1$th stand and the looper roll, $\xi(t)$ is the deviation of the inter-strand strip length with respect to L, g is the gravitational constant and H_1 is the distance between the actual passline and the looper pivot. Besides, the constituents of the system's dynamics are analyzed next:

Tension dynamics: The interstand tension dynamics is given by the strip stretch $L^{'}(\theta) - (L+\xi(t))$ and Young's modulus E (Zhong and Wang, 2011; Hesketh *et al.*, 1998)

$$\sigma(t) = E\left[\frac{L^{'}(\theta) - (L+\xi(t))}{L+\xi(t)}\right] \tag{5.81}$$

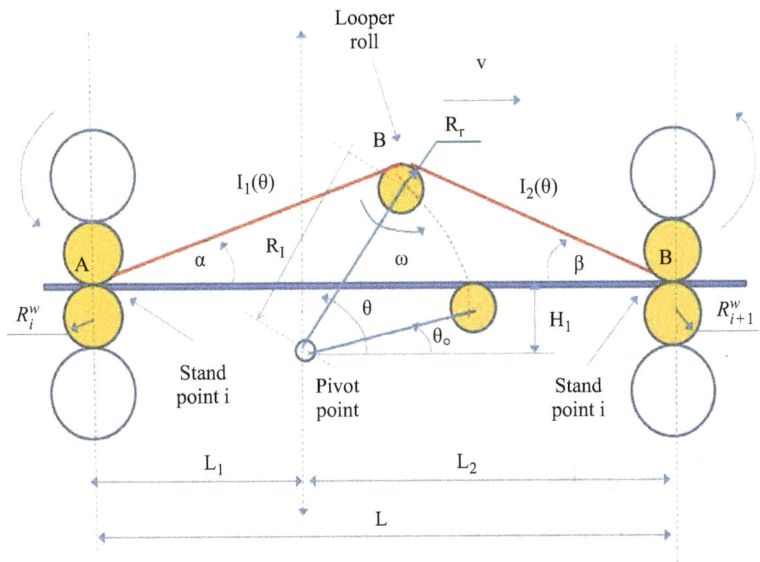

Figure 5.17 Diagram of the hot-steel rolling mill system

for $L^{'}(\theta)) > L + \xi(t)$, θ is the looper angle, L is the interstrand length, $L + \xi(t)$ is the accumulated material length, $v(t)$ is the mass flow difference of the strip between the stands. Actually $L^{'}(\theta)$ is the sum of the distances between the ith, $(i+1)$th stand and the looper roll. It holds that

$$
\begin{aligned}
& L^{'}(\theta) = l_1(\theta) + l_2(\theta), \quad \text{where} \\
& l_1(\theta) = \sqrt{(L_1 + R_l\cos(\theta))^2 + (R_l\sin(\theta) + R_r - H_1)^2} \\
& l_2(\theta) = \sqrt{(L_2 - R_l\cos(\theta))^2 + (R_l\sin(\theta) + R_r - H_1)^2}
\end{aligned}
\tag{5.82}
$$

where $\xi(t)$ is calculated as follows:

$$
\begin{aligned}
& \dot{\xi}(t) = v(t) + w_{\xi}(t) \\
& v(t) = v_s^i(t) - V_s^{i+1}(t) \\
& v_s^i(t) = (1 + S_f^i)V_R^i(t) \\
& V_s^{i+1}(t) = (1 - S_b^{i+1})V_R^{i+1}(t)
\end{aligned}
\tag{5.83}
$$

where $v(t)$ is the difference in the strip speed entering and leaving the interstand space, $v_s^i(t)$ is the strip speed leaving the upstream stand i, $V_s^{i+1}(t)$ is the strip speed exiting the downstream stand $(i+1)$, $V_R^i(t)$ is the linear speed of the i-th stand, $V_R^{i+1}(t)$ is the linear speed of the $(i+1)$th stand, S_f^i is the coefficient of forward slip which occurs between the rolls at the ith stand and the strip, S_b^{i+1} is the coefficient of backward slip which occurs between the rolls st the ith stand and the strip.

About the denominator of the relation that describes the strip's tension, that is (5.81) one can consider that the strip's elongation $\xi(t)$ is significantly smaller than L, consequently it holds that $L+\xi(t)\simeq L$. By differentiating (5.81) with respect to time one obtains (Zhong and Wang, 2011; Hesketh *et al.*, 1998)

$$\begin{aligned}\dot{\sigma}(t) &= \frac{E}{L}\left[\frac{d}{dt}L'(\theta) - \dot{\xi}(\theta)\right] \Rightarrow \\ \dot{\sigma}(t) &= \frac{E}{L}\{R_l[\sin(\theta+\beta) - \sin(\theta-\alpha)]\dot{\theta}(t) - (v(t) + w_g(t))\}\end{aligned} \tag{5.84}$$

where α is the upstream strip angle and β is the downstream strip angle. These angles are geometrically evaluated as (Zhong and Wang, 2011; Hesketh *et al.*, 1998)

$$\alpha = \tan^{-1}\left(\frac{[R_l\sin(\theta) - H_1 + R_r]}{[L_1 + R_l\cos(\theta)]}\right) \tag{5.85}$$

$$\beta = \tan^{-1}\left(\frac{[R_l\sin(\theta) - H_1 + R_r]}{[L_2 - R_l\cos(\theta)]}\right) \tag{5.86}$$

Looper dynamics: The application of Newton's law to the motion of the looper gives (Zhong and Wang, 2011; Hesketh *et al.*, 1998)

$$J\ddot{\theta}(t) = T_u(t) - T_{Load}(\theta) + w_\omega(t) \tag{5.87}$$

where J is the moment of inertia of the looper with respect to the pivoting point, $T_u(t)$ is the actuator torque on the looper, $T_{Load}(\theta)$ is the load torque of the looper, and $w_\omega(t)$ is the unmodelled dynamics (disturbance torque). It holds that (Zhong and Wang, 2011; Hesketh *et al.*, 1998)

$$T_{Load}(\theta) = T_\sigma(\theta) + T_s(\theta) + T_L(\theta) \tag{5.88}$$

where

$$\begin{aligned}T_\sigma(\theta) &= \sigma h{\cdot}w{\cdot}R_l[\sin(\theta+\beta) - \sin(\theta-\alpha)] \\ T_L(\theta) &= gM_LR_G\cos(\theta) \\ T_s(\theta) &\simeq 0.5g{\cdot}\rho{\cdot}L{\cdot}h{\cdot}w{\cdot}R_l{\cdot}\cos(\theta)\end{aligned} \tag{5.89}$$

The above relation means that the load torque of the looper $T_{Load}(\theta)$ is the sum of the load $T_\sigma(\theta)$ caused by the strip tension, the torque $T_s(\theta)$ caused by the strip weight and the torque $T_L(\theta)$ caused by the looper weight.

State-space model: The cumulative dynamics of the hot steel rolling mill is given by (5.84 and (5.87). Actually one has (Zhong and Wang, 2011; Hesketh *et al.*, 1998)

$$\dot{\theta}(t) = \omega(t) \tag{5.90}$$

$$\dot{\omega}(t) = \frac{1}{J}[T_u(t) - F_1(\theta) - \sigma A F_3(\theta) + w_\omega(t)] \tag{5.91}$$

$$\dot{\sigma}(t) = \frac{E}{L}[F_3(t)\omega(t) - (v(t) + w_\xi(t))] \tag{5.92}$$

where $A = h{\cdot}w$ and

$$F_1(\theta) = 0.5g{\cdot}\rho L{\cdot}h{\cdot}w{\cdot}R_l\cos(\theta) + gM_LR_G\cos(\theta)$$
$$F_3(\theta) = R_l[\sin(\theta+\beta) - \sin(\theta-\alpha)] \tag{5.93}$$

The control inputs of the system are $v(t)$ and $T_u(t)$, where $v(t)$ is the strip speed difference which is controlled by an automatic speed regulator, and $T_u(t)$ is the looper actuator torque which is also controlled by an external actuator.

By considering the following control loops for the speed v of the strip and for the torque T_u of the of the looper's actuator, one has

$$\dot{T}_u(t) = v_1 \quad v_1 = \dot{T}_{u,d}(t) - k_p^2(T_u(t) - T_{u.d}(t))$$
$$\dot{v}(t) = v_2 \qquad v_2 = \dot{v}_d(t) - k_p^1(v(t) - v_d(t)) \tag{5.94}$$

The control problem of the hot-steel rolling mill is thus formulated in two levels: (i) From (5.90) to (5.92) one computes the control inputs $v(t)$ and $T_u(t)$ that should be applied to the hot-steel roll mill system so as to assure that the angle of the looper and the tension on the strip will reach the desirable setpoints and (ii) the control inputs $v(t)$ and $T_u(t)$ become setpoints for the secondary control of (5.94). There, v_1 and v_2 stand for control currents/voltages that allow $T_u(t)$ and $v(t)$ to reach the values needed by the control loop of the hot-steel rolling mill.

Next, the dynamics of the hot-steel rolling mill is written in state-space form. By defining the following state variables: $x_1 = \theta$, $x_2 = \omega$ and $x_3 = \sigma$, and the control inputs: $u_1 = T_u(t)$, $u_2 = v(t)$, as well the outputs of the system: $y_1 = \theta$, $y_2 = \sigma$ one has

$$\begin{aligned}\dot{x}_1 &= x_2\\ \dot{x}_2 &= \frac{1}{J}[u_1 - F_1(x) - x_3AF_3(x) + w_\omega(t)]\\ \dot{x}_3 &= \frac{E}{L}[F_3(x)x_2 - (u_2 + w_\xi(t))]\end{aligned} \tag{5.95}$$

By omitting from the previous state-space description the additive-input disturbance terms $w_\omega(t)$ and $w_\xi(t)$ which are unknown and which are going to be compensated by the robustness of the control algorithm one has the description of the state-space model into the following matrix form:

$$\begin{pmatrix}\dot{x}_1\\ \dot{x}_2\\ \dot{x}_3\end{pmatrix} = \begin{pmatrix}x_2\\ \frac{1}{J}(-F_1(x) - x_3AF_3(x))\\ \frac{E}{L}(F_3(x)x_2)\end{pmatrix} + \begin{pmatrix}0 & 0\\ \frac{1}{J} & 0\\ 0 & -\frac{E}{L}\end{pmatrix}\begin{pmatrix}u_1\\ u_2\end{pmatrix} \tag{5.96}$$

Equivalently, the system can be written into the following concise state-space description

$$\dot{x} = f(x) + g(x)u \tag{5.97}$$

where $f(x) \in R^{3\times 1}$, $g(x) \in R^{3\times 2}$, $x \in R^{3\times 1}$ and $u \in R^{2\times 1}$.

5.3.3 Approximate linearization of the hot-steel rolling mill dynamics

The hot-steel rolling mill system dynamics undergoes approximate linearization around a temporary operating point (x^*, u^*) which is recomputed at each time-step of the control method. This operating point consists of the present value of the system's state vector x^* and of the last sampled value of the control inputs vector. Thus the initial nonlinear state-space model of the system given in (5.97) is now written in the linearized form

$$\dot{x} = Ax + Bu + \tilde{d} \tag{5.98}$$

where it holds that

$$A = \nabla_x[f(x) + g(x)u] \mid_{(x^*,u^*)} \Rightarrow A = \nabla_x[f(x)] \mid_{(x^*,u^*)} \tag{5.99}$$

$$B = \nabla_u[f(x) + g(x)u] \mid_{(x^*,u^*)} \Rightarrow B = g(x) \mid_{(x^*,u^*)} \tag{5.100}$$

The computation of the Jacobian matrix $\nabla_x[f(x)] \mid_{(x^*,u^*)}$ proceeds as follows:

About the first row of $\nabla_x[f(x)] \mid_{(x^*,u^*)}$ one has: $\frac{\partial f_1}{\partial x_1} = 0$, $\frac{\partial f_1}{\partial x_2} = 1$ and $\frac{\partial f_1}{\partial x_3} = 0$.

About the second row of $\nabla_x[f(x)] \mid_{(x^*,u^*)}$ one has: $\frac{\partial f_2}{\partial x_1} = -\frac{1}{J}\frac{\partial F_1(x)}{\partial x_1} - x_3 A\frac{\partial F_3(x)}{\partial x_1}$, $\frac{\partial f_2}{\partial x_2} = -\frac{1}{J}\frac{\partial F_1(x)}{\partial x_2} - x_3 A\frac{\partial F_3(x)}{\partial x_2}$ and $\frac{\partial f_2}{\partial x_3} = -\frac{1}{J}\frac{\partial F_1(x)}{\partial x_3} - AF_3(x) - x_3 A\frac{\partial F_3(x)}{\partial x_3}$

About the third row of $\nabla_x[f(x)] \mid_{(x^*,u^*)}$ one has: $\frac{\partial f_3}{\partial x_1} = \frac{\partial F_3}{\partial x_1}x_2$, $\frac{\partial f_3}{\partial x_2} = \frac{\partial F_3}{\partial x_2}x_2 + F_3(x)$ and $\frac{\partial f_3}{\partial x_3} = \frac{\partial F_3}{\partial x_3}x_2$

Next, the following partial derivatives are computed: $\frac{\partial F_1}{\partial x_1} = 0.5 \cdot g \cdot \rho \cdot L \cdot h \cdot w \cdot R_l(-\sin(x_1)) + gM_L R_G(-\sin(x_1))$, $\frac{\partial F_1}{\partial x_1} = 0$ and $\frac{\partial F_1}{\partial x_3} = 0$

and similarly: $\frac{\partial F_3}{\partial x_1} = R_l[\cos(x_1 + \beta)(1 + \frac{\partial \beta}{\partial x_1}) + \cos(x_1 - \alpha)(1 - \frac{\partial \alpha}{\partial x_1})]$, $\frac{\partial F_3}{\partial x_2} = 0$, $\frac{\partial F_3}{\partial x_3} = 0$.

Moreover, using (5.85) one finds that

$$\frac{\partial a}{\partial x_1} = \cos^2(\alpha)\frac{[R_l\cos(x_1)][L_1 + R_l\cos(x_1)] + [R_l\sin(x_1) - H_1 + R_r][R_l\sin(x_1)]}{[L_1 + R_l\cos(x_1)]^2} \tag{5.101}$$

while, using (5.86) one gets that

$$\frac{\partial \beta}{\partial x_1} = \cos^2(\beta)\frac{[R_l\cos(x_1)][L_2 - R_l\cos(x_1)] + [R_l\sin(x_1) - H_1 + R_r][R_l\sin(x_1)]}{[L_2 - R_l\cos(x_1)]^2} \tag{5.102}$$

5.3.4 Computation of the feedback control gains

For the linearized system given by (1.36) the cost function of (1.37) is defined, where coefficient r determines the penalization of the control input and weight coefficient ρ determines the reward of the disturbances' effects. It is assumed that (i) the energy that is transferred from the disturbances signal $\tilde{d}(t)$ is bounded, that is $\int_0^{\infty}\tilde{d}^T(t)\tilde{d}(t)dt < \infty$, (ii) matrices $[A, B]$ and $[A, L]$ are stabilizable and (iii) matrix $[A, C]$ is detectable. In the case of a tracking problem the optimal feedback control law is given by

$$u(t) = -Ke(t) \tag{5.103}$$

with $e = x - x_d$ to be the tracking error, and $K = \frac{1}{r}B^TP$ where P is a positive definite symmetric matrix. Matrix P is obtained from the solution of the Riccati equation

$$A^TP + PA + Q - P\left(\frac{2}{r}BB^T - \frac{1}{\rho^2}LL^T\right)P = 0 \tag{5.104}$$

where Q is a positive semi-definite symmetric matrix. The worst-case disturbance is given by

$$\tilde{d}(t) = \frac{1}{\rho^2}L^TPe(t) \tag{5.105}$$

The nonlinear optimal control loop for the hot-steel rolling mill's system is given in Figure 5.18. The global stability properties of the control method can be proven following the procedure presented in Section 1.1.2.5.

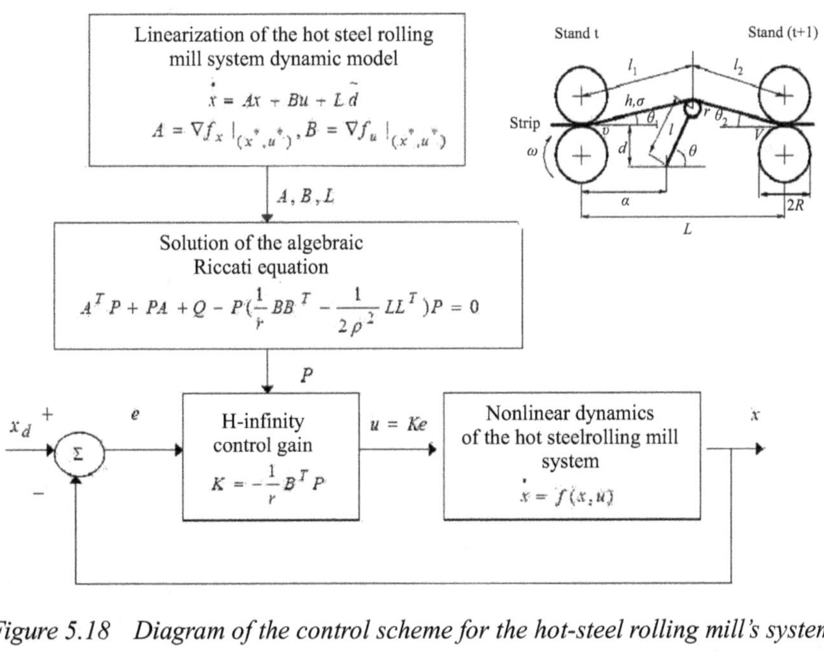

Figure 5.18 Diagram of the control scheme for the hot-steel rolling mill's system

5.3.5 Simulation tests

The performance of the nonlinear optimal control for the dynamic model of the hot-steel rolling-mill system was tested through simulation experiments. Nominal values for the parameters of the process are given in Zhong and Wang, 2011. The parameters' values which have been used in the simulation tests are as follows: R_l: 0.611 m, h: 0.1 m, w: 2.5 m, M_L: 1.8×10^3 kg, R_G: 0.2 m, r: 0.08 m, E: 1.2×10^{10} Pa, L: 5.8 m, l_1: 2.34 m, l_2: 3.46 m, ρ: 8.0×10^3 kg/m^3, g: 9.8 m/s^2, H_1: 0.195 m and J: 600 kg·m^2. The obtained results are depicted in Figures 5.19–5.25. The simulated parameters of the system are its state variables, that is $x_1=\theta$ which is the turn angle of the looper (rad), ω which is the angular speed of the looper (rad/s) and σ which is the strip's tension (MPa). The control inputs of the system are the looper's actuator torque u_1 (kNt·m) and the strip's speed difference u_2 (m/s). The disturbances that affect the control loop were due to modelling errors and parametric changes in its state-space description.

The control scheme is a state estimation-based one and its implementation requires measurement of only one state variable. The measured state variable is the turn angle $x_1=\theta$ of the looper. The rest of the state variables $x_2=\omega$ (angular speed of the looper) and $x_3=\sigma$ (strip's tension) are not measured directly but are estimated through the processing of the measurements of $x_1=\theta$ with the use of the H-infinity Kalman Filter. There is measurement noise in x_1 which is considered to be the only measurable state variable. The real value of the state vector elements of the system is shown in blue, the estimated value as provided by the H-infinity Kalman Filter is shown in green, while the associated reference setpoints are printed in red. The simulation tests show that fast and accurate convergence of the state variables to their

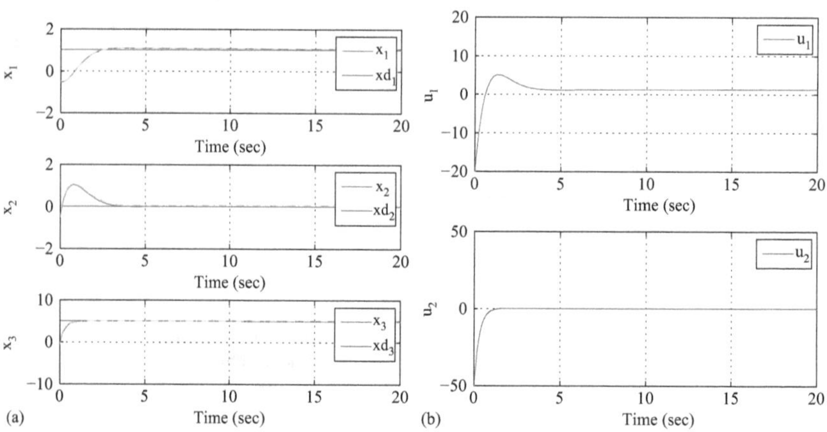

Figure 5.19 Test case 1: (a) tracking of reference setpoints by the state variables x_1 to x_3 of the hot-steel rolling mills system and (b) control inputs u_1 and u_2 applied to the hot-steel rolling mills system

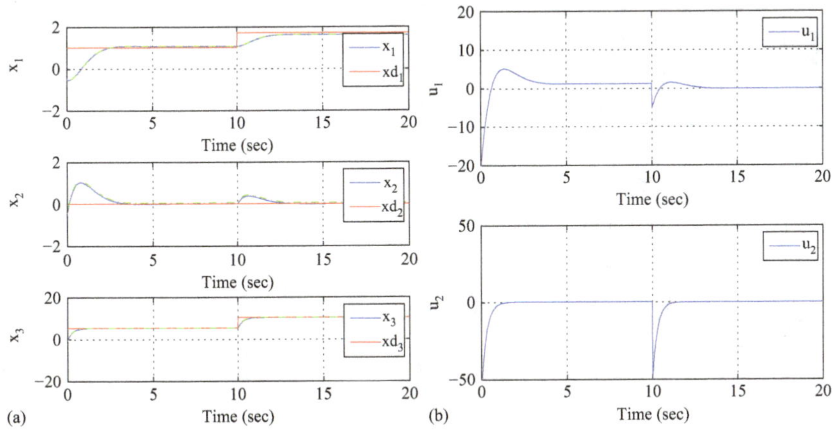

Figure 5.20 *Test case 2: (a) tracking of reference setpoints by the state variables x_1 to x_3 of the hot-steel rolling mills system and (b) control inputs u_1 and u_2 applied to the hot-steel rolling mills system*

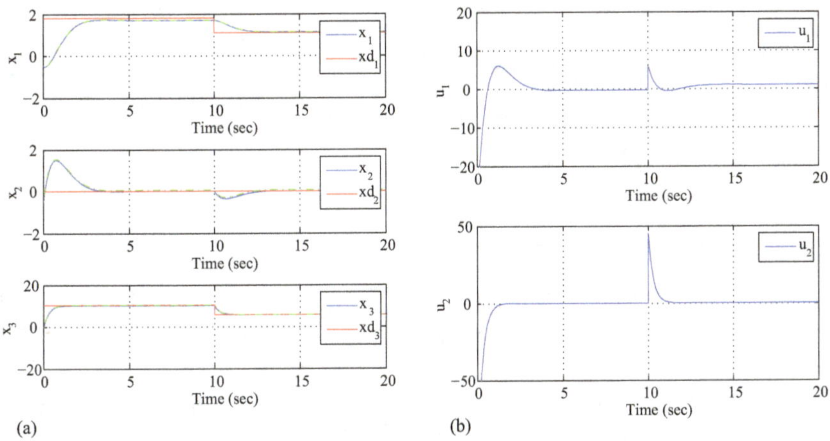

Figure 5.21 *Test case 3: (a) tracking of reference setpoints by the state variables x_1 to x_3 of the hot-steel rolling mills system and (b) control inputs u_1 and u_2 applied to the hot-steel rolling mills system*

setpoints was achieved, under moderate variations of the control inputs. The proposed control method not only achieves minimization of the tracking error for the state variables of the hot-steel rolling mill system, but it also minimizes the control effort or equivalently the cost of the electric power consumed by the rolling mill industrial units. Regarding the transient performance of the control method this is affected by

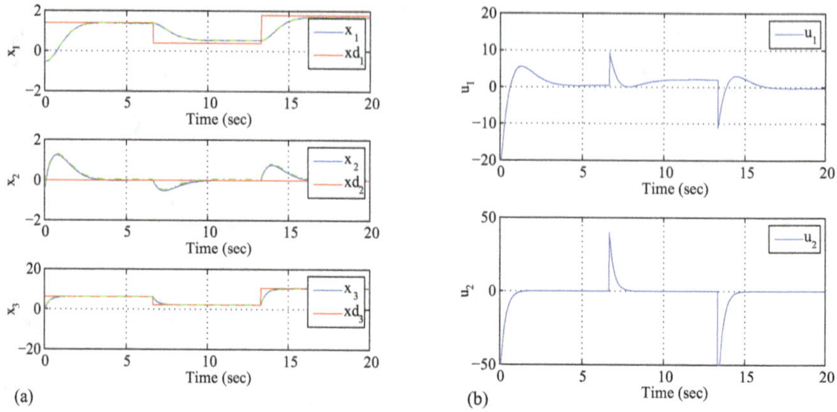

Figure 5.22 *Test case 4: (a) tracking of reference setpoints by the state variables x_1 to x_3 of the hot-steel rolling mills system and (b) control inputs u_1 and u_2 applied to the hot-steel rolling mills system*

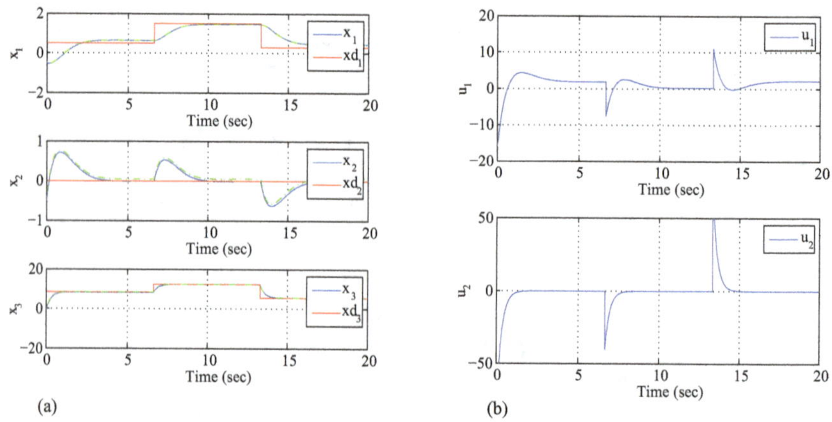

Figure 5.23 *Test case 5: (a) tracking of reference setpoints by the state variables x_1 to x_3 of the hot-steel rolling mills system and (b) control inputs u_1 and u_2 applied to the hot-steel rolling mills system*

the gains and parameters r, ρ and Q which appear in the Riccati equation of (5.104). Actually, the smallest value of the attenuation coefficient ρ for which a valid solution of the Riccati equation can be obtained is the one that provides the control loop with maximum robustness.

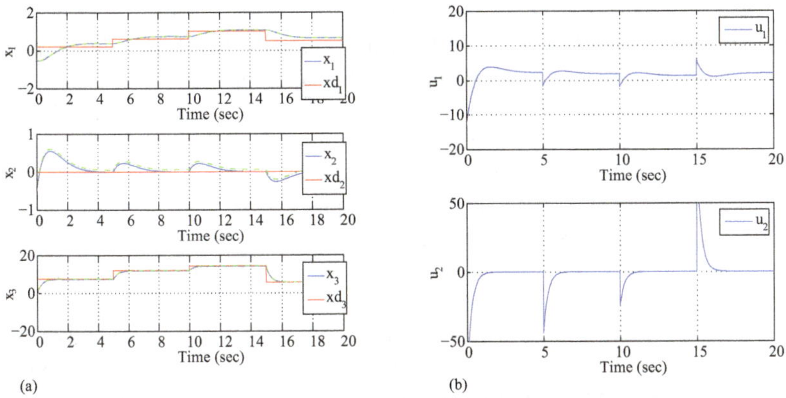

Figure 5.24 *Test case 6: (a) tracking of reference setpoints by the state variables x_1 to x_3 of the hot-steel rolling mills system and (b) control inputs u_1 and u_2 applied to the hot-steel rolling mills system*

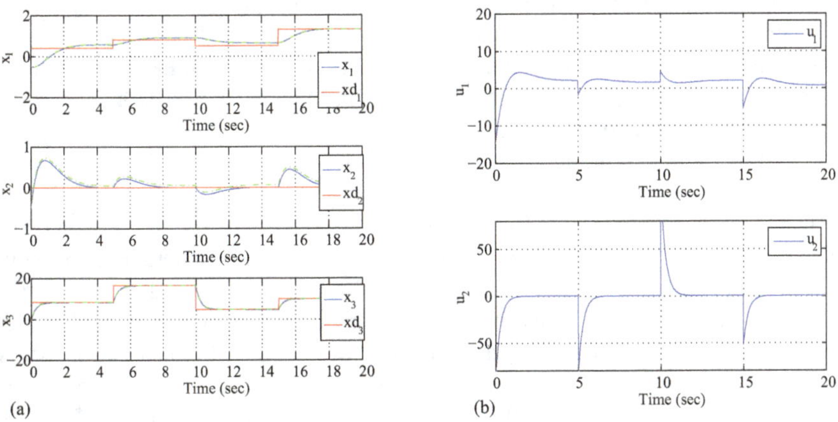

Figure 5.25 *Test case 7: (a) tracking of reference setpoints by the state variables x_1 to x_3 of the hot-steel rolling mills system and (b) control inputs u_1 and u_2 applied to the hot-steel rolling mills system*

5.4 Nonlinear control of paper mills

5.4.1 Outline

The mechanical pulping process is among the most energy intensive processes in the industry (Tian, Lu *et al.*, 2019; Henon, 2019; Luppi, Garcia *et al.*, 2010; Harinath *et al.*, 2011; Li, Zhou *et al.*, 2019). Paper is an important commodity and the demand for paper of various types and quantities grows steadily (Durocher and Higginson, 2019; Mercangoz and Doyle, 2004; Marshman *et al.*, 2010; Li, Zhang *et al.*, 2014; Austin *et al.*, 2010). The annual revenues of the paper industry are at the level of 500 billion USD, while the total produced quantity reaches the amount of 300 million tons. A pulp and paper mill with a production capacity of 300,000 tons per year has a construction cost of about 1 billion USD (Durocher and Higginson, 2019; Mercangoz and Doyle, 2004). Challenging objectives for the paper making industry are (i) to minimize energy consumption, (ii) to reach specific quality features for the produced paper despite the variability of the raw material and (iii) to minimize hazardous emissions and environmental risks (Elsinga, 2003; Li, Zhou, Wang *et al.*, 2019; Ding, Lejeune *et al.*, 2012; Ashfar *et al.*, 2011). Due to the strongly nonlinear and multivariable structure of the state-space model of the pulping process, control and stabilization of this process at setpoints that optimize the paper mills' functioning is a non-trivial task (Harinath *et al.*, 2013; Harinath *et al.*, 2011; Berg *et al.*, 2004; Jilkaistavaski, 2015). So far several approaches have been proposed for the control of the mechanical pulping process in the paper industry with the majority of them to consider linear models of the process. Often, there is a need for further elaboration on the global stability properties of the related control schemes (Wisnewshi and Doyle, 2001; Choi and Kwon, 2019; Tian, Lu *et al.*, 2016; Khanbaghi *et al.*, 2002). Energy efficiency and energy cost minimization in the paper industry remains a challenge that is directly related to the solution of the control problem for the mechanical pulping process in paper mills (Tian, Prakash *et al.*, 2020; Man, Li *et al.*, 2020; Sanaye *et al.*, 2020; Niu, Xi *et al.*, 2020).

In this section, a nonlinear optimal (H-infinity) control method is developed for the dynamic model of the mechanical pulping process in the paper industry (Rigatos and Busawon, 2018; Rigatos, Siano and Abbaszadeh, 2017; Rigatos, Siano *et al.*, 2015). First, the dynamic model of the pulping process undergoes approximate linearization around a temporary operating point which is updated at each iteration of the control algorithm. This operating point is defined by the present value of the process's state vector and by the last value of the control inputs vector that was applied to the process. The linearization relies on first-order Taylor series expansion and on the computation of the related Jacobian matrices (Rigatos and Tzafestas, 2007; Basseville and Nikiforov, 1993; Rigatos and Zhang, 2009). For the approximately linearized model of the pulping process, an optimal (H-infinity) stabilizing feedback controller is defined. The modelling error which is due to the truncation of higher-order terms in the Taylor series expansion is considered to a perturbation which is asymptotically compensated by the robustness of the control algorithm.

The proposed H-infinity controller provides a solution to the optimal control problem of the pulping process under model uncertainty and external perturbations. Actually, it represents the solution to a min–max differential game in which the controller tries to minimize a cost function that comprises a quadratic term of the state vector's tracking error, whereas the model uncertainty and external perturbation terms try to minimize this cost function. To compute the controller's feedback gains an algebraic Riccati equation is solved at each time-step of the control method (Rigatos, 2011; Rigatos, 2015; Rigatos, 2016b). The stability properties of the control scheme are proven through Lyapunov analysis. First, it is shown that the control loop satisfies the H-infinity tracking performance criterion, which signifies elevated robustness against model uncertainty and external perturbations (Rigatos and Busawon, 2018; Tousssaint, Basar and Bullo, 2000; Lublin and Athans, 1995). Moreover, under moderate conditions, the global asymptotic stability properties of the control method are proven. To perform feedback control without the need of measuring the entire state vector of the process, the H-infinity Kalman Filter is proposed as reliable state estimator. The proposed optimal (H-infinity) control method retains the advantages of typical linear optimal control, that is fast and accurate tracking of the reference setpoints, under moderate variations of the control inputs.

The section's control approach differs from all control methods that one could have considered for the nonlinear model of the mechanical pulping process in paper mills. A large part of the competitive control methods is of known suboptimal performance (for instance global linearization-based control methods, sliding-mode control, back-stepping control, and PID control) while the rest of the control methods are not of proven global stability (MPC, NMPC, or PID control). Besides, multiple models-based optimal control approaches require the definition of multiple operating points, and the solution of multiple Riccati equations or LMIs and this incurs a significant computation effort comparing to the section's nonlinear optimal control approach. Finally, about LQR control it can be stated that this method is completely unsuitable for use in the case of the nonlinear mechanical pulping process of paper mills because of not performing compensation of the modelling errors of the linearization procedure and consequently because of lacking a global stability proof.

5.4.2 Dynamic model of the mechanical pulping process in paper mills

5.4.2.1 The state-space model of the pulping process

The multi-stage mechanical pulping process comprises the following stages Tian, Lu *et al.*, 2019; Li, Zhou *et al.*, 2019: (i) wood chip pretreatment, (ii) wood chip refining and (iii) pulp refining. In the wood chip pretreatment stage, wood chips are screened to remove over and under-sized sections and next they are heated and steamed at a temperature of about 100°C. In the wood chip refining stage, there are typically two high consistency refiners, the primary and the secondary one, where the refined wood material is exposed again to steam flow.

After pretreatment, the wood chips are introduced into the inlet of the primary high consistency (HC) refiner by the cylindrical chip transfer screw feeder. Dilution water is usually fed into the inlet of the refiner to control the consistencies of the refining zone. Finally, the wood chips end at being broken down into fibers as they pass through the two rotating discs of the refiners. The pressure exerted by the discs that rotate in opposite directions frees fibers from lignine, thus releasing the mass that will be turned into paper pulp.

The diagram of the mechanical pulping process in paper mills is shown in Figure 5.26. The following state variables are defined: x_1 is the production rate (tons/day), x_2: is the primary motor load (MW), x_3 is the primary consistency indicator (%), x_4: is the secondary motor load (MW) and x_5: is the secondary consistency %. The control inputs of the pulping process are defined as follows: u_1 is the chip transfer screw speed (rpm), x_2: primary refiner plane gap (rpm), u_3: is the primary dilation flow rate (kg/s), u_4: is the secondary refiner plane gap (mm) and u_5: is the secondary dilution flow rate (kg/s).

The significance of the aforementioned state variables and control inputs is as follows (Tian, Lu *et al.*, 2019; Li, Zhou *et al.*, 2019):

x_1: is the production rate. The production rate is related to the wood chips feed-in rate. By changing the chip transfer screw speed, the production rate can be also modified. The production rate is also affected by parameters such as the quality of the wood species, the chip density and the moisture content. The production rate affects also the specific energy of the process and the pulp's quality.

x_2, (x_4): Primary (secondary) electric motor's load. The motor's load is a parameter that affects significantly the quality of the produced paper's pulp. As noted, the motor's load is modified by changing the plate's gap. Actually, the motor's load exhibits high sensitivity to changes of the plate's distance.

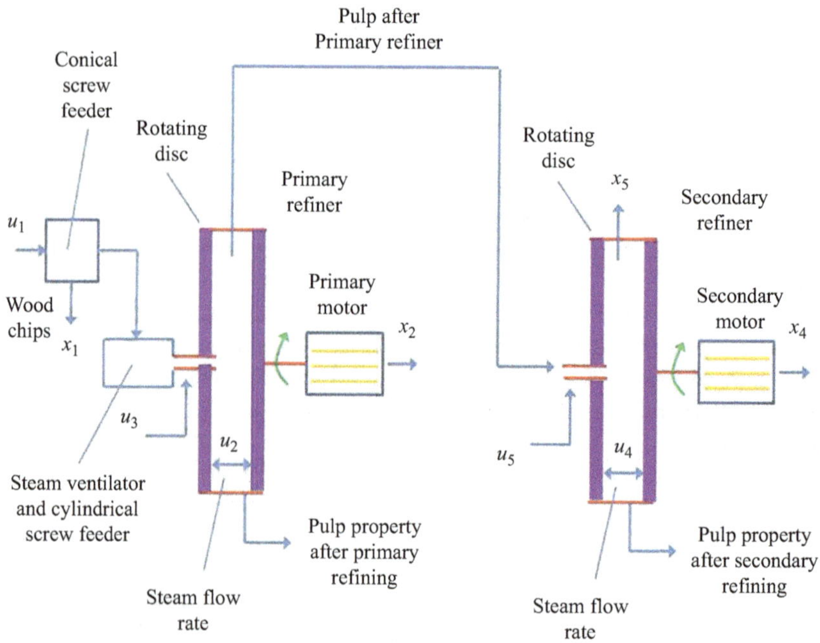

Figure 5.26 Diagram of the mechanical pulping process in paper mills

x_3, (x_5): Primary (secondary) consistency: For the constant value of the process's specific energy, the primary and secondary consistency have a significant impact in the quality of the produced paper's pulp. The efficient control of the process's primary (and secondary) consistency determines also the motor's load and the quality properties of the produced paper pulp.

The significance of the aforementioned control input variables is as follows (Tian, Lu *et al.*, 2019; Li, Zhou *et al.*, 2019):

u_1: is the chip transfer screw speed. The chip transfer screw for the primary refiner is the main manipulated variable that controls the flow of chips from the preheater at the pretreatment stage to the inlet of the primary refiner. Changes to the screw speed affect the flow of dry fibers to the refiner. The variable is also used to modify the pulp's production rate.

u_2, (u_4): primary (secondary) refiner plate gap. The plate gap is the distance between the plates of a refiner, that is of the two discs that squeeze wood chips while rotating in opposite directions. The plate gap is controlled by a mechanical load system. Variations in gap size affect the mechanical force exerted by plants on the wood chips and thus they also affect the loads of the two electric motors.

u_3, (u_5): primary (secondary) dilution flow rate. The consistency of the refining zone affects the pulp's properties. The added water modifies this consistency and finally changes the pulp's quality. Large variations in the dilution mater flow rate can even result into an unstable refining process. Thus the incoming water's flow at the refiners is among the control variables that determine the quality of the paper's pulp.

A description of the mechanical pulping process has been first formulated in discrete-time form (Tian, Lu *et al.*, 2019; Tian, Lu *et al.*, 2016):

$$\begin{aligned}
x_1(k+1) &= a_1x_1(k) + b_1k_ak_ps_cd_cu_1(k) \\
x_2(k+1) &= a_2x_2(k) + b_2\frac{k_{m_1}x_1(k)}{u_3(k)}(1 - e^{(-10u_2(k))}(c_1 - e_1u_3(k))) \\
x_3(k+1) &= a_2x_3(k) + b_3\frac{100x_1(k)}{x_1(k) + k_au_3(k) - k_{c_p}x_2(k)} \\
x_4(k+1) &= a_4x_4(k) + b_4\frac{k_{m_2}x_1(k)}{u_5(k)}(1 - e^{-(10u_4(k))})(c_2 - e_2u_4(k)) \\
x_5(k+1) &= a_5x_5(k) + b_5\frac{100x_1(k)}{\frac{x_1(k)}{0.01x_3(k)} + k_au_5(k) - k_{c_s}x_4(k)}
\end{aligned} \tag{5.106}$$

By using the first-order Euler approximation to describe the time-derivative and by defining the following normalized coefficients $\bar{a}_1 = \frac{a_1-1}{T_s}$, $\bar{b}_1 = \frac{b_1}{T_s}$, $\bar{a}_2 = \frac{a_2-1}{T_s}$,

$\bar{b}_2 = \frac{b_2}{T_s}$, $\bar{a}_3 = \frac{a_3-1}{T_s}$, $\bar{b}_3 = \frac{b_1}{T_s}$, $\bar{a}_4 = \frac{a_4-1}{T_s}$, $\bar{b}_4 = \frac{b_4}{T_s}$, and $\bar{a}_5 = \frac{a_5-1}{T_s}$, $\bar{b}_5 = \frac{b_1}{T_s}$, one can obtain a continuous-time state-space description of the dynamics of the mechanical pulping process in paper mills:

$$\begin{aligned}
x_1(t) &= \bar{a}_1 x_1(t) + \bar{b}_1 k_a k_p s_c d_c u_1(t) \\
x_2(t) &= \bar{a}_2 x_2(k) + \bar{b}_2 \frac{k_{m_1} x_1(t)}{u_3(t)} (1 - e^{(-10u_2(t))} (c_1 - e_1 u_3(t))) \\
x_3(t) &= \bar{a}_2 x_3(t) \bar{b}_3 \frac{100 x_1(t)}{x_1(t) + k_a u_3(t) - k_{c_p} x_2(t)} \\
x_4(t) &= \bar{a}_4 x_4(t) + \bar{b}_4 \frac{k_{m_2} x_1(t)}{u_5(t)} (1 - e^{-(10u_4(t))}) (c_2 - e_2 u_4(t)) \\
x_5(t) &= \bar{a}_5 x_5(t) + \bar{b}_5 \frac{100 x_1(t)}{\frac{x_1(t)}{0.01 x_3(t)} + k_a u_5(t) - k_{c_s} x_4(t)}
\end{aligned} \tag{5.107}$$

It is assumed that the control inputs u_3 (primary dilution flow rate) and u_5 (secondary dilution flow rate) are kept constant. Moreover, the following parameters and control inputs variables are considered in the individual rows of the state-space model: In the first row one defines $\bar{b}_1 k_a k_p s_c d_c = \tilde{b}_1$, $\bar{a}_1 = \tilde{a}_1$ and $u_1 = v_1$. In the second row of the state-space model one defines $\bar{b}_2 \frac{k_{m1}}{u_3}(c_1 - e_2 u_3) = \tilde{b}_2$, $e^{-10u_2} = v_2$, and $\bar{a}_2 = \tilde{a}_2$. In the third row of the state-space model one defines $\bar{b}_3 100 = \tilde{b}_3$, $k_a u_3 = q_1$ (constant) and $\bar{a}_3 = \tilde{a}_3$. In the fourth row of the state-space model one defines $\bar{b}_4 \frac{k_{m2}}{u_5} = \tilde{b}_4$, $(1 - e^{-10u_4})(c_2 - e_2 u_4) = c_2 - v_3$ where $v_3 = c_2 e^{-10u_4} - e_2 u_4 + e^{-10u_4} e_2 u_4$, and $\bar{a}_4 = \tilde{a}_4$. Finally, in the fifth row of the state-space model one defines $\bar{b}_5 = \tilde{b}_5$, $q_2 = 0.01 k_a u_5$ (constant), $q_3 = 0.01$ and $\bar{a}_5 = \tilde{a}_5$. Using the previous definitions, the state-space model of the pulping process can be written as:

$$\begin{aligned}
\dot{x}_1 &= \tilde{a}_1 x_1 + \tilde{b}_1 v_1 \\
\dot{x}_2 &= \tilde{a}_2 x_2 + \tilde{b}_2 x_1 (1 - v_2) \\
\dot{x}_3 &= \tilde{a}_3 x_3 + \tilde{b}_3 \frac{x_1}{(x_1 + q_1 - k_{c_p}) x_2} \\
\dot{x}_4 &= \tilde{a}_4 x_4 + \tilde{b}_4 x_1 (c_2 - v_3) \\
\dot{x}_5 &= \tilde{a}_5 x_5 + \tilde{b}_5 \frac{x_1}{x_1 + q_2 x_3 - q_3 k_{c_s} x_3 x_4}
\end{aligned} \tag{5.108}$$

By defining $x \in R^{5\times 1}$, $u = [v_1, v_2, v_3]^T \in R^{3\times 1}$, $f(x) \in R^{5\times 1}$ and $g(x) \in R^{5\times 3}$ the dynamic model of the process is written in the following concise form:

$$\dot{x} = f(x) + g(x)u \tag{5.109}$$

where vector $f(x)$ and matrix $g(x)$ are given by:

$$f(x) = \begin{pmatrix} \tilde{a}_1 x_1 \\ \tilde{a}_2 x_2 + \tilde{b}_2 x_1 \\ \tilde{a}_3 x_3 + \tilde{b}_3 \frac{x_1}{(x_1+q_1-k_{cp})x_2} \\ \tilde{a}_4 x_4 + \tilde{b}_4 x_1 c_2 \\ \tilde{a}_5 x_5 + \tilde{b}_5 \frac{x_1}{x_1+q_2 x_3 - q_3 k_{cs} x_3 x_4} \end{pmatrix} \quad g(x) = \begin{pmatrix} b_1 & 0 & 0 \\ 0 & -\tilde{b}_2 x_1 & 0 \\ 0 & 0 & 0 \\ 0 & 0 & -\tilde{b}_4 x_1 \\ 0 & 0 & 0 \end{pmatrix} \tag{5.110}$$

The measured outputs of the process are x_1 (production rate), x_2 (load at the primary motor) and x_4 (load at the secondary motor). The non-measurable state variables of the process, that is the primary consistency x_3 and the secondary consistency x_5 have to be estimated through a filtering procedure.

5.4.2.2 Parameters of the mechanical pulping process

Next, the main parameters of the mechanical pulping process are overviewed (Tian, Lu *et al.*, 2019; Li, Zhou *et al.*, 2019).

(a) The *production rate* P is measured in tons/day and is given by

$$P = k_a k_p s_c d_c R \tag{5.111}$$

where k_a and k_p (m^3/rpm) are constant parameters with values depending on the specific production units. Besides, s_c(%) is the chip solid content, d_c(kg/m^3) is the chip bulk density and R(rpm) is the chip transfer screw speed.

(b) The *motor load* (MW) is defined as

$$M = \frac{k_{m_i} P}{D_i}(1 - e^{-10G_i})(c_i - e_i G_i) \tag{5.112}$$

where $i = \{p, s\}$ with p to stand for primary and s to stand for secondary. It holds that M (MW) is the motor's load at the ith refiner, D (l/min) is the dilution mater flow rate, G_i (mm) is the gap distance, and c_i, e_i and k_{m_i} are parameters of each refiner.

(c) The *Consistency* of the refiners is defined as

$$C_P = \frac{100P}{P + k_a D_p - k_{e_p} M_p} \tag{5.113}$$

$$C_S = \frac{100P}{\frac{P}{0.01 C_p} + k_a D_s - k_{e_s} M_s} \tag{5.114}$$

where C_P and C_S denote the consistency for the primary and the secondary refiner, and k_s, k_{e_p} and k_{e_s} are the refiner parameters.

(d) The *Specific Energy* (MW/tons/day) is the energy consumed per ton of dry pulp, which is defined as the ratio between the motor load and the production rate. For the ith refiner and motor the specific energy (SE) is defined as

$$SE = \frac{\text{Motor load i}}{\text{Production rate}} \tag{5.115}$$

One can also define the Total Specific Energy (TSE)

$$TSE = \frac{\text{Total Motor load}}{\text{Production rate}} \tag{5.116}$$

5.4.2.3 Quality indicators in the mechanical pulping process

At present, the widely used technological indicators for evaluating pulp quality are the Canadian Standard Freeness (CSF), the long fiber content and the shive content. CSF is closely associated with the energy consumption of the pulping process. Another quality indicator is the pulp consistency which stands for the ratio of the mass of dry fibers over the sum of the total mass of dry fibers and of the water suspension. When this ratio is between 20% and 50% the pulp is considered to be of high consistency, while when it ranges between 3% and 5% the pulp is considered to be low consistency (Tian, Lu *et al.*, 2019; Li, Zhou *et al.*, 2019).

In general, wood consists of cellulose ~42%, hemicellulose ~27%, lignin ~28% and extractives ~3%. Lignin is a glue-like substance binding cellulose and semi-cellulose fibers throughout the wood's cellular structure. Pulp's quality is also determined by the so-called kappa number κ which indicates the residual lignin content in the wood pulp. The purpose of thermo-mechanical pulping is to remove lignin and semi-cellulose from the wood's mass, therefore pulp's production should have a low kappa number.

5.4.3 *Approximate linearization of the state-space model of the pulping process*

5.4.3.1 Approximately linearized model

The dynamic model of the pulping process of (5.109) undergoes approximate linearization around the temporary operating point (equilibrium) (x^*, u^*), where x^* is the present value of the process's state vector and u^* is the last value of the control inputs vector that was applied to the process. The linearization procedure gives

$$\dot{x} = Ax + Bv + \tilde{d} \tag{5.117}$$

where $\tilde{d}$ is the cumulative disturbances vector which comprises (i) the modelling error due to the truncation of higher-order terms in the Taylor series expansion, (ii) exogenous perturbations, and (iii) sensors noise of any distribution. Considering that matrix $g(x)$ consists of the column vectors $g(x)=[g_1(x)\,g_2(x)\,g_3(x)]$, matrices A and B are obtained from the computation of the system's Jacobians, that is

$$\begin{aligned} A &= \nabla_x f(x)\mid_{(x^*,u^*)})+\nabla_x g_1(x)v_1\mid_{(x^*,u^*)}) \\ &\quad +\nabla_x g_2(x)v_2\mid_{(x^*,u^*)})+\nabla_x g_3(x)v_3\mid_{(x^*,u^*)}) \\ B &= \nabla_u[f(x)+g(x)v]\mid_{(x^*,u^*)})\Rightarrow B=g(x\mid_{(x^*,u^*)})) \end{aligned} \tag{5.118}$$

Next, computation of the Jacobian matrix $\nabla_x f(x)\mid_{(x^*,u^*)})$ is performed:

About the first row of the Jacobian matrix $\nabla_x f(x)\mid_{(x^*,u^*)})$ one has: $\frac{\partial f_1}{\partial x_1}=\tilde{a}_1$, $\frac{\partial f_1}{\partial x_2}=0$, $\frac{\partial f_1}{\partial x_3}=0$, $\frac{\partial f_1}{\partial x_4}=0$ and $\frac{\partial f_1}{\partial x_5}=0$.

About the second row of the Jacobian matrix $\nabla_x f(x)\mid_{(x^*,u^*)})$ one has: $\frac{\partial f_2}{\partial x_1}=\tilde{b}_2$, $\frac{\partial f_2}{\partial x_2}=\tilde{a}_2$, $\frac{\partial f_2}{\partial x_3}=0$, $\frac{\partial f_2}{\partial x_4}=0$ and $\frac{\partial f_2}{\partial x_5}=0$.

About the third row of the Jacobian matrix $\nabla_x f(x)\mid_{(x^*,u^*)})$ one has: $\frac{\partial f_3}{\partial x_1}=\tilde{b}_3\frac{(q_1-k_{cp}x_2)}{(x_1+q_1-k_{cp}x_2)^2}$, $\frac{\partial f_3}{\partial x_2}=0$, $\frac{\partial f_3}{\partial x_3}=\tilde{a}_3$, $\frac{\partial f_3}{\partial x_4}=0$ and $\frac{\partial f_3}{\partial x_5}=0$.

About the fourth row of the Jacobian matrix $\nabla_x f(x)\mid_{(x^*,u^*)})$ one has: $\frac{\partial f_4}{\partial x_1}=\tilde{b}_4c_2$, $\frac{\partial f_4}{\partial x_2}=0$, $\frac{\partial f_4}{\partial x_3}=03$, $\frac{\partial f_4}{\partial x_4}=\tilde{s}_4$ and $\frac{\partial f_4}{\partial x_5}=0$.

About the fifth row of the Jacobian matrix $\nabla_x f(x)\mid_{(x^*,u^*)})$ one has: $\frac{\partial f_5}{\partial x_1}=\tilde{b}_5\frac{(q_2x_3-q_3k_{cs}x_4)}{(x_1+q_2x_3-q_3k_{cp}x_3x_4)^2}$, $\frac{\partial f_5}{\partial x_2}=0$, $\frac{\partial f_5}{\partial x_3}=\tilde{b}_5\frac{(-x_1(q_2-q_3k_{cs}x_4))}{(x_1+q_2x_3-q_3k_{cp}x_3x_4)^2}$, $\frac{\partial f_5}{\partial x_4}=\tilde{b}_5\frac{(q_3k_{cs}x_3)}{(x_1+q_2x_3-q_3k_{cp}x_3x_4)^2}$ and $\frac{\partial f_5}{\partial x_5}=0$.

The Jacobian matrix $\nabla_x g_1(x)\mid_{(x^*,u^*)})$ is $0\in R^{5\times 5}$.

Finally, computation of the Jacobian matrices $\nabla_x g_2(x)\mid_{(x^*,u^*)})$ and $\nabla_x g_3(x)\mid_{(x^*,u^*)})$ is performed:

$$\nabla_x g_2(x)\mid_{(x^*,u^*)})= \begin{pmatrix} 0 & 0 & 0 & 0 & 0 \\ -\tilde{n}_2 & 0 & 0 & 0 & 0 \\ 0 & 0 & 0 & 0 & 0 \\ 0 & 0 & 0 & 0 & 0 \\ 0 & 0 & 0 & 0 & 0 \end{pmatrix}\mid_{(x^*,u^*)}) \qquad \nabla_x g_3(x)\mid_{(x^*,u^*)})= \begin{pmatrix} 0 & 0 & 0 & 0 & 0 \\ 0 & 0 & 0 & 0 & 0 \\ 0 & 0 & 0 & 0 & 0 \\ 0 & 0 & 0 & -\tilde{n}_4 & 0 \\ 0 & 0 & 0 & 0 & 0 \end{pmatrix}\mid_{(x^*,u^*)}) \tag{5.119}$$

5.4.4 Stabilizing feedback control

After linearization around its current operating point, the dynamic model of the mechanical pulping process is written as

$$\dot{x}=Ax+Bu+d_1 \tag{5.120}$$

Parameter d_1 stands for the linearization error in the pulping process's dynamic model appearing previously in (5.120). The reference setpoints for the pulping process's state vector are denoted by $\mathbf{x_d} = [x_1^d, \ldots, x_5^d]$. Tracking of this trajectory is achieved after applying the control input u_d. At every time instant, the control input u_d is assumed to differ from the control input u appearing in (5.120) by an amount equal to Δu, that is $u_d = u + \Delta u$

$$\dot{x}_d = Ax_d + Bu_d + d_2 \tag{5.121}$$

The dynamics of the controlled system described in (5.120) can be also written as

$$\dot{x} = Ax + Bu + Bu_d - Bu_d + d_1 \tag{5.122}$$

and by denoting $d_3 = -Bu_d + d_1$ as an aggregate disturbance term one obtains

$$\dot{x} = Ax + Bu + Bu_d + d_3 \tag{5.123}$$

By subtracting (5.121) from (5.123) one has

$$\dot{x} - \dot{x}_d = A(x - x_d) + Bu + d_3 - d_2 \tag{5.124}$$

By denoting the tracking error as $e = x - x_d$ and the aggregate disturbance term as $\tilde{d} = d_3 - d_2$, the tracking error dynamics becomes

$$\dot{e} = Ae + Bu + \tilde{d} \tag{5.125}$$

For the approximately linearized model of the system a stabilizing feedback controller is developed. The controller has the form

$$u(t) = -Ke(t) \tag{5.126}$$

with $K = \frac{1}{r}B^T P$ where P is a positive definite symmetric matrix which is obtained from the solution of the Riccati equation (Rigatos and Busawon, 2018)

$$A^T P + PA + Q - P(\frac{2}{r}BB^T - \frac{1}{\rho^2}LL^T)P = 0 \tag{5.127}$$

where Q is a positive semi-definite symmetric matrix. The diagram of the considered control loop is depicted in Figure 5.27. The global stability properties of the control method can be proven following the procedure presented in Section 1.1.2.5.

5.4.5 Simulation tests

The performance of the proposed nonlinear optimal control method for the mechanical pulping process in paper industry is tested through simulation experiments. To compute the feedback gains of the controller, the algebraic Riccati equation appearing in (5.127) is iteratively solved at each time-step of the control method. The parameters of the state-space model of the pulping process were according to Refs. [1] and [21]. Considering a sampling period of $T_s = 0.01$ sec, indicative values for these parameters have been $\bar{a}_1 = \frac{a_1 - 1}{T_s} = -0.6$, $\bar{b}_1 = \frac{b_1}{T_s} = 0.1$, $\bar{a}_2 = \frac{a_2 - 1}{T_s} = -0.6$, $\bar{b}_2 = \frac{b_2}{T_s} = -0.1$,

$\bar{a}_3 = \frac{a_3-1}{T_s} = -0.6$, $\bar{b}_3 = \frac{b_1}{T_s} = -0.1$, $\bar{a}_4 = \frac{a_4-1}{T_s} = -0.6$, $\bar{b}_4 = \frac{b_4}{T_s} = -0.1$, and $\bar{a}_5 = \frac{a_5-1}{T_s} = -0.6$, $\bar{b}_5 = \frac{b_1}{T_s}0.1$. In the presented simulation tests the following values were given to the parameters r, ρ and Q which appear in the Riccati equation of the control method: $r = 0.001$, $\rho = 0.1$ and $Q = 40I_{5\times5}$ and $L_m = 10^{-4}I_{5\times5}$

The obtained results are shown in Figures 5.28–5.33 The real values of the state variables of the process are printed in blue, the estimated values which are provided by the H-infinity Kalman Filter are plotted in green, while the related setpoints are plotted in red. The values of the control inputs are shown in Figures 5.34–5.36. It can be noticed, that under the proposed nonlinear optimal control method, precise tracking of the reference setpoints was achieved. The state variables of the system converged fast to the targeted values while the variations of the control inputs remained moderate. Since the control algorithm provides the values of the control variables v_i, $i = 1, 2, 3$ one has to solve also a set of three algebraic equations given in Section 5.4.2 to find finally the initial control variables u_i, $i = 1, 2, 3$. By using the H-infinity Kalman Filter as a state estimator it was possible to implement state estimation-based control through the processing of the measurable state variable, that is the production rate x_1, the motor's load at the first refiner x_2 and the motor's load at the second refiner x_4.

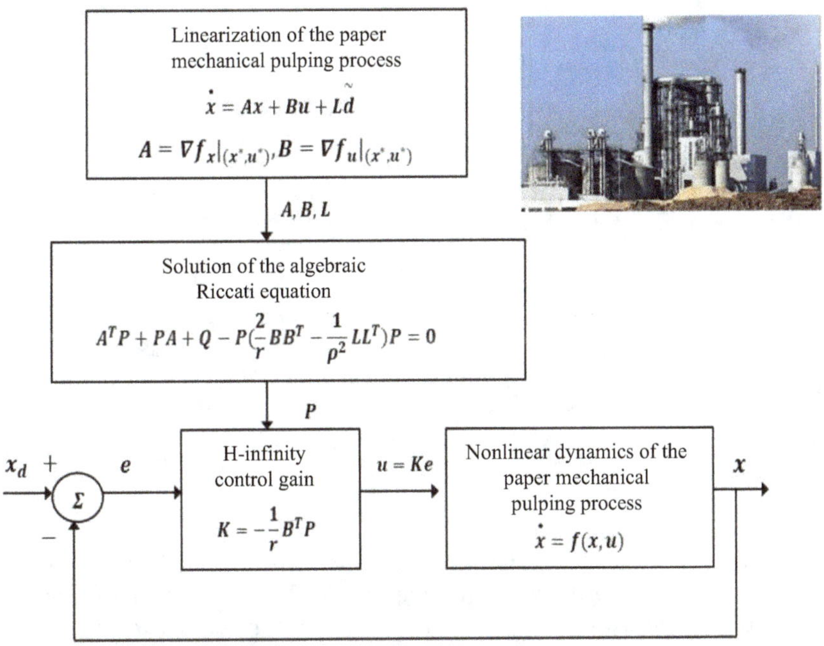

Figure 5.27 Diagram of the optimal (H-infinity) control loop for the mechanical pulping process in paper mills

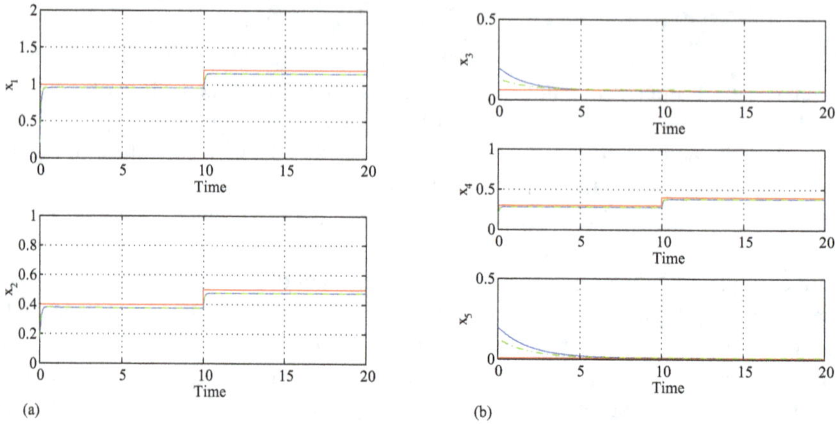

Figure 5.28 *Nonlinear optimal control of the mechanical pulping process in the paper industry when tracking setpoint 1: (a) convergence of state variables x_1 to x_2 to the related reference setpoints and (b) convergence of state variables x_3 to x_5 (blue lines) to the related reference setpoints*

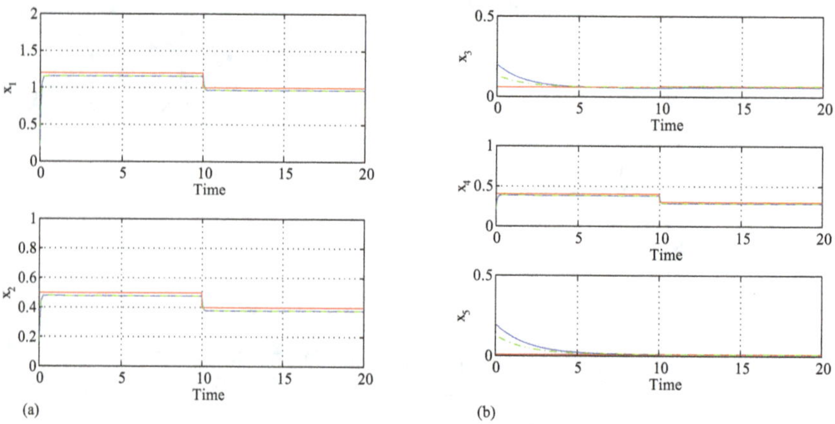

Figure 5.29 *Nonlinear optimal control of the mechanical pulping process in the paper industry when tracking setpoint 2: (a) convergence of state variables x_1 to x_2 to the related reference setpoints and (b) convergence of state variables x_3 to x_5 (blue lines) to the related reference setpoints*

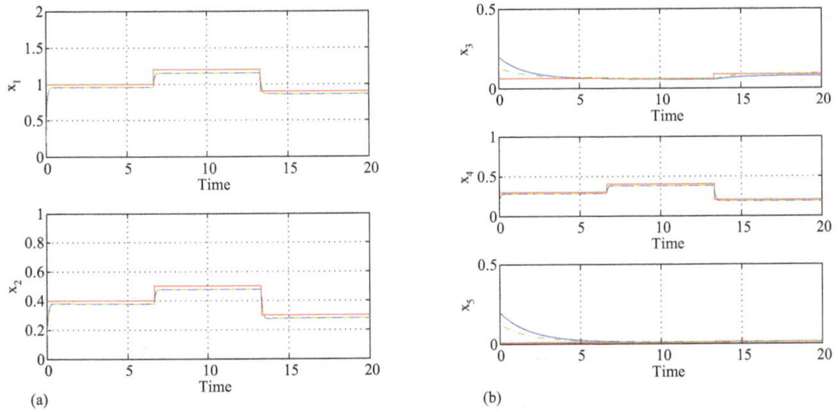

Figure 5.30 *Nonlinear optimal control of the mechanical pulping process in the paper industry when tracking setpoint 3: (a) convergence of state variables x_1 to x_2 to the related reference setpoints and (b) convergence of state variables x_3 to x_5 (blue lines) to the related reference setpoints*

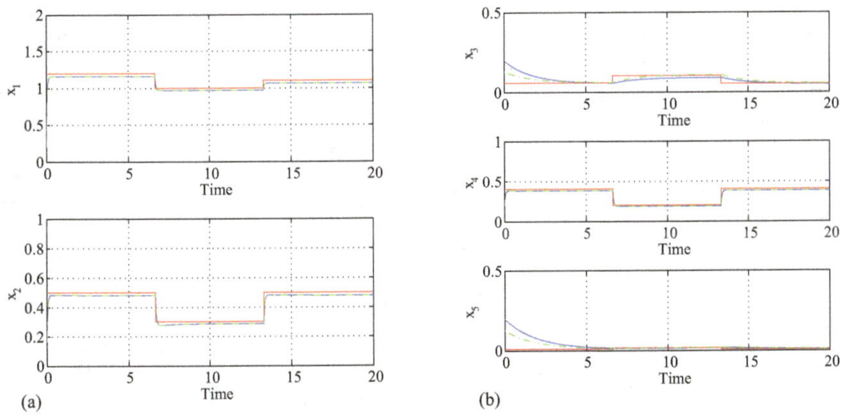

Figure 5.31 *Nonlinear optimal control of the mechanical pulping process in the paper industry when tracking setpoint 4: (a) convergence of state variables x_1 to x_2 to the related reference setpoints and (b) convergence of state variables x_3 to x_5 (blue lines) to the related reference setpoints*

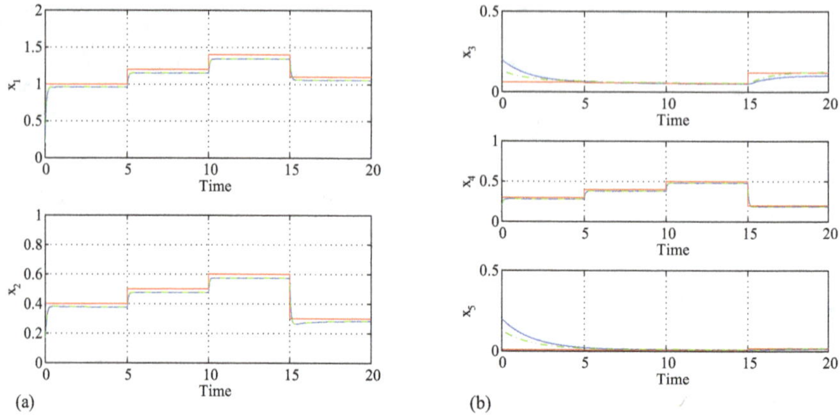

Figure 5.32 Nonlinear optimal control of the mechanical pulping process in the paper industry when tracking setpoint 5: (a) convergence of state variables x_1 to x_2 to the related reference setpoints and (b) convergence of state variables x_3 to x_5 (blue lines) to the related reference setpoints

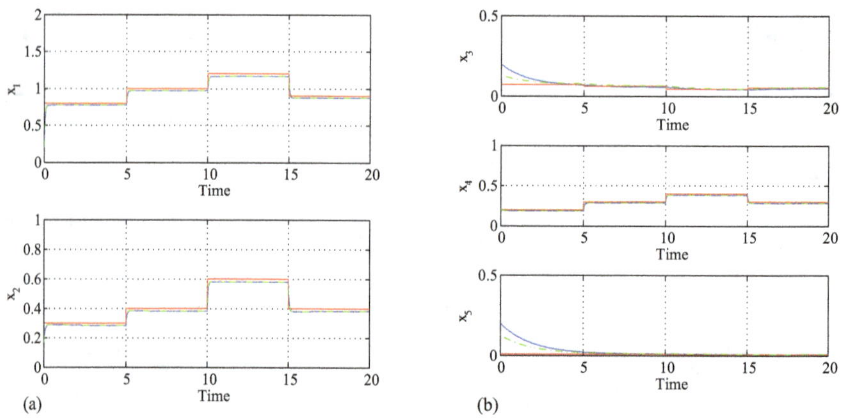

Figure 5.33 Nonlinear optimal control of the mechanical pulping process in the paper industry when tracking setpoint 6: (a) convergence of state variables x_1 to x_2 to the related reference setpoints and (b) convergence of state variables x_3 to x_5 (blue lines) to the related reference setpoints

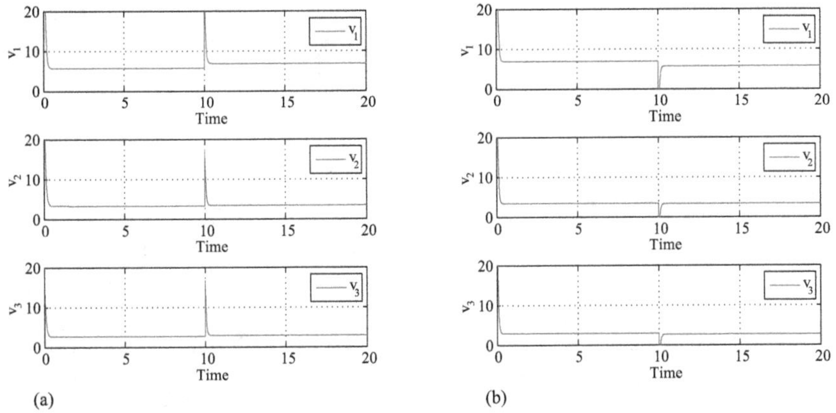

Figure 5.34 *(a) Nonlinear optimal control of the mechanical pulping process in the paper industry: variations of the control inputs v_1 to v_3 when tracking setpoint 1 and (b) variations of the control inputs v_1 to v_3 when tracking setpoint 2*

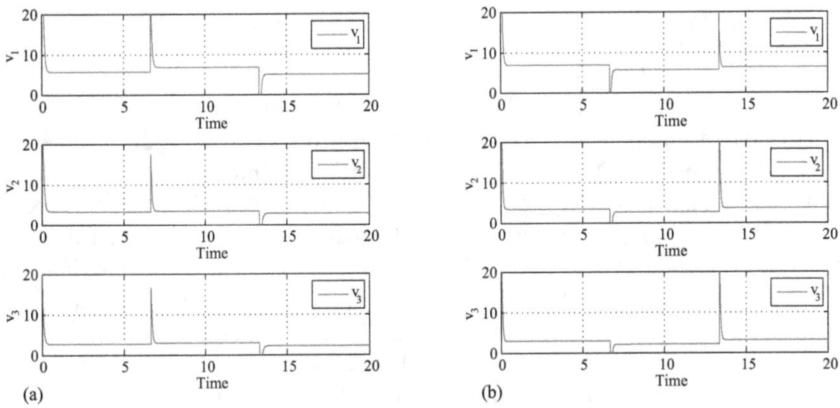

Figure 5.35 *(a) Nonlinear optimal control of the mechanical pulping process in the paper industry: variations of the control inputs v_1 to v_3 when tracking setpoint 3, and (b) variations of the control inputs v_1 to v_3 when tracking setpoint 4*

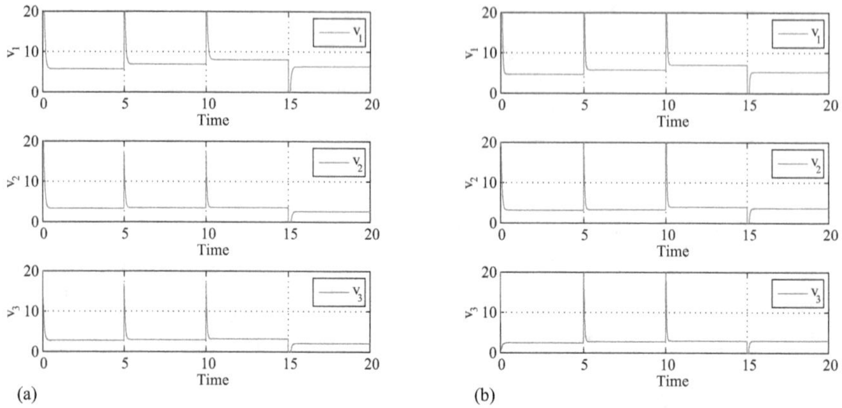

Figure 5.36 (a) Nonlinear optimal control of the mechanical pulping process in the paper industry: variations of the control inputs v_1 to v_3 when tracking setpoint 5 and (b) variations of the control inputs v_1 to v_3 when tracking setpoint 6

5.5 Nonlinear control of the injection moulding machine

5.5.1 Outline

The injection moulding process is nowadays the most widespread production method in the thermoplastics industry. It achieves controlled polymerization of raw material and the massive production of new plastic items (Froechlich *et al.*, 2020; Froechlich *et al.*, 2018; Lindert *et al.*, 2014). The injection moulding process takes place in injection moulding machines (Rigatos *et al.*, 2001; Stemmler *et al.*, 2019; Stemmler *et al.*, 2020). These machines have usually electro-hydraulic or servo-electric actuation (Tsai *et al.*, 2009; Hopmann *et al.*, 2021; Reiter *et al.*, 2014). The injection moulding process comprises two main stages, namely the injection stage, and the packing stage (Yang and Gao, 2000; Tan *et al.*, 2001; Huang, Tan *et al.*, 2004). In the injection stage, raw material drops from a storage tank called hopper into a metallic horizontal tube which is called barrel. There, the raw material is heated and melted. In the interior of the barrel, a rotating screw transfers the melted material forward to the opening of a cavity which is the mould that will give shape to the plastic products. When reaching the end of the barrel the screw injects the melted material in the mould up to the point that the cavity is filled. In the packing stage high pressure is initially maintained in the cavity until the material gets solidified. Cooling down follows for the plastic item and finally the mould opens and the well-formed and finished item is ejected. A new production cycle follows. There are several parameters of the injection moulding machine that affect the completeness of shape, the avoidance of deformations, the mechanical strength, the fine finishing and other properties defining the quality of the produced items (Yao, Gao *et al.*, 2008; Ogordayk and Martinsen, 2018; Dubay, 2002).

Among them, of primary importance is the temperature of the barrel and the pressure of the cavity, each one of them being controlled by a separate control loop (Huang and Lee, 2000; He *et al.*, 2003; Cho *et al.*, 2009). Energy dispersion is a major issue in the functioning of injection moulding machines. Actually, a significant part of energy is consumed by the machines' actuators (Lu, Zhang *et al.*, 2020; Abeykoon *et al.*, 2021). Consequently, minimization of energy losses is a challenge for the associated control loops (Meekers *et al.*, 2018; Zhao, Zhou *et al.*, 2014; Lafranche *et al.*, 2006).

The present section is concerned with the cavity pressure control problem in a servo-electric injection moulding machine. In this specific type of injection moulding machine actuation is provided to the screw through a DC motor and a gears-transmission system. Two pressure variables are monitored through dedicated pressure sensors, the pressure of the screw and the pressure of the cavity (Rigatos *et al.*, 2001; Stemmler *et al.*, 2020). Both of these variables are shown to be connected to the position and speed variables of the screw. This allows to associate finally the set-points of the pressure variables with the servo-actuator's turn speed setpoint. Next, to control the turn motion of the actuator and subsequently the position and the velocity of the screw a nonlinear optimal control method is developed (Rigatos and Busawon, 2018; Rigatos, 2016b). The differential flatness properties of the injection moulding machine are proven, thus allowing to solve the setpoints definition problem for the rest of the state variables of the servo-actuator. The dynamic model of the electrical actuator of the injection moulding process undergoes approximate linearization with the use of first-order Taylor series expansion and through the computation of the associated Jacobian matrices (Rigatos and Siano, 2015; Rigatos, Siano *et al.*, 2015). The linearization process takes place at each sampling instance around a temporary operating point which is defined by the present value of the state vector of the injection moulding machine and by the last value of the control inputs vector (Rigatos and Tzafestas, 2007; Basseville and Nikiforov, 1993; Rigatos and Zhang, 2009). The modelling error that is due to the truncation of higher-order terms from the Taylor series expansion is considered to be a perturbation which is asymptotically compensated by the robustness of the control method.

For the approximately linearized model of the injection moulding machine a stabilizing H-infinity (optimal) feedback controller is designed. Actually, this controller stands for the solution of the nonlinear optimal control problem for the injection moulding process under model uncertainty and external perturbations. This controller implements a min–max differential game taking place between (i) the control inputs which try to minimize a cost function that contains a quadratic term of the state vector's tracking error and (ii) the model uncertainty and external disturbance terms which try to maximize this cost function. To select the stabilizing feedback gains of this controller an algebraic Riccati equation is repetitively solved at each time-step of the control algorithm (Rigatos, 2011; Rigatos, 2015). The global stability properties of the control loop are proven through Lyapunov analysis (Rigatos and Busawon, 2018; Rigatos, 2016b). First, it is proven that the control scheme satisfies the H-infinity tracking performance criterion which signifies elevated robustness against model uncertainty and exogenous perturbations. Moreover, under moderate conditions it is proven that the control loop is globally asymptotically stable. The method retains the advantages of linear optimal control, that is fast and accurate tracking of reference trajectories under moderate variations of the control inputs

The present section provides one of the few existing solutions to the nonlinear optimal control problem of injection moulding machines which is of proven global stability while also remaining computationally efficient (Rigatos and Busawon, 2018). Preceding results on the use of H-infinity control to nonlinear dynamical systems were limited to the case of affine-in-the-input systems with drift-only dynamics and considered that the control inputs gain matrix is never dependent on the values of the system's state vector. Moreover, in these approaches the linearization was performed around points of the desirable trajectory whereas in the present section's control method the linearization points are related to the value of the state vector at each sampling instant as well as with the last sampled value of the control inputs vector. The Riccati equation which has been proposed for computing the feedback gains of the controller is novel, so is the presented global stability proof through Lyapunov analysis.

5.5.2 Dynamic model of the injection moulding process

5.5.2.1 Main parameters of the injection moulding process

Plastics products are widely used in industrial and everyday living applications. Among such products, 70%–80% are manufactured by the injection moulding process (Lu, Zhang *et al.*, 2020; Abeykoon *et al.*, 2021). With the raise of plastics consumption, the demand for injection moulding machines has also grown significantly. The scale of the plastics industry worldwide is huge. In 2016, the European plastics industry comprised 60,000 companies and was employing 1.7 million people with annual sales reaching the amount of 350 bn Euros. Besides the market size of such machines was estimated to be 6.63 billion dollars in 2018 with a perspective to reach the amount of 10.40 billion dollars in 2025.

Injection moulding machines often exhibit low efficiency and high energy waste. The greatest part of energy losses comes from the actuators of these machines (hydraulic or servo-electric actuators) and is due to sub-optimality of their control loops and empirical selection of controllers parameters (Lu, Zhang *et al.*, 2020; Abeykoon *et al.*, 2021). It is estimated that injection moulding machines consume on an annual basis 185 GWh of electric energy. Therefore, it is a main challenge to achieve optimality in the control of such machines. This will allow to eliminate energy dispersion in the machines' functioning and to reduce significantly the plastics' production cost. The diagram of a typical injection moulding machine is shown in Figure 5.37.

The main parameters of the injection moulding process are shown in Figure 5.38. These are (i) the pressure p_c and the temperature T_c of the cavity, the cavity's volume V_c, the mass m_c of the polymer in the cavity, (ii) the pressure p_s and the temperature T_s of the barrel (screw), the barrel's volume V_s, the mass m_s of the polymer in the barrel and (iii) the position x, velocity v and acceleration $\dot{v}$ of the screw, and (iv) the turn angle θ, the angular speed ω and the rotor's current of the screw's servo-actuator (Rigatos *et al.*, 2001; Stemmler *et al.*, 2020).

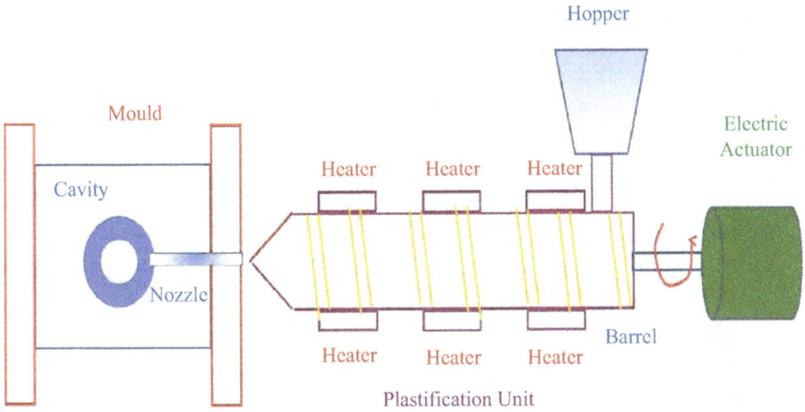

Figure 5.37 Diagram of the injection moulding machine

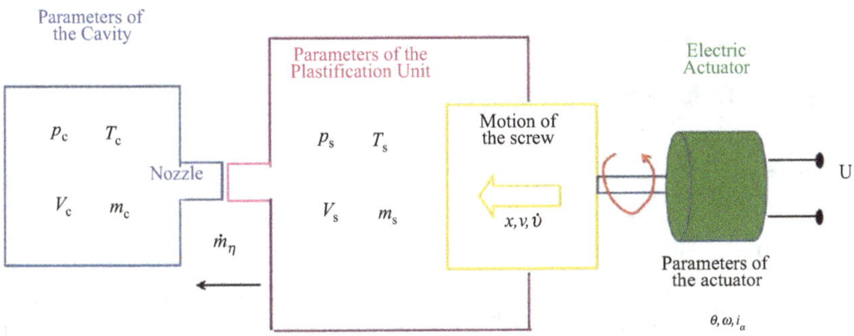

Figure 5.38 Parameters of the cavity and the plastification unit of the injection moulding machine

The dynamic model of the injection moulding process is described by a state-space model of the form given below (Stemmler *et al.*, 2019; Stemmler *et al.*, 2020):

$$\dot{x} = f(x, u) \tag{5.128}$$

with the associated state vector to be defined as $x = [x, v, \dot{v}, p_s, p_c]^T$, where x is the position of the screw v is the velocity of the screw, $\dot{v}$ is the acceleration of the screw, p_s is the pressure of the barrel and p_c is the pressure of the cavity. Equivalently, the dynamics of the linear motion of the screw can be substituted by the dynamics of the DC servo-motor that provides actuation to the screw. In such an approach the state vector of the injection moulding machine is defined as $x = [\theta, \omega, i_a, p_s, p_c]^T$ where θ is the turn angle of the motor, ω is the angular speed of the motor and i_a is the current at the rotor of the DC motor.

The state-space of the process is written in the following form (Stemmler *et al.*, 2019; Stemmler *et al.*, 2020): $f_1(x,u)$ if $p_{c,far} \leq p_{switch}$ and $f_2(x,u)$ if $p_{c,far} > p_{switch}$. Function $f_1(x,u)$ is valid until the cavity is fully filled. This is indicated by the value of the cavity pressure measured near the end of the flow path. The latter variable is denoted as $p_{c,far}$. When the value of pressure variable $p_{c,far}$ exceeds threshold variable p_{switch} it is concluded that the cavity has been filled with polymer and packing stage of the process begins. In the latter functioning mode, the dynamics of the process is described by function $f_2(x,u)$.

In the injection stage of the process it holds that (Stemmler *et al.*, 2019; Stemmler *et al.*, 2020)

$$\begin{pmatrix} \dot{p}_s \\ \dot{p}_c \end{pmatrix} = \begin{pmatrix} \frac{\beta_s}{\theta}(-\omega - \frac{\pi R^4}{8L\eta}(p_s - p_c)) \\ -a_2 p_s + k_s p_s^* \end{pmatrix} \text{ if } p_{c,far} \leq p_{switch} \tag{5.129}$$

The delayed pressure of the barrel (screw) is defined as

$$p_s^* = \begin{cases} p_s(t - T_{d,2}) \text{ for } t > T_{d,2} \\ 0 \text{ otherwise} \end{cases} \tag{5.130}$$

By applying Taylor series expansion in the above relation one obtains

$$p_s^* = \begin{cases} p_s(t) - T_{d,2}\dot{p}_s(t) \text{ for } t > T_{d,2} \\ 0 \text{ otherwise} \end{cases} \tag{5.131}$$

where $p_s(t) = x_4$ and $\dot{p}_s(t) = \dot{x}_4$, therefore

$$p_s^* = \begin{cases} p_s - T_{d,2}\dot{p}_s \text{ for } t > T_{d,2} \\ 0 \text{ otherwise} \end{cases} \tag{5.132}$$

and by substituting the relation about state variable $\dot{x}_4$ one obtains (Stemmler *et al.*, 2019; Stemmler *et al.*, 2020):

$$p_s^* = \begin{cases} p_s - T_{d,2}\frac{\beta_s}{\theta}\left[-\omega - \frac{\pi R^4}{8L\eta}(p_s - p_c)\right] \text{ for } t > T_{d,2} \\ 0 \text{ otherwise} \end{cases} \tag{5.133}$$

where β_s is constant, R is the radius of the nozzle, L is the length of the nozzle and η is a viscosity coefficient. Thus, the pressure variables in the injection phase of the process become

$$\begin{pmatrix} \dot{p}_s \\ \dot{p}_c \end{pmatrix} = \begin{pmatrix} \frac{\beta_s}{\theta}(-\omega - \frac{\pi R^4}{8L\eta}(p_s - p_c)) \\ -a_2 p_c + k_s\left\{p_s - T_{d,2}\frac{\beta_s}{\theta}\left[-\omega - \frac{\pi R^4}{8L\eta}(p_s - p_c)\right]\right\} \end{pmatrix} \begin{matrix} \text{if } p_{c,far} \leq p_{switch} \\ \text{and } t > T_{d,2} \end{matrix} \tag{5.134}$$

while it holds that

$$\begin{pmatrix} \dot{p}_s \\ \dot{p}_c \end{pmatrix} = \begin{pmatrix} \frac{\beta_s}{\theta}\left(-\omega - \frac{\pi R^4}{8L\eta}(p_s - p_c)\right) \\ -a_2 p_c \end{pmatrix} \quad \text{if } p_{c,far} \leq p_{switch} \text{ and } t \leq T_{d,2} \tag{5.135}$$

About the packing stage of the injection moulding process, which begins once the filling of the cavity has been completed it holds that (Stemmler *et al.*, 2019; Stemmler *et al.*, 2020):

$$\begin{pmatrix} \dot{p}_s \\ \dot{p}_c \end{pmatrix} = \begin{pmatrix} \frac{\beta_s}{\theta}\left(-\omega - \frac{\pi R^4}{8L\eta}(p_s - p_c)\right) \\ -\frac{\beta_c}{V_0}(p_s - p_c)\frac{\pi R^4}{8L\eta} \end{pmatrix} \quad \text{if } p_{c,far} > p_{switch} \tag{5.136}$$

The above relations connecting the pressure of the barrel (screw's tube) p_s and the pressure of the cavity p_c to the angular speed of the DC motor can be used to define setpoints for the motor's turn speed.

5.5.2.2 Actuator of the injection moulding machine

The screw which feeds the cavity with melted polymer receives actuation from a DC motor. The model of the DC motor is given by (Rigatos, 2011)

$$\dot{x}_m = f_m(x) + g_m(x)u \tag{5.137}$$

where $x_m = [\theta, \dot{\theta}, i_a]^T$ with $x_m \in R^{3\times 1}$ and θ to be the turn angle of the rotor, $\dot{\theta}$ is the angular speed of the motor, and i_a is the rotor's current. Besides $f_m(x) \in R^{3\times 1}$, $g_m(x) \in R^{3\times 1}$, and $u \in R$. About vectors $f_m(x)$ and $g(_m x)$ it holds that Rigatos, 2011

$$f_m(x) = \begin{pmatrix} x_2 \\ k_1 x_2 + k_2 x_3 + k_3 x_3^2 + k_4 T_1 \\ k_5 x_2 + k_6 x_2 x_3 + k_7 x_3 \end{pmatrix} \quad g_m(x) = \begin{pmatrix} 0 \\ 0 \\ k_8 \end{pmatrix} \tag{5.138}$$

where $k_1 = -\frac{F}{J}$ $k_2 = \frac{A}{J}$, $\frac{k_3 = B}{J}$, $k_4 = -\frac{A}{J}$, $k_5 = -\frac{A}{L}$, $k_6 = -\frac{B}{L}$, $k_7 = -\frac{R}{L}$, k_8, A, B are constants, L is the induction of the rotor, J is the rotor's inertia and F is a friction coefficient. Finally, T_1 is the external load that the servo-actuation may have to compensate for, which in the case of the injection moulding machine can be due to the accumulation of melted polymer in the front part of the screw during the injection stage.

5.5.2.3 Differential flatness properties of the actuator

The model of the servo-actuator of the injection moulding machine is differentially flat, with flat out output $y = x_1 = \theta$, which is the turn angle of the rotor. From the first row of the servo-motor's model one has

$$x_2 = \dot{x}_1 \Rightarrow x_2 = h_2(y, \dot{y}) \tag{5.139}$$

which signifies that state variable x_2 is a differential function of the motor's flat outputs. From the second row of the state-space model of the DC actuator and after considering that the load T_l is a function of the turn angle of the rotor, that is $T_l(\theta)$ one has

$$\begin{aligned}\dot{x}_2 &= k_1x_2 + k_2x_3 + k_3x_3^2 + k_3T_1(x_1) \Rightarrow \\ k_2{x_3}^2 &+ k_3x_3 + [k_1x_2 + k_4T_1(x_1) - \dot{x}_2] = 0\end{aligned} \tag{5.140}$$

This is a binomial with respect to x_3 and the associated solutions are

$$x_3 = \frac{-k_2 \pm \sqrt{|k_2^2 - 4k_3[k_1x_2 + k_4T_1(x_1) - \ddot{x}_1]|}}{2k_3} \tag{5.141}$$

and by retaining the largest of the above-noted solutions one has that x_3 is also a differential function of the flat outputs of the system. Finally, from the last row of the state-space model of the DC motor one has

$$u = \frac{1}{k_3}[\dot{x}_3 - k_5x_2 - k_6x_2x_3 - k_7x_3] \tag{5.142}$$

Since x_2, x_3 are differential functions of the flat outputs of the motor one obtains that the control input u is also a differential function of the system's flat output.

In conclusion, all state variables and control inputs of the DC servo-motor that provides actuation to the screw are differential functions of the flat outputs of the system, and this actuator is a differentially flat system. The differential flatness property can be used for defining a setpoint for the rotor's current i_a after knowing the setpoints for the rotor's turn angle and angular velocity.

5.5.2.4 Selection of setpoints

The pressure variable of the barrel p_s and the pressure variable of the cavity p_c are given the state variables notation x_4 and x_5, respectively. Thus the previous relation of (5.134) becomes

$$\begin{pmatrix}\dot{x}_4 \\ \dot{x}_5\end{pmatrix} = \begin{pmatrix} \frac{\beta_s}{x_1}\left(-x_2 - \frac{\pi R^4}{8L\eta}(x_4 - x_5)\right) \\ -a_2x_5 + k_s\left\{x_4 - T_{d,2}\frac{\beta_s}{x_1}\left[-x_2 - \frac{\pi R^4}{8L\eta}(x_4 - x_5)\right]\right\}\end{pmatrix} \quad \begin{matrix}\text{if } p_{c,far} \leq p_{switch} \\ \text{and } t > T_{d,2}\end{matrix} \tag{5.143}$$

By denoting the setpoints which are related with the state variables of the process as $x_i^d, i = 1, \ldots, 5$ the following relation holds

$$\begin{pmatrix}\dot{x}_4^d \\ \dot{x}_5^d\end{pmatrix} = \begin{pmatrix} \frac{\beta_s}{x_1^d}\left(-x_2^d - \frac{\pi R^4}{8L\eta}(x_4^d - x_5^d)\right) \\ -a_2x_5^d + k_s\left\{x_4^d - T_{d,2}\frac{\beta_s}{x_1^d}\left[-x_2^d - \frac{\pi R^4}{8L\eta}(x_4^d - x_5^d)\right]\right\}\end{pmatrix} \quad \begin{matrix}\text{if } p_{c,far} \leq p_{switch} \\ \text{and } t > T_{d,2}\end{matrix} \tag{5.144}$$

Using (5.144) and by considering that setpoints x_4^d and x_5^d are piece-wise constant and thus $\dot{x}_4^d = 0$ and $\dot{x}_5^d = 0$ one gets

$$\begin{pmatrix}0\\0\end{pmatrix} = \begin{pmatrix} \frac{\beta_s}{x_1^d}\left(-x_2^d - \frac{\pi R^4}{8L\eta}(x_4^d - x_5^d)\right) \\ -a_2 x_5^d + k_s\left\{x_4^d - T_{d,2}\frac{\beta_s}{x_1^d}\left[-x_2^d - \frac{\pi R^4}{8L\eta}(x_4^d - x_5^d)\right]\right\}\end{pmatrix} \quad \begin{matrix}\text{if } p_{c,far} \leq p_{switch} \\ \text{and } t > T_{d,2}\end{matrix} \tag{5.145}$$

which after intermediate operations gives

$$\begin{pmatrix} -\frac{\pi R^4}{8L\eta} & \frac{\pi R^4}{8L\eta} \\ K_s + K_s T_{d,2}\frac{\beta_s}{x_1^d}\frac{\pi R^4}{8L\eta} & -a_2 - K_s T_{d,2}\frac{\beta_s}{x_1^d}\frac{\pi R^4}{8L\eta}\end{pmatrix}\begin{pmatrix}x_4^d\\x_5^d\end{pmatrix} = \begin{pmatrix}x_2^d \\ -K_s T_{d,2}\frac{\beta_s}{x_1^d}x_2^d\end{pmatrix} \tag{5.146}$$

while solving with respect to x_4^d and x_5^d one obtains

$$\begin{pmatrix}x_4^d\\x_5^d\end{pmatrix} = \begin{pmatrix} -\frac{\pi R^4}{8L\eta} & \frac{\pi R^4}{8L\eta} \\ K_s + K_s T_{d,2}\frac{\beta_s}{x_1^d}\frac{\pi R^4}{8L\eta} & -a_2 - K_s T_{d,2}\frac{\beta_s}{x_1^d}\frac{\pi R^4}{8L\eta}\end{pmatrix}^{-1}\begin{pmatrix}x_2^d \\ -K_s T_{d,2}\frac{\beta_s}{x_1^d}x_2^d\end{pmatrix} \tag{5.147}$$

It can be concluded that the pressure setpoints for the screw's tube and the cavity are dependent on the velocity setpoints for the DC electric actuator of the injection moulding machine. Inversely, by knowing the pressure setpoints for the screw's tube and the cavity one can designated the functioning of the DC motor at the associated turn speed setpoints.

5.5.3 Stable feedback control of the injection moulding process

5.5.3.1 Approximate linearization of the Injection Moulding Machine

The dynamic model of the injection moulding process undergoes approximate linearization around the temporary operating point (x^*, u^*) at each sampling instance, where x^* is the present value of the system's state vector and u^* is the last sampled value of the control inputs vector. The linearization relies on first-order Taylor series expansion and on the computation of the associated Jacobian matrices. The linearization process gives

$$\dot{x} = Ax + Bu + \tilde{d} \tag{5.148}$$

where $\tilde{d}$ is the cumulative disturbances vector which comprises (a) the modelling error due to the truncation of higher-order terms in the Taylor series expansion, (b) exogenous perturbations and (c) sensors measurement noise of any distribution. Matrices A and B are the Jacobian matrices of the system which are computed as follows:

$$A = \nabla_x[f(x) + g(x)u] \mid_{(x^*,u^*)} \Rightarrow A = \nabla_x f(x) \mid_{(x^*,u^*)} \tag{5.149}$$

$$B = \nabla_x u[f(x) + g(x)u] \mid_{(x^*,u^*)} \Rightarrow B = g(x) \mid_{(x^*,u^*)} \tag{5.150}$$

The computation of the Jacobian matrix $A = \nabla_x f(x) \mid_{(x^*,u^*)}$ proceeds as follows:

First row of the Jacobian matrix $A = \nabla_x f(x) \mid_{(x^*,u^*)}$: $\frac{\partial f_1}{\partial x_1} = 0$, $\frac{\partial f_1}{\partial x_2} = 1$, $\frac{\partial f_1}{\partial x_3} = 0$

Second row of the Jacobian matrix $A = \nabla_x f(x) \mid_{(x^*,u^*)}$: $\frac{\partial f_2}{\partial x_1} = k_4 \frac{\partial T_1}{\partial x_1}$, $\frac{\partial f_2}{\partial x_2} = k_1$, $\frac{\partial f_2}{\partial x_3} = k_2 + 2k_3x_3$

Third row of the Jacobian matrix $A = \nabla_x f(x) \mid_{(x^*,u^*)}$: $\frac{\partial f_3}{\partial x_1} = 0$, $\frac{\partial f_3}{\partial x_2} = k_5 + k_6x_3$, $\frac{\partial f_3}{\partial x_3} = k_6x_2 + k_7$

5.5.3.2 Stabilizing feedback control

After linearization around its current operating point, the dynamic model of the injection moulding machine is written as

$$\dot{x} = Ax + Bu + d_1 \tag{5.151}$$

Parameter d_1 stands for the linearization error in the injection moulding machine's dynamic model appearing previously in (5.151). The reference setpoints for the injection moulding machine's state vector are denoted by $\mathbf{x_d} = [x_1^d, \ldots, x_5^d]$. Tracking of this trajectory is achieved after applying the control input u_d. At every time instant the control input u_d is assumed to differ from the control input u appearing in (5.151) by an amount equal to Δu, that is $u_d = u + \Delta u$

$$\dot{x}_d = Ax_d + Bu_d + d_2 \tag{5.152}$$

where d_2 is the linearization error at points of the desirable trajectory. The dynamics of the controlled system described in (5.151) can be also written as

$$\dot{x} = Ax + Bu + Bu_d - Bu_d + d_1 \tag{5.153}$$

and by denoting $d_3 = -Bu_d + d_1$ as an aggregate disturbance term one obtains

$$\dot{x} = Ax + Bu + Bu_d + d_3 \tag{5.154}$$

By subtracting (5.152) from (5.154) one has

$$\dot{x} - \dot{x}_d = A(x - x_d) + Bu + d_3 - d_2 \tag{5.155}$$

By denoting the tracking error as $e = x - x_d$ and the aggregate disturbance term as $\tilde{d} = d_3 - d_2$, the tracking error dynamics becomes

$$\dot{e} = Ae + Bu + \tilde{d} \tag{5.156}$$

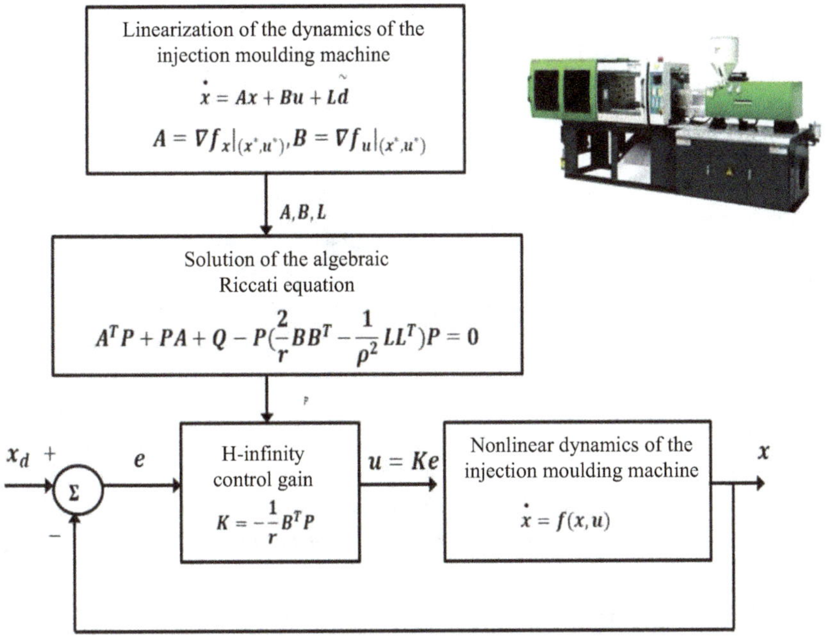

Figure 5.39 Diagram of the control scheme for the injection moulding machine

For the approximately linearized model of the system a stabilizing feedback controller is developed. The controller has the form

$$u(t) = -Ke(t) \tag{5.157}$$

with $K = \frac{1}{r}B^T P$ where P is a positive definite symmetric matrix which is obtained from the solution of the Riccati equation (Rigatos and Busawon, 2018)

$$A^T P + PA + Q - P\left(\frac{2}{r}BB^T - \frac{1}{\rho^2}LL^T\right)P = 0 \tag{5.158}$$

where Q is a positive semi-definite symmetric matrix. The diagram of the considered control loop for the injection moulding machine is depicted in Figure 5.39. The global stability properties of the control method can be proven following the procedure presented in Section 1.1.2.5.

5.5.4 Simulation tests

The performance of the nonlinear optimal control scheme for the servo-electric injection moulding machine has been confirmed through simulation experiments. The obtained results are shown in Figure 5.40–5.45. To implement the nonlinear optimal control method, the algebraic Riccati equation of Eq. (5.158) had to be repetitively

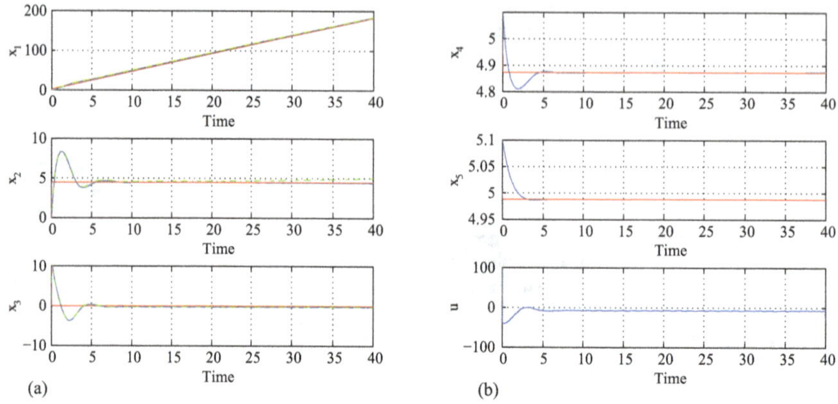

Figure 5.40 Control of the injection moulding machine in the case of cavity pressure setpoint 1 (a) convergence of state variables x_1, x_2, x_3 to their reference setpoints and (b) convergence of state variables x_4 and x_5 to their reference setpoints and variations of the control inputs u

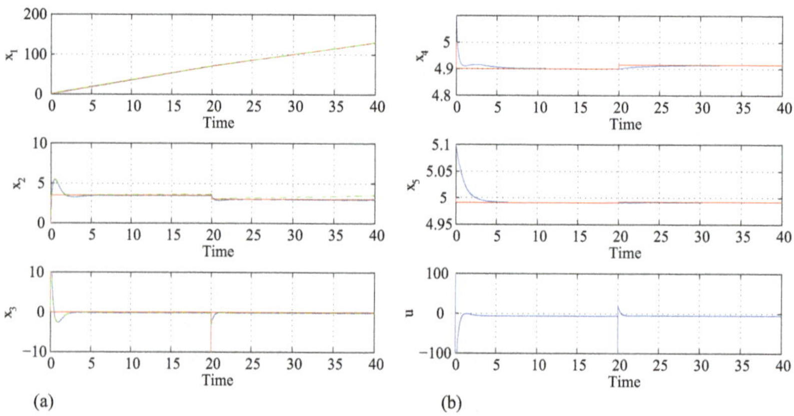

Figure 5.41 Control of the injection moulding machine in the case of cavity pressure setpoint 2 (a) convergence of state variables x_1, x_2, x_3 to their reference setpoints and (b) convergence of state variables x_4 and x_5 to their reference setpoints and variations of the control inputs u

solved at each time-step of the control algorithm. It can be observed, that under variable setpoints for the machine's cavity pressure and consequently under variable setpoints for the servo-actuator, the reference signals' tracking error for the state variables of the injection moulding machine was eliminated. Besides, the variations of the control inputs remained moderate. The latter is a significant result for reducing the energy consumption of the injection moulding process. Actually, it is only the solution

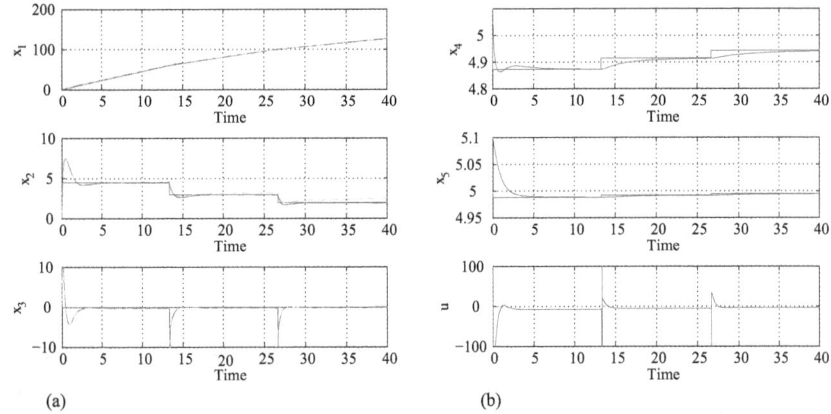

Figure 5.42 Control of the injection moulding machine in the case of cavity pressure setpoint 3 (a) convergence of state variables x_1, x_2, x_3 to their reference setpoints and (b) convergence of state variables x_4 and x_5 to their reference setpoints and variations of the control inputs u

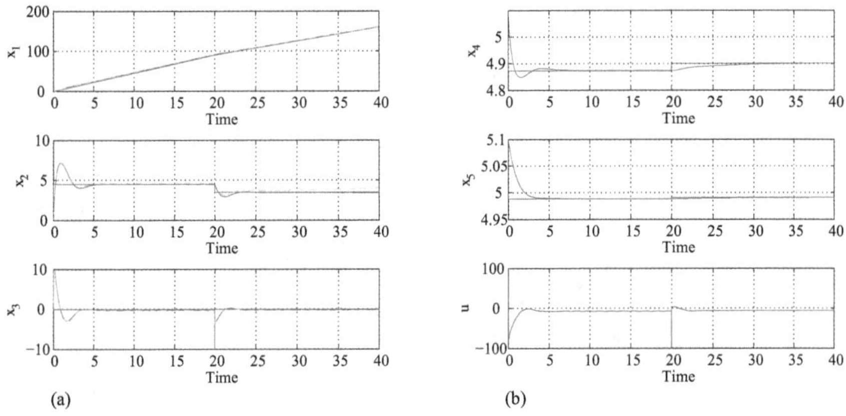

Figure 5.43 Control of the injection moulding machine in the case of cavity pressure setpoint 4 (a) convergence of state variables x_1, x_2, x_3 to their reference setpoints and (b) convergence of state variables x_4 and x_5 to their reference setpoints and variations of the control inputs u

of the nonlinear optimal control problem that allows for restraining the variations of the machine's control inputs. On the other side, all other control methods one can consider for these machines are suboptimal. In these methods the minimization of the state vector's tracking error does not ensure that excessive values for the control signals of the machine will be avoided.

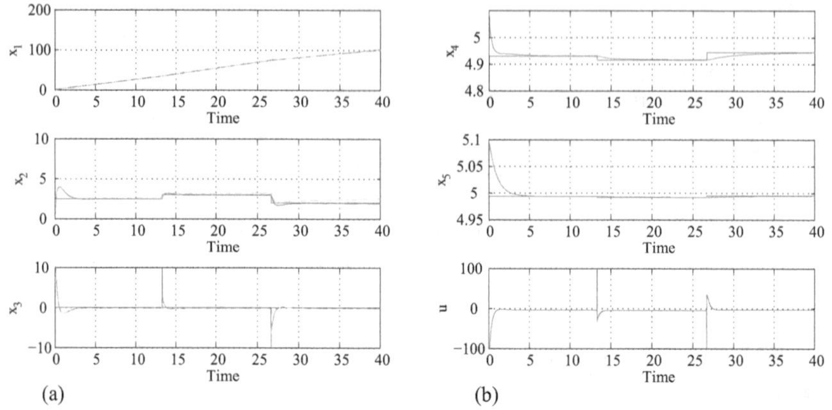

Figure 5.44 *Control of the injection moulding machine in the case of cavity pressure setpoint 5 (a) convergence of state variables* x_1, x_2, x_3 *to their reference setpoints and (b) convergence of state variables* x_4 *and* x_5 *to their reference setpoints and variations of the control inputs u*

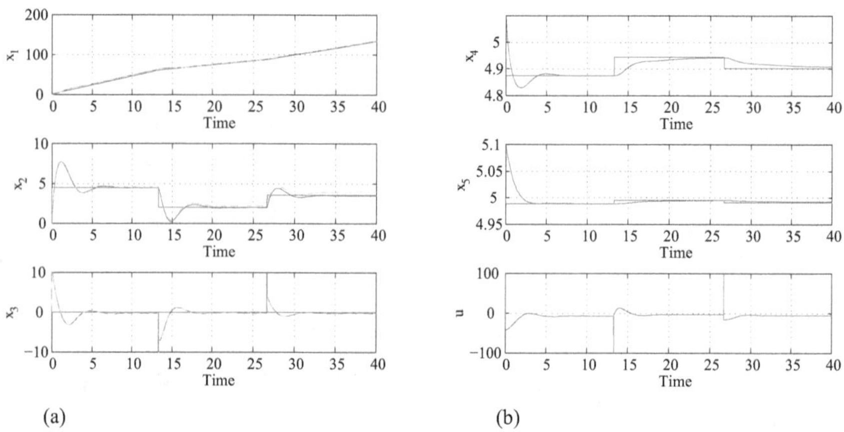

Figure 5.45 *Control of the injection moulding machine in the case of cavity pressure setpoint 6 (a) convergence of state variables* x_1, x_2, x_3 *to their reference setpoints and (b) convergence of state variables* x_4 *and* x_5 *to their reference setpoints and variations of the control inputs u*

5.6 Nonlinear control of the slosh-container system dynamics

5.6.1 Outline

Sloshing is the uncontrolled motion of a liquid in a half-filled container (tank) (Thakar *et al.*, 2017; Bandyopadhyay *et al.*, 2009). Sloshing can appear because of acceleration of the container (Kurode *et al.*, 2013; Kurode *et al.*, 2009; Thakar *et al.*, 2016; Mishra and Kurode, 2014). Sloshing is unavoidable when executing certain tasks (for instance the maneuvers performed by vehicles, vessels and spacecrafts that incorporate fuel tanks) (Hamaguchi, 2019; Kimura *et al.*, 2003; Zang and Huang, 2016; Yano and Terashima, 2001). Sloshing is undesirable because it can have several dangerous consequences (Vreeburg, 2008; Mazamanyan and Ayoubi, 2018; Miera, Wang *et al.*, 2012). In large liquid cargos transported by long vehicles or by vessels, sloshing can provoke hazardous overturns. A typical example is sloshing of liquified gas in the compartments of LNG vessels which can excite undesirable vibrations for the entire ship's structure. In spacecrafts uncontrolled sloshing can provoke additional accelerations, which affect the guidance system and can even lead to instability. Several failures in space missions have been attributed to sloshing effects. Furthermore, sloshing of the fuel that circulates in the interior of aircraft wings can generate undesirable oscillations too.

So far there have been many attempts to solve the control problem for the slosh-container system. In several cases simplified linear models of this system are used to design stabilizing feedback controllers. There have been several approaches based on robust control concepts such as sliding-mode control and H-infinity control (Yano and Terashima, 2003; Rabu, Bandouin *et al.*, 2012; Leiten *et al.*, 2019; Zhang and Yang, 2019). Additionally, Lyapunov theory-based methods in the design of controllers for the control problem of the slosh-container system can be also found. The implementation of sloshing control requires the processing of real-time sensor measurements, or if such measurements are not directly available they have to be estimated with the use of state-observers or filtering methods (Hervas and Reyhanoglou, 2013; Hervas and Reyhanoglou, 2014). Often, the dynamics of the slosh-container system is represented by an equivalent cart-pendulum model (Reyhanoglou and Hervas, 2013; Moriello *et al.*, 2018; Moriello *et al.*, 2017). Apart from the above noted active control methods, there exist passive control methods that try to mitigate the liquid's sloshing without direct measurements of the sloshing state. In all control approaches, the objective is to eliminate the oscillations of the transported liquid in the minimal time while also avoiding abrupt displacements of the container (Soyyaadi and Soltani, 2018; Sira-Ramirez and Fliess, 2002).

In the present section, a nonlinear optimal (H-infinity) control method is developed for the dynamic model of the slosh-container system (Rigatos and Busawon, 2018; Rigatos, 2016b; Rigatos, Siano *et al.*, 2015). This system is first represented by an equivalent cart-pendulum model. Next, using Euler-Lagrange analysis the related state-space model is obtained. This state-space description, being a highly nonlinear and underactuated one, undergoes approximate linearization with the use of first-order Taylor series expansion and through the computation of the related Jacobian matrices (Rigatos and Tzafestas, 2007; Basseville and Nikiforov, 1993; Rigatos and Zhang, 2009). The linearization takes place at each time-step of the control algorithm, around a time-varying operating point which is defined by the present value of the system's state vector and by the last value of the control inputs vector. The modelling error that is due to the truncation of higher-order terms in the Taylor series expansion is considered to be a perturbation that is asymptotically compensated by the robustness of the control algorithm. For the approximately linearized dynamics of the slosh-container system a stabilizing H-infinity feedback controller is designed.

The proposed H-infinity controller stands for the solution of the optimal control problem for the slosh-container dynamics under model uncertainty and external perturbations. Actually, this H-infinity controller represents a min–max differential game taking place between (i) the control inputs which try to minimize a cost function that comprises a quadratic term of the state vector's error and (ii) the model uncertainty and exogenous perturbations that try to maximize this cost function. For the selection of the controller's gains an algebraic Riccati equation is repetitively solved at each iteration of the control algorithm (Rigatos, 2011; Rigatos, 2015). The stability properties of the feedback control scheme are proven through Lyapunov analysis. First, it is proven that the related control loop satisfies the H-infinity tracking performance criterion, which signifies robustness against modelling imprecision and external disturbances (Rigatos and Busawon, 2018; Tousssaint, Basar and Bullo, 2000). Besides, under moderate conditions it is proven that the control loop is globally asymptotically stable. Moreover, to implement state estimation-based control without the need to measure the entire state vector of the slosh-container system, the H-infinity Kalman Filter is introduced as a reliable state estimator (Rigatos and Busawon, 2018; Rigatos, 2015). The proposed nonlinear optimal (H-infinity) control method retains the advantages of the typical (linear) optimal control, that is fast and accurate convergence to reference setpoints with minimum variations of the control inputs.

5.6.2 Dynamic model of the slosh-container system

The diagram of the slosh-container system and of the equivalent pendulum system is depicted in Figure 5.46. To analyze the dynamics of this system, an equivalent description of it is used, consisting of an oscillating pendulum that is mounted on a moving basis (Kurode *et al.*, 2013; Kurode *et al.*, 2009; Thakar *et al.*, 2016; Mishra and Kurode, 2014). Actually, the rigid mass M_r of the container and of its liquid content is taken to be equal to the mass of the moving basis of the pendulum. Besides, the slosh-mass of the liquid m is taken to be equal to the mass of the pendulum. The main parameters of the initial slosh-container system are defined as follows: M_r is the total rigid mass, including the mass of the tank, the mass of the base plate and the mass of the liquid that does not slosh. The slosh mash is denoted as m. The total mass of the system is denoted as M and is the sum of M_r and m. Angle ϕ defines the raise of the level of the slosh mass with reference to the top surface of the liquid that does not slosh. The force that causes the translational motion of the slosh-container system is denoted as f.

The main parameters of the equivalent pendulum system are defined as follows (Kurode *et al.*, 2013; Kurode *et al.*, 2009; Thakar *et al.*, 2016; Mishra and Kurode, 2014): M_r is the rigid mass of the moving base of the pendulum, m is the mass of the pendulum, M is the total mass of the pendulum and of its moving basis, that is the sum $M = M_r + m$. ϕ is the turn angle of the pendulum that is the angle of the oscillating slosh mass with respect to the vertical reference axis, l is the length of the pendulum, and f is the force that causes the rotational motion of the pendulum and the

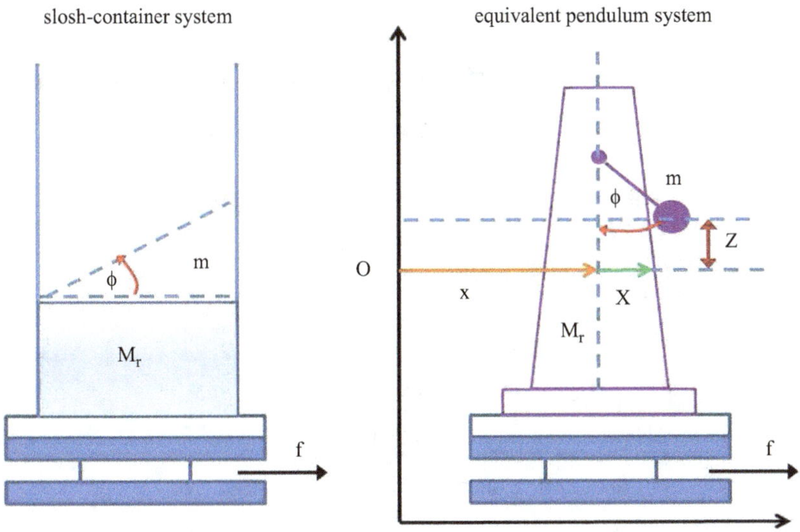

Figure 5.46 Diagram of the slosh-container system and of the equivalent pendulum system

translational motion of its moving basis. Besides, by defining the inertial reference frame *OXZ* one denotes as x the displacement of the rigid mass M_r along the OX axis and as X the displacement of the pendulum's mass also along the OX axis. Finally, the displacement of the pendulum's mass along the OZ axis is denoted as Z.

To define the equations of the motion of the slosh-container system, or the equations of motion of the equivalent pendulum system on a moving basis, Euler–Lagrange analysis is used. First, it is noted that the position of the rigid mass M_r along the OX axis is X, while the position of mass m along the OX axis $x+X$ or $x+l\sin(\phi)$. Besides the position of mass m along the OZ axis is Z or $l(1-\cos(\phi))$.

The kinetic energy of mass M_r (rigid mass) is

$$K_r = \frac{1}{2}M_r\dot{X}^2 \tag{5.159}$$

The kinetic energy of the pendulum (slosh mass) is

$$K_s = \frac{1}{2}m(\dot{x}+\dot{X})^2 + \frac{1}{2}m\dot{Z}^2 \tag{5.160}$$

where $X = l\sin(\phi)$ and $Z = l(1-\cos(\phi))$, thus

$$K_s = \frac{1}{2}m(\dot{x}+l\cos(\phi)\dot{\phi})^2 + \frac{1}{2}m(l\sin(\phi)\dot{\phi})^2 \tag{5.161}$$

Thus, the total kinetic energy of the system is

$$K = K_r + K_s \Rightarrow K = \frac{1}{2}M_r\dot{x}^2 + \frac{1}{2}m(\dot{x}^2 + l^2\cos^2(\phi)\dot{\phi}^2 + 2\dot{x}l\cos(\phi)\dot{\phi}) \\ \Rightarrow K = \frac{1}{2}M\dot{x}^2 + \frac{1}{2}ml^2\dot{\phi}^2 + m\dot{x}l\cos(\phi)\dot{\phi} \tag{5.162}$$

The potential energy of mass M_r (rigid mass) is

$$P_r = 0 \tag{5.163}$$

The potential energy of the pendulum (slosh mass) is

$$P_s = mgl(1 - \cos(\phi)) \tag{5.164}$$

The total potential energy of the system is

$$P = P_r + P_s \Rightarrow P = mgl(1 - \cos(\phi)) \tag{5.165}$$

The Lagrangian of the slosh-container system is

$$L = K - P \Rightarrow \\ L = \frac{1}{2}M\dot{x}^2 + \frac{1}{2}ml^2\dot{\phi}^2 + m\dot{x}l\cos(\phi)\dot{\phi} - mgl(1 - \cos(\phi)) \tag{5.166}$$

The motion of the system is defined by the following two Euler–Lagrange equations:

$$\frac{\partial}{\partial t}\frac{\partial L}{\partial \dot{x}} - \frac{\partial L}{\partial x} = f \tag{5.167}$$

$$\frac{\partial}{\partial t}\frac{\partial L}{\partial \dot{\phi}} - \frac{\partial L}{\partial \phi} = -c\dot{\phi} \tag{5.168}$$

where f is a force causing the displacement of the mobile basis of the pendulum along the OX axis and $c\dot{\phi}$ is a friction force opposing to the rotational motion of the pendulum.

It holds that $\frac{\partial}{\partial t}\frac{\partial L}{\partial \dot{x}} = M\ddot{x} - mlsing(\phi)\dot{\phi}^2 + ml\cos(\phi)\ddot{\phi}$ and $\frac{\partial L}{\partial x} = 0$. Thus (5.167) gives

$$M\ddot{x} + ml\cos(\phi)\ddot{\phi} - ml\sin(\phi)\dot{\phi}^2 = f \tag{5.169}$$

It holds that $\frac{\partial}{\partial t}\frac{\partial L}{\partial \dot{\phi}} = ml^2\ddot{\phi} + m\ddot{x}l\cos(\phi) - m\dot{x}l\sin(\phi)\dot{\phi}$ and $\frac{\partial L}{\partial \phi} = -m\dot{x}l\sin(\phi)\dot{\phi} - mgl\sin(\phi)$. Thus (5.168) gives

$$ml\cos(\phi)\ddot{x} + ml^2\ddot{\phi} + c\dot{\phi} + mgl\sin(\phi) = 0 \tag{5.170}$$

Consequently, the dynamic model of the slosh-container system is described by the following two equations

$$M\ddot{x} + ml\cos(\phi)\ddot{\phi} - ml\sin(\phi)\dot{\phi}^2 = f \\ ml\cos(\phi)\ddot{x} + ml^2\ddot{\phi} + c\dot{\phi} + mgl\sin(\phi) = 0 \tag{5.171}$$

The dynamic model of (5.171) can be also written in matrix form

$$\begin{pmatrix} M & ml\cos(\phi) \\ ml\cos(\phi) & ml^2 \end{pmatrix}\begin{pmatrix} \ddot{x} \\ \ddot{\phi} \end{pmatrix} - \begin{pmatrix} ml\dot{\phi}^2\sin(\phi) \\ -c\dot{\phi} - mgl\sin(\phi) \end{pmatrix} + \begin{pmatrix} -1 \\ 0 \end{pmatrix} f = \begin{pmatrix} 0 \\ 0 \end{pmatrix} \tag{5.172}$$

By defining vector $\mathbf{x} = [x\ \phi]^T$, the state-space model of (5.172) can be written in the concise form

$$D(\mathbf{x})\ddot{\mathbf{x}} + C(\mathbf{x}, \dot{\mathbf{x}}) + Nf = \mathbf{0} \tag{5.173}$$

where matrices $D(\mathbf{x})$ (inertia matrix), $C(\mathbf{x}$ (coriolis matrix) and N (control inputs gain matrix) are defined as

$$D = \begin{pmatrix} M & Ml\cos(\phi) \\ ml\cos(\phi) & ml^2 \end{pmatrix} \quad C = -\begin{pmatrix} ml\dot{\phi}\sin(\phi) \\ -c\dot{\phi} - mgl\sin(\phi) \end{pmatrix} \quad N = -\begin{pmatrix} 1 \\ 0 \end{pmatrix} \tag{5.174}$$

The inverse of matrix D is given by

$$D(\mathbf{x})^{-1} = \frac{1}{detD}\begin{pmatrix} ml^2 & -ml\cos(\phi) \\ -ml\cos(\phi) & M \end{pmatrix} \tag{5.175}$$

with determinant $detD = ml^2(M - m\cos^2(\phi))$. It holds that

$$D(\mathbf{x})^{-1}C(\mathbf{x}, \dot{\mathbf{x}}) = \begin{pmatrix} \dfrac{m^2l^3\dot{\phi}^2\sin(\phi) + ml\cos(\phi)(c\dot{\phi} + mgl\sin(\phi))}{ml^2(M - m\cos^2(\phi))} \\ \\ \dfrac{-m^2l^2\dot{\phi}^2\sin(\phi)\cos(\phi) - M(c\dot{\phi} + mgl\sin(\phi))}{ml^2(M - m\cos^2(\phi))} \end{pmatrix} \tag{5.176}$$

and also

$$D(\mathbf{x})^{-1}N = \begin{pmatrix} \dfrac{1}{(M - m\cos^2(\phi))} \\ \\ -\dfrac{\cos(\phi)}{l(M - m\cos^2(\phi))} \end{pmatrix} \tag{5.177}$$

Thus, the dynamic model of the slosh-container system becomes

$$\ddot{\mathbf{x}} = -D(\mathbf{x})^{-1}C(\mathbf{x}, \dot{\mathbf{x}}) - D(\mathbf{x})^{-1}Nf \tag{5.178}$$

or analytically

$$\begin{aligned}
\ddot{x} &= \frac{m^2 l^3 \dot{\phi}^2 \sin(\phi) + ml\cos(\phi)(c\dot{\phi} + mgl\sin(\phi))}{ml^2(M - m\cos^2(\phi))} \\
&\quad + \frac{1}{(M - m\cos^2(\phi))} f \\
\ddot{\phi} &= \frac{-m^2 l^2 \dot{\phi}^2 \sin(\phi)\cos(\phi) - M(c\dot{\phi} + mgl\sin(\phi))}{ml^2(M - m\cos^2(\phi))} \\
&\quad - \frac{\cos(\phi)}{l(M - m\cos^2(\phi))} f
\end{aligned} \tag{5.179}$$

Moreover, using the notation of state variables $x_1 = x, x_2 = \dot{x}, x_3 = \phi$ and $x_4 = \dot{\phi}$ one obtains the following state-space description for the slosh-container system

$$\begin{aligned}
\dot{x}_1 &= x_2 \\
\dot{x}_2 &= \frac{m^2 l^3 x_4^2 \sin(x_3) + ml\cos(x_3)(cx_4 + mgl\sin(x_3))}{ml^2(M - m\cos^2(x_3))} + \frac{1}{(M - m\cos^2(x_3))} f \\
\dot{x}_3 &= x_4 \\
\dot{x}_4 &= \frac{-m^2 l^2 x_4^2 \sin(x_3)\cos(x_3) - M(cx_4 + mgl\sin(x_3))}{ml^2(M - m\cos^2(x_3))} - \frac{\cos(x_3)}{l(M - m\cos^2(x_3))} f
\end{aligned} \tag{5.180}$$

The previous state-space model can be written in the affine-in-the-input underactuated form

$$\dot{x} = f(x) + g(x)u \tag{5.181}$$

where $x \in R^{4\times1}$, $f(x) \in R^{4\times1}$, $g(x) \in R^{4\times1}$ and $u \in R$, and vector fields $f(x)$ and $g(x)$ are given by

$$f(x) = \qquad\qquad g(x) =$$

$$\begin{pmatrix} x_2 \\ \dfrac{m^2 l^3 x_4^2 \sin(x_3) + ml\cos(x_3)(cx_4 + mgl\sin(x_3))}{ml^2(M - m\cos^2(x_3))} \\ x_4 \\ \dfrac{-m^2 l^2 x_4^2 \sin(x_3)\cos(x_3) - M(cx_4 + mgl\sin(x_3))}{ml^2(M - m\cos^2(x_3))} \end{pmatrix} \begin{pmatrix} 0 \\ \dfrac{1}{(M - m\cos^2(x_3))} \\ 0 \\ \dfrac{\cos(x_3)}{l(M - m\cos^2(x_3))} \end{pmatrix} \tag{5.182}$$

5.6.3 Approximate linearization of the model of the slosh-container system

5.6.3.1 Approximately linearized model

The dynamic model of the slosh-container system undergoes approximate linearization around the temporary operating point (x^*, u^*) which is updated at each time-step of the control algorithm, where x^* is the present value of the system's state vector and u^* is the last sampled value of the control inputs vector. The approximately linearized model of the system is written as

$$\dot{x} = Ax + Bu + \tilde{d} \tag{5.183}$$

where $\tilde{d}$ is the cumulative disturbances vector comprising (i) the modelling error due to the truncation of higher-order terms in the Taylor-series expansion, (ii) exogenous perturbations and (iii) sensor noises of any distribution. Matrices A and B are related with the system's Jacobians that is

$$\begin{aligned} &A = \nabla_x[f(x) + g(x)u] \mid_{(x^*,u*)} \Rightarrow \\ &A = \nabla_x[f(x)] \mid_{(x^*,u*)} + \nabla_x[g(x)]u \mid_{(x^*,u*)} \end{aligned} \tag{5.184}$$

$$\begin{aligned} &B = \nabla_u[f(x) + g(x)u] \mid_{(x^*,u*)} \Rightarrow \\ &B = g(x) \mid_{(x^*,u*)} \end{aligned} \tag{5.185}$$

The computation of the Jacobian matrix $\nabla_x f(x) \mid_{(x^*,u^*)}$ is performed as follows:

First row of the Jacobian matrix $\nabla_x f(x) \mid_{(x^*,u^*)}$: $\frac{\partial f_1}{\partial x_1} = 0$, $\frac{\partial f_1}{\partial x_2} = 0$, $\frac{\partial f_1}{\partial x_3} = 0$ and $\frac{\partial f_1}{\partial x_4} = 0$.

Second row of the Jacobian matrix $\nabla_x f(x) \mid_{(x^*,u^*)}$: $\frac{\partial f_2}{\partial x_1} = 0$, $\frac{\partial f_2}{\partial x_2} = 0$, $\frac{\partial f_2}{\partial x_3} = \frac{\frac{\partial f_{2n}}{\partial x_3} f_{2d} - f_{2n} \frac{\partial f_{2d}}{\partial x_3}}{f_{2d}^2}$, where $f_{2n} = m^2 l^3 x_4^2 \sin(x_3) + ml\cos(x_3)(cx_4 + mgl\sin(x_3))$, $f_{2d} = ml^2(M - m\cos^2(x_3))$ and $\frac{\partial f_{2n}}{\partial x_3} = m^2 l^3 x_4^2 \cos(x_3) - ml\sin(x_3)(cx_4 + mgl\sin(x_3)) + ml\cos(x_3)(-mgl\cos(x_3))$, $\frac{\partial f_{2d}}{\partial x_3} = m^2 l^2 (2m\cos(x_3)\sin(x_3))$.

and also $\frac{\partial f_2}{\partial x_4} = \frac{m^2 l^2 2x_4 - \sin(x_3) + ml\cos(x_3)c}{ml^2(M - m\cos^2(x_3))}$

Third row of the Jacobian matrix $\nabla_x f(x) \mid_{(x^*,u^*)}$: $\frac{\partial f_3}{\partial x_1} = 0$, $\frac{\partial f_3}{\partial x_2} = 0$, $\frac{\partial f_3}{\partial x_3} = 0$ and $\frac{\partial f_3}{\partial x_4} = 0$.

Fourth row of the Jacobian matrix $\nabla_x f(x) \mid_{(x^*,u^*)}$: $\frac{\partial f_4}{\partial x_1} = 0$, $\frac{\partial f_4}{\partial x_2} = 0$,

$\frac{\partial f_4}{\partial x_3} = \frac{\frac{\partial f_{4n}}{\partial x_3} f_{4d} - f_{4n} \frac{\partial f_{4d}}{\partial x_3}}{f_{4d}^2}$, where $f_{4n} = -m^2 l^2 x_4^2 \sin(x_3)\cos(x_3) - M(cx_4 + mgl\sin(x_3))$, $f_{4d} = ml^2(M - m\cos^2(x_3))$ and $\frac{\partial f_{4n}}{\partial x_3} = -ml^2 x_4^2 \sin(x_3)\cos(x_3) + M(cx_4 + mgl\sin(x_3))$, $\frac{\partial f_{4d}}{\partial x_3} = ml^2(M - m\cos^2(x_3))$.

and also $\frac{\partial f_4}{\partial x_4} = \frac{-m^2 l^2 2x_4 \sin(x_3)\cos(x_3) - Mc}{ml^2(M - m\cos^2(x_3))}$.

The computation of the Jacobian matrix $\nabla_x g(x) \mid_{(x^*,u^*)}$ is performed as follows:

First row of the Jacobian matrix $\nabla_x g(x) \mid_{(x^*,u^*)}$: $\frac{\partial g_1}{\partial x_1} = 0$, $\frac{\partial g_1}{\partial x_2} = 0$, $\frac{\partial g_1}{\partial x_3} = 0$ and $\frac{\partial g_1}{\partial x_4} = 0$.

Second row of the Jacobian matrix $\nabla_x g(x) \mid_{(x^*,u^*)}$: $\frac{\partial g_2}{\partial x_1} = 0$, $\frac{\partial g_2}{\partial x_2} = 0$, $\frac{\partial g_2}{\partial x_3} = \frac{(2m\cos(x_3)\sin(x_3))}{(M - m\cos^2(x_3))^2}$, $\frac{\partial g_2}{\partial x_4} = 0$.

Third row of the Jacobian matrix $\nabla_x g(x) \mid_{(x^*,u^*)}$: $\frac{\partial g_3}{\partial x_1} = 0$, $\frac{\partial g_3}{\partial x_2} = 0$, $\frac{\partial g_3}{\partial x_3} = 0$ and $\frac{\partial g_3}{\partial x_4} = 0$.

Fourth row of the Jacobian matrix $\nabla_x g(x) \mid_{(x^*,u^*)}$: $\frac{\partial g_4}{\partial x_1} = 0$, $\frac{\partial g_4}{\partial x_2} = 0$, $\frac{\partial g_4}{\partial x_4} = 0$ and also $\frac{\partial g_4}{\partial x_3} = \frac{\sin(x_3) l (M - m\cos^2(x_3)) - \cos(x_3)(2ml\cos(x_3)\sin(x_3))}{[l(M - m\cos^2(x_3))]^2}$.

5.6.3.2 Stabilizing feedback control

After linearization around its current operating point (x^*, u^*), the slosh-container system's dynamic model is written as (Rigatos and Busawon, 2018)

$$\dot{x} = Ax + Bu + d_1 \tag{5.186}$$

Parameter d_1 stands for the linearization error in the slosh-container system's model appearing previously in (5.186). The reference setpoints for the slosh-container system's state vector are denoted by $\mathbf{x_d} = [x_1^d, \ldots, x_5^d]$. Tracking of this trajectory is achieved after applying the control input u_d. At every time instant,the control input u_d is assumed to differ from the control input u appearing in (5.186) by an amount equal to Δu, that is $u_d = u + \Delta u$

$$\dot{x}_d = Ax_d + Bu_d + d_2 \tag{5.187}$$

The dynamics of the controlled system described in (5.186) can be also written as

$$\dot{x} = Ax + Bu + Bu_d - Bu_d + d_1 \tag{5.188}$$

and by denoting $d_3 = -Bu_d + d_1$ as an aggregate disturbance term one obtains

$$\dot{x} = Ax + Bu + Bu_d + d_3 \tag{5.189}$$

By subtracting (5.187) from (5.189) one has

$$\dot{x} - \dot{x}_d = A(x - x_d) + Bu + d_3 - d_2 \tag{5.190}$$

By denoting the tracking error as $e = x - x_d$ and the aggregate disturbance term as $\tilde{d} = d_3 - d_2$, the tracking error dynamics becomes

$$\dot{e} = Ae + Bu + \tilde{d} \tag{5.191}$$

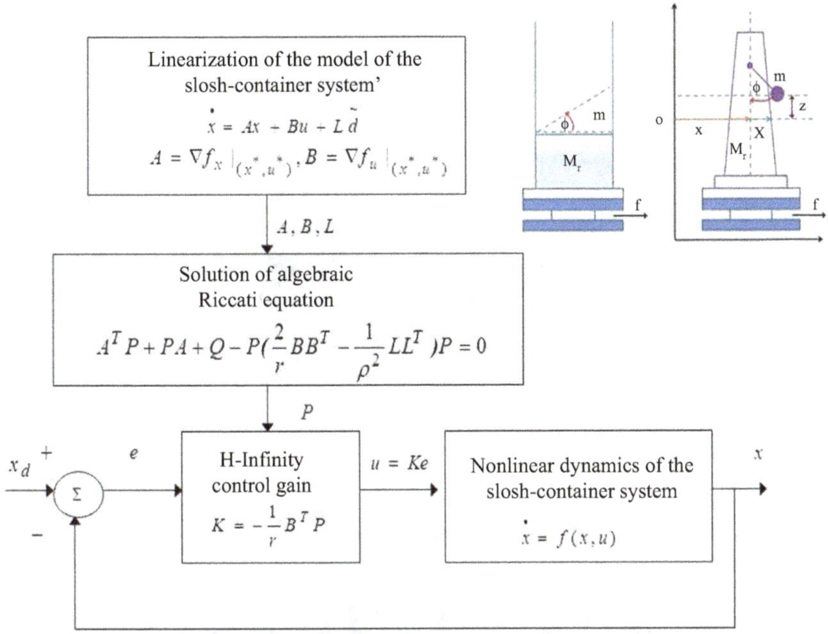

Figure 5.47 Diagram of the slosh-container system's nonlinear optimal (H-infinity) control loop

For the approximately linearized model of the system, a stabilizing feedback controller is developed. The controller has the form

$$u(t) = -Ke(t) \tag{5.192}$$

with $K = \frac{1}{r}B^T P$ where P is a positive definite symmetric matrix which is obtained from the solution of the Riccati equation (Rigatos and Busawon, 2018)

$$A^T P + PA + Q - P\left(\frac{2}{r}BB^T - \frac{1}{\rho^2}LL^T\right)P = 0 \tag{5.193}$$

where Q is a positive semi-definite symmetric matrix. The diagram of the considered control loop is depicted in Figure 5.47. The global stability properties of the control method can be proven following the procedure presented in Section 1.1.2.5.

5.6.4 Simulation tests

The performance of the proposed nonlinear optimal (H-infinity) control scheme for the model of the slosh-container system can be further evaluated through simulation experiments. To compute the controller's feedback gain the algebraic Riccati equation appearing in (5.193) has been solved at each time-step of the control method.

The obtained results are depicted in Figures 5.48–5.55. It is noted that a state estimation-based control scheme has been applied with the use of the H-infinity Kalman Filter. The real values of the state variables are printed in blue, the estimated values provided by the H-infinity Kalman Filter are plotted in green, while the

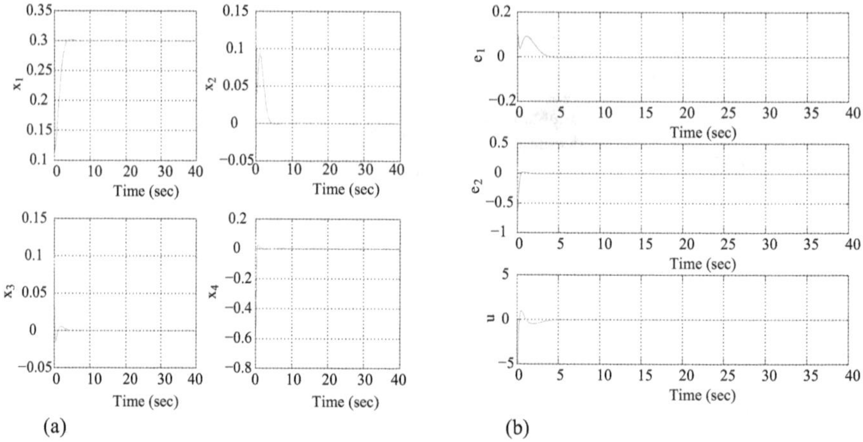

Figure 5.48 *(a) Tracking of reference setpoint 1 (red lines) by the state variables x_1 to x_4 of the slosh-container system (blue lines) and state estimates provided by the H_∞ KF (green lines) and (b) convergence of the tracking errors e_1 of position x and e_3 of angle ϕ (red lines) and variations of the control input u (blue line)*

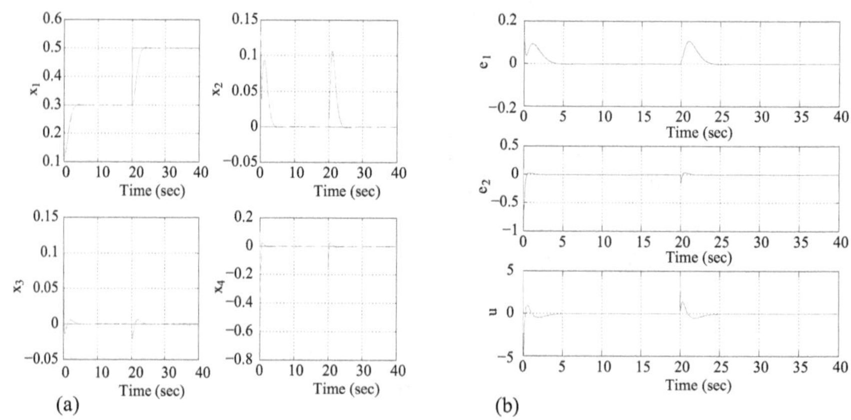

Figure 5.49 *(a) Tracking of reference setpoint 2 (red lines) by the state variables x_1 to x_4 of the slosh-container system (blue lines) and state estimates provided by the H_∞ KF (green lines) and (b) convergence of the tracking errors e_1 of position x and e_3 of angle ϕ (red lines) and variations of the control input u (blue line)*

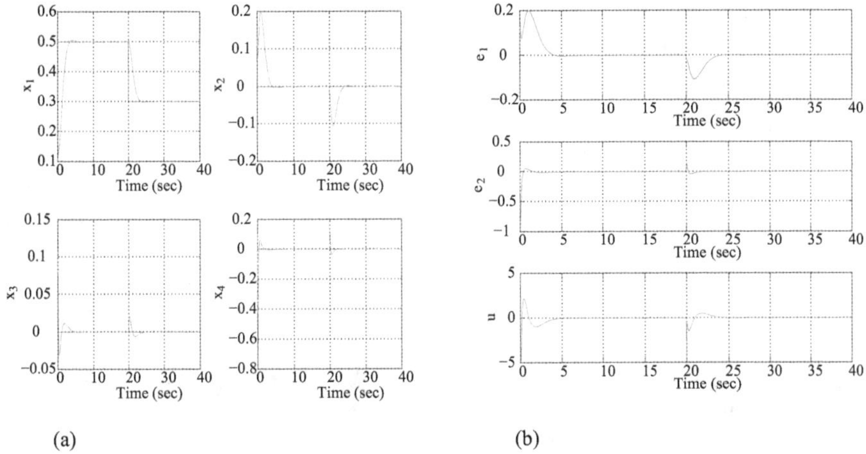

Figure 5.50 *(a) Tracking of reference setpoint 3 (red lines) by the state variables x_1 to x_4 of the slosh-container system (blue lines) and state estimates provided by the H_∞ KF (green lines) and (b) convergence of the tracking errors e_1 of position x and e_3 of angle ϕ (red lines) and variations of the control input u (blue line)*

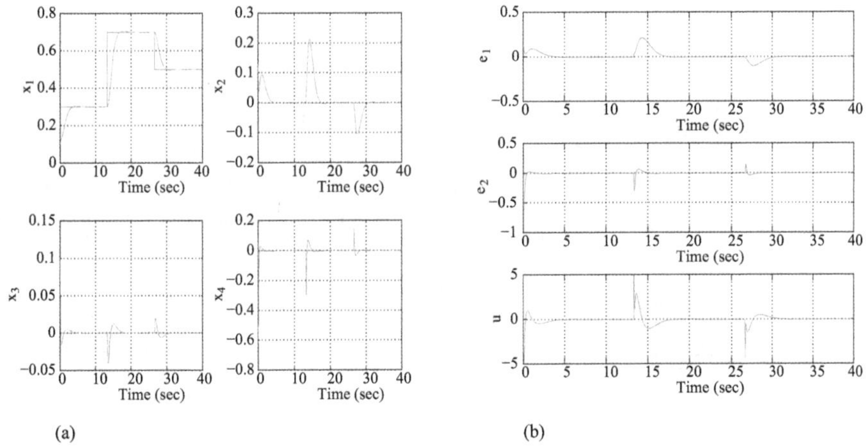

Figure 5.51 *(a) Tracking of reference setpoint 4 (red lines) by the state variables x_1 to x_4 of the slosh-container system (blue lines) and state estimates provided by the H_∞ KF (green lines) and (b) convergence of the tracking errors e_1 of position x and e_3 of angle ϕ (red lines) and variations of the control input u (blue line)*

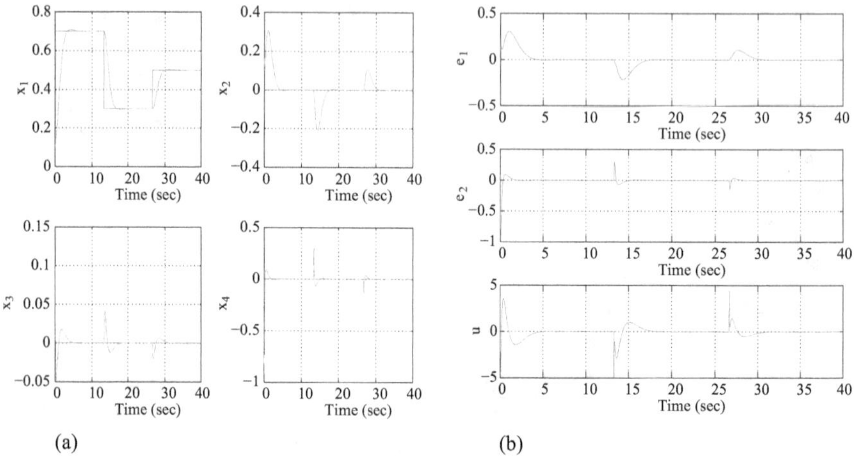

Figure 5.52 *(a) Tracking of reference setpoint 5 (red lines) by the state variables* x_1 *to* x_4 *of the slosh-container system (blue lines) and state estimates provided by the* H_∞ *KF (green lines) and (b) convergence of the tracking errors* e_1 *of position x and* e_3 *of angle* ϕ *(red lines) and variations of the control input u (blue line)*

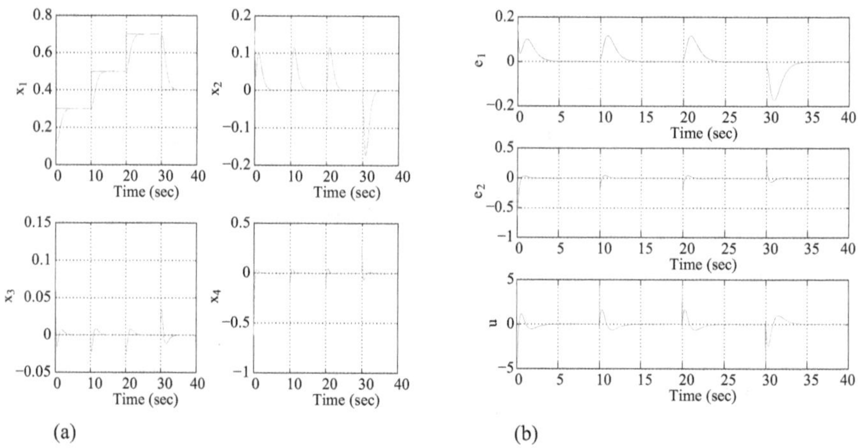

Figure 5.53 *(a) Tracking of reference setpoint 6 (red lines) by the state variables* x_1 *to* x_4 *of the slosh-container system (blue lines) and state estimates provided by the* H_∞ *KF (green lines) and (b) convergence of the tracking errors* e_1 *of position x and* e_3 *of angle* ϕ *(red lines) and variations of the control input u (blue line)*

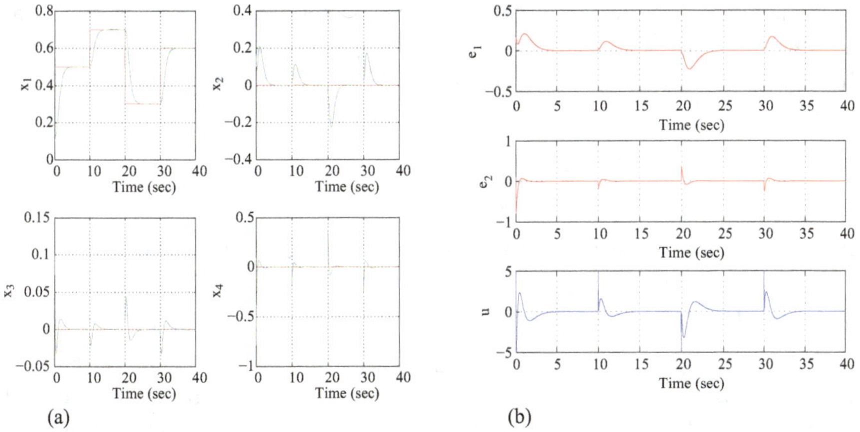

Figure 5.54 (a) Tracking of reference setpoint 7 (red lines) by the state variables x_1 to x_4 of the slosh-container system (blue lines) and state estimates provided by the H_∞ KF (green lines) and (b) convergence of the tracking errors e_1 of position x and e_3 of angle ϕ (red lines) and variations of the control input u (blue line)

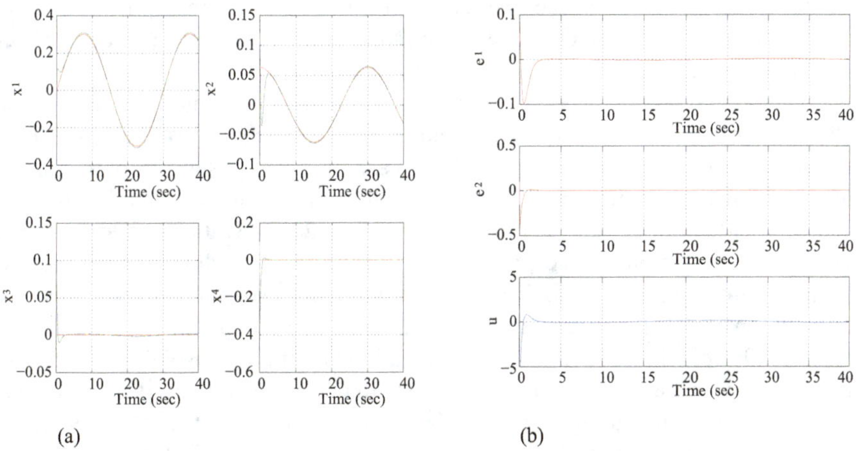

Figure 5.55 (a) Tracking of reference setpoint 8 (red lines) by the state variables x_1 to x_4 of the slosh-container system (blue lines) and state estimates provided by the H_∞ KF (green lines) and (b) convergence of the tracking errors e_1 of position x and e_3 of angle ϕ (red lines) and variations of the control input u (blue line)

associated setpoints are depicted in red. It can be noticed, that in all cases, fast and accurate tracking of the reference setpoints was achieved under moderate variations of the control inputs. Besides, the method exhibits robustness to parametric uncertainty, for instance imprecision in the value of slosh mass m.

5.7 Nonlinear control of micro-satellites' attitude dynamics

5.7.1 Introduction

Attitude control for satellites is important for the precise orientation of theirs and consequently for assuring the reliable transmission of communication data or the accomplishment of image acquisition, navigation, geodesic and surveillance tasks (Shi, Kinkaid *et al.*, 2016; Jiang *et al.*, 2008; Lee and Singh, 2011). During the operation of a satellite, attitude control is applied to compensate for various space–environmental disturbances, as well as for inner moment-of-inertia variations (Horri *et al.*, 2012; Patel *et al.*, 2009; Hu, Xiao *et al.*, 2019; Han *et al.*, 2014). The attitude determination and control subsystem (ADCS) consists of various sensors and actuators which provide information about attitude and finally stabilize the satellite at the desirable orientation (Rodrigues and Murillo, 2019; Han and Pechev, 2007; Ahmed and Kerrigan, 2014). Control of the satellite's attitude can be accomplished by using as control inputs electromagnetic actuation, solar-powered actuation, reaction wheels as well as thrusters.

The present section is particularly focusing on micro-satellites. These have usually mass that is less than 100 kg and thus their manufacturing is less costly, takes shorter time and has less failure risks. Micro-satellites have a significant role in navigation, earth observation, atmosphere data collection and tele-communications (Wu and Saif, 2005; Singh and Iyer, 1989; Yang and Sun, 2002). Unlike satellites of larger dimensions and weight that comprise flexible parts, micro-satellites can by described by the dynamics of a rigid body. Three-axis attitude control of micro-satellites requires the solution of a multi-variable and highly nonlinear control problem (Xiao, Cao and Ram, 2019; Doroshin, 2017; Yoshimoto *et al.*, 2011). Elimination of the satellite's orientation error is the primary objective of an attitude control system. Besides, another important feature is robustness against the disturbances that may affect the satellite's control loop (Giuseppi *et al.*, 2019; Petersen, Leve *et al.*, 2017; Cao *et al.*, 2013). There have been several attempts to solve the satellite's attitude stabilization problem with the use of linear and nonlinear control techniques (Show *et al.*, 2014; Xue, Tan *et al.*, 2012; Liang, Wang *et al.*, 2019). Although significant developments have been witnessed in nonlinear controllers design for the attitude control problem, this research area is still open.

In this section, a nonlinear optimal (H-infinity) controller is designed for the stabilization of the attitude of micro-satellites (Rigatos and Busawon, 2018; Rigatos, 2016b; Rigatos, Siano and Cecati, 2015; Rigatos, Siano *et al.*, 2015).

At the first stage, the attitude's dynamic model undergoes approximate linearization around a temporary operating point which is updated at each time-step of the control method. This operating point is defined by the present value of the state vector and by the last sampled value of the control inputs vector. The linearization procedure takes place with the use of first-order Taylor series expansion and through the computation of the related Jacobian matrices (Rigatos and Tzafestas, 2007; Basseville and Nikiforov, 1993; Rigatos and Zhang, 2009). For the approximately linearized model of the satellite, an H-infinity feedback controller is designed.

This controller can provide a solution to the satellite's optimal control problem under model uncertainty and external perturbations. Actually a min–max differential game is implemented between (i) the controller which tries to minimize a cost function that comprises a quadratic term of the state vector's tracking error and between (ii) the disturbances and model uncertainty which try to maximize this cost function. To select the controller's feedback gains an algebraic Riccati equation is repetitively solved at each iteration of the control algorithm (Rigatos, 2011; Rigatos, 2015). The stability properties of the satellite's control loop are solved with the use of Lyapunov analysis. First, it is proven that the control scheme satisfies the H-infinity tracking performance criterion, which signifies elevated robustness against model uncertainty and external perturbations (Rigatos and Busawon, 2018; Tousssaint, Basar and Bullo, 2000). Moreover, global asymptotic stability for the control loop is proven (Rigatos and Busawon, 2018). To implement state estimation-based control for the satellite's model, the H-infinity Kalman Filter has been used as a robust state estimator (Rigatos and Busawon, 2018; Gibbs, 2012). The proposed control method retains the advantages of the typical (linear) optimal control, that is fast and accurate tracking of the reference setpoints under moderate variations of the control inputs (Rigatos and Busawon, 2018).

Control and estimation for the satellite's nonlinear attitude dynamics remains a challenging research topic (Zhang, Zeng *et al.* 2020; Pirat, Anderson *et al.*, 2020; Chen, Sun *et al.*, 2020; Castaldi *et al.*, 2020; Zhang, Zhang *et al.* 2019; Sutherland *et al.*, 2019). The present section (a) provides one of the few existing solutions to the nonlinear optimal control problem of the satellite's attitude stabilization which is of proven global stability while also remaining computationally efficient and (b) applies a genuine approach for implementing H-infinity control on the satellite's nonlinear attitude dynamics. Preceding results on the use of H-infinity control to nonlinear dynamical systems were limited to the case of affine-in-the-input systems with drift-only dynamics and considered that the control inputs gain matrix is not dependent on the values of the system's state vector. Moreover, in these approaches the linearization was performed around points of the desirable trajectory whereas in the present section's control method the linearization points are related with the value of the state vector at each sampling instant as well as with the last sampled value of the control inputs vector. The Riccati equation which has been proposed for computing the feedback gains of the controller is novel, so is the presented global stability proof through Lyapunov analysis.

5.7.2 Dynamic model of the micro-satellite attitude system

The diagram of the micro-satellite attitude system and the related body-fixed and inertial reference frames are depicted in Figure 5.56. Typical motion equations of the satellite are given in the following concise state-space model (Wu and Saif, 2005; Singh and Iyer, 1989:

$$\dot{x} = f(x) + g(x)u + T_d \tag{5.194}$$

where $x = [\theta_1, \theta_2, \theta_3, \omega_1, \omega_2, \omega_3]^T$ with $\theta_1 = \phi, \theta_2 = \theta$ and $\theta_3 = \psi$ to stand for the roll, pitch and yaw angles of the satellite in the body-fixed reference frame, and $\omega_1 = \omega_x$, $\omega_2 = \omega_y$ and $\omega_3 = \omega_z$ to stand for the angular velocities of the satellite about the axes of the inertial reference frames.

In Figure 5.56 the turn angles of the micro-satellite around the axes of the body-fixed reference frame are denoted as ϕ, θ and ψ respectively and are depicted in this diagram with curved blue arrows, while the associated angular velocities are denoted as $\dot{\phi}$, $\dot{\theta}$ and $\dot{\psi}$ and are collinear to the axes of the body-fixed frame (that is $\dot{\phi}$ is collinear to axis B_1, $\dot{\theta}$ is collinear to axis B_2 and $\dot{\psi}$ is collinear to axis B_3). In the earth-fixed (inertial) reference frame the turn angles of the micro-satellite around the axes of the inertial reference frame are denoted as θ_x, θ_y and θ_z respectively and are

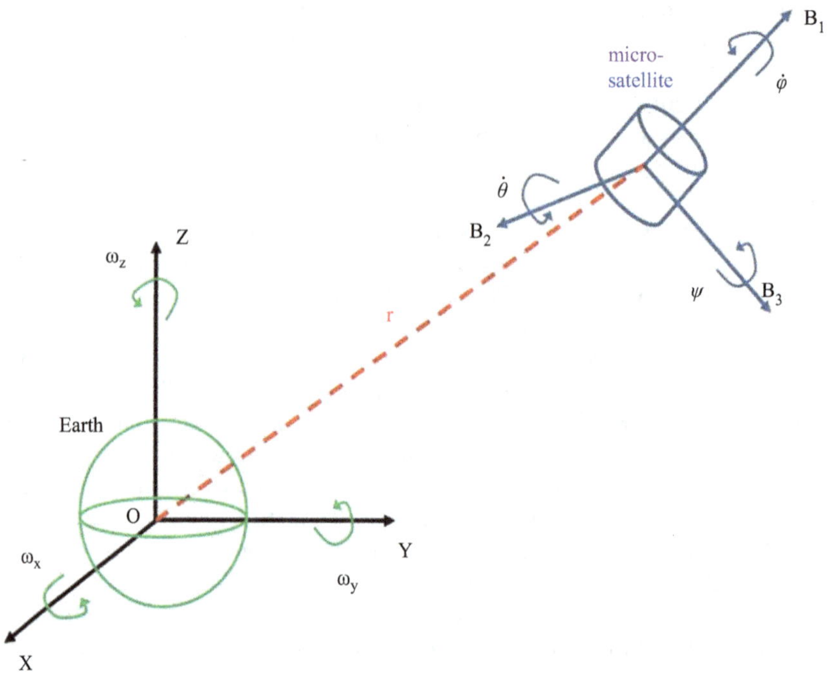

Figure 5.56 Diagram of the micro-satellite attitude system and the related body-fixed and inertial reference frames

depicted in this diagram with the curved green arrows, while the associated angular velocities are denoted as ω_x, ω_y and ω_z and are collinear to the axes of the earth-fixed frame (that is ω_x is collinear to axis Ox, ω_y is collinear to axis O_y and ω_z is collinear to axis Oz).

The control inputs vector u of the satellite comprises the torques which are exerted on the satellite and which can be provided by its actuators. It holds that $u = [u_1, u_2, u_3]^T = [T_x, T_y.T_z]^T$, while in the case of underactuation some of the elements of the control inputs vector can be 0. Vector $T_d = [0, 0, 0, T_{d_1}/I_1, T_{d_2}/I_2, T_{d_3}/I_3]^T$ represents the disturbance torques that affect the rotational motion of the satellite about the axes of the inertial reference frames. It has been considered that the inertial matrix of the satellite is $I = diag[I_1, I_2, I_3]^T$. The orbital rate of the satellite is denoted as ω_0 (rad/sec). The state variables and key parameters of the model are outlined in nomenclature Table 5.1. About vector fields $f(x)$ and $g(x)$ it holds that (Wu and Saif, 2005; Singh and Iyer, 1989)

$$f(x) = \begin{pmatrix} \dfrac{\cos(\theta_3)\omega_2 - \sin(\theta_3)\omega_3}{\cos(\theta_2)} - \omega_0 \\ \sin(\theta_3)\omega_2 + \cos(\theta_3)\omega_3 \\ \omega_1 + \dfrac{\sin(\theta_2)}{\cos(\theta_2)}[\sin(\theta_3)\omega_3 - \cos(\theta_3)\omega_2] \\ \dfrac{I_2 - I_3}{I_1}(\omega_2\omega_3 - 3\omega_0^2\zeta_2\zeta_3) \\ \dfrac{I_3 - I_1}{I_2}(\omega_3\omega_1 - 3\omega_0^2\zeta_3\zeta_1) \\ \dfrac{I_1 - I_2}{I_3}(\omega_1\omega_2 - 3\omega_0^2\zeta_1\zeta_2) \end{pmatrix} \quad g(x) = \begin{pmatrix} 0 & 0 & 0 \\ 0 & 0 & 0 \\ 0 & 0 & 0 \\ \frac{1}{I_1} & 0 & 0 \\ 0 & \frac{1}{I_2} & 0 \\ 0 & 0 & \frac{1}{I_3} \end{pmatrix} \tag{5.195}$$

Table 5.1 Parameters of the satellite's attitude dynamic model

Parameter	Definition
θ_1	turn angle for rotation around axis B_1 of the satellite-fixed frame
θ_2	turn angle for rotation around axis B_2 of the satellite-fixed frame
θ_3	turn angle for rotation around axis B_3 of the satellite-fixed frame
ω_x	angular speed for rotation around axis Ox of the earth-fixed frame
ω_y	angular speed for rotation around axis Oy of the earth-fixed frame
ω_z	angular speed for rotation around axis Oz of the earth-fixed frame
I_x	moment of inertia for rotation around the Ox axis of the earth-fixed frame
I_y	moment of inertia for rotation around the Oy axis of the earth-fixed frame
I_z	moment of inertia for rotation around the Oz axis of the earth-fixed frame
ω_0	orbital rate of the satellite

Moreover, about vector $\zeta = [\zeta_1, \zeta_2, \zeta_3]^T$ it holds that

$$\zeta = \begin{pmatrix} -\sin(\theta_1)\cos(\theta_2) \\ \cos(\theta_1)\sin(\theta_3) - \sin(\theta_1)\sin(\theta_2)\cos(\theta_3) \\ \cos(\theta_1)\cos(\theta_3) - \sin(\theta_1)\sin(\theta_2)\sin(\theta_3) \end{pmatrix} \tag{5.196}$$

The main parameters of the micro-satellite are given in the following (Wu and Saif, 2005; Singh and Iyer, 1989): I_1 is the principal moment of inertia about the x-axis and an indicative value for it is 18.4 (kg·m^2), I_2 is the principal moment of inertia about the y-axis and an indicative value for it is 18.4 (kg·m^2), I_3 is the principal moment of inertia about the z-axis and an indicative value for it is 6.8 (kg·m^2), and finally ω_0 is the orbital rate of the satellite and an indicative value for it 0.0011(rad/sec).

Next, using the state vector notation $x = [\theta_1, \theta_2, \theta_3, \omega_1, \omega_2, \omega_3]^T$, that is $x = [x_1, x_2, x_3, x_4, x_5, x_6]^T$ one has the following state-space description for the micro-satellite

$$\dot{x} = f(x) + g(x)u + T_d \tag{5.197}$$

$$f(x) = \begin{pmatrix} \frac{\cos(x_3)x_5 - \sin(x_3)x_6}{\cos(x_2)} - \omega_0 \\ \sin(x_3)x_5 + \cos(x_3)x_6 \\ x_4 + \frac{\sin(x_2)}{\cos(x_2)}[\sin(x_3)x_6 - \cos(x_3)x_5] \\ \frac{I_2 - I_3}{I_1}(x_5x_6 - 3\omega_0^2\zeta_2\zeta_3) \\ \frac{I_3 - I_1}{I_2}(x_6x_4 - 3\omega_0^2\zeta_3\zeta_1) \\ \frac{I_1 - I_2}{I_3}(x_4x_5 - 3\omega_0^2\zeta_1\zeta_2) \end{pmatrix} \quad g(x) = \begin{pmatrix} 0 & 0 & 0 \\ 0 & 0 & 0 \\ 0 & 0 & 0 \\ \frac{1}{I_1} & 0 & 0 \\ 0 & \frac{1}{I_2} & 0 \\ 0 & 0 & \frac{1}{I_3} \end{pmatrix} \tag{5.198}$$

while, about vector $\zeta = [\zeta_1, \zeta_2, \zeta_3]^T$ it holds that

$$\zeta = \begin{pmatrix} -\sin(x_1)\cos(x_2) \\ \cos(x_1)\sin(x_3) - \sin(x_1)\sin(x_2)\cos(x_3) \\ \cos(x_1)\cos(x_3) - \sin(x_1)\sin(x_2)\sin(x_3) \end{pmatrix} \tag{5.199}$$

5.7.3 Approximate linearization of the satellite's state-space model

5.7.3.1 Approximately linearized model

The state-space model of the satellite's attitude dynamics undergoes approximate linearization around the temporary operating point (x^*, u^*), where x^* is the present value of the system's state vector, while u^* is the last sampled value of the control inputs vector. This setpoint is updated at each iteration of the control algorithm. From the initial nonlinear state-space model $\dot{x} = f(x) + g(x)u$ one arrives at the equivalent linearized model

$$\dot{x} = Ax + Bu + \tilde{d} \tag{5.200}$$

where $\tilde{d}$ is the cumulative disturbances vector that comprises modelling errors due to truncation of higher-order terms in the Taylor series expansion, external perturbations, as well as measurement noises of any distribution. Matrices A and B are computed from the Jacobians of the state-space model

$$A = \nabla_x[f(x) + g(x)u] \mid_{(x^*,u^*)} \Rightarrow A = \nabla_x[f(x)] \mid_{(x^*,u^*)} \tag{5.201}$$

$$B = \nabla_u[f(x) + g(x)u] \mid_{(x^*,u^*)} \Rightarrow B = g(x) \mid_{(x^*,u^*)} \tag{5.202}$$

5.7.3.2 Computation of Jacobian matrices

Next, the Jacobian matrix $A = \nabla_x[f(x)] \mid_{(x^*,u^*)}$ is computed:

First row of the Jacobian matrix $A = \nabla_x[f(x)] \mid_{(x^*,u^*)}$: $\frac{\partial f_1}{\partial x_1} = 0$, $\frac{\partial f_1}{\partial x_2} = \frac{[\cos(x_3)x_5 - \sin(x_3)x_6]\sin(x_2)}{[\cos(x_2)]^2}$, $\frac{\partial f_1}{\partial x_3} = \frac{[-\sin(x_3)x_5 - \cos(x_3)x_6]}{\cos(x_2)}$, $\frac{\partial f_1}{\partial x_4} = 0$, $\frac{\partial f_1}{\partial x_5} = \frac{\cos(x_3)}{\cos(x_2)}$, $\frac{\partial f_1}{\partial x_6} = -\frac{\sin(x_3)}{\cos(x_2)}$.

Second row of the Jacobian matrix $A = \nabla_x[f(x)] \mid_{(x^*,u^*)}$: $\frac{\partial f_2}{\partial x_1} = 0$, $\frac{\partial f_2}{\partial x_2} = 0$, $\frac{\partial f_2}{\partial x_3} = \cos(x_3)x_5 - \sin(x_3)x_6$, $\frac{\partial f_2}{\partial x_4} = 0$, $\frac{\partial f_2}{\partial x_5} = \sin(x_3)$, $\frac{\partial f_2}{\partial x_6} = \cos(x_3)$.

Third row of the Jacobian matrix $A = \nabla_x[f(x)] \mid_{(x^*,u^*)}$: $\frac{\partial f_3}{\partial x_1} = 0$, $\frac{\partial f_3}{\partial x_2} = \frac{[\sin(x_3)x_6 - \cos(x_3)x_5]}{[\cos^2(x(k,2))]}$, $\frac{\partial f_3}{\partial x_3} = \frac{\sin(x_2)}{\cos(x_2)}[\cos(x_3)x_6 - \sin(x_3)x_5]$, $\frac{\partial f_3}{\partial x_4} = 1$, $\frac{\partial f_3}{\partial x_5} = -\frac{\sin(x_2)}{\cos(x_2)}$, $\frac{\partial f_3}{\partial x_6} = \frac{\sin(x_2)}{\cos(x_2)}\sin(x_3)$.

Fourth row of the Jacobian matrix $A = \nabla_x[f(x)] \mid_{(x^*,u^*)}$: $\frac{\partial f_4}{\partial x_i} = \frac{I_2 - I_3}{I_1}(-3\omega_0^2)[\frac{\partial \zeta_2(x)}{\partial x_i}\zeta_3(x) + \zeta_2(x)\frac{\partial \zeta_3(x)}{\partial x_i}]$, for $i = 1, 2, 3, 4$, $\frac{\partial f_4}{\partial x_5} = \frac{I_2 - I_3}{I_1}(x_6 - 3\omega_0^2)[\frac{\partial \zeta_2(x)}{\partial x_5}\zeta_3(x) + \zeta_2(x)\frac{\partial \zeta_3(x)}{\partial x_5}])$, $\frac{\partial f_4}{\partial x_6} = \frac{I_2 - I_3}{I_1}(x_5 - 3\omega_0^2)[\frac{\partial \zeta_2(x)}{\partial x_6}\zeta_3(x) + \zeta_2(x)\frac{\partial \zeta_3(x)}{\partial x_6}])$

Fifth row of the Jacobian matrix $A = \nabla_x[f(x)] \mid_{(x^*,u^*)}$: $\frac{\partial f_5}{\partial x_i} = \frac{I_3 - I_1}{I_2}(-3\omega_0^2)[\frac{\partial \zeta_3(x)}{\partial x_i}\zeta_1(x) + \zeta_3(x)\frac{\partial \zeta_1(x)}{\partial x_i}]$, for $i = 1, 2, 3, 5$, $\frac{\partial f_5}{\partial x_4} = \frac{I_3 - I_1}{I_2}(x_6 - 3\omega_0^2)[\frac{\partial \zeta_3(x)}{\partial x_4}\zeta_1(x) + \zeta_3(x)\frac{\partial \zeta_1(x)}{\partial x_4}])$, $\frac{\partial f_5}{\partial x_6} = \frac{I_3 - I_1}{I_2}(x_4 - 3\omega_0^2)[\frac{\partial \zeta_3(x)}{\partial x_6}\zeta_1(x) + \zeta_3(x)\frac{\partial \zeta_1(x)}{\partial x_6}])$

Sixth row of the Jacobian matrix $A = \nabla_x[f(x)] \mid_{(x^*,u^*)}$: $\frac{\partial f_6}{\partial x_i} = \frac{I_1 - I_2}{I_3}(-3\omega_0^2)[\frac{\partial \zeta_1(x)}{\partial x_i}\zeta_2(x) + \zeta_1(x)\frac{\partial \zeta_2(x)}{\partial x_i}]$, for $i = 1, 2, 3, 6$, $\frac{\partial f_6}{\partial x_4} = \frac{I_1 - I_2}{I_3}(x_6 - 3\omega_0^2)[\frac{\partial \zeta_1(x)}{\partial x_4}\zeta_2(x) + \zeta_1(x)\frac{\partial \zeta_2(x)}{\partial x_4}])$, $\frac{\partial f_6}{\partial x_5} = \frac{I_1 - I_2}{I_3}(x_4 - 3\omega_0^2)[\frac{\partial \zeta_1(x)}{\partial x_5}\zeta_2(x) + \zeta_1(x)\frac{\partial \zeta_2(x)}{\partial x_5}])$

Next, the following partial derivatives are computed: $\frac{\partial \zeta_1(x)}{\partial x_i}$ $i = 1, 2, \ldots, 6$, $\frac{\partial \zeta_2(x)}{\partial x_i}$ $i = 1, 2, \cdots, 6$ and $\frac{\partial \zeta_3(x)}{\partial x_i}$ $i = 1, 2, \cdots, 6$

For $\frac{\partial\zeta_1(x)}{\partial x_i}$: $\frac{\partial\zeta_1(x)}{\partial x_1}=-\cos(x_1)\cos(x_2)$, $\frac{\partial\zeta_1(x)}{\partial x_2}=\sin(x_1)\sin(x_2)$, $\frac{\partial\zeta_1(x)}{\partial x_3}=0$, $\frac{\partial\zeta_1(x)}{\partial x_4}=0$, $\frac{\partial\zeta_1(x)}{\partial x_5}=0$ and $\frac{\partial\zeta_1(x)}{\partial x_6}=0$.

For $\frac{\partial\zeta_2(x)}{\partial x_i}$: $\frac{\partial\zeta_2(x)}{\partial x_i}-\sin(x_1)\sin(x_3)-\cos(x_1)\sin(x_2)\cos(x_3)$, $\frac{\partial\zeta_2(x)}{\partial x_2}=-\sin(x_1)\cos(x_2)$ $\cos(x_3)$, $\frac{\partial\zeta_2(x)}{\partial x_3}=\cos(x_1)\cos(x_3)+\sin(x_1)\sin(x_2)\sin(x_3)$, $\frac{\partial\zeta_2(x)}{\partial x_4}=0$, $\frac{\partial\zeta_2(x)}{\partial x_5}=0$, $\frac{\partial\zeta_2(x)}{\partial x_6}=0$

For $\frac{\partial\zeta_3(x)}{\partial x_i}$: $\frac{\partial\zeta_3(x)}{\partial x_1}=-\cos(x_1)\sin(x_2)\sin(x_3)$, $\frac{\partial\zeta_3(x)}{\partial x_2}=-\sin(x_1)\cos(x_2)\sin(x_3)$, $\frac{\partial\zeta_1(x)}{\partial x_3}=$ $-\cos(x_1)\sin(x_3)-\sin(x_1)\sin(x_2)\cos(x_3)$, $\frac{\partial\zeta_3(x)}{\partial x_4}=0$, $\frac{\partial\zeta_3(x)}{\partial x_5}=0$ and $\frac{\partial\zeta_3(x)}{\partial x_6}=0$.

5.7.3.3 Differential flatness properties of the satellite's attitude dynamics

The state-space model of the satellite comprises the following state equations:

$$\dot{x}_1=\frac{\cos(x_3)x_5-\sin(x_3)x_6}{\cos(x_2)}-\omega_0 \tag{5.203}$$

$$\dot{x}_2=\sin(x_3)x_5+\cos(x_3)x_6 \tag{5.204}$$

$$\dot{x}_3=x_4+\frac{\sin(x_2)}{\cos(x_2)}[\sin(x_3)x_6-\cos(x_3)x_5] \tag{5.205}$$

$$\dot{x}_4=\frac{I_2-I_3}{I_1}(x_5x_6-3\omega_0^2\zeta_2\zeta_3)+\frac{1}{I_1}u_1 \tag{5.206}$$

$$\dot{x}_5=\frac{I_3-I_1}{I_2}(x_6x_4-3\omega_0^2\zeta_3\zeta_1)+\frac{1}{I_1}u_2 \tag{5.207}$$

$$\dot{x}_6=\frac{I_1-I_2}{I_3}(x_4x_5-3\omega_0^2\zeta_1\zeta_2)+\frac{1}{I_1}u_3 \tag{5.208}$$

It can be proven that this model is differentially flat, with a flat outputs vector that is given by $y=[x_1,x_2,x_3]^T$. The differential flatness property signifies that: (i) all state variables and control inputs of the system can be written as differential functions of its flat outputs and (ii) the flat outputs are differentially independent, which means that they are not connected through a relation in the form of a homogeneous differential equation (Rigatos and Busawon, 2018; Rigatos, 2015; Levine, 2009).

From (5.203) and (5.204) and by solving with respect to x_5 and x_6 one gets:

$$\begin{aligned} x_5&=(\dot{x}_1+\omega_0)\cos(x_2)\cos(x_3)+\dot{x}_2\sin(x_3)\\ x_6&=-(\dot{x}_1+\omega_0)\cos(x_2)\cos(x_3)+\dot{x}_2\cos(x_3)\end{aligned} \tag{5.209}$$

Thus, state variables x_5 and x_6 are differential functions of the system's flat outputs. Moreover, from (5.205) one gets

$$x_4=\dot{x}_3-\frac{\sin(x_2)}{\cos(x_2)}[\sin(x_3)x_6-\cos(x_3)x_5] \tag{5.210}$$

Consequently, state variable x_4 is also a differential function of the system's flat outputs. Additionally, from (5.206), (5.207) and (5.208) one has:

$$\begin{aligned} u_1 &= I_1 \left\{ \dot{x}_4 - \frac{I_2 - I_3}{I_1}(x_5 x_6 - 3\omega_0^2 \zeta_2 \zeta_3) \right\} \\ u_2 &= I_2 \left\{ \dot{x}_5 - \frac{I_3 - I_1}{I_2}(x_6 x_4 - 3\omega_0^2 \zeta_3 \zeta_1) \right\} \\ u_3 &= I_3 \left\{ \dot{x}_6 - \frac{I_1 - I_2}{I_3}(x_4 x_5 - 3\omega_0^2 \zeta_1 \zeta_2) \right\} \end{aligned} \quad (5.211)$$

where $\zeta_1(x)$, $\zeta_2(x)$ and $\zeta_3(x)$ are functions of the system's flat outputs $y_1 = x_1$, $y_2 = x_2$ and $y_3 = x_3$. Consequently, the control inputs u_1, u_2 and u_3 are also differential functions of the system's flat outputs. As a result of the above, the satellite's attitude dynamics is a differentially flat system.

By proving the differential flatness properties of the satellite's attitude model, one can also solve the setpoints definition problem for the associated control loop. Actually, one can define setpoints for the flat outputs x_1, x_2 and x_3, and next with the use of (5.209) and (5.210) can also define setpoints for state variables x_4, x_5 and x_6.

5.7.3.4 Stabilizing feedback control

After linearization around its current operating point, the micro-satellite's attitude dynamic model is written as (Rigatos and Busawon, 2018)

$$\dot{x} = Ax + Bu + d_1 \quad (5.212)$$

Parameter d_1 stands for the linearization error in the micro-satellite's attitude dynamic model appearing previously in (5.212). The reference setpoints for the micro-satellite's attitude state vector are denoted by $\mathbf{x_d} = [x_1^d, \ldots, x_6^d]$. Tracking of this trajectory is achieved after applying the control input u_d. At every time instant the control input u_d is assumed to differ from the control input u appearing in (5.212) by an amount equal to Δu, that is $u^* = u + \Delta u$

$$\dot{x}_d = Ax_d + Bu_d + d_2 \quad (5.213)$$

The dynamics of the controlled system described in (5.212) can be also written as

$$\dot{x} = Ax + Bu + Bu_d - Bu_d + d_1 \quad (5.214)$$

and by denoting $d_3 = -Bu_d + d_1$ as an aggregate disturbance term one obtains

$$\dot{x} = Ax + Bu + Bu_d + d_3 \quad (5.215)$$

By subtracting (5.213) from (5.215) one has

$$\dot{x} - \dot{x}_d = A(x - x_d) + Bu + d_3 - d_2 \quad (5.216)$$

By denoting the tracking error as $e = x - x_d$ and the aggregate disturbance term as $\tilde{d} = d_3 - d_2$, the tracking error dynamics becomes

$$\dot{e} = Ae + Bu + \tilde{d} \quad (5.217)$$

For the approximately linearized model of the system a stabilizing feedback controller is developed. The controller has the form

$$u(t) = -Ke(t) \tag{5.218}$$

with $K = \frac{1}{r}B^T P$ where P is a positive definite symmetric matrix which is obtained from the solution of the Riccati equation (Rigatos and Busawon, 2018)

$$A^T P + PA + Q - P\left(\frac{2}{r}BB^T - \frac{1}{\rho^2}LL^T\right)P = 0 \tag{5.219}$$

where Q is a positive semi-definite symmetric matrix. The diagram of the considered control loop is depicted in Figure 5.57. The global stability properties of the control method can be proven following the procedure presented in Section 1.1.2.5.

It is also noteworthy that the solution of the H-infinity feedback control problem for the satellite's attitude dynamics and the computation of the worst-case disturbance that this controller can sustain, comes from superposition of Bellman's optimality principle when considering that the process is affected by two separate inputs (i) the control input u (ii) the cumulative disturbance input $\tilde{d}(t)$. Solving the optimal control problem for u that is for the minimum variation (optimal) control input that achieves

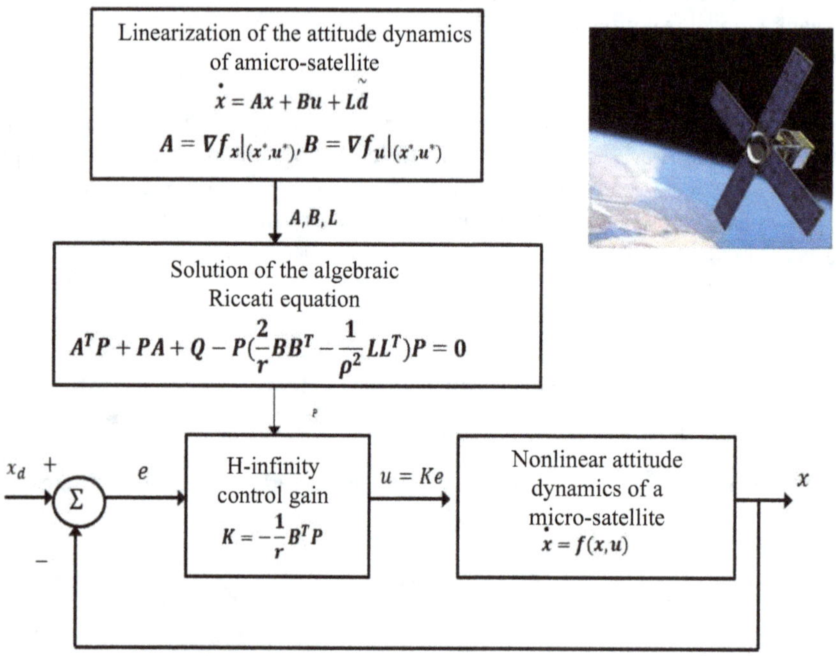

Figure 5.57 Diagram of the satellite's attitude nonlinear optimal (H-infinity) control loop

elimination of the state vector's tracking error gives $u = -\frac{1}{r}B^T Pe$. Equivalently, solving the optimal control problem for $\tilde{d}$, that is for the worst-case disturbance that the control loop can sustain gives $\tilde{d} = \frac{1}{\rho^2}L^T Pe$.

5.7.4 Simulation tests

The performance of the proposed nonlinear optimal control method for the satellite's attitude has been tested through simulation experiments. To compute the feedback gains of the controller, the algebraic Riccati equation appearing in (5.219) is iteratively solved at each time-step of the control method. The reference setpoints have been defined by exploiting the differential flatness properties of the state-space model of the satellite. Out of the 6 state variables of this state-space model 3 have been defined as flat outputs. The rest of the state variables of the satellite's model can be written as differential functions of the system's flat outputs. Consequently, by defining reference setpoints only for the flat outputs of the system (that is state variables x_1, x_2 and x_3) and by using the differential functions which connect the rest of the state variables to the flat outputs, one can define reference setpoints for the remaining state variables of the satellite which are not flat outputs (that is state variables x_4, x_5 and x_6).

The real values of the state variables of the satellite's attitude are printed in blue, the estimated values which are provided by the H-infinity Kalman Filter are plotted in green, while the related setpoints are plotted in red. The obtained results are given in Figures 5.58–5.71. It can be noticed, that under the proposed nonlinear optimal control method, precise tracking of the reference setpoints was achieved. The state

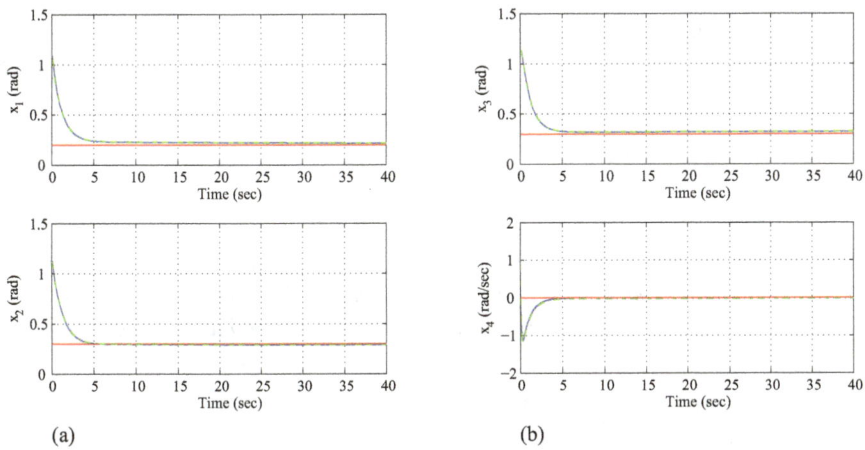

Figure 5.58 Tracking of setpoint 1 by the satellite's attitude control system: (a) convergence of state variables x_1 and x_2 to the reference setpoints and (b) convergence of state variables x_3 and x_4 to the reference setpoints

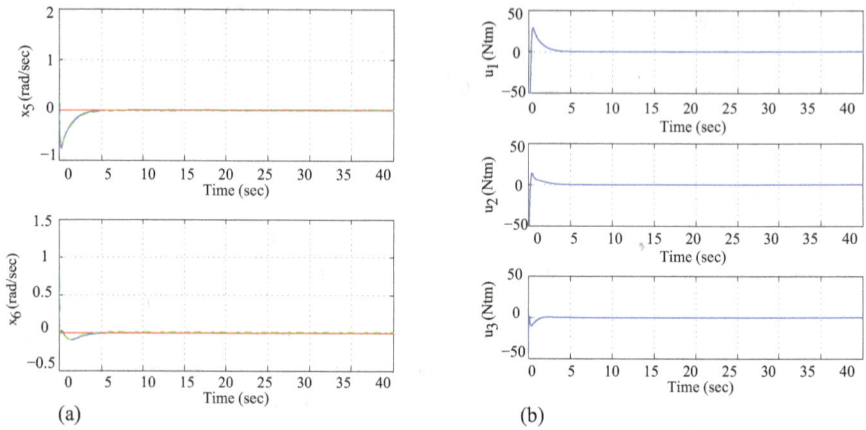

Figure 5.59 Tracking of setpoint 1 by the satellite's attitude control system: (a) convergence of state variables x_5 and x_6 to the reference setpoints and (b) variation of the control inputs u_1 to u_3

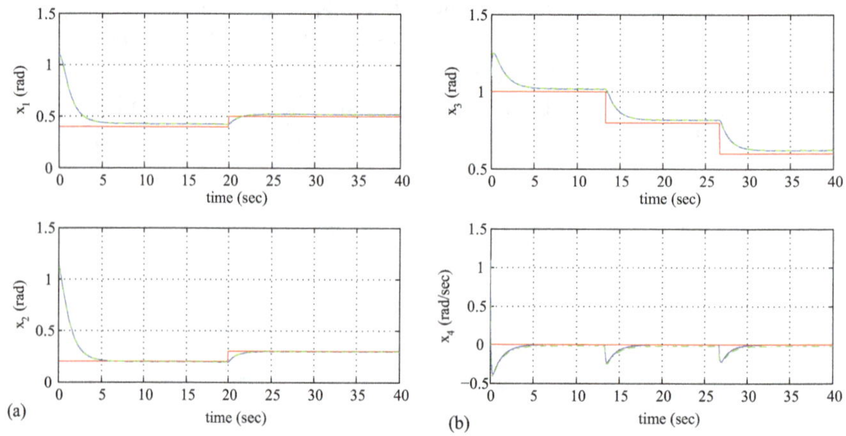

Figure 5.60 Tracking of setpoint 2 by the satellite's attitude control system: (a) convergence of state variables x_1 and x_2 to the reference setpoints and (b) convergence of state variables x_3 and x_4 to the reference setpoints

variables of the system converged fast to the targeted values while the variations of the control inputs remained moderate.

With the use of the H-infinity Kalman Filter, state estimation-based control has been finally implemented for the stabilization of the satellite's attitude. The measured state variables where the roll, pitch and yaw rotation angles of the body-fixed reference

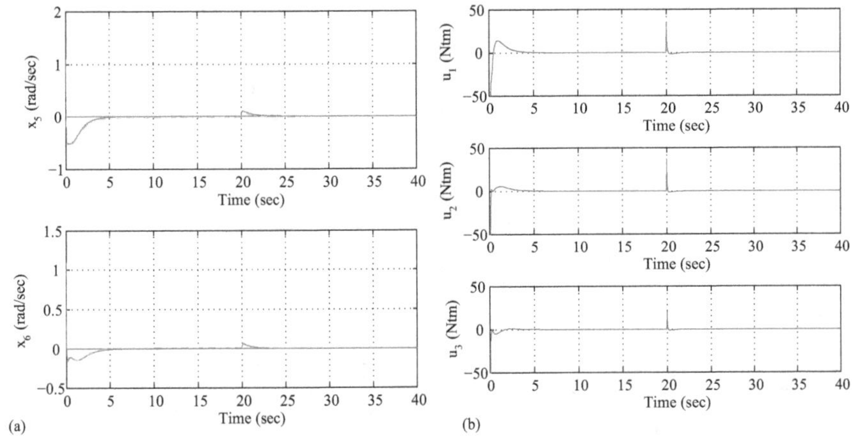

Figure 5.61 Tracking of setpoint 2 by the satellite's attitude control system: (a) convergence of state variables x_5 and x_6 to the reference setpoints and (b) variation of the control inputs u_1 to u_3

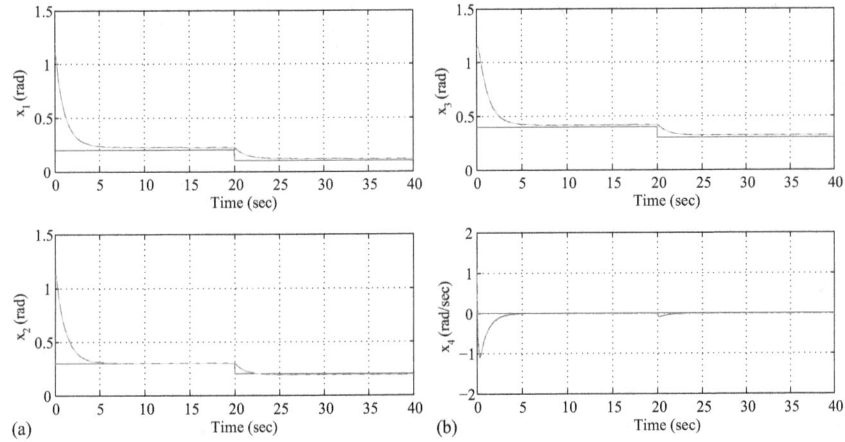

Figure 5.62 Tracking of setpoint 3 by the satellite's attitude control system: (a) convergence of state variables x_1 and x_2 to the reference setpoints and (b) convergence of state variables x_3 and x_4 to the reference setpoints

frame ($x_1 = \phi$, $x_2 = \theta$, and $x_3 = \psi$), while the rest of the state vector elements that is the angular velocities of the inertial reference frame ($x_4 = \omega_1$, $x_5 = \omega_2$ and $x_6 = \omega_3$) have been estimated through the recursion of the H-infinity Kalman Filter. the transient performance of the control algorithm depends on the parameters r, ρ and Q which appear in the above noted algebraic Riccati equation of (1.50). Small

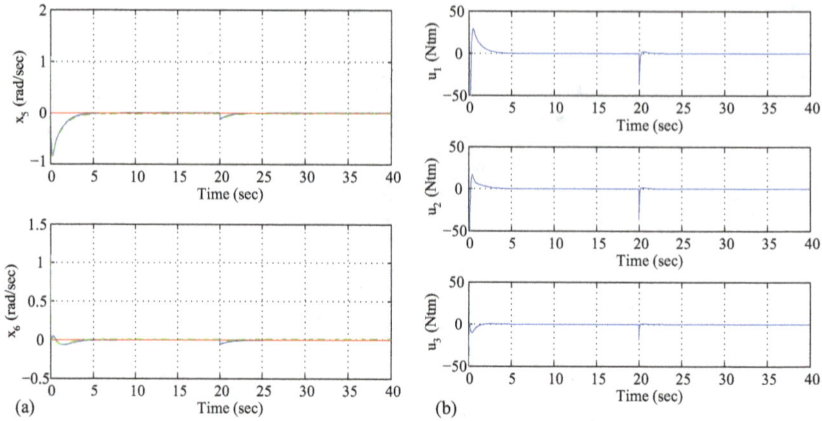

Figure 5.63 Tracking of setpoint 3 by the satellite's attitude control system: (a) convergence of state variables x_5 and x_6 to the reference setpoints and (b) variation of the control inputs u_1 to u_3

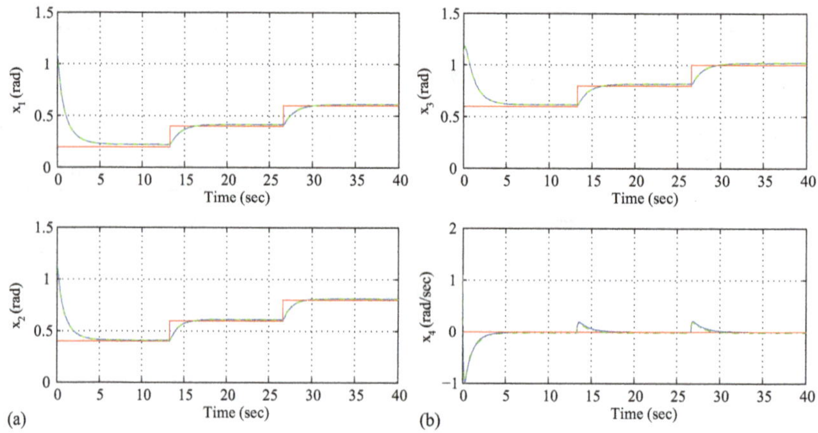

Figure 5.64 Tracking of setpoint 4 by the satellite's attitude control system: (a) convergence of state variables x_1 and x_2 to the reference setpoints and (b) convergence of state variables x_3 and x_4 to the reference setpoints

values of r minimize the tracking error while relatively large values of Q allow for fast convergence to the reference setpoints. Besides, the smallest value of ρ for which an acceptable solution P is obtained from the Riccati equation of (1.50) is the one that provides the control loop with maximum robustness.

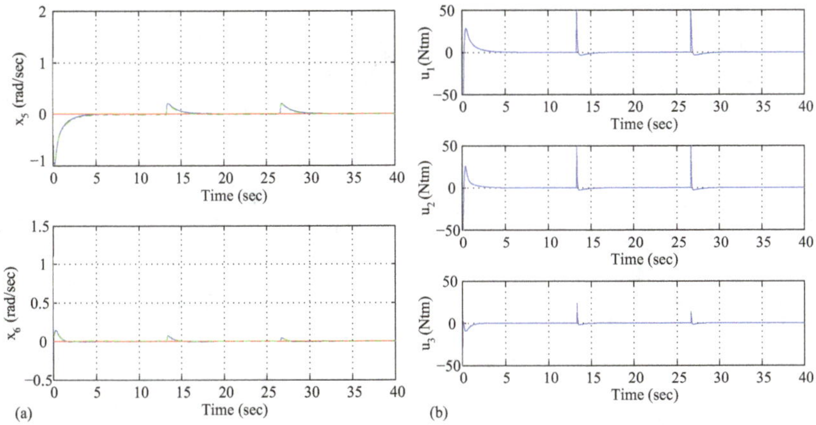

Figure 5.65 *Tracking of setpoint 4 by the satellite's attitude control system: (a) convergence of state variables x_5 and x_6 to the reference setpoints and (b) variation of the control inputs u_1 to u_3*

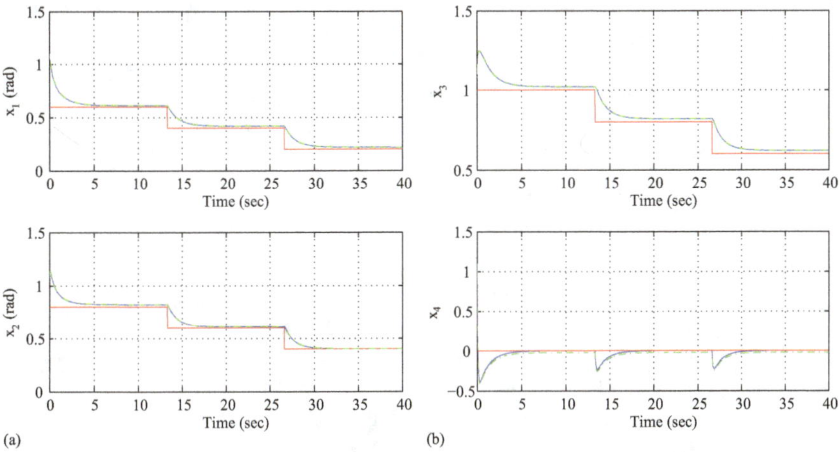

Figure 5.66 *Tracking of setpoint 5 by the satellite's attitude control system: (a) convergence of state variables x_1 and x_2 to the reference setpoints and (b) convergence of state variables x_3 and x_4 to the reference setpoints*

5.8 Nonlinear control of the industrial crystallization process

5.8.1 Outline

Control of the crystallization process is important for the fertilizers, food and pharmaceutical industry (Mesbah, Huesman *et al.*, 2011; Hours *et al.*, 2016; Yang and

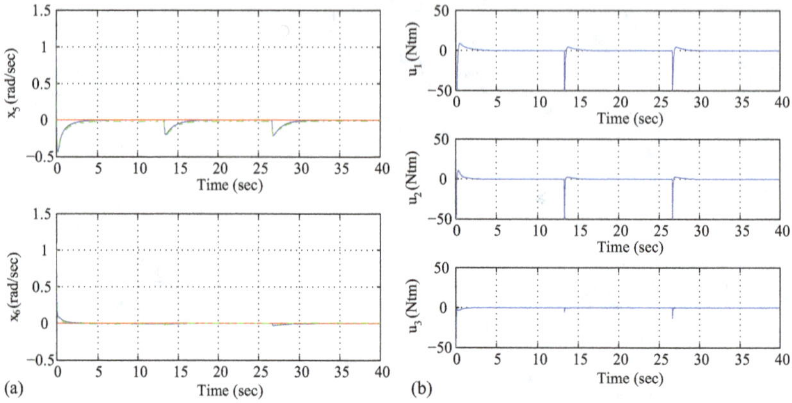

Figure 5.67 Tracking of setpoint 5 by the satellite's attitude control system: (a) convergence of state variables x_5 and x_6 to the reference setpoints and (b) variation of the control inputs u_1 to u_3

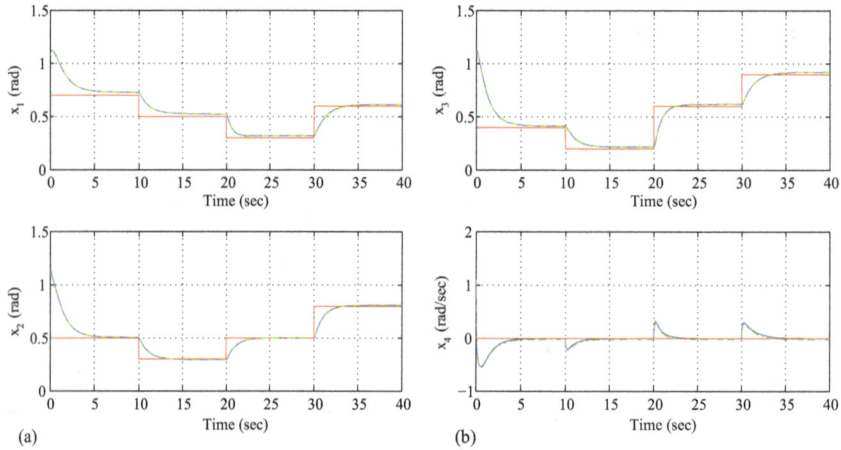

Figure 5.68 Tracking of setpoint 6 by the satellite's attitude control system: (a) convergence of state variables x_1 and x_2 to the reference setpoints and (b) convergence of state variables x_3 and x_4 to the reference setpoints

Nagy, 2015; Damout *et al.*, 2011). Efficient control of crystallization determines the quality and the cost of production (Nagy and Braatz, 2003; Liu, Hao *et al.*, 2019). Due to nonlinearities in the dynamics of the crystallization process and due to limitations in the number of the control inputs, the solution of the related control problem is a non-trivial task (Nagy and Braatz, 2003; Liu, Hao *et al.*, 2019; Mebsah, Nagy *et al.*, 2012; Nagy, Fujiwara *et al.*, 2008; Fujiwara *et al.*, 2005). The use of optimal

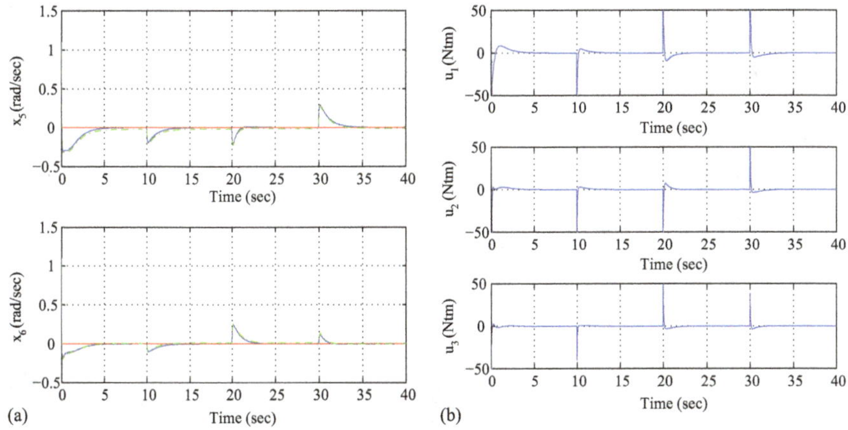

Figure 5.69 Tracking of setpoint 6 by the satellite's attitude control system: (a) convergence of state variables x_5 and x_6 to the reference setpoints and (b) variation of the control inputs u_1 to u_3

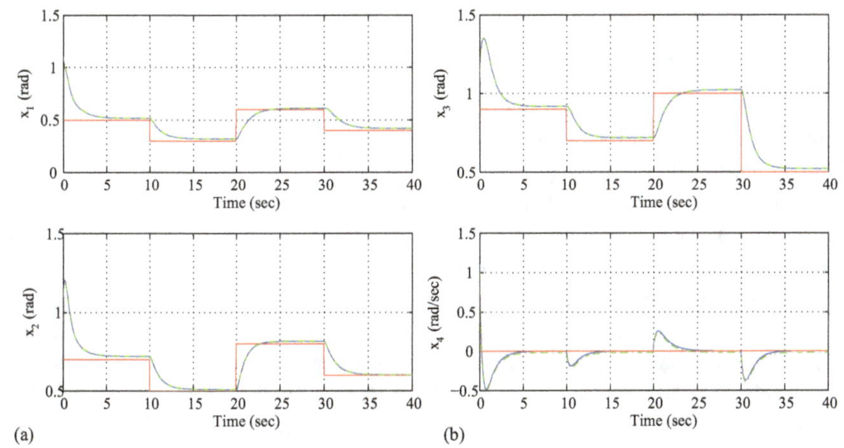

Figure 5.70 Tracking of setpoint 7 by the satellite's attitude control system: (a) convergence of state variables x_1 and x_2 to the reference setpoints and (b) convergence of state variables x_3 and x_4 to the reference setpoints

control in this process allows for reducing energy consumption and for making production cost-efficient. However, application to the crystallization process of popular approaches of industrial optimal control, such as Model Predictive Control (MPC) and Nonlinear Model Predictive Control (NMPC) can be hindered by stability problems (Gao, Aceval *et al.*, 2017; Aydin *et al.*, 2008; Damout *et al.*, 2010; Bradford *et al.*,

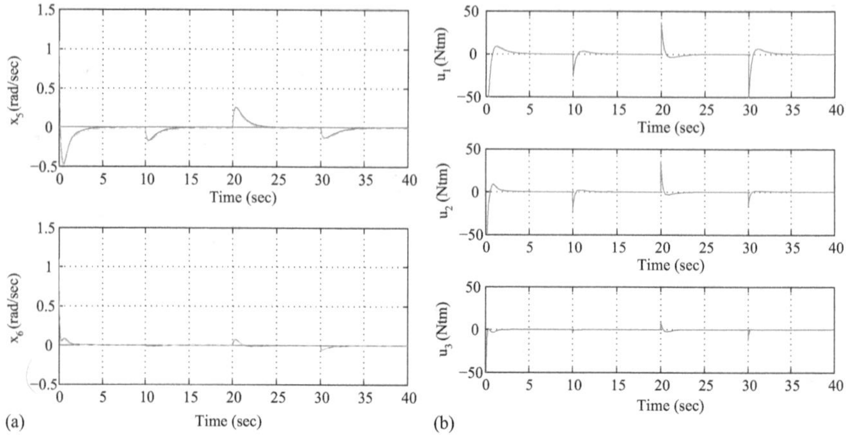

Figure 5.71 Tracking of setpoint 7 by the satellite's attitude control system: (a) convergence of state variables x_5 and x_6 to the reference setpoints and (b) variation of the control inputs u_1 to u_3

2019). Approaches based on other nonlinear control methods try also to prove global stability and to minimize the process's energy costs (Hermanto *et al.*, 2007; Mesbaj, Paulson *et al.*, 2015; Quinynen *et al.*, 2014). Other issues that one has to handle in the design of controllers for the crystallization process are model uncertainties and the unavailability of measurements for specific state variables of this system (Lovets *et al.*, 2015; Bakin *et al.*, 2006; Mesbah, Lardlast *et al.*, 2010; Nagy and Braatz, 2012; Nagy, Fujiwara *et al.*, 2008).

A concise formulation of the optimal control problem for the industrial crystallization process has been given in Rigatos, Cuccurullo *et al.*, 2019). In this section, a nonlinear optimal control approach is developed for the problem of control and stabilization of the industrial crystallization process for the production of fertilizers (Rigatos and Busawon, 2018; Rigatos, 2016b; Rigatos, 2015). First, the dynamic model of the crystallization process undergoes approximate linearization around a temporary operating point which is updated at each iteration of the control algorithm (Rigatos and Siano, 2015; Rigatos, Siano *et al.*, 2015). The linearization point is defined by the present value of the system's state vector and by the last sampled value of the control inputs vector. The linearization relies on first-order Taylor series expansion and on the computation of the associated Jacobian matrices (Rigatos and Tzafestas, 2007; Basseville and Nikiforov, 1993; Tousssaint, Basar and Bullo, 2000; Rigatos and Zhang, 2009). The approximation error which is due to the truncation of higher-order terms in the Taylor series expansion is considered as a perturbation that is finally compensated by the robustness of the control algorithm. For the linearized model of the process a stabilizing H-infinity feedback controller is designed.

The H-infinity controller stands for the solution of the optimal control problem for the crystallization process under model uncertainty and external perturbations. Actually, it represents a min–max differential game in which the control inputs try to minimize a cost function that comprises a quadratic term of the state vector's tracking error, whereas the model uncertainty and the external perturbation terms try to maximize this cost function. To select the controller's feedback gains an algebraic Riccati equation is solved at each time-step of the control method (Rigatos, 2011; Rigatos, 2013; Rigatos, 2017). The stability properties of the control scheme are proven through Lyapunov analysis. At a first stage, it is proven that the H-infinity tracking performance criterion is satisfied. This signifies elevated robustness against model uncertainty and perturbations. Next, under moderate conditions, the global asymptotic stability of the control method is proven. To implement state estimation-based control for the crystallization process without the need to measure the entire state vector of the system, the H-infinity Kalman Filter is used as a robust state estimator (Rigatos and Busawon, 2018; Rigatos, 2016b; Rigatos, 2015).

Comparing to other control schemes that one could have considered for the crystallization process, the proposed nonlinear optimal (H-infinity) control method exhibits significant advantages. In particular, (i) unlike global linearization-based control schemes (for instance Lie algebra-based control and differential flatness theory-based control), the section's method does not make use of state variables transformations (diffeomorphisms). Besides, it is applied directly to the initial non-linear model of the crystallization process, thus avoiding inverse transformations that may come against singularity problems, (ii) unlike popular approaches being used in industry for solving optimal control problems, such as MPC and NMPC, the section's method is of proven global stability. Actually, MPC is a linear control method which if applied to the nonlinear dynamics of the crystallization process will result into the control loop's destabilization. Moreover, the performance of NMPC is not of assured performance either. Actually, the convergence of this method's iterative search for an optimum depends on initialization and on the selection of certain parameters' values, (iii) unlike sliding mode control, the section's control method, does not require the empirical and intuitive selection of sliding surfaces and avoids unnecessary chattering of the control inputs, (iv) unlike backstepping control, the section's control method does not require the system's state-space model to be found a-priori or to be transformed into the triangular (backstepping integral) form, (v) unlike PID control, the section's control method avoids the heuristic selection of parameters, is of proven global stability and functions reliably at changes of operating points or under external perturbations and (vi) unlike multiple local models-based control the nonlinear optimal control method presented in this section uses only one linearization point and needs the solution of only one Riccati equation so as to compute the stabilizing feedback gains of the controller. It is widely reckoned that the off-line definition of multiple linearization points and local linear models that may cover the entire state-space of a nonlinear dynamical system comes against dimensionality problems. Besides, the related need for solving multiple Riccati equations or LMIs reveals that control approaches based on multiple linearization points devour computational power. Consequently, in terms of computation load the proposed control method for the crystallization process is much more efficient.

Control and estimation for the nonlinear dynamics of the crystallization process remains a challenging research topic (Geyyer *et al.*, 2015; Komaraju and Chiu, 2012; Ucchedou *et al.*, 2011). The present section (a) provides the only existing solution to the nonlinear optimal control problem of the crystallization process which is of proven global stability while also remaining computationally efficient and (b) applies a genuine approach for implementing H-infinity control on the nonlinear dynamics of the crystallization process. Preceding results on the use of H-infinity control to nonlinear dynamical systems were limited to the case of affine-in-the-input systems with drift-only dynamics and considered that the control inputs gain matrix is not dependent on the values of the system's state vector. Moreover, in these approaches the linearization was performed around points of the desirable trajectory whereas in the present section's control method the linearization points are related with the value of the state vector at each sampling instant as well as with the last sampled value of the control inputs vector. The Riccati equation which has been proposed for computing the feedback gains of the controller is novel, so is the presented global stability proof through Lyapunov analysis.

5.8.2 Dynamic model of the industrial crystallization process

Without loss of generality, the ammonium sulphate ($(NH_4)_2SO_4$)—water crystallizing system will be considered (Mesbah, Huesman *et al.*, 2011). Ammonium sulphate is a chemical substance that is widely used in agriculture as fertilizer. It is also used for biochemical applications, as for instance proteins purification. The related industrial crystallization process is defined by: (i) the population balance equation, (ii) the concentration balance equation and (iii) kinetic relations. The method of moments is applied to turn the population balance equation into a set of ordinary differential equations. The method of moments allows to calculate the relevant properties of the total crystals population. Actually, one has the following differential equation:

$$\frac{dm_i}{dt} = O^i B_o + iG_{mi} - \frac{m_i Q_P}{V} \tag{5.220}$$

with initial conditions $m_i(t_o) = m_{i,o}$, $i = 0, 1, 2, \ldots, 4$. Indicative parameters of this differential equation are defined as follows (Mesbah, Huesman *et al.*, 2011: m_i is the ith moment of the crystal size distribution m^i/m^3, G is the crystal growth rate m/sec, B_o is the total nucleation rate #/m^3/s, V is the crystallization volume m^3, Q_p is the unclassified product removal flow rate m^3/s. The four leading moments have physical significance: m_0 is the total number of crystals, m_1 is the total length of crystals, m_2 is proportional to the total surface area of the crystal, m_3 is proportional to the total crystal's volume, m_4 designates the mean crystal's size as the ratio of m_4 to m_3.

The particles' formation is attributed to secondary nucleation from crystal surfaces, since it is often the most dominant nucleation mechanism. The empirical relations about the total nucleation rate and size-independent crystal growth rate are:

$$\begin{aligned} B_o &= K_b m_3 G \\ G &= K_g(C - C^*)^g \end{aligned} \tag{5.221}$$

where K_b is the nucleation rate constant, K_g is the crystal growth rate constant, g is the crystal growth rate exponent, C is the solute concentration and C^* is the saturation concentration. Typical values of the model's parameters are given in the following table (Mesbah, Huesman *et al.*, 2011): g : 1.0, C^* : 0.46 (kg)*solute*/(kg)solution, ρ_c : 1767.35 (kg/m^3), ρ_L : 1248.93 (kg/m^3), H_c : 60.75 (kJ/kg), H_L : 69.86 (kJ/kg), H_v : 2.59$\cdot 10^3$ (kJ/kg), K_b: 1.02$\cdot 10^{14}$ $\#/m^4$, K_g: 7.50$\cdot 10^{-5}$ m^3/s, Q_p : 1.73$\cdot 10^{-6}$m/s, V: 7.50$\cdot 10^{-2}$ m^2, K_v : 0.43.

The evaporative crystallizer has isothermal operation and this allows for simplifying the mass and energy balance equations to a single equation for the solute concentration (Mesbah, Huesman *et al.*, 2011)

$$\frac{dC}{dt} = \frac{Q_P(C - C^*)/V + 3K_v G m_2(k_1 + C)}{1 - K_v m_g} + \frac{K_2 H_{in}}{1 - K_v m_g} \tag{5.222}$$

with initial condition $C(t_0) = C_0$, while the following coefficients are defined

$$K_1 = \frac{H_v C^*}{H_v - H_L}\left(\frac{\rho_C}{\rho_L} - 1 + \frac{\rho_L H_L - \rho_C H_c}{\rho_L H_v}\right) - \frac{\rho_C}{\rho_L} \tag{5.223}$$

$$K_2 = \frac{C^*}{V \cdot \rho_L(H_c - H_L)} \tag{5.224}$$

where K_v is the crystal volumetric shape factor, H_{in} is the heat input to the crystallizer (kW), ρ_L is the density of the saturated solution (kg/m^3) ρ_C is the density of the crystals (kg/m^3), H_L is the solution's enthalpy (kJ/kg), H_C is the crystal's specific enthalpy (kJ/kg), H_w is the vapor's specific enthalpy (kJ/Kg).

Next, the nonlinear dynamics of the crystallization process is given by (Mesbah, Huesman *et al.*, 2011)

$$f(x) = \begin{pmatrix} K_b m_3 K_g(C - C^*)^g - \dfrac{m_0 Q_P}{V} \\ K_g(C - C^*)^g m_0 - \dfrac{m_1 Q_P}{V} \\ 2K_g(C - C^*)^g m_1 - \dfrac{m_2 Q_P}{V} \\ 3K_g(C - C^*)^g m_2 - \dfrac{m_3 Q_P}{V} \\ 4K_g(C - C^*)^g m_3 - \dfrac{m_4 Q_P}{V} \\ \dfrac{Q_P(C - C^*)/V + 3K_v K_g(C - C^*)^g(k_1 + C)}{1 - K_v m_3} \end{pmatrix} \quad g(x) = \begin{pmatrix} 0 \\ 0 \\ 0 \\ 0 \\ 0 \\ \dfrac{K_2}{1 - K_v m_3} \end{pmatrix} \tag{5.225}$$

while the output measurements vector of the system is $H(x) = [m_0, m_2, m_4, C]^T$.

$$\begin{aligned} \dot{x} &= f(x) + g(x)u \\ y_i &= h_i(x)\ i = 1, 2, \cdots, 4 \end{aligned} \tag{5.226}$$

The state vector of the system is $x = [m_0, m_1, m_2, m_3, m_4, C]^T$, that is $x \in R^{6\times 1}$. The control input is the heat power provided to the system $u = H_{in}$, that is $u \in R$. This is regulated through the opening of a heating/cooling water valve. Moreover, using the state variables notation, vector fields $f(x) \in R^{6\times 1}$ and $g(x) \in R^{6\times 1}$ are defined as follows:

$$f(x) = \begin{pmatrix} K_b x_4 K_g (x_6 - C^*)^g - \frac{x_1 Q_P}{V} \\ K_g (x_6 - C^*)^g x_1 - \frac{x_2 Q_P}{V} \\ 2K_g (x_6 - C^*)^g x_2 - \frac{x_3 Q_P}{V} \\ 3K_g (x_6 - C^*)^g x_3 - \frac{x_4 Q_P}{V} \\ 4K_g (x_6 - C^*)^g x_4 - \frac{x_5 Q_P}{V} \\ \frac{Q_P (x_6 - C^*)/V + 3K_v K_g (x_6 - C^*)^g (k_1 + C)}{1 - K_v x_4} \end{pmatrix} \quad g(x) = \begin{pmatrix} 0 \\ 0 \\ 0 \\ 0 \\ 0 \\ \frac{K_2}{1 - K_v x_4} \end{pmatrix} \tag{5.227}$$

5.8.3 Approximate linearization of the dynamics of the crystallization process

5.8.3.1 Computation of the system's Jacobian matrices

The dynamic model of the crystallizat4ion process was shown to be

$$\begin{aligned} \dot{x} &= f(x) + g(x)u \\ y &= h(x) \end{aligned} \tag{5.228}$$

where $f(x) \in R^{6\times 1}$, $g(x) \in R^{6\times 1}$ and $h(x) \in R^{4\times 1}$. The model undergoes approximate linearization around the temporary operating point (x^*, u^*), where x^* is the present value of the system's state vector and u^* is the last sampled value of the control inputs vector. The linearization procedure gives

$$\dot{x} = Ax + Bu + \tilde{d} \tag{5.229}$$

where

$$A = \nabla_x [f(x) + g(x)u] \mid_{(x^*, u^*)} \Rightarrow A = \nabla_x f(x) \mid_{(x^*, u^*)} + \nabla_x g(x) u \mid_{(x^*, u^*)} \tag{5.230}$$

$$B = \nabla_u [f(x) + g(x)u] \mid_{(x^*, u^*)} \Rightarrow B = g(x) \mid_{(x^*, u^*)} \tag{5.231}$$

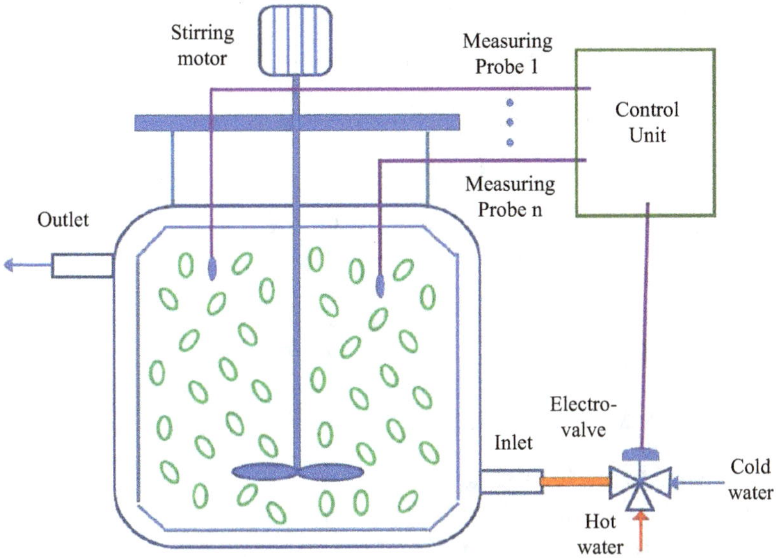

Figure 5.72 Diagram of industrial crystallizer

while $\tilde{d}$ is a vector denoting the modelling error due to the truncation of higher-order terms in the Taylor series expansion, as well as the cumulative effect of external perturbations. About the computation of matrix A the following Jacobian matrices are considered:

1st row of the Jacobian matrix $\nabla_x f(x)$: $\frac{\partial f_1}{\partial x_1} = -\frac{Q_P}{V}$, $\frac{\partial f_1}{\partial x_2} = 0$, $\frac{\partial f_1}{\partial x_3} = 0$, $\frac{\partial f_1}{\partial x_4} = K_b K_g (x_6 - C^*)^g$, $\frac{\partial f_1}{\partial x_5} = 0$, $\frac{\partial f_1}{\partial x_6} = K_b x_4 K_g g (x_6 - C^*)^{g-1}$.

2nd row of the Jacobian matrix $\nabla_x f(x)$: $\frac{\partial f_2}{\partial x_1} = K_g (x_6 - C^*)^g$, $\frac{\partial f_2}{\partial x_2} = -\frac{Q_P}{V}$, $\frac{\partial f_2}{\partial x_3} = 0$, $\frac{\partial f_2}{\partial x_4} = 0$, $\frac{\partial f_2}{\partial x_5} = 0$, $\frac{\partial f_2}{\partial x_6} = K_g g (x_6 - C^*)^{g-1} x_1$

3rd row of the Jacobian matrix $\nabla_x f(x)$: $\frac{\partial f_3}{\partial x_1} = 0$, $\frac{\partial f_3}{\partial x_2} = 2K_g (x_6 - C^*)^g$, $\frac{\partial f_3}{\partial x_3} = -\frac{Q_p}{V}$, $\frac{\partial f_3}{\partial x_4} = 0$, $\frac{\partial f_3}{\partial x_5} = 0$, $\frac{\partial f_3}{\partial x_6} = 2K_g g (x_6 - C^*)^{g-1} x_2$

4th row of the Jacobian matrix $\nabla_x f(x)$: $\frac{\partial f_4}{\partial x_4} = 0$, $\frac{\partial f_4}{\partial x_2} = 0$, $\frac{\partial f_4}{\partial x_3} = 3K_g (x_6 - C^*)$, $\frac{\partial f_4}{\partial x_4} = -\frac{Q_p}{V}$, $\frac{\partial f_4}{\partial x_5} = 0$, $\frac{\partial f_4}{\partial x_6} = 3K_g g (x_6 - C^*)^{g-1} x_3$

5th row of the Jacobian matrix $\nabla_x f(x)$: $\frac{\partial f_5}{\partial x_1} = 0$, $\frac{\partial f_5}{\partial x_2} = 0$, $\frac{\partial f_5}{\partial x_3} = 0$, $\frac{\partial f_5}{\partial x_4} = 4K_g (x_6 - C^*)^g$, $\frac{\partial f_5}{\partial x_5} = -\frac{Q_p}{V}$, $\frac{\partial f_5}{\partial x_6} = 4K_g g (x_6 - C^*)^{g-1} x_4$

6th row of the Jacobian matrix $\nabla_x f(x)$: $\frac{\partial f_6}{\partial x_1}=0$, $\frac{\partial f_6}{\partial x_2}=0$, $\frac{\partial f_6}{\partial x_3}=$ $=3K_vK_g(x_6-C^*)(k_1+x_6)/[1-K_vx_4]$, $\frac{\partial f_6}{\partial x_4}=\frac{-K_v\{[Q_p*(C^*-x_6)/V]+3K_vK_g(x_6-C^*)x_3(K_1+x_6)\}}{[1-K_vx_4]^2}$, $\frac{\partial f_6}{\partial x_5}=0$, $\frac{\partial f_6}{\partial x_6}=\frac{-\frac{Q_p}{V}+3K_vK_gx_3(K_1+x_6)+3K_vK_g(x_6-C^*)x_3}{[1-K_vx_4]}$.

About the Jacobian matrix $\nabla_x g(x)$ one has

$$\nabla_x g(x)=\begin{pmatrix} 0 & 0 & 0 & 0 & 0 & 0 \\ 0 & 0 & 0 & 0 & 0 & 0 \\ 0 & 0 & 0 & 0 & 0 & 0 \\ 0 & 0 & 0 & 0 & 0 & 0 \\ 0 & 0 & 0 & 0 & 0 & 0 \\ 0 & 0 & 0 & \frac{K_2K_v}{[1-K_vx_4]^2} & 0 & 0 \end{pmatrix} \tag{5.232}$$

5.8.3.2 Stabilizing feedback control

After linearization around its current operating point, the dynamic model of the crystallization process is written as

$$\dot{x}=Ax+Bu+d_1 \tag{5.233}$$

Parameter d_1 stands for the linearization error in the crystallization process's dynamic model appearing previously in (5.233). The reference setpoints for the crystallization process's state vector are denoted by $\mathbf{x_d}=[x_1^d,\cdots,x_6^d]$. Tracking of this trajectory is achieved after applying the control input u_d. At every time instant the control input u_d is assumed to differ from the control input u appearing in (5.233) by an amount equal to Δu, that is $u_d=u+\Delta u$

$$\dot{x}_d=Ax_d+Bu_d+d_2 \tag{5.234}$$

where d_2 is the linearization error at points of the desirable trajectory. The dynamics of the controlled system described in (5.233) can be also written as

$$\dot{x}=Ax+Bu+Bu_d-Bu_d+d_1 \tag{5.235}$$

and by denoting $d_3=-Bu_d+d_1$ as an aggregate disturbance term one obtains

$$\dot{x}=Ax+Bu+Bu_d+d_3 \tag{5.236}$$

By subtracting (5.234) from (5.236) one has

$$\dot{x}-\dot{x}_d=A(x-x_d)+Bu+d_3-d_2 \tag{5.237}$$

By denoting the tracking error as $e=x-x_d$ and the aggregate disturbance term as $\tilde{d}=d_3-d_2$, the tracking error dynamics becomes

$$\dot{e}=Ae+Bu+\tilde{d} \tag{5.238}$$

For the approximately linearized model of the system a stabilizing feedback controller is developed. The controller has the form

$$u(t)=-Ke(t) \tag{5.239}$$

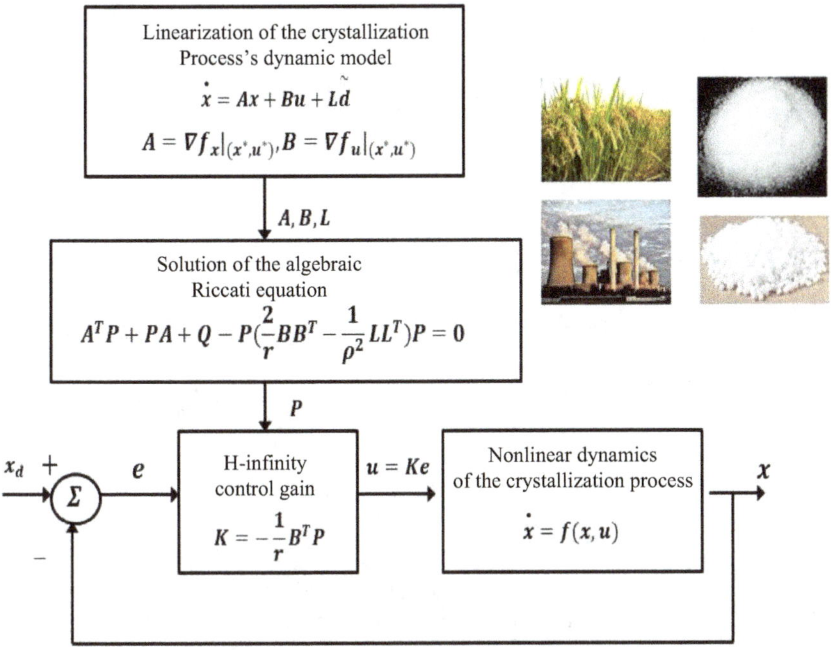

Figure 5.73 Diagram of the control scheme for the industrial crystallization process

with $K = \frac{1}{r}B^TP$ where P is a positive definite symmetric matrix which is obtained from the solution of the Riccati equation (Rigatos and Busawon, 2018)

$$A^TP + PA + Q - P\left(\frac{2}{r}BB^T - \frac{1}{\rho^2}LL^T\right)P = 0 \tag{5.240}$$

where Q is a positive semi-definite symmetric matrix. The diagram of the considered control loop for the industrial crystallization process is depicted in Figure 5.73. The global stability properties of the control method can be proven following the procedure presented in Section 1.1.2.5.

5.8.4 Simulation tests

The performance of the considered nonlinear optimal control method for the industrial crystallization process and for the production of fertilizers has been further confirmed through simulation experiments. The obtained results are depicted in Figure 5.74–5.79. The control scheme has been implemented through partial measurement of the system's state vector, while the H-infinity Kalman Filter has been used to estimate the non-measurable state variables (state variables x_1, x_2, x_3 and x_5 were measured, while state variables x_4 and x_6 were estimated). In the provided diagrams, the real value of

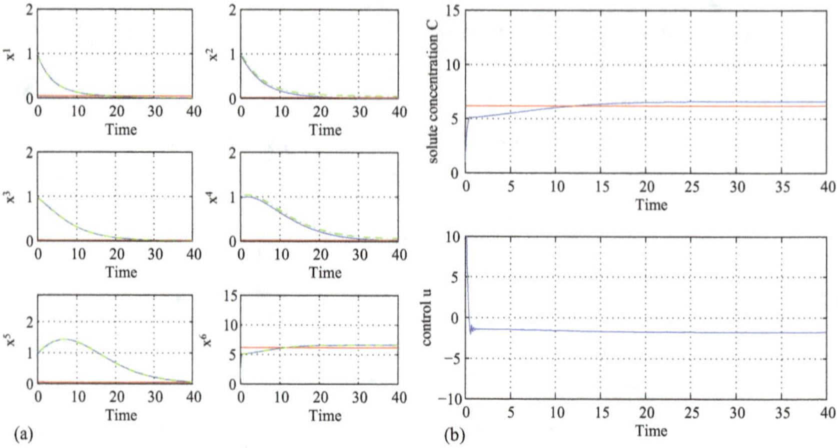

Figure 5.74 Tracking performance of the crystallization process control in case of setpoint 1: (a) convergence of the state variables x_1 to x_6 to the reference setpoints and (b) state variable $x_6 = C$ (solute concentration) and control input u (opening of the heating/cooling valve) applied to the industrial crystallization process

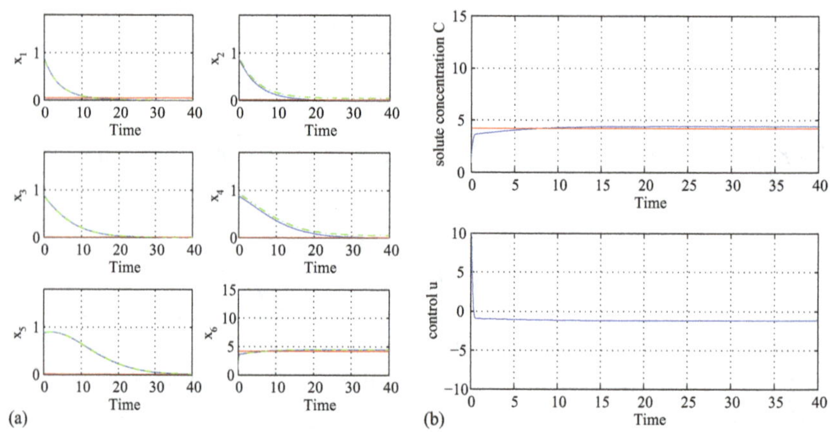

Figure 5.75 Tracking performance of the crystallization process control in case of setpoint 2: (a) convergence of the state variables x_1 to x_6 to the reference setpoints and (b) state variable $x_6 = C$ (solute concentration) and control input u (opening of the heating/cooling valve) applied to the industrial crystallization process

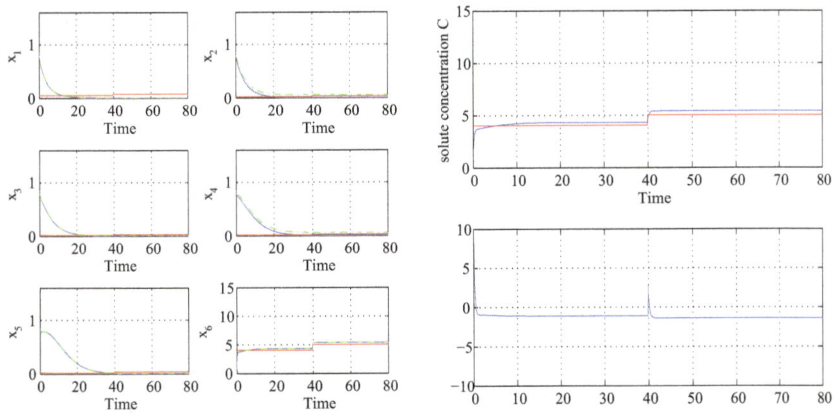

Figure 5.76 Tracking performance of the crystallization process control in case of setpoint 3: (a) convergence of the state variables x_1 to x_6 to the reference setpoints and (b) state variable $x_6 = C$ (solute concentration) and control input u (opening of the heating/cooling valve) applied to the industrial crystallization process

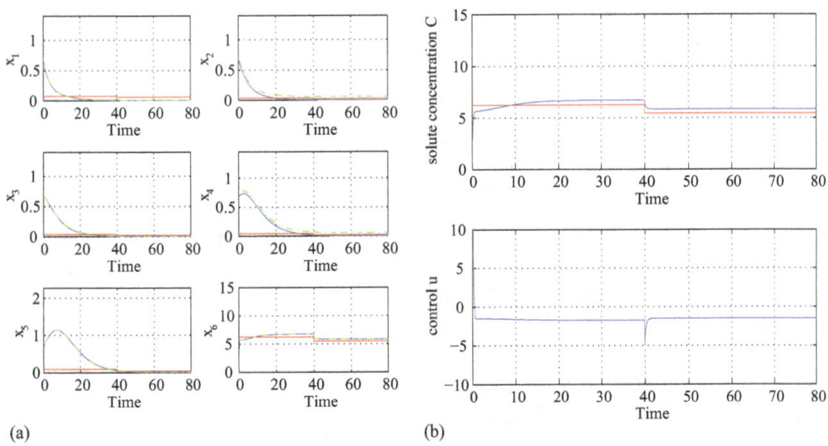

Figure 5.77 Tracking performance of the crystallization process control in case of setpoint 4: (a) convergence of the state variables x_1 to x_6 to the reference setpoints and (b) state variable $x_6 = C$ (solute concentration) and control input u (opening of the heating/cooling valve) applied to the industrial crystallization process

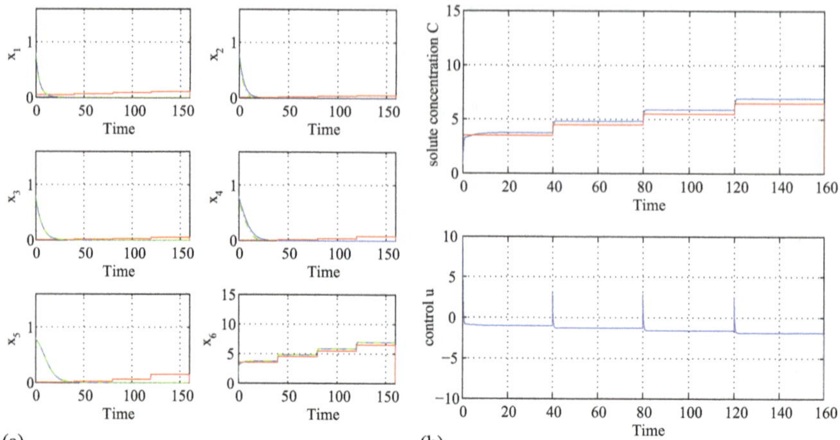

(a) (b)

Figure 5.78 Tracking performance of the crystallization process control in case of setpoint 5: (a) convergence of the state variables x_1 to x_6 to the reference setpoints and (b) state variable $x_6 = C$ (solute concentration) and control input u (opening of the heating/cooling valve) applied to the industrial crystallization process

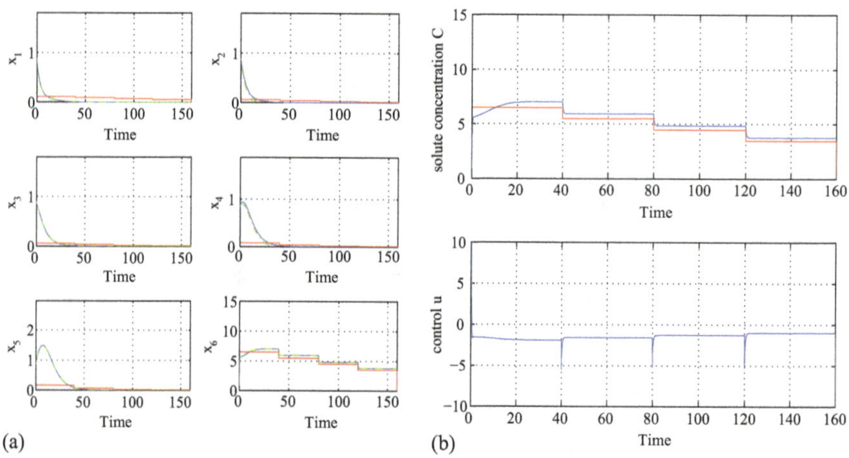

(a) (b)

Figure 5.79 Tracking performance of the crystallization process control in case of setpoint 6: (a) convergence of the state variables x_1 to x_6 to the reference setpoints and (b) state variable $x_6 = C$ (solute concentration) and control input u (opening of the heating/cooling valve) applied to the industrial crystallization process

the state variables is printed in blue, the estimated value is plotted in green while the associated setpoints are printed in red. It can be observed that, in all cases, fast and accurate tracking of the reference setpoints has been achieved under moderate variations of the control inputs. The transient performance of the control method relies on the selection of the gains and matrices r, ρ and Q that appear in the controller's algebraic Riccati equation. The robustness of the control method depends on the selection of ρ. Actually, the smallest value of this parameter that assures a solution for the algebraic Riccati equation of (5.240) is the one that provides maximum robustness to the control loop.

Chapter 6

Control based on global linearisation for industrial and PDE systems

Abstract

The chapter analyses methods and presents results on control and estimation based on global linearisation for industrial and PDE systems: (i) control of a robotic exoskeleton subject to time-delays, (ii) adaptive control of synchronous reluctance machines, (iii) control of a mobile robotic manipulator (iv) state-of-charge estimation in EVs with a Kalman Filter-based disturbance observer, (v) control of nonlinear wave PDE dynamics, (vi) control of PDE diffusion dynamics for bandwidth allocation in internet routes, (vii) control of diffusion PDE for data flow management in communication networks, (viii) control of the diffusion PDE in Li-ion batteries, (ix) control of the diffusion PDE in industrial assets' management, (x) estimation of PDE dynamics of the highway traffic and (xi) estimation of the PDE dynamics of a cable-suspended bridge.

6.1 Control of a robotic exoskeleton subject to time-delays

6.1.1 Outline

The problem of reliable functioning of robotic systems under unmodelled time-delayed disturbance forces and torques is of significant difficulty and has been the subject of extensive research (Rigatos and Busawon, 2018; Rigatos, 2015). In the case of robotic exoskeletons, compensation of disturbance forces and torques is often pursued with disturbance observers . In (Meng, Liu *et al.*, 2015) an overview is given about force and torque control techniques in rehabilitation robots. In (Long and Peng, 2021) a 2-link lower-limb robotic exoskeleton is controlled with a sliding-mode controller, while an extended state observer is used to estimate disturbances. In (Zhang, Yang *et al.*, 2018) a 5-DOF upper limb robotic exoskeleton is considered and sliding mode control is applied jointly with an extended state observer which estimates disturbances. In (Chen, Du *et al.*, 2019) a 2-DOF lower-limb robotic exoskeleton is considered on which sliding-mode control and an extended state observer are applied, aiming at compensating for exogenous perturbations In (Yang, Ma *et al.*, 2019) a 5-DOF upper limb robotic exoskeleton is considered and sliding mode control is used jointly with an extended state observer which estimates disturbances. In (Li, Su, Wang *et al.*, 2015) adaptive fuzzy control jointly with a disturbance estimator are used in a 5-DOF robotic exoskeleton. In (Brajimi *et al.*, 2021) an adaptive control scheme jointly with a disturbance observer are used in the 7-DOF model of an upper-limb robotic exoskeleton to compensate for the effects of time-delayed perturbations. In (Wang, Wang and Tian, 2011) a backstepping controller is used jointly with a disturbance observer to eliminate the perturbations' impact in the model of a 2-DOF lower-limb robotic exoskeleton. In (Masoud, Mattson *et al.*, 2020) a 4-DOF upper-link robotic exoskeleton is considered and a feedback controller designed in the s-frequency domain is used jointly with a disturbance estimator to ensure reliable functioning. In (Hun *et al.*, 2020) a 2-DOF lower-limb robotic exoskeleton is analysed and a PD feedback controller is applied to it jointly with an extended state observer which allows to estimate and compensate for lumped disturbances. In (Just, Ozen *et al.*, 2015) an inverse dynamics controller is used jointly with a disturbance observer in the model of a 7-DOF upper-limb rehabilitation exoskeleton. In (Sudo, Yap *et al.*, 2019) a 3-DOF lower limb robotic exoskeleton is examined and an inverse dynamics controller is applied to it jointly with a dual Unscented Kalman Filter which performs both states and parameters estimation.

The use of disturbance observers in robotic exoskeletons is advantageous because it is computationally simple and ensures fast and precise estimation of unknown forces and torques. In (Ahma *et al.*, 2018) a 2-DOF lower-limb robotic exoskeleton is analysed in which a fractional-order sliding-mode controller is applied together with a time-delays estimation scheme. In (Saadatzi, Long *et al.*, 2018) a 2-DOF wrist rehabilitation exoskeleton is considered in which a PD feedback controller is applied jointly with an estimator of contact torques. In (Chan *et al.*, 2020) a 3-DOF robotic exoskeleton is examined in which adaptive compliance control is used jointly with parameters' identification. In (Liang and Hsiao, 2020) an inverse dynamics controller is developed together with a disturbance observer which allows for joint torques' estimation in a 2-DOF lower-limb rehabilitation exoskeleton. In (Zhang, Yang *et al.*, 2020) a 5-DOF lower-limb robotic exoskeleton is considered in which a sliding-mode controller is applied together with a disturbance observer. In (Wu, Chen *et al.*, 2018) a neural network is used as torques estimator in an upper-limb soft rehabilitation exoskeleton. In (Ugurlu *et al.*, 2015) an inverse dynamics controller is used in a 2-DOF upper-limb robotic exoskeleton jointly with a disturbance observer which allows for estimating and compensating for perturbation torques. In (Vantili *et al.*, 2018) a 6-DOF lower-limb robotic exoskeleton is controlled with the use of PD-type controllers while joint torques are computed through an optimisation procedure related to energy variables of the robotic mechanism, Finally, in (Wu, Huang *et al.*, 2014) a knee rehabilitation exoskeleton is considered in which sliding-mode control is applied together with a nonlinear disturbance observer.

In the present section, a flatness-based controller is developed jointly with a flatness-based disturbance observer for a 2-link lower-limb robotic exoskeleton which is subject to time-delayed torques. First, it is proven that the dynamic model of the 2-link robotic exoskeleton is differentially flat. This property means that all state variables and the control inputs of the robot can be written as differential functions of a subset of its state vector elements which are the flat outputs of the exoskeleton (Rudolph, 2003; Sira-Ramirez and Agrawal, 2014; Levine, 2011). This property signifies also that the exoskeleton's dynamic model can be transformed into the input–output linearised form, or equivalently into the linear canonical (Brunovsky) form (Villagra *et al.*, 2007; Boouden *et al.*, 2011). The latter description of the dynamics of this robotic system is both controllable and observable. This allows for solving the stabilisation and feedback control problem with the use of linear feedback control methods, such as the pole-placement (eigenvalues assignment) method. At the same time, it allows for solving optimally the state estimation problem of the robot with Kalman Filtering (Rigatos and Tzafestas, 2007, Basseville and Nikiforov, 1993, Rigatos and Zhang, 2009). The implementation of this filtering method on the flatness-based equivalent linearised model of the robotic exoskeleton and the use of flatness-based inverse transformations to obtain state estimates for the initial nonlinear state-space description of the robot is also known as derivative-free nonlinear Kalman Filtering (Rigatos, 2012c).

In the disturbance-free case, the use of the above-noted flatness-based controller jointly with the flatness-based implementation of Kalman Filtering would suffice to ensure stabilisation of the exoskeleton and precise tracking of setpoints by its joints through the feedback of the estimated state vector (Rigatos and Busawon, 2018; Rigatos, 2015). However, the functioning of the exoskeleton comes against perturbations which are due to contact forces exerted at its free-end. The Jacobian matrix of the kinematic model of the exoskeleton shows how disturbance torques are in-turn developed at the joints of this robotic mechanism. These disturbance forces and torques may also be subject to time-delays. Both the mathematical description of these disturbance forces and torques and the associated time-delays are unknown thus making their compensation by the exoskeleton's controller be a nontrivial problem. To treat this shortcoming, the section proposes the redesign of the aforementioned Kalman Filter as a disturbance observer that will allow for estimating in real-time the perturbation torques which affect the robot's joints (Rigatos and Busawon, 2018; Rigatos, 2015). Using the estimates of these disturbance torques, additional terms can be included in the flatness-based controller of the exoskeleton thus finally achieving the complete elimination of the disturbances' impact and the precise tracking of setpoints by the exoskeleton's state variables.

6.1.2 Dynamic model of the robotic exoskeleton

6.1.2.1 State-space description of the robotic exoskeleton

A two-link lower-limb robotic exoskeleton is considered, as shown in Figure 6.1. The dynamic model of the robotic exoskeleton can be also written in the concise form (Rigatos, Abbaszadeh *et al.*, 2020)

$$M(\theta)\ddot{\theta} + C(\theta,\dot{\theta}) + G(\theta) = \tau \tag{6.1}$$

where the state-vector is $\theta = [\theta_1, \theta_2]^T$, the joints' torques vector is $\tau = [\tau_1, \tau_2]^T$, while the inertia matrix $M(\theta)$, the Coriolis matrix $C(\theta,\dot{\theta})$ and the gravitational forces matrix $G(\theta)$ are defined as follows (Rigatos, Abbaszadeh *et al.*, 2020):

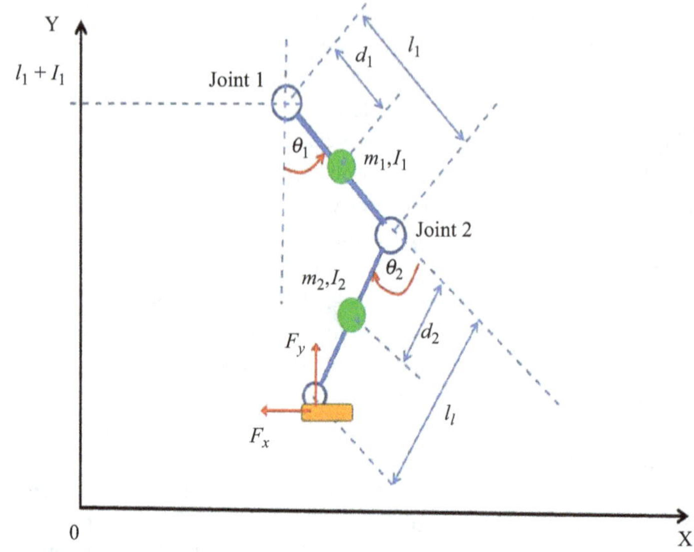

Figure 6.1 Diagram of the two-link lower-limb robotic exoskeleton with forces exerted at its free end

$$M(\theta) = \begin{pmatrix} m_1 d_1^2 + I_1 + m_2 l_1^2 + m_2 d_1^2 + 2m_2 l_1 d_2 \cos(\theta_2) + I_2 & -m_2 d_2^2 - m_2 l_1 d_2 \cos(\theta_2) - I_2 \\ -m_2 d_2^2 - m_2 l_1 d_2 \cos(\theta_2) - I_2 & m_2 d_2^2 + I_2 \end{pmatrix} \tag{6.2}$$

$$C(\theta, \dot{\theta}) = \begin{pmatrix} -m_2 l_1 d_2 \sin(\theta_2)\dot{\theta}_2(2\dot{\theta}_1 - \dot{\theta}_2) \\ m_2 l_1 d_2 \sin(\theta_2)\dot{\theta}_1^2 \end{pmatrix}$$
$$G(\theta) = \begin{pmatrix} (m_1 g d_1 + m_2 g l_1)\sin(\theta_1) + m_2 g d_2 \sin(\theta_1 - \theta_2) \\ m_2 g d_2 \sin(\theta_1 - \theta_2) \end{pmatrix} \tag{6.3}$$

Using that the symmetric inertia matrix $M(\theta)$ is

$$M = \begin{pmatrix} M_{11} & M_{12} \\ M_{21} & M_{22} \end{pmatrix} \tag{6.4}$$

and that the determinant of $M(\theta)$ is $detM = M_{11}M_{22} - M_{12}^2$, the inverse of $M(\theta)$ is

$$M^{-1} = \begin{pmatrix} \frac{M_{22}}{M_{11}M_{22} - M_{12}^2} & \frac{-M_{12}}{M_{11}M_{22} - M_{12}^2} \\ \frac{-M_{21}}{M_{11}M_{22} - M_{12}^2} & \frac{M_{11}}{M_{11}M_{22} - M_{12}^2} \end{pmatrix} \tag{6.5}$$

Thus the dynamic model of the exoskeleton robot is written as

$$\ddot{\theta} = -M^{-1}(\theta)[C(\theta, \dot{\theta}) + G(\theta)] + M^{-1}(\theta)\tau \tag{6.6}$$

By performing operations between the individual matrices one arrives at the state-space description

$$\begin{pmatrix}\ddot{\theta}_1\\ \ddot{\theta}_2\end{pmatrix}=\begin{pmatrix}\frac{-M_{22}(C_1+G_1)+M_{12}(C_2+G_2)}{M_{11}M_{22}-M_{12}^2}\\ \frac{M_{12}(C_1+G_1)-M_{11}(C_2+G_2)}{M_{11}M_{22}-M_{12}^2}\end{pmatrix}$$
$$+\begin{pmatrix}\frac{M_{22}}{M_{11}M_{22}-M_{12}^2} & \frac{-M_{12}}{M_{11}M_{22}-M_{12}^2}\\ \frac{-M_{21}}{M_{11}M_{22}-M_{12}^2} & \frac{M_{11}}{M_{11}M_{22}-M_{12}^2}\end{pmatrix}\begin{pmatrix}\tau_1\\ \tau_2\end{pmatrix} \tag{6.7}$$

The state variables of the exoskeleton are $x_1=\theta_1$, $x_2=\dot{\theta}_1]$, $x_3=\theta_2$ and $x_4=\dot{\theta}_2$. In case that no contact forces are exerted on the free-end of the robotic exoskeleton the associated state-space model is given by

$$\dot{x}_1=x_2 \tag{6.8}$$

$$\dot{x}_2=\frac{-M_{22}(C_1+G_1)+M_{12}(C_2+G_2)}{M_{11}M_{22}-M_{12}^2}+\frac{M_{22}}{M_{11}M_{22}-M_{12}^2}\tau_1+\frac{M_{12}}{M_{11}M_{22}-M_{12}^2}\tau_2 \tag{6.9}$$

$$\dot{x}_3=x_4 \tag{6.10}$$

$$\dot{x}_4=\frac{M_{12}(C_1+G_1)-M_{11}(C_2+G_2)}{M_{11}M_{22}-M_{12}^2}-\frac{M_{21}}{M_{11}M_{22}-M_{12}^2}\tau_1+\frac{M_{11}}{M_{11}M_{22}-M_{12}^2}\tau_2 \tag{6.11}$$

With reference to the diagram of Figure 6.1 the coordinates of the free-end of the two-link robotic exoskeleton are

$$\begin{aligned}x&=x_0+l_1\sin(\theta_1)+l_2\sin(\theta_1-\theta_2)\\ y&=l_1+l_2-l_1\cos(\theta_1)-l_2\cos(\theta_1-\theta_2)\end{aligned} \tag{6.12}$$

The associated Jacobian matrix, having as elements the partial derivatives of x and y with respect to the joint angles of the robotic exoskeleton θ_1 and θ_2 are:

$$J=\begin{pmatrix}l_1\cos(\theta_1)+l_2\cos(\theta_1-\theta_2) & -l_2\cos(\theta_1-\theta_2)\\ l_1\sin(\theta_1)-l_2\sin(\theta_1-\theta_2) & -l_2\sin(\theta_1-\theta_2)\end{pmatrix} \tag{6.13}$$

It is considered that the following forces are exerted on the free-end of the exoskeleton

$$\begin{aligned}F_x&=k_{11}\theta_1(t-\tau_{11})+k_{21}\theta_2(t-\tau_{21})\\ F_y&=k_{12}\theta_1(t-\tau_{21})+k_{22}\theta_2(t-\tau_{22})\end{aligned} \tag{6.14}$$

that is the forces are expressed as the sum of the products of the time-delayed turn angles of the joints with elasticity coefficients. The torques which are transferred to the joints of the exoskeleton are given by $T = J^T F$ where $F = [F_x, F_y]^T$. Both the elasticity coefficients $k_{11}, k_{21}, k_{12}, k_{22}$ and the time-delay coefficients $\tau_{11}, \tau_{21}, \tau_{12}, \tau_{22}$ are taken to be unknown.

The torques which are transferred to the joints of the robotic exoskeleton are given by

$$J^T F = \begin{pmatrix} l_1\cos(\theta_1) + l_2\cos(\theta_1 - \theta_2) & -l_2\cos(\theta_1 - \theta_2) \\ l_1\sin(\theta_1) - l_2\sin(\theta_1 - \theta_2) & -l_2\sin(\theta_1 - \theta_2) \end{pmatrix}^T \cdot \begin{pmatrix} k_{11}\theta_1(t - \tau_{11}) + k_{21}\theta_2(t - \tau_{21}) \\ k_{12}\theta_1(t - \tau_{21}) + k_{22}\theta_2(t - \tau_{22}) \end{pmatrix} \tag{6.15}$$

By approximating the time-delayed terms of the contact forces through Taylor series expansion, that is

$$\begin{aligned} F_x &= k_{11}\theta_1(t) - k_{11}\tau_{11}\dot{\theta}_1(t) + k_{21}\theta_2(t) - k_{21}\tau_{21}\dot{\theta}_2(t) \\ F_y &= k_{12}\theta_1(t) - k_{12}\tau_{12}\dot{\theta}_2(t) + k_{22}\theta_2(t) - k_{22}\tau_{22}\dot{\theta}_2(t) \end{aligned} \tag{6.16}$$

the torques which are transferred to the joints of the robotic exoskeleton are given by

$$J^T F = \begin{pmatrix} l_1\cos(\theta_1) + l_2\cos(\theta_1 - \theta_2) & -l_2\cos(\theta_1 - \theta_2) \\ l_1\sin(\theta_1) - l_2\sin(\theta_1 - \theta_2) & -l_2\sin(\theta_1 - \theta_2) \end{pmatrix}^T \cdot \begin{pmatrix} k_{11}\theta_1(t) - k_{11}\tau_{11}\dot{\theta}_1(t) + k_{21}\theta_2(t) - k_{21}\tau_{21}\dot{\theta}_2(t) \\ k_{12}\theta_1(t) - k_{12}\tau_{12}\dot{\theta}_2(t) + k_{22}\theta_2(t) - k_{22}\tau_{22}\dot{\theta}_2(t) \end{pmatrix} \tag{6.17}$$

or equivalently using the state variables notation $x_1 = \theta_1$, $x_2 = \dot{\theta}_1$, $x_3 = \theta_3$, $x_4 = \dot{\theta}_4$ one obtains

$$J^T F = \begin{pmatrix} l_1\cos(x_1) + l_2\cos(x_1 - x_3) & l_1\sin(x_1) - l_2\sin(x_1 - x_3) \\ -l_2\cos(x_1 - x_3) & -l_2\sin(x_1 - x_3) \end{pmatrix} \cdot \begin{pmatrix} k_{11}x_1 - k_{11}\tau_{11}x_2 + k_{21}x_3 - k_{21}\tau_{21}x_4 \\ k_{12}x_1 - k_{12}\tau_{12}x_2 + k_{22}x_3 - k_{22}\tau_{22}x_4 \end{pmatrix} \tag{6.18}$$

By denoting the torques which are transferred to the joints of the robotic exoskeleton as

$$\begin{aligned} \tilde{\tau}_1 = {} & [l_1\cos(x_1) + l_2\cos(x_1 - x_3)][k_{11}x_1 - k_{11}\tau_{11}x_2 + k_{21}x_3 - k_{21}\tau_{21}x_4] \\ & + [l_1\sin(x_1) - l_2\sin(x_1 - x_3)][k_{12}x_1 - k_{12}\tau_{12}x_2 + k_{22}x_3 - k_{22}\tau_{22}x_4] \end{aligned} \tag{6.19}$$

$$\begin{aligned} \tilde{\tau}_2 = {} & [-l_2\cos(x_1 - x_3)][k_{11}x_1 - k_{11}\tau_{11}x_2 + k_{21}x_3 - k_{21}\tau_{21}x_4] \\ & + [-l_2\sin(x_1 - x_3)][k_{12}x_1 - k_{12}\tau_{12}x_2 + k_{22}x_3k_{22}\tau_{22}x_4] \end{aligned} \tag{6.20}$$

the state-space model of the robotic exoskeleton is rewritten as

$$\dot{x}_1 = x_2 \tag{6.21}$$

$$\dot{x}_2 = \frac{-M_{22}(C_1+G_1)+M_{12}(C_2+G_2)}{M_{11}M_{22}-M_{12}^2} + \frac{M_{22}}{M_{11}M_{22}-M_{12}^2}(\tau_1+\tilde{\tau}_1) - \frac{M_{12}}{M_{11}M_{22}-M_{12}^2}(\tau_2+\tilde{\tau}_2) \tag{6.22}$$

$$\dot{x}_3 = x_4 \tag{6.23}$$

$$\dot{x}_4 = \frac{M_{12}(C_1+G_1)-M_{11}(C_2+G_2)}{M_{11}M_{22}-M_{12}^2} - \frac{M_{21}}{M_{11}M_{22}-M_{12}^2}(\tau_1+\tilde{\tau}_1) - \frac{M_{11}}{M_{11}M_{22}-M_{12}^2}(\tau_2+\tilde{\tau}_2) \tag{6.24}$$

Thus, the model of the robotic exoskeleton under the effects of time-delayed external forces and torques is written in the concise form

$$\dot{x} = f(x) + g(x)u \tag{6.25}$$

where $x \in R^{4\times1}$, $f(x) \in R^{4\times1}$, $g(x) \in R^{4\times2}$ and $u \in R^{2\times1}$, with the drift vector $f(x)$ to be given by

$$f(x) = \begin{pmatrix} x_2 \\ \frac{-M_{22}(C_1+G_1)+M_{12}(C_2+G_2)}{M_{11}M_{22}-M_{12}^2} + \frac{M_{22}}{M_{11}M_{22}-M_{12}^2}\tilde{\tau}_1 + \frac{M_{12}}{M_{11}M_{22}-M_{12}^2}\tilde{\tau}_2 \\ x_4 \\ \frac{M_{12}(C_1+G_1)-M_{11}(C_2+G_2)}{M_{11}M_{22}-M_{12}^2} - \frac{M_{21}}{M_{11}M_{22}-M_{12}^2}\tilde{\tau}_1 + \frac{M_{11}}{M_{11}M_{22}-M_{12}^2}\tilde{\tau}_2 \end{pmatrix} \tag{6.26}$$

while the control inputs gain matrix $g(x)$ is given by

$$g(x) = \begin{pmatrix} 0 & 0 \\ \frac{M_{22}}{M_{11}M_{22}-M_{12}^2} & \frac{M_{12}}{M_{11}M_{22}-M_{12}^2} \\ 0 & 0 \\ -\frac{M_{21}}{M_{11}M_{22}-M_{12}^2} & \frac{M_{11}}{M_{11}M_{22}-M_{12}^2} \end{pmatrix} \tag{6.27}$$

Equivalently the state-space model of the robotic exoskeleton under time-delayed external forces is given by

$$\begin{aligned} \ddot{x}_1 &= f_2(x) + g_{21}(x)u_1 + g_{22}(x)u_2 \\ \ddot{x}_3 &= f_4(x) + g_{41}(x)u_1 + g_{42}(x)u_2 \end{aligned} \tag{6.28}$$

6.1.2.2 Stabilising feedback control for the robotic exoskeleton

In (6.30) the following transformed control inputs are defined

$$\begin{aligned} v_1 &= f_2(x) + g_{21}(x)u_1 + g_{22}(x)u_2 \\ v_2 &= f_4(x) + g_{41}(x)u_1 + g_{42}(x)u_2 \end{aligned} \tag{6.29}$$

Then, the dynamic model of the exoskeleton is written as

$$\ddot{x}_1 = v_1 \quad \ddot{x}_3 = v_2 \tag{6.30}$$

The dynamic model of the exoskeleton robot can be also written in the canonical (Brunovsky) form

$$\begin{pmatrix} \dot{x}_1 \\ \dot{x}_2 \\ \dot{x}_3 \\ \dot{x}_4 \end{pmatrix} = \begin{pmatrix} 0 & 1 & 0 & 0 \\ 0 & 0 & 0 & 0 \\ 0 & 0 & 0 & 1 \\ 0 & 0 & 0 & 0 \end{pmatrix} \begin{pmatrix} x_1 \\ x_2 \\ x_3 \\ x_4 \end{pmatrix} + \begin{pmatrix} 0 & 0 \\ 1 & 0 \\ 0 & 0 \\ 0 & 1 \end{pmatrix} \begin{pmatrix} v_1 \\ v_2 \end{pmatrix} \tag{6.31}$$

and the associated measurement equation is

$$\begin{pmatrix} x_1^m \\ x_3^m \end{pmatrix} = \begin{pmatrix} 1 & 0 & 0 & 0 \\ 0 & 0 & 1 & 0 \end{pmatrix} \begin{pmatrix} x_1 \\ x_2 \\ x_3 \\ x_4 \end{pmatrix} \tag{6.32}$$

The stabilising feedback control is computed as follows

$$\begin{aligned} v_1 &= \ddot{x}_1^d - \tilde{k}_{11}(\dot{x}_1 - \dot{x}_1^d) + \tilde{k}_{12}(x_1 - x_1^d) \\ v_2 &= \ddot{x}_3^d - \tilde{k}_{13}(\dot{x}_3 - \dot{x}_3^d) + \tilde{k}_{23}(x_3 - x_3^d) \end{aligned} \tag{6.33}$$

and by denoting the tracking error variables as $e_1 = x_1 - x_1^d$ and $e_3 = x_3 - x_3^d$ one has the tracking error dynamics

$$\begin{aligned} \ddot{e}_1 + \tilde{k}_{11}\dot{e}_1 + \tilde{k}_{12}e_1 &= 0 \\ \ddot{e}_3 + \tilde{k}_{13}\dot{e}_3 + \tilde{k}_{23}e_3 &= 0 \end{aligned} \tag{6.34}$$

Through suitable selection of the feedback gains $\tilde{k}_{11}, \tilde{k}_{21}, \tilde{k}_{13}$ and $\tilde{k}_{23}$, it can be assured that the associated characteristic polynomials are Hurwitz stable

$$\begin{aligned} p_1(s) &= s^2 + \tilde{k}_{11}s + \tilde{k}_{12} = 0 \\ p_2(s) &= s^2 + \tilde{k}_{13}s + \tilde{k}_{23} = 0 \end{aligned} \tag{6.35}$$

This ensures that the tracking error variables will converge asymptotically to 0, that is

$$\begin{aligned} &\lim_{t\to\infty} e_1(t) = 0 && \lim_{t\to\infty} e_3(t) = 0 \\ &\lim_{t\to\infty} x_1(t) = x_1^d(t) && \lim_{t\to\infty} x_3(t) = x_3^d(t) \end{aligned} \tag{6.36}$$

The control inputs which are applied to the initial nonlinear model of the robotic exoskeleton are given by

$$\begin{pmatrix} v_1 \\ v_2 \end{pmatrix} = \begin{pmatrix} g_{11}(x) & g_{12}(x) \\ g_{21}(x) & g_{22}(x) \end{pmatrix}^{-1} \begin{pmatrix} v_1 - f_1(x) \\ v_2 - f_2(x) \end{pmatrix} \tag{6.37}$$

6.1.2.3 Differential flatness properties of the model of the robotic exoskeleton

It will be proven that the dynamic model of the robotic exoskeleton under time-delayed external forces is a differentially flat system with flat outputs $y_1 = x_1$ and $y_2 = x_3$ (Rigatos and Busawon, 2018; Rigatos, 2015). The dynamic model of the robotic exoskeleton has been given in the form

$$\begin{aligned}\ddot{x}_1 &= f_2(x) + g_{21}(x)u_1 + g_{22}(x)u_2\\ \ddot{x}_3 &= f_4(x) + g_{41}(x)u_1 + g_{42}(x)u_2\end{aligned} \tag{6.38}$$

The state vector of the robotic exoskeleton is re-defined as $z = [z_1, z_2, z_3, z_4]^T = [x_1, \dot{x}_1, x_3, \dot{x}_3]^T$, thus the associated dynamic model can be written as

$$\begin{aligned}\dot{z}_1 &= z_2\\ \dot{z}_2 &= f_2(z) + g_{21}(z)u_1 + g_{22}(z)u_2\\ \dot{z}_3 &= z_4\\ \dot{z}_4 &= f_4(z) + g_{41}(z)u_1 + g_{42}(z)u_2\end{aligned} \tag{6.39}$$

It can be confirmed that the robotic exoskeleton is a differentially flat system with flat outputs $y_1 = z_1$ and $y_2 = z_3$. This signifies that all state variables of the robot and its control inputs can be written as differential functions of the system's flat outputs, (ii) the state variables and their derivatives are differentially independent, which means that they are not connected between them through a relation in the form of an homogeneous ordinary differential equation.

From the first and third row of the state-space model one has

$$z_2 = \dot{z}_1 \quad z_4 = \dot{z}_3 \tag{6.40}$$

which signifies that state variables z_2 and z_4 are differential functions of the system's flat outputs. Moreover, from the second and fourth rows of the state-space model one has a system of two equations with respect to u_1 and u_2 which gives

$$\begin{pmatrix}u_1\\ u_2\end{pmatrix} = \begin{pmatrix}g_{21}(z) & g_{22}(z)\\ g_{41}(z) & g_{42}(z)\end{pmatrix}^{-1} \begin{pmatrix}\ddot{z}_1 - f_1(z)\\ \ddot{z}_3 - f_3(z)\end{pmatrix} \tag{6.41}$$

which signifies that the control inputs u_1 and u_2 are also differential functions of the system's flat outputs. Consequently, the robotic exoskeleton is a differentially flat system. By defining setpoints z_1^d and z_3^d it is straightforward to compute setpoints z_2^d and z_4^d as well.

6.1.3 *Estimation of perturbations with the use of a disturbance observer*

In the dynamic model of the exoskeleton robot, the disturbance terms which are due to the time-delayed external forces are described as

$$\tilde{d}_1 = \frac{M_{22}}{M_{11}M_{22} - M_{12}^2}\tilde{\tau}_1 - \frac{M_{12}}{M_{11}M_{22} - M_{12}^2}\tilde{\tau}_2 \tag{6.42}$$

$$\tilde{d}_2 = -\frac{M_{21}}{M_{11}M_{22} - M_{12}^2}\tilde{\tau}_1 + \frac{M_{11}}{M_{11}M_{22} - M_{12}^2}\tilde{\tau}_2 \tag{6.43}$$

The new drift vector $\tilde{f}(x)$ in the model of the exoskeleton is defined as

$$\tilde{f}(x) = \begin{pmatrix} x_2 \\ \dfrac{-M_{22}(C_1 + G_1) + M_{12}(C_2 + G_2)}{M_{11}M_{22} - M_{12}^2} \\ x_4 \\ \dfrac{M_{12}(C_1 + G_1) - M_{11}(C_2 + G_2)}{M_{11}M_{22} - M_{12}^2} \end{pmatrix} \tag{6.44}$$

Thus, the dynamic model of the exoskeleton is written as

$$\begin{aligned} \ddot{x}_1 &= \tilde{f}_2(x) + g_{21}(x)u_1 + g_{22}(x)u_2 + \tilde{d}_1 \\ \ddot{x}_3 &= \tilde{f}_4(x) + g_{41}(x)u_1 + g_{42}(x)u_2 + \tilde{d}_2 \end{aligned} \tag{6.45}$$

Next, by considering that the dynamics of the disturbance terms are given by the second-order time-derivative of $\tilde{d}_1$, and $\tilde{d}_2$, one has

$$\dot{\tilde{d}}_1 = f_{\tilde{d}_1} \quad \dot{\tilde{d}}_2 = f_{\tilde{d}_2} \tag{6.46}$$

This assumption is reasonable because every function can be described by its nth order derivative and the associated initial conditions. If the estimation is to be carried out with the use of a convergent filtering method then knowledge of initial conditions becomes obsolete. By considering the disturbance terms $\tilde{d}_1$ and $\tilde{d}_2$ and their time-derivatives as additional state variables one arrives at an extended state-space description of the robotic exoskeleton. Actually, the following state variables are defined in an extended state-space description of this robotic system: $z_1 = x_1$, $z_2 = \dot{x}_1$, $z_3 = x_3$, $z_4 = \dot{x}_3$, $z_5 = \tilde{d}_1$, $z_6 = \dot{\tilde{d}}_1$, $z_7 = \tilde{d}_2$ and $z_8 = \dot{\tilde{d}}_2$. Thus, the extended state-space model of the robotic exoskeleton is written as:

$$\begin{pmatrix} \dot{z}_1 \\ \dot{z}_2 \\ \dot{z}_3 \\ \dot{z}_4 \\ \dot{z}_5 \\ \dot{z}_6 \\ \dot{z}_7 \\ \dot{z}_8 \end{pmatrix} = \begin{pmatrix} 0 & 1 & 0 & 0 & 0 & 0 & 0 & 0 & 0 \\ 0 & 0 & 0 & 0 & 1 & 0 & 0 & 0 \\ 0 & 0 & 0 & 1 & 0 & 0 & 0 & 0 \\ 0 & 0 & 0 & 0 & 0 & 0 & 1 & 0 \\ 0 & 0 & 0 & 0 & 0 & 1 & 0 & 0 \\ 0 & 0 & 0 & 0 & 0 & 0 & 0 & 0 \\ 0 & 0 & 0 & 0 & 0 & 0 & 0 & 1 \\ 0 & 0 & 0 & 0 & 0 & 0 & 0 & 0 \end{pmatrix} \begin{pmatrix} z_1 \\ z_2 \\ z_3 \\ z_4 \\ z_5 \\ z_6 \\ z_7 \\ z_8 \end{pmatrix} + \begin{pmatrix} 0 & 0 & 0 & 0 \\ 1 & 0 & 0 & 0 \\ 0 & 0 & 0 & 0 \\ 0 & 1 & 0 & 0 \\ 0 & 0 & 0 & 0 \\ 0 & 0 & 1 & 0 \\ 0 & 0 & 0 & 0 \\ 0 & 0 & 0 & 1 \end{pmatrix} \begin{pmatrix} v_1 \\ v_2 \\ f_{\tilde{d}_1} \\ f_{\tilde{d}_2} \end{pmatrix} \tag{6.47}$$

and the associated measurement equation becomes

$$\begin{pmatrix} z_1^m \\ z_3^m \end{pmatrix} = \begin{pmatrix} 1 & 0 & 0 & 0 & 0 & 0 & 0 & 0 \\ 0 & 0 & 1 & 0 & 0 & 0 & 0 & 0 \end{pmatrix} \begin{pmatrix} z_1 \\ z_2 \\ z_3 \\ z_4 \\ z_5 \\ z_6 \\ z_7 \\ z_8 \end{pmatrix} \tag{6.48}$$

The extended state-space model of the robotic exoskeleton is concisely written as:

$$\begin{aligned} \dot{z} &= A_e z + B_e v + w \\ z^m &= C_e Z + v \end{aligned} \tag{6.49}$$

where w is the vector of process noise and v is the vector of measurement noise. Next, the non-measurable state variables of the robotic exoskeleton, as well as the disturbance terms $\tilde{d}_1$ and $\tilde{d}_2$ can be estimated with the use of a disturbance observer, which is written first in the following continuous-time form

$$\begin{aligned} \dot{\hat{z}} &= A_{e,o}\hat{z} + B_{e,o}v_e + K_f(z^m - \hat{z}^m) \\ \hat{z}^m &= C_{e,o}\hat{z} \end{aligned} \tag{6.50}$$

where matrices $A_{e,o}$, $B_{e,o}$ and $C_{e,o}$ are defined as follows:

$$A_{e,o} = \begin{pmatrix} 0 & 1 & 0 & 0 & 0 & 0 & 0 & 0 \\ 0 & 0 & 0 & 0 & 1 & 0 & 0 & 0 \\ 0 & 0 & 0 & 1 & 0 & 0 & 0 & 0 \\ 0 & 0 & 0 & 0 & 0 & 0 & 1 & 0 \\ 0 & 0 & 0 & 0 & 0 & 1 & 0 & 0 \\ 0 & 0 & 0 & 0 & 0 & 0 & 0 & 0 \\ 0 & 0 & 0 & 0 & 0 & 0 & 0 & 1 \\ 0 & 0 & 0 & 0 & 0 & 0 & 0 & 0 \end{pmatrix} \quad B_{e,o} = \begin{pmatrix} 0 & 0 \\ 1 & 0 \\ 0 & 0 \\ 0 & 1 \\ 0 & 0 \\ 0 & 0 \\ 0 & 0 \\ 0 & 0 \end{pmatrix} \quad C_{e,o}^T = \begin{pmatrix} 1 & 0 \\ 0 & 0 \\ 0 & 1 \\ 0 & 0 \\ 0 & 0 \\ 0 & 0 \\ 0 & 0 \\ 0 & 0 \end{pmatrix} \tag{6.51}$$

The observer's gain K_f is computed through Kalman Filter's recursion after expressing (6.50) in discrete-time form. To this end, matrices $A_{e,o}$, $B_{e,o}$ and C_e, are substituted by their discrete-time equivalents A_d, B_d, C_d after using common discretisation methods. The Kalman Filter's recursion consists of a measurement-update stage and of a time-update stage:

measurement update:

$$\begin{aligned} K_f(k) &= P^-(k)C_d^T[C_dP^-(k)C_d^T + R]^{-1} \\ \hat{z}(k) &= \hat{z}^-(k) + K_f(k)[z_m - \hat{z}_m] \\ P(k) &= P^-(k) - K_f(k)C_dP^-(k) \end{aligned} \tag{6.52}$$

time update:

$$\begin{aligned} P^-(k+1) &= A_dP(k)A_d^T + Q \\ \hat{z}^-(k+1) &= A_d\hat{z}(k) + B_dv(k) \end{aligned} \tag{6.53}$$

In the equations of the Kalman Filter's recursion P is the a-posteriori state vector's error covariance matrix (after receiving the output measurements at the kth sampling interval), and P^- is the a-priori state vector's error covariance matrix (before receiving the output measurements at the kth sampling interval). Besides Q denotes the process noise covariance matrix, while R is the measurement noise covariance matrix. Once the estimates $\hat{\tilde{d}}_1$ and $\hat{\tilde{d}}_2$ of the disturbance terms $\tilde{d}_1$ and $\tilde{d}_2$ are obtained the stabilising feedback control becomes

$$\begin{aligned} v_1 &= \ddot{x}_1^d - \tilde{k}_{11}(\dot{x}_1 - \dot{x}_1^d) - \tilde{k}_{21}(x_1 - \tilde{x}_1^d) - \hat{\tilde{d}}_1 \\ v_2 &= \ddot{x}_3^d - \tilde{k}_{12}(\dot{x}_3 - \dot{x}_3^d) - \tilde{k}_{23}(x_3 - \tilde{x}_3^d) - \hat{\tilde{d}}_3 \end{aligned} \tag{6.54}$$

The control inputs which are applied to the initial nonlinear model of the robotic exoskeleton are given by

$$\begin{pmatrix} v_1 \\ v_2 \end{pmatrix} = \begin{pmatrix} g_{11}(x) & g_{12}(x) \\ g_{21}(x) & g_{22}(x) \end{pmatrix}^{-1} \begin{pmatrix} v_1 - \tilde{f}_1(x) \\ v_2 - \tilde{f}_2(x) \end{pmatrix} \tag{6.55}$$

6.1.4 Simulation tests

The performance of the proposed flatness-bases control scheme has been tested in the dynamic model of the previously analysed lower-limb 2-link robotic exoskeleton. The use of a Kalman Filter-based disturbance observer has allowed to estimate time-delayed disturbance torques that were applied on the robot's joints. Using these estimates, additional control terms have been included in the controller of the robotic exoskeleton, thus allowing finally to compensate for the perturbations' effects. The measured state vector elements where the two turn angles of the joints that is $x_1 = \theta_1$ and $x_3 = \theta_2$. The obtained results are depicted in Figures 6.2 to 6.9.

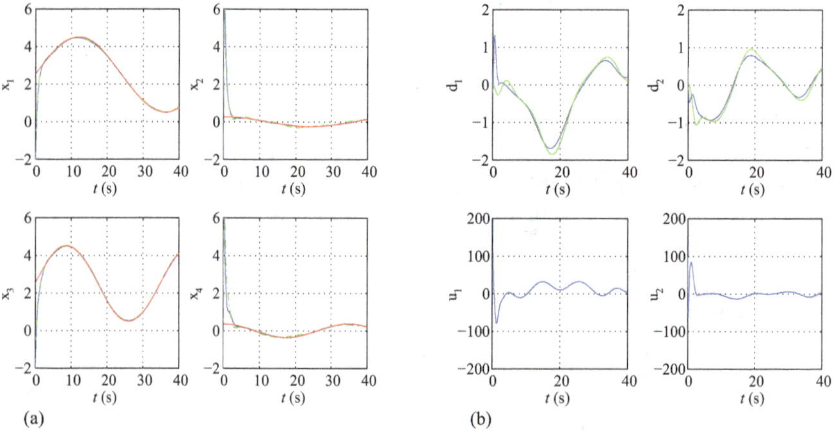

Figure 6.2 Tracking of setpoint 1 by the 2-link robotic exoskeleton: (a) convergence of state variables x_1 to x_4 to the reference setpoints and (b) top diagrams: Kalman Filter-based estimation of disturbance torques $\tilde{d}_1$ and $\tilde{d}_2$, bottom diagrams: control inputs (torques) u_1 and u_2

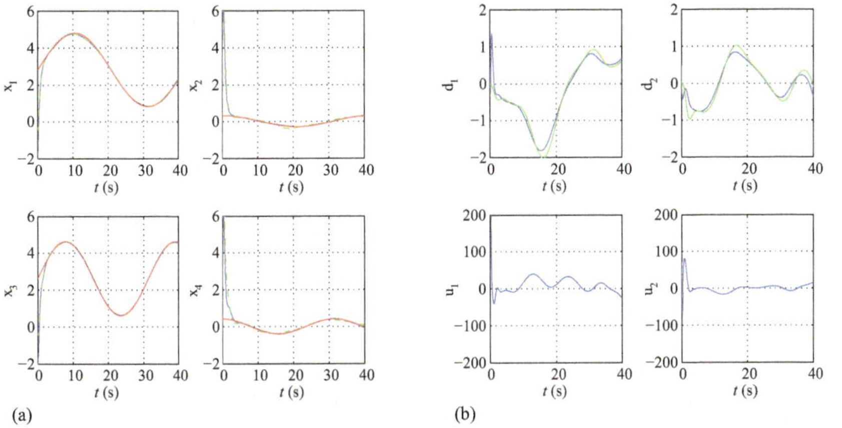

Figure 6.3 Tracking of setpoint 2 by the 2-link robotic exoskeleton: (a) convergence of state variables x_1 to x_4 to the reference setpoints, (b) top diagrams: Kalman Filter-based estimation of disturbance torques $\tilde{d}_1$ and $\tilde{d}_2$, bottom diagrams: control inputs (torques) u_1 and u_2

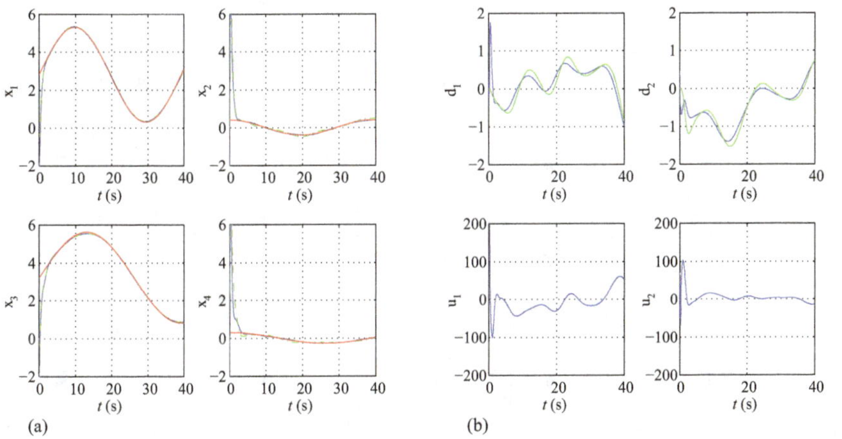

Figure 6.4 Tracking of setpoint 3 by the 2-link robotic exoskeleton: (a) convergence of state variables x_1 to x_4 to the reference setpoints and (b) top diagrams: Kalman Filter-based estimation of disturbance torques $\tilde{d}_1$ and $\tilde{d}_2$, bottom diagrams: control inputs (torques) u_1 and u_2

The advantages of using a global linearisation-based control method for the dynamic model of the robotic exoskeleton are outlined as follows: (i) the transformation that is performed on the robotic system's state-space model is an exact one and does not introduce any modelling errors, (ii) by expressing the dynamic

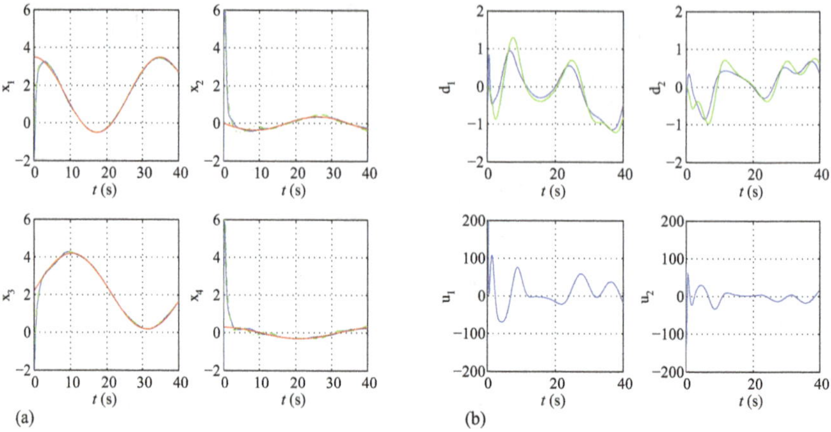

Figure 6.5 Tracking of setpoint 4 by the 2-link robotic exoskeleton: (a) convergence of state variables x_1 to x_4 to the reference setpoints and (b) top diagrams: Kalman Filter-based estimation of disturbance torques $\tilde{d}_1$ and $\tilde{d}_2$, bottom diagrams: control inputs (torques) u_1 and u_2

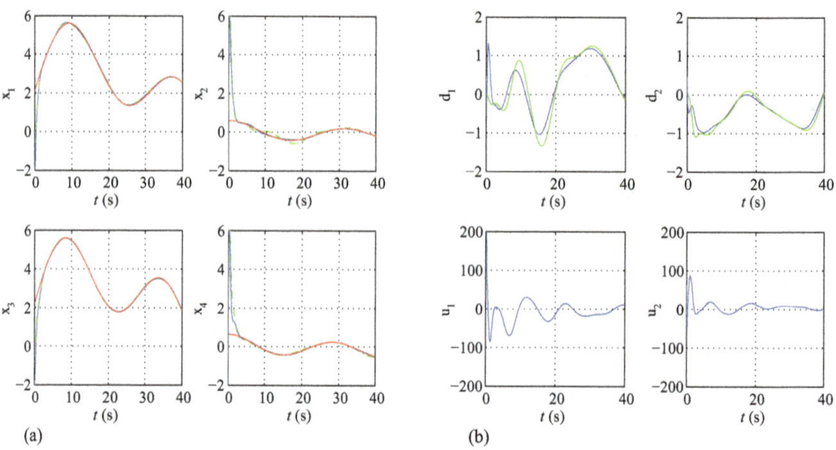

Figure 6.6 Tracking of setpoint 5 by the 2-link robotic exoskeleton: (a) convergence of state variables x_1 to x_4 to the reference setpoints and (b) top diagrams: Kalman Filter-based estimation of disturbance torques $\tilde{d}_1$ and $\tilde{d}_2$, bottom diagrams: control inputs (torques) u_1 and u_2

model of the robotic exoskeleton into the linear canonical form it is assured that the separation principle holds and that the design of the control problem can be solved independently from the design of the state-observer, (iii) by using the Kalman Filter as a disturbance observer the estimation and compensation of perturbation terms that

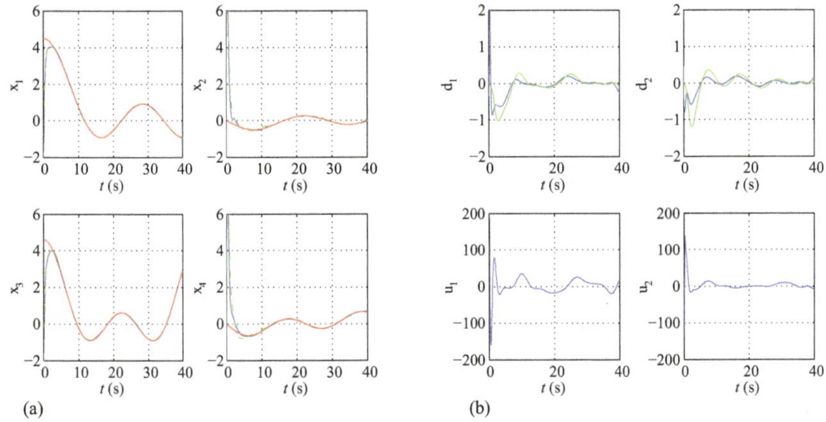

Figure 6.7 Tracking of setpoint 6 by the 2-link robotic exoskeleton: (a) convergence of state variables x_1 to x_4 to the reference setpoints and (b) top diagrams: Kalman Filter-based estimation of disturbance torques $\tilde{d}_1$ and $\tilde{d}_2$, bottom diagrams: control inputs (torques) u_1 and u_2

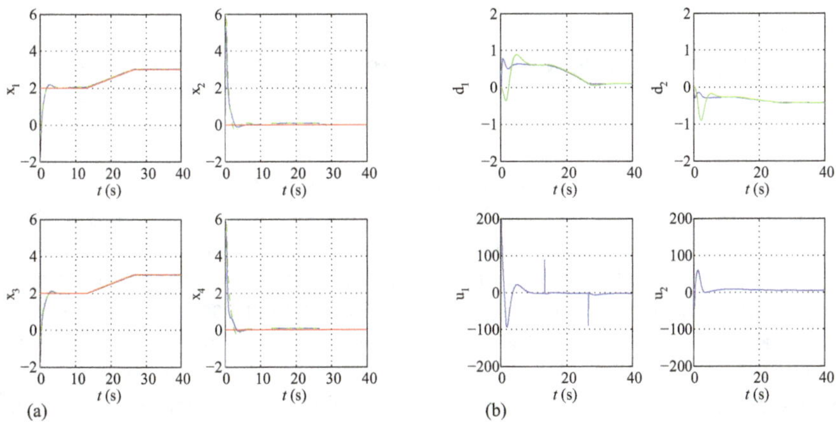

Figure 6.8 Tracking of setpoint 7 by the 2-link robotic exoskeleton (a) convergence of state variables x_1 to x_4 to the reference setpoints and (b) top diagrams: Kalman Filter-based estimation of disturbance torques $\tilde{d}_1$ and $\tilde{d}_2$, bottom diagrams: control inputs (torques) u_1 and u_2

affect the exoskleton's model is achieved and thus the robustness of the control scheme is improved. (iv) The robustness properties of the control method are equivalent to those of LQG (Linear Quadratic Gaussian) control, (v) by using the Kalman Filter as a disturbance observer, optimality in the estimation of the perturbation torques is assured.

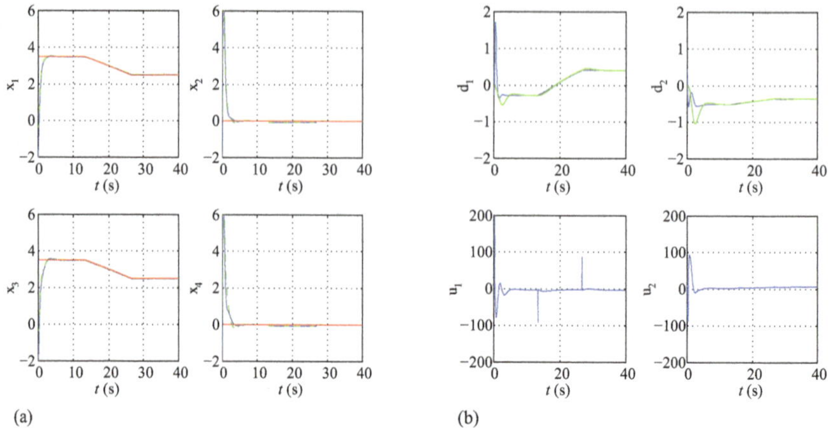

Figure 6.9 Tracking of setpoint 8 by the 2-link robotic exoskeleton (a) convergence of state variables x_1 to x_4 to the reference setpoints and (b) top diagrams: Kalman Filter-based estimation of disturbance torques $\tilde{d}_1$ and $\tilde{d}_2$, bottom diagrams: control inputs (torques) u_1 and u_2

6.2 Adaptive control of synchronous reluctance machines

6.2.1 Outline

The section proposes a nonlinear adaptive fuzzy control method being based on differential flatness theory, in an aim to implement model-free control for synchronous reluctance machines, while also assuring global asymptotic stability for the control loop. The use of Synchronous Reluctance Machines (SRMs) either as power generators or as electric motors exhibits significant advantages but also comes against several technical problems, related to the control of their nonlinear dynamics. Synchronous reluctance machines do not have windings at the rotor. Comparing to permanent magnet synchronous machines and to induction machines, the synchronous reluctance machines appear to have improved power and torque characteristics and reduced rate of failures (Willer and Kamper, 2014; Lyshevski, 2002; Rigatos, Siano *et al.*, 2018; Lin, Huang *et al.*, 2019; Truong *et al.*, 2015; Wai and Liu, 2013). A comparison of synchronous reluctance machines against permanent magnet synchronous machines shows that SRMs have a simpler mechanical structure and are of reduced manufacturing cost. Besides, when functioning under harsh conditions or in high speed the SRMs are more reliable (Barcaro *et al.*, 2012; Park, Kalev *et al.*, 2009; Obe, 2010; Hofmann *et al.*, 2004). The control of synchronous reluctance machines is a non-trivial problem due to the fact that their dynamic model is a nonlinear and multivariable one. For the control of synchronous reluctance machines many approaches have been based on linearisation procedures (Lee, Kang *et al.*, 1999; Zarchi, Soltani *et al.*, 2010; Shyu *et al.*, 2000; Betz *et al.*, 1993; Chiang and Cheng, 2004). Moreover, one can note several results on sensorless control of such machines (Ghaderi and Hanamoto, 2011; Wai and Liu, 2012; Wei, Liu and Lin, 2011; Senjiya *et al.*, 2001; Chen, Liu et. al., 2004).

In the present section, the dynamic model of the Synchronous Reluctance Machine undergoes first global linearisation through a state-space transformation that is based on differential flatness theory (Levine, 2011; Rudolph, 2003; Sira-Ramirez and Agrawal, 2014). At the first stage it is proven that the dynamic model of the synchronous reluctance machine is differentially flat which implies that all its state variables and its control inputs can be expressed as differential functions of specific state vector elements that constitute the flat output. By proving that the dynamic model of the synchronous reluctance machine is differentially flat, one can assure its transformation into the linear canonical (Brunovsky) form (Rigatos, 2011; Rigatos, 2015; Rigatos, 2016b). By exploiting differential flatness properties, the machine's dynamics are transformed into an input–output linearised and decoupled form (Levine, 2009; Menhour, d'Andrea-Novel *et al.*, 2014). This finally is written as a state-space model of the Brunovsky canonical form (Rouchon, 2005; Boouden *et al.*, 2011). Using the latter linear state-space form of the system the solution of the feedback control and state-estimation problems become possible. To stabilise the system's dynamics and to eliminate the tracking error of the state variables with respect to reference setpoints, a linear feedback controller is designed, using the pole-placement concept.

However, the dynamic model of the Synchronous Reluctance Machine is practically unknown and the values of its parameters are imprecise. Actually, it is assumed that only the order of the SRM's state-space model is known while its parameters and the specific form of its state-space equations are completely unknown. To handle this an indirect adaptive neurofuzzy controller is developed. The design of the adaptive controller takes advantage of the fact that in the linearised equivalent model of the Synchronous Reluctance Machine the model uncertainty has the form of unknown functions that appear as part of the system's transformed control inputs. This unknown part of the system's dynamics is identified with the use of neurofuzzy approximators. The estimates of the SRM's dynamics are used by a state feedback controller thus establishing an indirect adaptive control scheme. The learning procedure for these estimators is defined by the requirement to assure negative definiteness for the system's Lyapunov function (Rigatos, 2011; Rigatos, 2015; Rigatos, 2016b). Another issue that arises is that the state vector that is fed into the controller is partially measurable, and the missing state vector elements have to be estimated with the use of a state observer. By including such an observer in the control loop an output feedback-based control scheme is established. The computation of the gains of this state estimation-based adaptive control scheme required the simultaneous solution of two algebraic Riccatic equations (one for the controller and one for the observer). The stability of the control method is proven through Lyapunov analysis. Global asymptotic stability is finally demonstrated for the SRM's adaptive control loop.

6.2.2 Dynamic model of the synchronous reluctance machines

A state-space description of the Synchronous Reluctance machine can be found in (Lee, Kang *et al.*, 1999). The state vector of the machine is defined as $[x_1, x_2, x_3, x_4]^T = [\theta, \omega, i_d, i_q]^T$ while the control inputs vector consists of two elements denoting control voltage components expressed in the synchronously rotating dq reference frame. The state-space equations of this model are obtained from the application of Kirchhoff's voltage equations as shown in Figure 6.10

$$\begin{aligned}\dot{x}_1 &= x_2\\ \dot{x}_2 &= \frac{3P(L_d - L_q)}{4J_m}x_3x_4 - \frac{B_m}{J_m}x_2 - \frac{P}{2J_m}T_L\\ \dot{x}_3 &= -\frac{R_s}{L_d}x_3 + \frac{L_q}{L_d}x_2x_4 + \frac{1}{L_d}v_d\\ \dot{x}_4 &= -\frac{R_s}{L_q}x_4 - \frac{L_d}{L_q}x_2x_3 + \frac{1}{L_q}v_q\end{aligned} \tag{6.56}$$

Thus one obtains the state-space description

$$\dot{x} = f(x) + g(x)u \tag{6.57}$$

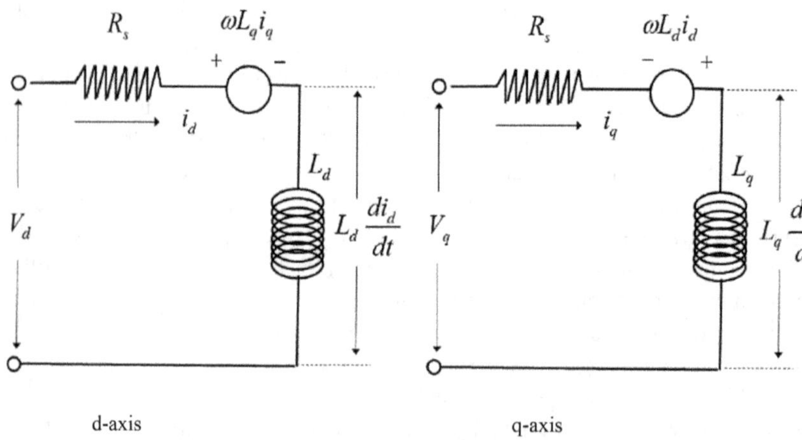

Figure 6.10 Equivalent circuits for the synchronous reluctance machine (a) on the d direct axis and (b) on the q quadrature axis

where $f(x) \in R^{4\times 1}$ and $g(x) \in R^{4\times 2}$ are given by

$$f(x) = \begin{pmatrix} x_2 \\ \frac{3P(L_d - L_q)}{4J_m} x_3 x_4 - \frac{B_m}{J} x_2 - \frac{P}{2J} T_L \\ -\frac{R_s}{L_d} x_3 + \frac{L_q}{L_d} x_2 x_4 \\ -\frac{R_s}{L_q} x_4 - \frac{L_d}{L_q} x_2 x_3 \end{pmatrix} \quad g(x) = \begin{pmatrix} 0 & 0 \\ 0 & 0 \\ \frac{1}{L_d} & 0 \\ 0 & \frac{1}{L_q} \end{pmatrix} \tag{6.58}$$

where R_s is the stator's resistance, L_d is the direct axis inductance, L_q is the quadrature axis inductance, P is the number of poles of the machine, B_m is a friction coefficient and T_L is the externally applied mechanical torque which in the following will be considered to be unknown and will be treated as a disturbance term. The electromagnetic torque (Figure 6.11) that is developed by the synchronous reluctance machine is given by

$$\begin{aligned} &T_e = K(\psi_d i_q - \psi_q i_d) \Rightarrow T_e = \frac{3}{2}\frac{P}{2}(i_d L_d i_q - i_q L_q i_d) \Rightarrow \\ &T_e = \frac{3P}{4}(L_d - L_q) i_d i_q \end{aligned} \tag{6.59}$$

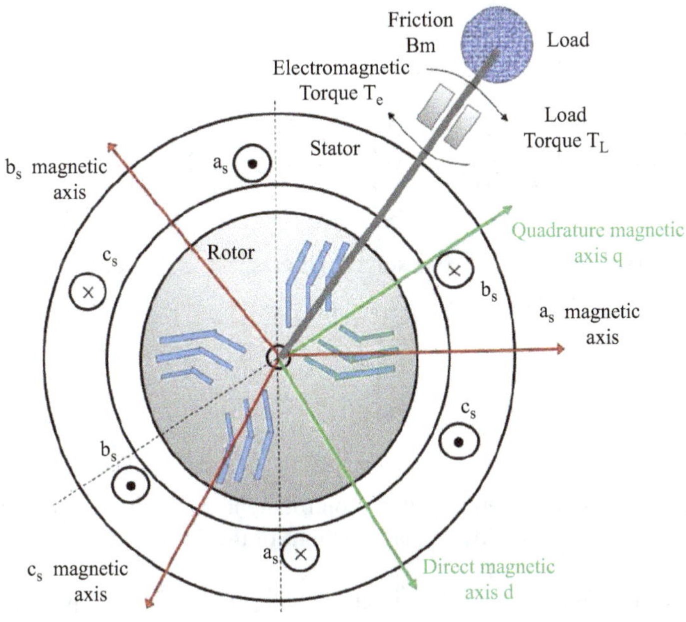

Figure 6.11 Diagram and reference axes for the synchronous reluctance machine

6.2.3 Differential flatness of the synchronous reluctance machine

As noted above in (6.56), the dynamic model of the synchronous reluctance machine comprises four differential equations. The flat output of the system is chosen to be $Y=[y_1,y_2]=[x_1,x_3]=[\theta,i_d]$.

From the first row of (6.56) one has that

$$x_2=\dot{x}_1 \Rightarrow x_2=\dot{y}_1 \Rightarrow y_2=f_2(Y,\dot{Y}) \tag{6.60}$$

From the second row of (6.56), and considering the external load T_L as a disturbance one has that

$$\begin{aligned} \dot{x}_2 &= \frac{3P(L_d-L_q)}{4J_m}x_3x_4-\frac{B_m}{J_m}x_2 \Rightarrow \\ x_4 &= \frac{\dot{x}_2+\frac{B_m}{J_m}x_2}{\frac{3P(L_d-L_q)}{4J_m}x_3} \Rightarrow x_4=\frac{\ddot{y}_1+\frac{B_m}{J_m}y_1}{\frac{3P(L_d-L_q)}{4J_m}y_2} \Rightarrow \\ x_4 &= f_4(Y,\dot{Y}) \end{aligned} \tag{6.61}$$

From the third row of (6.56), one solves with respect to the control input v_d. Thus, the following relation is obtained

$$v_d=L_d\left\{\dot{x}_3+\frac{R_s}{L_d}x_3-\frac{L_q}{L_d}x_2x_4\right\} \Rightarrow v_d=f_a(Y,\dot{Y}) \tag{6.62}$$

From the fourth row of (6.56), one solves with respect to the control input v_q. Thus, the following relation is obtained

$$v_q=L_q\left\{\dot{x}_4+\frac{R_s}{L_q}x_4-\frac{L_d}{L_q}x_2x_3\right\} \Rightarrow v_q=f_b(Y,\dot{Y}) \tag{6.63}$$

Consequently, all state variables and the control inputs of the synchronous reluctance machine can be written as differential functions of the flat output $Y=[y_1,y_2]$. Thus, the system is differentially flat.

As a consequence of its differential flatness property and of the Lie-Backlund equivalence the system is first written into an input–output linearised form, while at a second stage it can be written into the linear canonical (Brunovsky) state-space form.

From the second row of (6.56), and by considering the mechanical torque T_l as a disturbance the differentiation with respect to time gives

$$\ddot{x}_2 = \frac{3P(L_d - L_q)}{4J_m}\dot{x}_3 x_4 + \frac{3P(L_d - L_q)}{4J_m}x_3\dot{x}_4 - \frac{B_m}{J_m}\dot{x}_2 - \frac{P}{2J_m}\dot{T}_L \Rightarrow$$

$$\ddot{x}_2 = \left\{\frac{3P(L_d - L_q)}{4J_m}\left[-\frac{R_s}{L_d}x_3 - \frac{L_q}{L_d}x_2x_4\right]x_4 + \frac{3P(L_d - L_q)}{4J_m}x_3\left[-\frac{R_s}{L_q}x_4 - \frac{L_d}{L_q}x_2x_3\right]\right.$$
$$\left. -\frac{B_m}{J_m}\left[\frac{3P(L_d - L_q)}{4J_m}x_3x_4 - \frac{B_m}{J_m}x_2\right] + \frac{3P(L_d - L_q)}{4J_m}\frac{1}{L_q}x_4v_d + \frac{3P(L_d - L_q)}{4J_m}\frac{1}{L_q}x_3v_q\right\}$$
$$+\frac{B_m}{J_m}\frac{P}{2J_m}T_L - \frac{P}{2J_m}\dot{T}_L \tag{6.64}$$

By defining the aggregate disturbance term $\tilde{d}_1 = \frac{B_m}{J_m}\frac{P}{2J_m}T_L - \frac{P}{2J_m}\dot{T}_L$ and by denoting the following functions

$$f_1(Y,\dot{Y}) = \left\{\frac{3P(L_d - L_q)}{4J_m}\left[-\frac{R_s}{L_d}x_3 - \frac{L_q}{L_d}x_2x_4\right]x_4\right.$$
$$\left. +\frac{3P(L_d - L_q)}{4J_m}x_3\left[-\frac{R_s}{L_q}x_4 - \frac{L_d}{L_q}x_2x_3\right] - \frac{B_m}{J_m}\left[\frac{3P(L_d - L_q)}{4J_m}x_3x_4 - \frac{B_m}{J_m}x_2\right]\right. \tag{6.65}$$

$$g_{11}(Y,\dot{Y}) = \frac{3P(L_d - L_q)}{4J_m}\frac{1}{L_q}x_4 \quad g_{21}(Y,\dot{Y}) = \frac{3P(L_d - L_q)}{4J_m}\frac{1}{L_q}x_3 \tag{6.66}$$

then (6.64) is written in the concise form:

$$\ddot{x}_2 = f_1(y,\dot{y}) + g_{11}(Y,\dot{Y})u_1 + g_{21}(Y,\dot{Y})u_2 + \tilde{d}_1 \tag{6.67}$$

Equivalently, the third row of (6.56) can be written in the concise form

$$\dot{x}_3 = f_2(y,\dot{y}) + g_{12}(Y,\dot{Y})u_1 + g_{22}(Y,\dot{Y})u_2 + \tilde{d}_2 \tag{6.68}$$

where functions $f_2(Y,\dot{Y})$, $g_12(Y,\dot{Y})$ and $g_{22}(Y,\dot{Y})$ are defined as follows:

$$f_2(Y,\dot{Y}) = -\frac{R_s}{L_d}x_3 + \frac{L_q}{L_d}x_2x_4 \tag{6.69}$$

$$g_{12}(Y,\dot{Y}) = \frac{1}{L_d} \quad g_{22}(Y,\dot{Y}) = 0 \tag{6.70}$$

and $\tilde{d}_2$ is an additive disturbance input. Thus, one obtains the following description about the system's dynamics:

$$\begin{aligned} y_1^{(3)} &= f_1(Y,\dot{Y}) + g_{11}(Y,\dot{Y})u_1 + g_{12}(Y,\dot{Y})u_2 + \tilde{d}_1 \\ \dot{y}_2 &= f_2(Y,\dot{Y}) + g_{12}(Y,\dot{Y})u_1 + g_{22}(Y,\dot{Y})u_2 + \tilde{d}_2 \end{aligned} \tag{6.71}$$

By defining the new control inputs for the system

$$\begin{aligned} v_1 &= f_1(Y,\dot{Y}) + g_{11}(Y,\dot{Y})u_1 + g_{21}(Y,\dot{Y})u_2 + \tilde{d}_1 \\ v_2 &= f_2(Y,\dot{Y}) + g_{12}(Y,\dot{Y})u_1 + g_{22}(Y,\dot{Y})u_2 + \tilde{d}_2 \end{aligned} \tag{6.72}$$

the input–output linearised description of the system can be written as

$$\begin{aligned} y_1^{(3)} &= v_1 \\ \dot{y}_2 &= v_2 \end{aligned} \tag{6.73}$$

Next, by denoting the new state variables for the system $z = [z_1, z_2, z_3, z_4] = [y_1, \dot{y}_1, \ddot{y}_1, y_2]$ one arrives at the following state-space description:

$$\begin{pmatrix} \dot{z}_1 \\ \dot{z}_2 \\ \dot{z}_3 \\ \dot{z}_4 \end{pmatrix} = \begin{pmatrix} 0 & 1 & 0 & 0 \\ 0 & 0 & 1 & 0 \\ 0 & 0 & 0 & 0 \\ 0 & 0 & 0 & 0 \end{pmatrix} \begin{pmatrix} z_1 \\ z_2 \\ z_3 \\ z_4 \end{pmatrix} + \begin{pmatrix} 0 & 0 \\ 0 & 0 \\ 1 & 0 \\ 0 & 1 \end{pmatrix} \begin{pmatrix} v_1 \\ v_2 \end{pmatrix} \tag{6.74}$$

The stabilising feedback control for the input–output linearised form of the system is computed from (6.73) as follows:

$$\begin{aligned} v_1 &= y_{1,d}^{(3)} - k_1^1(\ddot{y}_1 - \ddot{y}_{1,d}) - k_2^1(\dot{y}_1 - \dot{y}_{1,d}) - k_3^1(y_1 - y_{1,d}) - \tilde{d}_1 \\ v_2 &= \dot{y}_{2,d} - k_1^2(y_2 - y_{2,d}) - \tilde{d}_2 \end{aligned} \tag{6.75}$$

The stabilising control input that is actually exerted on the synchronous reluctance motor is given by:

$$\begin{pmatrix} u_1 \\ u_2 \end{pmatrix} = \begin{pmatrix} g_{11} & g_{12} \\ g_{21} & g_{22} \end{pmatrix}^{-1} \cdot \left\{ \begin{pmatrix} v_1 \\ v_2 \end{pmatrix} - \begin{pmatrix} f_1(Y,\dot{Y}) \\ f_2(Y,\dot{Y}) \end{pmatrix} \right\} \tag{6.76}$$

After the application of the feedback control law of (6.75) and (6.76) to the input–output linearised dynamics of the synchronous reluctance motor that is given in (6.73), and after defining the tracking error variables $e_1 = y_1 - y_{1,d}$, $e_2 = y_2 - y_{2,d}$ the tracking error dynamics becomes:

$$\begin{aligned} e_1^{(3)} + k_1^1\ddot{e}_1 + k_2^1\dot{e} + k_3^1 e &= 0 \\ \dot{e}_2 + k_1^2 e_2 &= 0 \end{aligned} \tag{6.77}$$

By selecting the feedback gains k_i^1, $i = 1, 2, 3$ and k_1^2 such that the characteristic polynomials of the two differential equations appearing in (6.77) to be Hurwitz stable one has the tracking error gets asymptotically eliminated, that is $\lim_{t\to\infty} e_1(t) = 0$ and $\lim_{t\to\infty} e_2(t) = 0$.

6.2.4 *Flatness-based adaptive neurofuzzy control*

6.2.4.1 Transformation of MIMO nonlinear systems into the Brunovsky form

Following the previous procedure, that is after defining the flat outputs of the initial MIMO nonlinear model of the SRM, and after expressing the system's state variables and control inputs as functions of the flat output and of the associated derivatives, the SRM is transformed into the Brunovsky canonical form of (6.73), or equivalently into the form (Rigatos, 2011; Rigatos, 2015):

$$
\begin{aligned}
&\dot{x}_1 = x_2 \\
&\dot{x}_2 = x_3 \\
&\cdots \\
&\dot{x}_{r_1-1} = x_{r_1} \\
&\dot{x}_{r_1} = f_1(x) + \sum_{j=1}^{p} g_{1_j}(x)u_j + d_1 \qquad\qquad y_1 = x_1 \\
&\qquad\qquad\qquad\qquad\qquad\qquad\qquad\qquad y_2 = x_{r_1-1} \\
&\qquad\qquad\qquad\qquad\qquad\qquad\qquad\qquad \cdots \\
&\dot{x}_{r_1+1} = x_{r_1+2} \qquad\qquad\qquad\qquad\quad\; y_p = x_{n-r_p+1} \\
&\dot{x}_{r_1+2} = x_{r_1+3} \\
&\cdots \\
&\dot{x}_{p-1} = x_p \\
&\dot{x}_p = f_p(x) + \sum_{j=1}^{p} g_{p_j}(x)u_j + d_p
\end{aligned}
\tag{6.78}
$$

where $x = [x_1, \cdots, x_n]^T$ is the state vector of the transformed system (according to the differential flatness formulation), $u = [u_1, \cdots, u_p]^T$ is the set of control inputs, $y = [y_1, \ldots, y_p]^T$ is the output vector, f_i are the drift functions and $g_{i,j}$, $i, j = 1, 2, \ldots, p$ are smooth functions corresponding to the control input gains, while d_j is a variable associated to external disturbances. In holds that $r_1 + r_2 + \ldots + r_p = n$. Having written the initial nonlinear system into the canonical (Brunovsky) form it holds

$$
y_i^{(r_i)} = f_i(x) + \sum_{j=1}^{p} g_{ij}(x)u_j + d_j \tag{6.79}
$$

Equivalently, in vector form, one has the following description for the system dynamics

$$
y^{(r)} = f(x) + g(x)u + d \tag{6.80}
$$

where the following vectors and matrices are be defined

$$
\begin{aligned}
&y^{(r)} = [y_1^{(r_1)}, \ldots, y_p^{(r_p)}] \\
&f(x) = [f_1(x), \ldots, f_p(x)]^T \\
&g(x) = [g_1(x), \ldots, g_p(x)] \\
&\mathit{with}\; g_i(x) = [g_{1i}(x), \ldots, g_{pi}(x)]^T \\
&A = diag[A_1, \ldots, A_p], \;\; B = diag[B_1, \ldots, B_p] \\
&C^T = diag[C_1, \ldots, C_p], \;\; d = [d_1, \ldots, d_p]^T
\end{aligned}
\tag{6.81}
$$

where matrix A has the MIMO canonical form, i.e. with elements

$$A_i = \begin{pmatrix} 0 & 1 & \cdots & 0 \\ 0 & 0 & \cdots & 0 \\ \vdots & \vdots & \cdots & \vdots \\ 0 & 0 & \cdots & 1 \\ 0 & 0 & \cdots & 0 \end{pmatrix}_{r_i \times r_i} \quad B_i = \begin{pmatrix} 0 \\ 0 \\ \cdots \\ 0 \\ 1 \end{pmatrix}_{r_i \times 1} \quad C_i^T = \begin{pmatrix} 1 \\ 0 \\ \cdots \\ 0 \\ 0 \end{pmatrix}_{r_i \times 1} \tag{6.82}$$

Thus, (6.79) can be written in state-space form

$$\begin{aligned} \dot{x} &= Ax + B[f(x) + g(x)u + \tilde{d}] \\ y &= C^T x \end{aligned} \tag{6.83}$$

which can be also written in the equivalent form:

$$\begin{aligned} \dot{x} &= Ax + Bv + B\tilde{d} \\ y &= C^T x \end{aligned} \tag{6.84}$$

where $v = f(x) + g(x)u$. The reference setpoints for the system's outputs $y_1, \ldots, y_p$ are denoted as $y_{1m}, \ldots, y_{pm}$, thus for the associated tracking errors it holds

$$\begin{aligned} e_1 &= y_1 - y_{1m} \\ e_2 &= y_2 - y_{2m} \\ &\cdots \\ e_p &= y_p - y_{pm} \end{aligned} \tag{6.85}$$

The error vector of the outputs of the transformed MIMO system is denoted as

$$\begin{aligned} E_1 &= [e_1, \ldots, e_p]^T \\ y_m &= [y_{1m}, \ldots, y_{pm}]^T \\ &\cdots \\ y_m^{(r)} &= [y_{1m}^{(r)}, \ldots, y_{pm}^{(r)}]^T \end{aligned} \tag{6.86}$$

where $y_{im}^{(r)}$ denotes the rth order derivative of the ith reference output of the MIMO dynamical system. Thus, one can also define the following vectors: (i) a vector containing the state variables of the system and the associated derivatives, (ii) a vector containing the reference outputs of the system and the associated derivatives

$$x = [x_1, \ldots, x_1^{r_1 - 1}, \ldots, x_p, \ldots, x_p^{r_p - 1}]^T \tag{6.87}$$

$$Y_m = [y_{1m}, \ldots, y_{1m}^{r_1 - 1}, \ldots, y_{pm}, \ldots, y_{pm}^{r_p - 1}]^T \tag{6.88}$$

while in a similar manner one can define a vector containing the tracking error of the system's outputs and the associated derivatives

$$e = Y_m - x = [e_1, \ldots, e_1^{r_1 - 1}, \ldots, e_p, \ldots, e_p^{r_p - 1}]^T \tag{6.89}$$

It is assumed that matrix $g(x)$ is a nonsingular one, i.e. $g^{-1}(x)$ exists and is bounded for all $x \in U_x$, where $U_x \subset R^n$ is a compact set. In any case, the problem of singularities in matrix $g(x)$ can be handled by appropriately modifying the state feedback-based control input.

The objective of the adaptive neurofuzzy controller, denoted as $u = u(x, e|\theta)$ is: all the signals involved in the controller's design are bounded and it holds that $\lim_{t\to\infty} e = 0$, (ii) the H_∞ tracking performance criterion is achieved for a prescribed attenuation level.

In the presence of non-gaussian disturbances w_d, successful tracking of the reference signal is denoted by the H_∞ criterion (Rigatos, 2011; Rigatos, 2015):

$$\int_0^T e^T Q e dt \leq \rho^2 \int_0^T {w_d}^T w_d dt \tag{6.90}$$

where ρ is the attenuation level and corresponds to the maximum singular value of the transfer function $G(s)$ of the linearised model associated to (6.83) and (6.84).

6.2.4.2 Control law

The control signal of the MIMO nonlinear SRM model which has been transformed into the Brunovsky form as described by (6.84) contains the unknown nonlinear functions $f(x)$ and $g(x)$. In case that the complete state vector x is measurable these unknown functions can be approximated by

$$\begin{aligned} \hat{f}(x|\theta_f) &= \Phi_f(x)\theta_f \\ \hat{g}(x|\theta_g) &= \Phi_g(x)\theta_g \end{aligned} \tag{6.91}$$

where

$$\Phi_f(x) = (\xi_f^1(x), \xi_f^2(x), \dots \xi_f^n(x))^T \tag{6.92}$$

with $\xi_f^i(x), \imath = 1, \dots, n$ being the vector of kernel functions (e.g. normalised fuzzy Gaussian membership functions), where

$$\xi_f^i(x) = (\phi_f^{i,1}(x), \phi_f^{i,2}(x), \dots, \phi_f^{i,N}(x)) \tag{6.93}$$

thus giving

$$\Phi_f(x) = \begin{pmatrix} \phi_f^{1,1}(x) & \phi_f^{1,2}(x) & \cdots & \phi_f^{1,N}(x) \\ \phi_f^{2,1}(x) & \phi_f^{2,2}(x) & \cdots & \phi_f^{2,N}(x) \\ \cdots & \cdots & \cdots & \cdots \\ \phi_f^{n,1}(x) & \phi_f^{n,2}(x) & \cdots & \phi_f^{n,N}(x) \end{pmatrix} \tag{6.94}$$

while the weights vector is defined as

$${\theta_f}^T = \left(\theta_f^1, \theta_f^2, \dots \theta_f^N\right) \tag{6.95}$$

$j = 1, \dots, N$ is the number of basis functions that is used to approximate the components of function f which are denoted as $i = 1, \dots, n$. Thus, one obtains the relation of (6.91), i.e. $\hat{f}(x|\theta_f) = \Phi_f(x)\theta_f$.

In a similar manner, for the approximation of function g one has

$$\Phi_g(x) = \left(\xi_g^1(x), \xi_g^2(x), \dots \xi_g^N(x)\right)^T \tag{6.96}$$

with $\xi_g^i(x), \iota = 1, \ldots, N$ being the vector of kernel functions (e.g. normalised fuzzy Gaussian membership functions), where

$$\xi_g^i(x) = (\phi_g^{i,1}(x), \phi_g^{i,2}(x), \ldots, \phi_g^{i,N}(x)) \tag{6.97}$$

thus giving

$$\Phi_g(x) = \begin{pmatrix} \phi_g^{1,1}(x) & \phi_g^{1,2}(x) & \cdots & \phi_g^{1,N}(x) \\ \phi_g^{2,1}(x) & \phi_g^{2,2}(x) & \cdots & \phi_g^{2,N}(x) \\ \cdots & \cdots & \cdots & \cdots \\ \phi_g^{n,1}(x) & \phi_g^{n,2}(x) & \cdots & \phi_g^{n,N}(x) \end{pmatrix} \tag{6.98}$$

while the weights vector is defined as

$$\theta_g = (\theta_g^1, \theta_g^2, \ldots, \theta_g^p) \tag{6.99}$$

where the components of matrix θ_g are defined as

$$\theta_g^j = (\theta_{g1}^j, \theta_{g2}^j, \ldots \theta_{gN}^j)^T \tag{6.100}$$

$j = 1, \cdots, p$ is the number of basis functions that is used to approximate the components of function g which are denoted as $i = 1, \ldots, n$. Thus one obtains about matrix $\theta_g \in R^{N \times p}$

$$\theta_g = \begin{pmatrix} \theta_{g1}^1 & \theta_{g1}^2 & \cdots & \theta_{g1}^p \\ \theta_{g2}^1 & \theta_{g2}^2 & \cdots & \theta_{g2}^p \\ \cdots & \cdots & \cdots & \cdots \\ \theta_{gN}^1 & \theta_{gN}^2 & \cdots & \theta_{gN}^p \end{pmatrix} \tag{6.101}$$

It holds that

$$g = \begin{pmatrix} g_1 \\ g_2 \\ \cdots \\ g_n \end{pmatrix} = \begin{pmatrix} g_1^1 & g_1^2 & \cdots & g_1^p \\ g_2^1 & g_2^2 & \cdots & g_2^p \\ \cdots & \cdots & \cdots & \cdots \\ g_n^1 & g_n^2 & \cdots & g_n^p \end{pmatrix} \tag{6.102}$$

Using the above, one finally has the relation of (6.91), i.e. $\hat{g}(x|\theta_g) = \Phi_g(x)\theta_g$. If the state variables of the system are available for measurement then a state-feedback control law can be formulated as

$$u = \hat{g}^{-1}(x|\theta_g)[-\hat{f}(x|\theta_f) + y_m^{(r)} - K^T e + u_c] \tag{6.103}$$

where $\hat{f}(x|\theta_f)$ and $\hat{g}(x|\theta_g)$ are neurofuzzy models to approximate $f(x)$ and $g(x)$, respectively. u_c is a supplementary control term, e.g. an H_∞ control term that is used to compensate for the effects of modelling inaccuracies and external disturbances. Moreover, K^T is the feedback gain matrix that assures that the characteristic polynomial of matrix $A - BK^T$ will be a Hurwitz one.

6.2.4.3 Estimation of the state vector

The control of the system described by (6.80) becomes more complicated when the state vector x of the SRM is not directly measurable and has to be reconstructed through a state observer. The following definitions are used

- error of the state vector $e = x - x_m$
- error of the estimated state vector $\hat{e} = \hat{x} - x_m$
- observation error $\tilde{e} = e - \hat{e} = (x - x_m) - (\hat{x} - x_m)$

When an observer is used to reconstruct the state vector, the control law of (6.103) is written as

$$u = \hat{g}^{-1}(\hat{x}|\theta_g)[-\hat{f}(\hat{x}|\theta_f) + y_m^{(r)} - K^T\hat{e} + u_c] \tag{6.104}$$

Applying (6.104) to the nonlinear system described by (6.80) results into $y^{(r)} = f(x) + g(x)\hat{g}^{-1}(\hat{x})[-\hat{f}(\hat{x}) + y_m^{(r)} - K^T\hat{e} + u_c] + d \Rightarrow y^{(r)} = f(x) + [g(x) - \hat{g}(\hat{x}) + \hat{g}(\hat{x})]\hat{g}^{-1}(\hat{x})[-\hat{f}(\hat{x}) + y_m^{(r)} - K^T\hat{e} + u_c] + d \Rightarrow y^{(r)} = [f(x) - \hat{f}(\hat{x})] + [g(x) - \hat{g}(\hat{x})]u + y_m^{(r)} - K^T\hat{e} + u_c + d$.

It holds $e = x - x_m \Rightarrow y^{(r)} = e^{(r)} + y_m^{(r)}$. Substituting $y^{(r)}$ in the above equation gives

$$\begin{aligned} e^{(r)} + y_m^{(r)} = y_m^{(r)} - K^T\hat{e} + u_c + [f(x) - \hat{f}(\hat{x})] \\ + [g(x) - \hat{g}(\hat{x})]u + d \end{aligned} \tag{6.105}$$

and equivalently

$$\begin{aligned} &\dot{e} = Ae - BK^T\hat{e} + Bu_c + B\{[f(x) - \hat{f}(\hat{x})] \\ &e^{(r)} + y_m^{(r)} = +[g(x) - \hat{g}(\hat{x})]u + \tilde{d}\} \end{aligned} \tag{6.106}$$

$$e_1 = C^T e \tag{6.107}$$

where $e = [e^1, e^2, \ldots, e^p]^T$ with $e^i = [e_i, \dot{e}_i, \ddot{e}_i, \cdots, e_i^{r_i-1}]^T$, $i = 1, 2, \ldots, p$ and also $\hat{e} = [\hat{e}^1, \hat{e}^2, \ldots, \hat{e}^p]^T$ with $\hat{e}^i = [\hat{e}_i, \dot{\hat{e}}_i, \ddot{\hat{e}}_i, \ldots, \hat{e}_i^{r_i-1}]^T$, $i = 1, 2, \cdots, p$. Matrices A,B and C have been defined in (6.82).

A state observer is designed according to (6.106) and (6.107) and is given by Rigatos, 2015:

$$\dot{\hat{e}} = A\hat{e} - BK^T\hat{e} + K_o[e_1 - C^T\hat{e}] \tag{6.108}$$

$$\hat{e}_1 = C^T\hat{e} \tag{6.109}$$

The observer's feedback gain matrix is denoted as $K \in R^{n\times p}$, and is computed according to the requirement for asymptotic elimination of the state vector's estimation error (Rigatos, Siano and Zervos, 2014; Rigatos, 2012c). The observation gain matrix is denoted as $K_o \in R^{p\times n}$ and its elements are selected so as to assure the asymptotic elimination of the observation error.

6.2.5 Application of flatness-based adaptive neurofuzzy control to the SRM

6.2.5.1 Tracking error dynamics under feedback control

If the state vector in the linearised description of the SRM of (6.73) is written as $x=[x_1,x_2,x_3,x_4]^T=[\theta,\omega,\dot{\omega},i_d]^T$, it holds that

$$\begin{aligned} x_1^{(3)} &= f_1(x)+g_1(x)u \\ \dot{x}_4 &= f_2(x)+g_2(x)u \end{aligned} \tag{6.110}$$

or equivalently

$$\begin{aligned} \dot{x}_1 &= x_2 \\ \dot{x}_2 &= x_3 \\ \dot{x}_3 &= f_1(x)+g_1(x)u \\ \dot{x}_4 &= f_2(x)+g_2(x)u \end{aligned} \tag{6.111}$$

Next, taking into account also the effects of additive disturbances the dynamic model becomes

$$\begin{aligned} x_1^{(3)} &= f_1(x,t)+g_1(x,t)u+d_1 \\ \dot{x}_4 &= f_2(x,t)+g_2(x,t)u+d_2 \end{aligned} \tag{6.112}$$

$$\begin{pmatrix} x_1^{(3)} \\ \dot{x}_4 \end{pmatrix} = \begin{pmatrix} f_1(x,t) \\ f_2(x,t) \end{pmatrix} + \begin{pmatrix} g_1(x,t) \\ g_2(x,t) \end{pmatrix} u + \begin{pmatrix} d_1 \\ d_2 \end{pmatrix} \tag{6.113}$$

The following control input is defined

$$u = \begin{pmatrix} \hat{g}_1(x,t) \\ \hat{g}_2(x,t) \end{pmatrix}^{-1} \left\{ \begin{pmatrix} x_{1,d}^{(3)} \\ \dot{x}_{4,d} \end{pmatrix} - \begin{pmatrix} \hat{f}_1(x,t) \\ \hat{f}_2(x,t) \end{pmatrix} - \begin{pmatrix} K_1^T \\ K_2^T \end{pmatrix} e + \begin{pmatrix} u_{c_1} \\ u_{c_2} \end{pmatrix} \right\} \tag{6.114}$$

where $[u_{c_1}\ u_{c_2}]^T$ is a robust control term that is used for the compensation of the model's uncertainties as well as of the external disturbances and $K_i^T=[k_1^i,k_2^i,\ldots,k_{n-1}^i,k_n^i]$. Substituting (6.114) into (6.113) the closed-loop tracking error dynamics is obtained

$$\begin{aligned} \begin{pmatrix} x_1^{(3)} \\ \dot{x}_3 \end{pmatrix} &= \begin{pmatrix} f_1(x,t) \\ f_2(x,t) \end{pmatrix} + \begin{pmatrix} g_1(x,t) \\ g_2(x,t) \end{pmatrix} \begin{pmatrix} \hat{g}_1(x,t) \\ \hat{g}_2(x,t) \end{pmatrix}^{-1} \\ &\cdot \left\{ \begin{pmatrix} x_{1,d}^{(3)} \\ \dot{x}_{4,d} \end{pmatrix} - \begin{pmatrix} \hat{f}_1(x,t) \\ \hat{f}_2(x,t) \end{pmatrix} - \begin{pmatrix} K_1^T \\ K_2^T \end{pmatrix} e + \begin{pmatrix} u_{c_1} \\ u_{c_2} \end{pmatrix} \right\} + \begin{pmatrix} d_1 \\ d_2 \end{pmatrix} \end{aligned} \tag{6.115}$$

Equation (6.115) can now be written as

$$\begin{aligned} \begin{pmatrix} x_1^{(3)} \\ \dot{x}_4 \end{pmatrix} &= \begin{pmatrix} f_1(x,t) \\ f_2(x,t) \end{pmatrix} + \left\{ \begin{pmatrix} g_1(x,t)-\hat{g}_1(x,t) \\ g_2(x,t)-\hat{g}_2(x,t) \end{pmatrix} + \begin{pmatrix} \hat{g}_1(x,t) \\ \hat{g}_2(x,t) \end{pmatrix} \right\} \begin{pmatrix} \hat{g}_1(x,t) \\ \hat{g}_2(x,t) \end{pmatrix}^{-1} \\ &\cdot \left\{ \begin{pmatrix} x_{1,d}^{(3)} \\ \dot{x}_{4,d} \end{pmatrix} - \begin{pmatrix} \hat{f}_1(x,t) \\ \hat{f}_2(x,t) \end{pmatrix} - \begin{pmatrix} K_1^T \\ K_2^T \end{pmatrix} e + \begin{pmatrix} u_{c_1} \\ u_{c_2} \end{pmatrix} \right\} + \begin{pmatrix} d_1 \\ d_2 \end{pmatrix} \end{aligned} \tag{6.116}$$

and using (6.114) this results into

$$\begin{pmatrix} e_1^{(3)} \\ \dot{e}_4 \end{pmatrix} = \begin{pmatrix} f_1(x,t) - \hat{f}_1(x,t) \\ f_2(x,t) - \hat{f}_2(x,t) \end{pmatrix} + \begin{pmatrix} g_1(x,t) - \hat{g}_1(x,t) \\ g_2(x,t) - \hat{g}_2(x,t) \end{pmatrix} u - \begin{pmatrix} K_1^T \\ K_2^T \end{pmatrix} e + \begin{pmatrix} u_{c_1} \\ u_{c_2} \end{pmatrix} + \begin{pmatrix} d_1 \\ d_2 \end{pmatrix} \tag{6.117}$$

The following description for the approximation error is defined

$$w = \begin{pmatrix} f_1(x,t) - \hat{f}_1(x,t) \\ f_2(x,t) - \hat{f}_2(x,t) \end{pmatrix} + \begin{pmatrix} g_1(x,t) - \hat{g}_1(x,t) \\ g_2(x,t) - \hat{g}_2(x,t) \end{pmatrix} u \tag{6.118}$$

Moreover, the following matrices are defined

$$A = \begin{pmatrix} 0 & 1 & 0 & 0 \\ 0 & 0 & 1 & 0 \\ 0 & 0 & 0 & 0 \\ 0 & 0 & 0 & 0 \end{pmatrix}, \quad B = \begin{pmatrix} 0 & 0 \\ 0 & 0 \\ 1 & 0 \\ 0 & 1 \end{pmatrix} \tag{6.119}$$

$$K^T = \begin{pmatrix} K_1^1 & K_2^1 & K_3^1 & K_4^1 \\ K_1^2 & K_2^2 & K_3^2 & K_4^2 \end{pmatrix}$$

Using matrices A, B, K^T, (6.117) is written in the following form

$$\dot{e} = (A - BK^T)e + Bu_c + B\left\{ \begin{pmatrix} f_1(x,t) - \hat{f}_1(x,t) \\ f_2(x,t) - \hat{f}_2(x,t) \end{pmatrix} + \begin{pmatrix} g_1(x,t) - \hat{g}_1(x,t) \\ g_2(x,t) - \hat{g}_2(x,t) \end{pmatrix} u + \tilde{d} \right\} \tag{6.120}$$

When the estimated state vector $\hat{x}$ is used in the feedback control loop, equivalently to (6.106) one has

$$\dot{e} = Ae - BK^T\hat{e} + Bu_c + B\left\{ \begin{pmatrix} f_1(x,t) - \hat{f}_1(\hat{x},t) \\ f_2(x,t) - \hat{f}_2(\hat{x},t) \end{pmatrix} + \begin{pmatrix} g_1(x,t) - \hat{g}_1(\hat{x},t) \\ g_2(x,t) - \hat{g}_2(\hat{x},t) \end{pmatrix} u + \tilde{d} \right\} \tag{6.121}$$

and considering that the approximation error w is now denoted as

$$w = \begin{pmatrix} f_1(x,t) - \hat{f}_1(\hat{x},t) \\ f_2(x,t) - \hat{f}_2(\hat{x},t) \end{pmatrix} + \begin{pmatrix} g_1(x,t) - \hat{g}_1(\hat{x},t) \\ g_2(x,t) - \hat{g}_2(\hat{x},t) \end{pmatrix} u \tag{6.122}$$

Equation (6.121) can be also written as

$$\dot{e} = Ae - BK^T\hat{e} + Bu_c + Bw + B\tilde{d} \tag{6.123}$$

The associated state observer will be described again by (6.108) and (6.109).

6.2.5.2 Dynamics of the observation error

The observation error is defined as $\tilde{e} = e - \hat{e} = x - \hat{x}$. Subtracting (6.108) from (6.106) as well as (6.109) from (6.107) one gets

$$\begin{aligned} \dot{e} - \dot{\hat{e}} &= A(e-\hat{e}) + Bu_c + B\{[f(x,t) - \hat{f}(\hat{x},t)] \\ &\quad + [g(x,t) - \hat{g}(\hat{x},t)]u + \tilde{d}\} - K_o C^T(e-\hat{e}) \\ e_1 - \hat{e}_1 &= C^T(e-\hat{e}) \end{aligned}$$

or equivalently

$$\begin{aligned} \dot{\tilde{e}} &= A\tilde{e} + Bu_c + B\{[f(x,t) - \hat{f}(\hat{x},t)] \\ &\quad + [g(x,t) - \hat{g}(\hat{x},t)]u + \tilde{d}\} - K_o C^T\tilde{e} \\ \tilde{e}_1 &= C^T\tilde{e} \end{aligned}$$

which can be written as

$$\begin{aligned} \dot{\tilde{e}} = (A - K_o C^T)\tilde{e} + Bu_c + B\{[f(x,t) - \hat{f}(\hat{x},t)] \\ + [g(x,t) - \hat{g}(\hat{x},t)]u + \tilde{d}\} \end{aligned} \tag{6.124}$$

$$\tilde{e}_1 = C^T\tilde{e} \tag{6.125}$$

or equivalently, it can be written as

$$\dot{\tilde{e}} = (A - K_o C^T)\tilde{e} + Bu_c + Bw + \tilde{d}\} \tag{6.126}$$

$$\tilde{e}_1 = C^T\tilde{e} \tag{6.127}$$

6.2.5.3 Approximation of functions $f(x,t)$ and $g(x,t)$

Next, the following neurofuzzy approximators of the unknown system dynamics are defined

$$\hat{f}(\hat{x}) = \begin{pmatrix} \hat{f}_1(\hat{x}|\theta_f)\ \hat{x}\in R^{4\times1}\ \hat{f}_1(\hat{x}|\theta_f) \ \in \ R^{1\times1} \\ \hat{f}_2(\hat{x}|\theta_f)\ \hat{x}\in R^{4\times1}\ \hat{f}_2(\hat{x}|\theta_f) \ \in \ R^{1\times1} \end{pmatrix} \tag{6.128}$$

with kernel functions

$$\phi_f^{i,j}(\hat{x}) = \frac{\prod_{j=1}^{n}\mu_{A_j}^{i}(\hat{x}_j)}{\sum_{i=1}^{N}\prod_{j=1}^{n}\mu_{A_j}^{i}(\hat{x}_j)} \tag{6.129}$$

where $l = 1,2$, $\hat{x}$ is the estimate of the state vector and $\mu_{A_j^i}(\hat{x})$ is the ith membership function associated with the lth hidden layer node (fuzzy rule), as shown in Figure 6.12. Similarly, the following approximators of the unknown system dynamics are defined

$$\hat{g}(\hat{x}) = \begin{pmatrix} \hat{g}_1(\hat{x}|\theta_g)\ \hat{x}\in R^{4\times1}\ \hat{g}_1(\hat{x}|\theta_g) \ \in \ R^{1\times2} \\ \hat{g}_2(\hat{x}|\theta_g)\ \hat{x}\in R^{4\times1}\ \hat{g}_2(\hat{x}|\theta_g) \ \in \ R^{1\times2} \end{pmatrix} \tag{6.130}$$

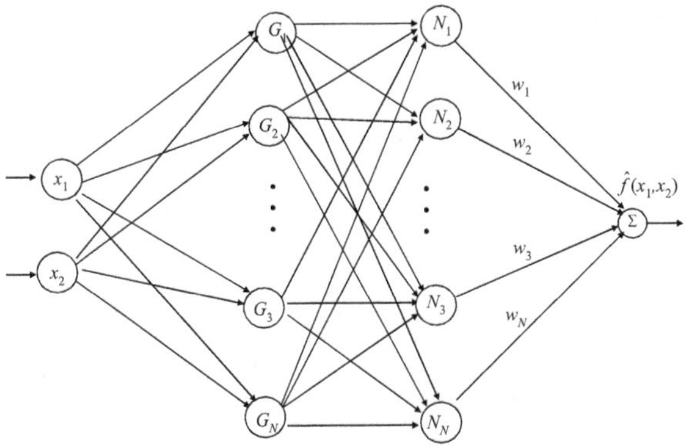

Figure 6.12 Neurofuzzy approximator used for estimating the unknown system dynamics

The values of the weights that result in optimal approximation are

$$\begin{aligned} \theta_f^* &= arg\, min_{\theta_f \in M_{\theta_f}} [sup_{\hat{x} \in U_{\hat{x}}} (f(x) - \hat{f}(\hat{x}|\theta_f))] \\ \theta_g^* &= arg\, min_{\theta_g \in M_{\theta_g}} [sup_{\hat{x} \in U_{\hat{x}}} (g(x) - \hat{g}(\hat{x}|\theta_g))] \end{aligned} \tag{6.131}$$

where the variation ranges for the weights are defined as

$$\begin{aligned} M_{\theta_f} &= \{\theta_f \in R^h : \ ||\theta_f|| \leq m_{\theta_f}\} \\ M_{\theta_g} &= \{\theta_g \in R^h : \ ||\theta_g|| \leq m_{\theta_g}\} \end{aligned} \tag{6.132}$$

The value of the approximation error defined in (6.118) that corresponds to the optimal values of the weights vectors θ_f^* and θ_g^* one has

$$w = \left(f(x,t) - \hat{f}(\hat{x}|\theta_f^*)\right) + \left(g(x,t) - \hat{g}(\hat{x}|\theta_g^*)\right)u \tag{6.133}$$

which is next written as

$$\begin{aligned} w = &\left(f(x,t) - \hat{f}(\hat{x}|\theta_f) + \hat{f}(\hat{x}|\theta_f) - \hat{f}(\hat{x}|\theta_f^*)\right) \\ &+ \left(g(x,t) - \hat{g}(\hat{x}|\theta_g) + \hat{g}(\hat{x}|\theta_g) - \hat{g}(\hat{x}|\theta_g^*)\right)u \end{aligned} \tag{6.134}$$

which can be also written in the following form

$$w = \ \left(w_a + w_b\right) \tag{6.135}$$

where

$$w_a = \{[f(x,t) - \hat{f}(\hat{x}|\theta_f)] + [g(x,t) - \hat{g}(\hat{x}|\theta_g)]\}u \tag{6.136}$$

$$w_b = \{[\hat{f}(\hat{x}|\theta_f) - \hat{f}(\hat{x}|\theta_f^*)] + [\hat{g}(\hat{x},\theta_g) - \hat{g}(\hat{x}|\theta_g^*)]\}u \tag{6.137}$$

Moreover, the following weights' error vectors are defined

$$\begin{aligned} \tilde{\theta}_f &= \theta_f - \theta_f^* \\ \tilde{\theta}_g &= \theta_g - \theta_g^* \end{aligned} \tag{6.138}$$

It is noted that both neural networks and neurofuzzy networks are function approximators using non-orthogonal kernel functions. This is to point out the difference to other approximators using orthogonal basis functions as for instance Fourier series expansions. The distinctive features for neurofuzzy networks are (i) they use as kernel functions in their hidden layer nodes some typical membership functions met in fuzzy logic (for instance Gaussian functions) and (ii) the outputs of the hidden layer nodes are normalised so as to sum up to 1 (this is a property of most fuzzy membership functions). On the other hand in neural networks, (i) it is possible to have several types of membership functions (e.g. wavelet functions, Gauss-Hermite polynomials, Bezier-Bernstein polynomials, B-splines) and (ii) the outputs of the hidden layer nodes are not necessarily normalised. The learning (weights update) in both neural networks and neurofuzzy networks is performed with gradient algorithms of first or higher order.

6.2.6 Lyapunov stability analysis

6.2.6.1 Design of the Lyapunov function

The adaptation law of the neurofuzzy approximators weights θ_f and θ_g as well as the equation of the supervisory control term u_c are derived from the requirement for negative definiteness of the Lyapunov function

$$V = \frac{1}{2}\hat{e}^T P_1 \hat{e} + \frac{1}{2}\tilde{e}^T P_2 \tilde{e} + \frac{1}{2\gamma_1}\tilde{\theta}_f^T \tilde{\theta}_f + \frac{1}{2\gamma_2} tr[\tilde{\theta}_g^T \tilde{\theta}_g] \tag{6.139}$$

The selection of the Lyapunov function is based on the following principle of indirect adaptive control $\hat{e}: \lim_{t\to\infty}\hat{x}(t) = x_d(t)$ and $\tilde{e}: \lim_{t\to\infty}\hat{x}(t) = x(t)$. This yields $\lim_{t\to\infty} x(t) = x_d(t)$. Substituting (6.108), (6.109) and (6.124), (6.125) into (6.294) and differentiating results into

$$\begin{aligned}\dot{V} &= \frac{1}{2}\dot{\hat{e}}^T P_1 \hat{e} + \frac{1}{2}\hat{e}^T P_1 \dot{\hat{e}} + \frac{1}{2}\dot{\tilde{e}}^T P_2 \tilde{e} + \frac{1}{2}\tilde{e}^T P_2 \dot{\tilde{e}} \\ &+ \frac{1}{\gamma_1}\dot{\tilde{\theta}}_f^T \tilde{\theta}_f + \frac{1}{\gamma_2} tr[\dot{\tilde{\theta}}_g^T \tilde{\theta}_g] \Rightarrow\end{aligned} \tag{6.140}$$

$$\dot{V} = \tfrac{1}{2}\{(A - BK^T)\hat{e} + K_o C^T \tilde{e}\}^T P_1 \hat{e} + \tfrac{1}{2}\hat{e}^T P_1\{(A - BK^T)\hat{e} + K_o C^T \tilde{e}\} + \tfrac{1}{2}\{(A - K_o C^T)\tilde{e} + Bu_c + B\tilde{d} + Bw\}^T P_2 \tilde{e} + \tfrac{1}{2}\tilde{e}^T P_2\{(A - K_o C^T)\tilde{e} + Bu_c + B\tilde{d} + Bw\} + \tfrac{1}{\gamma_1}\dot{\tilde{\theta}}_f^T \tilde{\theta}_f + \tfrac{1}{\gamma_2} tr[\dot{\tilde{\theta}}_g^T \tilde{\theta}_g] \Rightarrow.$$

$$\dot{V} = \tfrac{1}{2}\{\hat{e}^T (A - BK^T)^T + \tilde{e}^T C K_o^T\} P_1 \hat{e} + \tfrac{1}{2}\hat{e}^T P_1\{(A - BK^T)\hat{e} + K_o C^T \tilde{e}\} + \tfrac{1}{2}\{\tilde{e}^T (A - K_o C^T)^T + u_c^T B^T + w^T B^T + \tilde{d}^T B^T\} P_2 \tilde{e} + \tfrac{1}{2}\tilde{e}^T P_2\{(A - K_o C^T)\tilde{e} + Bu_c + Bw + B\tilde{d}\} + \tfrac{1}{\gamma_1}\dot{\tilde{\theta}}_f^T \tilde{\theta}_f + \tfrac{1}{\gamma_2} tr[\dot{\tilde{\theta}}_g^T \tilde{\theta}_g] \Rightarrow$$

$$\dot{V} = \tfrac{1}{2}\hat{e}^T (A - BK^T)^T P_1 \hat{e} + \tfrac{1}{2}\tilde{e}^T C K_o^T P_1 \hat{e} + \tfrac{1}{2}\hat{e}^T P_1 (A - BK^T)\hat{e} + \tfrac{1}{2}\hat{e}^T P_1 K_o C^T \tilde{e} + \tfrac{1}{2}\tilde{e}^T (A - K_o C^T)^T P_2 \tilde{e} + \tfrac{1}{2}(u_c^T + w^T + \tilde{d}^T) B^T P_2 \tilde{e} + \tfrac{1}{2}\tilde{e}^T P_2 (A - K_o C^T)\tilde{e} + \tfrac{1}{2}\tilde{e}^T P_2 B(u_c + w + \tilde{d}) + \tfrac{1}{\gamma_1}\dot{\tilde{\theta}}_f^T \tilde{\theta}_f + \tfrac{1}{\gamma_2} tr[\dot{\tilde{\theta}}_g^T \tilde{\theta}_g]$$

Assumption 1: For given positive definite matrices Q_1 and Q_2 there exist positive definite matrices P_1 and P_2, which are the solution of the following Riccati equations (Rigatos, 2011; Rigatos, 2015)

$$(A - BK^T)^T P_1 + P_1(A - BK^T) + Q_1 = 0 \tag{6.141}$$

$$\begin{aligned}&(A - K_oC^T)^T P_2 + P_2(A - K_oC^T)\\ &-P_2B(\frac{2}{r} - \frac{1}{\rho^2})B^T P_2 + Q_2 = 0\end{aligned} \tag{6.142}$$

The conditions given in (6.141) to (6.142) are related to the requirement that the systems described by (6.108), (6.109) and (6.124), (6.125) are strictly positive real. Substituting (6.141) to (6.142) into $\dot{V}$ yields

$$\dot{V} = \tfrac{1}{2}\hat{e}^T\{(A - BK^T)^T P_1 + P_1(A - BK^T)\}\hat{e} + \tilde{e}^T CK_o^T P_1\hat{e} + \tfrac{1}{2}\tilde{e}^T\{(A - K_oC^T)^T P_2 + P_2(A - K_oC^T)\}\tilde{e} + \tilde{e}^T P_2B(u_c + w + \tilde{d}) + \tfrac{1}{\gamma_1}\dot{\tilde{\theta}}_f^T\tilde{\theta}_f + \tfrac{1}{\gamma_2}tr[\dot{\tilde{\theta}}_g^T\tilde{\theta}_g]$$

that is

$$\dot{V} = -\tfrac{1}{2}\hat{e}^T Q_1\hat{e} + \tilde{e}^T CK_o^T P_1\hat{e} - \tfrac{1}{2}\tilde{e}^T\{Q_2 - P_2B(\tfrac{2}{r} - \tfrac{1}{\rho^2})B^T P_2\}\tilde{e} + \tilde{e}^T P_2B(u_c + w + \tilde{d}) + \tfrac{1}{\gamma_1}\dot{\tilde{\theta}}_f^T\tilde{\theta}_f + \tfrac{1}{\gamma_2}tr[\dot{\tilde{\theta}}_g^T\tilde{\theta}_g]$$

The supervisory control u_c is decomposed in two terms, u_a and u_b.

- Control term u_a is given by

$$u_a = -\frac{1}{r}\tilde{e}^T P_2B + \Delta u_a \tag{6.143}$$

  where assuming that the measurable elements of vector $\tilde{e}$ are $\{\tilde{e}_1, \tilde{e}_3, \ldots, \tilde{e}_k\}$, the term Δu_a is such that

$$-\frac{1}{r}\tilde{e}^T P_2B + \Delta u_a = -\frac{1}{r}\begin{pmatrix} p_{11}\tilde{e}_1 + p_{13}\tilde{e}_3 + \cdots + p_{1k}\tilde{e}_k \\ p_{13}\tilde{e}_1 + p_{33}\tilde{e}_3 + \cdots + p_{3k}\tilde{e}_k \\ \cdots\cdots\cdots \\ p_{1k}\tilde{e}_1 + p_{3k}\tilde{e}_3 + \cdots + p_{kk}\tilde{e}_k \end{pmatrix} \tag{6.144}$$

- Control term u_b is given by

$$u_b = -[(P_2B)^T(P_2B)]^{-1}(P_2B)^T CK_o^T P_1\hat{e} \tag{6.145}$$

- u_a is an H_∞ control used for the compensation of the approximation error w and the additive disturbance $\tilde{d}$. Its first component $-\frac{1}{r}\tilde{e}^T P_2B$ has been chosen so as to compensate for the term $\frac{1}{r}\tilde{e}^T P_2BB^T P_2\tilde{e}$, which appears in the previously computed function about $\dot{V}$. By including also the second component Δu_a, one has that u_a is computed based on the feedback of only the measurable variables $\{\tilde{e}_1, \tilde{e}_3, \ldots, \tilde{e}_k\}$, out of the complete vector $\tilde{e} = [\tilde{e}_1, \tilde{e}_2, \ldots, \tilde{e}_n]$. Equation (6.143) is finally rewritten as $u_a = -\frac{1}{r}\tilde{e}^T P_2B + \Delta u_a$.

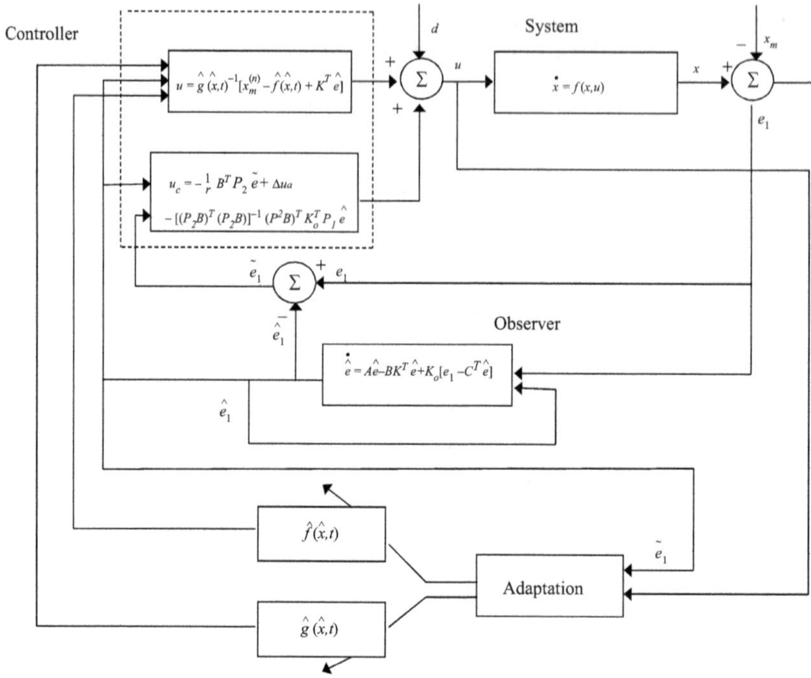

Figure 6.13 Flatness-based adaptive neurofuzzy control for the synchronous reluctance machine

- u_b is a control used for the compensation of the observation error (the control term u_b has been chosen so as to satisfy the condition $\tilde{e}^T P_2 B u_b = -\tilde{e}^T C K_o^T P_1 \hat{e}$).

The control scheme is depicted in Figure 6.13

Substituting (6.143) and (6.145) in $\dot{V}$ and assuming that (6.141) and (6.142) hold, one gets

$$\begin{aligned} \dot{V} = &-\frac{1}{2}\hat{e}^T Q_1 \hat{e} + \tilde{e}^T C K_o^T P_1 \hat{e} - \frac{1}{2}\tilde{e}^T Q_2 \tilde{e} \\ &+ \frac{1}{r}\tilde{e}^T P_2 B B^T P_2 \tilde{e} - \frac{1}{2\rho^2}\tilde{e}^T P_2 B B^T P_2 \tilde{e} \\ &+ \tilde{e}^T P_2 B u_a + \tilde{e}^T P_2 B u_b + \tilde{e}^T P_2 B(w + \tilde{d}) \\ &+ \frac{1}{\gamma_1}\dot{\tilde{\theta}}_f^T \tilde{\theta}_f + \frac{1}{\gamma_2} tr[\dot{\tilde{\theta}}_g^T \tilde{\theta}_g] \end{aligned} \tag{6.146}$$

or equivalently,

$$\begin{aligned} \dot{V} = &-\frac{1}{2}\hat{e}^T Q_1 \hat{e} - \frac{1}{2}\tilde{e}^T Q_2 \tilde{e} - \frac{1}{2\rho^2}\tilde{e}^T P_2 B B^T P_2 \tilde{e} \\ &+ \tilde{e}^T P_2 B(w + \tilde{d} + \Delta u_a) + \frac{1}{\gamma_1}\dot{\tilde{\theta}}_f^T \tilde{\theta}_f + \frac{1}{\gamma_2} tr[\dot{\tilde{\theta}}_g^T \tilde{\theta}_g] \end{aligned} \tag{6.147}$$

It holds that $\dot{\tilde{\theta}}_f = \dot{\theta}_f - \dot{\theta}_f^* = \dot{\theta}_f$ and $\dot{\tilde{\theta}}_g = \dot{\theta}_g - \dot{\theta}_g^* = \dot{\theta}_g$. The following weight adaptation laws are considered:

$$\begin{aligned}\dot{\theta}_f &= -\gamma_1 \Phi(\hat{x})^T B^T P_2 \tilde{e} \\ \dot{\theta}_g &= -\gamma_2 \Phi(\hat{x})^T B^T P_2 \tilde{e} u^T\end{aligned} \tag{6.148}$$

where assuming N fuzzy rules and associated kernel functions the matrices dimensions are $\theta_f \in R^{N\times 1}$, $\theta_g \in R^{N\times 2}$, $\Phi(x) \in R^{2\times N}$, $B \in R^{4\times 2}$, $P \in R^{4\times 4}$ and $\tilde{e} \in R^{4\times 1}$.

The update of θ_f is a gradient type algorithm (Rigatos and Tzafestas, 2007; Basseville and Nikiforov, 1993; Rigatos and Zhang, 2009). The update of θ_g is also a gradient type algorithm, where u_c implicitly tunes the adaptation gain γ_2. Substituting (6.148) in $\dot{V}$ gives

$$\begin{aligned}\dot{V} = &-\tfrac{1}{2}\hat{e}^T Q_1 \hat{e} - \tfrac{1}{2}\tilde{e}^T Q_2 \tilde{e} - \tfrac{1}{2\rho^2}\tilde{e}^T P_2 B B^T P_2 \tilde{e} + B^T P_2 \tilde{e}(w + d + \Delta u_a) \\ &+ \tfrac{1}{\gamma_1}(-\gamma_1)\tilde{e}^T P_2 B \Phi(\hat{x})(\theta_f - \theta_f^*) + \tfrac{1}{\gamma_2}(-\gamma_2) tr[u \tilde{e}^T P_2 B \Phi(\hat{x})(\theta_g - \theta_g^*)]\end{aligned}$$

or equivalently

$$\begin{aligned}\dot{V} = &-\tfrac{1}{2}\hat{e}^T Q_1 \hat{e} - \tfrac{1}{2}\tilde{e}^T Q_2 \tilde{e} - \tfrac{1}{2\rho^2}\tilde{e}^T P_2 B B^T P_2 \tilde{e} + B^T P_2 \tilde{e}(w + \tilde{d} + \Delta u_a) \\ &+ \tfrac{1}{\gamma_1}(-\gamma_1)\tilde{e}^T P_2 B \Phi(\hat{x})(\theta_f - \theta_f^*) + \tfrac{1}{\gamma_2}(-\gamma_2) tr[u \tilde{e}^T P_2 B(\hat{g}(\hat{x}|\theta_g) - \hat{g}(\hat{x}|\theta_g^*)]\end{aligned}$$

Taking into account that $u \in R^{2\times 1}$ and $\tilde{e}^T P B(\hat{g}(x|\theta_g) - \hat{g}(x|\theta_g^*)) \in R^{1\times 2}$ it holds

$$\begin{aligned}\dot{V} = &-\frac{1}{2}\hat{e}^T Q_1 \hat{e} - \frac{1}{2}\tilde{e}^T Q_2 \tilde{e} - \frac{1}{2\rho^2}\tilde{e}^T P_2 B B^T P_2 \tilde{e} + B^T P_2 \tilde{e}(w + \tilde{d} + \Delta u_a) \\ &+ \frac{1}{\gamma_1}(-\gamma_1)\tilde{e}^T P_2 B \Phi(\hat{x})(\theta_f - \theta_f^*) + \frac{1}{\gamma_2}(-\gamma_2) tr[\tilde{e}^T P_2 B(\hat{g}(\hat{x}|\theta_g) - \hat{g}(\hat{x}|\theta_g^*))u]\end{aligned}$$

Since $\tilde{e}^T P_2 B(\hat{g}(\hat{x}|\theta_g) - \hat{g}(\hat{x}|\theta_g^*))u \in R^{1\times 1}$ it holds

$$\begin{aligned}&tr(\tilde{e}^T P_2 B(\hat{g}(x|\theta_g) - \hat{g}(x|\theta_g^*)u) \\ &= \tilde{e}^T P_2 B(\hat{g}(x|\theta_g) - \hat{g}(x|\theta_g^*))u\end{aligned} \tag{6.149}$$

Therefore, one finally obtains

$$\begin{aligned}\dot{V} = &-\tfrac{1}{2}\hat{e}^T Q_1 \hat{e} - \tfrac{1}{2}\tilde{e}^T Q_2 \tilde{e} - \tfrac{1}{2\rho^2}\tilde{e}^T P_2 B B^T P_2 \tilde{e} + B^T P_2 \tilde{e}(w + \tilde{d} + \Delta u_a) \\ &+ \frac{1}{\gamma_1}(-\gamma_1)\tilde{e}^T P_2 B \Phi(\hat{x})(\theta_f - \theta_f^*) + \frac{1}{\gamma_2}(-\gamma_2)\tilde{e}^T P_2 B(\hat{g}(\hat{x}|\theta_g) - \hat{g}(\hat{x}|\theta_g^*))u\end{aligned}$$

Next, the following approximation error is defined

$$w_\alpha = [\hat{f}(\hat{x}|\theta_f^*) - \hat{f}(\hat{x}|\theta_f)] + [\hat{g}(\hat{x}|\theta_g^*) - \hat{g}(\hat{x}|\theta_g)]u \tag{6.150}$$

Thus, one obtains

$$\begin{aligned}\dot{V} = &-\frac{1}{2}\hat{e}^T Q_1 \hat{e} - \frac{1}{2}\tilde{e}^T Q_2 \tilde{e} - \frac{1}{2\rho^2}\tilde{e}^T P_2 B B^T P_2 \tilde{e} + \\ &+ B^T P_2 \tilde{e}(w + \tilde{d}) + \tilde{e}^T P_2 B w_\alpha\end{aligned} \tag{6.151}$$

Denoting the aggregate approximation error and disturbances vector as

$$w_1 = w + \tilde{d} + w_\alpha + \Delta u_a \tag{6.152}$$

the derivative of the Lyapunov function becomes

$$\dot{V} = -\frac{1}{2}\hat{e}^T Q_1 \hat{e} - \frac{1}{2}\tilde{e}^T Q_2 \tilde{e} - \frac{1}{2\rho^2}\tilde{e}^T P_2 B B^T P_2 \tilde{e} + \tilde{e}^T P_2 B w_1 \tag{6.153}$$

which in turn is written as

$$\begin{aligned}\dot{V} = &-\frac{1}{2}\hat{e}^T Q_1 \hat{e} - \frac{1}{2}\tilde{e}^T Q_2 \tilde{e} - \frac{1}{2\rho^2}\tilde{e}^T P_2 B B^T P_2 \tilde{e} \\ &+ \frac{1}{2}\tilde{e}^T P B w_1 + \frac{1}{2} w_1^T B^T P_2 \tilde{e}\end{aligned} \tag{6.154}$$

Lemma: The following inequality holds

$$\frac{1}{2}\tilde{e}^T P_2 B w_1 + \frac{1}{2} w_1^T B^T P_2 \tilde{e} - \frac{1}{2\rho^2}\tilde{e}^T P_2 B B^T P_2 \tilde{e} \le \frac{1}{2}\rho^2 w_1^T w_1 \tag{6.155}$$

Proof: The binomial $(\rho a - \frac{1}{\rho} b)^2 \ge 0$ is considered. Expanding the left part of the above inequality one gets

$$\begin{aligned}&\rho^2 a^2 + \frac{1}{\rho^2} b^2 - 2ab \ge 0 \Rightarrow \frac{1}{2}\rho^2 a^2 + \frac{1}{2\rho^2} b^2 - ab \ge 0 \Rightarrow \\ &ab - \frac{1}{2\rho^2} b^2 \le \frac{1}{2}\rho^2 a^2 \Rightarrow \frac{1}{2} ab + \frac{1}{2} ab - \frac{1}{2\rho^2} b^2 \le \frac{1}{2}\rho^2 a^2\end{aligned} \tag{6.156}$$

The following substitutions are carried out: $a = w_1$ and $b = \tilde{e}^T P_2 B$ and the previous relation becomes

$$\frac{1}{2} w_1^T B^T P_2 \tilde{e} + \frac{1}{2}\tilde{e}^T P_2 B w_1 - \frac{1}{2\rho^2}\tilde{e}^T P_2 B B^T P_2 \tilde{e} \le \frac{1}{2}\rho^2 w_1^T w_1 \tag{6.157}$$

The above relation is used in $\dot{V}$, and the right part of the associated inequality is enforced

$$\dot{V} \le -\frac{1}{2}\hat{e}^T Q_1 \hat{e} - \frac{1}{2}\tilde{e}^T Q_2 \tilde{e} + \frac{1}{2}\rho^2 w_1^T w_1 \tag{6.158}$$

Thus, (6.158) can be written as

$$\dot{V} \le -\frac{1}{2} E^T Q E + \frac{1}{2}\rho^2 w_1^T w_1 \tag{6.159}$$

where

$$E = \begin{pmatrix}\hat{e} \\ \tilde{e}\end{pmatrix}, \quad Q = \begin{pmatrix} Q_1 & 0 \\ 0 & Q_2\end{pmatrix} = diag[Q_1, Q_2] \tag{6.160}$$

Hence, the H_∞ performance criterion is derived. For ρ sufficiently small (6.158) will be true and the H_∞ tracking criterion will be satisfied. In that case, the integration of $\dot{V}$ from 0 to T gives

$$\begin{aligned}&\int_0^T \dot{V}(t)dt \leq -\frac{1}{2}\int_0^T ||E||^2 dt + \frac{1}{2}\rho^2\int_0^T ||w_1||^2 dt \Rightarrow \\ &2V(T) - 2V(0) \leq -\int_0^T ||E||_Q^2 dt + \rho^2\int_0^T ||w_1||^2 dt \Rightarrow \\ &2V(T) + \int_0^T ||E||_Q^2 dt \leq 2V(0) + \rho^2\int_0^T ||w_1||^2 dt\end{aligned} \tag{6.161}$$

It is assumed that there exists a positive constant $M_w > 0$ such that $\int_0^\infty ||w_1||^2 dt \leq M_w$. Therefore for the integral $\int_0^T ||E||_Q^2 dt$ one gets

$$\int_0^\infty ||E||_Q^2 dt \leq 2V(0) + \rho^2 M_w \tag{6.162}$$

Thus, the integral $\int_0^\infty ||E||_Q^2 dt$ is bounded and according to Barbalat's Lemma

$$\begin{aligned}&\lim_{t\to\infty} E(t) = 0 \Rightarrow \\ &\lim_{t\to\infty} \hat{e}(t) = 0 \\ &\lim_{t\to\infty} \tilde{e}(t) = 0\end{aligned} \tag{6.163}$$

6.2.7 Simulation tests

The flatness-based adaptive neurofuzzy control scheme for the Synchronous Reluctance Machines, using only output feedback, has been evaluated through simulation experiments. The turn speed of the rotor ω was considered to be measurable. The obtained experimental results are presented in Figures 6.14 to 6.18. It can be noticed

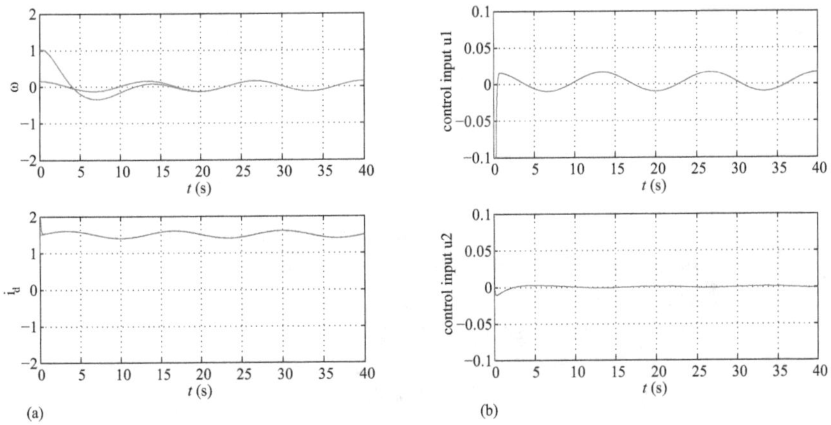

Figure 6.14 Tracking of setpoint 1 (red lines) from state variables of the SRM (blue lines): (a) state variables $x_1 = \omega$ and $x_4 = i_d$ of the SRM (blue lines) and convergence to reference setpoints (red lines) and (b) control inputs of the SRM u_1 and u_2.

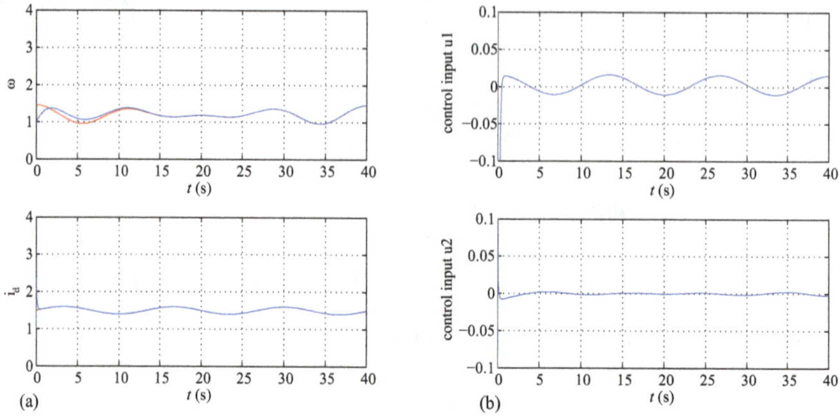

Figure 6.15 Tracking of setpoint 2 (red lines) from state variables of the SRM (blue lines): (a) state variables $x_1 = \omega$ and $x_4 = i_d$ of the SRM (blue lines) and convergence to reference setpoints (red lines) and (b) control inputs of the SRM u_1 and u_2

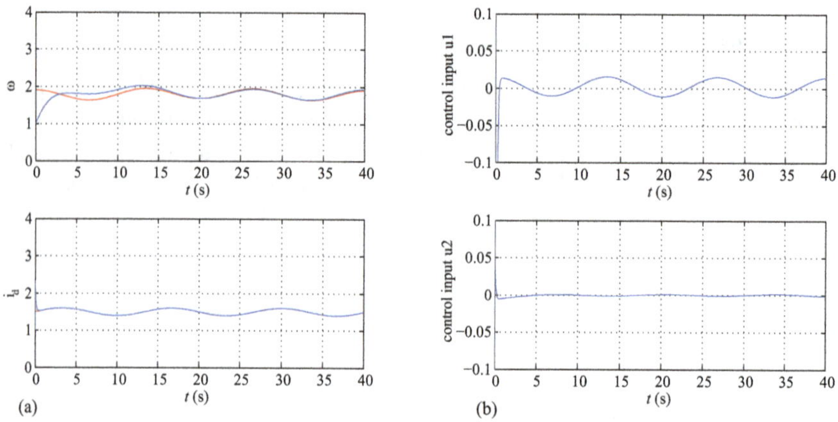

Figure 6.16 Tracking of setpoint 3 (red lines) from state variables of the SRM (blue lines): (a) state variables $x_1 = \omega$ and $x_4 = i_d$ of the SRM (blue lines) and convergence to reference setpoints (red lines) and (b) control inputs of the SRM u_1 and u_2

that the proposed control scheme succeeded in all cases fast and accurate tracking of the reference setpoints.

The treated control problem, that is (i) simultaneous estimation of the unknown dynamics of the Synchronous Reluctance Machine and estimation of the machine's

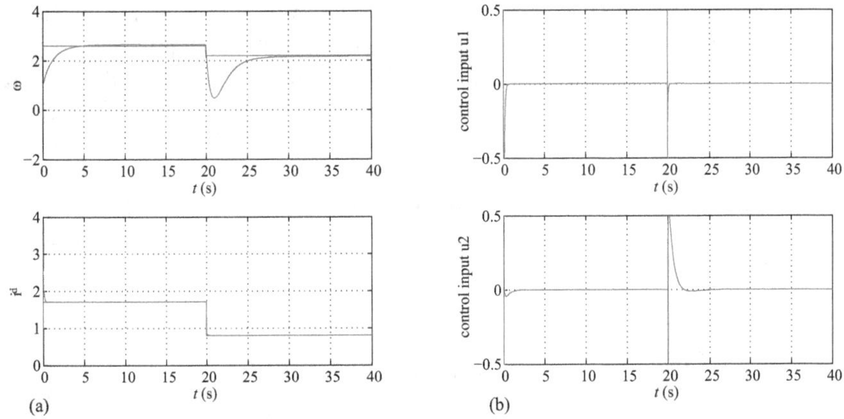

Figure 6.17 *Tracking of setpoint 4 (red lines) from state variables of the SRM (blue lines): (a) state variables $x_1 = \omega$ and $x_4 = i_d$ of the SRM (blue lines) and convergence to reference setpoints (red lines); (b) control inputs of the SRM u_1 and u_2*

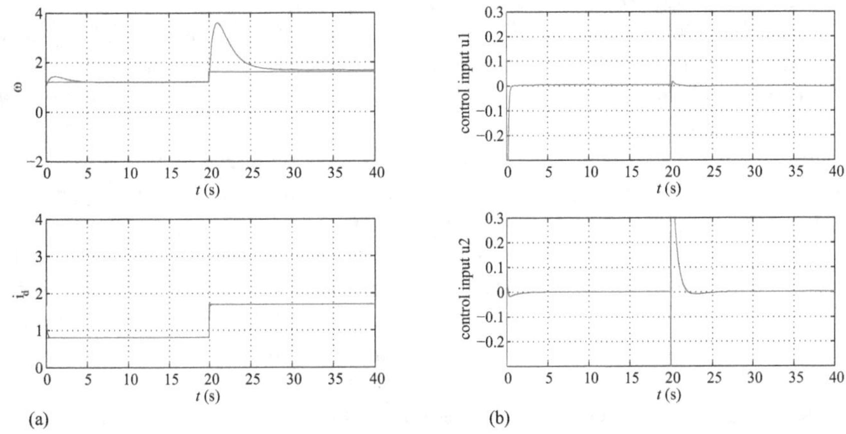

Figure 6.18 *Tracking of setpoint 5 (red lines) from state variables of the SRM (blue lines): (a) state variables $x_1 = \omega$ and $x_4 = i_d$ of the SRM (blue lines) and convergence to reference setpoints (red lines); (b) control inputs of the SRM u_1 and u_2*

state vector and (ii) use of this information by an adaptive control scheme, is of elevated difficulty. The provided Lyapunov analysis has come to confirm the stability properties of the control loop. The presented adaptive control method of the SRM is a model-free one. This means that no prior knowledge about the dynamic model of the machine is needed. Moreover, no knowledge is needed about the mechanical torque exerted on the machine's rotor. Through the identification procedure that was analysed in Section 6.2.5 it becomes possible to estimate in real-time the unknown dynamical model of the synchronous reluctance machine and subsequently to use this information in the control loop. Besides, this adaptive control method does not need to measure the complete state vector of the SRM and is based only on output feedback. The analysed adaptive neurofuzzy control approach exhibits robustness to external perturbation thus assuring the reliable functioning of the Synchronous Reluctance Machine under harsh conditions.

6.3 Control of a mobile robotic manipulator

6.3.1 Outline

Mobile manipulators are widely used in several industrial and human assisting tasks. For instance they can be used in pick and placement tasks and for carrying objects, in assembling, in painting, spraying, harvesting, for patrolling and defence purposes, as well as for providing services to the elderly and the disabled (Li, Tan, Ge *et al.*, 2009; Dai and Liu, 2017; Abeygunawardhana and Murakami, 2010; Andaluz *et al.*, 2015; Li, Tan, Ge *et al.*, 2008a). Dexterity and accuracy in the handling of objects as well as in the maneuvers performed by the mobile manipulators are dependant on the efficiency of the related control algorithms (Rigatos and Busawon, 2018; Boyle *et al.*, 2003; Kocemarek *et al.*, 2017; Koraye and Nekao, 2016). There are several results on nonlinear control approaches for robotic vehicles and mobile manipulators (Rigatos, 2011; Rigatos, 2015; Li, Tan, Ge *et al.*, 2008b; Li, Ge and Ming, 2016; Najjaran and Goldenberg, 2007). In particular, the application of sliding-mode and backstepping methods can be hindered by the need to transform previously the dynamic model of mobile manipulators into canonical or triangular state-space forms. One can also note results on robust and adaptive control schemes for mobile manipulators which aim at compensating for model uncertainty and disturbances in these robotic systems (Xu, Zhao *et al.*, 2009; Souzanchi *et al.*, 2017; Wu, Wang and Dong, 2014; Park, Park *et al.*, 2018; Mozur and Kulawik, 2006). There are also findings on global linearisation-based control schemes for mobile manipulators, as for instance in the case of flatness-based control (Tang, Miller *et al.*, 2011; Morales, Sira-Ramirez *et al.*, 2014; Levine, 2011; Fliess and Mounier, 1999; Sira-Ramirez and Agrawal, 2014; Villagra *et al.*, 2007). Apart from motion control and the end-effector's positioning problem for mobile manipulators, compliance tasks and joint position and force control problems for the end-effector have been also analysed (Galicki, 2016; Lin and Goldberg, 2002; Mai and Wang, 2014; Li, Yang and Li, 2010; Liu and Liu, 2009). The development of functional mobile manipulators is completed with the solution of the related motion planning and trajectory generation problems.

In the present section, a differential flatness theory-based approach is developed for solving the control and state estimation problems in mobile manipulators under parametric model uncertainty and external disturbances (Rigatos and Busawon, 2018;

Rigatos, 2011; Rigatos, 2015). First, the dynamic model of the mobile manipulator, comprising a four-wheel vehicle and a two-DOF robotic manipulator, is obtained through the application of the Euler–Lagrange analysis. It is proven that all state variables and the control inputs of the dynamic model can be written as differential functions of a subset of its state-vector elements, the so-called flat outputs of the system. Besides, it is shown that the flat outputs of the system are differentially independent, meaning that both these variables and their derivatives are not connected through a relation in the form of a linear differential equation. These come to confirm that the dynamic model of the mobile manipulator is a differentially flat one. By proving the differential flatness of the mobile manipulator it is confirmed that (i) it can be transformed into an equivalent input–output linearised form, (ii) it can be written in the canonical (Brunovsky) state-space form.

6.3.2 Dynamic model of the mobile manipulator

To obtain the dynamic model of the mobile manipulator (Figure 6.19) the Lagrangian functions of both the robotic manipulator and of the wheeled platform are computed first. The mass of the wheeled platformed is denoted by M and its moment of inertia for rotation around the vertical axis is denoted as I_z. The mass of the first link of the robotic manipulator is m_1 and the associated moment of inertia (for rotation around its center of gravity) is I_1.The mass of the second link of the robotic manipulator is m_2 and the associated moment of inertia (for rotation around its center of gravity) is I_2. The inertial reference frame of the system is denoted as $O_1X_1Y_1Z_1$ while the body-fixed reference frame is denoted as $O_MX_MY_MZ_M$. The angle between the transversal

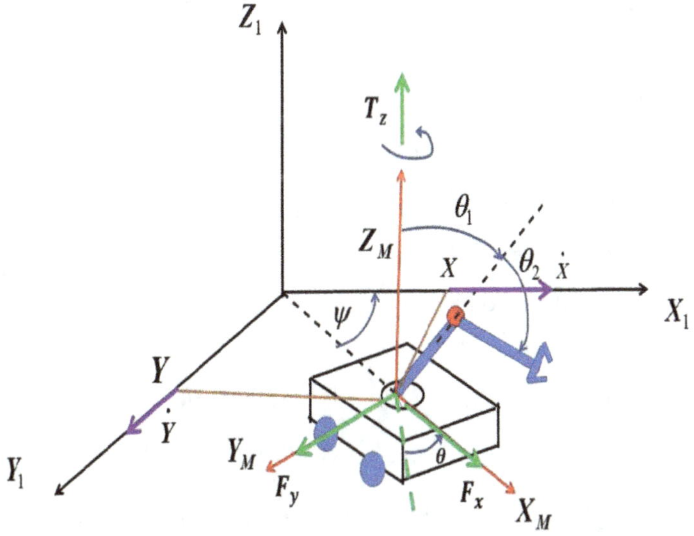

Figure 6.19 Diagram of the mobile manipulator together with the body-fixed and the inertial reference frames

axis of the vehicle and the OX_1 axis is denoted as ψ. The turn angle of the steering wheels of the vehicle with respect to its transversal axis is denoted as θ. The turn angles of the joints of the robotic manipulator are denoted as θ_1 and θ_2 respectively.

Computation of the Lagrangian of the robotic manipulator: About link 1 it holds

$$K_1 = \frac{1}{2}m_1(l_{c_1}\dot{\theta}_1)^2 + \frac{1}{2}I_1\dot{\theta}_1^2 + \frac{1}{2}m_1(V_x^2 + V_y^2) + \frac{1}{2}m_1(l_{c_1}\dot{\psi})^2 + \frac{1}{2}I_1\dot{\psi}^2 \quad (6.164)$$

The potential energy of the first link of the manipulator is given by

$$P_1 = m_1 g l_{c_1}\sin(\theta_1) \quad (6.165)$$

About link 2 it holds

$$\begin{aligned} K_2 &= \frac{1}{2}v_{c_2}v_{c_2}^T + \frac{1}{2}I_2(\dot{\theta}_1^2 + \dot{\theta}_2^2) \\ &+ \frac{1}{2}m_2(V_x^2 + V_y^2) + \frac{1}{2}m_2\{[l_1 + l_{c_2}\cos(\theta_2)]\dot{\psi}\}^2 + \frac{1}{2}I_2\dot{\psi}^2 \end{aligned} \quad (6.166)$$

where the velocity of its center of gravity is $v_{c_2} = [\dot{x}_{c_2}, \dot{y}_{c_2}]$ and after intermediate operations one gets

$$v_{c_2}{v_{c_2}}^T = l_1^2\dot{\theta}_1^2 + l_{c_1}^2(\dot{\theta}_1 + \dot{\theta}_2)^2 + 2l_1l_{c_2}\cos(\theta_1)\dot{\theta}_1(\dot{\theta}_1 + \dot{\theta}_2)$$

Thus the kinetic energy of the second link is given by

$$\begin{aligned} K_2 &= \frac{1}{2}m_2[l_1^2\dot{\theta}_1^2 + l_{c_1}^2(\dot{\theta}_1 + \dot{\theta}_2)^2 + 2l_1l_{c_2}\cos(\theta_1)\dot{\theta}_1(\dot{\theta}_1 + \dot{\theta}_2)] \\ &+ \frac{1}{2}I_2(\dot{\theta}_1^2 + \dot{\theta}_2^2) + \frac{1}{2}m_x(V_x^2 + V_y^2) + \frac{1}{2}m_2\{[l_1 + l_{c_2}\cos(\theta_2)]\dot{\psi}\}^2 + \frac{1}{2}I_2\dot{\psi}^2 \end{aligned} \quad (6.167)$$

The potential energy of the second link of the manipulator is given by

$$P_2 = m_2 g(l_1\sin(\theta_1)) + l_{c2}\sin(\theta_1 + \theta_2) \quad (6.168)$$

The Lagrangian of the robotic manipulator is given by:

$$L_r = K_1 + K_2 - P_1 - P_2 \quad (6.169)$$

while the detailed description of the Lagrangian is

$$\begin{aligned} L_r &= \frac{1}{2}m_1(l_{c_1}\dot{\theta}_1)^2 + \frac{1}{2}I_1\dot{\theta}_1^2 + \frac{1}{2}m_1(V_x^2 + V_y^2) + \frac{1}{2}m_1(l_{c_1}\dot{\psi})^2 + \frac{1}{2}I_1\dot{\psi}^2 \\ &\frac{1}{2}m_2[l_1^2\dot{\theta}_1^2 + l_{c_1}^2(\dot{\theta}_1 + \dot{\theta}_2)^2 + 2l_1l_{c_2}\cos(\theta_1)\dot{\theta}_1(\dot{\theta}_1 + \dot{\theta}_2)] \\ &+ \frac{1}{2}I_2(\dot{\theta}_1^2 + \dot{\theta}_2^2) + \frac{1}{2}m_x(V_x^2 + V_y^2) + \frac{1}{2}m_2\{[l_1 + l_{c_2}\cos(\theta_2)]\dot{\psi}\}^2 + \frac{1}{2}I_2\dot{\psi}^2 \\ &- [m_1 g l_{c_1}\sin(\theta_1)] - [m_2 g(l_1\sin(\theta_1)) + l_{c2}\sin(\theta_1 + \theta_2) \end{aligned} \quad (6.170)$$

Computation of the Lagrangian of the robotic vehicle: The kinetic energy of the vehicle is

$$K_v = \frac{1}{2}M(V_x^2 + V_y^2) + \frac{1}{2}I_z\dot{\psi}^2 \tag{6.171}$$

The vehicle's velocity is initially expressed in the body-fixed reference frame $O_MX_MY_MZ_m$ (Figure 6.19) and is given by the vector $[V_x, V_y]$. When the vehicle's velocity is written in the inertial reference frame $O_1X_1Y_1Z_1$, then it is described by the vector $[\dot{X}, \dot{Y}]$. It holds that

$$\begin{aligned} V_x &= \cos(\psi)\dot{X} - \sin(\psi)\dot{Y} \\ V_y &= \sin(\psi)\dot{X} + \cos(\psi)\dot{Y} \end{aligned} \tag{6.172}$$

It holds that

$$V_x^2 + V_y^2 = \dot{X}^2 + \dot{Y}^2 \tag{6.173}$$

Consequently, the kinetic energy of the vehicle is also written as

$$K_v = \frac{1}{2}M(\dot{X}^2 + \dot{Y}^2) + \frac{1}{2}I_z\dot{\psi}^2 \tag{6.174}$$

The potential energy of the vehicle is taken to be zero, considering that the vertical distance between its center of gravity and the ground is negligible.

$$P_v = 0 \tag{6.175}$$

Thus, about the Lagrangian of the vehicle one has:

$$L_v = K_v - P_v \tag{6.176}$$

and after using the previous relations the Lagrangian of the vehicle becomes

$$L_v = \frac{1}{2}M(\dot{X}^2 + \dot{Y}^2) + \frac{1}{2}I_Z\dot{\psi}^2 \tag{6.177}$$

The aggregate Lagrangian of the mobile manipulator is

$$L = L_v + L_r \tag{6.178}$$

Next, the state vector for the mobile manipulator is defined as

$$x = [X, \dot{X}, Y, \dot{Y}, \psi, \dot{\psi}, \theta_1, \dot{\theta}_1, \theta_2, \dot{\theta}_2]^T$$

The control inputs which are exerted on the model of the mobile manipulator are as follows: (i) force F which is generated by the vehicle's electric motor, (ii) Force F_b which is generated by the vehicle's brakes, (ii) torques T_1 and T_2 which are generated by the actuators of the manipulator.

The force giving propulsion to the vehicle is F. It is considered that the vector of force F forms an angle θ with the longitudinal axis of the vehicle (angle of the steering wheels), and that the angle between this axis and the O_1X_1 axis of the inertial reference system is ψ (Figure 6.20). Moreover, it is considered that a breaking force F_b is exerted at the rear wheels of the vehicle, which is aligned with the longitudinal

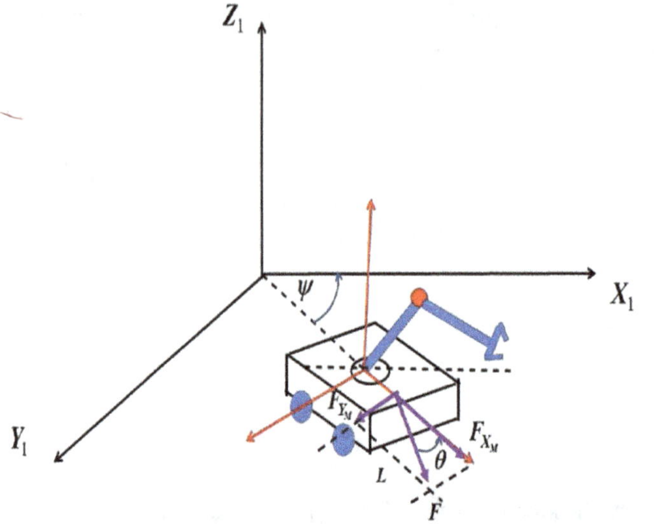

Figure 6.20 Diagram of the forces and torques exerted on the robotic vehicle

axis of the vehicle. Thus, one has that the aggregate force exerted on the vehicle along the O_1X_1 axis is $F_X = F\cos(\psi - \theta) - F_b\cos(\psi)$. Equivalently, the aggregate force that is exerted on the vehicle along the O_1Y_1 axis is $F_Y = F\sin(\psi - \theta) - F_b\sin(\psi)$. Moreover, considering that the distance between the front wheels axis of the vehicle and its transversal axis is L, one has that the torque that causes rotation of the vehicle around the O_MZ_M axis is $T_z = F\sin(\psi - \theta)L$.

By applying the Euler-Lagrange principle to the model of the mobile manipulator one has:

$$\frac{\partial}{\partial t}\frac{\partial L}{\partial \dot{x}} - \frac{\partial L}{\partial x} = F_X \tag{6.179}$$

$$\frac{\partial}{\partial t}\frac{\partial L}{\partial \dot{y}} - \frac{\partial L}{\partial y} = F_Y \tag{6.180}$$

$$\frac{\partial}{\partial t}\frac{\partial L}{\partial \dot{\psi}} - \frac{\partial L}{\partial \psi} = T_Z \tag{6.181}$$

$$\frac{\partial}{\partial t}\frac{\partial L}{\partial \dot{\theta}_1} - \frac{\partial L}{\partial \theta_1} = T_1 \tag{6.182}$$

$$\frac{\partial}{\partial t}\frac{\partial L}{\partial \dot{\theta}_2} - \frac{\partial L}{\partial \theta_2} = T_2 \tag{6.183}$$

The aggregate Lagrangian for the model of the mobile manipulator is

$$\begin{aligned}L = &\frac{1}{2}m_1(l_{c_1}\dot{\theta}_1)^2 + \frac{1}{2}I_1\dot{\theta}_1^2 + \frac{1}{2}m_1(V_x^2 + V_y^2) + \frac{1}{2}m_1(l_{c_1}\dot{\psi})^2 + \frac{1}{2}I_1\dot{\psi}^2 \\ &\frac{1}{2}m_2[l_1^2\dot{\theta}_1^2 + l_{c_1}^2(\dot{\theta}_1 + \dot{\theta}_2)^2 + 2l_1l_{c_2}\cos(\theta_1)\dot{\theta}_1(\dot{\theta}_1 + \dot{\theta}_2)] \\ &+ \frac{1}{2}I_2(\dot{\theta}_1^2 + \dot{\theta}_2^2) + \frac{1}{2}m_x(V_x^2 + V_y^2) + \frac{1}{2}m_2\{[l_1 + l_{c_2}\cos(\theta_2)]\dot{\psi}\}^2 + \frac{1}{2}I_2\dot{\psi}^2 \\ &- [m_1gl_{c_1}\sin(\theta_1)] - [m_2g(l_1\sin(\theta_1)) + l_{c2}\sin(\theta_1 + \theta_2)] \\ &+ \frac{1}{2}M(\dot{X}^2 + \dot{Y}^2) + \frac{1}{2}I_Z\dot{\psi}^2\end{aligned} \tag{6.184}$$

From the Lagrangian equation given in (6.179) one has:

$$\ddot{X} = \frac{1}{m_1 + m_2 + M}F_X \tag{6.185}$$

$$\ddot{Y} = \frac{1}{m_1 + m_2 + M}F_Y \tag{6.186}$$

$$\ddot{\psi} = \frac{1}{I_Z + I_1 + I_2 + m_1l_{c_1}^2 + m_2(l_1 + l_{c_2}\cos(\theta_2))^2}T_Z \tag{6.187}$$

Moreover, about the dynamics of the robotic manipulator one obtains

$$D_{11}\ddot{\theta}_1^2 + D_{12}\ddot{\theta}_2^2 + h_1(\theta, \dot{\theta}) + g_1(\theta) = T_1 \tag{6.188}$$

where $D_{11}(\theta) = m_1l_{c_1}^2 + I_1 + m_1l_1^2 + m_2l_{c_2}^2 + 2m_2l_1l_{c_2}\cos(\theta_1) + I_2, D_{12}(\theta) = m_2l_{c_2}^2 + m_1l_1l_{c_2}\cos(\theta_1)\ h_1(\theta, \dot{\theta}) = -m_2l_1l_{c_2}\sin(\theta_1)\dot{\theta}_1^2$,
and $g_1(\theta) = -m_1gl_{c_1}\cos(\theta_1) - m_2g[l_1\cos(\theta_1) + l_{c_2}\cos(\theta_1 + \theta_2)$.

and also

$$D_{21}\ddot{\theta}_1^2 + D_{22}\ddot{\theta}_2^2 + h_2(\theta, \dot{\theta} + g_2(\theta)) = T_2 \tag{6.189}$$

where $D_{21}(\theta) = m_2l_{c_1}^2 + m_2l_1l_{c_2}\cos(\theta_1), D_{22}(\theta) = m_2l_{c_2}^2 + I_2$,
$h_2(\theta.\dot{\theta}) = -m_2l_1l_{c_2}\sin(\theta_1)\dot{\theta}_1^2 + m_2(l_1 + l_{c_2}\cos(\theta_2))(l_2\sin(\theta_2))\dot{\psi}^2$,
and $g_2(\theta) = -m_2gl_{c_2}\cos(\theta_1 + \theta_2)$.

Thus, about the manipulator attached to the mobile platform one arrives at the following state-space description:

$$\begin{pmatrix} D_{11} & D_{12} \\ D_{21} & D_{22} \end{pmatrix}\begin{pmatrix} \ddot{\theta}_1 \\ \ddot{\theta}_2 \end{pmatrix} + \begin{pmatrix} h_1(\theta, \dot{\theta}) \\ h_2(\theta, \dot{\theta}) \end{pmatrix} + \begin{pmatrix} g_1(\theta) \\ g_2(\theta) \end{pmatrix} = \begin{pmatrix} T_1 \\ T_2 \end{pmatrix} \tag{6.190}$$

By inverting matrix $D(\theta)$, the state-space description of the robotic manipulator can be written as

$$\begin{aligned}\begin{pmatrix}\ddot{\theta}_1\\ \ddot{\theta}_2\end{pmatrix} &= -\frac{1}{D_{11}D_{22}-D_{12}D_{21}}\begin{pmatrix}D_{22} & -D_{12}\\ -D_{21} & D_{11}\end{pmatrix}\begin{pmatrix}h_1(\theta,\dot{\theta})\\ h_2(\theta,\dot{\theta})\end{pmatrix}\\ &-\frac{1}{D_{11}D_{22}-D_{12}D_{21}}\begin{pmatrix}D_{22} & -D_{12}\\ -D_{21} & D_{11}\end{pmatrix}\begin{pmatrix}g_1(\theta)\\ g_2(\theta)\end{pmatrix}\\ &+\frac{1}{D_{11}D_{22}-D_{12}D_{21}}\begin{pmatrix}D_{22} & -D_{12}\\ -D_{21} & D_{11}\end{pmatrix}\begin{pmatrix}T_1\\ T_2\end{pmatrix}\end{aligned} \tag{6.191}$$

The dynamics of the robotic manipulator can be also brought to the following form:

$$\begin{aligned}\begin{pmatrix}\ddot{\theta}_1\\ \ddot{\theta}_2\end{pmatrix} &= \begin{pmatrix}-\frac{D_{22}h_1-D_{21}h_2}{(D_{11}D_{22}-D_{12}D_{21})}-\frac{D_{22}g_1-D_{21}g_2}{(D_{11}D_{22}-D_{12}D_{21})}\\ -\frac{-D_{12}h_1+D_{11}h_2}{(D_{11}D_{22}-D_{12}D_{21})}-\frac{-D_{12}g_1+D_{11}g_2}{(D_{11}D_{22}-D_{12}D_{21})}\end{pmatrix}\\ &+\begin{pmatrix}\frac{D_{22}}{(D_{11}D_{22}-D_{12}D_{21})} & \frac{-D_{21}}{(D_{11}D_{22}-D_{12}D_{21})}\\ \frac{-D_{12}}{(D_{11}D_{22}-D_{12}D_{21})} & \frac{D_{11}}{(D_{11}D_{22}-D_{12}D_{21})}\end{pmatrix}\begin{pmatrix}T_1\\ T_2\end{pmatrix}\end{aligned} \tag{6.192}$$

Consequently, the state-space model of the mobile manipulator becomes:

$$\begin{pmatrix}\ddot{X}\\ \ddot{Y}\\ \ddot{\psi}\\ \ddot{\theta}_1\\ \ddot{\theta}_2\end{pmatrix} = \begin{pmatrix}\tilde{F}_1\\ \tilde{F}_2\\ \tilde{F}_3\\ \tilde{F}_4\\ \tilde{F}_5\end{pmatrix} + \begin{pmatrix}g_{11} & 0 & 0 & 0 & 0\\ 0 & g_{22} & 0 & 0 & 0\\ 0 & 0 & g_{33} & 0 & 0\\ 0 & 0 & 0 & g_{44} & g_{45}\\ 0 & 0 & 0 & g_{54} & g_{55}\end{pmatrix}\begin{pmatrix}F_X\\ F_Y\\ T_Z\\ T_1\\ T_2\end{pmatrix} \tag{6.193}$$

where $\tilde{F}_1 = 0$, $\tilde{F}_2 = 0$, $\tilde{F}_3 = 0$, $\tilde{F}_4 = -\frac{D_{22}h_1-D_{21}h_2}{(D_{11}D_{22}-D_{12}D_{21})} - \frac{D_{22}g_1-D_{21}g_2}{(D_{11}D_{22}-D_{12}D_{21})}$, $\tilde{F}_5 = -\frac{-D_{12}h_1+D_{11}h_2}{(D_{11}D_{22}-D_{12}D_{21})} - \frac{-D_{12}g_1+D_{11}g_2}{(D_{11}D_{22}-D_{12}D_{21})}$,

while it also holds that $g_{11} = \frac{1}{m_1+m_2+M}$, $g_{22} = \frac{1}{m_1+m_2+M}$, $g_{33} = \frac{1}{I_Z+I_1+I_2+m_1l_{c_1}^2+m_2(l_1+l_{c_2}\cos(\theta_2))^2}$, $g_{44} = \frac{D_{22}}{(D_{11}D_{22}-D_{12}D_{21})}$, $g_{45} = \frac{-D_{21}}{(D_{11}D_{22}-D_{12}D_{21})}$, $g_{54} = \frac{-D_{12}}{(D_{11}D_{22}-D_{12}D_{21})}$, and $g_{55} = \frac{D_{11}}{(D_{11}D_{22}-D_{12}D_{21})}$.

6.3.3 *Differential flatness properties of the model of the mobile manipulator*

It will be proven that the dynamic model of the mobile manipulator is a differentially flat one, which implies that all its state variables and its control inputs can be expressed as differential functions of a specific subset of its state vector elements which are known as flat outputs of the system. The following state variables of the mobile manipulator are defined: $x_1 = X, x_2 = \dot{X}, x_3 = Y, x_4 = \dot{Y}, x_5 = \psi, x_6 = \dot{\psi}, x_7 = \theta_1$,

$x_8 = \dot{\theta}_1$, $x_9 = \theta_2$, $x_{10} = \dot{\theta}_2$. Additionally, the control inputs of the robotic system are taken to be $u_1 = F_X$, $u_2 = F_Y$, $u_3 = T_Z$, $u_4 = T_1$ and $u_5 = T_2$. Thus, the following state-space description is obtained:

$$\begin{pmatrix} \dot{x}_1 \\ \dot{x}_2 \\ \dot{x}_3 \\ \dot{x}_4 \\ \dot{x}_5 \\ \dot{x}_6 \\ \dot{x}_7 \\ \dot{x}_8 \\ \dot{x}_9 \\ \dot{x}_{10} \end{pmatrix} = \begin{pmatrix} x_2 \\ \tilde{F}_1 \\ x_4 \\ \tilde{F}_2 \\ x_6 \\ \tilde{F}_3 \\ x_8 \\ \tilde{F}_4 \\ x_{10} \\ \tilde{F}_5 \end{pmatrix} + \begin{pmatrix} 0 & 0 & 0 & 0 & 0 \\ g_{11} & 0 & 0 & 0 & 0 \\ 0 & 0 & 0 & 0 & 0 \\ 0 & g_{22} & 0 & 0 & 0 \\ 0 & 0 & 0 & 0 & 0 \\ 0 & 0 & g_{33} & 0 & 0 \\ 0 & 0 & 0 & 0 & 0 \\ 0 & 0 & 0 & g_{44} & g_{45} \\ 0 & 0 & 0 & 0 & 0 \\ 0 & 0 & 0 & g_{54} & g_{55} \end{pmatrix} \begin{pmatrix} u_1 \\ u_2 \\ u_3 \\ u_4 \\ u_5 \end{pmatrix} \tag{6.194}$$

The vector of the system's flat outputs is $x_f = [x_1, x_3, x_5, x_7, x_9]^T$. It holds that $x_2 = \dot{x}_1$, $x_4 = \dot{x}_3$, $x_6 = \dot{x}_5$, $x_8 = \dot{x}_7$ and $x_{10} = \dot{x}_9$. Therefore, all state vector elements of the mobile manipulator can be written as differential functions of the flat output vector. Moreover, from the second row of (6.194) one has

$$u_1 = \frac{1}{g_{11}}\dot{x}_2 - \tilde{F}_1 \Rightarrow u_1 = q_1(x_f, \dot{x}_f) \tag{6.195}$$

From the fourth row of (6.194) one has

$$u_2 = \frac{1}{g_{22}}\dot{x}_4 - \tilde{F}_2 \Rightarrow u_2 = q_2(x_f, \dot{x}_f) \tag{6.196}$$

From the sixth row of (6.194) one has

$$u_3 = \frac{1}{g_{33}}\dot{x}_6 - \tilde{F}_3 \Rightarrow u_3 = q_3(x_f, \dot{x}_f) \tag{6.197}$$

From the eight and tenth rows of (6.194) one has

$$\begin{pmatrix} u_5 \\ u_6 \end{pmatrix} = D\begin{pmatrix} \dot{x}_8 \\ \dot{x}_{10} \end{pmatrix} + \begin{pmatrix} h_1 \\ h_2 \end{pmatrix} + \begin{pmatrix} g_1 \\ g_2 \end{pmatrix} \Rightarrow \begin{matrix} u_4 = q_4(x_f, \dot{x}_f) \\ u_5 = q_5(x_f, \dot{x}_f) \end{matrix} \tag{6.198}$$

Consequently, the control inputs of the mobile manipulator are also written as differential functions of the flat outputs vector and the system is a differentially flat one.

6.3.4 Design of a flatness-based controller for the mobile manipulator

The dynamic model of the mobile manipulator has been given in (6.194). The following control inputs are applied:

$$u_1 = \frac{1}{g_{11}}[\ddot{x}_1^d - \tilde{F}_1 + k_1^1(\dot{x}_1 - \dot{x}_1^d) + k_2^1(x_1 - x_1^d)] \tag{6.199}$$

$$u_2 = \frac{1}{g_{22}}[\ddot{x}_3^d - \tilde{F}_2 + k_1^2(\dot{x}_3 - \dot{x}_3^d) + k_2^2(x_3 - x_3^d)] \tag{6.200}$$

$$u_3 = \frac{1}{g_{33}}[\ddot{x}_5^d - \tilde{F}_3 + k_1^3(\dot{x}_5 - \dot{x}_5^d) + k_2^3(x_5 - x_5^d)] \tag{6.201}$$

$$\begin{pmatrix} u_4 \\ u_5 \end{pmatrix} = \begin{pmatrix} g_{44} & g_{45} \\ g_{54} & g_{55} \end{pmatrix}^{-1} \left\{ \begin{pmatrix} \ddot{x}_7^d \\ \ddot{x}_9^d \end{pmatrix} - \begin{pmatrix} \tilde{F}_4 \\ \tilde{F}_5 \end{pmatrix} + \begin{pmatrix} k_1^4(\dot{x}_7 - \dot{x}_7^d) + k_2^4(x_7 - x_7^d) \\ k_1^5(\dot{x}_9 - \dot{x}_9^d) + k_2^5(x_9 - x_9^d) \end{pmatrix} \right\} \tag{6.202}$$

By applying the previous control inputs to the state-space model of (6.194) one gets:

$$\begin{aligned}
&(\ddot{x}_1 - \ddot{x}_1^d) + k_1^1(\dot{x}_1 - \dot{x}_1^d) + k_2^1(x_1 - x_1^d) = 0 \\
&(\ddot{x}_3 - \ddot{x}_3^d) + k_1^2(\dot{x}_3 - \dot{x}_3^d) + k_2^2(x_3 - x_3^d) = 0 \\
&(\ddot{x}_5 - \ddot{x}_5^d) + k_1^3(\dot{x}_5 - \dot{x}_5^d) + k_2^3(x_5 - x_5^d) = 0 \\
&(\ddot{x}_7 - \ddot{x}_7^d) + k_1^4(\dot{x}_7 - \dot{x}_7^d) + k_2^4(x_7 - x_7^d) = 0 \\
&(\ddot{x}_9 - \ddot{x}_9^d) + k_1^5(\dot{x}_9 - \dot{x}_9^d) + k_2^5(x_9 - x_9^d) = 0
\end{aligned} \tag{6.203}$$

By defining the state variables' tracking error as $e_i = x_i - x_i^d$, $i = 1, 2, \cdots, 5$ one obtains the tracking error dynamics through the following equations:

$$\begin{aligned}
&\ddot{e}_1 + k_1^1\dot{e}_1 + k_2^1 e_1 = 0 \quad \ddot{e}_2 + k_1^2\dot{e}_2 + k_2^2 e_2 = 0 \\
&\ddot{e}_3 + k_1^3\dot{e}_3 + k_2^3 e_3 = 0 \quad \ddot{e}_4 + k_1^4\dot{e}_4 + k_2^4 e_4 = 0 \\
&\ddot{e}_5 + k_1^5\dot{e}_5 + k_2^5 e_1 = 0
\end{aligned} \tag{6.204}$$

Next, by selecting the feedback gains (k_1^i, k_2^i), $i = 1, 2, \cdots, 5$ such that the characteristic polynomials which are associated with the aforementioned differential equations to be Hurwitz stable, one has that

$$\begin{aligned}
&\lim_{t\to\infty} e_i = 0, \;\; i = 1, 2, \cdots, 5 \Rightarrow \\
&\lim_{t\to\infty} x_i = x_i^d, \;\; i = 1, 2, \cdots, 5
\end{aligned} \tag{6.205}$$

6.3.5 Design of a flatness-based disturbances estimator

Next, the dynamic model of the mobile manipulator is considered to be affected by additive input disturbances:

$$\begin{pmatrix} \dot{x}_1 \\ \dot{x}_2 \\ \dot{x}_3 \\ \dot{x}_4 \\ \dot{x}_5 \\ \dot{x}_6 \\ \dot{x}_7 \\ \dot{x}_8 \\ \dot{x}_9 \\ \dot{x}_{10} \end{pmatrix} = \begin{pmatrix} x_2 \\ \tilde{F}_1 \\ x_4 \\ \tilde{F}_2 \\ x_6 \\ \tilde{F}_3 \\ x_8 \\ \tilde{F}_4 \\ x_{10} \\ \tilde{F}_5 \end{pmatrix} + \begin{pmatrix} 0 & 0 & 0 & 0 & 0 \\ g_{11} & 0 & 0 & 0 & 0 \\ 0 & 0 & 0 & 0 & 0 \\ 0 & g_{22} & 0 & 0 & 0 \\ 0 & 0 & 0 & 0 & 0 \\ 0 & 0 & g_{33} & 0 & 0 \\ 0 & 0 & 0 & 0 & 0 \\ 0 & 0 & 0 & g_{44} & g_{45} \\ 0 & 0 & 0 & 0 & 0 \\ 0 & 0 & 0 & g_{54} & g_{55} \end{pmatrix} \left[\begin{pmatrix} u_1 \\ u_2 \\ u_3 \\ u_4 \\ u_5 \end{pmatrix} + \begin{pmatrix} d_1 \\ d_2 \\ d_3 \\ d_4 \\ d_5 \end{pmatrix} \right] \tag{6.206}$$

Using the previous state-space model the following virtual control inputs are defined: $v_1 = \tilde{F}_1 + g_{11}u_1$, $\;v_2 = \tilde{F}_2 + g_{22}u_2$, $\;v_3 = \tilde{F}_3 + g_{33}u_3$, $\;v_4 = \tilde{F}_4 + g_{44}u_4 + g_{45}u_5$.

Equivalently, the dynamics of the mobile manipulator under additive input disturbances is written as: $\ddot{x}_1 = v_1 + d_1$, $\ddot{x}_2 = v_2 + d_2$, $\ddot{x}_3 = v_3 + d_3$, $\ddot{x}_4 = v_4 + d_4$, $\ddot{x}_5 = v_5 + d_5$.

The disturbances terms d_i, $i = 1, 2, \ldots, 5$ are considered to be described by the associated 2nd order time-derivative and by the related initial conditions. This is reasonable, because every function f can be equivalently represented by its nth order time-derivative $f^{(n)}(t)$ and by the related initial conditions $f^{(i)}(0)$, $i = 1, 2, \ldots, n-1$. However, since the estimation of such functions is going to be performed with the use of Kalman Filtering, prior knowledge about initial conditions becomes obsolete.

The second-order derivative of the aforementioned disturbance inputs is given by $\ddot{d}_i = f_{d_i}$, $i = 1, 2, \ldots, 5$. Next, the disturbance variables and their time-derivatives are denoted as additional state variables of the model of the mobile manipulator. This results into the following state-space description of the system: $z_1 = x_1, z_2 = \dot{x}_1, z_3 = x_2, z_4 = \dot{x}_2, z_5 = x_3, z_6 = \dot{x}_3, z_7 = x_4, z_8 = \dot{x}_4, z_9 = x_5, z_{10} = \dot{x}_5, z_{11} = d_1, z_{12} = \dot{d}_1, z_{13} = d_2, z_{14} = \dot{d}_2, z_{15} = d_3, z_{16} = \dot{d}_3, z_{17} = d_4, z_{18} = \dot{d}_4$ and $z_{19} = d_5, z_{20} = \dot{d}_5$.

The model of the mobile manipulator can be written in the following concise state-space form:

$$\begin{aligned} \dot{z} &= Az + B\tilde{v} \\ z_m &= Cz \end{aligned} \tag{6.207}$$

where vector $\tilde{v} = [u_1, u_2, u_3, u_4, u_5, f_{d_1}, f_{d_2}, f_{d_3}, f_{d_4}, f_{d_5}]^T$ while matrices $A \in R^{20\times 20}$, $B \in R^{20\times 10}$ and $C \in R^{5\times 20}$ are given by

$$A = \begin{pmatrix}
0&1&0&0&0&0&0&0&0&0&0&0&0&0&0&0&0&0&0&0\\
0&0&0&0&0&0&0&0&0&0&1&0&0&0&0&0&0&0&0&0\\
0&0&0&1&0&0&0&0&0&0&0&0&0&0&0&0&0&0&0&0\\
0&0&0&0&0&0&0&0&0&0&0&0&1&0&0&0&0&0&0&0\\
0&0&0&0&0&1&0&0&0&0&0&0&0&0&0&0&0&0&0&0\\
0&0&0&0&0&0&0&0&0&0&0&0&0&0&1&0&0&0&0&0\\
0&0&0&0&0&0&0&1&0&0&0&0&0&0&0&0&0&0&0&0\\
0&0&0&0&0&0&0&0&0&0&0&0&0&0&0&0&1&0&0&0\\
0&0&0&0&0&0&0&0&0&1&0&0&0&0&0&0&0&0&0&0\\
0&0&0&0&0&0&0&0&0&0&0&0&0&0&0&0&0&0&1&0\\
0&0&0&0&0&0&0&0&0&0&0&1&0&0&0&0&0&0&0&0\\
0&0&0&0&0&0&0&0&0&0&0&0&0&0&0&0&0&0&0&0\\
0&0&0&0&0&0&0&0&0&0&0&0&0&1&0&0&0&0&0&0\\
0&0&0&0&0&0&0&0&0&0&0&0&0&0&0&0&0&0&0&0\\
0&0&0&0&0&0&0&0&0&0&0&0&0&0&0&1&0&0&0&0\\
0&0&0&0&0&0&0&0&0&0&0&0&0&0&0&0&0&0&0&0\\
0&0&0&0&0&0&0&0&0&0&0&0&0&0&0&0&0&1&0&0\\
0&0&0&0&0&0&0&0&0&0&0&0&0&0&0&0&0&0&0&0\\
0&0&0&0&0&0&0&0&0&0&0&0&0&0&0&0&0&0&0&1\\
0&0&0&0&0&0&0&0&0&0&0&0&0&0&0&0&0&0&0&0
\end{pmatrix} \quad B = \begin{pmatrix}
0&0&0&0&0&0&0&0&0&0\\
1&0&0&0&0&0&0&0&0&0\\
0&0&0&0&0&0&0&0&0&0\\
0&1&0&0&0&0&0&0&0&0\\
0&0&0&0&0&0&0&0&0&0\\
0&0&1&0&0&0&0&0&0&0\\
0&0&0&0&0&0&0&0&0&0\\
0&0&0&1&0&0&0&0&0&0\\
0&0&0&0&0&0&0&0&0&0\\
0&0&0&0&1&0&0&0&0&0\\
0&0&0&0&0&0&0&0&0&0\\
0&0&0&0&0&1&0&0&0&0\\
0&0&0&0&0&0&0&0&0&0\\
0&0&0&0&0&0&1&0&0&0\\
0&0&0&0&0&0&0&0&0&0\\
0&0&0&0&0&0&0&1&0&0\\
0&0&0&0&0&0&0&0&0&0\\
0&0&0&0&0&0&0&0&0&1\\
0&0&0&0&0&0&0&0&0&0
\end{pmatrix} \tag{6.208}$$

$$C = \begin{pmatrix} 1&0&0&0&0&0&0&0&0&0&0&0&0&0&0&0&0&0&0&0 \\ 0&0&1&0&0&0&0&0&0&0&0&0&0&0&0&0&0&0&0&0 \\ 0&0&0&0&1&0&0&0&0&0&0&0&0&0&0&0&0&0&0&0 \\ 0&0&0&0&0&0&1&0&0&0&0&0&0&0&0&0&0&0&0&0 \\ 0&0&0&0&0&0&0&0&1&0&0&0&0&0&0&0&0&0&0&0 \end{pmatrix} \tag{6.209}$$

The associated state-vector and disturbances estimation problem can be solved by considering a disturbance observer in the form:

$$\begin{aligned} \hat{\dot{z}} &= A_o\hat{z} + B_o v + K_f[z_m - \hat{z}_m] \\ \hat{z}_m &= C_o\hat{z} \end{aligned} \tag{6.210}$$

where $A_o = A$, $B_o = B$ and B_o is obtained from B after zeroing all elements in its last 5 columns. The estimator's gain K_f is computed by applying the Kalman Filter's recursion. To this end, matrices A_o, B_o and C_o are substituted by their discrete-time equivalents A_d, B_d, C_d after using common discretisation methods. The Kalman Filter's recursion consists of a measurement-update stage and of a time-update stage:

measurement update:

$$\begin{aligned} K_f(k) &= P^-(k)C_d^T[C_dP^-(k)C_d^T + R]^{-1} \\ \hat{z}(k) &= \hat{z}^-(k) + K_f(k)[z_m - \hat{z}_m] \\ P(k) &= P^-(k) - K_f(k)C_dP^-(k) \end{aligned} \tag{6.211}$$

time update:

$$\begin{aligned} P^-(k+1) &= A_dP(k)A_d^T + Q \\ \hat{z}^-(k+1) &= A_d\hat{z}(k) + B_dv(k) \end{aligned} \tag{6.212}$$

By identifying the unknown disturbance terms d_i, $i - 1, 2, \ldots, 5$ the control system of the robotic unit is modified as follows:

$$u_1 = \frac{1}{g_{11}}[\ddot{x}_1^d - \tilde{F}_1 + k_1^1(\dot{x}_1 - \dot{x}_1^d) + k_2^1(x_1 - x_1^d) - \hat{d}_1] \tag{6.213}$$

$$u_2 = \frac{1}{g_{22}}[\ddot{x}_3^d - \tilde{F}_2 + k_1^2(\dot{x}_3 - \dot{x}_3^d) + k_2^2(x_3 - x_3^d)] \tag{6.214}$$

$$u_3 = \frac{1}{g_{33}}[\ddot{x}_5^d - \tilde{F}_3 + k_1^3(\dot{x}_5 - \dot{x}_5^d) + k_2^3(x_5 - x_5^d) - \hat{d}_3] \tag{6.215}$$

$$\begin{aligned} \begin{pmatrix} u_4 \\ u_5 \end{pmatrix} = \begin{pmatrix} g_{44} & g_{45} \\ g_{54} & g_{55} \end{pmatrix}^{-1} & \left\{ \begin{pmatrix} \ddot{x}_7^d \\ \ddot{x}_9^d \end{pmatrix} - \begin{pmatrix} \tilde{F}_4 \\ \tilde{F}_5 \end{pmatrix} \right. \\ & \left. + \begin{pmatrix} k_1^4(\dot{x}_7 - \dot{x}_7^d) + k_2^4(x_7 - x_7^d) \\ k_1^5(\dot{x}_9 - \dot{x}_9^d) + k_2^5(x_9 - x_9^d) \end{pmatrix} - \begin{pmatrix} \hat{d}_4 \\ \hat{d}_5 \end{pmatrix} \right\} \end{aligned} \tag{6.216}$$

6.3.6 Simulation tests

The performance of the proposed differential flatness theory-based scheme for control and state estimation in mobile manipulators has been confirmed through simulation experiments. The derivative-free nonlinear Kalman Filter has been redesigned as a disturbances estimator, thus allowing for simultaneous estimation of both the non-measurable state variables of the robotic system and of the additive perturbation terms that were affecting it. The measured state vector elements where $x_1 = X$ that is the displacement of the robotic vehicle along the X axis, $x_2 = Y$ that is the displacement of the robotic vehicle along the Y axis, ψ that is the rotation angle of the vehicle around the Z axis, θ_1 the turn angle of the first joint of the robotic manipulator, and θ_2 that is the turn angle of the second joint of the robotic manipulator. The obtained results are depicted in Figures 6.21–6.36. The real value of the state variables of the robotic system is printed in blue, the estimated value is plotted in green while the associated reference setpoints are shown in red.

The advantages of using a global linearisation-based control method for the dynamic model of the mobile manipulator comprising a four-wheels ground vehicle and a two-DOF robotic manipulator are outlined as follows: (i) the transformation that is performed on the robotic system's state-space model is an exact one and does not introduce any modelling errors (ii) by expressing the dynamic model of the mobile manipulator into the linear canonical form it is assured that the separation principle holds and that the design of the control problem can be solved independently from the design of the state-observer, (iii) by using the Kalman Filter as a disturbance observer the estimation and compensation of perturbation terms that affect the mobile manipulator's model is achieved and thus the robustness of the control scheme is improved

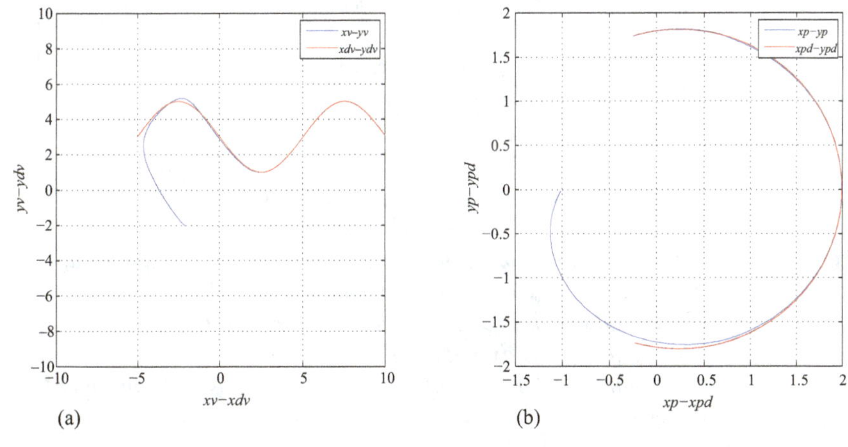

Figure 6.21 Test 1: (a) Tracking of the reference path (red line) by the robotic vehicle of the mobile manipulator (blue line) and (b) tracking of the reference path (red line) by the end-effector of the robotic arm of the mobile manipulator (blue line)

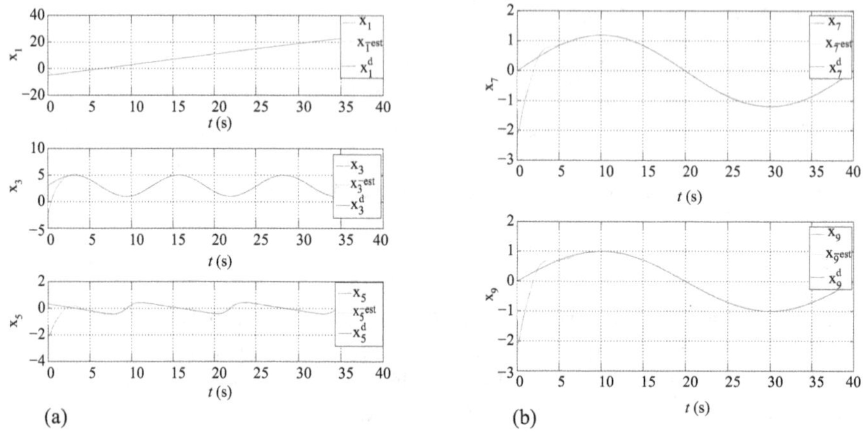

Figure 6.22 Test 1: (a) Convergence of state variables $x_1 = X$, $x_3 = Y$ and $x_5 = \psi$ of the robotic vehicle of the mobile manipulator to their reference setpoints and (b) converge of state variables $x_7 = \theta_1$, and $x_9 = \theta_2$ of the robotic arm of the mobile manipulator to their reference setpoints

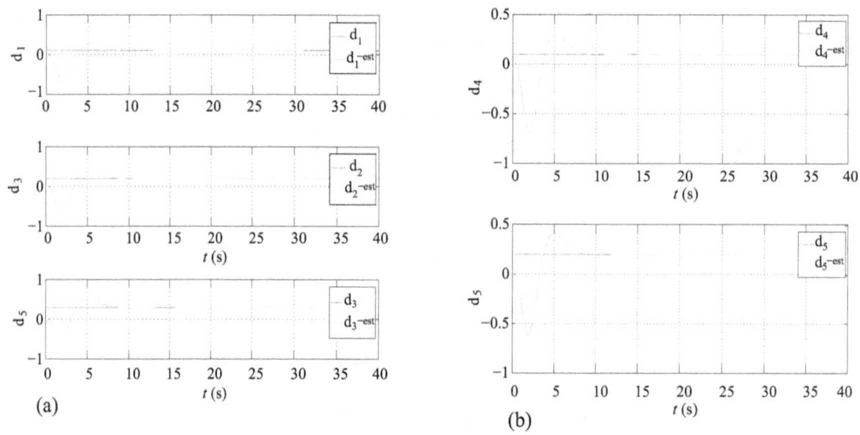

Figure 6.23 Test 1: (a) Estimation of the disturbance inputs d_1, d_2 and d_3 affecting the robotic vehicle of the mobile manipulator. with the use of a Kalman Filter-based disturbance observer and (b) estimation of the disturbance inputs d_4 and d_5 affecting the robotic arm of the mobile manipulator. with the use of a Kalman Filter-based disturbance observer

(iv) The robustness properties of the control method are equivalent to those of LQG (Linear Quadratic Gaussian) control and (v) by using the Kalman Filter as a disturbance observer it is assured that the optimality of the estimation performed by the Kalman Filter is retained.

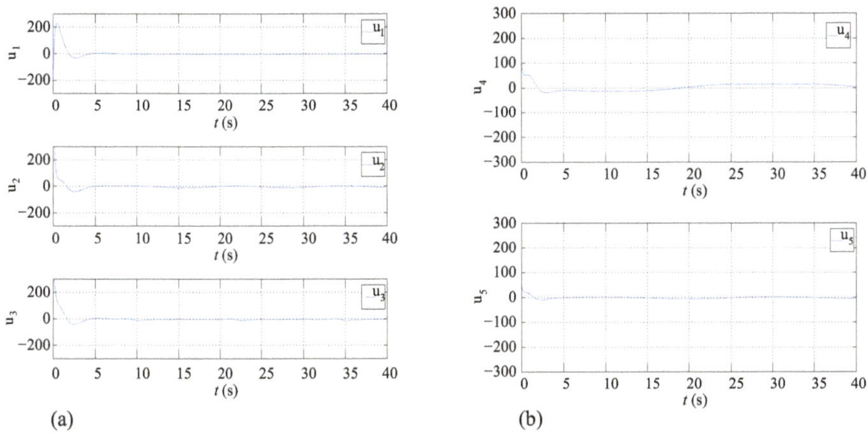

Figure 6.24 *Test 1: (a) Control inputs u_1, u_2 and u_3 applied to the robotic vehicle of the mobile manipulator and (b) control inputs u_4 and u_5 applied to the robotic arm of the mobile manipulator*

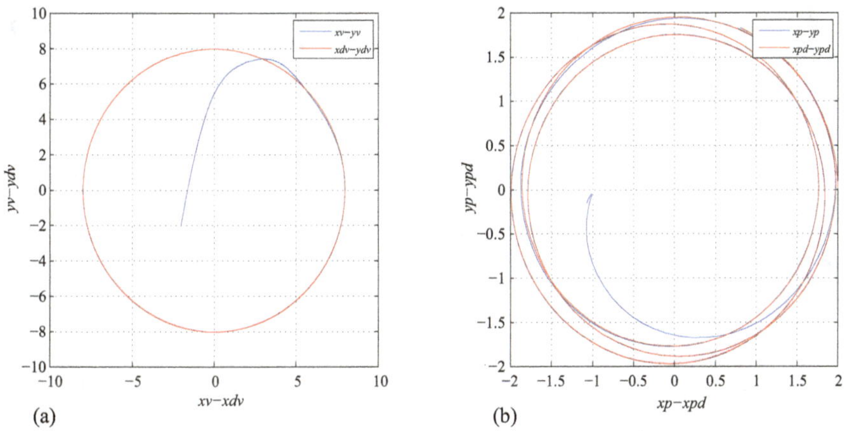

Figure 6.25 *Test 2: (a) Tracking of the reference path (red line) by the robotic vehicle of the mobile manipulator (blue line) and (b) tracking of reference path (red line) by the end-effector of the robotic arm of the mobile manipulator (blue line)*

It is noted that by finding control inputs u_1, u_2 and u_3 the previously analysed procedure computed actually the forces F_X, F_Y and the torque T_Z which are exerted on the mobile robot and which define the motion performed by the mobile manipulator. To find also the real control inputs applied on the mobile robot, that is the propulsion

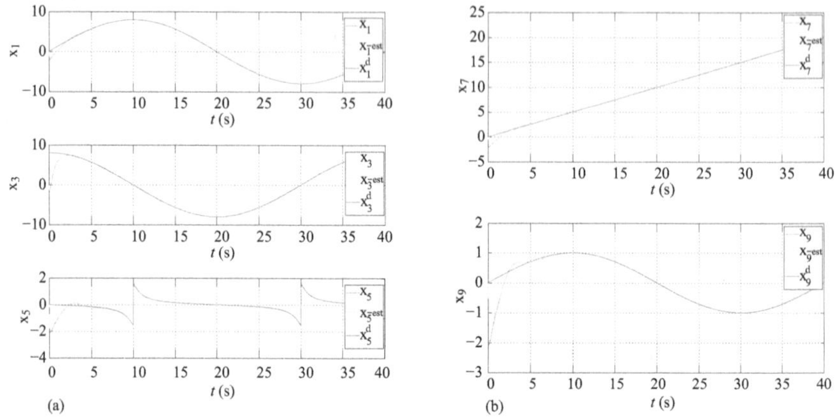

Figure 6.26 Test 2: (a) Convergence of state variables $x_1 = X$, $x_3 = Y$ and $x_5 = \psi$ of the robotic vehicle of the mobile manipulator to their reference setpoints and (b) converge of state variables $x_7 = \theta_1$, and $x_9 = \theta_2$ of the robotic arm of the mobile manipulator to their reference setpoints

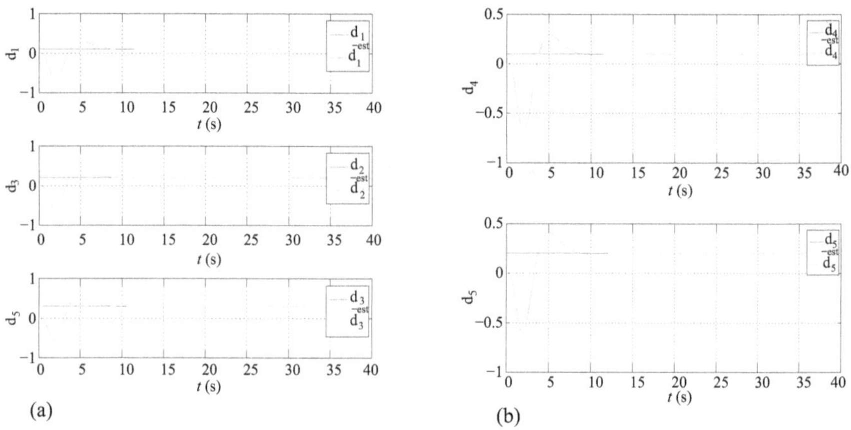

Figure 6.27 Test 2: (a) Estimation of the disturbance inputs d_1, d_2 and d_3 affecting the robotic vehicle of the mobile manipulator with the use of a Kalman Filter-based disturbance observer and (b) estimation of the disturbance inputs d_4 and d_5 affecting the robotic arm of the mobile manipulator with the use of a Kalman Filter-based disturbance observer

force F of its motor, the turn angle θ of its steering wheels and the braking force F_b applied to the rear wheels of the vehicle, the following relations are used:

$$\begin{aligned} F_X &= F\cos(\psi - \theta) - F_b\cos(\psi) \\ F_Y &= F\sin(\psi - \theta) - F_b\sin(\psi) \\ T_Z &= F\sin(\psi - \theta)L \end{aligned} \tag{6.217}$$

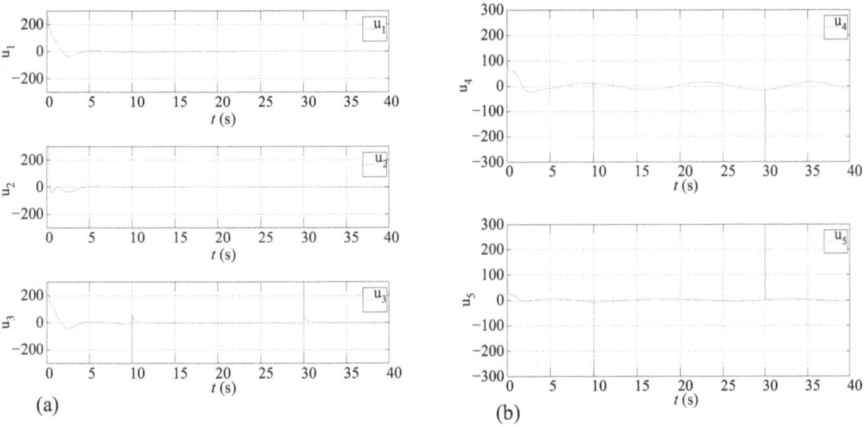

Figure 6.28 *Test 2: (a) Control inputs u_1, u_2 and u_3 applied to the robotic vehicle of the mobile manipulator and (b) control inputs u_4 and u_5 applied to the robotic arm of the mobile manipulator*

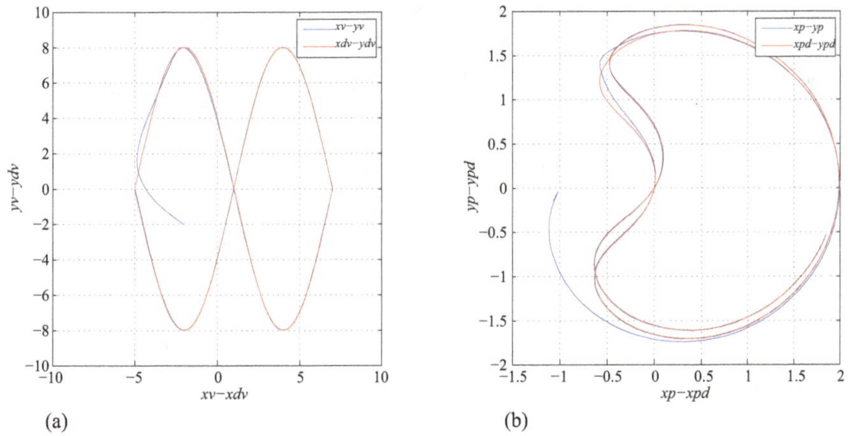

Figure 6.29 *Test 3: (a) Tracking of reference path (red line) by the robotic vehicle of the mobile manipulator (blue line) and (b) tracking of reference path (red line) by the end-effector of the robotic arm of the mobile manipulator (blue line)*

For the linearised state-space representation of the robotic system, both the solution of its control and state estimation problem becomes possible. Actually, to solve the control problem one can apply pole-placement methods or optimal control approaches on the equivalent linearised description of the system. Moreover, to solve the associated state estimation problem a filtering method under the name of Derivative-free

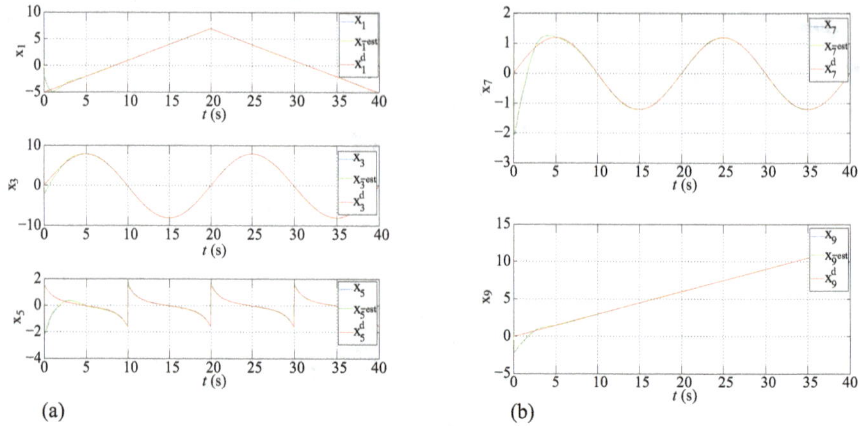

Figure 6.30 *Test 3: (a) convergence of state variables $x_1 = X$, $x_3 = Y$ and $x_5 = \psi$ of the robotic vehicle of the mobile manipulator to their reference setpoints and (b) converge of state variables $x_7 = \theta_1$, and $x_9 = \theta_2$ of the robotic arm of the mobile manipulator to their reference setpoints*

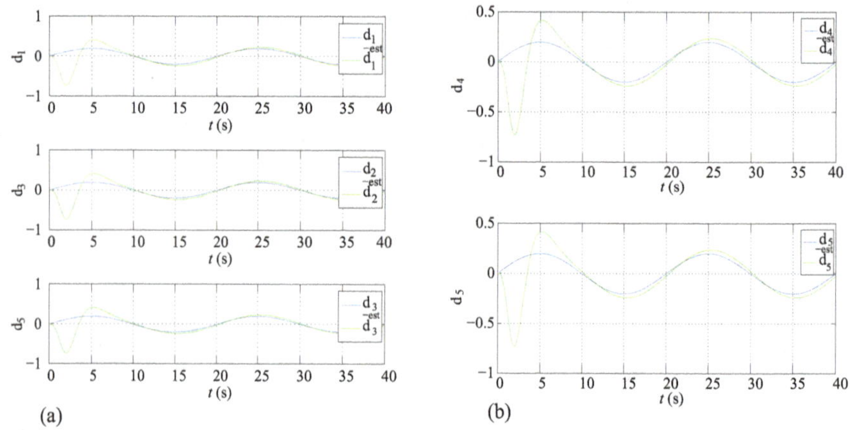

Figure 6.31 *Test 3: (a) Estimation of the disturbance inputs d_1, d_2 and d_3 affecting the robotic vehicle of the mobile manipulator. with the use of a Kalman Filter-based disturbance observer and (b) estimation of the disturbance inputs d_4 and d_5 affecting the robotic arm of the mobile manipulator. with the use of a Kalman Filter-based disturbance observer*

nonlinear Kalman Filter can be used (Rigatos and Tzafestas, 2007; Basseville and Nikiforov, 1993; Rigatos and Zhang, 2009). This filtering approach consists of the standard Kalman Filter recursion on the equivalent linearised description of the system and of an inverse transformation that provides estimates for the state variables

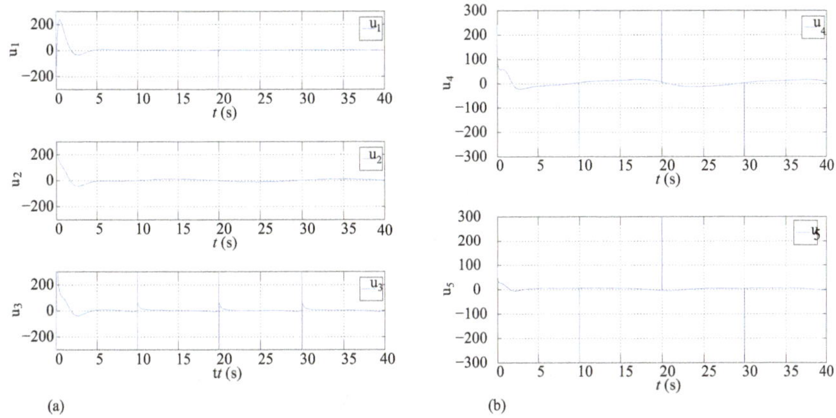

Figure 6.32 *Test 3: (a) Control inputs u_1, u_2 and u_3 applied to the robotic vehicle of the mobile manipulator and (b) control inputs u_4 and u_5 applied to the robotic arm of the mobile manipulator*

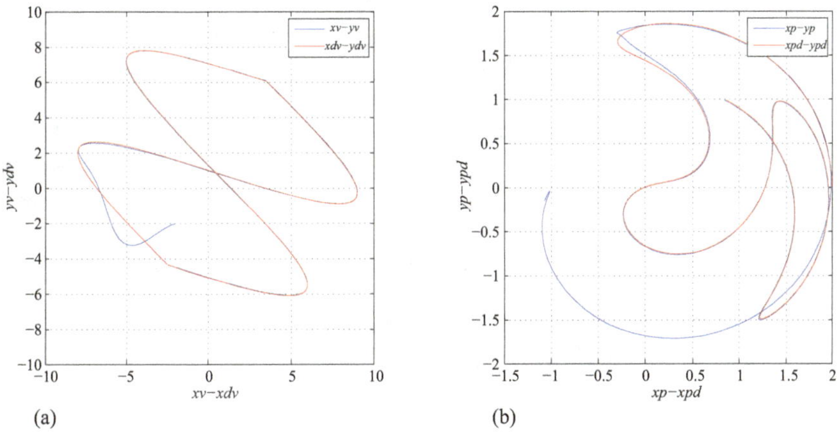

Figure 6.33 *Test 4: (a) Tracking of reference path (red line) by the robotic vehicle of the mobile manipulator (blue line) and (b) tracking of reference path (red line) by the end-effector of the robotic arm on the mobile manipulator (blue line)*

of the initial nonlinear model of the mobile manipulator. Moreover, by redesigning the aforementioned Kalman Filter as a disturbance observer, one can also estimate in real-time and compensate for additive input disturbances that affect the mobile manipulator. To this end, the state vector of the robotic system is extended by considering as additional state variables the disturbance inputs and their time derivatives.

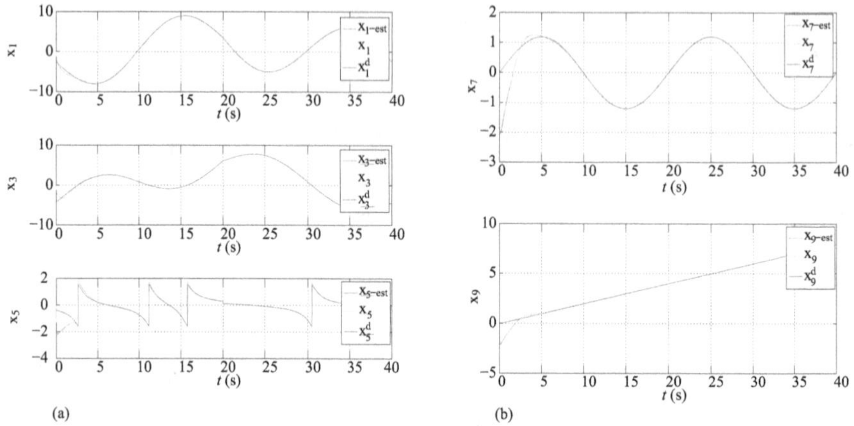

Figure 6.34 Test 4: (a) Convergence of state variables $x_1 = X$, $x_3 = Y$ and $x_5 = \psi$ of the robotic vehicle of the mobile manipulator to their reference setpoints and (b) converge of state variables $x_7 = \theta_1$, and $x_9 = \theta_2$ of the robotic arm of the mobile manipulator to their reference setpoints

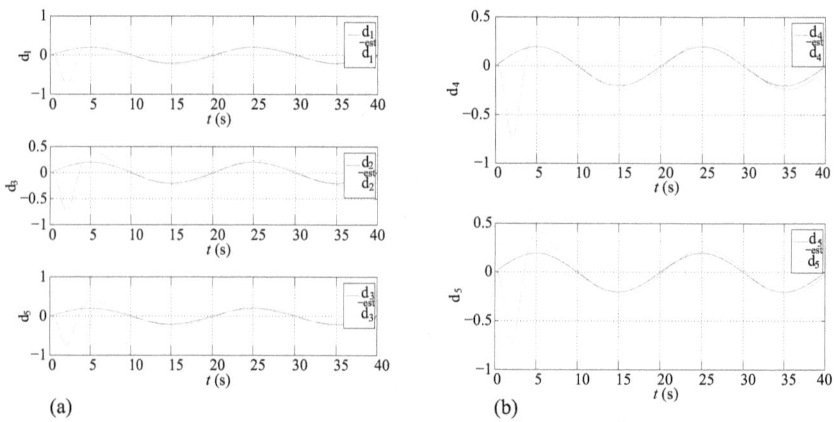

Figure 6.35 Test 4: (a) Estimation of the disturbance inputs d_1, d_2 and d_3 affecting the robotic vehicle of the mobile manipulator with the use of a Kalman Filter-based disturbance observer and (b) estimation of the disturbance inputs d_4 and d_5 affecting the robotic arm of the mobile manipulator with the use of a Kalman Filter-based disturbance observer

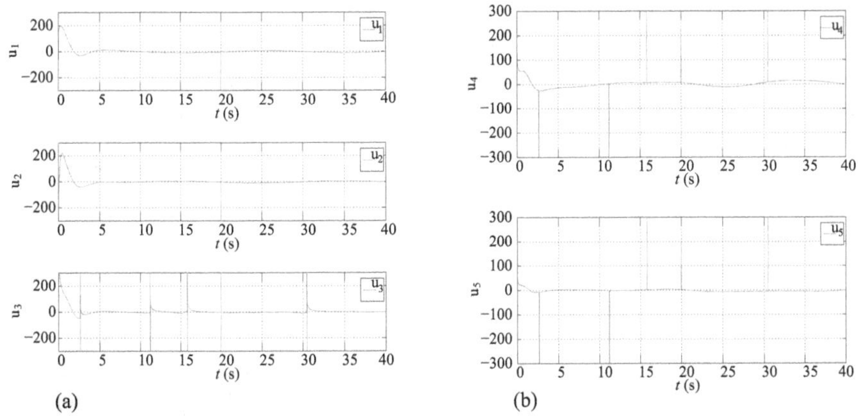

Figure 6.36 Test 4: (a) Control inputs u_1, u_2 and u_3 applied to the robotic vehicle of the mobile manipulator and (b) control inputs u_4 and u_5 applied to the robotic arm of the mobile manipulator

6.4 State of charge estimation in EVs with a KF-based disturbance observer

6.4.1 Outline

Estimation of the state-of-charge in the batteries of electric vehicles is important for achieving the vehicle's reliable functioning, and for optimising the vehicles' charging procedure. Actually, by knowing precisely the batteries state-of-charge, one can make the need for recharging be less frequent and can minimise the time that electric vehicles have to stay in charging lots. Monitoring of the state-of-charge is also significant for the autonomous functioning of unmanned vehicles (such as mobile robots, UAVs or AUVs), or for biomedical applications (Li, Liang, Liu *et al.*, 2017; Zhang, Li *et al.*, 2017; Zheng and Fang, 2015; Sin, Zhou *et al.*, 2016; Wang, Cao *et al.*, 2007; Gregedal *et al.*, 2017). Estimation of the state-of-charge is advantageous comparing to the monitoring of a battery's electric charge through the integration of its output current. This is because the measurement of the current variable can be subject to errors that are accumulated through numerical integration. Besides, the previous procedure for approximating the battery's state-of-charge requires knowledge of initial conditions. By using an erroneous value of initial conditions additional error can be introduced (Rigatos, 2011; Rigatos, 2015; Rigatos, 2016b; Basseville and Nikiforov, 1993; Rigatos and Zhang, 2009).

Among estimation techniques for computing the battery's state-of-charge one can primarily distinguish the model-free and the model-based ones. Model-free techniques assume that there is no prior knowledge about the dynamic model of the battery's electric circuit and that this model can be identified with black-box type estimators such as neural networks. The method relies on the processing of raw (input–output) data from the battery's circuit in the least squares sense (Chen, Ouyang *et al.*, 2018; Liu, Mendoza *et al.*, 2017). In model-based estimation techniques one makes use of Kirchhoff's current and voltage laws to develop the battery's dynamic model after assigning to the elements of the battery's circuit (resistors, capacitors and voltage sources) their nominal values (Song, Lin *et al.*, 2017; Zhao, Duncan *et al.*, 2017; Fridholm *et al.*, 2014; Chen, Balance *et al.*, 2000; Cortesao *et al.*, 2005; Gupta and Malley, 2011). One can note several results on the use of nonlinear observers aiming at estimating the battery's state-of-charge through current or voltage measurements (Dey and Ayalev, 2014; Rauh *et al.*, 2013; Muhammad, Rafique *et al.*, 2017). Equivalently, one can include the aforementioned model in the recursion of the Kalman Filter. Among Kalman filtering techniques, the Extended Kalman Filter appears to be particularly popular (Dai, Wai *et al.*, 2012; Lee, Dai *et al.*, 2018; Fang, Zhang *et al.*, 2017; Wang, Fang *et al.*, 2017; Chen, Fu *et al.*, 2013; Xiong, Gong *et al.*, 2013). Additionally, elaborated Kalman Filter schemes such as the H-infinity Kalman Filter and the Unscented Kalman Filter have been applied for the solving the state-of-charge estimation problem in batteries (Peng, Chen *et al.*, 2017; Partovibakhsh and Liu, 2017; Parakkadavath and Bhikkaji, 2017). Moreover, to monitor the battery's state-of-charge under non-Gaussian measurement noise, estimation with the use of the Particle Filter has been proposed (Wang, Zhang and Chen, 2015; Cheng, Wang *et al.*, 2017).

The present section aims at solving the problem of a battery's state-of-charge estimation with Kalman Filtering, despite uncertainty or parametric variations in the battery's dynamic model. Changes in the parameters of the Li-ion battery's circuit or external perturbations are considered to take the form of additive disturbances distorting the circuit's current. They can be the result of the batteries' aging or the result of the functioning of the battery under harsh conditions (e.g. temperature variations). To accomplish robust estimation, the Kalman Filter is re-designed as a disturbance observer (Rigatos, 2016b). The state-vector of the battery is extended by including in it, as additional elements, the disturbance variables and their derivatives. This filtering approach allows to identify simultaneously (i) the state vector elements of the battery's model, comprising also the state-of-charge, (ii) the perturbations that affect the electric circuit. The only measured variable is the output voltage of the battery. As it has been confirmed through simulation experiments, the Kalman Filter-based disturbance observer is computationally fast and capable of identifying perturbation inputs that vary in time.

6.4.2 Dynamic model of the battery

The electric circuit of the battery is shown in Figure 6.37. The R_D, C_D and R_T, C_T circuits are used to describe the polarisation of the battery during charging and discharging, caused by transfer, diffusion and other factors (Dai, Wai *et al.*, 2012). U_{oc} describes the open-circuit voltage. The value of U_{oc} is directly related to the state-of-charge of the battery. R_o is the internal resistance of the battery. The state-of-charge of the battery is related to the battery's current I_B through the following integration:

$$SOC = SOC_{init} - \int \frac{\eta(I_B - I_{sd})dt}{C_{nom}} \tag{6.218}$$

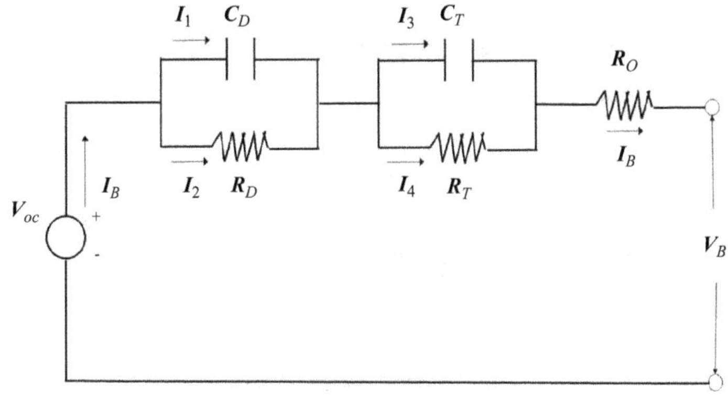

Figure 6.37 Diagram of the battery's electric circuit

where SOC_{init} is the initial value of the state-of-charge, I_B is the working current of the battery and I_{sd} is the self-discharge current. Moreover, η is a coulombic coefficient and C_{nom} is the battery's nominal capacity. By considering that the self-discharge current is negligible one finally gets

$$SOC = SOC_{init} - \int \frac{\eta I_B dt}{C_{nom}} \tag{6.219}$$

With reference to Figure 6.37, the state-space description of the battery's dynamics is given by the following equations:

$$\begin{aligned} \frac{d}{dt} SOC_{avg} &= -\frac{\eta}{C_{nom}} I_B \\ \frac{d}{dt} V_D &= -\frac{1}{R_D C_D} V_D + \frac{1}{C_D} I_B \\ \frac{d}{dt} V_T &= -\frac{1}{R_T C_T} V_T + \frac{1}{C_T} I_B \end{aligned} \tag{6.220}$$

where the measurement equation of the system is given by

$$V_B = \bar{V}_{oc} SOC_{avg} - I_B R_o - V_D - V_T \tag{6.221}$$

It can be considered that current I_B is constant subject only to perturbation terms:

$$\frac{d}{dt} I_B = \tilde{d} \tag{6.222}$$

An elaborated model can consider the function of the voltage source V_{oc} to include nonlinear terms of the state-of-charge variable SOC_{avg}, in polynomial approximation sense. Here, the voltage source V_{oc} is considered to vary with the state-of-charge of the battery as follows:

$$V_{oc} = \bar{V}_{oc} SOC_{avg} \tag{6.223}$$

Defining the state vector of the system as $x_1 = SOC_{avg}$, $x_2 = V_D$, $x_3 = V_T$, $x_4 = I_B$ one obtains the following state-space description of the system:

$$\begin{pmatrix}\dot{x}_1\\ \dot{x}_2\\ \dot{x}_3\\ \dot{x}_4\end{pmatrix} = \begin{pmatrix} 0 & 0 & 0 & \frac{-\eta}{C_{\max}} \\ 0 & -\dfrac{1}{R_D C_D} & 0 & \dfrac{1}{C_D} \\ 0 & 0 & -\dfrac{1}{R_T C_T} & \dfrac{1}{C_T} \\ 0 & 0 & 0 & 0 \end{pmatrix} \begin{pmatrix} x_1\\ x_2\\ x_3\\ x_4\end{pmatrix} + \begin{pmatrix}0\\0\\0\\ \tilde{d}\end{pmatrix} \tag{6.224}$$

$$y = \begin{pmatrix} V_{oc} & -1 & -1 & -R_o\end{pmatrix} \begin{pmatrix} x_1\\ x_2\\ x_3\\ x_4\end{pmatrix} \tag{6.225}$$

Next, the state-space model of the system is extended by considering as additional state variables, the disturbance term and their derivatives affecting the current variable I_B. This gives $x_5 = \tilde{d}$ and $x_6 = \dot{\tilde{d}}$. The disturbance term is taken to be described by its second-order derivative $\ddot{\tilde{d}} = f_d$ and the associated initial conditions. This is reasonable because every function can be represented by its nth order derivative and the related initial conditions. However, since the disturbance term is finally going to be estimated with the use of Kalman Filtering, knowledge of these initial conditions becomes obsolete.

Using the extended state-space description of the EV battery's one has

$$\begin{pmatrix}\dot{x}_1\\ \dot{x}_2\\ \dot{x}_3\\ \dot{x}_4\\ \dot{x}_5\\ \dot{x}_6\end{pmatrix} = \begin{pmatrix} 0 & 0 & 0 & \frac{-\eta}{C_{\max}} & 0 & 0 \\ 0 & -\dfrac{1}{R_D C_D} & 0 & \dfrac{1}{C_D} & 0 & 0 \\ 0 & 0 & -\dfrac{1}{R_T C_T} & \dfrac{1}{C_T} & 0 & 0 \\ 0 & 0 & 0 & 0 & 1 & 0 \\ 0 & 0 & 0 & 0 & 0 & 1 \\ 0 & 0 & 0 & 0 & 0 & 0 \end{pmatrix} \begin{pmatrix} x_1\\ x_2\\ x_3\\ x_4\\ x_5\\ x_6\end{pmatrix} + \begin{pmatrix}0\\0\\0\\0\\0\\1\end{pmatrix} f_d \tag{6.226}$$

$$y = \begin{pmatrix} V_{oc} & -1 & -1 & -R_o & 0 & 0\end{pmatrix} \begin{pmatrix} x_1\\ x_2\\ x_3\\ x_4\\ x_5\\ x_6\end{pmatrix} \tag{6.227}$$

The extended state-space description is written in the concise form

$$\begin{aligned} \dot{x} &= Ax + Bu + w \\ y &= Cx + v \end{aligned} \tag{6.228}$$

6.4.3 Kalman Filter-based disturbance observer

A disturbance observer can be derived for the battery's charging model in the form

$$\begin{aligned} \dot{\hat{x}} &= A_o\hat{x} + B_o u + K_f[C_o x - C_o\hat{x}] \\ \hat{y} &= C_o\hat{x} \end{aligned} \tag{6.229}$$

where $A_o = A$, $B_o = O_{6\times 1}$ and $C_o = C$. The computation of the gain of the disturbances' estimator can be performed with the use of the Kalman Filter recursion.

The previously defined matrices A_o, B_o and C_o are discretised using common discretisation methods. This provides matrices A_d, B_d and C_d. Matrices Q and R denote the process and measurement noise covariance matrices. Matrix P is the state-vector error covariance matrix. The Kalman Filter comprises a measurement update and a time-update stage, which is given below:

measurement update:

$$\begin{aligned} K_f(k) &= P^-(k)C_d^T[C_dP^-(k)C_d^T + R]^{-1} \\ \hat{x}(k) &= \hat{x}^-(k) + K_f(k)[y(k) - \hat{y}(k)] \\ P(k) &= P^-(k) - K_f(k)C_dP^-(k) \end{aligned} \tag{6.230}$$

time update:

$$\begin{aligned} P^-(k+1) &= A_dP(k)A_d^T + Q \\ \hat{x}^-(k+1) &= A_d\hat{x}(k) + B_dv(k) \end{aligned} \tag{6.231}$$

A generic diagram of a Kalman Filter-based disturbance observer is given in Figure 6.38.

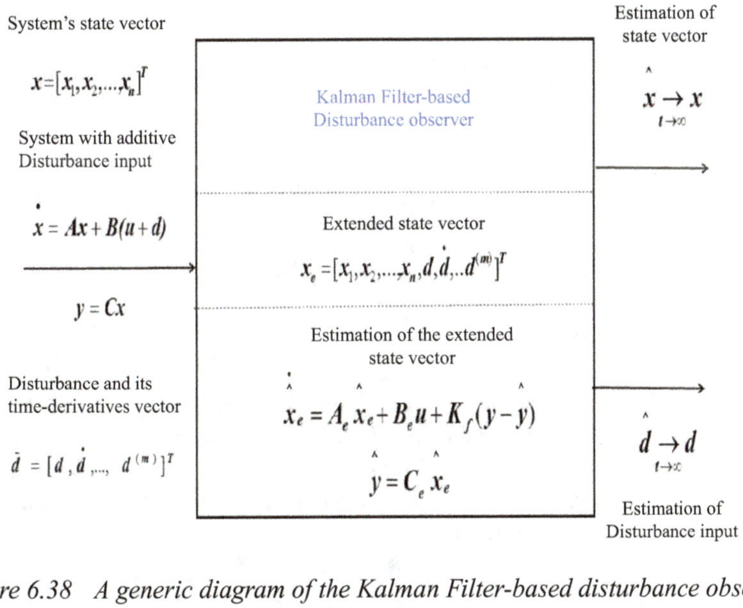

Figure 6.38 A generic diagram of the Kalman Filter-based disturbance observer

6.4.4 Simulation tests

The performance of the proposed Kalman Filtering approach for estimation of the state-of-charge of electric vehicles' batteries was tested through simulation experiments. For the implementation of the Kalman Filter algorithm, a measurement of the output voltage of the battery had to be sampled at each sampling period. The process noise covariance matrix was taken to be a diagonal matrix of dimension $Q \in^{6\times 6}$ and with diagonal elements equal to 10^{-4}. Due to measuring only one output from the battery's model, the measurement noise covariance matrix R was a scalar with value 10^{-3}.

The measurement equation of the system was given in (6.227). As noted, an elaborated model can consider the mathematical function of the voltage source V_{oc} to include nonlinear terms of the state-of-charge variable SOC_{avg}, in a polynomial approximation sense. This would result into a relation of the form $V_{oc} = a_o + a_1 SOC_{avg} + a_2 SOC_{avg}^2 + \cdots + a_m SOC_{avg}^m$. In such a case, in the recursion of the Kalman Filter one would have to use the Jacobian matrix J_γ in place of the measurement matrix C. The obtained results are depicted in Figures 6.39 to 6.43. It is confirmed that the Kalman Filter provides accurate estimates of the state variables, as well as accurate estimates of the system's disturbance terms.

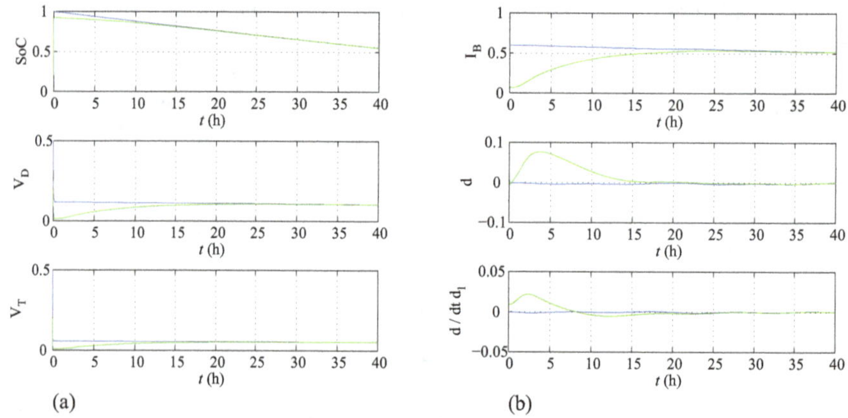

Figure 6.39 Test case 1: (a) State variables x_i, $i = 1, 2, 3$ (blue lines) of the battery's charging dynamics and their estimates (green lines) provided by the Kalman Filter-based disturbance observer and (b) state variables x_i, $i = 4, 5, 6$ (blue lines) of the battery's charging dynamics and their estimates (green lines) provided by the Kalman Filter-based disturbance observer

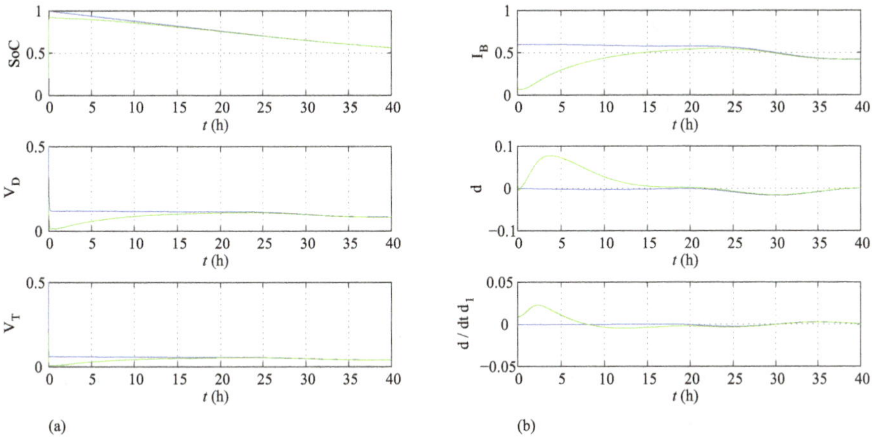

Figure 6.40 *Test case 2: (a) State variables x_i, $i = 1, 2, 3$ (blue lines) of the battery's charging dynamics and their estimates (green lines) provided by the Kalman Filter-based disturbance observer and (b) state variables x_i, $i = 4, 5, 6$ (blue lines) of the battery's charging dynamics and their estimates (green lines) provided by the Kalman Filter-based disturbance observer*

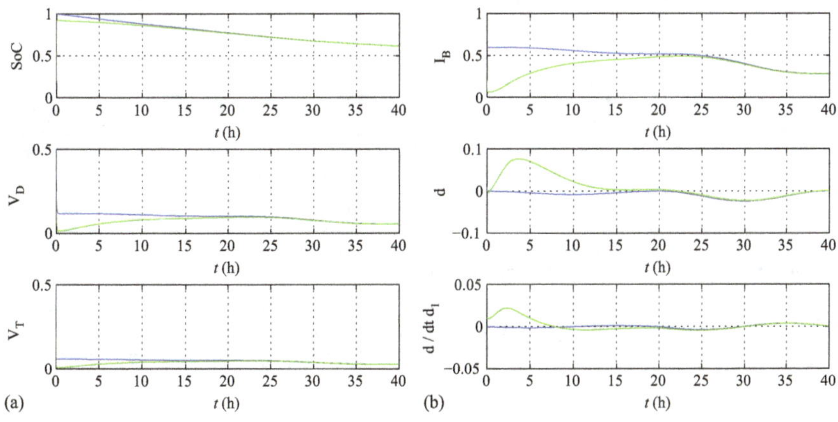

Figure 6.41 *Test case 3: (a) State variables x_i, $i = 1, 2, 3$ (blue lines) of the battery's charging dynamics and their estimates (green lines) provided by the Kalman Filter-based disturbance observer and (b) state variables x_i, $i = 4, 5, 6$ (blue lines) of the battery's charging dynamics and their estimates (green lines) provided by the Kalman Filter-based disturbance observer*

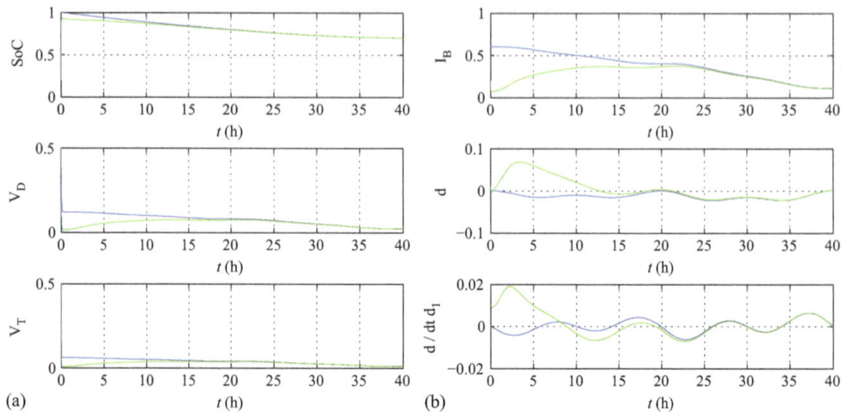

Figure 6.42 *Test case 4: (a) State variables x_i, $i = 1, 2, 3$ (blue lines) of the battery's charging dynamics and their estimates (green lines) provided by the Kalman Filter-based disturbance observer and (b) state variables x_i, $i = 4, 5, 6$ (blue lines) of the battery's charging dynamics and their estimates (green lines) provided by the Kalman Filter-based disturbance observer*

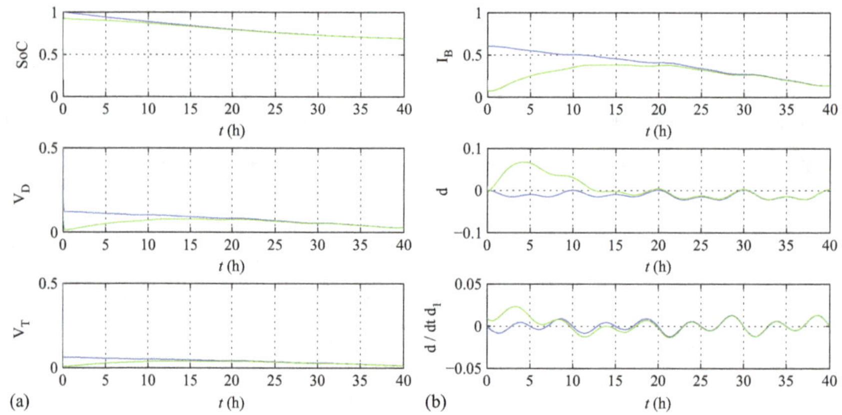

Figure 6.43 *Test case 5: (a) State variables x_i, $i = 1, 2, 3$ (blue lines) of the battery's charging dynamics and their estimates (green lines) provided by the Kalman Filter-based disturbance observer and (b) state variables x_i, $i = 4, 5, 6$ (blue lines) of the battery's charging dynamics and their estimates (green lines) provided by the Kalman Filter-based disturbance observer*

6.5 Control of nonlinear wave PDE dynamics

6.5.1 Outline

Control of distributed parameter systems described by nonlinear partial differential equations is a nontrivial problem of high importance for several applications (Rigatos, 2011; Rigatos, 2013; Rudolph, 2003; Rigatos, 2015). For instance, parabolic differential equations are met in diffusion phenomena that appear in heat transfer, chemical processes as well as in quantum mechanics (Maidi and Corriu, 2014; Zwart *et al.*, 2011; Winkler, Krause *et al.*, 2009). Moreover, nonlinear hyperbolic differential equations mostly describe wave-type phenomena. These are met in communication systems (transmission lines, optical fibers and electromagnetic waves propagation), in electronics (Josephson junctions), in fluid flow models, in structural engineering (dynamic analysis of buildings under seismic waves, mechanical structures subjected to vibrations, pendulum chains), in biomedical systems (voltage propagation and variations in neuron's membrane), etc. (Saadatpour and Levi, 2013; Tang and Zhang, 2010; Basseville *et al.*, 1993; Gerdres *et al.*, 2006; Dubljevic, 2009; Ning and Yan, 2010; Krstic, 2010; Krstic, Guo *et al.*, 2008). Additionally, elliptic partial differential equations are met in problems of vibrating structures, acoustic cavities and radiation waves. The section treats the problem of boundary control of nonlinear distributed parameter systems, which means that the boundary conditions are used as control inputs to modify the PDE dynamics. Boundary control or lumped-input control of nonlinear distributed parameter systems are problems of elevated difficulty and many remarkable results on this topic exist (Woitteneck and Mounier, 2010; Mounier, Rudolph *et al.*, 2010; Fliess and Mounier, 2002; Boussaada *et al.*, 2013). If the PDE system is decomposed into an equivalent set of ordinary nonlinear differential equations, then controlling such a set of dynamical subsystems by the boundary conditions implies underactuation (actually one has to control the different degrees of freedom associated with the set of nonlinear ODEs by using only a couple of control inputs).

The control approach followed in this section is as follows. First, by implementing a semi-discretisation procedure as in the numerical solution of PDEs a set of equivalent nonlinear ordinary differential equations is obtained (Pinsky, 1991; Gerdts *et al.*, 2008; Guo and Billings, 2007; Kröner, 2011; Winler and Lohmann, 2010; Dragonescu and Soane, 2013). Next, it is shown that the system of the nonlinear ODEs is a differentially flat one. This means that all its state variables and the control inputs can be written as differential functions of one single algebraic variable which is the flat output (Sira-Ramirez and Agrawal, 2014; Levine, 2009; Boouden *et al.*, 2011; Fliess and Mounier, 1999; Levine, 2011). Moreover, by examining independently each nonlinear ODE it is shown that this stands again for a differentially flat system, for which a virtual control input can be computed as in the case of flatness-based control for the trivial system. The virtual control input is chosen such that the ODE subsystem dynamics is linearised and the tracking error is eliminated. The boundary condition that appears in the nonlinear ODE subsystem that comprises the last rows of the state-space description, stands for the aggregate control input. The computation of the boundary control input uses recursively all virtual control inputs mentioned above, moving from the last ODE subsystem to the first one. Thus, by tracing the rows of the state-space model backwards, at each iteration of the control algorithm, one can finally obtain the control input that should be applied to the nonlinear PDE system so as to assure that all its state vector elements will converge to the desirable setpoints. By analyzing the dynamics of the closed-loop system that results from the application of the aforementioned control method, asymptotic stability is confirmed.

6.5.2 Transformation of the PDE model into a set of nonlinear ODEs

The following nonlinear wave equation is considered

$$\frac{\partial^2\phi}{\partial t^2} = k\frac{\partial^2\phi}{\partial x^2} + f(\phi) \tag{6.232}$$

where $u(x,t)$ is assumed to be the external control input. Using the approximation for the partial derivative

$$\frac{\partial^2\phi}{\partial x^2} \simeq = \frac{\phi_{i+1} - 2\phi_i + \phi_{i-1}}{\Delta x^2} \tag{6.233}$$

and considering spatial measurements of variable ϕ along axis x at points $x_0 + i\Delta x,\ i = 1, 2, \ldots, N$ one has

$$\frac{\partial^2\phi_i}{\partial t^2} = \frac{K}{\Delta x^2}\phi_{i+1} - \frac{2K}{\Delta x^2}\phi_i + \frac{K}{\Delta x^2}\phi_{i-1} + f(\phi_i) \tag{6.234}$$

By considering the associated samples of ϕ given by $\phi_0, \phi_1, \ldots, \phi_N, \phi_{N+1}$ one has

$$\begin{aligned}
\frac{\partial^2\phi_1}{\partial t^2} &= \frac{K}{\Delta x^2}\phi_2 - \frac{2K}{\Delta x^2}\phi_1 + \frac{K}{\Delta x^2}\phi_0 + f(\phi_1) \\
\frac{\partial^2\phi_2}{\partial t^2} &= \frac{K}{\Delta x^2}\phi_3 - \frac{2K}{\Delta x^2}\phi_2 + \frac{K}{\Delta x^2}\phi_1 + f(\phi_2) \\
\frac{\partial^2\phi_3}{\partial t^2} &= \frac{K}{\Delta x^2}\phi_4 - \frac{2K}{\Delta x^2}\phi_3 + \frac{K}{\Delta x^2}\phi_2 + f(\phi_3) \\
&\cdots \\
\frac{\partial^2\phi_{N-1}}{\partial t^2} &= \frac{K}{\Delta x^2}\phi_N - \frac{2K}{\Delta x^2}\phi_{N-1} + \frac{K}{\Delta x^2}\phi_{N-2} + f(\phi_{N-1}) \\
\frac{\partial^2\phi_N}{\partial t^2} &= \frac{K}{\Delta x^2}\phi_{N+1} - \frac{2K}{\Delta x^2}\phi_N + \frac{K}{\Delta x^2}\phi_{N-1} + f(\phi_N)
\end{aligned} \tag{6.235}$$

By defining the following state vector

$$x^T = (\phi_1, \phi_2, \ldots, \phi_N) \tag{6.236}$$

one obtains the following state-space description

$$\begin{aligned}
\ddot{x}_1 &= \frac{K}{\Delta x^2}x_2 - \frac{2K}{\Delta x^2}x_1 + \frac{K}{\Delta x^2}\phi_0 + f(x_1) \\
\ddot{x}_2 &= \frac{K}{\Delta x^2}x_3 - \frac{2K}{\Delta x^2}x_2 + \frac{K}{\Delta x^2}x_1 + f(x_2) \\
\ddot{x}_3 &= \frac{K}{\Delta x^2}x_4 - \frac{2K}{\Delta x^2}x_3 + \frac{K}{\Delta x^2}x_2 + f(x_3) \\
&\cdots \\
\ddot{x}_{N-1} &= \frac{K}{\Delta x^2}x_N - \frac{2K}{\Delta x^2}x_{N-1} + \frac{K}{\Delta x^2}x_{N-2} + f(x_{N-1}) \\
\ddot{x}_N &= \frac{K}{\Delta x^2}\phi_{N+1} - \frac{2K}{\Delta x^2}x_N + \frac{K}{\Delta x^2}x_{N-1} + f(x_N)
\end{aligned} \tag{6.237}$$

Next, the following state variables are defined

$$\begin{aligned}
y_{1,i} &= x_i \\
y_{2,i} &= \dot{x}_i
\end{aligned} \tag{6.238}$$

and the state-space description of the system becomes as follows

$$\begin{aligned}
\dot{y}_{1,1} &= y_{2,1} \\
\dot{y}_{2,1} &= \frac{K}{\Delta x^2}y_{1,2} - \frac{2K}{\Delta x^2}y_{1,1} + \frac{K}{\Delta x^2}\phi_0 + f(y_{1,1}) \\
\dot{y}_{1,2} &= y_{2,2} \\
\dot{y}_{2,2} &= \frac{K}{\Delta x^2}y_{1,3} - \frac{2K}{\Delta x^2}y_{1,2} + \frac{K}{\Delta x^2}y_{1,1} + f(y_{1,2}) \\
\dot{y}_{1,3} &= y_{2,3} \\
\dot{y}_{2,3} &= \frac{K}{\Delta x^2}y_{1,4} - \frac{2K}{\Delta x^2}y_{1,3} + \frac{K}{\Delta x^2}y_{1,2} + f(y_{1,3}) \\
&\cdots \\
&\cdots \\
\dot{y}_{1,N-1} &= y_{2,N-1} \\
\dot{y}_{2,N-1} &= \frac{K}{\Delta x^2}y_{1,N} - \frac{2K}{\Delta x^2}y_{1,N-1} + \frac{K}{\Delta x^2}y_{1,N-2} + f(y_{1,N-1}) \\
\dot{y}_{1,N} &= y_{2,N} \\
\dot{y}_{2,N} &= \frac{K}{\Delta x^2}\phi_{N+1} - \frac{2K}{\Delta x^2}y_{1,N} + \frac{K}{\Delta x}y_{1,N-1} + f(y_{1,N})
\end{aligned} \tag{6.239}$$

The dynamical system described in (6.430) is a differentially flat one with flat output defined as the vector $\tilde{y} = [y_{1,1}, y_{1,2}, \ldots, y_{1,N}]$. Indeed all state variables can be written as functions of this flat output and its derivatives.

6.5.3 Differential flatness of the nonlinear PDE model

It will be also shown that the state-space description of the nonlinear wave PDE dynamics, which has as control input only the boundary condition ϕ_{N+1} is a differentially flat one. One has

$$\begin{aligned}\dot{y}_{1,1} &= y_{2,1}\\ \dot{y}_{2,1} &= -\frac{2K}{\Delta x^2}y_{1,1} + \frac{K}{\Delta x^2}\phi_0 + f(y_{1,1}) + \frac{K}{\Delta x^2}y_{1,2}\end{aligned} \quad (6.240)$$

$$\begin{aligned}\dot{y}_{1,2} &= y_{2,2}\\ \dot{y}_{2,2} &= -\frac{2K}{\Delta x^2}y_{1,2} + \frac{K}{\Delta x^2}y_{1,1} + f(y_{1,2}) + \frac{K}{\Delta x^2}y_{1,3}\end{aligned} \quad (6.241)$$

$$\begin{aligned}\dot{y}_{1,3} &= y_{2,3}\\ \dot{y}_{2,3} &= -\frac{2K}{\Delta x^2}y_{1,3} + \frac{K}{\Delta x^2}y_{1,2} + f(y_{1,3}) + \frac{K}{\Delta x^2}y_{1,4}\end{aligned} \quad (6.242)$$

$\cdots\cdots$

$$\begin{aligned}\dot{y}_{1,i} &= y_{2,i}\\ \dot{y}_{2,i} &= -\frac{2K}{\Delta x^2}y_{1,i} + \frac{K}{\Delta x^2}y_{1,i-1} + f(y_{1,i}) + \frac{K}{\Delta x^2}y_{1,i+1}\end{aligned} \quad (6.243)$$

$\cdots\cdots$

$$\begin{aligned}\dot{y}_{1,N-1} &= y_{2,N-1}\\ \dot{y}_{2,N-1} &= -\frac{2K}{\Delta x^2}y_{1,N-1} + \frac{K}{\Delta x^2}y_{1,N-2} + f(y_{1,N-1}) +\\ &+\frac{K}{\Delta x^2}y_{1,N}\end{aligned} \quad (6.244)$$

$$\begin{aligned}\dot{y}_{1,N} &= y_{2,N}\\ \dot{y}_{2,N} &= -\frac{2K}{\Delta x^2}y_{1,N} + \frac{K}{\Delta x^2}y_{1,N-1} + f(y_{1,N}) + \frac{K}{\Delta x^2}\phi_{N+1}\end{aligned} \quad (6.245)$$

The flat output is considered to be the state variable $y_{1,1}$, which is denoted as $y = y_{1,1}$. Next, it is shown that all state variables $\alpha_i = y_{1,i+1}$ which stand also for virtual control inputs of the system, can be written as functions of the flat output $y = y_{1,1}$.

From (6.443) and considering that ϕ_0 is constant one obtains

$$\begin{aligned}y_{i,2} = \alpha_1 &= \frac{1}{K/\Delta x^2}\left[\ddot{y}_{1,1} + \frac{2K}{\Delta x^2}y_{1,1} - \frac{K}{\Delta x^2}\phi_0 - f(y_{1,1})\right]\\ &\Rightarrow y_{1,2} = h_1(y, \dot{y}, \ddot{y}, \ldots)\end{aligned} \quad (6.246)$$

From (6.444) one obtains

$$\begin{aligned}y_{1,3} = \alpha_2 &= \frac{1}{K/\Delta x^2}\left[\ddot{y}_{1,2} + \frac{2K}{\Delta x^2}y_{1,2} - \frac{K}{\Delta x^2}y_{1,1} - f(y_{1,2})\right]\\ &\Rightarrow y_{1,3} = h_2(y, \dot{y}, \ddot{y}, \ldots)\end{aligned} \quad (6.247)$$

From (6.445) one obtains

$$y_{1,4} = \alpha_3 = \frac{1}{K/\Delta x^2}\left[\ddot{y}_{1,3} + \frac{2K}{\Delta x^2}y_{1,3} - \frac{K}{\Delta x^2}y_{1,2} - f(y_{1,3})\right] \\ \Rightarrow y_{1,4} = h_3(y, \dot{y}, \ddot{y}, \ldots) \tag{6.248}$$

Continuing in a similar manner, from (6.446) one obtains

$$y_{1,i+1} = \alpha_i = \frac{1}{K/\Delta x^2}[\ddot{y}_{1,i} + \frac{2K}{\Delta x^2}y_{1,i} - \frac{K}{\Delta x^2}y_{1,i-1} - f(y_{1,i})] \\ \Rightarrow y_{1,i+1} = h_i(y, \dot{y}, \ddot{y}, \ldots) \tag{6.249}$$

and following a similar procedure, from (6.447) one gets

$$y_{1,N} = \alpha_{N-1} = \frac{1}{K/\Delta x^2}\left[\ddot{y}_{1,N-1} + \frac{2K}{\Delta x^2}y_{1,N-1} \right. \\ \left. - \frac{K}{\Delta x^2}y_{1,N-2} - f(y_{1,N-1})\right] \Rightarrow y_{1,N} = h_{N-1}(y, \dot{y}, \ddot{y}, \ldots) \tag{6.250}$$

Finally, from (6.448) one gets

$$\phi_{N+1} = \alpha_{N-1} = \frac{1}{K/\Delta x^2}\left[\ddot{y}_{1,N} + \frac{2K}{\Delta x^2}y_{1,N} - \frac{K}{\Delta x^2}y_{1,N-1} - f(y_{1,N})\right] \\ \Rightarrow \phi_{1,N+1} = h_N(y, \dot{y}, \ddot{y}, \ldots) \tag{6.251}$$

The above procedure confirms that all state variables of the model

$$\begin{array}{llll} y_{1,1}, y_{2,1} & y_{1,2}, y_{2,2} & y_{1,3}, y_{2,3} & \cdots \\ y_{1,i}, y_{2,i} & \cdots & y_{1,N-1}, y_{2,N-1} & y_{1,N}, y_{2,N} \end{array} \tag{6.252}$$

and the control input which is the boundary condition ϕ_{N+1} can be written as functions of the flat output $y = y_{1,1}$ and of the flat output's derivatives. Consequently, the state-space model of the PDE dynamics is differentially flat.

Additionally, one can consider decomposition of the PDE state-space equation into submodels, where at each submodel the virtual control input is $\alpha_i = y_{1,i+1}$

$$\dot{y}_{1,i} = y_{2,i} \\ \dot{y}_{2,i} = -\frac{2K}{\Delta x^2}y_{1,i} + \frac{K}{\Delta x^2}y_{1,i-1} + f(y_{1,i}) + \frac{K}{\Delta x^2}y_{1,i+1} \\ \cdots\ \cdots \tag{6.253}$$

and with local flat output $y_{1,i}$. Thus one can confirm that all such subsystems $i = 1, 2, \ldots, N$ are differentially flat. Next, one can compute the virtual inputs which are applied to each subsystem.

6.5.4 Computation of a boundary conditions-based feedback control law

To implement boundary feedback control in the system of (6.443) to (6.448), the nonlinear wave-PDE model is rewritten as follows:

$$\ddot{y}_{1,1} = -\frac{2K}{\Delta x^2}y_{1,1} + \frac{K}{\Delta x^2}\phi_0 + f(y_{1,1}) + \frac{K}{\Delta x^2}y_{1,2} \tag{6.254}$$

$$\ddot{y}_{1,2} = -\frac{2K}{\Delta x^2}y_{1,2} + \frac{K}{\Delta x^2}y_{1,1} + f(y_{1,2}) + \frac{K}{\Delta x^2}y_{1,3} \tag{6.255}$$

$$\ddot{y}_{1,3} = -\frac{2K}{\Delta x^2}y_{1,3} + \frac{K}{\Delta x^2}y_{1,2} + f(y_{1,3}) + \frac{K}{\Delta x^2}y_{1,4} \tag{6.256}$$

$\cdots\ \cdots$

$$\ddot{y}_{1,i} = -\frac{2K}{\Delta x^2}y_{1,i} + \frac{K}{\Delta x^2}y_{1,i-1} + f(y_{1,i}) + \frac{K}{\Delta x^2}y_{1,i+1} \tag{6.257}$$

$\cdots\ \cdots$

$$\ddot{y}_{1,N-1} = -\frac{2K}{\Delta x^2}y_{1,N-1} + \frac{K}{\Delta x^2}y_{1,N-2} + f(y_{1,N-1}) + \frac{K}{\Delta x^2}y_{1,N} \tag{6.258}$$

$$\ddot{y}_{1,N} = -\frac{2K}{\Delta x^2}y_{1,N} + \frac{K}{\Delta x^2}y_{1,N-1} + f(y_{1,N}) + \frac{K}{\Delta x^2}\phi_{N+1} \tag{6.259}$$

The boundary condition ϕ_0 is assumed to be known and remains steady. The boundary condition ϕ_{N+1} stands for the control input. The feedback control law is designed as follows:

$$\begin{aligned}\alpha_1 = y_{1,2}^* = \frac{1}{(K/\Delta x^2)}\Big[&\ddot{y}_{1,1}^d - k_{d,1}(\dot{y}_{1,1} - \dot{y}_{1,1}^d)\\ &- k_{p,1}(y_{1,1} - y_{1,1}^d) + \frac{2K}{\Delta x^2}y_{1,1} - \frac{K}{\Delta x^2}\phi_0 - f(y_{1,1})\Big]\end{aligned} \tag{6.260}$$

$$\begin{aligned}\alpha_2 = y_{1,3}^* = \frac{1}{(K/\Delta x^2)}\Big[&\ddot{y}_{1,2}^d - k_{d,2}(\dot{y}_{1,2} - \dot{y}_{1,2}^d)\\ &-k_{p,2}(y_{1,2} - y_{1,2}^d) + \frac{2K}{\Delta x^2}y_{1,2} - \frac{K}{\Delta x^2}y_{1,1} - f(y_{1,2})\Big] \Rightarrow\\ \alpha_2 = \frac{1}{(K/\Delta x^2)}\Big[&\ddot{\alpha}_1 - k_{d,2}(\dot{y}_{1,2} - \dot{\alpha}_1)\\ &-k_{p,2}(y_{1,2} - \alpha_1) + \frac{2K}{\Delta x^2}y_{1,2} - \frac{K}{\Delta x^2}y_{1,1} - f(y_{1,2})\Big]\end{aligned} \tag{6.261}$$

$$\begin{aligned}\alpha_3 = y_{1,4}^* = \frac{1}{(K/\Delta x^2)}\Big[&\ddot{y}_{1,3}^d - k_{d,3}(\dot{y}_{1,3} - \dot{y}_{1,3}^d)\\ &- k_{p,3}(y_{1,3} - y_{1,3}^d) + \frac{2K}{\Delta x^2}y_{1,3} - \frac{K}{\Delta x^2}y_{1,2} - f(y_{1,3})\Big] \Rightarrow\\ \alpha_3 = \frac{1}{(K/\Delta x^2)}\Big[&\ddot{\alpha}_2 - k_{d,3}(\dot{y}_{1,3} - \dot{\alpha}_2)\\ &- k_{p,3}(y_{1,3} - \alpha_2) + \frac{2K}{\Delta x^2}y_{1,3} - \frac{K}{\Delta x^2}y_{1,2} - f(y_{1,3})\Big]\end{aligned} \tag{6.262}$$

and continuing in a similar manner

$$\begin{aligned}\alpha_i = y_{1,i+1}^* = & \frac{1}{(K/\Delta x^2)}\Bigg[\ddot{y}_{1,i}^d - k_{d,i}(\dot{y}_{1,i} - \dot{y}_{1,i}^d) \\ & -k_{p,i}(y_{1,i} - y_{1,i}^d) + \frac{2K}{\Delta x^2}y_{1,i} - \frac{K}{\Delta x^2}y_{1,i-1} - f(y_{1,i})\Bigg] \Rightarrow \\ \alpha_i = & \frac{1}{(K/\Delta x^2)}\Bigg[\ddot{\alpha}_{i-1} - k_{d,i}(\dot{y}_{1,i} - \dot{\alpha}_{i-1}) \\ & - k_{p,i}(y_{1,i} - \alpha_{i-1}) + \frac{2K}{\Delta x^2}y_{1,i} - \frac{K}{\Delta x^2}y_{1,i-1} - f(y_{1,i})\Bigg]\end{aligned} \tag{6.263}$$

Following this procedure one arrives to compute the control inputs which are associated with the last two rows of the state-space model

$$\begin{aligned}\alpha_{N-1} = y_{1,N}^* = & \frac{1}{(K/\Delta x^2)}\Bigg[\ddot{y}_{1,N-1}^d - k_{d,N-1}(\dot{y}_{1,N-1} - \dot{y}_{1,N-1}^d) \\ & - k_{p,N-1}(y_{1,N-1} - y_{1,N-1}^d) + \frac{2K}{\Delta x^2}y_{1,N-1} \\ & - \frac{K}{\Delta x^2}y_{1,N-2} - f(y_{1,N-1})\Bigg] \Rightarrow \\ \alpha_{N-1} = & \frac{1}{(K/\Delta x^2)}\Bigg[\ddot{\alpha}_{N-2} - k_{d,N-1}(\dot{y}_{1,N-1} - \dot{\alpha}_{N-2}) - k_{p,N-1}(y_{1,N-1} - \alpha_{N-2}) \\ & + \frac{2K}{\Delta x^2}y_{1,N-1} - \frac{K}{\Delta x^2}y_{1,N-2} - f(y_{1,N-1})\Bigg]\end{aligned} \tag{6.264}$$

$$\begin{aligned}\phi_{N+1} = & \frac{1}{(K/\Delta x^2)}\Bigg[\ddot{y}_{1,N}^d - k_{d,N}(\dot{y}_{1,N} - \dot{y}_{1,N}^d) \\ & - k_{p,N}(y_{1,N} - y_{1,N}^d) + \frac{2K}{\Delta x^2}y_{1,N} - \frac{K}{\Delta x^2}y_{1,N-1} - f(y_{1,N})\Bigg] \\ \Rightarrow \alpha_N = & \frac{1}{(K/\Delta x^2)}\Bigg[\ddot{\alpha}_{N-1} - k_{d,N}(\dot{y}_{1,N} - \dot{\alpha}_{N-1}) \\ & - k_{p,N}(y_{1,N} - \alpha_{N-1}) + \frac{2K}{\Delta x^2}y_{1,N} - \frac{K}{\Delta x^2}y_{1,N-1} - f(y_{1,N})\Bigg]\end{aligned} \tag{6.265}$$

Consequently, the computation of the aggregate control input ϕ_{N+1} which is applied the PDE model is performed backwards, by substituting recursively into ϕ_{N+1} the virtual control inputs a_{N-1}, $a_{N-2}, \ldots, a_i s, \ldots, a_2, a_1$. It is noted that the concept of flatness-based control implementation in cascading-loops has been already used in lumped parameter systems and the above results show that it can also be applied effectively to distributed parameter systems (Rigatos and Siano, 2012a; Rigatos and Siano, 2012b).

6.5.5 Closed-loop dynamics

By substituting (6.464) into (6.259) of the state-space model of the PDE dynamics, and using the definition $y_{1,N} - \alpha_{N-1} = z_N$ one has

$$\ddot{y}_{1,N} = \ddot{\alpha}_{N-1} - k_{d,N}(\dot{y}_N - \dot{\alpha}_{N-1}) - k_{p,N}(y_{1,N} - \alpha_{N-1}) \Rightarrow (\ddot{y}_{1,N} - \ddot{\alpha}_{N-1}) + k_{d,N}(\dot{y}_{1,N} - \dot{\alpha}_{N-1}) + k_{p,N}(y_{1,N} - \alpha_{N-1}) = 0 \Rightarrow \ddot{z}_N + k_{d,N}\dot{z}_N + k_{p,N}z_N = 0$$

By substituting (6.465) into (6.258) of the state-space model of the PDE dynamics, and using the definition $y_{1,N-1} - \alpha_{N-2} = z_{N-1}$ one has

$$\ddot{y}_{1,N-1} = \ddot{\alpha}_{N-2} - k_{d,N-1}(\dot{y}_{N-1} - \dot{\alpha}_{N-2}) - k_{p,N-1}(y_{1,N-1} - \alpha_{N-2}) \Rightarrow (\ddot{y}_{1,N-1} - \ddot{\alpha}_{N-2}) + k_{d,N-1}(\dot{y}_{1,N-1} - \dot{\alpha}_{N-2}) + k_{p,N-1}(y_{1,N-1} - \alpha_{N-2}) = 0 \Rightarrow \ddot{z}_{N-1} + k_{d,N-1}\dot{z}_{N-1} + k_{p,N-1}z_{N-1} = 0$$

Moving backwards, and by substituting (6.467) into (6.257) of the state-space model of the PDE dynamics, and using the definition $y_{1,i} - \alpha_{i-1} = z_i$ one has

$$\ddot{y}_{1,i} = \ddot{\alpha}_{i-1} - k_{d,i}(\dot{y}_i - \dot{\alpha}_{i-1}) - k_{p,i}(y_{1,i} - \alpha_{i-1}) \Rightarrow (\ddot{y}_{1,i} - \ddot{\alpha}_{i-1}) + k_{d,i}(\dot{y}_{1,i} - \dot{\alpha}_{i-1}) + k_{p,i}(y_{1,i} - \alpha_{i-1}) = 0 \Rightarrow \ddot{z}_i + k_{d,i}\dot{z}_i + k_{p,i}z_i = 0$$

Similarly, continuing with the rest of the equations of the state-space model and by substituting (6.468) into (6.256), while also using the definition $y_{1,3} - \alpha_2 = z_3$ one has

$$\ddot{y}_{1,3} = \ddot{\alpha}_2 - k_{d,3}(\dot{y}_3 - \dot{\alpha}_2) - k_{p,3}(y_{1,3} - \alpha_2) \Rightarrow (\ddot{y}_{1,3} - \ddot{\alpha}_2) + k_{d,3}(\dot{y}_3 - \dot{\alpha}_2) + k_{p,3}(y_{1,3} - \alpha_2) = 0 \Rightarrow \ddot{z}_3 + k_{d,3}\dot{z}_3 + k_{p,3}z_3 = 0$$

Equivalently, by substituting (6.469) into (6.255), and using the definition $y_{1,2} - \alpha_1 = z_2$ one has

$$\ddot{y}_{1,2} = \ddot{\alpha}_1 - k_{d,2}(\dot{y}_2 - \dot{\alpha}_1) - k_{p,2}(y_{1,2} - \alpha_1) \Rightarrow (\ddot{y}_{1,2} - \ddot{\alpha}_1) + k_{d,2}(\dot{y}_2 - \dot{\alpha}_1) + k_{p,2}(y_{1,2} - \alpha_1) = 0 \Rightarrow \ddot{z}_2 + k_{d,2}\dot{z}_2 + k_{p,2}z_2 = 0$$

Finally, by substituting (6.470) into (6.254), and using the definition $y_{1,1} - y_{1,1}^d = z_1$ one obtains

$$\ddot{y}_{1,1} = \ddot{y}_{1,1}^d - k_{d,1}(\dot{y}_{1,1} - \dot{y}_{1,1}^d) - k_{p,1}(y_{1,1} - y_{1,1}^d) \Rightarrow (\ddot{y}_{1,1} - \ddot{y}_{1,1}) + k_{d,1}(\dot{y}_{1,1} - \dot{y}_{1,1}^d) + k_{p,1}(y_{1,1} - y_{1,1}^d) = 0 \Rightarrow \ddot{z}_1 + k_{d,1}\dot{z}_1 + k_{p,1}z_1 = 0$$

Thus, the dynamics of the closed-loop system becomes

$$\begin{aligned}
&\ddot{z}_1 + k_{d,1}\dot{z}_1 + k_{p,1}z_1 = 0\\
&\ddot{z}_2 + k_{d,2}\dot{z}_2 + k_{p,2}z_2 = 0\\
&\ddot{z}_3 + k_{d,3}\dot{z}_3 + k_{p,3}z_3 = 0\\
&\cdots\ \cdots
\end{aligned}$$

$$\begin{aligned} &\ddot{z}_i + k_{d,i}\dot{z}_i + k_{p,i}z_i = 0 \\ &\cdots\ \cdots \\ &\ddot{z}_{N-1} + k_{d,N-1}\dot{z}_{N-1} + k_{p,N-1}z_{N-1} = 0 \\ &\ddot{z}_N + k_{d,N}\dot{z}_N + k_{p,N}z_N = 0 \end{aligned} \quad (6.266)$$

The dynamics of the closed-loop system can be also written in matrix form

$$\ddot{\tilde{Z}} + K_D\dot{\tilde{Z}} + K_P\tilde{Z} = 0 \quad (6.267)$$

where $\tilde{Z} = [z_1, z_2, z_3, \ldots, z_i, \ldots, z_{N-1}, z_N]^T$,
$K_D = \text{diag}[k_{d,1}, k_{d,2}, k_{d,3}, \ldots, k_{d,i}, \ldots, k_{d,N-1}, k_{d,N}]$ and
$K_P = \text{diag}[k_{p,1}, k_{p,2}, k_{p,3}, \ldots, k_{p,i}, \ldots, k_{p,N-1}, k_{p,N}]$.

After suitable selection of the coefficients $k_{d,i}$, $k_{P,i}$, $i = 1, 2, \ldots, N$ such that the polynomials $p(s) = s^2 + k_{d,i}s + k_{p,i}$ to be Hurwitz ones, it can be assured that $\lim_{t\to\infty} z_i(t) = 0 \ \forall\ i = 1, \ldots, N$ and that the closed-loop system is asymptotically stable.

Moreover, to prove asympotic stability for the closed-loop system, the following Lyapunov function can be used

$$V = \sum_{i=1}^{N} \frac{k_{p,i}}{2} z_i^2 + \sum_{i=1}^{N} \frac{1}{2}\dot{z}_i^2 \quad (6.268)$$

The derivative of this Lyapunov function with respect to time is

$$\begin{aligned} &\dot{V} = \sum_{i=1}^{N} k_{p,i}z_i\dot{z}_i + \sum_{i=1}^{N} \frac{1}{2}\dot{z}_i\ddot{z}_i \Rightarrow \\ &\dot{V} = \sum_{i=1}^{N} k_{p,i}z_i\dot{z}_i + \sum_{i=1}^{N} \dot{z}_i(-k_{d,i}\dot{z}_i - k_{p_i}z_i) \Rightarrow \\ &\dot{V} = \sum_{i=1}^{N} k_{p,i}z_i\dot{z}_i - \sum_{i=1}^{N} k_{d,i}\dot{z}_i^2 - \sum_{i=1}^{N} k_{p,i}z_i\dot{z}_i \Rightarrow \\ &\dot{V} = -\sum_{i=1}^{N} k_{d,i}\dot{z}_i^2 < 0 \end{aligned} \quad (6.269)$$

Thus, it is proven again that the closed-loop system is globally asymptotically stable.

6.5.6 Simulation tests

Simulation examples about the proposed control method for distributed parameter systems are provided for the case of the nonlinear wave PDE that was given in (6.423). The obtained results are depicted in Figures 6.45–6.48. The spatial discretisation of the PDE model consisted of $N = 25$ points. The sampling period was chosen to be $T_s = 0.01$sec. The boundary condition ϕ_0 of the PDE was taken to be known and constant. The boundary condition ϕ_{N+1} served as the control input and was computed as each iteration of the control algorithm according to the procedure described in Section 6.7.5.

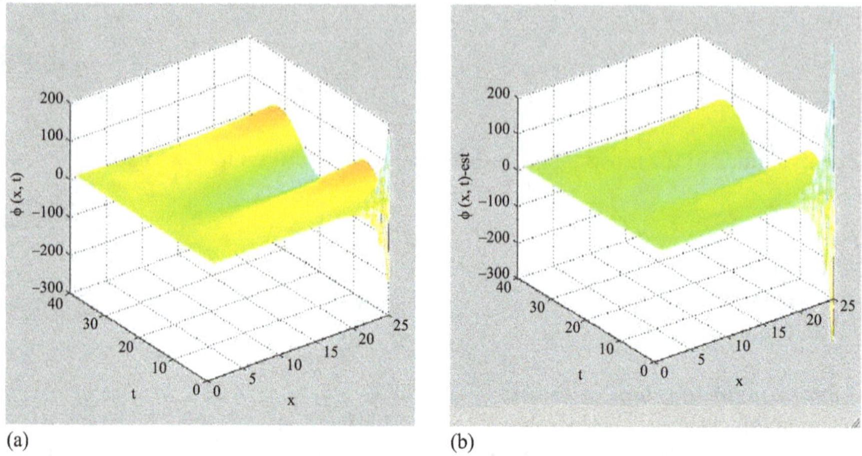

Figure 6.44 (a) Plot of the 2D-wave function after applying boundary control and (b) plot of the reference 2D-wave function

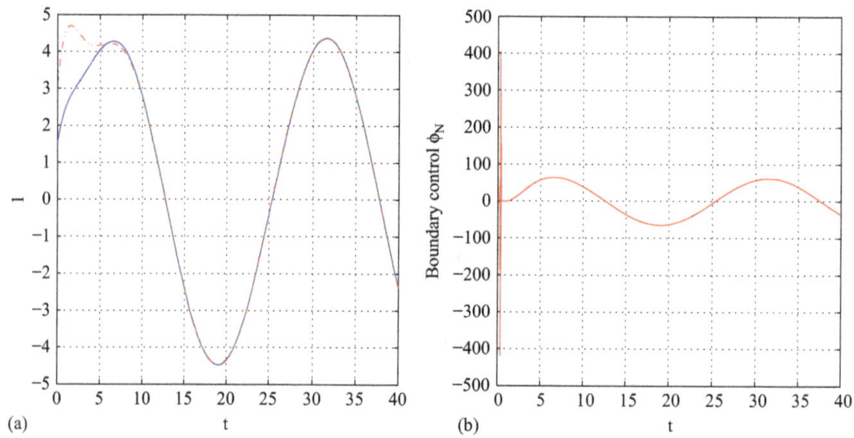

Figure 6.45 (a) Variation in time of the solution of the PDE at grid point x_1 (blue line) and associated reference setpoint (red line) and (b) control input exerted to the PDE system through the boundary condition ϕ_{N+1}

The numerical simulation experiments have confirmed the theoretical findings of this section. It has been shown that by applying the proposed control method, the nonlinear wave PDE dynamics can be modified so as to converge to the desirable reference profile. The control input that succeeds this, changes smoothly and has a moderate range of variation. The accuracy of tracking of the reference setpoints was

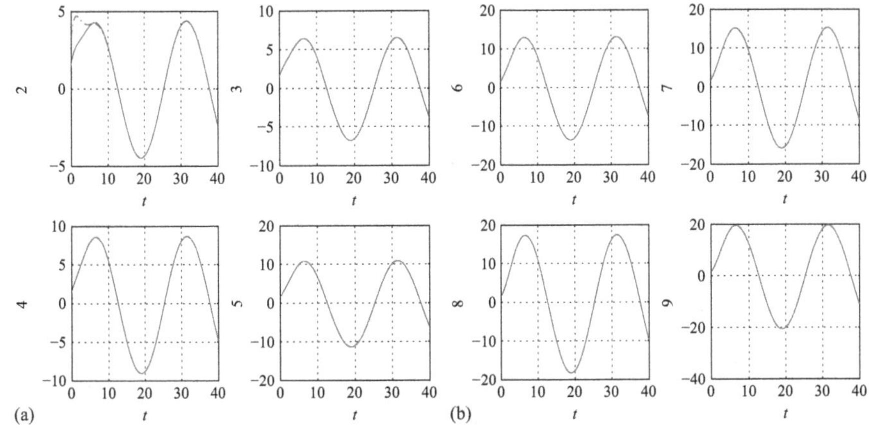

Figure 6.46 (a) Variation in time of the solution of the PDE at grid points x_2 to x_5 (blue line) and associated reference setpoint (red line) and (b) variation in time of the solution of the PDE at grid points x_6 to x_9 (blue line) and associated reference setpoint (red line)

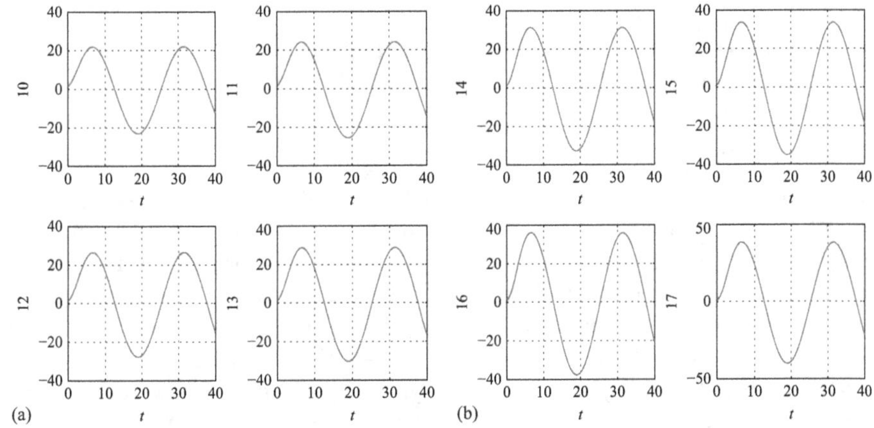

Figure 6.47 (a) Variation in time of the solution of the PDE at grid points x_{10} to x_{13} (blue line) and associated reference setpoint (red line) and (b) variation in time of the solution of the PDE at grid points x_{14} to x_{17} (blue line) and associated reference setpoint (red line)

indeed satisfactory. It is noted that the application of the proposed nonlinear PDE feedback control method is not limited to the case of nonlinear wave dynamics but can be also extended to several other types of PDEs of the parabolic, hyperbolic and elliptic type and to various types of nonlinear PDEs.

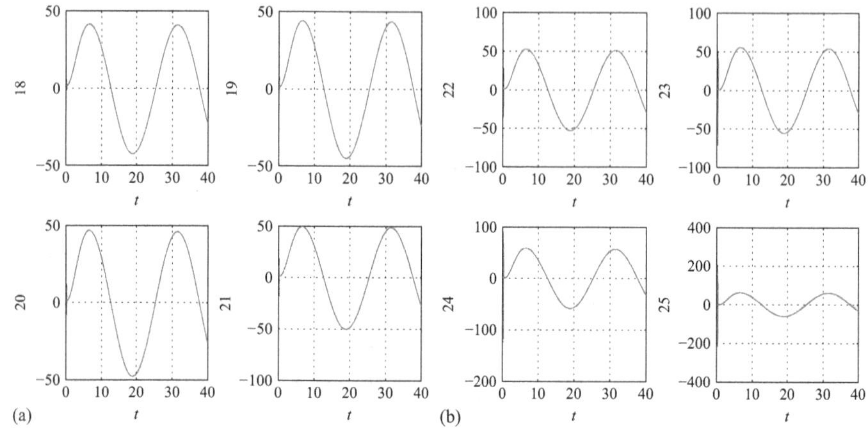

Figure 6.48 (a) Variation in time of the solution of the PDE at grid points x_{18} to x_{21} (blue line) and associated reference setpoint (red line) and (b) variation in time of the solution of the PDE at grid points x_{22} to x_{25} (blue line) and associated reference setpoint (red line)

6.6 Control of data-flow PDE for bandwidth allocation in internet routes

6.6.1 Outline

This section proposes a PDE control method based on differential flatness theory that is addressed to the problem of bandwidth allocation for internet routes. The provided results give also a solution to the generic congestion control problem for communication networks. The problem of bandwidth allocation and congestion control in communication networks has been analysed in (Paganini, Tang *et al.*, 2012; Ferragut and Paganini, 2016; Paganini, Zubellin *et al.*, 2015). Partial Differential Equation (PDE) models describing spatiotemporal dynamics of data flow in communication networks can be found in (Leung *et al.*, 1994; Paganini and Ferragut, 2012; Ule and Boucherie, 2003; Kelly and Williams, 2004). Moreover, indicative results on the problem of congestion control in communication networks have been given in (Paganini and Ferragut, 2014a; Paganini and Ferragut, 2014b; Ben-Fredj *et al.*, 2001). Applications of congestion control to internet connections management can be found in (Massey and White, 1994; Paganini, Ferragut *et al.*, 2013; Ferragut and Paganini, 2012).

In this section, it is proven first that the data flow along internet routes has spatiotemporal dynamics. Actually, it is shown that the number of tasks exceeding in data size a threshold related to the servicing capability of the route and with the allocated bandwidth for this route follows a partial differential equation. The first dimension of this PDE is time while the second dimension of this PDE is the aforementioned threshold parameter. The question that arises next is if such a PDE can be controlled by an external input in the form of a boundary condition. This signifies the controlled bandwidth allocation for each route of the internet connections and the ability to control congestion at these routes after making the size of unprocessed data converge to desirable levels (setpoints).

By applying semi-discretisation and the finite differences method, the PDE describing the data-flow is decomposed into a set of ODEs and thus a state-space description for the data flow is obtained. It is proven that this state-space model satisfies differential flatness properties, with its first state variable to stand for the flat-output. The properties of differentially flat systems have been analysed in (Rudolph, 2003; Sira-Ramirez and Agrawal, 2014; Fliess and Mounier, 1999; Rigatos, 2013; Rigatos, 2016b). Transformation of differentially flat dynamical systems into the canonical form has been demonstrated in (Levine, 2011; Rouchon, 2005; Laroche *et al.*, 2007; Boouden *et al.*, 2011). Modelling and control methods for distributed parameter systems have been analysed in (Salberg *et al.*, 2010; Chauvin, 2011; Pinsky, 1991; Rigatos, 2016a). Flatness-based control for PDE systems has been given in (Woitteneck and Mounier, 2010; Fliess and Mounier, 1999; Boussaada *et al.*, 2013; Rouchon, Martin *et al.*, 2014). Flatness-based control of PDE systems through boundary conditions has been demonstrated in (Rigatos, 2015; Rigatos, Siano and Melkikh, 2016; Mounier, Rudolph *et al.*, 2010).

It is shown that the ith row of the aforementioned state-space model is a differentially flat subsystem with flat out the associated state variable i and virtual control input state variable $i+1$. By considering the differential flatness properties of these subsystems it is possible to design stabilising feedback controllers based on the previous virtual control inputs and this procedure is followed for all rows of the state-space model $i = 1, 2, \ldots, N-1$. Finally, from the last row of the state-space description, one can compute in the form of a boundary condition the control input that should be actually applied to the PDE model of the data flow so as to control routes' congestion.

The real control input that is applied to the PDE model contains recursively all virtual control inputs associated with the previous $N-1$ rows of the system. The asymptotic stability properties of the control scheme are proven by confirming analytically that the tracking error for each state variable of the model will converge to zero. Moreover, the asymptotic stability features of the control loop are confirmed through Lyapunov analysis. The significance of the proposed boundary control method for the data flow PDE is as follows: the boundary condition is determined by the maximum value of the threshold variable designating the capability of the route to service tasks and files up to a certain size. By changing this upper bound one can also control congestion levels in the route and actually one can achieve at every time instant that the number of the unprocessed tasks (files) follows closely desirable reference values.

Unlike the section's control approach which is of proven global stability, a large part of research in the areas of bandwidth allocation and congestion control for communication networks arrives at empirical results, without any stability guarantee.Actually, for bandwidth allocation in the routes of communication networks one can distinguish approaches (i) making use of simplified ODE models about the data flow (Beggia *et al.*, 2005; Greyb and Orlawska, 2015; Alpcan *et al.*, 2005; Yaeihu, Mazumder *et al.*, 2009), (ii) making use of model-free heuristic control methods (Silva *et al.*, 2014; Ryu, Rumpand *et al.*, 2003; Lee and Hou, 2000; Divakaran *et al.*, 2014), (iii) making use of PDE control methods. The present section has a novel and substantial contribution in the research area of control of the data flow and congestion management in communication networks. Both the proposed PDE control method and the provided stability analysis are genuine findings. Although the concepts of semi-discretisation and the finite differences method have been previously used for describing a PDE through an equivalent model of ODEs, the proposed control method for boundary control of the bandwidth allocation PDE through successive loops stands for a genuine research result.

6.6.2 PDE of the internet flow per route

The PDE of the data flow per route (Figure 6.49) is given by (Paganini, Tang *et al.*, 2012; Ferragut and Paganini, 2016; Paganini, Zubellin *et al.*, 2015)

$$\frac{\partial F_m(t,\sigma)}{\partial t} = \frac{\partial F_m(t,\sigma)}{\partial \sigma}\frac{\phi_m(t)}{Z_m(t)} + \lambda_m G_m(\sigma) \tag{6.270}$$

where $F_m(t,\sigma)$, $t \geq 0$ and $\sigma \geq 0$ is the number of the mth route files at time t with residual size larger than σ. $F_m(t,\sigma)$ is a finer descriptor of the system than $Z_m(t)$, where actually it holds that $Z_m(t) = F_m(t,0)$. Moreover, $G_m(\sigma) = e^{-\sigma t}$ is an exponential size distribution about incoming to the route files that are of size larger than σ and λ_m is the files arrival rate.

A justification for the previous PDE about a data flow along the internet routes is provided next Paganini, Tang *et al.*, 2012; Ferragut and Paganini, 2016; Paganini, Zubellin *et al.*, 2015). At time instant $t + \Delta t$ tasks that have residual file size at least σ come from two sources:

(i) data that arrive in the time interval between t and $t + \Delta t$, having residual file size that exceeds σ. Their arrival rate is defined as λ_m, thus the aggregate volume of data associated with these arrivals is $\lambda_m dt G_m(\sigma)$.

(ii) files already existing in the queue at time instant t, having a residual file size given by $\sigma + \frac{\phi_m(t)}{Z_m(t)}dt$. The service rate for data in queue is $\frac{\phi_m(t)}{Z_m(t)}$.

It holds that (Paganini, Tang *et al.*, 2012; Ferragut and Paganini, 2016; Paganini, Zubellin *et al.*, 2015)

$$F_m(t+dt,\sigma) = F_m(t,\sigma + \frac{\phi_m(t)}{Z_m(t)}dt) + \lambda_m G_m(\sigma)dt \tag{6.271}$$

After subtracting $F_m(\sigma,t)$ from both sides of (6.271) and by dividing with dt one obtains

$$\frac{F_m(t+dt,\sigma) - F_m(t,\sigma)}{dt} = \frac{F_m(t,\sigma + \frac{\phi_m(t)}{Z_m(t)}dt) - F_m(t,\sigma)}{dt} + \lambda_m G_m(\sigma) \tag{6.272}$$

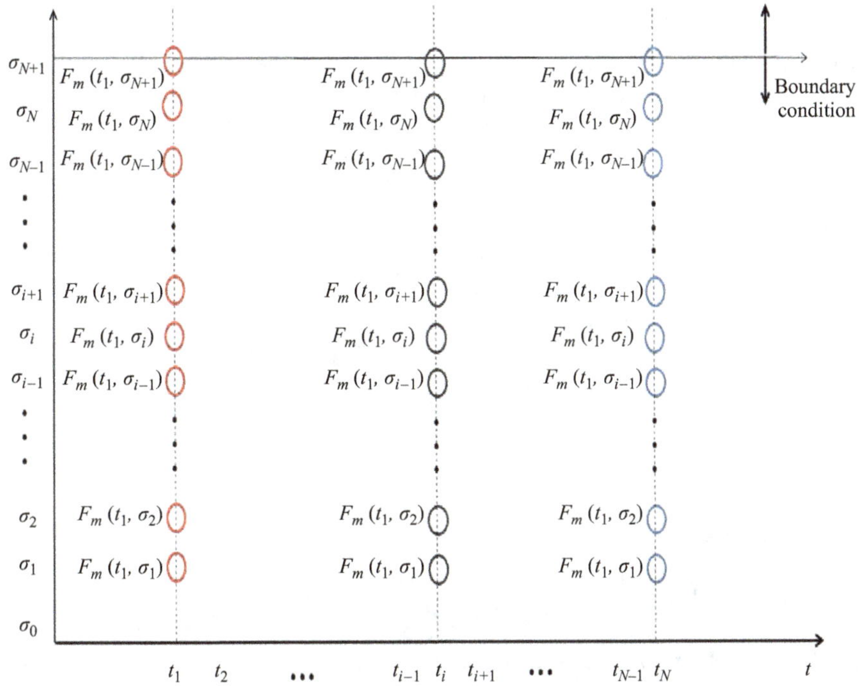

Figure 6.49 Spatiotemporal dynamics of data flow in the communication network routes: $F_m(t, \sigma)$ is the number of tasks (files) exceeding threshold σ at time instant t that cannot be serviced by the route

By denoting $\Delta\sigma = \frac{\phi_m(t)}{Z_m(t)}dt$, one has about the fraction appearing in the right part of (6.272) (Paganini, Tang *et al.*, 2012; Ferragut and Paganini, 2016; Paganini, Zubellin *et al.*, 2015)

$$\frac{\partial F_m(t, \sigma + d\sigma)}{\partial t} = \frac{\partial F_m(t, \sigma + \Delta\sigma)}{\partial \sigma} \cdot \frac{\partial \sigma}{dt} \tag{6.273}$$

and by applying Taylor series expansion one arrives at the initial PDE about the flow of data in internet routes, as given in (6.270)

$$\frac{\partial F_m(t, \sigma)}{\partial t} = \frac{\partial F_m(t, \sigma)}{\partial \sigma} \frac{\phi_m(t)}{Z_m(t)} + \lambda_m G_m(\sigma) \tag{6.274}$$

6.6.3 Data flow model

The following grid-partitioning along the σ axis in the previous PDE model is used: $[\sigma_0, \sigma_1, \sigma_2, \ldots, \sigma_{N-1}, \sigma_N, \sigma_{N+1}]$. By applying semi-discretisation and the finite differences method one obtains for the grid points σ_i, $i = 1, 2 \ldots, N$ (Rigatos, 2015)

$$\frac{\partial F_m(t,\sigma_i)}{\partial t} = \frac{F_m(t,\sigma_{i+1}) - F_m(t,\sigma_i)}{\Delta\sigma}\frac{\phi_m(t)}{Z_m(t)} + \lambda_m G_m(\sigma_i) \tag{6.275}$$

Thus, at the aforementioned grid points, the partial differential equation of the data flow is decomposed into a set of ordinary differential equations of the form

$$\begin{aligned}
\dot{F}_m(t,\sigma_1) &= \frac{F_m(t,\sigma_2) - F_m(t,\sigma_1)}{\Delta\sigma}\frac{\phi_m(t)}{Z_m(t)} + \lambda_m G_m(\sigma_1) \\
\dot{F}_m(t,\sigma_2) &= \frac{F_m(t,\sigma_3) - F_m(t,\sigma_2)}{\Delta\sigma}\frac{\phi_m(t)}{Z_m(t)} + \lambda_m G_m(\sigma_2) \\
&\cdots\cdots\cdots \\
&\cdots\cdots\cdots \\
\dot{F}_m(t,\sigma_{N-1}) &= \frac{F_m(t,\sigma_N) - F_m(t,\sigma_{N-1})}{\Delta\sigma}\frac{\phi_m(t)}{Z_m(t)} + \lambda_m G_m(\sigma_{N-1}) \\
\dot{F}_m(t,\sigma_N) &= \frac{F_m(t,\sigma_{N+1}) - F_m(t,\sigma_N)}{\Delta\sigma}\frac{\phi_m(t)}{Z_m(t)} + \lambda_m G_m(\sigma_N)
\end{aligned} \tag{6.276}$$

Next, by denoting $q(t) = \frac{\phi_{mt}}{Z_m(t)}$ and also by using the notation $y_i = F_m(t,\sigma_i)$ one can define the state vector $Y = [y_1, y_2, \ldots, y_{N_1}, y_N]$. This results in the following state-space description:

$$\begin{aligned}
\dot{y}_1 &= \frac{y_2 - y_1}{\Delta\sigma}q(t) + \lambda_m G_m(\sigma_1) \\
\dot{y}_2 &= \frac{y_3 - y_2}{\Delta\sigma}q(t) + \lambda_m G_m(\sigma_2) \\
&\cdots\cdots \\
\dot{y}_{N-1} &= \frac{y_N - y_{N-1}}{\Delta\sigma}q(t) + \lambda_m G_m(\sigma_{N-1}) \\
\dot{y}_N &= \frac{y_{N+1} - y_N}{\Delta\sigma}q(t) + \lambda_m G_m(\sigma_N)
\end{aligned} \tag{6.277}$$

The previous state-space model is also written as

$$\begin{aligned}
\dot{y}_1 &= -\frac{y_1}{\Delta\sigma}q(t) + \lambda_m G_m(\sigma_1) + \frac{1}{\Delta\sigma}y_2 \\
\dot{y}_2 &= -\frac{y_2}{\Delta\sigma}q(t) + \lambda_m G_m(\sigma_2) + \frac{1}{\Delta\sigma}y_3 \\
&\cdots\cdots \\
\dot{y}_{N-1} &= -\frac{y_{N-1}}{\Delta\sigma}q(t) + \lambda_m G_m(\sigma_{N-1}) + \frac{1}{\Delta\sigma}y_N \\
\dot{y}_N &= -\frac{y_N}{\Delta\sigma}q(t) + \lambda_m G_m(\sigma_N) + \frac{1}{\Delta\sigma}y_{N+1}
\end{aligned} \tag{6.278}$$

The boundary condition $u = y_{N+1}$ is considered to be the control input to the data-flow model.

6.6.4 Differential flatness of the data flow model

First differential flatness is proven for the entire state-space model of the data flow PDE that has been described in (6.278). From the first row of (6.278) one has that

$$\begin{aligned} y_2 &= \Delta\sigma\left[\dot{y}_1 + \frac{y_1}{\Delta\sigma}q(t) - \lambda_m G_m(\sigma_1)\right] \\ &\Rightarrow y_2 = f_2(y_1, \dot{y}_1) \end{aligned} \tag{6.279}$$

From the second row of Eq, (6.278) one has that

$$\begin{aligned} y_3 &= \Delta\sigma\left[\dot{y}_2 + \frac{y_2}{\Delta\sigma}q(t) - \lambda_m G_m(\sigma_2)\right] \\ &\Rightarrow y_3 = f_3(y_2, \dot{y}_2) = f_3(y_1, \dot{y}_1, \ddot{y}_1) \end{aligned} \tag{6.280}$$

Continuing in the same manner, from the $N-1$th row of (6.278) one has that

$$\begin{aligned} y_N &= \Delta\sigma\left[\dot{y}_{N-1} + \frac{y_{N-1}}{\Delta\sigma}q(t) - \lambda_m G_m(\sigma_{N-1})\right] \\ &\Rightarrow y_n = f_N(y_{N-1}, \dot{y}_{N-1}) = f_N(y_1, \dot{y}_1, \ddot{y}_1, \ldots, y_1^{(N-1)}) \end{aligned} \tag{6.281}$$

Finally, from the Nth row of (6.278) one has that

$$\begin{aligned} u = y_{N+1} &= \Delta\sigma\left[\dot{y}_N + \frac{y_N}{\Delta\sigma}q(t) - \lambda_m G_m(\sigma_N)\right] \\ &\Rightarrow u = f_{N+1}(y_N, \dot{y}_N) = f_{N+1}(y_1, \dot{y}_1, \ddot{y}_1, \ldots, y_1^{(N)}) \end{aligned} \tag{6.282}$$

Consequently, all state variables and the control inputs of the data flow model can be written as differential functions of the flat output and its derivatives. Thus, the PDE data flow-model is a differentially flat one with flat output $y_1 = F_m(t, \sigma_1)$.

Besides, differential flatness can be also shown in a different manner, and by considering the ith row of (6.278) as an independent subsystem, where the output is y_i and the virtual control input of this subsystem is $v_i = y_{i+1}$ $i = 1, 2, \ldots, N$. By defining the vector of the virtual control inputs $[v_1, v_2, \ldots, v_N, u] = [y_2, y_3, \ldots, y_N]$, one has

$$\begin{aligned} \dot{y}_1 &= -\frac{y_1}{\Delta\sigma}q(t) + \lambda_m G_m(\sigma_1) + \frac{1}{\Delta\sigma}v_1 \\ \dot{y}_2 &= -\frac{y_2}{\Delta\sigma}q(t) + \lambda_m G_m(\sigma_2) + \frac{1}{\Delta\sigma}v_2 \\ &\cdots\cdots \\ \dot{y}_{N-1} &= -\frac{y_{N-1}}{\Delta\sigma}q(t) + \lambda_m G_m(\sigma_{N-1}) + \frac{1}{\Delta\sigma}v_{N-1} \\ \dot{y}_N &= -\frac{y_N}{\Delta\sigma}q(t) + \lambda_m G_m(\sigma_N) + \frac{1}{\Delta\sigma}u \end{aligned} \tag{6.283}$$

Taking that the first row of (6.283) describes an independent subsystem with flat output y_1 one has that the virtual control input v_1 is a differential function of y_1.

Similarly, taking that the second row of (6.283) describes an independent subsystem with flat output y_2 one has that the virtual control input v_2 is a differential function of y_2.

Continuing in the same manner and, taking that the $N-1$th row of (6.283) describes an independent subsystem with flat output y_{N-1} one has that the virtual control input v_{N-1} is a differential function of y_{N-1}.

Finally, by considering the last row of (6.283) as an independent subsystem, with flat output y_N, one has that u is a differential function of y_N.

6.6.5 *Flatness-based control for the data-flow model*

For the state-space description of the data-flow PDE that was given in (6.283) one can obtain a stabilising flatness-based controller. From the first row of (6.283) the stabilising virtual control input of the first subsystem is given by

$$v_1 = y_2^* = \Delta\sigma\left[\dot{y}_1^* + \frac{y_1}{\Delta\sigma}q(t) - \lambda_m G_m(\sigma_1) - K_{p_1}(y_1 - y_1^*)\right] \tag{6.284}$$

The virtual control input $v_1 = y_2^*$ becomes the reference setpoint for the subsystem described by the second row of (6.283). Thus, the stabilising virtual control input of the first subsystem is given by

$$v_2 = y_3^* = \Delta\sigma\left[\dot{y}_2^* + \frac{y_2}{\Delta\sigma}q(t) - \lambda_m G_m(\sigma_2) - K_{p_2}(y_2 - y_2^*)\right] \tag{6.285}$$

Continuing in a similar manner for the rest of the rows of the state-space model of (6.283), the virtual control input $v_{N-1} = y_N^*$ becomes the reference setpoint for the $N-1$ subsystem. Thus, the stabilising virtual control input of the $N-1$ subsystem is given by

$$v_{N-1} = y_N^* = \Delta\sigma\left[\dot{y}_{N-1}^* + \frac{y_{N-1}}{\Delta\sigma}q(t) - \lambda_m G_m(\sigma_{N-1}) - K_{p_{N-1}}(y_{N-1} - y_{N-1}^*)\right] \tag{6.286}$$

Finally, the real control input $u = y_{N+1}^*$ is applied to the Nth subsystem. Thus, the stabilising virtual control input of the Nth subsystem is given by

$$u = \Delta\sigma[\dot{y}_N^* + \frac{y_N}{\Delta\sigma}q(t) - \lambda_m G_m(\sigma_N) - K_{p_N}(y_N - y_N^*)] \tag{6.287}$$

6.6.6 *Stability analysis for the data-flow control loop*

6.6.6.1 Proof of elimination of the tracking error

By substituting the control input given in (6.287) into the Nth row of (6.283), and for $K_{p_N} > 0$ one obtains

$$\begin{aligned}&(\dot{y}_N - \dot{y}_N^*) + K_{p_N}(y_N - y_N^*) = 0 \Rightarrow \\ &\dot{e}_N + K_{p_N}e_N = 0 \Rightarrow \lim_{t\to\infty} e_N = 0\end{aligned} \tag{6.288}$$

By assuring that $y_N \to y_N^*$ and by substituting the control input given in (6.286) into the $N-1$th row of (6.283), and for $K_{p_{N-1}} > 0$ one obtains

$$\begin{aligned}&(\dot{y}_{N-1} - \dot{y}_{N-1}^*) + K_{p_{N-1}}(y_{N-1} - y_{N-1}^*) = 0 \Rightarrow \\ &\dot{e}_{N-1} + K_{p_{N-1}}e_{N-1} = 0 \Rightarrow \lim_{t\to\infty} e_{N-1} = 0\end{aligned} \tag{6.289}$$

Next, the rows of the state-space model of (6.283) are traced backwards. Following the previous procedure one arrives at the second row of the states-space model. By assuring that $y_3 \to y_3^*$ and by substituting the control input given in (6.285) into the second row of (6.283), and for $K_{p_2} > 0$ one obtains

$$
\begin{aligned}
&(\dot{y}_2 - \dot{y}_2^*) + K_{p_2}(y_2 - y_2^*) = 0 \Rightarrow \\
&\dot{e}_2 + K_{p_2} e_2 = 0 \Rightarrow \lim_{t\to\infty} e_2 = 0
\end{aligned}
\tag{6.290}
$$

Finally, by assuring that $y_2 \to y_2^*$ and by substituting the control input given in (6.284) into the first row of (6.283), and for $K_{p_1} > 0$ one obtains

$$
\begin{aligned}
&(\dot{y}_1 - \dot{y}_1^*) + K_{p_1}(y_1 - y_1^*) = 0 \Rightarrow \\
&\dot{e}_1 + K_{p_1} e_1 = 0 \Rightarrow \lim_{t\to\infty} e_1 = 0
\end{aligned}
\tag{6.291}
$$

According to the above, it is proven that the tracking error for all state variables of the data-flow model gets asymptotically eliminated. The tracking error dynamics of (6.291) to (6.288) can be also written in the matrix form

$$
\begin{pmatrix} \dot{e}_1 \\ \dot{e}_2 \\ \cdots \\ \cdots \\ \dot{e}_N \end{pmatrix} = \begin{pmatrix} -K_{p_1} & 0 & \cdots & 0 & 0 \\ 0 & -K_{p_2} & \cdots & 0 & 0 \\ \cdots & \cdots & \cdots & \cdots & \cdots \\ \cdots & \cdots & \cdots & \cdots & \cdots \\ 0 & 0 & \cdots & -K_{p_{N-1}} & 0 \\ 0 & 0 & \cdots & 0 & K_{p_N} \end{pmatrix} \begin{pmatrix} e_1 \\ e_2 \\ \cdots \\ \cdots \\ e_N \end{pmatrix}
\tag{6.292}
$$

where coefficients K_{p_i} $i = 1, 2, \ldots, N$ are positive and the elimination of the state variables' tracking error is confirmed.

6.6.6.2 Lyapunov stability analysis

The global asymptotic stability for the PDE model of the data-flow along the communication network routes can be also shown through Lyapunov analysis. From (6.291) to (6.288) it has been shown that the tracking error dynamics for the state variables of the model is

$$
\begin{aligned}
&\dot{e}_1 + K_{p_1} e_1 = 0 \\
&\dot{e}_2 + K_{p_2} e_2 = 0 \\
&\cdots \\
&\cdots \\
&\dot{e}_{N_1} + K_{p_{N-1}} e_{N-1} = 0 \\
&\dot{e}_N + K_{p_N} e_N = 0
\end{aligned}
\tag{6.293}
$$

The following Lyapunov function is defined

$$
V = \frac{1}{2}\sum_{i=1}^{N} e_i^2
\tag{6.294}
$$

By differentiating (6.294) with respect to time one obtains the proof about the global asymptotic stability of the control scheme

$$\dot{V} = \frac{1}{2}\sum_{i=1}^{N} 2e_i\dot{e}_i \Rightarrow \dot{V}\sum_{i=1}^{N} e_i(-K_{p_i}e_i) \Rightarrow$$
$$\dot{V} = \sum_{i=1}^{N} -K_{p_i}e_i^2 \Rightarrow \dot{V} < 0 \tag{6.295}$$

6.6.7 Simulation tests

The performance of the proposed differential flatness theory-based PDE control scheme of the data flow in routes of a communication network was confirmed through simulation experiments. The boundary condition $F_m(t, \sigma_{N+1})$ has been used as a control input to the PDE system. A spatial grid comprising $N = 10$ σ_i points has been considered. The obtained results are depicted in Figures 6.50 to 6.55. The tracking performance of the control method was very satisfactory since all state variables of the PDE model were made to converge to their reference setpoints. The variations of the control input remained smooth.

Unlike a large number of data-flow control methods and congestion control methods in communication networks, the proposed control scheme was of proven stability. The boundary condition $F_m(t, \sigma_N)$ which serves as a control input is determined by the maximum value of the threshold variable σ_N which in turn defines the size beyond

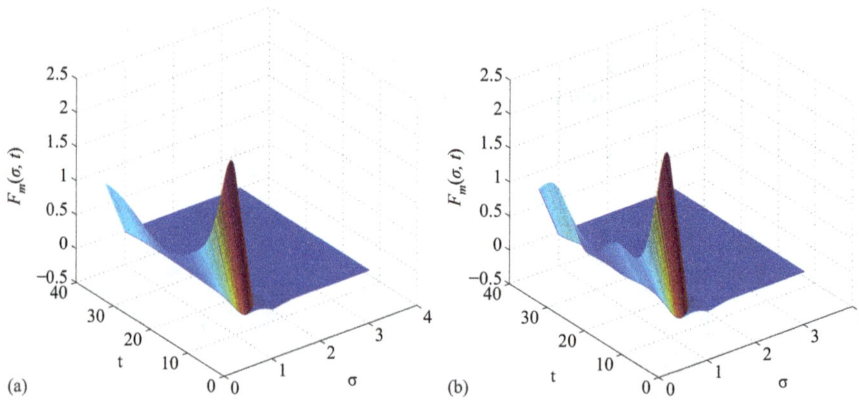

Figure 6.50 Variation of the congestion function $F_m(t, \sigma)$ which denotes the number of m-th route files at time instant t with residual size larger than σ (a) in the case of congestion profile 1 and (b) in the case of congestion profile 2

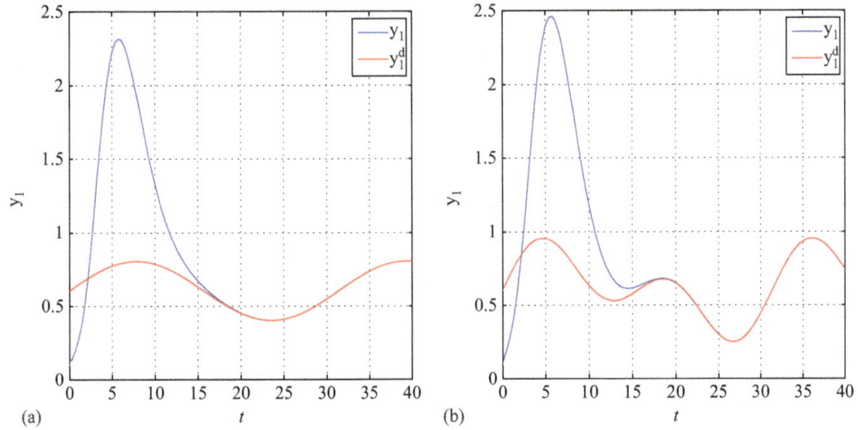

Figure 6.51 Variation of the congestion function $F_m(t, \sigma_1)$ at the initial grid point σ_1 (a) in the case of congestion profile 1 and (b) in the case of congestion profile 2

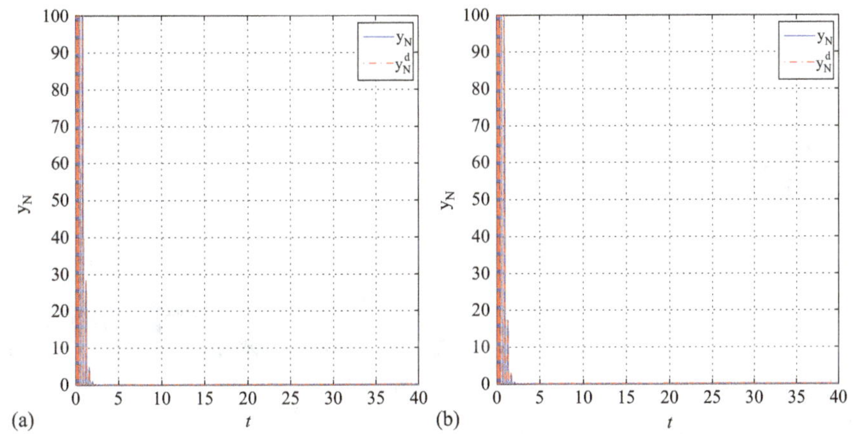

Figure 6.52 Variation of the congestion function $F_m(t, \sigma_N)$ at the final grid point σ_N (a) in the case of congestion profile 1 and (b) in the case of congestion profile 2

which a task (file) in the route cannot be serviced at a specific time instant. By changing this upper bound one can also control congestion levels in the route and actually one can achieve at every step of the control method that the number of the unprocessed tasks (files) follows closely desirable reference values.

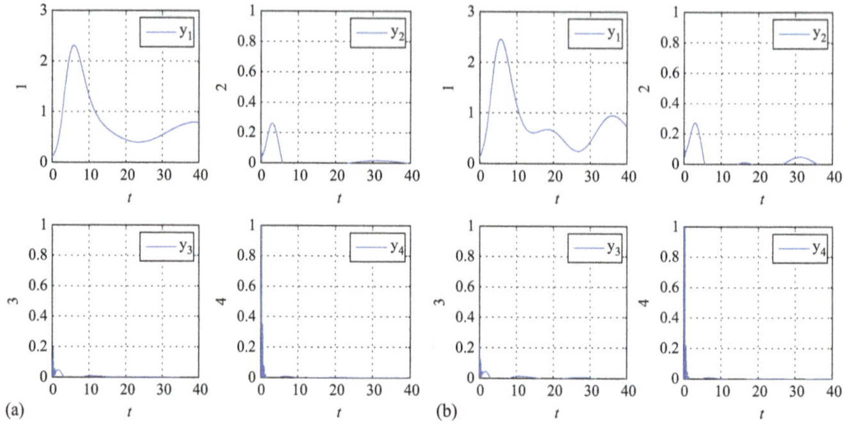

Figure 6.53 Variation of the congestion function $F_m(t, \sigma_N)$ at grid points σ_1 to σ_4 (a) in the case of congestion profile 1 and (b) in the case of congestion profile 2

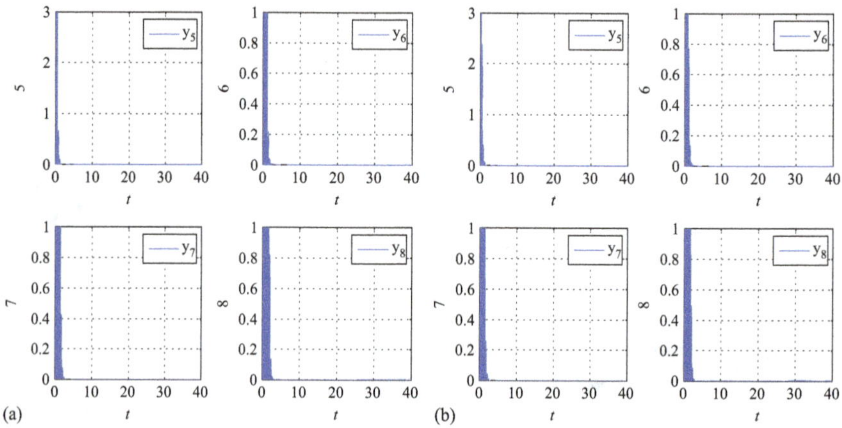

Figure 6.54 Variation of the congestion function $F_m(t, \sigma_N)$ at grid points σ_5 to σ_8 (a) in the case of congestion profile 1 and (b) in the case of congestion profile 2

The PDE model of data flow (number of unprocessed data packets exceeding a specific size) has been given in (Paganini and Ferragut, 2014a; Paganini and Ferragut, 2014b). However, in none of the aforementioned references has a solution of the related PDE control problem been attempted. Consequently, the present section has a unique contribution because it solves the control problem for the PDE model that describes data flow in the communication network's routes. This is a significant

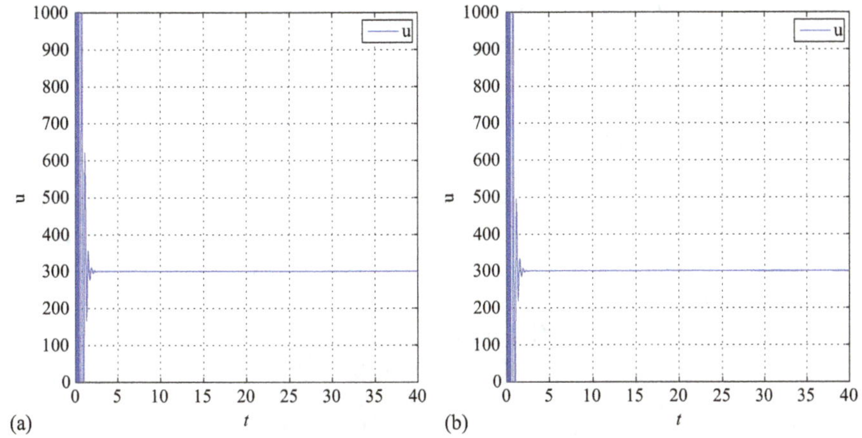

Figure 6.55 Variation of the control input $u = \sigma_{N+1}$ (boundary condition) which determines the congestion levels at the internet routes (a) in the case of congestion profile 1 and (b) in the case of congestion profile 2

advancement in research on congestion control and bandwidth allocation in communication networks, because the majority of the attempts to treat this problem (i) either use simplified descriptions of the data flow dynamics in the form of ODEs, (ii) or rely on heuristics and empirical schemes which lack stability proof and which are not of generic use.

6.7 Diffusion PDE control of data flow in communication networks

6.7.1 Outline

Modelling of data packets flow in telecom networks with the use of diffusion PDEs has been a topic of research for several years (Takeno *et al.*, 2012; Czachorski *et al.*, 2010; Carofiglio *et al.*, 2006; Takeno *et al.*, 2008). Actually, diffusion PDE approximations have been used to describe (i) the congestion of data packets in telecommunication and computer networks (Czachorski *et al.*, 2006; Mukherjee and Strikwerde, 1993; Takeno and Aida, 2004; Sumi, Takeno *et al.*, 2011), (ii) the dynamics of queues in networks comprising multiple servers (Daley *et al.*, 2007; Czachorski *et al.*, 2009), (iii) the dynamics of data-flow in high-speed communication networks such as ATM networks (Takeno and Aida, 2005; Takeno, Aida and Kirubayashi, 2004; Takeno, Muranaka *et al.*, 2006; Takeno and Aida, 2003; Ren and Kobayashi, 1997) and (iv) the number of consecutive hops that data packets in WSN need until they reach the final node (Gelenbe, 2007; Czarchoski *et al.*, 1995; Czachorski *et al.*, 2008). Although diffusion approximations offer a compact and concise representation of data propagation in the aforementioned networks, their feedback control remains a difficult problem (Rigatos, 2015; Rigatos, 2016b; Rigatos, Siano and Melkikh, 2016). The diffusion PDE describing data flow is a nonlinear one and exhibits spatiotemporal dynamics, thus evolving simultaneously along with the space and time axis. For the solution of the PDE control problem one can consider the decomposition of the PDE system into an equivalent set of ordinary nonlinear differential equations. However control of such a set of dynamical subsystems through boundary conditions is a non-trivial problem because of missing a sufficient number of control inputs (actually one has to control the different degrees of freedom associated with the set of nonlinear ODEs by using only one control input).

The present section proposes a method for boundary control of the diffusion PDE that appears in communication networks. Actually, the control of this PDE signifies the control of data packets flow and congestion control at the nodes and links of the communication network. First, the diffusion model of data flow in the communication network is introduced. The index used to count the nodes of the communication or computer network is taken to vary by infinitesimally small steps, and is finally substituted by the continuous spatial variable x. Using the principle of the conservation of data packets flow at each point x, the dynamics of the data packets concentration is shown to be described by a diffusion PDE (Rigatos, 2015; Rigatos, 2016b; Rigatos, Siano and Melkikh, 2016). The main variable of this PDE model is the concentration $\rho(x,t)$ of packets at point x of the network and at time instant t. By applying semi-discretisation and the finite differences approach in the representation of the partial derivatives appearing in the PDE, a state-space description for the diffusion dynamics of the data packets flow is finally obtained (Pinsky, 1991).

Next, by applying a per-row decomposition of the PDE's state-space model, the latter is shown to be described by an equivalent set of nonlinear ordinary differential equations. Moreover, it is shown that each one of these differential equations stands for a differentially flat system (Rudolph, 2003; Sira-Ramirez and Agrawal, 2014; Levine, 2011; Rouchon, 2005; Boouden *et al.*, 2011; Chauvin, 2011). Results on boundary feedback control of PDE systems, with the use of differential flatness theory, can be found in (Woitteneck and Mounier, 2010; Salberg *et al.*, 2010; Fliess and Mounier, 1999; Rouchon, Martin *et al.*, 2014; Mounier, Rudolph *et al.*, 2010). In the present section, the state variable that is associated with each row of the state-space description is taken to be the flat output while the state variable of the adjacent row (in ascending order) is taken to be a virtual control input. The condition for the elimination of the output tracking error for each individual ODE subsystem provides the targeted value for its virtual control input. The boundary condition that appears in the nonlinear ODE subsystem that comprises the last row of the state-space description, stands for the aggregate control input. The computation of the boundary control input uses recursively all virtual control inputs mentioned above, moving from the last ODE system to the first one (Rigatos, 2015). Thus, by tracing forward and backwards the rows of the state-space model backwards, at each iteration of the control algorithm, one can finally obtain the control input that should be applied to the diffusion PDE system so as to assure that all its state vector elements will converge to the desirable setpoints. By analyzing the dynamics of the closed-loop system that results from the application of the aforementioned control method, asymptotic stability is confirmed.

6.7.2 *Model of diffusion describing data flow in the communication network*

Diffusion PDEs are met in several forms of communication and computer systems, for instance (i) in models of congestion of data packets in telecommunication and computer networks, (ii) in queueing taking place at multiple servers (where it describes the evolution of the number of tasks waiting to be processed by the individual servers), (iii) in models of data-flow in high-speed communication networks and (iv) in wireless sensor networks (where it describes the number of hops that data packets have to perform before reaching their destination).

In this section, a communication network comprising N information processing nodes is considered. The associated diffusion model describes the number of data packets concentrated at each node of the network at every time instant. The continuous space-time representation of the diffusion process is obtained after replacing index i for the ith node with the continuous spatial variable x which means that the variation step of i now becomes infinitesimally small (Takeno *et al.*, 2012; Takeno and Aida, 2005).

The concentration of packets at x at time instant t is denoted as $\rho(x,t)$. The data packets' flow at point x and time instant t is given by

$$J(x,t) = r(x,t) - D\frac{\partial \rho(x,t)}{\partial x} \tag{6.296}$$

where $J(x,t)$ is the number of packets per unit of time leaving from x, $r(x,t)$ denotes the number of data packets per unit of time arriving at x and the term $D\frac{\partial \rho(x,t)}{\partial x}$ denotes the number of packets diffused around x. Coefficient D is taken to be inversely proportional to the time delay appearing in the propagation of data packets between the network's nodes. The appearance of a bottleneck in the network's link is related with a change in this propagation delay coefficient.

From the condition about the conservation of the number of packets at point x one has

$$\frac{\partial \rho(x,t)}{\partial t} + \frac{\partial J(x,t)}{\partial x} = 0 \tag{6.297}$$

Thus, the spatio-temporal evolution of the data packets under processing at x and at time instance t, that is $\rho(x,t)$ is given by the following diffusion PDE (Fokker–Planck PDE)

$$\frac{\partial \rho(x,t)}{\partial t} = -\frac{\partial r(x,t)}{\partial x} + D\frac{\partial^2 \rho(x,t)}{\partial x^2} \tag{6.298}$$

The communication network is depicted in Figure 6.56.

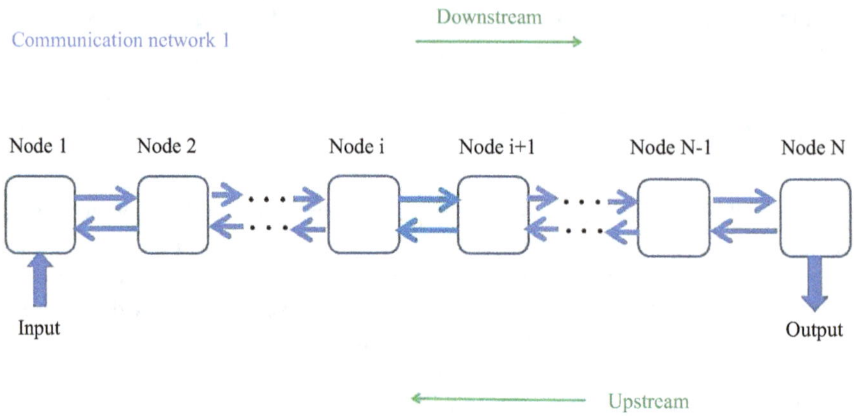

Figure 6.56 Communication network consisting of multiple datapackets processing nodes

6.7.3 Transformation of the Fokker-Planck PDE into a set of nonlinear ODEs

In the previously given Fokker-Planck PDE, the following substitutions are carried out: $r = -\alpha x\rho(x,t)$ and $D = \frac{1}{2}b^2$. Therefore, one arrives at the following form of the Fokker-Planck PDE

$$\frac{\partial \rho}{\partial t} = \frac{1}{2}b^2\frac{\partial^2 \rho}{\partial x^2} + \frac{\partial}{\partial x}[\alpha x\rho] \tag{6.299}$$

which in turn is written as

$$\frac{\partial \rho}{\partial t} = \frac{1}{2}b^2\frac{\partial^2 \rho}{\partial x^2} + \alpha\rho + \alpha x\frac{\partial}{\partial x}\rho \tag{6.300}$$

which can be also written in the form

$$\frac{\partial \rho}{\partial t} = \frac{1}{2}b^2\nabla^2\rho + \alpha\rho + \alpha x\nabla\rho \tag{6.301}$$

A grid of N points is considered, that is $\{x_1, x_2, \cdots, x_{N-1}, x_N\}$ which are placed at equal distances on the x axis. At the points of spatial discretisation it holds

$$\frac{\partial \rho_i}{\partial t} = \frac{1}{2}b^2\frac{\rho_{i+1} - 2\rho_i + \rho_{i-1}}{\Delta x^2} + \alpha\rho_i + \alpha x_i\frac{\rho_{i+1} - \rho_i}{\Delta x} \tag{6.302}$$

where $\rho_0 = \rho(x_0, t)$ and $\rho_{N+1} = \rho(x_{N+1}, t)$ are taken to be boundary conditions. For the ith ODE and at sampling point x_i along the x axis one has

$$\frac{\partial \rho_i}{\partial t} = \frac{1}{2}b^2\frac{\rho_{i-1}}{\Delta x^2} - \frac{1}{2}b^2\frac{2\rho_i}{\Delta x^2} + \frac{1}{2}b^2\frac{\rho_{i+1}}{\Delta x^2} + \alpha\rho_i + \alpha x_i\Delta x\frac{\rho_{i+1} - \rho_i}{\Delta x^2} \tag{6.303}$$

Equation (6.440) is also written in the form

$$\frac{\partial\rho_i}{\partial t}=\frac{1}{2}b^2\frac{\rho_{i-1}}{\Delta x^2}-\frac{1}{2}b^2\frac{2\rho_i}{\Delta x^2}+\left[\frac{1}{2}b^2+\alpha x_i\Delta x\right]\frac{\rho_{i+1}}{\Delta x^2}+\alpha\rho_i-\alpha x_i\Delta x\frac{\rho_i}{\Delta x^2} \tag{6.304}$$

By denoting $K_1=\frac{1}{2}b^2$, $K_2=\frac{1}{2}b^2+\alpha x_i\Delta x$ and $f(\rho_i)=\alpha\rho_i-\alpha x_i\Delta x\frac{\rho_i}{\Delta x^2}$ one obtains the following description for (6.441)

$$\frac{\partial\rho_i}{\partial t}=\frac{K_1}{\Delta x^2}\rho_{i-1}-\frac{2K_1}{\Delta x^2}\rho_i+\frac{K_2}{\Delta x^2}\rho_{i+1}+f(\rho_i) \tag{6.305}$$

6.7.4 Differential flatness of the Fokker–Planck PDE model

Next, the following state vector is defined for the PDE model:

$$\tilde{Y}=[y_{1,1},y_{1,2},\dots,y_{1,i},\dots,y_{1,N-1},y_{1,N}],$$

where $y_{1,1}=\rho_1$, $y_{1,2}=\rho_2,\dots,y_{1,i}=\rho_i,\dots,y_{1,N-1}=\rho_{N-1}$ and $y_{1,N}=\rho_N$. It will be shown that the state-space description of the nonlinear PDE dynamics , which has as control input only the boundary condition ρ_0 is differentially flat. One has

$$\dot{y}_{1,1}=-\frac{2K_1}{\Delta x^2}y_{1,1}+\frac{K_1}{\Delta x^2}\phi_0+f(y_{1,1})+\frac{K_2}{\Delta x^2}y_{1,2} \tag{6.306}$$

$$\dot{y}_{1,2}=-\frac{2K_1}{\Delta x^2}y_{1,2}+\frac{K_1}{\Delta x^2}y_{1,1}+f(y_{1,2})+\frac{K_2}{\Delta x^2}y_{1,3} \tag{6.307}$$

$$\dot{y}_{1,3}=-\frac{2K_1}{\Delta x^2}y_{1,3}+\frac{K_1}{\Delta x^2}y_{1,2}+f(y_{1,3})+\frac{K_2}{\Delta x^2}y_{1,4} \tag{6.308}$$

$\cdots\ \cdots$

$$\dot{y}_{1,i}=-\frac{2K_1}{\Delta x^2}y_{1,i}+\frac{K_1}{\Delta x^2}y_{1,i-1}+f(y_{1,i})+\frac{K_2}{\Delta x^2}y_{1,i+1} \tag{6.309}$$

$\cdots\ \cdots$

$$\dot{y}_{1,N-1}=-\frac{2K_1}{\Delta x^2}y_{1,N-1}+\frac{K_1}{\Delta x^2}y_{1,N-2}+f(y_{1,N-1})+\frac{K_2}{\Delta x^2}y_{1,N} \tag{6.310}$$

$$\dot{y}_{1,N}=-\frac{2K_1}{\Delta x^2}y_{1,N}+\frac{K_1}{\Delta x^2}y_{1,N-1}+f(y_{1,N})+\frac{K_2}{\Delta x^2}\rho_{N+1} \tag{6.311}$$

The flat output is considered to be the state variable $y_{1,N}$, which is denoted as $y=y_{1,N}$. Next, it is shown that all state variables which stand also for virtual control inputs of the system $\alpha_i=y_{1,i+1}$, can be written as functions of the flat output $y=y_{1,N}$. From (6.448) and considering that ρ_{N+1} is constant one obtains

$$\begin{aligned}y_{1,N-1}=\alpha_1&=\frac{1}{K_1/\Delta x^2}\left[\dot{y}_{1,N}+\frac{2K_1}{\Delta x^2}y_{1,N}-\frac{K_2}{\Delta x^2}\rho_{N+1}-f(y_{1,N})\right]\\ &\Rightarrow y_{1,N-1}=h_1(y,\dot{y},\cdots)\end{aligned} \tag{6.312}$$

and following a similar procedure, from (6.447) one gets

$$\begin{aligned} y_{1,N-2} = \alpha_2 = & \frac{1}{K_1/\Delta x^2}\left[\dot{y}_{1,N-1} + \frac{2K_1}{\Delta x^2}y_{1,N-1} - \frac{K_2}{\Delta x^2}y_{1,N} - f(y_{1,N-1})\right] \\ & \Rightarrow y_{1,N-2} = h_2(y,\dot{y},\dots) \end{aligned} \tag{6.313}$$

Continuing in a similar manner, from (6.446) one obtains

$$\begin{aligned} y_{1,i-1} = \alpha_{N-i+1} = & \frac{1}{K_1/\Delta x^2}\left[\dot{y}_{1,i} + \frac{2K_1}{\Delta x^2}y_{1,i} - \frac{K_2}{\Delta x^2}y_{1,i+1} - f(y_{1,i})\right] \\ & \Rightarrow y_{1,i-1} = h_{N-i+1}(y,\dot{y},\dots) \end{aligned} \tag{6.314}$$

From (6.445) one obtains

$$\begin{aligned} y_{1,2} = \alpha_{N-2} = & \frac{1}{K_1/\Delta x^2}\left[\dot{y}_{1,3} + \frac{2K_1}{\Delta x^2}y_{1,3} - \frac{K_2}{\Delta x^2}y_{1,4} - f(y_{1,3})\right] \\ & \Rightarrow y_{1,2} = h_{N-2}(y,\dot{y},\dots) \end{aligned} \tag{6.315}$$

From (6.444) one obtains

$$\begin{aligned} y_{1,1} = \alpha_{N-1} = & \frac{1}{K_1/\Delta x^2}\left[\dot{y}_{1,2} + \frac{2K_1}{\Delta x^2}y_{1,2} - \frac{K_2}{\Delta x^2}y_{1,3} - f(y_{1,2})\right] \\ & \Rightarrow y_{1,1} = h_{N-1}(y,\dot{y},\dots) \end{aligned} \tag{6.316}$$

Finally, from (6.443) one obtains

$$\begin{aligned} \rho_0 = \alpha_N = & \frac{1}{K_1/\Delta x^2}\left[\dot{y}_{1,1} + \frac{2K_1}{\Delta x^2}y_{1,1} - \frac{K_2}{\Delta x^2}y_{1,2} - f(y_{1,1})\right] \\ & \Rightarrow \rho_0 = h_N(y,\dot{y},\dots) \end{aligned} \tag{6.317}$$

The above procedure confirms that all state variables of the model

$$\begin{matrix} y_{1,1} & y_{1,2} & y_{1,3} & \cdots \\ y_{1,i} & \cdots & y_{1,N-1} & y_{1,N} \end{matrix} \tag{6.318}$$

and the control input which is the boundary condition ρ_0 can be written as functions of the flat output $y = y_{1,N}$ and of the flat output's derivatives. Consequently, the state-space model of the PDE dynamics is differentially flat.

Additionally, one can consider decomposition of the PDE state-space equation into submodels, where at each submodel the virtual control input is $\alpha_i = y_{1,i+1}$

$$\begin{aligned} \dot{y}_{1,i} = & -\frac{2K_1}{\Delta x^2}y_{1,i} + \frac{K_1}{\Delta x^2}y_{1,i-1} + f(y_{1,i}) + \frac{K_2}{\Delta x^2}y_{1,i+1} \\ & \cdots\ \cdots \end{aligned} \tag{6.319}$$

and with local flat output $y_{1,i}$ one can confirm that all such subsystems $i = 1, 2, \dots, N$ are differentially flat. Next, one can compute the virtual inputs which should be applied to each subsystem.

6.7.5 Computation of a boundary conditions-based feedback control law

To implement boundary feedback control in the system of (6.443) to 6.448), the nonlinear diffusion-PDE model is rewritten as follows:

$$\dot{y}_{1,N} = -\frac{2K_1}{\Delta x^2}y_{1,N} + f(y_{1,N}) + \frac{K_2}{\Delta x^2}\rho_{N+1} + \frac{K_1}{\Delta x^2}y_{1,N-1} \tag{6.320}$$

$$\dot{y}_{1,N-1} = -\frac{2K_1}{\Delta x^2}y_{1,N-1} + f(y_{1,N-1}) + \frac{K_2}{\Delta x^2}y_{1,N} + \frac{K_1}{\Delta x^2}y_{1,N-2} \tag{6.321}$$

$$\dot{y}_{1,N-2} = -\frac{2K_1}{\Delta x^2}y_{1,N-2} + f(y_{1,N-2}) + \frac{K_2}{\Delta x^2}y_{1,N-1} + \frac{K_1}{\Delta x^2}y_{1,N-1} \tag{6.322}$$

$\cdots\ \cdots$

$$\dot{y}_{1,i} = -\frac{2K_1}{\Delta x^2}y_{1,i} + f(y_{1,i}) + \frac{K_2}{\Delta x^2}y_{1,i+1} + \frac{K_1}{\Delta x^2}y_{1,i-1} \tag{6.323}$$

$\cdots\ \cdots$

$$\dot{y}_{1,3} = -\frac{2K_1}{\Delta x^2}y_{1,3} + f(y_{1,3}) + \frac{K_2}{\Delta x^2}y_{1,4} + \frac{K_1}{\Delta x^2}y_{1,2} \tag{6.324}$$

$\cdots\ \cdots$

$$\dot{y}_{1,2} = -\frac{2K_1}{\Delta x^2}y_{1,2} + f(y_{1,2}) + \frac{K_2}{\Delta x^2}y_{1,3} + \frac{K_1}{\Delta x^2}y_{1,1} \tag{6.325}$$

$$\dot{y}_{1,1} = -\frac{2K_1}{\Delta x^2}y_{1,1} + f(y_{1,1}) + \frac{K_2}{\Delta x^2}y_{1,2} + \frac{K_1}{\Delta x^2}\phi_0 \tag{6.326}$$

The boundary condition ρ_{N+1} is assumed to be known and remains steady. The boundary condition ρ_0 stands for the control input. The feedback control law is designed as follows:

$$\begin{aligned}\alpha_1 = y^*_{1,N-1} = \frac{1}{(K_1/\Delta x^2)}\Big[&\dot{y}^d_{1,N} - k_{p,1}(y_{1,N} - y^d_{1,N}) \\ &+ \frac{2K_1}{\Delta x^2}y_{1,N} - f(y_{1,N}) - \frac{K_2}{\Delta x^2}\rho_{N+1}\Big]\end{aligned} \tag{6.327}$$

$$\begin{aligned}\alpha_2 = y^*_{1,N-2} = \frac{1}{(K_1/\Delta x^2)}\Big[&\dot{y}^d_{1,N-1} - k_{p,2}(y_{1,N-1} - y^d_{1,N-1}) \\ &+ \frac{2K_1}{\Delta x^2}y_{1,N-1} - f(y_{1,N-1}) - \frac{K_2}{\Delta x^2}y_N\Big] \Rightarrow\end{aligned} \tag{6.328}$$

$$\alpha_2 = \frac{1}{(K_1/\Delta x^2)}\left[\dot{\alpha}_1 - k_{p,2}(y_{1,N-1} - \alpha_1) + \frac{2K_1}{\Delta x^2}y_{1,N-1} - f(y_{1,N-1}) - \frac{K_2}{\Delta x^2}y_N\right]$$

$$\alpha_3 = y^*_{1,N-3} = \frac{1}{(K_1/\Delta x^2)}\left[\dot{y}^d_{1,N-2} - k_{p,3}(y_{1,N-2} - y^d_{1,N-2}) + \frac{2K_1}{\Delta x^2}y_{1,N-2} - f(y_{1,N-2}) - \frac{K_2}{\Delta x^2}y_{N-1}\right] \Rightarrow$$

$$\alpha_3 = \frac{1}{(K_1/\Delta x^2)}\left[\dot{\alpha}_2 - k_{p,3}(y_{1,N-2} - \alpha_2) + \frac{2K_1}{\Delta x^2}y_{1,N-2} - f(y_{1,N-2}) - \frac{K_2}{\Delta x^2}y_{N-1}\right] \tag{6.329}$$

and continuing in a similar manner

$$\begin{aligned}\alpha_i = y^*_{1,N-i} &= \frac{1}{(K_1/\Delta x^2)}\left[\dot{y}^d_{1,N-i+1} - k_{p,i}(y_{1,N-i+1} - y^d_{1,N-i+1}) + \frac{2K_1}{\Delta x^2}y_{1,N-i+1} - f(y_{1,N-i+1}) - \frac{K_2}{\Delta x^2}y_{N-i+2}\right] \\ \Rightarrow \alpha_i &= \frac{1}{(K_1/\Delta x^2)}[\dot{\alpha}_{i-1} - k_{p,i}(y_{1,N-i+1} - \alpha_{i-1}) + \frac{2K_1}{\Delta x^2}y_{1,N-i+1} - f(y_{1,N-i+1}) - \frac{K_2}{\Delta x^2}y_{N-i+2}]\end{aligned} \tag{6.330}$$

Following this procedure one arrives to compute the control inputs which are associated with the last two rows of the state-space model

$$\alpha_{N-2} = y^*_{1,2} = \frac{1}{(K_1/\Delta x^2)}\left[\dot{y}^d_{1,3} - k_{p,N-1}(y_{1,3} - y^d_{1,3}) + \frac{2K_1}{\Delta x^2}y_{1,3} - f(y_{1,3}) - \frac{K_2}{\Delta x^2}y_{1,4}\right] \Rightarrow$$

$$\alpha_{N-2} = \frac{1}{(K_1/\Delta x^2)}\left[\dot{\alpha}_{N-3} - k_{p,N-2}(y_{1,3} - \alpha_{N-3}) + \frac{2K_1}{\Delta x^2}y_{1,3} - f(y_{1,3}) - \frac{K_2}{\Delta x^2}y_{1,4}\right] \tag{6.331}$$

$$\alpha_{N-1} = y^*_{1,1} = \frac{1}{(K_1/\Delta x^2)}\left[\dot{y}^d_{1,2} - k_{p,N-1}(y_{1,2} - y^d_{1,2}) + \frac{2K_1}{\Delta x^2}y_{1,2} - f(y_{1,2}) - \frac{K_2}{\Delta x^2}y_{1,3}\right] \Rightarrow$$

$$\alpha_{N-1} = \frac{1}{(K_1/\Delta x^2)}\left[\dot{\alpha}_{N-2} - k_{p,N-1}(y_{1,2} - \alpha_{N-2}) + \frac{2K_1}{\Delta x^2}y_{1,2} - f(y_{1,2}) - \frac{K_2}{\Delta x^2}y_{1,3}\right] \tag{6.332}$$

and finally

$$\begin{aligned}\alpha_N = \phi_0 = &\frac{1}{(K_1/\Delta x^2)}\left[\dot{y}_{1,1}^d - k_{p,N}(y_{1,1} - y_{1,1}^d)\right.\\ &\left. + \frac{2K_1}{\Delta x^2}y_{1,1} - f(y_{1,1}) - \frac{K_2}{\Delta x^2}y_{1,2}\right] \Rightarrow\end{aligned}$$

$$\alpha_N = \frac{1}{(K_1/\Delta x^2)}\left[\dot{\alpha}_{N-1} - k_{p,N}(y_{1,1} - \alpha_{N-1}) + \frac{2K_1}{\Delta x^2}y_{1,1} - f(y_{1,1}) - \frac{K_2}{\Delta x^2}y_{1,2}\right] \tag{6.333}$$

Consequently, the computation of the control input $\alpha_N = \rho_0$ which is applied to the PDE model is performed backwards, by substituting recursively into ρ_0 the virtual control inputs $a_{N-1}, a_{N-2},\ldots,a_i,\ldots,a_2,a_1$.

6.7.6 Closed-loop dynamics

By substituting (6.470) into (6.463) of the state-space model of the PDE dynamics, and using the definition $y_{1,1} - y_{1,1}^d = z_1$ one has

$$\begin{aligned}&\dot{y}_{1,1} = \dot{y}_{1,1}^d - k_{p,1}(y_{1,1} - y_{1,1}^d)\Rightarrow\\ &(\dot{y}_{1,1} - \dot{y}_{1,1}^d) + k_{p,1}(y_{1,1} - y_{1,1}^d) = 0\Rightarrow\\ &\dot{z}_1 + k_{p,1}z_1 = 0\end{aligned} \tag{6.334}$$

Equivalently, by substituting (6.469) into (6.462), and using the definition $y_{1,2} - \alpha_1 = z_2$ one has

$$\begin{aligned}&\dot{y}_{1,2} = \dot{\alpha}_1 - k_{p,2}(y_{1,2} - \alpha_1)\Rightarrow\\ &(\dot{y}_{1,2} - \dot{\alpha}_1) + k_{p,2}(y_{1,2} - \alpha_1) = 0\Rightarrow\\ &\dot{z}_2 + k_{p,2}z_2 = 0\end{aligned} \tag{6.335}$$

Similarly, continuing with the rest of the equations of the state-space model and by substituting (6.468) into (6.461), while also using the definition $y_{1,3} - \alpha_2 = z_3$ one has

$$\begin{aligned}&\dot{y}_{1,3} = \dot{\alpha}_2 - k_{p,3}(y_{1,3} - \alpha_2)\Rightarrow\\ &(\dot{y}_{1,3} - \dot{\alpha}_2) + k_{p,3}(y_{1,3} - \alpha_2) = 0\Rightarrow\\ &\ddot{z}_3 + k_{p,3}z_3 = 0\end{aligned} \tag{6.336}$$

Moving backwards, and by substituting (6.467) into (6.460) of the state-space model of the PDE dynamics, and using the definition $y_{1,i} - \alpha_{i-1} = z_i$ one has

$$\begin{aligned}&\dot{y}_{1,i} = \dot{\alpha}_{i-1} - k_{p,i}(y_{1,i} - \alpha_{i-1})\Rightarrow\\ &(\dot{y}_{1,i} - \dot{\alpha}_{i-1}) + k_{p,i}(y_{1,i} - \alpha_{i-1}) = 0\Rightarrow\\ &\dot{z}_i + k_{p,i}z_i = 0\end{aligned} \tag{6.337}$$

By substituting (6.465) into (6.458) of the state-space model of the PDE dynamics, and using the definition $y_{1,N-1} - \alpha_{N-2} = z_{N-1}$ one has

$$\begin{aligned}
&\dot{y}_{1,N-1} = \dot{\alpha}_{N-2} - k_{p,N-1}(y_{1,N-1} - \alpha_{N-2}) \Rightarrow \\
&(\dot{y}_{1,N-1} - \dot{\alpha}_{N-2}) + k_{p,N-1}(y_{1,N-1} - \alpha_{N-2}) = 0 \Rightarrow \\
&\dot{z}_{N-1} + k_{p,N-1} z_{N-1} = 0
\end{aligned} \tag{6.338}$$

Finally, by substituting (6.464) into (6.458), and using the definition $y_{1,N} - \alpha_{N-1} = z_N$ one obtains

$$\begin{aligned}
&\dot{y}_{1,N} = \dot{\alpha}_{N-1} - k_{p,N}(y_{1,N} - \alpha_{N-1}) \Rightarrow \\
&(\dot{y}_{1,N} - \dot{\alpha}_{N-1}) + k_{p,N}(y_{1,N} - \alpha_{N-1}) = 0 \Rightarrow \\
&\dot{z}_N + k_{p,N} z_N = 0
\end{aligned} \tag{6.339}$$

Thus, the dynamics of the closed-loop system becomes

$$\begin{aligned}
&\dot{z}_1 + k_{p,1} z_1 = 0 \\
&\dot{z}_2 + k_{p,2} z_2 = 0 \\
&\dot{z}_3 + k_{p,3} z_3 = 0 \\
&\cdots\ \cdots \\
&\dot{z}_i + k_{p,i} z_i = 0 \\
&\cdots\ \cdots \\
&\dot{z}_{N-1} + k_{p,N-1} z_{N-1} = 0 \\
&\dot{z}_N + k_{p,N} z_N = 0
\end{aligned} \tag{6.340}$$

The dynamics of the closed-loop system can be also written in matrix form

$$\dot{\tilde{Z}} + K_P \tilde{Z} = 0 \tag{6.341}$$

where $\tilde{Z} = [z_1, z_2, z_3, \ldots, z_i, \ldots, z_{N-1}, z_N]$,and
$K_P = \text{diag}[k_{p,1}, k_{p,2}, k_{p,3}, \ldots, k_{p,i}, \ldots, k_{p,N-1}, k_{p,N}]$.

After suitable selection of the coefficients $k_{P,i}$, $i = 1, 2, \ldots, N$ such that the monomials $p(s) = s + k_{p,i}$ to have a negative root, it can be assured that $\lim_{t \to \infty} z_i(t) = 0$ and that the closed-loop system is asymptotically stable.

Moreover, to prove asymptotic stability for the closed-loop system, the following Lyapunov function can be used

$$V_L = \sum_{i=1}^{N} \frac{1}{2} z_i^2 \tag{6.342}$$

The derivative of this Lyapunov function with respect to time is given by

$$\dot{V}_L = \sum_{i=1}^{N} z_i \dot{z}_i \Rightarrow \dot{V}_L = \sum_{i=1}^{N} z_i(-k_{p_i} z_i) \Rightarrow$$
$$\dot{V}_L = -\sum_{i=1}^{N} k_{p,i} z_i^2 < 0 \tag{6.343}$$

Thus, it is proven again that the closed-loop system is globally asymptotically stable.

6.7.7 Simulation tests

Simulation examples about the proposed control method for distributed parameter systems are provided for the case of the Fokker–Planck PDE that was given in (6.362). The obtained results are depicted in Figures 6.57–6.61. The spatial discretisation of the PDE model consisted of $N = 25$ points. The sampling period was chosen to be $T_s = 0.01$sec. A normalised scale has been used for the $\rho(x, t)$ axis. The boundary condition ρ_{N+1} of the PDE was taken to be known and constant. The boundary condition ρ_0 served as the control input and was computed as each iteration of the control algorithm according to the procedure described in Section 6.7.5.

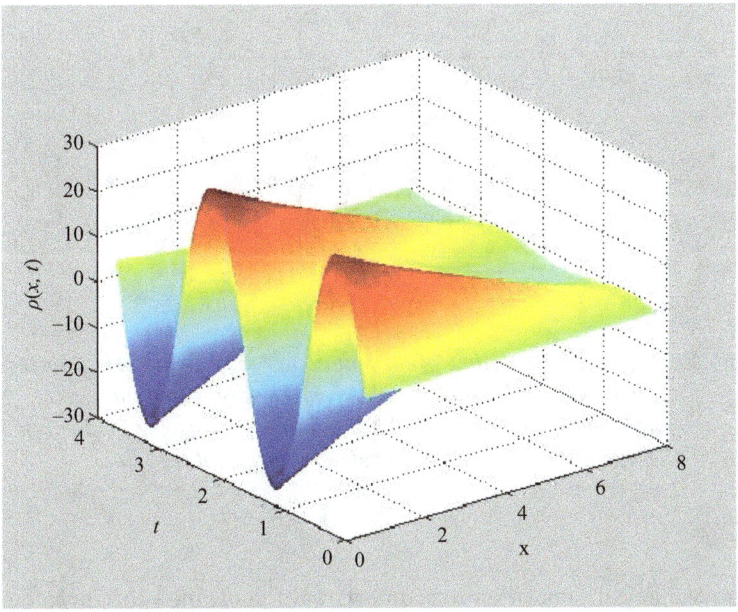

Figure 6.57 Fokker–Planck PDE under boundary feedback control

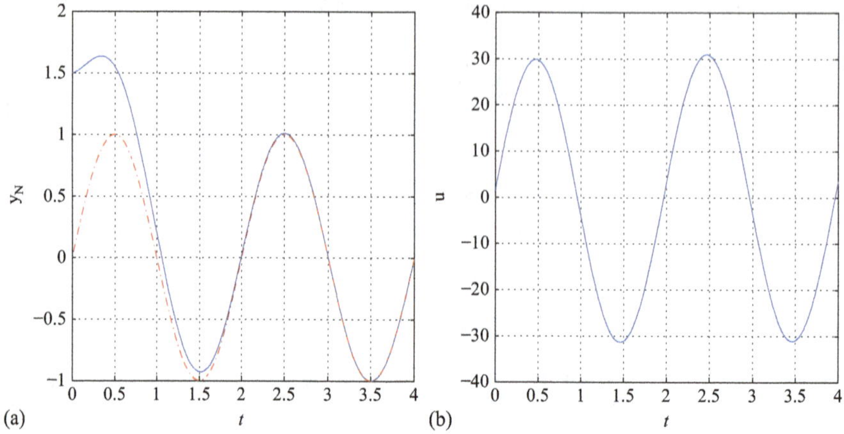

Figure 6.58 (a) tracking of reference setpoint by the value of the FP PDE at the final grid point x_N and (b) control input

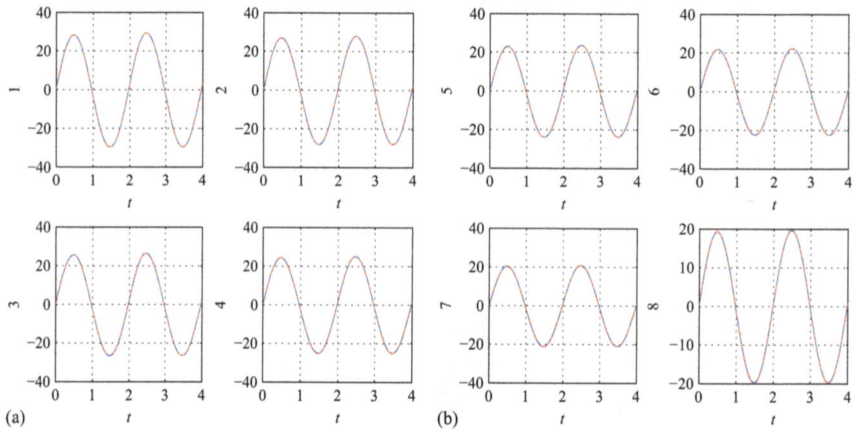

Figure 6.59 (a) Variation in time of the solution of the PDE at grid points x_1 to x_4 (blue line) and associated reference setpoint (red line) and (b) variation in time of the solution of the PDE at grid points x_5 to x_8 (blue line) and associated reference setpoint (red line)

The numerical simulation experiments have confirmed the theoretical findings of this section. It has been shown that by applying the proposed control method, the Fokker–Planck PDE dynamics can be modified so as to converge to the desirable reference profile. The control input that succeeds this changes smoothly

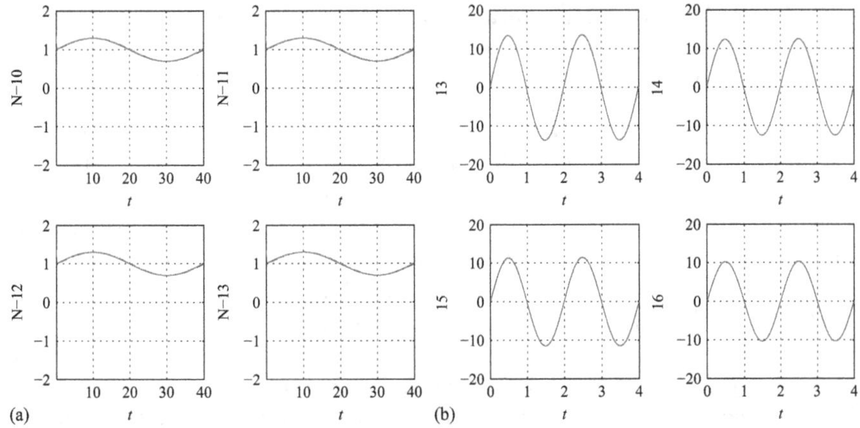

Figure 6.60 *(a) Variation in time of the solution of the PDE at grid points x_9 to x_{12} (blue line) and associated reference setpoint (red line) and (b) variation in time of the solution of the PDE at grid points x_{13} to x_{16} (blue line) and associated reference setpoint (red line)*

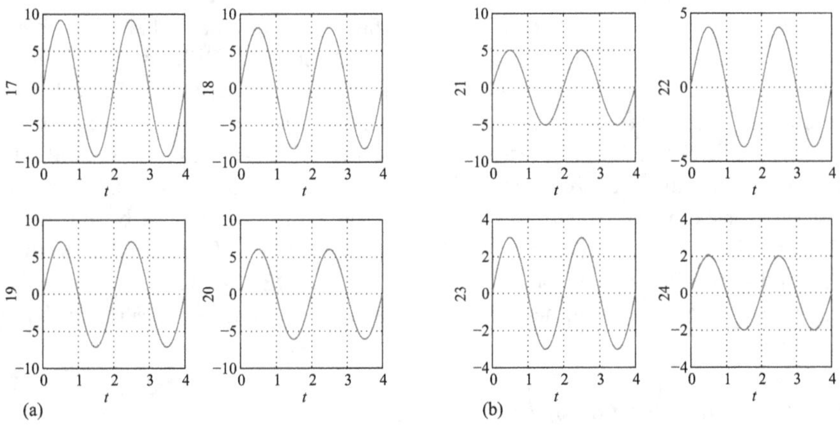

Figure 6.61 *(a) Variation in time of the solution of the PDE at grid points x_{17} to x_{20} (blue line) and associated reference setpoint (red line) and (b) variation in time of the solution of the PDE at grid points x_{21} to x_{24} (blue line) and associated reference setpoint (red line)*

and has a moderate range of variation. The accuracy of tracking of the reference setpoints was quite satisfactory. It is noted that the application of the proposed Fokker–Planck PDE feedback control method is also applicable to other diffusion phenomena.

6.8 Control of the diffusion PDE in Li-ion batteries

6.8.1 Outline

The need for more efficient power storage has led to the development of new control and estimation methods of the batteries' dynamics which are not based on lumped parameter models but which make use of distributed parameter (PDE) models. Specific interest has been shown in Lithium-ion batteries because these exhibit higher power density than other types of batteries, low self-discharge and prolonged life cycle (Klein *et al.*, 2013; Corno *et al.*, 2015; Smith, Rahn *et al.*, 2010; Smith, Rahn *et al.*, 2007). The distributed parameter models of Li-ion batteries comprise several coupled PDEs which describe the diffusion of Li ions at the solid electrodes as well as their diffusion in the electrolyte (Chaturvedi *et al.*, 2010; Smith, 2010; Zou, Mansic *et al.*, 2014; Klein *et al.*, 2010). Control of these diffusion phenomena, by applying external voltage or currents can result in making the battery dynamics follow specific charge or discharge patterns as well to keep its functioning within desirable charge levels. Obviously, this can further improve the performance indexes of Li-ion batteries.

At the first stage this section develops a solution to the control problem of diffusion dynamics of Li-ion batteries. It is proven that this diffusion PDE dynamics satisfies differential flatness properties and this makes possible to design a stabilising feedback controller. The present section applied semi-discretisation for the numerical representation of the PDE dynamics, which is an approach also implemented in (Olivier and Sedoglavic, 2001; Kröner, 2011; Guo and Billings, 2007; Pinsky, 1991; Utz, Meurer *et al.*, 2011). Methods for feedback stabilisation of systems with nonlinear PDE dynamics have been a flourishing research subject in the last years (Balogh and Kristic, 2002; Smyshlyaev and Krstic, 2010; Boscovic *et al.*, 2002; Liu, 2003). In particular, feedback control of diffusion-type (parabolic) PDEs has been a subject of extensive research and several remarkable results have been produced (Maidi and Corriu, 2014; Zwart *et al.*, 2011; Woitteneck and Mounier, 2010; Mounier, Rudolph *et al.*, 2010). For the control of diffusion PDEs boundary and distributed control methods have been developed (Fliess and Mounier, 2002; Winkler, Krause *et al.*, 2009).

6.8.2 Diffusion PDE in Li-ion batteries

The conservation of Li in a single spherical active material is described by Fick's law of diffusion (Smith, Rahn *et al.*, 2007)

$$\frac{\partial c_s}{\partial t} = \frac{D_s}{r^2}\frac{\partial}{\partial r}\left(r^2\frac{\partial c_s}{\partial r}\right) \tag{6.344}$$

with boundary conditions

$$\frac{\partial c_s}{\partial r}\,|_{r=0} = 0 \tag{6.345}$$

$$D_s\frac{\partial c_s}{\partial r}\,|_{r=R_s} = -\frac{j^{Li}}{\alpha_s F} \tag{6.346}$$

In (6.344) to (6.346) c_s denotes the concentration of Li and the subscript s stands for the solid phase. Moreover, D_s is a solid-phase diffusion coefficient, j^{Li} is the volumetric rate of electrochemical reaction at the psection surface with $j^{Li} > 0$ to denote ion discharge, a_s is the specific interfacial surface area, and F is Faraday's constant (96.487 C/mol.) For spherical active material psections of radius R_s occupying electrode volume fraction equal to ε_0, the interfacial surface area is $\alpha_s = 3\varepsilon_0/R_s$.

Equations (6.344) to (6.346) are applied on a continuum basis across each electrode, thus giving to the solid-phase concentration of Li the form of a 2D spatial variation $c_s(x, r, t)$, where x is the psection's position, r is the radial position within a psection and t is time. The electrochemical model depends only on concentration on a psection's surface $c_{s,e} = c_s(x, R_s, t)$. where the subscript s denotes the solid condition and the subscript e denotes the electrolyte condition (Figure 6.62).

Conservation of Li in the electrolyte phase gives (Smith, Rahn *et al.*, 2007)

$$\frac{\partial(\varepsilon_0 c_e)}{\partial t} = \frac{\partial}{\partial x}\left(D_e^{\text{eff}}\frac{\partial}{\partial x}c_e\right) + \frac{(1-t_+^0)}{F}j^{Li} \tag{6.347}$$

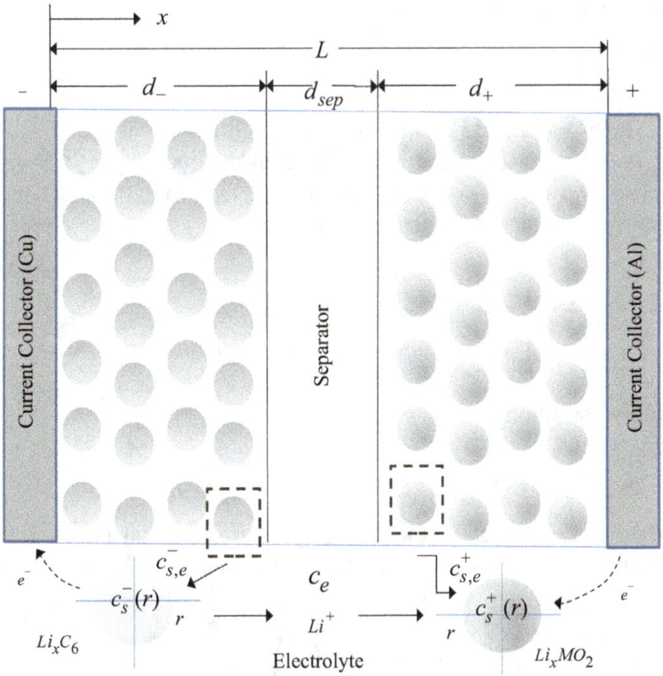

Figure 6.62 Diffusion of Lithium ions in a Li-ion battery (Smith, Rahn et al., 2007)

with boundary conditions

$$\frac{\partial c_e}{\partial x}\,|_{x=0} = 0 \tag{6.348}$$

$$\frac{\partial c_e}{\partial x}\,|_{x=L} = 0 \tag{6.349}$$

where $c_e(x,t)$ is Li concentration in the electrolyte phase, ε_e is the electrolyte phase volume fraction, t_+^0 is the transference number of Li^+ with respect to the velocity of the solvent. The effective diffusion coefficient is calculated from a reference coefficient using the Bruggeman relation $D_e^{\text{eff}} = D_e\varepsilon_e^p$ that accounts for the path that the Li^+ ions follow through the porous media. Moreover, it assumed that t_+^0 is constant.

Charge conservation in the solid phase of each electrode is described by Ohm's law

$$\frac{\partial}{\partial x}(\sigma^{\text{eff}})\frac{\partial}{\partial x}\phi_s - j^{Li} = 0 \tag{6.350}$$

with boundary conditions at the current collectors to be proportional to the applied current

$$-\sigma_-^{\text{eff}}\frac{\partial \phi_s}{\partial x}\,|_{x=0} = \sigma_+^{\text{eff}}\frac{\partial \phi_s}{\partial x}\,|_{x=L} = \frac{I}{A} \tag{6.351}$$

while the electronic current at the separator is zero

$$\frac{\partial \phi_s}{\partial x}\,|_{x=\delta_-} = \frac{\partial \phi_s}{\partial x}\,|_{x=\delta_- + \delta_{sep}} = 0 \tag{6.352}$$

In (6.350), (6.351) and (6.352) $\phi_s(x,t)$ is the potential, σ^{eff} is the effective conductivity, and is evaluated from the active material conductivity $\sigma^{\text{eff}} = \sigma\varepsilon_s$, A is the electrode plate area, and $I(t)$ is the applied current, following the sign convention that a positive current discharges the battery.

Electrolyte phase charge conservation gives

$$\frac{\partial}{\partial x}\left(\kappa^{\text{eff}}\frac{\partial}{\partial x}\phi_e\right) + \frac{\partial}{\partial x}\left(\kappa_D^{\text{eff}}\frac{\partial}{\partial x}lnc_e\right) + j^{Li} = 0 \tag{6.353}$$

with boundary conditions

$$\frac{\partial \phi_e}{\partial x}\,|_{x=0} = \frac{\partial \phi_e}{\partial x}\,|_{x=L} = 0 \tag{6.354}$$

In (6.353) and (6.354), $\phi_e(x,t)$ is the electrolyte phase potential and κ^{eff} is the effective ionic conductivity calculated from the Bruggeman relation $\kappa^{\text{eff}} = \kappa\varepsilon_e^D$. The effective diffusion conductivity is given by

$$\kappa_D^{\text{eff}} = \frac{2RT\kappa^{\text{eff}}}{F}(t_+^0 - 1)\left(1 + \frac{d\cdot\ln(f_\perp)}{d\cdot\ln(c_e)}\right) \tag{6.355}$$

where R is the universal gas constant (8.3143J/molK), T is the temperature and $f_\perp$ is the activity coefficient which is taken to be constant.

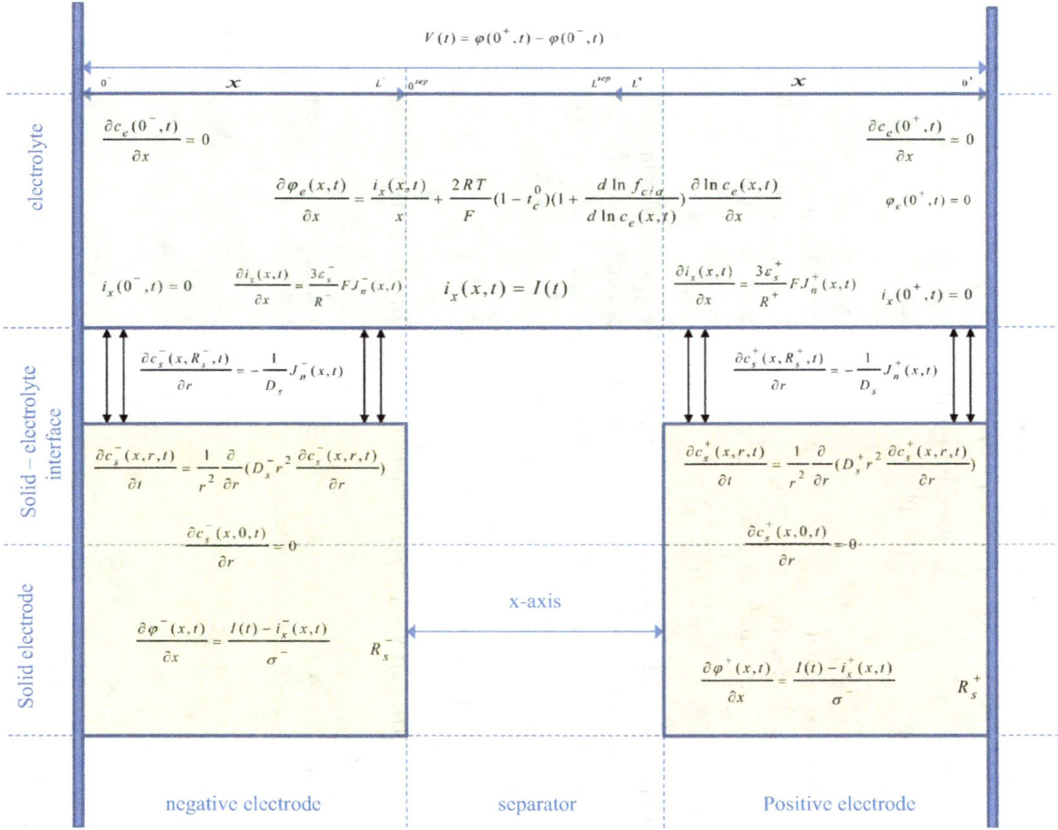

Figure 6.63 Diagram of the Li-ions diffusion dynamics in the battery (Klein et al., 2013)

The battery's dynamics consists of four PDEs which describe (i) the lithium concentration at the electrode $c_s(x,t)$ given in (6.344), (ii) the lithium concentration at the electrolyte $c_e(x,t)$ given in (6.347), (iii) the potential at the electrode $\phi_s(x,t)$ given in (6.350) and (iv) the potential at the electrolyte $\phi_e(x,t)$ given in (6.353). The aggregate PDE dynamics of the Li-ions battery is shown in Figure 6.63. These PDEs are coupled by the Butler–Volmer electrochemical kinetic expression

$$j^{Li} = \alpha_s i_0 \left\{ \exp\left[\frac{\alpha_a F}{RT}\eta\right] - \exp\left[-\frac{\alpha_c F}{RT}\eta\right]\right\} \tag{6.356}$$

where η is the difference between the solid and the electrolytic phase potential minus the thermodynamic equilibrium potential U. Thus one has

$$\eta = \phi_s - \phi_e - U \tag{6.357}$$

The equilibrium potential $U(c_{s,e})$ is evaluated as a function of the solid phase concentration at the psection surface. In (6.356) the exchange current density i_0 is a function of ion concentrations both at the solid electrode and at the electrolyte

$$i_0 = k(c_e)^{\alpha_0}(c_{s\max} - c_{s,0})^{\alpha_0}(c_{s,e})^{\alpha_c} \tag{6.358}$$

where k is a kinetic rate constant, while α_0 and α_c are the anodic and cathodic transfer coefficients. The current $I(t)$ is provided by the boundary conditions of (6.351). Then the voltage across the cell terminals is given by

$$V(t) = \phi_s(L,t) - \phi_s(0,t) - \frac{R_I}{A}I(t) \tag{6.359}$$

where R_I is an empirical contact resistance.

6.8.3 Modeling in state-space form of the of the Li-ions diffusion PDE

In the previously given diffusion PDE of (6.344), which describes the concentration of Li ions on the solid electrode one has

$$\frac{\partial c_s}{\partial t} = \frac{D_s}{r^2}\frac{\partial}{\partial r}\left(r^2\frac{\partial c_s}{\partial r}\right) \tag{6.360}$$

which in turn is written as

$$\frac{\partial c_s}{\partial t} = \frac{2D_s}{r}\frac{\partial c_s}{\partial r} + D_s\frac{\partial^2 c_s}{\partial r^2} \tag{6.361}$$

which can be also written in the form

$$\frac{\partial c_s}{\partial t} = D_s\nabla^2 c_s + \frac{2D_s}{r}\nabla c_s \tag{6.362}$$

A grid of N points is considered, that is $\{r_1, r_2, \ldots, r_{N-1}, r_N\}$ which are placed at equal distances on the r axis. At the points of spatial discretisation it holds

$$\frac{\partial c_s}{\partial t} = D_s\frac{c_{s,i+1} - 2c_{s,i} + c_{s,i-1}}{\Delta r^2} + \frac{2D_s}{r}\frac{c_{s,i+1} - c_{s,i}}{\Delta r} \tag{6.363}$$

where the boundary conditions are $c_s(r,t)|_{r=0} = 0$ and $c_s(r,t)|_{r=R_s}$. The second boundary conditions are taken to act on the system as a control input.

For the ith ODE one has at sampling point r_i along the r axis one has

$$\frac{\partial c_s}{\partial t} = \frac{Ds}{\Delta r^2}c_{s,i+1} - \frac{2Ds}{\Delta r^2}c_{s,i} + \frac{Ds}{\Delta r^2}c_{s,i-1} + \frac{2D_s}{r}\frac{c_{s,i+1}}{\Delta r} - \frac{2D_s}{r}\frac{c_{s,i}}{\Delta r} \tag{6.364}$$

Equation (6.440) is also written in the form

$$\frac{\partial c_s}{\partial t} = \frac{Ds}{\Delta r^2}c_{s,i+1} - \frac{2Ds}{\Delta r^2}c_{s,i} + \frac{Ds}{\Delta r^2}c_{s,i-1} + \frac{2D_s}{r}\left[\frac{c_{s,i+1} - c_{s,i}}{\Delta r}\right] \tag{6.365}$$

By denoting $K_1 = D_s$, $K_2 = -2D_s$ and $f(c_{s,i}) = \frac{2D_s}{r}[\frac{c_{s,i+1}-c_{s,i}}{\Delta r}]$ or equivalently $f(c_{s,i}) = \frac{\partial c_s}{\partial r}|_{r=r_i}$ one obtains the following description for (6.441)

$$\frac{\partial c_s}{\partial t} = \frac{K_1}{\Delta r^2}c_{s,i-1} + \frac{K_2}{\Delta r^2}c_{s,i} + \frac{K_1}{\Delta r^2}c_{s,i+1} + f(c_{s,i}) \tag{6.366}$$

6.8.4 Differential flatness of the battery's PDE diffusion model

Next, the following state vector is defined for the PDE model:

$$\tilde{Y} = [y_{1,1}, y_{1,2}, \ldots, y_{1,i}, \ldots, y_{1,N-1}, y_{1,N}],$$

where $y_{1,1} = c_{s,1}$, $y_{1,2} = c_{s.2}, \ldots, y_{1,i} = c_{s,i}, \ldots, y_{1,N-1} = c_{s,N-1}$ and $y_{1,N} = c_{s,N}$. It will be shown that the state-space description of the nonlinear PDE dynamics, which has as control input only the boundary condition $c_{s,N+1}$ is differentially flat. One has

$$\dot{y}_{1,1} = \frac{K_1}{\Delta r^2}y_{1,0} + \frac{K_2}{\Delta r^2}y_{1,1} + \frac{K_1}{\Delta r^2}y_{1,2} + f(y_{1,1}) \tag{6.367}$$

$$\dot{y}_{1,2} = \frac{K_1}{\Delta r^2}y_{1,1} + \frac{K_2}{\Delta r^2}y_{1,2} + \frac{K_1}{\Delta r^2}y_{1,3} + f(y_{1,2}) \tag{6.368}$$

$$\dot{y}_{1,3} = \frac{K_1}{\Delta r^2}y_{1,2} + \frac{K_2}{\Delta r^2}y_{1,3} + \frac{K_1}{\Delta r^2}y_{1,4} + f(y_{1,3}) \tag{6.369}$$

$\cdots\ \cdots$

$$\dot{y}_{1,i} = \frac{K_1}{\Delta r^2}y_{1,i-1} + \frac{K_2}{\Delta r^2}y_{1,i} + \frac{K_1}{\Delta r^2}y_{1,i+1} + f(y_{1,i}) \tag{6.370}$$

$\cdots\ \cdots$

$$\dot{y}_{1,N-1} = \frac{K_1}{\Delta r^2}y_{1,N-2} + \frac{K_2}{\Delta r^2}y_{1,N-1} + \frac{K_1}{\Delta r^2}y_{1,N} + f(y_{1,N-1}) \tag{6.371}$$

$$\dot{y}_{1,N} = \frac{K_1}{\Delta r^2}y_{1,N-1} + \frac{K_2}{\Delta r^2}y_{1,N} + \frac{K_1}{\Delta r^2}y_{1,N+1} + f(y_{1,N}) \tag{6.372}$$

Next, it is considered that outside the grid of the N points the value of c_s becomes 0, therefore $y_{1,0} = 0$ while $y_{1,N+1}$ acts as control input. The flat output is considered to be the state variable $y_{1,1}$, which is denoted as $y = y_{1,1}$. Next, it is shown that all state variables which stand also for virtual control inputs of the system $\alpha_i = y_{1,i-1}$, can be written as functions of the flat output $y = y_{1,1}$. From (6.367) and considering that $y_{1,0} = 0$ one obtains

$$\begin{aligned} \alpha_1 = y_{1,2} = & \frac{1}{K_1/\Delta r^2}\left[\dot{y}_{1,1} - \frac{K_1}{\Delta r^2}y_{1,0} - \frac{K_2}{\Delta r^2}y_{1,1} - f(y_{1,1})\right] \\ \Rightarrow & y_{1,2} = h_1(y, \dot{y}, \ldots) \end{aligned} \tag{6.373}$$

and following a similar procedure, from (6.368) one gets

$$\begin{aligned} \alpha_2 = y_{1,3} = & \frac{1}{K_1/\Delta r^2}\left[\dot{y}_{1,2} - \frac{K_1}{\Delta r^2}y_{1,1} - \frac{K_2}{\Delta r^2}y_{1,2} - f(y_{1,2})\right] \\ \Rightarrow & \alpha_2 = y_{1,3} = h_2(y, \dot{y}, \ldots) \end{aligned} \tag{6.374}$$

Continuing in a similar manner, from (6.369) one obtains

$$\begin{aligned} \alpha_3 = y_{1,4} = & \frac{1}{K_1/\Delta r^2}\left[\dot{y}_{1,3} - \frac{K_1}{\Delta r^2}y_{1,2} - \frac{K_2}{\Delta r^2}y_{1,3} - f(y_{1,3})\right] \\ \Rightarrow & \alpha_3 = y_{1,4} = h_3(y, \dot{y}, \ldots) \end{aligned} \tag{6.375}$$

From (6.370) one obtains

$$\begin{aligned} \alpha_i = y_{1,i+1} = & \frac{1}{K_1/\Delta r^2}\left[\dot{y}_{1,i} - \frac{K_1}{\Delta r^2}y_{1,i-1} - \frac{K_2}{\Delta r^2}y_{1,i} - f(y_{1,i})\right] \\ \Rightarrow & \alpha_i = y_{1,i+1} = h_i(y, \dot{y}, \ldots) \end{aligned} \tag{6.376}$$

From (6.371) one obtains

$$\begin{aligned} \alpha_{N-1} = y_{1,N-2} = & \frac{1}{K_1/\Delta r^2}\left[\dot{y}_{1,N-1} - \frac{K_2}{\Delta r^2}y_{1,N-1} - \frac{K_1}{\Delta r^2}y_{1,N} - f(y_{1,N-1})\right] \\ \Rightarrow & \alpha_{N-1} = y_{1,N-2} = h_{N-1}(y, \dot{y}, \ldots) \end{aligned} \tag{6.377}$$

Finally, from (6.372) one obtains

$$\begin{aligned} \alpha_N = y_{1,N+1} = & \frac{1}{K_1/\Delta r^2}\left[\dot{y}_{1,N} - \frac{K_1}{\Delta r^2}y_{1,N-1} - \frac{K_2}{\Delta r^2}y_{1,N} - f(y_{1,N})\right] \\ \Rightarrow & \alpha_N = y_{1,N+1} = h_N(y, \dot{y}, \ldots) \end{aligned} \tag{6.378}$$

The above procedure confirms that all state variables of the model

$$\begin{matrix} y_{1,1} & y_{1,2} & y_{1,3} & \cdots \\ y_{1,i} & \cdots & y_{1,N-1} & y_{1,N} \end{matrix} \tag{6.379}$$

and the control input which is the boundary condition $y_{1,N+1}$ can be written as functions of the flat output $y = y_{1,1}$ and of the flat output's derivatives. Consequently, the state-space model of the PDE dynamics is differentially flat.

Additionally, one can consider the decomposition of the PDE state-space equation into submodels, where at each submodel the virtual control input is $\alpha_i = y_{1,i+1}$

$$\dot{y}_{1,i} = \frac{K_1}{\Delta r^2}y_{1,i-1} + \frac{K_2}{\Delta r^2}y_{1,i} + f(y_{1,i}) + \frac{K_1}{\Delta r^2}y_{1,i+1} \tag{6.380}$$

and with local flat output $y_{1,i}$ one can confirm that all such subsystems $i = 1, 2, \ldots, N$ are differentially flat. Next, one can compute the virtual inputs which should be applied to each subsystem.

6.8.5 Computation of a boundary conditions-based feedback control law

To implement boundary feedback control, the feedback control input for (6.367) is computed first. Following this procedure one arrives to compute the control inputs which are associated with the last row of the state-space model. This uses recursively all virtual control inputs associated with the virtual control inputs of the previous $N-1$ row of the state-space description of the system.

$$\alpha_1 = y_{1,2}^* = \frac{1}{K_1/\Delta r^2}\left[\dot{y}_{1,1}^* - \frac{K_1}{\Delta r^2}y_{1,0} - \frac{K_2}{\Delta r^2}y_{1,1} - f(y_{1,1}) - K_{p_1}(y_{1,1} - y_{1,1}^*)\right] \tag{6.381}$$

$$\alpha_2 = y_{1,3}^* = \frac{1}{K_1/\Delta r^2}\left[\dot{y}_{1,2}^* - \frac{K_1}{\Delta r^2}y_{1,1} - \frac{K_2}{\Delta r^2}y_{1,2} - f(y_{1,2}) - K_{p_2}(y_{1,2} - y_{1,2}^*)\right] \Rightarrow$$

$$\alpha_2 = y_{1,3}^* = \frac{1}{K_1/\Delta r^2}\left[\dot{\alpha}_1^* - \frac{K_1}{\Delta r^2}y_{1,1} - \frac{K_2}{\Delta r^2}y_{1,2} - f(y_{1,2}) - K_{p_2}(y_{1,2} - \alpha_1)\right] \tag{6.382}$$

$$\alpha_3 = y_{1,4}^* = \frac{1}{K_1/\Delta r^2}\left[\dot{y}_{1,3}^* - \frac{K_1}{\Delta r^2}y_{1,2} - \frac{K_2}{\Delta r^2}y_{1,3} - f(y_{1,3}) - K_{p_3}(y_{1,3} - y_{1,3}^*)\right] \Rightarrow$$

$$\alpha_3 = y_{1,4}^* = \frac{1}{K_1/\Delta r^2}\left[\dot{\alpha}_{1,3}^* - \frac{K_1}{\Delta r^2}y_{1,2} - \frac{K_2}{\Delta r^2}y_{1,3} - f(y_{1,3}) - K_{p_3}(y_{1,i} - \alpha_2)\right] \tag{6.383}$$

while advancing forward one has

$$\alpha_i = y_{1,i+1}^* = \frac{1}{K_1/\Delta r^2}\left[\dot{y}_{1,i}^* - \frac{K_1}{\Delta r^2}y_{1,i-1} - \frac{K_2}{\Delta r^2}y_{1,i} - f(y_{1,i}) - K_{p_i}(y_{1,i} - y_{1,i}^*)\right] \Rightarrow$$

$$\alpha_i = y_{1,i+1}^* = \frac{1}{K_1/\Delta r^2}\left[\dot{\alpha}_{1,i-1}^* - \frac{K_1}{\Delta r^2}y_{1,i-1} - \frac{K_2}{\Delta r^2}y_{1,i} - f(y_{1,i}) - K_{p_i}(y_{1,i} - \alpha_{i-1})\right] \tag{6.384}$$

and continuing in a similar manner

$$\begin{aligned}\alpha_{N-2} = y_{1,N-1}^* &= \frac{1}{K_1/\Delta r^2}\left[\dot{y}_{1,N-2}^* - \frac{K_1}{\Delta r^2}y_{1,N-3}\right.\\ &\left. - \frac{K_2}{\Delta r^2}y_{1,N-2} - f(y_{1,n-2}) - K_{p_{N-2}}(y_{1,N-2} - y_{1,N-2}^*)\right] \Rightarrow\\ \alpha_{N-2} = y_{1,N-1}^* &= \frac{1}{K_1/\Delta r^2}\left[\dot{\alpha}_{1,N-3}^* - \frac{K_1}{\Delta r^2}y_{1,N-3}\right.\\ &\left. - \frac{K_2}{\Delta r^2}y_{1,N-2} - f(y_{1,n-2}) - K_{p_{N-2}}(y_{1,N-2} - \alpha_{N-3})\right]\end{aligned} \tag{6.385}$$

$$\alpha_{N-1} = y_{1,N}^* = \frac{1}{K_1/\Delta r^2}\left[\dot{y}_{1,N-1}^* - \frac{K_1}{\Delta r^2}y_{1,N-2} - \frac{K_2}{\Delta r^2}y_{1,N-1} - f(y_{1,N-1}) - K_{p_N}(y_{1,N-1} - y_{1,N-1}^*)\right] \Rightarrow$$
$$\alpha_{N-1} = y_{1,N}^* = \frac{1}{K_1/\Delta r^2}\left[\dot{\alpha}_{1,N-2}^* - \frac{K_1}{\Delta r^2}y_{1,N} - f(y_{1,N-2}) - \frac{K_2}{\Delta r^2}y_{1,N-1} - K_{p_N}(y_{1,N-1} - \alpha_{N-2})\right] \tag{6.386}$$

and finally

$$\alpha_N = y_{1,N+1} = \frac{1}{K_1/\Delta r^2}\left[\dot{y}_{1,N}^* - \frac{K_1}{\Delta r^2}y_{1,i-1} - \frac{K_2}{\Delta r^2}y_{1,N} - K_{p_N}(y_{1,N} - y_{1,N}^*)\right] \Rightarrow$$
$$\alpha_N = y_{1,N+1} = \frac{1}{K_1/\Delta r^2}\left[\dot{\alpha}_{N-1} - \frac{K_1}{\Delta r^2}y_{1,N-1} - \frac{K_2}{\Delta r^2}y_{1,N} - K_{p_N}(y_{1,N} - \alpha_{N-1})\right] \tag{6.387}$$

Consequently, the computation of the control input $\alpha_N = y_{1,N+1}$ which is exerted on the PDE model is performed backwards, by substituting recursively into α_N the virtual control inputs $\alpha_{N-1}, \alpha_{N-2}, \ldots, \alpha_i, \ldots, \alpha_2, \alpha_1$.

The control input is the boundary condition $y_{1,N+1} = c_s(r,t)\,|_{r=R_s}$. From (6.346) and (6.356) it can be noticed that the current $I(t)$ is the variable that implicitly defines the control input. The measurable output can be obtained from the boundary condition of (6.346) $\frac{\partial c_s(r,t)}{\partial r}|_{r=R_s} = -\frac{1}{D_s}j^{Li}(x,t)$. Variable c_s is indirectly related to the voltage $V(t)$.

6.8.6 Closed-loop dynamics

By substituting (6.464) into (6.372) of the state-space model of the PDE dynamics, and using the definition $y_{1,N} - y_{1,N}^* = y_{1,N} - \alpha_{N-1} = z_N$ one has

$$\begin{aligned}&\dot{y}_{1,N} = \dot{\alpha}_{N-1} - k_{p_N}(y_{1,N} - \alpha_{N-1}) \Rightarrow\\&(\dot{y}_{1,N} - \dot{\alpha}_{N-1} - k_{p_N}(y_{1,N} - \alpha_{N-1}) = 0 \Rightarrow\\&\dot{z}_N + k_{p_N}z_N = 0\end{aligned} \tag{6.388}$$

Equivalently, by substituting (6.465) into (6.371), considering that $(y_{1,N} \rightarrow \alpha_{N-1})$ and by using the definition $y_{1,N-1} - y_{1,N-1}^* = y_{1,N-1} - \alpha_{N-2} = z_{N-1}$ one has

$$\begin{aligned}&\dot{y}_{1,N-1} = \dot{\alpha}_{N-2} - k_{p_{N-1}}(y_{1,N-1} - \alpha_{N-2}) \Rightarrow\\&(\dot{y}_{1,N-1} - \dot{\alpha}_{N-2} - k_{p_{N-1}}(y_{1,N-1} - \alpha_{N-2}) = 0 \Rightarrow\\&\dot{z}_{N-1} + k_{p_{N-1}}z_{N-1} = 0\end{aligned} \tag{6.389}$$

Similarly, continuing with the rest of the equations of the state-space model and by substituting (6.467) into (6.370), considering that $(y_{1,i+1} \rightarrow \alpha_i)$ while also using the definition $y_{1,i} - y^*_{1,i} = y_{1,i} - \alpha_{i-1} = z_i$ one has

$$\begin{aligned} &\dot{y}_{1,i} = \dot{\alpha}_{i-1} - k_{p_i}(y_{1,i} - \alpha_{i-1}) \Rightarrow \\ &(\dot{y}_{1,i} - \dot{\alpha}_{i-1} - k_{p_i}(y_{1,i} - \alpha_{i-1}) = 0 \Rightarrow \\ &\dot{z}_i + k_{p_i} z_i = 0 \end{aligned} \tag{6.390}$$

Moving backwards, and by substituting (6.469) into (6.368) of the state-space model of the PDE dynamics, considering that $(y_{1,3} \rightarrow \alpha_2)$ while also using the definition $y_{1,2} - y^*_{1,2} = y_{1,2} - \alpha_1 = z_2$ one has

$$\begin{aligned} &\dot{y}_{1,2} = \dot{\alpha}_1 - k_{p_2}(y_{1,2} - \alpha_1) \Rightarrow \\ &(\dot{y}_{1,2} - \dot{\alpha}_1 - k_{p_2}(y_{1,2} - \alpha_1) = 0 \Rightarrow \\ &\dot{z}_2 + k_{p_2} z_2 = 0 \end{aligned} \tag{6.391}$$

Finally, by substituting (6.470) into (6.367), and using the definition $y_{1,1} - y^*_{1,1} = z_1$ one obtains

$$\begin{aligned} &\dot{y}_{1,N} = \dot{\alpha}_{N-1} - k_{p,N}(y_{1,N} - \alpha_{N-1}) \Rightarrow \\ &(\dot{y}_{1,N} - \dot{\alpha}_{N-1}) + k_{p,N}(y_{1,N} - \alpha_{N-1}) = 0 \Rightarrow \\ &\dot{z}_N + k_{p,N} z_N = 0 \end{aligned} \tag{6.392}$$

Thus, the dynamics of the closed-loop system asymptotically becomes

$$\begin{aligned} &\dot{z}_1 + k_{p,1} z_1 = 0 \\ &\dot{z}_2 + k_{p,2} z_2 = 0 \\ &\dot{z}_3 + k_{p,3} z_3 = 0 \\ &\cdots \cdots \\ &\dot{z}_i + k_{p,i} z_i = 0 \\ &\cdots \cdots \\ &\dot{z}_{N-1} + k_{p,N-1} z_{N-1} = 0 \\ &\dot{z}_N + k_{p,N} z_N = 0 \end{aligned} \tag{6.393}$$

The dynamics of the closed-loop system can be also written in matrix form

$$\dot{\tilde{Z}} + K_P \tilde{Z} = 0 \tag{6.394}$$

where $\tilde{Z} = [z_1, z_2, z_3, \ldots, z_i, \ldots, z_{N-1}, z_N]$,and
$K_P = \text{diag}[k_{p,1}, k_{p,2}, k_{p,3}, \ldots, k_{p,i}, \ldots, k_{p,N-1}, k_{p,N}]$.
After suitable selection of the coefficients $k_{P,i}$, $i = 1, 2, \cdots, N$ such that the monomials $p(s) = s + k_{p,i}$ to have a negative root, it can be assured that $\lim_{t \rightarrow \infty} z_i(t) = 0$ and that the closed-loop system is asymptotically stable.

Moreover, to prove asymptotic stability for the closed-loop system, the following Lyapunov function can be used

$$V_L = \sum_{i=1}^{N} \frac{1}{2} z_i^2 \tag{6.395}$$

The derivative of this Lyapunov function with respect to time is given by

$$\dot{V}_L = \sum_{i=1}^{N} z_i \dot{z}_i \Rightarrow \dot{V}_L = \sum_{i=1}^{N} z_i(-k_{p_i} z_i) \Rightarrow$$
$$\dot{V}_L = -\sum_{i=1}^{N} k_{p,i} z_i^2 < 0 \tag{6.396}$$

Thus, it is proven again that the closed-loop system is globally asymptotically stable.

6.8.7 State estimation for the PDE diffusion model

The measurable output of the PDE system is considered to be only $y_{1,1} = c_{s,1}$. The rest of the state variables of the model that is $y_{1,2} = c_{s,2}$, $y_{1,3} = c_{s,3}$, …, $y_{1,i} = c_{s,i}$, …, $y_{1,N-1} = c_{s,N-1}$ and $y_{1,N} = c_{s,N}$ have to be reconstructed with the use of a state estimator. As explained before, the system's dynamics consists of the ODEs given in (6.367) to (6.372). This is rewritten as

$$\dot{y}_{1,1} = \Delta m_1 + \frac{K_1}{\Delta r^2} y_{1,2}$$
$$\text{where } \Delta m_1 = \frac{K_1}{\Delta r^2} y_{1,0} + \frac{K_2}{\Delta r^2} y_{1,1} + f(y_{1,1}) \tag{6.397}$$

$$\dot{y}_{1,2} = \Delta m_2 + \frac{K_1}{\Delta r^2} y_{1,3}$$
$$\text{where } \Delta m_2 = \frac{K_1}{\Delta r^2} y_{1,1} + \frac{K_2}{\Delta r^2} y_{1,2} + f(y_{1,2}) \tag{6.398}$$

$$\dot{y}_{1,3} = \Delta m_2 + \frac{K_1}{\Delta r^2} y_{1,4}$$
$$\text{where } \Delta m_3 = \frac{K_1}{\Delta r^2} y_{1,2} + \frac{K_2}{\Delta r^2} y_{1,3} + f(y_{1,3}) \tag{6.399}$$

$$\cdots\ \cdots\ \cdots$$

$$\dot{y}_{1,i} = \Delta m_i + \frac{K_1}{\Delta r^2} y_{1,i+1}$$
$$\text{where } \Delta m_i = \frac{K_1}{\Delta r^2} y_{1,i-1} + \frac{K_2}{\Delta r^2} y_{1,i} + f(y_{1,i}) \tag{6.400}$$

$$\cdots\ \cdots\ \cdots$$

$$\dot{y}_{1,N-1} = \Delta m_{N-1} + \frac{K_1}{\Delta r^2} y_{1,N}$$
$$\text{where } \Delta m_{N-1} = \frac{K_1}{\Delta r^2} y_{1,N-2} + \frac{K_2}{\Delta r^2} y_{1,N-1} + f(y_{1,N-1}) \tag{6.401}$$

$$\dot{y}_{1,N} = \Delta m_N + u$$
$$\text{where } \Delta m_N = \frac{K_2}{\Delta r^2} y_{1,N} + \frac{K_1}{\Delta r^2} y_{1,N+1} + f(y_{1,N}) \text{ and } u = \frac{K_1}{\Delta r^2} y_{1,N-1} \tag{6.402}$$

By deriving the expression about $y_{1,1}$ given in the first row of (6.397) with respect to time one obtains

$$\begin{aligned}
\ddot{y}_{1,1} &= \Delta\dot{m}_1 + \frac{K_1}{\Delta r^2}\dot{y}_{1,2} \Rightarrow \\
\ddot{y}_{1,1} &= \Delta\dot{m}_1 + \frac{K_1}{\Delta r^2}\left(\Delta m_1 + \frac{K_1}{\Delta r^2}y_{1,3}\right) \Rightarrow \\
\ddot{y}_{1,1} &= \Delta\dot{m}_1 + \frac{K_1}{\Delta r^2}\Delta m_1 + \left(\frac{K_1}{\Delta r^2}\right)^2 y_{1,3}
\end{aligned} \tag{6.403}$$

By deriving again with respect to time one gets

$$\begin{aligned}
\frac{d^3}{dt^3}y_{1,1} &= \Delta\ddot{m}_1 + \frac{K_1}{\Delta r^2}\Delta\dot{m}_1 + \left(\frac{K_1}{\Delta r^2}\right)^2\left[\Delta m_3 + \frac{K_1}{\Delta r^2}y_4\right] \Rightarrow \\
\frac{d^3}{dt^3}y_{1,1} &= \Delta\ddot{m}_1 + \frac{K_1}{\Delta r^2}\Delta\dot{m}_1 + \left(\frac{K_1}{\Delta r^2}\right)^2\Delta m_3 + \left(\frac{K_1}{\Delta r^2}\right)^3 y_4
\end{aligned} \tag{6.404}$$

Continuing in a similar manner one obtains

$$\frac{d^i}{dt^i}y_{1,1} = \sum_{k=1}^{i}\left(\frac{K_1}{\Delta r^2}\right)^{k-1}\frac{d^{i-k}}{dt^{i-k}}\Delta m_k + \left(\frac{K_1}{\Delta r^2}\right)^{i+1} y_{i+1} \tag{6.405}$$

By advancing with the computation of higher-order time derivatives for $y_{1,1}$ one gets

$$\frac{d^{N-1}}{dt^{N-1}}y_{1,1} = \sum_{k=1}^{N-1}\left(\frac{K_1}{\Delta r^2}\right)^{k-1}\frac{d^{N-1-k}}{dt^{N-1-k}}\Delta m_k + \left(\frac{K_1}{\Delta r^2}\right)^{N} y_N \tag{6.406}$$

By computing one more derivative of $y_{1,1}$ with respect to time the following relation is obtained

$$\begin{aligned}
\frac{d^N}{dt^N}y_{1,1} &= \sum_{k=1}^{N-1}\left(\frac{K_1}{\Delta r^2}\right)^{k-1}\frac{d^{N-k}}{dt^{N-k}}\Delta m_k + \left(\frac{K_1}{\Delta r^2}\right)^{N}[\Delta m_N + u] \Rightarrow \\
\frac{d^N}{dt^N}y_{1,1} &= \sum_{k=1}^{N-1}\left(\frac{K_1}{\Delta r^2}\right)^{k-1}\frac{d^{N-k}}{dt^{N-k}}\Delta m_k + \left(\frac{K_1}{\Delta r^2}\right)^{N}\Delta m_N + \left(\frac{K_1}{\Delta r^2}\right)^{N} u
\end{aligned} \tag{6.407}$$

This is an input–output linearised description of the system which is a consequence of the differential flatness properties of the diffusion process (since the system is differentially flat it can be transformed to the trivial canonical form). Equation (6.407) can be also written as

$$\begin{aligned}
&y_{1,1}^{(N)} = \tilde{f}(y_{1,1},\dot{y}_{1,1},\ddot{y}_{1,1},\ldots) + \tilde{g}(y_{1,1},\dot{y}_{1,1},\ddot{y}_{1,1},\ldots)u \\
&\text{where } \tilde{f}(y_{1,1},\dot{y}_{1,1},\ddot{y}_{1,1},\ldots) = \sum_{k=1}^{N-1}\left(\frac{K_1}{\Delta r^2}\right)^{k-1}\frac{d^{N-k}}{dt^{N-k}}\Delta m_k + \left(\frac{K_1}{\Delta r^2}\right)^{N}\Delta m_N \\
&\text{and } \tilde{g}(y_{1,1},\dot{y}_{1,1},\ddot{y}_{1,1},\ldots) = \left(\frac{K_1}{\Delta r^2}\right)^{N}
\end{aligned} \tag{6.408}$$

For this description of the system, the estimation problem can be solved by applying the derivative-free nonlinear Kalman Filter. To this end, the following state variables are defined: $z_1 = y_{1,1}$, $z_2 = \dot{y}_{1,1}$, $z_3 = \ddot{y}_{1,1}, \ldots, z_i = y_{1,1}^{(i-1)}, \ldots, z_{N-1} = y_{1,1}^{(N-2)}$ and $z_N = y_{1,1}^{(N-1)}$.

For the input–output linearised model of the PDE one gets a state-space description of the form

$$\begin{pmatrix} \dot{z}_1 \\ \dot{z}_2 \\ \dot{z}_3 \\ \cdots \\ \dot{z}_i \\ \cdots \\ \dot{z}_{N-1} \\ \dot{z}_N \end{pmatrix} = \begin{pmatrix} 0 & 1 & 0 & 0 & \cdots & 0 & 0 & \cdots & 0 & 0 \\ 0 & 0 & 1 & 0 & \cdots & 0 & 0 & \cdots & 0 & 0 \\ 0 & 0 & 0 & 1 & \cdots & 0 & 0 & \cdots & 0 & 0 \\ \cdots & & & & \cdots & & \cdots & & & \cdots \\ 0 & 0 & 0 & 0 & \cdots & 0 & 1 & \cdots & 0 & 0 \\ \cdots & & & & \cdots & & \cdots & & & \cdots \\ 0 & 0 & 0 & 0 & \cdots & 0 & 0 & \cdots & 0 & 1 \\ 0 & 0 & 0 & 0 & \cdots & 0 & 0 & \cdots & 0 & 0 \end{pmatrix} \cdot \begin{pmatrix} z_1 \\ z_2 \\ z_3 \\ \cdots \\ z_i \\ \cdots \\ z_{N-1} \\ z_N \end{pmatrix} + \begin{pmatrix} 0 \\ 0 \\ 0 \\ \cdots \\ 0 \\ \cdots \\ 0 \\ 1 \end{pmatrix} (\tilde{f} + \tilde{g}u)$$

$$z^{\text{meas}} = \begin{pmatrix} 1 & 0 & 0 & 0 & \cdots & 0 & 0 & \cdots & 0 & 0 \end{pmatrix} Z \tag{6.409}$$

where $Z = [z_1,\ z_2,\ z_3,\ \ldots\ z_i,\ \ldots\ z_{N-1},\ z_N]^T$ and $v = \tilde{f} + \tilde{g}u$. The state estimation problem for the input–output linearised description of the system is solved with the use of the Derivative-free nonlinear Kalman Filter. The above continuous-time state-space description of the system takes the form

$$\begin{aligned} \dot{z} &= Az + Bv \\ z^{\text{meas}} &= CZ \end{aligned} \tag{6.410}$$

The state-space model of (6.410) is turned into discrete-time form after applying common discretisation methods. The discrete-time form of matrices A, B, C is denoted as A_d, B_d and C_d respectively. The Derivative-free nonlinear Kalman Filter is given by

$$\dot{\hat{z}} = A\hat{z} + B\hat{v} + K_f(z^{\text{meas}} - \hat{z}) \tag{6.411}$$

where $\hat{v} = \tilde{\hat{f}} + \tilde{\hat{g}}u$. The filter's gain K_f is computed through the following recursion

Measurement update:

$$\begin{aligned} K_f(k) &= P^-(k)C_d^T[C_dP^-(k)C_d^T + R(k)]^{-1} \\ \hat{z}^-(k) &= \hat{z}(k) + K_f[z^{\text{meas}} - C\hat{Z}] \\ P(k) &= P^-(k) - K(k)A_dP^-(k) \end{aligned} \tag{6.412}$$

Time update:

$$\begin{aligned} P^-(k+1) &= A_dP(k)A_d^T + Q(k) \\ \hat{z}^-(k+1) &= A_d\hat{z}(k) + B_d\hat{v}(k) \end{aligned} \tag{6.413}$$

Once estimation of $y_{1,1}$ and its derivatives is obtained from the above filtering algorithm it is possible to compute the estimates of all state vector elements $c_{s,i}$ $i = 1, 2, \ldots, N$ of the initial state-space model of the system given in (6.367) to (6.372). As explained in (6.454) to (6.450) the PDE model is differential flat and all state variables $c_{s,i}$ can be expressed as differential functions of the flat output $y_{1,1}$.

6.8.8 Simulation tests

Simulation examples about the proposed control method for distributed parameter systems are provided for the case of the lithium-ions diffusion PDE that was given in (6.344). The obtained results are depicted in Figures 6.64 to 6.66. The spatial discretisation of the PDE model consisted of $N = 15$ points. The sampling period was chosen to be $T_s = 0.01$s. The boundary conditions of the PDE model were $c_s(r,t)\mid_{r=0}= 0$, while $c_s(r,t)\mid_{r=R_s}$ served as the control input and was computed as each iteration of the control algorithm according to the procedure described in Section 6.8.5.

The numerical simulation experiments have confirmed the theoretical findings of this section. It has been shown that by applying the proposed control method, the Li-ions diffusion PDE dynamics can be modified so as to converge to the desirable reference profile. The control input that achieve these changes smoothly and has a moderate range of variation. The accuracy of tracking of the reference setpoints was quite satisfactory.

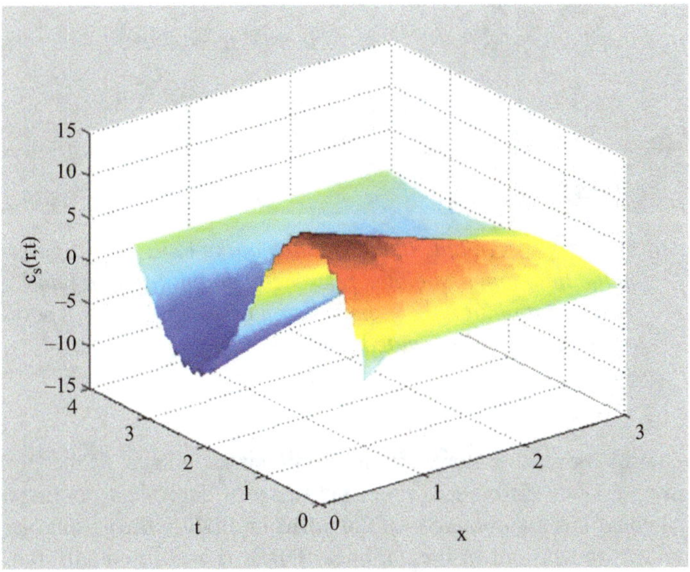

Figure 6.64 Controlled PDE diffusion in the Li-ion battery

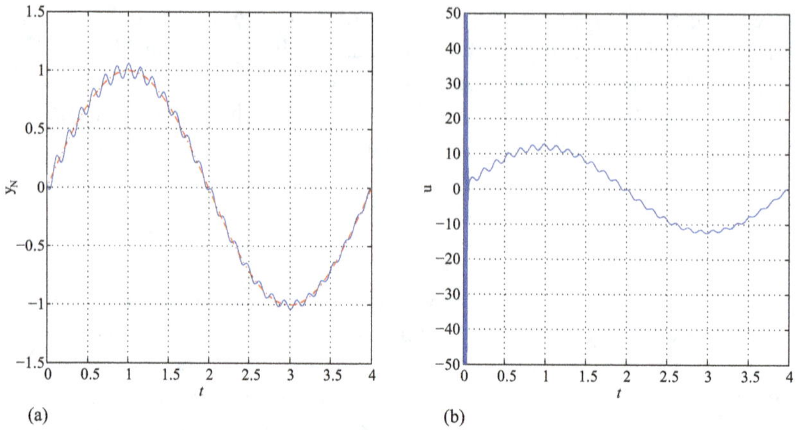

Figure 6.65 *(a) Tracking of reference setpoint by the value of the Li-ions PDE at the final grid point x_N and (b) control input*

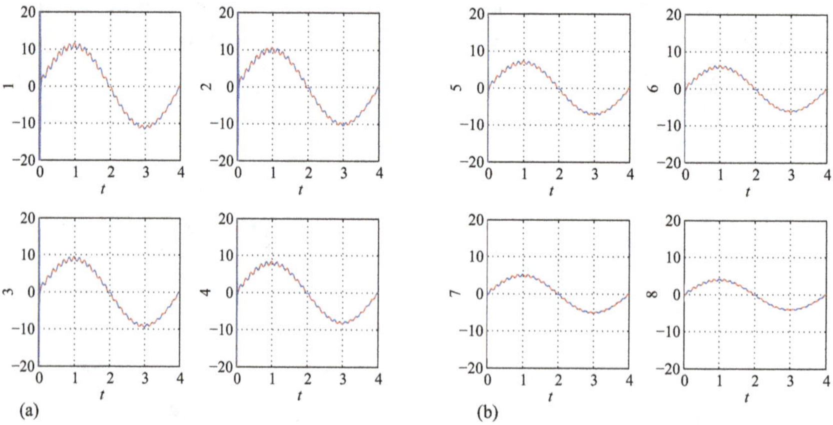

Figure 6.66 *(a) Variation in time of the solution of the PDE at grid points x_1 to x_4 (blue line) and associated reference setpoint (red line) and (b) variation in time of the solution of the PDE at grid points x_5 to x_8 (blue line) and associated reference setpoint (red line)*

The control approach followed in this section is as follows. At the first stage, by implementing a finite differences procedure a set of equivalent nonlinear ordinary differential equations are obtained for the diffusion PDE in the Li-ions model. Next it is shown that the system of the nonlinear ODEs is a differentially flat one. This means that all its state variables and the control inputs can be written as differential functions of one single algebraic variable which is the flat output (Boouden *et al.*, 2011; Levine, 2009; Levine, 2011; Rudolph, 2003; Rigatos, 2015; Sira-Ramirez and Agrawal, 2014). By defining specific state variables as virtual control inputs it is shown that each row of the state-space model is a differentially flat subsystem for which a feedback control law can be found that achieves the elimination of the output's tracking error. From the last row of the state-space model, the control input that is finally exerted on the PDE system is computed and this contains recursively all virtual control inputs of the previous rows of the state-space description. Therefore, by tracing the rows of the state-space model backwards on can calculate the control input that should be applied to the PDE system so as to succeed that all its state vector elements track the reference setpoints.

At a second stage, and by exploiting differential flatness properties the state estimation problem for the diffusion PDE model in Li-ions batteries can be solved which consequently means that feedback control can be implemented with the use of a small number of measurements. It is shown that, being differentially flat, the state-space model of the PDE can be transformed into the canonical form and that state-estimation can be performed with the use of the Derivative-free nonlinear Kalman Filter. This filter consists of the Kalman Filter recursion, applied to the canonical form state-space description of the PDE and of an inverse transformation based on differential flatness theory which provides estimates for the initial state variables of the diffusion model (Rigatos, 2011; Rigatos, 2013; Rigatos, 2015; Basseville and Nikiforov, 1993). By performing state estimation it is possible to reconstruct the entire PDE dynamics after taking measurements at a small number of points in the diffusion's spatial grid.

6.9 Control of the diffusion PDE in financial assets' management

6.9.1 Outline

The section develops a differential flatness theory-based approach for feedback control and stabilisation of the valuation PDE process of financial assets, such as stock loans (Rigatos, 2017). Aiming at alleviating the liquidity stress that several companies face, banks and other financial entities or institutions provide stock loans (Chen, Xu *et al.*, 2015; Avolin, 2002; Zhang and Zhou, 2009; Lu and Putri, 2015). Stock loans allow for increasing liquidity for a company without selling its stock (Xin and Zhou, 2007; Fan *et al.*, 2018; Dai and Xu, 2011). The two parties of a stock loan are the borrower (e.g. the company) who initially possesses the stock and the lender (e.g. the bank) to whom the stock is transferred after providing to the borrower a loan (Grasselli and Velez, 2013; Liang and Wu, 2012). The lender can become the owner of the stock in case that the borrower fails to pay back his loan (Liang, Wu *et al.*, 2010; Prager and Zhang, 2016; Cai and Sun, 2014). In such a loan agreement, the borrower may retrieve the stock at any time before or on the maturity, where maturity is the agreed time for paying back the loan (Liu and Xu, 2010; WingWang *et al.*, 2012). The value of the stock loan is the difference between the cumulative value of the stock minus the accumulated interest according to the agreed interest rate (Shi, Zhang *et al.*, 2017; Zhang, Liu *et al.*, 2017). The valuation problem of stock loans is a non-trivial one because the model of a stock-loan's price is complicated and is described either in the form of a stochastic differential equation or in the form of a partial differential equation (Wong and Wong, 2012; Cai and Sun, 2014; Zhang and Zhou, 2009; Huszar and Prado, 2018; Wong and Wong, 2013).

In the present section, the value of the stock loan is shown to vary in accordance to a partial differential equation (PDE) that follows the concept of the Black-Scholes PDE,for options pricing (Pascucci *et al.*, 2013; Hu and Gan, 2018; Aregui *et al.*, 2017). In the Black-Scholes PDE the price of the option is the function of an underlying security index and time, whereas in the stock-loan PDE the value of the loan is a function of the stock and of time. The boundary condition of the stock-loan PDE, at the early exit time or at maturity is shown to be dependent on interest rates and on dividend rates. The section demonstrates that feedback control can be applied to the stock loan PDE, which is equally effective to the PDE control applied so far to other dynamical systems (Bensoussan *et al.*, 2006; Fliess and Join, 2012; Basseville and Nikiforov, 1993; Forsyth and Labahn, 2007). Previous results on optimal control of PDEs with potential use in the problem of price stabilisation of financial assets can be found in (Zheng, 2015; Gao, Xiao *et al.*, 2016; Liu, Zhang *et al.*, 2016; Zheng, 2016). By applying semi-discretisation and the finite differences method, a state-space description of the stock-loan PDE is obtained consisting of a set of nonlinear differential equations (Platen and Heath, 2008; Pascucci, 2011; Orlov *et al.*, 2019; Olivier and Sedoglavic, 2001; Pinsky, 1991). It is demonstrated that this state-space description is differentially flat, which signifies that all its state variables and control inputs can be written as differential functions of specific state variables of it, which are known as flat outputs of the financial system (Rigatos, 2015; Sira-Ramirez and Agrawal, 2014; Levine, 2009). Actually, it is shown that the stock-loan PDE state-space model is brought into the triangular (backstepping integral) form, for which the differential flatness properties have been proven and confirmed. Moreover, in this section a novel proof of the differential flatness properties of the stock loan PDE and of the associated state-space model is given. By examining independently each row of the state-space model it is shown that this stands again for a differentially flat system, for which a virtual control input can be computed as in the case of flatness-based control for the trivial system (Mounier, Rudolph *et al.*, 2010; Fliess and Mounier, 2002; Woitteneck and Mounier, 2010). The virtual control input is chosen such that the ODE subsystem dynamics is linearised and the tracking error is eliminated. The boundary condition that appears in the nonlinear ODE subsystem that comprises the last row of the state-space description, stands for the aggregate control input.

To develop the stabilising feedback control law for the stock-loan PDE, the aforementioned virtual control inputs are processed recursively, moving from the last row of the state-space model to the first one. Consequently, it can be stated that the proposed flatness-based control is implemented in successive (cascading) loops. At each iteration of the control algorithm, and by tracing the rows of the state-space model backwards, one can finally find the control input that needs to be applied to the stock loan PDE system for achieving that all its state vector elements will converge to the desirable setpoints. In addition to the above, one can prove the stability of the control loop in two manners. First, convergence to zero is proven for the tracking error of all subsystems into which the PDE's state-space model is decomposed. Second, with the use of Lyapunov analysis it is reconfirmed that this control scheme is asymptotically stable.

The section has a meaningful contribution in the area of control and financial engineering since it shows that through boundary control the dynamics of the stock-loans PDE (and equivalently of other assets and equities) can be stabilised and can be made to converge to desirable reference levels. The majority of the results in this area analyse the valuation of stock-loans under constant boundary conditions and possibly after considering only variations in volatility and interest rates. Unlike such results, in the present section it is proven that globally stable control of the stock-loans PDE can be achieved after using the associated boundary conditions as control inputs. The section's results allow for considering tools and policies that assure stability in the stock-loans market.

6.9.2 Dynamic model of stock-loans valuation

Considering that in a stock loan the negotiated asset is stock, the value of the loan depends on the value of stock, where the latter is expected to vary in time according to a Black-Scholes PDE. The stock loan variation resembles one of American call options. During the stock loan contract, if the stock value increases, the borrower can repay the loan so as to retrieve the stock and can profit from the elevated value of the stock. If the stock price falls the lender can exit the loan by returning the stock to the borrower and can also claim for return of the amount that had been transferred to the borrower. This limits the loss risk for the lender.

As in the case of option pricing and its dependence on an underlying security index, in the case of a stock loan the loan's value depends on the underlying value of the stock. This is described by the following stochastic differential equation (Lu and Putri, 2015)

$$dS = (r - \delta)Sdt + \sigma SdW_t \tag{6.414}$$

where dW_t is a Wiener process for $t\in[0, \infty)$, $r\geq 0$ is the interest rate, δ is the continuous dividend yield and σ is the volatility of the stock. It is assumed that at time instant $t = 0$ a client borrows an amount q from the lender at an interest rate equal to γ. The value of the stock at the initial time instant is taken to be S_0 and pays also a service fee to the lender equal to c. Then, at any time $t\geq 0$ the borrower owes the amount $qe^{-\gamma t}$.

By employing Ito's lemma in the previous stochastic differential equation, and by using the arbitrage-free opportunity principal the value V of the stock-loan is shown to vary with respect to the stock's value S according to the following partial differential equation (Lu and Putri, 2015):

$$\frac{\partial V}{\partial t} + \frac{1}{2}\sigma^2 S^2 \frac{\partial^2 V}{\partial S^2} + (r - \delta)S\frac{\partial V}{\partial S} - rV = 0 \tag{6.415}$$

with boundary condition $V(0,t)=0$ which signifies that when the stock price $S=0$ has no value, the stock loan has no value either. Equation (2) bares resemblance to the Black-Scholes PDE used in option pricing. However, in (2) the terminal condition of the pay-off at the maturity depends on how the dividend of the stocks is distributed between the borrower and the lender.

It is possible that the early exit of the borrower or the lender from the stock loan takes place. This provides the stock-loan PDE with additional boundary conditions at $S=S_f(t)$ (optimal exit price) and makes the stock-loan evaluation problem a free (moving) boundary value problem, which is equivalent to an American option evaluation problem.

Next, taking into account different forms of the terminal boundary condition, the stock-loan PDE model can be formulated as follows (Lu and Putri, 2015):

a. The lender collects dividend before redemption:

If the lender collects the stock's dividend and pays it back to the borrower at maturity time or upon early exit, then the value of the stock loan at time t is the difference between the stock price S and the accumulated loan amount due to interest rates $qe^{\gamma t}$. Thus the value of the loan at the exit or maturity time is given by $V(S,t)=\max(S-qe^{\gamma t},0)$. The PDE of the stock-loan is formulated as follows:

$$\frac{\partial V_1}{\partial t}+\frac{1}{2}\sigma^2 S_1^2\frac{\partial^2 V_1}{\partial S_1^2}+(r-\delta)S\frac{\partial V_1}{\partial S_1}-rV_1=0 \tag{6.416}$$

with boundary conditions: $V_1(0,t)=0$, $V_1(S,T)=\max(S_1-qe^{\gamma T},0)$, $V_1(S_{f1}(t),t)=S_{f1}(t)-qe^{\gamma t}$, $\frac{\partial V_1}{\partial S_1}(S_{f1}(t),t)=1$

b. *The dividend is re-invested before redemption*:

In such a case there is direct reinvestment of the dividend after this is paid to the lender. This will be returned to the borrower upon redemption, that is when the borrower pays-off the loan. The evolution of the stock price due to acquiring the dividend is

$$e^{\delta t}S=S_0(rt+\sigma)W_t-\frac{1}{2}\sigma^2 t \tag{6.417}$$

Thus, in the case of dividend reinvestment the lender owes to the borrower the difference between the earnings from the dividend and the accumulated debt due to interest rates, that is

$$\max(e^{\delta t}S-qe^{\gamma t},0)\ t\in[0,T] \tag{6.418}$$

It is assumed that $S_2=e^{\delta t}S$ denotes the revenues to be generated by reinvesting the dividend. Then, the stock-loan PDE is formulated as follows

$$\frac{\partial V_2}{\partial t}+\frac{1}{2}\sigma^2 S_2^2\frac{\partial^2 V_2}{\partial S_2^2}+rS_2\frac{\partial V_2}{\partial S_2}-rV_2=0 \tag{6.419}$$

with boundary conditions: $V_2(0,t)=0$, $V_2(S_2,T)=\max(S_2-qe^{\gamma T},0)$, $V_2(S_{f2}(t),t)=S_{f2}(t)-qe^{\gamma t}$, $\frac{\partial V_2}{\partial S_2}(S_{f2}(t),t)=1$

c. The dividend is returned to the borrower:

The dividend's payment to the borrower is taken to be continuous, and during the time interval du the borrower receives back $\delta S_u e^{rt-u}du,\ 0<u<t<T$. Thus at the time instant t the borrower receives the following amount

$$I_t=\int_{u=0}^{u=t}\delta S_u e^{rt-u}du \tag{6.420}$$

Then, the value of the stock-loan becomes

$$\max(S-qe^{\gamma t},0)+I_t\ t\in[0,T] \tag{6.421}$$

Next, by defining $H(S_3,t)=V_3(S,I,t)-I$, a modified PDE for the stock loan's value is obtained:

$$\frac{\partial H}{\partial t}+\frac{1}{2}\sigma^2 S_3^2\frac{\partial^2 H}{\partial S_3^2}+(r-\delta)S_3\frac{\partial H}{\partial S_3}-rH=0 \tag{6.422}$$

with boundary conditions: $H(0,t)=0$, $H(S_3,T)=\max(S_3-qe^{\gamma T},0)$, $H(S_{f3}(t),t)=S_{f3}(t)-qe^{\gamma t}$, $\frac{\partial H}{\partial S_3}(S_{f3}(t),t)=1$.

6.9.3 Transformation of the stock-loan PDE into a set of nonlinear ODEs

6.9.3.1 Decomposition of the PDE model into a set of equivalent ODEs

The stock-loan valuation PDE described above is a diffusion PDE and the solution of the associated boundary control problem will be obtained in the context of feedback control for diffusion PDEs. To this end, the following nonlinear diffusion equation is considered

$$\frac{\partial\phi}{\partial t}=K\frac{\partial^2\phi}{\partial x^2}+f(\phi) \tag{6.423}$$

Using the approximation for the partial derivative

$$\frac{\partial^2\phi}{\partial x^2}\simeq\frac{\phi_{i+1}-2\phi_i+\phi_{i-1}}{\Delta x^2} \tag{6.424}$$

and considering spatial measurements of variable ϕ along axis x at points $x_0+i\Delta x,\ i=1,2,\ldots,N$ one has

$$\frac{\partial\phi_i}{\partial t}=\frac{K}{\Delta x^2}\phi_{i+1}-\frac{2K}{\Delta x^2}\phi_i+\frac{K}{\Delta x^2}\phi_{i-1}+f(\phi_i) \tag{6.425}$$

By considering the associated samples of ϕ given by $\phi_0, \phi_1, \ldots, \phi_N, \phi_{N+1}$ one has

$$\begin{aligned}
\frac{\partial \phi_1}{\partial t} &= \frac{K}{\Delta x^2}\phi_2 - \frac{2K}{\Delta x^2}\phi_1 + \frac{K}{\Delta x^2}\phi_0 + f(\phi_1) \\
\frac{\partial \phi_2}{\partial t} &= \frac{K}{\Delta x^2}\phi_3 - \frac{2K}{\Delta x^2}\phi_2 + \frac{K}{\Delta x^2}\phi_1 + f(\phi_2) \\
\frac{\partial \phi_3}{\partial t} &= \frac{K}{\Delta x^2}\phi_4 - \frac{2K}{\Delta x^2}\phi_3 + \frac{K}{\Delta x^2}\phi_2 + f(\phi_3) \\
&\cdots \\
\frac{\partial \phi_{N-1}}{\partial t} &= \frac{K}{\Delta x^2}\phi_N - \frac{2K}{\Delta x^2}\phi_{N-1} + \frac{K}{\Delta x^2}\phi_{N-2} + f(\phi_{N-1}) \\
\frac{\partial \phi_N}{\partial t} &= \frac{K}{\Delta x^2}\phi_{N+1} - \frac{2K}{\Delta x^2}\phi_N + \frac{K}{\Delta x^2}\phi_{N-1} + f(\phi_N)
\end{aligned} \tag{6.426}$$

By defining the following state vector

$$x^T = (\phi_1, \phi_2, \ldots, \phi_N) \tag{6.427}$$

one obtains the following state-space description

$$\begin{aligned}
\dot{x}_1 &= \frac{K}{\Delta x^2}x_2 - \frac{2K}{\Delta x^2}x_1 + \frac{K}{\Delta x^2}\phi_0 + f(x_1) \\
\dot{x}_2 &= \frac{K}{\Delta x^2}x_3 - \frac{2K}{\Delta x^2}x_2 + \frac{K}{\Delta x^2}x_1 + f(x_2) \\
\dot{x}_3 &= \frac{K}{\Delta x^2}x_4 - \frac{2K}{\Delta x^2}x_3 + \frac{K}{\Delta x^2}x_2 + f(x_3) \\
&\cdots \\
\dot{x}_{N-1} &= \frac{K}{\Delta x^2}x_N - \frac{2K}{\Delta x^2}x_{N-1} + \frac{K}{\Delta x^2}x_{N-2} + f(x_{N-1}) \\
\dot{x}_N &= \frac{K}{\Delta x^2}\phi_{N+1} - \frac{2K}{\Delta x^2}x_N + \frac{K}{\Delta x^2}x_{N-1} + f(x_N)
\end{aligned} \tag{6.428}$$

Next, the following state variables are defined

$$y_{1,i} = x_i \tag{6.429}$$

and the state-space description of the system becomes as follows

$$
\begin{aligned}
\dot{y}_{1,1} &= \frac{K}{\Delta x^2}y_{1,2} - \frac{2K}{\Delta x^2}y_{1,1} + \frac{K}{\Delta x^2}\phi_0 + f(y_{1,1}) \\
\dot{y}_{1,2} &= \frac{K}{\Delta x^2}y_{1,3} - \frac{2K}{\Delta x^2}y_{1,2} + \frac{K}{\Delta x^2}y_{1,1} + f(y_{1,2}) \\
\dot{y}_{1,3} &= \frac{K}{\Delta x^2}y_{1,4} - \frac{2K}{\Delta x^2}y_{1,3} + \frac{K}{\Delta x^2}y_{1,2} + f(y_{1,3}) \\
&\cdots \\
&\cdots \\
\dot{y}_{1,N-1} &= \frac{K}{\Delta x^2}y_{1,N} - \frac{2K}{\Delta x^2}y_{1,N-1} + \frac{K}{\Delta x^2}y_{1,N-2} + f(y_{1,N-1}) \\
\dot{y}_{1,N} &= \frac{K}{\Delta x^2}\phi_{N+1} - \frac{2K}{\Delta x^2}y_{1,N} + \frac{K}{\Delta x}y_{1,N-1} + f(y_{1,N})
\end{aligned}
\tag{6.430}
$$

The dynamical system described in (6.430) is a differentially flat one with flat output defined as the vector $\tilde{y} = [y_{1,1}, y_{1,2}, \ldots, y_{1,N}]$. Indeed all state variables can be written as functions of the flat output and its derivatives.

Denoting $a = \frac{K}{Dx^2}$ and $b = -\frac{2K}{Dx^2}$, the initial description of the system is rewritten as follows

$$
\begin{aligned}
\begin{pmatrix} \dot{y}_{1,1} \\ \dot{y}_{1,2} \\ \cdots \\ \dot{y}_{1,N-1} \\ \dot{y}_{1,N} \end{pmatrix} &= \begin{pmatrix} b & a & 0 & 0 & \cdots & 0 & 0 & 0 \\ a & b & a & 0 & \cdots & 0 & 0 & 0 \\ 0 & a & b & a & \cdots & 0 & 0 & 0 \\ \cdots & \cdots & \cdots & \cdots & \cdots & & & \\ 0 & 0 & 0 & 0 & \cdots & a & b & a \\ 0 & 0 & 0 & 0 & \cdots & 0 & a & b \end{pmatrix} \begin{pmatrix} y_{1,1} \\ y_{1,2} \\ \cdots \\ y_{1,N-1} \\ y_{1,N} \end{pmatrix} \\
&+ \begin{pmatrix} 1 & 0 & 0 & \cdots & 0 & 0 \\ 0 & 1 & 0 & \cdots & 0 & 0 \\ 0 & 0 & 1 & \cdots & 0 & 0 \\ \cdots & \cdots & \cdots & \cdots & & \\ 0 & 0 & 0 & \cdots & 1 & 0 \\ 0 & 0 & 0 & \cdots & 0 & 1 \end{pmatrix} \begin{pmatrix} v_1 \\ v_2 \\ v_3 \\ \cdots \\ v_{N-1} \\ v_N \end{pmatrix}
\end{aligned}
\tag{6.431}
$$

The associated control inputs are defined as

$$
\begin{aligned}
v_1 &= \frac{K}{\Delta x^2}\phi_0 + f(y_{1,1}) \\
v_2 &= f(y_{1,2}) \\
v_3 &= f(y_{1,3}) \\
&\cdots \\
v_{N-1} &= f(y_{1,N-1}) \\
v_N &= \frac{K}{\Delta x^2}\phi_{N+1} + f(y_{1,N})
\end{aligned}
\tag{6.432}
$$

By selecting measurements from a subset of points x_j $j\in[1,2,\ldots,m]$, the associated observation (measurement) equation becomes

$$\begin{pmatrix} z_1 \\ z_2 \\ \cdots \\ z_m \end{pmatrix} = \begin{pmatrix} 1 & 0 & 0 & \cdots & 0 & 0 \\ \cdots & \cdots & \cdots & \cdots & & \\ 0 & 0 & 0 & \cdots & 0 & 1 \end{pmatrix} \begin{pmatrix} y_{1,1} \\ y_{1,2} \\ \cdots \\ y_{1,N} \end{pmatrix} \tag{6.433}$$

Thus, in matrix form one has the following state-space description of the system

$$\begin{aligned} \dot{\tilde{y}} &= A\tilde{y} + Bv \\ \tilde{z} &= C\tilde{y} \end{aligned} \tag{6.434}$$

6.9.3.2 Modeling in state-space form of the stock-loan PDE

Next, the following nonlinear stock-loans PDE is considered:

$$\frac{\partial V(t,S)}{\partial t} + (r_t - \delta_t)S\frac{\partial V(t,S)}{\partial S} + \frac{1}{2}\sigma_t^2 S^2 \frac{\partial^2 V(t,S)}{\partial S^2} - r_t V(t,S) = 0 \tag{6.435}$$

It is considered that the volatility σ_t has a nonlinear dependence on V_{ss}, that is

$$\sigma_t = \sigma(V_{ss}) \tag{6.436}$$

where $V_{ss} = \frac{\partial^2 V(t,s)}{\partial S^2}$ and σ is a nonlinear function. Again, a grid of N points is considered, that is $\{s_1, s_2, \cdots, s_{N-1}, s_N\}$ which are placed at equal distances on the S axis. At the points of spatial discretisation it holds

$$\begin{aligned} \frac{\partial V(t,s_i)}{\partial t} &= r_t V(t,s_i) - (r_t - \delta_t)s_i \frac{\partial V(t,s_i)}{\partial S} \\ &\quad - \frac{1}{2}\sigma_t^2(V_{ss})s_i^2 \frac{\partial^2 V(t,s_i)}{\partial S^2} \end{aligned} \tag{6.437}$$

where $i = 1, 2, \ldots, N$. The following state vector is defined

$$\begin{aligned} \tilde{V} &= [V(t,s_1), V(t,s_2), V(t,s_3), \ldots, V(t,s_{N-1}), V(t,s_N)] \text{ or} \\ \tilde{V} &= [V_1, V_2, V_3, \ldots, V_{N-1}, V_N] \end{aligned} \tag{6.438}$$

thus one has

$$\begin{aligned}
\frac{\partial V_1}{\partial t} &= r_t V_1 - (r_t - \delta_t) s_1 \frac{V_2 - V_1}{\Delta S} - \frac{1}{2}\sigma_t^2 (V_{ss}) s_1^2 \frac{V_2 - 2V_1 + V_0}{\Delta S^2} \\
\frac{\partial V_2}{\partial t} &= r_t V_2 - (r_t - \delta_t) s_2 \frac{V_3 - V_2}{\Delta S} - \frac{1}{2}\sigma_t^2 (V_{ss}) s_2^2 \frac{V_3 - 2V_2 + V_1}{\Delta S^2} \\
\frac{\partial V_3}{\partial t} &= r_t V_3 - (r_t - \delta_t) s_3 \frac{V_4 - V_3}{\Delta S} - \frac{1}{2}\sigma_t^2 (V_{ss}) s_3^2 \frac{V_4 - 2V_3 + V_2}{\Delta S^2} \\
&\cdots \\
&\cdots \\
\frac{\partial V_{N-1}}{\partial t} &= r_t V_{N-1} - (r_t - \delta_t) s_{N-1} \frac{V_N - V_{N-1}}{\Delta S} \\
&\quad - \frac{1}{2}\sigma_t^2 (V_{ss}) s_{N-1}^2 \frac{V_N - 2V_{N-1} + V_{N-2}}{\Delta S^2} \\
\frac{\partial V_N}{\partial t} &= r_t V_N - (r_t - \delta_t) s_N \frac{V_{N+1} - V_N}{\Delta S} - \frac{1}{2}\sigma_t^2 (V_{ss}) s_N^2 \frac{V_{N+1} - 2V_N + V_{N-1}}{\Delta S^2}
\end{aligned} \tag{6.439}$$

For the ith ODE one has at sampling point S_i along the S axis one has

$$\begin{aligned}
\frac{\partial V_i}{\partial t} &= r_t V_i - (r_t - \delta_t) s_i \frac{V_{i+1} - V_i}{\Delta S} - \frac{1}{2}\sigma_t^2 (V_{ss}) s_i^2 \frac{V_{i+1}}{\Delta S^2} \\
&\quad + \sigma_t^2 (V_{ss}) s_i^2 \frac{V_i}{\Delta S^2} - \frac{1}{2}\sigma_t^2 (V_{ss}) s_i^2 \frac{V_{i-1}}{\Delta S^2}
\end{aligned} \tag{6.440}$$

Equation (6.440) is also written in the form

$$\begin{aligned}
\frac{\partial V_i}{\partial t} &= -\frac{1}{2}\frac{\sigma_t^2 (V_{ss})}{\Delta S^2} s_i^2 V_{i-1} + \frac{\sigma_t^2 (V_{ss})}{\Delta S^2} s_i^2 V_i \\
&\quad - \left[\frac{(r_t - \delta_t) s_i}{\Delta S} + \frac{1}{2}\frac{\sigma_t^2 (V_{ss})}{\Delta S^2} s_i^2] V_{i+1} + [r_t + \frac{(r_t - \delta_t) s_i}{\Delta S} \right] V_i
\end{aligned} \tag{6.441}$$

Next, by denoting $K_1 = -\sigma_t^2 (V_{ss}) s_i^2$, $K_2 = -[2(r_t - \delta_t) s_i \Delta S + \sigma_t^2 (V_{ss}) s_i^2]$ and $f(V_i) = [r_t + \frac{(r_t - \delta_t) s_i}{\Delta S}] V_i$ one obtains the following description for (6.441)

$$\frac{\partial V_i}{\partial t} = \frac{K_1}{2\Delta S^2} V_{i-1} - \frac{K_1}{\Delta S^2} V_i + \frac{K_2}{2\Delta S^2} V_{i+1} + f(V_i) \tag{6.442}$$

6.9.4 Differential flatness of the stock-loan PDE model

Next, the following state vector and the associated state variables are defined for the stock-loan PDE model of (6.439): $\tilde{Y} = [y_{1,1}, y_{1,2}, \ldots, y_{1,i}, \ldots, y_{1,N-1}, y_{1,N}]$, where $y_{1,1} = V_1$, $y_{1,2} = V_2, \ldots, y_{1,i} = V_i, \ldots, y_{1,N-1} = V_{N-1}$ and $y_{1,N} = V_N$. It will be shown that the state-space description of the nonlinear PDE dynamics , which has as control input only the boundary condition ϕ_0 is a differentially flat one. One has

$$\dot{y}_{1,1} = -\frac{K_1}{\Delta S^2} y_{1,1} + \frac{K_1}{2\Delta S^2} \phi_0 + f(y_{1,1}) + \frac{K_2}{2\Delta S^2} y_{1,2} \tag{6.443}$$

$$\dot{y}_{1,2} = -\frac{K_1}{\Delta S^2}y_{1,2} + \frac{K_1}{2\Delta S^2}y_{1,1} + f(y_{1,2}) + \frac{K_2}{2\Delta S^2}y_{1,3} \tag{6.444}$$

$$\dot{y}_{1,3} = -\frac{K_1}{\Delta S^2}y_{1,3} + \frac{K_1}{2\Delta S^2}y_{1,2} + f(y_{1,3}) + \frac{K_2}{2\Delta S^2}y_{1,4} \tag{6.445}$$

$\cdots\ \cdots$

$$\dot{y}_{1,i} = \frac{K_1}{\Delta S^2}y_{1,i} + \frac{K_1}{\Delta S^2}y_{1,i-1} + f(y_{1,i}) + \frac{K_2}{2\Delta S^2}y_{1,i+1} \tag{6.446}$$

$\cdots\ \cdots$

$$\begin{aligned}\dot{y}_{1,N-1} = &-\frac{K_1}{\Delta S^2}y_{1,N-1} + \frac{K_1}{2\Delta S^2}y_{1,N-2} \\ &+ f(y_{1,N-1}) + \frac{K_2}{2\Delta S^2}y_{1,N}\end{aligned} \tag{6.447}$$

$$\begin{aligned}\dot{y}_{1,N} = &-\frac{K_1}{\Delta S^2}y_{1,N} + \frac{K_1}{2\Delta S^2}y_{1,N-1} \\ &+ f(y_{1,N}) + \frac{K_2}{2\Delta S^2}\phi_{N+1}\end{aligned} \tag{6.448}$$

The flat output is considered to be the state variable $y_{1,N}$, which is denoted as $y = y_{1,N}$. Next, it is shown that all state variables which stand also for virtual control inputs of the system $\alpha_i = y_{1,N-i}$, can be written as functions of the flat output $y = y_{1,N}$.

From (6.448) and considering that ϕ_{N+1} is constant one obtains

$$\begin{aligned}y_{1,N-1} = \alpha_1 = &\frac{1}{K_1/2\Delta S^2}\left[\dot{y}_{1,N} + \frac{K_1}{\Delta S^2}y_{1,N} - \frac{K_2}{2\Delta S^2}\phi_{1,N+1} - f(y_{1,N})\right] \\ &\Rightarrow y_{1,N-1} = h_1(y, \dot{y}, \ldots)\end{aligned} \tag{6.449}$$

and following a similar procedure, from (6.447) one gets

$$\begin{aligned}y_{1,N-2} = \alpha_2 = &\frac{1}{K_1/2\Delta S^2}\left[\dot{y}_{1,N-1} + \frac{K_1}{\Delta S^2}y_{1,N-1} - \frac{K_2}{2\Delta S^2}y_{1,N} - f(y_{1,N-1})\right] \\ &\Rightarrow y_{1,N-2} = h_2(y, \dot{y}, \ldots)\end{aligned} \tag{6.450}$$

Continuing in a similar manner, from (6.446) one obtains

$$\begin{aligned}y_{1,i-1} = \alpha_{N-i+1} = &\frac{1}{K_1/2\Delta S^2}\left[\dot{y}_{1,i} + \frac{K_1}{\Delta S^2}y_{1,i} - \frac{K_2}{2\Delta S^2}y_{1,i+1} - f(y_{1,i})\right] \\ &\Rightarrow y_{1,i-1} = h_{N-i+1}(y, \dot{y}, \ldots)\end{aligned} \tag{6.451}$$

From (6.445) one obtains

$$\begin{aligned}y_{1,2} = \alpha_{N-2} = &\frac{1}{K_1/2\Delta S^2}\left[\dot{y}_{1,3} + \frac{K_1}{\Delta S^2}y_{1,3} - \frac{K_2}{\Delta S^2}y_{1,4} - f(y_{1,3})\right] \\ &\Rightarrow y_{1,2} = h_{N-2}(y, \dot{y}, \ldots)\end{aligned} \tag{6.452}$$

From (6.444) one obtains

$$\begin{aligned}y_{1,1} = \alpha_{N-1} = &\frac{1}{K_1/2\Delta S^2}\left[\dot{y}_{1,2} + \frac{K_1}{\Delta S^2}y_{1,2} - \frac{K_2}{\Delta S^2}y_{1,3} - f(y_{1,2})\right] \\ &\Rightarrow y_{1,1} = h_{N-1}(y, \dot{y}, \ldots)\end{aligned} \tag{6.453}$$

Finally, from (6.443) one obtains

$$\begin{aligned}\phi_0 = \alpha_N &= \frac{1}{K_1/2\Delta S^2}\left[\dot{y}_{1,1} + \frac{K_1}{\Delta S^2}y_{1,1} - \frac{K_2}{\Delta S^2}y_{1,2} - f(y_{1,1})\right] \\ &\Rightarrow \phi_0 = h_N(y, \dot{y}, \dots)\end{aligned} \tag{6.454}$$

The above procedure confirms that all state variables of the model

$$\begin{array}{llll} y_{1,1} & y_{1,2} & y_{1,3} & \cdots \\ y_{1,i} & \cdots & y_{1,N-1} & y_{1,N} \end{array} \tag{6.455}$$

and the control input which is the boundary condition ϕ_0 can be written as functions of the flat output $y = y_{1,N}$ and of the flat output's derivatives. Consequently, the state-space model of the PDE dynamics is differentially flat.

Additionally, one can consider decomposition of the PDE state-space equation into submodels, where at each submodel the virtual control input is $\alpha_i = y_{1,N-i}$

$$\begin{aligned}\dot{y}_{1,i} &= -\frac{K_1}{\Delta x^2}y_{1,i} + \frac{K_1}{2\Delta x^2}y_{1,i-1} + f(y_{1,i}) + \frac{K_2}{2\Delta x^2}y_{1,i+1} \\ &\cdots\cdots\end{aligned} \tag{6.456}$$

and with local flat output $y_{1,i}$ one can confirm that all such subsystems $i = 1, 2, \dots, N$ are differentially flat. Next, one can compute the virtual inputs which are applied to each subsystem.

6.9.5 Computation of a boundary conditions-based feedback control law

To implement boundary feedback control in the system of (6.443) to (6.448), the nonlinear wave-PDE model is rewritten as follows:

$$\dot{y}_{1,N} = -\frac{K_1}{\Delta S^2}y_{1,N} + f(y_{1,N}) + \frac{K_2}{2\Delta S^2}\phi_{N+1} + \frac{K_1}{2\Delta S^2}y_{1,N-1} \tag{6.457}$$

$$\dot{y}_{1,N-1} = -\frac{K_1}{\Delta S^2}y_{1,N-1} + f(y_{1,N-1}) + \frac{K_2}{2\Delta S^2}y_{1,N} + \frac{K_1}{2\Delta S^2}y_{1,N-2} \tag{6.458}$$

$$\begin{aligned}\dot{y}_{1,N-2} &= -\frac{K_1}{\Delta S^2}y_{1,N-2} + f(y_{1,N-2}) + \frac{K_2}{2\Delta S^2}y_{1,N-1} + \frac{K_1}{2\Delta S^2}y_{1,N-1} \\ &\cdots\cdots\end{aligned} \tag{6.459}$$

$$\begin{aligned}\dot{y}_{1,i} &= -\frac{K_1}{\Delta S^2}y_{1,i} + f(y_{1,i}) + \frac{K_2}{2\Delta S^2}y_{1,i+1} + \frac{K_1}{2\Delta S^2}y_{1,i-1} \\ &\cdots\cdots\end{aligned} \tag{6.460}$$

$$\begin{aligned}\dot{y}_{1,3} &= -\frac{K_1}{\Delta S^2}y_{1,3} + f(y_{1,3}) + \frac{K_2}{2\Delta S^2}y_{1,4} + \frac{K_1}{2\Delta S^2}y_{1,2} \\ &\cdots\cdots\end{aligned} \tag{6.461}$$

$$\dot{y}_{1,2} = -\frac{K_1}{\Delta S^2}y_{1,2} + f(y_{1,2}) + \frac{K_2}{2\Delta S^2}y_{1,3} + \frac{K_1}{2\Delta S^2}y_{1,1} \tag{6.462}$$

$$\dot{y}_{1,1} = -\frac{K_1}{\Delta S^2}y_{1,1} + f(y_{1,1}) + \frac{K_2}{2\Delta S^2}y_{1,2} + \frac{K_1}{2\Delta S^2}\phi_0 \tag{6.463}$$

The boundary condition ϕ_{N+1} is assumed to be known and remains steady. The boundary condition ϕ_0 stands for the control input (for example, it can be dependent on the interest rate). The feedback control law is designed as follows:

$$\begin{aligned}\alpha_1 = y^*_{1,N-1} = &\frac{1}{(K_1/2\Delta S^2)}\cdot\Big[\dot{y}^d_{1,N} - k_{p,1}(y_{1,N} - y^d_{1,N})\\ &+\frac{K_1}{2\Delta S^2}y_{1,N} - f(y_{1,N}) - \frac{K_2}{2\Delta S^2}\phi_{N+1}\Big]\end{aligned} \tag{6.464}$$

$$\begin{aligned}\alpha_2 = y^*_{1,N-2} = &\frac{1}{(K_1/2\Delta S^2)}\cdot\Big[\dot{y}^d_{1,N-1} - k_{p,2}(y_{1,N-1} - y^d_{1,N-1})\\ &+\frac{K_1}{2\Delta S^2}y_{1,N-1} - f(y_{1,N-1}) - \frac{K_2}{2\Delta S^2}y_N\Big] \Rightarrow\\ \alpha_2 = &\frac{1}{(K_1/2\Delta S^2)}\Big[\dot{\alpha}_1 - k_{p,2}(y_{1,N-1} - \alpha_1)\\ &+\frac{K_1}{2\Delta S^2}y_{1,N-1} - f(y_{1,N-1}) - \frac{K_2}{2\Delta S^2}y_N\Big]\end{aligned} \tag{6.465}$$

$$\begin{aligned}\alpha_3 = y^*_{1,N-3} = &\frac{1}{(K_1/2\Delta S^2)}\cdot\Big[\dot{y}^d_{1,N-2} - k_{p,3}(y_{1,N-2} - y^d_{1,N-2})\\ &+\frac{K_1}{2\Delta S^2}y_{1,N-2} - f(y_{1,N-2}) - \frac{K_2}{2\Delta S^2}y_{N-1}\Big] \Rightarrow\\ \alpha_3 = &\frac{1}{(K_1/2\Delta S^2)}[\dot{\alpha}_2 - k_{p,3}(y_{1,N-2} - \alpha_2)\\ &+\frac{K_1}{2\Delta S^2}y_{1,N-2} - f(y_{1,N-2}) - \frac{K_2}{2\Delta S^2}y_{N-1}]\end{aligned} \tag{6.466}$$

and continuing in a similar manner

$$\begin{aligned}\alpha_i = y^*_{1,N-i} = &\frac{1}{(K_1/2\Delta S^2)}\cdot\Big[\dot{y}^d_{1,N-i+1} - k_{p,i}(y_{1,N-i+1} - y^d_{1,N-i+1})\\ &+\frac{K_1}{2\Delta S^2}y_{1,N-i+1} - f(y_{1,N-i+1}) - \frac{K_2}{2\Delta S^2}y_{N-i+2}\Big] \Rightarrow\\ \alpha_i = &\frac{1}{(K_1/2\Delta S^2)}\Big[\dot{\alpha}_{i-1} - k_{p,i}(y_{1,N-i+1} - \alpha_{i-1})\\ &+\frac{K_1}{2\Delta S^2}y_{1,N-i+1} - f(y_{1,N-i+1}) - \frac{K_2}{2\Delta S^2}y_{N-i+2}\Big]\end{aligned} \tag{6.467}$$

Following this procedure one arrives to compute the control inputs which are associated with the last two rows of the state-space model

$$\begin{aligned}\alpha_{N-2} = y^*_{1,2} = &\frac{1}{(K_1/2\Delta S^2)}\cdot\Big[\dot{y}^d_{1,3} - k_{p,N-1}(y_{1,3} - y^d_{1,3})\\ &+\frac{K_1}{2\Delta S^2}y_{1,3} - f(y_{1,3}) - \frac{K_2}{2\Delta S^2}y_{1,4}\Big] \Rightarrow\\ \alpha_{N-2} = &\frac{1}{(K_1/\Delta S^2)}\Big[\dot{\alpha}_{N-3} - k_{p,N-2}(y_{1,3} - \alpha_{N-3})\\ &+\frac{K_1}{2\Delta S^2}y_{1,3} - f(y_{1,3}) - \frac{K_2}{2\Delta S^2}y_{1,4}\Big]\end{aligned} \tag{6.468}$$

$$\begin{aligned}\alpha_{N-1} = y_{1,1}^{*} &= \frac{1}{(K_1/2\Delta S^2)}\cdot\left[\dot{y}_{1,2}^{d} - k_{p,N-1}(y_{1,2} - y_{1,2}^{d})\right.\\ &\left.+\frac{K_1}{2\Delta S^2}y_{1,2} - f(y_{1,2}) - \frac{K_2}{2\Delta S^2}y_{1,3}\right]\Rightarrow\\ \alpha_{N-1} &= \frac{1}{(K_1/\Delta S^2)}\left[\dot{\alpha}_{N-2} - k_{p,N-1}(y_{1,2} - \alpha_{N-2})\right.\\ &\left.+\frac{K_1}{2\Delta S^2}y_{1,2} - f(y_{1,2}) - \frac{K_2}{2\Delta S^2}y_{1,3}\right]\end{aligned} \tag{6.469}$$

and finally

$$\begin{aligned}\alpha_{N} = \phi_0 &= \frac{1}{(K_1/2\Delta S^2)}\cdot\left[\dot{y}_{1,1}^{d} - k_{p,N}(y_{1,1} - y_{1,1}^{d})\right.\\ &\left.+\frac{K_1}{2\Delta S^2}y_{1,1} - f(y_{1,1}) - \frac{K_2}{2\Delta S^2}y_{1,2}\right]\Rightarrow\\ \alpha_{N} &= \frac{1}{(K_1/2\Delta S^2)}\left[\dot{\alpha}_{N-1} - k_{p,N}(y_{1,1} - \alpha_{N-1})\right.\\ &\left.+\frac{K_1}{2\Delta S^2}y_{1,1} - f(y_{1,1}) - \frac{K_2}{2\Delta S^2}y_{1,2}\right]\end{aligned} \tag{6.470}$$

Consequently, the computation of the aggregate control input $\alpha_N = \phi_0$ which is applied to the PDE model is performed backwards, by substituting recursively into ϕ_0 the virtual control inputs a_{N-1}, a_{N-2},. . .,$a_i s$,. . .,a_2,a_1.

6.9.6 Closed-loop dynamics

By substituting (6.470) into (6.463) of the state-space model of the PDE dynamics, and using the definition $y_{1,1} - y_{1,1}^{d} = z_1$ one has

$$\begin{aligned}&\dot{y}_{1,1} = \dot{y}_{1,1}^{d} - k_{p,1}(y_{1,1} - y_{1,1}^{d})\Rightarrow\\ &(\dot{y}_{1,1} - \dot{y}_{1,1}) + k_{p,1}(y_{1,1} - y_{1,1}^{d}) = 0\Rightarrow\\ &\dot{z}_1 + k_{p,1}z_1 = 0\end{aligned} \tag{6.471}$$

Equivalently, by substituting (6.469) into (6.462), and using the definition $y_{1,2} - \alpha_1 = z_2$ one has

$$\begin{aligned}&\dot{y}_{1,2} = \dot{\alpha}_1 - k_{p,2}(y_{1,2} - \alpha_1)\Rightarrow\\ &(\dot{y}_{1,2} - \dot{\alpha}_1) + k_{p,2}(y_{1,2} - \alpha_1) = 0\Rightarrow\\ &\dot{z}_2 + k_{d,2}\dot{z}_2 + k_{p,2}z_2 = 0\end{aligned} \tag{6.472}$$

Similarly, continuing with the rest of the equations of the state-space model and by substituting (6.468) into (6.461), while also using the definition $y_{1,3} - \alpha_2 = z_3$ one has

$$\begin{aligned}&\dot{y}_{1,3} = \dot{\alpha}_2 - k_{p,3}(y_{1,3} - \alpha_2)\Rightarrow\\ &(\dot{y}_{1,3} - \dot{\alpha}_2) + k_{p,3}(y_{1,3} - \alpha_2) = 0\Rightarrow\\ &\ddot{z}_3 + k_{p,3}z_3 = 0\end{aligned} \tag{6.473}$$

Moving backwards, and by substituting (6.467) into (6.460) of the state-space model of the PDE dynamics, and using the definition $y_{1,i} - \alpha_{i-1} = z_i$ one has

$$\begin{aligned}&\dot{y}_{1,i} = \dot{\alpha}_{i-1} - k_{p,i}(y_{1,i} - \alpha_{i-1})\Rightarrow \\ &(\dot{y}_{1,i} - \dot{\alpha}_{i-1}) + k_{p,i}(y_{1,i} - \alpha_{i-1}) = 0\Rightarrow \\ &\dot{z}_i + k_{p,i}z_i = 0\end{aligned} \tag{6.474}$$

By substituting (6.465) into (6.458) of the state-space model of the PDE dynamics, and using the definition $y_{1,N-1} - \alpha_{N-2} = z_{N-1}$ one has

$$\begin{aligned}&\dot{y}_{1,N-1} = \dot{\alpha}_{N-2} - k_{p,N-1}(y_{1,N-1} - \alpha_{N-2})\Rightarrow \\ &(\dot{y}_{1,N-1} - \dot{\alpha}_{N-2}) + k_{p,N-1}(y_{1,N-1} - \alpha_{N-2}) = 0\Rightarrow \\ &\dot{z}_{N-1} + k_{p,N-1}z_{N-1} = 0\end{aligned} \tag{6.475}$$

Finally, by substituting (6.464) into (6.458), and using the definition $y_{1,N} - \alpha_{N-1} = z_N$ one obtains

$$\begin{aligned}&\dot{y}_{1,N} = \dot{\alpha}_{N-1} - k_{p,N}(y_{1,N} - \alpha_{N-1})\Rightarrow \\ &(\dot{y}_{1,N} - \dot{\alpha}_{N-1}) + k_{p,N}(y_{1,N} - \alpha_{N-1}) = 0\Rightarrow \\ &\dot{z}_N + k_{p,N}z_N = 0\end{aligned} \tag{6.476}$$

Thus, the dynamics of the closed-loop system becomes

$$\begin{aligned}&\dot{z}_1 + k_{p,1}z_1 = 0 \\ &\dot{z}_2 + k_{p,2}z_2 = 0 \\ &\dot{z}_3 + k_{p,3}z_3 = 0 \\ &\cdots\ \cdots \\ &\dot{z}_i + k_{p,i}z_i = 0 \\ &\cdots\ \cdots \\ &\dot{z}_{N-1} + k_{p,N-1}z_{N-1} = 0 \\ &\dot{z}_N + k_{p,N}z_N = 0\end{aligned} \tag{6.477}$$

The dynamics of the closed-loop system can be also written in matrix form

$$\dot{\tilde{Z}} + K_P\tilde{Z} = 0 \tag{6.478}$$

where $\tilde{Z} = [z_1, z_2, z_3, \ldots, z_i, \ldots, z_{N-1}, z_N]$,and
$K_P = \text{diag}[k_{p,1}, k_{p,2}, k_{p,3}, \ldots, k_{p,i}, \ldots, k_{p,N-1}, k_{p,N}]$.

After suitable selection of the coefficients $k_{P,i}$, $i = 1, 2, \ldots, N$ such that the monomials $p(s) = s + k_{p,i}$ to have negative roots, it can be assured that $\lim_{t\to\infty}z_i(t) = 0$ and that the closed-loop system is asymptotically stable.

Moreover, to prove asymptotic stability for the closed-loop system, the following Lyapunov function can be used

$$V = \sum_{i=1}^{N}\frac{1}{2}z_i^2 \tag{6.479}$$

The derivative of this Lyapunov function with respect to time is

$$\begin{aligned}
\dot{V} &= \sum_{i=1}^{N} \frac{1}{2} 2 z_i \dot{z}_i \Rightarrow \dot{V} = \sum_{i=1}^{N} z_i(-k_{p_i} z_i) \Rightarrow \\
\dot{V} &= -\sum_{i=1}^{N} k_{p,i} z_i^2 < 0
\end{aligned} \tag{6.480}$$

Thus, it is proven again that the closed-loop system is globally asymptotically stable.

6.9.7 Simulation tests

Indicative results about the performance of the proposed control method for the stabilisation of the stock-loan PDE are shown in the diagrams given in Figures 6.67–6.71. The proposed control method assures that the value of the state variables of the stock-loan PDE will track accurately the reference setpoints. The spatial discretisation of the PDE model consisted of $N = 12$ points. The sampling period was chosen to be T_s=0.01 sec. Other parameters of the model were $r_t = 0.1$, $\delta = 0.05$ and $\sigma = 0.5$. The boundary condition V_{N+1} served as the control input, while the boundary condition V0 was set to zero.

The numerical simulation experiments have confirmed the theoretical findings of this section. It has been shown that by applying the proposed control method, the stock-loan PDE dynamics can be modified so as to converge to the desirable reference profile. The control input that achieves this changes smoothly and has a moderate range of variation. The accuracy of tracking of the reference setpoints was quite satisfactory. The proposed method shows that stabilisation of financial

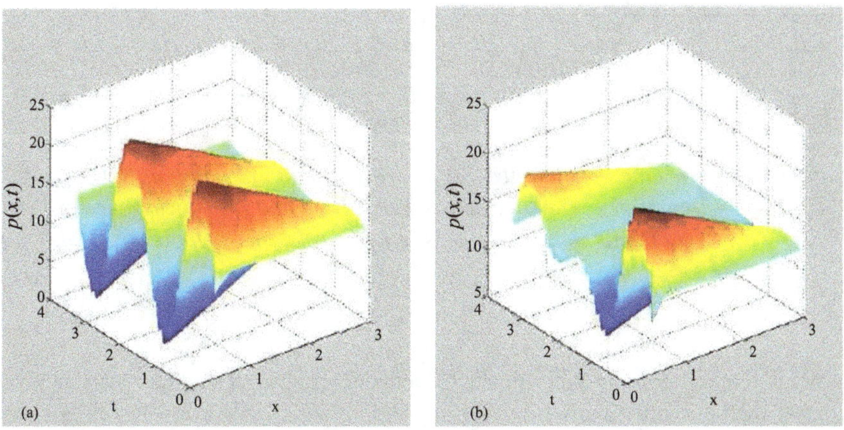

Figure 6.67 (a) Tracking of setpoint 1: plot of the 2D-stock loan PDE function after applying boundary control and (b) tracking of setpoint 2: plot of the 2D-stock loan PDE function after applying boundary control

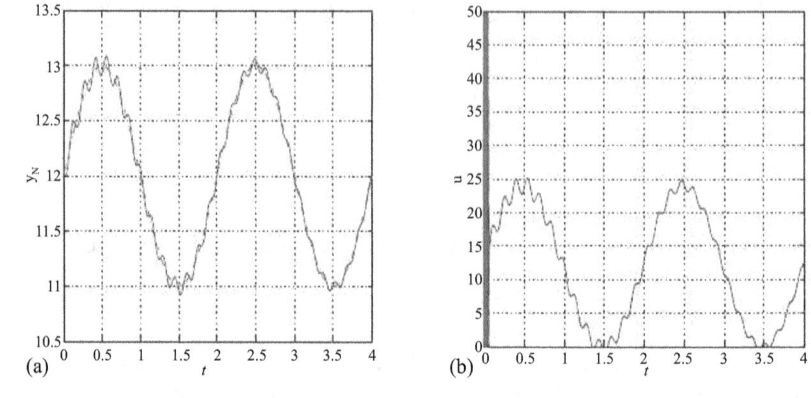

Figure 6.68 Tracking of setpoint 1: (a) convergence of the final grid point y_N (blue line) to the reference setpoint (red line) and (b) variation of the boundary control input

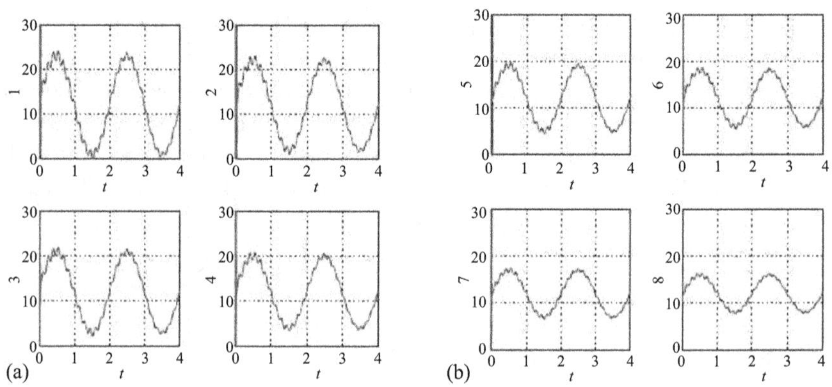

Figure 6.69 Tracking of setpoint 1: (a) convergence of the grid points y_1 to y_4 (blue line) to the reference setpoint (red line) and (b) convergence of the grid points y_5 to y_8 (blue line) to the reference setpoint (red line)

systems dynamics is possible through feedback control. The previously analysed control method confirms that by varying the boundary condition which is related to the exit time or the maturity of the loan's agreement one can make the stock loan's value track and converge to specific reference values. Implicitly the boundary condition can be determined by the interest and dividend rates.

The results of the section demonstrate that by applying feedback control based on differential flatness theory the spatiotemporal dynamics of the stock-loans PDE can

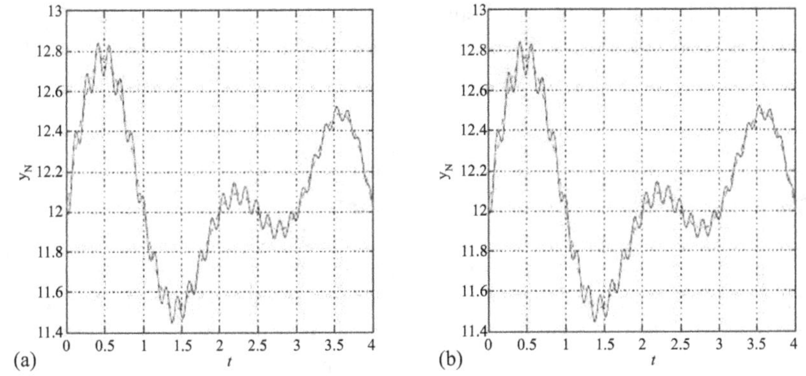

Figure 6.70 *Tracking of setpoint 2: (a) convergence of the final grid point y_N (blue line) to the reference setpoint (red line) and (b) variation of the boundary control input*

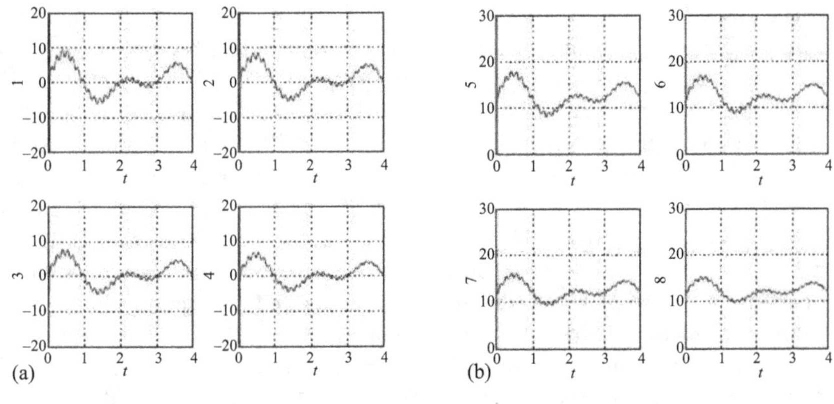

Figure 6.71 *Tracking of setpoint 2: (a) convergence of the grid points y_1 to y_4 (blue line) to the reference setpoint (red line) and (b) convergence of the grid points y_5 to y_8 (blue line) to the reference setpoint (red line)*

be stabilised and actually the price of stock loans can be made to follow specific reference values. The tracking error for all state variables of the related state-space model is asymptotically eliminated. The section has a meaningful contribution towards showing that it is feasible to control the value of several tradeable assets in the area of financial engineering. Such assets take the form of options, bonds, swaps, futures and forward contracts, stock-loans, mortgages, etc. The dynamics of the price of such loans are usually described by variants of the Black-Scholes PDE. Such is the case of stock-loans' pricing, which has been analysed in the submitted manuscript. The section's results confirmed that technically it is possible to stabilise or modify the value of stock-loans.

6.10 Estimation of PDE dynamics of the highway traffic

6.10.1 Outline

It is important to obtain estimates of highway traffic, in real-time, through the processing of measurements from a limited number of sensors (such as induction loops, cameras or GPS devices mounted on vehicles) because this enables better monitoring of the road's condition, the detection of events that disrupt normal flow of vehicles and the take up of measures that improve the road's safety. To this end, it is necessary to develop filtering techniques that will be able to estimate the traffic's dynamics through the processing of information that is provided by spatially distributed and heterogeneous sensors of the type that was described above (Wamg, Vam Schuppen *et al.*, 2014; Yuan, van Lint *et al.*, 2012; Kong *et al.*, 2009; Perera, Oliveira *et al.*, 2012).

The spatiotemporal dynamics of traffic in highways is usually described with the use of partial differential equations, such as the Lighthill–Whitham–Richards PDE or the Payne–Whitham PDE (Thai and Bayen, 2015; Blandin *et al.*, 2012). By applying spatial discretisation to such PDEs, these are turned into equivalent descriptions consisting of sets of nonlinear ordinary differential equations. For this latter description state estimation can be performed with the use of nonlinear filtering methods, among which the Extended Kalman Filter (EKF) is frequently used (Van Hinsbergen *et al.*, 2012; Work, Tassavainen *et al.*, 2008; Mihaylova *et al.*, 2012; Sun and Work, 2014). However, it is widely reckoned that the EKF suffers from cumulative computational errors which are due to the approximate linearisation it performs to the nonlinear model of the monitored system (Rigatos, 2014). Thus, in the case of highway traffic monitoring, the EKF can result into inaccurate estimates of the traffic's dynamics while in certain cases instability of the filtering algorithm may also appear. Therefore, it is significant to develop nonlinear filtering methods that will be free from numerical errors and will provide highly accurate estimates of the highway traffic.

In this section, the Payne–Whitham PDE model is considered and by applying semi-discretisation and the finite differences method it is turned into an equivalent description which consists of a set of coupled nonlinear ordinary differential equations. Thus, a state-space description is obtained. Next, it is proven that this state-space model of the Payne–Whitham PDE is a differentially flat one. This means that all its state variables and its control inputs can be expressed as differential functions of a specific algebraic variable which is the system's flat output (Rudolph, 2003; Sira-Ramirez and Agrawal, 2014; Levine, 2011; Fliess and Mounier, 1999; Rouchon, 2005; Laroche *et al.*, 2007; Boouden *et al.*, 2011). The differential flatness properties of the traffic model indicate also that it is possible to transform it into a linear canonical form, for which state estimation can be performed with the use of the standard

Kalman Filter recursion. Unlike EKF, the Kalman Filter estimation based on the transformed linearised model of the PDE does not require the computation of Jacobian matrices and partial derivatives and thus is known as derivative-free nonlinear Kalman Filter (Rigatos, 2011; Rigatos, 2013; Rigatos, 2015). Actually, the transformation of the Payne–Whitham PDE into the aforementioned linear canonical state-space form is an exact linearisation that is free of numerical approximation errors. Consequently, estimation of the traffic's dynamics with the use of the Derivative-free nonlinear Kalman Filter is numerically more accurate and computationally more robust than the estimation which is performed by the Extended Kalman Filter.

By applying the previously noted spatial discretisation method in the Payne–Whitham PDE model of traffic flow, this PDE model is transformed into the Brunovsky canonical state-space form, for which the solution of the state-estimation (filtering) problem becomes possible. By performing reconstruction of the entire PDE state-space vector out of a number of limited number of measurements of state-vector elements one can monitor traffic flow along a highway without the need to install a large number of sensors. In this manner, it is also possible to detect the occurrence of abnormal variations in the traffic flow, which can be due to traffic jams or accidents. Thus, the solution of the state-estimation problem for the traffic flow PDE model can be also exploited for highways condition monitoring.

6.10.2 Traffic modeling with the use of PDEs

6.10.2.1 Scalar models of the traffic flow

Macroscopic data modeling considers that the motion of vehicles in highways is a continuous-time phenomenon and has been inspired by hydrodynamics. For estimating traffic dynamics, both one-dimensional and higher dimensional PDE models have been proposed. Classical 1D PDE models of traffic consider the traffic state at a point x and at time instant t to be represented by the density function $\rho(x,t)$ (Thai and Bayen, 2015; Blandin *et al.*, 2012).

A typical state equation used to model the evolution of the density function $\rho(x,t)$ is the LWR (Lighthill–Whitham–Richards) PDE, which expresses the conservation of vehicles on road links:

$$\frac{\partial\rho}{\partial t}+\frac{\partial Q(\rho)}{\partial x}=0 \qquad (6.481)$$

where $Q()$ is a flux function, which denotes the flux of vehicles associated with density ρ, at a stationary state. The initial conditions for the PDE are ρ_0. The flux depends on the density and on the mean value of the vehicles according to the relation

$$Q(\rho)=q=\rho v=\rho V(\rho) \qquad (6.482)$$

where v is the vehicles' flow speed and $V(\rho)$ is a nonlinear function to be defined next. Several parametric flux functions have been proposed. One such a function is the so-called Greenshields flux function, which expresses a linear relationship between density and speed:

$$Q(\rho) = v_{\max}\rho\left(1 - \frac{\rho}{\rho_{\max}}\right) \tag{6.483}$$

and thus

$$V(\rho) = v_{\max}\left(1 - \frac{\rho}{\rho_{\max}}\right) \tag{6.484}$$

while $v_{\max}$ denotes the free-flow speed and $\rho_{\max}$ the jam density. One can note also the Newell–Daganzo flux function or triangular flux function, which is a piecewise linear function of the density, with different slopes in the free-flow and in congestion:

$$Q(\rho) = \begin{cases} \rho v_{\max} \text{ if } \rho \in [0, \rho_c] \\ \rho_c v_{\max}\dfrac{\rho_{\max} - \rho}{\rho_{\max} - \rho_c} \text{ if } \rho \in [\rho_c, \rho_{\max}] \end{cases} \tag{6.485}$$

where ρ_c denotes the critical density, that is the density at which the traffic flow becomes maximal. In congestion, the speed exhibits a wave dynamics denoted by $w = v_{\max}\rho_c/(\rho_c - \rho_{\max})$. Variations on a flux function based on an exponential relation between density and flow, parameterised by α, can be also found in the form:

$$Q(\rho) = \rho v_{\max} \exp\left(-\frac{1}{\alpha}(\frac{\rho}{\rho_c})^n\right) \tag{6.486}$$

There is also the so-called Godunov flux which is defined as follows:

$$Q_G(\rho_l, \rho_\tau) = \begin{cases} Q(\rho_l) & \text{if } \rho_\tau \leq \rho_l < \rho_c \\ Q(\rho_c) & \text{if } \rho_\tau \leq \rho_c < \rho_l \\ Q(\rho_\tau) & \text{if } \rho_c \leq \rho_\tau < \rho_l \\ \min\left(Q(\rho_l, \rho_\tau)\right) & \text{if } \rho_l \leq \rho_\tau \end{cases} \tag{6.487}$$

6.10.2.2 Non-scalar models of the traffic flow

In the case of non-scalar models of the traffic flow, additional degrees of freedom are included in the model of the traffic's PDEs. One such description is the Payne–Whitham model

$$\begin{aligned} &\frac{\partial \rho}{\partial t} + \frac{\partial q}{\partial x} = 0 \\ &\frac{\partial v}{\partial t} + v\frac{\partial v}{\partial x} + \frac{c_0^2}{\rho}\frac{\partial \rho}{\partial x} = \frac{V(\rho) - v}{\tau} \end{aligned} \tag{6.488}$$

Using a set of n equidistant points x_i $i = 1, \cdots, n$ along the x-axis, discretisation of the model of (6.488) is performed as follows:

$$\begin{aligned}
\frac{\partial \rho_i}{\partial t} &= \frac{q_i - q_{i-1}}{\Delta x} \\
\frac{\partial v_i}{\partial t} &= -v_i \frac{v_i - v_{i-1}}{\Delta x} - \frac{c_0^2}{\rho_i}\frac{\rho_i - \rho_{i-1}}{\Delta x} + \frac{V(\rho_i) - v_i}{\tau} \\
q_i &= \rho_i v_i
\end{aligned} \tag{6.489}$$

The associated boundary conditions of the PDE are: $\rho(x_0, t)$ and $v(x_0, t)$.

6.10.3 Estimation of the Payne–Whitham model using Extended Kalman Filter

Nonlinear observers, among which the Extended Kalman Filter, have been used in several cases for state estimation in distributed parameters systems described by nonlinear PDEs (Salberg *et al.*, 2010; Chauvin, 2011). To apply Extended Kalman Filtering for traffic monitoring, the x-axis of the Payne–Whitham model is discretised at $n + 1$ equidistant points, which are denoted as $[x_0, x_2, x_2, \ldots, x_{n-1}, x_n]$ (Pinsky, 1991). The boundary conditions for the model are denoted as $\rho(x_0) = \rho_0$ and $v(x_0) = v_0$.

For point x_1 it holds

$$\begin{aligned}
\frac{\partial \rho(x_1)}{\partial t} &= -\frac{\rho(x_1)u(x_1) - \rho_0 u_0}{\Delta x} \\
\frac{\partial v(x_1)}{\partial t} &= -\frac{v(x_1)v(x_1) - v_0}{\Delta x} - \frac{c_0^2}{\rho(x_1)}\frac{\rho(x_1) - \rho_0}{\Delta x}
\end{aligned} \tag{6.490}$$

while for points x_i, $i = 2, 3, \ldots, n-1, n$ one has

$$\begin{aligned}
\frac{\partial \rho(x_i)}{\partial t} &= -\frac{\rho(x_i)u(x_i) - \rho(x_{i-1})u(x_{i-1})}{\Delta x} \\
\frac{\partial v(x_i)}{\partial t} &= -v(x_i)\frac{v(x_i) - v(x_{i-1})}{\Delta x} - \frac{c_0^2}{\rho(x_i)}\frac{\rho(x_i) - \rho(x_{i-1})}{\Delta x} + \frac{V(x_i) - v_i}{\tau}
\end{aligned} \tag{6.491}$$

Next, the following state variables are defined: $z_{i,1} = \rho(x_i)$ and $z_{i,2} = v(x_i)$. Then the Payne–Whitham model becomes

At point x_1:

$$\begin{aligned}
\frac{\partial z_{1,1}}{\partial t} &= -\frac{z_{1,1}z_{1,2} - \rho_0 v_0}{\Delta x} \\
\frac{\partial z_{1,2}}{\partial t} &= -z_{1,2}\frac{z_{1,2} - v_0}{\Delta x} - \frac{c_0^2}{z_{1,1}}\frac{z_{1,1} - \rho_0}{\Delta x} + \frac{V(z_{1,1}) - z_{1,2}}{\tau}
\end{aligned} \tag{6.492}$$

At point x_2:

$$\begin{aligned}
\frac{\partial z_{2,1}}{\partial t} &= -\frac{z_{2,1}z_{2,2} - z_{1,1}z_{1,2}}{\Delta x} \\
\frac{\partial z_{1,2}}{\partial t} &= -z_{1,2}\frac{z_{2,2} - z_{1,2}}{\Delta x} - \frac{c_0^2}{z_{2,1}}\frac{z_{2,1} - z_{1,1}}{\Delta x} + \frac{V(z_{2,1}) - z_{2,2}}{\tau}
\end{aligned} \tag{6.493}$$

At point x_3:

$$\begin{aligned}
\frac{\partial z_{3,1}}{\partial t} &= -\frac{z_{3,1}z_{3,2} - z_{2,1}z_{2,2}}{\Delta x} \\
\frac{\partial z_{3,2}}{\partial t} &= -z_{3,2}\frac{z_{3,2} - z_{3,2}}{\Delta x} - \frac{c_0^2}{z_{3,1}}\frac{z_{3,1} - z_{2,1}}{\Delta x} + \frac{V(z_{3,1}) - z_{3,2}}{\tau}
\end{aligned} \tag{6.494}$$

$\cdots \quad \cdots$

At point x_{n-1}:

$$\begin{aligned}
\frac{\partial z_{n-1,1}}{\partial t} &= -\frac{z_{n-1,1}z_{n-1,2} - z_{n-2,1}z_{n-2,2}}{\Delta x} \\
\frac{\partial z_{n-1,2}}{\partial t} &= -z_{n-1,2}\frac{z_{n-1,2} - z_{n-2,2}}{\Delta x} - \frac{c_0^2}{z_{n-1,1}}\frac{z_{n-1,1} - z_{n-2,1}}{\Delta x} + \frac{V(z_{n-1,1}) - z_{n-1,2}}{\tau}
\end{aligned} \tag{6.495}$$

At point x_n:

$$\begin{aligned}
\frac{\partial z_{n,1}}{\partial t} &= -\frac{z_{n,1}z_{n,2} - z_{n-1,1}z_{n-1,2}}{\Delta x} \\
\frac{\partial z_{n,2}}{\partial t} &= -z_{n,2}\frac{z_{n,2} - z_{n-1,2}}{\Delta x} - \frac{c_0^2}{z_{n,1}}\frac{z_{n,1} - z_{n-1,1}}{\Delta x} + \frac{V(z_{n,1}) - z_{n,2}}{\tau}
\end{aligned} \tag{6.496}$$

Next, the following functions are defined

$$\begin{aligned}
f_{11} &= -\frac{\rho(x_1)u(x_1) - \rho_0 u_0}{\Delta x} \\
f_{12} &= -z_{1,2}\frac{z_{1,2} - v_0}{\Delta x} - \frac{c_0^2}{z_{1,1}}\frac{z_{1,1} - \rho_0}{\Delta x} + \frac{V(z_{1,1}) - z_{1,2}}{\tau} \\
f_{21} &= -\frac{z_{2,1}z_{2,2} - z_{1,1}z_{1,2}}{\Delta x} \\
f_{22} &= -z_{1,2}\frac{z_{2,2} - z_{1,2}}{\Delta x} - \frac{c_0^2}{z_{2,1}}\frac{z_{2,1} - z_{1,1}}{\Delta x} + \frac{V(z_{2,1}) - z_{2,2}}{\tau} \\
f_{31} &= -\frac{z_{3,1}z_{3,2} - z_{2,1}z_{2,2}}{\Delta x} \\
f_{32} &= -z_{3,2}\frac{z_{3,2} - z_{3,2}}{\Delta x} - \frac{c_0^2}{z_{3,1}}\frac{z_{3,1} - z_{2,1}}{\Delta x} + \frac{V(z_{3,1}) - z_{3,2}}{\tau} \\
&\cdots \\
f_{n-1,1} &= -\frac{z_{n-1,1}z_{n-1,2} - z_{n-2,1}z_{n-2,2}}{\Delta x} \\
f_{n-2,2} &= -z_{n-1,2}\frac{z_{n-1,2} - z_{n-2,2}}{\Delta x} - \frac{c_0^2}{z_{n-1,1}}\frac{z_{n-1,1} - z_{n-2,1}}{\Delta x} + \frac{V(z_{n-1,1}) - z_{n-1,2}}{\tau} \\
f_{n,1} &= -\frac{z_{n,1}z_{n,2} - z_{n-1,1}z_{n-1,2}}{\Delta x} \\
f_{n,2} &= -z_{n,2}\frac{z_{n,2} - z_{n-1,2}}{\Delta x} - \frac{c_0^2}{z_{n,1}}\frac{z_{n,1} - z_{n-1,1}}{\Delta x} + \frac{V(z_{n,1}) - z_{n,2}}{\tau}
\end{aligned} \tag{6.497}$$

and the state-space description of the system becomes

$$\begin{aligned} &\dot{z}_{1,1}=f_{1,1} && \dot{z}_{1,2}=f_{1,2} \\ &\dot{z}_{2,1}=f_{2,1} && \dot{z}_{2,2}=f_{2,2} \\ &\dot{z}_{3,1}=f_{3,1} && \dot{z}_{3,2}=f_{3,2} \\ &\cdots && \cdots \\ &\dot{z}_{n-1,1}=f_{n-1,1} && \dot{z}_{n-1,2}=f_{n-1,2} \\ &\dot{z}_{n,1}=f_{n,1} && \dot{z}_{n,2}=f_{n,2} \end{aligned} \tag{6.498}$$

The state vector of the system is written as

$$Z=[z_{1,1},z_{1,2},z_{2,1},z_{2,2},z_{3,1},z_{3,2},\ldots\cdots z_{n-1,1},z_{n-1,2},z_{n,1},z_{n,2}]^T.$$

The measurements vector is denoted as $Z^m=[z_{1,1},z_{2,1},z_{3,1},\ldots z_{n-1,1},z_{n,1}]^T$. Thus, measurements are considered to be received according to the following measurement equation

$$Z^m=\gamma(Z) \ \text{ or}$$

$$Z^m=\begin{pmatrix} 1 & 0 & 0 & 0 & 0 & 0 & \cdots & 0 & 0 & 0 & 0 \\ 0 & 0 & 1 & 0 & 0 & 0 & \cdots & 0 & 0 & 0 & 0 \\ 0 & 0 & 0 & 0 & 1 & 0 & \cdots & 0 & 0 & 0 & 0 \\ \cdots & & \cdots & & \cdots & & \cdots & & \cdots & & \cdots \\ 0 & 0 & 0 & 0 & 0 & 0 & \cdots & 1 & 0 & 0 & 0 \\ 0 & 0 & 0 & 0 & 0 & 0 & \cdots & 0 & 0 & 1 & 0 \end{pmatrix} Z \tag{6.499}$$

The Jacobian matrix $J_zf(Z)$ is defined as:

$$J_zf(Z)=$$
$$\begin{bmatrix} \frac{\partial f_{1,1}}{\partial z_{1,1}} & \frac{\partial f_{1,1}}{\partial z_{1,2}} & \frac{\partial f_{1,1}}{\partial z_{2,1}} & \frac{\partial f_{1,1}}{\partial z_{2,2}} & \frac{\partial f_{1,1}}{\partial z_{3,1}} & \frac{\partial f_{1,1}}{\partial z_{3,2}} & \cdots & \frac{\partial f_{1,1}}{\partial z_{n-1,1}} & \frac{\partial f_{1,1}}{\partial z_{n-1,2}} & \frac{\partial f_{1,1}}{\partial z_{n,1}} & \frac{\partial f_{1,1}}{\partial z_{n,2}} \\ \frac{\partial f_{1,2}}{\partial z_{1,1}} & \frac{\partial f_{1,2}}{\partial z_{1,2}} & \frac{\partial f_{1,2}}{\partial z_{2,1}} & \frac{\partial f_{1,2}}{\partial z_{2,2}} & \frac{\partial f_{1,2}}{\partial z_{3,1}} & \frac{\partial f_{1,2}}{\partial z_{3,2}} & \cdots & \frac{\partial f_{1,2}}{\partial z_{n-1,1}} & \frac{\partial f_{1,2}}{\partial z_{n-1,2}} & \frac{\partial f_{1,2}}{\partial z_{n,1}} & \frac{\partial f_{2,1}}{\partial z_{n,2}} \\ \frac{\partial f_{2,1}}{\partial z_{1,1}} & \frac{\partial f_{2,1}}{\partial z_{1,2}} & \frac{\partial f_{2,1}}{\partial z_{2,1}} & \frac{\partial f_{2,1}}{\partial z_{2,2}} & \frac{\partial f_{2,1}}{\partial z_{3,1}} & \frac{\partial f_{2,1}}{\partial z_{3,2}} & \cdots & \frac{\partial f_{2,1}}{\partial z_{n-1,1}} & \frac{\partial f_{2,1}}{\partial z_{n-1,2}} & \frac{\partial f_{2,1}}{\partial z_{n,1}} & \frac{\partial f_{2,2}}{\partial z_{n,2}} \\ \frac{\partial f_{2,2}}{\partial z_{1,1}} & \frac{\partial f_{2,2}}{\partial z_{1,2}} & \frac{\partial f_{2,2}}{\partial z_{2,1}} & \frac{\partial f_{2,2}}{\partial z_{2,2}} & \frac{\partial f_{2,2}}{\partial z_{3,1}} & \frac{\partial f_{2,2}}{\partial z_{3,2}} & \cdots & \frac{\partial f_{2,2}}{\partial z_{n-1,1}} & \frac{\partial f_{2,2}}{\partial z_{n-1,2}} & \frac{\partial f_{2,2}}{\partial z_{n,1}} & \frac{\partial f_{2,2}}{\partial z_{n,2}} \\ \frac{\partial f_{3,1}}{\partial z_{1,1}} & \frac{\partial f_{3,1}}{\partial z_{1,2}} & \frac{\partial f_{3,1}}{\partial z_{2,1}} & \frac{\partial f_{3,1}}{\partial z_{2,2}} & \frac{\partial f_{3,1}}{\partial z_{3,1}} & \frac{\partial f_{3,1}}{\partial z_{3,2}} & \cdots & \frac{\partial f_{3,1}}{\partial z_{n-1,1}} & \frac{\partial f_{3,1}}{\partial z_{n-1,2}} & \frac{\partial f_{3,1}}{\partial z_{n,1}} & \frac{\partial f_{3,1}}{\partial z_{n,2}} \\ \frac{\partial f_{3,1}}{\partial z_{1,1}} & \frac{\partial f_{3,2}}{\partial z_{1,2}} & \frac{\partial f_{3,2}}{\partial z_{2,1}} & \frac{\partial f_{3,2}}{\partial z_{2,2}} & \frac{\partial f_{3,2}}{\partial z_{3,1}} & \frac{\partial f_{3,2}}{\partial z_{3,2}} & \cdots & \frac{\partial f_{3,2}}{\partial z_{n-1,1}} & \frac{\partial f_{3,2}}{\partial z_{n-1,2}} & \frac{\partial f_{3,2}}{\partial z_{n,1}} & \frac{\partial f_{3,2}}{\partial z_{n,2}} \\ \cdots & \cdots & \cdots & \cdots & \cdots & \cdots & \cdots & \cdots & \cdots & \cdots & \cdots \\ \frac{\partial f_{n-1,1}}{\partial z_{1,1}} & \frac{\partial f_{n-1,1}}{\partial z_{1,2}} & \frac{\partial f_{n-1,1}}{\partial z_{2,1}} & \frac{\partial f_{n-1,1}}{\partial z_{2,2}} & \frac{\partial f_{n-1,1}}{\partial z_{3,1}} & \frac{\partial f_{n-1,1}}{\partial z_{3,2}} & \cdots & \frac{\partial f_{n-1,1}}{\partial z_{n-1,1}} & \frac{\partial f_{n-1,1}}{\partial z_{n-1,2}} & \frac{\partial f_{n-1,1}}{\partial z_{n,1}} & \frac{\partial f_{n-1,1}}{\partial z_{n,2}} \\ \frac{\partial f_{n-1,2}}{\partial z_{1,1}} & \frac{\partial f_{n-1,2}}{\partial z_{1,2}} & \frac{\partial f_{n-1,2}}{\partial z_{2,1}} & \frac{\partial f_{n-1,2}}{\partial z_{2,2}} & \frac{\partial f_{n-1,2}}{\partial z_{3,1}} & \frac{\partial f_{n-1,2}}{\partial z_{3,2}} & \cdots & \frac{\partial f_{n-1,2}}{\partial z_{n-1,1}} & \frac{\partial f_{n-1,2}}{\partial z_{n-1,2}} & \frac{\partial f_{n-1,2}}{\partial z_{n,1}} & \frac{\partial f_{n-1,2}}{\partial z_{n,2}} \\ \frac{\partial f_{n,1}}{\partial z_{1,1}} & \frac{\partial f_{n,1}}{\partial z_{1,2}} & \frac{\partial f_{n,1}}{\partial z_{2,1}} & \frac{\partial f_{n,1}}{\partial z_{2,2}} & \frac{\partial f_{n,1}}{\partial z_{3,1}} & \frac{\partial f_{n,1}}{\partial z_{3,2}} & \cdots & \frac{\partial f_{n,1}}{\partial z_{n-1,1}} & \frac{\partial f_{n,1}}{\partial z_{n-1,2}} & \frac{\partial f_{n,1}}{\partial z_{n,1}} & \frac{\partial f_{n,1}}{\partial z_{n,2}} \\ \frac{\partial f_{n,2}}{\partial z_{1,1}} & \frac{\partial f_{n,2}}{\partial z_{1,2}} & \frac{\partial f_{n,2}}{\partial z_{2,1}} & \frac{\partial f_{n,2}}{\partial z_{2,2}} & \frac{\partial f_{n,2}}{\partial z_{3,1}} & \frac{\partial f_{n,2}}{\partial z_{3,2}} & \cdots & \frac{\partial f_{n,2}}{\partial z_{n-1,1}} & \frac{\partial f_{n,2}}{\partial z_{n-1,2}} & \frac{\partial f_{n,2}}{\partial z_{n,1}} & \frac{\partial f_{n,2}}{\partial z_{n,2}} \end{bmatrix}$$

About $J_z\gamma(Z)$ one has

$$J_z\gamma(Z) = \begin{pmatrix} 1\,0\,0\,0\,0\,0 \cdots 0\,0\,0\,0 \\ 0\,0\,1\,0\,0\,0 \cdots 0\,0\,0\,0 \\ 0\,0\,0\,0\,1\,0 \cdots 0\,0\,0\,0 \\ \cdots\cdots\cdots\cdots\cdots\cdots \\ 0\,0\,0\,0\,0\,0 \cdots 1\,0\,0\,0 \\ 0\,0\,0\,0\,0\,0 \cdots 0\,0\,1\,0 \end{pmatrix} \tag{6.500}$$

The EKF recursion is as follows: First, the time update is considered: by $\hat{Z}(k)$ the estimation of the state vector at time instant k is denoted. Given initial conditions $\hat{Z}^-(0)$ and $P^-(0)$ the recursion proceeds as (Rigatos and Tzafestas, 2007; Basseville and Nikiforov, 1993; Rigatos and Zhang, 2009:

Measurement update. Acquire $z(k)$ and compute:

$$\begin{aligned} &K(k) = P^-(k)J_\gamma^T(\hat{x}^-(k))\cdot[J_\gamma(\hat{x}^-(k))P^-(k)J_\gamma^T(\hat{x}^-(k)) + R(k)]^{-1} \\ &\hat{Z}(k) = \hat{Z}^-(k) + K(k)[Z^m(k) - \gamma(\hat{Z}^-(k))] \\ &P(k) = P^-(k) - K(k)J_\gamma(\hat{Z}^-(k))P^-(k) \end{aligned} \tag{6.501}$$

Time update. Compute:

$$\begin{aligned} &P^-(k+1) = J_\phi(\hat{Z}(k))P(k)J_\phi^T(\hat{Z}(k)) + Q(k) \\ &\hat{Z}^-(k+1) = \phi(\hat{Z}(k)) + L(k)u(k) \end{aligned} \tag{6.502}$$

Thus, to implement Extended Kalman Filtering for the state-space description of the Payne–Whitham system the Jacobian matrices $J_zf(Z)$ and $J_z\gamma(Z)$ have to be computed. Moreover, since the EKF is implemented in discrete-time the Jacobian matrix $J_zf(Z)$ should be finally substituted by $I + T_sJ_zf(Z)$. About $J_zf(Z)$ one has

For the first couple of rows of the state-space description of the Payne–Whitham model given in (6.498) one has

$$\begin{aligned} &\frac{\partial f_{1,1}}{\partial z_{1,1}} = -\frac{z_{1,2}}{\Delta x}\ \frac{\partial f_{1,1}}{\partial z_{1,2}} = -\frac{z_{1,1}}{\Delta x}\ \frac{\partial f_{1,1}}{\partial z_{2,1}} = 0\ \frac{\partial f_{1,1}}{\partial z_{2,2}} = 0\ \frac{\partial f_{1,1}}{\partial z_{3,1}} = 0\ \frac{\partial f_{1,1}}{\partial z_{3,2}} = 0 \\ &\cdots\ \frac{\partial f_{1,1}}{\partial z_{n-1,1}} = 0\ \frac{\partial f_{1,1}}{\partial z_{n-1,2}} = 0\ \frac{\partial f_{1,1}}{\partial z_{n,1}} = 0\ \frac{\partial f_{1,1}}{\partial z_{n,2}} = 0 \end{aligned} \tag{6.503}$$

and also

$$\begin{aligned} &\frac{\partial f_{1,2}}{\partial z_{1,1}} = -\frac{c_0^2}{z_{1,1}}\frac{z_{1,1}-\rho_0}{\Delta x} - \frac{c_0^2}{z_{1,1}}\frac{1}{\Delta x} \\ &\quad + \frac{v_{\max}}{\tau}\exp\left(-\frac{1}{\alpha}\left(\frac{z_{1,1}}{\rho_c}^2\right)\right)\left(-\left(\frac{z_{1,1}}{\rho_c}\right)^{\alpha-1}\right)\ \frac{\partial f_{1,2}}{\partial z_{1,2}} = -\frac{2z_{1,2}}{\Delta x} - \frac{1}{\tau} \\ &\frac{\partial f_{1,2}}{\partial z_{2,1}} = 0\ \frac{\partial f_{1,2}}{\partial z_{2,2}} = 0\ \frac{\partial f_{1,2}}{\partial z_{3,1}} = 0\ \frac{\partial f_{1,2}}{\partial z_{3,2}} = 0 \cdots\ \frac{\partial f_{1,2}}{\partial z_{n-1,1}} = 0\ \frac{\partial f_{1,2}}{\partial z_{n-1,2}} = 0\ \frac{\partial f_{1,2}}{\partial z_{n,1}} = 0\ \frac{\partial f_{1,2}}{\partial z_{n,2}} = 0 \end{aligned} \tag{6.504}$$

For the ith couple of rows of the state-space description of the Payne–Whitham model given in (6.498) one has

$$\begin{aligned}&\frac{\partial f_{i,1}}{\partial z_{1,1}}=0\ \frac{\partial f_{i,1}}{\partial z_{1,2}}=0\ \cdots\ \frac{\partial f_{i,1}}{\partial z_{i-1,1}}=\frac{z_{i-1,2}}{\Delta x}\ \frac{\partial f_{i,1}}{\partial z_{i-1,2}}=\frac{z_{i-1,1}}{\Delta x}\\&\frac{\partial f_{i,1}}{\partial z_{i,1}}=-\frac{z_{i,2}}{\Delta x}\ \frac{\partial f_{i,1}}{\partial z_{i,2}}=-\frac{z_{i,1}}{\Delta x}\cdots\ \frac{\partial f_{i,1}}{\partial z_{n,1}}=0\ \frac{\partial f_{i,1}}{\partial z_{n,2}}=0\end{aligned} \tag{6.505}$$

and also

$$\begin{aligned}&\frac{\partial f_{i,2}}{\partial z_{1,1}}=0\ \frac{\partial f_{i,2}}{\partial z_{1,2}}=0\ \cdots\ \frac{\partial f_{i,2}}{\partial z_{i-1,1}}=\frac{c_0^2}{z_{1,1}}\frac{1}{\Delta x}\ \frac{\partial f_{i,2}}{\partial z_{i-1,2}}=\frac{z_{i,2}}{\Delta x}\\&\frac{\partial f_{i,2}}{\partial z_{i,1}}=\frac{c_0^2}{z_{1,1}}\frac{z_{i,1}-z_{i-1,1}}{\Delta x}-\frac{c_0^2}{z_{1,1}}\frac{1}{\Delta x}+\frac{v_{\max}}{c}\exp\left(-\frac{1}{a}\left(\frac{z_{i,1}}{\rho_c}\right)^2\right)\left(-\left(\frac{z_{i,1}}{\rho_c}\right)^{a-1}\right)\\&\frac{\partial f_{i,2}}{\partial z_{i,2}}=-\frac{2z_{i,2}}{\Delta x}-\frac{1}{c}\cdots\ \frac{\partial f_{i,2}}{\partial z_{n,1}}=0\ \frac{\partial f_{i,2}}{\partial z_{n,2}}=0\end{aligned} \tag{6.506}$$

6.10.4 *Estimation of Payne–Whitham PDE with the derivative-free KF*

6.10.4.1 Differential flatness of the Payne–Whitham PDE

Differential flatness theory has been applied for solving several problems of control and estimation in distributed parameter systems (Woitteneck and Mounier, 2010; Mounier, Rudolph *et al.*, 2010; Fliess and Mounier, 1999; Boussaada *et al.*, 2013, Rouchon, Martin *et al.*, 2014). In the case of the Payme–Whitham traffic model, the flat output of the system is taken to be the vector

$$Z_f=[z_{1,1},z_{2,1},z_{3,1},\ldots,z_{n-1,1},z_{n,1}]^T \tag{6.507}$$

It can be proven that all state variables of the system can be written as functions of the flat output and its derivatives. It has been shown that for the 1-st couple of equations on the Payne–Whitham model it holds

$$\begin{aligned}\frac{\partial z_{1,1}}{\partial t}&=-\frac{z_{1,1}z_{1,2}-\rho_0 v_0}{\Delta x}\\\frac{\partial z_{1,2}}{\partial t}&=-z_{1,2}\frac{z_{1,2}-v_0}{\Delta x}-\frac{c_0^2}{z_{i,1}}\frac{z_{i,1}-\rho_0}{\Delta x}+\frac{V(z_{i,1})-z_{i,2}}{\tau}\end{aligned} \tag{6.508}$$

By solving the first row of the above equation with respect to $z_{1,2}$ one finally gets

$$z_{1,2}=\frac{(\Delta x\dot{z}_{1,1}-\rho_0 v_0)}{-z_{1,1}} \tag{6.509}$$

and using that $z_{1,1}$ is by definition a component of the flat output vector as well as that the boundary conditions ρ_0 and v_0 are constants, one gets that $z_{1,2}$ is a function of the flat output and its derivatives, or equivalently $z_{1,2} = f_1(Z_f, \dot{Z}_f)$.

For the 2-nd couple of equations of the Payne–Whitham model it holds

$$\begin{aligned}\frac{\partial z_{2,1}}{\partial t} &= -\frac{z_{2,1}z_{2,2} - z_{1,1}z1,2}{\Delta x} \\ \frac{\partial z_{2,2}}{\partial t} &= -z_{2,2}\frac{z_{2,2} - z_{1,2}}{\Delta x} - \frac{c_0^2}{z_{2,1}}\frac{z_{2,1} - z_{1,1}}{\Delta x} + \frac{V(z_{2,1}) - z_{2,2}}{\tau}\end{aligned} \tag{6.510}$$

By solving the first row of the above equation with respect to $z_{2,2}$ one finally gets

$$z_{2,2} = \frac{(\Delta x\dot{z}_{2,1} - z_{1,1}z_{1,2})}{-z_{2,1}} \tag{6.511}$$

and using that $z_{2,1}$ and $z_{1,1}$ are by definition components of the flat output vector as well as that $z_{1,2}$ is a differential function of the flat output, one gets that $z_{2,2}$ is a function of the flat output and its derivatives, or equivalently $z_{2,2} = f_2(Z_f, \dot{Z}_f)$.

For the 3-rd couple of equations on the Payne–Whitham model it holds

$$\begin{aligned}\frac{\partial z_{3,1}}{\partial t} &= -\frac{z_{3,1}z_{3,2} - z_{2,1}z_{2,2}}{\Delta x} \\ \frac{\partial z_{3,2}}{\partial t} &= -z_{3,2}\frac{z_{3,2} - z_{2,2}}{\Delta x} - \frac{c_0^2}{z_{3,1}}\frac{z_{3,1} - z_{2,1}}{\Delta x} + \frac{V(z_{3,1}) - z_{3,2}}{\tau}\end{aligned} \tag{6.512}$$

By solving the first row of the above equation with respect to $z_{3,2}$ one finally gets

$$z_{3,2} = \frac{(\Delta x\dot{z}_{3,1} - z_{2,1}z_{2,2})}{-z_{3,1}} \tag{6.513}$$

and using that $z_{3,1}$ and $z_{2,1}$ are by definition components of the flat output vector as well as that $z_{2,2}$ is a differential function of the flat output, one gets that $z_{3,2}$ is a function of the flat output and its derivatives, or equivalently $z_{3,2} = f_3(Z_f, \dot{Z}_f)$.

Continuing in a similar manner one has that for the $n-1$th couple of equations on the Payne–Whitham model it holds

$$\begin{aligned}\frac{\partial z_{n-1,1}}{\partial t} &= -\frac{z_{n-1,1}z_{3,2} - z_{n-2,1}zn-2,2}{\Delta x} \\ \frac{\partial z_{n-1,2}}{\partial t} &= -z_{n-1,2}\frac{z_{n-1,2} - z_{n-2,2}}{\Delta x} - \frac{c_0^2}{z_{n-1,1}}\frac{z_{n-1,1} - z_{n-2,1}}{\Delta x} + \frac{V(z_{n-1,1}) - z_{n-1,2}}{\tau}\end{aligned} \tag{6.514}$$

By solving the first row of the above equation with respect to $z_{n-1,2}$ one finally gets

$$z_{n-1,2} = \frac{(\Delta x\dot{z}_{n-1,1} - z_{n-2,1}z_{n-2,2})}{-z_{n-1,1}} \tag{6.515}$$

and using that $z_{n-1,1}$ and $z_{n-2,1}$ are by definition components of the flat output vector as well as that $z_{n-2,2}$ is a differential function of the flat output, one gets that $z_{n-1,2}$ is a function of the flat output and its derivatives, or equivalently $z_{n-1,2} = f_{n-1}(Z_f, \dot{Z}_f)$.

Finally, for the nth couple of equations on the Payne–Whitham model it holds

$$\begin{aligned} \frac{\partial z_{n,1}}{\partial t} &= -\frac{z_{n,1}z_{3,2} - z_{n-1,1}z_{n-1,2}}{\Delta x} \\ \frac{\partial z_{n,2}}{\partial t} &= -z_{n,2}\frac{z_{n,2} - z_{n-1,2}}{\Delta x} - \frac{c_0^2}{z_{n,1}}\frac{z_{n,1} - z_{n-1,1}}{\Delta x} + \frac{V(z_{n,1}) - z_{n,2}}{\tau} \end{aligned} \tag{6.516}$$

By solving the first row of the above equation with respect to $z_{n,2}$ one finally gets

$$z_{n,2} = \frac{(\Delta x\dot{z}_{n,1} - z_{n-1,1}z_{n-1,2})}{-z_{n,1}} \tag{6.517}$$

and using that $z_{n,1}$ and $z_{n-1,1}$ are by definition components of the flat output vector as well as that $z_{n-1,2}$ is a differential function of the flat output, one gets that $z_{n,2}$ is a function of the flat output and its derivatives, or equivalently $z_{n,2} = f_n(Z_f, \dot{Z}_f)$.

According to the above, all state variables of the Payne–Whitham model can be written as differential functions of the previously defined flat output Z_f. Consequently, the Payne–Whitham model is a differentially flat one.

6.10.5 Derivative-free nonlinear Kalman Filter for the Payne–Whitham PDE

The Payne–Whitham PDE model is rewritten as:

$$\begin{aligned} \frac{\partial z_{1,1}}{\partial t} &= -\frac{z_{1,1}z_{1,2}}{\Delta x} + d_1 \\ \frac{\partial z_{1,2}}{\partial t} &= v_{1,2} \end{aligned} \tag{6.518}$$

where the disturbance term d_1 and the control input v_1 are defined as

$$\begin{aligned} d_1 &= \frac{\rho_0 v_0}{\Delta x} \\ v_1 &= -z_{1,2}\frac{z_{1,2} - v_0}{\Delta x} - \frac{c_0^2}{z_{1,1}}\frac{z_{1,1} - \rho_0}{\Delta x} + \frac{V(z_{1,1}) - z_{1,2}}{\tau} \end{aligned} \tag{6.519}$$

At point x_2:

$$\begin{aligned} \frac{\partial z_{2,1}}{\partial t} &= -\frac{z_{2,1}z_{2,2} - z_{1,1}z_{1,2}}{\Delta x} \\ \frac{\partial z_{1,2}}{\partial t} &= v_2 \end{aligned} \tag{6.520}$$

where the control input v_2 are defined as

$$v_2 = -z_{1,2}\frac{z_{2,2} - z_{1,2}}{\Delta x} - \frac{c_0^2}{z_{2,1}}\frac{z_{2,1} - z_{1,1}}{\Delta x} + \frac{V(z_{2,1}) - z_{2,2}}{\tau} \tag{6.521}$$

At point x_3:

$$\begin{aligned}
\frac{\partial z_{3,1}}{\partial t} &= -\frac{z_{3,1}z_{3,2} - z_{2,1}z_{2,2}}{\Delta x} \\
\frac{\partial z_{3,2}}{\partial t} &= v_3 \\
\cdots & \cdots
\end{aligned} \tag{6.522}$$

where the control input v_3 is defined as

$$v_3 = -z_{3,2}\frac{z_{3,2} - z_{3,2}}{\Delta x} - \frac{c_0^2}{z_{3,1}}\frac{z_{3,1} - z_{2,1}}{\Delta x} + \frac{V(z_{3,1}) - z_{3,2}}{\tau} \tag{6.523}$$

$$\cdots \cdots \cdots$$

At point x_{n-1}:

$$\begin{aligned}
\frac{\partial z_{n-1,1}}{\partial t} &= -\frac{z_{n-1,1}z_{n-1,2} - z_{n-2,1}z_{n-2,2}}{\Delta x} \\
\frac{\partial z_{n-1,2}}{\partial t} &= v_{n-1}
\end{aligned} \tag{6.524}$$

where the control inputs v_{n-1} is defined as

$$v_{n-1} = -z_{n-1,2}\frac{z_{n-1,2} - z_{n-2,2}}{\Delta x} - \frac{c_0^2}{z_{n-1,1}}\frac{z_{n-1,1} - z_{n-2,1}}{\Delta x} + \frac{V(z_{n-1,1}) - z_{n-1,2}}{\tau} \tag{6.525}$$

At point x_n:

$$\begin{aligned}
\frac{\partial z_{n,1}}{\partial t} &= -\frac{z_{n,1}z_{n,2} - z_{n-1,1}z_{n-1,2}}{\Delta x} \\
\frac{\partial z_{n,2}}{\partial t} &= v_n
\end{aligned} \tag{6.526}$$

Thus, one arrives at a linearised description for the Payne–Whitman state-space model

$$\dot{Z} = AZ + BV + D \tag{6.527}$$

where the state vector is defined as

$$Z = [z_{1,1}, z_{1,2}, z_{2,1}, z_{2,2}, \ldots, z_{n,1}, z_{n,2}]^T \tag{6.528}$$

the control inputs v_n is defined as

$$v_n = -z_{n,2}\frac{z_{n,2} - z_{n-1,2}}{\Delta x} - \frac{c_0^2}{z_{n,1}}\frac{z_{n,1} - z_{n-1,1}}{\Delta x} + \frac{V(z_{n,1}) - z_{n,2}}{\tau} \tag{6.529}$$

the disturbances vector D describes the effects of initial conditions to this model and is given by

$$D = [\rho_0 v_0/\Delta x, 0, 0, 0, 0, 0, \ldots, 0, 0, 0, 0]^T \tag{6.530}$$

while matrices A and B are defined as

$$A = \begin{bmatrix} 0 & -\frac{z_{1,1}}{\Delta x} & 0 & 0 & 0 & 0 & \cdots & 0 & 0 & 0 & 0 & 0 & 0 \\ 0 & 0 & 0 & 0 & 0 & 0 & \cdots & 0 & 0 & 0 & 0 & 0 & 0 \\ 0 & \frac{z_{1,1}}{\Delta x} & 0 & -\frac{z_{2,1}}{\Delta x} & 0 & 0 & \cdots & 0 & 0 & 0 & 0 & 0 & 0 \\ 0 & 0 & 0 & 0 & 0 & 0 & \cdots & 0 & 0 & 0 & 0 & 0 & 0 \\ 0 & 0 & 0 & \frac{z_{2,1}}{\Delta x} & 0 & -\frac{z_{3,1}}{\Delta x} & \cdots & 0 & 0 & 0 & 0 & 0 & 0 \\ 0 & 0 & 0 & 0 & 0 & 0 & \cdots & 0 & 0 & 0 & 0 & 0 & 0 \\ \cdots & \cdots & \cdots & \cdots & \cdots & \cdots & \cdots & \cdots & \cdots & \cdots & \cdots & \cdots & \cdots \\ 0 & 0 & 0 & 0 & 0 & 0 & \cdots & 0 & \frac{z_{n-2,1}}{\Delta x} & 0 & -\frac{z_{n-1,1}}{\Delta x} & 0 & 0 \\ 0 & 0 & 0 & 0 & 0 & 0 & \cdots & 0 & 0 & 0 & 0 & 0 & 0 \\ 0 & 0 & 0 & 0 & 0 & 0 & \cdots & 0 & 0 & 0 & \frac{z_{n-1,1}}{\Delta x} & 0 & -\frac{z_{n,1}}{\Delta x} \\ 0 & 0 & 0 & 0 & 0 & 0 & \cdots & 0 & 0 & 0 & 0 & 0 & 0 \end{bmatrix}$$

$$B = \begin{bmatrix} 0 & 0 & 0 & \cdots & 0 & 0 \\ 1 & 0 & 0 & \cdots & 0 & 0 \\ 0 & 0 & 0 & \cdots & 0 & 0 \\ 0 & 1 & 0 & \cdots & 0 & 0 \\ 0 & 0 & 0 & \cdots & 0 & 0 \\ 0 & 0 & 1 & \cdots & 0 & 0 \\ \cdots & \cdots & \cdots & & & \\ 0 & 0 & 0 & \cdots & 0 & 0 \\ 0 & 0 & 0 & \cdots & 1 & 0 \\ 0 & 0 & 0 & \cdots & 0 & 0 \\ 0 & 0 & 0 & \cdots & 0 & 1 \end{bmatrix}$$

For the linearised equivalent model of the Peyne–Whitham PDE, and after substituting matrices A, B, C by their discrete-time equivalents A_d, B_d and C_D respectively, one can perform state estimation using the Kalman Filter recursion. Finally, the Kalman filter equations are:

measurement update:

$$\begin{aligned} K(k) &= P^{-}(k){C_d}^T[C_d \cdot P^{-}(k){C_d}^T + R]^{-1} \\ \hat{Z}(k) &= \hat{Z}^{-}(k) + K(k)[Z^m(k) - C_d\hat{Z}^{-}(k)] \\ P(k) &= P^{-}(k) - K(k)C_dP^{-}(k) \end{aligned} \tag{6.531}$$

time update:

$$\begin{aligned} P^{-}(k+1) &= A_d(k)P(k){A_d}^T(k) + Q(k) \\ \hat{Z}^{-}(k+1) &= A_d(k)\hat{Z}(k) + B_d(k)u(k) \end{aligned} \tag{6.532}$$

6.10.5.1 A Brunovsky canonical form for the PDE model

Elaborating on the previous results is also possible to show that the Payne–Whitham PDE model can be also written in the linear canonical Brunovsky form. The first couple of equations of the discretised Payne–Whitham PDE is written as:

$$\begin{aligned}\dot{z}_{1,1} &= -\frac{z_{1,1}z_{1,2} - \rho_0 v_0}{\Delta x} \\ \dot{z}_{1,2} &= -z_{1,2}\frac{z_{1,2} - v_0}{\Delta x} - \frac{c_0^2}{z_{1,1}}\frac{z_{1,1} - \rho_0}{\Delta x} + \frac{V(z_{1,1} = z_{1,2})}{\tau}\end{aligned} \tag{6.533}$$

By deriving the first row of (6.533) with respect to time one gets

$$\ddot{z}_{1,1} = \frac{\dot{z}_{1,1}z_{1,2}}{\Delta x} - \frac{z_{1,1}\dot{z}_{1,2}}{\Delta x} \tag{6.534}$$

By substituting the first and the second row of (6.533) into (6.534) one gets

$$\ddot{z}_{1,1} = \left[-\frac{z_{1,1}z_{1,2} - \rho_0 v_0}{\Delta x}\right]\frac{z_{1,2}}{\Delta x} - \frac{z_{1,1}}{\Delta x}\left[-\frac{\dot{z}_{1,1}z_{1,2}}{\Delta x} + \frac{z_{1,1}\dot{z}_{1,2}}{\Delta x}\right] \tag{6.535}$$

where as proven above $z_{1,2} = f_1(Z_f, \dot{Z}_f)$. Next, one can define the right part of 6.535 as the new control input $\tilde{v}_1(Z_f, \dot{Z}_f)$, which gives

$$\ddot{z}_{1,1} = \tilde{v}_1 \tag{6.536}$$

In a similar manner, for the ith couple of equations of the discretised Payne–Whitham PDE description one has

$$\begin{aligned}\dot{z}_{i,1} &= -\frac{z_{i,1}z_{i,2} - z_{i-1,1}z_{i-1,2}}{\Delta x} \\ \dot{z}_{i,2} &= -z_{i,2}\frac{z_{i,2} - z_{i-1,1}}{\Delta x} - \frac{c_0^2}{z_{i,1}}\frac{z_{i,1} - \rho_0}{\Delta x} + \frac{V(z_{i,1} = z_{i,2})}{\tau}\end{aligned} \tag{6.537}$$

By deriving the first row of (6.537) with respect to time one obtains

$$\ddot{z}_{i,1} = -\frac{\dot{z}_{i,1}z_{i,2}}{\Delta x} - \frac{z_{i,1}\dot{z}_{i,2}}{\Delta x} + \frac{\dot{z}_{i-1,1}z_{i-1,2}}{\Delta x} + \frac{z_{i-1,1}\dot{z}_{i-1,2}}{\Delta x} \tag{6.538}$$

By substituting the first and the second row of (6.537) into (6.538) one gets

$$\begin{aligned}\ddot{z}_{i,1} = &\left[-\frac{z_{1,1}z_{1,2} - z_{i-1,1}z_{i-1,2}}{\Delta x}\right]\frac{z_{i,2}}{\Delta x} \\ &- \frac{z_{i,1}}{\Delta x}\left[-z_{i,2}\frac{z_{i,2} - z_{i-1,2}}{\Delta x} - \frac{c_0^2}{z_{i,1}}\frac{z_{i,1} - z_{i-1,1}}{\Delta x} + \frac{V(z_{i,1}) - z_{i,2}}{\tau}\right] \\ &- \left[-\frac{z_{i-1,1}z_{i-1,2} - z_{i-2,1}z_{i-2,2}}{\Delta x}\right]\frac{z_{i-1,1}}{\Delta x} \\ &+ \frac{z_{i-1,1}}{\Delta x}\left[-z_{i-1,2}\frac{z_{i-1,2} - z_{i-2,2}}{\Delta x} - \frac{c_0^2}{z_{i-1,1}}\frac{z_{i-1,1} - z_{i-2,1}}{\Delta x} + \frac{V(z_{i-1,1}) - z_{i-1,2}}{\tau}\right]\end{aligned} \tag{6.539}$$

where $z_{i,1}$, $z_{i-1,1}$ are elements of the flat output vector and $z_{i-1,2} = f_{i-1}(Z_f, \dot{Z}_f)$, while also $z_{i-2,1} = f_i(Z_f, \dot{Z}_f)$. Next, one can define the right part of (6.539) as the new control input $\tilde{v}_i(Z_f, \dot{Z}_f)$, which gives

$$\ddot{z}_{i,1} = \tilde{v}_i \tag{6.540}$$

for $i = 1, 2, \ldots, n$. In this manner, the state-space description of the Payne–Whitham PDE is transformed into the equivalent description

$$\begin{array}{lll} \ddot{z}_{1,1} = \tilde{v}_1 & \ddot{z}_{2,1} = \tilde{v}_3 & \ddot{z}_{3,1} = \tilde{v}_3 \\ \cdots & \ddot{z}_{n-1,1} = \tilde{v}_{n-1} & \ddot{z}_{n,1} = \tilde{v}_{n-1} \end{array} \tag{6.541}$$

which stands for the linear canonical Brunovsky form of the system.

6.10.6 Simulation tests

The efficiency of the proposed nonlinear Kalman filtering scheme for the estimation of traffic dynamics described by the Payne–Whitman PDE model has been confirmed through simulation experiments. The sampling period was taken to be $T_s = 0.01$ s. The x-axis (corresponding to the traffic lane) of the vehicles' spatiotemporal variables $\rho(x, t)$ and $v(x, t)$ was divided into $n = 50$ equidistant points. Estimates of the vehicles' density $\hat{\rho}(x, t)$ and of the vehicles' flow speed $\hat{v}(x, t)$ were obtained with the use of the Derivative free nonlinear Kalman Filter, through the processing of measurements of $\rho(x_i, t)$ at specific points of the grid. The obtained results are first depicted in Figures 6.72 and 6.73, where the real and estimated values of $\rho(x, t)$ and $v(x, t)$ are given. Moreover, in Figures 6.74, 6.75 and 6.76 one can see the convergence of the estimated state variables of the vehicles PDE dynamics to the associated real

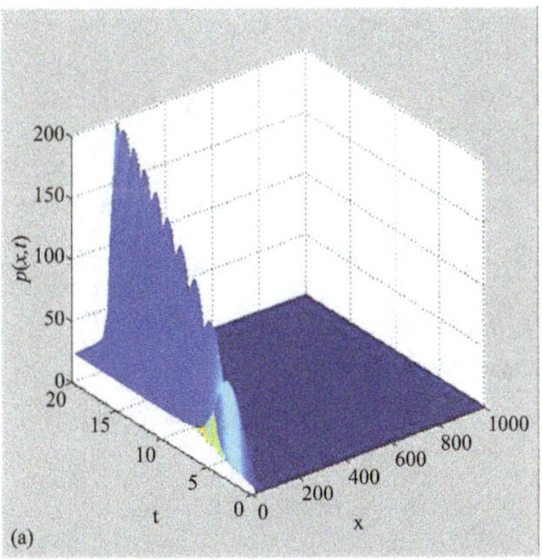

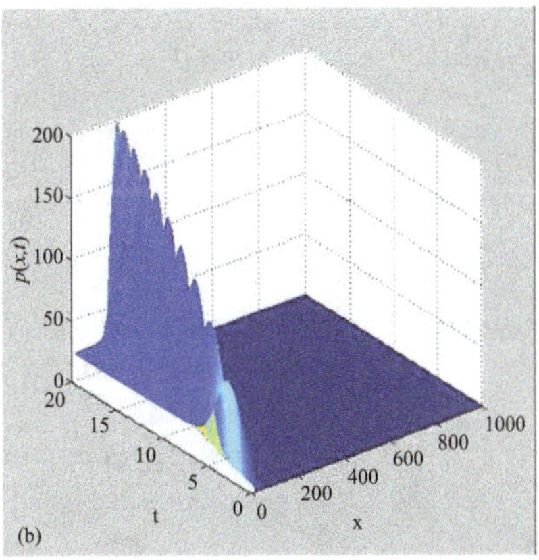

Figure 6.72 (a) Spatiotemporal variation of the vehicles' density $\rho(x, t)$ (state variable $z_{i,1}$, $i = 1, \ldots, n$) and (b) estimation of the spatiotemporal variation of the vehicles' density $\hat{\rho}(x, t)$ provided by the Derivative-free nonlinear Kalman Filter

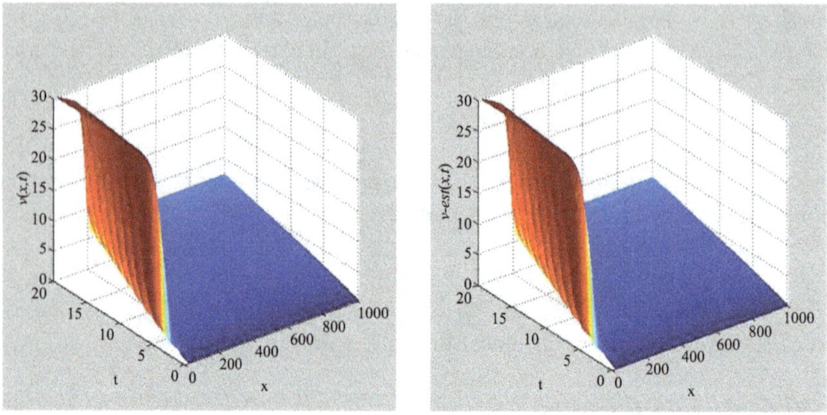

Figure 6.73 *(a) Spatiotemporal variation of the vehicles' flow speed* $v(x,t)$ *(state variable* $z_{i,2}$, $i = 1,\ldots,n$*) and (b) estimation of the spatiotemporal variation of the vehicles' flow speed* $\hat{v}(x,t)$ *provided by the Derivative-free nonlinear Kalman Filter*

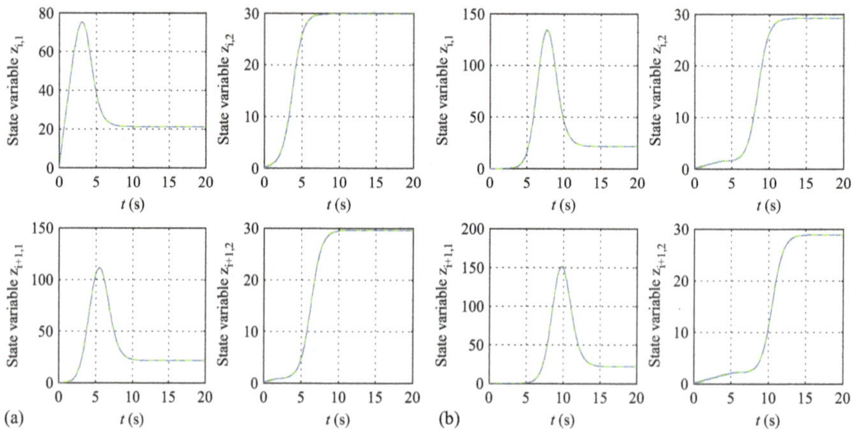

Figure 6.74 *State variables (blue line) and their estimation provided by the derivative free nonlinear Kalman Filter (green line) and (a)* $z_{1,1}, z_{1,2}$, *at grid point 1 and* $z_{2,1}, z_{2,2}$ *at grid point 2 (b)* $z_{3,1}, z_{3,2}$, *at grid point 3 and* $z_{4,1}, z_{4,2}$ *at grid point 4*

state variables. Results are given for the first 12 grid points of the spatiotemporal representation of the system.

Having proven differential flatness of the highway PDE traffic model the solution of the associated state estimation problem has become also possible. Actually, it has been shown that state estimation can be performed in this model with the use of

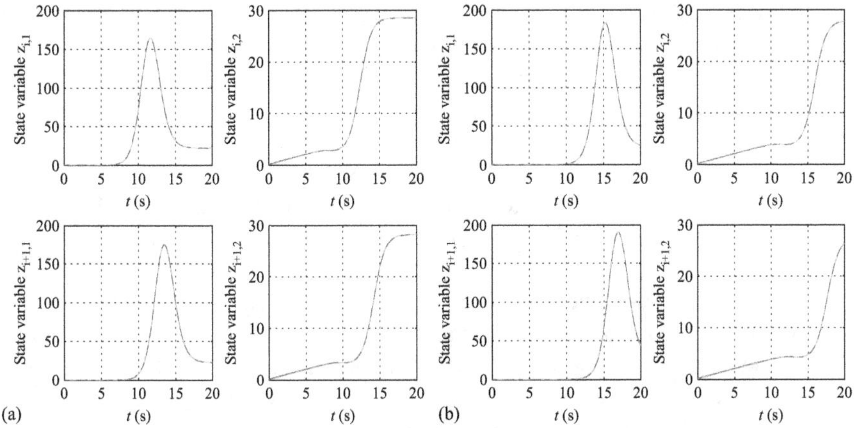

Figure 6.75 State variables (blue line) and their estimation provided by the Derivative free nonlinear Kalman Filter (green line) (a) $z_{5,1}, z_{5,2}$, *at grid point 5 and* $z_{6,1}, z_{6,2}$ *at grid point 6 and (b)* $z_{7,1}, z_{7,2}$, *at grid point 7 and* $z_{8,1}, z_{8,2}$ *at grid point 8*

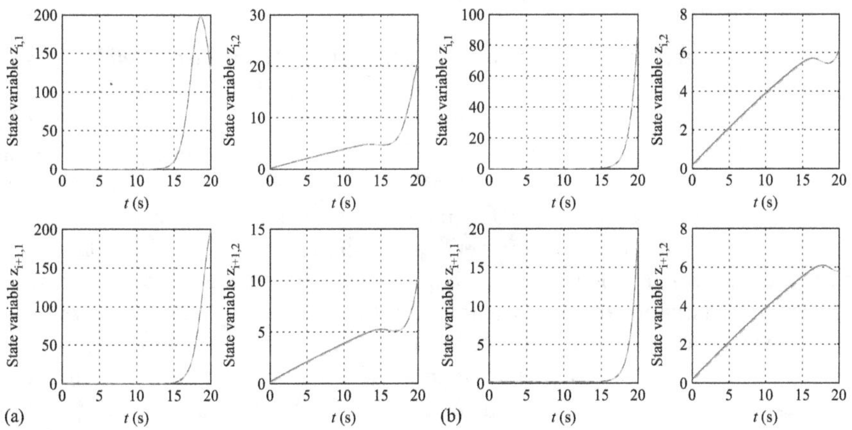

Figure 6.76 State variables (blue line) and their estimation provided by the Derivative free nonlinear Kalman Filter (green line) (a) $z_{9,1}, z_{9,2}$, *at grid point 9 and* $z_{10,1}, z_{10,2}$ *at grid point 10 and (b)* $z_{11,1}, z_{11,2}$, *at grid point 11 and* $z_{12,1}, z_{12,2}$ *at grid point 12*

differential flatness theory. The estimation method that achieves this is the Derivative-free nonlinear Kalman Filter. Comparing to other estimation methods, such as the Extended Kalman Filter, the new filtering approach is of improved accuracy because it avoids approximate linearisation in the state-space model of the PDE that describes the traffic flow. Besides, it is computationally faster because it does not require the iterative computation of Jacobian matrices.

From the simulation experiments, it can be observed that the Derivative-free nonlinear Kalman Filter was capable of computing fast and with high accuracy the estimates of the non-measurable state variables of the PDE traffic model. This is important for the assessment of the condition of highways in real-time, because it helps to develop a low-cost monitoring infrastructure (Rigatos, Clare and Blackmore, 2012). Actually, to extract information about the highway's condition there is no need to deploy a sensor network that comprises a large number of sensors (cameras, induction loops, GPS devices, etc.). It suffices to use a smaller number of suitably placed sensors (so as to maintain the system's observability and detectability) that will provide input to the Derivative-free nonlinear Kalman Filter. At a second stage, the filter computes the critical variables of the highway, such as $\rho(x,t)$ and $v(x,t)$ even at points where no sensor measurements are available.

6.11 Estimation of the PDE dynamics of a cable-suspended bridge

6.11.1 Outline

Condition monitoring of concrete bridges and particularly of suspended bridges is a non-trivial problem (Jaksic *et al.*, 2014; Pakrashi *et al.*, 2010; Gonzalez *et al.*, 2012). Bridges are exposed to degradation and failures due to aging, corrosion, environmental factors and materials stressing under harsh operating conditions (ElHattab *et al.*, 2016; Antolin *et al.*, 2013; Pakrashi *et al.*, 2019; Ding, Kong *et al.*, 2010). Visual inspection of a bridge is labor intensive, expensive and tedious while being often unreliable and inadequate for identifying incipient damages in such a large-scale structure. Failures in bridges set in risk human lives while having also a negative financial impact. Disruption of the traffic that goes through bridges can affect severely the economic activities in entire regions while the restoration cost can be too high to manage. So far there have been several attempts for condition monitoring of suspended bridges, however the fault detection and isolation problem in this critical infrastructure has not been solved in a conclusive manner (Li, Xu, Wu *et al.*, 2010; Zhang, Chen *et al.*, 2018; Rigatos, Zervos *et al.*, 2019; Zhu, Lin *et al.*, 2018). In the last years, the interaction between bridges and moving vehicles has emerged as a condition monitoring approach (Hester and Golzalez, 2012; Tsao *et al.*, 2001; Wei and Lin, 2007; Karoumi, 2000). The dynamic behavior of suspended bridges subjected to moving vehicles is complicated and comprises both ordinary differential equations (for the vehicle's vibrations) and partial differential equations (for the bridge's spatiotemporal deformation) (Yang, Zhang *et al.*, 2018; Ahmed *et al.*, 1998; Pi and Ouyang, 2017; DeBrunner *et al.*, 2004). By monitoring the bridge-vehicle interaction and by processing the related sensor measurements one can achieve fault detection and isolation (Majumder and Manohar, 2014; Kohn and Dyke, 2007; Gong and Cheung, 2006). Primarily, it is aimed to achieve early detection of failures, and to take up timely repair actions (Zhang, Basseville and Benveniste, 1994; Rigatos, Siano *et al.*, 2009; Rigatos, 2015; Isermann, 2006). Moreover, it is aimed to reduce false alarms rate so as to avoid unnecessary interruptions of vehicles' traffic (Basseville and Nikiforov, 1993; Benveniste *et al.*, 1987; Basseville *et al.*, 1996; Zhang, 1997). To implement efficient fault detection and isolation in bridges it is necessary to use precise estimation and statistical decision-making methods (Zhang, Basseville and Benveniste, 1998; Zhang and Benveniste, 1993; Rigatos and Zhang, 2009).

In the present section, a Kalman Filtering approach based on differential flatness theory (derivative-free nonlinear Kalman Filter) is used jointly with elaborated statistical decision-making tools so as to provide an effective solution to the condition monitoring problem of suspended bridges (Rigatos, 2015). There is particular difficulty in solving the fault detection and isolation problem for suspended bridges because such structures exhibit a PDE dynamics where common model-based fault diagnosis tools cannot be directly applied (Jaksic *et al.*, 2014; Pakrashi *et al.*, 2010; Gonzalez *et al.*, 2012). It is considered that the bridge's dynamics receive mechanical excitation by a vehicle that moves over it at a constant speed. Next, the joint dynamic model of the interaction between the suspended bridge and the moving vehicle is obtained. The vehicle is modeled as a quarter-car suspension system, comprising two masses and exhibiting two degrees of freedom (Rigatos, 2015). The suspended-bridge is modeled with the use of the Euler-Bernoulli flexible beam PDE. This designates that the bridge exhibits spatiotemporal deformations (Yang, Zhang *et al.*, 2018; Ahmed *et al.*, 1998; Pi and Ouyang, 2017; DeBrunner *et al.*, 2004). Next, by applying semi-discretisation and the finite differences approach, and considering a grid of N equidistant points on the bridge's longitudinal axis, the PDE that describes the spatiotemporal dynamics of the bridge is decomposed into a set of N ordinary differential equations (Rigatos, Zervos *et al.*, 2019; Rigatos, 2015). Thus, a state-space model of the bridge and vehicle interaction is obtained comprising $N+2$ ordinary differential equations. Equivalently, one can obtain a matrix description of this state-space model of dimension $2N+4$. Moreover, by denoting transformed control inputs, this state-space model is finally written in the canonical Brunovsky form which is known to be observable (Rigatos, Zervos *et al.*, 2019; Rigatos, 2015).

The implementation stages of the proposed fault detection and isolation methods have been as follows. First, by comparing the outputs of the Kalman Filter with the real-time outputs of the bridge and moving vehicle dynamical system the residuals sequence is generated. The known optimality properties of Kalman Filter's estimation assure that even when using a limited number of sensor measurements, one can obtain minimum variance estimates of the entire state vector of the suspended bridge and moving vehicle dynamic model (Rigatos, 2015). Next, it is demonstrated that the sum of the squares of the residuals vectors weighted by the inverse of their covariance matrix stands for a stochastic variable which follows the χ^2 distribution. Thus, by exploiting the 96% or the 98% confidence intervals of the χ^2 distribution one can define fault thresholds which show in an almost infallible manner, whether the bridge retains its structural integrity or whether it has been subjected to a fault. Actually, as long as the above noted statistical test remains within these confidence intervals and close to a value that is equal to the total number of the monitored outputs of the bridge-vehicle system (that is up to $N+2$) one can conclude that no damage has affected the bridge's structure. By applying, the proposed statistical tests into subspaces of the state-space model of the bridge-vehicle interaction one can also perform fault isolation, thus finding the specific part in the bridge's structure that has undergone a damage.

6.11.2 Dynamic model of the suspended-bridge and vehicle interaction

The diagram of the suspended-bridge is shown in Figure 6.77. The bridge exhibits both oscillations when subjected to external excitation, as well as deformations. This is represented in the spatiotemporal variable for the displacement $y_b(x,t)$ of each individual segment of the bridge from the equilibrium position. The dynamic model of the suspended bridge is that of the Euler–Bernoulli flexible beam and is given by the following PDE (Jaksic *et al.*, 2014; Pakrashi *et al.*, 2010; Gonzalez *et al.*, 2012):

$$M\frac{\partial^2 y_b}{\partial t^2} + EI\frac{\partial^4 y_b}{\partial x^4} + c_b\frac{\partial y_b}{\partial t} = -Ky_b^+ + F(x) + W(x) \tag{6.542}$$

The boundary conditions of this PDE are $y_b(0,t)=0, y_b(L,t)=0$ and $\frac{\partial^2 y_b}{\partial^2 x}(0,t)=0$, $\frac{\partial^2 y_b}{\partial^2 x}(L,t)=0$. The main parameters of the PDE equation of the suspended bridge are defined as follows: M is the mass of the bridge per unit length, E is Young's modulus, I is the moment of inertia of the cross-section part of the bridge, c_b is a damping coefficient, K is the stiffness of the cables, $y_b^+ = \max(y_b, 0)$ provides the condition for non-zero tension from the cables, $w(x)$ is the weight per unit length of

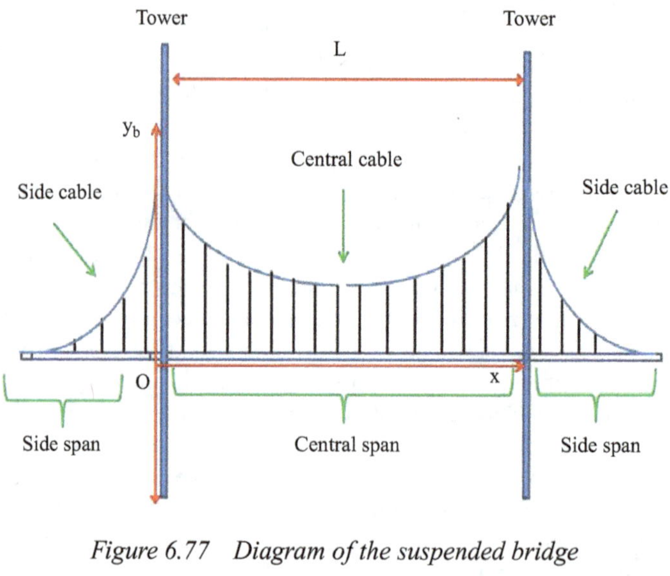

Figure 6.77 Diagram of the suspended bridge

the bridge, $f(x,t)$ is the external excitation applied to the bridge (for instance because of the moving vehicle), and L is the length of the central span of the bridge.

The model of the quarter vehicle's suspension is shown in Figure 6.78. The parameters of this model are as follows: m_1 is the sprung mass, y_1 is the displacement of the sprung mass with respect to the standing-still position, k_1 is the elasticity of the spring element between the sprung and the unsprung mass, c_1 is the damping of the absorber between the sprung and the unsprung mass, m_2 is the unsprung mass, k_2 is the elasticity of the spring element between the unsprung mass and the road's surface, c_2 is the damping of the absorber element between the unsprung mass and the road surface, and y_2 is the displacement of the unsprung mass with respect to the standing still position.

The equations of the dynamics of the quarter-vehicle suspension model are (Rigatos, 2015):

$$\begin{aligned} &m_1\ddot{y}_1 + c_1(\dot{y}_1 - \dot{y}_2) + k_1(y_1 - y_2) = 0 \\ &m_2\ddot{y}_2 + c_1(\dot{y}_2 - \dot{y}_1) + k_1(y_2 - y_1) + c_2(\dot{y}_2 - \dot{y}_b) \\ &\quad + k_2(y_2 - y_b) = \bar{F}(t) - (m_1 + m_2)g \end{aligned} \tag{6.543}$$

$\bar{F}(t)$ is the interaction force with the road surface. An elaborated model for this force is $\bar{F}(t) = (m_1 + m_2)g + m_1\ddot{y}_1 + m_2\ddot{y}_2$. A simplified model for $\bar{F}(t))$ is $F(t) =$

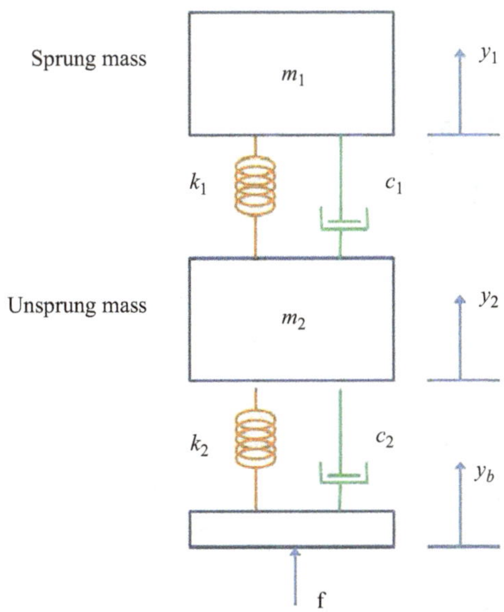

Figure 6.78 Diagram of the quarter-vehicle and suspension system

$(m_1 + m_2)g$. which stands for the moving force model. Using the Latter description for $\bar{F}(t)$ one obtains

$$\begin{aligned} &m_1\ddot{y}_1 + c_1(\dot{y}_1 - \dot{y}_2) + k_1(y_1 - y_2) = 0 \\ &m_2\ddot{y}_2 + c_1(\dot{y}_2 - \dot{y}_1) + k_1(y_2 - y_1) + c_2(\dot{y}_2 - \dot{y}_b) + k_2(y_2 - y_b) = 0 \end{aligned} \tag{6.544}$$

Next, to find the dynamic model of the bridge under the moving load of the vehicle, one has to substitute in (6.542) the force $F(t)$ which is due to the interaction of the vehicle with the bridge's surface. This is

$$F(x) = -c_2(\dot{y}_b) - \dot{y}_2 + k_2(y_b - y_2) + (m_1 + m_2)g \tag{6.545}$$

The diagram of the interaction between the moving vehicle and the bridge is shown in Figure 6.79

Weight function $W(x)$ is taken to be equal to $W(x) = Mg$, where M is the mass of the ith structural element of the bridge $i = 1, 2, \ldots, N$. Thus the dynamic model of the bridge becomes

$$\begin{aligned} &M\frac{\partial^2 y_b}{\partial t^2} + EI\frac{\partial^4 y_b}{\partial x^4} + c_b\frac{\partial y_b}{\partial t} + [c_2(\dot{y}_b - \dot{y}_2) + k_2(y_b - y_2)] \,|_{\delta(x-vt)} \\ &= -Ky_b^+(x) + Mg + [(m_1 + m_2)g] \mid \delta(x - vt) \end{aligned} \tag{6.546}$$

where $\delta(x - vt)$ is a Dirac delta function denoting that the effect of spring-type or absorber-type forces on the bridge due to its interaction with the vehicle, as well as the effect of the vehicle's weight appears at point x when $x = vt$ where v is the velocity of the vehicle.

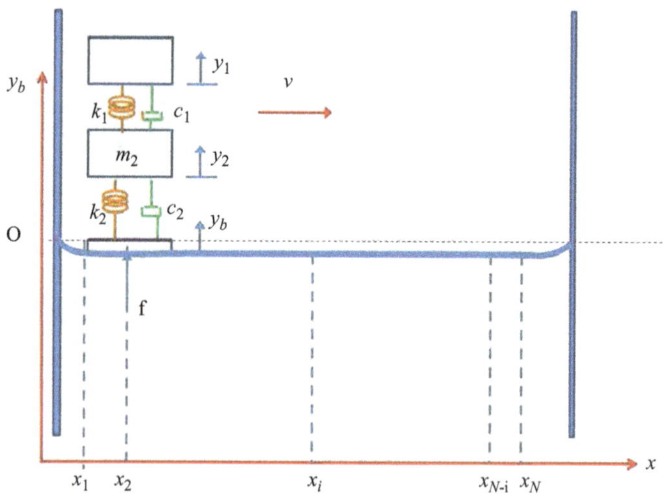

Figure 6.79 Diagram of the interaction between the moving vehicle and the suspended bridge

Next, to compute the partial derivative $\frac{\partial^4 y_b}{\partial x^4}$, along a grid partition of the Ox axis the finite-differences concept is used (Rigatos, 2015). Actually, it holds that

$$\frac{\partial^4 y_b}{\partial x^4}\mid_{x=x_i}=\frac{y_b(x_{i+4})-4y_b(x_{i+3})+6y_b(x_{i+2})-4y_b(x_{i+1})+y_b(x_i)}{\Delta x^4} \tag{6.547}$$

Thus, using the semi-discretisation concept, the dynamic model of the bridge becomes

$$\begin{aligned} &M\frac{\partial^2 y_b}{\partial t^2}+EI\frac{y_b(x_{i+4})-4y_b(x_{i+3})+6y_b(x_{i+2})-4y_b(x_{i+1})+y_b(x_i)}{\Delta x^4}\\ &+c_b\frac{\partial y_b}{\partial t}+[c_2(\dot{y}_b-\dot{y}_2)+k_2(y_b-y_2)]\mid_{\delta(x=x_i)}\\ &=-Ky_b^+ +Mg+[(m_1+m_2)g]\mid_{\delta(x=x_i)} \end{aligned} \tag{6.548}$$

Considering a grid partitioning of N-points along the Ox axis one can define the following notation for the associate values of $y_b(x,t)$ (Rigatos, 2015)

$$\begin{aligned} &x=[y_b(x_1),y_b(x_2),\dots,y_b(x_i),\dots,y_b(x_{N-1},y_b(x_N))]\Rightarrow\\ &x=[y_b(1),y_b(2),\dots,y_b(i),\dots,y_b(N-1,y_b(N))] \end{aligned} \tag{6.549}$$

At point x_1 (i=1) the dynamic model becomes

$$\begin{aligned} &M\ddot{y}_b(1)+EI\frac{y_b(5)-4y_b(4)+6y_b(3)-4y_b(2)+y_b(1)}{\Delta x^4}+c_b\dot{y}_b(1)\\ &+[c_2(\dot{y}_b(1)-\dot{y}_2)+k_2(y_b(1)-y_2)]\mid_{\delta(1}=-Ky_b^+(1)\\ &+Mg+[(m_1+m_2)g]\mid_{\delta(1)} \end{aligned} \tag{6.550}$$

At point x_2 (i=2) the dynamic model becomes

$$\begin{aligned} &M\ddot{y}_b(2)+EI\frac{y_b(6)-4y_b(5)+6y_b(4)-4y_b(3)+y_b(2)}{\Delta x^4}+c_b\dot{y}_b(2)\\ &+[c_2(\dot{y}_b(2)-\dot{y}_2)+k_2(y_b(2)-y_2)]\mid_{\delta(2}=-Ky_b^+(2)\\ &+Mg+[(m_1+m_2)g]\mid_{\delta(2)} \end{aligned} \tag{6.551}$$

At point x_3 (i=3) the dynamic model becomes

$$\begin{aligned} &M\ddot{y}_b(3)+EI\frac{y_b(7)-4y_b(6)+6y_b(5)-4y_b(4)+y_b(3)}{\Delta x^4}+c_b\dot{y}_b(3)\\ &+[c_2(\dot{y}_b(3)-\dot{y}_2)+k_2(y_b(3)-y_2)]\mid_{\delta(3}=-Ky_b^+(3)\\ &+Mg+[(m_1+m_2)g]\mid_{\delta(3)} \end{aligned} \tag{6.552}$$

$$\begin{aligned} &\dots \quad \dots \quad \dots\\ &\dots \quad \dots \quad \dots \end{aligned}$$

At point x_i (i) the dynamic model becomes

$$\begin{aligned} &M\ddot{y}_b(i)+EI\frac{y_b(i+4)-4y_b(i+3)+6y_b(i+2)-4y_b(i+1)+y_b(i)}{\Delta x^4}+c_b\dot{y}_b(i)\\ &+[c_2(\dot{y}_b(i)-\dot{y}_2)+k_2(y_b(i)-y_2)]\mid_{\delta(i}=-Ky_b^+(i)\\ &+Mg+[(m_1+m_2)g]\mid_{\delta(i)}\\ &\dots \quad \dots \quad \dots\\ &\dots \quad \dots \quad \dots \end{aligned} \tag{6.553}$$

At point x_{N-4} (i=N-4) the dynamic model becomes

$$\begin{aligned}
&M\ddot{y}_b(N-4)+EI\frac{y_b(N)-4y_b(N-1)+6y_b(N-2)-4y_b(N-3)+y_b(N-4)}{\Delta x^4}\\
&\quad+c_b\dot{y}_b(N-4)+[c_2(\dot{y}_b(N-4)-\dot{y}_2)+k_2(y_b(N-4)-y_2)]\mid_{\delta(N-4}\\
&\quad=-Ky_b^+(N-4)+Mg+[(m_1+m_2)g]\mid_{\delta(N-4)}
\end{aligned} \tag{6.554}$$

At point x_{N-3} (i=N-3) the dynamic model becomes

$$\begin{aligned}
&M\ddot{y}_b(N-3)+EI\frac{y_b(N+1)-4y_b(N)+6y_b(N-1)-4y_b(N-2)+y_b(N-3)}{\Delta x^4}\\
&\quad+c_b\dot{y}_b(N-3)+[c_2(\dot{y}_b(N-3)-\dot{y}_2)+k_2(y_b(N-3)-y_2)]\mid_{\delta(N-3}\\
&\quad=-Ky_b^+(N-3)+Mg+[(m_1+m_2)g]\mid_{\delta(N-3)}
\end{aligned} \tag{6.555}$$

At point x_{N-2} (i=N-2) the dynamic model becomes

$$\begin{aligned}
&M\ddot{y}_b(N-2)+EI\frac{y_b(N+2)-4y_b(N+1)+6y_b(N)-4y_b(N-1)+y_b(N-2)}{\Delta x^4}\\
&\quad+c_b\dot{y}_b(N-2)+[c_2(\dot{y}_b(N-2)-\dot{y}_2)+k_2(y_b(N-2)-y_2)]\mid_{\delta(N-2}\\
&\quad=-Ky_b^+(N-2)+Mg+[(m_1+m_2)g]\mid_{\delta(N-2)}
\end{aligned} \tag{6.556}$$

At point x_{N-1} (i=N-1) the dynamic model becomes

$$\begin{aligned}
&M\ddot{y}_b(N-1)+EI\frac{y_b(N+3)-4y_b(N+2)+6y_b(N+1)-4y_b(N)+y_b(N-1)}{\Delta x^4}\\
&\quad+c_b\dot{y}_b(N-1)+[c_2(\dot{y}_b(N-1)-\dot{y}_2)+k_2(y_b(N-1)-y_2)]\mid_{\delta(N-1}\\
&\quad=-Ky_b^+(N-1)+Mg+[(m_1+m_2)g]\mid_{\delta(N-1)}
\end{aligned} \tag{6.557}$$

At point x_N (i=N) the dynamic model becomes

$$\begin{aligned}
&M\ddot{y}_b(N)+EI\frac{y_b(N+4)-4y_b(N+3)+6y_b(N+2)-4y_b(N+1)+y_b(N)}{\Delta x^4}\\
&\quad+c_b\dot{y}_b(N)+[c_2(\dot{y}_b(N)-\dot{y}_2)+k_2(y_b(N)-y_2)]\mid_{\delta(N}\\
&\quad=-Ky_b^+(N)+Mg+[(m_1+m_2)g]\mid_{\delta(N)}
\end{aligned} \tag{6.558}$$

where $\forall i>N$, that is $i=N+1,\ldots N+4$ it holds that $y_b(i)=0$.

Next, one defines the following state variables

$$
\begin{array}{cccc}
x_1 = y_1 & x_2 = \dot{y}_1 & x_3 = y_2 & x_4 = \dot{y}_2 \\
x_5 = y_b(1) & x_6 = \dot{y}_b(1) & x_7 = y_b(2) & x_8 = \dot{y}_b(2) \\
x_9 = y_b(3) & x_{10} = \dot{y}_b(3) & x_{11} = y_b(4) & x_{12} = \dot{y}_b(4) \\
x_{13} = y_b(5) & x_{14} = \dot{y}_b(5) & x_{15} = y_b(6) & x_{16} = \dot{y}_b(6) \\
\cdots & \cdots & \cdots & \cdots \\
\cdots & \cdots & \cdots & \cdots \\
x_{2i+3} = y_b(i) & x_{2i+4} = \dot{y}_b(i) & x_{2i+5} = y_b(i+1) & x_{2i+6} = \dot{y}_b(i+1) \\
\cdots & \cdots & \cdots & \cdots \\
\cdots & \cdots & \cdots & \cdots \\
x_{2N-3} = y_b(N-3) & x_{2N-2} = \dot{y}_b(N-3) & x_{2N-1} = y_b(N-2) & x_{2N} = \dot{y}_b(N-2) \\
x_{2N+1} = y_b(N-1) & x_{2N+2} = \dot{y}_b(N-1) & x_{2N+3} = y_b(N) & x_{2N+4} = \dot{y}_b(N)
\end{array}
$$

Thus, the state-space model of the bridge–vehicle interaction, when using a N-point partition of the longitudinal axis of the bridge, has $2N+4$ state variables. Next, using the previous state variables notation, the dynamics of the bridge–vehicle system is written as

$$
\begin{aligned}
\dot{x}_1 &= x_2 \\
\dot{x}_2 &= -\frac{c_1}{m_1}(x_2 - x_4) - \frac{k_1}{m1}(x_1 - x_3) = 0 \\
\dot{x}_3 &= x_4 \\
\dot{x}_4 &= -\frac{c_1}{m_2}(x_4 - x_2) - \frac{k_1}{m2}(x_3 - x_1) - \frac{c_2}{m_2}(x_4 - x_{2i+4}) - \frac{k_2}{m2}(x_3 - x_{2i+3}) \\
\dot{x}_5 &= x_6 \\
\dot{x}_6 &= -\frac{EI}{M}\left[\frac{x_{13} - 4x_{11} + 6x_9 - 4x_7 + x_5}{\delta x^4}\right] + \frac{c_b}{M}x_6 \\
&\quad + \frac{1}{M}[c_2(x_6 - x_4) + k_2(x_5 - x_3)]\mid_{\delta(1)} - \frac{K}{M}y_b^+(1) + g + \frac{[(m_1 + m2)]}{M}g\mid_{\delta(1)} \\
\dot{x}_7 &= x_8 \\
\dot{x}_8 &= -\frac{EI}{M}\left[\frac{x_{15} - 4x_{13} + 6x_{11} - 4x_9 + x_7}{\delta x^4}\right] + \frac{c_b}{M}x_8 \\
&\quad + \frac{1}{M}[c_2(x_8 - x_4) + k_2(x_7 - x_3)]\mid_{\delta(2)} - \frac{K}{M}y_b^+(2) + g + \frac{[(m_1 + m2)]}{M}g\mid_{\delta(2)} \\
\dot{x}_9 &= x_{10} \\
\dot{x}_{10} &= -\frac{EI}{M}\left[\frac{x_{17} - 4x_{15} + 6x_{13} - 4x_{11} + x_9}{\delta x^4}\right] + \frac{c_b}{M}x_{10} \\
&\quad + \frac{1}{M}[c_2(x_{10} - x_4) + k_2(x_9 - x_3)]\mid_{\delta(3)} - \frac{K}{M}y_b^+(3) + g + \frac{[(m_1 + m2)]}{M}g\mid_{\delta(3)}
\end{aligned}
\tag{6.559}
$$

$$\cdots \quad \cdots \quad \cdots$$
$$\cdots \quad \cdots \quad \cdots$$

$$\begin{aligned}
\dot{x}_{2i+3} &= x_{2i+4} \\
\dot{x}_{2i+4} &= -\frac{EI}{M}\left[\frac{x_{2i+11} - 4x_{2i+9} + 6x_{2i+7} - 4x_{2i+5} + x_{2i+3}}{\delta x^4}\right] + \frac{c_b}{M}x_{2i+4} \\
&+\frac{1}{M}[c_2(x_{2i+4} - x_4) + k_2(x_{2i+3} - x_3)] \mid_{\delta(i)} - \frac{K}{M}y_b^+(i) + g + \frac{[(m_1 + m2)]}{M}g \mid_{\delta(i)}
\end{aligned} \tag{6.560}$$

$$\cdots \quad \cdots \quad \cdots$$
$$\cdots \quad \cdots \quad \cdots$$

$$\begin{aligned}
\dot{x}_{2N+1} &= x_{2N+2} \\
\dot{x}_{2N=2} &= -\frac{EI}{M}\left[\frac{x_{2N+9} - 4x_{2N+7} + 6x_{2N+5} - 4x_{2N+3} + x_{2N+1}}{\delta x^4}\right] + \frac{c_b}{M}x_{2N+2} \\
&+ \frac{1}{M}[c_2(x_{2N+2} - x_4) + k_2(x_{2N+1} - x_3)] \mid_{\delta(N-1)} - \frac{K}{M}y_b^+(N-1) \\
&+ g + \frac{[(m_1 + m2)]}{M}g \mid_{\delta(N-1)} \\
\dot{x}_{2N+3} &= x_{2N+4} \\
\dot{x}_{2N=4} &= -\frac{EI}{M}\left[\frac{x_{2N+11} - 4x_{2N+9} + 6x_{2N+7} - 4x_{2N+5} + x_{2N+3}}{\delta x^4}\right] + \frac{c_b}{M}x_{2N+4} \\
&+ \frac{1}{M}[c_2(x_{2N+4} - x_4) + k_2(x_{2N+3} - x_3)] \mid_{\delta(N)} - \frac{K}{M}y_b^+(N) \\
&+ g + \frac{[(m_1 + m2)]}{M}g \mid_{\delta(N)}
\end{aligned} \tag{6.561}$$

6.11.3 Kalman Filtering for state-estimation in the bridge and vehicle system

6.11.3.1 State-space model of the bridge and vehicle system in the canonical matrix form

For the first two rows of the state-space model the following transformed control inputs are defined

$$\begin{aligned}
v_1 &= -\frac{c_1}{m_1}(x_2 - x_4) - \frac{k_1}{m_1}(x_1 - x_3) \\
v_2 &= -\frac{c_1}{m_2}(x_4 - x_2) - \frac{k_1}{m_2}(x_3 - x_1) - \frac{c_2}{m_2}(x_4 - x_{2i+4}) - \frac{k_2}{m_2}(x_3 - x_{2i+3})
\end{aligned} \tag{6.562}$$

Next, by defining the new control input v_{i+2} for $i = 3, 4, \ldots, N$ at the $(2i+4)$th row of the state-space model as

$$\begin{aligned}
v_{i+2} = &-\frac{EI}{M}\left[\frac{x_{2i+11} - 4x_{2i+9} + 6x_{2i+7} - 4x_{2i+5} + x_{2i+3}}{\delta x^4}\right] + \frac{c_b}{M}x_{2i+4} \\
&+ \frac{1}{M}[c_2(x_{2i+4} - x_4) + k_2(x_{2i+3} - x_3)] \mid_{\delta(i)} - \frac{K}{M}y_b^+(i) \\
&+ g + \frac{[(m_1 + m2)}{M}g] \mid_{\delta(i)}
\end{aligned} \tag{6.563}$$

the state-space model of the bridge-vehicle system becones

$$\begin{array}{ll}
\dot{x}_1 = x_2 & \dot{x}_2 = v_1 \\
\dot{x}_3 = x_4 & \dot{x}_4 = v_2 \\
\dot{x}_5 = x_6 & \dot{x}_6 = v_3 \\
\dot{x}_7 = x_8 & \dot{x}_8 = v_4 \\
\dot{x}_9 = x_{10} & \dot{x}_{10} = v_6 \\
\cdots & \cdots \\
\cdots & \cdots \\
\dot{x}_{2i+3} = x_{2i+4} & \dot{x}_{2i+4} = v_{i+2} \\
\cdots & \cdots \\
\cdots & \cdots \\
\dot{x}_{2N+1} = x_{2N+2} & \dot{x}_{2N+2} = v_{N+1} \\
\dot{x}_{2N+3} = x_{2N+4} & \dot{x}_{2N+4} = v_{N+2}
\end{array} \tag{6.564}$$

Consequently, the state-space model of the bridge-vehicle interaction model can be written in the following canonical form:

$$\begin{pmatrix} \dot{x}_1 \\ \dot{x}_2 \\ \dot{x}_3 \\ \dot{x}_4 \\ \dot{x}_5 \\ \dot{x}_6 \\ \cdots \\ \cdots \\ \dot{x}_{2i+3} \\ \dot{x}_{2i+4} \\ \cdots \\ \cdots \\ \dot{x}_{2N+1} \\ \dot{x}_{2N+2} \\ \dot{x}_{2N+3} \\ \dot{x}_{2N+4} \end{pmatrix} = \begin{pmatrix} 0\,1\,0\,0\,0\,0 \cdots 0\,0 \cdots 0\,0\,0\,0 \\ 0\,0\,0\,0\,0\,0 \cdots 0\,0 \cdots 0\,0\,0\,0 \\ 0\,0\,0\,1\,0\,0 \cdots 0\,0 \cdots 0\,0\,0\,0 \\ 0\,0\,0\,0\,0\,0 \cdots 0\,0 \cdots 0\,0\,0\,0 \\ \cdots\ \cdots\ \cdots\ \cdots\ \cdots \\ \cdots\ \cdots\ \cdots\ \cdots\ \cdots \\ 0\,0\,0\,0\,0\,0 \cdots 0\,1 \cdots 0\,0\,0\,0 \\ 0\,0\,0\,0\,0\,0 \cdots 0\,0 \cdots 0\,0\,0\,0 \\ \cdots\ \cdots\ \cdots\ \cdots\ \cdots \\ \cdots\ \cdots\ \cdots\ \cdots\ \cdots \\ 0\,0\,0\,0\,0\,0 \cdots 0\,0 \cdots 0\,1\,0\,0 \\ 0\,0\,0\,0\,0\,0 \cdots 0\,0 \cdots 0\,0\,0\,0 \\ 0\,0\,0\,0\,0\,0 \cdots 0\,0 \cdots 0\,0\,1\,0 \\ 0\,0\,0\,0\,0\,0 \cdots 0\,0 \cdots 0\,0\,0\,0 \end{pmatrix} \begin{pmatrix} x_1 \\ x_2 \\ x_3 \\ x_4 \\ x_5 \\ x_6 \\ \cdots \\ \cdots \\ x_{2i+3} \\ x_{2i+4} \\ \cdots \\ \cdots \\ x_{2N+1} \\ x_{2N+2} \\ x_{2N+3} \\ x_{2N+4} \end{pmatrix}$$

$$
+\begin{pmatrix}
0\,0\,0\cdots 0 \cdots 0\,0 \\
1\,0\,0\cdots 0 \cdots 0\,0 \\
0\,0\,0\cdots 0 \cdots 0\,0 \\
0\,1\,0\cdots 0 \cdots 0\,0 \\
0\,0\,0\cdots 0 \cdots 0\,0 \\
0\,0\,1\cdots 0 \cdots 0\,0 \\
\cdots \cdots \cdots \cdots \\
\cdots \cdots \cdots \cdots \\
0\,0\,0\cdots 0 \cdots 0\,0 \\
0\,0\,0\cdots 1 \cdots 0\,0 \\
\cdots \cdots \cdots \cdots \\
\cdots \cdots \cdots \cdots \\
0\,0\,0\cdots 1 \cdots 0\,0 \\
0\,0\,0\cdots 1 \cdots 1\,0 \\
0\,0\,0\cdots 1 \cdots 0\,0 \\
0\,0\,0\cdots 1 \cdots 0\,1
\end{pmatrix}
\begin{pmatrix}
v_1 \\ v_2 \\ v_3 \\ \cdots \\ \cdots \\ v_{i+1} \\ \cdots \\ \cdots \\ v_{N+1} \\ v_{N+2}
\end{pmatrix}
$$

or, in concise form the state-space model of the bridge-vehicle interaction is written as

$$\dot{x} = Ax + Bv \tag{6.565}$$

where $x \in R^{(2N+4)\times 1}$, $A \in R^{(2N+4)\times(2N+4)}$, $B \in R^{(2N+4)\times(N+2)}$, and $v \in R^{(N+2)\times 1}$. The measurement equation of the system is written as

$$
\begin{pmatrix}
y_1 \\ y_2 \\ y_3 \\ \cdots \\ \cdots \\ y_i \\ \cdots \\ \cdots \\ y_{N+1} \\ y_{N+2}
\end{pmatrix}
=
\begin{pmatrix}
1\,0\,0\,0\,0\,0 \cdots 0\,0 \cdots 0\,0\,0\,0 \\
0\,0\,1\,0\,0\,0 \cdots 0\,0 \cdots 0\,0\,0\,0 \\
0\,0\,0\,0\,1\,0 \cdots 0\,0 \cdots 0\,0\,0\,0 \\
\cdots \cdots \cdots \cdots \\
\cdots \cdots \cdots \cdots \\
0\,0\,1\,0\,0\,0 \cdots 1\,0 \cdots 0\,0\,0\,0 \\
\cdots \cdots \cdots \cdots \\
\cdots \cdots \cdots \cdots \\
0\,0\,0\,0\,0\,0 \cdots 0\,0 \cdots 1\,0\,0\,0 \\
0\,0\,0\,0\,0\,0 \cdots 0\,0 \cdots 0\,0\,1\,0
\end{pmatrix}
\begin{pmatrix}
x_1 \\ x_2 \\ x_3 \\ x_4 \\ x_5 \\ x_6 \\ \cdots \\ \cdots \\ x_{2i+3} \\ x_{2i+4} \\ \cdots \\ \cdots \\ x_{2N+1} \\ x_{2N+2} \\ x_{2N+3} \\ x_{2N+4}
\end{pmatrix}
\tag{6.566}
$$

where in concise form, the measurement equation is written as

$$y = Cx \tag{6.567}$$

where $y\in R^{(N+2)\times 1}$, $C\in R^{(N+2)\times 2N+4}$ and $x\in R^{(2N+4)\times 1}$. The bridge-vehicle system, described in (6.565) and (6.567) is differential flat and its flat output is

$$Y = [x_1, x_3, x_5, \ldots, x_{2i+3}, \ldots, x_{2N+1}, x_{2N+3}]^T \tag{6.568}$$

Actually, all state variables and the control inputs of this system can be expressed as differential functions of the flat outputs defined in (6.568) (Rigatos, 2015).

6.11.3.2 State-estimation with the use of the Kalman Filter

It has been confirmed that the bridge-vehicle interaction system can be written in the linear canonical form. This was expected after having proven that the system is differentially flat (Rigatos, 2015). The system was written in the linear canonical (Brunovsky) form which consisted of the following state equations

$$\begin{aligned} \dot{x} &= Ax + Bv \\ y &= Cx \end{aligned} \tag{6.569}$$

$x\in R^{(2N+4)\times 1}$, $v\in R^{(N+2)\times 1}$, $A\in R^{(2N+4)\times(2N+4)}$, $B\in R^{(2N+4)\times(N+2)}$, $y\in R^{(N+2)\times 1}$ and $B\in R^{(N+2)\times(2N+4)}$. To solve the related state-estimation problem the Kalman Filter is applied to the bridge-vehicle system of (6.570). The use of the Kalman Filter in nonlinear or in PDE-based dynamical systems, after transforming such systems in the linear canonical (Brunovsky) form is known to be a new nonlinear filtering method under the name *Derivative-free nonlinear Kalman Filter* (Rigatos, 2015).

The Kalman Filter's estimator is implemented in discrete-time, therefore, the state-space model of (6.570) has to be brought into a discrete-time form. This is done, by applying discretisation transformations, such as the Tustin transform. The discretised state-space model of the system is written in the form

$$\begin{aligned} x(k+1) &= A_d x(k) + B_d v(k) \\ y(k) &= C_d x(k) \end{aligned} \tag{6.570}$$

where A_d, B_d and C_d are the discrete-time equivalents of matrices A, B and C, respectively. Finally, the Kalman Filter's recursion consists of a *measurement update part* and of a *time-update part.*

measurement update part:

$$\begin{aligned} K(k) &= P^-(k)C_d^T[C_d P^-(k)C_d^T + R]^{-1} \\ \hat{x}(k) &= \hat{x}^-(k) + K(k)[y(k) - C_d x(k)] \\ P(k) &= P^-(k) - K(k)C_d P^-(k) \end{aligned} \tag{6.571}$$

time-update part:

$$\begin{aligned} P^-(k+1) &= A_d P(k) A_d^T + Q \\ \hat{x}^-(k+1) &= A_d \hat{x}(k) + B_d v(k) \end{aligned} \tag{6.572}$$

where R and Q are the related measurement noise covariance matrix and process noise covariance matrix, respectively. One can perform condition monitoring for the bridge-vehicle structure using statistical processing of the residuals of the derivative-free nonlinear Kalman Filter. Actually, the residuals' vector is defined as the difference, at each sampling instance, between the real outputs vector of the system y and the estimated outputs vector $\hat{y}$, and is given by

$$e(k) = y(k) - \hat{y}(k) \ \in R^{(N+2)\times 1} \tag{6.573}$$

Next, after defining the mean value of the residuals vector as $\bar{e}(k)$, the related covariance matrix is computed

$$Cov_e = E[(e(k) - \bar{e}(k))(e(k) - \bar{e}(k))^T] \in R^{(N+2)\times(N+2)} \tag{6.574}$$

It holds that the stochastic variable which is defined by the sum of the squares of the residuals' vector, weighted by the inverse of the associated covariance matrix, and normalised through the division with the dimension of the residuals' vector follows a χ^2 distribution of dimension that is equal to the number of the measured outputs of the system, that is up to $(N+2)$

$$t_{\chi^2} = \frac{1}{M}\sum_{k=1}^{M} e^T(k)Cov_e e(k) \sim \chi^2 \tag{6.575}$$

By exploiting the confidence intervals of the χ^2 distribution, at a confidence level of 96% or 98%, one can define fault thresholds that reveal in a credible manner the existence of faults in the bridge's structure.

6.11.4 Statistical fault diagnosis using the Kalman Filter

6.11.4.1 Fault detection

The residuals' sequence, that is the differences between (i) the real outputs of the system that describes the interaction between the bridge and the moving vehicle and (ii) the outputs estimated by the Kalman Filter (Figure 6.80) is a discrete error process e_k with dimension $m\times 1$ (here $m = N$ is the dimension of the output measurements vector). Actually, it is a zero-mean Gaussian white-noise process with covariance given by $E_k = Cov_e$.

A conclusion can be stated based on a measure of certainty, that the system that describes the interaction between the bridge and the moving vehicle has not been subjected to a fault. To this end, the following *normalised error square* (NES) is defined (Rigatos, 2015)

$$\varepsilon_k = e_k^T E_k^{-1} e_k \tag{6.576}$$

The sum of this normalised residuals' square follows a χ^2 distribution, with a number of degrees of freedom that is equal to the dimension of the residuals' vector. The form of the χ^2 distribution for various degrees of freedom is shown in Figure 6.81. An appropriate test for the normalised error sum is to numerically show that the

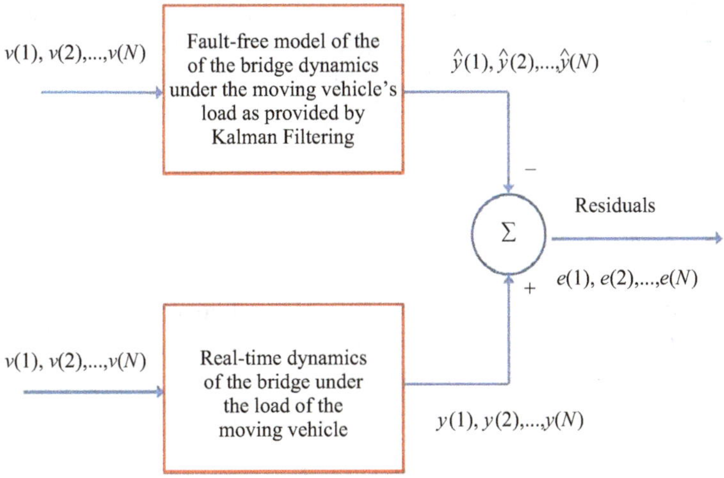

Figure 6.80 Residuals' generation for the bridge-vehicle interaction, with the use of Kalman Filtering

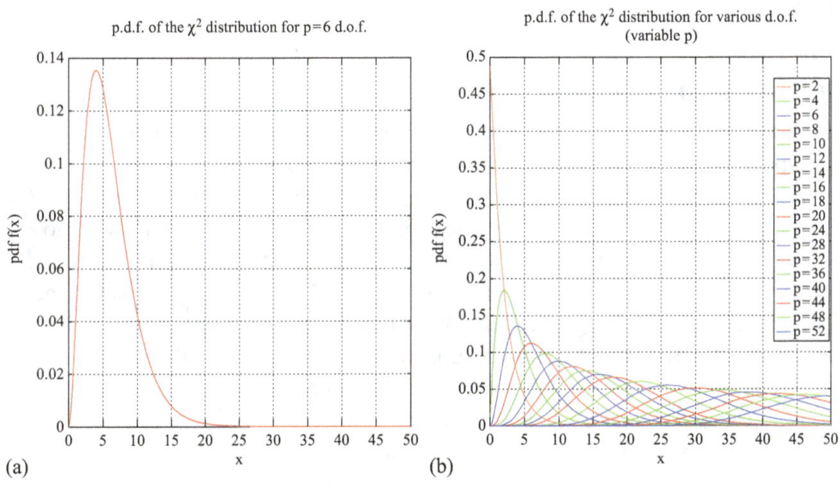

Figure 6.81 (a) Probability density function of the χ^2 distribution for $p = 6$ degrees of freedom and (b) probability density function of the distribution for several values of the degrees of freedom (variable p)

following condition is met within a level of confidence (according to the properties of the χ^2 distribution)

$$E\{\varepsilon_k\} = m \tag{6.577}$$

This can be achieved using statistical hypothesis testing, which are associated with confidence intervals. A 95% confidence interval is frequently applied, which is specified using $100(1-a)$ with $a=0.05$. Actually, a two-sided probability region is considered cutting-off two end tails of 2.5% each. For M runs the normalised error square that is obtained is given by

$$\bar{\varepsilon}_k = \frac{1}{M}\sum_{i=1}^{M}\varepsilon_k(i) = \frac{1}{M}\sum_{i=1}^{M} e_k^T(i)E_k^{-1}(i)e_k(i) \tag{6.578}$$

where ε_i stands for the ith run at time t_k. Then $M\bar{\varepsilon}_k$ will follow a χ^2 density with Mm degrees of freedom. This condition can be checked using a χ^2 test. The hypothesis holds, if the following condition is satisfied

$$\bar{\varepsilon}_k \in [\zeta_1, \zeta_2] \tag{6.579}$$

where ζ_1 and ζ_2 are derived from the tail probabilities of the χ^2 density. For example, for $m=20$ (dimension of the measurements vector) and $M=100$ (total number of the output vector's samples) one has $\chi^2_{Mm}(0.025)=1878$ and $\chi^2_{Mm}(0.975)=2126$. Using that $M=100$ one obtains $\zeta_1=\chi^2_{Mm}(0.025)/M=18.78$ and $\zeta_2=\chi^2_{Mm}(0.975)/M=21.26$.

6.11.4.2 Fault isolation

By applying the statistical test into the individual components of the system that describes the interaction between the bridge and the moving vehicle, it is also possible to find out the specific component that has been subjected to a fault (Rigatos, 2015). For a bridge-vehicle system of n parameters suspected for change one has to carry out n χ^2 statistical change detection tests, where each test is applied to the subset that comprises parameters $i-1$, i and $i+1$, $i=1,2,\ldots,n$. Actually, out of the n χ^2 statistical change detection tests, the one that exhibits the highest score are those that identify the parameter that has been subjected to change.

In the case of multiple faults, one can identify the subset of parameters that has been subjected to change by applying the χ^2 statistical change detection test according to a combinatorial sequence. This means that

$$\binom{n}{k} = \frac{n}{k!(n-k)!} \tag{6.580}$$

tests have to take place, for all clusters in the system that describe the interaction between the bridge and the moving vehicle, that finally comprise $n, n-1, n-2, \ldots,$ 2, 1 parameters. Again the χ^2 tests that give the highest scores indicate the parameters which are most likely to have been subjected to change.

As a whole, the concept of the proposed fault detection and isolation method is simple. The sum of the squares of the residuals' vector, weighted by the inverse of the residuals' covariance matrix, stands for a stochastic variable that follows the χ^2 distribution. Actually, this is a multi-dimensional χ^2 distribution and the number of its degrees of freedom is equal to the dimension of the residuals' vector. For the grid partition over the longitudinal axis of the bridge, 198 points provided measurable outputs. Thus, the residuals' vector is of dimension 198 and the number of degrees of freedom of the associated χ^2 distribution is also 198. Next, from the properties of the χ^2 distribution, the mean value of the aforementioned stochastic variable in the fault-free case should be also equal to 198. However, due to having sensor measurements subject to noise, the value of the statistical test in the fault-free case will not be precisely equal to 198 but it may vary within a small range around this value. This range is determined by the confidence intervals of the χ^2 distribution. For a probability of 98% to get a value of the stochastic variable about 198, the associated confidence interval is given by the lower bound $L = 196.55$ and by the upper bound $U = 199.45$. Consequently, as long as the statistical test provides an indication that the aforementioned stochastic variable is in the interval $[L, U]$ the functioning of the bridge-vehicle interaction system can be concluded to be free of faults. On the other side, when the bounds of the previously given confidence interval are exceeded it can be concluded that the bridge-vehicle interaction system has been subject to a fault. Finally, by performing the statistical test into subspaces of the bridge-vehicle interaction state-space model, where each subspace is associated with different components, one can also achieve fault isolation. This signifies that the specific component that has caused the malfunctioning of the bridge can be identified.

6.11.5 Simulation tests

The performance of the proposed fault detection and isolation method for suspended bridges, using Kalman Filtering and the χ^2 distribution properties of this filter's residuals has been tested through simulation experiments. A model of 2-DOF quarter-vehicle suspension was taken to represent the dynamics of the vehicle that is moving on the bridge. Besides, the grid partitioning was considered along the longitudinal axis of the bridge consisting of $N = 200$ points. Out of them, 198 points were considered to be providing measurable outputs. Considering a bridge with a span equal to 200 m, a sampling period of $T_s = 0.01$ and that the vehicle moves at a constant speed of 72km/h, it can be used that its load is exerted on a new grid point i every 5 sampling periods.

As already noted the output of the model (Kalman Filter) is compared against the measurements which are obtained from the real bridge-vehicle interaction system. Thus, the residuals' sequence is generated. Next, it is shown that when the squares of the residuals' vectors are multiplied by the inverse of the model's (Kalman Filter's) covariance matrix and summed up, a new stochastic variable is generated which follows the χ^2 distribution with n degrees of freedom where n is the dimension of the residuals' vector. This stochastic variable stands for a χ^2 statistical test. Consequently, if the values of this variable are compared against the confidence intervals of the χ^2 distribution one can infer with a high degree of certainty whether the structural integrity of the bridge is preserved or whether a fault has occurred. The confidence intervals of the χ^2 distribution allow to define with precision the ranges for the appearance of a failure in the bridge's structure. Obviously, fault detection and

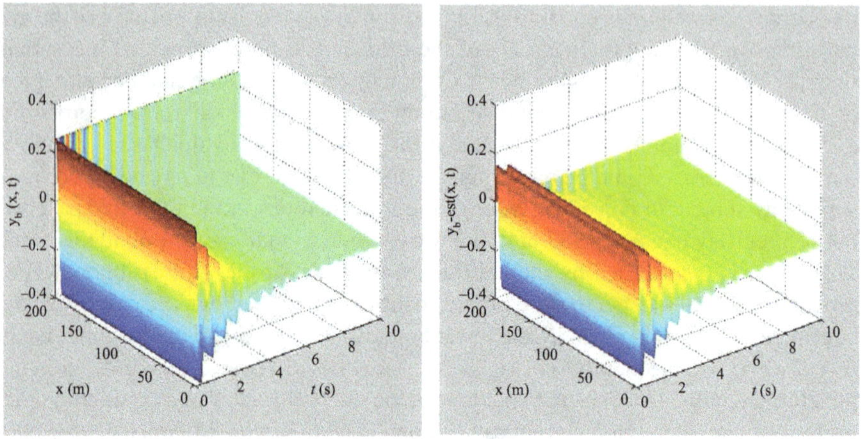

(a) (b)

Figure 6.82 Spatiotemporal vibrations of the suspended bridge under the load of the moving vehicle (a) as monitored by sensors in real-time and (b) as estimated by the Kalman Filter

isolation for the bridge-vehicle interaction systems cannot be performed with the use of metrics such as the RMSE, because in such a case the definition of fault thresholds comes from intuition and heuristics. Consequently, the RMSE of the Kalman Filter's residuals can lead to either overlooking of incipient faults in the bridge's structure, or to the launching of false alarms. As far as the p-value test is concerned this can be performed on the stochastic value which is obtained from the residuals' data, after using the Central Limit Theorem (CLT). The p-value test is related to the probability, of obtaining a value of the test statistic $z = (\bar{x} - \mu)/(\sigma/\sqrt{n})$ that annihilates the null hypothesis (where $\bar{x}$ is the average value, $\mu = 0$ is the sample's mean, s is the sample standard deviation, and n is the sample's size.). This is equivalent to saying that one uses the $a\%$ confidence intervals of the χ^2 distribution (for instance with a to be 96% or 98%) because in the latter case the probability for an erroneous conclusion from the χ^2 test is $1 - a\%$ (for instance 4% or 2% respectively). Consequently, there is no benefit from using a p-value test for finding faults in the bridge-vehicle interaction system. On the other side, the p-value test is not efficient in processing the multi-dimensional residuals vector of the Kalman Filter, whereas the χ^2 test is. One has to apply the p-value test for the sequence of the individual state variables of the residuals vector of the Kalman Filter, whereas in the χ^2 test the sequence of the entire residuals vector can be processed in batch mode. Additionally, it is noted that the application of the p-value test, for instance in the case that the test variable comes from CLT, requires to know the mean value and the variance of the i.i.d. processed data distribution (μ and σ) whereas the χ^2 test does not.

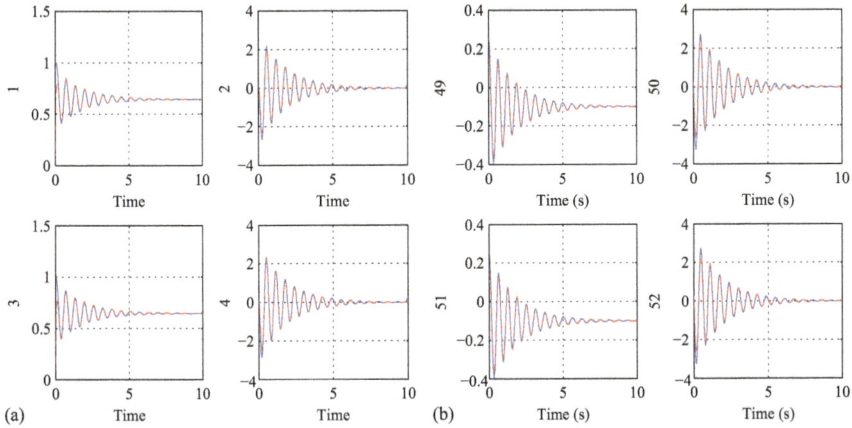

Figure 6.83 (a) Variation of state variables x_1 to x_4 (blue lines) of the suspended bridge and vehicle system (moving vehicle) and Kalman Filter-based estimates (dashed red lines) and (b) variation of state variables x_{49} to x_{52} (blue lines) of the suspended bridge and vehicle system (bridge's structure) and Kalman Filter-based estimates (dashed red lines)

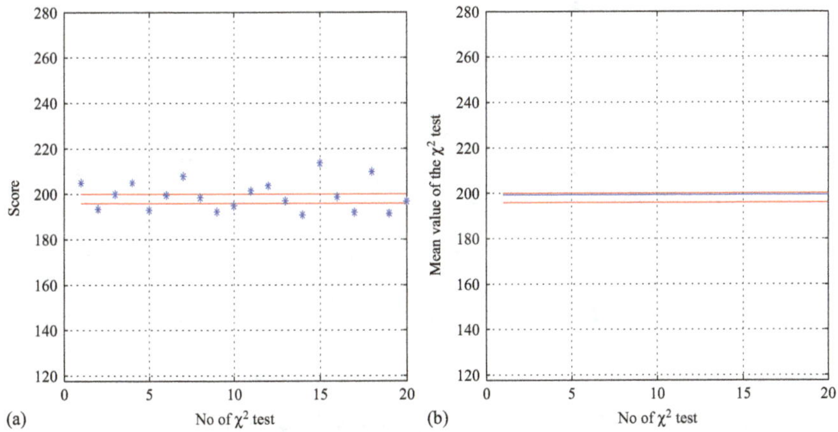

Figure 6.84 Test case 1: (a) Values of successive χ^2 tests were performed for the suspended bridge and vehicle system when no fault exists in the structure and (b) mean value (blue line) of the successive χ^2 tests performed for the suspended bridge and vehicle system when no fault exists in the structure

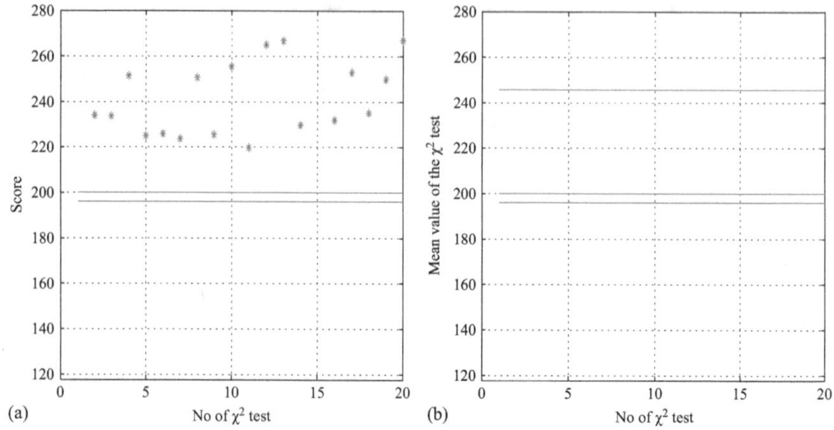

Figure 6.85 *Test case 2: (a) values of successive χ^2 tests performed for the suspended bridge and vehicle system when change (0.1%) has taken place in the stiffness of the cables and (b) mean value (blue line) of the successive χ^2 tests performed for the suspended bridge and vehicle system when change (0.1%) has taken place in the stiffness of the cables*

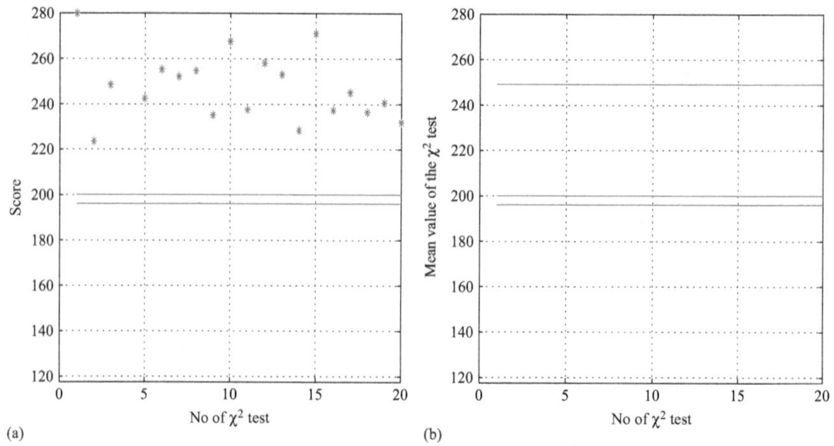

Figure 6.86 *Test case 3: (a) Values of successive χ^2 tests performed for the suspended bridge and vehicle system when change (0.1%) has taken place in the mass of the concrete segments of the bridge's deck and (b) mean value (blue line) of the successive χ^2 tests performed for the suspended bridge and vehicle system when change (0.1%) has taken place in the mass of the concrete segments of the bridge's deck*

Epilogue

The monograph has presented advances in the control of nonlinear and PDE dynamical systems and their use in robotic manipulators, autonomous vehicles, mechatronic systems, electric power systems, power electronics, industrial production processes, and in specific cyber-physical systems such as telecommunication and computer networks. The focus of the monograph has been on industrial systems control. However, throughout the related research it has been confirmed that the analysed control methods can be also used in non-industrial systems such as several biosystems and financial systems. For instance, the fine performance of these control methods has been also shown in the problem of vaccination policy control for the abatement of the spreading of epidemics, in the control of the acute inflammatory response, in the optimized infusion of medication, as well as in the synchronization problem of biological oscillators. Besides, the fine performance of these control methods has been shown in the control problem of coupled economic growth models under time-delays and of endogenous economic growth models. The above demonstrates the potential for using the aforementioned control approaches not only in areas of technological interest but also in domains of wider societal impact. The research domain is indeed wide and there is ongoing work which is expected to give additional important results in the forthcoming years.

Glossary

AC/DC	Alternating Current/Direct Current
ADCS	Attitude Determination and Control Subsystem
AUV	Autonomous Underwater Vessel
BLDC	Brushless DC Motor
CLT	Central Limit Theorem
DC/DC	Direct Current/Direct Current
DFIG	Doubly-Fed Induction Generator
DOF	Degrees of Freedom
EKF	Extended Kalman Filter
EMF	Electromagnetic Force
H_∞	Control: H-infinity Control
H_∞	Kalman Filter: H-infinity Kalman Filter
HESG	Hybrid Excited Synchronous Generator
HESM	Hybrid Excited Synchronous Machine
HEV	Hybrid Electric Vehicle
HSV	Hypersonic Vehicle
KF	Kalman Filter
LMI	Linear Matrix Inequality
LPV	Linear Parameter Varying system
LQR	Linear Quadratic Regulator
LQG	Linear Quadratic Gaussian
MIMO	Multi-Input Multi-Output
MPC	Model Predictive Control
NES	Normalized Error Square
NMPC	Nonlinear Model Predictive Control
ODE	Ordinary Differential Equation
PDE	Partial Differential Equation
PID	Proportional Derivative Integral
PLI	Power Line Inspection Robot
PMBLDC	Permanent Magnet Brushless Direct Current motor
PMLSG	Permanent Magent Linear Synchronous Generator
PMSG	Permanent Magnet Synchronous Generator
PMSM	Permanent Magnet Synchronous Motor
PV	Photovoltaic Unit
PWM	Pulse Width Modulation
RTAC	Rotational-Translational Actuator

SDRE	State-Dependent Riccati Equation
SRG	Synchronous Reluctance Generator
SRM	Synchronous Reluctance Machine
SISO	Single-Input Single-Output
SMC	Sliding Mode Control
SoC	State-of-Charge
TORA	Tortional Oscillator with Rotational Actuator
UGV	Unmanned Ground Vehicle
USV	Unmanned Surface Vessel
VSC	Voltage Source Converter
VSI	Voltage Source Inverter
VTOL	Vertical Take-Off and Landing Aircraft
WSN	Wireless Sensor Networks
UAV	Unmanned Aerial Vehicle

References

[1] A. Abdessameud and A. Tayebi, Global trajectory tracking control of VTOL UAVs without linear velocity measurements, Automatica, Elsevier, vol. 45, pp. 2876–2882, 2009.

[2] C. Abeykoon, A. McMillan and B.K. Nguyen, Energy efficiency in extrusion-related polymer processing: A review of state-of-the-art and potential efficiency improvements, Renewable and Sustainable Energy Reviews, Elsevier, vol. 147, pp. 111219–111231, 2021.

[3] P.K.W. Abeygunawardhana and T. Murakami, Vibration suppression of two-wheel mobile manipulator using resonance-ratio-control-based null-space control, IEEE Transactions on Industrial Electronics, vol. 57, no. 12, pp. 4137–4146, 2010.

[4] M. Abroshan, M. Taeibat, A. Goodarzi and A. Khajepour, Automatic steering control in tractor semi-trailer vehicles for low-speed maneuverability enhancement, Proceedings of the Institution of Mechanical Engineers, Part K: Journal of Multibody dynamics, vol. 23, no. 1, pp. 83–102, 2016.

[5] F. Abry, X. Brun, S. Sesmant, E. Bidaux and C. Ducat, Electromagnetic cylinder backstepping position controller design with real-time closed-loop stiffness and damping timing, IEEE Transactions on Control Systems Technology, vol. 24, no. 2, pp. 541–552, 2016.

[6] A. Accetta and M. Pucci, Energy management system in DC microgrid of smart ships: Main Gen-Set fuel consumption minimization and fault compensation, IEEE Transactions on industry Applications, vol. 55, no. 5, pp. 2097–3113, 2019.

[7] J.A. Acosta, C.R. de Cos and A. Ollero, A robust decentralized strategy for multi-task control of unmanned aerial systems: Application on under-actuated aerial manipulator, 2016 International Conference on Unmanned Aircraft Systems, IEEE ICUAS 2016, Arlington, Virginia, June 2016.

[8] J.A. Acosta, M.I. Sanchez and A. Ollero, Robust control of underactuated aerial manipulators via IDA-PBC, IEEE CDC 2014, 53rd IEEE Conference on Decision and Control, Los Angeles, California, USA, December 2014

[9] R. Adlgostar, H. Azimian and D. Tughirad, Robust H_∞, H_2/H_∞ controller for rotational translational actuator, Proceedings of the 2006 International Conference on Control Applications, Munich, Germany, October 2006.

[10] R.S. Aeashloo, M. Salahifer, L. Rame and V. Sala, A robust predictive current controller for healthy and open-circuit faulty conditions of five-phase BLDC drives applicable for wind generators and electric vehicles, Energy Conversion and Management, Elsevier, vol. 82, pp. 437–447, 2018.

[11] P. Afshar, M. Brown, J. Maciejowski and H. Wang, Distributed robust multi-objective optimization of interconnected processes energy efficiency case study in papermaking, IEEE Transactions on Neural Networks, vol. 22, no. 12, pp. 2324–2338, 2011.

[12] F. Aghili, Ripple suppression of BLDC motors with finite observer/amplifier bandwidth at high velocity, IEEE Transactions on Control Systems Technology, vol. 19, no. 2, pp. 391–397, 2011.

[13] L.T. Aguilar, I.M. Boiko, L.M. Fridman and L.B. Freidovich, Generating oscillations in inertia wheel pendulum via two-relay controller, International Journal of Robust and Nonlinear Control, J. Wiley, vol. 22, no. 3, pp. 318–330, 2012.

[14] S. Ahmed and E.C. Kerrigan, Suboptimal predictive control for satellite detumbling, Journal of Guidance, Control and Dynamics, vol. 37, no. 3, pp. 850–859, 2014.

[15] M. Ahmed, L. Meegahapola, A. Vahidnia and M. Datta, Stability and control aspects of microgrid architectures – A comprehensive review, IEEE Access, vol. 8, pp. 144730–147766, 2020.

[16] N.U. Ahmed and H. Harbi, Mathematical analysis of dynamic models of suspension bridges, SIAM Journal of Applied Mathematics, vol. 58, no. 3, pp. 853–874, 1998.

[17] S. Ahma, H. Wang and Y. Tian, Model-free control using time-delay estimation and fractional-order nonsingular fast terminal sliding-mode for uncertain over-limb exoskeleton, Journal of Vibration and Control, Sage Publications, vol. 24, no. 22, pp. 5273–5290, 2018.

[18] A. Ailon, Simple tracking controllers for autonomous VTOL aircraft with bounded inputs, IEEE Transactions on Automatic Control, vol. 55, no. 3, pp. 737–743, 2010.

[19] N. Ajjanaramvat and M. Parnichkan, Trajectory tracking using only learning LQR with adaptive learning method of a leg-exoskeleton for disorder gait rehabilitation, Mechatronics, Elsevier, vol. 51, pp. 85–96, 2018.

[20] S.A. Ali, A. Christen, S. Begg and N. Langlois, Continuous-discrete time observer design for state and disturbance estimator of electro-hydraulic actuator systems, IEEE Transactions on Industrial Electronics, vol. 63, no. 7, pp. 4314–4324, 2016.

[21] M.E.M. Ali, A.M. Kassem and M.F.A. Hmed, Modeling and control of a mini hybrid hydro matrix / wind in micro-grid applications, IEEE Access, vol. 8, pp. 170843–170852, 2020.

[22] N.A. Ali-Hussain, S.S. Azhar Ali, M. Ovinis, M.R. Aeshad and U.M. Al-Saggaf, Underactuated coupled nonlinear adaptive control synthesis using

U-Model for multivariable Unmanned Marine Robots, IEEE Access, vol. 8, pp. 1851–1865, 2020.

[23] G. Allibert, M.D. Hua, S. Krupinsky and T. Hamel, Pipeline following by visual serving for autonomous underwater vehicles, Control Engineering Practice, Elsevier, vol. 82, pp. 151–160, 2019.

[24] T. Alpcan, T. Basar and R. Tempo, Randomized algorithms for stability and robustness analysis of high-speed communication networks, IEEE Transactions on Neural Networks, vol. 16, no. 5, pp. 11229–11241, 2005.

[25] M. Alramlawi, A. Cabush, E. Mohagheghi and P. Li, Optimal operation of PV-battery-diesel microgrid for industrial loads under grid blackouts, IEEE EEEIC 2018, IEEE 2018 International Conference on Environment and Electrical Engineering, Palermo, Italy, June 2018.

[26] A. Alyshev, N. Dudarenko and V. Melnikov, Parametric identification of reaction wheel pendulum with adaptive control, Cybernetics and Physics, vol. 7, no. 2, pp. 57–65, 2018.

[27] H. Alwi and C. Edwards, Sliding-mode fault tolerant control of an octorotor using linear parameter varying-based schemes, IET Control Theory and Applications, vol. 9, no. 4, pp. 618–636, 2015.

[28] H. Alwi and C. Edwards, Fault tolerant control of octorotor using sliding-mode control allocation, Proceedings of the EuroGNC 2013, 2nd CEAS Specialist Conference on Guidance, Navigation and Control, Delft, The Netherlands, April 2013.

[29] H. Alwi, M.T. Hamayan and C. Edwards, An integral sliding-mode fault tolerant control scheme for an octorotor using fixed control allocation, 13th IEEE Workshop on Variable Structure Systems, IEEE VSS 2014, Nantes, France, July 2014.

[30] H.H. Aly, Dynamic modelling and control of the tidal current turbine using DFIG and DDPMSG for power system stability analysis, Electric Power and Energy Systems, Elsevier, vol. 83, pp. 525–540, 2016.

[31] Y. Ait-Gougen, R. Ibtiouen, O. Touhami, J.P. Louis and M. Gabsi, Inverse modeling and pulsating torque minimization of salient pole non-sinusoidal synchronous machines, Electric Power Systems Research, Elsevier, vol. 78, pp. 88–96, 2008.

[32] R. Ali, P. Yunfeng, M. Touseef Iqbal, R. Ul Amin, M.O. Zahid and O.I. Khan, Adaptive backstepping sliding-mode control of coaxial octorotor unmanned aerial vehicle, IEEE Access, vol. 7, pp. 27526–27534, 2019.

[33] O. Ameur, P. Massion, G. Schorletti, X. Brun and M. Smaoui, Lyapunov stability analysis of switching controllers in presence of sliding modes and parametric uncertainties with applications to pneumatic systems, IEEE Transactions on Control Systems Technology, vol. 24, no. 6, pp. 1953–1964, 2016.

[34] H. An, Q. Wu, H. Xin and C. Wang, Control of a time-varying hypersonic vehicle model subject to inlet unstart condition, Journal of the Franklin Institute, Elsevier, vol. 355, pp. 4164–4197, 2018.

[35] S. Andary, A. Chemori and S. Krut, Estimation-based disturbance rejection in control for limit cycle generation on inertia wheel inverted pendulum testbed, IEEE IROS 2005, IEEE International Conference on Intelligent Robots an Systems, Edmonton, Canada, October 2009.

[36] S. Andary, A. Chemori and S. Krut, Stable limit cycle generation for underactuated mechanical systems, Application: Inertia wheel inverted pendulum, IEEE IROS 2008, IEEE Conference on Intelligent Robots and Systems, Nice, France, September 2008.

[37] V.H. Andaluz, F. Roberi, L. Salinas, J.M. Toibeiro and R. Carelli, Passivity-based visual feedback control with dynamic compensation of mobile manipulators: Stability and L_2-gain performance analysis, Robotics and Autonomous Systems, Elsevier, vol. 66, pp. 64–74, 2015.

[38] P. Antolin, N. Zhang, J.M. Goicolea, M. Xia, M.A. Astiz and J. Oliva, Coordination of nonlinear wheel and contact forces for dynamic vehicle-bridge interaction in high-speed railways, Journal of Sound and Vibration. Elsevier, vol. 332, pp. 1231–1251, 2013.

[39] B. Anton and A. Floresccu, Design and Development of Series-Hybrid Automotive Powertrains, IEEE Access, vol. 8, pp. 226026–226041, 2020.

[40] I. Aregui, B. Salvador and C. Vasquez, PDE models and numerical methods for total value adjustment in European and American options with counterparty risk, Applied Mathematics and Computation, Elsevier, vol. 308, pp. 31–33, 2017.

[41] H. Armghem, M. Young, A. Armghem, N. Ali, M.O. Wang and I. Ahmed, Design of integral terminal sliding-mode controller for the hybrid AC/DC microgrids involving renewables and energy storage systems, Electric Power and Energy Systems, Elsevier, vol. 199, pp. 105857–105871, 2020.

[42] L.T. Aschepkov, D.V. Dolgy, T. Kim, and R.P. Agarwal, Optimal Control, Springer, 2017.

[43] S. Audrey, A. Chemori, M. Benoit and J. Sallantin, A dual model-free control of underactuated mechanical systems, application to the inertia wheel inverted pendulum, IEEE ACC 2012, American Control Conference, Montreal, Canada, pp. 1029–1034, June 2012.

[44] P. Austim, M. McEwan, J. Mack, J. Godsal, J. Tyler and J. Maciejowski, Optimization of fuel usage and steam availability in the power and steam plant of a paper mill, 1st IFAC Conference on Control Methodologies and Technology for Energy Efficiency, IFAC CMTEE 2010, Vilamura, Portugal, March 2010.

[45] J.M. Avis, S.G. Nersesov, R. Nathan, H. Ashrafiwan and K.R. Muske, A comparison study of nonlinear control techniques for the RTAC system, Nonlinear Analysis: Real World Applications, Elsevier, vol. 11, pp. 2647–2658, 2010.

[46] J.M. Avis, N.G. Nersesov and R. Nathan, Energy-based hybrid control for the RTAC system: Experimental results, IEEE ACC 2008, IEEE 2008 American Control Conference, June 2008, Seattle, Washington, DC, USA, 2008.

[47] J.M. Avis, N.G. Nersesov and R. Nathan, Decentralized energy-based hybrid control for the multi-RTAC system, IEEE ACC 2009, IEEE 2009 American Control Conference, June 2009, St. Louis, Missouri, USA, 2009.

[48] G.D. Avolio, The market for borrowing stock, Journal of Financial Economics, Elsevier, vol. 66, pp. 271–306, 2002.

[49] M. Ayaz and E. Mese, A permanent magnet alternator with increased power capability for hybrid electric vehicle applications, Electric Power Systems Research, Elsevier, vol. 33, pp. 292–303, 2016.

[50] E. Aydin, D. Bonvin and K, Sundmacher, Computationally efficient NMPC for batch and semi-batch processes using parsimonious input parametrization, Journal of Process Control, Elsevier, vol. 18, pp. 399–407, 2008.

[51] P.D. Ba, S.G. Lee, S. Back, J. Kim and H.K. Lee, Balancing and translational control of a Ball Segway that a human can ride, 2016 16th International Conference on Control, Automation and Systems, IEEE ICCAS 2016, Gyeangju, Korea, October 2016.

[52] H.R. Baghase, M. Mirsalim, G.B. Gharehpetian and H.A. Talebi, A decentralized power management and sliding-mode control strategy for hybrid AC/DC microgrids including renewable energy resources, IEEE Transactions on Industrial Informatics, 2021.

[53] T. Bakin, S. Ochman, G. Ferotte and H. Hammouri, Nonlinear observer of crystal-size distribution during batch crystallization, AiChe Journal, John Wiley, vol. 50, no. 6, pp. 2188–2197, 2006.

[54] A. Baldini, R. Felicetti, A. Freddi, S. Longhi and A. Monteriu, Dynamic surface control of multirotor vehicles, IEEE ICUAS 2018, IEEE 2018 International Conference on Unmanned Aircraft Systems, Dallas, Texas, USA, June 2018.

[55] A. Balogh and M. Kristic, Infinite dimensional backstepping style feedback transformations for a heat equation with an arbitrary level of instability, European Journal of Control, vol. 8, pp. 165–175, 2002.

[56] B. Bandyopadhyay, P.S. Gandhi and S. Kurode, Sliding-mode observer-based sliding-mode controller for slosh-free motion through PID scheme, IEEE Transactions on Industrial Electronics, vol. 56, no. 9, pp. 3432–3442, 2009.

[57] M. Barcaro, N. Bianchi and F. Magnussen, Permanent magnet optimization in permanent-magnet assisted Synchronous Reluctance Motor for a wide constant power speed range, IEEE Transactions on Industrial Electronics, vol. 59, no. 6, pp. 2495–2502, 2012.

[58] M. Basin, P. Yu and Y.B. Stessel, Hypersonic vehicle adaptive sliding-mode control using finite and fixed-time observers, IEEE Transactions on Industrial Electronics, vol. 65, no. 1, pp. 930–941, 2018.

[59] M. Basseville and I. Nikiforov, Detection of Abrupt Changes: Theory and Applications, Prentice-Hall, 1993.

[60] M. Basseville, A. Benveniste, B. Gach-Devauchelle, *et al.*, In situ damage monitoring in vibration mechanics: Diagnostics and Predictive

Maintenance, Mechanical Systems and Signal Processing, vol. 7, no. 5, pp. 401–423, 1993.

[61] M. Basseville, A. Benveniste and Q. Zhang, Surveillance d' installations industrielles: démarche générale et conception de l' algorithmique, IRISA Publication Interne no. 1010, 1996.

[62] F. Bauer, C.H. Hackl, K.M. Smedley and R.M. Kennel, Multicopter with series connected propeller drives, IEEE Transactions on Control Systems Technology, vol. 26, no. 2, pp. 563–574, 2016.

[63] A. Becker and T. Bretl, Approximate steering of a plate-ball system under bounded model perturbation using ensemble control, IEEE IROS 2012, IEEE/RSJ International Conference on Intelligent Robots and System, Vilamoura, Portugal, October 2012.

[64] G. Beggia, P. Camanda and B.S. Mascolo, Feedback-based bandwidth allocation with call admission control for providing delay guarantees in IEEE 802.11e networks, Computer Communications, Elsevier, vol. 28, no. 3, pp. 325–337, 2005.

[65] R. Behrad, G.M. Bone, G. Ashby and E. Li, Design, implementation and control of an improved hybrid pneumatic-electric actuator for robot arms, IEEE Access, vol. 7, pp. 14699–14714, 2019.

[66] H. Beikzadeh and G. Liu, Trajectory tracking of quadrotor flying manipulators using L_1 adaptive control, Journal of the Franklin Institute, Elsevier, 2018.

[67] S. Benelghali, M. Benbouzid and J.F. Charpentier, Modelisation et commande d'une hydrolienne equipée d'une génératrice asynchrone double alimentation, European Journal of Electrical Engineering, Lavoisier, vol. 13, no. 2, pp. 161–178, 2010.

[68] S. Ben-Fredj, S. Ouweslati-Boulahia and J.W. Roberts, Measurement-based admission-control for elastic traffic, in Proceedings of the ITC17, September 2001, Slavador da Bahia, Brazil, 2001.

[69] A. Bensoussan, G.D. Prato, M.C. Delfour and S.K. Mitter, Representation and Control of Infinite Dimensional Systems, Birkhäuser, Boston, MA, 2006.

[70] A. Benveniste, M. Basseville, and G. Moustakides, The asymptotic local approach to change detection and model validation, IEEE Transactions on Automatic Control, vol. 32, no. 7, pp. 583–592, 1987.

[71] D. Berg and A. Karlstrom, Approach to 3×3 decoupling and control of thermo-mechanical pulp refiners, IEEE IAS 2004, IEEE Advanced Process Control Workshop, 2004.

[72] L.D. Berkovitz and N.G. Medhin, Nonlinear Optimal Control Theory, Chapman and Hall/CRC Press, 2012.

[73] M. Bermudez, M.R. Arahal, M. Duran and I.G. Prieto, Model predictive control of six-phase electric drives including ARX disturbance estimator, IEEE Transactions on Industrial Electronics, vol. 68, no. 1, pp. 81–91, 2021.

[74] C.M. Best, M.T. Gillespie, P. Hyatt, L. Rapert, V. Sherrad and M.D. Killpack, A new soft robot control method: Using model predictive control

for a pneumatically actuated humanoid, IEEE Robotics and Automation Magazine, vol. 23, no. 3, pp. 75–84, 2016.

[75] R.E. Betz, R. Lagerquist, M. Jovanovic, T.J. Miller and R. Middleton, Control of synchronous reluctance machines, IEEE Transactions on Industry Applications, vol. 29, no. 6, pp. 1110–1122, 1993.

[76] S. Bezzaoucha, B. Marx, D. Maquin, and J. Ragot, State and multiplicative sensor fault estimation for nonlinear systems, 2013 Conference on Control and Fault-Tolerant Systems (SysTol), Nice, France, October 2013.

[77] S. Blandin, A. Couque, A. Bayen and D. Work, On sequential data assimilation for scalar macroscopic traffic flow models, Physica D, Elsevier, vol. 241, pp. 1421–1440, 2012.

[78] M. Bock and A. Kugi, Real-time nonlinear model predictive path-following control of a laboratory tower crane, IEEE Transactions on Control Systems Technology, vol. 22, no. 4, pp. 1461–1463, 2014.

[79] E. Boje, Attitude and position estimation for a power line inspection robot, Proceedings of the 7th IFAC Symposium on Mechatronic Systems, IFAC Mechatronics 2016, Loughborough, UK, September 2016.

[80] R. Bojec, B. Pevec and D. Mijovec, Optimal control of brushless PM motor in parallel hybrid propulsion system, Mechatronics, Elsevier, vol. 20, pp. 464–473, 2010.

[81] I. Boldea, L. Tutelea, L. Parsu and D. Dorell, Automotive electric propulsion systems with reduced or no permanent magnets: An overview, IEEE Transactions on Industrial Electronics, vol. 81, no. 10, pp. 5696–5705, 2014.

[82] I. Boldea, Starter/alternator systems for HEV and their control: A review, KIFE International Transactions, vol. 48, no. 4, pp. 157–189, 2004.

[83] G.M. Bone, M. Xue and J. Flett, Position control of hybrid pneumatic-electric actuators using discrete-valued model predictive control, Mechatronics, Elsevier, vol. 25, pp. 1–10, 2015.

[84] F. Borrelli, A. Bemporad and M. Morari, Predictive Control for Linear and Hybrid Systems, Cambridge University Press, 2017.

[85] D.M. Boskovic, M. Krstic and W.J. Liu, Boundary control of an unstable heat equation via measurement of domain averaged temperature, IEEE Transactions on Automatic Control, vol. 46, pp. 2022–2028, 2002.

[86] M. Bouakou and R. Channa, Stabilization and fault detection for Takagi–Sugeno system Application to a bioreactor, Proceedings of the 7th International Conference on Systems and Control, Valencia, Spain, October 2018.

[87] I. Boussaada, A. Cela, H. Mounier and S.I. Niculescu, Control of Drilling Vibrations: A time-delay system-based approach, 11th Workshop on Time Delay Systems, France, February 2013.

[88] S. Bououden, D. Boutat, G. Zheng, J.P. Barbot and F. Kratz, A triangular canonical form for a class of 0-flat nonlinear systems, International Journal of Control, Taylor and Francis, vol. 84, no. 2, pp. 261–269, 2011.

[89] F. Bourbonnais, P. Bigras and H.A. Boner, Minimum-time trajectory planning and control of a pick-and-place five-bar parallel robot, IEEE/ASME Transactions on Mechatronics, vol. 20, no. 2, pp. 740–749, 2015.

[90] M. Bouri and D. Thomasset, Sliding mode control of an electropneumatic actuator using an integral switching surface, IEEE Transactions on Control Systems Technology, vol. 9, no. 2, pp. 968–975, 2001.

[91] Y. Bouzid, H. Siguerdidjane and Y. Bestaoui, Hirearchical autopilot design based on immersion and invariance and nonlinear internal model tracking controllers for autonomous system, 4th IFAC International Conference on Intelligent Control and Automation Sciences (ICONS 2016), June 2016, Reims, France.

[92] Y. Bouzid, H. Siguerdidjane and Y. Bestaoui, Improved 3D trajectory tracking by nonlinear internal-model-feedback linearization control strategy for autonomous systems, Part of special issue: 6th IFAC Symposium on System Structure and Control SSSC 2016: Istanbul, Turkey, 22–24 June 2016.

[93] B. Boyle, J.Y. Fourquet and M. Renaud, Manipulability of wheeled mobile manipulators: Application to motion generation, Intl. Journal of Robotic Research, J. Wiley, vol. 22, no. 7-8, pp. 565–581, 2003.

[94] E. Bradford and L. Imsland, Output feedback stochastic nonlinear model predictive control for batch processes, Computers and Chemical Engineering, Elsevier, vol. 126, pp. 434–450, 2019.

[95] B. Brajimi, H. DrisellL, K. El-Bojairimi, M. Saad and A. Brahimi, Novel adaptive impedance control for exoskeleton robot for rehabilization using a using a nonlinear time-delay observer, ISA Transactions, Elsevier, vol. 108, pp. 381–392, 2021.

[96] F. Bu, H. Liu, W. Huang, *et al.*, Induction machine-based starter/generator systems, IEEE Industrial Electronics Magazine, pp. 4–19, March, 2020.

[97] F. Bullo and A.D. Lewis, Geometric Control of Mechanical Systems: Modeling Analysis and Design for Simple Mechanical Control Systems, Springer, 2010.

[98] A. Bravo-Doddoli and L.C. Garcia-Naranjo, The dynamics of an articulated N-trailer vehicle, Regular and Chaotic Dynamics, Springer, vol. 20, no. 5, pp. 497–517, 2015.

[99] D.J. Braun, F. Petit, F. Huber, *et al.*, Robots driven by compliant actuators: Optimal control under actuation constraints, IEEE Transactions on Robotics, vol. 29, no. 5, pp. 1085–1101, 2013.

[100] A. Brezina, M. Tkacik, T. Tkacik and S. Jadlovska, Upgrade of the ball and plate laboratory model, IFAC PDEs 2019, 16th IFAC Conference on Programmable Devices and Embedded Systems, High Tatras, Slovakia, October 2019.

[101] T. Broomhead, C. Menzie, P. Hield, R. Shakhar and M. Brear, Economic model predictive control and applications for diesel generators, IEEE Transactions on Control Systems Technology, vol. 25, no. 2, pp. 388–400, 2017.

[102] M. Brucker and P. Heidtmann, Flatness-based control of a rotary vane actuator, Proceedings of 16th International Conference on Mathematical Theory of Networks and Systems, IEEE MTNS 2004, Leuven, Belgium, July 2004.

[103] D. de Brun and P.P.J. van den Bosch, Modelling and control of a double articulated vehicle with four steerable axes, IEEE ACC 1999, Proceedings of the 1999 American Control Conference, San Diego, California, June 1999.

[104] G. Buandonno and A. de Luca, Efficient computation of inverse dynamics and feedback linearization for VSA-based robots, IEEE Robotics and Automation Letters, vol. 1, no. 2, pp. 908–915, 2018.

[105] A. Burch, C. Weise and J. Reger, Application of higher-order sliding-modes to a ball and plate system, IEEE VSS 2018, IEEE 2018 15th Workshop on Variable Structure Systems, Graz, Austria, July 2018.

[106] T. Burg and D. Dawson, Additional notes on the TORA example: A filtering approach to eliminate velocity measurements, IEEE Transactions on Control Systems Technology, vol. 3, no. 5, pp. 520–527, 1997.

[107] L.G. Bushnell, D.M. Tilbury and S.S. Sastry, Steering three-input nonholonomic systems: The fire-truck example, International Journal of Robotic Research, Sage Publications, vol. 14, no. 4, pp. 366–381, 1995.

[108] L. Bushnell, D. Tilbury and S. Sastry, Steering three-input chained form nonholonomic systems using sinusoids: The fire-truck example, Proceedings of the European Control Conference, Groningen, The Netherlands, June 1993.

[109] F. Caccavale, G. Giglio, G. Muscio and F. Pierri, Adaptive control for UAVs equipped with a robotic arm, Proceedings of the IFAC World Congress, Cape Town, South Africa, August 2014.

[110] R. Cai, R. Zheng, M. Liu and M. Li, Robust control of PMSM using Geometric model reduction and μ-synthesis, IEEE Transactions on Industrial Electronics, vol. 65, no. 1, pp. 498–509, 2018.

[111] J. Cao, S.Q. Xie and R. Das, MIMO sliding-mode controller for gait exoskeleton driven by pneumatic muscles, IEEE Transactions on Control Systems Technology, vol. 26, no. 1, pp. 274–281, 2018.

[112] Y. Cao and J. Huang, Neural-network-based nonlinear model predictive tracking control of a pneumatic muscle actuator-driven exoskeleton, IEEE/CAA Journal of Automatica Sinica, vol. 7, no. 6, pp. 1478–1488, 2020.

[113] L. Cao, X.Q. Chen, and T. Sheng, Fault tolerant small satellite attitude control using adaptive non-singular terminal sliding modes, Advances in Space Research, Elsevier, vol. 51, pp. 2374–2393, 2013.

[114] C. Cai, J. Lu and Z. Li, Kinematic analysis and control algorithm for the ballbot, IEEE Access, vol. 7, pp. 38314–38312, 2017.

[115] N. Cai and L. Sun, Valuation of stock loans with jump risk, Journal of Economic Dynamics and Control, Elsevier, vol. 40, pp. 213-241, 2014.

[116] L. Cappelli, F. Marignetti, G. Mattiazzo, *et al.*, Linear tubular permanent-magnet generators for the inertial sea wave energy converter, IEEE Transactions on Industry Applications, vol. 50, no. 3, pp. 1817–1829, 2014.

[117] G. Carofiglio, R. Gaeta, M. Garetto, P. Giacone, E. Leonard and M. Sereno, A fluid-diffusion approach for modelling P2P systems, Proceedings of

the 14th IEEE International Sympon Modelling, Analysis and Simulation of Computer and Telecommunication Systems, IEEE MASCOTS 2006, Monterey, California, September 2006.

[118] D. Casagrande, A. Astolfi and T. Parisini, Switching-driving Lyapunov function and the stabilization of the ball-and-plate system, IEEE Transactions on Automatic Control, vol. 54, no. 8, pp. 1881–1887, 2009.

[119] P. Castaldi, N. Mimmo and S. Simani, LEO satellite active FTC with aerodynamic disturbance decoupled fault diagnosis, European Journal of Control, Elsevier, vol. 53, pp. 76–94, 2020.

[120] K. Cavanagh, J.A. Bolk and K. Turitsyn, Transient stability guarantees for ad-hoc DC microgrids, IEEE Control Systems Letters, vol. 2, no. 1, pp. 139–145, 2018.

[121] A. Chalaga and F. Plestan, High-order sliding-mode control with predefined convergence time for electropneumatic actuator, IEEE Transactions on Control Systems Technology, 2021.

[122] F. Celani, Output regulation of the TORA benchmark via rotational position feedback, Automatica, Elsevier, vol. 47, pp. 584–590, 2011.

[123] S.B. Chabane, M. Alamir, M. Fiacchini, R. Riah, T. Kovaltchouk and S. Bacha, Electricity grid connection of a tidal farm: An active power control framework constrained to grid code requirements, IEEE Transaction on Sustainable Energy, vol. 9, no. 4, pp. 1948–1956, 2018.

[124] H. Chadir, H. Ouadi and F. Giri, Output feedback control of wind energy conversion system with hybrid excitation synchronous generator, 1st IFAC Conference on Modelling, Identification and Control of Nonlinear Systems, IEEE MICNON 2015, Saint Petersburg, Russia, June 2015.

[125] M. Chadli, S. Aouaouda, H.R. Karimi and P. Shi,. Robust fault tolerant tracking controller design for a VTOL aircraft, Journal of the Franklin Institute, vol. 350, pp. 2627–2645, 2013.

[126] J. Chan, Y Huang, X. Guo, S. Zhou and L. Jia, Parameter identification and adaptive compliant control of rehabilitation exoskeleton based on multiple sensors, Measurement, Elsevier, vol. 159, pp. 107765–107777, 2020.

[127] L. Chang, Y. Lin, Z.G. Hou, M. Tan, T. Huang and W.J. Zhang, Adaptive tracking control of hybrid machines: A closed-chain five-bar mechanism case, IEEE/ASME Transactions on Mechatronics, vol. 16, no. 6, pp. 1155–1163, 2011.

[128] W. Chang, G. Yang, J. Yu, Z. Liang, L. Cheng and C. Zhou, Development of a power line inspection robot with hybrid operation modes, IEEE/RSJ 2017, IEEE 2017 International Conference on Intelligent Robots and Systems, September 2017, Vancouver, Canada.

[129] L. Chang, Y. Liu, Z.G. Hou, M. Tan, J. Huang and W.J. Zhang, Integrated design of machine body and control algorithm for improving the robustness of a closed-chain five-bar machine, IEEE/ASME Transactions on Mechatronics, vol. 17, no. 3, pp. 587–592, 2012.

[130] T.K. Chau, S.S. Yu, T. Fernando, and H.C. Iu, Senior demand-side regulation provision from industrial loads integrated with solar PV panels and energy

storage system for ancillary services, IEEE Transactions on Industrial Informatics, vol. 14, no. 11, pp. 5038–5049, 2018.

[131] N.A. Chaturvedi, R. Klein, J. Christensen, J. Ahmed and A. Kojic, Algorithms for advanced battery-management systems, IEEE Control Systems Magazine, pp. 49–68, June 2010.

[132] J. Chauvin, Observer Design for a Class of Wave Equations Driven by an Unknown Periodic Input, 18th World Congress, Milano, Italy, September 2011.

[133] J. Chebhi, F. Defey, Y. Briere and A. Derouaz-Pepin, Novel model-based control mixing strategy for a coaxial push-pull multirotor, IEEE Robotics and Automation Letters, vol. 5, no. 2, pp. 485–492, 2020.

[134] M.A.M. Cheema, J.E. Fletcher, M. Furshodnia and M.F. Rahman, Sliding-mode based combined speed and direct thrust force control of linear permanent magnet synchronous motors with first-order plus integral sliding condition, IEEE Transactions on Power Electronics, vol. 34, no. 3, pp. 2526–2538, 2019.

[135] W. Chen, L. Xu and S.P. Zhu, Stock loan valuation under a stochastic interest rate model, Computers and Mathematics with Applications, Elsevier, vol. 70, pp. 1757–1771, 2015.

[136] H. Chen, T. Tang, N. Aot-Ahmed, M.H. Benbouzid, M. Machmoum and M.H. Zaom, Attraction, challenge and current status of marine current energy, IEEE Access, vol. 6, pp. 12665–12695, 2018.

[137] Z.H. Chen, Y. Wang, P. Ouyang, J. Huang and W.J. Zhang, A novel iteration-based controller for hybrid machine systems for trajectory tracking at the end-effector level, Robotica, Cambridge University Press, vol. 23, pp. 317–324, 2011.

[138] G.P. Chen, Z.J. Du, L. He, J.Q. Wang, D.M. Wu and W. Deng, Active disturbance rejection with fast terminal sliding-mode control for a lower-limb exoskeleton in swing phase, IEEE Access, vol. 7, pp. 72343–72357, 2019.

[139] Q. Chen, W. Cheng, L. Guo and J. Fortner, A pure neural network controller for double-pendulum crane anti-sway control: Based on Lyapunov stability theory, Asian Journal of Control, J. Wiley, pp. 1–12, 2019.

[140] H. Chen and N. Sun, Nonlinear control of underactuated systems subject to both actuated and unactuated state constraints with experimental verification, IEEE Transactions on Industrial Electronics, vol. 67, no. 9, pp. 7702–7714, 2020.

[141] H. Chen, Y. Fang and N. Sun, A swing constraint MPC algorithm for underactuated overhead cranes, IEEE/ASME Transactions on Mechatronics, vol. 21, no. 5, pp. 2543–2544, 2016.

[142] S. Chen, Z. Chen and B. Yao, Precision cascade force control of multi-DOF hydraulic leg exoskeleton, IEEE Access vol. 6, pp. 8574–8583, 2018.

[143] Z. Chen, Z. Li and C.L.P. Chen, Disturbance observer-based fuzzy control of uncertain MIMO mechanical systems with input nonlinearities and its application to robotic exoskeleton, IEEE Transactions on Cybernetics, vol. 47, no. 4, pp. 984–994, 2012.

[144] L. Chen, G. Liu, H. Dong, S. Hoang and S. Dion, Position control on center of mass for power line inspection robot, IEEE CCC 2016, Proceedings of the 35th Chinese Control Conference, Chengdu, China, July 2016.

[145] Y. Chen and B. Guo, Sliding-mode fault-tolerant tracking control for a single-link flexible-joint manipulator system, IEEE Access, vol. 7, pp. 83046–83055, 2019.

[146] Y. Chen, A.W. Peterson, C. Zhang and M. Ahmadian, A simulation-based comparative study on lateral characteristics of trucks with double and triple trailers, International Journal of Vehicle Safety, Inderscience, vol. 11, no. 2, pp. 136–157, 2019.

[147] L. Chen, C. Yang and W. Yan, Adaptive neural network control of underactuated surface vessels with guaranteed transient performance: Theory and experimental results, IEEE Transactions on Industrial Electronics, vol. 67, no. 5, pp. 4024–4035, 2020.

[148] X. Chen, Z. Liu, J. Zhang, D. Zhou and J. Dong, Adaptive sliding-mode path-following control system of the underactuated USV under the influence of ocean currents, IEEE Journal of Systems Engineering and Electronics, vol. 29, no. 6, pp. 1271–1283, 2018.

[149] Z. Chen, Y. Zhang, Y. Nie, J. Tang and S. Zhu, Adaptive sliding-mode control design for nonlinear unmanned surface vessel using RBFNN and Disturbance Observer, IEEE Access, vol. 8, pp. 45457–45467, 2020.

[150] F. Chen, J. Niu and G. Jiang, Nonlinear fault-tolerant control for hypersonic flight vehicle, IEEE Access, vol. 6, pp. 25427–25436, 2018.

[151] J. Chen, W. Yao, Y. Ren, R. Wang, L. Zhang and L. Jiang, Nonlinear adaptive speed control of a Permanent Magnet Synchronous Motor: A perturbation estimation approach, Control Engineering Practice, Elsevier, vol. 85, pp. 163–175, 2019.

[152] K. Chen, M. Kamezaki, T. Katano, *et al.*, Compound locomotion control system combining crawling and walking for multi-crawler multi-arm robot to adapt unstructured and unknown terrain, Robomech Journal, Springer, vol. 5, no. 2, pp. 1–17, 2018.

[153] X. Chen, R. Sun, M. Liu and D. Song, Two-stage exogenous Kalman Filter for time-varying fault estimation of satellites attitude control system, Journal of the Franklin Institute, Elsevier, vol. 357, pp. 2354–2370, 2020.

[154] X. Chen and G. Liu, Sensorless optimal commutation steady speed control method for a non ideal back EMF BLDC motor drive system including buck converter, IEEE Transactions on Industrial Electronics, vol. 67, no. 7, pp. 6147–6157, 2020.

[155] S. Chen, W. Sun, K. Wang, G. Liu and L. Zhu, Sensorless high-recession position correction strategy for a 100KW@20.000 rpm BLDC motor with low stator inductance, IEEE Transactions on Industrial Informatics, vol. 14, no. 10, pp. 4297–4299, 2018.

[156] L. Chen, Z. Laio and X. Ma, Nonlinear model predictive control for heavy-duty hybrid vehicles using random power prediction method, IEEE Access, vol. 8, pp. 202819–202835, 2020.

[157] C.F. Chen, Z.J. Du, L. He, J.O. Wang, D.M. Wu and W. Dong, Active disturbance rejection with fast terminal sliding-mode control for a lower-limb exoskeleton in swing-phase, IEEE Access, vol. 7, pp. 72343–72357, 2019.

[158] G.G. Chen, T.H. Liu, M.T. Lin and C.A. Tai, Position control of a sensorless synchronous reluctance motor, IEEE Transactions on Industrial Electronics, vol. 51, no. 1, pp. 15–25, 2004.

[159] J. Chen, Q. Ouyang, C. Xu and H. Su, Neural network-based state of charge observer design for Lithium-ion batteries, IEEE Transactions on Control Systems Technology, vol. 26, no. 1, pp. 313–320, 2018.

[160] W.H. Chen, D.J. Ballance, P.J. Gawthrop and J.O. Reilly, A nonlinear disturbance observer for robotic manipulators, IEEE Transactions on Industrial Electronics, vol. 47, no. 4, pp. 932–938, 2000.

[161] C. Chen, R. Xiong and W. Shen, A lithium-ion battery-in-the-loop approach to test and validate multiscale dual H-infinity filters for state-of-charge and capacity estimation, IEEE Transactions on Power Electronics, vol. 33, no. 1, pp. 332–342, 2017.

[162] Y. Chen, C. Zhang, X. Zhang and H. Hu, Autonomous flight control for multi-motor UAVs at low altitude, IEEE Access, vol. 7, pp. 42614–42623, 2019.

[163] Z. Chen, Y. Fu and C.C. Mi, State of charge estimation of lithium-ion batteries in electric drive vehicles using Extended Kalman Filtering, IEEE Transactions on Vehicular Technology, vol. 62, no. 3, pp. 1020–1030, 2013.

[164] L. Cheng, Z.G. Hou, M. Tan and W.J. Zhang, Tracking control of a closed-chain five-bar robot with two degrees of freedom by integration of an approximation-based approach and mechanical design, IEEE Transactions on Systems, Man and Cybernetics – Part B: Cybernetics, vol. 42, no. 5, pp. 1470–1479, 2012.

[165] X. Cheng, S. Wang and L. Yao, Li-ion battery pack state-of-charge estimation disturbed by colored noises, ICAE 2016, The 8th International Conference on Applied Energy, Energy Procedia, Elsevier, vol. 105, pp. 4104–4109, 2017.

[166] S. Chenicherc, J.P. Cassar and K.M. Pekpe, Fault detection and Isolation from an identified MIMO Takagi-Sugeno model of a bioreactor, Proceedings of the 7th IFAC Symposium on Fault Detection, Supervision and Safety of Technical Processes Barcelona, Spain, June–July, 2009.

[167] T. Chevet, M. Makarov, C. Stoica-Maniu, I. Hinostroza and P. Tarascon, State estimation of an octorotor with unknown inputs: Applications to radar imaging, IEEE ICSTCC 2017, IEEE 21st International Conference on Systems Theory, Control and Computing, Romania, 2017.

[168] J.W.T. Cheung and Y.S. Huang, Modelling and control of a 2-DOF planar parallel manipulator for semiconductor packaging systems, IEEE/ASME AIM 2005, IEEE/ASME International Conference on Advanced Intelligent Mechatronics, July 2005.

[169] H.K. Chiang and C.H. Cheng, Design and implementation of a grey sliding-mode controller for synchronous reluctance motor drive, Control Engineering Practice, Elsevier, vol. 17, pp. 155–163, 2004.

[170] G. Chiaverini, G. Oriolo and A. Maciejewski, Redundant robots, Springer Handbooks on Robotics, 2nd ed., pp. 221–242, 2016.

[171] L. Chikh, P. Poignet, F. Pierrot and M. Michelin, A generalized predictive force controller for electropneumatic cylinder, 8th IFAC Symposium on Nonlinear Control Systems, University of Bologna, Italy, September 2010.

[172] L. Chikh, P. Poignet, M. Michelin and F. Pierrot, Commande predictive robuste des vérins electropneumatiques en cascade position-pression, CIFA 2010, Conférence Internationale Francophone d' Automatique, Nancy, France, 2010.

[173] F. Chishti, S. Murshid and B. Singh, Robust normalized mixed-norm adaptive control scheme for PQ improvement at PCC of a remotely located wind solar PV-BES microgrid, IEEE Transactions on Industrial Informatics, vol. 16, no. 3, pp. 1708–1723, 2020.

[174] C.H. Chiu and W.J. Wang, Implementation of a ball inverted pendulum with omnidirectional moving ability using a robust, fuzzy control design, ISA Transactions, Elsevier, vol. 86, pp. 287–298, 2019.

[175] S.H. Cho, S. Backlede and S. Helluser, Position tracking control of a clamp-cylinder for energy-saving injection moulding machines with electric-hydraulic drives, Proceedings of the Institution of Mechanical Engineers, Part K: Part I: Journal of Systems and Control Engineering, Sage Publications, vol. 223, pp. 479–492, 2009.

[176] L.S. Choi, J.A. Rossiter and P.J. Fleming, Looper and tension control in hot rolling mills: A survey, Journal of Process Control Elsevier, vol. 17, pp. 509–521, 2007.

[177] H.K. Choi and J.S. Kwon, Multi-scale modelling and model-based feedback control of pulp digester, IEEE ACC 2019, IEEE 2019 American Control Conference, Philadelphia, USA, July 2019.

[178] H. Choukri Lamraoui and Z. Qidan, Path following control of fully actuated autonomous underwater vehicle in presence of fast varying disturbances, Applied Ocean Research, Elsevier, vol. 88, pp. 40–46, 2019.

[179] D. Chwa, Fuzzy adaptive output feedback tracking control of VTOL aircraft with uncertain input coupling and input-dependent disturbances, IEEE Transactions on Fuzzy Systems, vol. 23, no. 5, pp. 1505–1518, 2015.

[180] A. Colbio-Vega, J. de Leon-Morales, L. Fridman, O. Sala-Pena and M.T. Mata-Jimenez, Robust excitation control design using sliding-mode technique for multi-machine power systems, Electric Power Systems Research, Elsevier, vol. 78, pp. 1627–1634, 2008.

[181] J.T. Cook, L.E. Ray and J.H. Lever, Dynamics model for mobility optimization and control of off-road tractor convoys, IEEE ACC 2016, IEEE 2016 American Control Conference, Boston, Massachusetts, June 2016.

[182] M. Corno, N. Bhatt, S.M. Savaresi and M. Verhaegen, Electrochemical model-based state of charge estimation in Li-ion Cells, IEEE Transactions on Control Systems Technology, vol. 23, no. 1, pp. 112–127, 2015.

[183] R. Cortesao, J. Park and O. Khatib, Real-time adaptive control for haptic telemanipulation with Kalman Active Observers, IEEE Transactions on Robotics, vol. 22, no. 5, pp. 987–999, 2005.

[184] C.R. de Cos, J.A. Acosta and A. Ollero, Relation-pose optimization for robust and nonlinear control of unmanned aerial manipulators, IEEE ICUAS 2017, IEEE 2017 International Conference on Unmanned Aerial Systems, Miami, Florida, June 2017.

[185] R.J. Coverly, D.E. Zlotnik and J.R. Forbes, Saturated Control of Flexible-Joint Manipulator using a Hammerstein Strictly Positive Real Compensator, Robotica, Cambridge University Press, vol. 34, no. 6, pp. 1367–1382, 2016.

[186] J.L. Crassidis and J.L. Junkins, Optimal Estimation of Dynamic Systems, 2nd ed., CRC Press, 2012.

[187] M. Cucuzzella, R. Lazzani, S. Trip, S. Rosti and C. Sandati and A. Ferrara, Sliding-mode voltage control of boost converters in DC microgrids, Control Engineering Practice, Elsevier, vol. 73, pp. 161–170, 2018.

[188] T. Czachorski and F. Pekergin, Diffusion approximation on a modelling tool in congestion control and performance evaluation, 2nd International Working Conference Performance Modelling and Evaluation of Heterogeneous Networks HET-NETs 2004, West Yorkshire, UK, July 26–28, 2004.

[189] T. Czachorski, K. Grochia, T. Nycz and F. Pekergin, A diffusion approximation model for wireless networks based on IEEE 802.11 standard, Computer Communications, Elsevier, vol. 33, pp. 586–592, 2010.

[190] T. Czachorski, J. Fourneau, T. Nycz and F. Pekergin, Diffusion approximation model of multi-server stations with losses, Electronic Notes in Theoretical Computer Science, Elsevier, vol. 232, pp. 125–143, 2009.

[191] T. Czachorski, J. Fourneau and F. Pekergin, Diffusion model to study non-stationary traffic and cell losses, in: Performance Modelling and Evaluation of ATM Networks, Springer, 1995.

[192] T. Czachorshki, K.G. Grochia and F. Peregrin, Un modèle d' approximation de diffusion pour la distribution du temps d' acheminement des paquets dans les réseaux des serveurs, Colloque Francophone sur l' Ingénierie des Protocols (CFIP), Mars 2008, Les Ares, France.

[193] G.B. Dai and Y.C. Liu, Distributed coordination and cooperation control for networked mobile manipulators, IEEE Transactions on Industrial Electronics, vol. 64, no. 6, pp. 5065–5074, 2017.

[194] M. Dai and Z.Q. Xu, Optimal redeeming strategy of stock loans with finite maturity, Journal of Mathematical Finance, J. Wiley, vol. 21, no. 4, pp. 775–793, 2011.

[195] H. Dai, X. Wai, Z. Sun, J. Wang and W. Gu, Online cell SoC estimation of Li-ion battery packs using a dual-time-scale Kalman Filtering for EV applications, Applied Energy, Elsevier, vol. 95, pp. 227–257, 2012.

[196] D. Daley, C. Constantinou and S. Quighy, A Fokker-Planck equation method predicting buffer occupancy in a single queue, Computer Networks, Elsevier, vol. 51, pp. 2198–2216, 2007.

[197] C. Damout, M. Benne, L. Boillereaux, G. Grordin Perez and C.B. Chabriat, Multivariable linearizing control of an industrial sugar crystallization process, Journal of Process control, Elsevier, vo. 21, pp. 46–54, 2011.

[198] J. Dannehl and F.W. Fuchs, Flatness-based control of an induction machine fed via voltage source inverter – Concept, control design and performance analysis, IECON 2006, IEEE Conference on Industrial Electronics, pp. 5125–5130, 2006.

[199] T.T. Dang, L. Prévond, H.B. Ahmed and B. Multon, Sizing optimization of tubular linear induction generator and its possible application in high acceleration free-piston sterling microcogeneration, IEEE Transactions on Industry Applications, IEEE Transactions on Industry Applications, vol. 51, no. 5, pp. 3716–3733, 2015.

[200] M. Dansoko, H. NKwawo, F. Floret, *et al.*, Marine-turbine system directly connected to electrical grid: Experimental implementations using a nonlinear and robust control, Ocean Engineering, Elsevier, vol. 149, pp. 260–267, 2018.

[201] M. Dansoko, H. NKwawo, B. Diourté, F. Floret, R. Goma and G. Kenné, Decentralized sliding-mode control for marine-turbines connected to grid, 11th IFAC International Workshop on Adaptation and Learning in Control and Signal Processing, Caen, France, July 2013.

[202] M. Dansoko, H. NKwawo, B. Diourté, F. Floret, R. Goma, and G. Kenné, Robust multi-variable sliding-mode control design for generator excitation of marine turbine in multi-machine configuration, Electrical Power and Energy Systems, Elsevier, vol. 63, pp. 423–428, 2014.

[203] R.I. Davis and R.D. Lorenz, Engine torque ripple cancellation with integrated starter/alternator in a hybrid electric vehicle: Implementation and control, IEEE Transactions on Industry Applications, vol. 99, no. 6, pp. 1765–1774, 2003.

[204] S. Dey and B. Ayalev, Nonlinear observer designs for state-of-charge estimation of lithium-ion batteries, 2014 American Control Conference, IEEE ACC 2014, June 2014, Portland, Oregon, USA.

[205] A. De, A. Stewart-Height and D.E. Koditchek, Task-based control and design of a BLDC actuator for robotics, IEEE Robotics and Automation Letters, vol. 4, no. 3, pp. 2393–2400, 2019.

[206] P. Debenest, M. Guarnieri, K. Takita, *et al.*, Expliner – robot for inspection of transmission lines, IEEE ICRA 2008, IEEE 20198 International Conference on Robotics and Automation, Pasadena, California, USA, May 2008.

[207] D. Debono and M. Bugeja, Application of sliding-mode control to the ball and plate problem, IEEE ICINCO 2015, 2015 12th International Conference on Informatics in Control, Automation and Robotics, Colmar, France, July, 2015.

[208] V. DeBrunner, M. Ta and D. Zhou, On the design of structural vibration control systems for highway bridges, IEEE 2004 Intl. Vehicle Symposium, Parma, Italy, June 2004.

[209] V. Delli Colli, P. Cancelliere, F. Marignetti, R. di Stefano and M. Scarano, A tubular-generator drive for wave energy conversion, IEEE Transactions on Industrial Electronics, vol. 17, no. 4, pp. 1152–1159, 2006.

[210] T. Devos and J. Levine, A flatness-based nonlinear predictive approach for crane control, IFAC Workshop on Nonlinear Model Predictive Control for Fast Systems, October 2006, Grenoble, France.

[211] S. Dian, L. Chen, S. Hoang, M. Pu and J. Liu, Dynamic balance control based on an adaptive gain-scheduled backstepping scheme for power-line inspection robots, IEEE/CAA Journal of Automatica Sinica, vol. 6, no. 1, pp. 198–208, 2019.

[212] S. Dian, C. Lin, S. Hoang, T. Zhao and J. Tan, Gain-scheduled dynamic surface control for a class of underactuated mechanical systems using neural network disturbance observer, Neurocomputing, Elsevier, vol. 275, pp. 1998–2008, 2018.

[213] P. Di Barba, M.E. Mognaschi, M. Borislawski, R. Porlka, P. Paplicki and M. Wardach, Hybrid excited synchronous machine with flux control possibility, International Journal of Applied Electromagnetics and Mechanics, IOS Press, vol. 50, pp. 1–4, 2015.

[214] T.X. Dinh, T.D. Thien, T.H.V. Ahn and K.K. Ahn, Disturbance observer-based finite-time trajectory tracking control for a 3-DOF hydraulic manipulator including actuator dynamics, IEEE Access, vol. 6, pp. 36798–36809, 2018.

[215] W.H. Ding, H. Deng, Y.M. Xin and X.G. Duan, Tracking control of electro-hydraulic servo multi-closed-chain mechanisms with the use of an approximate nonlinear internal model, Control Engineering Practice, Elsevier, vol. 58, pp. 225–241, 2017.

[216] W. Ding and J. Yao, Extended State Observer-based adaptive control of electrohydraulic servomechanisms without velocity measurement, IEEE/ASME Transactions on Mechatronics, vol. 25, no. 3, pp. 1151–1161, 2020.

[217] X. Ding, Y. He, J. Ren and T. Sun, A comparative study of control algorithms for active trailer steering systems of articulated heavy vehicles, IEEE ACC 2012, IEEE 2012 American Control Conference, Montreal, Canada, June, 2012.

[218] F. Ding, C. Lejeune, A. Paulin and L. Laperriére, Wood chip fiber flow rate control and TMP refining energy savings, 2012 IEEE International Conference on Instrumentation and Measurement Technology, Grez, Austria, July 2012.

[219] Y. Ding, S. Kong, J. Huang and X. Xie, FEM analysis of vehicle-bridge vibration considering bridge deck's vibration, IEEE 2010 International Conference on Mechanic Automation and Control Engineering, Wuhan, China, June 2010.

[220] T.X. Dinh, T.D. Thien, T.H.V. Ahn and K.K. Ahn, Disturbance observer-based finite-time trajectory tracking control for a 3-DOF hydraulic manipulator including actuator dynamics, IEEE Access, vol. 6, pp. 36798–36809, 2018.

[221] A.M. Disqah, A. Maheri, K. Busawon and A. Kamjoo, A multivariable optimal energy strategy for standalone DC microgrids, IEEE Transactions on Power Systems, vol. 30, no. 5, pp. 2278–2287, 2015.
[222] D.M. Divakaran, M. Gurusamy and M. Sellamuthu, Bandwidth allocation with differential pricing for flexible demand in data center networks, Computer Networks, Elsevier, vol. 73, pp. 84–97, 2014.
[223] A. Dragonescu and A.M. Soane, Multigrid solution of a distributed optimal control problem constrained by the Stokes equations, Applied Mathematics and Computation, Elsevier, vol. 219, pp. 5622–5634, 2013.
[224] H. Duan, Y. Tian and G. Wang, Trajectory tracking control of the ball and plate system based on auto-disturbance rejection controller, Proceeding of the 7th Asian Control Conference, Hong Kong, China, August 2009.
[225] S. Dubljevic, Optimal boundary control of Kuramoto-Sivasjinsky equation, 2009 American Control Conference, St. Louis, MO, USA, June 2009.
[226] K.D. Do, Z.P. Jiang and J. Pan, On global tracking control of a VTOL aircraft without velocity measurements, IEEE Transactions on Automatic Control, vol. 48, no. 12, pp. 2212–2217, 2003.
[227] J. Dong, J. Shi, C. Liu and T. Yu, Research on pneumatic polishing force control system based on a high-speed on/off with PWM controlling, Robotics and Computer Integrated Manufacturing, Elsevier, vol. 70, pp. 102133–102147, 2021.
[228] C. Dong, X. Liu and Q. Wang, Barrier Lyapunov function-based adaptive finite-time control for hypersonic flight vehicles with state constraints, ISA Transactions, Elsevier, 2019.
[229] M. Dong, C. Liu and G. Li, Robust fault diagnosis based on nonlinear model of hydraulic gauge control system on rolling mill, IEEE Transactions on Control Systems Technology, vol. 18, no. 2, pp. 510–515, 2010.
[230] A.V. Doroshin, Attitude dynamics of gyrostat-satellites under control by magnetic actuators at small perturbations, Communications in Nonlinear Science and Numerical Simulation, Elsevier, vol. 49, pp. 159–175, 2017.
[231] L. Dou, P. Su, Q. Zong and Z. Ding, Fuzzy disturbance observer-based dynamic surface control for air-breathing hypersonic vehicle with variable geometry inlets, IET Control Theory and Applications, vol. 12, no. 1, pp. 10–19, 2018.
[232] L. Dou, J. Guo, Q. Zong and Z. Ding, Modelling and switching control of air-breathing hypersonic vehicle with variable geometry inlet, Journal of the Franklin Institute, Elsevier, vol. 355, pp. 6904–6926, 2018.
[233] R. Dubay, Self-optimizing MPC of melt temperature in injection moulding, ISA Transactions, Elsevier, vol. 41, pp. 91–94, 2002.
[234] C. Damout, M. Benne, L. Boillereaux, G. Grordin Perez and C.B. Chabriat, NMPC of an industrial crystallization process using model-based observers, Journal of Industrial and Engineering Chemistry, Elsevier, vol. 16, pp. 708–716, 2010.
[235] M.J. Duran, I. Gonzalez Prieto, M. Bermudez, F. Barrero, H. Guzman and M.R. Arahal, Optimal fault-tolerant control of six-phase induction motor

drives with parallel converters, IEEE Transactions on Industrial Electronics, vol. 63, no. 1, pp. 629–640, 2016.

[236] D.B. Durocher and M. Higginson, Thermomechanical pulp mill energy upgrade: New technology enhances energy efficiency to reduce operating costs, IEEE Transactions on Industry Applications, vol. 25, no.5, pp. 57–67, 2019.

[237] A.A. Eaied, M. Muda, A. Aderibole, M.A. Hosano, H. Zeineldin, and E.T. El-Saasdany, Stability evaluation of AC/DC microgrids considering bidirectional power flow through the interlinking converters, IEEE Access, vol. 8, pp. 43876–43889, 2021.

[238] K. Ebrahimi and M. Navmar, Port-Hamiltonian control of a brachiating robot via generalized canonical transformations, IEEE ACC 2016, IEEE 2016 Automatic Control Conference, Boston, Massachusetts, July 2016.

[239] F. Ejaz, M.T. Hamayun, S. Houssein, *et al.*, An adaptive sliding-mode actuator fault tolerant control scheme for octorotor system, Journal of Advanced Robotic Systems, Sage Publications, March–April 2019, pp. 1–17, 2019.

[240] R. Ekstrom, B. Ekergard and M. Leijan, Electrical damping of linear generators for wave energy conversion, Renewable and Sustainable Energy Reviews, Elsevier, vol. 42, pp. 116–128, 2015.

[241] M. Elsinga, TMP optimization using multivariable analysis, IEEE Transactions on Industry Applications, vol. 3, pp. 893-898, 2003.

[242] A. ElHattab, N. Uddin and E.O. Brien, Drive by bridge damage monitoring using bridge displacement profile difference, Journal of Civil Structural Health Monitoring, Springer, vol. 6, no. 5, pp. 839–850, 2016.

[243] A. Emadi, Y.J. Lee and K. Rajeshkara, Power electronics and motor drives in electric, hybrid electric and plugin hybrid electric vehicles, IEEE Transactions on Industrial Electronics, vol. 55, no. 6, pp. 2237–2245, 2008.

[244] T.E. Englert and K. Graicher, Nonlinear model predictive torque control of OMSMs for high performance applications, Control Engineering Practice, Elsevier, vol. 81, pp. 43–54, 2018.

[245] H.N. Esfahani, R. Szlapczynski and H. Gaemi, High-performance super-twisting sliding-mode control of a maritime autonomous surface ship (MASS) using ADP-based adaptive gain and time-delay estimation, Ocean Engineering, Elsevier, vol. 191, pp. 106526–106544, 2019.

[246] E. Espina, C. Burgos-Mellada, R. Cardenas-Dobson, M. Martinez-Gomez and D. Saez, Distributed control strategies for microgrids: An overview, IEEE Access, vol. 8, pp. 193412–193448, 2020.

[247] I. El Makrini, C. Rodriguez-Guerrero, D. Lefeber and B. Vanderborght, The variable boundary layer sliding-mode control: A safe and performant control for compliant joint manipulators, IEEE Robotics and Automation Letters, vol. 2, no. 1, pp. 187–192, 2017.

[248] D. Endo, Y. Okada, K. Nagatani, and K. Yoshida, Path following control for tracked vehicles based on slip-compensating odometry, IEEE IROS 2007, The IEEE 2007 Conference on Intelligent Robots and Systems, San Diego, California, December 2007.

[249] S. Enev, Feedback linearization control of the inertia wheel pendulum, Cybernetics and Information Technologies, De Gruyter, vol. 14, no. 3, pp. 96–100, 2014.

[250] J. Escareno, J. Castillo, W. Abassi, G. Flores and K. Camarillo, Navigation strategy in flight retrieving and transportation for a rotorcraft MAV, 4th Workshop on Research, Education and Development of Unmanned Aerial Systems, REF-UAS 2017, Linkoping, Sweden, October 2017.

[251] A. Estrada and F. Plestan, Second-order sliding-mode output feedback control with switching gains – application to the control of a pneumatic actuator, Journal of the Franklin Institute, Elsevier, vol. 351, pp. 2335–2355, 2014.

[252] E. Fabregas, J. Chacon, S. Dormido-Canto, G. Farias and S. Dormido, Virtual laboratory of the ball and plate system, IFAC Workshop on Internet-Based Control Education, Brescia, Italy, November 2015.

[253] O. Fall, J.F. Charpentier, N.K. Nguyen and P. Letellier, Maximum torque per ampere control strategy of a 5-phase PM Generator in healthy and faulty modes for tidal marine turbine application, IPEC 2014, International Power Electronics Conference 2014, Shanghai, China, November 2014.

[254] X. Fan, N. Zhang and S. Teng, Trajectory planning and tracking of ball and plate system using hierarchical fuzzy control scheme, Fuzzy Sets and Systems, Elsevier, vol. 144, pp. 297–312, 2004.

[255] Y. Fan, A. Mu and T. Ma, Modelling and control of a hybrid wind-tidal turbine with hydraulic accumulator, Energy, Elsevier, vol. 182, pp. 188–199, 2016.

[256] X. Fan, S. Singh, F. Oppolzer, E. Nettleton, R. Hennessy, A. Lowe, and H. Durrant-Whyte, Integrated Planning and Control of Large Tracked Vehicles in Open Terrain, 2010 IEEE International Conference on Robotics and Automation, Anchorage, Alaska, USA, May 2010.

[257] C. Fan, K. Xiong and S. Chen, Stock-loan valuation based on the finite-moment log-stable process, Computers and Mathematics with Applications, Elsevier, vol. 75, pp. 374–387, 2018.

[258] Y. Fang, P. Wang, N. Sun, and X. Zhang, Dynamics analysis and nonlinear control of an offshore boom crane, IEEE Transactions on Industrial Electronics, vol. 61, no. 1, pp. 414–427, 2014.

[259] Y. Fang and P. Wang, Advanced nonlinear control of an offshore boom crane, IEEE ACC 2012, Proceedings of the 2012 American Control Conference, Montreal, Canada, June 2012.

[260] S. Fang, B. Gou, Y. Wang, Y. Xu, C. Shang and H. Wang, Optimal hierarchical management of shipboard multi-battery energy storage system using a data-driven degradation model, IEEE Transactions on Transportation Electrification, vol. 5, no. 4, pp. 1306–1318, 2019.

[261] Y. Fang, P. Wang, N. Sun and Y. Zhang, Dynamics analysis and nonlinear control of an offshore boom crane, IEEE Transactions on Industrial Electronics, vol. 61, no. 1, pp. 414–427, 2014.

[262] Y. Fang, E. Zergeroglou, M.S. de Queiroz and D.M. Dawson, Global output feedback control of dynamically positioned surface vessels: An adaptive control approach, Mechatronics, Elsevier, vol. 14, pp. 341–356, 2004.

[263] H. Fang, Y. Zhang, M. Liu and W. Shen, State-of-charge estimation of lithium-ion battery based on an improved Kalman Filter, Proceedings of the 2017 IEEE 21st International Conference on Computer Supported Cooperative Work in Design, 2017.

[264] M. Farza, M.M. Saad, M.L. Fall, O. Gehan and K. Busawon, Continuous-discrete time observers for a class of MIMO nonlinear systems, IEEE Transactions on Automatic Control, vol. 59, no. 4, pp. 1060–1065, 2013.

[265] S. Farzen, A.P. He, E. Davies and J. Rogers, Feedback motion planning and control of brachiation robots traversing flexible cables, IEEE ACC 2019, IEEE ACC 2019 American Control Conference, Philadelphia, USA, July 2019.

[266] J. Felez, C. Carcia-Sanchez and J.A. Lozano, Control design for an articulated track with autonomous driving in an electrified highway, IEEE Access, vol. 6, pp. 60171–60186, 2018.

[267] A. Ferragut and F. Paganini, Content dynamics in P2P networks from queuing and fluid perspectives, ACM ITC 2012, Proceedings of the 24th International Teletraffic Congress, Krakow, Poland, September 2012.

[268] A. Ferragut and F. Paganini, Fluid models of population and download progress in P2P networks, IEEE Transactions on Control of Network Systems, vol. 3, no. 1, pp. 34–45, 2016.

[269] N. Fischer, D. Hughes, P. Walters, E.M. Schwartz and W.E. Dixon, Nonlinear RISE-based control of an autonomous underwater vehicle, IEEE Transactions on Robotics, vol. 30, no. 4, pp. 845–853, 2014.

[270] C. Fletcher, C. Manzie and M. Good, Trailer steering: An Australian research perspective and application for by-wire control, 9th International Symposium on Heavy Vehicle Weights and Dimensions, Pennsylvania, USA, June 2006.

[271] M. Fliess, J. Levine, P. Martin and P. Rouchon, A Lie-Backlünd approach to equivalence and flatness of nonlinear systems, IEEE Transactions on Automatic Control, vol. 44, no. 5, pp. 922–937, 1999.

[272] M. Fliess and H. Mounier. An algebraic framework for infinite-dimensional linear systems, Proceedings of the International School on Automatic Control of Lille, "Control of Distributed Parameter Systems: Theory and Applications", Grenoble, France, 2002.

[273] M. Fliess and H. Mounier, Tracking control and π-freeness of infinite dimensional linear systems, In: G. Picci and D.S. Gilliam Eds., Dynamical Systems, Control, Coding and Computer Vision, vol. 258, pp. 41–68, *Birkhaüser*, 1999.

[274] M. Fliess and C. Join, Preliminary remarks on option pricing and dynamic hedging. IEEE ICSCS 2012, 1st International Conference on Systems and Computer Science, Lille, Villeneuve d' Ascq, France, 2012.

[275] P.A. Forsyth and C. Labahn, Numerical methods for controlled Hamilton-Jacobi-Bellman PDE in finance. Journal of Computational Finance, vol. 11, no. 2, pp. 1, 2007.

[276] F. Forte, R. Naldi, A. Macchelli and L. Marconi, Impendence control of an aerial manipulator, IEEE ACC 2012, IEEE 2012 American Control Conference, June 2012, Montreal, Canada.

[277] A. Frances-Roger, A. Anvan-Moghadelam, E. Rodriguez-Diaz, J.C. Vasquez, J.M. Guerrero and J. Uceda, Dynamic assessment of COTS converters-based DC microgrid power systems in electric ships, IEEE Transactions in Industrial Informatics, vol. 14, no. 12, pp. 5518–5529, 2018.

[278] E. Franco, Adaptive IDA-PBC for underactuated mechanical systems with constant disturbances, International Journal of Adaptive Control and Signal Processing, John Wiley, vol. 33, pp. 1–15, 2019.

[279] B. Fridholm, M. Nilsson and T. Wik, Robustness comparison of battery state of charge observers for automotive applications, Proceedings of the 19th IFAC World Congress, Cape Town, South Africa, August 2014.

[280] C. Frochlich, W. Kemmetmuller and A. Kugi, Model-predictive control of servo-pump driven injection moulding machines, IEEE Transactions on Control Systems Technology, vol. 28, no. 5, pp. 1665–1680, 2020.

[281] C. Froehlich, W. Kemmetmuller and A. Kugi, Control-oriented modelling of servo-pump driven injection moulding machines with filling and packing phase, Mathematical and Computer Modelling of Dynamical Systems, Taylor and Francis, vol. 24, no. 5, pp. 451–474, 2018.

[282] S. Fu, Z. Sui, H. Wang and Y. Tian, Controllability of the nonlinear underactuated ball and plate system, IEEE 2009 International Conference on Mechatronics and Automation, Changchun, China, August 2009.

[283] S. Fuchshumer, K. Schlacher and G. Keinstel, A non-circular arc roll gap model for control applications in steel rolling mills, IEEE CCA 2005, Proceedings of the IEEE 2005 Conference on Control Applications, Toronto, Canada, August 2005.

[284] M. Fujiwara, Z.K. Nagy, J.W. Chen and R.D. Baartz, First principles and direct design approaches for the control pharmaceutical crystallization, Journal of Process Control, Elsevier, vol. 15, pp. 493–504, 2005.

[285] R. Furlan, F.A. Cuzzola and T. Parisini, Friction compensation in the interstand looper of hot-strip mills: A sliding-mode control approach, Control Engineering Practice, Elsevier, vol. 16, pp. 214–224, 2008.

[286] M. Galicki, Path-constrained control of a redundant manipulator in a task space, Robotics and Autonomous Systems, Elsevier, vol. 34, pp. 234–243, 2006.

[287] M. Galicki, Inverse-free control of robotic manipulator in a task space, Robotics and Autonomous Systems, Elsevier, vol. 62, pp. 131–141, 2014.

[288] M. Galicki, Fimite-time trajectory tracking control in a task-space of robotic manipulator, Automatica, Elsevier, vol. 67, pp. 165–170, 2016.

[289] M. Galicki, An adaptive regulator for robotic manipulators in the task space, IEEE Transactions on Automatic Control, vol. 53, no. 4, pp. 1058–1063, 2008.

[290] M. Galicki, Real-time constrained trajectory generation of mobile manipulators, Robotics and Autonomous Systems, Elsevier, vol. 78, pp. 49–62, 2016.

[291] A. Galkina, I. Gafer and K. Schlacher, Model predictive control and linear programming of the strip infeed in hot rolling mills, IFAC 2017 World Congress, Toulouse, France, July 2017.

[292] G.S. Galloway, V.M. Catterson, C. Love, A. Rabb and T. Fay, Modelling and interpretation of tidal turbine vibration through weighted least squares regression, IEEE Transactions on Systems, Man and Cybernetics: Systems, vol. 50, no. 4, pp. 1252–1259, 2017.

[293] B. Gao and J. Zhao, Dynamical modelling and controllability analysis of an underactuated 2-dimensional TORA system on a slope, Journal of Advanced Robotic Systems, Sage Publications, vol. 11, pp. 70, 2014.

[294] Q. Gao, Y. Liu and D. Jiang, Nonlinear control method of electro-hydraulic system driving two-DOF robot arm with output constraint, IOP Conference Series on Materials Science and Engineering, vol. 715, pp. 012090, 2020.

[295] G. Gao, Y. Hen, H. Yu and J. Qin, Spatial path following control of underactuated AUV with multiple uncertainties and input saturation, IEEE Access, vol. 7, pp. 98014–98022, 2019.

[296] H. Gao, J. Zhang and W. Tang, Offset-free trajectory tracking control for hypersonic vehicle under external disturbance and parametric uncertainty, Journal of the Franklin Institute, Elsevier, vol. 355, pp. 997–1017, 2018.

[297] Z. Gao, J. Guo, J. Zhou and J. Chang, Robust tracking for hypersonic re-entry vehicles via disturbance estimation-triggered control, IEEE Transactions on Aerospace and Electronic Systems, vol. 56, no. 2, pp. 1279–1289, 2019.

[298] Y. Gao, J. Lin, Z. Wang and L. Wan, Interval type-2 FNN-based quantized tracking control hypersonic flight vehicles with prescribed performance, IEEE Transactions on Systems, Man and Cybernetics: Systems, vol. 51, no. 3, pp. 1981–1993, 2019.

[299] Z. Gao, S. Li and Y. Zhang, Distributed model predictive control for secondary voltage of the inverter-based microgrid, IEEE CCC 2017, Proceedings of the 36th Chinese Control Conference, July 2017, Dalian, China.

[300] J. Gao, L. Dai, W. Zhang, S. Huang and X. Wu, Multi-interval efficiency design optimization for permanent magnet synchronous generators used in hybrid electric special vehicles, IEEE Transactions on Industrial Electronics, 2021.

[301] Z.L.C. Gao, H. Xiao and Y. Kao, Non-fragile observer-based H-infinity control for uncertain neutral-type systems via sliding mode technique, Asian Journal of Control, John Wiley, vol. 19, no. 2, 2016.

[302] Y. Gao, D. Aceval, Z. Nagy and C.D. Laird, Real-time feasible multi-objective optimization based nonlinear model predictive control of particle size and shape in a batch crystallization process, Control Engineering Practice, Elsevier, vol. 69, pp. 1–8, 2017.

[303] L. Guo and S.A. Billings, State-space reconstruction and spatio-temporal prediction of lattice dynamical systems, IEEE Transactions on Automatic Control, vol. 52, no. 4, pp. 622–632, 2007.

[304] R.A. Garcia-Garcia and M. Arias-Montiel, Linear controllers for the NXT ballbot with parameter variations using linear matrix inequalities, IEEE Control Systems Magazine, June 2012, pp. 121–136.

[305] F. Garcia-Torres, C. Bordons and M.A. Ridao, Optimal economic schedule for a network of microgrids with hybrid energy storage system using distributed model predictive control, IEEE Transactions on Industrial Electronic, vol. 66, no. 3, pp. 1919–1929, 2019.

[306] G. Garimella and M. Kobilarov, Towards model-predictive control for aerial pick and place, IEEE ICRA 2015, IEEE 2015 International Conference On Robotics And Automation, Seattle, USA, July 2013.

[307] H. Ge, Y. Miao, B. bilgin, B. Nahid-Mobarakeh and A. Emadi, Speed-range extended maximum torque per ampere control for PM drives considering inverter and motor nonlinearities, IEEE Transactions on Power Electronics, vol. 32, no. 9, pp. 7151–7158, 2017.

[308] E.J.M. Geddes and I. Postlethwaite, Improvements in product quality in tandem cold rolling using robust multivariable control, IEEE Transactions on Control Systems Technology, vol. 6, no. 2, pp. 257–269, 1998.

[309] E. Gelenbe, A diffusion model for packet travel time in a random multihop medium, IEEE ACM Transactions on Sensor Networks, vol. 3, no. 2, 2007.

[310] Y. Geng, Z. Lai, Y. Li, D. Wang, R. Chen and P. Zhang, Sensorless fault-tolerant control strategy of six-phase induction machine based on harmonic suppression and sliding-mode observers, IEEE Access, vol. 7, pp. 110986–110104, 2019.

[311] M. Gerdes, G. Greif and H.J. Peich, Numerical optimal control of the wave equation: Optimal boundary control of a string to rest in finite time, Proceedings of the 5th Mathmod Conference, Vienna, Austria, February 2006.

[312] M. Gerdts, G. Greif and H. J. Pesch, Numerical optimal control of the wave equation: Optimal boundary control of a string to rest in finite time, Mathematics and Computers in Simulation, Elsevier, vol. 79, no. 4, pp. 1020–1032, 2008.

[313] R. Geyyer, A. Kienle and S. Palis, Robust control of continuous crystallization process, IFAC MICNON 2015, 1st IFAC Conference on Modelling, Identification and Control of Nonlinear Systems, St. Petersburg, Russia, June 2015.

[314] A. Ghaderi and T. Hanamoto, Wide-speed-range sensorless vector control of synchronous reluctance motors based on extended programmable cascaded low-pass filters, IEEE Transactions on Industrial Electronics, vol. 58, no. 6, pp. 2322–2333, 2011.

[315] M.G. Ghiasi, M.A. Golkar and A. Hajizadeh, Lyapunov-based distributed fuzzy-sliding mode control for building integrated DC microgrid with plug-in electric vehicle, IEEE Access Journal, vol. 5, pp. 7746–7752, 2014.

[316] F.H. Ghorbel, O. Chetelat, R. Gunawardana and R. Longchamp, Modeling and setpoint control of closed-chain mechanism: Theory and experiment, IEEE Transactions on Control Systems Technology, vol. 8, no. 5, pp. 801–815, 2000.

[317] M. Giacomelli, F. Padula, L. Simoni and A. Visioli, Simplified input-output inversion control of a double-pendulum overhead crane for residual oscillations reduction, Mechatronics, Elsevier, vol. 56, pp. 37–47, 2018.

[318] B.P. Gibbs, Advanced Kalman Filtering, Least Squares and Modelling: A Practical Handbook, J. Wiley, 2011.

[319] A. Girin and F. Plestan, A new experimental test bench for a high performance double electropneumatic actuator system, IEEE ACC 2009, IEEE 2009 American Control Conference, St. Louis, Missouri, USA, June 2009.

[320] A. Girin, F. Plestan, X. Brun and A. Glumineau, Higher-order sliding-mode controllers of an electropneumatic actuator: Application to an aeronautic benchmark, IEEE Transactions on Control Systems Technology, vol. 17, no. 3, pp. 633–647, 2009.

[321] A. Giuseppi, A. Pietrabissa, S. Cilione and L. Galvani, Feedback linearization-based satellite attitude control with a life-support device without communications, Control Engineering Practice, Elsevier, vol. 90, pp. 221–230, 2019.

[322] M. Gokasan, S. Bogosyan and D.J. Goering, Sliding-mode based powertrain control for efficiency improvement in hybrid-electric vehicles, IEEE Transactions on Power Electronics, vol. 21, pp. 779–790, 2006.

[323] M.W. Gomes and A.L. Ruino, Five-link 2D brachiating ape model with life-like zero-energy-cost function, Journal ofTheoretical Biology, Elsevier, vol. 257, no. 3, pp. 265–278, 2005.

[324] P.F.C. Goncalves, S.M.A. Cruz and A.M.S. Mendes, Disturbance-observer based predictive current control of six-phase PMSMs for the mitigation of steady-state errors and current harmonics, IEEE Transactions on Industrial Electronics, 2021.

[325] L. Gong and M.S. Cheung, Computer simulation of dynamic interaction between vehicle and long span box girder bridges, Tsinghua Science and Technology, vol. 13, no. S1, pp. 71–77, 2006.

[326] A. Gonzalez, E.J. O' Brien and P. Mc Getrick, Identification of damping in a bridge using a moving instrumented vehicle, Journal of Sound and Vibration, Elsevier, vol. 331, pp. 4115–4131, 2012.

[327] G. Grabmair, K. Schlacher and A. Kugi, Geometric energy-based analysis and controller design of hydraulic actuators applied in rolling mills, IEEE ECC 2003, European Control Conference 2003, Cambridge, UK, September 2003.

[328] O. Gregedal, P.J.S. Vie and M. Molins, Battery modelling and Kalman Filter-based state-of-charge estimation for a race car application, IEEE ICNSC 2017, IEEE 14th International Conference on Networking, Sensing and Control, Calabria, Italy, August 2017.

[329] G. Grenier and J.P. Louis, Modelling for control of non-sinewave permanent magnet synchronous drives by extending Park's transformation, Mathematics and Computers in Simulation, Elsevier, vol. 38, pp. 445–452, 1995.

[330] S. Greyb and P. Orlawska, Congestion feedback control for computer networks with bandwidth estimation, IEEE MMAR 2015, 20th International Conference On Methods in Automation and Robotics, Mielzydroje, Poland, August 2015.

[331] H. Gritli and S. Belghith, Robust feedback control of the underactuated inertia wheel inverted pendulum under parametric uncertainties and subject to external disturbances: LMI formulation, Journal of the Franklin Institute, Elsevier, vol. 355, pp. 9150–9191, 2018.

[332] L. Grüne and Jü Pannek, Nonlinear Model Predictive Control: Theory and Algorithms, Springer, 2017.

[333] C. Guan and S. Pan, Nonlinear adaptive robust control of single-rod electro-hydraulic actuator with unknown nonlinear parameters, IEEE Transactions on Control Systems Technology, vol. 16, no. 3, pp. 434–445, 2008.

[334] Y. Guan, J.C.V. Quintero, J.M. Guerrero, J.M. Wang and W. Feng, Frequency stability of hierarchically controlled hybrid photovoltaic–battery–hydropower microgrids, IEEE Transactions on Industry Applications, vol.51, no. 6, pp. 4729–4742, 2015.

[335] J. Guerrero, J. Torres, V. Creuze and A. Chemori, Observation-based nonlinear proportional derivative control for robust trajectory tracking for autonomous underwater vehicles, IEEE Journal of Oceanic Engineering, 2019.

[336] L. Guevara, M. Torres-Torriti and F.A. Cheein, Collision-free navigation of N-trailer vehicles with motion constraints, IEEE RoMoCo 2019, Proceedings of the 12th International Workshop on Robot Motion and Control, Poznan, Poland, July 2019.

[337] L. Guewara, R.P. Rocha and F.A. Cheein, Improving the manual harvesting operation efficiency by coordinating a fleet of N-trailer vehicles, Computers and Electronics in Agriculture, Elsevier, vol. 195, pp. 106103–106113, 2021.

[338] L. Guevara, M. Michalek and F.A. Cheein, Headland turning algorithmization for autonomous N-trailer vehicles in agricultural scenarios, Computers and Electronics in Agriculture, vol. 175, pp. 105541–105553, 2020.

[339] Q. Guo, J. Yin, T. Yu and D. Jiang, Saturated adaptive control of an electro-hydraulic actuator with parametric uncertainty and load disturbance, IEEE Transactions on Industrial Electronics, vol. 67, no. 10, pp. 7930–7941, 2017.

[340] W. Guo and D. Liu, Nonlinear dynamic surface control the underactuated translational nonlinear oscillator with rotating actuator system, IEEE Access, vol. 7, pp. 11844–11853, 2019.

[341] L. Guo, X. Mo and Y. Song, Dynamic modelling and adaptive controller design for a wire-moving robot, IEEE CCC 2015, Proceedings of the 34th Chinese Control Conference Hangzhou, China, July 2015.

[342] Q. Guo, Y. Zhang, B.G. Keller and S.W. Wu, State-constrained control of single-rod electrohydraulic actuator with parametric uncertainty and load disturbance, IEEE Transactions on Control Systems Technology, vol. 26, no. 6, pp. 2942–2950, 2018.

[343] Q. Guo, Q. Wang and Y. Liu, Anti-windup control of an electrohydraulic system with load disturbance and modeling uncertainty, IEEE Transactions on Industrial Informatics, vol. 14, no. 7, pp. 3097–3108, 2018.

[344] Q. Guo, P. Sun, J.M. Yin, T. Yu and D. Jiang, Parametric adaptive estimation and backstepping control of electro-hydraulic actuator with delayed memory filter, ISA Transactions, Elsevier, vol. 62, pp. 202–214, 2016.

[345] K. Guo, J. Wei, J. Fang, R. Feng and X. Wang, Position tracking control of electro-hydraulic actuator based on extended disturbance observer, Mechatronics, Elsevier, vol. 27, pp. 47–56, 2016.

[346] Q. Guo, Q. Wang and X. Li, Finite-time convergence control of electro-hydraulic velocity servo-system under uncertain parameter and external load, IEEE Transactions on Industrial Electronics, vol. 68, no. 6, pp. 4513–4523, 2013.

[347] J. Guo, G. Zhang, P. Wu, Z. Zhao, T. Wang and W. Yan, Model predictive visual servoing of fully-actuated underwater vehicles with a sliding-mode disturbance observer, IEEE Access, vol. 7, pp. 25516–25527, 2019.

[348] Y. Guo, L. Yu and J. Xu, Robust finite-time trajectory tracking control of wheeled mobile robots with parametric uncertainties and disturbances, Journal of Systems Science and Complexity, Springer, vol. 32, pp. 1358–1374, 2019.

[349] L. Guo and S.A. Billings, State-space reconstruction and spatio-temporal prediction of lattice dynamical systems, IEEE Transactions on Automatic Control, vol. 52, no.4, pp. 622–632, 2007.

[350] A. Gupta and M.K.O. Malley, Disturbance-observer-based force estimation for haptic feedback, ASME Journal of Dynamical Systems Measurement and Control, vol. 133, no. 1, 2011.

[351] M.R.Grasselli and C.G. Velez, Stock loans in incomplete markets, Applied Mathematical Finance, Taylor and Francis, vol. 20, no. 2, pp. 118–136, 2013.

[352] B.C. Gruenwald, T. Yucelen, K.M. Dogan and J.A. Muse, An adaptive architecture for control of uncertain dynamical systems with unknown actuator bandwidth, IFAC NAASS 2018, IFAC Workshop of Networked and Autonomous Space Systems, Santa Fe, New Mexico, USA, June 2018.

[353] M. Grun and O. Konigorski, Observer-based method for joint torque estimation in active orthosis, MathMod 2012, 7th Vienna International Conference on Mathematical Modelling, Vienna, Austria, February 2012.

[354] N.K. Haddad, A. Chemori and S. Belghith, Robustness enhancing of IDA-PBC controller for underactuated mechanical systems: Theory and real-time experiments, International Journal of Control, Taylor and Francis, vol. 91, pp. 2157–2672, 2016.

[355] V. Hagenmeyer, S. Streif and M. Zeitz, Flatness-based feedforward and feedback linearization of the ball and plate lab experiment, IFAC NOLCOS 2004, 6th IFAC Conference on Nonlinear Control Systems, Stuttgart, Germany, September 2004.

[356] V. Hagenmeyer and E. Delaleau, Robustness analysis of exact feedforward linearization based on differential flatness, Automatica, Elsevier, vol. 39, pp. 1941–1946, 2003.

[357] V. Hagenmeyer and E. Delaleau, Robustness analysis with respect to exogenous perturbations for flatness-based exact feedforward linearization, IEEE Transactions on Automation Control, vol. 55, no.3, pp. 727–731, 2010.

[358] M. Hamaguchi, Damping control of sloshing in liquid container in cart with active reducer: The case of a curved path on a horizontal plane, IEEE Transactions on Mechatronics, vol. 24, no. 1, pp. 361–372, 2019.

[359] C. Han and A.N. Pechev, Underactuated satellite attitude control with two parallel CMGs, IEEE 2007 International Conference on Control and Automation, Guangzhou, China, May 2007.

[360] C. Han, J. Guo and A. Pechev, Nonlinear H_∞-based underactuated attitude control for small satellites with two reaction wheels, Acta Astronautica, Elsevier, vol. 194, pp. 159–172, 2014.

[361] R. Han, M. Tucci, R. Solopetro, J.M. Guerrero and G. Ferrani-Trecante, Plug-and-play design of current controllers for grid-feeding converters in DC microgrids, IEEE ASCC 2017, IEEE 11th Asian Control Conference, Australia, December 2017.

[362] M.R.J. Harandi, S.A. Khlilpour, H.D. Taghirad and J.G. Romero, Adaptive control of parallel robots with uncertain kinematics and dynamics, Mechanical Systems and Signal Processing, Elsevier, vol. 157, pp. 107693–107705, 2021.

[363] E. Harinath, L.T. Biegler and G.A. Dumont, Control and optimization strategies for thermo-mechanical pulping processes: Nonlinear model predictive control, Journal of Process Control, Elsevier, vol. 21, pp. 519–528, 2011.

[364] E. Harinath, L.T. Biegler and G.A. Dumont, Predictive optimal control for thermomechanical pulping processes with multi-stage low consistency refining, Journal of Process Control, Elsevier, vol. 23, pp. 2001–2011, 2013.

[365] E. Harinath, L.T. Biegler and G.A. Dumont, Advanced step nonlinear model predictive control for two-stage thermomechanical pulping processes, IEEE ACC 2016, IEEE 2016 American Control Conference, Boston, USA, June 2016.

[366] A. Haseltalab and R.R. Negenborn, Model predictive maneuvering control and energy management for all-electric autonomous ships, Applied Energy, Elsevier, vol. 251, pp. 113308–113334, 2014.

[367] A. Haseltalab, M.A. Botto, and R.R. Negenborn, Model predictive DC voltage control for all-electric ships, Control Engineering Practice, Elsevier, vol. 90, pp. 133–147, 2019.

[368] A. Haseltalab, M.A. Botto and R.R. Negenborn, On-board voltage regulation for all-electric DC-ships, Proceedings IFAC Conference on Control Applications in Marine Systems, Robotics and Vehicles, IFAC CAMS 2018, Opatija, Croatia, September 2018.

[369] T. Hashimoto, Y. Yoshioto and T. Ohtshaka, Receding horizon control for hot-strip mill cooling systems, IEEE Transactions on Mechatronics, vol. 18, no. 3, pp. 998–1005, 2012.

[370] T. Haus, M. Orsag and S. Bogdan, A concept of non-tilting multirotor UAV based on moving mass control, IEEE ICUAS 2017, IEEE 2017 International

Conference on Unmanned Aircraft Systems, June 2017, Miami, Florida, USA.

[371] J. Hauser, S. Shastry and G. Meyer, Nonlinear control design for slightly non-minimum phase systems: Application to V/STOL aircraft, Automatica, Elsevier, vol. 28, no. 4, pp. 665–679, 1992.

[372] Z. He, W. Wang, H. Ruan, Y. Yao and X. Li, A two-wheel load balance control strategy for an HVTL inspection robot based on second-order sliding-mode, Industrial Robot, vol. 46, no. 1, pp. 83–92, 2019.

[373] S. He, Y. Li, G. Zhou, *et al.*, Digital collaborative development of a reliable auxiliary electric drive system for transportation: From dual three-phase PMSM to control algorithm, IEEE Access, vol. 8, pp. 178755–178769, 2020.

[374] P. Heidtmann and M. Brucker, Nonlinear modelling and tracking control of a hydraulic rotary vane actuator, PAMM Proceeding on Applied Mathematics and Mechanics, vol. 5, pp. 161–162, 2005.

[375] B. Heilan, Z. Chen and B. Yao, Precise motion control of a servomotor pump direct-drive electrohydrauic system with a nonlinear pump flow mapping, IEEE Transactions on Industrial Electronics, vol. 67pp. 9638–9648, 2020.

[376] J. Henikl, W. Kemmetmuller, T. Meurer and A. Kugi, Infinite dimensional decentralized damping control of large-scale manipulators with hydraulic actuation, Automatica, Elsevier, vol. 63, pp. 101–115, 2016.

[377] C. Henon, Installing, Operating and maintaining DC motors in a paper-mill environment, IEEE Industry Applications Magazine, pp. 25–33, December 2019.

[378] P.M. La Hera, U. Mettin, S. Westenberg and A.S. Shiriaev, Modelling and control of hydraulic rotatory actuators used in forestry cranes, IEEE ICRA 2009, IEEE 2009 International Conference on Robotics and Automation, Kobe, Japan, May 2009.

[379] P. La Hera, L. Freidovich, A. Shiriaev and U. Mettin, Traversing from point-to-point along a straight line with a ballbot, 8th IFAC Symposium on Nonlinear Control Systems, University of Bologna, Italy, September 2010.

[380] B. Hérisé, T. Hamel, R. Mahony and F.X. Russotto, Landing a VTOL unmanned aerial vehicle on a moving platform using optical flow, IEEE Transactions on Robotics, vol. 28, no. 1, pp. 77–89, 2012.

[381] L. Herrera, R. Hernandez and F. Jurada, Control and extended Kalman Filter-based estimation of a ballbot robotic system, 2018 IEEE Mexican Congress in Robotics, IEEE ComRob 2018, Ensenada, Mexico, September 2018.

[382] J.M. Hervas and M. Reyhanoglu, Observer-based nonlinear optimal control of slosh in liquid-container transfer via a PPR robot, IEEE ICCAS 2013, IEEE 2013 13th International Conference on Control, Automation and Systems, Gwengju, Korea, October 2013.

[383] J.M. Hervas and M. Reyhanoglu, Observer-based nonlinear control of space vehicles with multi-mass fuel slosh dynamics, IEEE ISIE 2014, IEEE 2014

International Symposium on Industrial Electronics, Istanbul, Turkey, June 2014.

[384] T. Hesketh, Y.A. Jiang, D.J. Clemens, D.H. Butler and R. van der Loon, Controller design for hot-strip finishing mills, IEEE Transactions on Control Systems Technology, vol. 6, no. 2, pp. 208–219, 1998.

[385] D. Hester and A. Golzalez, A wavelet-based damage detection algorithm based on bridge acceleration response to a vehicle, Mechanical Systems and Signal Processing, vol. 28, pp. 145–166, 2012.

[386] N. Herzig, R. Moreau, T. Redarce, F. Abry and X. Brun, Nonlinear position and stiffness backstepping controller for a two-degrees of freedom pneumatic robot, Control Engineering Practice, Elsevier, vol. 73, pp. 26–39, 2018.

[387] B. Heymann, J.F. Bonnans, P, Martinan, F. Silva, F. Lanas and J. Jimenez, Continuous optimal control approaches to microgrid energy management, Energy Systems, vol. 9, no. 1, pp. 54–77, 2018.

[388] A. Hilderbrandt, R. Neumann and o. Sawodny, Optimal system design of SISO-servopneumatic position drives, IEEE Transactions on Control Systems Technology, vol. 18, no.1, pp. 35–44, 2010.

[389] Q.D. Hoang, V.A. Rosas-Cervantes, S.G. Lee, J.S. Weon, J.H. Choi and Y.H. Kwon, Robust finite-time convergence control mechanism for high-precision tracking in a hybrid fluid power actuator, IEEE Access, vol. 8, pp. 196775–196790, 2020.

[390] T. Ho, K. Suzuki, M. Tsume, R. Tasaki, T. Miyashi and K. Terashima, A switched optimal control approach to reduce transferring time, energy consumption and residual vibration of payload's skew rotation in crane systems, Control Engineering Practice, Elsevier, vol. 84, pp. 247, 2019.

[391] Q.K. Ho and C.B. Pham, Study on inertia wheel pendulum applied to self-balancing electric motorcycle, IEEE GTSD 2018, 4th IEEE International Conference on Green Technology and Sustainable Development, Ho Chi Minh City, Vietnam, November 2018.

[392] N. Hodgins, O. Keysan, A.S. McDonald and M.A. Mueller, Design and testing of a linear generator for wave-energy applications, IEEE Transactions on Industrial Electronics, vol. 57, no. 5, pp. 2094–2103, 2012.

[393] S. Hodgsan, M.Q. Le, M. Tavakoli, and M.T. Pham, Improved tracking and switching performance of an electro-pneumatic positioning system, Mechatronics, Elsevier, vol. 22, pp. 1–12, 2012.

[394] H.F. Hofmann, S.R. Sanders and A. El-Antably, Stator-flux-oriented vector control of synchronous reluctance machines with maximized efficiency, IEEE Transactions on Industrial Electronics, vol. 51, no. 5, pp. 1066–1072, 2004.

[395] C. Hopmmann, C.E. Kahve, H. Schmitz and M. Roebig, A comparison of predictive control strategies for a highly segmented injection mould tempering, Polymer, Elsevier, vol. 219, pp. 123494–123501, 2021.

[396] N.M. Horri, P. Palmer and M. Roberts, Gain-scheduled inverse optimal satellite attitude control, IEEE Transactions of Aerospace and Electronic Systems, vol. 48, no. 3, pp. 2437–2457, 2012.

[397] Z. Hou, R. Li, W. Wu, H. Qu and Z. Hou, Kinematic analysis and experimental verification of a controllable five-bar parallel kinematic manipulator, IEEE 2015 International Conference on Control, Automation and Information Systems, Changsu, China, October 2015.

[398] Z. Hou, Z. Song, Hofmann and J. Sun, Adaptive model-predictive control for hybrid energy storage energy management in all-electric ship microgrids, Energy Conversion and Management, vol. 198, pp. 111929–111939, 2019.

[399] J. Hou, Z. Sang, H.F. Hofmann and J. Sun, Control strategy for battery-flywheel hybrid energy storage in electric shipboard microgrids, IEEE Transactions on Industrial Informatics, vol. 17, no. 2, pp. 1088–1100, 2021.

[400] H. Hou, X. Nian, J. Chen and D. Xiao, Decentralized coordinated control of elastic web winding systems without tension sensor, ISA Transactions, Elsevier, vol. 80, pp. 350–359, 2018.

[401] J.H. Hours, H. Shukla and C.N. Jones, A parametric augmented Lagrangian algorithm for real-time economic NMPC, 2016 European Control Conference, Aalborg, Denmark, July 2016.

[402] C. Hu, R. Wang, F. Yan and N. Chen, Robust composite nonlinear feedback path-following control for underactuated surface vessels with desired-heading amendment, IEEE Transactions on Industrial Electronics, vol. 69, no. 10, pp. 6386–6395, 2016.

[403] X. Hu, B. Xu and C. Hu, Robust adaptive fuzzy control for HSV with parameter uncertainty and unmodelled dynamics, IEEE Transactions on Industrial Electronics, vol. 65, no. 11, pp. 8851–8860, 2018.

[404] S. Hu, M. Xiong, Z. Liang and X. He, Torque distributed control strategy for the dual three-phase PMSM in hybrid energy storage system application, IEEE Transactions on Industrial Electronics, vol. 67, no. 4, pp. 2544–2553, 2020.

[405] Y. Hu, X. Song, W. Cao and B. Ji, New SR drive with integrated charging capacity for plug-in hybrid electric vehicle (PHEVs), IEEE Transactions on Industrial Electronics, vol. 61, no. 10, pp. 5722–5731, 2014.

[406] J. Hu and S. Gan, High-order methods for Black-Scholes PDE, Computers and Mathematics with Applications, Elsevier, vol. 75, no. 7, pp. 2259–2270, 2018.

[407] S. Hun, H. Wang and Y. Tian, A linear discrete-time extended state observer-based intelligent PD controller for a 12DOFs lower-limb exoskeleton LLF-RePA, Mechanical Systems and Signal Processing, Elsevier, vol. 138, pp. 106547–106561, 2020.

[408] X. He, X. Wang and Z. Feng, Nonlinear modelling of electro-hydraulic servo injection moulding machine including asymmetric cylinder, Proceedings of the IEEE ACC 2003 American Control Conference, Denver, Colorado, USA, June 2003.

[409] M.W. Hermanto, R.D. Braatz and M.S. Chiu, Optimal control of polymorphic transformations in batch pharmaceutical crystallization, 16th IEEE International Conference on Control Applications, Singapore, October 2007.

[410] Q. Hu, L. Xiao and C. Wang, Adaptive fault-tolerant attitude tracking control for spacecraft with time-varying inertia uncertainties, Chinese Journal of Aeronautics, Elsevier, vol. 32, no. 3, pp. 674–687, 2019.

[411] M.D. Hua, T. Humel, P. Morin and C. Samson, Introduction to feedback control of underactuated VTOL vehicles, IEEE Control Systems Magazine, pp. 61–75, 2013.

[412] M.D. Hua, T. Humel, P. Morin and C. Samson, A control approach for thrust propelled underactuated vehicles and its application to VTOL drones, IEEE Transactions on Automatic Control, vol. 54, no. 8, pp. 1837–1853, 2009.

[413] M.D. Hua, T. Humel, P. Morin and C. Samson, Control of VTOL vehicles with thrust-tilting augmentation, Automatica, Elsevier, vol. 52, pp. 1–7, 2015.

[414] X. Huang, H.K. Khalil and Y. Song, Regulation of nonminimum-phase nonlinear systems using slow integrators and high-gain feedback, IEEE Transactions on Automatic Control, vol. 64, no. 2, pp. 640–653, 2019.

[415] B. Huang, Z. Li, X. Wu, A. Ajoudani, A. Bicchi and J. Liu, Coordination control of a dual-arm exoskeleton robot using human impedance transfer skills, IEEE Transactions on Systems, Man and Cybernetics: Systems, vol. 49, no. 5, pp. 954–963, 2019.

[416] J. Huang, H. An, Y. Yang, G. Yu and H. Ma, Model predictive trajectory tracking control of electrohydraulic actuator in legged robot with multi-scale online estimation, IEEE Access, vol. 6, pp. 95918–05933, 2020.

[417] J. Huang, H. An, L. Liang, Q. Wei and H. Ma, A data-driven multi-scale online joint estimation of states and parameters for electrohydraulic actuation of legged robot, IEEE Access, vol. 8, pp. 06886–96903, 2020.

[418] S.J. Huang and T.H. Lee, Fuzzy logic controller for a retrofitted closed-loop injection moulding machine, Proceedings of the Institutions of Mechanical Engineers Part I: Journal of Systems and Control Engineering, Sage Publications, vol. 214, pp. 9–22, 2000.

[419] G. Huang, D. Zhang, H. Tang, L. Kong and S. Song, Analysis and control for a new reconfigurable parallel mechanism, International Journal of Advanced Robotic Systems, Sage Publications, pp. 1–20, September–October 2020.

[420] W. Huang, Y. Zhao, Y. Ye and W. Xie, State feedback control for stabilization of the ball and beam system, Proceedings of the 38th Chinese Control Conference, Guangzhou, China, July 2019.

[421] H. Huang, M. Gang, Y. Zhang, S. Sharma and D. Xu, A new guidance law for trajectory tracking of an underactuated unmanned surface vehicle with parameter perturbations, Ocean Engineering, Elsevier, vol. 175, pp. 217–22, 2019.

[422] S.N. Huang, K.K. Tan and T.H. Lee, Neural network-based predictive learning control of ram velocity in injection moulding, IEEE Transactions on Systems, Man and Cybernetics, Part C: Applications and Reviews, vol. 34, no. 3, pp. 363–368, 2004.

[423] O. Huber, V. Acory, B. Brogliato and F. Plestan, Implicit discrete-time twisting controller without numerical chattering: Analysis and experimental results, Control Engineering Practice, Elsevier, vol. 73, pp. 26–39, 2018.

[424] J. Hun and F. Liu, Positioning control research on ball and plate system based on Kalman Filter, IEEE 2016 6th International Conference on Intelligent Human-Machine Systems and Cybernetics, Hangzhou, China, August 2016.

[425] L.C. Hung, H.P. Lin and H.Y. Chung, Design of self-tuning fuzzy sliding-mode control for TORA system, Expert Systems with Applications, vol. 332, pp. 201–212, 2007.

[426] Z.R. Huszar and M.P. Prado, An analysis of over-the-counter and centralized stock lending markets, Journal of Financial Markets, Elsevier, 2018.

[427] S.H. Hyan, Y. Taniai, K. Hianum, K. Yosunaga and H. Mizui, Overpressure compensation for hydraulic hybrid servo booster applied to hydraulic manipulators, IEEE Robotics and Automation Letters, vol. 4, no. 2, pp. 942–949, 2020.

[428] J.G. Iossaqui, J.F. Camino and D.E. Zampieri, A nonlinear control design for tracked robots with longitudinal slip, Proceedings of the 18th World Congress The International Federation of Automatic Control, Milano, Italy, September 2011.

[429] A. Iovine, S. Benamane, G. Damm, E. De Santis and M.D. di Benedetto, Nonlinear control of a DC microgrid for the integration of photovoltaic panels, IEEE Transactions on Automation Science and Engineering, vol. 14, no. 2, pp. 524–535, 2015.

[430] A. Iovine, T. Rigaut, G. Damm, E. de Santis and M.D. de Benedetto, Power management for a DC MicroGrid integrating renewables and storages, Control Engineering Practice, Elsevier, vol. 85, pp. 59–79, 2019.

[431] A. Iovine, M. Jimenez Carrizosa, G. Damm and P. Alou, Nonlinear control for DC MicroGrids enabling efficient renewable power integration and ancillary services for AC grids, IEEE Transactions on Power Systems, 2019.

[432] R. Isermann, Fault-Diagnosis Systems: An Introduction from Fault Detection to Fault Tolerance, Springer, 2006.

[433] R. Iriarte, L.T. Aguilar and L. Fridman, Second-order sliding-mode tracking controller for inertia wheel pendulum, Journal of the Franklin Institute, Elsevier, vol. 350, pp. 92–106, 2013.

[434] A. Isidori, Nonlinear Control Systems, 3rd ed. New York: Springer-Verlag, 1995.

[435] R.M.T. Raja Ismail, N.D. That and Q.P. Ha, Modelling and robust trajectory following for offshore boom crane systems, Automation in Construction, Elsevier, vol. 39, pp. 179–187, 2015.

[436] M. Jabbari Asl, T. Nasikaya and M. Kawanishui, Neural network-based bounded control of robotic exoskeletons without velocity measurements, Control Engineering Practice, Elsevier, vol. 80, pp. 94–104, 2018.

[437] M. Jafarinasab, S. Sirouspour and E. Dyer, Model-based motion control of a robotic manipulator with a flying multirotor base, IEEE/ASME Transactions on Mechatronics, vol. 24, no. 5, pp. 2328–2340, 2019.

[438] M.J. Jahrami, A.I. Maswood and K.S. Tseng, Design and evaluation of a new converter control strategy for near-shore tidal turbines, IEEE Transactions on Industrial Electronics, vol. 60, no. 12, pp. 5648–5659, 2013.

[439] V. Jaksic, A.O. Connor and V. Pakrashi, Damage detection and calibration from beam-moving oscillator interaction employing surface roughness, Journal of Sound and Vibration, Elsevier, vol. 333, pp. 3917–3920, 2014.

[440] M. Jama and A. Wahyudie, Online damping strategy for controlling heaving wave energy converters using three-phase bridge boost rectifiers, IEEE Access, vol. 5, pp. 7682–7651, 2017.

[441] C. Jammazi, Finite-time partial stabilizability of chained systems, Comptes Rendus de l' Academie des Sciences de Paris, Section I, vol. 346, pp. 975–980, 2008.

[442] E. Jamshidpour, P. Poure, B. Nahid-Mobarakeh and S. Saadate, Differential flatness-based control of a stand-alone solar PV energy generating system, IEEE CCA 2015, IEEE Conference on Control Applications, Part of the 2015 Multi-Conference on Systems and Control, September 2015, Sydney, Australia.

[443] A.M. Jassour and M. Farrokhi, Adaptive neuropredictive control for redundant robot manipulators in presence of static and dynamic obstacles: A Lyapunov-based approach, International Journal of Adaptive Control and Signal Processing, vol. 28, no. 3–5, pp. 386–411, 2014.

[444] N. Jiang, S. Zhang, D. Guo, D. Zhang and J. Xu, Experimental study of event-based neural network control of parallel manipulator, Mechatronics, Elsevier, vol. 75, pp. 102514–102523, 2021.

[445] W. Jiang, Y. Yan, L. Yu, M. Peng, H. Li and W. Chen, Research on dual-arm coordination motion control strategy for power cable mobile robot, Transactions of the Institute of Measurement and Control, vol. 41, no. 11, pp. 3235–3247, 2019.

[446] P. Jiang, S. Huang, J. Xiang and M.Z.Q. Chen, Iteratively successive projection: A novel continuous approach for the task-based control of redundant robots, IEEE Access, vol. 7, pp. 25347–25358, 2019.

[447] T. Jiang, K. Khorasani and S. Tafazoli, Parameter estimation-based fault detection, isolation and recovery for nonlinear satellite model, IEEE Transactions on Control Systems Technology, vol. 16, no. 4, pp. 799–808, 2008.

[448] J. Jiao, L. Sun, W. Kang, Y. Zhang, Y. Qiao and C. Yuan, A sliding parameter estimation method based on UKF for agricultural tracked robot, IEEE ICSAI 2014, IEEE 2nd International Conference on Systems and Informatics, Shanghai, China, November 2014.

[449] H. Jilkaistavaski, Model predictive control technology in paper quality control: A case study of a system update, 91st Annual Meeting of the Pulp and Paper Technical Association of Canada, Montreal Canada, February 2005.

[450] A.E. Jimenez-Cano, J. Martin, G. Heredia, A. Ollero and R. Cano, Control of an aerial robot with multi-link arm for assembly tasks, IEEE ICRA 2013, IEEE 2013 International Conference on Robotics and Automation, Karlsruhe, Germany, October 2013.

[451] J. Jin and Y. Zhang, G2-type SRMPC scheme for synchronous manipulation of two redundant robot arms, IEEE Transactions on Cybernetics, vol. 45, no. 2, pp. 153–164, 2015.

[452] Z. Jin, G. Salligui, R. Cuzner, L. Meng, J.C. Vasquez and J.M. Guerrero, Next generation shopboard DC power systems, IEEE Electrification Magazine, pp. 45–57, 2016.

[453] Z. Jin, L. Meng, J.M. Guerrero and R. Han, Hierarchical and control design for a shipboard power system with DC distribution and energy storage aboard future more-electric ships, IEEE Transactions on Industrial Informatics, vol. 14, no. 2, pp. 709–720, 2018.

[454] J. Joy and S. Ushakumari, Performance comparison of canonical switching cell with SPWM and SVPWM fed sensorless PMBLDC motor drive under conventional and fuzzy logic controllers, Journal of the Franklin Institute, Elsevier, vol. 34, pp. 5996–6032, 2011.

[455] B. Jujnovich and D. Cebon, Comparative performance of semi-trailer steering systems, 7th International Symposium on Heavy Vehicle Weights and Dimensions, Delft, The Netherlands, June 2002.

[456] F. Just, O. Ozen, F. Bosch, *et al.*, Exoskeleton transparency: Feed-forward compensation vs. disturbance observer, Automatisierungstechnik, De Gruyter, vol. 66, no. 12, pp. 1014–1026, 2018.

[457] C. Kaddissi, J.P. Kenné and M. Saad, Indirect adaptive control of an electrohydraulic servo system based on nonlinear backstepping. IEEE Transactions on Mechatronics, vol. 16, no. 6, pp. 1171–1177, 2011.

[458] J. Kalmari, J. Backman and A. Visala, Coordinated motion of an hydraulic forestry crane and vehicle using nonlinear model predictive control, Computers and Electronics in Agriculture, Elsevier, vol. 133, pp. 119–127, 2017.

[459] M.B. Kang and J.H. Wa, Adaptive control of 5-DOF upper limb exoskeleton with improved safety, ISA Transactions, Elsevier vol. 52, pp. 844–852, 2013.

[460] N. Kant and R. Mukherjee, Implicit dynamics and control of the inertia-wheel pendulum, IEEE Robotics and Automation Letters, vol. 3, no. 4, pp. 3208–3215, 2018.

[461] N. Kant, R. Mukherjee and H.K. Khalil, Swing-up of the inertia wheel pendulum using impulsive torques, IEEE CDC 2017, IEEE 56th Annual Conference on Decision and Control, Melbourne, Australia, December 2017.

[462] A. Karami, H. Sadeghian, M. Keshmiri and G. Oriolo, Force orientation and position control in redundant manipulators in prioritized scheme with null space compliance, Control Engineering Practice, Elsevier, vol. 85, pp. 23–33, 2019.

[463] M.A. Kardan, M.H. Asemani, A. Khayutian, *et al.*, Improved stabilization of nonlinear DC microgrids: Cubature Kalman Filter approach, IEEE Transactions on Industry Applications, pp. 1–9, 2019.

[464] M. Karpenko and N. Sepehri, Hardware-in-the-loop simulator for research on fault tolerant control of electrohydraulic actuators in a flight control application, Mechatronics, Elsevier, vol. 19, pp. 1067–1077, 2009.

[465] J. Karttunen, S. Kallio, P. Peltaniemi and P. Silventainen, Current harmonic compensator in dual three-phase PMSM using a disturbance observer, IEEE Transactions on Industrial Electronics, vol. 63, no. 1, pp. 583–594, 2016.

[466] R. Karoumi, Modelling of cable-stayed bridges for analysis of traffic induced vibrations, IMAC XVIII Conference on Structural Dynamics, San Antonio, Texas, USA, February 2000.

[467] A. Kassem, H. Haddad and C. Albitar, Comparison between different methods of control of ball and plate system with 6DOF Stewart Platform, 1st IFAC Conference on Modelling, Identification and Control of Nonlinear Systems MICNON 2015, St. Petersburg, Russia, June 2015.

[468] S. Kayastha, L. Shi, J. Katapitiya and G. Pearce, Nonlinear model predictive control of a planar three-link space manipulator, IEEE ASCC 2017, 11th Asian Control Conference, Gold Coast, Australia, December 2017.

[469] S. Kawamoto, S. Morimoto and S. Tanaka, Nonlinear equilibrium point control for nonlinear benchmark problem (RTAC model), 5th IFAC International Conference on Nonlinear Control Systems, St. Petersburg, Russia, July 2001.

[470] F.P. Kelly and R.J. Williams, Fluid model for a network operating under a fair bandwidth-sharing policy, The Annals of Applied Probability, vol. 14, no. 3, pp. 1055–1083, 2004.

[471] A. Keymasi-Khalaji, M. Rahimi Bidgali and S.A. Moosavian, Non-model-based control for a wheeled mobile robot towing two trailers, Proceedings of the Institute of Mechanical Engineers – Part K: Journal of Multi-body Dynamics, Springer, vol. 229, no.1, pp. 97–108, 2014.

[472] N. Khalid and A.Y. Memon, Output feedback stabilization of an inertia wheel pendulum using sliding mode control, UKACC 10th International Conference on Control, Loughborough, UK, July 2014.

[473] A. Khalifa, M. Fanni and T. Namerikawa, MPC and DOb-based robust optimal control of a new quadrotor manipulation system. ECC 2016, European Control Conference 2016, Aalborg, Denmark, June 2016.

[474] H.K. Khalil, Nonlinear Systems, 2nd Edition, Prentice Hall, 1996.

[475] K.R. Khan and M.D. Sueaz Miah, Fault-tolerant BLDC motor-driven pump for fluids with unknown specific gravity: An experimental approach, IEEE Access, vol. 8, pp. 30160–30173, 2020.

[476] M.H. Khooban, M.H. Ghisornejad, H. Farizadeh, A. Masoudian and J. Dragisevic, A new nonlinear hybrid control approach for DC-DC converters in zero-emission ferry ships, IEEE Transactions on Power Electronics, vol. 35, no. 6, pp. 5832–5841, 2020.

[477] M. Khanbaghi, R.P. Mulhame and M. Perrier, Optimal white water and broke recirculation policies in paper mills via jump linear quadratic control, IEEE Transactions on Control Systems Technology, vol. 10, pp. 578–588, 2002.

[478] P. Kshirsagar and R. Krishnan, High-efficiency current excitation strategy for variable-speed nonsinusoidal back-EMF PMSM machines, IEEE Transactions on Industry Applications, vol. 48, no. 6, pp. 1975–1890, 2012.

[479] D. Kim and Y. Park, Tracking control in xy-plane of an offshore boom crane, Journal of Vibration and Control, Sage Publications, vol. 23, no. 3, pp. 469–485, 2015.

[480] W. Kim, D. Won, D. Shin and C. Chung, Output feedback nonlinear control for electro-hydraulic systems, Mechatronics, vol. 22, pp. 766–777, 2012.

[481] W. Kim, D. Shin, D. Won and C. Chung, Disturbance-observer-based position tracking controller in the presence of biased sinusoidal disturbance for electrohydraulic actuators, IEEE Transactions on Control Systems Technology, vol. 11, no. 6, pp. 2290–2298, 2013.

[482] J. Kim, W. Choi and J. Lee, Discrete-time delay control for hydraulic excavator motion control with terminal sliding-mode control, Mechatronics, Elsevier, vol. 60, pp. 15–25, 2019.

[483] C. Kim, J.H. Chang and D. Hong, Coordination control of an active deburring tool, Robotics and Computer Integrated Manufacturing, Elsevier, vol. 24, pp. 462–471, 2008.

[484] Y. Kim, Two-time scale control of flexible-joint robots with an improved slow model, IEEE Transactions on Industrial Electronics, vol. 65, no. 4, pp. 3317–3325, 2018.

[485] B.S. Kim and J.B. Song, Design and control of a variable stiffness actuator, based on adjustable moment arm, IEEE Transactions on Robotics, vol. 28, no. 5, pp. 1145–1151, 2012.

[486] S.K. Kim, Robust adaptive speed regulator with self-tuning law for surface-mounted permanent magnet synchronous motor, Control Engineering Practice, Elsevier, vol. 61, pp. 55–71, 2017.

[487] D.E. Kim and D.C. Lee, Feedback linearization control of three-phase UPS inverter systems, IEEE Transactions on Industrial Electronics, vol. 57, no. 3, pp. 963–968, 2010.

[488] H.W. Kim, M.J. Youn, K.Y. Cho and H.S. Kim, Nonlinear estimation and cooperation of PWM VSI for PMSM under resistance and flux linkage uncertainty, IEEE Transactions on Control Systems Technology, vol. 14, no. 4, pp. 589–602, 2006.

[489] D.E. Kim, and D.C. Lee, Feedback linearization control of three-phase UPS inverter systems, IEEE Transactions on Industrial Electronics, vol. 57, no. 3, pp. 963–968, 2010.

[490] S. Kimura, M. Hamaguchi and T. Taniguchi, Damping control of liquid container by a carrier with dual swing type active vibration reducer, IEEE SICE 2002, IEEE 2002 41st Annual SICE Conference, Osaka, Japan, April 2003.

[491] R. Klein, N.A. Chaturvedi, J. Christensen, J. Ahmed, R. Findeisen and A. Kojic, Electrochemical model-based observer design for a lithium-ion

battery, IEEE Transactions on Control Systems Technology, vol. 21, no. 2, pp. 289–301, 2013.
[492] R. Klein, N.A. Chaturvedi, J. Christensen, J. Ahmed, R. Findeisen and A. Kojic, State estimation of a reduced electrochemical model of a lithium-ion battery, 2010 American Control Conference, Baltimore, MD, USA, June 2010.
[493] M. Kocemarek, W. Domski and A. Mazur, Position-force control of mobile manipulator – Nonadaptive and adaptive case, Archives of Control Sciences, De Gruyter, vol. 27, no. 4, pp. 487–503, 2017.
[494] B.H. Kohn and S.J. Dyke, Structural health monitoring for flexible bridge structures using correlation and sensitivity of model data, Computers and Structures, Elsevier, vol. 85, pp. 117–130, 2007.
[495] J. Koivumaki and J. Mattila, Stability-generated force-sensorless contact force-motion control of heavy duty hydraulic manipulators, IEEE Transactions on Robotixs, vol. 31, no. 4, pp. 916–936, 2015.
[496] J. Koivumaki, W.H. Zhu and J. Mattila, Energy-efficient and high-precision control of hydraulic robots, Control Engineering Practice, Elsevier, vol. 85, pp. 176–193, 2019.
[497] J. Koivumaki and J. Mattila, High performance nonlinear motion/force controller design for redundant hydraulic construction crane automation, Automation in Construction, Elsevier, vol. 51, pp. 59–97, 2015.
[498] K. Kolano, Improved sensor control method for BLDC motors, IEEE Access, vol. 7, pp. 186159–186177, 2019.
[499] V.K. Komaraju and M.S. Chiu, An integrated approach for C-control of antisolvent crystallization process, 8th IFAC Symposium on Advanced Control of Chemical Processes, Singapore, July 2012.
[500] T. Konrad and D. Abei, A flatness-based control strategy for multirotors in industrial applications, 12th IFAC Symposium on Robot Control SYROCO 2018 Budapest, Hungary, 27–30 August 2018.
[501] Q.J. Kong, Z. Li, Y. Chen and Y. Liu, An approach to urban traffic state estimation by fusing multi-source information, IEEE Transactions on Intelligent Transportation Systems, vol. 10, no. 3, pp. 499–511, 2009.
[502] M.H. Koraye and S.R. Nekao, The SDRE control of mobile-base cooperative manipulators: Collision-free path planning and moving obstacle avoidance, Robotics and Autonomous Systems, Elsevier, vol.88, pp. 86–105, 2016.
[503] A. Kröner, Adaptive finite element methods for optimal control second order hyberbolic equations, Computational methods in applied Mathematics, vol. 11, no. 2, pp. 214–240, 2011.
[504] M. Krstic, Adaptive control of an anti-stable wave PDE, Dynamics of Continuous, Discrete and Impulsive Systems – Series A: Mathematical Analysis, vol. 17, pp. 853–882, 2010.
[505] M. Krstic, B.Z. Guo. A. Balogh and A. Smyshlyaev, Output-feedback stabilization of an unstable wave equation, Automatica, Elsevier, vol. 44, pp. 63–74, 2008.

[506] K. Kural, P. Hatzidimitris, N. van de Wouw, I. Besselink and H. Nijmeijer, Active trailer steering control for high-capacity vehicle combinations, IEEE Transactions on Intelligent Vehicles, vol. 2, no. 4, pp. 251–265, 2017.

[507] S. Kurode, S.K. Spurgeon, B. Badyopadhyay and P.S. Gandhi, Sliding-mode control for slosh-free motion using a nonlinear sliding surface, IEEE Transactions on Mechatronics, vol. 18, no. 2, pp. 714–724, 2013.

[508] S. Kurode, S. Spurgeon, B. Badyopadhyay and P.S. Gandhi, Sliding-mode control for slosh-free motion using nonlinear sliding surface, ECC 2009, European Control Conference 2009, Budapest, Hungary, August 2009.

[509] N. Kumar, V. Panwar, N. Sukovanam, S.P. Sharma and J.H. Borm, Neural network-based nonlinear tracking control of kinematically redundant robot manipulator, Mathematics and Computer Modelling, Elsevier, vol. 57, pp. 1889–1901, 2001.

[510] R. Kumar and B. Singh, BLDC motor driver solar PV array-fed water pumping system employing zero converter, IEEE Transactions on Industry Applications, vol. 52, no. 3, pp. 2315–2322, 2016.

[511] N.K. Kwon, B.Y. Park and P. Park, LPV controller design with multiple parameters for the nonlinear RTAC system, IEEE ASCC 2013, IEEE 9th Asian Control Conference, Istanbul, Turkey, June 2013.

[512] E. Lafranche, L. Cilleruelo, M. Ryekebusch and P.P. Krawczak, A novel adaptive process control for injection moulding of BMC and CIC polyester compounds, Journal of Composites and Advanced Materials, Elsevier, vol. 14, no. 5, pp. 1–11, 2006.

[513] S. Lagrouche, F. Shakil Ahmed and A. Mehmood, Pressure and direction observer-based backstepping control for a VGT pneumatic actuator, IEEE Transactions on Control Systems Technology, vol. 72, no. 2, 2014.

[514] S. Lagrouche, M. Smaoui, F. Plestan and X. Brun, Higher-order sliding-mode control based on optimal approach of an electropneumatic actuator, International Journal of Control, Taylor and Francis, vol. 77, no. 2, pp. 119–131, 2006.

[515] W. Lan and B.M. Chen, Explicit constructions of global stabilization control laws for a class of nonminimum phase nonlinear systems, IEEE 2006 9th International Conference on Control, Automation, Robotics and Vision, Singapore, December 2006.

[516] F.A. Lara-Molina, D. Dumur and E.H. Koroishi, Uncertainty analysis of a 2-DOF planar parallel robot by means of fuzzy dynamic approach, IEEE LARS/SBR 2016, IEEE 2016 8th Latin American Robotic Symposium and 4th Brazilian Robotics Symposium Recife, Brazil, October 2016.

[517] F.A. Lara-Molina, E.H. Koroishi and D. Dumur, Design criteria of a 2-DOF planar parallel manipulator with flexible joints, IEEE Induscon 2016, Proceedings of the 12th IEEE International Conference on Industry Applications, Curitiba, Brazil, November 2014.

[518] B. Laroche, P. Martin, and N. Petit, Commande par platitude: Equations différentielles ordinaires et aux derivées partielles, Ecole Nationale Supérieure des Techniques Avancées, Paris, 2007.

[519] B. Laroche, P. Martin and P. Rouchon, Motion planning of the heat equation, International Journal of Robust and Nonlinear Control, J. Wiley, vol. 40, no. 8, pp. 629–643, 2000.

[520] F. Lamiraux, J.P. Laumond, C. van Geem, G. Boutannet and G. Raust, Trailer-truck trajectory optimization: The transportation of components for the Airbus A380, IEEE Robotics and Automation Magazine, pp. 14–21, March 2005.

[521] N. Lashkari, M. Biglarbegian and S.Y. Yang, Development of a new robust controller with velocity estimator for docked mobile robots: Theory and applications, IEEE/ASME Transactions on Mechatronics, vol. 23, no. 3, pp. 1287–1298, 2017.

[522] N. Lashkari, M. Biglarbegian and S.Y. Yang, Backstepping tracking control design for a tractor robot pulling multiple trailers, IEEE ACC 2018, IEEE 2018 American Control Conference, Wisconsin, Milwaukee USA, June 2018.

[523] O.M. Le, M.T. Pham, M. Tavakoli, R. Moreau, J.P. Simon and T. Redurce, Bilateral control of nonlinear pneumatic teleoperation system with solenoid valves, IEEE Transactions on Control Systems Technology, vol. 21, no. 4, pp. 1463–1471, 2013.

[524] T.D. Le, H.J. Kung and Q.V. Doan, A method for optimal kinematic design of five-bar planar parallel manipulators, IEEE 2013 International Conference on Control, Automation and Information Systems, Nha Trang, Vietnam, November 2013.

[525] T.D. Le, H.J. Kang an Y.S. Soh, Chattering-free neuro-sliding mode control of 2-DOF planar parallel manipulators, International Journal of Advanced Robotic Systems, Sage Publications, vol. 10, pp. 1–15, 2013.

[526] T.D. Le, H. Kung, Y.S. Soh and Y.S. Ro, A online self-gain tuning method using neural networks for nonlinear PD computed torque controller of a 2-DOF parallel manipulator, Neurocomputing, Elsevier, vol. 166, pp. 53–61, 2013.

[527] P.A. Leban, J.D. Gonzalez, G.G. Parker and W. Zhao, Inverse kinematic control of a dual crane system experiencing base motion, IEEE Transactions on Control Systems Technology, vol. 23, no. 1, pp. 331–340, 2015.

[528] S.J. Lee amd C.L. Hou, A neural fuzzy system for congestion control in ATM networks, IEEE Transactions on Systems, Man and Cybernetics – Part B: Cybernetics, vol. 30, no. 1, pp. 2–9, 2000.

[529] C.H. Lee and S.K. Chung, Experimental implementation of nonlinear TORA system and adaptive backstepping controller design, Neural Computing and Applications, vol. 22, no. 4, pp. 785–800, 2012.

[530] W. Lee and W.K. Chung, Disturbance observer-based compliance control of electro-hydraulic actuators with backdrivability, IEEE Robotics and Automation Letters, vol. 4, no. 2, pp. 1722–1730, 2019.

[531] K.W. Lee and S.N. Singh, Attractive manifold-based adaptive solar attitude control of satellites in elliptic orbits, Acta Astronautica, Elsevier, vol. 68, pp. 185–196, 2011.

[532] H.D. Lee, S.J. Kang, and S.K. Sul, Efficiently optimized direct torque control of synchronous reluctance motor using feedback linearization, IEEE Transactions on Industrial Electronics, vol. 46, no. 1, pp. 192–198, 1999.

[533] K.T. Lee, M.J. Dai and C.C. Chuang, Temperature compensated model for lithium-ion batteries with extended Kalman Filter state-of-charge estimation for an implantable charger, IEEE Transactions on Industrial Electronics, vol. 65, no. 1, pp. 589–596, 2018.

[534] N. Leiten, A. Priel and O. Yaniv, Robust spacecraft attitude control under multi-tank sloshing disturbances during orbit correction maneuvers, IEEE MED 2019, IEEE 2019 27th Mediterranean Conference on Control and Automation, Akko, Israel, July 2019.

[535] E. Lenz, D.J. Pagano and V. Stramosk, Nonlinear control applied to a DC-DC power converter and the load sharing problem in a DC microgrid, 19th IFAC World Congress, Cape Town, Africa, August 2014.

[536] E. Lenz and D.J. Pagano, Nonlinear control of bidirectional power converter in a DC microgrid, 9th IFAC Symposium on Nonlinear Control Systems, Toulouse, France, September 2013.

[537] K.K. Leung, W.A. Massey and W. White, Traffic models for wireless communication networks, IEEE Journal of Selected Areas in Communications, vol. 12, no. 8, pp. 1353–1364, 1994.

[538] J. Levine, Analysis and Control of Nonlinear Systems: A Flatness-Based Approach, Springer, 2009.

[539] J. Lévine, On necessary and sufficient conditions for differential flatness, Applicable Algebra in Engineering, Communications and Computing, Springer, vol. 22, no. 1, pp. 47–90, 2011.

[540] C.A. Lightcap and S.A. Banks, An extended Kalman Filter for real-time estimation and control of a rigid-link flexible-joint manipulator, IEEE Transactions on Control Systems Technology, vol. 18, no. 1, pp. 91–101, 2010.

[541] S.O. Lindert, R. Reindl and K. Schlacher, Identification and control of an injection moulding machine, 19th IFAC World Congress, Proceedings of the 19th World Congress of the International Federation of Automatic Control, Cape Town, South Africa, August 2014.

[542] A.S. Lofimejani, M.T. Masouleh and A. Kalhor, Trajectory tracking control of a pneumatically actuated 6-DOF Gough-Stewart parallel robot using backstepping sliding-mode controller and geometry-based quasi-forward kinematic model, Robotics and Computer Integrated Manufacturing, Elsevier, vol. 54, pp. 96–114, 2016.

[543] M. Li, W. Guo, R. Lin, C. Wu and L. Moro, An efficient motion generation method for redundant humanoid robot arm based on the intrinsic principles of human arm motion, International Journal of Humanoid Robotics, World Scientific, vol. 15, no. 6, pp. 1850026, 2018.

[544] Z. Li and R. Yuan, Autonomous inspection robot for power transmission lines maintenance while operating on the overhead ground wires, Journal of Advanced Robotic Systems, Sage Publications, vol. 7, no. 4, pp. 107–113, 2010.

[545] Z. Li, Z. Bai and S. Xu, Design and operating mode analysis of hybrid actuation system based on EHA/SHA, IET Journal of Engineering, vol. 018, no. 13, pp. 385–391, 2018.

[546] S. Li and W. Wang, Adaptive robust H_∞ control for double support balance systems, Information Sciences, Elsevier, 2019.

[547] L. Li, L. Xin, X. Luo and Z. Wang, Compliance control using hydraulic heavy-duty manipulator, IEEE Transactions on Industrial Informatics, vol. 15, no. 2, pp. 1193–1202, 2019.

[548] X. Li, Y. Pan, G. Chen and H. Yu, Continuous tracking control for a compliant actuator with two-stage stiffness, IEEE Transactions on Automation Science and Engineering, vol. 15, no. 1, pp. 57–66, 2018.

[549] M. Li, C. Guo and H. Yu, Filtered extended state observer-based line-of-sight guidance for path following for unmanned surface vehicles with unknown dynamics and disturbances, IEEE Access, vol. 7, pp. 178401–178412, 2019.

[550] Z. Li, L. Zhang, H. Liu, Z. Zuo and C. Liu, Nonlinear robust control of tail-sitter aircrafts in flight-mode transitions, Aerospace Science and Technology, Elsevier, vol. 81, pp. 348–361, 2018.

[551] X. Li and G. Li, Novel fuzzy approximation control scheme for flexible air-breathing hypersonic vehicles with non-affine dynamics and amplitude and rate constraints, IEEE Access, vol. 7, pp. 73602-3616, 2019.

[552] S. Li, C. Xia and X. Zhou, Disturbance rejection control method for permanent magnet synchronous motor speed-regulation system, Mechatronics, Elsevier, vol. 22, pp. 706–714, 2012.

[553] Z. Li, X. Jing, B. Sun and J. Yu, Autonomous navigation of a tracked mobile robot with novel passive bio-inspired suspension, IEEE/ASME Transactions on Mechatronics, vol. 14, no. 5, pp. 528–538, 2009.

[554] Z. Li, Y. Xu, S. Feng, X. Zhang and X. Feng, Robust coordination of a hybrid AC/DC multi-energy ship microgrid with flexible voyage and thermal loads, IEEE Transactions on Smart Grids, vol. 11, no. 4, pp. 2782–2793, 2020.

[555] J. Li, F. Liu, Y. Chen, *et al.*, Resilience control of shipboard DC power systems, IEEE Transactions on Power Systems, vol. 37, no. 8, pp. 6675–6685, 2018.

[556] P. Li, T. Guo, P. Zhou, J. Yang and Y. Liu, Nonlinear coordinated control of parallel bidirectional power converters in AC/DC hybrid microgrid, Electric Power and Energy System, Elsevier, vol. 122, pp. 106208–106216, 2020.

[557] P. Li, T. Guo, X. Han, *et al.*, The optimal decentralized coordinated control method based on the H_∞ performance index for AC/DC hybrid microgrid, Electric Power and Energy Systems, Elsevier, vol. 125, pp. 106442–106451, 2021.

[558] P. Li, T. Guo, Y. Li, *et al.*, An adaptive coordinated optimal control method for parallel bidirectional converters in AC/DC hybrid microgrid, Electric Power and Energy Systems, Elsevier, vol. 126, pp. 106596–106605, 2021.

[559] Z. Li, Y. Xu, S. Feng, X. Zhang and X. Feng, Robust coordination of a hybrid AC/DC multi-energy ship microgrid with flexible voyage and thermal loads, IEEE Transactions on Smart Grids, vol. 11, no. 4, pp. 2782–2793, 2020.

[560] M. Li, P. Zhou, H. Wang and T. Choi, Nonlinear multi-objective MPC-based optimal operation of a high consistency refining system in papermaking, IEEE Transactions on Systems, Man and Cybernetics: Systems, 2019.

[561] G. Li and F. Janabi-Sharifi, Fuzzy looperless tension control for hot strip rolling, Fuzzy Sets and Systems, Elsevier, vol. 160, pp. 521–536, 2009.

[562] J. Li, Y. Zhang and H. Liu, Simulation of multiple effect evaporators for energy-saving studies in a chemical pulp mill, IEEE WCICA 2014, IEEE 11th World Congress on Intelligent Control and Automation, Shenyang, China, July 2014.

[563] M. Li, P. Zhou, H. Wang and T. Choi, Geometric analysis-based double closed-loop iterative learning control of output PDF shaping of fiber length distribution in refining process, IEEE Transactions on Industrial Electronics, vol. 66, no. 9, pp. 7229–7238, 2019.

[564] Q. Li, Y.L. Xu, D.J. Wu and Z.W. Chen, Computer-aided nonlinear vehicle-bridge interaction analysis, Journal of Vibration and Control, Elsevier, vol. 10, no. 12, pp. 1791–1816, 2010.

[565] Z. Li, C.Y. Su, L. Wang, Z. Chen and T. Choi, Nonlinear disturbance observer-based control design for a robotic exoskeleton incorporating fuzzy approximate, IEEE Transactions on Industrial Electronics, vol. 62, no. 9, pp. 5763–5775, 2015.

[566] Z. Li, P.Y. Tan, S.S. Ge, M. Adams and W.S. Wijesoma, Robust adaptive control of cooperative mobile manipulators with relative motion, IEEE Transactions on Systems, Man and Cybernetics – Part B: Cybernetics, vol. 37, no.1, pp. 103–116, 2009.

[567] Z. Li, P.Y. Tan, S.S. Ge, M. Adams and W.S. Wijesoma, Adaptive robust output-feedback motion/force control of electrically driven nonholonomic mobile manipulators, IEEE Transactions on Control Systems Technology, vol. 16, no. 6, pp. 1308–1315, 2008.

[568] Z. Li, P.Y. Tan, S.S. Ge, M. Adams and W.S. Wijesoma, Robust adaptive control of uncertain force/motion constrained nonholonomic mobile manipulators, Automatica, Elsevier, vol. 44, pp. 776–784, 2008.

[569] Z. Li, S.S. Ge and A. Ming, Adaptive robust motion/force control of holonomic-constrained nonholonomic mobile manipulators, IEEE Transactions on Systems, Man and Cybernetics: Part B- Cybernetics, vol. 37, no. 3, pp. 607–616, 2016.

[570] Z. Li, Y. Yang and J. Li, Adaptive motion/force control of mobile underactuated manipulators with dynamics uncertainty by dynamic coupling and output feedback, IEEE Transactions on Control Systems Technology, vol. 18, no. 5, pp. 1068–1079, 2010.

[571] W. Li, L. Liang, W. Liu and W. Xu, State of charge estimation of lithium-ion batteries using a discrete-time nonlinear observer, IEEE Transactions on Industrial Electronics, vol. 64, no. 11, pp. 8557–8565, 2017.

[572] L. Li, L. Xie, X. Luo and Z. Wang, Compliance control using hydraulic heavy-duty manipulator, IEEE Transactions on Industrial Informatics, vol. 15, no. 2, pp. 1193–1204, 2019.

[573] X. Liang, Y. Wan, C. Zhang, Y. Kou, Q. Xin and W. Yi, Robust position control of hydraulic manipulators using time-delay estimation and non-singular fast terminal sliding-mode, Proceedings Of the IMeche Part I: Journal of Systems and Control Engineering, Sage Publications, vol. 232, no. 1, pp. 50–61, 2018.

[574] X. Liang, P. Zhang, Y. Feng, H. Lin and W. He, Nonlinear control for aerial transportation systems with double-pendulum swing effects, IEEE Transactions on Industrial Electronics, pp. 1–10, 2021.

[575] L.H. Liang, L.Y. Wang and J.F. Wang, Energy-saving nonlinear friction compensation method for direct drive volume control flange-type rotary-vane steering gear, IEEE Access, vol. 7, pp. 148351–148362, 2019.

[576] X. Liang, X. Qu, Y. Hou, Y. Li and R. Zhang, Distributed coordinated tracking control of multiple unmanned surface vehicles under complex marine environments, Ocean Engineering, Elsevier, vol. 205, pp. 107328–107336, 2020.

[577] X. Liang, X. Qu, N. Wang, Y. Li and R. Zhang, Swarm control with collision avoidance for multiple underactuated surface vehicles, Ocean Engineering, Elsevier, vol. 191, pp. 106616–106625, 2019.

[578] X. Liang, G. Wang, C. Hu and C. Dong, Observer-based H-infinity fault tolerant attitude control for satellite with actuator sensor faults, Journal of Aerospace Science and Technology, Elsevier, vol. 95, 2019.

[579] X. Liang, Y. Wan, C. Zhang, Y. Kim, O. Xin and W. Yi, Robust position control of hydraulic manipulators using time-delay estimation and non-singular fast terminal sliding-mode, Proceedings of the IMehE, Journal of Systems and Control Engineering, vol. 232, no. 1, pp. 50–61, 2018.

[580] Z. Liang and W. Wu, Variational inequalities in stock-loan models, Journal of Optimization in Engineering, Springer, vol. 13, pp. 459–470, 2012.

[581] Z. Liang, W. Wu and S. Jiang, Stock loan with automatic termination clause, cap and growth, Computers and Mathematics with Applications, Elsevier, vol. 60, pp. 3160–3176, 2010.

[582] C. Liang and T. Hsiao, Admittance control of power exoskeletons based on joint torque estimation, IEEE Access vol. 8, pp. 94404–94414, 2020.

[583] S.C. Liao, Y.W. Tu, K.S. Yang, and M.T. Ho, Balance and tracking control of ball and plate systems, Proceedings of the 17th World Congress The International Federation of Automatic Control Seoul, Korea, July 6–11, 2008.

[584] J. Lie, D. Shin, H. Ryu, D. Lee and D.H. Shin, Fault tolerant adaptive control using time-delay control scheme under motor faults of octocopter, IEEE CSC 2018, Proceedings of the 7th IEEE Conference on Systems and Control, Valencia, Spain, October 2018.

[585] C. Lie, K.T. Chau and J.Z. Jiang, A permanent-magnet hybrid brushless integrated starter-generator for hybrid electric vehicles, IEEE Transactions on Industrial Electronics, vol. 57, no. 12, pp. 4255–4265, 2010.

[586] S. Lin and A. Goldenberg, Robust damping control of mobile manipulators, IEEE Transactions on Systems, Man and Cybernetics: Part B – Cybernetics, vol. 32, no. 1, pp. 126–132, 2002.

[587] F.J. Lin, M.S. Huang, S.G. Chen and C.W. Hau, Intelligent maximum torque per ampere tracking control of synchronous reluctance motor using recurrent legendre fuzzy neural network, IEEE Transactions on Power Electronics, vol. 34, no. 12, pp. 12080–12094, 2019.

[588] C.Y. Lin and Z.H. Yang, TRBR: Flight body posture compensation for transverse ricochetal brachiation robot, Mechatronics, Elsevier, vol. 65, pp. 102307, 2020.

[589] F.J. Lin, Y.C. Hung and M.T. Tsui, Fault tolerant control for six-phase PMSM drive system via intelligent complementary sliding-mode control using TSK-FNN-AMF, IEEE Transactions on Industrial Electronics, vol. 60, no. 112. pp. 5747–5762, 2013.

[590] F.J. Lin, S.G. Chen and I.F. Sun, Adaptive-backstepping control of a six-phase PMSM using functional link radial basis function network uncertainty observer, Asian Journal of Control, John Wiley, vol. 19, no. 6, pp. 2255–2269, 2017.

[591] F.J. Lin, Y.C. Huang, J.C. Hwang and M.T. Tsai, Fault-tolerant control of a six-phase motor drive system using a Takagi-Sugeno-Kung type fuzzy neural network with asymmetric membership function, IEEE Transactions on Power Electronics, vol. 28, no. 7, pp. 3557–3572, 2013.

[592] W. Lin, K. Rattanamongkhonkun and R. Pongvuthithum, L_gV -Type adaptive controllers for uncertain non-affine systems and application to a DC-microgrid with PV and battery, IEEE Transactions on Automatic Control, vol. 64, no. 5, pp. 2182–2189, 2019.

[593] J. Linares-Flores, C. Garcia-Rodriquez, H. Sira-Ramirez and O.D. Ramirez-Cardenas, Robust backstepping tracking controller for low-speed PMSM positioning system: Design, analysis and implementation, IEEE Transactions on Industrial Informatics, vol. 11, no. 5, pp. 1130–1141, 2015.

[594] V. Lippiello and F. Ruggiero, Cartesian impedance control of a UAV with a robotic arm, IFAC SYROCO 2012, 10th IFAC Symposium on Robot Control, Dubrovnik, Croatia, September 2012.

[595] V. Lippiello and F. Ruggiero, Cartesian impedance control of a UAV with a robotic arm, 10th IFAC Symposium on Robot Control, Dubrovnik, Croatia, September 2012.

[596] D.A. Lizarraga, P. Morin and C. Samson, Exponential stabilization of certain configurations of the general N-trailer system, INRIA Research Report No. 3412, April 1998.

[597] W. Liu, R. Guon, L. Yuan, Z. Li and M. Tong, Position feedback dynamic surface control for pneumatic actuator position servo system, Systems Science and Control Engineering, Taylor and Francis, vol. 6, no. 1, pp. 388–397, 2018.

[598] D. Liu and G. Yan, Motion control of an underactuated brachiation robot based on Lyapunov direct methods, IOP Conference Series: Materials Science and Engineering, vol. 768, no. 042040, 2020.

[599] D. Liu, Y. Tian and H. Duan, Ball and plate control system based on sliding-mode control and uncertain items observe compensation, 2009 IEEE International Conference on Intelligent Computing and Intelligent Systems, Shanghai, China, November 2009.

[600] Z.W. Liu, H. Hou and Y.W. Wang, Formation containment control of multiple underactuated vessels with sampling communication via hierarchical sliding-mode approach, ISA Transactions, Elsevier, 2020.

[601] S. Liu, X. Guo and L. Zhanng, Robust adaptive backstepping sliding-mode control for six-phase permanent magnet synchronous motor using recurrent wavelet fuzzy neural network, IEEE Access, vol. 5, pp. 14502–14515, 2017.

[602] Y. Liu and G. Liu, Track-stair interaction analysis and online tipover prediction for a self-reconfigurable tracked mobile robot climbing stairs, IEEE / ASME Transactions on Mechatronics, vol. 14, no. 9, pp. 528–538, 2009.

[603] Y. Liu and G. Liu, Modeling of tracked mobile manipulators with consideration of track-terrain and vehicle-manipulator interactions, Robotics and Autonomous Systems, Elsevier, vol. 57, pp. 1065–1074, 2009.

[604] Y. Liu and G. Liu, Interaction analysis and online tip-over avoidance for a reconfigurable tracked mobile modular manipulator negotiating slopes, IEEE/ASME Transactions on Mechatronics, vol. 15, no. 4, pp. 623–635, 2010.

[605] G. Liu, X. Chen, X. Zhou and S. Zhang, Sensorless commutation deviation correction of brushless DC motor with three-phase asymmetry back-EMF, IEEE Transactions on Industrial Electronics, vol. 67, no. 7, pp. 6158–6167, 2020.

[606] Y. Liu, W. Xu, J. Zhu and F. Blaabjerg, Sensorless control of standalone brushless doubly-fed induction generator feeding unbalanced loads in a ship shaft power generation system, IEEE Transactions on Industrial Electronics, vol. 66, no. 1, pp. 739–749, 2019.

[607] W.J. Liu, Boundary stabilization of an unstable heat equation, SIAM Journal on Control and Optimization, vol. 42, pp. 1033–1043, 2003.

[608] B. Liu, System deployment and decentralized control of islanded AC microgrid without communication facility, IEEE Journal of Modern Power Systems and Clean Energy, vol. 7, no. 4, pp. 913–922, 2019.

[609] C. Liu, T. Chai, H. Wang, S. Zhou, Data-based hybrid tension estimation and faults diagnosis of cold rolling continuous annealing processes, IEEE Transactions on Neural Networks, vol. 22, no. 11, pp. 2284–2295, 2011.

[610] G. Liu and Y. Xu, Capped stock loans, Computers and Mathematics and Application, Elsevier, vol. 9, pp. 3548–3558, 2010.

[611] X. Liu, K. Zhang, S. Li, S. Fei and H. Wei, Optimal control of switching times in switched stochastic systems, Asian Journal of Control, J. Wiley, vol. 17, no. 5, 2016.

[612] T. Liu, S. Hao, D. Li, W.H. Chen and Q.G. Wang, Predictor-based disturbance rejection control for sampled systems with input delay, IEEE Transactions on Control Systems Technology, vol. 27, no. 2, pp. 772–780, 2019.

[613] Y. Liu and G. Liu, Modelling of tracked mobile manipulators with consideration of truck-terrain and vehicle-manipulator interactions, Robotics and Autonomous Systems, Elsevier, vol. 57, pp. 1065–1074, 2009.

[614] J. Liu, S. Mendoza and H.K. Fathy, Total least squares state-of-charge estimation for lithium-ion batteries: An efficient moving horizon estimation approach, Proceedings of the 20th IFAC World Congress, Toulouse, France, July 2017.

[615] D. Liu, Q. Wei, D. Wang and X. Yang, Hongliang Li, Adaptive Dynamic Programming with Applications in Optimal Control, Springer, 2017.

[616] O. Ljungquist and D. Aschill, A predictive path following controller for multi-steered articulated vehicles, IFAC 21st World Congress, Berlin, Germany, July 2020.

[617] J. Löber, Optimal Trajectory Tracking of Nonlinear Dynamical Systems, Springer, 2017.

[618] Y. Long and Y. Peng, Extended state observer-based nonlinear terminal sliding-mode control with feedforward with feedforward compensation for lower extremity exoskeleton, IEEE Access, vol. pp. 1–10, 2021.

[619] G. Lou, W. Gu, W. Sheng, X. Song and F. Gao, Distributed model predictive secondary voltage control of isolated microgrids with feedback linearization, IEEE Access, vol. 6, pp. 50169–50178, 2018.

[620] M. di Lovets, L. Boillereaux, G. Conte, G. Gialini and C.H. Moog, Experimentation of melting kinetics control in a convective food thawing process, IEEE Transactions on Control Systems Technology, vol. 13, no. 5, pp. 826–831, 2015.

[621] B. Lu, Y. Feng and N. Sun, Nonlinear control for underactuated multi-input cranes: Modeling theoretical design and hardware experiments, Control Engineering Practice, Elsevier, vol. 76, pp. 123–132, 2018.

[622] B. Lu, Y. Fang and N. Sun, Enhanced-coupling adaptive control for double-pendulum overhead cranes with payload hoisting and learning, Automatica, Elsevier, vol. 101, pp. 241–251, 2019.

[623] B. Lu, Y. Fang, N. Sun and X. Wang, Anti-swing control of offshore boom cranes with ship roll disturbances, IEEE Transactions on Control Systems Technology, vol. 26, no. 2, pp. 740–747, 2018.

[624] B. Lu, Y. Fang and N. Sun, Global stabilization of inertia wheel systems with a novel sliding-mode based strategy, IEEE International Workshop on Variable Structure Systems, Nanjing, China, June 2016.

[625] R. Lu, Z. Li, C.Y. Su and A. Xue, Development and learning control of a human limb with a rehabilitation exoskeleton, IEEE Transactions on Industrial Electronics, vol. 61, no. 7, pp. 3776–3785, 2014.

[626] Z. Lu, K. Sekiyama, T. Aoyama, Y. Hosegawa, T. Kobayashi and T. Fukuda, Energetically efficient ladder descent motion with internal stress and body

motion optimized for a multi-locomotion robot, IEEE Transactions on Industrial Electronics, vol. 62, no. 8, pp. 4972–4984, 2015.

[627] L. Lu and B. Yao, Energy-saving adaptive robust control of a hydraulic manipulator using five cartridge valves with an accumulator, IEEE Transactions on Industrial Electronics, vol. 61,no. 12, pp. 7046–7055, 2014.

[628] H. Lu, J. Li, R. Qu, D. Ye and Y. Lu, Fault-tolerant predictive control of six-phase PMSM drives based on pulse-width-modulation, IEEE Transactions on Industrial Electronics, vol. 86, no. 7, pp. 4992–5003, 2019.

[629] Q. Lu and Y. Ye, Design and analysis of tubular linear PM generator, IEEE Transactions on Magnetics, vol. 45, no. 10, pp. 4716–4719, 2009.

[630] W. Lu, Y. Wu, D. Zhu, D. Wang and M. Yu, A new short-time high-overload BLDC driving system based on electronic flywheel and time-division switching control, Mechatronics, Elsevier, vol. 69, pp. 102385–102396, 2020.

[631] H. Lu, X. Zhang, L. Quan and H. Zhang, Research on energy consumption of injection moulding machine driven by five different types of electro-hydraulic power units, Journal of Cleaner Production, Elsevier, vol. 242, pp. 118355–118365, 2020.

[632] X. Lu and E.R.M. Putri, Semi-analytic valuation of stock-loans with finite memory, Communications in Nonlinear Science and Numerical Simulation, Elsevier, vol. 27, pp. 206–215, 2015.

[633] L. Lublin and M. Athans, An experimental comparison of and designs for interferometer testbed, Lectures Notes in Control and Information Sciences: Feedback Control, Nonlinear Systems and Complexity, (Francis B. and Tannenbaum A., eds.), Springer, pp. 150–172, 1995.

[634] E. Lucet and A. Micaelli, Stabilization of a road-train of articulated vehicles, Robotics and Autonomous systems, Elsevier, vol. 114, pp. 106–123, 2019.

[635] D. Lumni, A. Santamariti-Navarro, R. Rossi, P. Rocco, L. Bascetta and T. Andrade-Cetto, Nonlinear model predictive control for aerial manipulation, IEEE ICUAS 2017, IEEE 2017 International Conference on Unmanned Aerial Systems, Miami, Florida, June 2017.

[636] P.A. Luppi, M.P. Garcia and M.S. Basualdo, Improvements on model predictive control for a pulp mill process, IFAC DYCOPS 2010, 9th International Symposium on Dynamics and Control of Process Systems, Leuven, Belgium, July 2010.

[637] L. Lyn, Z. Chen and B. Yao, Development of pump and valves combined hydraulic system for both high tracking precision and high energy efficiency, IEEE Transactions on Industrial Electronics, vol. 66, no. 9, pp. 7169–7178, 2019.

[638] S.E. Lyshevksi, Nonlinear modelling and robust control of synchronous reluctance motors, Energy Conversion and Management, vol. 43, pp. 523–536, 2002.

[639] Y. Luo and C. Liu, Model predictive control for a six-phase PMSM motor with a reduced-dimension cost function, IEEE Transactions on Industrial Electronics, vol. 67, no. 2, pp. 969–979, 2020.

[640] Y. Luo and C. Liu, Pre and post-fault tolerant operation of a six-phase PMSM motor using FCS-MPC without controller reconfiguration, IEEE Transactions on Vehicular Technology, vol. 68, no. 1, pp. 254–269, 2019.

[641] X. Lyu, J. Zhou, H. Gui, Z. Li, S. Shen and F. Zhang, Disturbance observer-based hovering control of quadrotor tail-sitter VTOL UAVs using H_{∞} synthesis, IEEE Robotics and Automation Letters, vol. 3, no. 4, pp. 2910–2917, 2018.

[642] G. Lv, M. Zhou and R.D. Gregg, A discussion of past and ongoing work on the design and control of highly back-drivable lower-limb exoskeleton, IEEE Control Systems Magazine, pp. 88–113, December 2018.

[643] J. Ma, J.G. Zhu and H. Schock, Adaptive control of a pneumatic valve actuator for an internal combustion engine, IEEE Transactions on Control Systems Technology, vol. 19, no. 4, pp. 730–743, 2013.

[644] Z. Ma, J. Hu, J. Feng and A. Liu, Diving adaptive position tracking control for underactuated vehicles, IEEE Access, vol. 7, pp. 24602–24611, 2019.

[645] Y. Maddahi, S. Liao, W.K. Fung and N. Sepehri, Position referenced force augmentation in teleoperated hydraulic manipulators operating under delayed and lossy networks: A pilot survey, Robotics and Autonomous Systems, Elsevier, vol. 83, pp. 231–242, 2016.

[646] T. Madani, B. Doachi and K. Djouani, Non-singular terminal sliding-mode controller: Application to an actuated exoskeleton, Mechatronics, Elsevier, vol. 53, pp. 136–145, 2016.

[647] S. Madanzadeh, A. Abedini, A. Radur and J.S. Roi, Application of quadratic linearization state feedback control with hysteresis, ISA Transactions, Elsevier, 2020.

[648] S.K. Maddugari, V.B. Borghate, S. Sabyasachi and R.R. Karasani, A linear generator-based wave power plant model using reliable multilevel inverters, IEEE Transactions on Industry Applications, vol. 55, no. 3, pp. 2964–2972, 2019.

[649] A. El Magri, F. Giri, A. Abouloifa and F.Z. Chaoui, Robust control of synchronous motor through AC/DC/AC converters, Control Engineering Practice, Elsevier, vol. 18, pp. 540–553, 2010.

[650] W. Ma, T. Li, X. Zhang, J. Fan, Y. Yan and J. Yang, The effect of the parameters of PI control for hydraulic looper control in hot rolling mill, IOP Conference Series: Materials Science and Engineering, vol. 381, 012063, 2018.

[651] T. Mai and Y. Wang, Adaptive force/motion control system based on recurrent fuzzy wavelet CMAC neural networks for condenser cleaning crawler-type mobile manipulator robot, IEEE Transactions on Control Systems Technology, vol. 22, no. 5, pp. 1973–1982, 2014.

[652] D. Mahindrakar, R. Banawar and M. Reyhanoglu, Controllability and point-to-point control of 3-DOF planar horizontal underactuated manipulators, International Journal of Control, Taylor and Francis, vol. 78, no. 1, pp. 1–18, 2005.

[653] M.A. Mahmud, H.R. Pota, M. Alden and M.J. Hossain, Partial feedback linearizing excitation controller for multi-machine power systems to

improve transient stability, IEEE Transactions on Power Systems, vol. 29, no. 2, pp. 561–571, 2014.

[654] A. Maidi and J.P. Corriou, Distributed control of nonlinear diffusion systems by input-output linearization, International Journal of Robust and Nonlinear Control, vol. 26, pp. 389–405, 2014.

[655] L. Majumder and C.S. Manohar, Nonlinear reduced models for beam damage detection using data on moving oscillator-beam interaction, Computers and Structures, Elsevier, vol. 82, pp. 301–314, 2014.

[656] C.D. Makavita, S.G. Jayasingh, H.D. Mguyen and D. Ranmurthugala, Experimental study of command governor adaptive control for unmanned underwater vehicles, IEEE Transactions on Control Systems Technology, vol. 27, no. 1, pp. 332–345, 2019.

[657] Y. Man, J. Li, M. Hong and Y. Han, Energy transition for the low-carbon pulp and paper industry in China, Renewable and Sustainable Energy Reviews, Elsevier, vol. 131, pp. 1–14, 2020.

[658] M.M. Mandani, M.H. Khooban, A. Masoudian and T. Dragisevic, Model predictive control of DC-DC converter to mitigate the effects of pulsed power loads in Naval DC microgrids, IEEE Transactions on Industrial Electronics, vol. 66, no. 7, pp. 5678–5685, 2017.

[659] S. Mane, M. Mejani, F. Kazi and N. Singh, Improving life-time of fuel cell in hybrid energy management system by Lure-Lyapunov-based control formulation, IEEE Transactions on Industrial Electronics, vol. 64, no. 8, pp. 6671–6678, 2017.

[660] A. Maqsood and K. Corzine, Integration of Z-source breakers into zonal DC ship power system microgrids, IEEE Journal of Emerging and Selected Topics in Power Electronics, vol. 5, no. 1, pp. 269–277, 2017.

[661] M.A. Mardani, M.H. Khooban, A. Masoudian and T. Dragisevic, Model predictive control of DC-DC converters to mitigate the effects of pulsed power loads in novel DC microgrids, IEEE Transactions on Industrial Electronics, pp. 1–10, 2019.

[662] A. Marks, J.F. Whidborne and J. Yamamoto, Control allocation for fault tolerant control of a VTOL octorotor, IEEE UKACC 2012, IEEE UK Automatic Control Conference, Cardiff, UK, September 2012.

[663] D.J. Marshman, T. Chmelyk, M.S. Sidhu, R.B. Gopulani and G.A. Dumont, Energy optimization in a pulp and paper-mill cogeneration facility, IEEE ACC 2010, IEEE 2010 American Control Conference, Baltimore, MD USA, July 2010.

[664] Ph. Martin and P. Rouchon, Systèmes plats: planification et suivi des trajectoires, Journées X-UPS, École des Mines de Paris, Centre Automatique et Systèmes, Mai, 1999.

[665] P. Martin, S. Devasia and B. Paden, A differential look at output tracking: Control of VTOL aircraft, Automatica, Elsevier, vol. 32, no. 1, pp. 101–107, 1996.

[666] J.L. Martinez, A. Mandow, J. Morales, S. Pedraza and A. Garcia-Cerezo, Approximating kinematics for tracked mobile robots, The International

Journal of Robotics Research, J. Wiley, vol. 24, no. 10, pp. 867–878, 2005.

[667] N. Masoud, P. Mattson, C. Smith and M. Isaksson, On stability and performance of disturbance observer-based dynamic load torque compensation for assistive exoskeleton: A hybrid approach, Mechatronics, Elsevier, vol. 63, pp. 102373–102387, 2020.

[668] W.A. Massey and W. White, A stochastic model to capture space and time dynamics in wireless communication systems, Probability in the Engineering and Informational Sciences, vol. 8, pp. 541–569, 1994.

[669] J. Mayr, F. Spanlang and Gattringer, Mechatronic design of a self-balancing three-dimensional inertia wheel pendulum, Mechatronics, Elsevier, vol. 350, pp. 92–106, 2013.

[670] L. Mazamanyan and M.A. Ayoubi, Fuzzy attitude control of spacecraft with fuel sloshing via linear matrix inequalities, IEEE Transactions on Aerospace and Electronic Systems, vol. 34, no. 5, pp. 2526–2536, 2018.

[671] S. Mefoued, S. Mohammad and Y. Amirat, Toward movement restoration of knee joint using robust control of powered orthosis, IEEE Transactions on Control Systems Technology, vol. 31, no. 6, pp. 2156–2168, 2013.

[672] A. Meghdari, S.H. Lavasani, M. Norouzi and M.S.R. Mousavi, Minimum control effort trajectory planning and tracking of the CEDRA brachiation robot, Robotica, Cambridge University Press, vol. 31, no. 7, pp. 1119–1129, 2013.

[673] A. Meghdari, S.M.H. Lavasani, M.S.R. Mousavi and M. Norouzi, Optimal trajectory planning for brachiation robot on ladder with irregular branches, ASME 2011 International Design Engineering Technical Conference & Computers and Information in Engineering Conference, Washington DC, USA, August 2011.

[674] S. Mekri, M. Shafir-Khah, P. Siano, M. Moullem, M. Mokhtari and J.P.S. Catalao, Contribution of tidal power generation system for damping interarea oscillation, Energy Conversion and Management, Elsevier, vol. 132, pp. 136–146, 2017.

[675] O. Menendez, F.A. Auat Cheein, M. Perez and S. Kouro, Robotics in power systems: Enabling a more reliable and safe grid, IEEE Industrial Electronics Magazine, vol. 11, no. 2, pp. 22–34, 2017.

[676] W. Menezes de Oliveira and W.F. Lages, MPC applied motion control of an underactuated brachiation robots, IEEE ETFA 2006, IEEE 2006 Conference on Emerging Technologies and Factory Automation, Prague, Czech Republic, September 2006.

[677] V. Menezes de Oliveira and W.F. Lages, Control of a brachiating robot for inspection of aerial power lines, IEEE 2000 1st International Conference on Applied Robotics for the Power Industry, Montreal, Canada, October 2010.

[678] V. Menezes de Oliveira and W.F. Lages, Control of a brachiation robot with a single underactuated joint using nonlinear model predictive control, IFAC SSSC 2007, IFAC 3rd Symposium on Structure and Control, Foz de Iguessa, Brazil, October 2007.

[679] V. Menezes de Oliveria and W.F. Lages, Linear predictive control of a brachiation robot, IEEE 2006 Canadian Conference on Electrical and Computer Engineering, Ottawa, Canada, May 2006.

[680] V. Menezes de Oliveira and W.F. Lages, Predictive control of an underactuated brachiation robot, 8th IFAC Symposium on Robot Control, Bologna, Italy, September 2006.

[681] V. Menezes de Oliveira and W.F. Lages, Real-time predictive control of a brachiation robot, IEEE ETFA 2009, IEEE 2009 Conference on Emerging Technologies and Factory Automation, Mallorca, Spain, September 2009.

[682] L. Menhour, B. d'Andréa-Novel, M. Fliess and H. Mounier, Coupled nonlinear vehicle control: Flatness-based setting with algebraic estimation techniques, Control Engineering Practice, vol. 2, pp. 135–146, 2014.

[683] M. Mercangoz and F.J. Doyle, Model-based control in the pulp and paper industry, IEEE Control Systems Magazine, pp. 30–39, August 2006.

[684] X. Meng, Y. He, F. Gu, *et al.*, Design and implementation of rotor aerial manipulator system. IEEE 2016 International Conference on Robotics and Biomimetics, Qingdao, China, December 2016.

[685] L. Meekers, P. Rafalo and A. Rochman, Analysis of process parameters affecting energy consumption in plastic injection moulding, Procedia CIRP, Elsevier, vol. 69, pp. 242–347, 2018.

[686] A.R. Merheb, H. Noura and F. Bateman, Active fault tolerant control of Octorotor UAV using dynamic control allocation, IEEE ICIUS 2014, The 2014 International Conference on Intelligent Unmanned systems, Montreal, Canada, 2014.

[687] A. Mebsah, Z.K. Nagy, A.E. Huesman, J.J.M. Kramer and P.M.J. van der Hof, Nonlinear model-based control of a semi-industrial batch crystallizer using a population balance modelling framework, IEEE Transactions on Control Systems Technology, vol. 20, no. 5, pp. 1188–1203, 2012.

[688] O. Menendez, F. Auat Cheein, M. Perez and S. Koura, Robotics in power systems, IEEE Industrial Electronics Magazine, vol. 11, no.2, pp. 22–34, 2017.

[689] W. Meng, Q. Liu, Z. Zhou, Q. Ai, B. Sheng, and S. Xie, Recent development of mechanisms and control strategies for robot-assisted lower limb rehabilitation, Mechatronics, Elsevier, vol. 31, pp. 132–145, 2015.

[690] A. Mesbah, J.A. Paulson, R. Lakerveld and R.D. Braatz, Plant-wide model predictive control for a continuous pharmaceutical process, 2015 American Control Conference, July 2015, Chicago, USA.

[691] A. Mesbah, J. Lardlast, A.e.M. Huesman, *et al.*, A model-based control framework for industrial batch crystallization processes, Chemical Engineering Research and Design, Elsevier, vol. 88, pp. 1223–1233, 2010.

[692] A. Mesbah, A.E.M. Huesman, H.J.M. Kramer and P.M.J. van der Hof, A comparison of nonlinear observers for output feedback model-based control of seeded batch crystallization processes, Journal of Process Control, Elsevier, vol. 21, pp. 662–666, 2011.

[693] Q. Miao and D. Cebon, Path-following control based-on ground-watching navigation, IEEE Transactions on Intelligent Transportation Systems, vol. 19, no. 8, pp. 2592–2602, 2012.

[694] M.M. Michalek, Modular approach to compact low-speed kinematic modelling of multi-articulated urban buses for motion algorithmization purposes, 2019 IEEE Intelligent Vehicle Symposium, Paris, France, June 2019.

[695] M.M. Michalek, Agile maneuvering with intelligent articulated vehicles: A control perspective, 10th IFAC Symposium on Intelligent Autonomous Vehicles, IFAC IAV 2019, Gdansk Poland, July 2019.

[696] M.M. Michalek, Trailer maneuverability in N-trailer structures, IEEE Robotics and Automation Letters, vol. 5, no. 4, pp. 5105–5112, 2020.

[697] M.M. Michalek, B. Patkowski and T. Gawron, Modular kinematic modelling of articulated buses, IEEE Transactions on Vehicular Technology, vol. 69, no. 8, pp. 8381–8394, 2020.

[698] L. Michel, S. Selvarajan, M. Ghanes, F. Plestan, Y. Aoustin and J.P. Barbot, An experimental investigation of discretized homogeneous differentiators: Pneumatic actuator case, IEEE Journal of Emerging and Selected Topics in Industrial Electronics, 2021.

[699] S. Miera, C.Z. Wang, J.N. Reddy and B.C. Khoo, A 3D fully coupled analysis of nonlinear sloshing with ship motion, Ocean Engineering, Elsevier, vol. 19, pp. 1–13, 2012.

[700] L. Mihaylova, A. Heygi, A. Gning and R.K. Boel, Parallelized particle and Gaussian sum particle filters for large-scale freeway traffic systems, IEEE Transactions on Intelligent Transportation Systems, vol. 13, no. 1, pp. 36–48, 2012.

[701] H.A. Mintsa, R. Venugopal, J.P. Kenné and C. Belleau, Feedback linearization-based position control of an electrohydraulic servosystem with supply pressure uncertainty, IEEE Transactions on Control Systems Technology, vol. 20, no. 4, pp. 1092–1099, 2012.

[702] B.J.E. Misgeld, L. Hewing, L. Liu and S. Leanhardt, Closed-loop positive real optimal control of variable stiffness actuators, Control Engineering Practice, Elsevier, vol. 82, pp. 142–150, 2019.

[703] J.P. Mishra and S.R. Kurode, Robust output-feedback control for container-slash system using variable gain super-twisting algorithm, IEEE VSS 2014, IEEE 13th Workshop on Variable Structure Systems, Nantes, France, July 2014.

[704] S. Mohammad, W. Huo, J. Huang, H. Rifai and Y. Amirat, Nonlinear disturbance observer-based sliding-mode control of a human-driven knee-joint orthosis, Robotics and Autonomous Systems, Elsevier, vol. 75, pp. 41–49, 2016.

[705] A. Mohanty and B. Yao, Indirect adaptive robust control of hydraulic manipulators with accurate parameter estimates, IEEE Transactions on Control Systems Technology, vol. 19, no. 3, pp. 567–576, 2011.

[706] A. Mohanty and B. Yao, Integrated direct/indirect adaptive robust control of hydraulic manipulators with valve deadband, IEEE Transactions on Mechatronics, vol. 16, no. 6, pp. 707–716, 2011.

[707] A. Mohanty, M. Viswandya, P.K. Rey and S. Mohanty, Reactive power control optimization of hybrid offshore tidal turbine with system uncertainties,

Journal of Ocean Engineering and Science, Elsevier, vol. 1, pp. 256–267, 2016.

[708] A. Monzur and M. Kulawik, Adaptive trajectory tracking for doubly-fed nonholonomic mobile manipulator – 3 pendulum on mobile platform of (2.0) Class,8th IFAC Symposium on Robot Control, Syroco2 2006, Bologna, Italy, September 2006.

[709] O.D. Montoya and W. Gil-Gonzalez, Nonlinear analysis and control of a reaction wheel pendulum: Lyapunov-based approach, Journal of Engineering Science and Technology, Elsevier, 2019.

[710] M. Mooref, M. Saadat and G. Vossoughi, Mechatronic design and position control of a novel ball and plate system, 16th Mediterranean Conference on Control and Automation, Ajaccio, France, June 2008.

[711] R. Morales, H. Sira-Ramirez and J.A. Somolinos, Robust control of under-actuated wheeled mobile manipulators using GPI disturbance-observers, Multibody System Dynamics, Springer, vol. 32, no. 4, pp. 511–533, 2014.

[712] J. Morales, J.L. Martinez, A. Mandow and A.J. Garcia-Cerezo, Steering the last trailer as a virtual tractor for reversing vehicles with passive on- and off-axle hitches, IEEE Transactions on Industrial Electronics, vol. 60, no. 12, pp. 5729–5736, 2013.

[713] M. Morawiec, The adaptive backstepping control of permanent magnet synchronous motor supplied by current source inverter, IEEE Transactions on Industrial Informatics, vol. 9, no. 2, pp. 1947–1055, 2013.

[714] M.A. Moreno Armendaviz, C.A. Perez-Oliveira, F. Ortiz-Rodriguez and E. Rubio, Indirect hierarchical FCMAC control for the ball and plate system, Neurocomputing, Elsevier, vol. 73, pp. 2454–2463, 2010.

[715] V. Moria, F. Cazaurang and P. Vernis, Flatness-based hypersonic re-entry guidance of a lifting-body vehicle, Control Engineering Practice, Elsevier, vol. 17, pp. 586–596, 2009.

[716] L. Moriello, K. Biagiotti, C. Melchiorri and A. Paoli, Manipulating liquids with robots: A sloshing-free solution, Control Engineering Practice, Elsevier, vol. 78, pp. 129–141, 2018.

[717] L. Moriello, K. Biagiotti, C. Melchiorri and A. Paoli, Control of liquid handling robotic systems: A feed-forward approach to suppress sloshing, IEEE ICRA 2017, IEEE 2017 International Conference on Robotics and Automation, Singapore, July 2017.

[718] P. Morin and C. Samson, Trajectory tracking for nonholonomic systems: Theoretical background and applications, Research Report INRIA No 6464, March 2008.

[719] P. Morin and C. Samson, Transverse function control of a class of non-invariant driftless systems: Applications to vehicles with trailers, IEEE Conference on Decision and Control, IEEE CDC 2008, Proceedings of the 47th IEEE Conference on Decision and Control, Cancun, Mexico, December 2008.

[720] P. Morin and C. Samson, Stabilization of trajectories for systems on Lie groups: Application to the rolling sphere, IFAC WC 2008, Proceedings of the 17th IFAC World Congress, Seoul, Korea, July 2008.

[721] P. Morin and C. Samson, Control of nonholonomic mobile robots based on the transverse function approach, IEEE Transactions on Robotics, vol. 25, no. 5, pp. 1058–1076, 2009.

[722] P. Morin and C. Samson, Feedback control of the general two-trailer system with the transverse function approach, INRIA Research Report No 7938, pp. 1–27, 2012.

[723] M. Morlock, N. Meyer, M.A. Pick and S. Seifreid, Real-time trajectory tracking control of a parallel robot with flexible links, Mechanism and Machine Theory, Elsevier, vol. 158, pp. 104220–104236, 2021.

[724] F.L. Moro, Balancing while executing computing reaching tasks: An attractor-based whole body motion control system using gravitational stiffness, International Journal of Humanoid Robotics, World Scientific, vol. 13, no. 1, pp. 1550046, 2016.

[725] N. Morozovsky and F. Bewley, SkySweeper: A low DOF dynamic high wire robot, IEEE Robotics and Automation Letters, vol. 4, no. 2, pp. 1993–1998, 2019.

[726] H. Mounier, Document de synthèse pour l' habilitation à diriger des recherches, Université de Paris XI, Juillet, 2005.

[727] H. Mounier, J. Rudolph and F. Woittenneck, Boundary value problems and convolutional systems over rings of ultradistributions, Advances in the theory of Control, Signal and Systems with Physical Modelling, Lecture Notes in Control an Information Sciences, Springer, pp. 179–188, 2010.

[728] H. Mounier and J. Rudolph, Trajectory tracking for π-flat nonlinear delay systems with a motor example, In: A. Isidori, F. Lamnabhi-Lagarrigue and W. Respondek, editors, Nonlinear control in the year 2000, vol.1, Lecture Notes in Control and Inform. Sci., vol. 258, pp. 339–352, Springer, 2001.

[729] S. Muhammad, M.U. Rafique, S. Li, Z. Shao, Q. Wang and N. Guan, A robust algorithm for state-of-charge estimation with gain optimization, IEEE Transactions on Industrial Informatics, vol. 13, no. 6, pp. 2983–2994, 2017.

[730] A. Mukherjee and J. Strikwerde, Analysis of dynamic congestion control protocols: A Fokker-Planck approach, IEEE ACM 1993, Proceedings of IEEE SIGCOMM 1993 International Conference, Symposium Communications Architectures Protocols, San Francisco, CA, September 1993.

[731] A. Muller and T. Hufnagel, Model-based control of redundantly actuated parallel manipulators in redundant conditions, Robotic and Autonomous Systems, Elsevier, vol. 60, pp. 563–571, 2012.

[732] G. Munoz-Gomez, A.V. Alanis and J. Rivera, Nested high-order sliding-mode controller applied on a Brushless Direct Current Motor, IFAC MICNON 2018, 2nd IFAC Conference of Modelling, Identification and Control of Nonlinear Systems, Guadalajara, Mexico, June 2018.

[733] M. Nachtane, M. Tarfaoui, A.El. Moumen and D. Saifaoui, Damage protection of horizontal axis marine current turbines under hydrodynamic, hydrostatic and impact loads, Composite Structures, Elsevier, vol. 170, pp. 146–157, 2017.

[734] U. Nagarajan, G. Kantor and R. Hollis, Trajectory planning and control of an underactuated dynamically stable single spherical wheeled mobile

robot, 2009 IEEE International Conference on Robotics and Automation, IEEE ICRA 2009, Kobe, Japan, May 2009.

[735] U. Nagarajan, G. Kantor and R. Hollis, Integrated motion planning and control of graceful balancing mobile robots, International Journal of Robotics Research, Sage Publications, vol. 32, no. 9–10, pp. 1005–1029, 2013.

[736] Z.K. Nagy and R.D. Braatz, Advances and new directions in crystallization control, Annual Review of Chemical and Bimolecular Engineering, vol. 3 pp. 55–76, 2012.

[737] Z.K. Nagy, M. Fujiwara, J.W. Chen and R.D. Baartz, Comparative performance of concentration and temperature controlled batch crystallization, Journal of Process Control, Elsevier, vol. 18, pp. 399–407, 2008.

[738] Z.K. Nagy and R.D. Braatz, Worst case and distributional robustness analysis of finite-time control trajectories for nonlinear distributed parameter systems, IEEE Transactions on Control Systems Technology, vol. 11, no. 3, pp. 694–704, 2003.

[739] B. Nahid-Mobarakeh, F. Meibody-Tubur and F.M. Sargos, Back EMF estimation-based sensorless control of PMSM: Robustness with respect to measurement errors and inverter irregularities, IEEE Transactions on Industry Applications, vol. 43, no. 2, pp. 485–494, 2007.

[740] Z.K. Nagy, M. Fujiwara and R.D. Baetz, Modelling and control of combined cooling and antisolvent crystallization process, Journal of Process control, Elsevier, vol. 18, pp. 356–364, 2008.

[741] H. Najjaran and A. Goldenberg, Real-time motion planning of an autonomous mobile manipulator using fuzzy adaptive Kalman Filter, Robotics and Autonomous Systems, Elsevier, vol. 55, pp. 96–106, 2007.

[742] J. Nakanishi, T. Fukuda and D.E. Koditschek, A brachiating robot controller, IEEE Transactions on Robotics and Automation, vol. 16, no. 2,pp. 109–124, 2000.

[743] J. Nakanishi and S. Vijayakumar, Exploiting passive dynamics with variable stiffness actuation in robot brachiation, IEEE RSS 2012, IEEE 2012 Robotics Science and Systems Conference, Sydney, Australia, July 2012.

[744] J. Nakanishi, R. Cory, M. Minstry, J. Peters and S. School, Operational space control: A theoretical and experimental comparison, International Journal of Robotics Research, vol. 27, no. 6, pp. 737–757, 2008.

[745] R. Naldi, M. Falci, R.G. Sanfelice and L. Marconi, Robust global trajectory tracking for underactuated VTOL aerial vehicles using inner-outer loop control paradigms, IEEE Transactions on Automatic Control, vol. 62, no. 1, pp. 97–112, 2017.

[746] B. Narayan Rath and D. Subudhi, Online extreme learning algorithm based identification and nonlinear model predictive controller for way-point tracking application of an autonomous underwater vehicle, IET Cognitive Computation and Systems, vol. 1, no. 3, pp. 61–71, 2019.

[747] I. Nasiruddin, K.R. Niazi and G. Sharma, Study on dynamic participation of wind turbines in AGC of power system, Electric Power Component and System Journal, Taylor & Francis, vol. 43, no. 1, pp. 44–55, 2014.

[748] F. Nejabbatkhah, V.W. Li and H. Tian, Power quality control of smart hybrid AC/DC microgrids: An overview, IEEE Access, vol. 7, pp. 52295–52318, 2019.

[749] J. Neupert, E. Arnold, K. Schneider and O. Sawodny, Tracking and anti-swing control for boom cranes, Control Engineering Practice, Elsevier, vol. 18, pp. 31–44, 2010.

[750] K.D. Nguyen and D. Liu, Robust control of a brachiating robot, IEEE IROS 2017, IEEE 2017 International Conference on Intelligent Robots and Systems, Vancouver, Canada, September 2017.

[751] K.D. Nguyen and D. Liu, Gibbon-inspired robust asymmetric brachiation along an upward slope, International Journal of Control, Automation and Systems, Springer, vol. 17, no. 10, pp. 2647–2654, 2019.

[752] M.N. Nguyen, D.T. Tran and K.K. Ahn, Robust position and vibration control of an electrohydraulic series elastic manipulator against disturbance generated by a variable stiffness actuator, Mechatronics, Elsevier, vol. 52, pp. 22–35, 2018.

[753] H.T. Nguyen and J.W. Jung, Finite control set model predictive control to guarantee stability and robustness for surface-mounted PM synchronous motors, IEEE Transactions on Industrial Electronics, vol. 65, no. 11, pp. 8510–8519, 2018.

[754] Z. Ni and Y. He, Design and validation of a robust active trailer steering system for multi-trailer articulated heavy vehicles, Vehicle System Dynamics, Taylor and Francis, 2018.

[755] D. Nicolis, F. Allewi and P. Rocco, Operational space model predictive sliding-mode control for redundant manipulators, IEEE Transactions on Robotics, vol. 36, no. 4, pp. 1348–1355, 2020.

[756] Z. Nie, X. Xiao, R. McMahon, P. Clifton, Y. Wu and R. Shao, Emulation and control methods for direct drive linear wave energy converters, IEEE Transactions on Industrial Informatics, vol. 9, no. 2, pp. 790–798, 2013.

[757] P. Niermeyer, V.S. Akkinejpalli, M. Pak, F. Holzapfel and B. Lohmann, Geometric path following control for multirotor vehicles using nonlinear model predictive control and 3D spline paths, IEEE ICUAS 2016, IEEE 2016 International Conference on Unmanned Aircraft Systems, Arlington, Virginia, USA, June 2016.

[758] Z.H. Ning and Q.X. Yan, Stabilization of the wave equation with variable coefficients and a delay in dissipative boundary feedback, Journal of Mathematical Analysis and Applications, Elsevier, vol. 367, pp. 167–173, 2010.

[759] C. Niu, H. Shen, J. Li and Y. Yi, Disturbance observer-based L_1 robust tracking control for single-link flexible-joint manipulator systems, IET Journal of Engineering, vol. 2019, no. 22, pp. 8390–8394, 2019.

[760] G. Niu, X. Yi, C. Chen, *et al.*, A novel effluent quality predicting model based on genetic-deep belief network algorithm for cleaner production in a full-scale paper-making wastewater treatment, Journal of Cleaner Production, Elsevier, vol. 265, pp. 1–10, 2020.

[761] I. Noh, S. Wan and Y.J. Jang, Non-interactive looper and strip tension control for hot finishing mill using nonlinear disturbance observer, ISIJ International, vol. 52, no. 6, pp. 1092–1100, 2012.

[762] N. Noroozi, S. Trip and R. Geiselhart, Model predictive control of DC microgrids: Current sharing and voltage regulation, IFAC NecSys 2018, 7th IFAC Workshop on Distributed Estimation and Control in Networked Systems, Groningen, The Netherlands, August 2018.

[763] G. Notomista, Y. Emam and N.M. Egersted, The Slothbot: A novel design for a wire-traversing robot, IEEE Robotics and Automation Letters, vol. 4, no. 2, pp. 1993–1998, 2019.

[764] Q.H. Njo, W.P. Nguyen, C.N. Nguyen, T.H. Tram and Q.P. Ho, Fuzzy sliding-mode control of an offshore container crane, Ocean Engineering, Elsevier, vol. 140, pp. 125–134, 2017.

[765] E. Obe, Steady-state performance of a line-start synchronous reluctance motor with capacitive assistance, Electric Power Systems Research Elsevier, vol. 80, pp. 1240–1246, 2010.

[766] O. Ogordayk and K. Martinsen, Modelling and control for thermoplastics injection moulding: A review, Procedia CIRP, Elsevier, vol. 67, pp. 380–385, 2018.

[767] Y.J. Oh, J.S. Park, B.J. Hyon and J. Lee, Novel control strategy of wave energy converter using linear permanent magnet synchronous generator, IEEE Transactions on Applied Superconductivity, vol. 28, no. 3, pp. 1–5, 2018.

[768] R. Olfati-Saber, Global configuration stabilization for the VTOL aircraft with strong input coupling, IEEE Transactions on Automatic Control, vol. 7, no. 11, pp. 1949–1952, 2002.

[769] F. Olivier and A. Sedoglavic, A generalization of flatness to nonlinear systems of partial differential equations: Application to the control of a flexible rod, Proceedings of the 5th IFAC Symposium on Nonlinear Control Systems, Saint-Petersburg, July 2001.

[770] K. Omkar, K.B. Karthikeyan, R. Srimathi, *et al.*, A performance analysis of tidal turbine conversion system based on control strategies, Energy Procedia, Elsevier, vol. 160, pp. 526–533, 2019.

[771] K. Omoto, M. Dei and T. Ohtsuka, Integrated optimization of climbing locomotion for a humanoid robot, 11th IFAC Symposium on Nonlinear Control Systems, IFAC NOLCOS 2019, Vienna, Austria, September 2019.

[772] G.A. Orcajo, J.D. Rodriguez, P.G. Ardura, *et al.*, Dynamic estimation of electrical demand in hot rolling mills, IEEE Transactions on Industry Applications, vol. 53, no. 3, pp. 2714–2723, 2016.

[773] G.A. Orcajo, J.D. Rodriguez, J.M. Cano, *et al.*, Retrofit of a hot rolling mill plant with three-level active front end drives, IEEE Transactions on Industry Applications, vol. 54, no. 3, pp. 2954–2964, 2018.

[774] P. Ordaz and A. Poznyak, The Furuta's pendulum stabilization without the use of a mathematical model: Attractive ellipsoid method with KL-adaptation, IEEE CDC 2012, IEEE 2012 51st IEEE Conference on Decision and Control, Maui, Hawaii, USA, December 2012.

[775] P. Ordaz and A. Poznyak, Adaptive-Robust stabilization of the Furuta's Pendulum Via Attractive Ellipsoid Method, Journal of Dynamic System, Measurement and Control, vol. 138, no. 2, pp. 021005, 2015.

[776] G. Oriolo, M. Cefalo and M. Venditelli, Repeatable motion planning for redundant robots over cyclic tasks, IEEE Transactions on Robotics, vol. 33, no. 5, pp. 1170–1187, 2017.

[777] G. Oriolo and M. Venditelli, A framework for the stabilization of general nonholonomic systems with an application to the plate-ball mechanism, IEEE Transactions on Robotics, vol. 21, no. 2, pp. 162–175, 2005.

[778] Y. Orlov, S. Chakrabarty, D. Zhao and S.K. Spurgeon, Sliding mode observer design for a parabolic PDE in the presence of unknown inputs, Asian Journal of Control, J. Wiley, vol. 21, no. 1, pp. 224–235, 2019

[779] F. Ornelas-Tellez, J.J. Rico-Melgoza, E. Espinosa-Juarez and E.N. Sanchez, Optimal and robust control in DC microgrids, IEEE Transactions on Smart Grid, vol. 9, no. 6, pp. 5543–5553, 2018.

[780] M. Orsag, C. Corpela, S. Bogdan and P. Oh, Dexterous aerial robots in mobile manipulation using unmanned aerial systems, IEEE Transactions on Robotics, vol. 33, no. 6, pp. 1453–1476, 2017.

[781] H. Ouyang, J. Wang, G. Zhang, L. Mei and X. Deng, Novel adaptive hierarchical sliding-mode control for trajectory tracking and load sway rejection in double-pendulum overhead cranes, IEEE Access, vol. 7, pp. 10353–10362, 2017.

[782] H. Ouyang, J. Hu, G. Zhang, L. Mei and X. Deng, Sliding-mode based trajectory tracking and load sway suppression control for double-pendulum overhead cranes, IEEE Access, vol. 7, pp. 4371–4375, 2019.

[783] H. Ouyang, Z. Tian, L. Yu and G. Zhang, Motion planning approach for payload swing reduction in tower-cranes with double-pendulum effect, Journal of the Franklin Institute, Elsevier, vol. 352, pp. 8290–8320, 2020.

[784] H. Ouyang, Z. Tian, L. Yu and G. Zhang, Adaptive tracking controller design for double-pendulum tower cranes, Mechanism and Machine Theory, Elsevier, vol. 153, pp. 103980, 2020.

[785] H. Ouyang, Z. Tian, L. Yu and G. Zhang, Load swing rejection for double-pendulum tower cranes using energy-shaping-based control with actuator output limitation, ISA Transactions, Elsevier, vol. 101, pp. 246–255, 2020.

[786] H. Ouyang, Z. Tian, L. Yu and G. Zhang, Motion planning approach for payload swing reduction in tower cranes with double-pendulum effect, Journal of the Franklin Institute, Elsevier, vol. 357, pp. 8299–8320, 2020.

[787] H. Ouyang, X. Xu and G. Zhang, Energy-shaping-based nonlinear controller design for rotary cranes with double-pendulum effect considering actuator saturation, Automation in Construction, Elsevier, vol. 111, pp. 103054, 2020.

[788] H. Ouyang, X. Xu and G. Zhang, Boom motion trajectory generation approach for load sway rejection in rotary cranes considering double pendulum effect, Journal of Measurement and Control, Sage Publications, pp. 1–11, 2020.

[789] P.R. Ouyang. Q. Li, W.J. Zhang and L.C. Guo, Design, modelling and control of a hybrid machine system, Mechatronics, Elsevier, vol. 14, pp. 1197–1217, 2004.

[790] P.R. Ouyang, W.J. Zhang and R. Moazl, Experimental study of a hybrid actuation system, IEEE/ASME 2008 International Conference on Advanced Intelligent Mechatronics, Xi'an, China, July 2006.

[791] P.R. Ouyang, H.M. Kong, W.H. Yue and D.S. Liu, Revisiting hybrid five-bar mechanism position domain control application, Proceedings of the IEEE 2014 International Conference on Information and Automation, Hailan, China, July 2014.

[792] P.R. Ouyang, M. Kuang, S. Jayasinghe, W.H. Yue and H.M. Kang, Adaptive PD-SMC in position domain for contour tracking of hybrid actuated system, Proceedings of the IEEE 2017 International Conference on Mechatronics and Automation, Tokamatsu, Japan, August 2017.

[793] K. Ozaki, T. Oheshuka, K. Fujimoto, A. Kitamura and M. Nakayama, Nonlinear receding horizon control of a tandem cold mill in acceleration and deceleration conditions, IEEE ICROS-SICE 2009, Fukuoka, Japan, August 2009.

[794] F. Paganini and A. Ferragut, PDE models for population and residuals work applied to peer-to-peer networks, 46th Annual Conference on Information Sciences and Systems, Princeton, USA, Mar. 2012.

[795] F. Paganini, A. Tang, A. Ferragut and L. Andrew, Network stability under alpha fair bandwidth allocation with general file size distribution, IEEE Transactions on Automatic Control, vol. 57, no. 3, pp. 579–591, 2012.

[796] F. Paganini. H. Zubelliu and A. Ferragut, Trading-off efficiency reciprocity in wireless peer-to-peer file sharing, WiOpt2015, 19th IEEE International Symposium on Modelling and Optimization in Mobile AdHoc and Wireless Networks, Paris, May 2015.

[797] F. Paganini and A. Ferragut, Monotonicity and global stability in bounded dynamics of content-sharing networks, 53rd IEEE Conference on Decision and Control, December 15–17, 2014, Los Angeles, California, USA.

[798] F. Paganini and A. Ferragut, Monotonicity and global stability download dynamics of content sharing networks, IEEE CDC 2014, Los Angeles, USA, December 2014.

[799] F. Paganini, A. Ferragut and H. Zubelliu, Dynamics of heterogeneous peer-to-peer, 2013 IEEE 52nd Annual Conference on Decision and Control, Florence, Italy, December 2013.

[800] V. Pakrashi, A. O'Connor and B. Basu, Effect of tuned mass damper on the interaction of a quarter car model with a damage bridge, Structure and Infrastructure Engineering, Taylor and Francis, vol. 6, no. 4, pp. 409–421, 2010.

[801] V. Pakrashi, A. O'Connor and B. Basu, Critical speed variations for damaged bridges, ICASP 2013, 13th International Conference on Applications of Statistics and Probability in Civil Engineering, Seoul, South Korea, May 2019.

[802] Y. Pan, H. Wang, X. Li and H. Yu, Adaptive command-filtered backstepping control of robot arms with compliant actuators, IEEE Transactions on Control Systems Technology, vol. 26, no. 3, pp. 1149–1157, 2018.

[803] S. Parakkadavath and B. Bhikkaji, Identification if the nonlinear dynamics and state of charge estimation of a $LiFePO_4$ battery using constrained Unscented Kalman Filter, Proceedings of the 20th IFAC World Congress, Toulouse, France, July 2017.

[804] J.Y. Park, J.K. Lee, B.H. Cho and K.Y. Oh, An inspection robot for live-line suspension insulator strings in 345 kV power lines, IEEE Transactions on Power Delivery, vol. 27, no. 2, pp. 632–639, 2012.

[805] B.S. Park and S.J. Yoo, An error transformation approach connectivity preserving and collision-avoidance formation tracking of networked uncertain underactuated surface vessels, IEEE Transactions on Cybernetics, vol. 49, no. 8, pp. 2955–2966, 2019.

[806] H. Park, I. Noh and S. Wen, Non-interactive control of hot rolling mill using nonlinear disturbance observers, IEEE ICROS-SICE 2003, Fukuoka, Japan, August 2009.

[807] H. Park, D. Jung and S. Won, Based on nonlinear disturbance observer approach non-interactive control for hot rolling mill – experimental validation, 13th IFAC Symposium on Automation and Mining, Mineral and Metal Processing, Cape Town, South Africa, August 2010.

[808] J.D. Park, C. Kalev and H.F. Hofmann, Control of high-speed solid-rotor synchronous reluctance motor / generator for flywheel-based uninterruptible power supplies, IEEE Transactions on Industrial Electronics, vol. 55, no. 8, pp. 3038–3046, 2009.

[809] B.S. Park, J.B. Park and Y.H. Choi, Adaptive observer-based trajectory tracking control of nonholonomic mobile robots, International Journal of Control, Automation and Systems, vol. 9, no. 3, pp. 534–541, 2018.

[810] M. Partovibakhsh and G. Liu, An adaptive Unscented Kalman Filtering approach for online estimation of model parameters and state-of-charge of lithium-ion batteries for autonomous mobile robots, IEEE Transactions on Control Systems Technology, vol. 23, no. 4, pp. 357–363, 2017.

[811] D. Pazderki, K. Kozlowski and D.K. Waskoswicz, Control of a unicycle-like robot with trailers using transverse function approach, Bulletin of the Polish Academy of Sciences: Technical Sciences, vol. 60, no. 3, pp. 537–579, 2012.

[812] A. Pascucci, PDE and Martingale Methods in Option Pricing, Springer, 2011.

[813] A. Pascucci, M. Suarez-Taboada and C. Vasquez, Mathematical analysis and numerical methods for a PDE model of a stock-loan pricing problem, Journal of Mathematical Analysis and Applications, Elsevier, vol. 403, pp. 38–58, 2013.

[814] W. Paszkowiack, T. Bartkowiak and M. Pelic, Kinematic model of multiple trailers on a tractor system for production logistics applications, Archives of Mechanical Technology and Materials, DeGruyter, vol. 39, pp. 16–25, 2019.

[815] T.R. Patel, K.D. Kumar and K. Behdinan, Variable structure control for satellite attitude stabilization on elliptic orbits suing solar radiation pressure, Acta Astronautica, Elsevier, vol. 64, pp. 359–373, 2009.

[816] A. Pavlov, B. Janssen, N. van de Wouw and H. Nijmeijer, Experimental output regulation for a nonlinear benchmark system, IEEE Transactions on Control Systems Technology, vol. 15, no. 4, pp. 786–793, 2007.

[817] L.M. Pedersen and B. Wittenmark, Multivariable controller design for a hot rolling mill, IEEE Transactions on Control Systems Technology, vol. 6, no. 3, pp. 304–312, 1998.

[818] A.U. Pehlivan, J.P. Lasey and M.K. O'Malley, Minimal assist-as-needed controller for upper limb robotic rehabilitation, IEEE Transactions on Robotics, vol. 32, no. 1, pp. 113–124, 2016.

[819] Z. Peng, J. Wang and J. Wang, Constrained control of autonomous underwater vehicles based on command optimization and disturbance estimation, IEEE Transactions on Industrial Electronics, vol. 66, no. 5, pp. 3627–3635, 2019.

[820] J. Peng, Z. Yang, Y. Wang, F. Zhang and Y. Liu, Robust adaptive motion/force control scheme for crawler-type mobile manipulator with nonholonomic constraint based on sliding mode control approach, ISA Transactions, Elsevier, vol. 92, pp. 166–179, 2019.

[821] X. Peng, K. Guo and Z. Geng, Full state tracking and formation control for under-actuated VTOL UAVs, IEEE Access, vol.7, pp. 3755–3766, 2019.

[822] S. Peng, C. Chen, H. Shi and Z. Yao, State of charge estimation of battery energy storage systems based on adaptive unscented Kalman Filter with a noise statistics estimator, IEEE Access: Special Section on Battery Energy Storage and Management Systems, vol. 5, pp. 13202–13212, 2017.

[823] X. Penz, M. Jiu, L. He, X. Yu and Y. Lu, Fuzzy sliding-mode control based on longitudinal force estimation for electromechanical braking system using BDC motor, CES Transaction on Electrical Machines and Systems, vol. 2, no. 1, pp. 142–151, 2018.

[824] L.P. Perera, P. Oliveira and C.G. Soares, Maritime traffic monitoring based on vessel detection, tracking, state estimation and trajectory prediction, IEEE Transactions on Intelligent Transportation Systems, vol. 13, no. 3, pp. 1188–1200, 2012.

[825] C.D. Petersen, F. Leve and J. Kolmanovsky, Model predictive control of an underactuated spacecraft with two reaction wheels, Journal of Guidance, Control and Dynamics, vol. 40, no. 2, pp. 320–333, 2017.

[826] P.B. Petrovic and N.A. Lukic, Configuration-based compliance control of kinematically redundant robot arm, Part I: Theoretical Framework, Transactions of the Faculty of Mechanical Engineering, University of Belgrade, February 2017.

[827] A. Pfeffer, T. Glück, F. Schausberger and A. Kugi, Control and estimation strategies for pneumatic drives with partial position information, Mechatronics, Elsevier, vol. 50, pp. 259–270, 2018.

[828] D.B. Pham, J. Kim and S.G. Lee, Combined control with sliding-mode and partial feedback linearization for a spatial ridable ballbot, Mechanical Systems and Signal Processing, Elsevier, vol. 128, pp. 531–550, 2019.

[829] D.B. Pham, H. Kim, J. Kim and S.G. Lee, Balancing and transferring control of a ball Segway using a double-loop approach, IEEE Control Systems Magazine, vol. 38, no. 2, pp. 15–37, 2018.

[830] D.B. Pham and S.G. Lee, Hierarchical sliding-mode control for a two-dimensional ball Segway that is a class of an second-order underactuated system, Journal of Vibration and Control, Sage Publications, vol. 28, no. 1, pp. 72–83, 2019.

[831] Y. Pi and H. Ouyang, Lyapunov-based boundary control of a multi-span bridge subjected to moving masses, Journal of Vibration and Control, Sage Publications, vol. 13, no. 14, pp. 1–14, 2017.

[832] M. Pinsky, Partial differential equations and boundary value problems, Prentice-Hall, 1991.

[833] C. Pirat, F. Anderson, R. Walker and V. Gass, H_∞ and μ synthesis for nanosatellites rendezvous and docking, IEEE Transactions on Control Systems Technology, vol. 25 no. 5, pp. 1050–158, 2020.

[834] M. Pisaturo and A. Senatore, Improvement of traction through mMPC in linear vehicle dynamics based on electrohydraulic dry clutch, Journal of Intelligent Industrial Systems, Springer, vol. 1, pp. 153–161, 2015.

[835] E. Platen and D. Heath, A Benchmark Approach to Quantitative Finance. Springer, 2008.

[836] F. Plestan, Y. Shtseel, V. Brégeault and A. Poznyak, Sliding-mode control with gain adaptation: Application to an electropneumatic actuator, Control Engineering Practice, Elsevier, vol. 21, pp. 679–688, 2013.

[837] H. Polinder, B.C. Mecrow, A.G. Jack, P. Dieckinson and M.A. Mueller, Conventional and TFPM linear generators for direct-drive wave energy conversion, IEEE Transactions on Energy Conversion, vol. 20, no. 2, pp. 260–267, 2005.

[838] V. Potkonjak, M. Popovic, M. Lazarevic and J. Sinanovic, Redundancy problem in writing: From human to anthropomorphic robot arm, IEEE Transactions on Systems, Man and Cybernetics – Part B: Cybernetics, vol. 28, no. 6, pp. 790–805, 1998.

[839] N. Pouliot, P.L. Richard and S. Montabault, LineScout Technology opens the way to robotic inspection and maintenance of high-voltage power lines, IEEE Power and Energy Technology Systems Journal, vol. 2, no. 1, pp. 1–10, 2015.

[840] N. Pouliot and S. Montamabault, Geometric Design of the LineScout, a Teleoperated Robot for Power Line Inspection and Maintenance, 200 IEEE Intl. Conference on Robotics and Automation, Pasadena, California, USA, May 2009.

[841] M. Pourbehzadi, T. Niknam, J. Aghael, G. Mokryani, M. Shafie-Khah and J.P.S. Catalao, Optimal operation of hybrid AC/DC microgrids under uncertainty of renewable energy resources: A comprehensive review, Electric Power and Energy Systems, Elsevier, vol. 129, pp. 139–159, 2019.

[842] A. Pozo and M. Maggiore, A class of position controllers for underactuated VTOL vehicles, IEEE Transactions on Automatic Control, vol. 59, no. 9, pp. 2580–2585, 2014.

[843] D. Prager and Q. Zhang, Valuation of stable loans under a Markov chain model, Journal of Systems Science and Complexity, Springer, vol. 29, pp. 171–186, 2016.

[844] J. Prudell, M. Stoddard, E. Amon, T.K.A. Brekken and A. von Jouanne, A permanent-magnet tubular linear generator for ocean wave energy conversion, IEEE Transactions on Industry Applications, vol. 46, no.6, pp. 2392–2400, 2010.

[845] D. Qian, S. Tong and S. Lee, Fuzzy logic-based control of payloads subjected to double-pendulum motion in overhead cranes, Automation in Construction, Elsevier, vol. 65, pp. 133–143, 2016.

[846] Y. Qian and Y. Fang, Switching logic-based nonlinear feedback control of offshore ship-mounted tower-cranes: A disturbance observer-based approach, IEEE Transactions on Automation Science and Engineering, vol. 16, no. 3, pp. 1125–1136, 2018.

[847] L. Qiao and W. Zhang, Adaptive non-singular integral terminal sliding-mode tracking control for autonomous underwater vehicles, IET Control Theory and Applications, vol. 11, no. 8, pp. 1293–1306, 2017.

[848] H. Qin, C. Li, Y. Sun, X. Li, Y. Du and Z. Ding, Finite-time trajectory tracking control of unmanned surface vessel with error constraints and input saturation, Journal of the Franklin Institute, Elsevier, vol. 357, no. 16, pp. 11472–11495, 2020.

[849] B. Qiu, G. Wang and Y. Fan, Predictor LOS-based trajectory linearization control for path following of underactuated unmanned surface vehicles with input saturation, Ocean Engineering, Elsevier, vol. 214, pp. 107874–107885, 2020.

[850] Q. Quan and K.Y. Cai, Additive-state-decomposition-based tracking control for TORA benchmark, Journal of Sound and Vibration, Elsevier, vol. 332, pp. 4829–4841, 2013.

[851] R. Quinynen, B. Houske, M. Vallero, *et al.*, Symmetric algorithmic differentiation based exact Hessian SQP method and software for economic MPC, 53rd IEEE Conference on Decision and Control, Lo Angeles, California, December 2014.

[852] B. Rabu, L. Bandouin, C. Prieur and D. Arzelier, Simultaneous H_∞ vibration control of fluid/plate system via reduced-order controller, IEEE Transactions on Control Systems Technology, vol. 20, no. 3, pp. 700–711, 2012.

[853] G. Raghavendra Gopinath and P.D. Shyama, Sensorless control of PMSM using adaptively tuned SCKF, IET Journal of Engineering, vol. 2019, no. 12, pp. 4304–4308, 2019.

[854] R.M.T. Raja Ismail and Q.P. Ha, Trajectory tracking control for offshore boom cranes using higher-order sliding modes, Proceedings of the 2013 International Symposium on Automation and Robotics in Construction, pp. 996–1001, Montreal, Canada, August 2014.

[855] K.S. Rama Rao, T. Sunderan and M. Refat Adiris, Performance and design optimization of two model-based wave energy permanent magnet linear generators, Renewable Energy, Elsevier, vol. 101, pp. 196–203, 2017.

[856] R. Ramirez-Villalobos, L.T. Aguilar and L.N. Coria, Sensorless H_∞ speed-tracking synthesis for surface mount permanent magnet synchronous motor, ISA Transactions, Elsevier, vol. 67, pp. 140–150, 2017.

[857] K. Rangavajhula and H.S.J. Tsao, Command-steering and command-steering-based optimal control of an articulated system for tractor track following, Proceedings of Institute of Mechanical Engineering, Part J, Journal of Automobile Engineering, vol. 222, pp. 935–955, 2008.

[858] H.V. Rangel, A.G. Loukianov, B. Castillo-Toledo and E. Bayio-Corrochano, Sliding control variation for ball-plate system stabilization, IEEE VSS 2014, IEEE 13th Workshop on Variable Structure Systems, Nantes, France, July, 2014.

[859] P. Ranjan, G. Wrat, M. Bhala, S.K. Mishara and J. Das, A novel approach for the energy recovery and position control of a hybrid hydraulic excavator, ISA Transactions, Elsevier, vol. 99, pp. 387–402, 2020.

[860] A. Rauh, S.S. Butt and H. Aschemann, Nonlinear state observers and extended Kalman Filters for battery systems, International Journal of Applied Mathematics and Computer Science, De Gruyter, vol. 23, no. 3, pp. 539–556, 2013.

[861] S. Rawanasamy, G. Wu and K. Sreenath, Dynamically feasible motion planning through partial differential flatness, 2014 IEEE International Conference on Robotics Science and Systems, IEEE RSS 2014, Berkeley, California, July 2014.

[862] H. Reiter, S. Stemler, C. Hopmann, A. Ressmann and D. Abel, Model-predictive control of cavity pressure in the injection molding process, 19th IFAC World Congress, Cape Town, South African August 2014.

[863] H. Ren, G. Deng, B. Hou, S. Wang and G. Zhou, Finite-time command filtered backstepping algorithm-based pitch angle tracking control for wind-turbine pitch systems, IEEE Access, vol. 5, pp. 135514–135534, 2019.

[864] Q. Ren and H. Kobayashi, Diffusion approximation modelling for Markov modulated bursty traffic and its application to bandwidth allocation in ATM networks, IEEE Journal of Selected Areas in Communications, vol. 16, no. 3, pp. 679–691, 1997.

[865] M. Reyhanoglu and J.R. Hervas, Nonlinear modelling and control of slosh in liquid container transfer via a PPR robot, Communication in Nonlinear Science and Numerical Simulation, Elsevier, vol. 18, pp. 1481–1490, 2013.

[866] A. Riani, T. Madani, A. Benallegue and K. Djouanni, Adaptive integral terminal sliding-mode control for upper-limb rehabilitation exoskeleton, Control Engineering Practice, Elsevier, vol. 75, pp. 108–117, 2018.

[867] G. Rigatos, G. Vagelatos and S. Tzafestas, Incremental fuzzy supervisory controller design for optimizing the injection moulding process, Expert Systems with Applications, Elsevier, vol. 20, no. 2, pp. 207–216, 2001.

[868] G.G. Rigatos and S.G. Tzafestas, Extended Kalman Filtering for fuzzy modelling and multi-sensor fusion, Mathematical and Computer Modelling of Dynamical Systems, Taylor & Francis, vol. 13, pp. 251–266, 2007.
[869] G. Rigatos and Q. Zhang, Fuzzy model validation using the local statistical approach, Fuzzy Sets and Systems, Elsevier, vol. 60, no. 7, pp. 882–904, 2009.
[870] G.G. Rigatos, P. Siano and A. Piccolo, A neural network-based approach for early detection of cascading events in electric power systems, IET Journal on Generation Transmission and Distribution, vol.3, no. 7, pp. 650–665, 2009.
[871] G.G. Rigatos, Extended Kalman and Particle Filtering for sensor fusion in motion control of mobile robots, Mathematics and Computers in Simulation, Elsevier, vol. 81, no. 3, pp. 590–607, 2010.
[872] G. Rigatos, Modelling and Control for Intelligent Industrial Systems: Adaptive Algorithms in Robotics and Industrial Engineering, Springer, 2011.
[873] G.G. Rigatos, Adaptive fuzzy control of MIMO dynamical systems using differential flatness theory, IEEE ISIE 2012, 21st International Symposium on Industrial Electronics, Hangzhou, China, May 2012.
[874] G.G. Rigatos, Adaptive fuzzy control for non-linear dynamical systems based on differential flatness theory, IET Control Theory and Applications, vol. 6, no. 17, pp. 2644–2656, 2012.
[875] G.G. Rigatos, A derivative-free Kalman Filtering approach to state estimation-based control of nonlinear dynamical systems, IEEE Transactions on Industrial Electronics, vol. 59, no. 10, pp. 3987–3997, 2012.
[876] G. Rigatos, Advanced Models of Neural Networks: Nonlinear Dynamics and Stochasticity of Biological Neurons, Springer, 2013.
[877] G.G. Rigatos, A chaotic communication system of improved performance based on the derivative-free nonlinear Kalman Filter, International Journal of Systems Science, Taylor and Francis, 2014.
[878] G. Rigatos, Nonlinear Control and Filtering Using Differential Flatness Theory: Applications to Electromechanical Systems, Springer, 2015.
[879] G. Rigatos, Control of the nonlinear wave-type dynamics using the derivative-free nonlinear Kalman Filter, Journal of Intelligent Industrial Systems, vol. 2, no. 2, pp. 119–132, 2016.
[880] G. Rigatos, Intelligent Renewable Energy Systems: Modelling and Control, Springer, 2016.
[881] G. Rigatos, State-space Approaches for Modelling and Control in Financial Engineering: Systems Theory and Machine Learning Methods, Springer, 2017.
[882] G. Rigatos and K. Busawon, Robotic Manipulators and Vehicles: Control, Estimation and Filtering. Springer, 2018.
[883] G. Rigatos and E. Karapanou, Advances in Applied Nonlinear Optimal Control, Cambridge Scholars Publishing, 2020.
[884] G.G. Rigatos and P. Siano, Design of robust electric power system stabilizers using Kharitonov's theorem, Mathematics and Computers in Simulation, Elsevier, vol. 81, no.1, pp. 181–191, 2011.

[885] G. Rigatos and P. Siano, Sensorless nonlinear control of induction motors using Unscented Kalman Filtering, IEEE IECON 2012, 38th International Conference of the Industrial Electronics Society, Montreal, Canada, October 2012.

[886] G. Rigatos and P. Siano, DFIG control using Differential flatness theory and Extended Kalman Filtering, IFAC INCOM 2012, 14th IFAC International Conference on Information Control Problems in Manufacturing, Bucharest, Romania, May 2012.

[887] G. Rigatos and P. Siano, A new nonlinear H-infinity feedback control approach to the problem of autonomous robot navigation, Journal of Intelligent Industrial Systems, Springer, vol. 1, no. 3, pp. 179–186, 2015.

[888] G. Rigatos, D. Clare and S. Blackmore, Automata-based modeling and synthesis for programmable logic controllers, IFAC International Conference, Dynamics and Control in Agriculture and Food Processing, 2012.

[889] G. Rigatos, P. Siano and N. Zervos, Sensorless Control of Distributed Power Generators With the Derivative-free Nonlinear Kalman Filter, IEEE Transactions on Industrial Electronics, vol. 61, no. 11, pp. 6369–6382, 2014.

[890] G. Rigatos, P. Siano and C. Cecati, A new non-linear H-infinity feedback control approach for three-phase voltage source converters, Electric Power Components and Systems, Taylor and Francis, vol. 44, no. 3, pp. 302–312, 2015.

[891] G. Rigatos, P. Siano, P. Wira and F. Profumo, Nonlinear H-infinity feedback control for asynchronous motors of electric trains, Journal of Intelligent Industrial Systems, Springer, vol. 1, pp. 85–98, 2015.

[892] G. Rigatos, P. Siano and A. Melkikh, Differential flatness properties and feedback control of the Fokker-Planck PDE, Journal of Intelligent Industrial Systems, Springer, vol. 2, pp. 233–241, 2016.

[893] G. Rigatos, P. Siano and M. Abbaszadeh, Nonlinear H-infinity control for 4-DOF underactuated overhead cranes, Transactions of the Institute of Measurement and Control, Sage Publications, vol. 40, no. 7, pp. 2364–2377, 2018.

[894] G. Rigatos, P. Siano, M. Jovanovic, S. Ademi, P. Wira and Z. Tir, Nonlinear optimal control for Synchronous Reluctance Machines, IEEE CPE-PowerEng 2017, 2017 11th IEEE International Conference on Compatibility, Power Electronics and Power Engineering, Cadiz, Spain, May 2017.

[895] G. Rigatos, G. Cuccurullo, P. Siano, M. Hamida and M. Abbaszadeh, Nonlinear optimal control for the industrial crystallization process, Proceedings of the ICNAAM 2019, 17th International Conference on Numerical Analysis and Applied Mathematics, Rhodes, Greece, September 2019.

[896] G. Rigatos, K. Busawon, J. Pomares and M. Abbaszadeh, Nonlinear optimal control for the wheeled inverted pendulum, Robotica, Cambridge University Press, vol. 38, no. 1, pp. 29–47, 2019.

[897] G. Rigatos, P. Siano, F. Marignetti and I. Gros, A nonlinear optimal control approach for PM Linear Synchronous Motors, IEEE INDIN 2018, IEEE

16th International Conference on Industrial Informatics, Porto, Portugal, July 2018.

[898] G. Rigatos, P. Siano, P. Wira and A. Munoz, Flatness-based Adaptive Control of Synchronous Reluctance Machines with Output Feedback, 2018 IEEE EFEA 2018, 5th International Symposium on Environment-Friendly Energies and Applications, Rome, Italy, September 2018.

[899] G. Rigatos, K. Busawon, J. Pomares and M. Abbaszadeh, Nonlinear optimal control for the wheeled inverted pendulum, Robotica, vol. 38, no. 1, pp. 29–47, 2020.

AQ1

[900] G Rigatos, M Abbaszadeh, J Pomares, P Wira, A nonlinear optimal control approach for a lower-limb robotic exoskeleton, World Scientific, vol. 17, no. 5, pp. 2050018, 2020.

[901] G. Rigatos, P. Siano, P. Wira, K. Busawon an R. Binns, A nonlinear H-infinity control approach for autonomous truck and trailer systems, Unmanned Systems, World Scientific, vol. 98, no. 1, pp. 49–69, 2019.

[902] G. Rigatos, N. Zervos, M. Abbaszadeh, J. Pomares and P. Wira, Nonlinear optimal control for multi-DOF electro-hydraulic robotic manipulators, IET Cyber-systems and Robotics, vol. 2, no. 2, pp. 96–106, 2020.

[903] G. Rigatos, M. Abbaszadeh, J. Pomares, and P. Wira, A nonlinear optimal control approach for a lower-limb robotic exoskeleton, World Scientific, vol. 17, no. 5, pp. 2050018, 2020.

[904] G. Rigatos, N. Zervos, D. Serpanos, V. Siadimas, P. Siano and M. Abbaszadeh, Fault diagnosis of gas turbine power units with the derivative-free nonlinear Kalman Filter, Electric Power Systems Research, Elsevier, vol. 174, 105518, 2019.

[905] P. Ritzen, E. Roebroek, N. van de Wouw, Z.P. Jiang and H. Nijmeijer, Trailer steering control of a tractor-trailer robot, IEEE Transactions on Control Systems Technology, vol. 24, no. 4, pp. 1240–1252, 2016.

[906] R. Rodrigues, A. Murillo, R. Vilela-Lopes and L.C. Cadelh de Souza, Hardware in the loop simulation for model predictive control applied to satellite attitude control, IEEE Access, vol. 7, pp. 157401–157416, 2019.

[907] C.A. Rodriguez, P. Ponce and A. Molina, ANFIS and MPC controllers for a reconfigurable linear limb exoskeleton, Soft Computing, Springer, vol. 21, no. 3, pp. 571–584, 2017.

[908] A. Rosales, G. Scaglia, V. Mut and F. di Sciascio, Controller design by means of numeric methods for a benchmark problem: RTAC (Rotational Translational Actuator), IEEE CERMA 2006, IEEE Conference on Electronics, Robotics and Automotive Mechanics, Cuernavaca, Mexico, September 2006.

[909] J.A. Rossiter and P. Fleming, Robust constrained predictive controllers for hot rolling mills: Disturbance uncertainty case. IMech Part I, Journal of Systems and Control Engineering, vol. 222, pp. 132–153, 2008.

[910] P. Rouchon, Flatness-based control of oscillators, ZAMM Journal of Applied Mathematics and Mechanics, vol. 85, no. 6, pp. 411–421, 2005.

[911] P. Rouchon, P. Martin and L. Rosier, Null controllability of the heat equation using flatness, Automatica, Elsevier, vol. 50, no. 12, pp. 3067–3076, 2014.

[912] R. Rout, R. Cui and Z. Han, Modified line-of-sight guidance law with adaptive neural network control of underactuated marine vehicle with state and input constraints, IEEE Transactions on Control Systems Technology, vol. 28, no. 5, pp. 1902–1914, 2020.

[913] K. Rsetam, Z. Cao and Z. Man, Cascaded extended state observer-based sliding-mode control for underactuated flexible-joint robot, IEEE Transactions on Industrial Electronics, vol. 67, no. 12, pp. 10822–10832, 2020.

[914] M. Ruderman, Compensation of nonlinear torsion in flexible-joint robots: Comparison of two approaches, IEEE Transactions on Industrial Electronics, vol. 83, no. 9, pp. 5744–5751, 2016.

[915] J. Rudolph, Flatness-based control of Distributed Parameter Systems: Examples and Computer Exercises from Various Technological Domains, Shaker Verlag, Aachen, 2003.

[916] F. Rugierro, M.A. Trajillo, R Cano, *et al.*, A multi-layer control for multi-rotor UAVs equipped with a servo robot arm, IEEE ICRA 2015, IEEE 2015 International Conference on Robotics and Automation, Seattle USA, May 2015.

[917] M. Ryalat and D.S. Laila, A simplified IDA-PBC design for underactuated mechanical systems with applications, European Journal of Control, Elsevier, vol. 27, pp. 1–16, 2016.

[918] S. Ryu, C. Rumpand and C. Qiao, Advances in internet congestion control, IEEE Communication Surveys, vol. 5, no. 1, pp. 28–39, 2003.

[919] M. Saadatzi, D.C. Long and O. Celik, Comparison of human-robot interaction torque estimation method in a wrist rehabilitation exoskeleton, Journal of Intelligent and Robotic Systems, Springer, vol. 94, pp. 565–581, 2018.

[920] A. Saadatpour and M. Levi, Travelling waves in chains of pendulum, Physica D, vol. 244, pp. 68–73, 2013.

[921] H. Sadeghian, L. Villani, H. Keshmiri and B. Siciliano, Task-space control of robot manipulators with null-space compliance, IEEE Transactions on Robotics, vol. 30, no. 2, pp. 2014.

[922] P. Saeed, P.D. Lawrence, D.G. Lowe, *et al.*, An autonomous excavator with vision-based track slippage control, IEEE Transactions on Control Systems Technology, vol. 13, no. 1, pp. 67–84, 2004.

[923] O. Saeed, A. Wahyudie, T.B. Susilo and H. Shareef, Simple resonance circuit to improve electrical power conversion in a two-sided planar permanent magnet linear generator for wave energy converters, IEEE Access, vol. 5, pp. 18654–18664, 2017.

[924] H. Saied, B. Lussier, I. Fantoni, H. Shraim and C. Francis, Passive fault-tolerant control of an octorotor using super-twisting algorithm: Theory and experiments, IEEE SysTol 2016, IEEE 3rd Conference on Control and Fault-tolerant Systems, Barcelona, Spain, September 2016.

[925] J.C. Salazar, A. Sanjuan, F. Nejjari and R. Sarrate, Health-aware control of an octorotor UAV system based on actuator reliability, IEEE CoDIT 2017, 2017 4th International Conference on Control, Decision and Information Technologies, April 2017, Barcelona, Spain.

[926] S. Sakain and T. Tsuji, Development of friction-free controller for electro-hydrostatic actuator using feedback modulator and disturbance observer, Robomech, Springer, vol. 4, no. 1, pp. 1–10, 2017.
[927] S.A. Salberg, P.S. Maybeck and M.E. Oxley, Infinite-dimensional sampled-data Kalman Filtering and stochastic heat equation, 49th IEEE Conference on Decision and Control, Atlanta, Georgia, USA, December 2010.
[928] M. Salehifar, M. Moreno-Eguilaz, G. Putrus and P. Baras, Simplified fault-tolerant finite control set model predictive control of a five-phase inverter supplying BLDC motor in electric vehicles, Electric Power Systems Research, Elsevier, vol. 132, pp. 56–66, 2016.
[929] M. Salehifar and M. Moeno-Equilaz, Fault diagnosis and fault tolerant finite control set-model predictive control of multi-phase voltage-source inverter supplying BLDC motor, ISA Transactions, Elsevier, vol. 60, pp. 145–153, 2018.
[930] S. Sanaye, N. Khakpaay, and A. Chitsaz, Thermo-economic and environmental multi-objective optimization of a novel arranged biomass-fueled gas engine and backpressure steam turbine combined system for pulp and paper mills, Sustainable Energy Technologies and Assessments, Elsevier, vol. 40, pp. 1–18, 2020.
[931] E.N. Sanchez and F. Ornelas-Tellez, Discrete-time Inverse Optimal Control for Nonlinear Systems, CRC Press, 2017.
[932] B. Sanchez, C. Cuvas, P. Ordaz, O. Santos-Sanchez and A. Poznyak, Observer for a class of nonlinear systems with unmatched uncertainties: Joint attractive ellipsoid and sliding mode concepts, IEEE Transactions on Industrial Electronics, vol. 67, no. 7, pp. 5677–5686, 2020.
[933] J. Sandeval, P. Vieyes and G. Poisson. Generalized framework for control of redundant manipulators in robot-assisted minimal invasive surgery, IRBM Journal (Innovation and Research in Biomedical Engineering), Elsevier, vol. 39, pp. 160–166, 2018.
[934] V. Santibanez, R. Kelly and J. Sandoval, Control of the inertia wheel pendulum by bounded torques, IEEE CDC 2005, 44th IEEE Conference on Decision and Control, Seville, Spain, December 2005.
[935] T. Sasaki and K. Kawashima, Remote control of backhoe at construction site with a pneumatic robot system, Automation in Construction, Elsevier, vol. 17, pp. 907–914, 2008.
[936] E.H. Satel, P. Acharya and K.D. Nguyen, Brachiating Robot Analysis and Design, 2019 IEEE International Conference on Electro Information Technology (EIT), Brookings, USA, May 2019.
[937] A.C. Satici, F. Ruggiero, V. Lippiello and B. Siciliano, Intrinsic Euler-Lagrange dynamics and control analysis of the ballbot, 2016 American Control Conference, IEEE ACC 2016, Boston, Massachusetts, June 2016.
[938] A. Satici, A. Donaire and B. Siciliano, Intrinsic dynamics and total energy-shaping control of the ballbot system, International Journal of Control, Taylor and Francis, vol. 90, no. 12, pp. 2734–2747, 2017.

[939] K. Satpathi, A. Ukil, S.S. Nag, J. Poou and M.A. Zagradnik, DC marine power system: Transient behavior and fault management aspects, IEEE Transactions on Industrial Informatics, vol. 15, no. 4, pp. 1911–1924, 2019.

[940] D. Sbarbaro-Hofer, D. Neumerkel and K. Hune, Neural control of a steel rolling mill, IEEE Control Systems Magazine, vol. 13, no. 3, pp. 69–78, 1993.

[941] J. Scerri, G.S. Dordevic and D. Todorovic, Modelling and control of a reaction wheel pendulum with visual feedback, IEEE ICCAD 2017, Hammamet, Tunisia, June 2017.

[942] F. Schausberger, A. Steinback and A. Kugi, Feedback control of the contour shape in heavy-plate hot rolling, IEEE Transactions on Control Systems Technology, vol. 26, no. 3, pp. 842–852, 2018.

[943] K.H. Seok and Y.S. Kim, A state-of-the-art of power transmission for maintenance robots, Journal of Electrical Engineering Technology, The Korean Institute of Electrical Engineers, vol. 11, no. 5, pp. 1412–1422, 2016.

[944] J. Seron, J.L. Martinez, A. Mandow, A.J. Reina, J. Morales, and A.J. Garcia-Cerezo, Automation of the arm-aided climbing maneuver for tracked mobile manipulators, IEEE Transactions on Industrial Electronics, vol. 61, no. 7, pp. 3638–3647, 2014.

[945] W. Shang and S. Cong, Nonlinear adaptive task-space control for a 2-DOF redundantly actuated parallel manipulator, Nonlinear Dynamics, Springer, vol. 59, pp. 61–72, 2020.

[946] W. Shang, S. Cong and Y. Ge, Adaptive computed torque control for a parallel manipulator with redundant actuation, Robotica, Cambridge University Press, vol. 30, pp. 457–466, 2011.

[947] W. Shang and S. Cong, Motion control of parallel manipulators using acceleration feedback, IEEE Transactions on Control Systems Technology, vol. 22, no. 1, pp. 314–321, 2014.

[948] G. Sharma, I. Nasiruddin, K.R. Niazi and R.C. Bansal, Robust automatic generation control regulators for a two-area power system interconnected via AC/DC tie-lines considering new structures of matrix Q, IET-Generation Transmission and Distribution, vol. 10, no. 14, pp. 3570–3579, 2016.

[949] G. Sharma, I. Nasiruddin and K.R. Niazi, Optimal AGC of asynchronous power systems using output feedback control strategy with dynamic participation of wind turbines, Electric Power Component and System Journal, Taylor & Francis, vol. 43, no. 4, pp. 384–398, 2015.

[950] H. Shen, X. Zhang, A.B. Alhassan, J. Gup and H. Xu, Research on the coupling model for inspection robot to depart from high-voltage transmission line, IEEE 2018 International Conference on Sensing, Diagnostics, Prognostics and Control, IEEE SDPC 2018, Xian, China, August 2018.

[951] M. Shomin and R. Hollis, Differentially flat trajectory generation for a dynamically stable mobile robot, 2013 IEEE International Conference on Robotics and Automation, IEEE ICRA 2013, Karlsruhe, Germany, May 2013.

[952] M. Shamin, J. Fodicci and R. Hollis, Sit-to-stand assistance with a balancing mobile robot, 2015 IEEE International Conference on Robotics and Automation, IEEE ICRA 2015, Seattle, Washington, USA, May 2015.

[953] C. Shen, Y. Shi, and B. Buckham, Path-following control of an AUV: A multiobjective model predictive control approach, IEEE Transactions on Control Systems Technology, vol. 27, no. 3, pp. 1334–1343, 2019.

[954] C. Shen, Y. Shi and B. Buckham, Trajectory tracking control of an autonomous underwater vehicle using Lyapunov-based model predictive control, IEEE Transactions on Industrial Electronics, vol. 65, no. 7, pp. 5796–5805, 2018.

[955] C. Shen, B. Buckham and Y. Shi, Modified C/GMRES algorithm for fast nonlinear model predictive tracking control of AUVs, IEEE Transactions on Control Systems Technology, vol. 25, no. 5, pp. 1896–1905, 2017.

[956] L. Shi, N. Kinkaid and J. Katupitiya, Robust control for satellite attitude regulation during on-orbit assembly, IEEE Transactions on Aerospace Science and Electronic Systems, vol. 52, no. 1, pp. 49–59, 2016.

[957] C. Shi, X. Wang, S. Wang, J. Wang and M.T. Tomovic, Adaptive decoupling synchronous control of dissimilar redundant actuator system for large civil aircraft, Aerospace Science and Technology, Elsevier, vol. 47, pp. 114–124, 2015.

[958] H. Schulte and K. Guelton, Descriptor modelling control of two-link pneumatic robot manipulator: A T-S multimodel approach, Nonlinear Analysis: Hybrid Systems, Elsevier, vol. 3, no. 2, pp. 124–132, 2009.

[959] M.R. Sirouspour and S.E. Salcudean, Nonlinear control of hydraulic robots, IEEE Transactions on Robotics and Automation, vol. 17, no. 2, pp. 173–182, 2001.

[960] Y. Shtessel, M. Taleb and F. Plestan, A novel adaptive-gain super-twisting sliding-mode controller: Methodology and application, Automatica, Elsevier, vol. 48, pp. 759–769, 2012.

[961] D. Schnidele, R. Prabel and H. Aschemann, Nonlinear model-predictive control of an electropneumatic clutch for truck applications, IFAC Mathmod 2012, IFAC 7th International Conference on Mathematical Modelling, Vienna, Austria, 2012.

[962] B. Sebastian and P. Ben-Tzvi, Active disturbance rejection control for handling slip in tracked vehicle locomotion, ASME Journal of Mechanisms and Robotics, vol. 11, pp. 021003, 1–12, 2019.

[963] T. Senjiya, T. Shingaki and K. Uezato, Sensorless vector control of synchronous reluctance motors with disturbance torque observer, IEEE Transactions on Industrial Electronics, vol. 48, no. 2, pp. 402–407, 2001.

[964] L. Shang, Z. Yan, J. Li, A. Qiu and H. Zhang, Canonical residual based incipient fault detection method for industrial process, IEEE CCDC 2020, Chinese Control and Decision Conference, Hefei, China, August 2020.

[965] E. Sharifi and C.L. Damaren, Nonlinear optimal approach to spacecraft attitude control using magnetic and impulsive actuations, Journal of Guidance, Control and Dynamics, AIAA Journals, vol. 43, no. 6, pp. 1154–1163, 2020.

[966] R. Sharma, S. Kewat and B. Singh, Robust 3IMPL control algorithm for power management of SyRG/PV/BES-based distributed islanded microgrid, IEEE Transactions on Industrial Electronics, vol. 65, no. 10, pp. 7765–7772, 2019.

[967] W. Sheng, H. Chen and N. Xi, Navigating a miniature crawler robot for engineered structure inspection, IEEE Transactions on Automation Science and Engineering, vol. 5, no. 2, pp. 368–375, 2008.

[968] P. Shi, J. Li, D. Zhou, B. Liu and D. Han, Nonlinear torsional vibrational dynamics of rolling mill's drive system under spindle parametric excitation, Proceedings of the 10th World Congress on Intelligent Control and Automation, July 2012, Beijing, China.

[969] G. Shi, Z. Zhang and Y. Sheng, Valuation of stock loan under uncertain mean-reverting stock model, Journal of Intelligent and Fuzzy Systems, IOS Press, vol. 33, no. 3, pp. 1355–1361, 2017.

[970] M.S. Shiruai Bouroujani, G.R Arab Markadeh and J. Soltani, Torque ripple reduction of brushless DC motor based on adaptive input-output feedback linearization, ISA Transactions, Elsevier, vol. 70, pp. 502–511, 2017.

[971] L.L. Show, J.C. Juang and Y.W. Jan, An LMI-based nonlinear attitude control approach, IEEE Transactions on Control Systems Technology, vol. 11, no. 1, pp. 73–83, 2014.

[972] K.K. Shyu, C.K. Lai and Y.W. Tsai, Optimal position control of synchronous reluctance motor via totally invariance variable structure control, IEE Proceedings on Control Theory and Applications, vol. 147, no.7, pp. 28–36, 2000.

[973] S. Siad, A. Malkawi, G. Damma, L. Lopes and L. Dold, Nonlinear control of a DC MicroGrid for the integration of distributed generation based on different time scales, Electric Power and Energy Systems, Elsevier, vol. 111, pp. 93–100, 2018.

[974] A.P. Silva, S. Barleigh, C.M. Hirata and K. Obraczks, A survey on congestion control for delay and disruption tolerant networks, Ad-hoc Networks, Elsevier, vol. 25, Part B, pp. 480–494, 2015.

[975] R. da Silveira Castro, J. Vieira Flores, A. Tergolina Salton, and L.F. Alves Pereira, A comparative analysis of repetitive and resonant controllers to a servo-vision ball and plate system, Proceedings of the 19th World Congress The International Federation of Automatic Control Cape Town, South Africa, August 24–29, 2014.

[976] D. Simon, A game theory approach to constrained minimax state estimation. IEEE Transactions on Signal Processing, vol. 54, no. 2, pp. 405–412, 2006.

[977] R. Singh, R. Sharma and S. Kewat, Robust control strategies for RG-PV and wind-based islanded microgrid, IEEE Transactions on Industrial Electronics, vol. 68, no. 4, pp. 3137–3147, 2021.

[978] S. Singh and A. Iyer, Nonlinear decoupling sliding-mode control and attitude control of spacecraft, IEEE Transactions on Aerospace and Electronic Letters, vol. 25, no. 5, pp. 621–633, 1989.

[979] H. Sira-Ramirez and M. Fliess, A flatness-based generalized PI control approach to liquid sloshing regulation in a moving container, IEEE ACC 2002, IEEE 2002 American Control Conference, Anchorage, Alaska, May 2002.

[980] H. Sira-Ramirez and S. Agrawal, Differentially Flat Systems, Marcel Dekker, New York, 2004.

[981] J.J. Slotine and W. Li, Applied Nonlinear Control, Prentice Hall, 1991.

[982] M. Smaoui, X. Brun and D. Thomasset, Systematic control of an electropneumatic system: Integrator backstepping and sliding-mode control, IEEE Transactions on Control Systems Technology, IEEE Transactions on Control Systems Technology, vol. 14, no. 5, pp. 905–913, pp. 457-467, 2006.

[983] M. Smaoui, X. Brun and D. Thomasset, A study on tracking position control of an electropneumatic system using backstepping design, Control Engineering Practice, Elsevier, vol. 14, pp. 923–933, 2006.

[984] K. Smith, Electrochemical control of lithium-ion batteries, IEEE Control Systems Magazine, vol 30, no. 2, pp. 18–25, 2010.

[985] K.A. Smith, C.D. Rahn and C.Y. Wang, Control oriented 1D electrochemical model of lithium-ion battery, Energy Conversion and Management, Elsevier, vol. 48, pp. 2565–2578, 2007.

[986] K. Smith, C. Rahn and C.Y. Wang, Model-based electrochemical estimation and constraint management for pulse operation of lithium-ion batteries, IEEE Transactions on Control Systems Technology, vol. 10, no. 18, pp. 654–665, 2010.

[987] A. Smyshlyaev and M. Krstic, Adaptive Control of Parabolic PDEs, Princeton University Press, 2010.

[988] J.J. Slotine, Applied Nonlinear Control, Prentice-Hall, 1991.

[989] A.A. Sofiane, Sample-data observer-based intersample output predictor for electrohydraulic actuators, ISA Transactions, Elsevier, vol. 58, pp. 421–433, 2015.

[990] J. Son, S. Zhou, C. Sankavaram, X. Du and Y. Zhang, Remaining useful life prediction based on noisy condition monitoring signals using constrained Kalman filter, Reliability Engineering & System Safety, Elsevier, vol. 152, pp. 38–50, 2016.

[991] F. Sonleitner, R. Shu and R.L. Hollis, The mechanics and control of leaning to lift heavy objects with a dynamically stable mobile robot, 2019 IEEE International Conference on Robotics and Automation, IEEE ICRA 2019, Montreal, Canada, May 2019.

[992] F. Song and J. Cameron, Evaluating the dynamics of tension mooring supported tidal turbines, 12th European Wave and Tidal Energy Conferences, University of College Cork, Ireland, August 2017.

[993] Q. Song, L. Wang, and J. Chen, A decentralized energy management strategy for a fuel cell-battery hybrid electric vehicle based on composite control, IEEE Transactions on Industrial Electronics, vol. 68, no. 7, pp. 5486–5496, 2020.

[994] Y. Song, W. Lin, H. Li, Y. Zhou, Z. Huang and F. Jiang, Robust and accurate state-of-char estimation for lithium-ion batteries using generalized extended state observer, 2017 IEEE International Conference on Systems, Man and Cybernetics, Baniff, Canada, October 2017.

[995] M. Souzanchi, A. Arab, M.R. Akbarzadeh and M.M. Fetch, Robust impedance control of uncertain mobile manipulators using time-delay compensation, IEEE Transactions on Control Systems Technology, vol. 26, no. 6, pp. 1942–1953, 2018.

[996] J. Sowman, D.S. Laila and S. Longo, Real-time approximate explicit Nonlinear Model Predictive Control for the swing-up of a reaction wheel pendulum, IEEE CDC 2015, IEEE 54th Annual Conference on Decision and Control, Osaka, Japan, December 2015.

[997] H. Soyyaadi and A. Soltani, Modelling and control for cooperative transport of a slung fluid container using quadrotors, Chinese Journal of Aeronautics, Elsevier, vol. 31, no. 2, pp. 262–272, 2018.

[998] M.W. Spong, P. Corke and R. Lozano, Nonlinear control of the reaction wheel pendulum, Automatica, Elsevier, vol. 37, pp. 1845–1851, 2001.

[999] A. Steinboeck, G. Muhlberger and A. Kugi, Control of strip tension in a rolling mill based on loopers and impedance control, Control of strip tension in a rolling mill base on loopers and impendance control, IFAC 2014 World Congress, Cape Town, South Africa, August 2014.

[1000] S. Stemmler, M. Ay, M. Vukovic, D Abel, J. Heinisch and C. Hopmann, Cross-phase model-based predictive cavity pressure control in injection moulding, IEEE CCTA 2019, IEEE 2019 Conference on Control Technology and Applications, Hong Kong, China, August 2019.

[1001] S. Stemmler, M. Vukovic, M. Ay, *et al.*, Quality control of injection moulding based on norm-optimal iterative learning cavity pressure control, 21st IFAC World Congress, Proceedings of the 21st Word Congress of the International Federation of Automatic Control, Berlin, Germany, July 2020.

[1002] N. Su, Y. Fang, H. Chen, B. Lu and Y. Fu, Slew/Translation positioning and swing suppression for 4-DOF tower cranes with parametric uncertainties: Design and hardware experimentation, IEEE Transactions on Industrial Electronics, vol. 63, no. 10, pp. 6407–6418, 2016.

[1003] B. Su, T. Wang, S. Kuang and J. Wang, Effects of physical parameters on dynamic behavior of a ballbot-like robot, International Journal of Robotics and Automation, vol. 29, no. 1, pp. 105–115, 2014.

[1004] B. Su and Y. Gang, Euler-Lagrangian modeling and exact trajectory following controller of a ballbot-like robot, IEEE 2017 International Conference on Robotics and Biomimetics, IEEE ICRB 2107, Macau, China, December 2017.

[1005] S. Su and Y. Lin, Robust output tracking control of VTOL aircraft without velocity measurements, Proceedings of the 18th IFAC World Congress, Milan, Italy, September 2011.

[1006] B. Su, T. Wang, S. Kuang and J. Wang, Effects of physical parameters on dynamic behavior of a ballbot-like robot, International Journal of Robotics and Automation, vol. 29, no. 1, pp. 105–115, 2014.

[1007] A. Suarez, A.E. Jimenez-Cano, V.M. Vega, G. Heredia, A. Rodriguez-Castano and A. Ollero, Design of a lightweight dual-arm system for aerial manipulation, Mechatronics, Elsevier, vol. 50, pp. 30–44, 2018.

[1008] F. Sudo, H.J. Yap, R.A.R. Ghazilla and N. Ahmad, Exoskeleton robot control for synchronous walking assistance in repetitive manual handling works based on dual Unscented Kalman Filter, PLoS One, vol. 13, no. 7, pp. e0200193, 2019.

[1009] R.A. Suarez Fernandez, D. Grande, A. Martins, L. Bascetta, S. Dominguez and C. Rossi, Modelling and control of underwater mine explorer robot UX-1, IEEE Access, vol. 7, pp. 39452–39465, 2019.

[1010] A. Suarez, G. Heredia and A. Ollero, Physical virtual impendance control in ultra lightweight and compliant dual-arm aerial manipulators, IEEE Robotics and Automation Letters. vol. 3, pp. 2553–2560. 2018.

[1011] T. Sumi, C. Takeno, K. Ishida and M. Aida, Implementation and verification of autonomous decentralized flow control based on local interaction, IEEE ISADS 2011, 10th International Symposium on Autonomous Decentralized Systems, Tokyo, Japan, November 2011.

[1012] Y. Sun and D.B. Work, A distributed local Kalman consensus filter for traffic estimation, 53rd IEEE Conference on Decision and Control, Los Angeles, California, December 2014.

[1013] N. Sun, Y. Wu, X. Liang and Y. Fang, Nonlinear state transportation control for double-pendulum shipboard cranes with ship-motion-induced disturbances, IEEE Transactions on Industrial Electronics, vol. 66, no. 12, pp. 96467–9479, 2019.

[1014] N. Sun, Y. Wu, Y. Fang and H. Chen, Non-linear anti-swing control for crane systems with double-pendulum swing effects and uncertain parameters: Design and experiment, IEEE Transactions on Automation Science and Engineering, vol. 15, no. 3, pp. 1413–1422, 2018.

[1015] N. Sun, Y. Fu, T. Yang, J. Zhang, Y. Feng and X. Xin, Nonlinear matrix control of complicated dual rotary crane systems without velocity feedback: Design, analysis and hardware experiment, IEEE Transactions on Automation Science and Engineering, vol. 17, no. 2, pp. 1017–1029, 2020.

[1016] N. Sun, Y. Fang, H. Chen and B. Lu, Amplitude-saturated nonlinear output feedback antiswing control for underactuated cranes with double-pendulum cargo dynamics, IEEE Transactions on Industrial Electronics, vol. 64, no. 3, pp. 2135–2146, 2017.

[1017] Y.G. Sun, H.Y. Qiang, J. Xu and D.S. Dong, The nonlinear dynamics and anti-sway tracking control for offshore boom crane on a mobile harbor, Journal of Marine Science and Technology, Springer, vol. 25, no. 6. pp. 656–665, 2017.

[1018] N. Sun, Y. Wu, Y. Fang, H. Chen, and B. Lu, Nonlinear continuous global stabilization control for underactuated RTAC systems: Design, analysis, and experimentation, IEEE/ASME Transactions on Mechatronics, vol. 22, no. 2, pp. 1104–1115, 2017.

[1019] L. Sun, W. Yu, M. Yang and J. Liu, Position control for flexible-joint robot based on online gravity compensation with vibration suppression, IEEE Transactions on Industrial Electronics, vol. 65, no. 6, pp. 4840–4848, 2018.

[1020] J. Sun, Z. Pu and J. Yi, Conditional disturbance negation-based source disturbance rejection control for hypersonic vehicles, Control Engineering Practice, Elsevier, vol. 84, pp. 159–191, 2019.

[1021] J. Sun, J. Yi, Z. Pu and Z. Liu, Adaptive fuzzy non-smooth backstepping output feedback control for hypersonic vehicles with finite-time convergence, IEEE Transactions on Fuzzy Systems, vol. 28, no. 10, pp. 2320–2334, 2019.

[1022] N. Sun, Y. Fu, T. Yang, J. Zhang, Y. Fang and Xin Xin, Nonlinear motion control of complicated dual rotary crane systems without velocity feedback: Design, analysis, and hardware experiments, IEEE Transactions on Automation Science and Engineering, vol. 17, no. 2, pp. 1017–1029, 2020.

[1023] R. Sutherland, I. Kolmanovsky and A.R. Girard, Attitude control of a 2U CubeSat by magnetic and air drag torques, IEEE Transactions on Control Systems Technology, vol. 27, no. 3, pp. 1047–1059, 2019.

[1024] T. Suzuki, Y. Hayashi, H. Kabune, and N. Ito, Pulse width modulation control algorithm for a six-phase PMSM: Reducing the current in the inverter capacitor and current sensing with resistors, IEEE Transactions on Industrial Electronics, vol. 86, no. 6, pp. 4240–4249, 2019.

[1025] G. Tadman, Dissipative design, lossless dynamics and the nonlinear TORA benchmark example, IEEE Transactions on Control Systems Technology, vol. 9, no. 2, pp. 191–198, 2001.

[1026] S. Tafazoli, S.F. Salcudean, K. Hashtradi-Zaad and P.D. Lawrence, Impedance control of a teleoperated excavator, IEEE Transactions on Control Systems Technology, vol. 9, no. 5, pp. 355–369, 2002.

[1027] E. Tahoumi, F. Plestan, M. Ghanes and J.P. Barbot, New robust control schemes based on both linear and sliding-mode approaches: Design and application to an electropneumatic actuator, IEEE Transactions on Control Systems Technology, vol. 29, no. 2, pp. 818–825, 2021.

[1028] R. Takai, L. Yang and N. Noguchi, Development of a crawler-type robot tractor using RTK-GPS and IMU, Engineering in Agriculture, Environment and Food, Elsevier, vol. 7, pp. 143–147, 2014.

[1029] C. Takeno, M. Aida, M. Murata and M. Imase, Proposal for autonomous decentralized structure formation based on local interaction and back-diffusion potential, IEICE Transactions on Communications, vol. E95-8, pp. 1529–1538, 2012.

[1030] C. Takeno, K. Sugiyama and M. Aida, Parameter design for diffusion-type autonomous decentralized flow control, IEICE Transactions on Communications, vol. E91-B, no. 9, pp. 2828–2837, 2008.

[1031] C. Takeno and M. Aida, Diffusion-type autonomous decentralized flow control for open networks, IEEE ICDCS 2004, 24th International Conference on Distributed Computing Systems, Tokyo, Japan, August 2004.

[1032] C. Takeno and M. Aida, Diffusion-type autonomous decentralized flow control for end-to-end flow in high-speed networks, IEICE Transactions on Communications, vol. E88-B, no. 4, pp. 1559–1567, 2005.

[1033] C. Takeno, M. Aida and S. Kirubayashi, Autonomous decentralized flow control in high-speed networks with inhomogeneous configurations, IEICE Transactions on Communications, vol. E.87-B, no. 6, pp. 1551–1560, 2004.

[1034] C. Takeno, K. Muranaka, K. Sugiyama and M. Aida, Mutual complementarity between diffusion-type flow control and TCP, IEICE Transactions on Communications, vol. E89-B, no. 10, pp. 2850–2859, 2006.

[1035] C. Takeno and M. Aida, Stability and adaptability of autonomous decentralized flow controlling high-speed networks, IEICE Transactions on Communications, vol. E86-B, no. 10, pp. 2882–2889, 2003.

[1036] S.E. Talole, J.P. Kolhe and S.N. Phadke, Extended-state-observer-based control of flexible-joint system with experimental validation, IEEE Transactions on Industrial Electronics, vol. 57, no. 4, pp. 1411–1419, 2010.

[1037] K. Tan, S.N. Huang an X. Jiang, Adaptive control of ram velocity for the injection moulding machine, IEEE Transactions on Control Systems Technology, vol. 9, no. 4, pp. 663–672, 2001.

[1038] H. Tang and Q. Zhang. An efficient numerical inverse scattering algorithm for generalized Zakharov-Shabat equations with two potential functions, INRIA Research Report, <inria-00447358>, 2010.

[1039] C.P. Tang, P.T. Miller, V.N. Kravi, J.C. Rya and S.K. Agrawal, Differential-flatness-based planning and control of a wheeled mobile manipulator – Theory and experiments, IEEE Transactions on Mechatronics, vol. 16, no. 5, pp. 768–773, 2011.

[1040] X. Tao, J. Yi, Z. Pu and T. Xiang, State-estimator-integrated robust adaptive tracking control for flexible air-breathing hypersonic vehicle with noisy measurements, IEEE Transactions on Instrumentation and Measurement, vol. 68, no. 11, pp. 4285–4299, 2019.

[1041] X. Tao, J. Yi, Z. Pu and T. Xiang, Robust adaptive tracking control for hypersonic vehicle based on interval type-2 fuzzy logic system and small gain approach, IEEE Transactions on Cybernetics, vol. 51, no. 5, pp. 2504–2517, 2019.

[1042] W. Tao, Y. Ou, and H. Feng, Research on dynamics and stability in the stairs-climbing of a tracked mobile robot, International Journal of Advanced Robotic Systems, Sage Publications, vol. 9, no. 146, pp. 1–9, 2012.

[1043] A. Tayebi and S. McGilvray, Attitude stabilization of a VTOL quadrotor aircraft, IEEE Transactions on Control Systems Technology, vol. 14, no. 3, pp. 562–571, 2006.

[1044] C.J. Taylor and D. Robertson, State-dependent control of a hydraulically actuated nuclear decommissioning robot, Control Engineering Practice, Elsevier, vol. 21, pp. 1716–1725, 2013.

[1045] E. Tedeschi and M. Molinas, Tunable control strategy for wave energy converters with limited power takeoff rating, IEEE Transactions on Industrial Electronics, vol. 59, no. 10, pp. 3838–3847, 2012.

[1046] I. Temiz, J. Leijon, B. Ekergard and C. Bostrom, Economic aspects of latching control for a wave energy converter with a direct drive linear generator power take-off, Renewable Energy, Elsevier, vol. 128, pp. 57–67, 2018.

[1047] D.V. Thach and S.G. Lee, LQG control design for a coupled ballbot dynamical system, 2018 18th International Conference on Control Automation and Systems, IEEE ICCAS 2018, GangWon, Korea, 2018.

[1048] D.V. Thach, P.B. Ba and S.G. Lee, Design of an energy-based controller for a 2D ball segway, 2017 17th International Conference on Control, Automation and Systems, IEEE ICCAS 2017, Jeju, Korea, October 2017.

[1049] J. Thai and A.M. Bayen, State estimation for polyhedral hybrid systems and application to the Godunov scheme for highway traffic estimation, IEEE Transactions on Automatic Control, vol. 60, no. 2, pp. 311–326, 2015.

[1050] N. Thatte, J. Shah and H. Geyer, Robust and adaptive lower limb prosthetic stance control with extended Kalman Filter-based gait phase estimation, IEEE Robotics and Automation Letters, vol. 4, no. 4, pp. 3129–3137, 2019.

[1051] H.D. Tho and K. Terashima, Robust control design of payload's skew rotation in a boom crane system, IEEE Transactions on Control Systems Technology, vol. 27, no. 4, pp. 1606–1621, 2019.

[1052] P. Thounthong, S. Sikkabet, N. Poonay, *et al.*, Nonlinear differential flatness-based speed/torque control with state observers of permanent magnet synchronous motor drives, IEEE Transactions on Industry Applications, vol. 56, no. 5, pp. 2874–2884, 2018.

[1053] Z. Tian, L. Yu, H. Ouyang and G. Zhang, Transportation and swing-reduction for double-pendulum tower cranes using partial enhanced coupling nonlinear controller with initial saturation, ISA Transactions, Elsevier, 2021.

[1054] B. Tian, O.T. An and M. Molinas, 'High-frequency injection-based sensorless control for a general five-phase BLDC motor incorporating system delay and phase resistance, IEEE Access, vol. 7, pp. 162682–162873, 2019.

[1055] H. Tian, X. Wen and Y.W, Li, A harmonic compensation approach for interlinking voltage source converter in hybrid AC-DC microgrids with low switching frequency, IEEE Journal of Power and Energy Systems, vol. 4, no. 1, pp. 39–48, 2018.

[1056] H. Tian, Q. Lu, R.B. Gopulani and V.M. Zavala, Multiobjective economic model predictive control of mechanical pulping processes, IEEE CDC 2016, IEEE 2016 55th Conference on Decision and Control, Las Vegas, USA, December 2016.

[1057] H. Tian, Q. Lu, R.B. Gopaluni, V.M. Zavala and J.A. Olson, An economic model predictive control framework for mechanical pulping processes, Control Engineering Practice, Elsevier, vol. 85, pp. 100–109, 2019.

[1058] W. Tian, J. Prakash, V.M. Zavala, J.A. Olson, and R. Bhushan Gopaluni, A tractable approximation for stochastic MPC and application to mechanical pulping processes, Computers and Chemical Engineering, Elsevier, vol. 44, no. 1–10, 2020.

[1059] D. Tilbury, O. Sordalen, L. Bushnell and S. Sastry, A multi-steering trailer system: Conversion into chained form using dynamic feedback, Preprints of the 4th IFAC Symposium on Robot Control, Capri, Italy, September 1994.

[1060] P.S. Thakar, B. Bandyopadhyay and P.S. Gandhi, Sliding-mode control for underactuated slosh-container system using collocated and non-collocated linearization, 2016 IEEE Indian Control Conference, Indian Institute of Technology, Hyderabad, India, January 2016.

[1061] P.S. Thakar, P.K. Trivedi, B. Badhyopadhyay and P.S. Gandhi, A new nonlinear control for asymptotic stabilization of a class of underactuated systems: An implementation to slosh-container problem, IEEE Transactions on Mechatronics, vol. 22, no. 2, pp. 1082–1092, 2017.

[1062] N. Tom, Confirmation of a nonlinear model predictive control strategy applied to a permanent magnet linear generator for wave energy conversion, Proceedings of the 33rd ASME International Conference on Ocean Offshore and Arctic Engineering, San Francisco, California, USA, June 2014.

[1063] N. Tom and R.W. Yeung, Experimental confirmation of nonlinear model predictive control applied offline to a permanent magnet linear generator for ocean-wave energy conversion, IEEE Transactions on Oceanic Engineering, vol. 41, no. 2, pp. 281–295, 2016.

[1064] M. Tong, M. Cheng, S. Wang and W. Hua, An on-board two-stage integrated fast battery charger for EVs based on the five-phase hybrid excitation flux-switching machine, IEEE Transactions on Industrial Electronics, vol. 89, no. 2, pp.1780–1790, 2021.

[1065] C.C. Tsai, C.H. Chiang, F.C. Tai and K.S. Hwang, Adaptive RFWCMAC cooperative formation control or multiple ballbots incorporated with coupling dynamics, 2015 International Conference on Informative and Cybernetics for Computational Social Systems, IEEE ICCSS 2015, Chengdu, China, August 2015.

[1066] L.A. Tuan, H.M. Cuang, S.G. Lee, L.C. Nho and K. Moon, Nonlinear feedback control of boom crane mounted on elastic foundation with the flexibility of suspended cable, Journal of Vibration and Control, Sage Publications vol. 22, no. 13, pp. 3067–3078, 2014.

[1067] W. Tuvayonond and M. Parnichkun, Position control of a pneumatic surgical robot using PSO-based 2-DOF H_∞ loop shaping structural controller, Mechatronics, Elsevier, vol. 43, pp. 40–55, 2017.

[1068] A. Toedtheide, T. Lilge and S. Haddadin, Antagonistic impedance control for pneumatically actuated robot joints, IEEE Robotics and Automation Letters, vol. 1, no. 1, pp. 161–169, 2016.

[1069] G.J. Toussaint, T. Basar and F. Bullo, H_∞ optimal tracking control techniques for nonlinear underactuated systems, Proceedings Of the IEEE CDC 2000, 39th IEEE Conference on Decision and Control, Sydney Australia, 2000.

[1070] J. Trachte, F. Gonzalez and A. McFadyen, Nonlinear model predictive control for a multi-rotor with heavy slug load, IEEE ICUAS 2014, IEEE 2014 International Conference on Unmanned Aircraft Systems, May 2016, Orlando, Florida, USA.

[1071] M.S. Trivedi and R.V. Kumar Keshri, Evaluation of predictive current control techniques for PMBLDC motor in stationary plane, IEEE Access, vol. 8, pp. 46217–46228, 2020.

[1072] M. Trumic, K. Jovanovic and A. Fagiolini, Decoupled nonlinear adaptive control of position and stiffness for pneumatic soft robots, International Journal of Robotics Research, Sage Publications, vol. 40, no. 1, pp. 277–295, 2020.

[1073] P.M. Truong, D. Flieller, N.K. Nguyen, J. Mercklé and M.D. Dat, Optimal efficiency control of synchronous reluctance motor-based ANN considering cross magnetic saturation and iron loss, IEEE IECON 2015, IEEE Industrial Electronics Conference, Yokohama, Japan, November 2015.

[1074] C.C. Tsai, S.M. Hsieh and H.E. Kao, Mechatronic design and injection speed control of an ultra high-speed plastic injection moulding machine, Mechatronics, Elsevier, vol. 19, pp. 142–155, 2009.

[1075] T.C. Tsao, C.A. Tom, A. Pesterev, B. Youngsnd and L.A. Bergman, Control oriented formulation for structures interacting with moving loads, IEEE ACC 2001, IEEE 2001 American Control Conference, Arlington, Virginia, June 2001.

[1076] M. Tucci, S. Riverso and G. Ferrari Trecante, Line independent plug an play controllers for voltage stabilization in DC microgrids, IEEE Transactions on Control Systems Technology, vol. 25, no. 3, pp. 1115–1123, 2018.

[1077] B. Ucchedou, K. Zhang, H. Hammouri and G. Fevotte, Design of a CSD observer during batch cooling crystallization dealing with uncertain nucleation parameters, 18th IFAC World Congress, Milano, Italy, September 2011.

[1078] B. Ugurlu, N. Nishimura, K. Hyodo, M. Kawanishi and T. Narikoyo, Proof of concept for robot-aided upper-limb rehabilitation using disturbance observers, IEEE Transactions on Human-Machine Systems, vol. 45, no. 1, pp. 110–118, 2015.

[1079] A. Ule and R.J. Boucherie, On the distribution of calls in a wireless network driven by fluid traffic, European Journal of Operational Research, vol. 147, pp. 146–155, 2003.

[1080] T. Utz, T. Meurer and A. Kugi, Trajectory planning for two-dimensional quasi-linear parabolic PDE based on finite difference semi-discretization, 18th IFAC World Congress, Milano, Italy, September 2011.

[1081] N. Vafamand, S. Yousefizadeh, M.H. Khooban, J.D. Bendtsen and T. Dragicevic, Adaptive TS fuzzy-based MPC for DC microgrids with dynamic CPLs: Nonlinear power observer approach, IEEE Systems Journal, 2019.

[1082] R.M. Valle, P.M. de Almeida, G.A. Fogli, A.A. Ferreira and P.G. Barbosa, Simple and effective digital control of a variable-speed low inductance BLDC motor drive, IEEE Access, vol.8, pp. 13240–13250, 2020.

[1083] M. Van Damme, B. Vanderborght, R. Van Ham, B. Verrelst, F. Daersden and B. Lefeber, Sliding-mode control of a 2-DOF planar pneumatic manipulator, Journal of Dynamical Systems, Measurement and Control, vol. 131, no. 2, pp. 021013, 2009.

[1084] C.P. Van Hinsbergen, T. Schreiter, F.S. Zuurbier, J.W.C. van Lint and H.J. van Zuylen, Localized extended Kalman Filter for scalable real-time traffic state estimation, IEEE Transactions on Intelligent Transportation Systems, vol. 13, no. 1, pp. 385–394, 2012.

[1085] N. Van de Wouw, P. Ritzen, E. Roebroek, Z.P. Jiang and H. Nijmeijer, Active trailer steering for robotic tractor-trailer combinations, IEEE CDC 2015, IEEE 54th Conference on Decision and Control, Osaka, Japan, December 2015.

[1086] T. Van Vu, D. Gonsoulin, F. Diaz, C.S. Edrington and T. El-Mezyani, Predictive control for energy management in ship power systems under high-power ramp-rate loads, IEEE Transactions on Energy Conversion, vol. 32, no. 2, pp. 788–799, 2017.

[1087] J. Vantili, C. Giraldi, E. Aertbelien, F. de Groote, and J.J. de Schutter, Estimating contact forces and moments for walking robots and exoskeletons using complementary energy methods, IEEE Robotics and Automation Letters, vol. 3, no. 4, pp. 3410–3417, 2018.

[1088] J. Varghese, V.M. Akhil, P.K. Rajendrakumar and K.S. Sirandan, A rotary pneumatic actuator for the actuation of the exoskeleton knee joint, Theoretical and Applied Mechanics Letters, Elsevier, vol. 7, pp. 222–230, 2012.

[1089] G. Vaughan, D. Kim and W. Singhose, Control of tower cranes with double-pendulum payload dynamics, IEEE Transactions on Control Systems Technology, vol. 18, no. 6, pp. 1345–1358, 2010.

[1090] R. Vermaak and M.J. Kampen, Design aspects of a novel topology air-cored permanent magnet linear generator for direct drive wave energy converters, IEEE Transactions on Industrial Electronics, vol. 50, no. 5, pp. 2104–2115, 2012.

[1091] J. Villagra, B. d'Andrea-Novel, H. Mounier and M. Pengov, Flatness-based vehicle steering control strategy with SDRE feedback gains tuned via a sensitivity approach, IEEE Transactions on Control Systems Technology, vol. 15, pp. 554–565, 2007.

[1092] B. Voidya, M. Shomin, R. Hollis and G. Kantor, Operation of the ballbot on slopes and with center-of-mass offset, Proceedings of the IEEE International Conference on Robotics and Automation, IEEE ICRA 2015, Seattle, USA, May 2015.

[1093] J.P.B. Vreeburg, Spacecraft maneuvers and slosh control, IEEE Control Systems Magazine, vol. 25, no. 3, pp. 12–16, 2008.

[1094] R.T. Vu, N.K. Nguyen, E. Semail, and T.J. dos Santos Moraes, Torque optimisation of seven-phase BLDC machines in normal and degraded modes with constraints on currents and voltages, IET Journal of Engineering, vol. 2019, no. 17, pp. 3818–3824, 2014.

[1095] M.Y. Wai and T.H. Liu, Design and implementation of an online tuning adaptive controller for synchronous reluctance motor drives, IEEE Transactions on Industrial Electronics, vol. 60,no. 9, pp. 3644–3657, 2013.

[1096] H. Wang, B. Yan, Z. Pu, *et al.*, Optimal design and stress analysis of the transmission line inspection robot along the ground line, IET Journal of Engineering (Proceedings of the 1th International Conference on AC and DC Power Transmission, IET ACDC 2018), vol. 16, pp. 2088–3091, 2019.

[1097] L. Wang and H. Wang, A survey on insulator inspection robot for power transmission lines, IEEE CARPI 2016, IEE 2016 4th International Conference on Applied Robotics for the Power Industry, Jinan, China, November 2016.

[1098] H. Wang, F. Zhang, Y. Jiang, G. Liu and X. Peng, Development of an inspection robot for 500kV EHV Power Transmission Lines, IEEE/RSJ 2010, IEEE 2010 International Conference on Intelligent Robots and Systems, Taipei, Taiwan, October 2010.

[1099] W. Wang, H. Chi, S. Zhao and Z. Du, A control method for hydraulic manipulators in automatic emulsion filling, Automation in Construction, Elsevier, vol. 91, pp. 92–99, 2018.

[1100] H. Wang, Y. Pan, S. Li and H. Yu, Robust sliding-mode control for robots driven by compliant actuators, IEEE Transactions on Control Systems Technology, vol. 27, no. 3, pp. 1259–1266, 2019.

[1101] Z. Wang, B. Wang, L. Xu and Q. Xie, Feedback-added pseudoinverse type balanced minimization scheme for kinematic control of redundant robot manipulators, IEEE Access, vol. 7, pp. 23806–23815, 2017.

[1102] X. Wang, J. Taghia and J. Katupitiya, Robust model predictive control for part tracking of a tracked vehicle with steerable trailer in the presence of slip, 5th IFAC Conference on Steering Control and Automation Technologies, Agricontrol 2016, Seattle USA, August 2018.

[1103] Y. Wang, M. Sun, Z. Wang, Z. Liu and Z. Chen, A novel disturbance-observer based friction compensation scheme for ball and plate system, ISA Transactions, Elsevier, vol. 53, pp. 671–678, 2014.

[1104] H. Wang, Y. Tian, S. Fu and Z. Sui, Nonlinear control for output regulation of ball and plate system, Proceedings of the 27th Chinese Control Conference, Kumming Yunnan, China, July 2008.

[1105] A. Wang, X. Li and X. Cao, Backstepping-based robust H_∞ tracking controller design for ball and plate system, IEEE CCC, 2019, IEEE 2019 38th Chinese Control Conference, GuangZhou, China, July 2014.

[1106] N. Wang, Z. Sun, J. Yin, Z. Zou and S.F. Su, Fuzzy unknown observer-based robust adaptive path-following control of underactuated surface vehicles subject to multiple unknowns, Ocean Engineering, Elsevier, vol. 176, pp. 57–64, 2019.

[1107] N. Wang and S.F. Su, Finite-time unknown observer-based interactive trajectory tracking control of asymmetric underactuated surface vehicles,

IEEE Transactions on Control Systems Technology, vol. 29, no. 2, pp. 794–803, 2021.

[1108] N. Wang and H.R. Karimi, Successive waypoints tracking of an underactuated surface vessel, IEEE Transactions on Industrial Informatics, vol. 16, no. 2, pp. 898–908, 2020.

[1109] X. Wang, Takeoff/landing control based on acceleration measurements for VTOL aircraft, Journal of the Franklin Institute, Elsevier, vol. 350, pp. 3045–3063, 2013.

[1110] X. Wang, J. Liu and K.Y. Cai, Tracking control for a velocity – Sensorless VTOL aircraft with delayed outputs, Automatica, Elsevier, vol. 45, pp. 2876–2882, 2009.

[1111] B. Wang, Y. Zhang, J.C. Ponsart and D. Theilliol, Fault-tolerant adaptive control allocation for unmanned multirotor helicopters, 20th IFAC World Congress, Toulouse, France, July 2017.

[1112] Z. Wang, W. Bao and H. Li, Second-order dynamic sliding-mode control for non-minimum phase underactuated hypersonic vehicles, IEEE Transactions on Industrial Electronics, vol. 64, no. 4, pp. 3105–3112, 2017.

[1113] Y. Wang, T. Chao, S. Wang and M. Yang, Trajectory tracking control based on differential flatness, Proceedings of the 35th Chinese Control Conference, Chengdu, China, July 2014.

[1114] W. Wang, Z. Hou, S. Shan and L. Chen, Optimal periodic control of hypersonic cruise vehicle: Trajectory features, IEEE Access, vol. 4, pp. 1–16, 2016.

[1115] Y. Wang, X. Yang and H. Yan, Reliable fuzzy tracking control of near-space hypersonic vehicle using aperiodic measurement information, IEEE Transactions on Industrial Electronics, vol. 66, no. 2, pp. 9439–9447, 2019.

[1116] Z. Wang, J. Chen, M. Cheng and K.T. Chan, Field-oriented control and direct torque control for paralleled VSI-fed PMSM drives with variable switching frequencies, IEEE Transactions on Power Electronics, vol. 31, no. 3, pp. 2417–2428, 2016.

[1117] Z. Wang, J. Chen, M. Cheng and Y. Cheng, Fault-tolerant control of paralleled voltage source inverter-Fed PMSM drives, IEEE Transactions on Industrial Electronics, vol. 62, no. 8, pp. 4749–4760, 2015.

[1118] Z. Wang, Y. Zhang, Z. Zou and M. Cheng, Position sensorless control of interleaved VSI-fed PMSM drive with extended Kalman Filter, IEEE Transactions on Magnetics, vol. 18, no. 11, pp. 3688–3694, 2017.

[1119] Y. Wang, Y. Xu and J. Zou, Sliding-mode sensorless control of PMSM with inverter nonlinearity compensation, IEEE Transactions on Power Electronics, vol. 34, no. 10, pp. 10206–10220, 2019.

[1120] W. Wang, Y. Fan, S. Chen and Q. Zhang, Finite control set model predictive current control of a five-phase PMSM with virtual voltage vectors and adaptive control set, CES Transactions on Electrical Machines and Systems, vol. 2, no. 1, pp. 136–141, 2018.

[1121] L. Wang, J. Isberg and E. Tedeschi, Review of control strategies for wave energy conversion systems and their validation: The wave-to-wire approach, Renewable and Sustainable Energy Reviews, Elsevier, vol. 81, pp. 966–979, 2018.

[1122] L. Wang, Z.Q. Zhu, B. Hang and L.M. Gong, A commutation error compensation strategy for high-speed brushless DC drive based on adaline filters, IEEE Transactions on Industrial Electronics, 2020.

[1123] Q. Wang, S. Niu and X. Luo, A novel hybrid dual-PM machine excited by AC with DC bias for electric vehicle propulsion, IEEE Transactions on Industrial Electronics, vol. 64, no. 9, pp. 6908–6919, 2017.

[1124] L. Wang, X. Fu and M.C. Wang, Operational control of a hybrid coupled interlinking converter for hybrid AC / low voltage DC microgrids, IEEE Transactions on Industrial Electronics, vol. 68, no. 8, pp. 7158–7168, 2021.

[1125] P. Wang, J. Li and Y. Hu, Frequency control of isolated wind-diesel microgrid power system in double equivalent-input-disturbance controllers, IEEE Access, vol. 7, pp. 105617–105626, 2019.

[1126] Y. Wang, J.H. van Schuppen and J. Vrancken, Prediction of traffic flow at the boundary of a motorway network, IEEE Transactions on Intelligent Transportation Systems, vol. 15. no. 1, pp. 214–227, 2014.

[1127] Y. Wang, H. Wang and Y. Tian, Nonlinear disturbance observer-based flexible boundary prescribed performance control for a lower limb exoskeleton, International Journal of Systems Science, Taylor and Francis, vol. 52, no. 15, pp. 3176–3189, 2021.

[1128] J. Wang, B. Cao, O. Chen and F. Wang, Combined state of charge estimation for electric vehicle battery pack, Control Engineering Practice, Elsevier, vol. 15, p. 1569–1576, 2007.

[1129] Y. Wang, H. Fang, L. Zhou and T. Wada, Revisiting the state-of-charge estimation for lithium-ion batteries: A methodological investigation of the extended Kalman Filter approach, IEEE Control Systems Magazine, pp. 73–96, 2017.

[1130] Y. Wang, C. Zhang and Z. Chen, A method for state-of-charge estimation of $LiFePO_4$ batteries at dynamic currents and temperatures using particle filter, Journal of Power Sources, Elsevier, vol. 279, pp. 306–311, 2015.

[1131] K. Watanabe, Y. Ueda, J. Nagai and S. Maeyama, Stabilization of a fire-truck robot by an invariant manifold theory, Energy Procedia, Elsevier, vol. 41, pp. 1099–1104, 2012.

[1132] K. Watanabe, T. Yamamoto, K. Izumi and S. Maeyama, Underactuated control for nonholonomic mobile robots by using double integrator model and invariant manifold theory, The 2010 IEEE/RSJ International Conference on Intelligent Robots and Systems, Taipei, Taiwan, October 2010.

[1133] H. Wei, C. Shen and Y. Shi, Distributed Lyapunov-based model predictive formation tracking control for autonomous underwater vehicles subject to disturbances, IEEE Transactions on Systems, Man and Cybernetics: Systems, vol. 51, no. 5, pp. 5198–5208, 2020.

[1134] W. Wei and L. Lin, Vibration control of vehicle-bridge dynamic interactive system, IEEE CCC 2007, IEEE 2007 26th Chinese Control Conference, Zhangjiajie, Hunan, China, 2007.

[1135] M.Y. Wei and T.H. Liu, A high-performance sensorless position control system of a synchronous reluctance motor using dual current; slope estimation techniques, IEEE Transactions on Industrial Electronics, vol. 50, no. 9, pp. 3411–3426, 2012.

[1136] M.Y. Wei, T.H. Liu and C.K. Lin, Design and implementation of a passivity-based controller for sensorless synchronous reluctance motor drive systems, IET Electric Power Applications, vol. 5, no. 4, pp. 335–349, 2011.

[1137] Q. Wei, R. Song, B. Li, and X. Lin, Self-learning Optimal Control of Nonlinear Systems: Adaptive Dynamic Programming Approach, Springer, 2016.

[1138] R. Werner, G. Kormann and S. Mueller, Systematic model-based path tracking control of actively steered implements in Simulation and experiments, 4th IFAC Conference on Modelling and Control in Agriculture, Horticulture and Post Harvest Industry, Espoo, Finland, August 2013.

[1139] W.T. Willet and M.J. Kamper, Variable-gear EV reluctance synchronous motors drives: An evaluation of rotor structures for position sensorless control, IEEE Transactions on Industrial Electronics, vol. 61, no. 10, pp. 5732–5740, 2014.

[1140] T. WingWang and H. Ying Wang, Stochastic volatility asymptotics of stock loans: Valuation and optimal stopping, Journal of Mathematical Analysis and Applications, Elsevier, vol. 394, pp. 337–346, 2012.

[1141] F.J. Winkler, I. Krause and B. Lohmann, Flatness-based control of a continuous furnace, 18th International Conference on Control Applications, Part of 2009 IEEE Multi-Conference on Systems and Control, Saint Petersburg, Russia, July 2009.

[1142] H. Wind, A. Renner, S. Schmit, S. Albrecht and O. Sawodny, Comparison of joint angle, velocity and acceleration estimates for hydraulically actuated manipulators to a novel dynamical approach, Control Engineering Practice, Elsevier, vol. 91, pp. 10411, 2019.

[1143] F.J. Winkler and B. Lohmann, Design of a decoupling controller structure for first order hyperbolic PDEs with distributed control action, 2010 American Control Conference, Baltimore, MD USA, July 2010.

[1144] F. Winkler, I. Krause and B. Lohmann, Flatness-based control of a continuous furnace, 18th International Conference on Control Applications, Part of 2009 IEEE Multi-Conference on Systems and Control, Saint Petersburg, Russia, July 2009.

[1145] P.A. Wisnewski and F.J. Doyle, Model-based predictive control studies for a continuous pulp digester, IEEE Transactions on Control Systems Technology, vol. 9, no. 3, pp. 435–444, 2001.

[1146] J. Woolfrey, W. Lu and D. Liu, A control method for joint-torque minimization of redundant manipulators handling large external forces, Journal of Intelligent and Robotic Systems, Springer, vol. 93, pp. 3–16, 2019.

[1147] F. Woitteneck and H. Mounier, Controllability of networks of spatially one-dimensional second order PDEs – An algebraic approach, SIAM Journal of Control and Optimization, vol. 48, no. 6, pp. 3882–3902, 2010.

[1148] T.W. Wong and H.Y. Wong, Stochastic volatility asymptotics of stock loans: Valuation and optimal stopping, Journal of Mathematical Analysis and Applications, Elsevier, vol. 394, pp. 337–346, 2012.

[1149] T.W. Wong and H.Y. Wong, Valuation of stock loans using exponential phase-type Levy models, Applied Mathematics and Computation, Elsevier, vol. 222, pp. 275–289, 2013.

[1150] D.B. Work, O.P. Tassavainen, S. Blandin, A.M. Bayen, T. Iwuchukwa and K. Tracton, An Ensemble Kalman Filtering approach to highway traffic estimation using GPS enabled mobile devices, 47th IEEE Conference on Decision and Control, Cancun, Mexico, December 2008.

[1151] J. Wu, G. Wang, H. Zhou and K. Sun, Study on electromechanical performance of steering of the electric articulated tracked vehicle, Journal of Mechanical Science and Technology, vol. 33, no. 7, pp. 3171–3185, 2019.

[1152] Q. Wu, X. Wang, L. Hua and M. Xia, Improved time-optimal anti-swing control system based on low-gain filter for double-pendulum crane system with distributed mass beam, Mechanical Systems and Signal Processing, Elsevier, vol. 151, pp. 107444, 2021.

[1153] T.S. Wu, M. Karkoub, W.S. Yu, C.T. Chen, M.G. Her and K.W. Wu, Anti-sway tracking control of tower cranes with delayed uncertainty using a robust adaptive fuzzy control, Fuzzy Sets and Systems, Elsevier, vol. 290, pp. 118–137, 2016.

[1154] Y. Wu, N. Sun, Y. Fang and D. Liang, An increased nonlinear coupling motion controller for underactuated multi-TORA systems: Theoretical design and hardware experimentation, IEEE Transactions on Systems, Man and Cybernetics: Systems, vol. 49, no. 6, pp. 1186–1193, 2010.

[1155] T. Wu, D. Gui, D. Hu and C. Du, Adaptive fuzzy sliding-mode control for translational oscillator with rotating actuator: A fuzzy model, IEEE Access vol.6, pp. 55861–55869, 2018.

[1156] W. Wu, M. Huang and X. Gu, Underactuated control of bionic ape robot based on the energy pumping method and big damping condition turn-back angle feedback, Robotics and Autonomous Systems, Elsevier, vol. 100, pp. 119–131, 2015.

[1157] F.X. Wu, W.J. Zhang, Q. Li, P.R. Ouyang and Z.X. Zhou, Control of hybrid machines with 2-DOF for trajectory tracking problem, IEEE Transactions on Control Systems Technology, vol. 13, no. 2, pp. 338–342, 2005.

[1158] H.N. Wu, S. Feng, S.Y. Liu and L. Guo, Disturbance observer-based robust mixed H_2/H_∞ fuzzy tracking control for hypersonic vehicles, Fuzzy Sets and Systems, Elsevier, vol. 306, pp. 116–136, 2017.

[1159] Y.J. Wu, J.X. Zuo and L.H. Sun, Adaptive terminal sliding-mode control for hypersonic flight vehicles with strictly lower convex function based nonlinear disturbance observer, ISA Transactions, Elsevier, vol. 71, pp. 215–226, 2017.

[1160] B. Wu, Z. Shi, F. Sun and W. He, Barrier Lyapunov function-based learning control of hypersonic flight vehicle with AOA constraint and actuator faults, IEEE Transactions on Cybernetics, vol. 49, no. 3, pp. 1047–1057, 2019.

[1161] F. Wu, X.P. Zhang, P. Ju and M.J. Sterling, Optimal control for AWS-based wave energy conversion system, IEEE Transactions on Power Systems, vol. 24, no. 4, pp. 1747–1755, 2009.

[1162] R. Wu, X.P. Zhang, P. Ju and M.J.H. Sterling, Modelling and control of AWS-based wave energy conversion system integrated into power grid, IEEE Transactions on Power Grid, vol. 23, no. 3, pp. 1196–1205, 2008.

[1163] Q. Wu and M. Saif, Neural adaptive observer-based fault detection and identification for satellite attitude control systems, IEEE ACC 2005, IEEE 2005 American Control Conference, Portland, Oregon, USA, July 2005.

[1164] Q. Wu, B. Chen and H. Wu, Neural network-enhanced torque estimation control of a soft wearable exoskeleton for elbow assistance, Mechatronics, Elsevier, vol. 63, pp. 102279–102288, 2018.

[1165] J. Wu, J. Huang, Y. Wang, and K. Xing, Nonlinear disturbance observer-based dynamic surface control for trajectory tracking of pneumatic muscle system, IEEE Transactions on Control Systems Technology, vol. 23, no. 2, pp. 440–455, 2014.

[1166] X. Wu, Y.Wang and X. Dong, Robust adaptive sliding-mode control of condense-cleaning mobile manipulator using fuzzy wavelet neural network, Fuzzy Sets and Systems, Elsevier, vol. 235, pp. 62–82, 2014.

[1167] R. Xia, Q. Wu and S. Shao, Disturbance observer-based optimal flight control of near space vehicle with external disturbance, Transactions of the Institute of Measurement and Control, Sage Publications, vol. 42, no. 2, pp. 272–284, 2019.

[1168] B. Xian, M.S. de Queiroz, D. Dawson and I. Walker, Task-based tracking control of redundant robot manipulators via quaternion feedback, IEEE Transactions on Robotics and Automation, vol. 20, no. 1, pp. 160–167, 2004.

[1169] L. Xian, L. Zhang, F. Gao and J. Qian, Robust fault-tolerant synergetic control for dual three-phase PMSM drives considering speed sensor faults, IEEE Access, vol. 8, pp. 78912–78922, 2020.

[1170] X. Xiang, C. Yu, Q. Zhang, P.A. Wilson and G. Xu, Maneuvering-based actuation evaluation of an AUV with control surfaces and through body thrusters, Applied Ocean Research, Elsevier, vol. 96, pp. 102046–102052, 2020.

[1171] X. Xiang, L. Lapierre and B. Jouvencel, Smooth transition of AUV motion control: From fully actuated to under-actuated configuration, Robotics and Autonomous Systems, Elsevier, vol, 67, pp. 14–22, 2015.

[1172] X. Xiang, G. Yu and Q. Zhang, Robust fuzzy 3D path following for autonomous underwater vehicle subject to uncertainties, Computers and Operations Research, Elsevier, vol. 84, pp. 165–177, 2017.

[1173] Z.X. Xiao, M.M. Li, H.W. Fang, *et al.*, Operation control for improving energy efficiency of shipboard microgrid including bow thrusters and hybrid energy storage IEEE Transactions on Transportation Electrification, vol. 6, no. 2, pp. 856–868, 2020.

[1174] B. Xiao, L. Cao and D. Ram, Attitude exponential stabilization control of rigid bodies via disturbance observer, IEEE Transactions on Systems, Man and Cybernetics: Systems, 2019.

[1175] F. Xie, W. Shung, B. Zhang, S. Cong and Z. Li, High-precision tracking control of cable-driven parallel robots using robust synchronization, IEEE Transactions on Industrial Informatics, vol. 17, no. 4, pp. 2488–2499, 2021.

[1176] G. Xie, K. Lu, S.K. Dwivedi, J.R. Rasholm and F. Blaabjerg, Minimum-voltage vector injection method for sensorless control of PMSM for low-speed operations, IEEE Transactions on Power Electronics, vol. 31, no. 2, pp. 1785–1794, 2016.

[1177] J. Xin and X.Y. Zhou, Stock loans, Journal of Mathematical Finance, J. Wiley, vol. 17, no. 2,pp. 307–317, 2007.

[1178] H. Xing and W. Chen, Walking beam furnace temperature control using fuzzy feed-forward method in steel rolling mills, IEEE CCC 2013, Proceedings of the 32nd Chinese Control Conference, Xian, China, July 2013.

[1179] R. Xiong, X. Gong, C.C. Mi and F. Sun, A robust state-of-charge estimation for multiple types of lithium-ion batteries using adaptive extended Kalman Filter, Journal of Power Sources, Elsevier, vol. 243, pp. 805–816, 2013.

[1180] K. Xiu, H. Xing, L. Ding, H. Gao, G. Liu and Z. Deng, Virtual decomposition-based modeling for multi-DOF manipulator with flexible joints, IEEE Access, vol. 7, pp. 91582–91592, 2019.

[1181] P. Xiong, X. Lai and M. Wu, Position control strategy based on energy attenuation for planar three-link underactuated manipulators, Proceedings of the 35th Chinese Control Conference, Chengdu, China, July 2016.

[1182] R. Xu, D. Tang, L. Hun, H. Huang and D. Xie. Robust finite-time attitude tracking control of a CMG-based AUV with unknown disturbances and input saturation, IEEE Access, vol. 7, pp. 56409–56423, 2019.

[1183] B. Xu, D. Wang, Y. Zhang and Z. Shi, DOB-based neural control of flexible hypersonic flight vehicle considering wind effects, IEEE Transactions on Industrial Electronics, vol. 64, no. 11, pp. 8676–8685, 2017.

[1184] B. Xu, D. Yang, Z. Shi, Y. Pan, B. Chen and F. Sun, Online recorded data-based composite neural control of strict-feedback systems with application to hypersonic flight dynamics, IEEE Transactions on Neural Networks and Learning Systems, vol. 29, no. 8, pp. 3839–3849, 2018.

[1185] B. Xu and Y. Pan, Global neural dynamic surface tracking control of strict feedback systems with application to hypersonic flight vehicle, IEEE Transactions on Neural Networks and Learning Systems, vol. 26, no. 10, pp. 2563–2575, 2015.

[1186] B. Xu, Z. Shi, F. Sun and W. He, Barrier Lyapunov function-based learning control of hypersonic flight vehicle with angle of attack constraints, IEEE Transactions on Cybernetics, vol. 49, no. 3, pp. 1047–1057, 2019.

[1187] J. Xu, B. Zhang, H. Fang and H. Guo, Guaranteeing the fault-tolerant performance of aerospace multi-phase permanent magnet motors system: An adaptive robust speed control approach, CES Transactions on Electrical Machines and Systems, vol. 4, no. 2, pp. 114–122, 2020.

[1188] X. Xu, Q. Liu, C. Zhang and Z. Zeng, Prescribed performance controller design for DC converter system with constant power loads in DC microgrids, IEEE Transactions on Systems, Man and Cybernetics: Systems, vol. 50, no. 11, pp. 4339–4348, 2018.

[1189] Q. Xu, W. Li, Y. Lin, H. Liu and Y. Gu, Investigation of the performance of a stand-alone horizontal axis tidal current turbine based on in-site experiment, Ocean Engineering, Elsevier, vol. 113, pp. 111–120, 2016.

[1190] D. Xu, D. Zhao, J. Yi and X. Tan, Trajectory tracking control of omni-directional wheeled mobile manipulators: Robust neural network-based sliding-mode approach, IEEE Transactions on Systems, Man and Cybernetics: Part B – Cybernetics, vol. 39, no. 3, pp. 788–799, 2009.

[1191] C. Xue, S. Tan and H. Liu, An improved iterative sliding-mode control for tethered satellite, Proceedings of the 36th Chinese Control Conference, Dalian, China, July 2012.

[1192] M. Yaiehu, R. Mazumder and C. Rosenborg, A game-theoretic framework for bandwidth allocation and pricing in broadband networks, IEEE ACM Transactions on Networking, vol. 8, no. 5, pp. 667–678, 2009.

[1193] H. Yamaguchi, M. Mori and A. Kawakami, Control of a five-axle three-steering coupled-vehicle system and its experimental validation, Proceedings of the 13th IFAC World Congress, Milan Italy, September 2011.

[1194] G. Yamauchi, K. Nagatani, Hashimoto and K. Fujino, Slip-compensated odometry for tracked vehicle on loose and weak slope, RoboMech Journal, Springer, vol. 4, no. 27, pp. 1–11, 2017.

[1195] G. Yamauchi, D. Suzuki and K. Nagatani, Online slip parameter estimation for tracked vehicle odometry on loose slope, IEEE SSSR 2016. IEEE 2016 International Symposium on Safety, Security, and Rescue Robotics (SSRR) EPFL, Lausanne, Switzerland, October 2016.

[1196] Y. Yamukawa, Brachiaton motion by a 2-DOF brachiating robot with hook-shaped end effectors, Bulletin of the JSME, Mechanical Engineering Letters, vol. 4, pp. 1–9, 2018.

[1197] X. Yan, A. Estrada and F. Plestan, Adaptive pulse output feedback controller based on second-order sliding mode technology and applications, IEEE Transactions on Control Systems Technology, vol. 24, no. 6, pp. 2233–2242, 2016.

[1198] Z. Yan, X. Lai, O. Meng and M. Wu, A novel robust control method for motion control of uncertain single-link flexible-joint manipulator, IEEE Transactions on Systems, Man and Cybernetics, 2020.

[1199] Y. Yan, C. Zhang, A. Narayan, J, Yang, S. Li and H. Yu, Generalized dynamic predictive control for nonparametric uncertain systems with application to series elastic actuators, IEEE Transactions on Industrial Informatics, vol. 14, no. 11, pp. 4829–4840, 2018.

[1200] Z. Yan, Z. Yang, I. Zhang, J. Zhou, A. Jiang and X. Du, Trajectory tracking control of UUV based on backstepping sliding-mode with fuzzy switching gain in diving plane, IEEE Access, vol. 7, pp. 166788–166795, 2019.

[1201] X.G. Yan, S.K. Spurgeon, and C. Edwards, Variable Structure Control of Complex Systems: Analysis and Design, Springer, 2017.

[1202] P. Yang, X. Ma, J. Wang, G. Zhang, Y. Zhang and L. Chan, Disturbance observer-based sliding-mode control of a 5-DOF upper-limb exoskeleton robot, IEEE Access, vol. 7, pp. 6283–62839, 2019.

[1203] L. Yang, J. Fan, Y. Liu, E. Li, J. Pang and Z. Liang, A review on state-of-the-art power line inspection technologies, IEEE Transactions on Instrumentation and Measurement, vol. 69, no. 12, pp. 9350–9365, 2020.

[1204] G. Yang, E. Li, C. Fan, W. Lei and Z. Liang, Modelling and control of a bi-brachiate inspection robot for power transmission lines, IEEE ICMA 2010, IEEE 2010 International Conference on Mechatronics and Automation, Xian, China, October 2010.

[1205] G. Yang, E. Li, C. Fan, W. Lei and Z. Liang, Adaptive fuzzy logic-based inspection robot for high voltage power transmission lines, IEEE ICMA 2010, IEEE 2010 International Conference on Mechatronics and Automation, Xian, China, October 2010.

[1206] D. Yang, Z. Feng, R. Sha and X. Ren, Robust control of a class of under-actuated mechanical systems with model uncertainty, International Journal of Control, Taylor and Francis, vol. 92, no. 7, pp. 1567–1579, 2018.

[1207] G. Yang, J. Yao, G. Le and D. Ma, Asymptotic output tracking control of electrohydraulic systems with unmatched disturbances, IET Control Theory and Applications, vol. 10, no. 19, pp. 2543–2551, 2016.

[1208] G. Yang and J. Yao, Output feedback control of electrohydraulic servo actuators with matched and mismatched disturbances rejection, Journal of the Franklin Institute, Elsevier, vol. 356, pp. 3152–3179, 2019.

[1209] T. Yang, Y. Guo, Y. Zhou and S. Wen, Joint communication and control for small underactuated USV based on mobile computing technology, IEEE Access, vol. 7, pp. 160610–160622, 2019.

[1210] N. Yang, B. Nahid-Mobarakeh, F. Gao, D. Poire, A. Miraoui and W. Liu, Modelling and stability analysis of multi-time scale DC microgrid, Electric Power Systems Research, vol. 140, pp. 906–916, 2016.

[1211] T. Yang, N. Sun, H. Chen, and Y. Fang, Neural network-based adaptive antiswing control of an underactuated ship-mounted crane with roll motions and input dead zones, IEEE Transactions on Neural Networks and Learning Systems, vol. 31, no. 3, pp. 901–914, 2019.

[1212] R. Yang, X. Yang, W. Huang1and, S. Zhang, Energy management of the power-split hybrid electric city bus based on the Stochastic model predictive control, IEEE Access, vol. 9, pp. 2055–3071, 2020.

[1213] C. Yang, M. Zha, W. Wang and K. Liu, Changle Xiang, Efficient energy management strategy for hybrid electric vehicles/plug-in hybrid electric vehicles: Review and recent advances under intelligent transportation system, IET Intelligent Transport Systems, vol. 14, no. 7, pp. 7040711, 2020.

[1214] P. Yang, Y. Xia, M. Yu, W. Wei and Y. Peng, A decentralized coordination control method for parallel bidirectional power converters in a hybrid AC/DC microgrid, IEEE Transactions on Industrial Electronics, vol. 66, no. 6, pp. 6217–6228, 2018.

[1215] Y. Yang and X. Zhang, An optimized variable structure control for hydraulic AGC of cold rolling mill, IEEE ICIT 2009, IEEE International Conference on Industrial Technology, Gippsland, Australia, February 2009.

[1216] Y. Yang, and T. Gao, Adaptive control of the filling velocity of thermoplastics injection molding, Control Engineering Practice, Elsevier, vol. 8, pp. 1285–1296, 2000.

[1217] B. Yang, Y. He, J. Han and G. Liu, Modelling and control of rotor-flying multi-joint manipulator, Proceedings of the 19th IFAC World Congress, Cape Town, South Africa, August 2014.

[1218] Y.B. Yang, B. Zhang, Y. Qian and Y. Wu, Contact-point response for model identification of bridges by a moving test vehicle, International Journal of Structural Stability and Dynamics, vol. 18, no. 65, pp. 11–24, 2018.

[1219] K. Yano and K. Terashima, Robust liquid container transfer control for complete sloshing suppression, IEEE Transactions on Control Systems Technology, vol. 9, no. 5, pp. 483–493, 2001.

[1220] L.D. Yang and Y.P. Sun, Mixed H_2/H_∞ state feedback design for micro-satellite attitude control, Control Engineering Practice, Elsevier, vol. 10, pp. 951–970, 2002.

[1221] Y. Yang and Z.K. Nagy, Application of nonlinear model predictive control to continuous crystallization systems, IEEE ACC 2015, 2015 American Control Conference, Chicago, Illinois, USA, July 2015.

[1222] P. Yang, X. Ma, J. Wang, G. Zhang, Y. Zhang and L. Chen, Disturbance observer-based terminal sliding-mode control of a 5-DOF upper-limb exoskeleton robot, IEEE Access, vol. 7, pp. 62813–62889, 2019.

[1223] K. Yano and K. Terashima, Sliding suppression control of liquid transfer systems considering a 3D transfer path, IEEE Transactions on Mechatronics, vol. 10, no. 1, pp. 8–16, 2003.

[1224] C. Yao, J. Krieglstein and K. Janschek, Modelling and sliding-mode control of a fully actuated multirotor with tilted propellers, 12th IFAC Symposium on Robot Control SYROCO 2018: Budapest, Hungary, Aug. 2018.

[1225] K. Yao, F. Gao and F. Allgower, Barrel temperature control during operation transition in injection moulding, Control Engineering Practice, Elsevier, vol. 16, pp. 1253–1264, 2008.

[1226] H. Ye, H. Wang and H. Wang, Stabilization of a PVTOL aircraft and an inertia wheel pendulum using saturation techniques, IEEE Transactions on Control Systems Technology, vol. 15, no. 6, pp. 1143–1150, 2007.

[1227] L. Ye, Q. Zong, J.L. Crassides and B. Tian, Output-redefinition-based dynamic inversion control for a non-minimum phase hypersonic vehicle, IEEE Transactions on Industrial Electronics, vol. 65, no. 4, pp. 3447–3457, 2018.

[1228] S.K. Yildiz, J.F. Forbes, B. Huang, *et al.*, Dynamic modelling and simulation of a hot-strip finishing mill, Applied Mathematical Modelling, Elsevier, vol. 33, pp. 3208–3225, 2003.

[1229] S.K. Yildiz, B. Huang and J.F. Forbes, Dynamics and variance control of hot mill loopers, Control Engineering Practice, Elsevier, vol. 16, pp. 89–100, 2008.

[1230] F.C. Yin, D.H. Zhang and Y.C. Zhang, Dynamic modelling and rolling data analysis of the tandem hot rolling process, Simulation, Transaction of the Society for Modelling and Simulation, vol. 93, no. 4, pp. 307–321, 2017.

[1231] X. Ying, J. Yan, C. Hun and X. Guan, Trajectory tracking control of autonomous underwater vehicle with unknown parameter and external disturbances, IEEE Transactions on Systems, Man and Cybernetics: Systems, vol. 51, no. 2, pp. 1054–1063, 2019.

[1232] X. Yin, M. Lei and H. Pan, Direct optimal power extraction control for a tidal turbine system based on fuzzy power tuning, Ocean Engineering, Elsevier, vol. 170, pp. 426–439, 2018.

[1233] C. Yong, X. Jiao, L. Li, Y. Zhang, L. Zhang and J. Song, Electromechanical coupling driving control for single-shaft parallel hybrid powertrain, Science China: Technological Sciences, Springer, vol. 57, pp. 541–549, 2014.

[1234] K. Yoo and W. Chung, Pushing motion control of N passive off-hooked trailers by a car-like mobile robot, 2010 IEEE International Conference on Robotics and Automation, IEEE ICRA 2010, Anchorage, Alaska, May 2010.

[1235] Y. Yoon, E.N. Johnson and L. Ren, Autonomous flight control for multirotors by a simple input-output linearization with nested saturation, IEEE CCTA 2018, IEEE 2018 Conference on Control Technology and Applications, Copenhagen, Denmark, August 2018.

[1236] Y. Yoshimoto, T. Matsuno and S. Hokamoto, Position and attitude control of an underactuated satellite with constant thrust, AIAA Guidance, Navigation, and Control Conference, Portland, Oregon, USA, August 2011.

[1237] M.Z. Youssef, Design and performance of a cost-effective BLDC drive for water-pump applications, IEEE Transactions on Industrial Electronics, vol. 62, no. 5, pp. 3277–3285, 2015.

[1238] S. Yousefizadeh, J.D. Bedtsen, N. Vafamand, M.H. Khooban, T. Dragisevic and F.Blaabjerg, EKF-based predictive stabilization of shipboard DC microgrids with uncertain time-varying load, IEEE Journal of Emerging and Selected Topics in Power Electronics, vol. 7, no. 2, pp. 901–909, 2019.

[1239] J. Yuan, F. Sun and Y. Huang, Trajectory generation and tracking control for double-steering tractor-trailer mobile robots with on-axle hitching, IEEE Transactions on Industrial Electronics, vol. 60, no. 12, pp. 7665–7677, 2015.

[1240] S. Yu, H. Lee, W. Kim and C. Han, Development of an underactuated exoskeleton for effective walking and load-carrying assist, Advanced Robotics, Taylor and Francis, vol. 30, no. 8, pp. 535–551, 2016.

[1241] N. Yu, J. Song and G. Ren, Model-based adaptive command filtering control of an electrohydraulic actuator with input saturation and friction, IEEE Access, vol. 5, pp. 48252–48263, 2020.

[1242] H. Yu, S. Huang, G. Chen and Z. Gao, Human-robot interaction control of rehabilitation robots with series elastic actuators, IEEE Transactions on Robotics, vol. 31, no. 5, pp. 1089–1100, 2015.

[1243] H. Yu, Modelling and control of hybrid machine systems: A five-bar mechanism case, International Journal of Automation and Computing, vol. 9, pp. 235–243, 2006.

[1244] C. Yu, X. Xiang, P.A. Wilson and Q. Zhang, Guidance-error-based robot adaptive control for bottom following of a flight-style AUV with saturated actuator dynamics, IEEE Transactions on Cybernetics, vol. 50, no. 5, pp. 1887–1899, 2019.

[1245] Q. Yu, R. Xiong and C. Lin, Online estimation of state-of-charge based on H-infinity and Unscented Kalman Filters for Lithium-Ion batteries, Proceedings Of the ICAE 2016, The 8th International Conference on Applied Energy, Energy Procedia, Elsevier, vol. 105, pp. 2791–2796, 2017.

[1246] J. Yuan, S. Yang and J. Cai, Consistent path planning for on-axle hitching multisteering trailer systems, IEEE Transactions on Industrial Electronics, vol. 65, no. 2, pp. 9625–9634, 2018.

[1247] L. Yuan, M.L. Chen, J.O. Shen and F. Xiao, Current harmonics elimination control method for six-phase PM synchronous motor drives, ISA Transactions, Elsevier, vol. 59, pp. 443–449, 2016.

[1248] Y. Yuan, Z. Wang, L. Guo and H. Liu, Barrier Lyapunov functions-based adaptive fault tolerant control for flexible hypersonic flight vehicles with full state constraints, IEEE Transactions on Systems, Man and Cybernetics: Systems, vol. 50, no. 9, pp, 3391–3400, 2018.

[1249] Y. Yuan, J.W.C. van Lint, R.E. Wilson, F. van Wageningen-Kessels and S.P. Hoogendoorn, Real-time Lagrangian Traffic state estimator for freeways, IEEE Transactions on Intelligent Transportation Systems, vol. 13, no. 1, pp. 59–70, 2012.

[1250] X. Yue, H. Wang and J. Jiamg, A novel 110 kV power line inspection robot and its climbing stability analysis, International Journal of Advanced Robotic Systems, Sage Publications, pp. 1–10, May–June 2017.

[1251] B. Yuksel, S. Mahboubi, C. Secchi, H.H. Buhthoff and A. Franchi, Design, identification and experimental testing of a light-weight flexible-joint arm for aerial physical interaction, IEEE ICRA 2015, IEEE 2015 International Conference on Robotics and Automation, May 2015, Seattle, USA.

[1252] B. Yuksel, G. Buondonno, and A. Franchi, Differential flatness and control of protocentric aerial manipulators with any number of arms and mixed rigid / flexible joints, IEEE IROS 2016, IEEE 2016 International Conference on Intelligent Robots and Systems, Deajeon, South Korea, October 2016.

[1253] B. Yuksel, N. Staub and A. Franchi, Aerial robots with rigid/elastic joint arms: Single joint controllability study and preliminary experiments, IEEE IROS 2016, IEEE 2016 International Conference on Intelligent Robots and Systems, Daejeon, South Korea, October 2016.

[1254] J.N. Yun and J.B. Sun, Design of a disturbance observer for a two-link manipulator with flexible-joints, IEEE Transactions on Control Systems Technology, vol. 22, no. 2, pp. 809–815, 2014.

[1255] R. Yung, M. Schofied, N. Zhao and A. Emadi, Dual three-phase permanent magnet synchronous machine investigation for battery electric vehicle power-trains, IET Journal of Engineering, vol. 2019, no. 17, pp. 3981–3985, 2019.

[1256] C. Zacarrias Ferreira, R. Cardoso, M.E. Mendosa Mesa and J.P. Julca Avila, Controlling tracking trajectory of a robotic vehicle for inspection of underwater structures, Ocean Engineering, Elsevier, vol. 149, pp. 373–382, 2018.

[1257] Q. Zang and J. Huang, Dynamics and control of three-dimensional slosh in a moving rectangular liquid container undergoing planar excitations, IEEE Transactions on Industrial Electronics, vol. 68, no. 4, pp. 2308–2317, 2016.

[1258] H.A. Zarchi, J. Soltani and G. A. Markadeh, Adaptive input-output feedback linearization-based torque control of synchronous reluctance motor without mechanical sensor, IEEE Transactions on Industrial Electronics, vol. 57, no.1, pp. 375–384, 2010.

[1259] S. Zeglache, H. Mekki, A. Boufouerra and A. Djerioui, Actuator fault tolerant control using adaptive RBFNN fuzzy sliding mode controller for co-axial octorotor UAV, ISA Transactions, Elsevier, vol. 80, pp. 267–278, 2018.

[1260] M. Zerar, A. Zolghadri and F. Cazaurang, Coupled linear parameter varying and flatness-based approach for space re-entry vehicles guidance, IET Control Theory and Applications, vol. 3, no. 8, pp. 1081–1092, 2009.

[1261] M. Zhang, Y. Zhang, B. Ji, C. Ma and X. Chang, Modeling and energy-based sway reduction control for tower-crane systems with double-pendulum and spherical pendulum effects, Journal of Measurement and Control, Sage Publications, vol. 53, no. 1–2. pp. 14–150, 2019.

[1262] X. Zhang, X. Huang and H. Lu, Finite-time attractivity of manifold based immersion and invariance control, IEEE CCC 2015, Proceedings of the 34th Chinese Control Conference, Hangzhou, China, July 2016.

[1263] G. Zhang, P. Young, J. Wang, and J. Sun, Multivariable finite-time control for a 5-DOF upper-limb exoskeleton based on linear extended observers, IEEE Access, vol. 6, pp. 43213–43221, 2018.

[1264] F. Zhang, G. Liu, L. Fang and H. Wang, Estimation of battery state of charge worth H_∞ observer: Applied to a robot for inspecting power transmission lines, IEEE Transactions on Industrial Electronics, vol. 59, no. 2, pp. 1086–1097, 2012.

[1265] Y. Zhang, S. Li, J. Zou and A.H. Khan, A passivity-based approach for kinematic control of manipulators with constraints, IEEE Transactions on Industrial Informatics, vol. 16, no. 5, pp. 3029–3038, 2020.

[1266] Y. Zhang, X. Liu, M. Luo and C. Yang, MPC-based 3D trajectory tracking for an autonomous underwater vehicle with constraints in complex ocean environments, Ocean Engineering, vol. 189, pp. 106309, 2019.

[1267] S. Zhang, Q. Wang and C. Dong, Extended state observed-based control for generic hypersonic vehicles with non-affine-in-control character, ISA Transactions, Elsevier, vol. 80, pp. 127–136, 2018.

[1268] S. Zhang, Q. Wang, G. Yang and M. Zhang, Anti-disturbance backstepping control for air-breathing hypersonic vehicles based on extended state observer, ISA Transactions, Elsevier, vol. 92, pp. 84–93, 2019.

[1269] Y. Zhang, D. Xu and L. Huang, Generalized multiple-vector-based model predictive control for PMSM drives, IEEE Transactions on Industrial Electronics, vol. 65, no. 12, pp. 9356–9366, 2018.

[1270] X. Zhang, L. Zhang and Y. Zhang, Model predictive current control for PMSM drives with parameter robustness improvement, IEEE Transactions on Power Electronics, vol. 34, no. 2, 2019.

[1271] Z. Zhang, D. Yue and C.X. Dou, DMPC-based coordinated voltage control for integrated hybrid energy system, IEEE Transactions on Industrial Informatics, vol. 17, no. 10, pp. 6786–6797, 2021.

[1272] Bo Zhang, F. Xu, J. Zhang and T. Shen, Real-time control algorithm for minimising energy consumption in parallel hybrid electric vehicles, IET Electrical Systems in Transportation, vol. 10, no. 4, pp. 331–340, 2020.

[1273] Q.X. Zhang, C. Zhou, D. Wang and Z. Cui, A research based on the energy consumption of a balanced cold rolling mill production scheduling optimization, IEEE CCC 2014, Proceedings of the 32nd Chinese Control Conference, Nanjing, China, July 2014.

[1274] T. Zhang and J. Yang, Nonlinear dynamics and robust control of sloshing in a tank, Journal of Vibration and Control, Sage Publications, vol. 25, no. 1, pp. 132–140, 2019.

[1275] Z. Zhang, W. Liu and J. Ding, Valuation of stock loan under uncertain environment, Soft Computing, Springer, vol. 22, pp. 5663–5669, 2017.

[1276] Q. Zhang and X.Y. Zhou, Valuation of stock loans with regime switching, SIAM Journal on Control and Optimization, vol. 48, no. 3, pp. 1229–1250, 2009.

[1277] C. Zhang, K. Li, J. Deng and S. Song, Improved real-time state-of-charge estimation of $LiFePO_4$ battery based on a novel thermoelectric model, IEEE Transactions on Industrial Electronics, vol. 64, no. 1, pp. 654–663, 2017.

[1278] H. Zheng, J. Wu, W. Wu, and Y. Wang, Integrated motion and powertrain predictive control of intelligent fuel cell/battery hybrid vehicles, IEEE Transactions on Industrial Informatics, vol. 16, no. 5, 2020.

[1279] P. Zhao, H. Zhou, Y. He, K. Cai and J. Fu, A non-destructive online method for monitoring the injection moulding process by collecting and analyzing machine running data, International Journal of Advanced Manufacturing Technology, Springer, vol. 70, pp. 765–777, 2014.

[1280] J. Zheng, Optimal control of multi-dimensional modified Swift-Hohenberg equation, International Journal of Control, vol. 88, no. 10, pp. 2117–2125, 2015.

[1281] J Zheng, The bang-bang principle for time-optimal control of the Kuramoto-Sivashinsky-KdV equation with internal control, International Journal of Robust and Nonlinear Control, vol. 26, pp. 1667–1685, 2016.

[1282] X. Zheng and H. Fang, An integrated unscented Kalman Filter and relevance vector regression approach for lithium-ion battery remaining useful life and short-term capacity prediction, Reliability Engineering & System Safety, Elsevier, vol. 144, pp. 74–82, 2015.

[1283] D. Zhenye and J. Zhang, An active disturbance rejection controller design for ball and plate system, IEEE CCC 2015, IEEE 34th Chinese Control Conference, Huangzhou, China, July, 2015.

[1284] Z. Zhong and J. Wang, Looper-tension almost disturbance decoupling control for hot strip finishing mill based on feedback linearization, IEEE Transactions on Industrial Electronics, vol. 58, no. 8, pp. 3668–3679, 2011.

[1285] Y. Zhonglin, Z. Xiaoming and B. Leping, Investigation of the stabilization method for DC microgrid based on the loop counteraction technique, IEEE CAC 2015, IEEE 2015 Chinese Automation Congress, Wuhan, China, January 2016.

[1286] L. Zhou, J. Sun, H. Yang and T. Wang, Position control of a rodless cylinder in pneumatic servo with actuator saturation, ISA Transactions, Elsevier, vol. 90, pp. 235–243, 2019.

[1287] Y. Zhou, X. Qi, A. Incecik, Y. Ma and Z. Li, Broken lines path-following algorithm for a water-jet propulsion USV with disturbance uncertainties, Ocean Engineering, Elsevier, vol. 21, pp. 107118–107128, 2020.

[1288] Y. Zhou and G. Chen, Predictive DTC strategy with fault-tolerant function for six-phase and three-phase PMSM series-connected drive system, IEEE Transactions on Industrial Electronics, vol. 65, no. 11, pp. 9101–9112, 2018.

[1289] L. Zhou, Z. Wang, Y.H. Mok, R. Hun and Y. Hou, Steering stability control for four-motor distributed drive high speed tracked vehicles, IEEE Access, vol. 8, pp. 94968–94983, 2020.

[1290] X. Zhou, X. Chen, C. Peng and Y. Zhou, High performance nonsalient sensorless BLDC motor control strategy from standstill to high speed, IEEE Transactions on Industrial Informatics, vol. 14, no. 10, pp. 4365–4375, 2018.

[1291] Z.Q. Zhou and S. Cai, Hybrid excited permanent magnet machines for electric and hybrid electric vehicles, CES Transactions on Electrical Machines and Systems, vol. 3, no. 3, pp. 233–247, 2019.

[1292] X. Zhoxin, Z. Tiamli, L. Huaimin, J.M. Guerrero, C.L. Su amd J.C. Vasquez, Coordinated control of a hybrid electric ferry shipboard microgrid, IEEE Transactions on Transportation Electrification, vol. 5, no. 3, pp. 828–839, 2019.

[1293] B. Zhu and X. Xia, Nonlinear trajectory tracking control for heavy-haul trains, IFAC MICNON 2015, 1st IFAC Conference on Modelling, Identification and Control of Nonlinear Systems, St. Petersburg, Russia, June 2015.

[1294] W. Zhuang, S. Li, X. Zhang, *et al.*, A survey of powertrain configuration studies on hybrid electric vehicles, Applied Energy, Elsevier, vol. 226, pp. 114553–114569, 2020.

[1295] W. Zhuang, X. Zhang, G. Yin, H. Peng and L. Wang, Mode shift schedule and control strategy design of multimode hybrid powertrain, IEEE Transactions on Control Systems Technology, vol. 28, no. 3, pp. 804–815, 2020.

[1296] J. Zhang, G. Zeng, Y. Guo and J. Wang, Underactuated microsatellites electromagnetic docking control using disturbance observer-based sliding-mode controllers, IEEE Access, vol. 8, pp. 124476–124485, 2020.

[1297] X. Zhang, Z. Zhang, Z. Lu and W. Liao, Optimal path-planning-based finite-time control for agile CubeSat Attitude Maneuvers, IEEE Access, vol. 7, pp. 102180–102197, 2019.

[1298] B. Zhang, Y. Chen, Y. Wu and Y.B. Yang, An efficient means for damage detection of bridges using the contact-point response of a moving test vehicle, Journal of Sound and Vibration, Elsevier, vol. 419, pp. 158–172, 2018.

[1299] Q.Zhang, M. Basseville and A. Benveniste, Early warning of slight changes in systems, Special Issue on Statistical Methods in Signal Processing and Control, Automatica, Elsevier, vol. 30, no. 1, pp. 95–113, 1994.

[1300] Q. Zhang, Fault detection and isolation with nonlinear black-box models, Proceedings of the IFAC SYSID 1997, IFAC Symposium on System Identification, Kitakyushu, Japan, 1997.

[1301] Q. Zhang, M. Basseville and A. Benveniste, Fault detection and isolation in nonlinear dynamic systems: A combined input-output and local approach, Automatica, Elsevier, vol. 34, no. 11, pp. 1359–1373, 1998.

[1302] Q. Zhang and A. Benveniste, Wavelet networks, IEEE Transactions on Neural Networks, vol. 3, no. 6, pp. 869–898, 1993.

[1303] G. Zhang, P. Yang, J. Wang and J. Sun, Multivariable finite-time control of 5-DOF upper limb exoskeleton based on linear extended state observer, IEEE Access, vol. 6, pp. 43213–43221, 2018.

[1304] G.W. Zhang, P. Yang, J. Wang, J.F. Sun and Y. Zhang, Integrated observer-based fixed-time control with backstepping method for exoskeleton robot, International Journal of Automation and Computing, Springer, vol. 17, no. 1, pp. 71–82, 2020.

[1305] J. Zhang, D. Gu, C. Deng and B. Wen, Robust and adaptive backstepping control for hexacopter UAV, IEEE Access, vol. 7, pp. 163502–163514, 2019.

[1306] S. Zhao, R. Duncan and D.A. Howey, Observability analysis and state estimation of lithium-ion batteries in the presence of sensor biases, IEEE Transactions on Control Systems Technology, vol. 25, no. 1, pp. 326–333, 2017.

[1307] S. Zhu, D. Wang, Q. Shen and E.K. Poh, Satellite attitude stabilization control with actuator faults, Journal of Guidance, Control and Dynamics, AIAA Journals, vol. 40, no. 5, pp. 1309–1318, 2017.

[1308] Z.Q. Zhu, S.S. Lin, L. Huang and S.Y. Zhu, Damage identification of supporting structures with a moving sensory system, Journal of Sound and Vibration, Elsevier, vol. 415, pp. 111–127, 2018.

[1309] B. Zi, J. Cao and Z. Zhu, Dynamic simulation of hybrid-driven planar five-bar parallel mechanism based on SimMechanics and tracking control,

International Journal of Advanced Robotic Systems, Sage Publications, vol. 5, no. 4, pp. 28–33, 2011.

[1310] T. Zilic, D. Pavkovic and D. Zorc, Modelling and control for a pneumatically actuated inverted pendulum, ISA Transactions, Elsevier, vol. 48, pp. 327–345, 2009.

[1311] Z. Ziye, L. Haiou, C. Huiyan, H. Jiaming, and G. Hongming, Kinematics-aware model predictive control for autonomous high-speed tracked vehicles under the off-road conditions, Mechanical Systems and Signal Processing, Elsevier, vol. 123, pp. 333–350, 2019.

[1312] T. Zou, J. Angeles and F. Hassani, Dynamic modeling and trajectory tracking control of unmanned tracked vehicles, Robotic and Autonomous Systems, Elsevier, vol. 110, pp. 102–111, 2018.

[1313] M. Zou, H. Bai, Y. Wang and S. Yu, Mechanical design of a self-adaptive transformable tracked robot for cable tunnel inspection, Proceedings of 2016 IEEE International Conference on Mechatronics and Automation, Harbin, China, August 2016.

[1314] J. Zou, M. Zhao, Q. Wang, J. Zou and G. Ku, Development and analysis of tubular transverse flux machine with permanent magnet excitation, IEEE Transactions on Industrial Electronics, vol. 59, pp. 2198–2207, 2012.

[1315] C. Zou, C. Mansic and S. Anway, Control-oriented modelling of a lithium-ion battery for fast charging, 19th IFAC World Congress, Cape Town, South Africa, pp. 3912–3917, August 2014.

[1316] H. Zwart, Y. Le Gorrec and B. Moschke, Linking hyperbolic and parabolic PDEs, 2011 50th IEEE Conference on Decision and Control and European Control Conference, CDC-ECC, Orlando, Florida, USA, December 2011.

[1317] Y.H. Zweiri, L.D. Seneviratne and K. Althoefer, Modelling and control of an unmanned excavator vehicle, modelling and control of an unmanned excavator vehicle, Proceedings Of the IMeche, Part I: Journal of Systems and Control Engineering, Sage Publications, vol. 227, pp. 259–274, 2016.

Index

www.ingramcontent.com/pod-product-compliance
Lightning Source LLC
LaVergne TN
LVHW020326220926
848484LV00001BB/266
* 9 7 8 1 8 3 9 5 3 4 2 6 3 *